Principles of Life • Correlation to College Board's AP® Biology Curriculum Framework

BIG IDEA 1: The process of evolution drives the diversity and unity of life.

ESSENTIAL KNOWLEDGE	CHAPTERS/SECTIONS	ILLUSTRATIVE EXAMPLES COVERED
1.A.1 Natural selection is a major mechanism of evolution	1.4, 15.1-7, 42.4	• Graphical analysis of allele frequencies in a population • Application of Hardy-Weinberg Equation
1.A.2 Natural selection acts on phenotypic variations in populations	1.3, 1.4, 9.3, 10.1, 15.1-7, 24.4, 27.3, 43.3	• DDT resistance in insects • Artificial selection • Loss of genetic diversity within a crop species • Overuse of antibiotics • Sickle cell anemia
1.A.3 Evolutionary change is also by random processes	15.2, 15.5	*No illustrative examples listed in Curriculum Framework.*
1.A.4 Biological evolution is supported by scientific evidence from many disciplines, including mathematics.	15.1, 15.3, 16.1-4, 18.1-3	• Graphical analyses of allele frequencies in a population • Analysis of sequence data sets • Analysis of phylogenetic trees • Construction of phylogenetic trees based on sequence data
1.B.1 Organisms share many conserved core processes and features that evolved and are widely distributed among organisms today.	1.1, 16.1-4, 19.1, 32.2	• Cytoskeleton (a network of structural proteins that facilitate cell movement, morphological integrity, organelle transport) • Membrane-bound organelles (mitochondria and/or chloroplasts) • Linear chromosomes • Endomembrane systems, including the nuclear envelope
1.B.2 Phylogenetic trees and cladograms are graphical representations (models) of evolutionary history that can be tested.	1.1, 16.1-16.4, 19.1, 20.2, 21.1, 22.3, 23.1, 23.6, 23.7, 32.2, 41.4	• Number of heart chambers in animals• Absence of legs in some sea mammals • Opposable thumbs
1.C.1 Speciation and extinction have occurred throughout the Earth's history.	17.1-4, 18.1-3, 43.2, 44.5	• Five major extinctions • Human impact on ecosystems and species extinction rates
1.C.2 Speciation may occur when two populations become reproductively isolated from each other.	17.2, 17.3, 41.4	*No illustrative examples listed in Curriculum Framework.*
1.C.3 Populations of organisms continue to evolve.	1.5, 10.5, 15 Opening and Q&A, 15.1-7; 17.4, 43.3	• Chemical resistance • Emergent diseases • Observed directional phenotypic change in a population • A eukaryotic example that describes evolution of a structure or process
1.D.1 There are several hypotheses about the natural origin of life on Earth, each with supporting scientific evidence.	1.1, 2 Opening and Q&A, 6.1	*No illustrative examples listed in Curriculum Framework.*
1.D.2 Scientific evidence from many different disciplines supports models of the origin of life.	1.1, 2 Opening and Q&A, 3.3, 4 Q&A, 6.1	*No illustrative examples listed in Curriculum Framework.*

BIG IDEA 2: Biological systems utilize free energy and molecular building blocks to grow, to reproduce, and to maintain dynamic homeostasis.

ESSENTIAL KNOWLEDGE	CHAPTERS/SECTIONS	ILLUSTRATIVE EXAMPLES COVERED
2.A.1 All living systems require constant input of free energy.	2.5, 6.1, 6.2, 6.3, 6.4, 6.5, 30.1, 42.3, 44.3	• Krebs cycle • Glycolysis • Calvin cycle • Fermentation • Seasonal reproduction in animals and plants • Ectothermy • Endothermy • Life history strategy • Change in primary production affects higher trophic levels • Change in each trophic level affects higher trophic levels

BIG IDEA 2 (continued)		
ESSENTIAL KNOWLEDGE	CHAPTERS/SECTIONS	ILLUSTRATIVE EXAMPLES COVERED
2.A.2 Organisms capture and store free energy for use in biological processes.	6.1, 6.2, 6.3, 30.1, 43.4, 45.2	• $NADP^+$ in photosynthesis • Oxygen in cellular respiration
2.A.3 Organisms must exchange matter with the environment to grow, reproduce, and maintain organization.	2.2, 25.1, 29.4, 30.1, 30.4, 31.2. 32.4, 42.3, 45.2, 45.3	• Cohesion, adhesion, high specific heat, universal solvent • Root hairs • Cells of villi • Cells of alveoli • Microvilli
2.B.1 Cell membranes are selectively permeable due to their structure.	5.1, 34.2, 36.3, 36.5	*No illustrative examples listed in Curriculum Framework.*
2.B.2 Growth and dynamic homeostasis are maintained by the constant movement of molecules across membranes.	5.2, 5.3, 5.4, 29.1	• Glucose transport • Na^+/K^+ transport
2.B.3 Eukaryotic cells maintain internal membranes that partition the cell into specialized regions.	4.3	• Endoplasmic reticulum • Mitochondria • Chloroplasts • Golgi • Nuclear envelope
2.C.1 Organisms use negative feedback mechanisms to maintain their internal environments and respond to external environmental changes.	7.2, 7.3, 11.1, 11.2, 28.3, 29.3, 35.3, 37.2, 27.4, 30.5	• Operons in gene regulation • Plants and water limitations • Cell cycle checkpoints • Temperature regulation in animals • Plant responses to water limitation • Lactation in mammals • Onset of labor • Ripening of fruit • Diabetes mellitus • Dehydration in response to decreased ADH • Blood clotting
2.C.2 Organisms respond to changes in their external environments.	27.2, 29.3, 35, 40.4, 41.3	• Photoperiodism in plants • Behavioral thermoregulation • Hibernation and migration in animals • Circadian rhythms • Shivering and sweating in humans
2.D.1 All biological systems from cells and organisms to populations, communities, and ecosystems are affected by complex biotic and abiotic interactions involving exchange of matter and free energy	29.1, 29.3, 29.6, 31.2, 41.1, 41.2, 42.1-6, 43.1-4, 44.1-6	• Cell density • Population density • Biofilms • Temperature • Water availability • Symbiosis • Predator-prey relationships • Water and nutrient availability • Availability of nesting sites • Food chains and food webs • Species diversity
2.D.2 Homeostatic mechanisms reflect both common ancestry and divergence due to adaptation in different environments.	29.4, 30.2, 31.2, 32.2, 36.2	• Gas exchange in aquatic and terrestrial plants • Digestive mechanisms in animals • Respiratory systems of aquatic and terrestrial animals • Nitrogeneous waste production in animals • Excretory systems in animals • Circulatory systems in animals • Thermoregulation in animals (countercurrent)
2.D.3 Biological systems are affected by disruptions to their dynamic homeostasis.	28.3, 30.1-3, 32.1-6, 39.1-5, 42.4, 43.4, 44.5	• Plant responses to toxins, water stress and salinity • Immune response • Human impact • Invasive species • Fires • Water limitation • Salination • Dehydration • Physiological responses to toxic substances
2.D.4 Plants and animals have a variety of chemical defenses against infections that affect dynamic homeostasis.	28.1, 39.1-5	• Plant defenses against pathogens • Animal nonspecific defenses and specific defenses • Mammalian cellular and humoral immunity, antibodies
2.E.1 Timing and coordination of specific events are necessary for the normal development of an organism, and these events are regulated by a variety of mechanisms.	14.1-14.3, 24.1, 26.1, 39.2	• Morphogenesis of fingers and toes • *C. elegans* development • Flower development • Immune function
2.E.2 Timing and coordination of physiological events are regulated by multiple mechanisms.	19 Opening and Q&A, 19.3, 26.4, 29.3, 34.4, 35.1, 40.4	• Quorum sensing in bacteria • Circadian rhythms • Seasonal responses such as hibernation, estivation, and migration • Release and reaction to pheromones
2.E.3 Timing and coordination of behavior are regulated by various mechanisms and are important in natural selection.	22.3, 29.3, 40.1, 43.2	• Quorum sensing in bacteria • Fruiting body formation in fungi • Hibernation • Migration • Niche and Resource Partitioning

BIG IDEA 3: Living systems store, retrieve, transmit, and respond to information essential to life processes.		
ESSENTIAL KNOWLEDGE	CHAPTERS/SECTIONS	ILLUSTRATIVE EXAMPLES COVERED
3.A.1 DNA, and in some cases RNA, is the primary source of heritable information.	3.1-4, 9.1-3, 10.1-5, 13.1-4	• Poly A tail • GTP cap • Excision of introns • Enzymes • Transport by proteins • Synthesis • Degradation • GM foods • Transgenic animals • Cloned animals • Pharmaceuticals • Electrophoresis • Plasmid-based transformation • Polymerase chain reaction
3.A.2 In eukaryotes, heritable information is passed to the next generation via processes that include the cell cycle and mitosis, or meiosis plus fertilization.	7.1-4, 37.1-3	• Mitosis-promoting factor • Cancer and cell cycle control
3.A.3 The chromosomal basis of inheritance provides an understanding of the pattern of passage (transmission) of genes from parent to offspring.	8.1-3, 12.4	• Down syndrome • X-linked color blindness • Sickle cell anemia • Civic issues
3.A.4 The inheritance pattern of many traits cannot be explained by simple Mendelian genetics.	8.3, 9.3, 35.4	• Sex-linked genes • The Y chromosome carries few genes • In mammals and flies, females are XX and males are XY
3.B.1 Gene regulation results in differential gene expression, leading to cell specialization.	11.1-4	• Promoter • Terminator • Enhancers
3.B.2 A variety of intercellular and intracellular signal transmissions mediate gene expression.	5.5, 14.3, 26.1, 26.2, 26.3, 35.4, 38.3	• Morphogens stimulate development • Cytokines regulate gene expression • HOX genes and development • Seed germination and gibberellin
3.C.1 Changes in genotype can result in changes in phenotype.	7.4, 9.3	• Antibiotic resistance mutations • Sickle cell disorder and heterozygote advantage
3.C.2 Biological systems have multiple processes that increase genetic variation.	7.4, 8.4, 9.2	*No illustrative examples listed in Curriculum Framework.*
3.C.3 Viral replication results in genetic variation, and viral infection can introduce genetic variation into the hosts.	8.4, 9.1, 12.3	• Transposons
3.D.1 Cell communication processes share common features that reflect a shared evolutionary history.	5.5, 5.6, 9.2	• Epinephrine stimulation of glycogen breakdown • DNA repair mechanisms
3.D.2 Cells communicate with each other through direct contact with other cells or from a distance via chemical signaling.	4.5, 14.3, 28.1, 34.3, 35.1-2, 39.4, 39.5, 40.4	• Immune cells interact • Plasmodesmata between plant cells • Plant immune response • Morphogens and embryonic development • Neurotransmitters • Insulin • Quorum sensing in bacteria • Thyroid hormone • Testosterone • Estrogen
3.D.3 Signal transduction pathways link signal reception with cellular response.	5.5, 5.6	• G-protein linked receptors • Ligand gated ion channels • Receptor tyrosine kinases • Second messengers
3.D.4 Changes in signal transduction pathways can alter cellular response.	5.6, 35, 37.1-2	• Diabetes • Effects of neurotoxins • Drugs
3.E.1 Individuals can act on information and communicate it to others.	28.2, 35.2, 40.1-6	• Fight or flight response• Predator warnings • Colony behavior • Herbivory responses • Coloration • Parent-offspring interactions • Territorial marking • Plant-plant interactions in herbivory • Courtship and mating behaviors • Bee dances • Bird songs
3.E.2 Animals have nervous systems that detect external and internal signals, transmit and integrate information, and produce responses.	34.1-4, 33.1-3	• Acetylcholine • Epinephrine • Dopamine • Serotonin • GABA • Hearing • Muscle movement • Abstract thought • Neurohormone production • Forebrain, midbrain and hindbrain • Right and left cerebral hemispheres

BIG IDEA 4: Biological systems interact, and these systems and their interactions possess complex properties.		
ESSENTIAL KNOWLEDGE	CHAPTERS/SECTIONS	ILLUSTRATIVE EXAMPLES COVERED
4.A.1 The subcomponents of biological molecules and their sequence determine the properties of that molecule.	3.1-4, 9.1	*No illustrative examples listed in Curriculum Framework.*
4.A.2 The structure and function of subcellular components, and their interactions, provide essential cellular processes.	4.3, 4.4, 6.2, 6.5	*No illustrative examples listed in Curriculum Framework.*
4.A.3 Interactions between external stimuli and regulated gene expression result in specialization of cells, tissues and organs.	14.2, 14.3, 38.1	*No illustrative examples listed in Curriculum Framework.*
4.A.4 Organisms exhibit complex properties due to interactions between their constituent parts.	24.1-3, 30.4, 33.1, 33.3	• Plant vascular and leaf • Root, stem and leaf • Kidney and bladder • Respiratory and circulatory • Nervous and muscular • Stomach and small intestines
4.A.5 Communities are composed of populations of organisms that interact in complex ways.	42.1-4, 43.1-4, 44.1-6	• Predator-prey relationship • Symbiotic relationship • Graphical representation of field data • Introduction of species • Global climate change models
4.A.6 Interactions among living systems and with their environment result in the movement of matter and energy.	42.2, 44.3, 45.3, 45.2	*No illustrative examples listed in Curriculum Framework.*
4.B.1 Interactions between molecules affect their structure and function.	3.3, 3.4	*No illustrative examples listed in Curriculum Framework.*
4.B.2 Cooperative interactions within organisms promote efficiency in the use of energy and matter.	29.1, 30.5, 31.2, 34.4, 33.2, 41.1	• Exchange of gases • Circulation of fluids • Digestion of food • Excretion of wastes • Bacterial community in the rumen • Bacterial community in the gut
4.B.3 Interactions between and within populations influence patterns of species distribution and abundance.	22.3, 43.1, 42.1, 42.4, 42.5, 43.1-4, 44.1-6	• Loss of keystone species • Dutch elm disease
4.B.4 Distribution of local and global ecosystems changes over time.	41.1-4, 44.2, 45.5	• Continental drift • Impacts of human land use • Effects of introduced species • Volcanic eruption • Impacts of climate change
4.C.1 Variation in molecular units provides cells with a wider range of functions.	5.1, 6.5, 39.4	• Phospholipids in membranes • MHC proteins • Chlorophylls • Molecular diversity in antibodies
4.C.2 Environmental factors influence the expression of the genotype in an organism.	8.2, 11.1-3, 30.1	• Height and weight in humans • Effect of adding lactose to a Lac^+ bacterial culture • Darker fur in cooler regions of the body
4.C.3 The level of variation in a population affects population dynamics.	15.2-4, 22.3, 28.1, 28.3	• Wheat rust • Prairie chickens
4.C.4 The diversity of species within an ecosystem may influence the stability of the ecosystem.	41.5, 44.4, 45.5	*No illustrative examples listed in Curriculum Framework.*

Principles of Life

SECOND EDITION

for the AP® course

David M. Hillis
University of Texas at Austin

David Sadava
Emeritus, The Claremont Colleges

Richard W. Hill
Michigan State University

Mary V. Price
Emerita, University of California, Riverside

SINAUER ASSOCIATES

bfw FREEMAN

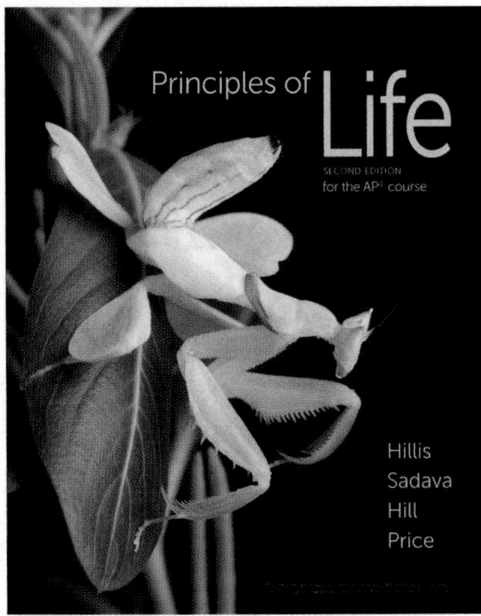

About the Cover

A juvenile pink orchid mantis (*Hymenopus coronatus*) looks, at first glance, like an orchid flower. Its abdomen, head, and four walking legs look like the petals of the flower, and the small black dot at the posterior tip of the abdomen resembles a small fly investigating the flower. This mimicry is advantageous to the mantid for two reasons. The mantis is concealed from potential predators, which mistake the mantis for a flower. At the same time, insects looking for nectar become prey for the mantis, which captures visiting insects with its front pair of toothed, grasping legs. As a result of these advantages, natural selection favored the evolution of this spectacular example of an insect that resembles an orchid flower. © Ch'ien Lee/Minden Pictures.

Principles of Life, Second Edition, for the AP® course

ISBN 978-1-4641-5641-0

Address editorial correspondence to:
Sinauer Associates, Inc., P.O. Box 407, Sunderland, MA 01375 U.S.A.

www.sinauer.com
publish@sinauer.com

Address orders to:
MPS / W.H. Freeman & Co., Order Dept., 16365 James Madison Highway, U.S. Route 15, Gordonsville, VA 22942 U.S.A. or call 1-888-330-8477
Examination copy information: highschool.bfwpub.com/pol2e or 1-800-446-8923

Library of Congress Cataloging-in-Publication Data
Hillis, David M., 1958-
Principles of life / David M. Hillis, University of Texas at Austin, David Sadava, Emeritus, The Claremont Colleges, Richard W. Hill, Michigan State University, Mary V. Price, Emerita, University of California, Riverside. -- Second edition, high school edition.
pages cm
1. Biology. I. Title.
QH308.2.P75 2014
570--dc23

2014021713

Printed in U.S.A.
Third Printing May 2016
RR Donnelley

To all our students. You have taught us, too, and inspired us to write this book.

The Authors

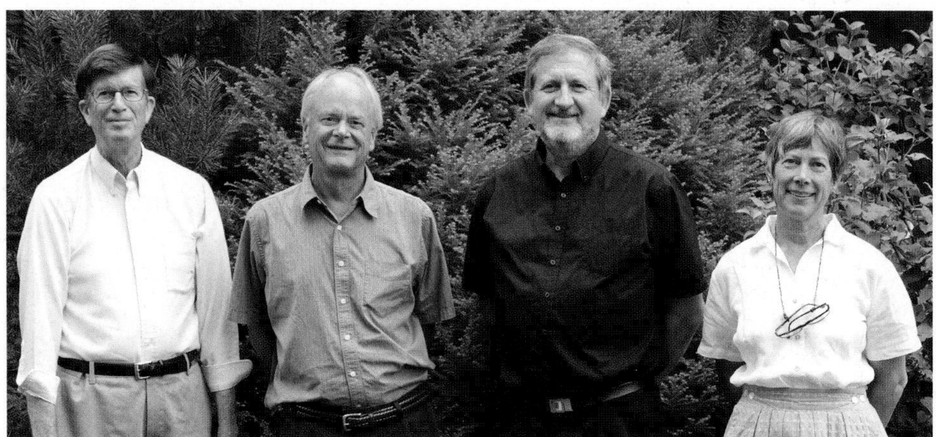

David Sadava, Richard W. Hill, David M. Hillis, and Mary V. Price

David M. Hillis is the Alfred W. Roark Centennial Professor in Integrative Biology at the University of Texas at Austin, where he also has directed the Center for Computational Biology and Bioinformatics and the School of Biological Sciences. Dr. Hillis has taught courses in introductory biology, genetics, evolution, systematics, and biodiversity. He has been elected to the National Academy of Sciences and the American Academy of Arts and Sciences, awarded a John D. and Catherine T. MacArthur Fellowship, and has served as President of the Society for the Study of Evolution and of the Society of Systematic Biologists. He served on the National Research Council committee that wrote the report BIO 2010: Transforming Undergraduate Biology Education for Research Biologists, and currently serves on the Executive Committee of the National Academies Scientific Teaching Alliance. His research interests span much of evolutionary biology, including experimental studies of evolving viruses, empirical studies of natural molecular evolution, applications of phylogenetics, analyses of biodiversity, and evolutionary modeling. He is particularly interested in teaching and research about the practical applications of evolutionary biology.

David Sadava is the Pritzker Family Foundation Professor of Biology, Emeritus, at the Keck Science Center of Claremont McKenna, Pitzer, and Scripps, three of The Claremont Colleges. In addition, he is Adjunct Professor of Cancer Cell Biology at the City of Hope Medical Center. Twice winner of the Huntoon Award for superior teaching, Dr. Sadava has taught courses in introductory biology, biotechnology, biochemistry, cell biology, molecular biology, plant biology, and cancer biology. In addition to *Life: The Science of Biology*, he is the author or coauthor of books on cell biology and on plants, genes, and crop biotechnology. His research has resulted in many papers coauthored with his students, on topics ranging from plant biochemistry to pharmacology of narcotic analgesics to human genetic diseases. For the past 15 years, he has investigated multi-drug resistance in human small-cell lung carcinoma cells with a view to understanding and overcoming this clinical challenge. At the City of Hope, his current work focuses on new anti-cancer agents from plants.

Richard W. Hill is Professor in the Department of Zoology at Michigan State University and a frequent Guest Investigator at Woods Hole Oceanographic Institution. He received his Ph.D. in Zoology from the University of Michigan. Apart from Sinauer Associates' editions of *Animal Physiology*, Dr. Hill has authored two other books on the subject (the second with Gordon Wyse), as well as numerous articles for scientific journals, encyclopedias, and edited volumes. Among the awards he has received are the Outstanding Faculty Award (Michigan State University Senior Class Council) and election as Fellow of the American Association for the Advancement of Science. He was a U.S. Senior Fulbright Scholar in 2000–2001. He has taught courses in environmental physiology and marine biology, and for the last five years has led student trips to the Galápagos Islands. His research interests include: temperature regulation and energetics in birds and mammals, especially neonates; and environmental physiology of marine tertiary sulfonium and quaternary ammonium compounds, especially in the contexts of biogeochemistry and animal–algal symbioses.

Mary V. Price is Professor of Biology, Emerita, at the University of California, Riverside and Adjunct Professor in the School of Natural Resources and the Environment at the University of Arizona. In "retirement," she continues to teach and study, having learned the joy and art of scientific discovery as an undergraduate student at Vassar College and doctoral student at the University of Arizona. Dr. Price has mentored and published with independent-research students and has developed and taught general biology and ecology courses from introductory (majors and nonmajors) to graduate levels. She has particularly enjoyed leading field classes in the arid regions of North America and Australia, and the tropical forests of Central America, Africa, and Madagascar. Dr. Price's research focuses on understanding the ecology of North American deserts and mountains. She has asked why so many desert rodents can coexist, how best to conserve endangered kangaroo rat species, how pollinators and herbivores influence floral evolution and plant population dynamics, and how climate change affects ecological systems.

Brief Table of Contents

Preface

If you are reading this preface and are a high school student, you are probably enrolled in an Advanced Placement® Biology class. Taking AP® Biology is an exciting and rewarding experience because the curriculum is inherently interesting and challenging. Today's AP® Biology course not only teaches you about how your cells and body function, but more importantly, it helps you learn how to design experiments and to ask appropriate questions about how science works. In contrast to old-school biology courses that emphasized memorization, AP® Biology now encourages you to experience biology and learn about the big ideas through many fun, challenging, and enlightening hands-on investigations. Further, AP® Biology will help you to develop conceptual understanding of the discipline and a grounding in science skills that will prove useful throughout your college career and beyond.

If you are like most AP® Biology students, you like a challenge and love to learn. You may well find the AP® Biology curriculum is unlike most classes you have had before. It will require you to think conceptually. It will emphasize how processes work, how they are regulated, and how data is interpreted. You will be asked to apply your knowledge to new and unique situations. *Principles of Life*, Second Edition, and its companion online and print resources have been developed to help you succeed in your quest to understand biology, to prepare for subsequent courses, and to do well on the AP® Biology exam.

Principles of Life, Second Edition, emphasizes the mastery of major concepts in biology through active learning, problem solving in realistic scenarios, and understanding rather than memorization. It embodies the modern, innovative, and inquiry-based approach to teaching biology that was described in an important report called *Vision and Change in Undergraduate Biology Education: A Call to Action* (funded by the American Association for the Advancement of Science and supported by the National Science Foundation, 2011). This report identified five "core concepts for biological literacy" that should be integrated throughout the college biology curriculum. These core concepts center on the themes of:

- evolution,
- the relationship between structure and function,
- information flow, exchange, and storage,
- pathways and transformations of energy and matter, and
- biological systems.

At about the same time that the *Vision and Change* report published, the College Board was redesigning the Advanced Placement® Biology course with the same objectives, resulting in the *AP® Biology Curriculum Framework* (College Board, 2011). For AP® Biology, the five core concepts from *Vision and Change* were restated as the four "Big Ideas":

- **Big Idea 1:** The process of evolution drives the diversity and unity of life.
- **Big Idea 2:** Biological systems utilize free energy and molecular building blocks to grow, to reproduce, and to maintain dynamic homeostasis.
- **Big Idea 3:** Living systems store, retrieve, transmit, and respond to information essential to life processes.
- **Big Idea 4:** Biological systems interact, and these systems and their interactions possess complex properties.

In the Second Edition of *Principles of Life*, the authors have worked to ensure that these Big Ideas are stressed and reinforced throughout the text, problems, media links, and other activities. To help you build bridges between different portions of the course and areas of knowledge, the authors have provided **Links** throughout the book. Using these Links will help you see, for example, that information learned about molecular or cell biology is connected to topics in evolution, diversity, physiology, and ecology.

In addition to encouraging a focus on core concepts, the AP® Biology framework directs students to cultivate certain core competencies to become successful scientists. These competencies are expressed as the seven "Science Practices" in the *AP® Biology Curriculum Framework*:

- **Science Practice 1:** Use representations and models to communicate scientific phenomena and solve scientific problems.
- **Science Practice 2:** Use mathematics appropriately.
- **Science Practice 3:** Engage in scientific questioning to extend thinking or to guide investigations within the context of the AP course.
- **Science Practice 4:** Plan and implement data collection strategies appropriate to a particular scientific question.
- **Science Practice 5:** Perform data analysis and evaluation of evidence.
- **Science Practice 6:** Work with scientific explanations and theories.
- **Science Practice 7:** Connect and relate knowledge across various scales, concepts, and representations in and across domains.

There are numerous opportunities for you to practice these skills throughout the Second Edition of *Principles of Life*. Every chapter contains **Apply the Concept** exercises that give you practice working with data and help reinforce concepts central to each chapter. These problems tie in with **Making Sense of Data: A Statistics Primer** (Appendix B), which provides extra support on understanding why and how biologists draw conclusions from biological data, and thus helps you develop quantitative reasoning skills. The authors have also added more online **Animated Tutorials** and **Activities**, which include opportunities to use modeling and simulation modules to further reinforce an understanding of concepts. By engaging in these activities, you will learn the importance of biological concepts and analyses for addressing societal issues and challenges. The Second Edition also includes links (as a QR code or a URL) to online videos in the new **Media Clips**, giving you an opportunity to see and appreciate the relevance and excitement of biology.

Another valuable element of this textbook is that nearly every chapter incorporates **Investigation** figures that let you see *how* we know *what* we know. Each presents a Hypothesis, Method, Results, and Conclusion. Most of the Investigation figures now have an **Analyze the Data** section that has an extracted subset of data from the published experiment. You are asked to work with these data and to apply basic statistical approaches to understand the results and draw conclusions. Original references and extensive resources are found online for each Investigation. Moreover, the authors have expanded opportunities for you to apply what you've learned by using real data and examples, and have better integrated and explained the concepts of statistical analysis of data. Additionally, to help you understand research tools and how those tools are used in biology, **Research Tools** boxes explain major tools, including laboratory, computational, and field methods. Each chapter begins with an application of a major concept—a story that illustrates a social, medical, scientific, or historical context for the material. Each of these vignettes ends with an open-ended question you can keep in mind as you read and study the rest of the chapter. At the close of the chapter, we return to the question to show how information presented throughout the chapter illuminates the question and helps provide an answer. By pondering these questions as you study, you are learning to think like a scientist. Throughout every chapter, **Checkpoints** are designed to help you self-evaluate their understanding of the material.

In addition to your textbook, there are numerous resources offered both online and in print by the publisher. Learn more about them by going to highschool.bfwpub.com/pol2e. One companion resource for this text is *Strive for a 5: Preparing for the AP® Biology Exam* that accompanies this text, written by John Lepri and I, who are experienced leaders in AP® Biology. We wrote this guide to give you insight into how best to study for the AP® Biology exam while developing a rich understanding of biology.

Each chapter in *Strive for a 5* begins with a synopsis of the textbook chapter. This synopsis points out which Big Ideas apply to the chapter's content, and helps you navigate the chapter efficiently to focus on the most important features in the chapter. We frequently reproduce figures from the textbook to reinforce concepts and to test comprehension. In addition, each section of the chapter is briefly reviewed before you are presented with questions. The majority of these questions are conceptual, requiring you to apply your knowledge. Finally, each chapter concludes with one or two questions, written in the style that you'll encounter on the AP® Biology exam, that tie content to the seven Science Practices. These questions run the gamut from short and long free response questions to grid-in items. In addition, we've developed two full-length, AP®-style exams at the end of the guide and an introduction to the Science Practices begins the guide—making this the ultimate prep tool for the AP® Biology exam.

I sincerely hope that as a student you will come to love and enjoy learning about biology. Teaching biology has been my passion for over thirty years. Whether it is evolution, cellular biology, ecology, genetics, physiology, or one of the other new emerging fields in biotechnology, I know that you will find some element of this textbook especially fascinating and I hope that in the future you might make it your passion. Have a great year and the best of luck to you in your studies!

FRANKLIN BELL

MERCERSBURG ACADEMY,
MERCERSBURG, PA

Advisors and Reviewers

Special Contributions

Many people contributed to the creation of the Second Edition of *Principles of Life*. However, two individuals deserve special mention for their contributions. Susan D. Hill did a masterful job in writing Chapter 38 on Animal Development. Nickolas Waser worked extensively with Mary Price on the Ecology section (Part 7), and was otherwise intimately involved in discussions of the book's planning and execution.

Many People to Thank

In addition to the many biologists listed on these pages who provided formal reviews, each of us benefitted enormously from personal contacts with colleagues who helped us resolve issues and made critical suggestions for new material. They are: Walter Arnold, University of Veterinary Medicine (Vienna); Harry Greene, Cornell University; Will Petry, University of California, Irvine; David Sleboda, Brown University; Thomas Ruf, University of Veterinary Medicine (Vienna); Andrew Zanella, The Claremont Colleges; Edward McCabe, University of Colorado and the March of Dimes Foundation; and Frank Price, Utica College.

Between-editions Reviewers

Rebecca Achterman, Western Washington University
William Anderson, Harvard University
Felicitas Avendano, Grand View University
Dave Bailey, St. Norbert College
Annalisa Berta, San Diego State University
Lori Boies, St. Mary's University
Nicole Bournias-Vardiabasis, California State University, Santa Bernardino
Elizabeth Braker, Occidental College
Cheryl Burrell, Wake Forest University
Robert Cabin, Brevard College
David Carroll, Florida Institute of Technology
Karen Champ, College of Central Florida
Shelton Charles, Forsyth Technical Community College
Sixue Chen, University of Florida
Doug Darnowski, Indiana University Southeast
Maria Davis, Olivet College
Weston Dulaney, Nashville State Community College
Laurie Eberhardt, Valparaiso University
Ingeborg Eley, Hudson Valley Community College
Lewis Feldman, University of California, Berkeley
Austin Francis, Jr., Armstrong Atlantic State University
Jason Garvon, Lake Superior State University
Patricia Geppert, The University of Texas at San Antonio
Nicole Gerlanc, Frederick Community College
Arundhati Ghosh, University of Pittsburgh
Suzanne Gollery, Sierra Nevada College
Elizabeth Good, University of Illinois, Urbana-Champaign
Eileen Gregory, Rollins College
Susan Hester, The University of Arizona
Samantha Hilber, University of Florida
Jill Holliday, University of Florida
Brian Hurdle, Frederick Community College
Sanjeeda Jafar, Morgan State University
Jerald Johnson, Brigham Young University
Susan Jorstad, The University of Arizona
Erin Keen-Rhineheart, Susquehanna University
Stephen Kilpatrick, University of Pittsburgh
Michael Koban, Morgan State University
Daniel Kueh, Emory University
Kristen Lennon, Frostburg State University
Jun Liu, Cornell University
James Malcolm, The University of Redlands
Steve Marcus, The University of Alabama
Nilo Marin, Broward College
Erin Martin, University of Portland
Robert Minckley, University of Rochester
Alexey Nikitin, Grand Valley State University
David Oppenheimer, University of Florida
Aswini Pai, St. Lawrence University
Crima Pogge, City College of San Francisco
Mary Preuss, Webster University
Frank Price, Hamilton College
David Puthoff, Frostburg State University
Sabiha Rahman, University of Ottawa
Nancy Rice, Western Kentucky University
Lori Rose, Hill College
Paul Schulte, University of Nevada, Las Vegas
Jon Shenker, Florida Institute of Technology
Cara Shillington, Eastern Michigan
Miles Silman, Wake Forest University
Thomas Terleph, Sacred Heart University
Mark Thogerson, Grand Valley State University
Elethia Tillman, Spelman College
Mike Troyan, Pennsylvania State University
Ximena Valderrama, Ramapo College of New Jersey
Sebastian Vélez, Worcester State University
Alexander Wait, Missouri State University
Katherine Warpeha, University of Illinois, Chicago
Ted Weinert, The University of Arizona
Erika Whitney, Morgan State University
Michelle Wien, Bryn Mawr College

Accuracy Reviewers

Brian Antonsen, Marshall University
Michael Baltzley, Western Oregon University
Alexa Bely, University of Maryland
Smriti Bhotika, University of Florida
Greta Bolin, Tarrant County College
Katherine Boss-Williams, Emory University
Michelle Boone, Miami University of Ohio
Jackie Brittingham, Simpson College
W. Randy Brooks, Florida Atlantic University
Justin Brown, Simpson College
Deborah Cardenas, Collin College
Trevor Caughlin, University of Florida
Shelton Charles, Forsyth Technical Community College
Donna Charley-Johnson, University of Wisconsin, Oshkosh
Kelly Jean Craig, Colorado Mesa University

Kathryn Craven, Armstrong Atlantic State University
Noelle Cutter, Molloy College
Nancy Dalman, University of North Georgia
Doug Darnowski, Indiana University Southeast
Sandra L. Davis, University of Indianapolis
Ann Marie Davison, Kwantlen Polytechnic University
James Demastes, University of Northern Iowa
Jesse Dillon, California State University, Long Beach
Wes Dulaney, Nashville State Community College
Michele DuRand, Memorial University
Todd Egan, Elmira College
Gordon Fain, The University of California, Los Angeles
Victor Fet, Marshall University
Jeffrey Firestone, University of Maryland
Michael Fultz, Morehead State University
Mark Garcia, Collin College
Deborah Gelman, Pace University
Emily Gillespie, Marshall University
Christopher Gilson, Emory University
Richard Gonzalez-Diaz, Seminole State College of Florida
Harry Greene, Cornell University
Shala Hankison, Ohio Wesleyan University
Amy Hark, Muhlenberg College
Phillip Harris, The University of Alabama
Julie Havens, Armstrong Atlantic State University
Marshal Hedin, San Diego State University
Anna Hicks, Memorial University
Nicole Huber, University of Colorado, Colorado Springs
Amanda Hyde, Greenfield Community College
Erika Iyengar, Muhlenberg College
Laura Jackson, University of Northern Iowa
Jerald Johnson, Brigham Young University
Carly Jordan, The George Washington University
Matthew Kayser, University of Pennsylvania Medical School
Kim Lackey, University of Alabama, Tuscaloosa
Marc Lajeunesse, University of South Florida
Evan Lampert, University of North Georgia
Charlie Lawrence, Viterbo University
Eric Liebgold, Salisbury University
Clark Lindgren, Grinnell College
Jessica Lucas, Santa Clara University
Lisa Lyons, Florida State University
C. Smoot Major, University of Southern Alabama
Marlee Marsh, Columbia College
Megan McNulty, The University of Chicago
Kyle McQuade, Colorado Mesa University
Susan T. Meiers, Western Illinois University
Tamra Mendelson, University of Maryland, Baltimore County
Anthony Metcalf, California State University, San Bernadino
Clara Moore, Franklin and Marshall College
Jeanelle Morgan, University of North Georgia
Jacqueline Bee Nesbit, University of New Orleans
Marcia Newcomer, Louisiana State University
Richard Niesenbaum, Muhlenberg College
John Ophus, University of Northern Iowa
Rebecca Orr, Collin College
Cassandra Osborne, Western State Colorado University
Andrew Palmer, Florida Institute of Technology
Karla Passalacqua, Emory University
Lisa Petrella, Marquette University
Eric Ribbens, Western Illinois University
George Robinson, University at Albany, State University of New York
Nicole M. Roy, Sacred Heart University
Aaron Schrey, Armstrong Atlantic State University
James Schulte, Clarkson University
Sarah Shannon, Indiana University
Frederick Sheldon, Louisiana State University
Bin Shuai, Wichita State University
Nancy Solomon, Miami University of Ohio
Sarah Spinette, Rhode Island College
Robert Sterner, University of Minnesota, Twin Cities
Oliver Sturm, Rhodes College
Timothy Sullivan, Armstrong Atlantic State University
Thomas Terleph, Sacred Heart University
Dustin Vale-Cruz, Springfield College
Kenneth van Golen, University of Delaware
Christine Weilhoefer, University of Portland
Robert Wise, University of Wisconsin, Oshkosh
MaryJo A. Witz, Monroe Community College
John Yoder, University of Alabama, Tuscaloosa
Ximena Valderrrama, Ramapo College of New Jersey

Media and Supplements Authors and Reviewers

Juliann Aukema, University of California, Santa Barbara
Edward Awad, Vanier College
Roberta Batorsky, Rowan University
Alexa Bely, University of Maryland
Joe Bruseo, Holyoke Community College
Shannon Compton, University of Massachusetts, Amherst
Douglas Darnowski, Indiana University Southeast
Ann Marie Davison, Kwantlen Polytechnic University
Donna Francis, University of Massachusetts, Amherst
Emily Gillespie, Marshall University
Richard Gonzalez-Diaz, Seminole State University
Carol Hand, science writer
Phillip Harris, University of Alabama
Laurel Hester, Keuka College
Peggy Hill, science writer
Jerry Johnson, Brigham Young University
Norman Johnson, University of Massachusetts, Amherst
Stephen Kilpatrick, University of Pittsburgh
Kim Lackey, University of Alabama
Marlee Marsh, Columbia College
Betty McGuire, Cornell University
Stephanie Moeckel-Cole, University of Massachusetts, Amherst
Paul Nolan, The Citadel
Kristine Nowak, Kennesaw State University
Don O'Malley, Northeastern University
Lesley Evans Ogden, science writer
Sherry Ogg, Johns Hopkins University
Barbara Peckarsky, University of Wisconsin, Madison
Sarah Shannon, Indiana University
Leo Shapiro, University of Maryland
Molly Solanki, science writer
Nancy Solomon, Miami University
Meredith Stafford, Johns Hopkins University
John Townsend-Mehler, Montana State University
Mary Tyler, University of Maine
Robert Wise, University of Wisconsin, Oshkosh

for *Principles of Life*, SECOND EDITION

LaunchPad (**highschool.bfwpub.com/launchpad/pol2e**) is the easy-to-use online platform that integrates the e-Book, all of the student and instructor media resources, and assessment functions into a unified interface. In addition to a wealth of course management, communication, organization, and gradebook features, LaunchPad includes the following resources.

e-Book

A complete online version of the textbook, the e-Book is fully integrated into LaunchPad and includes media resources, in-text links to all glossary entries (with audio pronunciations), and flexible notes and highlighting features. In addition, instructors can easily hide chapters or sections that they don't cover in their course, rearrange the order of chapters and sections, and add their own content directly into the e-Book.

*LEARNING*Curve

LearningCurve is a powerful adaptive quizzing system with a game-like format to engage students. Rather than simply answering a fixed set of questions, students answer dynamically selected questions to progress toward a target level of understanding. At any point, students can view a report (with links to e-Book sections and media resources) of how well they are performing in each topic, to help them focus on problem areas.

Student Resources

INTERACTIVE SUMMARIES. For each chapter, these dynamic summaries combine a review of important concepts with links to all of the key figures, Activities, and Animated Tutorials.

ANIMATED TUTORIALS. In-depth animations and simulations present complex topics in a clear, easy-to-follow format that includes a brief quiz.

MEDIA CLIPS. New for the Second Edition, these short, engaging video clips depict fascinating examples of some of the many organisms, processes, and phenomena discussed in the textbook.

ACTIVITIES. A range of interactive activities helps students learn and review key facts and concepts through labeling diagrams, identifying steps in processes, and matching concepts.

LECTURE NOTEBOOK. Available online as PDF files, the Lecture Notebook includes all of the textbook's figures and tables, with space for note-taking.

BIONEWS FROM *SCIENTIFIC AMERICAN*. BioNews makes it easy for instructors to bring the dynamic nature of the biological sciences and up-to-the-minute currency into their course via an automatically updated news feed.

BIONAVIGATOR. A unique visual way to explore all of the Animated Tutorials and Activities across the various levels of biological inquiry—from the global scale down to the molecular scale.

ANALYZE THE DATA. Online versions of the Analyze the Data exercises included in many of the Investigation figures in the textbook.

FLASHCARDS AND KEY TERMS. An ideal way for students to learn and review the extensive terminology of introductory biology, featuring a review mode and a quiz mode.

INVESTIGATION LINKS. An overview of the experiments featured in each of the textbook's Investigation figures, with links to the original paper(s), related research or applications that followed, and additional information related to the experiment.

GLOSSARY. The full glossary, with audio pronunciations for all terms.

TREE OF LIFE. An interactive version of the Tree of Life from Appendix A, with links to a wealth of information on each group listed.

MATH FOR LIFE. A collection of mathematical shortcuts and references to help students with the quantitative skills they need in the biology laboratory.

SURVIVAL SKILLS. A guide to more effective study habits, including time management, note-taking, highlighting, and exam preparation.

Assessment Resources

SUMMATIVE QUIZZES. The pre-built summative quizzes assess overall student understanding of each chapter, and provide instructors with data on class and individual-student comprehension of chapter material.

LEARNINGCURVE. Reports provide instructors with instant information on student performance, broken down by individual section.

Strive for a 5: Preparing for the AP® Biology Exam

Franklin Bell, Mercersburg Academy, Mercersburg, PA
John Lepri, University of North Carolina, Greensboro

(ISBN 1-4641-8652-9)

Strive for a 5 is a study guide and test preparation workbook for use throughout the AP® course. Following the textbook chapter by chapter, it reinforces the book's key concepts and focuses on the Big Ideas and Learning Objectives from the AP® Biology Curriculum. The workbook includes four main sections:

SCIENCE PRACTICES. Scientists share seven fundamental methods when performing their work. These include modeling, applying mathematics, posing questions, designing experiments, analyzing data, using explanations and theories, and connecting related knowledge across many domains. These "Science Practices" are enforced throughout the AP® Biology curriculum and are emphasized at the end of each *Strive* chapter in a section of the workbook that provides a thorough explanation about the importance and application of each method.

STUDY GUIDE. The study guide section is presented in chapters that follow the textbook. Each study chapter opens with an overview of the corresponding textbook material. The Big Ideas and Learning Objectives from the AP® Biology Curriculum that are covered in the chapter are clearly stated. The study guide presents each book concept and follows it with a number of free-response exercises. Emphasis on describing, explaining, and analyzing helps students master every concept in preparation for the exam.

TEST PREPARATION. This section of the workbook provides a description of the AP® Biology exam and includes strategies to help students prepare for and take the test. There are study tips for students to apply during the months of preparation, as well as strategies to use on test day. There's even a list of things to avoid while taking the test—things that can waste valuable time or lower a score.

FULL-LENGTH PRACTICE EXAMS. The workbook concludes with two full-length practice exams that are designed to look and feel just like the real AP® exam. (Answer keys are included.) Working through these exams under the appropriate time limits will give students practice—and can help them gain the confidence they need to walk into the exam room feeling fully prepared.

Companion Website

bcs.whfreeman.com/pol2e

The *Principles of Life* Companion Website is available free of charge to students. The site provides access **to all of the media referenced in the textbook links,** as well as a variety of additional resources, including flashcards, study ideas, and more.

Additional Media and e-Book Offerings

For information about additional resources that are available to support your book, please visit the catalog page: highschool.bfwpub.com/pol2e.

Each chapter introduces essential biological concepts and the science that led to our understanding of them. Chapters are designed to help you focus on what's

OPENING STORY & QUESTION

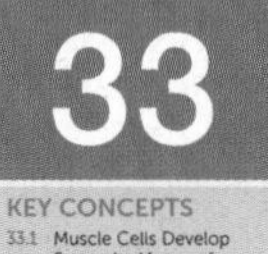

Muscle and Movement

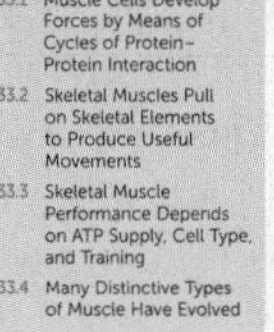

KEY CONCEPTS

33.1 Muscle Cells Develop Forces by Means of Cycles of Protein–Protein Interaction

33.2 Skeletal Muscles Pull on Skeletal Elements to Produce Useful Movements

33.3 Skeletal Muscle Performance Depends on ATP Supply, Cell Type, and Training

33.4 Many Distinctive Types of Muscle Have Evolved

A sphinx moth (*Hemaris thysbe*) uses its flight muscles to hover at a flower, collecting nectar with its long proboscis.

This moth, a type of sphinx or hawk moth, is sometimes mistaken for a hummingbird, accounting for its name, the hummingbird clearwing moth (*Hemaris thysbe*). Common in several parts of the United States and Canada, it feeds during daylight by hovering at flowers using rapid wingstrokes driven by the flight muscles in its thorax. It has frequency whining wingstrokes of mosquitoes. Today, engineers are studying insects to learn more about the aerodynamics of flight. Sphinx moths are of particular interest because they are powerful fliers that can quickly alternate between hovering and flying straight ahead at high speeds.

Muscle cells—one of the defining draw the interest of muscle physiologists because they are among the animals that have the highest frequencies of muscle contraction during flight while retaining this 1:1 ratio. Their wingstrokes can number more than 30 per second (30 Hz).

Per gram, insect flight muscle stands out as one of the tissues that attain the

Chapters begin with an **OPENING STORY** designed to show you how the biology relates to historical, medical, or social issues. Each story ends with an intriguing question.

The **ANSWER** comes at the chapter's conclusion, with references to relevant information and illustrations in the chapter.

Why is it likely that available space inside cells has limited the contents of contractile proteins and mitochondria in high-performance muscles?

You will find the answer to this question on page 697.

Why is it likely that available space inside cells has limited the contents of contractile proteins and mitochondria in high-performance muscles?

ANSWER **FIGURE 33.17** is a highly magnified image of an insect flight muscle cell, obtained by electron microscopy. The inside of the cell is filled almost completely by mitochondria and contractile proteins. Open space is thus a scarce resource. Put simply, a high-performance muscle cell needs as large a set of contractile proteins as possible and as many mitochondria as possible—meaning there is a sort of "competition" for space in which the amounts of contractile proteins and mitochondria are each limited by space shortage. If, over evolutionary time, natural selection started to favor larger numbers of contractile protein molecules, the contractile proteins could edge out some of the mitochondria—jeopardizing the ability of the contractile proteins to get enough ATP. If natural selection started to favor more mitochondria, the mitochondria would edge out contractile proteins—jeopardizing the ability of the cell to use the ATP it could produce. The fact that space is limited has resulted in a sort of compromise in the use of space inside a high-performance muscle cell.

KEY CONCEPTS & CHECKPOINTS

KEY CONCEPTS

33.1 Muscle Cells Develop Forces by Means of Cycles of Protein–Protein Interaction

33.2 Skeletal Muscles Pull on Skeletal Elements to Produce Useful Movements

33.3 Skeletal Muscle Performance Depends

CHECKpoint CONCEPT **33.1**

✓ Imagine planting your feet and trying to push through a concrete wall that's far too heavy to move. As you push, would you describe the associated muscles in your back, arms, and legs as contracting, shortening, lengthening, or a combination of these words? Explain.

✓ In a muscle fiber, how is force development aided by the interdigitated arrangement of actin and myosin filaments?

✓ Describe how the concentration of Ca^{2+} in the sarcoplasmic reticulum of a muscle cell changes before, during, and after the cell is excited by a nerve impulse (action potential).

KEY CONCEPTS begin each chapter.

CHECKpoints revisit the Key Concepts at the end of each section.

LINKS

to both partners. Mutualisms take many forms and involve many kinds of organisms. They also vary in how essential the interaction is to the partners.

LINK

We have seen several examples of mutualisms in this book, including interactions between mycorrhizal fungi and plants (see **Concepts 22.2 and 25.2**); between fungi, algae, and cyanobacteria in lichens (see **Concept 22.2**); and between corals and dinoflagellates (see **Concept 20.4**).

Competition, consumer–resource interactions, and mutualism all affect the fitness of both participants. The other two defined types of interactions affect only one of the participants.

In-text **LINKS** point you to additional discussion of a concept or key term elsewhere in the book.

APPLY THE CONCEPT

APPLY THE CONCEPT

Interactions within and among species affect population dynamics and species distributions

A leaf-cutter ant nest can be considered a community—an ecological system (see Concepts 1.2 and 41.1) in which the species are components that interact with one another. Major components of the system are shown as labeled boxes, and their interactions as arrows between those boxes.

Use the description of interactions within leaf-cutter ant nests in the opening story of this chapter to answer the following questions:

1. What is the sign of the following direct effects of each species on another?

Ants on fungus	Fungus on ants
Fungus on mold	Mold on fungus
Mold on bacteria	Bacteria on mold
Bacteria on ants	Ants on bacteria

2. Explain for each interaction the mechanism by which the fitness of interacting individuals is affected.

3. To which of the five categories of interspecific interactions does each pairwise interaction belong?

4. Explain how the spatial distribution of the green mold *Escovopsis* might affect the spatial distribution of leaf-cutter ant colonies.

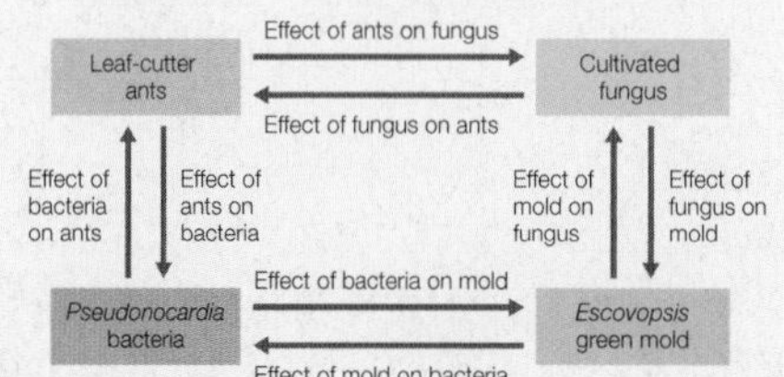

APPLY THE CONCEPT exercises ask you to use a concept in a real-world setting to interpret actual research data and draw your own conclusions.

important, and they offer a number of ways to analyze and review what you've read as you prepare for class or exams.

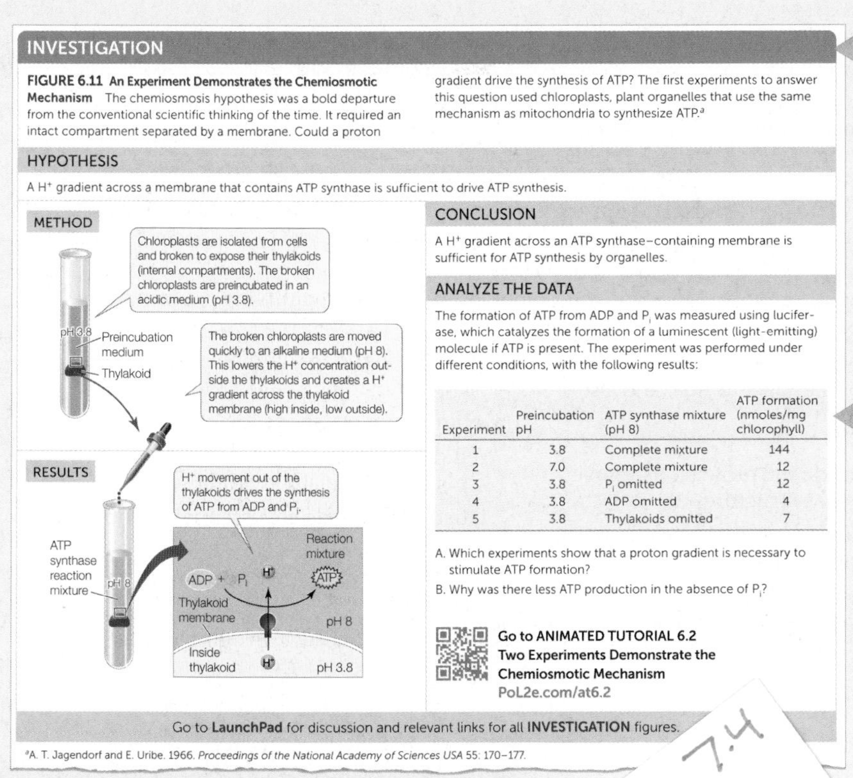

INVESTIGATION

INVESTIGATION figures emphasize the process of scientific inquiry to give you a realistic sense of how science is done. Each Investigation figure is organized in order of **hypothesis**, **method**, **results,** and **conclusion** and cites the original research paper(s).

ANALYZE THE DATA

Most **INVESTIGATION** figures are followed by **ANALYZE THE DATA** problems, which ask you to work with data from published biological research and make your own connections between observations, analyses, hypotheses, and conclusions.

INSTANT ACC

QUICK RESPONSE (QR) CODES and **DIRECT W**
grated into the text link you immediately to
clips, and activities. Just scan the code w
type the short Web address into any
(Free QR reader apps are availab

Go to MEDIA CLIP 23.7
Octopuses Can Pass through Small Openings
PoL2e.com/mc23.7

HELPFUL ART

Numbered **BALLOON C** ns make it easy to follow key processes step

MMARY

Three Domains That
mon Ancestor

omains, Bacteria and Archaea, are **prokaryotic**.
guished from Eukarya in several ways, including their
cleus and of membrane-enclosed organelles. **Review**
.1

karyotes are related to both Archaea and Bacteria and appear to have formed through endosymbiosis between members of these two lineages. The last common ancestor of all three domains probably lived about 3 billion years ago. **Review Figure 19.1 and ANIMATED TUTORIAL 19.1**

CHAPTER SUMMARIES provide a thorough review of chapter content, including key figures, and references to supporting online resources, including Animated Tutorials and Activities.

Table of Contents

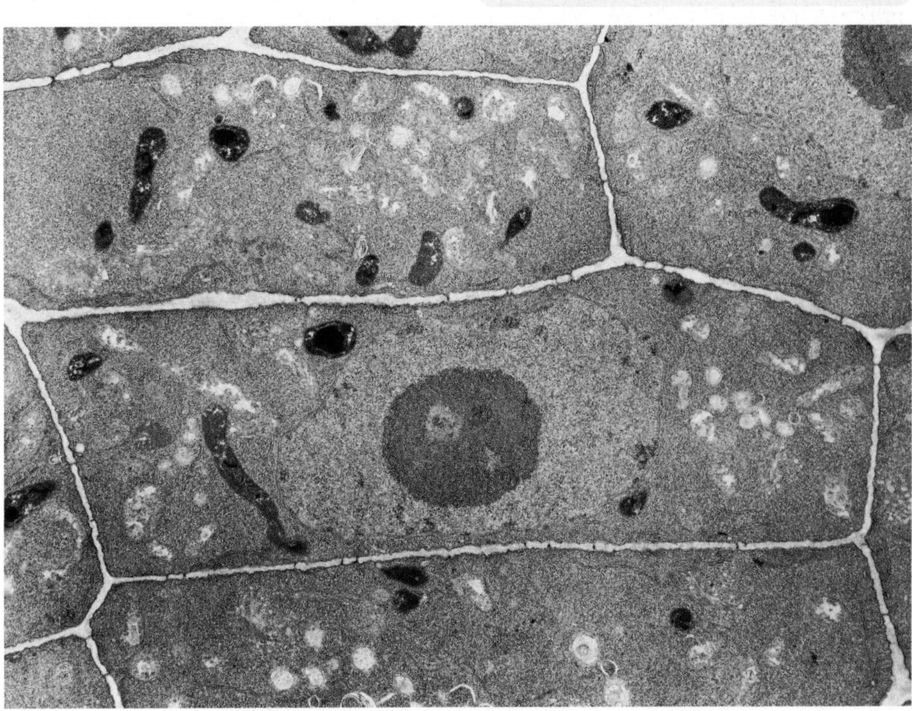

PART 2 Genetics

8 Inheritance, Genes, and Chromosomes 150

9 DNA and Its Role in Heredity 172

10 From DNA to Protein: Gene Expression 194

11 Regulation of Gene Expression 215

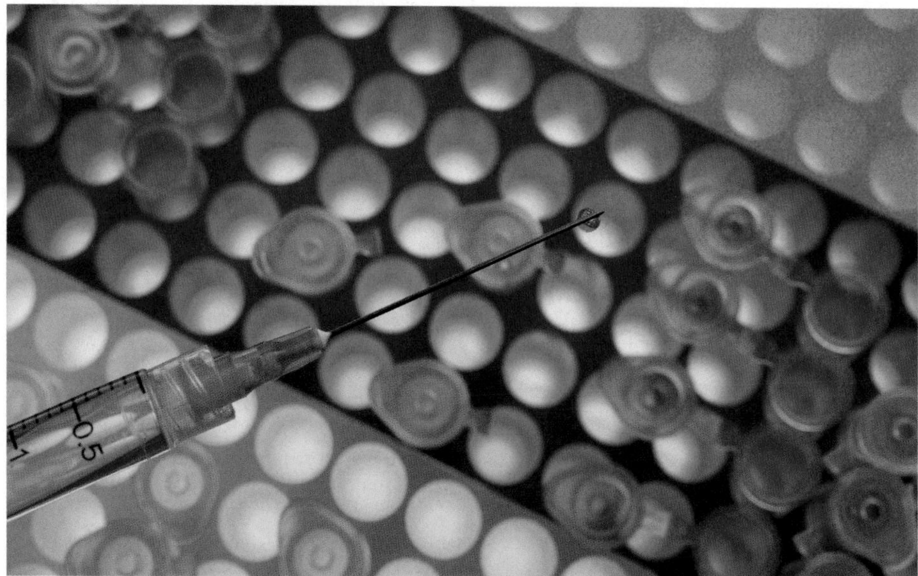

12 Genomes 234

13 Biotechnology 253

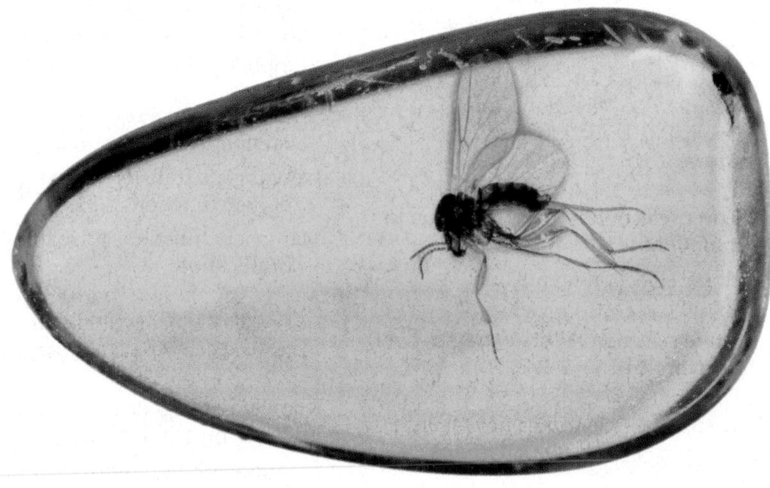

PART 4 Diversity

PART 6 Animal Form and Function

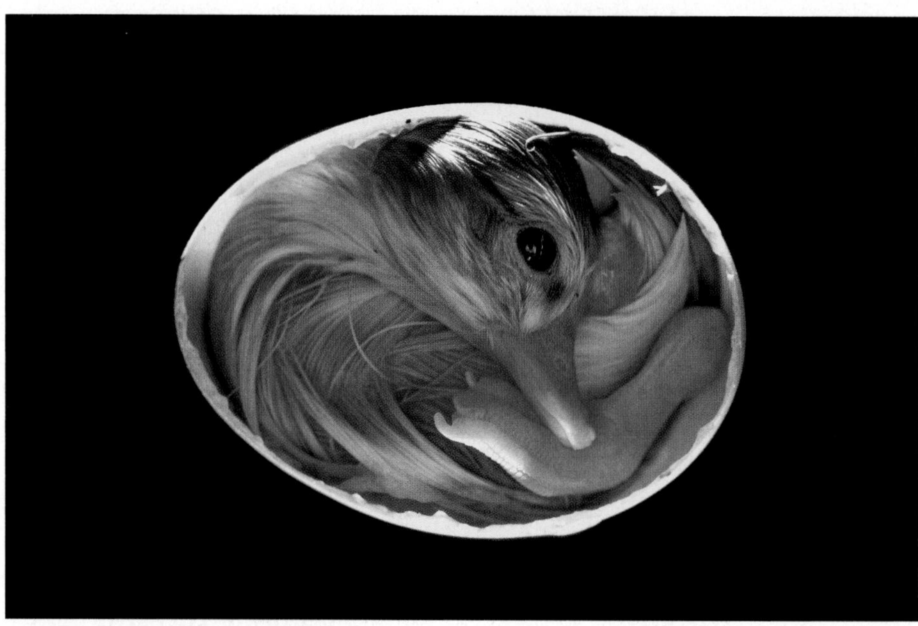

PART 7 Ecology

1 Principles of Life

KEY CONCEPTS

1.1 Living Organisms Share Common Aspects of Structure, Function, and Energy Flow

1.2 Life Depends on Organization and Energy

1.3 Genetic Systems Control the Flow, Exchange, Storage, and Use of Information

1.4 Evolution Explains the Diversity as Well as the Unity of Life

1.5 Science Is Based on Quantitative Observations, Experiments, and Reasoning

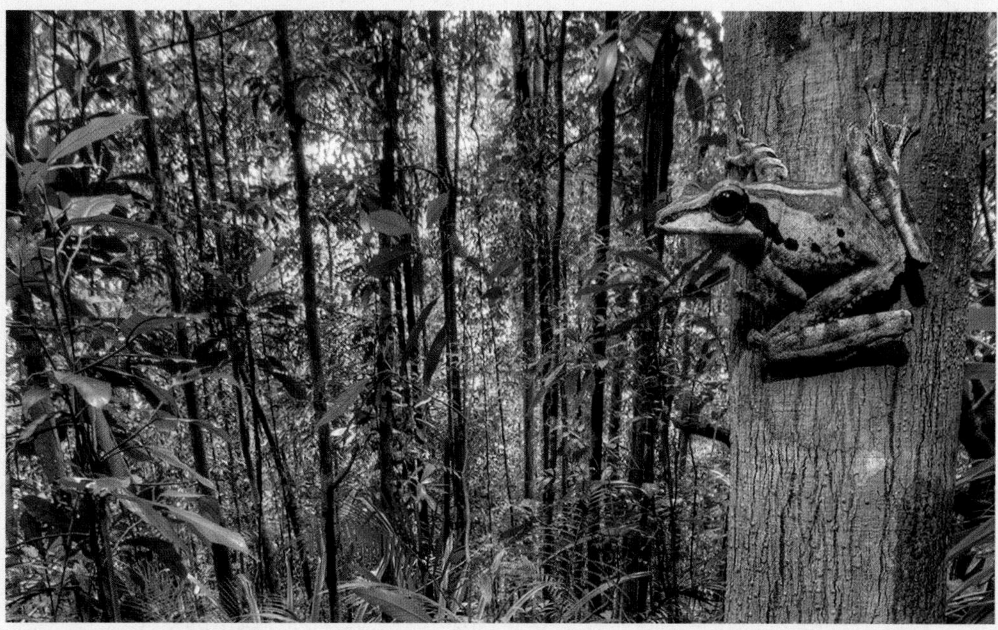

What principles of life are illustrated in this scene?

When you take a walk through the woods and fields or a park near your home, what do you see? Like most people, you probably notice the trees, colorful flowers, and some animals. But do you spend more than a little time thinking about how these living things survive, reproduce, interact with one another, or affect their environment? With the introduction to biology in this book, we would like to inspire you to ask questions about what life is, how living systems work, and how the living world came to be as we observe it today.

Biologists have amassed a huge amount of information about the living world, and some introductory biology classes focus on memorizing these details. In this book we take a different approach, focusing on the major principles of life that underlie everything in biology.

What do we mean by "principles of life"? Look at the photograph. Why is the view so overwhelmingly green? A fundamental principle of life, namely that all living organisms require energy to grow, move, reproduce, and maintain their bodies, can explain the color. Ultimately, most of that energy comes from the sun. The leaves of plants contain chlorophyll, a green pigment that captures energy from the sun and uses it to transform water and carbon dioxide into sugar and oxygen (in the process called photosynthesis). That sugar stores some of the energy from the sun in its chemical bonds. The plant, or other organisms that eat the plant, can then obtain energy by breaking down the sugar. The frog in the photo used energy to climb up the tree. That energy came from molecules in the bodies of insects eaten by the frog. The insects, in turn, had built up their bodies by ingesting tissues of plant leaves, which grew by using some of the sun's energy captured through photosynthesis. The frog, like the plants, is ultimately solar-powered, as is the human observer who took this photograph.

The photograph also illustrates other principles of biology. One is that living organisms often survive and thrive by interacting with one another in complex ways. You probably noticed the frog and the trees. But did you notice the patches of growth on the trunk of the tree? Most of those are lichens, a complex interaction between a fungus and a photosynthetic organism (in this case, a kind of alga). In lichens, the fungus and the alga depend on each other for survival. Many other organisms in this scene are too small to be seen, but they are critical components for keeping this living system functioning over time.

After reading this book, you should understand the main principles of life. You'll be able to describe how organisms capture and transform energy; pass genetic information to their offspring in reproduction; grow, develop, and behave; and interact with other organisms and with their physical environment. You will also have learned how this system of life on Earth evolved, and how it continues to change. May a walk in the park never be the same for you again!

CONCEPT 1.1 Living Organisms Share Common Aspects of Structure, Function, and Energy Flow

Biology is the scientific study of life, which encompasses all living things, or **organisms**. The living things we know about are all descended from a single-celled ancestor that lived on Earth almost 4 billion years ago. We can imagine that something with some similarities to life as we know it might have originated differently, perhaps on other planets. But the evidence suggests that all of life on Earth today has a single origin—a single common ancestor—and we consider all the organisms that descended from that common ancestor to be a part of life.

Life as we know it had a single origin

The overwhelming evidence for the common ancestry of life lies in the many characteristics that are shared among living organisms. Typically, living organisms

- are composed of a common set of chemical parts, such as nucleic acids (one example is DNA, which is the important molecule that carries our genetic information) and amino acids (the chemical building blocks that make up proteins), and similar structures, such as cells enclosed within membranes
- depend on intricate interactions among structurally complex parts to maintain the living state
- contain genetic information that uses a nearly universal code to specify how proteins are assembled
- convert molecules obtained from their environment into new biological molecules
- extract energy from the environment and use it to carry out life functions
- replicate their genetic information in the same manner when reproducing themselves
- share structural similarities among a fundamental set of genes
- evolve through gradual changes in their genetic information

Taken together, these shared characteristics logically lead to the conclusion that all life has a common ancestry, and that the diverse organisms that exist today originated from one life form. If life had multiple origins, there would be little reason to expect a nearly universal genetic code, or the similarities among many genes, or a common set of amino acids. If we were to discover something similar to life, such as a self-replicating system that originated independently on another planet, we would expect it to be fundamentally different in these aspects. It might be similar in some ways to life on Earth, such as using genetic information to reproduce. But we would not expect the details of its genetic code, for example, to be like ours.

The simple list of shared characteristics above, however, does not describe the incredible complexity and diversity of life. Some forms of life may not even display all of these characteristics all of the time. For example, the seed of a desert plant may exist for many years in a dormant state in which it doesn't extract energy from the environment, convert molecules, or reproduce. Yet the seed is alive.

And then there are viruses, which are not composed of cells and cannot carry out physiological functions on their own. Instead they use the cells they invade to perform these functions for them. Yet viruses contain genetic information, and they mutate and evolve. So even though viruses are not independent cellular organisms, their existence depends on cells, and there is strong evidence that viruses evolved from cellular life forms. For these reasons, most biologists consider viruses to be a part of life. But as viruses illustrate, the boundaries between "living" and "nonliving" are not always clear, and all biologists do not agree exactly on where we should draw the lines.

Major steps in the history of life are compatible with known physical and chemical processes

Geologists estimate that Earth formed between 4.6 and 4.5 billion years ago. At first the planet was not a very hospitable place. It was some 600 million years or more before the earliest life evolved. If we picture the history of Earth as a 30-day month, with each day representing about 150 million years, life first appeared somewhere toward the end of the first week (**FIGURE 1.1**).

How might life have arisen from nonliving matter? In thinking about this question, we must take into account that the young Earth's atmosphere, oceans, and climate all were very different than they are today. Biologists have conducted many experiments that simulate the conditions on early Earth. These experiments have confirmed that the formation of complex organic molecules under such conditions is possible, even probable.

The critical step for the evolution of life, however, was the appearance of **nucleic acids**—molecules that could reproduce themselves and also contain the information for the synthesis, or manufacture, of large molecules with complex but stable shapes. These large, complex molecules were proteins. Their shapes varied enough to enable them to participate in increasing numbers and kinds of chemical reactions with other molecules.

CELLULAR STRUCTURE EVOLVED IN THE COMMON ANCESTOR OF LIFE In the next big step in the origin of life, a membrane surrounded and enclosed complex proteins and other biological molecules, forming a tiny **cell**. This membrane kept the enclosed components separate from the surrounding external environment. Molecules called fatty acids played a critical role because these molecules form membrane-like films instead of dissolving in water. When agitated, these films can form hollow spheres, which could have enveloped assemblages of biological molecules. The creation of a cell interior, separate from the external environment, allowed the reactants and products of chemical reactions to be concentrated, opening up the possibility that those reactions could be integrated and controlled. This natural process of membrane formation likely resulted in the first cells with the ability to reproduce—the evolution of the first cellular organisms.

For more than 2 billion years after cells originated, every organism consisted of only one cell. These first organisms were **prokaryotes**, which are made up of single cells containing genetic material and other biochemical structures enclosed in a membrane (**FIGURE 1.2**). Vast numbers of their descendants, such as bacteria, exist in similar form today. Early prokaryotes were confined to the oceans, which had an abundance of complex molecules they could use as raw materials and sources of energy. The oceans also shielded them from the damaging effects of ultraviolet (UV) light, which was intense at that time because there was little or no oxygen (O_2) in the atmosphere, and for that reason, no protective ozone (O_3) layer in the upper atmosphere.

PHOTOSYNTHESIS ALLOWED LIVING ORGANISMS TO CAPTURE THE SUN'S ENERGY To fuel the chemical reactions inside them, the earliest prokaryotes took in molecules directly from their environment and broke down these small molecules to release and use the energy contained in their chemical bonds. Many modern prokaryotes still function this way, and very successfully.

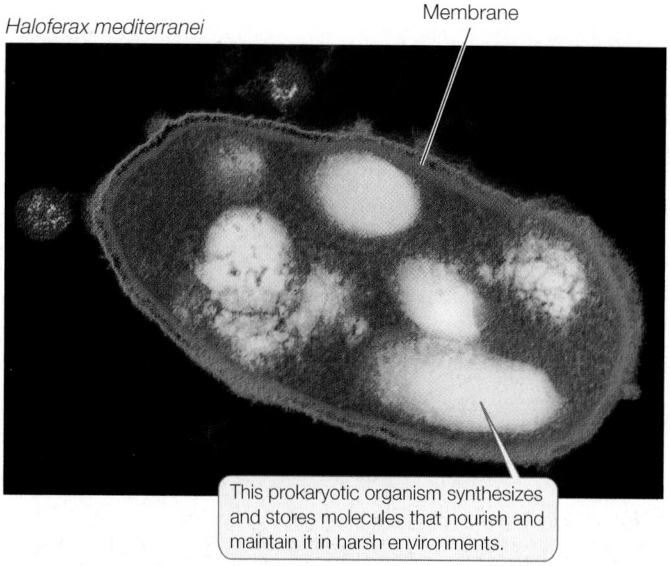

Haloferax mediterranei

Membrane

This prokaryotic organism synthesizes and stores molecules that nourish and maintain it in harsh environments.

FIGURE 1.2 The Basic Unit of Life Is the Cell The concentration of reactions within the enclosing membrane of a cell allowed the evolution of integrated organisms. Today all organisms, even the largest and most complex, are made up of cells. Single-celled organisms such as this one, however, remain the most abundant living organisms (in absolute numbers) on Earth.

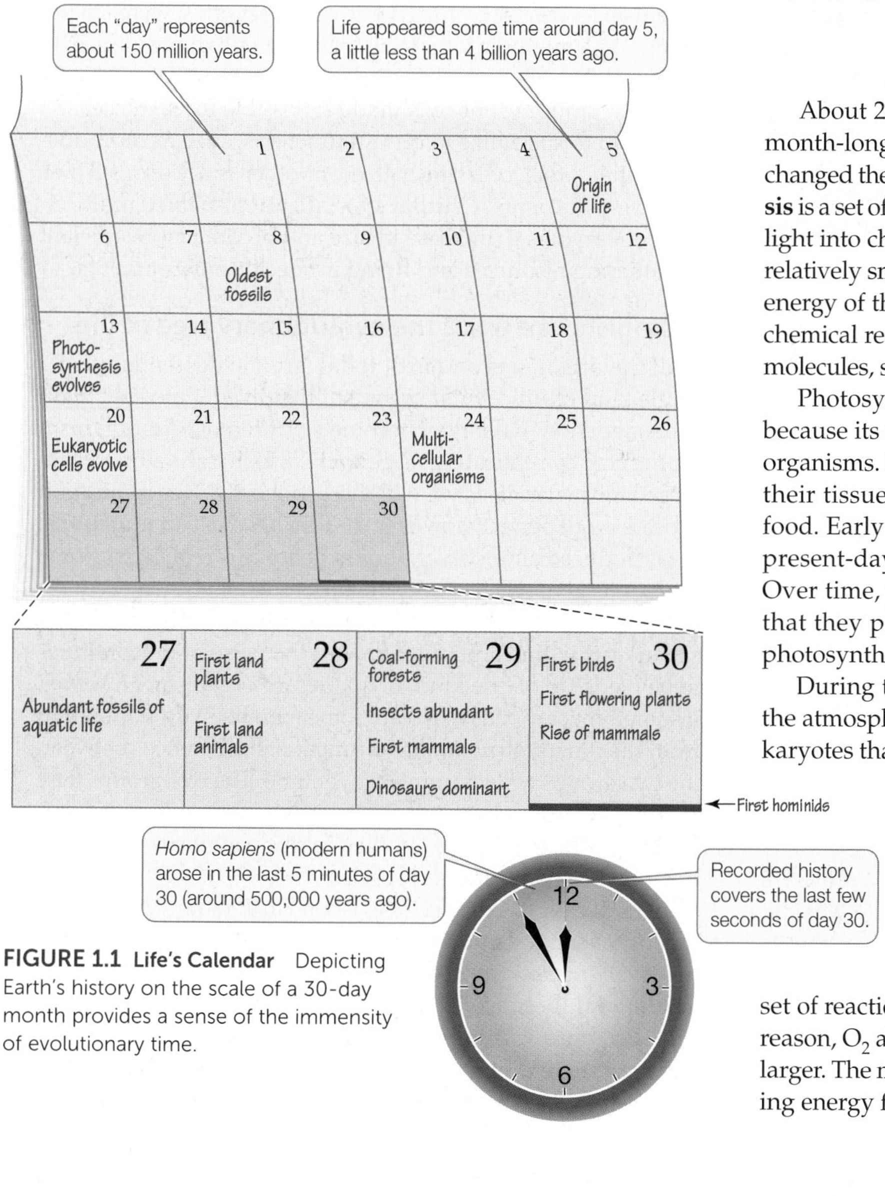

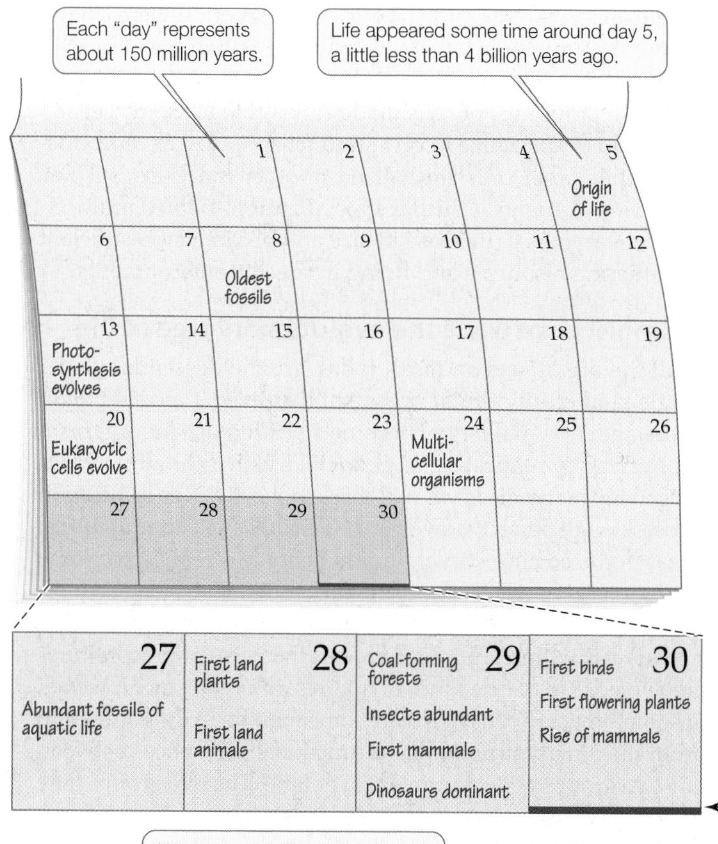

FIGURE 1.1 Life's Calendar Depicting Earth's history on the scale of a 30-day month provides a sense of the immensity of evolutionary time.

About 2.7 billion years ago, or on day 13 of our imaginary month-long calendar of life, the emergence of photosynthesis changed the nature of life on Earth (see Figure 1.1). **Photosynthesis** is a set of chemical reactions that transforms the energy of sunlight into chemical-bond energy of the sugar glucose and other relatively small biological molecules. In turn, the chemical-bond energy of these small molecules can be tapped to power other chemical reactions inside cells, including the synthesis of large molecules, such as proteins, that are the building blocks of cells.

Photosynthesis is the basis of much of life on Earth today because its energy-capturing processes provide food for other organisms. Photosynthetic organisms use solar energy to build their tissues, and then other organisms use those tissues as food. Early photosynthetic cells were probably similar to the present-day prokaryotes called cyanobacteria (**FIGURE 1.3**). Over time, photosynthetic prokaryotes became so abundant that they produced vast quantities of O_2 as a by-product of photosynthesis.

During the early eons of life on Earth, there was no O_2 in the atmosphere. In fact, O_2 was poisonous to many of the prokaryotes that lived at that time. But those organisms that tolerated O_2 were able to proliferate as O_2 slowly began to accumulate in the atmosphere. The presence of O_2 opened up vast new avenues of evolution. **Aerobic metabolism**, a set of chemical reactions that releases energy from life's molecules by using O_2, proved to be more efficient than **anaerobic metabolism**, a set of reactions that extracts energy without using O_2. For this reason, O_2 allowed organisms to live more intensely and grow larger. The majority of living organisms today use O_2 in extracting energy from molecules.

(A)

(B)

FIGURE 1.3 Photosynthetic Organisms Changed Earth's Atmosphere Cyanobacteria were the first photosynthetic organisms on Earth. (A) Colonies of cyanobacteria called stromatolites are known from the ancient fossil record. (B) Living stromatolites are still found in suitable environments on Earth today.

Oxygen in the atmosphere also made it possible for life to move onto land. For most of life's history, UV radiation falling on Earth's surface was so intense that it destroyed any living cell that was not well shielded by water. But as a result of photosynthesis, O_2 accumulated in the atmosphere for more than 2 billion years and gradually resulted in a layer of ozone in the upper atmosphere. By about 500 million years ago, or about day 28 on our imaginary calendar of life, the ozone layer was sufficiently dense and absorbed enough of the sun's UV radiation to make it possible for organisms to leave the protection of the water and live on land (see Figure 1.1).

EUKARYOTIC CELLS AROSE THROUGH ENDOSYMBIOSIS Another important, earlier step in the history of life was the evolution of cells with membrane-enclosed compartments called **organelles**. Organelles were—and are—important because specialized cellular functions could be performed inside them, separated from the rest of the cell. The first organelles probably appeared about 2.5 billion years after life first appeared on Earth, or about day 20 on Figure 1.1. One of these organelles, the **nucleus**, came to contain the cell's genetic information. The nucleus (Latin *nux*, "nut" or "core") gives these cells their name: **eukaryotes** (Greek *eu*, "true"; *karyon*, "kernel" or "core"). The eukaryotic cell is distinct from the cells of prokaryotes (*pro*, "before"), which lack nuclei and other internal compartments.

Some organelles are hypothesized to have originated by **endosymbiosis**, which means "living inside another" and may have occurred when larger cells ingested smaller ones. The **mitochondria** that release energy for use by a eukaryotic cell probably evolved from engulfed prokaryotic organisms. And **chloroplasts**—the organelles specialized to conduct photosynthesis in eukaryotic photosynthetic organisms—could have originated when larger eukaryotes ingested photosynthetic prokaryotes. If the larger cell failed to break down this intended food object, a partnership could have evolved in which the ingested prokaryote provided sugars from photosynthesis and the host cell provided a good environment for its smaller partner.

MULTICELLULARITY ALLOWED SPECIALIZATION OF TISSUES AND FUNCTIONS For the first few billion years of life, all organisms—whether prokaryotic or eukaryotic—were single-celled. At some point, the cells of some eukaryotes failed to separate after cell division and remained attached to each other. These groupings of cells made it possible for some cells in the group to specialize in certain functions, such as reproduction, while other cells specialized in other functions, such as absorbing nutrients. **Cellular specialization** enabled multicellular eukaryotes to increase in size and become more efficient at gathering resources and living in specific environments.

Biologists can trace the evolutionary tree of life

If all the organisms on Earth today are the descendants of a single kind of unicellular organism that lived almost 4 billion years ago, how have they become so different? An organism reproduces by replicating its **genome**, which is the sum total of its genetic material, as we will discuss shortly. This replication process is not perfect, however, and changes, called **mutations**, are introduced almost every time a genome is replicated. Some mutations give rise to structural and functional changes in organisms. As individuals mate with one another, these changes can spread within a population, but the population continues to be made up of one kind, or species, of organism. However, if something happens to isolate some members of a population from the others, structural and functional differences between the two groups will accumulate over time. The two groups may eventually differ enough that their members no longer regularly reproduce with one another. In this way the two populations become two different species.

Tens of millions of species exist on Earth today. Many times that number lived in the past but are now extinct. As biologists discover species, they give each one a scientific name called a **binomial** (because it is made up of two Latinized words). The first word identifies the species' genus—a group of species that share a recent common ancestor. The second word indicates the species. For example, the scientific name for the

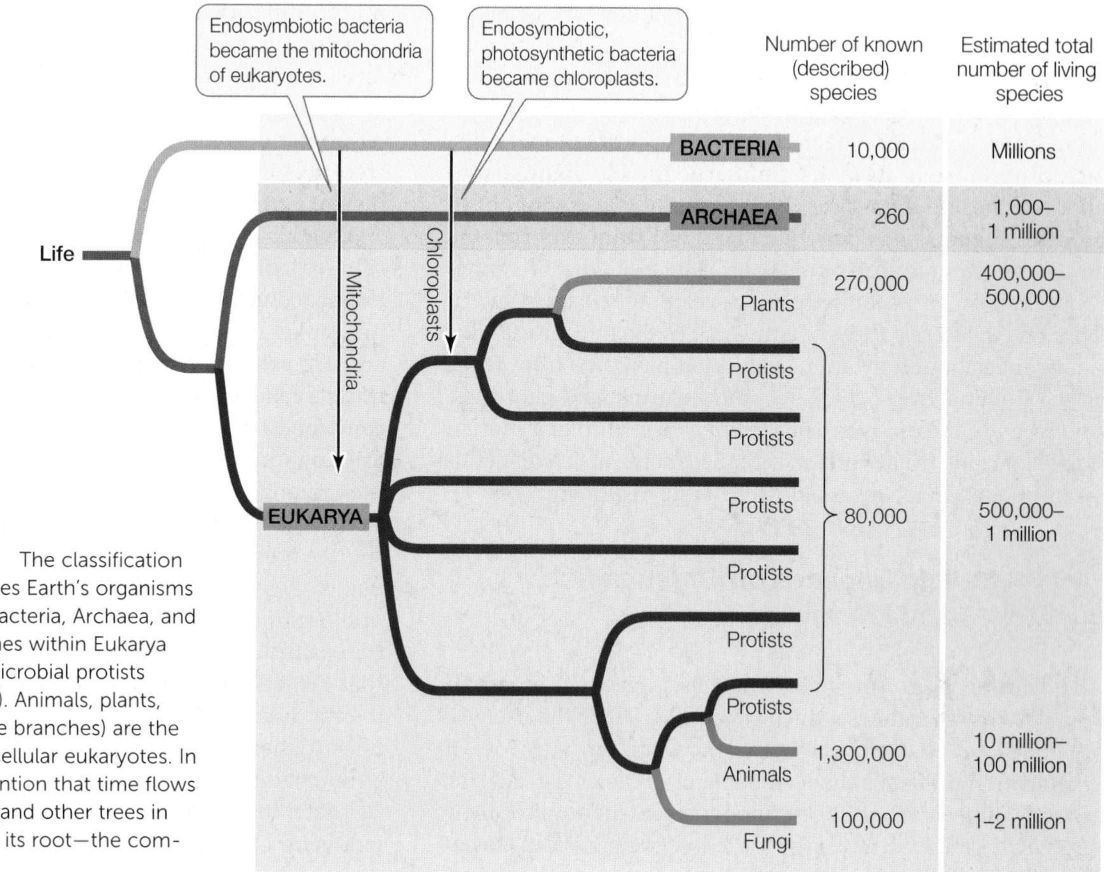

FIGURE 1.4 The Tree of Life The classification system used in this book divides Earth's organisms into three primary domains: Bacteria, Archaea, and Eukarya. The dark blue branches within Eukarya represent various groups of microbial protists (mostly unicellular eukaryotes). Animals, plants, and fungi (green and turquoise branches) are the most familiar groups of multicellular eukaryotes. In this book we adopt the convention that time flows from left to right, so this tree (and other trees in this book) lies on its side, with its root—the common ancestor—at the left.

human species is *Homo sapiens*: *Homo* is our genus and *sapiens* our species. *Homo* is Latin for "man," and *sapiens* is from the Latin word for "wise" or "rational." Our closest relatives in the genus *Homo* are the Neanderthals (*Homo neanderthalensis*), which are now extinct and are known only from fossil remains.

Much of biology is based on comparisons among species. Our ability to make relevant comparisons has improved greatly in recent decades as a result of our relatively newfound ability to study and compare the genomes of different species. We do this by sequencing a genome (in whole or in part), which means we can determine the order of the nucleotides that serve as the building blocks of the organism's DNA. Genome sequencing and other molecular techniques have allowed biologists to add a vast array of molecular evidence to existing evolutionary knowledge based on the fossil record. The result is the ongoing compilation of **phylogenetic trees** that document and diagram evolutionary relationships as part of an overarching **tree of life.** The broadest categories of this tree are shown in **FIGURE 1.4.** (The tree is expanded in Appendix A, and you can also explore the tree interactively online.)

Many details remain to be clarified, but the broad outlines of the tree of life have been determined. Its branching patterns are based on a rich array of evidence from fossils, structures, chemical processes, behavior, and molecular analyses of genomes. Molecular data in particular have been used to separate the tree into three major branches called **domains**: Archaea, Bacteria, and Eukarya. The organisms of each domain have been evolving separately from those in the other domains for more than a billion years. Note that all organisms that are alive today descended from common ancestors in the past. In other words, living species did not evolve from other species living today. Rather, all living organisms evolved from now-extinct common ancestors. For example, humans did not evolve from our close relatives, the chimpanzees, but humans and chimpanzees both evolved from a common (now extinct) ancestral species.

Organisms in the domains **Archaea** and **Bacteria** are single-celled prokaryotes. However, members of these two groups differ so fundamentally that they are thought to have separated into distinct evolutionary lineages very early. Species belonging to the third domain—**Eukarya**—have eukaryotic cells whose mitochondria and chloroplasts originated from endosymbioses with bacteria, as we have described.

Plants, fungi, and animals are examples of familiar multicellular eukaryotes. We know that multicellularity arose independently in each of these three multicellular groups because they are each most closely related to different groups of unicellular eukaryotes (commonly called protists), as you can see from the branching pattern of Figure 1.4.

Life's unity allows discoveries in biology to be generalized

Knowledge gained from investigations of one kind of organism can, with care, be generalized to other organisms because all life is related by descent from a common ancestor, shares a genetic code, and consists of similar molecular building blocks. Biologists use certain **model organisms** for research, knowing they can extend their findings to other organisms, including humans.

Our basic understanding of the chemical reactions in cells came from research on bacteria but is applicable to all cells, including those of humans. Similarly, the biochemistry of photosynthesis—the process by which plants use sunlight to produce sugars—was largely worked out from experiments on *Chlorella*, a unicellular green alga. Much of what we know about the genes that control plant development is the result of work on *Arabidopsis thaliana*, a member of the mustard family. Knowledge about how animals develop has come from work on sea urchins, frogs, chickens, roundworms, and fruit flies. And recently, the discovery of a major gene controlling human skin color came from work on zebrafish. Being able to generalize from model systems is a powerful tool in biology.

CONCEPT 1.2 Life Depends on Organization and Energy

All of life depends on organization. Physics gives us the second law of thermodynamics, which states that, left to themselves, organized entities tend to become more random. Any loss of organization threatens the well-being of organisms. Cells, for example, must combat the thermodynamic tendency for their molecules, structures, and systems to lose organization—to become disorganized. Energy is required to maintain organization. For this reason, cells require energy throughout their lives.

Organization is apparent in a hierarchy of levels from molecules to ecosystems

Cells synthesize, or manufacture, proteins and other complex molecules by assembling atoms into new, highly organized configurations. Such complex molecules give cells their structure and enable them to function. For example, a fatty acid molecule that the cell synthesizes may become part of a membrane that structures the inside of the cell by dividing it into compartments. Or a protein made by a cell may enable a specific chemical reaction to take place in the cell by helping start or speed up the reaction—that is, by acting as a catalyst for the reaction.

Organization is also essential for many cells to function together in a multicellular organism. As we have seen, multicellularity allows individual cells to specialize and depend on other cells for functions they themselves do not perform. But the different specialized cells also work together. For example, division of labor in a multicellular organism usually requires a circulatory system so that the functions of specialized cells in one part of the body are of use to cells in other, distant parts of the body.

Overall, a multicellular organism exhibits many hierarchical levels of organization (**FIGURE 1.5A**). Small molecules are organized into larger ones, such as DNA and proteins. Large molecules are organized into cells, and assemblages of differentiated cells are organized into **tissues**. For example, a single muscle cell cannot generate much force, but when many cells combine to form the tissue of a working muscle, considerable force and movement can be generated. Different tissue types are organized to form **organs** that accomplish specific functions. The heart, brain, and stomach are each constructed of several types of tissues, as are the roots, stems, and leaves of plants. Organs whose functions are interrelated can be grouped into **organ systems**; the esophagus, stomach, and intestines, for example, are all part of the digestive system. Because all these levels of organization are subject to the second law of thermodynamics, they all tend to degrade unless energy is applied to the system. This is why an organism must use energy to maintain its functions.

Matching the internal hierarchy of an individual organism is an external hierarchy in the larger biological world where organisms interact with their physical environment—an **ecological system**, often shortened to **ecosystem** (**FIGURE 1.5B**). Individual organisms interacting with their immediate

FIGURE 1.5 Life Consists of Organized Systems at a Hierarchy of Scales (A) The hierarchy of systems within a multicellular organism. DNA—a molecule—encodes the information for cells—a higher level of organization. Cells, in turn, are the components of still higher levels of organization: tissues, organs, and the organism itself. (B) Organisms interacting with their external environment form ecological systems on a hierarchy of scales. Individual organisms form the smallest ecological system. Individuals of a species form populations, which interact with other populations to form communities. Multiple communities in turn interact within landscapes at progressively larger scales until they include all the landscapes and organisms of Earth: the entire biosphere.

Go to ACTIVITY 1.1 The Hierarchy of Life
PoL2e.com/ac1.1

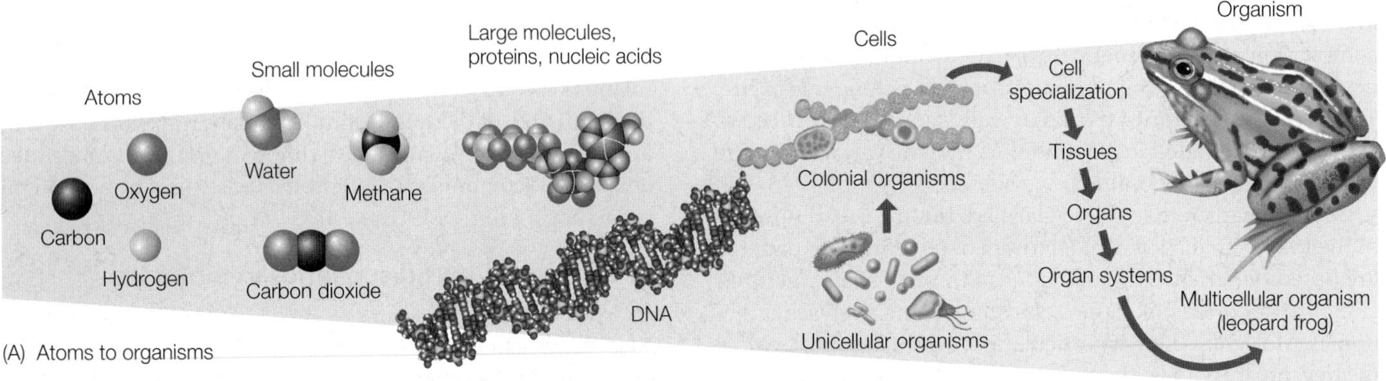

environment form the smallest ecological system. Groups of individuals of any one species live together and interact in **populations**, and populations of different species that live and interact in a single area form ecological **communities**. Multiple communities interact within **landscapes**. The landscape of the entire Earth and all its life is known as the **biosphere**.

But there are some important differences between biological systems at the organismal level and these larger scales. All the hierarchical levels of organization within an individual organism are encoded by its single genome, so that these levels generally interact harmoniously. By contrast, the external hierarchy of populations, communities, and landscapes involves interactions among multiple species with multiple genomes, so that interactions are not always harmonious. For example, individuals often prevent others of their own species from exploiting a necessary resource such as food, or they exploit members of their own or different species as food.

Each level of biological organization consists of systems

We have already discussed organ systems and ecological systems. More generally, a **system** is a set of interacting parts in which neither the parts nor the whole can be understood without taking into account the interactions. A simple biological system might consist of a few **components** (e.g., proteins, pools of nutrients, or organisms) and the **processes** by which the components interact (e.g., protein synthesis, nutrient metabolism, or grazing) (**FIGURE 1.6**).

Consider, for example, the system within a cell that synthesizes and controls the quantity of a particular protein, which we'll call Protein T (**FIGURE 1.7A**). The components of the system are the amino acids from which Protein T is made, Protein T, and the breakdown products of Protein T. The processes are the biochemical pathways that synthesize and break down Protein T. To understand how the cell controls the amount of Protein T, we must understand how all the other components and processes in this system function.

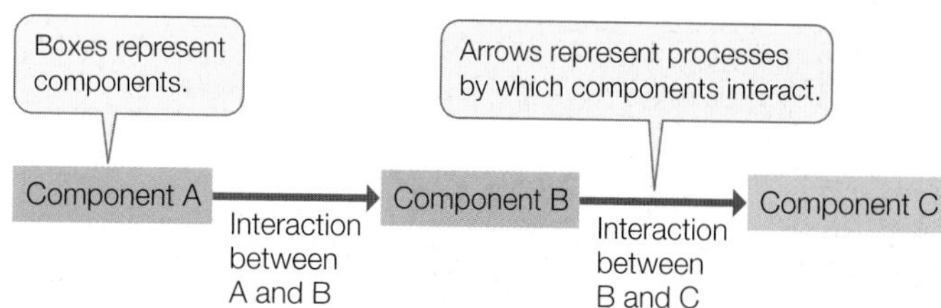

FIGURE 1.6 A Generalized System Systems in cells, whole organisms, and ecosystems can be represented with boxes and arrows.

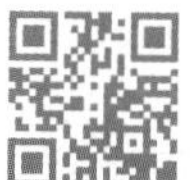

Go to ANIMATED TUTORIAL 1.1
System Simulation
PoL2e.com/at1.1

Systems are found at every level of biological organization. For example, our bodies have a physiological system that controls the amount of sodium (Na^+) in our body fluids (**FIGURE 1.7B**). Grass, voles, and predators (foxes and owls) are components of a community-level system (**FIGURE 1.7C**).

Biological systems are highly dynamic even as they maintain their essential organization

Given the central importance of organization, you might think that biological systems are inflexible and static. Actually, they are often incredibly dynamic—characterized by rapid flows of matter and energy. On average, for example, a cell in your body breaks down and rebuilds 2–3 percent of its protein molecules per day. Each day it also makes and uses more than 100,000 trillion (10^{14}) molecules of adenosine triphosphate (ATP), the molecule responsible for shuttling energy from sources to uses. Collectively, all the cells in your body liberate more than 90 grams of hydrogen every day from the foods they break down to obtain energy. Your cells also combine that hydrogen with oxygen (O_2) to make almost a liter of water every day.

This dynamic aspect of biological systems means that they constantly exchange energy and matter with their surroundings. For example, even after a single-celled or multicellular

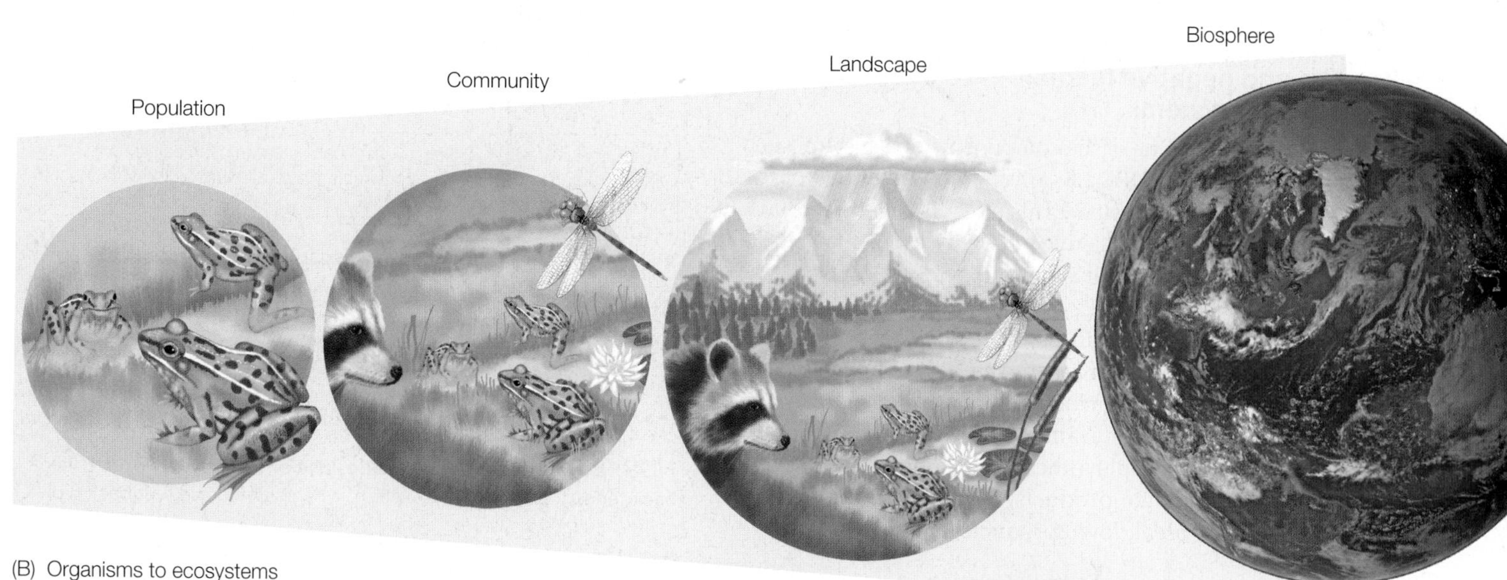

(B) Organisms to ecosystems

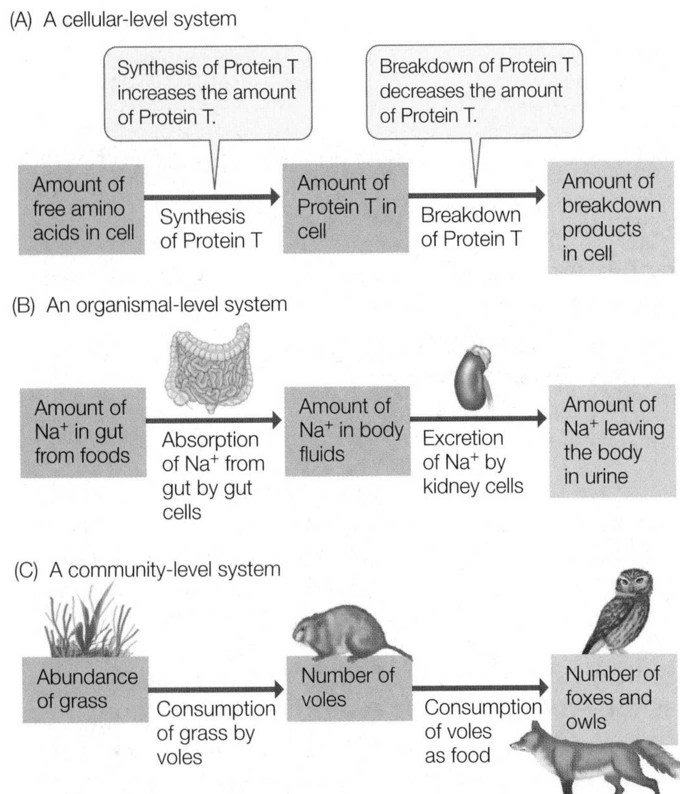

FIGURE 1.7 Organized Systems Exist at Many Levels (A) This cellular-level system synthesizes and breaks down a cell protein called Protein T. (B) This organismal-level system determines the amount (and thus the concentration) of sodium (Na^+) in the blood plasma and other extracellular body fluids of a human. (C) This community-level system helps determine the number of meadow voles (*Microtus pennsylvanicus*) in a field in the spring.

organism has reached maturity, most of its molecules are steadily replaced. In this ceaseless, dynamic process, atoms are lost from the cells in the organism to the surrounding soil, air, or water, and they are replaced with atoms from the soil, air, or water. Yet as the atomic building blocks of any particular cell come and go, the organization of the molecules, structures, and systems in the cell persists. This fact emphasizes the central importance of organization.

Positive and negative feedback are common in biological systems

Often, the amount of one of the components of a system, such as component C in **FIGURE 1.8**, affects the rate of one of the earlier processes in the system. This effect is called **feedback** and may be described as positive or negative. Feedback is often diagrammed simply with a line and symbol, but its actual mechanism may be complex.

Positive feedback occurs in a system when a product of the system *speeds up* an earlier process. The effect of positive feedback is to cause the product to be produced faster and faster. To return to one of our earlier examples, if the breakdown products of Protein T sped up synthesis of Protein T, this would lead to more breakdown products, then even more Protein T, then even more breakdown products, and so on. Positive feedback tends to destabilize a system, but destabilization can sometimes be advantageous, provided it is ultimately brought under control.

Negative feedback occurs when a product of a system *slows down* an earlier process in the system. Often, as the product increases in amount or concentration, it exerts more and more of a slowing effect. Negative feedback stabilizes the amount of the product in this way: if a high amount of the product accumulates, that accumulation tends to reduce further production of the product. For example, if an increase in the amount of breakdown products of Protein T slowed down synthesis of Protein T, this would lead to a decreased amount of breakdown products and a return to the previous rate of Protein T synthesis. Negative feedback is very common in **regulatory systems**, which are systems that tend to stabilize amounts or concentrations.

Systems analysis is a conceptual tool for understanding all levels of organization

Biologists today employ an approach known as **systems analysis** to understand how biological systems function. In systems analysis, we identify the parts or components of a biological system and specify the processes by which the components interact (see Figure 1.6). We may also be able to specify the *rates* of these interactions and how the rates are affected by feedback. What we can do then is analyze how the system will change through time. Will the amounts of different components increase or decrease, and how quickly, and how will this depend on the rates of the interactions? Will there be any stable balance, or equilibrium, that the system eventually reaches?

To do the analysis we write out mathematical equations that express the amounts of the different components and that include the processes and their rates. Expressed in words, such an equation for component B in Figure 1.6 has the following form:

The amount of B present at some time in the future =
the amount of B now + the amount of A converted into B
– the amount of B converted into C

We write out a similar equation for each component in the system.

We can analyze the relatively simple biological systems in Figure 1.7 by hand, but the analysis of larger systems quickly

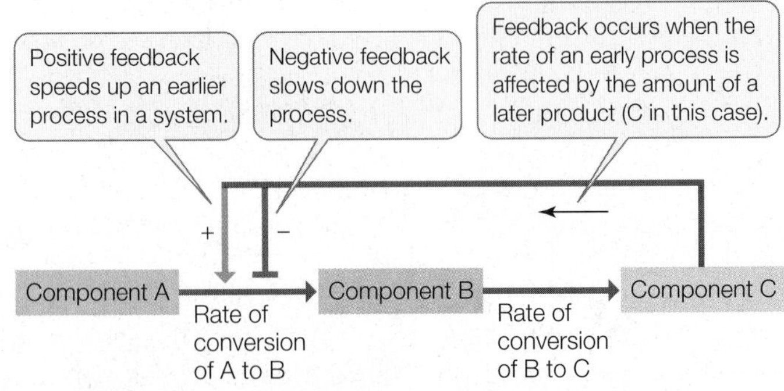

FIGURE 1.8 Feedback Can Be Positive or Negative Positive feedback tends to destabilize a system, whereas negative feedback typically stabilizes a system.

becomes very complicated and is typically carried out using computers. The approach, however, is the same: We express the rates of all processes as mathematical equations.

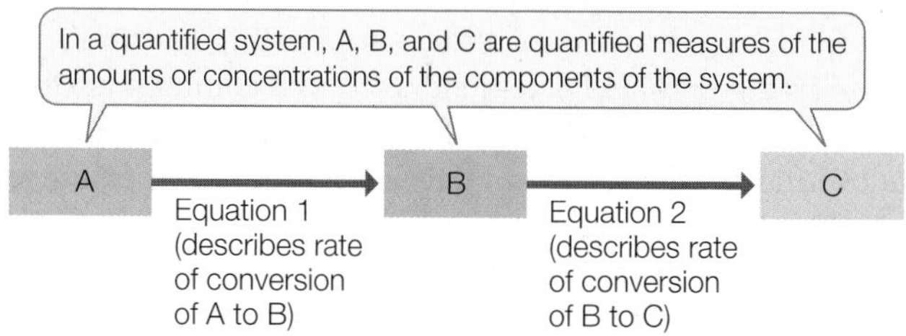

After this analysis is done, we have a **computational model** of the biological system. If the computational model is well grounded in factual knowledge of the biological system, the model will mimic the biological system.

An important use of computational models is prediction. For instance, if atmospheric temperature affects a biological system, we can use a computational model to develop a hypothetical prediction of the future behavior of the system in a warming world by adjusting the model to take into account the expected increases in temperature.

CONCEPT 1.3 Genetic Systems Control the Flow, Exchange, Storage, and Use of Information

The information required for an organism to function—the "blueprint" for its existence—is contained in the organism's genome, which as we noted earlier is the sum total of all the information encoded by its genes. The presence of genetic information and the processes by which organisms "decode" and use it to build the proteins that underlie a body's structure and function involve fundamental principles that we will discuss and expand on throughout the book, especially in Chapters 10–14.

Genomes encode the proteins that govern an organism's structure

Early in the chapter we noted the importance of self-replicating nucleic acids in the origin of life. Nucleic acid molecules contain long sequences of four subunits called **nucleotides**. The sequence of these nucleotides in **deoxyribonucleic acid**, or **DNA**, allows the organism to assemble **proteins**. Each **gene** is a specific segment of DNA whose sequence carries the information for building, or controlling the building of, one or more proteins (**FIGURE 1.9**). Proteins, in turn, are the molecules that govern the chemical reactions within cells and form much of an organism's structure. For these reasons, in biology we often say that genes "encode" proteins.

By analogy with a book, the nucleotides of DNA are like the letters of an alphabet. The sentences in the book are genes that encode proteins, which means that the genes provide instructions for making the proteins at a particular time or place. If you were to write out your own genome using four letters to represent the four DNA nucleotides, you would write more than 3 billion letters. Using the size type you are reading now, your genome would fill more than 1,000 books the size of this one.

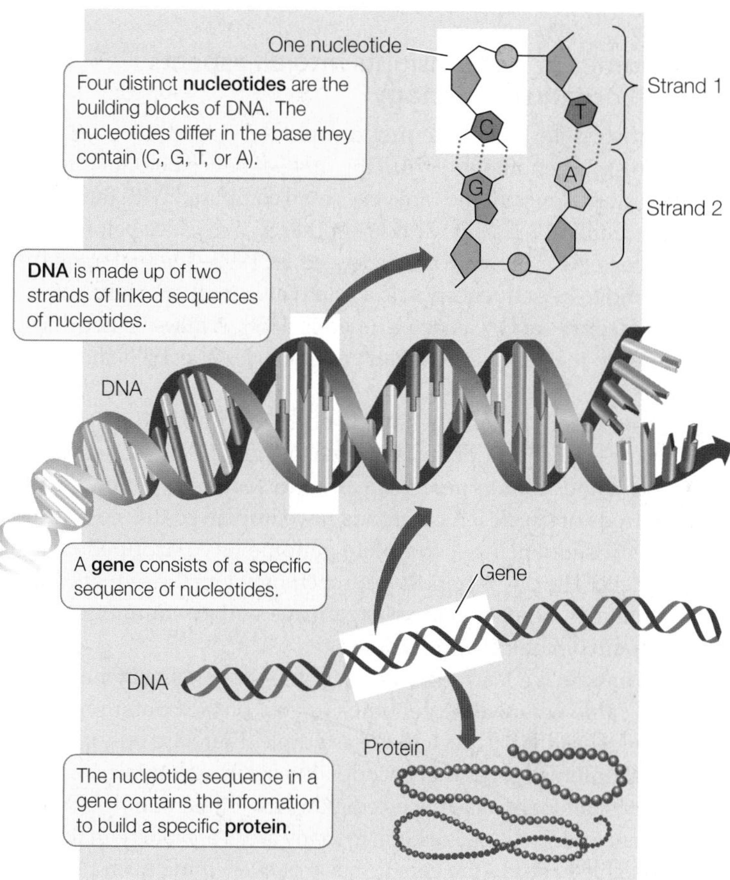

One nucleotide
Four distinct **nucleotides** are the building blocks of DNA. The nucleotides differ in the base they contain (C, G, T, or A).
C
T
G
A
Strand 1
Strand 2
DNA is made up of two strands of linked sequences of nucleotides.
DNA
A **gene** consists of a specific sequence of nucleotides.
Gene
DNA
Protein
The nucleotide sequence in a gene contains the information to build a specific **protein**.

FIGURE 1.9 DNA Is Life's Blueprint The instructions for life are contained in the sequences of nucleotides in DNA molecules. Specific DNA nucleotide sequences comprise genes. The average length of a single human gene is 27,000 nucleotides. The information in each gene provides the cell with the information it needs to manufacture molecules of a specific protein.

All the cells of a given multicellular organism contain the same genome, yet the different cells have different functions and form different proteins. For example, oxygen-carrying hemoglobin occurs in red blood cells, gut cells produce digestive proteins, and so on. Therefore different types of cells in an organism must express, or use, different parts of the genome. How any given cell controls which genes it expresses, or uses (and which genes it suppresses, or doesn't use), is a major focus of current biological research.

The genome of an organism contains thousands of genes. If mutations alter the nucleotide sequence of a gene, the protein that the gene encodes is often altered as well. Mutations may occur spontaneously, as happens when mistakes take place during replication of DNA. Mutations can also be caused by certain chemicals (such as those in cigarette smoke) and radiation (including UV radiation from the sun). Most mutations either are harmful or have no effect. Occasionally a mutation improves the functioning of the organism under the environmental conditions the individual encounters. Mutations are the raw material of evolution.

Genomes provide insights into all aspects of an organism's biology

Scientists determined the first complete DNA sequence of an organism's genome in 1976. This first sequence belonged to a virus, and viral genomes are very small compared with those of most cellular organisms. It was another two decades before the first bacterial genome was sequenced, in 1995. The first animal genome to be sequenced was a relatively small one—that of a roundworm—and was determined in 1998. A massive effort to sequence the complete human genome began in 1990 and finished 13 years later.

Since then, scientists have used the methods developed in these pioneering projects, as well as new DNA sequencing technologies that appear each year, to sequence genomes of hundreds of species. As methods have improved, the cost and time for sequencing a complete genome have dropped dramatically. The day is rapidly approaching when the sequencing of genomes from individual organisms will be commonplace for many biological applications.

What are we learning from genome sequencing? One surprise came when some genomes turned out to contain many fewer genes than expected. For example, there are only about 21,000 different genes that encode proteins in a human genome, but most biologists had expected many times that number. Gene sequence information is a boon to many areas of biology, making it possible to study the genetic basis of everything from physical structures to inherited diseases. Biologists can also compare genomes from many species to learn how and why one species differs from another. Such comparative genomic studies allow biologists to trace the evolution of genes through time and to document how particular changes in gene sequences result in changes in structure and function.

The vast amount of information being collected from genome studies has led to rapid development of the field of bioinformatics, the study of biological information. In this emerging field, biologists and computer scientists work together closely to develop new computational tools to organize, process, and study databases used in comparing genomes.

CONCEPT 1.4 Evolution Explains the Diversity as Well as the Unity of Life

Evolution—change in the genetic makeup of biological populations through time—is a major unifying principle of biology. Any process that can lead to changes in the frequencies of genes in a population from generation to generation is an evolutionary process. A common set of evolutionary processes is at work in populations of all organisms. The constant change that occurs in these populations gives rise to all the diversity we see in life. These two themes—unity and diversity—provide a framework for organizing and thinking about the evolution of life. The similarities of life allow us to make comparisons and predictions from one species to another, as we have discussed. The differences are what make biology such a rich and exciting field for investigation and discovery.

Natural selection is an important process of evolution

Charles Darwin compiled factual evidence for evolution in his 1859 book *On the Origin of Species*. Since then, biologists have gathered massive amounts of data supporting Darwin's idea that all living organisms descended from a common ancestor. Darwin also proposed one of the most important processes that produce evolutionary change. He argued that the differing survival and reproduction among individuals in a population, which he termed **natural selection**, could account for much of the evolution of life.

When Darwin proposed that living organisms descended from a common ancestor and are therefore related to one another, he did not have the advantage we have today of understanding the processes of genetic inheritance. Those processes, which we will cover in depth in Chapters 7–9, were not widely understood until the early 1900s. But he knew that offspring differed from their parents, even though they showed strong similarities. And he knew that any population of a plant or animal species displays variation.

Darwin himself bred pigeons, and he knew that if you select breeding pairs on the basis of some particular trait, then that trait is more likely to be present in their offspring than in the general population. He was well aware of how pigeon fanciers selected breeding pairs to produce offspring with unusual feather patterns, beak shapes, or body sizes. He realized that if humans could select for specific traits in organisms such as pigeons, a similar process could operate in nature. Darwin emphasized that human-imposed selection, which he called "artificial selection," has been practiced on crop plants and domesticated animals since the dawn of human civilization. In coining the term "natural selection," he argued that a similar process occurs in nature. But in nature, the "selection" occurs not by human choice but by the fact that some individuals contribute more offspring to future generations than others.

How does natural selection work? Darwin thought that differing probabilities of survival and reproductive success could account for evolutionary change. He reasoned that the reproductive capacity of plants and animals, if unchecked, would result in unlimited growth of populations, but we do not observe such growth in nature. In most species, only a small percentage of offspring survive to reproduce. For this reason, any trait will spread in the population if that trait gives an individual organism even a small increase in the probability that the individual will survive and reproduce.

Because organisms with certain traits survive and reproduce best under specific sets of conditions, natural selection leads to **adaptations**: structural, physiological, or behavioral traits that increase an organism's chances of surviving and reproducing in its environment. For example, remember the frog in the opening photograph of this chapter? Look at the frog's feet and notice that the frog's toes appear to be greatly expanded. These expanded toes would be especially obvious if you could compare them with the toes of frog species that do not live in trees. Expanded toes increase the ability of tree frogs to climb trees, which allows them to hunt insects for food in the treetops and to escape predators on the ground. For this

This ground-living frog walks across the ground using its short legs and peglike digits (toes).
Webbed rear feet are evident in this highly aquatic species of frog.
Dyscophus guineti
Pelophylax sp.
This tree frog has toe pads, which are adaptations for climbing.
A different tree frog species has extended webbing between the toes, which increases surface area and allows the frog to glide from tree to tree.
Phyllomedusa bicolor
Rhacophorus nigropalmatus

FIGURE 1.10 Adaptations to the Environment The limbs of frogs show adaptations to the different environments of each species.

Go to MEDIA CLIP 1.1 Wallace's Flying Frog PoL2e.com/mc1.1

reason the expanded toe pads of tree frogs are an adaptation to life in trees. **FIGURE 1.10** shows other adaptations in the limbs of frogs to different environments.

Biologists often think about two different kinds of explanations for adaptations. On the one hand, we can consider the immediate genetic, physiological, neurological, and developmental processes that explain how an adaptation works. We call these **proximate explanations**. For example, a proximate explanation for the toes of tree frogs might examine the physical structure of the toe pads and explain how expansion of the toe leads to greater adhesion to a substrate. Such an explanation tells us how the adaptation works, but it does not explain how tree frogs came to possess such toe pads. An **ultimate explanation**, on the other hand, concerns the processes that led to the evolution of toe pads in various groups of climbing frogs. Ultimate explanations involve comparison of variation within and among species and describe how a given trait affects an organism's chances for survival and reproduction.

Natural selection has been demonstrated in countless biological investigations, but it is not the only process that results in evolution, as we will explore in Chapters 15–18. An example of another evolutionary process is genetic drift, which refers to random changes in gene frequencies in a population because of chance events. As a result of the various evolutionary processes, all biological populations evolve through time. All the evolutionary processes operating over the long history of Earth have led to the remarkable diversity of life on our planet.

Evolution is a fact, as well as the basis for broader theory

The famous biologist Theodosius Dobzhansky once wrote that "Nothing in biology makes sense except in the light of evolution." Dobzhansky was emphasizing the need to include an evolutionary perspective and approach in all aspects of biological study. Everything in biology is a product of evolution, and biologists need a perspective of change and adaptation to fully understand biological systems.

You may have heard someone say that evolution is "just a theory," implying that there is some question about whether or not biological populations evolve. This is a common misunderstanding that originates in part from the different meanings of the word "theory" in everyday language and in science. In everyday speech, some people use the word "theory" to mean "hypothesis" or even—disparagingly—"a guess." In science, however, a **theory** is *a body of scientific work in which rigorously tested and well-established facts and principles are used to make predictions about the natural world.* In short, evolutionary theory is both (1) a body of knowledge supported by facts and (2) the resulting understanding of the various processes by which biological populations have changed and diversified over time, and by which Earth's populations continue to evolve.

We can observe and measure evolution directly, and many biologists conduct experiments on evolving populations. We constantly observe changes in the genetic makeup of populations over relatively short time periods. For example, every year health agencies need to produce new flu vaccines, because populations of influenza viruses evolve so quickly that last year's vaccines may not be effective against this year's populations of viruses. In addition, we can directly observe a record of the history of evolution in the fossil record over the almost unimaginably long periods of geological time. Exactly *how* biological populations change through time is something that is subject to testing and experimentation. The fact that biological populations evolve, however, is not disputed among biologists.

CONCEPT 1.5 Science Is Based on Quantitative Observations, Experiments, and Reasoning

Regardless of the many different tools and methods used in research, all scientific investigations are based on quantitative observation, experimentation, and reasoning. In each of these areas, scientists are guided by an established set of scientific methodological principles.

Observing and quantifying are important skills

Many biologists are motivated by their observations of the living world. Learning *what to observe* in nature is a skill that develops with experience in biology. An intimate understanding of the **natural history** of a group of organisms—how the organisms get their food, reproduce, behave, regulate their functions, and interact with other organisms—leads to better observations and prompts biologists to ask questions about those observations. The more a biologist knows about general principles of life, the more he or she is likely to gain new insights from observing nature.

Biologists have always observed the world around them, but today our ability to observe is greatly extended by technologies such as electron microscopes, rapid genome sequencing, magnetic resonance imaging, and global positioning satellites. These technologies allow us to observe everything from the distribution of molecules in the body (by using electron microscopes) to the daily movement of animals across continents and oceans (by using global positioning satellites).

Observation is a basic tool of biology, but as scientists we must also be able to **quantify** our observations—turn the observations into explicit counts or measures. Whether we are testing a new drug or mapping the migrations of whales, mathematical and statistical calculations are essential. For example, biologists once classified organisms entirely on the basis of qualitative descriptions of the physical differences among them. There was no way of objectively determining evolutionary relationships of organisms, and biologists had to depend on the fossil record for insight. Today our ability to quantify the molecular and physical differences among species, combined with explicit mathematical models of the evolutionary process, enables quantitative analyses of evolutionary history. These mathematical calculations, in turn, make it easier to compare all other aspects of the biology of different organisms.

Scientific methods combine observation, experimentation, and logic

Often, science textbooks describe "*the* scientific method," as if there is a single flow chart that all scientists follow. This view is an oversimplification. Such flow charts include much of what scientists do, but you should not conclude that scientists necessarily go through these steps in one prescribed, linear order.

Observations lead to questions, and scientists make additional observations and often do experiments to answer those questions. This approach, called the hypothesis–prediction method, has five steps: (1) making observations; (2) asking questions; (3) forming hypotheses, or tentative answers to the questions; (4) making predictions based on the hypotheses; and (5) testing the predictions by making additional observations or conducting experiments. These are the steps in traditional flow charts such as the one shown in **FIGURE 1.11**.

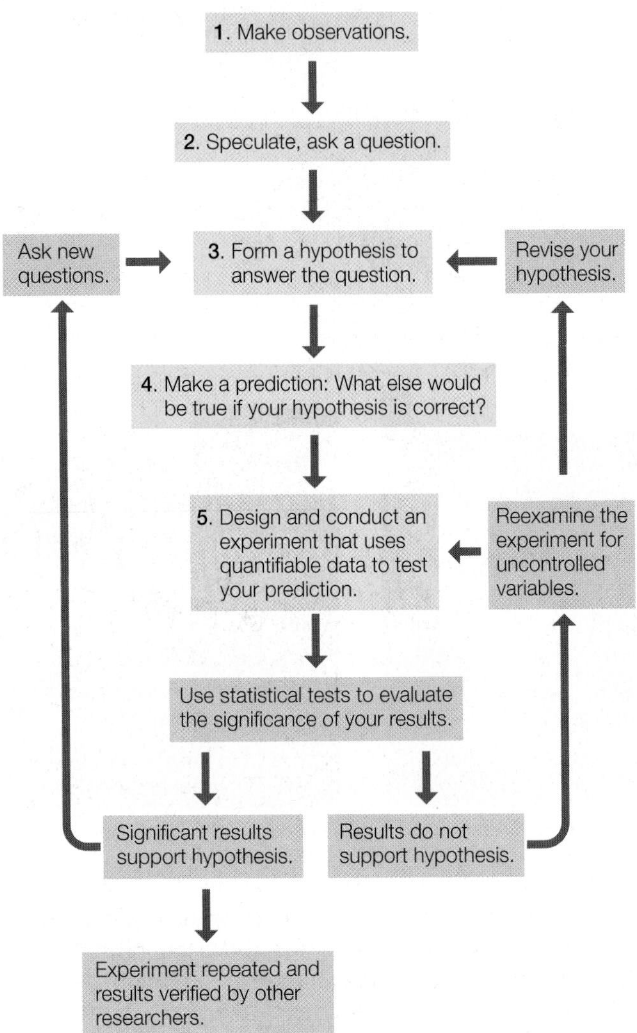

FIGURE 1.11 Scientific Methodology The process of observation, speculation and questioning, hypothesis formation, prediction, and experimentation is a cornerstone of modern science, although scientists may initiate their research at any of several different points.

Go to ANIMATED TUTORIAL 1.2 Using Scientific Methodology PoL2e.com/at1.2

Getting from questions to answers

Let's consider an example of how scientists start with a general question and work to find answers. Amphibians—such as the frog in the opening photograph of this chapter—have been around for a long time. They watched the dinosaurs come and go. But today scientists have observed that amphibian populations around the world are in dramatic decline. In quantitative terms, more than a third of the world's amphibian species are threatened with extinction. Why is this happening?

To answer big questions like this, biologists begin by sifting through what is already known to arrive at possible answers, or **hypotheses**. In the case of amphibians, biologists know that instances of recent population declines have been associated with various environmental changes such as loss of moist habitats, changing climate, pathogens, or increased quantities of environmental toxins. Tyrone Hayes, a biologist at the University of California at Berkeley, chose to test the hypothesis that frog populations have been adversely affected by agricultural insecticides and herbicides (weed killers). He did so even though several prior studies had shown that many of these chemicals tested at realistic concentrations do not kill amphibians.

Hayes focused on atrazine, the most widely used herbicide in the world and a common contaminant in fresh water. More than 70 million pounds of atrazine are applied to farmland in the United States every year, and it is used in at least 20 countries. Atrazine kills several types of weeds that can choke fields of important crops such as corn. The chemical is usually applied before weeds emerge in the spring—at the same time many amphibians are breeding and thousands of tadpoles swim in the ditches, ponds, and streams that receive runoff from farms.

In his laboratory, Hayes and his associates raised frog tadpoles in water containing no atrazine and in water with concentrations ranging from 0.01 parts per billion (ppb) up to 25 ppb. The U.S. Environmental Protection Agency considers environmental levels of atrazine of 10–20 ppb of no concern, and it considers 3 ppb a safe level in drinking water. Rainwater in Iowa has been measured to contain 40 ppb. In Switzerland, where the use of atrazine is illegal, the chemical has been measured at approximately 1 ppb in rainwater.

In the Hayes laboratory, an atrazine concentration as low as 0.1 ppb had a dramatic effect on tadpole development: it feminized the males. In some of the adult males that developed from these tadpoles, the vocal structures used in mating calls were smaller than normal, female sex organs developed, and eggs were found growing in the testes. In other studies, normal adult male frogs exposed to 25 ppb had a tenfold reduction in levels of the male sex hormone testosterone and did not produce sperm. You can imagine the disastrous effects these changes could have on the capacity of frogs to breed and reproduce.

But Hayes's experiments were performed in the laboratory, with a species of frog bred for laboratory use. Could atrazine be affecting frogs in nature? If so, then developmental abnormalities in natural frog populations should occur where atrazine is present in their environment. Hayes and his students traveled across the middle of North America, sampling water and collecting frogs to test this prediction. They analyzed the water for atrazine and examined the frogs. In the only site where atrazine was undetectable in the water, the frogs were normal. In all the other sites, male frogs had abnormalities of the sex organs.

Like other biologists, Hayes made observations. He then made predictions based on those observations, and designed and carried out experiments to test his predictions. Some of the conclusions from his experiments, described below, could have profound implications not only for amphibians but also for other animals, including humans.

Well-designed experiments have the potential to falsify hypotheses

Once predictions are made from a hypothesis, experiments can be designed to test those predictions. The most informative experiments have the ability to show that the prediction is wrong. If the prediction is wrong, the hypothesis must be modified or rejected.

There are two general types of experiments. Both compare data from different groups or samples. A **controlled experiment** changes, or manipulates, one or more of the factors being tested. A **comparative experiment** compares unmanipulated data gathered from different sources.

In a controlled experiment, we start with groups or samples that are as similar as possible. We predict on the basis of our hypothesis that some critical factor, or **variable**, has an effect on the phenomenon we are investigating. We devise some method to manipulate *only that variable* in an "experimental" group, and we compare the resulting data with data from an unmanipulated "control" group. If the predicted difference occurs, we then apply statistical tests to find the probability that the manipulation created the difference (as opposed to the difference being the result of random chance). **FIGURE 1.12** describes one of the many controlled experiments performed by the Hayes laboratory to quantify the effects of atrazine on male frogs.

The basis of controlled experiments is that one variable is manipulated while all others are held constant. The variable that is manipulated is called the independent variable (because an investigator can manipulate it independently of other considerations). The response that is measured is the dependent variable (because it is not manipulated directly by the investigator but is permitted to vary in ways that depend on the independent variable). A perfectly controlled experiment is not easy to design because biological variables are so interrelated that it is difficult to alter just one.

A comparative experiment starts with the prediction that there will be a difference among naturally existing samples or groups based on the hypothesis. In comparative experiments, we do not control any of the variables, and often we cannot even identify all the variables that are present. We simply gather and compare data from different naturally occurring sample groups.

When his controlled experiments indicated that atrazine indeed affects reproductive development in frogs, Hayes and his colleagues performed a comparative experiment. They collected frogs and water samples from eight widely separated sites across the United States and compared the percentages of abnormal frogs from environments with very different levels of atrazine (**FIGURE 1.13**). Of course, the sample sites differed in many ways besides the level of atrazine present.

The results of experiments frequently reveal that the situation is more complex than the hypothesis anticipated, thus raising new questions. There are no "final answers" in science. As a result, biologists often develop new questions, hypotheses, and experiments as they collect more data. The process of science is open-ended in this regard, and continued research leads to an ever-better understanding of the living world, with practical implications for agriculture, medicine, conservation of species, and other endeavors.

Statistical methods are essential scientific tools

Whether we do controlled or comparative experiments, at the end we have to decide whether there is a difference among the

INVESTIGATION

FIGURE 1.12 Controlled Experiments Manipulate a Variable The Hayes laboratory created controlled environments that differed only in the concentrations of atrazine in the water. Eggs from leopard frogs (*Rana pipiens*) raised specifically for laboratory use were allowed to hatch and the tadpoles were separated into experimental tanks containing water with different concentrations of atrazine.[a]

HYPOTHESIS

Exposure to atrazine during larval development causes abnormalities in the reproductive tissues of male frogs.

METHOD

1. Establish 9 tanks in which all attributes are held constant except the water's atrazine concentration. Establish 3 atrazine conditions (3 replicate tanks per condition): 0 ppb (control condition), 0.1 ppb, and 25 ppb.
2. Place *Rana pipiens* tadpoles from laboratory-reared eggs in the 9 tanks (30 tadpoles per tank).
3. When tadpoles have transitioned into adults, sacrifice the animals and evaluate their reproductive tissues.
4. Test for relationship between degree of atrazine exposure and the presence of abnormalities in the gonads (testes) of male frogs.

RESULTS

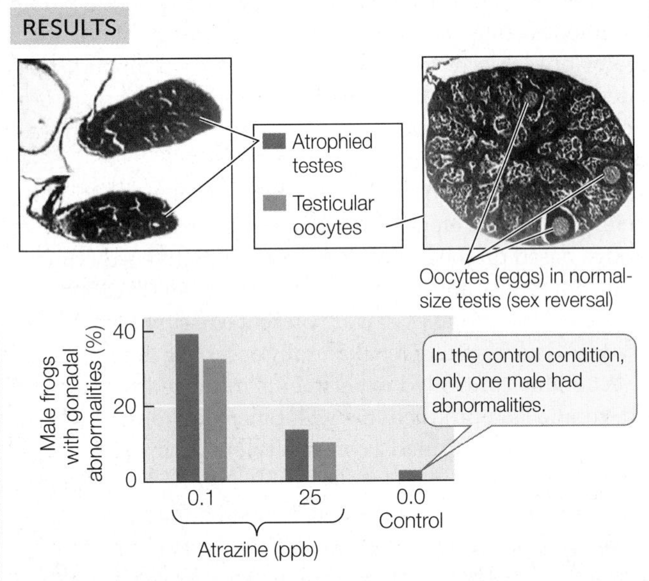

CONCLUSION

Exposure to atrazine at concentrations as low as 0.1 ppb induces abnormalities in the gonads of male frogs. The effect is not proportional to the level of exposure.

Go to **LaunchPad** for discussion and relevant links for all **INVESTIGATION** figures.

[a]T. Hayes et al. 2003. *Environmental Health Perspectives III*: 568–575.

samples, individuals, groups, or populations in the study. How do we decide whether a measured difference is enough to support or reject a hypothesis? In other words, how do we decide in an unbiased, objective way whether the measured difference is meaningful, or significant?

Statistical significance refers to the extent to which a result is unlikely to be due to chance alone. Scientists use statistics because they recognize that variation is always present in any set of measurements. Statistical tests calculate the probability that the differences observed in an experiment could be due to random variation. The results of statistical tests are therefore probabilities. Many statistical tests start with a **null hypothesis**—the premise that any observed differences are simply the result of random differences that arise from drawing two samples from the same population. When scientists collect quantified observations, or **data**, they apply statistical methods to those data to calculate the likelihood that the null hypothesis is correct.

More specifically, statistical methods tell us the probability of obtaining a particular result by chance alone, even if the samples being tested are drawn from the same population. As scientists, we need to eliminate, insofar as possible, the possibility that any differences seen are simply due to chance variation in the samples. Appendix B in this book is a short primer on statistical methods that you can refer to as you analyze data that will be presented throughout the text.

Not all forms of inquiry into nature are scientific

Science is a human endeavor that is bounded by certain standards of practice. Other areas of scholarship share with science the practice of making observations and asking questions, but scientists are distinguished by what they do with their observations and how they answer their questions. Data, subjected to appropriate statistical analysis, are critical in testing hypotheses. Science is the most powerful approach humans have devised for learning about the world and how it works.

Scientific explanations for natural processes are objective and reliable because the hypotheses proposed *must be testable* and *must have the potential of being rejected* by direct observations and experiments. Scientists must clearly describe the methods they use to test hypotheses so that other scientists can repeat their experiments to see if they get the same results. Not all experiments are repeated, but surprising or controversial results are always subjected to independent verification. Scientists worldwide share this process of testing and rejecting hypotheses, contributing to a common body of scientific knowledge.

If you understand the methods of science, you can distinguish science from non-science. Art, music, and literature all contribute to the quality of human life, but they are not science. They do not use scientific methods to establish what is fact. Religion is not science, although religions have historically attempted to explain natural events ranging from unusual weather patterns to crop failures to human diseases. Most such phenomena that at one time were mysterious can now be explained in terms of scientific principles. Fundamental tenets of religious faith, such as the existence of a supreme deity or deities, cannot be confirmed or refuted by experimentation and for this reason are outside the realm of science.

The power of science derives from the uncompromising objectivity and absolute dependence on evidence that comes from reproducible and quantifiable observations. A religious or spiritual explanation of a natural phenomenon may be

INVESTIGATION

FIGURE 1.13 Comparative Experiments Look for Differences among Groups To see whether the presence of atrazine correlates with testicular abnormalities in male frogs, the Hayes lab collected frogs and water samples from different locations around the U.S. The analysis that followed was "blind," meaning that the frogs and water samples were coded so that experimenters working with each specimen did not know which site the specimen came from.[a]

HYPOTHESIS

Presence of the herbicide atrazine in environmental water correlates with gonadal abnormalities in frog populations.

METHOD

1. Based on commercial sales of atrazine, select 4 sites (sites 1–4) less likely and 4 sites (sites 5–8) more likely to be contaminated with atrazine.
2. Visit all sites in the spring (i.e., when frogs have transitioned from tadpoles into adults); collect frogs and water samples.
3. In the laboratory, sacrifice frogs and examine their reproductive tissues, documenting abnormalities.
4. Analyze the water samples for atrazine concentration (the sample for site 7 was not tested).
5. Quantify and correlate the incidence of reproductive abnormalities with environmental atrazine concentrations.

RESULTS

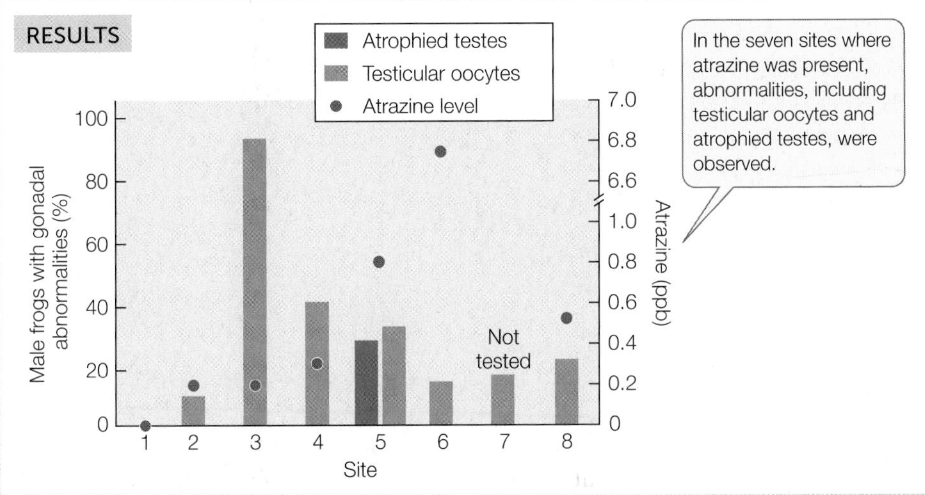

CONCLUSION

Reproductive abnormalities exist in frogs from environments in which aqueous atrazine concentration is 0.2 ppb or above. The frequency of abnormalities does not appear to be proportional to atrazine concentration at the time of transition to adulthood.

Go to **LaunchPad** for discussion and relevant links for all **INVESTIGATION** figures.

[a]T. Hayes et al. 2002. *Nature* 419: 895–896.

coherent and satisfying for the person holding that view, but it is not testable and therefore it is not science. To invoke a supernatural explanation (such as a "creator" or "intelligent designer" with no known bounds) is to depart from the world of science. Science does not say that untestable religious beliefs are necessarily wrong, just that they are not something that we can address using scientific methods.

Science describes how the world works. It is silent on the question of how the world "ought to be." Many scientific advances that contribute to human welfare also raise major ethical issues. Recent developments in genetics and developmental biology may enable us to select the sex of our children, to use stem cells to repair our bodies, and to modify the human genome. Scientific knowledge allows us to do these things, but science cannot tell us whether or not we should do so, or if we choose to do them, how we should regulate them. Such questions are as crucial to human society as the science itself. A responsible scientist does not lose sight of these questions or neglect the contributions of the humanities in attempting to come to grips with them.

Consider the big themes of biology as you read this book

You will see evolution and the other fundamental principles of life introduced in this chapter at work in each part of this book. In Part 1 you will learn about the molecular organization of life. We will discuss the origin of life, the energy in atoms and molecules, and how proteins and nucleic acids became the self-replicating cellular systems of life. Part 2 will describe how these self-replicating systems work and the genetic principles that explain heredity and mutation, which are the basis of evolution. In Part 3 we will describe the processes of evolution and go into detail about how evolution works. Part 4 will examine the products of evolution: the vast diversity of life and the many different ways organisms solve common problems such as how to reproduce, defend themselves, and obtain nutrients. Parts 5 and 6 will explore the physiological adaptations that allow plants and animals to survive and function in a wide range of physical environments. Finally, in Part 7 we will discuss these environments and the integration of individual organisms, populations, and communities into the interrelated ecological systems of Earth.

You may enjoy returning to this chapter occasionally as the course progresses. The brief explanations we have given here should become more meaningful as you read about the facts and phenomena that underlie the principles. Our knowledge of the "facts" of biology, however, is not based just on reading, contemplation, or discussion, although all of these activities are important. Scientific knowledge is based on active and always-ongoing research.

PART 1

Cells

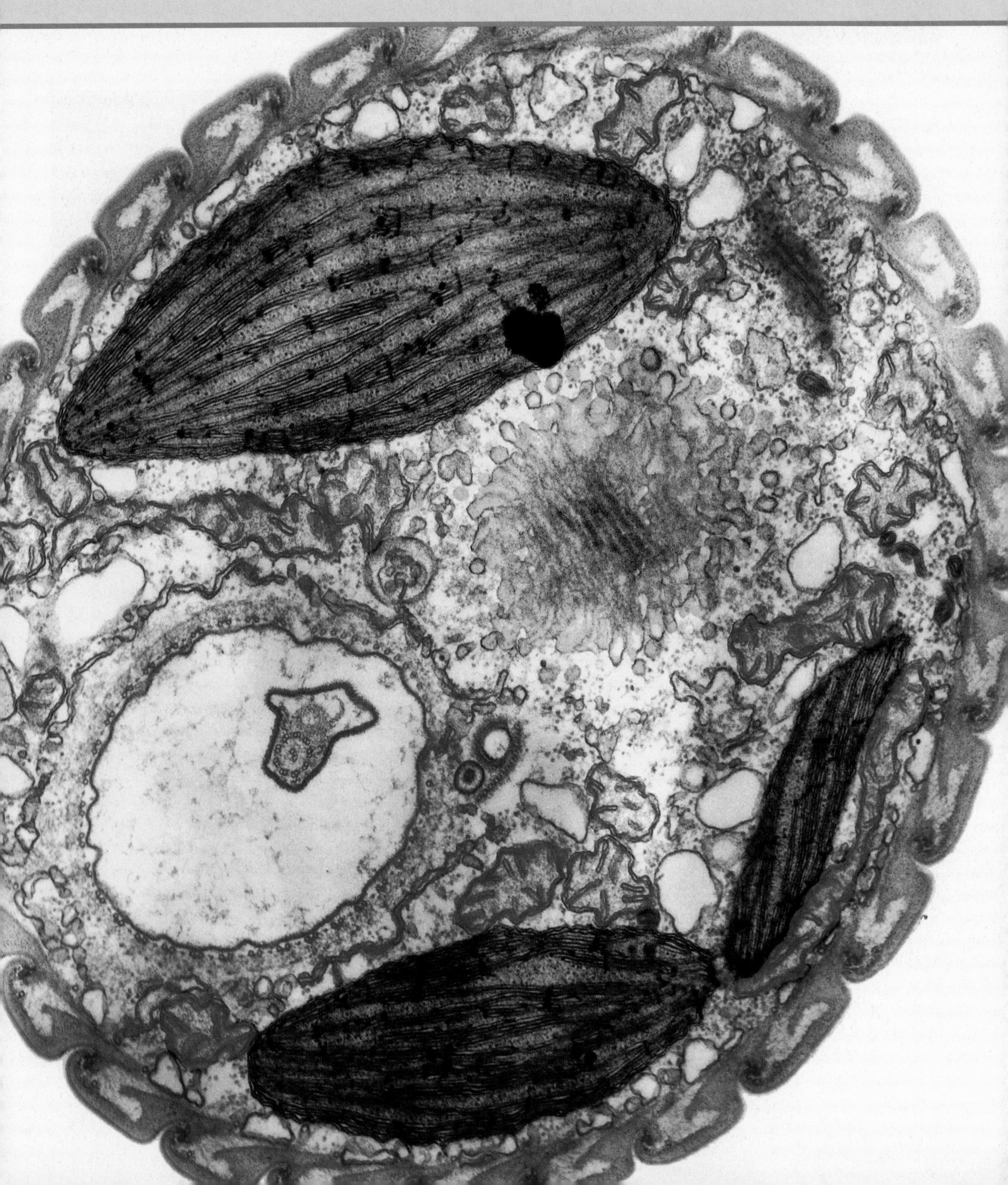

2 The Chemistry and Energy of Life

KEY CONCEPTS

Polar ice caps, as shown here, have been observed on Mars for a long time, but recent evidence also shows water at the milder mid-latitudes of Mars. Finding water on Mars may indicate that it is or was hospitable to life.

A major discovery of biology was that living things are composed of the same chemical elements as the vast nonliving portion of the universe. This mechanistic view—that life is chemically based and obeys the universal laws of chemistry and physics—is relatively new in human history. Until the nineteenth century, many scientists thought that a "vital force," distinct from the forces governing the inanimate world, was responsible for life. Many people still assume that such a vital force exists. However, the mechanistic view of life has led to great advances in biological science, and it underpins many of the applications of biology to medicine and agriculture. We use a mechanistic view throughout this book.

Among the most abundant chemical elements in the universe are hydrogen and oxygen, and life as we know it requires the presence of these elements as water (H_2O). Water makes up about 70 percent of the bodies of most organisms, and those that live on land have evolved elaborate ways to retain the water in their bodies. Aquatic organisms do not need these water-retention mechanisms; thus biologists think that life originated in a watery environment.

Life has been found in some surprising places, often in extreme conditions. There are organisms living in hot springs at temperatures above the boiling point of water, 5 kilometers below Earth's surface, at the bottom of the ocean, and in extremely acid or salty conditions.

Life has been found even below the Antarctic ice, a finding especially relevant to the search for life beyond Earth. Some moons that orbit Jupiter and Saturn have ice on their surfaces, overlaying oceans of water much larger than the oceans on Earth. The subsurface water on these moons may be maintained as a liquid in large part because ice is a good insulator and traps heat generated by tidal forces between the moons and their planets. Europa is a moon of Jupiter that is slightly smaller than Earth's moon. It has an atmosphere composed primarily of oxygen, and scientists speculate that Europa's subsurface oceans could harbor life, just as the Antarctic ice lakes do. Mars also has ice on its surface, with some evidence of subsurface water. Two of NASA's highest priorities are to launch missions to Mars and Europa to bring back samples from these lakes to Earth.

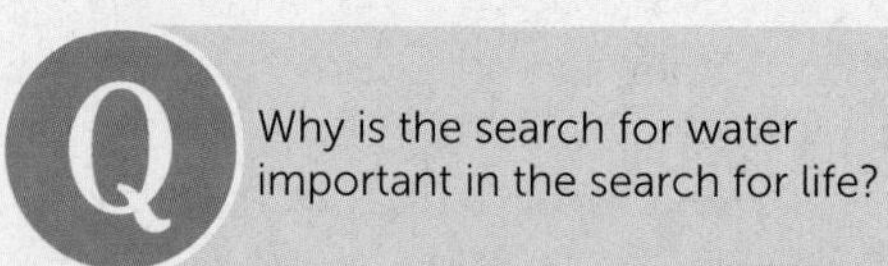

Why is the search for water important in the search for life?

You will find the answer to this question on page 34.

CONCEPT 2.1 Atomic Structure Is the Basis for Life's Chemistry

Living and nonliving matter is composed of **atoms**. Each atom consists of a dense, positively charged **nucleus**, with one or more negatively charged **electrons** moving around it. The nucleus contains one or more positively charged **protons**, and may contain one or more **neutrons** with no electrical charge:

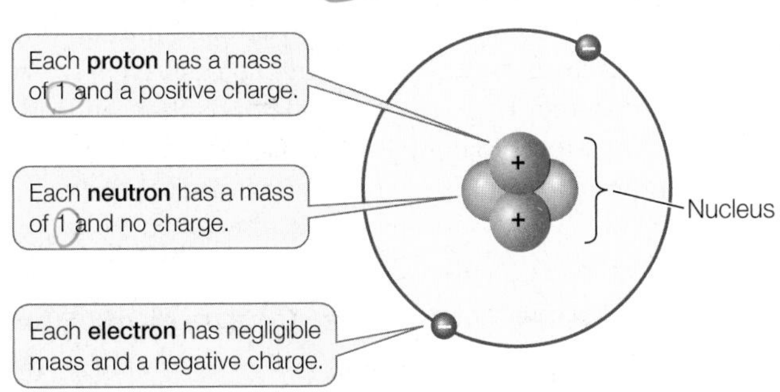
Each proton has a mass of 1 and a positive charge.
Each neutron has a mass of 1 and no charge.
Each electron has negligible mass and a negative charge.
Nucleus

Charges that are different (+/–) attract each other, whereas charges that are alike (+/+, –/–) repel one another. Most atoms are electrically neutral because the number of electrons in an atom equals the number of protons.

The standard unit of measure for the mass of an atom (atomic mass) is the dalton—named after the English chemist John Dalton. A single proton or neutron has a mass of about 1 dalton (Da), which is 1.7×10^{-24} grams, but an electron is even tinier, at 0.0005 Da (9×10^{-28} g). Because the mass of an electron is only about 1/2,000th of the mass of a proton or neutron, the contribution of electrons to the mass of an atom can usually be ignored when chemical measurements and calculations are made.

An element consists of only one kind of atom

An **element** is a pure substance that contains only one kind of atom. The element hydrogen consists only of hydrogen atoms, the element gold only of gold atoms. The atoms of each element have characteristics and properties that distinguish them from the atoms of other elements.

There are 94 elements in nature, and at least another 24 have been made in physics laboratories. Most of the 94 natural elements have been detected in living organisms, but just a few predominate. About 98 percent of the mass of every living organism (bacterium, turnip, or human) is composed of just six elements:

Carbon (symbol C)	Hydrogen (H)	Nitrogen (N)
Oxygen (O)	Phosphorus (P)	Sulfur (S)

The chemistry of these six elements will be our primary concern in this chapter, but other elements found in living organisms are important as well. Sodium and potassium, for example, are essential for nerve function; calcium can act as a biological signal; iodine is a component of a human hormone; and magnesium is bound to chlorophyll in green plants.

The physical and chemical (reactive) properties of atoms depend on the numbers of protons, neutrons, and electrons they contain. The atoms of an element differ from those of other elements by the number of protons in their nuclei. The number of protons is called the **atomic number**, and it is unique to and characteristic of each element. A carbon atom has six protons and thus an atomic number of 6; the atomic number of oxygen is 8. For electrical neutrality, each atom has the same number of electrons as protons, so a carbon atom has six electrons and an oxygen atom has eight.

Along with a definitive number of protons, every element except hydrogen has one or more neutrons. The **mass number** of an atom is the total number of protons and neutrons in its nucleus. The number of neutrons may vary among atoms of a particular element. For example, carbon atoms with six, seven, and eight neutrons are found in nature. These variants are referred to as **isotopes**. The most common carbon isotope has six neutrons and a mass number of 12, and is referred to as carbon-12 (often written ^{12}C). The most common oxygen isotope (^{16}O) has eight protons and eight neutrons, and a mass number of 16.

Go to MEDIA CLIP 2.1
The Elements Song
PoL2e.com/mc2.1

Electrons determine how an atom will react

The **Bohr model** for atomic structure (see diagram in previous column) provides a concept of an atom that is largely empty space, with a central nucleus surrounded by electrons in orbits, or **electron shells**, at various distances from the nucleus. This model is much like our solar system, with planets orbiting around the sun. Although highly oversimplified (you will learn about the reality of atomic structure in physical chemistry courses), the Bohr model is useful for describing how atoms behave. Specifically, *the behaviors of electrons determine whether a chemical bond will form and what shape the bond will have*. These are two key properties for determining biological changes and structure.

In the Bohr model, each electron shell is a certain distance from the nucleus. Since electrons are negatively charged and protons are positive, an electron needs energy to escape from the attraction of the nucleus. The farther away an electron shell is from the nucleus, the more energy the electron must have. We will return to this topic when we discuss biological energetics in Chapter 6.

The electron shells, in order of their distance from the nucleus, can be filled with electrons as follows:

- First shell: up to 2 electrons
- Second shell: up to 8 electrons
- Third shell: up to 18 electrons
- Fourth and subsequent shells: up to 32 electrons

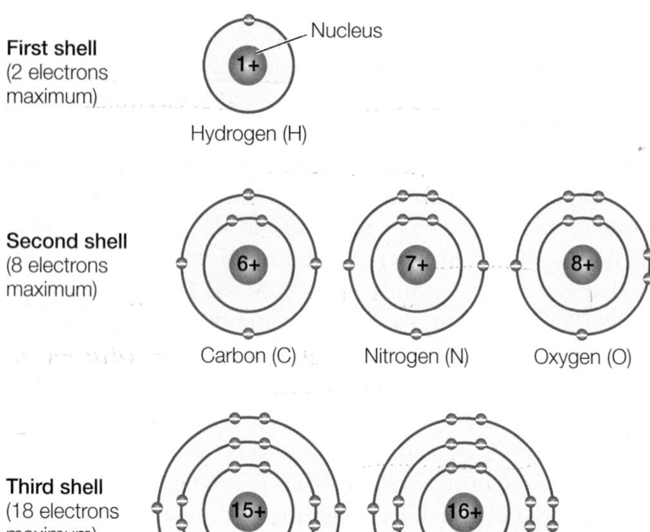

FIGURE 2.1 Electron Shells Each shell can hold a specific maximum number of electrons and must be filled before electrons can occupy the next shell. The energy level of an electron is higher in a shell farther from the nucleus. An atom with fewer than eight electrons in its outermost shell (or two in the case of hydrogen) can react (bond) with other atoms.

Electrons fill shells closest to the nucleus before occupying shells farther from the nucleus. **FIGURE 2.1** illustrates the electron shell configurations for the six major elements found in living systems.

For elements with atomic numbers between 6 and 20 there is a chemical rule of thumb called the **octet rule**, which states that an atom will lose, gain, or share electrons in order to achieve a stable configuration of eight electrons in its outermost shell. Oxygen, for example, which has six electrons in its outermost shell, will undergo chemical reactions to gain two electrons. When atoms share electrons, they form stable associations called **molecules**. Most atoms in biologically important molecules—for example, carbon and nitrogen—follow the octet rule. However, very small atoms such as hydrogen (with one proton and one electron) tend to gain, lose, or share electrons such that their single shell contains its maximum number of two electrons.

CHECKpoint CONCEPT 2.1

- ✓ What is the arrangement of protons, neutrons, and electrons in an atom?
- ✓ Sketch the electron shell configuration of a sodium atom (symbol Na), which has 11 protons. According to the octet rule, what would be the simplest way for a sodium atom to achieve electron stability?
- ✓ Many elements have isotopes, which are rare variants of the element with additional neutrons in the nucleus. Deuterium is an isotope of hydrogen that has one neutron (normal hydrogen has no neutrons).

  Does the neutron change the chemical reactivity of deuterium, compared with normal hydrogen? Explain why or why not.

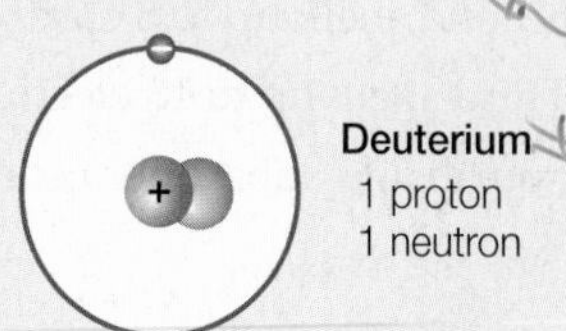

We have introduced the basic unit of matter that makes up all living organisms—the atom. We have discussed the tendency of atoms to attain stable configurations of electrons: a single shell of two electrons in the case of hydrogen, and an outer shell of eight electrons in the case of larger atoms. Next we will describe the different types of chemical bonds that can lead to stability, joining atoms together into molecular structures with different properties.

CONCEPT 2.2 Atoms Interact and Form Molecules

A **chemical bond** is an attractive force that links two atoms together in a molecule. There are several kinds of chemical bonds (**TABLE 2.1**; see p. 22). In this section we will begin with covalent bonds, the strong bonds that result from the sharing of electrons. Next we will consider weaker interactions, including hydrogen bonds, which are enormously important to biology. We will then examine ionic attractions, which form when an atom gains or loses electrons to achieve stability. Finally, we will see how atoms are bonded to make functional groups—groups of atoms that give important properties to biological molecules.

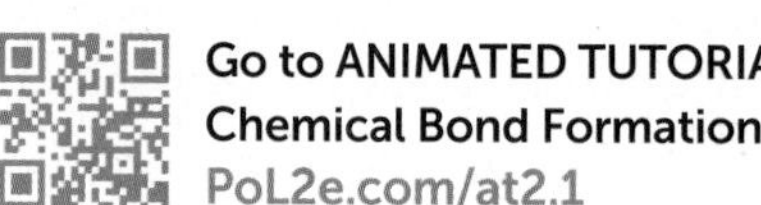
Go to ANIMATED TUTORIAL 2.1
Chemical Bond Formation
PoL2e.com/at2.1

Covalent bonds consist of shared pairs of electrons

A **covalent bond** forms when two atoms attain stable electron numbers in their outermost shells by *sharing* one or more pairs of electrons. In this case, each atom contributes one member of each electron pair. Consider two hydrogen atoms coming close together, each with just one electron in its single shell (**FIGURE 2.2**). When the electrons pair up, a stable association is formed, and this links the two hydrogen atoms in a covalent bond, forming the molecule H_2.

Let's see how covalent bonds are formed in the somewhat more complicated methane molecule (CH_4). The carbon atom has six electrons: two electrons fill its inner shell, and four electrons are in its outer shell. Because of the octet rule, carbon is most stable when it shares electrons with four other atoms—*it can form four covalent bonds* (**FIGURE 2.3A**). Methane forms when an atom of carbon reacts with four hydrogen atoms. As a result of electron sharing, the outer shell of the carbon atom is now filled with eight electrons—a stable configuration. The single shell of each hydrogen atom is also filled. Four covalent bonds—four shared electron pairs—hold methane together.

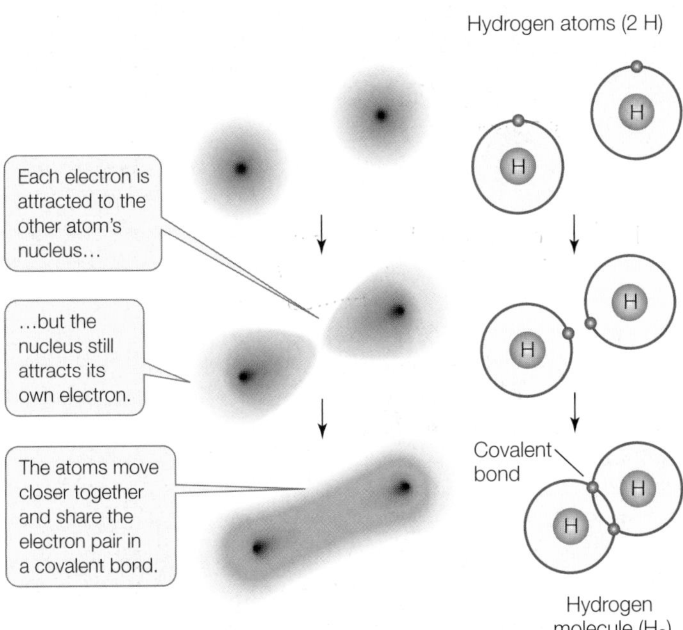
Hydrogen atoms (2 H)
Each electron is attracted to the other atom's nucleus...
...but the nucleus still attracts its own electron.
The atoms move closer together and share the electron pair in a covalent bond.
Covalent bond
Hydrogen molecule (H_2)

FIGURE 2.2 Electrons Are Shared in Covalent Bonds Two hydrogen atoms can combine to form a hydrogen molecule. A covalent bond forms when the electron shells of the two atoms overlap in an energetically stable manner.

FIGURE 2.3B shows several different ways to represent the molecular structure of methane. **TABLE 2.2** shows the covalent bonding capacities of some biologically important elements.

The properties of molecules are influenced by the characteristics of their covalent bonds. Four important aspects of covalent bonds are orientation, strength and stability, multiple covalent bonds, and the degree of sharing of electrons.

ORIENTATION For a given pair of elements, such as carbon bonded to hydrogen, the length of the covalent bond is always the same. And for a given atom within a molecule, the angle of each covalent bond with respect to the others is generally the same. This is true regardless of the type of larger molecule that contains the atom. For example, the four covalent bonds formed by the carbon atom in methane are always distributed in space so that the bonded hydrogens point to the corners of a regular tetrahedron, with the carbon in the center (see Figure 2.3B). Even when carbon is bonded to four atoms other than hydrogen, this three-dimensional orientation is more or less maintained. As you will see, the orientations of covalent bonds in space give molecules their three-dimensional geometry, and the shapes of molecules contribute to their biological functions.

STRENGTH AND STABILITY Covalent bonds are very strong (see Table 2.1), meaning it takes a lot of energy to break them. At the temperatures at which life exists, the covalent bonds of biological molecules are quite stable, as are their three-dimensional structures. However, this stability does not rule out change, as we will discover.

MULTIPLE COVALENT BONDS As shown in Figure 2.3B, covalent bonds can be represented by lines between the chemical symbols for the linked atoms:

- A *single bond* involves the sharing of a single pair of electrons (for example, H—H or C—H).

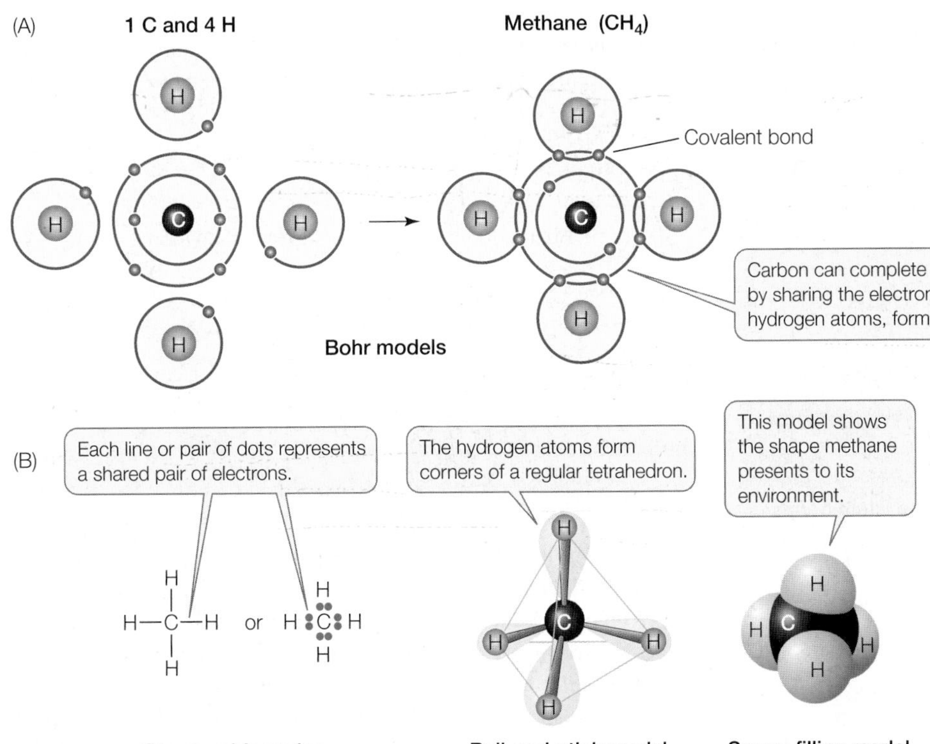
(A)
1 C and 4 H
Methane (CH_4)
Covalent bond
Carbon can complete its outer shell by sharing the electrons of four hydrogen atoms, forming methane.
Bohr models
(B)
Each line or pair of dots represents a shared pair of electrons.
The hydrogen atoms form corners of a regular tetrahedron.
This model shows the shape methane presents to its environment.
Structural formulas
Ball-and-stick model
Space-filling model

FIGURE 2.3 Covalent Bonding (A) Bohr models showing the formation of covalent bonds in methane, whose molecular formula is CH_4. Electrons are shown in shells around the nuclei. (B) Three additional ways of representing the structure of methane. The ball-and-stick and the space-filling models show the spatial orientations of the bonds. The space-filling model indicates the overall shape and surface of the molecule. In the chapters that follow, different conventions will be used to depict molecules. Bear in mind that these are models to illustrate certain properties and are not the most accurate portrayals of reality.

TABLE 2.1 Chemical Bonds and Interactions

Name	Basis of interaction	Structure	Bond energy[a]
Ionic attraction	Attraction of opposite charges	$-N^{+}H_3 \cdots {}^{-}O-C(=O)-$	3–7
Covalent bond	Sharing of electron pairs	$-N(H)-C(=O)-$	50–110
Hydrogen bond	Attraction between H (δ+) and a strongly electronegative atom	$-N(H)-H^{\delta+}\cdots O^{\delta-}=C-$	3–7
Hydrophobic interaction	Interaction of nonpolar substances in the presence of polar substances (especially water)	$-CH_2-CH_3$ … $C-C$ (nonpolar groups enclosed)	1–2
van der Waals interaction	Interaction of electrons of nonpolar substances	$H-H \cdots CH_4$	1

[a]Bond energy is the amount of energy (Kcal/mol) needed to separate two bonded or interacting atoms under physiological conditions.

- A *double bond* involves the sharing of four electrons (two pairs; C═C).
- *Triple bonds*—six shared electrons—are rare, but there is one in nitrogen gas (N≡N), which is the major component of the air we breathe.

UNEQUAL SHARING OF ELECTRONS If two atoms of the same element are covalently bonded, there is an equal sharing of the pair(s) of electrons in their outermost shells. However, when the two atoms are different elements, the sharing is not necessarily equal. One nucleus may exert a greater attractive force on the electron pair than the other nucleus, so that the pair tends to be closer to that atom.

The attractive force that an atomic nucleus exerts on electrons in a covalent bond is called its **electronegativity**. The electronegativity of a nucleus depends on how many positive charges it has (nuclei with more protons are more positive and thus more attractive to electrons) and on the distance between the electrons in the bond and the nucleus (the closer the electrons, the greater the electronegative pull). **TABLE 2.3** shows the electronegativities of some elements important in biological systems. (Electronegativity is calculated to produce a dimensionless quantity, meaning that it has no unit of measurement such as for length, time, mass, etc.)

TABLE 2.2 Covalent Bonding Capacities of Some Biologically Important Elements

Element	Usual number of covalent bonds
Hydrogen (H)	1
Oxygen (O)	2
Sulfur (S)	2
Nitrogen (N)	3
Carbon (C)	4
Phosphorus (P)	5

If two atoms are 0.5 or less apart in electronegativity, they will share electrons equally in what is called a nonpolar covalent bond. Two oxygen atoms, for example, each with an electronegativity of 3.4, will share electrons equally. So will two hydrogen atoms (each with an electronegativity of 2.2). But when hydrogen bonds with oxygen to form water, the electrons involved are *unequally shared*: they tend to be nearer to the oxygen nucleus because it is more electronegative than hydrogen. When electrons are drawn to one nucleus more than to the other, the result is a **polar covalent bond**:

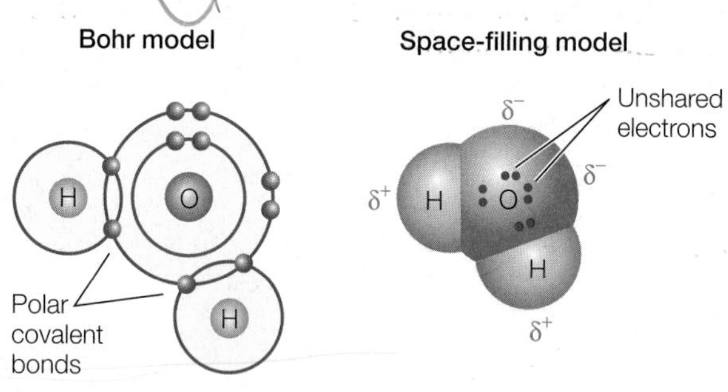

Because of this unequal sharing of electrons, the oxygen end of the bond has a slightly negative charge (symbolized by δ⁻ and spoken of as "delta negative," meaning a partial unit of charge), and the hydrogen end has a slightly positive charge (δ⁺). The bond is *polar* because these opposite charges are separated at the two ends, or poles, of the bond. The partial charges that result from polar covalent bonds produce polar molecules or

TABLE 2.3 Some Electronegativities

Element	Electronegativity
Oxygen (O)	3.4
Chlorine (Cl)	3.2
Nitrogen (N)	3.0
Carbon (C)	2.6
Phosphorus (P)	2.2
Hydrogen (H)	2.2
Sodium (Na)	0.9
Potassium (K)	0.8

polar regions of large molecules. Polar bonds within molecules greatly influence the interactions they have with other polar molecules. The polarity of the water molecule has significant effects on its physical properties and chemical reactivity, as we will see in later chapters.

Hydrogen bonds may form within or between molecules with polar covalent bonds

In liquid water, the negatively charged oxygen (δ^-) atom of one water molecule is attracted to the positively charged hydrogen (δ^+) atoms of other water molecules (**FIGURE 2.4A**). The bond resulting from this attraction is called a **hydrogen bond** (see Table 2.1). These bonds are not restricted to water molecules. A hydrogen bond may also form between a strongly electronegative atom and a hydrogen atom that is covalently bonded to another electronegative atom (oxygen or nitrogen), as shown in **FIGURE 2.4B**.

A hydrogen bond is much weaker than a covalent bond (see Table 2.1). Although individual hydrogen bonds are weak, many of them can form within one molecule or between two molecules. In these cases, the hydrogen bonds together have considerable strength and can greatly influence the structure and properties of the substances. Hydrogen bonds play important roles in determining and maintaining the three-dimensional shapes of giant molecules such as DNA and proteins (see Chapter 3).

Hydrogen bonding also contributes to several properties of water that have great significance for life. As we will discuss later in this Concept, water plays a vital role in living systems as a **solvent**: a liquid in which other molecules dissolve. Other important properties of water are its heat capacity, cohesion, adhesion, and surface tension.

HEAT CAPACITY At any given time in liquid water, a water molecule forms an average of 3.4 hydrogen bonds (dotted red lines below) with other water molecules:

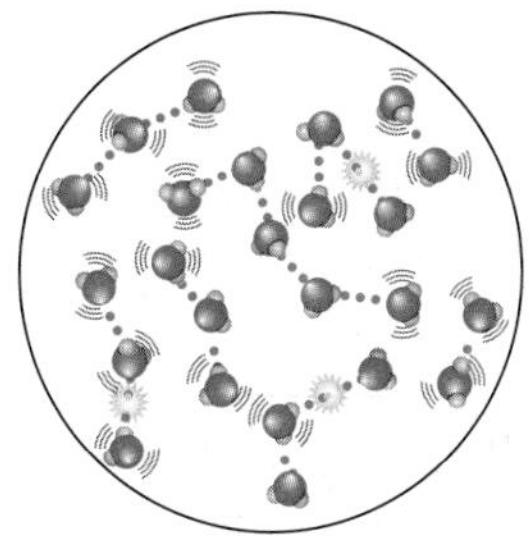

Liquid water

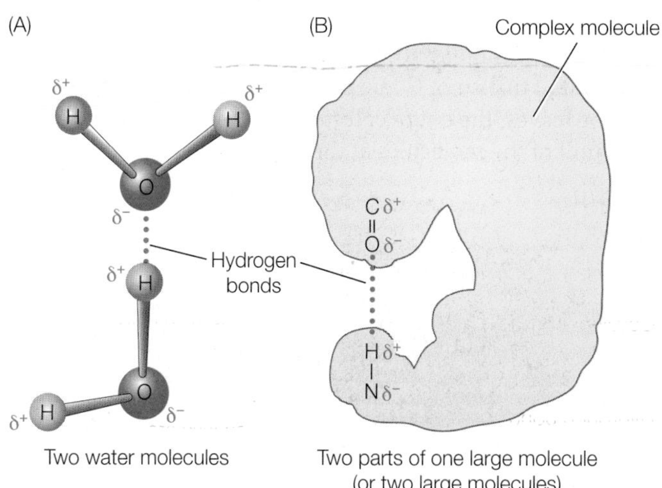

FIGURE 2.4 Hydrogen Bonds Can Form between or within Molecules (A) A hydrogen bond forms between two molecules because of the attraction between an atom with a partial negative charge on one molecule and a hydrogen with a partial positive charge on a second molecule. (B) Hydrogen bonds can form between different parts of the same large molecule.

These multiple hydrogen bonds contribute to the high **heat capacity** of water. Raising the temperature of liquid water takes a lot of heat, because much of the heat energy is used to break the hydrogen bonds that hold the liquid together (indicated by the yellow energy bursts in the liquid water diagram in the previous column). Think of what happens when you apply heat to a pan of water on the stove: it takes a while for the water to begin boiling. The same happens with a living organism—the large amount of water in living tissues shields the organism from fluctuations in environmental temperature.

Hydrogen bonding also gives water a high **heat of vaporization**. This means that a lot of heat is required to change water from its liquid to its gaseous state, in the process of evaporation. Once again, much of the heat energy is used to break the many hydrogen bonds between the water molecules. This heat must be absorbed from the environment in contact with the water. Evaporation thus has a cooling effect on the environment—whether a leaf, a forest, or an entire land mass. This effect explains why sweating cools the human body: as sweat evaporates from the skin, it absorbs some of the adjacent body heat.

LINK

Evaporation is important in the physiology of both plants and animals; see **Concepts 25.3 and 29.4**

COHESION, ADHESION, AND SURFACE TENSION The numerous hydrogen bonds that give water a high heat capacity and high heat of vaporization also explain the cohesive strength of liquid water. This cohesive strength, or **cohesion**, is defined as the capacity of water molecules to resist coming apart from one another when placed under tension. Hydrogen bonding between liquid water molecules and the solid surfaces surrounding them also allows for **adhesion** between the water and the solid surfaces. Together, cohesion and adhesion permit narrow columns of liquid water to move from the roots to the leaves of tall trees (see Concept 25.3). When water evaporates from the leaves, the entire column moves upward in response to the pull of the molecules at the top.

Cohesion causes water droplets to round up.
Adhesion allows water to cling to solid surfaces.

The surface of liquid water exposed to air is difficult to puncture because the water molecules at the surface are hydrogen-bonded to other molecules below them. This **surface tension** permits a container to be filled with water above its rim without overflowing, and it permits spiders to walk on the surface of a pond.

Polar and nonpolar substances: Each interacts best with its own kind

Just as water molecules can interact with one another through hydrogen bonds, any polar molecule can interact with any other polar molecule through the weak (δ^+ to δ^-) attractions of hydrogen bonds. Polar molecules interact with water in this way and are called **hydrophilic** ("water-loving"). In aqueous (watery) solutions, these molecules become separated and surrounded by water molecules (**FIGURE 2.5A**). Thus, water functions as a solvent for polar molecules.

Because they do not have partial charges, nonpolar molecules do not interact with water in this way. Instead, these molecules tend to aggregate with one another. This allows more hydrogen bonds to form between water molecules. Therefore, nonpolar molecules are known as **hydrophobic** ("water-hating"), and the interactions between them are called hydrophobic interactions (**FIGURE 2.5B**). Hydrophobic substances do not really "hate" water—they can form weak interactions with it, since the electronegativities of the atoms in many nonpolar bonds (e.g., C—H bonds) are not exactly the same. But these interactions are far weaker than the hydrogen bonds between the water molecules, so the nonpolar substances tend to aggregate.

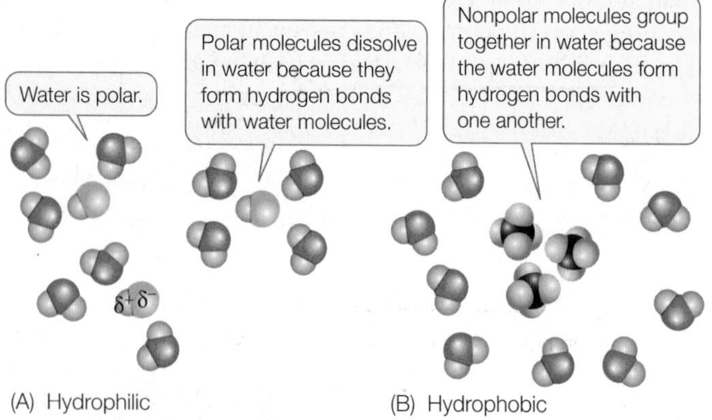

FIGURE 2.5 Hydrophilic and Hydrophobic (A) Molecules with polar covalent bonds are attracted to polar water (they are hydrophilic). (B) Molecules with nonpolar covalent bonds show greater attraction to one another than to water (they are hydrophobic). The color convention in the models shown here (gray, H; red, O; black, C; green, F) is often used.

APPLY THE CONCEPT

Atoms interact and form molecules

The concepts of chemical bonding and electronegativity (see Table 2.2) allow us to predict whether a molecule will be polar or nonpolar, and how it will interact with water. Typically, a difference in electronegativity greater than 0.5 will result in polarity. For each of the bonds below, indicate:

1. Whether the bond is polar or nonpolar
2. If polar, which is the δ^+ end
3. How a molecule with the bond will interact with water (hydrophilic or hydrophobic)

N—H   C—H   C=O   C—N
O—H   C—C   H—H   O—P

Ionic attractions form between anions and cations

When one interacting atom is much more electronegative than the other (see Table 2.3), a complete transfer of one or more electrons may occur. Consider sodium (11 protons) and chlorine (17 protons). A sodium atom has only one electron in its outermost shell; this condition is unstable. A chlorine atom has seven electrons in its outermost shell—another unstable condition. The most straightforward way for both atoms to achieve stability is to transfer an electron from sodium's outermost shell to that of chlorine (**FIGURE 2.6**). This reaction makes the two atoms more stable because they both have eight electrons in their outer shells. The result is two ions.

An **ion** is an electrically charged particle that forms when an atom gains or loses one or more electrons:

- The sodium ion (Na^+) in our example has a charge of +1 because it has one less electron than it has protons. The outermost electron shell of the sodium ion is stable because it has eight electrons. Positively charged ions are called **cations**.
- The chloride ion (Cl^-) has a charge of –1 because it has one more electron than it has protons. This additional electron gives Cl^- a stable outermost shell with eight electrons. Negatively charged ions are called **anions**.

Ionic attractions form as a result of the electrical attraction between ions bearing opposite charges. Ionic attractions result in stable crystalline structures that are often referred to as ionic compounds or salts. An example is sodium chloride (NaCl; table salt), where cations and anions are held together by ionic attractions. Ionic attractions in salt crystals may be stronger, but attractions between ions in solution, as occur in living systems, are typically weak (see Table 2.1).

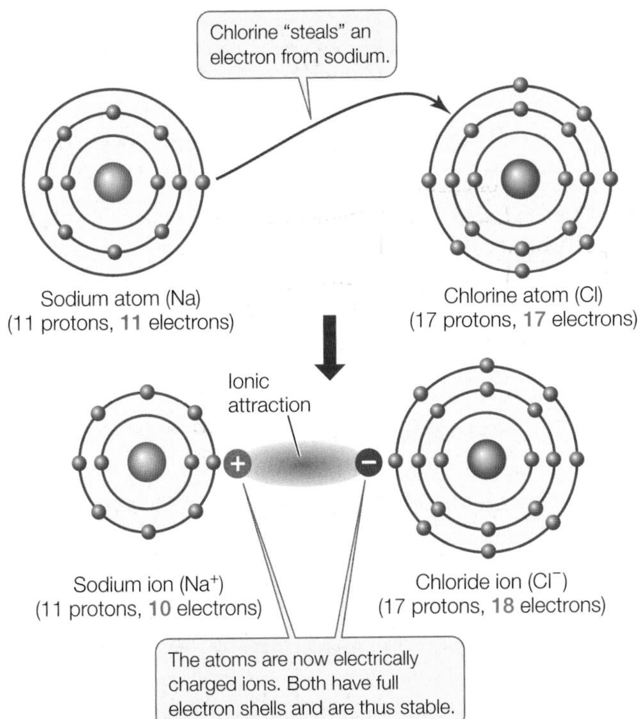

FIGURE 2.6 Ionic Attraction between Sodium and Chlorine
When a sodium atom reacts with a chlorine atom, the chlorine fills its outermost shell by "stealing" an electron from the sodium. In so doing, the chlorine atom becomes a negatively charged chloride ion (Cl^-). With one less electron, the sodium atom becomes a positively charged sodium ion (Na^+).

Given that living organisms consist of about 70 percent water, most biological processes occur in the presence of water. Because ionic attractions are weak, salts dissolve in water; the ions separate from one another and become surrounded by water molecules. The water molecules are oriented with their negative poles nearest to the cations and their positive poles nearest to the anions:

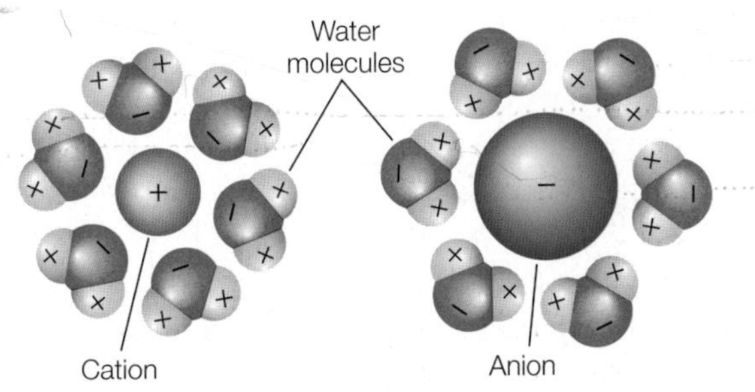

Functional groups confer specific properties to biological molecules

Certain small groups of atoms, called **functional groups**, are consistently found together in very different biological molecules. You will encounter several functional groups repeatedly in your study of biology (**FIGURE 2.7**). Each functional group has specific chemical properties (for example, polarity), and when attached to a larger molecule, it gives those properties to the larger molecule. The consistent chemical behavior

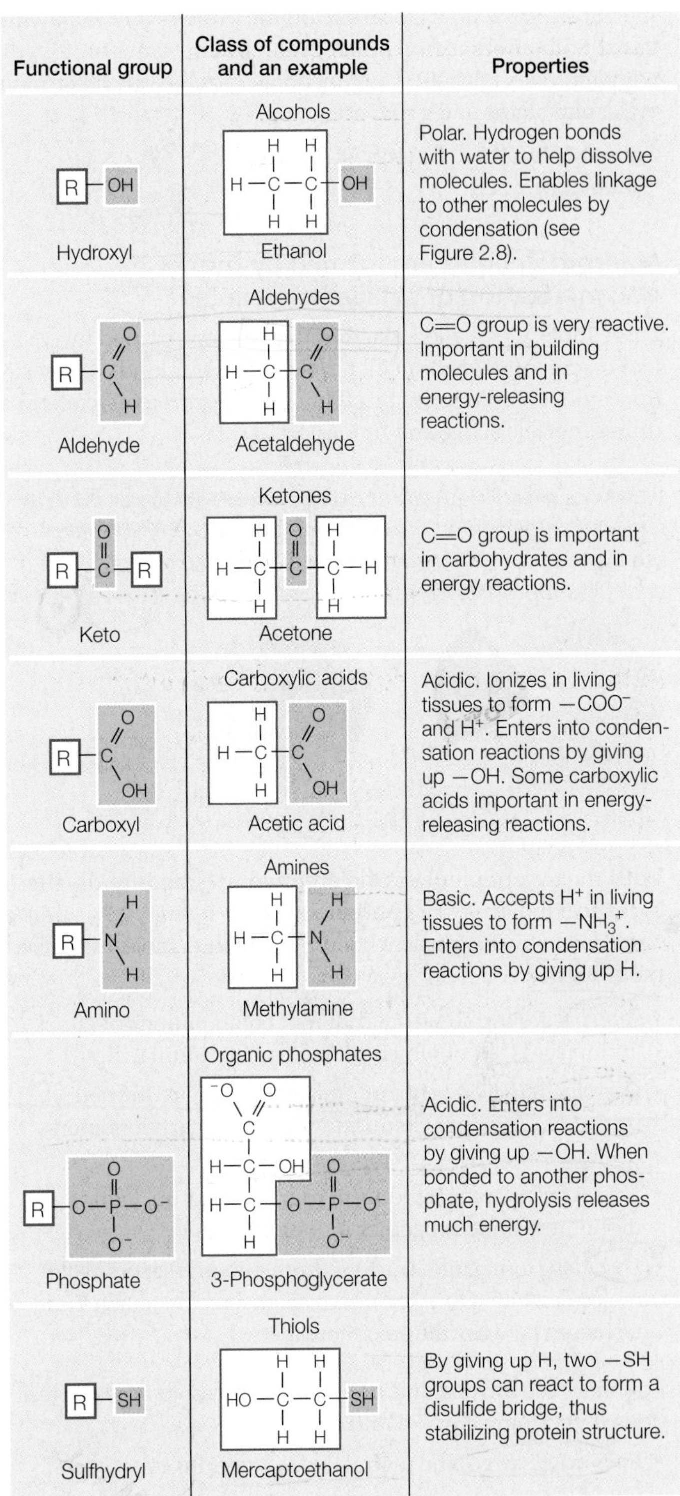

Functional group	Class of compounds and an example	Properties
R—OH Hydroxyl	Alcohols Ethanol	Polar. Hydrogen bonds with water to help dissolve molecules. Enables linkage to other molecules by condensation (see Figure 2.8).
R—CHO Aldehyde	Aldehydes Acetaldehyde	C=O group is very reactive. Important in building molecules and in energy-releasing reactions.
R—CO—R Keto	Ketones Acetone	C=O group is important in carbohydrates and in energy reactions.
R—COOH Carboxyl	Carboxylic acids Acetic acid	Acidic. Ionizes in living tissues to form —COO⁻ and H^+. Enters into condensation reactions by giving up —OH. Some carboxylic acids important in energy-releasing reactions.
R—NH_2 Amino	Amines Methylamine	Basic. Accepts H^+ in living tissues to form $-NH_3^+$. Enters into condensation reactions by giving up H.
R—O—PO_3^{2-} Phosphate	Organic phosphates 3-Phosphoglycerate	Acidic. Enters into condensation reactions by giving up —OH. When bonded to another phosphate, hydrolysis releases much energy.
R—SH Sulfhydryl	Thiols Mercaptoethanol	By giving up H, two —SH groups can react to form a disulfide bridge, thus stabilizing protein structure.

FIGURE 2.7 Functional Groups Important to Living Systems
Highlighted in yellow are the seven functional groups most commonly found in biological molecules. "R" represents the rest of the molecule.

of functional groups helps us understand the properties of the molecules that contain them.

Biological molecules often contain many different functional groups. A single large protein may contain hydrophobic, polar, and charged functional groups. Each group gives a different specific property to its local site on the protein, and it may interact with another functional group on the same protein or with another molecule. The functional groups thus determine molecular shape and reactivity.

Go to ACTIVITY 2.1 Functional Groups
PoL2e.com/ac2.1

Macromolecules are formed by the polymerization of smaller molecules

Large molecules, called **macromolecules**, are formed by covalent linkages between smaller molecules. Four kinds of macromolecules are characteristic of living things: proteins, carbohydrates, nucleic acids, and lipids.

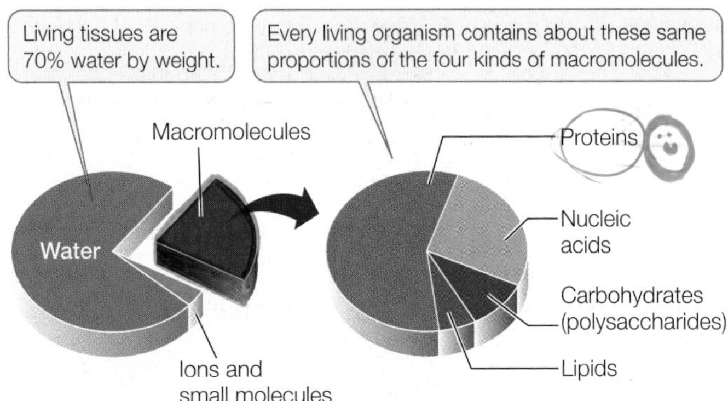

With the exception of small carbohydrates and lipids, these biological molecules are **polymers** (*poly*, "many"; *mer*, "unit") constructed by the covalent bonding of smaller molecules called **monomers**.

- *Proteins* are formed from different combinations of 20 amino acids, all of which share chemical similarities.
- *Carbohydrates* can be giant molecules, and are formed by linking together chemically similar sugar monomers (monosaccharides) to form polysaccharides.
- *Nucleic acids* are formed from four kinds of nucleotide monomers linked together in long chains.
- *Lipids* also form large structures from a limited set of smaller molecules, but in this case noncovalent forces maintain the interactions between the lipid monomers.

Polymers are formed and broken down by two types of reactions involving water (**FIGURE 2.8**):

- In **condensation**, the removal of water links monomers together.
- In **hydrolysis**, the addition of water breaks a polymer into monomers.

How the macromolecules function and interact with other molecules depends on the properties of the functional groups in their monomers.

Go to ANIMATED TUTORIAL 2.2
Macromolecules: Carbohydrates and Lipids
PoL2e.com/at2.2

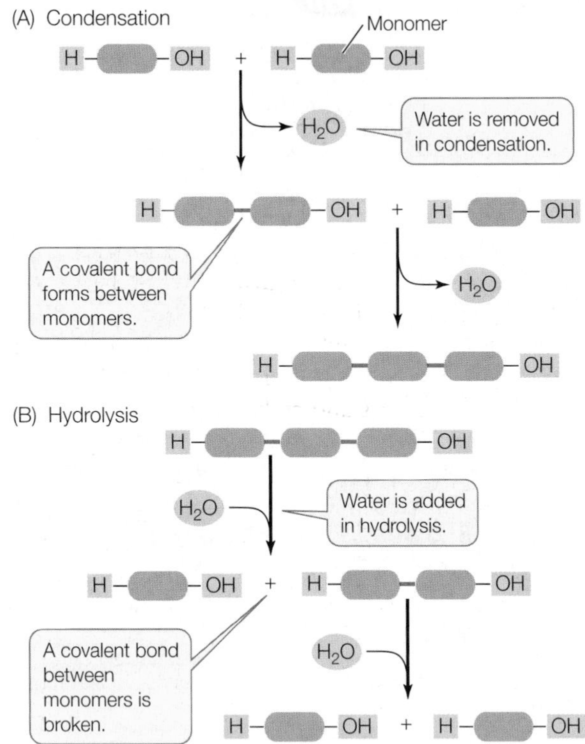

FIGURE 2.8 Condensation and Hydrolysis of Polymers
(A) Condensation reactions link monomers into polymers and produce water. (B) Hydrolysis reactions break polymers into individual monomers and consume water.

CHECKpoint CONCEPT 2.2

✓ How do differences in electronegativity result in the unequal sharing of electrons in polar molecules?

✓ Some functional groups (see Figure 2.7) can either donate or accept hydrogen bonds with other molecules, acting either as a donor (like the oxygen in a water molecule) or as an acceptor (like the hydrogens in a water molecule). For each of the following, is it an H-bond donor, acceptor, both, or neither?
a. Aldehyde
b. Amino
c. Hydroxyl

✓ Here is the structure of the molecule glycine:

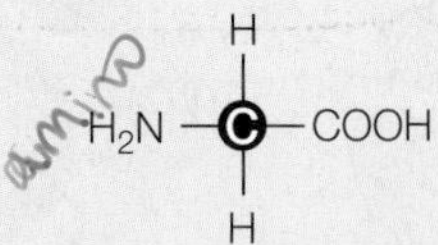

a. What are the functional groups on this molecule? What is the R group to which they are attached? Is the R group hydrophilic or hydrophobic? Explain.
b. Draw two glycine molecules and show how they can be linked by a condensation reaction.

✓ The boiling point (the temperature at which a liquid vaporizes) of water (H_2O) is 100°C, whereas the boiling point of methane (CH_4) is –161°C. Explain this difference in terms of hydrogen bonding between molecules.

We will begin our discussion of the molecules of life with carbohydrates, as they exemplify many of the chemical principles we have outlined so far.

CONCEPT 2.3 Carbohydrates Consist of Sugar Molecules

Carbohydrates are a large group of molecules that all have similar atomic compositions but differ greatly in size, chemical properties, and biological functions. Carbohydrates usually have the general formula $C_mH_{2n}O_n$, where *m* and *n* represent numbers. This makes them appear to be hydrates of carbon [associations between water molecules and carbon in the ratio $C_m(H_2O)_n$], hence their name. However, carbohydrates are not really "hydrates" because the water molecules are not intact. Rather, the linked carbon atoms are bonded with hydrogen atoms (—H) and hydroxyl groups (—OH), the components of water. Carbohydrates have four major biochemical roles:

- They are a source of stored energy that can be released in a form that organisms can use.
- They are used to transport stored energy within complex organisms.
- They function as structural molecules that give many organisms their shapes.
- They serve as recognition or signaling molecules that can trigger specific biological responses.

Some carbohydrates are relatively small, such as the simple sugars (for example, glucose) that are the primary energy source for many organisms. Others are large polymers of simple sugars; an example is starch, which is stored in seeds.

Monosaccharides are simple sugars

Monosaccharides (*mono*, "one") are relatively simple molecules with up to seven carbon atoms. They differ in their arrangements of carbon, hydrogen, and oxygen atoms (**FIGURE 2.9**).

Pentoses (*pente*, "five") are five-carbon sugars. Two pentoses are of particular biological importance: the backbones of the nucleic acids RNA and DNA contain ribose and deoxyribose, respectively.

LINK
For a description of the nucleic acids RNA and DNA see **Concept 3.1**

The hexoses (*hex*, "six") all have the formula $C_6H_{12}O_6$. They include glucose, fructose (so named because it was first found in fruits), mannose, and galactose.

Go to ACTIVITY 2.2 Forms of Glucose
PoL2e.com/ac2.2

Glycosidic linkages bond monosaccharides

The disaccharides, oligosaccharides, and polysaccharides are all constructed from monosaccharides that are covalently bonded by condensation reactions that form **glycosidic linkages**. A single glycosidic linkage between two monosaccharides forms a **disaccharide**. For example, sucrose—common table sugar—is a major disaccharide formed in plants from a glucose molecule and a fructose molecule (see Figure 2.9 for complete structures):

Glucose + Fructose → (Formation of linkage, H_2O) → Sucrose

Another disaccharide is maltose, which is formed from two glucose units and is a product of starch digestion. Maltose is an important carbohydrate for making beer.

Oligosaccharides contain several monosaccharides bound together by glycosidic linkages. Many oligosaccharides have additional functional groups, which give them special properties. Oligosaccharides are often covalently bonded to proteins and lipids on the outer surfaces of cells, where they serve as recognition signals. For example, the different human blood groups (the ABO blood types) get their specificity from oligosaccharide chains.

Polysaccharides store energy and provide structural materials

Polysaccharides are large polymers of monosaccharides connected by glycosidic linkages (**FIGURE 2.10**). Polysaccharides are not necessarily linear chains of monomers. Each monomer unit has several sites that are capable of forming glycosidic linkages, and thus branched molecules are possible.

FIGURE 2.9 Monosaccharides Monosaccharides are made up of varying numbers of carbons. Many have the same kind and number of atoms, but the atoms are arranged differently.

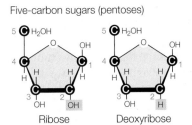

Five-carbon sugars (pentoses)

Ribose

Deoxyribose

Ribose and deoxyribose each have five carbons, but very different chemical properties and biological roles.

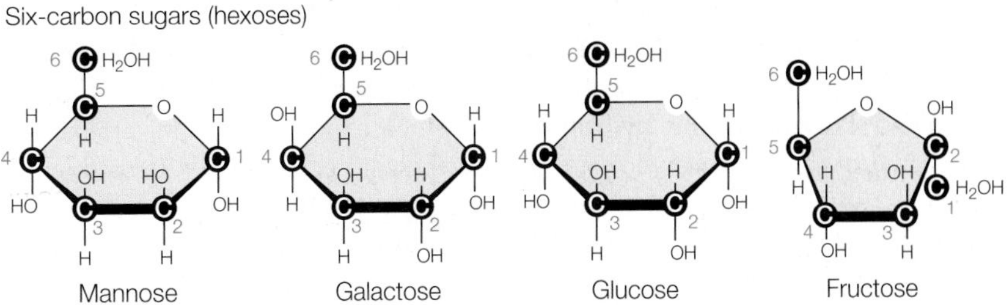

These hexoses all have the formula $C_6H_{12}O_6$, but each has distinct biochemical properties.

(A) Molecular structure

Cellulose

Hydrogen bonding to other cellulose molecules can occur at these points.

Cellulose is an unbranched polymer of glucose with linkages that are chemically very stable.

Starch and glycogen

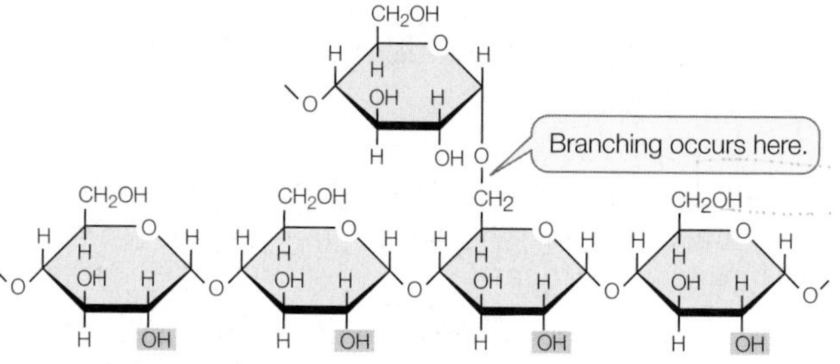

Glycogen and starch are polymers of glucose, with branching at carbon 6 (see Figure 2.9).

(B) Macromolecular structure

Linear (cellulose)

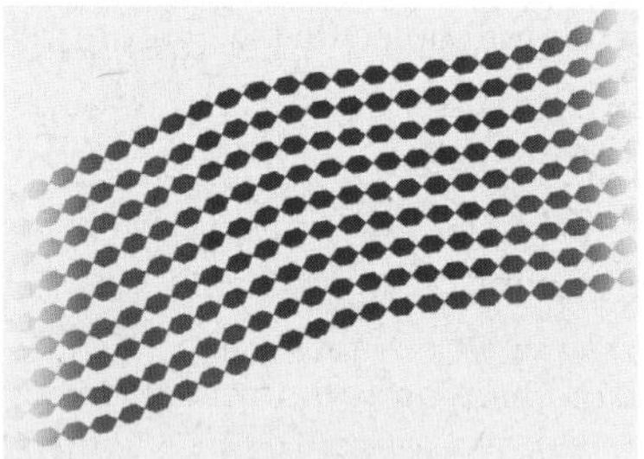

Parallel cellulose molecules form hydrogen bonds, resulting in thin fibrils.

Branched (starch)

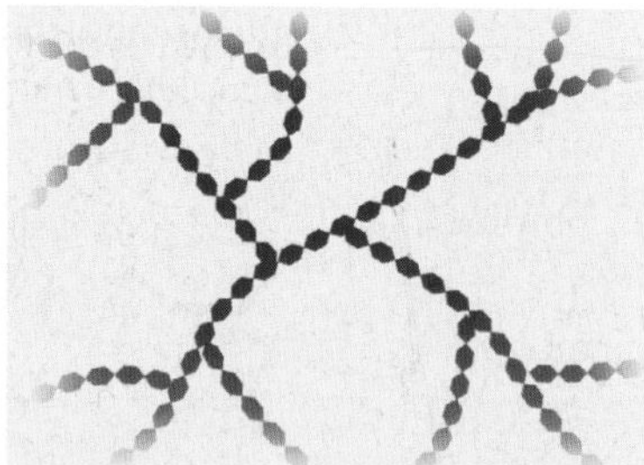

Branching limits the number of hydrogen bonds that can form in starch molecules, making starch less compact than cellulose.

Highly branched (glycogen)

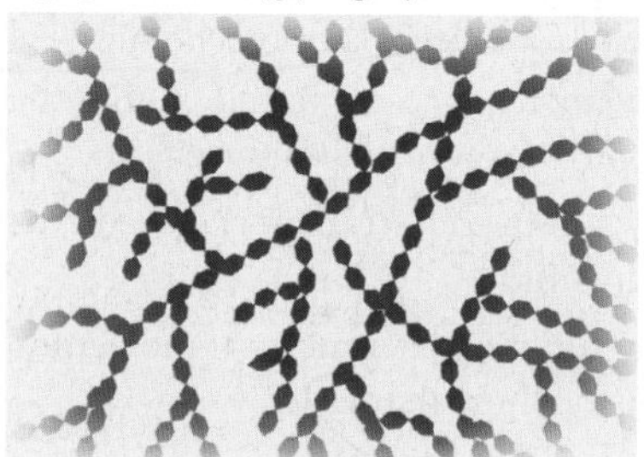

The high amount of branching in glycogen makes its solid deposits more compact than starch.

(C) Polysaccharides in cells

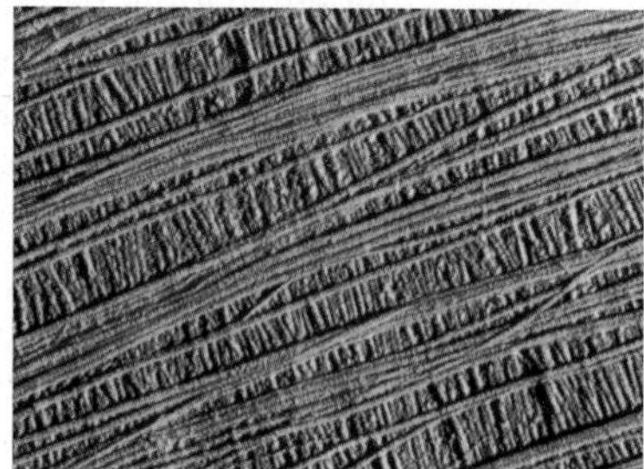

Layers of cellulose fibrils, as seen in this scanning electron micrograph, give plant cell walls great strength. 0.1 μm

Within these potato cells, starch deposits have a granular shape. 15 μm

The dark clumps in this electron micrograph are glycogen deposits in a monkey liver cell. 50 μm

FIGURE 2.10 Polysaccharides Cellulose, starch, and glycogen are all composed of long chains of glucose but with different levels of branching and compaction.

Starches comprise a family of giant molecules that are all polysaccharides of glucose. The different starches can be distinguished by the amount of branching in their polymers. Starch is the principal energy storage compound of plants.

Glycogen is a water-insoluble, highly branched polymer of glucose that is the major energy storage molecule in mammals. It is produced in the liver and transported to the muscles. Both glycogen and starch are readily hydrolyzed into glucose monomers, which in turn can be broken down to liberate their stored energy.

If glucose is the major source of fuel, why store it in the form of starch or glycogen? The reason is that 1,000 glucose molecules would exert 1,000 times the osmotic pressure of a single glycogen molecule, causing water to enter the cells (see Concept 5.2). If it were not for polysaccharides, many organisms would expend a lot of energy expelling excess water from their cells.

As the predominant component of plant cell walls, cellulose is by far the most abundant carbon-containing (organic) compound on Earth. Like starch and glycogen, cellulose is a polysaccharide of glucose, but its glycosidic linkages are arranged in such a way that it is a much more stable molecule. Whereas starch is easily broken down by chemicals or enzymes to supply glucose for energy-producing reactions, cellulose is an excellent structural material that can withstand harsh environmental conditions without substantial change.

LINK

Most animals cannot digest (hydrolyze) cellulose; **Chapter 30** describes adaptations in some animals to use cellulose as an energy source

CHECKpoint CONCEPT 2.3

- ✓ Draw the chemical structure of a disaccharide formed from two glucose monosaccharides.
- ✓ Examine the glucose molecule shown in Figure 2.9. What are the functional groups on the molecule?
- ✓ Notice the large number of hydrogen bonding groups present in the linear structure of cellulose (see Figure 2.10A). Why is this structure so strong?
- ✓ Some sugars have other functional groups in addition to those typically present. Draw the structure of the amino sugar glucosamine, which has an amino group bonded at carbon 2 of glucose. Would this molecule be more or less polar than glucose? Explain why.

We have seen that carbohydrate structure is an example of the monomer–polymer theme in biology. Now we will turn to lipids, which are unusual among the four classes of biological macromolecules in that they are not, strictly speaking, polymers.

CONCEPT 2.4 Lipids Are Hydrophobic Molecules

Lipids—commonly called fats and oils—are hydrocarbons (composed of C and H atoms) that are insoluble in water because of their many nonpolar covalent bonds. As you have seen, nonpolar molecules are hydrophobic and preferentially aggregate together, away from polar water (see Figure 2.5). When nonpolar hydrocarbons are sufficiently close together, weak but additive **van der Waals interactions** (see Table 2.1) hold them together. The huge macromolecular aggregations that can form are not polymers in a strict chemical sense, because the individual lipid molecules are not covalently bonded.

Lipids play several roles in living organisms, including the following:

- They store energy in the C—C and C—H bonds.
- They play important structural roles in cell membranes and on body surfaces, largely because their nonpolar nature makes them essentially insoluble in water.
- Fat in animal bodies serves as thermal insulation.

Fats and oils are triglycerides

The most common units of lipids are **triglycerides**, also known as simple lipids. Triglycerides that are solid at room temperature (around 20°C) are called fats; those that are liquid at room temperature are called oils. A triglyceride contains three fatty acid molecules and one glycerol molecule. **Glycerol** is a small molecule with three hydroxyl (—OH) groups; thus it is an alcohol. A **fatty acid** consists of a long nonpolar hydrocarbon chain attached to the polar carboxyl (—COOH) group, and it is therefore a carboxylic acid. The long hydrocarbon chain is very hydrophobic because of its abundant C—H and C—C bonds.

Synthesis of a triglyceride involves three condensation reactions (**FIGURE 2.11**). The resulting molecule has very little polarity and is extremely hydrophobic. That is why fats and oils do not mix with water but float on top of it in separate globules or layers. The three fatty acids in a single triglyceride molecule need not all have the same hydrocarbon chain length or structure; some may be saturated fatty acids, while others may be unsaturated:

- In a **saturated fatty acid**, all the bonds between the carbon atoms in the hydrocarbon chain are single; there are no double bonds. That is, all the available bonds are saturated with hydrogen atoms (**FIGURE 2.12A**). These fatty acid molecules are relatively rigid and straight, and they pack together tightly, like pencils in a box.
- In an **unsaturated fatty acid**, the hydrocarbon chain contains one or more double bonds. Linoleic acid is an example of a polyunsaturated fatty acid that has two double bonds near the middle of the hydrocarbon chain, causing kinks in the chain (**FIGURE 2.12B**). Such kinks prevent the unsaturated molecules from packing together tightly.

The kinks in fatty acid molecules are important in determining the fluidity and melting point of the lipid. The triglycerides of animal fats tend to have many long-chain saturated fatty acids, which pack tightly together; these fats are usually solid at room temperature and have a high melting point. The triglycerides of plants, such as corn oil, tend to have short or

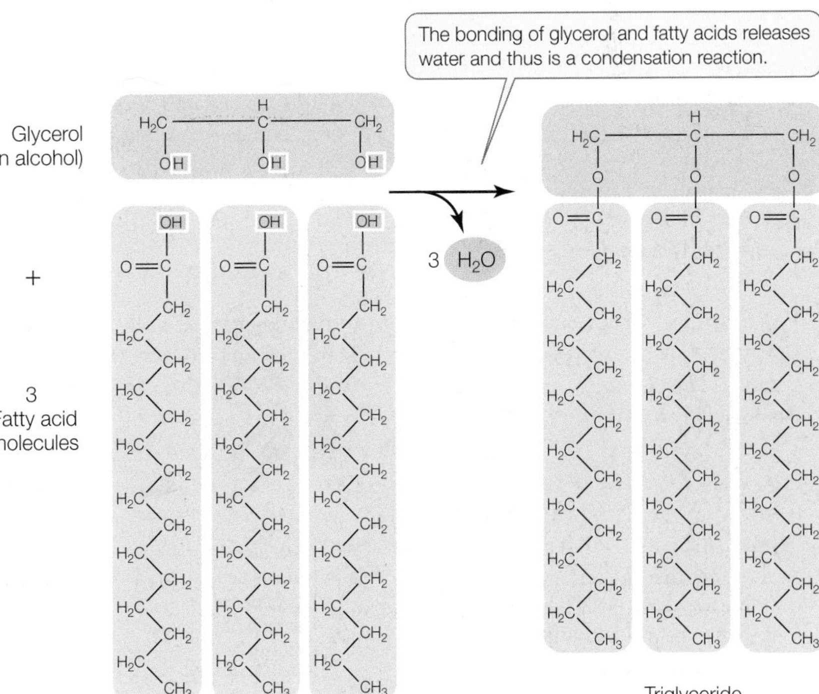

FIGURE 2.11 Synthesis of a Triglyceride In living things, the reaction that forms a triglyceride is more complex than the single step shown here.

unsaturated fatty acids. Because of their kinks, these fatty acids pack together poorly, have low melting points, and are usually liquid at room temperature.

Fats and oils are excellent storehouses for chemical energy. As you will see in Chapter 6, energy is released when these compounds are broken down into smaller molecules. The released energy can be used by an organism for other purposes, such as movement or the synthesis of complex molecules. On a per weight basis, broken-down lipids yield more than twice as much energy as degraded carbohydrates.

Phospholipids form biological membranes

We have mentioned the hydrophobic nature of the many C—C and C—H bonds in a fatty acid. But what about the carboxyl functional group at the end of the molecule? When it ionizes and forms COO^-, it is strongly hydrophilic. So a fatty acid has two opposing chemical properties: a hydrophilic end and a long hydrophobic tail. A molecule that is partly hydrophilic and partly hydrophobic is said to be **amphipathic**.

In triglycerides, a glycerol molecule is bonded to three fatty acid chains and the resulting molecule is entirely hydrophobic.

Phospholipids are like triglycerides in that they contain fatty acids bound to glycerol. However, in phospholipids, a phosphate-containing compound replaces one of the fatty acids, giving these molecules amphipathic properties (**FIGURE 2.13A**). The phosphate functional group (there are several different kinds in different phospholipids) has a negative electric charge, so this portion of the molecule is hydrophilic, attracting polar water molecules. But the two fatty acids are hydrophobic, so they tend to avoid water and aggregate together or with other hydrophobic substances.

In an aqueous environment, phospholipids line up in such a way that the nonpolar, hydrophobic "tails" pack tightly together and the phosphate-containing "heads" face outward, where they interact with water. The phospholipids thus form a **bilayer**: a sheet two molecules thick, with water excluded from the core (**FIGURE 2.13B**). Although no covalent bonds link individual lipids in these large aggregations, such stable aggregations form readily in aqueous conditions. Biological membranes have this kind of **phospholipid bilayer** structure, and we will devote Chapter 5 to their biological functions.

CHECKpoint CONCEPT 2.4

- ✓ What is the difference between fats and oils?
- ✓ Why are phospholipids amphipathic, and how does this result in a lipid bilayer membrane?
- ✓ If fatty acids are carefully put onto the surface of water, they form a single molecular layer. If the mixture is then shaken vigorously, the fatty acids will form round structures called micelles. Explain these observations.

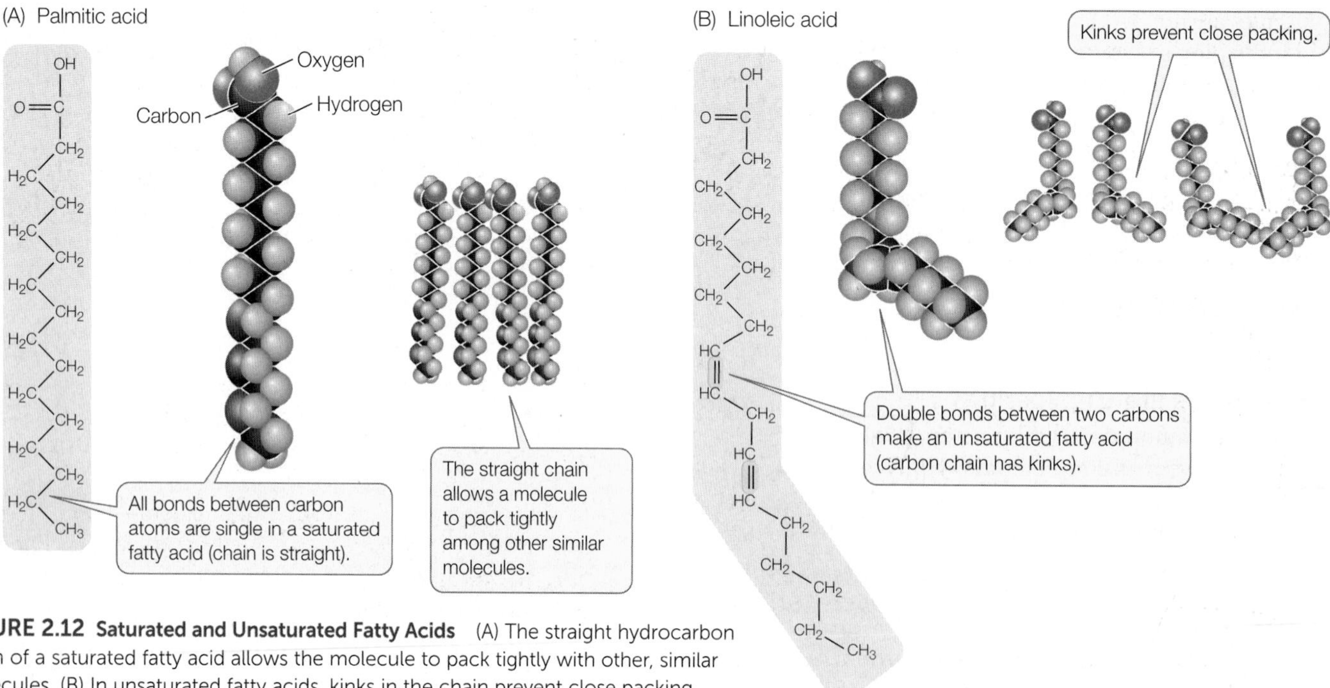

FIGURE 2.12 Saturated and Unsaturated Fatty Acids (A) The straight hydrocarbon chain of a saturated fatty acid allows the molecule to pack tightly with other, similar molecules. (B) In unsaturated fatty acids, kinks in the chain prevent close packing.

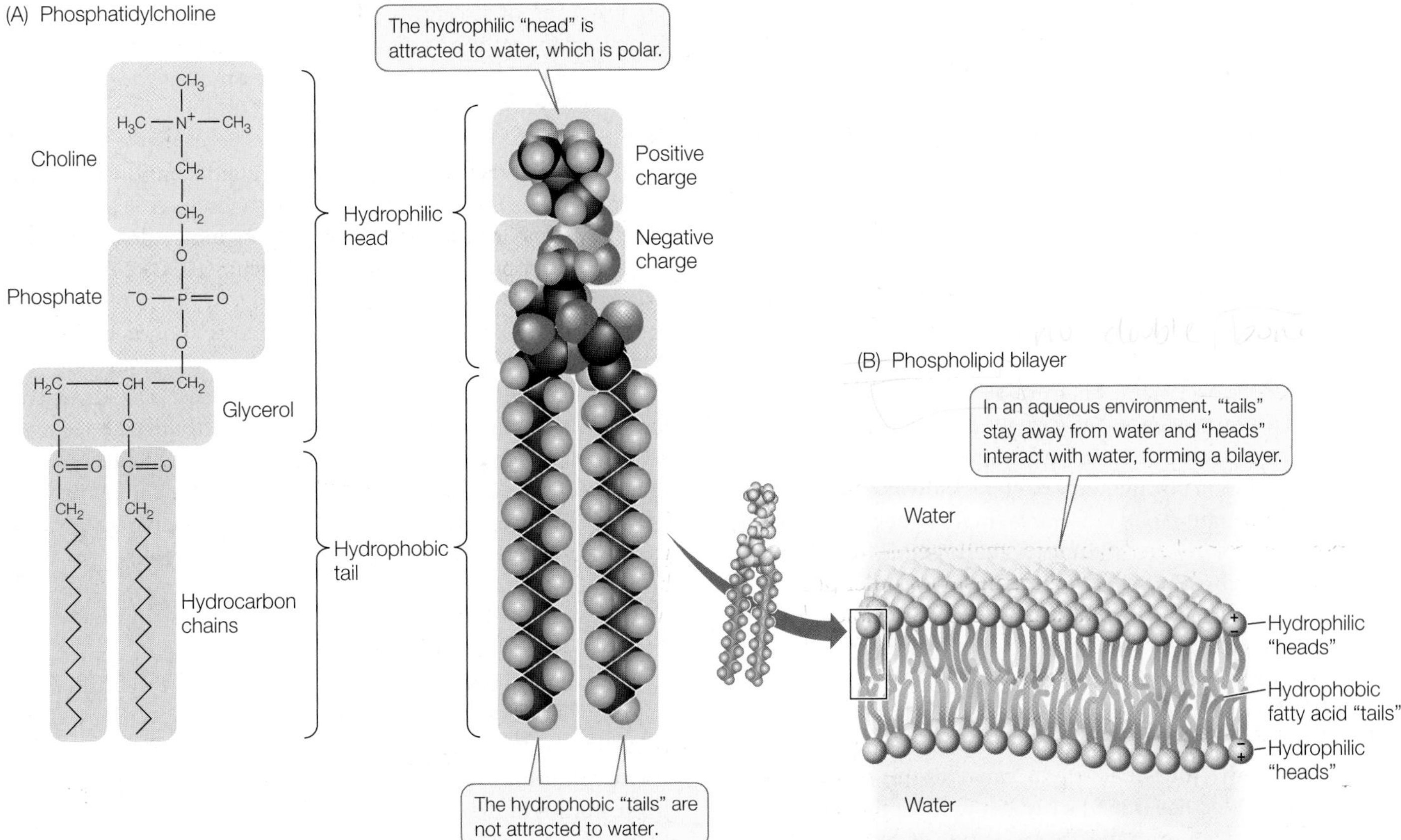

FIGURE 2.13 Phospholipids (A) Phosphatidylcholine (lecithin) is an example of a phospholipid molecule. In other phospholipids, the amino acid serine, the sugar alcohol inositol, or another compound replaces choline. (B) In an aqueous environment, hydrophobic interactions bring the "tails" of phospholipids together in the interior of a bilayer. The hydrophilic "heads" face outward on both sides of the bilayer, where they interact with the surrounding water molecules.

Molecules such as carbohydrates and lipids are not always stable in living systems. Rather, a hallmark of life is its *ability to transform molecules*. This involves making and breaking covalent bonds, as some atoms are removed and others are attached. As part of our introduction to biochemical concepts, we will now turn to these processes of chemical change.

CONCEPT 2.5 Biochemical Changes Involve Energy

A **chemical reaction** occurs when atoms have sufficient energy to combine, or to change their bonding partners. Consider the hydrolysis of the disaccharide sucrose to its component monomers, glucose and fructose (see p. 27 for the chemical structures). We can express this reaction using a chemical equation:

$$\underset{(C_{12}H_{22}O_{11})}{\text{Sucrose}} + H_2O \rightarrow \underset{(C_6H_{12}O_6)}{\text{glucose}} + \underset{(C_6H_{12}O_6)}{\text{fructose}}$$

In this equation, sucrose and water are the **reactants**, and glucose and fructose are the **products**. The reaction proceeds as some bonds in the reactants are broken and new bonds form to make the products. Electrons and protons are transferred from one reactant to the other to form the products. The products of this reaction have different properties from those of the reactants. Chemical reactions involve *changes in energy*; for example, the energy contained in the chemical bonds of sucrose and water (the reactants) is greater than the energy in the bonds of the two products, glucose and fructose.

What is energy? Physicists define it as the capacity to do work, which occurs when a force operates on an object over a distance. But in biochemistry, it is more useful to consider energy as *the capacity for change*. In biochemical reactions, energy changes are usually associated with changes in the chemical composition and properties of molecules. Energy comes in many forms: chemical, electrical, heat, light, and mechanical. But all forms of energy can be considered as one of two basic types:

- **Potential energy** is the energy of state or position—that is, stored energy. It can be stored in many forms: in chemical bonds, as a concentration gradient, or even as an electric charge imbalance.
- **Kinetic energy** is the energy of movement—that is, the type of energy that does work, that makes things change. For example, heat causes molecular motions and can even break chemical bonds.

Potential energy can be converted into kinetic energy and vice versa, and the form that the energy takes can be converted.

Think of reading this book: light energy is converted to chemical energy in your eyes, and then is converted to electrical energy in the nerve cells that carry messages to your brain. When you decide to turn a page, the electrical and chemical energy of nerves and muscles are converted to mechanical energy for movement of your hand and arm.

Metabolism involves reactions that store and release energy

The sum total of all the chemical reactions occurring in a biological system at a given time is called **metabolism**. Metabolic reactions involve energy changes, in which energy is either stored in, or released from, chemical bonds. In general, the formation of a bond releases energy, whereas the breaking of a bond requires an input of energy. More energy is released in the formation of a stronger, more stable (lower energy) bond than in the formation of a less stable (higher energy) one. Conversely, the breaking of a stronger bond consumes more energy than the breaking of a weaker bond. A chemical reaction will occur spontaneously if the total energy consumed by breaking bonds in the reactants is *less than* the total energy released by forming bonds in the products.

Anabolic reactions (collectively called anabolism) link simple molecules to form more complex molecules. Anabolic reactions require an input of energy because strong bonds within the smaller molecules must be broken to form the more complex molecules. For example, the formation of sucrose requires the breaking of strong O—H bonds in glucose and fructose. Reactions that require an input of energy are called endergonic or endothermic (**FIGURE 2.14A**). The energy used in anabolic reactions is stored in the newly formed (higher energy) chemical bonds.

Catabolic reactions (collectively called catabolism) break down complex molecules into simpler ones and release the energy that was used to make the complex molecules. Chemists call such reactions exergonic or exothermic (**FIGURE 2.14B**). For example, when sucrose is hydrolyzed, energy is released by the formation of more stable (lower energy) bonds within the monosaccharides.

Catabolic and anabolic reactions are often linked. The energy released in catabolic reactions is often used to drive anabolic reactions—that is, to do biological work. For example, the energy released by the breakdown of glucose (catabolism) is used to drive anabolic reactions such as the synthesis of triglycerides. That is why fat accumulates if you eat food in excess of your energy needs.

Biochemical changes obey physical laws

Recall from the opening of this chapter that we described the mechanistic view of life, whereby living systems obey the same rules that govern the nonliving world. The **laws of thermodynamics** (thermo, "energy"; dynamics, "change") were derived from studies of the fundamental properties of energy, and the ways energy interacts with matter. These laws apply to all

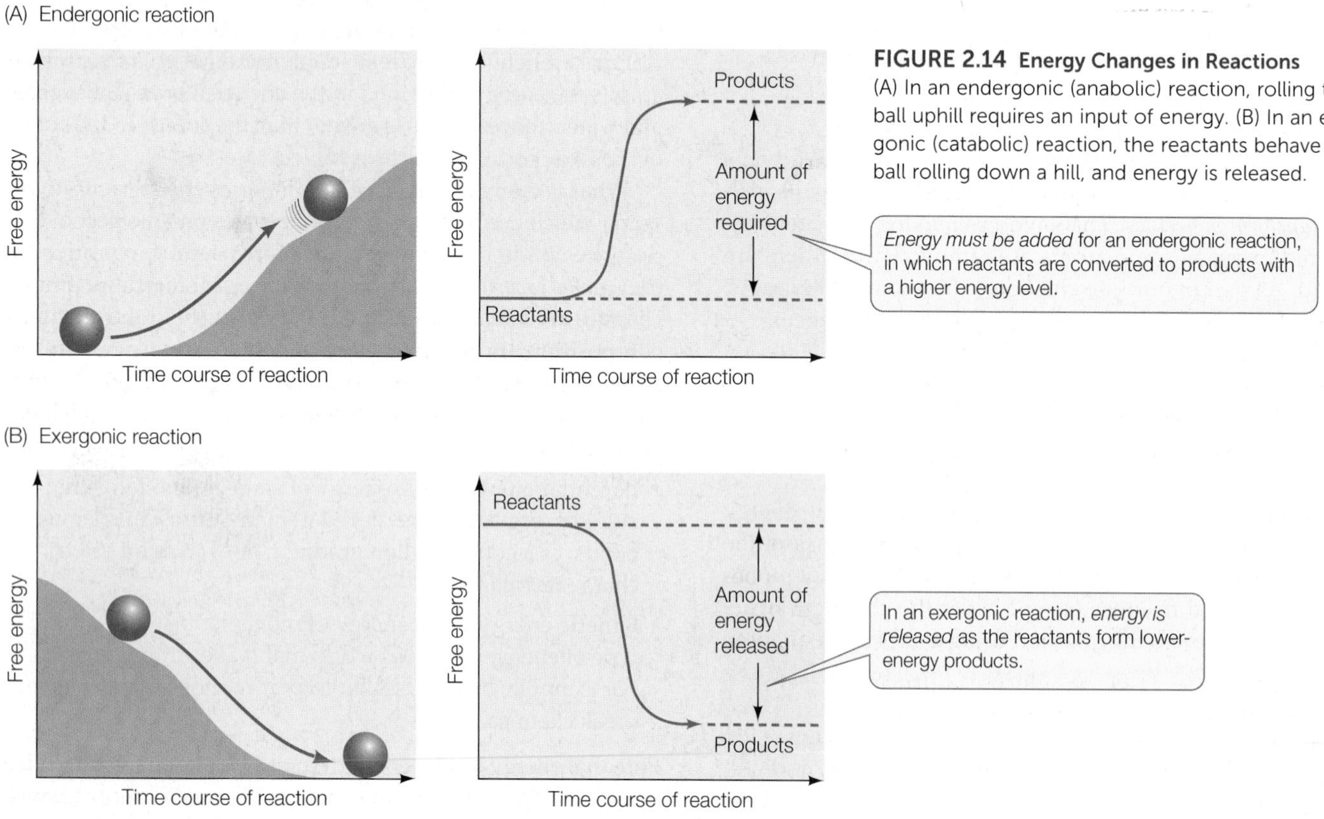

FIGURE 2.14 Energy Changes in Reactions (A) In an endergonic (anabolic) reaction, rolling the ball uphill requires an input of energy. (B) In an exergonic (catabolic) reaction, the reactants behave like a ball rolling down a hill, and energy is released.

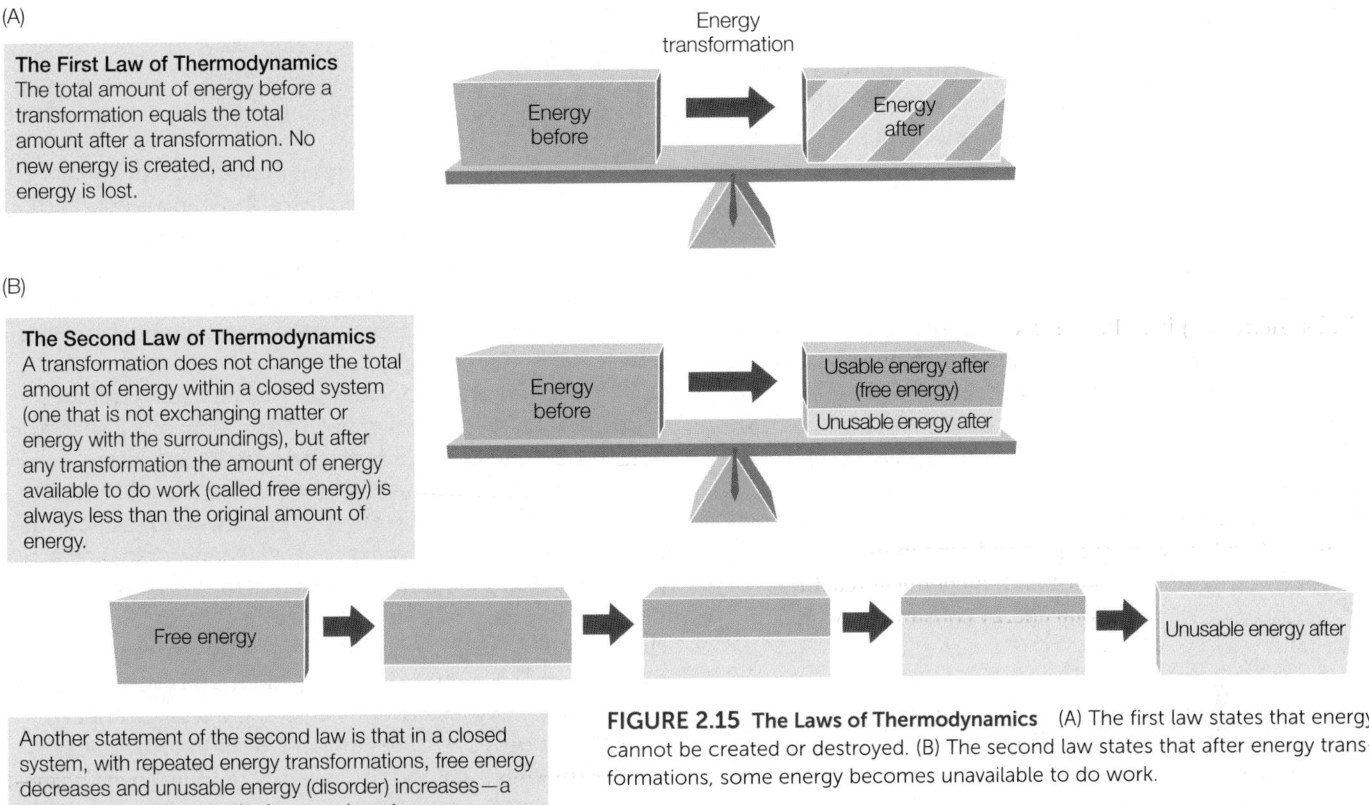

(A)
The First Law of Thermodynamics
The total amount of energy before a transformation equals the total amount after a transformation. No new energy is created, and no energy is lost.
Energy transformation
Energy before
Energy after
(B)
The Second Law of Thermodynamics
A transformation does not change the total amount of energy within a closed system (one that is not exchanging matter or energy with the surroundings), but after any transformation the amount of energy available to do work (called free energy) is always less than the original amount of energy.
Energy before
Usable energy after (free energy)
Unusable energy after
Free energy
Unusable energy after
Another statement of the second law is that in a closed system, with repeated energy transformations, free energy decreases and unusable energy (disorder) increases—a phenomenon known as the increase in entropy.

FIGURE 2.15 The Laws of Thermodynamics (A) The first law states that energy cannot be created or destroyed. (B) The second law states that after energy transformations, some energy becomes unavailable to do work.

matter and all energy transformations in the universe. Their application to living systems helps us understand how organisms and cells harvest and transform energy to sustain life.

The first law of thermodynamics: Energy is neither created nor destroyed. The first law of thermodynamics states that in any conversion, energy is neither created nor destroyed. Another way of stating this is that the total energy before and after an energy conversion is the same (**FIGURE 2.15A**). [Similarly, matter is also conserved: in the hydrolysis of sucrose (see p. 31), there are 12 carbons, 24 hydrogens, and 12 oxygens on both sides of the equation.]

Although the total amount of energy is conserved, chemical reactions involve changes in the amount of (potential) energy stored in chemical bonds. If energy is released during the reaction, it is available to do work—for example, to drive another chemical reaction. In general, reactions that release energy (catabolic, or exergonic reactions) can occur spontaneously.

The second law of thermodynamics: Useful energy tends to decrease. Although energy cannot be created or destroyed, the second law of thermodynamics implies that when energy is converted from one form to another, some of that energy becomes unavailable for doing work (**FIGURE 2.15B**). In other words, no physical process or chemical reaction is 100 percent efficient; some of the released energy is lost in a form associated with disorder. Think of disorder as a kind of randomness caused by the thermal motion of particles; this energy is so dispersed that it is unusable. **Entropy** is a measure of the disorder in a system.

If a chemical reaction increases entropy, its products are more disordered or random than its reactants. The disorder in a solution of glucose and fructose is greater than that in a solution of sucrose, where the glycosidic bond between the two monosaccharides prevents free movement. Conversely, if there are fewer products and they are more restrained in their movements than the reactants, the disorder is reduced. But this requires an energy input to achieve.

The second law of thermodynamics predicts that, as a result of energy transformations, disorder tends to increase; some energy is always lost to random thermal motion (entropy). Chemical changes, physical changes, and biological processes all tend to increase entropy (see Figure 2.15B), and this tendency gives direction to these processes. Changes in entropy are mathematically related to changes in free energy (the energy available to do work), and thus the second law helps explain why some reactions proceed in one direction rather than another.

How does the second law of thermodynamics apply to organisms? Consider the human body, with its highly organized tissues and organs composed of large, complex molecules. This level of complexity appears to be in conflict with the second law, but for two reasons, it is not. First, the construction of complex molecules also generates disorder. The anabolic reactions needed to construct 1 kg of an animal body require the catabolism of about 10 kg of food. So metabolism creates far more disorder (more energy is lost to entropy) than the amount of order stored in flesh. Second, life requires a constant input of energy to maintain order. Without this energy, the complex structures of living systems would break down. Because energy is used to generate and maintain order, and biological processes cause an overall increase in entropy, there is no conflict with the second law of thermodynamics.

CHECKpoint CONCEPT 2.5

- ✓ When you eat a candy bar and then decide to go for a walk, energy transformations take place. Beginning with the food energy in the candy bar, describe the forms of energy used and the changes in energy that occur as you decide to walk and as you do the walking.
- ✓ What is the difference between anabolism and catabolism? Between endergonic and exergonic reactions?
- ✓ Predict whether these situations are endergonic or exergonic, and explain your reasoning:
  a. The formation of a phospholipid bilayer membrane
  b. Turning on a TV set

APPLY THE CONCEPT

Biochemical changes involve energy

Chemical reactions in living systems involve changes in energy. These can be expressed as changes in available energy, called free energy (designated *G*, for Gibbs—the scientist who first described this parameter). The overall direction of a spontaneous chemical reaction is from higher to lower free energy. In other words, if the $G_{reactants}$ is greater than the $G_{products}$ (negative ΔG; the Greek letter delta stands for "change in" or "difference"), the reaction will be spontaneous; it will tend to go in the direction from reactants to product and release free energy in the process. Reactions where the $G_{reactants}$ is less than the $G_{products}$ (positive ΔG) will occur only if additional free energy is supplied.

The table shows some reactions and the absolute values of their associated free energy changes, $|\Delta G|$ (the vertical lines indicate absolute value).

REACTION	REACTANTS	PRODUCTS	$\lvert\Delta G\rvert$
Hydrolysis of sucrose	Sucrose + H_2O	Glucose + fructose	7.0
Triglyceride attachment	Glycerol + fatty acid	Monoglyceride	3.5

1. For each reaction, would you expect ΔG to be positive or negative?
2. Which reactions will be spontaneous? Explain your answer.

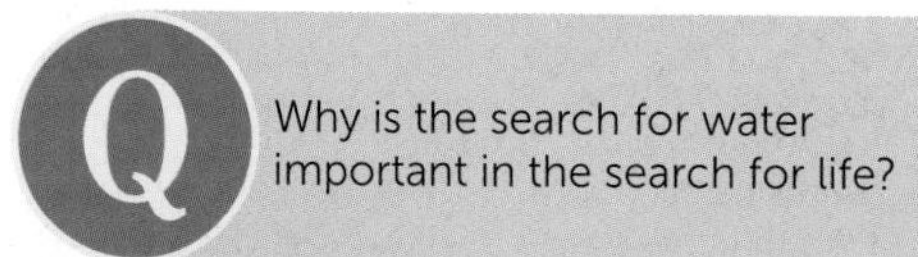

Why is the search for water important in the search for life?

ANSWER You have seen throughout this chapter that water is essential for the chemistry of life. Water is composed of two of the most abundant elements (Concept 2.1). Water is a polar molecule (Concept 2.2), which allows biologically important polar molecules such as monosaccharides (Concept 2.3) to dissolve in water. Because they are hydrophobic, lipids interact with water to form important biological structures (Concept 2.4). Water molecules participate directly in the formation and breakdown of polymers (Concept 2.2). In short, all of the processes of life as we know it require water.

In the opening essay of this chapter, we described recent evidence for the presence of water on other bodies in our solar system. Could this water harbor life, now or in the past? One way to investigate this possibility is to study how life on Earth may have originated in an aqueous environment. Geological evidence suggests that Earth was formed about 4.5 billion years ago and that life arose about 3.8 billion years ago. During the time when life originated, there was apparently little oxygen gas (O_2) in the atmosphere. In the 1950s, Stanley Miller and Harold Urey at the University of Chicago set up an experimental "atmosphere" containing various gases thought to be present in Earth's early atmosphere. Among them were ammonia (NH_3), hydrogen (H_2), methane (CH_4), and water vapor (H_2O). Miller and Urey passed an electric spark over the mixture to simulate lightning, providing a source of energy for covalent bond formation. Then they cooled the system so the gases would condense and collect in a watery solution, or "ocean" (**FIGURE 2.16**). Note that water was essential for this experiment as a source of oxygen atoms.

After several days of continuous operation, the system contained numerous complex molecules, including amino acids, the building blocks of proteins. In later experiments the researchers added other gases, such as carbon dioxide (CO_2), nitrogen (N_2), and sulfur dioxide (SO_2). This resulted in the formation of functional groups such as carboxylic acids, fatty acids, and many three- to six-carbon sugars. Taken together, these data suggest a plausible mechanism for the formation of life's chemicals in the aqueous environment of early Earth.

INVESTIGATION

FIGURE 2.16 Miller and Urey Synthesized Prebiotic Molecules in an Experimental Atmosphere With an increased understanding of the atmospheric conditions that existed on primitive Earth, the researchers devised an experiment to see if these conditions could lead to the formation of organic molecules.[a]

HYPOTHESIS

Organic chemical compounds can be generated under conditions similar to those that existed in the atmosphere of primitive Earth.

METHOD

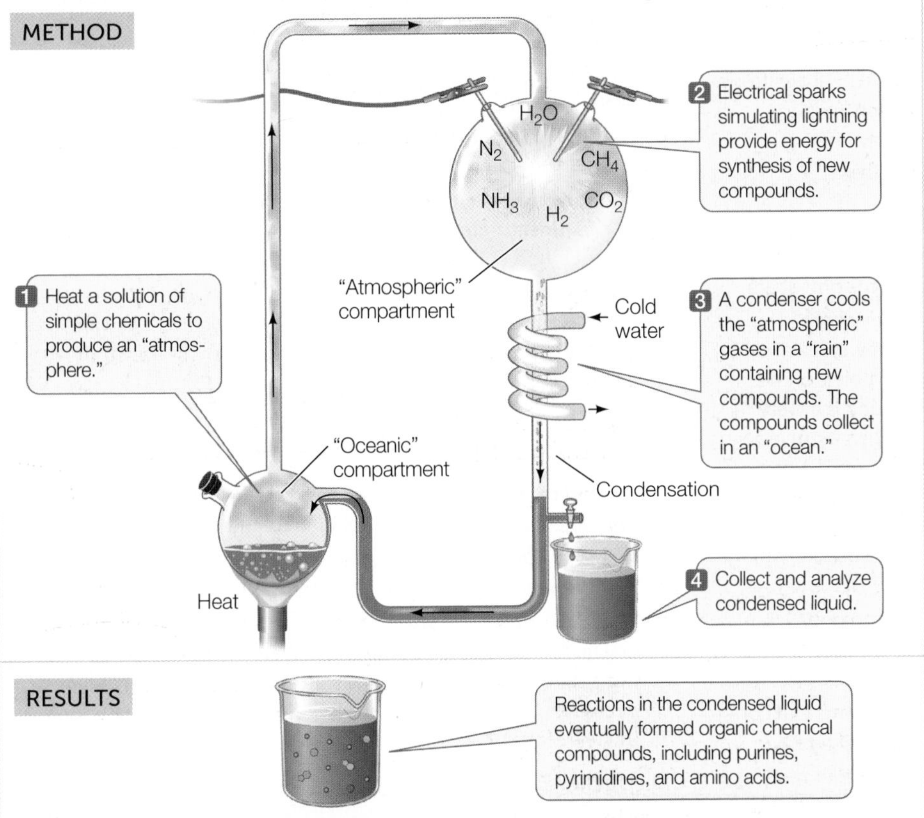

RESULTS

Reactions in the condensed liquid eventually formed organic chemical compounds, including purines, pyrimidines, and amino acids.

CONCLUSION

The chemical building blocks of life could have been generated in the probable atmosphere of early Earth.

ANALYZE THE DATA

The following data show the amount of energy impinging on Earth in different forms.

Source	Energy (cal/[cm^2 x yr])
Total radiation from sun	260,000
Ultraviolet light	199
Wavelength <250 nm	570
Wavelength <200 nm	85
Wavelength <150 nm	3.5
Electric discharges	4
Cosmic rays	0.0015
Radioactivity	0.8
Volcanoes	0.13

A. Only a small fraction of the sun's energy is ultraviolet light (less than 250 nm). What is the rest of the solar energy?

B. The molecules CH_4, H_2O, NH_3, and CO_2 absorb light at wavelengths less than 200 nm. What fraction of total solar radiation is in this range?

C. Instead of electric discharges, what other sources of energy could be used in these experiments?

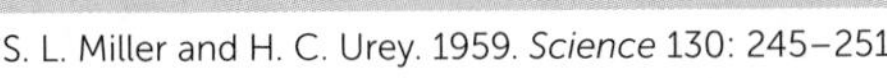

Go to **LaunchPad** for discussion and relevant links for all **INVESTIGATION** figures.

[a]S. L. Miller and H. C. Urey. 1959. *Science* 130: 245–251.

Go to ANIMATED TUTORIAL 2.3 Synthesis of Prebiotic Molecules
PoL2e.com/at2.3

SUMMARY

CONCEPT 2.1 Atomic Structure Is the Basis for Life's Chemistry

- Matter is composed of atoms. Each **atom** consists of a positively charged **nucleus** made up of **protons** and **neutrons**, surrounded by **electrons** bearing negative charges.
- The number of protons in the nucleus defines an **element**. There are many elements in the universe, but only a few of them (C, H, O, P, N, and S) make up the bulk of living organisms.
- Electrons are distributed in **electron shells** at varying energy levels away from the nucleus. The first shell can have a maximum of 2 electrons; the second shell, 8 electrons; the third shell, 18 electrons; and subsequent shells, 32 electrons. **Review Figure 2.1**

CONCEPT 2.2 Atoms Interact and Form Molecules

- A **chemical bond** is an attractive force that links two atoms together in a molecule. **Review ANIMATED TUTORIAL 2.1**
- A **covalent bond** is a strong bond formed when two atoms share one or more pairs of electrons. **Review Figures 2.2 and 2.3**
- When two atoms of unequal electronegativity bond with each other, a **polar covalent bond** is formed. The two ends, or poles, of the bond have partial charges (δ^+ or δ^-). A **hydrogen bond** is a weak electrical attraction that forms between a δ^+ hydrogen atom in one molecule and a δ^- atom in another molecule (or in another part of a large molecule). Hydrogen bonds are abundant in water. **Review Figure 2.4**
- **Ions** are electrically charged particles that form when atoms gain or lose one or more electrons in order to form more stable electron configurations. **Anions** and **cations** are negatively and positively charged ions, respectively. **Ionic attractions** form when ions with opposite charges attract. **Review Figure 2.6**
- **Functional groups** are covalently bonded groups of atoms that confer specific properties to biological molecules. **Review Figure 2.7**
- **Macromolecules** are formed by polymerization of smaller molecules call **monomers**. **Review Figure 2.8, ANIMATED TUTORIAL 2.2, and ACTIVITY 2.1**

CONCEPT 2.3 Carbohydrates Consist of Sugar Molecules

- **Carbohydrates** contain carbon bonded to hydrogen and oxygen.
- **Monosaccharides** include pentoses (with five carbons) and hexoses (with six carbons). **Review Figure 2.9 and ACTIVITY 2.2**
- Glycosidic linkages are covalent bonds between saccharides. **Disaccharides** such as sucrose each contain two monosaccharides, whereas **polysaccharides** such as starch and cellulose contain long chains of monomers. **Review Figure 2.10**

CONCEPT 2.4 Lipids Are Hydrophobic Molecules

- Fats and oils are **triglycerides**, composed of three **fatty acids** covalently linked to glycerol. **Review Figure 2.11**
- **Saturated fatty acids** have hydrocarbon chains with no double bonds. **Unsaturated fatty acids** contain double bonds in their hydrocarbon chains. **Review Figure 2.12**
- **Phospholipids** contain two fatty acids and a hydrophilic, phosphate-containing polar group attached to glycerol. They are **amphipathic**, with both polar and nonpolar ends. They form into a structural bilayer in water. **Review Figure 2.13**

CONCEPT 2.5 Biochemical Changes Involve Energy

- A **chemical reaction** occurs when atoms have sufficient energy to combine or to change their bonding partners.
- **Anabolic reactions** require energy and are endergonic. **Catabolic reactions** release energy and are exergonic. **Review Figure 2.14**
- The **laws of thermodynamics** govern biochemical reactions. The first law states that in any transformation, energy is neither created nor destroyed. The second law states that useful energy tends to decrease. In other words, **entropy** (disorder) tends to increase. **Review Figure 2.15**

See **ANIMATED TUTORIAL** 2.3

Go to the Interactive Summary to review key figures, Animated Tutorials, and Activities
PoL2e.com/is2

Go to LaunchPad at **macmillanhighered.com/launchpad** for additional resources, including LearningCurve Quizzes, Flashcards, and many other study and review resources.

3 Nucleic Acids, Proteins, and Enzymes

KEY CONCEPTS

3.1 Nucleic Acids Are Informational Macromolecules

3.2 Proteins Are Polymers with Important Structural and Metabolic Roles

3.3 Some Proteins Act as Enzymes to Speed up Biochemical Reactions

3.4 Regulation of Metabolism Occurs by Regulation of Enzymes

The bark of the willow tree (*Salix alba*) was the original source of salicylic acid, later modified to aspirin.

Despite suffering from the "ague," the Reverend Edward Stone went walking in the English countryside. Feverish, tired, with aching muscles and joints, he came across a willow tree. Although apparently unaware that many ancient healers used willow bark extracts to reduce fever, the clergyman knew of the tradition of natural remedies for various diseases. The willow reminded him of the bitter extracts from the bark of South American trees then being sold (at high prices) to treat fevers. Removing some willow bark, Stone sucked on it and found it did indeed taste bitter—and that it relieved his symptoms.

Later he gathered a pound of willow bark and ground it into a powder, which he gave to about 50 people who complained of pain; all said they felt better. Stone reported the results of this "clinical test" in a letter to the Royal Society, England's most respected scientific body. Stone had discovered the main source of salicylic acid, the basis of the most widely used drug in the world. The date of his letter (which still exists) was April 25, 1763.

The chemical structure of salicylic acid (named for *Salix*, the willow genus) was worked out about 70 years later, and soon chemists could synthesize it in the laboratory. Although the compound alleviated pain, its acidity irritated the digestive system. In the late 1890s, the German chemical company Bayer synthesized a milder yet equally effective form, acetylsalicylic acid, which it marketed as aspirin. The new medicine's success launched Bayer to world prominence as a pharmaceutical company, a position it maintains today.

In the 1960s and 1970s, aspirin use declined when two alternative medications, acetaminophen (Tylenol) and ibuprofen (Motrin and Advil), became widely available. But over this same time, clinical studies revealed a new use for aspirin: it is an effective anticoagulant, shown to prevent heart attacks and strokes caused by blood clots. Today many people take a daily low dose of aspirin as a preventive agent against clotting disorders.

Fever, joint pain, headache, blood clots. What do these symptoms have in common? They all are mediated by fatty acid products called prostaglandins and molecules derived from them. Salicylic acid blocks the synthesis of the primary prostaglandin. The biochemical mechanism by which aspirin works was described in 1971. As we will see, an understanding of this mechanism requires an understanding of protein and enzyme function—two subjects of this chapter.

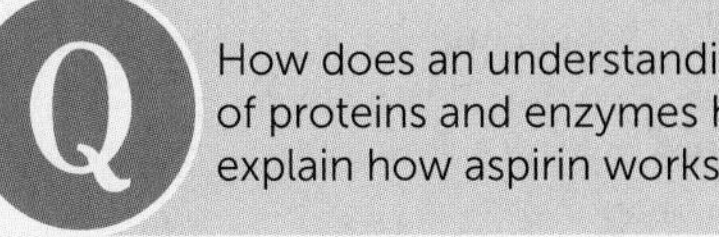

Q How does an understanding of proteins and enzymes help explain how aspirin works?

You will find the answer to this question on page 57.

CONCEPT 3.1 Nucleic Acids Are Informational Macromolecules

Nucleic acids are polymers that store, transmit, and express hereditary (genetic) information. This information is encoded in the sequences of monomers that make up nucleic acids. There are two types of nucleic acids: **DNA** (*d*eoxyribo*n*ucleic *a*cid) and **RNA** (*r*ibo*n*ucleic *a*cid). DNA stores and transmits genetic information. Through RNA intermediates, the information encoded in DNA is used to specify the amino acid sequences of proteins. As you will see later in this chapter, proteins are essential for both metabolism and structure. Certain specialized RNA molecules also play roles in metabolism. Ultimately, *nucleic acids and the proteins encoded by them determine the metabolic functions of an organism.*

Nucleotides are the building blocks of nucleic acids

Nucleic acids are polymers composed of monomers called nucleotides. A **nucleotide** consists of three components: a nitrogen-containing **base**, a pentose sugar, and one to three phosphate groups (**FIGURE 3.1**). Molecules consisting of a pentose sugar and a base—but no phosphate group—are called nucleosides. The nucleotides that make up nucleic acids contain just one phosphate group—they are nucleoside monophosphates.

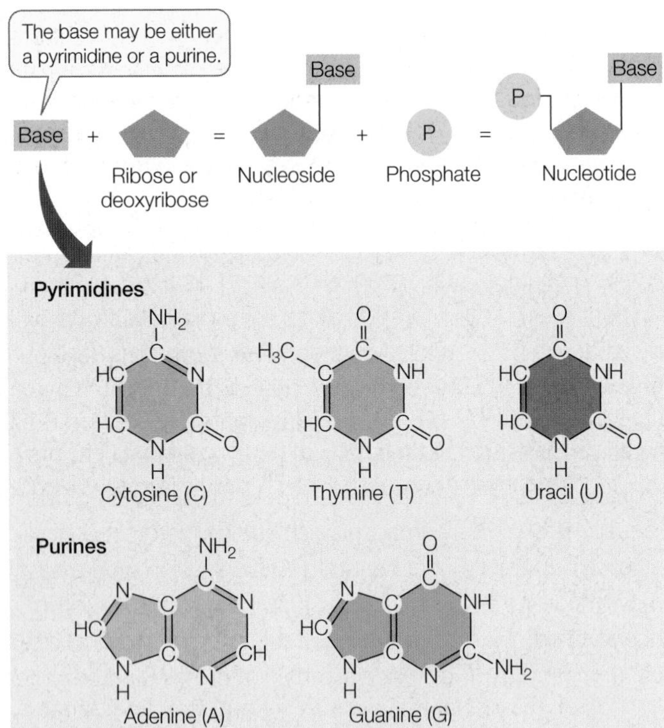

FIGURE 3.1 Nucleotides Have Three Components Nucleotide monomers are the building blocks of DNA and RNA polymers. Nucleotides may have one to three phosphate groups; those in DNA and RNA have one.

Go to ACTIVITY 3.1 Nucleic Acid Building Blocks PoL2e.com/ac3.1

TABLE 3.1 Distinguishing RNA from DNA

Nucleic acid	Sugar	Bases	Strands
RNA	Ribose	Adenine Cytosine Guanine Uracil	Single
DNA	Deoxyribose	Adenine Cytosine Guanine Thymine	Double

The bases of the nucleic acids take one of two chemical forms: a six-membered single-ring structure called a **pyrimidine**, or a fused double-ring structure called a **purine** (see Figure 3.1). In DNA, the pentose sugar is **deoxyribose**, which differs from the **ribose** found in RNA by the absence of one oxygen atom (see Figure 2.9).

During the formation of a nucleic acid, new nucleotides are added to an existing chain one at a time. The pentose sugar in the last nucleotide of the existing chain and the phosphate on the new nucleotide undergo a condensation reaction (see Figure 2.8) and the resulting linkage is called a **phosphodiester bond**. The phosphate on the new nucleotide is attached to the 5′ (5 prime) carbon atom of its sugar, and the bond occurs between it and the 3′ (3 prime) carbon on the last sugar of the existing chain. Because each nucleotide is added to the 3′ carbon of the last sugar, nucleic acids are said to *grow in the 5′ to 3′ direction* (**FIGURE 3.2**).

Nucleic acids can be oligonucleotides, with a few to about 20 nucleotide monomers, or longer polynucleotides:

- *Oligonucleotides* include RNA molecules that function as "primers" to begin the duplication of DNA; RNA molecules that regulate the expression of genes; and synthetic DNA molecules used for amplifying and analyzing other, longer nucleotide sequences.
- *Polynucleotides*, more commonly referred to as nucleic acids, include DNA and most RNA. Polynucleotides can be very long, and indeed are the longest polymers in the living world. Some DNA molecules in humans contain hundreds of millions of nucleotides.

Base pairing occurs in both DNA and RNA

In addition to differing in their sugar groups, DNA and RNA also differ in their bases and general structures (**TABLE 3.1**). Four bases are found in DNA: **adenine** (**A**), **cytosine** (**C**), **guanine** (**G**), and **thymine** (**T**). RNA also contains adenine, cytosine, and guanine, but the fourth base in RNA is **uracil** (**U**) rather than thymine. The lack of a hydroxyl group at the 2′ position of the deoxyribose sugar in DNA makes the structure of DNA less flexible than that of RNA. As we describe below, DNA is composed of two polynucleotide strands whereas RNA is usually single-stranded. However, a long RNA can fold up on itself, forming a variety of structures.

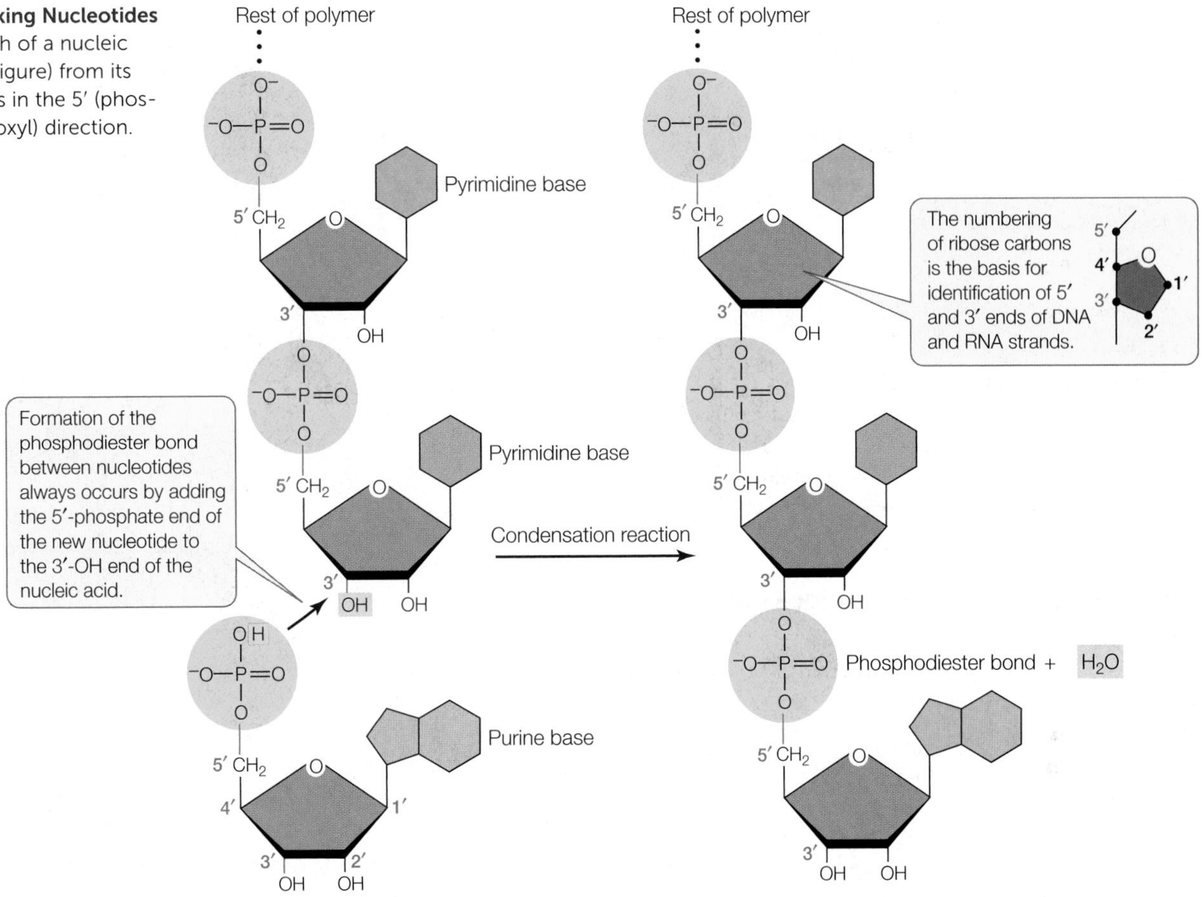

FIGURE 3.2 Linking Nucleotides Together Growth of a nucleic acid (RNA in this figure) from its monomers occurs in the 5′ (phosphate) to 3′ (hydroxyl) direction.

Formation of the phosphodiester bond between nucleotides always occurs by adding the 5′-phosphate end of the new nucleotide to the 3′-OH end of the nucleic acid.

The numbering of ribose carbons is the basis for identification of 5′ and 3′ ends of DNA and RNA strands.

The key to understanding the structure and function of both DNA and RNA is the principle of **complementary base pairing**. In DNA, adenine and thymine always pair (A-T), and cytosine and guanine always pair (C-G):

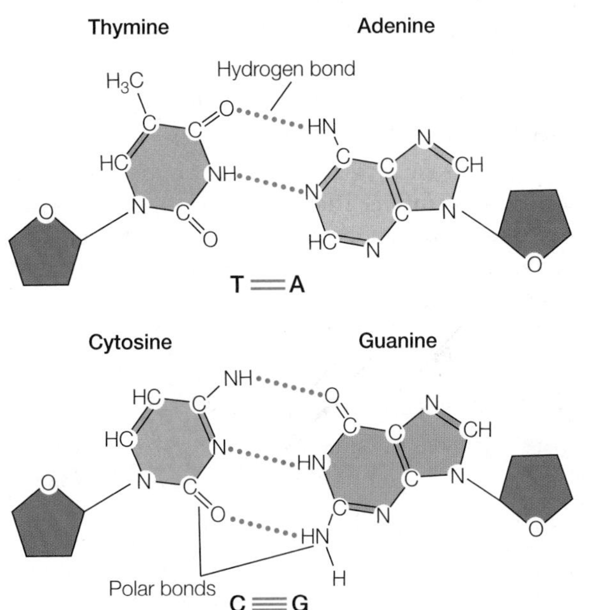

In RNA, the base pairs are A-U and C-G. Base pairs are held together primarily by hydrogen bonds. As you can see, there are polar C═O and N—H covalent bonds in the nucleotide bases (see Concept 2.2 for a discussion of polar covalent bonds). Hydrogen bonds form between the partial negative charge (δ^-) on an oxygen or nitrogen atom of one base, and the partial positive charge (δ^+) on a hydrogen atom of another base. Complementary base pairing occurs because the arrangements of polar bonds in the nucleotide bases favor the pairing of bases as they occur (C with G, and A with U or T).

Individual hydrogen bonds are relatively weak, but there are so many of them in DNA and RNA that collectively they provide a considerable force of attraction. However, this attraction is not as strong as that provided by multiple covalent bonds. This means that base pairs are relatively easy to separate with a modest input of energy. As you will see in Chapters 9 and 10, the breaking and making of hydrogen bonds in nucleic acids is vital to their roles in living systems. Let's now look in a little more detail at the structures of RNA and DNA.

RNA Usually, RNA is single-stranded (**FIGURE 3.3A**). However, many single-stranded RNA molecules fold up into three-dimensional structures, because of hydrogen bonding between nucleotides in separate portions of the molecules (**FIGURE 3.3B**). An RNA strand can also fold back on itself to form a double-stranded helix. This results in a three-dimensional surface for the bonding and recognition of other molecules. It is important to realize that this folding occurs by complementary base pairing, and the structure is thus determined by the particular order of bases in the RNA molecule.

DNA Usually, DNA is double-stranded; that is, it consists of two separate polynucleotide strands of the same length

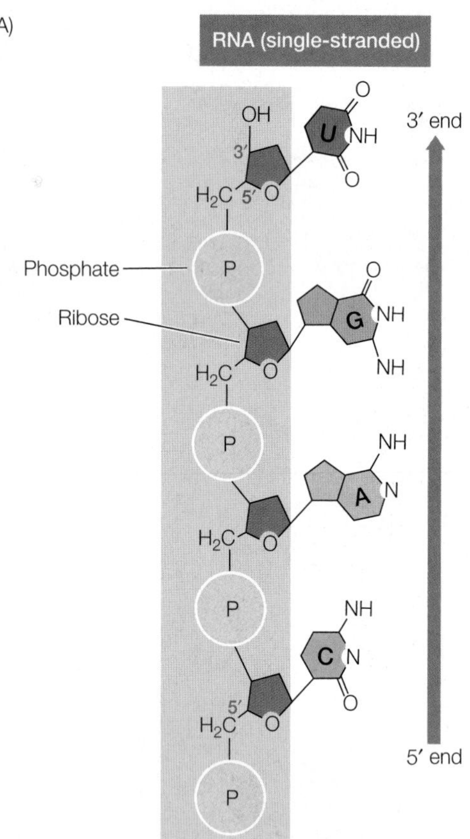

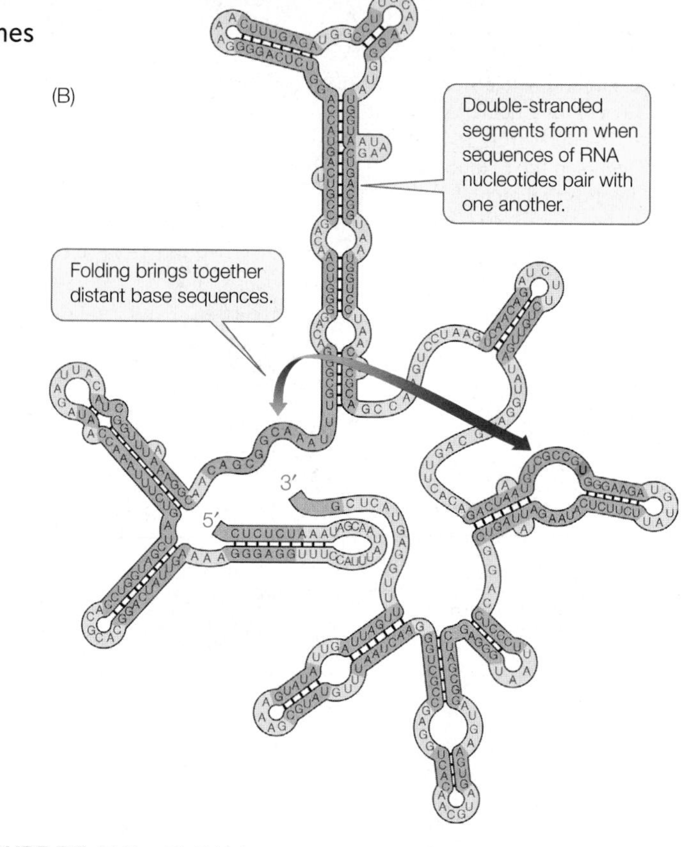

FIGURE 3.3 RNA (A) RNA is usually a single strand. (B) When a single-stranded RNA folds back on itself, hydrogen bonds between complementary sequences can stabilize it into a three-dimensional shape with distinct surface characteristics.

(**FIGURE 3.4A**). The two polynucleotide strands are antiparallel: they run in opposite directions so that their 5′ ends are at opposite ends of the double-stranded molecule. In contrast to RNA's diversity in three-dimensional structure, DNA is remarkably uniform. The A-T and G-C base pairs are about the same size (each is a purine paired with a pyrimidine), and the two polynucleotide strands form a "ladder" that twists into a double helix (**FIGURE 3.4B**). The sugar–phosphate groups form the sides of the ladder, and the bases with their hydrogen bonds form the rungs on the inside. The double helix is almost always right-handed:

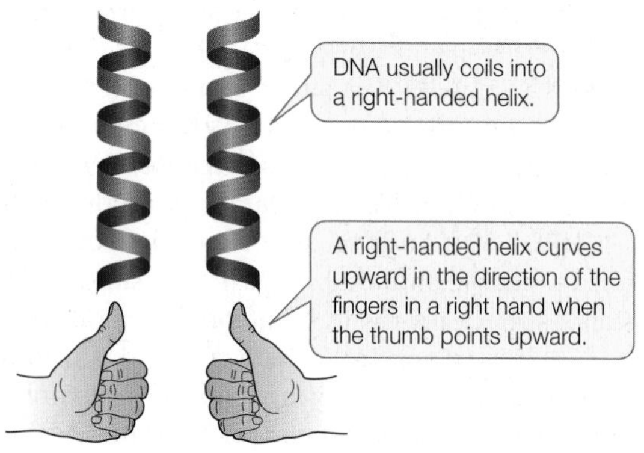

DNA usually coils into a right-handed helix.
A right-handed helix curves upward in the direction of the fingers in a right hand when the thumb points upward.

Go to ACTIVITY 3.2 DNA Structure
PoL2e.com/ac3.2

DNA carries information and is expressed through RNA

DNA is a purely informational molecule. The information is encoded in the sequence of bases carried in its strands. For example, the information encoded in the sequence TCAGCA is different from the information in the sequence CCAGCA. DNA has two functions in terms of information:

- DNA can be reproduced precisely by **DNA replication**. DNA is replicated by polymerization using an existing strand as a base-pairing template.
- Some DNA sequences can be copied into RNA, in a process called **transcription**. The nucleotide sequences in most RNA molecules can then be used to specify sequences of amino acids in proteins (polypeptides). This process is called **translation**.

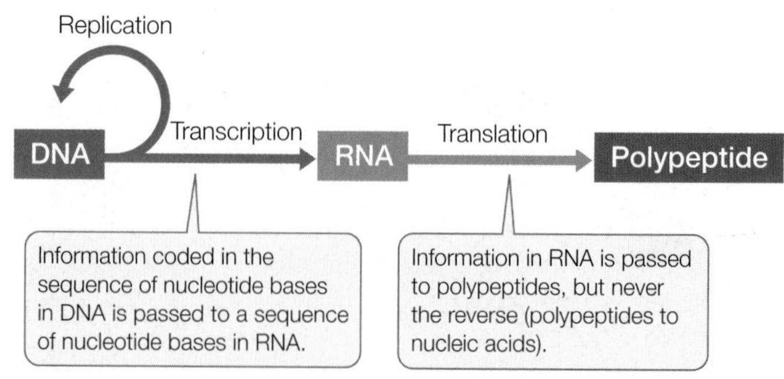

The details of these important processes are described in Chapters 9 and 10, but it is important to realize several things at this point:

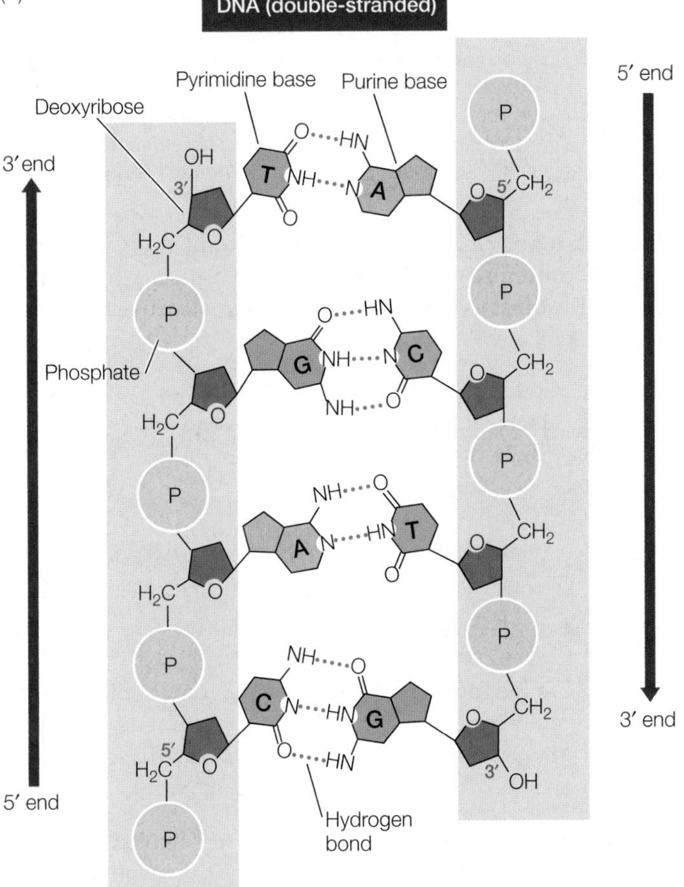

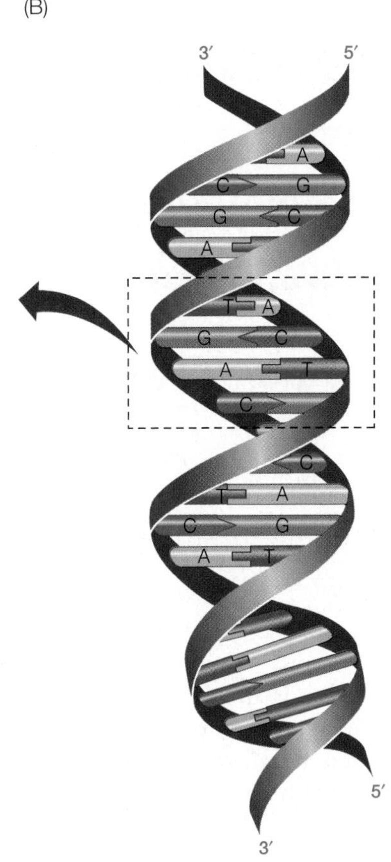

FIGURE 3.4 DNA (A) DNA usually consists of two strands running in opposite directions that are held together by base pairing between purines and pyrimidines opposite one another on the two strands. (B) The two antiparallel strands in a DNA molecule are twisted into a double helix.

- *DNA replication and transcription depend on the base pairing properties of nucleic acids.* In both replication and transcription, the hydrogen bonds between two DNA strands are broken, so that complementary base pairing can occur between an existing DNA strand and a newly forming strand of DNA or RNA. The resulting new DNA or RNA strand is *complementary to* the existing DNA template strand. Recall that the hydrogen-bonded base pairs are A-T and G-C in DNA and A-U and G-C in RNA. Now, consider this double-stranded DNA region:

  5′-TCAGCA-3′

  3′-AGTCGT-5′

  Transcription of the lower strand will result in a single strand of RNA with the sequence 5′-UCAGCA-3′. Can you figure out what RNA sequence the top strand would produce?
- *DNA replication usually involves the entire DNA molecule.* Since DNA holds essential information, it must be replicated completely so that each new cell or new organism receives a complete set of DNA from its parent (**FIGURE 3.5A**).
- *Gene expression is the transcription and translation of specific DNA sequences.* Sequences of DNA that encode specific proteins and are transcribed into RNA are called **genes** (**FIGURE 3.5B**). The complete set of DNA in a living organism is called its **genome**. However, not all of the information in the genome is needed at all times and in all tissues. For example, in humans, the gene that encodes the major protein in hair (keratin) is expressed only in skin cells. The genetic information in the keratin-encoding gene is transcribed into RNA and then translated into the protein keratin. In other tissues such as the muscles, the keratin gene is not transcribed, but other genes are—for example, the genes that encode proteins present in muscles but not in skin.

The DNA base sequence reveals evolutionary relationships

Because DNA carries hereditary information from one generation to the next, a theoretical series of DNA molecules stretches back through the lineage of every organism to the beginning of biological evolution on Earth, about 3.8 billion years ago. The genomes of organisms gradually accumulate changes in their DNA base sequences over evolutionary time. Therefore closely related living species should have more similar base sequences than species that are more distantly related.

Over the past two decades there have been remarkable developments in technologies for determining the order of nucleotides in DNA molecules (DNA sequencing), and in computer technologies to analyze these sequences. These

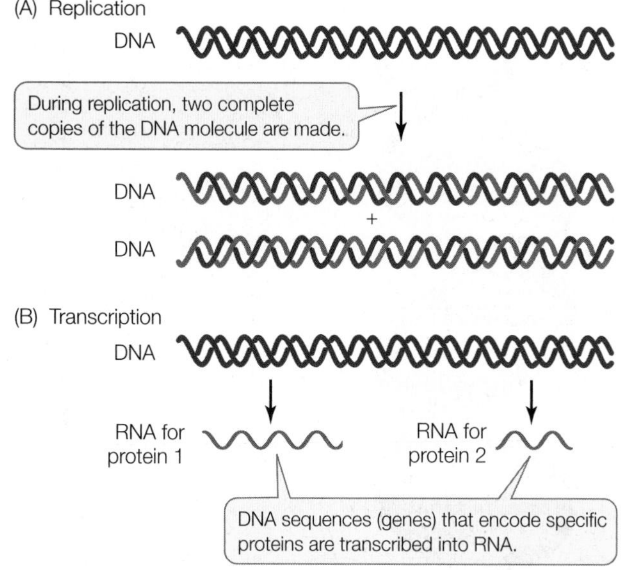

FIGURE 3.5 DNA Replication and Transcription DNA is completely replicated during cell reproduction (A), but it is only partially transcribed (B). In transcription, the DNA code is copied to RNA. The sequence of the latter determines the amino acid sequence of a protein. Transcription of the genes for many different proteins is activated at different times and, in multicellular organisms, in different cells of the body.

advances have enabled scientists to determine the entire DNA base sequences of whole organisms, including the human genome, which contains about 3 billion base pairs. These studies have confirmed many of the evolutionary relationships that were inferred from more traditional comparisons of body structure, biochemistry, and physiology. Traditional comparisons had indicated that the closest living relative of humans (*Homo sapiens*) is the chimpanzee (genus *Pan*). In fact, the chimpanzee genome shares nearly 99 percent of its DNA base sequence with the human genome. Increasingly, scientists turn to DNA analyses to figure out evolutionary relationships when other comparisons are not possible or are not conclusive. For example, DNA studies revealed a close relationship between starlings and mockingbirds that was not expected on the basis of their anatomy or behavior.

LINK

For more on the use of DNA sequences to reconstruct the evolutionary history of life, see **Concept 16.2**

CHECKpoint CONCEPT 3.1

- ✓ List the key differences between DNA and RNA and between purines and pyrimidines.
- ✓ What are the differences between DNA replication and transcription?
- ✓ If one strand of a DNA molecule has the sequence 5′-TTCCGGAT-3′, what is the sequence of the other strand of DNA? If RNA is transcribed from the 5′-TTCCGGAT-3′ strand, what would be its sequence? And if RNA is transcribed from the other DNA strand, what would be its sequence? (Note that it is conventional to write these sequences with the 5′ end on the left.)
- ✓ How can DNA molecules be so diverse when they appear to be structurally similar?

Nucleic acids are largely informational molecules that encode proteins. We will now turn to a discussion of proteins—the most structurally and functionally diverse class of macromolecules.

Go to ANIMATED TUTORIAL 3.1
Macromolecules: Nucleic Acids and Proteins
PoL2e.com/at3.1

CONCEPT 3.2 Proteins Are Polymers with Important Structural and Metabolic Roles

Proteins are the fourth and final type of biological macromolecule we will discuss, and in terms of structural diversity and function, they are at the top of the list. Here are some of the major functions of proteins in living organisms:

- *Enzymes* are catalytic molecules that speed up biochemical reactions. Most enzymes are proteins (some are RNA molecules).
- *Defensive proteins* such as antibodies recognize and respond to substances or particles that invade the organism from the environment.
- *Hormonal and regulatory proteins* such as insulin control physiological processes.
- *Receptor proteins* receive and respond to molecular signals from inside and outside the organism.
- *Storage proteins* store chemical building blocks—amino acids—for later use.
- *Structural proteins* such as collagen provide physical stability and enable movement.
- *Transport proteins* such as hemoglobin carry substances within the organism.
- *Genetic regulatory proteins* (transcription factors) regulate when, how, and to what extent a gene is expressed.

Clearly, the biochemistry of proteins warrants our attention!

Amino acids are the building blocks of proteins

As we noted in Chapter 2, **proteins** are polymers made up of monomers called **amino acids**. As their name suggests, the amino acids all contain two functional groups: the nitrogen-containing amino group and the (acidic) carboxyl group.

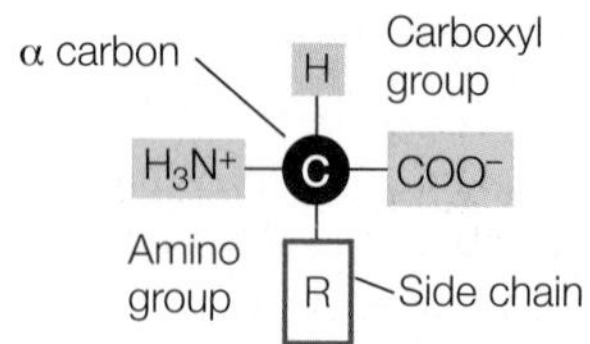

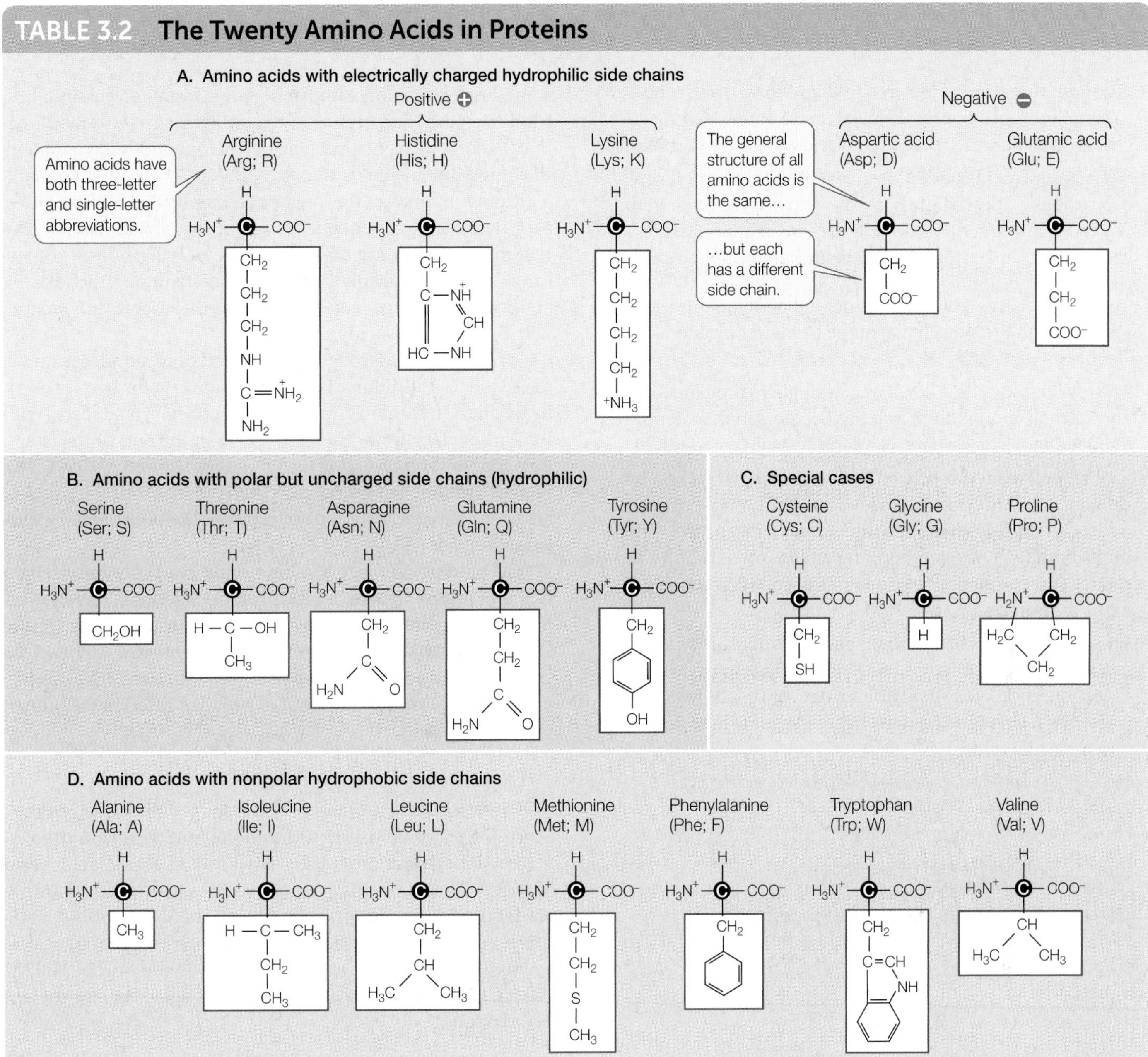

The amino and carboxyl groups shown in the diagram are charged. How does this happen? Under the conditions that exist in most living systems, the carboxyl group releases an H^+ (a cation), leaving the rest of the group as an anion:

$$—COOH \rightarrow —COO^- + H^+$$

From your studies of chemistry, you may recognize the carboxyl group as an acid. Conversely, under the same conditions the amino group tends to form a bond with H^+:

$$—NH_2 + H^+ \rightarrow —NH_3^+$$

Your chemistry knowledge should tell you that the amino group is a base.

The central carbon atom of an amino acid—the α (alpha) carbon—has four available electrons for covalent bonding. In all amino acids, two of the electrons are occupied by the two functional groups noted above, and a third is occupied by a hydrogen atom. The fourth bonding electron is shared with a group that differs in each amino acid. This is often referred to as the **R group**, or **side chain**, and is designated by the letter **R**. Each amino acid is identified by its R group.

Go to ACTIVITY 3.3 Features of Amino Acids
PoL2e.com/ac3.3

There are hundreds of amino acids known in nature, and many of these occur in plants. But *only 20 amino acids* (listed in **TABLE 3.2**) *occur extensively in the proteins of all organisms*. These 20 amino acids can be grouped according to the properties conferred by their side chains (R groups):

- Five amino acids have electrically charged side chains (+1 or –1), attract water (are hydrophilic), and attract oppositely charged ions of all sorts.

- Five amino acids have polar side chains (δ^+, δ^-) and tend to form hydrogen bonds with water and other polar or charged substances. These amino acids are also hydrophilic.
- Seven amino acids have side chains that are nonpolar hydrocarbons or very slightly modified hydrocarbons. In the watery environment of the cell, these hydrophobic side chains may cluster together in the interior of the protein.

Three amino acids—glycine, proline, and cysteine—are special cases, although the side chains of the former two generally are hydrophobic:

- The glycine side chain consists of a single hydrogen atom and is small enough to fit into tight corners in the interior of a protein molecule, where a larger side chain could not fit.
- Proline possesses a modified amino group that lacks a hydrogen atom and instead forms a covalent bond with the hydrocarbon side chain, resulting in a ring structure. This limits both its hydrogen-bonding ability and its ability to rotate. Thus proline often functions to stabilize bends or loops in proteins.
- The cysteine side chain, which has a terminal —SH group, can react with another cysteine side chain to form a covalent bond called a **disulfide bridge**, or disulfide bond (—S—S—). Disulfide bridges help determine how a protein molecule folds.

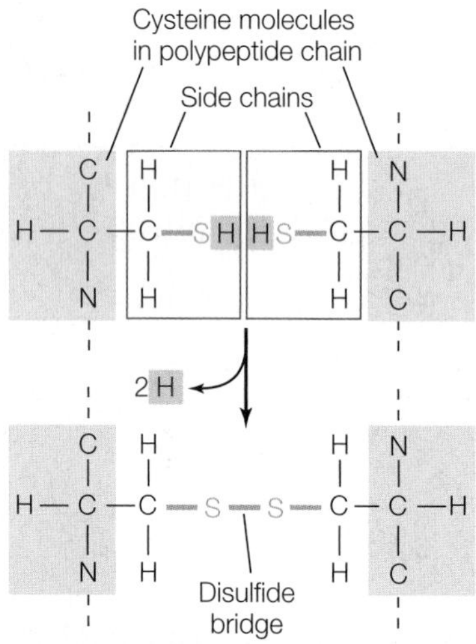

In the reaction shown above, the two —SH groups each lose a hydrogen atom (a proton and an electron) and become **oxidized**. The —SH group, which carries the extra electron and proton, is in its **reduced** state.

LINK

In addition to their role in protein structure, oxidation and reduction reactions are important in cellular metabolism; see **Concept 6.1**

Amino acids are linked together by peptide bonds

Amino acids can form short polymers of 20 or fewer amino acids, called **oligopeptides** or simply **peptides**. These include some hormones and other molecules involved in signaling from one part of an organism to another. Even with their relatively short chains of amino acids, oligopeptides have distinctive three-dimensional structures.

More common are the longer polymers called **polypeptides**, each with a unique sequence of amino acids. A functional protein may be made up of one or more polypeptides. Proteins range in size from small ones such as insulin, which has 51 amino acids, to huge molecules such as the muscle protein titin, with 34,350 amino acids.

Like nucleic acids, oligopeptides and polypeptides form via the sequential addition of new amino acids to the ends of existing chains. The amino group of the new amino acid reacts with the carboxyl group of the amino acid at the end of the chain. This condensation reaction forms a **peptide bond** (**FIGURE 3.6**). Note that there is directionality here, just as with the nucleic acids. In this case, *polymerization takes place in the amino to carboxyl direction*.

The precise sequence of amino acids in a polypeptide chain is the **primary structure** of a protein. Scientists have determined the primary structures of many proteins. The single-letter abbreviations for amino acids (see Table 3.2) are used to record the amino acid sequences of proteins. Here, for example, are the first 20 amino acids (out of a total of 1,827) in the human protein sucrase:

MARKKFSGLEISLIVLFVIV

The theoretical number of different proteins is enormous. Since there are 20 different amino acids, there could be 20 × 20 = 400 distinct dipeptides (two linked amino acids) and 20 × 20 × 20 = 8,000 different tripeptides (three linked amino acids). So for even a small polypeptide of 100 amino acids there are 20^{100} possible sequences, each with its own distinctive primary structure. How large is the number 20^{100}? Physicists tell us there aren't that many electrons in the entire universe.

Higher-level protein structure is determined by primary structure

The primary structure of a protein is established by covalent bonds, but higher levels of structure are determined largely by weaker forces, including hydrogen bonds and hydrophobic and hydrophilic interactions. Follow **FIGURE 3.7** as we describe how a protein chain becomes a three-dimensional structure.

SECONDARY STRUCTURE A protein's **secondary structure** consists of regular, repeated spatial patterns in different regions of a polypeptide chain. There are two basic types of secondary structure, both determined by hydrogen bonding between the amino acids that make up the primary structure:

- The α (alpha) **helix** is a right-handed coil that turns in the same direction as a standard wood screw (see Figure 3.7B). The R groups extend outward from the peptide backbone of the helix. The coiling results from hydrogen bonds that form between the N—H group on one amino acid and the C=O group on another within the same turn of the helix.

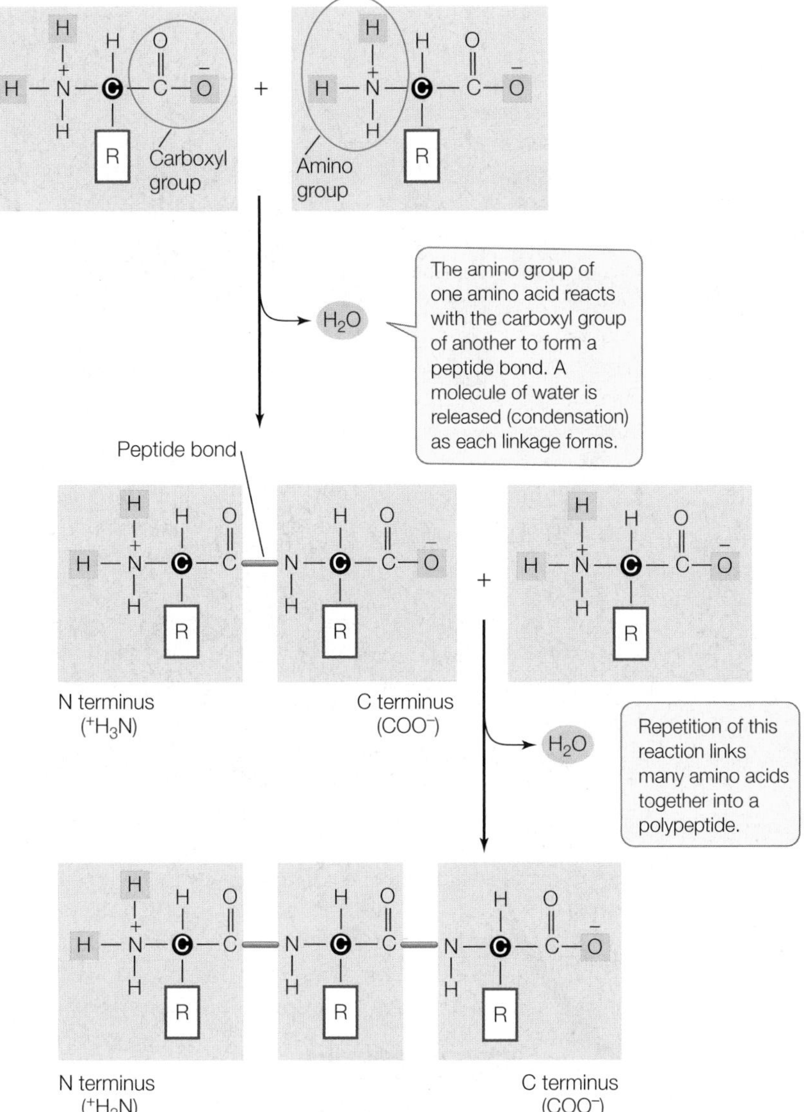

FIGURE 3.6 Formation of a Peptide Bond In living things, the reaction leading to a peptide bond has many intermediate steps, but the reactants and products are the same as those shown in this simplified diagram.

- The β (**beta**) **pleated sheet** is formed from two or more sequences of amino acids that are extended and aligned. The sheet is stabilized by hydrogen bonds between the N—H groups and the C=O groups on the two chains (see Figure 3.7C). A β pleated sheet may form between separate polypeptide chains or between different regions of a single polypeptide chain that is bent back on itself. Many proteins contain both α helices and β pleated sheets in different regions of the same polypeptide chain.

TERTIARY STRUCTURE In many proteins, the polypeptide chain is bent at specific sites and then folded back and forth, resulting in **tertiary structure** (see Figure 3.7D). Tertiary structure results in the polypeptide's definitive three-dimensional shape, including a buried interior as well as a surface that is exposed to the environment. The protein's exposed outer surfaces present functional groups capable of interacting with other molecules in the cell. These molecules might be other proteins or smaller chemical reactants (as in enzymes; see below).

Whereas hydrogen bonding between the N—H and C=O groups within and between chains is responsible for a protein's secondary structure, it is the interactions between R groups—the amino acid side chains—that determine tertiary structure (**FIGURE 3.8**):

- *Covalent disulfide bridges* can form between specific cysteine side chains, holding a folded polypeptide together.
- *Hydrogen bonds* between side chains also stabilize folds in proteins.
- *Hydrophobic side chains* can aggregate together in the interior of a protein, away from water, folding the polypeptide in the process.
- *van der Waals interactions* can stabilize close associations between hydrophobic side chains.
- *Ionic interactions* can form between positively and negatively charged side chains, forming "salt bridges" between amino acids. Ionic interactions can also be buried deep within a protein, away from water. These interactions occur between positively and negatively charged amino acids, for example arginine (which has a positively charged R group) and glutamic acid (which has a negatively charged R group):

Arg—C(NH_2)=NH_2^+ ^-O—C(=O)—Glu

LINK
To review the strong and weak interactions that can occur between atoms, see **Concept 2.2**

A complete description of a protein's tertiary structure would specify the location of every atom in the molecule in three-dimensional space, relative to all the other atoms. Many such descriptions are available, including one for the human protein sucrase (**FIGURE 3.9**).

Remember that both secondary and tertiary structure derive from primary structure. If a protein is heated slowly, the heat energy will disrupt only the weaker interactions, causing the secondary and tertiary structure to break down. The protein is then said to be **denatured**. Chemical treatments can also be used to denature proteins. In many cases a denatured protein can return to its normal tertiary structure when it cools or the denaturing

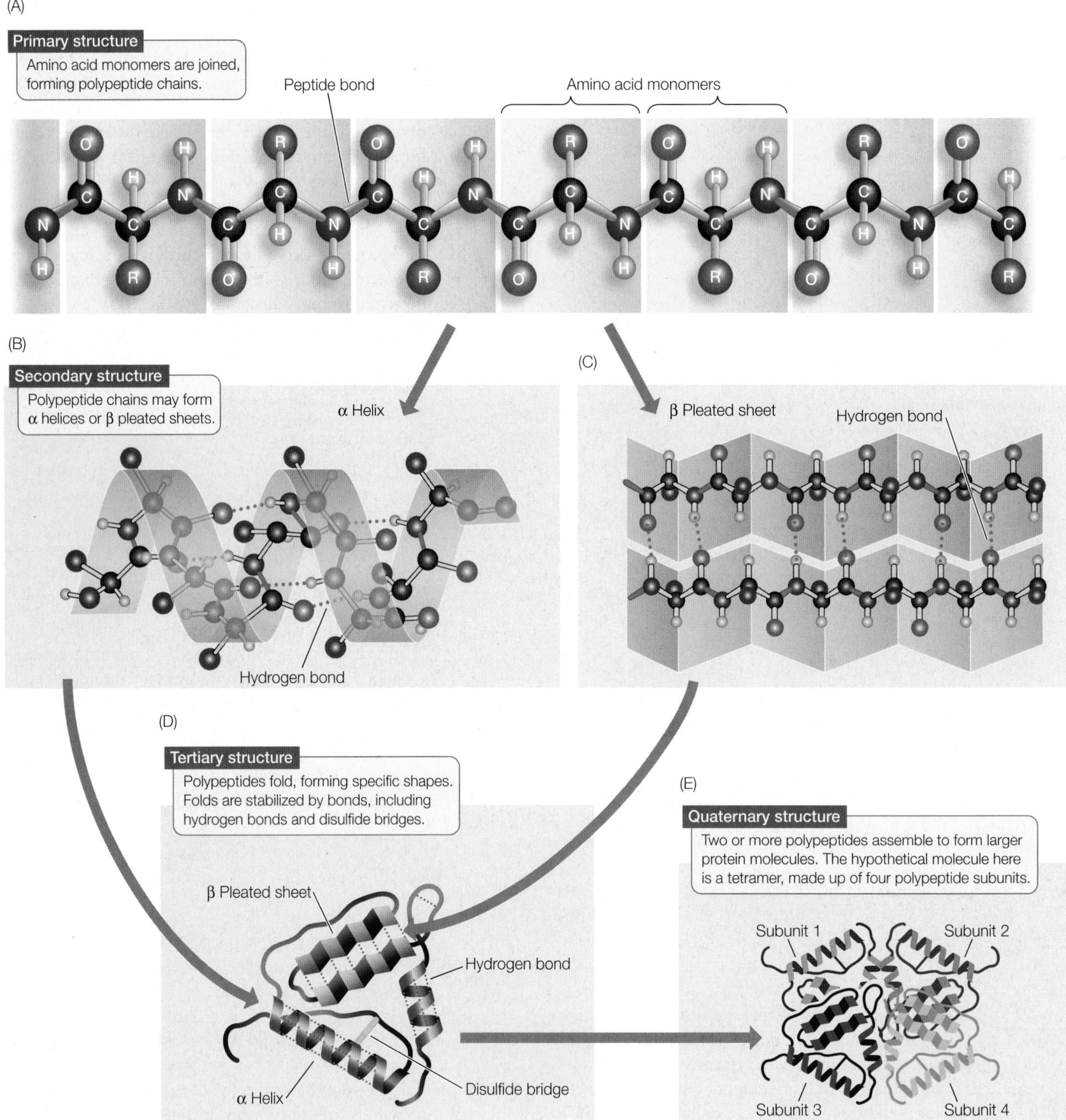
(A)
Primary structure
Amino acid monomers are joined, forming polypeptide chains.
Peptide bond
Amino acid monomers
(B)
Secondary structure
Polypeptide chains may form α helices or β pleated sheets.
α Helix
Hydrogen bond
(C)
β Pleated sheet
Hydrogen bond
(D)
Tertiary structure
Polypeptides fold, forming specific shapes. Folds are stabilized by bonds, including hydrogen bonds and disulfide bridges.
β Pleated sheet
Hydrogen bond
α Helix
Disulfide bridge
(E)
Quaternary structure
Two or more polypeptides assemble to form larger protein molecules. The hypothetical molecule here is a tetramer, made up of four polypeptide subunits.
Subunit 1
Subunit 2
Subunit 3
Subunit 4

FIGURE 3.7 The Four Levels of Protein Structure The primary structure (A) of a protein determines what its secondary (B and C), tertiary (D), and quaternary (E) structures will be.

chemicals are removed, demonstrating that all the information needed to specify the protein's unique shape is contained in its primary structure. This fact was first shown by biochemist Christian Anfinsen for the protein ribonuclease (**FIGURE 3.10**).

QUATERNARY STRUCTURE Many functional proteins contain two or more polypeptide chains, called subunits, each folded into its own unique tertiary structure. The protein's **quaternary structure** results from the ways in which these

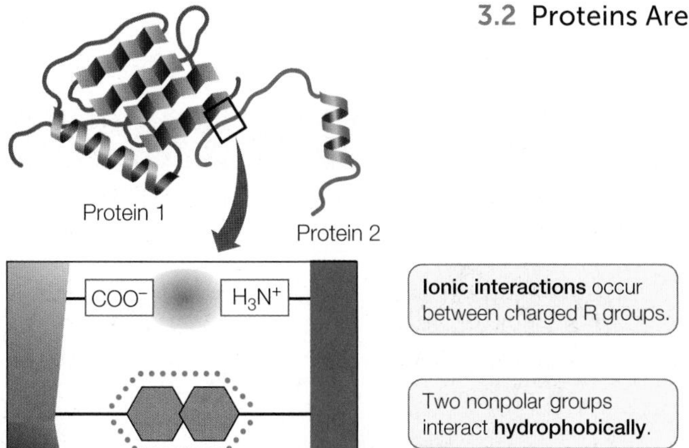

FIGURE 3.8 Noncovalent Interactions between Proteins and Other Molecules Noncovalent interactions allow a protein (brown) to bind tightly to another protein (blue) with specific properties. Noncovalent interactions also allow regions within a single protein to interact with one another.

subunits bind together and interact (see Figure 3.7E). Hemoglobin is an example of a protein with multiple subunits:

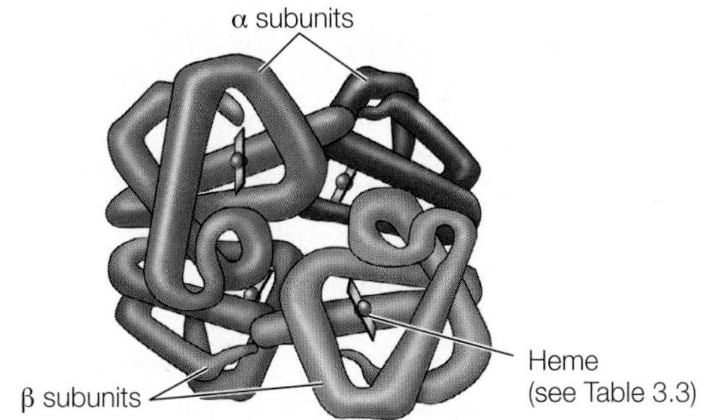

Hydrophobic interactions, hydrogen bonds, and ionic interactions all help hold the four subunits together to form a hemoglobin macromolecule. The weak nature of these forces permits small changes in the quaternary structure to aid the protein's function—which is to carry oxygen in red blood cells. As hemoglobin binds one O_2 molecule, the four subunits shift their relative positions slightly, changing the quaternary structure. Ionic interactions are broken, exposing buried side chains that enhance the binding of additional O_2 molecules. The quaternary structure changes again when hemoglobin releases its O_2 molecules to the cells of the body.

Protein structure can change

The environment that surrounds a protein, as well as interactions with other molecules, can change protein structure.

ENVIRONMENT Various conditions can alter the weak, noncovalent interactions that hold proteins together in their secondary, tertiary, and quaternary structures:

- *Increases in temperature* cause more rapid molecular movements and thus can break hydrogen bonds and hydrophobic interactions.

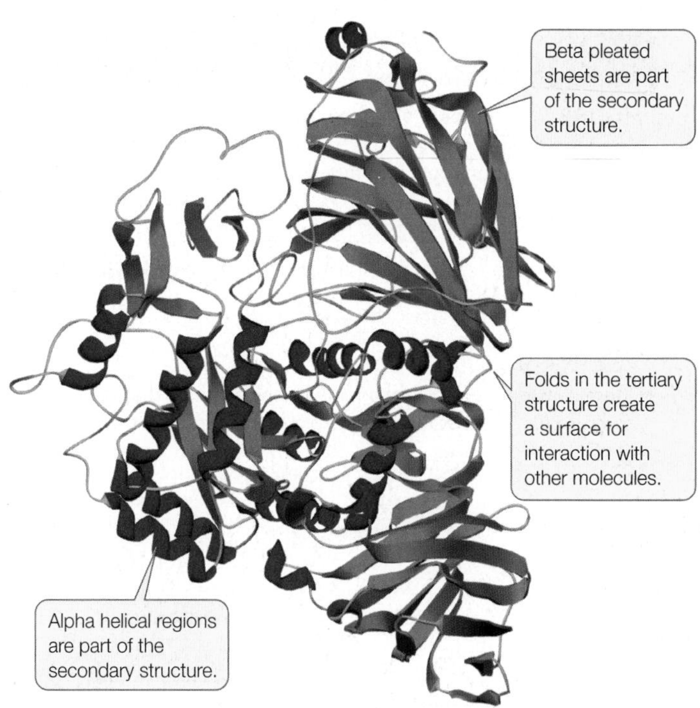

Beta pleated sheets are part of the secondary structure.
Folds in the tertiary structure create a surface for interaction with other molecules.
Alpha helical regions are part of the secondary structure.

FIGURE 3.9 The Structure of a Protein Sucrase has a specific three-dimensional structure, determined by its primary structure. Sucrase plays a role in digestion in humans.

- *Alterations in the concentration of H^+* (pH) in the solution surrounding the protein can change the patterns of ionization of the exposed carboxyl and amino groups. This can disrupt the patterns of ionic attractions and repulsions.
- *High concentrations of polar substances* such as urea can disrupt the hydrogen bonding that is crucial to protein structure.
- *Nonpolar substances* may also denature a protein in cases where hydrophobic groups are essential for maintaining the protein's structure.

Go to MEDIA CLIP 3.1
Protein Structures in 3D
PoL2e.com/mc3.1

Denaturation can be irreversible when amino acids that were buried in the interior of the protein become exposed at the surface, or vice versa, causing a new structure to form, or causing different molecules to bind to the protein. Boiling an egg denatures its proteins and is, as you know, not reversible.

MOLECULAR INTERACTIONS Proteins do not exist in isolation. Within a living organism, a protein may interact with other proteins, other kinds of macromolecules, or a variety of smaller molecules. These interactions are reminiscent of the interactions that make up quaternary structure (see above). If a polypeptide comes into contact with another molecule, R groups on its surface may form weak interactions (such as hydrogen bonds or ionic interactions) with groups on the surface

INVESTIGATION

FIGURE 3.10 Primary Structure Specifies Tertiary Structure Using the protein ribonuclease, Christian Anfinsen showed that proteins spontaneously fold into functionally correct three-dimensional configurations.[a] As long as the primary structure is not disrupted, the information for correct folding (under the right conditions) is retained.

HYPOTHESIS

Under controlled conditions that simulate the normal cellular environment, a denatured protein can refold into a functional three-dimensional structure.

METHOD

Chemically denature a functional ribonuclease, so that only its primary structure (i.e., an unfolded polypeptide chain) remains.

RESULTS

When the disruptive agents are removed, three-dimensional structure is restored and the protein once again is functional.

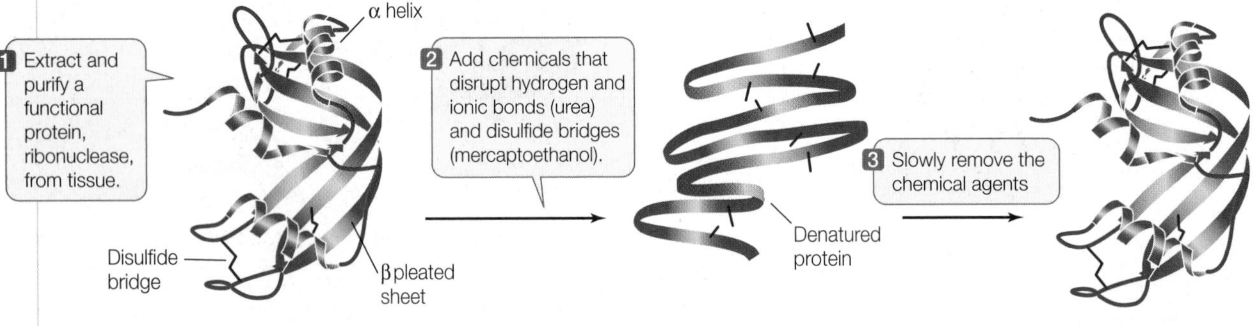

CONCLUSION

In normal cellular conditions, the primary structure of a protein specifies how it folds into a functional, three-dimensional structure.

ANALYZE THE DATA

Initially, disulfide bridges (S—S) in ribonuclease were eliminated because the sulfur atoms in cysteine were reduced (—SH). At time 0, reoxidation began and at various times, the amount of disulfide bridge re-formation (blue circles) and the function of ribonuclease (enzyme activity; red circles) were measured by chemical methods. Here are the data:

A. At what time did disulfide bridges begin to form?

B. At what time did enzyme activity begin to appear?

C. Explain the difference between your answers for the times of (A) and (B).

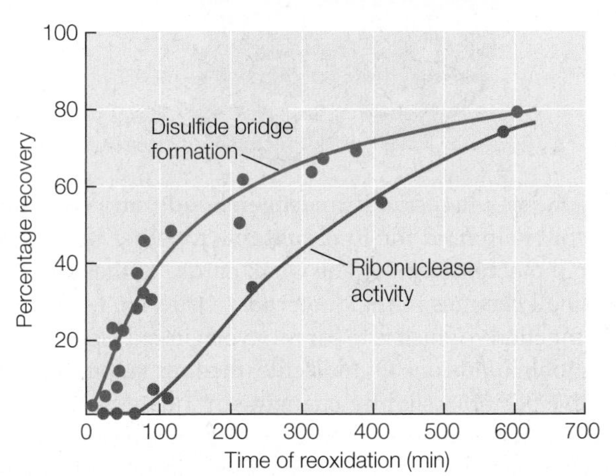

Go to **LaunchPad** for discussion and relevant links for all **INVESTIGATION** figures.

[a]C. B. Anfinsen et al. 1961. *Proceedings of the National Academy of Sciences USA*. 47: 1309–1314.

of the other molecule. This may disrupt some of the interactions between R groups within the polypeptide, causing it to undergo a change in shape (**FIGURE 3.11A**).

The structure of a protein can also be modified by the covalent bonding of a chemical group to the side chain of one or more of its amino acids. The chemical modification of just one amino acid can alter the shape and function of a protein. An example is the addition of a charged phosphate group to a relatively nonpolar R group. This can cause the amino acid to become more hydrophilic and to move to the outer surface of the protein, altering the shape of the protein in the region near the amino acid (**FIGURE 3.11B**).

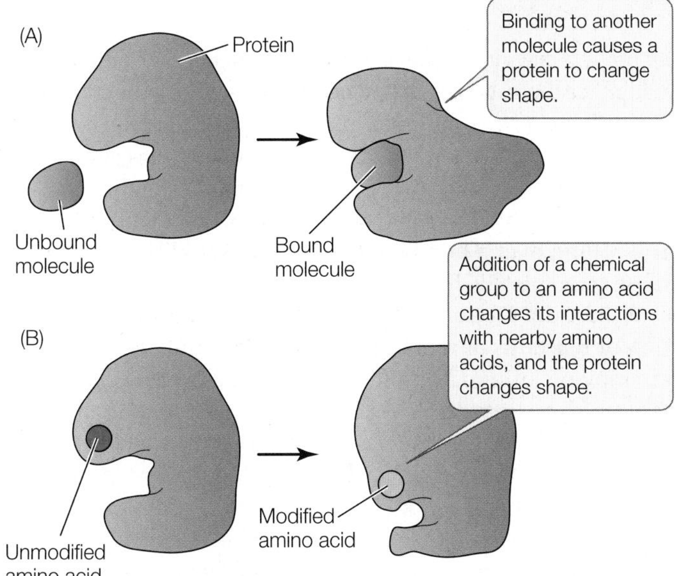

FIGURE 3.11 Protein Structure Can Change Proteins can change their tertiary structure when they bind to other molecules (A) or are modified chemically (B).

CHECKpoint CONCEPT 3.2

- ✓ Sketch the peptide bonding of the two amino acids glycine and leucine (in that order). Now add a third amino acid, alanine, in the position it would have if added within a biological system. What is the directionality of this process?
- ✓ Examine the structure of sucrase (see Figure 3.9). Where in the protein might you expect to find the following amino acids: valine, proline, glutamic acid, and threonine? Explain your answers.
- ✓ Detergents disrupt hydrophobic interactions by coating hydrophobic molecules with a molecule that has a hydrophilic surface. When hemoglobin is treated with a detergent, the four polypeptide chains separate and become random coils. Explain these observations.
- ✓ Several small molecules interact with a protein. The chemical groups on the small molecules interact with specific amino acids as shown in the table below. Fill in the table to show the types of noncovalent interactions that occur between the small molecules and the amino acids.

SMALL MOLECULE CHEMICAL GROUP	AMINO ACID IN PROTEIN	TYPE OF INTERACTION (HYDROGEN BOND; IONIC INTERACTION; HYDROPHOBIC INTERACTION)
$-NH_3^+$	Aspartic acid	
$-CH_3$	Isoleucine	
$-OH$	Glutamine	

We have discussed the remarkable diversity in protein structures. These structures carry functional groups (on exposed amino acid side chains) that can interact with other molecules. In the next section we will see how these interactions can result in catalysis, the speeding up of biochemical reactions.

APPLY THE CONCEPT

Proteins are polymers with important structural and metabolic roles

Biological systems contain "supermolecular complexes" (for example, the ribosome; see Chapter 4), which are composed of individual molecules of RNA and protein that fit together noncovalently. These complexes can be split apart with detergents that disrupt hydrophobic interactions. Based on the concepts discussed in this chapter, fill in the table below to indicate which of the observations are characteristic of RNA, which are characteristic of protein, and which are characteristic of both. Explain your answers.

OBSERVATION	CHARACTERISTIC OF: PROTEIN	RNA
Has three-dimensional (3-D) structure		
3-D structure destroyed by heat		
Monomers connected by N—C bonds		
Contains sulfur atoms		
Contains phosphorus atoms		

CONCEPT 3.3 Some Proteins Act as Enzymes to Speed up Biochemical Reactions

In Chapter 2 we introduced the concepts of biological energetics. We showed that some metabolic reactions are exergonic and some are endergonic, and that biochemistry obeys the laws of thermodynamics (see Figures 2.14 and 2.15). Knowing whether energy is supplied or released in a particular reaction tells us whether the reaction *can* occur in a living system. But it does not tell us *how fast* the reaction will occur.

Living systems depend on reactions that occur spontaneously. But without help, most of these reactions would proceed at such slow rates that an organism could not survive. The role of a **catalyst** is to speed up a reaction without itself being permanently altered. A catalyst does not cause a reaction to occur, but it increases the rate of the reaction. This is an important point: *No catalyst makes a reaction occur that would not proceed without it.*

Biological catalysts are called **enzymes**; for example, the synthesis of prostaglandin (see the opening story) is catalyzed by an enzyme (cyclooxygenase). Most enzymes are proteins, but a few important enzymes are RNA molecules called ribozymes. An enzyme can bind the reactants in a chemical reaction and participate in the reaction itself. However, this participation does not permanently change the enzyme. At the end of the reaction, the enzyme is unchanged and available to catalyze additional, similar reactions.

An energy barrier must be overcome to speed up a reaction

An exergonic reaction releases **free energy** (G), which is the amount of energy in a system that is available to do work. For example, the free energy released in an exergonic reaction can be used by the cell to drive an endergonic reaction, or it can be converted to mechanical energy for movement (see Figure 6.1). But without a catalyst, a reaction will usually take place very slowly. This is because there is an energy barrier between the reactants and the products. Think about the hydrolysis of sucrose, which we described in Concept 2.5.

$$\text{Sucrose} + H_2O \rightarrow \text{glucose} + \text{fructose}$$

In humans, this reaction is part of the process of digestion. The reaction is exergonic, but even if water is abundant, the sucrose molecule will only rarely bind the H atom and the –OH group in the water molecule at the appropriate locations to break the covalent bond between the glucose and fructose—*unless there is an input of energy to initiate the reaction*. Such an input of energy will place the sucrose into a reactive mode called the **transition state**. The energy input required for sucrose to reach this state is called the **activation energy** (E_a). Once the transition state is reached, the reaction can proceed spontaneously with a release of free energy (ΔG is negative) (**FIGURE 3.12A**). The image of a ball rolling over a bump and then down a hill helps illustrate these concepts (**FIGURE 3.12B**).

Where does the activation energy come from? In any collection of reactants at room or body temperature, the molecules are moving around. Recall from Chapter 2 that the energy the molecules possess due to this motion is called kinetic energy. A few molecules are moving fast enough that their kinetic energy can overcome the energy barrier; they enter the transition state and react. So the reaction takes place—but very slowly. If the system is heated, all the reactant molecules have more kinetic energy, and the reaction speeds up. You have probably used this technique in the chemistry laboratory.

Adding enough heat to increase the average kinetic energy of the molecules would not work in a living system, however. Such a nonspecific approach would accelerate all reactions, including destructive ones such as the denaturation of proteins.

An enzyme lowers the activation energy for a reaction by enabling the reactants to come together and react more easily; the reactants need lower amounts of kinetic energy to enter their transition states (**FIGURE 3.13**). In this way, an enzyme can change the rate of a reaction substantially. For example, if a molecule of sucrose just sits in solution, hydrolysis may take hundreds of years. But with the enzyme sucrase present, the same reaction occurs in 1 second! Typically, an enzyme-catalyzed reaction proceeds 10^3 to 10^8 times faster than the uncatalyzed reaction, and the enzyme converts 100 to 1,000 substrate molecules into product per second.

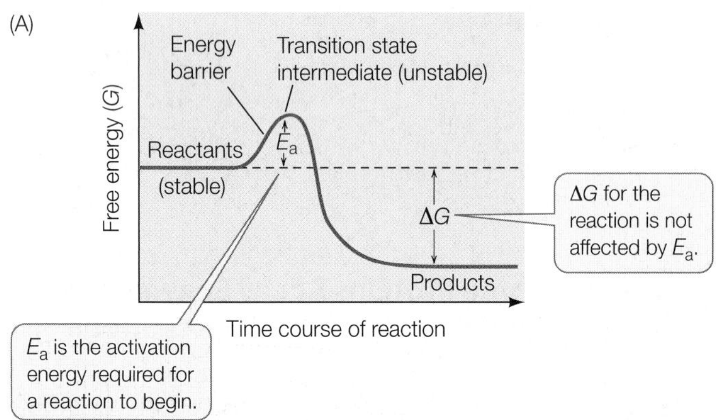

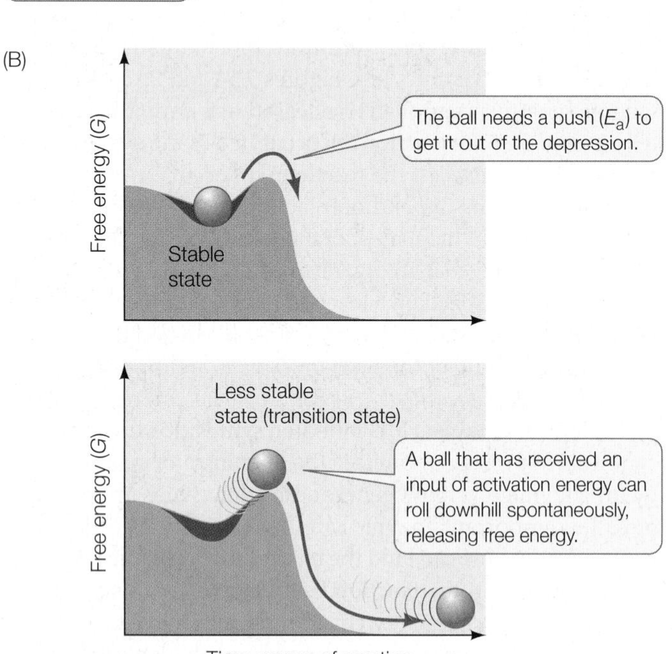

FIGURE 3.12 Activation Energy Initiates Reactions (A) In any chemical reaction, an initial stable state must become less stable before change is possible. (B) A ball on a hillside provides a physical analogy to the biochemical principle graphed in A. Although these graphs show an exergonic reaction, activation energy is needed for endergonic reactions as well.

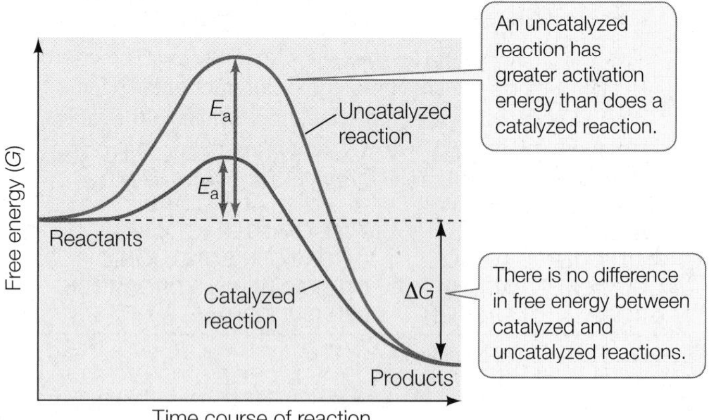

FIGURE 3.13 Enzymes Lower the Energy Barrier The activation energy (E_a) is lower in an enzyme-catalyzed reaction than in an uncatalyzed reaction, but the free energy released is the same with or without catalysis. A lower activation energy means the reaction will take place at a faster rate.

Go to ACTIVITY 3.4 Free Energy Changes
PoL2e.com/ac3.4

Enzymes bind specific reactants at their active sites

Catalysts increase the rates of chemical reactions. Most nonbiological catalysts are nonspecific. For example, powdered platinum catalyzes virtually any reaction in which molecular hydrogen (H_2) is a reactant. In contrast, most biological catalysts are highly specific. An enzyme usually recognizes and binds to only one or a few closely related reactants, and it catalyzes only a single chemical reaction.

In an enzyme-catalyzed reaction, the reactants are called **substrates**. Substrate molecules bind to a particular site on the enzyme, called the **active site**, where catalysis takes place (**FIGURE 3.14**). The specificity of an enzyme results from the exact three-dimensional shape (also called conformation) and chemical properties of its active site. Only a narrow range of substrates, with specific shapes, functional groups, and chemical properties, can fit properly and bind to the active site. The names of enzymes reflect their functions and often end with the suffix "ase." For example, the enzyme sucrase catalyzes the hydrolysis of sucrose, and we write the reaction as follows:

$$\text{Sucrose} + H_2O \xrightarrow{\text{Sucrase}} \text{glucose} + \text{fructose}$$

The binding of a substrate (S) to the active site of an enzyme (E) produces an **enzyme–substrate complex** (**ES**) that is held together by one or more means, such as hydrogen bonding, ionic attraction, or temporary covalent bonding. The enzyme–substrate complex gives rise to product (P) and free enzyme:

$$E + S \rightarrow ES \rightarrow E + P$$

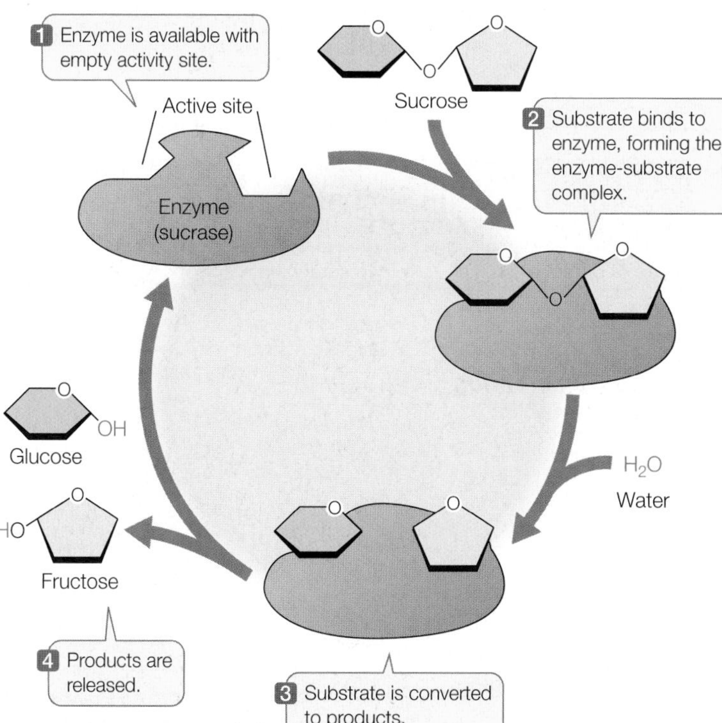

FIGURE 3.14 Enzyme Action Sucrase catalyzes the hydrolysis of sucrose. After the reaction, the enzyme is unchanged and is ready to accept another substrate molecule.

(As we have seen in the case of sucrase, a single enzyme-catalyzed reaction may involve multiple substrates and/or products.) The free enzyme (E) is in the same chemical form at the end of the reaction as at the beginning. While bound to the substrate(s), it may change chemically, but by the end of the reaction it has been restored to its initial form and is ready to catalyze the same reaction again (see Figure 3.14).

HOW ENZYMES WORK During and after the formation of the enzyme–substrate complex, chemical interactions occur. These interactions contribute directly to the breaking of old bonds and the formation of new ones. In catalyzing a reaction, an enzyme may use one or more mechanisms:

- *Inducing strain*: Once the substrate has bound to the active site, the enzyme causes bonds in the substrate to stretch, putting it in an unstable transition state:

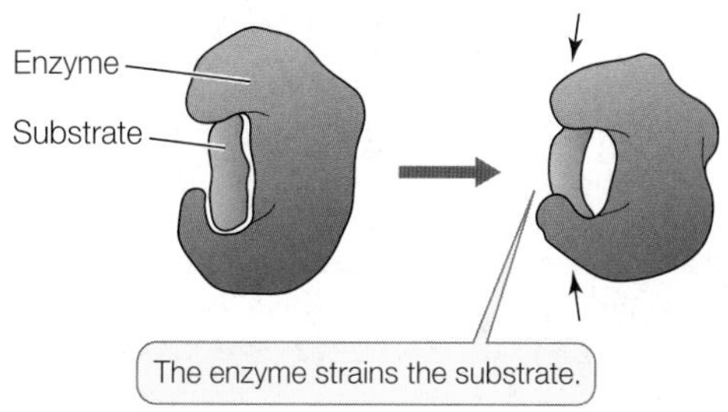

- *Substrate orientation:* When free in solution, substrates are moving from place to place randomly while at the same time vibrating, rotating, and tumbling. They only rarely have the proper orientation to react when they collide. The enzyme lowers the activation energy needed to start the reaction, by bringing together specific atoms so that bonds can form.
- *Adding chemical groups:* The side chains (R groups) of an enzyme's amino acids may be directly involved in the reaction. For example, in acid–base catalysis, the acidic or basic side chains of the amino acids in the active site transfer H^+ ions to or from the substrate, destabilizing a covalent bond in the substrate and permitting the bond to break.

The active site is usually only a small part of the enzyme protein. But its three-dimensional structure is so specific that it binds only one or a few related substrates. The binding of the substrate to the active site depends on the same relatively weak forces that maintain the tertiary structure of the enzyme: hydrogen bonds, the attraction and repulsion of charged groups, and hydrophobic interactions. Scientists used to think of substrate binding as being similar to a lock and key fitting together. Actually, for most enzymes and substrates the relationship is more like a baseball and a catcher's mitt: the substrate first binds, and then the active site changes slightly to make the binding tight. **FIGURE 3.15** illustrates this "induced fit" phenomenon. (We introduced the concept of protein structure changes earlier; see Figure 3.11.)

Induced fit at least partly explains why enzymes are so large. The rest of the macromolecule has at least three roles:

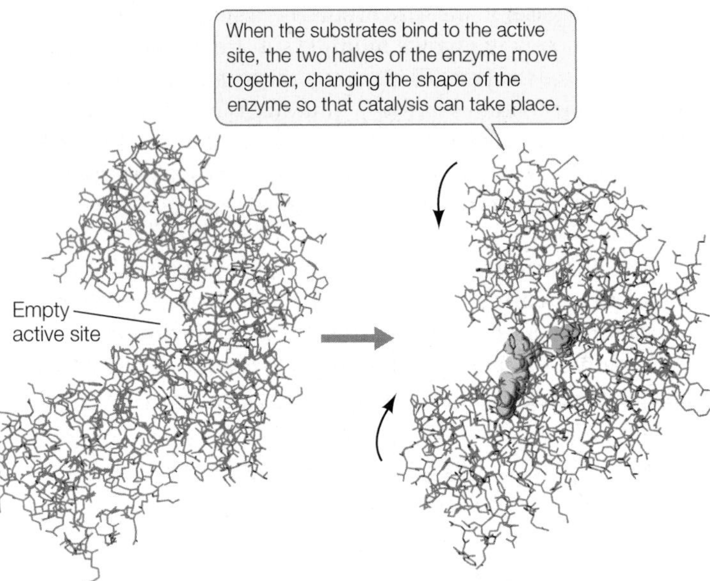
When the substrates bind to the active site, the two halves of the enzyme move together, changing the shape of the enzyme so that catalysis can take place.
Empty active site

FIGURE 3.15 Some Enzymes Change Shape When Substrate Binds to Them Shape changes result in an induced fit between enzyme and substrate, improving the catalytic ability of the enzyme. Induced fit can be observed in the enzyme hexokinase, seen here with and without its substrates, glucose (green) and ATP (yellow).

- It provides a framework so the amino acids of the active site are properly positioned in relation to the substrate(s).
- It participates in the changes in protein shape and structure that result in induced fit.
- It provides binding sites for regulatory molecules (as we will discuss in Concept 3.4).

NONPROTEIN PARTNERS FOR ENZYMES Some enzymes require ions or other molecules in order to function. These molecules are referred to as **cofactors**, and they can be grouped into three categories (**TABLE 3.3**):

- *Metal ions* such as copper, zinc, and iron bind to certain enzymes and participate in the enzyme-catalyzed reactions. For example, the cofactor zinc binds to the enzyme alcohol dehydrogenase, which catalyzes the breakdown of toxic alcohol.
- A *coenzyme* is a relatively small, carbon-containing (organic) molecule that is required for the action of one or more enzymes. It binds to the active site of the enzyme, adds or removes a chemical group from the substrate, and then separates from the enzyme to participate in other reactions. A coenzyme differs from a substrate in that it can participate in many different reactions with different enzymes.
- *Prosthetic groups* are organic molecules that are permanently bound to their enzymes. An example is a flavin nucleotide, which binds to succinate dehydrogenase, an important enzyme in energy metabolism.

RATE OF REACTION The rate of an uncatalyzed reaction is directly proportional to the concentration of the substrate. The higher the concentration, the more reactions per unit of time.

As we have seen, the addition of the appropriate enzyme speeds up the reaction, but it also changes the shape of the plot of rate versus substrate concentration (**FIGURE 3.16**). For a given concentration of enzyme, the rate of the enzyme-catalyzed reaction initially increases as the substrate concentration increases from zero, but then it levels off.

Why does this happen? The concentration of an enzyme is usually much lower than that of its substrate and does not change as substrate concentration changes. When all the enzyme molecules are bound to substrate molecules, the enzyme is working at its maximum rate. Under these conditions the active sites are said to be saturated.

The maximum rate of a catalyzed reaction can be used to measure how efficient the enzyme is—that is, how many molecules of substrate are converted into product by an individual enzyme molecule per unit of time, when there is an excess of substrate present. This turnover number ranges from 1 molecule every second for sucrase to an amazing 40 million molecules per second for the liver enzyme catalase.

CHECKpoint CONCEPT 3.3

- ✓ Explain how the structure of an enzyme makes that enzyme specific.
- ✓ What is activation energy? How does an enzyme lower the activation energy needed to start a reaction?
- ✓ Compare coenzymes with substrates. How do they work together in enzyme catalysis?
- ✓ Compare the state of an enzyme active site at a low substrate concentration and at a high substrate concentration. How does this affect the rate of the reaction?

TABLE 3.3 Some Examples of Enzyme Cofactors

Type of cofactor	Role in catalyzed reactions
METAL IONS	
Iron (Fe^{2+} or Fe^{3+})	Oxidation/reduction
Copper (Cu^{+} or Cu^{2+})	Oxidation/reduction
Zinc (Zn^{2+})	Helps bind NAD
COENZYMES	
Biotin	Carries $-COO^-$
Coenzyme A	Carries $-CO-CH_3$
NAD	Carries electrons
FAD	Carries electrons
ATP	Provides/extracts energy
PROSTHETIC GROUPS	
Heme	Binds ions, O_2, and electrons; contains iron cofactor
Flavin	Binds electrons
Retinal	Converts light energy

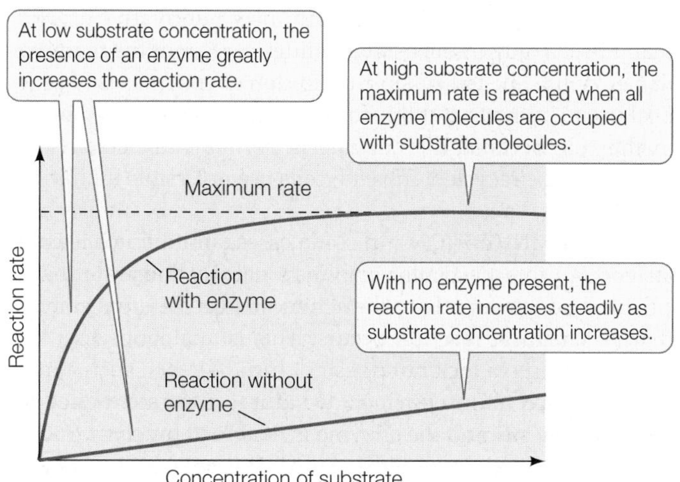

FIGURE 3.16 Catalyzed Reactions Reach a Maximum Rate Because there is usually less enzyme than substrate present, the reaction rate levels off when the enzyme becomes saturated.

Now that you understand more about how enzymes function, let's see how different enzymes work in the metabolism of living organisms.

CONCEPT 3.4 Regulation of Metabolism Occurs by Regulation of Enzymes

The enzyme-catalyzed reactions we have been discussing often operate within **metabolic pathways** in which the product of one reaction is a substrate for the next. For example, the pathway for the catabolism of sucrose begins with sucrase and ends many reactions later with the production of CO_2 and H_2O. Energy is released along the way. Each step of this catabolic pathway is catalyzed by a specific enzyme:

$$\text{Sucrose} + H_2O \xrightarrow{\text{Sucrase}} \text{glucose} + \text{fructose} \longrightarrow$$

$$\xrightarrow{\text{Many enzymes}} \longrightarrow\longrightarrow\longrightarrow\longrightarrow CO_2 + H_2O$$

Other enzymes participate in anabolic pathways, which produce relatively complex molecules from simpler ones. A typical cell contains hundreds of enzymes that participate in many interconnecting metabolic pathways, forming a metabolic system (**FIGURE 3.17**). Consider a single molecule in the midst of this map:

- There may be two or more enzyme-catalyzed reactions affecting it: either making it or metabolizing it.
- Other pathways affect the concentrations of the substrates and products of these reactions.
- Each enzyme-catalyzed reaction has its own rate, depending on these concentrations.

Clearly, every component of this complex system is affected by every other component, making it difficult to predict what would happen if one or more components were altered. In the new field of **systems biology**, scientists describe mathematically the components of metabolic systems—the concentrations of all the reactants and the rates of the reactions—and use computer algorithms to make predictions about what would happen if a component of the system were altered (see Concept 1.2, pp. 8–9).

Cells need to maintain stable internal conditions, including constant levels of certain metabolites. In addition, cells need to regulate their metabolic pathways to respond to changes, either within the organism or in its environment. One way a cell can regulate its metabolism is to control the *amount* of an enzyme. For example, the product of a metabolic pathway may be available from the cell's environment in adequate amounts. In this case, it would be energetically wasteful for the cell to continue making large proteins (as most enzymes are) that it doesn't need. For this reason, cells often have the ability to turn off the synthesis of certain enzymes.

LINK

The amount of an enzyme is controlled by regulating the expression of gene(s), a topic covered in **Chapter 11**

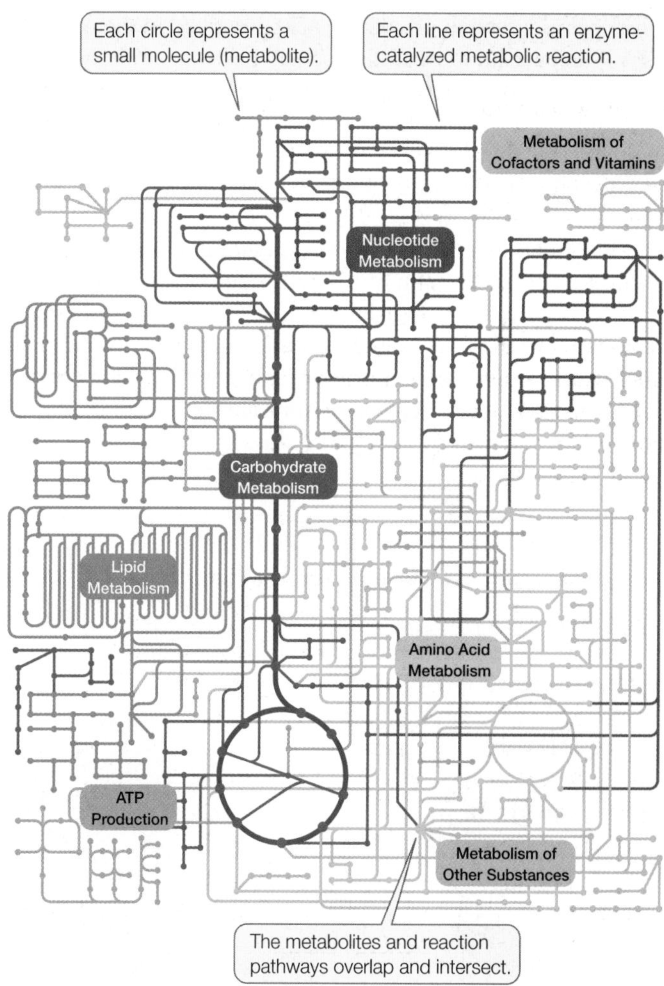

FIGURE 3.17 A Biochemical System The complex interactions of metabolic pathways can be studied using the tools of systems biology. Enzymes are a major element controlling these pathways.

The consequences of *too little* enzyme can be significant. For example, in humans sucrase is important in digestion. In rare cases, infants are born with a congenital sucrase deficiency and the pathway that begins with sucrose is essentially blocked. If these infants ingest foods containing sucrose, the sucrose accumulates rather than being catabolized, and the infant gets diarrhea and stomach cramps. In some cases this leads to slower growth. This deficiency can be treated by limiting sucrose consumption or taking tablets that contain sucrase at every meal.

Cells can also maintain stable internal conditions by regulating the *activity* of enzymes. An enzyme protein may be present continuously, but it may be active or inactive depending on the needs of the cell. Synthesizing and breaking down enzymes takes time, whereas regulating enzyme activity allows cells to fine-tune metabolism relatively quickly in response to changes in the environment. In this section, we will describe how enzyme regulation occurs.

Enzymes can be regulated by inhibitors

Various chemical inhibitors can bind to enzymes, slowing down the rates of the reactions they catalyze. Some inhibitors occur naturally in cells; others can be made in laboratories. Naturally occurring inhibitors regulate metabolism; artificial ones (such as the improved version of salicylic acid described in the opening story) can be used to treat disease, kill pests, or study how enzymes work. In some cases the inhibitor binds the enzyme irreversibly, and the enzyme becomes permanently inactivated. In other cases the inhibitor has reversible effects; it can separate from the enzyme, allowing the enzyme to function fully as before.

IRREVERSIBLE INHIBITION If an inhibitor covalently binds to an amino acid side chain at the active site of an enzyme, the enzyme is permanently inactivated because it cannot interact with its substrate. An example of an irreversible inhibitor is DIPF (diisopropyl phosphorofluoridate), which irreversibly inhibits acetylcholinesterase, an important enzyme that functions in the nervous system. DIPF does so by reacting with a hydroxyl group on a serine in the active site (**FIGURE 3.18**).

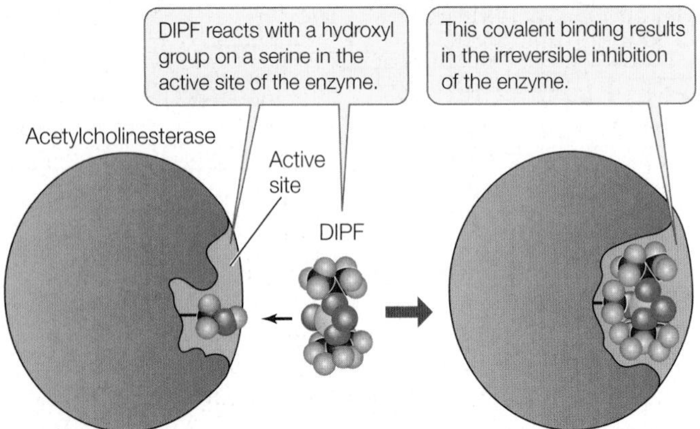

FIGURE 3.18 Irreversible Inhibition DIPF forms a stable covalent bond with the amino acid serine at the active site of the enzyme acetylcholinesterase, thus irreversibly disabling the enzyme.

The widely used insecticide malathion is a derivative of DIPF that inhibits only insect acetylcholinesterase, not the mammalian enzyme. The irreversible inhibition of enzymes is of practical use to humans, but this form of regulation is not common in the cell, because the enzyme is permanently inactivated and cannot be recycled. Instead, cells use reversible inhibition.

REVERSIBLE INHIBITION In some cases, an inhibitor is similar enough to a particular enzyme's natural substrate that it can bind noncovalently to the active site, yet different enough that no chemical reaction occurs. This is analogous to a key that inserts into a lock but does not turn it. When such a molecule is bound to the enzyme, the natural substrate cannot enter the active site and the enzyme is unable to function. Such a molecule is called a **competitive inhibitor** because it competes with the natural substrate for the active site (**FIGURE 3.19A**). Many drugs are competitive inhibitors of enzyme targets. For example, methotrexate is a drug designed with a structure similar to the metabolite dihydrofolate. The latter is converted by an enzyme to a substance essential to cell division. Acting as a competitive inhibitor of the enzyme, methotrexate blocks cell division and is used in cancer therapy. Competitive inhibition is reversible. When the concentration of the competitive inhibitor is reduced, the active site is less likely to be occupied by the inhibitor, and the enzyme regains activity.

A **noncompetitive inhibitor** binds to an enzyme at a site distinct from the active site. This binding causes a change in the shape (the conformation) of the enzyme, altering its activity (**FIGURE 3.19B**). The active site may no longer bind the substrate, or if it does, the rate of product formation may be reduced. Like competitive inhibitors, noncompetitive inhibitors can become unbound, so their effects are reversible.

An allosteric enzyme is regulated by changes in its shape

Noncompetitive inhibition is an example of allostery (*allo*, "different"; *stereos*, "shape"). **Allosteric regulation** occurs when a non-substrate molecule binds or modifies a site other than the active site of an enzyme. The site bound by the non-substrate molecule is called the allosteric site. This binding induces the enzyme to change its conformation, altering the chemical attraction (affinity) of the active site for the substrate. As a result, the rate of the reaction is changed.

An allosteric site may be modified by either noncovalent or covalent binding:

- *Noncovalent binding*: A regulatory molecule may bind noncovalently to an allosteric site, causing the enzyme to change shape. This noncovalent binding is reversible, and may result in the inactivation of an enzyme (see Figure 3.19B) or the activation of a formerly inactive enzyme (**FIGURE 3.20A**).
- *Covalent binding*: Some allosteric sites can be modified by the covalent binding of a molecule or chemical group. For example, an amino acid residue can be covalently modified by the addition of a phosphate group, in a process called phosphorylation (**FIGURE 3.20B**). If this occurs in a hydro-

(A) Competitive inhibition

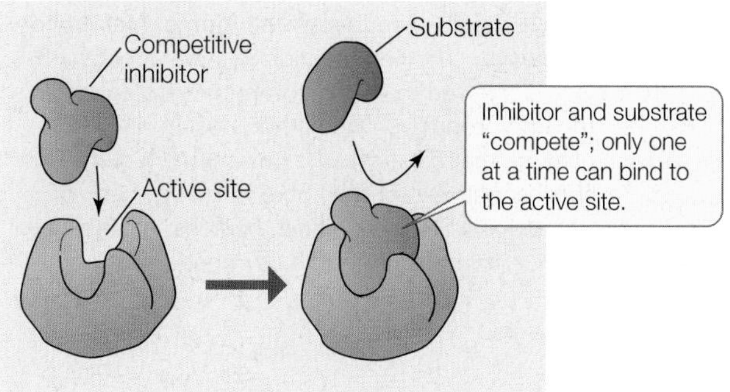

(B) Noncompetitive inhibition

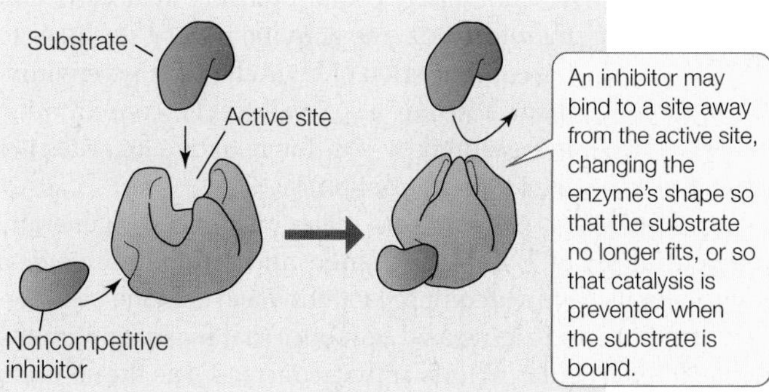

FIGURE 3.19 Reversible Inhibition (A) A competitive inhibitor binds temporarily to the active site of an enzyme. (B) A noncompetitive inhibitor binds temporarily to the enzyme at a site away from the active site. In both cases, the enzyme's function is disabled for only as long as the inhibitor remains bound.

Go to ANIMATED TUTORIAL 3.2
Enzyme Catalysis
PoL2e.com/at3.2

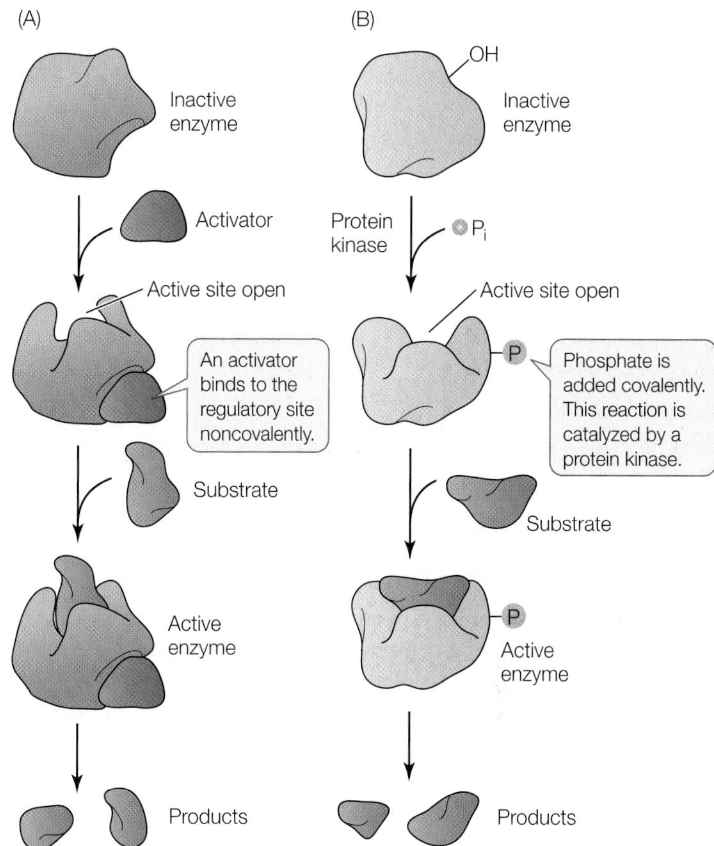

FIGURE 3.20 Allosteric Regulation of Enzyme Activity
(A) Noncovalent binding of a regulator (in this case an activator) can cause an enzyme to change shape and expose an active site. B) Enzymes can also be activated by covalent modification, in this case phosphorylation. Note that allosteric regulation can be negative as well, with the active site becoming hidden.

Go to ANIMATED TUTORIAL 3.3
Allosteric Regulation of Enzymes
PoL2e.com/at3.3

phobic region of the enzyme, it makes that region hydrophilic, because phosphate carries a negative charge. The protein twists, and this can expose or hide the active site. Protein phosphorylation is an extremely important mechanism by which cells regulate many different enzymes and other proteins. It is a reversible process: a class of enzymes called protein kinases catalyze the addition of phosphate groups to proteins, whereas protein phosphatases remove phosphate groups from proteins. Humans have hundreds of different protein kinases and phosphatases. We will return to the exact functions of these proteins many times in this book.

LINK

Protein kinases are of particular importance in intracellular signaling pathways (see **Concepts 5.5 and 5.6**) and in the control of cell reproduction (see **Concept 7.3**)

Some metabolic pathways can be controlled by feedback inhibition

A metabolic pathway typically involves a starting material, various intermediate products, and an end product that is used for some purpose by the cell. In each pathway there are a number of reactions, each forming an intermediate product and each catalyzed by a different enzyme. In many pathways the first step is the commitment step, meaning that once this enzyme-catalyzed reaction occurs, the "ball is rolling," and the other reactions happen in sequence, leading to the end product. But as we pointed out earlier, it is energetically wasteful for the cell to make something it does not need.

One way to regulate a metabolic pathway is by having the final product inhibit the enzyme that catalyzes the commitment step (**FIGURE 3.21**). When the end product is present at a high concentration, some of it binds to a site on the commitment step enzyme, thereby causing it to become inactive. The end product may bind to the active site on the enzyme (as a competitive inhibitor) or an allosteric site (as a noncompetitive inhibitor). This mechanism is known as **feedback inhibition** or end-product inhibition. We will describe many other examples of such inhibition in later chapters.

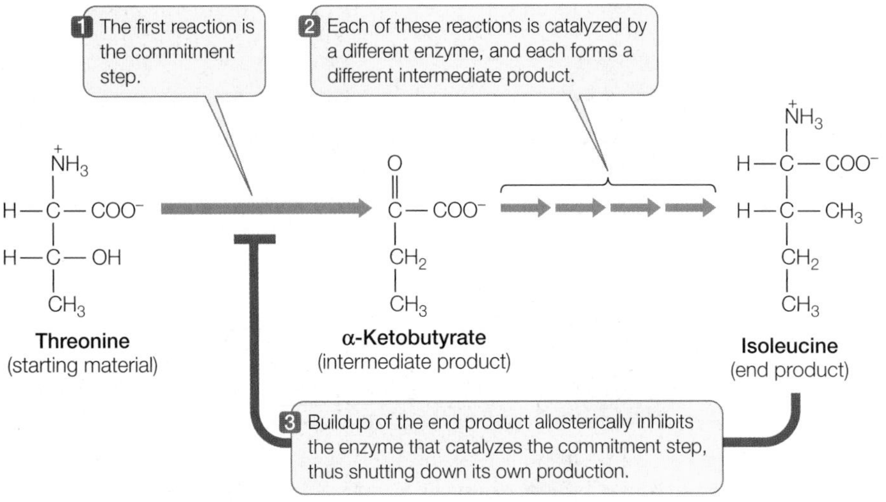
1 The first reaction is the commitment step.
2 Each of these reactions is catalyzed by a different enzyme, and each forms a different intermediate product.
Threonine
(starting material)
α-Ketobutyrate
(intermediate product)
Isoleucine
(end product)
3 Buildup of the end product allosterically inhibits the enzyme that catalyzes the commitment step, thus shutting down its own production.

FIGURE 3.21 Feedback Inhibition of Metabolic Pathways The first reaction in a metabolic pathway is referred to as the commitment step. Often the end product of the pathway can inhibit the enzyme that catalyzes the commitment step. The specific pathway shown here is the synthesis of isoleucine from threonine in bacteria. It is typical of many enzyme-catalyzed biosynthetic pathways.

Enzymes are affected by their environment

As we have seen, the specificity and activity of an enzyme depend on its three-dimensional structure, and this in turn depends on weak forces such as hydrogen bonds (see Figure 3.7). In living systems, two environmental factors can change protein structure and thereby enzyme activity.

pH We introduced the concept of acids and bases when we discussed amino acids. Some amino acids have side chains that are acidic or basic (see Table 3.2). That is, they either generate H^+ and become anions, or attract H^+ and become cations. These reactions are often reversible. For example:

$$\text{Glutamic acid—COOH} \rightleftharpoons \text{glutamic acid—COO}^- + H^+$$

The ionic form of this amino acid (right) is far more hydrophilic than the nonionized form (left).

From your studies of chemistry, you may recall the law of mass action or Le Chatelier's principle. In this case the law implies that the higher the H^+ concentration in the solution, the more the reaction will be driven to the left (forming more of the nonionized form of glutamic acid). Therefore changes in the H^+ concentration can alter how hydrophobic some regions of a protein are and thus affect its shape. To generalize, protein tertiary structure, and therefore enzyme activity, is very sensitive to the concentration of H^+ in the aqueous environment. You may also recall that H^+ concentration is measured by pH (the negative logarithm of the H^+ concentration).

Although the water inside cells is generally at a neutral pH of 7, this can change, and different biological environments have different pH values. Each enzyme has a tertiary structure and amino acid sequence that make it optimally active at a particular pH. Its activity decreases as the solution is made more acidic or more basic than this ideal (optimal) pH (**FIGURE 3.22A**). As an example, consider the human digestive system (see Concept 30.4). The pH inside the human stomach is highly acidic, about pH 1.5. Pepsin, an enzyme that is active in the stomach, has a pH optimum near 2. Many enzymes that hydrolyze macromolecules in the intestine, such as proteases, have pH optima in the neutral range. So when food enters the small intestine, a buffer (bicarbonate) is secreted into the intestine to raise the pH to 6.5. This allows the hydrolytic enzymes to be active and digest the food.

TEMPERATURE In general, warming increases the rate of a chemical reaction because a greater proportion of the reactant molecules have enough kinetic energy to provide the activation energy for the reaction. Enzyme-catalyzed reactions are no different (**FIGURE 3.22B**). However, temperatures that are

APPLY THE CONCEPT

Regulation of metabolism occurs by regulation of enzymes

The concept of enzymes as biological catalysts has many applications. In a pile of clothes in your garage, you notice there are bacteria growing on some socks made of this synthetic polymer:

$$[\text{—CO—(CH}_2)_4\text{—CO—NH—(CH}_2)_4\text{—NH—}]_n$$

You make a protein extract from the bacteria and isolate what you think is an enzyme that can cleave the monomers from the polymer. You also synthesize the dipeptide glycine-glycine (see Table 3.2) to test as a possible inhibitor of the enzyme. The table shows the results from several of your experiments.

EXPERIMENT	CONDITION	RATE OF POLYMER CLEAVAGE
1	No enzyme	0.505
2	Enzyme	825.0
3	Enzyme pre-boiled at 100°C	0.520
4	Enzyme + dipeptide	0.495
5	Enzyme + RNA	799.0

1. Explain the results of each experiment.
2. How might the dipeptide work? How would you test your hypothesis?

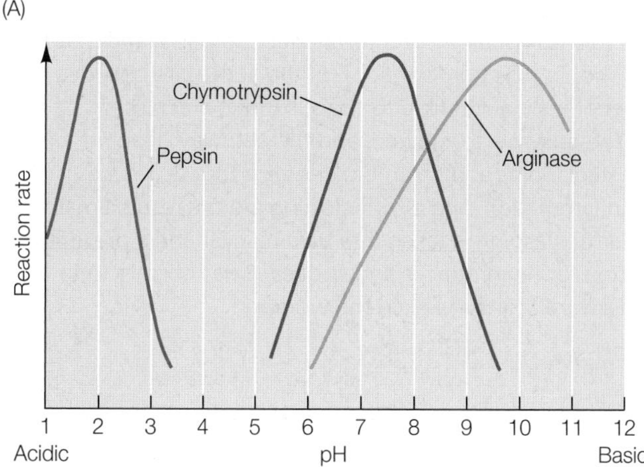

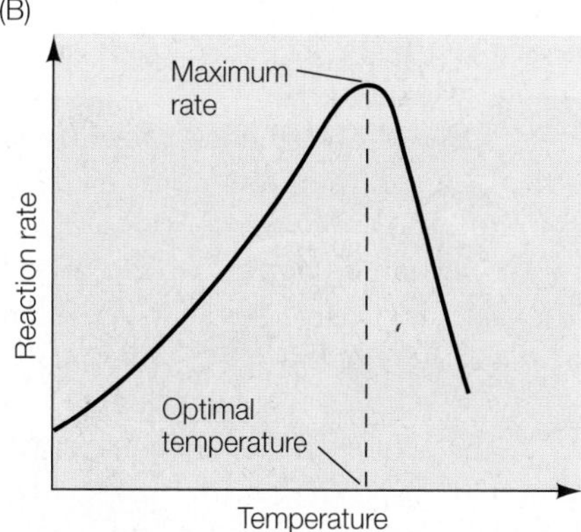

FIGURE 3.22 Enzyme Activity Is Affected by the Environment (A) The activity curve for each enzyme peaks at its optimal pH. For example, pepsin is active in the acidic environment of the stomach, whereas chymotrypsin is active in the neutral environment of the small intestine, and arginase is active in a basic environment. (B) Similarly, there is an optimal temperature for each enzyme. At higher temperatures the enzyme becomes denatured and inactive; this explains why the activity curve falls off abruptly at temperatures that are above optimal.

too high inactivate enzymes, because at high temperatures the polypetides vibrate and twist so rapidly that some of their noncovalent bonds break. When an enzyme's tertiary structure is changed by heat, the enzyme can no longer function. Some enzymes denature at temperatures only slightly above that of the human body, but a few are stable even at the boiling point (or freezing point) of water. All enzymes have an optimal temperature for activity.

Individual organisms adapt to changes in the environment in many ways, one of which is based on groups of enzymes called isozymes, which catalyze the same reaction but have different chemical compositions and physical properties. Different isozymes within a given group may have different optimal temperatures. The rainbow trout, for example, has several isozymes of the enzyme acetylcholinesterase. If a rainbow trout is transferred from warm water to near-freezing water (2°C), the fish produces a different isozyme of acetylcholinesterase. The new isozyme has a lower optimal temperature, allowing the fish's nervous system to perform normally in the colder water.

In general, enzymes adapted to warm temperatures do not denature at those temperatures because their tertiary structures are held together largely by covalent bonds such as disulfide bridges, instead of the more heat-sensitive weak chemical interactions.

CHECKpoint CONCEPT 3.4

- ✓ Explain and give examples of irreversible and reversible enzyme inhibitors.
- ✓ The amino acid glutamic acid (see Table 3.2) is at the active site of an enzyme. Normally the enzyme is active at pH 7. At pH 4 (higher concentration of H^+), the enzyme is inactive. Explain these observations.
- ✓ An enzyme is subject to allosteric regulation. How would you design an inhibitor of the enzyme that was competitive? Noncompetitive? Irreversible?
- ✓ Some organisms thrive at pH 2; other organisms thrive at a temperature of 65°C. Yet mammals cannot tolerate either environment in their tissues. Explain.

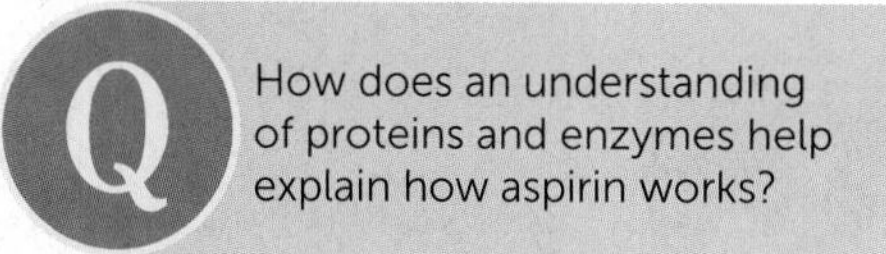

Q
How does an understanding of proteins and enzymes help explain how aspirin works?

ANSWER The mechanism by which aspirin works exemplifies many of the concepts introduced in this chapter. Robert Vane showed that aspirin binds to a protein with a specific three-dimensional structure (Concept 3.2). This protein is cyclooxygenase, an enzyme (Concept 3.3) that catalyzes the commitment step in a metabolic pathway (Concept 3.4). Aspirin acts as an irreversible inhibitor of cyclooxygenase (Concept 3.4). Follow the description below carefully, as it illustrates these important concepts.

Cyclooxygenase catalyzes the conversion of a fatty acid with 20 carbon atoms, arachidonic acid, to a structure with a ring (thus the "cyclo" in the name of the enzyme). O_2 is a substrate (thus the "oxygen"; **FIGURE 3.23**). The product of

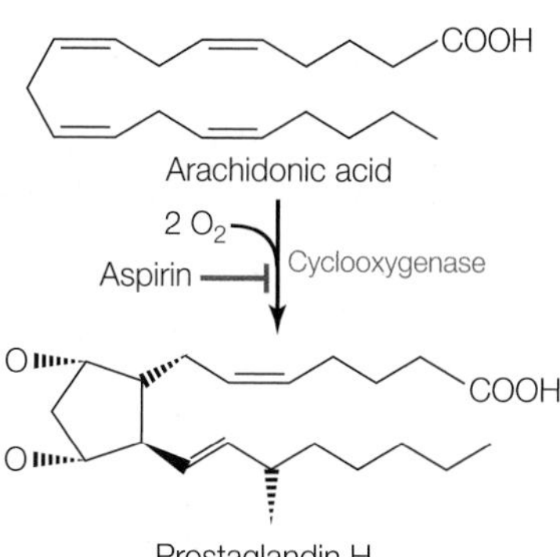

FIGURE 3.23 Aspirin: An Enzyme Inhibitor Aspirin inhibits a key enzyme in the metabolic pathways leading to inflammation and blood clotting.

this reaction (prostaglandin H_2) is the starting material for biochemical pathways that produce two types of molecules:

- prostaglandins, which are involved in inflammation and pain, and
- thromboxanes, which stimulate blood clotting and constriction of blood vessels.

Aspirin binds and reacts with a serine residue within the active site of cyclooxygenase. As a result of this binding, an acetyl group is transferred to the exposed hydroxyl group of the serine residue (**FIGURE 3.24**):

$$\text{(Cyclooxygenase)–serine–OH} \longrightarrow \text{(cyclooxygenase)–serine–O–CH}_2\text{–CH}_3$$

This covalent modification changes the exposed, polar serine to a less polar molecule, and it becomes slightly more hydrophobic. The conformation of the active site changes and becomes inaccessible to the substrate, arachidonic acid. The enzyme is inhibited, and the pathways leading to prostaglandins and thromboxanes are shut down. Less pain, inflammation, and blood clotting are the result. Small wonder that aspirin is taken as a pain reliever and a preventive medicine for heart attacks and strokes. It has come a long way from Edward Stone's walk in the woods

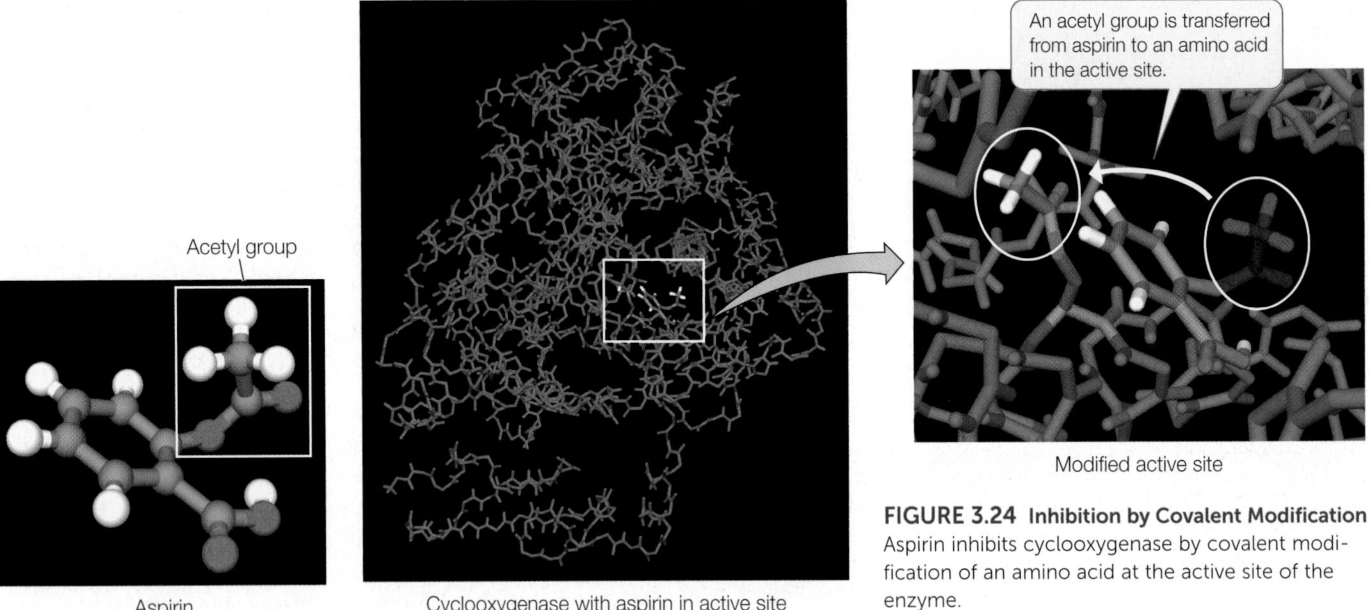
Acetyl group
Aspirin
Cyclooxygenase with aspirin in active site
An acetyl group is transferred from aspirin to an amino acid in the active site.
Modified active site

FIGURE 3.24 Inhibition by Covalent Modification Aspirin inhibits cyclooxygenase by covalent modification of an amino acid at the active site of the enzyme.

SUMMARY

CONCEPT 3.1 Nucleic Acids Are Informational Macromolecules

- The **nucleic acids**—**DNA** and **RNA**—are used mainly to store, transmit, and express hereditary (genetic) information.
- Nucleic acids are polymers of nucleotides. A **nucleotide** consists of one to three phosphate groups, a pentose sugar (**ribose** in RNA and **deoxyribose** in DNA), and a nitrogen-containing **base. Review Figure 3.1 and ACTIVITY 3.1**
- In DNA, the nucleotide bases are **adenine** (**A**), **guanine** (**G**), **cytosine** (**C**), and **thymine** (**T**). **Uracil** (**U**) replaces thymine in RNA. The nucleotides are joined by **phosphodiester bonds** between the sugar of one and the phosphate of the next. RNA is usually single-stranded, whereas DNA is double-stranded. **Review Figure 3.2**
- **Complementary base pairing**, based on hydrogen bonds between A and T, A and U, and G and C, occurs in RNA and DNA. In RNA the hydrogen bonds result in a folded molecule; in DNA the hydrogen bonds connect two antiparallel strands into a double helix. **Review Figures 3.3 and 3.4 and ACTIVITY 3.2**
- DNA is expressed as RNA in the process of **transcription**. RNA can then specify the amino acid sequence of a protein in the process of **translation**.

See **ANIMATED TUTORIAL** 3.1

CONCEPT 3.2 Proteins Are Polymers with Important Structural and Metabolic Roles

- The functions of proteins include support, protection, catalysis, transport, defense, regulation, storage, and movement.
- **Amino acids** are the monomers from which polymeric proteins are made by **peptide bonds**. There are 20 different amino acids in proteins, each distinguished by a **side chain** (**R group**) that confers specific properties. **Review Table 3.2 and ACTIVITY 3.3**
- The **primary structure** of a protein is the sequence of amino acids in the polypeptide chain. This chain is folded into a **secondary structure**, which in different parts of the protein may take the form of an **α helix** or a **β pleated sheet**. **Review Figure 3.7**
- **Disulfide bridges** and noncovalent interactions between amino acids cause polypeptide chains to fold into three-dimensional **tertiary structures**. Multiple polypeptides can interact to form **quaternary structures**. A protein's unique shape and chemical structure allow it to bind specifically to other molecules.
- Heat and certain chemicals can result in a protein becoming **denatured**, which involves the loss of tertiary or secondary structure. **Review Figure 3.10**

CONCEPT 3.3 Some Proteins Act as Enzymes to Speed up Biochemical Reactions

- A chemical reaction must overcome an energy barrier to get started. An **enzyme** is a catalyst that affects the rate of a biological reaction by lowering the **activation energy** needed to initiate the reaction. **Review Figure 3.13 and ACTIVITY 3.4**
- A **substrate** binds to the enzyme's **active site**—the site of catalysis—forming an **enzyme–substrate complex**. Enzymes are highly specific for their substrates.
- At the active site, a substrate enters its **transition state**, and the reaction proceeds.
- Substrate binding causes many enzymes to change shape, exposing their active site(s) and allowing catalysis. **Review Figure 3.15**
- Some enzymes require nonprotein "partners" called **cofactors** to carry out catalysis. **Review Table 3.3**
- Substrate concentration affects the rate of an enzyme-catalyzed reaction. At the maximum rate, the enzyme is saturated with substrate. **Review Figure 3.16**

CONCEPT 3.4 Regulation of Metabolism Occurs by Regulation of Enzymes

- Metabolism is organized into pathways in which the product of one reaction is a substrate for the next reaction. A specific enzyme catalyzes each reaction in the pathway.
- Metabolic pathways are integrated into a biochemical system. **Systems biology** is a way to study how biochemical systems behave. **Review Figure 3.17**
- Enzyme activity is subject to regulation. Some inhibitors bind irreversibly to enzymes. Other inhibitors bind reversibly. **Review Figures 3.18 and 3.19 and ANIMATED TUTORIAL 3.2**
- In **allosteric regulation**, a molecule binds to a site on the enzyme other than the active site. This changes the overall structure of the enzyme (including that of its active site) and results in either activation or inhibition of the enzyme's catalytic activity. **Review Figure 3.20 and ANIMATED TUTORIAL 3.3**
- The end product of a metabolic pathway may inhibit an enzyme that catalyzes the "commitment step" of that pathway. This is called **feedback inhibition**. **Review Figure 3.21**
- Environmental pH and temperature affect enzyme activity. **Review Figure 3.22**

Go to the Interactive Summary to review key figures, Animated Tutorials, and Activities PoL2e.com/is3

Go to LaunchPad at **macmillanhighered.com/launchpad** for additional resources, including LearningCurve Quizzes, Flashcards, and many other study and review resources.

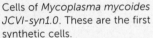

4 Cells: The Working Units of Life

KEY CONCEPTS

4.1 Cells Provide Compartments for Biochemical Reactions

4.2 Prokaryotic Cells Do Not Have a Nucleus

4.3 Eukaryotic Cells Have a Nucleus and Other Membrane-Bound Compartments

4.4 The Cytoskeleton Provides Strength and Movement

4.5 Extracellular Structures Provide Support and Protection for Cells and Tissues

Cells of *Mycoplasma mycoides JCVI-syn1.0*. These are the first synthetic cells.

In 1818, a 21-year-old London writer, Mary Shelley, published a novel that shocked a society in the midst of the Industrial Revolution. In Shelley's story, Dr. Victor Frankenstein discovers how to use electricity to reanimate dead creatures. Collecting body parts from graves and medical labs, the fictional doctor assembles them into a huge 8-foot-tall body and uses his secret method to bring it to life. The results are disastrous, and the novel became a cautionary tale about the limits of science.

Almost 200 years later, in 2010, biologists Craig Venter and Hamilton Smith also gave new life to an "empty shell." In this case, the "shell" was a cell of the tiny bacterium *Mycoplasma capricolum*, into which the scientists inserted a complete new set of genetic material, bringing a new organism to life. The scientists used a computer to design an artificial DNA sequence that had all the genes necessary for bacterial life, plus some unique sequences. Then they went into the chemistry lab and made the DNA from individual nucleotides. They inserted this synthetic genome (similar to the genome of the closely related bacterium *Mycoplasma mycoides*) into the host bacterium, where it replaced the host bacterium's normal DNA. The new DNA directed the cell to perform all the biochemical characteristics of life, including cell reproduction. Eventually, all of the cell's original proteins and RNAs were replaced with proteins and RNAs encoded by the new genome. Since the new genome had some distinctive DNA sequences devised by the scientists, these experiments resulted in an entirely new organism, called *Mycoplasma mycoides JCVI-syn1.0*.

Why did Venter and Smith need to start with a preexisting cell? The chemical reactions of life (metabolism, polymerization, and replication) cannot occur in a dilute aqueous environment; it would be too unlikely for reactants and enzymes to collide with one another. Life requires compartments that bring together and concentrate the molecules involved in these events, which ultimately are directed by the DNA genome.

After about 30 cell divisions, the cells of the new organism no longer had any of the original cell's proteins or small molecules. The cells had used substances in the environment to synthesize their own small and large molecules. They were truly individuals of a new organism, whose "parent" was a synthetic DNA molecule!

The practical aim of this research is to create cells with new capabilities, such as synthesizing clean-burning fuels. But it also puts cells into broader focus as the basic units of biological structure and function.

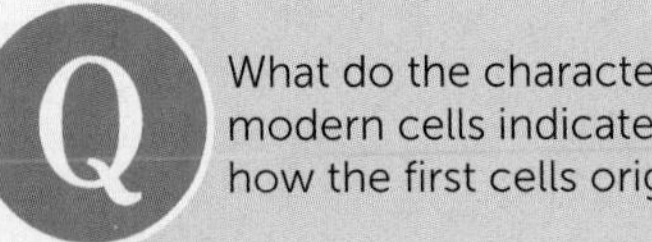

Q What do the characteristics of modern cells indicate about how the first cells originated?

You will find the answer to this question on page 80.

CONCEPT 4.1 Cells Provide Compartments for Biochemical Reactions

Cells contain water and other small and large molecules, which we examined in Chapters 2 and 3. Each cell contains at least 10,000 different types of molecules, most of them present in many copies. Cells use these molecules to transform matter and energy, to respond to their environments, and to reproduce. As we mentioned in the opening story, these biological processes would not be possible outside the enclosure of a cell.

The **cell theory**, developed in the nineteenth century, recognizes this basic fact about life. It was the first unifying principle of biology and has three critical components:

- Cells are the fundamental units of life.
- All living organisms are composed of cells.
- All cells come from preexisting cells.

Cell theory has two important conceptual implications:

- *Studying cell biology is in some sense the same as studying life.* The principles that underlie the functions of a single bacterial cell are similar to those governing the approximately 60 trillion cells in an adult human.
- *Life is continuous.* All those human cells came from a single cell, a zygote (or fertilized egg). The zygote was formed when two cells fused: a sperm from the father and an egg from the mother. The cells of the parents' bodies were all derived from their parents, and so on back through the generations—all the way back to the evolution of the first living cells.

Cell size can be limited by the surface area-to-volume ratio

Most cells are tiny. Their diameters range from about 1 to 100 micrometers (**FIGURE 4.1**). There are some exceptions: the eggs of birds are single cells that are, relatively speaking, enormous, and individual cells of several types of algae and bacteria are large enough to be viewed with the unaided eye.

Small cell size is a practical necessity arising from the decrease in the **surface area-to-volume ratio** of any object as it increases in size. As an object increases in volume, its surface area also increases, but not as quickly (**FIGURE 4.2**). This phenomenon has biological significance for two reasons:

- The *volume* of a cell determines the amount of metabolic activity it carries out per unit of time.
- The *surface area* of a cell determines the amount of substances that can enter it from the outside environment, and the amount of waste products that can exit to the environment.

As a living cell grows larger, its metabolic activity, and thus its need for resources and its rate of waste production, increases faster than its surface area. In addition, substances must move from one location to another within the cell; the smaller the cell, the more easily this is accomplished. The large surface area-to-volume ratio represented by the many small cells of a

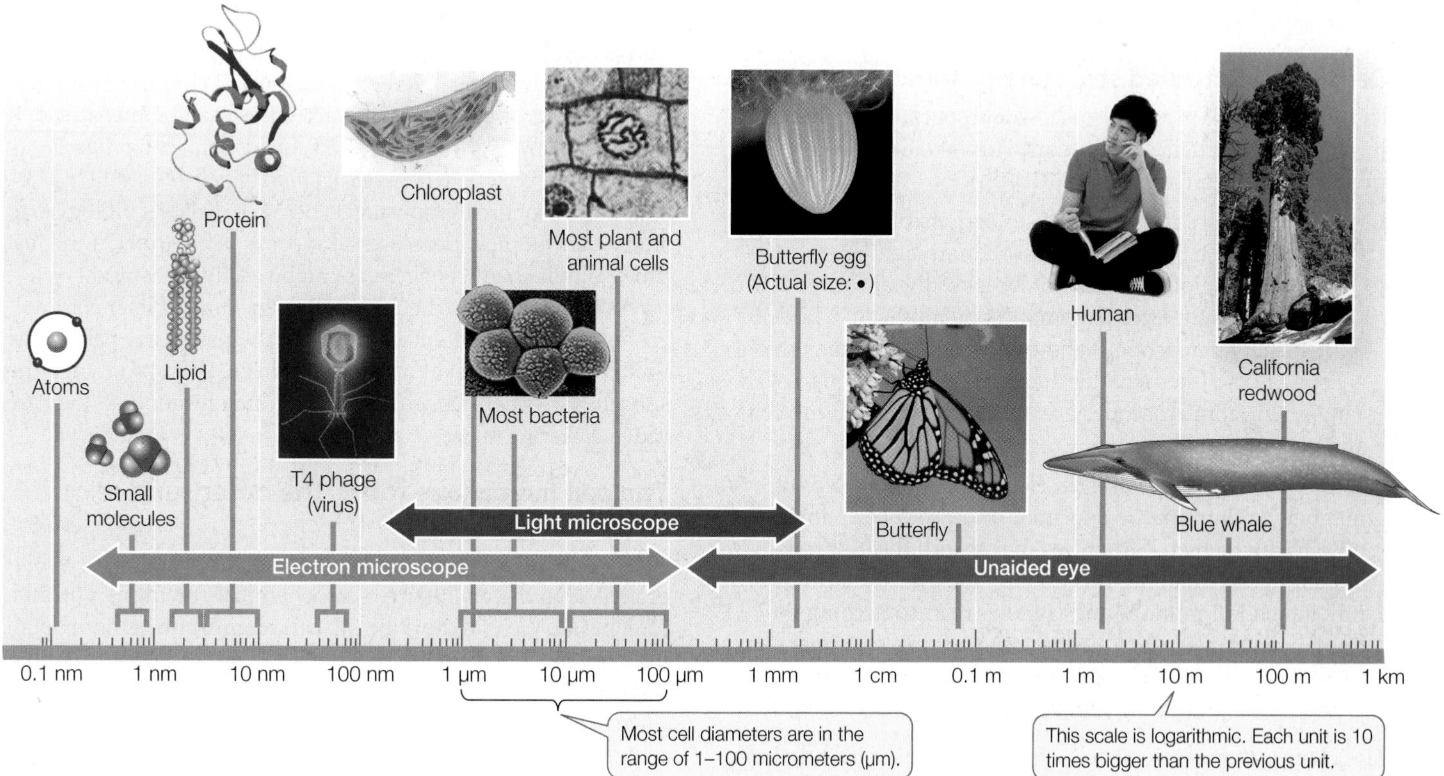

FIGURE 4.1 The Scale of Life This logarithmic scale shows the relative sizes of molecules, cells, and multicellular organisms.

Go to ACTIVITY 4.1 The Scale of Life
PoL2e.com/ac4.1

Larger surface area compared with volume.

Smaller surface area compared with volume.

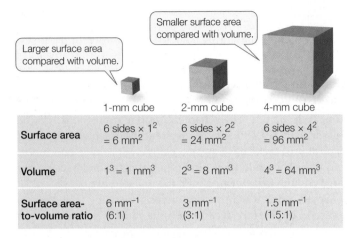

	1-mm cube	2-mm cube	4-mm cube
Surface area	6 sides × 1^2 = 6 mm^2	6 sides × 2^2 = 24 mm^2	6 sides × 4^2 = 96 mm^2
Volume	1^3 = 1 mm^3	2^3 = 8 mm^3	4^3 = 64 mm^3
Surface area-to-volume ratio	6 mm^{-1} (6:1)	3 mm^{-1} (3:1)	1.5 mm^{-1} (1.5:1)

FIGURE 4.2 Why Cells Are Small As an object grows larger, its volume increases more rapidly than its surface area. Cells must maintain a large surface area-to-volume ratio in order to function. This explains why multicellular organisms must be composed of many small cells rather than a few large ones.

RESEARCH TOOLS

FIGURE 4.3 Microscopy Light and electron microscopes are used to examine cell structures.

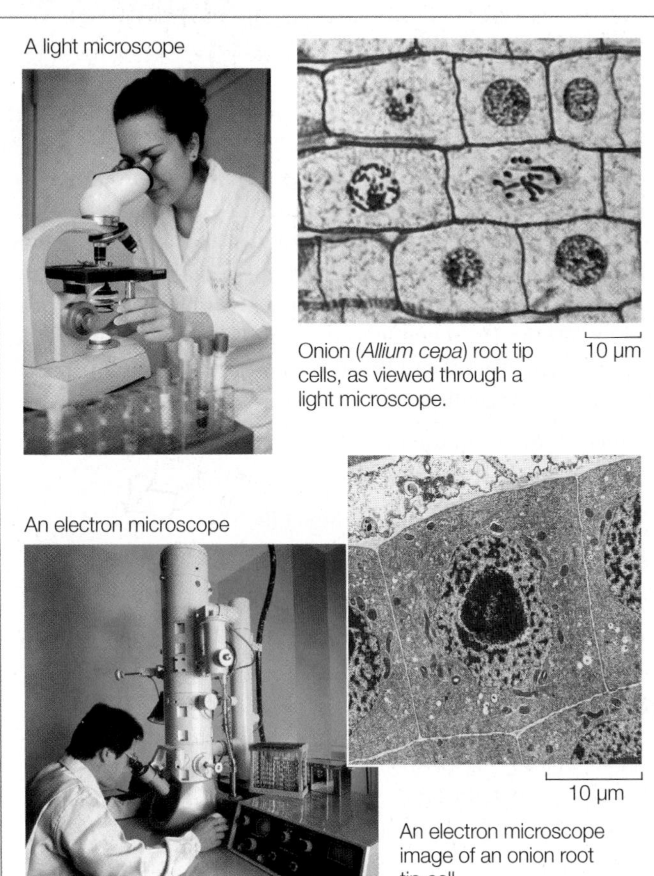
A light microscope
Onion (*Allium cepa*) root tip cells, as viewed through a light microscope.
10 µm
An electron microscope
10 µm
An electron microscope image of an onion root tip cell.

multicellular organism enables it to carry out the many different functions required for survival.

However, for some cell types this general argument does not hold. Many cells are not shaped like cubes, but rather have an irregular shape; as they get larger, they can still have adequate exchange of materials with the environment by increasing their surface area by folds of their cell membrane. In other cases, cells can be quite large without greatly increased membrane surface. For instance, in a giraffe, nerve cells can be several meters long. In these cases, the rate of exchange of materials across the cell membrane must be increased, since the surface area is not adequately larger.

Cells can be studied structurally and chemically

The small sizes of most cells necessitate special instruments to study them and their constituents. For visualizing cells, there are two types of microscopes (**FIGURE 4.3**):

- *A light microscope* uses glass lenses and visible light to form images. Structures that absorb more light are seen as darker than regions that do not, because the absorbed light does not reach the eye. The smallest detail that can be seen with such a microscope is about 0.2 µm in diameter, which is about 1,000 times smaller than an object the human eye can see. Light microscopes are used to visualize living cells and general cell structure.
- *An electron microscope* uses an electron beam focused by magnets to illuminate a specimen and produce an image on a TV-like screen. Structures that absorb the electrons appear darker than regions that do not. The size limit is 0.1 nm, which is 2 million times smaller than something the human eye can see. For electron microscopy, specimens must be preserved and stained using toxic heavy metals, so living cells cannot be visualized this way.

The chemical analysis of cells usually begins with breaking them open to make a cell-free extract. This can be done physically, using a blender or other homogenizing machinery, or by placing the cell in a chemical environment where it swells and bursts (see Figure 5.3). In either case, the resulting extract can be analyzed in terms of its composition and chemical reactions. For example, specific enzyme activities may be measured. If conditions are right in this test tube system, the *properties of the cell-free extract are the same as those inside the cell*. This last statement is of great importance, because it allows biologists to study the chemical processes that occur inside cells in the test tube, so that chemical changes can be easily measured.

A cell's internal structures and even some of its macromolecules can be separated according to their sizes in a centrifuge that spins the tubes at a high speed (**FIGURE 4.4**). Once the subcellular structures are separated from one another, they are much easier to study.

The cell membrane forms the outer surface of every cell

As we will discuss at the end of this chapter, a key to the origin of cells was the enclosure of biochemical functions within a membrane. We will describe the **cell membrane** in more detail in Chapter 5, but for now, note that it consists of a phospholipid bilayer with proteins (see Figure 2.13B). Unless stained for light microscopy, the very thin (7 nm) cell membrane is visible only with electron microscopy. It has several important roles:

- The cell membrane acts as a *selectively permeable barrier*, preventing some substances from crossing it while permitting other substances to enter and leave the cell. In doing

RESEARCH TOOLS

FIGURE 4.4 Centrifugation Structures within cells can be separated from one another on the basis of size and density, and the isolated structures can then be analyzed chemically.

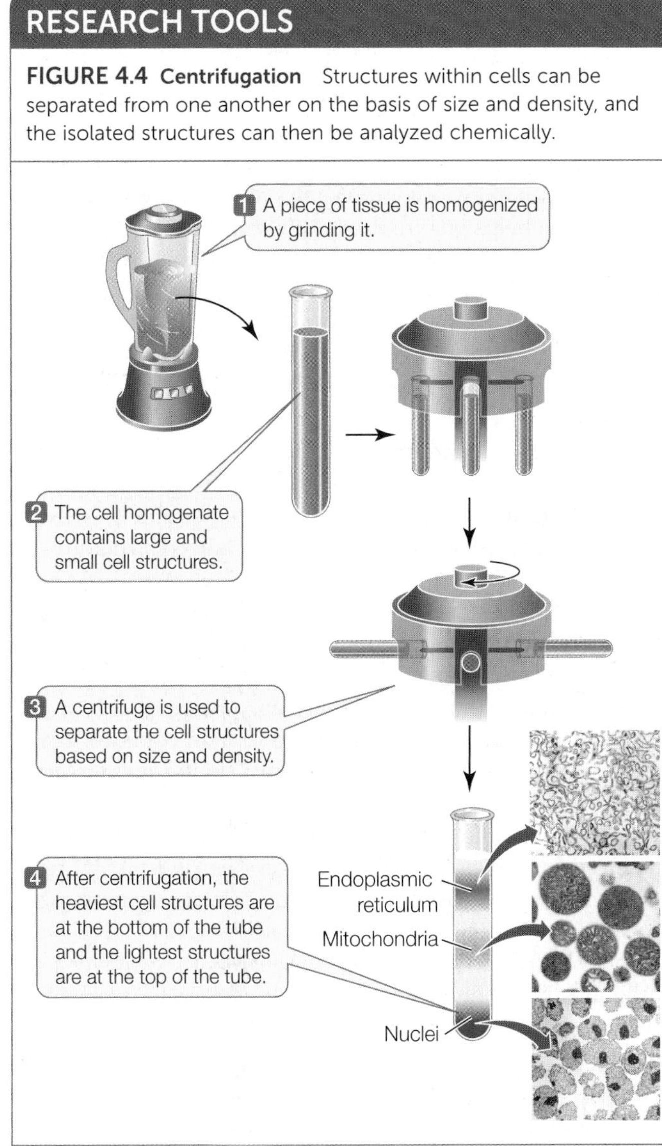

1 A piece of tissue is homogenized by grinding it.
2 The cell homogenate contains large and small cell structures.
3 A centrifuge is used to separate the cell structures based on size and density.
4 After centrifugation, the heaviest cell structures are at the bottom of the tube and the lightest structures are at the top of the tube.
Endoplasmic reticulum
Mitochondria
Nuclei

so, it allows the cell to maintain a stable internal environment that is distinct from the surrounding environment. This explains why a red blood cell contains the pigmented molecule hemoglobin but the surrounding blood plasma does not.

- As the cell's boundary with the outside environment, the cell membrane is important in *communicating* with adjacent cells and receiving signals from the environment.
- The cell membrane often has proteins protruding from it that are responsible for *binding and adhering* to adjacent cells or to a surface. Thus the cell membrane plays an important structural role and contributes to cell shape.

Cells are classified as either prokaryotic or eukaryotic

Biologists classify all living things into three domains: Archaea, Bacteria, and Eukarya. The organisms in Archaea and Bacteria are collectively called **prokaryotes** because they have in common a prokaryotic cellular organization. A prokaryotic cell typically does not have membrane-enclosed internal compartments; in particular, it does not have a nucleus.

Eukaryotic cell organization is found in members of the domain Eukarya—the **eukaryotes**—which includes the protists (a diverse group of microorganisms), plants, fungi, and animals. In contrast to the prokaryotes, eukaryotes contain membrane-enclosed compartments called **organelles** where specific metabolic functions occur. The most notable of these is the cell **nucleus**, where most of the cell's DNA is located and where gene expression begins:

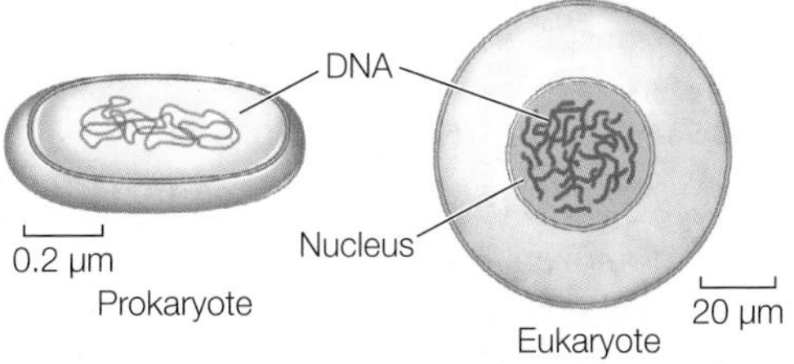

Just as a cell is an enclosed compartment that separates its contents from the surrounding environment, each organelle provides a compartment that separates certain molecules and biochemical reactions from the rest of the cell. This impressive "division of labor" provides possibilities for regulation and efficiency that were important in the evolution of complex organisms.

LINK

Eukaryotes arose from prokaryotes by endosymbiosis; see **Concept 20.1**

CHECKpoint CONCEPT 4.1

- ✓ In considering the origin of cells, why do biologists focus on the origin of the cell membrane?
- ✓ If a cell has a cube shape that is 500 μm on a side, what is its surface area-to-volume ratio? If the surface area-to-volume ratio should be more than 0.1 μm^{-1} for optimal cell function, would dividing this cell into 1 million individual cells (also cubes) meet this standard?
- ✓ What is the surface area-to-volume ratio of a giraffe's nerve cell? [For simplicity, assume that it is tubular (a cylinder), with a length of 3 m and a diameter of 5 μm.]
- ✓ What evolutionary advantages does a eukaryotic cell have compared with a prokaryotic cell?

This section has introduced two structural themes in cell architecture: prokaryotic and eukaryotic. We'll turn now to the organization of prokaryotic cells.

CONCEPT 4.2 Prokaryotic Cells Do Not Have a Nucleus

In terms of sheer numbers and diversity, prokaryotes are the most successful organisms on Earth. As we generalize about the features of these cells, bear in mind that there are vast numbers of prokaryotic species, and that the Bacteria and Archaea can be distinguished from one another in numerous ways. These differences, and the vast diversity of organisms in these two domains, are the subject of Chapter 19.

Prokaryotic cells, with diameters or lengths in the range of 1–10 μm, are generally smaller than eukaryotic cells, whose diameters or lengths are usually in the range of 10–100 μm. While some prokaryotes exist as single cells, other types form chains or small clusters of cells, and in some cases certain cells in a group perform specialized functions.

Prokaryotic cells share certain features

All prokaryotes have the same basic structure (**FIGURE 4.5**):

- The cell membrane encloses the cell, separating its interior from the external environment, and regulates the traffic of materials into and out of the cell.
- The **nucleoid** is a region in the cell where the DNA is located. DNA is the hereditary material that controls cell growth, maintenance, and reproduction (see Chapter 3).
- The rest of the material inside the cell is called the **cytoplasm**. The cytoplasm consists of a liquid component, the cytosol, and a variety of insoluble filaments and particles, the most abundant of which are ribosomes (see below).
- The **cytosol** consists mostly of water containing dissolved ions, small molecules, and soluble macromolecules such as proteins.
- **Ribosomes** are complexes of RNA and proteins that are about 25 nm in diameter. They can be visualized only with the electron microscope. They are the sites of protein synthesis, where the information encoded by nucleic acids directs the sequential linking of amino acids to form proteins.

The cytoplasm is not a static region. Rather, the substances in this environment are in constant motion. For example, a typical protein moves around the entire cell within a minute, and it collides with many other molecules along the way. This constant motion helps ensure that biochemical reactions proceed at sufficient rates to meet the needs of the cell. Prokaryotes may look simple, but in reality they are functionally complex, carrying out thousands of biochemical reactions.

Specialized features are found in some prokaryotes

As they evolved, some prokaryotes developed specialized structures that gave them a selective advantage in their particular environments. These cells were better able to survive and reproduce than cells lacking the specialized structures.

CELL WALLS Most prokaryotes have a **cell wall** located outside the cell membrane. The rigidity of the cell wall supports the cell and determines its shape. The cell walls of most bacteria, but not those of archaea, contain peptidoglycan, a polymer of carbohydrates that is linked at regular intervals to short peptides. Cross-linking among these peptides results in a single giant molecule that surrounds the entire cell. In some bacteria, another layer, the outer membrane (a polysaccharide-rich phospholipid membrane), encloses the peptidoglycan layer (see Figure 4.5). Unlike the cell membrane, this outer membrane is relatively permeable, allowing the movement of molecules across it.

Enclosing the cell wall of some bacteria is a slimy layer composed mostly of polysaccharides, referred to as the capsule. In some cases these capsules protect the bacteria from attack by white blood cells in the animals they infect. Capsules also help keep the cells from drying out, and sometimes they help bacteria attach to other cells. Many prokaryotes produce no capsule, and those that do have capsules can survive even if they lose

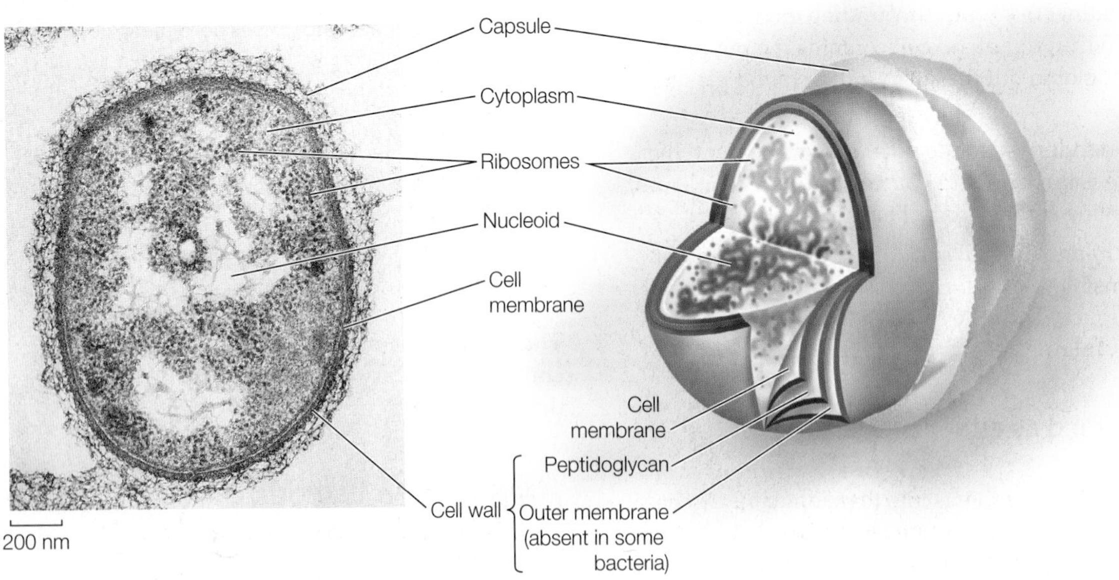

FIGURE 4.5 A Prokaryotic Cell The bacterium *Pseudomonas aeruginosa* illustrates the typical structures shared by all prokaryotic cells. This bacterium also has a protective outer membrane and a capsule, which are not present in all prokaryotes.

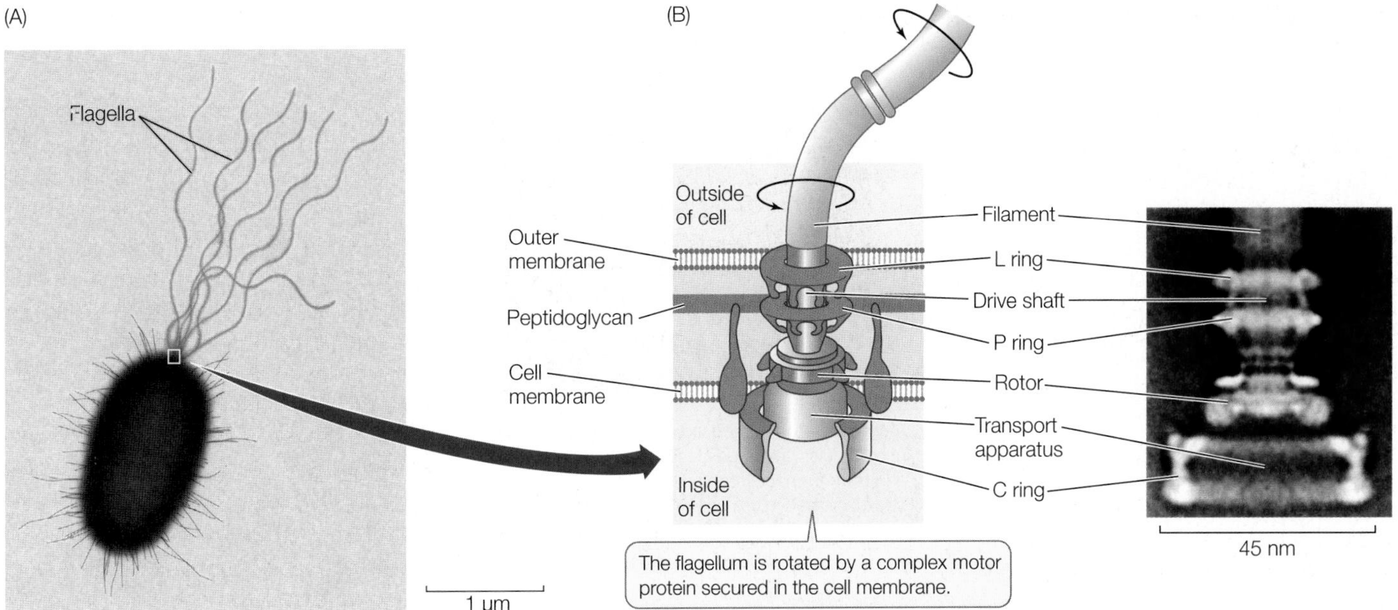

FIGURE 4.6 Prokaryotic Flagella (A) Flagella contribute to the movement and adhesion of prokaryotic cells. (B) Complex protein ring structures anchored in the cell membrane form a motor unit that rotates the flagellum and propels the cell.

them, so the capsule is not essential to prokaryotic life. Some strains of the bacterium *Streptococcus pneumoniae* have a capsule and can cause pneumonia in humans and other mammals; however, non-encapsulated strains do not cause the disease.

INTERNAL MEMBRANES Some groups of bacteria—including the cyanobacteria—carry out photosynthesis: they use energy from the sun to convert carbon dioxide and water into carbohydrates. These bacteria have an **internal membrane** system that contains molecules needed for photosynthesis. The development of photosynthesis, which requires membranes, was an important event in the early evolution of life on Earth. Other prokaryotes have internal membrane folds that are attached to the cell membrane. These folds may function in cell division or in various energy-releasing reactions.

A bacterium with enclosed compartments would have several evolutionary advantages. Chemicals could be concentrated within particular regions of the cell, allowing chemical reactions to proceed more efficiently. Certain biochemical activities could be segregated within compartments with more favorable conditions for those reactions, such as a different pH from the rest of the cell.

FLAGELLA Some cells swim by using appendages called **flagella**, which sometimes look like tiny corkscrews (**FIGURE 4.6A**). These movements are important in allowing the cells to swim toward food, for example. In bacteria, the filament of the flagellum is made of a protein called flagellin. A complex motor protein (see Concept 4.4) spins each flagellum on its axis like a propeller, driving the cell along. This motor protein is anchored to the cell membrane and, in some bacteria, to the outer membrane of the cell wall (**FIGURE 4.6B**). We know that flagella cause the motion of cells because if they are removed, the cells do not move. Flagellar motion can result in impressive speeds. The prokaryote *Methanocaldococcus* has been clocked at over 400 μm (500 body lengths) per second. For a car, this would mean 4,000 miles per hour.

CYTOSKELETON The **cytoskeleton** is the collective name for filaments made up of polymers of monomer subunits that play roles in cell division or in maintaining the shapes of cells. One such protein forms helical structures that extend down the lengths of rod-shaped bacterial cells, helping maintain their shapes. This protein is similar to actin in eukaryotic cells, which we will discuss next.

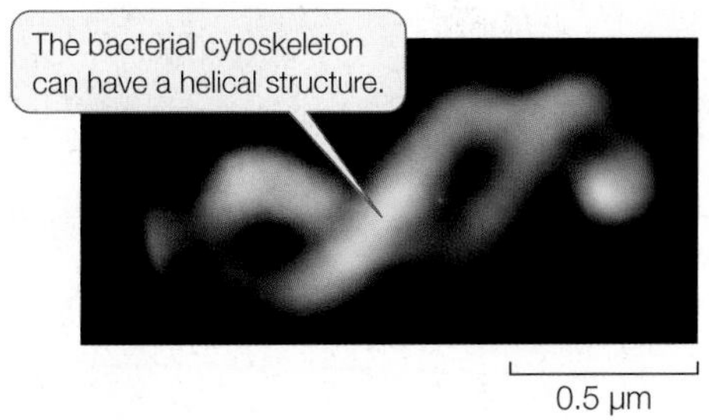

The bacterial cytoskeleton can have a helical structure.
0.5 μm

CHECKpoint CONCEPT 4.2

- ✓ Compare the structures and functions of bacterial cell walls with those of bacterial cytoskeletons.
- ✓ What is the evolutionary advantage of bacteria that have flagella over bacteria that do not?

The prokaryotic cell is one of two broad types of cells recognized in cell biology. The other is the eukaryotic cell. Eukaryotic cells, and multicellular eukaryotic organisms, are more structurally and functionally complex than prokaryotic cells.

CONCEPT 4.3 Eukaryotic Cells Have a Nucleus and Other Membrane-Bound Compartments

Like prokaryotic cells, eukaryotic cells have a cell membrane, cytoplasm, and ribosomes. But as we noted earlier in this chapter, eukaryotic cells also have organelles within the cytoplasm

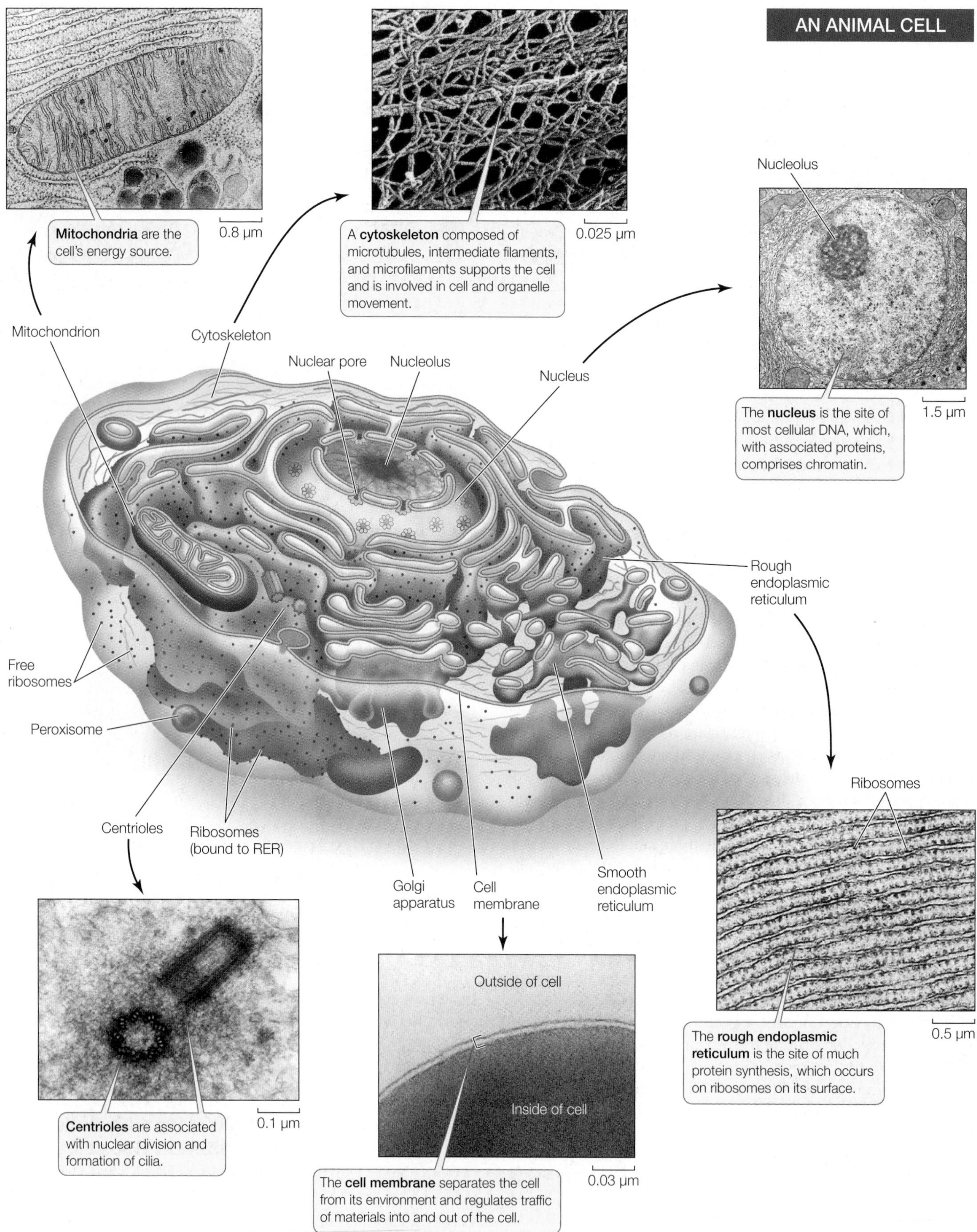
AN ANIMAL CELL
0.8 µm
Mitochondria are the cell's energy source.
A cytoskeleton composed of microtubules, intermediate filaments, and microfilaments supports the cell and is involved in cell and organelle movement.
0.025 µm
Nucleolus
The nucleus is the site of most cellular DNA, which, with associated proteins, comprises chromatin.
1.5 µm
Mitochondrion
Cytoskeleton
Nuclear pore
Nucleolus
Nucleus
Rough endoplasmic reticulum
Free ribosomes
Peroxisome
Centrioles
Ribosomes (bound to RER)
Golgi apparatus
Cell membrane
Smooth endoplasmic reticulum
Ribosomes
The rough endoplasmic reticulum is the site of much protein synthesis, which occurs on ribosomes on its surface.
0.5 µm
Centrioles are associated with nuclear division and formation of cilia.
0.1 µm
Outside of cell
Inside of cell
The cell membrane separates the cell from its environment and regulates traffic of materials into and out of the cell.
0.03 µm

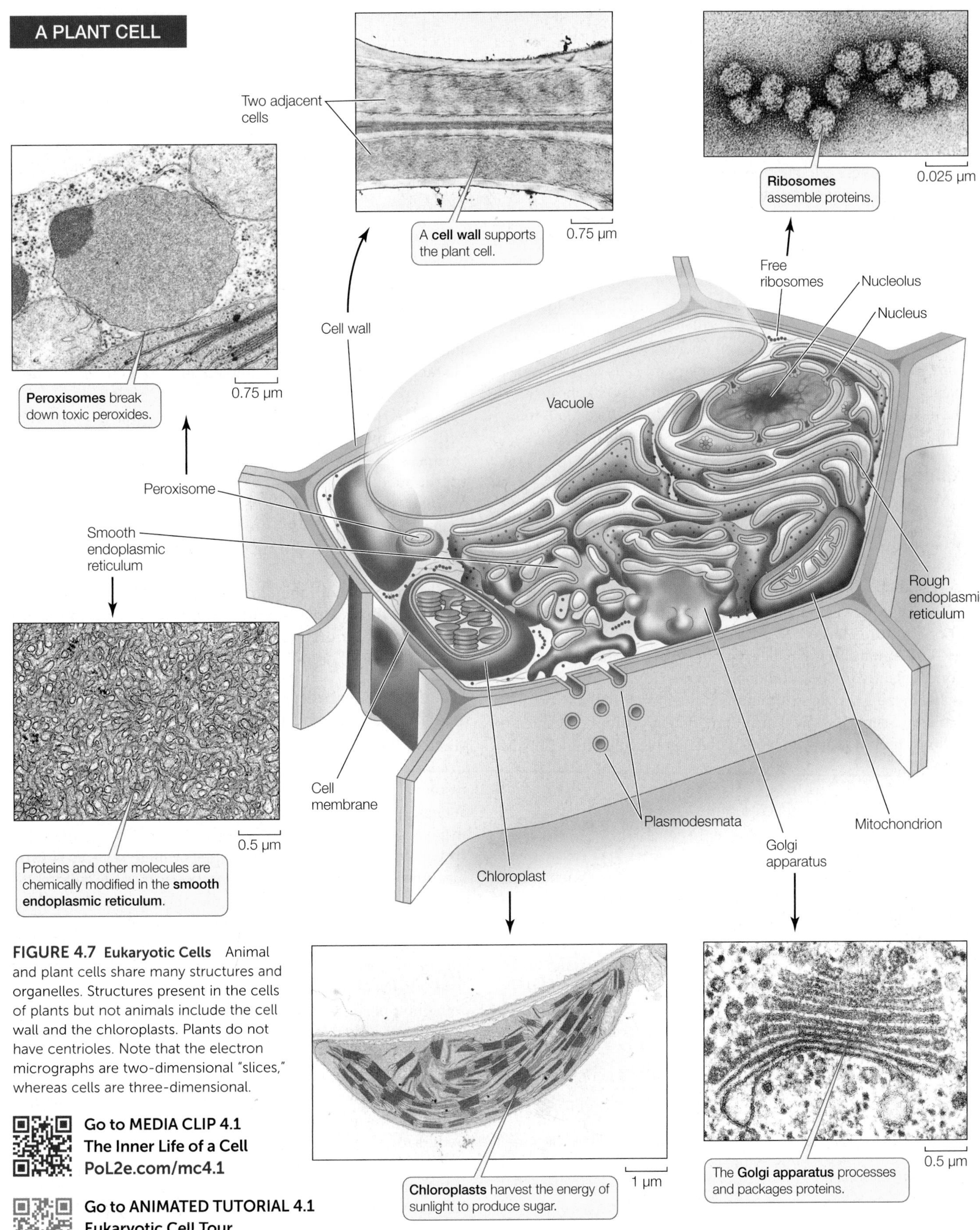

FIGURE 4.7 Eukaryotic Cells Animal and plant cells share many structures and organelles. Structures present in the cells of plants but not animals include the cell wall and the chloroplasts. Plants do not have centrioles. Note that the electron micrographs are two-dimensional "slices," whereas cells are three-dimensional.

Go to MEDIA CLIP 4.1
The Inner Life of a Cell
PoL2e.com/mc4.1

Go to ANIMATED TUTORIAL 4.1
Eukaryotic Cell Tour
PoL2e.com/at4.1

whose interiors are separated from the cytosol by membranes (FIGURE 4.7).

Compartmentalization is the key to eukaryotic cell function

Each type of organelle has a specific role in the cell. Some organelles have been characterized as factories that make specific products. Others are like power plants that take in energy in one form and convert it into a more useful form. In addition, eukaryotic cells have some structures that are analogous to those seen in prokaryotes. For example, they have a cytoskeleton composed of protein fibers, and outside the cell membrane, an extracellular matrix.

When animal and plant cells are examined using electron microscopy, they have many organelles and structures in common—the most obvious is the cell nucleus. But they also have some differences. For example, many plant cells have chloroplasts that are colored green by the pigment used in photosynthesis. Figure 4.7 shows diagrams of an animal and a plant cell, with electron micrographs of some of the subcellular structures.

Ribosomes are factories for protein synthesis

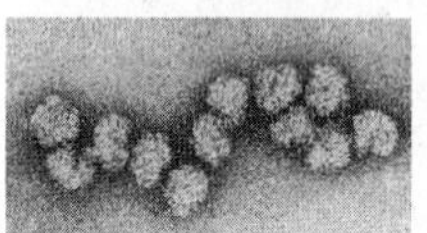

Ribosomes translate the nucleotide sequence of a messenger RNA molecule into a polypeptide chain. The ribosomes of both prokaryotes and eukaryotes consist of one larger and one smaller subunit; the sizes of the subunits differ between the two cell types. Each subunit consists of one to three large RNA molecules called ribosomal RNA (rRNA) and multiple smaller protein molecules that are bound noncovalently to one another and to the rRNA. The ribosome is an amazingly precise structure. If the individual macromolecules are separated by disruption of their hydrophobic interactions, they will spontaneously reassemble into a functional complex.

LINK

Protein synthesis is described in more detail in **Concept 10.4**

Ribosomes are not membrane-enclosed compartments. In prokaryotic cells, ribosomes float freely in the cytoplasm. In eukaryotic cells they are found in the cytoplasm, where they may be free or attached to the surface of the endoplasmic reticulum (a membrane-enclosed organelle; see below), and also inside certain organelles—namely the mitochondria and the chloroplasts.

The nucleus contains most of the cell's DNA

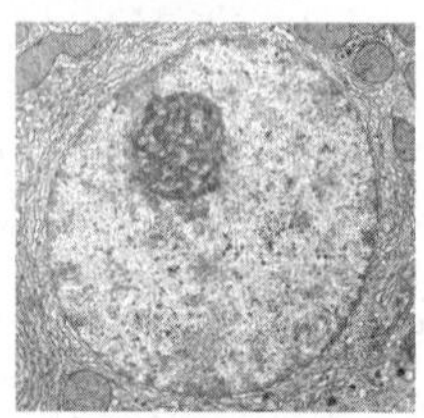

As we noted in Chapter 3, hereditary information is stored in the sequence of nucleotides in DNA molecules. In eukaryotic cells, most of the DNA is in the nucleus. Most cells have a single nucleus, and it is usually the largest organelle; at 5 μm in diameter, the nucleus is substantially larger than most prokaryotic cells. The nucleus has several functions:

- It is the location of the DNA and the site of DNA replication.
- It is where DNA is transcribed into RNA (see Concept 3.1).
- It contains the **nucleolus**, a region where ribosomes begin to assemble from RNA and proteins.

As you can see in Figure 4.7, the nucleus is enclosed by not one but two membranes: two lipid bilayers that together form the nuclear envelope. Functionally, this barrier separates DNA transcription (which occurs in the nucleus) from translation (which occurs in the cytoplasm). The two membranes of the nuclear envelope are perforated by thousands of nuclear pores, each measuring approximately 9 nm in diameter, which connect the interior of the nucleus to the cytoplasm. The pores regulate traffic between these two cellular compartments by allowing some molecules to enter or leave the nucleus and by blocking others. This allows the nucleus to regulate its information-processing functions.

Inside the nucleus, each DNA molecule is combined with proteins to form exceedingly long, thin threads called **chromosomes**. Different eukaryotic organisms have different numbers of chromosomes (ranging from two in one kind of Australian ant to hundreds in some plants). These DNA–protein complexes, which are also called **chromatin**, become much more compact during cell division, as you will see in Concept 7.2.

The outer membrane of the nuclear envelope folds outward into the cytoplasm and is continuous with the membrane of another organelle, the endoplasmic reticulum (see Figure 4.7).

The endomembrane system is a group of interrelated organelles

Much of the volume of many eukaryotic cells is taken up by an extensive **endomembrane system**. This interconnected system of membrane-enclosed compartments includes the nuclear envelope, endoplasmic reticulum, Golgi apparatus, and lysosomes, which are derived from the Golgi apparatus. Tiny, membrane-surrounded droplets called **vesicles** shuttle substances between the various components of the endomembrane system, as well as the cell membrane (**FIGURE 4.8**). In drawings and electron micrographs this system appears static, but in the living cell, membranes and the materials they contain are in constant motion. Membrane components have been observed to shift from one organelle to another within the endomembrane system. This suggests that all these membranes must be functionally related.

ENDOPLASMIC RETICULUM Electron micrographs of eukaryotic cells reveal networks of interconnected membranes branching throughout the cytoplasm, forming tubes and flattened sacs about 1 μm across. These membranes are collectively called the **endoplasmic reticulum**, or **ER**. The interior compartment (lumen) of the ER is separate and distinct from the surrounding cytoplasm (see Figure 4.8). The ER can enclose up to 10 percent of the interior volume of the cell, and its extensive folding results in a surface area many times greater

Rough endoplasmic reticulum is studded with ribosomes that are sites for protein synthesis. They produce its rough appearance.

Nucleus

Cytosol

1 Protein-containing vesicles from the endoplasmic reticulum transfer substances to the *cis* region of the Golgi apparatus.

Lumen

Cisterna

2 The Golgi apparatus chemically modifies proteins in its lumen...

cis region

3 ...and "targets" them to the correct destinations.

Medial region

trans region

Proteins for use within the cell

Cell membrane

Proteins for use outside the cell

Outside of cell

Smooth endoplasmic reticulum is a site for lipid and steroid synthesis, glycogen degradation, chemical modification of toxins, and calcium ion storage.

FIGURE 4.8 The Endomembrane System Membranes of the nucleus, endoplasmic reticulum (ER), and Golgi apparatus form a network that is connected by vesicles. Parts of the membrane move between these organelles. Membrane synthesized in the smooth ER becomes sequentially part of the rough ER, then the Golgi apparatus, then vesicles formed from the Golgi apparatus. These vesicles may eventually fuse with, and become part of, the cell membrane.

Go to ANIMATED TUTORIAL 4.2
The Golgi Apparatus
PoL2e.com/at4.2

APPLY THE CONCEPT

Eukaryotic cells have a nucleus and other membrane-bound compartments

Imagine that a group of scientists wanted to trace the path that the enzyme lipase follows between its site of synthesis and its final destination within a liver cell. First, the scientists exposed cultured liver cells to radioactive amino acids for 3 minutes. The amino acids entered the cells and became incorporated into all proteins synthesized during that time period. Then the radioactive amino acids were removed, so any proteins synthesized subsequently were *not* radioactive. At 5-minute intervals after the brief exposure to radioactive amino acids, some of the cells were broken open and fractionated to separate the organelles, as shown in Figure 4.4. An antibody that binds specifically to lipase was used to measure how much radioactive lipase was in each organelle at each time point (see Concept 39.4 for information on antibodies). The table shows the results.

	PERCENTAGE OF RADIOACTIVE LIPASE			
TIME (MIN)	ER LUMEN	GOLGI APPARATUS	LYSOSOMES	RIBOSOMES
5	5	0	0	95
10	25	10	0	65
15	75	20	5	0
20	25	55	20	0
25	0	65	35	0
30	0	25	75	0
35	0	0	100	0

1. Plot percentage radioactive lipase versus time for each organelle. What can you conclude about the pathway of lipase in the cell after it is synthesized?
2. Lipase breaks down lipids. Why is its organelle destination appropriate (see Figure 4.9)?

than that of the cell membrane. There are two types of ER: rough and smooth.

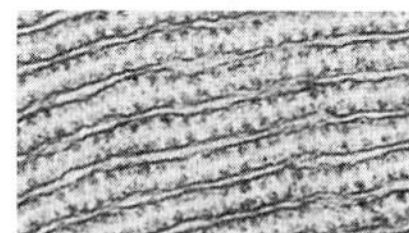

The **rough endoplasmic reticulum** (**RER**) is called "rough" because of the many ribosomes attached to the outer surface of the membrane, giving it a rough appearance in electron micrographs. These ribosomes are not permanently attached to the ER but become attached when they begin synthesizing proteins destined for modification within the RER:

- A protein enters the RER only if it contains a specific short sequence of amino acids that signals the ribosome to attach to the RER (see Concept 10.5).
- Once inside the RER, proteins are chemically modified to induce their three-dimensional functional shape and to chemically "tag" them for delivery to specific cellular destinations.
- The RER participates in transporting these proteins to other locations in the cell. The proteins are transported in vesicles that pinch off from the ER. All secreted proteins pass through the RER.
- Most membrane-bound proteins are made on the RER.

A polypetide that is synthesized on the RER surface is transported across the membrane and into the lumen while it is being translated. Once inside, it undergoes several changes, including the formation of disulfide bridges and folding into its tertiary structure. Many proteins are covalently linked to carbohydrate groups, thus becoming **glycoproteins**. These carbohydrate groups often have roles in recognition. For example, the carbohydrate groups on some secreted glycoproteins play roles in recognition and interactions between cells. Other carbohydrate groups "tag" their proteins for transfer to specific cellular locations. This "addressing" system is very important for ensuring that proteins arrive at their correct destinations. For example, the enzymes within the lysosomes (see below) are highly destructive and could destroy the cell if they were released into the cytosol.

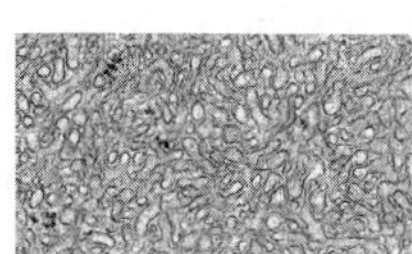

The **smooth endoplasmic reticulum** (**SER**) is connected to portions of the RER but lacks ribosomes and is more tubular (less like flattened sacs) than the RER. The SER has four important roles:

- It is responsible for the chemical modification of small molecules taken in by the cell that may be toxic to the cell. These modifications make the targeted molecules more polar, so they are more water-soluble and easily removed.
- It is the site for glycogen degradation in animal cells.
- It is the site where lipids and steroids are synthesized.
- It stores calcium ions, which when released trigger a number of cell responses, including muscle contraction.

LINK

The role of the SER in muscle contraction is described in **Concept 33.1**

Cells that synthesize a lot of protein for export are usually packed with RER. Examples include glandular cells that secrete digestive enzymes and white blood cells that secrete antibodies. In contrast, cells that carry out less protein synthesis (such as storage cells) contain less RER. Liver cells, which modify molecules (including toxins) that enter the body from the digestive system, have abundant SER.

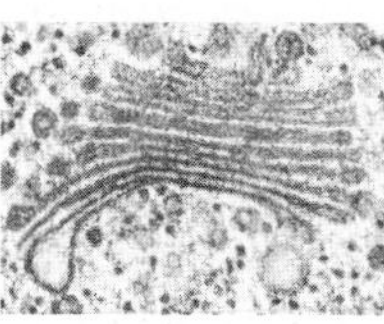

GOLGI APPARATUS The **Golgi apparatus** is named after its discoverer, Camillo Golgi. It has two components: flattened membranous sacs called cisternae (singular cisterna), which are piled up like saucers in a stack about 1 μm thick, and small membrane-enclosed vesicles (see Figure 4.8). There can be many of these stacks in a cell.

When protein-containing vesicles from the RER fuse with the Golgi apparatus membrane, the proteins are released into the lumen of a Golgi apparatus cisterna, where they may be further modified. The Golgi apparatus has several roles:

- It concentrates, packages, and sorts proteins before they are sent to their cellular or extracellular destinations.
- It adds some carbohydrates to proteins.
- It is where some polysaccharides for the plant cell wall are synthesized.

The cisternae of the Golgi apparatus have three functionally distinct regions: the *cis* region lies nearest to the nucleus or a patch of RER, the *trans* region lies closest to the cell membrane, and the medial region lies in between (see Figure 4.8). (The terms *cis*, *trans*, and medial derive from Latin words meaning "on the same side," "on the opposite side," and "in the middle," respectively.) These three parts of the Golgi apparatus contain different enzymes and perform different functions.

Protein-containing vesicles from the ER fuse with the *cis* membrane of the Golgi apparatus. Other vesicles may transport proteins from one cisterna to the next, although it appears that some proteins move between cisterna through tiny channels. Vesicles budding off from the *trans* region carry their contents away from the Golgi apparatus. These vesicles go to the cell membrane or to the lysosome.

LYSOSOMES The **primary lysosomes** originate from the Golgi apparatus. They contain hydrolases (digestive enzymes), and they are the sites where macromolecules—proteins, polysaccharides, nucleic acids, and lipids—are hydrolyzed into their monomers (see Chapter 2):

$$R_1\text{-}R_2 \text{ (linked monomers)} + H_2O \rightarrow R_1\text{-}OH + R_2\text{-}H$$

For example,

$$\text{Polypeptide} + H_2O \rightarrow \text{amino acids}$$

A lysosome is about 0.5 μm in diameter; it is surrounded by a single membrane and has a densely staining, featureless interior (**FIGURE 4.9**). There may be dozens of lysosomes in a cell.

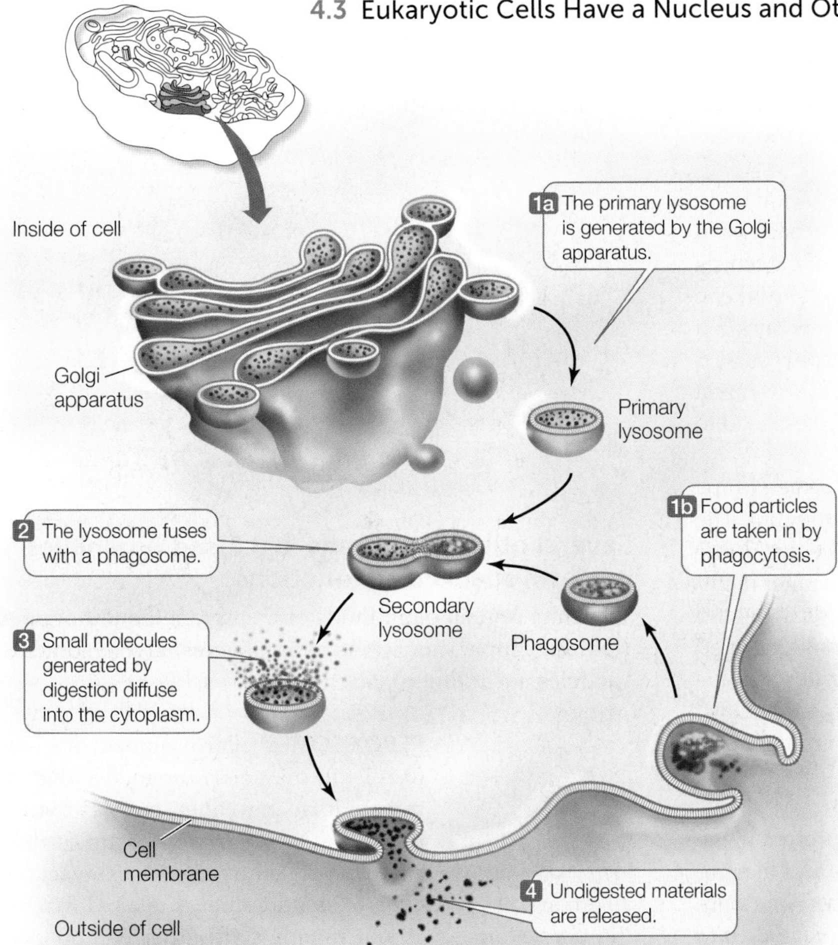

Inside of cell
1a The primary lysosome is generated by the Golgi apparatus.
Golgi apparatus
Primary lysosome
1b Food particles are taken in by phagocytosis.
2 The lysosome fuses with a phagosome.
Secondary lysosome
Phagosome
3 Small molecules generated by digestion diffuse into the cytoplasm.
Cell membrane
4 Undigested materials are released.
Outside of cell

FIGURE 4.9 Lysosomes Isolate Digestive Enzymes from the Cytoplasm Lysosomes are sites for the hydrolysis of material taken into the cell by phagocytosis.

Go to ACTIVITY 4.2 Lysosomal Digestion PoL2e.com/ac4.2

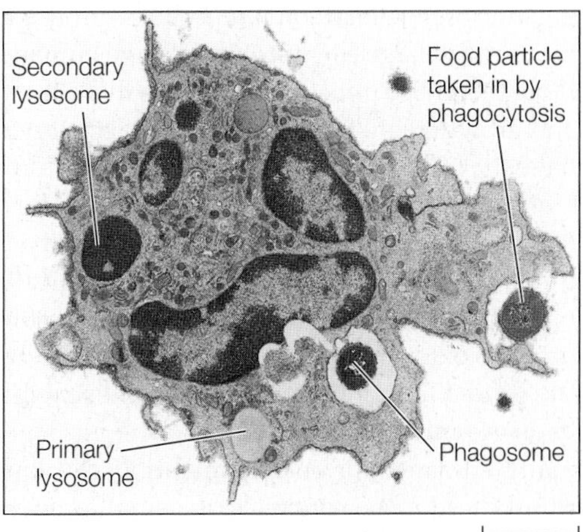

Some macromolecules that are hydrolyzed in lysosomes enter from the environment outside the cell by a process called **phagocytosis** (*phago*, "eat"; *cytosis*, "cellular"). In this process, a pocket forms in the cell membrane and then deepens and encloses material from outside the cell. The pocket becomes a small vesicle containing macromolecules (e.g., proteins), called a **phagosome**, which breaks free of the cell membrane to move into the cytoplasm. The phagosome fuses with a primary lysosome to form a **secondary lysosome**, in which hydrolysis occurs. The products of digestion (e.g., amino acids) pass through the membrane of the lysosome, providing monomers for other cellular processes. The "used" secondary lysosome, now containing undigested particles, then moves to the cell membrane, fuses with it, and releases the undigested contents to the environment.

Phagocytes are specialized cells whose major role is to take in and break down materials; they are found in nearly all animals and many protists. However, lysosomes are active even in cells that do not perform phagocytosis. All cells continually break down some of their components and replace them with new ones. The programmed destruction of cell components is called **autophagy**, and lysosomes are where the cell breaks down its own materials, even entire organelles, hydrolyzing their constituents.

How important is autophagy? An entire class of human diseases called lysosomal storage diseases occur when lysosomes fail to digest internal components; these diseases are often very harmful or fatal. For example, Tay-Sachs disease occurs when a particular lipid called a ganglioside is not broken down in the lysosomes and instead accumulates in brain cells and damages them. In the most common form of this disease, a baby starts exhibiting neurological symptoms and becomes blind, deaf, and unable to swallow after six months of age. Death occurs before age four.

Plant cells do not appear to contain lysosomes, but the central vacuole of a plant cell (which we will describe below) may function in an equivalent capacity because it, like lysosomes, contains many digestive enzymes.

Some organelles transform energy

A cell requires energy to make the molecules it needs for activities such as growth, reproduction, responsiveness, and movement. Mitochondria (found in all eukaryotic cells) harvest chemical energy, whereas chloroplasts (found in plants and other photosynthetic cells) harvest energy from sunlight.

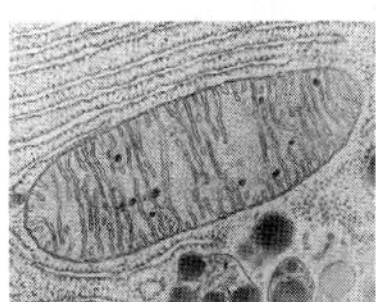

MITOCHONDRIA In eukaryotic cells, the breakdown of energy-rich molecules such as the monosaccharide glucose begins in the cytosol. The molecules that result from this partial degradation enter the **mitochondrion** (plural *mitochondria*), whose primary function is to harvest the chemical energy of those molecules in a form the cell can use, namely the energy-rich nucleotide ATP (adenosine triphosphate). We will discuss these energy-harvesting processes in Chapter 6.

A typical mitochondrion is somewhat less than 1.5 μm in diameter and 2–8 μm in length—about the size of many bacteria. It contains some DNA and can divide independently of the central nucleus. The number of mitochondria per cell ranges from one gigantic organelle in some unicellular protists to a few hundred thousand in large egg cells. An average human liver cell contains more than 1,000 mitochondria. Cells that are active in movement and growth require the most chemical energy, and these tend to have the most mitochondria per unit of volume.

Mitochondria have two membranes. The outer membrane has large pores, and most substances can pass through it. The inner membrane separates the biochemical processes of the mitochondrion from the surrounding cytosol. The inner membrane is extensively folded into structures called cristae, and the fluid-filled region inside the inner membrane is referred to as the mitochondrial matrix. The mitochondrion contains many enzymes for energy metabolism, as well as DNA and ribosomes for the synthesis of a small proportion of the mitochondrial proteins.

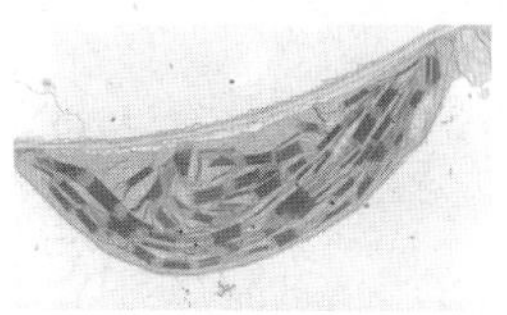

PLASTIDS Plastids are present in the cells of plants and algae, and like mitochondria, they can divide autonomously. Plastids can differentiate into a variety of organelles, some of which are used for the storage of pigments (as in flowers), carbohydrates (as in potatoes), lipids, or proteins. An important type of plastid is the **chloroplast**, which contains the green pigment chlorophyll and is the site of photosynthesis (see Concepts 6.5 and 6.6). Photosynthesis is an anabolic process that converts light energy into the chemical energy contained in bonds between the atoms of carbohydrates.

A chloroplast is enclosed within two membranes. In addition, it contains a series of internal membranes that look like stacks of flat, hollow discs, called **thylakoids**. Each stack of thylakoids is called a granum (plural grana). Light energy is converted to chemical energy on the thylakoid membranes. The aqueous fluid surrounding the thylakoids is called the stroma, and it is there that carbohydrates are synthesized. Like the mitochondrial matrix, the chloroplast stroma contains ribosomes and DNA, which are used to synthesize some of the chloroplast proteins.

Other types of plastids have functions different from those of chloroplasts. Chromoplasts make and store red, yellow, and orange pigments, especially in flowers and fruits:

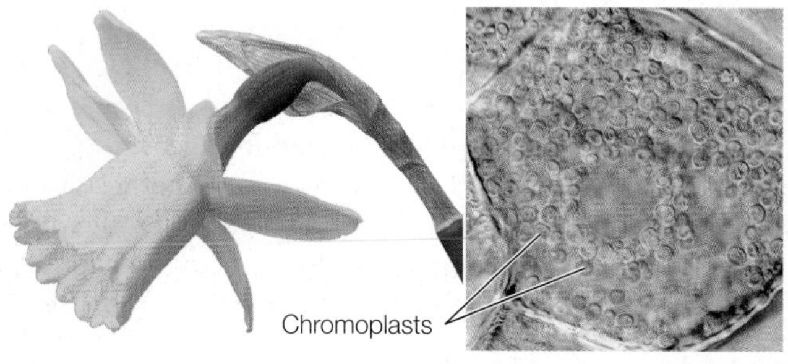

Leucoplasts are storage organelles for macromolecules such as starch:

Several other membrane-enclosed organelles perform specialized functions

There are several other kinds of membrane-bound organelles with specialized functions: peroxisomes, glyoxysomes, and vacuoles, including contractile vacuoles.

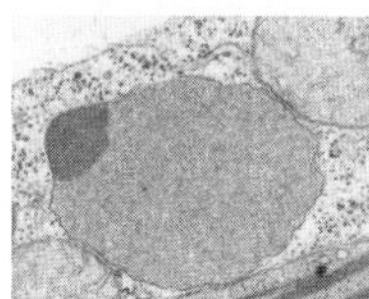

PEROXISOMES **Peroxisomes** are small (0.2–1 μm diameter) organelles that accumulate toxic peroxides, such as hydrogen peroxide (H_2O_2), which occur as the by-products of some biochemical reactions in all eukaryotes. These peroxides are safely broken down inside the peroxisomes without mixing with other components of the cell. A peroxisome has a single membrane and a granular interior containing specialized enzymes.

GLYOXYSOMES **Glyoxysomes** are found only in plants. They are most abundant in young plants, which have many in each cell, and are the locations where stored lipids are converted into carbohydrates for transport to growing cells.

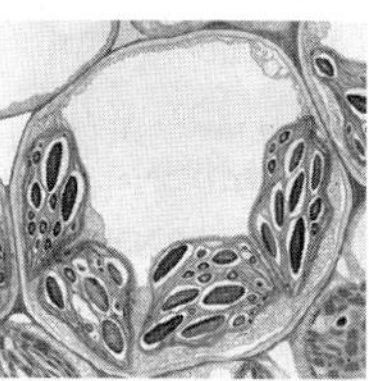

VACUOLES **Vacuoles** occur in many eukaryotic cells, but particularly those of plants and fungi (see Figure 4.7). There can be one large vacuole or many small ones in a cell. Plant vacuoles have several functions:

- *Storage*: Like all cells, plant cells produce a variety of toxic by-products and waste products. Plants store many of these in vacuoles. Because they are poisonous or distasteful, these stored materials deter some animals from eating the plants, and may thus contribute to the plants' defenses and survival.
- *Structure*: In many plant cells, enormous vacuoles take up more than 90 percent of the cell volume and grow as the cell grows. The presence of dissolved substances in the vacuole causes water to enter it from the cytoplasm (which in turn takes up water from outside the cell), making the vacuole swell like a water-filled balloon. (This osmotic effect is illustrated in Figure 5.3.) The plant cell wall resists the swelling, causing the cell to stiffen from the increase in water pressure. This pressure is called turgor pressure, and it helps support the plant.

- *Reproduction*: Some pigments in the petals and fruits of flowering plants are contained in vacuoles. These pigments—the red, purple, and blue anthocyanins—are visual cues that help attract animals, which assist in pollination and seed dispersal.
- *Catabolism*: In the seeds of some plants, the vacuoles contain enzymes that hydrolyze stored seed proteins into monomers. The developing plant seedling uses these monomers as building blocks and sources of energy.

Many freshwater protists have contractile vacuoles. Their function is to get rid of the excess water that rushes into the cell because of the imbalance in solute concentration between the interior of the cell and its freshwater environment. The contractile vacuole enlarges as water enters, and then abruptly contracts, forcing the water out of the cell through a special pore structure.

CHECKpoint CONCEPT 4.3

✓ Make a table that summarizes eukaryotic cell organelles with regard to typical size, numbers per cell, and functions.

✓ What are some functions of the cell nucleus? What are the advantages of confining these functions within the nucleus, separated from the cytoplasm?

✓ Compare the structural and functional differences between rough and smooth endoplasmic reticulum.

✓ In I-cell disease, an enzyme in the endomembrane system that normally adds phosphorylated sugar groups to proteins is lacking, and the proteins are not targeted to the lysosomes as they would be in normal cells. The "I" stands for inclusion bodies that appear in the cells. What do you think these inclusions are, and why do they accumulate?

So far, we have discussed numerous membrane-enclosed organelles. Now we'll turn to a group of cytoplasmic structures that do not directly involve membranes.

CONCEPT 4.4 The Cytoskeleton Provides Strength and Movement

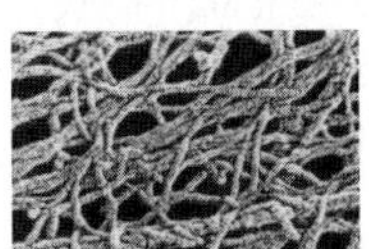

The interior of the cell has a meshwork of protein filaments. Each type of filament is a polymer, made up of monomers that are proteins (which in turn are polymers of amino acids). This cytoskeleton fills several important roles:

- It supports the cell and maintains its shape.
- It holds cell organelles and other particles in position within the cell.
- It moves organelles and other particles around within the cell.
- It is involved with movements of the cytoplasm called cytoplasmic streaming.
- It interacts with extracellular structures, helping anchor the cell in place.

There are three components of the eukaryotic cytoskeleton: microfilaments (smallest diameter), intermediate filaments, and microtubules (largest diameter). These filaments have very different functions.

Microfilaments are made of actin

Microfilaments (**FIGURE 4.10A**) are usually in bundles. Each filament is about 7 nm in diameter and up to several micrometers long. Microfilaments have two major roles:

- They help the entire cell or parts of the cell to move.
- They determine and stabilize cell shape.

Microfilaments are assembled from **actin** monomers that attach to the filament at one end (the "plus end") and detach at the other (the "minus end"). In an intact filament, assembly and detachment are in equilibrium. But sometimes the filaments can shorten (more detachment) or lengthen (more assembly):

$$\text{Actin polymer (filament)} \rightleftharpoons \text{actin monomers}$$

This property of dynamic instability is a hallmark of the cytoskeleton. Portions of it can be made and broken down rather quickly, depending on cell function. Actin-associated proteins work at both ends of the filament to catalyze assembly and disassembly.

In the muscle cells of animals, actin filaments are associated with the motor protein myosin, and the interactions of these two proteins account for the contraction of muscles. A **motor protein** (or molecular motor) is any protein that causes movement within a cell. In non-muscle cells, actin filaments are associated with localized changes in cell shape. For example, microfilaments are involved in the flowing movement of the cytoplasm called cytoplasmic streaming, and in the "pinching" contractions that divide an animal cell into two daughter cells. Microfilaments are also involved in the formation of cellular extensions called pseudopodia (*pseudo*, "false"; *podia*, "feet") that enable some cells (such as *Amoeba*, see Figure 4.14) to move.

Intermediate filaments are diverse and stable

There are at least 50 different kinds of **intermediate filaments** (**FIGURE 4.10B**), many of them specific to just a few cell types. They generally fall into six molecular classes based on amino acid sequence. One of these classes consists of the fibrous keratin proteins, which are also found in hair and fingernails. The intermediate filaments are tough, ropelike protein assemblages 8–12 nm in diameter. Intermediate filaments are more permanent than the other two types of filaments and do not show dynamic instability. Intermediate filaments have two major structural functions:

- They anchor cell structures in place. In some cells, intermediate filaments radiate from the nuclear envelope and help

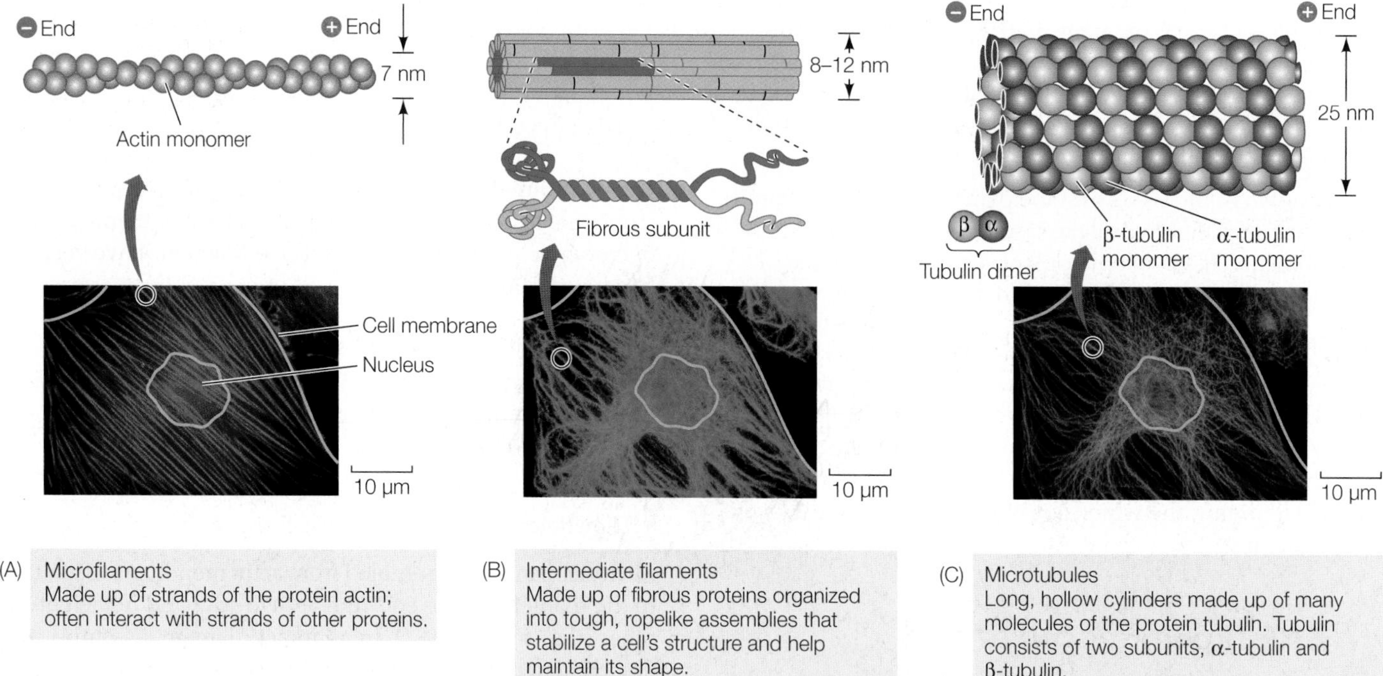

− End
+ End
7 nm
Actin monomer
Cell membrane
Nucleus
10 μm
8–12 nm
Fibrous subunit
10 μm
− End
+ End
25 nm
β α
Tubulin dimer
β-tubulin monomer
α-tubulin monomer
10 μm
(A) Microfilaments
Made up of strands of the protein actin; often interact with strands of other proteins.
(B) Intermediate filaments
Made up of fibrous proteins organized into tough, ropelike assemblies that stabilize a cell's structure and help maintain its shape.
(C) Microtubules
Long, hollow cylinders made up of many molecules of the protein tubulin. Tubulin consists of two subunits, α-tubulin and β-tubulin.

FIGURE 4.10 The Cytoskeleton Three highly visible and important structural components of the cytoskeleton are shown here in detail. Specific stains were used to visualize them in a single cell. These structures maintain and reinforce cell shape and contribute to cell movement.

maintain the positions of the nucleus and other organelles in the cell.

- They resist tension. For example, they maintain rigidity in body surface tissues by extending through the cytoplasm and connecting specialized membrane structures called desmosomes (see Figure 4.18).

Microtubules are the thickest elements of the cytoskeleton

Microtubules (**FIGURE 4.10C**) are long, hollow, unbranched cylinders about 25 nm in diameter and up to several micrometers long. Microtubules have two roles:

- They form a rigid internal skeleton for some cells or cell regions.
- They act as a framework along which motor proteins can move structures within the cell.

Microtubules are assembled from dimers of the protein **tubulin**. The dimers consist of one molecule each of α-tubulin and β-tubulin. Thirteen chains of tubulin dimers surround the hollow microtubule. Like microfilaments, microtubules show dynamic instability, with plus and minus ends and associated proteins.

$$\text{Microtubule} \rightleftharpoons \text{tubulin dimers}$$

Tubulin polymerization results in a rigid structure, and tubulin depolymerization leads to its collapse. Microtubules often form an interior skeleton for projections that come out of the cell membrane, such as cilia and flagella (see below).

LINK

Microtubules are important components of the spindle, which separates chromosomes during cell division; see **Concept 7.2**

Cilia and flagella provide mobility

Microtubules internally line movable cell appendages: the **cilia** (singular *cilium*; **FIGURE 4.11**) and the flagella (see Concept 4.2). Many eukaryotic cells have one or both of these appendages, which are projections of the cell membrane lined with microtubules and their associated proteins:

- Cilia are only 0.25 μm in length. They are present by the hundreds and move stiffly either to propel a cell (for example, in protists) or to move fluid over a stationary cell (as in the human respiratory system).
- Flagella are much longer—100 to 200 μm—and occur singly or in pairs. They can push or pull the cell through its aqueous environment.

The microtubules that line cilia and flagella do more than just make them rigid. Microtubules and their associated proteins are responsible for the movement of these organelles by bending.

In cross section, a typical cilium or eukaryotic flagellum is surrounded by the cell membrane and contains a "9 + 2" array of microtubules. As Figure 4.11B shows, nine fused pairs of microtubules—called doublets—form an outer cylinder, and one pair of unfused microtubules runs up the center. Each doublet is connected to the center of the structure by a radial spoke. This structure is essential to the bending motions of both cilia and flagella. How does this bending occur?

The motion of cilia and flagella results from the sliding of the microtubule doublets past one another. This sliding is driven

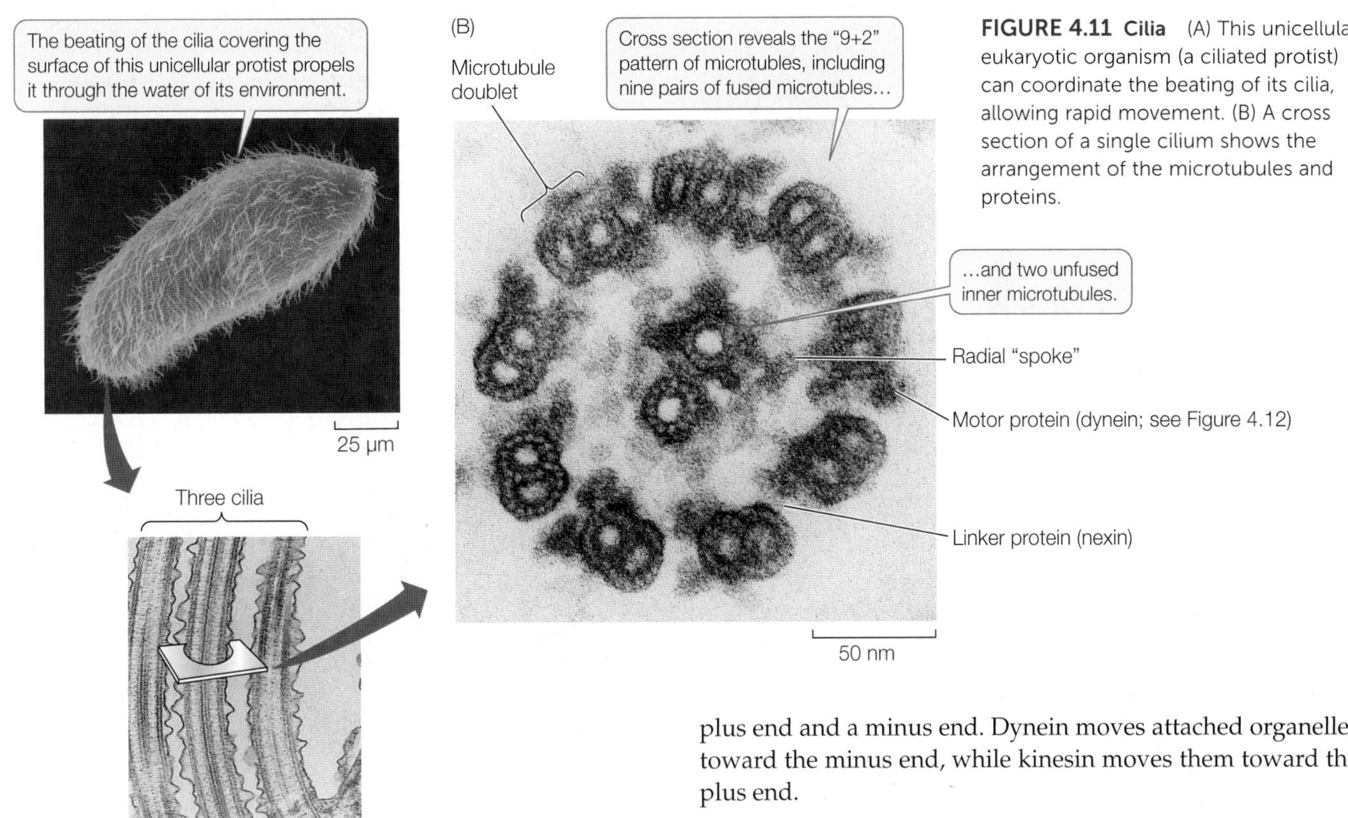
(A)
The beating of the cilia covering the surface of this unicellular protist propels it through the water of its environment.
25 μm
Three cilia
250 nm
(B)
Microtubule doublet
Cross section reveals the "9+2" pattern of microtubles, including nine pairs of fused microtubles...
...and two unfused inner microtubules.
Radial "spoke"
Motor protein (dynein; see Figure 4.12)
Linker protein (nexin)
50 nm

FIGURE 4.11 Cilia (A) This unicellular eukaryotic organism (a ciliated protist) can coordinate the beating of its cilia, allowing rapid movement. (B) A cross section of a single cilium shows the arrangement of the microtubules and proteins.

by a motor protein called dynein, which can change its three-dimensional shape. (All motor proteins work by undergoing reversible shape changes powered by energy from ATP hydrolysis.) Dynein molecules bind between two neighboring microtubule doublets. As the dynein molecules change shape, they move the doublets past one another (**FIGURE 4.12**). Another protein, nexin, can cross-link the doublets and prevent them from sliding past one another; in this case, the cilium bends.

Other motor proteins, including kinesin, carry protein-laden vesicles or other organelles from one part of the cell to another (**FIGURE 4.13**). These proteins bind to the organelle and "walk" it along a microtubule by a repeated series of shape changes. A slightly different form of dynein from the one that moves cilia also performs this function. Recall that microtubules are directional, with a plus end and a minus end. Dynein moves attached organelles toward the minus end, while kinesin moves them toward the plus end.

Biologists manipulate living systems to establish cause and effect

How do we know that the structural fibers of the cytoskeleton can achieve the dynamic functions described above? We can observe an individual structure under the microscope, and we can note that living cells containing that structure can perform a particular function. Such simultaneous observations suggest that the structure may carry out that function, but *mere correlation does not establish cause and effect*. For example, light microscopy of living cells reveals the movement of the cytoplasm within the cell. The observed presence of cytoskeletal components *suggests, but does not prove*, their role in this process. Scientists seek to understand the specific relationship between a

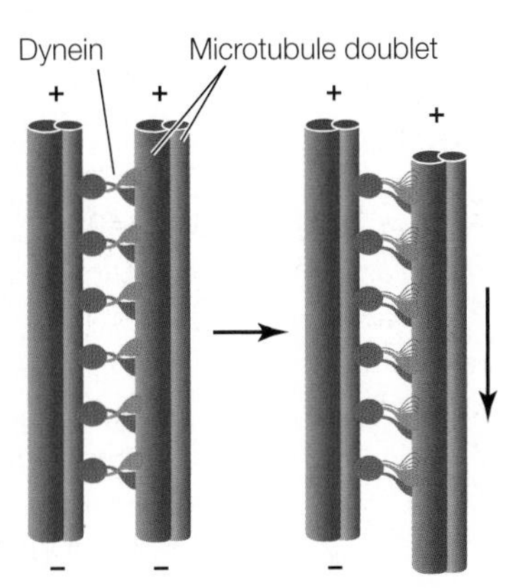
Dynein
Microtubule doublet
In isolated cilia without nexin cross-links, movement of dynein motor proteins causes microtubule doublets to slide past one another.

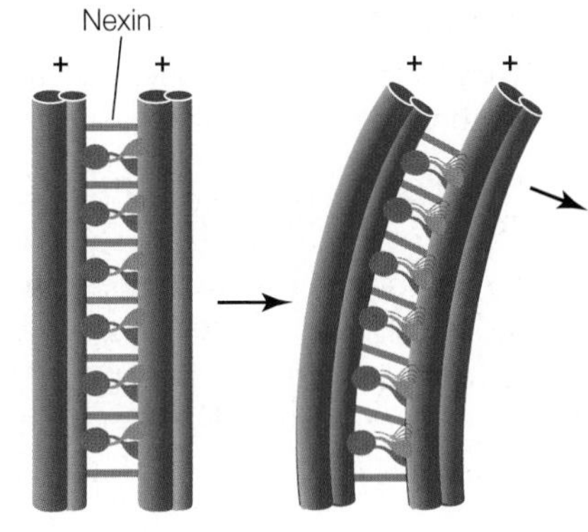
Nexin
When nexin is present to cross-link the doublets, they cannot slide and the force generated by dynein movement causes the cilium to bend.

FIGURE 4.12 A Motor Protein Moves Microtubules in Cilia and Flagella The motor protein dynein causes microtubule doublets to slide past one another. If the protein nexin is present to anchor the microtubule doublets together, the flagellum or cilium bends.

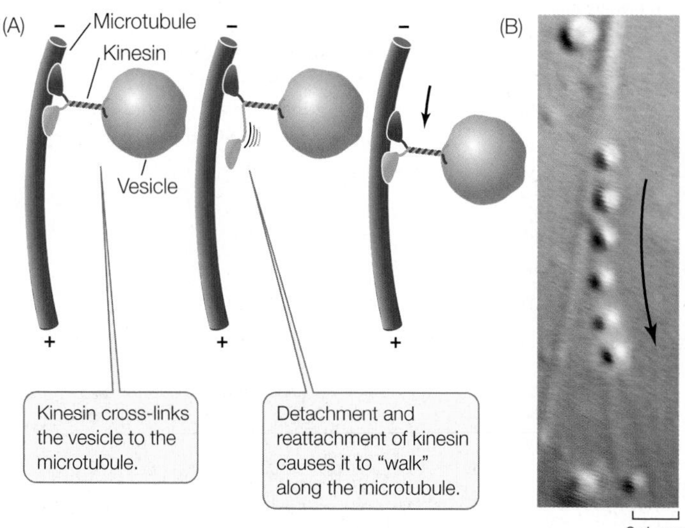
(A)
Microtubule
Kinesin
Vesicle
(B)
Kinesin cross-links the vesicle to the microtubule.
Detachment and reattachment of kinesin causes it to "walk" along the microtubule.
0.1 μm

FIGURE 4.13 A Motor Protein Drives Vesicles along Microtubules (A) Kinesin delivers vesicles or organelles to various locations in the cell by moving along microtubule "railroad tracks." (B) The process is seen by time-lapse photography at half-second intervals in the protist *Dictyostelium*.

structure or molecule ("A") and a function ("B"). Two manipulative approaches are commonly used in cell biology:

- *Inhibition*: Use a drug that inhibits A, and see if B still occurs. If B does not occur, then A is probably a causative factor for B. **FIGURE 4.14** shows an experiment in which an inhibitor is used to demonstrate cause and effect in the case of microfilaments and cell movement.
- *Mutation*: Examine a cell or organism that lacks the gene (or genes) for A, and see if B still occurs. If it does not, then A is probably a causative factor for B. You will see many examples of this experimental approach later in this book.

CHECKpoint CONCEPT 4.4

- ✓ Make a table that compares the three major components of the cytoskeleton with regard to composition, structure, and function.
- ✓ The neuron (nerve cell) has a long extension called an axon. Molecules made in the cell's main body must travel a long distance to reach the end of the axon. The axon is lined with microtubules. Explain how motor proteins, vesicles, and microtubules move these molecules along the axon.
- ✓ In a dividing cell, the chromosomes become very compact, and then the duplicated sets of chromosomes move along microtubules to opposite ends of the cell. How would you use an inhibitor to show that microtubules are essential for this chromosomal separation? What control treatments would you suggest?

All cells interact with their environments, and many eukaryotic cells are part of multicellular organisms and must interact with

INVESTIGATION

FIGURE 4.14 The Role of Microfilaments in Cell Movement: Showing Cause and Effect in Biology In test tubes, the drug cytochalasin B prevents microfilament formation from monomeric precursors. This led to the question: Will the drug work like this in living cells and inhibit the movement of *Amoeba*[a]?

HYPOTHESIS

Amoeboid cell movements are caused by microfilaments.

METHOD

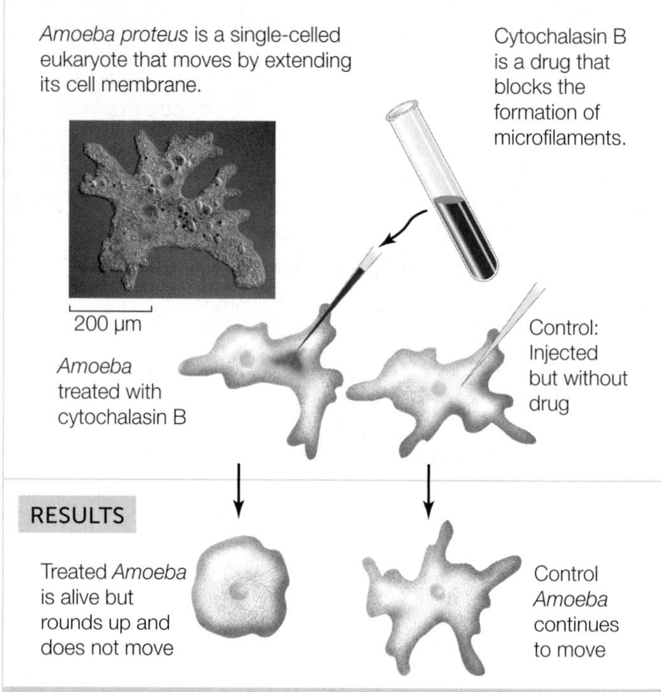
Amoeba proteus is a single-celled eukaryote that moves by extending its cell membrane.
Cytochalasin B is a drug that blocks the formation of microfilaments.
200 μm
Amoeba treated with cytochalasin B
Control: Injected but without drug
RESULTS
Treated *Amoeba* is alive but rounds up and does not move
Control *Amoeba* continues to move

CONCLUSION

Microfilaments are essential for amoeboid cell movement.

ANALYZE THE DATA

Several important controls were done to validate the conclusions of this experiment. The experiment was repeated in the presence of cycloheximide, which inhibits new protein synthesis, and colchicine, which inhibits the polymerization of microtubules. Here are the results:

Condition	Rounded cells (%)
No drug	3
Cytochalasin B	95
Colchicine	4
Cycloheximide	3
Cycloheximide + cytochalasin B	94

A. Explain the reasoning behind each condition. Why were the controls important?

B. Interpret the results of this experiment. What can you conclude about movements in *Amoeba* and the cytoskeleton?

Go to **LaunchPad** for discussion and relevant links for all **INVESTIGATION** figures.

[a]T. D. Pollard and R. R Weihing. 1974. *CRC Critical Reviews in Biochemistry* 2: 1–65.

other cells. The cell membrane (which we will discuss in detail in Chapter 5) plays a crucial role in these interactions, but other structures outside that membrane are involved as well. We will now turn to these extracellular structures in animals and plants.

CONCEPT 4.5 Extracellular Structures Provide Support and Protection for Cells and Tissues

In Chapter 5 we will look at the role of the cell membrane in cell communication. Although the cell membrane is the functional barrier between the inside and the outside of a cell, cells produce molecules and secrete them to the outside of the cell membrane. There these molecules form structures that play essential roles in protecting, supporting, or attaching cells to each other. Because they are outside the cell membrane, these structures are said to be "extracellular." In eukaryotes, these structures are made up of two components:

- A fibrous macromolecule
- A gel-like medium in which the fibers are embedded

The plant cell wall is an extracellular structure

The plant cell wall is a semirigid structure outside the cell membrane (**FIGURE 4.15**). The fibrous component is the polysaccharide cellulose (see Figure 2.10), and the gel-like matrix contains extensively cross-linked polysaccharides and proteins. The wall has three major roles:

- It provides support for the cell and limits the volume of a mature cell by remaining rigid.
- It acts as a barrier to infection by fungi and other organisms that can cause plant diseases.
- It contributes to plant form by controlling the direction of cell expansion during growth and development.

Because of their thick cell walls, plant cells viewed under a light microscope appear to be entirely isolated from one another. But electron microscopy reveals that this is not the case. The cytoplasms of adjacent plant cells are connected by numerous cell membrane–lined channels called **plasmodesmata**. These are about 20–40 nm in diameter and extend through the cell walls (see Figure 4.7). Plasmodesmata allow water, ions, small molecules, hormones, and even some RNA and protein molecules to move between connected cells. In this way, energy-rich molecules such as sugars can be shared among cells, and plant hormones can affect growth at sites far from where they were synthesized. This intercellular communication integrates a plant organ composed of thousands of cells.

The extracellular matrix supports tissue functions in animals

Animal cells lack the semirigid wall that is characteristic of plant cells, but many animal cells are surrounded by, or in contact with, an **extracellular matrix** (**FIGURE 4.16**). The fibrous component of the extracellular matrix is the protein **collagen**, and the gel-like medium consists of **proteoglycans**, which are glycoproteins with long carbohydrate side chains. A third group of proteins links the collagen and the proteoglycan matrix together.

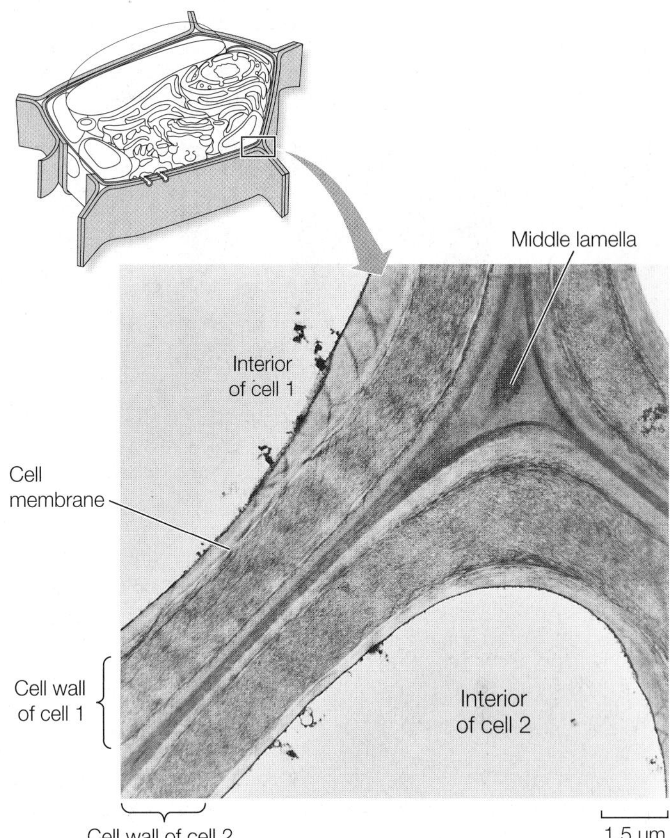

FIGURE 4.15 The Plant Cell Wall The semirigid cell wall provides support for plant cells. It is composed of cellulose fibers embedded in a matrix of polysaccharides and proteins.

The extracellular matrices of animal cells have several roles:

- They hold cells together in tissues.
- They contribute to the physical properties of cartilage, skin, and other tissues. For example, the mineral component of bone is laid down on an organized extracellular matrix.
- They help filter materials passing between different tissues. This is especially important in the kidney.
- They help orient cell movements during embryonic development and during tissue repair.

Proteins connect the cell's cell membrane to the extracellular matrix. These proteins (for example, integrin) span the cell membrane and have two binding sites: one on the interior of the cell, usually to microfilaments in the cytoplasm just below the cell surface, and the other to collagen in the extracellular matrix. These binding sites are noncovalent and reversible. When a cell moves its location in an organism, the first step is

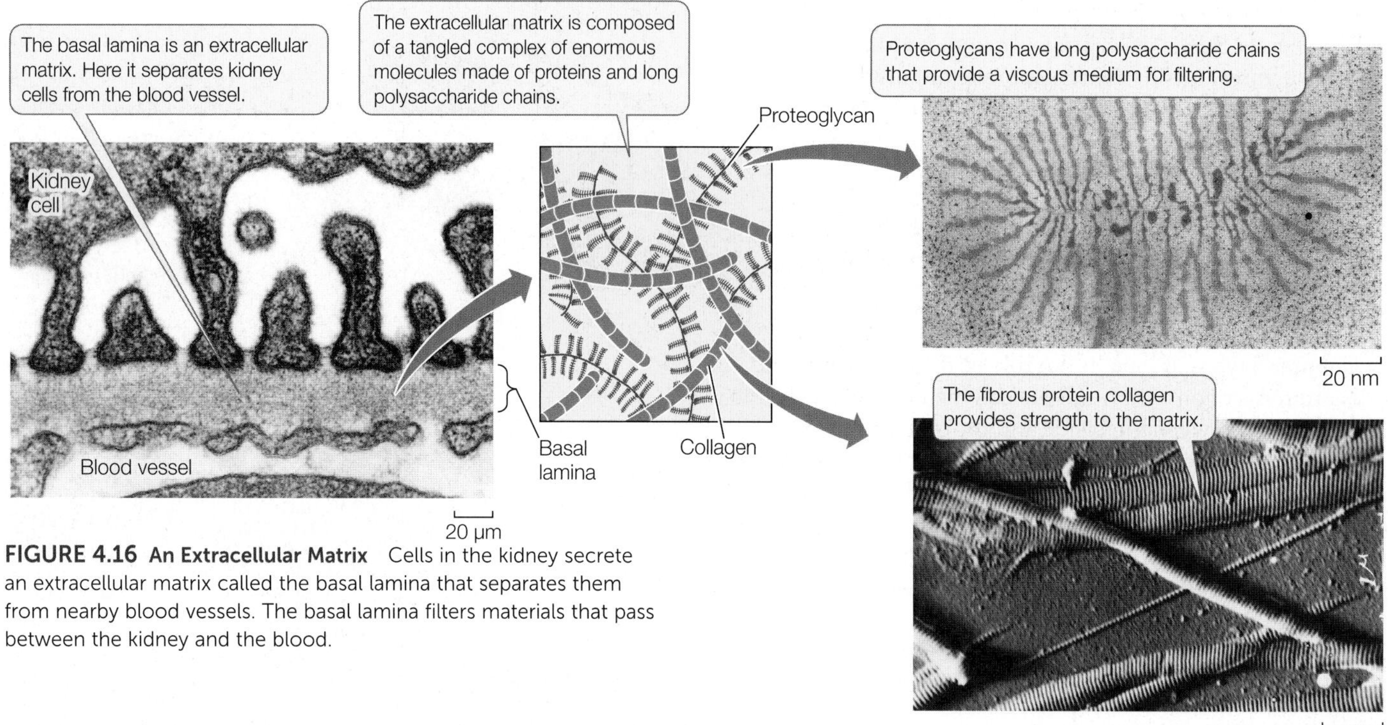

The basal lamina is an extracellular matrix. Here it separates kidney cells from the blood vessel.
The extracellular matrix is composed of a tangled complex of enormous molecules made of proteins and long polysaccharide chains.
Proteoglycans have long polysaccharide chains that provide a viscous medium for filtering.
Proteoglycan
Kidney cell
Blood vessel
Basal lamina
Collagen
20 μm
20 nm
The fibrous protein collagen provides strength to the matrix.
100 nm

FIGURE 4.16 An Extracellular Matrix Cells in the kidney secrete an extracellular matrix called the basal lamina that separates them from nearby blood vessels. The basal lamina filters materials that pass between the kidney and the blood.

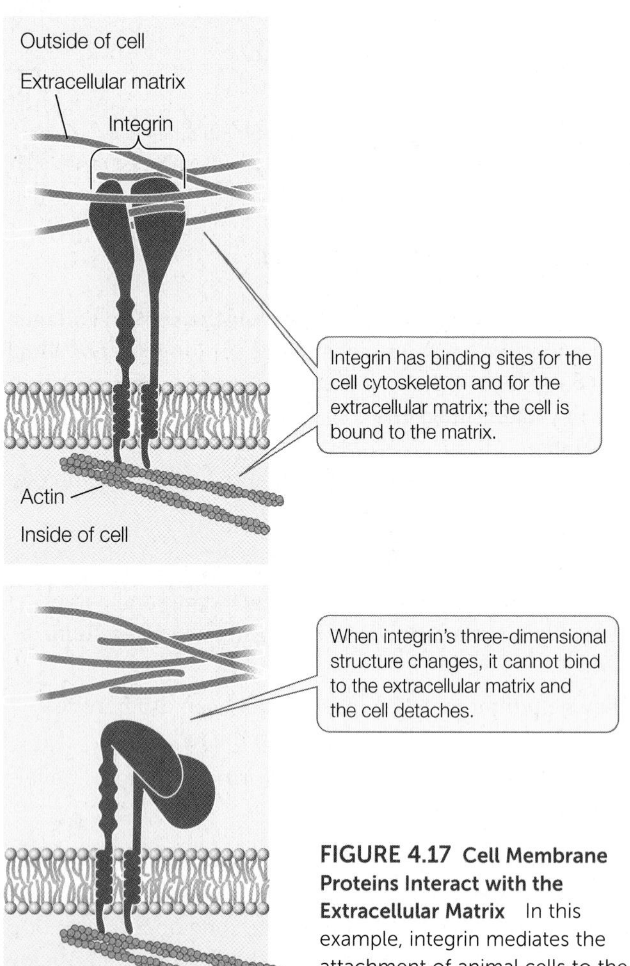

Outside of cell
Extracellular matrix
Integrin
Integrin has binding sites for the cell cytoskeleton and for the extracellular matrix; the cell is bound to the matrix.
Actin
Inside of cell
When integrin's three-dimensional structure changes, it cannot bind to the extracellular matrix and the cell detaches.

FIGURE 4.17 Cell Membrane Proteins Interact with the Extracellular Matrix In this example, integrin mediates the attachment of animal cells to the extracellular matrix.

for integrin to change its three-dimensional structure so that it detaches from the collagen (**FIGURE 4.17**).

Cell junctions connect adjacent cells

In a multicellular animal, specialized structures protrude from adjacent cells to "glue" them together. These **cell junctions** are most evident in electron micrographs of epithelial tissues, which are layers of cells that line body cavities or cover body surfaces (examples are skin and the lining of the windpipe leading to the lungs). These surfaces are often exposed to environmental factors that might disrupt the integrity of the tissues, so it is particularly important that their cells stick together tightly. There are three types of junctions (**FIGURE 4.18**):

- *Tight junctions* prevent substances from moving through spaces between cells. For example, the epithelium of the urinary bladder contains tight junctions to prevent urine from leaking out into the body.
- *Desmosomes* hold adjacent cells together with stable protein connections, but materials can still move around in the extracellular matrix. This provides mechanical stability for tissues such as skin that receive physical stress.
- *Gap junctions* are like plant plasmodesmata: they are channels that run between membrane pores in adjacent cells, allowing substances to pass between cells. In the heart, for example, gap junctions allow the rapid spread of electric current mediated by ions so the heart muscle cells can beat in unison.

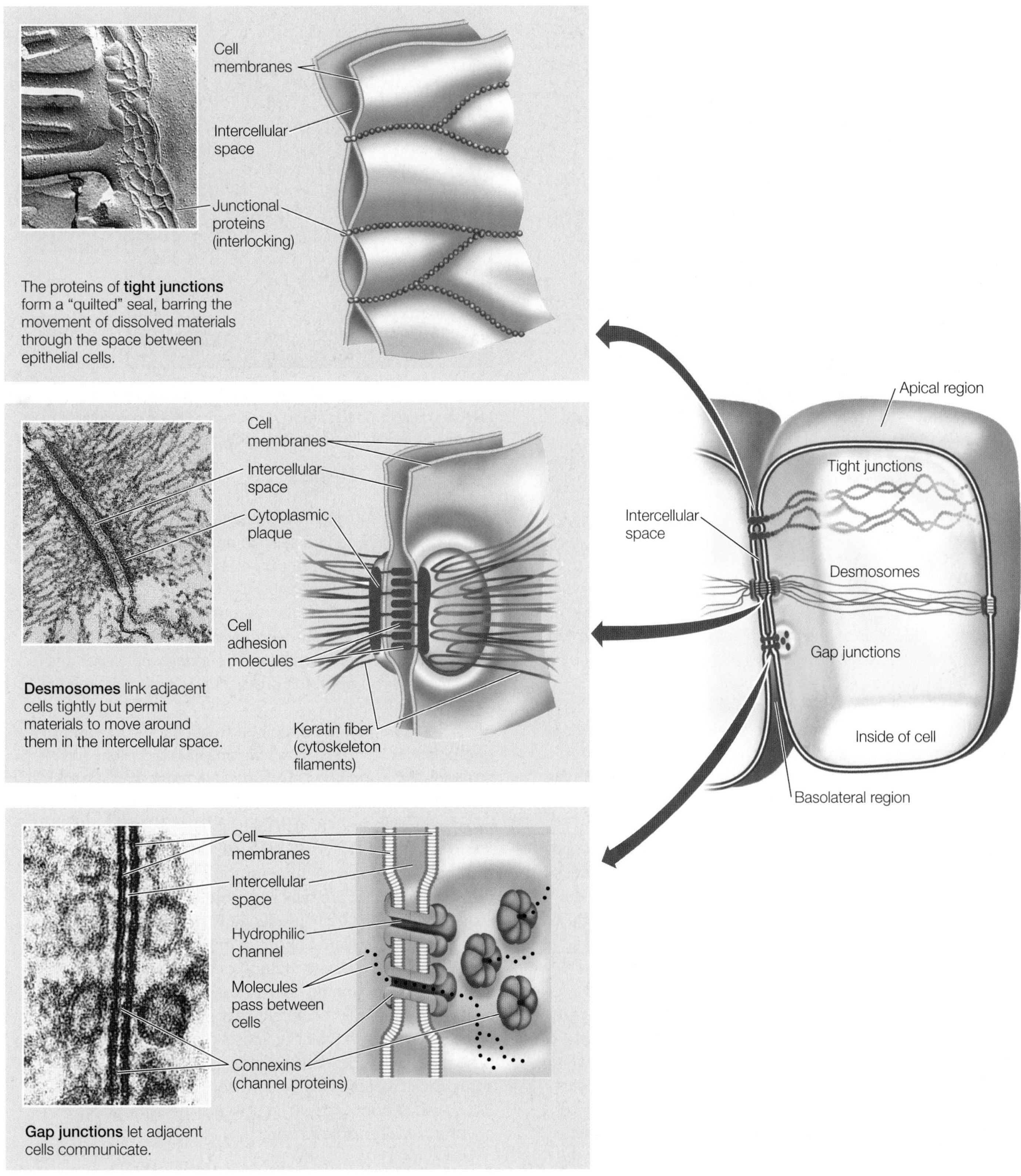

FIGURE 4.18 Junctions Link Animal Cells Although all three types of junctions are shown in the cell at right, they don't necessarily all occur in the same cell.

Go to ACTIVITY 4.3 Animal Cell Junctions
PoL2e.com/ac4.3

CHECKpoint CONCEPT 4.5

- ✓ Compare the fibrous and gel-like components of the extracellular matrices of plant and animal cells.
- ✓ What kinds of cell junctions would you expect to find, and why, in the following situations?
  a. In the digestive system, where material must pass through cells and not go through the extracellular material, to get from the intestine to the blood vessels.
  b. In a small animal, where a chemical signal passes rapidly through cells to go from the head to the tail.
  c. In the lining of the intestine, where cells in the lining are constantly jostled by the churning of the underlying muscle.
- ✓ When cancer spreads from its primary location to other parts of the body (a process called metastasis), tumor cells detach from their original location and then reattach at a different location. How would the integrin–collagen system be involved in this process?

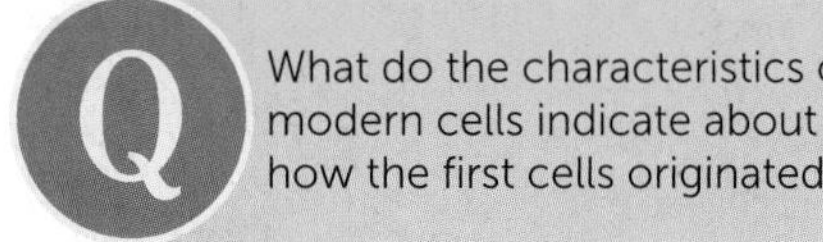

What do the characteristics of modern cells indicate about how the first cells originated?

ANSWER Ideas about how the first cells may have formed focus on two questions: how and when. As to how cells could arise from a chemical-rich environment, most biologists assume that a cell membrane formed first and was necessary to provide a compartment for the chemical transformations of life to occur, separated from the environment (Concept 4.1). Biologists also assume that the first cells were relatively simple prokaryotes (Concept 4.2), without the organelles that define eukaryotic cells (Concept 4.3).

Jack Szostak, a Nobel laureate at Harvard University, builds synthetic cell models that give insights into the origin of cells. He and his colleagues make small membrane-lined droplets by putting fatty acids into water and then shaking the mixture. The lipids form water-filled droplets, each surrounded by a lipid bilayer "membrane" (see Figure 2.13). With water (and other molecules of the scientists' choosing) trapped inside, these spheres have many properties characteristic of modern cells—so many that they have been called **protocells** (**FIGURE 4.19**). For example, the membrane barrier determines what goes in and out of a protocell, by excluding macromolecules like RNA but allowing smaller molecules such as nucleotides to pass through. Moreover, RNA inside the protocell can act as a catalyst, replicating itself from nucleotides that enter the protocell. The spheres are somewhat unstable, and under the microscope they can be seen to grow, elongate, and break, a possible precursor of more precise cell division.

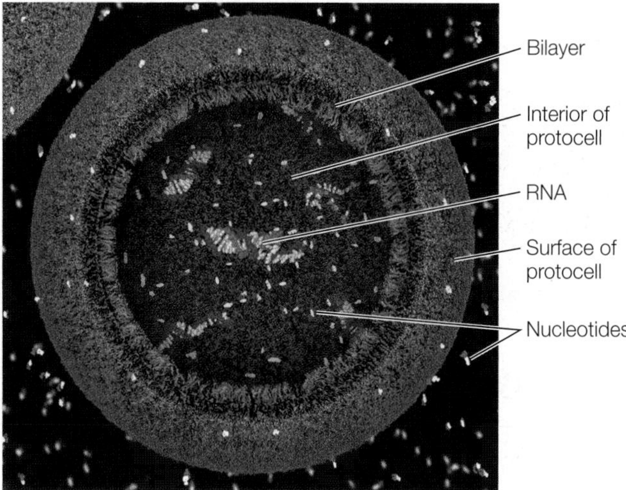

FIGURE 4.19 A Protocell A protocell can be made in the lab and can carry out some functions of modern cells—in particular, it provides a compartment for biochemical reactions.

Is this really a cell, possibly like the one where life started? Certainly not: it cannot fully reproduce itself, and its capacities for metabolism are limited. But by providing a compartment for biochemical reactions with a boundary that separates it from the environment, the protocell is a model for the first cell.

When did the first cells on Earth appear? According to geologists, Earth is about 4.5 billion years old. Heat and atmospheric conditions precluded life for at least a half-billion years after Earth formed. The oldest fossils of multicellular organisms date from about 1.2 billion years ago.

In all probability, life began with single-celled organisms resembling modern bacteria. Unfortunately, such cells lack the structures that are typically preserved in fossils, and so they die without a trace. Recently, however, geochemist and paleontologist William Schopf at the University of California, Los Angeles used a new method of microscopy called confocal laser scanning microscopy, combined with chemical analyses, to identify fossil cells that are about 800 million years old. Some of these look like Szostak's protocells. These were probably not the first cells, as there is chemical evidence in some rocks that life was present about 3.8 billion years ago. But so far, Schopf's fossilized cells are the oldest cells that anyone has been able to find.

SUMMARY

CONCEPT 4.1 Cells Provide Compartments for Biochemical Reactions

See **ACTIVITY 4.1**

- Cell theory states that the cell is the fundamental unit of biological structure and function.
- Cells are small because a cell's surface area must be large compared with its volume to accommodate exchanges between the cell and its environment. **Review Figure 4.2**
- All cells are enclosed by a selectively permeable **cell membrane** that separates their contents from the external environment.

CONCEPT 4.2 Prokaryotic Cells Do Not Have a Nucleus

- Prokaryotic cells usually have no internal compartments, but have a **nucleoid** containing DNA, and a **cytoplasm** containing **cytosol**, **ribosomes** (the sites of protein synthesis), proteins, and small molecules. Many have an extracellular **cell wall**. **Review Figure 4.5**
- Some prokaryotes have folded membranes, for example photosynthetic membranes, and some have **flagella** for motility. **Review Figure 4.6**

CONCEPT 4.3 Eukaryotic Cells Have a Nucleus and Other Membrane-Bound Compartments

- Eukaryotic cells contain many membrane-enclosed **organelles** that compartmentalize their biochemical functions. **Review Figure 4.7 and ANIMATED TUTORIAL 4.1**
- The **nucleus** contains most of the cell's DNA.
- The **endomembrane system**—consisting of the nuclear envelope, **endoplasmic reticulum**, **Golgi apparatus**, and **lysosomes**—is a series of interrelated compartments enclosed by membranes. It segregates proteins and modifies them. Lysosomes contain many digestive enzymes. **Review Figures 4.8 and 4.9, ANIMATED TUTORIAL 4.2, and ACTIVITY 4.2**
- **Mitochondria** and **chloroplasts** are semiautonomous organelles that process energy.
- A **vacuole** is prominent in many plant cells. It is a membrane-enclosed compartment full of water and dissolved substances.

CONCEPT 4.4 The Cytoskeleton Provides Strength and Movement

- The **microfilaments**, **intermediate filaments**, and **microtubules** of the **cytoskeleton** provide the cell with shape, strength, and movement. **Review Figure 4.10**
- Microfilaments and microtubles have dynamic instability and can grow or shrink in length rapidly.
- **Cilia** and **flagella** are microtubule-lined extensions of the cell membrane that produce movements of cells or their surrounding fluid medium. **Review Figures 4.11 and 4.12**
- Motor proteins move cellular components, such as **vesicles**, around the cell by "walking" them along the microtubules. **Review Figure 4.13**
- Biologists establish cause-and-effect relationships by manipulating biological systems. **Review Figure 4.14**

CONCEPT 4.5 Extracellular Structures Provide Support and Protection for Cells and Tissues

- The plant cell wall consists principally of cellulose. Cell walls are pierced by **plasmodesmata** that join the cytoplasms of adjacent cells. **Review Figure 4.15**
- In animals, the **extracellular matrix** consists of different kinds of proteins, including **collagen** and **proteoglycans**. Integrins connect the cell cytoplasm with the extracellular matrix. **Review Figures 4.16 and 4.17**
- Specialized **cell junctions** connect cells in animal tissues. These include tight junctions, desmosomes, and gap junctions. Gap junctions are involved in intercellular communication. **Review Figure 4.18 and ACTIVITY 4.3**

Go to the Interactive Summary to review key figures, Animated Tutorials, and Activities
PoL2e.com/is4

Go to LaunchPad at **macmillanhighered.com/launchpad** for additional resources, including LearningCurve Quizzes, Flashcards, and many other study and review resources.

5 Cell Membranes and Signaling

KEY CONCEPTS

5.1 Biological Membranes Have a Common Structure and Are Fluid

5.2 Passive Transport across Membranes Requires No Input of Energy

5.3 Active Transport Moves Solutes against Their Concentration Gradients

5.4 Large Molecules Cross Membranes via Vesicles

5.5 The Membrane Plays a Key Role in a Cell's Response to Environmental Signals

5.6 Signal Transduction Allows the Cell to Respond to Its Environment

Many people rely on caffeine to wake themselves up and to keep their minds alert.

If you are like most people, you consume a significant amount of caffeine every day. In fact, more than 90 percent of North Americans and Europeans drink coffee or tea to get their "caffeine fix." Coffee and tea plants contain caffeine as a defense against the insects that eat them. Caffeine acts as an insecticide in plant parts that are particularly vulnerable to insect attacks, such as seeds, young seedlings, and leaves. But it is not toxic to humans.

Legend has it that about 5,000 years ago, a Chinese emperor found out by accident that a pleasant beverage could be made by boiling tea leaves. About 1,000 years ago, monks living in what is now Ethiopia found that roasting coffee seeds (also called "beans") gave a similarly pleasant effect and that the beverage kept them awake during long periods of prayer. Caffeine is now the most widely consumed psychoactive molecule in the world, but unlike other psychoactive drugs, it is not subject to government regulation.

Most people know from personal experience what caffeine does to the body: because it keeps us awake, it obviously affects the brain. In fact, it is often given to premature babies in the hospital nursery when they stop breathing. But it also affects other parts of the body—for example, it increases urination and speeds up the heart. How does this molecule work?

The key to understanding caffeine's action is to understand how it interacts with the cell membrane. In Chapter 4 we introduced the concept of the membrane as a structural boundary between the inside of a cell and the surrounding environment. The cell membrane physically separates the cell cytoplasm from its surroundings and helps maintain chemical differences between these two environments. The same can be said of the membranes that surround cell organelles, separating them from the cytoplasmic environment.

When caffeine arrives at a cell in the body, it first encounters the cell membrane. The properties of this membrane determine whether and how the cell will react to caffeine. Will it cross the membrane boundary and enter the cell? What determines whether it crosses the membrane? If it does not, how can caffeine's interactions with membrane components lead to changes in cell function?

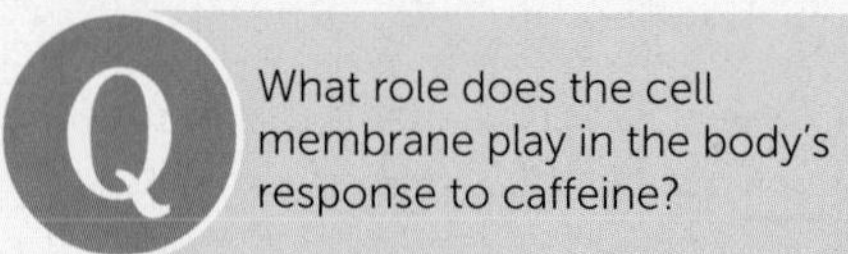

Q
What role does the cell membrane play in the body's response to caffeine?

You will find the answer to this question on page 102.

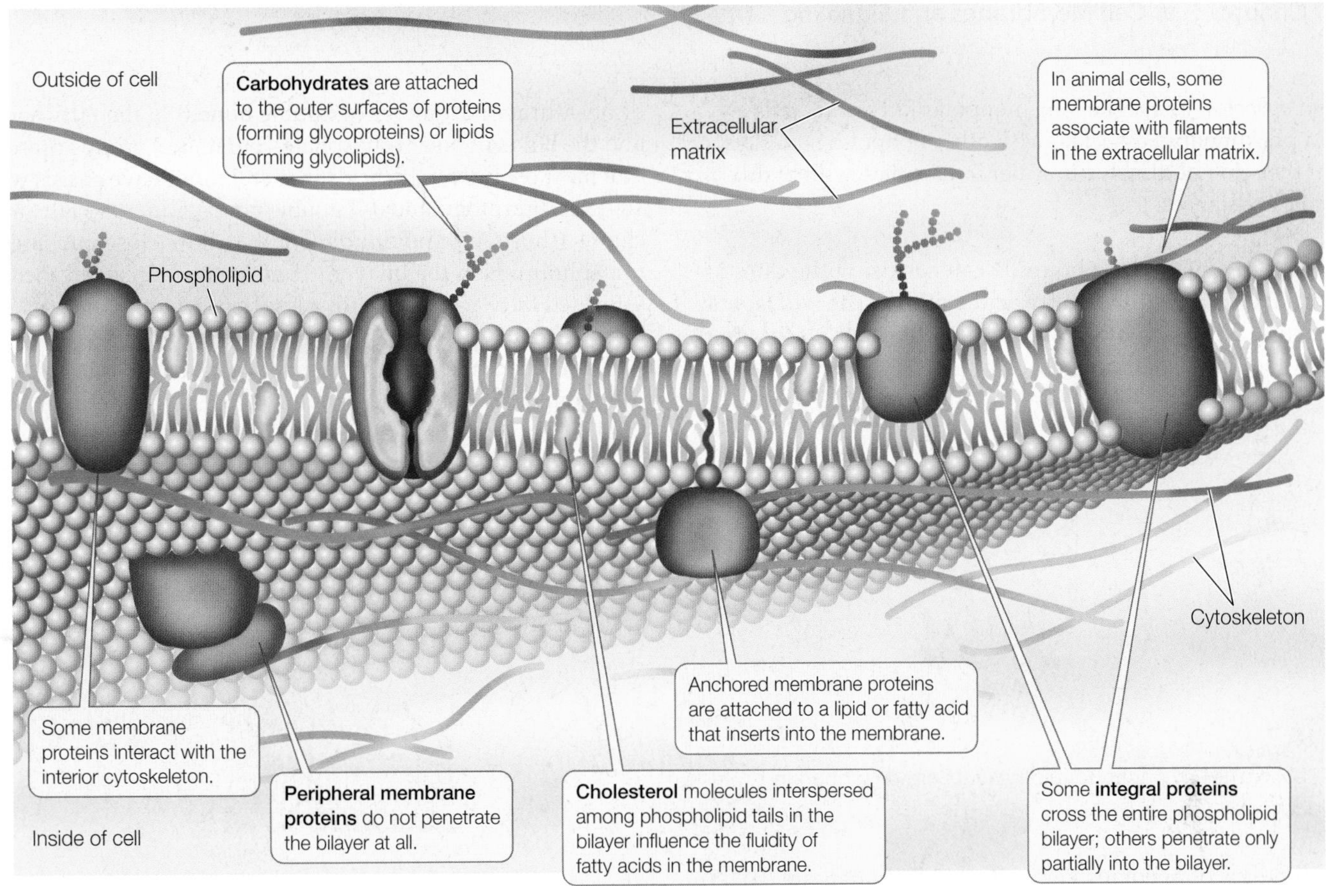

Outside of cell
Carbohydrates are attached to the outer surfaces of proteins (forming glycoproteins) or lipids (forming glycolipids).
Extracellular matrix
In animal cells, some membrane proteins associate with filaments in the extracellular matrix.
Phospholipid
Cytoskeleton
Anchored membrane proteins are attached to a lipid or fatty acid that inserts into the membrane.
Some membrane proteins interact with the interior cytoskeleton.
Inside of cell
Peripheral membrane proteins do not penetrate the bilayer at all.
Cholesterol molecules interspersed among phospholipid tails in the bilayer influence the fluidity of fatty acids in the membrane.
Some integral proteins cross the entire phospholipid bilayer; others penetrate only partially into the bilayer.

FIGURE 5.1 Membrane Structure The general molecular structure of biological membranes is a continuous phospholipid bilayer in which proteins are embedded. The phospholipid bilayer separates two aqueous regions, the external environment outside the cell and the cell cytoplasm.

Go to ACTIVITY 5.1 Membrane Molecular Structure PoL2e.com/ac5.1

CONCEPT 5.1 Biological Membranes Have a Common Structure and Are Fluid

The evolution of cellular life required the presence of a boundary, a way to separate the inside of the cell from the surrounding environment. This need was—and still is—fulfilled by biological membranes. Like many of the basic processes of life at the cellular level, the functions of membranes are carried out by a molecular structure shared by most organisms.

A biological membrane's structure and functions are determined by the chemical properties of its constituents: lipids, proteins, and carbohydrates. An important concept that emerges from our consideration of these molecules is polarity and how polarity influences the way a molecule interacts with water. Recall from Concept 2.2 that some compounds are polar and hydrophilic ("water-loving"), whereas others are nonpolar and hydrophobic ("water-hating"), and that a phospholipid has both polar and nonpolar regions. The nonpolar regions of phospholipids and membrane proteins interact to form an insoluble barrier. The phospholipid bilayer serves as a lipid "lake" in which a variety of proteins "float" (**FIGURE 5.1**). This general design is known as the **fluid mosaic model**.

Membranes contain a wide array of proteins, most of which are noncovalently embedded in the phospholipid bilayer. These proteins are held within the membrane by their hydrophobic regions (also called their hydrophobic "domains"). The proteins' hydrophilic regions are exposed to the watery conditions on one or both sides of the bilayer. Membrane proteins have three major functions: some move materials through the membrane, others are involved in intercellular recognition and adhesion, while others receive chemical signals from the cell's external environment.

The carbohydrates associated with membranes are attached to either lipids or protein molecules. They are generally located on the outside of the cell, where they interact with substances in the external environment. Like some membrane proteins, carbohydrates are crucial for recognizing specific molecules, such as those on the surfaces of adjacent cells.

Each membrane has constituents that are suitable for the specialized functions of the cell or organelle it surrounds. As you read about the different molecules in membranes, keep in mind that some membranes contain many protein molecules, others are lipid-rich, others have significant amounts of cholesterol, and still others are rich in carbohydrates.

Lipids form the hydrophobic core of the membrane

The lipids in biological membranes are usually **phospholipids**, with hydrophilic and hydrophobic regions:

- *Hydrophilic regions.* The phosphorus-containing "head" of a phospholipid is electrically charged and therefore associates with polar water molecules.

- *Hydrophobic regions.* The long, nonpolar fatty acid "tails" of a phospholipid associate with other nonpolar materials, but they do not dissolve in water or associate with hydrophilic substances.

One way in which phospholipids can coexist with water is to form a bilayer, with the fatty acid "tails" of the two layers interacting with each other, and the polar "heads" facing the outside, aqueous environment:

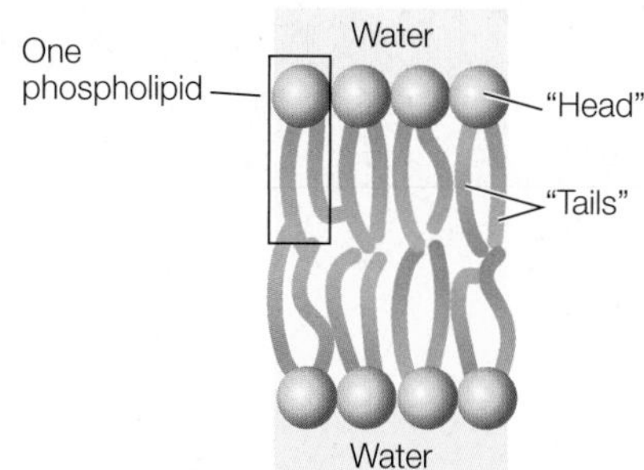

LINK

The properties of phospholipid bilayers are described in **Concept 2.4**

The thickness of a biological membrane is about 8 nm (0.008 μm), which is twice the length of a typical phospholipid. The page you are reading is about 3,000 times thicker than this.

As we noted in Chapter 4, it is possible in the laboratory to make artificial bilayers with the same organization as natural membranes. Small holes in such bilayers seal themselves spontaneously. The capacity of phospholipids to associate with one another and maintain a bilayer organization helps biological membranes fuse during vesicle formation, phagocytosis, and related processes (see Concept 4.3, especially Figure 4.9).

Although biological membranes all share a similar structure, there are many different kinds of phospholipids, and membranes from different cells or organelles may differ greatly in their lipid composition. Phospholipids can differ in terms of fatty acid chain length (number of carbon atoms), degree of unsaturation (number of double bonds) in the fatty acids, and the kinds of polar (phosphate-containing) groups present. The most common fatty acids in membranes have chains with 16–18 carbon atoms and 0–2 double bonds. Saturated fatty acid chains (those with no double bonds) allow close packing of phospholipids in the bilayer, whereas the "kinks" in the unsaturated fatty acids make for a less dense, more fluid packing (see Figure 2.12).

Up to 25 percent of the lipid content of an animal cell's cell membrane may be the steroid cholesterol. Steroids are a family of carbon compounds that have multiple linked rings. Cholesterol plays an important role in modulating membrane fluidity (see below); other steroids function as hormones (see Concept 35.2). The hydroxyl group (—OH) on the cholesterol molecule interacts with the polar heads of the phospholipids, while the nonpolar rings insert among the fatty acid chains of the membrane:

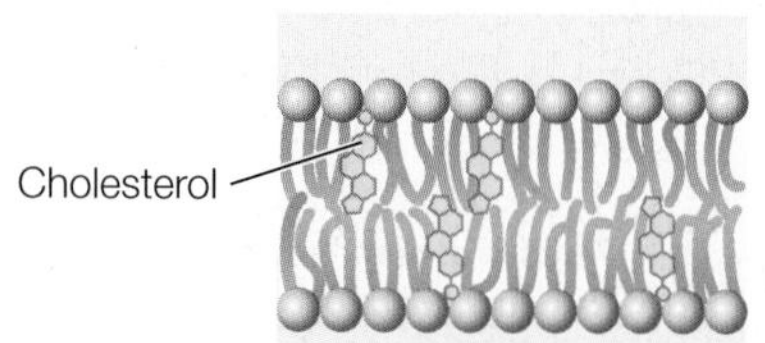

Cholesterol

The fatty acids of the phospholipids make the membrane somewhat fluid—about as fluid as olive oil. This fluidity permits some molecules to move laterally within the plane of the membrane. A given phospholipid molecule in the cell membrane can travel from one end of the cell to the other in a little more than 1 second!

APPLY THE CONCEPT

Biological membranes have a common structure and are fluid

The membrane lipids of a cell can be labeled with a fluorescent tag so the entire surface of the cell will glow evenly under ultraviolet light. If a strong laser light is then shone on a tiny region of the cell, that region gets bleached (the strong light destroys the fluorescent tag) and there is a "hole" in the cell surface fluorescence (though not an actual hole in the cell's membrane). After the laser is turned off, the hole gradually fills in with fluorescent lipids that diffuse in from other parts of the membrane. The time it takes for the "hole" to disappear is a measure of membrane fluidity. The table shows some data for cells with altered membrane compositions.[a] Explain the effect of each alteration.

CONDITION	TIME (sec) FOR "HOLE" TO BECOME FLUORESCENT
No alteration	65
Decreased length of fatty acid chains	38
Increased cholesterol	88
Increased percentage of unsaturated fatty acids	42
Increased membrane protein content	90

[a] Adapted from E. Wu et al. 1977. *Biochemistry* 16: 3936–3941.

Membrane fluidity is affected by several factors, two of which are particularly important:

- *Lipid composition.* Cholesterol and long-chain, saturated fatty acids pack tightly together, resulting in less fluid membranes. Unsaturated fatty acids or those with shorter chains tend to increase membrane fluidity. Some anesthetics are nonpolar and act by inserting into cell membranes. They reduce the fluidity of nerve cell membranes and thereby decrease nerve activity.
- *Temperature.* Membrane fluidity declines under cold conditions because molecules move more slowly at lower temperatures. For example, when your fingers get numb after contact with ice, it is due to a reduction in membrane fluidity in the nerve cells. To address this problem, some organisms simply change the lipid composition of their membranes when their environment gets cold, replacing saturated with unsaturated fatty acids and using fatty acids with shorter chains. These changes play a role in the survival of plants, bacteria, and hibernating animals during the winter.

While phospholipid molecules can easily move laterally within a membrane, it is rare for a phospholipid in one half of the bilayer to spontaneously flip over to the other side. For that to happen, the polar part of the molecule would have to move through the hydrophobic interior of the membrane. Since spontaneous flip-flops are rare, the inner and outer halves of the bilayer may be quite different in the kinds of phospholipids they contain.

Go to ANIMATED TUTORIAL 5.1
Lipid Bilayer Composition
PoL2e.com/at5.1

Proteins are important components of membranes

All biological membranes contain proteins. Typically, cell membranes have about 1 protein molecule for every 25 phospholipid molecules. This ratio varies depending on membrane function. In the inner membrane of the mitochondrion, which is specialized for energy processing, there is 1 protein for every 5 lipids. By contrast, myelin—a membrane that encloses portions of some neurons (nerve cells) and acts as an electrical insulator—has only 1 protein for every 70 lipids.

Recall from Table 3.2 that some amino acids contain nonpolar, hydrophobic R groups, whereas others contain polar (or charged), hydrophilic R groups. The arrangement of these amino acids in a membrane protein determines whether the membrane protein will insert into the nonpolar lipid bilayer and how it will be positioned. There are two general types of membrane proteins:

- **Peripheral membrane proteins** lack exposed hydrophobic groups and are not embedded in the bilayer. Instead, they have polar or charged regions that interact with exposed parts of integral membrane proteins, or with the polar heads of phospholipid molecules (see Figure 5.1).
- **Integral membrane proteins** are at least partly embedded in the phospholipid bilayer. Like phospholipids, these proteins have both hydrophilic and hydrophobic regions:

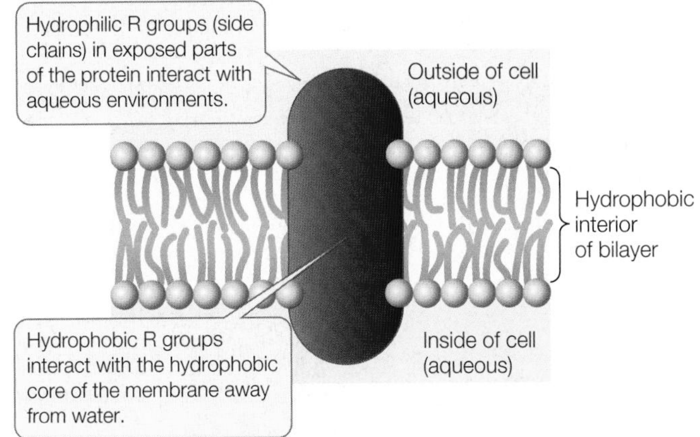

Hydrophilic R groups (side chains) in exposed parts of the protein interact with aqueous environments.
Outside of cell (aqueous)
Hydrophobic interior of bilayer
Hydrophobic R groups interact with the hydrophobic core of the membrane away from water.
Inside of cell (aqueous)

Membrane proteins and lipids generally interact only noncovalently. The polar ends of proteins can interact with the polar ends of lipids, and the nonpolar regions of both molecules can interact hydrophobically. However, some membrane proteins have fatty acids or other lipid groups covalently attached to them. These are referred to as anchored membrane proteins, because it is their hydrophobic lipid components that anchor them in the phospholipid bilayer (see Figure 5.1).

Proteins are asymmetrically distributed on the inner and outer surfaces of membranes. An integral membrane protein that extends all the way through the phospholipid bilayer and protrudes on both sides is known as a **transmembrane protein**. In addition to one or more transmembrane domains (regions) that extend through the bilayer, such a protein may have domains with other specific functions on the inner and outer sides of the membrane. Transmembrane proteins are always oriented the same way—domains with specific functions inside or outside the cell are always found on the correct side of the membrane. Peripheral membrane proteins are located on one side of the membrane or the other. This asymmetrical arrangement gives the two surfaces of the membrane different properties. As we will soon see, these differences have great functional significance.

Like lipids, some membrane proteins move relatively freely within the phospholipid bilayer. Cell fusion experiments illustrate this migration dramatically. When two cells fuse, a single continuous membrane forms around both cells, and some proteins from each cell distribute themselves uniformly around this membrane (**FIGURE 5.2**).

Although some proteins are free to migrate throughout the membrane, others appear to be contained within specific regions. These membrane regions are like a corral of horses on a farm: the horses are free to move around within the fenced area but not outside it. For example, a muscle cell protein that recognizes a chemical signal from a neuron is normally found only within a specific region of the cell membrane, where the neuron meets the muscle cell.

How does this happen? Proteins inside the cell can restrict the movement of proteins within a membrane. Components

INVESTIGATION

FIGURE 5.2 Rapid Diffusion of Membrane Proteins A human cell can be fused to a mouse cell in the laboratory, forming a single large cell (heterokaryon). This phenomenon was used to test whether membrane proteins can diffuse independently in the plane of the cell membrane.[a]

HYPOTHESIS

Proteins embedded in a membrane can diffuse freely within the membrane.

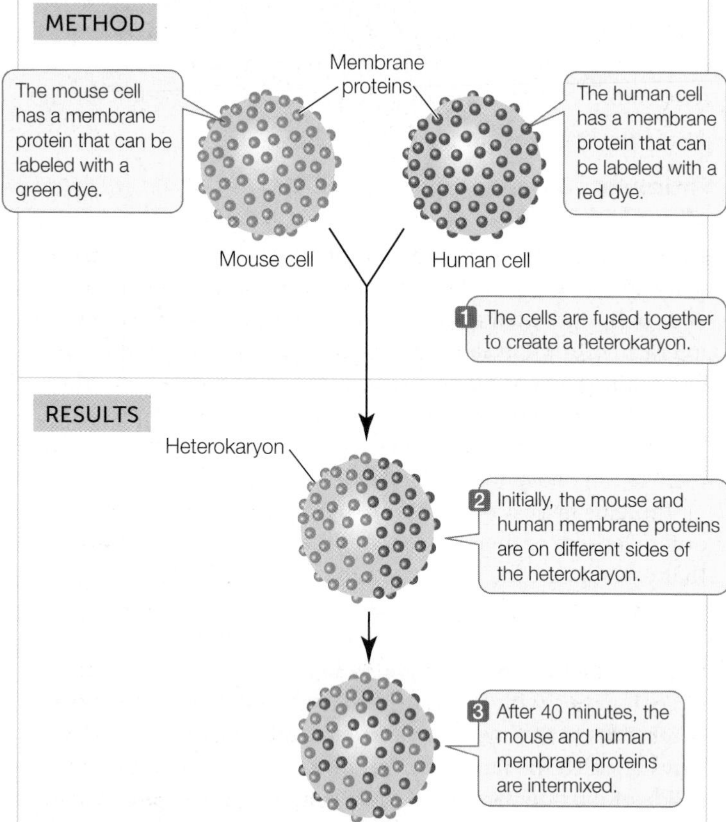

METHOD
Membrane proteins
The mouse cell has a membrane protein that can be labeled with a green dye.
The human cell has a membrane protein that can be labeled with a red dye.
Mouse cell
Human cell
1 The cells are fused together to create a heterokaryon.
RESULTS
Heterokaryon
2 Initially, the mouse and human membrane proteins are on different sides of the heterokaryon.
3 After 40 minutes, the mouse and human membrane proteins are intermixed.

CONCLUSION

Membrane proteins can diffuse rapidly in the plane of the membrane.

ANALYZE THE DATA

The experiment was repeated at various temperatures with the following results:

Temperature (°C)	Cells with mixed proteins (%)
0	0
15	8
20	42
26	77

Plot these data on a graph of Percentage Mixed vs. Temperature. Explain these data, relating the results to the concepts of diffusion and membrane fluidity.

Go to **LaunchPad** for discussion and relevant links for all **INVESTIGATION** figures.

[a]L. Frye and M. Edidin. 1970. *Journal of Cell Science* 7: 319–335.

of the cytoskeleton may be attached to membrane proteins protruding into the cytoplasm (see Figure 5.1). The stability of the cytoskeleton may thus restrict the movement of attached membrane proteins.

Cell membrane carbohydrates are recognition sites

In addition to lipids and proteins, the cell membrane contains carbohydrates. The carbohydrates are located on the outer surface of the cell membrane and may be covalently bonded to lipids or to proteins:

- A **glycolipid** consists of a carbohydrate covalently bonded to a lipid.
- A **glycoprotein** consists of one or more short carbohydrate chains covalently bonded to a protein. The bound carbohydrates are oligosaccharides, usually not exceeding 15 monosaccharide units in length (see Concept 2.3).
- A **proteoglycan** is a protein with even more carbohydrate molecules attached to it, and the carbohydrate chains are often longer than in glycoproteins.

These chains of monosaccharides can generate a large diversity of linear or branched structures. These diverse structures play roles in cell communication and cell adhesion. For example, the carbohydrates on some glycolipids change when cells become cancerous. This change may allow white blood cells to target cancer cells for destruction.

LINK

The activities of white blood cells are described in **Chapter 39**

Cells can stick together (adhere) due to interactions between similar carbohydrates on the outer surfaces of two cells, or between a carbohydrate on one cell and a membrane protein on another cell. Or, two proteins can interact directly:

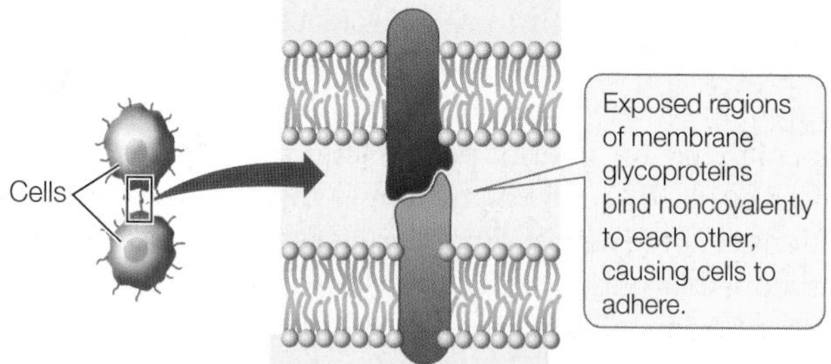

Cells
Exposed regions of membrane glycoproteins bind noncovalently to each other, causing cells to adhere.

Cell adhesion occurs in all kinds of multicellular organisms.

Membranes are constantly changing

Membranes in eukaryotic cells are constantly forming, transforming from one type to another, fusing with one another, and breaking down. As we discussed in Concept 4.3, fragments of membrane move (in the form of vesicles) from the endoplasmic reticulum to the Golgi apparatus, and from the Golgi apparatus to the cell membrane (see Figure 4.8). Secondary lysosomes form when primary lysosomes from the Golgi apparatus fuse with phagosomes from the cell membrane (see Figure 4.9).

Because membranes interconvert so readily, we might expect all subcellular membranes to be chemically identical. However, that is not the case: there are major chemical differences among the membranes of even a single cell. Membranes are changed chemically when they form parts of certain organelles. In the Golgi apparatus, for example, the membranes of the *cis* face closely resemble those of the endoplasmic reticulum in chemical composition, but those of the *trans* face are more similar to the cell membrane.

CHECKpoint CONCEPT 5.1

- ✓ What are the differences between peripheral and integral membrane proteins?
- ✓ A membrane protein has the following amino acid sequence (see Table 3.2 for abbreviations):

  EWDRHDFESGPTFIWLIWLVLAVLFLLLWAVLRPGKYKDKHE

  Considering the R groups on the amino acids, predict the region of the protein that will be embedded within the membrane.
- ✓ What is the evidence for membrane fluidity?
- ✓ If the cells of certain sponges are separated, they reaggregate because of binding between their membrane-associated proteoglycans. What would happen if the same experiment were conducted with cells treated to remove cell surface carbohydrates?

Now that you understand the structure of biological membranes, let's see how their components function. In the sections that follow, we will focus on the cell membrane. We'll look at how the cell membrane regulates the passage of substances that enter or leave a cell. Bear in mind that these principles also apply to the membranes that surround organelles.

CONCEPT 5.2 Passive Transport across Membranes Requires No Input of Energy

An important property of all life is the ability to regulate the internal composition of a cell, distinguishing it from the surrounding environment. Biological membranes allow some substances, but not others, to pass through them. This characteristic of membranes is called **selective permeability**. If a membrane is permeable to a particular substance, that substance can simply diffuse (as we describe below) across the membrane from a region of higher concentration to a region of lower concentration. However, some substances must be transported across membranes, and this process is facilitated by specialized membrane proteins.

There are two fundamentally different processes by which substances cross biological membranes:

- The processes of **passive transport** do not require direct inputs of metabolic energy to drive them. In general, passive transport occurs when a substance moves from the side of the membrane where its concentration is higher to the side where its concentration is lower. In other words, the substance moves *down its concentration gradient*. Passive transport can also occur when an ion is transported across a membrane to reduce charge differences between the two sides of the membrane.
- The processes of **active transport** require the input of metabolic energy because they involve the movement of substances *against their concentration gradients.* Even if the membrane is permeable to a substance, that substance must be actively transported from the side where its concentration is lower to the side where it is higher.

This section focuses on passive transport across the membrane. Passive transport can occur by simple diffusion through the phospholipid bilayer, or it can be facilitated by channel proteins or carrier proteins.

Simple diffusion takes place through the phospholipid bilayer

In a solution, there is a tendency for all of the components to be evenly distributed. You can see this when a drop of ink is allowed to fall into a gelatin suspension (a "gel"). Initially the pigment molecules are very concentrated, but they will move about at random, slowly spreading until the intensity of color is exactly the same throughout the gel:

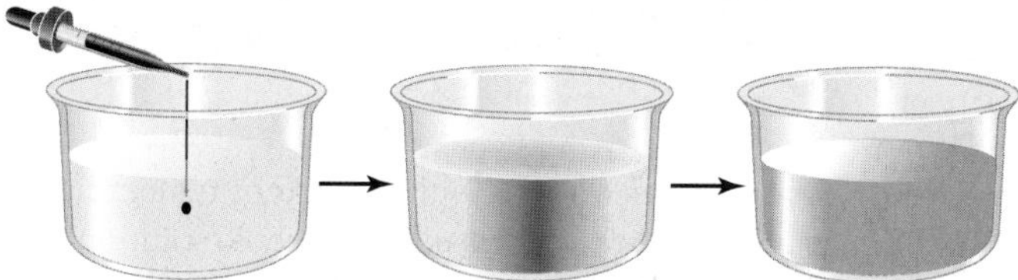

A solution in which the solute molecules are uniformly distributed is said to be at equilibrium. This does not mean the molecules have stopped moving; it just means they are moving in such a way that their overall distribution does not change.

Diffusion is the process of random movement toward a state of equilibrium. In effect, it is a net movement from regions of greater concentration to regions of lesser concentration. Diffusion is generally a *very slow process in living tissues* except over short distances, especially when we consider the gel-like consistency of the cell cytoplasm. For example, it would take about 3 years for a molecule of oxygen gas (O_2) to diffuse from the human lung to a cell at the fingertip! So it is not surprising that as plants and animals evolved and became larger and multicellular, those with circulatory systems to distribute vital molecules such as O_2 had a distinct advantage over organisms relying on simple diffusion.

How fast a substance diffuses depends on three factors:

- The *diameter* of the molecules or ions: smaller molecules diffuse faster.
- The *temperature* of the solution: higher temperatures lead to faster diffusion because the heat provides more energy for movement.
- The *concentration gradient* in the system—that is, the change in solute concentration with distance in a given direction. The greater the concentration gradient, the more rapidly a substance diffuses.

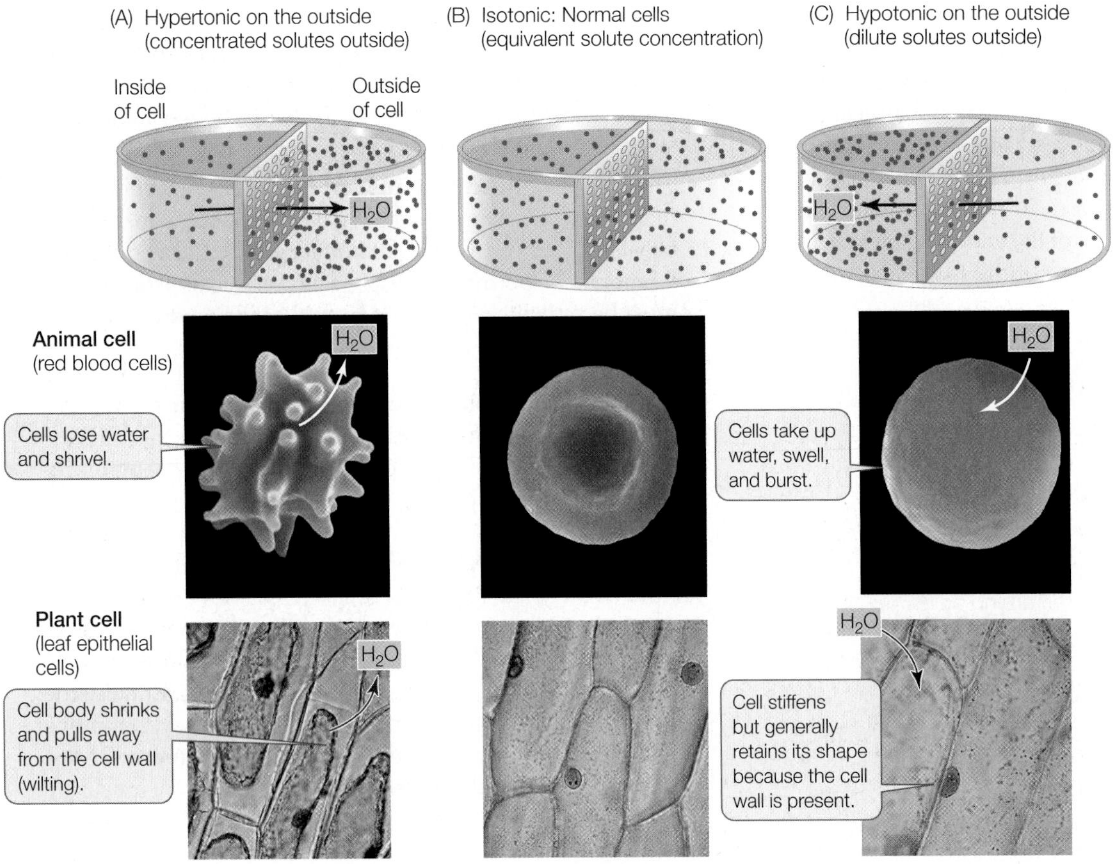

FIGURE 5.3 Osmosis Can Modify the Shapes of Cells
(A) In a solution that is hypertonic to the cytoplasm of a plant or animal cell, water flows out of the cell. (B) In a solution that is isotonic with the cytoplasm, the cell maintains a consistent, characteristic shape because there is no net movement of water into or out of the cell. (C) In a solution that is hypotonic to the cytoplasm, water enters the cell. An animal cell will swell and may burst under these conditions; a plant cell will not swell too much because of its rigid cell wall.

What does this mean for a cell surrounded by a membrane? The cytoplasm is largely a water-based (aqueous) solution, and so is the surrounding environment. In a complex solution (one with many different solutes), the diffusion of each solute depends only on its own concentration, not on the concentrations of other solutes. So one might expect a substance with a higher concentration inside the cell to diffuse out, and one with a higher concentration outside the cell to diffuse in. Indeed, some small molecules can pass through the phospholipid bilayer of the membrane by **simple diffusion**. Gases, including oxygen and carbon dioxide, can cross membranes this way. Small nonpolar and uncharged molecules can enter the membrane readily and pass through it. The more lipid-soluble the molecule is, the more rapidly it diffuses through the lipid bilayer.

In contrast, electrically charged or polar (hydrophilic) molecules, such as amino acids, sugars, ions, and water, do not pass readily through a membrane because they are not soluble in the hydrophobic interior of the lipid bilayer. However, as we discuss below, specialized proteins facilitate the transport of these molecules across membranes.

Osmosis is the diffusion of water across membranes

Water molecules pass through specialized channels in membranes (see below) by a diffusion process called **osmosis**. This process depends on the relative concentrations of water molecules on both sides of the membrane. In a particular solution, the higher the *total* solute concentration, the lower the concentration of water molecules. **Osmotic pressure** is defined as the pressure that needs to be applied to a solution to prevent the flow of water across a membrane by osmosis. This pressure is proportional to the total concentration of solutes in the solution—the more dissolved solutes there are, the fewer water molecules there are, and so water moves across the membrane and into the solution. The equation for osmotic pressure (symbolized by the Greek capital letter *pi*, Π) due to water is

$$\Pi = cRT$$

where c is total solute concentration, R is the gas constant, and T is the absolute temperature. In thermodynamic terms, the higher concentration of a substance in a compartment on one side of a membrane represents stored energy.

Consider a situation where a membrane separating two different solutions allows water, *but not solutes*, to pass through. The water molecules will move across the membrane toward the solution with the higher solute concentration and the lower concentration of water molecules.

Here we are referring to the *net* movement of water. Since it is so abundant, water is constantly moving (through channel proteins) across the cell membrane, into and out of cells. But if there is a concentration difference between the two sides of the membrane, the overall movement will be greater in one direction or the other.

Three terms are used to compare the solute concentrations of two solutions separated by a membrane:

- A **hypertonic** solution has a higher solute concentration than the other solution (**FIGURE 5.3A**).

- **Isotonic** solutions have equal solute concentrations (**FIGURE 5.3B**).
- A **hypotonic** solution has a lower solute concentration than the other solution (**FIGURE 5.3C**).

The concentration of solutes in the environment determines the direction of osmosis in all animal cells. A red blood cell takes up water from a solution that is hypotonic to the cell's contents. If this happens, the cell bursts because its cell membrane cannot withstand the pressure created by the water entry and the resultant swelling (see Figure 5.3C). Conversely, the cell shrinks if the solution surrounding it is hypertonic to its contents (see Figure 5.3A). The integrity of blood cells is absolutely dependent on the maintenance of a constant solute concentration in the surrounding blood plasma—the plasma must be isotonic to the blood cells. Regulation of the solute concentrations of body fluids is thus an important process for organisms without cell walls.

In contrast to animal cells, the cells of plants, archaea, bacteria, fungi, and some protists have cell walls that limit their volumes and keep them from bursting. Cells with sturdy walls take up a limited amount of water, and in so doing they build up internal pressure against the cell wall, which prevents further water from entering. This pressure within the cell is called **turgor pressure**; it keeps the green parts of plants upright and is the driving force for enlargement of plant cells (see Concept 25.3). It is a normal and essential component of plant growth. If enough water leaves the cells, turgor pressure drops and the plant wilts. Turgor pressure reaches about 100 pounds per square inch (0.7 kg/cm^2), which is greater than the pressure in auto tires (about 35 pounds per square inch).

LINK

The roles of osmosis in plant physiological processes are described in **Concept 25.3**. Excretion in animals also involves osmosis; see **Concept 36.1**

Diffusion may be aided by channel proteins

As we saw earlier, polar or charged substances such as water, amino acids, sugars, and ions do not readily diffuse across membranes. But they can cross the hydrophobic phospholipid bilayer passively (that is, without the input of energy) in one of two ways, depending on the substance:

- **Channel proteins** are integral membrane proteins that form channels across the membrane through which certain substances can pass.
- Some substances can bind to membrane proteins called **carrier proteins** that speed up their diffusion through the phospholipid bilayer.

Both of these processes are forms of **facilitated diffusion**. The substances diffuse according to their concentration gradients, but their diffusion is made easier by channel or carrier proteins. Particular channel or carrier proteins allow diffusion both into and out of a cell or organelle. In other words, they can operate in both directions.

We will focus here on two examples of channel proteins and discuss carrier proteins in the next section.

ION CHANNELS The best-studied channel proteins are the **ion channels**. As you will see in later chapters, the movement of ions across membranes is important in many biological processes, including ATP production within the mitochondria, the electrical activity of the nervous system, and the opening of pores in plant leaves to allow gas exchange with the environment. Several types of ion channels have been identified, each of them specific for a particular ion. All of them show the same basic structure of a hydrophilic pore that allows a particular ion to move through it.

Just as a fence may have a gate that can be opened or closed, most ion channels are "gated": they can be opened or closed to ion passage. A **gated channel** opens when a stimulus causes a change in the three-dimensional shape of the channel. In some cases, this stimulus is the binding of a chemical signal, or **ligand**. Channels controlled in this way are called ligand-gated channels (**FIGURE 5.4**). In contrast, a voltage-gated channel is stimulated to open or close by a change in the voltage (electrical charge difference) across the membrane (see Figure 34.5).

AQUAPORINS Water crosses membranes at a much faster rate than would be expected if it simply diffused through the phospholipid bilayer. One way water does this is by "hitchhiking" with some ions, such as Na^+, as they pass through ion channels. Up to 12 water molecules may coat an ion as it

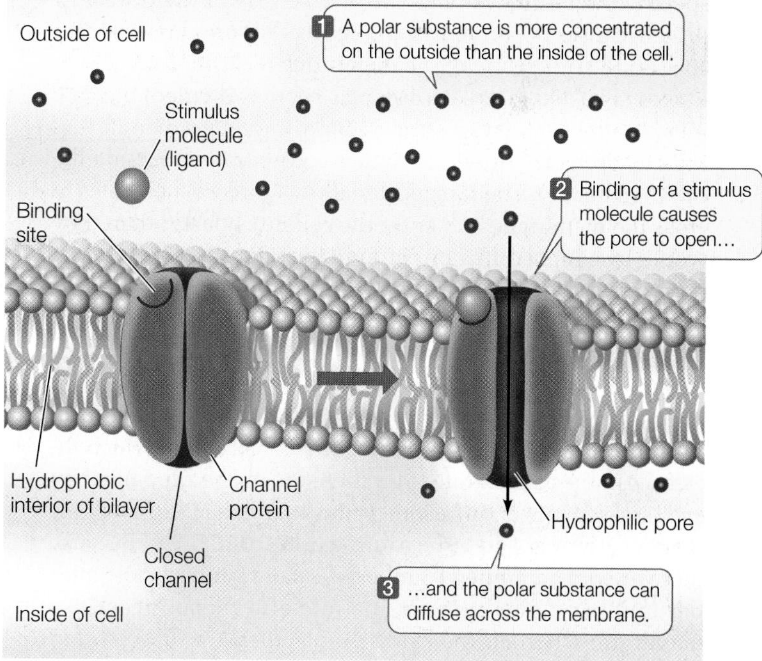
Outside of cell
1 A polar substance is more concentrated on the outside than the inside of the cell.
Stimulus molecule (ligand)
Binding site
2 Binding of a stimulus molecule causes the pore to open...
Hydrophobic interior of bilayer
Channel protein
Closed channel
Hydrophilic pore
Inside of cell
3 ...and the polar substance can diffuse across the membrane.

FIGURE 5.4 A Ligand-Gated Channel Protein Opens in Response to a Stimulus The channel protein is anchored in the lipid bilayer by the nonpolar (hydrophobic) amino acids exposed on the protein's surface. The protein changes its three-dimensional shape when a stimulus molecule (ligand) binds to it, opening a pore lined with polar amino acids. This allows hydrophilic, polar substances to pass through.

traverses a channel. But there is an even faster way for water to cross membranes. Plants and some animal cells (such as red blood cells and kidney cells) have membrane channels called **aquaporins**. These specific channels allow large amounts of water to move down its concentration gradient, as you will see when we discuss water relations in plants (see Chapter 25) and animals (see Chapter 36).

Aquaporins were first identified by Peter Agre at Duke University. He noticed a membrane protein that was present in red blood cells, kidney cells, and plant cells, all of which show rapid diffusion of water across their membranes. To test the idea that the membrane protein might be a water channel, Agre injected egg cells (oocytes) with the mRNA for the protein. The injected cells produced the protein and inserted it into their membranes. An oocyte membrane does not normally permit much diffusion of water. However, the injected oocytes began swelling immediately after being transferred to a hypotonic solution, indicating the rapid diffusion of water into the cells (**FIGURE 5.5**).

Carrier proteins aid diffusion by binding substances

Another kind of facilitated diffusion involves the actual binding of the transported substance to a membrane protein called a carrier protein. Carrier proteins transport polar molecules such as sugars and amino acids.

Glucose is the major energy source for most mammalian cells, and they require a great deal of it. Their membranes contain a carrier protein—the glucose transporter—that facilitates glucose uptake into the cell. Binding of glucose to a specific three-dimensional site on one side of the transport protein causes the protein to change its shape and release glucose on the other side of the membrane (**FIGURE 5.6A**). Since glucose is usually broken down as soon as it enters the cell, there is almost always a strong concentration gradient favoring glucose entry (that is, a higher concentration outside the cell than inside). The transporter allows glucose molecules to cross the membrane and enter the cell much faster than they would by simple diffusion through the bilayer. This rapid entry is necessary to ensure that the cell receives enough glucose for its energy needs.

Transport by carrier proteins is different from simple diffusion. In both processes, the rate of movement depends on the concentration gradient across the membrane. However, in carrier-mediated transport, a point is reached at which increases in the concentration gradient are not accompanied by an increased rate of diffusion. At this point, the facilitated diffusion system is said to be saturated (**FIGURE 5.6B**). Because there are only a limited number of carrier protein molecules per unit of membrane area, the rate of diffusion reaches a maximum when all the carrier molecules are fully loaded with solute molecules. This situation is similar to that of enzyme saturation (see Figure 3.16).

Go to ANIMATED TUTORIAL 5.2
Passive Transport
PoL2e.com/at5.2

INVESTIGATION

FIGURE 5.5 Aquaporins Increase Membrane Permeability to Water A protein was isolated from the membranes of cells in which water diffuses rapidly across the membranes. When mRNA encoding the protein was inserted into and translated in oocytes, which do not normally have the protein, the water permeability of the oocytes was greatly increased.[a]

HYPOTHESIS

Aquaporin increases membrane permeability to water.

METHOD

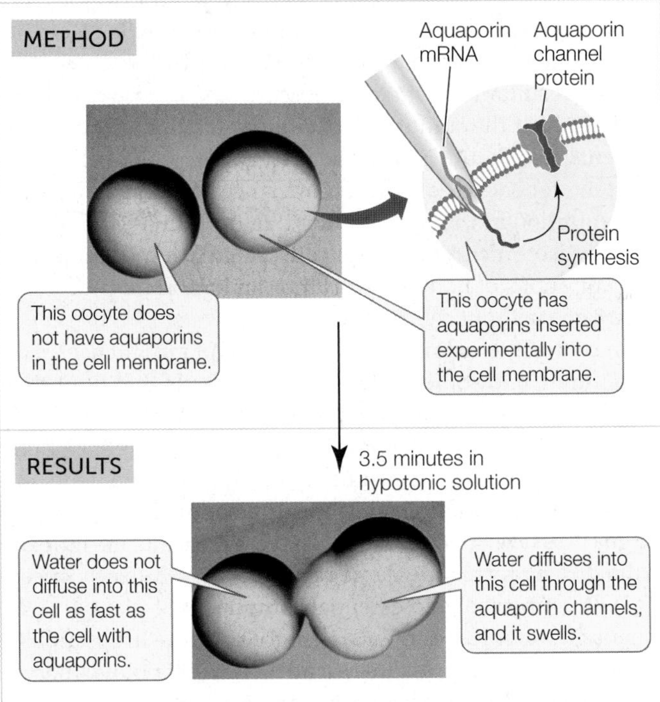

Aquaporin mRNA
Aquaporin channel protein
Protein synthesis
This oocyte does not have aquaporins in the cell membrane.
This oocyte has aquaporins inserted experimentally into the cell membrane.
RESULTS
3.5 minutes in hypotonic solution
Water does not diffuse into this cell as fast as the cell with aquaporins.
Water diffuses into this cell through the aquaporin channels, and it swells.

CONCLUSION

Aquaporin increases the rate of water diffusion across the cell membrane.

ANALYZE THE DATA

Oocytes were injected with aquaporin mRNA (red circles) or a solution without mRNA (blue circles). Water permeability was tested by incubating the oocytes in hypotonic solution and measuring cell volume. After time X in the upper curve, intact oocytes were not visible:

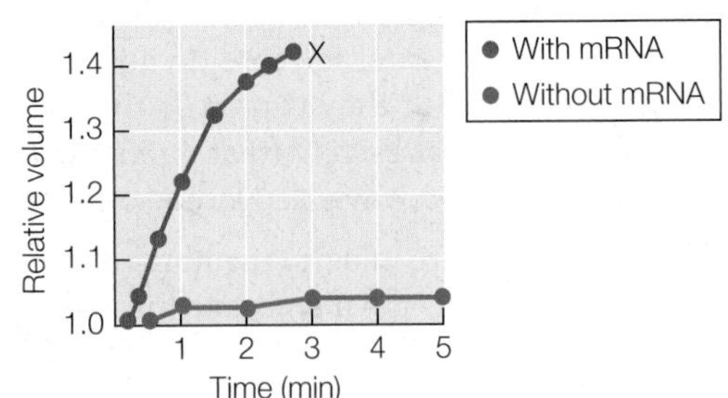

A. Why did the cells with aquaporin mRNA increase in volume?

B. What happened at time X?

C. Calculate the relative rates (volume increase per minute) of swelling in the control and experimental curves. What does this show about the effectiveness of mRNA injection?

Go to **LaunchPad** for discussion and relevant links for all **INVESTIGATION** figures.

[a]G. M. Preston et al. 1992. *Science* 256: 385–387.

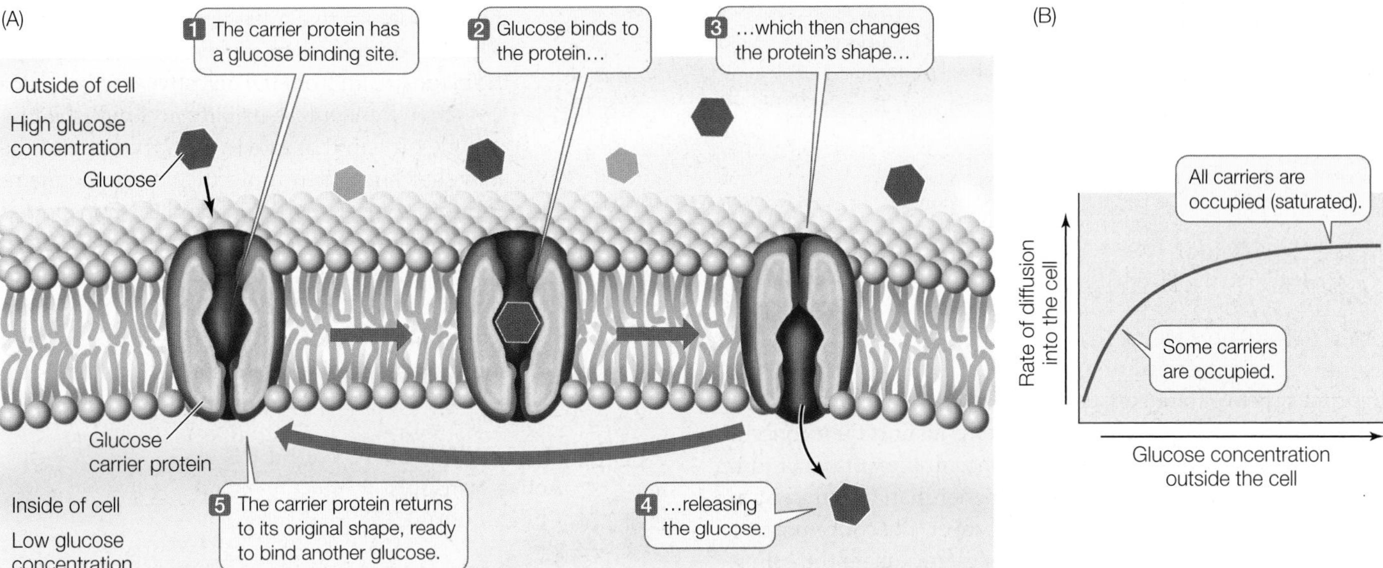
(A)
1 The carrier protein has a glucose binding site.
2 Glucose binds to the protein...
3 ...which then changes the protein's shape...
Outside of cell
High glucose concentration
Glucose
Glucose carrier protein
Inside of cell
Low glucose concentration
5 The carrier protein returns to its original shape, ready to bind another glucose.
4 ...releasing the glucose.
(B)
All carriers are occupied (saturated).
Some carriers are occupied.
Rate of diffusion into the cell
Glucose concentration outside the cell

FIGURE 5.6 A Carrier Protein Facilitates Diffusion The glucose transporter is a carrier protein that allows glucose to enter the cell at a faster rate than would be possible by simple diffusion. (A) The transporter binds to glucose, and as it does so, it changes shape, releasing the glucose into the cell cytoplasm. (B) The graph shows the rate of glucose entry via a carrier versus the concentration of glucose outside the cell. As the glucose concentration increases, the rate of diffusion increases until the point at which all the available transporters are being used (the system is saturated).

CHECKpoint CONCEPT 5.2

- ✓ What properties of a substance determine whether, and how fast, it will diffuse across a membrane?
- ✓ Compare the process of facilitated diffusion through a channel and by a carrier protein. Which might be faster, and why?
- ✓ After celery is stored in an open, dry container in the refrigerator for two days, it is wilted. However, immersing the cut stalk in water for a few hours restores the integrity of the celery. How?

Diffusion tends to equalize the concentrations of substances between the outsides and insides of cells or organelles. However, one hallmark of a living thing is that it can have an internal composition quite different from that of its environment. To achieve this, a cell must sometimes move substances *against their concentration gradients*. This process requires work—the input of energy—and is known as active transport.

CONCEPT 5.3 Active Transport Moves Solutes against Their Concentration Gradients

In many biological situations, there is a different concentration of a particular ion or small molecule inside compared with outside a cell. In these cases, the concentration imbalance is maintained by a protein in the cell membrane that moves the substance against its concentration gradient. This is called active transport, and because it is acting "against the normal flow," it requires the expenditure of energy. Often the energy source is the nucleotide adenosine triphosphate (ATP). In eukaryotes, ATP is produced in the mitochondria and plastids, and it has chemical energy stored in its terminal phosphate bond. This energy is released when ATP is converted to adenosine diphosphate (ADP) in a hydrolysis reaction that breaks the bond between the terminal phosphate and the rest of the molecule.

LINK

You will find more details about how ATP functions as an energy shuttle in cells in **Concept 6.1**

The differences between diffusion and active transport are summarized in **TABLE 5.1**. In many cases of simple and facilitated diffusion, ions or molecules can move down their concentration gradients in either direction across the cell membrane. In contrast, *active transport is directional*, and moves a substance either into or out of a cell or organelle, depending on the transport protein's function. As in facilitated diffusion, there is usually a specific carrier protein for each substance that is transported.

Different energy sources distinguish different active transport systems

There are two basic types of active transport:

- **Primary active transport** involves the direct hydrolysis of ATP, which provides the energy required for transport.
- **Secondary active transport** does not use ATP directly. Instead, its energy is supplied by an ion concentration gradient or an electrical gradient, established by primary active transport. This transport system uses the energy of ATP indirectly to set up the gradient.

TABLE 5.1 Membrane Transport Mechanisms

	Simple diffusion	Facilitated diffusion (channel or carrier protein)	Active transport
Cellular energy required?	No	No	Yes
Driving force	Concentration gradient	Concentration gradient	ATP hydrolysis (against concentration gradient)
Membrane protein required?	No	Yes	Yes
Specificity	No	Yes	Yes

In primary active transport, energy released by the hydrolysis of ATP drives the movement of specific ions against their concentration gradients. For example, the concentration of potassium ions (K^+) inside a cell is often much higher than the concentration in the fluid bathing the cell. However, the concentration of sodium ions (Na^+) is often much higher outside the cell. A protein in the cell membrane pumps Na^+ out of the cell and K^+ into the cell against these concentration gradients, ensuring that the gradients are maintained. This **sodium–potassium (Na^+–K^+) pump** is an integral membrane glycoprotein that is found in all animal cells. It breaks down a molecule of ATP to ADP and a free phosphate ion (P_i) and uses the released energy to bring two K^+ ions into the cell, and export three Na^+ ions (**FIGURE 5.7**).

In secondary active transport, the movement of a substance against its concentration gradient is accomplished using energy "regained" by letting ions move across the membrane *down* their concentration gradients. For example, once the Na^+–K^+ pump establishes a concentration gradient of sodium ions, the passive diffusion of some Na^+ back into a cell can provide energy for the secondary active transport of glucose into the cell. This occurs when glucose is absorbed into the bloodstream from the digestive tract. Secondary active transport is usually accomplished by a single protein that moves both the ion and the actively transported molecule across the membrane. In some cases, the ion and the transported molecule move in opposite directions, whereas in others they move in the same direction (as for glucose and Na^+ in the digestive tract). Secondary active transport aids in the uptake of amino acids and sugars, which are essential raw materials for cell maintenance and growth.

Go to ANIMATED TUTORIAL 5.3
Active Transport
PoL2e.com/at5.3

CHECKpoint CONCEPT 5.3

- ✓ Why is energy required for active transport?
- ✓ The drug ouabain inhibits the activity of the Na^+–K^+ pump. A nerve cell is incubated in ouabain. Make a table in which you predict what would happen to the concentrations of Na^+ and K^+ inside and outside the cell, as a result of the action of ouabain.
- ✓ How would you use experiments to distinguish between the following two ways for glucose to enter a cell: (1) facilitated diffusion via a carrier protein and (2) secondary active transport?

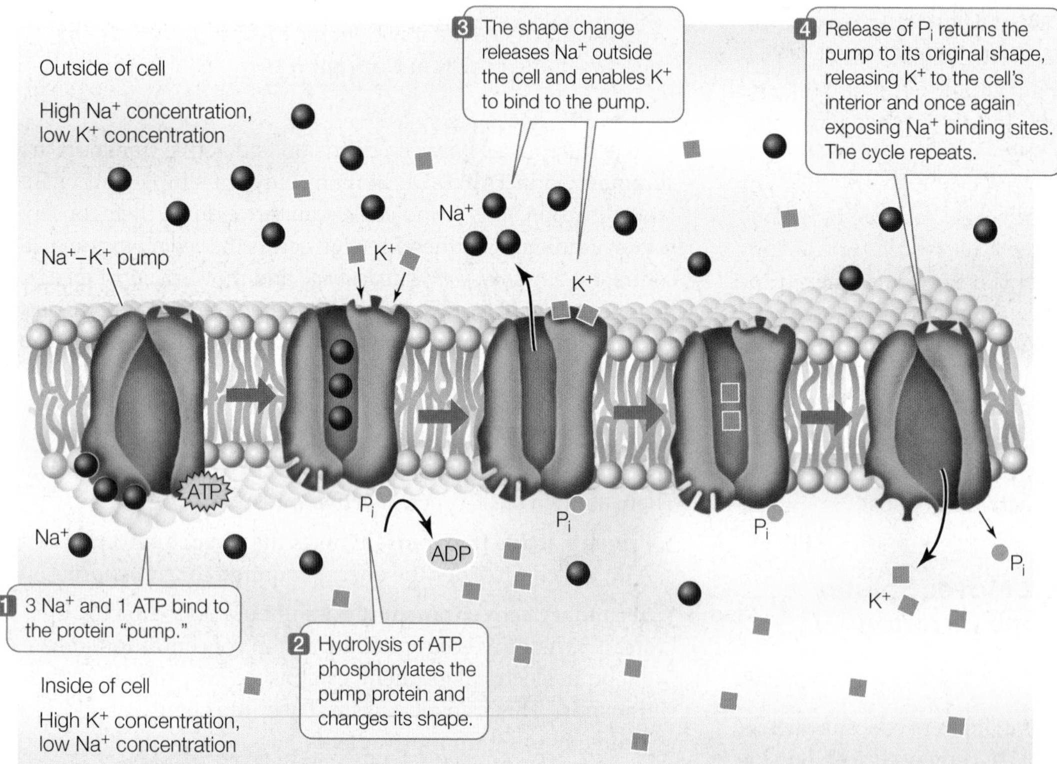

FIGURE 5.7 Primary Active Transport: The Sodium–Potassium Pump In active transport, energy is used to move a solute against its concentration gradient. Here, energy from ATP is used to move Na^+ and K^+ against their concentration gradients.

We have examined a number of passive and active ways by which ions and small molecules can enter and leave cells. But what about large molecules such as proteins? Many proteins are so large that they diffuse very slowly, and their bulk makes it difficult for them to pass through the phospholipid bilayer. It takes a completely different mechanism to move intact large molecules across membranes.

CONCEPT 5.4 Large Molecules Cross Membranes via Vesicles

Macromolecules such as proteins, polysaccharides, and nucleic acids are simply too large and too charged or polar to pass through biological membranes. This is a fortunate property—cellular integrity depends on containing these macromolecules in specific locations. However, cells must sometimes take up or **secrete** (release to the external environment) intact large molecules. This is done via vesicles, and the general terms for the mechanisms by which cells secrete and take up large molecules or particles are exocytosis and endocytosis (**FIGURE 5.8**).

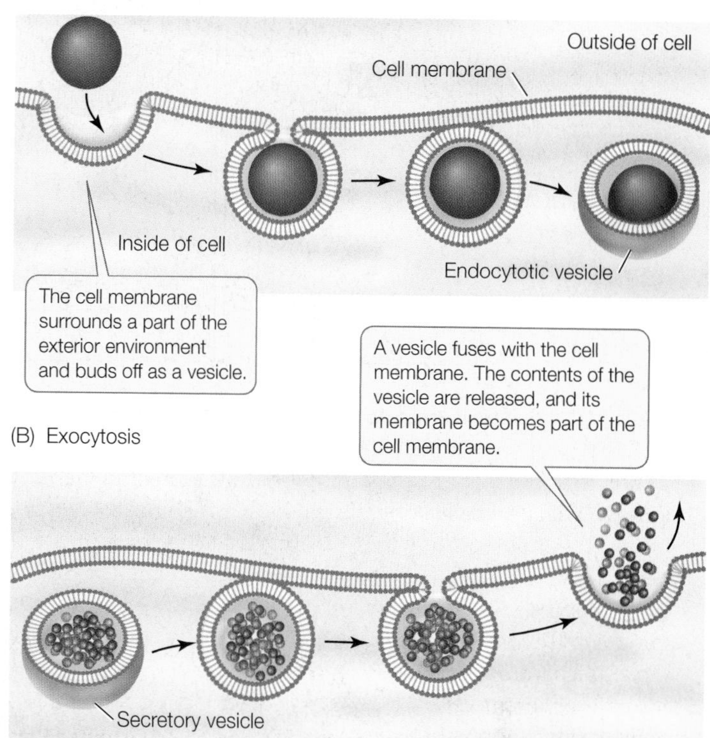
(A) Endocytosis
Outside of cell
Cell membrane
Inside of cell
Endocytotic vesicle
The cell membrane surrounds a part of the exterior environment and buds off as a vesicle.
A vesicle fuses with the cell membrane. The contents of the vesicle are released, and its membrane becomes part of the cell membrane.
(B) Exocytosis
Secretory vesicle

FIGURE 5.8 Endocytosis and Exocytosis Eukaryotic cells use endocytosis (A) and exocytosis (B) to take up and release large molecules and particles. Even small cells can be engulfed via endocytosis.

Go to MEDIA CLIP 5.1
An Amoeba Eats by Phagocytosis
PoL2e.com/mc5.1

Exocytosis moves materials out of the cell

Exocytosis is the process by which materials packaged in vesicles are secreted from the cell (see Figure 5.8B). When the vesicle membrane fuses with the cell membrane, an opening is made to the outside of the cell. The contents of the vesicle are released into the environment, and the vesicle membrane is smoothly incorporated into the cell membrane.

In Chapter 4 we encountered exocytosis as the last step in the processing of material engulfed by phagocytosis—the release of undigested materials back to the extracellular environment (see Figure 4.9). Secreted proteins are also transported out of the cell via exocytosis. The proteins are folded and modified in the endoplasmic reticulum and then transported in vesicles to the Golgi apparatus, where they may be further modified. Finally, the proteins are packaged in new vesicles for secretion (see Figure 4.8).

Exocytosis is important in the secretion of many types of substances, including digestive enzymes from the pancreas, neurotransmitters from neurons, and materials for the construction of the plant cell wall. You will encounter these processes in later chapters.

Macromolecules and particles enter the cell by endocytosis

Endocytosis is a general term for a group of processes that bring small molecules, macromolecules, large particles, and even small cells into eukaryotic cells (see Figure 5.8A). The cell membrane invaginates (folds inward), forming a small pocket around materials from the environment. The pocket deepens, forming a vesicle. This vesicle separates from the cell membrane and migrates with its contents to the cell's interior.

Endocytosis often depends on **receptors** (see Concept 5.5), which are proteins that bind to specific molecules (their ligands) and then set off specific cellular responses. In endocytosis, the receptors are integral membrane proteins located on the extracellular surface of the cell membrane. Vesicle formation results in the internalization of both the receptor and its ligand, along with other substances present near the site of invagination.

There are three broad types of endocytosis: phagocytosis, pinocytosis, and receptor endocytosis:

- In **phagocytosis** ("cellular eating"), receptors in the cell membrane recognize a specific ligand on the surface of a large particle or even an entire cell. The binding of the ligand to the receptor causes the phagocytic cell to engulf the particle or other cell. Phagocytosis is restricted to specialized cells; for example, unicellular protists use phagocytosis for feeding, and some white blood cells use phagocytosis to engulf foreign cells and substances. The food vesicle (phagosome) that forms usually fuses with a lysosome, where the vesicle's contents are digested.

LINK
Review the discussion of phagocytosis in **Concept 4.3**

- Vesicles also form in **pinocytosis** ("cellular drinking"). However, in this case the vesicles bring fluids and dissolved substances, including proteins, into the cell. Pinocytosis is relatively nonspecific regarding what it brings into the cell. For example, pinocytosis goes on constantly in

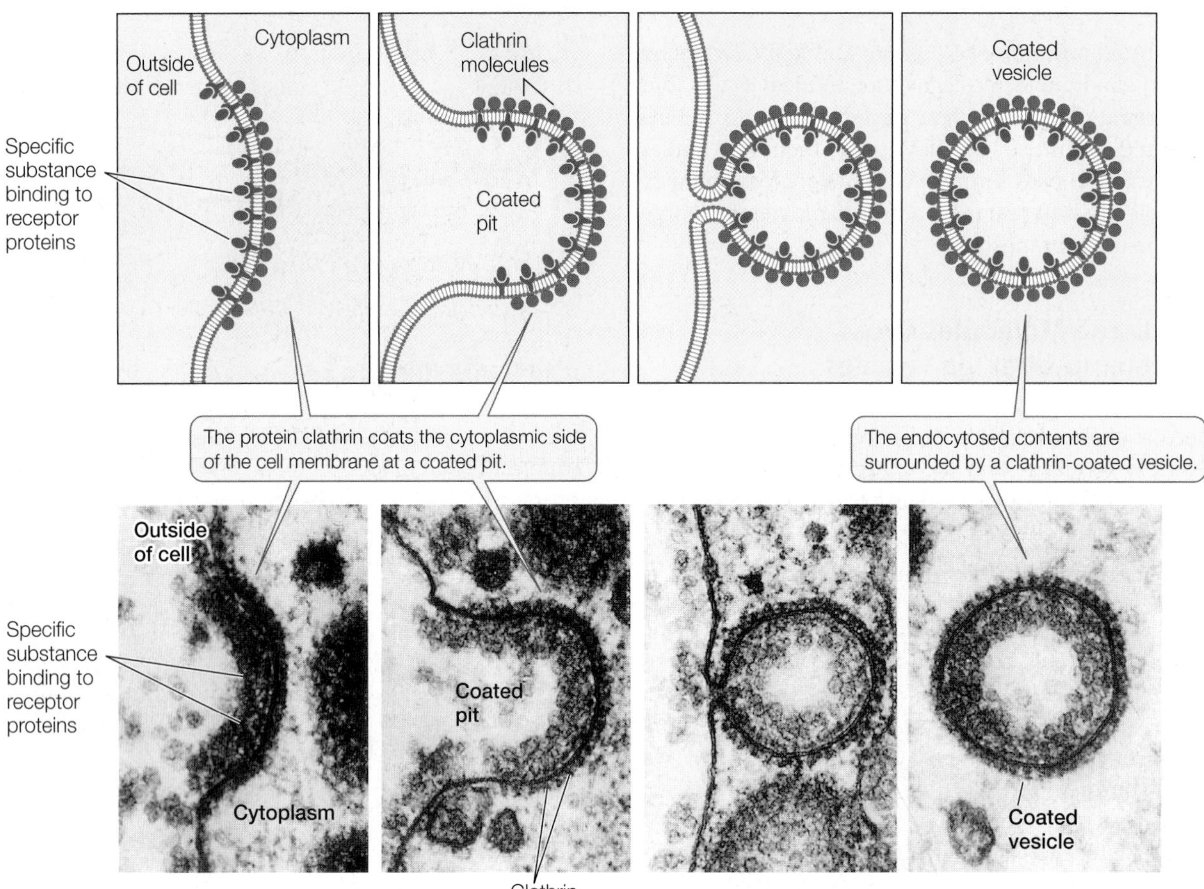

FIGURE 5.9 Receptor Endocytosis The receptor proteins in a coated pit bind specific macromolecules, which are then carried into the cell by a coated vesicle.

the endothelium—the single layer of cells that separates a blood capillary from the surrounding tissue. Pinocytosis allows cells of the endothelium to rapidly acquire fluids and dissolved solutes from the blood.

- **Receptor endocytosis** (also called receptor-mediated endocytosis) is a mechanism for bringing specific large molecules, recognized by specific receptors, into the cell. In recent years it has become clear that receptor endocytosis also plays an important role in cell signaling, which we will discuss in Concepts 5.5 and 5.6. Put simply, receptor endocytosis allows cells to control their internal processes by controlling the location and abundance of each type of receptor on the cell membrane.

Let's take a closer look at the process of receptor endocytosis.

Receptor endocytosis often involves coated vesicles

In receptor endocytosis, the receptors are often located at particular regions, called coated pits, on the extracellular surface of the cell membrane. These pits form slight depressions in the cell membrane, and their cytoplasmic surfaces are coated by another protein, often clathrin. The uptake process is similar to that in phagocytosis. The clathrin (or other protein) molecules strengthen and stabilize the vesicle (**FIGURE 5.9**).

Once inside the cell, the vesicle loses its clathrin coat and fuses with a membrane-enclosed compartment called an **endosome**, where the ligands, receptors, and other substances in the vesicle are separated and sorted. Some of these components are transferred to the lysosome for degradation, while others may be transferred back to the cell membrane. Thus a receptor may be recycled to the cell membrane or degraded in the lysosome, and as we mentioned above, this is an important mechanism by which the cell controls the abundance of each kind of receptor at its surface.

Receptor endocytosis is the way cholesterol is taken up by most mammalian cells. Cholesterol and triglycerides, which have low solubility in water, are packaged by liver cells into lipoprotein particles. Most of the cholesterol is packaged into low-density lipoproteins (LDLs) and circulated via the bloodstream. When a particular cell requires cholesterol, it produces LDL receptors, which are inserted into the cell membrane. The receptors diffuse laterally through the membrane until they become associated with clathrin-coated pits. LDLs bind to the receptors and are taken into the cell via receptor endocytosis. After separation from the receptors, the LDL particles are transferred to the lysosome, where they are broken down and the cholesterol made available for use by the cell.

APPLY THE CONCEPT

Some substances require energy to cross the membrane

The liver plays several vital metabolic roles, including protein synthesis, detoxification, and the production of substances necessary for digestion. Liver cells are in contact with the blood and exchange a variety of substances with the blood plasma (the noncellular part of blood). Below is a list of observations about the relative concentrations of various molecules in a liver cell cytoplasm and in the blood plasma. Explain each observation in terms of membrane permeability and transport mechanisms.

1. The concentration of serum albumin, a blood protein synthesized in the liver, is much higher in the plasma.
2. The concentration of RNA is much higher in the cytoplasm.
3. The concentration of Na^+ is lower in the cytoplasm.
4. The concentration of water is equal in the plasma and the cytoplasm.
5. The concentration of low-density lipoproteins is higher in the cytoplasm.
6. The concentration of glucose is equal in the plasma and the cytoplasm.
7. If K^+ enters the plasma, its concentration rapidly equalizes between the plasma and the cytoplasm.

In healthy individuals, the liver takes up unused LDLs for recycling. People with the inherited disease familial hypercholesterolemia have a defective LDL receptor in their livers. This prevents receptor endocytosis of LDLs in the liver, resulting in dangerously high levels of cholesterol in the blood. The cholesterol builds up in the arteries that nourish the heart and causes heart attacks. In extreme cases where only the defective receptor is present, children and teenagers can have severe cardiovascular disease.

Receptor endocytosis also plays an important role in cell signaling, which we will discuss in the following concepts.

Go to ANIMATED TUTORIAL 5.4
Endocytosis and Exocytosis
PoL2e.com/at5.4

CHECKpoint CONCEPT 5.4

✓ What is the difference between phagocytosis and pinocytosis?

✓ Would a small molecule such as an amino acid enter a cell by receptor endocytosis?

We have just introduced the concept of a membrane-bound receptor, which is a key factor in a cell's interaction with its environment. Let's look more closely at receptors and how they respond to signals.

CONCEPT 5.5 The Membrane Plays a Key Role in a Cell's Response to Environmental Signals

A hallmark of living cells is their ability to process information from their environments. We can think of this information in terms of **cell signaling**. In this context, the signal may be a physical stimulus such as light or heat, or a chemical such as a hormone. A chemical signal may also be referred to as a **ligand**: a molecule that binds to a receptor (see Concept 5.4). The mere presence of a signal, however, does not mean a particular cell will respond to it. In order to respond, the cell must have a specific receptor that can detect the signal. Once the signal activates its receptor, it sets off a **signal transduction pathway**, a sequence of molecular events and chemical reactions within a cell that lead to the cell's response to the signal. This ability of cells to sense and respond to signals in the environment is key to the maintenance of stable intracellular conditions, a theme that recurs throughout this book.

Cells are exposed to many signals and may have different responses

Inside a large multicellular animal, chemical signals made by the body itself reach a target cell by local diffusion or by circulation within the blood. These signals are usually in tiny concentrations (as low as 10^{-10} M) and differ in their sources and mode of delivery (**FIGURE 5.10**):

- **Autocrine signals** affect the same cell that releases them. For example, many tumor cells reproduce uncontrollably because they self-stimulate cell division by making their own division signals.
- **Paracrine signals** diffuse to and affect nearby cells. An example is a neurotransmitter made by a nerve cell that diffuses to an adjacent cell and stimulates it.
- **Juxtacrine signaling** requires direct contact between the signaling and the responding cell, and usually involves interaction between signaling molecules bound to the surfaces of the two cells.
- Signals that travel through the circulatory systems of animals or the vascular systems of plants to reach receptors on distant cells are generally called **hormones**.

Chemical signals do not always come from within the multicellular organism—some come from the external environment. For example, specific molecules produced by pathogenic (disease-causing) organisms trigger signal transduction pathways in plants, leading to defense responses.

For the information from a signal to be transmitted to a cell, the target cell must be able to sense the signal and respond to it. In a multicellular animal, all the cells may receive chemical signals that are circulated in the blood, but most body cells are not capable of responding to every signal. *Only the cells with the necessary receptors can respond.*

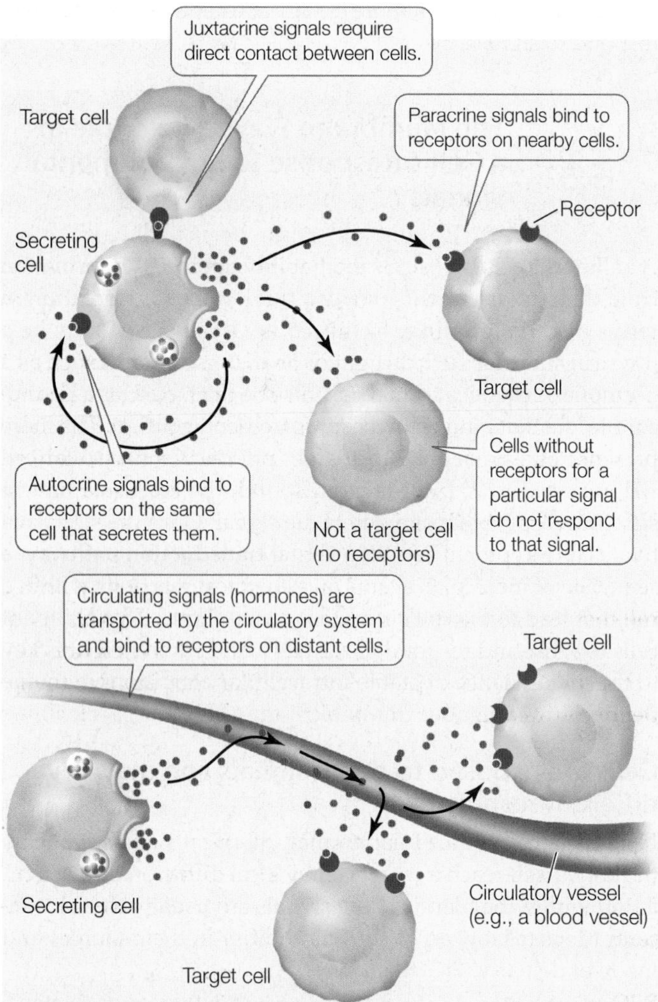

FIGURE 5.10 Chemical Signaling Concepts A signal molecule can act on the cell that produces it, on a nearby cell, or be transported by the organism's circulatory system to a distant target cell.

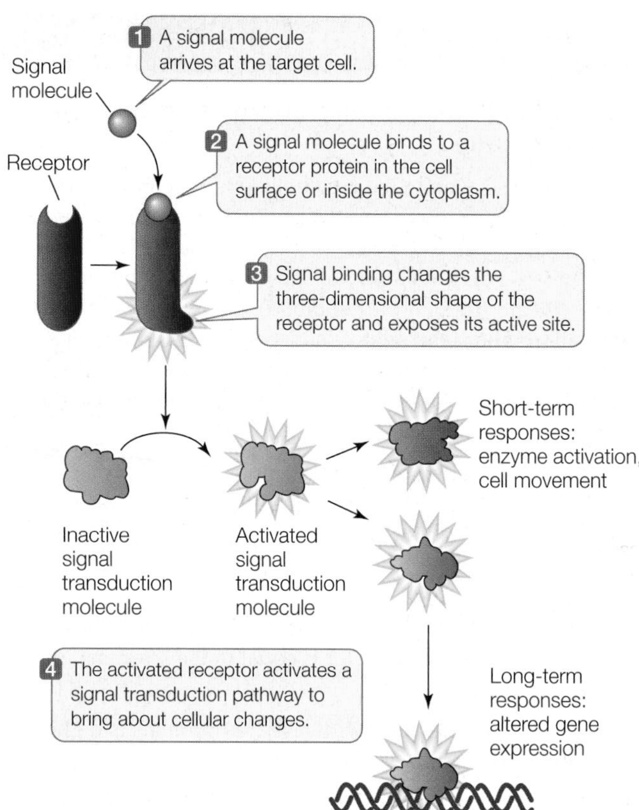

FIGURE 5.11 Signal Transduction Concepts This general pathway is common to many cells and situations. The ultimate cellular responses are either short-term or long-term.

Typically, a signal transduction pathway involves a signal, a receptor, and a response (**FIGURE 5.11**). These pathways vary in their details, but they commonly include allosteric regulation. Recall that **allosteric regulation** involves an alteration in the three-dimensional shape of a protein as a result of the binding of another molecule at a site other than the protein's active site (see Figure 3.20). You saw an example of allosteric regulation earlier in this chapter when we considered a ligand-gated channel, which opens (changes shape) after binding to another molecule (see Figure 5.4).

A signal transduction pathway may end in a response that is short-term, such as the activation of an enzyme, or long-term, such as an alteration in gene expression.

LINK

Gene expression—the transcription of specific DNA sequences and the translation of these sequences into proteins—is described in **Chapter 10**

Receptors can be classified by location and function

Chemical signals (ligands) are quite variable, but they can be divided into two groups based on whether or not they can diffuse through membranes. Physical signals such as light and sound also vary in their ability to penetrate particular cells and tissues. Accordingly, we can classify a receptor by its location in the cell, which largely depends on the nature of its ligand:

- **Intracellular receptors** are located inside the cell. Small or nonpolar ligands can diffuse across the phospholipid bilayer of the cell membrane and enter the cell. Estrogen, for example, is a lipid-soluble steroid hormone that can easily diffuse across the cell membrane; it binds to a receptor inside the cell.
- **Membrane receptors** are located on the cell surface. Large or polar ligands cannot cross the lipid bilayer. Insulin, for example, is a protein hormone that cannot diffuse through the cell membrane. Instead, it binds to a transmembrane receptor with an extracellular ligand-binding domain.

Many receptors are associated with the cell membrane

Membrane receptors are found on the surfaces of cells, and they respond to signals from outside the cell. If the signal is a

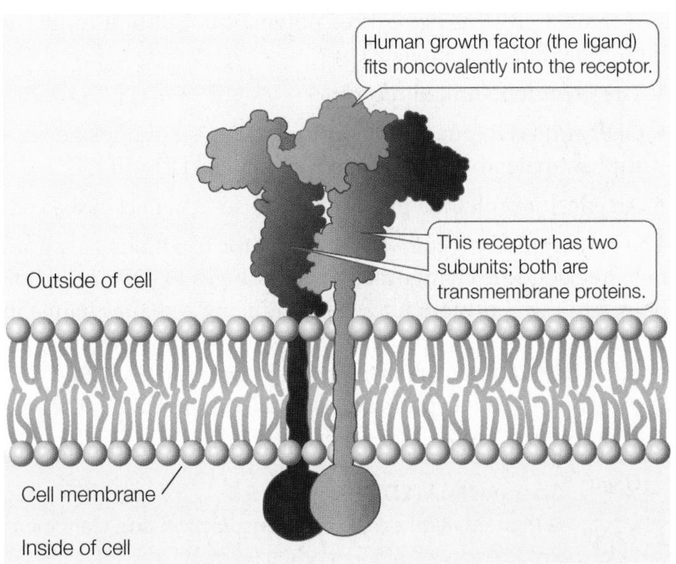

FIGURE 5.12 A Signal Binds to Its Receptor Human growth factor fits into its membrane-bound receptor (a protein with two subunits) and binds to it noncovalently.

chemical ligand, it fits into a three-dimensional site on its corresponding receptor protein (**FIGURE 5.12**). In many cases the receptor has a catalytic domain and functions as an enzyme, with its active site on the cytoplasmic side of the membrane. The ligand acts as an allosteric regulator, exposing the active site of the catalytic domain. The ligand does not contribute further to the cellular response; its role is purely to "knock on the door." (This is in sharp contrast to the enzyme–substrate interactions we described in Concept 3.3. The whole purpose of those interactions is to change substrates into useful products.)

Ligands (L) bind to their receptors (R) noncovalently and reversibly, according to chemistry's law of mass action:

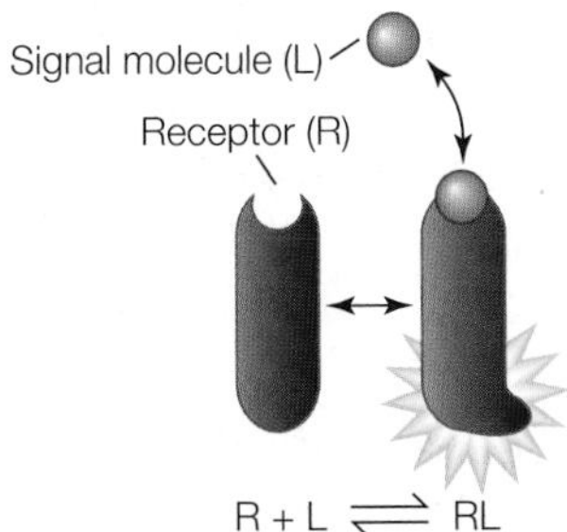

Reversibility is important because if the ligand were never released, the receptor would be continuously stimulated. In most cases, the cell needs to *stop* responding to a signal after the appropriate response has occurred. For example, if a signal transduction pathway results in the production of a particular protein, the cell needs to stop producing the protein when enough of it has been made. Nevertheless, for most ligand–receptor complexes, the equilibrium point is far to the right in the above reaction—that is, binding is favored, even at low ligand concentrations.

As noted above, receptors are removed from the cell membrane by endocytosis and either degraded inside the cell or recycled back to the membrane.

An inhibitor (or antagonist) can also bind to a receptor protein, preventing the binding of the normal ligand. This is analogous to the competitive inhibition of enzymes (see Concept 3.4). There are both natural and artificial antagonists of receptor binding. For example, many substances that alter human behavior (such as caffeine; see the opening story) bind to specific receptors in the brain and prevent the binding of the receptors' specific ligands.

In complex eukaryotes such as mammals and higher plants, there are three well-studied categories of cell membrane receptors, which are grouped according to their activities: ion channels, protein kinase receptors, and G protein–linked receptors. Because you will see these receptors several times later in this book, we describe them in some detail here.

ION CHANNEL RECEPTORS As described in Concept 5.2, the cell membranes of many cells contain ligand-gated channels for ions such as Na^+, K^+, Ca^{2+}, or Cl^- (see Figure 5.4). These proteins are receptors because their functioning depends on ligand binding. An example is the acetylcholine receptor, a ligand-gated sodium channel located in the cell membranes of skeletal muscle cells. Acetylcholine is a neurotransmitter—a chemical signal released from nerve cells. Opening of the channel allows Na^+, which is more concentrated outside the cell than inside, to diffuse into the cell. This initiates a series of events that result in muscle contraction (see Figure 34.9).

PROTEIN KINASE RECEPTORS Like ligand-gated channel receptors, protein kinase receptors change shape upon ligand binding. But in this case, the new conformation exposes or activates a catalytic domain on the cytoplasmic side of the transmembrane protein that has **protein kinase** activity—it modifies specific target proteins in the cell by adding phosphate groups to them. In general, protein kinases catalyze the following reaction:

$$\text{ATP} + \text{protein} \xrightarrow{\text{Protein kinase}} \text{ADP} + \text{phosphorylated protein}$$

This reaction results in the covalent modification (phosphorylation) of the target protein, thereby changing its activity (see Figure 3.20B). Protein kinases are extraordinarily important in biological signaling: about 1 human gene in 50 is a protein kinase gene, and there is an even higher proportion of such genes in some plants.

An example of a protein kinase receptor is the receptor for the hormone insulin (**FIGURE 5.13**). The activation of this receptor results in the phosphorylation of target proteins, which then bring about the cell's response, which includes the insertion of glucose transport proteins into the cell membrane.

It should be noted that not all protein kinases are receptors—many function in later steps of signal transduction pathways, as we will discuss in Concept 5.6.

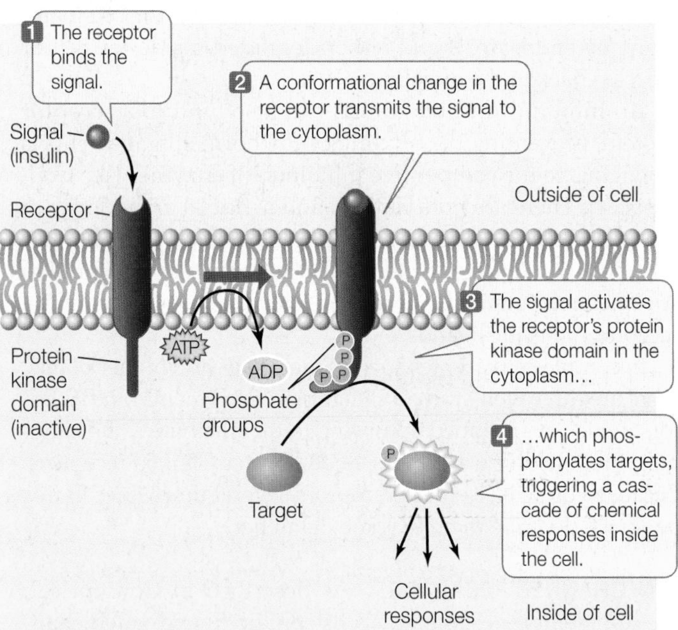

FIGURE 5.13 A Protein Kinase Receptor The mammalian hormone insulin binds to a protein kinase receptor on the outside surface of the cell and initiates a response.

G PROTEIN–LINKED RECEPTORS A third category of eukaryotic cell membrane receptors is the family of **G protein–linked receptors**. In this case, ligand binding on the extracellular domain of the receptor exposes a site on the cytoplasmic side that can bind to a mobile membrane protein called a **G protein**. The G protein is partially inserted in the lipid bilayer and partially exposed on the cytoplasmic surface of the membrane.

Many G proteins have three polypeptide subunits and can bind three different molecules:

- The G protein–linked receptor
- GDP and GTP (guanosine diphosphate and guanosine triphosphate; these are nucleotides, like ADP and ATP)
- An effector protein (a protein that causes an effect in the cell)

The activated G protein–linked receptor functions as a guanine nucleotide exchange factor. It exchanges a GDP nucleotide bound to the G protein for a GTP, inducing a shape change in the G protein. The G protein then activates the effector protein, leading to downstream signal amplification (**FIGURE 5.14**). G protein–linked receptors are especially important in the sensory systems of animals (see Concept 34.4).

Go to ANIMATED TUTORIAL 5.5
G Protein–Linked Signal Transduction and Cancer
PoL2e.com/at5.5

CHECKpoint CONCEPT 5.5

✓ Name three major steps in cell signaling that were discussed in this concept.

✓ What are the differences and similarities between ion channel receptors and G protein–linked receptors?

✓ If an intact cell is treated to remove cell surface proteins, will the cell be able to receive any environmental signals? Explain.

When a signal activates a receptor, a signal transduction pathway ensues. This often involves multiple steps, and it leads to one or more specific cellular responses. We will discuss these processes next.

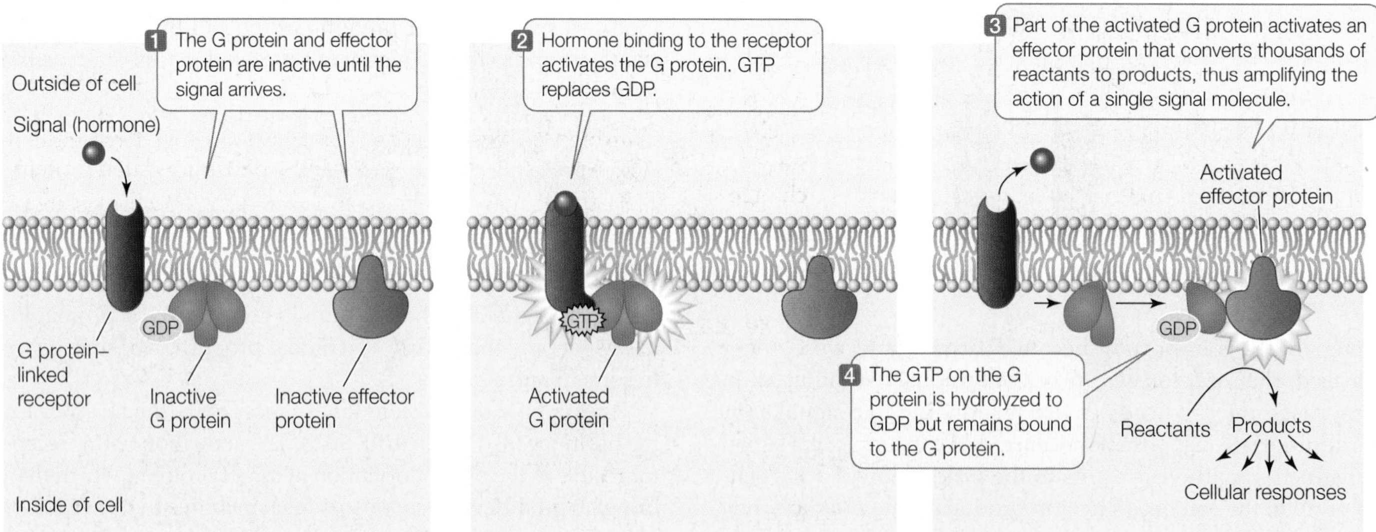

FIGURE 5.14 A G Protein–Linked Receptor The G protein is an intermediary between the receptor and an effector protein.

CONCEPT 5.6 Signal Transduction Allows the Cell to Respond to Its Environment

As we mentioned in Concept 5.5, a signal may be a chemical ligand or a physical stimulus such as light or heat. Its effect is to activate a specific receptor, leading to a cellular response that is brought about by a signal transduction pathway. Typically, signaling at the cell membrane initiates a cascade (or series) of events in the cell. Proteins interact with other proteins until the final responses are achieved. Through such a cascade, an initial signal can be both *amplified* and *distributed* to cause several different responses.

Before we discuss how signals are amplified and distributed by signal transduction pathways, let's look at some of the cellular responses that can result from cell signaling.

Cell functions change in response to environmental signals

The activation of a receptor by a signal, and the subsequent transduction and amplification of the signal, ultimately leads to changes in cell function. There are many ways in which a cell might respond, some of which we mention here:

- *Opening of ion channels* changes the balance of ion concentrations between the outside of the cell membrane and its interior (see Figure 5.4). As you will see in Chapter 34, this results in a change in the electrical potential across the membrane, with important consequences in nerve and muscle cells.
- Many signal transduction pathways lead to *alterations in gene expression*. The expression of some genes may be switched on (upregulated), whereas others may be switched off (downregulated). This alters the abundance of the proteins (often enzymes) encoded by the genes, thus changing cell function. You will see many examples that highlight the importance of gene regulation throughout this book.
- A third kind of response involves the *alteration of enzyme activities*. An example is the activation of specific enzymes in liver cells exposed to the hormone epinephrine, which we discuss below. An alteration in enzyme activity is a much more rapid response than one that involves a change in gene expression.

LINK

The different types of enzyme regulation are discussed in **Concept 3.4**

The same signal can lead to different responses in different types of cells. For example, in heart muscle cells, the hormone epinephrine *activates* a signal transduction cascade that results in glucose mobilization for energy and muscle contraction. However, in the smooth muscle cells that line the digestive tract, epinephrine stimulates a pathway that *inhibits* a target enzyme, allowing the muscle cells to relax. This increases the diameter of the blood vessels, allowing more nutrients to be carried from the digestive system to the rest of the body. Heart and digestive tract muscle cells respond differently to the same signal—epinephrine—because the signal transduction pathways stimulated by epinephrine are different in the different cell types. Let's take a closer look at the mechanism by which cells amplify and transduce signals to bring about these responses.

Second messengers can stimulate signal transduction

Often there is a small molecule intermediary between the activated receptor and the cascade of events that ensues. In a series of clever experiments, Earl Sutherland and his colleagues at Case Western Reserve University discovered that a small, water-soluble chemical could mediate cytoplasmic events initiated by a cell membrane receptor. The researchers were investigating the activation of the liver enzyme glycogen phosphorylase by the hormone epinephrine (also called adrenaline)—the "fight-or-flight" hormone (see Concept 35.2). The enzyme is activated when an animal faces life-threatening conditions and needs energy fast for the fight-or-flight response. Glycogen phosphorylase catalyzes the breakdown of glycogen stored in the liver so that the resulting glucose molecules can be released to the blood (see Figure 30.16). The enzyme is present in the liver cell cytoplasm but is inactive in the absence of epinephrine.

The researchers found that epinephrine could activate glycogen phosphorylase in liver cells that had been broken open, but only if the entire cell contents, including cell membrane fragments, were present. Under these conditions epinephrine was bound to the cell membrane fragments, but the active phosphorylase was in the solution. The researchers hypothesized that there must be a second "messenger" that transmits the epinephrine signal (the "first messenger") from the cell membrane to the phosphorylase in the cytoplasm. They investigated this by separating cell membrane fragments from the cytoplasmic fractions of broken liver cells and following the sequence of steps described in **FIGURE 5.15**. This experiment confirmed the existence of a second messenger, later identified as **cyclic AMP** (**cAMP**; **FIGURE 5.16**).

A second messenger is a small molecule that brings about later steps in a signal transduction pathway. Second messengers do not have enzymatic activity themselves; rather, they act to regulate target enzymes by binding to them noncovalently. Whereas receptor binding is highly specific, second messengers allow a cell to respond to a single event at the cell membrane with *many events inside the cell*—in other words, the second messenger *distributes* the initial signal. Second messengers also serve to *amplify* the signal—for example, the binding of a single epinephrine molecule leads to the production of many molecules of cAMP. In turn, cAMP activates many enzyme targets by binding to them noncovalently. In the case of epinephrine and the liver cell, glycogen phosphorylase is just one of several enzymes that are activated.

INVESTIGATION

FIGURE 5.15 The Discovery of a Second Messenger Glycogen phosphorylase is activated in liver cells after epinephrine binds to a membrane receptor. Sutherland and his colleagues observed that this activation could occur in a test tube only if fragments of the cell membrane were present. They designed experiments to show that a second messenger caused the activation of glycogen phosphorylase.[a]

HYPOTHESIS

A second messenger mediates between receptor activation at the cell membrane and enzyme activation in the cytoplasm.

METHOD

Liver
1 Liver tissue is homogenized and separated into cell membrane and cytoplasm fractions.
Cytoplasm contains inactive glycogen phosphorylase
Membranes contain epinephrine receptors
2 The hormone epinephrine is added to the membranes and allowed to incubate along with the substrate for synthesis of a second messenger.
3 The membranes are removed by centrifugation, leaving only the solution in which they were incubated.
4 Drops of membrane-free solution are added to the cytoplasm.

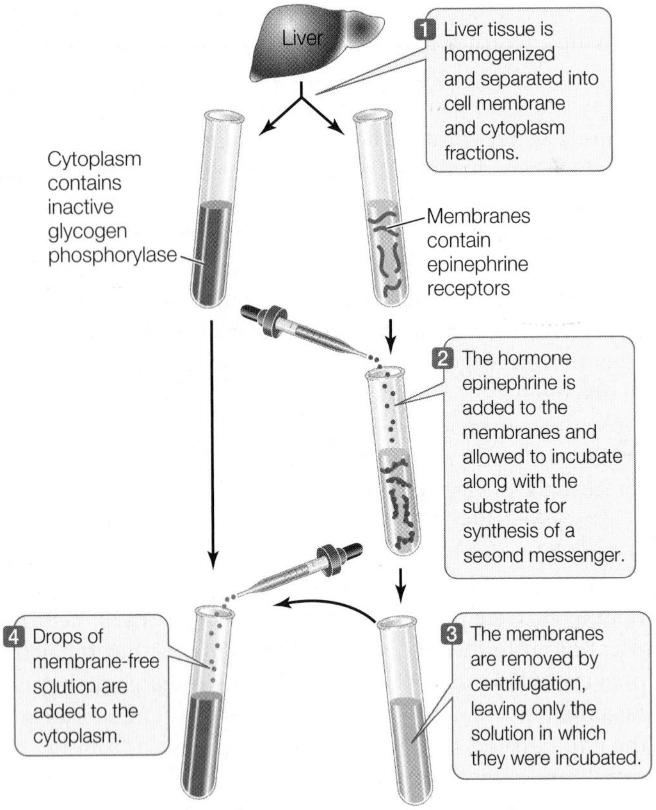

RESULTS

Active glycogen phosphorylase is present in the cytoplasm.

CONCLUSION

A soluble second messenger, produced by hormone-activated membranes, is present in the solution and activates enzymes in the cytoplasm.

ANALYZE THE DATA

The experiment was repeated under various conditions with the following results:

Condition	Enzyme activity (units)
Homogenate	0.4
Homogenate + epinephrine	2.5
Cytoplasm fraction	0.2
Cytoplasm + epinephrine	0.4
Membranes + epinephrine	0.4
Cytoplasm + membranes + epinephrine	2.0

A. What do these data show?

B. Propose an experiment to show that the factor that activates the enzyme is stable on heating (and therefore probably not a protein) and give predicted data.

C. Propose an experiment to show that cAMP can replace the membrane fraction and hormone treatment and give predicted data.

Go to **LaunchPad** for discussion and relevant links for all **INVESTIGATION** figures.

[a]T. W. Rall et al. 1957. *Journal of Biological Chemistry* 224: 463–475.

A signaling cascade involves enzyme regulation and signal amplification

Signal transduction pathways often involve multiple sequential steps, in which particular enzymes are either activated or inhibited by other enzymes in the pathway. For example, a protein kinase adds a phosphate group to a target protein, and this covalent change alters the protein's conformation and activates or inhibits its function. Cyclic AMP binds noncovalently to a target protein, and this changes the protein's shape, activating or inhibiting its function. In the case of activation, a previously inaccessible active site is exposed, and the target protein goes on to perform a new cellular role.

A good example of a signaling cascade is the G protein–mediated protein kinase pathway stimulated by epinephrine in liver cells (**FIGURE 5.17**). Binding of epinephrine to the membrane receptor results in the activation of a G protein, followed by the production of cAMP, which activates a key signaling molecule, the enzyme protein kinase A. In turn, protein kinase A phosphorylates two other enzymes, with opposite effects:

- *Inhibition*. Glycogen synthase, which catalyzes the joining of glucose molecules to form the energy-storing molecule glycogen, is inactivated when a phosphate group is added

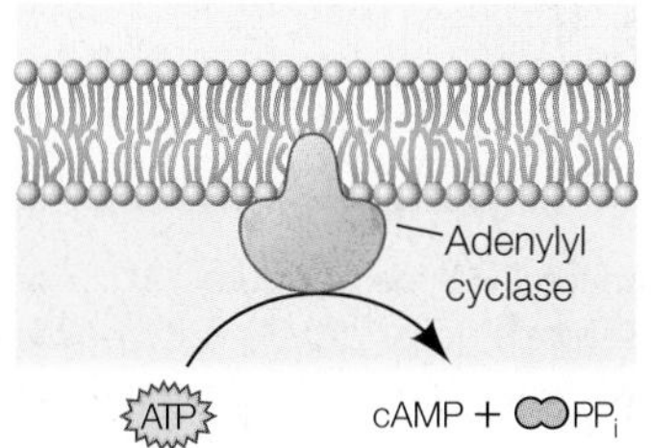

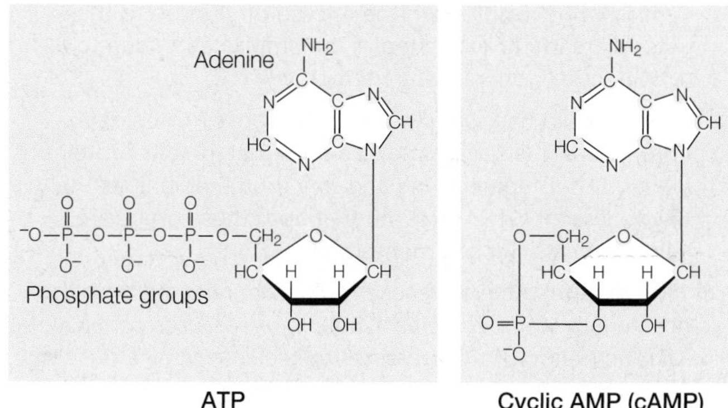

FIGURE 5.16 The Formation of Cyclic AMP The formation of cAMP from ATP is catalyzed by adenylyl cyclase, an enzyme that is activated by G proteins.

1 Epinephrine

Outside of cell

Epinephrine receptor

Activated G protein subunit

Cell membrane

1 GTP

Activated adenylyl cyclase

ATP

20 cAMP

1 Phosphorylation, induced by epinephrine binding, *inactivates* glycogen synthase, preventing glucose from being stored as glycogen.

Inactive protein kinase A

20 Active protein kinase A

Active glycogen synthase

Inactive glycogen synthase

Inactive phosphorylase kinase

2 The protein kinase cascade amplifies the signal. Here, for every molecule of epinephrine bound, 20 molecules of cAMP are made, each of which activates a molecule of protein kinase A.

100 Active phosphorylase kinase

Inactive glycogen phosphorylase

1,000 Active glycogen phosphorylase

Glycogen

10,000 Glucose 1-phosphate

3 Phosphorylation *activates* glycogen phosphorylase, releasing stored glucose molecules from glycogen.

Glucose

Inside of cell

10,000 Blood glucose

4 Release of glucose by carrier transport fuels "fight-or-flight" response.

Outside of cell

to it by protein kinase A. Thus the epinephrine signal *prevents glucose from being stored* in glycogen (see Figure 5.17, step 1).

- *Activation*. Phosphorylase kinase is activated when a phosphate group is added to it. It is part of a cascade of reactions that ultimately leads to the activation of glycogen phosphorylase, another key enzyme in glucose metabolism. This enzyme results in the liberation of glucose molecules from glycogen (see Figure 5.17, steps 2 and 3).

An important consequence of having multiple steps in a signal transduction cascade is that the signal is amplified with each step. The amplification of the signal in the pathway illustrated in Figure 5.17 is impressive. Each molecule of epinephrine that arrives at the cell membrane ultimately results in 10,000 molecules of blood glucose:

1	molecule of epinephrine bound to the membrane activates
1	molecule of adenylyl cyclase, which produces
20	molecules of cAMP, which activate
20	molecules of protein kinase A, which activate
100	molecules of phosphorylase kinase, which activate
1,000	molecules of glycogen phosphorylase, which produce
10,000	molecules of glucose 1-phosphate, which produce
10,000	molecules of blood glucose

Signal transduction is highly regulated

Signal transduction is a temporary event in the cell, and gets "turned off" once the cell has responded. We have already discussed the turnover of cell surface receptors by endocytosis. In addition, there are enzymes that convert

FIGURE 5.17 A Cascade of Reactions Leads to Altered Enzyme Activity Liver cells respond to epinephrine by activating G proteins, which in turn activate the synthesis of the second messenger cAMP. Cyclic AMP initiates a protein kinase cascade, greatly amplifying the epinephrine signal, as indicated by the blue numbers. The cascade both inhibits the conversion of glucose to glycogen and stimulates the release of previously stored glucose.

Go to ANIMATED TUTORIAL 5.6
Signal Transduction Pathway
PoL2e.com/at5.6

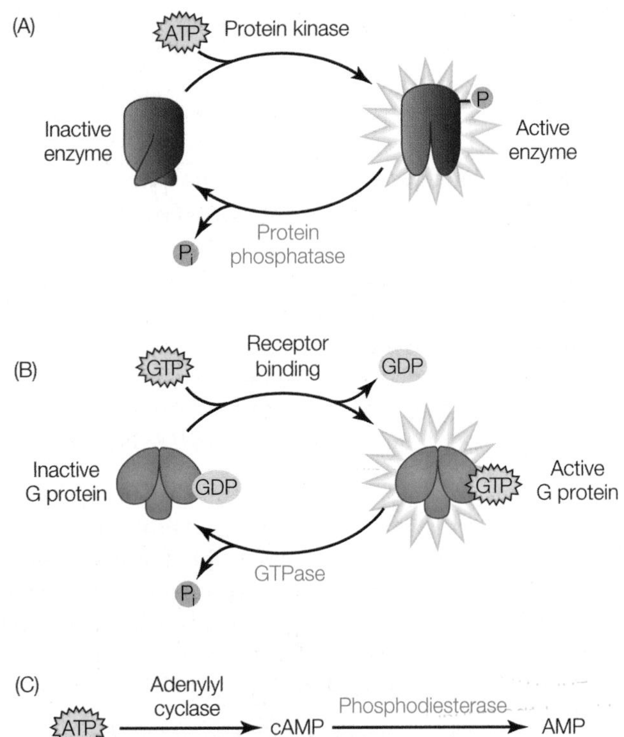

FIGURE 5.18 Signal Transduction Regulatory Mechanisms Some signals lead to the production of active signal transduction molecules such as (A) protein kinases, (B) G proteins, and (C) cAMP. Other enzymes (red type) inactivate or remove these active molecules.

signal transduction molecules back to their inactive precursors. For example, protein phosphatases remove phosphate groups from target proteins, thus reversing the effects of protein kinases (**FIGURE 5.18A**). G proteins have GTPase activity, which removes a phosphate group from GTP, converting it to GDP (**FIGURE 5.18B**). Cyclic AMP is converted back to AMP by the enzyme phosphodiesterase (**FIGURE 5.18C**). The balance between the activities of these regulating enzymes and the signaling enzymes themselves is what determines the ultimate cellular response to a signal. Cells can alter this balance in several ways, including:

- Synthesis or breakdown of the enzymes involved
- Activation or inhibition of the enzymes by other molecules (see Concept 3.4)

A great deal has been learned about signal transduction pathways and cellular responses in the past two decades, and there is still much to learn. As biologists tease apart specific pathways, they find that many of them are interconnected: one pathway may be switched on by a particular signal or molecule, and another may be switched off. In this chapter we have concentrated on signaling pathways that occur in animal cells. However, signal transduction pathways are important in the functioning of all living organisms.

CHECKpoint CONCEPT 5.6

- ✓ Compare "first messengers" (e.g., hormones) with "second messengers" (e.g., cAMP) with regard to their chemical nature, where and when they are made, and their activity.
- ✓ Outline the steps in the amplification of signaling by epinephrine, resulting in the release of glucose to the bloodstream. At each step, is the amplification due to a covalent or noncovalent interaction?
- ✓ What would happen to a liver cell exposed to epinephrine and at the same time to a drug that inhibits protein kinase A? To epinephrine and to a drug that inhibits the hydrolysis of GTP? (Assume that both these drugs are able to cross the cell membrane.)
- ✓ The disease cholera is caused by a toxin released from the bacterium *Vibrio cholerae*. Cholera toxin causes continuous activation of a G protein at the cell membrane of cells lining the intestine. This in turn results in continuous activation of adenylyl cyclase. As a result, there is continuous release of Na^+ from the intestine, followed by massive outflow of water, resulting in severe diarrhea, dehydration, and if untreated, death. How does cholera toxin work on the second messenger system, and what is the normal role of that second messenger in the intestine cell membrane?

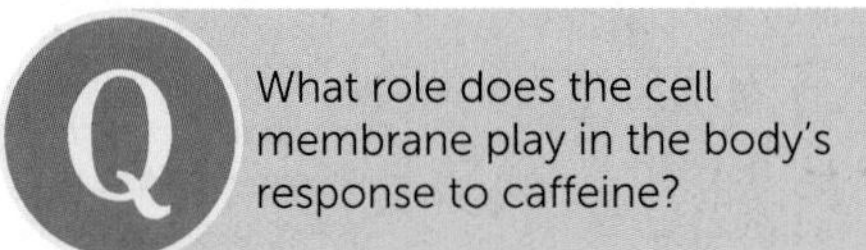

Q
What role does the cell membrane play in the body's response to caffeine?

ANSWER Caffeine has many effects on the body, but the most noticeable is that it keeps us awake. The caffeine molecule is somewhat large and polar, and it is unlikely to diffuse through the nonpolar lipids of the cell membrane (Concept 5.2). Instead, it binds to receptors on the surfaces of nerve cells in the brain (Concept 5.5).

The nucleoside adenosine (adenine attached to a five-carbon sugar) accumulates in the brain when a person is under stress or has prolonged mental activity. When it binds to a specific receptor in the brain, adenosine sets in motion a signal transduction pathway (Concept 5.6) that results in reduced brain activity, which usually means drowsiness. This membrane-associated signaling by adenosine has evolved as a protective mechanism against the adverse effects of stress.

Caffeine has a three-dimensional structure similar to that of adenosine and is able to bind to the adenosine receptor (**FIGURE 5.19**). Because its binding does not activate the receptor, caffeine functions as an antagonist of adenosine signaling, with the result that the brain stays active and the person remains alert.

When we discussed the interaction between a ligand and its receptor, we noted that this is a reversible, noncovalent interaction. In time, after drinking coffee or tea, the caffeine molecules come off the adenosine receptors in the brain,

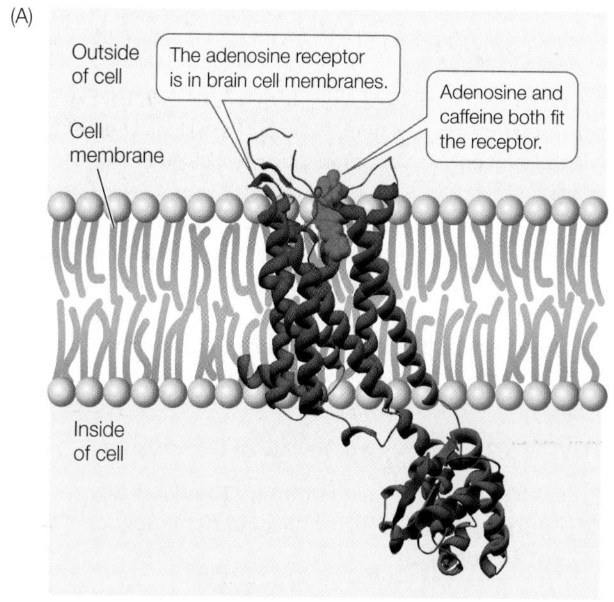
(A)
Outside of cell
The adenosine receptor is in brain cell membranes.
Adenosine and caffeine both fit the receptor.
Cell membrane
Inside of cell

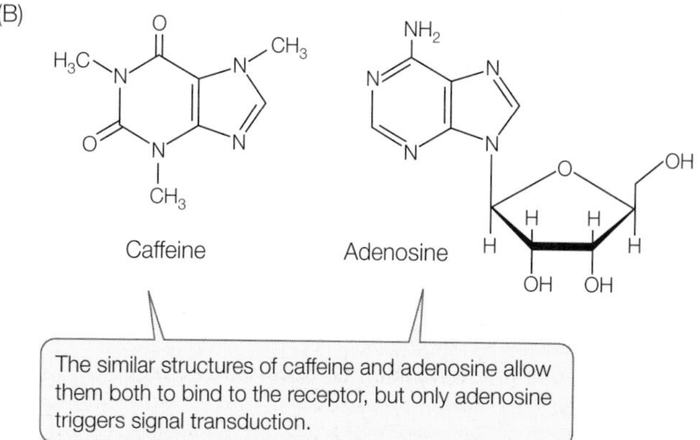
(B)
Caffeine
Adenosine
The similar structures of caffeine and adenosine allow them both to bind to the receptor, but only adenosine triggers signal transduction.

FIGURE 5.19 Caffeine and the Cell Membrane (A) The adenosine 2A receptor is present in the human brain, where it is involved in inhibiting arousal. (B) Adenosine is the normal ligand for the receptor. Caffeine has a structure similar to that of adenosine and can act as an antagonist that binds the receptor and prevents its normal functioning.

allowing adenosine to bind once again. Otherwise, coffee drinkers might never get to sleep!

In addition to competing with adenosine for a membrane receptor, caffeine blocks the enzyme cAMP phosphodiesterase. This enzyme acts in signal transduction (Concept 5.6) to break down the second messenger cAMP. Looking at the signal transduction pathway in Figure 5.17, can you explain how caffeine augments the fight-or-flight response, which includes an increase in blood sugar and increased heartbeat?

SUMMARY

CONCEPT 5.1 Biological Membranes Have a Common Structure and Are Fluid

- Biological membranes consist of lipids, proteins, and carbohydrates. The **fluid mosaic model** of membrane structure describes a **phospholipid** bilayer in which proteins can move about within the plane of the membrane.
- The two layers of a membrane may have different properties because of their different phospholipid compositions, exposed domains of **integral membrane proteins**, and **peripheral membrane proteins**. **Transmembrane proteins** span the membrane. **Review Figure 5.1, ACTIVITY 5.1, and ANIMATED TUTORIAL 5.1**

CONCEPT 5.2 Passive Transport across Membranes Requires No Input of Energy

- Membranes exhibit **selective permeability** that regulates which substances can pass through them.
- A substance can diffuse passively across a membrane by one of two processes: **simple diffusion** through the phospholipid bilayer or **facilitated diffusion**, either through a channel created by a **channel protein** or by means of a **carrier protein**. In both cases, molecules diffuse down their concentration gradients. **Review Figure 5.4 and ANIMATED TUTORIAL 5.2**
- In **osmosis**, water diffuses from a region of higher water concentration to a region of lower water concentration, largely through membrane channels called **aquaporins**. Ions diffuse across membranes through **ion channels**. **Review Figures 5.3 and 5.5**
- Carrier proteins bind to polar molecules such as sugars and amino acids and transport them across the membrane. **Review Figure 5.6**

CONCEPT 5.3 Active Transport Moves Solutes against Their Concentration Gradients

- **Active transport** requires the use of chemical energy to move substances across membranes against their concentration gradients. The **sodium–potassium (Na^+–K^+) pump** uses energy released from the hydrolysis of ATP. **Review Figure 5.7 and ANIMATED TUTORIAL 5.3**

CONCEPT 5.4 Large Molecules Cross Membranes via Vesicles

- **Endocytosis** is the transport of molecules, large particles, and small cells into eukaryotic cells via the invagination of the cell membrane and the formation of vesicles. **Review Figure 5.8A**
- In **receptor endocytosis**, a specific receptor on the cell membrane binds to a particular macromolecule that is to be transported into the cell. **Review Figure 5.9 and ANIMATED TUTORIAL 5.4**
- In **exocytosis**, materials in vesicles are secreted from the cell when the vesicles fuse with the cell membrane. **Review Figure 5.8B**

CONCEPT 5.5 The Membrane Plays a Key Role in a Cell's Response to Environmental Signals

- Cells receive many signals from the physical environment and from other cells. Chemical signals are often at very low concentrations. **Review Figure 5.10**
- A **signal transduction pathway** involves the interaction of a signal (often a chemical **ligand**) with a receptor; the transduction and amplification of the signal via a series of steps within the cell; and a cellular response. The response may be short-term or long-term. **Review Figure 5.11**

(continued)

SUMMARY (continued)

- Cells respond to signals only if they have specific receptor proteins that can be activated by those signals. Many receptors are located at the cell membrane. They include ion channels, **protein kinases**, and **G protein–linked receptors**. **Review Figures 5.13 and 5.14 and ANIMATED TUTORIAL 5.5**

CONCEPT 5.6 Signal Transduction Allows the Cell to Respond to Its Environment

- A cascade of events, one following another, occurs after a receptor is activated by a signal.
- Often, a soluble second messenger conveys signaling information from the primary messenger (ligand) at the membrane to downstream signaling molecules in the cytoplasm. **Cyclic AMP** (**cAMP**) is an important second messenger. **Review Figure 5.16**
- Activated enzymes may in turn activate other enzymes in a signal transduction pathway, leading to impressive amplification of a signal. **Review Figure 5.17 and ANIMATED TUTORIAL 5.6**
- Protein kinases covalently add phosphate groups to target proteins; cAMP binds target proteins noncovalently. Both kinds of binding change the target protein's conformation to expose or hide its active site.
- Signal transduction can be regulated in several ways. The balance between the activation and inactivation of the molecules involved determines the ultimate cellular response to a signal. **Review Figure 5.18**
- The cellular responses to signals may include the opening of ion channels, changes in gene expression, or the alteration of enzyme activities.

See **ACTIVITY 5.2** for a concept review of this chapter.

Go to the Interactive Summary to review key figures, Animated Tutorials, and Activities PoL2e.com/is5

Go to LaunchPad at **macmillanhighered.com/launchpad** for additional resources, including LearningCurve Quizzes, Flashcards, and many other study and review resources.

6 Pathways that Harvest and Store Chemical Energy

KEY CONCEPTS

6.1 ATP and Reduced Coenzymes Play Important Roles in Biological Energy Metabolism

6.2 Carbohydrate Catabolism in the Presence of Oxygen Releases a Large Amount of Energy

6.3 Carbohydrate Catabolism in the Absence of Oxygen Releases a Small Amount of Energy

6.4 Catabolic and Anabolic Pathways Are Integrated

6.5 During Photosynthesis, Light Energy Is Converted to Chemical Energy

6.6 Photosynthetic Organisms Use Chemical Energy to Convert CO_2 to Carbohydrates

An Old Brew Carvings from more than 4,000 years ago in ancient Egypt show barley being crushed and mixed with water (left), then put into closed vessels (center) where airless conditions are suitable for the production of alcohol by yeast cells residing on the vessels' walls. The beer is then ready for consumption (right).

Agriculture was a key invention in the development of human civilizations. The planting and harvesting of seeds began about 10,000 years ago. One of the first plants to be turned into a reliable crop was barley, and one of the first uses of barley was to brew beer. Living in what is now Iraq, ancient Sumerians learned that partly germinated and then mashed-up barley seeds, stored under the right conditions, could produce a potent and pleasant drink. An ancient king, Hammurabi, laid down the oldest known laws regarding an alcoholic beverage: the daily beer ration was 2 liters for a normal worker, 3 liters for a civil servant, and 5 liters for a high priest. Alcoholic beverages were not just a diversion to these people; their health depended on them. Drinking water from rivers and ponds caused diseases, and whatever caused these diseases was not present in liquids containing alcohol.

Early chemists and biologists were interested in how mashed barley seeds (or grapes, in the case of wine) were transformed into alcoholic beverages. By the nineteenth century there were two theories. Chemists claimed that these transformations were simply chemical reactions, not some special property of the plant material. Biologists, armed with their microscopes and cell theory (see Chapter 4), said that the barley and grape extracts were converted to beer and wine by living cells.

The great French scientist Louis Pasteur tackled the question in the 1860s, responding to a challenge posed by a group of distillers who wanted to use sugar beets to produce alcohol. Pasteur found that (1) nothing happened to beet mash unless microscopic yeast cells were present; (2) in the presence of fresh air, yeast cells grew vigorously on the mash, and bubbles of CO_2 were formed; and (3) without fresh air, the yeast grew slowly, less CO_2 was produced, and alcohol was formed. So the biologists were right: living cells produced alcohol from ground-up, sugary extracts. Later, biochemists broke open yeast cells and unraveled the sequence of chemical transformations from sugar to alcohol. It turned out that the chemists were right too: the production of alcohol involves a series of chemical reactions that require energy transfers. The flow of energy in living systems (such as yeast cells) involves the same chemical principles as energy flow in the inanimate world.

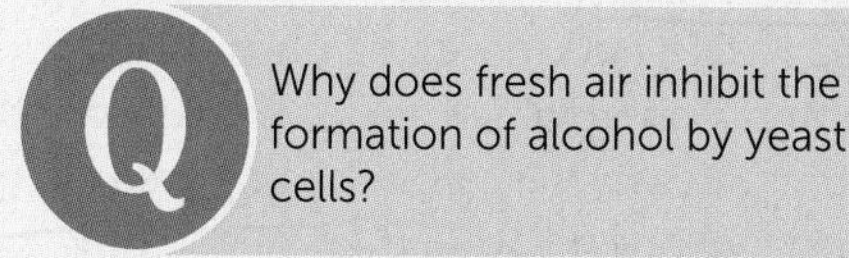

Q Why does fresh air inhibit the formation of alcohol by yeast cells?

You will find the answer to this question on page 126.

CONCEPT 6.1

ATP and Reduced Coenzymes Play Important Roles in Biological Energy Metabolism

In Chapters 2 and 3 we introduced the general concepts of energy, enzymes, and metabolism. Energy is stored in the chemical bonds of molecules, and it can be released and transformed by the metabolic pathways of living cells. There are five general principles governing metabolic pathways:

- A complex chemical transformation occurs in a series of separate, intermediate reactions that form a metabolic pathway.
- Each reaction is catalyzed by a specific enzyme.
- Most metabolic pathways are similar in all organisms, from bacteria to plants to humans.
- In eukaryotes, many metabolic pathways are compartmentalized, with certain reactions occurring inside specific organelles.
- Each metabolic pathway is controlled by key enzymes that can be inhibited or activated, thereby determining how fast the reactions will go.

Chemical energy available to do work is termed free energy (G). According to the laws of thermodynamics, a biochemical reaction may change the *form* of energy but not the net *amount*. A biochemical reaction is exergonic if it releases energy from the reactants, or endergonic if energy must be added to the reactants.

LINK

You can review the principles of energy transformations in **Concept 2.5**

In the chemistry lab, energy can be released or added in the form of heat. But in cells, energy-transforming reactions are often coupled; that is, an energy-releasing (exergonic) reaction is coupled in time and location to an energy-requiring (endergonic) reaction. Two widely used coupling molecules are the coenzymes ATP and NADH.

ATP hydrolysis releases energy

Cells use adenosine triphosphate (ATP) as a kind of "energy currency." Just as it is more effective, efficient, and convenient for you to trade money for a lunch than to trade your actual labor, it is useful for cells to have a currency for transferring energy between different reactions and cell processes. Some of the energy that is released in exergonic reactions is captured in chemical bonds when ATP is formed from adenosine diphosphate (ADP) and inorganic phosphate (hydrogen phosphate; commonly abbreviated as P_i). The ATP can then be hydrolyzed at other sites in the cell, releasing free energy to drive endergonic reactions (**FIGURE 6.1**).

An active cell requires the production of millions of molecules of ATP per second to drive its biochemical machinery.

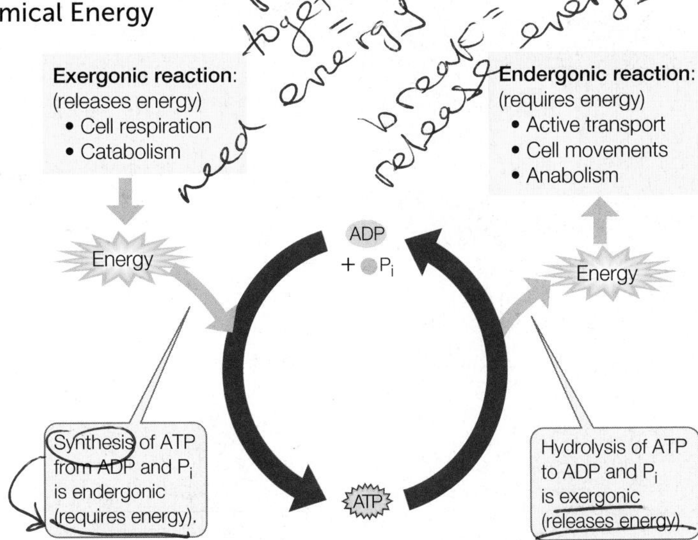

FIGURE 6.1 The Concept of Coupling Reactions Some exergonic cellular reactions are coupled with the formation of ATP from ADP and P_i (an endergonic reaction). The cell can later couple the (exergonic) hydrolysis of ATP with endergonic cellular processes.

Go to ACTIVITY 6.1 ATP and Coupled Reactions PoL2e.com/ac6.1

You are already familiar with some of the activities in the cell that require free energy derived from the hydrolysis of ATP:

- Active transport across a membrane (Concept 5.3)
- Condensation reactions that use enzymes to form polymers (Concept 2.2)
- Motor proteins that move vesicles along microtubules (Concept 4.4)

An ATP molecule consists of the nitrogen-containing base adenine bonded to ribose (a sugar), which is attached to a sequence of three phosphate groups (**FIGURE 6.2**). The hydrolysis

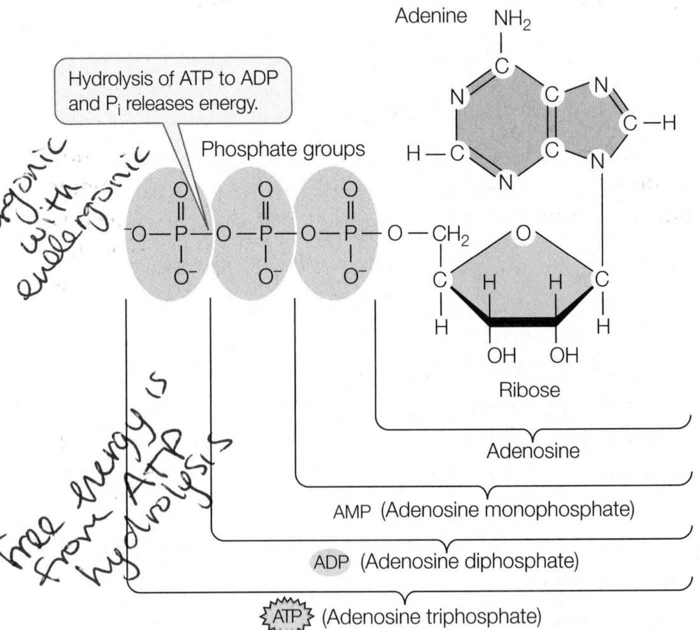

FIGURE 6.2 ATP ATP is built by the addition of terminal phosphate groups onto the nucleoside adenosine.

of a molecule of ATP yields free energy, ADP, and the inorganic phosphate ion (P_i):

$$ATP + H_2O \rightarrow ADP + P_i + \text{free energy}$$

The important property of this reaction is that it is exergonic, releasing free energy. Under standard laboratory conditions, the change in free energy for this reaction (ΔG) is about −7.3 kcal/mol (−30 kJ/mol). Recall that a negative change in free energy means that the product molecules (in this case, ADP and P_i) have less free energy than the reactants (ATP and H_2O), so the change is negative. A molecule of ATP can also be hydrolyzed to adenosine monophosphate (AMP) and a pyrophosphate ion ($P_2O_7^{4-}$; commonly abbreviated as PP_i). In this case, additional energy may be released by the subsequent conversion of PP_i to two molecules of P_i.

Energy is released as a result of ATP hydrolysis because the P—O bonds in a free hydrogen phosphate (P_i) molecule are stronger and more stable than the relatively weak P—O bonds (called phosphoanhydride bonds) between the phosphate groups in ATP. (Phosphoanhydride bonds are often denoted by wavy lines in chemical structures, as highlighted below). Recall that in general, stable bond formation is an exergonic process, whereas breaking bonds requires an input of energy. In this case, the amount of energy released by the formation of a new P—O bond in the P_i molecule is greater than the energy needed to break the phosphoanhydride bond.

Adenine—ribose—O—P(=O)(O⁻)—O~P(=O)(O⁻)—O~P(=O)(O⁻)—O⁻ + H_2O

ATP

Phosphoanhydride bond

↓

Adenine—ribose—O—P(=O)(O⁻)—O~P(=O)(O⁻)—OH + HO—P(=O)(O⁻)—O⁻

ADP P_i

In some reactions, ATP is formed by substrate-level phosphorylation—the enzyme-mediated direct transfer of phosphate from another molecule (the substrate) to ADP. This is the case for some reactions of glycolysis, as we will see in Concept 6.2. But most of the ATP in living cells is formed by oxidative phosphorylation, which we will discuss shortly.

Redox reactions transfer electrons and energy

Another way of transferring energy in chemical reactions is to transfer electrons. A reaction in which one substance transfers one or more electrons to another substance is called a reduction–oxidation reaction, or **redox** reaction.

- **Reduction** is the gain of one or more electrons by an atom, ion, or molecule.
- **Oxidation** is the loss of one or more electrons.

Oxidation and reduction *always occur together*: as one chemical is oxidized, the electrons it loses are transferred to another

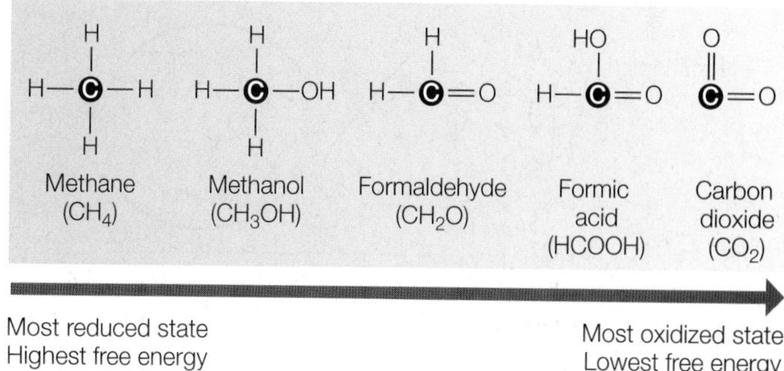

FIGURE 6.3 Oxidation, Reduction, and Energy The more oxidized a carbon atom is, the less free energy it has.

chemical, reducing it. Thus some molecules are called oxidizing agents and others are reducing agents:

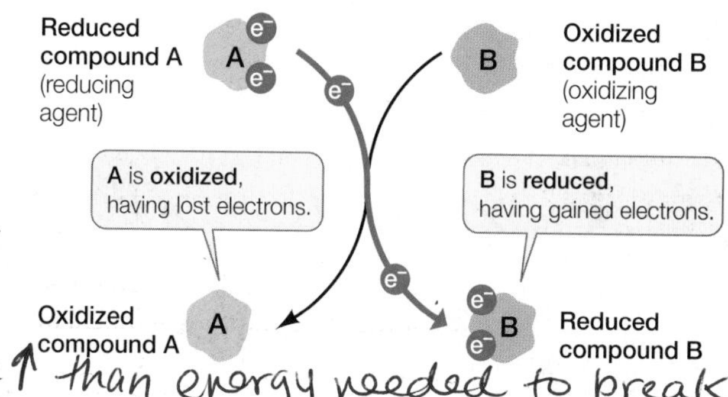

Although oxidation and reduction are defined in terms of traffic in electrons, it is often helpful to think in terms of the gain or loss of hydrogen atoms. Transfers of hydrogen atoms involve transfers of electrons ($H = H^+ + e^-$). So when a molecule loses a hydrogen atom, it becomes oxidized.

In general, the more reduced a molecule is, the more energy is stored in its covalent bonds (**FIGURE 6.3**). Indeed, highly reduced molecules can be used as energy sources; for example, methane and methanol can be burned as fuel. However, oxidized molecules such as CO_2 cannot be used as sources of energy. In a redox reaction, some energy is transferred from the reducing agent to the reduced product. Some energy remains in the reducing agent (now oxidized), and some is lost to entropy.

Cells use the coenzyme nicotinamide adenine dinucleotide (NAD) as an electron carrier in redox reactions (**FIGURE 6.4**). NAD exists in two chemically distinct forms, one oxidized (NAD^+) and the other reduced (NADH). The reduction reaction

$$NAD^+ + H^+ + 2\,e^- \rightarrow NADH$$

involves the transfer of a proton (the hydrogen ion, H^+) and two electrons, which are released by an accompanying oxidation reaction.

The reduction of NAD^+ is highly endergonic, and within the cell, the electrons do not remain with NADH. Oxygen is highly electronegative and readily accepts electrons from the reduced NADH molecule. The oxidation of NADH by O_2 (which occurs in several steps):

$$NADH + H^+ + \tfrac{1}{2}\,O_2 \rightarrow NAD^+ + H_2O$$

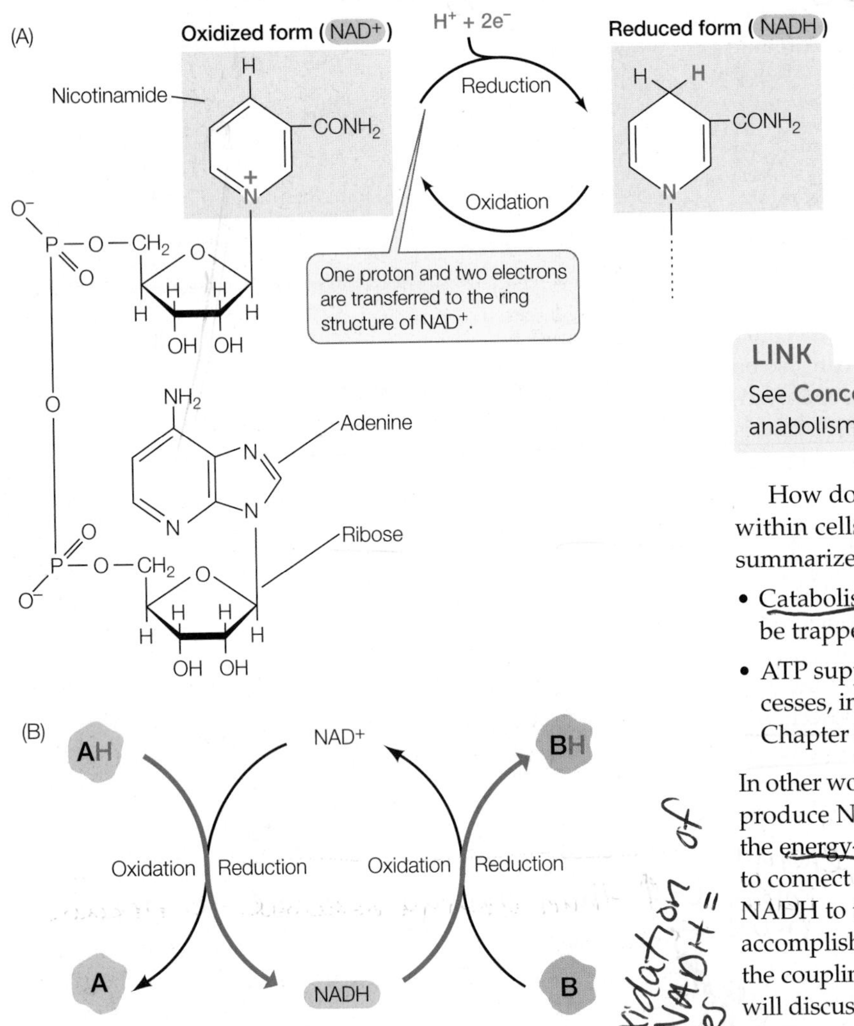

FIGURE 6.4 NAD^+/NADH Is an Electron Carrier in Redox Reactions (A) NAD^+ is an important electron acceptor in redox reactions, and its reduced form, NADH, is an important energy intermediary in cells. The unshaded portion of the molecule (left) remains unchanged by the redox reaction. (B) Coupling of redox reactions using NAD^+/NADH.

is highly exergonic, releasing energy with a ΔG of –52.4 kcal/mol (–219 kJ/mol). Note that the oxidizing agent appears here as "½ O_2" instead of "O." This notation emphasizes that it is molecular oxygen (O_2) that acts as the oxidizing agent. This is clearer if the molecules of the reaction above are doubled:

$$2\ NADH + 2\ H^+ + O_2 \rightarrow 2\ NAD^+ + 2\ H_2O$$

Because the oxidation of NADH releases more energy than the hydrolysis of ATP, NADH can be thought of as a larger package of free energy than ATP. NAD^+ is a common electron carrier in cells, but not the only one. Others include flavin adenine dinucleotide (FAD), which also transfers electrons during glucose metabolism (see Concept 6.2), and nicotinamide adenine dinucleotide phosphate ($NADP^+$), which is used in photosynthesis (see Concept 6.5).

The processes of NADH oxidation and ATP production are coupled

In order to carry out the many metabolic processes needed to sustain life, cells release and reuse the energy contained in chemical bonds. The energy-coupling coenzymes (in particular, ATP and NADH) play vital roles in the transfer of energy between cellular reactions that release energy (catabolism) and those that require energy (including anabolism).

LINK

See **Concept 2.5** to review the principles of catabolism and anabolism

How do these coenzymes participate in the flow of energy within cells? The release and reuse of cellular energy can be summarized as follows:

- Catabolism releases energy by oxidation; this energy can be trapped by the reduction of coenzymes such as NAD^+.
- ATP supplies the energy for many energy-requiring processes, including anabolism. For example, as we noted in Chapter 5, active transport requires ATP.

In other words, most of the energy-releasing reactions in the cell produce NADH (or similar reduced coenzymes), but most of the energy-consuming reactions require ATP. Cells need a way to connect the two coenzymes; that is, to transfer energy from NADH to the phosphoanhydride bond of ATP. This transfer is accomplished in a process called oxidative phosphorylation—the coupling of NADH oxidation to the production of ATP. We will discuss the mechanisms of this process in Concept 6.2.

CHECKpoint CONCEPT 6.1

- For each of the reactions
  a. $C_6H_{12}O_6 + 6\ O_2 \rightarrow 6\ CO_2 + 6\ H_2O$
  b. $6\ CO_2 + 6\ H_2O \rightarrow C_6H_{12}O_6 + 6\ O_2$
  which reactants get oxidized and which get reduced?
- What kinds of coenzymes might be involved in the following reactions? Explain your answer.
  a. Glucose → glucose 6-phosphate
  b. Fatty acid → $CO_2 + H_2O$
- A typical, active young man requires 2,800 kilocalories of food energy a day to fuel metabolism, movement, active transport, etc. The energy stored in the third phosphodiester bond of ATP is 0.0145 kcal/gram.
  a. If the energy from the man's food were all stored as ATP, how much ATP would be produced each day from ADP and P_i?
  b. The man actually has about 50 grams of ATP. What does this mean in terms of ATP hydrolysis and synthesis?

In this concept we saw that both ATP and NADH function as energy-coupling coenzymes, which are used by cells to store and transfer energy. We will now look at how cells capture energy from the catabolism of glucose to produce NADH, and then transfer this energy from NADH to ATP.

CONCEPT 6.2 Carbohydrate Catabolism in the Presence of Oxygen Releases a Large Amount of Energy

Cellular respiration is the set of metabolic reactions used by cells to harvest energy from food. Energy is released when reduced organic molecules, with many C—C and C—H bonds, are oxidized to CO_2. We will consider in detail only the oxidation (catabolism) of carbohydrates, but bear in mind that cells also obtain energy from the catabolism of other molecules, such as lipids.

The chemical energy released from the complete oxidation of glucose to CO_2 is considerable:

$$\text{Glucose} + 6\,O_2 \rightarrow 6\,CO_2 + 6\,H_2O + \text{energy (686 kcal per mole of glucose)}$$

In a chemistry lab, where sugar is "burned" in the presence of O_2, this energy is all lost as heat. In the cell, some of the released energy (234 kcal/mol; 34% of the total) is trapped as ATP. The efficiency of this energy-trapping process is impressive, even when compared with motors that humans have devised. The cell can achieve this through the general principles that govern metabolism listed in Concept 6.1. Most notably, the oxidation occurs in a series of small steps (**FIGURE 6.5**).

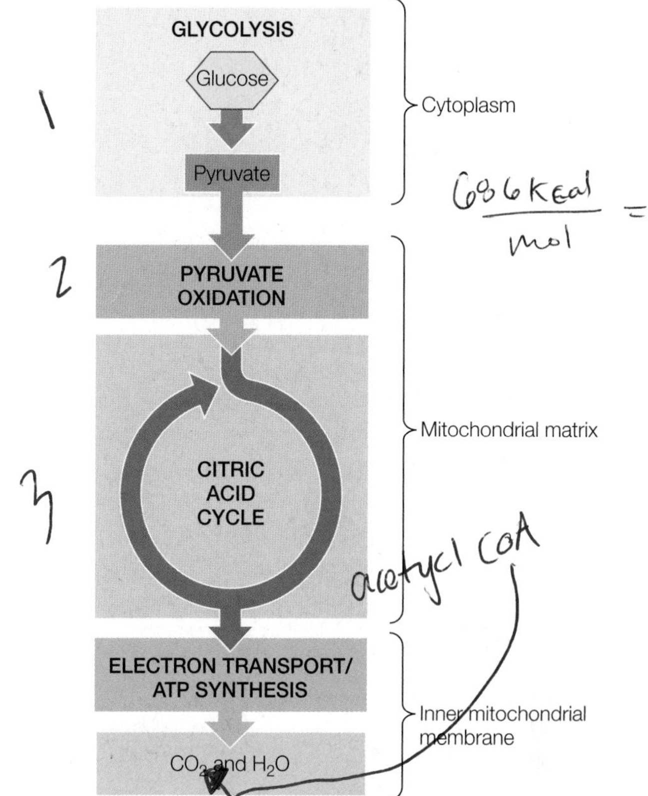

FIGURE 6.6 Energy-Releasing Metabolic Pathways The catabolism of glucose under aerobic conditions occurs in three sequential metabolic pathways: glycolysis, pyruvate oxidation, and the citric acid cycle. The reduced coenzymes are then oxidized by the respiratory chain, and ATP is made.

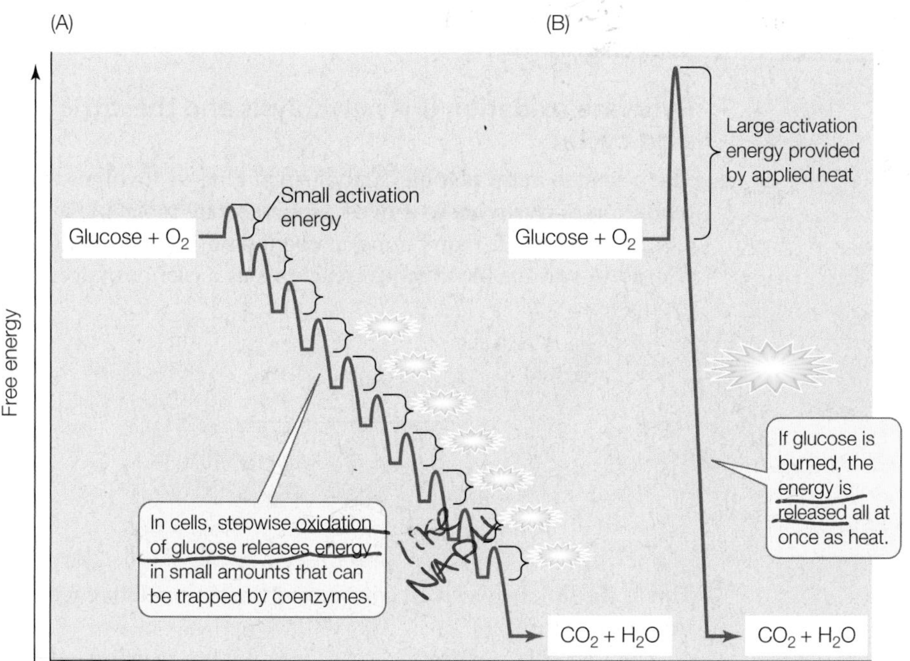

FIGURE 6.5 Energy Metabolism Occurs in Small Steps (A) In living systems, glucose is oxidized via a series of steps, releasing small amounts of energy that can be efficiently trapped by coenzymes. (B) Glucose that is burned releases its energy as heat in one big step.

In the catabolism of glucose under **aerobic** conditions (in the presence of O_2), the small steps can be grouped into three linked biochemical pathways (**FIGURE 6.6**):

- In **glycolysis**, the six-carbon monosaccharide glucose is converted into two three-carbon molecules of pyruvate.
- In **pyruvate oxidation**, two three-carbon molecules of pyruvate are oxidized to two two-carbon molecules of acetyl CoA and two molecules of CO_2.
- In the **citric acid cycle**, two two-carbon molecules of acetyl CoA are oxidized to four molecules of CO_2.

In glycolysis, glucose is partially oxidized and some energy is released

Glycolysis takes place in the cytosol and involves ten enzyme-catalyzed reactions. During glycolysis, some of the C—H bonds in the glucose molecule are oxidized, releasing some stored energy. The final products are two molecules of pyruvate (the anion of pyruvic acid), two molecules of ATP, and

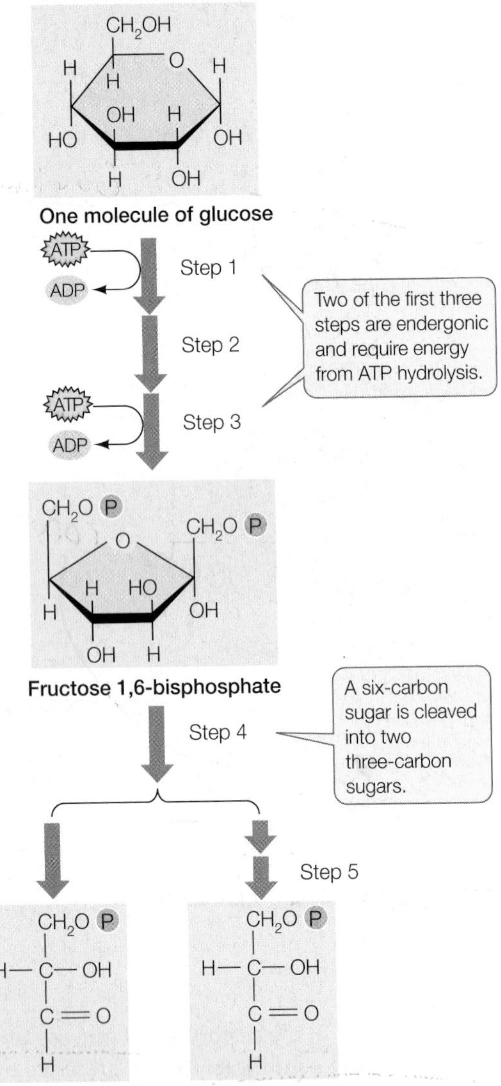

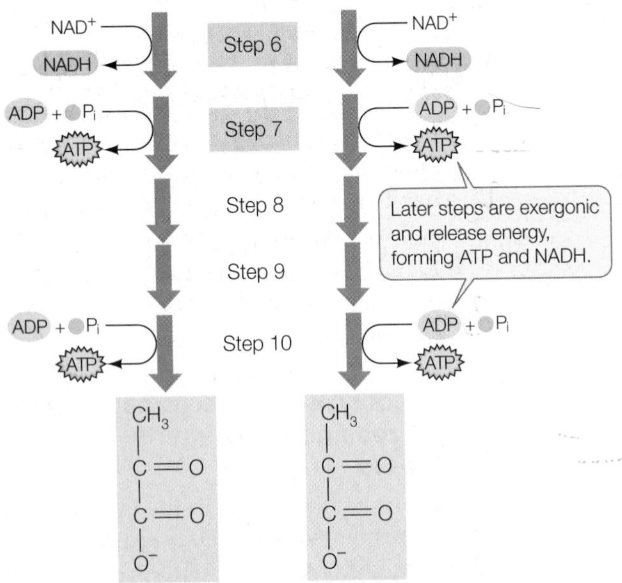

FIGURE 6.7 Glycolysis Converts Glucose into Pyruvate Glucose is converted to pyruvate in ten enzyme-catalyzed steps. Along the way, energy is released to form ATP and NADH.

two molecules of NADH. Glycolysis can be divided into two stages: the initial energy-investing reactions that consume chemical energy stored in ATP, and the energy-harvesting reactions that produce ATP and NADH (**FIGURE 6.7**).

To help you understand the process without getting into extensive detail, we will focus on two consecutive reactions in this pathway (steps 6 and 7 in Figure 6.7).

Step 6: Glyceraldehyde 3-phosphate dehydrogenase (Glyceraldehyde 3-phosphate → 1,3-Bisphosphoglycerate; P_i, NAD^+ → NADH)

Step 7: Phosphoglycerate kinase (1,3-Bisphosphoglycerate → 3-Phosphoglycerate; ADP → ATP)

These are examples of two types of reactions that occur repeatedly in glycolysis and in many other metabolic pathways:

- *Oxidation–reduction*: In the exergonic reaction of step 6, more than 50 kcal/mol of energy are released in the oxidation of glyceraldehyde 3-phosphate. (Look at the first carbon atom, highlighted in blue, where an H is replaced by an O.) The energy is trapped via the reduction of NAD^+ to NADH.
- *Substrate-level phosphorylation*: The second reaction in this series is also exergonic, but in this case less energy is released. It is enough to transfer a phosphate from the substrate (1,3-bisphosphoglycerate) to ADP, forming ATP.

The end product of glycolysis, pyruvate, is somewhat more oxidized than glucose. In the presence of O_2, further oxidation can occur. In prokaryotes these subsequent reactions take place in the cytosol, but in eukaryotes they take place in the mitochondrial matrix.

Pyruvate oxidation links glycolysis and the citric acid cycle

The next step in the aerobic catabolism of glucose involves the oxidation of pyruvate to a two-carbon acetate molecule and CO_2. The acetate is then bound to **coenzyme A (CoA)**, which is used in various biochemical reactions as a carrier of acetyl groups:

Pyruvate → Acetyl CoA (Coenzyme A; NAD^+ → NADH, CO_2)

This is the link between glycolysis and further oxidative reactions (see Figure 6.6).

The formation of acetyl CoA is a multistep reaction catalyzed by the pyruvate dehydrogenase complex, which contains 60 individual proteins and 5 different coenzymes. The overall reaction is exergonic, and one molecule of NAD^+ is reduced.

The main role of acetyl CoA is to donate its acetyl group to the four-carbon compound oxaloacetate, forming the six-carbon molecule citrate (the anion of citric acid). This initiates the citric acid cycle, one of life's most important energy-harvesting pathways.

The citric acid cycle completes the oxidation of glucose to CO_2

Acetyl CoA is the starting point for the citric acid cycle. This pathway of eight reactions completely oxidizes the two-carbon acetyl group to two molecules of CO_2. The free energy released from these reactions is captured by ADP and the electron carriers NAD^+ and FAD (**FIGURE 6.8**). This is a cycle because the starting material, oxaloacetate, is regenerated in the last step and is ready to accept another acetate group from acetyl CoA. The citric acid cycle operates twice for each glucose molecule that enters glycolysis (once for each pyruvate that enters the mitochondrion).

Let's focus on the final reaction of the cycle (step 8 in Figure 6.8) as an example of the kind of reaction that occurs:

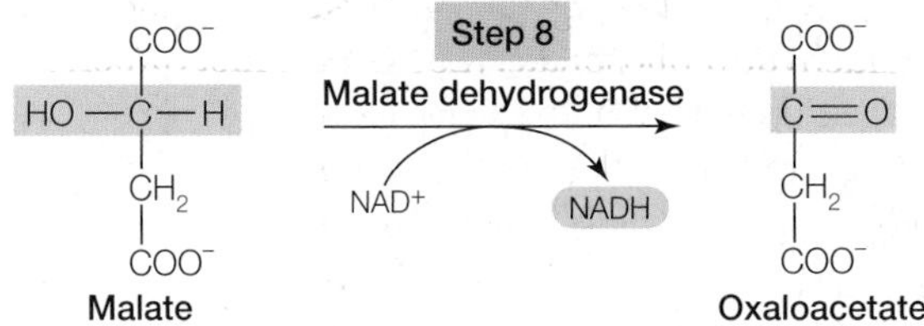

This oxidation reaction (see the blue-highlighted carbon atom) is exergonic, and the released energy is trapped by NAD^+, forming NADH. With four such reactions ($FADH_2$ is a reduced coenzyme similar to NADH), the citric acid cycle harvests a great deal of chemical energy from the oxidation of acetyl CoA.

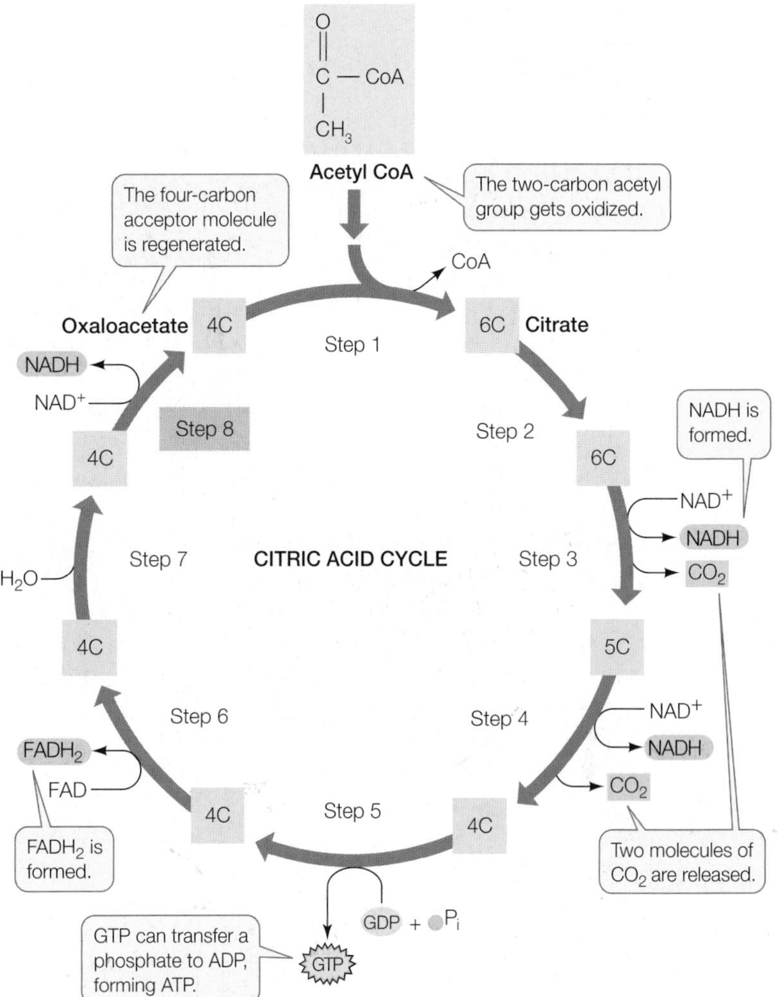

FIGURE 6.8 The Citric Acid Cycle Also called the Krebs cycle for its discoverer, Hans Krebs, the citric acid cycle involves eight steps and fully oxidizes acetyl CoA to CO_2.

Go to ACTIVITY 6.2 The Citric Acid Cycle
PoL2e.com/ac6.2

Energy is transferred from NADH to ATP by oxidative phosphorylation

As we mentioned in Concept 6.1, energy-consuming processes in the cell use ATP as their source of energy. In order to fully use the energy harvested in catabolism, cells need to transfer energy from NADH (and $FADH_2$) to the phosphoanhydride bond of ATP. In eukaryotic mitochondria, this transfer is accomplished by **oxidative phosphorylation**: NADH oxidation is used to actively transport protons (H^+ ions) across the inner mitochondrial membrane, resulting in a proton gradient across the membrane. The diffusion of protons back across the membrane is then used to drive the synthesis of ATP. (In prokaryotes, oxidative phosphorylation takes place at the cell membrane.)

First let's examine how the oxidation of NADH and $FADH_2$ leads to the production of the proton gradient. For example, when NADH is reoxidized to NAD^+, O_2 is reduced to H_2O:

$$NADH + H^+ + \frac{1}{2} O_2 \rightarrow NAD^+ + H_2O$$

This does not happen in a single step. Rather, there is a series of redox electron carrier proteins called the **respiratory chain** embedded in the inner membrane of the mitochondrion (**FIGURE 6.9**). The electrons from the oxidation of NADH and $FADH_2$ pass from one carrier to the next in the chain in a process called **electron transport**. The oxidation reactions are exergonic, and they release energy that is used to actively transport H^+ ions across the membrane.

An important aspect of this process is that an oxidation reaction is always coupled with a reduction. When NADH is oxidized to NAD^+, the corresponding reduction reaction is the formation of water from O_2:

$$2\,H^+ + 2\,e^- + \frac{1}{2} O_2 \rightarrow H_2O$$

So the *key role of O_2 in cells*—the reason we breathe and have a blood system to deliver O_2 to tissues—*is to act as an electron acceptor and become reduced.*

Chemiosmosis uses the proton gradient to generate ATP

In addition to the electron transport carriers, the inner mitochondrial membrane contains an enzyme called **ATP synthase** (**FIGURE 6.10A**). This enzyme uses the H^+ gradient to drive the

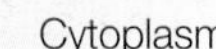

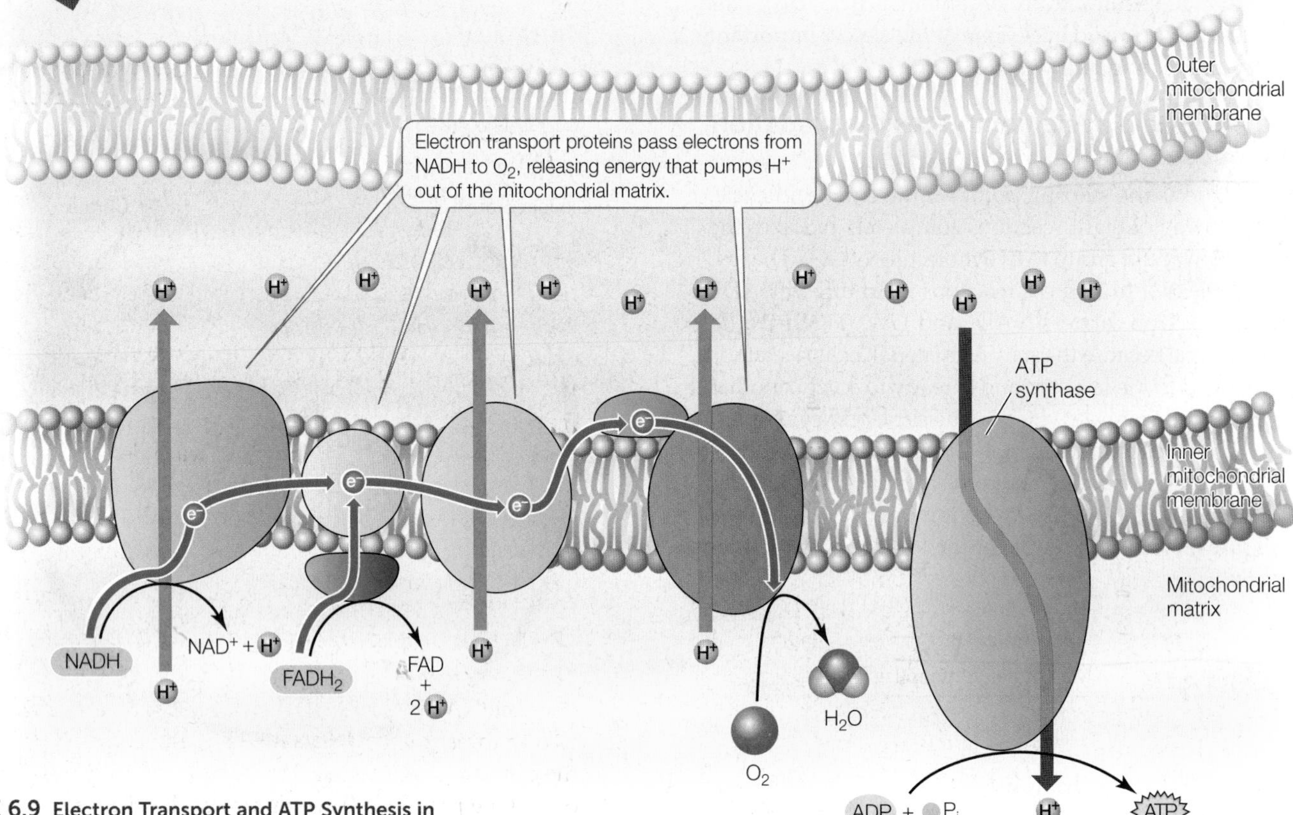

FIGURE 6.9 Electron Transport and ATP Synthesis in Mitochondria As electrons pass through the protein complexes of the respiratory chain, protons are pumped from the mitochondrial matrix into the intermembrane space. As the protons return to the matrix through ATP synthase, ATP is formed.

Go to ANIMATED TUTORIAL 6.1
Electron Transport and ATP Synthesis
PoL2e.com/at6.1

Go to ACTIVITY 6.3 Respiratory Chain
PoL2e.com/ac6.3

synthesis of ATP via a mechanism called **chemiosmosis**—the movement of ions across a semipermeable barrier from a region of higher concentration to a region of lower concentration. Chemiosmosis relies on concepts covered in earlier chapters:

- If the concentration of a substance is greater on one side of a membrane than the other, the substance will tend to diffuse across the membrane to its region of lower concentration (see Concept 5.2).
- If a membrane blocks this diffusion, the substance at the higher concentration has potential energy, which can be converted to other forms of energy (see Concept 2.5).
- Because the interior of a membrane is nonpolar, protons (H^+) cannot readily diffuse across the membrane, but they can cross the membrane through the ATP synthase enzyme. ATP synthase converts the potential energy of the proton gradient (called the proton motive force) into the chemical energy in ATP.

ATP synthase is a molecular motor composed of two parts: the F_o unit, which is a transmembrane domain that functions as the H^+ channel; and the F_1 unit, which contains the active sites for ATP synthesis (**FIGURE 6.10B**). The F_1 unit consists of six subunits (three each of two polypeptide chains), arranged like the segments of an orange around a central polypeptide. The potential energy set up by the proton gradient drives the passage of protons through the ring of polypeptides that make up the F_o component. This ring rotates as the protons pass through the membrane, causing part of the F_1 unit to rotate as well. ADP and P_i bind to active sites that become exposed on the F_1 unit as it rotates, and ATP is made.

Go to MEDIA CLIP 6.1
ATP Synthase in Motion
PoL2e.com/mc6.1

The structure and function of ATP synthase enzymes are shared by living organisms as diverse as bacteria and humans. These enzymes make ATP at rates of up to 100 molecules per second. In all organisms, these molecular motors rely on protein gradients across membranes:

- In prokaryotes, the gradient is set up across the cell membrane, using energy from various sources.
- In eukaryotes, chemiosmosis occurs in the mitochondria and the chloroplasts.

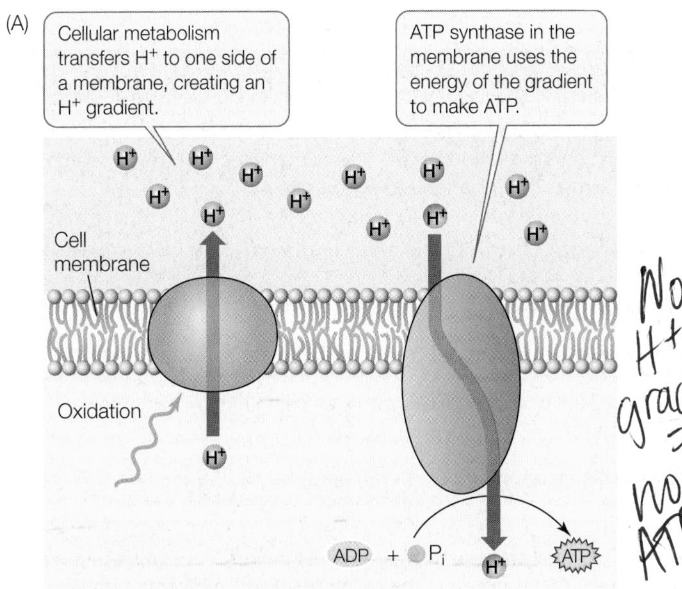

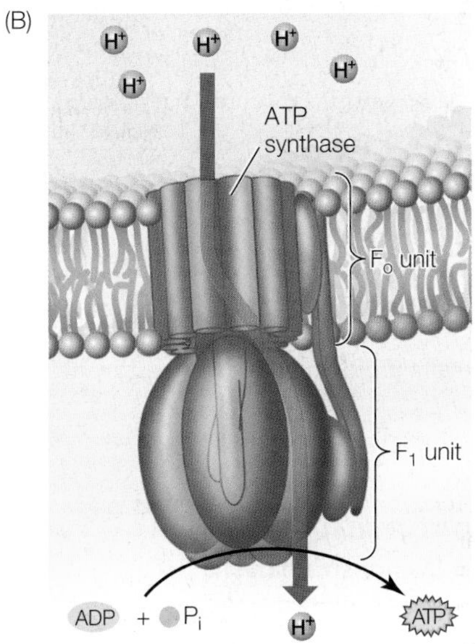

FIGURE 6.10 Chemiosmosis (A) If a cell can generate a proton (H^+) gradient across a membrane, the potential energy resulting from the concentration gradient can be used by a membrane-spanning enzyme to make ATP. (B) ATP synthase has a membrane-embedded channel for H^+ diffusion and a motor that turns, releasing some energy to produce ATP.

- As we have just seen, the H^+ gradient in mitochondria is set up across the inner mitochondrial membrane, using energy released by the oxidation of NADH and $FADH_2$.
- In chloroplasts, the H^+ gradient is set up across the thylakoid membrane using energy from light (see Concept 6.5). In this case, the reduced molecule is $NADP^+$, a relative of NAD^+.

Despite these differences in detail, the mechanism of chemiosmosis is similar in almost all forms of life.

Chemiosmosis can be demonstrated experimentally

If chloroplasts or mitochondria are isolated from cells and put in a test tube, a proton gradient can be introduced artificially. This artificial gradient drives ATP synthesis (**FIGURE 6.11**), but only if ATP synthase, ADP, inorganic phosphate, and the membrane are present.

What happens if the H^+ gradient is destroyed by the presence of a membrane channel that is always open to protons? ATP cannot be made, but the oxidation of NADH still occurs and O_2 is reduced, releasing considerable energy. The released energy forms heat instead of being used to make ATP. In newborn human infants, a membrane protein appropriately called uncoupling protein 1 disrupts the H^+ gradient in fat cell mitochondria, and this results in the release of heat. Because infants lack body hair, this process helps keep them warm.

A popular weight loss drug in the 1930s was the synthetic uncoupler molecule dinitrophenol. There were claims of dramatic weight loss when the drug was administered to obese patients. Unfortunately, the heat that was released caused fatally high fevers, and the effective dose and fatal dose were quite close. The use of this drug was discontinued in 1938, but the general strategy of using an uncoupler for weight loss remains a subject of research.

Oxidative phosphorylation and chemiosmosis yield a lot of ATP

For each NADH (or $FADH_2$) that begins the respiratory chain, two to three (let's say 2.5) ATP molecules are formed under the conditions in the cell. Thus the four molecules of reduced coenzyme produced by each turn of the citric acid cycle yield about ten (4 × 2.5) molecules of ATP. Two molecules of acetyl CoA are produced from each glucose, so the total is about 20 ATPs per molecule of glucose. Add to this the NADH produced by glycolysis and pyruvate oxidation, and the ATP formed by substrate-level phosphorylation during glycolysis and the citric acid cycle, and the total is about *32 molecules of ATP produced per fully oxidized glucose.*

The vital role of O_2 is now clear: most of the ATP produced in cellular respiration is formed by oxidative phosphorylation—the process of transferring electrons from NADH to O_2, resulting in the reoxidation of NADH to NAD^+. The accumulation of atmospheric O_2 as a result of photosynthesis by ancient microorganisms (see Concept 18.2) set the stage for the evolution of oxidative phosphorylation; organisms that could exploit the O_2 would have had a selective advantage.

Nevertheless, many microorganisms still thrive where O_2 is scarce. These anaerobic bacteria and archaea use alternative electron acceptors in their natural environments. For instance, the bacterium *Geobacter metallireducens* typically lives in sediments under streams or ponds, and uses metal ions as terminal electron acceptors. For example:

$$Fe^{3+} + e^- \text{ (from electron transport)} \rightarrow Fe^{2+}$$

This bacterium can also use radioactive uranium ions as electron acceptors. In the process, the uranium is converted from a soluble to an insoluble form, making *Geobacter* of potential use in environmental cleanup. The bacterium can convert uranium in contaminated water into a form that accumulates in the sediment instead and can be more readily removed.

INVESTIGATION

FIGURE 6.11 An Experiment Demonstrates the Chemiosmotic Mechanism The chemiosmosis hypothesis was a bold departure from the conventional scientific thinking of the time. It required an intact compartment separated by a membrane. Could a proton gradient drive the synthesis of ATP? The first experiments to answer this question used chloroplasts, plant organelles that use the same mechanism as mitochondria to synthesize ATP.[a]

HYPOTHESIS

A H^+ gradient across a membrane that contains ATP synthase is sufficient to drive ATP synthesis.

METHOD

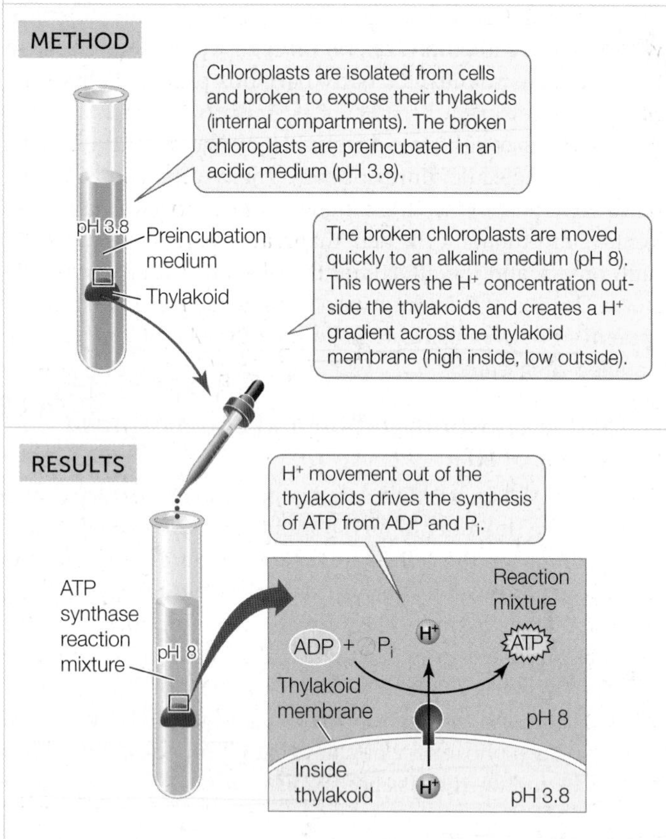
Chloroplasts are isolated from cells and broken to expose their thylakoids (internal compartments). The broken chloroplasts are preincubated in an acidic medium (pH 3.8).
pH 3.8
Preincubation medium
Thylakoid
The broken chloroplasts are moved quickly to an alkaline medium (pH 8). This lowers the H⁺ concentration outside the thylakoids and creates a H⁺ gradient across the thylakoid membrane (high inside, low outside).
RESULTS
H⁺ movement out of the thylakoids drives the synthesis of ATP from ADP and Pᵢ.
ATP synthase reaction mixture
pH 8
Reaction mixture
ADP + Pᵢ
H⁺
ATP
Thylakoid membrane
pH 8
Inside thylakoid
H⁺
pH 3.8

CONCLUSION

A H^+ gradient across an ATP synthase–containing membrane is sufficient for ATP synthesis by organelles.

ANALYZE THE DATA

The formation of ATP from ADP and P_i was measured using luciferase, which catalyzes the formation of a luminescent (light-emitting) molecule if ATP is present. The experiment was performed under different conditions, with the following results:

Experiment	Preincubation pH	ATP synthase mixture (pH 8)	ATP formation (nmoles/mg chlorophyll)
1	3.8	Complete mixture	144
2	7.0	Complete mixture	12
3	3.8	P_i omitted	12
4	3.8	ADP omitted	4
5	3.8	Thylakoids omitted	7

A. Which experiments show that a proton gradient is necessary to stimulate ATP formation?

B. Why was there less ATP production in the absence of P_i?

Go to ANIMATED TUTORIAL 6.2
Two Experiments Demonstrate the Chemiosmotic Mechanism
PoL2e.com/at6.2

Go to **LaunchPad** for discussion and relevant links for all **INVESTIGATION** figures.

[a]A. T. Jagendorf and E. Uribe. 1966. *Proceedings of the National Academy of Sciences USA* 55: 170–177.

APPLY THE CONCEPT

Carbohydrate catabolism in the presence of oxygen releases a large amount of energy

The following reaction occurs in the citric acid cycle:

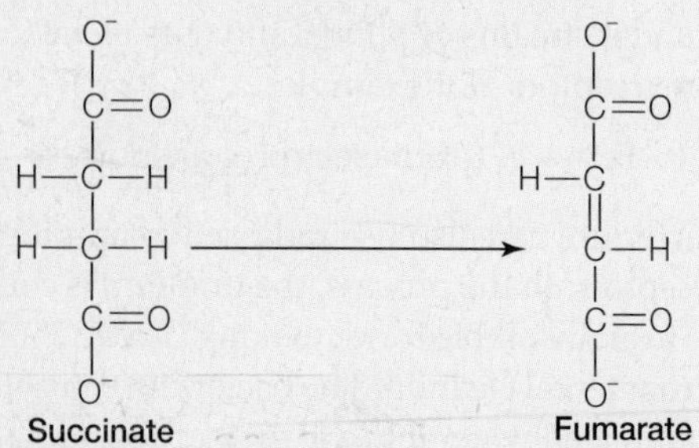
Succinate
Fumarate

Answer each of the following questions, and explain your answers:

1. Is this reaction an oxidation or reduction?
2. Is the reaction exergonic or endergonic?
3. This reaction requires a coenzyme. What kind of coenzyme?
4. What happens to the fumarate after the reaction is completed?
5. What happens to the coenzyme after the reaction is completed?

CHECKpoint CONCEPT 6.2

- ✓ Draw the molecules of glucose and pyruvate and compare them with respect to their oxidation/reduction states. What does this mean in terms of available chemical bond energy?
- ✓ Compare the energy yields (in terms of ATP per glucose) from glycolysis and from the combination of pyruvate oxidation and the citric acid cycle.
- ✓ High levels of ammonia (NH_3) are toxic to mammals. One reason is that ammonia bonds with α-ketoglutarate (the substrate for step 4 in Figure 6.8), forming the amino acid glutamate. This removes α-ketoglutarate from the citric acid cycle. Explain the consequence of this for the cell.
- ✓ Hibernating animals usually have low rates of metabolism. Periodically, they have episodes of arousal, where they rapidly increase body temperature. During arousal, they synthesize a membrane protein not normally present in their cells. What might that protein do?

We have seen that a large amount of energy is released when carbohydrates are catabolized in the presence of O_2. If O_2 is absent, the yield of ATP is much lower. We will now turn to this situation.

CONCEPT 6.3 Carbohydrate Catabolism in the Absence of Oxygen Releases a Small Amount of Energy

In the absence of O_2—that is, when conditions are **anaerobic**—the respiratory chain cannot operate. (The exceptions, as we noted earlier, are the respiratory chains of anaerobic microbes adapted to use terminal electron acceptors other than oxygen.) Without an alternative, the NADH produced by glycolysis would not be reoxidized and glycolysis would stop, because there would be no NAD^+ for step 6 of glycolysis (see Figure 6.7). To solve this problem, organisms use **fermentation** to reoxidize the NADH, thus allowing glycolysis to continue (**FIGURE 6.12**).

Like glycolysis, fermentation pathways occur in the cytoplasm. There are many different types of fermentation used by different organisms, but all operate to regenerate NAD^+. The consequence of this is that the NADH made during glycolysis is not available for reoxidation by the respiratory chain to form ATP. Therefore the overall yield of ATP from fermentation is restricted to the ATP made in glycolysis (two ATP per glucose).

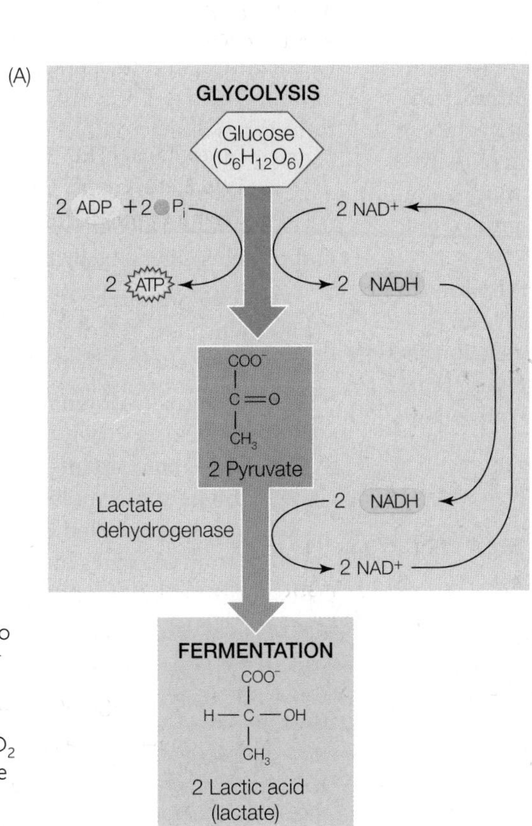

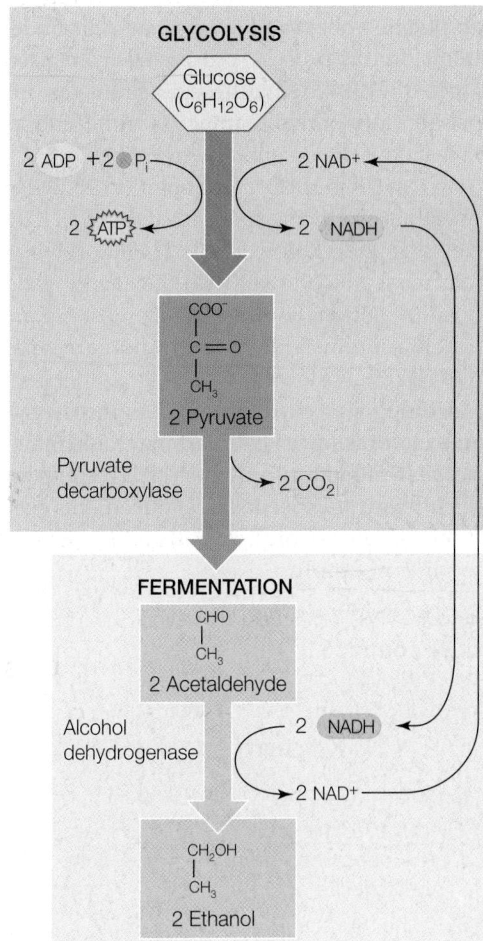

FIGURE 6.12 Fermentation (A) In lactic acid fermentation, NADH is used to reduce pyruvate to lactic acid, thus regenerating NAD^+ to keep glycolysis operating. (B) In alcoholic fermentation, pyruvate is converted to acetaldehyde, and CO_2 is released. NADH is used to reduce acetaldehyde to ethanol, again regenerating NAD^+ for glycolysis.

Go to ACTIVITY 6.4 Glycolysis and Fermentation PoL2e.com/ac6.4

Two fermentation pathways are found in a wide variety of organisms:

- Lactic acid fermentation, whose end product is lactic acid (lactate)
- Alcoholic fermentation, whose end product is ethyl alcohol (ethanol)

In **lactic acid fermentation**, pyruvate serves as the electron acceptor and lactate is the product (see Figure 6.12A). This process takes place in many microorganisms and complex organisms, including more complex plants and vertebrates. A notable example of lactic acid fermentation occurs in vertebrate muscle tissue. Usually, vertebrates get their energy for muscle contractions aerobically, with the circulatory system supplying O_2 to muscles. This is almost always adequate for small vertebrates, which explains why birds can fly long distances without resting. But in larger vertebrates such as humans, the circulatory system is not up to the task of delivering enough O_2 when the need is great, such as during a long sprint. At this point, the muscle cells break down glycogen (a stored polysaccharide) and undergo lactic acid fermentation. The process is reversible; lactate is converted back to pyruvate once O_2 is available again.

Alcoholic fermentation takes place in certain yeasts (eukaryotic microbes) and some plant cells under anaerobic conditions. In this process, pyruvate is converted to ethanol (see Figure 6.12B). We saw these reactions (as did Pasteur) in the opening story of this chapter. As with lactic acid fermentation, the reactions are essentially reversible.

By recycling NAD^+, fermentation allows glycolysis to continue, thus producing small amounts of ATP through substrate-level phosphorylation. The net yield of two ATPs per glucose molecule is much lower than the energy yield from oxidative phosphorylation. For this reason, most organisms that rely on fermentation instead of respiration are small microbes that grow relatively slowly.

Although its yield of ATP per glucose is generally low, in some circumstances cellular anaerobic metabolism can produce an adequate supply of ATP if the enzymatic reactions in the pathway are speeded up. Indeed, this occurs in vertebrate muscle cells (see Concept 33.3) and in some cancer cells where O_2 is in low supply.

Go to ACTIVITY 6.5 Energy Levels
PoL2e.com/ac6.5

CHECKpoint CONCEPT 6.3

- ✓ Why is replenishing NAD^+ crucial to cellular metabolism?
- ✓ Compare the sources and total energy yield in terms of ATP per glucose in human cells in the presence versus the absence of O_2.
- ✓ Conditions can become anaerobic in a heart muscle cell during a heart attack, because of the inadequate supply of blood. If O_2 is restored, what will happen to the lactate produced by the heart muscle?

You have seen how cells harvest chemical energy in cellular respiration. Now we will see how that energy moves through other metabolic pathways in the cell.

CONCEPT 6.4 Catabolic and Anabolic Pathways Are Integrated

Metabolic transformations are a hallmark of life. The pathways we have seen thus far in the chapter, including glycolysis and the citric acid cycle, do not operate in isolation. Rather, there is an interchange of molecules into and out of these pathways, to and from the metabolic pathways for the synthesis and breakdown of amino acids, nucleotides, fatty acids, and other building blocks of life. Carbon skeletons (a term describing molecules with covalently linked carbon atoms) can enter catabolic pathways and be oxidized to release their energy, or they can enter anabolic pathways to be used in the formation of the macromolecules that are the major constituents of the cell. These relationships are summarized in **FIGURE 6.13** and comprise a metabolic system (see Figure 3.17). The inputs and outputs of this system can be described, and predictions can be made regarding what would happen if the concentration of one component changes.

Catabolism and anabolism are linked

A hamburger or veggie burger on a bun contains three major sources of carbon skeletons: carbohydrates, mostly in the form of starch (a polysaccharide); lipids, mostly as triglycerides (three fatty acids attached to glycerol); and proteins (polymers of amino acids). Look at Figure 6.13 to see how each of these three types of macromolecules can be hydrolyzed and used in catabolism or anabolism.

CATABOLIC INTERCONVERSIONS Polysaccharides, lipids, and proteins can all be broken down to provide energy:

- *Polysaccharides* can be hydrolyzed to glucose. Glucose then passes through glycolysis, pyruvate oxidation, and the respiratory chain, where its energy is captured in ATP.
- *Lipids* are broken down into their constituents—glycerol and fatty acids. Glycerol is converted into dihydroxyacetone phosphate, an intermediate in glycolysis. Fatty acids are highly reduced molecules that are converted to acetyl CoA in a process called β-oxidation, catalyzed by a series of oxidation enzymes inside the mitochondrion. For example, the β-oxidation of a C_{16} (16-carbon) fatty acid occurs in several steps:

$$C_{16}\text{ fatty acid} + \text{CoA} \rightarrow C_{16}\text{ fatty acyl CoA}$$

$$C_{16}\text{ fatty acyl CoA} + \text{CoA} \rightarrow C_{14}\text{ fatty acyl CoA} + \text{acetyl CoA}$$

$$\text{(Repeat 6 times)} \rightarrow 8\text{ acetyl CoA}$$

The acetyl CoA can then enter the citric acid cycle and be catabolized to CO_2.

- *Proteins* are hydrolyzed to their amino acid building blocks. The 20 different amino acids feed into glycolysis

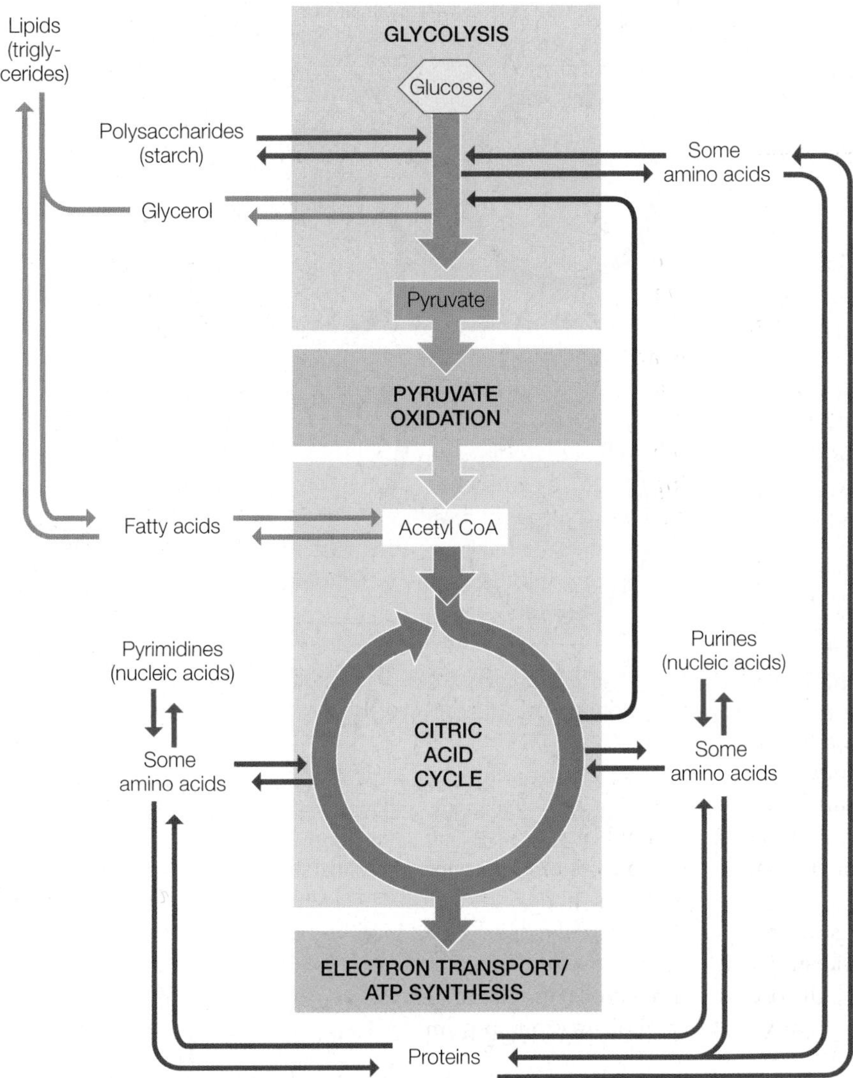

FIGURE 6.13 Relationships among the Major Metabolic Pathways of the Cell Note the central positions of glycolysis and the citric acid cycle in this system of metabolic pathways. Also note that many of the pathways can operate essentially in reverse.

or the citric acid cycle at different points. For example, the amino acid glutamate is converted into α-ketoglutarate, an intermediate in the citric acid cycle.

ANABOLIC INTERCONVERSIONS Many catabolic pathways can operate essentially in reverse, with some modifications. Glycolytic and citric acid cycle intermediates, instead of being oxidized to form CO_2, can be reduced and used to form glucose in a process called **gluconeogenesis** (which means "new formation of glucose"). Likewise, acetyl CoA can be used to form fatty acids. The most common fatty acids have even numbers of carbons: 14, 16, or 18. These are formed by the addition of two-carbon acetyl CoA "units" one at a time until the appropriate chain length is reached.

Some intermediates in the citric acid cycle are reactants in pathways that synthesize important components of nucleic acids. For example, α-ketoglutarate and oxaloacetate are starting points for purines and pyrimidines, respectively.

Catabolism and anabolism are integrated into a system

A carbon atom from a protein in your burger can end up in DNA, fat, or CO_2, among other fates. How does the organism "decide" which metabolic pathways to follow, in which cells? With hundreds of enzymes and all the possible interconversions, you might expect that the cellular concentrations of various biochemical molecules would fluctuate widely. Remarkably, the levels of these substances in what is called the metabolic pool—the sum total of all the biochemical molecules in a cell—are usually quite constant. Metabolic changes in the cell are a bit like the changes in traffic patterns in a city: if an accident blocks traffic on a major road, drivers take alternate routes, where the traffic volume consequently changes.

Consider what happens to the starch in your burger bun. In the digestive system, starch is hydrolyzed to glucose, which enters the blood. If it is needed, the glucose is distributed to the rest of the body. But if there is already enough glucose in the blood to supply the body's needs, the excess glucose is converted into glycogen and stored in the liver. If not enough glucose is supplied by food, glycogen is broken down, or other molecules are used to make glucose by gluconeogenesis. The end result is that the level of glucose in the blood is remarkably constant. How does the body accomplish this?

Metabolic enzymes (including those of glycolysis, the citric acid cycle, and the respiratory chain) are subject to regulation, and often the regulatory mechanisms involve allosteric effects. An example is feedback inhibition, illustrated in Figure 3.21. In a metabolic pathway, a high concentration of the final product can inhibit the enzyme that catalyzes the commitment step. For example, the product of glucose catabolism, ATP, feedback inhibits key enzymes in both glycolysis and the citric acid cycle. Sometimes feedback involves the product of one pathway speeding up reactions in another pathway. Feedback regulation generally occurs rapidly, affecting a pathway within minutes.

LINK
Review the discussion of enzyme regulation in **Concept 3.4**

The rate of a biochemical reaction can also be controlled by reducing or increasing the number of enzyme molecules present relative to substrate (see Figure 11.6). This can be done by altering the transcription of the genes that encode the enzymes. These events take time, and typically the effects on metabolism will take days to appear. For example, excess levels of glucose and other dietary factors can lead to increased transcription of the gene for fatty acid synthase, a key enzyme in the synthesis of fatty acids. Excess citrate produced by the citric acid cycle is broken down to acetyl CoA, which in turn is used in fatty acid synthesis. This is one reason why people accumulate fat after eating too much. The fatty acids may be catabolized later to produce more acetyl CoA.

LINK

The regulation of gene transcription is described in **Chapter 11**

ATP and reduced coenzymes link catabolism, anabolism, and photosynthesis

Thus far in this chapter we have discussed the major catabolic pathway called cellular respiration, in which a reduced molecule, the carbohydrate glucose, is oxidized, often all the way to CO_2:

$$\text{Glucose} + 6\ O_2 \rightarrow 6\ CO_2 + 6\ H_2O + \text{chemical energy}$$

We have also seen how the energy derived from cellular respiration is transferred between catabolic and anabolic pathways (depending on a cell's needs), and that the coenzymes ATP and NADH play key roles in these energy transfers.

But where does the energy for all these processes come from? As we will see in the next two concepts, almost all living organisms on Earth ultimately derive their energy from the sun (exceptions include organisms that derive energy from geothermal vents deep within oceans). In addition to cellular respiration, green plants, algae, some protists, and some prokaryotes also carry out photosynthesis—a major anabolic pathway that converts light energy into chemical energy in the form of carbohydrates:

$$CO_2 + H_2O + \text{light energy} \rightarrow O_2 + \text{carbohydrates}$$

Just as catabolism and anabolism are linked, cellular respiration and photosynthesis are also linked—not only by their reactants and products (O_2, CO_2, and carbohydrates), but also by the energy "currency" of ATP and reduced coenzymes (**FIGURE 6.14**). The pathways of cellular respiration and photosynthesis can occur in the same cell. For example, cells in green plants often carry out both photosynthesis and cellular respiration simultaneously.

More often, however, cells do not carry out photosynthesis for themselves. Even in a green plant, a root cell is not photosynthetic and relies on carbohydrates transported from the leaf to carry out cellular respiration. Humans do not carry out photosynthesis anywhere in their bodies; they rely on carbohydrates obtained in their diet (ultimately derived from photosynthesis) to carry out cellular respiration, which provides the chemical energy for their bodies' activities, such as active transport and anabolism.

Go to ACTIVITY 6.6 Regulation of Energy Pathways
PoL2e.com/ac6.6

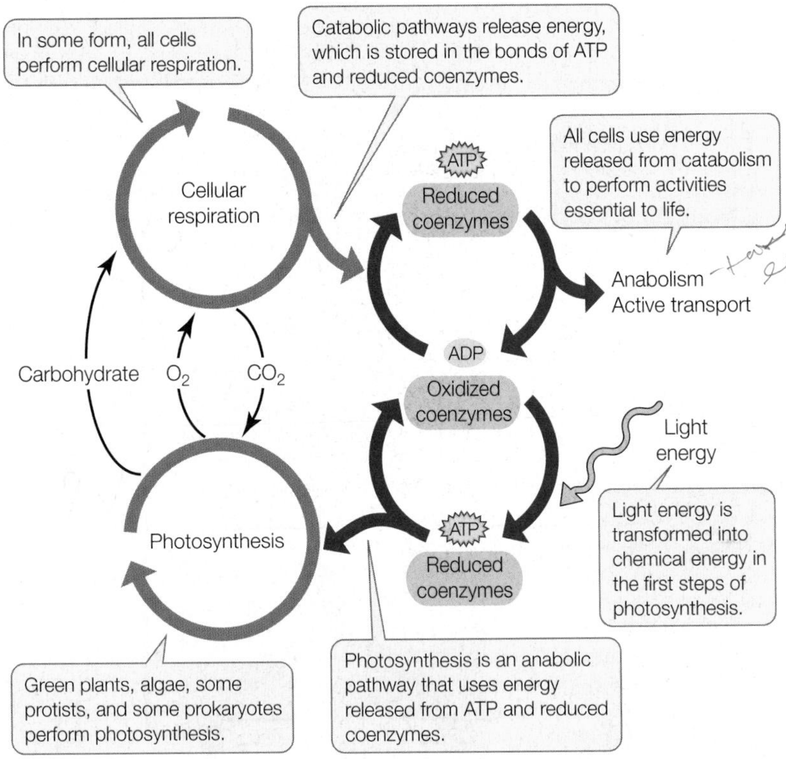

FIGURE 6.14 ATP, Reduced Coenzymes, and Metabolism The major pathways of energy metabolism, cellular respiration and photosynthesis, are related by their use of energy-transferring substrates, ATP and reduced coenzymes. Note that the net result of these pathways is to convert light energy into chemical energy to fuel the processes of life.

CHECKpoint CONCEPT 6.4

- ✓ Give examples of the catabolic conversion of a lipid and the anabolic conversion of a protein.
- ✓ Trace the biochemical pathway by which a carbon atom from a starch molecule in rice eaten today can end up in a muscle protein tomorrow.
- ✓ Describe what might happen if there were no mechanisms for modulating the level of acetyl CoA.

We have seen how cellular respiration allows organisms to harvest chemical energy from organic molecules. For the rest of the chapter, we'll look at how plants and other photosynthetic organisms produce these organic molecules using energy from light.

CONCEPT 6.5 During Photosynthesis, Light Energy Is Converted to Chemical Energy

The energy released by catabolic pathways in almost all organisms, including animals, plants, and prokaryotes, ultimately comes from the sun. **Photosynthesis** (literally, "synthesis from light") is an anabolic process by which the energy of sunlight is captured and used to convert carbon dioxide (CO_2) and water

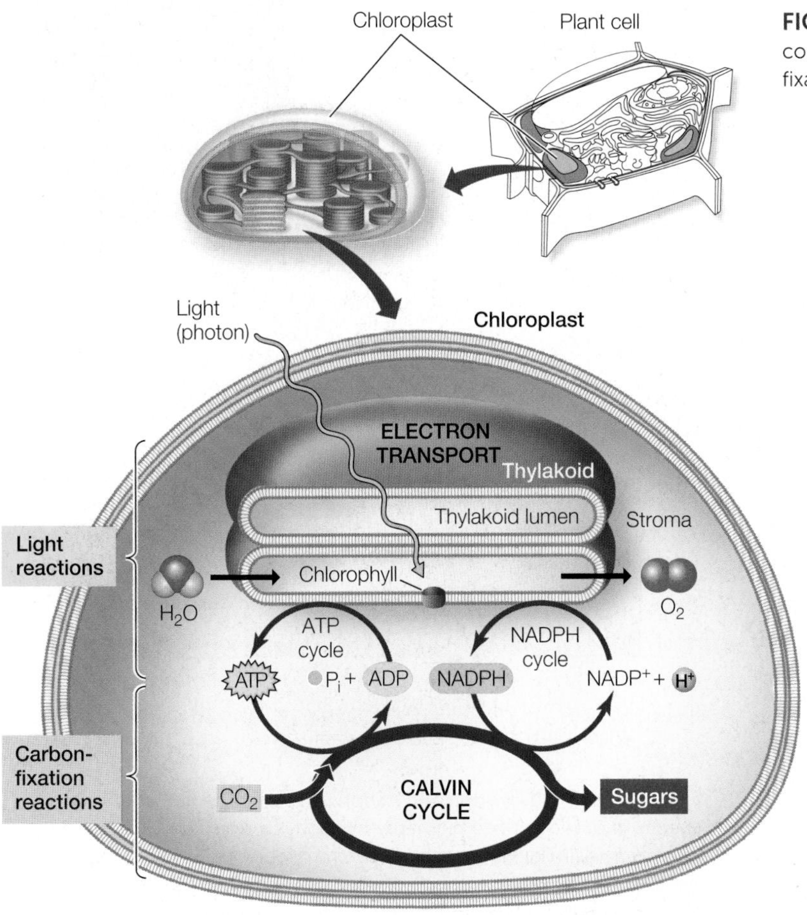

FIGURE 6.15 An Overview of Photosynthesis Photosynthesis consists of two pathways: the light reactions and the carbon-fixation reactions. In eukaryotes, these occur in the chloroplast.

(H_2O) into carbohydrates (which we represent as a six-carbon sugar, $C_6H_{12}O_6$) and oxygen gas (O_2):

$$6\,CO_2 + 6\,H_2O \rightarrow C_6H_{12}O_6 + 6\,O_2$$

This equation shows a highly endergonic reaction. The net outcome is the reverse of the general equation for glucose catabolism that we discussed in Concept 6.2. Many of the molecular processes of photosynthesis are similar to those for glucose catabolism. For example, both processes involve redox reactions, electron transport, and chemiosmosis. However, the details of photosynthesis are quite different.

Photosynthesis involves two pathways (**FIGURE 6.15**):

- The **light reactions** convert light energy into chemical energy in the form of ATP and the reduced electron carrier NADPH. This molecule is similar to NADH (see Figure 6.4A) but with an additional phosphate group attached to the sugar of its adenosine.
- The **carbon-fixation reactions** do not use light directly, but instead use the ATP and NADPH made by the light reactions, along with CO_2, to produce carbohydrates.

Both the light reactions and the carbon-fixation reactions stop in the dark because ATP synthesis and $NADP^+$ reduction require light. In photosynthetic prokaryotes (e.g., cyanobacteria), the light reactions take place on internal membranes and the carbon-fixation reactions occur in the cytosol. In plants, which will be our focus here, both pathways proceed within the chloroplast, but they occur in different parts of that organelle (see Figure 6.15).

Light energy is absorbed by chlorophyll and other pigments

Light is a form of energy that can be converted to other forms, such as heat or chemical energy. It is helpful here to discuss light in terms of its photochemistry and photobiology.

PHOTOCHEMISTRY Light is a form of **electromagnetic radiation**. Electromagnetic radiation is propagated in waves, and the amount of energy in the radiation is inversely proportional to its **wavelength**—the shorter the wavelength, the greater the energy. The visible portion of the electromagnetic spectrum (**FIGURE 6.16**) encompasses a wide range of wavelengths and energy levels. In addition to traveling in waves, light also behaves as particles, packets of light energy called **photons**, which have no mass. In plants and other photosynthetic organisms, receptive molecules absorb photons in order to harvest their energy for biological processes. These receptive molecules absorb only specific wavelengths of light—photons with specific amounts of energy.

When a photon meets a molecule, one of three things can happen:

- The photon may bounce off the molecule—it may be *scattered* or *reflected.*
- The photon may pass through the molecule—it may be *transmitted.*
- The photon may be *absorbed* by the molecule, adding energy to the molecule.

Neither of the first two outcomes causes any change in the molecule. However, in the case of absorption, the photon disappears and its energy is absorbed by the molecule. The photon's *energy* cannot disappear, because according to the first law of thermodynamics, energy is neither created nor destroyed. When the molecule acquires the energy of the photon, it is raised from a ground state (with lower energy) to an excited state (with higher energy):

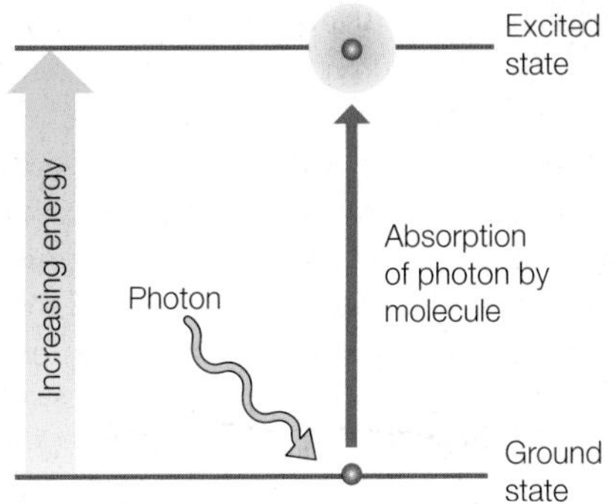

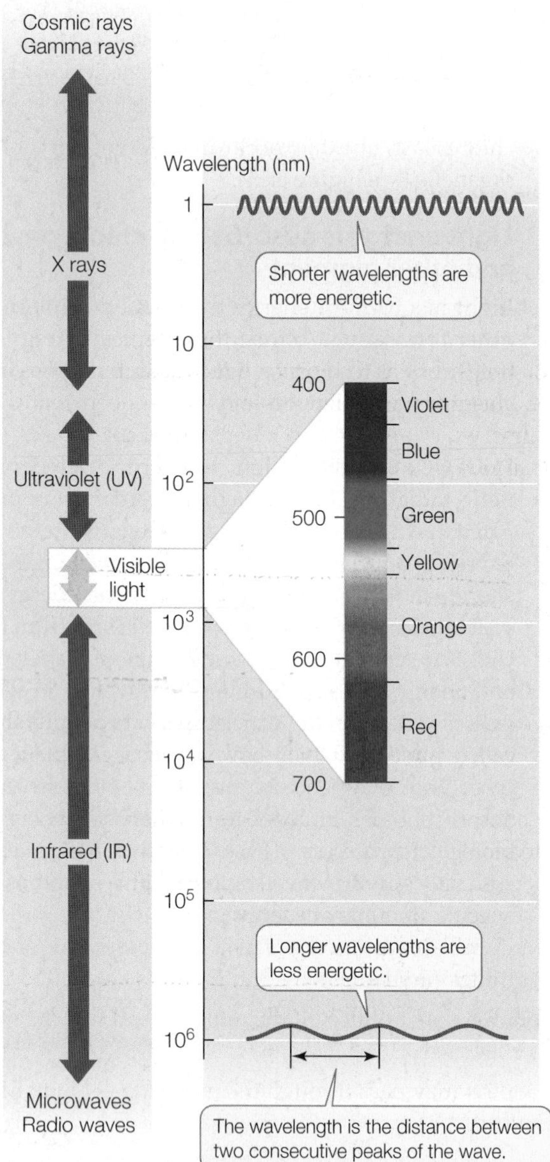

FIGURE 6.16 The Electromagnetic Spectrum The portion of the electromagnetic spectrum that is visible to humans as light is shown in detail at the right.

The difference in free energy between the molecule's excited state and its ground state is approximately equal to the free energy of the absorbed photon (a small amount of energy is lost to entropy). The increase in energy boosts one of the electrons within the molecule into a shell farther from its nucleus; this electron is now held less firmly, making the molecule unstable and more chemically reactive.

PHOTOBIOLOGY Each type of molecule absorbs light at specific, characteristic wavelengths. Molecules that absorb wavelengths in the visible spectrum are called **pigments**.

When a beam of white light (containing all the wavelengths of visible light) falls on a pigment, certain wavelengths are absorbed. The remaining wavelengths are scattered or transmitted and make the pigment appear to us as colored. For example, the pigment **chlorophyll** absorbs both blue and red light, and we see the remaining light, which is primarily green. If we

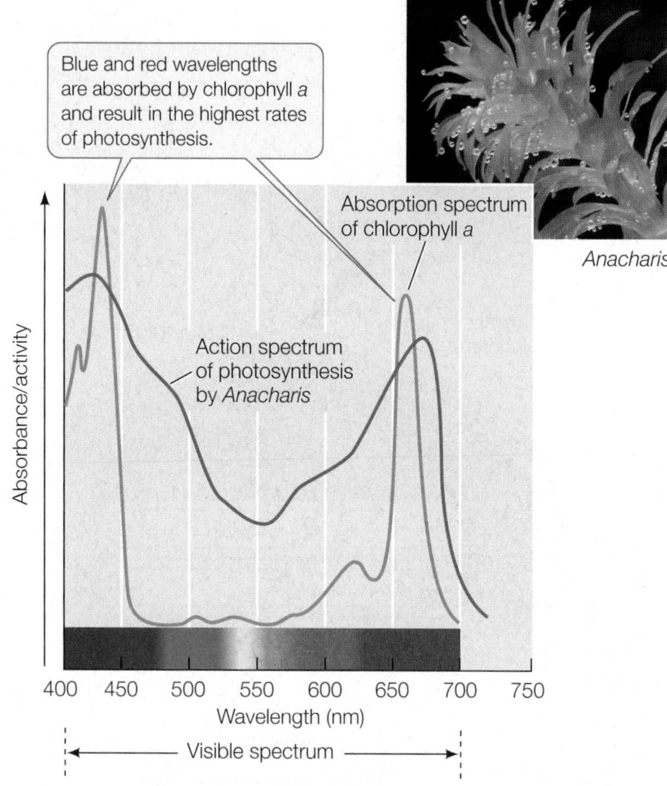

FIGURE 6.17 Absorption and Action Spectra The absorption spectrum of the purified pigment chlorophyll *a* from the aquatic plant *Anacharis* is similar to the action spectrum, obtained when different wavelengths of light are shone on the intact plant and the rate of photosynthesis is measured. In the thicker leaves of land plants, the action spectra show less of a dip in the green region (500–650 nm).

plot light absorbed by a purified pigment against wavelength, the result is an **absorption spectrum** for that pigment.

In contrast to the absorption spectrum, an **action spectrum** is a plot of the biological activity of an organism against the wavelengths of light to which it is exposed. An action spectrum can be determined as follows:

1. Place the organism (for example, a water plant with thin leaves) in a closed container.
2. Expose it to light of a certain wavelength for a period of time.
3. Measure the rate of photosynthesis by the amount of O_2 released.
4. Repeat with light of other wavelengths.

FIGURE 6.17 shows the absorption spectrum of the pigment chlorophyll *a*, which was isolated from the leaves of *Elodea* (also known as *Anacharis*), a common aquarium plant. Also shown is the action spectrum for photosynthetic activity by the same plant. A comparison of the two spectra shows that the wavelengths at which photosynthesis is highest are the same wavelengths at which chlorophyll *a* absorbs light.

In plants, two chlorophylls absorb light energy to drive the light reactions: chlorophyll *a* and chlorophyll *b*. These two molecules differ only slightly in their molecular structures. Both have a complex ring structure, similar to that of the heme group of hemoglobin, with a magnesium ion at the center (**FIGURE 6.18**).

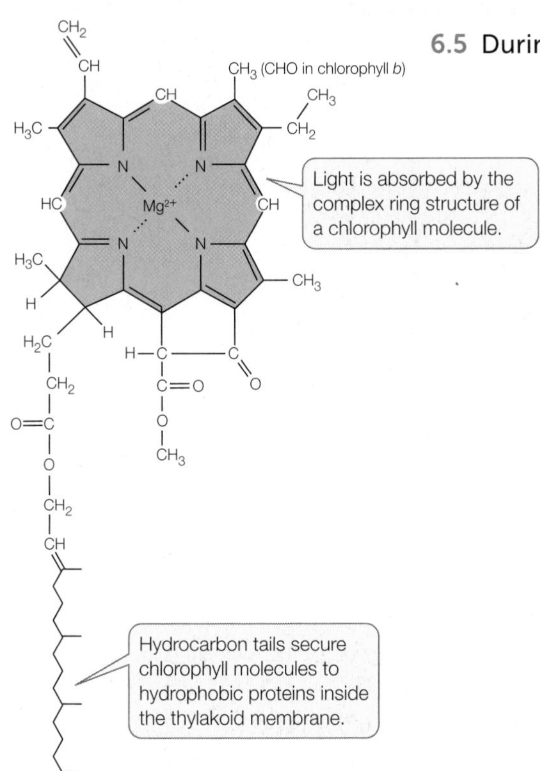

FIGURE 6.18 The Molecular Structure of Chlorophyll Chlorophyll consists of a complex ring structure (green) with a magnesium ion at its center, plus a hydrocarbon "tail." The tail anchors chlorophyll molecules to integral membrane proteins in the thylakoid membrane. Chlorophyll *a* and chlorophyll *b* are identical except for the replacement of a methyl group ($—CH_3$) with an aldehyde group (—CHO), shown on the upper right side of the ring structure.

A long hydrocarbon "tail" anchors the chlorophyll molecule to integral proteins in the thylakoid membrane of the chloroplast.

As mentioned above, the chlorophylls absorb blue and red light, which are near the two ends of the visible spectrum (see Figure 6.17). In addition, plants possess accessory pigments that absorb photons intermediate in energy between the red and the blue wavelengths, and then transfer a portion of that energy to the chlorophylls. Among these accessory pigments are carotenoids such as β-carotene, which absorb photons in the blue and blue-green wavelengths and appear deep yellow. The phycobilins, which are found in red algae and in cyanobacteria, absorb various yellow-green, yellow, and orange wavelengths.

LINK

Some plant pigments act as sensors that regulate growth and development; see **Concept 26.4**

Light absorption results in photochemical change

The pigments in photosynthetic organisms are arranged into energy-absorbing antenna systems, also called **light-harvesting complexes** (**FIGURE 6.19A**). These form part of a large multiprotein complex called a **photosystem**, where light energy is converted into chemical energy (**FIGURE 6.19B**). The photosystem spans the thylakoid membrane and consists of multiple antenna systems with their associated pigment molecules, all surrounding a **reaction center**.

When chlorophyll absorbs light, it enters an excited state. This is an unstable situation, and the chlorophyll rapidly returns to its ground state, releasing most of the absorbed energy.

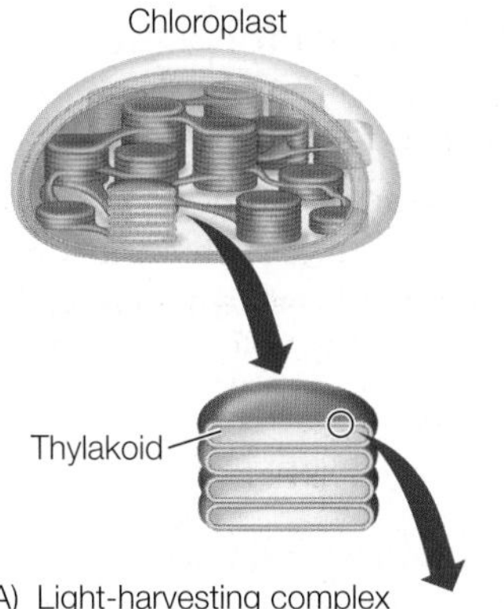

FIGURE 6.19 Photosystem Organization (A) The molecular structure of a single light-harvesting complex shows the polypeptide in brown with three helices that span the thylakoid membrane. Pigment molecules (carotenoids and chlorophylls *a* and *b*) are bound to the polypeptide. (B) Light-harvesting complexes are organized into large photosystems that span the thylakoid membrane and have a centrally placed reaction center. The light-harvesting chlorophylls absorb light and pass the energy on to a chlorophyll in the reaction center.

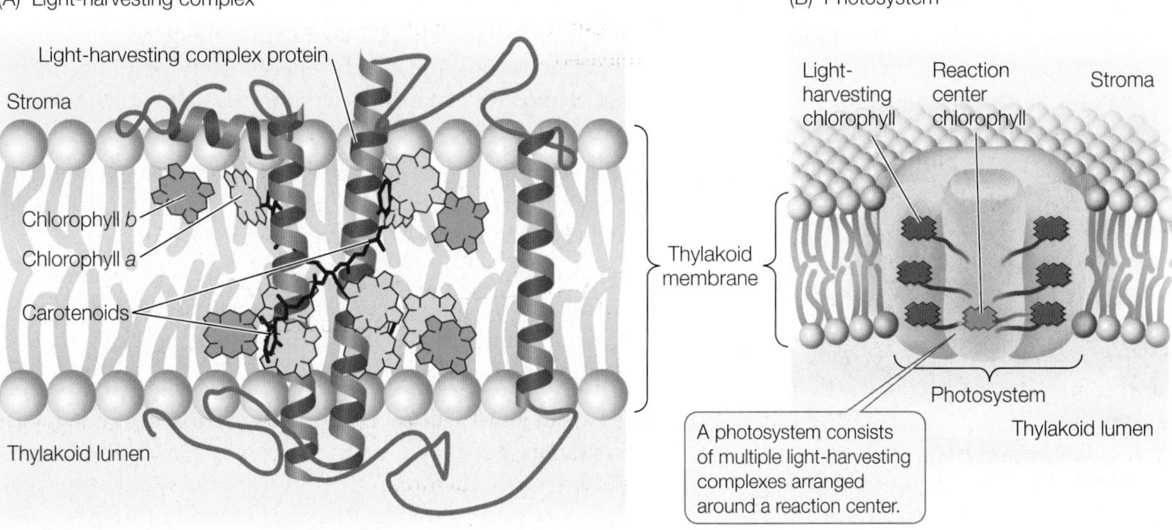

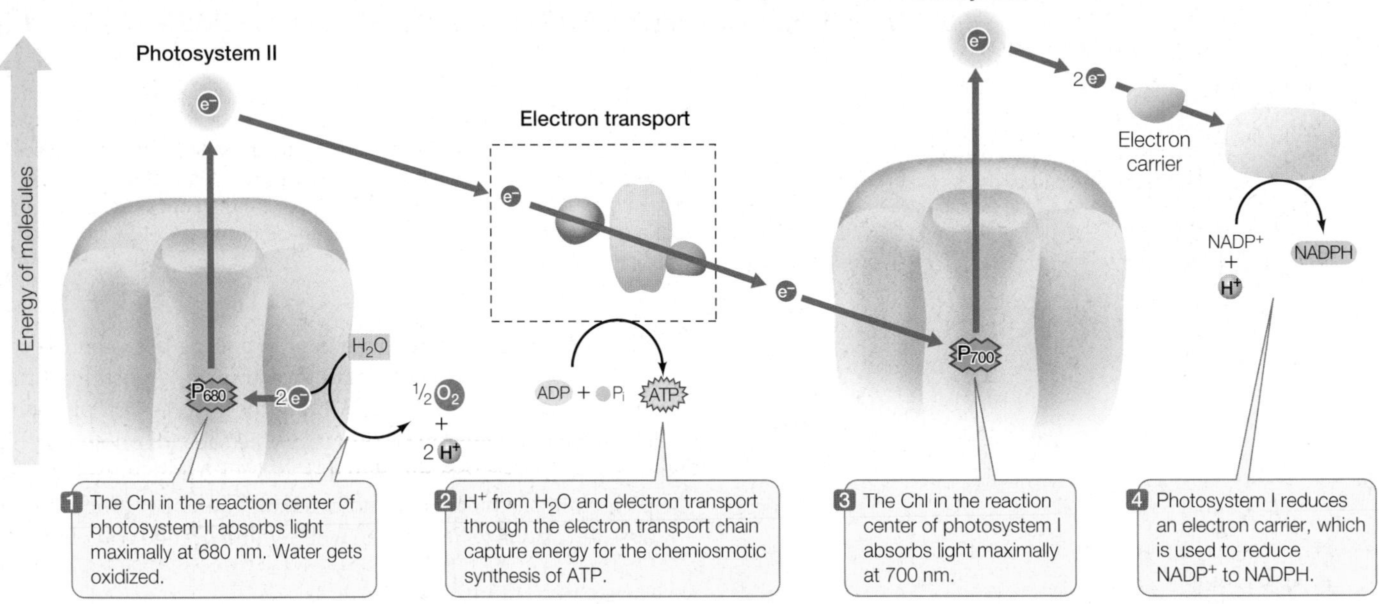

FIGURE 6.20 Noncyclic Electron Transport Uses Two Photosystems As chlorophyll molecules in the reaction centers of photosystems I and II absorb light energy, they pass electrons into a series of redox reactions, ultimately producing NADPH and ATP. The term "Z scheme" describes the path (blue arrows) of electrons as they travel through the two photosystems. In this scheme the vertical positions represent the energy levels of the molecules in the electron transport system.

Go to ANIMATED TUTORIAL 6.3
Photophosphorylation
PoL2e.com/at6.3

This is an extremely rapid process—measured in picoseconds (trillionths of a second)! For most chlorophyll molecules embedded in the thylakoid membrane, the released energy is absorbed by other, adjacent chlorophyll molecules. The energy eventually arrives at a ground-state chlorophyll molecule at the reaction center (symbolized by Chl), which absorbs the energy and becomes excited (Chl*). But when the reaction center chlorophyll returns to the ground state, something very different occurs. *The reaction center converts the absorbed light energy into chemical energy*. The chlorophyll molecule in the reaction center absorbs sufficient energy that it actually *gives up its excited electron to a chemical acceptor*:

$$\text{Chl}^* + \text{acceptor} \rightarrow \text{Chl}^+ + \text{acceptor}^-$$

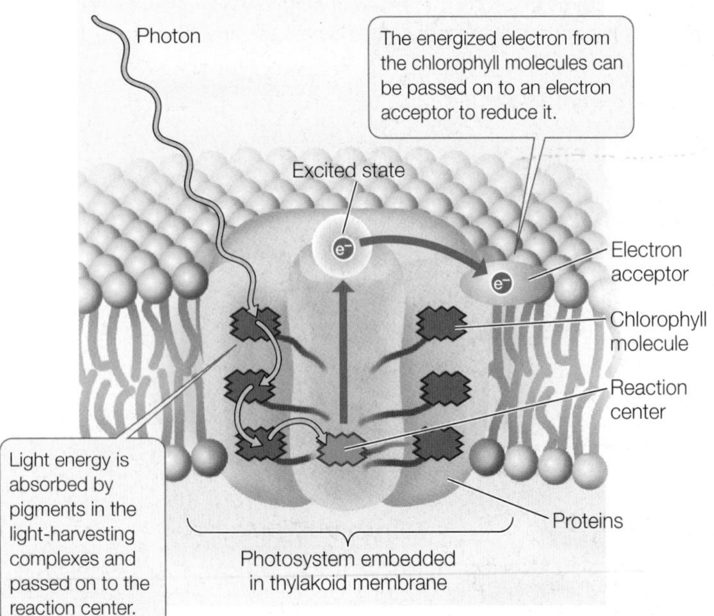

Photon
The energized electron from the chlorophyll molecules can be passed on to an electron acceptor to reduce it.
Excited state
Electron acceptor
Chlorophyll molecule
Reaction center
Proteins
Light energy is absorbed by pigments in the light-harvesting complexes and passed on to the reaction center.
Photosystem embedded in thylakoid membrane

This, then, is the first consequence of light absorption by chlorophyll: *the reaction center chlorophyll (Chl*) loses its excited electron in a redox reaction and becomes Chl⁺*. As a result of this transfer of an electron, the chlorophyll gets oxidized, while the acceptor molecule is reduced.

Reduction leads to ATP and NADPH formation

The electron acceptor that is reduced by Chl* is the first of a chain of electron carriers in the thylakoid membrane. Electrons are passed from one carrier to another in an energetically "downhill" series of reductions and oxidations. Thus the thylakoid membrane has an electron transport system similar to the respiratory chain of mitochondria (see Concept 6.2). The final electron acceptor is $NADP^+$, which gets reduced:

$$NADP^+ + H^+ + 2\,e^- \rightarrow NADPH$$

As in mitochondria, ATP is produced chemiosmotically during the process of electron transport (a process called photophosphorylation). **FIGURE 6.20** shows the series of electron transport reactions that use the energy from light to generate NADPH and ATP. There are two photosystems, each with its own reaction center:

- **Photosystem I** (containing the "P_{700}" chlorophylls at its reaction center) absorbs light energy at 700 nm and passes an excited electron to $NADP^+$, reducing it to NADPH.
- **Photosystem II** (with "P_{680}" chlorophylls at its reaction center) absorbs light energy at 680 nm, oxidizes water molecules, and initiates the electron transport chain that produces ATP.

Let's look in more detail at these photosystems, beginning with photosystem II.

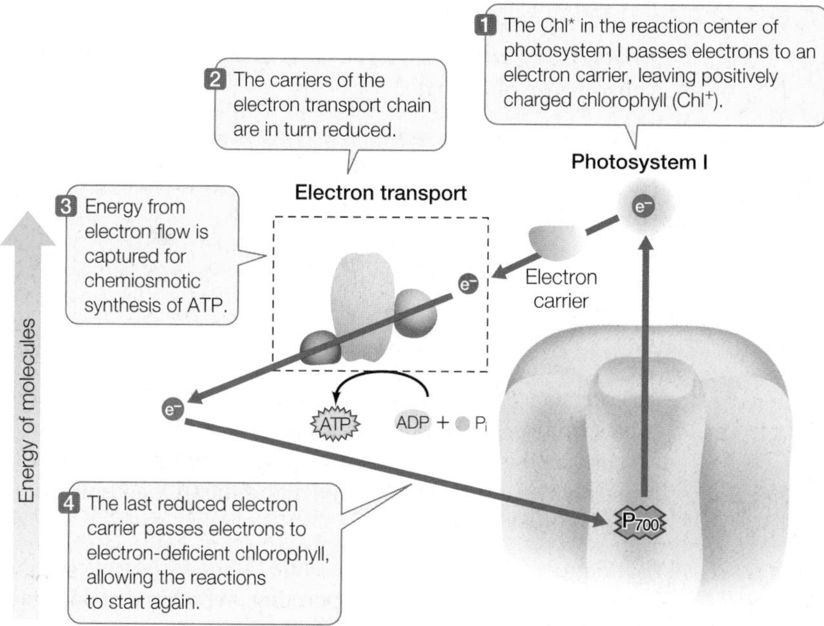

1 The Chl* in the reaction center of photosystem I passes electrons to an electron carrier, leaving positively charged chlorophyll (Chl+).
2 The carriers of the electron transport chain are in turn reduced.
Photosystem I
Electron transport
3 Energy from electron flow is captured for chemiosmotic synthesis of ATP.
e−
Electron carrier
Energy of molecules
ATP
ADP + Pi
4 The last reduced electron carrier passes electrons to electron-deficient chlorophyll, allowing the reactions to start again.
P700

FIGURE 6.21 Cyclic Electron Transport Traps Light Energy as ATP Cyclic electron transport produces ATP but no NADPH.

PHOTOSYSTEM II After an excited chlorophyll in the reaction center (Chl*) gives up its energetic electron to reduce a chemical acceptor molecule, the chlorophyll lacks an electron and is very unstable. It has a strong tendency to "grab" an electron from another molecule to replace the one it lost—in chemical terms, it is a strong oxidizing agent. The replenishing electrons come from water, splitting the H—O—H bonds:

$$H_2O \rightarrow \tfrac{1}{2}\,O_2 + 2\,H^+ + 2\,e^-$$

$$2\,e^- + 2\,Chl^+ \rightarrow 2\,Chl$$

$$\text{Overall: } 2\,Chl^* + H_2O \rightarrow 2\,Chl + 2\,H^+ + \tfrac{1}{2}\,O_2$$

The source of the O_2 produced by photosynthesis is H_2O.

Go to ANIMATED TUTORIAL 6.4 The Source of the Oxygen Produced by Photosynthesis
PoL2e.com/at6.4

Back to the electron acceptor in the electron transport system: the energetic electrons are passed through a series of thylakoid membrane–bound carriers to a final acceptor at a lower energy level. As in the mitochondrion, a proton gradient is generated and is used by ATP synthase to store energy in the bonds of ATP.

PHOTOSYSTEM I In photosystem I, an excited electron from the Chl* at the reaction center reduces an acceptor. The oxidized chlorophyll (Chl^+) now "grabs" an electron, but in this case the electron comes from the last carrier in the electron transport system. This links the two photosystems chemically. They are also linked spatially, with the two photosystems in the thylakoid membrane. The energetic electrons from photosystem I pass through several molecules and end up reducing $NADP^+$ to NADPH.

Next in the process of harvesting light energy to produce carbohydrates is the series of carbon-fixation reactions. These reactions require more ATP than NADPH. If the pathway we just described—the linear, or noncyclic, pathway—were the only set of light reactions operating, there might not be sufficient ATP for carbon fixation. **Cyclic electron transport** makes up for this imbalance. This pathway uses only photosystem I and produces ATP but not NADPH; it is cyclic because the electrons flow from the reaction center of photosystem I, through the electron transport chain, and then back to photosystem I (**FIGURE 6.21**).

APPLY THE CONCEPT

During photosynthesis, light energy is converted to chemical energy

The key role of water in supplying electrons for reduction of light-activated chlorophyll in the light reactions and in the release of O_2 to the atmosphere in the process of energy conversion has been investigated using isotopes of oxygen. The ^{18}O isotope is heavier than normal oxygen (^{16}O), and a mass spectrometer can be used to detect the difference. Green plant cells were exposed to light, water, and CO_2. (The CO_2 was supplied as the bicarbonate ion HCO_3^-, which forms CO_2 when dissolved in water.) In the first experiment, some of the oxygen atoms in the water molecules were ^{18}O ($H_2{}^{18}O$), while CO_2 had the normal form of oxygen ($C^{16}O_2$). In the second experiment, the situation was reversed, with $H_2{}^{16}O$ and $C^{18}O_2$ being supplied to the cells. After 2 hours of photosynthesis, the ratio of ^{18}O to ^{16}O was measured in the O_2 produced by the cells.[a]

EXPERIMENT	ISOTOPE RATIO		
	H_2O	CO_2	O_2
1	0.85	0.31	0.84
2	0.20	0.50	0.20

1. In experiment 1, was the isotopic ratio of O_2 more similar to that of H_2O or CO_2?
2. What about experiment 2? What can you conclude?

[a] S. Ruben et al. 1941. *Journal of the American Chemical Society* 63: 877–879.

CHECKpoint CONCEPT 6.5

- ✓ What are the reactants and products of the light reactions of photosynthesis?
- ✓ Chlorophyll absorbs light of blue and red wavelengths, but leaves also absorb some light at other wavelengths. Explain why.
- ✓ Trace the flow of electrons in noncyclic electron transport in the chloroplast and compare it with that of cyclic electron transport.
- ✓ Write equations for the production of the following in photosynthesis, and indicate whether they are oxidations, reductions, or neither: Chl*; O_2; ATP; NADPH.

We have seen how photosystems I and II absorb light energy, which ultimately ends up as chemical energy in ATP and NADPH. Let's look now at how these two energy-rich molecules are used in the carbon-fixation reactions to reduce CO_2 and thereby form carbohydrates.

CONCEPT 6.6 Photosynthetic Organisms Use Chemical Energy to Convert CO_2 to Carbohydrates

The energy in ATP and NADPH is used in the carbon-fixation reactions to "fix" CO_2 into a reduced form and convert it to carbohydrates. Most CO_2 fixation occurs only in the light, when ATP and NADPH are being generated. The metabolic pathway occurs in the stroma, or central region, of the chloroplast (see Figure 6.15) and is called the **Calvin cycle** after one of its discoverers, Melvin Calvin.

Like all biochemical pathways, each reaction in the Calvin cycle is catalyzed by a specific enzyme. The cycle is composed of three distinct processes (**FIGURE 6.22**):

- *Fixation of CO_2.* The initial reaction of the Calvin cycle adds the one-carbon CO_2 to an acceptor molecule, the five-carbon ribulose 1,5-bisphosphate (RuBP). The immediate product is a six-carbon molecule, which quickly breaks down into two three-carbon molecules called 3-phosphoglycerate (3PG; **FIGURE 6.23**). The enzyme that catalyzes this reaction, **ribulose bisphosphate carboxylase/oxygenase (rubisco)**, is rather sluggish as enzymes go. It typically catalyzes two to three fixation reactions per second. Because of this, plants need a lot of rubisco to perform enough photosynthesis to satisfy the needs of growth and metabolism. Rubisco constitutes about half of all the protein in a leaf, and it is probably the most abundant protein in the world!
- *Reduction of 3PG to form glyceraldehyde 3-phosphate.* This series of reactions involves a phosphorylation (using the high-energy phosphate from an ATP made in the light reactions) and a reduction (using an NADPH made in the light reactions). The product is **glyceraldehyde 3-phosphate (G3P)**, which is a three-carbon sugar phosphate, also called triose phosphate:

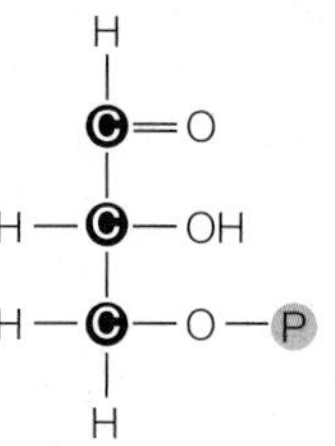

Glyceraldehyde 3-phosphate (G3P)

- *Regeneration of the CO_2 acceptor, RuBP.* Most of the G3P ends up as ribulose monophosphate (RuMP), and ATP is used to convert this compound into RuBP. So for every "turn" of the Calvin cycle, one CO_2 is fixed and the CO_2 acceptor is regenerated.

What happens to the extra G3P made by the Calvin cycle (see Figure 6.22)? It has two fates, depending on the time of day and the needs of different parts of the plant:

- Some of the extra G3P is exported out of the chloroplast to the cytosol, where it is converted to hexoses (glucose and fructose). This is the familiar $C_6H_{12}O_6$ from the general equation for photosynthesis. These molecules may be catabolized for energy in mitochondria as part of cellular respiration; used as carbon skeletons for the synthesis of amino acids and other molecules (see Figure 6.13); or converted to sucrose, which is transported out of the leaf to other organs in the plant.
- Late in the day when glucose has accumulated inside the chloroplast, the glucose units are linked to form the polysaccharide starch. This storage carbohydrate can then be drawn on during the night so that the photosynthetic tissues can continue to export sucrose to the rest of the plant, even when photosynthesis is not taking place. In addition, starch is abundant in nonphotosynthetic organs such as roots, underground stems, and seeds, where it provides a ready supply of glucose to fuel cellular activities, including plant growth.

The products of the Calvin cycle are of crucial importance to Earth's entire biosphere. The C—H covalent bonds generated by this cycle store almost all of the energy for life on Earth. Photosynthetic organisms, which are also called **autotrophs** ("self-feeders"), release most of this energy in cellular respiration, and use it to support their own growth, development, and reproduction. But plants are also the source of energy for other organisms. Much plant matter ends up being consumed by **heterotrophs** ("other-feeders"), including humans and other animals, which cannot photosynthesize. Heterotrophs depend on autotrophs for chemical energy, which they harvest via cellular respiration. In addition, many heterotrophs rely on plants to produce other molecules (such as vitamins) that the heterotrophs cannot synthesize for themselves.

LINK

Concept 44.3 describes how the energy captured by autotrophs flows through ecological communities

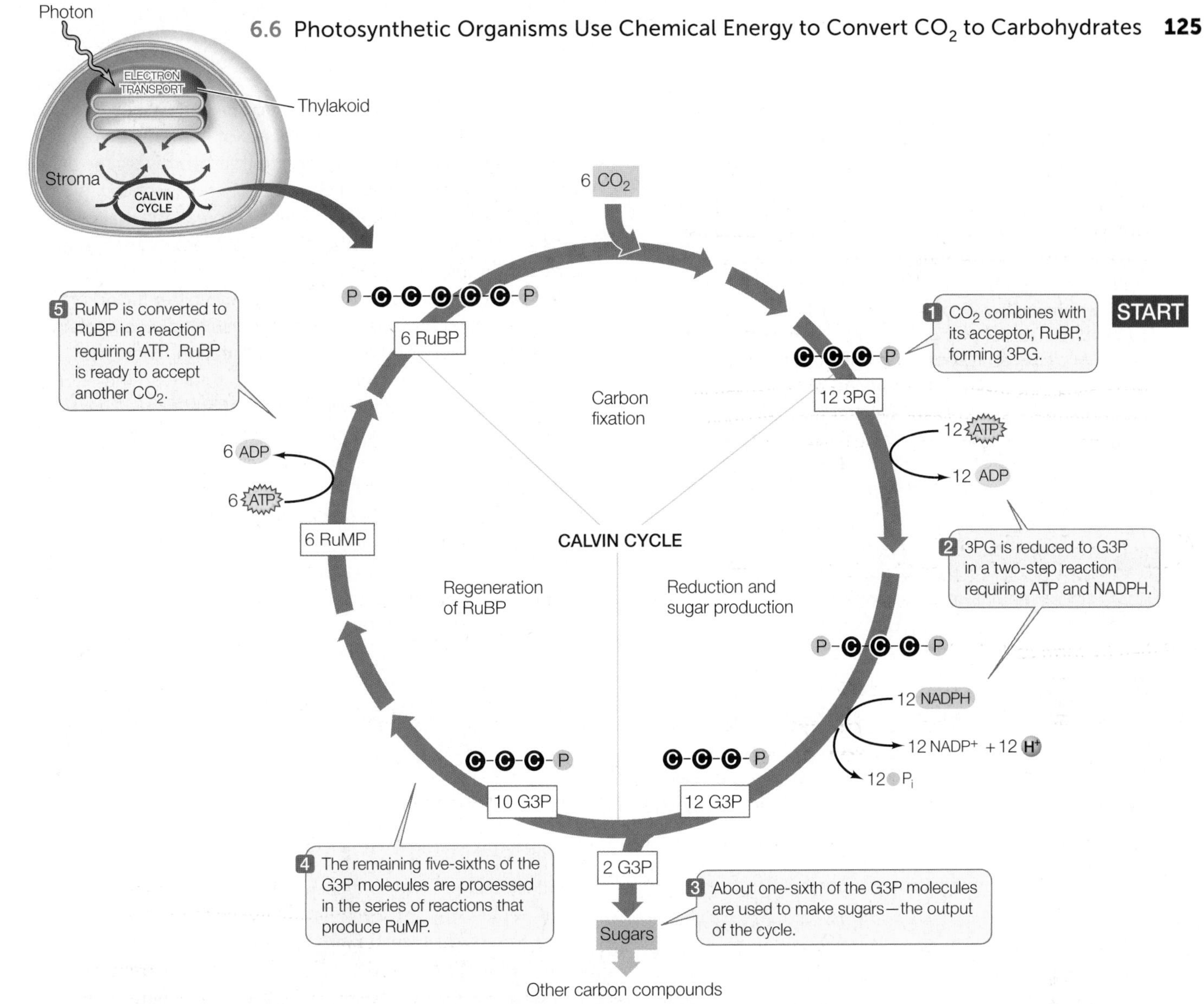

Photon
ELECTRON TRANSPORT
Thylakoid
Stroma
CALVIN CYCLE
$6\ CO_2$
START
1 CO_2 combines with its acceptor, RuBP, forming 3PG.
6 RuBP
12 3PG
Carbon fixation
12 ATP
12 ADP
2 3PG is reduced to G3P in a two-step reaction requiring ATP and NADPH.
CALVIN CYCLE
Regeneration of RuBP
Reduction and sugar production
12 NADPH
12 $NADP^+$ + 12 H^+
12 P_i
12 G3P
10 G3P
2 G3P
Sugars
Other carbon compounds
3 About one-sixth of the G3P molecules are used to make sugars—the output of the cycle.
4 The remaining five-sixths of the G3P molecules are processed in the series of reactions that produce RuMP.
6 RuMP
6 ATP
6 ADP
5 RuMP is converted to RuBP in a reaction requiring ATP. RuBP is ready to accept another CO_2.

FIGURE 6.22 The Calvin Cycle The Calvin cycle uses the ATP and NADPH generated in the light reactions to produce G3P from CO_2. The G3P is used as a starting material for the production of glucose and other carbohydrates. Six turns of the cycle are needed to produce one molecule of the hexose glucose.

Go to ACTIVITY 6.7 The Calvin Cycle
PoL2e.com/ac6.7

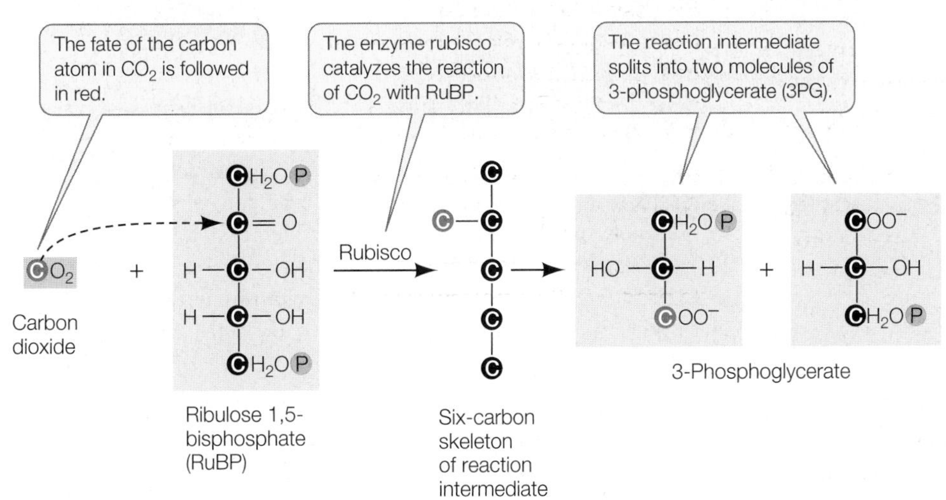

FIGURE 6.23 RuBP Is the Carbon Dioxide Acceptor The enzyme rubisco adds CO_2 to the five-carbon compound RuBP. The resulting six-carbon compound immediately splits into two molecules of 3PG.

Go to ANIMATED TUTORIAL 6.5
Tracing the Pathway of CO_2
PoL2e.com/at6.5

CHECKpoint CONCEPT 6.6

- ✓ What are the three processes of the Calvin cycle?
- ✓ If green plant cells are incubated in the presence of CO_2 molecules containing radioactive carbon atoms, the fate of the carbon atoms can be followed. In an experiment, radioactive CO_2 was given for 1 minute to plant cells, and then the cells were examined after 1, 5, 10, 20, and 30 minutes. The following molecules were labeled with radioactive carbon at some point(s): glucose, glyceraldehyde 3-phosphate, glycine (an amino acid), 3-phosphoglycerate, ribulose 1,5-bisphosphate, and sucrose. List these molecules in the order in which they first become labeled.

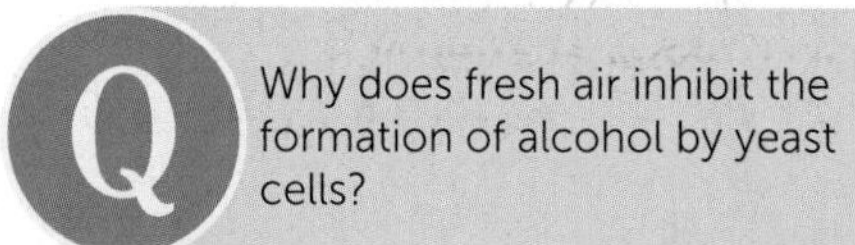

Q
Why does fresh air inhibit the formation of alcohol by yeast cells?

ANSWER Armed with our knowledge of metabolism, we can now explain Pasteur's observations of beet sugar and alcohol:

1. *Nothing happened to beet mash unless microscopic yeast cells were present.* Beet sugar is a product of photosynthetic CO_2 fixation (Concept 6.6). Catabolism of carbohydrates is a cellular activity, characteristic of virtually all living things (Concepts 6.2 and 6.3).
2. *In the presence of fresh air, yeast cells grew vigorously on the mash, and bubbles of CO_2 were formed.* Carbohydrate catabolism under aerobic conditions yields a large amount of ATP (Concepts 6.1 and 6.2), which can be used to fuel anabolic pathways for growth (Concept 6.4). Under aerobic conditions, glucose is fully oxidized to CO_2 (Concept 6.2).
3. *Without fresh air, the yeast grew slowly, less CO_2 was produced, and alcohol was formed.* Alcoholic fermentation occurs in anaerobic conditions, and yields far less ATP (for growth) than aerobic metabolism; a small amount of CO_2 is produced (Concept 6.3).

As we noted in the opening story, the use by humans of yeasts for fermentation has a long history (**FIGURE 6.24**).

Beer comes from the fermentation of barley seeds. These are soaked to begin the process of germination, which includes the induction of an enzyme that hydrolyzes the starch stored in the seed. The resulting disaccharide, maltose, is what the yeast cells use for energy—they first break the maltose down to its glucose monomers. At a cool temperature and under anaerobic conditions, yeast produces alcohol. An herb called hops is added to impart a distinctive bitter flavor.

Wine grapes are crushed, and the resulting juice contains sugars that are used by yeast for fermentation. There are yeasts growing on the skins of grapes naturally, so the original winemakers did not add yeast. More recently, special yeast strains have been used to control the fermentation process. The longer this process is allowed to proceed, the less sugar remains (resulting in a less sweet taste) and the higher the alcohol content. After fermentation the wine is stored in wooden casks, typically at cool temperatures. During this time hundreds of molecular transformations occur, giving different wines distinctive properties.

Bread is made from ground-up plant seeds (flour), which contain abundant starch. Moisture activates enzymes inside the seed that catabolize the starch to monosaccharides. These monosaccharides are used by bread yeast in fermentation reactions that result in CO_2 production, and the resulting gas bubbles cause the complex flour mixture to rise.

FIGURE 6.24 Products of Glucose Metabolism Beer, wine, and bread are all made using fermentation reactions in yeast cells.

SUMMARY

CONCEPT 6.1 ATP and Reduced Coenzymes Play Important Roles in Biological Energy Metabolism

- Metabolism is carried out in small steps and involves coenzymes as carriers of chemical energy.
- Adenosine triphosphate (ATP) serves as "energy currency" in the cell. Hydrolysis of ATP releases a large amount of free energy. **Review Figure 6.1 and ACTIVITY 6.1**
- In **oxidation**, a material loses electrons by transfer to another material, which thereby undergoes **reduction**. Such **redox** reactions transfer large amounts of energy.
- The coenzyme nicotinamide adenine dinucleotide (NAD) is a key electron carrier in biological redox reactions. It exists in two forms, one oxidized (NAD^+) and the other reduced (NADH). **Review Figure 6.4**

CONCEPT 6.2 Carbohydrate Catabolism in the Presence of Oxygen Releases a Large Amount of Energy

- The sequential pathways of aerobic glucose catabolism are **glycolysis**, **pyruvate oxidation**, and the **citric acid cycle**. **Review Figure 6.6**
- In glycolysis, a series of ten enzyme-catalyzed reactions in the cell cytoplasm converts glucose to two molecules of pyruvate. Energy is released and captured as ATP and NADH. **Review Figure 6.7**
- The next pathway, pyruvate oxidation, links glycolysis to the citric acid cycle. Pyruvate oxidation converts pyruvate into the two-carbon molecule acetyl CoA.
- In the citric acid cycle, a series of eight enzyme-catalyzed reactions fully oxidizes acetyl CoA to CO_2. Much energy is released, and most is used to form NADH. **Review Figure 6.8 and ACTIVITY 6.2**
- The energy in NADH is used to make ATP via a series of **electron transport** carriers and chemiosmosis. **Review Figure 6.9, ANIMATED TUTORIAL 6.1, and ACTIVITY 6.3**
- In **oxidative phosphorylation**, ATP is formed with the energy derived from the reoxidation of reduced coenzymes. This depends on the process of **chemiosmosis**, in which a proton gradient across a membrane powers ATP formation. This occurs at the cell membrane in prokaryotes, and in the mitochondria and chloroplasts in eukaryotes. **Review Figures 6.10 and 6.11 and ANIMATED TUTORIAL 6.2**

CONCEPT 6.3 Carbohydrate Catabolism in the Absence of Oxygen Releases a Small Amount of Energy

- In the absence of O_2, glycolysis is followed by **fermentation**. Together, these pathways partially oxidize pyruvate and generate the end products lactic acid or ethanol. In the process, NAD^+ is regenerated from NADH so that glycolysis can continue, thus generating a small amount of ATP. **Review Figure 6.12 and ACTIVITY 6.4**
- For each molecule of glucose used, fermentation yields 2 molecules of ATP. In contrast, glycolysis, pyruvate oxidation, the citric acid cycle, and oxidative phosphorylation yield up to 32 molecules of ATP per molecule of glucose. **Review ACTIVITY 6.5**

CONCEPT 6.4 Catabolic and Anabolic Pathways Are Integrated

- The catabolic pathways for the breakdown of carbohydrates, lipids, and proteins feed into the energy-harvesting metabolic pathways. **Review Figure 6.13**
- Anabolic pathways use intermediate components of the energy-harvesting pathways to synthesize fatty acids, amino acids, and other essential building blocks.
- The formation of glucose from intermediates of glycolysis and the citric acid cycle is called **gluconeogenesis**.
- The enzymes of glycolysis and the citric acid cycle are regulated by various mechanisms, including allosteric regulation. Excess acetyl CoA is diverted into fatty acid synthesis. **Review ACTIVITY 6.6**

CONCEPT 6.5 During Photosynthesis, Light Energy Is Converted to Chemical Energy

- The **light reactions** of photosynthesis convert light energy into chemical energy. They produce ATP and reduce $NADP^+$ to NADPH. **Review Figure 6.15**
- Light is a form of **electromagnetic radiation**. It is emitted in particle-like packets called **photons** but has wavelike properties. Molecules that absorb light in the visible spectrum are called **pigments**. Photosynthetic organisms have several pigments, most notably **chlorophylls**. **Review Figures 6.16–6.18**
- The absorption of a photon puts a chlorophyll molecule into an excited state that has more energy than its ground state. This energy can be transferred via other chlorophylls to one in the **reaction center** of a photosystem. **Review Figure 6.19**
- An excited chlorophyll can act as a reducing agent, transferring excited electrons to other molecules. Oxidized chlorophyll regains electrons by the splitting of H_2O.
- In the thylakoid membrane of the chloroplast, **photosystems I** and **II** and a noncyclic electron transport system produce ATP via oxidative phosphorylation. NADPH and O_2 are also produced. **Review Figure 6.20 and ANIMATED TUTORIALS 6.3 and 6.4**
- **Cyclic electron transport** uses only photosystem I and produces only ATP. **Review Figure 6.21**

CONCEPT 6.6 Photosynthetic Organisms Use Chemical Energy to Convert CO_2 to Carbohydrates

- The **Calvin cycle** makes carbohydrates from CO_2. The cycle consists of three processes: fixation of CO_2, reduction and sugar production, and regeneration of RuBP. **Review Figure 6.22 and ACTIVITY 6.7**
- RuBP is the initial CO_2 acceptor, and 3PG is the first stable product of CO_2 fixation. The enzyme **rubisco** catalyzes the reaction of CO_2 and RuBP to form 3PG. **Review Figure 6.23 and ANIMATED TUTORIAL 6.5**
- ATP and NADPH formed by the light reactions are used to fuel the reduction of 3PG to form **glyceraldehyde 3-phosphate (G3P)**—a starting material for the synthesis of glucose and other carbohydrates.

Go to the Interactive Summary to review key figures, Animated Tutorials, and Activities
PoL2e.com/is6

Go to LaunchPad at **macmillanhighered.com/launchpad** for additional resources, including LearningCurve Quizzes, Flashcards, and many other study and review resources.

PART 2
Genetics

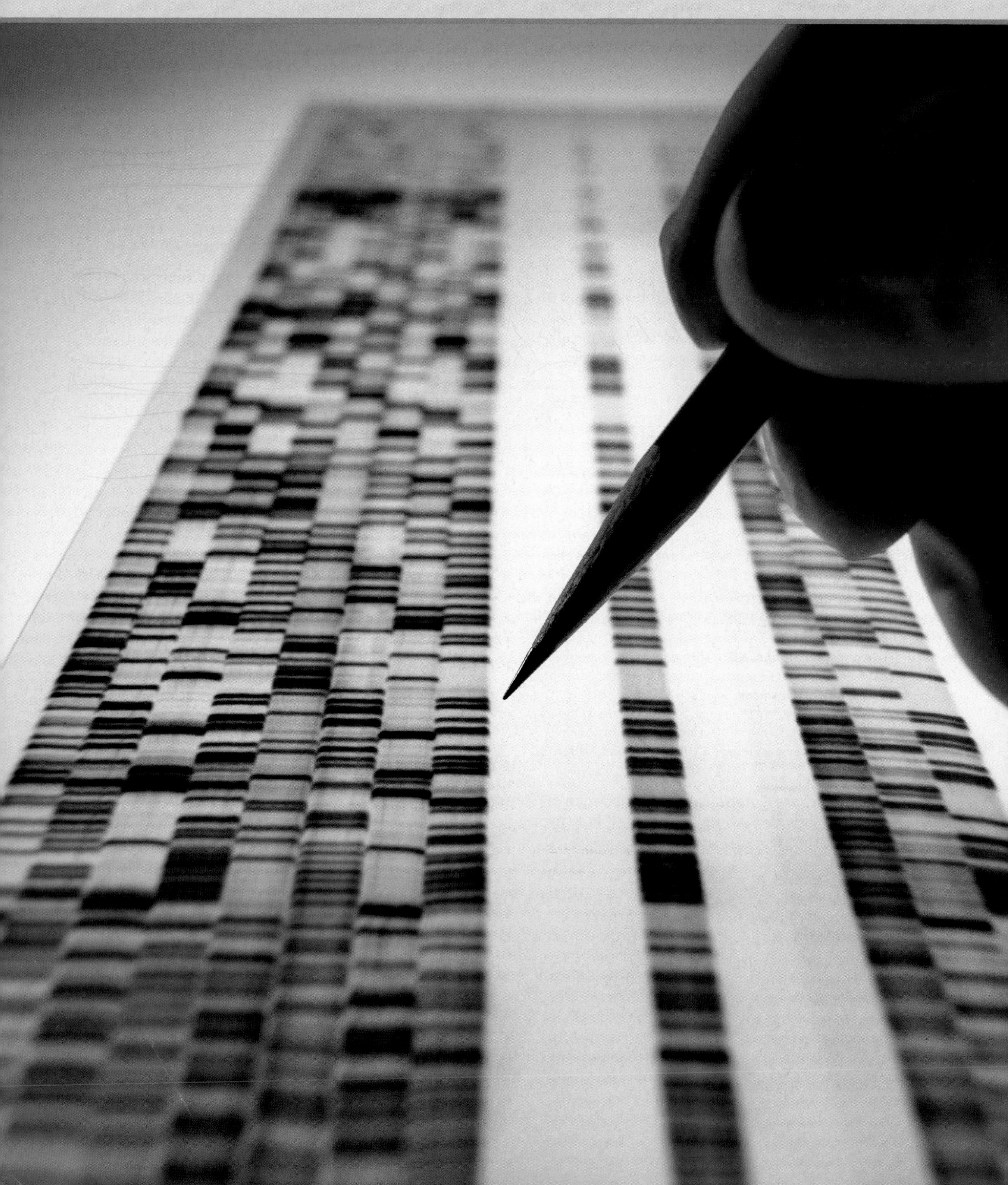

7 The Cell Cycle and Cell Division

KEY CONCEPTS

7.1 Different Life Cycles Use Different Modes of Cell Reproduction

7.2 Both Binary Fission and Mitosis Produce Genetically Identical Cells

7.3 Cell Reproduction Is Under Precise Control

7.4 Meiosis Halves the Nuclear Chromosome Content and Generates Diversity

7.5 Programmed Cell Death Is a Necessary Process in Living Organisms

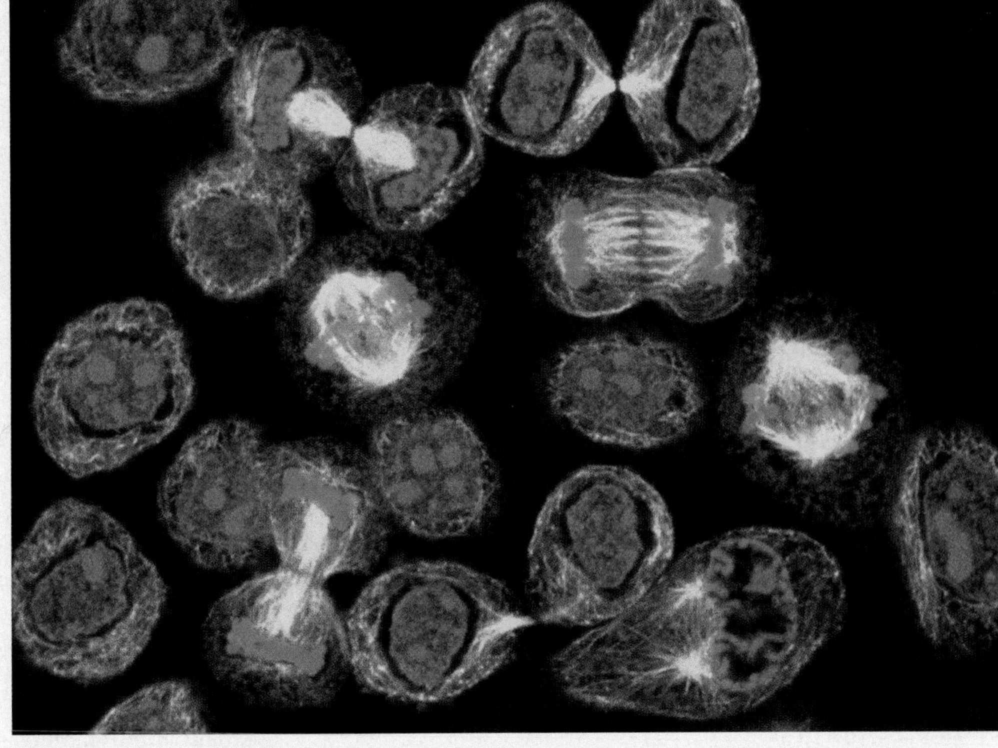

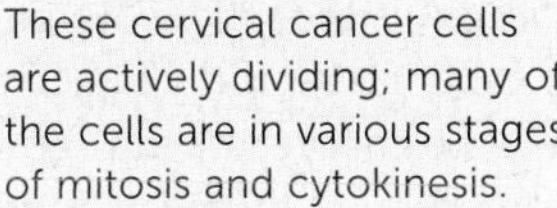

These cervical cancer cells are actively dividing; many of the cells are in various stages of mitosis and cytokinesis.

Ruth felt healthy and was surprised when she was called back to her physician's office a week after her annual checkup. "Your lab report indicates you have early cervical cancer," said the doctor. "I ordered a follow-up test, and it came back positive. At some point, you were infected with HPV."

Ruth felt numb as soon as she heard the word "cancer." Her mother had died of breast cancer in the previous year. The doctor's statement about HPV—human papillomavirus—did not register in her consciousness. Sensing Ruth's discomfort, the doctor quickly reassured her that the cancer had been caught at an early stage and that a simple surgical procedure would remove it. Two weeks later, the cancer was removed and Ruth remains cancer-free. She was fortunate that her annual medical exam included a Papanicolau (Pap) test, in which the cells lining the cervix are examined for abnormalities. Since they were begun almost 50 years ago in Europe, Pap tests have resulted in the detection and removal of millions of early cervical cancers, and the death rate from this potentially lethal disease has plummeted.

The role of HPV in causing most cervical cancers was discovered only recently. The German physician Harald Zur-Hausen was awarded the Nobel Prize in 2008 for this discovery, and it has led to a vaccine to prevent future infections. There are many different types of HPV, and many of the ones that infect humans cause warts, which are small, rough growths on the skin. The types of HPV that infect tissues at the cervix get there by sexual transmission, and this is a common infection.

When HPV arrives at the tissues lining the cervix, it has one of two fates. Most of the time it enters the cells and turns them into HPV factories, releasing a lot of HPV particles into the mucus outside the uterus. These viruses can infect another person during a sexual encounter. But in some cases the virus follows a different, more sinister path. The viral DNA becomes incorporated into the DNA of the cervical cells, and the cells are stimulated to reproduce.

Cell reproduction in healthy humans is tightly controlled by a variety of mechanisms, but the virus-infected cells lose these controls. Understanding how cell division is controlled is clearly an important subject for the development of cancer treatments. But cell division is not just important in medicine. It underlies the growth, development, and reproduction of all organisms.

How does infection with HPV result in uncontrolled cell reproduction?

You will find the answer to this question on page 148.

CONCEPT 7.1 Different Life Cycles Use Different Modes of Cell Reproduction

In Chapter 4 we described cells as the basic compartments of life, where biological processes are separated from the external environment. Cells are also essential for biological reproduction.

The life span of an organism from birth to death is intimately linked to cell reproduction, which is commonly referred to as **cell division**: a process by which a parent cell duplicates its genetic material and then divides into two similar cells. Cell division plays important roles in the growth and repair of tissues in multicellular organisms, as well as in the reproduction of all organisms (**FIGURE 7.1**). Although the details vary widely, organisms have two basic strategies for reproducing themselves: asexual reproduction and sexual reproduction. These two strategies make use of different types of cell division.

Asexual reproduction by binary fission or mitosis results in genetic constancy

Asexual reproduction is a rapid and effective means of making new individuals, and it is common in nature. The offspring resulting from asexual reproduction are **clones** of the parent organism—they are genetically identical (or virtually identical) to each other and the parent. Any genetic variations among the parent and offspring are due to changes called mutations, which are alterations in DNA sequence caused by environmental factors or errors in DNA replication (see Concept 9.3). This small amount of variation contrasts with the extensive variation possible in sexually reproducing organisms, as we will see later in this chapter.

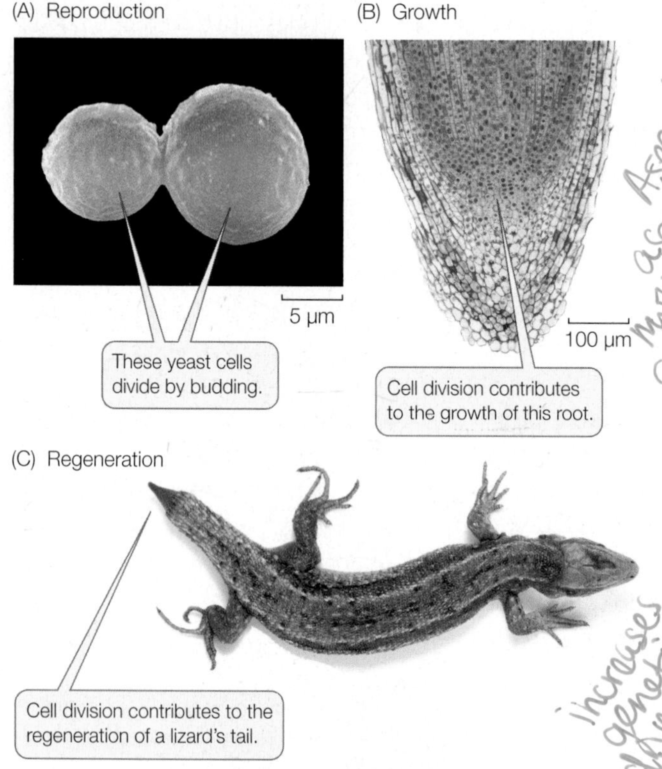

FIGURE 7.1 The Importance of Cell Division Cell division is the basis for (A) reproduction, (B) growth, and (C) repair and regeneration of tissues.

FIGURE 7.2 Asexual Reproduction on a Large Scale This forest of aspens in Utah's Wasatch Mountains arose via asexual reproduction. Genetically, all these trees are virtually identical.

In most cases, single-celled prokaryotes reproduce by binary fission, an asexual process that we will discuss in Concept 7.2. A cell of the bacterium *Escherichia coli* is the whole organism, so when it divides to form two new cells, it is reproducing. Similarly, single-celled eukaryotes (such as fission yeast) can reproduce asexually through mitosis followed by cytokinesis, processes that also produce two genetically identical cells (see Concept 7.2).

Many multicellular eukaryotes, including fungi and plants, can also reproduce asexually. Perhaps the most dramatic example of this is a forest containing thousands of aspen trees (*Populus tremuloides*) in the Wasatch Mountains of Utah (**FIGURE 7.2**). DNA analyses have shown that these trees are clones—they are virtually identical genetically. Aspen can reproduce sexually, with male and female plants, but in many aspen stands all the trees are the same sex and reproduction is asexual. An extensive root system spreads through the soil, and at intervals stems form and grow into new trees.

Sexual reproduction by meiosis results in genetic diversity

Sexual reproduction involves the fusion of two specialized cells called **gametes**, and can result in offspring with considerable genetic variation. In many diploid organisms, the gametes form by **meiosis**—a process of cell division (described in Concept 7.4) resulting in daughter cells with only half the genetic material of the original cell. During meiosis, the genetic material is randomly separated and reorganized so that the daughter cells differ genetically from one another. Because of this genetic variation, some offspring of sexual reproduction may be better adapted than others to survive and reproduce in a particular environment. Meiosis thus increases genetic diversity, which is the raw material for natural selection and evolution.

As we described in Chapter 4, the DNA in eukaryotic cells is organized into multiple structures called chromosomes. Each chromosome consists of a double-stranded molecule of DNA and associated proteins. In multicellular organisms, the body cells that are *not* specialized for reproduction are called **somatic cells.** In many familiar organisms, including most vascular plants and animals, the somatic cells each contain two sets of chromosomes, and the chromosomes occur in pairs called **homologous pairs.** One chromosome of each pair comes from the organism's female parent, and the other comes from its male parent. For example, in humans with 46 chromosomes, 23 come from the mother and 23 from the father, with, for example, a chromosome 1 from each parent, and so on.

The two chromosomes in a homologous pair (called homologs of one another) bear corresponding, though not identical, genetic information. For example, a homologous pair of chromosomes in a plant may carry different versions of a gene that controls seed shape. One homolog may carry the version for wrinkled seeds, while the other may carry the version for smooth seeds.

LINK

The inheritance of characteristics such as seed shape is discussed in **Concept 8.1**

Gametes contain only a single set of chromosomes—that is, one homolog from each pair. The number of chromosomes in a gamete is denoted by n, and the cell is said to be **haploid.** During sexual reproduction, two haploid gametes fuse to form a **zygote** in a process called **fertilization.** The zygote thus has two sets of chromosomes, just as the somatic cells do. The chromosome number in the zygote is denoted by $2n$, and the cells are said to be **diploid.**

In many familiar organisms the zygote divides by mitosis, producing a new, mature organism with diploid somatic cells (as we have just described). But some other organisms have haploid stages in their life cycles.

Sexual life cycles are diverse

All sexual life cycles involve meiosis to produce haploid cells. In some cases, gametes develop immediately after meiosis. In others, each haploid cell divides and develops into a haploid organism—the haploid stage of the life cycle—that eventually produces gametes by mitosis. The fusion of gametes—fertilization—results in a zygote and begins the diploid stage of the life cycle. Since the origin of sexual reproduction, evolution has generated many different versions of the sexual life cycle. **FIGURE 7.3**

FIGURE 7.3 Sexual Life Cycles Involve Fertilization and Meiosis In sexual reproduction, haploid (n) cells or organisms alternate with diploid ($2n$) cells or organisms.

Go to ACTIVITY 7.1 Sexual Life Cycle PoL2e.com/ac7.1

Bread mold (*Rhizopus stolonifer*) (haploid organism)

Lady fern (*Athyrium felix-femina*) (diploid sporophyte)

Nyala (*Tragelaphus angasii*) (diploid organism)

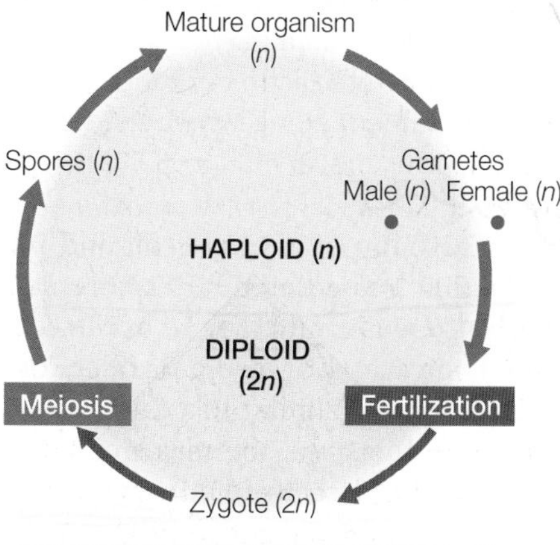

In the **haplontic life cycle**, the mature organism is haploid and the zygote is the only diploid stage. Found in most protists, fungi, and some green algae.

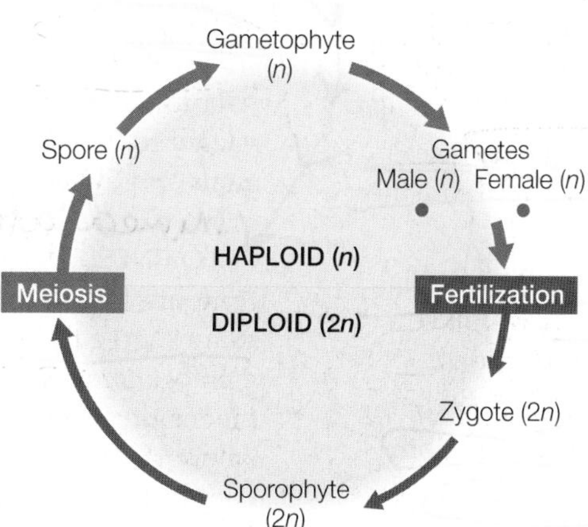

In **alternation of generations**, the organism passes through haploid and diploid stages that are both multicellular. Found in most plants and some fungi.

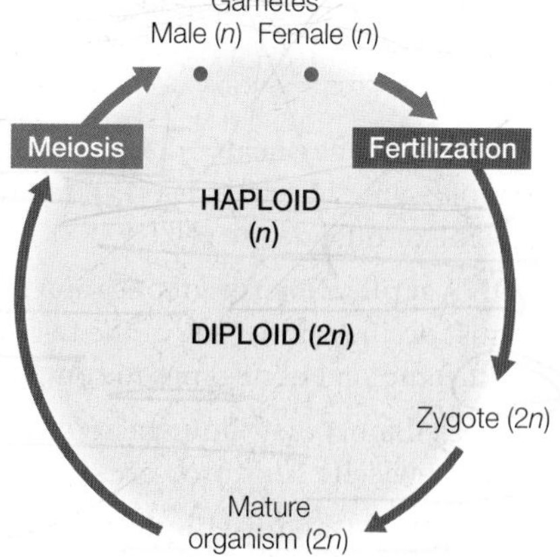

In the **diplontic life cycle**, the organism is diploid and the gametes are the only haploid stage. Found in animals, brown algae, and some fungi.

presents three examples. The life cycles of a variety of organisms will be described in detail in Part 4. For now we will focus on the role of sexual reproduction in generating diversity among individuals.

The essence of sexual reproduction is the *random selection of half of the diploid chromosome set* to make a haploid gamete, followed by fusion of two haploid gametes from separate parents to produce a diploid cell. As we will see later in this chapter, further diversity is introduced by events that take place during meiosis. All of these steps contribute to a shuffling of genetic information in the population, so that no two individuals have exactly the same genetic constitution. The diversity provided by sexual reproduction opens up enormous opportunities for evolution.

CHECKpoint CONCEPT 7.1

- ✓ In terms of the genetic composition of offspring, what is the difference between sexual and asexual reproduction?
- ✓ Discuss the advantages of sexual versus asexual reproduction in terms of evolution. Could evolution proceed without sexual reproduction? Explain your answer.

We have briefly mentioned the different types of cell division and the roles they play in the life cycles of organisms. Now let's look in more detail at the processes of cell division, starting with binary fission and mitosis.

CONCEPT 7.2 Both Binary Fission and Mitosis Produce Genetically Identical Cells

Cell division by either binary fission or mitosis produces two genetically identical cells. This is the basis of asexual reproduction in single-celled organisms: prokaryotes reproduce by binary fission, and single-celled eukaryotes reproduce by mitosis. In multicellular organisms, mitosis is a way to build tissues and organs during development and to repair damaged tissues once development is complete.

In order for any cell to divide, the following events must occur:

- There must be one or more **reproductive signals**. These signals initiate cell division and may originate from either inside or outside the cell.
- **DNA replication** (i.e., replication of the genetic material) must occur so that each of the two new cells will have a full complement of genes to complete cell functions.
- The cell must distribute the replicated DNA to each of the two new cells. This process is called **DNA segregation**.
- The cytoplasm must divide to form the two new cells, each surrounded by a cell membrane and a cell wall in organisms that have one. This process is called **cytokinesis**.

Let's see how these events occur during the processes of binary fission in prokaryotes and mitosis in eukaryotes.

Prokaryotes divide by binary fission

In prokaryotes, cell division results in the reproduction of the entire single-celled organism. The cell grows in size, replicates its DNA, and then separates the cytoplasm and DNA into two new cells by a process called **binary fission**.

REPRODUCTIVE SIGNALS External factors such as environmental conditions and nutrient concentrations are common reproductive signals for prokaryotes. For example, the bacterium *Bacillus subtilis* can divide every 30 minutes under ideal conditions. But when nutrients in its environment are low, it stops dividing. It then resumes dividing when conditions improve.

DNA REPLICATION In most prokaryotic cells, almost all of the genetic information is carried on one single chromosome. In many cases the ends of the single DNA molecule are covalently joined, making the chromosome circular. Two regions of the prokaryotic chromosome play functional roles in cell reproduction:

- *ori*: the site where replication of the circular chromosome starts (the *ori*gin of replication)
- *ter*: the site where replication ends (the *ter*minus of replication)

Chromosome replication takes place as the DNA is threaded through a "replication complex" of proteins near the center of the cell. Replication begins at the *ori* site and moves toward the *ter* site. When replication is complete, the two daughter DNA molecules separate and segregate from one another at opposite ends of the cell. In rapidly dividing prokaryotes, DNA replication occupies the entire time between cell divisions.

DNA SEGREGATION Replication begins near the center of the cell, and as it proceeds, the *ori* regions move toward opposite ends of the cell (**FIGURE 7.4**). DNA sequences adjacent to the *ori* region bind proteins that are essential for this segregation. This is an active process, since the binding proteins hydrolyze ATP. Components of the prokaryotic cytoskeleton are involved in the segregation process. In particular, a bacterial protein that is structurally related to actin but functionally related to tubulin provides a filament along which the *ori* regions and their associated proteins move.

CYTOKINESIS The actual division of a single cell and its contents into two cells begins immediately after chromosome segregation. Initially, there is a pinching in of the cell membrane caused by the contraction of a ring of fibers on the inside surface of the membrane (similar to a drawstring on shorts being tightened). In this case, the major component of these fibers is structurally similar to eukaryotic tubulin (which makes up microtubules), but its function is analogous to that of actin in the contractile ring of an animal cell (see below). As the membrane pinches in, new cell wall materials are deposited, which finally separate the two cells.

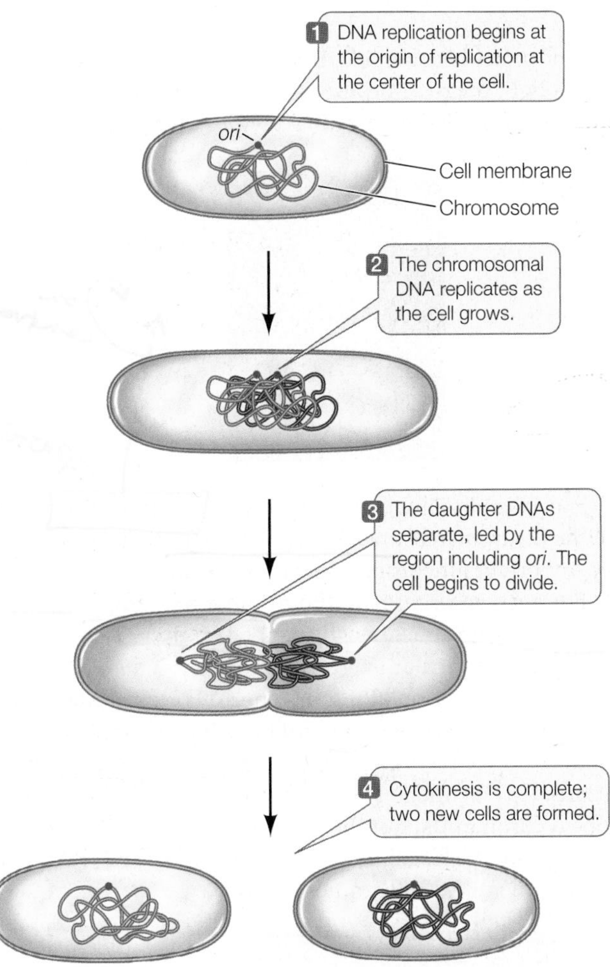

1 DNA replication begins at the origin of replication at the center of the cell.
ori
Cell membrane
Chromosome
2 The chromosomal DNA replicates as the cell grows.
3 The daughter DNAs separate, led by the region including ori. The cell begins to divide.
4 Cytokinesis is complete; two new cells are formed.

FIGURE 7.4 Prokaryotic Cell Division: Binary Fission The process of cell division in a bacterium involves DNA replication, DNA segregation, and cytokinesis.

Eukaryotic cells divide by mitosis followed by cytokinesis

As in prokaryotes, cell division in eukaryotes entails reproductive signals, DNA replication, DNA segregation, and cytokinesis. Some of the details, however, are quite different:

- *Reproductive signals*. Unlike prokaryotes, eukaryotic cells do not constantly divide whenever environmental conditions are adequate. In fact, most cells in a multicellular organism are specialized and do not divide. In a eukaryotic organism, the signals for cell division are usually not related to the environment of a single cell, but to the function of the entire organism. We will discuss the signals that control eukaryotic cell division in Concept 7.3.
- *DNA replication*. Unlike prokaryotes, eukaryotes have more than one chromosome. But the replication of each eukaryotic DNA molecule is similar to replication in prokaryotes, in that it is achieved by threading the long strands through replication complexes (see Concept 9.2). DNA replication occurs only during a specific stage of the cell cycle.
- *DNA segregation*. This is much more complicated than in prokaryotes, because first, there is a nuclear envelope, and second, there are multiple chromosomes. When a cell divides, one copy of each chromosome must end up in each of the two new cells—for example, each new somatic cell in a human will have all 46 chromosomes. In eukaryotes, the pairs of newly replicated chromosomes are initially attached to one another. They become highly condensed, and then the pairs are pulled apart before segregating into two new nuclei. The cytoskeleton is involved in this process.
- *Cytokinesis*. The process of cytokinesis in plant cells (which have cell walls) is different than in animal cells (which do not have cell walls). We describe both processes below.

LINK

The cytoskeleton is crucial for cell division. Review the description of the cytoskeleton and its molecular components in **Concept 4.4**

These events occur within the context of the **cell cycle**: the period from one cell division to the next. In eukaryotes, the cell cycle can be divided into several stages (**FIGURE 7.5**):

- **Mitosis** is the set of processes in which the chromosomes become condensed and then segregate into two new nuclei.

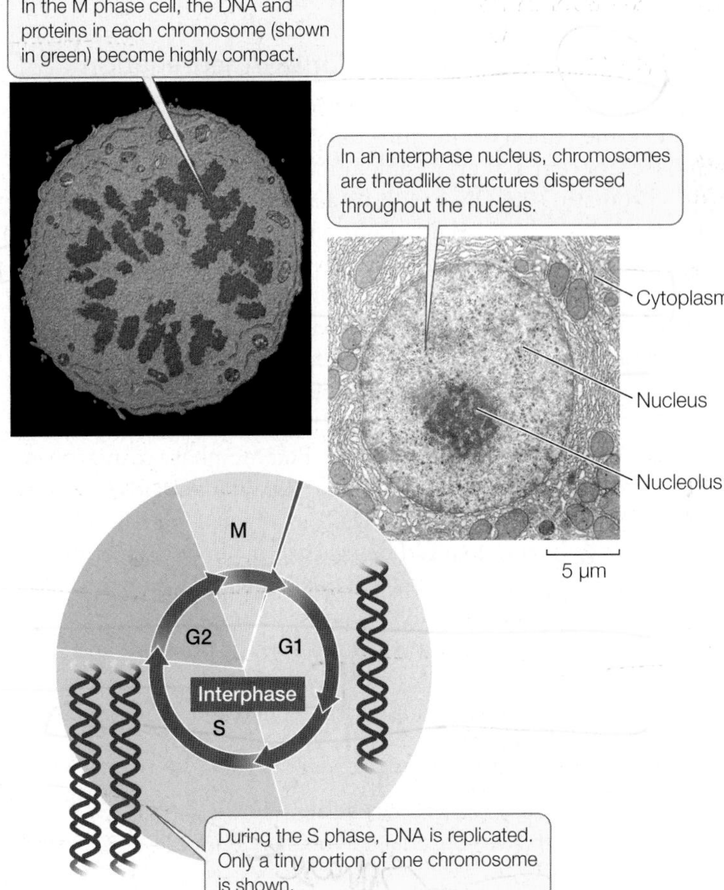

In the M phase cell, the DNA and proteins in each chromosome (shown in green) become highly compact.
In an interphase nucleus, chromosomes are threadlike structures dispersed throughout the nucleus.
Cytoplasm
Nucleus
Nucleolus
5 μm
M
G2
G1
Interphase
S
During the S phase, DNA is replicated. Only a tiny portion of one chromosome is shown.

FIGURE 7.5 The Phases of the Eukaryotic Cell Cycle The eukaryotic cell cycle has several phases. DNA in the interphase nucleus is diffuse and becomes compacted as mitosis begins.

- Cytokinesis usually follows immediately after mitosis, and these two stages—mitosis and cytokinesis—are referred to as **M phase**.
- M phase is followed by a much longer period called **interphase**, when the cell nucleus is visible and typical cell functions occur—including DNA replication in cells that are preparing to divide.

Interphase has three subphases called G1, S, and G2 (the G stands for gap). **G1** is quite variable, and a cell may spend a long time in this phase carrying out its specialized functions. The cell's DNA is replicated during **S** phase (S for synthesis). During **G2**, the cell makes preparations for mitosis—for example, by synthesizing components of the microtubules that will move the segregating chromosomes to opposite ends of the dividing cell.

In mitosis, *a single nucleus gives rise to two daughter nuclei that each contain the same number of chromosomes as the parent nucleus.* Although mitosis is a continuous process in which each event flows smoothly into the next, it is convenient to subdivide it into a series of stages: prophase, prometaphase, metaphase, anaphase, and telophase. Next we will look at these stages in more detail.

Go to ANIMATED TUTORIAL 7.1
Mitosis
PoL2e.com/at7.1

Prophase sets the stage for DNA segregation

During interphase, only the nuclear envelope and the nucleolus (the region of the nucleus where ribosomes are formed; see Concept 4.3) are visible under the light microscope. The chromatin (the DNA with its associated proteins) is not yet condensed, and individual chromosomes cannot be discerned. The appearance of the nucleus changes as the cell enters **prophase**—the beginning of mitosis. Here we describe three structures that appear during prophase and contribute to the orderly segregation of the replicated DNA: the condensed chromosomes, the reoriented centrosomes, and the spindle.

CONDENSED CHROMOSOMES Before S phase of interphase, each chromosome contains one very long double-stranded DNA molecule. If all of the DNA in a typical human cell were put end to end, it would be nearly 2 meters long. Yet the nucleus is only 5 μm (0.000005 m) in diameter. So even during interphase, eukaryotic DNA is packaged in a highly organized way. The DNA is wound around specific proteins, and other proteins coat the DNA coils. During prophase the chromosomes become much more tightly coiled and condensed.

LINK
DNA "packaging" and the specialized proteins that accomplish it are described in detail in **Concept 11.3**

After DNA replication, each chromosome has *two* DNA molecules, known as **sister chromatids**. Until they are separated during anaphase (see below), the chromatids are held together at a region called the **centromere**. During prophase the chromosomes become so compact that they can be seen clearly with a light microscope after staining with special dyes. They are even more clearly visualized with an electron microscope:

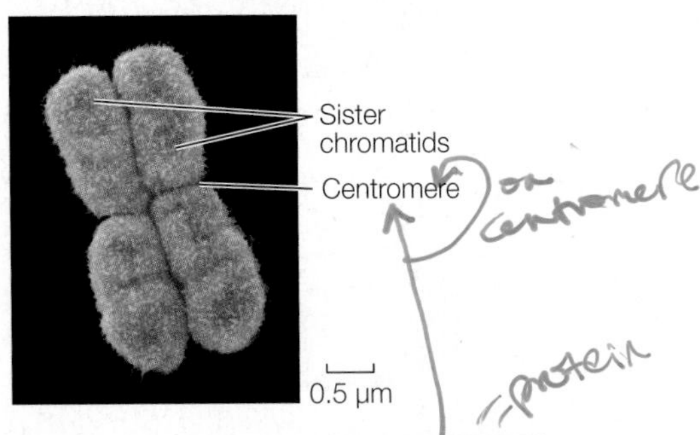

Specialized protein structures called **kinetochores** assemble on the centromeres, one on each chromatid. These structures are important for chromosome movement.

For a given organism, the number and sizes of the condensed chromosomes constitute the **karyotype**. Each chromosome has a particular length, and the centromere is located at a particular position along its length. For example, humans have 46 chromosomes (23 homologous pairs) that can be distinguished from one another by their sizes and centromere positions. (In the image below, each chromosome is composed of two chromatids, but the individual chromatids cannot be distinguished. The colors are from dyes that were used to help identify the different chromosomes.)

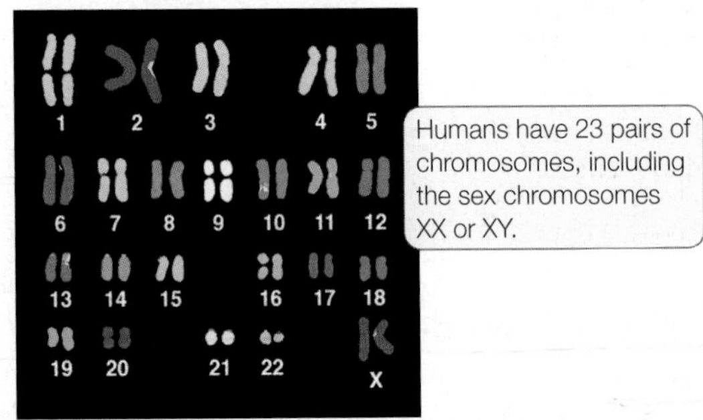

In the past, karyotype analysis was used as a way to identify organisms, and this method is still used to detect chromosomal abnormalities in humans. However, DNA sequence analysis is now much more commonly used to identify individuals and to classify related organisms (see Concept 16.2).

REORIENTED CENTROSOMES Before the spindle apparatus forms (see below), its orientation is determined. In many cells this is accomplished by the **centrosome** ("central body"), an organelle in the cytoplasm near the nucleus. The centrosome consists of a pair of **centrioles**, each one a hollow tube formed by nine triplets of microtubules. During S phase the centrosome becomes duplicated, and at the G2–M transition, the two centrosomes separate from one another, moving to opposite sides of the nucleus. Eventually these identify "poles" toward which chromosomes move during segregation.

The positions of the centrosomes determine the plane at which the cell divides; therefore they determine the spatial relationship between the two new cells. This relationship may be of little consequence to single free-living cells such as yeasts, but it is important for development in a multicellular organism. For example, during the development of an embryo, the daughter cells from some divisions must be positioned correctly to receive signals to form new tissues. Plant cells lack centrosomes, but distinct microtubule organizing centers at each end of the cell play the same role.

SPINDLE Each of the two centrosomes, when positioned on opposite sides of the nucleus, serves as a pole toward which the chromosomes move. Tubulin dimers from around the centrosomes aggregate into microtubules that extend from the poles into the middle region of the cell. Together these microtubules make up a **spindle**. The spindle forms during prophase and prometaphase, when the nuclear envelope breaks down. The microtubules are initially unstable, constantly forming and falling apart, until they contact kinetochores or microtubules from the other half-spindle and become more stable.

There are three types of microtubules in the spindle:

- *Polar microtubules* overlap in the middle region of the cell and keep the two poles apart.
- *Astral microtubules* interact with proteins attached to the cell membrane, and also assist in keeping the poles apart.
- *Kinetochore microtubules* attach to the kinetochores on the chromosomes. The two sister chromatids in each chromosome become attached to kinetochore microtubules from opposite sides of the cell. This ensures that the two chromatids will move to opposite poles.

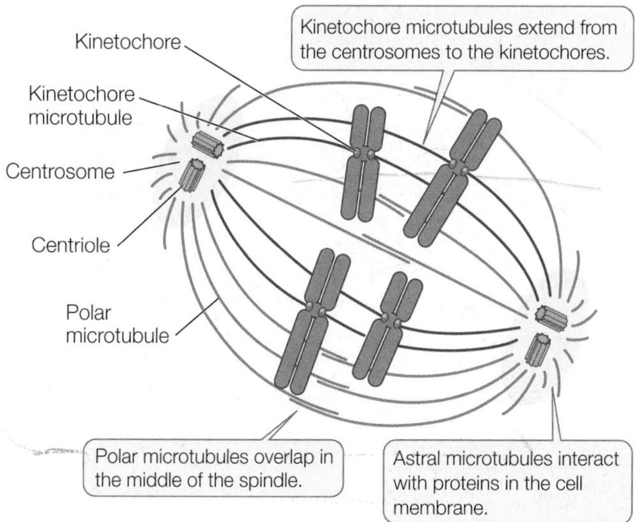

Separation of the chromatids and movement of the **daughter chromosomes** (which the sister chromatids become after separation) is the central feature of mitosis. It accomplishes the DNA segregation that is needed for cell division and completion of the cell cycle. *Note the difference between chromatids and chromosomes:*

- Chromatids share a centromere.
- Chromosomes have their own centromere.

Go to ACTIVITY 7.2 The Mitotic Spindle
PoL2e.com/ac7.2

Chromosome separation and movement are highly organized

During the next three phases of mitosis—prometaphase, metaphase, and anaphase—dramatic changes take place in the cell and the chromosomes (FIGURE 7.6):

- In **prometaphase** the nuclear envelope breaks down and the compacted chromosomes, each consisting of two chromatids, attach to the kinetochore microtubules.
- In **metaphase** the chromosomes line up at the midline of the cell (the equatorial position).
- In **anaphase** the chromatids separate, and the daughter chromosomes move away from each other toward the poles.

The separation of chromatids into daughter chromosomes occurs at the beginning of anaphase. The migration of the daughter chromosomes to the poles of the cell is a highly organized, active process. Two mechanisms operate to move the chromosomes along. First, the kinetochores contain molecular motor proteins, including kinesin and dynein (see Concept 4.4), which use energy from ATP hydrolysis to move the chromosomes along the microtubules. Second, the kinetochore microtubules shorten from the poles, drawing the chromosomes toward the poles.

Telophase occurs after the chromosomes have separated and is the last phase of mitosis. During this period, a nuclear envelope forms around each set of new chromosomes, nucleoli appear, and the chromosomes become less compact. The spindle also disappears at this stage. As a result, there are two new nuclei in a single cell.

Cytokinesis is the division of the cytoplasm

Mitosis refers only to the division of the nucleus. Cytokinesis, the division of the cell's cytoplasm, is the final stage of cell reproduction. This process occurs differently in plants and animals.

ANIMAL CELLS Cytokinesis usually begins with a furrowing of the cell membrane, as if an invisible thread were cinching the cytoplasm between the two nuclei (FIGURE 7.7A). This contractile ring is composed of microfilaments of actin and myosin, which form a ring on the cytoplasmic surface of the cell membrane. These two proteins interact to produce a contraction (just as they do in muscles; described in Concept 33.1), pinching the cell in two. The microfilaments assemble rapidly from actin monomers that are present in the interphase cytoskeleton. Their assembly is controlled by calcium ions (commonly used in cellular signaling) that are released from storage sites in the center of the cell.

PLANT CELLS In plant cells, the cytoplasm divides differently because plants have cell walls. As the spindle breaks down after mitosis, vesicles derived from the Golgi apparatus appear along the plane of cell division, roughly midway between the two daughter nuclei. The vesicles are propelled

Interphase

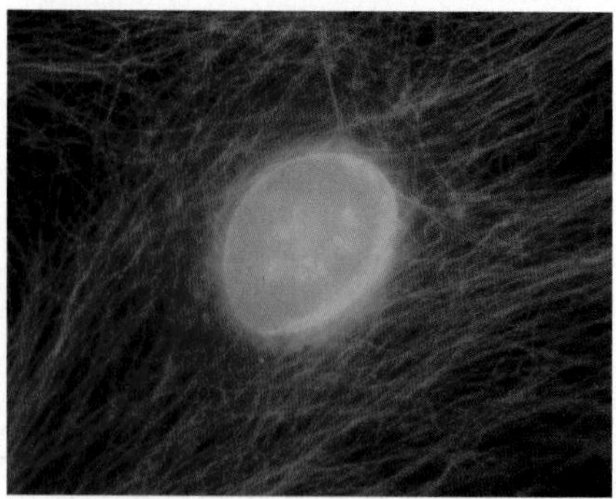

Prophase

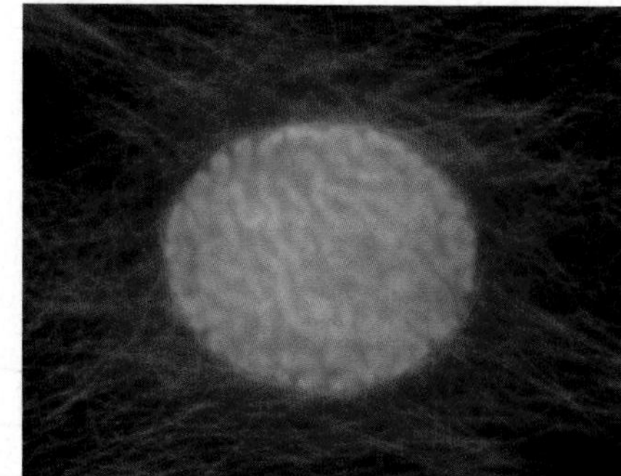

Prometaphase

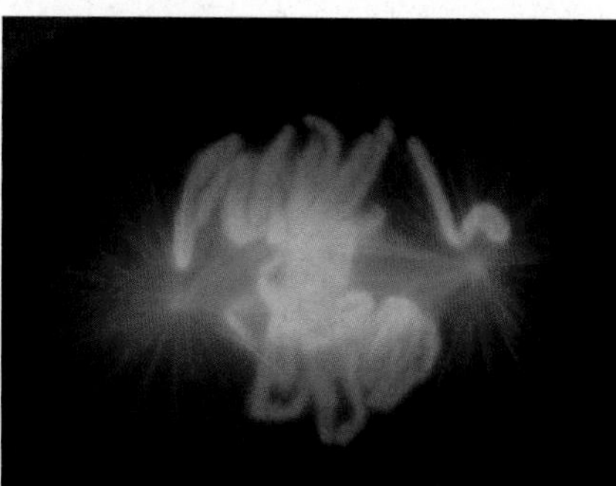

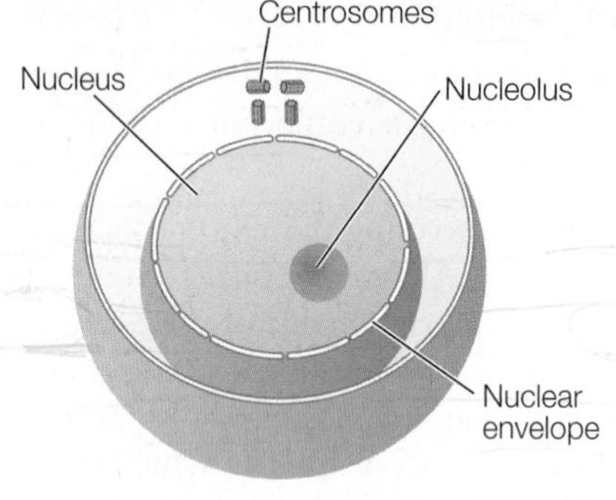

1 During S phase of interphase, the nucleus replicates its DNA and centrosomes.

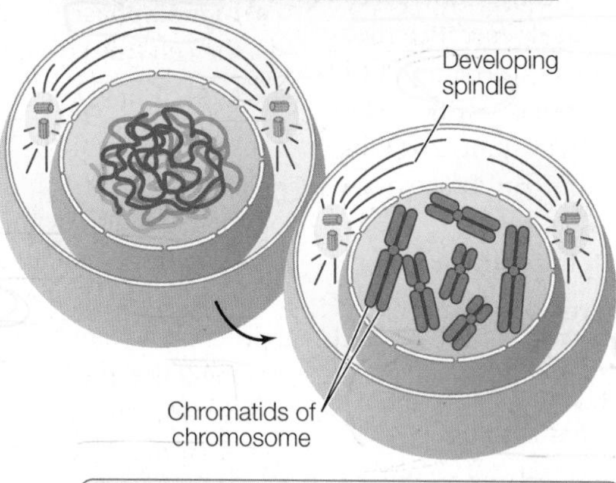

2 The chromatin coils and supercoils, becoming more and more compact and condensing into visible chromosomes. The chromosomes consist of identical, paired sister chromatids. Centrosomes move to opposite poles.

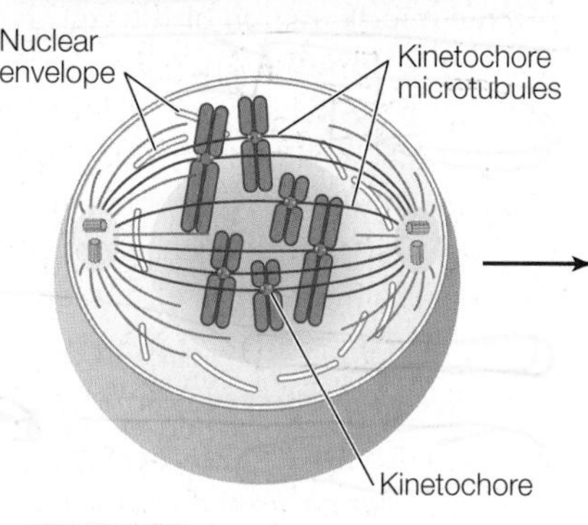

3 The nuclear envelope breaks down. Kinetochore microtubules appear and connect the kinetochores to the poles.

FIGURE 7.6 The Phases of Mitosis Mitosis results in two new nuclei which are genetically identical to each other and to the nucleus from which they were formed. In the micrographs, the green dye stains microtubules (and thus the spindle); the blue dye stains the chromosomes. The chromosomes in the diagrams are stylized to emphasize the fates of the individual chromatids.

Go to ACTIVITY 7.3 Images of Mitosis
PoL2e.com/ac7.3

Go to MEDIA CLIP 7.1
Mitosis: Live and Up Close
PoL2e.com/mc7.1

along microtubules by the motor protein kinesin and fuse to form a new cell membrane. At the same time they contribute their contents to a cell plate, which is the beginning of a new cell wall between the two daughter cells (**FIGURE 7.7B**).

Following cytokinesis, each daughter cell contains all the components of a complete cell. A precise distribution of chromosomes is ensured by mitosis. In contrast, organelles such as mitochondria and chloroplasts are not necessarily distributed equally, although at least one of each must be present in each daughter cell. The *orientation* of cell division is important in development (see above), but there does not appear to be a precise mechanism for the distribution of the cytoplasmic contents.

TABLE 7.1 summarizes the major events of the eukaryotic cell cycle.

TABLE 7.1 Summary of Eukaryotic Cell Cycle Events

Phase	Events
INTERPHASE	
G1	Growth; specialized cell functions
S	DNA replication
G2	Spindle synthesis begins; preparation for mitosis
MITOSIS	
Prophase	Condensation of chromosomes; spindle assembly
Prometaphase	Nuclear envelope breakdown; chromosome attachment to spindle
Metaphase	Alignment of chromosomes at equatorial plate
Anaphase	Separation of chromatids; migration to poles
Telophase	Chromosomes decondense; nuclear envelope re-forms
Cytokinesis	Cell separation; cell membrane and/or cell wall formation

Metaphase

Equatorial (metaphase) plate

4 The centromere/kinetochore complexes become aligned in a plane, which is often at the cell's equator.

Anaphase

Daughter chromosomes

5 The paired sister chromatids separate, and the new daughter chromosomes begin to move toward the poles.

Telophase

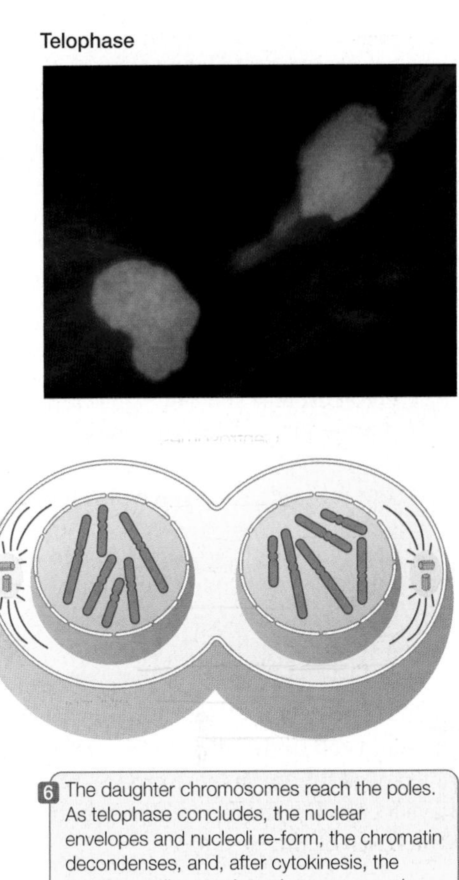

6 The daughter chromosomes reach the poles. As telophase concludes, the nuclear envelopes and nucleoli re-form, the chromatin decondenses, and, after cytokinesis, the daughter cells enter interphase once again.

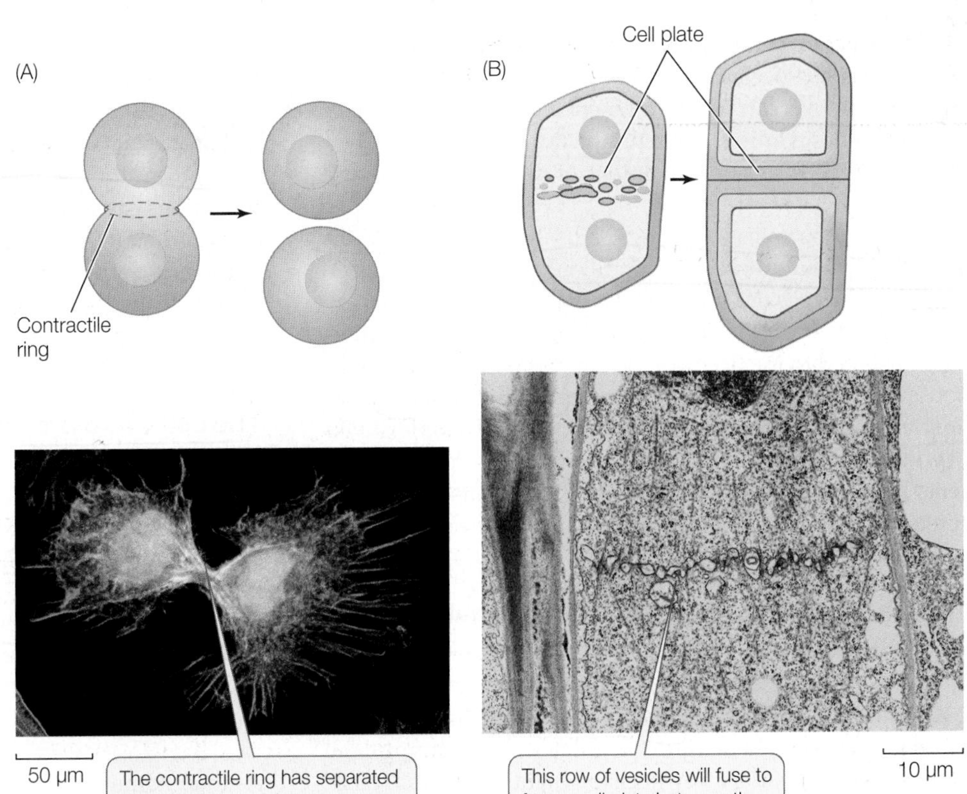
(A)
Contractile ring
(B)
Cell plate
50 µm
The contractile ring has separated the cytoplasms of these two daughter cells.
This row of vesicles will fuse to form a cell plate between the cell above and the cell below.
10 µm

FIGURE 7.7 Cytokinesis Differs in Animal and Plant Cells (A) A HeLa cell (a type of human cancer cell) undergoing cytokinesis. In this flurorescence micrograph, nuclei are yellow, mitochondria are red, and actin filaments are green. (B) An electron micrograph of a plant cell in late telophase. Plant cells divide differently than animal cells because they have cell walls.

CHECKpoint CONCEPT 7.2

- ✓ How does the mitotic spindle ensure that each daughter cell receives a full complement of the genetic material in the cell nucleus?
- ✓ Compare the cell cycles of prokaryotes and eukaryotes with regard to reproductive signals for initiation, how many chromosomes are present, and how the replicated DNA segregates.
- ✓ Sketch the five stages of mitosis for a diploid organism with four chromosomes (two pairs). Clearly label chromosomes and chromatids and note the number of double-stranded DNA molecules in each structure at each stage.
- ✓ The drug cytochalasin B blocks the assembly and function of microfilaments. What would you expect to happen in dividing animal cells treated with this drug after telophase but before cytokinesis?

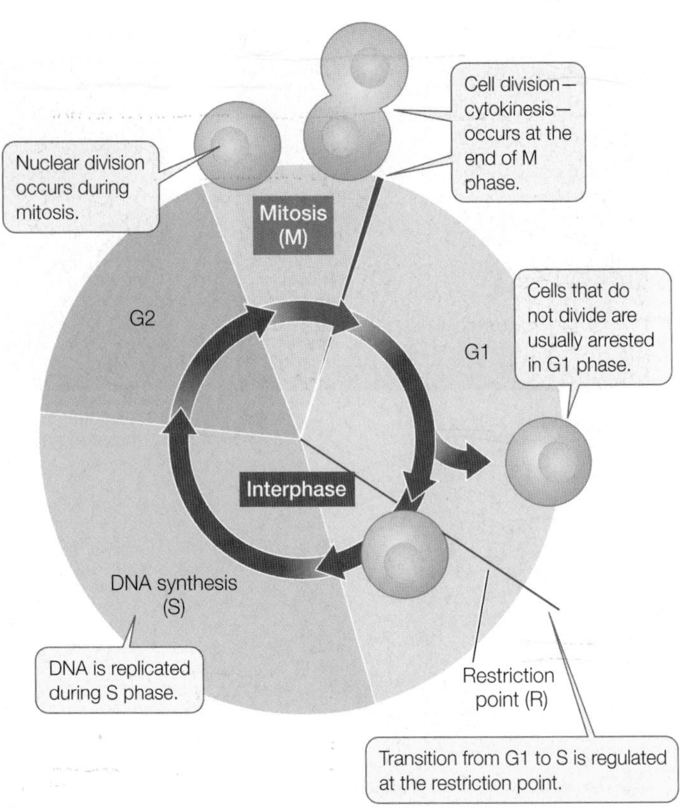

FIGURE 7.8 The Eukaryotic Cell Cycle The cell cycle consists of a mitotic (M) phase, during which mitosis and cytokinesis take place, and a long period of growth known as interphase. Interphase has three subphases (G1, S, and G2) in cells that divide.

Light microscopists have studied the dramatic events of mitosis and cell division since the 1880s, and we now have detailed descriptions of these processes. More recently, biologists have focused on the mechanisms controlling cell reproduction, which will be our next topic.

CONCEPT 7.3 Cell Reproduction Is Under Precise Control

Cell reproduction cannot go on continuously and indefinitely. If a single-celled species had no control over its reproduction, it would soon overrun its environment and starve to death. In a multicellular organism, cell reproduction must be controlled to maintain the forms and functions of different parts of the body.

Unlike prokaryotes, eukaryotic cells do not constantly divide whenever environmental conditions are adequate. In fact, the specialized cells of a multicellular eukaryotic organism may seldom or never divide. The signals for eukaryotic cell division are related to the needs of the entire organism. Mammals produce a variety of substances called **growth factors** that stimulate cell division and differentiation. For example, if you cut yourself and bleed, a blood clot eventually forms. Cell fragments called platelets in the blood vessels surrounding the clot secrete various growth factors that stimulate nearby cells to divide and heal the wound.

The eukaryotic cell division cycle is regulated internally

As we discussed in Concept 7.2, the eukaryotic cell cycle can be divided into four stages: G1, S, G2, and M. Progression through these phases is tightly regulated. For example, the **G1–S transition** marks a key decision point for the cell: passing this point (called R, the restriction point) usually means the cell will proceed with the rest of the cell cycle and divide (**FIGURE 7.8**).

What events cause a cell to enter S phase or M phase? A first indication that there were substances controlling these transitions came from cell fusion experiments. For example, an experiment involving the fusion of mammalian cells at G1 phase and S phase showed that a cell in S phase produces a substance that activates DNA replication (**FIGURE 7.9**). Similar experiments pointed to a molecular activator for entry into M phase.

The cell cycle is controlled by cyclin-dependent kinases

The molecular activators revealed by the cell fusion experiments turned out to be protein kinases, a class of enzymes that are common in cell signaling (see Concept 5.5). The kinases involved in cell cycle regulation are called **cyclin-dependent kinases (CDKs)**. They catalyze the phosphorylation of target proteins that regulate the cell cycle:

$$\text{Cell cycle regulator} + \text{ATP} \xrightarrow{\text{CDK}} \text{cell cycle regulator-P} + \text{ADP}$$

As their name implies, CDKs are activated by binding to the protein **cyclin**. This binding changes the shape of a CDK such

INVESTIGATION

FIGURE 7.9 Regulation of the Cell Cycle Nuclei in G1 do not undergo DNA replication, but nuclei in S phase do. To determine if there is some signal in the S cells that stimulates G1 cells to replicate their DNA, cells in G1 and S phases were induced to fuse, creating cells with both G1 and S properties.[a]

HYPOTHESIS

A cell in S phase contains an activator of DNA replication.

METHOD

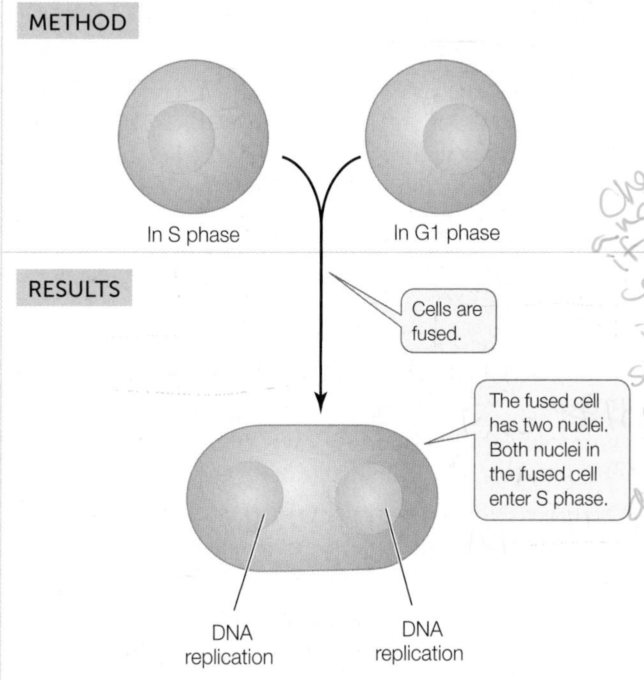

RESULTS

CONCLUSION

The S phase cell contains a substance that diffuses to the G1 nucleus and activates DNA replication.

ANALYZE THE DATA

The researchers used mammalian cells undergoing the cell cycle synchronously. Radioactive labeling and microscopy were used to determine which nuclei were synthesizing DNA; only nuclei that were synthesizing DNA became labeled:

Type of cells	Cells with labeled nuclei/total cells
Unfused G1	6/300
Unfused S	435/500
Fused G1 and S cells	17*/19

*Both nuclei labeled

A. What were the percentages of cells in S phase in each of the three experiments?

B. What does this mean in terms of control of the cell cycle?

Go to **LaunchPad** for discussion and relevant links for all **INVESTIGATION** figures.

[a]P. N. Rao and R. T. Johnson. 1970. *Nature* 225: 159–164.

that its active site is exposed, and is an example of allosteric regulation (see Concept 3.4):

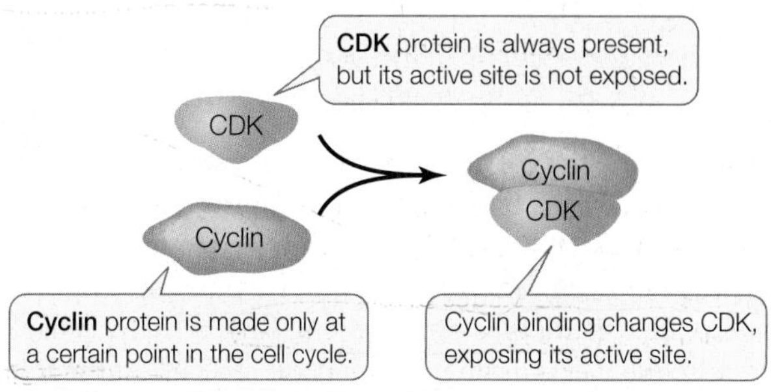

Several different CDKs function at specific stages of the cell cycle, called **cell cycle checkpoints**. At these points, signaling pathways regulate the progress of the cell cycle. For example, if DNA is substantially damaged by radiation or toxic chemicals, the cell may be prevented from successfully completing the cell cycle. So the damage to DNA is repaired before the cycle proceeds. There are three checkpoints during interphase and one during mitosis:

- G1 checkpoint is triggered by DNA damage.
- S checkpoint is triggered by incomplete replication or DNA damage.
- G2 checkpoint is triggered by DNA damage.
- M checkpoint is triggered by a chromosome that fails to attach to the spindle.

Each CDK has its own cyclin to activate it, and the cyclin is made only at the right time. After the CDK acts, the cyclin is broken down by a protease (**FIGURE 7.10**). So a key event controlling the transition from one cell cycle phase to the next is the

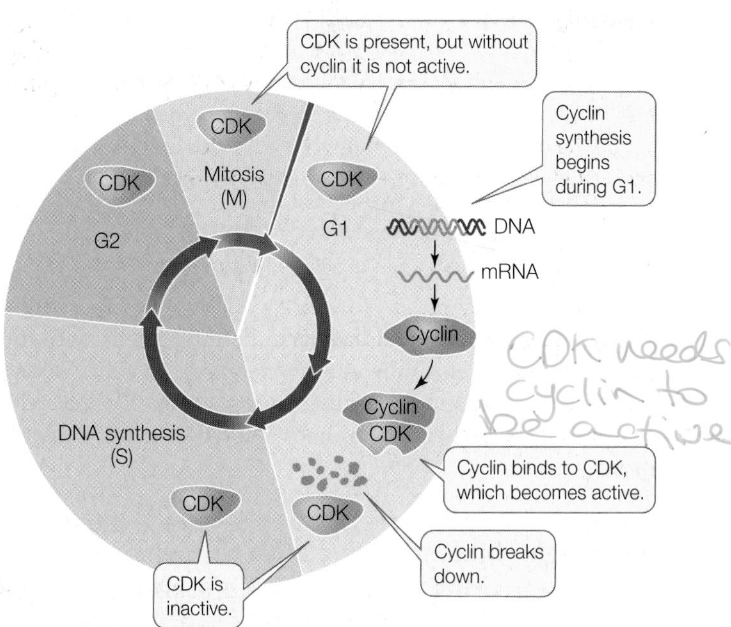

FIGURE 7.10 Cyclins Are Transient in the Cell Cycle Cyclins are made at a particular time and then break down. In this case, the cyclin is present during G1 and activates a CDK at that time.

synthesis and subsequent breakdown of a cyclin. Cyclins are synthesized in response to various molecular signals, including growth factors. This starts a chain reaction:

Growth factor → cyclin synthesis → CDK activation → cell cycle events

To illustrate the concept of cell cycle control by a particular cyclin–CDK complex, let's take a look at the complex that controls the R point at the G1–S transition (see Figure 7.8).

The G1–S cyclin–CDK catalyzes the phosphorylation of a protein called retinoblastoma protein (RB). In many cells, RB or a protein like it acts as an *inhibitor of the cell cycle* at the R point. To begin S phase, a cell must overcome the RB block. Here is where the G1–S cyclin–CDK comes in: it catalyzes the addition of a phosphate to RB. This causes a change in the three-dimensional structure of RB, thereby inactivating it. With RB out of the way, the cell cycle can proceed. To summarize:

$$\underset{\text{(active: blocks cell cycle)}}{\text{RB + ATP}} \xrightarrow{\text{G1–S cyclin–CDK}} \underset{\text{(inactive: allows cell cycle)}}{\text{RB-P + ADP}}$$

Now we can be more specific about the chain of events involved in growth factor stimulation of cell division: the specific cyclin whose synthesis is activated is the one that allosterically activates the CDK that phosphorylates RB, and this allows the cell cycle to exit G1 and begin DNA replication in S phase. This example illustrates how regulation of the cell cycle involves a number of cellular processes that we have examined in this and other chapters: signal transduction (see Chapter 5), gene expression and protein synthesis (see Chapter 3), and cell division.

CHECKpoint CONCEPT 7.3

✓ Draw a diagram and describe the events that occur in the four stages of the eukaryotic cell cycle (M phase, G1, S, and G2).

✓ Cultures of eukaryotic cells can be synchronized, so they are all at the same phase of the cell cycle at the same time. If you examined a culture at the beginning of G1, would the CDK that acts at the R point be present? Would it be active? Would its cyclin be present? What would your answers be if the culture were at the R point?

Binary fission and mitosis result in daughter cells with the same number of chromosomes as their parent cells. Sexual reproduction, however, requires a process of cell division in which the number of chromosomes is halved. We'll look at this process next.

CONCEPT 7.4 Meiosis Halves the Nuclear Chromosome Content and Generates Diversity

In Concept 7.1 we described the role and importance of meiosis in sexual reproduction. Now we will see how the orderly and precise generation of haploid cells is accomplished.

Meiosis consists of *two* nuclear divisions that reduce the number of chromosomes to the haploid number. Although the nucleus divides twice during meiosis, the DNA is replicated only once. Unlike the products of mitosis, the haploid cells produced by meiosis are genetically different from one another and from the parent cell. **FIGURE 7.11** compares the two processes.

To understand the process of meiosis and its specific details, it is useful to keep in mind the overall functions that meiosis has evolved to serve:

- To reduce the chromosome number from diploid to haploid
- To ensure that each of the haploid products has a complete set of chromosomes
- To generate genetic diversity among the products (gametes)

The events of meiosis are illustrated in **FIGURE 7.12**. In the rest of this section we will discuss some of the key features that distinguish meiosis from mitosis.

Go to ANIMATED TUTORIAL 7.2
Meiosis
PoL2e.com/at7.2

Meiotic division reduces the chromosome number

As noted above, meiosis consists of two nuclear divisions, meiosis I and meiosis II. Two unique features characterize **meiosis I**:

- *Homologous chromosomes come together and line up* along their entire lengths. No such pairing occurs in mitosis.
- *The homologous chromosome pairs separate*, but the individual chromosomes, each consisting of two sister chromatids, remain intact. (The chromatids will separate during meiosis II.)

Like mitosis, meiosis I is preceded by an interphase with an S phase, during which each chromosome is replicated. As a result, each chromosome consists of two sister chromatids. At the end of meiosis I two nuclei form, each with half of the original chromosomes (one member of each homologous pair). Since the centromeres did not separate, these chromosomes are still double—composed of two sister chromatids. The sister chromatids are separated during **meiosis II**, which is *not* preceded by DNA replication. As a result, the products of meiosis I and II are four cells, each containing the haploid number of chromosomes. But *these four cells are not genetically identical.*

Crossing over and independent assortment generate diversity

A diploid organism has two sets of chromosomes ($2n$): one set derived from its male parent, the other from its female parent. As the organism grows and develops, its cells undergo mitotic

FIGURE 7.11 Mitosis and Meiosis: A Comparison Meiosis involves two cell divisions, the first of which is very different from the single division of mitosis. Meiosis II is similar to mitosis, in that the centromeres separate during anaphase, allowing the chromatids of the two homologous pairs to separate into four daughter chromosomes that are genetically distinct from the parental chromosomes. ▶

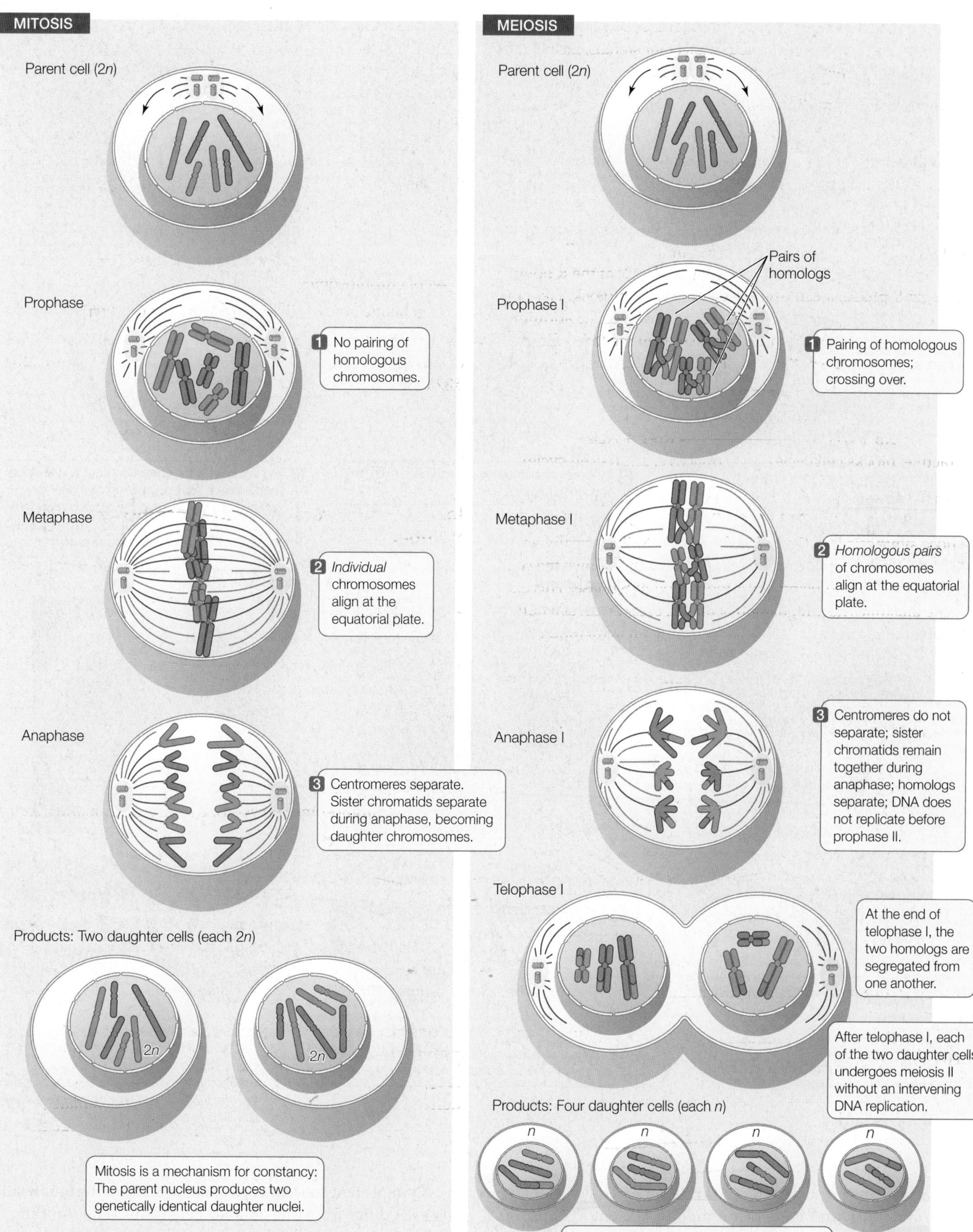
MITOSIS
Parent cell (2n)
Prophase
1 No pairing of homologous chromosomes.
Metaphase
2 Individual chromosomes align at the equatorial plate.
Anaphase
3 Centromeres separate. Sister chromatids separate during anaphase, becoming daughter chromosomes.
Products: Two daughter cells (each 2n)
2n
2n
Mitosis is a mechanism for constancy: The parent nucleus produces two genetically identical daughter nuclei.
MEIOSIS
Parent cell (2n)
Pairs of homologs
Prophase I
1 Pairing of homologous chromosomes; crossing over.
Metaphase I
2 Homologous pairs of chromosomes align at the equatorial plate.
Anaphase I
3 Centromeres do not separate; sister chromatids remain together during anaphase; homologs separate; DNA does not replicate before prophase II.
Telophase I
At the end of telophase I, the two homologs are segregated from one another.
After telophase I, each of the two daughter cells undergoes meiosis II without an intervening DNA replication.
Products: Four daughter cells (each n)
n
n
n
n
Meiosis II produces four haploid daughter cells that are genetically distinct. Meiosis is thus a mechanism for generating diversity.

MEIOSIS I

Early prophase I

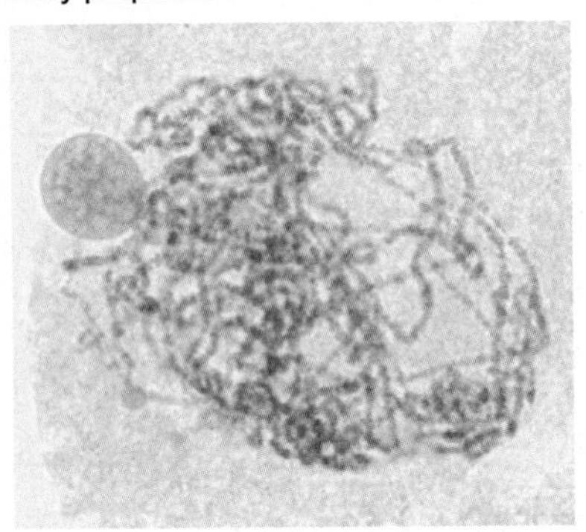

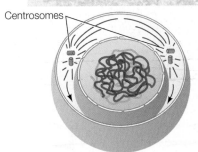

Centrosomes

1 The chromatin begins to condense following interphase.

Mid-prophase I

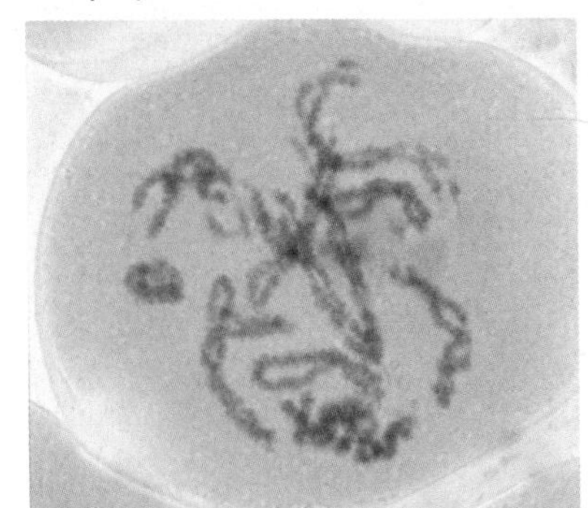

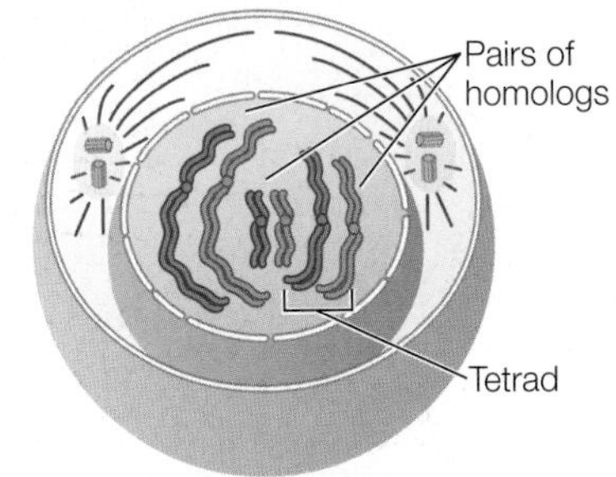

2 Synapsis aligns homologs, and chromosomes condense further.

Late prophase I–Prometaphase

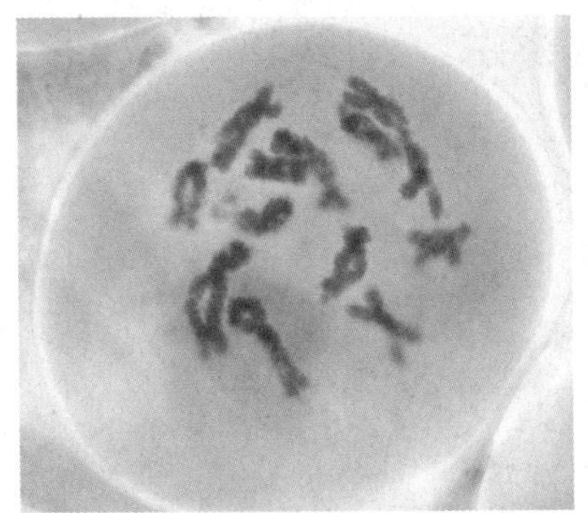

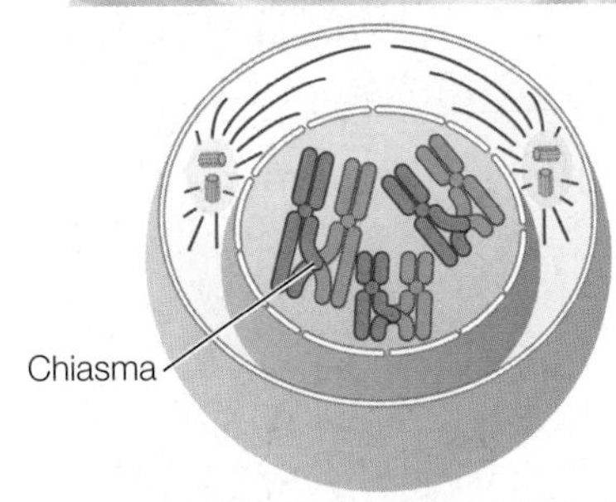

3 The chromosomes continue to coil and shorten. The chiasmata reflect crossing over, the exchange of genetic material between nonsister chromatids in a homologous pair. In prometaphase the nuclear envelope breaks down.

MEIOSIS II

Prophase II

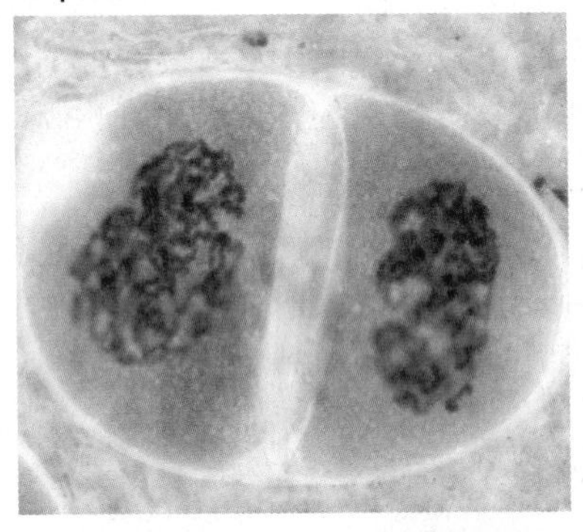

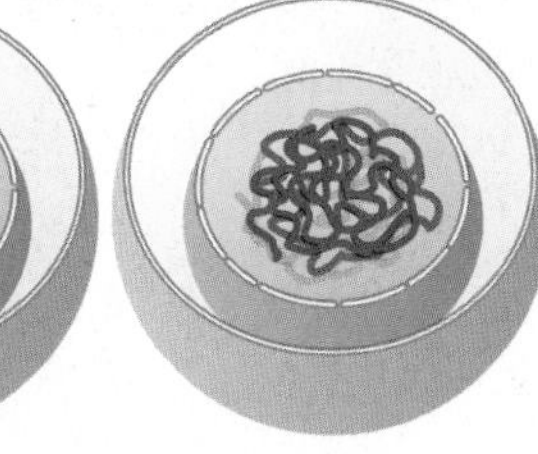

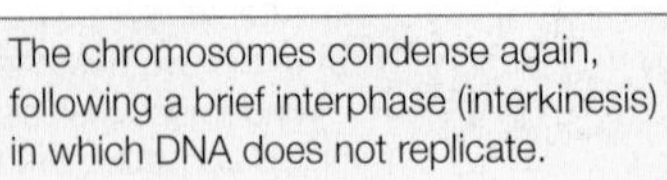

7 The chromosomes condense again, following a brief interphase (interkinesis) in which DNA does not replicate.

Metaphase II

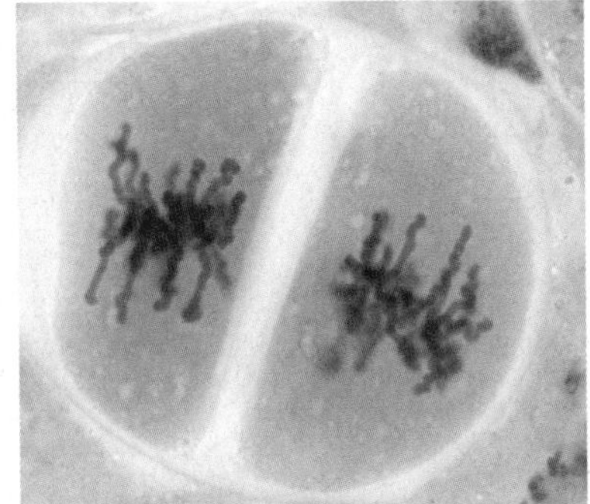

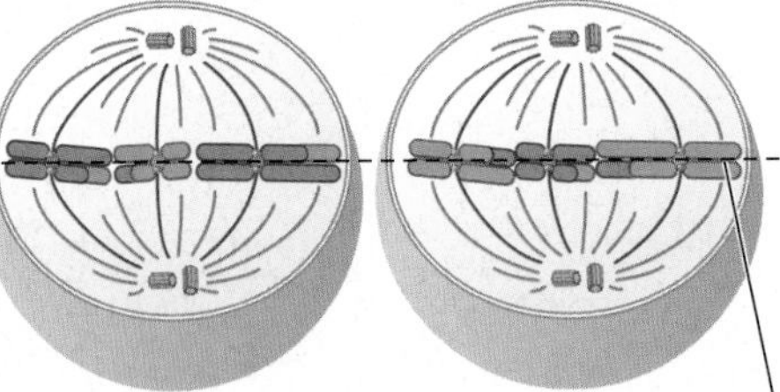

8 The centromeres of the paired chromatids line up across the equatorial plates of each cell.

Anaphase II

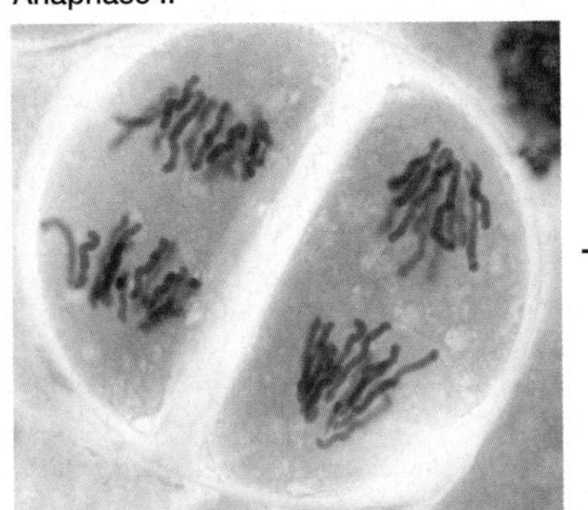

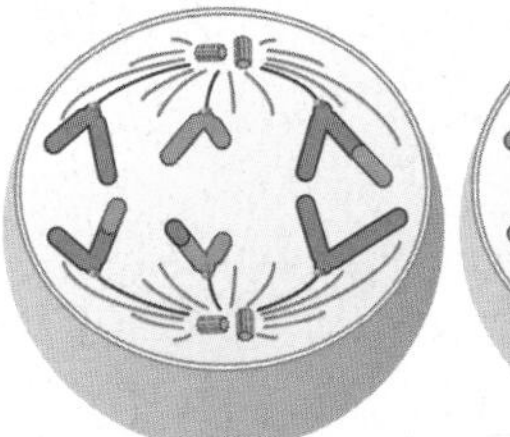

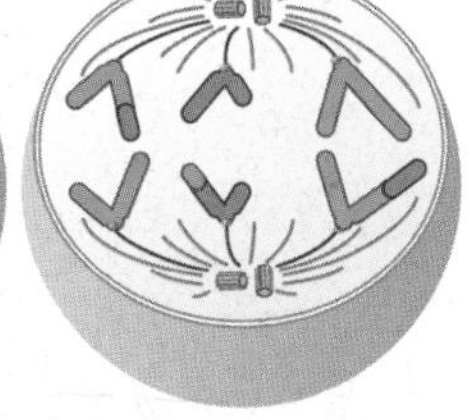

9 The chromatids finally separate, becoming chromosomes in their own right, and are pulled to opposite poles. Because of crossing over and independent assortment, each new cell will have a different genetic makeup.

divisions. In mitosis, each chromosome behaves independently of its homolog, and its two chromatids are sent to opposite poles during anaphase. Each daughter nucleus ends up with an identical set of $2n$ chromosomes. In meiosis, things are very different (see Figure 7.11).

An important consequence of meiosis is that the four resulting cells differ from one another genetically. The shuffling of genetic material occurs by two processes: crossing over and independent assortment.

MEIOSIS I (continued)

Metaphase I

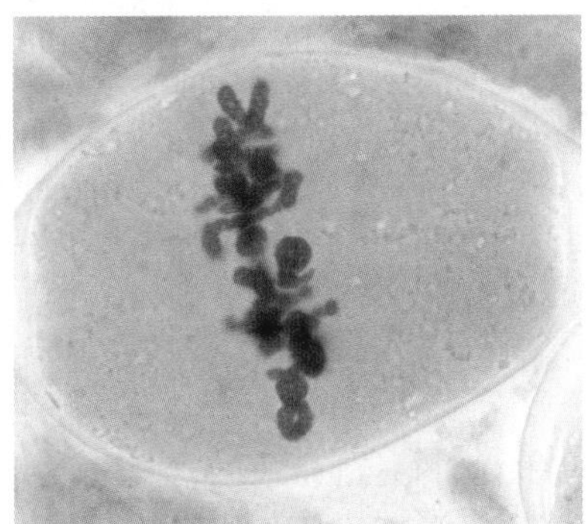

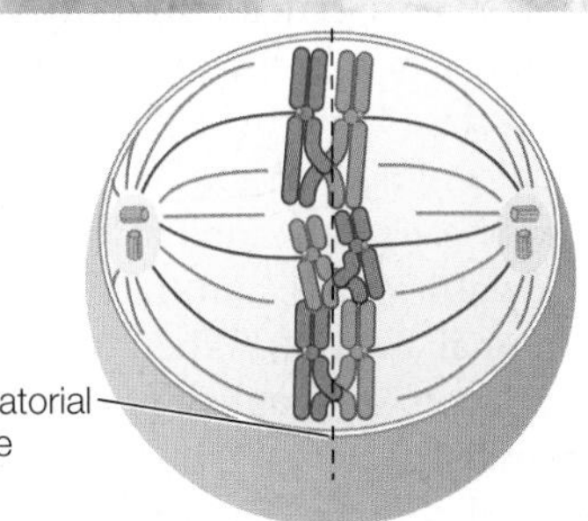

4 The homologous pairs line up on the equatorial (metaphase) plate.

Anaphase I

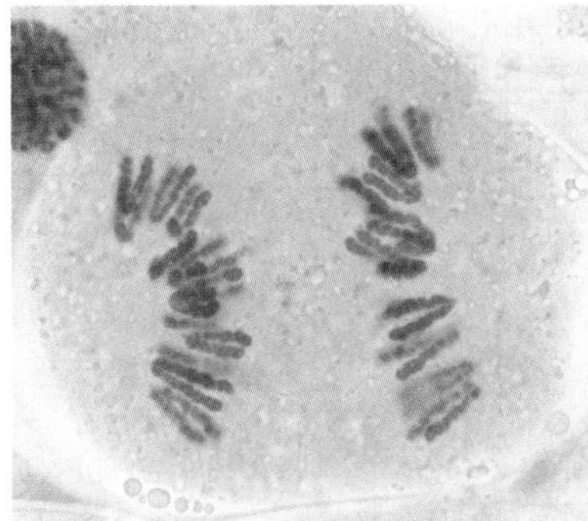

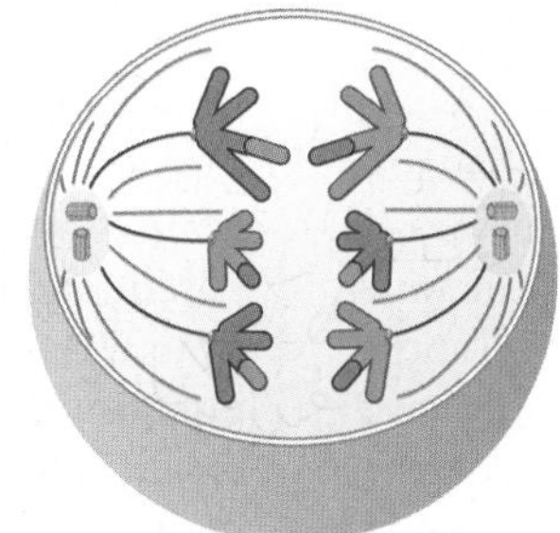

5 The homologous chromosomes (each with two chromatids) move to opposite poles of the cell.

Telophase I

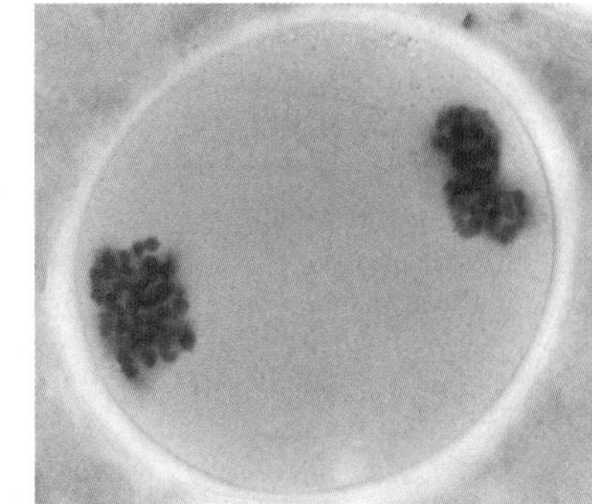

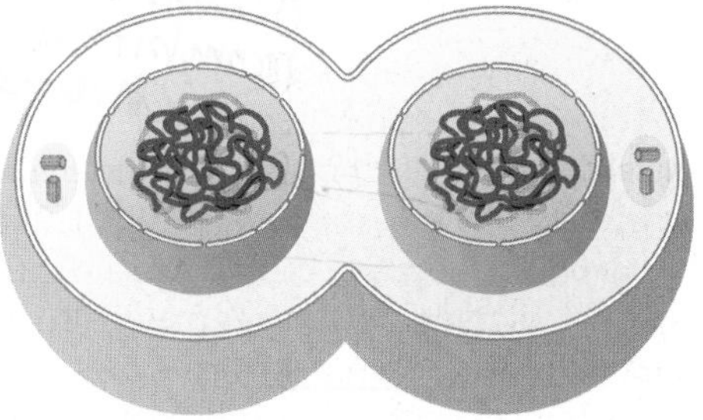

6 The chromosomes gather into nuclei, and the original cell divides.

MEIOSIS II (continued)

Telophase II

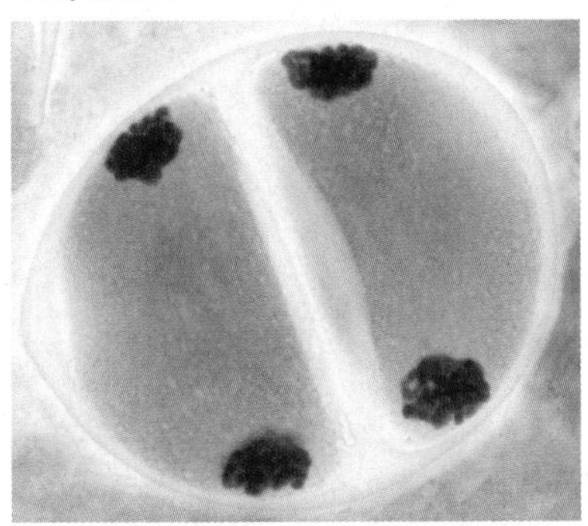

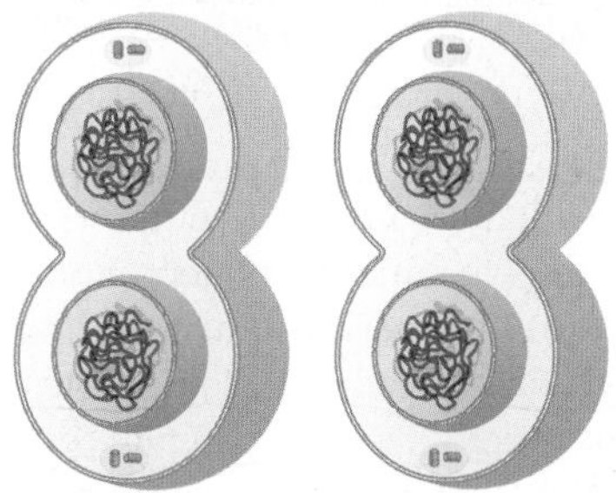

10 The chromosomes gather into nuclei, and the cells divide.

Products

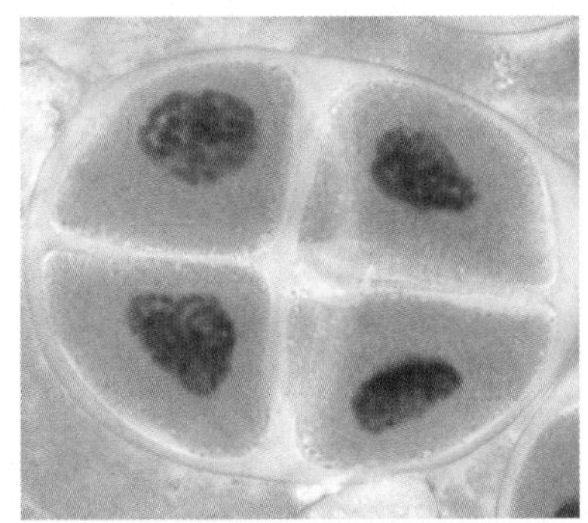

11 Each of the four cells has a nucleus with a haploid number of chromosomes.

FIGURE 7.12 Meiosis: Generating Haploid Cells In meiosis, two sets of chromosomes are divided among four daughter cells, each of which has half as many chromosomes as the original cell. The four haploid cells are the result of two successive nuclear divisions. The micrographs show meiosis in the male reproductive organ of a lily; the diagrams show the corresponding phases in an animal cell. (For instructional purposes, the chromosomes from one parent of the original organism are colored blue and those from the other parent are red.)

Go to ACTIVITY 7.4 Images of Meiosis
PoL2e.com/ac7.4

CROSSING OVER Meiosis I begins with a long prophase I (the first three panels of Figure 7.12), during which the chromosomes change markedly. The homologous chromosomes pair by adhering along their lengths in a process called synapsis. (This does not happen in mitosis.) This pairing process lasts from prophase I to the end of metaphase I. The four chromatids of each pair of homologous chromosomes form a **tetrad**, or bivalent. For example, in a human cell at the end of prophase I there are 23 tetrads, each consisting of four chromatids. The four chromatids come from the two partners in each homologous pair of chromosomes.

Throughout prophase I and metaphase I, the chromatin continues to coil and compact and the chromosomes become more condensed. At a certain point, the homologous chromosome pairs appear to repel each other, especially near the centromeres, but they remain attached. The X-shaped attachment points are called **chiasmata** (singular chiasma, "cross"):

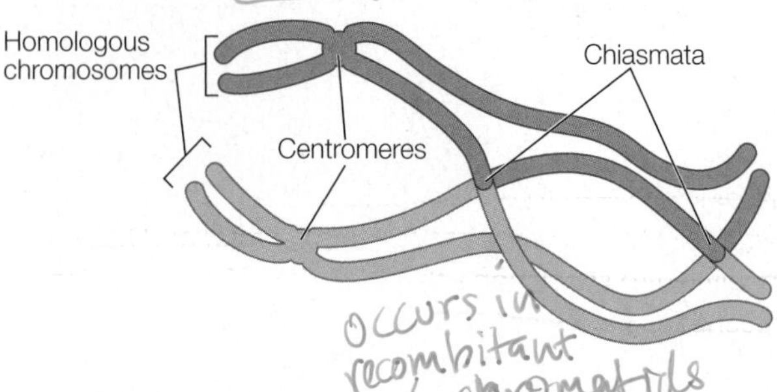

A chiasma is a point where genetic material is exchanged between nonsister chromatids on homologous chromosomes—a process called **crossing over** (FIGURE 7.13). Any of the four chromatids in the tetrad can participate in this exchange, and a single chromatid can exchange material at more than one point along its length. Crossing over occurs shortly after synapsis begins, but chiasmata do not become visible until later, when the homologs are repelling each other. Crossing over results in **recombinant** chromatids, and it increases genetic variation among the products of meiosis by reshuffling genetic information between homologous chromosome pairs. In Concept 8.3 we will explore further the genetic consequences of crossing over.

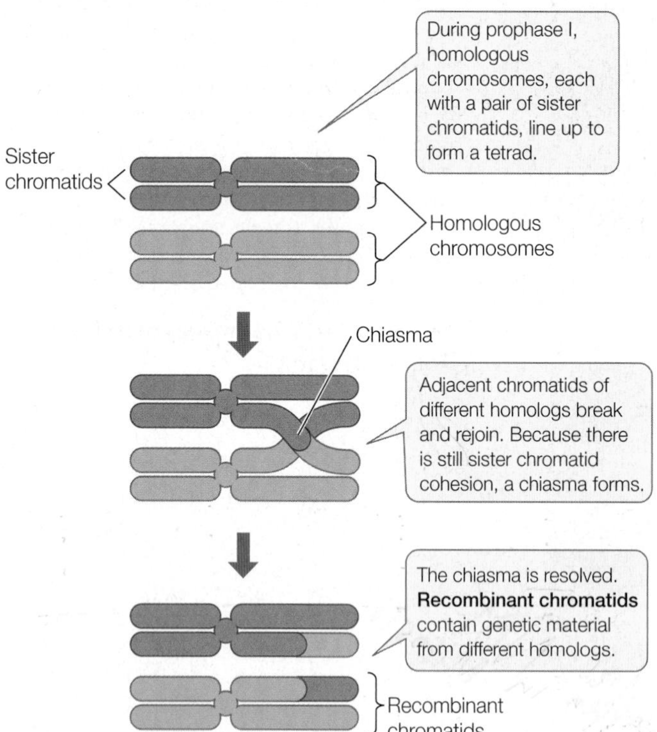

FIGURE 7.13 Crossing Over Forms Genetically Diverse Chromosomes The exchange of genetic material by crossing over results in new combinations of genetic information on the recombinant chromosomes. The two different colors distinguish the chromosomes contributed by the male and female parents of the organism whose cell is undergoing meiosis.

Mitosis seldom takes more than an hour or two, but meiosis can take *much* longer. In human males, the cells in the testis that undergo meiosis take about a week for prophase I and about a month for the entire meiotic cycle. In females, prophase I begins long before a woman's birth, during her early fetal development. Meiosis continues as much as decades later, during the monthly ovarian cycle, and is completed only after fertilization.

INDEPENDENT ASSORTMENT In addition to crossing over, meiosis provides a second source of genetic diversity. It is a matter of chance which member of a homologous pair goes to which daughter cell at anaphase I. For example, consider a diploid organism with two pairs of homologous chromosomes (pairs 1 and 2). One member of each pair came from the male parent of the organism (paternal 1 and 2), and the other came from the female parent (maternal 1 and 2). When cells in this organism undergo meiosis, a particular daughter nucleus could receive paternal 1 and maternal 2, paternal 2 and maternal 1, both maternal, or both paternal chromosomes. It all depends on how the homologous pairs line up at metaphase I. This phenomenon is called **independent assortment**.

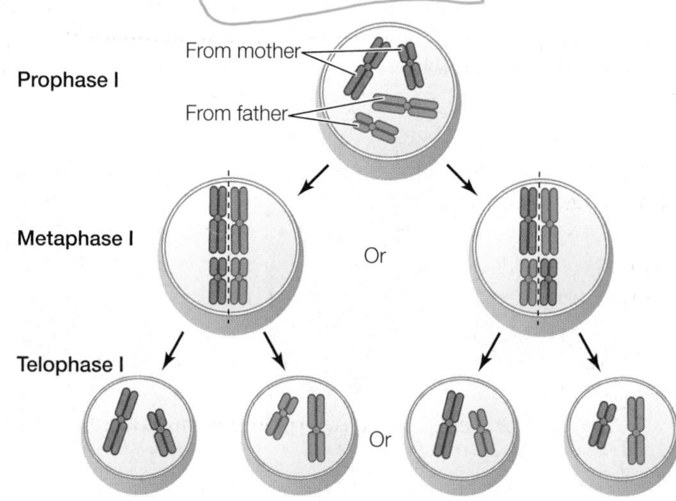

Note that of the four possible outcomes in the figure above, only two daughter nuclei receive either all maternal or all paternal chromosomes (apart from material exchanged by crossing over). The greater the number of chromosomes, the lower the probability of reestablishing the original parental combinations, and thus the greater the potential for genetic diversity. Most species of diploid organisms have more than two pairs of chromosomes. In humans, with 23 chromosome pairs, 2^{23} (8,388,608) different combinations of maternal and paternal chromosomes can be produced just by the mechanism of independent assortment! Taking into account the extra genetic shuffling afforded by crossing over, the number of possible combinations is virtually infinite. Crossing over and independent assortment, along with the processes that result in mutations, provide the genetic diversity needed for evolution by natural selection.

We have seen how meiosis I is fundamentally different from mitosis. However, meiosis II is similar to mitosis in that it involves the separation of chromatids into daughter nuclei (see steps 7–11 in Figure 7.12). The final products of meiosis I and meiosis II are four haploid daughter cells, each with one set (n) of chromosomes.

Meiotic errors lead to abnormal chromosome structures and numbers

Meiosis is a complex process, and things occasionally go wrong. For example, chromosomes may break, homologs may fail to separate at anaphase I, or chromatids may fail to separate at anaphase II. The gametes formed from meiotic errors carry abnormal chromosomes, and when abnormal chromosomes take part in fertilization, the consequences for offspring can be significant.

NONDISJUNCTION Occasionally a homologous chromosome pair fails to separate (fails to "disjoin") at anaphase I, or a pair of chromatids fail to separate at anaphase II. This failure to separate is referred to as **nondisjunction**. If a chromosome pair fails to separate at anaphase I, two of the four daughter nuclei will each end up with both members of that homologous pair, and the other two will have neither member of the pair. If nondisjunction occurs at anaphase II, only two of the four daughter nuclei will be affected: one will have an extra chromosome and the other will have one less than the full complement of chromosomes.

Using humans as an example, if during anaphase I the two homologs of chromosome 10 fail to separate, half the gametes will have two copies of chromosome 10, with a total of 24 chromosomes instead of 23. If one of these gametes fuses with a normal gamete during fertilization, the zygote will have 47 (23 + 24) chromosomes, with three copies of chromosome 10. The condition of having an abnormal number of chromosomes is called **aneuploidy**; having one extra chromosome is called trisomy, and missing one chromosome is called monosomy (**FIGURE 7.14**).

For reasons that are unclear, aneuploidy is a common and harmful condition in humans. About 10–30 percent of all conceptions show aneuploidy, but most of the embryos that develop from such zygotes do not survive to birth, and those that do often die before the age of 1 year. At least one-fifth of all recognized human pregnancies are spontaneously terminated (miscarried) during the first 2 months, largely because of trisomies and monosomies. The actual proportion of spontaneously terminated pregnancies is certainly higher, because the earliest ones often go unrecognized. The most common form of aneuploidy in humans is trisomy 16 (three copies of chromosome 16), but almost none of these embryos survive to birth. Among the few aneuploidies that allow survival is Down syndrome—trisomy 21. Such individuals generally have intellectual disabilities but can lead long and productive lives.

POLYPLOIDY Most organisms are either diploid (for example, most animals) or haploid (for example, most fungi). Under some circumstances, triploid ($3n$), tetraploid ($4n$), or higher-order **polyploid** nuclei may form. This can occur in a variety of ways. For example, there could be an extra round of DNA replication preceding meiosis, or there could be no spindle formed in meiosis II. Polyploidy occurs naturally in some animal species and in many plants.

LINK

Polyploidy can lead to reproductive isolation (the inability of two individuals to produce fertile offspring) and has probably led to speciation—the evolution of new species—as described in **Concept 17.3**

A diploid nucleus can undergo normal meiosis because there are two sets of chromosomes to make up homologous pairs, which separate during anaphase I. Similarly, a tetraploid nucleus has an even number of each kind of chromosome, so each chromosome can pair with its homolog. However, a triploid nucleus cannot undergo normal meiosis because one-third of the chromosomes would lack partners. Polyploidy has implications for agriculture, particularly in the production of hybrid plants. For example, ploidy (the number of chromosomes in the nucleus) must be taken into account in wheat breeding because

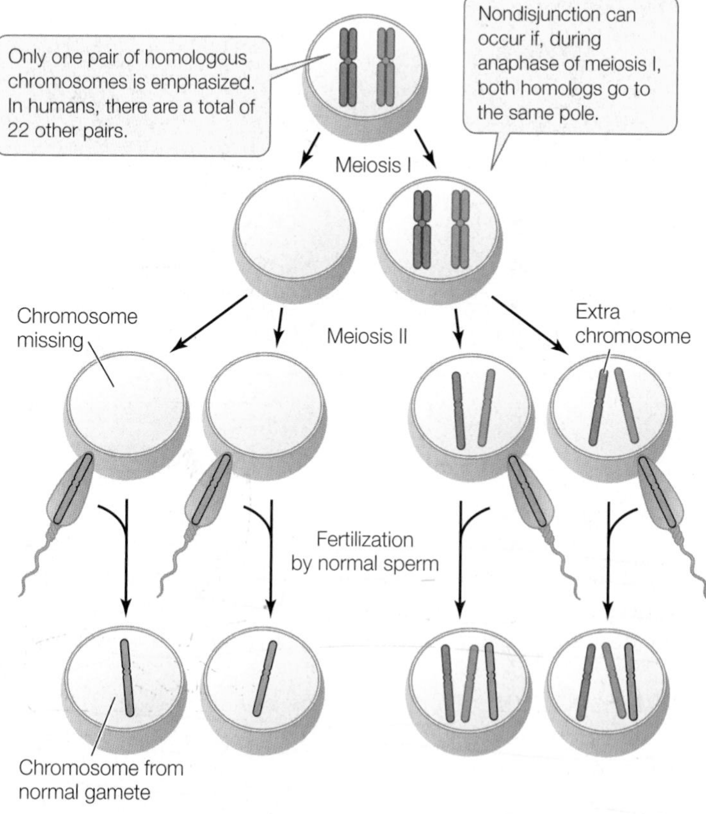

FIGURE 7.14 Nondisjunction Leads to Aneuploidy Nondisjunction, shown here occurring in meiosis I, results in aneuploidy: one or more chromosomes are either lacking or present in excess. Generally, aneuploidy is lethal to a developing embryo.

APPLY THE CONCEPT

Meiosis halves the nuclear chromosome content and generates diversity

Cells from a diploid organism ($2n = 4$) are shown undergoing division in the diagrams. For each diagram, indicate the type of cell division (mitosis or meiosis), the phase of division, and any special condition that is depicted.

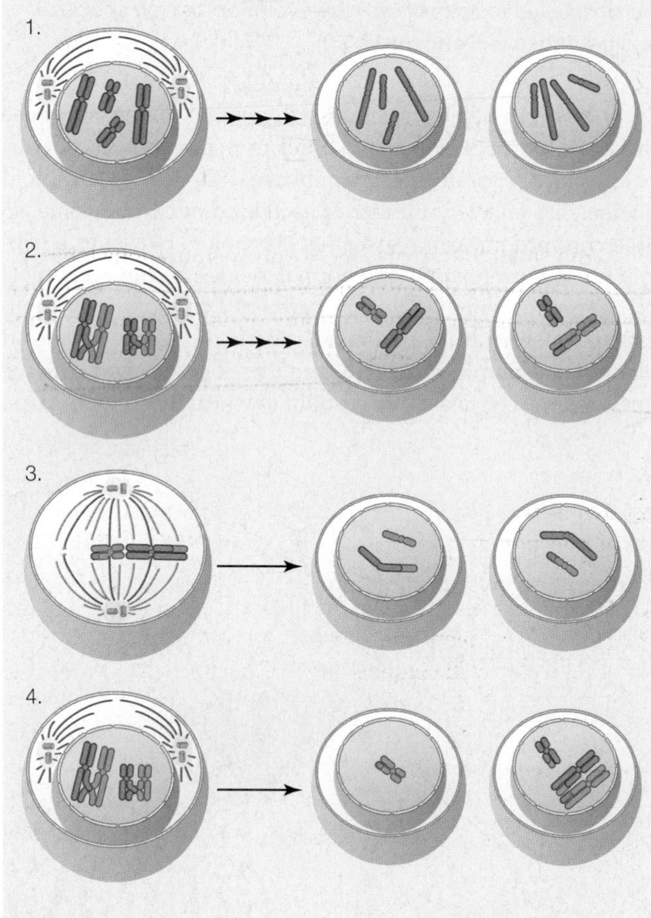

there are diploid, tetraploid, and hexaploid wheat varieties. Polyploidy can be a desirable trait in crops and ornamental plants because it often leads to more robust plants with larger flowers, fruits, and seeds. In addition, triploid fruit varieties are desirable because they are infertile and therefore seedless.

TRANSLOCATION During crossing over in meiosis I, chromatids from homologous chromosome pairs break and rejoin. Occasionally this can happen between *non-homologous chromosomes*. The result is a **translocation**, and these are quite common, even in mitotic cells. As we will point out in our discussion of gene expression and its regulation in Chapters 10 and 11, the location of genes relative to other DNA sequences is important, and translocations can have profound effects on gene expression.

An example of a translocation known to occur in humans is a swap of material between chromosomes 9 and 22:

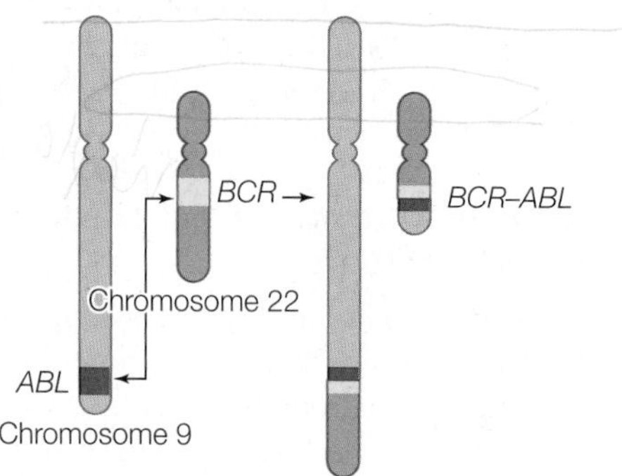

In this case, part of the *BCR* gene sequence on chromosome 22 comes to lie adjacent to part of the *ABL* gene sequence, which was translocated from chromosome 9. If the translocation occurs in a mitotic cell forming white blood cells, the result of this combination is a form of leukemia, a cancer of white blood cells.

A translocation that occurs during meiosis may be carried on the gametes that result and passed on to offspring at fertilization.

CHECKpoint CONCEPT 7.4

- ✓ How do crossing over and independent assortment during meiosis result in daughter nuclei that differ genetically?
- ✓ What are the differences between meiosis and mitosis?
- ✓ A vertebrate animal has a diploid number of 6. How many chromosomes are present in the following cells: A gamete? A gamete with monosomy of chromosome 2? A skin cell? A sperm cell at meiotic anaphase II?

An essential role of cell division in complex eukaryotes is to replace cells that die. What causes cells to die?

CONCEPT 7.5 Programmed Cell Death Is a Necessary Process in Living Organisms

Cells die in one of two ways. The first type of cell death, **necrosis**, occurs when cells are damaged by mechanical means or toxins, or are starved of oxygen or nutrients. These cells often swell up and burst, releasing their contents into the extracellular environment. This process often results in inflammation (see Concept 39.1).

More typically, cell death is due to **apoptosis** (Greek, "falling apart"). Apoptosis is a genetically programmed series of events that result in cell death. Why would a cell initiate apoptosis, which is essentially cell suicide? In animals, there are two possible reasons:

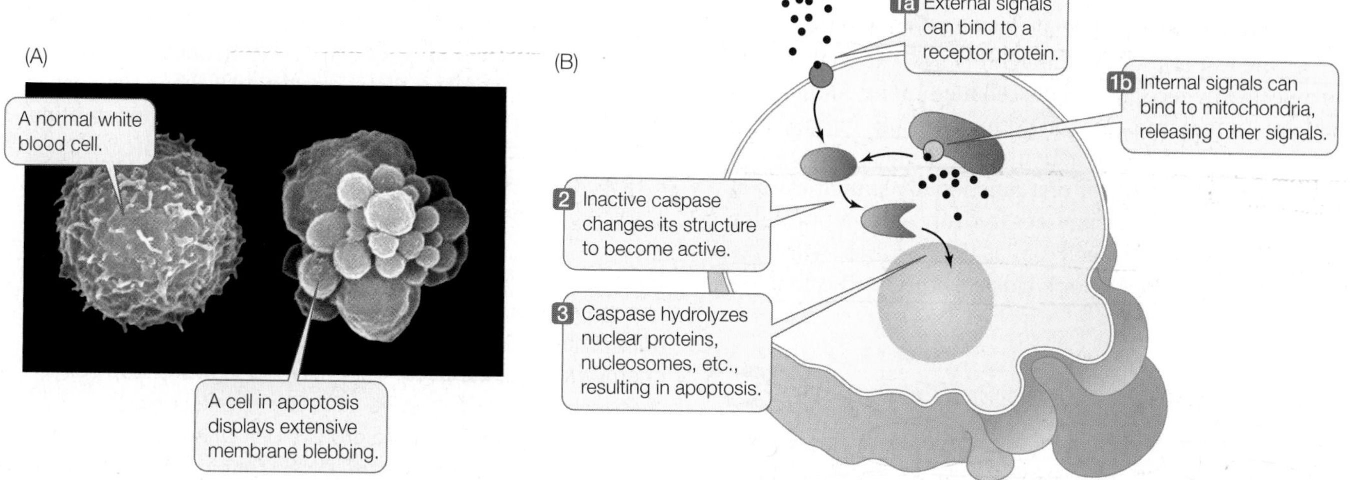
(A)
A normal white blood cell.
A cell in apoptosis displays extensive membrane blebbing.
(B)
1a External signals can bind to a receptor protein.
1b Internal signals can bind to mitochondria, releasing other signals.
2 Inactive caspase changes its structure to become active.
3 Caspase hydrolyzes nuclear proteins, nucleosomes, etc., resulting in apoptosis.

FIGURE 7.15 Apoptosis: Programmed Cell Death (A) Many cells are programmed to "self-destruct" when they are no longer needed, or when they have lived long enough to accumulate a burden of DNA damage that might harm the organism. (B) Both external and internal signals stimulate caspases (or similar enzymes in plants), which break down specific cell constituents, resulting in apoptosis.

- *The cell is no longer needed by the organism.* For example, before birth, a human fetus has weblike hands, with connective tissue between the fingers. As development proceeds, this unneeded tissue disappears as the cells undergo apoptosis in response to specific signals.
- *The longer cells live, the more prone they are to genetic damage that could lead to cancer.* This is especially true of epithelial cells on the surface of an organism, which may be exposed to radiation or toxic substances. Such cells normally die after only days or weeks and are replaced by new cells.

The events of apoptosis are similar in many organisms. The cell becomes detached from its neighbors, it hydrolyzes its DNA into small fragments, and forms membranous lobes, or "blebs," that break up into cell fragments (**FIGURE 7.15A**). In a remarkable example of the economy of nature, the surrounding living cells usually ingest the remains of the dead cell by phagocytosis. The remains are digested in the lysosomes, and the digestion products are recycled.

Apoptosis is also used by plant cells in an important defense mechanism called the hypersensitive response. Plants can protect themselves from disease by undergoing apoptosis at the site of infection by a fungus or bacterium. With no living tissue to grow in, the invading organism is not able to spread to other parts of the plant. Because of their rigid cell walls, plant cells do not form blebs the way animal cells do. Instead, they digest their own cell contents in the vacuole and then release the digested components into the vascular system.

Despite these differences between plant and animal cells, they share many of the signal transduction pathways that lead to apoptosis. Like the cell division cycle, programmed cell death is controlled by signals, which may come from inside

APPLY THE CONCEPT

Programmed cell death is a necessary process in living organisms

The DNA content of an individual cell can be measured by applying a DNA-specific dye to the cell and then passing it through an instrument that measures the staining intensity. A new drug was tested on a population of rapidly dividing tumor cells, and the DNA contents of the treated cells were analyzed and compared with those of untreated cells:[a]

DYE INTENSITY	% UNTREATED CELLS	% TREATED CELLS
<10	0	20
10	10	5
20	55	60
30	5	5
40	30	10

1. Plot percentage of cells versus dye intensity for the untreated and treated cells.
2. Explain the data for the untreated cells. Which cells are in G1? What do the data indicate about how much time cells spend in G1 relative to other phases?
3. Explain the data for treated cells and compare them with untreated cells. At what stage of the cell cycle do you think the new drug acts?

[a] Author's own, unpublished data.

or outside the cell. Internal signals may be linked to the age of the cell or the recognition of damaged DNA. External signals can be detected by receptors in the cell membrane, and in turn they activate signal transduction pathways. Both internal and external signals lead to the activation of a class of enzymes called **caspases** in animals or of a functionally similar class of enzymes in plants. These enzymes hydrolyze target proteins in a cascade of events. The cell dies as the caspases hydrolyze proteins of the nuclear envelope, nucleosomes, and cell membrane (**FIGURE 7.15B**).

CHECKpoint CONCEPT 7.5

- ✓ What are some differences between apoptosis and necrosis?
- ✓ Give examples of situations in which apoptosis occurs in animals and in plants.
- ✓ In the worm *Caenorhabditis elegans* the fertilized egg divides by mitosis to produce 1,090 somatic cells. But the adult worm has only 959 cells. What happens to the 131 other cells formed during worm embryo development? What might happen if the 131 cells did not undergo this process?

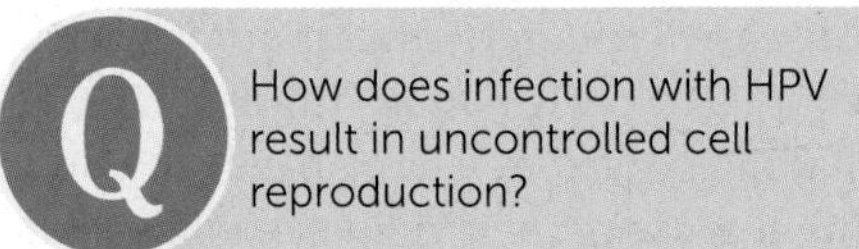

Q
How does infection with HPV result in uncontrolled cell reproduction?

ANSWER Human papillomavirus (HPV) stimulates the cell cycle when it infects tissues lining the cervix. It does this by "hijacking" the regulatory mechanisms that control the cell cycle (Concept 7.3). There are two types of proteins that regulate the cell cycle:

- **Oncogene** proteins are positive regulators of the cell cycle in cancer cells. They are derived from normal positive regulators that have become mutated to be overly active, or that are present in excess, and they stimulate the cancer cells to divide more often than normal cells. An example of an oncogene protein is the growth factor receptor in a breast cancer cell (**FIGURE 7.16A**). Normal breast cells have relatively low numbers of the growth factor receptor *h*uman *e*pidermal growth factor *r*eceptor 2 (HER2). So the growth factor does not normally find many receptors with which to bind and initiate cell division. In about 25 percent of breast cancers, a DNA change results in the increased production of HER2. This results in positive stimulation of the cell cycle, and a rapid proliferation of cells with the altered DNA.
- **Tumor suppressors** are negative regulators of the cell cycle in normal cells, but in cancer cells they are inactive. An example is the retinoblastoma (RB) protein that acts at R (the restriction point) in G1 (see Figure 7.8). When RB is active the cell cycle does not proceed, but it is inactive in cancer cells, allowing the cell cycle to occur (**FIGURE 7.16B**). *This is where HPV hijacks the system*. When it infects cells lining the cervix, HPV causes the synthesis of a protein called E7, which has a three-dimensional shape that just fits into the protein-binding site of RB, thereby inactivating it. With no active RB to prevent it, cell division proceeds. Uncontrolled cell reproduction is a hallmark of cancer—and so cervical cancer begins.

Most tumors are treated by surgery. But when a tumor has spread from its original site (a common occurrence, unfortunately), surgery does not cure it. Instead, drugs—chemotherapy—are used. Generally, these drugs stop cell division by targeting specific cell cycle events (Concepts 7.2 and 7.3). For example, some drugs block DNA replication (e.g., 5-fluorouracil); others damage DNA, stopping the cells at G2 (e.g., etoposide); and still others prevent the normal functioning of the mitotic spindle (e.g., paclitaxel). Many of these drugs do not kill the cell, but they cause the cell cycle to stop, and the damaged cell is stimulated to undergo apoptosis (Concept 7.5).

A major problem with these treatments is that they target normal cells as well as the tumor cells. They are toxic to

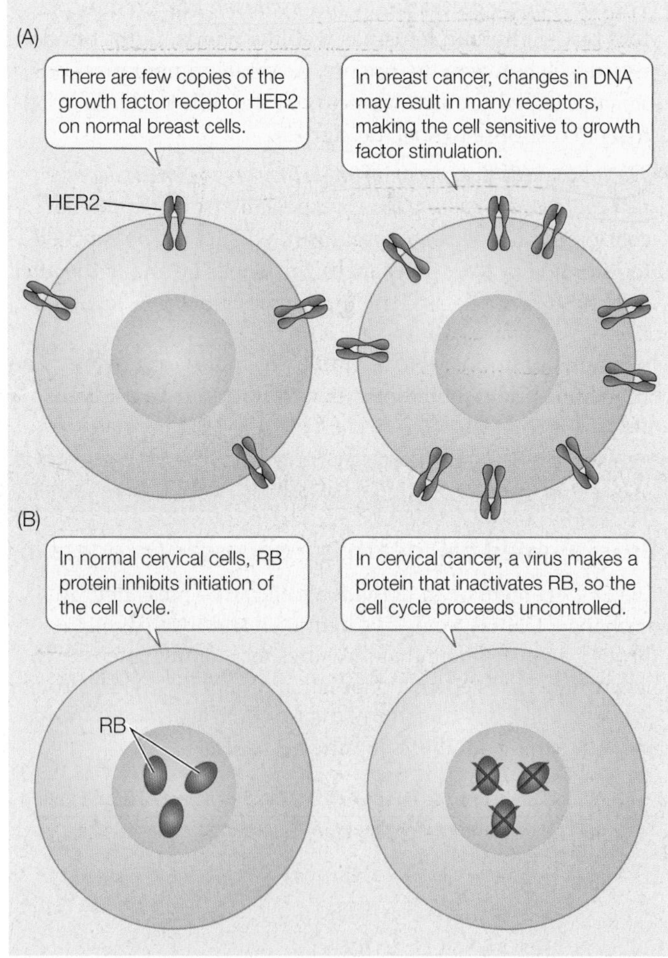

FIGURE 7.16 Molecular Changes Regulate the Cell Cycle in Cancer Cells In cancer cells, oncogene proteins become active (A) and tumor suppressor proteins become inactive (B).

tissues with large populations of normally dividing cells such as those in the intestine, skin, and bone marrow (producing blood cells). There is an ongoing search for better and more specific drugs. For example, a drug has been identified that affects the protein produced as a result of the translocation between chromosomes 9 and 22 (Concept 7.4). The drug is rather specific and has been very successful at treating leukemia caused by this translocation.

In this chapter we examined the cell cycle and cell division by binary fission and mitosis. We have seen how the normal cell cycle is disrupted in cancer. We also examined meiosis and the production of haploid cells in sexual life cycles. In the coming chapters we will examine heredity, genes, and DNA. In Concept 8.1 we will discuss Gregor Mendel's studies of heredity and how the enormous power of his discoveries founded the science of genetics.

SUMMARY

CONCEPT 7.1 Different Life Cycles Use Different Modes of Cell Reproduction

- **Cell division** is necessary for the reproduction, growth, and repair of organisms. **Review Figure 7.1**
- **Asexual reproduction** produces **clones**, new organisms that are virtually identical genetically to the parent. Any genetic variation is the result of mutations.
- In **sexual reproduction**, two **haploid** gametes—usually one from each parent—unite in **fertilization** to form a genetically unique **diploid zygote**. There are many different sexual life cycles that can be haplontic, diplontic, or involve alternation of generations. **Review Figure 7.3 and ACTIVITY 7.1**
- Diploid cells contain **homologous pairs** of chromosomes. In sexually reproducing organisms, certain cells undergo **meiosis**, a process of cell division in which the chromosome number is halved. Each of the haploid daughter cells contains one member of each homologous pair of chromosomes.

CONCEPT 7.2 Both Binary Fission and Mitosis Produce Genetically Identical Cells

- Cell division must be initiated by a reproductive signal. Before a cell can divide, the genetic material (DNA) must undergo **replication** and **segregation** to separate portions of the cell. **Cytokinesis** then divides the cytoplasm into two cells.
- In prokaryotes, most cellular DNA is a single molecule, usually in the form of a circular chromosome. Prokaryotes reproduce by **binary fission**. **Review Figure 7.4**
- During most of the eukaryotic cell cycle, the cell is in **interphase**, which is divided into three subphases: **G1**, **S**, and **G2**. DNA is replicated during S phase. **Mitosis** (**M phase**) and cytokinesis follow. **Review Figure 7.5 and ANIMATED TUTORIAL 7.1**
- In mitosis, a single nucleus gives rise to two nuclei that are genetically identical to each other and to the parent nucleus.
- At mitosis, the replicated chromosomes, called **sister chromatids**, are held together at the **centromere**. Each chromatid contains one double-stranded DNA molecule. During mitosis, sister chromatids line up at the equatorial plate and attach to the **spindle**. **Review ACTIVITY 7.2**
- Mitosis can be divided into several phases called **prophase**, **prometaphase**, **metaphase**, **anaphase**, and **telophase**. **Review Figure 7.6 and ACTIVITY 7.3**
- Nuclear division is usually followed by cytokinesis. Animal cell cytoplasms divide via a contractile ring made up of actin microfilaments. In plant cells, cytokinesis is accomplished by vesicles that fuse to form a cell plate. **Review Figure 7.7**

CONCEPT 7.3 Cell Reproduction Is Under Precise Control

- Interactions between **cyclins** and **CDKs** regulate the passage of cells through checkpoints in the cell cycle. External controls such as **growth factors** can stimulate the cell to begin a division cycle. **Review Figure 7.10**

CONCEPT 7.4 Meiosis Halves the Nuclear Chromosome Content and Generates Diversity

- Meiosis consists of two nuclear divisions, **meiosis I** and **meiosis II**, which collectively reduce the chromosome number from diploid to haploid. Meiosis results in four genetically diverse haploid cells, often gametes. **Review ANIMATED TUTORIAL 7.2**
- In meiosis I, entire chromosomes, each with two chromatids, migrate to the poles. In meiosis II, the sister chromatids separate. **Review Figures 7.11 and 7.12 and ACTIVITY 7.4**
- During prophase I, homologous chromosomes undergo synapsis to form pairs in a **tetrad**. Chromatids can form junctions called chiasmata, and genetic material may be exchanged between the two homologs by **crossing over**. **Review Figure 7.13**
- Both crossing over during prophase I and **independent assortment** of the homologs as they separate during anaphase I ensure that gametes are genetically diverse.
- Meiotic errors can result in abnormal numbers of chromosomes in the resulting gametes and offspring. **Review Figure 7.14**

CONCEPT 7.5 Programmed Cell Death Is a Necessary Process in Living Organisms

- A cell may die by **necrosis**, or it may self-destruct by **apoptosis**, a genetically programmed series of events that includes the fragmentation of the cell's nuclear DNA.
- Apoptosis is regulated by both external and internal signals. These signals result in activation of a class of enzymes called **caspases** that hydrolyze proteins in the cell. **Review Figure 7.15**

Go to the Interactive Summary to review key figures, Animated Tutorials, and Activities
PoL2e.com/is7

Go to LaunchPad at **macmillanhighered.com/launchpad** for additional resources, including LearningCurve Quizzes, Flashcards, and many other study and review resources.

8 Inheritance, Genes, and Chromosomes

KEY CONCEPTS

8.1 Genes Are Particulate and Are Inherited According to Mendel's Laws

8.2 Alleles and Genes Interact to Produce Phenotypes

8.3 Genes Are Carried on Chromosomes

8.4 Prokaryotes Can Exchange Genetic Material

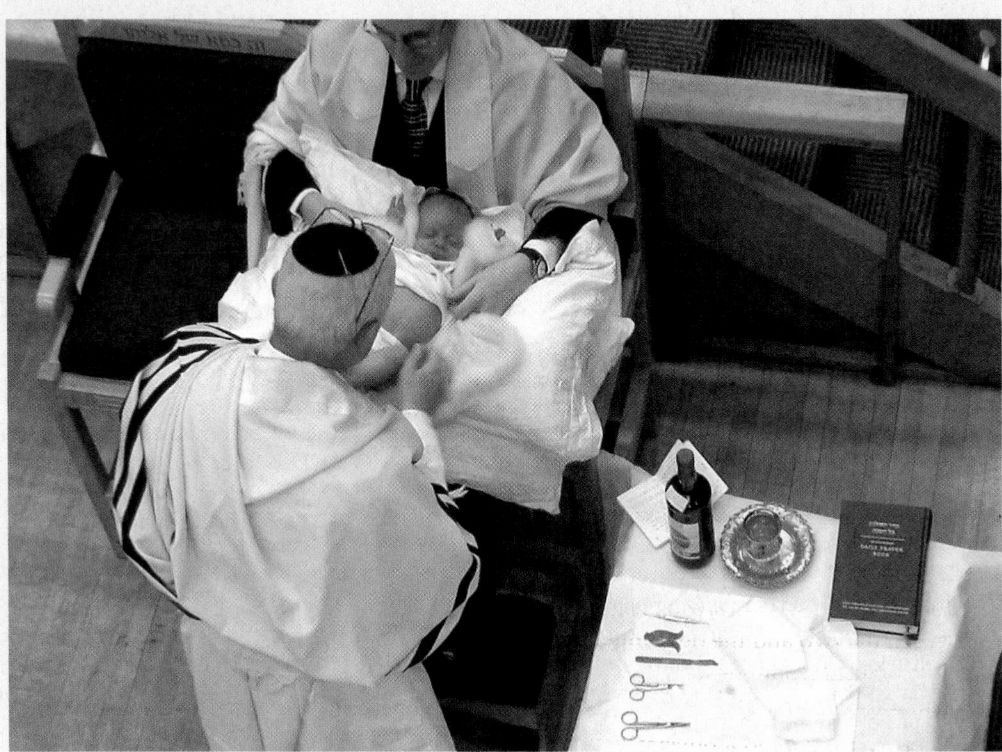

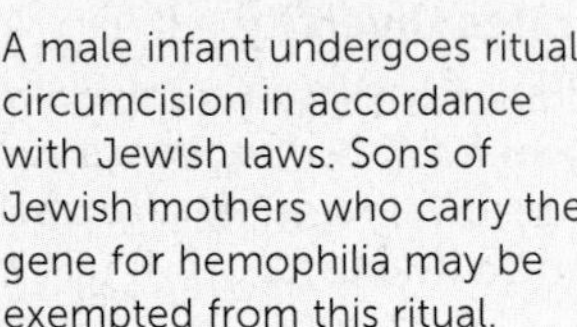

A male infant undergoes ritual circumcision in accordance with Jewish laws. Sons of Jewish mothers who carry the gene for hemophilia may be exempted from this ritual.

In the Middle Eastern desert 1,800 years ago, a rabbi faced a dilemma. A Jewish woman had given birth to a son. As required by Jewish custom, the mother brought her 8-day-old son to the rabbi for ritual penile circumcision. The rabbi knew that the woman's two previous sons had bled to death when their foreskins were cut. Yet the biblical requirement remained: unless he was circumcised, the boy could not be counted among those with whom their God had made a solemn covenant. After consultation with other rabbis, the religious leaders decided to exempt this third son.

Almost 1,000 years later, in the twelfth century, the physician and biblical commentator Moses Maimonides reviewed this and other cases in the rabbinical literature and stated that in such instances the third son should not be circumcised. Furthermore, the exemption should apply whether the mother's son was "from her first husband or from her second husband." The bleeding disorder, he reasoned, was clearly carried by the mother and passed on to her sons. In all cases, the parents did not show any evidence of having the disease.

Without any knowledge of modern concepts of genes and genetics, the rabbis had linked a human disease with a pattern of inheritance. We now have a name for the disease: it is hemophilia, which affects about 18,000 people in the United States, almost all of them males. The bleeding disorder is due to the absence of a specific protein that is crucial for the formation of blood clots. When a person who does not have hemophilia gets a cut there is usually some bleeding, but soon a clot forms and prevents further bleeding. In the case of hemophilia, no clot forms and the bleeding can continue until the person dies. Indeed, well into the twentieth century the slightest accident could be lethal for such a person. Internal bleeding is also an extremely serious problem for people with this disease, and permanent joint damage due to bleeding in the joints is a common problem for untreated patients.

Treatment of hemophilia by injection of clotting factor into the bloodstream is now possible because the proteins can be isolated from donated blood or made in the laboratory using biotechnological techniques. An issue has been whether people who suffer from hemophilia should receive injections of clotting factor all the time as a preventive measure (an expensive proposition), or only when the factor is needed. Based on reductions in joint damage in children treated by the former (preventive) approach, recent studies have concluded that this approach is best.

How is hemophilia inherited, and why is it most frequent in males?

You will find the answer to this question on page 169.

CONCEPT 8.1 Genes Are Particulate and Are Inherited According to Mendel's Laws

Genetics, the field of biology concerned with inheritance, has a long history. There is good evidence that people were deliberately breeding animals (horses) and plants (the date palm tree) for desirable characteristics as long as 5,000 years ago. The general idea was to examine the natural variation among the individuals of a species and "breed the best to the best and hope for the best." This was a hit-or-miss method—sometimes the resulting offspring had all the good characteristics of the parents, but often they did not.

By the mid-nineteenth century, two hypotheses had emerged to explain the results of breeding experiments:

- The hypothesis of *blending inheritance* proposed that gametes contained hereditary determinants (what we now call genes) that blended when the gametes fused during fertilization. Like inks of different colors, the two different determinants lost their individuality after blending and could never be separated. For example, if a plant that made smooth, round seeds was mated (crossed) with a plant that made wrinkled seeds, the offspring would be intermediate between the two and the determinants for the two parental characteristics would be lost.
- The hypothesis of *particulate inheritance* proposed that each determinant had a physically distinct nature; when gametes fused in fertilization, the determinants remained intact. According to this theory, if a plant that made round seeds was crossed with a plant that made wrinkled seeds, the offspring (no matter the shape of their seeds) would still contain the determinants for the two characteristics.

The story of how these competing hypotheses were tested provides a great example of how the scientific method can be used to support one theory and reject another. In the following sections we will look in detail at experiments performed in the 1860s by an Austrian monk and scientist, Gregor Mendel, whose work clearly supported the particulate hypothesis.

Mendel used the scientific method to test his hypotheses

After entering the priesthood at a monastery in Brno, in what is now the Czech Republic, Gregor Mendel was sent to the University of Vienna, where he studied biology, physics, and mathematics. He returned to the monastery in 1853 to teach. The abbot in charge had set up a small plot of land to do experiments with plants and encouraged Mendel to continue with them. Over seven years, Mendel made crosses with many thousands of plants. Analysis of his meticulously gathered data suggested to him that inheritance was due to particulate factors.

Mendel presented his theories in two public lectures in 1865 and a detailed written publication in 1866, but his work was ignored by mainstream scientists until 1900. By that time, the discovery of chromosomes had suggested to biologists that genes might be carried on chromosomes. When they read Mendel's work on particulate inheritance, the biologists connected the dots between genes and chromosomes.

Mendel chose to study the common garden pea because it is easily cultivated and is amenable to manipulation and controlled crosses. Pea flowers have both male and female sex organs—stamens and pistils, which produce gametes that are contained within the pollen and ovules, respectively:

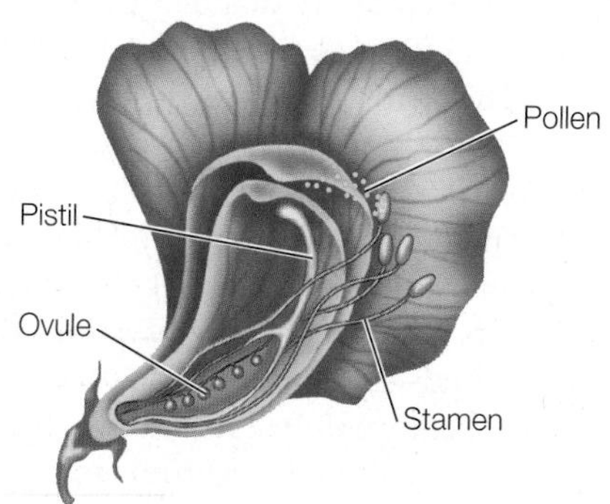

Pea flowers normally self-fertilize. However, the male organs can be removed from a flower so that it can be manually fertilized with pollen from a different flower.

There are many varieties of pea plants with easily recognizable characteristics. A **character** is an observable physical feature, such as seed shape. A **trait** is a particular form of a character, such as round or wrinkled seeds. Mendel worked with seven pairs of varieties with contrasting traits for characters such as seed shape, seed color, and flower color. These varieties were true-breeding: that is, when he crossed a plant that produced wrinkled seeds with another of the same variety, all of the offspring plants produced wrinkled seeds.

As we will see, Mendel developed a set of hypotheses to explain the inheritance of particular pea traits, and then designed crossing experiments to test his hypotheses. He performed his crosses in the following manner:

- He removed the stamens to emasculate flowers of one parental variety so that it couldn't self-fertilize. Then he collected pollen from another parental variety and placed it on the pistils of the emasculated flowers. The plants providing and receiving the pollen were the **parental generation**, designated **P**. The pollen provider was the male parent, and the pollen receiver was the female parent.
- In due course, seeds formed and were planted. The seeds and the resulting new plants constituted the **first filial generation**, or $\mathbf{F_1}$. (The word "filial," from the Latin *filius*, "son," refers to the relationship between offspring and parents.) Mendel examined each F_1 plant to see which traits it bore and then recorded the number of F_1 plants expressing each trait.
- In some experiments the F_1 plants were allowed to self-pollinate and produce a **second filial generation**, the $\mathbf{F_2}$. Again, each F_2 plant was characterized and counted.

Go to MEDIA CLIP 8.1
Mendel's Discoveries
PoL2e.com/mc8.1

Mendel's first experiments involved monohybrid crosses

The term "hybrid" refers to the offspring of crosses between organisms differing in one or more characters. In Mendel's first experiments, he crossed parental (P) varieties with contrasting traits for a single character, producing monohybrids (from the Greek *monos*, "single") in the F_1 generation. He subsequently planted the F_1 seeds and allowed the resulting plants to self-pollinate to produce the F_2 generation. This technique is referred to as a **monohybrid cross**.

Mendel performed the same experiment for seven pea characters. His method is illustrated in **FIGURE 8.1**, using seed shape as an example. When he crossed a strain that made round seeds with one that made wrinkled seeds, the F_1 seeds were round—it was as if the wrinkled seed trait had disappeared completely. However, when F_1 plants were allowed to self-pollinate to produce F_2 seeds, about one-fourth of the seeds were wrinkled. These observations were key to distinguishing the two theories noted above:

- The F_1 offspring were *not a blend* of the two traits of the parents. Only one of the traits was present (in this case, round seeds).
- Some F_2 offspring had wrinkled seeds. The trait *did not disappear*.

These observations led to a rejection of the blending theory of inheritance and provided support for the particulate theory. We now know that hereditary determinants are not actually "particulate," but they are physically distinct entities: sequences of DNA carried on chromosomes (see Concept 8.3).

All seven crosses between varieties with contrasting traits gave the same kind of data (see Figure 8.1). In the F_1 generation only one of the two traits was seen, but the other one reappeared in about one-fourth of the offspring in the F_2 generation. Mendel called the trait that appeared in the F_1 and was more abundant in the F_2 the **dominant** trait, and the other trait **recessive**.

Mendel went on to expand the particulate hypothesis. He proposed that hereditary determinants—what we know today as genes, though Mendel did not use that term—occur in pairs and segregate (separate) from one another during the formation of gametes. He concluded that each pea plant has two copies of the gene for each character (such as seed shape), one inherited from each parent. We now use the term **diploid** to describe the state of having two copies of each gene; **haploids** have just one copy (see Concept 7.1).

Mendel concluded that while each gamete contains one copy of each gene (i.e., is haploid), the resulting zygote contains two copies (is diploid), because it is produced by the fusion of two gametes. Furthermore, different traits arise because there can be *different forms of a gene*—now called **alleles**—for a particular character. For example, Mendel studied two alleles for seed shape: one that caused round seeds and the other causing wrinkled seeds.

INVESTIGATION

FIGURE 8.1 Mendel's Monohybrid Experiments Mendel performed crosses with pea plants and carefully analyzed the outcomes to show that genetic determinants are particulate.[a]

HYPOTHESIS

When two strains of peas with contrasting traits are bred, their characteristics are irreversibly blended in succeeding generations.

METHOD

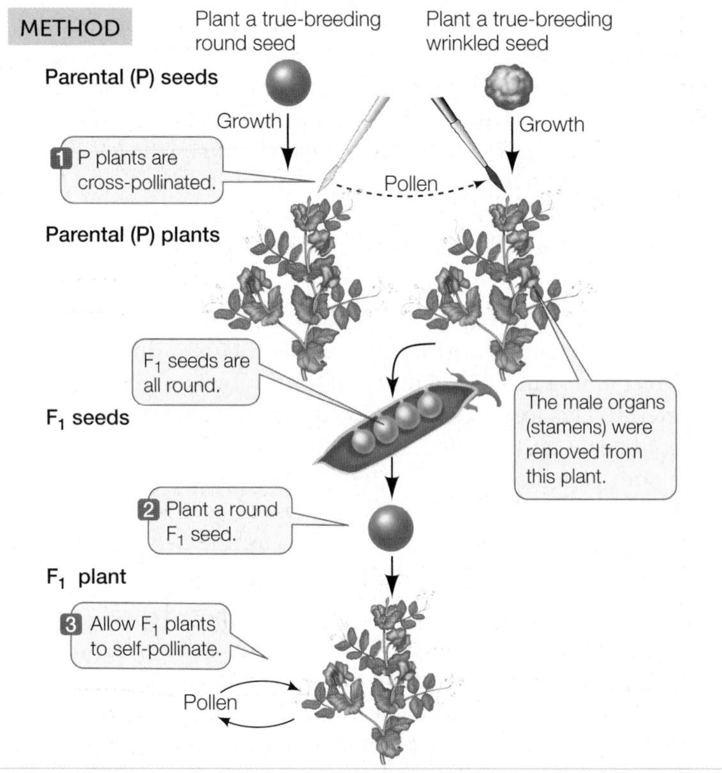
Plant a true-breeding round seed
Plant a true-breeding wrinkled seed
Parental (P) seeds
Growth
Growth
1 P plants are cross-pollinated.
Pollen
Parental (P) plants
F1 seeds are all round.
F1 seeds
The male organs (stamens) were removed from this plant.
2 Plant a round F1 seed.
F1 plant
3 Allow F1 plants to self-pollinate.
Pollen

RESULTS

F$_2$ seeds from F$_1$ plant

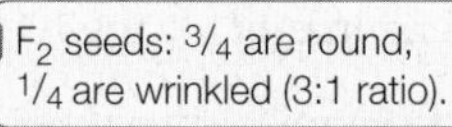
4 F2 seeds: 3/4 are round, 1/4 are wrinkled (3:1 ratio).

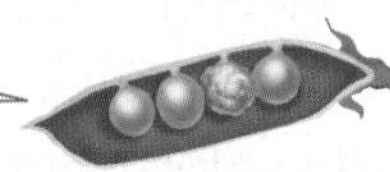

CONCLUSION

The hypothesis is rejected. There is no irreversible blending of characteristics, and a recessive trait can reappear in succeeding generations.

ANALYZE THE DATA

The table gives Mendel's data—the number of offspring showing each trait—for the F_2 from crosses between P generation plants with contrasting traits:

Characteristic	Dominant	Recessive
Seed shape	5,474 round	1,850 wrinkled
Seed color	6,022 yellow	2,001 green
Flower color	705 purple	224 white
Pod color	428 green	152 yellow
Stem height	787 tall	277 short

A. Calculate the phenotypic ratio of dominant:recessive in the F_2 offspring.

B. What can you conclude about the behavior of alleles during gamete formation in a plant that is heterozygous for a trait?

C. Perform a chi-square test to evaluate the statistical significance of these data (refer to Appendix B).

Go to **LaunchPad** for discussion and relevant links for all **INVESTIGATION** figures.

[a]See www.mendelweb.org/Mendel.plain.html

- An organism that is **homozygous** for a gene has two alleles that are the same (for example, two copies of the allele for round seeds).
- An organism that is **heterozygous** for a gene has two different alleles (for example, one allele for round seeds and one allele for wrinkled seeds).

In a heterozygote, one of the two alleles may be dominant (such as round, *R*) and the other recessive (wrinkled, *r*). By convention, we designate dominant alleles with uppercase letters and recessive alleles with lowercase letters.

- The physical appearance of an organism is its **phenotype**.
- The phenotype is the result of the **genotype**, or genetic constitution, of the organism.

Round seeds and wrinkled seeds are two phenotypes resulting from three possible genotypes: the wrinkled seed phenotype is produced by the genotype *rr*, whereas the round seed phenotype is produced by either of the genotypes *RR* or *Rr* (because the *R* allele is dominant to the *r* allele).

Mendel's first law states that the two copies of a gene segregate

How do Mendel's theories explain the proportions of traits seen in the F_1 and F_2 generations of his monohybrid crosses? Mendel's first law—the **law of segregation**—states that *when any individual produces gametes, the two copies of a gene separate, so each gamete receives only one copy*. Thus gametes from a parent with the *RR* genotype will all be *R*; gametes from an *rr* parent will all be *r*; and the progeny derived from a cross between these parents will all be *Rr*, producing seeds with a round phenotype (**FIGURE 8.2**).

Now let's consider the composition of the F_2 generation. Because the alleles segregate, half of the gametes produced by the F_1 generation will have the *R* allele and the other half will have the *r* allele. What genotypes are produced when these gametes fuse to form the next (F_2) generation? We can predict the allele combinations that will result from a cross using a **Punnett square**, a method devised in 1905 by the British geneticist Reginald Punnett. This device ensures that we consider all possible random combinations of gametes when calculating expected genotype frequencies. A Punnett square looks like this:

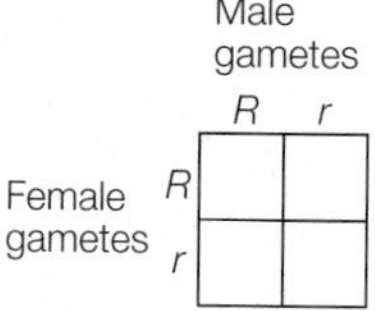

All possible male gamete (haploid sperm) genotypes are shown along the top and all possible female gamete (haploid egg) genotypes along the left side. The grid is completed by filling in each square with the diploid genotype that can be generated from each combination of gametes. In this example, to fill in the top right square, we put in the *R* from the female gamete (the egg cell) and the *r* from the male gamete (the sperm cell in the pollen tube), yielding *Rr*.

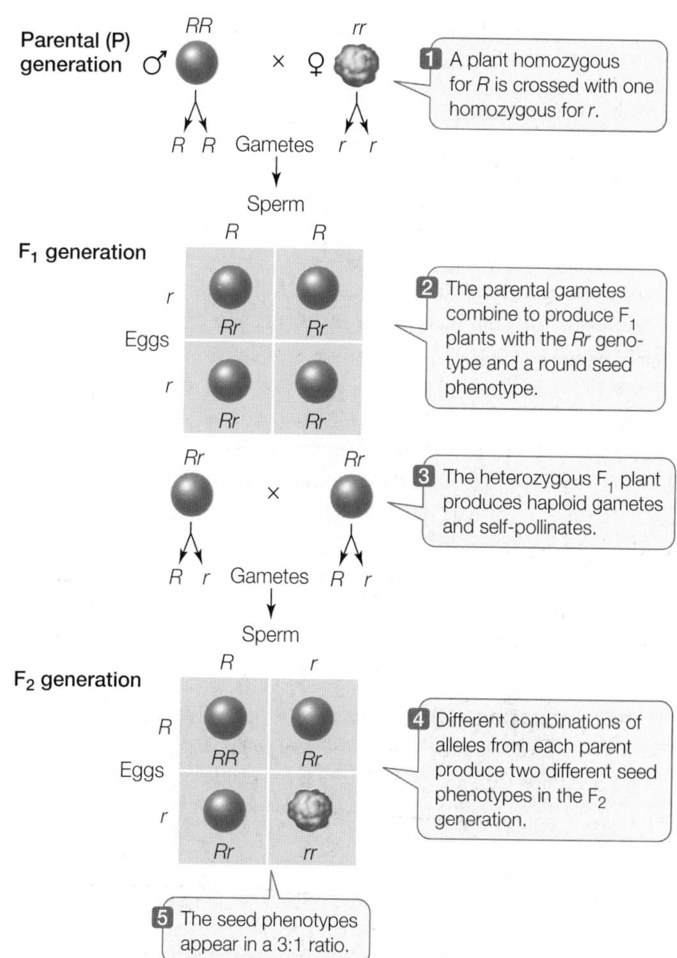

FIGURE 8.2 Mendel's Explanation of Inheritance Mendel concluded that inheritance depends on discrete factors from each parent that do not blend in the offspring.

Once the Punnett square is filled in, we can readily see that there are four possible combinations of alleles in the F_2 generation: *RR*, *Rr*, *rR*, and *rr* (see Figure 8.2). Since *R* is dominant, there are three ways to get round-seeded plants in the F_2 generation (*RR*, *Rr*, or *rR*), but only one way to get a plant with wrinkled seeds (*rr*). Therefore we predict a 3:1 ratio of these phenotypes in the F_2 generation, remarkably close to the values Mendel found experimentally for all seven of the traits he compared.

Mendel did not live to see his theories placed on a sound physical footing with the discoveries of chromosomes and DNA. Genes are now known to be relatively short sequences of DNA (usually a few thousand base pairs in length) found on the much longer DNA molecules that make up chromosomes (which are often millions of base pairs long). Today we can picture the different alleles of a gene segregating as chromosomes separate during meiosis I (**FIGURE 8.3**).

LINK

You can review the process of meiosis and the separation of chromosomes in **Concept 7.4**

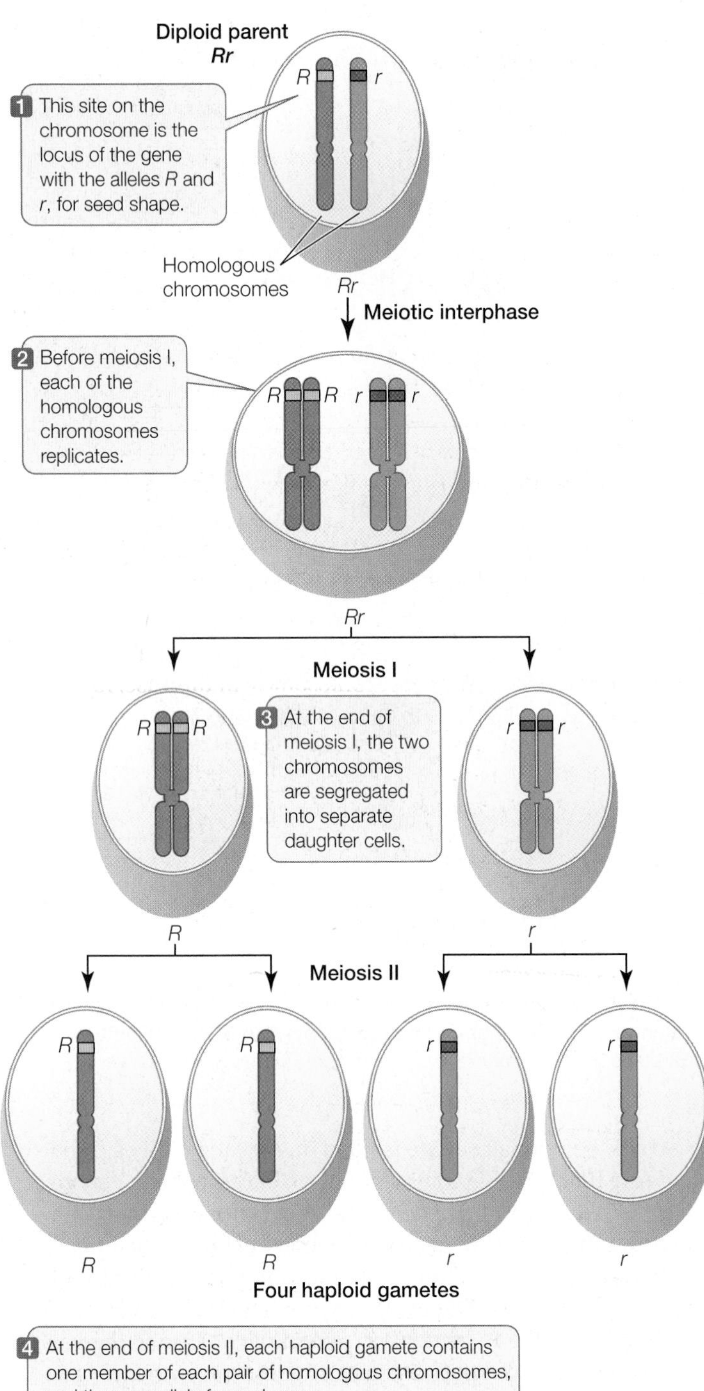
Diploid parent
Rr
R
r
1 This site on the chromosome is the locus of the gene with the alleles R and r, for seed shape.
Homologous chromosomes
Rr
Meiotic interphase
2 Before meiosis I, each of the homologous chromosomes replicates.
R R r r
Rr
Meiosis I
R R
3 At the end of meiosis I, the two chromosomes are segregated into separate daughter cells.
r r
R
r
Meiosis II
R R r r
R R r r
Four haploid gametes
4 At the end of meiosis II, each haploid gamete contains one member of each pair of homologous chromosomes, and thus one allele for each gene.

FIGURE 8.3 Meiosis Accounts for the Segregation of Alleles Although Mendel had no knowledge of chromosomes or meiosis, we now know that a pair of alleles resides on homologous chromosomes, and that those alleles segregate during meiosis.

Genes determine phenotypes mostly by producing proteins with particular functions, such as enzymes. So in many cases a dominant gene is expressed (transcribed and translated; see Concept 3.1) to produce a functional protein, whereas a recessive gene is mutated so that it is no longer expressed, or it encodes a mutant protein that is nonfunctional. For example, the molecular nature of the wrinkled pea seed phenotype is the absence of an enzyme called starch branching enzyme 1 (SBE1), which is essential for starch synthesis. With less starch, the developing seed has more sucrose, which causes an inflow of water by osmosis. When the seed matures, this water is lost, leaving a shrunken, wrinkled seed.

Mendel verified his hypotheses by performing test crosses

As mentioned above, Mendel arrived at his laws of inheritance by developing a series of hypotheses and then designing experiments to test them. One such hypothesis was that there are two possible allele combinations (*RR* or *Rr*) for seeds with the round phenotype. Mendel verified this hypothesis by performing test crosses with F_1 seeds derived from a variety of other crosses. A **test cross** is used to determine whether an individual showing a dominant trait is homozygous or heterozygous. The individual in question is crossed with an individual that is homozygous for the recessive trait—an easy individual to identify, because all individuals with the recessive phenotype are homozygous for that trait.

The recessive homozygote for the seed shape gene has wrinkled seeds and the genotype *rr*. The individual being tested may be described initially as *R_* because we do not yet know the identity of the second allele. We can predict two possible results:

- If the individual being tested is homozygous dominant (*RR*), all offspring of the test cross will be *Rr* and show the dominant trait (round seeds) (**FIGURE 8.4, LEFT**).
- If the individual being tested is heterozygous (*Rr*), then approximately half of the offspring of the test cross will be heterozygous and show the dominant trait (*Rr*), and the other half will be homozygous for the recessive trait (*rr*) (**FIGURE 8.4, RIGHT**).

Mendel obtained results consistent with both of these predictions; thus his hypothesis accurately predicted the results of his test crosses.

Go to ACTIVITY 8.1 Homozygous or Heterozygous? PoL2e.com/ac8.1

Mendel's second law states that copies of different genes assort independently

Consider an organism that is heterozygous for two genes (*RrYy*). In this example, the dominant *R* and *Y* alleles came from one true-breeding parent, and the recessive *r* and *y* alleles came from the other true-breeding parent. When this organism produces gametes, do the *R* and *Y* alleles always go together in one gamete, and *r* and *y* alleles in another? Or can a single gamete receive one recessive and one dominant allele (*R* and *y* or *r* and *Y*)?

Mendel performed another series of experiments to answer these questions. He began with peas that differed in *two* characters: seed shape and seed color. One parental variety produced only round, yellow seeds (*RRYY*), and the other produced only wrinkled, green ones (*rryy*). A cross between these two varieties produced an F_1 generation in which all the plants were *RrYy*.

INVESTIGATION

FIGURE 8.4 Homozygous or Heterozygous? An individual with a dominant phenotype may have either a homozygous or a heterozygous genotype. The test cross determines which.[a]

HYPOTHESIS

The progeny of a test cross can reveal whether a parental organism is homozygous or heterozygous.

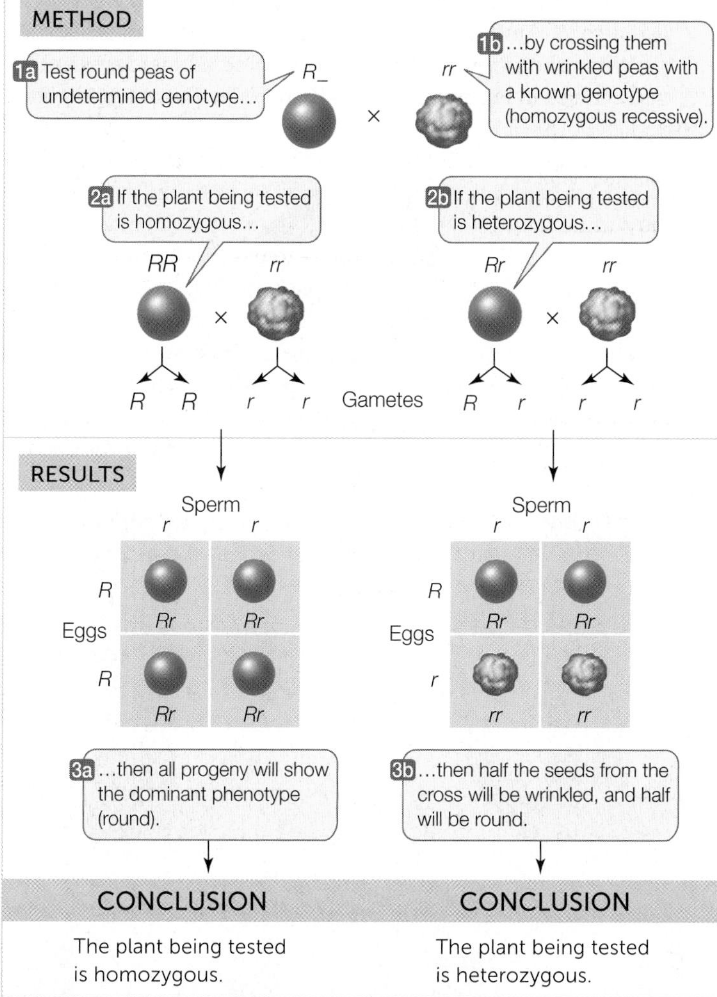

ANALYZE THE DATA

A tall plant was crossed with a short plant to produce three F_1 plants (A, B, and C). Each of the F_1 plants was test crossed, with the following results:

F_1	Tall progeny	Short progeny
A	34	0
B	18	15
C	21	0

A. For each test cross, what were the genotypes and phenotypes of both parents?

B. Choose and perform a statistical test from Appendix B to evaluate the significance of these data.

Go to **LaunchPad** for discussion and relevant links for all **INVESTIGATION** figures.

[a]See www.mendelweb.org/Mendel.plain.html

Because the *R* and *Y* alleles were dominant, the F_1 seeds were all round and yellow.

Mendel continued this experiment into the F_2 generation by performing a **dihybrid cross**, a cross between individuals that are identical double heterozygotes. In this case he simply allowed the F_1 plants, which were all double heterozygotes, to self-pollinate (**FIGURE 8.5**). Depending on whether the alleles of the two genes are inherited together or separately, there are two possible outcomes, as Mendel saw:

- The alleles could maintain the associations they had in the parental generation—they could be near each other (linked) on the same chromosome. If this were the case, the F_1 plants would produce two types of gametes (*RY* and *ry*). The F_2 progeny resulting from self-pollination of these F_1 plants would consist of two phenotypes: round yellow and wrinkled green in the ratio of 3:1, just as in the monohybrid cross.
- The segregation of *R* from *r* could be independent of the segregation of *Y* from *y*—the two genes could be unlinked, on different chromosomes. In this case, four

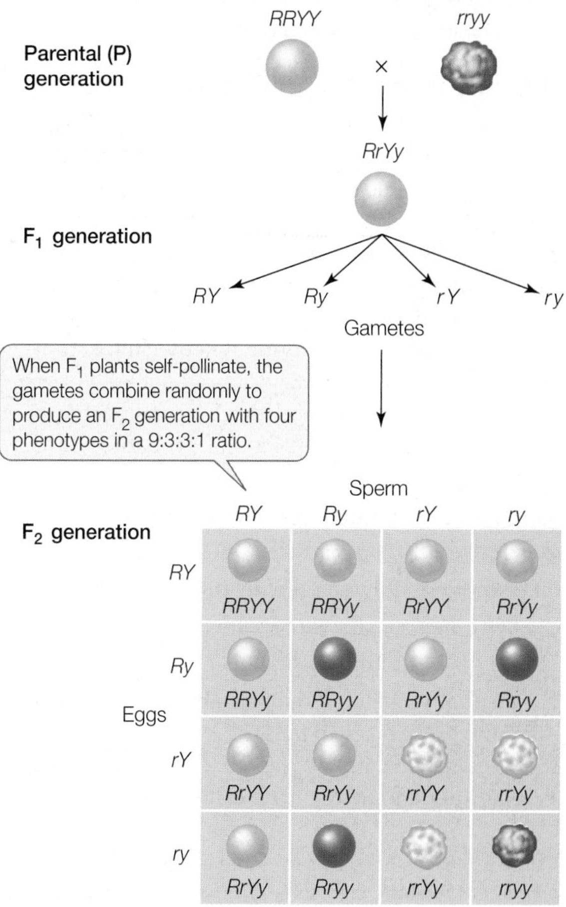

FIGURE 8.5 Independent Assortment The 16 possible combinations of gametes in this dihybrid cross result in nine different genotypes. Because *R* and *Y* are dominant over *r* and *y*, respectively, the nine genotypes result in four phenotypes in a ratio of 9:3:3:1. These results show that the two genes segregate independently.

kinds of gametes would be produced in equal numbers: *RY*, *Ry*, *rY*, and *ry*. When these gametes combine at random, they should produce an F_2 having nine different genotypes. The nine genotypes would produce four phenotypes (round yellow, round green, wrinkled yellow, wrinkled green). Putting these possibilities into a Punnett square, we can predict that these four phenotypes just mentioned would occur in a ratio of 9:3:3:1.

Mendel's dihybrid crosses supported the second prediction: four different phenotypes appeared in the F_2 generation in a ratio of about 9:3:3:1 (see Figure 8.5). On the basis of such experiments, Mendel proposed his second law, the **law of independent assortment**: *alleles of different genes assort independently of one another during gamete formation*. In the example above, the segregation of the *R* and *r* alleles is independent of the segregation of the *Y* and *y* alleles. As you will see in Concept 8.3, this is not as universal as the law of segregation because it does not apply to genes located near one another on the same chromosome. However, it is correct to say that *chromosomes segregate independently* during the formation of gametes, and so do any two genes located on separate chromosome pairs (**FIGURE 8.6**).

Probability is used to predict inheritance

One key to Mendel's success was his use of large sample sizes. By counting many progeny from each cross, he observed clear patterns that allowed him to formulate his theories. After his work became widely recognized, geneticists began using simple probability calculations to predict the ratios of genotypes and phenotypes in the progeny of a given cross or mating. They use statistics to determine whether the actual results match the prediction.

You can think of probabilities by considering a coin toss. The basic conventions of probability are simple:

- If an event is absolutely certain to happen, its probability is 1.
- If it cannot possibly happen, its probability is 0.
- All other events have a probability between 0 and 1.

There are two possible outcomes of a coin toss, and both are equally likely, so the probability of heads is ½—as is the probability of tails.

If two coins (say a penny and a dime) are tossed, each acts independently of the other (**FIGURE 8.7**). What is the probability of both coins coming up heads? In half of the tosses, the penny comes up heads, and in half of that fraction, the dime comes up heads. The probability of both coins coming up heads is ½ × ½ = ¼. In general, *the probability of two independent outcomes occurring together is found by multiplying the two individual probabilities*. This can be applied to a monohybrid cross (see Figure 8.2). After the self-pollination of an *Rr* F_1 plant, the probability that an F_2 plant will have the genotype *RR* is ½ × ½ = ¼, because the chance that the sperm will have the genotype *R* is ½, and the chance that the egg will have the genotype *R* is also ½. Similarly, the probability of *rr* offspring is also ¼.

What about the probability of getting a heterozygote? As you can see in Figures 8.2 and 8.7, there are *two* ways to get an

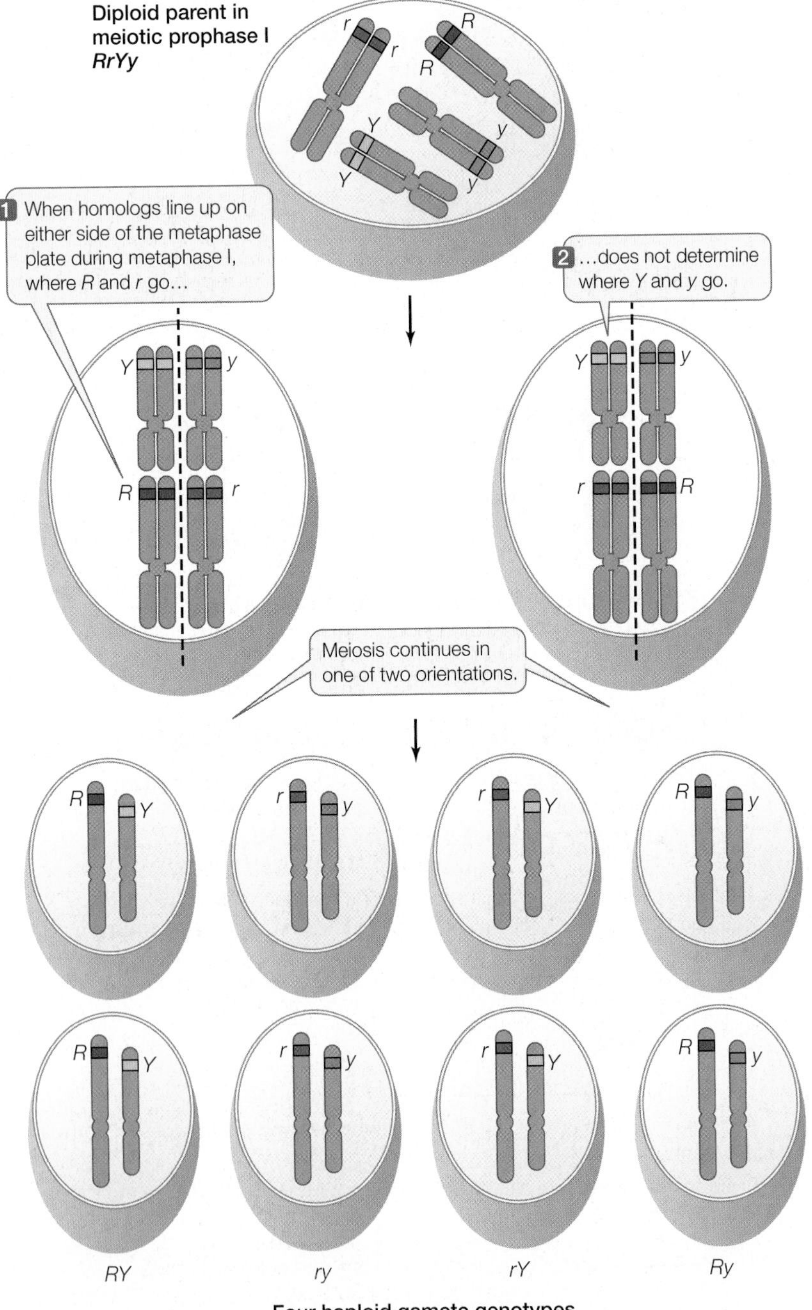

FIGURE 8.6 Meiosis Accounts for Independent Assortment of Alleles We now know that copies of genes on different chromosomes are segregated independently during metaphase I of meiosis. Thus a parent of genotype *RrYy* can form gametes with four different genotypes.

Go to ANIMATED TUTORIAL 8.1
Independent Assortment of Alleles
PoL2e.com/at8.1

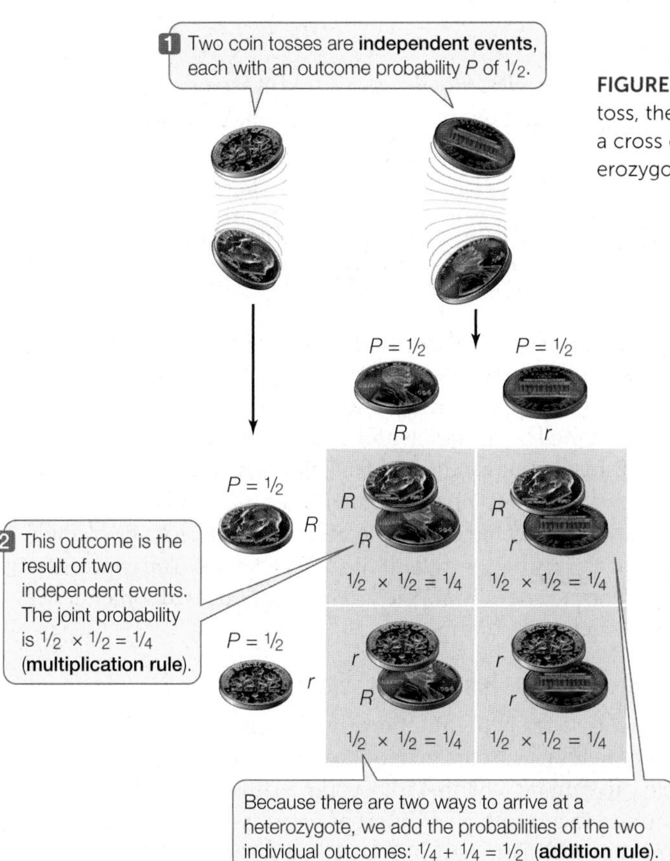
1 Two coin tosses are **independent events**, each with an outcome probability P of 1/2.
P = 1/2
P = 1/2
R
r
P = 1/2
R
r
R
R
R
r
1/2 × 1/2 = 1/4
1/2 × 1/2 = 1/4
P = 1/2
r
r
R
r
r
1/2 × 1/2 = 1/4
1/2 × 1/2 = 1/4
2 This outcome is the result of two independent events. The joint probability is 1/2 × 1/2 = 1/4 (**multiplication rule**).
Because there are two ways to arrive at a heterozygote, we add the probabilities of the two individual outcomes: 1/4 + 1/4 = 1/2 (**addition rule**).

FIGURE 8.7 Using Probability Calculations in Genetics Like the results of a coin toss, the probability of any given combination of alleles appearing in the offspring of a cross can be obtained by multiplying the probabilities of each event. Since a heterozygote can be formed in two ways, these two probabilities are added together.

Rr plant or a head and a tail in a coin toss. In the case of the seed shape gene, the *R* allele can come from a sperm and the *r* from an egg (probability ¼). Or the *R* allele could come from the egg and the *r* from the sperm (probability ¼). *The probability of an event that can occur in two or more different ways is the sum of the individual probabilities of those ways.* Thus the probability that an F_2 plant will be a heterozygote is equal to the sum of the probabilities of the two ways of forming a heterozygote: ¼ + ¼ = ½.

Mendel's laws can be observed in human pedigrees

Mendel developed his theories by performing many planned crosses and counting many offspring. This approach is not possible with humans, so human geneticists rely on **pedigrees**: family trees that show the occurrence of inherited phenotypes in several generations of related individuals (**FIGURE 8.8**).

Because humans have relatively few offspring, human pedigrees do not show the clear proportions of phenotypes that Mendel saw in his pea plants. For example, when a man and a woman who are both heterozygous for a recessive allele (say, *Aa*) have children together, each child has a ¼ probability of

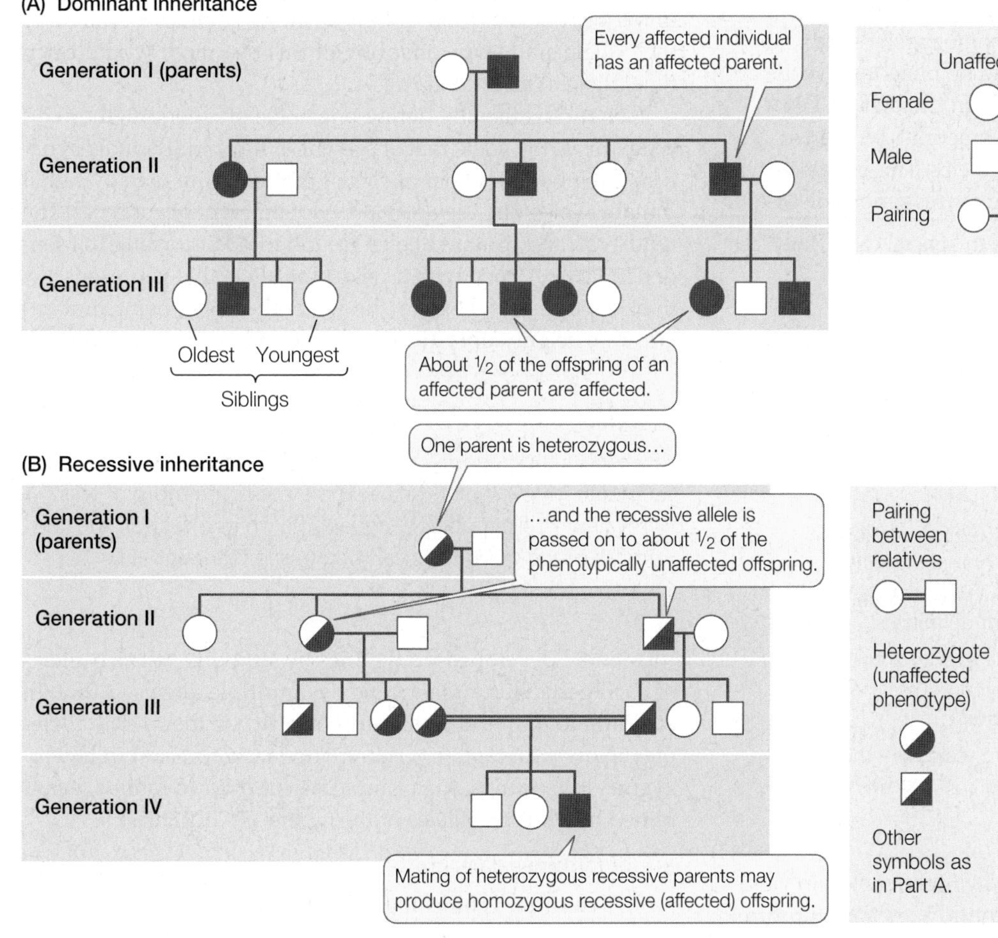
(A) Dominant inheritance
Generation I (parents)
Generation II
Generation III
Every affected individual has an affected parent.
Unaffected
Affected
Female
Male
Pairing
Oldest
Youngest
Siblings
About 1/2 of the offspring of an affected parent are affected.
(B) Recessive inheritance
One parent is heterozygous…
Generation I (parents)
Generation II
Generation III
Generation IV
…and the recessive allele is passed on to about 1/2 of the phenotypically unaffected offspring.
Pairing between relatives
Heterozygote (unaffected phenotype)
Other symbols as in Part A.
Mating of heterozygous recessive parents may produce homozygous recessive (affected) offspring.

FIGURE 8.8 Pedigree Analysis and Inheritance (A) This pedigree represents a family affected by Huntington's disease, which results from a rare dominant allele. Everyone who inherits this allele is potentially affected. (B) The family in this pedigree carries the allele for albinism, a recessive trait. Because the trait is recessive, heterozygotes do not have the albino phenotype, but they can pass the allele on to their offspring. Affected persons must inherit the allele from two heterozygous parents, or (rarely) from one homozygous recessive and one heterozygous parent, or (very rarely) two homozygous recessive parents. In this family, the heterozygous parents in generation III are cousins; however, the same result would occur if the parents were unrelated.

being a recessive homozygote (*aa*). But the offspring of a single couple are likely to be too few to reliably show the one-fourth proportion. In a family with only two children, for example, both could easily be *aa* (or *Aa*, or *AA*).

Despite this limitation, pedigrees do show inheritance patterns that can provide information about the allele(s) controlling a particular phenotype. For example, it is useful to know whether a particular rare allele that causes an abnormal phenotype is dominant or recessive. Figure 8.8A is a pedigree showing the pattern of inheritance of a rare dominant allele. The following are the key features to look for in such a pedigree:

- Every person with the abnormal phenotype (affected) has an affected parent.
- Either all (in the case of a homozygous parent) or about half (in the case of a heterozygous parent) of the offspring are affected. In Figure 8.8A, each of the affected individuals in generations II and III is heterozygous for the rare dominant allele. This is clear because each person has inherited a recessive allele from his or her unaffected parent.

Compare this pattern with the one shown in Figure 8.8B, which is typical for the inheritance of a rare recessive allele:

- Affected people most often have two unaffected parents.
- Only a small proportion of people are affected: about one-fourth of children whose parents are both heterozygotes. Such parents are **heterozygous carriers** of the recessive allele. Unfortunately, it is not possible to use the pedigree alone to determine who else in the family may be a heterozygous carrier (that is, unless they parent a homozygous child), because the allele is recessive and shows no phenotype in the heterozygote. However, as we will see in Chapter 12, DNA sequencing or other techniques can sometimes be used to identify carriers of specific recessive alleles. This is especially important in cases where a rare allele causes serious disease.

Other patterns of inheritance can arise in special cases, such as sex-linked characters (see Concept 8.3).

Go to ANIMATED TUTORIAL 8.2
Pedigree Analysis Simulation
PoL2e.com/at8.2

CHECKpoint CONCEPT 8.1

- ✓ What are the differences between genes and alleles? Between homozygous and heterozygous conditions? Between genotype and phenotype? Between Mendel's definition of a gene and the current definition?
- ✓ In a monohybrid cross, how do the events of meiosis explain Mendel's first law? In a dihybrid cross, how does meiosis explain Mendel's second law?
- ✓ Using the cross shown in Figure 8.5, calculate the probability that an F_2 seed will be round and yellow.

The laws of inheritance as articulated by Mendel remain valid today, and his discoveries laid the groundwork for all future studies of genetics. However, as we will see next, the relationship of one gene to one phenotype can be complicated by interactions among alleles and among genes.

CONCEPT 8.2 Alleles and Genes Interact to Produce Phenotypes

Phenotypes do not always follow the simple patterns of inheritance shown by the pairs of alleles for seed color and seed shape in peas, described in Concept 8.1. Existing alleles are subject to change by mutation and can give rise to new alleles—in fact, a single gene can have many alleles. In addition, alleles do not always show simple dominant–recessive relationships. A single allele may have multiple phenotypic effects, and a single character may be controlled by multiple genes. The expression of a gene is generally affected by interactions with other genes and with the environment.

New alleles arise by mutation

Genes are subject to **mutations**, which are rare, stable, and inherited changes in the genetic material. In other words, an allele can mutate (change) to become a different allele (this can happen in a number of different ways, as will be detailed in Concept 9.3). For example, we can envision that at one time all pea plants made round seeds and had the seed shape allele *R*. At some point, a mutation in *R* resulted in a new allele, *r* (wrinkled seeds). If this mutation was present in a cell that underwent meiosis, some of the resulting gametes would carry the *r* allele, and some offspring of this pea plant would carry the *r* allele in all of their cells.

Geneticists usually define one allele of a gene as the **wild type**; this is the allele that is present in most individuals in nature ("the wild"). Other alleles of that gene are usually called mutant alleles, and they may produce different phenotypes. The wild-type and mutant alleles are inherited according to Mendelian laws. A gene with a wild-type allele that is present less than 99 percent of the time (the rest of the alleles being mutant) is said to be **polymorphic** (Greek *poly*, "many"; *morph*, "form").

LINK

By producing phenotypic variety, mutations provide raw material for evolution. An allele may become more or less prevalent in a population, depending on its effect on the fitness of the individuals carrying it; see **Concept 15.3**

Mendel developed his theories by studying just two alleles of each gene. But often a gene has multiple alleles (although any diploid individual will carry only two of them). The alleles may show a hierarchy of dominance when present in heterozygous individuals. An example is coat color in rabbits, determined by multiple alleles of the *C* gene (**FIGURE 8.9**):

- *C* determines dark gray.
- c^{chd} determines chinchilla, a lighter gray.

Possible genotypes	CC, Cc^{chd}, Cc^{h}, Cc	$c^{chd}c^{chd}$, $c^{chd}c$	$c^{h}c^{h}$, $c^{h}c$	cc
Phenotype	Dark gray	Chinchilla	Point restricted	Albino

FIGURE 8.9 Multiple Alleles for Coat Color in Rabbits These photographs show the phenotypes conferred by four alleles of the *C* gene for coat color in rabbits. Different combinations of two alleles give different coat colors and pigment distributions.

- c^h determines Himalayan (point restricted).
- c determines albino, no pigment.

The hierarchy of dominance for these alleles is $C > c^{chd} > c^h > c$. Any rabbit with the *C* allele (paired with itself or another allele) is dark gray, and a *cc* rabbit is albino. Intermediate colors result from different allele combinations, as shown in Figure 8.9. As this example illustrates, multiple alleles can increase the number of possible phenotypes.

Dominance is not always complete

Many genes have alleles that are neither dominant nor recessive to one another. Instead, the heterozygotes have an intermediate phenotype, in a situation called **incomplete dominance**. For example, if a true-breeding red snapdragon is crossed with a white one, all the F_1 flowers are an intermediate pink. Such cases appear to support the old blending theory of inheritance. However, further crosses indicate that this apparent blending can still be explained in terms of Mendelian genetics (**FIGURE 8.10**). The red and white snapdragon alleles have not disappeared, as those colors reappear in the F_2 generation.

Sometimes two alleles of a gene both produce their phenotypes when present in a heterozygote—a phenomenon called **codominance**. An example is the ABO blood group in humans. The gene *I* encodes an enzyme involved in the attachment of sugars to a glycoprotein present on the surfaces of red blood cells. There are three alleles of the gene: I^A, I^B, and I^O. The I^A and I^B alleles both encode active enzymes, but the enzymes attach different sugars to the glycoprotein. The I^O allele does not encode an active enzyme, so no sugar is attached at that position on the glycoprotein. When two different alleles (e.g., I^A and I^B) are present, both alleles are expressed (both enzymes are

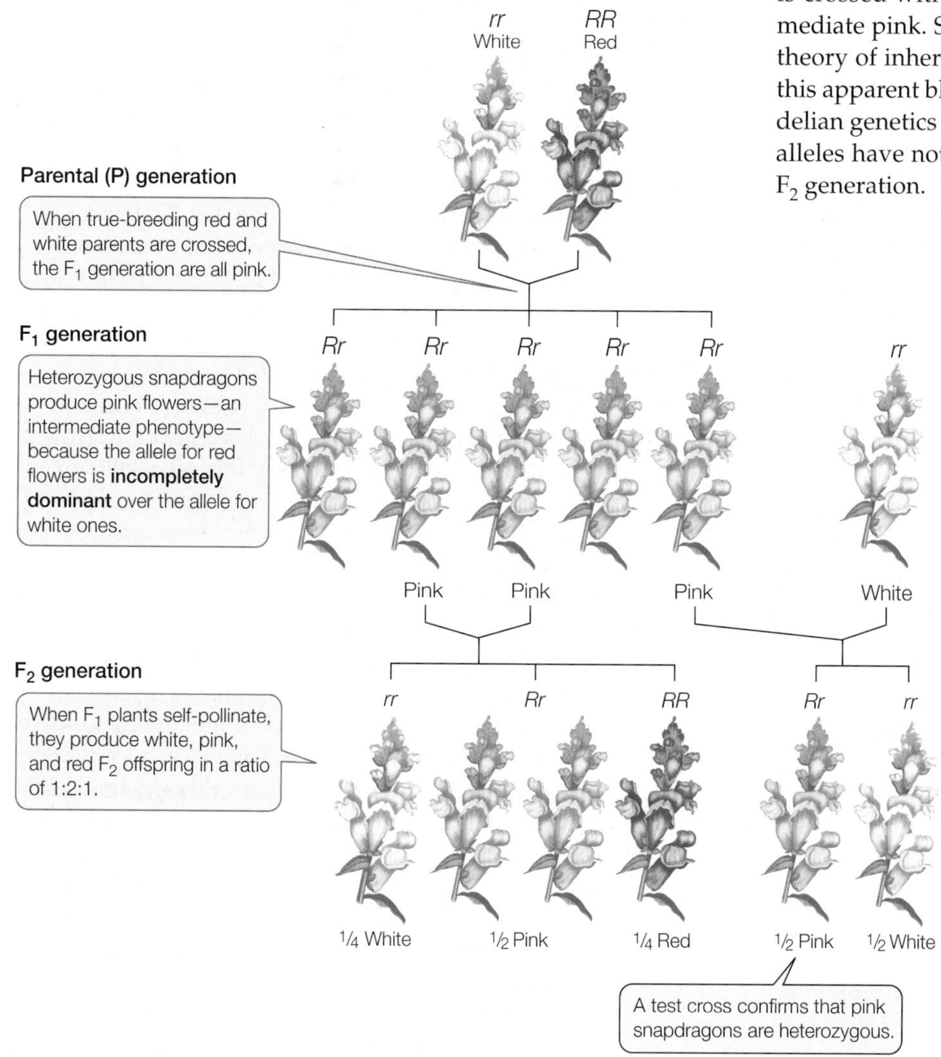

rr
White
RR
Red
Parental (P) generation
When true-breeding red and white parents are crossed, the F1 generation are all pink.
F1 generation
Heterozygous snapdragons produce pink flowers—an intermediate phenotype—because the allele for red flowers is incompletely dominant over the allele for white ones.
Rr
Rr
Rr
Rr
Rr
rr
Pink
Pink
Pink
White
F2 generation
When F1 plants self-pollinate, they produce white, pink, and red F2 offspring in a ratio of 1:2:1.
rr
Rr
RR
Rr
rr
1/4 White
1/2 Pink
1/4 Red
1/2 Pink
1/2 White
A test cross confirms that pink snapdragons are heterozygous.

FIGURE 8.10 Incomplete Dominance Follows Mendel's Laws An intermediate phenotype can occur in heterozygotes when neither allele is dominant. The heterozygous phenotype (here, pink flowers) may give the appearance of a blended trait, but the traits of the parental generation reappear in their original forms in succeeding generations, as predicted by Mendel's laws of inheritance.

made, so both types of glycoproteins are made). The A and B glycoproteins are antigenic: if a red blood cell with the A glycoprotein on its surface gets into the bloodstream of a person who lacks the I^A allele, the recipient produces antibodies against the "nonself" cells (**FIGURE 8.11**). While the A and B glycoproteins are antigenic in people who do not have the I^A or I^B alleles, respectively, the O glycoprotein does not provoke antibody production. This makes people who are I^OI^O good blood donors.

LINK

Reactions against A or B glycoproteins are an adaptive immune response, as described in **Concept 39.3**

Genes interact when they are expressed

Epistasis occurs when the phenotypic expression of one gene is affected by another gene. For example, two genes (*B* and *E*) determine coat color in Labrador retrievers:

- Allele *B* (black pigment) is dominant to *b* (brown).
- Allele *E* (pigment deposition in hair) is dominant to *e* (no deposition, so hair is yellow).

An *EE* or *Ee* dog with *BB* or *Bb* is black; one with *bb* is brown. A dog with *ee* is yellow regardless of whether *B* or *b* alleles are present. Clearly, gene *E* determines the phenotypic expression of gene *B*, and is therefore epistatic to *B* (**FIGURE 8.12**).

Perhaps the most dramatic example of interacting genes or alleles is **hybrid vigor** (or heterosis). In 1876, Charles Darwin reported that when he crossed two different genetic varieties of corn, the offspring were 25 percent taller than either of the parent strains. Darwin's observation was largely ignored for the next 30 years. In 1908, George Shull "rediscovered" this idea, reporting that not just plant height but the weight of the corn grain produced was dramatically higher in the offspring:

Agricultural scientists took note, and Shull's paper had a lasting impact on the field of applied genetics. The cultivation of hybrid corn spread rapidly, and the practice of hybridization is now used for many other agricultural crops and animals. For example, beef cattle strains that are crossbred are larger and live longer than cattle bred within their own genetic strains.

What determines the "vigor" in hybrid vigor? A phenotype such as the amount of grain that a variety of corn produces in a given environment is determined by many genes and their alleles. Put more generally, *most complex phenotypes are determined by multiple genes*. Traits conferred by multiple genes are usually **quantitative traits** that need to be *measured* rather than assessed qualitatively. For example, grain yields must be measured, whereas a character such as Mendel's pea seed color is an either-or quality that can be assessed by eye. The genetic basis of hybrid vigor is not well understood, but presumably the combinations of alleles and their products from two different varieties can interact to produce more vigorous offspring.

The environment affects gene action

The phenotype of an individual does not always result from its genotype alone. *Genotype and environment often interact to determine the phenotype of an organism*. It is especially important to remember this fact in the era of genome sequencing (see Chapter 12). When the sequence of the human genome was completed in 2003, it was hailed as the "book of life," and public expectations of the benefits gained from this knowledge were (and are) high. But this kind of "genetic determinism"—the idea that an organism's genome sequence determines all of its phenotype—is wrong. Common knowledge tells us that environmental variables such as light, temperature, and nutrition can affect the phenotypic expression of a genotype. For example, in humans body weight is determined not only by multiple genes but also by nutrition and activity.

A common example of the interaction of genes and environment involves "point restriction" coat patterns found in Siamese cats and certain rabbit breeds. These animals carry the c^h allele (see Figure 8.9), a mutant version of the *C* gene that controls coat

Red blood cell type	Genotype	Blood group that body rejects	Reaction to added antibodies: Anti-A	Reaction to added antibodies: Anti-B
A	I^AI^A or I^AI^O	B and AB		
B	I^BI^B or I^BI^O	A and AB		
AB	I^AI^B	None		
O	I^OI^O	A, B, and AB		

Red blood cells that do not react with antibody remain evenly dispersed.

Red blood cells that react with antibody clump together (speckled appearance).

FIGURE 8.11 ABO Blood Groups Are Important in Transfusions The table shows the results of mixing red blood cells of types A, B, AB, and O with serum containing anti-A or anti-B antibodies. As you look down the two columns on the right, note that each of the types, when mixed separately with anti-A and with anti-B, gives a unique pair of results. This is the basic method by which blood is typed.

APPLY THE CONCEPT

Alleles and genes interact to produce phenotypes

1. In the genetic cross *AaBbCcDdEE* × *AaBBCcDdEe* where all the genes are on separate chromosomes, what fraction of the offspring will be heterozygous for all of these genes?
2. In squash plants, fruit color is determined by one gene (*W*: white, dominant to *w*: yellow) and fruit shape by another gene on a different chromosome (*D*: disk shape, dominant to *d*: round). What were the genotypes of the parents for each cross in the table at right?
3. In chickens, when the dominant alleles of the genes for rose comb (*R*) and pea comb (*A*) are present together, the bird has a walnut comb. Birds that are homozygous recessive for both genes (i.e., genotype *ra*) have a single comb. A rose-combed bird mated with a walnut-combed bird, and the phenotypes of the 8 offspring were:

   3/8 walnut : 3/8 rose : 1/8 pea : 1/8 single.

   What were the genotypes of the parents?

	NUMBER OF OFFSPRING			
PARENTS	WHITE, DISK	WHITE, ROUND	YELLOW, DISK	YELLOW, ROUND
White, round × White, round	0	52	0	16
White, disk × White, round	62	58	18	20
White, disk × White, disk	176	60	54	21
White, disk × Yellow, round	22	24	21	22

color. As a result of this mutation, the enzyme encoded by the gene is inactive at temperatures above a certain point—usually 35°C. The animals maintain a body temperature above this, so their body fur is light. However, their extremities—feet, ears, nose, and tail—are cooler (around 25°C), and the fur on these regions is dark:

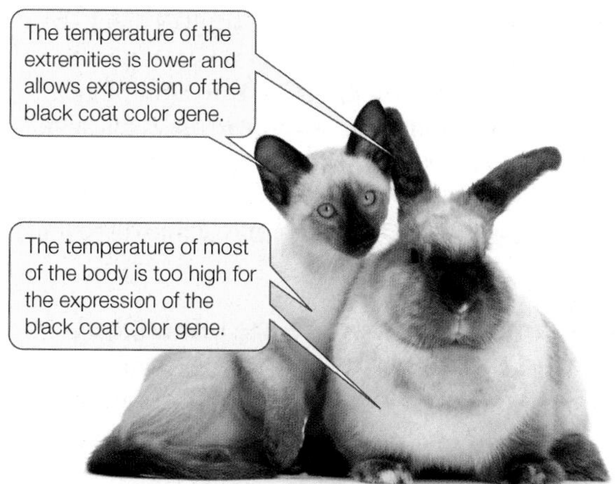

The temperature of the extremities is lower and allows expression of the black coat color gene.
The temperature of most of the body is too high for the expression of the black coat color gene.

A simple experiment shows that the dark fur is temperature-dependent. If a patch of white fur on a point-restricted animal is shaved off and an ice pack is placed on the skin where the patch was, the fur that grows in is dark. This indicates that although the allele for dark fur was expressed all

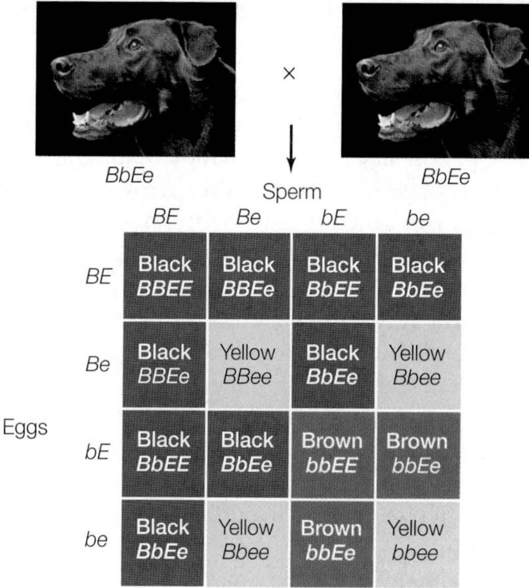

FIGURE 8.12 Genes Interact Epistatically Epistasis occurs when one gene alters the phenotypic effect of another gene. In Labrador retrievers, the *E* gene determines the expression of the *B* gene.

along, the environment inhibited the activity of the mutant enzyme. These animals are all white at birth because the extremities are kept warm in the mother's womb.

Two parameters describe the effects of genes and environment on phenotype:

- **Penetrance** is the proportion of individuals in a group with a given genotype that actually show the expected phenotype. For example, many people who inherit a mutant allele of the gene *BRCA1* develop breast cancer in their lifetimes. But for reasons that are not yet clear and must involve other genes and/or the environment, some people with the mutation do not develop breast cancer. So the *BRCA1* mutation is said to be incompletely penetrant.
- **Expressivity** is the degree to which a genotype is expressed in an individual. For example, a woman with the *BRCA1* allele may develop both breast and ovarian cancer as part of the phenotype, but another woman with the same mutation may only get breast cancer. So the mutation is said to have variable expressivity.

In a population of an organism, there is often wide variability with regard to a particular phenotype. Consider, for example, height in humans:

Human height is a quantitative trait. Many genes contribute to height, and the interactions between these genes are complex. But the environment also contributes to variation in height; for example, some people have better nutrition than others, and this can affect their growth. The **heritability** of a character is the relative contribution of genetic versus environmental factors to the variation in that character in a particular population. Typically, heritability varies from 0 to 1. For example, the heritability of human height varies from about 0.65 to 0.8, depending on the population studied. In a population where heritability is 0.65, 65 percent of the variation in height is due to genetic factors, while the remaining 35 percent is due to environmental effects. It is important to note that heritability estimates apply to variations within populations. They cannot be used to estimate the contribution of genetics to particular characters in an individual.

Heritability estimates are important for breeders of plants and animals because they provide information about whether it is more worthwhile to modify the environment or do genetic crosses to improve a phenotype.

CHECKpoint CONCEPT 8.2

- ✓ What is the difference between incomplete dominance and codominance?
- ✓ A point-restricted rabbit (see Figure 8.9) was mated with a chinchilla rabbit. The two offspring were albino and chinchilla. What were the genotypes of the parents?
- ✓ If a dominant allele of one gene, *A*, is necessary for hearing in humans, and the dominant allele of another gene, *B*, results in deafness regardless of the presence of other genes, what fraction of offspring in a pairing of *AaBb* and *Aabb* individuals will be deaf?
- ✓ Give an example from your own experience of a genotype whose expression is affected by the environment.

So far we have considered genes that obey Mendel's law of independent assortment. But many genes are inherited together, with multiple genes transmitted as one unit. This apparent anomaly can be explained by the presence of multiple genes on a single chromosome.

CONCEPT 8.3 Genes Are Carried on Chromosomes

Genes are parts of chromosomes. More specifically, a gene is a sequence of DNA that resides at a particular site on a chromosome called a **locus** (plural *loci*). You have seen how the behavior of chromosomes during meiosis can explain Mendel's laws of segregation (see Figure 8.3) and independent assortment (see Figure 8.6). However, the **genetic linkage** of genes on a single chromosome alters their pattern of inheritance.

Genetic linkage was first discovered in the fruit fly *Drosophila melanogaster*. This animal is an attractive experimental subject because it is small, easily bred, and has a short generation time (from fertilized egg to reproducing adult). In fact, the fruit fly has been a model organism for experimental genetics for more than a century. In Concept 8.1 we saw how Mendel successfully applied the scientific method to arrive at his laws of inheritance. Now we will examine the work of Thomas Hunt Morgan, who worked at Columbia University early in the twentieth century and used a similar approach to discover genetic linkage.

Genes on the same chromosome are linked, but can be separated by crossing over in meiosis

Some of the crosses Morgan performed with fruit flies yielded phenotypic ratios that were not in accordance with those predicted by Mendel's law of independent assortment. Morgan did a test cross between flies with two known genotypes: *BbVgvg* and *bbvgvg*. The *B* and *Vg* genes control two characters, body color and wing shape:

- *B* (wild-type gray body) is dominant over *b* (black body).
- *Vg* (wild-type wing) is dominant over *vg* (vestigial, or very small, wing).

INVESTIGATION

FIGURE 8.13 Some Alleles Do Not Assort Independently Morgan's studies showed that the genes for body color and wing size in *Drosophila* are linked, so that their alleles do not assort independently.[a]

HYPOTHESIS

Alleles for different characteristics always assort independently.

METHOD

Parent (P)

BbVgvg Wild type (gray body, normal wings) ♀ × *bbvgvg* (black body, vestigial wings) ♂

RESULTS

These are the results expected from Mendel's second law (independent assortment)...

F_1

Genotypes	*BbVgvg* Gray normal	*bbvgvg* Black vestigial	*Bbvgvg* Gray vestigial	*bbVgvg* Black normal
Expected frequencies	575	575	575	575
Observed frequencies (number of individuals)	965	944	206	185
	Parental phenotypes		Recombinant phenotypes	

...but the actual results were inconsistent with the law.

CONCLUSION

The hypothesis is rejected. These two genes do not assort independently, but are linked (on the same chromosome).

Go to **LaunchPad** for discussion and relevant links for all **INVESTIGATION** figures.

[a]T. H. Morgan. 1912. *Science* 36: 718–720

Go to ANIMATED TUTORIAL 8.3
Alleles That Do Not Assort Independently
PoL2e.com/at8.3

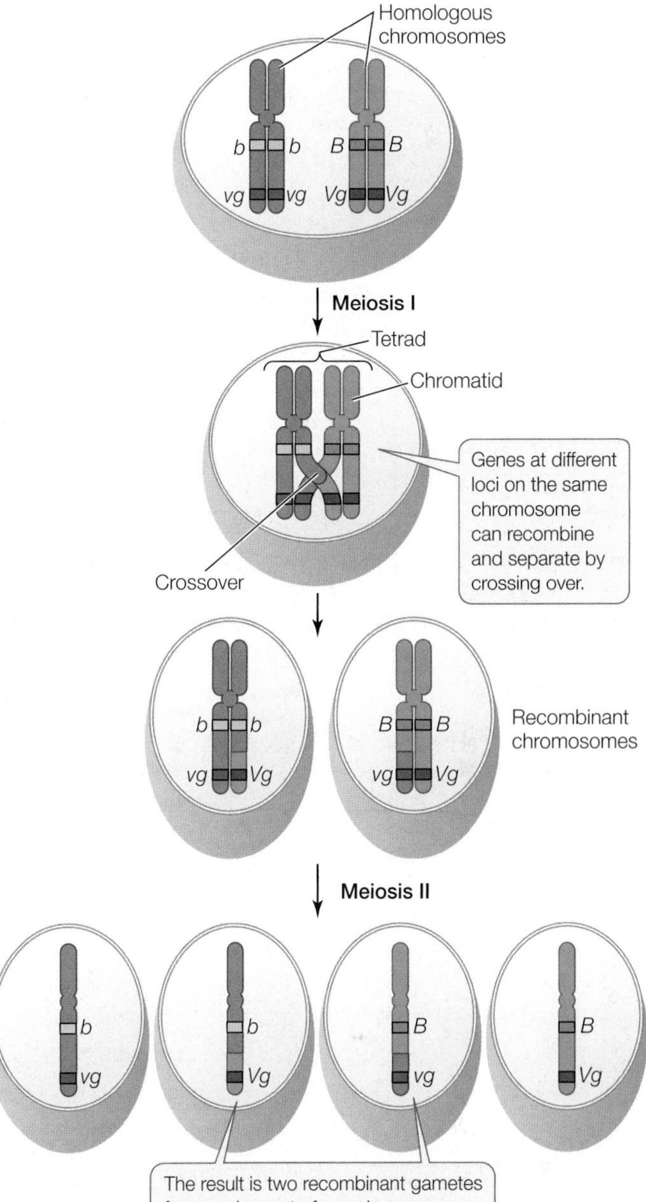

FIGURE 8.14 Crossing Over Results in Genetic Recombination Recombination accounts for why linked alleles are not always inherited together. Alleles at different loci on the same chromosome can be recombined by crossing over and then being separated from one another. Such recombination occurs during prophase I of meiosis.

Morgan expected to see four phenotypes in a ratio of 1:1:1:1, but that is not what he observed. The body color gene and the wing size gene did not assort independently; instead, they were frequently inherited together, and most of the progeny showed one or the other of the parental phenotypes (**FIGURE 8.13**).

These results became understandable when Morgan considered the possibility that the two loci were *linked* on the same chromosome. Such genes would not be able to assort independently as predicted by Mendel's second law. In this case, the test cross offspring might be expected to have only the parental phenotypes (gray flies with normal wings or black flies with vestigial wings) in a 1:1 ratio. If linkage were absolute, we would see *only* these two types of progeny. However, this did not happen. Why did some of Morgan's flies show phenotypes different from their parents?

Some of Morgan's flies displayed **recombinant** phenotypes because two homologous chromosomes can physically exchange corresponding segments during prophase I of meiosis—that is, by crossing over (**FIGURE 8.14**; see also Figure 7.13). Each exchange event involves two of the four chromatids in a tetrad—one from each member of the homologous pair—and can occur at any point along the length of the chromosome. The chromosome segments are exchanged reciprocally, so both chromatids become recombinant (that is, each chromatid ends up with genes from both of the organism's parents).

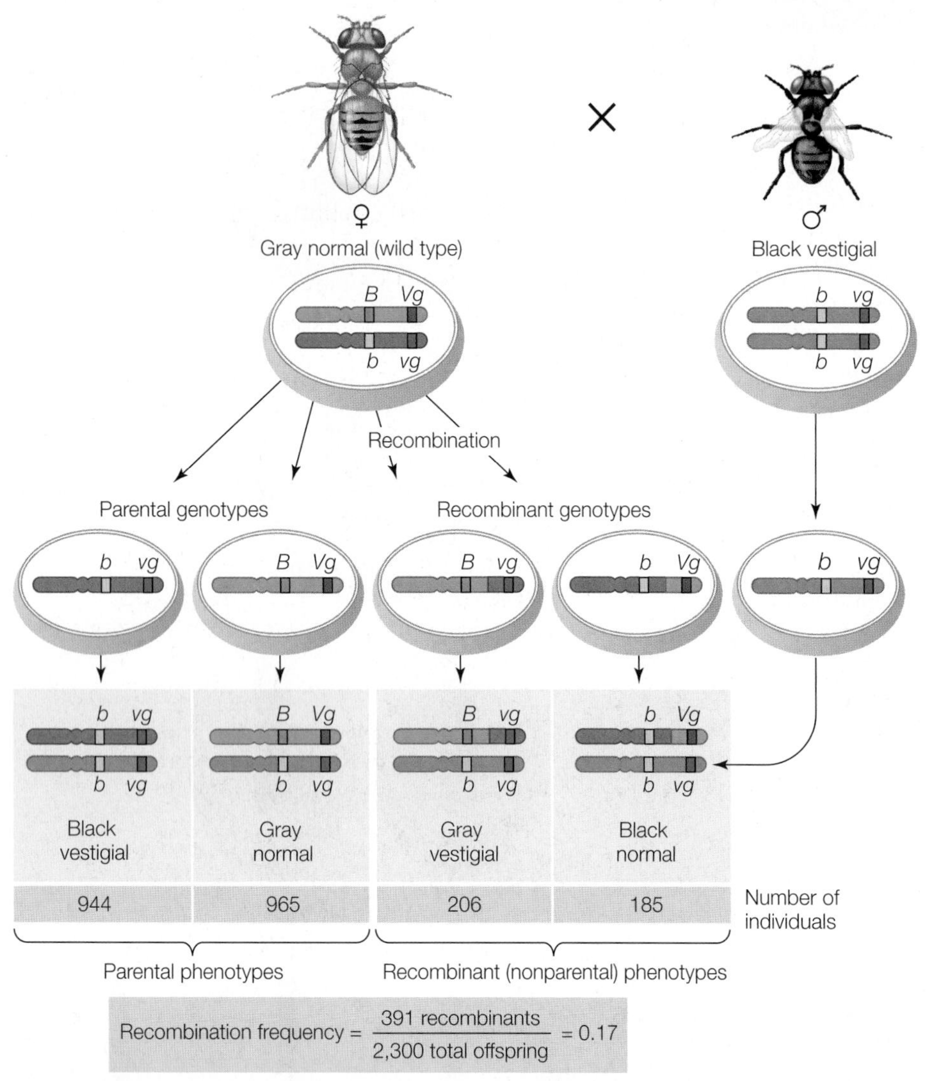

FIGURE 8.15 **Recombination Frequencies** The frequency of recombinant offspring (those with a phenotype different from either parent) can be calculated.

As a result of a crossing over event between two linked genes, not all the progeny of a cross have the parental phenotypes. Recombinant offspring appear as well, generally in proportions related to the **recombination frequency** between the two genes, which is calculated by dividing the number of recombinant progeny by the total number of progeny (**FIGURE 8.15**). *Recombination frequencies are greater for loci that are farther apart on the chromosome* than for loci that are closer together because crossing over is more likely to occur between genes that are far apart. By calculating recombination frequencies, geneticists can infer the locations of genes along a chromosome and generate a genetic map. Below is a map showing five genes on a fruit fly chromosome. It was constructed using the recombination frequencies generated by test crosses involving various pairs of the genes:

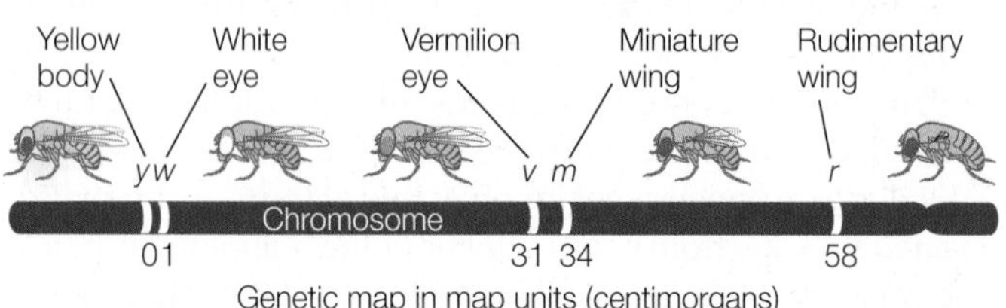

The recombination frequency between *y* and *w* is low, so they are close together on the map. Recombination between *y* and *v* is more frequent, so they are farther apart. The recombination frequencies are converted to map units (also called centimorgans, cM); one map unit is equivalent to an average recombination frequency of 0.01 (one percent).

The era of gene sequencing has made mapping less important in some areas of genetics research. However, mapping is still one way to verify that a particular DNA sequence corresponds with a particular phenotype. The phenomenon of linkage has allowed biologists to isolate genes and to create genetic markers that are linked to important genes, making it easy to identify individuals carrying particular alleles. This is particularly important in breeding new crops and animals for agriculture, and for identifying humans carrying medically significant mutations.

Linkage is also revealed by studies of the X and Y chromosomes

The fruit fly genome has four pairs of chromosomes: in three pairs, the chromosomes are similar in size to one another and are called **autosomes**. The fourth pair has two chromosomes of different sizes. These determine the sex of the fly and are called the **sex chromosomes**, shown here in green:

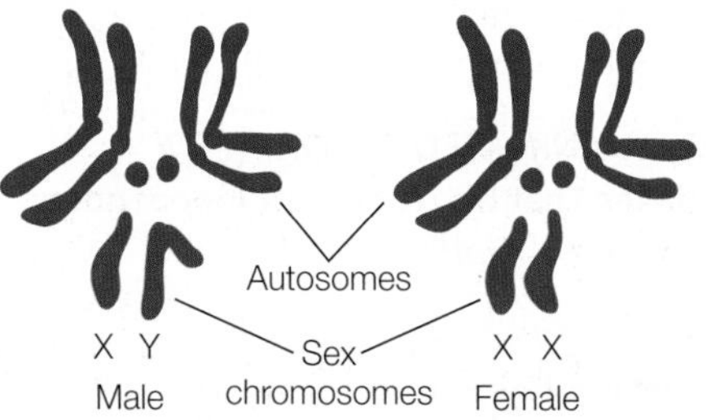

Note that the female fly has two X chromosomes and that the male has only one, the other being the Y chromosome: females are XX and males are XY. It turns out that in addition to being different sizes, *many genes on the X chromosome are not present on the Y*. This means that males have only one copy of these genes. The X chromosome was one of the first to have specific genes assigned to it.

Morgan identified a gene that controls eye color in *Drosophila*. The wild-type allele of the gene confers red eyes, whereas a recessive mutant allele confers white eyes. Morgan's experimental crosses with flies carrying the mutant allele demonstrated that this eye color locus is on the X chromosome. If we abbreviate the eye color alleles as *R* (red eyes) and *r* (white eyes), the presence of the alleles on the X chromosome is designated by X^R and X^r.

Morgan crossed a homozygous red-eyed female (X^RX^R) with a white-eyed male. The male is designated X^rY because the Y does not carry any allele for this gene. (Any gene that is present as a single copy in a diploid organism is called **hemizygous**. A male will express all the alleles of his one X chromosome, whether or not they are dominant.) All the sons and daughters from this cross had red eyes, because red (*R*) is dominant over white (*r*) and all the progeny had inherited a wild-type X chromosome (X^R) from their mother (**FIGURE 8.16A**). Note that this phenotypic outcome would have occurred even if the *R* gene had been present on an autosome rather than a sex chromosome. In that case, the male would have been homozygous recessive—*rr*.

When Morgan performed the reciprocal cross, in which a white-eyed female (X^rX^r) was mated with a red-eyed male (X^RY), the results were unexpected: *all the sons were white-eyed and all the daughters were red-eyed* (**FIGURE 8.16B**). The sons from the reciprocal cross inherited their only X chromosome from their white-eyed mother and were therefore hemizygous for the white allele. The daughters, however, got an X chromosome bearing the white allele from their mother and an X chromosome bearing the red allele from their father; therefore they were red-eyed heterozygotes. When these heterozygous females were mated with red-eyed males, half their sons had white eyes but all their daughters had red eyes. Together, these results showed that eye color was carried on the X chromosome and not on the Y.

These and other experiments led to the term **sex-linked inheritance**: inheritance of a gene that is carried on a sex chromosome. (This term is misleading because "sex-linked" inheritance is not really linked to the sex of an organism—after all,

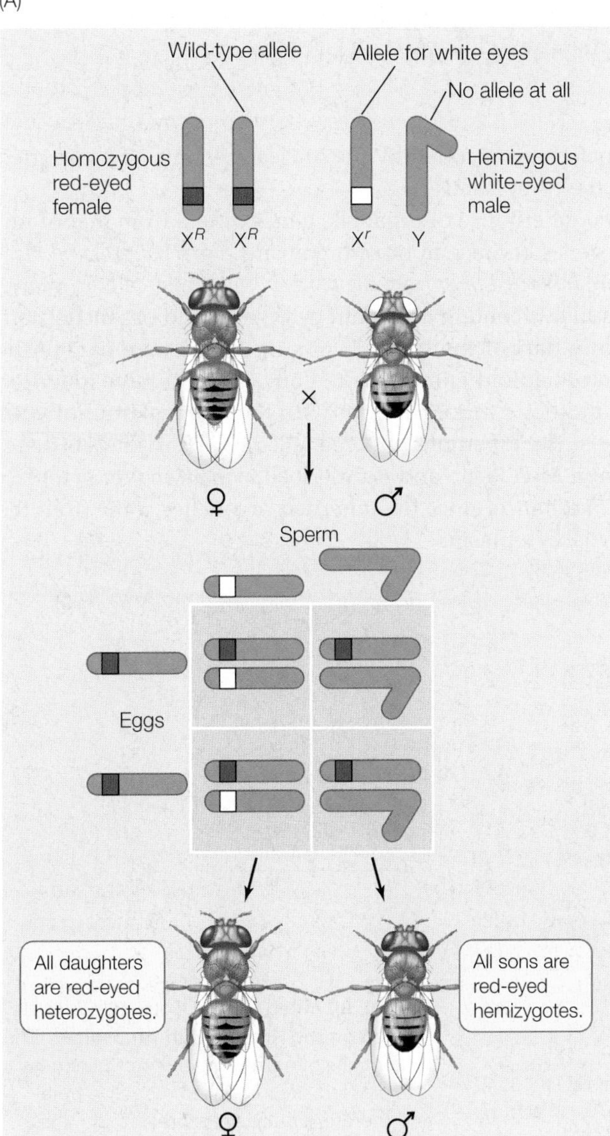

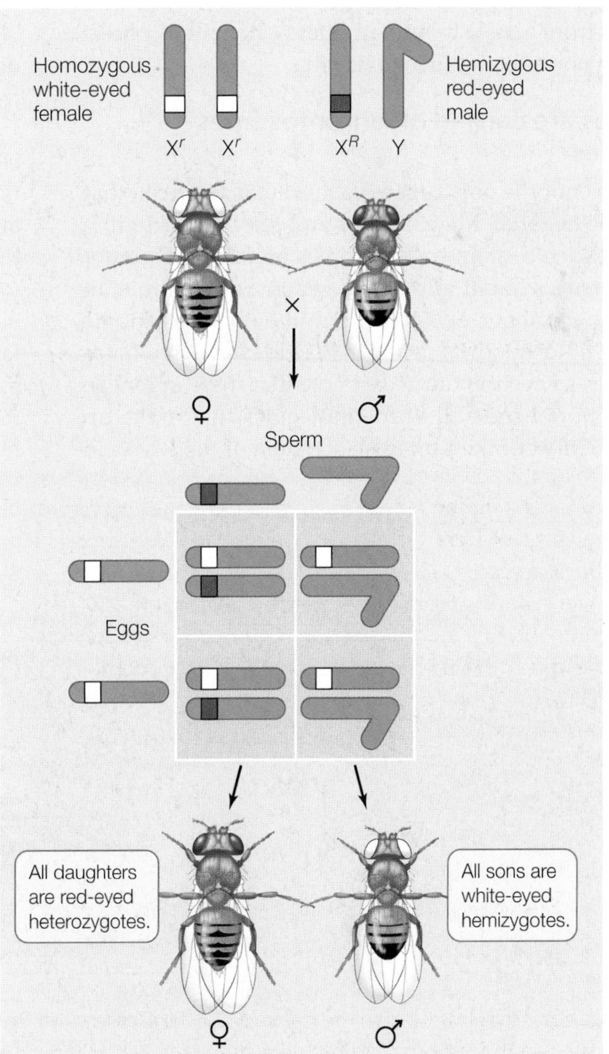

FIGURE 8.16 A Gene for Eye Color Is Carried on the *Drosophila* X Chromosome Morgan demonstrated that a mutant allele that causes white eyes in *Drosophila* is carried on the X chromosome. Note that in this case, the reciprocal crosses do not have the same results.

both males and females carry X chromosomes.) In mammals, the X chromosome is larger and carries more genes than the Y. For this reason, most examples of sex-linked inheritance involve genes that are carried on the X chromosome.

Many sexually reproducing species, including humans, have sex chromosomes. As in fruit flies, human males are XY, females are XX, and relatively few of the genes that are present on the X chromosome are present on the Y. Pedigree analyses of X-linked recessive phenotypes like the one in **FIGURE 8.17** reveal the following patterns (compare with the pedigrees of non-X-linked phenotypes in Figure 8.8):

- The phenotype appears much more often in males than in females, because only one copy of the allele is needed for its expression in males, whereas two copies must be present in females.
- A male with the mutation can pass it on only to his daughters; all his sons get his Y chromosome.
- Daughters who receive one X-linked mutation are heterozygous carriers. They are phenotypically normal, but they can pass the mutant allele to their sons or daughters. On average, half their children will inherit the mutant allele since half of their X chromosomes carry the normal allele.
- The mutant phenotype can skip a generation if the mutation passes from a male to his daughter (who will be phenotypically normal) and then to her son.

Some genes are carried on chromosomes in organelles

The nucleus is not the only organelle in a eukaryotic cell that carries genetic material. Mitochondria and plastids (including chloroplasts) each contain several copies of a small chromosome that carries a small number of genes. For example, in humans there are about 21,000 genes in the nuclear genome and 37 in the mitochondrial genome. Plastids have five times as many genes as mitochondria. But note that these organelle genomes do not encode all of the molecules that make up the organelle. Most of the proteins and some of the RNAs in organelles are encoded by the nuclear genome and imported from the cytoplasm.

The inheritance of organelle genes differs from that of nuclear genes because in most organisms, *mitochondria and plastids are inherited only from the mother*. Most egg cells in plants and animals contain abundant cytoplasm and organelles, but the only part of the sperm that survives to take part in the union of haploid gametes is the nucleus. You have inherited your mother's mitochondria (with their genes) but not your father's. The inheritance of organelles and their genes is therefore non-Mendelian and is described as maternal or cytoplasmic inheritance, since the inherited organelles come from the maternal cytoplasm.

APPLY THE CONCEPT

Genes are carried on chromosomes

The pedigree shows the inheritance pattern of a rare mutant phenotype in humans, congenital cataract (filled-in symbols).

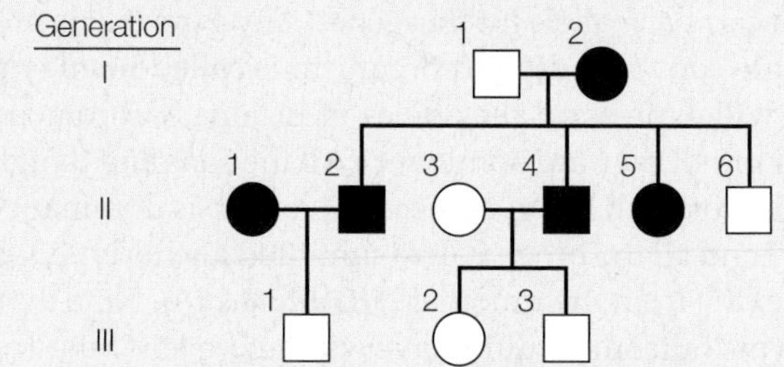

1. Are cataracts inherited as autosomal dominant? Autosomal recessive? Sex-linked dominant? Sex-linked recessive?
2. Person #5 in the second generation marries a man who does not have cataracts. Two of their four children, a boy and a girl, develop cataracts. What is the probability that their next child will be a girl with cataracts?

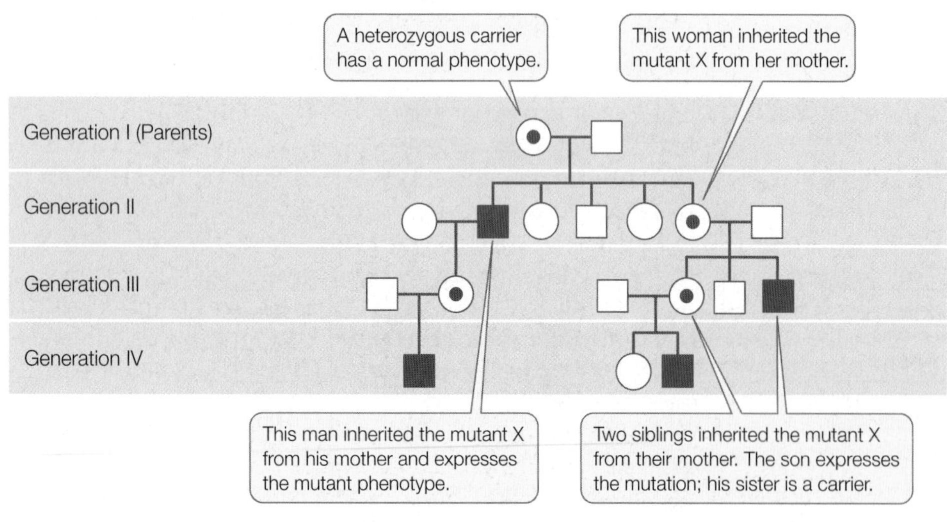

FIGURE 8.17 Red–Green Color Blindness Is Carried on the Human X Chromosome The mutant allele for red–green color blindness is expressed as an X-linked recessive trait, and therefore is always expressed in males when they carry that allele.

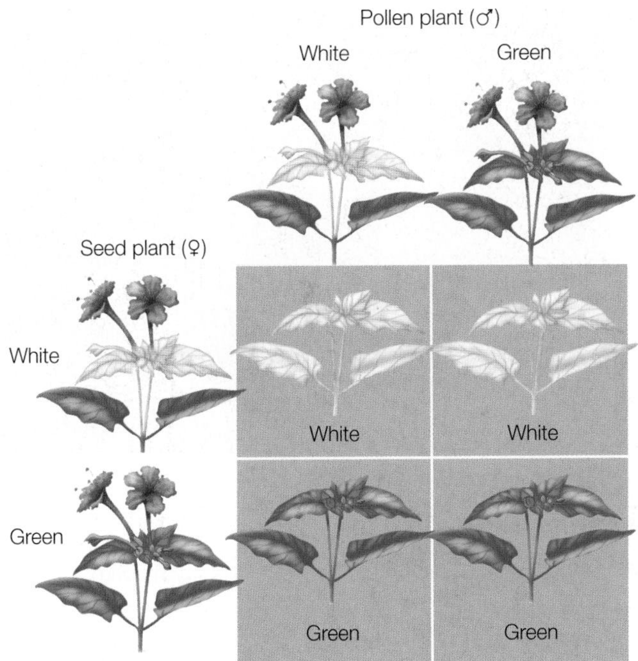

FIGURE 8.18 Cytoplasmic Inheritance In four o'clock plants, leaf color is inherited through the female plant only. In the parent plant with some white leaves, the white leaf color is caused by a chloroplast mutation that occurred during the life of the plant; the leaves that formed before the mutation occurred are green. When this plant is used as a pollen donor in a cross with an all green plant, the offspring are all green. But when the same plant is used as an egg donor, the offspring inherit the mutation cytoplasmically and are entirely white.

Some of the genes carried by cytoplasmic organelles are important for organelle assembly and function, and mutations of these genes can have profound effects on the organism. For example, in plants, certain plastid gene mutations affect the proteins that assemble chlorophyll molecules into photosystems (see Figure 6.19). These mutations result in a phenotype that is essentially white instead of green. **FIGURE 8.18** illustrates the cytoplasmic inheritance of a mutant plastid gene that confers white color to leaves.

CHECKpoint CONCEPT 8.3

- ✓ Describe the differences in patterns of inheritance between a gene present in the nucleus and a gene present in the mitochondria.
- ✓ Explain the concept of linkage. If you performed a test cross with a fruit fly that is heterozygous for two genes, how would you conclude that the two genes are linked?
- ✓ Red–green color blindness is inherited as a sex-linked recessive. Two parents with normal color vision have a child who is red–green color-blind. Is the child a boy or girl? Draw a pedigree of the family. Draw a Punnett square to show gametes and offspring with regard to the X and Y chromosomes and the normal and color-blind alleles.

Like eukaryotes, prokaryotes contain genes that determine their phenotypes. Sexual reproduction in eukaryotes involves two sets of chromosomes and meiosis, giving rise to haploid gametes. Let's look next at how reproduction and inheritance differ in prokaryotes, which are haploid.

CONCEPT 8.4 Prokaryotes Can Exchange Genetic Material

As described in Concept 4.2, prokaryotic cells lack nuclei; they contain their genetic material mostly as single chromosomes in central regions of their cells. Prokaryotes reproduce asexually by binary fission, a process that gives rise to progeny that are virtually identical genetically (see Concept 7.2). That is, the offspring of cell reproduction in prokaryotes constitute a clone.

How, then, do prokaryotes evolve? Mutations occur in prokaryotes just as they do in eukaryotes, and the resulting new alleles increase genetic diversity. You might expect, therefore, that there is no way for individuals of these organisms to exchange genes as happens in sexual reproduction. It turns out, however, that prokaryotes *do* have a way of transferring genes between cells. This transfer of genes from one individual organism to another without sexual reproduction is called **horizontal** or **lateral gene transfer** to distinguish it from vertical gene transfer (gene transfer from parent to offspring). Along with mutation, this process generates genetic diversity among prokaryotes.

LINK

The evolutionary consequences of lateral gene transfer and its role in identifying and classifying bacterial species are discussed in **Concepts 15.6** and **19.2**

Bacteria exchange genes by conjugation

To illustrate genetic exchange in bacteria, let's consider two strains of the bacterium *E. coli* with different alleles for each of six genes (which code for enzymes in a biochemical pathway that synthesizes a certain small molecule). The two strains have the following genotypes (remember that bacteria are haploid):

ABCdef and *abcDEF*

where capital letters stand for wild-type alleles that encode functional gene products and lowercase letters stand for mutant alleles that encode defective gene products. Neither strain is able to synthesize the small molecule because neither has a full complement of wild-type genes.

When the two strains are grown together in the laboratory, most of the cells produce clones. That is, almost all of the cells that grow have either genotype *ABCdef* or genotype *abcDEF*. However, out of millions of bacteria, a few occur that have the genotype *ABCDEF*.

Unlike the original strains, cells of genotype *ABCDEF* are able to synthesize the small molecule. How could these completely wild-type bacteria arise? One possibility is mutation: in the *abcDEF* bacteria, the *a* allele could have mutated to *A*, the *b* allele to *B*, and the *c* allele to *C*. The problem with this explanation is that a mutation at any particular point in an organism's DNA sequence is a very rare event (about 1 in a million). The probability of all three events occurring in the same cell is extremely low—much lower than the observed rate of appearance of cells with genotype *ABCDEF*. So the mutant cells must have acquired wild-type genes some other way—and this turns out to be the transfer of DNA between cells.

Electron microscopy shows that gene transfers between bacteria can happen via physical contact between the cells (**FIGURE 8.19A**). Contact is initiated by a thin projection called a **sex pilus** (plural *pili*) that extends from one cell (the donor), attaches to another cell (the recipient), and draws the two cells together. Genetic material can then pass from the donor to the recipient through a thin cytoplasmic bridge called a conjugation tube. There is no reciprocal transfer of DNA from the recipient to the donor. This process is referred to as **bacterial conjugation**.

Once the donor DNA is inside the recipient cell, it can recombine with the recipient cell's genome. In much the same way that chromosomes pair up in prophase I of meiosis, the donor DNA can line up beside its homologous genes in the recipient, and crossing over can occur. Gene(s) from the donor can become integrated into the genome of the recipient, thus changing the recipient's genetic constitution (**FIGURE 8.19B**). In general, about half the transferred genes become integrated in this way. When the recipient cells proliferate, the integrated donor genes are passed on to all progeny cells, and the other transferred genes are lost. If the new combination of alleles is advantageous, the progeny of the recipient cell may be able to proliferate faster than the original strains.

Plasmids transfer genes between bacteria

In addition to their main chromosome, many bacteria harbor additional smaller, circular DNA molecules called **plasmids** that replicate independently inside the cell. Plasmids typically contain at most a few dozen genes, which may fall into one of several categories, including:

- Genes for unusual metabolic capacities, such as the ability to break down hydrocarbons. Bacteria carrying these plasmids can be used to clean up oil spills.
- Genes for antibiotic resistance. Plasmids carrying such genes are called R factors, and since they can be transferred between bacteria via conjugation, they are a major threat to human health.

Plasmids can move between cells during conjugation, thereby transferring new genes to the recipient bacterium (**FIGURE 8.20**). A single strand of the donor plasmid is transferred to the recipient; synthesis of complementary DNA strands results in two complete copies of the plasmid, one in the donor and one in the recipient. Because plasmids can replicate independently of the main chromosome, they do not need to recombine with the main chromosome to add their genes to the recipient cell's genome.

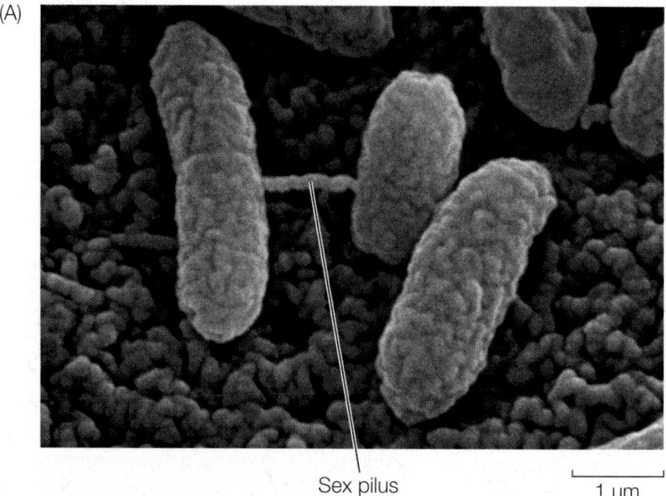

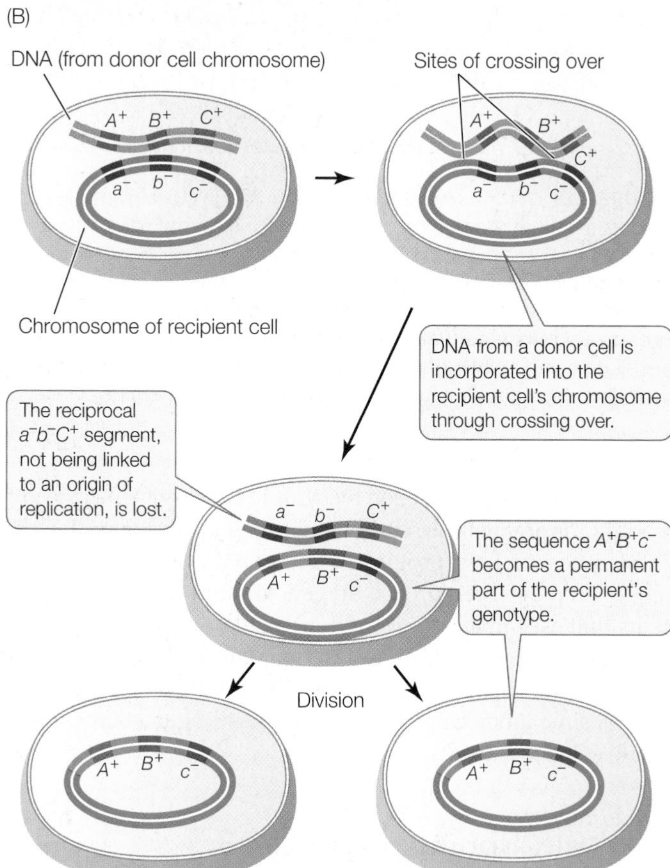

FIGURE 8.19 Bacterial Conjugation and Recombination (A) A sex pilus draws two bacteria into close contact, so that a cytoplasmic bridge (conjugation tube) can form. DNA is transferred from the donor cell to the recipient cell via the conjugation tube. (B) DNA from a donor cell can become incorporated into a recipient cell's chromosome through crossing over.

The evolution of drug-resistant bacteria is a major public health problem

Until the twentieth century, bacterial infections were a major scourge of humanity. With the discovery of antibiotics

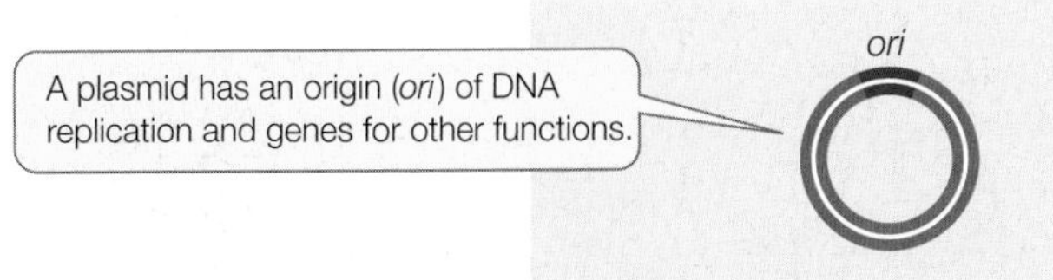

ori
A plasmid has an origin (ori) of DNA replication and genes for other functions.

FIGURE 8.20 Gene Transfer by Plasmids When plasmids enter a cell via conjugation, their genes can be expressed in the recipient cell.

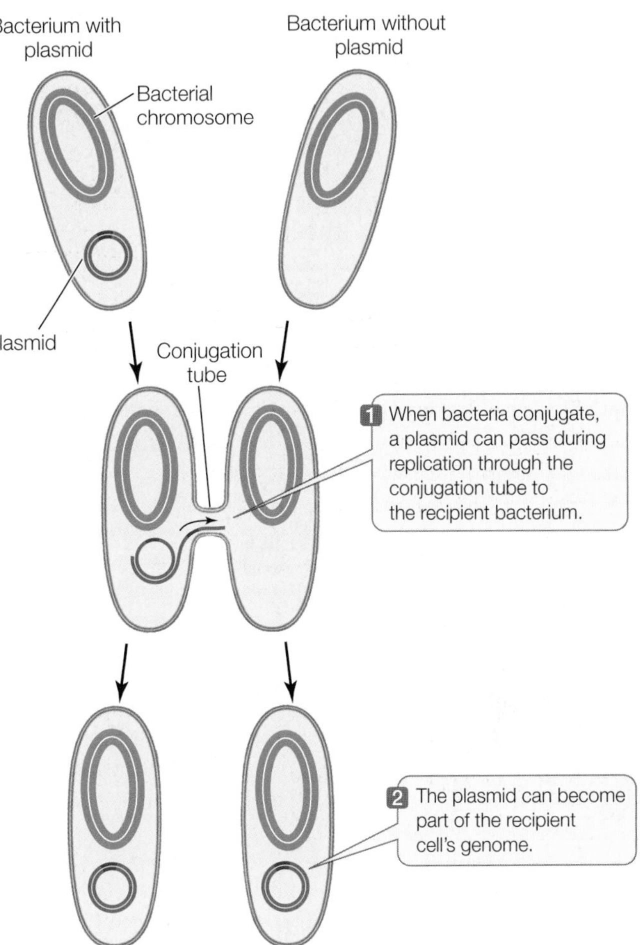

(particularly penicillin, which prevents the assembly of the bacterial call wall), many lethal infections were kept at bay. But over time some bacteria acquired mutations that rendered them resistant to penicillin. These bacteria had a selective advantage when faced with penicillin, and as penicillin use increased, the resistant bacteria became widespread. So scientists designed chemical variants, such as methicillin and vancomycin, to attack penicillin-resistant bacteria. With the development of strains of bacteria resistant to these antibiotics as well, the "arms race" has continued. The latest weapon against resistant bacteria is a group of antibiotics called carbapenems (for example, colistin), but bacteria resistant to these antibiotics are beginning to appear. Some resistance genes are carried on plasmids and are transferred between bacterial species by conjugation. This poses a major public health problem worldwide, since the new antibiotics are currently the last line of defense against lethal infections.

CHECKpoint CONCEPT 8.4

- ✓ How does recombination occur in prokaryotes?
- ✓ What is the evolutionary advantage of recombination in prokaryotes?
- ✓ What are the differences between recombination after conjugation in prokaryotes and recombination during meiosis in eukaryotes?

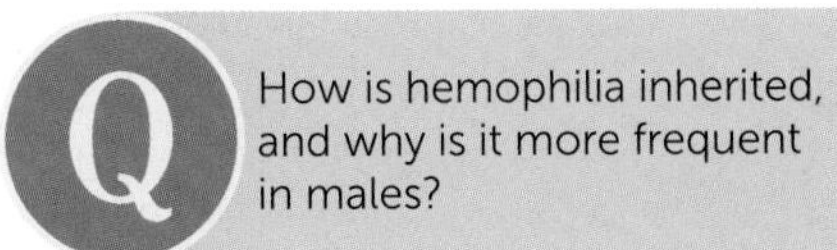

How is hemophilia inherited, and why is it more frequent in males?

ANSWER The ancient rabbis in the opening story were dealing with male babies that bled to death when they were cut. We know this as the blood-clotting disease hemophilia. The mutant allele of the gene coding for a blood clotting factor missing in hemophilia must be recessive, because the babies' parents did not suffer from the disease (Concept 8.1). The rabbis noted that the disease occurred in boys, and any relatives with the disease were males on the mother's side of the family. This is because the gene for the clotting factor is carried on the human X chromosome and its inheritance is sex-linked. Females have two X chromosomes, so even if they receive a mutant X chromosome from their mother, the second X chromosome, inherited from their father, will usually provide sufficient functional clotting factor. Males, however, have a single X chromosome, always inherited from their mother (Concept 8.3). If they receive their mother's recessive mutant chromosome (a 50–50 probability), they cannot produce clotting factor and thus suffer from hemophilia.

Hemophilia played a role in the history of modern Europe. England's Queen Victoria, who ruled for much of the nineteenth century, had nine children, one of whom, Leopold, had hemophilia and died after a minor accident at age 31. The queen did not have the disease, nor was it present in any of her forbears, so it is probable that a mutation arose spontaneously on one of Victoria's X chromosomes or that of her father (Concept 8.2). Three of Victoria's grandchildren had hemophilia, indicating that their mothers were carriers of the mutant allele. The mutation was thus introduced into the royal families of Spain, Germany, and Russia (**FIGURE 8.21**). While there are many descendants of Queen Victoria still living (including Queen Elizabeth II), none have hemophilia. It remains possible, however, that some of Victoria's female descendants are carriers.

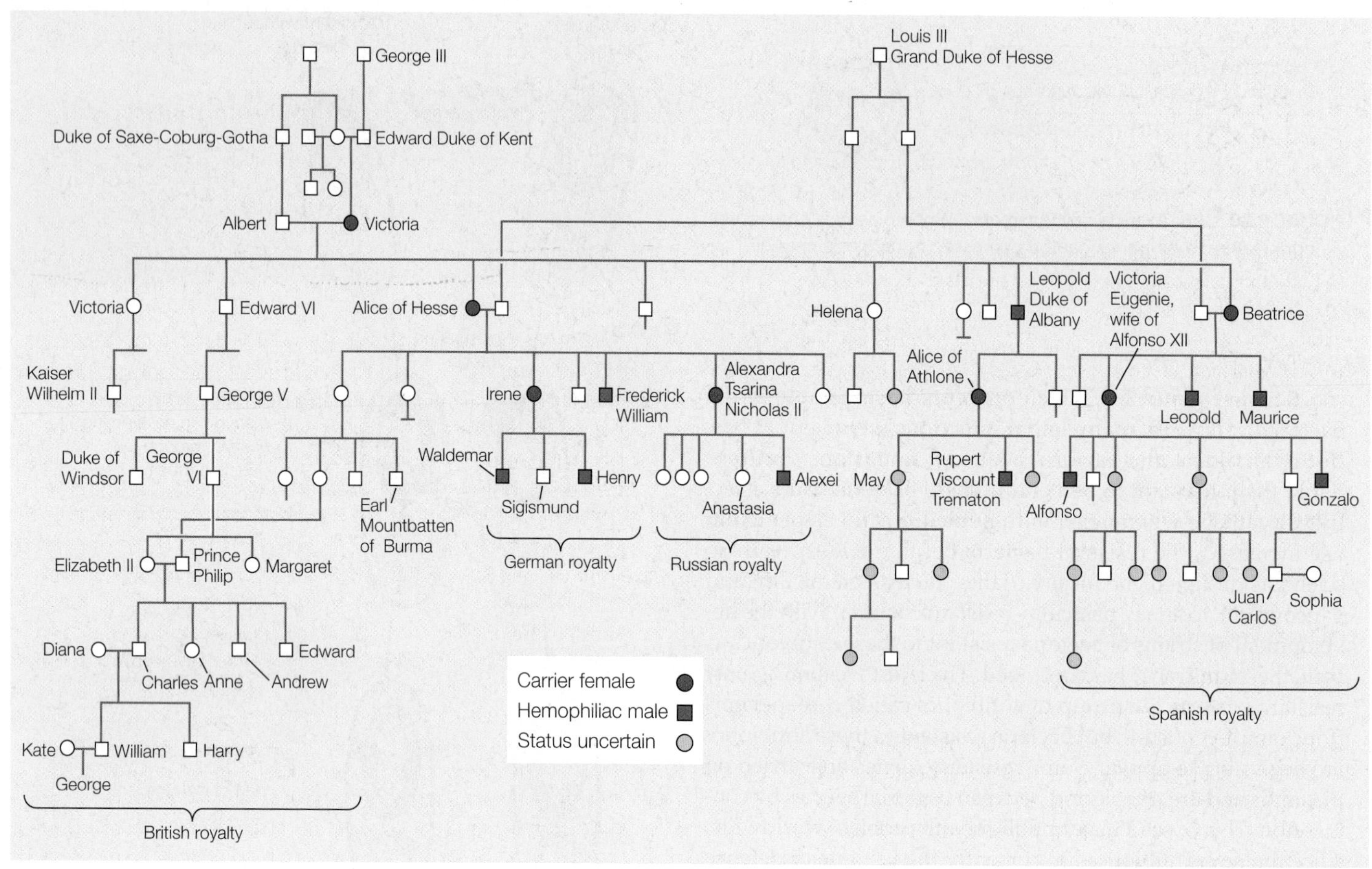

FIGURE 8.21 Sex Linkage in Royal Families of Europe England's Queen Victoria passed an X chromosome carrying the mutant allele for hemophilia to three of her children.

SUMMARY

CONCEPT 8.1 Genes Are Particulate and Are Inherited According to Mendel's Laws

- Mendel's experiments on pea plants supported the particulate theory of inheritance stating that discrete units (now called genes) are responsible for the inheritance of specific traits. **Review Figure 8.1**
- Mendel's first law, the **law of segregation**, states that when any individual produces gametes, the two copies of a gene separate, so that each gamete receives only one member of the pair. **Review Figures 8.2 and 8.3**
- Mendel used a **test cross** to find out if an individual with a dominant phenotype was homozygous or heterozygous for that phenotype. **Review Figure 8.4 and ACTIVITY 8.1**
- Mendel's use of **dihybrid crosses** to study the inheritance of two characters led to his second law: the **law of independent assortment**. The independent assortment of genes in meiosis leads to novel combinations of phenotypes. **Review Figures 8.5 and 8.6 and ANIMATED TUTORIAL 8.1**
- **Pedigree** analysis can determine whether an allele is **dominant** or **recessive**. **Review Figure 8.8 and ANIMATED TUTORIAL 8.2**

CONCEPT 8.2 Alleles and Genes Interact to Produce Phenotypes

- New alleles arise by random **mutation**. Many genes have multiple alleles. A **wild-type** allele gives rise to the predominant form of a trait. When the wild-type allele is present at a locus less than 99 percent of the time, the locus is said to be **polymorphic**.
- In **incomplete dominance**, neither of two alleles is dominant. The heterozygous phenotype is intermediate between the homozygous phenotypes. **Review Figure 8.10**
- **Codominance** exists when two alleles at a locus produce two different phenotypes that both appear in heterozygotes. **Review Figure 8.11**
- In **epistasis**, one gene affects the expression of another. **Review Figure 8.12**
- Environmental conditions can affect the expression of a genotype.
- **Heritability** is the relative contribution of genetic versus environmental factors to the variation in a character in a population.

CONCEPT 8.3 Genes Are Carried on Chromosomes

- Each chromosome carries many genes, and the genes on a single chromosome show **genetic linkage**. **Review Figure 8.13 and ANIMATED TUTORIAL 8.3**
- Genes on the same chromosome can recombine by crossing over. The resulting **recombinant** chromosomes have new combinations of alleles. **Review Figure 8.14**
- **Recombination frequencies** can be used to generate a genetic map of a chromosome. **Review Figure 8.15**
- In fruit flies and mammals, the X chromosome carries many genes, but the Y chromosome has only a few. Males have only one allele (are **hemizygous**) for X-linked genes, so recessive sex-linked mutations are expressed phenotypically more often in males than in females. Females may be unaffected **heterozygous carriers** of such alleles. **Review Figure 8.16**
- Some genes are present on the chromosomes of organelles such as plastids and mitochondria. In many organisms, cytoplasmic genes are inherited only from the mother because the male gamete contributes only its nucleus (i.e., no cytoplasm) to the zygote at fertilization. **Review Figure 8.18**

CONCEPT 8.4 Prokaryotes Can Exchange Genetic Material

- Prokaryotes reproduce asexually but can transfer genes from one cell to another in a process called **bacterial conjugation**. **Review Figure 8.19**
- **Plasmids** are small, extra DNA molecules in bacteria that carry genes involved in important metabolic processes. Plasmids can be transmitted from one cell to another. **Review Figure 8.20**

See **ACTIVITIES 8.2** and **8.3** for a concept review of this chapter.

Go to the Interactive Summary to review key figures, Animated Tutorials, and Activities PoL2e.com/is8

Go to LaunchPad at **macmillanhighered.com/launchpad** for additional resources, including LearningCurve Quizzes, Flashcards, and many other study and review resources.

9 DNA and Its Role in Heredity

KEY CONCEPTS

9.1 DNA Structure Reflects Its Role as the Genetic Material

9.2 DNA Replicates Semiconservatively

9.3 Mutations Are Heritable Changes in DNA

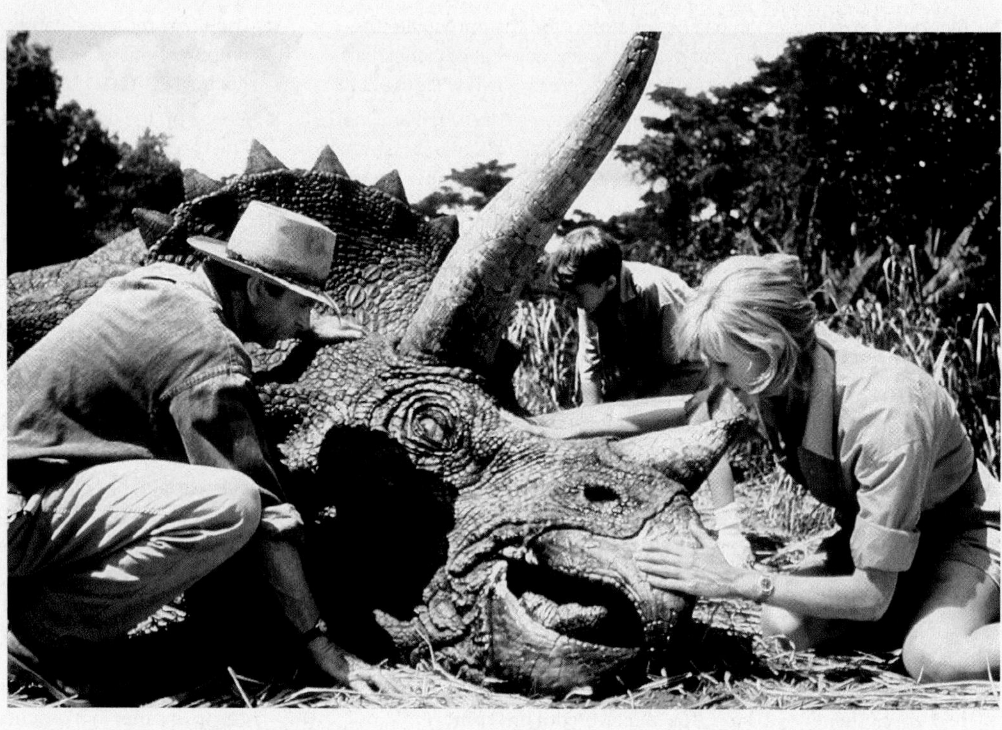

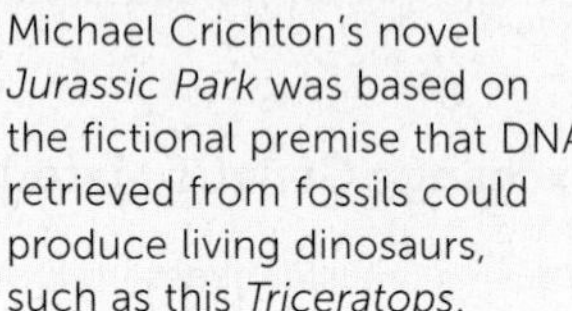

Michael Crichton's novel *Jurassic Park* was based on the fictional premise that DNA retrieved from fossils could produce living dinosaurs, such as this *Triceratops*.

Jurassic Park, in both its literary and film incarnations, featured a fictional theme park populated with live dinosaurs. In the story, scientists isolated DNA from dinosaur blood extracted from the digestive tracts of fossil insects. The insects supposedly sucked the reptiles' blood immediately before being preserved in amber (fossilized tree resin). According to the novel, this DNA could be manipulated to produce living individuals of long-extinct organisms such as velociraptors and the famous *Tyrannosaurus rex*.

The late Michael Crichton got the idea for his novel from an actual scientific paper in which the authors cracked open amber that was 40 million years old and extracted DNA from a fossilized bee that had been trapped inside. Other scientists had reported on ancient DNA from amber-trapped termites and gnats. Then several reports emerged of DNA from 80-million-year-old dinosaur bones. Unfortunately, upon additional study, these "preserved" DNAs turned out to be contamination—either from microorganisms living in the surrounding soil or even from the scientists studying the samples. In fact, one of the supposed dinosaur DNAs turned out to be from the human Y chromosome.

It is unlikely that any long DNA polymers would survive over millions of years. The oldest fossilized insects in amber are about 40 million years old, and the dinosaurs died out about 65 million years ago. Nevertheless, the huge success of Crichton's book brought ancient DNA to the attention of millions of people, including biologists who study the evolution of life on Earth. DNA samples have been isolated from the remains of entire ecosystems of organisms preserved for many thousands of years in permafrost. With improved methods for DNA analysis, large portions of these organisms' genomes are being sequenced.

Methods to replicate tiny amounts of DNA and keep it from contamination have improved, and attention has turned to ancient human DNA. For example, DNA samples have been studied from people whose bodies were preserved in ice, such as the "Tyrolean Iceman" who died in the Austrian Alps 5,300 years ago. There is even a Neandertal Genome Project to analyze the DNA from preserved specimens of *Homo neanderthalensis*, a species that lived in Europe at the same time as early humans, between 350,000 and 30,000 years ago.

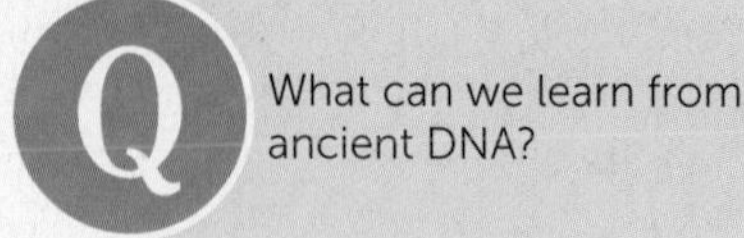

Q What can we learn from ancient DNA?

You will find the answer to this question on page 192.

CONCEPT 9.1 DNA Structure Reflects Its Role as the Genetic Material

In Chapter 8 we described Mendel's experiments in the 1860s demonstrating that genes are physically distinct entities, and other work in the early twentieth century showing that groups of genes are linked together. By the early twentieth century, a "chromosomal theory of inheritance" had been developed, proposing that Mendel's genes are present in the chromosomes of the cell nucleus. This theory came partly from observations of sea urchins: it was shown that an entire set of chromosomes must be present for a sea urchin embryo to grow and develop. Scientists also observed that homologous chromosomes are paired during meiosis, that crossing over occurs during meiosis I (see Figure 8.14), and that the chromosome pairs separate independently at anaphase I. Thus the behavior of chromosomes accounted for Mendel's laws of segregation and independent assortment, as well as the later discoveries of linkage and recombination.

We now turn to the actual chemical nature of genes, beginning with the evidence that DNA is the carrier of heritable information. Scientists used two types of evidence to show that DNA is the genetic material: circumstantial and experimental. We will provide examples of both types.

Circumstantial evidence suggested that DNA is the genetic material

Early observations pointed to the possibility that DNA is the genetic material. Scientists found that DNA:

- is present in the cell nucleus and in condensed chromosomes
- doubles during S phase of the cell cycle
- is twice as abundant in the diploid cells as in the haploid cells of a given organism.

Let's look at some of these lines of evidence.

DNA IN THE NUCLEUS DNA was first isolated in 1868 by the young Swiss researcher–physician Friedrich Miescher, who isolated cell nuclei from white blood cells in pus from the bandages of wounded soldiers. When he treated these nuclei chemically, a fibrous substance came out of solution. He called it "nuclein" and found that it contained the elements C, H, O, N, and P. With no evidence except for finding it in the nucleus, Miescher boldly proposed that nuclein was the genetic material. His supervising professor was so astounded by Miescher's work that he repeated it himself in the laboratory, and finally allowed his student to publish it in a scientific journal.

DNA IN THE CHROMOSOMES In the early twentieth century dyes were developed that react specifically with DNA, only showing color when they bind to it. This allowed individual cells to be examined for the location and amount of DNA they contained. When dividing cells were stained with such a dye, only the chromosomes were stained:

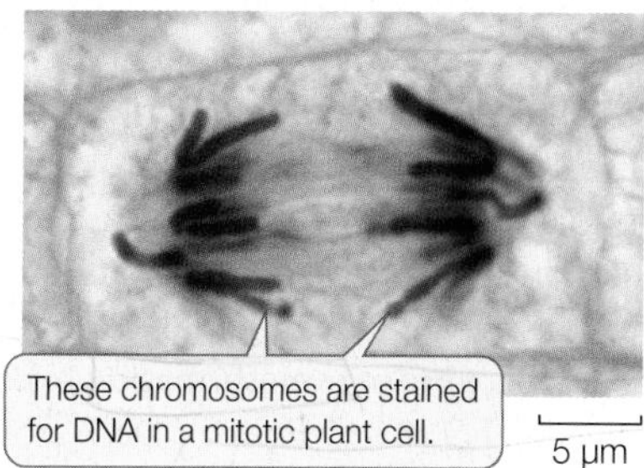

These chromosomes are stained for DNA in a mitotic plant cell.
5 µm

DNA AMOUNTS The amount of dye binding to DNA, and hence the intensity of color observed, was directly related to the amount of DNA present: the greater the intensity, the more DNA. This allowed scientists to analyze DNA amounts in individual cells during the cell cycle (see Concept 7.2).

When a population of actively dividing cells was stained with dye, the amount of DNA in each cell could be quantified by passing the cells one by one through an instrument called a flow cytometer. In general, two populations of cells were seen: most cells were in G1 and contained half the amount of DNA that was in the remaining cells, which were in S, G2, or M (**FIGURE 9.1**). Such staining experiments confirmed two other predictions for DNA as the genetic material:

- Virtually all nondividing somatic cells of a particular organism have the same amount of nuclear DNA. This amount varies from species to species.
- Similar experiments showed that after meiosis, gametes have half the amount of nuclear DNA as somatic cells.

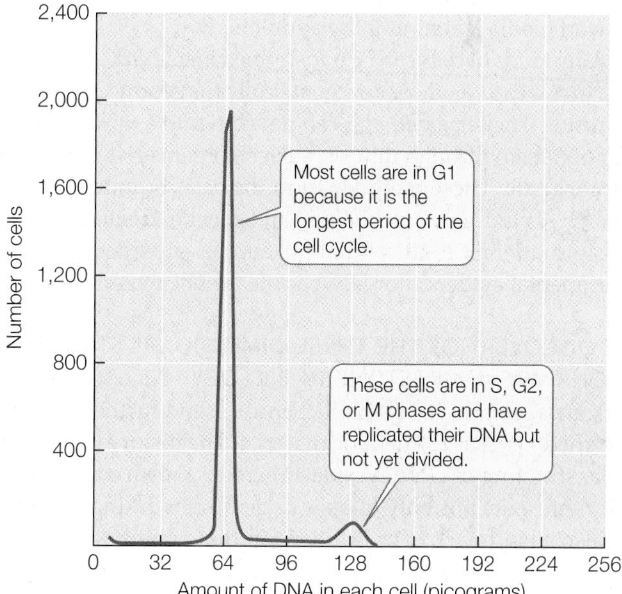

FIGURE 9.1 DNA in the Cell Cycle When dividing cells are stained and analyzed by flow cytometry, there are two populations in terms of DNA content, seen as two peaks in the above graph.

Experimental evidence confirmed that DNA is the genetic material

Circumstantial evidence can show correlations between two phenomena. However, *scientists rely on experiments to provide evidence of a cause-and-effect relationship*. Chromosomes in eukaryotic cells contain DNA, but they also contain proteins that are bound to DNA. Therefore it was difficult to rule out the possibility that genetic information might be carried in proteins. In order to confirm that DNA was the genetic material, biologists used model organisms such as bacteria in **transformation** experiments. They found, for example, that the addition of DNA from one strain of bacterium could genetically transform another strain:

Bacterium strain A + strain B DNA → bacterium strain B

Viruses provided another system to explore this question. Many viruses, including **bacteriophage** (viruses that infect bacteria), are composed of DNA and only one or a few kinds of protein. When a bacteriophage infects a bacterium, it takes about 20 minutes for the virus to hijack the bacterium's metabolic capabilities and turn the bacterium into a virus factory. Minutes later, the bacterium is dead and hundreds of viruses are released.

The transition from bacterium to virus producer is a change in the genetic program of the bacterial cell, resulting in a change of phenotype. Experiments showed that *only the viral DNA* is injected into the cell during infection (**FIGURE 9.2**). Since the viral DNA genetically transformed the bacteria, this was further evidence that DNA and not protein is the genetic material.

The transformation of mammalian cells with a gene for antibiotic resistance provided another model system for showing that DNA is the genetic material (**FIGURE 9.3**). When cultured mammalian cells were treated with DNA containing a gene for resistance to the antibiotic neomycin, the cells were able to grow on media containing the antibiotic.

Many kinds of cells can be transformed in this way—even egg cells. In this case, a whole new genetically transformed organism can result. The fertilized egg can develop into a new multicellular organism through mitosis; such an organism is referred to as **transgenic**. These methods form the basis of much applied research, including biotechnology and genetic engineering. The transformation of multicellular eukaryotes provides powerful experimental evidence for DNA as the genetic material.

THE DISCOVERY OF THE THREE-DIMENSIONAL STRUCTURE OF DNA WAS A MILESTONE IN BIOLOGY Mendel showed that genes are physically distinct entities, and further research identified DNA as the genetic material. The history of how the actual structure of DNA was deciphered is worth considering, as it represents not only talented scientists working together, but also a landmark in our understanding of biology.

By the mid-twentieth century, the chemical makeup of DNA, as a polymer made up of nucleotide monomers, had been known for several decades. In determining the structure of DNA, scientists hoped to answer two additional questions:

- How is DNA replicated between cell divisions?
- How does it direct the synthesis of specific proteins?

They were eventually able to answer both questions. The structure of DNA was deciphered only after many types of experimental evidence were considered together.

X-RAY CRYSTALLOGRAPHY PROVIDED CLUES TO DNA'S STRUCTURE The most crucial evidence was obtained using X-ray crystallography. Some chemical substances, when they are isolated and purified, can be made to form crystals. The positions of atoms in a crystallized substance can be inferred from the diffraction pattern of X rays passing through the substance (**FIGURE 9.4A**). The structure of DNA would not have been characterized without the crystallographs prepared in the early 1950s by the English chemist Rosalind Franklin (**FIGURE 9.4B**). Franklin's work, in turn, depended on the success of the English biophysicist Maurice Wilkins, who prepared samples containing very uniformly oriented DNA fibers. These fibers and the crystallographs Franklin prepared from them suggested a spiral or helical molecule.

THE NUCLEOTIDE COMPOSITION OF DNA WAS KNOWN The chemical composition of DNA also provided important clues

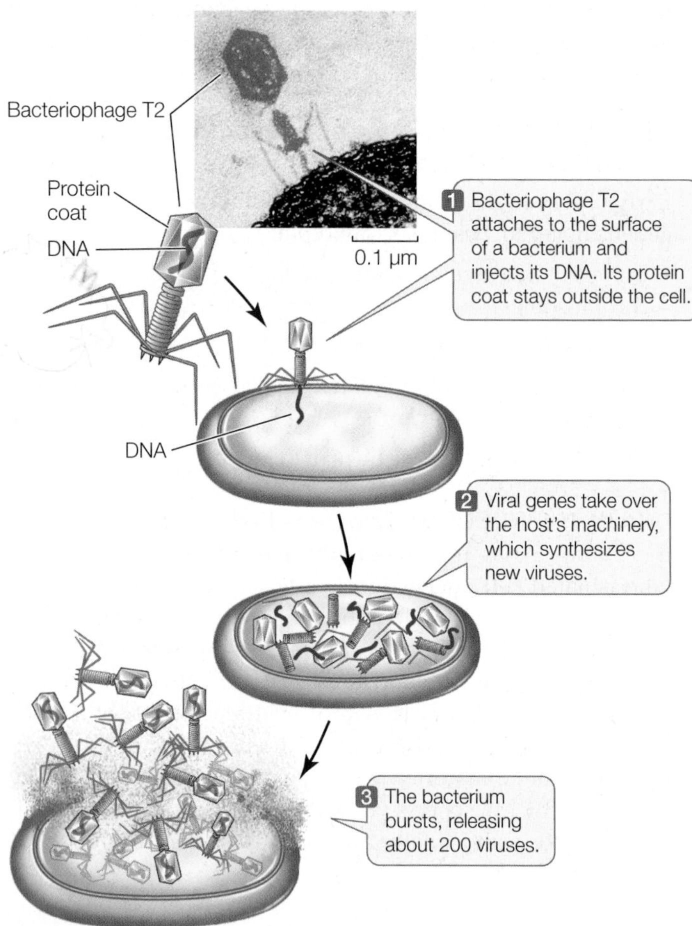

Bacteriophage T2
Protein coat
DNA
0.1 µm
1 Bacteriophage T2 attaches to the surface of a bacterium and injects its DNA. Its protein coat stays outside the cell.
DNA
2 Viral genes take over the host's machinery, which synthesizes new viruses.
3 The bacterium bursts, releasing about 200 viruses.

FIGURE 9.2 Viral DNA and Not Protein Enters Host Cells
Bacteriophage T2 infects *E. coli* and depends on the bacterium to produce new viruses. The bacteriophage consists entirely of DNA contained within a protein coat. When the virus infects an *E. coli* cell, its DNA, but not its protein coat, is injected into the host bacterium.

INVESTIGATION

FIGURE 9.3 Transformation of Eukaryotic Cells A DNA molecule can be treated chemically so that it is taken up from a solution by mammalian cells. The inclusion of an antibiotic resistance gene shows that the cells have been genetically transformed by the DNA.[a]

HYPOTHESIS

DNA can transform eukaryotic cells.

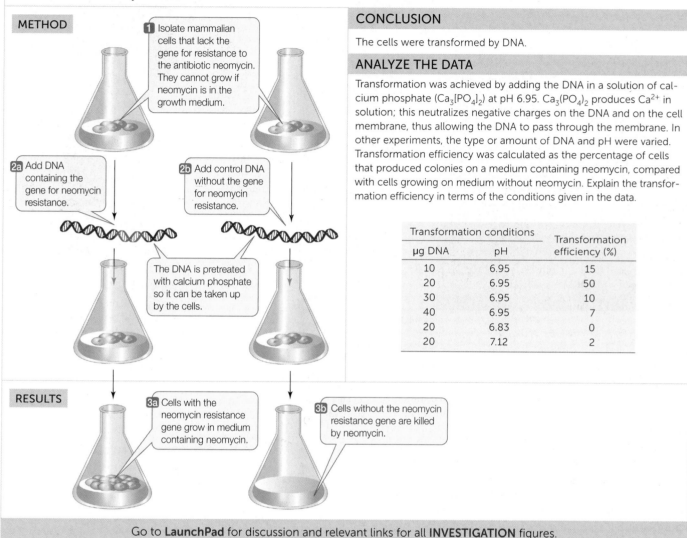

METHOD

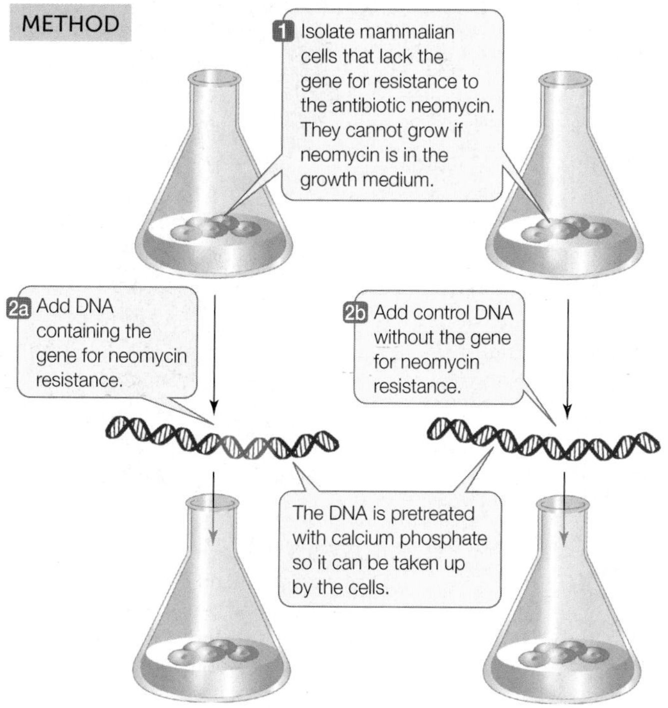

1 Isolate mammalian cells that lack the gene for resistance to the antibiotic neomycin. They cannot grow if neomycin is in the growth medium.

2a Add DNA containing the gene for neomycin resistance.

2b Add control DNA without the gene for neomycin resistance.

The DNA is pretreated with calcium phosphate so it can be taken up by the cells.

RESULTS

3a Cells with the neomycin resistance gene grow in medium containing neomycin.

3b Cells without the neomycin resistance gene are killed by neomycin.

CONCLUSION

The cells were transformed by DNA.

ANALYZE THE DATA

Transformation was achieved by adding the DNA in a solution of calcium phosphate ($Ca_3[PO_4]_2$) at pH 6.95. $Ca_3(PO_4)_2$ produces Ca^{2+} in solution; this neutralizes negative charges on the DNA and on the cell membrane, thus allowing the DNA to pass through the membrane. In other experiments, the type or amount of DNA and pH were varied. Transformation efficiency was calculated as the percentage of cells that produced colonies on a medium containing neomycin, compared with cells growing on medium without neomycin. Explain the transformation efficiency in terms of the conditions given in the data.

Transformation conditions		Transformation efficiency (%)
μg DNA	pH	
10	6.95	15
20	6.95	50
30	6.95	10
40	6.95	7
20	6.83	0
20	7.12	2

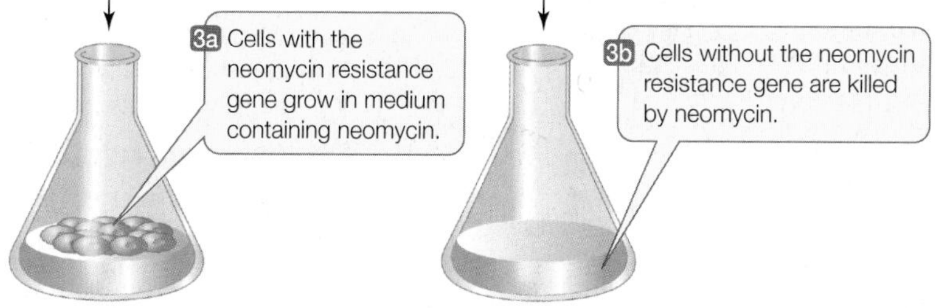

Go to **LaunchPad** for discussion and relevant links for all **INVESTIGATION** figures.

[a]C. Chen and H. Okayama. 1987. *Molecular and Cellular Biology* 7: 2745–2752.

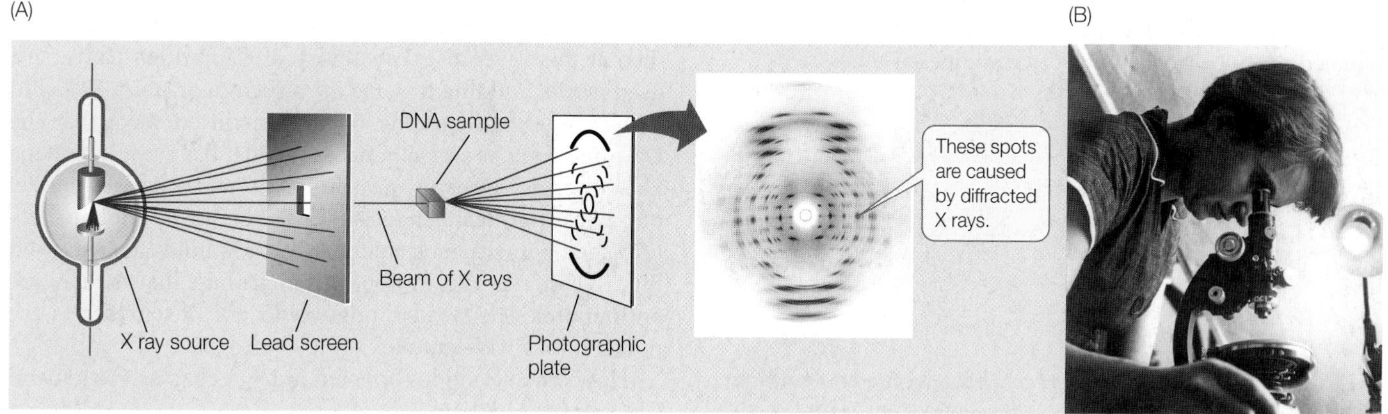

FIGURE 9.4 X-Ray Crystallography Helped Reveal the Structure of DNA (A) The positions of atoms in a crystallized chemical substance can be inferred by the pattern of diffraction of X rays passed through it. The pattern of DNA is both highly regular and repetitive. (B) Rosalind Franklin's crystallographs helped other scientists visualize the helical structure of the DNA molecule.

FIGURE 9.5 DNA Is a Double Helix
(A) Francis Crick (right) and James Watson (left) proposed that the DNA molecule has a double-helical structure. (B) Biochemists can now pinpoint the position of every atom in a DNA molecule and have verified the essential features of the Watson–Crick model. Two sugar–phosphate chains provide the outer backbones of the double helix, with hydrogen-bonded nitrogenous bases forming "rungs" within the structure.

Go to MEDIA CLIP 9.1
Discovery of the Double Helix
PoL2e.com/mc9.1

to its structure. Biochemists knew that DNA is a polymer of nucleotides. Each of these nucleotides consists of a molecule of the sugar deoxyribose, a phosphate group, and a nitrogen-containing base (see Figure 3.1). The only differences among the four nucleotides of DNA are their nitrogenous bases: the purines adenine (A) and guanine (G), and the pyrimidines cytosine (C) and thymine (T) (see Figure 3.1).

In 1950, biochemist Erwin Chargaff at Columbia University reported an important observation. He and his colleagues had found that DNA samples from many different species—and from different sources within a single organism—exhibited certain regularities. The following rule held for each sample: the amount of adenine equaled the amount of thymine (A = T), and the amount of guanine equaled the amount of cytosine (G = C). As a result, the total abundance of purines (A + G) equaled the total abundance of pyrimidines (T + C):

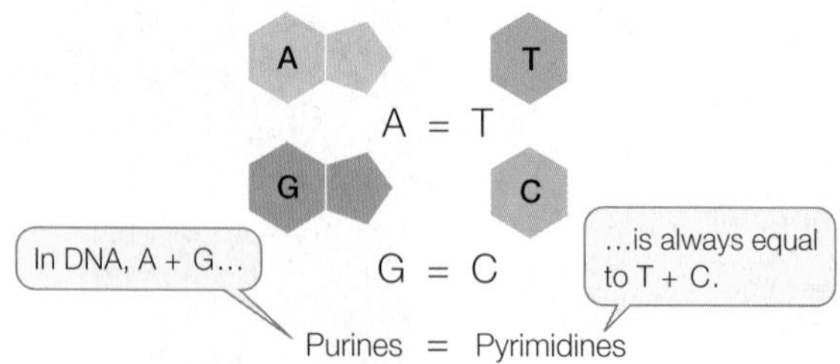

The structure of DNA could not have been worked out without this observation, now known as Chargaff's rule.

WATSON AND CRICK DESCRIBED THE DOUBLE HELIX Chemical model building is the assembly of three-dimensional structures using known relative molecular dimensions and known bond angles. The English physicist Francis Crick and the American geneticist James D. Watson (**FIGURE 9.5A**), both then at the Cavendish Laboratory of Cambridge University, used model building to solve the structure of DNA. Rosalind Franklin's crystallography results convinced them that the DNA molecule must be **helical**—it must have a spiral shape like a spring. Density measurements and previous model building experiments suggested that there are two polynucleotide chains in the molecule. Modeling studies also showed that the strands run in opposite directions, that is, they are **antiparallel**. The two strands would not fit together in the model if they were parallel.

How are nucleotides oriented in DNA chains? Watson and Crick suggested that:

- the nucleotide bases are on the interior of the two strands, with a sugar–phosphate backbone on the outside. The strands would not fit together otherwise:

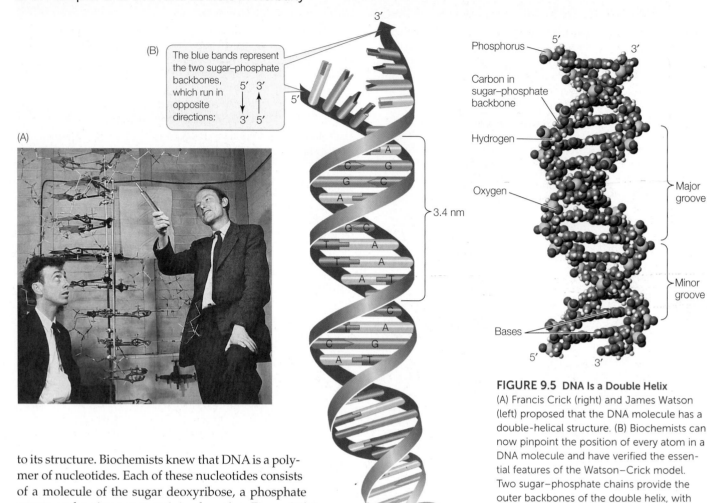

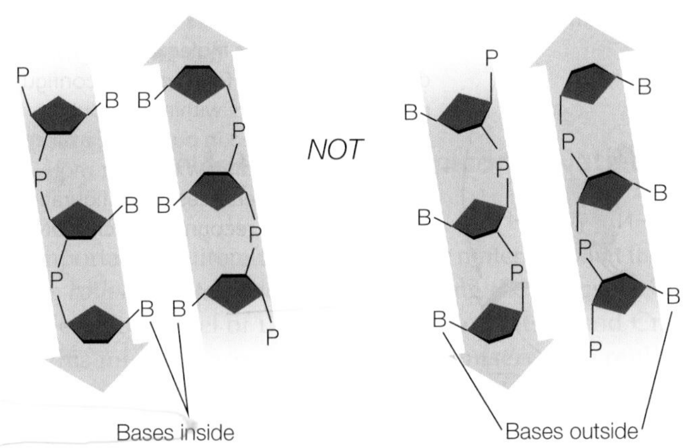

- to satisfy Chargaff's rule (purines = pyrimidines), a purine on one strand is always paired with a pyrimidine on the opposite strand. These **base pairs** (A-T and G-C) have the same width down the double helix, a uniformity shown by X-ray diffraction:

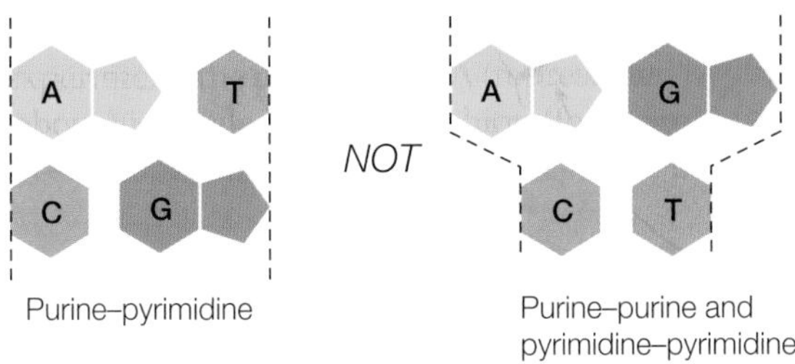

In late February of 1953, Crick and Watson built a model out of tin that established the general structure of DNA. This structure explained all the known chemical properties of DNA, and it opened the door to understanding its biological functions. There have been minor amendments to that first published structure, but its principal features remain unchanged.

Four key features define DNA structure

Four features summarize the molecular architecture of the DNA molecule (**FIGURE 9.5B**; also review Figure 3.4):

- *DNA is a double-stranded helix of uniform diameter*. The sugar–phosphate backbones of the two chains form a coil on the outside of the helix, and the nitrogenous bases point toward the center. The chains are held together by two chemical forces: hydrogen bonding between the bases

and van der Waals forces between adjacent bases on the same strand. When the base rings come near one another, they tend to stack like poker chips because of these weak attractions.

- *The two DNA strands are antiparallel*. The backbone of each strand contains repeating units of the monosaccharide (sugar) deoxyribose:

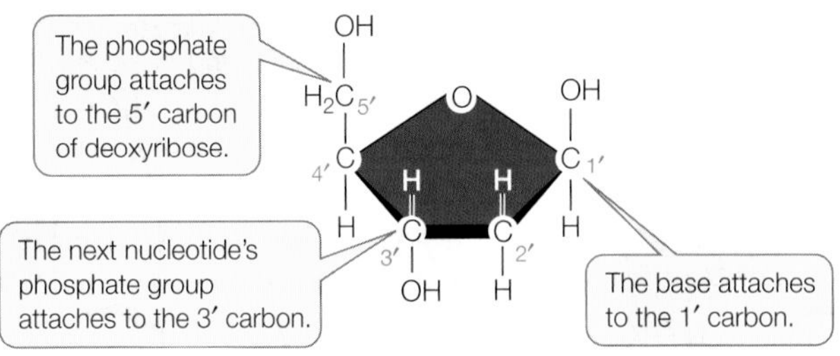

The number followed by the prime sign (′) designates the position of a carbon atom in the sugar. In the sugar–phosphate backbone of DNA, the phosphate groups are connected to the 5′ carbon of one deoxyribose molecule and to the 3′ carbon of the next, linking the two sugar molecules together. Thus the two ends of the polynucleotide chain are different. The **5′ end** of the chain is a free (not connected to another nucleotide) phosphate group. The **3′ end** of the chain is a free 3′ hydroxyl (OH) group. In the double helix of DNA, the 5′ end of one strand is paired with the 3′ end of the other strand, and vice versa (see Figure 3.4).

- *In DNA, the outer edges of the nitrogenous bases are exposed in the major and minor grooves*. These grooves exist because the helices formed by the backbones of the two DNA strands are not evenly spaced relative to one another (see Figure 9.5B). **FIGURE 9.6** shows the four possible configurations of the flat, hydrogen-bonded base pairs within the major and minor grooves. The exposed outer edges of the base pairs are accessible for additional hydrogen bonding. Notice that the arrangements of unpaired atoms and groups differ in the A-T base pairs compared with the G-C base pairs. Thus the *surfaces of the A-T and G-C base pairs are chemically distinct*, allowing other molecules, such as proteins, to recognize specific base-pair sequences and bind to them. The atoms and groups in the major groove are more accessible, and tend to bind other molecules more frequently, than those in the minor groove. *This binding of proteins to specific base-pair sequences is the key to protein–DNA interactions*, which are necessary for the replication and expression of the genetic information in DNA.

- *The DNA double helix is right-handed*. Hold your right hand with the thumb pointing up (see Figure 9.10B). Imagine the curve of the helix following the direction of your fingers as it winds upward, and you have the idea.

While DNA usually forms a right-handed helix, it can sometimes be found as a much less stable left-handed helix. So-called Z-DNA ("zig-zag DNA") does not have major and minor grooves and is more elongated and less compact than normal DNA. Z-DNA appears to form in regions of DNA that are being actively transcribed, and it may play a role in stabilizing the DNA during transcription.

The double-helical structure of DNA is essential to its function

The genetic material performs four important functions, and the DNA structure proposed by Watson and Crick was elegantly suited to three of them.

RESEARCH TOOLS

FIGURE 9.15 The Polymerase Chain Reaction The steps in this cyclic process are repeated many times to produce millions of identical copies of a DNA fragment. This makes enough DNA for chemical analysis and genetic manipulations.

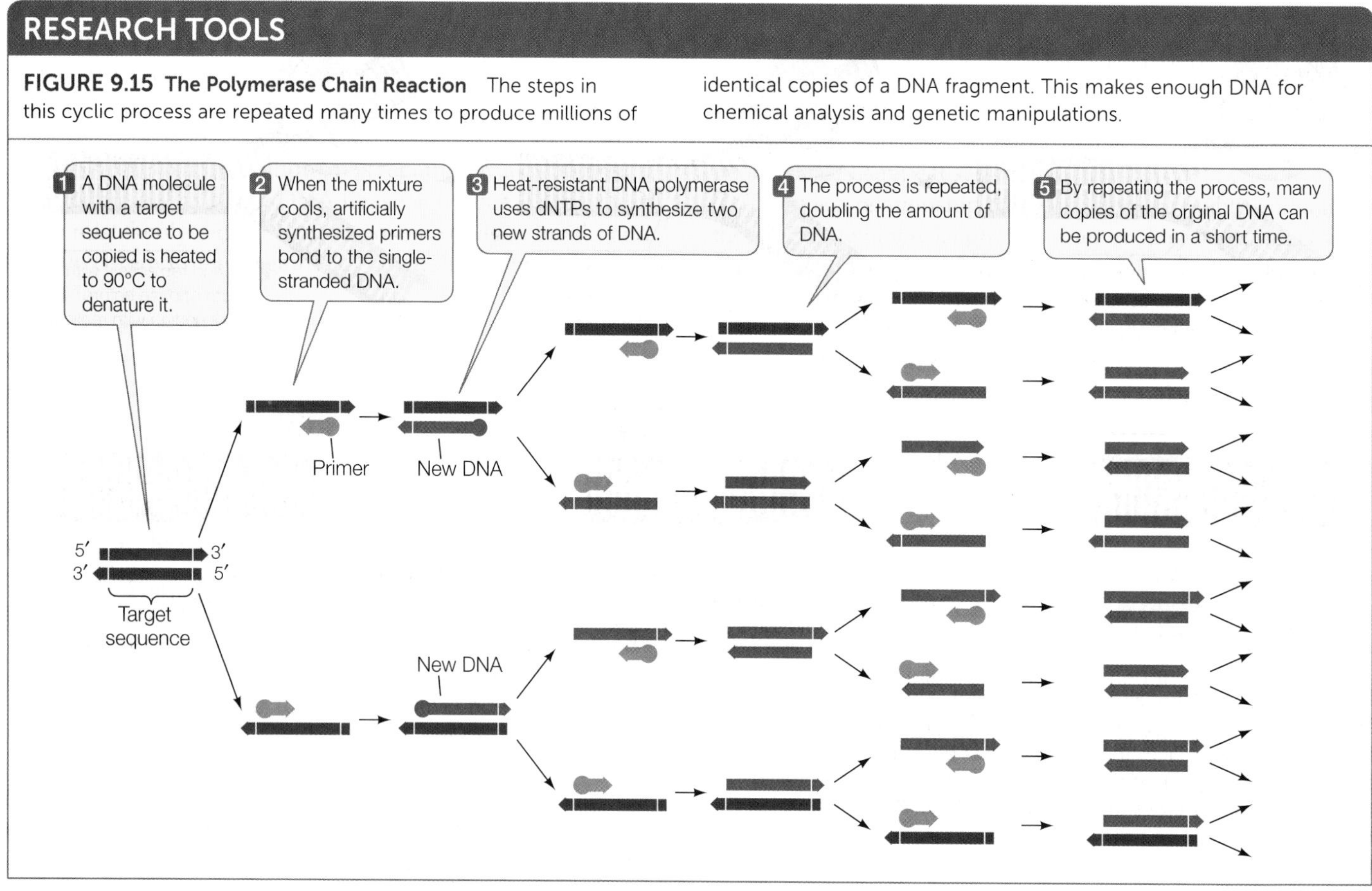
1 A DNA molecule with a target sequence to be copied is heated to 90°C to denature it.
2 When the mixture cools, artificially synthesized primers bond to the single-stranded DNA.
3 Heat-resistant DNA polymerase uses dNTPs to synthesize two new strands of DNA.
4 The process is repeated, doubling the amount of DNA.
5 By repeating the process, many copies of the original DNA can be produced in a short time.
Primer
New DNA
5′
3′
3′
5′
Target sequence
New DNA

CHECKpoint CONCEPT 9.2

- ✓ What is semiconservative DNA replication?
- ✓ Why does the leading strand in DNA replicate continuously and the lagging strand discontinuously?
- ✓ Cells from older people have shorter telomeres than cells from younger people. How might this relate to aging?
- ✓ If you have a small amount of a large chromosome of 20 million bp and want to amplify a short sequence of 1,000 bp, how would you do it? Explain the role of primers in this process.

We have now described (1) the lines of evidence for DNA as the genetic material and (2) the precise replication of DNA during cell division. A less obvious requirement for DNA as the genetic material is its ability to mutate. Mutation creates variability in DNA, which is the raw material for evolution.

CONCEPT 9.3 Mutations Are Heritable Changes in DNA

In Chapter 8 we described mutations as stable and inherited changes in the genetic material. A mutation may result in a new allele of a gene, and different alleles may produce different phenotypes (for example, pea plants with wrinkled seeds versus round seeds). With reference to the chemical nature of genes, we can state that *mutations are changes in the nucleotide sequence of DNA that are passed on from one cell or organism to another.*

Mutations occur by a variety of processes. For example, in Concept 9.2 we described how DNA polymerases can make errors. Repair systems such as proofreading are in place to correct them, but some errors escape being corrected and are passed on to daughter cells.

Mutations in multicellular organisms can be divided into two types:

- **Somatic mutations** occur in the somatic (body) cells of a multicellular organism. These mutations are passed on to the daughter cells during mitosis, and in turn to the offspring of those cells. For example, a mutation in a single skin cell could result in a patch of skin cells that all have the same mutation. However, somatic mutations are not passed on to sexually produced offspring. (Exceptions occur in plants, where germline cells can arise from somatic cells and thus pass on somatic mutations.)
- **Germline mutations** occur in the cells of the germ line—the specialized cells that give rise to gametes (the eggs and sperm of sexual reproduction). A gamete with the mutation passes it on to a new organism at fertilization.

In either case, the mutations may or may not affect the phenotype.

Mutations can have various phenotypic effects

An organism's genome is the total DNA sequence present in all of its chromosomes (or in its single chromosome, in the case of prokaryotes). Depending on the organism, it can consist of millions or billions of base pairs of DNA. Most genomes include both genes and regions of DNA that are not expressed.

As we discussed briefly in Concept 3.1, gene expression involves the transcription of DNA into RNA, followed by the translation of the RNA into a polypeptide. (Some RNAs are not translated but have catalytic or other roles in the cell.) We will discuss gene expression in much more detail in Chapter 10. For now we will look at some of the ways that mutations can affect gene expression and phenotypes. It is also the case that many mutations have no effects on phenotypes.

Mutations are often discussed in terms of their effects on protein-coding genes and their functions (**FIGURE 9.16**):

- **Silent mutations** do not affect gene function (see Figure 9.16B). They can be mutations in DNA that is not expressed, or mutations within an expressed region that do not have any effect on the encoded protein. *Most mutations in large genomes are silent.*

LINK

Silent mutations are a source of neutral alleles that can be acted upon by the mechanisms of evolution, as described in **Concept 15.2**

- **Loss-of-function mutations** can result in either the loss of expression of a gene or in the production of a nonfunctional protein or RNA. Some loss-of-function mutations prevent a gene from being transcribed or cause transcription to terminate too soon. In other cases the gene is transcribed and translated, but the resulting protein no longer works as a structural protein or enzyme (as illustrated in Figure 9.16C). Loss-of-function mutations almost always show recessive inheritance in a diploid organism, because the presence of one wild-type allele usually results in sufficient functional protein for the cell. For example, the wrinkled seed phenotype studied by Mendel (see Figure 8.1) is due to a recessive loss-of-function mutation in the gene for starch branching enzyme 1 (*SBE1*). Even in plants with only one copy of the wild-type allele, there is enough SBE1 enzyme to produce the wild-type round phenotype.
- **Gain-of-function mutations** lead to a protein with an altered function (see Figure 9.16D). This kind of mutation usually shows dominant inheritance, because the presence of the wild-type allele does not prevent the mutant allele from functioning. This type of mutation is common in cancer. For example, a receptor for a growth factor normally requires binding of the growth factor (the ligand) to activate the cell division cycle. Some cancers are caused by mutations in genes that encode these receptors so that the receptors are "always on," even in the absence of their ligands. This leads to the unrestrained cell proliferation that is characteristic of cancer cells.

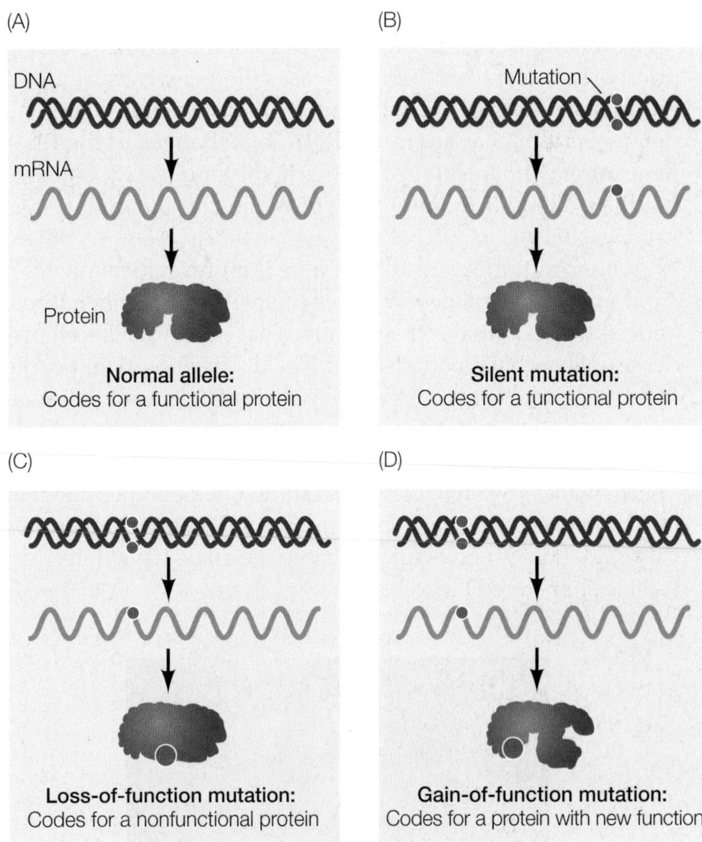

FIGURE 9.16 Mutation and Phenotype Mutations may or may not affect the function and phenotype of a protein.

- **Conditional mutations** cause their phenotypes only under certain *restrictive* conditions. The wild-type phenotype is expressed under other, *permissive* conditions. Many conditional mutants are temperature-sensitive; that is, they show the altered phenotype only at a certain temperature. For example, in Chapter 8 (see p. 161) we described a temperature-sensitive mutation that affects coat color in rabbits and cats. In warmer parts of the body, the mutant protein is inactive, resulting in pale fur. In the cooler extremities of the body, the protein is active, producing dark fur.

All mutations are alterations in the nucleotide sequence of DNA. They can be small-scale mutations that alter only one or a few nucleotides, or they can be large-scale mutations in which entire segments of DNA are rearranged, duplicated, or irretrievably lost. Next we will consider small-scale mutations, in particular, point mutations.

Point mutations are changes in single nucleotides

A **point mutation** is the addition or subtraction of a single nucleotide base, or the substitution of one base for another. Point mutations can arise because of errors in DNA replication that are not corrected during proofreading, or they may be caused by environmental **mutagens**: substances that cause mutations, such as radiation or certain chemicals.

Some point mutations that occur within genes are loss-of-function mutations because they prevent the gene from being properly transcribed. In other cases the gene is transcribed normally. A point mutation in the coding region of a gene may result in changes in the RNA, but changes in the RNA may or may not result in a change in the amino acid sequence of the protein. If the protein is not changed, the mutation is silent.

Other mutations result in altered amino acid sequences, and in some cases these changes can have drastic phenotypic effects. An example is the mutation that causes sickle-cell disease, a heritable blood disorder. The disease occurs in people who carry two copies of the sickle allele of the gene for human β-globin (a subunit of hemoglobin, the protein in human blood that carries oxygen; see p. 47). The sickle allele differs from the normal allele by one base pair, resulting in a polypeptide that differs by one amino acid from the normal protein. Individuals who are homozygous for this recessive allele have defective, sickle-shaped red blood cells:

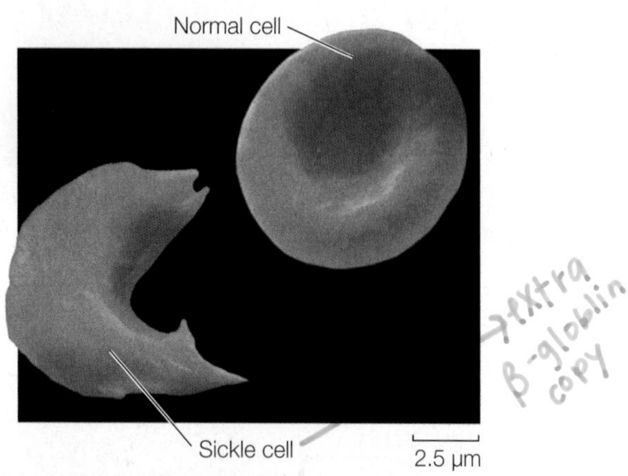

Go to MEDIA CLIP 9.2
Sickle Cells: Deformed by a Mutation
PoL2e.com/mc9.2

The deformed cells tend to block narrow capillaries, which results in tissue damage.

Not all changes in the amino acid sequence of a protein affect its function. For example, a hydrophilic amino acid may be substituted for another hydrophilic amino acid, so that the shape of the protein is unchanged. Or a mutation might result in a protein that has reduced efficiency but is not completely inactivated. Individuals homozygous for a point mutation of this type may show no change in phenotype if enough of the protein's function is retained.

LINK

Review **Concept 3.2** to better understand how a mutation affecting amino acid sequence (the primary protein structure) can affect higher levels of protein structure and thus the protein phenotype

In some cases, gain-of-function point mutations occur. An example is a class of mutations in the human gene *TP53*, which encodes the tumor suppressor protein p53. The p53 protein normally functions to inhibit the cell cycle, but certain mutations cause the protein to promote the cell cycle and prevent programmed cell death. So a p53 protein mutated in this way has a gain of oncogenic (cancer-causing) function.

Chromosomal mutations are extensive changes in the genetic material

In addition to point mutations there are other kinds of mutations that affect longer sequences of DNA. The most dramatic changes that can occur in the genetic material are **chromosomal mutations**. Whole chromosomes can break and rejoin, grossly disrupting the sequences of genes. There are four types of chromosomal mutations: deletions, duplications, inversions, and translocations. This kind of severe damage to chromosomes can result from mutagens or from drastic errors in chromosome replication. Like point mutations, chromosome mutations provide new combinations of genes and genetic diversity important to evolution by natural selection.

- **Deletions** result in the removal of part of the genetic material (**FIGURE 9.17A**). Their consequences can be severe or even fatal. It is easy to imagine one mechanism that could

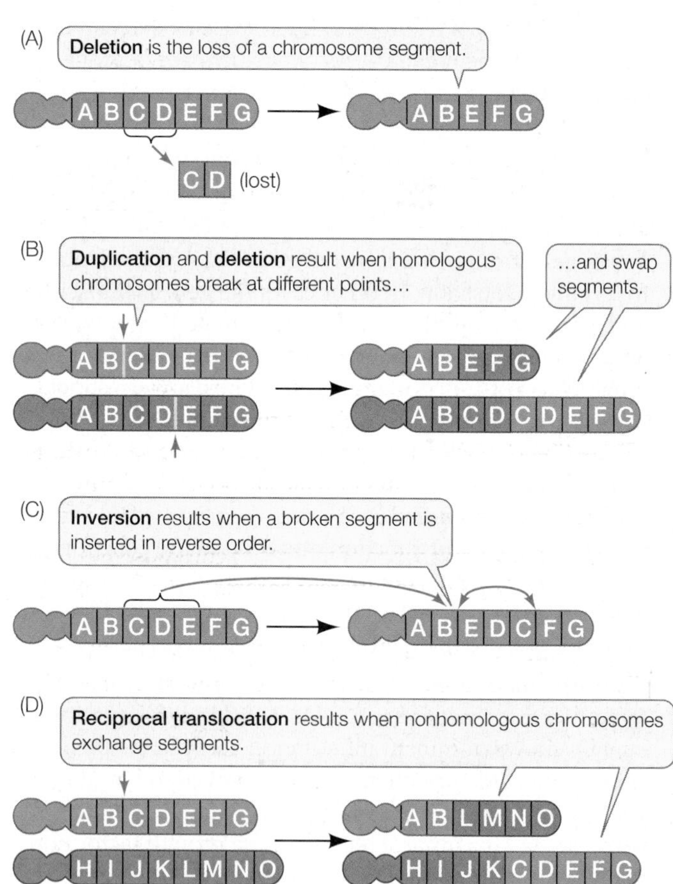

FIGURE 9.17 Chromosomal Mutations Chromosomes may break during replication, and parts of chromosomes may then rejoin incorrectly. The letters on these illustrations represent large chromosomal segments containing anywhere from zero to hundreds or thousands of genes.

APPLY THE CONCEPT

Mutations are heritable changes in DNA

Nitrosamines (R–N–NO_2) are potent mutagens. They can be formed from the reaction of nitrites (R–NO_2) with amino groups in proteins. So there are concerns about using nitrites to preserve meats, which contain amino groups. An experiment was performed to test the effect of vitamin C (ascorbate) on mutagenesis (induction of mutations) caused by meats cured with nitrites. Bacterial cells were incubated with cured meat extracts in the presence or absence of ascorbate. The rates of mutation (number of mutant bacteria per total bacteria) are shown in the table.

MEAT EXTRACT (μg/ml)	AMOUNT OF ASCORBATE (μg/ml)	RATE OF MUTATION ($\times 10^{-5}$)
0	0	2
0	50	2
10	0	5
10	50	2
20	0	14
20	50	2
30	0	35
30	50	2

1. What did the experiment with no extract and no ascorbate show?
2. What did the experiments with increasing amounts of extract and no ascorbate show?
3. What was the effect of ascorbate on the mutation rate?
4. In the bacterium tested, the wild-type DNA had the sequence 5′-ACTTAT-3′, and the mutated strain had the sequence 5′-ATTTAT-3′. What does this tell you about the nature of the mutation? Outline the steps in mutagenesis, noting DNA replication(s). (Hint: see Figure 9.18.)

produce a deletion: a DNA molecule might break at two points and the two end pieces might rejoin, leaving out the DNA between the breaks.

- **Duplications** can be produced at the same time as deletions (**FIGURE 9.17B**). A duplication would arise if homologous chromosomes broke at different positions and then reconnected to the wrong partners. One of the two chromosomes produced by this mechanism would lack a segment of DNA (it would have a deletion), and the other would have two copies (a duplication) of the segment that was deleted from the first chromosome.
- **Inversions** can also result from the breaking and rejoining of chromosomes. A segment of DNA may be removed and reinserted into the same location in the chromosome, but "flipped" end over end so that it runs in the opposite direction (**FIGURE 9.17C**). If either break site occurs within a gene, it is likely to cause a loss-of-function mutation in that gene.
- **Translocations** result when segments of chromosomes break off and become joined to different chromosomes. Translocations may involve reciprocal exchanges of chromosome segments, as in **FIGURE 9.17D**. Translocations often lead to duplications and deletions and may result in sterility if normal chromosome pairing cannot occur during meiosis.

LINK

Mobile DNA elements called transposons are another source of mutation; see **Figure 12.7**

Mutations can be spontaneous or induced

When thinking about the causes of mutations, it is useful to distinguish between mutations that are spontaneous and those that are induced.

Spontaneous mutations are permanent changes in the genetic material that occur without any outside influence. In other words, they occur simply because cellular processes are imperfect. Spontaneous mutations may occur by several mechanisms:

- *DNA polymerase can make errors in replication.* Most of these errors are repaired by the proofreading function of the replication complex, but some errors escape detection and become permanent.
- *The four nucleotide bases of DNA have alternate structures that affect base pairing.* Each nucleotide can exist in two different forms (called tautomers), one of which is common and one rare. When a base temporarily forms its rare tautomer, it can pair with the wrong base (**FIGURE 9.18A,C**).
- *Bases in DNA may change because of spontaneous chemical reactions.* One such reaction is the deamination (conversion of an amino group to a keto group) in cytosine to form the base uracil, which pairs with A rather than G. Usually these errors are repaired, but since the repair mechanism is not perfect, the altered nucleotide will sometimes remain and cause a permanent base change after replication.
- *Meiosis is not perfect.* Sometimes errors occur during the complex process of meiosis. This can result in nondisjunction and aneuploidy (see Concept 7.4) or chromosomal breakage and rejoining (discussed above).
- *Gene sequences can be disrupted.* Random chromosome breakage and rejoining can produce deletions, duplications, inversions, or translocations.

Induced mutations occur when some agent from outside the cell—a mutagen—causes a permanent change in the DNA sequence:

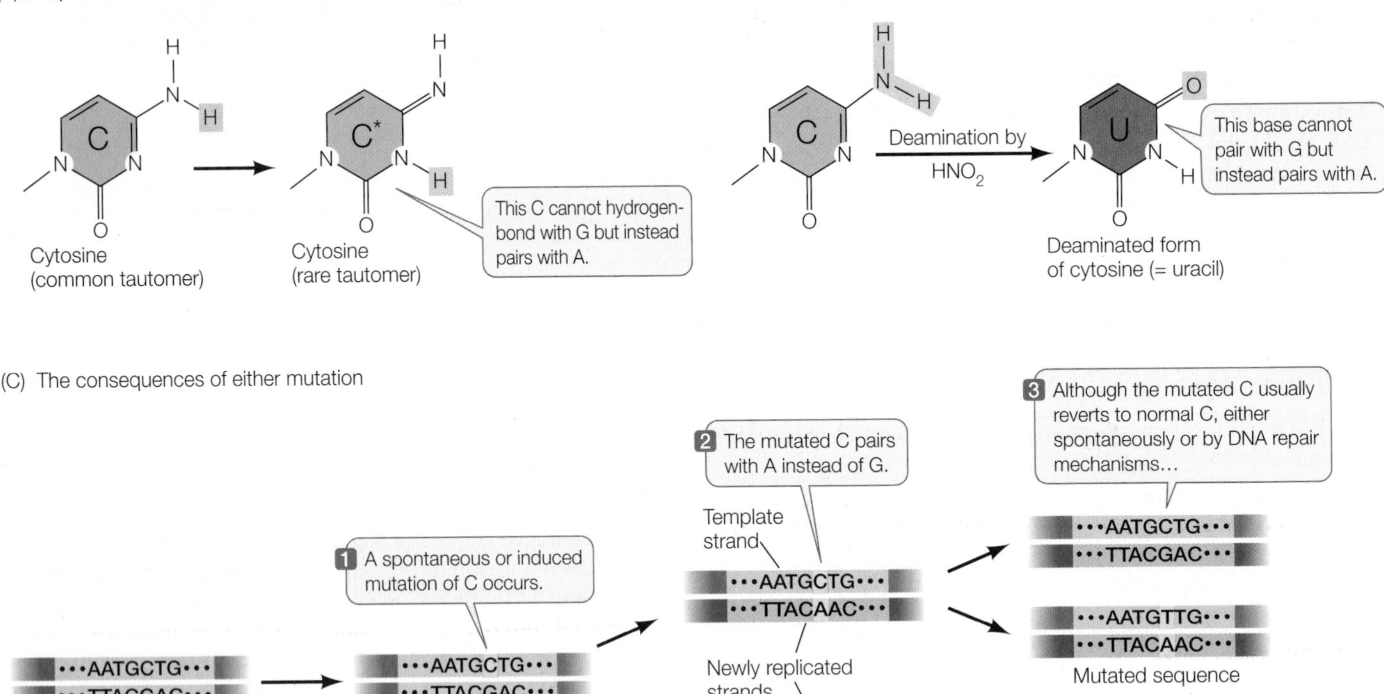

FIGURE 9.18 Spontaneous and Induced Mutations (A) Each of the nitrogenous bases exists in both a common (prevalent) form and a rare form. When a base spontaneously switches to its rare tautomer, it can pair with a different base. (B) Mutagens such as nitrous acid can induce changes in the bases. (C) The results of both spontaneous and induced mutations are permanent changes in the DNA sequence following replication.

- *Some chemicals alter the nucleotide bases.* For example, nitrous acid (HNO_2) reacts with cytosine and converts it to uracil by deamination (**FIGURE 9.18B**). This alteration has the same result as spontaneous deamination: instead of a G, DNA polymerase inserts an A (see Figure 9.18C).
- *Some chemicals add groups to the bases.* An example is benzopyrene, a component of cigarette smoke that adds a large chemical group to guanine, making it unavailable for base pairing. When DNA polymerase reaches such a modified guanine, it inserts any one of the four bases, resulting in a high frequency of mutations.

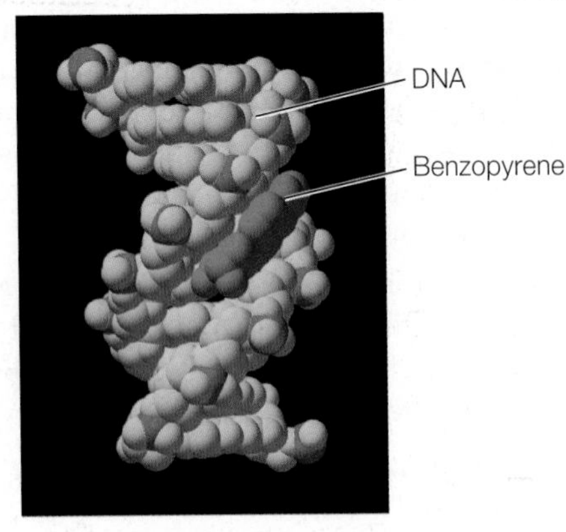

- *Radiation damages the genetic material.* Radiation can damage DNA in three ways. First, ionizing radiation (including X rays, gamma rays, and particles emitted by unstable isotopes) can detach electrons from atoms or molecules and produce highly reactive chemicals called free radicals. Free radicals can change bases in DNA to forms that are not recognized by DNA polymerase. Second, ionizing radiation can also break the sugar–phosphate backbone of DNA, causing chromosomal abnormalities. And third, ultraviolet radiation (from the sun or a tanning lamp) can cause thymine bases to form covalent bonds with adjacent thymines. This, too, plays havoc with DNA replication by distorting the double helix, and can result in a mutation.

Some base pairs are more vulnerable than others to mutation

DNA sequencing has revealed that mutations occur most often at certain base pairs. These "hotspots" are often located where cytosine has been methylated to 5-methylcytosine. (Methylation of DNA is a normal process in the regulation of chromatin structure and gene expression; see Concept 11.3.)

As we discussed above, unmethylated cytosine can lose its amino group to form uracil (see Figure 9.18B). This error is usually repaired because uracil (which normally occurs in

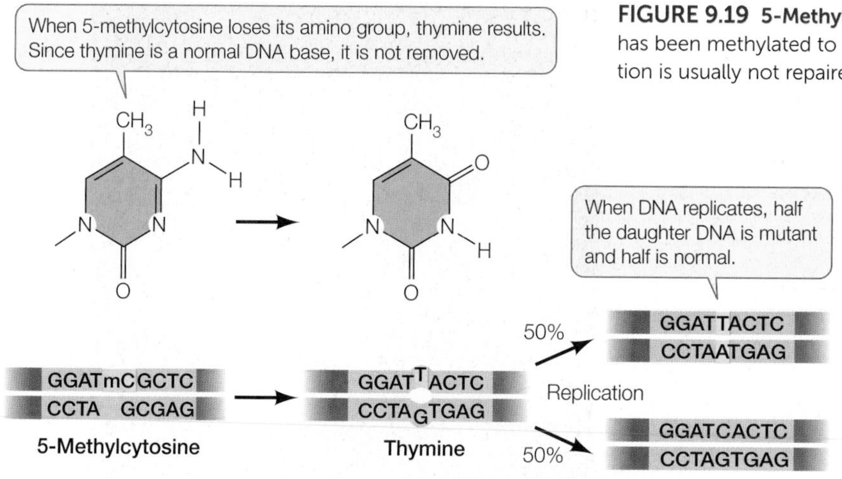
When 5-methylcytosine loses its amino group, thymine results. Since thymine is a normal DNA base, it is not removed.
When DNA replicates, half the daughter DNA is mutant and half is normal.
GGATmCGCTC
CCTA GCGAG
5-Methylcytosine
GGATTACTC
CCTAGTGAG
Thymine
Replication
50%
GGATTACTC
CCTAATGAG
50%
GGATCACTC
CCTAGTGAG

FIGURE 9.19 5-Methylcytosine Is a "Hotspot" for Mutations If a cytosine in DNA has been methylated to 5-methylcytosine and then becomes deaminated, the mutation is usually not repaired, and a C-G base pair is replaced with a T-A base pair.

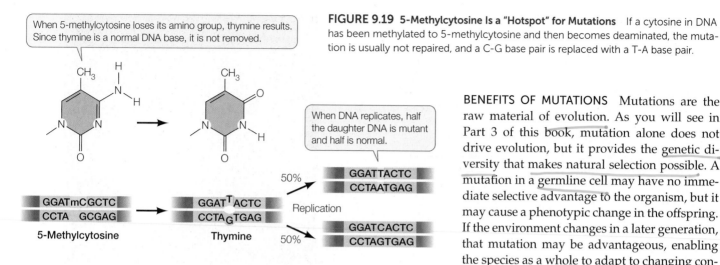

RNA) is recognized by the repair mechanism as an inappropriate component of DNA. When 5-methylcytosine loses its amino group, however, the product is thymine, a natural base in DNA. The DNA repair mechanism ignores the thymine, and in this case, half of the new DNA molecules contain A-T base pairs instead of the original G-C pair (**FIGURE 9.19**). Thus the frequency of mutation is much greater in regions containing 5-methylcytosine.

Mutagens can be natural or artificial

Many people associate mutagens with materials made by humans, but there are also many naturally occurring mutagens. Plants (and to a lesser extent animals) make thousands of small molecules that serve a range of purposes, including defense against pathogens (see Concept 28.1). Some of these are mutagenic and potentially carcinogenic. An example of a naturally occurring mutagen is aflatoxin, which is made by many species of the mold *Aspergillus*. When mammals ingest the mold, the aflatoxin is converted into a product that binds to guanine and causes mutations, just as benzopyrene from cigarette smoke does.

Radiation can be human-made or natural. Some of the isotopes made in nuclear reactors and nuclear bomb explosions are certainly harmful—as was shown by the increased mutation rates in survivors of the atom bombs dropped on Japan in 1945. A certain amount of radiation comes from space in the form of cosmic radiation, and as mentioned above, natural ultraviolet radiation in sunlight also causes mutations. There is a well-established link between excessive exposure to the sun in light-skinned people and skin cancer.

Biochemists have estimated how much DNA damage occurs in the human genome under normal circumstances: among the haploid genome's 3.2 billion base pairs, there are about 16,000 DNA-damaging events per cell per day, of which 80 percent are repaired.

Mutations have both benefits and costs

What is the overall effect of mutation? For a species as a whole, the evolutionary benefits are clear. But there are costs as well as benefits for individual organisms.

BENEFITS OF MUTATIONS Mutations are the raw material of evolution. As you will see in Part 3 of this book, mutation alone does not drive evolution, but it provides the genetic diversity that makes natural selection possible. A mutation in a germline cell may have no immediate selective advantage to the organism, but it may cause a phenotypic change in the offspring. If the environment changes in a later generation, that mutation may be advantageous, enabling the species as a whole to adapt to changing conditions. A mutation in a somatic cell can sometimes benefit the individual organism, particularly if it occurs in a stem cell that produces a large number of offspring cells.

COSTS OF MUTATIONS Mutations can be harmful if they result in the loss of function of genes (and their protein products) or other DNA sequences that are needed for survival. A harmful mutation in a germline cell may be inherited in heterozygous form by the organism's descendants. If two individuals carrying the mutation mate, some of the offspring may be homozygous for the mutation. In their extreme form, such mutations produce phenotypes that are lethal. Lethal mutations can kill an organism during early development, or the organism may die before it matures and reproduces.

In Chapter 7 we described how mutations in somatic cells can lead to cancer. Typically these are mutations in oncogenes that result in the stimulation of cell division, or mutations in tumor suppressor genes that result in a lack of inhibition of cell division. These mutations can occur spontaneously or they can be induced.

We attempt to minimize our exposure to mutagens

Spontaneous mutagenesis is not in our control, but we can certainly try to avoid mutagenic substances and radiation. Not surprisingly, many things that cause cancer (carcinogens) are also mutagens. A good example is benzopyrene (discussed above), which is found in coal tar, car exhaust fumes, and charbroiled foods, as well as in cigarette smoke.

A major public policy goal is to minimize the effects of both human-made (anthropogenic) and natural mutagens on human health. For example, the Montreal Protocol (the only international environmental agreement signed and adhered to by all nations) bans chlorofluorocarbons (CFCs) and other substances that deplete the ozone layer in Earth's upper atmosphere. The ozone layer screens out ultraviolet radiation from the sun—radiation that can cause somatic mutations that lead to skin cancer. Similarly, bans on cigarette smoking have rapidly spread throughout the world. Cigarette smoking causes cancer because of the increased exposure of lung and other cells to benzopyrene and other carcinogens.

CHECKpoint CONCEPT 9.3

- ✓ What are the differences between a germline mutation and a somatic mutation?
- ✓ Describe how the same base—cytosine—can be mutated spontaneously or by a mutagen. In which case would the mutation be efficiently repaired?
- ✓ The *Bar eyes* mutation in fruit flies was found in genetic crosses to be inherited as an autosomal (non-X chromosome) dominant. Would you expect the mutant allele to be a loss-of-function or a gain-of-function mutation? Explain your answer.

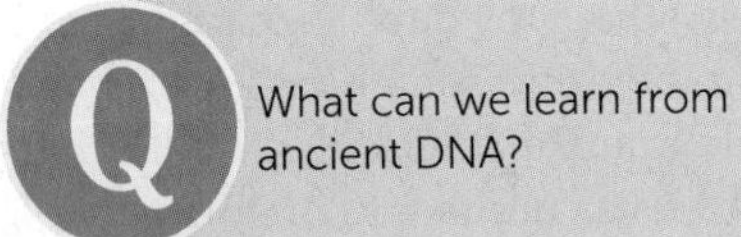

What can we learn from ancient DNA?

ANSWER As we described in the opening story of this chapter, studying the DNA of fossils is a challenge. DNA is a large polymer that is rather stable (Concept 9.1). However, during the formation of fossils, most of the soft tissues and cells are broken down and their contents consumed by organisms in the environment. So it is not surprising that research on ancient DNA has been focused on places where DNA is likely to be preserved intact, such as frozen specimens and the interior of bones.

The polymerase chain reaction (PCR; Concept 9.2) has been invaluable in amplifying tiny amounts of ancient DNA to be sequenced. However, even a single molecule of contaminating DNA can ruin the experiment, because it too will be amplified. This has been a major challenge in studies of ancient DNA related to humans.

The Neandertal Genome Project involves an international team of scientists who are extracting DNA from the bones of skeletons of Neanderthals who lived in Europe more than 50,000 years ago, amplifying it by PCR, and then examining the genes. The entire DNA sequence has been completed. It is more than 99 percent identical to our human DNA, justifying the classification of Neanderthals as part of the same genus, *Homo*.

Comparisons of humans and Neanderthals with regard to specific genes and mutations (Concept 9.3) is ongoing and has already shown several interesting facts. For example:

- The gene *MC1R* is involved in skin and hair pigmentation. A point mutation found in Neanderthals but in not humans caused lower activity of the MC1R protein when it was induced in cell cultures. Such lower activity of MC1R is known to result in fair skin and red hair in humans. So it appears that at least some Neanderthals may have had pale skin and red hair (**FIGURE 9.20**).
- The gene *FOXP2* is involved in vocalization in many organisms, including birds and mammals. Mutations in this gene result in severe speech impairment in humans. The Neanderthal *FOXP2* gene is identical to that of humans, whereas that of chimpanzees is slightly different. This has led to speculation that Neanderthals may have been capable of speech.

FIGURE 9.20 A Neanderthal Child This reconstruction of a Neanderthal child who lived about 60,000 years ago was made using bones recovered at Gibraltar, as well as phenotypic projections made from DNA analyses.

While the two genome sequences are very similar, there are differences in many point mutations and larger chromosomal arrangements (Concept 9.3). There are distinctive "human" DNA sequences and also distinctive "Neanderthal" sequences. Some Neanderthal sequences have been found in modern humans, indicating that ancient humans and Neanderthals may have interbred. Such reconstructions of ancient DNAs and their phenotypic expression are being repeated for many other species. These studies underline the universality of DNA as the genetic material.

SUMMARY

CONCEPT 9.1 DNA Structure Reflects Its Role as the Genetic Material

- Circumstantial evidence for DNA as the genetic material includes its presence in the nucleus, its doubling during S phase of the mitotic cell cycle, and its injection into host cells by viruses. **Review Figures 9.1 and 9.2**
- Experimental evidence for DNA as the genetic material is provided by the **transformation** of one genotype into another by adding DNA. **Review Figure 9.3**
- In DNA, the amount of A equals the amount of T and the amount of G equals the amount of C. This observation, along with X-ray crystallography data, helped Watson and Crick unravel the **helical** structure of DNA. **Review Figures 9.4 and 9.5**

CONCEPT 9.2 DNA Replicates Semiconservatively

- DNA exhibits **semiconservative replication**. Each parent strand acts as a **template** for the synthesis of a new strand; thus the two replicated DNA molecules each contain one parent strand (the template) and one newly synthesized strand. **Review ANIMATED TUTORIALS 9.1 and 9.2**
- In DNA replication, the enzyme **DNA polymerase** catalyzes the addition of nucleotides to the 3′ end of each new strand. **Review Figure 9.7**
- Replication proceeds in both directions from the **origin of replication**. The parent DNA molecule unwinds to form a **replication fork**. **Review Figure 9.8 and ANIMATED TUTORIAL 9.3**
- **Primase** catalyzes the synthesis of a short RNA **primer** to which nucleotides are added by DNA polymerase. **Review Figure 9.9 and ACTIVITY 9.1**
- The **leading strand** is synthesized continuously. The **lagging strand** is synthesized in pieces called **Okazaki fragments**. The fragments are joined together by **DNA ligase**. **Review Figures 9.11 and 9.12 and ANIMATED TUTORIAL 9.4**
- Eukaryotic chromosomes have repetitive sequences at each end called **telomeres**. DNA replication leaves a short, unreplicated sequence at the 5′ end of each new DNA strand. Unless the enzyme **telomerase** is present, the sequence is removed. After multiple cell cycles the telomeres shorten, leading to chromosome instability and cell death. **Review Figure 9.13**
- DNA polymerases make errors, which can be repaired by proofreading and mismatch repair. **Review Figure 9.14**
- The **polymerase chain reaction**, or **PCR** technique uses DNA polymerase to make multiple copies of DNA in the laboratory. **Review Figure 9.15 and ANIMATED TUTORIAL 9.5**

CONCEPT 9.3 Mutations Are Heritable Changes in DNA

- **Somatic mutations** occur in the body cells of an individual and are passed on to daughter cells during mitosis. Only **germline mutations** (mutations in the cells that give rise to gametes) can be passed on to sexually produced offspring.
- **Point mutations** are alterations in single base pairs of DNA. **Silent mutations** can occur in genes or nontranscribed regions and do not affect the amino acid sequences of proteins. A mutation in a protein-coding region can lead to an alteration in the amino acid sequence of the protein. **Review Figure 9.16**
- Chromosomal mutations (**deletions**, **duplications**, **inversions**, and **translocations**) involve large regions of chromosomes. **Review Figure 9.17**
- **Spontaneous mutations** occur because of instabilities in DNA or chromosomes. **Induced mutations** occur when a mutagen damages DNA. **Review Figure 9.18**

Go to the Interactive Summary to review key figures, Animated Tutorials, and Activities PoL2e.com/is9

Go to LaunchPad at **macmillanhighered.com/launchpad** for additional resources, including LearningCurve Quizzes, Flashcards, and many other study and review resources.

10 From DNA to Protein: Gene Expression

KEY CONCEPTS

Although these *Staphylococcus aureus* cells look like normal bacteria, they have genes for resistance to multiple antibiotics and are difficult to eradicate. Antibiotic-resistant bacteria pose an ever increasing challenge to public health.

Humans have more prokaryotic cells on and in their bodies than they have eukaryotic cells of their own. Among the billions of bacteria that inhabit the skin and noses of many people is *Staphylococcus aureus*. Healthy people can carry this bacterium without symptoms, but sometimes, especially when the immune system has been weakened by age or disease, *S. aureus* can cause major skin infections and may even enter the body through the nose or a wound site. In these cases, much more serious infections of organs such as the heart and lungs can occur, and may result in death.

Until recently, most *S. aureus* infections were successfully treated with penicillin and related drugs, including methicillin. These antibiotics bind and inactivate several related enzymes (called penicillin-binding proteins) that are involved in the assembly of bacterial cell walls. Bacteria treated with these antibiotics have defective cell walls, and because of this, new cells cannot survive after cell division. Unfortunately, some *S. aureus* strains have acquired mutant versions of a penicillin-binding protein that can catalyze the assembly of cell walls in the presence of the antibiotics, thus conferring antibiotic resistance to these strains. The mutant penicillin-binding protein has an altered shape that doesn't bind the antibiotics. This protein is encoded by the *mecA* gene, which can be passed from one bacterium to another by bacterial conjugation (see Concept 8.4). At a more general level, the mutant phenotype demonstrates that *a gene is expressed as a protein*.

By the late 1990s these bacterial strains were being called "superbugs," with the formal name "methicillin-resistant *S. aureus*," or MRSA. The first decade of the new millennium saw a dramatic rise in MRSA infections. At first, most cases occurred in hospitals and nursing homes, but more recently MRSA has occurred in communities as well. Resistant strains have a selective advantage because of the extensive use of antibiotics in health care. With close to 100,000 serious MRSA infections and 20,000 deaths in the United States each year, more people are dying from MRSA than from AIDS.

MRSA can be treated if detected early. Antibiotics such as tetracycline, which targets bacterial protein synthesis, can be effective in some strains. But there is reasonable concern that MRSA may become resistant to these antibiotics as well.

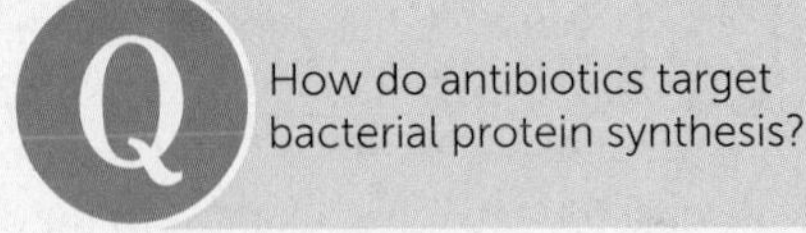

How do antibiotics target bacterial protein synthesis?

You will find the answer to this question on page 213.

CONCEPT 10.1 Genetics Shows That Genes Code for Proteins

Following Mendel's definition of the gene as a physically distinct entity (see Concept 8.1), biologists identified the genetic material as DNA (see Concept 9.1). In this chapter we will show that in most cases, genes code for proteins, and it is proteins that determine phenotypes. The connection between protein and phenotype was made before it was known that DNA is the genetic material.

Observations in humans led to the proposal that genes determine enzymes

The identification of a gene product as a protein began with a mutation. In the early twentieth century, the English physician Archibald Garrod saw several children with a rare disease. One symptom was that the urine turned dark brown or black in air, and for this reason the disease was named alkaptonuria ("black urine").

Garrod noticed that the disease was most common in children whose parents were first cousins. Mendelian genetics had just been "rediscovered," and Garrod realized that because first cousins can inherit some alleles that are the same from their shared grandparents, their children are more likely than others to inherit a rare mutant allele from both parents—and therefore are more likely to be homozygous recessive for rare genetic conditions (see Figure 8.8B). Garrod proposed that alkaptonuria was a phenotype caused by a recessive mutant allele.

LINK

You can review the inheritance patterns of recessive alleles in **Concept 8.1**; see especially **Figure 8.8**

Garrod took the analysis a step further by identifying the biochemical abnormality in the affected children. He isolated from them an unusual substance, homogentisic acid, which accumulated in the blood, joints (where it crystallized and caused severe pain), and urine (where it turned black when exposed to air).

Enzymes as biological catalysts had just been discovered, and Garrod proposed that in healthy individuals, homogentisic acid might be broken down to a harmless product by an enzyme:

Normal allele
↓
Active enzyme
↓
Homogentisic acid ⟶ harmless product

Garrod speculated that the synthesis of the active enzyme is determined by the dominant wild-type allele of the gene that was mutated in alkaptonuria patients. These and other studies led him to correlate one gene to one enzyme, and to coin the term "inborn error of metabolism" to describe this kind of genetically determined biochemical disease.

But Garrod's hypothesis needed direct confirmation by the identification of the specific enzyme and the specific gene mutation involved. In 1958 the enzyme was identified as homogentisic acid oxidase, which breaks down homogentisic acid to a harmless product, just as Garrod predicted (**FIGURE 10.1**). The specific DNA mutation leading to alkaptonuria was described in 1996.

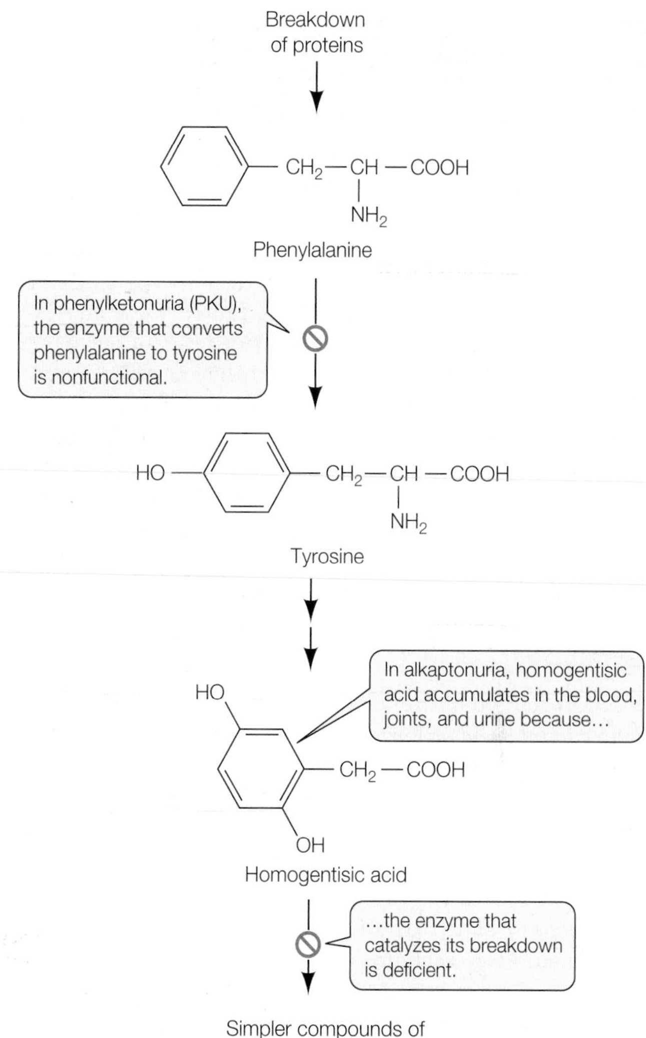

FIGURE 10.1 Metabolic Diseases and Enzymes Both phenylketonuria and alkaptonuria are caused by abnormalities in specific enzymes in a pathway that breaks down proteins.

Homogentisic acid is part of a biochemical pathway that catabolizes proteins, with the amino acids phenylalanine and tyrosine as intermediate products. Phenylketonuria, another genetic disease involving the same pathway, was discovered several decades after Garrod did his work. In phenylketonuria, the enzyme that converts phenylalanine to tyrosine is nonfunctional (see Figure 10.1). If left untreated, this disease leads to significant intellectual disability. Fortunately, the accumulation of phenylalanine can be easily detected in the blood of a newborn infant, and if the child consumes a diet low in proteins containing phenylalanine, intellectual disability is avoided.

The concept of the gene has changed over time

The phenotypic expression of mutations underlying alkaptonuria and phenylketonuria led to the "one gene–one enzyme" hypothesis. Once it was known that proteins (including enzymes) are polymers of amino acids, and that the sequence

APPLY THE CONCEPT

Genetics shows that genes code for proteins

Wild-type bacteria can synthesize the amino acid tryptophan (T) using a biochemical pathway that begins with chorismate (C) and involves four intermediates that we will call D, E, F, and G. Bacterial strains with mutant alleles for the five enzymes (1–5) involved in this pathway cannot synthesize tryptophan, and it must be supplied as a nutrient in the growth medium. The table gives the phenotypes of five mutant strains, each of which has a mutation in a gene for a different enzyme of the five. A "+" means the strain grew when the indicated compound was added to the medium, and a "0" means the strain did not grow. Based on these data, order the compounds (C, D, E, F, G, and T) and the enzymes (1, 2, 3, 4, and 5) in a biochemical pathway, as in the (incorrect) depiction here:

$$C \xrightarrow{1} D \xrightarrow{2} E \xrightarrow{3} F \xrightarrow{4} G \xrightarrow{5} T$$

Hint: Mutant strain 5 will not grow if any compound other than T is supplied, so it must carry a loss-of-function mutation in the enzyme that transforms another compound (either C, D, E, F, or G) into T. Thus enzyme 5 is the final enzyme in the pathway.

MUTANT STRAIN	ADDITION TO THE MEDIUM					
	C	D	E	F	G	T
1	0	0	0	0	+	+
2	0	+	+	0	+	+
3	0	+	0	0	+	+
4	0	+	+	+	+	+
5	0	0	0	0	0	+

of amino acids determines protein function, it became clear that *a mutant phenotype arises from a change in the protein's amino acid sequence*. However, scientists soon realized that the one gene–one enzyme (or one protein) hypothesis was an oversimplification. Once again, studies of human mutations were a key to this realization.

In humans, the oxygen-carrying protein hemoglobin has a quaternary structure of four polypeptide chains—two α-chains and two β-chains (see Concept 3.2, p. 47). Concept 9.3 introduced sickle-cell disease, which is caused by a point mutation in the gene for β-globin and is inherited as an autosomal recessive (carried on an autosome rather than a sex chromosome). In sickle-cell disease, one of the 147 amino acids in the β-globin chain is abnormal: at position 6, the normal glutamic acid has been replaced by valine. This replacement changes the charge of the protein (glutamic acid is negatively charged and valine is neutral), causing it to form long, needlelike aggregates in the red blood cells. The phenotypic result is anemia, an impaired ability of the blood to carry oxygen.

Because hemoglobin is easy to isolate and study, its variations in the human population have been extensively documented (**FIGURE 10.2**). Hundreds of single amino acid alterations in β-globin have been reported. For example, at the same position that is mutated in sickle-cell disease (resulting in hemoglobin S), the normal glutamic acid may be replaced by lysine, causing hemoglobin C disease. In this case, the resulting anemia is usually not severe. Many alterations of hemoglobin do not affect the protein's function. That is fortunate, because about 5 percent of all humans are carriers for one of these variants.

Studies of proteins that are made up of multiple polypeptides (such as hemoglobin) resulted in a modification of the one gene–one enzyme hypothesis. Scientists began to think of the relationship as **one gene–one polypeptide**, which remains a powerful and useful concept today. However, as you will see in this and later chapters, we are learning that this, too, is an oversimplification.

LINK

Exceptions to the one gene–one polypeptide relationship include the alternative splicing of RNA, which can produce multiple functional polypeptides from a single gene; see **Concept 11.4**

Mutations such as those that cause alkaptonuria and phenylketonuria result in alterations in amino acid sequences. But not all genes code for polypeptides. As we will see below and in Chapter 11, some DNA sequences are transcribed into RNA molecules that are *not* translated into polypeptides, but instead have other functions. Like all other DNA sequences, these RNA genes are subject to mutations, which may or may not affect the functions of the RNAs they produce.

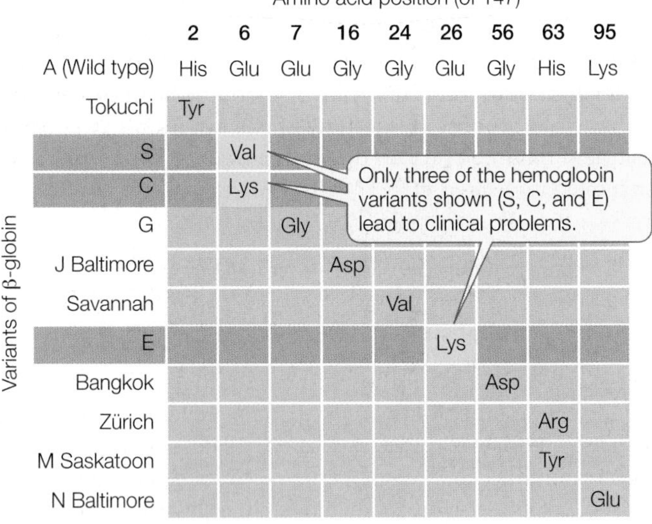

FIGURE 10.2 Gene Mutations and Amino Acid Changes Each of these mutant alleles (e.g., hemoglobin Tokuchi) codes for a β-globin polypeptide with an alteration in one of its 147 amino acids (e.g., a change from His to Tyr at position 2 of the polypeptide).

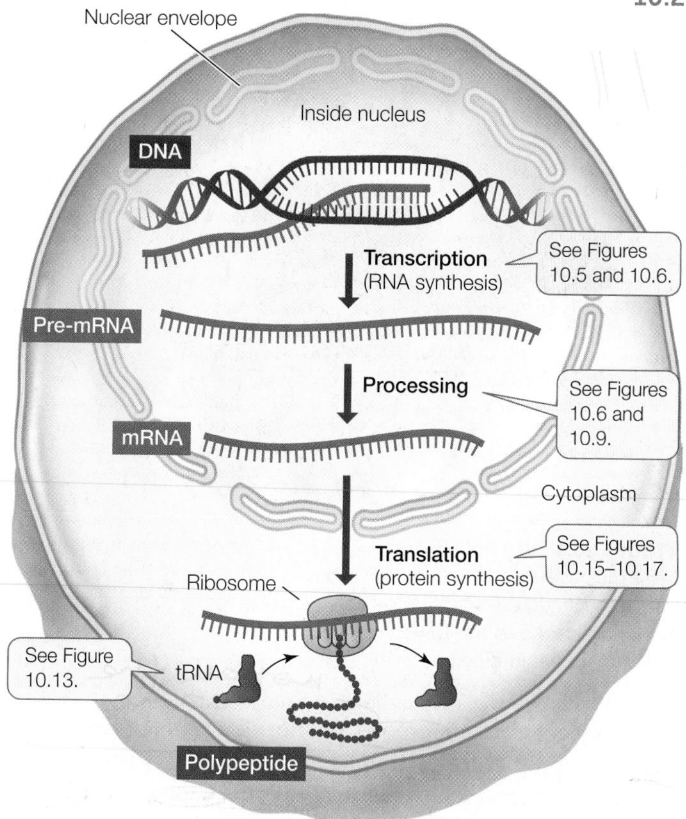

FIGURE 10.3 From Gene to Protein This diagram summarizes the processes of gene expression in eukaryotes.

Go to ACTIVITY 10.1 Eukaryotic Gene Expression
PoL2e.com/ac10.1

Our understanding of genes and how they are expressed has increased dramatically over the past 60 years, since Watson and Crick first worked out the structure of DNA (see Concept 9.1). Let's begin our discussion of gene expression with an overview of the kinds of RNA molecules that are involved in transcription and translation.

Genes are expressed via transcription and translation

Molecular biology is the study of nucleic acids and proteins, and it often focuses on gene expression. As we described briefly in Chapter 3, genes are expressed as RNAs, many of which are translated into proteins. This process involves two steps:

- During **transcription**, the information in a DNA sequence (a gene) is copied into a complementary RNA sequence.
- During **translation**, this RNA sequence is used to create the amino acid sequence of a polypeptide.

DNA —Transcription→ RNA —Translation→ Polypeptide

Information coded in the DNA provides instructions for assembling nucleotide bases into RNA.

Information coded in the RNA provides instructions for assembling amino acids into polypeptides.

Here we will consider three types of RNA with regard to their roles in protein synthesis:

- *Messenger RNA and transcription*: When a particular gene is expressed, the two strands of DNA unwind and separate into a **coding strand** and a **template strand**. The template strand is then transcribed to produce an RNA strand by complementary base pairing. The RNA strand is then modified to produce a **messenger RNA (mRNA)**. In eukaryotic cells, the mRNA is processed in the nucleus and then moves to the cytoplasm, where it is translated into a polypeptide (**FIGURE 10.3**). The nucleotide sequence of the mRNA determines the ordered sequence of amino acids in the polypeptide chain, which is built by a ribosome.
- *Ribosomal RNA and translation:* The **ribosome** is essentially a protein synthesis factory with multiple proteins and several **ribosomal RNAs (rRNAs)**. One of the rRNAs catalyzes peptide bond formation between amino acids, to form a polypeptide.
- *Transfer RNA mediates between mRNA and protein:* A third kind of RNA called **transfer RNA (tRNA)** can both bind a specific amino acid and recognize a specific sequence of nucleotides in mRNA, by complementary base pairing (A with U, and G with C). It is the tRNA that recognizes which amino acid should be added next to a growing polypeptide chain (see Figure 10.3).

In Chapter 11 we will consider other RNAs, which play roles in the regulation of gene expression.

CHECKpoint CONCEPT 10.1

- ✓ What is the difference between the "one gene–one protein" and "one gene–one polypeptide" hypotheses?
- ✓ What is the difference between gene transcription and translation?
- ✓ Could a person inherit homozygous recessive alleles for both alkaptonuria and phenylketonuria? If so, what would the symptoms be?
- ✓ Defining phenotype as the presence of a polypeptide chain of a particular amino acid sequence, would you expect the Zürich variant of β-globin (see Figure 10.2) to be inherited as a dominant, recessive, or codominant? Explain your answer.

In this section we have shown how the connection between genes and phenotypes can be understood in terms of DNA and proteins. We will now turn to some details of the process of gene expression, which is at the heart of what genes do.

CONCEPT 10.2 DNA Expression Begins with Its Transcription to RNA

Transcription—the formation of a specific RNA sequence from a specific DNA sequence—requires several key components:

- A DNA template for complementary base pairing
- The four ribonucleoside triphosphates (ATP, GTP, CTP, and UTP) to act as substrates
- An RNA polymerase enzyme

The same transcription process is responsible for the synthesis of mRNA, tRNA, and rRNA. Like mRNA, tRNA and rRNA are encoded by specific genes; their important roles in protein synthesis will be described in Concepts 10.3 and 10.4. There are also other kinds of RNA in the cell, with functions other than protein synthesis.

LINK

Small nuclear RNAs are involved in processing mRNA after it is transcribed, and microRNAs play important roles in stimulating or inhibiting gene expression; see **Concept 11.4**

RNA polymerases share common features

RNA polymerases from both prokaryotes and eukaryotes catalyze the synthesis of RNA from the DNA template. There is only one kind of RNA polymerase in bacteria and archaea, whereas there are several kinds in eukaryotes. However, they all share a common structure (**FIGURE 10.4**). Like DNA polymerases, RNA polymerases are processive; that is, a single enzyme–template binding event results in the polymerization of hundreds of RNA nucleotides. But unlike DNA polymerases, RNA polymerases do not require a primer.

Transcription occurs in three steps

Transcription can be divided into three distinct processes: initiation, elongation, and termination. You can follow these processes in **FIGURE 10.5**.

INITIATION Transcription begins with initiation, which requires a **promoter**, a special DNA sequence to which the RNA polymerase binds very tightly (see Figure 10.5A). Promoters are control sequences that "tell" the RNA polymerase two crucial things:

- Where to start transcription
- Which of the two DNA strands to transcribe and under what conditions

The promoter has a nucleotide sequence that can be "read" in a particular direction and orients the RNA polymerase, thus "aiming" it at the correct strand to use as a template. Part of each promoter is the **transcription initiation site**, where transcription begins. Groups of nucleotides lying "upstream" from the initiation site (5′ on the coding strand and 3′ on the template strand) are bound by other proteins, which help the RNA polymerase bind. These other proteins, called sigma factors and transcription factors, help determine which genes are expressed at a particular time in a particular cell.

LINK

The roles of sigma factors and transcription factors are described in **Concepts 11.1 and 11.2**

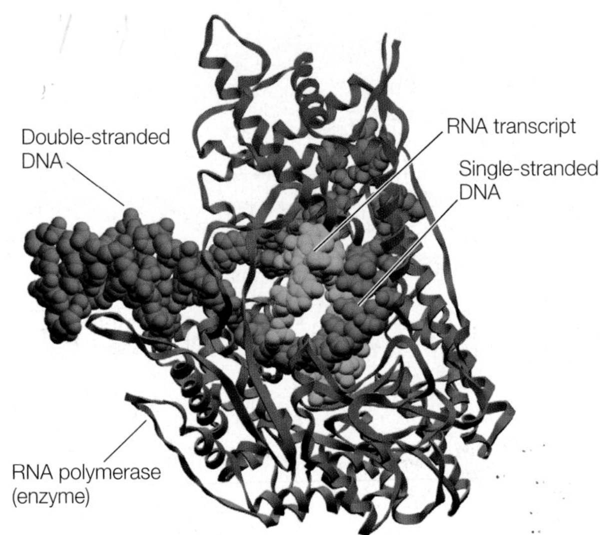

FIGURE 10.4 RNA Polymerase This enzyme from a virus is smaller than most other RNA polymerases, but its active site is similar to those of bacterial and eukaryotic RNA polymerases. The enzyme, shown in brown, is bound to the DNA it is transcribing (blue). The RNA transcript is shown in green.

Every gene has a promoter, but not all promoters are identical; some promoters are more effective at transcription initiation than others. Furthermore, bacteria, archaea, and eukaryotes differ in the details of transcription initiation. Despite these variations, the basic mechanisms of initiation are the same throughout the living world and provide further evidence of the biochemical unity of life on Earth.

ELONGATION Once RNA polymerase has bound to the promoter, it begins the process of **elongation** (see Figure 10.5B). RNA polymerase unwinds the DNA about 13 base pairs at a time and reads the template strand in the 3′-to-5′ direction. Like DNA polymerase, RNA polymerase adds new nucleotides to the 3′ end of the growing strand, beginning with the first nucleotide at the transcription initiation site. Thus the first nucleotide in the new RNA forms its 5′ end, and the RNA transcript is antiparallel to the DNA template strand.

RNA polymerase adds new nucleotides to the RNA molecule by complementary base pairing with nucleotides in the template strand of the DNA. This process is similar to DNA replication except that the base uracil (rather than thymine) in the RNA molecule is paired with adenine in the DNA molecule. In a mechanism very similar to that used by DNA polymerase, the RNA polymerase uses the ribonucleoside triphosphates ATP, UTP, GTP, and CTP as substrates and catalyzes the formation of phosphodiester bonds between them, releasing pyrophosphate in the process (see Figure 9.7). As transcription progresses, the two DNA strands rewind and the RNA grows as a single-stranded molecule (see Figure 10.5B).

Like DNA polymerases, RNA polymerases and associated proteins have mechanisms for proofreading during transcription, but these mechanisms are not as efficient as those for DNA. Transcriptional errors occur at rates of about 1 for every 10^4 to 10^5 bases. Because many copies of RNA are made, however, and because they often have relatively short life spans, these errors are not as potentially harmful as mutations in DNA.

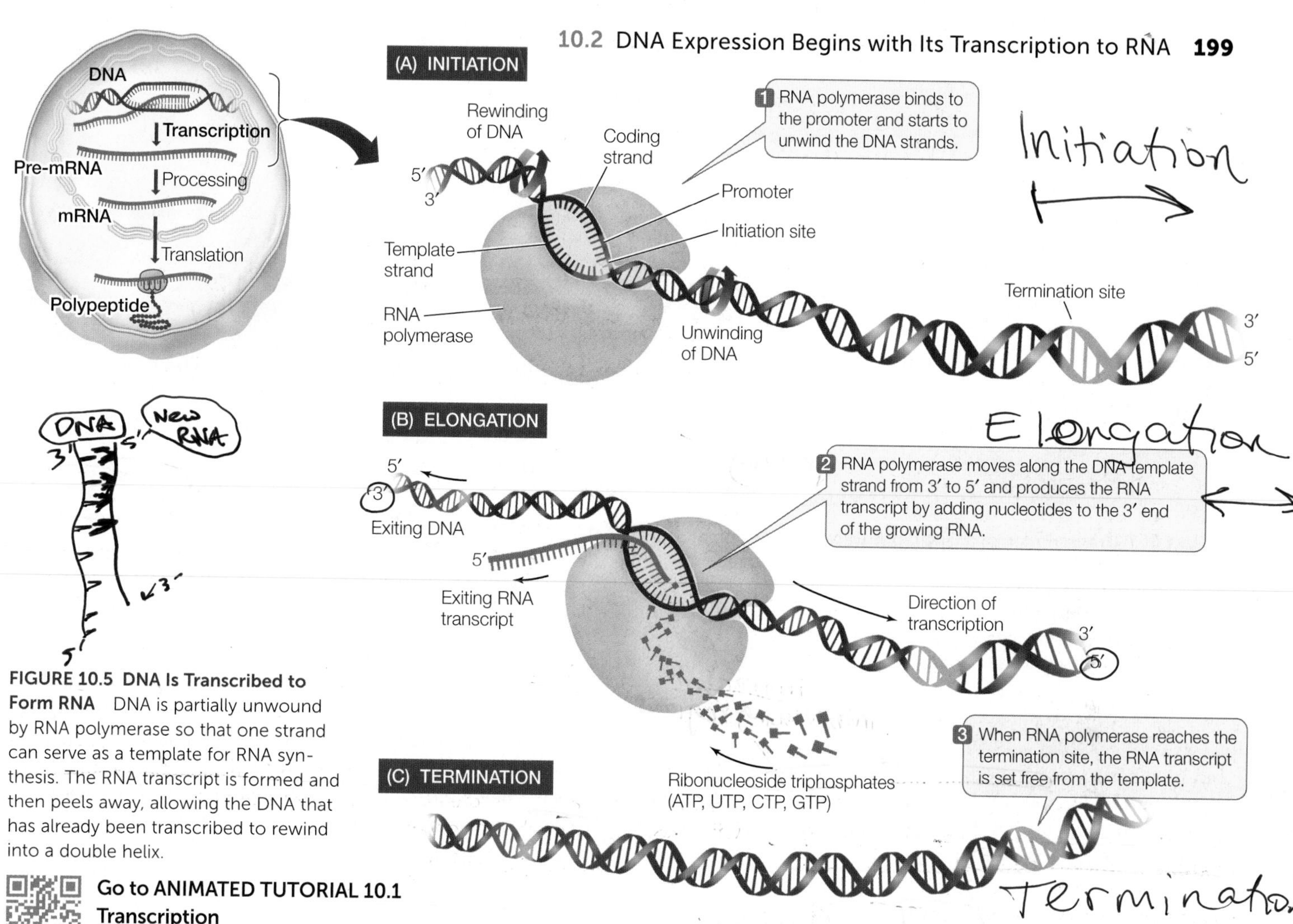

FIGURE 10.5 DNA Is Transcribed to Form RNA DNA is partially unwound by RNA polymerase so that one strand can serve as a template for RNA synthesis. The RNA transcript is formed and then peels away, allowing the DNA that has already been transcribed to rewind into a double helix.

Go to ANIMATED TUTORIAL 10.1
Transcription
PoL2e.com/at10.1

TERMINATION Just as initiation sites in the DNA template strand specify the starting point for transcription, particular base sequences specify its **termination** (see Figure 10.5C). The mechanisms of termination are complex and vary among different genes and organisms. In eukaryotes, multiple proteins are involved in recognizing the transcription termination site and separating the newly formed RNA strand from the DNA template and the RNA polymerase.

Eukaryotic coding regions are often interrupted by introns

Coding regions are sequences within a DNA molecule that are eventually translated as proteins. The coding region on the DNA template strand is transcribed into a complementary mRNA molecule, which has the same base sequence (with U's instead of T's) as the DNA coding strand. In prokaryotes, most of the genomic DNA is made up of coding regions, and the mRNAs are usually co-linear with them. That is, the mRNA sequence (e.g., 5′-AUGAUAGCCCC....) can be found in the DNA coding strand (e.g., 5′-ATGATAGCCCC....) without interruptions. In eukaryotes the situation is often different (**TABLE 10.1**).

A diagram of the structure and transcription of a typical eukaryotic gene is shown in **FIGURE 10.6**. In prokaryotes and viruses several adjacent genes sometimes share one promoter,

TABLE 10.1 Differences between Prokaryotic and Eukaryotic Gene Expression

Characteristic	Prokaryotes	Eukaryotes
Transcription and translation occurrence	At the same time in the cytoplasm	Transcription in the nucleus, then translation in the cytoplasm
Gene structure	DNA sequence usually not interrupted by introns	Transcribed regions often interrupted by noncoding introns
Modification of mRNA after initial transcription but before translation	Usually none	Introns spliced out; 5′ cap and 3′ poly A tail added

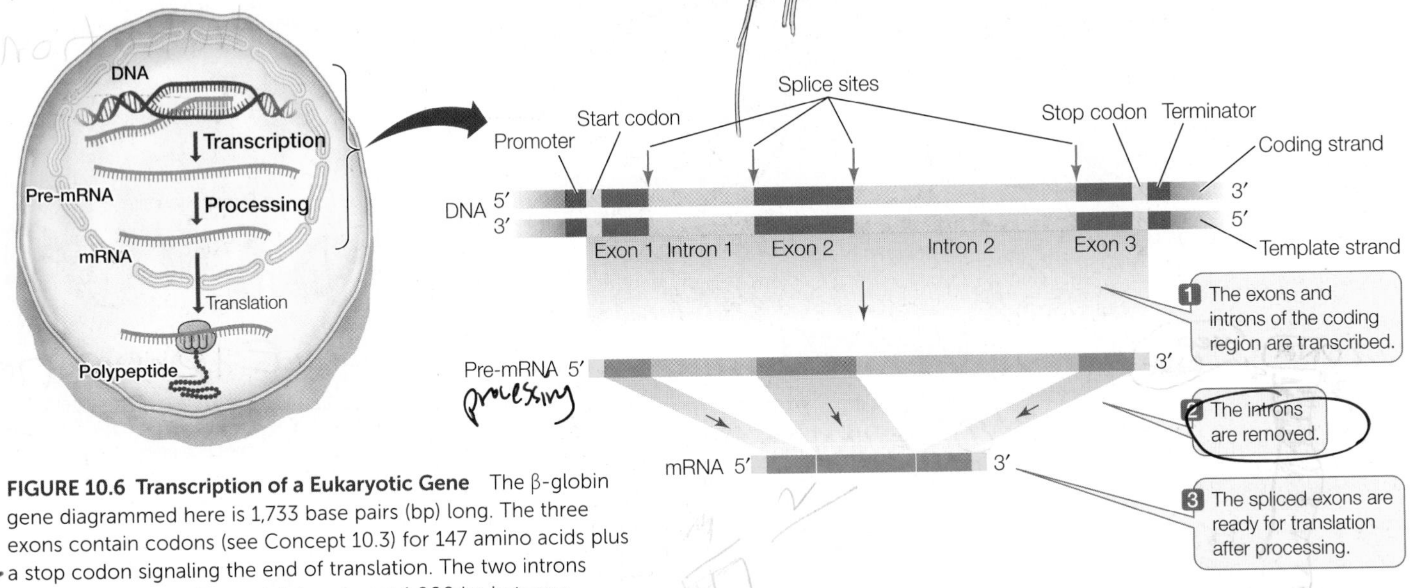

FIGURE 10.6 Transcription of a Eukaryotic Gene The β-globin gene diagrammed here is 1,733 base pairs (bp) long. The three exons contain codons (see Concept 10.3) for 147 amino acids plus a stop codon signaling the end of translation. The two introns (noncoding sequences containing almost 1,000 bp between them) are initially transcribed but then are spliced out of the pre-mRNA transcript.

but in eukaryotes each gene has its own promoter. And while the coding region of a prokaryotic gene is usually continuous (with no interruptions), a eukaryotic gene may contain noncoding sequences called **introns** (*int*ervening *regions*) that interrupt the coding region. The transcribed regions that are interspersed with the introns are called **exons** (*ex*pressed *regions*). Both introns and exons appear in the primary mRNA transcript, called the **precursor RNA**, or **pre-mRNA**, but the introns are removed by the time the mature mRNA leaves the nucleus. Pre-mRNA processing involves cutting introns out of the pre-mRNA transcript and splicing together the exon transcripts (see Figure 10.6). If this seems surprising to you, you are in good company. For scientists who were familiar with prokaryotic genes and gene expression, the discovery of introns in eukaryotic genes was entirely unexpected.

How can we locate introns within a eukaryotic gene? One way is by **nucleic acid hybridization**, the method that originally revealed the existence of introns. This method, outlined in **FIGURE 10.7**, has been crucial for studying the relationship between eukaryotic genes and their transcripts and is widely used in many applications. It involves two steps:

- The DNA to be analyzed is denatured by heat to break the hydrogen bonds between the base pairs and separate the two strands.
- A single-stranded nucleic acid from another source (called a **probe**) is incubated with the denatured DNA. If the probe has a base sequence complementary to the target DNA, a probe–target double helix forms by hydrogen bonding between the bases. Because the two strands are from different sources, the resulting double-stranded molecule is called a hybrid.

Biologists used this technique to examine the β-globin gene, which encodes a hemoglobin subunit (**FIGURE 10.8**). The researchers first denatured DNA containing the gene by heating it slowly, then used previously isolated β-globin mRNA as a probe. They were able to view the hybridized molecules using electron microscopy. As expected, the mRNA bound to the template DNA by complementary base pairing. The researchers expected to obtain a linear (1:1) matchup of the mRNA to the

RESEARCH TOOLS

FIGURE 10.7 Nucleic Acid Hybridization Base pairing permits the detection of a sequence that is complementary to the probe.

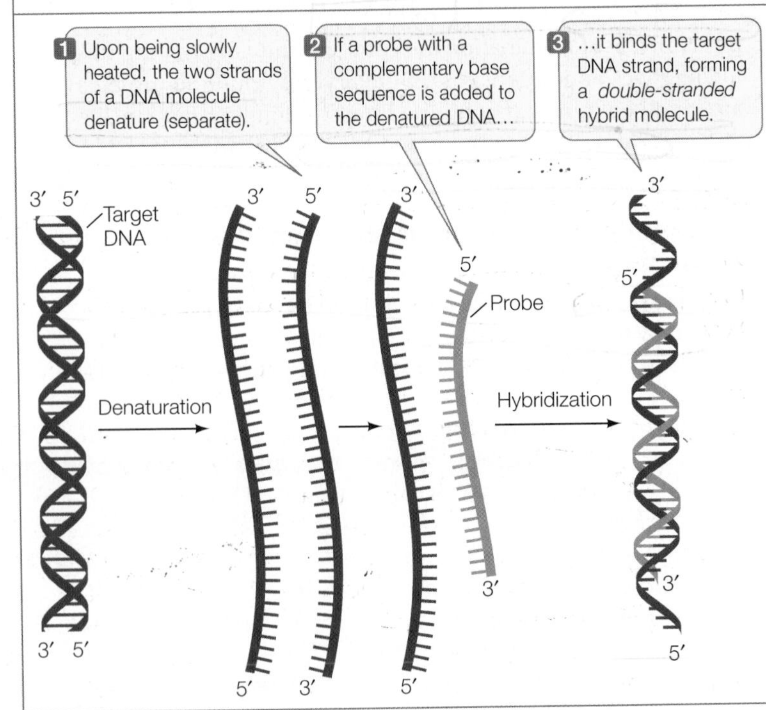

1 Upon being slowly heated, the two strands of a DNA molecule denature (separate).
2 If a probe with a complementary base sequence is added to the denatured DNA...
3 ...it binds the target DNA strand, forming a *double-stranded* hybrid molecule.
3′ 5′
Target DNA
3′ 5′
Denaturation
3′ 5′
5′ 3′
3′
5′
Probe
5′
3′
Hybridization
3′
5′
3′
5′

INVESTIGATION

FIGURE 10.8 Demonstrating the Existence of Introns When an mRNA transcript of the β-globin gene was hybridized with the double-stranded DNA of that gene, the introns in the DNA "looped out." This demonstrated that the coding region of a eukaryotic gene can contain noncoding DNA that is not present in the mature mRNA transcript.[a]

HYPOTHESIS

All regions within the coding sequence of a gene end up in its mRNA.

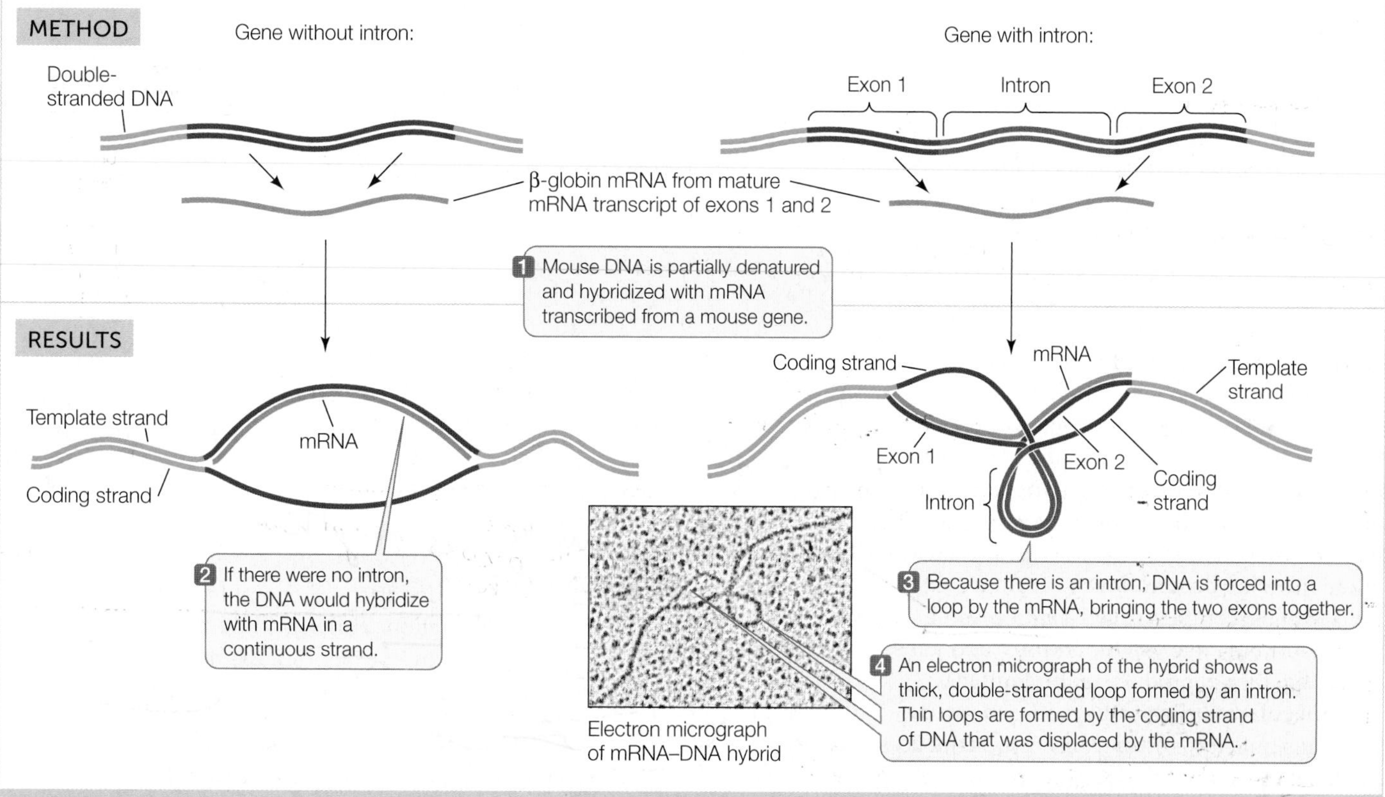
METHOD
Gene without intron:
Double-stranded DNA
Gene with intron:
Exon 1
Intron
Exon 2
β-globin mRNA from mature mRNA transcript of exons 1 and 2
1 Mouse DNA is partially denatured and hybridized with mRNA transcribed from a mouse gene.
RESULTS
Template strand
mRNA
Coding strand
2 If there were no intron, the DNA would hybridize with mRNA in a continuous strand.
Coding strand
mRNA
Template strand
Exon 1
Exon 2
Intron
Coding strand
3 Because there is an intron, DNA is forced into a loop by the mRNA, bringing the two exons together.
Electron micrograph of mRNA–DNA hybrid
4 An electron micrograph of the hybrid shows a thick, double-stranded loop formed by an intron. Thin loops are formed by the coding strand of DNA that was displaced by the mRNA.

CONCLUSION

The DNA contains noncoding regions within the genes that are not present in the mature mRNA.

Go to **LaunchPad** for discussion and relevant links for all **INVESTIGATION** figures.

[a]S. M. Tilghman et al. 1978. *Proceedings of the National Academy of Sciences USA* 75: 725–729.

coding DNA. That expectation was only partially met. There were indeed stretches of RNA–DNA hybrid, but some *unexpected looped structures* were also visible. These loops turned out to be introns, stretches of DNA that did not have complementary base sequences on the mature mRNA.

When pre-mRNA was used instead of mature mRNA to hybridize to the DNA, there was complete hybridization with *no loops*, revealing that the introns were part of the pre-mRNA transcript. Somewhere on the path from primary transcript (pre-mRNA) to mature mRNA, the introns had been removed, and the exons had been spliced together. We will examine this splicing process in the next section.

Introns *interrupt, but do not scramble,* the DNA sequence of a gene. The base sequences of the exons in the template strand, if joined and taken in order, form a continuous sequence that is complementary to that of the mature mRNA. Most (but not all) eukaryotic genes contain introns, and in rare cases, introns are also found in prokaryotes. The largest human gene encodes a muscle protein called titin; it has 363 exons, which together code for 38,138 amino acids. Can you deduce how many introns the titin gene has?

Eukaryotic gene transcripts are processed before translation

The primary transcript of a eukaryotic gene is modified in several ways before it leaves the nucleus: introns are removed, and both ends of the pre-mRNA are chemically modified.

SPLICING TO REMOVE INTRONS After the pre-mRNA is made, its introns must be removed. If this did not happen, the extra nucleotides in the mRNA would be translated at the ribosome and a nonfunctional protein would result. A process called **RNA splicing** removes the introns and splices the exons together.

Primary mRNA transcript

5′ 5′ Exon 5′ Splice site Intron 3′ Splice site 3′ Exon 3′

snRNP snRNP

Spliceosome

5′ 3′

5′ 3′

5′ OH 3′

5′ 5′ Exon 3′ Exon 3′

Mature mRNA

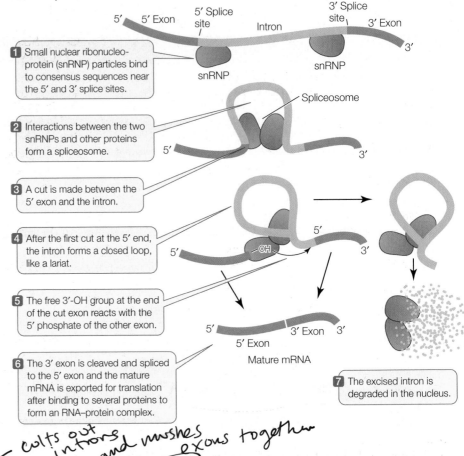

1 Small nuclear ribonucleoprotein (snRNP) particles bind to consensus sequences near the 5′ and 3′ splice sites.

2 Interactions between the two snRNPs and other proteins form a spliceosome.

3 A cut is made between the 5′ exon and the intron.

4 After the first cut at the 5′ end, the intron forms a closed loop, like a lariat.

5 The free 3′-OH group at the end of the cut exon reacts with the 5′ phosphate of the other exon.

6 The 3′ exon is cleaved and spliced to the 5′ exon and the mature mRNA is exported for translation after binding to several proteins to form an RNA–protein complex.

7 The excised intron is degraded in the nucleus.

FIGURE 10.9 The Spliceosome: An RNA Splicing Machine Small nuclear ribonucleoprotein particles (snRNPs) bind to consensus sequences bordering the introns on pre-mRNA transcripts. Other proteins then bind, forming a large complex called a spliceosome. This structure determines the exact position of each cut in the pre-mRNA with great precision.

Go to ANIMATED TUTORIAL 10.2 RNA Splicing
PoL2e.com/at10.2

At the boundaries between introns and exons are **consensus sequences**—short stretches of DNA that appear with little variation ("consensus") in many different genes. As soon as the pre-mRNA is transcribed, these consensus sequences are bound by several **small nuclear ribonucleoprotein particles (snRNPs)** (**FIGURE 10.9**). The RNA in one of the snRNPs has a stretch of bases complementary to the consensus sequence at the 5′ exon–intron boundary, and it binds to the pre-mRNA by complementary base pairing. Another snRNP binds to the pre-mRNA near the 3′ intron–exon boundary, and then other proteins accumulate to form a large RNA–protein complex called a **spliceosome**. This complex cuts the pre-mRNA, releases the introns, and joins the ends of the exons together to produce mature mRNA.

Molecular studies of human genetic diseases have provided insights into intron consensus sequences and the splicing machinery. For example, people with the genetic disease beta thalassemia have a defect in the production of one of the hemoglobin subunits. These people suffer from severe anemia because they have an inadequate supply of red blood cells. In some cases, the genetic mutation that causes the disease occurs at an intron consensus sequence in the β-globin gene. Consequently, the β-globin pre-mRNA cannot be spliced correctly, and a defective β-globin mRNA is made. This finding offers another example of how biologists can *use mutations to elucidate cause-and-effect relationships*.

MODIFICATION AT BOTH ENDS While the pre-mRNA is still in the nucleus it undergoes two processing steps, one at each end of the molecule:

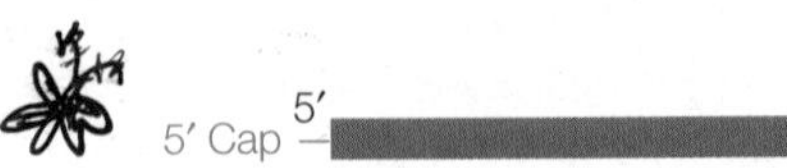

- A **5′ cap** (or G cap) is added to the 5′ end of the pre-mRNA as it is transcribed. The 5′ cap is a chemically modified molecule of guanosine triphosphate (GTP). It facilitates the binding of mRNA to the ribosome for translation, and it protects the mRNA from being digested by ribonucleases (enzymes that break down RNAs).
- A **poly A tail** is added to the 3′ end of the pre-mRNA at the end of transcription. This sequence of 100–300 adenine nucleotides assists in the export of the mRNA from the nucleus and is also important for mRNA stability.

CHECKpoint CONCEPT 10.2

- Part of a DNA template strand has the sequence 5′-ATGGTGTACG-3′. What will be the sequence of the RNA transcribed from this DNA? (Be careful to specify the 5′ and 3′ ends.)
- What would be the consequences of the following?
  a. A mutation of a promoter sequence such that the promoter is deleted
  b. A mutation of the gene that encodes RNA polymerase, such that the polymerase is not made
  c. Deletion of intron consensus sequences from a gene
- Refer to the experiment shown in Figure 10.8. What would the result have been if there were five exons and four introns? Sketch what this would look like in an electron micrograph.

The transcription of a gene to produce mRNA is only the first step in gene expression. The next step in the pathway from DNA to RNA to protein is translation, the subject of Concepts 10.3 and 10.4. First we will discuss the genetic code, which enables the base sequence in an mRNA to be translated into a specific amino acid sequence in the resulting polypeptide. Then we will look in more detail at the process of translation.

CONCEPT 10.3 The Genetic Code in RNA Is Translated into the Amino Acid Sequences of Proteins

The translation of the nucleotide sequence of an mRNA into the amino acid sequence of a polypeptide occurs at the ribosome. In prokaryotes, transcription and translation are coupled: there is no nucleus, and ribosomes often bind to an mRNA as it is being transcribed in the cytoplasm. In eukaryotes, the nuclear envelope separates the locations of mRNA production and translation, the latter occurring at ribosomes in the cytoplasm. In both cases, the key event is the decoding of one chemical "language" (the nucleotide sequence) into another (the amino acid sequence).

The information for protein synthesis lies in the genetic code

The genetic information in an mRNA molecule is a series of sequential, nonoverlapping three-letter "words" called **codons**. Each codon specifies a particular amino acid. The "letters" are three adjacent nucleotide bases in the mRNA polynucleotide chain. Each codon in the mRNA is complementary to the corresponding triplet of bases in the template strand of the DNA molecule from which it was transcribed. The genetic code relates codons to their specific amino acids.

CHARACTERISTICS OF THE GENETIC CODE Molecular biologists "broke" the genetic code in the early 1960s. The problem they addressed was perplexing: how could 20 different amino acids be specified using only four nucleotide bases (A, U, G, and C)? A triplet code with three-letter codons was considered likely because it was the shortest sequence with enough possible variations to encode all 20 amino acids. With four available bases, a triplet codon has $4 \times 4 \times 4 = 64$ variations.

Marshall Nirenberg and Heinrich Matthaei, at the U.S. National Institutes of Health, made the first decoding breakthrough in 1961 when they realized they could use a simple artificial polynucleotide instead of a complex natural mRNA as a template for polypeptide synthesis in a test tube. They could then identify the polypeptide the artificial messenger encoded. This led to the identification of the first two codons, as described in **FIGURE 10.10**.

Go to ANIMATED TUTORIAL 10.3 Deciphering the Genetic Code
PoL2e.com/at10.3

Other scientists later found that an artificial mRNA only three nucleotides long—amounting to one codon—could bind to a ribosome, and that the resulting complex could bind to a corresponding tRNA carrying a specific amino acid. Thus, for example, a simple UUU mRNA caused the tRNA carrying phenylalanine to bind to the ribosome. After this discovery, the complete deciphering of the genetic code was relatively simple.

The complete genetic code is shown in **FIGURE 10.11**. Notice that there are many more codons than there are different amino acids in proteins. All possible combinations of the four available bases give 64 (4^3) different three-letter codons, yet these codons determine only 20 amino acids. AUG, which codes for

INVESTIGATION

FIGURE 10.10 Deciphering the Genetic Code Nirenberg and Matthaei used a test tube protein synthesis system to determine the amino acids specified by synthetic mRNAs of known compositions.[a]

HYPOTHESIS

An artificial mRNA containing only one repeating base will direct the synthesis of a protein containing only one repeating amino acid.

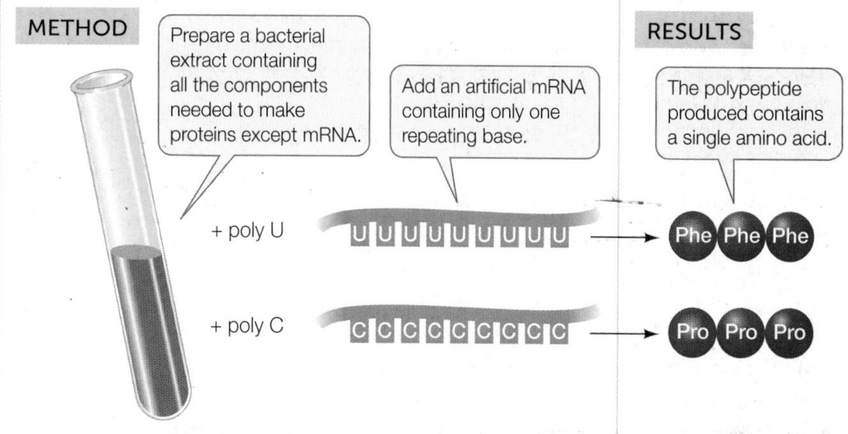

CONCLUSION

Poly U contains codons for phenylalanine only.
Poly C contains codons for proline only.

ANALYZE THE DATA

Poly U, an artificial mRNA, was added to a test tube with all other components for protein synthesis ("Complete system"). Other test tubes differed from the complete system as indicated in the table. Samples were tested for radioactive phenylalanine incorporation with the results shown in the table. Explain the results for each of the conditions.

Condition	Radioactivity in polypeptide
Complete system	29,500
Minus poly U mRNA	70
Minus ribosomes	52
Minus ATP	83
Plus RNase (hydrolyzes RNA)	120
Plus DNase	27,600
Mixture of 5 radioactive amino acids minus phenylalanine	33

Go to **LaunchPad** for discussion and relevant links for all **INVESTIGATION** figures.

[a]M. W. Nirenberg and J. H. Matthaei. 1961. *Proceedings of the National Academy of Sciences USA* 47: 1588–1602.

First letter	Second letter: U	Second letter: C	Second letter: A	Second letter: G	Third letter
U	UUU UUC Phenyl-alanine UUA UUG Leucine	UCU UCC UCA UCG Serine	UAU UAC Tyrosine UAA Stop codon UAG Stop codon	UGU UGC Cysteine UGA Stop codon UGG Tryptophan	U C A G
C	CUU CUC CUA CUG Leucine	CCU CCC CCA CCG Proline	CAU CAC Histidine CAA CAG Glutamine	CGU CGC CGA CGG Arginine	U C A G
A	AUU AUC AUA Isoleucine AUG Methionine; start codon	ACU ACC ACA ACG Threonine	AAU AAC Asparagine AAA AAG Lysine	AGU AGC Serine AGA AGG Arginine	U C A G
G	GUU GUC GUA GUG Valine	GCU GCC GCA GCG Alanine	GAU GAC Aspartic acid GAA GAG Glutamic acid	GGU GGC GGA GGG Glycine	U C A G

FIGURE 10.11 The Genetic Code Genetic information is encoded in three-letter units—codons—that are read in the 5′-to-3′ direction on the mRNA. To decode a codon, find its first letter in the left column, then read across the top to its second letter, then read down the right column to its third letter. The amino acid the codon specifies is given in the corresponding row. For example, AUG codes for methionine, and GUA codes for valine.

Go to ACTIVITY 10.2 The Genetic Code
PoL2e.com/ac10.2

methionine, is also the **start codon**, the initiation signal for translation. The AUG codon is somewhat like the capitalized first word of a sentence, indicating how the sequence of words should be read. Three of the codons (UAA, UAG, and UGA) are **stop codons**, or termination signals for translation. When the translation machinery reaches one of these codons, translation stops and the polypeptide is released.

THE GENETIC CODE IS REDUNDANT BUT NOT AMBIGUOUS The 60 codons that are not start or stop codons are far more than enough to code for the other 19 amino acids—and indeed, there is more than one codon for almost all the amino acids. Thus we say that the genetic code contains redundancies. For example, leucine is represented by six different codons (see Figure 10.11). Only methionine and tryptophan are represented by just one codon each.

A *redundant* code should not be confused with an *ambiguous* code. If the code were ambiguous, a single codon could specify two (or more) different amino acids, and there would be doubt about which amino acid should be incorporated into a growing polypeptide chain. The genetic code is not ambiguous: a given amino acid may be encoded by more than one codon, but each codon encodes only one amino acid.

THE GENETIC CODE IS (NEARLY) UNIVERSAL The same genetic code is used by all the species on our planet. Thus the code must be an ancient one that has been maintained intact throughout the evolution of living organisms. Exceptions are known: within mitochondria and chloroplasts, the code differs slightly from that in prokaryotes and in the nuclei of eukaryotic cells; and in one group of protists, the codons UAA and UAG encode glutamine rather than functioning as stop codons. The significance of these differences is not yet clear. What is clear is that the exceptions are few.

The common genetic code unifies life, and indicates that all life came ultimately from a common ancestor. The genetic code probably originated early in the evolution of life. The common code also has profound implications for genetic engineering, as we will see in Chapter 13, since it means that the code for a human gene is the same as that for a bacterial gene. It is therefore impressive, but not surprising, that a human gene can be expressed in *Escherichia coli* via laboratory manipulations, since these cells speak the same "molecular language."

The codons shown in Figure 10.11 are for mRNA. The base sequence of the template DNA strand is complementary and antiparallel to these codons. Thus, for example, 3′-ACC-5′ in the template DNA strand corresponds to tryptophan (which is encoded by the mRNA codon 5′-UGG-3′). However, the coding DNA strand has the same sequence as the mRNA (but with T's

APPLY THE CONCEPT

The genetic code in RNA is translated into the amino acid sequences of proteins

The double-stranded DNA sequence for the coding region of a short peptide is:

```
5′-A T G T T T T C G A C G T G C G A T T G A-3′
3′-T A C A A A A G C T G C A C G C T A A C T-5′
   1       5         10        15        20
```

1. Which strand of DNA (top or bottom) is transcribed into mRNA? Explain.
2. What is the amino acid sequence of the peptide coded for by the DNA?
3. A mutant strain has a C-G base pair at position 5 instead of the T-A pair shown here. What is the amino acid sequence of the peptide? Explain.
4. A mutation at base pair 15 results in a peptide that is not full length. What point mutation would cause this? Explain.

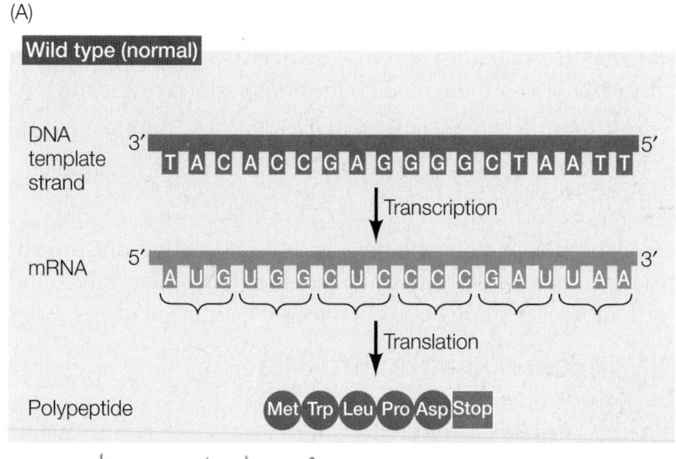

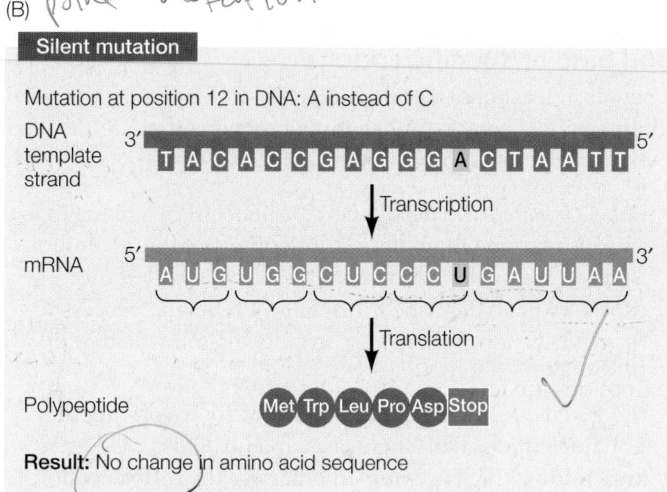

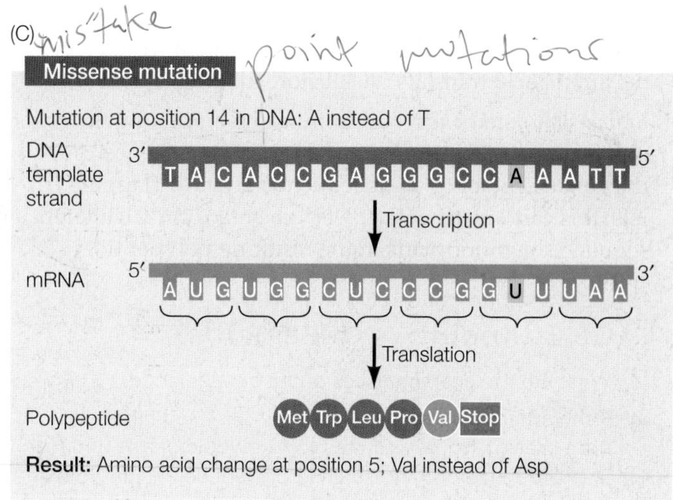

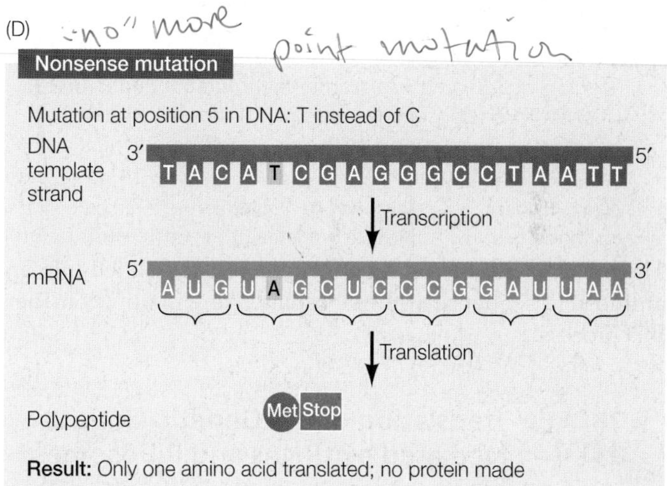

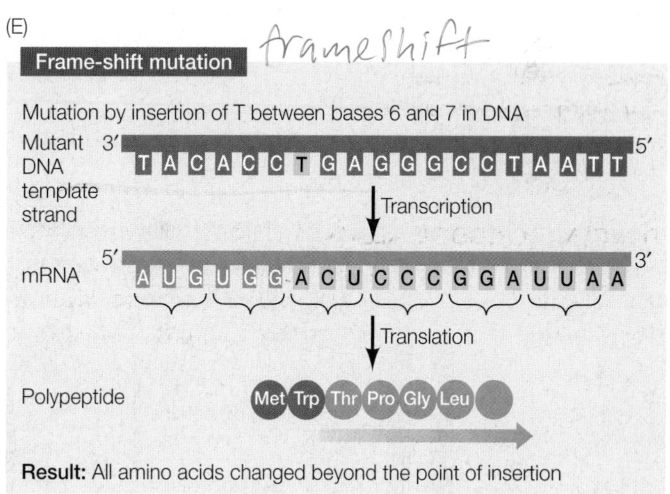

FIGURE 10.12 Mutations Changes in a coding region of DNA can have different effects on the protein the DNA encodes.

Go to ANIMATED TUTORIAL 10.4
Genetic Mutations Simulation
PoL2e.com/at10.4

instead of U's). By convention, gene sequences are usually shown as the sequence of the coding strand, beginning at the 5′ end.

Point mutations confirm the genetic code

Strong support for the assignment of codons in the genetic code comes from point mutations (changes in single nucleotides within a sequence). These have been studied in a wide variety of organisms. When point mutations within the coding region of a gene are compared with the amino acid sequences in the encoded polypeptide, they are consistent with the genetic code.

In Concept 9.3 we discussed mutations in terms of their effects on phenotypes. We can now define types of mutations in terms of their effects on polypeptide sequences (**FIGURE 10.12**):

- *Silent mutations* can occur because of the redundancy of the genetic code. For example, the codons CCG and CCU are both translated from mRNA as proline (Pro). So a change in the template strand of the DNA from 3′-GGC to 3′-GGA (a mutation from C to A) will not cause any change in amino acid sequence.
- *Missense mutations* result in a change in the amino acid sequence. For example, GAU in mRNA is translated as aspartic acid (Asp), whereas a mutation that results in GUU is translated as valine (Val).
- *Nonsense mutations* result in a premature stop codon. For example, the codon UGG is translated as the amino acid tryptophan (Trp). A DNA point mutation could convert this to the stop codon UAG, which acts as a translation termination

signal. If this occurred, the polypeptide chain would end at the amino acid translated just before the stop codon.

- *Frame-shift mutations* result from the insertion or deletion of one or more base pairs within the coding sequence. Since the genetic code is read as sequential, nonoverlapping triplets, this can cause new triplets to be read, and an altered sequence of amino acids in the resulting polypeptide.

CHECKpoint CONCEPT 10.3

- ✓ What are the characteristics of the genetic code?
- ✓ If the artificial mRNA UAUAUAUAUA... is used in a test tube protein synthesis system, what would be the amino acid sequence of the resulting polypeptide chain? Note that in this system translation can begin anywhere on the mRNA.
- ✓ A deletion of two consecutive base pairs in the coding region of DNA causes a frame-shift mutation. But a deletion of three consecutive base pairs causes the deletion of only one amino acid, with the rest of the polypeptide chain intact. Explain.

The mRNA with its coding information is translated into an amino acid sequence at the ribosome. We will now consider this process.

CONCEPT 10.4 Translation of the Genetic Code Is Mediated by tRNAs and Ribosomes

The translation of mRNA into proteins requires molecules that can link the information contained in each mRNA codon with a specific amino acid. That function is performed by a set of transfer RNAs (tRNAs). Two key events must take place to ensure that the protein made is the one specified by the mRNA:

- A tRNA must chemically read each mRNA codon correctly.
- The tRNA must deliver the amino acid that corresponds to the mRNA codon.

Once the tRNAs "decode" the mRNA and deliver the appropriate amino acids, components of the ribosome catalyze the formation of peptide bonds between the amino acids.

Go to ANIMATED TUTORIAL 10.5
Protein Synthesis
PoL2e.com/at10.5

Transfer RNAs carry specific amino acids and bind to specific codons

There is at least one specific tRNA molecule for each of the 20 amino acids. Each tRNA has three functions that are fulfilled by its structure and base sequence (**FIGURE 10.13**):

- *tRNAs bind to particular amino acids.* Each tRNA binds to a specific enzyme that attaches it to only 1 of the 20 amino acids. This covalent attachment is at the 3′ end of the tRNA. We will describe the details of this vital process in the next section. When it is carrying an amino acid, the tRNA is said to be "charged."
- *tRNAs bind to mRNA.* At about the midpoint on the tRNA polynucleotide chain there is a triplet of bases called the **anticodon**, which is complementary to the mRNA codon for the particular amino acid that the tRNA carries. Like the two strands of DNA, the codon and anticodon bind together via noncovalent hydrogen bonds. For example, the

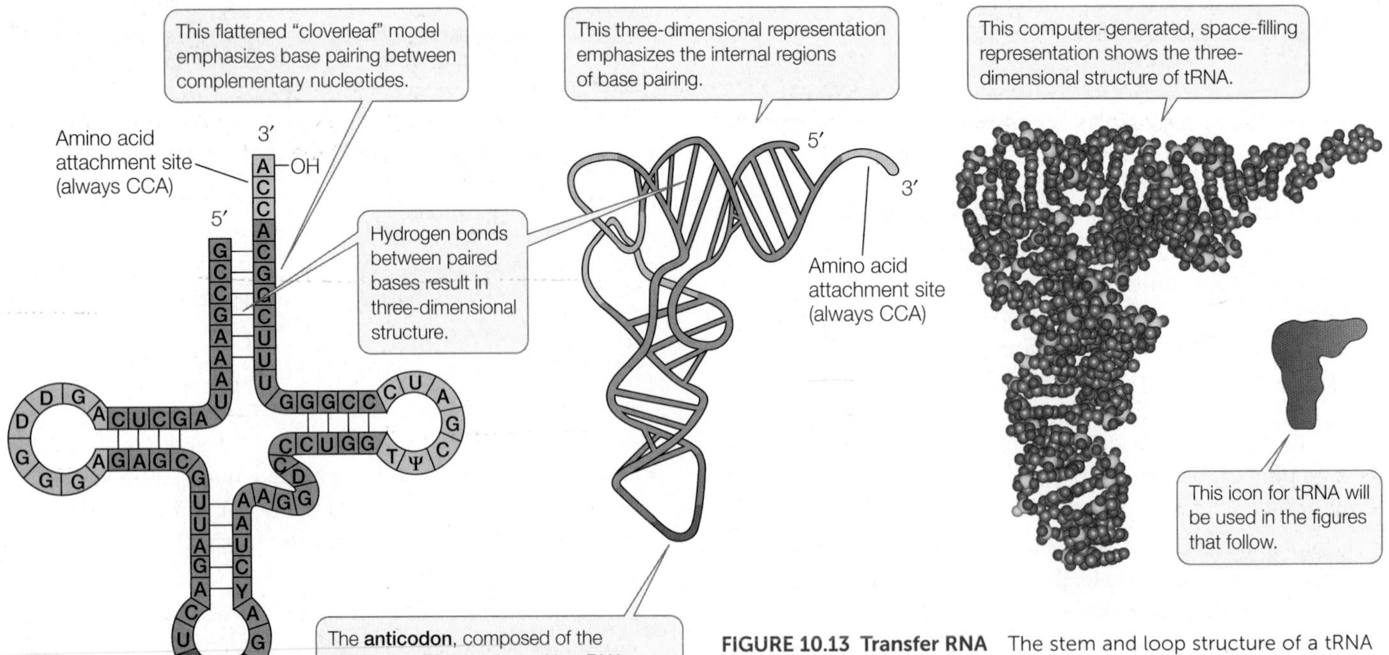

FIGURE 10.13 Transfer RNA The stem and loop structure of a tRNA molecule is well suited to its functions: binding to amino acids, associating with mRNA molecules, and interacting with ribosomes.

mRNA codon for arginine is 5′-CGG-3′, and the tRNA anticodon is 3′-GCC-5′.

- *tRNAs interact with ribosomes.* The ribosome has several sites on its surface that just fit the three-dimensional structure of a tRNA molecule. Interaction between the ribosome and the tRNA is noncovalent.

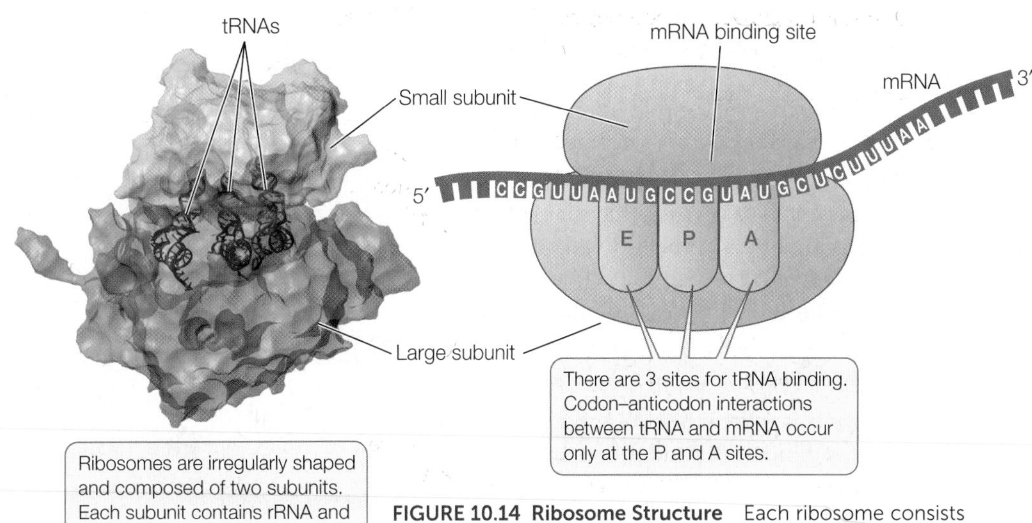

tRNAs
Small subunit
Large subunit
mRNA binding site
mRNA
5′
3′
CCGUUAAUGCCGUAUGCUCUUUAA
E
P
A
Ribosomes are irregularly shaped and composed of two subunits. Each subunit contains rRNA and numerous proteins.
There are 3 sites for tRNA binding. Codon–anticodon interactions between tRNA and mRNA occur only at the P and A sites.

FIGURE 10.14 Ribosome Structure Each ribosome consists of a large and a small subunit. The subunits remain separate when they are not in use for protein synthesis.

Recall that 61 different codons encode the 20 amino acids in proteins (see Figure 10.11). Does this mean that the cell must produce 61 different tRNA species, each with a different anticodon? No. The cell gets by with about two-thirds of that number of tRNA species because the specificity for the base at the 3′ end of the codon (and the 5′ end of the anticodon) is not always strictly observed. This phenomenon is called "wobble," and it occurs because in some cases unusual or modified nucleotide bases occur in the 5′ position of the anticodon. One such unusual base is inosine (I), which can pair with A, C, and U. For example, its presence allows three of the alanine codons—GCA, GCC, and GCU—to be recognized by the same tRNA (with the anticodon 3′-CGI-5′). Wobble occurs in some matches but not in others; of most importance, it does not allow the genetic code to be ambiguous. That is, *each mRNA codon binds to just one tRNA species, carrying a specific amino acid.*

Each tRNA is specifically attached to an amino acid

The charging of each tRNA with its correct amino acid is achieved by a family of enzymes known as aminoacyl-tRNA synthases. Each enzyme is specific for one amino acid and for its corresponding tRNA. The reaction uses the energy in ATP to form a high-energy bond between the amino acid and the tRNA:

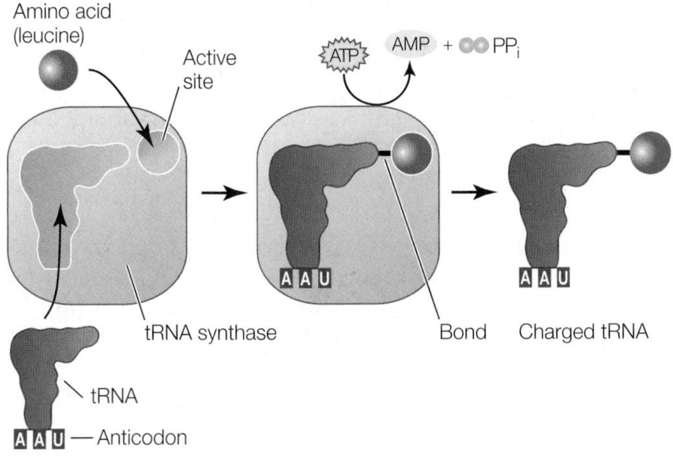

The energy in this bond is later used in the formation of peptide bonds between amino acids in a growing polypeptide chain.

Clearly, the specificity between the tRNA and its corresponding amino acid is extremely important. These two reactions, for example, are highly specific:

$$\text{Cysteine} + \text{tRNA}_{cys}\ (\text{anticodon ACA}) \xrightarrow{\text{Cys-tRNA synthase}} \text{Cys-tRNA}_{cys}$$

$$\text{Alanine} + \text{tRNA}_{ala}\ (\text{anticodon CGA}) \xrightarrow{\text{Ala-tRNA synthase}} \text{Ala-tRNA}_{ala}$$

A clever experiment by Seymour Benzer and his colleagues at Purdue University demonstrated the importance of this specificity. They took the Cys-tRNA$_{cys}$ molecule and chemically converted the cysteine into alanine, resulting in Ala-tRNA$_{cys}$. Which component—the amino acid or the tRNA—would be recognized when this hybrid charged tRNA was put into a protein-synthesizing system? The answer was the tRNA. Everywhere in the synthesized protein where cysteine was supposed to be, alanine appeared instead. The cysteine-specific tRNA had delivered its cargo (alanine) to every mRNA codon for cysteine. This experiment showed that the protein synthesis machinery recognizes the anticodon of the charged tRNA, not the amino acid attached to it.

Translation occurs at the ribosome

The ribosome is the molecular workbench where the translation of mRNA by tRNA is accomplished. All prokaryotic and eukaryotic ribosomes consist of two subunits (**FIGURE 10.14**). In eukaryotes, the large subunit consists of 3 different ribosomal RNA (rRNA) molecules and about 49 protein molecules arranged in a precise pattern. The small subunit consists of one rRNA molecule and about 33 proteins. These two subunits and several dozen other molecules interact noncovalently, fitting together like a jigsaw puzzle. If the hydrophobic interactions between the proteins and RNAs are disrupted, the ribosome falls apart, but it will reassemble perfectly when the disrupting agent is removed. When not active in the translation of mRNA, the ribosome exists as two separate subunits.

On the large subunit of the ribosome there are three sites to which a tRNA can bind, designated the A, P, and E sites (see Figure 10.14). The mRNA and ribosome move in relation to one another, and as they do so, a charged tRNA traverses these three sites in order:

- The *A (amino acid) site* is where the charged tRNA anticodon binds to the mRNA codon, thus lining up the correct amino acid to be added to the growing polypeptide chain.
- The *P (polypeptide) site* is where the tRNA adds its amino acid to the polypeptide chain.
- The *E (exit) site* is where the tRNA, having given up its amino acid, resides before being released from the ribosome and going back to the cytosol to pick up another amino acid and begin the process again.

The ribosome has a fidelity function, which ensures that a charged tRNA with the correct anticodon binds to the appropriate codon in the mRNA. When proper binding occurs, hydrogen bonds form between the three base pairs. The rRNA of the small ribosomal subunit plays a role in validating the three-base-pair match. Any tRNA that does *not* form hydrogen bonds with all three bases of the codon is ejected from the ribosome.

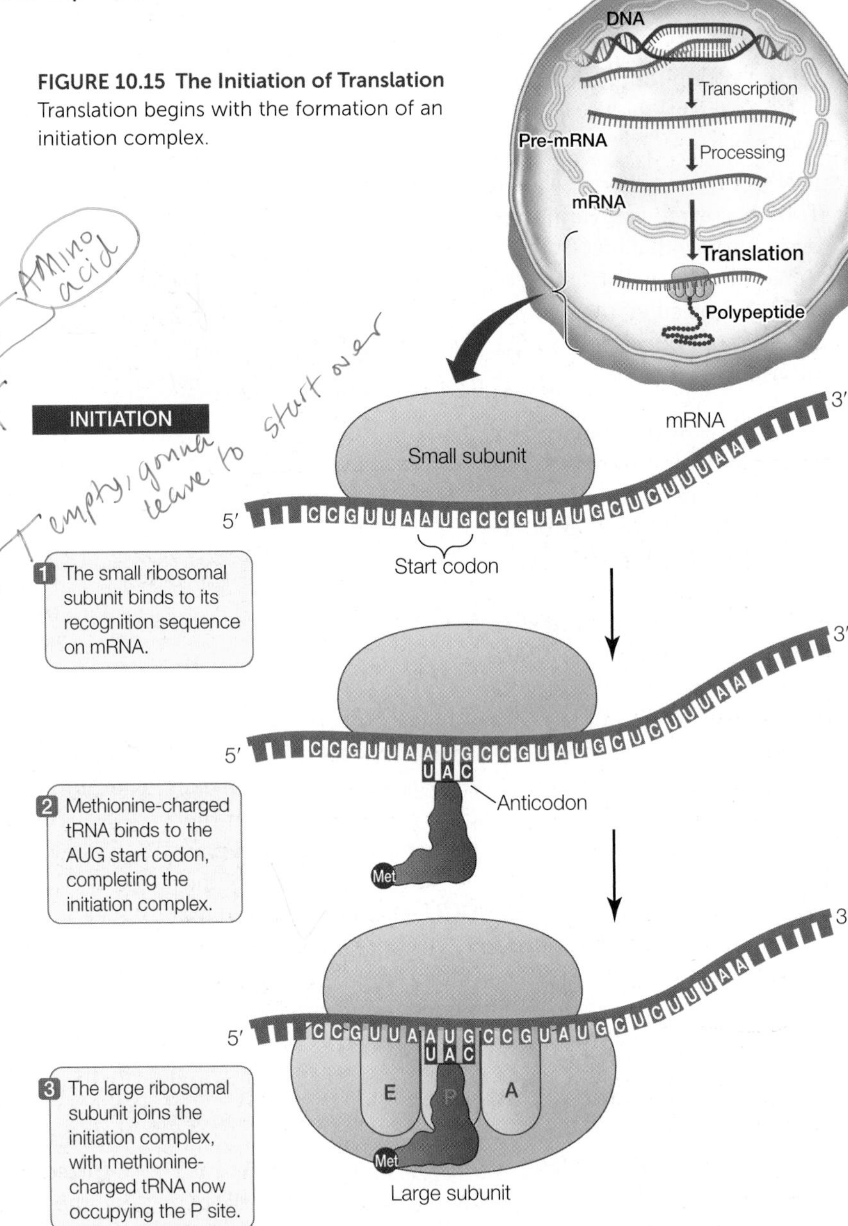

FIGURE 10.15 The Initiation of Translation Translation begins with the formation of an initiation complex.

Translation takes place in three steps

Like transcription, translation occurs in three steps: initiation, elongation, and termination.

INITIATION The **initiation complex** consists of a charged tRNA and a small ribosomal subunit, both bound to the mRNA (**FIGURE 10.15**). While different organisms have different ways to effect this binding, here is an example from a prokaryote: The rRNA of the small ribosomal subunit binds by base pairing to a complementary sequence on the mRNA, about 8 base pairs upstream of the translation start codon (AUG; see Figure 10.11). After binding, the mRNA and the small subunit are aligned in such a way that the start codon at the beginning of the coding sequence will be aligned with the P site on the large subunit:

mRNA 5′ ... A G G A G G ... AUG ... 3′

rRNA 3′ ... U C C U C C ... (P site) ... 5′

The anticodon of a methionine-charged tRNA binds to the start codon by complementary base pairing to complete the initiation complex. Thus the first amino acid in a new polypeptide chain is always methionine. (In bacteria, but not archaea, the first amino acid is a slightly modified form of methionine called formylmethionine.) However, not all mature proteins have methionine as their first amino acid. In many cases, the initial methionine is removed by an enzyme after translation.

After the methionine-charged tRNA has bound to the mRNA, the large subunit of the ribosome joins the complex. The methionine-charged tRNA lies in the P site of the large subunit, and the A site is aligned with the second mRNA codon. These ingredients—mRNA, two ribosomal subunits, and methionine-charged tRNA—are put together properly by a group of proteins called initiation factors.

ELONGATION A charged tRNA whose anticodon is complementary to the second codon of the mRNA now enters the open A site of the large ribosomal subunit (**FIGURE 10.16**). The large subunit then catalyzes two reactions:

- It breaks the bond between the methionine and its tRNA in the P site.
- It catalyzes the formation of a peptide bond between the methionine and the amino acid attached to the tRNA in the A site.

ELONGATION

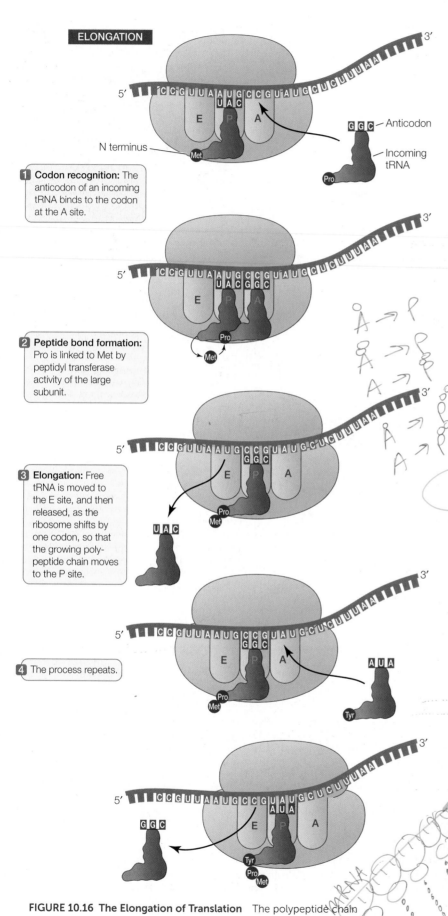

1 **Codon recognition:** The anticodon of an incoming tRNA binds to the codon at the A site.

2 **Peptide bond formation:** Pro is linked to Met by peptidyl transferase activity of the large subunit.

3 **Elongation:** Free tRNA is moved to the E site, and then released, as the ribosome shifts by one codon, so that the growing polypeptide chain moves to the P site.

4 The process repeats.

5′ CCGUUAAUGCCGUAUGCUCUUUAA 3′

UAC

GGC — Anticodon

Incoming tRNA

N terminus

Met

Pro

Tyr

AUA

E P A

FIGURE 10.16 The Elongation of Translation The polypeptide chain elongates as the mRNA is translated.

Because the large ribosomal subunit performs these two actions, it is said to have **peptidyl transferase** activity. The component with this activity is actually one of the rRNAs in the ribosome, so the catalyst is an example of a **ribozyme** (from *ribo*nucleic acid and en*zyme*).

Methionine thus becomes the amino (N) terminus of the new protein (recall that polypeptides grow in the amino to the carboxyl direction; see Concept 3.2). The second amino acid is now bound to methionine but remains attached to its tRNA at the A site.

After the first tRNA releases its methionine, the ribosome moves so that the first tRNA is at the E site. The tRNA then dissociates from the ribosome and returns to the cytosol to become charged with another methionine. The second tRNA, now bearing a dipeptide (a chain of two amino acids), is shifted to the P site as the ribosome moves one codon along the mRNA in the 5′-to-3′ direction (see Figure 10.16). These steps are repeated, and the polypeptide chain grows as each new amino acid is added.

LINK

You can review the structure and formation of peptide bonds in **Concept 3.2**, especially **Figure 3.6**

TERMINATION The elongation cycle terminates at the end of the coding sequence, which is marked by a stop codon: UAA, UAG, or UGA (**FIGURE 10.17**). When a stop codon enters the A site, it binds a protein release factor, which allows hydrolysis of the bond between the polypeptide chain and the tRNA in the P site. The newly completed polypeptide then separates from the ribosome.

TABLE 10.2 summarizes the nucleic acid signals for initiation and termination of transcription and translation.

Go to MEDIA CLIP 10.1
Protein Synthesis:
An Epic on a Cellular Level
PoL2e.com/mc10.1

Polysome formation increases the rate of protein synthesis

Several ribosomes can simultaneously translate a single mRNA molecule, producing multiple polypeptides at the same time. As soon as the first ribosome has moved far enough from the translation initiation site, a second initiation complex can form, then a third, and so on. An assemblage consisting of a strand of mRNA with its beadlike ribosomes and their growing polypeptide chains is called a **polyribosome**, or **polysome** (**FIGURE**

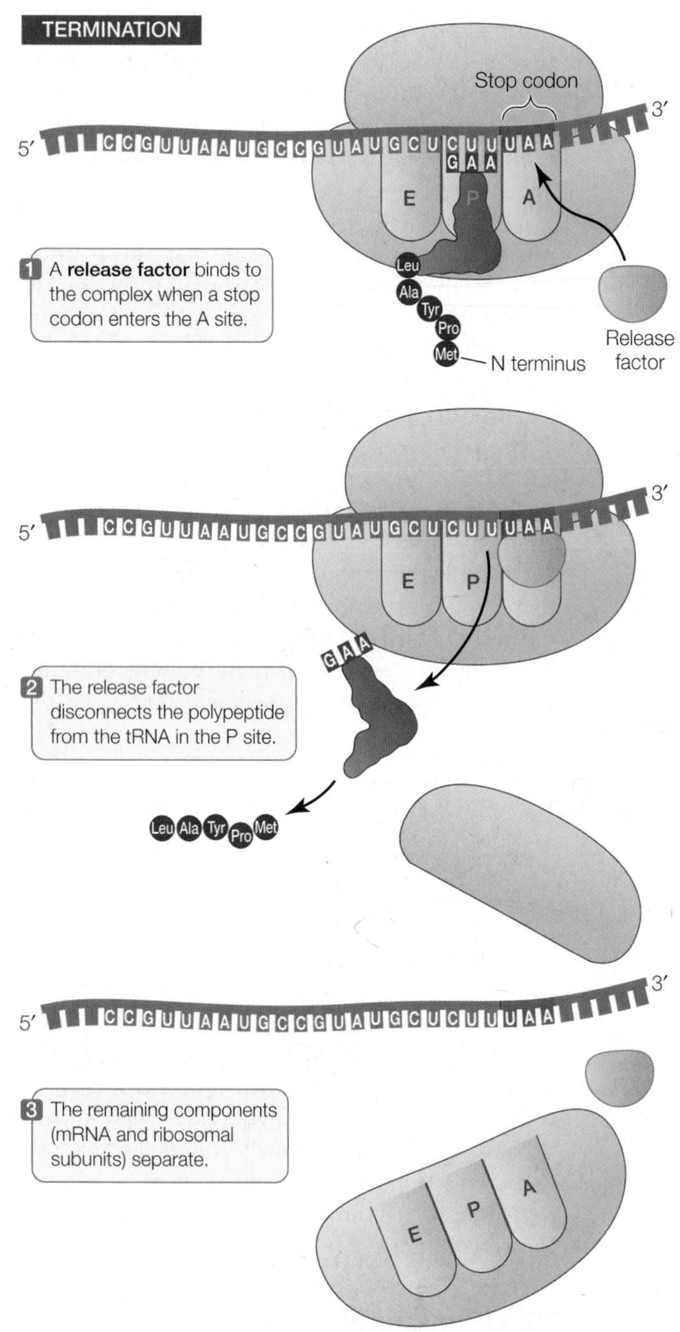

FIGURE 10.17 **The Termination of Translation** Translation terminates when the A site of the ribosome encounters a stop codon on the mRNA.

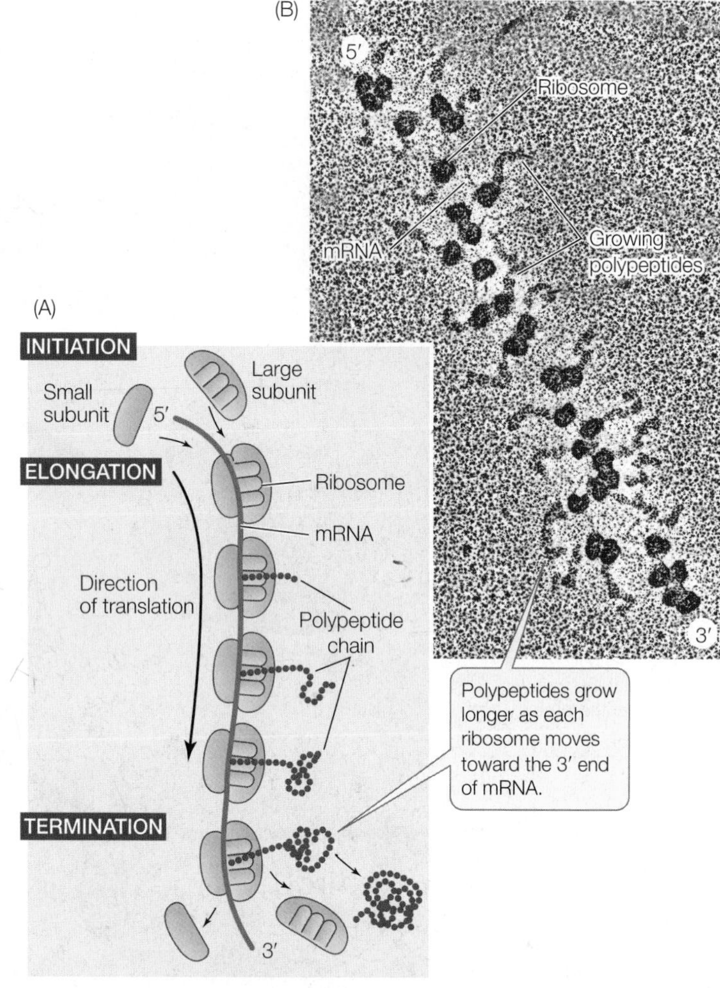

FIGURE 10.18 **A Polysome** (A) A polysome consists of multiple ribosomes and their growing polypeptide chains moving along an mRNA molecule. (B) An electron micrograph of a polysome.

10.18). Cells that are actively synthesizing proteins contain large numbers of polysomes and few free ribosomes or ribosomal subunits.

TABLE 10.2 Signals that Start and Stop Transcription and Translation

	Transcription	Translation
Initiation	Promoter DNA	AUG start codon in the mRNA
Termination	Terminator DNA	UAA, UAG, or UGA in the mRNA

CHECKpoint CONCEPT 10.4

- ✓ Describe the sequence of events in translation that involve tRNA, from charging with an amino acid to initiation, elongation, and termination.
- ✓ Imagine a polypeptide whose second amino acid is tryptophan. Sketch a ribosome with the mRNA and the first two tRNAs for this polypeptide, noting their positions in the A, P, and E sites.
- ✓ What would happen to polypeptides and cell function if a valine-tRNA synthase lost its specificity and attached any of the 20 amino acids to the 3′ end of the valine tRNA?

The process of protein synthesis usually does not end with translation. Proteins can undergo covalent modifications both during and after translation, with chemical groups being added or parts of the polypeptide chains removed. We now turn to these modifications.

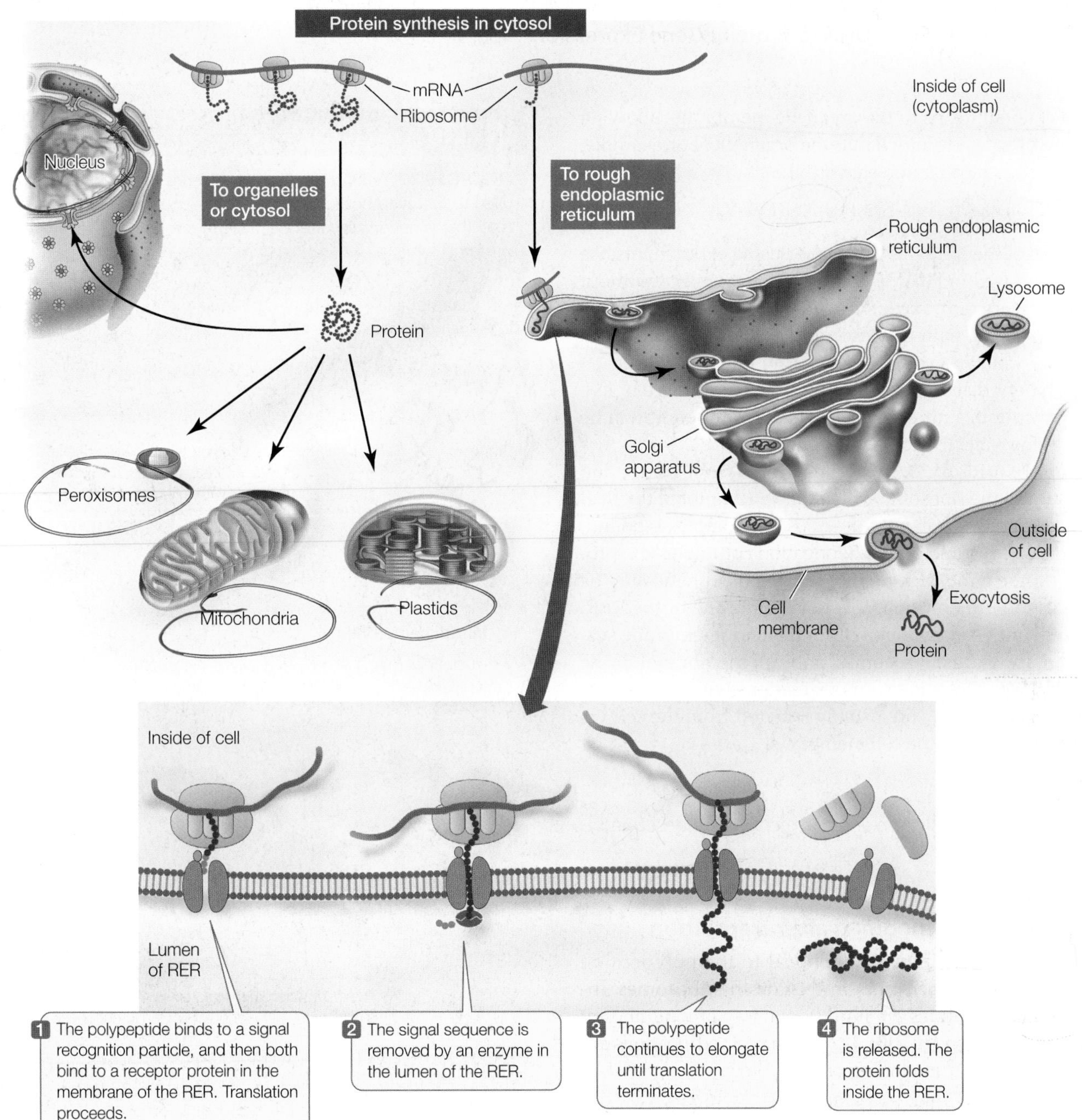
Protein synthesis in cytosol
mRNA
Ribosome
Nucleus
To organelles or cytosol
To rough endoplasmic reticulum
Inside of cell (cytoplasm)
Rough endoplasmic reticulum
Lysosome
Protein
Golgi apparatus
Peroxisomes
Mitochondria
Plastids
Outside of cell
Cell membrane
Exocytosis
Protein
Inside of cell
Lumen of RER
1 The polypeptide binds to a signal recognition particle, and then both bind to a receptor protein in the membrane of the RER. Translation proceeds.
2 The signal sequence is removed by an enzyme in the lumen of the RER.
3 The polypeptide continues to elongate until translation terminates.
4 The ribosome is released. The protein folds inside the RER.

FIGURE 10.19 Destinations for Newly Translated Polypeptides in a Eukaryotic Cell Signal sequences on newly synthesized polypeptides bind to specific receptor proteins on the outer membranes of the organelles to which they are directed. Once the protein has bound to it, the protein enters the organelle through a channel in the membrane.

CONCEPT 10.5 Proteins Are Modified after Translation

The site of a polypeptide's function in the cell may be far away from its point of synthesis at the ribosome. This is especially true for eukaryotes, where a polypeptide may be moved into an organelle. Furthermore, polypeptides are often modified by the addition of new chemical groups that contribute to the function of the mature protein. In this section we examine these posttranslational aspects of protein synthesis.

Signal sequences in proteins direct them to their cellular destinations

Protein synthesis always begins on free ribosomes floating in the cytoplasm, and the "default" location for a protein is the cytosol. As the polypeptide chain emerges from the ribosome it may simply fold into its three-dimensional shape and perform its cellular role in the cytosol. However, a newly formed polypeptide may contain a **signal sequence** (or signal peptide)—a short stretch of amino acids that indicates where in the cell the polypeptide belongs. Proteins destined for different locations have different signals.

In the absence of a signal sequence, the protein will remain in the same cellular compartment where it was synthesized. Some proteins, however, contain signal sequences that "target" them to the nucleus, mitochondria, plastids, or peroxisomes (**FIGURE 10.19, LEFT**). A signal sequence binds to a specific receptor protein at the surface of the organelle. Once it has bound, a channel forms in the organelle membrane, allowing the targeted protein to move into the organelle. For example, here is a nuclear localization signal (NLS):

-Pro-Pro-Lys-Lys-Lys-Arg-Lys-Val-

The function of the NLS was established using experiments like the one illustrated in **FIGURE 10.20**. Proteins with or without this peptide were introduced into cells and then located by labeling the proteins with fluorescent dyes. Only proteins with the nuclear localization signal were found in the nucleus.

If a polypeptide carries a particular signal sequence of five to ten hydrophobic amino acids at its N terminus, it will be directed to the rough endoplasmic reticulum (RER) for further processing (**FIGURE 10.19, RIGHT AND BOTTOM**). Translation will pause, and the ribosome will bind to a receptor at the RER membrane. Once the polypeptide–ribosome complex is bound, translation will resume, and as elongation continues, the protein will traverse the RER membrane. Such proteins may be retained in the lumen (the inside) or membrane of the RER, or they may move elsewhere within the endomembrane system (Golgi apparatus, lysosomes, and cell membrane). If the proteins lack specific signals for destinations within the endomembrane system, they are usually secreted from the cell via vesicles that fuse with the cell membrane.

LINK

The endomembrane system and its functions are described in **Concept 4.3**

INVESTIGATION

FIGURE 10.20 Testing the Signal A series of experiments were used to test whether a nuclear localization signal (NLS) sequence is all that is needed to direct a protein to the nucleus.[a]

HYPOTHESIS

A nuclear localization signal is necessary for importing a protein into the cell nucleus.

METHOD

1 A protein is introduced into the cytoplasm and labeled with a fluorescent dye.

RESULTS

Introduced protein:

Nucleoplasmin, a nuclear protein, with its NLS

Nucleoplasmin with the NLS removed

Pyruvate kinase, a cytoplasmic protein without an NLS

Pyruvate kinase with attached NLS

NLS

NLS

2 The distribution of the protein in the cell is observed with a fluorescence microscope.

CONCLUSION

An NLS is essential for nuclear protein import and will direct a normally cytoplasmic protein to the nucleus.

Go to **LaunchPad** for discussion and relevant links for all **INVESTIGATION** figures.

[a]C. Dingwall et al. 1988. *Journal of Cell Biology* 107: 841–849.

Many proteins are modified after translation

Most mature proteins are not identical to the polypeptide chains that are translated from mRNA on the ribosomes. Instead, most polypeptides are modified in any of a number of ways after translation (**FIGURE 10.21**). These modifications are essential to the final functioning of the protein.

- *Proteolysis* is the cutting of a polypeptide chain. For example, the ER signal sequence is cut off from the growing polypeptide chain as it enters the ER. Some mature proteins are actually made from polyproteins—long polypeptides containing the primary sequences of multiple distinct proteins—that are cut into final products by enzymes called proteases. Proteases are essential to some viruses, including human immunodeficiency virus (HIV), because the large viral polyprotein cannot fold properly unless it is cut. Certain drugs used to treat acquired immune deficiency syndrome (AIDS) work by inhibiting the HIV protease, thereby preventing the formation of proteins needed for viral reproduction.
- *Glycosylation* is the addition of carbohydrates to proteins to form glycoproteins. In both the ER and the Golgi apparatus, resident enzymes catalyze the addition of various oligosaccharides (short chains of monosaccharides; see Concept 2.3) to certain amino acid R groups on proteins. One such type of "sugar coating" is essential for directing proteins to lysosomes. Other types are important for protein conformation and for recognition functions at the cell surface. As we noted in Chapter 8, different chains of sugars added to red blood cell proteins determine an individual's blood type. In many cases the attached oligosaccharides help stabilize proteins, such as those in the extracellular matrix, and those in the storage vacuoles of plants.
- *Phosphorylation* is the addition of phosphate groups to proteins and is catalyzed by protein kinases. The charged phosphate groups change the conformation of the protein, often exposing the active site of an enzyme or the binding

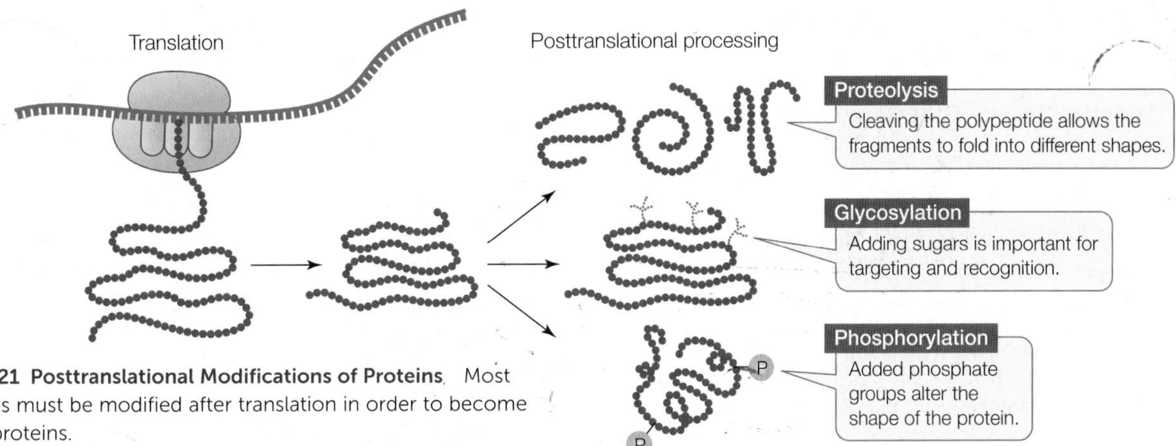

Translation
Posttranslational processing
Proteolysis
Cleaving the polypeptide allows the fragments to fold into different shapes.
Glycosylation
Adding sugars is important for targeting and recognition.
Phosphorylation
Added phosphate groups alter the shape of the protein.
P
P

FIGURE 10.21 Posttranslational Modifications of Proteins Most polypeptides must be modified after translation in order to become functional proteins.

site for another protein. Phosphorylation is especially important in cell signaling (see Concepts 5.5 and 5.6).

CHECKpoint CONCEPT 10.5

- ✓ Describe how signal sequences determine where a protein will go after it is made.
- ✓ What are some ways in which posttranslational modifications alter protein structure and function?
- ✓ Describe an experiment you would perform to test a proposed chloroplast-targeting signal sequence. Be specific about the type of cell and the proteins you would use. Describe the results you would expect if the sequence is indeed a chloroplast-targeting signal.

Q How do antibiotics target bacterial protein synthesis?

ANSWER Tetracyclines are antibiotics that are effective against some strains of MRSA and many other bacterial infections. They derive their name from the four hydrocarbon rings that are common to this family of molecules. Tetracyclines kill bacteria by interrupting translation (Concept 10.1). They do this by binding noncovalently to the small subunit of bacterial ribosomes (Concept 10.4), where binding changes ribosome structure such that charged tRNAs can no longer bind to the A site on the ribosome (**FIGURE 10.22**). The specificity of antibiotics for bacterial ribosomes comes from the fact that bacterial and eukaryotic ribosomes have different proteins and RNAs. The target protein for tetracyclines is not present in eukaryotic ribosomes, so these antibiotics disrupt translation in prokaryotes but not in eukaryotes, including humans.

Strains of MRSA with resistance to tetracyclines are emerging. The genes that confer resistance to this group of antibiotics are carried on mobile genetic elements such as plasmids, which can move at high frequencies between bacteria, by bacterial conjugation (see Concept 8.4). Some of the resistance genes encode proteins that transfer the tetracyclines out of the cell, whereas others encode proteins that prevent the antibiotics from binding to the ribosomes. These resistance genes present a major challenge, because MRSA can be lethal. To overcome resistance, new antibiotics are being developed and tried. The evolutionary race between genetically caused drug resistance and new therapies continues. In the meantime, health-care providers and the general public are being advised to take precautions to prevent the spread of MRSA.

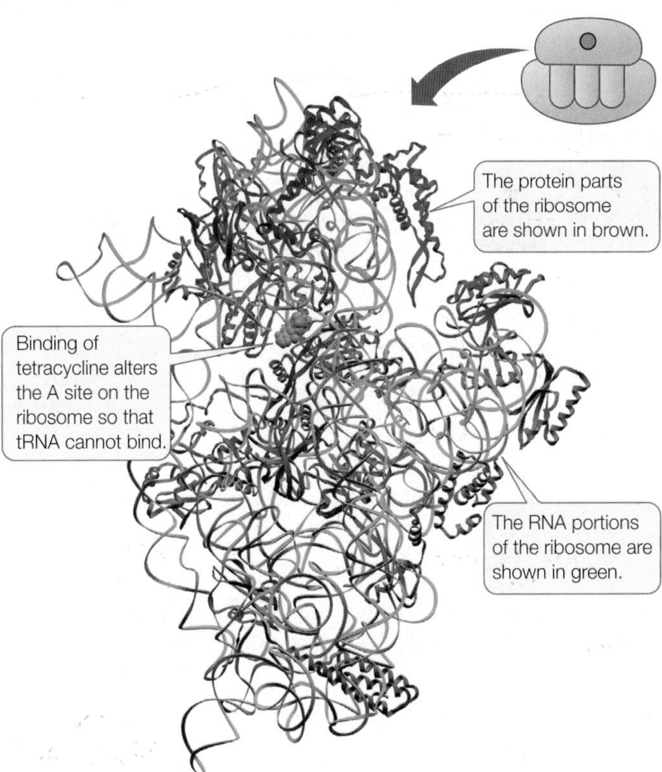

The protein parts of the ribosome are shown in brown.
Binding of tetracycline alters the A site on the ribosome so that tRNA cannot bind.
The RNA portions of the ribosome are shown in green.

FIGURE 10.22 An Antibiotic at the Ribosome The antibiotic tetracycline binds to the small subunit of bacterial ribosomes. This causes a change in the structure of the A site, preventing tRNAs from binding, and protein synthesis stops.

SUMMARY

CONCEPT 10.1 Genetics Shows That Genes Code for Proteins

- Studies of human genetic diseases such as alkaptonuria linked genes to proteins. **Review Figure 10.1**
- Hemoglobin abnormalities demonstrate that mutations can alter the sequence of amino acids in proteins. **Review Figure 10.2**
- Genes are expressed via transcription and translation. During **transcription**, the information in a gene is copied into a complementary RNA sequence. During **translation**, this RNA sequence is used to create the amino acid sequence of a polypeptide. **Review Figure 10.3 and ACTIVITY 10.1**
- The product of transcription is **messenger RNA (mRNA)**. **Transfer RNA (tRNA)** molecules translate the genetic information in the mRNA into a corresponding sequence of amino acids.
- **Ribosomal RNA (rRNA)** helps provide structure to the **ribosome** and acts as a ribozyme that catalyzes peptide bond formation between amino acids during protein synthesis.

CONCEPT 10.2 DNA Expression Begins with Its Transcription to RNA

- In a given gene, only one of the two strands of DNA (the template strand) acts as a template for transcription. **RNA polymerase** is the catalyst for transcription.
- RNA transcription from DNA proceeds in three steps: initiation, **elongation**, and **termination**. Initiation requires a **promoter** to which RNA polymerase binds. Elongation of the RNA molecule proceeds by the addition of nucleotides to the 3′ end of the molecule. **Review Figure 10.5 and ANIMATED TUTORIAL 10.1**
- After transcription, eukaryotic **pre-mRNA** is spliced to remove **introns**. **Review Figures 10.6 and 10.9 and ANIMATED TUTORIAL 10.2**
- Eukaryotic mRNA is also modified by the addition of a **5′ cap** and, at the 3′ end, a **poly A tail**.

CONCEPT 10.3 The Genetic Code in RNA Is Translated into the Amino Acid Sequences of Proteins

- Experiments involving synthetic mRNAs and protein synthesis in the test tube established the genetic code. **Review Figure 10.10 and ANIMATED TUTORIAL 10.3**
- The genetic code consists of triplets of mRNA nucleotide bases (**codons**) that correspond to 20 specific amino acids and to **start codons** and **stop codons**.
- The genetic code is redundant (an amino acid may be represented by more than one codon) but not ambiguous (no single codon represents more than one amino acid). **Review Figure 10.11 and ACTIVITY 10.2**
- Mutations in the coding regions of genes can be silent, missense, nonsense, or frame-shift mutations. **Review Figure 10.12 and ANIMATED TUTORIAL 10.4**

Review ANIMATED TUTORIAL 10.5

CONCEPT 10.4 Translation of the Genetic Code Is Mediated by tRNAs and Ribosomes

- Transfer RNA (tRNA) mediates between mRNA and amino acids during translation at the ribosome.
- Each tRNA species has an amino acid attachment site and an **anticodon** that is complementary to a specific mRNA codon. **Review Figure 10.13**
- A specific synthase enzyme charges each tRNA with its specific amino acid.
- Three sites on the large subunit of the ribosome interact with tRNA anticodons. The A site is where the charged tRNA anticodon binds to the mRNA codon. The P site is where the tRNA adds its amino acid to the growing polypeptide chain. The E site is where the tRNA is released. **Review Figure 10.14**
- Translation occurs in three steps: initiation, elongation, and termination. **Review Figures 10.15–10.17**
- In a **polyribosome**, or **polysome**, more than one ribosome moves along a strand of mRNA at one time. **Review Figure 10.18**

CONCEPT 10.5 Proteins Are Modified after Translation

- **Signal sequences** are short sequences of amino acids that direct polypeptides to their cellular destinations.
- These destinations include the nucleus and other organelles, which proteins enter after being recognized and bound by surface receptors.
- If a ribosome begins translating a polypeptide with an N-terminus RER signal sequence, it pauses and then resumes translation after attachment to a receptor in the RER membrane. **Review Figure 10.19**
- Posttranslational modifications of polypeptides include proteolysis, in which a polypeptide is cut into smaller fragments; glycosylation, in which sugars are added; and phosphorylation, in which phosphate groups are added. **Review Figure 10.21**

Go to the Interactive Summary to review key figures, Animated Tutorials, and Activities PoL2e.com/is10

Go to LaunchPad at **macmillanhighered.com/launchpad** for additional resources, including LearningCurve Quizzes, Flashcards, and many other study and review resources.

11 Regulation of Gene Expression

KEY CONCEPTS

11.1 Many Prokaryotic Genes Are Regulated in Operons

11.2 Eukaryotic Genes Are Regulated by Transcription Factors

11.3 Gene Expression Can Be Regulated via Epigenetic Changes to Chromatin

11.4 Eukaryotic Gene Expression Can Be Regulated after Transcription

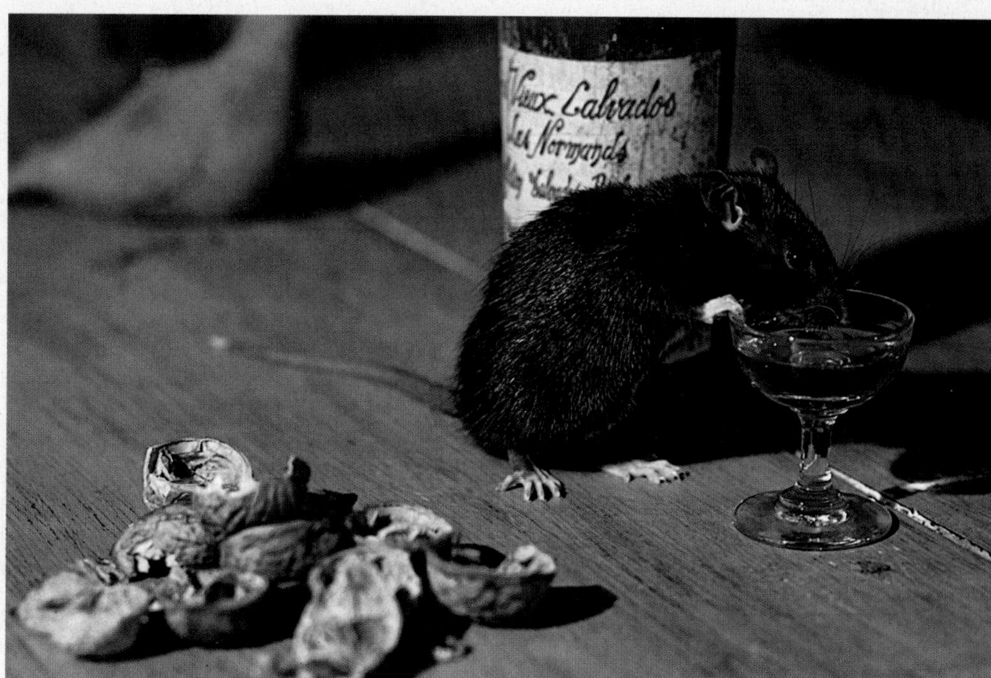

Some rats are genetically programmed to prefer alcohol over plain water.

Many people drink alcoholic beverages, but relatively few of them become addicted to alcohol (alcoholic). Alcoholism is characterized by a compulsion to consume alcohol, tolerance (increasing doses are needed for the same effect), and dependence (abrupt cessation of consumption leads to severe withdrawal symptoms). In most alcoholics, alcohol provides pleasant sensations (positive reinforcement) and alleviates unpleasant ones such as anxiety (negative reinforcement).

Alcoholism is a complex disease. Psychologists sometimes speak of "addictive personalities," and genetic studies indicate there may be inherited factors that predispose people to the disease. One approach to describing the genes involved is to study animal models of alcoholism at the molecular level. James Murphy at Indiana University has bred a genetic strain of alcoholic rats, called P rats, that prefer alcohol when given the choice of alcohol-containing or alcohol-free water. P rats show many of the symptoms of addiction, including compulsive drinking, tolerance, and withdrawal. These rats appear more anxious than wild-type rats, spending more time in a closed rather than an open environment. Drinking alcohol alters this behavior and seems to relieve their anxiety.

There may be a link between a particular protein and alcohol consumption. CREB (*c*yclic AMP *r*esponse *e*lement *b*inding protein) is abundant in the brain and regulates the expression of hundreds of genes that are important in metabolism. CREB becomes activated when it is phosphorylated by the enzyme protein kinase A, which in turn is activated by the second messenger cyclic AMP. In an effort to understand the molecular basis of alcoholism and anxiety, neuroscientist Subhash Pandey and his colleagues at the University of Illinois compared CREB levels in the brains of P rats and wild-type rats. They found that P rats have inherently lower levels of CREB in certain parts of the brain. When these rats consumed alcohol, the total levels of CREB did not increase, but the levels of phosphorylated CREB did. It is the phosphorylated version of CREB that regulates gene transcription.

The prospect that CREB, a molecule that regulates gene expression, is a key element in the genetic propensity for alcoholism is important because it begins to explain the molecular nature of a complex behavioral disease. Such understanding may permit more effective treatment of alcohol abuse, or even its prevention. And for our purpose here, it underscores the importance of the regulation of gene expression in biological processes.

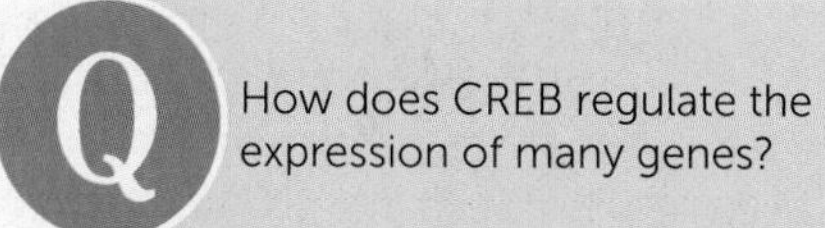

Q How does CREB regulate the expression of many genes?

You will find the answer to this question on page 232.

CONCEPT 11.1 Many Prokaryotic Genes Are Regulated in Operons

In Chapter 10 we introduced the concepts of gene expression. DNA is initially expressed as RNA, and in many cases the RNA is then translated into protein by the ribosome. Throughout this book we describe instances where gene expression is altered so that the level of protein produced from a particular gene is altered. Such changes are influenced by environmental conditions and the developmental stage of the cell or organism. Here are a few examples:

- In Chapter 5: When an extracellular signal binds to its receptor on a eukaryotic cell, it sets in motion a signal transduction pathway that may end with some genes being activated (their expression switched on) or others being repressed (their expression switched off).
- In Chapter 7: During the cell cycle, cyclins are synthesized only at specific stages. The genes for cyclins are inactive at other stages in the cycle.
- In Chapter 9: When a virus infects a host cell, it can "hijack" the host gene expression machinery and divert it to viral gene expression.

These and other examples indicate that gene expression is precisely regulated.

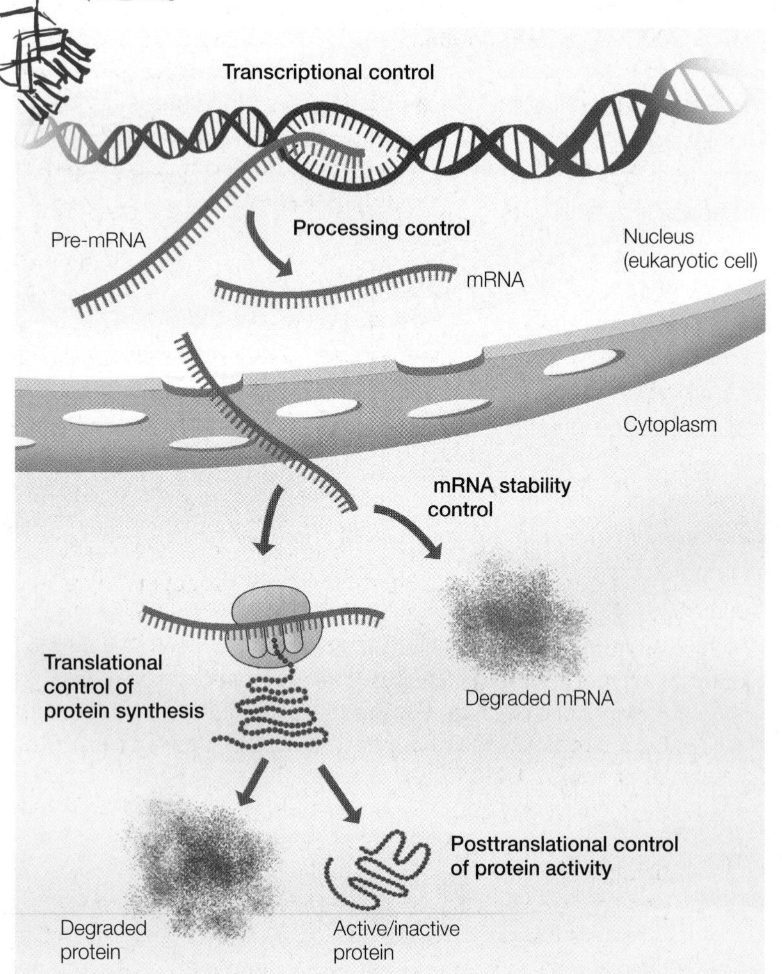

Genes are subject to positive and negative regulation

At every step of the way from DNA to protein, gene expression can be regulated (**FIGURE 11.1**). As we proceed through this chapter, you will see examples of gene regulation at the transcriptional, posttranscriptional, translational, and posttranslational levels. An important form of gene regulation is at the level of transcription.

Gene expression begins at the **promoter**, a region of DNA containing the site where RNA polymerase binds to initiate transcription. As we mentioned above, not all genes are active (being transcribed) at a given time. Two types of regulatory proteins—called **transcription factors**—control whether or not a gene is active: repressors and activators. These proteins bind to specific DNA sequences at or near the promoter (**FIGURE 11.2**):

- In negative regulation, a **repressor** binds a specific site in or near the promoter to prevent transcription.
- In positive regulation, the binding of an **activator** stimulates transcription.

You will see these mechanisms, or combinations of them, as we examine the regulation of prokaryotic, eukaryotic, and viral genes. Let's begin by looking at the regulation of gene expression in prokaryotes.

LINK

You can review the processes of transcription in **Concept 10.2**

Regulating gene transcription is a system that conserves energy

Prokaryotes conserve energy and resources by making certain proteins only when they are needed. Because their environments can change abruptly, prokaryotes have evolved mechanisms to rapidly alter the expression levels of certain genes when conditions warrant. An example is the bacterium *Escherichia coli*, which normally inhabits the intestines of humans and other mammals. *E. coli* must be able to adjust to sudden changes in its chemical environment as the foods consumed by its host change (for example, from a meal containing glucose at one time to one containing lactose at another). In many cases *E. coli* responds to such changes by changing the transcription of its genes. To illustrate this we will look at the regulation of the pathway for lactose catabolism in *E. coli*.

FIGURE 11.1 Potential Points for the Regulation of Gene Expression In a eukaryotic cell, gene expression can be regulated before transcription, during transcription, after transcription but before translation, at translation, or after translation.

Go to ACTIVITY 11.1 Eukaryotic Gene Expression Control Points
PoL2e.com/ac11.1

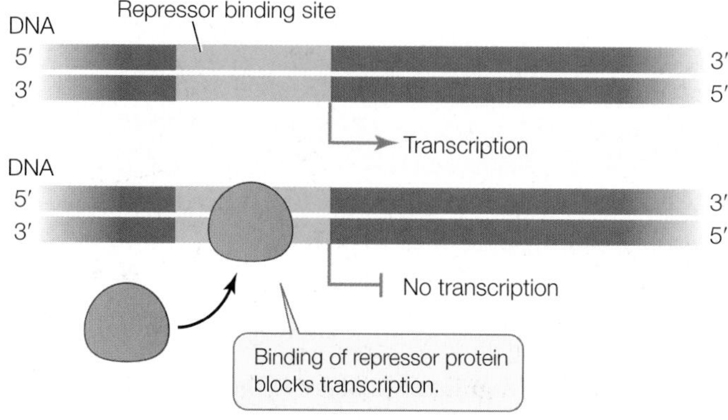

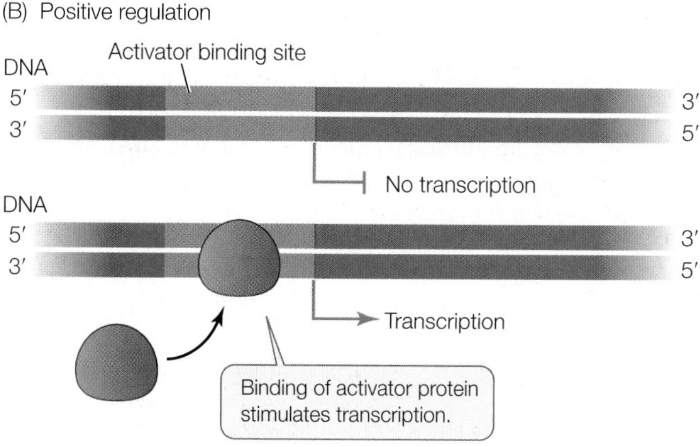

FIGURE 11.2 Positive and Negative Regulation Transcription factors regulate gene expression by binding to DNA and (A) repressing or (B) activating transcription by RNA polymerase.

Lactose is a β-galactoside—a disaccharide containing galactose linked to glucose. Three proteins are involved in the initial uptake and metabolism of lactose by *E. coli*:

- β-galactoside permease is a carrier protein in the bacterial cell membrane that moves the sugar into the cell.
- β-galactosidase is an enzyme that hydrolyzes lactose to glucose and galactose.
- β-galactoside transacetylase transfers acetyl groups from acetyl CoA to certain β-galactosides. Its role in the metabolism of lactose is not clear.

When *E. coli* is grown on a medium that contains glucose but no lactose, the basal (uninduced) levels of these three proteins are extremely low—only a few molecules per cell. But if the cells are transferred to a medium with lactose as the predominant sugar, they promptly begin making all three proteins after a short lag period, and within 10 minutes there are about 3,000 of each of these proteins per cell (the induced level):

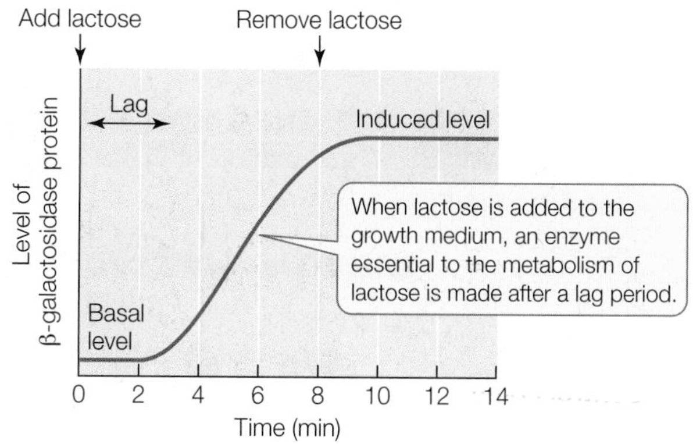

What causes this dramatic increase? A clue comes from measuring the concentration of mRNA for β-galactosidase. After lactose is added to the medium, the mRNA level increases *before* the level of β-galactosidase protein begins to rise:

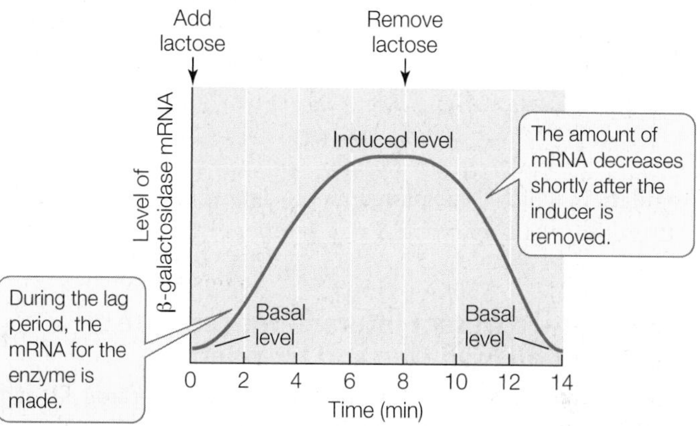

The mRNA is produced during the lag phase and then is translated into protein. The high mRNA level depends on the presence of lactose, because if lactose is removed, the mRNA level goes down. The response of the bacteria to lactose is clearly at the level of transcription. (The level of β-galactosidase protein does not go down immediately after the inducer is removed because the protein is more stable than the mRNA.)

Compounds that stimulate the transcription of specific genes are called **inducers**, and genes that can be activated by inducers are called **inducible genes**. In contrast, some other genes are expressed most of the time at a constant rate; these are called **constitutive genes**. The lactose-metabolizing proteins in *E. coli* are encoded by inducible genes. When lactose first enters the cell, some of it is converted to a similar molecule called allolactose. Allolactose is the inducer that switches on the expression of the genes for the lactose-metabolizing proteins.

Operons are units of transcriptional regulation in prokaryotes

The genes that encode the three proteins for processing lactose in *E. coli* lie adjacent to one another on the *E. coli* chromosome. This arrangement—which is common for functionally related genes in prokaryotes—is no coincidence: the genes share a single promoter, and their DNA is transcribed into a single, continuous molecule of mRNA that contains the coding regions for the three proteins. Because this particular mRNA governs the synthesis of all three lactose-metabolizing enzymes, either all or none of these enzymes are made at any particular time.

A cluster of genes with a single promoter is called an **operon**, and the operon that encodes the three lactose-metabolizing enzymes in *E. coli* is called the *lac* operon. The *lac* operon promoter can be very efficient (the maximum rate of mRNA synthesis can be high), but its activity can be reduced when the enzymes are not needed. This example of transcriptional

lac Operon
DNA
P_i — Gene *i* promoter
i — Gene for repressor protein
P_{lac} — *lac* operon promoter
o — Operator
z — β-galactosidase gene
y — β-galactoside permease gene
a — β-galactoside transacetylase gene

FIGURE 11.3 The *lac* Operon of *E. coli* The *lac* operon is a segment of DNA that includes a promoter, an operator, and the three genes that code for lactose-metabolizing enzymes. In reality, the coding sequences (genes) are much longer than the short regulatory sequences.

regulation, which we explore in more detail below, was worked out in the 1960s by Nobel Prize winners François Jacob and Jacques Monod.

Operator–repressor interactions regulate transcription in the *lac* and *trp* operons

The *lac* operon has a DNA sequence called an operator, which is near the promoter and controls transcription of the *lac* genes (**FIGURE 11.3**). An **operator** is a repressor-binding site that can bind very tightly with a repressor protein (see Figure 11.2A). Repressors play different roles in different operons:

- An inducible operon is turned off unless needed.
- A repressible operon is turned on unless not needed.

In the case of the inducible *lac* operon, a repressor protein prevents transcription until the *lac*-encoded proteins are needed. In contrast, the *trp* operon (described below) is a repressible operon that is turned off by a repressor only under particular circumstances.

lac OPERON As we described above, the *lac* operon is not transcribed at high levels unless a β-galactoside such as lactose is the predominant sugar available in the cell's environment. A repressor protein normally binds to the operator, preventing RNA polymerase from binding and thereby blocking transcription. When lactose is present, the repressor detaches from the operator, allowing RNA polymerase to bind to the promoter and start transcribing the *lac* genes (**FIGURE 11.4**).

The key to this regulatory system is the repressor protein. Expressed from a constitutive gene (one that is always active), the repressor is always present in the cell in adequate amounts to occupy the operator and keep the operon turned off. The repressor has a recognition site for the DNA sequence in the operator, and it binds very tightly. However, it also has an allosteric binding site for the inducer. When the inducer (allolactose) binds to the repressor, the repressor changes shape so that it can no longer bind DNA.

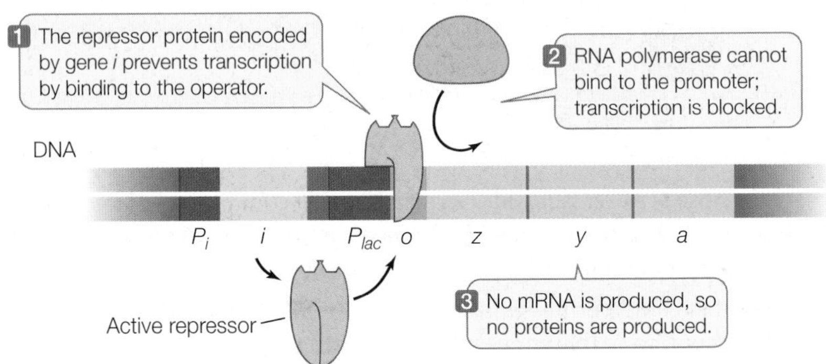

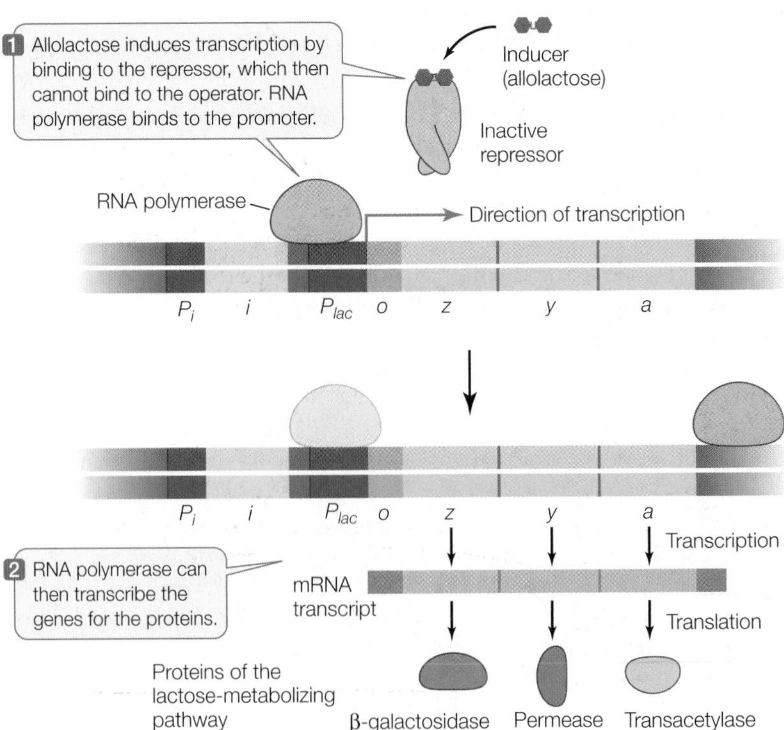

FIGURE 11.4 The *lac* Operon: An Inducible System Allolactose (the inducer) leads to synthesis of the proteins in the lactose-metabolizing pathway by binding to the repressor protein and preventing its binding to the operator.

Go to ANIMATED TUTORIAL 11.1
The *lac* Operon
PoL2e.com/at11.1

The gene for the lac repressor (gene *i* in Figure 11.3) is located upstream of the *lac* operon on the *E. coli* chromosome. The *lac i* gene is referred to as a regulatory gene because it encodes a regulatory protein (a transcription factor). In contrast, a **structural gene** is any gene that encodes a protein that is not directly involved in gene regulation. The three genes that encode the lactose-metabolizing enzymes are structural genes.

***trp* OPERON** Like an inducible operon, a repressible operon is switched off when its repressor is bound to its operator. However, in this case the repressor binds to the DNA only in the presence of a **corepressor**. The corepressor is a molecule that binds to the repressor, causing it to change shape and bind to the operator, thereby inhibiting transcription. An example is the operon whose structural genes catalyze the synthesis of the amino acid tryptophan:

$$\text{Precursor molecules} \xrightarrow{\text{Five enzyme-catalyzed reactions}} \text{Tryptophan}$$

When tryptophan is present in adequate concentrations, it is energy-efficient for the cell to stop making the enzymes for tryptophan synthesis. Therefore tryptophan itself functions as a corepressor: tryptophan binds to the repressor of the *trp* operon, causing the repressor to bind to the *trp* operator to prevent transcription of the enzymes in the pathway (**FIGURE 11.5**).

To summarize the differences between these two regulatory systems:

- In *inducible* systems, the substrate of a metabolic pathway (the inducer) interacts with a transcription factor (the repressor), rendering the repressor incapable of binding to the operator and thus allowing transcription.
- In *repressible* systems, a product of a metabolic pathway (the corepressor) binds to the repressor protein, which is then able to bind to the operator and block transcription.

In general, inducible systems control catabolic pathways (which are turned on only when the substrate is available), whereas repressible systems control anabolic pathways (which are turned on until the concentration of the product becomes sufficient).

LINK
You can review catabolic and anabolic reactions in **Concept 2.5**

In both of the systems described above, the regulatory protein is a repressor that functions by binding to the operator. Transcription in prokaryotes can also be regulated by activator proteins that bind to DNA sequences at or near the promoter and promote transcription (see Figure 11.2). Like repressors, activators can regulate both inducible and repressible systems. Furthermore, many genes and operons are controlled by more than one regulatory mechanism. We will discuss transcription factors in more detail in Concept 11.2.

We have now seen two basic systems for regulating a metabolic pathway. In Concept 3.4 we described the allosteric

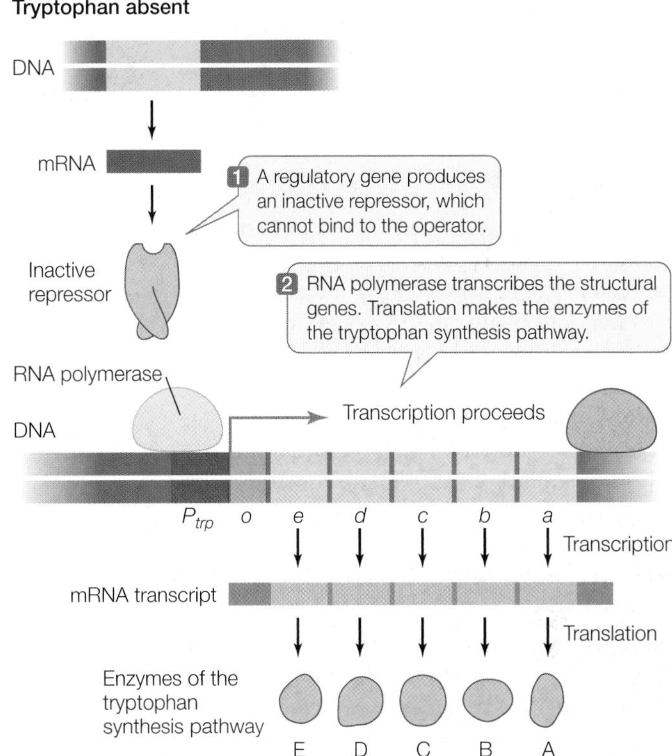

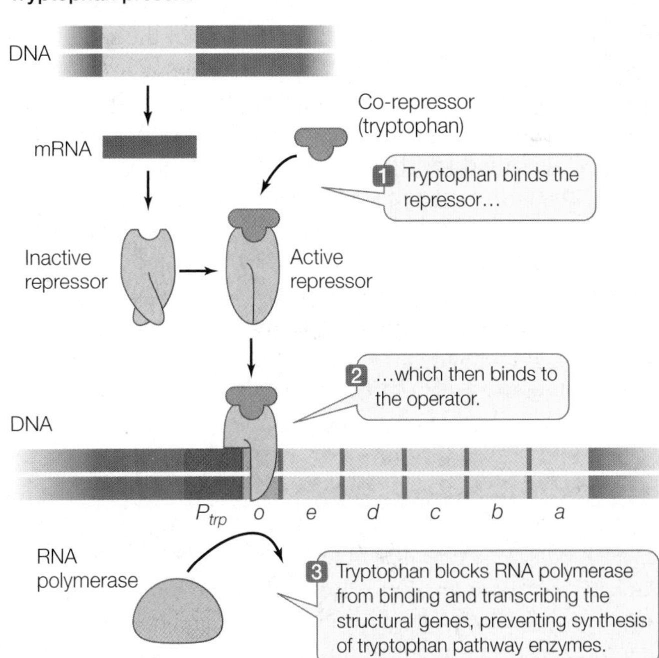

FIGURE 11.5 The *trp* Operon: A Repressible System Because tryptophan activates an otherwise inactive repressor, it is called a corepressor.

Go to ANIMATED TUTORIAL 11.2
The *trp* Operon
PoL2e.com/at11.2

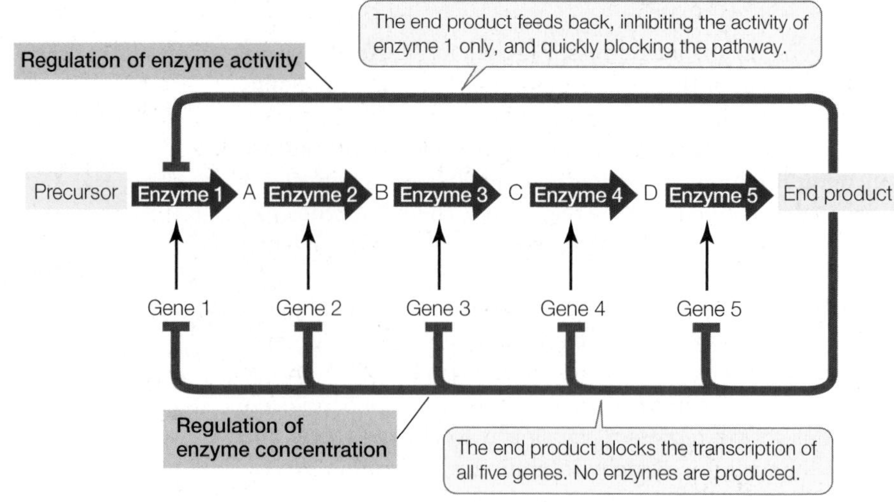

FIGURE 11.6 Systems to Regulate a Metabolic Pathway Feedback from the end product of a metabolic pathway can block enzyme activity (allosteric regulation), or it can stop the transcription of genes that code for the enzymes in the pathway (transcriptional regulation).

regulation of enzyme activity—a mechanism that allows rapid fine-tuning of metabolism. The regulation of transcription is slower but results in greater savings of energy and resources. Protein synthesis is a highly endergonic process; synthesizing mRNA, charging tRNA, and moving the ribosomes along mRNA all require large amounts of energy. **FIGURE 11.6** compares allosteric and transcriptional regulation.

RNA polymerase can be directed to a class of promoters

As noted above and in Chapter 10, RNA polymerase binds to specific DNA sequences at the promoter to initiate transcription. We have just described how repressor proteins can physically block RNA polymerase binding. However, there are other proteins in prokaryotes called **sigma factors** that can bind to RNA polymerase and direct the polymerase to specific promoters.

Genes that encode proteins with related functions may be at different locations in the genome but have the same promoter sequence. This allows them to be expressed at the same time and under the same physiological conditions. For example, some bacteria stop growing when nutrients in their environment are depleted. When this happens, they adopt an alternative lifestyle called sporulation—they reduce their metabolic activity and form a tough spore coat (see Concept 19.2). This process involves the sequential expression of specific classes of genes. Each member of a gene class has a common promoter sequence, and RNA polymerase is directed to the promoter in each case by a specific sigma factor. As we will see in Concept 11.2, this form of global gene regulation by proteins binding to RNA polymerase is also common in eukaryotes.

Viruses use gene regulation strategies to hijack host cells

The immunologist Sir Peter Medawar once described a virus as "a piece of bad news wrapped in protein." As we described in Concept 9.1, a **virus** injects its genetic material into a host cell, and in many cases it turns that cell into a virus factory:

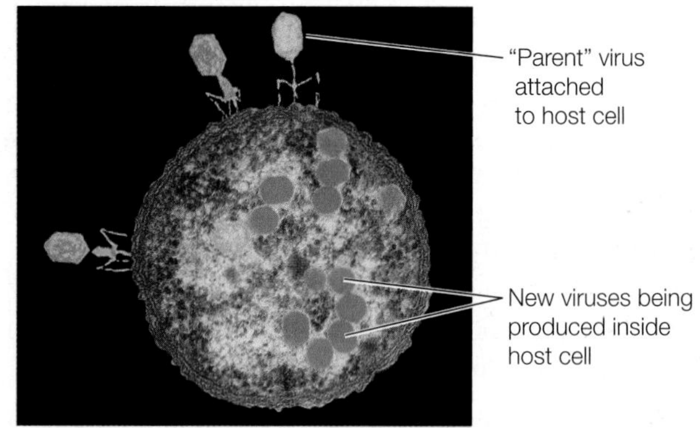

This involves a radical change in gene expression for the host cell, and can result in the death of the cell when new viral particles are released.

Viruses are not cells and do not carry out many of the processes characteristic of life. They are dependent on living

APPLY THE CONCEPT

Many prokaryotic genes are regulated in operons, which include regulatory DNA sequences

Genetic mutations are useful in analyzing the control of gene expression. In the *lac* operon of *E. coli* (see Figures 11.3 and 11.4), gene *i* codes for the repressor protein, P_{lac} is the promoter, *o* is the operator, and *z* is the first structural gene. The superscript "+" designates the wild type; superscript "–" means mutant. Fill in the table, describing the level of transcription in different genetic and environmental conditions. (The first line of the table has been filled in as an example.)

	z TRANSCRIPTION LEVEL	
GENOTYPE	LACTOSE PRESENT	LACTOSE ABSENT
$i^+ P_{lac}^+ o^+ z^+$ (wild type)	High	Low
$i^- P_{lac}^+ o^+ z^+$		
$i^+ P_{lac}^+ o^+ z^-$		
$i^+ P_{lac}^- o^+ z^+$		
$i^+ P_{lac}^+ o^- z^+$		

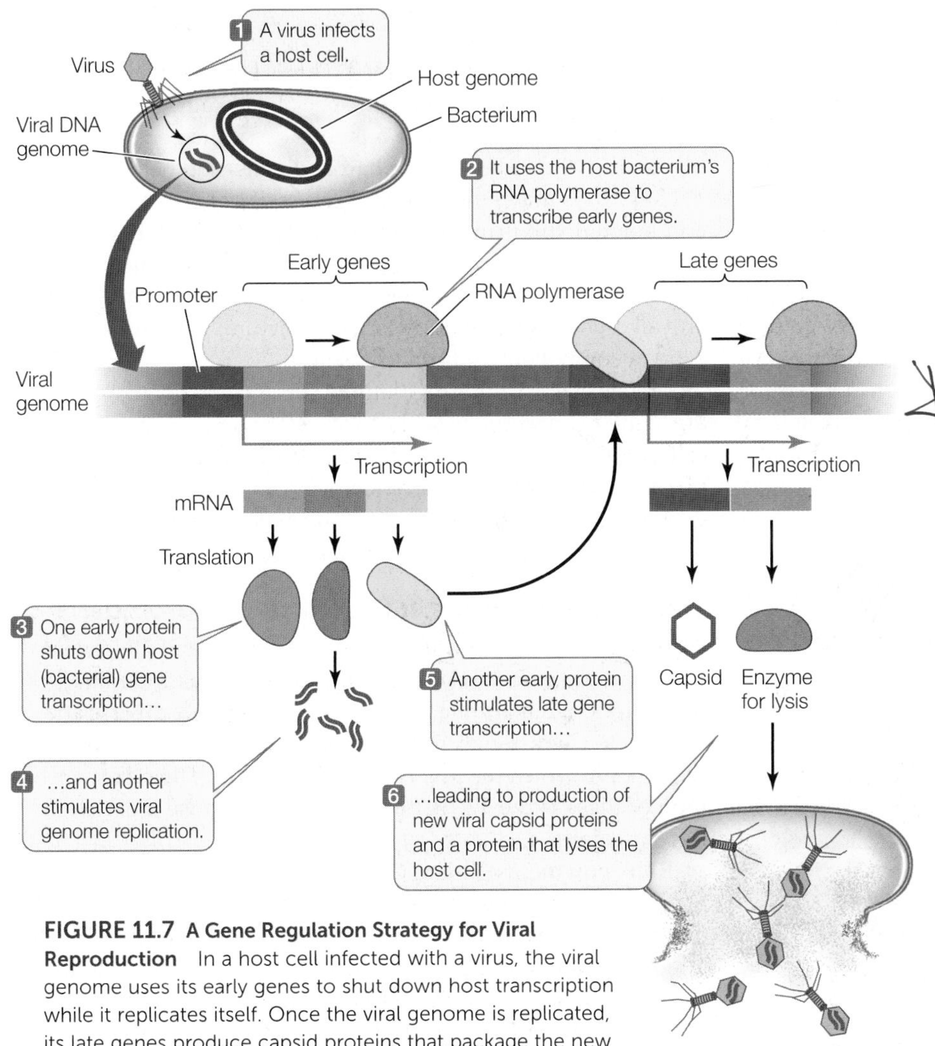

FIGURE 11.7 A Gene Regulation Strategy for Viral Reproduction In a host cell infected with a virus, the viral genome uses its early genes to shut down host transcription while it replicates itself. Once the viral genome is replicated, its late genes produce capsid proteins that package the new genomes, and other proteins that lyse the host cell.

cells to reproduce. Unlike living cells, not all viruses use double-stranded DNA as the genetic material that is contained within the viral particle and transmitted from one generation to the next. The viral genome may consist of double-stranded DNA, single-stranded DNA, or double- or single-stranded RNA. But whether the genetic material is DNA or RNA, the viral genome takes over the host's protein synthetic machinery within minutes of entering the cell.

Typically, the host cell immediately begins to produce new viral particles (virions), which are released as the cell breaks open, or lyses. This type of prokaryotic viral life cycle is called **lytic**. Some viral life cycles also include a **lysogenic** or dormant phase. In this case the viral genome becomes incorporated into the host cell genome and is replicated along with the host genome. The virus may survive in this way for many host cell generations. Sooner or later, an environmental signal can cause the host cell to begin producing virions—at which point the viral reproductive cycle enters the lytic phase.

LINK

The different types of viruses are described in **Concept 19.4**

FIGURE 11.7 illustrates molecular events in the lytic life cycle of T4, a typical double-stranded DNA bacteriophage (phage, or bacterial virus). At the molecular level, the lytic cycle has two stages, early and late:

- The viral genome contains a promoter that binds host RNA polymerase. In the early stage, viral genes that lie adjacent to this promoter are transcribed. These early genes encode proteins that shut down expression of host genes, stimulate viral genome replication, and activate the transcription of viral late genes. The host genes are shut down by a posttranscriptional mechanism: a virus-encoded enzyme degrades the host RNA before it can be translated. Another viral nuclease digests the host's chromosome, providing nucleotides for the synthesis of many copies of the viral genome. These processes can occur within a few minutes after the virus first infects the cell.
- In the late stage, viral late genes are transcribed; they encode the viral capsid proteins and enzymes that lyse the host cell to release the new virions.

Under ideal conditions, this entire process—from binding and infection to release of new phage—can be completed in only half an hour.

CHECKpoint CONCEPT 11.1

- ✓ What is the difference between positive and negative regulation of gene expression?
- ✓ Describe the molecular conditions at the *lac* operon promoter in the presence and absence of lactose.
- ✓ Describe the molecular events at the *trp* operon promoter in the presence and absence of tryptophan.
- ✓ If the *lac* repressor gene were mutated so that the allosteric site on the encoded protein no longer bound allolactose, what would be the effect on transcription of the *lac* operon? What about a similar mutation in the *trp* repressor gene?
- ✓ What would be the effect of a mutation in the gene that encodes RNA polymerase so that it does not bind to the late gene promoter of bacteriophage T4?

Studies of bacteria and bacteriophage provide a basic understanding of the mechanisms that regulate gene expression and of the roles of regulatory proteins in both positive and negative regulation. We will now turn to the control of gene expression in eukaryotes. You will see both negative and positive control of transcription, as well as posttranscriptional mechanisms of regulation.

CONCEPT 11.2 Eukaryotic Genes Are Regulated by Transcription Factors

As we mentioned in Concept 11.1, gene expression can be regulated at several different points in the process of transcribing a gene and translating the mRNA into a protein (see Figure 11.1). In this concept we will describe the mechanisms that result in the selective transcription of specific eukaryotic genes. As in prokaryotes, eukaryotic cells must precisely regulate the expression of their genes. Some genes are constitutive (expressed in most tissues most of the time), whereas others are inducible and expressed only when needed. This is especially important in multicellular organisms with specialized cells and tissues. For example, virtually all of our cells carry the genes encoding keratin (the protein in our hair and nails) and hemoglobin. Yet keratin is made only by epithelial cells such as skin cells, and hemoglobin is made only by developing red blood cells. In contrast, all human cells express the genes that encode enzymes needed for basic metabolic activities (such as glycolysis), and all cells must synthesize certain structural proteins such as actin (a component of the cytoskeleton).

The mechanisms for regulating transcription in eukaryotes are similar conceptually to those of prokaryotes. Both types of cells use DNA–protein interactions to bring about negative and positive control of gene expression. However, there are significant differences, which generally reflect the greater complexity of eukaryotic organisms (**TABLE 11.1**).

Transcription factors act at eukaryotic promoters

As in bacteria, a eukaryotic promoter is a region of DNA near the 5′ end of a gene where RNA polymerase binds and initiates transcription. Eukaryotic promoters are extremely diverse and difficult to characterize, but they each contain a core promoter sequence to which the RNA polymerase binds. The most common of these is the **TATA box**—so called because it is rich in A-T base pairs.

RNA polymerase II is the polymerase that transcribes the protein-coding genes in eukaryotes. It cannot bind to the promoter and initiate transcription by itself. Rather, it does so only after various **general transcription factors** have bound to the core promoter. General transcription factors bind to most promoters and are distinct from transcription factors that have specific regulatory effects only at certain promoters or classes of promoters. **FIGURE 11.8** illustrates the assembly of the resulting transcription complex at a promoter containing a TATA box. First, the protein TFIID ("TF" stands for transcription factor) binds to the TATA box. Binding of TFIID changes both its own shape and that of the DNA, presenting a new surface that attracts the binding of other transcription factors. RNA polymerase II binds only after several other proteins have bound to the complex.

The core promoter sequence is bound by general transcription factors that are needed for the expression of all RNA polymerase II–transcribed genes. Other sequences that are (usually) found in or near promoter regions are specific to only a few genes and are recognized by specific transcription factors. These transcription factors may be positive

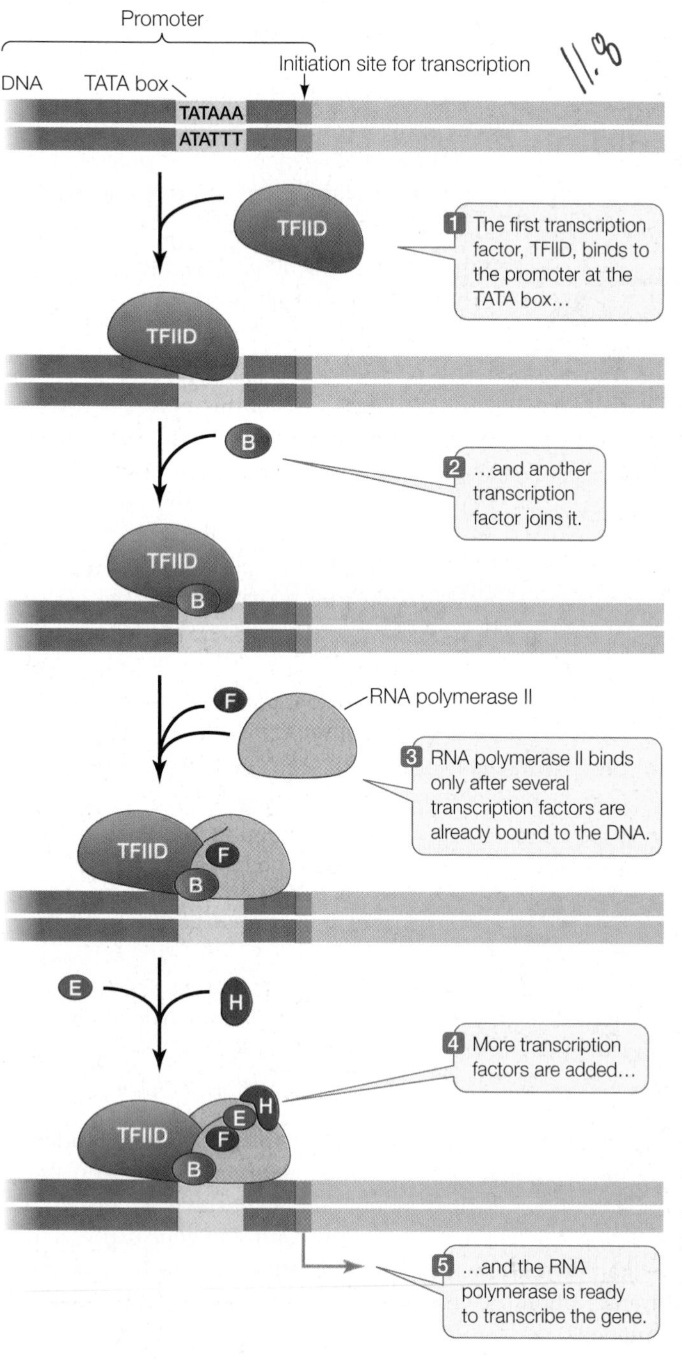

Promoter
DNA
TATA box
Initiation site for transcription
11.8
TATAAA
ATATTT
TFIID
1 The first transcription factor, TFIID, binds to the promoter at the TATA box…
B
2 …and another transcription factor joins it.
F
RNA polymerase II
3 RNA polymerase II binds only after several transcription factors are already bound to the DNA.
E
H
4 More transcription factors are added…
5 …and the RNA polymerase is ready to transcribe the gene.

FIGURE 11.8 The Initiation of Transcription in Eukaryotes Apart from TFIID, which binds to the TATA box, each transcription factor in this transcription complex has binding sites only for the other proteins in the complex, and does not bind directly to DNA. B, E, F, and H are general transcription factors.

Go to ANIMATED TUTORIAL 11.3
Initiation of Transcription
PoL2e.com/at11.3

TABLE 11.1 Transcription in Bacteria and Eukaryotes

Characteristic	Bacteria	Eukaryotes
Locations of functionally related genes	Often clustered in operons	Often distant from one another with separate promoters
RNA polymerases	One	Three:
		I: transcribes rRNA
		II: transcribes mRNA
		III: transcribes tRNA and small RNAs
Promoters and other regulatory sequences	Few	Many
Initiation of transcription	Binding of RNA polymerase	Binding of many proteins, including RNA polymerase, to promoter

regulators (activators) or negative regulators (repressors) of transcription:

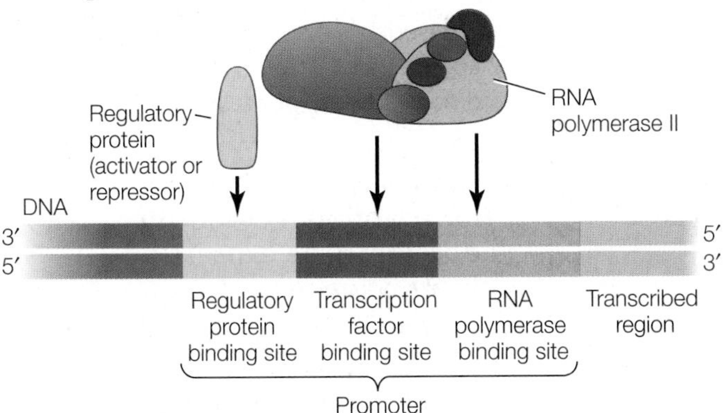

Such transcription factors may be present only in certain cell types, or they may be present in all cells but activated by specific signals. DNA sequences that bind activators are called enhancers, and those that bind repressors are called silencers. Some enhancers and silencers occur near the core promoter, and others can be as far as 20,000 base pairs away. When the activators or repressors bind to these DNA sequences, they interact with the RNA polymerase complex, causing the DNA to bend. Often many such binding proteins are involved, and the *combination* of factors present determines whether transcription is initiated. With about 2,000 different transcription factors in humans, there are many possibilities for regulation.

How do transcription factors recognize a specific nucleotide sequence in DNA? To answer this question, let's look at a specific example. NFATs (*n*uclear *f*actors of *a*ctivated *T* cells) are a group of transcription factors that control the expression of genes essential for the immune response (see Chapter 39). NFAT proteins bind to a 12-bp recognition sequence near the promoters of these genes, with the sequence CGAGGAAAATTG (**FIGURE 11.9**). Recall that there are atoms in the bases of DNA that are available for hydrogen bonding but are not involved in base pairing (see Figure 9.6). These atoms are important in the interactions between an NFAT and the DNA. In addition, there are hydrophobic interactions between the rings in the DNA bases and some amino acid R groups in the protein. As for an enzyme and its substrate (see Concept 3.3), there is an induced fit between the NFAT and the DNA, such that the protein undergoes a conformational change after binding begins.

The base sequence of a binding site on DNA determines the arrangement of chemical groups available for hydrogen bonding and hydrophobic interactions with DNA-binding proteins; this is the basis of the specificity of DNA–protein interactions.

The expression of transcription factors underlies cell differentiation

During the development of a complex organism from fertilized egg to adult, cells become more and more differentiated (specialized). Differentiation is brought about in many cases by changes in gene expression, resulting from the activation (and inactivation) of transcription factors. We will discuss this topic in more detail in Chapter 14. For now, remember that virtually all differentiated cells contain the entire genome, and that their specific characteristics arise from differential gene expression.

Currently there is great interest in cellular therapy: providing new, functional cells to patients who have diseases that involve the degeneration of certain cell types. An example is Alzheimer's disease, which involves the degeneration of neurons in the brain. Because of the possibility of immune system rejection (see Chapter 39), it would be optimal if patients could receive their own cells, modified in some way to be functional. Since specialized functions are under the control of transcription factors, turning readily available cells into a particular

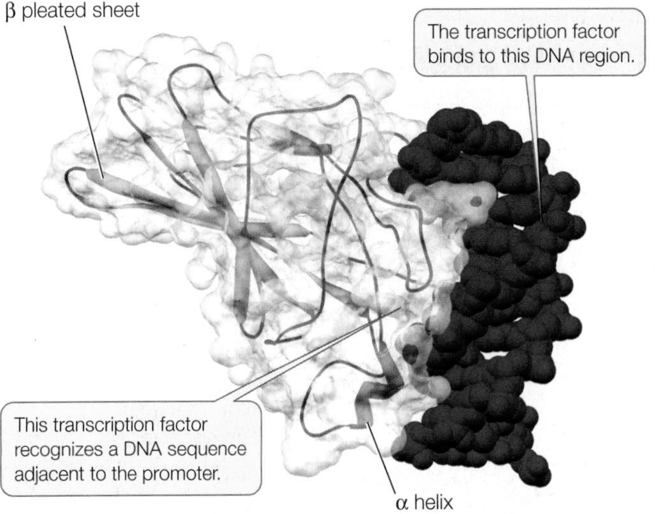

FIGURE 11.9 A Transcription Factor Protein Binds to DNA The transcription factor NFAT activates genes for the immune response by binding to a specific DNA sequence near the promoters of those genes.

INVESTIGATION

FIGURE 11.10 Expression of Specific Transcription Factors Turns Fibroblasts into Neurons Fibroblasts are cells that secrete abundant extracellular matrix and contribute to the structural integrity of organs. Neurons are highly specialized cells in the nervous system. Marius Wernig and his colleagues performed a series of experiments to find out whether expressing neuronal transcription factors in fibroblasts would be sufficient to cause the fibroblasts to become neurons.[a]

HYPOTHESIS

Expression of neuron-specific transcription factors in fibroblasts will turn the latter into neurons.

METHOD

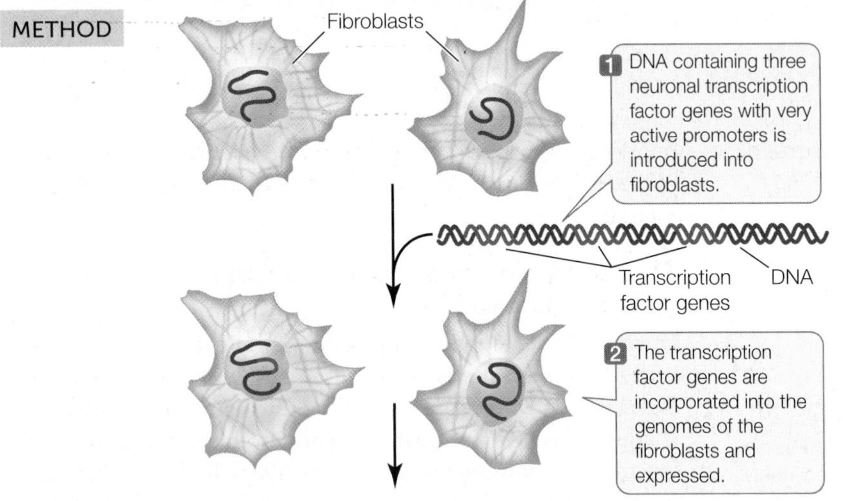

RESULTS

After 6 days, the fibroblasts develop into functional neurons, which form characteristic synapses with one another.

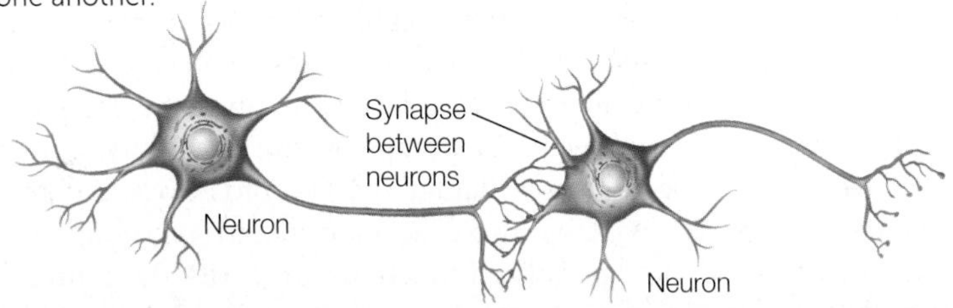

CONCLUSION

The expression of just three transcription factors is sufficient to transform a fibroblast into a neuron.

ANALYZE THE DATA

Fibroblasts are active in cell division; neurons are not. In addition to morphology, the lack of cell division was used as a criterion to show that the transformed cells were neurons. The rate of cell division in the transformed cells was measured by the incorporation of the labeled nucleotide bromodeoxyuridine (BrdU) into DNA. The percentage of labeled (hence dividing) cells is shown in the graph.

A. Was cell division stopped in the transformed cells? Explain your answer.

B. The error bars are standard deviations. What statistical test would you use to show whether the difference between the two cell populations was significant? See Appendix B for a Statistics Primer.

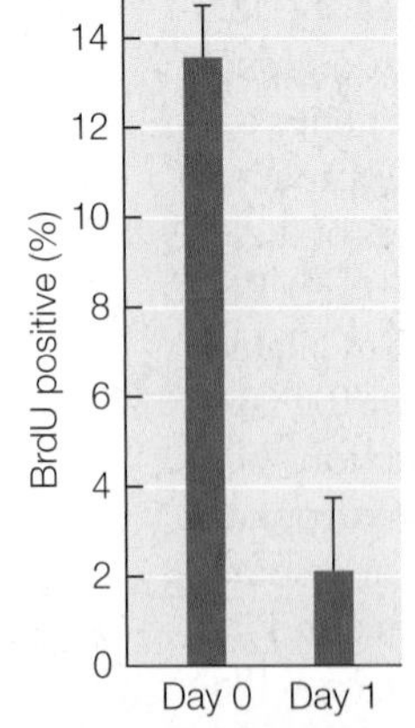

Go to **LaunchPad** for discussion and relevant links for all **INVESTIGATION** figures.

[a]T. Vierbuchen et al. 2010. *Nature* 463: 1035–1041.

desired cell type might be achieved by altering transcription factor expression. Marius Wernig and his colleagues at Stanford University have made important progress toward this goal (**FIGURE 11.10**). They took skin fibroblasts from mice and manipulated the expression of transcription factors in the cells to change them into neurons. By repeating their experiments on human fibroblasts, they have brought cellular therapy closer to reality.

LINK

The basis for rejection of nonself cells is described in **Concept 39.5**. Additional approaches used in cellular therapy are discussed in **Concept 14.1**

Transcription factors can coordinate the expression of sets of genes

We have seen that prokaryotes can coordinate the regulation of several genes by arranging them in an operon. In addition, bacteria can coordinate the expression of groups of genes using sigma factors, which guide RNA polymerase to particular classes of promoters. This latter mechanism is also used in eukaryotes to coordinately regulate genes that may be far apart, even on different chromosomes. The expression of genes can be coordinated if they share regulatory sequences that bind the same transcription factors.

This type of coordination is used by organisms to respond to stress—for example, by plants in response to drought. Under conditions of drought stress, a plant must simultaneously synthesize numerous proteins whose genes are scattered throughout the genome. The synthesis of these proteins comprises the stress response. To coordinate expression, each of these genes has a specific regulatory sequence near its promoter called the dehydration response element (DRE). In response to drought, a transcription factor changes so that it binds to this element and stimulates mRNA synthesis (**FIGURE 11.11**). The dehydration response proteins not only help the plant conserve water but also protect the plant against freezing or excess salt in the soil. This finding has considerable importance for agriculture because crops are often grown under less than optimal conditions.

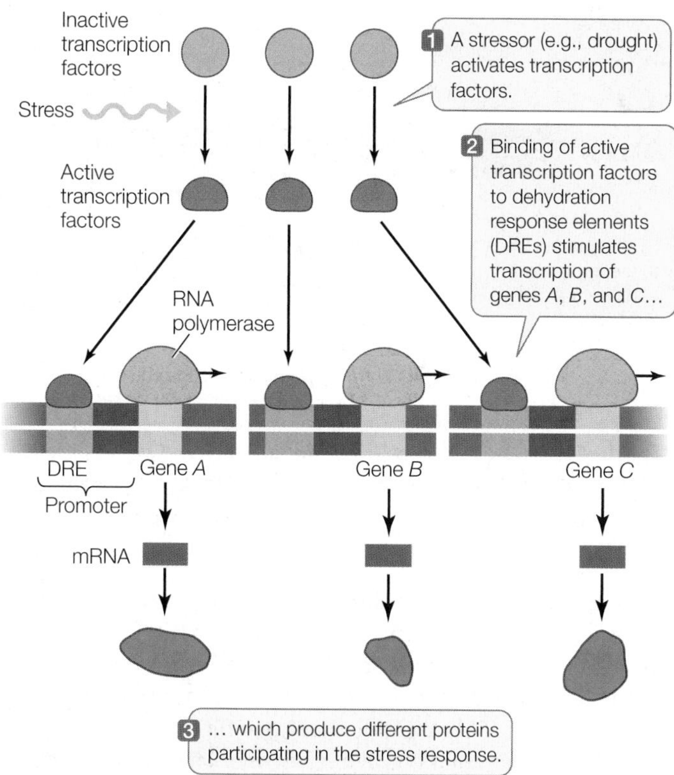
Inactive transcription factors
Stress
1 A stressor (e.g., drought) activates transcription factors.
Active transcription factors
2 Binding of active transcription factors to dehydration response elements (DREs) stimulates transcription of genes *A*, *B*, and *C*…
RNA polymerase
DRE
Gene *A*
Gene *B*
Gene *C*
Promoter
mRNA
3 … which produce different proteins participating in the stress response.

FIGURE 11.11 Coordinating Gene Expression A single environmental signal, such as drought stress, activates a transcription factor that acts on many genes.

LINK

Dehydration response proteins are one of many adaptations to drought stress found among the plants; see **Concept 28.3**

Eukaryotic viruses can have complex life cycles

Eukaryotes are susceptible to infections by various kinds of viruses that have a variety of life cycle strategies. These viral life cycles can be very efficient—for example, the poliovirus completes its life cycle (from infection to release of new particles) in 4–6 hours, and each dying host cell can release up to 10,000 new particles. Compare this with the 24-hour cell cycle typical of dividing human cells.

The viruses that infect eukaryotic cells may have genomes of single- or double-stranded DNA or RNA. Some viral life cycles can be quite complex. As an example, we focus here on **human immunodeficiency virus** (**HIV**), the infective agent that causes acquired immunodeficiency syndrome (AIDS) in humans. HIV typically infects only cells of the immune system that express a surface receptor called CD4. The virion is enclosed within a phospholipid membrane derived from its previous host cell. Proteins in the membrane are involved in the infection of new host cells, which HIV enters by direct fusion of the viral envelope with the host's cell membrane (**FIGURE 11.12**).

HIV is a retrovirus: its genome is single-stranded RNA, and it carries within the virion an enzyme called reverse transcriptase. Shortly after infection, the **reverse transcriptase** makes a DNA strand that is complementary to the RNA, while at the same time degrading the RNA and making a second DNA strand that is complementary to the first. The resulting double-stranded DNA becomes integrated into the host's chromosome. The integrated viral DNA is called a provirus.

The provirus resides permanently in the host chromosome and can remain in an inactive state for years. During this time transcription of the viral DNA is initiated, but host cell proteins called termination factors prevent the RNA from elongating, and transcription is terminated prematurely (**FIGURE 11.13A**). Under

FIGURE 11.12 The Reproductive Cycle of HIV This retrovirus enters a host cell via fusion of its envelope with the host's cell membrane. Reverse transcription of retroviral RNA then produces a complementary DNA that becomes inserted into the host's genome. The inserted viral DNA directs the synthesis of new virus particles.

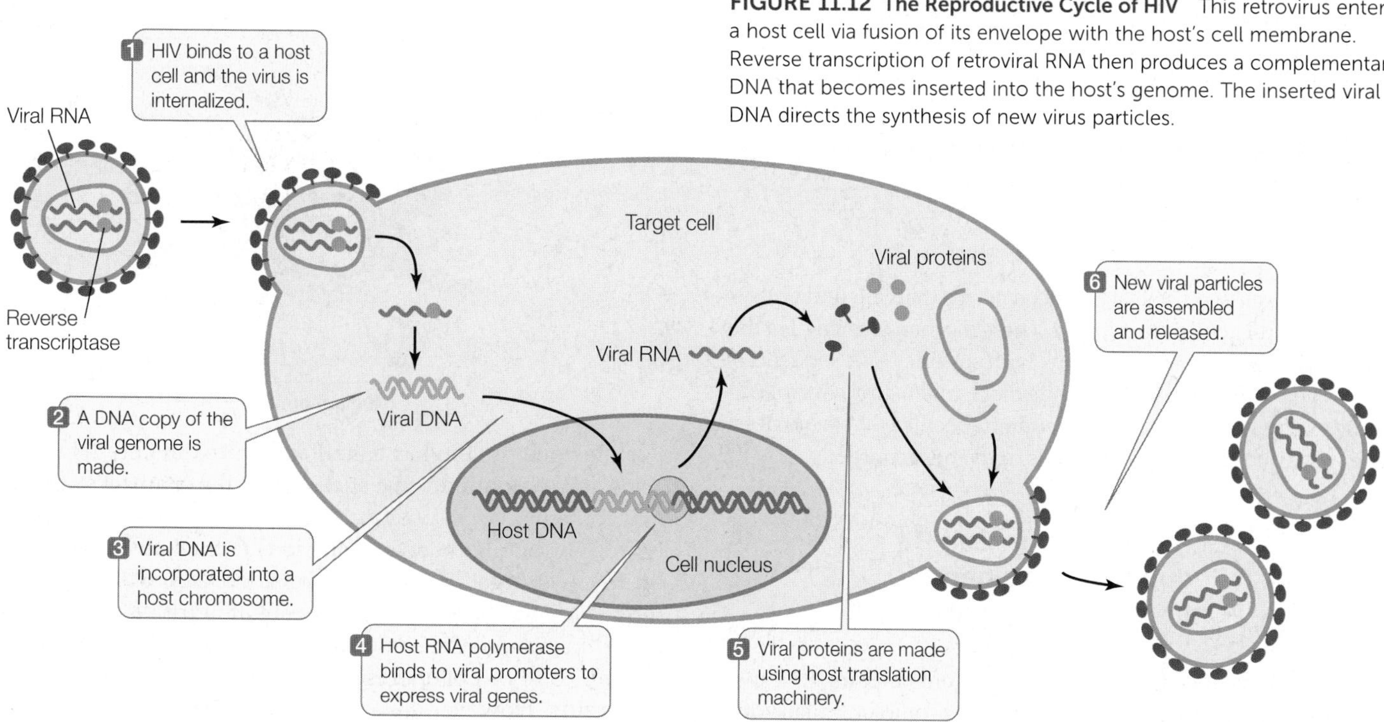
1 HIV binds to a host cell and the virus is internalized.
Viral RNA
Reverse transcriptase
Target cell
Viral proteins
6 New viral particles are assembled and released.
Viral RNA
Viral DNA
2 A DNA copy of the viral genome is made.
Host DNA
Cell nucleus
3 Viral DNA is incorporated into a host chromosome.
4 Host RNA polymerase binds to viral promoters to express viral genes.
5 Viral proteins are made using host translation machinery.

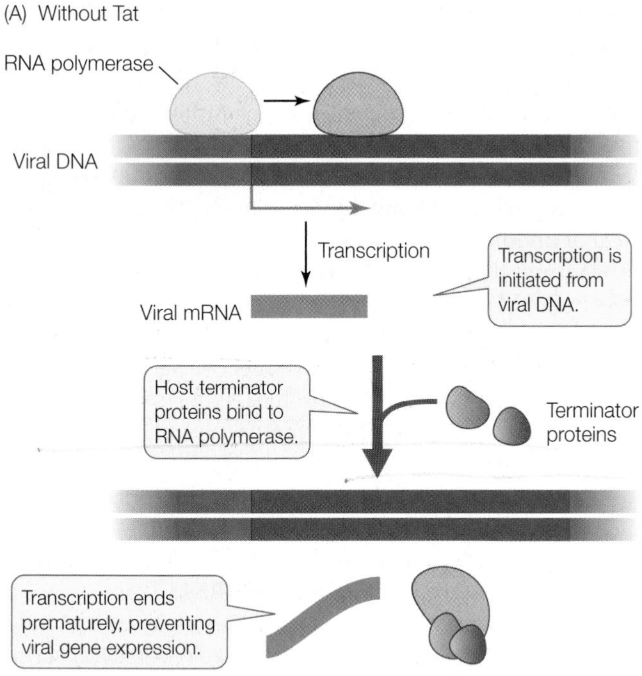

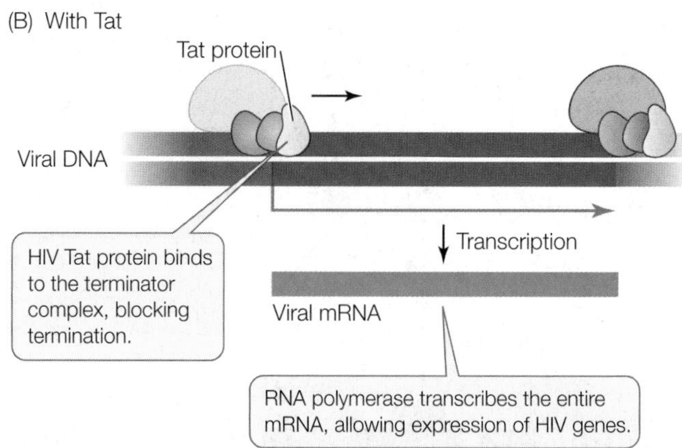

FIGURE 11.13 Regulation of Transcription by HIV The Tat protein acts as an antiterminator, allowing transcription of the HIV genome.

some circumstances, the level of transcription initiation increases and some viral RNA is made. One of the viral genes encodes a protein called Tat (*Trans*a*c*tivator of *t*ranscription), which binds to the 5′ end of the viral RNA. As a result of Tat binding, the production of full-length viral RNA is dramatically increased (**FIGURE 11.13B**), and the rest of the viral reproductive cycle is able to proceed. It was only after the discovery of this mechanism in HIV and similar viruses that researchers found that many eukaryotic genes are regulated at the level of transcription elongation.

CHECKpoint CONCEPT 11.2

- ✓ How do transcription factors regulate gene expression? How do the roles of transcription factors compare with the roles of proteins that regulate prokaryotic operons?
- ✓ What would be the effect of the inhibition of reverse transcriptase on infection of a cell with HIV?

We have discussed some of the mechanisms that cells and viruses use to control gene transcription. These mechanisms involve the interaction of regulatory proteins with specific DNA sequences. However, eukaryotes have other mechanisms for controlling gene expression that do not depend on specific DNA sequences. We will discuss these mechanisms in the next concept.

CONCEPT 11.3 Gene Expression Can Be Regulated via Epigenetic Changes to Chromatin

So far we have focused on regulatory events that involve specific DNA sequences at or near a gene's promoter. Eukaryotic cells are also able to regulate transcription via reversible, non-sequence-specific alterations to either the DNA or the chromosomal proteins that package the DNA in the nucleus. These alterations can be passed on to daughter cells after mitosis or meiosis. They are called **epigenetic** changes to distinguish them from mutations, which involve irreversible changes to the DNA's base sequence (see Concept 9.3).

Modification of histone proteins affects chromatin structure and transcription

Epigenetic gene regulation can occur via the alteration of chromatin structure, or **chromatin remodeling**. Large amounts of DNA (nearly 2 meters in humans!) are packed within the nucleus (which has a diameter of about 5 μm). The basic unit of DNA packaging in eukaryotes is the nucleosome, a core of positively charged **histone** proteins around which DNA is wound:

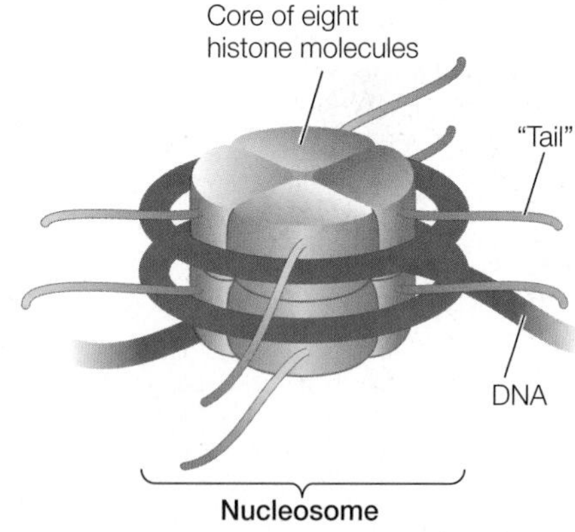

Each histone protein has a "tail" of approximately 20 amino acids at its N terminus that sticks out of the compact structure and contains certain positively charged amino acids, notably lysine. Ordinarily there is strong ionic attraction between the positively charged histone proteins and DNA, which is negatively charged because of its phosphate groups. Because of this attraction, nucleosomes can make DNA physically inaccessible to RNA polymerase and the rest of the transcription apparatus. However, a variety of transcription factors and

RNA polymerase can bind enzymes called **histone acetyltransferases**, which add acetyl groups to these positively charged amino acids and neutralize their charges:

$$\cdots\text{NH}-\text{CH}[(\text{CH}_2)_3\text{NH}_3^+]-\text{C}(=\text{O})\cdots + \text{CoA}-\text{S}-\text{C}(=\text{O})-\text{CH}_3 \rightarrow \cdots\text{NH}-\text{CH}[(\text{CH}_2)_3\text{HN}-\text{C}(=\text{O})-\text{CH}_3]-\text{C}(=\text{O})\cdots + \text{CoA}-\text{SH}$$

Lysine in histone; Acetyl CoA; Acetyl-lysine; CoA

Reducing positive charges on the histone tails reduces the affinity of the histones for the DNA, loosening the compact nucleosome (**FIGURE 11.14**). The majority of histone acetylation is found near gene promoters, but acetylated histones are also found throughout the transcribed regions of genes. Thus histone acetylation promotes both transcription initiation and elongation. Histone acetylases can be recruited to promoters by transcription factors. An example is CREB, the transcription factor that is associated with addiction (see the opening story of this chapter). CREB binds specific acetyltransferases that participate in the activation of CREB-responsive genes.

Another class of chromatin remodeling proteins, **histone deacetylases**, can remove the acetyl groups from histones and thereby *repress* transcription. Histones can also be modified in other ways, including methylation (the addition of a methyl group) and phosphorylation (the addition of a phosphate group). Histone methylation can contribute to either the activation or repression of gene expression, depending on which lysine residue is methylated. Histone phosphorylation is involved in chromosome condensation during mitosis and meiosis, as well as affecting gene regulation. All of these effects are reversible, and so the transcriptional activity of a eukaryotic gene may be determined by varying patterns of histone modification.

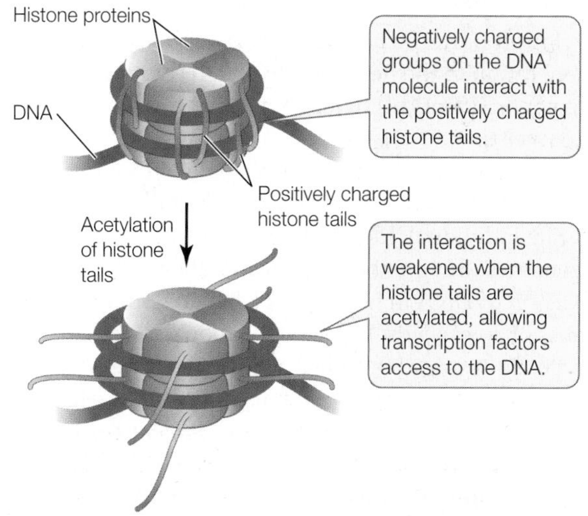

FIGURE 11.14 Epigenetic Remodeling of Chromatin for Transcription Initiation of transcription requires that nucleosomes change their structure, becoming less compact. This chromatin remodeling makes DNA accessible to the transcription complex (see Figure 11.8).

DNA methylation affects transcription

Depending on the organism, from 1 to 5 percent of cytosines in the DNA are chemically modified by the addition of a methyl group (—CH_3), to form 5-methylcytosine (**FIGURE 11.15**). This

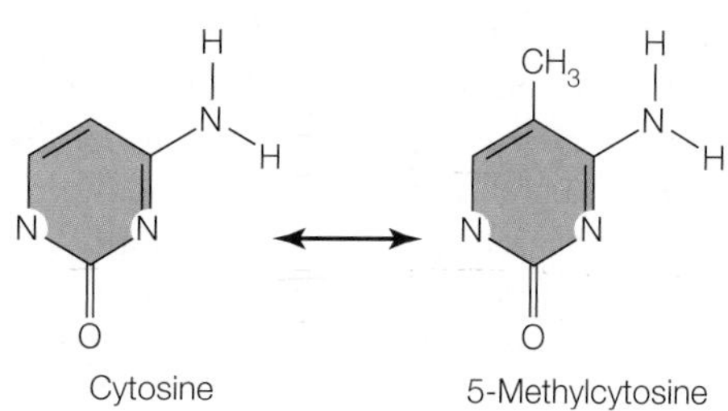

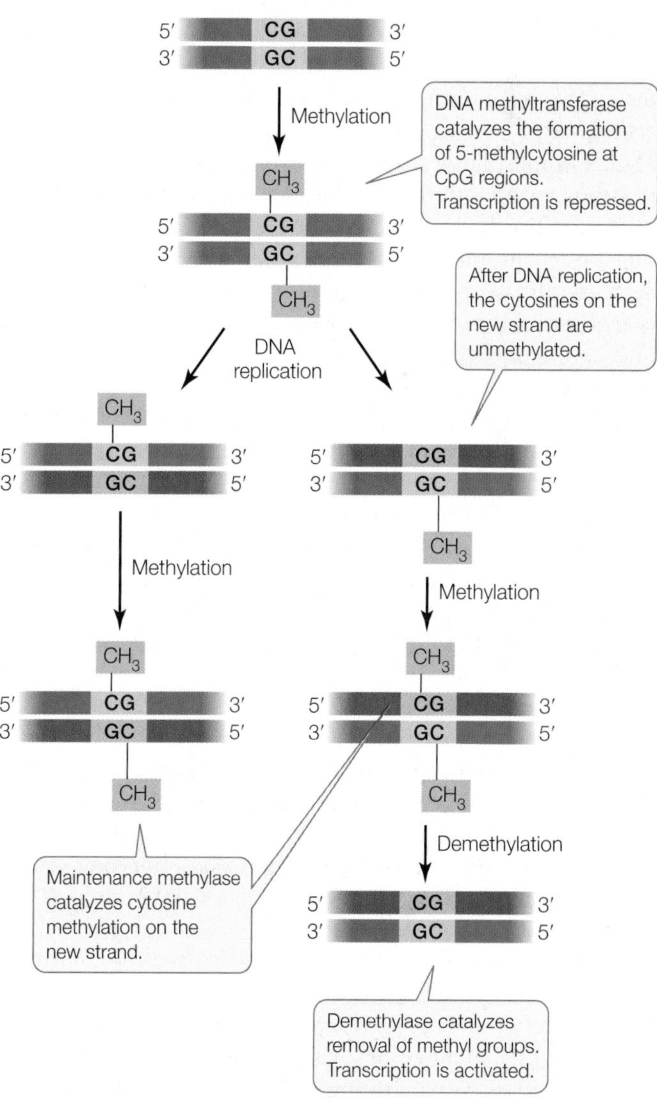

FIGURE 11.15 DNA Methylation: An Epigenetic Change The reversible formation of 5-methylcytosine in DNA can alter the rate of transcription.

covalent addition is catalyzed by the enzyme **DNA methyltransferase** and, in mammals, usually occurs on cytosines (C) that are adjacent to guanines (G). Virtually all of these CpG ("p" is for the phosphate in the DNA backbone) sites are methylated, apart from those found in and near transcriptionally active promoters. Promoters usually contain regions of DNA that are rich in CpG sites, and such regions are called **CpG islands**.

Methylated DNA binds specific proteins that are involved in the repression of transcription; thus heavily methylated genes tend to be inactive (silenced). Whereas histone acetylation/deacetylation (see above) are dynamic processes resulting in short-term changes in gene expression, DNA methylation is usually a stable, long-term silencing mechanism. When DNA is replicated, a **maintenance methyltransferase** catalyzes the formation of 5-methylcytosine in the new DNA strands. However, the pattern of cytosine methylation can also be altered, because methylation is reversible: a third enzyme, appropriately called **demethylase**, catalyzes the removal of the methyl group from cytosine (see Figure 11.15). In ways that are not fully understood, the enzymes involved in histone modification and DNA methylation/demethylation interact to ensure that genes whose products are needed in the cell are kept unmethylated, and their associated histones acetylated.

Sometimes, large stretches of DNA or almost entire chromosomes are methylated. Under a microscope, two kinds of chromatin can be distinguished in the stained interphase nucleus: euchromatin and heterochromatin. The **euchromatin** appears diffuse and stains lightly; it contains the DNA that is transcribed into mRNA. **Heterochromatin** is condensed and stains darkly; any genes it contains are generally not transcribed.

A dramatic example of heterochromatin is the X chromosome in female mammals. A normal female mammal has two X chromosomes, whereas a normal male has an X and a Y (see Concept 8.3). Because each female cell has two copies of each X chromosome gene, the female should have the potential to produce twice as much of each protein product as the male. Nevertheless, for 75 percent of the genes on the X chromosome, the total amount of mRNA produced is generally the same in males and in females. How does this happen?

In the early female embryo, one copy of X becomes heterochromatic and transcriptionally inactive in each cell, and the same X remains inactive in all of that cell's descendants. In a given female embryo cell, the "choice" of which X to inactivate is random. Recall that one X in a female comes from her father and one from her mother. Thus in one embryonic cell the paternal X might be inactivated, but in a neighboring cell the maternal X might be inactivated.

A familiar example of a phenotype caused by X chromosome inactivation is the tortoiseshell cat. In cats, two alleles of an X-linked gene that contributes to coat color are orange (X^B) and black (X^b). In the embryo of a heterozygous female (X^BX^b), there is random X-inactivation such that in some cells only the orange allele is expressed, while in others it is the black allele. These cells then form the progenitors of patches of skin in the animal.

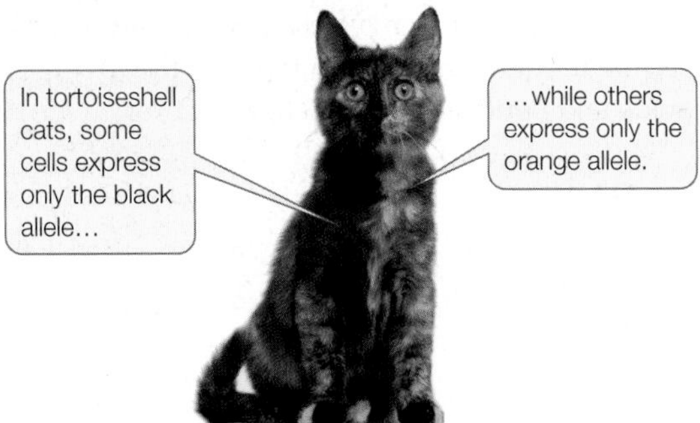

The inactive X is identifiable within the nucleus as a heterochromatic Barr body (named for its discoverer, Murray Barr):

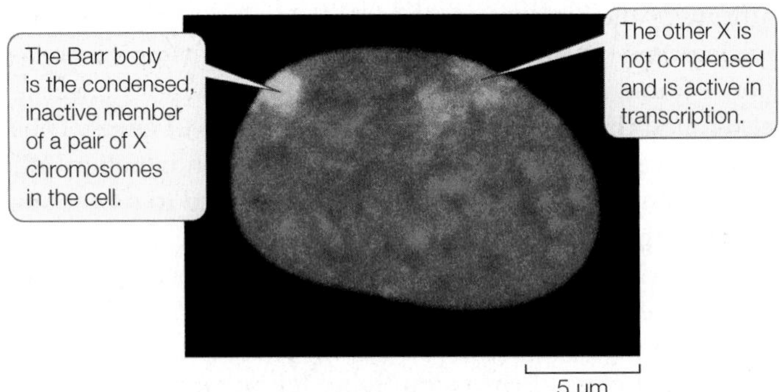

In this micrograph, both X chromosomes are stained yellow-green; the Barr body, which consists of heavily methylated DNA, is more condensed than the active X and therefore appears brighter.

Having the right "dosage" of transcriptionally active genes is important. This is illustrated by the fact that aneuploidy—an unusual number of a particular chromosome—is a harmful condition that often results in embryo death (see Concept 7.4).

Epigenetic changes can be induced by the environment

Methylation patterns are stable and can be passed on from one generation to the next. However, a recent study of human monozygotic (identical) twins shows that these patterns can be altered over time.

Monozygotic twins come from a single fertilized egg that divides to produce two separate cells; each of these develops into a separate individual. Identical twins thus have generally identical genomes. But are they identical in their "epigenomes"? A comparison of DNA in hundreds of such twin pairs shows that in tissues of 3-year-olds, DNA methylation patterns are virtually the same. But by age 50—when twins have usually been living apart and in different environments for decades—the patterns are quite different, and different genes are expressed. This indicates that the *environment plays an important role in epigenetic modifications*, and therefore in the regulation of genes that these modifications affect.

Go to MEDIA CLIP 11.1
The Surprising Epigenetics of Identical Twins
PoL2e.com/mc11.1

What factors in the environment lead to epigenetic changes? Chemicals such as tobacco smoke and dietary components such as folic acid can affect DNA methylation patterns. Another factor might be stress: when mice are put in a stressful situation, genes that are involved in important brain pathways become heavily methylated (and transcriptionally inactive). Treatment of the stressed mice with an antidepressant drug reverses these changes.

LINK

Stressful experiences can have lifelong effects on behavior patterns; see **Concept 40.2**

DNA methylation can result in genomic imprinting

In mammals, specific patterns of methylation develop for each sex during gamete formation. This happens in two stages: first, the existing methyl groups are removed from the 5′-methylcytosines by a demethylase, and then a DNA methylase adds methyl groups to a new set of cytosines. When the gametes form, they carry this new pattern of methylation.

The DNA methylation pattern in male gametes (sperm) differs from that in female gametes (eggs) at about 200 genes in the mammalian genome. That is, a given gene in this group may be methylated in eggs but unmethylated in sperm. In this case the offspring would inherit a maternal gene that is transcriptionally inactive (methylated) and a paternal gene that is transcriptionally active (demethylated). This is called **genomic imprinting**.

Imprinting of specific genes occurs primarily in mammals and flowering plants. Most imprinted genes are involved with embryonic development. An embryo must have both the paternally and maternally imprinted gene patterns to develop properly. In fact, attempts to make an embryo that has chromosomes from only one sex (for example, by chemically treating an egg cell to double its chromosomes) usually fail. So imprinting has an important lesson for genetics: *males and females may be the same genetically (except for the X and Y chromosomes), but they differ epigenetically.*

CHECKpoint CONCEPT 11.3

- ✓ What is the difference between epigenetic regulation and gene regulation by transcription factors?
- ✓ How can a DNA methylation pattern be inherited?
- ✓ In colorectal cancer, some tumor suppressor genes are inactive. This is an important factor resulting in uncontrolled cell division. Two of the possible explanations for the inactive genes are: (1) a mutation in the coding region, resulting in an inactive protein, and (2) epigenetic silencing at the promoter of the gene, resulting in reduced transcription. How would you investigate each of these possibilities?

Thus far we have examined transcriptional gene regulation in viruses, prokaryotes, and eukaryotes. In the final concept we will focus on the posttranscriptional mechanisms for regulating gene expression in eukaryotes.

CONCEPT 11.4 Eukaryotic Gene Expression Can Be Regulated after Transcription

Gene expression involves transcription and then translation. So far we have described how eukaryotic gene expression is regulated at the transcriptional level. But as Figure 11.1 shows, there are many points at which regulation can occur after the initial gene transcript is made.

Different mRNAs can be made from the same gene by alternative splicing

Most primary mRNA transcripts in eukaryotes contain several introns (see Figure 10.6). We have seen how the splicing mechanism recognizes the boundaries between exons and introns. What would happen if the β-globin pre-mRNA, which has two introns, were spliced from the start of the first intron to the end of the second? The middle exon would be spliced out along with the two introns. An entirely new protein (certainly not a β-globin) would be made, and the functions of normal β-globin would be lost. Such **alternative splicing** can be a deliberate mechanism for generating a family of different proteins with different activities and functions from a single gene (**FIGURE 11.16**).

Two examples of this mechanism are found in HIV and in the fruit fly (*Drosophila*):

- The HIV genome (see Figure 11.12) encodes nine proteins but is transcribed as a single pre-mRNA. Most of the nine proteins are then generated by alternative splicing of this pre-mRNA.
- In *Drosophila*, sex is determined by the *Sxl* gene. This gene has four exons, which we will designate 1, 2, 3, and 4. In the female embryo, splicing generates two active forms of the Sxl protein, containing exons 1 and 2, and 1, 2, and 4. However, in the male embryo, the protein contains all four exons (1, 2, 3, and 4) and is inactive.

Before the human genome was sequenced, most scientists estimated that they would find between 80,000 and 150,000 protein-coding genes. You can imagine their surprise when the actual sequence revealed only about 21,000 genes! In fact, there are many more human mRNAs than there are human genes, and most of this variation comes from alternative splicing. Indeed, recent surveys show that more than 80 percent of all human genes are alternatively spliced.

Alternative splicing may be a key to the differences in levels of complexity among organisms. For example, although humans and chimpanzees have similar-sized genomes, there is more alternative splicing in the human brain than in the brain of a chimpanzee.

MicroRNAs are important regulators of gene expression

As we will discuss in Concept 12.3, only a small fraction of the genome in most plants and animals codes for proteins. Some of the genome encodes ribosomal RNA and transfer RNAs, but until recently biologists thought that the rest of the genome

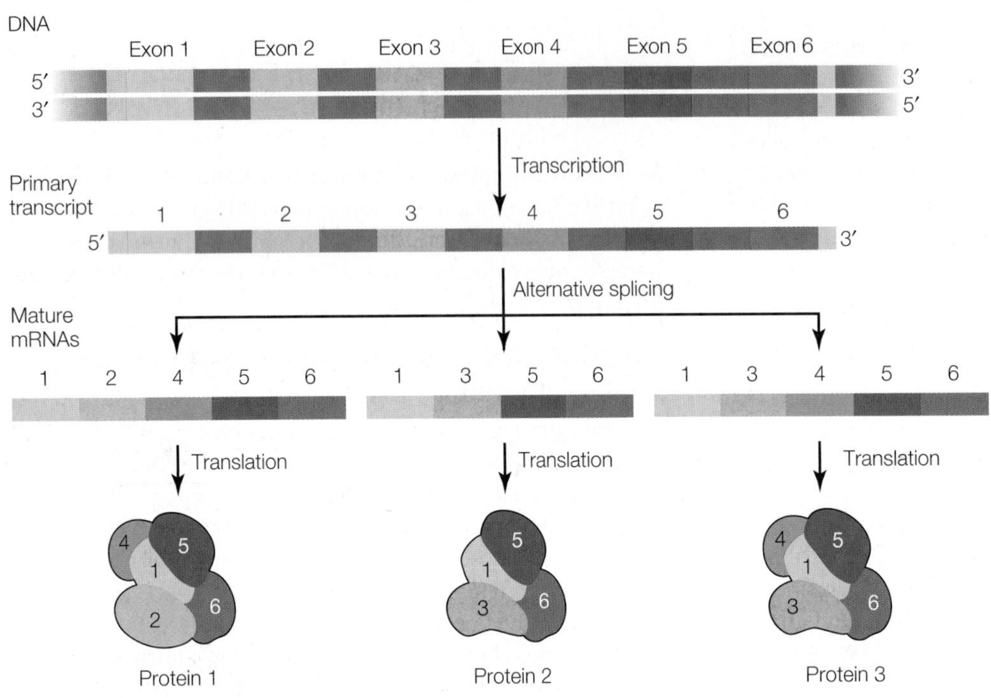

FIGURE 11.16 Alternative Splicing Results in Different Mature mRNAs and Proteins Pre-mRNA can be spliced differently in different tissues, resulting in different proteins.

was not transcribed; some even called it "junk." Recent investigations, however, have shown that some of these noncoding regions are transcribed into tiny RNA molecules called **microRNA (miRNA)**.

The first miRNA sequences were found in the worm *Caenorhabditis elegans*. This model organism, which has been studied extensively by developmental biologists, goes through several larval stages. Victor Ambros at the University of Massachusetts found *lin* mutations (named for abnormal cell *lin*eage) in two genes that had different effects on progress through these stages:

- *lin-14* mutations cause the larvae to skip the first stage and go straight to the second stage. Thus the gene's normal role is to facilitate events of the first larval stage.
- *lin-4* mutations cause certain cells in later larval stages to repeat a pattern of development normally observed in the first larval stage. It is as if the cells were stuck in that first stage. So the normal role of this gene is to *negatively regulate lin-14*, turning off its expression so the cells can progress to the next stage.

Not surprisingly, further investigation showed that *lin-14* encodes a transcription factor that affects the transcription of genes involved in larval cell progression. It was originally expected that *lin-4*, the negative regulator, would encode a protein that downregulates genes activated by the LIN-14 protein. But this turned out to be incorrect. Instead, *lin-4* encodes a 22-base miRNA that inhibits *lin-14* expression *posttranscriptionally* by binding to its mRNA.

Thousands of miRNAs in a variety of eukaryotes have now been described. Each miRNA is about 22 nucleotides long and usually has dozens of mRNA targets. MicroRNAs are transcribed as longer precursors that fold into double-stranded RNA molecules, which then are processed through a series of steps into single-stranded miRNAs. A protein complex guides the miRNA to its target mRNA, where translation is inhibited and the mRNA is degraded (**FIGURE 11.17**). The remarkable conservation of this gene-silencing mechanism, which is found in most eukaryotes, indicates that it is evolutionarily ancient and biologically important.

Translation of mRNA can be regulated

The amount of a protein in a cell is not determined simply by the amount of its mRNA. For example, in yeast cells only about a third of the genes show clear correlations in the amounts of mRNA and protein; in these cases, more mRNA leads to more protein. For two-thirds of the genes there is no apparent relationship between the two—there may be lots of mRNA and little or no protein, or lots of protein and little mRNA. The concentrations of these proteins must therefore be determined by factors acting after the mRNA is made. Cells do this in two major ways: by regulating the translation of mRNA or by altering how long proteins persist in the cell.

There are three known ways in which the translation of mRNA can be regulated:

- *Inhibition of translation with miRNAs.* This was discussed in the last section (see above).
- *Modification of the 5′ cap.* As noted in Concept 10.2, an mRNA usually has a chemically modified molecule of guanosine triphosphate (GTP) at its 5′ end. An mRNA that is capped with an unmodified GTP molecule is not translated. For example, stored mRNAs in the egg cells of the tobacco hornworm moth are capped with unmodified GTP molecules and are not translated. After the egg is fertilized, however, the caps are modified, allowing the mRNA to be translated to produce the proteins needed for early embryonic development.

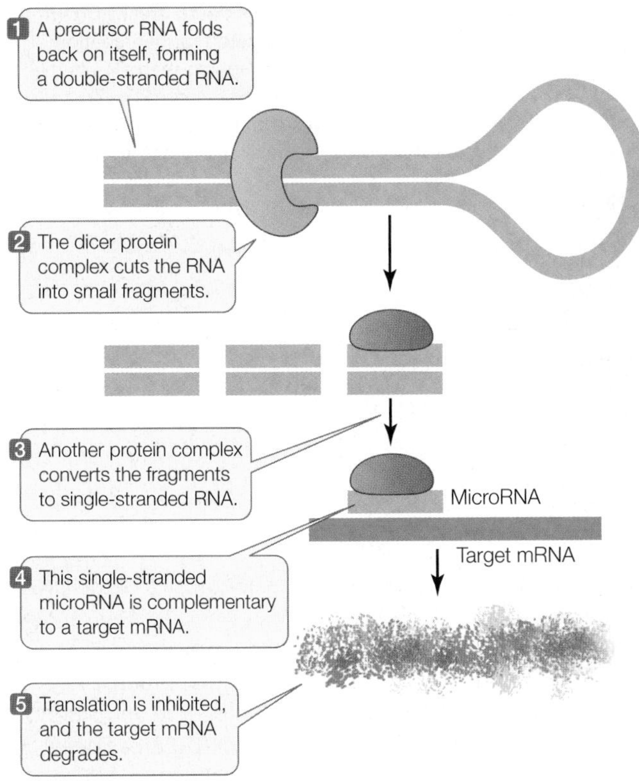

1 A precursor RNA folds back on itself, forming a double-stranded RNA.
2 The dicer protein complex cuts the RNA into small fragments.
3 Another protein complex converts the fragments to single-stranded RNA.
MicroRNA
Target mRNA
4 This single-stranded microRNA is complementary to a target mRNA.
5 Translation is inhibited, and the target mRNA degrades.

FIGURE 11.17 mRNA Degradation Caused by MicroRNAs
MicroRNAs inhibit the translation of specific mRNAs by causing their premature degradation.

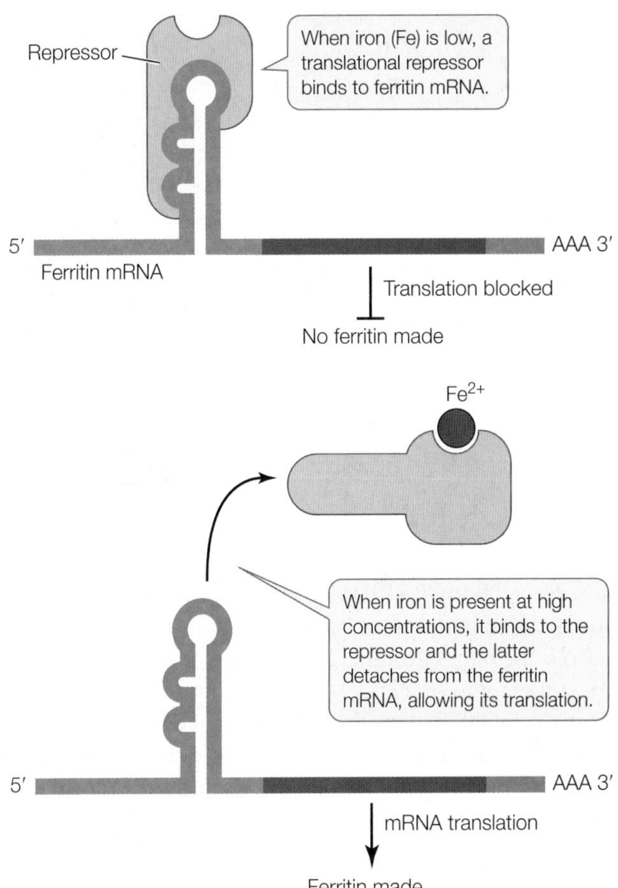

FIGURE 11.18 A Repressor of Translation Binding of a translational repressor to mRNA blocks the mRNA from associating with the ribosome. The repressor can be removed from the mRNA via allosteric regulation.

- *Translational repressor proteins.* Such proteins block translation by binding to mRNAs and preventing their attachment to the ribosome. For example, in mammalian cells the rate of translation of the protein ferritin increases rapidly when the level of free iron ions (Fe^{2+}) increases in the cell. Iron is an essential nutrient, but the free ions can be toxic to the cell; ferritin binds the ions and stores them in a safe but accessible form. The amount of ferritin mRNA in the cell remains constant, but when the iron level is low, a **translational repressor** binds to the ferritin mRNA and prevents its translation. When the iron level rises, some of the excess Fe^{2+} ions bind to the repressor and alter its three-dimensional structure, causing the repressor to detach from the mRNA and allowing translation to proceed (**FIGURE 11.18**).

Protein stability can be regulated

The protein content of any cell at a given time is a function of both protein synthesis and protein degradation. Certain proteins can be targeted for destruction in a chain of events that begins when an enzyme attaches a 76–amino acid protein called **ubiquitin** (so named because it is ubiquitous, or widespread) to a lysine residue of the protein to be destroyed. Other ubiquitins then attach to the primary one, forming a polyubiquitin chain. The protein–polyubiquitin complex then binds to a huge protein complex called a **proteasome** (from *protease*

APPLY THE CONCEPT

Eukaryotic gene expression can be regulated transcriptionally and posttranscriptionally

The enzyme HMG CoA reductase (HR) catalyzes an initial step in the synthesis of cholesterol. The table shows the HR levels in liver cells following various treatments. Explain the results of each of the treatments 1–5.

TREATMENT	AMOUNT OF HR PROTEIN
1. Actinomycin D, a drug that inhibits RNA polymerase II	Reduced
2. Suberoylanilide hydroxamic acid, a histone deacetylase inhibitor	Increased
3. Bortezomib, a proteasome inhibitor	Increased
4. High level of cholesterol	Reduced
5. Azacytidine, inhibitor of DNA methylation	Increased

1 A protein is targeted for breakdown.
2 An enzyme attaches ubiquitin to the protein...
3 ...which is recognized by a proteasome.
4 Ubiquitin is released and recycled.
5 The proteasome hydrolyzes the target protein.
Ubiquitin
Proteasome

FIGURE 11.19 A Proteasome Breaks Down Proteins Proteins targeted for degradation are bound by ubiquitin, which then directs the targeted protein to a proteasome. The proteasome is a complex structure where proteins are digested by several powerful proteases.

and *soma*, "body"; **FIGURE 11.19**). Upon entering the proteasome, the polyubiquitin is removed and ATP energy is used to unfold the target protein. Three different proteases then digest the protein into small peptides and amino acids. You may recall from Chapter 7 that cyclins are proteins that regulate the activities of key enzymes at specific points in the cell cycle. Cyclins must be broken down at just the right time, and this is done by proteasomes.

CHECKpoint CONCEPT 11.4

- ✓ How can a single gene transcribed into a single pre-mRNA code for several different proteins?
- ✓ Compare inhibition of translation by miRNA with inhibition by a repressor.
- ✓ You are studying the enzyme protease in germinating seeds. You find that protease activity increases tenfold after treatment of the seeds with a hormone, gibberellic acid. How would you show that this increase is due to:
  a. release of a translational repressor by the hormone?
  b. allosteric inhibition of a transcriptional repressor by the hormone?

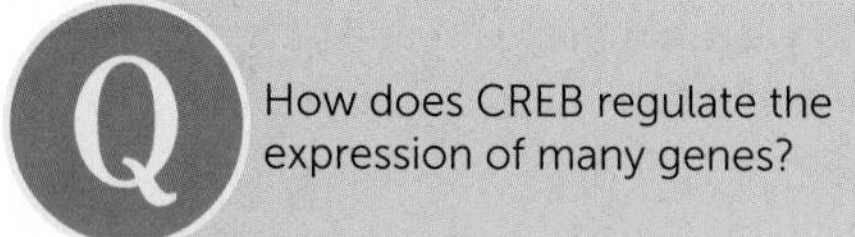

Q
How does CREB regulate the expression of many genes?

ANSWER CREB is a family of several closely related transcription factors that can either activate or repress gene expression (Concept 11.1). They bind to the *cAMP* response element (*CRE*), a short DNA sequence (GACGTCA) that is found in the promoter regions of many genes (Concept 11.2). CREB proteins have a "leucine zipper" structure that consists of two parallel α helices rich in the amino acid leucine, and fingerlike extensions that fit into the major groove of the DNA double helix (**FIGURE 11.20**).

CREB binding and regulation of gene expression are essential for a number of processes in several organs, including the brain. In addition to its role in drug and alcohol addiction (described in the opening story), CREB has been strongly implicated in long-term memory. Animals with mutations that result in a lack of active CREB can learn a maze test, but they don't remember it later. Similar effects are seen if CREB activity is blocked right after a task is learned. When animals learn a task, imaging studies reveal *CRE*-containing genes becoming active in the hippocampus, a region of the brain that is involved in long-term memory (see Concept 34.5). Thus CREB provides an insight into the molecular biology of memory, linking learning to the regulation of gene expression.

FIGURE 11.20 An Explanation for Alcoholism? The transcription factor CREB binds to DNA and activates the promoters of genes involved in addictive behaviors.

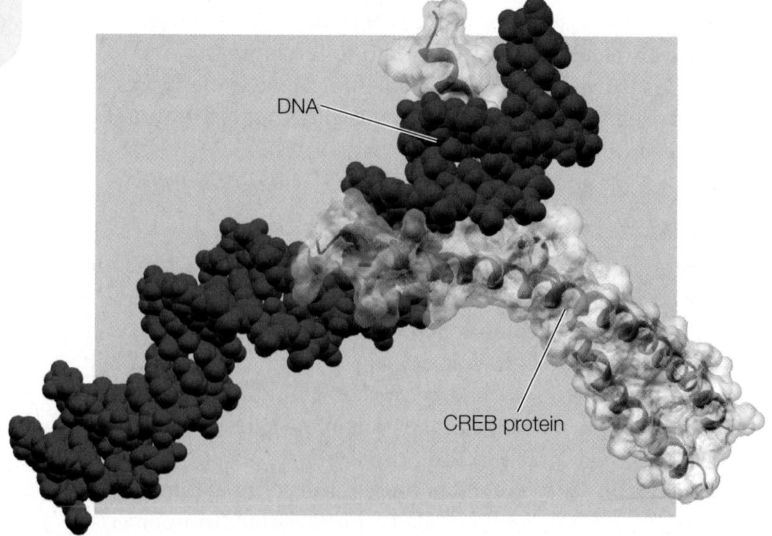

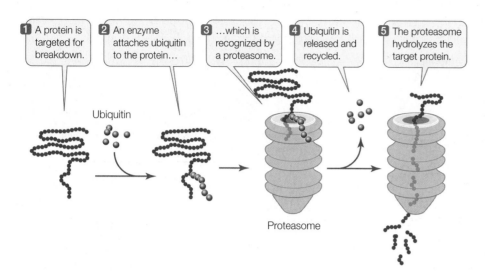

SUMMARY

CONCEPT 11.1 Many Prokaryotic Genes Are Regulated in Operons

- Gene expression can be regulated at the levels of transcription, RNA processing, translation, or posttranslation. **Review Figure 11.1 and ACTIVITY 11.1**
- Some genes are always expressed (**constitutive genes**), whereas others are expressed only at certain times and in certain cells (**inducible genes**).
- **Transcription factors** are regulatory proteins that bind DNA and regulate gene expression. **Activators** positively regulate gene expression. **Repressors** negatively regulate gene expression. **Review Figure 11.2**
- In prokaryotes, several genes can be part of a single transcriptional unit called an **operon**, which consists of a **promoter**, an **operator**, and two or more **structural genes**. **Review Figure 11.3**
- An inducible operon is turned off unless its expression is needed, whereas a repressible operon is turned on unless its expression is not needed. When an operon is turned off, it has a repressor protein bound to its operator, preventing transcription.
- The *lac* operon is an example of an inducible system, whereas the *trp* operon is an example of a repressible system. **Review Figures 11.4 and 11.5 and ANIMATED TUTORIALS 11.1 and 11.2**
- A metabolic pathway can be regulated either by allosteric regulation of an enzyme or by regulation of enzyme synthesis. **Review Figure 11.6**
- **Sigma factors** direct RNA polymerase to specific promoters in prokaryotes.
- **Viruses** provide examples of gene regulation as they convert the host cell into a virus factory. **Review Figure 11.7**

CONCEPT 11.2 Eukaryotic Genes Are Regulated by Transcription Factors

- Eukaryotic gene expression is regulated both during and after transcription.
- **General transcription factors** bind to the core promoter sequences of protein-coding genes and direct RNA polymerase II to the promoter. **Review Figure 11.8 and ANIMATED TUTORIAL 11.3**
- Specific transcription factors (activators and repressors) bind to specific DNA elements near the promoter and affect the rate of transcription initiation. **Review Figure 11.9**

CONCEPT 11.3 Gene Expression Can Be Regulated via Epigenetic Changes to Chromatin

- The term **epigenetic** refers to changes in gene expression that do not involve changes in DNA sequences.
- **Chromatin remodeling** via the modification of **histone** proteins in nucleosomes also affects transcription. **Review Figure 11.14**
- Methylation of cytosines in DNA generally inhibits transcription. **Review Figure 11.15**
- Epigenetic changes can be induced by the environment and can be inherited.

CONCEPT 11.4 Eukaryotic Gene Expression Can Be Regulated after Transcription

- **Alternative splicing** of pre-mRNA can produce different proteins. **Review Figure 11.16**
- A **microRNA** (**miRNA**) is a small noncoding RNA that inhibits the translation of specific mRNAs by causing their premature degradation. **Review Figure 11.17**
- The translation of mRNA to proteins can be regulated by **translational repressors**. **Review Figure 11.18**
- A **proteasome** can break down proteins, thus affecting protein longevity. **Review Figure 11.19**

See **ACTIVITY 11.2** for a concept review of this chapter.

Go to the Interactive Summary to review key figures, Animated Tutorials, and Activities
PoL2e.com/is11

Go to LaunchPad at **macmillanhighered.com/launchpad** for additional resources, including LearningCurve Quizzes, Flashcards, and many other study and review resources.

12 Genomes

KEY CONCEPTS

12.1 There Are Powerful Methods for Sequencing Genomes and Analyzing Gene Products

12.2 Prokaryotic Genomes Are Small, Compact, and Diverse

12.3 Eukaryotic Genomes Are Large and Complex

12.4 The Human Genome Sequence Has Many Applications

The Papillon and the Great Dane are the same species—*Canis lupus familiaris*—yet they show great variation in size. Genome sequencing has provided insights into how size is controlled by genes.

Canis lupus familiaris, the dog, was domesticated by humans from the gray wolf more than 10,000 years ago. There are several kinds of wolves, and they all look more or less the same. Not so with "man's best friend." The American Kennel Club recognizes about 155 different breeds, varying greatly in size, shape, coat color, hair length, and even behavior. For example, an adult Chihuahua weighs just 1.5 kilograms, whereas a Scottish Deerhound weighs 70 kilograms. No other mammalian species shows such large phenotypic variation. Furthermore, we know of hundreds of genetic diseases in dogs, and many of these diseases have counterparts in humans. Biologists are curious about the molecular basis of canine phenotypic variation, and they view dogs as models for studying genetic diseases. For these reasons, the Dog Genome Project began in the late 1990s. Since then the sequences of several dog genomes have been published.

Two dogs—a boxer and a poodle—were the first of their species to have their entire genomes sequenced. The dog genome contains 2.8 billion base pairs of DNA in 39 pairs of chromosomes. There are 22,000 protein-coding genes, most of them with close counterparts in other mammals, including humans. The entire genome sequence made it easy to create a map of genetic markers—specific nucleotides or short sequences of DNA at particular locations on the genome that differ among individual dogs or breeds.

Genetic markers are being used to map the locations of genes that control particular traits. To do this, scientists must extract DNA from many individual dogs that vary in just one or a few characters. Taking samples of cells for DNA isolation is relatively easy: a cotton swab is swept over the inside of the dog's cheek. As one scientist conducting genomic analyses of dogs said, the dogs "didn't care, especially if they were going to get a treat or if there was a tennis ball in our other hand."

The molecular methods used to analyze dogs have been applied to many other animals as well as to plants of economic and social importance to humans. And of course the human genome itself has been sequenced and is being studied intensively.

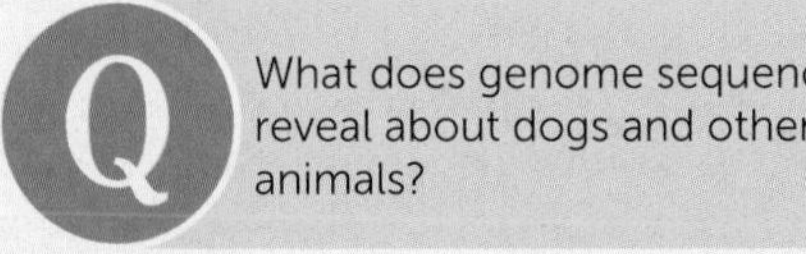

Q
What does genome sequencing reveal about dogs and other animals?

You will find the answer to this question on page 251.

CONCEPT 12.1 There Are Powerful Methods for Sequencing Genomes and Analyzing Gene Products

Genome sequencing involves determining the nucleotide base sequence of the entire genome of an organism. For a prokaryote with a single chromosome, the genome sequence is one continuous string of base pairs (bp). In a eukaryote, there are separate sequences for each chromosome. Scientists can use this genomic information in several ways:

- The genomes of different species can be compared to find out how they differ at the DNA level, and this can be used to trace evolutionary relationships.
- The sequences of individuals within a species and even tissues within an organism can be compared to identify mutations that affect particular phenotypes.
- The sequence information can be used to identify particular traits, such as genes associated with diseases.

The notion of sequencing the entire genome of a complex organism was not contemplated until 1986. The Nobel laureate Renato Dulbecco and others proposed at that time that the world scientific community be mobilized to undertake the sequencing of the entire human genome. One motive was to detect DNA damage in people who had survived the atomic bomb attacks and been exposed to radiation in Japan during the Second World War. But in order to detect changes in the human genome, scientists first needed to know its normal sequence.

The result was the publicly funded **Human Genome Project**, an enormous undertaking that was successfully completed in 2003. This effort was aided and complemented by privately funded groups. The project benefited from the development of many new methods that were first used in the sequencing of smaller genomes—those of prokaryotes and simple eukaryotes, the model organisms you are familiar with from studies in genetics and cell biology. Many of these methods are still applied widely, and powerful new methods for sequencing genomes have emerged. These are complemented by new ways to examine phenotypic diversity in a cell's proteins and in the metabolic products of the cell's enzymes.

Go to ANIMATED TUTORIAL 12.1
Sequencing the Genome
PoL2e.com/at12.1

Methods have been developed to rapidly sequence DNA

Many prokaryotes have a single chromosome, whereas eukaryotes have many. Because of their differing sizes, chromosomes can be separated from one another, identified, and experimentally manipulated. It might seem that the most straightforward way to sequence a chromosome would be to start at one end and simply sequence the DNA molecule one nucleotide at a time. The task is somewhat simplified because only one of the two strands needs to be sequenced, the other being complementary. However, this large-polymer approach is not practical. Using current methods, only several hundred bp can be sequenced at a time, whereas a human chromosome (for example) may be hundreds of millions of bp long.

As you will see, the key to determining genome sequences is to perform many sequencing reactions simultaneously, after first breaking the DNA up into millions of small, overlapping fragments.

In the 1970s, Frederick Sanger and his colleagues invented a way to sequence DNA by using chemically modified nucleotides that were originally developed to stop cell division in cancer. Variations of this method were used to obtain the first human genome sequence as well as those of several model organisms. However, it was relatively slow, expensive, and labor-intensive. The first decade of the new millennium saw the development of faster and less expensive methods, often referred to under the general term **high-throughput sequencing**. These methods use miniaturization techniques first developed for the electronics industry, as well as the principles of DNA replication and the polymerase chain reaction (PCR).

LINK

You can review the processes of DNA replication and PCR in **Concept 9.2**

High-throughput sequencing methods are rapidly evolving. Just one of the many approaches is outlined here and illustrated in **FIGURE 12.1**. First the DNA is prepared for sequencing:

1. A large molecule of DNA is cut into small fragments of about 100 bp each. This can be done physically, using mechanical forces to shear (break up) the DNA, or by using enzymes that hydrolyze the phosphodiester bonds between nucleotides at intervals in the DNA backbone.
2. The DNA is denatured by heat, breaking the hydrogen bonds that hold the two strands together. Each single strand acts as a template for the synthesis of new, complementary DNA.
3. Short, synthetic adapter sequences (oligonucleotides) are attached to each end of each fragment, and the fragments are attached to a solid support. The support can be a microbead or a flat surface.
4. Primers complementary to the adapters are used in PCR reactions to produce many (approximately 1,000) copies of each DNA fragment. The multiple copies at a single location allow for easy detection of added nucleotides during the sequencing steps.

Once the DNA has been attached to a solid substrate and amplified, it is ready to be used as a template for sequencing (see Figure 12.1B):

1. At the beginning of each sequencing cycle, the DNA fragments are heated to denature them. A universal primer, DNA polymerase, and the four deoxyribonucleoside triphosphates (dNTPs: dATP, dGTP, dCTP, and dTTP)

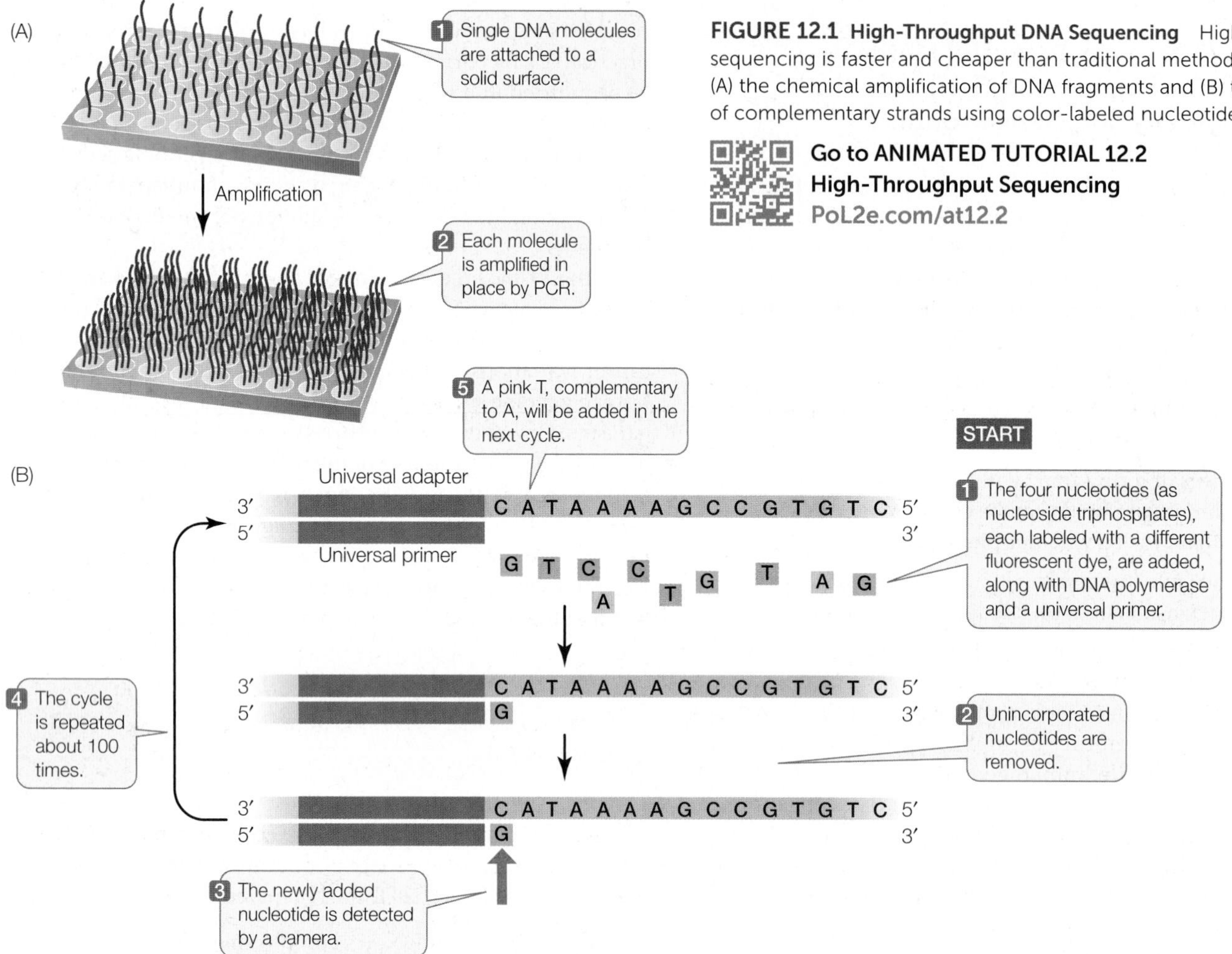
(A)
1 Single DNA molecules are attached to a solid surface.
Amplification
2 Each molecule is amplified in place by PCR.
(B)
5 A pink T, complementary to A, will be added in the next cycle.
START
Universal adapter
3′ C A T A A A A G C C G T G T C 5′
5′ 3′
Universal primer
G T C C A T G T A G
1 The four nucleotides (as nucleoside triphosphates), each labeled with a different fluorescent dye, are added, along with DNA polymerase and a universal primer.
4 The cycle is repeated about 100 times.
2 Unincorporated nucleotides are removed.
3 The newly added nucleotide is detected by a camera.

FIGURE 12.1 High-Throughput DNA Sequencing High-throughput sequencing is faster and cheaper than traditional methods. It involves (A) the chemical amplification of DNA fragments and (B) the synthesis of complementary strands using color-labeled nucleotides.

Go to ANIMATED TUTORIAL 12.2
High-Throughput Sequencing
PoL2e.com/at12.2

are added. Each of the four nucleotides (i.e., the dNTPs) is tagged with a different fluorescent dye. The universal primer is complementary to the adapter sequence at one end of each DNA fragment.

2. The DNA sequencing reaction is set up so that only one nucleotide at a time is added to the new DNA strand, which is complementary to the template strand. After each addition, the unincorporated nucleotides are removed.
3. The fluorescence of the new nucleotide at each location is detected with a camera. The color of the fluorescence indicates which of the four nucleotides was added.
4. The fluorescent tag is removed from the nucleotide that is already attached, and then the sequencing cycle is repeated. Images are captured after each nucleotide is added. The series of colors at each location indicates the sequence of nucleotides in the growing DNA strand at that location.

The power of this method derives from several factors:

- It is fully automated and miniaturized.
- Millions of different fragments are sequenced at the same time.
- It is an inexpensive way to sequence large genomes. For example, at the time of this writing, a complete human genome could be sequenced in a few days for several thousand dollars. In contrast, the Human Genome Project took 13 years and $2.7 billion to sequence one genome!

The technology used to sequence millions of short DNA fragments is only half the story, however. Once these sequences have been determined, the problem becomes how to put them together. In other words, how are they arranged in the chromosomes from which they came? Imagine if you cut out every word in this book (there are about 500,000 of them), put them on a table, and tried to arrange them in their original order! The enormous task of determining DNA sequences is possible because the original DNA fragments are *overlapping*.

Let's illustrate the process using a single 10-bp DNA molecule. (This is a double-stranded molecule, but for convenience we show only the sequence of the noncoding strand.) The molecule is cut three ways using three different enzymes. The first cut generates the fragments:

TG, ATG, and CCTAC

The second cut of the same molecule generates the fragments:

AT, GCC, and TACTG

The third cut results in:

CTG, CTA, and ATGC

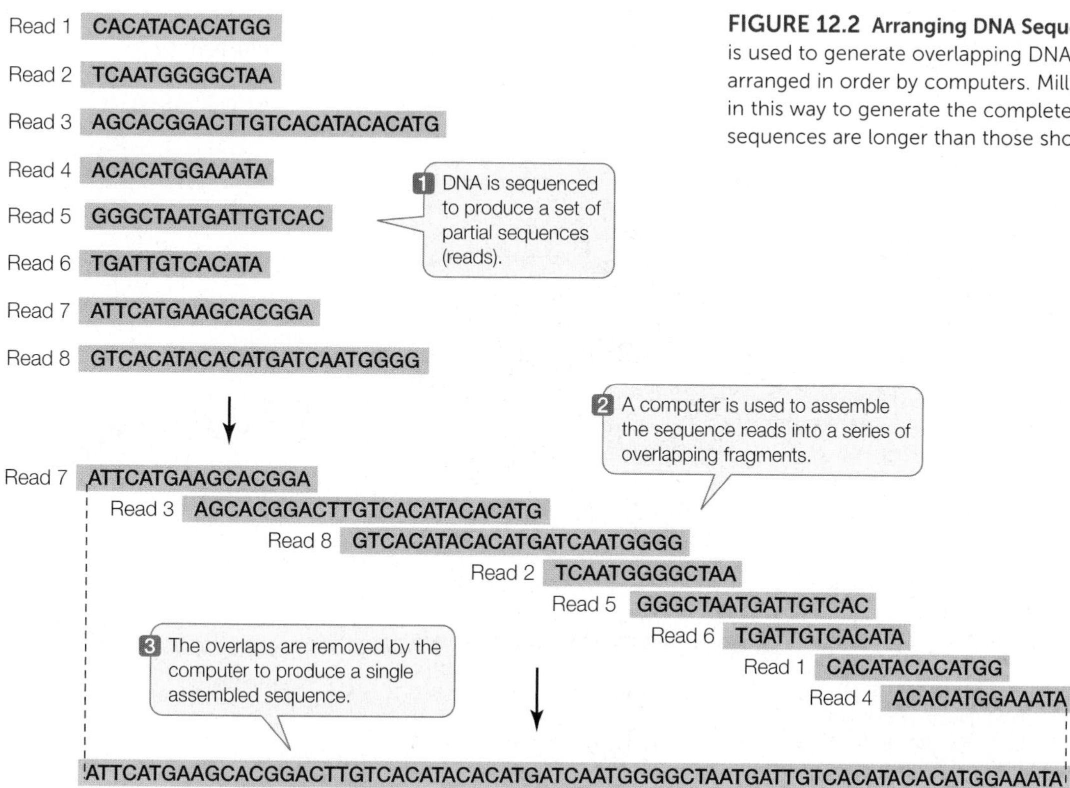

FIGURE 12.2 Arranging DNA Sequences A series of different cuts is used to generate overlapping DNA fragments. Their sequences are arranged in order by computers. Millions of sequences are arranged in this way to generate the complete sequence of a genome. Actual sequences are longer than those shown here.

Can you put the fragments in the correct order? (The answer is ATGCCTACTG.) For genome sequencing, the fragments are called "reads" (**FIGURE 12.2**). Of course, the problem of ordering 2.5 million fragments of 100 bp from human chromosome 1 (246 million bp) is more challenging than our 10-bp example above. The field of **bioinformatics** was developed to analyze DNA sequences using complex mathematics and computer programs. The cost of genome sequencing and analysis has gone down rapidly:

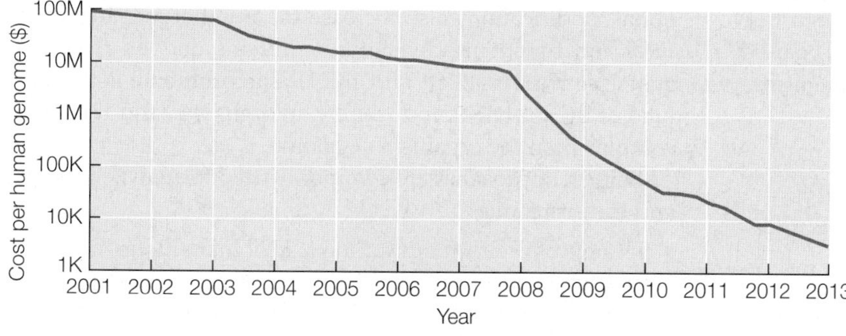

Genome sequences yield several kinds of information

New genome sequences are being published at an accelerating pace, creating a torrent of biological information. This information is used in two related fields of research, both focused on studying genomes. In **functional genomics**, biologists use sequence information to identify the functions of various parts of genomes (**FIGURE 12.3**). These parts include:

- **Open reading frames**, which are sequences of DNA that contain no stop codons, and thus may encode parts of proteins. An open reading frame that begins with a start codon or an intron consensus sequence (boundary between exon and intron) and ends with a stop codon or an intron consensus sequence may be an exon, which encodes part or all of a polypeptide. See Concepts 10.2 and 10.3 to review introns, exons, and the genetic code.
    - Regulatory sequences, such as promoters and terminators for transcription. These are identified by their proximity to open reading frames and because they contain consensus sequences for the binding of specific transcription factors and RNA polymerase.
    - Regions of DNA that have regulatory sequences at each end and one or more open reading frames; these may be protein-coding genes. The amino acid sequence of a protein can be deduced by applying the genetic code (see Figure 10.11) to the DNA sequences of the open reading frames within the gene. A major goal of functional genomics is to identify, and understand the function of, every protein-coding gene in each genome.
- RNA genes, including genes for rRNA, tRNA, and miRNA (see Concept 11.4).
- Other noncoding sequences that can be classified into various categories, including centromeric regions (see Concept 7.2), telomeric regions (see Concept 9.2), transposons (see Concept 12.2), and other repetitive sequences.

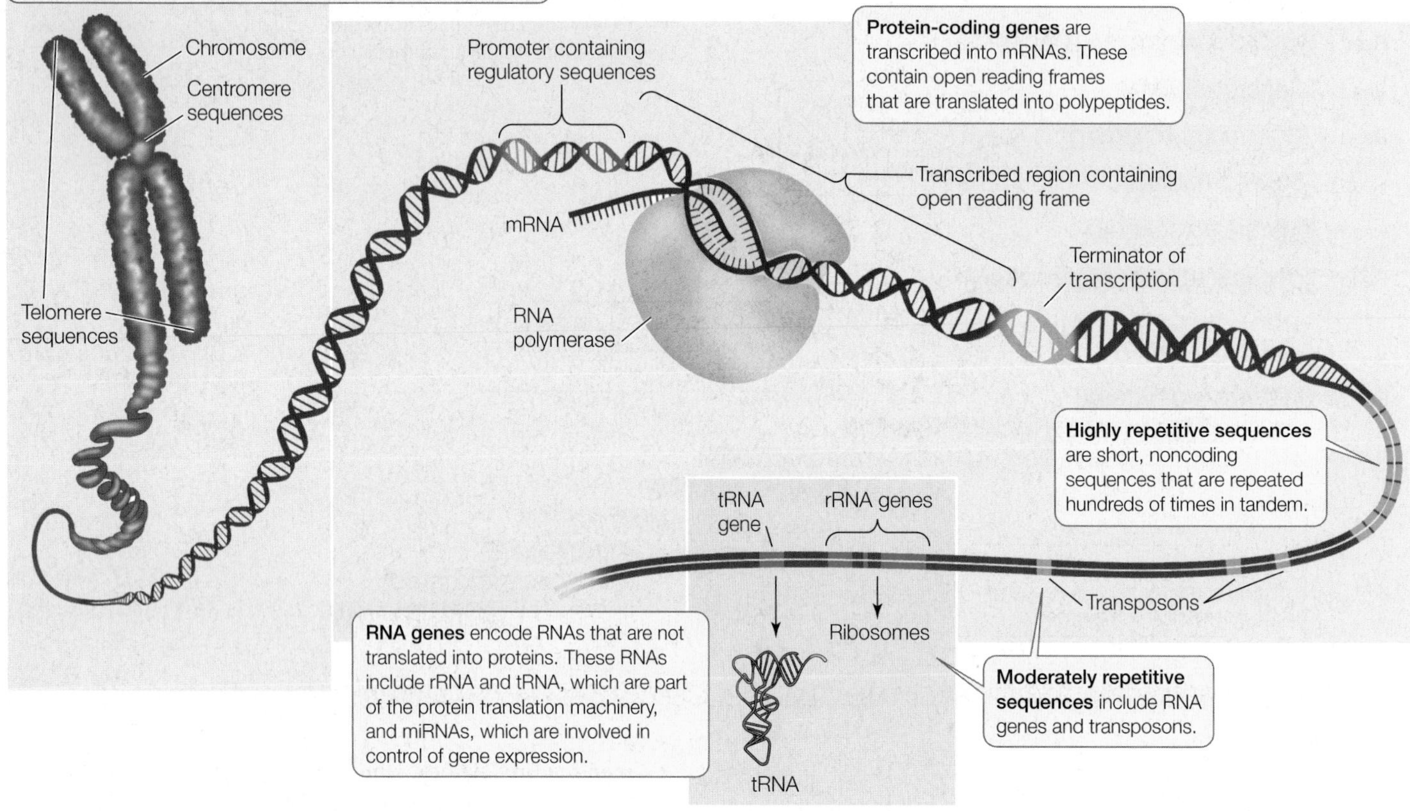
A **chromosome** has a single DNA molecule with specialized DNA sequences for the initiation of DNA replication, for spindle interactions in mitosis (centromeres), and for maintaining the integrity of the ends (telomeres).
Chromosome
Centromere sequences
Telomere sequences
Promoter containing regulatory sequences
Protein-coding genes are transcribed into mRNAs. These contain open reading frames that are translated into polypeptides.
Transcribed region containing open reading frame
mRNA
Terminator of transcription
RNA polymerase
Highly repetitive sequences are short, noncoding sequences that are repeated hundreds of times in tandem.
tRNA gene
rRNA genes
Transposons
Ribosomes
RNA genes encode RNAs that are not translated into proteins. These RNAs include rRNA and tRNA, which are part of the protein translation machinery, and miRNAs, which are involved in control of gene expression.
Moderately repetitive sequences include RNA genes and transposons.
tRNA

FIGURE 12.3 The Genomic Book of Life Genome sequences contain many features, some of which are summarized in this overview. Sifting through all the information contained in a genome sequence can help us understand how an organism functions and what its evolutionary history might be.

Functional regions in a newly described genomic sequence can be identified by searching DNA databases for similar or identical sequences in other organisms. There are now massive databases (accessible online) containing DNA sequences and their known or possible functions.

Sequence information is also used in **comparative genomics**: the comparison of a newly sequenced genome (or parts thereof) with sequences from other organisms. This can provide further information about the functions of sequences and can be used to trace evolutionary relationships among different organisms.

LINK

The application of genome sequencing to reconstructing phylogenies (evolutionary trees) is described in **Concept 16.2**

Phenotypes can be analyzed using proteomics and metabolomics

"The human genome is the book of life." Statements like this were common when the human genome sequence was first revealed. They reflect the concept of genetic determinism—the idea that a person's phenotype is determined solely by his or her genotype. But is an organism just a product of gene expression? We know that it is not. The proteins and small molecules present in any cell at a given point in time reflect not just gene expression but modifications caused by the intracellular and extracellular environments. Two new fields have emerged to complement genomics and take a more complete snapshot of a cell or organism: proteomics and metabolomics.

PROTEOMICS Many genes encode more than a single protein (**FIGURE 12.4A**). As we described in Concept 11.4, alternative splicing leads to different combinations of exons in the mature mRNAs transcribed from a single gene. Posttranslational modifications also increase the number and the structural and functional diversity of proteins derived from one gene (see Figures 12.4A and 10.21). The **proteome** is the sum total of the proteins produced by an organism, and it is more complex than the organism's genome.

Several approaches are commonly used to analyze proteins and the proteome:

- Because of their unique amino acid compositions (primary structures), most proteins have unique combinations of electric charge and size. On the basis of these two properties, they can be separated by two-dimensional gel electrophoresis (**FIGURE 12.4B**).
- Once they have been isolated, individual proteins can be analyzed by mass spectrometry. This technique uses electromagnets to identify molecules by the masses of their atoms, and it can also be used to determine the structures of molecules.
- Antibodies can also be used to isolate specific proteins, or to detect the proteins in cells or tissues.

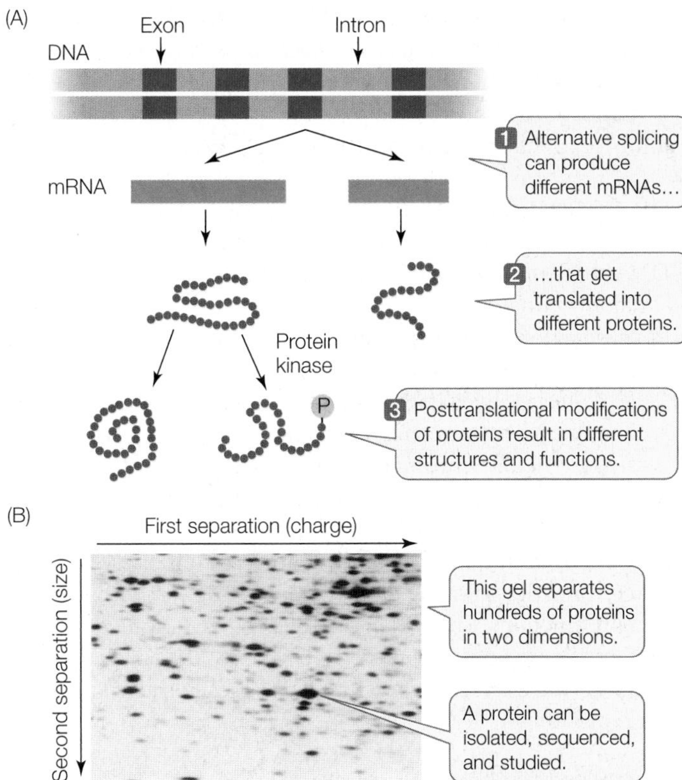

FIGURE 12.4 Proteomics (A) A single gene can code for multiple proteins. (B) Some of a cell's proteins can be separated on the basis of charge and size by two-dimensional gel electrophoresis. The two separations can distinguish some proteins from one another. Further analysis of each spot by mass spectrometry identifies different proteins.

Whereas genomics seeks to describe the genome and its expression, **proteomics** seeks to identify and characterize all of the expressed proteins. Its ultimate aim is just as ambitious as that of genomics. Comparisons of proteomes among organisms have revealed common sets of proteins that can be categorized into groups with similar amino acid sequences and functions. Often these share three-dimensional structural regions called domains (for example, the heme-binding domain of hemoglobin). While a particular organism may have many unique proteins, those proteins are often just unique combinations of domains that exist in proteins of other organisms. This reshuffling of the genetic deck is a key to evolution.

METABOLOMICS Studying genes and proteins gives a limited picture of what is going on in a cell. But as we have seen, both gene function and protein function are affected by a cell's internal and external environments. Many proteins are enzymes, and their activities affect the concentrations of their substrates and products. So as the proteome changes, so do the abundances of these (often small) molecules, called metabolites. The **metabolome** is the complete set of small molecules present in a cell, tissue, or organism. These include:

- *Primary metabolites* that are involved in normal processes, such as intermediates in pathways such as glycolysis. This category also includes hormones and other signaling molecules.
- *Secondary metabolites*, which are often unique to particular organisms or groups of organisms. They are often involved in special responses to the environment. Examples are antibiotics made by microbes, and the many chemicals made by plants that are used in defense against pathogens and herbivores.

Metabolomics aims to describe the metabolic profile of a tissue or organism under particular environmental conditions. Measuring metabolites involves sophisticated analytical instruments. If you have studied organic or analytical chemistry, you may be familiar with gas chromatography and high-performance liquid chromatography, which are used to separate molecules with different chemical properties. Mass spectrometry and nuclear magnetic resonance spectroscopy are used to identify molecules. These measurements result in "chemical snapshots" of cells or organisms, which can be related to physiological states.

Plant biologists are far ahead of medical researchers in the field of metabolomics. Tens of thousands of secondary metabolites, many of them made in response to environmental challenges, have been identified in plants. The metabolomes of agriculturally important plants are being described, and this information may be important in optimizing plant growth for food production.

LINK
Some of the many secondary metabolites made by plants are discussed in **Concept 28.2**

Taken together, the genome, proteome, and metabolome can move biologists toward a more comprehensive picture of an organism's genotype and phenotype (**FIGURE 12.5**).

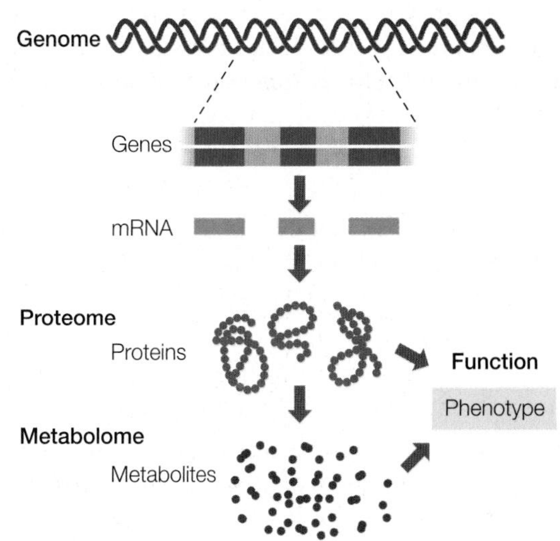

FIGURE 12.5 Genomics, Proteomics, and Metabolomics A combination of these approaches can give more comprehensive information about genotypes and phenotypes.

CHECKpoint CONCEPT 12.1

✓ Using a table, compare genomics, proteomics, and metabolomics with regard to the methods used and results obtained.

✓ A DNA molecule is cut into the following fragments that are sequenced: AGTTT, TAGG, CGAT, and CCT. The same molecule is cut in a different way to produce TTCGA, TCCT, AGT, GG, and TA. A third cut produces TTCGAT, CCTT, AGG, and AGT. What is the sequence of this DNA?

✓ If you were designing a computer program to recognize important sequences in DNA, what types of sequences would you include? What would you learn from finding these sequences in the DNA? Be as specific as you can.

The first cellular genomes to be fully sequenced were those of prokaryotes. In the next concept we will discuss these relatively small, compact genomes.

CONCEPT 12.2 Prokaryotic Genomes Are Small, Compact, and Diverse

When DNA sequencing became possible in the late 1970s, the first life forms to be sequenced were the simplest viruses. The sequences quickly provided new information about how viruses infect their hosts and reproduce. The next genomes to be fully sequenced were those of prokaryotes. We now have genome sequences for many microorganisms, to the great benefit of microbiology and medicine.

Prokaryotic genomes are compact

In 1995 a team led by Craig Venter and Hamilton Smith published the first complete genomic sequence of a free-living cellular organism, the bacterium *Haemophilus influenzae*. Many more prokaryotic sequences have followed. These sequences reveal not only how prokaryotic genes are organized to perform different cellular functions, but also how certain specialized functions of particular organisms are carried out.

There are several notable features of bacterial and archaeal genomes:

- They are relatively small. Prokaryotic genomes range in size from about 160,000 to 12 million bp and are usually organized into a single circular chromosome.
- They are compact. Typically, more than 85 percent of the DNA consists of protein-coding regions or RNA genes, with only short sequences between genes.
- Their genes usually do not contain introns. An exception is the rRNA and tRNA genes of archaea, which are frequently interrupted by introns.
- In addition to the main chromosome, they often carry smaller, circular DNA molecules called plasmids, which may be transferred between cells (see Concept 8.4).

Beyond these broad similarities, there is great diversity among these single-celled organisms, reflecting the huge variety of the environments where they are found.

Let's look in more detail at a few prokaryotic genomes in terms of functional and comparative genomics.

FUNCTIONAL GENOMICS As mentioned above, functional genomics is a biological discipline that assigns functions to DNA sequences. This field is less than 20 years old but is now a major occupation of biologists. You can see the various functions encoded by the genomes of three prokaryotes (in this case, all bacteria) in **TABLE 12.1**.

H. influenzae lives in the upper respiratory tracts of humans and can cause ear infections and (more seriously) meningitis. Its single circular chromosome has 1,830,138 bp. In addition to its origin of replication and the RNA genes, this bacterial chromosome has 1,727 open reading frames.

When this sequence was first announced, only 1,007 (58 percent) of the open reading frames encoded proteins with known functions. Since then scientists have identified the role of almost every protein encoded by the *H. influenzae* genome. All of the major biochemical pathways and molecular functions are represented.

COMPARATIVE GENOMICS Soon after the sequence of *H. influenzae* was announced, the smaller *Mycoplasma genitalium* (580,073 bp) and the larger *E. coli* (4,639,221 bp) genomic sequences were completed. Thus began the new era of comparative genomics. Scientists can identify genes that are present in one bacterium and missing in another, allowing them to relate these genes to bacterial function.

For example, *E. coli* has more genes than *H. influenzae* in each of the functional groups listed in Table 12.1. This suggests

TABLE 12.1 Gene Functions in Three Bacteria

	Number of genes in:		
Category	***E. coli***	***H. influenzae***	***M. genitalium***
Total protein-coding genes	4,288	1,727	482
Biosynthesis of amino acids	131	68	1
Biosynthesis of cofactors	103	54	5
Biosynthesis of nucleotides	58	53	19
Cell envelope proteins	237	84	17
Energy metabolism	243	112	31
Intermediary metabolism	188	30	6
Lipid metabolism	48	25	6
DNA replication, recombination, and repair	115	87	32
Protein folding	9	6	7
Regulatory proteins	178	64	7
Transcription	55	27	12
Translation	182	141	101
Uptake of molecules from the environment	427	123	34

that there may be more biochemical pathways in *E. coli* than in *H. influenzae*. *M. genitalium* lacks most of the enzymes needed to synthesize amino acids, which *E. coli* and *H. influenzae* both possess (see Table 12.1). This finding reveals that *M. genitalium* must obtain its amino acids from its environment (usually the human urogenital tract). Furthermore, *E. coli* has dozens of genes for regulatory proteins that encode transcriptional activators or repressors; *M. genitalium* has only seven such genes. This suggests that the biochemical flexibility of *M. genitalium* is limited by its relative lack of control over gene expression.

Metagenomics reveals the diversity of viruses and prokaryotic organisms

If you take a microbiology laboratory course, you will learn how to identify various prokaryotes on the basis of their growth in lab cultures. Microorganisms can be identified by their nutritional requirements or the conditions under which they will grow (such as aerobic versus anaerobic). For example, staphylococci are a group of bacteria that inhabit skin and nasal passages. Unlike many bacteria, staphylococci can use the sugar alcohol mannitol as an energy source and thus can grow on a special medium containing mannitol. Often a dye is included in the medium, which changes color if the bacteria are pathogenic (disease-causing). Such culture methods have been the mainstay of microbial identification for more than a century and are still useful and important. However, scientists can now use PCR and modern DNA analysis techniques to analyze microbes *without* culturing them in the laboratory.

In 1985 Norman Pace, then at Indiana University, came up with the idea of isolating DNA directly from environmental samples. He used PCR to amplify specific sequences from the samples to determine whether particular microbes were present. The PCR products were sequenced to explore their diversity. The term **metagenomics** was coined to describe this approach of analyzing genes without isolating the intact organism. It is now possible to do DNA sequencing with samples from almost any environment. The sequences can be used to detect the presence of previously unidentified organisms as well as known microbes (**FIGURE 12.6**). For example:

- Sequencing of DNA from 200 liters of seawater indicated that it contained 5,000 different viruses and 2,000 different bacteria, many of which had not been described previously.
- Samples from the intestines of chickens and turkeys from different flocks have led to the identification of viral causes of serious diseases in these domesticated birds.
- Water runoff from a mine contaminated with toxic chemicals contained many new species of prokaryotes thriving in this apparently inhospitable environment. Some of these organisms exhibited metabolic pathways that were previously unknown to biologists. These organisms and their capabilities may be useful in cleaning up pollutants from the water.
- Gut samples from 124 Europeans revealed that each person harbored at least 160 species of bacteria (constituting their gut microflora or microbiome). Many of these species were found in all of the individuals, but the presence of other bacteria varied from person to person. Such variations in gut microflora may be associated with obesity or bowel diseases.

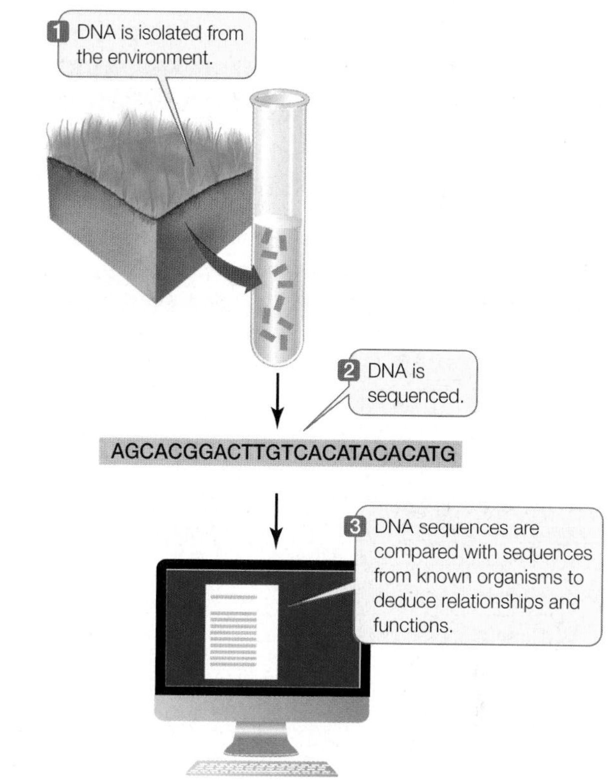

FIGURE 12.6 Metagenomics Microbial DNA extracted from the environment can be amplified and sequenced directly. This has led to the description of many new genes and species.

LINK

For more on the complex microbial ecosystem inside the human gut, see **Figures 19.20 and 41.1**

These and other discoveries are truly extraordinary and potentially very important. It is estimated that 90 percent of the microbial world has been invisible to biologists, in part because the cells could not be grown in the laboratory. These organisms are only now being revealed by metagenomics. Entirely new ecosystems of bacteria and viruses are being discovered in which, for example, one species produces a molecule that another metabolizes. It is hard to overemphasize the importance of such an increase in our knowledge of the hidden world of microbes. This new knowledge underscores the remarkable diversity among prokaryotic organisms, and will further our understanding of natural ecological processes. Furthermore, it has the potential to help us find better ways to manage environmental catastrophes such as oil spills, or to remove toxic heavy metals from soil and water.

Some sequences of DNA can move about the genome

Genome sequencing allowed scientists to study more broadly a class of DNA sequences that had been discovered by geneticists

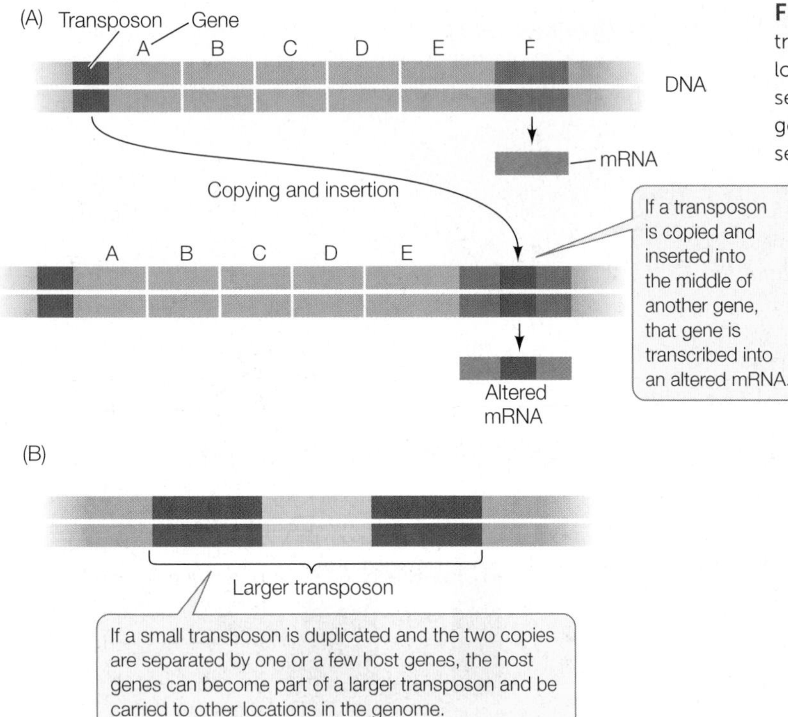

FIGURE 12.7 DNA Sequences That Move Transposons (or transposable elements) are DNA sequences that move from one location to another. (A) In one method of transposition, the DNA sequence is replicated and the copy inserts elsewhere in the genome. (B) Transposons can evolve to carry additional genomic sequences.

decades earlier. Segments of DNA called **transposons** (or transposable elements) can move from place to place in the genome and can even move from one piece of DNA (such as a chromosome) to another (such as a plasmid) in the same cell. A transposon might be at one location in the genome of one *E. coli* cell, and at a different location in another cell. The insertion of this movable DNA sequence from elsewhere in the genome into the middle of a protein-coding gene disrupts that gene (**FIGURE 12.7A**). Any mRNA expressed from the disrupted gene will have the extra sequence, and the protein will be abnormal. Consequently transposons can produce significant phenotypic effects by inactivating genes.

Transposons are often short sequences of 1,000–2,000 bp and are found at many sites in prokaryotic genomes. The mechanisms that allow them to move vary. For example, the transposon may be replicated, and then the copy inserted into another site in the genome. Or the transposon might splice out of one location and move to another location.

If a transposon becomes duplicated, with two copies separated by one or a few genes, the result may be a single larger transposon (up to about 5,000 bp). In this case, the additional genes can be carried to different locations in the genome (**FIGURE 12.7B**). Some of these transposons carry genes for antibiotic resistance. We will discuss transposons again in Concept 12.3.

Will defining the genes required for cellular life lead to artificial life?

When the genomes of prokaryotes and eukaryotes are compared, a striking conclusion arises: certain genes are present in all organisms (universal genes). There are also some (nearly) universal gene segments that are present in many genes in many organisms. One example is a sequence encoding an ATP binding site, which is a domain found in many proteins. These findings suggest that there is some ancient, minimal set of DNA sequences common to all cells. One way to identify these sequences is to look for them in computer analyses of sequenced genomes.

Another way to define the minimal genome is to take an organism with a simple genome and deliberately mutate one gene at a time to see what happens. *M. genitalium* has one of the smallest known genomes, with only 482 protein-coding genes. Even so, some of its genes are dispensable under some circumstances. For example, it has genes for metabolizing both glucose and fructose, but it can survive in the laboratory on a medium containing only one of these sugars. Under such circumstances, the bacterium doesn't need the genes for metabolizing the other sugar.

What about other genes? Researchers addressed this question using transposons as mutagens. When transposons in the bacterium were activated, they inserted themselves into genes at random, mutating and inactivating the genes (**FIGURE 12.8**). The mutated bacteria were tested for growth and survival, and DNA from interesting mutants was sequenced to find out which genes were mutated. The astonishing results of these studies suggested that only 382 of the 482 *M. genitalium* protein-coding genes were needed for survival in the laboratory!

One application of the research might be to design organisms with specific uses. The next step toward that goal is to create an artificial genome and insert it into bacterial cells. As we described in the opening story of Chapter 4, this was recently accomplished, using a synthetic genome based on that of the bacterium *Mycoplasma mycoides*. This research has promise for making organisms with novel functions, such as the synthesis of plastics polymers or the ability to break down environmental pollutants.

CHECKpoint CONCEPT **12.2**

- ✓ What are the characteristics of most prokaryotic genomes?
- ✓ Examine Table 12.1 and Figure 12.8. What gene functions would you predict are nonessential for *M. genitalium* as determined by transposon-mediated inactivation?
- ✓ You want to isolate a prokaryote that can live on discarded Styrofoam cups. Such an organism might live in a landfill where ground-up cups are discarded. How would you use metagenomics to identify such a bacterium?
- ✓ How would you show that the prokaryote's ability to live on Styrofoam is essential, and that it cannot live in another environment?

INVESTIGATION

FIGURE 12.8 Using Transposon Mutagenesis to Determine the Minimal Genome *Mycoplasma genitalium* has one of the smallest known genomes of any prokaryote. But are all of its genes essential to life? By inactivating the genes one by one, scientists determined which of them are essential for the cell's survival. This research may lead to the construction of artificial cells with customized genomes, designed to perform functions such as degrading oil and making plastics.[a]

HYPOTHESIS

Only some of the genes in a bacterial genome are essential for cell survival.

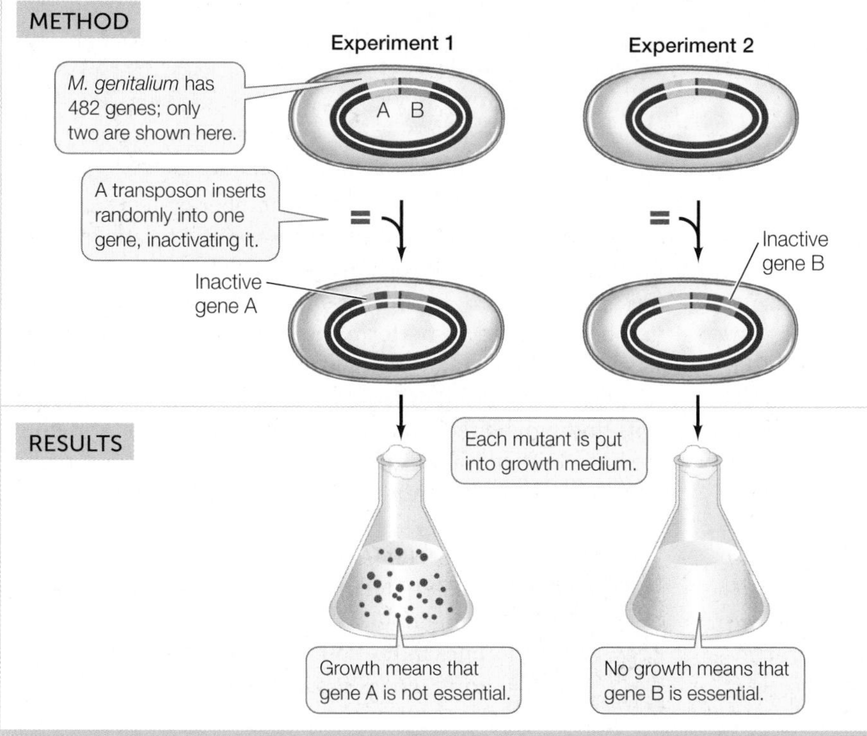

METHOD
Experiment 1
Experiment 2
M. genitalium has 482 genes; only two are shown here.
A B
A transposon inserts randomly into one gene, inactivating it.
Inactive gene B
Inactive gene A
RESULTS
Each mutant is put into growth medium.
Growth means that gene A is not essential.
No growth means that gene B is essential.

CONCLUSION

If each gene is inactivated in turn, a "minimal essential genome" can be determined.

ANALYZE THE DATA

The growth of *M. genitalium* strains with insertions in genes (intragenic regions) was compared with the growth of strains with insertions in noncoding (intergenic) regions of the genome:

Type of insertion	Number of strains with insertions	Number of strains that grew
Intragenic	482	100
Intergenic	199	184

A. Explain these data in terms of genes essential for growth and survival. Are all of the genes in *M. genitalium* essential for growth? If not, how many are essential? Why did some of the insertions in intergenic regions prevent growth?

B. If a transposon inserts into the following regions of genes, there might be no effect on phenotype. Explain why in each case:
 i. near the 3′ end of the coding region
 ii. within a gene coding for rRNA

How does this affect your answer to the first question?

Go to **LaunchPad** for discussion and relevant links for all **INVESTIGATION** figures.

[a]C. Hutchison et al. 1999. *Science* 286: 2165–2169. J. I. Glass et al. 2006. *Proceedings of the National Academy of Sciences USA* 103: 425–430.

The methods used to sequence and analyze prokaryotic genomes have also been applied to eukaryotic genomes, which we will examine next.

CONCEPT 12.3 Eukaryotic Genomes Are Large and Complex

As genomes have been sequenced and described, a number of major differences have emerged between eukaryotic and prokaryotic genomes:

- *Eukaryotic genomes are larger than those of prokaryotes*, and they have more protein-coding genes. This difference is not surprising given that multicellular organisms have many cell types with specialized functions. As we saw above, one of the simplest prokaryotes, *Mycoplasma*, has several hundred protein-coding genes in a genome of about 0.5 million bp. A rice plant, in contrast, has about 40,383 protein-coding genes.
- *Eukaryotic genomes have more regulatory sequences*—and many more regulatory proteins—than prokaryotic genomes. The greater complexity of eukaryotes requires much more regulation, which is evident in the many points of control associated with the expression of eukaryotic genes (see Concepts 11.2–11.4 and Figure 11.1).
- *Much of eukaryotic DNA does not encode proteins*. Distributed throughout many eukaryotic genomes are various kinds of DNA sequences that are not transcribed into mRNA. Some of these sequences are genes for functional RNAs, such as rRNA, tRNA, and miRNA. Others are introns or regulatory sequences. In addition, eukaryotic genomes contain various kinds of repeated sequences.

Model organisms reveal many characteristics of eukaryotic genomes

Most of our information about eukaryotic genomes has come from model organisms that have been studied extensively. These include the yeast *Saccharomyces cerevisiae*, the nematode (roundworm) *Caenorhabditis elegans*, the fruit fly *Drosophila melanogaster*, and the plant *Arabidopsis thaliana* (thale cress). Model organisms have been chosen

APPLY THE CONCEPT

Eukaryotic genomes are large and complex

Repetitive DNA sequences can be classified by nucleic acid hybridization (see Figure 10.7). A genome is initially cut into 300-bp fragments, and these are heated to denature the DNA. If the solution is cooled, the DNA strands will form hydrogen bonds and reassociate into double-stranded structures. If there are many copies of a DNA sequence in the solution (repetitive DNA), it will find its complementary sequence and reassociate faster than if there are only a few copies. The table shows typical results from equal amounts of DNA from three species.

	PERCENTAGE OF DNA REASSOCIATED		
REASSOCIATION TIME (MIN)	*E. COLI*	YEAST	MOUSE
1	0	3	10
10	0	17	35
100	100	100	100

1. Why do yeast and mouse DNAs reassociate faster than *E. coli* DNA?
2. Would you expect human DNA to reassociate faster or slower than yeast DNA?

because they are relatively easy to grow and study in a laboratory, their genetics are well studied, and they exhibit characteristics that represent a larger group of organisms. **TABLE 12.2** shows some characteristics of the genomes of these organisms.

YEAST: THE BASIC EUKARYOTIC CELL MODEL Yeasts are single-celled eukaryotes. Like other eukaryotes, they have membrane-enclosed organelles. They can live as either haploid or diploid organisms, and this is usually determined by environmental conditions: under adverse conditions the diploid cells will undergo meiosis and make spores. Whereas the prokaryote *E. coli* has a single circular chromosome and 4,288 protein-coding genes, *Saccharomyces cerevisiae* has 16 linear chromosomes and 6,275 protein-coding genes. The most striking difference between the yeast genome and that of *E. coli* is in the number of genes involved in secretion or targeting proteins to specific locations within the cell: yeast has 430 such genes; *E. coli* has only 35. Both of these single-celled organisms appear to use about the same number of genes to perform the basic functions of cell survival. It is the compartmentalization of the eukaryotic yeast cell into organelles that requires it to have many more genes. This finding is direct, quantitative confirmation of something we have known for a century: the eukaryotic cell is structurally and functionally more complex than the prokaryotic cell.

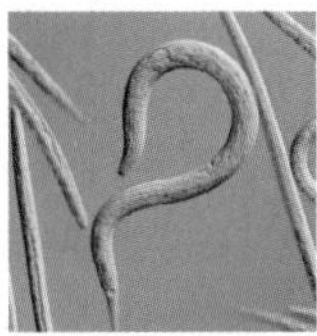

THE NEMATODE: UNDERSTANDING CELL DIFFERENTIATION The 1-millimeter-long nematode *Caenorhabditis elegans* normally lives in the soil. It can also live in the laboratory, where it has become a favorite model

TABLE 12.2 Representative Sequenced Genomes

Organism	Haploid genome size (Mb)[a]	Number of protein-coding genes	Percent of genome that codes for proteins	Notable attributes
BACTERIA				
Mycoplasma genitalium	0.58	482	88	Minimal genome
Haemophilus influenzae	1.83	1,727	89	
Escherichia coli	4.6	4,288	88	Well-studied enteric bacterium
YEASTS				Targeting; cell organelles
Saccharomyces cerevisiae	12.2	6,275	70	
Schizosaccharomyces pombe	13.8	4,824	60	
PLANTS				Photosynthesis; cell walls
Arabidopsis thaliana	125	27,416	25	Small plant genome
Oryza sativa (rice)	420	40,838	12	Water tolerance for roots
Glycine max (soybean)	973	46,430	7	Lipid synthesis, storage
ANIMALS				
Caenorhabditis elegans (nematode)	100	20,470	25	Tissue formation
Drosophila melanogaster (fruit fly)	140	13,733	13	Embryonic development
Homo sapiens (human)	3,200	~21,000	1.2	Language

[a]Mb = millions of base pairs

organism of developmental biologists (see Chapter 14). The nematode has a transparent body of about 1,000 somatic cells. It develops over 3 days from a fertilized egg to an adult worm that has a nervous system, digests food, and reproduces sexually. Its genome is 8 times larger than that of yeast and has about 3.3 times as many protein-coding genes. Many of these extra genes encode proteins needed for cell differentiation, for intercellular communication, and for holding cells together to form tissues.

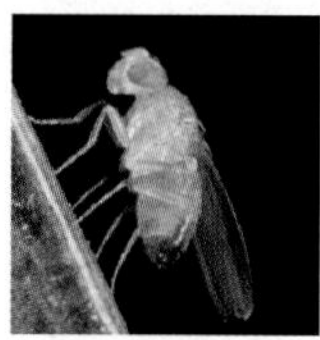

DROSOPHILA MELANOGASTER: UNDERSTANDING GENETICS AND DEVELOPMENT The fruit fly *Drosophila melanogaster* is a famous model organism. Studies of fruit fly genetics resulted in the formulation of many basic principles of genetics (see Concept 8.3). The fruit fly is a much larger organism than *C. elegans* (it has ten times as many cells), and it is much more complex: it undergoes complicated developmental transformations from egg to larva to pupa to adult. These differences are reflected in the fruit fly genome, which has many genes encoding transcription factors needed for complex embryonic development (you will study some of these in Chapter 14). In general, the fruit fly genome has a distribution of coding sequence functions quite similar to those of many other complex eukaryotes (**FIGURE 12.9**).

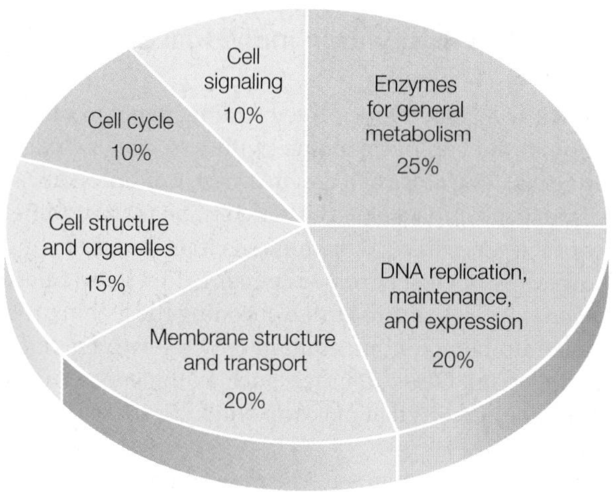

FIGURE 12.9 Functions of the Eukaryotic Genome The distribution of gene functions in *Drosophila melanogaster* shows a pattern that is typical of that of many complex organisms.

ARABIDOPSIS: STUDYING THE GENOMES OF PLANTS About 250,000 species of flowering plants dominate the land and fresh water. Although there is generally more interest in the plants we use for food and fiber, scientists first sequenced the genome of a simpler flowering plant with a relatively small genome. *Arabidopsis thaliana*, thale cress, is a member of the mustard family and has long been a favorite model organism of plant biologists. It is small (hundreds could grow and reproduce in the space occupied by this page) and easy to manipulate. Its genome has about 27,400 protein-coding genes, but many of these are duplicates and probably originated by chromosomal rearrangements. When these duplicate genes are subtracted from the total, about 15,000 unique genes are left—similar to the gene numbers found in fruit flies and nematodes. Indeed, many of the genes found in these animals have **orthologs**—genes that are derived from a common ancestral gene—in *Arabidopsis* and other plants, supporting the idea that plants and animals have a common ancestor.

Arabidopsis has some genes, however, that are unique to plants. These include genes involved in photosynthesis, in the transport of water throughout the plant, in the assembly of the cell wall, in the uptake and metabolism of inorganic substances from the environment, and in the synthesis of specific molecules used for defense against microbes and herbivores. These plant defense molecules may be a major reason why the numbers of protein-coding genes in some plants are higher than in many animals. Plants cannot escape their enemies or other adverse conditions as animals can, and they must cope with situations where they are. So they make tens of thousands of molecules to help them fight their enemies and adapt to their changing environments (see Chapter 28).

These plant-specific genes are also found in the genomes of other plants, including rice (*Oryza sativa*), the first major crop plant to be fully sequenced. Rice is the world's most important crop—it is a staple in the diets of 3 billion people. The larger genome of rice has a set of genes remarkably similar to that of *Arabidopsis*. The genome of the poplar tree *Populus trichocarpa* was also sequenced, to gain insight into the potential for this rapidly growing tree to be used as a source of fuel. Several more plant genomes have now been sequenced. Comparisons among diverse flowering plant species (including *Arabidopsis*, rice, poplar, and sorghum) suggest that about 12,000 protein-coding genes are shared among all flowering plants (**FIGURE 12.10**). These may comprise the basic plant genome.

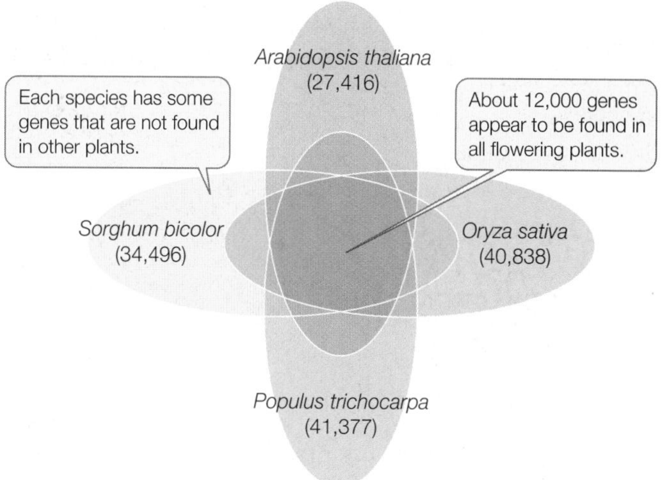

FIGURE 12.10 Plant Genomes Four plant genomes share a common set of approximately 12,000 genes that may comprise the "minimal" plant genome. The total numbers of protein-coding genes are shown in parentheses.

Gene families exist within individual eukaryotic organisms

About half of all eukaryotic protein-coding genes exist as only one copy in the haploid genome (with two alleles in diploid somatic cells). The other half are present in multiple copies that arose from gene duplications. Over evolutionary time, different copies of the genes have undergone separate mutations, giving rise to groups of closely related genes called **gene families.** Some gene families, such as those encoding the globin proteins that make up hemoglobin, contain only a few members in a single organism. Other families, such as the genes encoding the immunoglobulins that make up antibodies, have hundreds of members.

Within a single organism, the genes in a family are usually slightly different from one another. As long as at least one member encodes a functional protein, the other members may mutate in ways that change the functions of the proteins they encode. For evolution, the availability of multiple copies of a gene allows for selection of mutations that provide advantages under certain circumstances. If a mutated gene is useful, it may be selected for in succeeding generations. If the mutated gene is a total loss, the functional copy is still there to carry out its role.

As an example, let's look at the gene family encoding the globins in vertebrates. These proteins are found in hemoglobin and myoglobin (an oxygen-binding protein present in muscle). The globin genes all arose long ago from a single common ancestral gene (see Figure 15.23). In humans there are three functional members of the α-globin cluster and five in the β-globin cluster (**FIGURE 12.11**). Each hemoglobin molecule in an adult human is a tetramer containing two identical α-globin subunits, two identical β-globin subunits, and four heme pigments (see Chapter 3, p. 47).

During human development, different genes of the globin gene cluster are expressed at different times and in different tissues. This differential gene expression has great physiological significance. For example, the hemoglobin in the human fetus contains γ-globin, which binds O_2 more tightly than adult hemoglobin does. This specialized form of hemoglobin ensures that in the placenta, O_2 is transferred from the mother's blood to the developing fetus's blood. Just before birth the liver stops synthesizing fetal hemoglobin and the bone marrow cells take over, making the adult forms (two α and two β). Thus hemoglobins with different binding affinities for O_2 are provided at different stages of human development.

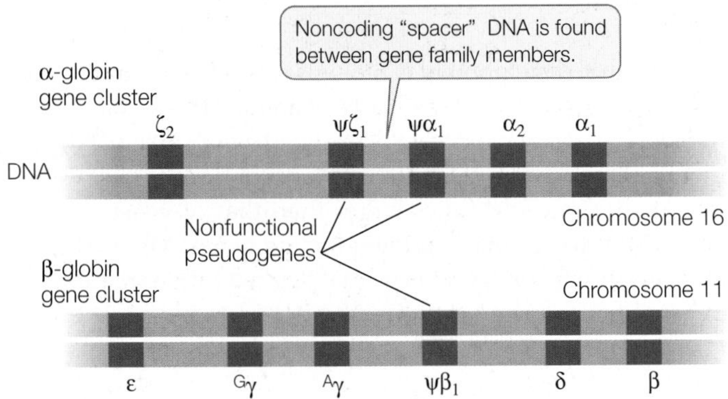

FIGURE 12.11 The Globin Gene Family The α-globin and β-globin clusters of the human globin gene family are located on different chromosomes. The genes of each cluster are separated by noncoding "spacer" DNA. The nonfunctional pseudogenes are indicated by the Greek letter psi (ψ). The γ gene has two variants, $^{A}\gamma$ and $^{G}\gamma$.

LINK

Gene duplication plays a role in the evolution of new protein functions, as described in **Concept 15.6**

In addition to genes that encode functional proteins, many gene families include nonfunctional **pseudogenes**, which are designated with the Greek letter psi (ψ; see Figure 12.11). These pseudogenes result from mutations that cause a loss of function rather than an enhanced or new function. The DNA sequence of a pseudogene may not differ greatly from that of other family members. It may simply lack a promoter, for example, and thus fail to be transcribed. Or it may lack a recognition site needed for the removal of an intron, so that the transcript it makes is not correctly processed into a useful mature mRNA. In some gene families pseudogenes outnumber functional genes. Because some members of the family are functional, there appears to be little selection pressure to preserve the functions of these pseudogenes.

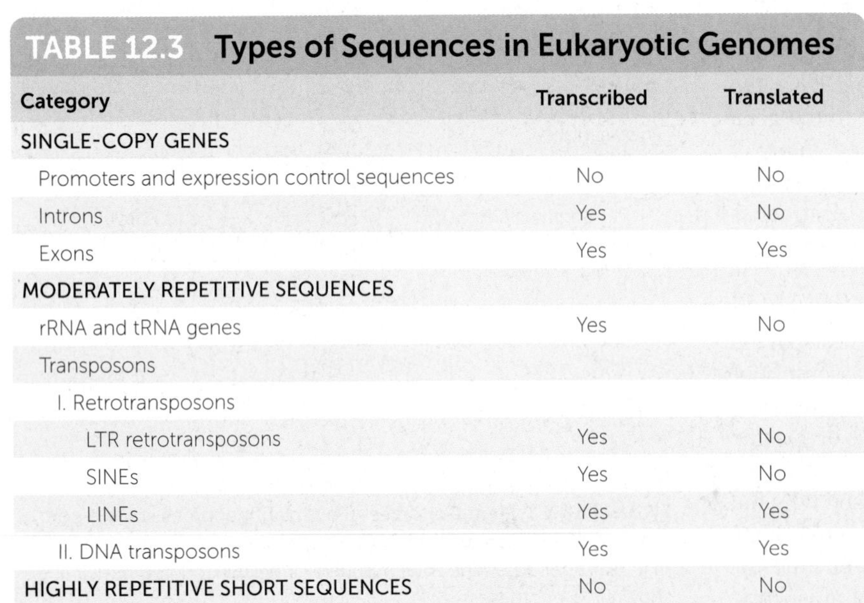

TABLE 12.3 Types of Sequences in Eukaryotic Genomes

Category	Transcribed	Translated
SINGLE-COPY GENES		
Promoters and expression control sequences	No	No
Introns	Yes	No
Exons	Yes	Yes
MODERATELY REPETITIVE SEQUENCES		
rRNA and tRNA genes	Yes	No
Transposons		
I. Retrotransposons		
LTR retrotransposons	Yes	No
SINEs	Yes	No
LINEs	Yes	Yes
II. DNA transposons	Yes	Yes
HIGHLY REPETITIVE SHORT SEQUENCES	No	No

Eukaryotic genomes contain many repetitive sequences

Eukaryotic genomes contain numerous repetitive DNA sequences that do not code for polypeptides. There are highly repetitive sequences and moderately repetitive sequences, which include rRNA genes, tRNA genes, and transposons (**TABLE 12.3**).

Highly repetitive sequences are short (less than 100 bp) sequences that are repeated thousands of times in tandem (side-by-side) arrangements in the genome. They are not transcribed. Their proportion in eukaryotic genomes varies, from 10 percent in humans to about half the genome in some species of fruit flies. Often they are associated with heterochromatin, the densely packed, largely transcriptionally inactive part of the genome (see Concept 11.3). Other highly repetitive sequences are scattered around the genome. For example, **short tandem repeats** (**STRs**) of 1–5 bp can be repeated up to 100 times at a particular chromosomal location. The copy number of an STR at a particular location varies among individuals and is inherited.

Moderately repetitive sequences are repeated 10–1,000 times in the eukaryotic genome. These sequences include the genes that are transcribed to produce tRNAs and rRNAs, which are used in protein synthesis. The cell makes tRNAs and rRNAs constantly, but even at the maximum rate of transcription, single copies of the tRNA and rRNA genes would be inadequate to supply the large amounts of these molecules needed by most cells. Thus the genome has multiple copies of these genes, in clusters containing transcribed regions (with introns) and nontranscribed "spacers" between the genes. Here, for example, is a region that encodes multiple sets of rRNA genes (in dark blue):

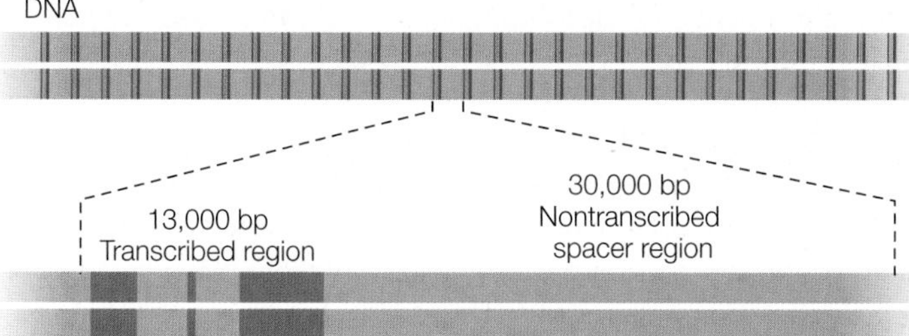

Most moderately repetitive sequences are not stably integrated into the genome but instead are transposons (see Figure 12.7). Transposons make up more than 40 percent of the human genome. There are two main types of transposons in eukaryotes: retrotransposons (Class I transposons) and DNA transposons (Class II; see Table 12.3).

Retrotransposons make RNA copies of themselves, which are then copied back into DNA before insertion at new locations in the genome. They are divided into two categories:

- *LTR retrotransposons* have long terminal repeats (LTRs) of DNA sequence (100–5,000 bp) at each end. LTR retrotransposons constitute about 8 percent of the human genome.
- *Non-LTR retrotransposons* do not have LTR sequences at their ends. They are further divided into two subcategories: SINEs and LINEs. SINEs (short *in*terspersed *e*lements) are up to 500 bp long and are transcribed but not translated. There are about 1.5 million of them scattered over the human genome, making up about 15 percent of the total DNA content. A single type, the 300-bp *Alu* element, accounts for 11 percent of the human genome; it is present in a million copies. LINEs (long *in*terspersed *e*lements) are up to 7,000 bp long, and some are transcribed and translated into proteins. They constitute about 17 percent of the human genome.

DNA transposons do not use RNA intermediates. Like some prokaryotic transposons, they are excised from the original location and become inserted at a new location without being replicated.

With so much of the genome made up of transposons, they must have a role in addition to just replicating themselves ("selfish DNA"). One possibly important function occurs when such a sequence moves to a new location within the coding region of a gene and causes mutation by disrupting it. Mutations are the raw material of evolution by natural selection. In most cases, the ability to move about the genome has been suppressed, so transposons are stable and do not move. Some occur within introns, where they may affect alternative splicing of pre-mRNA, causing phenotypic diversity. Others are at or near gene regulatory sequences such as promoters, where they can also affect gene expression.

CHECKpoint CONCEPT 12.3

- ✓ Compare the general properties of the genomes of prokaryotes and eukaryotes.
- ✓ Does the size of a genome determine how much information it contains? Explain in terms of repetitive sequences and protein-coding genes.
- ✓ What is the evolutionary role of eukaryotic gene families?
- ✓ During transposition, an adjacent gene is sometimes transposed along with a retrotransposon. What would be the consequence of making a new copy of this gene at a new location in the genome?

The analysis of eukaryotic genomes has resulted in an enormous amount of useful information, as we have seen. In the next concept we will look more closely at the human genome.

CONCEPT 12.4 The Human Genome Sequence Has Many Applications

During the first decade of this millennium the haploid genomes of more than ten individuals were sequenced and published. With the rapid development of new sequencing technologies, the time is approaching when a human genome can be sequenced for less than $1,000.

The human genome sequence held some surprises

The following are just some of the interesting facts we have learned about the human genome:

- Among the 3.2 billion bp in the haploid human genome, there are about 21,000 protein-coding genes. This was a surprise. Before sequencing began, the diversity of human proteins suggested there would be 80,000–150,000 genes. The actual number—not many more than in a nematode—means that posttranscriptional mechanisms (such as alternative splicing) must account for the observed number of proteins in humans. It turns out that most human genes encode multiple proteins.

- The average protein-coding gene spans 27,000 bp, and virtually all genes have many introns. Gene sizes vary greatly, from about 1,000 to 2.4 million bp. Variation in gene size is to be expected given that human proteins vary in size, from about 100 to 5,000 amino acids per polypeptide chain.
- More than 50 percent of the genome is made up of transposons and other repetitive sequences. Most transposons are inactive most of the time.
- About 75 percent of the genome is transcribed at some point in some cells. This result came from a recent analysis of genome expression in human cells in culture. A typical specialized cell only transcribes about 25 percent of its genome, so most transcripts are cell-type specific. Many transcripts are noncoding RNAs involved in regulating gene expression.
- Most of the genome (at least 99 percent) is the same in all people. Despite this apparent homogeneity, there are, of course, many individual differences. Current estimates suggest that each haploid genome has variations in about 3.3 million single nucleotide polymorphisms (SNPs; see below), as well as short repeated sequences that are variable in repeat number.
- An individual's genome changes over time, in specific sets of cells, as new mutations occur. These changes can be important if they affect cell function. For example, a mutation in a gene that blocks cell division may result in reduced expression of that gene, and cancer can then occur.

Comparisons among sequenced genomes from prokaryotes and eukaryotes have revealed some of the evolutionary relationships among genes. Some genes are present in both prokaryotes and eukaryotes; others are only in eukaryotes; still others are only in animals, or only in vertebrates (**FIGURE 12.12**).

More comparative genomics is possible now that the genomes of other primates, including all the great apes, have been sequenced. Chimpanzees are our closest living relatives, sharing nearly 99 percent of our DNA sequence. Gorillas and orangutans are next closest, with genomes that are about 98 percent and 97 percent similar to ours. Researchers have identified about 500 protein-coding genes that have undergone accelerated evolution in humans, chimpanzees, and gorillas, including genes involved in hearing and brain development. Further analyses of these sequences may reveal genes that distinguish us from other apes and that "make humans human."

Go to MEDIA CLIP 12.1
A Big Surprise from Genomics
PoL2e.com/mc12.1

Human genomics has potential benefits in medicine

Complex phenotypes are determined not by single genes but by multiple genes interacting with the environment. A single disease-causing allele, such as those associated with

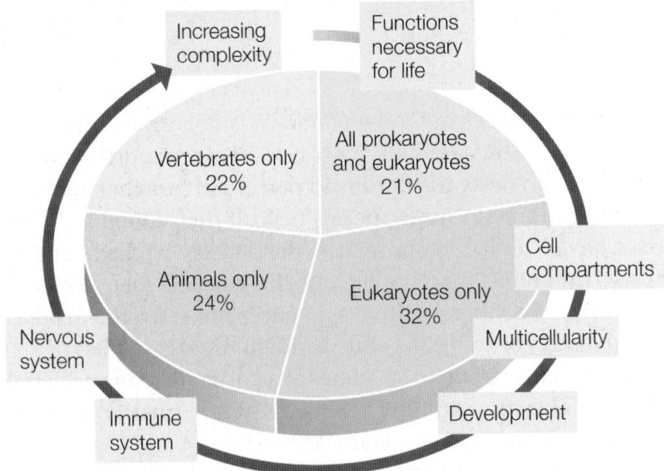

FIGURE 12.12 Evolution of the Genome A comparison of the human and other genomes has revealed how genes with new functions have been added over the course of evolution. Each percentage number refers to genes in the human genome. Thus 21 percent of human genes have orthologs in prokaryotes and other eukaryotes, 32 percent of human genes occur in all other eukaryotes, and so on.

phenylketonuria and sickle-cell anemia (see Concept 9.3), does not exist for such common disorders as diabetes, heart disease, and Alzheimer's disease. To understand the genetic bases of these diseases, biologists are now using rapid genotyping technologies to create "haplotype maps" that can be used to identify multiple genes involved in disease.

HAPLOTYPE MAPPING Haplotype maps are based on **single nucleotide polymorphisms (SNPs)**—DNA sequence variations that involve single nucleotides. SNPs (pronounced "snips") arise as point mutations (see Concept 9.3). Because of these mutations, a single nucleotide in a homologous DNA sequence may vary among individuals or even between alleles in a single individual. Biologists use SNPs to create genetic maps of organisms, to classify organisms and species, and to identify individual organisms carrying specific alleles.

The SNPs that differ among individuals are not necessarily inherited as independent alleles. Rather, a set of SNPs that are close together on a chromosome is inherited as a unit (the SNPs are tightly linked). A piece of chromosome with a set of linked SNPs is called a **haplotype**. You can think of the haplotype as a sentence and the SNP as a word in the sentence. Analyses of haplotypes in humans from all over the world have thus far identified 500,000 common variations.

GENOTYPING TECHNOLOGY AND PERSONAL GENOMICS New technologies are continually being developed to analyze thousands or millions of SNPs in the genomes of individuals. Such technologies include high-throughput sequencing methods and DNA microarrays, which depend on hybridization to identify specific SNPs.

A **DNA microarray** is a grid of microscopic spots of oligonucleotides (short DNA sequences) arrayed on a solid surface. It can be "probed" with a complex mixture of DNA or RNA; if the mixture contains a sequence that is complementary to one of the oligonucleotides, the sequence will hybridize to that spot. Colored fluorescent dyes are used to detect spots that hybridize with components of the probe mixture. The specific pattern of fluorescent dots reveals the haplotypes of the individual from whom the DNA came. For example, a microarray of 500,000 SNP-containing oligonucleotides has been used to analyze DNA from thousands of people to find out which SNPs are linked to genes associated with specific diseases. The aim is to identify particular alleles that contribute (along with particular alleles of other genes) to each complex disease (**FIGURE 12.13**). The amount of data from 500,000 SNPs and thousands of people with thousands of medical records is prodigious. With so much natural variation, statistical measures of association between a haplotype and a disease need to be very rigorous.

These association tests have revealed haplotypes or alleles that are associated with modestly increased risks for such diseases as breast cancer, diabetes, arthritis, obesity, and coronary heart disease. For example, 12 SNPs are associated with increased incidence of heart attacks, and if considered together, these SNPs can be used to identify individuals who are at increased risk. Indeed, the predictive value of this genetic test is greater than the widely used test for elevated blood cholesterol level. Private companies now offer to scan a human genome for SNP alleles, and the price for this service keeps getting lower. However, at this point it is unclear what a person without symptoms should do with the information, since multiple genes, environmental influences, and epigenetic effects all contribute to the development of these diseases.

PHARMACOGENOMICS Genetic variation can affect how an individual responds to a particular drug. For example, consider an enzyme in the liver that catalyzes the following reaction:

$$\text{Active drug} \xrightarrow{\text{Enzyme A}} \text{less active drug}$$

A mutation in the gene that encodes this enzyme may make the enzyme less active and reduce the rate at which the active drug is modified to a less active form. For a given dose of the drug, a person with the mutation would have more active drug in his or her bloodstream than a person without the mutation. So the dose of the drug needed for the same effect would be lower for this person.

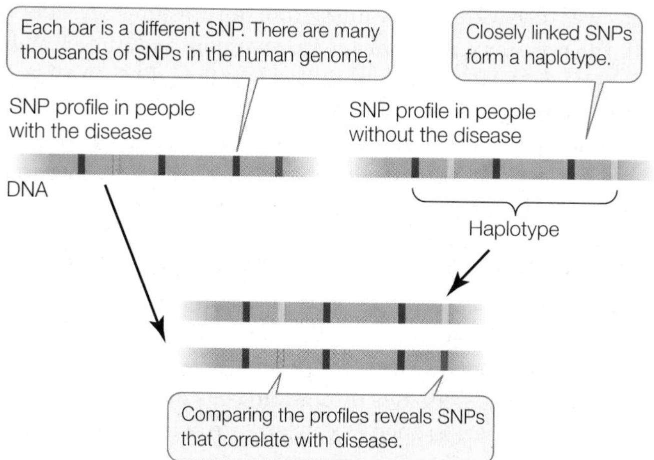

FIGURE 12.13 SNP Genotyping and Disease Scanning the genomes of people with and without particular diseases reveals correlations between SNPs and complex diseases.

Now consider a different case, in which a liver enzyme is needed to make the drug active:

$$\text{Inactive drug} \xrightarrow{\text{Enzyme B}} \text{active drug}$$

A person homozygous for a mutation in the gene encoding this enzyme would not be affected by the drug, since the activating enzyme is not present.

The study of how an individual's genome affects his or her response to drugs or other agents is called **pharmacogenomics**. This type of analysis makes it possible to predict whether or not a drug will be effective. The objective is to personalize drug treatment so that a physician can know in advance whether an individual will benefit from a particular drug (**FIGURE 12.14**). This approach might also be used to reduce the incidence of adverse drug reactions in individuals who metabolize particular drugs slowly.

PROTEOMICS Comparisons of the proteomes of humans and other eukaryotic organisms have revealed a common set of proteins that can be categorized into groups (families) with similar amino acid sequences and similar functions. Forty-six percent of the yeast proteome, 43 percent of the worm proteome, and 61 percent of the fruit fly proteome are shared by the human proteome. Functional analyses indicate that this set of 1,300 proteins provides the basic metabolic functions of a eukaryotic cell—including glycolysis, the citric acid cycle, membrane transport, protein synthesis and targeting, and DNA replication. Of course, these are not the only human

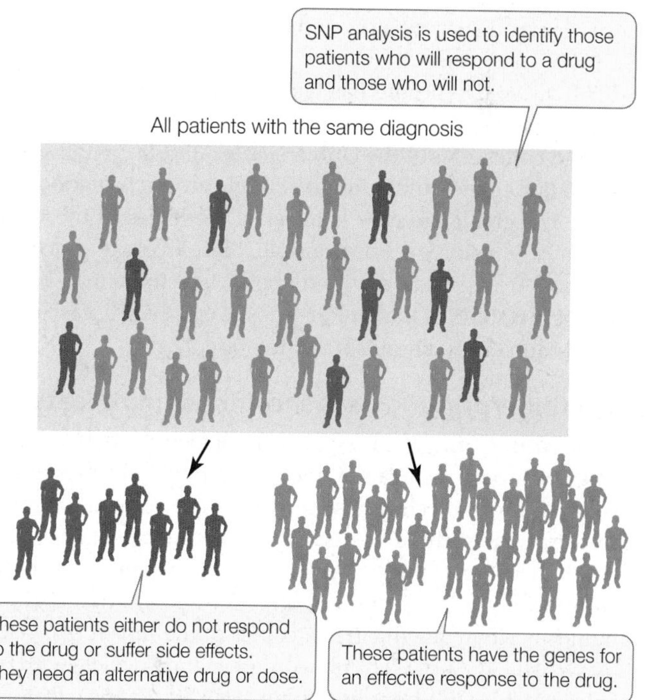

FIGURE 12.14 Pharmacogenomics Correlations between genotypes and responses to drugs will help physicians develop more personalized medical care. SNP analysis is used to identify people who will respond to a drug and those who will not. The different colors indicate individuals with different SNPs.

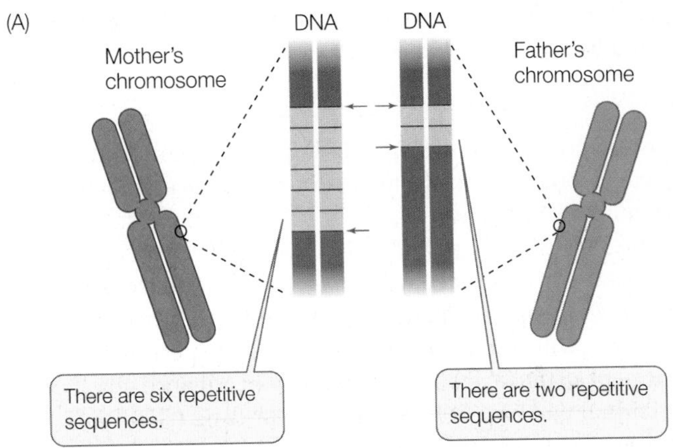

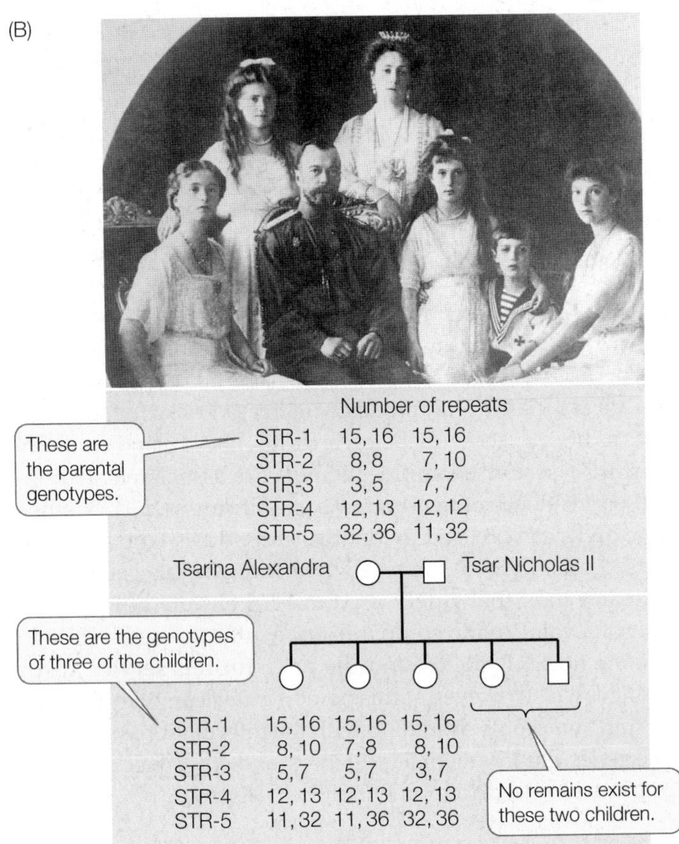

FIGURE 12.15 DNA Fingerprinting (A) A short tandem repeat (STR) can occur in a specific, inherited pattern. (B) STR analyses were used to determine that bony remains were from one family, and other evidence pointed to the Russian tsar and his family.

proteins. There are many more, which presumably distinguish us as *human* eukaryotic organisms.

There is considerable interest in using proteomics in the diagnosis of diseases. For diseases caused by a single gene, examining the single protein involved is possible (e.g., the hemoglobins; see Figure 10.2). But for more complex diseases such as diabetes and cancer, many genes and proteins may be involved, and the *pattern* of proteins made in a particular tissue at a certain time might indicate the presence or likelihood of the disease. Proteomic analyses of tissues (including blood) are being developed to provide early warnings of the presence of particular diseases before symptoms occur.

METABOLOMICS There has been some progress in defining the human metabolome. A database created by David Wishart and colleagues at the University of Alberta contains more than 40,000 entries, including metabolic products of foods and drugs. The challenge now is to relate the levels of these substances to physiology. For example, high levels of glucose in the blood are associated with diabetes, and there may be patterns of metabolites that are diagnostic of other diseases. This could aid in early diagnosis and treatment.

DNA fingerprinting uses short tandem repeats

As noted in Concept 12.3, short tandem repeats (STRs) are blocks of 1–5 bp that can be repeated up to 100 times at particular locations on chromosomes. Since the number of repeats can vary widely, there are usually numerous alleles for a particular STR. For example, at a particular location on human chromosome 15 there might be an STR of AGG. An individual might inherit an allele with six copies of the repeat (AGGAGGAGGAGGAGGAGG) from her mother and an allele with two copies (AGGAGG) from her father (**FIGURE 12.15A**). PCR can be used to amplify DNA fragments containing these repeat sequences, and the number of copies of the repeat can be determined by sizing the DNA fragments.

DNA fingerprinting refers to a group of techniques used to identify particular individuals by their DNA; the most common of these techniques involves STR analysis. When several different STR loci are analyzed, an individual's unique pattern becomes apparent. The U.S. Federal Bureau of Investigation uses 13 STR loci in its Combined DNA Index System (CODIS) database.

DNA fingerprinting is used to resolve questions of paternity, and in forensics (crime investigations) to identify criminals. It also has other uses—for example, it can help in analyses of historical events. In 1918 the Russian tsar Nicholas II and his wife and five children were killed during the Communist revolution. A report that the bodies had been burned was not questioned until 1991, when a shallow grave with several skeletons was discovered a few miles from the presumed execution site. The remains were from a man, a woman, and three female children, and STR analysis indicated that they were all related to one another (**FIGURE 12.15B**). These STR patterns were also related to living descendants of the tsar. The accuracy and specificity of this method gave historical and cultural closure to a major event of the twentieth century.

Genome sequencing is at the leading edge of medicine

With the rapid development of better and better ways to sequence genomes cheaply, we are on the verge of a new era in medicine. The genomes of random cells of people, as well as cells from diseases such as cancers, are being sequenced in two ways:

- Sequencing of the entire genome
- Sequencing of protein-coding exons only

APPLY THE CONCEPT

The human genome sequence has many applications

It is the year 2025. You are taking care of a patient who is worried that he may have an early stage of kidney cancer. His mother died from this disease.

1. Assume that the SNPs linked to genes involved in the development of this type of cancer have been identified. How would you determine if this man has a genetic predisposition for developing kidney cancer? Explain how you would do the analysis.
2. How might you develop a metabolomic profile for kidney cancer and then use it to determine whether your patient has kidney cancer?
3. If the patient is diagnosed with the cancer, how would you use pharmacogenomics to choose the right medications to treat his tumor?

The information these methods provide goes far beyond what SNP and STR genotyping can do. For example:

- The genome sequence of a tumor included mutations in specific genes that drive tumor formation. The proteins encoded by these genes are targets for specific therapy for that tumor.
- The genome sequence of an individual with a genetic disease has led to the identification of the causative gene.

CHECKpoint CONCEPT 12.4

- ✓ The average human gene spans 27,000 bp. The average human polypeptide has 300 amino acids. Explain.
- ✓ What is a haplotype with regard to an STR? How can haplotypes be used to relate DNA to a phenotype?
- ✓ A person has a rare allele for an STR (STR-1) that has a frequency in the population of 1 percent (0.01). The same person has allele frequencies for other STRs as follows: STR-2, 0.005; STR-3, 0.01; STR-4, 0.05; STR-5, 0.01. What is the probability that an individual will have all of these alleles? What does this mean in terms of identifying an individual with this genotype?

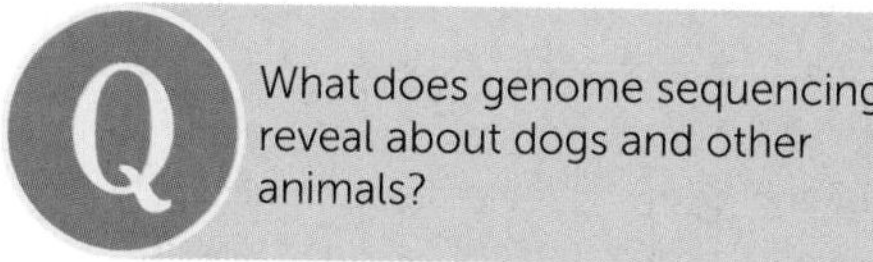

Q
What does genome sequencing reveal about dogs and other animals?

ANSWER In the opening story of this chapter we described how genome sequencing is being applied to breeds of dogs. For example, high-throughput sequencing methods (Concept 12.1) have allowed biologists to collect data on genes that control body size. This has led to the identification of a SNP (Concept 12.4) in the gene for *i*nsulin-like *g*rowth *f*actor 1 (IGF-1) that is important in determining size. Large breeds have an allele that encodes an active IGF-1, and small breeds have a different allele that encodes a less active version of the protein. In humans, IGF-1 mediates the overall effects of growth hormone, and people with a mutation in the IGF-1 gene have short stature.

Another gene important to phenotypic variation is found in whippets, sleek dogs that run fast and are often raced. A mutation in the gene for myostatin, a protein that inhibits the overdevelopment of muscles, results in a whippet that is more muscular and runs faster (**FIGURE 12.16A**). Comparative genomics shows that this gene is important in other animals as well. In Belgian Blue cattle, individuals homozygous for a particular SNP in the myostatin gene have huge muscles

(A)

(B)

FIGURE 12.16 Muscular Gene (A) These dogs are both whippets, but the muscle-bound dog (left) has a mutation in a gene that normally limits muscle buildup. (B) A similar mutation in Belgian Blue cattle also leads to overgrown muscles.

(**FIGURE 12.16B**). There is interest in applying this knowledge to humans. For example, in muscular dystrophy the skeletal muscles waste away, and blocking myostatin could be useful in keeping muscles robust. Athletes anxious to have bulkier muscles have also been focusing on this gene and its protein product.

Inevitably, some scientists have set up companies to test dogs for genetic variations using DNA supplied by dog owners and breeders. The black dog rescued from the pound that looks like a Labrador retriever may turn out to be a German pointer. Some traditional breeders frown on this practice, but others say it will bring more joy (and prestige) to owners.

SUMMARY

CONCEPT 12.1 There Are Powerful Methods for Sequencing Genomes and Analyzing Gene Products

- To sequence a genome, the chromosomes are cut into overlapping fragments, which are sequenced. Then the fragment sequences are lined up to assemble the DNA sequence of the chromosome. **See ANIMATED TUTORIAL 12.1**
- **High-throughput sequencing** involves attaching short, single-stranded DNA fragments to a solid surface. A primer and DNA polymerase are added, and tagged nucleotides are detected by a camera as they are added to the complementary DNA strand. Many sequences can be done in parallel. **Review Figure 12.1 and ANIMATED TUTORIAL 12.2**
- The analysis of DNA sequences is done by computer. Genomic sequences include protein-coding genes, RNA genes, regulatory sequences, and repeated sequences. **Review Figure 12.3**
- The **proteome** is the total protein content of an organism. It can be analyzed using chemical methods that separate and identify proteins. These include two-dimensional electrophoresis, mass spectrometry, and techniques involving antibodies. **Review Figure 12.4**
- The **metabolome** is the total content of small molecules in a tissue under particular conditions. These molecules include intermediates in metabolism, hormones and other signaling molecules, and secondary metabolites. **Review Figure 12.5**

CONCEPT 12.2 Prokaryotic Genomes Are Small, Compact, and Diverse

- Prokaryotic genomes have been studied using **functional genomics** to determine the roles of various parts of the genome, including the protein-coding genes. **Comparative genomics** is used to compare sequences among organisms. **Review Table 12.1**
- **Metagenomics** is the identification of DNA sequences in environmental samples without first isolating, growing, and identifying the organisms. **Review Figure 12.6**
- **Transposons** are sequences of DNA that can move about the genome. **Review Figure 12.7**
- Transposon mutagenesis can be used to inactivate genes one by one. Then the organism can be tested for survival. In this way, functionally important genes can be identified. **Review Figure 12.8**

CONCEPT 12.3 Eukaryotic Genomes Are Large and Complex

- Sequences from model organisms have highlighted some common features of eukaryotic genomes. In addition, there are specialized genes such as those for cellular compartmentalization, development, and features unique to plants. **Review Figures 12.9 and 12.10 and Table 12.2**
- Some genes exist as members of **gene families**. Proteins may be made from these closely related genes at different times and in different tissues. **Review Figure 12.11**
- Eukaryotic genomes contain various kinds of repeated sequences. **Review Table 12.3**

CONCEPT 12.4 The Human Genome Sequence Has Many Applications

- The haploid human genome has 3.2 billion bp.
- Only 1.5 percent of the genome codes for proteins; much of the rest consists of repeated sequences.
- Most of the genome is transcribed at some point in some cells.
- Virtually all human genes have introns, and alternative splicing leads to the production of more than one protein per gene.
- Genotyping using **single nucleotide polymorphisms** (**SNPs**) can be used to correlate variations in the genome with diseases or drug sensitivity. It may lead to personalized medicine. **Review Figure 12.13**
- **Pharmacogenomics** is the analysis of how a person's genetic makeup affects his or her drug metabolism. **Review Figure 12.14**
- Short tandem repeats (STRs) are DNA sequences that are variable in length. They can be used to identify individuals. **Review Figure 12.15**

See **ACTIVITY 12.1** for a concept review of this chapter.

Go to the Interactive Summary to review key figures, Animated Tutorials, and Activities PoL2e.com/is12

Go to LaunchPad at **macmillanhighered.com/launchpad** for additional resources, including LearningCurve Quizzes, Flashcards, and many other study and review resources.

13 Biotechnology

KEY CONCEPTS

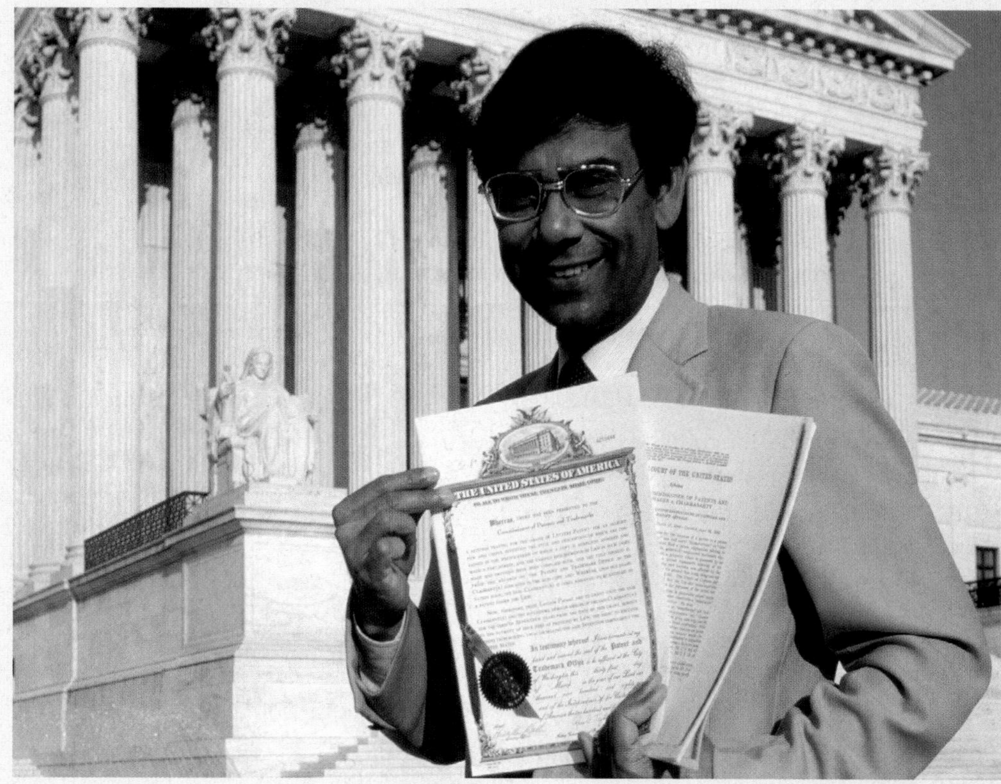

Ananda Chakrabarty received the first patent for a genetically modified organism, a bacterium that breaks down crude oil.

The United Nations defines **biotechnology** as "any technological application that uses biological systems, living organisms, or derivatives thereof to make or modify products or processes." This definition encompasses major human activities such as brewing beer (see Chapter 6) and the domestication of animals and plants (see Chapter 12). More recently, biotechnology has become associated with the genetic modification of microorganisms for the production of particular substances, and of a variety of plants and animals used in agriculture.

Industrial biotechnology began in England in 1917, during the First World War. The production of cordite, an explosive used to propel a bullet or shell to its target, required the solvent acetone, $(CH_3)_2CO$. But most acetone was manufactured by England's enemy, Germany. A microbiologist at the University of Manchester, Chaim Weizmann, found that if the bacterium *Clostridium acetylbutylicum* was grown using starch as an energy source, it produced abundant quantities of acetone. The British government set up a factory to grow large vats of these bacteria, and the cordite shortage was solved.

The contemporary era of biotechnology as a major industry dates from June 16, 1980, on the steps of the U.S. Supreme Court. In this case, scientists were studying bacteria not for their ability to make something, but to break it down. Many bacteria have genes that code for unusual enzymes and biochemical pathways, and they can use all sorts of substances as nutrients, including pollutants. Scientists have identified these organisms simply by mixing polluted soil with water and seeing what grows. In 1971, Ananda Chakrabarty at the General Electric Research Center in New York used genetic crosses to develop a single strain of the bacterium *Pseudomonas* that carried genes for the breakdown of various hydrocarbons in oil. He and his company applied for a patent to legally protect their discovery and profit from it. Nine years later, in a landmark case, the U.S. Supreme Court ruled that "a live, human-made microorganism is patentable" under the U.S. Constitution.

The 1980 Supreme Court ruling came at a time when new laboratory methods were being developed to insert specific DNA sequences into organisms by recombinant DNA technology. The resulting flood of patents for DNA sequences and genetically modified organisms, some of them developed to improve the environment, continues to this day and was the subject of another Supreme Court ruling in 2013.

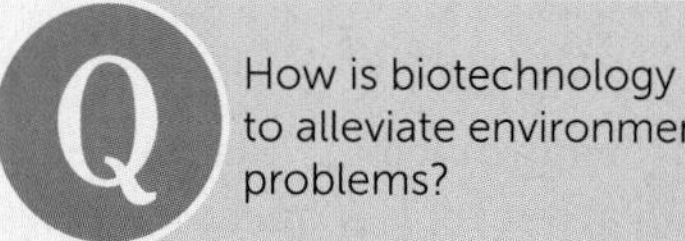

How is biotechnology used to alleviate environmental problems?

You will find the answer to this question on page 271.

CONCEPT 13.1 Recombinant DNA Can Be Made in the Laboratory

Biotechnology began with the use of organisms with genetic capabilities that occur in nature. For example, existing biochemical pathways in yeast are used to make alcohol, and as noted in the opening story, bacteria can be grown in large quantities to make acetone and other industrial chemicals. More recently, it has become possible to genetically modify organisms with genes from other, distantly related organisms, to create new combinations of genes that would not otherwise occur in the same cell. This technology involves the use of **recombinant DNA**: a single DNA molecule containing DNA sequences from two or more sources. Before the invention of PCR (polymerase chain reaction; see Concept 13.3 and also Concept 9.2), biologists relied on natural molecules and processes to manipulate DNA in the laboratory. Three key tools were, and still are, widely used:

- Restriction enzymes for cutting DNA into pieces (fragments) that can be manipulated
- Gel electrophoresis for the analysis and purification of DNA fragments
- DNA ligase for joining DNA fragments together in novel combinations

Restriction enzymes cleave DNA at specific sequences

All organisms, including bacteria, must have ways of dealing with their enemies. As we described in Concept 9.1, bacteria are attacked by viruses called bacteriophage (or phage, for short). These viruses inject their genetic material into the host cells and turn them into virus-producing factories, eventually killing the cells. Some bacteria defend themselves against such invasions by producing **restriction enzymes** (also known as restriction endonucleases), which cut double-stranded DNA molecules—such as those injected by bacteriophage—into smaller, noninfectious fragments (**FIGURE 13.1**). These enzymes break the bonds of the DNA backbone between the 3′ hydroxyl group of one nucleotide and the 5′ phosphate group of the next nucleotide:

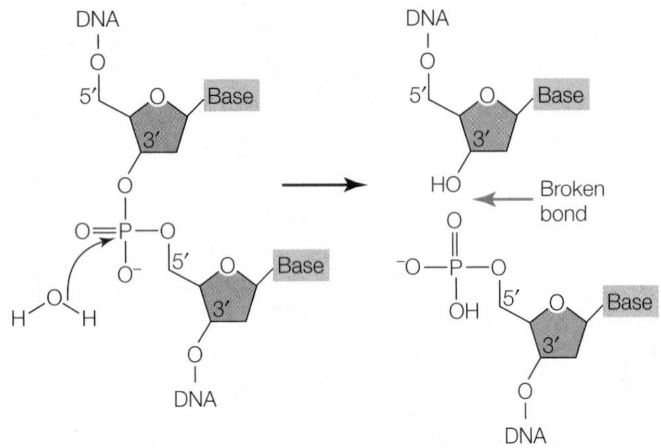

There are many different restriction enzymes, each of which cleaves DNA at a specific sequence of bases called a **restriction**

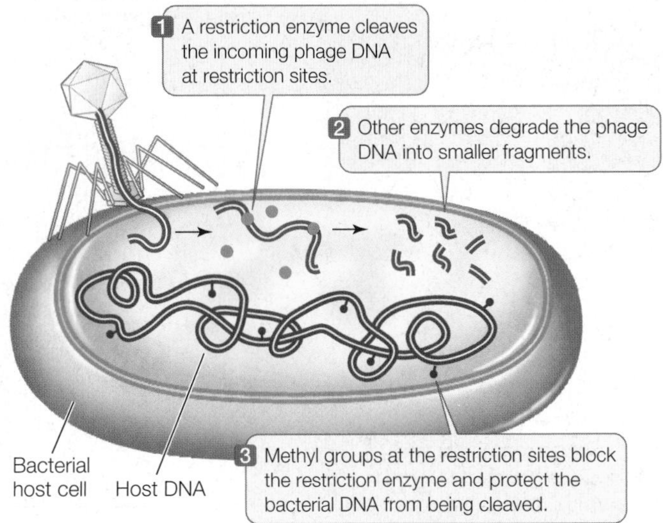

1 A restriction enzyme cleaves the incoming phage DNA at restriction sites.
2 Other enzymes degrade the phage DNA into smaller fragments.
3 Methyl groups at the restriction sites block the restriction enzyme and protect the bacterial DNA from being cleaved.
Bacterial host cell
Host DNA

FIGURE 13.1 Bacteria Fight Invading Viruses by Making Restriction Enzymes

site or recognition sequence. Hundreds of these enzymes have been purified from various microorganisms and can be used to cut DNA in the laboratory (by setting up a "restriction digest"). Most restriction sites are four to six base pairs (bp) long, and restriction enzymes catalyze hydrolysis of both strands of DNA. For example, below are three 6-bp sequences, each of which is recognized by a different restriction enzyme. The enzymes cleave the DNA at the sites indicated by the red arrows:

*Bam*HI
5′··· G↓G A T C C ··· 3′
3′··· C C T A G↑G ··· 5′

*Hind*III
5′··· A↓A G C T T ··· 3′
3′··· T T C G A↑A ··· 5′

*Eco*RI
5′··· G↓A A T T C ··· 3′
3′··· C T T A A↑G ··· 5′

The names of these enzymes reflect their organism of origin. *Eco*RI, for example, was found in *E. coli*, and *Bam*HI comes from the bacterium *Bacillus amyloliquifaciens*. Note that in each of these restriction sites both strands have the same sequence when read from their 5′ ends. This is similar to the word "racecar," which is the same when read in either direction. A word that is the same when read in either direction is called a palindrome. Restriction enzymes have two identical active sites on two subunits, which cleave the two strands simultaneously:

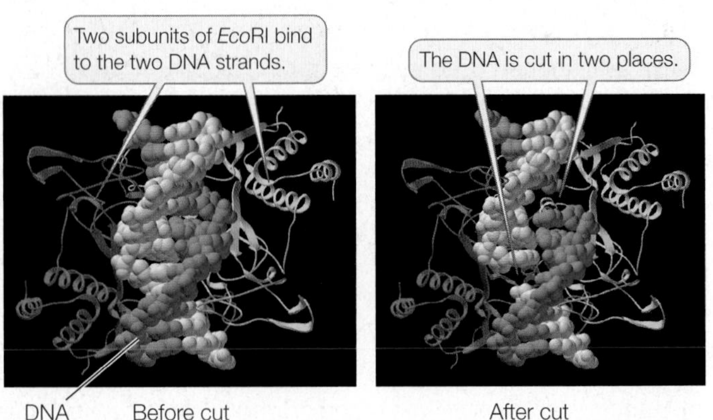

Two subunits of EcoRI bind to the two DNA strands.
The DNA is cut in two places.
DNA
Before cut
After cut

RESEARCH TOOLS

FIGURE 13.2 Separating Fragments of DNA by Gel Electrophoresis A mixture of DNA fragments is placed in a gel, and an electric field is applied across the gel. The negatively charged DNA moves toward the positive end of the field, with smaller molecules moving faster (and farther) than larger ones. After minutes to hours for separation, the electric power is shut off and the separated fragments can be analyzed.

1 A gel is made up of agarose polymer suspended in a buffer. It sits in a chamber between two electrodes.
2 Depressions in the gel (wells) are filled with DNA solutions.
Gel
Buffer solution
DNA solution
Enzyme 1
Enzyme 2
Enzymes 1 + 2
A B
C D
A E D
3 Restriction enzyme 1 cuts the DNA once, resulting in fragments A and B.
4 Restriction enzyme 2 cuts the DNA once, at a different restriction sequence.
5 If both restriction enzymes are used, two cuts are made in the DNA.
1 2 1+2
6 After enzyme incubation, each sample is loaded into one well in the gel.
Longer fragments
Shorter fragments
7 As fragments of DNA move toward the positive electrode, shorter fragments move faster (and therefore farther) than longer fragments.

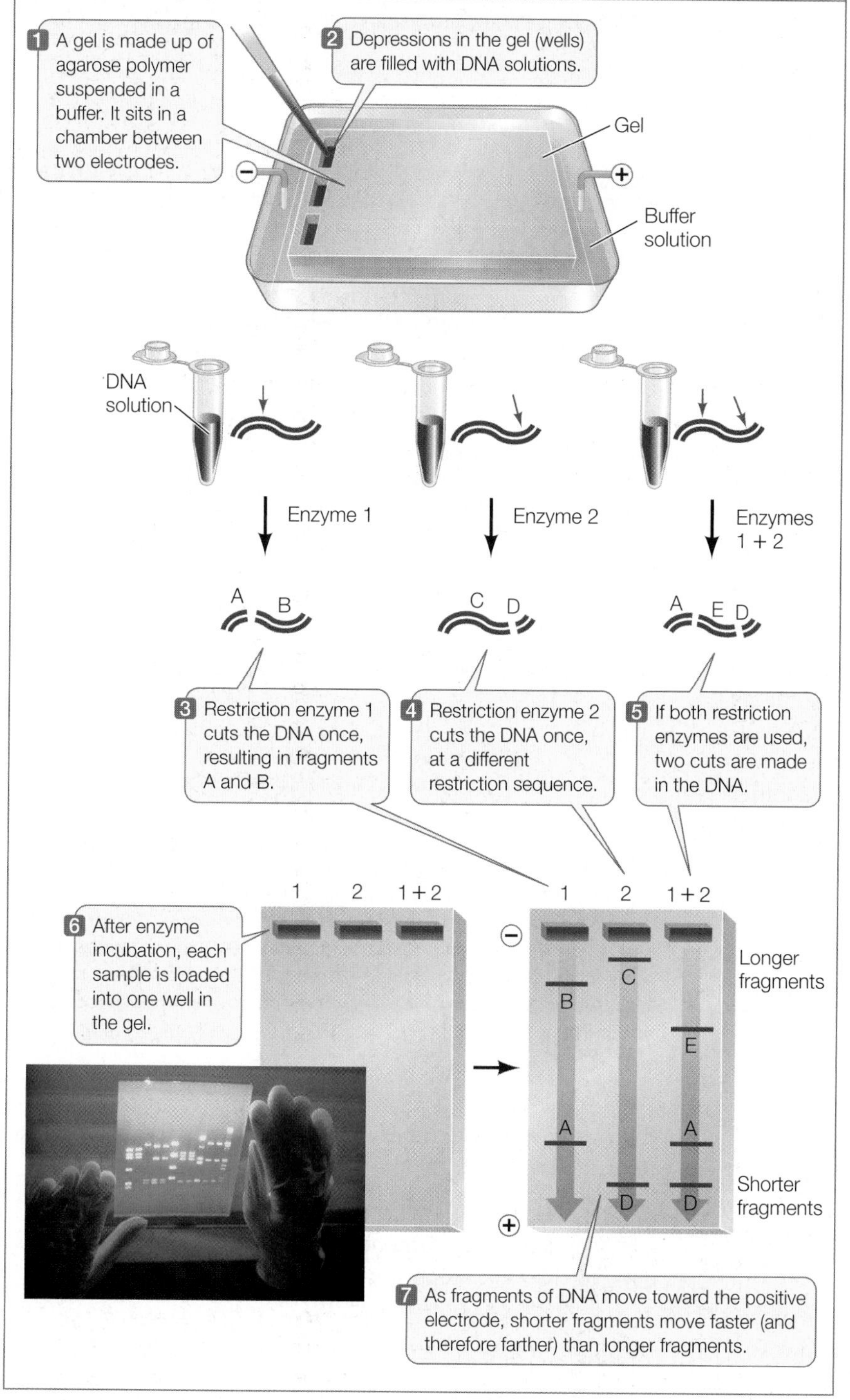

Note also that these enzymes cut the two DNA strands in such a way that there will be a short sequence of single-stranded DNA at each cut end. Many restriction enzymes create these single-stranded overhangs, which are referred to as "sticky ends" because they are able to form hydrogen bonds with complementary sequences on other DNA molecules. Other restriction enzymes make cuts directly opposite one another on the two DNA strands, creating "blunt ends."

As shown in Figure 13.1, the restriction enzymes may cleave host bacterial DNA. To prevent cleavage, a "stop sign" in the form of a methyl (—CH_3) group can be placed on restriction sites. This process involves specific DNA methyltransferases. The restriction enzymes do not recognize or cut the methylated restriction sites in the host's DNA. But unmethylated phage DNA is efficiently recognized and cleaved.

The *Eco*RI restriction site recognition sequence occurs, on average, about once in every 4,000 bp in a typical prokaryotic genome, or about once per four prokaryotic genes. So *Eco*RI can chop a large piece of DNA into smaller pieces containing, on average, just a few genes. When *Eco*RI is used in the laboratory to cut a small genome such as that of a virus with, say, 50,000 bp, only a few fragments are obtained. For a huge eukaryotic chromosome with tens of millions of bp, a very large number of fragments are created.

Of course, "on average" does not mean that the enzyme cuts all stretches of DNA at regular intervals. For example, the *Eco*RI restriction site does not occur even once in the 40,000 bp of the T7 phage genome—a fact that is crucial to the survival of this virus, since its host is *E. coli*. Fortunately for *E. coli*, the *Eco*RI restriction site does appear in the DNA of other bacteriophage.

Go to MEDIA CLIP 13.1
Striking Views of Recombinant DNA Being Made
PoL2e.com/mc13.1

Gel electrophoresis separates DNA fragments

After a sample of DNA has been cut with a restriction enzyme, the fragments can be separated from each other to determine the number of fragments and the size (in bp) of each fragment. In this way an individual fragment can be identified, and it can then be purified for further analysis or for use in an experiment.

A convenient way to separate or purify DNA fragments is by **gel electrophoresis**. Samples containing the fragments are placed in wells at one end of a semisolid gel (usually made of agarose or polyacrylamide polymers), and an electric field is applied to the gel (**FIGURE 13.2**). Because of its phosphate groups, DNA is

APPLY THE CONCEPT

Recombinant DNA can be made in the laboratory

The specificity of restriction enzyme recognition can be used to detect mutations. For example, the enzyme *Mst*II cuts DNA at CCTNAGG, where N is any base. Around the sixth codon in the β-globin gene is the sequence CCTGAGG. There are two additional *Mst*II sites on either side of the sixth codon, such that when *Mst*II is used to cut human DNA in this region, two fragments of 1.15 and 0.20 kilobases are obtained (1 kilobase = 1,000 bp). The sickle allele of the β-globin gene causes sickle-cell anemia when it occurs in the homozygous state. In this allele, the sequence around the sixth codon is mutated to CCTGTGG.

1. Identify the point mutation that led to the sickle allele.
2. What fragment(s) would result when *Mst*II is used to cut DNA with the sickle allele?
3. Sketch the patterns of cuts with normal and sickle alleles on a gel.
4. Could an individual have both patterns? Explain.
5. How can this information be used to make a DNA test for these alleles?

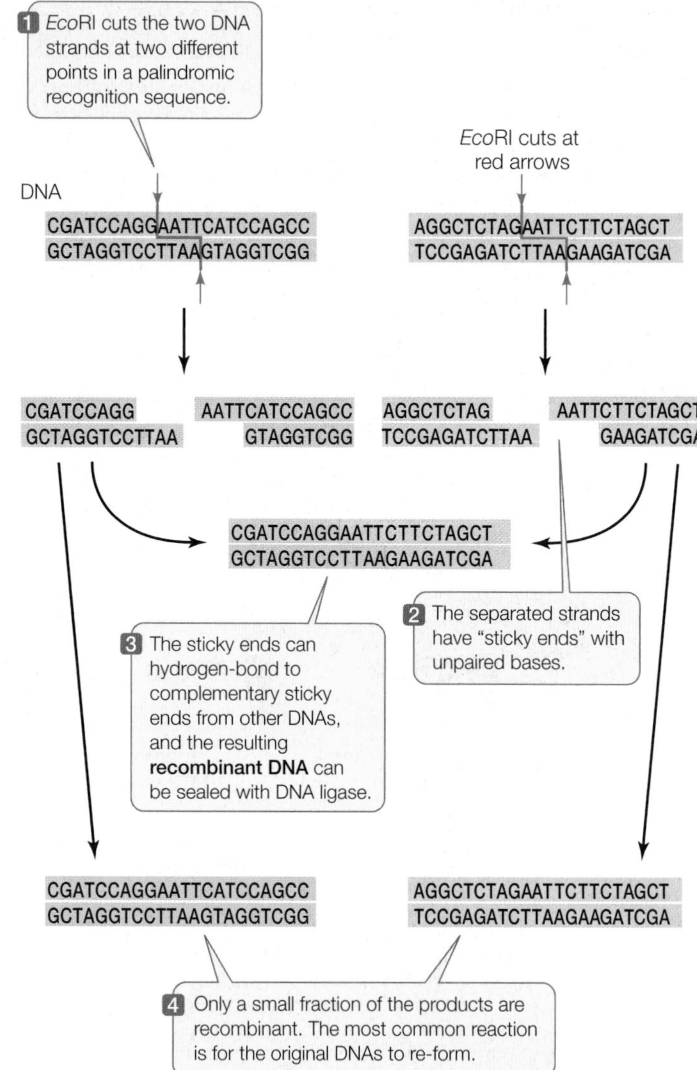

FIGURE 13.3 Cutting and Joining DNA Many restriction enzymes (*Eco*RI is shown here) make staggered cuts in DNA. *Eco*RI can be used to cut two different DNA molecules (blue and orange). The exposed bases can hydrogen-bond with complementary exposed bases on other DNA fragments, forming recombinant DNA. DNA ligase stabilizes the recombinant molecule by forming covalent phosphodiester bonds in the DNA backbone.

negatively charged at neutral pH. Therefore, because opposite charges attract, the DNA fragments move through the gel toward the positive end of the field. Because the spaces between the polymers of the gel are small, small DNA molecules can move through the gel faster than larger ones. Thus DNA fragments of different sizes separate from one another and can be detected with a dye. This gives us three types of information about a DNA sample:

- *The number of fragments*. The number of fragments produced by digestion of a DNA sample with a given restriction enzyme depends on how many times that enzyme's restriction site occurs in the sample. Thus gel electrophoresis can provide some information about the presence of specific DNA sequences in the DNA sample.
- *The sizes of the fragments*. DNA fragments of known size are often placed in one well of the gel to provide a standard for comparison. This tells us how large the DNA fragments in the other wells are. By comparing the fragment sizes obtained with two or more restriction enzymes, the locations of their recognition sequences relative to one another can be worked out (mapped).
- *The relative abundance of a fragment*. In many experiments, the investigator is interested in how much DNA is present. The relative intensity of a band produced by a specific fragment can indicate the amount of that fragment.

After separation on the gel, a slice of gel containing the desired DNA fragment (identified by its size) can be cut out and then be purified by one of a variety of methods. This fragment can then be analyzed to determine its sequence or used to make recombinant DNA.

Go to ANIMATED TUTORIAL 13.1 Separating Fragments of DNA by Gel Electrophoresis
PoL2e.com/at13.1

LINK

Determining and then comparing DNA sequences of the same gene from different species provides information about evolutionary relationships; see **Concepts 16.2 and 16.3**

Recombinant DNA can be made from DNA fragments

Another enzyme that is involved in DNA metabolism in cells is **DNA ligase**, which catalyzes the joining of DNA fragments by making phosphodiester bonds between them. This is the enzyme that joins Okazaki fragments during DNA replication (see Concept 9.2). With restriction enzymes (which *break* bonds) and DNA ligase (which *makes* bonds), scientists can cut DNA into fragments and then join the fragments together in new combinations as recombinant DNA. As shown in **FIGURE 13.3**, two fragments with complementary sticky ends first join by hydrogen bonding, and then the DNA ligase forms a phosphodiester bond in each strand, making a single intact DNA molecule.

In the early 1970s, Stanley Cohen and Herbert Boyer wondered whether recombinant DNA could be a functional carrier of genetic information. They used restriction enzymes to cut sequences from two *E. coli* plasmids (small circular DNAs; see Concept 8.4) containing different antibiotic resistance genes. Then they used DNA ligase to join the fragments together. The resulting plasmid, when inserted into new *E. coli* cells, gave those cells resistance to both antibiotics (**FIGURE 13.4**). A new era of biotechnology was born.

INVESTIGATION

FIGURE 13.4 Recombinant DNA With the discovery of restriction enzymes and DNA ligase, it became possible to combine DNA fragments from different sources in the laboratory. But would such "recombinant DNA" be functional when inserted into a living cell? The results of this experiment completely changed the scope of genetic research, increasing our knowledge of gene structure and function, and ushering in the new field of biotechnology.[a]

HYPOTHESIS

Biologically functional recombinant plasmids can be made in the laboratory.

METHOD

E. coli plasmids carrying a gene for resistance to either the antibiotic kanamycin (*kan*[r]) or tetracycline (*tet*[r]) are cut with a restriction enzyme.

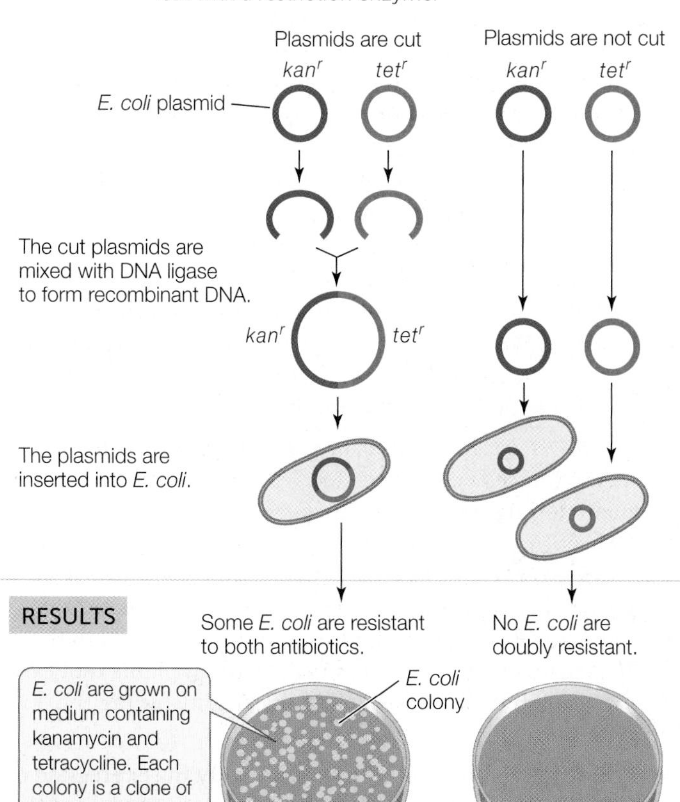

Plasmids are cut
Plasmids are not cut
kan^r
tet^r
E. coli plasmid
The cut plasmids are mixed with DNA ligase to form recombinant DNA.
The plasmids are inserted into E. coli.
RESULTS
Some E. coli are resistant to both antibiotics.
No E. coli are doubly resistant.
E. coli are grown on medium containing kanamycin and tetracycline. Each colony is a clone of millions of cells, all derived from a single bacterium.
E. coli colony

CONCLUSION

Two DNA fragments with different genes can be joined to make a recombinant DNA molecule, and the resulting DNA is functional.

ANALYZE THE DATA

Two plasmids were used in this study: pSC101 had a gene for resistance to tetracycline and pSC102 had a gene for resistance to kanamycin. Equal quantities of the plasmids—either intact, cut with *Eco*RI, or cut with *Eco*RI and then sealed with DNA ligase—were mixed and incubated with antibiotic-sensitive *E. coli*. The *E. coli* were then grown on various combinations of the antibiotics. Here are the results:

	Number of resistant colonies		
DNA treatment	Tetracycline only	Kanamycin only	Both antibiotics
None	200,000	100,000	200
*Eco*RI cut	10,000	1,100	70
*Eco*RI, then ligase	12,000	1,300	570

A. Did treatment with *Eco*RI affect the transformation efficiency? Explain.

B. Did treatment with DNA ligase affect the transformation efficiency of each cut plasmid? Which quantitative data support your answer?

C. How did doubly antibiotic-resistant bacteria arise in the "none" treatment? (Hint: see Concept 9.3.)

D. Did the *Eco*RI followed by ligase treatment increase the appearance of doubly antibiotic-resistant bacteria? What data support your answer?

Go to **LaunchPad** for discussion and relevant links for all **INVESTIGATION** figures.

[a]S. N. Cohen et al. 1973. *Proceedings of the National Academy of Sciences USA* 70: 3240–3244.

CHECKpoint CONCEPT 13.1

✓ Using diagrams of the chemical structure of DNA (see Concepts 3.1 and 9.2), compare the actions of a restriction enzyme and DNA ligase.

✓ How is recombinant DNA technology different from genetic recombination that occurs in meiosis?

✓ A DNA molecule of 12,000 bp (12 kilobase, or kb) is cut by restriction enzymes and analyzed by gel electrophoresis as shown in the table. The 2-kb band from the double digest (Enzymes A + B) is twice as intense as the 2-kb bands from the single digests.

CONDITION	SIZES OF FRAGMENTS (kb)
Enzyme A	2, 10
Enzyme B	2, 10
Enzymes A + B	2, 8

a. Sketch the results of the three cuts on an electrophoresis gel.

b. Indicate on a linear map where each enzyme cuts the DNA.

✓ DNA from two different sources is cut with two different restriction enzymes. Fragments from the two sources are then joined together. DNA from source A is cut with *Spe*I. Which of the four restriction enzymes shown below *Spe*I could be used to cut DNA from source B, such that it could join to a fragment from source A?

*Spe*I
5′··· A↓C T A G T ··· 3′
3′··· T G A T C↑A ··· 5′

*Hin*dIII
5′··· A↓A G C T T ··· 3′
3′··· T T C G A↑A ··· 5′

*Xba*I
5′··· T↓C T A G A ··· 3′
3′··· A G A T C↑T ··· 5′

*Eco*RV
5′··· G A T↓A T C ··· 3′
3′··· C T A↑T A G ··· 5′

*Xho*I
5′··· C↓T C G A G ··· 3′
3′··· G A G C T↑C ··· 5′

With the tools described in this concept—restriction enzymes, gel electrophoresis, and DNA ligase—scientists can cut and rejoin different DNA molecules from any and all sources, including artificially synthesized DNA sequences. In the next concept we will examine some of the ways these recombinant DNA molecules are used.

CONCEPT 13.2 DNA Can Genetically Transform Cells and Organisms

One goal of recombinant DNA technology is to **clone**—produce many identical copies of—a particular DNA sequence. We have seen the term "clone" used in the context of whole cells or organisms that are genetically identical to one another (see Concept 7.1). A gene can be cloned by inserting it into a bacterial cell such as *E. coli*. The bacterium is allowed to reproduce and multiply into millions of identical cells, all carrying copies of the gene. Cloning might be done for sequence analysis, to produce a protein product in quantity, or as a step toward creating an organism with a new phenotype.

The process of inserting recombinant DNA into host cells is called **transformation**, or **transfection** if the host cells are derived from an animal. A host cell or organism that contains recombinant DNA is described as **transgenic**. Later in this chapter we will encounter many examples of transgenic cells and organisms, including yeast, mice, rice plants, and even cattle.

Various methods are used to create transgenic cells. Generally, only a few of the cells that are exposed to the recombinant DNA actually become transformed with it. In order to grow only the transgenic cells, **selectable marker** genes, such as genes that confer resistance to antibiotics, are often included as part of the recombinant DNA molecule. Antibiotic resistance genes were the markers used in Cohen and Boyer's experiment (see Figure 13.4).

Genes can be inserted into prokaryotic or eukaryotic cells

In theory, any cell or organism can act as a host for the introduction of recombinant DNA. Most research has been done using model organisms:

- *Bacteria* are easily grown and manipulated in the laboratory. Much of their molecular biology is known, especially for well-studied bacteria such as *E. coli*. Furthermore, bacteria contain plasmids, which are easily manipulated to carry recombinant DNA into the cell. Because the processes of transcription and translation proceed differently in prokaryotes than they do in eukaryotes, however, bacteria might not be suitable as hosts to express eukaryotic genes.
- *Yeasts* such as *Saccharomyces* are commonly used as eukaryotic hosts for recombinant DNA studies. The advantages of using yeasts include rapid cell division (a life cycle completed in 2 hours), ease of growth in the laboratory, and a relatively small genome size. In addition, yeasts have most of the characteristics of other eukaryotes, except for those characteristics involved in multicellularity.
- *Plant cells* are good hosts, because even fully differentiated plant cells can be treated with hormones that make them dedifferentiate into unspecialized stem cells (see Concept 13.3). The unspecialized cells can be transformed with recombinant DNA and then studied in culture, or grown into new plants. There are also methods for making whole transgenic plants without going through the cell culture step. These methods result in plants that carry the recombinant DNA in all their cells, including the germline cells.
- *Cultured animal cells* can be used to study the expression of human or animal genes, for example for medical purposes. Whole transgenic animals can also be created.

A variety of methods are used to insert recombinant DNA into host cells

Methods for inserting DNA into host cells vary. The cells may be chemically treated to make their outer membranes more permeable, and then mixed with the DNA so it can diffuse into the cells. Another approach is called electroporation: a short electric shock is used to create temporary pores in the membranes through which the DNA can enter. Viruses can be altered so that they carry or insert recombinant DNA into cells. A common method for transforming plants involves a specific bacterium that inserts DNA into cells of some plants. Transgenic animals can be produced by injecting recombinant DNA into the nuclei of fertilized eggs. There are even "gene guns," which "shoot" the host cells with tiny particles carrying the DNA.

The challenge of inserting new DNA into a cell lies not just in getting it into the host cell, but in getting it to replicate as the host cell divides. DNA polymerase does not bind to just any sequence. If the new DNA is to be replicated, it must become part of a segment of DNA that contains an origin of replication (see Concept 9.2). Such a DNA molecule is called a replicon, or replication unit.

There are two general ways in which the newly introduced DNA can become part of a replicon within the host cell:

- It may be inserted into a host chromosome. Although the site of insertion is usually random, this is nevertheless a common method of integrating new genes into host cells.
- It can enter the host cell as part of a carrier DNA sequence, called a **vector**, and can either integrate into the host chromosome or have its own origin of DNA replication.

Several types of vectors are used to get DNA into cells.

PLASMIDS AS VECTORS As we described in Concept 8.4, plasmids are small, circular DNA molecules that replicate autonomously in many prokaryotic cells. A number of characteristics make plasmids useful as transformation vectors:

- Plasmids are relatively small (an *E. coli* plasmid usually has 2,000–6,000 bp) and are therefore easy to manipulate in the laboratory.
- A typical plasmid has one or more restriction enzyme recognition sequences that each occur only once in the plasmid sequence. These sites make it easy to insert additional DNA into the plasmid before it is used to transform host cells.
- Many plasmids contain genes that confer resistance to antibiotics and thus can serve as selectable markers.
- Plasmids have a bacterial origin of replication (*ori*) and can replicate independently of the host chromosome. It is not uncommon for a bacterial cell to contain hundreds of copies of a recombinant plasmid. For this reason, the power of bacterial transformation to amplify a gene is extraordinary. A 1-liter culture of bacteria harboring the human β-globin gene in a typical plasmid can have as many copies of that gene as there are cells in a typical adult human (10^{14}).

The plasmids used as vectors in the laboratory have been extensively altered to include convenient features: multiple cloning sites with 20 or more unique restriction enzyme sites for cloning purposes; origins of replication for a variety of host cells; and various kinds of reporter genes (see p. 260) and selectable marker genes. An example is pBR322, a plasmid used to transform *E. coli*:

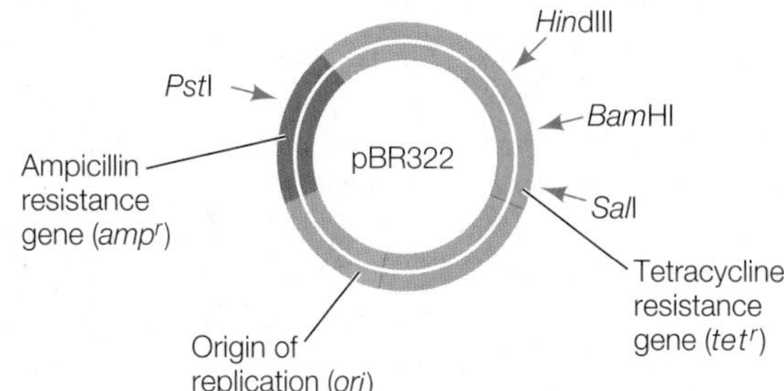

PLASMID VECTORS FOR PLANTS An important vector for carrying new DNA into many types of plants is a plasmid found in the bacterium *Agrobacterium tumefaciens*. This bacterium lives in the soil, infects plants, and causes a disease called crown gall, which is characterized by the presence of growths (or tumors) on the plant. *A. tumefaciens* contains a plasmid called Ti (for *t*umor-*i*nducing). When the bacterium infects a plant cell, a region of the Ti plasmid called the T DNA is inserted into the cell, where it becomes incorporated into one of the plant's chromosomes. The Ti plasmid carries the genes needed for this transfer and incorporation of the T DNA:

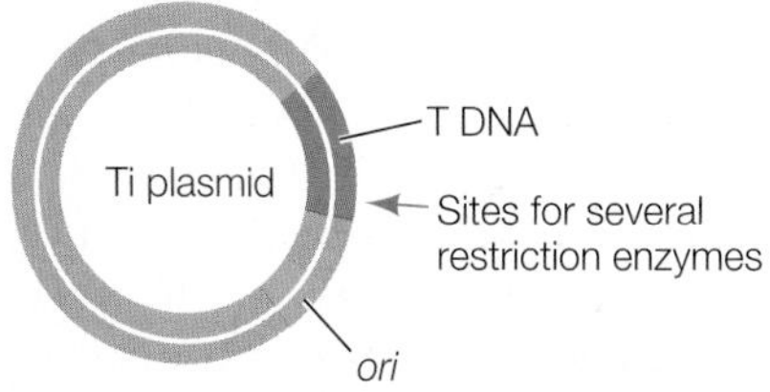

The T DNA carries genes that are expressed by the host cell, causing the growth of tumors and the production of specific sugars that the bacterium uses as sources of energy. Scientists have exploited this remarkable natural "genetic engineer" to insert foreign DNA into the genomes of plants. When used as a vector for plant transformation, the tumor-inducing and sugar-producing genes on the T DNA are removed and replaced with foreign DNA. The altered Ti plasmids are first used to transform *Agrobacterium* cells from which the original Ti plasmids have been removed. Then the *Agrobacterium* cells are used to infect plant cells.

VIRUSES AS VECTORS Constraints on plasmid replication limit the size of the new DNA that can be inserted into a plasmid to about 10,000 bp. Although many prokaryotic genes may be smaller than this, most eukaryotic genes—with their introns and extensive flanking sequences—are bigger. A vector that accommodates larger DNA inserts is needed for these genes.

Both prokaryotic and eukaryotic viruses are often used as vectors for eukaryotic DNA. Bacteriophage λ, which infects *E. coli*, has a DNA genome of about 45,000 bp; this is all that fits into the phage head. If the phage genes that cause the host cell to die and lyse—about 20,000 bp—are eliminated, the virus can still attach to a host cell and inject its DNA, but the host cell

APPLY THE CONCEPT

DNA can genetically transform cells and organisms

As shown in Figure 13.5, the β-galactosidase (*lacZ*) gene encodes an enzyme that can convert the colorless substrate X-gal into a bright blue product. A plasmid vector contains a modified version of the *lacZ* gene with multiple restriction sites within its coding sequence, and a gene for resistance (R) to the antibiotic ampicillin. This vector is used with an *E. coli* strain that carries no other functional *lacZ* gene and that is sensitive (S) to ampicillin. A biologist clones a wheat gene by inserting it into the multiple cloning site of the *lacZ* gene in this vector. As a control to ensure that the bacterial cells are viable, she also grows some cells that were not transformed with the ligation products. Remember that after a ligation reaction, only a few of the plasmids are recombinant. Fill in the table with the results of these transformations.

DNA TAKEN UP BY *E. COLI*	PHENOTYPE FOR AMPICILLIN (R OR S)	PHENOTYPE FOR X-GAL (BLUE OR WHITE)
None		
Plasmid only		
Recombinant plasmid		

will not die. The deleted 20,000 bp can be replaced with DNA from another organism. Because viruses infect cells naturally, they offer a great advantage over plasmids, which often require artificial means to coax them to enter host cells.

Reporter genes help select or identify host cells containing recombinant DNA

Even when a population of host cells interacts with an appropriate vector, only a small proportion of the cells actually take up the vector. Furthermore, the process of making recombinant DNA is far from perfect. After a ligation reaction, not all the vector copies contain the foreign DNA. How can we identify or select the host cells that contain the vector with foreign DNA?

As we described above, selectable markers such as antibiotic resistance genes can be used to select cells containing those genes. These cells can be selected because only cells carrying the antibiotic resistance gene can grow in the presence of that antibiotic. Selectable markers are one type of **reporter gene**, which is any gene whose expression is easily assayed. Other reporter genes code for proteins that can be detected visually. Two commonly used reporter genes are *lacZ* and the gene for green fluorescent protein:

- The β-galactosidase (*lacZ*) gene (see Figure 11.3) codes for an enzyme that can convert the white substrate X-gal into a bright blue product. Many cloning plasmids contain the *lacZ* gene, along with genes for antibiotic resistance. As shown in **FIGURE 13.5**, foreign DNA can be inserted into the *lacZ* gene, inactivating it. Bacteria transformed with the plasmid are selected on a solid medium containing X-gal and the appropriate antibiotic. Clones containing the recombinant plasmid cannot make β-galactosidase, and produce white colonies. Clones that contain the original plasmid with no insert express the *lacZ* gene and make blue colonies.
- Green fluorescent protein (GFP), which normally occurs in the jellyfish *Aequorea victoria*, emits green light when exposed to ultraviolet light. The gene for this protein is now widely used as a reporter gene (**FIGURE 13.6**).

LINK

The *lacZ* gene is part of the inducible *lac* operon in *E. coli*, which encodes proteins required for β-galactoside metabolism; see **Concept 11.1**

RESEARCH TOOLS

FIGURE 13.5 Selection for Recombinant DNA Selectable marker (reporter) genes are used by scientists to select for bacteria that have taken up a plasmid. In a typical experiment, most of the bacteria will not take up any DNA. Of those that do, only a small fraction will take up recombinant DNA.

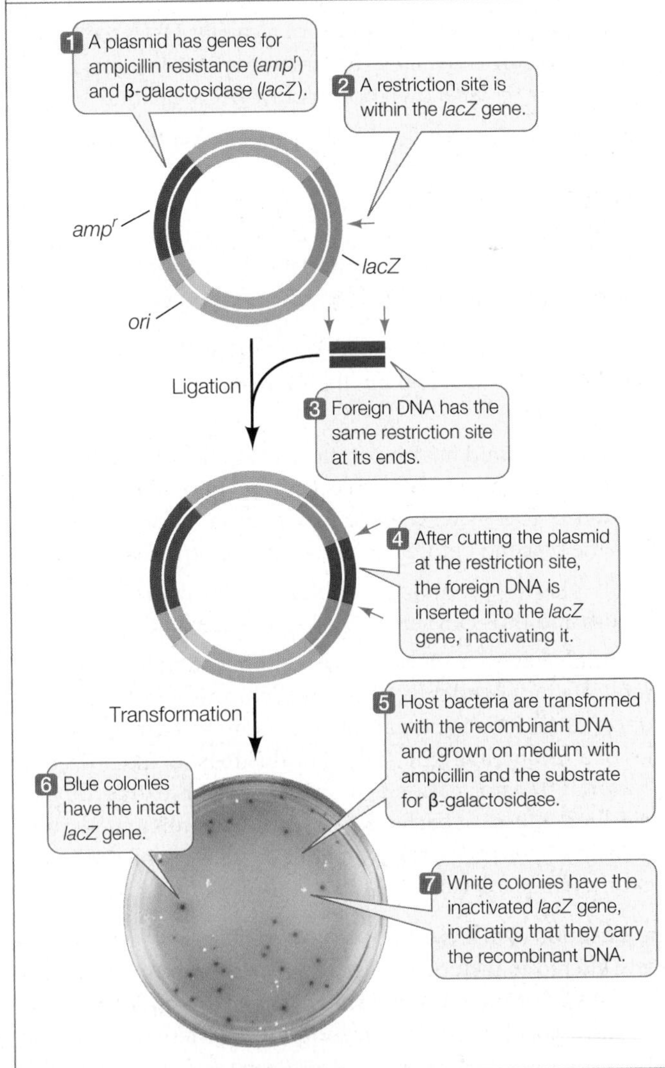

1 A plasmid has genes for ampicillin resistance (amp^r) and β-galactosidase (*lacZ*).
2 A restriction site is within the *lacZ* gene.
amp^r
lacZ
ori
Ligation
3 Foreign DNA has the same restriction site at its ends.
4 After cutting the plasmid at the restriction site, the foreign DNA is inserted into the *lacZ* gene, inactivating it.
Transformation
5 Host bacteria are transformed with the recombinant DNA and grown on medium with ampicillin and the substrate for β-galactosidase.
6 Blue colonies have the intact *lacZ* gene.
7 White colonies have the inactivated *lacZ* gene, indicating that they carry the recombinant DNA.

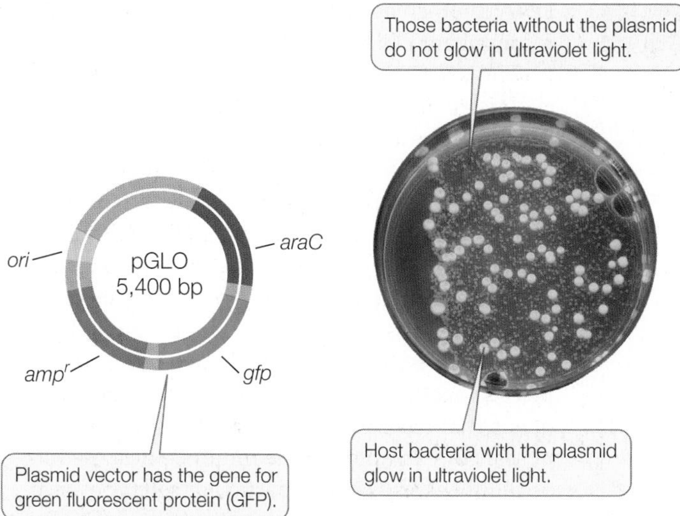
Those bacteria without the plasmid do not glow in ultraviolet light.
ori
pGLO
5,400 bp
araC
amp^r
gfp
Plasmid vector has the gene for green fluorescent protein (GFP).
Host bacteria with the plasmid glow in ultraviolet light.

FIGURE 13.6 Green Fluorescent Protein as a Reporter The presence of a plasmid with the gene for green fluorescent protein is readily apparent in transgenic cells because they glow under ultraviolet light.

Such reporters are not just used to select and identify cells carrying recombinant DNA. They can be attached to promoters in order to study how the promoters function under different conditions or in different tissues of a transgenic multicellular organism. Reporters can also be attached to other proteins, to study how and where those proteins become localized within eukaryotic cells.

CHECKpoint CONCEPT 13.2

- ✓ Outline the steps used to create recombinant DNA, transform host cells, and detect cells carrying the recombinant DNA.
- ✓ "Shuttle vectors" have the ability to transform both prokaryotic and eukaryotic cells. What sequences would you expect these vectors to have?
- ✓ What are the advantages of green fluorescent protein over antibiotic resistance as a marker on a plasmid for genetic transformation?
- ✓ Plasmid X has ampicillin and tetracycline resistance genes. The restriction enzyme *Eco*RI cleaves the plasmid once, within the tetracycline resistance gene. Plasmid B has a streptomycin resistance gene and one site for *Eco*RI cleavage that is not within the resistance gene. The two plasmids are cut with *Eco*RI and treated with DNA ligase. The mixture is used to transform an *E. coli* strain that is sensitive to the three antibiotics. Which antibiotic(s) would you add to the bacterial growth medium to select those bacteria carrying a recombinant plasmid?

We have described how DNA can be cut, inserted into a vector, and introduced into host cells. We have also seen how host cells carrying recombinant DNA can be identified. Now let's consider the sources of DNA used for cloning, as well as some molecular methods for manipulating gene expression.

CONCEPT 13.3 Genes Come from Various Sources and Can Be Manipulated

A major goal of molecular cloning experiments is to elucidate the functions of the DNA sequences and the proteins they encode. In this concept we look at ways that specific sequences can be identified and amplified. Millions of copies of a sequence are needed in order to study and manipulate the sequence in the laboratory. Cloned or amplified DNA can be used for various purposes, including the detection of expressed genes in specific cells and the artificial regulation of gene expression.

DNA fragments for cloning can come from several sources

The DNA fragments used in cloning procedures are obtained from a number of sources. In many cases the first step is to create a "library" of DNA fragments: a collection of clones that can be searched for the gene or genes of interest, or analyzed in other ways to learn more about the original source of the DNA fragments.

GENOMIC LIBRARIES A **genomic library** is a collection of DNA fragments that together comprise the genome of an organism. This is the starting point for some methods of genome sequencing. Restriction enzymes or other means, such as mechanical shearing, can be used to break chromosomes into smaller pieces (**FIGURE 13.7A**). Each fragment is inserted into a vector, which is then taken up by a host cell. Proliferation of a single transformed cell on a selective medium (such as for antibiotic resistance) produces a colony of recombinant cells, each of which harbors many copies of the same fragment of DNA. The colonies are grown by spreading the transformed cells over a solid culture medium in petri dishes (small circular plates), which are incubated at a suitable temperature for the host cells to grow.

A single petri dish can hold thousands of bacterial colonies and is easily screened for the presence of a particular DNA sequence. Colonies containing that sequence are identified by DNA hybridization using a probe labeled with complementary fluorescent or radioactive nucleotides. To do this, the petri dish with its bacterial colonies is duplicated, and then the bacteria on one of the plates are treated to expose the DNA for hybridization (see Figure 10.7).

cDNA A much smaller DNA library—one that includes only the genes transcribed in a particular tissue—can be made from **complementary DNA**, or **cDNA** (**FIGURE 13.7B**). This involves isolating mRNA from cells and making cDNA copies of that mRNA by complementary base pairing. The enzyme reverse transcriptase catalyzes this reaction. This collection of cDNAs from a particular tissue at a particular time is called a **cDNA library**, which is a "snapshot" of the transcription pattern of the cells in the sample. cDNA libraries have been invaluable for comparing gene expression in different tissues at different stages of development. For example, if cDNAs derived from developing red blood cells are examined, the

RESEARCH TOOLS

FIGURE 13.7 Constructing Libraries Intact genomic DNA is too large to be introduced into host cells. (A) A genomic library can be made by breaking the DNA into small fragments, incorporating the fragments into a vector, and then transforming host cells with the recombinant vectors. Each colony of cells contains many copies of a small part of the genome. (B) Similarly, there are many mRNAs in a cell. These can be copied into cDNAs and a library made from them. The DNA in these colonies can then be isolated for analysis.

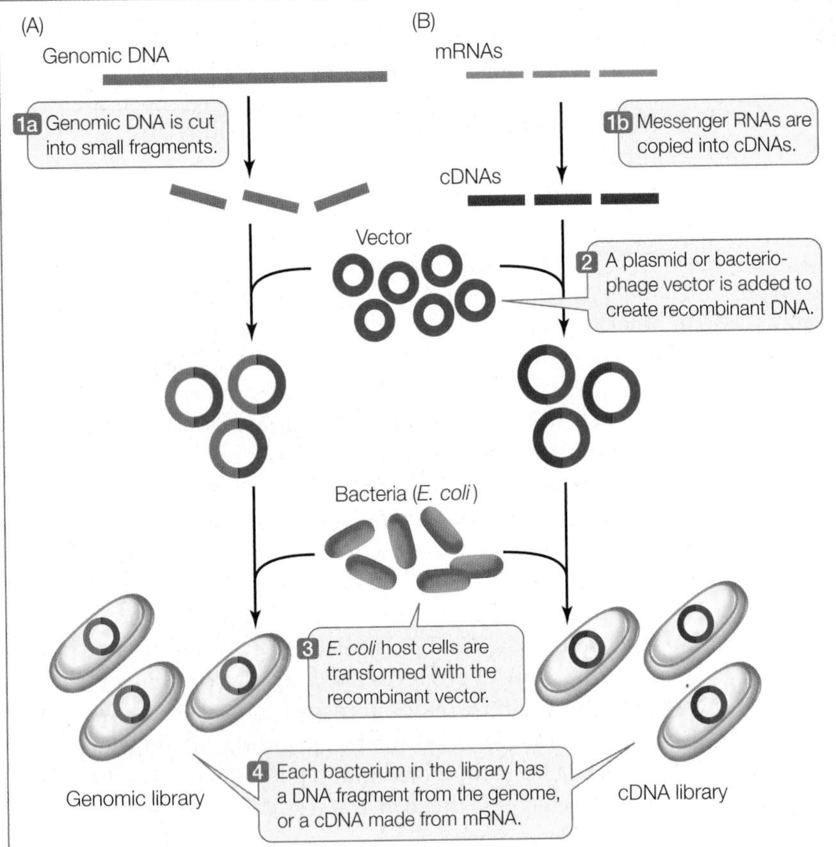
(A)
Genomic DNA
1a Genomic DNA is cut into small fragments.
(B)
mRNAs
1b Messenger RNAs are copied into cDNAs.
cDNAs
Vector
2 A plasmid or bacteriophage vector is added to create recombinant DNA.
Bacteria (E. coli)
3 E. coli host cells are transformed with the recombinant vector.
4 Each bacterium in the library has a DNA fragment from the genome, or a cDNA made from mRNA.
Genomic library
cDNA library

globin sequences (encoding the subunits of hemoglobin) are prominent. But a cDNA library derived from hair follicles does not contain those sequences.

Reverse transcriptase along with PCR (see below) can be used to create and amplify a specific cDNA sequence without the need to make a library. In this case, RNA is isolated from cells and then reverse transcriptase is used to make cDNA from the RNA. Then PCR is used to amplify a specific sequence directly from the cDNA. This method, called **RT-PCR**, is an invaluable tool for studies of the expression of particular genes in cells and organisms.

Synthetic DNA can be made in the laboratory

In Concept 9.2 (see Figure 9.15) we described the polymerase chain reaction (PCR), a method of amplifying DNA in a test tube. PCR can begin with just a single molecule of DNA, although larger quantities [in the picogram (10^{-12}) to microgram (10^{-6}) range] are more often used. Any fragment of DNA can be amplified as long as appropriate primers are available. This amplified DNA can then be inserted into a plasmid to create recombinant DNA, and cloned in host cells.

The artificial synthesis of DNA by organic chemistry methods is now fully automated. Synthetic oligonucleotides (single-stranded DNA fragments of 20–40 bp) are used as primers in PCR reactions. These primers can be designed to create short new sequences at the ends of the PCR products. This might be done to create a mutation in a recombinant gene, or to add restriction enzyme sites at the ends of the PCR product to aid in ligation reactions. Longer synthetic sequences can be pieced together to construct completely artificial genes that have been designed for specific purposes. For example, a gene might be designed to be highly expressed in a particular cell type, or to encode a highly active enzyme.

LINK

Synthetic DNA was used to create a novel bacterial genome to replace the genome in a host cell, resulting in a new bacterial species; see the opening story of **Chapter 4**

DNA sequences can be manipulated to study cause-and-effect relationships

Mutations that occur in nature have been important in demonstrating cause-and-effect relationships in biology. However, mutations in nature are rare events. Recombinant DNA technology allows us to ask "what if" questions by creating artificial gene constructs. Because synthetic DNA can be made with any desired sequence, it can be manipulated to create specific constructs or mutations, and the resulting phenotypes can be observed when the recombinant DNA is expressed in host cells. Such techniques have revealed thousands of cause-and-effect relationships.

One example involves the auxin response element, a short sequence of DNA that binds a specific transcription factor. This element is found in the promoters of plant genes that are switched on in the presence of the plant hormone auxin (see Concept 26.2). To study the role of the auxin response element in plants, scientists made an artificial promoter containing many copies of the element, and ligated the promoter to a reporter gene. The recombinant DNA was used to transform *Arabidopsis* plants. When the plants were treated with auxin, the reporter gene was switched on at very high levels (higher than those produced by a wild-type auxin-responsive promoter). This experiment helped show that the presence of the auxin response element (the "cause") results in gene expression in response to auxin (the "effect").

Genes can be inactivated by homologous recombination

Another way to understand a gene's function is to inactivate it so it is not transcribed and translated into a protein.

An example of this approach is the use of transposon mutagenesis in experiments designed to describe the minimal genome (see Figure 12.8). In animals, these "knockout" experiments often involve **homologous recombination** rather than transposon mutagenesis. As we saw in Chapter 8, recombination occurs when a pair of homologous chromosomes line up during meiosis. The chromosomes sometimes break and then rejoin in such a way that segments of the two chromosomes are exchanged. A key feature of homologous recombination is that it involves an exchange of DNA between molecules with identical, or nearly identical, sequences.

We will focus here on the technique used for mice (**FIGURE 13.8**). In order to knock out (inactivate) a target gene, the normal allele of the gene is inserted into a plasmid. Restriction enzymes are then used to insert a fragment containing a reporter gene or selectable marker into the middle of the normal gene. This addition of extra DNA disrupts the gene's coding region so that it no longer encodes a functional protein product.

Once the recombinant plasmid has been made, it is used to transfect mouse embryonic stem cells. A **stem cell** is an unspecialized cell that divides and differentiates into specialized cells. The gene sequences in the plasmid tend to line up with their homologous sequences in the mouse chromosome. If recombination occurs, the disrupted, inactive allele is "swapped" with the functional allele in the cell.

The knockout technique has been important in assessing the roles of many genes, and is especially valuable in studying human genetic diseases. Many such diseases (including phenylketonuria; see Concept 10.1) have knockout mouse models: mouse strains with similar diseases that were produced by homologous recombination. These models can be used to study the diseases and to test potential treatments.

RESEARCH TOOLS

FIGURE 13.8 Making a Knockout Mouse Animals carrying mutations are rare. Homologous recombination is used to replace a normal mouse gene with an inactivated copy of that gene, thus "knocking out" the gene. Discovering what happens to a mouse with an inactive gene tells us much about the normal role of that gene.

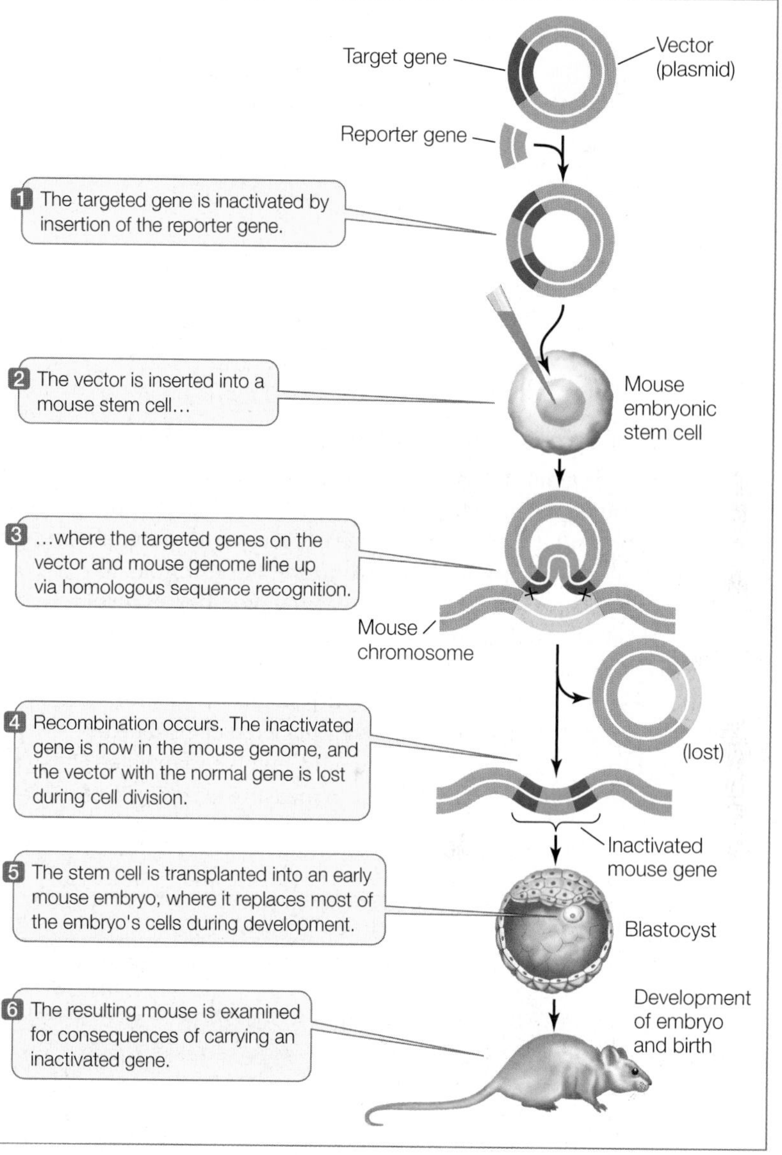

Complementary RNA can prevent the expression of specific genes

Another way to study the expression of a specific gene is to block the translation of its mRNA. This is yet another example of scientists imitating nature. As described in Concept 11.4, gene expression can be controlled in nature by the production of short, single-stranded RNA molecules (microRNAs or miRNAs) that inhibit the translation of target mRNA sequences. Many complex eukaryotes also produce small interfering RNAs (siRNAs), which are short (20–25 bp) double-stranded RNAs derived from much longer double-stranded RNA molecules. As in the production of miRNAs, these double-stranded siRNA molecules are processed into single-stranded molecules, and then each one is guided by a protein complex to a complementary region on an mRNA. The protein complex then catalyzes the breakdown of the targeted mRNA (**FIGURE 13.9**). These mechanisms for preventing mRNA translation are called **RNA interference (RNAi)**.

MicroRNAs and siRNAs are examples of **antisense RNA** because they bind by base pairing to the "sense" bases on the target mRNAs. siRNAs target specific mRNA molecules (from specific genes) because their sequences exactly match the target sequences in the mRNAs. By contrast, miRNAs do not match their targets perfectly, and therefore each one can reduce the expression of multiple, partially matching genes.

RNAi was discovered in the late 1990s, and since then scientists have used synthetic, single-stranded antisense RNAs and double-stranded siRNAs to inhibit the expression of known genes. This technique has been used extensively to block expression of specific genes in the laboratory, as well as in applied situations. For example, macular degeneration is an eye disease that results in near blindness when blood vessels proliferate in the eye. The signaling molecule that stimulates vessel

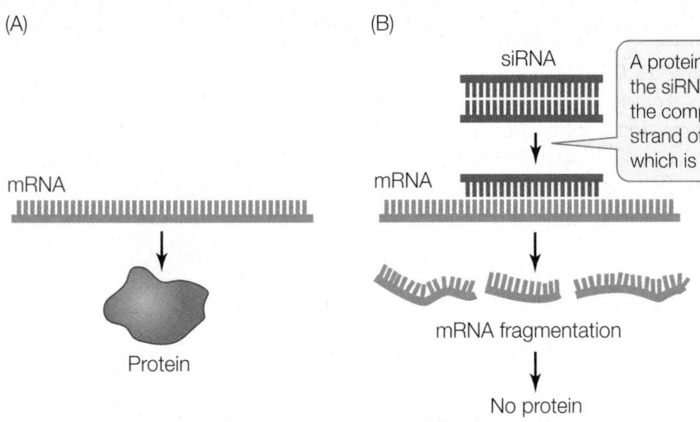
(A)
(B)
siRNA
A protein complex unwinds the siRNA and guides it to the complementary strand of the target mRNA, which is broken down.
mRNA
mRNA
Protein
mRNA fragmentation
No protein

FIGURE 13.9 Using siRNA to Block the Translation of mRNA (A) Normally an mRNA is translated to produce a protein. (B) Translation of a target mRNA can be prevented with a small interfering RNA (siRNA) that is complementary to part of the target mRNA.

proliferation is a growth factor. An RNAi-based therapy is being developed to target this growth factor's mRNA, and the therapy shows promise for stopping and even reversing the progress of the disease.

DNA microarrays reveal RNA expression patterns

The science of genomics faces two major quantitative realities. First, there are very large numbers of genes in eukaryotic genomes. Second, the pattern of gene expression in different tissues at different times is quite distinctive. For example, the cells of a skin cancer at its early stage may have a different set of mRNAs from those of normal skin cells and cells from a more advanced skin cancer.

To find such patterns, scientists could isolate mRNA from a cell and test for the presence of transcripts from each gene by hybridization or RT-PCR. But that would involve many steps and take a long time. It is far simpler to measure expression of every gene in one step. This is possible with DNA microarray technology, which provides large arrays of sequences for hybridization experiments.

A DNA microarray ("gene chip") contains a series of DNA sequences attached to a solid surface. The array is divided into a grid of microscopic spots, each containing thousands of copies of a particular oligonucleotide. A computer controls the addition of these oligonucleotide sequences in a predetermined pattern. Each oligonucleotide can hybridize with only one DNA or RNA sequence, and thus is a unique identifier of a gene. Many thousands of different oligonucleotides can be placed in a single microarray.

Microarrays can be used to examine patterns of gene expression in different tissues and under different conditions, and they can be used to identify individual organisms with particular mutations. You can visualize the concept of microarray analysis by following the example illustrated in **FIGURE 13.10**. Most women with breast cancer are treated with surgery to remove the tumor, and then treated with radiation soon afterward to kill cancer cells that the surgery may have missed. But a few cancer cells may survive in some patients, and these cells eventually form new tumors in the breast or elsewhere in the body. The challenge for physicians is to identify patients with surviving cancer cells so they can be treated aggressively with tumor-killing chemotherapy.

Scientists at the Netherlands Cancer Institute used medical records to identify patients whose cancer recurred or did not recur. They extracted mRNA from the patients' tumors and made cDNA from the samples. The cDNAs were hybridized to microarrays containing sequences derived from 1,000 human genes. The scientists found 70 genes whose expression differed dramatically between tumors from patients whose cancers recurred and tumors from patients whose cancers did not recur. From this information the Dutch group identified "gene expression signatures" that are useful in clinical decision-making: patients with a good prognosis can avoid unnecessary chemotherapy,

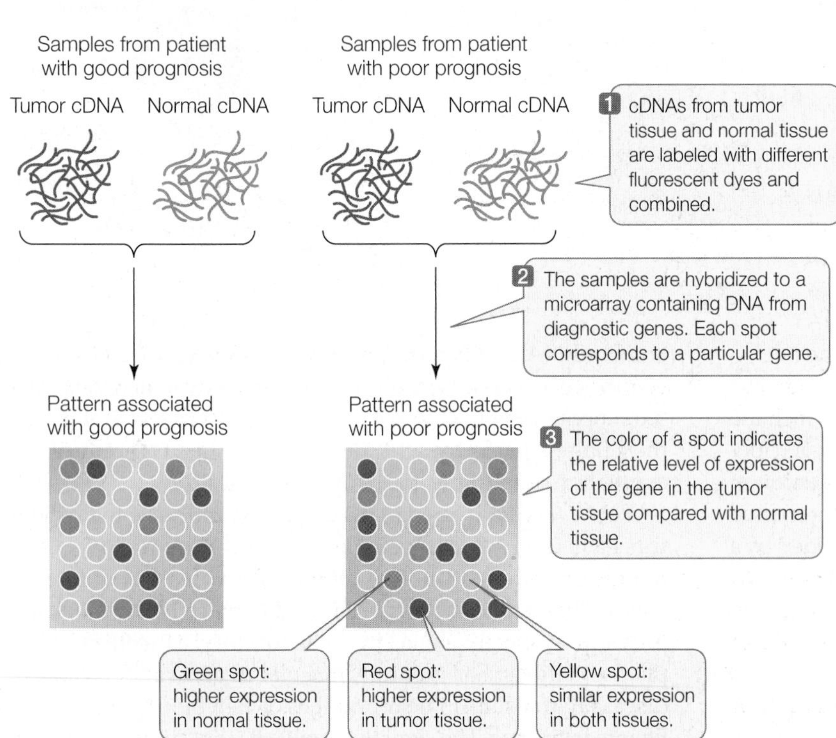
Samples from patient with good prognosis
Samples from patient with poor prognosis
Tumor cDNA
Normal cDNA
Tumor cDNA
Normal cDNA
1 cDNAs from tumor tissue and normal tissue are labeled with different fluorescent dyes and combined.
2 The samples are hybridized to a microarray containing DNA from diagnostic genes. Each spot corresponds to a particular gene.
Pattern associated with good prognosis
Pattern associated with poor prognosis
3 The color of a spot indicates the relative level of expression of the gene in the tumor tissue compared with normal tissue.
Green spot: higher expression in normal tissue.
Red spot: higher expression in tumor tissue.
Yellow spot: similar expression in both tissues.

FIGURE 13.10 Using DNA Microarrays for Clinical Decision-Making The pattern of expression of 70 genes in tumor tissues (the pattern of colored spots) indicates whether breast cancer is likely to recur. Actual arrays have more dots than shown here.

Go to ANIMATED TUTORIAL 13.2 DNA Chip Technology
PoL2e.com/at13.2

whereas those with a poor prognosis can receive more aggressive treatment.

CHECKpoint CONCEPT 13.3

- ✓ Outline the steps involved in knocking out a gene in a bacterium and in an animal. What are the uses for these methods?
- ✓ What are the differences between a genomic library and a cDNA library?
- ✓ You hypothesize that when a corn plant is infected with a fungus, a set of genes is turned on that fights the infection. How would you investigate this hypothesis using microarray technology?

We have now seen how recombinant DNA is made, how cells and organisms are transformed, and how gene expression can be manipulated. In the final concept we will look at some of the many applications of biotechnology.

CONCEPT 13.4 Biotechnology Has Wide Applications

In the opening story of this chapter we defined biotechnology as the use of cells or whole living organisms to make or modify materials or processes that are useful to people, such as foods, medicines, and chemicals. Bacteria and yeast cells can be transformed with almost any gene, and they can be induced to express that gene at high levels and to export the protein product out of their cells. This technology has turned these microbes into versatile factories for many important products. Today there is interest in producing nutritional supplements and pharmaceuticals in whole transgenic animals and plants and harvesting the products in large quantities—for example, from cow's milk or rice grains. Another goal is to produce animals and plants with improved characteristics, such as increased nutritional value or better tolerance of harsh environments. Key to this boom in biotechnology has been the development of specialized vectors that not only carry genes into animal and plant cells, but also make those cells express the genes at high levels.

Expression vectors can turn cells into protein factories

Many proteins that are potentially useful to humans come from eukaryotes. But if a eukaryotic gene is inserted into a typical plasmid and used to transform *E. coli*, none of the gene product will be made. Other key prokaryotic DNA sequences must be included with the gene. A bacterial promoter, a signal for transcription termination, and a special sequence that is necessary for ribosome binding on the mRNA must all be included in the transformation vector if the gene is to be expressed in the bacterial cell. In addition, the eukaryotic coding region must be made using cDNA so that it has no introns.

To solve this kind of problem, scientists make **expression vectors** that have all the characteristics of typical vectors, as well as the extra sequences needed for the foreign gene (also called a transgene) to be expressed in the host cell. For bacterial hosts, these additional sequences include the elements named above (**FIGURE 13.11**). For eukaryotes, they include the poly A addition sequence and a promoter that contains all the elements needed for expression in a eukaryotic cell. An expression vector can include various types of promoters and other features:

- An *inducible promoter*, which responds to a specific signal, can be included. For example, a promoter that responds to hormonal stimulation (e.g., a promoter containing the auxin response element described earlier) can be used so that the transgene will be expressed at high levels only when the hormone is added.
- A *tissue-specific promoter*, which is expressed only in a certain tissue at a certain time, can be used if localized expression is desired. For example, many seed proteins are expressed only in the plant embryo. Coupling a transgene to a seed-specific promoter will allow the gene to be expressed only in seeds and not, for example, in leaves.

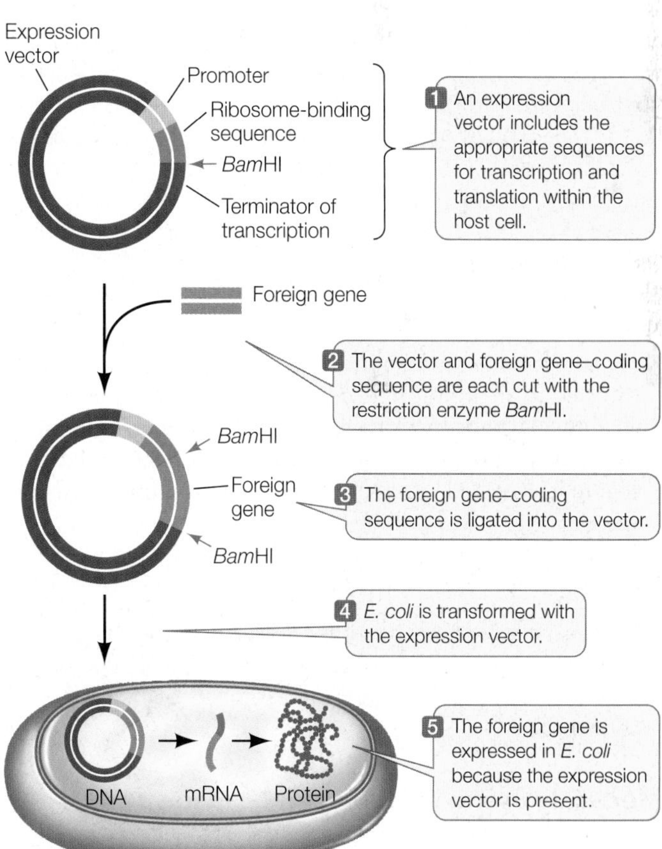

FIGURE 13.11 A Transgenic Cell Can Produce Large Amounts of the Transgene's Protein Product To be expressed in *E. coli*, a gene derived from a eukaryote requires bacterial sequences for transcription initiation (promoter), transcription termination, and ribosome binding. Expression vectors contain these additional sequences, enabling the eukaryotic protein to be synthesized in the prokaryotic cell.

Go to ACTIVITY 13.1 Expression Vectors
PoL2e.com/ac13.1

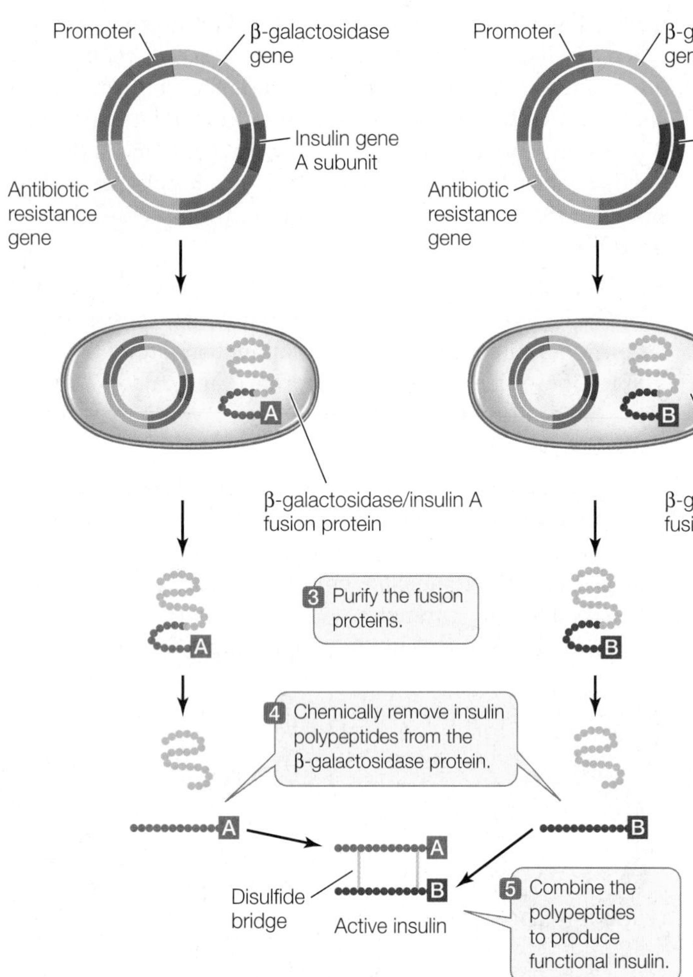

FIGURE 13.12 Human Insulin: From Gene to Drug Human insulin chains are made by recombinant DNA technology and then combined to produce the widely used drug.

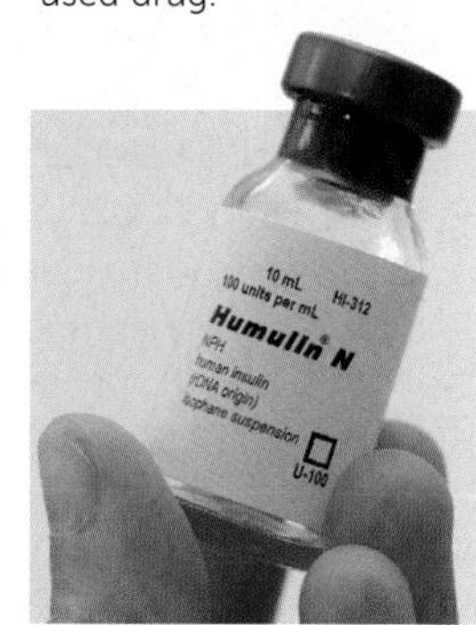

- *Signal sequences* can be added so that the gene product is directed to an appropriate destination. For example, when a protein is made by yeast or bacterial cells in a liquid medium, it is economical to include a signal directing the protein to be secreted into the extracellular medium for easier recovery.

LINK

You can review the mechanisms of transcription and mRNA processing in **Concept 10.2**, transcriptional regulation in **Concepts 11.2 and 11.3**, and signal sequences in **Concept 10.5**

Medically useful proteins can be made by biotechnology

Some medically useful products are being made by biotechnology (**TABLE 13.1**), and more are in various stages of development. Human insulin was the first medicine to be made using recombinant DNA, and provides a good illustration of a medical application of biotechnology. Insulin is essential for glucose uptake into cells. People with type I diabetes mellitus cannot make this hormone and must receive insulin injections. Before the advent of biotechnology, the insulin used for this purpose came from cattle and pigs.

Insulin is a protein of 51 amino acids and is made up of two polypeptide chains. The amino acid sequence of insulin from cattle differs from the human sequence by three amino acids, and the pig sequence differs from that of humans by just one amino acid. These differences are enough to cause an immune reaction in some patients who are diabetic, and these patients need to be treated with the human protein. Since the hormone is made in tiny amounts, harvesting enough from deceased persons is not practical. Recombinant DNA biotechnology solved this problem.

FIGURE 13.12 illustrates the strategy that was originally used to make human insulin in *E. coli* (today it is often made in yeast). The two insulin polypeptides were synthesized separately using an expression vector containing the gene for β-galactosidase (*lacZ*). Each insulin gene was inserted into the vector in such a way that it was induced, transcribed, and translated along with the β-galactosidase gene. After extraction and purification of the β-galactosidase–insulin fusion proteins, the insulin polypeptides were cleaved off by chemical treatment. The two insulin peptides were then combined to make a complete, functional human insulin molecule.

Before giving it to human patients, scientists had to be confident that the product made by biotechnology was functional human insulin. Several lines of evidence supported such confidence:

- The synthetic protein is the same size as human insulin.
- It has the same amino acid sequence.
- It has the same shape, as measured by physical techniques.
- It binds to the insulin receptor on cells and stimulates glucose uptake.

LINK

The crucial role of insulin in regulating glucose metabolism is detailed in **Concept 30.5**

Another way of making medically useful products in large amounts is **pharming**: the production of pharmaceuticals in farm animals or plants. For example, a gene encoding a useful

TABLE 13.1 Some Medically Useful Products of Biotechnology

Product	Use
Erythropoietin	Prevents anemia in patients undergoing kidney dialysis and cancer therapy
Colony-stimulating factor	Stimulates production of white blood cells in patients with cancer and AIDS
Bovine/porcine somatotropin	Stimulates growth and milk production in animals
Tissue plasminogen activator	Dissolves blood clots after heart attacks and strokes
Human growth hormone	Replaces missing hormone in people of short stature
Human insulin	Stimulates glucose uptake from blood in patients with type I diabetes mellitus
Factor VIII	Replaces clotting factor missing in patients with hemophilia A
Platelet-derived growth factor	Stimulates wound healing

protein might be placed next to the promoter of the gene that encodes lactoglobulin, an abundant milk protein. Transgenic animals carrying this recombinant DNA will secrete large amounts of the foreign protein into their milk. These natural "bioreactors" can produce abundant supplies of the protein, which can be separated easily from the other components of the milk (**FIGURE 13.13**).

Human growth hormone (hGH) is a protein made in the pituitary gland and has many effects, especially in growing children. People with hGH deficiencies have short stature as well as other abnormalities, a condition known as pituitary dwarfism. In the past they were treated with hGH isolated from the pituitary glands of dead people, but the supply was too limited to meet demand. Recombinant DNA technology was used to coax bacteria into making this protein, but the cost of treatment was high ($30,000 a year). In 2004, a team led by Daniel Salamone at the University of Buenos Aires produced a transgenic cow that secretes hGH in her milk. The yield is prodigious: a mere 15 such cows could meet the worldwide demand of children suffering from this type of dwarfism.

Plants and plant cells can be genetically transformed and induced to make proteins. Recently the first drug produced by plant biotechnology was approved by the U.S. government. The drug is an enzyme used to treat Gaucher's disease, an inherited disorder that affects the breakdown of certain lipids in lysosomes. More than 10,000 patients worldwide may soon be using this enzyme, made by transgenic carrot cells.

DNA manipulation is changing agriculture

The cultivation of plants and the husbanding of animals provide the world's oldest examples of biotechnology, dating back more than 10,000 years. Over the centuries, people have adapted crops and farm animals to their needs. Through selective breeding of these organisms, desirable characteristics such as large seeds, high fat content in milk, or resistance to disease have been selected for and improved.

The traditional way to improve crop plants and farm animals was to identify individuals with desirable phenotypes that existed as a result of natural variation. Through deliberate crosses, the genes responsible for the desirable traits could be introduced into widely used varieties or breeds. Despite some spectacular successes, such as the breeding of high-yielding varieties of wheat, rice, and hybrid corn, such deliberate crossing

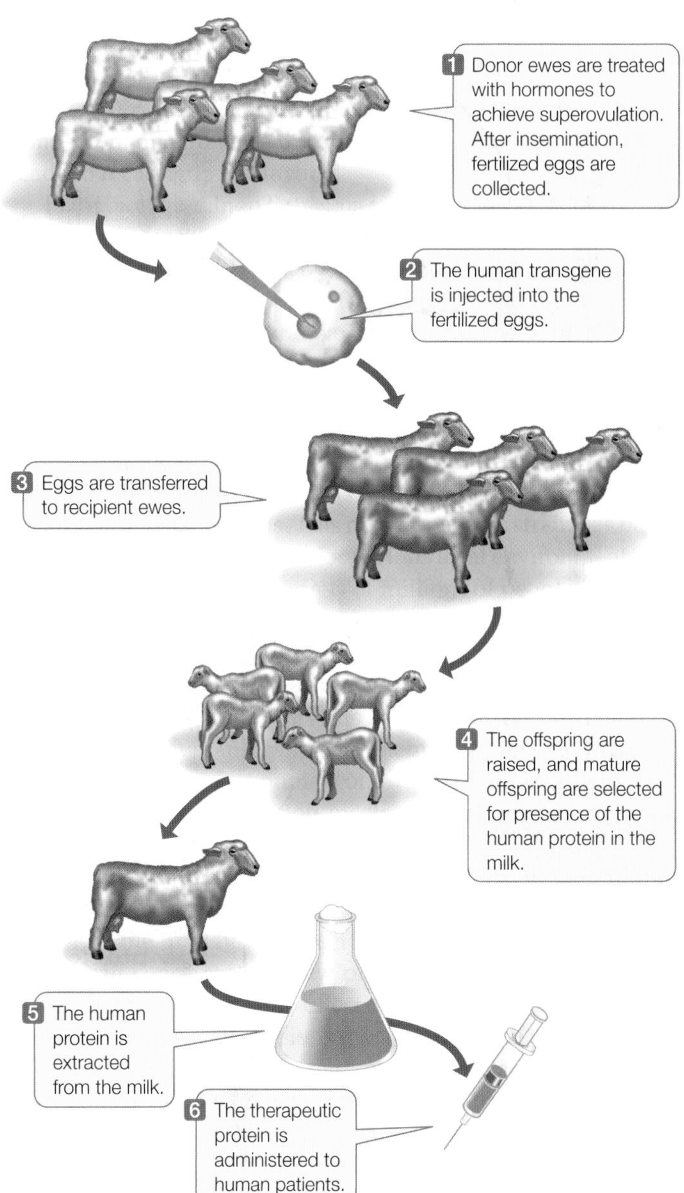

1 Donor ewes are treated with hormones to achieve superovulation. After insemination, fertilized eggs are collected.
2 The human transgene is injected into the fertilized eggs.
3 Eggs are transferred to recipient ewes.
4 The offspring are raised, and mature offspring are selected for presence of the human protein in the milk.
5 The human protein is extracted from the milk.
6 The therapeutic protein is administered to human patients.

FIGURE 13.13 Pharming An expression vector carrying a desired gene can be put into an animal egg, which is implanted into a surrogate mother. The transgenic offspring produce the new protein in their milk. The milk is easily harvested and the protein isolated, purified, and made clinically available for patients.

(A)
Transgenic plants
thrive in salty water.

(B)
Wild-type plants
wilt in salty water.

FIGURE 13.16 Salt-tolerant Tomato Plants Transgenic plants containing a gene for salt tolerance thrive in salty water (A), whereas plants without the transgene die (B). This technology may allow crops to be grown on salty soils.

(**FIGURE 13.16**). This finding raises the prospect of growing useful crops on previously unproductive soils.

LINK

Other stress adaptations of plants—some of which could be beneficial if introduced into crop plants—are discussed in **Concept 28.3**

The example described here illustrates what could become a fundamental shift in the relationship between crop plants and the environment. *Instead of manipulating the environment to suit the plant, biotechnology may allow us to adapt the plant to the environment.* As a result, some of the negative effects of agriculture, such as water pollution, could be lessened.

There is public concern about biotechnology

Concerns have been raised about the safety and wisdom of genetically modifying crops and other organisms. These concerns are centered on three claims:

- Genetic manipulation is an unnatural interference with nature.
- Genetically altered foods are unsafe to eat.
- Genetically altered crop plants are dangerous to the environment.

Advocates of biotechnology tend to agree with the first claim. However, they point out that all crops are unnatural in the sense that they come from artificially bred plants growing in a manipulated environment (a farmer's field). Recombinant DNA technology just adds another level of sophistication to these technologies.

To counter the concern about whether genetically engineered crops are safe for human consumption, biotechnology advocates point out that only single genes are added and that these genes are specific for plant function. For example, the *B. thuringiensis* toxin produced by transgenic plants has no effect on people. However, as plant biotechnology moves from adding genes that improve plant growth to adding genes that affect human nutrition, such concerns will become more pressing.

Various negative environmental impacts have been envisaged. There is concern about the possible "escape" of transgenes from crops to other species. If the gene for herbicide resistance, for example, were inadvertently transferred from a crop plant to a closely related weed, that weed could thrive in herbicide-treated areas. Another negative scenario is the possibility that increased use of an herbicide will select for weeds with naturally occurring mutations that make them resistant to that herbicide. This is indeed occurring. Widespread use of glyphosate on fields of glyphosate-resistant crops has resulted in the selection of rare mutations in weeds that make them resistant to glyphosate. To date more than ten resistant weed species have appeared in the U.S.

As we mentioned in the opening story, there are biotechnologically produced microorganisms that are able to break down components of crude oil. These have not been released into the environment because of the unknown effects that such organisms might have on natural ecosystems. However, these organisms potentially provide a way to rapidly clean up catastrophic oil spills.

Because of the potential benefits of biotechnology, most scientists believe that it is wise to proceed, albeit with caution.

CHECKpoint CONCEPT 13.4

- ✓ In addition to the coding sequence for a gene of interest, what other DNA sequences are required for the gene to be expressed in a different host?
- ✓ What is pharming, how is it done, and what are its advantages over more conventional biotechnology approaches?
- ✓ What are the advantages of using biotechnology for plant breeding compared with traditional methods?
- ✓ What are some concerns people might have about agricultural biotechnology?

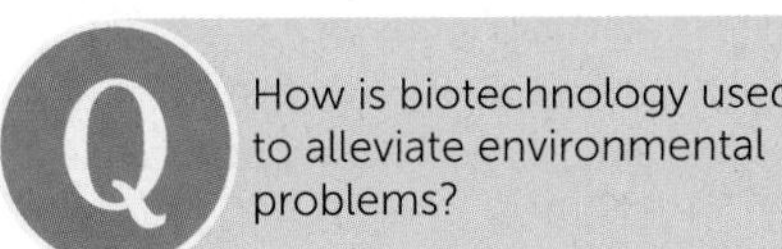

How is biotechnology used to alleviate environmental problems?

ANSWER Among the thousands of species of bacteria, there are many unique enzymes and biochemical pathways. New pathways are continually being discovered as new bacterial species are found (for example, by metagenomics; see Figure 12.6). Bacteria are natural recyclers, thriving on many types of nutrients—including what humans refer to as wastes.

Bioremediation is the use by humans of other organisms to remove contaminants from the environment. Two well-known examples of bioremediation are composting and wastewater treatment. Composting involves the use of bacteria and other microbes to break down large molecules, including carbon-rich polymers and proteins, in waste products such as wood chips, paper, straw, and kitchen scraps. For example, some species of bacteria make cellulase, an enzyme that hydrolyzes cellulose, a major component of plant cell walls and paper. Bacteria are used in wastewater treatment plants to break down human wastes, paper products, and household chemicals.

Bioremediation is also an attractive option for cleaning up oil spills. In 2010 the Deepwater Horizon oil rig in the Gulf of Mexico exploded, and crude oil began gushing out of the well below it. The oil flowed unabated for three months until the wellhead was finally capped; a total of 210 million gallons were released. The oil slick was visible from space—it damaged marine habitats and killed wildlife, washed up on beaches, and shut down fishing and tourism in the area. Efforts to remove the giant slick included collecting the oil, burning it, and dispersing it chemically. In addition, scientists identified bacteria living in the Gulf waters that were able to digest and break down components of the crude oil. Despite the clean-up efforts and the actions of these bacteria, much of the oil remains in the Gulf today.

FIGURE 13.17 The Spoils of War Massive oil spills occurred in Kuwait during the 1990–1991 Gulf War.

Oil spills can be remediated by encouraging the growth of natural microorganisms that digest components of crude oil. After the oil tanker *Exxon Valdez* ran aground near the Alaskan shore in 1989, nitrogen fertilizers were applied to nearby beaches to encourage the growth of oil-eating bacteria. Similar approaches have been tried in Kuwait, where the destruction of oil wells during the 1990–1991 Gulf War led to a massive release of oil (**FIGURE 13.17**). The success of such methods has been limited, however, because the naturally occurring bacteria digest only certain components of the crude oil, and because of technical difficulties in bringing the organisms into contact with the oil.

In the opening story we described how conventional genetic crosses were used to produce (and patent) bacteria that have the capacity to break down oil. Since that time, scientists have isolated the genes that encode the oil-degrading enzymes from such bacteria and cloned them into vectors (Concepts 13.2 and 13.3). This DNA has been used to transform several species of bacteria that live in environments where oil spills have occurred. However, as we mentioned in Concept 13.4, these genetically modified bacteria have not been released into the environment because of environmental concerns.

SUMMARY

CONCEPT 13.1 Recombinant DNA Can Be Made in the Laboratory

- **Biotechnology** is the use of living cells or their derivatives to make or modify materials and processes useful to people.
- **Restriction enzymes** make cuts in double-stranded DNA, creating fragments of various lengths.
- DNA fragments can be separated by size using **gel electrophoresis**. **Review Figure 13.2 and ANIMATED TUTORIAL 13.1**
- DNA fragments from different sources can be used to create **recombinant DNA** by joining them together using **DNA ligase**. **Review Figures 13.3 and 13.4**

CONCEPT 13.2 DNA Can Genetically Transform Cells and Organisms

- One goal of recombinant DNA technology is to **clone** a particular gene, either for analysis or to produce its protein product in quantity.
- Bacteria, yeasts, and cultured plant and animal cells are commonly used as hosts for recombinant DNA. The insertion of foreign DNA into host cells is called **transformation**, or if the host cells are derived from an animal, **transfection**.
- Various methods are used to get recombinant DNA into cells. These include chemical and physical treatments for plasmids and the use of viral **vectors**.
- **Selectable markers** such as genes for antibiotic resistance are used to select for host cells that have taken up a foreign gene. **Review Figure 13.5**
- **Reporter genes** (of which selectable markers are one type) are genetic markers with easily identifiable phenotypes. **Review Figure 13.6**

CONCEPT 13.3 Genes Come from Various Sources and Can Be Manipulated

- DNA fragments from a genome can be inserted into host cells to create a **genomic library**. A **cDNA library** is made by reverse transcribing mRNA to make cDNA. **Review Figure 13.7**
- Synthetic DNA containing any desired sequence can be made in the laboratory.
- Manipulating gene expression is one way to study the functions of particular genes.
- **Homologous recombination** is used to knock out a gene in a living organism. **Review Figure 13.8**
- Gene silencing techniques using miRNA or siRNA are used to prevent the translation of genes. **Review Figure 13.9**
- DNA microarray technology permits the screening of thousands of cDNA sequences at the same time. **Review Figure 13.10 and ANIMATED TUTORIAL 13.2**

CONCEPT 13.4 Biotechnology Has Wide Applications

- **Expression vectors** allow transgenes to be expressed in host cells. **Review Figure 13.11 and ACTIVITY 13.1**
- Recombinant DNA techniques have been used to make medically useful proteins. **Review Figure 13.12**
- **Pharming** is the use of transgenic plants or animals to produce pharmaceuticals. **Review Figure 13.13**
- Transgenic crop plants can be adapted to their environments, rather than vice versa. **Review Figure 13.16**
- There is public concern about the application of recombinant DNA technology to food production.

Go to the Interactive Summary to review key figures, Animated Tutorials, and Activities PoL2e.com/is13

Go to LaunchPad at **macmillanhighered.com/launchpad** for additional resources, including LearningCurve Quizzes, Flashcards, and many other study and review resources.

14 Genes, Development, and Evolution

KEY CONCEPTS

14.1 Development Involves Distinct but Overlapping Processes

14.2 Changes in Gene Expression Underlie Cell Fate Determination and Differentiation

14.3 Spatial Differences in Gene Expression Lead to Morphogenesis

14.4 Changes in Gene Expression Pathways Underlie the Evolution of Development

14.5 Developmental Genes Contribute to Species Evolution but Also Pose Constraints

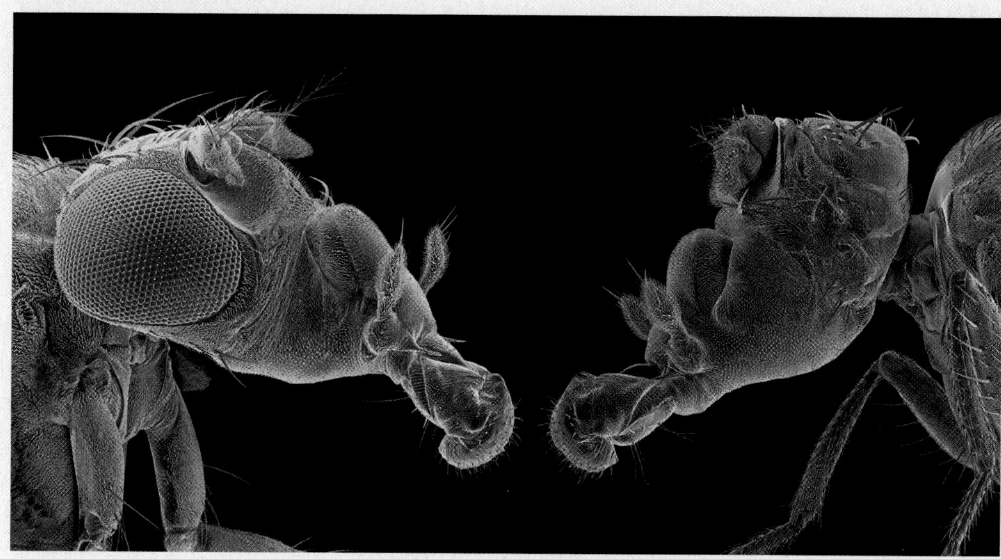

Scanning electron micrographs of a normal fruit fly (left) and an *eyeless* mutant fly of the same species (right).

Eyes are not essential for survival; many animals and all plants get by just fine without them. However, more than 90 percent of all animals *do* have eyes or some type of light-sensing organs, and having eyes can confer a selective advantage. About a dozen different kinds of eyes are found among the animals, including the camera-like eyes of humans and the compound eyes of insects, with their hundreds or thousands of individual units. In trying to understand how this variety came about, scientists—starting with Charles Darwin—proposed that eyes evolved independently many times in different animal groups, and that each improvement in the ability of eyes to gather light and form images conferred a selective advantage on their possessor.

Our understanding of the evolution of eyes remained at this level until 1915, when a mutant fruit fly without eyes was found and the gene involved, appropriately called *eyeless*, was mapped on one of the fly's chromosomes. This mutant fly remained a laboratory curiosity until the 1990s, when the Swiss developmental biologist Walter Gehring and his colleagues began looking for genes expressed in the fly embryo and found one that mapped to the *eyeless* locus. By transforming flies with recombinant DNA containing the *eyeless* gene, they showed that this gene's product controls the formation of the eye. Expression of the *eyeless* gene was then manipulated so that it was expressed in different parts of the body. In this way, eye structures were induced on legs, wings, and antennae of different flies. It did not matter where on the body the gene was expressed; if the gene was active, an eye developed there.

An even bigger surprise was in store when database searches revealed that the *eyeless* gene sequence was similar to that of *Pax6*, a mouse gene that, when mutated, leads to the development of very small eyes. Could the very different eyes of flies and mice result from simple variations on a common developmental theme? To test for functional similarity between the insect and mammalian genes, Gehring's team repeated their experiments on flies, but using the mouse *Pax6* gene instead of the fly *eyeless* gene. Once again, eyes developed. A gene whose expression normally leads to the development of a mammalian "camera" eye now led to the development of an insect's compound eye—a very different eye type. Thus a single gene product appears to function as a molecular switch that turns on eye development in diverse animals. Because of this and other recent findings, it now seems likely that all eye types evolved from a common origin very early in animal evolution.

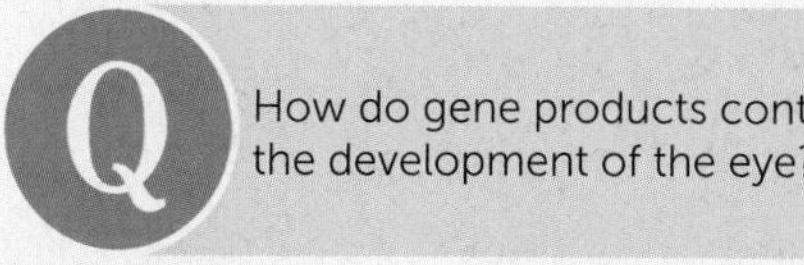

Q
How do gene products control the development of the eye?

You will find the answer to this question on page 294.

CONCEPT 14.1 Development Involves Distinct but Overlapping Processes

Development is the process by which a multicellular organism, beginning with a single cell, goes through a series of changes, taking on the successive forms that characterize its life cycle (**FIGURE 14.1**). After the egg is fertilized it is called a **zygote**, and in the earliest stages of development a plant or animal is called an **embryo**. Progress through a series of embryonic stages precedes emergence of the new, independent organism. Many organisms continue to develop throughout their lives, with development ceasing only at death.

Four key processes underlie development

The developmental changes an organism undergoes as it progresses from an embryo to mature adulthood involve four processes:

- **Determination** sets the developmental fate of a cell—what type of cell it will become—even before any characteristics of that cell type are observable. For example, in a developing mammalian embryo, as well as in some adult organs, there are mesenchymal stem cells that look unspecialized. But their fate to become muscle, fat, tendon, or other connective tissue cells has already been determined.
- **Differentiation** is the process by which different types of cells arise from less specialized cells, leading to cells with specific structures and functions. For example, mesenchymal stem cells differentiate to become the cells listed above.
- **Morphogenesis** (Greek for "origin of form") is the organization and spatial distribution of differentiated cells into the multicellular body and its organs. Morphogenesis can occur by cell division, cell expansion (especially in plants), cell movements, and apoptosis (programmed cell death).
- **Growth** is the increase in size of the body and its organs by cell division and cell expansion. Growth can occur by an increase in the number of cells or by the enlargement of existing cells. Growth continues throughout the individual's life in some organisms, but reaches a more or less stable end point in others.

All of these processes involve differential gene expression. The cells that arise from repeated mitoses in the early embryo may look the same superficially, but they soon begin to differ

FIGURE 14.1 Development Selected stages and processes of development from zygote to maturity are shown for an animal and for a plant. The blastula is a hollow sphere of cells; the gastrula has three cell layers (indicated by blue, red, and yellow).

Go to ACTIVITY 14.1 Stages of Development
PoL2e.com/ac14.1

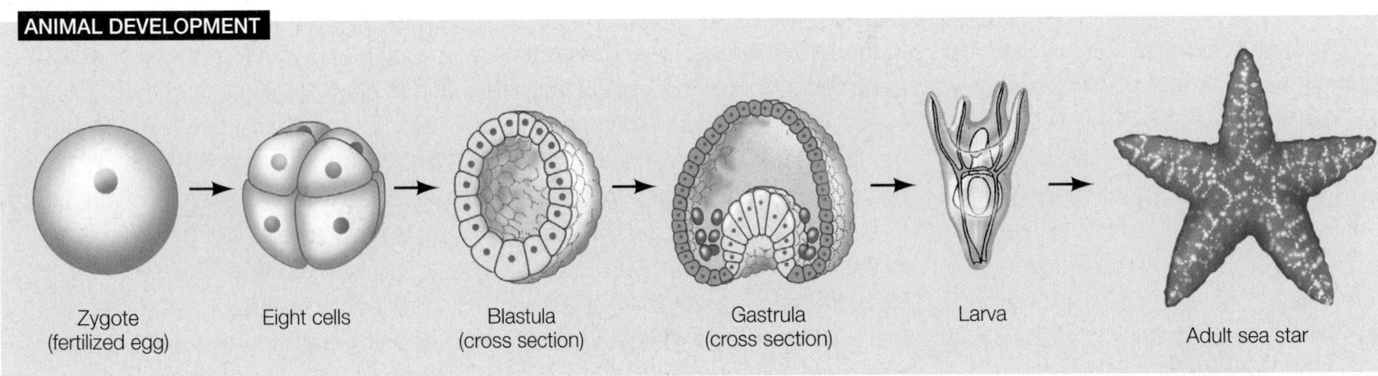

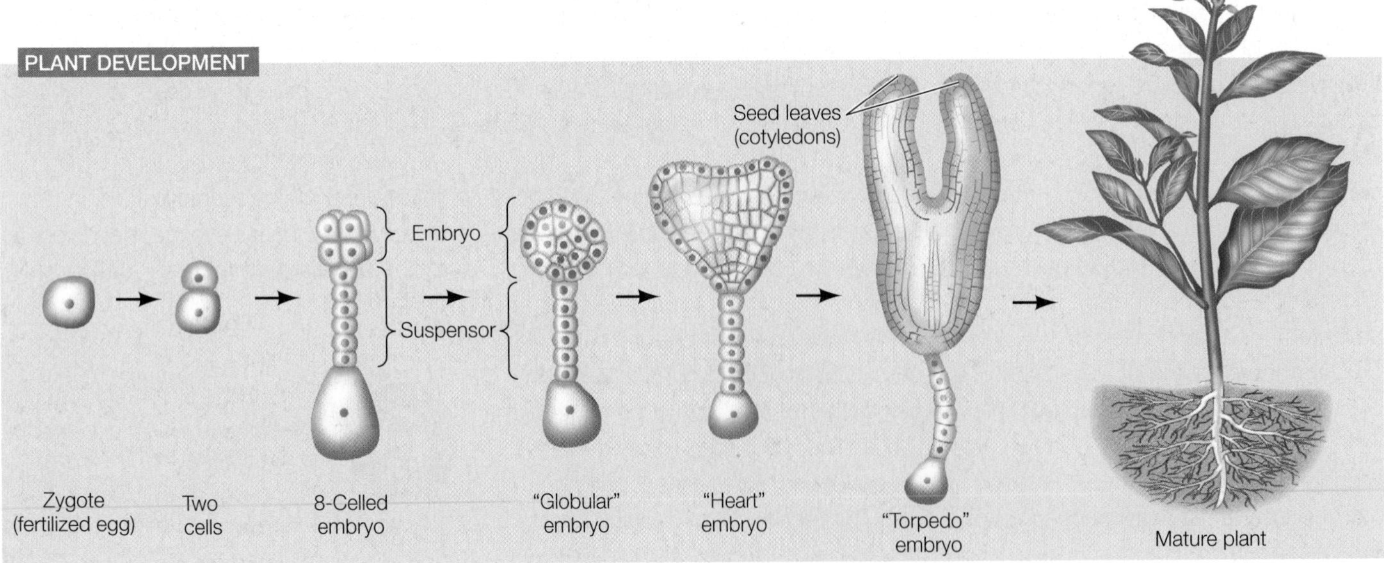

in terms of which genes they express. These processes also involve signals between cells. For example, within a developing embryo there are gradients of signaling molecules called morphogens that help determine cell fate and trigger cell differentiation.

Cell fates become progressively more restricted during development

A zygote is a single cell that gives rise to all the cells in the organism that will develop from it. As the zygote divides to form a multicellular embryo, each of the embryo's undifferentiated cells are destined to become part of a particular type of tissue—this is referred to as the **cell fate** of that undifferentiated cell.

When is the fate determined? At what point does a cell become committed to a particular fate and no other? One way to find out is to transplant cells from one embryo to a different region of a recipient embryo (**FIGURE 14.2**). A dye is used to mark the transplanted cells so that their subsequent development can be followed. The question is, will the transplanted cells adopt the differentiation pattern of their new surroundings, or will they continue on their own path, with their fate already sealed?

Experiments with frog embryos give an answer for that organism: If the donor tissue is from an early-stage embryo (blastula), it adopts the fate of its new surroundings; its fate *has not* been sealed. But if the donor tissue is from an older embryo (gastrula), it continues on its own path; its fate *has* been sealed. Determination is influenced by changes in gene expression as well as by the extracellular environment and is not something that is visible under the microscope—cells do not change their appearance when they become determined. Determination is followed by differentiation—the actual changes in biochemistry, structure, and function that result in cells of different types. *Determination is a commitment; the final realization of that commitment is differentiation.*

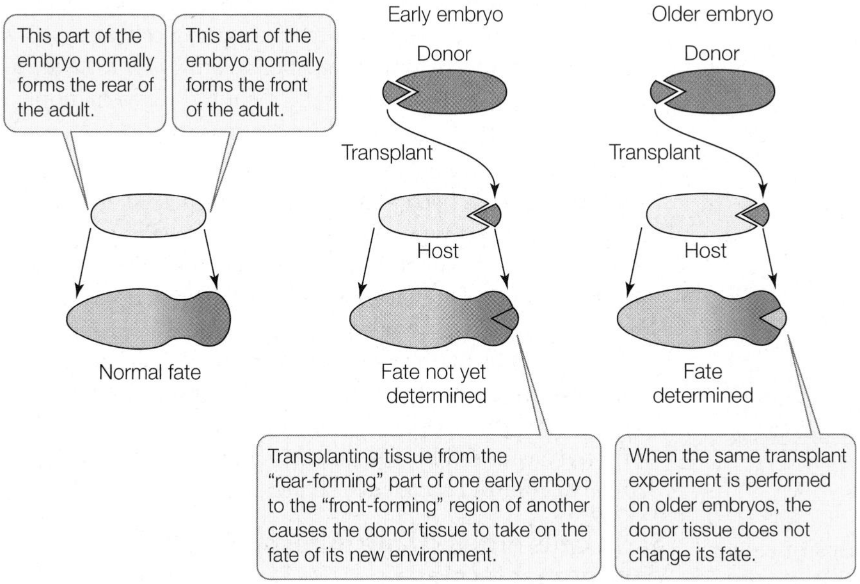

During animal development, cell fate becomes progressively restricted. This can be thought of in terms of **cell potency**, which is a cell's potential to differentiate into other cell types:

- The cells of an early embryo are **totipotent** (*toti*, "all"; *potent*, "capable"); they have the potential to differentiate into any cell type, including more embryonic cells.
- In later stages of the embryo, many cells are **pluripotent** (*pluri*, "many"); they have the potential to develop into most other cell types, but they cannot form new embryos.
- Through later developmental stages, including adulthood, certain stem cells are **multipotent**; they can differentiate into several different, related cell types. Mesenchymal stem cells (see above) are one kind of multipotent stem cell.
- Many cells in the mature organism are **unipotent**; they can produce only one cell type—their own.

Go to ANIMATED TUTORIAL 14.1
Cell Fates
PoL2e.com/at14.1

Cell differentiation is sometimes reversible

Once a cell's fate is determined, the cell differentiates. However, under the right experimental conditions, a determined or differentiated cell can become undetermined again. In some cases the cell can even become totipotent, meaning it is able to form the entire organism, with all of its differentiated cells. Normally this is a property of only the zygote or, in some cases, the first few cells of the early embryo.

PLANT CELL TOTIPOTENCY A carrot root cell normally faces a dark future. It cannot photosynthesize and generally does not give rise to new carrot plants. However, in 1958 Frederick Steward at Cornell University showed that if he isolated cells from a carrot root and maintained them in a suitable nutrient medium, he could induce them to dedifferentiate—to lose their differentiated characteristics. The cells could divide and give rise to masses of undifferentiated cells called calli (singular callus), which could be maintained in culture indefinitely. Furthermore, if they were provided with the right chemical cues, the cells could develop into embryos and eventually into complete new plants (**FIGURE 14.3**).

Since the new plants in Steward's experiments were genetically identical to the cells from which they came, they were clones of the original carrot plant. The ability to produce clones is evidence for the **genomic equivalence** of somatic (body) cells; that is, all somatic cells in a plant have a complete genome and thus have all the genetic information needed to become any cell in the plant.

Many types of cells from other plant species show similar behavior in the laboratory.

This part of the embryo normally forms the rear of the adult.
This part of the embryo normally forms the front of the adult.
Normal fate
Early embryo
Donor
Transplant
Host
Fate not yet determined
Older embryo
Donor
Transplant
Host
Fate determined
Transplanting tissue from the "rear-forming" part of one early embryo to the "front-forming" region of another causes the donor tissue to take on the fate of its new environment.
When the same transplant experiment is performed on older embryos, the donor tissue does not change its fate.

FIGURE 14.2 A Cell's Fate Is Determined in the Embryo Transplantation experiments using frog embryos show that the fate of cells is determined as the early embryo develops.

INVESTIGATION

FIGURE 14.3 Cloning a Plant When cells were removed from a plant and put into a medium with nutrients and hormones, they lost many of their specialized features—they dedifferentiated and became totipotent.[a]

HYPOTHESIS

Differentiated plant cells can be totipotent and can be induced to generate an entire new plant.

METHOD

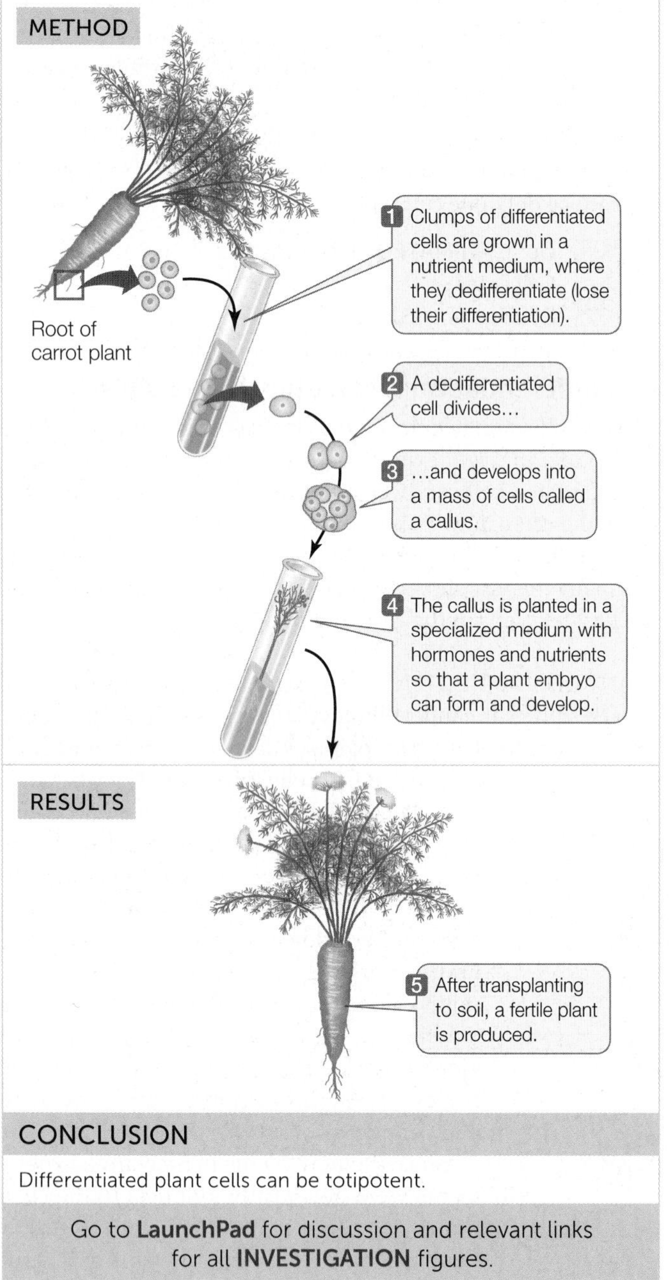
Root of
carrot plant
1 Clumps of differentiated cells are grown in a nutrient medium, where they dedifferentiate (lose their differentiation).
2 A dedifferentiated cell divides...
3 ...and develops into a mass of cells called a callus.
4 The callus is planted in a specialized medium with hormones and nutrients so that a plant embryo can form and develop.
RESULTS
5 After transplanting to soil, a fertile plant is produced.

CONCLUSION

Differentiated plant cells can be totipotent.

Go to **LaunchPad** for discussion and relevant links for all **INVESTIGATION** figures.

[a]F. C. Steward. 1958. *American Journal of Botany* 45: 705–709.

This ability to generate a whole plant from groups of cells or even a single cell has been invaluable in agriculture and forestry. For example, trees from planted forests are used in making paper, lumber, and other products. To replace the trees reliably, forestry companies regenerate new trees from the leaves of selected trees with desirable traits. The characteristics of these clones are more uniform and predictable than those of trees grown from seeds.

NUCLEAR TOTIPOTENCY IN ANIMALS Animal somatic cells cannot be manipulated as easily as plant cells can. Until recently, it was not possible to induce a cell from a fully developed animal to dedifferentiate and then redifferentiate into another cell type. However, nuclear transfer experiments have shown that the genetic information from a differentiated animal cell can be used to create cloned animals. The nucleus from an unfertilized egg is removed, forming an enucleated egg. A donor nucleus from a somatic cell is then introduced into the "empty" egg. If it is then stimulated to divide, the egg forms an embryo that can develop into an adult with the genetic composition of its nuclear donor. This is the basis of cloning animals. Dolly the sheep was the first experimentally produced mammalian clone, born in 1996 (**FIGURE 14.4**).

Many other animal species, including cats, dogs, horses, pigs, rabbits, and mice, have since been cloned by nuclear transfer. As in plants, the cloning of animals has shown that their differentiated cells have genomic equivalence. Cloning of animals has practical uses as well:

- Expansion of the numbers of valuable animals: One goal of the researchers who produced Dolly the sheep was to develop a method of cloning transgenic animals with useful phenotypes (see Concept 13.4). For example, a cow that was genetically engineered to make human growth hormone in milk has been cloned to produce two more cows that do the same thing. Only 15 such cows could supply the world's need for this protein, which is used to treat short stature that is due to growth hormone deficiency.
- Preservation of endangered species: The banteng, a relative of the cow, was the first endangered animal to be cloned and survive. The banteng was made using the enucleated egg from a cow, the nucleus from a banteng cell, and a cow surrogate mother. Cloning may be the only way to save endangered species with low rates of natural reproduction.
- Resurrection of extinct species. With the discoveries of intact DNA in fossils, the once-fictional idea of cloning an extinct species is becoming a possibility. In 2000, the last Pyrenian ibex—a type of mountain goat—died and the species became extinct. But in 2009, scientists used DNA from the dead animal to replace the DNA in a domestic goat egg and cloned a new ibex. Although this animal died shortly after birth, the resurrection of an extinct species was proven in principle. This has led to proposals to resurrect other extinct species, including the wooly mammoth and the Neanderthal. The genomes of both these species are available and have been sequenced.

Stem cells differentiate in response to environmental signals

The processes of development do not occur only in embryos. In adult plants, the growing regions at the tips of roots and stems

INVESTIGATION

FIGURE 14.4 Cloning a Mammal The experimental procedure described here was used to produce the first cloned mammal, a Dorset sheep named Dolly (shown on the left in the photo). As an adult, Dolly mated and subsequently gave birth to a normal offspring (the lamb on the right), thus proving the genetic viability of cloned mammals.[a]

HYPOTHESIS

Differentiated animal cells are totipotent.

METHOD

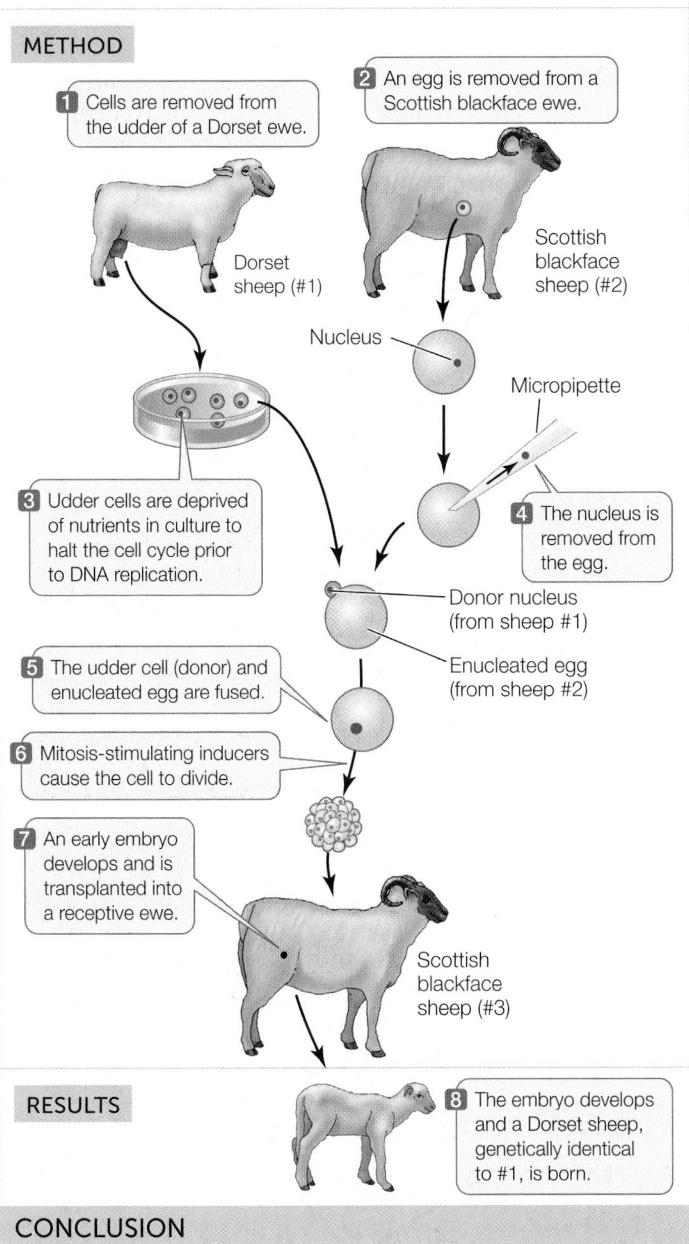

1 Cells are removed from the udder of a Dorset ewe.
2 An egg is removed from a Scottish blackface ewe.
Dorset sheep (#1)
Scottish blackface sheep (#2)
Nucleus
Micropipette
3 Udder cells are deprived of nutrients in culture to halt the cell cycle prior to DNA replication.
4 The nucleus is removed from the egg.
Donor nucleus (from sheep #1)
Enucleated egg (from sheep #2)
5 The udder cell (donor) and enucleated egg are fused.
6 Mitosis-stimulating inducers cause the cell to divide.
7 An early embryo develops and is transplanted into a receptive ewe.
Scottish blackface sheep (#3)
RESULTS
8 The embryo develops and a Dorset sheep, genetically identical to #1, is born.

CONCLUSION

Differentiated animal cells are totipotent in nuclear transplant experiments.

ANALYZE THE DATA

The team that cloned Dolly the sheep used a nucleus from a mammary epithelium (ME) cell. They also tried cloning by transplanting nuclei from fetal fibroblasts (FB) and embryos (EC), with the results shown in the table.

	Number of attempts that progressed to each stage		
Stage	ME	FB	EC
Egg fusions	277	172	385
Embryos transferred to recipients	29	34	72
Pregnancies	1	4	14
Live lambs	1	2	4

A. Calculate the percentage survival of eggs from fusion to birth. What can you conclude about the efficiency of cloning?

B. Compare the efficiencies of cloning using different nuclear donors. What can you conclude about the ability of nuclei at different stages to be totipotent?

C. What statistical test would you use to show whether the differences in A and B were significant (see Appendix B)?

Go to **LaunchPad** for discussion and relevant links for all **INVESTIGATION** figures.

[a]I. Wilmut et al. 1997. *Nature* 385: 810–813.

contain meristems, which are clusters of undifferentiated, rapidly dividing stem cells. These cells can differentiate into the 15–20 specialized cell types that make up roots, stems, leaves, and flowers. As you will see in Chapter 26, the plant body undergoes constant growth and renewal, with new organs forming often. (Think of flowers and leaves in the spring.)

In adult mammals, stem cells persist in many tissues, where they are used as a pool of cells that can differentiate and replace

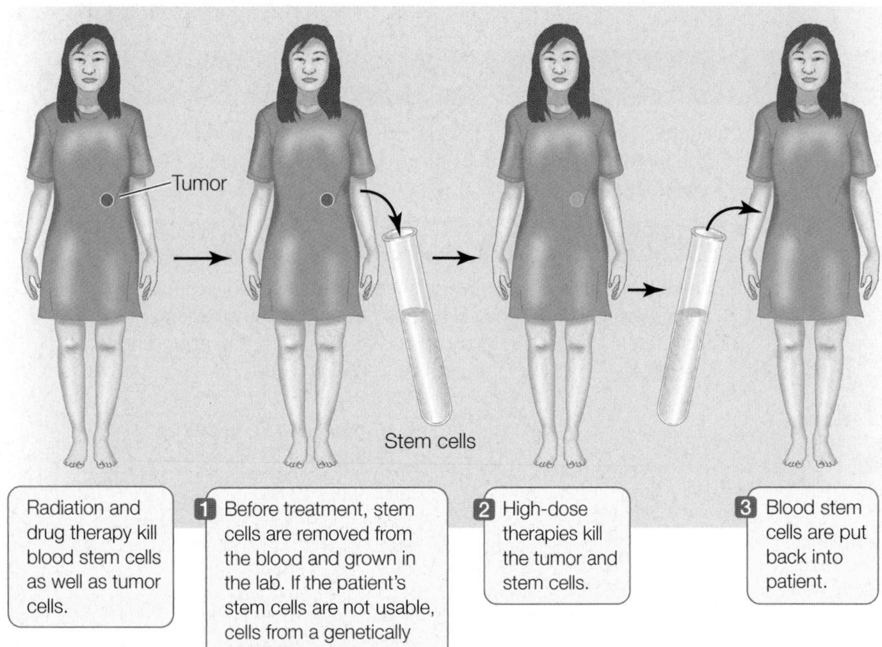

Tumor
Stem cells
Radiation and drug therapy kill blood stem cells as well as tumor cells.
1 Before treatment, stem cells are removed from the blood and grown in the lab. If the patient's stem cells are not usable, cells from a genetically related donor are used.
2 High-dose therapies kill the tumor and stem cells.
3 Blood stem cells are put back into patient.

FIGURE 14.5 Multipotent Stem Cells In hematopoietic stem cell transplantation, blood stem cells are used to replace stem cells destroyed by cancer therapy.

cells that are lost by "wear and tear" (necrosis) and programmed cell death (apoptosis). This is especially evident in tissues such as the skin, inner lining of the intestine, and blood. There are about 300 different cell types in a mammal.

MULTIPOTENT STEM CELLS Stem cells in particular mammalian tissues are multipotent, meaning they can form a limited repertoire of differentiated cells. For example, there are two types of multipotent stem cells in bone marrow. One type (called hematopoietic stem cells) produces the various kinds of red and white blood cells. The other type (mesenchymal stem cells) produces the cells that make bone and surrounding tissues, including muscle.

The proliferation and differentiation of multipotent stem cells is "on demand." Hematopoeitic stem cells in the bone marrow, for example, differentiate in response to specific signals. These signals can come from either adjacent bone marrow cells or from the circulating blood. This ability to differentiate is the basis of an important cancer therapy called hematopoietic stem cell transplantation (HSCT; **FIGURE 14.5**). Some treatments that kill cancer cells also kill other dividing cells, including the stem cells in the bone marrow of patients exposed to these treatments. To circumvent this problem, stem cells are harvested from the blood or bone marrow of the patient prior to treatment or from a donor; the cells are injected back into the patient after cancer treatment. Before the cells are harvested, the patient (or donor) receives injections of a growth factor that stimulates proliferation of the

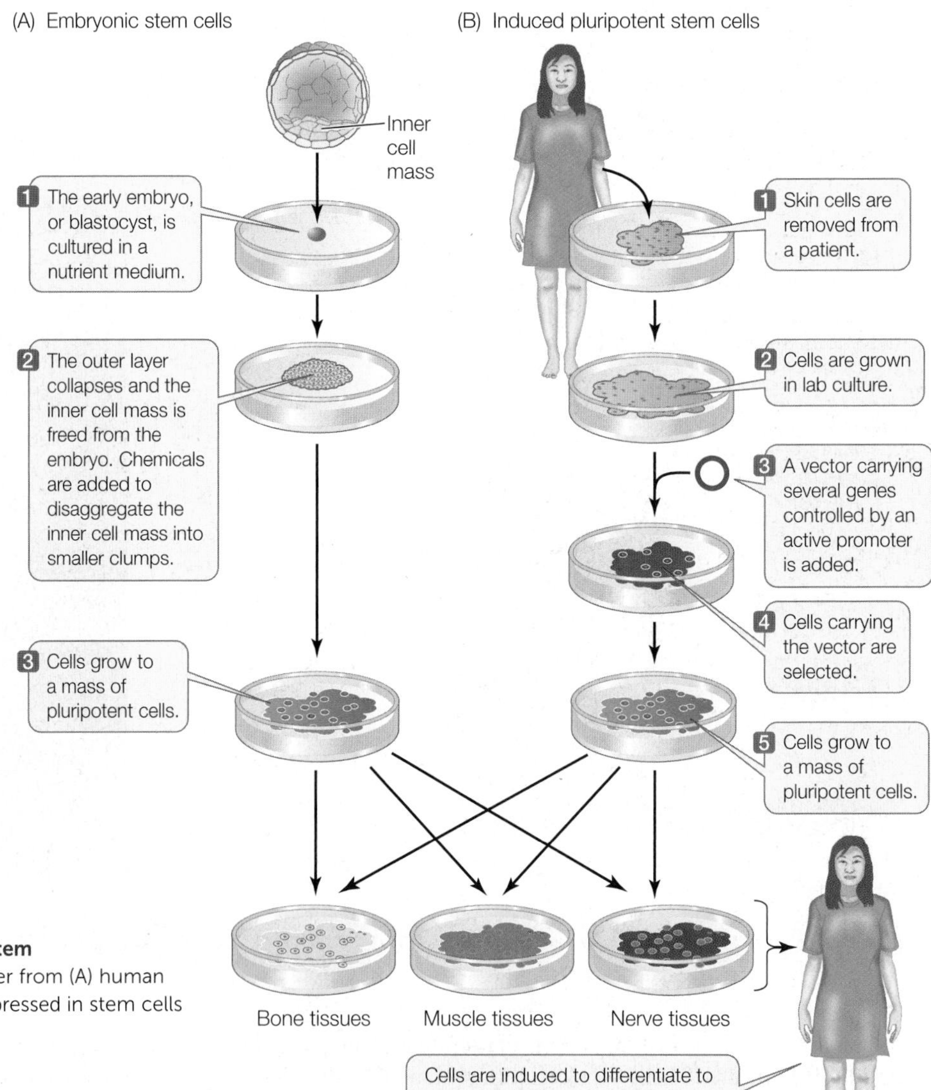

(A) Embryonic stem cells
(B) Induced pluripotent stem cells
Inner cell mass
1 The early embryo, or blastocyst, is cultured in a nutrient medium.
2 The outer layer collapses and the inner cell mass is freed from the embryo. Chemicals are added to disaggregate the inner cell mass into smaller clumps.
3 Cells grow to a mass of pluripotent cells.
1 Skin cells are removed from a patient.
2 Cells are grown in lab culture.
3 A vector carrying several genes controlled by an active promoter is added.
4 Cells carrying the vector are selected.
5 Cells grow to a mass of pluripotent cells.
Bone tissues
Muscle tissues
Nerve tissues
Cells are induced to differentiate to specialized cells and transplanted to patients as needed.

FIGURE 14.6 Two Ways to Obtain Pluripotent Stem Cells Pluripotent stem cells can be obtained either from (A) human embryos or (B) by adding genes that are highly expressed in stem cells to skin cells to transform them into stem cells.

Go to ANIMATED TUTORIAL 14.2
Embryonic Stem Cells
PoL2e.com/at14.2

hematopoietic stem cells. The stored stem cells retain their ability to differentiate in the bone marrow environment. By allowing the use of high doses of treatment to kill tumors, HSCT saves thousands of lives each year.

PLURIPOTENT STEM CELLS In mammals, totipotent stem cells that can individually give rise to an organism are found only in very early embryos. In both mice and humans, the last embryonic stage before differentiation occurs is called a blastocyst (the term for a mammalian blastula; see Figures 14.1 and 38.8). Although they cannot form an entire embryo, a group of cells in the blastocyst still retains the ability to form any cell type in the body; these cells are pluripotent. These **embryonic stem cells (ESCs)** can be removed from the blastocyst and grown in laboratory culture almost indefinitely if provided with the right conditions. They can also be induced to express appropriate genes and differentiate in a particular way if the right signal is provided (**FIGURE 14.6A**). For example, treatment of mouse ESCs with a derivative of vitamin A causes them to form neurons (nerve cells), whereas other growth factors induce them to form blood cells. Such experiments demonstrate both the cells' developmental potential and the roles of environmental signals. This finding raises the possibility of using ESC cultures as sources of differentiated cells to repair specific tissues, such as a damaged pancreas in diabetes, or a brain that malfunctions in Parkinson's disease.

ESCs can be harvested from human embryos conceived by in vitro ("under glass"—in the laboratory) fertilization, with the consent of the donors. Since more than one embryo is usually conceived in this procedure, embryos not used for reproduction might be available for embryonic stem cell isolation. These cells could then be grown in the laboratory and used as sources of tissues for transplantation into patients with tissue damage. There are two problems with this approach:

- Some people object to the destruction of human embryos for this purpose.
- The stem cells, and tissues derived from them, would provoke an immune response in a recipient (see Chapter 39).

Shinya Yamanaka and coworkers at Kyoto University in Japan developed another way to produce pluripotent stem cells that applies the concepts of gene expression and development (**FIGURE 14.6B**). Instead of extracting ESCs from blastocysts, they make **induced pluripotent stem cells (iPS cells)** from skin cells. This approach destroys no embryos and allows tissues to be made from skin cells of any individual, thus preventing an immune response. The scientists developed this method systematically (see Chapter 13 for more information on the techniques discussed here):

1. First, they used microarrays to compare the genes expressed in ESCs with those expressed in nonstem cells. They found several genes that were uniquely expressed at high levels in ESCs. These genes encode transcription factors believed to be essential to the undifferentiated state and function of stem cells. Recall that transcription factors are DNA binding proteins that regulate the expression of specific genes.
2. Next, they isolated the genes and inserted them into a vector for genetic transformation of skin cells. They found that the skin cells now expressed the newly added genes at high levels.
3. Finally, they showed that the transformed cells were pluripotent and could be induced to differentiate into many tissues—they had become iPS cells.

Yamanaka was awarded the Nobel Prize for his work. The ultimate aim is to use the cells for research and therapy in diseases.

CHECKpoint CONCEPT **14.1**

✓ Describe the four major processes of development.

✓ Not all the DNA in a cell is in the nucleus. What are the genetic differences between cloning in carrot plants and cloning in sheep? How would you show this?

✓ Identical twins are formed when a zygote divides once by mitosis and then each mitotic product forms an embryo. Are identical twins clones? Explain your answer.

Having considered the general principles of development, we will now turn to the mechanisms that govern developmental events. Not surprisingly, these mechanisms have been studied at the molecular level and involve changes in gene expression and the activities of specific proteins.

CONCEPT 14.2 Changes in Gene Expression Underlie Cell Fate Determination and Differentiation

Virtually every cell of an individual organism contains a complete copy of the organism's genome. Each cell, however, expresses only a subset of these DNA sequences. For example, certain cells in hair follicles produce keratin, the protein that makes up hair, whereas other cell types in the body do not. In Chapter 5 we discussed cell-signaling pathways, many of which result in changes in gene expression. In Chapter 11 we described several mechanisms by which cells control gene expression—by controlling transcription and translation, and by making posttranslational protein modifications. As we mentioned in Concept 14.1, all four processes of development—determination, differentiation, morphogenesis, and growth—involve changes in gene expression, and these changes often result from signaling between cells. In this concept we focus on the processes of cell fate determination and differentiation.

The most fundamental decisions in development are generally controlled at the level of transcription. Genes that determine cell fate and trigger differentiation (often by regulating the expression of other genes on other chromosomes) usually encode transcription factors. In some cases, a single transcription factor can cause a cell to differentiate in a certain way. In other cases, multiple interactions between genes and proteins set off a sequence of transcriptional events that leads to differential gene expression. There are two ways in which cell fate can be determined:

- by the asymmetrical distribution of cytoplasmic factors inside a cell, so that its two progeny cells receive unequal amounts of the factors, or
- by the differential exposure of two cells to an external signal (an inducer; see p. 282).

Cell fates can be determined by cytoplasmic polarity

An early event in development is often the establishment of axes that relate to the body plan of the organism. For example, an embryo may develop a distinct "top" and "bottom" corresponding to what will become opposite ends of the mature organism; such a difference is called **polarity**. Many examples of polarity are observed as development proceeds. Our heads are distinct from our rear ends, and the distal (far from the center) ends of our arms and legs (wrists, ankles, fingers, toes) differ from the proximal (near) ends (shoulders and hips).

Polarity may develop early; even within the egg, the yolk and other factors are often distributed asymmetrically. During early development in animals, polarity is specified by an "animal pole" at the top of the zygote and a "vegetal pole" at the bottom. This polarity can lead to the determination of cell fates at a very early stage of development. For example, sea urchin embryos can be bisected at the eight-cell stage in two different ways:

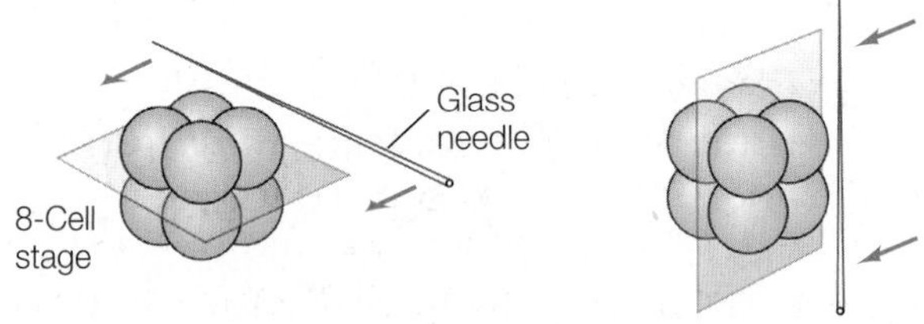

If the two halves (with four cells each) of these embryos are allowed to develop, the results are dramatically different for the two different cuts:

- For an embryo cut into a top half and a bottom half (left, above), the bottom half develops into a small sea urchin and the top half does not develop at all.
- For an embryo cut into two side halves (right, above), both halves develop into normal, though smaller, sea urchins.

These results indicate that the top and bottom halves of an eight-cell sea urchin embryo have already developed distinct fates. These observations led to the model of **cytoplasmic segregation**, which states that certain materials, called **cytoplasmic determinants**, are distributed unequally in the egg cytoplasm (**FIGURE 14.7**). During the early cell divisions of the embryo's development, the progeny cells receive unequal amounts of these determinants. Cytoplasmic determinants include specific transcription factors that promote differential gene expression in the two daughter cells. They also include small regulatory RNAs and mRNAs, which also contribute to differential gene expression. What accounts for the unequal distribution of these determinants?

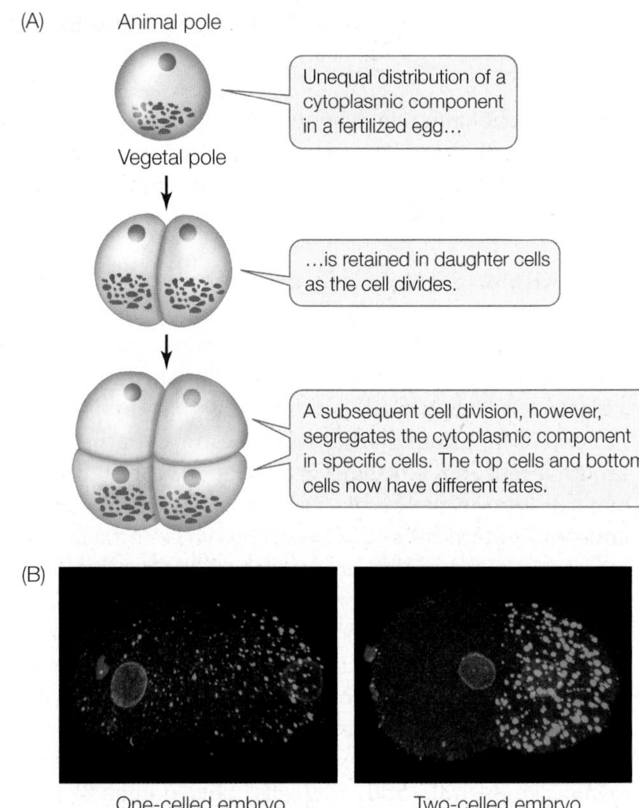

FIGURE 14.7 The Concept of Cytoplasmic Segregation (A) The unequal distribution of cytoplasmic determinants in a fertilized egg determines the fates of its descendants. (B) The zygote of the nematode worm *Caenorhabditis elegans* (left) shows an asymmetrical distribution of cytoplasmic particles (stained green). The progeny of the first cell division (right) receive unequal amounts of the particles.

Go to ANIMATED TUTORIAL 14.3
Early Asymmetry in the Embryo
PoL2e.com/at14.3

It turns out that the cytoskeleton contributes to the asymmetrical distribution of these determinants in the egg. Recall from Concept 4.4 that an important function of the microtubules and microfilaments in the cytoskeleton is to help move materials in the cell. Two properties allow these structures to accomplish this:

- Microtubules and microfilaments have polarity—they grow by adding subunits to the plus end.
- Cytoskeletal elements can bind specific proteins, which can be used in the transport of mRNA.

For example, in the sea urchin egg, a protein binds to both the growing (+) end of a microfilament and to an mRNA encoding a cytoplasmic determinant. As the microfilament grows toward one end of the cell, it carries the mRNA along with it. The asymmetrical distribution of the mRNA leads to asymmetrical distribution of the protein it encodes—a transcription factor.

Inducers passing from one cell to another can determine cell fates

The term "induction" has different meanings in different contexts. In biology it can be used broadly to refer to the initiation of, or cause of, a change or process. But in the context of cellular differentiation, it refers to the signaling events by which cells in a developing embryo communicate and influence one another's developmental fate. Induction involves chemical signals called **inducers** and the signal transduction pathways that are triggered by these signals. Exposure to different amounts of inductive signals can lead to differences in gene expression among cells in a developing organism.

LINK

Signal transduction pathways are described in **Concepts 5.5 and 5.6**

The nematode worm *Caenorhabditis elegans* was one of the first model eukaryotic organisms to have its entire genome sequenced (see Concept 12.3). This worm develops from fertilized egg to larva in only about 8 hours and reaches the adult stage in just 3.5 days. The process is easily observed using a low-magnification dissecting microscope because the body covering is transparent (**FIGURE 14.8A**). To illustrate the principles of induction, we focus here on the development of one part of the *C. elegans* body: the vulva (**FIGURE 14.8B**).

The adult nematode is hermaphroditic, containing both male and female reproductive organs. It lays eggs through a pore called the vulva on its ventral (lower) surface. During development, a single cell, called the anchor cell, induces the vulva to form from six cells on the worm's ventral surface. In this case there are two molecular signals, the primary inducer and the secondary inducer. Each of the six ventral cells has three possible fates: it may become a primary vulval precursor cell, a secondary vulval precursor cell, or simply become part of the worm's skin—an epidermal cell. You can follow the sequence of events in Figure 14.8B. The concentration of the primary inducer, LIN-3, is key: the anchor cell produces LIN-3, which diffuses out of the cell and forms a concentration gradient with respect to adjacent cells. The three cells that are closest to the anchor cell receive the most LIN-3 and become vulval precursor cells; cells slightly farther from the anchor cell receive less LIN-3 and become epidermal cells. A second induction event results in the two classes of vulval precursor cells: primary and secondary. Induction involves the activation or inactivation of specific sets of genes through

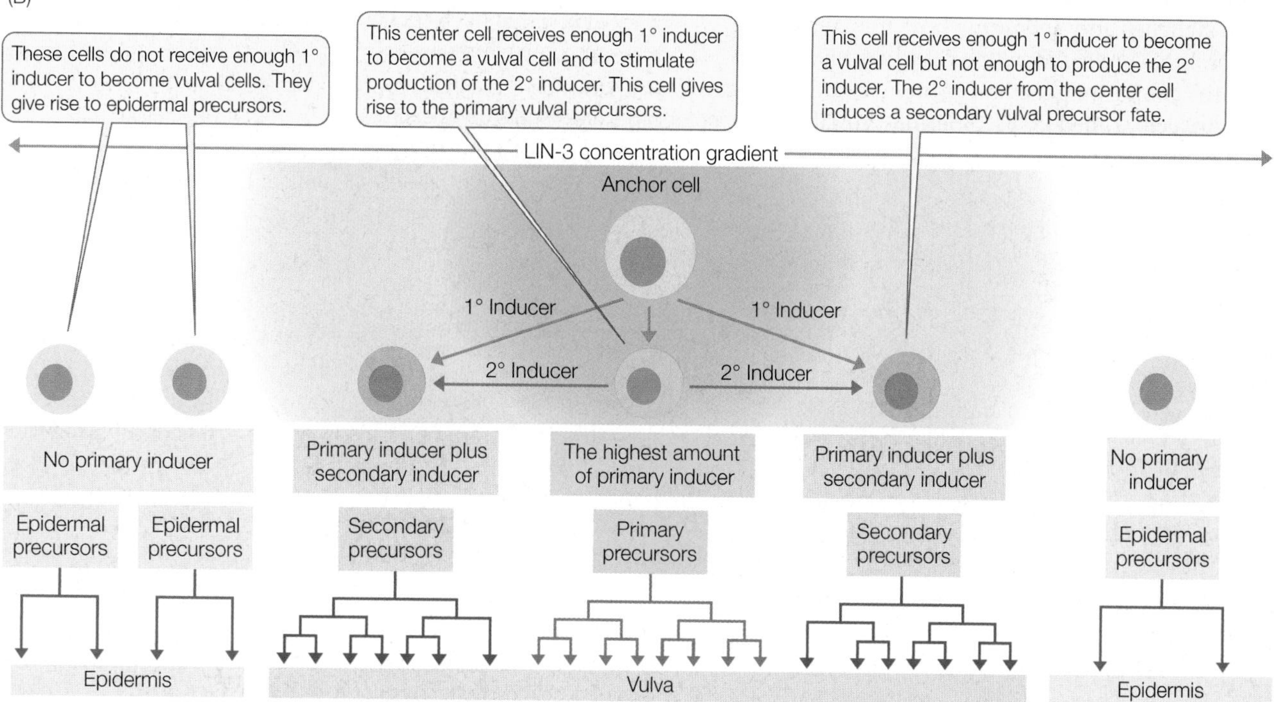
(A)
1 mm
Ovary
Intestine
Pharynx
Eggs
Vulva
Rectum
Anus
(B)
These cells do not receive enough 1° inducer to become vulval cells. They give rise to epidermal precursors.
This center cell receives enough 1° inducer to become a vulval cell and to stimulate production of the 2° inducer. This cell gives rise to the primary vulval precursors.
This cell receives enough 1° inducer to become a vulval cell but not enough to produce the 2° inducer. The 2° inducer from the center cell induces a secondary vulval precursor fate.
LIN-3 concentration gradient
Anchor cell
1° Inducer
1° Inducer
2° Inducer
2° Inducer
No primary inducer
Primary inducer plus secondary inducer
The highest amount of primary inducer
Primary inducer plus secondary inducer
No primary inducer
Epidermal precursors
Epidermal precursors
Secondary precursors
Primary precursors
Secondary precursors
Epidermal precursors
Epidermis
Vulva
Epidermis

FIGURE 14.8 Induction during Vulval Development in *Caenorhabditis elegans* (A) In the nematode *C. elegans* (shown in false color here), it has been possible to follow all of the cell divisions from the fertilized egg to the 959 cells found in the fully developed adult. (B) During vulval development, a molecule secreted by the anchor cell (the LIN-3 protein) acts as the primary (1°) inducer. The primary precursor cell (the one that received the highest concentration of LIN-3) then secretes a secondary (2°) inducer that acts on its neighbors. The gene expression patterns triggered by these molecular switches determine cell fates.

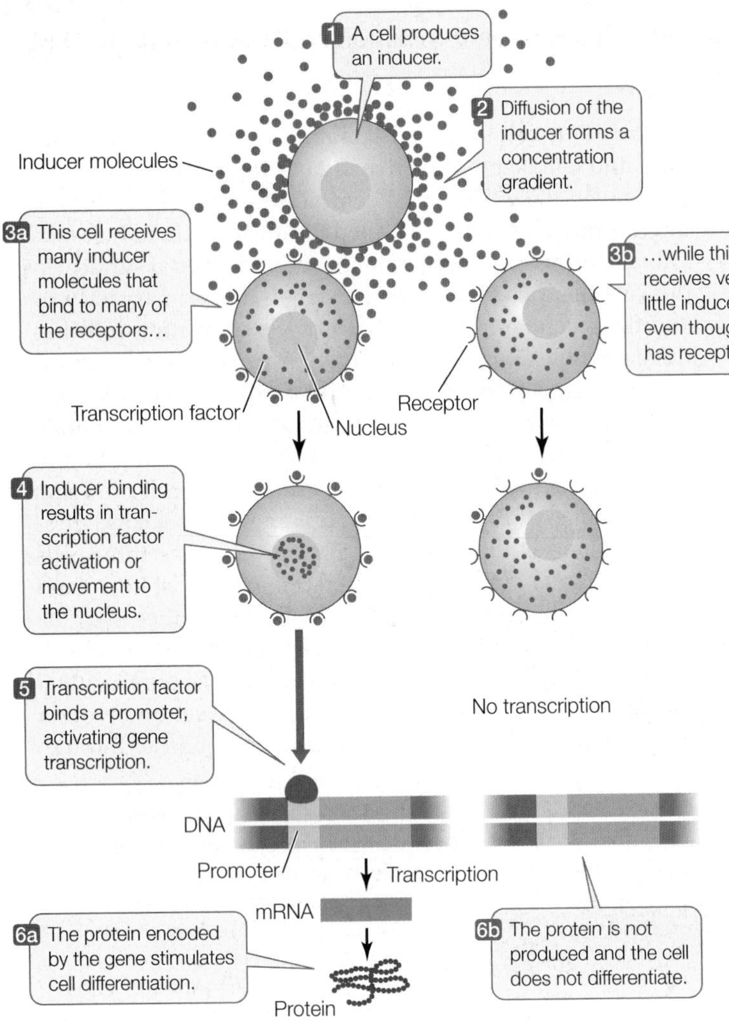

FIGURE 14.9 The Concept of Embryonic Induction The concentration of an inducer directly affects the degree to which a transcription factor is activated. The inducer acts by binding to a receptor on the target cell. This binding is followed by signal transduction involving transcription factor activation or movement from the cytoplasm to the nucleus. In the nucleus, the transcription factor acts to stimulate the expression of genes involved in cell differentiation.

signal transduction cascades in the responding cells (**FIGURE 14.9**).

This example from nematode development illustrates an important observation: *much of development is controlled by molecular switches that allow a cell to proceed down one of two alternative paths*. One challenge for developmental biologists is to find these switches and determine how they work. The primary inducer, LIN-3, released by the *C. elegans* anchor cell, is a growth factor similar in gene and protein sequence to a vertebrate growth factor called EGF (*e*pidermal *g*rowth *f*actor). LIN-3 binds to a receptor on the surfaces of vulval precursor cells, setting in motion a signal transduction cascade that results in increased transcription of the genes involved in the differentiation of vulval cells.

Differential gene transcription is a hallmark of cell differentiation

An important mechanism by which cells differentiate into specific cell types, with specific functions, is differential gene transcription. One well-studied example of cell differentiation is the conversion of undifferentiated muscle precursor cells (myoblasts) into the cells that make up muscle fibers (**FIGURE 14.10**). A key event in the commitment of these cells to become muscle is that they stop dividing. Indeed, in many parts of the embryo, *cell division and cell differentiation are mutually exclusive*. Cell signaling activates the gene for a transcription factor called **MyoD** (*myo*blast-*d*etermining gene). MyoD activates the gene for p21, an inhibitor of cyclin-dependent kinases (CDKs) that normally stimulate the cell cycle at G1 (see Figure 7.10). Expression of the *p21* gene causes the cell cycle to stop, and other transcription factors then enter the picture so that myoblasts can differentiate into muscle cells.

CHECKpoint CONCEPT 14.2

- ✓ What would be the effect on the embryo of injecting an inhibitor of microtubule polymerization into a fertilized sea urchin egg?
- ✓ Compare the internal and external stimuli that lead to differential gene expression in embryonic cells.
- ✓ What would be the consequences of a homozygous deletion mutation of *lin-3*, the gene that encodes LIN-3?

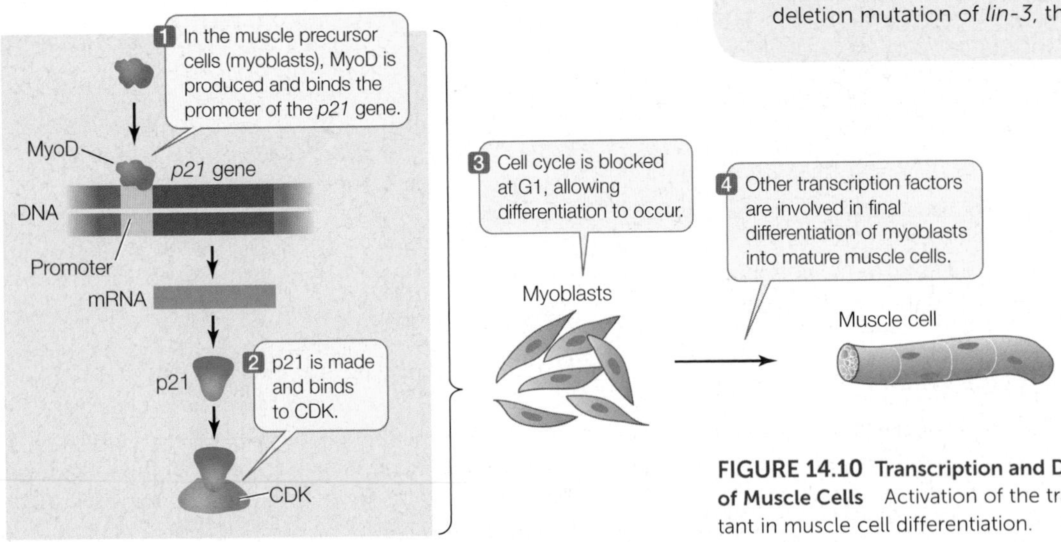

FIGURE 14.10 Transcription and Differentiation in the Formation of Muscle Cells Activation of the transcription factor MyoD is important in muscle cell differentiation.

Cytoplasmic polarity and inducers affect the expression of genes that determine cell fates. We will now look in more detail at how spatial differences in gene expression affect cell fate determination and the formation of tissues and organs.

CONCEPT 14.3 Spatial Differences in Gene Expression Lead to Morphogenesis

Pattern formation is the developmental process that results in the spatial organization of a tissue or organism. An example is the development of the vulva in *C. elegans* (see Figure 14.8). Pattern formation is inextricably linked to morphogenesis, the development of body form. Underlying both of these processes are spatial differences in gene expression, which determine whether, for example, a particular piece of tissue will become a leg or a wing or a flower petal. These instances where different genes are expressed in different places in the developing organism, in turn, depend on two cellular processes:

- The cells in the tissue must "know" where they are in relation to rest of the body.
- The cells must activate the pattern of gene expression that is appropriate for their location.

In the sections that follow, we will explore the mechanisms used by various organisms to direct pattern formation and morphogenesis.

Morphogen gradients provide positional information during development

During development, the key cellular question "What am I (or what will I be)?" is often answered in part by "Where am I?" Think of the cells in the developing nematode, which develop into different parts of the vulva depending on their positions relative to the anchor cell. The same is true for the cells between the digits of a developing hand and in different whorls of a developing flower. This spatial "sense" is called **positional information**.

Positional information often comes in the form of an inducer called a **morphogen**, which diffuses from one group of cells to surrounding cells, setting up a concentration gradient. There are two requirements for a signal to be considered a morphogen:

- It must specifically affect target cells.
- Different concentrations of the signal must cause different effects.

Developmental biologist Lewis Wolpert uses the "French flag model" to explain the action of morphogens (**FIGURE 14.11A**). This model can be applied to the differentiation of the vulva in *C. elegans* (see Figure 14.8), which relies on a gradient of LIN-3. Another example can be seen in the development of vertebrate limbs.

The vertebrate limb develops from a paddle-shaped limb bud (**FIGURE 14.11B**). The cells that develop into different digits must receive positional information; if they do not, the limb will not be organized properly (imagine a hand with only thumbs or only little fingers). How do the cells know where they are? A group of cells at the posterior base of the limb bud, just where it joins

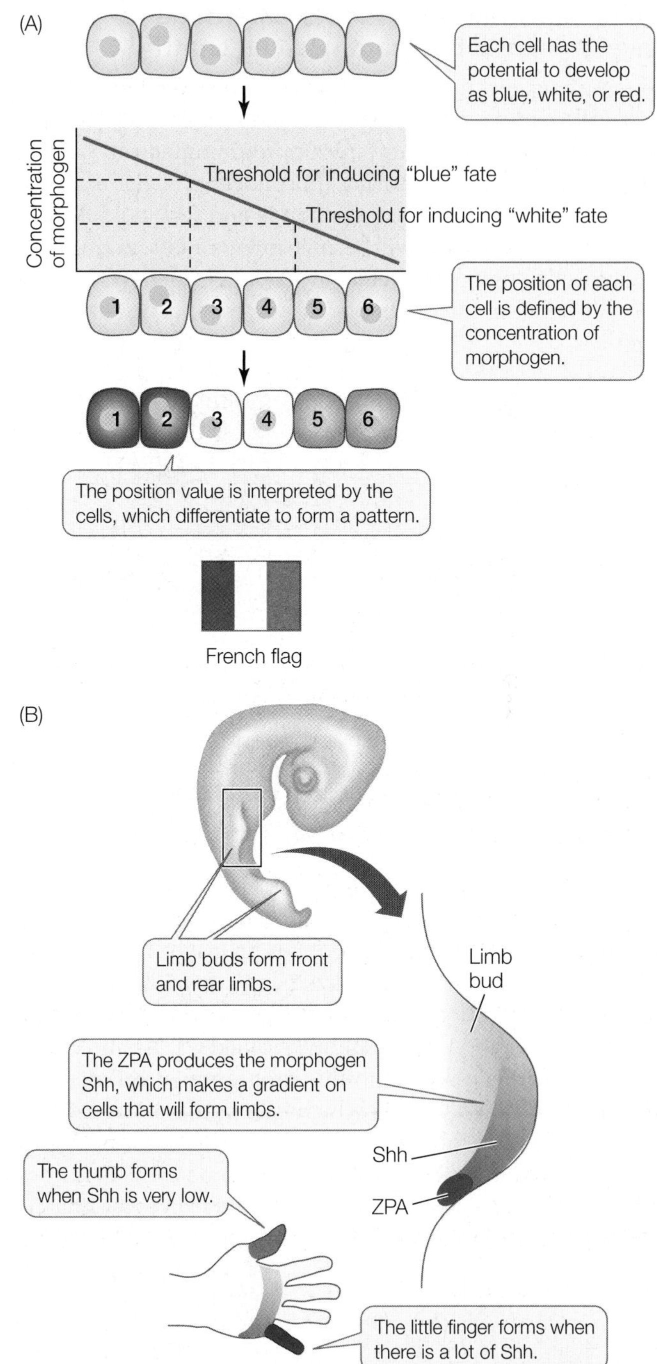

FIGURE 14.11 The French Flag Model (A) In the "French flag model," a concentration gradient of a diffusible morphogen signals each cell to specify its position. (B) The zone of polarizing activity (ZPA) in the limb bud of the embryo secretes the morphogen Sonic hedgehog (Shh). Cells in the bud form different digits depending on the concentration of Shh.

the body wall, is called the zone of polarizing activity (ZPA). The cells of the ZPA secrete a morphogen called Sonic *h*edge*h*og (Shh), which forms a gradient that determines the posterior–anterior (little finger to thumb) axis of the developing limb. The cells getting the highest dose of Shh form the little finger; those getting the lowest dose develop into the thumb. Recall the French flag model when considering the gradient of Shh.

Multiple proteins interact to determine developmental programmed cell death

You might expect morphogenesis to involve a lot of cell division, followed by differentiation—and it does. But what you might not expect is the amount of programmed cell death—apoptosis—that occurs during morphogenesis. For example, in an early human embryo, the hands and feet look like tiny paddles: the tissues that will become fingers and toes are linked by connective tissue. Between days 41 and 56 of development, the cells between the digits die, freeing the individual fingers and toes:

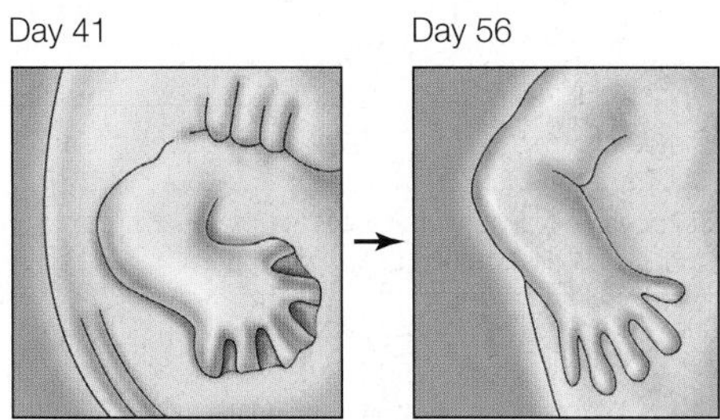

Many cells and structures form and then disappear during development, in processes involving apoptosis.

LINK

Concept 7.5 describes some of the cellular events of apoptosis

Model organisms have been very useful in studying the genes and proteins involved in apoptosis. For example, the nematode worm *C. elegans* produces precisely 1,090 somatic cells as it develops from a fertilized egg into an adult, but 131 of those cells die (leaving 959 cells in the adult worm). The sequential activation of two proteins called CED-4 and CED-3 (for *c*ell *d*eath) is essential to this programmed cell death. A third protein called CED-9, which is bound to the outside of the mitochondria, inhibits apoptosis in cells that are not programmed to die. In these cells, CED-9 binds CED-4 and prevents it from activating CED-3. If the cell receives a signal for apoptosis, CED-9 releases CED-4, which then activates CED-3:

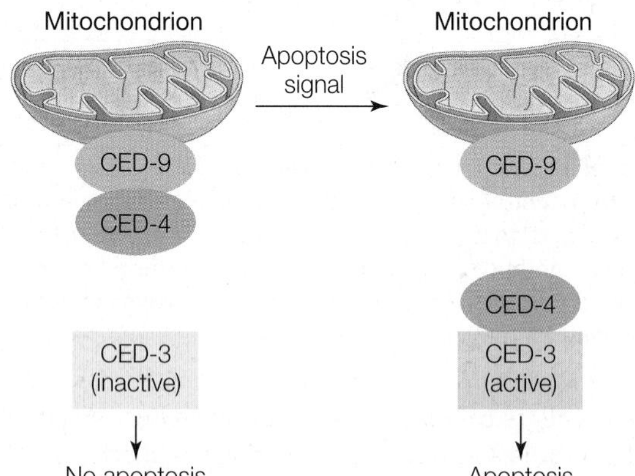

CED-3 is a caspase (a protease involved in apoptosis) that turns out to be similar to the caspase protein in humans. Several other proteins involved in the nematode apoptosis pathway (including CED-4 and CED-9) also have relatives in humans. So humans and *C. elegans*, two species separated by more than 600 million years of evolutionary history, have similar genes (encoding similar proteins) that control programmed cell death. The commonality of this pathway indicates its importance: most mutations in the genes that control this pathway are harmful and evolution selects against them. We will return to other examples of links between evolution and development in Concept 14.4.

Our example of apoptosis in the development of fingers and toes shows one of the many ways that the behavior of cells can give rise to body form during development. It also illustrates the two cellular processes underlying pattern formation: only cells in a particular place (between the digits) activate a specific pattern of gene expression (to trigger apoptosis).

Expression of transcription factor genes determines organ placement in plants

Like animals, plants have organs—for example, leaves and roots. Many plants form flowers, and many flowers are composed of four types of organs: sepals, petals, stamens (male reproductive organs), and carpels (female reproductive organs). These floral organs occur in concentric whorls (rings), with groups of each organ type encircling a central axis. The sepals are on the outside and the carpels are on the inside (**FIGURE 14.12A**).

In the model plant *Arabidopsis thaliana* (thale cress), flowers develop in a radial pattern around the top of the stem as it develops and elongates. At the shoot apex and in other parts of the plant where growth and differentiation occur (such as the root tip), there are groups of undifferentiated, rapidly dividing cells called **meristems** (see Concept 24.1). Each flower begins as a floral meristem of about 700 undifferentiated cells arranged in a dome, and the four whorls develop from this meristem. How is the identity of a particular whorl determined? Three classes of genes called **organ identity genes** encode proteins that act in combination to produce specific whorl features (**FIGURE 14.12B,C**):

- Genes in class A are expressed in whorls 1 and 2 (which form sepals and petals, respectively).
- Genes in class B are expressed in whorls 2 and 3 (which form petals and stamens).
- Genes in class C are expressed in whorls 3 and 4 (which form stamens and carpels).

APPLY THE CONCEPT

Gene expression and morphogenesis

Molecular biologists can attach genes to active promoters and insert them into cells. This results in higher than normal expression (overexpression) of the genes. What do you think would happen in each case if the four genes listed were to be overexpressed in the specified tissues? Explain your answers.

1. *ced-3* in embryonic neuron precursors of *C. elegans*
2. *myoD* in undifferentiated myoblasts
3. *Sonic hedgehog* in a chick limb bud
4. *LEAFY* in a leaf bud meristem of *Arabidopsis*

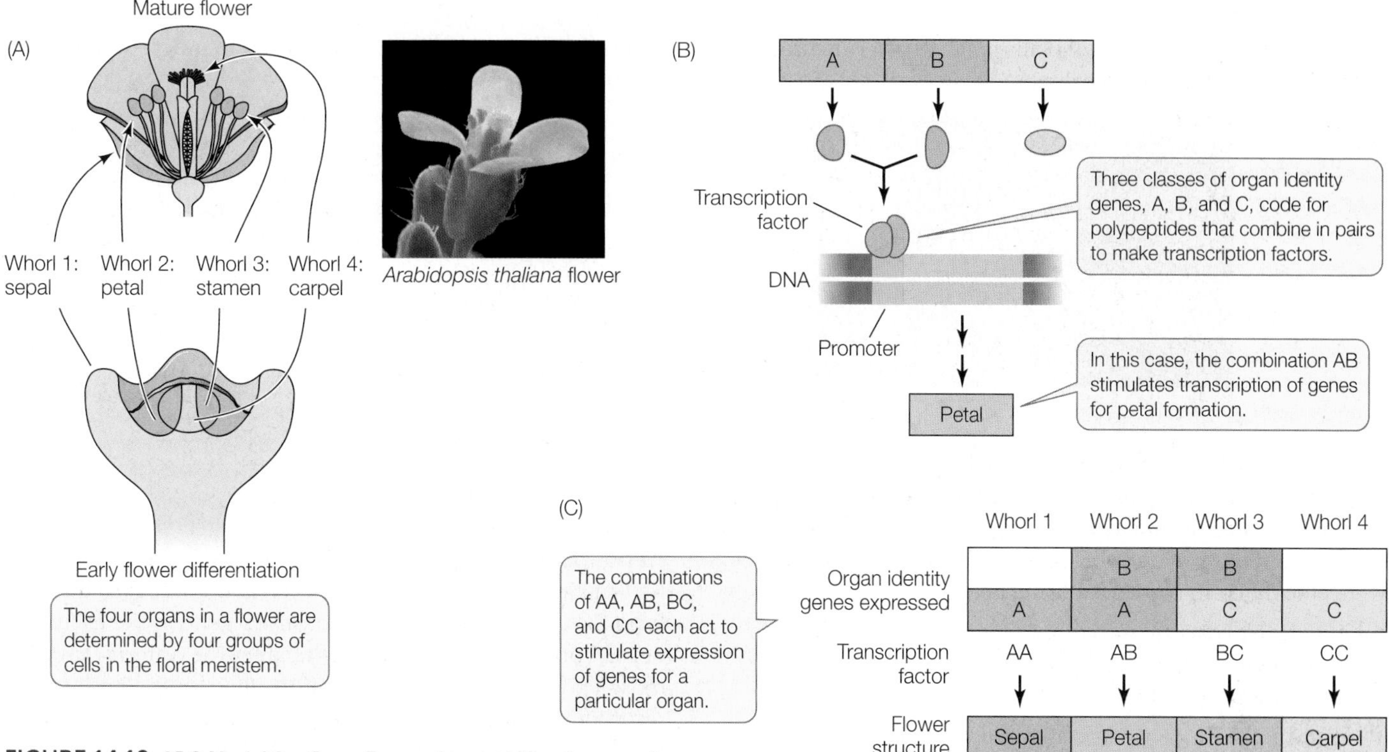

FIGURE 14.12 ABC Model for Gene Expression and Morphogenesis in *Arabidopsis thaliana* Flowers (A) The four organs of a flower—sepals (pink), petals (purple), stamens (green), and carpels (yellow)—grow in whorls that develop from the floral meristem. (B) Floral organs are determined by three classes of genes whose polypeptide products combine in pairs to form transcription factors. (C) Combinations of polypeptide subunits in transcription factors activate gene expression for specific organs.

These genes encode transcription factors that are active as dimers, that is, proteins with two polypeptide subunits. The composition of the dimer determines which genes the transcription factor activates. For example, a dimer made up of two class A monomers activates transcription of the genes that make sepals; a dimer made up of a class A monomer and a class B monomer results in petals, and so forth.

Two lines of experimental evidence support this model for floral organ determination:

- *Loss-of-function mutations*: for example, a mutation in a class A gene results in no sepals or petals.
- *Gain-of-function mutations*: for example, a promoter for a class C gene can be artificially coupled to a class A gene. In this case, the class A gene is expressed in all four whorls, resulting in only sepals and petals. In any organism, the replacement of one organ by another is called homeosis, and this type of mutation is a **homeotic mutation**.

Transcription of the floral organ identity genes is controlled by other gene products, including the LEAFY protein. Plants with loss-of-function mutations in the *LEAFY* gene make stems instead of flowers, with increased numbers of modified leaves called bracts. The wild-type LEAFY protein is a transcription factor that stimulates expression of the class A, B, and C genes so that they produce flowers. This finding has practical applications. It usually takes 6–20 years for a citrus tree to produce flowers and fruits. Scientists have made transgenic orange trees expressing the *LEAFY* gene coupled to a strongly expressed promoter. These trees flower and fruit years earlier than normal trees.

A cascade of transcription factors establishes body segmentation in the fruit fly

A major achievement in studies of developmental biology has been the ever-advancing description of how morphogens act in another model organism, the fruit fly *Drosophila melanogaster*. As you will see in Concept 14.4, the molecular events that underlie fruit fly development turn out to be similar to events that occur in many other organisms, including ourselves. So they merit examination in some detail.

The insect body is made up of segments that differ from one another. The adult fly has an anterior head (composed of several fused segments), three different thoracic segments, and eight abdominal segments at the posterior end. Each segment gives rise to different body parts: for example, antennae and eyes develop from head segments, wings from the thorax, and so on.

The life cycle of *Drosophila* from fertilized egg to adult takes about 2 weeks. The egg hatches into a larva, which then forms a pupa, which finally is transformed into the adult fly. By the time a larva appears—about 24 hours after fertilization—there are recognizable segments. The thoracic and abdominal segments all look similar, but *the fates of the cells to become different adult segments are already determined*.

As with other organisms, fertilization in *Drosophila* leads to a rapid series of mitoses. However, the first 12 nuclear divisions are not accompanied by cytokinesis. So a *multinucleate* embryo forms instead of a *multicellular* embryo. The nuclei are brightly stained in the micrographs below:

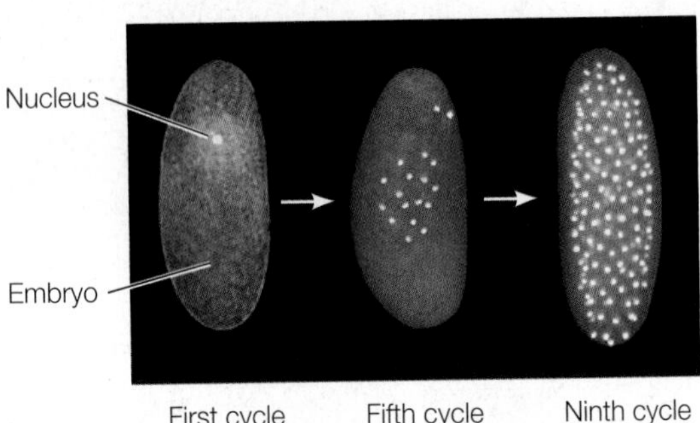

With no cell membranes to cross, morphogens can diffuse easily within the embryo. As you will see, many of these morphogens affect transcription in the cell nuclei. We focus here on cell fate determination events that occur in the first 24 hours, which were elucidated in *Drosophila* using genetics:

- First, developmental mutations were identified. For example, a mutant strain might produce larvae with two heads or missing certain segments.
- Second, each mutant was compared with wild-type flies, and the gene responsible for the developmental mistake, and its protein product (if appropriate), was isolated.
- Finally, experiments with the gene (making transgenic flies) and protein (injecting the protein into an egg or embryo) were done to confirm the proposed developmental pathway.

Together, these approaches revealed a sequential pattern (cascade) of gene expression that results in the determination of each segment within 24 hours after fertilization. Several classes of genes are involved:

- Maternal effect genes, which set up the major axes (anterior–posterior and dorsal–ventral) of the egg
- Segmentation genes, which determine the boundaries and polarity of each of the segments
- Hox genes, which determine what organ will be made at a given location

MATERNAL EFFECT GENES Like the eggs and early embryos of many other organisms (see Figure 14.7), *Drosophila* eggs and larvae are characterized by unevenly distributed cytoplasmic determinants. These molecular determinants, which include both mRNAs and proteins, are the products of specific **maternal effect genes**. These genes are transcribed in the cells of the mother's ovary, and the mRNAs are passed to the egg via cytoplasmic bridges. Two maternal effect genes, called *bicoid* and *nanos*, help determine the anterior–posterior axis of the egg. (The dorsal–ventral, or back–belly, axis is determined by other maternal effect genes that we will not describe here.)

The mRNAs for *bicoid* and *nanos* diffuse from the mother's cells into what will be the anterior (head) end of the egg. After fertilization, the *bicoid* mRNA is translated to produce Bicoid protein, a transcription factor that diffuses away from the anterior end, establishing a concentration gradient in the egg cytoplasm:

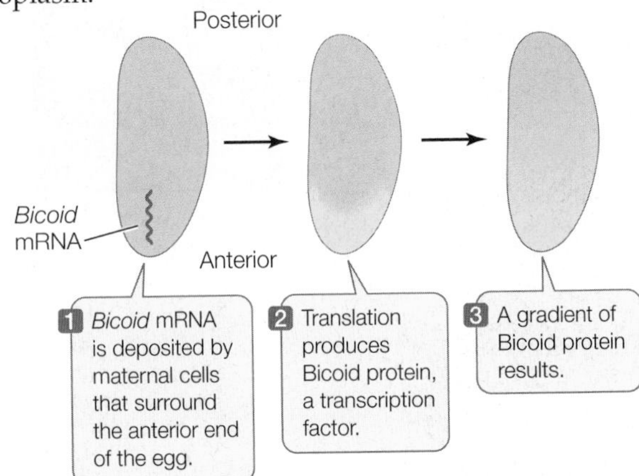

At this point, the egg is in its multinucleate stage.

Where it is present in sufficient concentration, Bicoid stimulates the transcription of the *hunchback* gene in the early embryo. Consequently, the nuclei nearest the anterior end are most active in the transcription of the *hunchback* gene, and the resulting gradient of Hunchback protein (itself a transcription factor) establishes the head, or anterior, region.

Meanwhile, the egg's cytoskeleton transports the *nanos* mRNA from the anterior end of the egg to the posterior (tail) end, where it is translated after fertilization. This results in a gradient of the Nanos protein, with the highest concentration at the posterior end. At that end, the Nanos protein inhibits the translation of *hunchback* mRNA, preventing accumulation of Hunchback protein. Thus the actions of both Bicoid and Nanos establish a Hunchback protein gradient, which determines the anterior and posterior ends of the embryo by influencing the gene expression patterns of the nuclei along the gradient.

The events involving *bicoid*, *nanos*, and *hunchback* begin before fertilization and continue after it, during the multinucleate stage, which lasts a few hours. At this stage the embryo looks like a bunch of indistinguishable nuclei under the light microscope. But the fates of the individual nuclei and the cells they will occupy have already begun to be determined. After the anterior and posterior ends have been established, the next step in pattern formation in fruit flies is the determination of segment number and locations.

SEGMENTATION GENES The number, boundaries, and polarity of the *Drosophila* larval segments are determined by proteins encoded by the **segmentation genes**. These genes are expressed when there are about 6,000 nuclei in the embryo (about 3 hours after fertilization). Three classes of segmentation genes act one after the other to regulate finer and finer details of the segmentation pattern (**FIGURE 14.13**):

- **Gap genes** organize broad areas along the anterior–posterior axis. Mutations in gap genes result in gaps in the body plan—the omission of several consecutive larval segments.
- **Pair rule genes** divide the embryo into units of two segments each. Mutations in pair rule genes result in embryos missing every other segment.

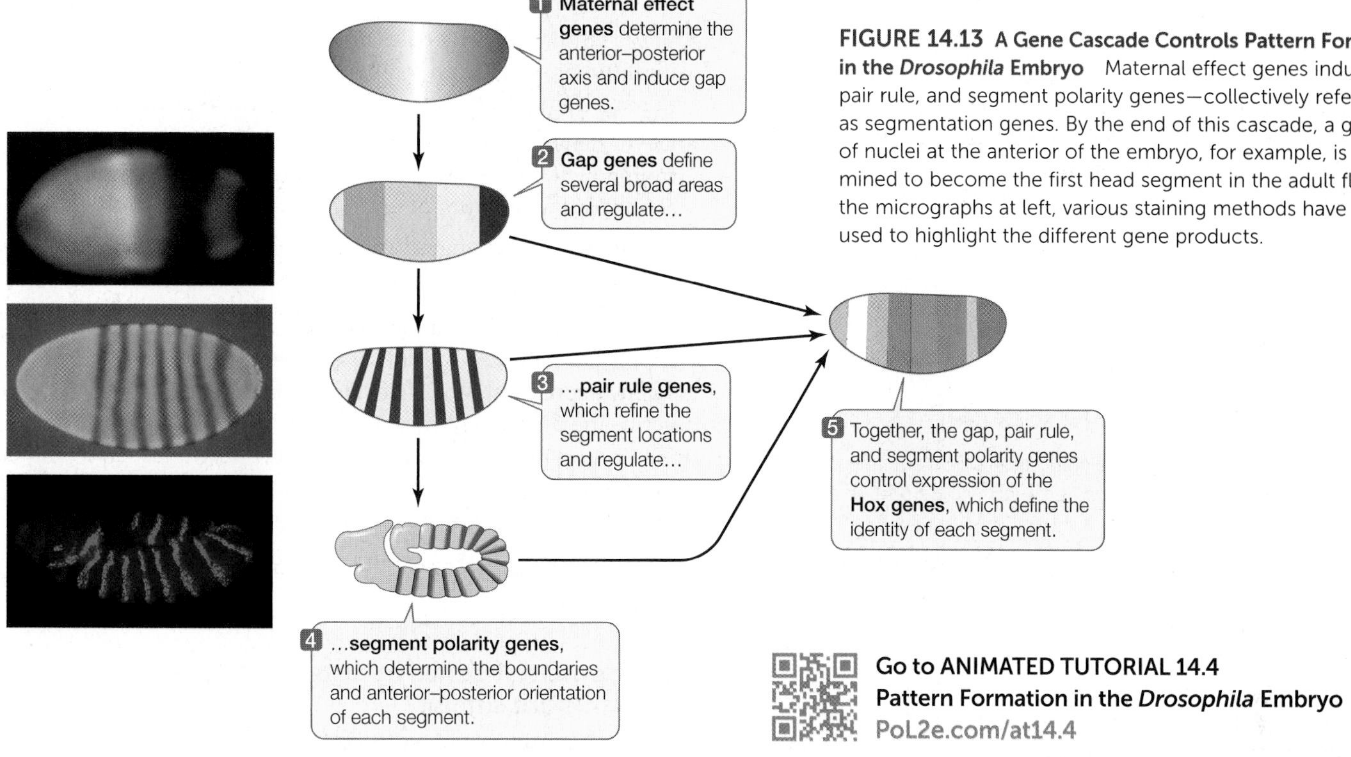
1 Maternal effect genes determine the anterior–posterior axis and induce gap genes.
2 Gap genes define several broad areas and regulate…
3 …pair rule genes, which refine the segment locations and regulate…
4 …segment polarity genes, which determine the boundaries and anterior–posterior orientation of each segment.
5 Together, the gap, pair rule, and segment polarity genes control expression of the Hox genes, which define the identity of each segment.

FIGURE 14.13 A Gene Cascade Controls Pattern Formation in the *Drosophila* Embryo Maternal effect genes induce gap, pair rule, and segment polarity genes—collectively referred to as segmentation genes. By the end of this cascade, a group of nuclei at the anterior of the embryo, for example, is determined to become the first head segment in the adult fly. In the micrographs at left, various staining methods have been used to highlight the different gene products.

Go to ANIMATED TUTORIAL 14.4
Pattern Formation in the *Drosophila* Embryo
PoL2e.com/at14.4

- **Segment polarity genes** determine the boundaries and anterior–posterior organization of the individual segments. Mutations in segment polarity genes can result in segments in which posterior structures are replaced by reversed (mirror-image) anterior structures.

By the end of this part of the cascade, nuclei throughout the embryo "know" which segment they will be part of in the adult fly. The next set of genes in the cascade determines the form and function of each segment.

Go to MEDIA CLIP 14.1
Spectacular Fly Development in 3D
PoL2e.com/mc14.1

HOX GENES **Hox genes** encode a family of transcription factors that are expressed in different combinations along the length of the embryo, and help determine cell fates within each segment. Expression of certain Hox genes leads to the development of antennae in the head segment, whereas other Hox genes are expressed in the thorax to make wings, and so on. Hox genes are homeotic genes that are shared by most animals, and they are functionally similar to the organ identity genes of plants (see Figure 14.12).

How do we know that the Hox genes determine segment identity? A clue comes from homeotic mutations in *Drosophila*. A gain-of-function mutation in the Hox gene *Antennapedia* causes legs to grow on the head in place of antennae (**FIGURE 14.14**). When another Hox gene, *bithorax*, is mutated, an extra pair of wings grows in a thoracic segment where wings do not normally occur. So the normal (wild-type) functions of the Hox genes must be to "tell" a segment what organ to form. Hox genes encode transcription factors and have a conserved 180-base-pair sequence called the **homeobox** (from which the

(A)

Antenna

(B)

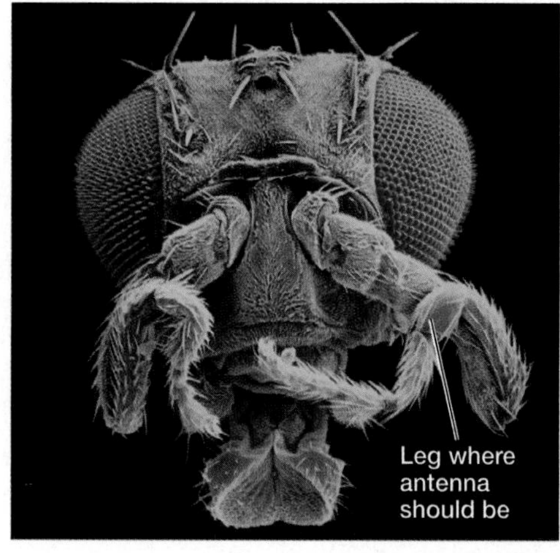
Leg where antenna should be

FIGURE 14.14 A Homeotic Mutation in *Drosophila* Mutations of the Hox genes cause body parts to form on inappropriate segments. (A) A wild-type fruit fly. (B) An *Antennapedia* mutant fruit fly. Mutations such as this reveal the normal role of the *Antennapedia* gene in determining segment function.

genes get their name). The homeobox encodes a 60-amino acid sequence called the homeodomain. The homeodomain recognizes and binds to a specific DNA sequence in the promoters of its target genes. As you will see in Concept 14.4, this domain is found in transcription factors that regulate development in many other animals with an anterior–posterior axis.

LINK

To review transcriptional regulation, see **Concept 11.2**

CHECKpoint CONCEPT 14.3

- ✓ Outline the steps that determine that a nucleus and cell in the developing *Drosophila* embryo will be part of an antenna.
- ✓ Compare the determination of organ identity in *Arabidopsis* and *Drosophila*.
- ✓ How does the "French flag model" apply to development in *Drosophila*?
- ✓ In the nematode nervous system, 302 neurons come from 405 precursors. How would you investigate the fate of the 103 "missing" cells? What gene(s) might be involved?

We have seen how positional information leads to changes in the expression of key developmental genes, which in turn control morphogenesis. It turns out that there are remarkable similarities in the genes used to guide development in diverse organisms, and this has led to a new way to look at the evolution of development.

CONCEPT 14.4 Changes in Gene Expression Pathways Underlie the Evolution of Development

The discovery of the genes that control the development of *Drosophila* provided biologists with tools to investigate the development of other organisms. For example, when scientists used homeobox DNA as a hybridization probe (see Figure 10.7) to search for similar genes elsewhere, they found that the homeobox is present in many genes in many other organisms. This, and other astounding discoveries that followed, showed a similarity in the molecular events underlying morphogenesis in organisms ranging from flies to fish to mammals. These results suggested that just as the forms of organisms evolved through descent with modification from a common ancestor, so did the molecular mechanisms that produce those forms. Biologists started to ask new questions about the interplay between evolutionary and developmental processes—a field of study called **evolutionary developmental biology** (**evo-devo**). The major findings of evo-devo are:

- Organisms share similar molecular mechanisms for development, including a "toolkit" of regulatory molecules that control the expression of genes.
- These regulatory molecules are able to act independently in different tissues and regions of the body, so that evolutionary change can occur in independent "modules."
- Developmental differences can arise from changes in the timing of action of a regulatory molecule, the location of its action, or the quantity of its action.
- Developmental changes can arise from environmental influences on developmental processes.

The development of a multicellular organism from a fertilized egg—a single cell—involves an intricate pattern of sequential gene expression. When developmental biologists began to describe the events responsible for the differentiation and controlled proliferation of cells and tissues at the molecular level, they found common regulatory genes and pathways in organisms that don't appear similar at all, such as fruit flies and mice.

Go to ACTIVITY 14.2 Concept Matching: Development PoL2e.com/ac14.2

Developmental genes in distantly related organisms are similar

Initially through hybridization with a homeobox probe, and then by genome sequencing and comparative genomics (see Concept 12.2), biologists have found that diverse animals share numerous molecular pathways that govern gene expression during development. For example, fruit fly homeotic genes such as *Antennapedia* and *bithorax* are similar to mouse (and human) genes that play similar developmental roles. This means that the positional information controlled by these genes has been conserved, even as the structures formed at each position

APPLY THE CONCEPT

Changes in gene expression pathways underlie the evolution of development

The control of eye formation during the development of many animals is under the control of a genetic switch involving a transcription factor. Partial DNA sequences for the control gene from two organisms are given below.

Mouse *Pax6* gene:
5′-GTA TCC AAC GGT TGT GTG AGT AAA ATT-3′

Fruit fly *eyeless* gene:
5′-GTA TCA AAT GGA TGT GTG AGC AAA ATT-3′

1. Calculate the percentage of identity (the percentage of bases that are identical) between the two DNA sequences.
2. Use the genetic code (see Figure 10.11) to determine the amino acid sequences (see Table 3.2) encoded by the two regions and calculate the percentage of identity between the two amino acid sequences.
3. The fruit fly and the mouse diverged from a common ancestor about 500 million years ago. Comment on your answers to 1 and 2 in terms of the evolution of developmental pathways.

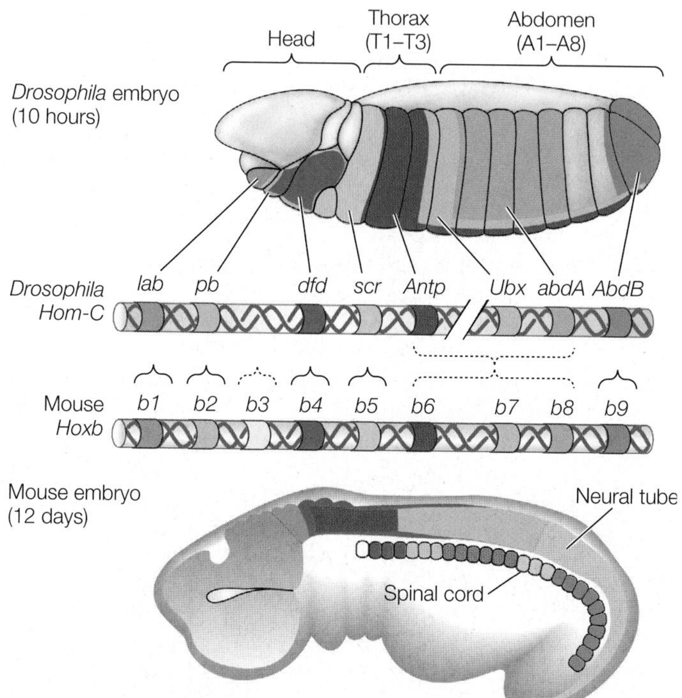

FIGURE 14.15 Regulatory Genes Show Similar Expression Patterns Similar genes encoding similar transcription factors are expressed in similar patterns along the anterior–posterior axis of both insects and vertebrates. Related genes and the locations of their expression are indicated by shared colors. The mouse (and human) Hox genes are actually present in multiple copies; this prevents a single mutation from having drastic effects.

have changed. Over the millions of years that have elapsed since these animals diverged from a common ancestor, the genes in question have mostly been maintained, suggesting that their functions are essential for animal development.

Remarkably, these genes are arranged along a chromosome in both the fruit fly and mouse *in the same order as they are expressed along the anterior–posterior axis of their embryos* (**FIGURE 14.15**). In the mouse and other vertebrates, these genes are switched on sequentially during embryogenesis. The Hox genes controlling the development of anterior structures (such as the head) are expressed earliest, and as a result, anterior structures develop earlier than posterior structures. The spatial organization of the Hox genes on the chromosome is important for the timing of expression of the genes.

These and other examples have led biologists to the idea that certain developmental mechanisms, controlled by specific DNA sequences, have been conserved over long periods during the evolution of multicellular organisms. These sequences comprise a **genetic toolkit**, the contents of which have been modified and reshuffled over the course of evolution to produce the remarkable diversity of plants, animals, and other organisms in the world today.

Genetic switches govern how the genetic toolkit is used

The genetic toolkit is also used to generate diverse structures within a single organism. Different structures can evolve within a single organism using a common set of genetic instructions because there are mechanisms called **genetic switches** (also called molecular switches) that control how the genetic toolkit is used. As we have seen, these mechanisms involve promoters and transcription factors. The signal cascades that converge on and operate these switches determine when and where genes will be turned on and off. Multiple switches control each gene by influencing its expression at different times and in different places. In this way, elements of the genetic toolkit can be involved in multiple developmental processes while still allowing individual modules to develop independently. For example, the morphogenesis of different flower organs is determined by different combinations of three classes of transcription factors (the ABC model; see Figure 14.12) acting at specific times and locations.

During evolution, changes in the functions of genetic switches have led to changes in the forms or functions of organisms. To illustrate this, let's look at the development of wings in *Drosophila* and other insects. *Drosophila* species are members of the insect group Diptera, which means "two wings"—that is, they have a single pair of wings, whereas most insects have two pairs of wings (i.e., four wings). In dipterans, the single pair of wings develops on the second thoracic segment, and a pair of balancing organs called halteres develops on the third thoracic segment. A critical difference between thoracic segments 2 and 3 is that the Hox gene *Ultrabithorax* (*Ubx*) is expressed in segment 3 but not in segment 2. The Ubx transcription factor represses the development of wings in dipterans (**FIGURE 14.16**). If the *Drosophila Ubx* gene is inactivated by mutation,

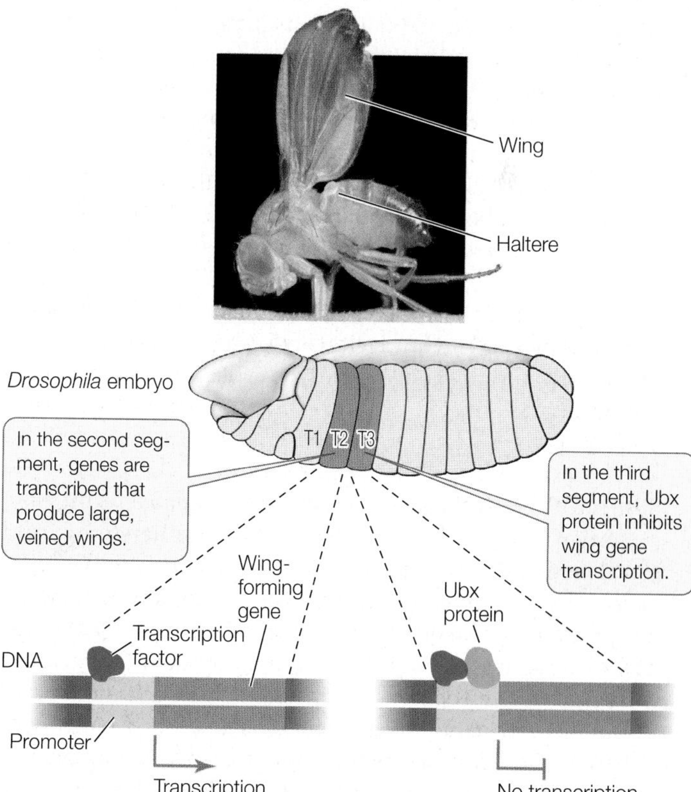

FIGURE 14.16 Segments Differentiate under Control of Genetic Switches The binding of a single protein, Ultrabithorax (Ubx), determines whether a thoracic segment in *Drosophila* produces full wings or halteres (balancers).

a second pair of wings forms in thoracic segment 3. In other insects such as butterflies, Ubx turns *on* the expression of wing-forming genes so that full hindwings develop. Therefore a simple genetic change in the effect of Ubx on genes that promote wing development results in a major morphological difference in the wings of flies and butterflies. This phenomenon—the same switch having different effects on target genes in different species—is important in evolution.

Modularity allows for differences in the pattern of gene expression among organisms

The modularity of development means that the molecular pathways for developmental processes such as organ formation operate independently from one another. For example, an *Antennapedia* mutant grows a leg where an antenna should be (see Figure 14.14), but all of the mutant's other organs develop normally and in their proper places. On an evolutionary time scale, modularity means that the timing and position of a particular developmental process can change without disrupting the whole organism.

Go to ANIMATED TUTORIAL 14.5
Modularity
PoL2e.com/at14.5

TIMING DIFFERENCES The genes regulating the development of a module may be expressed at different developmental stages or for different durations in different species, a phenomenon called **heterochrony**. An example is the evolution of the giraffe's neck. As in virtually all mammals (with the exception of manatees and sloths), there are seven vertebrae in the neck of the giraffe. So the giraffe did not get a longer neck than other mammals by adding vertebrae. Instead, each of the cervical (neck) vertebrae of the giraffe is much longer than those of other mammals (**FIGURE 14.17**). How does this happen?

Bones grow because of the proliferation of cartilage-producing cells called chondrocytes. Bone growth is stopped by a signal that results in death of the chondrocytes and the accumulation of calcium salts in the bone. In giraffes this signaling process is delayed in the cervical vertebrae, with the result that these vertebrae grow longer. Thus the evolution of longer necks occurred through *changes in the timing of expression* of the genes that control bone formation.

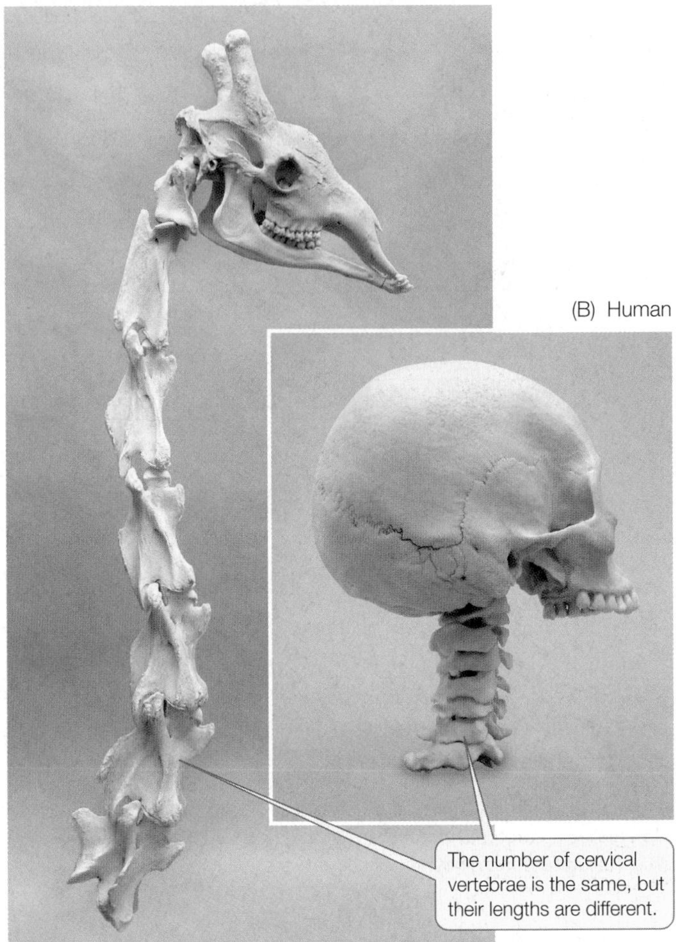

(A) Giraffe
(B) Human
The number of cervical vertebrae is the same, but their lengths are different.

FIGURE 14.17 Heterochrony in the Development of a Longer Neck There are seven vertebrae in the neck of the giraffe (left) and human (right; not to scale). But the vertebrae of the giraffe are much longer (25 cm compared with 1.5 cm) because during development, growth continues for a longer period of time. This timing difference is called heterochrony.

SPATIAL DIFFERENCES Changes in the spatial expression pattern of a developmental gene are known as **heterotopy** and can also result in evolutionary change. For example, the difference in foot webbing in ducks versus chickens is determined by an alteration in the spatial expression of a single gene. The feet of all bird embryos have webs of skin that connect their toes. This webbing is retained in adult ducks (and other aquatic birds) but not in adult chickens (and other non-aquatic birds). The loss of webbing is caused by a signaling protein called bone morphogenetic protein 4 (BMP4) that instructs the cells in the webbing to undergo apoptosis. The death of these cells destroys the webbing between the toes.

Embryonic duck and chicken hindlimbs both express the *BMP4* gene in the webbing between the toes, but they differ in expression of a gene called *Gremlin*, which encodes a BMP *inhibitor* protein (**FIGURE 14.18**). In ducks, but not chickens, the *Gremlin* gene is expressed in the webbing cells. The Gremlin protein inhibits the BMP4 protein from signaling for apoptosis, and the result is a webbed foot.

CHECKpoint CONCEPT 14.4

- ✓ Describe the major ideas of evolutionary developmental biology.
- ✓ What is the evidence that there was a common origin for the developmental pathways leading to segment identity in insects and the organization of the spinal cord in mice?
- ✓ What is the evidence that changes in the transcription of a single gene can lead to differences in morphogenesis between different regions of developing organ?
- ✓ Examine Figure 14.18 and the related text. If Gremlin protein were added to the webbed region between the developing toes of a chicken, what would be the result?

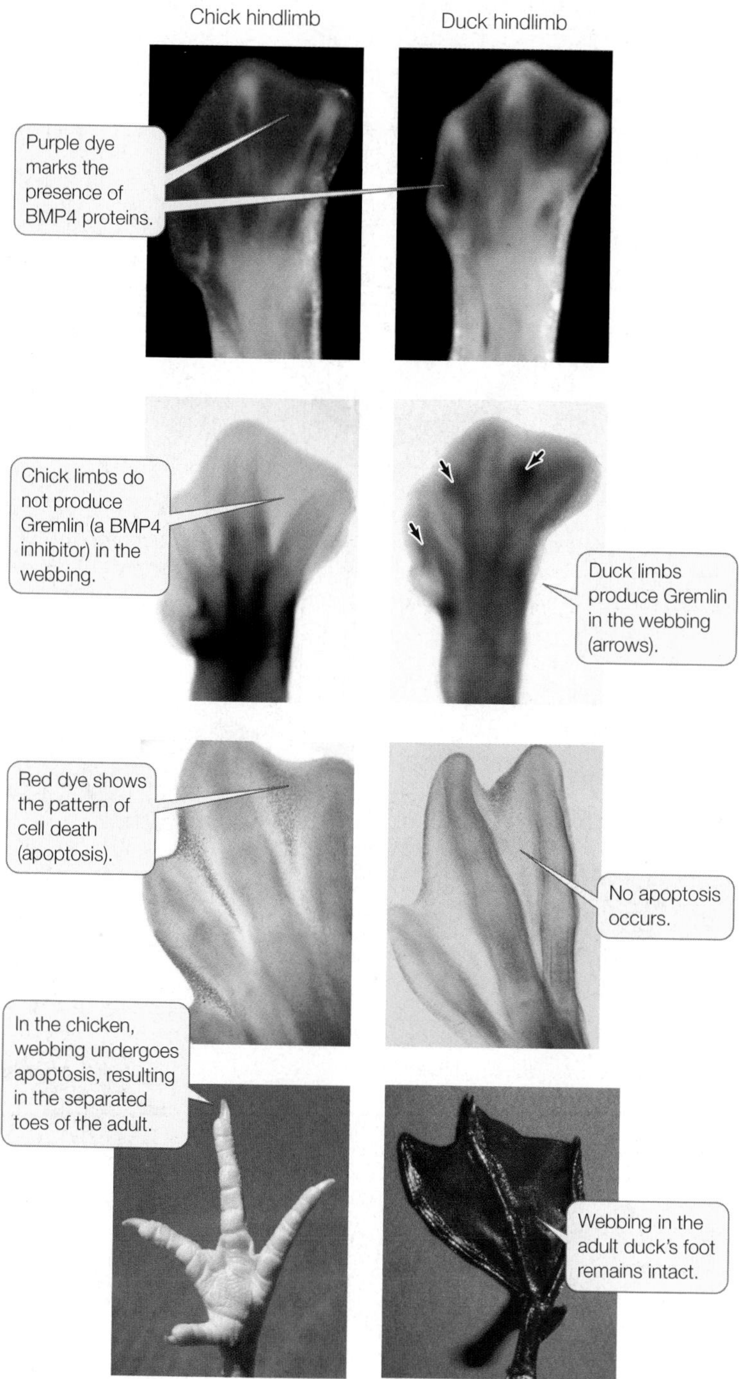

FIGURE 14.18 Changes in Gremlin Expression Correlate with Changes in Hindlimb Structure The left column of photos shows the development of a chicken's foot; the right column shows foot development in a duck. Gremlin protein in the webbing of the duck foot inhibits BMP4 signaling, thus preventing the embryonic webbing from undergoing apoptosis.

We have seen how the genetic toolkit guides morphogenesis in individual organisms, and how differences in genetic switches contribute to differences among species. In the next concept we will discuss further the roles that some of these same tools play in the evolution of new forms and new species.

CONCEPT 14.5 Developmental Genes Contribute to Species Evolution but Also Pose Constraints

The genetic switches that allow different structures to develop in different regions of an embryo can also give rise to major morphological differences among species. We have already seen examples of this: the difference in wing number between dipterans and other insects; the development of cervical vertebrae in giraffes versus other mammals; and the differences in Gremlin expression that determine whether a bird's foot will be webbed or not. Thus changes in the timing and position of a genetic switch can generate morphological variation, which then can be acted on by natural selection.

At the same time, the reliance of development on a genetic toolkit with a limited set of tools places constraints on how radically organisms can differ from one another. Four decades ago, the French geneticist François Jacob made the analogy that evolution works like a tinker, assembling new structures by *combining and modifying the available materials*, and not like an engineer, who is free to develop dramatically different designs (say, a jet engine to replace a propeller-driven engine). The evolution of morphology has not been governed by the appearance of radically new genes, but by modifications of existing genes and their regulatory pathways. Thus developmental genes and their expression constrain evolution in two major ways:

- Nearly all evolutionary innovations are modifications of previously existing structures.
- The genes that control development are highly conserved; that is, the regulatory genes usually change slowly over the course of evolution.

Mutations in developmental genes can cause major morphological changes

Sometimes a major developmental change is due to an alteration in the regulatory molecule itself rather than a change in where, when, or how much it is expressed. This is called **heterotypy** ("different type"). An excellent example of heterotypy is a gene that controls the number of legs in arthropods. Arthropods all have head, thoracic, and abdominal regions with variable numbers of segments. Insects such as *Drosophila* have three pairs of legs, one pair on each of their three thoracic segments, whereas centipedes have many legs on both thoracic and abdominal segments. All arthropods express a gene called *Distalless* (*Dll*) that controls segmental leg development. In insects, *Dll* expression is repressed in abdominal segments by the Hox gene *Ubx*. *Ubx* is expressed in the abdominal segments of all arthropods, but it has different effects in different species. In centipedes, Ubx protein *activates* expression of the *Dll* gene to promote the formation of legs. During the evolution of insects, a change in the *Ubx* gene

In the insect lineage, a mutation in the *Ubx* gene prevents legs from forming in the abdominal segments.

Insects

Fly

Mosquito

Butterfly

Moth

Beetle

Centipede

Shrimp

Spider

Onychophoran

Change in Ubx protein structure

Common ancestor

Insect

Most arthropod groups have legs growing from their abdominal segments.

Centipede

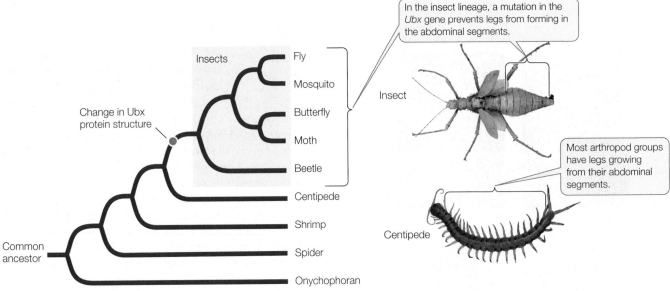

FIGURE 14.19 A Mutation in a Hox Gene Changed the Number of Legs in Insects In the insect lineage (blue box) of the arthropods, a change to the *Ubx* gene resulted in a protein that inhibits the *Dll* gene, which is required for legs to form. Because insects express this modified *Ubx* gene in their abdominal segments, no legs grow from these segments. Other arthropods, such as centipedes, do grow legs from their abdominal segments.

sequence resulted in a modified Ubx protein that *represses Dll* expression in abdominal segments. A phylogenetic tree of arthropods shows that this change in *Ubx* occurred in the ancestor of insects, at the same time that abdominal legs were lost (**FIGURE 14.19**).

LINK

Arthropod evolution and diversity are discussed in **Concept 23.4**

Evolution proceeds by changing what's already there

The features of organisms almost always evolve from preexisting features in their ancestors. New "wing genes" did not suddenly appear in birds and bats; instead, wings arose as modifications of existing structures (**FIGURE 14.20**). In vertebrates, the wings are modified limbs.

Although the wings of birds and bats look different, they are made from the same basic parts. Like limbs, wings have a common structure: a humerus that connects to the body; two longer bones, the radius and ulna, that project away from the humerus; and then metacarpals and phalanges (digits). During development these bones take on different lengths and

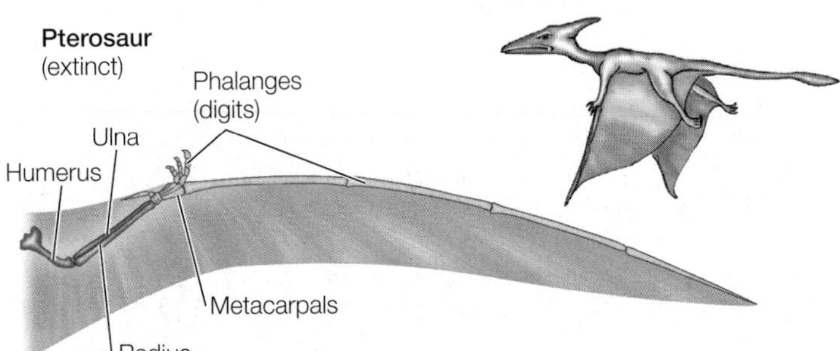

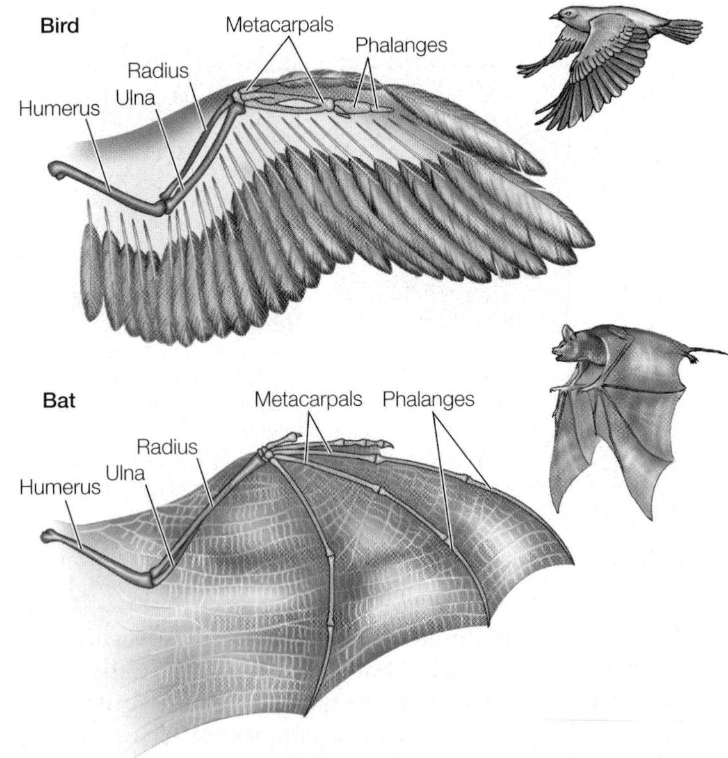

FIGURE 14.20 Wings Evolved Three Times in Vertebrates The wings of pterosaurs (the earliest flying vertebrates, which lived from 265 to 220 million years ago), birds, and bats are all modified forelimbs constructed from the same skeletal components. However, the components have different forms in the different groups of vertebrates.

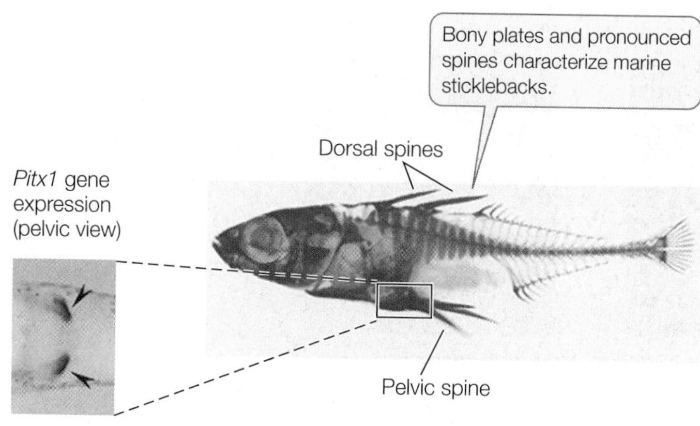
Bony plates and pronounced spines characterize marine sticklebacks.
Dorsal spines
Pitx1 gene expression (pelvic view)
Pelvic spine

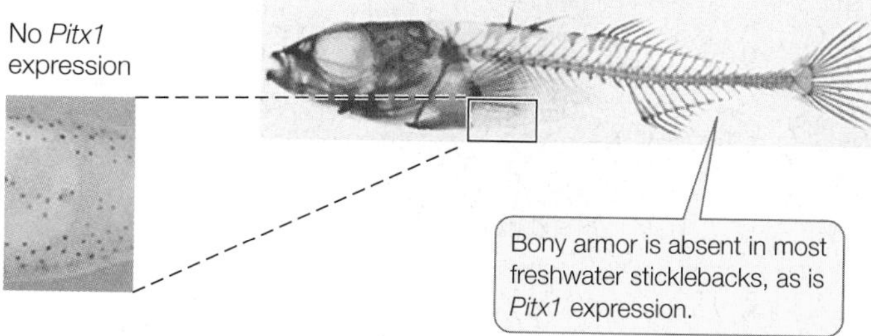
No Pitx1 expression
Bony armor is absent in most freshwater sticklebacks, as is Pitx1 expression.

FIGURE 14.21 Parallel Phenotypic Evolution in Sticklebacks
A developmental gene, *Pitx1*, encodes a transcription factor that stimulates the production of plates and spines. This gene is active in marine sticklebacks (indicated by arrowheads in inset at left) but is mutated and inactive in various freshwater populations of the fish. The fact that this mutation is found in geographically distant and isolated freshwater populations is evidence for parallel evolution.

weights in different organisms. For example, the phalanges are relatively short in birds and relatively long in bats. These differences arise from changes in the molecular mechanisms that control development, as we saw for cervical vertebrae in giraffes.

Developmental controls also influence how organisms lose structures. The ancestors of present-day snakes lost their forelimbs as a result of changes in the segmental expression of Hox genes. The snake lineage subsequently lost its hindlimbs by the loss of expression of the *Sonic hedgehog* gene in the limb bud tissue. But some snake species such as boas and pythons still have rudimentary pelvic bones and upper leg bones. Recall that Sonic hedgehog also functions as a morphogen in hand development (see p. 283). This is yet another example of how the same basic genetic tools are used in different ways in different species.

Conserved developmental genes can lead to parallel evolution

As we saw for the Hox genes, the nucleotide sequences of many developmental genes have been highly conserved throughout the evolution of multicellular organisms. In other words, these genes exist in similar form across a broad spectrum of species. The existence of these highly conserved genes makes it likely that similar traits will evolve repeatedly, especially among closely related species, in a phenomenon called **parallel phenotypic evolution**. A good example is provided by a small fish, the three-spined stickleback (*Gasterosteus aculeatus*: "bony stomach with spines").

Sticklebacks are widely distributed across the Atlantic and Pacific oceans and are also found in many freshwater lakes. Marine populations of this species spend most of their lives at sea but return to fresh water to breed. Members of freshwater populations live in lakes and never journey to salt water. Genetic evidence shows that freshwater populations have arisen independently from marine populations many times in different parts of the world, most recently at the end of the last ice age. Marine sticklebacks have structures that protect them from predators: well-developed pelvic bones with pelvic spines, and bony plates. In each of the separate freshwater populations, this body armor is greatly reduced, and dorsal and pelvic spines are much shorter or even lacking (**FIGURE 14.21**).

The difference between marine and freshwater sticklebacks is not induced by environmental conditions. Marine species that are reared in fresh water still grow spines. Not surprisingly, the difference is due to a gene that affects development. The *Pituitary homeobox transcription factor 1* (*Pitx1*) gene codes for a transcription factor that is normally expressed in regions of the developing embryo that form the head, trunk, tail, and pelvis of the marine stickleback. However, in independent freshwater populations from Canada, the United Kingdom, the United States, and Iceland, the gene is no longer expressed in the pelvis, and the spines do not develop. *This same gene sequence has evolved to produce similar phenotypic changes in several independent populations*, and is thus a good example of parallel evolution. What could be the common selective mechanism in these cases? Possibly, the decreased predation pressure in the freshwater environment allows for increased reproductive success in animals that invest less energy in the development of unnecessary protective structures.

CHECK**point** CONCEPT 14.5

- ✓ How have diverse body forms such as wings evolved by means of modifications in the functioning of existing genes?
- ✓ What would happen at the molecular and phenotypic levels if a *Ubx* gene from an adult insect replaced the *Ubx* gene in a fertilized insect egg?
- ✓ When several freshwater populations of stickleback fish were compared, the coding region of the *Pitx1* gene was identical to that found in marine populations. But in every case, the freshwater fish had mutations in *noncoding* regions of *Pitx1* that led to reduced expression. What might these noncoding-region mutations be?

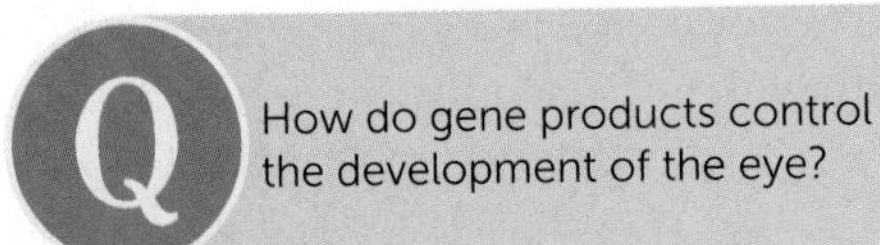

Q
How do gene products control the development of the eye?

ANSWER After reading this chapter, it should not surprise you that the product of both the fruit fly *eyeless* gene and the vertebrate *Pax6* gene is a transcription factor. This protein is produced in the front of the developing brain in a region called the neural plate. The result is a region called the "eye field," where an eye can develop. The separation of this single region into two eyes occurs when cells in the middle of the region produce Shh (Sonic hedgehog—a protein that is also involved in limb specification). Shh is a transcription factor that blocks the synthesis of Pax6, so where there is Shh, there is no eye development. Typically Shh is produced only in a central region of the neural plate. But in cave-dwelling fish, it occurs over a wider area of the eye field and adults have no eyes (**FIGURE 14.22**). If too little or no Shh is made, a single eye is formed; this occurs in the human disorder known as cyclopia.

(A) Adult Mexican tetras (*Astyanax mexicanus*)

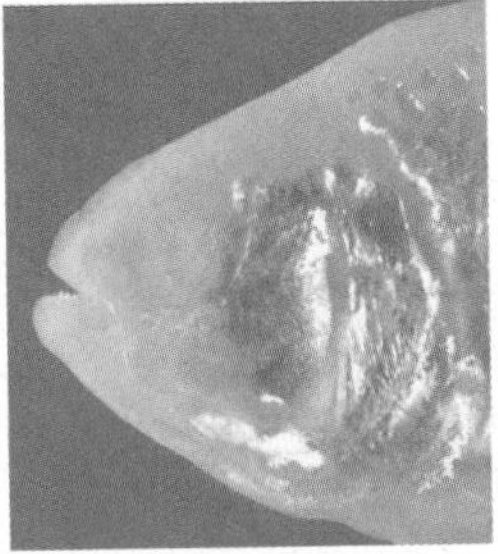

Surface-dwelling populations   Cave-dwelling populations

(B) Shh in embryonic eye region

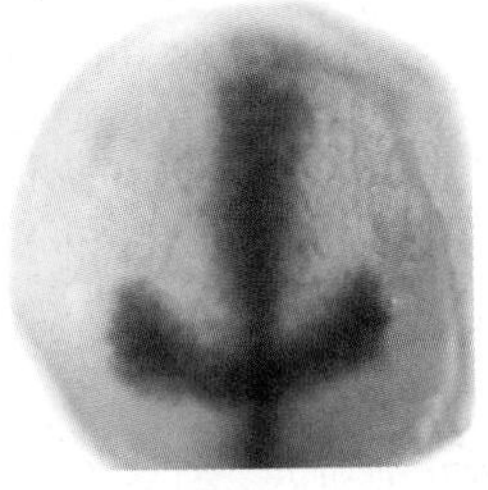
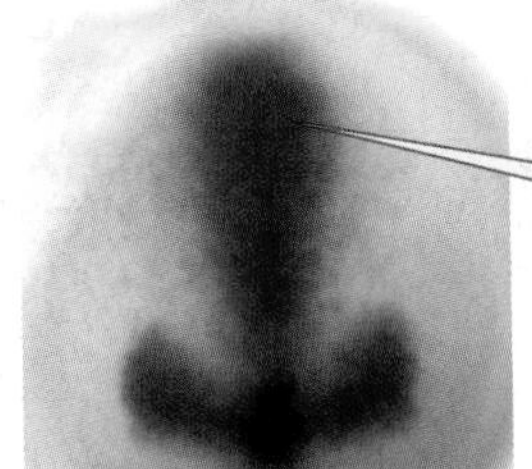

The area of Shh expression (dark areas) is broader in cave-dwelling populations than in surface-dwelling populations.

(C) Pax6 in embryonic eye region

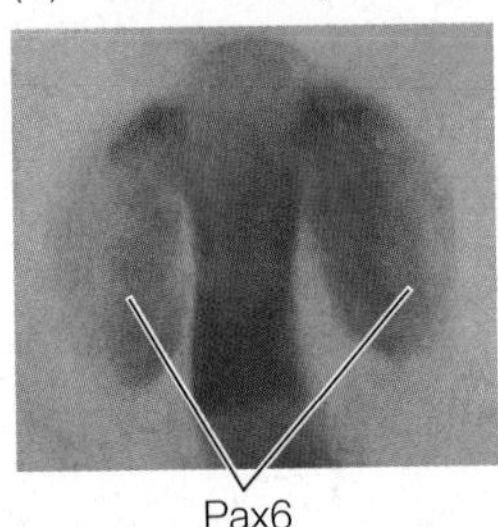

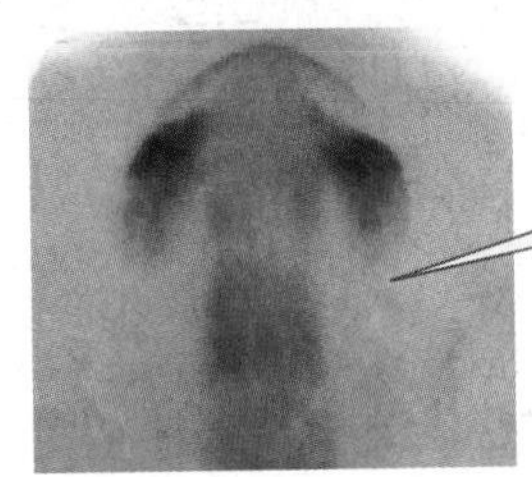

Shh prevents Pax6 expression and eye formation in cave-dwelling fish.

FIGURE 14.22 Inhibition of a Molecular Switch Results in No Eyes (A) In the Mexican tetra fish (*Astyanax mexicanus*), fish dwelling on the surface have two eyes (left), whereas those living in dark caves have no eyes (right). The difference results from overexpression of the *Shh* gene in cave-dwelling fish (B), which inhibits production of the molecular switch made by the *Pax6* gene (C).

SUMMARY

CONCEPT 14.1 Development Involves Distinct but Overlapping Processes

- A multicellular organism begins its development as an embryo, and several embryonic stages precede the production of an independent organism. **Review Figure 14.1 and ACTIVITY 14.1**
- The processes of development are **determination**, **differentiation**, **morphogenesis**, and **growth**.
- The zygote is **totipotent**; it is capable of producing an entire new organism, with every type of cell in the adult body. **Review ANIMATED TUTORIAL 14.1**
- The ability to create clones from differentiated cells demonstrates the principle of **genomic equivalence**. **Review Figures 14.3 and 14.4**
- **Multipotent** stem cells occur in the growing regions of many tissues in plants and animals. They constantly divide and form a pool of cells that can be used for differentiation to specialized cells. **Review Figure 14.5**
- **Pluripotent** stem cells can form every cell type of a mammal, but not an entire organism. They occur in the embryo and can be induced to form in the laboratory. They may have medical uses. **Review Figure 14.6 and ANIMATED TUTORIAL 14.2**

CONCEPT 14.2 Changes in Gene Expression Underlie Cell Fate Determination and Differentiation

- Differential gene expression results in cell differentiation. Transcription factors are especially important in regulating gene expression during differentiation.
- **Cytoplasmic segregation**—the unequal distribution of **cytoplasmic determinants** in the egg, zygote, or early embryo—can establish **polarity** and lead to cell fate determination. **Review Figure 14.7 and ANIMATED TUTORIAL 14.3**
- Induction is a process by which embryonic animal tissues direct the development of neighboring cells and tissues by secreting chemical signals called **inducers**. **Review Figure 14.9**

CONCEPT 14.3 Spatial Differences in Gene Expression Lead to Morphogenesis

- During development, selective elimination of cells by apoptosis results from the expression of specific genes.
- Both plants and animals use **positional information** in the form of a signal called a **morphogen** to stimulate cell determination. **Review Figure 14.11**
- In plants, **organ identity genes** encode polypeptides that associate to form transcription factors. These proteins determine the formation of flower organs. **Review Figure 14.12**
- In the fruit fly *Drosophila melanogaster*, a cascade of transcriptional activation sets up the axes of the embryo, the development of the segments, and the determination of cell fate in each segment. **Review Figure 14.13 and ANIMATED TUTORIAL 14.4**
- **Hox genes** determine cell fate in the embryos of many animals. The **homeobox** is a DNA sequence found in Hox genes and other genes that code for transcription factors. The sequence of amino acids encoded by the homeobox is called the homeodomain.

CONCEPT 14.4 Changes in Gene Expression Pathways Underlie the Evolution of Development

- **Evolutionary developmental biology** (**evo-devo**) is the modern study of the evolutionary aspects of development, and it focuses on molecular mechanisms.
- Hox genes have evolved from a common ancestor. **Review Figure 14.15 and ACTIVITY 14.2**
- Genes such as Hox genes underlie evolutionary changes in morphology that produce major differences in body forms.
- Evolutionary diversity is produced using a modest number of regulatory genes. **Review Figure 14.16**
- The transcription factors and chemical signals that govern pattern formation in the bodies of multicellular organisms, and the genes that encode them, can be thought of as a **genetic toolkit**.
- The bodies of developing and mature organisms are organized into self-contained units that can be modified independently in space and time. **Review ANIMATED TUTORIAL 14.5**
- Changes in **genetic switches** that determine where and when a set of genes will be expressed underlie the transformation of an individual from egg to adult.

CONCEPT 14.5 Developmental Genes Contribute to Species Evolution but Also Pose Constraints

- Evolutionary innovations are modifications of preexisting structures. **Review Figure 14.20**
- Because many genes that govern development have been highly conserved, similar traits are likely to evolve repeatedly, especially among closely related species. This process is called **parallel phenotypic evolution**. **Review Figure 14.21**

Go to the Interactive Summary to review key figures, Animated Tutorials, and Activities
PoL2e.com/is14

Go to LaunchPad at **macmillanhighered.com/launchpad** for additional resources, including LearningCurve Quizzes, Flashcards, and many other study and review resources.

PART 3
Evolution

15 Processes of Evolution

KEY CONCEPTS

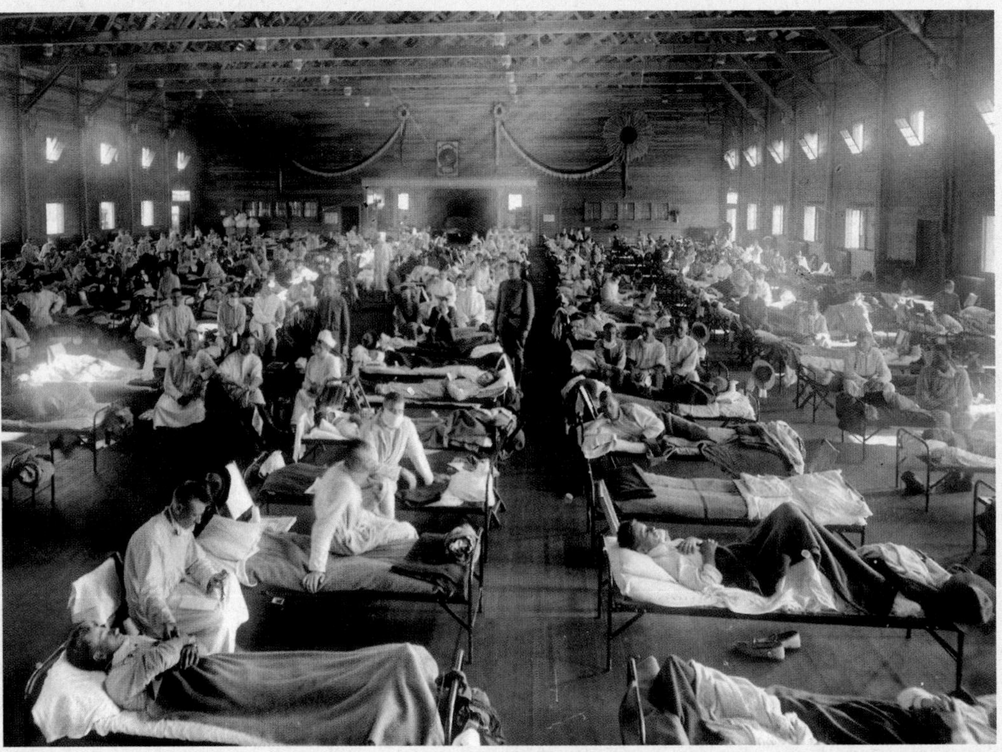

Flu victims are treated at a U.S. Army hospital in 1918.

On November 11, 1918, an armistice agreement signed in France signaled the end of World War I. But the death toll from four years of war was soon surpassed by the casualties of a massive influenza epidemic that began in the spring of 1918 among soldiers in a U.S. Army barracks. Over the next 18 months, this particular strain of flu virus spread across the globe, killing more than 50 million people worldwide—more than twice the number of World War I–related combat deaths.

The 1918–1919 pandemic was noteworthy because the death rate among young adults—who are usually less likely to die from influenza than are the elderly or the very young—was 20 times higher than in flu epidemics before or since. Why was that particular virus so deadly, especially to typically hardy individuals? The 1918 flu strain triggered an especially intense reaction in the human immune system. This overreaction meant that people with strong immune systems were likely to be more severely affected.

In most cases, however, our immune system helps us fight viruses; this response is the basis of vaccination. Since 1945, programs to administer flu vaccines have helped keep the number and severity of influenza outbreaks in check. Last year's vaccine, however, will probably not be effective against this year's virus. New strains of flu virus are evolving continuously, ensuring genetic variation in the population. If these viruses did not evolve, we would become resistant to them and annual vaccination would become unnecessary. But because the viruses do evolve, biologists must develop a new and different flu vaccine each year.

Vertebrate immune systems recognize proteins on the viral surface, and changes in these proteins mean that the virus can escape immune detection. Virus strains with the greatest number of changes to their surface proteins are most likely to avoid detection and infect their hosts, and thus have an advantage over other strains. Biologists can observe evolution in action by following changes in influenza virus proteins from year to year.

We learn a great deal about the processes of evolution by examining rapidly evolving organisms such as viruses, and these studies contribute to the development of evolutionary theory. Evolutionary theory, in turn, is put to practical uses, such as the development of better strategies for combating deadly diseases.

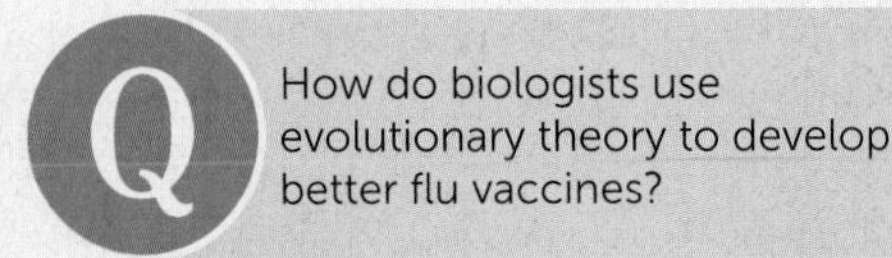

Q How do biologists use evolutionary theory to develop better flu vaccines?

You will find the answer to this question on page 322.

CONCEPT 15.1 Evolution Is Both Factual and the Basis of Broader Theory

All biological populations change in their genetic makeup over time. This change in the genetic composition of populations over time is called **evolution**. We can, and do, observe evolutionary change on a regular basis, both in laboratory experiments and in natural populations. We measure the rate at which new mutations arise, observe the spread of new genetic variants through a population, and see the effects of genetic change on the form and function of organisms. In the fossil record, we observe the long-term morphological changes (which are the result of underlying genetic changes) that have occurred among living organisms. These underlying changes in the genetic makeup of populations drive the origin and extinction of species and fuel the diversification of life.

In addition to observing and recording physical changes over evolutionary time, biologists have accumulated a large body of evidence about *how* these changes occur, and about *what* evolutionary changes have occurred in the past. The resulting understanding and application of the processes of evolutionary change to biological problems is known as **evolutionary theory**.

Evolutionary theory has many useful applications. We constantly apply it to the study and treatment of diseases. Evolutionary theory is critical to the development of better agricultural crops and practices, and to the development of industrial processes that produce new molecules with useful properties. At a more basic level, knowledge of evolutionary theory allows biologists to understand how life diversified. It also helps us make predictions about the biological world.

In everyday speech, people tend to use the word "theory" to mean an untested hypothesis, or even a guess. But evolutionary theory does not refer to any single hypothesis, and it certainly is not guesswork. The concept of evolutionary change among living organisms was present among a few scientists even before Charles Darwin so clearly described his observations, presented his conclusions, and articulated the premise of natural selection in his book *On the Origin of Species*. The rediscovery of Mendel's experiments and the subsequent establishment of the principles of genetic inheritance early in the 1900s set the stage for vast amounts of research. By the end of the twentieth century, findings from many fields of biology firmly upheld Darwin's basic premises about the common ancestry of life and the role of natural selection as an important process of evolution. Today a vast and rich array of geological, morphological, behavioral, and molecular data all support and expand the factual basis of evolution. Observations of fossils and natural populations are supported by experiments that demonstrate the basic operation of evolutionary processes.

When we refer to evolutionary theory, we are referring to our understanding of the processes that result in genetic changes in populations over time. We then apply that understanding to interpret the changes we observe in natural populations. We can directly observe the evolution of influenza viruses, but it is evolutionary theory that allows us to apply our observations to the task of developing more effective vaccines.

Several processes of evolutionary change are recognized, and the scientific community is continually using evolutionary theory to expand its understanding of how and when these processes apply to particular biological problems.

Go to MEDIA CLIP 15.1
Watching Evolution in Real Time
PoL2e.com/mc15.1

Darwin and Wallace introduced the idea of evolution by natural selection

In the early 1800s, it was not yet evident to many people that populations of living organisms evolve. But several biologists had suggested that the species living on Earth had changed over time—that is, that evolution had taken place. Jean-Baptiste Lamarck, for one, presented strong evidence for the fact of evolution in 1809, but his ideas about *how* it occurred were not convincing. At that time, no one had yet envisioned a viable process for evolution.

Charles Robert Darwin

In the 1820s, a young Charles Darwin became passionately interested in the subjects of geology (with its new sense of Earth's great age) and natural history (the scientific study of how different organisms function and carry out their lives in nature). Despite these interests, he planned, at his father's behest, to become a doctor. But surgery conducted without anesthesia nauseated Darwin, and he gave up medicine to study at Cambridge University for a career as a clergyman in the Church of England. Always more interested in science than in theology, he gravitated toward scientists on the faculty, especially the botanist John Henslow. In 1831, Henslow recommended Darwin for a position on HMS *Beagle*, a Royal Navy vessel that was preparing for a survey voyage around the world.

HMS *Beagle*

Whenever possible during the 5-year voyage (**FIGURE 15.1**), Darwin went ashore to study rocks and to observe and collect plants and animals. He noticed striking differences between the species he saw in South America and those of Europe. He

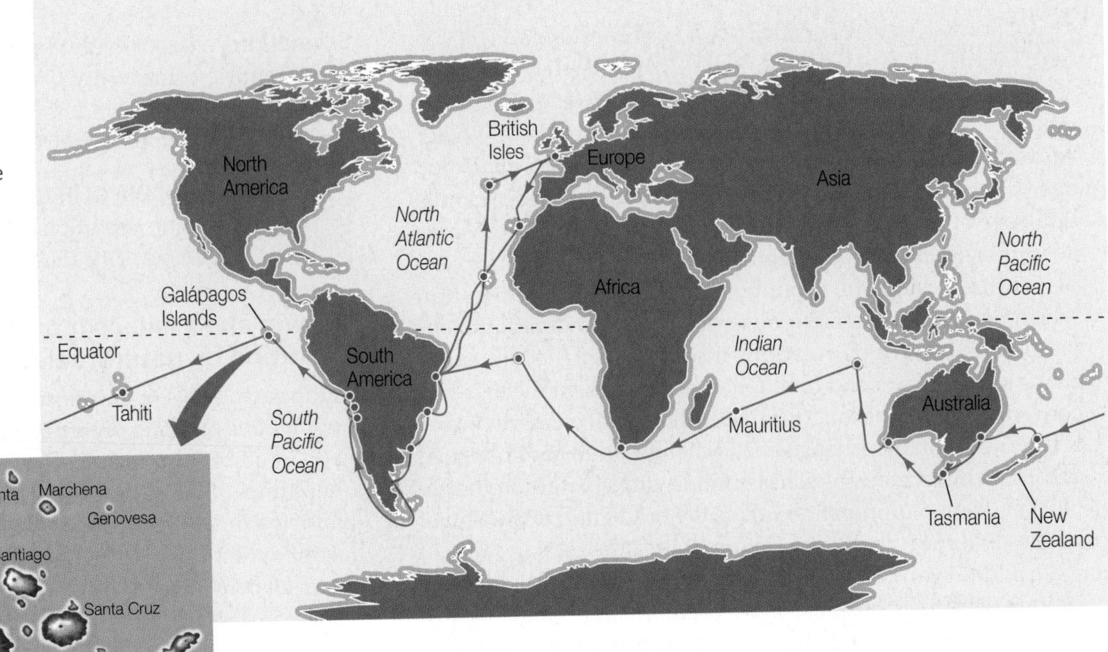

FIGURE 15.1 The Voyage of the *Beagle* The mission of HMS *Beagle* was to chart the oceans and collect oceanographic and biological information from around the world. The world map indicates the ship's path; the inset map shows the Galápagos Islands, whose organisms were an important source of Darwin's ideas on natural selection.

Go to ACTIVITY 15.1 Darwin's Voyage PoL2e.com/ac15.1

observed that the species of the temperate regions of South America (Argentina and Chile) were more similar to those of tropical South America (Brazil) than they were to temperate European species. When he explored the islands of the Galápagos archipelago west of Ecuador, he noted that most of the animals were endemic to the islands (that is, unique and found nowhere else), although they were similar to animals found on the mainland of South America. Darwin also observed that the fauna of the Galápagos differed from island to island. He postulated that some animals had come to the archipelago from mainland South America and had subsequently undergone different changes on each of the islands. He wondered what might account for these changes.

When he returned to England in 1836, Darwin continued to ponder his observations. His thoughts were strongly influenced by the geologist Charles Lyell, who had recently popularized the idea that Earth had been shaped by slow-acting forces that are still at work today. Darwin reasoned that similar thinking could be applied to the living world. Within a decade, he had developed the framework of an explanatory theory for evolutionary change based on three major propositions:

- Species are not immutable; they change over time.
- Divergent species share a common ancestor and have diverged from one another gradually over time (a concept Darwin termed **descent with modification**).
- Changes in species over time can be explained by **natural selection**: the increased survival and reproduction of some individuals compared with others, based on differences in their traits.

The first of these propositions was not unique to Darwin; several earlier authors had argued for the fact of evolution. A more revolutionary idea was his second proposition, that *divergent species are related to one another through common descent*. But Darwin is probably best known for his third proposition, that of natural selection.

Darwin realized that many more individuals of most species are born than survive to reproduce. He also knew that, although offspring usually resemble their parents, offspring are not identical to one another or to either parent. Finally, he was well aware of the fact that human breeders of plants and animals often selected their breeding stock based on the occurrence of particular traits. Over time, this selection resulted in dramatic changes in the appearance of the descendants of those plants or animals. In natural populations, wouldn't the individuals with the best chances of survival and reproduction be similarly "selected," and thus pass their traits on to the next generation? Darwin's simple but powerful idea was that nature did the selecting in natural populations on the basis of traits that resulted in greater survival and, eventually, greater likelihood of reproduction.

In 1844, Darwin wrote a long essay describing the role of natural selection as a process of evolution. But he was reluctant to publish it, preferring to assemble more evidence first. Darwin's hand was forced in 1858, when he received a letter and manuscript from another traveling English naturalist, Alfred Russel Wallace, who was studying the plants and animals of the Malay Archipelago. Wallace asked Darwin to evaluate his manuscript, which included an explanation of natural selection almost identical to Darwin's. Darwin was at first dismayed, believing Wallace to have preempted his idea. Parts of Darwin's 1844 essay, together with Wallace's manuscript, were presented to the Linnaean Society of London on July 1, 1858, thereby crediting both men for the idea of natural selection. Darwin then worked quickly to finish his own book, *On the Origin of Species*, which was published the following year.

Go to ANIMATED TUTORIAL 15.1 Natural Selection PoL2e.com/at15.1

Although Darwin and Wallace independently articulated the concept of natural selection, Darwin developed his ideas first. Furthermore, *On the Origin of Species* proved to be a stunning work of scholarship that provided exhaustive evidence from many fields supporting both the premise of evolution itself and the notion of natural selection as a process of evolution. Thus both concepts are more closely associated with Darwin than with Wallace.

The publication of *On the Origin of Species* in 1859 stirred considerable interest (and controversy) among scientists and the public alike. Scientists spent much of the rest of the nineteenth century amassing biological and paleontological data to test evolutionary ideas and document the history of life on Earth. By 1900, the fact of biological evolution (defined at that time as change in the physical characteristics of populations over time) was established beyond any reasonable doubt. But the *genetic* basis of evolutionary change was not yet understood.

Evolutionary theory has continued to develop over the past century

Shortly after 1900, several individuals rediscovered the work of Gregor Mendel (which had been published in 1866 but rarely read or cited), and the basic processes of genetic inheritance began to be unraveled. In the first decades of the twentieth century, Thomas Hunt Morgan's studies on fruit flies led to his discovery of the role of chromosomes in inheritance. In the 1920s and early 1930s, the major principles of population genetics were established, the genetic basis of new variation (i.e., mutations) began to be understood, and processes of evolution such as genetic drift were described (see Concept 15.2). This work set the stage for a "modern synthesis" of genetics and evolution that took place over the period 1936–1947. Some of the major contributors to this synthesis and a few of their books are listed in **FIGURE 15.2**.

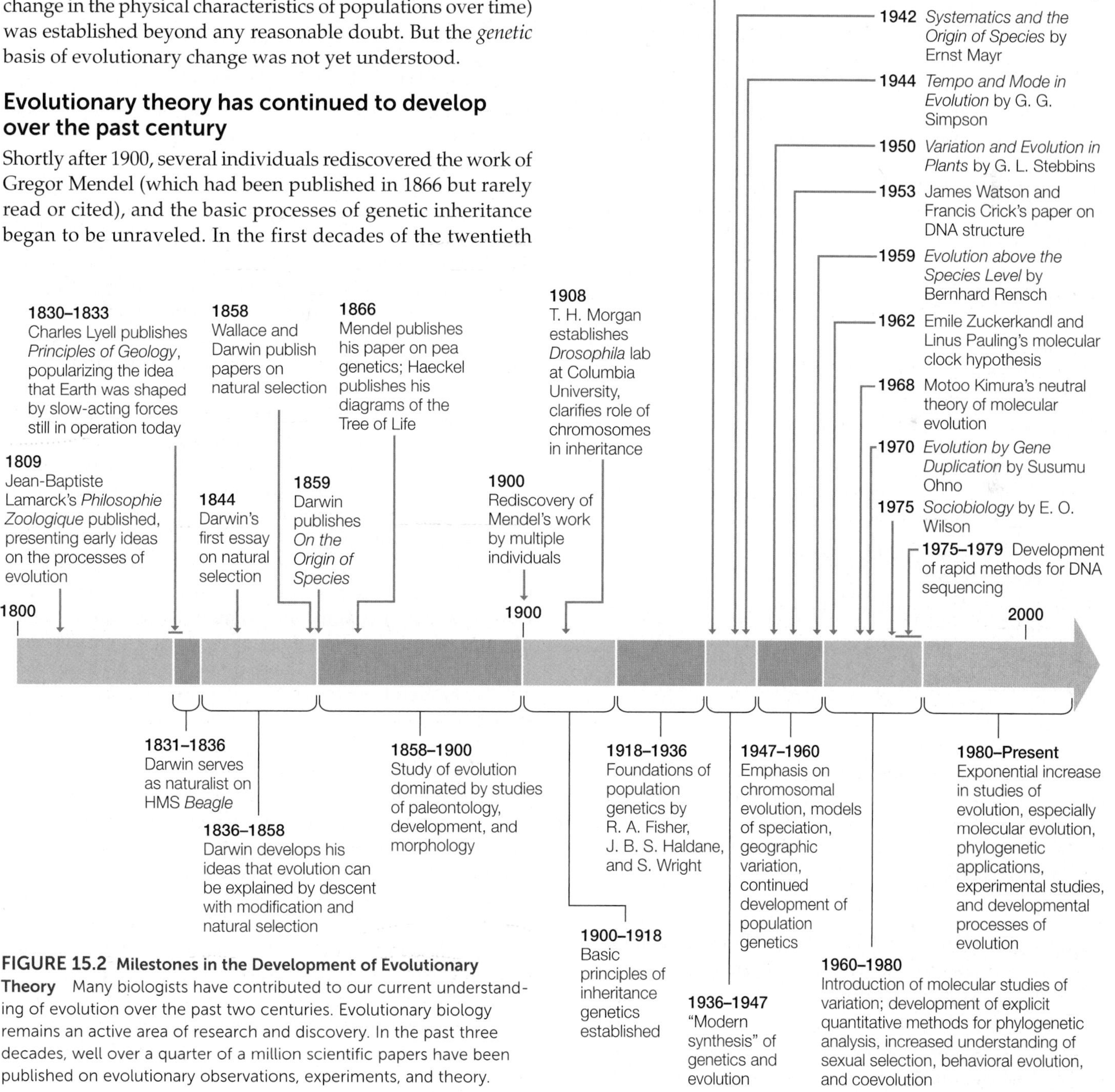

FIGURE 15.2 Milestones in the Development of Evolutionary Theory Many biologists have contributed to our current understanding of evolution over the past two centuries. Evolutionary biology remains an active area of research and discovery. In the past three decades, well over a quarter of a million scientific papers have been published on evolutionary observations, experiments, and theory.

Although chromosomes were understood to be the basis of genetic transmission in eukaryotes by the early 1900s, their molecular structure remained a mystery until soon after the modern synthesis. Then, in 1953, Watson and Crick published their paper on the structure of DNA, opening the door to our current detailed understanding of molecular evolutionary processes. By the 1960s, biologists could study and document changes in allele frequencies in populations over time (see Concept 15.3). Most of this early work necessarily focused on variants of proteins that differed within and between populations and species. Even though the molecular structure of DNA was known, it was not yet practical to sequence long stretches of DNA. Nonetheless, many important advances occurred in evolutionary theory during this time (see Figure 15.2), and these advances were not focused solely on a genetic understanding of evolution. E. O. Wilson's 1975 book *Sociobiology*, for example, invigorated studies of the evolution of behavior (a subject that had fascinated Darwin).

In the late 1970s, several techniques were developed that allowed the rapid sequencing of long stretches of DNA, which in turn allowed researchers to determine the amino acid sequences of proteins. This ability opened a new door for evolutionary biologists, who could now explore the structure of genes and proteins and document evolutionary changes within and between species in ways never before possible.

CHECKpoint CONCEPT 15.1

✓ How would you respond to someone who said that evolution was "just a theory"?

✓ Why do you think Darwin and Wallace formulated their ideas on natural selection at about the same time?

✓ Discuss the significance of each of the following scientific advances for evolutionary theory:

a. Elucidation of the principles of chromosomal inheritance

b. The discovery of DNA, its structure, and the universal genetic code

c. Technology that allows us to sequence long segments of DNA

Keep your discussion in mind as you continue reading this chapter.

Natural selection is not the only process that drives evolution, although the importance of natural selection to evolution has been confirmed in many thousands of scientific studies. In the next section we'll consider a more complete view of evolutionary processes and how they operate.

CONCEPT 15.2 Mutation, Selection, Gene Flow, Genetic Drift, and Nonrandom Mating Result in Evolution

The word "evolution" is often used in a general sense to mean simply "change," but in a biological context "evolution" refers specifically to change in the genetic makeup of populations over time. Developmental changes that occur in a single individual over the course of the life cycle are not the result of evolutionary change. Evolution is genetic change occurring in a **population**—a group of individuals of a single species that live and interbreed in a particular geographic area at the same time. It is important to remember that *individuals do not evolve; populations do.*

The premise of natural selection was one of Darwin's principal insights and has been demonstrated to be an important process of evolution, but natural selection does not act alone. Additional processes—gene flow, genetic drift, and nonrandom mating—affect the genetic makeup of populations over time. Before we consider how these processes change the frequencies of gene variants in a population, however, we need to understand how mutation brings such variants into existence.

Mutation generates genetic variation

The origin of novel genetic variation is mutation. As described in Concept 9.3, a mutation is any change in the nucleotide sequences of an organism's DNA. The process of DNA replication is not perfect, and some changes appear almost every time a genome is replicated. Mutations occur randomly with respect to an organism's needs; it is natural selection acting on this random variation that results in adaptation. Most mutations are either harmful to their bearers (deleterious mutations) or have no effect (neutral mutations). But a few mutations are beneficial, and even previously deleterious or neutral alleles may become advantageous if environmental conditions change. In addition, mutation can restore genetic variation that other evolutionary processes have removed. Thus mutation both creates and helps maintain genetic variation in populations.

Mutation rates can be high, as we saw in the case of the influenza viruses described at the opening of this chapter, but in many organisms the mutation rate is very low (on the order of 10^{-8} to 10^{-9} changes per base pair of DNA per generation). Even low overall mutation rates, however, create considerable genetic variation, because each of a large number of genes may change, and populations often contain large numbers of individuals. For example, if the probability of a point mutation (an addition, deletion, or substitution of a single base) were 10^{-9} per base pair per generation, then each human gamete—the DNA of which contains 3×10^9 base pairs—would average three new point mutations ($3 \times 10^9 \times 10^{-9} = 3$), and each zygote would carry an average of six new mutations. The current human population of about 7 billion people would thus be expected to carry about 42 billion new mutations (i.e., changes in the nucleotide sequences of their DNA that were not present one generation earlier). So even though the mutation rate in humans is low, human populations still contain enormous genetic variation on which other evolutionary processes can act.

As a result of mutation, different forms of a gene, known as **alleles**, may exist at a particular chromosomal locus. At any particular locus, a single diploid individual has no more than two of the alleles found in the population to which it belongs. The sum of all copies of all alleles at all loci found in a population constitutes its **gene pool** (**FIGURE 15.3**). (We can also refer

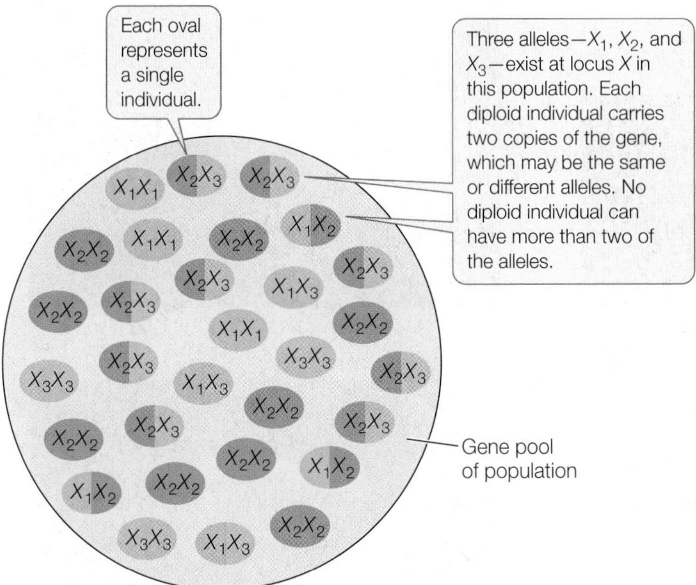

Each oval represents a single individual.
Three alleles—X_1, X_2, and X_3—exist at locus X in this population. Each diploid individual carries two copies of the gene, which may be the same or different alleles. No diploid individual can have more than two of the alleles.
Gene pool of population

FIGURE 15.3 A Gene Pool A gene pool is the sum of all the alleles found in a population or at a particular locus. This figure shows the gene pool for one locus, X. The allele frequencies in this case are 0.20 for X_1, 0.50 for X_2, and 0.30 for X_3 (see Figure 15.11).

to the gene pool for a particular chromosomal locus or loci.) The gene pool is the sum of the genetic variation in the population. The proportion of each allele in the gene pool is the allele frequency. Likewise, the proportion of each genotype among individuals in the population is the genotype frequency.

A simple experiment demonstrates how mutations accumulate in populations in a continuous, almost constant fashion over time (**FIGURE 15.4**). Lines of the bacterium *E. coli* were grown in the laboratory for 20,000 generations, and the genomes were sequenced from individuals in the experimental lines at least once every 5,000 generations. Over the experiment, the lines accumulated about 45 changes to their genomes, and these changes appeared at a fairly constant rate over time. All populations experience a similar accumulation of mutations over time (although the rate of change differs among species), and these changes provide the raw material for evolution.

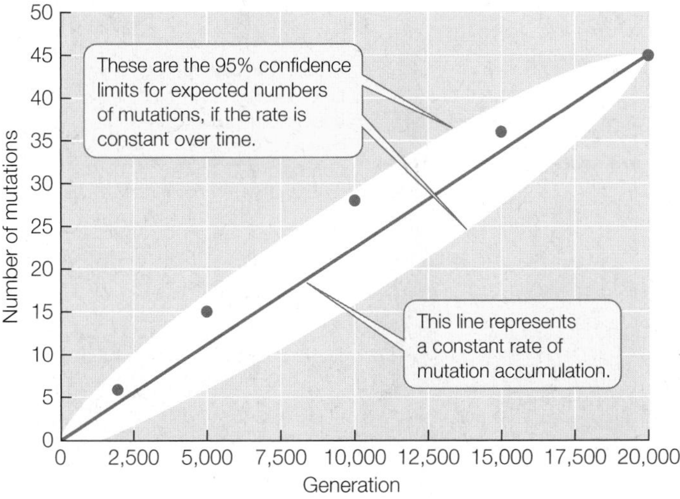

FIGURE 15.4 Mutations Accumulate Continuously An experimental lineage of the bacterium *Escherichia coli* was propagated in the laboratory for 20,000 generations. Genomes were sequenced from individuals sampled at various points during the experiment and were compared with the genome of the ancestral clone. Note that mutations accumulated at a relatively constant rate throughout the experiment.

LINK

Review the nature of alleles and genetic inheritance in **Concepts 8.1 and 8.2**

Selection on genetic variation leads to new phenotypes

As a result of mutation, the gene pools of nearly all populations contain variation for many traits. Selection that favors different traits can lead to many different lineages that descend from the same ancestor. For example, artificial selection on different traits in a single European species of wild mustard produced many important crop plants (**FIGURE 15.5**). Agriculturalists

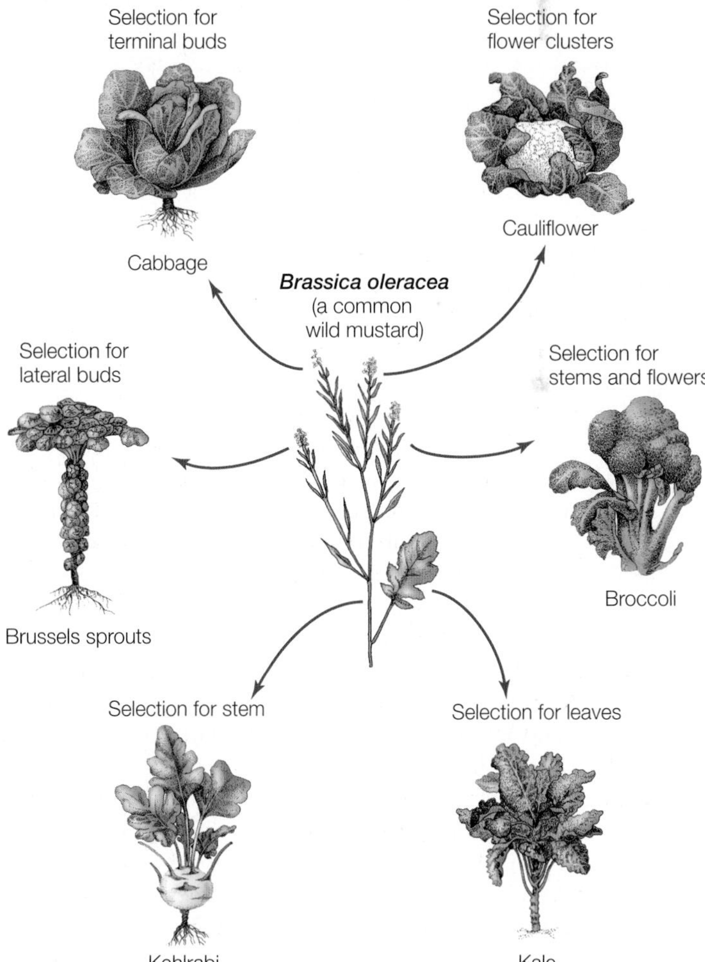

FIGURE 15.5 Many Vegetables from One Species All of the crop plants shown here derive from a single wild mustard species. European agriculturalists produced these crop species by selecting and breeding plants with unusually large buds, stems, leaves, or flowers. The results substantiate the vast amount of variation present in a gene pool.

FIGURE 15.6 Artificial Selection Charles Darwin raised pigeons as a hobby and noted similar forces at work in artificial and natural selection. The "fancy" pigeons shown here represent 3 of the more than 300 varieties derived from the wild rock pigeon (*Columba livia*; left) by artificial selection for character traits such as color and feather distribution.

were able to achieve these results because the original mustard population had genetic variation for the characteristics of interest (such as stem thickness or number of leaves).

Darwin compared this artificial selection, which was commonly practiced by animal and plant breeders, with natural selection that occurred in natural populations. Many of Darwin's observations on the nature of variation and selection came from domesticated plants and animals. Darwin bred pigeons and thus knew firsthand the astonishing diversity in color, size, form, and behavior that breeders could achieve (**FIGURE 15.6**). He recognized close parallels between selection by breeders and selection in nature. Whereas artificial selection resulted in traits that were preferred by the human breeders, natural selection resulted in traits that helped organisms survive and reproduce more effectively. In both cases, selection simply increased the frequency of the favored trait from one generation to the next.

Laboratory experiments also demonstrate the existence of considerable genetic variation in populations, and show how this variation can lead to evolution through selection. In one such experiment, investigators bred populations of the fruit fly *Drosophila melanogaster* with high or low numbers of bristles on their abdomens from an initial population with intermediate numbers of bristles. After 35 generations, all flies in both the high- and low-bristle lineages had bristle numbers that fell well outside the range found in the original population (**FIGURE 15.7**). Selection for high and low bristle numbers resulted in new combinations of the many different genes that were present in the original population, so that the phenotypic variation seen in subsequent generations fell outside the phenotypic variation seen in the original population.

Natural selection increases the frequency of beneficial mutations in populations

Darwin knew that far more individuals of most species are born than survive to reproduce. He also knew that, although offspring tend to resemble their parents, the offspring of most organisms are not identical either to their parents or to one another. He suggested that slight differences among individuals affect the chance that a given individual will survive and reproduce, which increases the frequency of the favored trait in the next generation. A favored trait that evolves through natural selection is known as an **adaptation**; this word is used to describe both the trait itself and the process that produces the trait.

Biologists regard an organism as being adapted to a particular environment when they can demonstrate that a slightly

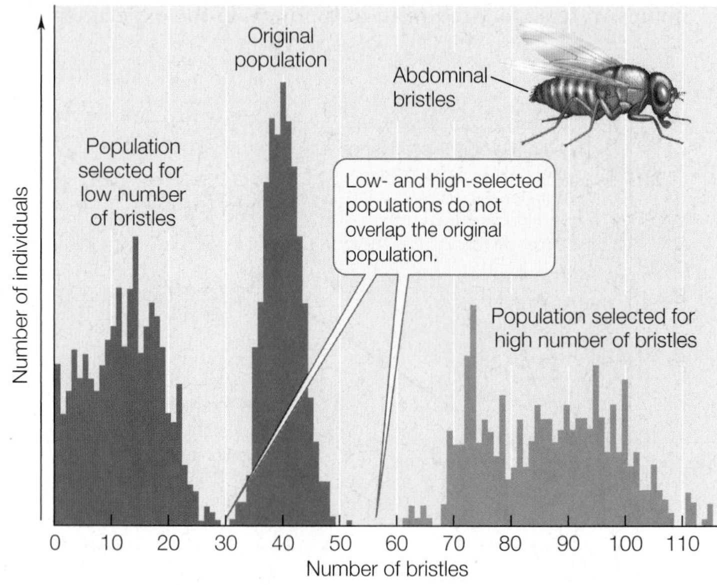

FIGURE 15.7 Artificial Selection Reveals Genetic Variation When investigators subjected *Drosophila melanogaster* to artificial selection for abdominal bristle number, that character evolved rapidly. The graph shows the number of flies with different numbers of bristles in the original population and after 35 generations of artificial selection. The bristle numbers of the selected lineages clearly diverged from those of the original population.

different organism is less likely to survive and reproduce in that environment. To understand adaptation, biologists compare the performances of individuals that differ in their traits.

Natural selection also acts to remove deleterious mutations from populations. Individuals with deleterious mutations are less likely to survive and reproduce, so they are less likely to pass their alleles on to the next generation.

Gene flow may change allele frequencies

Few populations are completely isolated from other populations of the same species. Migration of individuals and movements of gametes (in pollen, for example) between populations—a phenomenon called **gene flow**—can change allele frequencies in a population. If the arriving individuals survive and reproduce in their new location, they may add new alleles to the population's gene pool, or they may change the frequencies of alleles present in the original population.

LINK

If gene flow between two populations stops, those populations may diverge and become different species; see Concept 17.2

Genetic drift may cause large changes in small populations

In small populations, **genetic drift**—random changes in allele frequencies from one generation to the next—may produce large changes in allele frequencies over time. Harmful alleles may increase in frequency, and rare advantageous alleles may be lost. Even in large populations, genetic drift can influence the frequencies of neutral alleles (which do not affect the survival and reproductive rates of their bearers).

To illustrate the effects of genetic drift, suppose there are only two females in a small population of normally brown mice, and one of these females carries a newly arisen dominant allele that produces black fur. Even in the absence of any selection, it is unlikely that the two females will produce exactly the same number of offspring. Even if they do produce identical litter sizes and identical numbers of litters, chance events that have nothing to do with genetic characteristics are likely to result in differential mortality among their offspring. If each female produces one litter, but a flood envelops the black female's nest and kills all of her offspring, the novel allele could be lost from the population in just one generation. In contrast, if the brown female's litter is lost, then the frequency of the newly arisen allele (and phenotype) for black fur will rise dramatically in just one generation.

Genetic drift is especially potent when a population is reduced dramatically in size. Even populations that are normally large may occasionally pass through environmental events that only a small number of individuals survive, a situation known as a **population bottleneck**. The effect of genetic drift in such a situation is illustrated in **FIGURE 15.8**, in which red and yellow beans represent two alleles of a gene. Most of the beans in the small sample of the "population" that "survives" the bottleneck event are, just by chance, red, so the new population has a much higher frequency of red beans than the previous generation had. In a real population, the red and yellow allele frequencies would be described as having "drifted."

A population forced through a bottleneck is likely to lose much of its genetic variation. For example, when Europeans first arrived in North America, millions of greater prairie-chickens (*Tympanuchus cupido*) inhabited the midwestern prairies. As a result of hunting and habitat destruction by the new settlers, the Illinois population of this species plummeted from about 100 million birds in 1900 to fewer than 50 individuals in the 1990s. A comparison of DNA from birds collected in Illinois during the middle of the twentieth century with DNA from the surviving population in the 1990s showed that Illinois prairie-chickens have lost most of their genetic diversity. Loss of genetic variation in small populations is one of the problems facing biologists who attempt to protect endangered species.

Genetic drift can have similar effects when a few pioneering individuals colonize a new region. Because of its small size, the colonizing population is unlikely to possess all of the alleles found in the gene pool of its source population. The resulting change in genetic variation, called a **founder effect**, is equivalent to that in a large population reduced by a bottleneck.

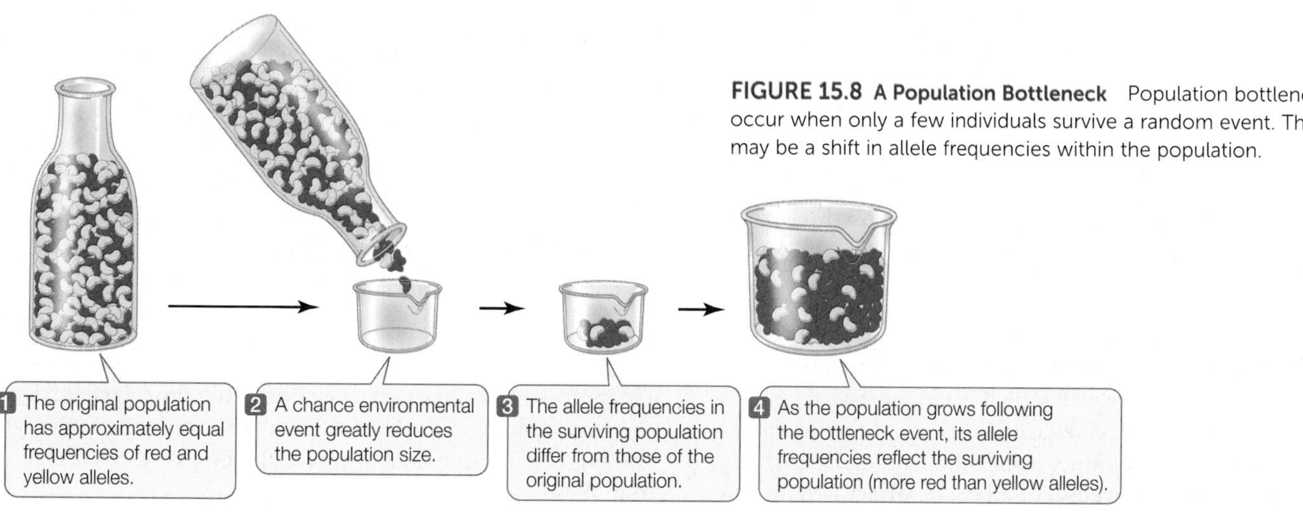

1 The original population has approximately equal frequencies of red and yellow alleles.
2 A chance environmental event greatly reduces the population size.
3 The allele frequencies in the surviving population differ from those of the original population.
4 As the population grows following the bottleneck event, its allele frequencies reflect the surviving population (more red than yellow alleles).

FIGURE 15.8 A Population Bottleneck Population bottlenecks occur when only a few individuals survive a random event. The result may be a shift in allele frequencies within the population.

Nonrandom mating can change genotype or allele frequencies

Mating patterns often alter genotype frequencies because the individuals in a population do not choose mates at random. For example, self-fertilization is common in many groups of organisms, especially plants. Any time individuals mate preferentially with other individuals of the same genotype (including themselves), homozygous genotypes will increase in frequency and heterozygous genotypes will decrease in frequency over time. The opposite effect (more heterozygotes, fewer homozygotes) is expected when individuals mate primarily or exclusively with individuals of different genotypes.

Nonrandom mating systems that do not affect the relative reproductive success of individuals produce changes in genotype frequencies but not in allele frequencies, and thus do not, by themselves, result in evolutionary change in a population. However, nonrandom mating systems that result in different reproductive success among individuals do produce allele frequency changes from one generation to the next. **Sexual selection** occurs when individuals of one sex mate preferentially with particular individuals of the opposite sex rather than at random.

Sexual selection was first suggested by Charles Darwin, who developed the idea to explain the evolution of conspicuous traits that would appear to inhibit survival, such as bright colors and elaborate courtship displays in males of many species. He hypothesized that these features either improved the ability of their bearers to compete for access to mates (intrasexual selection) or made their bearers more attractive to members of the opposite sex (intersexual selection). The concept of sexual selection was either ignored or questioned for many decades, but recent investigations have demonstrated its importance.

Darwin argued that while natural selection typically favors traits that enhance the survival of their bearers or their bearers' descendants, sexual selection is primarily about successful reproduction. An animal that survives but fails to reproduce makes no contribution to the next generation. Thus sexual selection may favor traits that enhance an individual's chances of reproduction even when these traits reduce its chances of survival. For example, females may be more likely to see or hear males with a given trait (and thus be more likely to mate with those males), even though the favored trait also increases the chances that the male will be seen or heard by a predator.

LINK

Some of the animal behaviors that have evolved in response to sexual selection are described in **Concepts 40.5 and 40.6**

One example of a trait that Darwin attributed to sexual selection is the remarkable tail of the male African long-tailed widowbird (*Euplectes progne*), which is longer than the bird's head and body combined (**FIGURE 15.9**). Male widowbirds normally select, and defend from other males, a territory where they perform courtship displays to attract females. To investigate whether sexual selection drove the evolution of widowbird tails, a biologist clipped the tails of some captured male widowbirds and lengthened the tails of others by gluing on additional feathers. He then cut and reglued the tail feathers of still other males, which served as controls. Both short- and long-tailed males successfully defended their display territories, indicating that a long tail does not confer an advantage in male–male competition. However, males with artificially elongated tails attracted about four times more females than did males with shortened tails (**FIGURE 15.10**). Thus males with long tails pass on their genes to more offspring than do males with short tails, which leads to the evolution of this unusual trait.

Euplectes progne

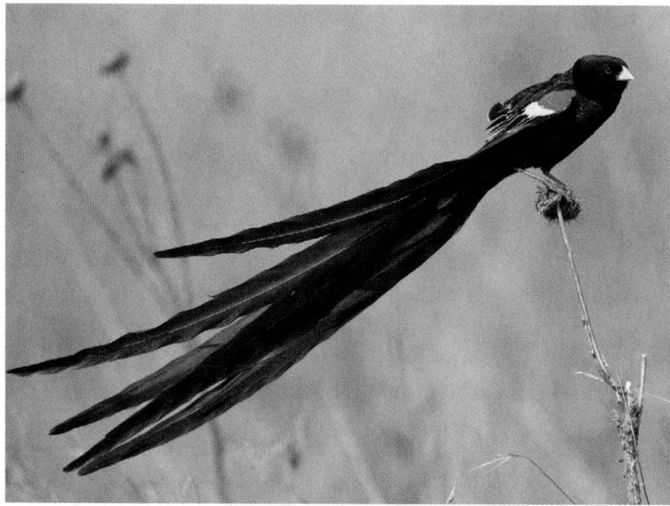

FIGURE 15.9 What Is the Advantage? The extensive tail of the male African long-tailed widowbird actually inhibits its ability to fly. Darwin attributed the evolution of this seemingly nonadaptive trait to sexual selection.

CHECKpoint CONCEPT 15.2

- ✓ How do deleterious, neutral, and beneficial mutations differ?
- ✓ Can you explain how natural selection results in an increase in the frequency of beneficial alleles in a population over time, and a decrease in the frequency of deleterious alleles?
- ✓ How can genetic drift cause large changes in small populations?
- ✓ How do self-fertilization and sexual selection differ in their expected effects on genotype and allele frequencies over time?

The processes of mutation, selection, gene flow, genetic drift, and nonrandom mating can all result in evolutionary change. We will consider next how evolutionary change that results from these processes is measured.

INVESTIGATION

FIGURE 15.10 Sexual Selection in Action Behavioral ecologist Malte Andersson tested Darwin's hypothesis that excessively long tails evolved in male widowbirds because female preference for longer-tailed males increased their mating and reproductive success.[a]

HYPOTHESIS

Female widowbirds prefer to mate with the male that displays the longest tail; longer-tailed males thus are favored by sexual selection because they will father more offspring.

METHOD

1. Capture males and artificially lengthen or shorten tails by cutting or gluing on feathers. In a control group, cut and replace tails to their normal length (to control for the effects of tail-cutting).
2. Release the males to establish their territories and mate.
3. Count the nests with eggs or young on each male's territory.

RESULTS

Male widowbirds with artificially shortened tails established and defended display sites sucessfully but fathered fewer offspring than did control or unmanipulated males. Males with artificially lengthened tails fathered the most offspring.

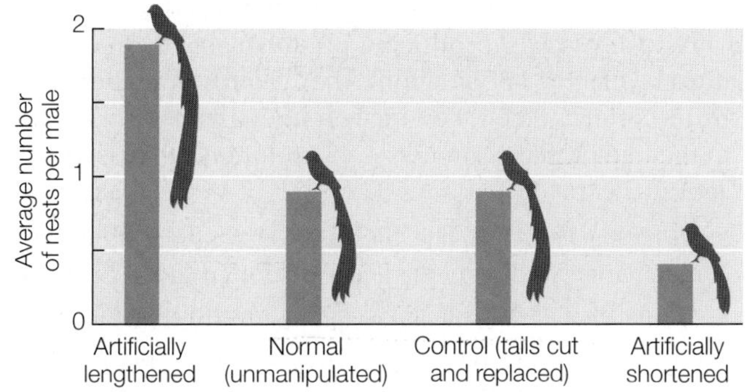

CONCLUSION

Sexual selection in *Euplectes progne* has favored the evolution of long tails in the male.

ANALYZE THE DATA

Are the differences plotted above significantly different? See Analyze the Data 15.1 in **LaunchPad** for a simple method to test the statistical significance of the differences using the following data.

	Number of nests per male		
Group	Shortened tail	Control	Elongated tail
1	0	0	2
2	0	0	2
3	2	3	5
4	1	2	4
5	0	1	2
6	0	1	2
7	0	1	0
8	0	0	0
9	1	0	0

Go to **LaunchPad** for discussion and relevant links for all **INVESTIGATION** figures.

[a]M. Andersson. 1982. *Nature* 299: 818–820.

CONCEPT 15.3 Evolution Can Be Measured by Changes in Allele Frequencies

Much of evolution occurs through gradual changes in the relative frequencies of different alleles in a population from one generation to the next. Major genetic changes can also be sudden, as happens when two formerly separated populations merge and hybridize, or when genes within a population are duplicated within the genome (see Concept 15.6). But in most cases, we measure evolution by looking at changes in allele and genotype frequencies in populations over time.

To measure allele frequencies in a population precisely, we would need to count every allele at every locus in every individual in the population. Fortunately, we do not need to make such complete measurements because we can reliably estimate allele frequencies for a given locus by counting alleles in a sample of individuals from the population. The sum of all allele frequencies at a locus is equal to 1, so measures of allele frequency range from 0 to 1.

Go to ANIMATED TUTORIAL 15.2
Genetic Drift Simulation
PoL2e.com/at15.2

An allele's frequency is calculated using the following formula:

$$p = \frac{\text{number of copies of the allele in the population}}{\text{total number of copies of all alleles in the population}}$$

If only two alleles (we'll call them A and a) for a given locus are found among the members of a diploid population, those alleles can combine to form three different genotypes: AA, Aa, and aa (see Figure 15.3). A population with more than one allele at a locus is said to be polymorphic ("many forms") at that locus. Applying the formula above, as shown in **FIGURE 15.11**, we can calculate the relative frequencies of alleles A and a in a population of N individuals as follows:

- Let N_{AA} be the number that are homozygous for the A allele (AA).
- Let N_{Aa} be the number that are heterozygous for the two alleles (Aa).
- Let N_{aa} be the number that are homozygous for the a allele (aa).

Note that $N_{AA} + N_{Aa} + N_{aa} = N$, the total number of individuals in the population, and that the total number of copies of both alleles present in the population is $2N$, because each individual is diploid. Each AA individual has two copies of the A allele, and each Aa individual has one copy of the A allele. Therefore the total number of A alleles in the population is $2N_{AA} + N_{Aa}$. Similarly, the total number of a alleles in the population is

RESEARCH TOOLS

FIGURE 15.11 Calculating Allele and Genotype Frequencies Allele and genotype frequencies for a gene locus with two alleles in the population can be calculated using the equations in panel 1. When the equations are applied to two populations (panel 2), we find that the frequencies of alleles *A* and *a* in the two populations are the same, but the alleles are distributed differently between heterozygous and homozygous genotypes.

1 In any population, where *N* is the total number of individuals in the population:

$$\text{Frequency of allele } A = p = \frac{2N_{AA} + N_{Aa}}{2N} \qquad \text{Frequency of allele } a = q = \frac{2N_{aa} + N_{Aa}}{2N}$$

Frequency of genotype $AA = N_{AA}/N$
Frequency of genotype $Aa = N_{Aa}/N$
Frequency of genotype $aa = N_{aa}/N$

2 Compute the allele and genotype frequencies for two separate populations of $N = 200$:

Population 1 (mostly homozygotes)	Population 2 (mostly heterozygotes)
$N_{AA} = 90$, $N_{Aa} = 40$, and $N_{aa} = 70$	$N_{AA} = 45$, $N_{Aa} = 130$, and $N_{aa} = 25$
$p = \frac{180 + 40}{400} = 0.55$	$p = \frac{90 + 130}{400} = 0.55$
$q = \frac{140 + 40}{400} = 0.45$	$q = \frac{50 + 130}{400} = 0.45$
Freq. *AA* = 90/200 = 0.45	Freq. *AA* = 45/200 = 0.225
Freq. *Aa* = 40/200 = 0.20	Freq. *Aa* = 130/200 = 0.65
Freq. *aa* = 70/200 = 0.35	Freq. *aa* = 25/200 = 0.125

$2N_{aa} + N_{Aa}$. If p represents the frequency of A, and q represents the frequency of a, then

$$p = \frac{2N_{AA} + N_{Aa}}{2N}$$

and

$$q = \frac{2N_{aa} + N_{Aa}}{2N}$$

The calculations in Figure 15.11 demonstrate two important points. First, notice that for each population, $p + q = 1$, which means that $q = 1 - p$. So when there are only two alleles at a given locus in a population, we can calculate the frequency of one allele and obtain the second allele's frequency by subtraction. If there is only one allele at a given locus in a population, its frequency is 1; the population is then monomorphic at that locus, and the allele is said to be **fixed.**

The second thing to notice is that population 1 (consisting mostly of homozygotes) and population 2 (consisting mostly of heterozygotes) have the same allele frequencies for *A* and *a*. Thus they have the same gene pool for this locus. Because the alleles in the gene pool are distributed differently among individuals, however, the genotype frequencies of the two populations differ.

The frequencies of the different alleles at each locus and the frequencies of the different genotypes in a population describe that population's **genetic structure**. Allele frequencies measure the amount of genetic variation in a population, whereas genotype frequencies show how a population's genetic variation is distributed among its members. Other measures, such as the proportion of loci that are polymorphic, are also used to measure variation in populations. With these measurements, it becomes possible to consider how the genetic structure of a population changes or remains the same over generations—that is, to measure evolutionary change.

Evolution will occur unless certain restrictive conditions exist

In 1908, the British mathematician Godfrey Hardy and the German physician Wilhelm Weinberg independently deduced the conditions that must prevail if the genetic structure of a population is to remain the same over time. If the conditions they identified do not exist, then evolution will occur. The resulting principle is known as **Hardy–Weinberg equilibrium**. Hardy–Weinberg equilibrium describes a model in which allele frequencies do not change across generations and genotype frequencies can be predicted from allele frequencies (**FIGURE 15.12**). The principles of Hardy–Weinberg equilibrium apply only to sexually reproducing organisms. Several conditions must be met for a population to be at Hardy–Weinberg equilibrium. Note that the following conditions correspond inversely to the five principal processes of evolution (discussed in Concept 15.2):

- *There is no mutation.* The alleles present in the population do not change, and no new alleles are added to the gene pool.
- *There is no selection among genotypes.* Individuals with different genotypes have equal probabilities of survival and equal rates of reproduction.
- *There is no gene flow.* There is no movement of individuals into or out of the population or reproductive contact with other populations.
- *Population size is infinite.* The larger a population, the smaller will be the effect of genetic drift.
- *Mating is random.* Individuals do not preferentially choose mates with certain genotypes.

If these idealized conditions hold, two major consequences follow. First, the frequencies of alleles at a locus remain constant from generation to generation. Second, following one generation of random mating, the genotype frequencies occur in the following proportions:

Genotype	*AA*	*Aa*	*aa*
Frequency	p^2	$2pq$	q^2

To understand why these consequences are important, start by considering a population that is *not* in Hardy–Weinberg equilibrium, such as generation I in Figure 15.12. This could occur, for example, if the initial population is founded by migrants from several other populations, thus violating the Hardy–Weinberg assumption of no gene flow. In this example,

Generation I (Founder population)

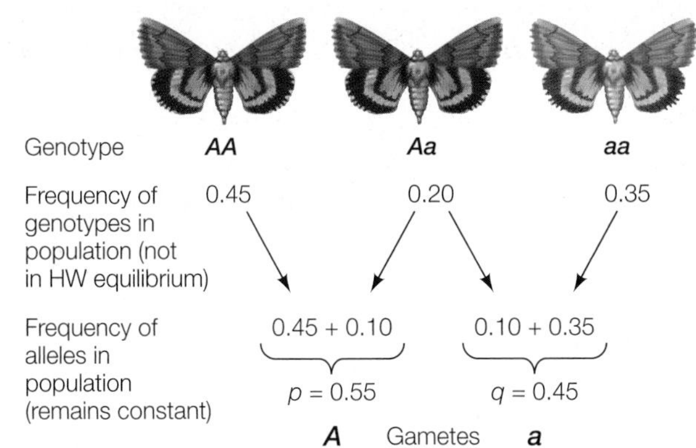

Generation II (Hardy–Weinberg equilibrium restored)

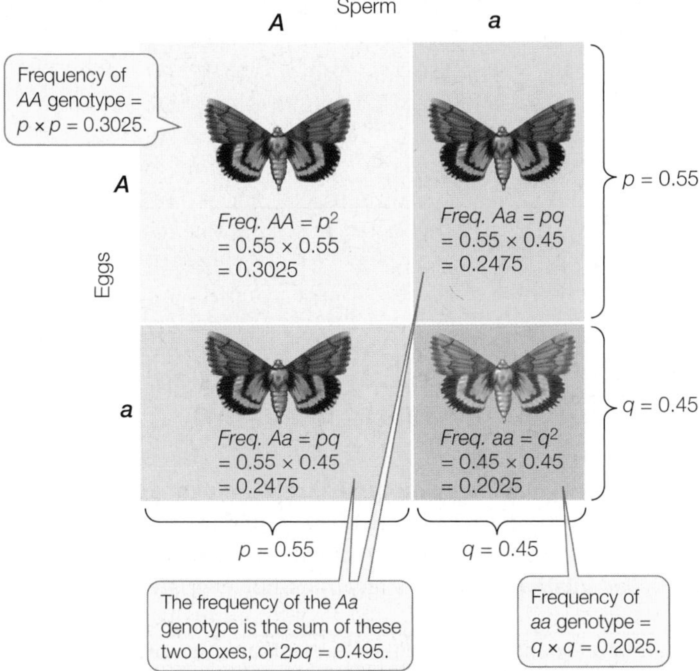

FIGURE 15.12 One Generation of Random Mating Restores Hardy–Weinberg Equilibrium Generation I of this population is made up of migrants from several source populations, and so is not in Hardy–Weinberg equilibrium. After one generation of random mating, the allele frequencies are unchanged, and the genotype frequencies return to Hardy–Weinberg expectations. The lengths of the sides of each rectangle are proportional to the allele frequencies in the population; the areas of the rectangles are proportional to the genotype frequencies.

generation I has more homozygous individuals and fewer heterozygous individuals than would be expected under Hardy–Weinberg equilibrium (a condition known as heterozygote deficiency).

Even with a starting population that is not in Hardy–Weinberg equilibrium, we can predict that after a single generation of random mating, and if the other Hardy–Weinberg assumptions are not violated, the allele frequencies will remain unchanged, but the genotype frequencies will return to Hardy–Weinberg expectations. Let's explore why this is true.

In generation I of Figure 15.12, the frequency of the *A* allele (p) is 0.55. Because we assume that individuals select mates at random, without regard to their genotype, gametes carrying *A* or *a* combine at random—that is, as predicted by the allele frequencies p and q. Thus in this example, the probability that a particular sperm or egg will bear an *A* allele is 0.55. In other words, 55 out of 100 randomly sampled sperm or eggs will bear an *A* allele. Because $q = 1 - p$, the probability that a sperm or egg will bear an *a* allele is $1 - 0.55 = 0.45$.

LINK

You may wish to review the discussion of probability and inheritance in **Concept 8.1**

To obtain the probability of two *A*-bearing gametes coming together at fertilization, we multiply the two independent probabilities of their occurrence:

$$p \times p = p^2 = (0.55)^2 = 0.3025$$

Therefore 0.3025, or 30.25 percent, of the offspring in generation II will have homozygous genotype *AA*. Similarly, the probability of two *a*-bearing gametes coming together is

$$q \times q = q^2 = (0.45)^2 = 0.2025$$

which means that 20.25 percent of generation II will have the *aa* genotype.

There are two ways of producing a heterozygote: an *A* sperm may combine with an *a* egg, the probability of which is $p \times q$; or an *a* sperm may combine with an *A* egg, the probability of which is $q \times p$. Consequently, the overall probability of obtaining a heterozygote is $2pq$, or 0.495. The frequencies of the *AA*, *Aa*, and *aa* genotypes in generation II of Figure 15.12 now meet Hardy–Weinberg expectations, and the frequencies of the two alleles (p and q) have not changed from generation I.

Under the assumptions of Hardy–Weinberg equilibrium, allele frequencies p and q remain constant from generation to generation. If Hardy–Weinberg assumptions are violated and the genotype frequencies in the parental generation are altered (say, by the loss of a large number of *AA* individuals from the population), then the allele frequencies in the next generation will be altered. However, based on the new allele frequencies, another generation of random mating will be sufficient to restore the genotype frequencies to Hardy–Weinberg equilibrium.

Go to ANIMATED TUTORIAL 15.3
Hardy–Weinberg Equilibrium
PoL2e.com/at15.3

Deviations from Hardy–Weinberg equilibrium show that evolution is occurring

You probably have realized that populations in nature never meet the stringent conditions necessary to be at Hardy–Weinberg equilibrium—which explains why all biological populations evolve. Why, then, is this model considered so

APPLY THE CONCEPT

Evolution can be measured by changes in allele frequencies

Imagine you have discovered a new population of curly-tailed lizards established on an island after immigrant lizards have arrived from several different source populations during a hurricane. You collect and tabulate genotype data (right) for the lactate dehydrogenase gene (*Ldh*) for each of the individual lizards. Use the table to answer the following questions.

1. Calculate the allele and genotype frequencies of *Ldh* in this newly founded population.
2. Is the population in Hardy–Weinberg equilibrium? If not, which genotypes are over- or underrepresented? Given the population's history, what is a likely explanation of your answer?
3. Under Hardy–Weinberg assumptions, what allele and genotype frequencies do you predict for the next generation?
4. Imagine that you are able to continue studying this population and determine the next generation's actual allele and genotype frequencies. What are some of the principal reasons you might expect the observed allele and genotype frequencies to differ from the Hardy–Weinberg expectations you calculated in question 3?

INDIVIDUAL NUMBER	SEX	INDIVIDUAL GENOTYPE FOR *Ldh*
1	Male	*Aa*
2	Male	*AA*
3	Female	*AA*
4	Male	*aa*
5	Female	*aa*
6	Female	*AA*
7	Male	*aa*
8	Male	*aa*
9	Female	*Aa*
10	Male	*AA*

important for the study of evolution? There are two reasons. First, the model is useful for predicting the approximate genotype frequencies of a population from its allele frequencies. Second—and crucially—the model allows biologists to evaluate which processes are acting on the evolution of a particular population. The specific patterns of deviation from Hardy–Weinberg equilibrium can help us identify the various processes of evolutionary change.

CHECKpoint CONCEPT 15.3

- ✓ Why is the concept of Hardy–Weinberg equilibrium important even though the assumptions on which it is based are never completely met in nature?
- ✓ Although the stringent assumptions of Hardy–Weinberg equilibrium are never met completely in real populations, the genotype frequencies of many populations do not deviate significantly from Hardy–Weinberg expectations. Can you explain why?
- ✓ Suppose you examine a population of toads breeding in a single pond and find that heterozygous genotypes at several different loci are present at significantly lower frequencies than predicted by Hardy–Weinberg equilibrium. What are some possible explanations?

Our discussion so far has focused on changes in allele frequencies at a single gene locus. Genes do not exist in isolation, however, but interact with one another (and with the environment) to produce an organism's phenotype. What effects can these interactions have on selection?

CONCEPT 15.4 Selection Can Be Stabilizing, Directional, or Disruptive

Until now, we have only discussed traits influenced by alleles at a single locus. Such traits are often distinguished by discrete qualities (black versus white, or smooth versus wrinkled) and so are called **qualitative traits**. Many traits, however, are influenced by alleles at more than one locus. Such traits are likely to show continuous quantitative variation rather than discrete qualitative variation, and so are known as **quantitative traits**. For example, the distribution of body sizes of individuals in a population, a trait that is influenced by genes at many loci as well as by the environment, is likely to resemble a continuous bell-shaped curve.

Natural selection can act on characters with quantitative variation in any one of several different ways, producing quite different results (**FIGURE 15.13**):

- **Stabilizing selection** preserves the average characteristics of a population by favoring average individuals.
- **Directional selection** changes the characteristics of a population by favoring individuals that vary in one direction from the mean of the population.
- **Disruptive selection** changes the characteristics of a population by favoring individuals that vary in both directions from the mean of the population.

Stabilizing selection reduces variation in populations

If the smallest and largest individuals in a population contribute fewer offspring to the next generation than do individuals

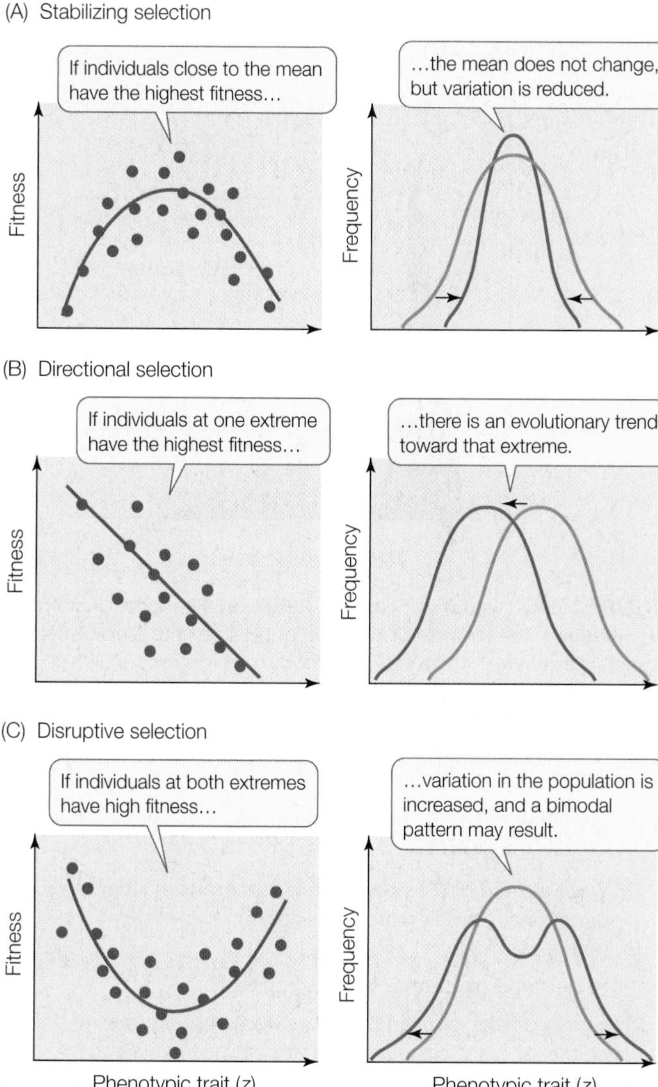

FIGURE 15.13 Natural Selection Can Operate in Several Ways The graphs in the left-hand column show the fitness of individuals with different phenotypes of the same trait. The graphs on the right show the distribution of the phenotypes in the population before (green) and after (blue) the influence of selection.

may result in favoring a particular genetic variant—referred to as **positive selection** for that variant. By favoring one phenotype over another, directional selection results in an increase of the frequencies of alleles that produce the favored phenotype (as with the surface proteins of influenza discussed in the opening of this chapter).

If directional selection operates over many generations, an evolutionary trend is seen in the population (see Figure 15.13B). Evolutionary trends often continue for many generations, but they can be reversed if the environment changes and different phenotypes are favored, or halted when an optimal phenotype is reached or trade-offs between different adaptational advantages oppose further change. The character then undergoes stabilizing selection.

The long horns of Texas Longhorn cattle (**FIGURE 15.15**) are an example of a trait that has evolved through directional selection. Texas Longhorns are descendants of cattle brought to the New World by Christopher Columbus, who picked up a few cattle in the Canary Islands and brought them to the island of Hispaniola in 1493. The cattle multiplied, and their descendants were taken to the mainland of Mexico. Spaniards exploring what would become Texas and the southwestern United States brought these cattle with them, some of which escaped and formed feral herds. Populations of feral cattle increased greatly over the next few hundred years, but there was heavy predation from bears, mountain lions, and wolves, especially on the young calves. Cows with longer horns were more successful in protecting their calves against attacks, and

closer to the average size, then stabilizing selection is operating on size (see Figure 15.13A). Stabilizing selection reduces variation in populations, but it does not change the mean. Natural selection frequently acts in this way, countering increases in variation brought about by sexual recombination, mutation, or gene flow. Rates of phenotypic change in many species are slow because natural selection is often stabilizing. Stabilizing selection operates, for example, on human birth weight. Babies who are lighter or heavier at birth than the population mean die at higher rates than babies whose weights are close to the mean (**FIGURE 15.14**). In discussions of specific genes, stabilizing selection is often called **purifying selection** because there is selection against any deleterious mutations to the usual gene sequence.

Directional selection favors one extreme

Directional selection is operating when individuals at one extreme of a character distribution contribute more offspring to the next generation than other individuals do, shifting the average value of that character in the population toward that extreme. In the case of a single gene locus, directional selection

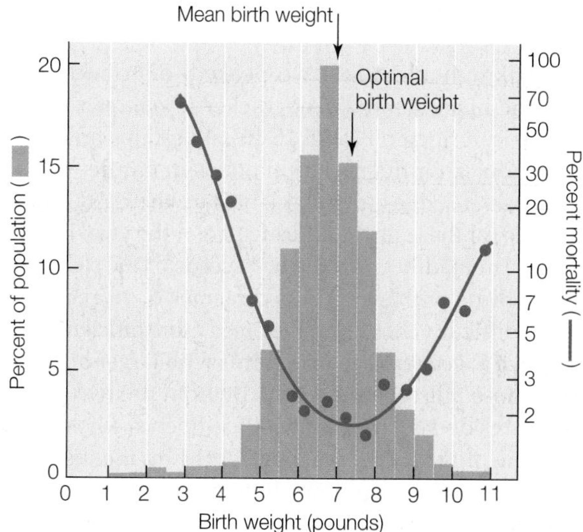

FIGURE 15.14 Human Birth Weight Is Influenced by Stabilizing Selection Babies that weigh more or less than average are more likely to die soon after birth than babies with weights close to the population mean.

FIGURE 15.15 Long Horns Are the Result of Directional Selection Long horns were advantageous for defending young calves from attacks by predators, so horn length increased in feral herds of Spanish cattle in the American Southwest between the early 1500s and the 1860s. The result was the familiar Texas Longhorn breed. This evolutionary trend has been maintained in modern times by ranchers practicing artificial selection.

over a few hundred years the average horn length in the feral herds increased considerably. In addition, the cattle evolved resistance to endemic diseases of the Southwest, as well as higher fecundity and longevity. Texas Longhorns often live and produce calves well into their twenties—about twice as long as many breeds of cattle that have been artificially selected by humans for traits such as high fat content or high milk production (which are examples of artificial directional selection).

Disruptive selection favors extremes over the mean

When disruptive selection operates, individuals at opposite extremes of a character distribution contribute more offspring to the next generation than do individuals close to the mean, which increases variation in the population (see Figure 15.13C).

The strikingly bimodal (two-peaked) distribution of bill sizes in the black-bellied seedcracker (*Pyrenestes ostrinus*), a West African finch (**FIGURE 15.16**), illustrates how disruptive selection can influence populations in nature. The seeds of two types of sedges (marsh plants) are the most abundant food source for these finches during part of the year. Birds with large bills can readily crack the hard seeds of the sedge *Scleria verrucosa*. Birds with small bills can crack *S. verrucosa* seeds only with difficulty; however, they feed more efficiently on the soft seeds of *S. goossensii* than do birds with larger bills. Young finches whose bills deviate markedly from the two predominant bill sizes do not survive as well as finches whose bills are close to one of the two sizes represented by the distribution peaks. Because there are few abundant food sources in the finches' environment, and because the seeds of the two sedges do not overlap in hardness, birds with intermediate-sized bills are less efficient in using either one of the species' principal food sources. Disruptive selection therefore maintains a bimodal bill size distribution.

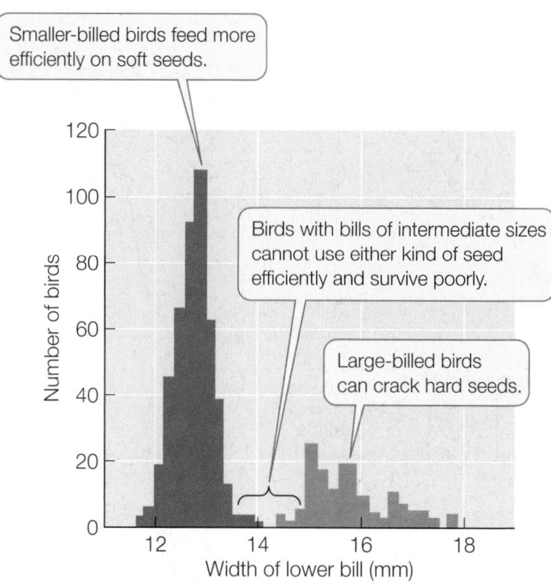

FIGURE 15.16 Disruptive Selection Results in a Bimodal Character Distribution The bimodal distribution of bill sizes in the black-bellied seedcracker of West Africa is a result of disruptive selection, which favors individuals with larger and smaller bill sizes over individuals with intermediate-sized bills.

CHECKpoint CONCEPT 15.4

- ✓ What are the different expected outcomes of stabilizing, directional, and disruptive selection?
- ✓ Why would you expect selection on human birth weight to be stabilizing rather than directional?
- ✓ Can you think of examples of extreme phenotypes in animal or plant populations that could be explained by directional selection?

Our discussion so far has largely focused on the evolution of phenotypes (what organisms look like and how they behave). We will now consider the specific mechanistic processes that operate at the level of genes and genomes.

CONCEPT 15.5 Genomes Reveal Both Neutral and Selective Processes of Evolution

Most natural populations harbor far more genetic variation than we would expect to find if genetic variation were influenced by natural selection alone. This discovery, combined with the knowledge that many mutations do not change molecular function, provided a major stimulus to the development of the field of molecular evolution.

To discuss the evolution of genes, we need to consider the specific types of mutations that are possible. A nucleotide substitution is a change in a single nucleotide in a DNA sequence (a type of point mutation). Many nucleotide substitutions have no effect on phenotype, even if the change occurs in a gene that encodes a protein, because most amino acids are specified by

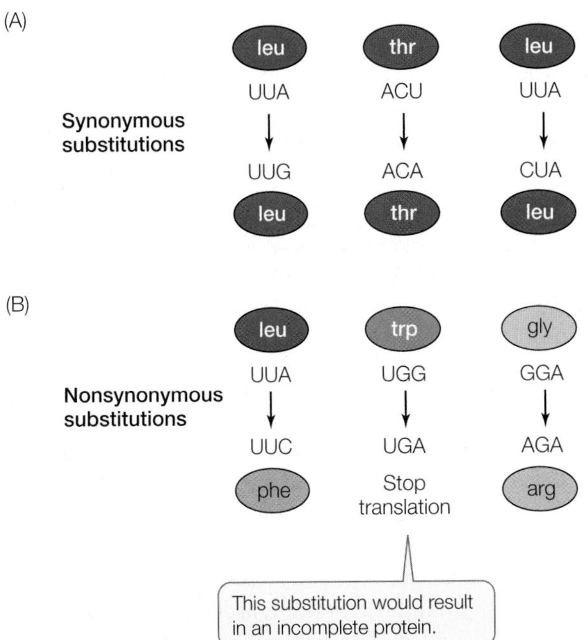

FIGURE 15.17 When One Nucleotide Changes (A) Synonymous substitutions do not change the amino acid specified and do not affect protein function. Such substitutions are less likely to be subject to natural selection, although they contribute greatly to the buildup of neutral genetic variation in a population. (B) Nonsynonymous substitutions do change the amino acid sequence and are likely to have an effect (often deleterious, but sometimes beneficial) on protein function. Such nucleotide substitutions are targets for natural selection.

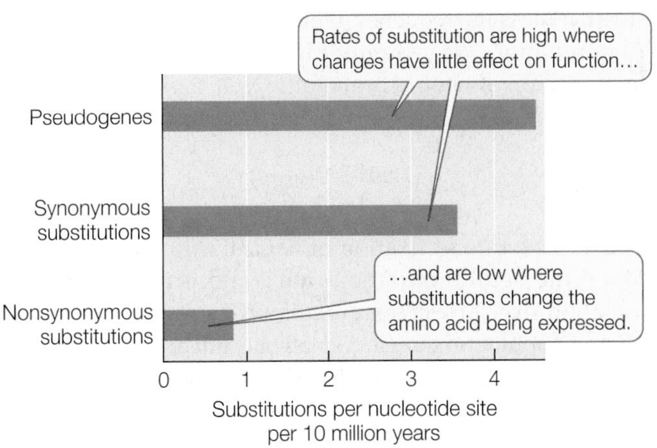

FIGURE 15.18 Rates of Substitution Differ Rates of nonsynonymous substitution are typically much lower than rates of synonymous substitution, and much lower than substitution rates in pseudogenes. This pattern reflects stronger stabilizing selection in functional genes than in pseudogenes.

more than one codon. A substitution that does not change the encoded amino acid is known as a **synonymous substitution** (also called a **silent substitution**; **FIGURE 15.17A**). Synonymous substitutions do not affect the functioning of a protein (although they may have other effects, such as changes in mRNA stability or translation rates) and are therefore less likely to be influenced by natural selection.

A nucleotide substitution that *does* change the amino acid sequence encoded by a gene is known as a **nonsynonymous substitution** (also called a **missense substitution**; **FIGURE 15.17B**). In general, nonsynonymous substitutions are likely to be deleterious to the organism. But not every amino acid replacement alters a protein's shape and charge (and hence its functional properties). Therefore some nonsynonymous substitutions are selectively neutral, or nearly so. A third possibility is that a nonsynonymous substitution alters a protein in a way that confers an advantage to the organism, and is therefore favored by natural selection.

LINK

The genetic code determines the amino acid that is encoded by each codon; see **Figure 10.11**

The rate of synonymous substitutions in most protein-coding genes is much higher than the rate of nonsynonymous substitutions. In other words, *substitution rates are highest at nucleotide positions that do not change the amino acid being expressed* (**FIGURE 15.18**). The rate of substitution is even higher in **pseudogenes**, which are copies of genes that are no longer functional.

Insertions, deletions, and rearrangements of DNA sequences are all mutations that may affect a larger portion of the gene or genome than do point mutations (see Concept 9.3). Insertions and deletions of nucleotides in a protein-coding sequence interrupt its reading frame, unless they occur in multiples of three nucleotides (the length of one codon). Rearrangements may merely change the order of whole genes along chromosomes, or they may rearrange functional domains among individual genes.

When biologists began to examine the details of genetic variation of populations, they soon discovered many gene variants that had little or no effect on function. This gave rise to new ideas about how these neutral variants arise and spread in populations.

Much of molecular evolution is neutral

Motoo Kimura proposed the neutral theory in 1968. He suggested that, at the molecular level, the majority of variants found in most populations are selectively neutral. That is, most gene variants confer neither an advantage nor a disadvantage on their bearers. Therefore these neutral variants must accumulate through genetic drift rather than through positive selection.

We saw in Concept 15.2 that genetic drift of existing gene variants tends to be greatest in small populations. However, the rate of fixation of new neutral mutations by genetic drift is independent of population size. To see why this is so, consider a population of size N and a neutral mutation rate of μ (mu) per gamete per generation at a particular locus. The number of new mutations would be, on average, $\mu \times 2N$, because $2N$ gene copies are available to mutate in a population of diploid organisms. The probability that a given mutation will be fixed by drift alone is its frequency, which equals $1/(2N)$ for a newly arisen mutation. We can multiply these two terms to get the rate of fixation of neutral mutations (m) in a given population of N individuals:

$$m = 2N\mu\frac{1}{2N}$$

Therefore the rate of fixation of neutral mutations depends only on the neutral mutation rate μ and is independent of population size. Any given mutation is more likely to appear in a large population than in a small one, but any mutation that does appear is more likely to become fixed in a small population. These two influences of population size cancel each other out. Therefore the rate of fixation of neutral mutations is equal to the mutation rate (i.e., $m = \mu$).

As long as the underlying mutation rate is constant, genes and proteins evolving in different populations should diverge from one another in neutral changes at a constant rate. The rate of evolution of particular genes and proteins is indeed often relatively constant over time, and therefore can be used as a "molecular clock" to calculate evolutionary divergence times between species (see Concept 16.3).

Neutral theory does not imply that most mutations have no effect on the individual organism, even though much of the genetic variation present in a population is the result of neutral evolution. Many mutations are never observed in populations because they are lethal or strongly detrimental, and the individuals that carry them are quickly removed from the population through natural selection. Similarly, because mutations that confer a selective advantage tend to be quickly fixed in populations, they also do not result in significant variation at the population level. Nonetheless, if we compare homologous proteins from different populations or species, some amino acid positions will remain constant under purifying selection, others will vary through neutral genetic drift, and still others will differ among species as a result of positive selection for change. How can these evolutionary processes be distinguished?

Go to MEDIA CLIP 15.2
The Ubiquitous Protein
PoL2e.com/mc15.2

Positive and purifying selection can be detected in the genome

As we have just seen, substitutions in a protein-coding gene can be either synonymous or nonsynonymous, depending on whether they change the resulting amino acid sequence of the protein. The relative rates of synonymous and nonsynonymous substitutions are expected to differ in regions of genes that are evolving neutrally, or evolving under positive selection for change, or staying unchanged under purifying selection.

- If a given amino acid in a protein can be one of many alternatives (without changing the protein's function), then an amino acid replacement is *neutral* with respect to the fitness of an organism. In this case, the rates of synonymous and nonsynonymous substitutions in the corresponding DNA sequences are expected to be very similar, so the ratio of the two rates should be close to 1.
- If a given amino acid position is under *positive selection* for change, the observed rate of nonsynonymous substitutions is expected to exceed the rate of synonymous substitutions in the corresponding DNA sequences.
- If a given amino acid position is under *purifying selection,* then the observed rate of synonymous substitutions is expected to be much higher than the rate of nonsynonymous substitutions in the corresponding DNA sequences.

The evolution of lysozyme illustrates how and why particular codons in a gene sequence might be under different modes of selection. The enzyme lysozyme is found in almost all animals. It is produced in the tears, saliva, and milk of mammals and in the albumen (whites) of bird eggs. Lysozyme digests the cell walls of bacteria, rupturing and killing them. As a result, it plays an important role as a first line of defense against invading bacteria. Most animals defend themselves against bacteria by digesting them, which is probably why most animals have lysozyme. Some animals also use lysozyme in the digestion of food.

Among mammals, a mode of digestion called foregut fermentation has evolved twice. In mammals with this mode of digestion, the foregut—consisting of part of the esophagus and/or stomach—has been converted into a chamber in which bacteria break down ingested plant matter by fermentation. Foregut fermenters can obtain nutrients from the otherwise indigestible cellulose that makes up a large proportion of plant tissue. Foregut fermentation evolved independently in ruminants (a group of hoofed mammals that includes cattle) and in certain leaf-eating monkeys, such as langurs. We know that these evolutionary events were independent because both langurs and ruminants have close relatives that are not foregut fermenters.

In both mammalian foregut-fermenting lineages, lysozyme has been modified to play a new, nondefensive role. The modified lysozyme enzyme ruptures some of the bacteria that live in the foregut, releasing nutrients metabolized by the bacteria, which the mammal then absorbs. How many changes in the lysozyme molecule were needed to allow it to perform this function amid the digestive enzymes and acidic conditions of the mammalian foregut? To answer this question, biologists compared the lysozyme-coding sequences in foregut fermenters with those in several of their nonfermenting relatives. They determined which amino acids differed and which were shared among the species (**FIGURE 15.19A**), as well as the rates of synonymous and nonsynonymous substitution in lysozyme genes across the evolutionary history of the sampled species.

For many of the amino acid positions of lysozyme, the rate of synonymous substitution in the corresponding gene sequence was much higher than the rate of nonsynonymous substitution. This observation indicates that many of the amino acids that make up lysozyme are evolving under purifying selection. In other words, there is selection against change in the lysozyme protein at these positions, and the encoded amino acids must therefore be critical for lysozyme function. At other

(A) *Semnopithecus* sp.

Bos taurus

(B) *Opisthocomus hoazin*

	Langur	Baboon	Human	Rat	Cattle	Horse
Langur		14	18	38	32	65
Baboon	0		14	33	39	65
Human	0	1		37	41	64
Rat	0	0	0		55	64
Cattle	5	0	0	0		71
Horse	0	0	0	0	1	

The lysozymes of langurs and cattle are convergent for 5 amino acid residues, indicative of the independent evolution of foregut fermentation in these two species.

FIGURE 15.19 Convergent Molecular Evolution of Lysozyme (A) The numbers of amino acid differences in the lysozymes of several pairs of mammals are shown above the diagonal line; the numbers of similarities that arose from convergence between species are shown below the diagonal. The two foregut-fermenting species (cattle and langur) share five convergent amino acid replacements related to this digestive adaptation. (B) The hoatzin—the only known foregut-fermenting bird species—has been evolving independently from mammals for hundreds of millions of years but has independently evolved modifications to lysozyme similar to those found in cattle and langurs.

positions, several different amino acids function equally well, and the corresponding codons have similar rates of synonymous and nonsynonymous substitution.

The most striking finding was that amino acid replacements in lysozyme happened at a much higher rate in the lineage leading to langurs than in any other primate lineage. The high rate of nonsynonymous substitution in the langur lysozyme gene shows that lysozyme went through a period of rapid change in adapting to the stomachs of langurs. Moreover, the lysozymes of langurs and cattle share five convergent amino

APPLY THE CONCEPT

Genomes reveal both neutral and selective processes of evolution

Analysis of synonymous and nonsynonymous substitutions in protein-coding genes can be used to detect neutral evolution, positive selection, and purifying selection. An investigator compared many gene sequences that encode the protein hemagglutinin (a surface protein of influenza virus) sampled over time, and collected the data at right.[a] Use the table to answer the following questions.

1. Which codon positions are likely evolving under positive selection? Why?
2. Which codon positions are likely evolving under purifying selection? Why?

(Hint: To calculate rates of each substitution type, you will need to consider the number of synonymous and nonsynonymous substitutions *relative to the number of possible substitutions of each type*. There are approximately three times as many possible nonsynonomous substitutions as there are synonymous substitutions.)

CODON POSITION	NUMBER OF SYNONYMOUS SUBSTITUTIONS IN CODON	NUMBER OF NONSYNONYMOUS SUBSTITUTIONS IN CODON
12	0	7
15	1	9
61	0	12
80	7	0
137	12	1
156	24	2
165	3	4
226	38	3

[a]R. M. Bush et al. 1999. *Molecular Biology and Evolution* 16: 1457–1465.

acid replacements, all of which lie on the surface of the lysozyme molecule, well away from the enzyme's active site. Several of these shared replacements are changes from arginine to lysine, which make the protein more resistant to degradation by the stomach enzyme pepsin. By understanding the functional significance of amino acid replacements, biologists can explain the observed changes in amino acid sequences in terms of changes in the functioning of the protein.

A large body of fossil, morphological, and molecular evidence shows that langurs and cattle do not share a recent common ancestor. However, langur and ruminant lysozymes share several amino acids that neither mammal shares with the lysozymes of its own closer relatives. The lysozymes of these two mammals have converged on some of the same amino acids despite their very different ancestry. The amino acids they share give these lysozymes the ability to lyse the bacteria that ferment plant material in the foregut.

The hoatzin, an unusual leaf-eating South American bird (**FIGURE 15.19B**) and the only known avian foregut fermenter, offers another remarkable example of the convergent evolution of lysozyme. Many birds have an enlarged esophageal chamber called a crop. The crop of the hoatzin contains lysozyme and bacteria and acts as a fermentation chamber. Many of the amino acid replacements that occurred in the adaptation of hoatzin lysozyme are identical to those that evolved in ruminants and langurs. Thus even though the hoatzin and foregut-fermenting mammals have not shared a common ancestor in hundreds of millions of years, similar adaptations have evolved in their lysozyme enzymes, enabling both groups to recover nutrients from fermenting bacteria.

Heterozygote advantage maintains polymorphic loci

In many cases, different alleles of a particular gene are advantageous under different environmental conditions. Most organisms, however, experience a wide diversity of environments. A night is dramatically different from the preceding day. A cold, cloudy day differs from a clear, hot one. Day length and temperature change seasonally. For many genes, a single allele is unlikely to perform well under all these conditions. In such situations, a heterozygous individual (with two different alleles) is likely to outperform individuals that are homozygous for either one of the alleles.

Colias butterflies of the Rocky Mountains live in environments where dawn temperatures often are too cold, and afternoon temperatures too hot, for the butterflies to fly. Populations of these butterflies are polymorphic for the gene that encodes

INVESTIGATION

FIGURE 15.20 A Heterozygote Mating Advantage Among butterflies of the genus *Colias*, males that are heterozygous for two alleles of the PGI enzyme can fly farther under a broader range of temperatures than males that are homozygous for either allele. Does this ability give heterozygous males a mating advantage?[a]

HYPOTHESIS

Heterozygous male *Colias* will have proportionally greater mating success than homozygous males.

METHOD

1. For each of two *Colias* species, capture butterflies in the field. In the laboratory, determine their genotypes and allow them to mate.
2. Determine the genotypes of the offspring, thus revealing paternity and mating success of the males.

RESULTS

For both species, the proportion of heterozygous males that mated successfully was higher than the proportion of all males seeking females ("flying").

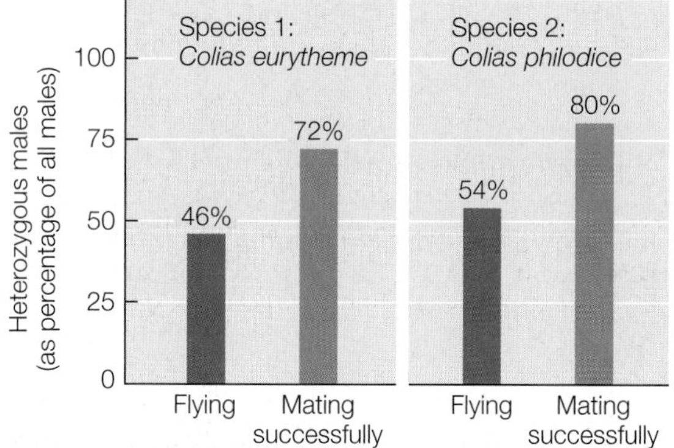

CONCLUSION

Heterozygous *Colias* males have a mating advantage over homozygous males.

ANALYZE THE DATA

Analyze these sampling data collected during the experiment (only one of several samples is shown for each species).

	All viable males*		Mating Males	
Species	Heterozygous/ total	Percent heterozygous	Heterozygous/ total	Percent heterozygous
C. philodice	32/74	43.2	31/50	62.0
C. eurytheme	44/92	47.8	45/59	76.3

*"Viable males" are all males captured flying with females (hence with the potential to mate).

A. Under the assumption that the proportions of each genotype (heterozygotes and homozygotes) of mating males are the same as the proportions seen among all viable males, calculate the number of mating males expected to be heterozygous and the number expected to be homozygous.

B. Use a chi-square test (see Appendix B) to evaluate the significance of the difference in your expected numbers in (A) and the observed percentages of heterozygous mating males. The critical value ($P = 0.05$) of the chi-square distribution with one degree of freedom is 3.841. Are the observed and expected numbers of heterozygotes and homozygotes among mating males significantly different in these samples?

Go to **LaunchPad** for discussion and relevant links for all **INVESTIGATION** figures.

[a]W. B. Watt et al. 1985. *Genetics* 109: 157–175.

phosphoglucose isomerase (PGI), an enzyme that influences how well an individual flies at different temperatures. Butterflies with certain PGI genotypes can fly better during the cold hours of early morning; those with other genotypes perform better during midday heat. The optimal body temperature for flight is 35°C–39°C, but some butterflies can fly with body temperatures as low as 29°C or as high as 40°C. Heat-tolerant genotypes are favored during spells of unusually hot weather; during spells of unusually cool weather, cold-tolerant genotypes are favored.

Heterozygous *Colias* butterflies can fly over a greater temperature range than homozygous individuals because they produce two different forms of PGI. This greater range of activity should give them an advantage in foraging and finding mates. A test of this prediction did find a mating advantage in heterozygous males, and further found that this mating advantage maintains the polymorphism in the population (**FIGURE 15.20**). The heterozygous condition can never become fixed in the population, however, because the offspring of two heterozygotes will always include both classes of homozygotes in addition to heterozygotes.

Genome size and organization also evolve

We know that genome size varies tremendously among organisms. Across broad taxonomic categories, there is some correlation between genome size and organismal complexity. The genome of the tiny bacterium *Mycoplasma genitalium* has only 470 genes. *Rickettsia prowazekii*, the bacterium that causes typhus, has 634 genes. *Homo sapiens*, by contrast, has about 21,000 protein-coding genes. **FIGURE 15.21** shows the number of genes from a sample of organisms whose genomes have been fully sequenced, arranged by their evolutionary relationships. As this figure reveals, however, a larger genome does not always indicate greater complexity (compare rice with the other plants, for example). It is not surprising that more complex genetic instructions are needed for building and maintaining a large, multicellular organism than a small, single-celled bacterium. What is surprising is that some organisms, such as lungfishes, some salamanders, and lilies, have about 40 times as much DNA as humans do (**FIGURE 15.22**). Structurally, a lungfish or a lily is not 40 times more complex than a human. So why does genome size vary so much?

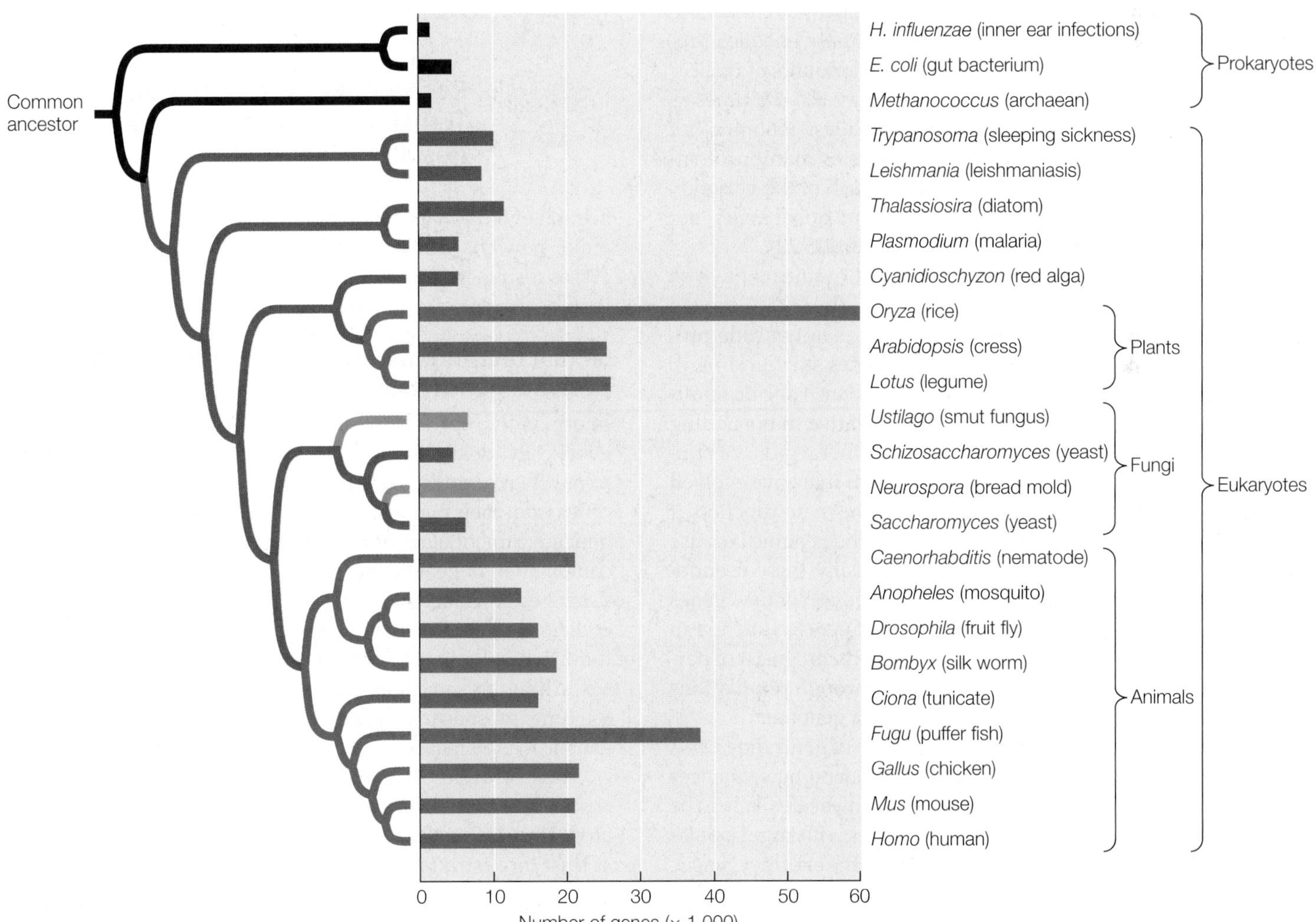

FIGURE 15.21 Evolution of Gene Number This figure shows the number of genes from a sample of organisms whose genomes have been fully sequenced, arranged by their evolutionary relationships. Bacteria and archaea (black branches) typically have fewer genes than most eukaryotes. Among eukaryotes, multicellular organisms with tissue organization (plants and animals; blue branches) have more genes than single-celled organisms (red branches) or multicellular organisms that lack pronounced tissue organization (green branches).

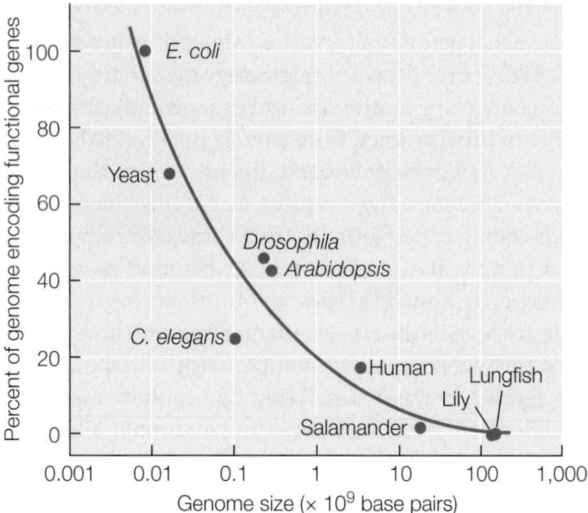

FIGURE 15.22 A Large Proportion of DNA Is Noncoding Most of the DNA of bacteria and yeasts encodes RNAs or proteins, but a large percentage of the DNA of multicellular species is noncoding.

Differences in genome size are not so great if we take into account only the portion of DNA that actually encodes proteins. The organisms with the largest total amounts of nuclear DNA (some ferns and flowering plants) have 80,000 times as much DNA as do the bacteria with the smallest genomes, but no species has more than about 100 times as many protein-coding genes as a bacterium. Therefore much of the variation in genome size lies not in the number of functional genes, but in the amount of noncoding DNA (see Figure 15.22).

Why do the cells of most eukaryotic organisms have so much noncoding DNA? Does this noncoding DNA have a function? Although some of this DNA does not encode proteins, it can alter the expression of the genes surrounding it. The degree or timing of gene expression can vary dramatically depending on the gene's position relative to noncoding sequences that regulate gene expression. Other regions of noncoding DNA consist of pseudogenes (regions that have evolved from functional genes, even though they have no function at present). Pseudogenes are often carried in the genome because the cost of doing so is very small. Occasionally, these pseudogenes become the raw material for the evolution of new genes with novel functions. Other noncoding sequences function in maintaining chromosomal structure. Still others consist of parasitic transposable elements that spread through populations because they reproduce faster than the host genome.

Another hypothesis is that the proportion of noncoding DNA is related primarily to population size. Noncoding sequences that are only slightly deleterious to the organism are likely to be purged by selection most efficiently in species with large population sizes. In species with small populations, the effects of genetic drift can overwhelm selection against noncoding sequences that have small deleterious consequences. Therefore selection against the accumulation of noncoding sequences is most effective in species with large populations, so such species (such as bacteria or yeasts) have relatively little noncoding DNA compared with species with small populations (see Figure 15.22).

CHECKpoint CONCEPT 15.5

- ✓ How can the ratio of synonymous to nonsynonymous substitutions be used to determine whether a particular gene is evolving neutrally, under positive selection, or under stabilizing selection?
- ✓ Why is the rate of fixation of neutral mutations independent of population size?
- ✓ Why do heterozygous individuals sometimes have an advantage over homozygous individuals?
- ✓ Why can a mutation that results in the replacement of one amino acid by another be a neutral event in some cases and in other cases be detrimental or beneficial? (Hint: Review the information about amino acids in Table 3.2 and the details of protein structure in Concept 3.2.)
- ✓ Postulate and contrast two hypotheses for the wide diversity of genome sizes among different organisms.

Most of our discussion so far has centered on changes in existing genes and phenotypes. Next we'll consider how new genes with novel functions arise in populations in the first place.

CONCEPT 15.6 Recombination, Lateral Gene Transfer, and Gene Duplication Can Result in New Features

Several evolutionary processes can result in the acquisition of major new characteristics in populations. Each of these processes results in larger and more rapid evolutionary changes than do single point mutations.

Sexual recombination amplifies the number of possible genotypes

In asexually reproducing organisms, each new individual is genetically identical to its parent unless there has been a mutation. When organisms reproduce sexually, however, offspring differ from their parents because of crossing over and independent assortment of chromosomes during meiosis, as well as the combination of genetic material from two different gametes, as described in Concept 7.4. Sexual recombination generates an endless variety of genotype combinations that increase the evolutionary potential of populations—a long-term advantage of sex. Although some species may reproduce asexually most of the time, most asexual species have some means of achieving genetic recombination.

The evolution of meiosis and sexual recombination was a crucial event in the history of life. Exactly how these processes arose is puzzling, however, because in the short term, sex has at least three striking disadvantages:

- Recombination breaks up adaptive combinations of genes.
- Sex reduces the rate at which females pass genes on to their offspring.
- Dividing offspring into separate genders greatly reduces the overall reproductive rate.

To see why this last disadvantage exists, consider an asexual female that produces the same number of offspring as a sexual female. Assume that both females produce two offspring, but that half of the sexual female's offspring are males. In the next (F_1) generation, then, each of the two asexual F_1 females will produce two more offspring—but there is only one sexual F_1 female to produce offspring. Thus the effective reproductive rate of the asexual lineage is twice that of the sexual lineage. The evolutionary problem is to identify the advantages of sex that can overcome such short-term disadvantages.

A number of hypotheses have been proposed to explain the existence of sex, none of which are mutually exclusive. One is that sexual recombination facilitates repair of damaged DNA, because breaks and other errors in DNA on one chromosome can be repaired by copying the intact sequence from the homologous chromosome.

Another advantage of sexual reproduction is that it permits the elimination of deleterious mutations through recombination followed by selection. As Concept 9.2 described, DNA replication is not perfect, and many replication errors result in lower fitness. Meiotic recombination distributes these deleterious mutations unequally among gametes. Sexual reproduction then produces some individuals with more deleterious mutations and some with fewer. The individuals with fewer deleterious mutations are more likely to survive. Therefore sexual reproduction allows natural selection to eliminate particular deleterious mutations from the population over time.

In asexual reproduction, deleterious mutations can be eliminated only by the death of the lineage or by a rare back mutation (that is, when a subsequent mutation returns a mutated sequence to its original DNA sequence). Hermann J. Muller noted that deleterious mutations in a non-recombining genome accumulate—"ratchet up"—at each replication. Mutations occur and are passed on each time a genome replicates, and these mutations accumulate with each subsequent generation. This accumulation of deleterious mutations in lineages that lack genetic recombination is known as **Muller's ratchet**.

Another explanation for the existence of sex is that the great variety of genetic combinations created in each generation can itself be advantageous. For example, genetic variation can be a defense against pathogens and parasites. Most pathogens and parasites have much shorter life cycles than their hosts and can rapidly evolve counter-adaptations to host defenses. Sexual recombination might give the host's defenses a chance to keep up.

Sexual recombination does not directly influence the frequencies of alleles. Rather, *it generates new combinations of alleles on which natural selection can act*. It expands variation in quantitative characters by creating new genotypes. That is why artificial selection for bristle number in *Drosophila* (see Figure 15.7) resulted in flies with either more or fewer bristles than were present in the flies in the initial population.

Lateral gene transfer can result in the gain of new functions

The tree of life is usually visualized as a branching diagram, with each lineage diverging into two (or more) lineages over time, from one common ancestor to the millions of species that are alive today. Ancestral lineages divide into descendant lineages, and it is those speciation events that the tree of life captures. However, there are also processes that result in **lateral gene transfer**—the horizontal movement of individual genes, organelles, or fragments of genomes from one lineage to another. Some species may pick up fragments of DNA directly from the environment. A virus may pick up some genes from one host and transfer them to a new host when the virus becomes integrated into the new host's genome. Hybridization between species also results in the lateral transfer of large numbers of genes.

Lateral gene transfer can be highly advantageous to the species that incorporates novel genes from a distant relative. Genes that confer antibiotic resistance, for example, are commonly transferred among different species of bacteria. Lateral gene transfer is another way, in addition to mutation and recombination, that species can increase their genetic variation.

The degree to which lateral gene transfer events occur in various parts of the tree of life is a matter of considerable current investigation and debate. Lateral gene transfer appears to be relatively uncommon among most eukaryote lineages, although the two major endosymbioses that gave rise to mitochondria and chloroplasts involved lateral transfers of entire bacterial genomes to the eukaryote lineage. Some groups of eukaryotes, most notably some plants, are subject to relatively high levels of hybridization among closely related species. Hybridization leads to the exchange of many genes among recently separated lineages of plants. The greatest degree of lateral transfer, however, occurs among bacteria. Many genes have been transferred repeatedly among bacteria, to the point that relationships and boundaries among species of bacteria are sometimes hard to decipher.

Many new functions arise following gene duplication

Gene duplication is yet another way that genomes can acquire new functions. When a gene is duplicated, one copy of that gene is potentially freed from having to perform its original function. The identical copies of a duplicated gene can have any one of four different fates:

- Both copies of the gene may retain their original function (which can result in a change in the amount of gene product that is produced by the organism).
- Both copies of the gene may retain the ability to produce the original gene product, but the expression of the genes may diverge in different tissues or at different times in development.
- One copy of the gene may be incapacitated by the accumulation of deleterious mutations and become a functionless pseudogene.
- One copy of the gene may retain its original function while the second copy changes and evolves a new function.

How often do gene duplications arise, and which of these four outcomes is most likely? Investigators have found that rates of gene duplication are fast enough for a yeast or *Drosophila* population to acquire several hundred duplicate genes over the course of a million years. They have also found that most of the duplicated genes that are still present in these organisms are very young. Many duplicated genes are lost from a genome within 10 million years—an eyeblink on an evolutionary time scale.

Many gene duplications affect only one or a few genes at a time, but in some cases entire genomes may be duplicated. When all the genes are duplicated, there are massive opportunities for new functions to evolve. That is exactly what seems to have happened during the course of vertebrate evolution. The genomes of the jawed vertebrates have four diploid sets of many major genes, which leads biologists to conclude that two genome-wide duplication events occurred in the ancestor of these species. These duplications allowed considerable specialization of individual vertebrate genes, many of which are now highly tissue-specific in their expression.

LINK

See **Concept 14.4** for a discussion of the role of duplicated Hox genes in vertebrate evolution

Several successive rounds of duplication and sequence evolution may result in a **gene family**, a group of homologous genes with related functions, often arrayed in tandem along a chromosome. An example of a group of genes related by gene duplication is the globin gene family (**FIGURE 15.23**). Comparisons of the amino acid sequences among globins strongly suggest that this family of proteins arose via gene duplications.

Hemoglobin is a tetramer (four-subunit molecule) consisting of two α-globin and two β-globin polypeptide chains. It carries oxygen in the blood. Myoglobin, a monomer, is the primary O_2 storage protein in muscle. Myoglobin's affinity for O_2 is much higher than that of hemoglobin, but hemoglobin has evolved to be more diversified in its role. Hemoglobin binds O_2 in the lungs or gills, where the O_2 concentration is relatively high, transports it to deep body tissues, where the O_2 concentration is low, and releases it in those tissues. With its more complex tetrameric structure, hemoglobin is able to carry four molecules of O_2, as well as hydrogen ions and carbon dioxide, in the blood. Hemoglobin and myoglobin are estimated to have arisen through gene duplication about 500 million years ago.

CHECKpoint CONCEPT 15.6

✓ What are some of the potential advantages of lateral gene transfer to the organisms that gain new genes by this process?

✓ Why is gene duplication considered important for long-term evolutionary change?

✓ Why is sexual reproduction so prevalent in nature, despite its having at least three short-term evolutionary disadvantages?

The development of evolutionary theory has helped reveal how biological molecules function, how genetic diversity is created

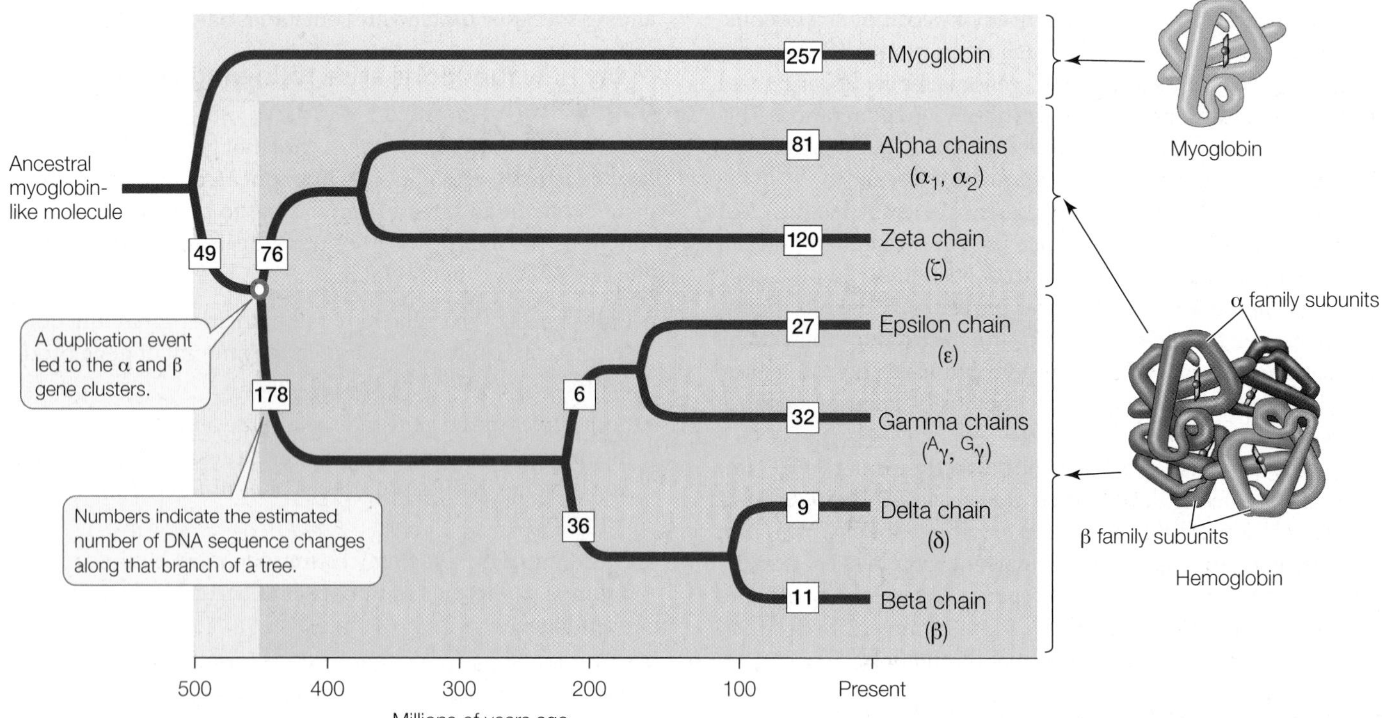

FIGURE 15.23 A Globin Family Gene Tree This gene tree suggests that the α-globin and β-globin gene clusters diverged about 450 million years ago (open circle), soon after the origin of the vertebrates.

Go to ACTIVITY 15.2 Gene Tree Construction PoL2e.com/ac15.2

and maintained, and how organisms develop new features. Next we will see how biologists put this theory into practice.

CONCEPT 15.7 Evolutionary Theory Has Practical Applications

Evolutionary theory has many practical applications across biology, and new ones are being developed every day. Here we'll discuss a few of these applications to fields such as agriculture, industry, and medicine.

Knowledge of gene evolution is used to study protein function

Earlier in this chapter we discussed some of the ways biologists can detect codons or genes that are under positive selection for change. These methods have greatly increased our understanding of the functions of many genes. Consider, for example, the gated sodium channel genes. Sodium channels have many functions, including the control of nerve impulses in the nervous system (see Concept 34.2). Sodium channels can become blocked when they bind certain toxins, one of which is the tetrodotoxin (TTX) present in puffer fishes and many other animals. A human who eats puffer fish tissues that contain TTX can become paralyzed and die because the toxin-blocked sodium channels prevent nerves and muscles from functioning properly.

But puffer fish themselves have sodium channels, so why doesn't the TTX in their system paralyze them? Nucleotide substitutions in the puffer fish genome have resulted in structural changes in the proteins that form the sodium channels, and those changes prevent TTX from binding to the channel pore. Several different substitutions that result in such resistance have evolved in the various duplicated sodium channel genes of the many species of puffer fish. Many other changes that have nothing to do with the evolution of tetrodotoxin resistance have occurred in these genes as well.

So how does what we have learned about the evolution of TTX-resistant sodium channels affect our lives? Mutations in human sodium channel genes are responsible for a number of neurological pathologies. By studying the function of sodium channels and understanding which changes have produced tetrodotoxin resistance, we are learning a great deal about how these crucial channels work and how various mutations affect them. Biologists do this by comparing rates of synonymous and nonsynonymous substitutions across sodium channel genes in various animals that have evolved TTX resistance. In a similar manner, molecular evolutionary principles are used to understand function and diversification of function in many other proteins.

In vitro evolution produces new molecules

Living organisms produce thousands of compounds that humans have found useful. The search for naturally occurring compounds that can be used for pharmaceutical, agricultural, or industrial purposes has been termed "bioprospecting." These compounds are the result of millions of years of molecular evolution across millions of species of living organisms. Yet biologists can imagine molecules that could have evolved but have not, in the absence of the right combination of selection pressures and opportunities.

For instance, we might want to find a molecule that binds a particular environmental contaminant so that the contaminant can be isolated and extracted from the environment. But if the contaminant is synthetic (not produced naturally), then it is unlikely that any living organism would have evolved a molecule with the function we desire. This problem was the inspiration for the field of **in vitro evolution**, in which new molecules are produced in the laboratory to perform novel and useful functions.

The principles of in vitro evolution are based on principles of molecular evolution that we have learned from the natural world. Consider a new RNA molecule that was produced in the laboratory using the principles of mutation and selection. The new molecule's intended function was to join two other RNA molecules (acting as a ribozyme with a function similar to that of the naturally occurring DNA ligase described in Concept 9.2, but for RNA molecules). The process started with a large pool of random RNA sequences (10^{15} different sequences, each about 300 nucleotides long), which were then selected for displaying any ligase activity (**FIGURE 15.24**). None were very effective ligases, but some were slightly better than others. The most functional of the ribozymes were selected and reverse-transcribed into cDNA (using the enzyme reverse transcriptase). The cDNA molecules were then amplified using the polymerase chain reaction (PCR; see Figure 9.15). PCR amplification is not perfect, and it introduced many new mutations into the pool of RNA sequences. These sequences were then transcribed back into RNA molecules using RNA polymerase, and the process was repeated.

The ligase activity of the RNAs evolved quickly; after ten rounds of in vitro evolution, it had increased by about 7 million times. Similar techniques have been used to create a wide variety of molecules with novel enzymatic and binding functions.

Evolutionary theory provides multiple benefits to agriculture

Well before humans had a clear understanding of evolution, they were selecting beneficial traits in the plants and animals they used for food. Modern agricultural practices have benefitted from a clearer understanding of evolutionary principles. Agriculturists have also used knowledge of evolutionary relationships and principles to incorporate beneficial genes into our food crops from many wild species.

Evolutionary theory has also proved important for understanding how to reduce the threats of pesticide and herbicide resistance. When farmers use the same pesticide over many seasons, the pests they are trying to kill gradually evolve resistance to the pesticide. Each year, a few pest individuals are slightly better at surviving in the presence of the pesticide, and those individuals produce most of the next generation of crop pests. Because their genes allow them to survive at a higher rate, and because they pass these resistant genes on to their offspring, pesticide resistance quickly evolves in the entire

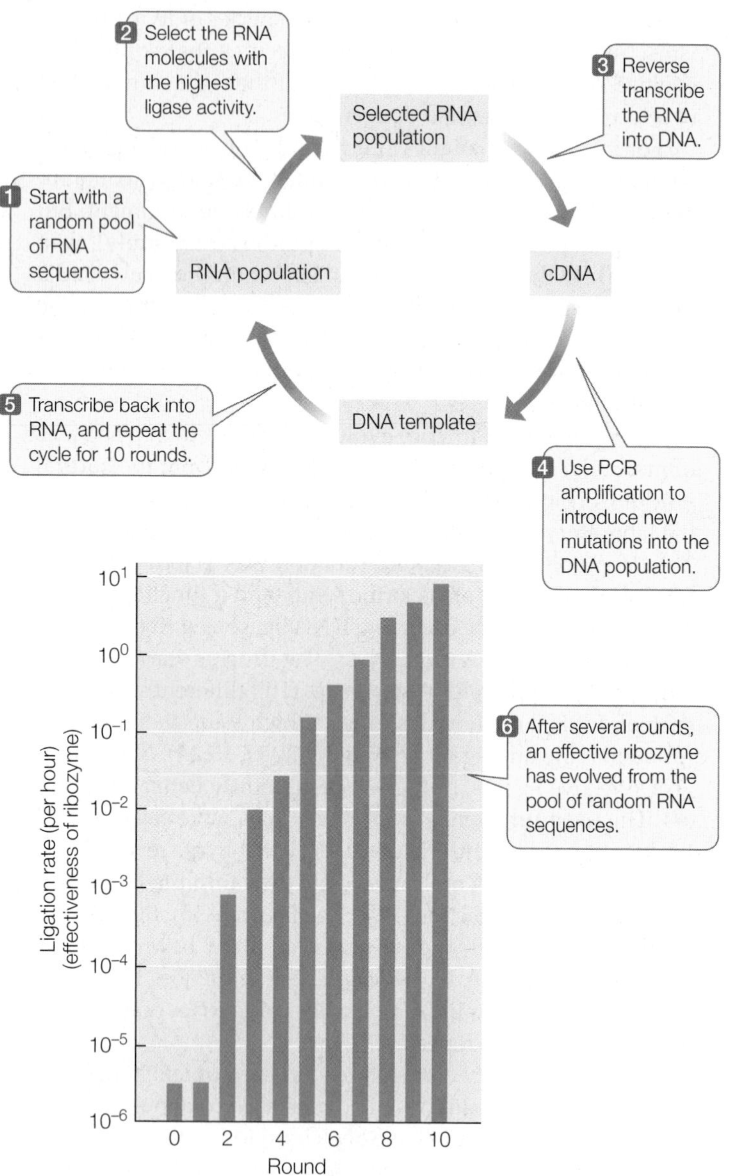

FIGURE 15.24 In Vitro Evolution Starting with a large pool of random RNA sequences, David Bartel and Jack Szostak of Massachusetts General Hospital produced a new ribozyme through rounds of mutation and selection for the ability to ligate (join) RNA sequences.

population. To combat this problem, evolutionary biologists have devised pesticide application and rotation schemes to reduce the rate of evolution of pesticide resistance, thus allowing farmers to use pesticides more effectively for longer periods of time.

Knowledge of molecular evolution is used to combat diseases

Many of the most problematic human diseases are caused by living, evolving organisms that present a moving target for modern medicine, as we described for influenza at the start of this chapter. The control of these and many other human diseases depends on techniques that can track the evolution of pathogenic organisms over time.

During the past century, transportation advances have allowed humans to move around the world with unprecedented speed and increasing frequency. Unfortunately, this mobility has increased the rate at which pathogens are transmitted among human populations, leading to the global emergence of many "new" diseases. Most of these emerging diseases are caused by viruses, and virtually all new viral diseases have been identified by evolutionary comparison of their genomes with those of known viruses. In recent years, rodent-borne hantaviruses have been identified as the source of widespread respiratory illnesses, and the virus that causes sudden acute respiratory syndrome (SARS) has been identified, as has its host, using evolutionary comparisons of genes. Studies of the origins, timing of emergence, and global diversity of many human pathogens (including HIV, the human immunodeficiency virus) depend on evolutionary principles and methods, as do efforts to develop effective vaccines against these pathogens.

At present, it is difficult to identify many common infections (the viral strains that cause "colds," for instance). As genomic databases increase, however, automated methods of sequencing and making evolutionary comparisons of sequences will allow us to identify and treat a much wider array of human (and other) diseases. Once biologists have collected genome data for enough infectious organisms, it will be possible to identify an infection by sequencing a portion of the pathogen's genome and comparing this sequence with other sequences on an evolutionary tree.

CHECKpoint CONCEPT 15.7

- ✓ How can gene evolution be used to study protein function?
- ✓ How are principles of evolutionary biology used to identify emerging diseases?
- ✓ What are the key elements of in vitro evolution, and how do these elements correspond to natural evolutionary processes?

The processes of evolution have produced a remarkable variety of organisms, some of which are adapted to most environments on Earth. In the next chapter we will describe how biologists study the evolutionary relationships across the great diversity of life.

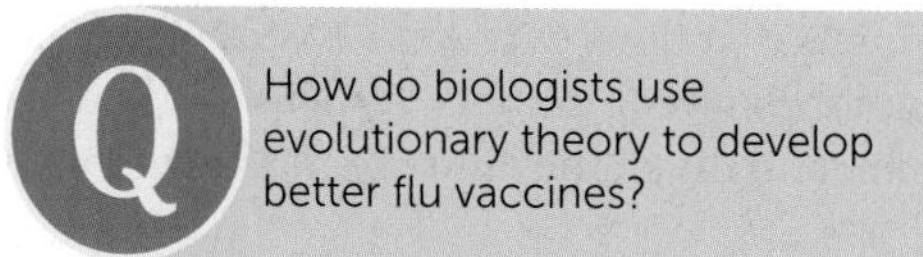

Q
How do biologists use evolutionary theory to develop better flu vaccines?

ANSWER Many different strains of influenza virus circulate among human populations and other vertebrate hosts each year, but only a few of those strains survive to leave descendants. Selection among these circulating influenza strains results in rapid evolution of the viral genome. One of the ways that influenza strains differ is in the configuration of

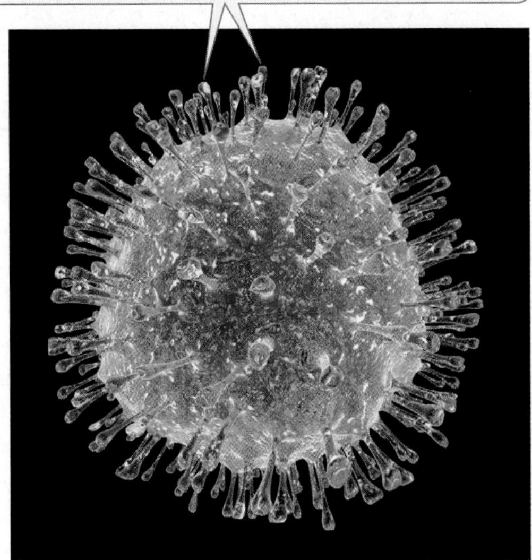

A vaccine stimulates our immune system to produce antibodies that recognize these proteins on the surface of the H1N1 virus.

FIGURE 15.25 Evolutionary Analysis of Surface Proteins Leads to Improved Flu Vaccines This computer-generated image is of the H1N1 virus that was the target of a 2009–2010 flu vaccine. Rapidly evolving surface proteins ("spikes" in this illustration) allow flu viruses to escape detection by the host's immune system. Analyzing the surface proteins among current strains of the virus can help biologists anticipate which strains are most likely to be the cause of future epidemics.

proteins on their surface. These surface proteins are the targets of recognition by the host immune system (**FIGURE 15.25**).

When changes occur in the surface proteins of an influenza virus, the host immune system may no longer detect the invading virus, so the virus is more likely to replicate successfully. The viral strains with the greatest number of changes to their surface proteins are most likely to escape detection by the host immune system, and are therefore most likely to spread among the host population and result in future flu epidemics. In other words, there is positive selection for change in the surface proteins of influenza.

By comparing the survival and proliferation rates of virus strains that have different gene sequences coding for their surface proteins, biologists can study adaptation of the viruses over time (Concept 15.2). If biologists can predict which of the currently circulating flu virus strains are most likely to escape host immune detection, then they can identify the strains that are most likely to be involved in upcoming influenza epidemics and can target those strains for vaccine production.

How can biologists make such predictions? By examining the ratio of synonymous to nonsynonymous substitutions in genes that encode viral surface proteins, biologists can detect which codon changes (i.e., mutations) are under positive selection (Concept 15.5). They can then assess which of the currently circulating flu strains show the greatest number of changes in these positively selected codons. It is these flu strains that are most likely to survive and lead to the flu epidemics of the future, so they are the best targets for new vaccines. This practical application of evolutionary theory leads to more effective flu vaccines—and thus fewer illnesses and influenza-related deaths each year.

SUMMARY

CONCEPT 15.1 Evolution Is Both Factual and the Basis of Broader Theory

- **Evolution** is genetic change in populations over time. Evolution can be observed directly in living populations as well as in the fossil record of life.
- **Evolutionary theory** refers to our understanding and application of the processes of evolutionary change.
- Charles Darwin in best known for his ideas on the common ancestry of divergent species and on **natural selection** as a process of evolution. **Review ANIMATED TUTORIAL 15.1 and ACTIVITY 15.1**
- Since Darwin's time, many biologists have contributed to the development of evolutionary theory, and rapid progress in our understanding continues today. **Review Figure 15.2**

CONCEPT 15.2 Mutation, Selection, Gene Flow, Genetic Drift, and Nonrandom Mating Result in Evolution

- Mutation produces new genetic variants (**alleles**).
- Within **populations**, natural selection acts to increase the frequency of beneficial alleles and decrease the frequency of deleterious alleles.
- **Adaptation** refers both to a trait that evolves through natural selection and to the process that produces such traits.
- Migration or mating of individuals between populations results in **gene flow**.
- **Genetic drift**—the random loss of individuals and the alleles they possess—may produce large changes in allele frequencies from one generation to the next and greatly reduce genetic variation.
- **Population bottlenecks** occur when only a few individuals survive a random event, resulting in a drastic shift in allele frequencies within the population and the loss of variation. Similarly, a population established by a small number of individuals colonizing a new region may lose variation via a **founder effect**. **Review Figure 15.8**
- Nonrandom mating may result in changes in genotype frequencies in a population.
- **Sexual selection** results from differential mating success of individuals based on their phenotype. **Review Figure 15.10**

(continued)

SUMMARY *(continued)*

CONCEPT 15.3 Evolution Can Be Measured by Changes in Allele Frequencies

- Allele frequencies measure the amount of genetic variation in a population. Genotype frequencies show how a population's genetic variation is distributed among its members. Together, allele and genotype frequencies describe a population's **genetic structure**. **Review Figure 15.11 and ANIMATED TUTORIAL 15.2**
- **Hardy–Weinberg equilibrium** predicts genotype frequencies from allele frequencies in the absence of evolution. Deviation from these frequencies indicates that evolutionary processes are at work. **Review Figure 15.12 and ANIMATED TUTORIAL 15.3**

CONCEPT 15.4 Selection Can Be Stabilizing, Directional, or Disruptive

- **Qualitative traits** differ by discrete qualities (e.g., black versus white) and often are determined by alleles of a single gene.
- **Quantitative traits** differ along a continuum (e.g., small to large size), and usually are influenced by variation at multiple genes.
- Natural selection can act on characters with quantitative variation in three different ways. **Review Figure 15.13**
- **Stabilizing selection** acts to reduce variation without changing the mean value of a trait. When applied to selection that maintains a particular genetic variant in a population, stabilizing selection is called **purifying selection**. **Review Figure 15.14**
- **Directional selection** acts to shift the mean value of a trait toward one extreme. When applied to selection for change at a single genetic locus, directional selection is called **positive selection**. **Review Figure 15.15**
- **Disruptive selection** favors both extremes of trait values, resulting in a bimodal character distribution. **Review Figure 15.16**

CONCEPT 15.5 Genomes Reveal Both Neutral and Selective Processes of Evolution

- **Nonsynonymous substitutions** of nucleotides result in amino acid replacements in proteins, but **synonymous substitutions** do not. **Review Figure 15.17**
- Rates of synonymous substitution are typically higher than rates of nonsynonymous substitution in protein-coding genes (a result of stabilizing selection). **Review Figure 15.18**
- Much of the change in nucleotide sequences over time is a result of neutral evolution. The rate of fixation of neutral mutations is independent of population size and is equal to the mutation rate.
- Positive selection for change in a protein-coding gene may be detected by a higher rate of nonsynonymous than synonymous substitution.
- Specific codons within a given gene sequence can be under different modes of selection. **Review Figure 15.20**
- The total size of genomes varies much more widely across multicellular organisms than does the number of functional genes. **Review Figures 15.21 and 15.22**
- Even though many noncoding regions of the genome may not have direct functions, these regions can affect the phenotype of an organism by influencing gene expression.
- Functionless **pseudogenes** can serve as the raw material for the evolution of new genes.

CONCEPT 15.6 Recombination, Lateral Gene Transfer, and Gene Duplication Can Result in New Features

- Despite its short-term disadvantages, sexual reproduction generates countless genotype combinations that increase genetic variation in populations.
- In the absence of genetic recombination (as in some asexual organisms), deleterious mutations accumulate with each replication—a phenomenon known as **Muller's ratchet**.
- **Lateral gene transfer** can result in the rapid acquisition of new functions from distantly related species.
- Gene duplications can result in increased production of the gene's product, in divergence of the duplicated genes' expression, in pseudogenes, or in new gene functions. Several rounds of gene duplication can give rise to multiple genes with related functions, known as a **gene family**. **Review Figure 15.23 and ACTIVITY 15.2**

CONCEPT 15.7 Evolutionary Theory Has Practical Applications

- Protein function can be studied by examining gene evolution. Detection of positive selection can be used to identify molecular changes that have resulted in functional changes.
- Agricultural applications of evolution include the development of new crop plants and domesticated animals, as well as a reduction in the rate of evolution of pesticide resistance.
- **In vitro evolution** is used to produce synthetic molecules with particular desired functions. **Review Figure 15.24**
- Many diseases are identified, studied, and combated through molecular evolutionary investigations.

Go to the Interactive Summary to review key figures, Animated Tutorials, and Activities
PoL2e.com/is15

Go to LaunchPad at **macmillanhighered.com/launchpad** for additional resources, including LearningCurve Quizzes, Flashcards, and many other study and review resources.

16 Reconstructing and Using Phylogenies

KEY CONCEPTS

- 16.1 All of Life Is Connected through Its Evolutionary History
- 16.2 Phylogeny Can Be Reconstructed from Traits of Organisms
- 16.3 Phylogeny Makes Biology Comparative and Predictive
- 16.4 Phylogeny Is the Basis of Biological Classification

The reef-building coral *Acropora millepora* shows cyan and red fluorescence. This photograph was taken under a fluorescent microscope that affects the colors we see. The colors are perceived differently by marine animals in their natural environment.

Green fluorescent protein (GFP) was discovered in 1962 when Osamu Shimomura, an organic chemist and marine biologist, led a team that was able to purify the protein from the tissues of the bioluminescent jellyfish *Aequorea victoria*. Some 30 years after GFP's initial discovery, Martin Chalfie had the idea (and the technology) to link the gene for GFP to other protein-coding genes, so that the expression of specific genes of interest could be visualized in glowing green within cells and tissues of living organisms (see Figure 13.6). This work was extended by Roger Tsien, who changed some of the amino acids within GFP to create proteins of several distinct colors. Different colored proteins meant that the expression of a number of different proteins could be visualized and studied in the same organism at the same time. These three scientists were awarded the 2008 Nobel Prize in Chemistry for the isolation and development of GFP for visualizing gene expression.

Tsien was able to produce different colored proteins, but he could not produce a *red* protein. This was frustrating; a red fluorescent protein would be particularly useful to biologists because red light penetrates tissues more easily than do other colors. Tsien's work stimulated Mikhail Matz to look for new fluorescent proteins in corals (which are relatives of the jellyfishes). Among the different species he studied, Matz found coral proteins that fluoresced in various shades of green, cyan (blue-green)—and red.

How had fluorescent red pigments evolved among the corals, given that the necessary molecular changes had eluded Tsien? To answer this question, Matz sequenced the genes of the fluorescent proteins and used these sequences to reconstruct the evolutionary history of the amino acid changes that produced different colors in different species of corals.

Matz's work showed that the ancestral fluorescent protein in corals was green, and that red fluorescent proteins evolved in a series of gradual steps. His analysis of evolutionary relationships allowed him to retrace these steps. Such an evolutionary history, as depicted in a tree of relationships among lineages, is called a phylogeny.

The evolution of many aspects of an organism's biology can be studied using phylogenetic methods. This information is used in all fields of biology to understand the structure, function, and behavior of organisms.

How are phylogenetic methods used to resurrect protein sequences from extinct organisms?

You will find the answer to this question on page 341.

CONCEPT 16.1 All of Life Is Connected through Its Evolutionary History

The sequencing of complete genomes from many diverse species has confirmed what biologists have long suspected: all of life is related through a common ancestor. The common ancestry of life explains why the general principles of biology apply to all organisms. Thus we can learn much about how the human genome works by studying the biology of model organisms because we share a common evolutionary history with those organisms. The evolutionary history of these relationships is known as **phylogeny**, and a **phylogenetic tree** is a diagrammatic reconstruction of that history.

Phylogenetic trees are commonly used to depict the evolutionary history of species, populations, and genes. For many years such trees have been constructed based on physical structures, behaviors, and biochemical attributes. Now, as genomes are sequenced for more and more organisms, biologists are able to reconstruct the history of life in ever greater detail.

In Chapter 15 we discussed why we expect populations of organisms to evolve over time. Such a series of ancestor and descendant populations forms a **lineage**, which we can depict as a line drawn on a time axis:

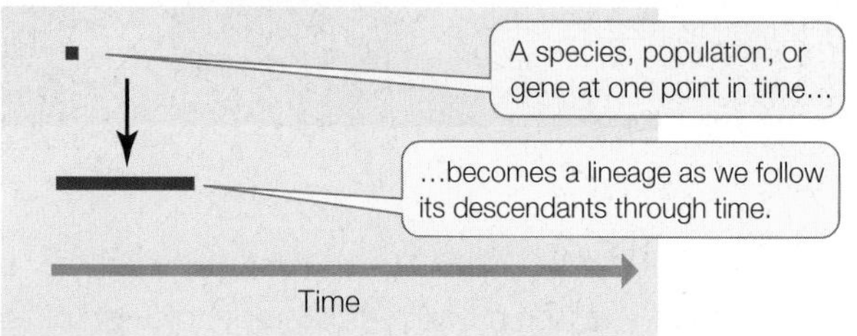

What happens when a single lineage divides into two? For example, a geographic barrier (such as a new mountain range) may divide an ancestral population into two descendant populations that no longer interact with one another. We depict such an event as a split, or **node**, in a phylogenetic tree. Each of the descendant populations gives rise to a new lineage, and as these independent lineages evolve, new traits arise in each:

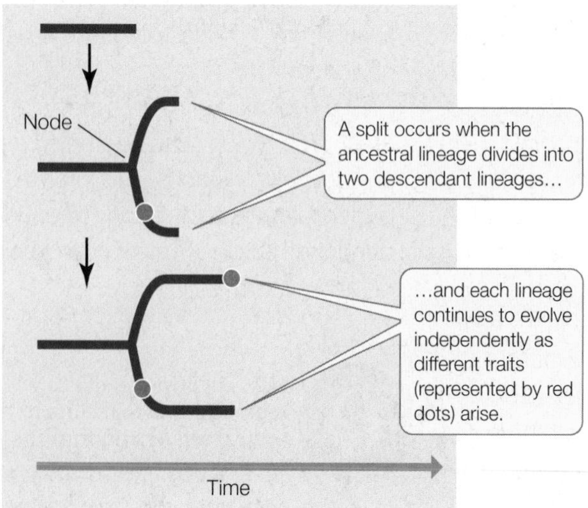

As the lineages continue to split over time, this history can be represented in the form of a branching tree that can be used to trace the evolutionary relationships from the ancient common ancestor of a group of species, through the various lineage splits, up to the present populations of the organisms:

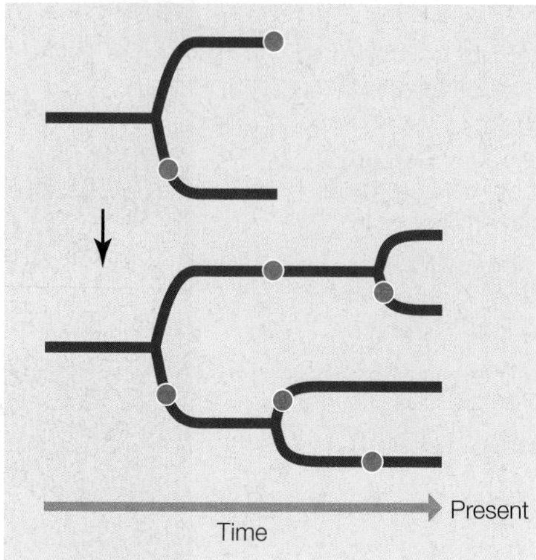

A phylogenetic tree may portray the evolutionary history of all life forms. Phylogenetic trees can also depict the history of a major evolutionary group (such as the insects) or of a much smaller group of closely related species. In some cases, phylogenetic trees are used to show the history of individuals, populations, or genes within a species. The common ancestor of all the organisms in the tree forms the **root** of the tree. The depictions of phylogenetic trees in this book are rooted at the left, with time flowing from left (earliest) to right (most recent):

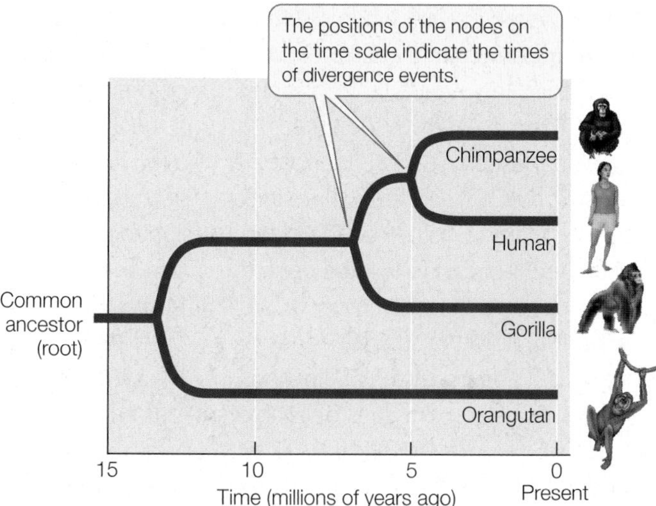

The timing of splitting events in lineages is shown by the position of nodes on a time axis. These splits represent events where one lineage diverged into two, such as a speciation event (for a tree of species), a gene duplication event (for a tree of genes), or a transmission event (for a tree of viral lineages transmitted through a host population). The time axis may have an explicit scale, or it may simply show the relative timing of divergence events.

In this book's illustrations, the order in which nodes are placed along the horizontal (time) axis has meaning, but the vertical distance between the branches does not. Vertical distances have been adjusted for legibility and clarity of presentation; they do not correlate with the degree of similarity or difference among groups. Note too that lineages can be rotated around nodes in the tree, so the vertical order of lineages is also largely arbitrary:

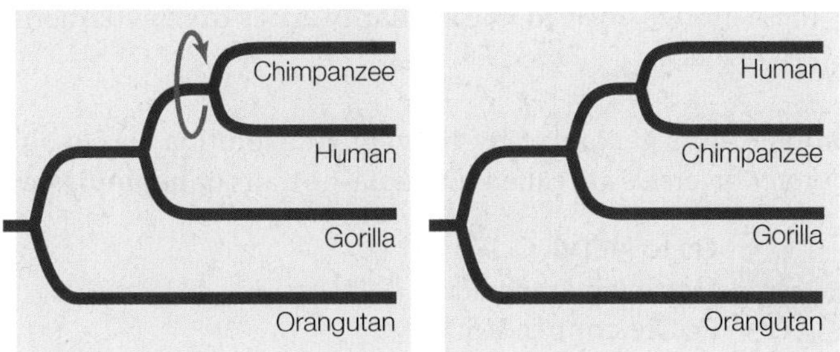

Any group of species that we designate with a name is a **taxon** (plural *taxa*). Examples of familiar taxa include humans, primates, mammals, and vertebrates; in this series, each taxon is also a member of the next, more inclusive taxon. Any taxon that consists of all the evolutionary descendants of a common ancestor is called a **clade**. Clades can be identified by picking any point on a phylogenetic tree and from that point tracing all the descendant lineages to the tips of the terminal branches (**FIGURE 16.1**). Two species that are each other's closest relatives are called **sister species**. Similarly, any two clades that are each other's closest relatives are **sister clades**.

Before the 1980s, phylogenetic trees tended to be seen only in the literature on evolutionary biology, especially in the area of **systematics**—the study and classification of biodiversity. But almost every journal in the life sciences published during the last few years contains phylogenetic trees. Trees are widely used in molecular biology, biomedicine, physiology, behavior, ecology, and virtually all other fields of biology. Why have phylogenetic studies become so widespread?

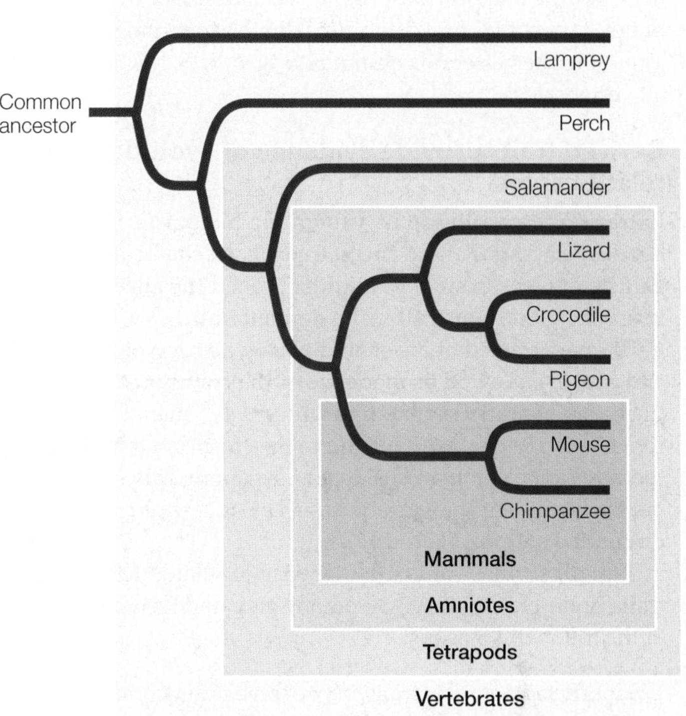

FIGURE 16.1 Clades Represent All the Descendants of a Common Ancestor All clades are subsets of larger clades, with all of life as the most inclusive taxon. In this example, the groups called mammals, amniotes, tetrapods, and vertebrates represent successively larger clades. Only a few species within each clade are represented on this tree.

Phylogenetic trees are the basis of comparative biology

In biology we study life at all levels of organization—from genes, cells, organisms, populations, and species to the major divisions of life. In most cases, however, no individual gene or organism (or other unit of study) is exactly like any other gene or organism that we investigate.

Consider the individuals in your biology class. We recognize each person as an individual human, but we know that no two are exactly alike. If we knew everyone's family tree in detail, the genetic similarity of any pair of students would be more predictable. We would find that more closely related students have many more traits in common (from the color of their hair to their susceptibility or resistance to diseases). Likewise, biologists use phylogenies to make comparisons and predictions about shared traits across genes, populations, and species.

The evolutionary relationships among species, as represented in the tree of life, form the basis for biological classification. Biologists estimate that there are tens of millions of species on Earth. So far, however, only about 1.8 million species have been classified—that is, formally described and named. New species are being discovered all the time and phylogenetic analyses are constantly reviewed and revised, so our knowledge of the tree of life is far from complete. Yet knowledge of evolutionary relationships is essential for making comparisons in biology, so biologists build phylogenies for groups of interest as the need arises. The tree of life's evolutionary framework allows us to make many predictions about the behavior, ecology, physiology, genetics, and morphology of species that have not yet been studied in detail.

When biologists compare species, they observe traits that differ within the group of interest and try to understand when these traits evolved. In many cases, investigators are interested in how the evolution of a trait relates to environmental conditions or selective pressures. For instance, scientists have used phylogenetic analyses to discover changes in the genome of human immunodeficiency viruses that result in resistance to particular drug treatments. The association of a particular genetic change in HIV with a particular treatment provides a hypothesis about the evolution of resistance that can be tested experimentally.

Any features shared by two or more species that have been inherited from a common ancestor are said to be **homologous**. Homologous features may be any heritable traits, including DNA sequences, protein structures, anatomical structures, and even some behavior patterns. For example, all living vertebrates have a vertebral column, as did the ancestral vertebrate. Therefore the vertebral column is judged to be homologous in all vertebrates.

Derived traits provide evidence of evolutionary relationships

In tracing the evolution of a character, biologists distinguish between ancestral and derived traits. Each character of an organism evolves from one condition, called the **ancestral trait**, to another condition, called the **derived trait**.

Derived traits that are shared among a group of organisms and are also viewed as evidence of the common ancestry of the group are called **synapomorphies** (*syn*, "shared"; *apo*, "derived"; *morph*, "form," referring to the "form" of a trait). Thus the vertebral column is considered a synapomorphy—a shared, derived trait—of the vertebrates. (The ancestral trait was an undivided supporting rod.)

Not all similar traits are evidence of relatedness. Similar traits in unrelated groups of organisms can develop for either of the following reasons:

- Superficially similar traits may evolve independently in different lineages, a phenomenon called **convergent evolution**. For example, although the *wing bones* of bats and birds are homologous, having been inherited from a common tetrapod ancestor, the *wings* of bats and birds are not homologous because they evolved independently from the forelimbs of different nonflying ancestors (**FIGURE 16.2**).

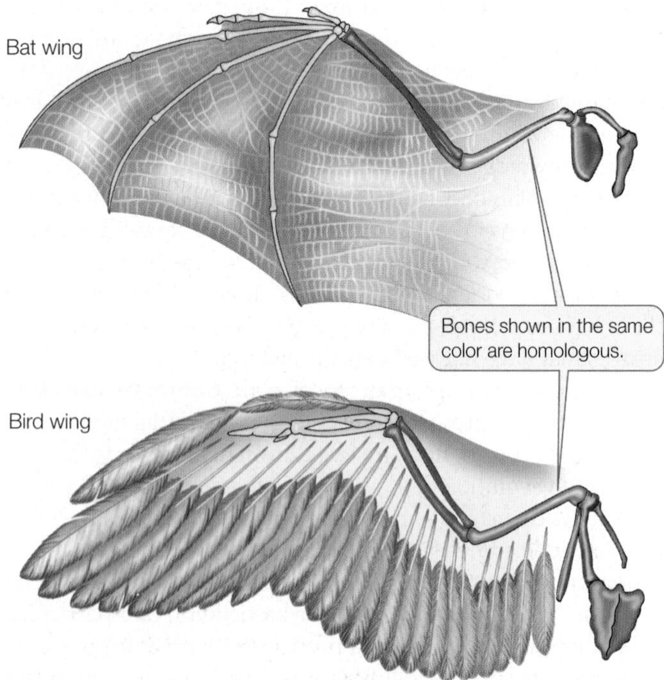
Bat wing
Bones shown in the same color are homologous.
Bird wing

FIGURE 16.2 The Bones Are Homologous, the Wings Are Not The supporting bone structures of both bat wings and bird wings are derived from a common tetrapod (four-limbed) ancestor and are thus homologous. However, the wings themselves—an adaptation for flight—evolved independently in the two groups.

- A character may revert from a derived state back to an ancestral state in an event called an **evolutionary reversal**. For example, the derived limbs of terrestrial tetrapods evolved from the ancestral fins of their aquatic ancestors. Then, within the mammals, the ancestors of modern cetaceans (whales and dolphins) returned to the ocean, and cetacean limbs evolved to once again resemble their ancestral state—fins. The superficial similarity of cetacean and fish fins does not suggest a close relationship between these groups. Instead, the similarity arises from evolutionary reversal.

Similar traits generated by convergent evolution and evolutionary reversals are called homoplastic traits or **homoplasies**.

Go to MEDIA CLIP 16.1
Morphing Arachnids
PoL2e.com/mc16.1

A particular trait may be ancestral or derived, depending on our point of reference. For example, all birds have feathers. We infer from this that feathers (which are highly modified scales) were present in the common ancestor of modern birds. Therefore we consider the presence of feathers to be an ancestral trait for any particular group of modern birds, such as the songbirds. However, feathers are not present in any other living animals. In reconstructing a phylogeny of all living vertebrates, the presence of feathers is a derived trait found only among birds, and thus is a synapomorphy of the birds.

CHECKpoint CONCEPT 16.1

- ✓ What biological processes can be represented in a phylogenetic tree?
- ✓ Why is it important to consider only homologous characters in constructing phylogenetic trees?
- ✓ What are some reasons that similar traits might arise independently in species that are only distantly related? Can you think of examples among familiar organisms?

Phylogenetic analyses of evolutionary history have become increasingly important to many types of biological research in recent years, and they are the basis for the comparative nature of biology. For the most part, however, evolutionary history cannot be observed directly. How, then, do biologists reconstruct the past?

CONCEPT 16.2 Phylogeny Can Be Reconstructed from Traits of Organisms

To illustrate how a phylogenetic tree is constructed, consider the eight vertebrate animals listed in **TABLE 16.1**: lamprey, perch, salamander, lizard, crocodile, pigeon, mouse, and chimpanzee. We will initially assume that any given derived trait arose only once during the evolution of these animals (that is,

TABLE 16.1 Eight Vertebrates and the Presence or Absence of Some Shared Derived Traits

	Derived trait							
Taxon	**Jaws**	**Lungs**	**Claws or nails**	**Gizzard**	**Feathers**	**Fur**	**Mammary glands**	**Keratinous scales**
Lamprey (outgroup)	–	–	–	–	–	–	–	–
Perch	+	–	–	–	–	–	–	–
Salamander	+	+	–	–	–	–	–	–
Lizard	+	+	+	–	–	–	–	+
Crocodile	+	+	+	+	–	–	–	+
Pigeon	+	+	+	+	+	–	–	+
Mouse	+	+	+	–	–	+	+	–
Chimpanzee	+	+	+	–	–	+	+	–

there has been no convergent evolution), and that no derived traits were lost from any of the descendant groups (there has been no evolutionary reversal). For simplicity, we have selected traits that are either present (+) or absent (–).

In a phylogenetic study, the group of organisms of primary interest is called the **ingroup**. As a point of reference, an ingroup is compared with an **outgroup**: a species or group that is closely related to the ingroup but is known to be phylogenetically outside it. In other words, the root of the tree is located between the ingroup and the outgroup. Any trait that is present in both the ingroup and the outgroup must have evolved before the origin of the ingroup and thus must be ancestral for the ingroup. In contrast, traits that are present in only some members of the ingroup must be derived traits within that ingroup. As we will see in Chapter 23, a group of jawless fishes called the lampreys is thought to have separated from the lineage leading to the other vertebrates before the jaw arose. Therefore we have included the lamprey as the outgroup for our analysis. Because derived traits are traits acquired by other members of the vertebrate lineage *after* they diverged from the outgroup, any trait that is present in both the lamprey and the other vertebrates is judged to be ancestral.

We begin by noting that the chimpanzee and mouse share two traits—mammary glands and fur—that are absent in both the outgroup and in the other species of the ingroup. Therefore we infer that mammary glands and fur are derived traits that evolved in a common ancestor of chimpanzees and mice after that lineage separated from the lineages leading to the other vertebrates. These characters are synapomorphies that unite chimpanzees and mice (as well as all other mammals, although we have not included other mammalian species in this example). By the same reasoning, we can infer that the other shared derived traits are synapomorphies for the various groups in which they are expressed. For instance, keratinous scales are a synapomorphy of the lizard, crocodile, and pigeon.

Table 16.1 also tells us that, among the animals in our ingroup, the pigeon has a unique trait: the presence of feathers. Feathers are a synapomorphy of birds and their extinct relatives. However, because we only have one bird in this example, the presence of feathers provides no clues concerning relationships among these eight species of vertebrates. However, gizzards are found in both birds and crocodiles, so this trait is evidence of a close relationship between birds and crocodilians.

By combining information about the various synapomorphies, we can construct a phylogenetic tree. We infer from our information that mice and chimpanzees—the only two animals that share fur and mammary glands—share a more recent common ancestor with each other than they do with pigeons and crocodiles. Otherwise we would need to assume that the ancestors of pigeons and crocodiles also had fur and mammary glands but subsequently lost them. There is no need to make these additional assumptions.

FIGURE 16.3 shows a phylogenetic tree for the vertebrates in Table 16.1, based on the shared derived traits we examined. This particular tree was easy to construct because it is based on a very small sample of traits, and the derived traits we examined evolved only once and were never lost after they appeared. Had we included a snake in the group, our analysis would not have been as straightforward. We would have needed to examine additional characters to determine that snakes evolved from a group of lizards that had limbs. In fact, the analysis of many characters shows that snakes evolved from burrowing lizards that became adapted to a subterranean existence.

Parsimony provides the simplest explanation for phylogenetic data

Typically, biologists construct phylogenetic trees using hundreds or thousands of traits. With larger data sets, we would expect to observe traits that have changed more than once, and thus would expect to see convergence and evolutionary reversal. How do we determine which traits are synapomorphies and which are homoplasies? One way is to invoke the principle of parsimony.

In its most general form, the **parsimony principle** states that the preferred explanation of observed data is the simplest

FIGURE 16.3 Inferring a Phylogenetic Tree This phylogenetic tree was constructed from the information given in Table 16.1 using the parsimony principle. Each clade in the tree is supported by at least one shared derived trait, or synapomorphy.

Go to ACTIVITY 16.1 Constructing a Phylogenetic Tree PoL2e.com/ac16.1

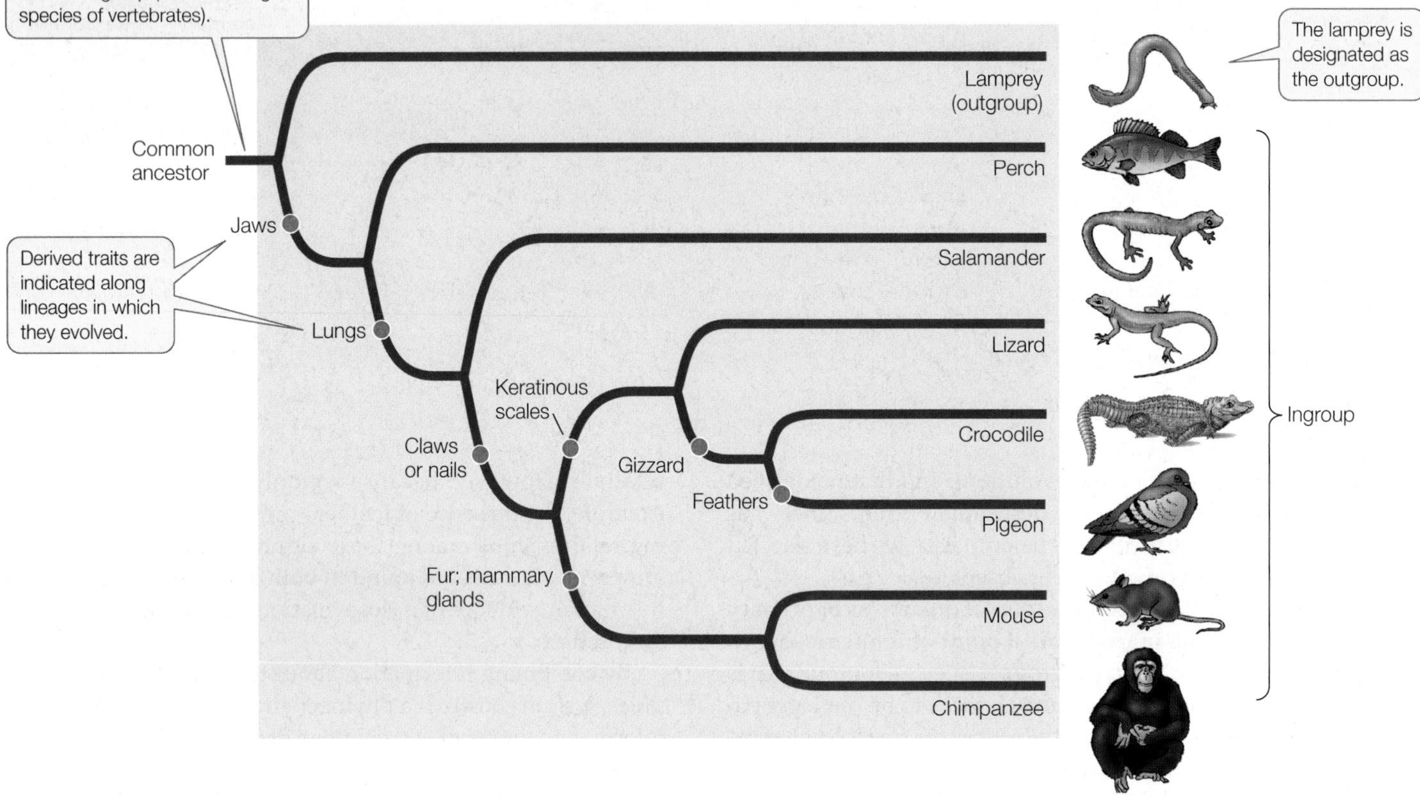

APPLY THE CONCEPT

Phylogeny can be reconstructed from traits of organisms

The matrix below supplies data for seven land plants and an outgroup (an aquatic plant known as a stonewort). Each trait is scored as either present (+) or absent (–) in each of the plants. Use this data matrix to reconstruct the phylogeny of land plants and answer the questions. See Activity 16.1 for help with constructing a phylogenetic tree.

1. Which two of these taxa are most closely related?
2. Plants that produce seeds are known as seed plants. What is the sister group to the seed plants among these taxa?
3. Which two traits evolved along the same branch of your reconstructed phylogeny?
4. Are there any homoplasies in your reconstructed phylogeny?

	TRAIT						
TAXON	PROTECTED EMBRYOS	TRUE ROOTS	PERSISTENTLY GREEN SPOROPHYTE	VASCULAR CELLS	STOMATA	MEGAPHYLLS (TRUE LEAVES)	SEEDS
Stonewort (outgroup)	–	–	–	–	–	–	–
Liverwort	+	–	–	–	–	–	–
Pine tree	+	+	+	+	+	+	+
Bracken fern	+	+	+	+	+	+	–
Club moss	+	+	+	+	+	–	–
Sphagnum moss	+	–	–	–	+	–	–
Hornwort	+	–	+	–	+	–	–
Sunflower	+	+	+	+	+	+	+

explanation. Applying the principle of parsimony to the reconstruction of phylogenies entails minimizing the number of evolutionary changes that need to be assumed over all characters in all groups in the tree. In other words, the best hypothesis under the parsimony principle is one that requires the fewest homoplasies. This application of parsimony is a specific case of a general principle of reasoning called Occam's razor: the best explanation is the one that best fits the data while making the fewest assumptions. More complicated explanations are accepted only when the evidence requires them. Phylogenetic trees represent our best estimates about evolutionary relationships, given our current knowledge. They are continually modified as additional evidence becomes available.

Phylogenies are reconstructed from many sources of data

Naturalists have constructed various forms of phylogenetic trees for more than 150 years. In fact, the only figure in the first edition of *On the Origin of Species* was a phylogenetic tree. Tree construction has been revolutionized, however, by the advent of computer software that allows us to consider far more data and analyze many more traits than could ever before be processed. Combining these advances in methodology with the massive comparative data sets being generated through studies of genomes, biologists are learning details about the tree of life at a remarkable pace (see Appendix A: The Tree of Life).

Any trait that is genetically determined, and therefore heritable, can be used in a phylogenetic analysis. Evolutionary relationships can be revealed through studies of morphology, development, the fossil record, behavioral traits, and molecular traits such as DNA and protein sequences. Let's take a closer look at the types of data used in modern phylogenetic analyses.

Go to ANIMATED TUTORIAL 16.1
Phylogeny and Molecular Evolution Simulation
PoL2e.com/at16.1

MORPHOLOGY An important source of phylogenetic information is **morphology**: the presence, size, shape, and other attributes of body parts. Since living organisms have been observed, depicted, and studied for millennia, we have a wealth of recorded morphological data as well as extensive museum and herbarium collections of organisms whose traits can be measured. New technological tools, such as the electron microscope and computed tomography (CT) scans, enable systematists to examine and analyze the structures of organisms at much finer scales than was formerly possible.

Most species are described and known primarily by their morphology, and morphology still provides the most comprehensive data set available for many taxa. The morphological features that are important for phylogenetic analysis are often specific to a particular group. For example, the presence, development, shape, and size of various features of the skeletal system are important in vertebrate phylogeny, whereas floral structures are important for studying the relationships among flowering plants.

Morphological approaches to phylogenetic analysis have some limitations, however. Some taxa exhibit little morphological diversity, despite great species diversity. For example, the phylogeny of the leopard frogs of North and Central America would be difficult to infer from morphological differences alone, because the many species look very similar, despite important differences in their behavior and physiology. At the other extreme, few morphological traits can be compared across distantly related species (earthworms and mammals, for example). Furthermore, some morphological variation has an environmental (rather than a genetic) basis and so must be excluded from phylogenetic analyses. An accurate phylogenetic analysis often requires information beyond that supplied by morphology.

DEVELOPMENT Similarities in developmental patterns may reveal evolutionary relationships. Some organisms exhibit similarities in early developmental stages only. The larvae of marine creatures called sea squirts, for example, have a flexible gelatinous rod in the back—the notochord—that disappears as the larvae develop into adults. All vertebrate animals also have a notochord at some time during their development (**FIGURE 16.4**). This shared structure is one of the reasons for inferring that sea squirts are more closely related to vertebrates than would be suspected if only adult sea squirts were examined.

LINK

For more on the role of developmental processes in evolution, see **Concepts 14.4 and 14.5**

PALEONTOLOGY The fossil record is another important source of information on evolutionary history. Fossils show us where and when organisms lived in the past and give us an idea of what they looked like. Fossils provide important evidence that helps us distinguish ancestral from derived traits. The fossil record can also reveal when lineages diverged and began their independent evolutionary histories. Furthermore, in groups with few species that have survived to the present, information on extinct species is often critical to an understanding of the large divergences among the surviving species. The fossil record has limitations, however. Few or no fossils have been found for some groups, and the fossil record for many groups is fragmentary.

BEHAVIOR Some behavioral traits are culturally transmitted and others are genetically inherited. If a particular behavior is culturally transmitted, it may not accurately reflect evolutionary relationships (but may nonetheless reflect cultural connections). Many bird songs, for instance, are learned and may be inappropriate traits for phylogenetic analysis. Frog calls, however, are genetically determined and appear to be acceptable sources of information for reconstructing phylogenies.

MOLECULAR DATA All heritable variation is encoded in DNA, and so the complete genome of an organism contains an

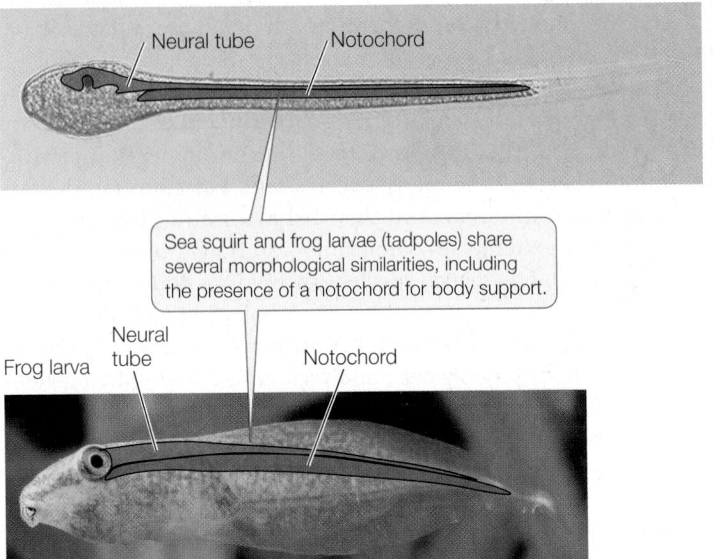

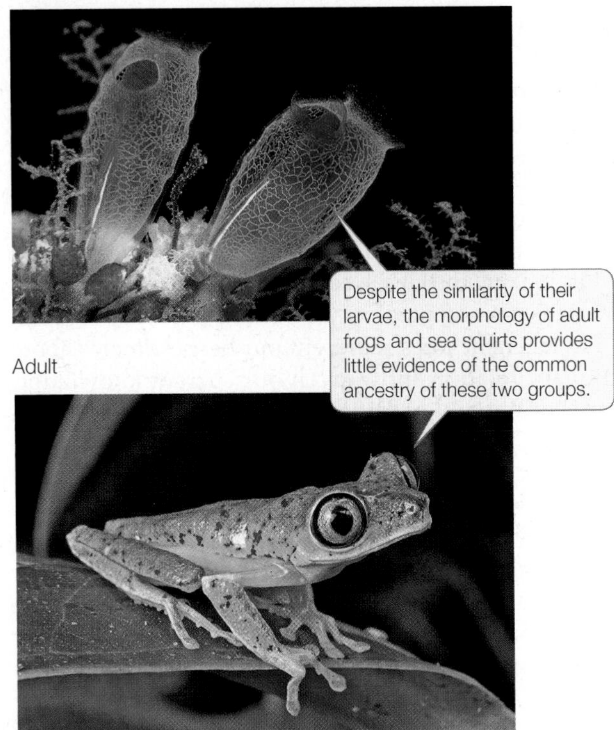

Adult
Adult
Despite the similarity of their larvae, the morphology of adult frogs and sea squirts provides little evidence of the common ancestry of these two groups.

FIGURE 16.4 The Chordate Connection Embryonic development can offer vital clues to evolutionary relationships, since larvae sometimes share similarities that are not apparent in the adults. An example is the notochord, a synapomorphy of the chordates (a taxonomic group that includes the sea squirts as well as vertebrates such as frogs). All chordates have a notochord during their early development. The notochord is lost in adult sea squirts, whereas in adult frogs—as in all vertebrates—the vertebral column replaces the notochord as the body's support structure.

enormous set of traits (the individual nucleotide bases of DNA) that can be used in phylogenetic analyses. In recent years, DNA sequences have become among the most widely used sources of data for constructing phylogenetic trees. Comparisons of nucleotide sequences are not limited to the DNA in the cell nucleus. Eukaryotes have genes in their mitochondria as well as in their nuclei. Plant cells also have genes in their chloroplasts. The chloroplast genome (cpDNA), which is used extensively in phylogenetic studies of plants, has changed slowly over evolutionary time, so it is often used to study relatively ancient phylogenetic relationships. Most animal mitochondrial DNA (mtDNA) has changed more rapidly, so mitochondrial genes are used to study evolutionary relationships among closely related animal species (the mitochondrial genes of plants evolve more slowly). Many nuclear gene sequences are also commonly analyzed, and now that entire genomes have been sequenced from many species, they too are used to construct phylogenetic trees. Information on gene products (such as the amino acid sequences of proteins) is also widely used for phylogenetic analyses.

Mathematical models expand the power of phylogenetic reconstruction

As biologists began to use DNA sequences to infer phylogenies in the 1970s and 1980s, they developed explicit mathematical models describing how DNA sequences change over time. These models account for multiple changes at a given position in a DNA sequence. They also take into account different rates of change at different positions in a gene, at different positions in a codon, and among different nucleotides. For example, transitions (changes between two purines or between two pyrimidines) are usually more likely than are transversions (changes between a purine and pyrimidine).

Mathematical models can be used to compute how a tree might evolve given the observed data. A **maximum likelihood** method will identify the tree that most likely produced the observed data, given the assumed model of evolutionary change. Maximum likelihood methods can be used for any kind of characters, but they are most often used with molecular data, for which explicit mathematical models of evolutionary change are easier to develop. The principal advantages to maximum likelihood analyses are that they incorporate more information about evolutionary change than do parsimony methods, and they are easier to treat in a statistical framework. The principal disadvantages are that they are computationally intensive and require explicit models of evolutionary change (which may not be available for some kinds of character change).

The accuracy of phylogenetic methods can be tested

How can we test the accuracy of phylogenetic methods? After all, phylogenetic trees represent reconstructions of past events, and many of these events occurred before any humans were around. To address this issue, biologists have conducted experiments both in living organisms and with computer simulations to test the effectiveness and accuracy of phylogenetic methods.

In one experiment designed to test the accuracy of phylogenetic analysis, a single viral culture of bacteriophage T7 was used as a starting point, and lineages were allowed to evolve from this ancestral virus in the laboratory (**FIGURE 16.5**). The initial culture was split into two separate lineages, one of which

INVESTIGATION

FIGURE 16.5 The Accuracy of Phylogenetic Analysis To test whether analysis of gene sequences can accurately reconstruct evolutionary history, we must have an unambiguously known phylogeny to compare against the reconstruction. Will the reconstruction match the observed phylogeny? See Animated Tutorial 16.1 for a simulation of this experiment.

HYPOTHESIS

A phylogenetic tree reconstructed from analysis of the DNA sequences of living organisms can accurately match the known evolutionary history of the organisms.

METHOD

In the laboratory, one group of investigators produced an experimental phylogeny of 9 viral lineages, enhancing the mutation rate to increase variation among the lineages.[a]

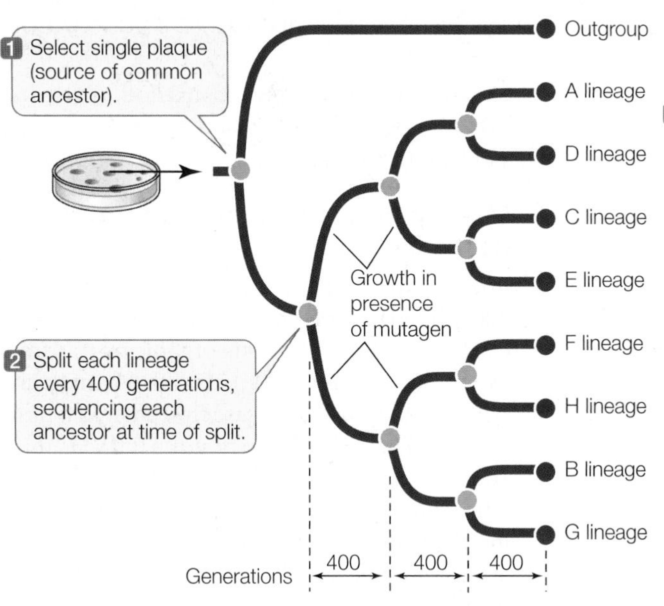

1 Select single plaque (source of common ancestor).
2 Split each lineage every 400 generations, sequencing each ancestor at time of split.
Growth in presence of mutagen
Outgroup
A lineage
D lineage
C lineage
E lineage
F lineage
H lineage
B lineage
G lineage
Generations
400
400
400
3 Present final genes (blue dots) to a second group of investigators who are unaware of the history of the lineages or the gene sequences of the ancestral viruses. These "blind" investigators then determine the sequences of the descendant genes and use these sequences to reconstruct the evolution of these lineages in the form of a phylogenetic tree.

RESULTS

The true phylogeny and ancestral DNA sequences were accurately reconstructed solely from the DNA sequences of the viruses at the tips of the tree.

CONCLUSION

Phylogenetic analysis of DNA sequences can accurately reconstruct evolutionary history.

ANALYZE THE DATA

The full DNA sequences for the T7 strains in this experiment are thousands of nucleotides long. The nucleotides ("characters") at 23 DNA positions are given in the table.[b] See Activity 16.1 for help with constructing a phylogenetic tree.

	Character at position																						
	1	2	3	4	5	6	7	8	9	10	11	12	13	14	15	16	17	18	19	20	21	22	23
Outgroup	C	C	G	G	G	C	C	T	C	C	T	C	G	A	C	C	G	G	C	A	C	G	G
A	T	C	G	G	G	C	C	C	C	C	C	C	A	A	C	C	G	A	T	A	C	A	A
B	C	C	G	G	G	T	C	C	C	T	C	C	G	A	T	T	A	G	C	G	T	G	G
C	C	C	G	G	G	C	C	C	T	C	C	T	A	A	C	C	G	G	T	A	C	A	A
D	T	C	A	G	G	C	C	C	C	C	C	C	A	A	C	C	G	A	T	A	C	A	A
E	C	T	G	G	G	C	C	C	C	C	C	T	A	A	C	C	G	G	T	A	C	A	A
F	C	T	G	A	A	C	C	C	C	C	C	C	G	A	C	T	G	G	C	G	C	G	G
G	C	C	G	G	G	T	T	C	C	T	C	C	G	A	T	T	A	G	C	G	C	G	G
H	C	C	G	G	A	C	C	C	C	C	C	C	G	C	C	T	G	G	C	G	C	G	G

A. Construct a phylogenetic tree from these DNA positions using the parsimony method. Use the outgroup to root your tree. Assume that all changes among nucleotides are equally likely.

B. Using your tree, reconstruct the DNA sequences of the ancestral lineages.

Go to **LaunchPad** for discussion and relevant links for all **INVESTIGATION** figures.

[a] D. M. Hillis et al. 1992. *Science* 255: 589–295.

[b] J. J. Bull et al. 1993. *Evolution* 47: 993–1007.

became the ingroup for analysis and the other of which became the outgroup for rooting the tree. The lineages in the ingroup were split in two after every 400 generations, and samples of the virus were saved for analysis at each branching point. The lineages were allowed to evolve until there were eight lineages in the ingroup. Mutagens were added to the viral cultures to increase the mutation rate so that the amount of change and the degree of homoplasy would be typical of the organisms analyzed in average phylogenetic analyses. The investigators then sequenced samples from the end points of the eight lineages, as well as from the ancestors at the branching points. They then gave the sequences from the end points of the lineages to other investigators to analyze, without revealing the known history of the lineages or the sequences of the ancestral viruses.

After the phylogenetic analysis was completed, the investigators asked two questions. Did phylogenetic methods reconstruct the known history correctly? And were the sequences of the ancestral viruses reconstructed accurately? The answer in both cases was yes. The branching order of the lineages was reconstructed exactly as it had occurred, more than 98 percent of the nucleotide positions of the ancestral viruses were reconstructed correctly, and 100 percent of the amino acid changes in the viral proteins were reconstructed correctly.

Go to ANIMATED TUTORIAL 16.2
Using Phylogenetic Analysis to Reconstruct Evolutionary History
PoL2e.com/at16.2

The experiment shown in Figure 16.5 demonstrated that phylogenetic analysis was accurate under the conditions tested, but it did not examine all possible conditions. Other experimental studies have taken other factors into account, such as the sensitivity of phylogenetic analysis to parallel selection and highly variable rates of evolutionary change. In addition, computer simulations based on evolutionary models have been used extensively to study the effectiveness of phylogenetic analysis. These studies have also confirmed the accuracy of phylogenetic methods and have been used to refine those methods and extend them to new applications.

CHECKpoint CONCEPT 16.2

- ✓ How is the parsimony principle used in reconstructing evolutionary history?
- ✓ Why is it useful to consider the entire life cycle when reconstructing an organism's evolutionary history?
- ✓ What are some comparative advantages and disadvantages of morphological and molecular approaches for reconstructing phylogenetic trees?
- ✓ Contrast experimental and computer simulation approaches for testing the accuracy of phylogenetic reconstructions of evolutionary history. Can you think of some aspects of phylogenetic accuracy that might be more practical to test using computer simulation rather than experimental studies of viruses?

Why do biologists expend the time and effort necessary to reconstruct phylogenies? Information about the evolutionary relationships among organisms is a useful source of data for scientists investigating a wide variety of biological questions. Next we will describe how phylogenetic trees are used to answer questions about the past, and to predict and compare traits of organisms in the present.

CONCEPT 16.3 Phylogeny Makes Biology Comparative and Predictive

Once a phylogeny is reconstructed, what do we do with it? What beyond an understanding of evolutionary history does phylogeny offer us?

Phylogenies are important for reconstructing past events

Reconstructing past events is important for understanding many biological processes. In the case of zoonotic diseases (diseases caused by infectious organisms transmitted to humans from another animal host), it is important to understand when, where, and how the disease first entered a human population. Human immunodeficiency virus (HIV) is the cause of such a zoonotic disease, acquired immunodeficiency syndrome, or AIDS. Phylogenetic analyses have become important for studying the transmission of viruses such as HIV. Phylogenies are also important for understanding the present global diversity of HIV and for determining the virus's origins in human populations. A broader phylogenetic analysis of immunodeficiency viruses shows that humans acquired these viruses from two different hosts: HIV-1 from chimpanzees, and HIV-2 from sooty mangabeys (**FIGURE 16.6**).

HIV-1 is the common form of the virus in human populations in central Africa, where chimpanzees are hunted for food, and HIV-2 is the common form in human populations in western Africa, where sooty mangabeys are hunted for food. Thus it seems likely that these viruses entered human populations through hunters who cut themselves while skinning chimpanzees and sooty mangabeys. The global pandemic of AIDS occurred when these infections in local African populations rapidly spread through human populations around the world.

In recent years, phylogenetic analysis has become important in forensic investigations that involve viral transmission events. For example, phylogenetic analysis was critical for a criminal investigation of a physician who was accused of purposefully injecting blood from one of his HIV-positive patients into his former girlfriend in an attempt to kill her. The phylogenetic analysis revealed that the HIV strains present in the girlfriend were a subset of those present in the physician's patient (**FIGURE 16.7**). Other evidence was needed, of course, to connect the physician to this purposeful transmission event, but the phylogenetic analysis was important to confirm the viral transmission event from the patient to the victim.

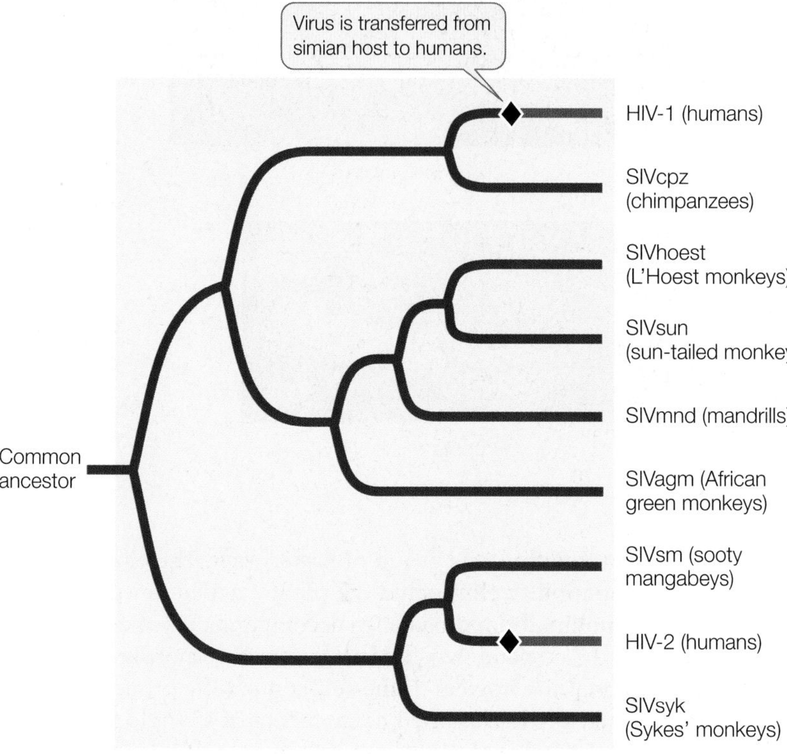

FIGURE 16.6 Phylogenetic Tree of Immunodeficiency Viruses
Immunodeficiency viruses have been transmitted to humans from two different simian hosts: HIV-1 from chimpanzees and HIV-2 from sooty mangabeys (the transmission events are marked by black diamonds). SIV stands for "simian immunodeficiency virus."

Phylogenies allow us to understand the evolution of complex traits

Male swordtails—a group of fishes in the genus *Xiphophorus*—have a long, colorful tail extension, and their reproductive success is closely associated with this appendage. Males with a long sword are more likely to mate successfully than are males with a short sword (an example of sexual selection; see Concept 15.2). Several explanations have been advanced for the evolution of this structure, including the hypothesis that the sword simply exploits a preexisting bias in the sensory system of the females. This sensory exploitation hypothesis suggests that female swordtails had a preference for males with long tails even before the tails evolved (perhaps because females assess the size of males by their total body length—including the tail—and prefer larger males).

To test the sensory exploitation hypothesis, phylogenetic analysis was used to identify the relatives of swordtails that had split most recently from their lineage before the evolution of swords. These closest relatives turned out to be fishes in the genus *Priapella*. Even though male *Priapella* do not normally have swords, when researchers attached artificial sword-like structures to the tails of male *Priapella*, female *Priapella* preferred those males. This result provided support for the hypothesis that female *Xiphophorus* had a preexisting sensory bias favoring tail extensions even before the trait evolved. Thus a long tail

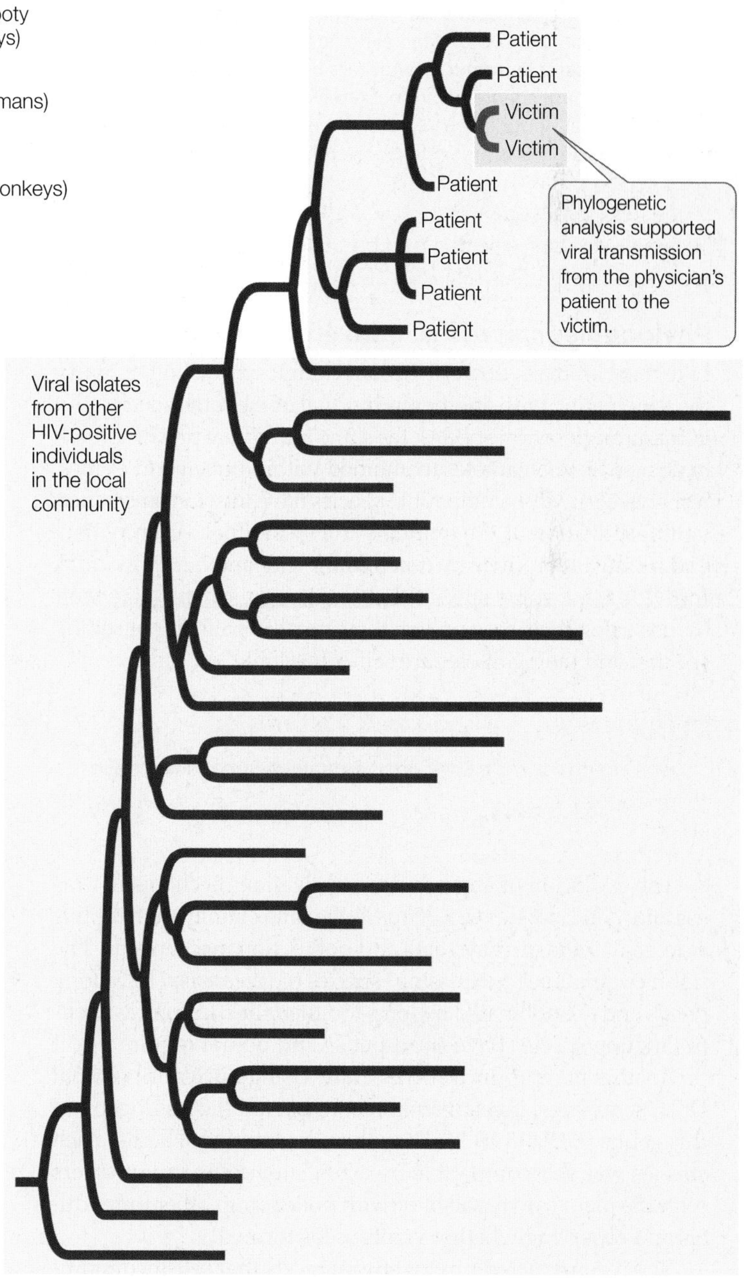

FIGURE 16.7 A Forensic Application of Phylogenetic Analysis
This phylogenetic analysis demonstrated that strains of HIV virus present in a victim (shown in red) were a phylogenetic subset of viruses isolated from a physician's patient (shown in blue). This analysis was part of the evidence used to show that the physician drew blood from his HIV-positive patient and injected it into the victim in an attempt to kill her. The physician was found guilty of attempted murder by the jury.

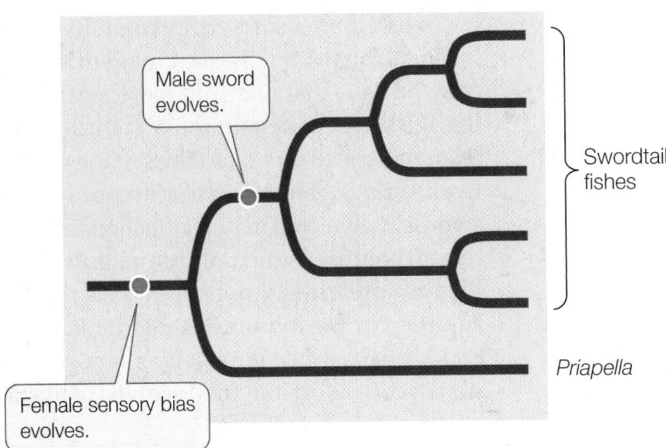

FIGURE 16.8 The Origin of a Sexually Selected Trait The long tail of male swordtail fishes (genus *Xiphophorus*) apparently evolved through sexual selection, with females mating preferentially with males with a longer "sword." Phylogenetic analysis reveals that the *Priapella* lineage split from the swordtails before the evolution of the sword. The independent finding that female *Priapella* prefer males with an artificial sword further supports the idea that this appendage evolved as a result of a preexisting preference in the females.

became a sexually selected trait because of the preexisting preference of the females (**FIGURE 16.8**).

Phylogenies can reveal convergent evolution

Like most animals, flowering plants (angiosperms) often reproduce by mating with another individual of the same species. But in many angiosperm species, the same individual produces both male and female gametes (contained within pollen and ovules, respectively). Self-incompatible species have mechanisms to prevent fertilization of the ovule by the individual's own pollen, and so must reproduce by outcrossing with another individual. Individuals of some species, however, regularly fertilize their ovules using their own pollen; they are self-fertilizing or selfing species, and their gametes are self-compatible.

LINK

Some mechanisms of self-incompatibility are discussed in **Concept 27.1**

The evolution of angiosperm fertilization mechanisms was examined in *Leptosiphon*, a genus in the phlox family that exhibits a diversity of mating systems and pollination mechanisms. The self-incompatible (outcrossing) species of *Leptosiphon* have long petals and are pollinated by long-tongued flies. In contrast, self-pollinating species have short petals and do not require insect pollinators to reproduce successfully. Using nuclear ribosomal DNA sequences, investigators reconstructed the phylogeny of this genus (**FIGURE 16.9**). They then determined whether each species was self-compatible by artificially pollinating flowers with the plant's own pollen or with pollen from other individuals and observing whether viable seeds formed.

The reconstructed phylogeny suggests that self-incompatibility is the ancestral state and that self-compatibility evolved three times within this group of *Leptosiphon*. The change to self-compatibility eliminated the plants' dependence on an outside pollinator and has been accompanied by the evolution of reduced petal size. Indeed, the striking morphological similarity of the flowers in the self-compatible groups once led to their being classified as members of a single species (*L. bicolor*). Phylogenetic analysis, however, shows them to be members of three distinct lineages. From this information we can infer that self-compatibility and its associated floral structure are convergent in the three independent lineages that had been called *L. bicolor*.

Ancestral states can be reconstructed

In addition to using phylogenetic methods to infer evolutionary relationships, biologists can use these techniques to reconstruct the morphology, behavior, or nucleotide and amino acid sequences of ancestral species (as was demonstrated for the ancestral sequence of bacteriophage T7 in Figure 16.5). In the opening of this chapter, we described how Mikhail Matz used phylogenetic analysis to reconstruct the sequence of changes in fluorescent proteins of corals to understand how red fluorescent proteins could be produced.

Reconstruction of ancient DNA sequences can also provide information about the biology of long-extinct organisms. For example, phylogenetic analysis was used to reconstruct an opsin protein in the ancestral archosaur (the most recent common ancestor of birds, dinosaurs, and crocodiles). Opsins are pigment proteins involved in vision; different opsins (with different amino acid sequences) are excited by different wavelengths of light. Knowledge of the opsin sequence in the ancestral archosaur would provide clues about the animal's visual capabilities and therefore about some of its probable behaviors. Investigators used phylogenetic analysis of opsin from living vertebrates to estimate the amino acid sequence of the pigment that existed in the ancestral archosaur. A protein with this same sequence was then constructed in the laboratory. The investigators tested the reconstructed opsin and found a significant shift toward the red end of the spectrum in the light sensitivity of this protein compared with that of most modern opsins. Modern species that exhibit similar sensitivity are adapted for nocturnal vision, so the investigators inferred that the ancestral

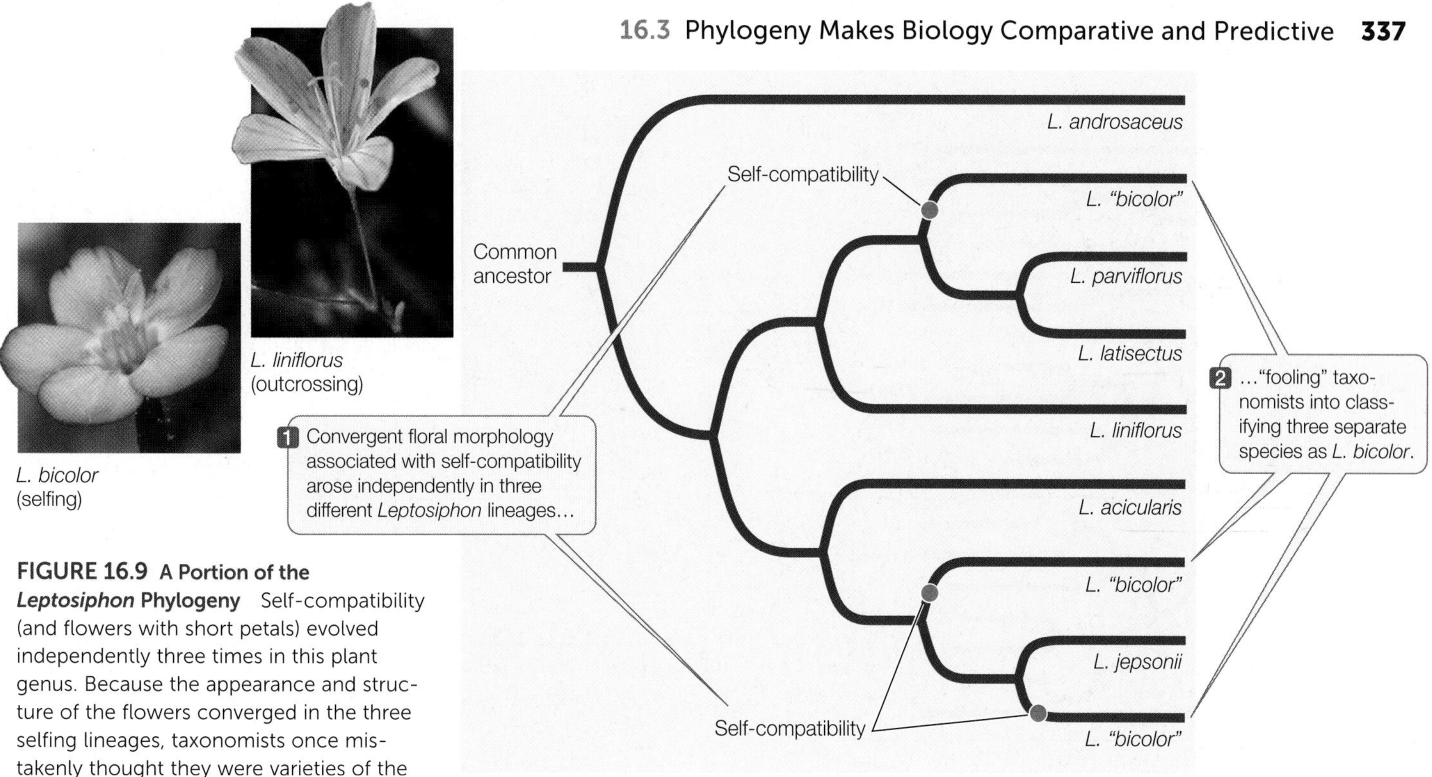

FIGURE 16.9 A Portion of the *Leptosiphon* Phylogeny Self-compatibility (and flowers with short petals) evolved independently three times in this plant genus. Because the appearance and structure of the flowers converged in the three selfing lineages, taxonomists once mistakenly thought they were varieties of the same species.

archosaur might have been active at night. Thus, reminiscent of the movie *Jurassic Park*, phylogenetic analyses are being used to reconstruct extinct species, one protein at a time.

Molecular clocks help date evolutionary events

For many applications, biologists want to know not only the order in which evolutionary lineages split but also the timing of those splits. In 1965, Emile Zuckerkandl and Linus Pauling hypothesized that rates of molecular change were constant enough that they could be used to predict evolutionary divergence times—an idea that has become known as the **molecular clock** hypothesis.

Of course, different genes evolve at different rates, and there are also differences in evolutionary rates among species related to differing generation times, environments, efficiencies of DNA repair systems, and other biological factors. Nonetheless, among closely related species, a given gene usually evolves at a reasonably constant rate. Therefore the protein encoded by the gene accumulates amino acid replacements at a relatively constant rate (**FIGURE 16.10**). A molecular clock uses the average rate at which a given gene or protein accumulates changes to gauge the time of divergence for a particular split in the phylogeny. Molecular clocks must be calibrated using independent data, such as the fossil record, known times of divergence, or biogeographic dates (e.g., the time of separations of continents). Using such calibrations, times of divergence have been estimated for many groups of species that have diverged over millions of years.

Molecular clocks are not only used to date ancient events; they are also used to study the timing of comparatively recent events. Most samples of HIV-1 have been collected from humans only since the early 1980s, although a few isolates from medical biopsies are available from as early as the 1950s. But biologists can use the observed changes in HIV-1 over the past several decades to project back to the common ancestor of all HIV-1 isolates, and estimate when HIV-1 first entered human populations from chimpanzees (**FIGURE 16.11**). This molecular clock was calibrated using the samples from the 1980s and 1990s, and then tested using the samples from the 1950s. As shown in Figure 16.11C, a sample from a 1959 biopsy is dated by molecular clock analysis at 1957 ± 10 years. Extrapolation back to

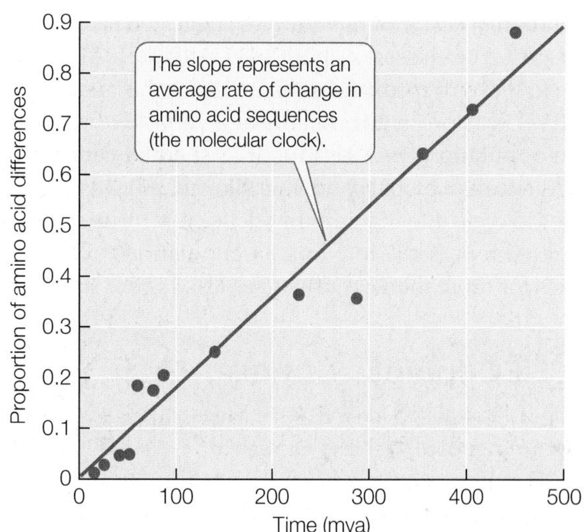

FIGURE 16.10 A Molecular Clock of the Protein Hemoglobin Amino acid replacements in hemoglobin have occurred at a relatively constant rate over nearly 500 million years of evolution. The graph shows the relationship between time of divergence and proportion of amino acid change for 13 pairs of vertebrate hemoglobin proteins. The average rate of change represents the molecular clock for hemoglobin in vertebrates.

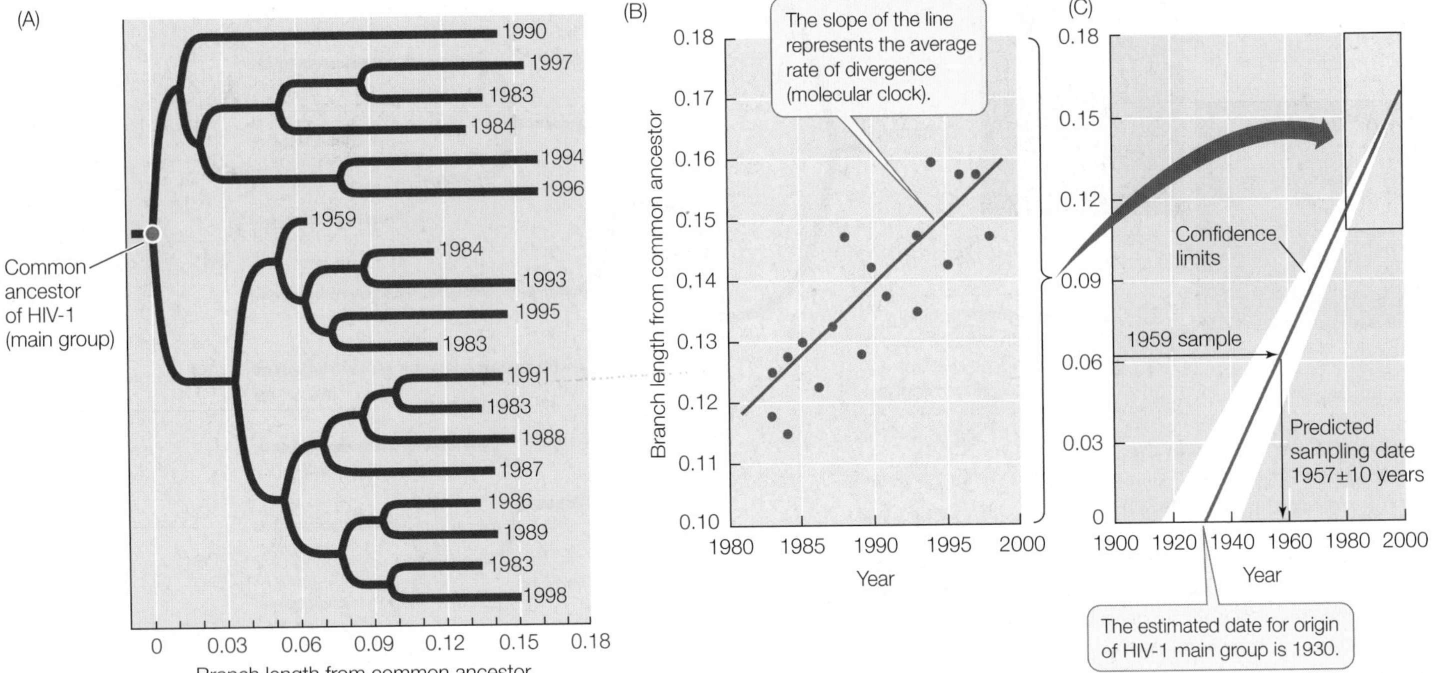

FIGURE 16.11 Dating the Origin of HIV-1 in Human Populations (A) A phylogenetic analysis of the main group of HIV-1 viruses. The dates indicate the years in which samples were taken. (For clarity, only a small fraction of the samples that were examined in the original study are shown.) (B) A plot of year of isolation versus genetic divergence from the common ancestor provides an average rate of divergence, or a molecular clock. (C) The molecular clock is used to date a sample taken in 1959 (as a test of the clock) and the unknown date of origin of the HIV-1 main group (about 1930). Branch length from a common ancestor represents the average number of substitutions per nucleotide.

the common ancestor of the samples suggested a date of origin for this group of viruses of about 1930. Although AIDS was unknown to Western medicine until the 1980s, this analysis shows that HIV-1 was present (probably at a very low frequency) in human populations in Africa for at least a half-century before its emergence as a global pandemic. Biologists have used similar analyses to conclude that immunodeficiency viruses have been transmitted repeatedly into human populations from multiple primates for more than a century (see also Figure 16.6).

CHECKpoint CONCEPT 16.3

- ✓ How can phylogenetic trees help determine the number of times a particular trait evolved?
- ✓ How does the reconstruction of ancestral traits help biologists explain the biology of extinct species?
- ✓ What is the importance of adding a time dimension to phylogenetic trees, and how do biologists accomplish this?

All of life is connected through evolutionary history, and the relationships among organisms provide a natural basis for making biological comparisons. For these reasons, biologists use phylogenetic relationships as the basis for organizing life into a coherent classification system.

CONCEPT 16.4 Phylogeny Is the Basis of Biological Classification

The biological classification system in widespread use today is derived from that developed by the Swedish biologist Carolus Linnaeus in the mid-1700s. Linnaeus developed a system of **binomial nomenclature**. Linnaeus gave each species two names, one identifying the species itself and the other the group of closely related species, or **genus** (plural, *genera*), to which it belongs. Optionally, the name of the taxonomist who first proposed the species name may be added at the end. Thus *Homo sapiens* Linnaeus is the name of the modern human species. *Homo* is the genus, *sapiens* identifies the particular species in the genus *Homo*, and Linnaeus is the person who proposed the name *Homo sapiens*.

You can think of *Homo* as equivalent to your surname and *sapiens* as equivalent to your first name. The first letter of the genus name is capitalized, and the specific name is lowercase. Both of these formal designations are italicized. Rather than repeating the name of a genus when it is used several times in the same discussion, biologists often spell it out only once and abbreviate it with the initial letter thereafter (e.g., *D. melanogaster* rather than *Drosophila melanogaster*).

As we noted earlier, any group of organisms that is treated as a unit in a biological classification system, such as all species in the genus *Drosophila*, or all insects, or all arthropods, is called a taxon. In the Linnaean system, species and genera are further grouped into a hierarchical system of higher taxonomic

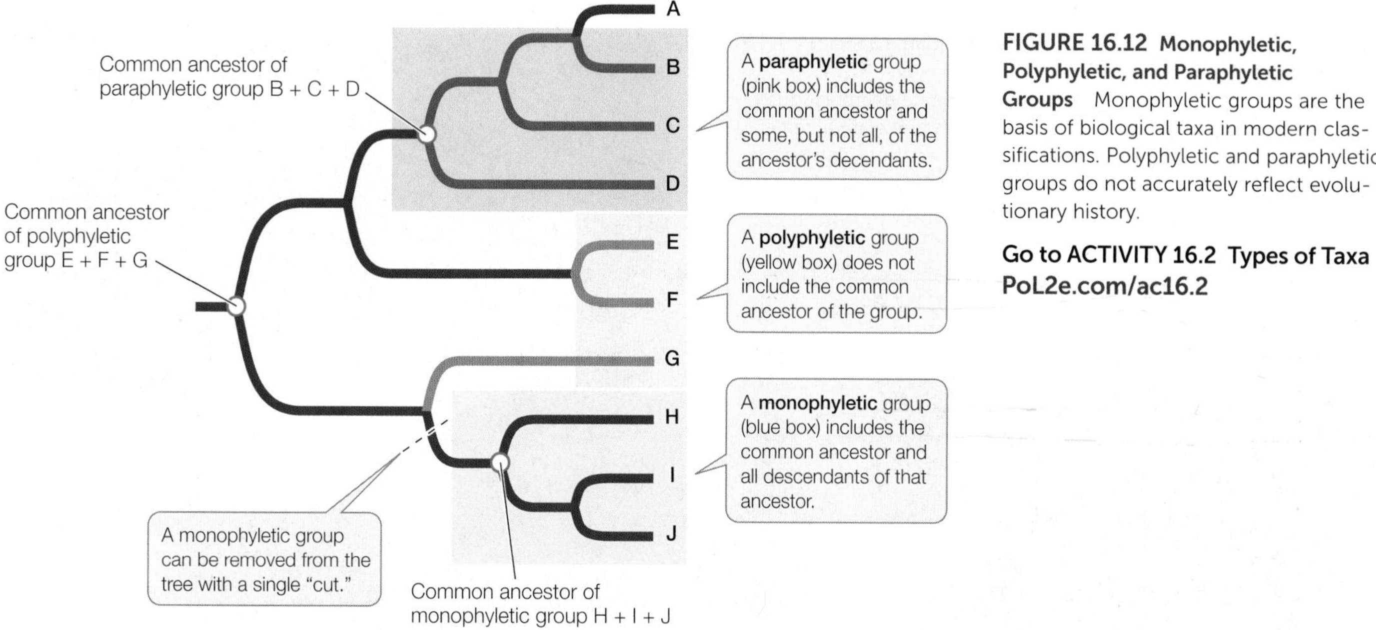

A
B
C
D
E
F
G
H
I
J
Common ancestor of paraphyletic group B + C + D
Common ancestor of polyphyletic group E + F + G
A paraphyletic group (pink box) includes the common ancestor and some, but not all, of the ancestor's decendants.
A polyphyletic group (yellow box) does not include the common ancestor of the group.
A monophyletic group (blue box) includes the common ancestor and all descendants of that ancestor.
A monophyletic group can be removed from the tree with a single "cut."
Common ancestor of monophyletic group H + I + J

FIGURE 16.12 Monophyletic, Polyphyletic, and Paraphyletic Groups Monophyletic groups are the basis of biological taxa in modern classifications. Polyphyletic and paraphyletic groups do not accurately reflect evolutionary history.

Go to ACTIVITY 16.2 Types of Taxa PoL2e.com/ac16.2

categories. The taxon above the genus in the Linnaean system is the **family**. The names of animal families end in the suffix "-idae." Thus Formicidae is the family that contains all ant species, and the family Hominidae contains humans and our recent fossil relatives, as well as our closest living relatives, the chimpanzees, gorillas, and orangutans. Family names are based on the name of a member genus; Formicidae is based on the genus *Formica*, and Hominidae is based on *Homo*. The same rules are used in classifying plants, except that the suffix "-aceae" is used for plant family names instead of "-idae." Thus Rosaceae is the family that includes the genus *Rosa* (roses) and its relatives.

In the Linnaean system, families are grouped into **orders**, orders into **classes**, and classes into **phyla** (singular *phylum*), and phyla into **kingdoms**. However, the ranking of taxa within Linnaean classification is subjective. Whether a particular taxon is considered, say, an order or a class is informative only with respect to the *relative* ranking of other related taxa. Although families are always grouped within orders, orders within classes, and so forth, there is nothing that makes a "family" in one group equivalent (in number of genera or in evolutionary age, for instance) to a "family" in another group.

Today the Linnaean terms above the genus level are used largely for convenience. Linnaeus recognized the overarching hierarchy of life, but he developed his system before evolutionary thought had become widespread. Biologists today recognize the tree of life as the basis for biological classification and often name taxa without placing them into the various Linnaean ranks. But regardless of whether they rank organisms into Linnaean categories or use unranked taxon names, modern biologists use evolutionary relationships as the basis for distinguishing, naming, and classifying biological groups.

Evolutionary history is the basis for modern biological classification

Today's biological classifications express the evolutionary relationships of organisms. Taxa are expected to be **monophyletic**, meaning that the taxon contains an ancestor and all descendants of that ancestor, and no other organisms. In other words, a monophyletic taxon is a historical group of related species, or a complete branch on the tree of life. As noted earlier, this is also the definition of a clade. A true monophyletic group can be removed from a phylogenetic tree by a single "cut" in the tree, as shown in **FIGURE 16.12**.

Note that there are many monophyletic groups on any phylogenetic tree, and that these groups are successively smaller subsets of larger monophyletic groups. This hierarchy of biological taxa, with all of life as the most inclusive taxon and many smaller taxa within larger taxa, down to the individual species, is the modern basis for biological classification.

Although biologists seek to describe and name only monophyletic taxa, the detailed phylogenetic information needed to do so is not always available. A group that does not include its common ancestor is **polyphyletic**. A group that does not include all the descendants of a common ancestor is referred to as **paraphyletic** (see Figure 16.12). Virtually all taxonomists now agree that polyphyletic and paraphyletic groups are inappropriate as taxonomic units because they do not correctly reflect evolutionary history. Some classifications still contain such groups because some organisms have not been evaluated phylogenetically. As mistakes in prior classifications are detected, taxonomic names are revised and polyphyletic and paraphyletic groups are eliminated from the classifications.

Several codes of biological nomenclature govern the use of scientific names

Several sets of explicit rules govern the use of scientific names. Biologists around the world follow these rules voluntarily to facilitate communication and dialogue. There may be dozens of common names for an organism in many different languages, and the same common name may refer to more than one species (**FIGURE 16.13**). The rules of biological nomenclature are designed so that there is only one correct scientific name for

(A) *Myrmecophaga tridactyla*

FIGURE 16.13 Same Common Name, Not the Same Species All three of these animals are known locally as anteaters. Unique scientific binomials allow biologists to communicate unambiguously about each species. (A) *Myrmecophaga tridactyla*, the giant anteater, searching for termites in Brazil. (B) *Tachyglossus aculeatus*, an echidna, is also called the spiny anteater. (C) *Orycteropus afer*, the aardvark, is also known as the Cape anteater.

(B) *Tachyglossus aculeatus*

(C) *Orycteropus afer*

APPLY THE CONCEPT

Phylogeny is the basis of biological classification

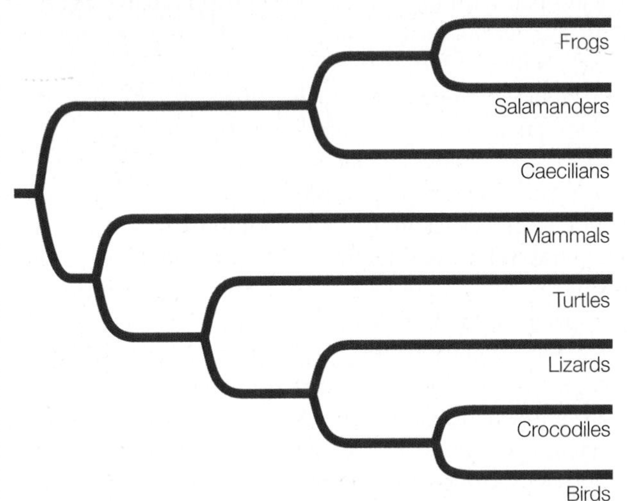

Classification One:

Named group	Included taxa
Amphibia	Frogs, salamanders, and caecilians
Mammalia	Mammals
Reptilia	Turtles, lizards, and crocodiles
Aves	Birds

Classification Two:

Named group	Included taxa
Amphibia	Frogs, salamanders, and caecilians
Mammalia	Mammals
Reptilia	Turtles, lizards, crocodiles, and birds

Classification Three:

Named group	Included taxa
Amphibia	Frogs, salamanders, and caecilians
Homothermia	Mammals and birds
Reptilia	Turtles, lizards, and crocodiles

Consider the above phylogeny and three possible classifications of the living taxa.

1. Which of these classifications contains a paraphyletic group?
2. Which of these classifications contains a polyphyletic group?
3. Which of these classifications is consistent with the goal of including only monophyletic groups in a biological classification?
4. Starting with the classification you named in Question 3, how many additional group names would you need to include all the clades shown in this phylogenetic tree?

any single recognized taxon, and (ideally) a given scientific name applies only to a single taxon (that is, each scientific name is unique). Sometimes the same species is named more than once (when more than one taxonomist has taken up the task). In these cases, the rules specify that the valid name is the first name that was proposed. If the same name is inadvertently given to two different species, then the species that was named second must be given a new name.

Because of the historical separation of the fields of zoology, botany (which originally included mycology, the study of fungi), and microbiology, different sets of taxonomic rules were developed for each of these groups. Yet another set of rules emerged later for classifying viruses. This separation of fields resulted in duplicated taxon names in groups governed by the different sets of rules. *Drosophila*, for example, is both a genus of fruit flies and a genus of fungi, and some species in both groups have identical names. Until recently these duplicated names caused little confusion, since traditionally biologists who studied fruit flies were unlikely to read the literature on fungi (and vice versa). Today, given the prevalence of large, universal biological databases (such as GenBank, which includes DNA sequences from across all life), it is increasingly important that each taxon have a unique and unambiguous name. Biologists are working on a universal code of nomenclature that can be applied to all organisms, so that every species will have a unique identifying name or registration number. This will assist efforts to build an online *Encyclopedia of Life* that links all the information for all the world's species.

CHECKpoint CONCEPT 16.4

- ✓ What is the difference between monophyletic, paraphyletic, and polyphyletic groups?
- ✓ Why do biologists prefer monophyletic groups in formal classifications?
- ✓ What advantages or disadvantages do you see to having separate sets of taxonomic rules for animals, plants, bacteria, and viruses?

Having described some of the mechanisms by which evolution occurs and how phylogenies can be used to study evolutionary relationships, we are now ready to consider the subject of speciation. Speciation is the process that leads to the splitting events (nodes) on the tree of life and eventually results in the millions of distinctive species that constitute Earth's biodiversity.

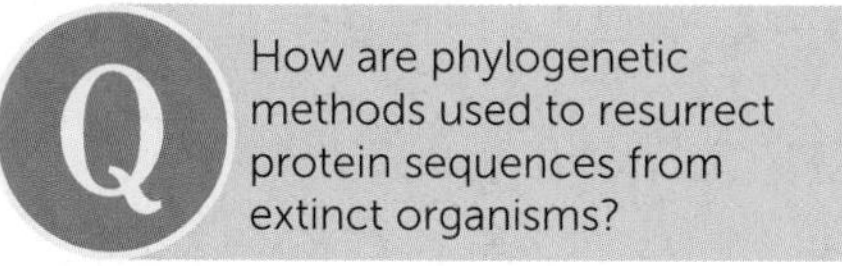

How are phylogenetic methods used to resurrect protein sequences from extinct organisms?

ANSWER Most genes and proteins of organisms that lived millions of years ago have decomposed in the fossil remains of these species. Nonetheless, sequences of many ancient genes and proteins can be reconstructed. Just as we can reconstruct the morphological features of a clade's ancestors, we can also reconstruct the ancestors' DNA and protein sequences—if we have enough information about the genomes of their descendants (Concept 16.3).

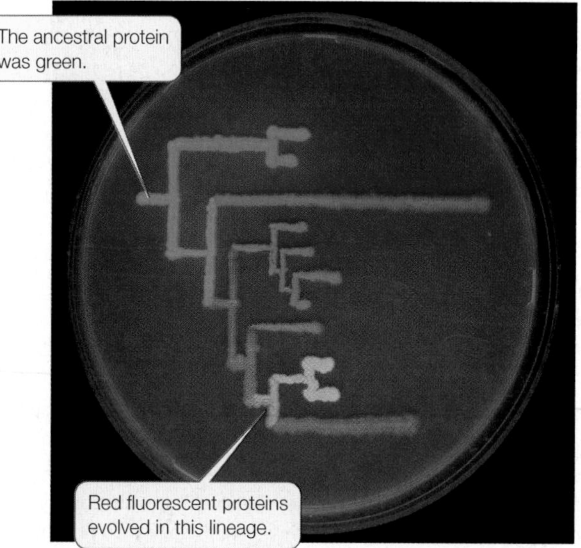

The ancestral protein was green.
Red fluorescent proteins evolved in this lineage.

FIGURE 16.14 Evolution of Fluorescent Proteins of Corals Mikhail Matz and his colleagues used phylogenetic analysis to reconstruct the sequences of extinct fluorescent proteins that were present in the ancestors of modern corals. They then expressed these proteins in bacteria and plated the bacteria in the form of a phylogenetic tree to show how the colors evolved over time.

Biologists have now reconstructed gene sequences from many species that have been extinct for millions of years. Using this information, a laboratory can reconstruct real proteins that correspond to the proteins that were present in long-extinct species. This is how Mikhail Matz and his colleagues were able to resurrect fluorescent proteins from the extinct ancestors of modern corals and then visualize the colors produced by these proteins in the laboratory and recreate the probable evolutionary path of the different pigments (**FIGURE 16.14**).

Biologists have even used phylogenetic analysis to reconstruct some protein sequences that were present in the common ancestor of life (Concept 16.1). These hypothetical protein sequences can then be resurrected into actual proteins in the laboratory. When biologists measured the temperature optima for these resurrected proteins, they found that the proteins functioned best in the range of 55°C–65°C. This analysis is consistent with hypotheses that life evolved in a high-temperature environment.

To reconstruct protein sequences from species that have been extinct for millions or even billions of years, biologists use detailed mathematical models that take into account much of what we have learned of molecular evolution (Concept 16.2). These models incorporate information on rates of replacement among different amino acid residues in proteins, information on different substitution rates among nucleotides, and changes in the rate of molecular evolution among the major lineages of life.

SUMMARY

CONCEPT 16.1 All of Life Is Connected through Its Evolutionary History

- **Phylogeny** is the history of descent of organisms from their common ancestor. Groups of evolutionarily related species are represented as related branches in a **phylogenetic tree**.
- A group of species that consists of a common ancestor and all its evolutionary descendants is called a **clade**. Named clades and species are called **taxa**. **Review Figure 16.1**
- **Homologies** are similar traits that have been inherited from a common ancestor.
- A trait that is shared by two or more taxa and is derived through evolution from a common ancestral form is called a **synapomorphy**.
- Similar traits may occur among species that do not result from common ancestry. **Convergent evolution** and **evolutionary reversals** can give rise to such traits, which are called **homoplasies**. **Review Figure 16.2**

CONCEPT 16.2 Phylogeny Can Be Reconstructed from Traits of Organisms

- Phylogenetic trees can be inferred from synapomorphies and using the **parsimony principle**. **Review Figure 16.3 and ACTIVITY 16.1**
- Sources of phylogenetic information include morphology, patterns of development, the fossil record, behavioral traits, and molecular traits such as DNA and protein sequences. **Review ANIMATED TUTORIAL 16.1**
- Phylogenetic trees can be inferred with **maximum likelihood** methods, which calculate the probability that a particular tree will have generated the observed data. **Review ANIMATED TUTORIAL 16.2**

CONCEPT 16.3 Phylogeny Makes Biology Comparative and Predictive

- Phylogenetic trees are used to reconstruct past events. **Review Figures 16.6 and 16.7**
- Biologists can use phylogenetic trees to reconstruct ancestral states. **Review Figure 16.8**
- Phylogenetic trees are used to reveal convergent evolution. **Review Figure 16.9**
- Phylogenetic trees may include estimates of times of divergence of lineages determined by **molecular clock** analysis. **Review Figures 16.10 and 16.11**

CONCEPT 16.4 Phylogeny Is the Basis of Biological Classification

- Taxonomists organize biological diversity on the basis of evolutionary history.
- Taxa in modern classifications are expected to be clades, or **monophyletic** groups. **Paraphyletic** and **polyphyletic** groups are not considered appropriate taxonomic units. **Review Figure 16.12 and ACTIVITY 16.2**
- Several sets of rules govern the use of scientific names, with the goal of providing unique and universal names for biological taxa.

Go to the Interactive Summary to review key figures, Animated Tutorials, and Activities PoL2e.com/is16

Go to LaunchPad at **macmillanhighered.com/launchpad** for additional resources, including LearningCurve Quizzes, Flashcards, and many other study and review resources.

17 Speciation

KEY CONCEPTS

This composite photograph shows several of the hundreds of species of diverse haplochromine cichlids that are endemic to Lake Malawi.

Not quite 2 million years ago, a tectonic split in the Great Rift Valley of East Africa led to the formation of Lake Malawi, which lies between the modern countries of Malawi, Tanzania, and Mozambique. A few fish species entered the new lake, including a type known as a haplochromine cichlid. Today the descendants of this early colonizer include nearly 1,000 species of haplochromine cichlids. All of them are endemic to Lake Malawi—they are found nowhere else. This vast array of cichlid species makes this the most diverse lake in the world in terms of its fish community. How did so many different species arise from a single ancestral species in less than 2 million years?

As we noted in Chapter 16, speciation is the process that produces the splits among lineages in the tree of life. Biologists have studied the history and timing of speciation events in Lake Malawi and have pieced together some of the processes that led to so many cichlid species. The earliest haplochromine cichlids to enter the new lake encountered diverse habitats in Lake Malawi, as some shores were rocky and others were sandy. Cichlid populations quickly adapted to these distinct habitat types. Fish in rocky habitats adapted to breeding and living in rocky conditions, and those in sandy habitats evolved specializations for life over sand. These changes resulted in an early speciation event.

Within each of these major habitat types, there were numerous opportunities for diet specialization. Various populations of cichlids became rock scrapers, bottom feeders, fish predators, scale biters, pelagic zooplankton eaters, and plant specialists. Each of these feeding specializations requires a different mouth morphology. The offspring of fish that bred with fish of similar morphology were more likely to survive than were fish with two very different parents. These differences in fitness led to the formation of many more new species, each adapted to a different feeding mode.

But still the Lake Malawi cichlids continued to diverge and form new species. Male cichlids compete for the attention of females through their bright body colors. Diversification of the body colors of males, and of the preferences of females for different body colors, led to many more new species of cichlids, each isolated from the other by their sexual preferences. Now biologists are studying the genomes of these Lake Malawi cichlids to understand the details of the genetic changes that have given rise to so many species over so little time.

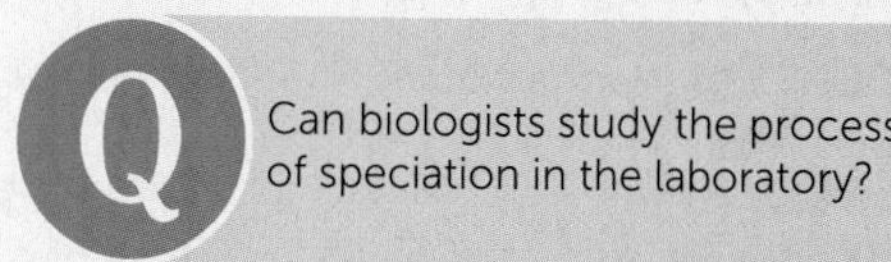

Q
Can biologists study the process of speciation in the laboratory?

You will find the answer to this question on page 356.

CONCEPT 17.1 Species Are Reproductively Isolated Lineages on the Tree of Life

Biological diversity does not vary in a smooth, incremental way. People have long recognized groups of similar organisms that mate with one another, and they have noticed that there are usually distinct morphological breaks between these groups. Groups of organisms that mate with one another are commonly called **species** (note that this is both the plural and singular form of the word). Species are the result of the process of **speciation**: the divergence of biological lineages and the emergence of reproductive isolation between lineages.

Although "species" is a useful and common term, its usage varies among biologists who are interested in different aspects of speciation. Different biologists think about species differently because they ask different questions: How can we recognize and identify species? How do new species arise? How do different species remain separate? Why do rates of speciation differ among groups of organisms? In answering these questions, biologists focus on different attributes of species, leading to several different ways of thinking about what species are and how they form. Most of the various **species concepts** proposed by biologists are simply different ways of approaching the question "What are species?" Let's compare three major classes of species concepts to contrast the way that biologists think about species.

We can recognize many species by their appearance

Someone who is knowledgeable about a group of organisms, such as birds or flowering plants, can usually distinguish the different species found in a particular area simply by looking at them. Standard field guides to birds, mammals, insects, and wildflowers are possible only because many species change little in appearance over large geographic distances (**FIGURE 17.1A**).

More than 250 years ago, Carolus Linnaeus developed the system of binomial nomenclature by which species are named today (see Concept 16.4). Linnaeus described and named thousands of species, but because he knew nothing about genetics or the mating behavior of the organisms he was naming, he classified them on the basis of their appearance alone. In other words, Linnaeus used a **morphological species concept**, a construct that assumes a species comprises individuals that look alike, and that individuals that do not look alike belong to different species. Although Linnaeus did not know it, the members of most of the groups he classified as species look alike because they share many alleles of genes that code for morphological features.

Using morphology to define species has limitations. Members of the same species do not always look alike. For example, males, females, and young individuals do not always resemble one another closely (**FIGURE 17.1B**). Furthermore, morphology is of little use in the case of cryptic species—instances in which two or more species are morphologically indistinguishable but do not interbreed (**FIGURE 17.2**). Biologists therefore cannot rely on appearance alone in determining whether individual organisms are members of the same or different species. Today biologists use several additional types of information—especially behavioral and genetic data—to differentiate species.

Reproductive isolation is key

The most important factor in the long-term isolation of sexually reproducing lineages from one another is the evolution of **reproductive isolation**, a state in which two groups of organisms can no longer exchange genes. If individuals of group "A" mate and reproduce only with one another, group "A" constitutes a distinct species within which genes recombine. In other words, group "A" is an independent evolutionary lineage—a separate branch on the tree of life.

Evolutionary biologist Ernst Mayr recognized the importance of reproductive isolation in maintaining species, and so he proposed the **biological species concept**: "*Species are groups of actually or potentially interbreeding natural populations which are reproductively isolated from other such groups.*" The phrase "actually or potentially" is an important element of this definition. "Actually" says that the individuals live in the same area and

(A)

Lophodytes cucullatus
Male, British Columbia

Lophodytes cucullatus
Male, New Mexico

(B)

Lophodytes cucullatus
Female

FIGURE 17.1 Members of the Same Species Look Alike—or Not (A) It is easy to identify these two male hooded mergansers as members of the same species, even though they were photographed thousands of miles apart in British Columbia and New Mexico, respectively. Despite their geographic separation, the two individuals are morphologically very similar. (B) Hooded mergansers are sexually dimorphic, which means the female's appearance is quite different from that of the male.

(A)

Hyla versicolor

(B)

Hyla chrysoscelis

FIGURE 17.2 Cryptic Species Look Alike but Do Not Interbreed These two species of gray treefrogs (*Hyla versicolor* and *H. chrysoscelis*) cannot be distinguished by their external morphology, but they do not interbreed even when they occupy the same geographic range. *H. versicolor* is a tetraploid species, whereas *H. chrysoscelis* is diploid. Although they look alike, the males have distinctive mating calls, and based on these calls, the females recognize and mate with males of their own species.

interbreed with one another. "Potentially" says that even though the individuals do not live in the same area, and therefore do not interbreed, other information suggests that they *would* do so if they were able to get together. This widely used species concept does not apply to organisms that reproduce asexually, and it is limited to a single point in evolutionary time.

The lineage approach takes a long-term view

Evolutionary biologists often think of species as branches on the tree of life. This idea can be termed a **lineage species concept**. In this framework for thinking about species, one species splits into two or more daughter species, which thereafter evolve as distinct lineages. A lineage concept allows biologists to consider species over evolutionary time.

A **lineage** is an ancestor–descendant series of populations followed over time. Each species has a history that starts with a speciation event by which one lineage on the tree is split into two, and ends either at extinction or at another speciation event, at which time the species produces two daughter species. The process of lineage splitting may be gradual, taking thousands of generations to complete. At the other extreme, an ancestral lineage may be split in two within a few generations (as happens with polyploidy, which we'll discuss in Concept 17.3). The gradual nature of some splitting events means that at a single point in time, the final outcome of the process may not be clear. In these cases, it may be difficult to predict whether the incipient species (lineages in an initial stage of development) will continue to diverge and become fully isolated from one another, or if they will merge again in the future.

The different species concepts are not mutually exclusive

Many named variants of these three major classes of species concepts exist. The various concepts are not entirely incompatible, however. Rather, they simply emphasize different aspects of species or speciation. The morphological species concept emphasizes the practical aspects of recognizing species, although it sometimes results in underestimation or overestimation of the actual number of species. Mayr's biological species concept emphasizes that reproductive isolation is what allows sexual species to evolve independently of one another. Lineage species concepts embrace the idea that sexual species are maintained by reproductive isolation, but extend the concept of a species as a lineage over evolutionary time. The species-as-lineage concept is also able to accommodate species that reproduce asexually.

Virtually all species exhibit some degree of genetic recombination among individuals, even if recombination events are relatively rare. Significant reproductive isolation between species is therefore necessary for lineages to remain distinct over evolutionary time. Furthermore, reproductive isolation is responsible for the morphological distinctiveness of most species, because mutations that result in morphological changes cannot spread between reproductively isolated species. Therefore no matter which species concept we emphasize, the evolution of reproductive isolation is important for understanding the origin of species.

CHECKpoint CONCEPT 17.1

- Why do different biologists emphasize different attributes of species in formulating species concepts?
- What makes reproductive isolation such an important component of each of the species concepts discussed here?
- Why is the biological species concept not applicable to asexually reproducing organisms? How do biologists consider the species status of asexual organisms?

Although Charles Darwin titled his groundbreaking book *On the Origin of Species*, in fact it included very little about speciation as we understand it today. Darwin devoted most of his attention to demonstrating that individual species are altered over time by natural selection. The following sections discuss the many aspects of speciation that biologists have learned about since Darwin's time.

CONCEPT 17.2 Speciation Is a Natural Consequence of Population Subdivision

Not all evolutionary changes result in new species. A single lineage may change over time without giving rise to a new species. Speciation requires the interruption of gene flow within a species whose members formerly exchanged genes. But if a genetic change prevents reproduction between individuals of a species, how can such a change spread through a species in the first place?

Incompatibilities between genes can produce reproductive isolation

If a new allele that causes reproductive incompatibility arises in a population, it cannot spread through the population because no other individuals will be reproductively compatible with the individual that carries the new allele. So how can one reproductively cohesive lineage ever split into two reproductively isolated species? Several early geneticists, including Theodosius Dobzhansky and Hermann Joseph Muller, developed a genetic model to explain this apparent conundrum (**FIGURE 17.3**).

The Dobzhansky–Muller model is quite simple. First, assume that a single ancestral population is subdivided into two daughter populations (by the formation of a new mountain range, for instance), which then evolve as independent lineages. In one of the descendant lineages, a new allele (*A*) arises and becomes fixed (see Figure 17.3). In the other population, another new allele (*B*) becomes fixed *at a different gene locus.* Neither new allele at either locus results in any loss of reproductive compatibility. However, the two new forms of these two different genes have never occurred together in the same individual or population. Recall that the products of many genes must work together in an organism. It is possible that the new protein forms encoded by the two new alleles will not be compatible with each other. If individuals from the two lineages come back together after these genetic changes, they may still be able to interbreed. However, the hybrid offspring may have a new combination of genes that is functionally inferior, or even lethal. This will not happen with all new combinations of genes, but over time, isolated lineages will accumulate many allele differences at many gene loci. Some combinations of these differentiated genes will not function well together in hybrids. Thus genetic incompatibility between the two isolated populations will develop over time.

Many empirical and experimental examples support the Dobzhansky–Muller model. This model works not only for pairs of individual genes but also for some kinds of chromosomal rearrangements. Bats of the genus *Rhogeessa*, for example, exhibit considerable variation in centric fusions of their chromosomes. The chromosomes of the various species contain the same basic chromosomal arms, but in some species two acrocentric (one-armed) chromosomes have fused at the centromere to form larger, metacentric (two-armed) chromosomes. A polymorphism in centric fusion causes few, if any, problems in meiosis because the respective chromosomes can still align and assort normally. Therefore a given centric fusion can become fixed in a lineage. However, if a *different* centric fusion becomes fixed in a second lineage, then hybrids between individuals of each lineage will not be able to produce normal gametes in meiosis (**FIGURE 17.4**). Most of the closely related species of *Rhogeessa* display different combinations of these centric fusions and are thereby reproductively isolated from one another.

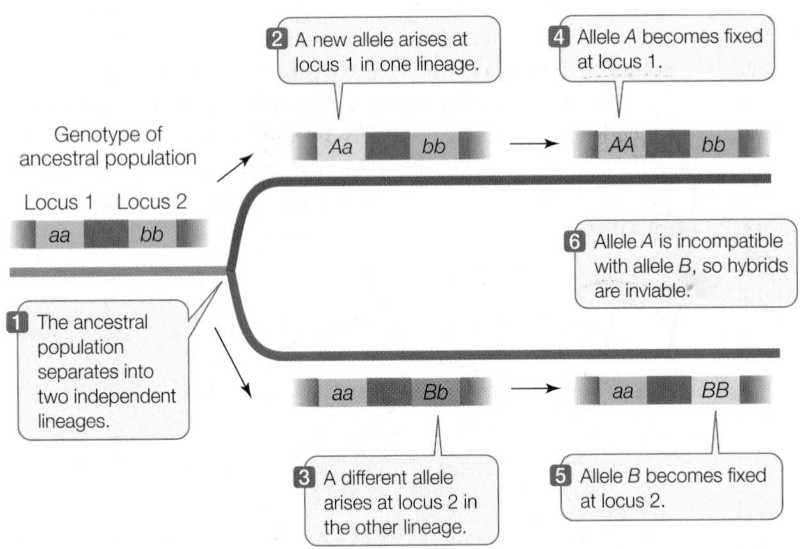

FIGURE 17.3 The Dobzhansky–Muller Model In this simple two-locus version of the model, two lineages from the same ancestral population become separated from each other and evolve independently. A new allele becomes fixed in each descendant lineage, but at two different genes. Neither of the new alleles is incompatible with the ancestral alleles, but the two new alleles in the two different genes are incompatible with each other. Thus the two descendant lineages are reproductively incompatible.

Go to ANIMATED TUTORIAL 17.1
Speciation Simulation
PoL2e.com/at17.1

Reproductive isolation develops with increasing genetic divergence

As pairs of species diverge genetically, they become increasingly reproductively isolated (**FIGURE 17.5**). Both the rate at which reproductive isolation develops and the mechanisms that produce it vary from group to group. Reproductive incompatibility has been shown to develop gradually in many groups of plants, animals, and fungi, reflecting the slow pace at which incompatible genes accumulate in each lineage. In some cases, complete reproductive isolation may take millions of years. In other cases (as with the chromosomal fusions of *Rhogeessa* described above), reproductive isolation can develop over just a few generations.

Partial reproductive isolation has evolved in strains of *Phlox drummondii* artificially isolated by humans. In 1835, Thomas Drummond, after whom this species of garden plant is named, collected seeds in Texas and distributed them to nurseries in Europe.

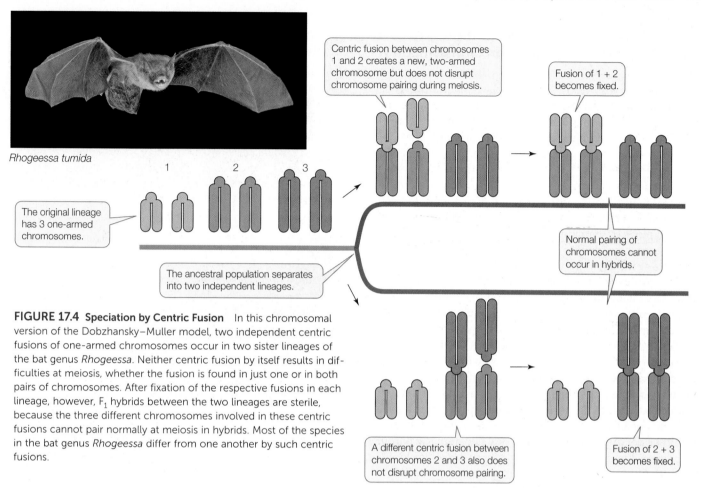

Rhogeessa tumida

1 2 3

The original lineage has 3 one-armed chromosomes.

The ancestral population separates into two independent lineages.

Centric fusion between chromosomes 1 and 2 creates a new, two-armed chromosome but does not disrupt chromosome pairing during meiosis.

Fusion of 1 + 2 becomes fixed.

Normal pairing of chromosomes cannot occur in hybrids.

A different centric fusion between chromosomes 2 and 3 also does not disrupt chromosome pairing.

Fusion of 2 + 3 becomes fixed.

FIGURE 17.4 Speciation by Centric Fusion In this chromosomal version of the Dobzhansky–Muller model, two independent centric fusions of one-armed chromosomes occur in two sister lineages of the bat genus *Rhogeessa*. Neither centric fusion by itself results in difficulties at meiosis, whether the fusion is found in just one or in both pairs of chromosomes. After fixation of the respective fusions in each lineage, however, F_1 hybrids between the two lineages are sterile, because the three different chromosomes involved in these centric fusions cannot pair normally at meiosis in hybrids. Most of the species in the bat genus *Rhogeessa* differ from one another by such centric fusions.

The European nurseries established more than 200 true-breeding strains of *P. drummondii* that differed in flower size, flower color, and plant growth form. The breeders did not select directly for reproductive incompatibility between strains, but in subsequent experiments in which strains were crossed and seed production was measured and compared, biologists found that reproductive compatibility between strains had been reduced by 14 to 50 percent, depending on the cross—even though the strains had been isolated from one another for less than two centuries.

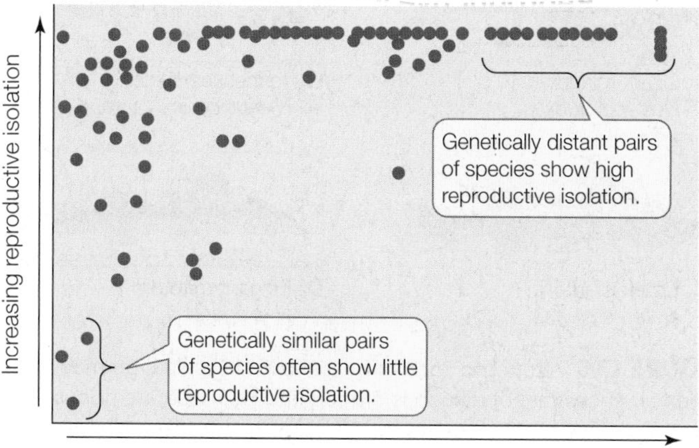

FIGURE 17.5 Reproductive Isolation Increases with Genetic Divergence Among pairs of *Drosophila* species, the more the species differ genetically, the greater their reproductive isolation from each other. Each dot represents a comparison of one species pair. Such positive relationships between genetic distance and reproductive isolation have been observed in many groups of plants, animals, and fungi.

CHECKpoint CONCEPT 17.2

- ✓ The Dobzhansky–Muller model suggests that divergence among alleles at *different* gene loci leads to genetic incompatibility between species. Why is genetic incompatibility between two alleles at the *same* locus considered less likely?
- ✓ Why do some combinations of chromosomal centric fusions cause problems in meiosis? Can you diagram what would happen at meiosis in a hybrid of the divergent lineages shown in Figure 17.4?
- ✓ Assume that the reproductive isolation seen in *Phlox* strains results from lethal combinations of incompatible alleles at several loci among the various strains. Given this assumption, why might the reproductive isolation seen among these strains be partial rather than complete?

We have now seen how splitting an ancestral population leads to genetic divergence in the two descendant lineages. Next we will consider ways in which the descendant lineages could have become separated in the first place.

CONCEPT 17.3 Speciation May Occur through Geographic Isolation or in Sympatry

Many scientists who study speciation have concentrated on geographic processes that result in the splitting of an ancestral species. Splitting the range of a species is one obvious way of achieving such a division, but it is not the only way.

Physical barriers give rise to allopatric speciation

Speciation that results when a population is divided by a physical barrier is known as **allopatric speciation** (Greek *allos*, "other"; *patria*, "homeland"). Allopatric speciation is thought to be the dominant mode of speciation in most groups of organisms. The physical barrier that divides the range of a species may be a body of water or a mountain range for terrestrial organisms, or dry land for aquatic organisms—in other words, any type of habitat that is inhospitable to the species. Such barriers can form when continents drift, sea levels rise and fall, glaciers advance and retreat, or climates change. The populations separated by such barriers are often, but not always, initially large. The lineages that descend from these founding populations evolve differences for a variety of reasons, including mutation, genetic drift, and adaptation to different environments in the two areas. As a result, many pairs of closely related **sister species**—species that are each other's closest relatives—may exist on opposite sides of the geographic barrier. An example of a physical geographical barrier that produced many pairs of sister species was the Pleistocene glaciation that isolated freshwater streams in the eastern highlands of the Appalachian Mountains from streams in the Ozark and Ouachita mountains about 20,000–80,000 years ago (**FIGURE 17.6**). This splitting event resulted in many parallel speciation events among isolated lineages of stream-dwelling organisms.

Allopatric speciation also may result when some members of a population cross an existing barrier and establish a new, isolated population. The 14 species of finches found in the Galápagos archipelago some 1,000 kilometers off the coast of Ecuador are the result of this process. Darwin's finches (as they are usually called, because Darwin was the first scientist to study them) arose in the Galápagos from a single South American finch species that colonized the islands. Today the Galápagos species differ strikingly from their closest mainland relative and from one another (**FIGURE 17.7**). The islands of the archipelago are sufficiently far apart that the finches move among them only infrequently. In addition, environmental conditions differ widely from island to island. Some islands are relatively flat and arid; others have forested mountain slopes. Over millions of years, finch lineages on the different islands have differentiated to the point that when occasional immigrants arrive from other islands, they either do not breed with the residents, or if they do, the resulting offspring do not survive as well as the offspring of established residents. The genetic distinctness of each finch species from the others and the genetic cohesiveness of the individual species are thus maintained.

Go to ANIMATED TUTORIAL 17.2
Founder Events and Allopatric Speciation
PoL2e.com/at17.2

(A) Pliocene

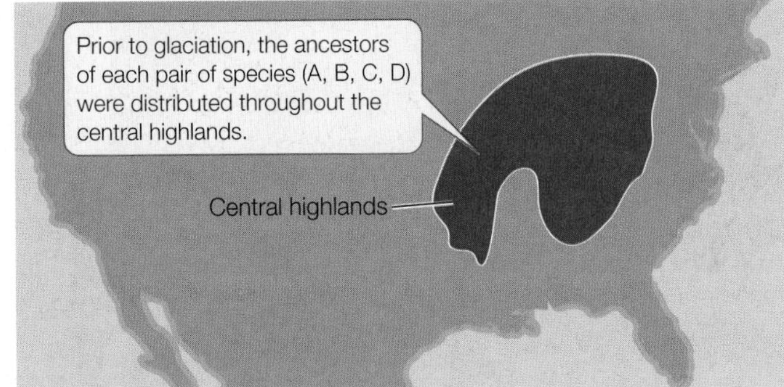

(B) Pleistocene

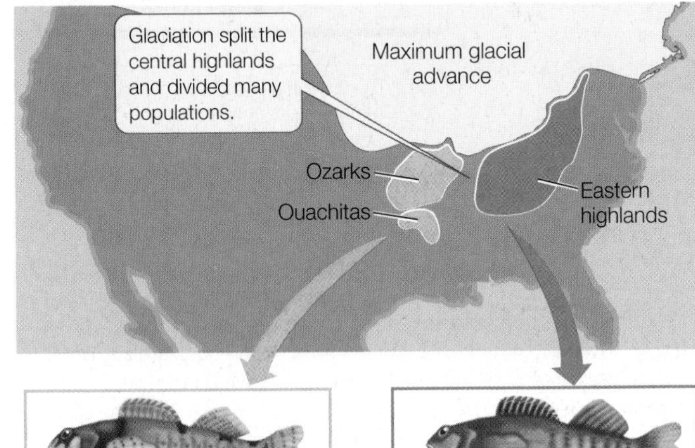

D_1 Ozark madtom *Noturus albater*

D_2 Elegant madtom *N. elegans*

FIGURE 17.6 Allopatric Speciation Allopatric speciation may result when an ancestral population is divided into two separate populations by a physical barrier, and the lineages that descend from these populations then diverge. (A) Many species of freshwater stream fishes were distributed throughout the central highlands of North America in the Pliocene epoch (about 3–5 million years ago). (B) During the Pleistocene, glaciers advanced and isolated fish populations in the Ozark and Ouachita mountains to the west from fish populations in the highlands to the east. Numerous species diverged as a result of this separation, including the ancestors of the four species pairs shown here.

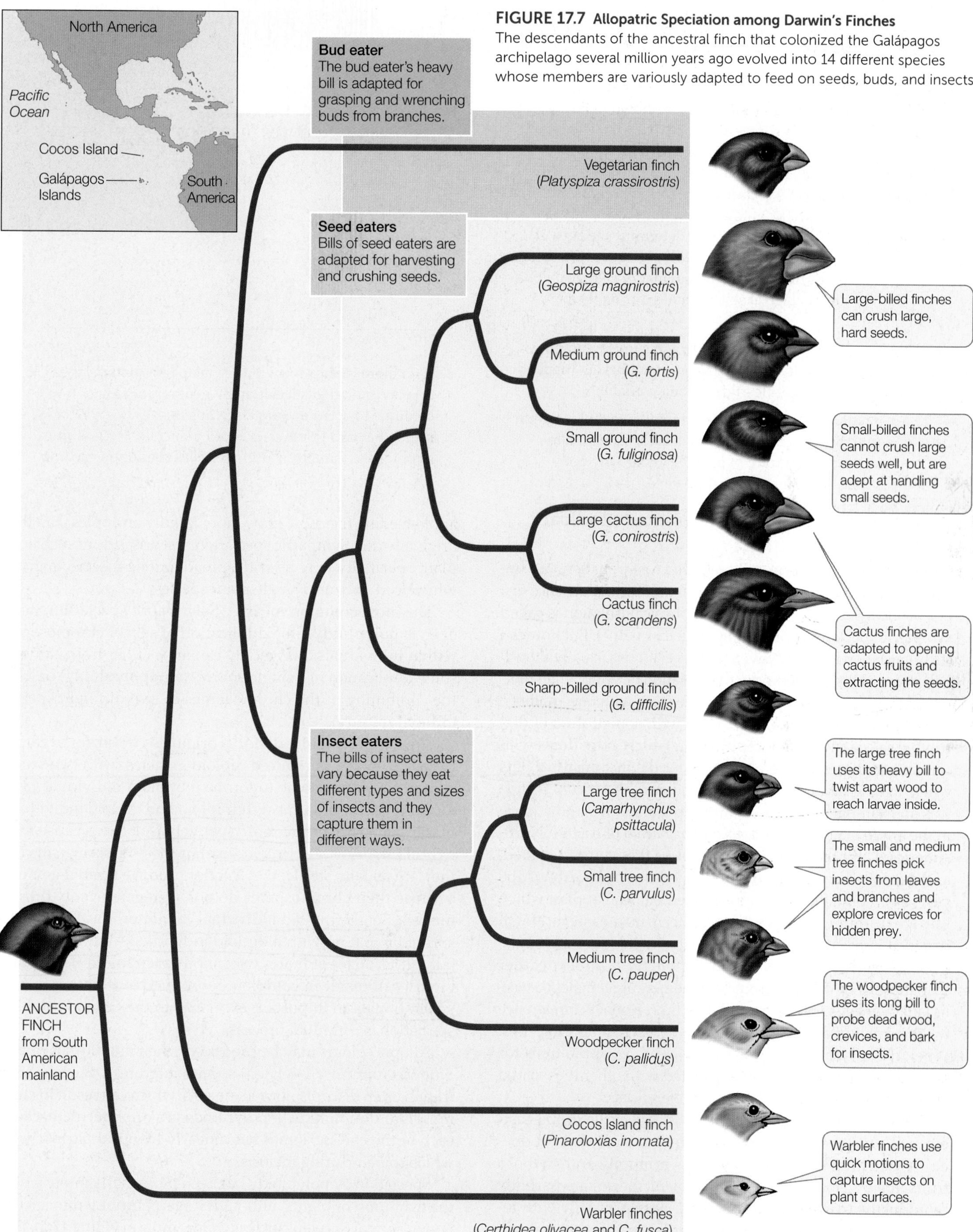

FIGURE 17.7 Allopatric Speciation among Darwin's Finches The descendants of the ancestral finch that colonized the Galápagos archipelago several million years ago evolved into 14 different species whose members are variously adapted to feed on seeds, buds, and insects.

APPLY THE CONCEPT

Speciation may occur through geographic isolation

The different species of Darwin's finches shown in the phylogeny in Figure 17.7 have all evolved on islands of the Galápagos archipelago within the past 3 million years. Molecular clock analysis (see Concept 16.3) has been used to determine the dates of the various speciation events in that phylogeny. Geological techniques for dating rock samples (see Concept 18.1) have been used to determine the ages of the various Galápagos islands. The table shows the number of species of Darwin's finches and the number of islands that have existed in the archipelago at several times during the past 4 million years.

TIME (mya)	NUMBER OF ISLANDS	NUMBER OF FINCH SPECIES
0.25	18	14
0.50	18	9
0.75	9	7
1.00	6	5
2.00	4	3
3.00	4	1
4.00	3	0

1. Plot the number of species of Darwin's finches and the number of islands in the Galápagos archipelago (dependent variables) against time (independent variable).
2. Are the data consistent with the hypothesis that isolation of populations on newly formed islands is related to speciation in this group of birds? Why or why not?
3. If no more islands form in the Galápagos archipelago, do you think that speciation by geographic isolation will continue to occur among Darwin's finches? Why or why not? What additional data could you collect to test your hypothesis (without waiting to see if speciation occurs)?

Sympatric speciation occurs without physical barriers

Geographic isolation is usually required for speciation, but under some circumstances speciation can occur in the absence of a physical barrier. Speciation without physical isolation is called **sympatric speciation** (Greek *sym*, "together with"). But how can such speciation happen? Given that speciation is usually a gradual process, how can reproductive isolation develop when individuals have frequent opportunities to mate with one another?

Sympatric speciation may occur with some forms of disruptive selection (see Concept 15.4) in which individuals with certain genotypes have a preference for distinct microhabitats where mating takes place. The opening story of African cichlids was an example. Another example appears to be taking place in the apple maggot fly (*Rhagoletis pomonella*) of eastern North America. Until the mid-1800s, *Rhagoletis* flies courted, mated, and deposited their eggs only on hawthorn fruits. About 150 years ago, some flies began to lay their eggs on apples, which European immigrants had introduced into eastern North America. Apple trees are closely related to hawthorns, but the smell of the fruits differs, and the apple fruits appear earlier than those of hawthorns. Some early-emerging female *Rhagoletis* laid their eggs on apples, and over time, a genetic preference for the smell of apples evolved among early-emerging flies. When the offspring of these flies sought out apple trees for mating and egg deposition, they mated with other flies reared on apples, which shared the same preferences.

Today the two groups of *Rhagoletis pomonella* in the eastern U.S. appear to be on the way to becoming distinct species. One group mates and lays eggs primarily on hawthorn fruits, the other on apples. The incipient species are partially reproductively isolated because they mate primarily with individuals raised on the same fruit and because they emerge from their pupae at different times of the year. In addition, the apple-feeding flies now grow more rapidly on apples than they originally did. Sympatric speciation that arises from such host-plant specificity may be widespread among insects, many of which feed only on a single plant species.

The most common means of sympatric speciation, however, is **polyploidy**—the duplication of sets of chromosomes within individuals. Polyploidy can arise either from chromosome duplication in a single species (**autopolyploidy**) or from the combining of the chromosomes of two different species (**allopolyploidy**).

An autopolyploid individual originates when, for example, two accidentally unreduced diploid gametes (with two sets of chromosomes) combine to form a tetraploid individual (with four sets of chromosomes). Tetraploid and diploid individuals of the same species are reproductively isolated because their hybrid offspring are triploid. Even if these offspring survive, they are usually sterile; they cannot produce normal gametes because their chromosomes do not segregate evenly during meiosis. So a tetraploid individual cannot produce fertile offspring by mating with a diploid individual—but it *can* do so if it self-fertilizes or mates with another tetraploid. Thus polyploidy can result in complete reproductive isolation in two generations—an important exception to the general rule that speciation is a gradual process.

Allopolyploids may be produced when individuals of two different (but closely related) species interbreed. Such hybridization often disrupts normal meiosis, which can result in chromosomal doubling. Allopolyploids are often fertile because each of the chromosomes has a nearly identical partner with which to pair during meiosis.

Speciation by polyploidy has been particularly important in the evolution of plants, although it has contributed to speciation in animals as well (such as the example in Figure 17.2). Botanists estimate that about 70 percent of flowering plant species and 95 percent of fern species are the result of recent polyploidization. Some of these species arose from hybridization between two species followed by chromosomal duplication and self-fertilization. Other species diverged from polyploid ancestors, so that the new species shared their ancestors' duplicated sets of chromosomes. New species arise by polyploidy more easily among plants than among animals because plants of many species can reproduce by self-fertilization. In addition, if polyploidy arises in several offspring of a single parent, the siblings can fertilize one another.

Go to ANIMATED TUTORIAL 17.3
Speciation Mechanisms
PoL2e.com/at17.3

CHECKpoint CONCEPT 17.3

- ✓ Explain how speciation via polyploidy can happen in only two generations.
- ✓ What are some obstacles to sympatric speciation?
- ✓ If allopatric speciation is the most prevalent mode of speciation, what do you predict about the geographic distributions of many closely related species? Does your answer differ for species that are sedentary versus species that are highly mobile?

Most populations separated by a physical barrier become reproductively isolated only slowly and gradually. If two incipient species once again come into contact with each other, what keeps them from merging back into a single species?

CONCEPT 17.4 Reproductive Isolation Is Reinforced When Diverging Species Come into Contact

As discussed in Concept 17.2, once a barrier to gene flow is established, reproductive isolation will begin to develop through genetic divergence. Over many generations, differences accumulate in the isolated lineages, reducing the probability that individuals from each lineage will mate successfully with one another when they come back into contact. In this way, reproductive isolation can evolve as a by-product of the genetic changes in the two diverging lineages.

Reproductive isolation may be incomplete when the incipient species come back into contact, however, in which case some hybridization will occur. If hybrid individuals are less fit than non-hybrids, selection favors parents that do not produce hybrid offspring. Under these conditions, selection results in strengthening, or **reinforcement**, of isolating mechanisms that prevent hybridization.

Mechanisms that prevent hybridization from occurring are called **prezygotic isolating mechanisms**. Mechanisms that reduce the fitness of hybrid offspring are called **postzygotic isolating mechanisms**. Postzygotic mechanisms result in selection against hybridization, which leads to the reinforcement of the prezygotic mechanisms.

Prezygotic isolating mechanisms prevent hybridization between species

Prezygotic isolating mechanisms, which come into play before fertilization, can prevent hybridization in several ways.

MECHANICAL ISOLATION Differences in the sizes and shapes of reproductive organs may prevent the union of gametes from different species. With animals, there may be a match between the shapes of the reproductive organs of males and females of the same species, so that reproduction between species with mismatched structures is not physically possible. In plants, mechanical isolation may involve a pollinator. For example, orchids of the genus *Ophrys* produce flowers that look and smell like the females of particular species of wasps (**FIGURE 17.8**). When a male wasp visits and attempts to mate with the flower (mistaking it for a female wasp of his species), his mating behavior results in the transfer of pollen to and from his body by appropriately configured anthers and stigmas on the flower. Insects that visit the flower but do not attempt to mate with it do not trigger the transfer of pollen between the insect and the flower.

TEMPORAL ISOLATION Many organisms have distinct mating seasons. If two closely related species breed at different times of the year (or different times of day), they may never have an opportunity to hybridize. For example, in sympatric populations of three closely related leopard frog species, each species breeds at a different time of year (**FIGURE 17.9**). Although there is some overlap in the breeding seasons, the opportunities for hybridization are minimal.

BEHAVIORAL ISOLATION Individuals may reject, or fail to recognize, individuals of other species as potential mating partners. For example, the mating calls of male frogs of related

FIGURE 17.8 Mechanical Isolation through Mimicry Many orchid species maintain reproductive isolation by means of flowers that look and smell like females of a specific bee or wasp species. Shown here are a fly orchid (*Ophrys insectifera*) and its pollinator, a male wasp of the genus *Argogorytes*.

ANSWER Although speciation usually takes thousands or millions of years, and although it is typically studied in natural settings such as Lake Malawi, some aspects of speciation can be studied and observed in controlled laboratory experiments. Most such experiments use organisms with short generation times, in which evolution is expected to be relatively rapid.

William Rice and George Salt conducted an experiment in which fruit flies were allowed to choose food sources in different habitats, where mating also took place. The habitats were vials in different parts of an experimental cage (**FIGURE 17.14**). The vials differed in three environmental factors: (1) light; (2) the direction (up or down) in which the fruit flies had to move to reach food; and (3) the concentrations of two aromatic chemicals, ethanol and acetaldehyde. In just 35 generations, the two groups of flies that chose the most divergent habitats had become reproductively isolated from each other, having evolved distinct preferences for the different habitats.

The experiment by Rice and Salt demonstrated an example of habitat isolation evolving to function as a prezygotic isolating mechanism. Even though the different habitats were in the same cage, and individual fruit flies were capable of flying from one habitat to the other, habitat preferences were inherited by offspring from their parents, and populations from the two divergent habitats did not interbreed. Similar habitat selection is thought to have resulted in the early split in cichlids that preferred the rocky versus the sandy shores of Lake Malawi.

In controlled experiments like this one, biologists can observe many aspects of the process of speciation directly.

FIGURE 17.14 Evolution in the Laboratory For their experiments on the evolution of prezygotic isolating mechanisms in *Drosophila melanogaster*, Rice and Salt built an elaborate system of varying habitats contained within vials inside a large fly enclosure. Some groups of flies developed preferences for widely divergent habitats and became reproductively isolated within 35 generations.

SUMMARY

CONCEPT 17.1 Species Are Reproductively Isolated Lineages on the Tree of Life

- **Speciation** is the process by which one species splits into two or more daughter species, which thereafter evolve as distinct lineages.
- The **morphological species concept** distinguishes species on the basis of physical similarities; it often underestimates or overestimates the actual number of reproductively isolated species.
- The **biological species concept** distinguishes species on the basis of **reproductive isolation**.
- **Lineage species concepts**, which recognize independent evolutionary lineages as species, allow biologists to consider species over evolutionary time.

CONCEPT 17.2 Speciation Is a Natural Consequence of Population Subdivision

- Genetic divergence results from the interruption of gene flow within a population.
- The Dobzhansky–Muller model describes how reproductive isolation between two descendant lineages can develop through the accumulation of incompatible genes or chromosomal arrangements. **Review Figures 17.3 and 17.4 and ANIMATED TUTORIAL 17.1**
- Reproductive isolation increases with increasing genetic divergence between populations. **Review Figure 17.5**

CONCEPT 17.3 Speciation May Occur through Geographic Isolation or in Sympatry

- **Allopatric speciation**, which results when populations are separated by a physical barrier, is the dominant mode of speciation. This type of speciation may follow founder events, in which some members of a population cross a barrier and found a new, isolated population. **Review Figures 17.6 and 17.7 and ANIMATED TUTORIAL 17.2**
- **Sympatric speciation** results when two species diverge in the absence of geographic isolation. It can result from disruptive selection in two or more distinct microhabitats.
- Sympatric speciation can occur within two generations via **polyploidy**, an increase in the number of chromosomes sets. Polyploidy may arise from chromosome duplications within a species (**autopolyploidy**) or from hybridization that combines the chromosomes of two species (**allopolyploidy**). **Review ANIMATED TUTORIAL 17.3**

CONCEPT 17.4 Reproductive Isolation Is Reinforced When Diverging Species Come into Contact

- **Prezygotic isolating mechanisms** prevent hybridization; **postzygotic isolating mechanisms** reduce the fitness of hybrids.
- Postzygotic isolating mechanisms lead to **reinforcement** of prezygotic isolating mechanisms by natural selection. **Review Figures 17.9, 17.10, and 17.12**
- **Hybrid zones** may persist between species with incomplete reproductive isolation. **Review Figure 17.13**

See **ACTIVITY 17.1** for a concept review of this chapter.

Go to the Interactive Summary to review key figures, Animated Tutorials, and Activities PoL2e.com/is17

Go to LaunchPad at **macmillanhighered.com/launchpad** for additional resources, including LearningCurve Quizzes, Flashcards, and many other study and review resources.

18 The History of Life on Earth

KEY CONCEPTS

18.1 Events in Earth's History Can Be Dated

18.2 Changes in Earth's Physical Environment Have Affected the Evolution of Life

18.3 Major Events in the Evolution of Life Can Be Read in the Fossil Record

Meganeuropsis permiana, shown in a reconstruction from fossils. Except for its size, this giant from the Permian period was similar to modern dragonflies (shown in the inset at the same scale).

Almost anyone who has spent time around freshwater ponds is familiar with dragonflies. Their bright colors and transparent wings stimulate our visual senses on bright summer afternoons as they fly about devouring mosquitoes, mating, and laying their eggs. The largest dragonflies alive today have wingspans that can be covered by a human hand. Three hundred million years ago, however, dragonflies such as *Meganeuropsis permiana* had wingspans of more than 70 centimeters—well over 2 feet, matching or exceeding the wingspans of many modern birds of prey. These dragonflies were the largest flying predators of their time.

No flying insects alive today are anywhere near this size. But during the Carboniferous and Permian geological periods, 350–250 million years ago, many groups of flying insects contained gigantic members. *Meganeuropsis* probably ate huge mayflies and other giant flying insects that shared their home in the Permian swamps. These enormous insects were themselves eaten by giant amphibians.

None of these insects or amphibians would be able to survive on Earth today. The oxygen concentrations in Earth's atmosphere were about 50 percent higher at that time compared to the present, and those high oxygen concentrations are thought to have been necessary to support giant insects and their huge amphibian predators.

Paleontologists have uncovered fossils of *Meganeuropsis permiana* in the rocks of Kansas. How do we know the age of these fossils, and how can we know how much oxygen that long-vanished atmosphere contained? The layering of the rocks allows us to tell their ages relative to one another, but it does not by itself indicate a given layer's absolute age.

One of the remarkable achievements of twentieth-century scientists was the development of sophisticated techniques that use the decay rates of various radioisotopes, the ratios of certain molecules in rocks and fossils, and changes in Earth's magnetic field to infer conditions and events in the remote past and to date them accurately. It is those methods that allow us to age the fossils of *Meganeuropsis* and to calculate the concentration of oxygen in Earth's atmosphere at the time.

Earth is about 4.5 billion years old, and life has existed on it for about 3.8 billion of those years. That means human civilizations have occupied Earth for less than 0.0003 percent of the history of life. Discovering what happened before humans were around is an ongoing and exciting area of science.

Can modern experiments test hypotheses about the evolutionary impact of ancient environmental changes?

You will find the answer to this question on page 374.

CONCEPT 18.1 Events in Earth's History Can Be Dated

Some evolutionary changes happen rapidly enough to be studied directly and manipulated experimentally. Plant and animal breeding by agriculturalists and evolution of resistance to pesticides are examples of rapid, short-term evolution that we saw in Chapter 15. Other evolutionary changes, such as the appearance of new species and evolutionary lineages, usually take place over much longer time scales.

To understand long-term patterns of evolutionary change, we must not only think in time scales spanning many millions of years, but also consider events and conditions very different from those we observe today. Earth of the distant past was so unlike our present Earth that it would seem like a foreign planet inhabited by strange organisms. The continents were not where they are now, and climates were sometimes dramatically different from those of today. We know this because much of Earth's history is recorded in its rocks.

We cannot tell the ages of rocks just by looking at them, but we can visually determine the ages of rocks *relative to one another*. The first person to formally recognize this fact was the seventeenth-century Danish physician Nicolaus Steno. Steno realized that in undisturbed **sedimentary rocks** (rocks formed by the accumulation of sediments), the oldest layers of rock, or **strata** (singular *stratum*), lie at the bottom, and successively higher strata are progressively younger.

Geologists subsequently combined Steno's insight with their observations of fossils contained in sedimentary rocks. They developed the following principles of **stratigraphy**:

- Fossils of similar organisms are found in widely separated places on Earth.
- Certain fossils are always found in younger strata, and certain other fossils are always found in older strata.
- Organisms found in younger strata are more similar to modern organisms than are those found in older strata.

These patterns revealed much about the relative ages of sedimentary rocks and the fossils they contain, as well as patterns in the evolution of life. But the geologists still could not tell how old particular rocks were. A method of dating rocks did not become available until after radioactivity was discovered at the beginning of the twentieth century.

Radioisotopes provide a way to date rocks

Radioactive isotopes of atoms—radioisotopes—decay in a predictable pattern over long periods. Over a specific time interval, known as a **half-life**, half of the atoms in a radioisotope decay to become a different, stable (nonradioactive) isotope (**FIGURE 18.1A**). The use of this knowledge to date fossils and rocks is known as **radiometric dating**.

To use a radioisotope to date a past event, we must know or estimate the concentration of that isotope at the time of that event, and we must know the radioisotope's half-life. In the case of carbon-14, a radioisotope of carbon, the production

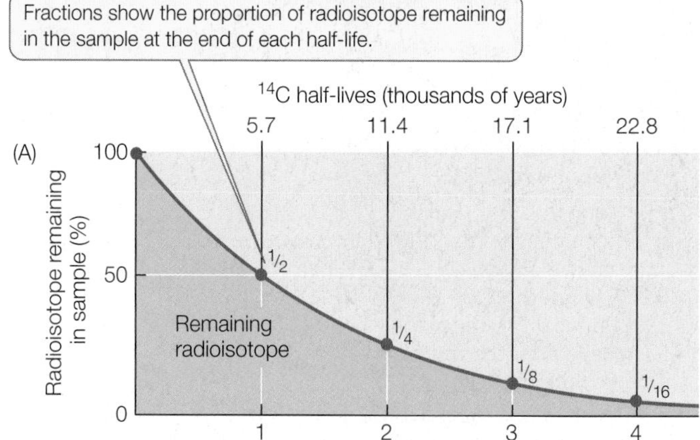

(B)

Radioisotope	Decay product	Half-life (years)	Useful dating range (years)
Carbon-14 (^{14}C)	Nitrogen-14 (^{14}N)	5,700	100 – 60,000
Uranium-234 (^{234}U)	Thorium-230 (^{230}Th)	80,000	10,000 – 500,000
Uranium-235 (^{235}U)	Lead-207 (^{207}Pb)	704 million	200,000 – 4.5 billion
Potassium-40 (^{40}K)	Argon-40 (^{40}Ar)	1.3 billion	10 million – 4.5 billion

FIGURE 18.1 Radioactive Isotopes Allow Us to Date Ancient Rocks The decay of radioactive isotopes into stable isotopes happens at a steady rate. A half-life is the time it takes for half of the remaining atoms to decay in this way. (A) The graph demonstrates the principle of half-life using carbon-14 (^{14}C) as an example. The half-life of ^{14}C is 5,700 years. (B) Examples of some radioisotopes with different characteristic half-lives that allow us to estimate the ages of many rocks.

of new carbon-14 (^{14}C) in the upper atmosphere—by the reaction of neutrons with nitrogen-14 (^{14}N, a stable isotope of nitrogen)—just balances the natural radioactive decay of ^{14}C into ^{14}N. Therefore the ratio of ^{14}C to the more common stable isotope of carbon, carbon-12 (^{12}C), is relatively constant in living organisms and in their environment. As soon as an organism dies, however, it ceases to exchange carbon compounds with its environment. Its decaying ^{14}C is no longer replenished, and the ratio of ^{14}C to ^{12}C in its remains decreases over time. Paleontologists can use the ratio of ^{14}C to ^{12}C in fossil material to date fossils that are less than 60,000 years old (and thus the sedimentary rocks that contain those fossils). If fossils are older than that, so little ^{14}C remains that the limits of detection using this particular isotope are reached.

Radiometric dating methods have been expanded and refined

Sedimentary rocks are formed from materials that existed for varying lengths of time before being weathered, fragmented, and transported, sometimes over long distances, to the site of their deposition. Therefore the radioisotopes in sedimentary rock do not contain reliable information about the date of its formation. Radiometric dating of rocks older than 60,000 years requires estimating radioisotope concentrations

APPLY THE CONCEPT

Events in Earth's history can be dated

Imagine you have been assigned the job of producing a geological map of volcanic rocks that were formed between 400 and 600 million years ago. You collect samples from ten sites (1–10 on the map at lower right). Then you can determine the ratio of ^{206}Pb to ^{238}U for each sample and use these ratios to estimate the ages of the rock samples, as given in the table below. Use the data to answer the following questions.

SITE	$^{206}PB/^{238}U$ RATIO	ESTIMATED AGE (mya)
1	0.076	474
2	0.077	479
3	0.069	431
4	0.081	505
5	0.076	474
6	0.070	435
7	0.089	550
8	0.080	500
9	0.079	495
10	0.077	479

1. Use the table shown here and Table 18.1 to assign each sample to a geological period.
2. Use these estimated ages and geological periods of the samples to mark rough boundaries between the geological periods among the sample locations on the map below.
3. If you wanted to refine the boundary between the Ordovician and Silurian on your map, which of three new sampling sites—*x*, *y*, or *z*—would you add to your analysis next?

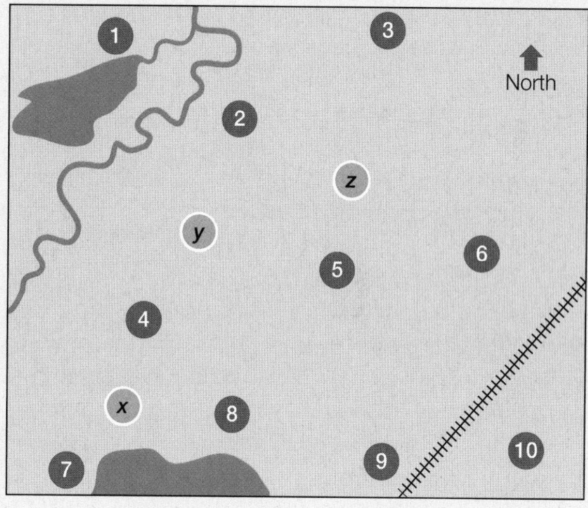

in **igneous** rocks, which are formed when molten material cools. To date sedimentary strata, geologists search for places where volcanic ash or lava flows have intruded into the sedimentary rock.

A preliminary estimate of the age of an igneous rock determines which radioisotopes can be used to date it (**FIGURE 18.1B**). The decay of potassium-40 (which has a half-life of 1.3 billion years) to argon-40, for example, has been used to date many of the ancient events in the evolution of life. Fossils in the adjacent sedimentary rock that are similar to those in other rocks of known ages provide additional clues to the rock's age.

Scientists have used several methods to construct a geological time scale

Radiometric dating of rocks, combined with fossil analysis, is the most powerful method of determining geological age. But in places where sedimentary rocks do not contain suitable igneous intrusions and few fossils are present, paleontologists turn to other dating methods.

One method, known as **paleomagnetic dating**, relates the ages of rocks to patterns in Earth's magnetism, which change over time. Earth's magnetic poles move and occasionally reverse themselves. Both sedimentary and igneous rocks preserve a record of Earth's magnetic field at the time they were formed, and that record can be used to determine the ages of those rocks. Other dating methods use information about continental drift, information about sea level changes, and molecular clocks.

LINK
Molecular clocks are described in **Concept 16.3**

Using all of these methods, geologists developed a **geological time scale** (**TABLE 18.1**). They divided the broad history of life into four **eons**. The Hadean eon refers to the time on Earth before life evolved. The early history of life occurred in the Archean eon, which ended about the time that photosynthetic organisms first appeared on Earth. Prokaryotic life diversified rapidly in the Proterozoic eon, and the first eukaryotes in the fossil record date from this time. These three eons are sometimes referred to collectively as Precambrian time, or simply the **Precambrian**. The Precambrian lasted for approximately 3.8 billion years and thus accounts for the vast majority of geological time. It was in the Phanerozoic eon, however—a mere 542-million-year time span—that multicellular eukaryotes rapidly diversified. To emphasize the events of the Phanerozoic, Table 18.1 shows the subdivision of this eon into eras and periods. The boundaries between these divisions of time are based

TABLE 18.1 Earth's Geological History

Eon	Era	Period	Onset	Major physical changes on Earth
Phanerozoic (~0.5 billion years long)	Cenozoic	Quaternary (Q)	2.6 mya	Cold/dry climate; repeated glaciations
		Tertiary (T)	65.5 mya	Continents near current positions; climate cools
	Mesozoic	Cretaceous (K)	145.5 mya	Laurasian continents attached to one another; Gondwana begins to drift apart; meteorite strikes near current Yucatán Peninsula at end of period
		Jurassic (J)	201.6 mya	Two large continents form: Laurasia (north) and Gondwana (south); climate warm
		Triassic (Tr)	251.0 mya	Pangaea begins to drift apart; hot/humid climate
	Paleozoic	Permian (P)	299 mya	Extensive lowland swamps; O_2 levels 50% higher than present; by end of period continents aggregate to form Pangaea, and O_2 levels drop rapidly
		Carboniferous (C)	359 mya	Climate cools; marked latitudinal climate gradients
		Devonian (D)	416 mya	Continents collide at end of period; one or more giant meteorites probably strike Earth
		Silurian (S)	444 mya	Sea levels rise; two large land masses emerge; hot/humid climate
		Ordovician (O)	488 mya	Massive glaciation; sea level drops 50 meters
		Cambrian (Ꞓ)	542 mya	Atmospheric O_2 levels approach current levels
Proterozoic	Collectively called the Precambrian (~4 billion years long)		2.5 bya	Atmospheric O_2 levels increase from negligible to about 18%; "snowball Earth" from about 750 to 580 mya
Archean			3.8 bya	Earth accumulates more atmosphere (still almost no O_2); meteorite impacts greatly reduced
Hadean			4.5–4.6 bya	Formation of Earth; cooling of Earth's surface; atmosphere contains almost no free O_2; oceans form; Earth under almost continuous bombardment from meteorites

Note: mya, million years ago; bya, billion years ago.

largely on the striking differences geologists observe in the assemblages of fossil organisms contained in successive strata. This geological record of life reveals a remarkable story of a world in which the continents and biological communities are constantly changing.

CHECKpoint CONCEPT 18.1

- ✓ What observations about fossils suggested to geologists that they could be used to determine the relative ages of rocks?
- ✓ Why are radioisotopes not measured directly from sedimentary rocks to determine their ages?
- ✓ Given the problems of dating sedimentary rocks using radioisotopes, what other methods can geologists use to date sedimentary rocks?

As geologists began to develop accurate ways to age Earth, they began to understand that Earth is far older than anyone had previously understood. During its 4.5-billion-year history, Earth has undergone massive physical changes. These changes have influenced the evolution of life, and life, in its turn, has influenced Earth's physical environment.

CONCEPT 18.2 Changes in Earth's Physical Environment Have Affected the Evolution of Life

As we saw in the previous section, the Phanerozoic eon has been notable for the rapid diversification of multicellular eukaryotes. But the diversity of multicellular organisms has not simply increased steadily through time. New species have arisen, and species have gone extinct, throughout the history of life. Let's consider some of the physical changes on Earth that have resulted in such dramatic changes in life's diversity.

The continents have not always been where they are today

The globes and maps that adorn our walls, shelves, and books give an impression of a static Earth. It would be easy for us to assume that the continents have always been where they are. But we would be wrong. The idea that Earth's land masses

Major events in the history of life

Humans evolve; many large mammals become extinct

Diversification of birds, mammals, flowering plants, and insects

Dinosaurs continue to diversify; mass extinction at end of period (~76% of species lost)

Diverse dinosaurs; radiation of ray-finned fishes; first fossils of flowering plants

Early dinosaurs; first mammals; marine invertebrates diversify; mass extinction at end of period (~65% of species lost)

Reptiles diversify; giant amphibians and flying insects present; mass extinction at end of period (~96% of species lost)

Extensive fern/horsetail/giant club moss forests; first reptiles; insects diversify

Jawed fishes diversify; first insects and amphibians; mass extinction at end of period (~75% of marine species lost)

Jawless fishes diversify; first jawed fishes; plants and animals colonize land

Mass extinction at end of period (~75% of species lost)

Rapid diversification of multicellular animals; diverse photosynthetic protists

Origin of photosynthesis, multicellular organisms, and eukaryotes

Origin of life; prokaryotes flourish

Life not yet present

have changed their positions over the millennia, and that they continue to do so, was first put forth in 1912 by the German meteorologist and geophysicist Alfred Wegener. His idea, known as **continental drift**, was initially met with skepticism and resistance. By the 1960s, however, physical evidence and increased understanding of **plate tectonics**—the geophysics of the movement of major land masses—had convinced virtually all geologists of the reality of Wegener's vision. Plate tectonics provided the geological mechanism that explained Wegener's hypothesis of continental drift.

Earth's crust consists of several solid plates. Thick continental and thinner oceanic plates overlie a viscous, malleable layer of Earth's mantle. Heat produced by radioactive decay deep in Earth's core sets up large-scale convection currents in the mantle. New crust is formed as mantle material rises between diverging plates, pushing them apart.

Where oceanic plates and continental plates converge, the thinner oceanic plate is forced underneath the thicker continental plate, a process known as **subduction**. Subduction results in volcanism and mountain building on the continental boundary (**FIGURE 18.2A**). For example, in the Pacific Northwest of North America, a series of volcanoes formed the Cascade mountain range as the Juan de Fuca oceanic plate has been subducted beneath a portion of the continental North American Plate (**FIGURE 18.2B**). When two oceanic plates collide, one is also subducted below the other, producing a deep oceanic trench and associated volcanic activity.

When two thick continental plates collide, neither plate is subducted. Instead, the plates push up against one another, forming high mountain chains. The highest mountain chain in the world, the Himalayas, was formed this way when the Indian Plate collided with the Eurasian Plate. When continental plates diverge, new crust forms in the intervening spaces, resulting in deep clefts called rift valleys in which large freshwater lakes typically form. The Great Rift Valley lakes of eastern Africa, including Lake Malawi (discussed at the opening of Chapter 17), were formed in this way.

Many physical conditions on Earth have oscillated in response to plate tectonic processes. We now know that the movement of the plates has sometimes brought continents together and at other times has pushed them apart, as seen in the maps across the top of Figure 18.12. The positions and sizes of the continents influence oceanic circulation patterns, global climates, and sea levels. Sea levels are influenced directly by plate tectonic processes (which can influence the depth of ocean basins) and indirectly by oceanic circulation patterns, which affect patterns of glaciation. As climates cool, glaciers form and tie up water over land masses; as climates warm, glaciers melt and release water.

Some of these dramatic changes in Earth's physical parameters resulted in **mass extinctions**, during which a large proportion of the species living at the time disappeared. These mass extinctions are the cause of the striking differences in fossil assemblages that geologists used to divide the units of the geological time scale. After each mass extinction, the diversity of life rebounded, but recovery took millions of years.

Go to ANIMATED TUTORIAL 18.1
Evolution of the Continents
PoL2e.com/at18.1

Earth's climate has shifted between hot and cold conditions

Through much of its history, Earth's climate was considerably warmer than it is today, and temperatures decreased more gradually toward the poles. At other times, Earth was colder than it is today. Rapid drops in sea levels throughout the history of Earth have resulted mainly from increased global glaciation (**FIGURE 18.3**). Many of these drops in sea levels were accompanied by mass extinctions—particularly of marine organisms, which could not survive the disappearance of the shallow seas that covered vast areas of the continental shelves.

Earth's cold periods were separated by long periods of milder climates. Because we are living in one of the colder periods, it is difficult for us to imagine the mild climates that were found at high latitudes during much of the history of life. The Quaternary period has been marked by a series of glacial advances, interspersed with warmer interglacial intervals during which the glaciers retreated.

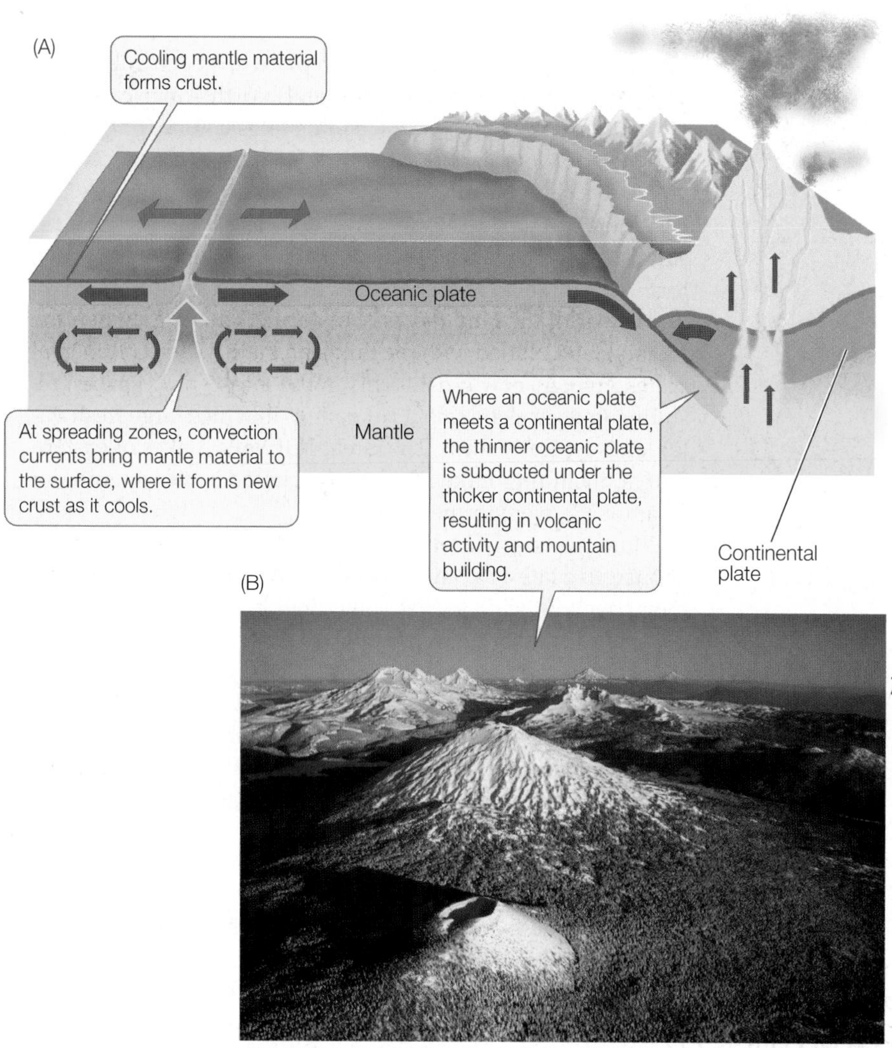
(A)
Cooling mantle material forms crust.
Oceanic plate
Mantle
At spreading zones, convection currents bring mantle material to the surface, where it forms new crust as it cools.
Where an oceanic plate meets a continental plate, the thinner oceanic plate is subducted under the thicker continental plate, resulting in volcanic activity and mountain building.
Continental plate
(B)

FIGURE 18.2 Plate Tectonics and Continental Drift (A) The heat of Earth's core generates convection currents in the viscous magma (the hot, molten rock of the mantle) underlying the oceanic and continental plates. Those currents push the continental plates, along with the land masses they carry, together or apart. Where plates collide, one may slide under the other, creating mountain ranges and often volcanoes. (B) The Cascade Range of the Pacific Northwest of North America is an example of a mountain range produced by subduction of an oceanic plate under a continental plate.

Go to MEDIA CLIP 18.1
Lava Flows and Magma Explosions
PoL2e.com/mc18.1

"Weather" refers to the daily events at a given location, such as individual storms and the high and low temperatures on a given day. "Climate" refers to long-term average expectations over the various seasons at a given location. Weather often changes rapidly, whereas climates typically change slowly. However, major climate shifts have taken place over periods as short as 5,000 to 10,000 years, primarily as a result of changes in Earth's orbit around the sun. A few climate shifts have been even more rapid. For example, during one Quaternary interglacial period, the ice-locked Antarctic Ocean became nearly ice-free in less than 100 years. Some climate changes have been so rapid that the extinctions caused by them appear to be nearly instantaneous in the fossil record. Such rapid changes are usually caused by sudden shifts in ocean currents.

We are currently living in a time of rapid climate change thought to be caused by a buildup of atmospheric CO_2, primarily from the burning of fossil fuels by human populations. We are reversing the energy transformations accrued in the burial and decomposition of organic material

FIGURE 18.3 Sea Levels Have Changed Repeatedly Rapid drops in sea levels are associated with periods of globally cooler temperatures and increased glaciation. Most mass extinctions of marine organisms have coincided with low sea levels.

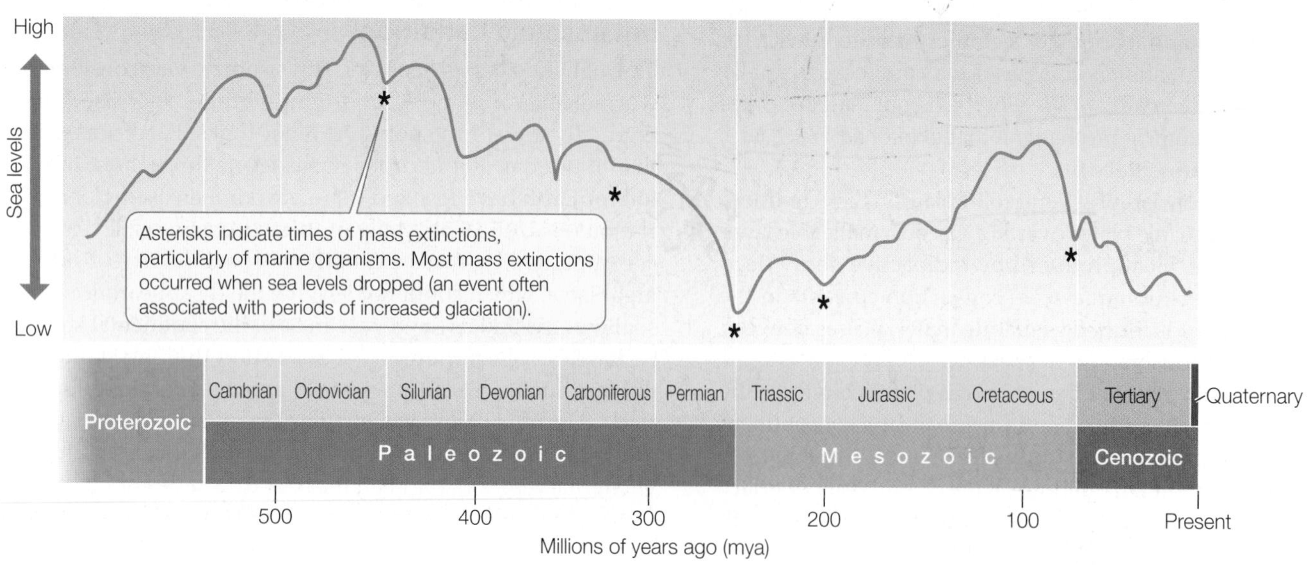
High
Sea levels
Low
Asterisks indicate times of mass extinctions, particularly of marine organisms. Most mass extinctions occurred when sea levels dropped (an event often associated with periods of increased glaciation).
Proterozoic
Cambrian
Ordovician
Silurian
Devonian
Carboniferous
Permian
Triassic
Jurassic
Cretaceous
Tertiary
Quaternary
Paleozoic
Mesozoic
Cenozoic
500
400
300
200
100
Present
Millions of years ago (mya)

that occurred (especially) in the Carboniferous, Permian, and Triassic, which gave rise to the fossil fuels we are using today. But we are burning these fuels over a few hundred years, rather than the many millions of years over which those deposits accumulated. The current rate of increase of atmospheric CO_2 is unprecedented in Earth's history. A doubling of the atmospheric CO_2 concentration—which may happen during the current century—is expected to increase the average temperature of Earth, change rainfall patterns, melt glaciers and ice caps, and raise sea levels.

LINK

The consequences of today's rapid climate changes are discussed in **Concept 45.5**

FIGURE 18.4 Volcanic Eruptions Can Cool Global Temperatures When Mount Pinatubo erupted in 1991, it increased the concentrations of sulfurous acid in high-altitude clouds, which temporarily lowered global temperatures by about 0.5°C.

Volcanoes have occasionally changed the history of life

Most volcanic eruptions produce only local or short-lived effects, but a few large volcanic eruptions have had major consequences for life. When Krakatau (a volcanic island in the Sunda Strait off Indonesia) erupted in 1883, it ejected more than 25 cubic kilometers of ash and rock, as well as large quantities of sulfur dioxide gas (SO_2). The SO_2 was ejected into the stratosphere and carried by high-altitude winds around the planet. Its presence led to high concentrations of sulfurous acid (H_2SO_3) in high-altitude clouds, creating a "parasol effect" so that less sunlight reached Earth's surface. Global temperatures dropped by 1.2°C in the year following the eruption, and global weather patterns showed strong effects for another 5 years. More recently, the eruption of Mount Pinatubo in the Philippines in 1991 (**FIGURE 18.4**) temporarily reduced global temperatures by about 0.5°C.

Although these individual volcanoes had only relatively short-term effects on global temperatures, they suggest that the simultaneous eruption of many volcanoes could have a much stronger effect on Earth's climate. What would cause many volcanoes to erupt at the same time? The collision of continents during the Permian period, about 275 million years ago (mya), formed a single, gigantic land mass and caused massive volcanic eruptions as the continental plates overrode one another (see Figure 18.2). Emissions from these eruptions blocked considerable sunlight, contributing to the advance of glaciers and a consequent drop in sea levels (see Figure 18.3). Thus volcanoes were probably responsible, at least in part, for the greatest mass extinction in Earth's history.

Extraterrestrial events have triggered changes on Earth

At least 30 meteorites of sizes between tennis and soccer balls strike Earth each year. Collisions with larger meteorites or comets are rare, but such collisions have probably been responsible for several mass extinctions. Several types of evidence tell us about these collisions. Their craters, and the dramatically disfigured rocks that result from their impact, are found in many places. Geologists have discovered compounds in these rocks that contain helium and argon with isotope ratios characteristic of meteorites, which are very different from the ratios found elsewhere on Earth.

A meteorite caused or contributed to a mass extinction at the end of the Cretaceous period (about 65.5 mya). The first clue that a meteorite was responsible came from the abnormally high concentrations of the element iridium found in a thin layer separating rocks deposited during the Cretaceous from rocks deposited during the Tertiary (**FIGURE 18.5**). Iridium is abundant in some meteorites, but it is exceedingly rare on Earth's surface. When scientists then discovered a circular crater 180 km in diameter buried beneath the northern coast of the Yucatán Peninsula of Mexico, they constructed the following scenario. When it collided with Earth, the meteorite released energy equivalent to that of 100 million megatons of high explosives, creating great tsunamis. A massive plume of debris rose into the atmosphere, spread around Earth, and descended. The descending debris heated the atmosphere to several hundred degrees and ignited massive fires. It also blocked the sun, preventing plants from photosynthesizing. The settling debris formed the iridium-rich layer. About a billion tons of soot with a composition matching that of smoke from forest fires was also deposited. These events had devastating effects on biodiversity. Many fossil species (including non-avian dinosaurs) that are found in Cretaceous rocks are not found in the overlying Tertiary rocks.

Oxygen concentrations in Earth's atmosphere have changed over time

As the continents have moved over Earth's surface, the world has experienced other physical changes, including large

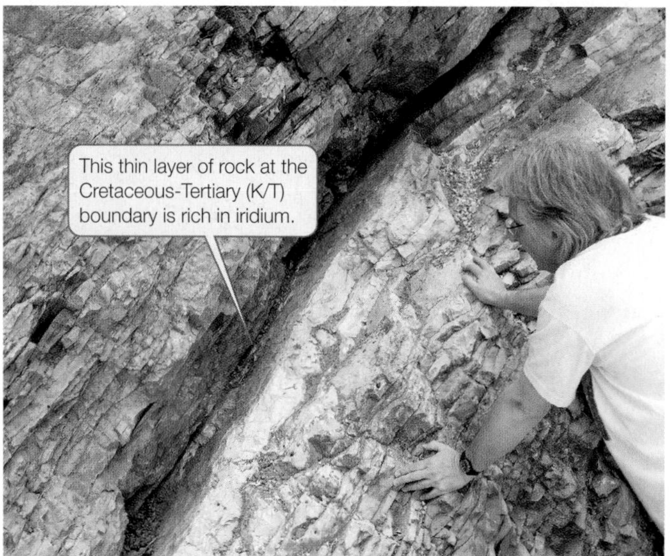

This thin layer of rock at the Cretaceous-Tertiary (K/T) boundary is rich in iridium.

FIGURE 18.5 Evidence of a Meteorite Impact The white layers of rock are Cretaceous in age, whereas the layers at the upper left were deposited in the Tertiary. Between the two is a thin, dark layer of clay that contains large amounts of iridium, a metal common in some meteorites but rare on Earth. Its high concentration in this sediment layer, deposited about 65.5 million years ago, suggests the impact of a large meteorite at that time.

increases and decreases in atmospheric oxygen concentrations. The atmosphere of early Earth probably contained little or no free oxygen gas (O_2). The increase in atmospheric O_2 came in two big steps more than a billion years apart.

The first step occurred about 2.5 billion years ago (bya), when the ancestors of modern cyanobacteria evolved the ability to use water as the source of hydrogen ions for photosynthesis. By chemically splitting H_2O, these bacteria generated O_2 as a waste product. They also made electrons available for reducing CO_2 to form the carbohydrate end-products of photosynthesis (see Concepts 6.5 and 6.6). The O_2 they produced dissolved in water and reacted with dissolved iron. The reaction product then precipitated as iron oxide, which accumulated in alternating layers of red and dark rock known as banded iron formations (**FIGURE 18.6A**). These formations provide evidence for the earliest photosynthetic organisms. As photosynthetic organisms continued to release O_2, oxygen gas began to accumulate in the atmosphere.

The second step occurred about a billion years later, when a cyanobacteria-like ancestor became symbiotic within eukaryote cells, leading to the evolution of chloroplasts in photosynthetic plants and other eukaryotes. One group of O_2-generating cyanobacteria formed rocklike structures called stromatolites, which are abundantly preserved in the fossil record (**FIGURE 18.6B**).

(A)

5 cm

(B)

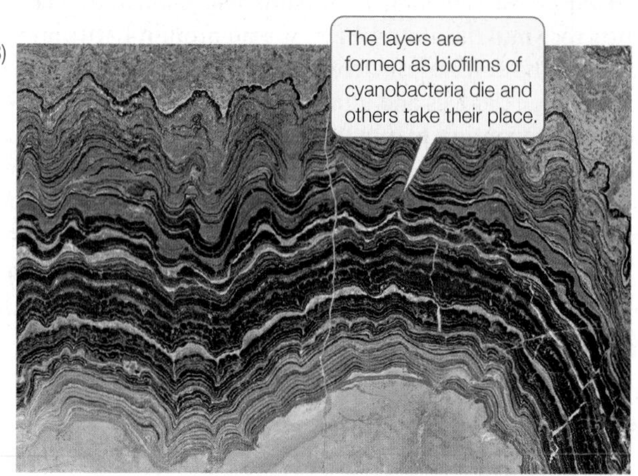

The layers are formed as biofilms of cyanobacteria die and others take their place.
5 cm

FIGURE 18.6 Evidence of Early Photosynthesis (A) The alternating red and dark layers in these "banded iron formations" resulted from a reaction between oxygen and dissolved iron, which produced iron oxide. The oxygen was produced by Earth's first photosynthetic organisms, beginning about 2.5 billion years ago. (B) A vertical section through a fossil stromatolite. Fossil stromatolites provide an early record of photosynthetic organisms. (C) These rocklike structures are living stromatolites that thrive in the very salty waters of Shark Bay in Western Australia.

(C)

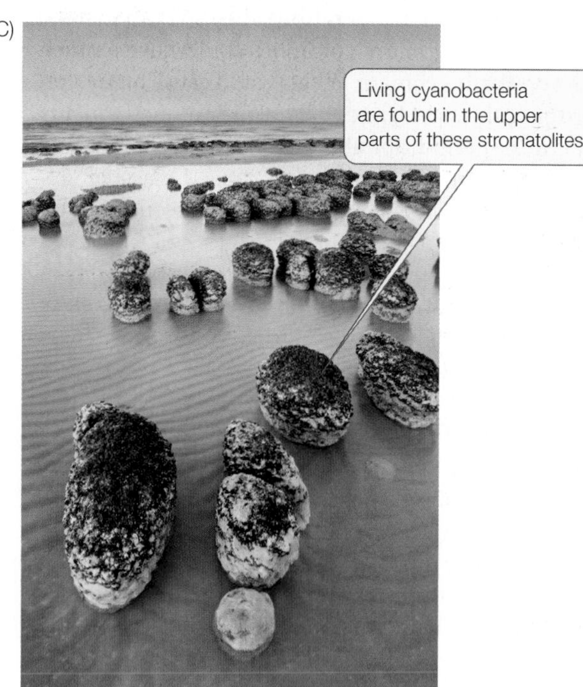

Living cyanobacteria are found in the upper parts of these stromatolites.

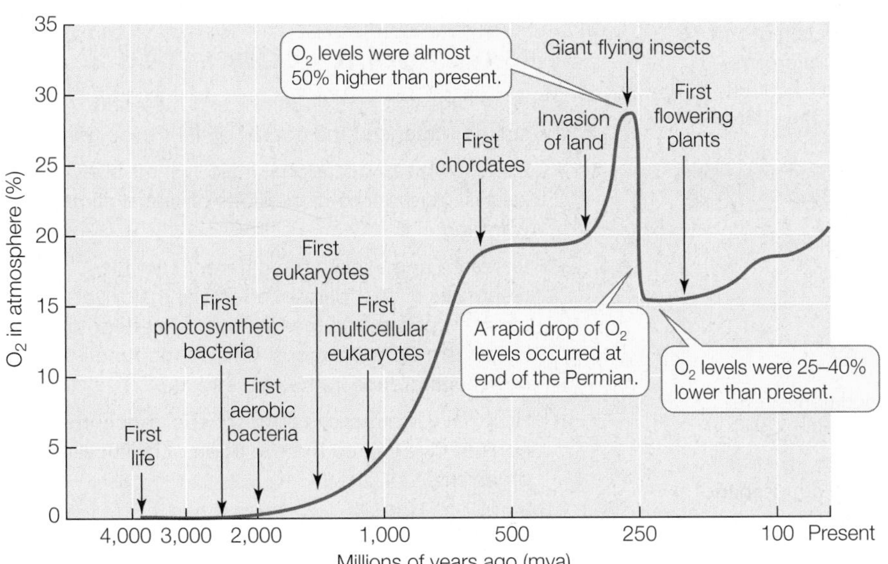

FIGURE 18.7 Atmospheric Oxygen Concentrations Have Changed Over Time Changes in atmospheric oxygen concentrations have strongly influenced, and been influenced by, the evolution of life. (Note that the horizontal axis of the graph is on a logarithmic scale.)

To this day, cyanobacteria still form stromatolites in a few very salty places (**FIGURE 18.6C**). Cyanobacteria liberated enough O_2 to open the way for the evolution of oxidation reactions as the energy source for the synthesis of ATP.

Thus the evolution of life irrevocably changed the physical nature of Earth. Those physical changes, in turn, influenced the evolution of life. When it first appeared in the atmosphere, O_2 was toxic to most of the anaerobic prokaryotes that inhabited Earth at the time. Over millennia, however, prokaryotes that evolved the ability to tolerate and use O_2 not only survived but gained the advantage. Aerobic metabolism proceeds more rapidly, and harvests energy more efficiently, than anaerobic metabolism. Organisms with aerobic metabolism replaced anaerobes in most of Earth's environments.

An atmosphere rich in O_2 also made possible larger and more complex organisms. Small single-celled aquatic organisms can obtain enough oxygen by simple diffusion even when dissolved oxygen concentrations in the water are very low. Larger single-celled organisms, however, have lower surface area-to-volume ratios. To obtain enough oxygen by simple diffusion, larger organisms must live in an environment with a relatively high oxygen concentration. Bacteria can thrive at 1 percent of the current oxygen concentration, but eukaryotic cells require levels that are at least 2–3 percent of the current concentration. For concentrations of dissolved oxygen in the oceans to have reached these levels, much higher atmospheric concentrations were needed.

Probably because it took many millions of years for Earth to develop an oxygenated atmosphere, only single-celled prokaryotes lived on Earth for more than 2 billion years. About 1.5 bya, atmospheric O_2 concentrations became high enough for larger eukaryotic cells to flourish (**FIGURE 18.7**). Further increases in atmospheric O_2 concentrations in the late Precambrian enabled several groups of multicellular organisms to evolve.

Oxygen concentrations increased again during the Carboniferous and Permian periods because of the evolution of large vascular plants. These plants lived in the expansive lowland swamps that existed at the time (see Table 18.1). Massive amounts of organic material were buried in these swamps as the plants died, leading to the formation of Earth's vast coal deposits. Because the buried organic material was not subject to oxidation as it decomposed, and because the living plants were producing large quantities of O_2, atmospheric O_2 increased to concentrations that have not been reached again in Earth's history (see Figure 18.7). As mentioned at the opening of this chapter, these high concentrations of atmospheric O_2 allowed the evolution of giant flying insects and amphibians that could not survive in today's atmosphere.

The drying of the lowland swamps at the end of the Permian reduced burial of organic matter as well as the production of O_2, so atmospheric O_2 concentrations dropped rapidly. Over the past 200 million years, with the diversification of flowering plants, O_2 concentrations have again increased, but not to the levels that characterized the Carboniferous and Permian periods.

Biologists have conducted experiments that demonstrate the changing selection pressures that can accompany changes in atmospheric O_2 concentrations. When fruit flies (*Drosophila*) are raised in hyperoxic conditions (i.e., with artificially increased atmospheric concentrations of O_2), they evolve larger body sizes in just a few generations (**FIGURE 18.8**). The present atmospheric O_2 concentrations appear to constrain body size in these flying insects, whereas increases in O_2 appear to relax those constraints. This experiment demonstrates that the stabilizing selection on body size at present O_2 concentrations can quickly switch to directional selection for a change in body size in response to a change in O_2 concentrations.

LINK

Stabilizing and directional selection are discussed and compared in **Concept 15.4**

INVESTIGATION

FIGURE 18.8 Atmospheric Oxygen Concentrations and Body Size in Insects C. Jaco Klok and his colleagues asked whether insects raised in hyperoxic conditions would evolve to be larger than their counterparts raised under today's atmospheric conditions. They raised strains of fruit flies (*Drosophila melanogaster*) under both conditions to test the effects of increased O_2 concentrations on the evolution of body size.[a]

HYPOTHESIS

In hyperoxic conditions, increased partial pressure of oxygen results in evolution of increased body size in flying insects.

METHOD

1. Separate a population of fruit flies into multiple lines.
2. Raise half the lines in current atmospheric (control) conditions; raise the other lines in hyperoxic (experimental) conditions. Continue all lines for seven generations.
3. Raise the F_8 individuals of all lines under identical (current) atmospheric conditions.
4. Weigh 50 flies from each of the replicate lines and test for statistical differences in body weight.

RESULTS

The average body mass of F_8 individuals of both sexes raised under hyperoxic conditions was significantly ($P < 0.001$) greater than that of individuals in the control lines. Error bars show 95% confidence intervals for the mean (see Appendix B).

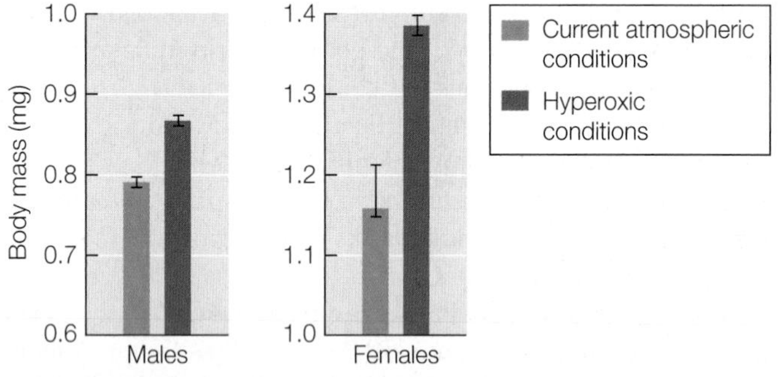

CONCLUSION

Increased O_2 concentrations led to evolution of larger body size in fruit flies, consistent with the trends seen among other flying insects in the fossil record.

ANALYZE THE DATA

The table shows the average body masses of the flies raised in hyperoxic conditions in the F_0 (i.e., before the first generation in hyperoxia), F_1, F_2, and F_8 generations.

	Average body mass (mg)	
Generation	Males	Females
F_0	0.732	1.179
F_1	0.847	1.189
F_2	0.848	1.254
F_8	0.878	1.392

A. Graph body mass versus generation for males and females.

B. Do the rates of evolution of larger body size appear to be constant throughout the experiment?

C. If you doubled the number of generations in the experiment, would you expect the increase in body mass seen under the hyperoxic conditions to double? Why or why not?

Go to **LaunchPad** for discussion and relevant links for all **INVESTIGATION** figures.

[a]C. J. Klok et al. 2009. *Journal of Evolutionary Biology* 22: 2496–2504.

CHECKpoint CONCEPT 18.2

✓ How have volcanic eruptions and meteorite strikes influenced the course of life's evolution?

✓ Explain why an occasional major winter blizzard is irrelevant to discussions of global climate warming.

✓ Research and explain the different climatic responses to the "parasol effect" of major volcanic eruptions and the "greenhouse effect" of long-term CO_2 emissions. What is the fundamental difference in the two effects?

✓ How have increases in atmospheric concentrations of O_2 affected the evolution of multicellular organisms?

The many dramatic physical events of Earth's history have influenced the nature and timing of evolutionary changes among Earth's living organisms. We will now look more closely at some of the major events that characterize the history of life on Earth.

APPLY THE CONCEPT

Changes in Earth's physical environment have affected the evolution of life

In the experiment shown in Figure 18.8, body mass of individuals in the experimental population of *Drosophila* increased (on average) about 2% per generation in the high-oxygen environment (although the rate of increase was not constant over the experiment). What rate of increase in body mass per generation would be sufficient to account for the giant dragonflies of the Permian?

1. Assume that the average rate of increase in dragonfly size during the Permian was much slower than the rate observed in the experiment in Figure 18.8. We'll assume that the actual rate of increase for dragonflies was just 0.01% per generation, rather than the 2% observed over a few generations for *Drosophila*. We'll assume further that dragonflies complete just one generation per year (as opposed to 40 or more generations for *Drosophila*). Starting with an average body mass of 1 gram, calculate the projected increase in body mass over 50,000 years.*
2. What percentage of the Permian period does 50,000 years represent? Use Table 18.1 for your calculation.
3. Given your calculations, do you think that increased oxygen concentrations during the Permian were sufficient to account for the evolution of giant dragonflies? Why or why not?

*This calculation is similar to computing compound interest for a savings account. Use the formula $W = S(1 + R)^N$, where W = the final mass, S = the starting mass (1 gram), R = the rate of increase per generation (0.0001 in this case), and N = the number of generations.

CONCEPT 18.3 Major Events in the Evolution of Life Can Be Read in the Fossil Record

How do we know about the physical changes in Earth's environment and their effects on the evolution of life? To reconstruct life's history, scientists rely heavily on the fossil record. As we have seen, geologists divided Earth's history into eons, eras, and periods based on their distinct fossil assemblages (see Table 18.1). Biologists refer to the assemblage of all organisms of all kinds living at a particular time or place as a **biota**. All of the plants living at a particular time or place are its **flora**, and all of the animals are its **fauna**.

About 300,000 species of fossil organisms have been described, and the number steadily grows. The number of named species, however, is only a tiny fraction of the species that have ever lived. We do not know how many species lived in the past, but we have ways of making reasonable estimates. Of the present-day biota, nearly 1.8 million species have been named. The actual number of living species is probably well over 10 million, and possibly much higher, because many species have not yet been discovered and described by biologists. So the number of described fossil species is only about 3 percent of the estimated minimum number of living species. Life has existed on Earth for about 3.8 billion years. Many species last only a few million years before undergoing speciation or going extinct. From this we know that Earth's biota must have turned over many times during geological history. So the total number of species that have lived over evolutionary time must vastly exceed the number living today. Why have only about 300,000 of these tens of millions of species been described from fossils to date?

Several processes contribute to the paucity of fossils

Only a tiny fraction of organisms ever become fossils, and only a tiny fraction of fossils are ever discovered by paleontologists. Most organisms live and die in oxygen-rich environments in which they quickly decompose. Organisms are not likely to become fossils unless they are transported by wind or water to sites that lack oxygen, where decomposition proceeds slowly or not at all. Furthermore, geological processes transform many rocks, destroying the fossils they contain, and many fossil-bearing rocks are deeply buried and inaccessible. Paleontologists have studied only a tiny fraction of the sites that contain fossils, although they find and describe many new ones every year.

The fossil record is most complete for marine animals that had hard skeletons (which resist decomposition). Among the nine major animal groups with hard-shelled members, approximately 200,000 species have been described from fossils—roughly twice the number of living marine species in these same groups. Paleontologists lean heavily on these groups in their interpretations of the evolution of life. Insects and spiders are also relatively well represented in the fossil record because they are numerically abundant and have hard exoskeletons (**FIGURE 18.9**). The fossil record, though incomplete, is good enough to document clearly the factual history of the evolution of life.

FIGURE 18.9 Insect Fossils Chunks of amber—fossilized tree resin—often contain insects that were preserved when they were trapped in the sticky resin.

By combining information about physical changes during Earth's history with evidence from the fossil record, scientists have composed portraits of what Earth and its inhabitants may have looked like at different times. We know in general where the continents were and how life changed over time, but many of the details are poorly known, especially for events in the more remote past.

Precambrian life was small and aquatic

Life first appeared on Earth about 3.8 bya (**FIGURE 18.10**). The fossil record of organisms that lived prior to the Phanerozoic is fragmentary, but it is good enough to establish that the total number of species and individuals increased dramatically in the late Precambrian.

For most of its history, life was confined to the oceans, and all organisms were small. For more than 3 billion years, all organisms lived in shallow seas. These seas slowly began to teem with microscopic prokaryotes. After the first eukaryotes appeared about 1.5 billion years ago, during the Proterozoic, unicellular eukaryotes and small multicellular animals fed on the microorganisms. Small floating organisms, known collectively as **plankton**, were strained from the water and eaten by slightly larger filter-feeding animals. Other animals ingested sediments on the seafloor and digested the remains of organisms within them. But it still took nearly a billion years before eukaryotes began to diversify rapidly into the many different morphological forms that we know today.

What limited the diversity of multicellular eukaryotes (in terms of their size and shape) for much of their early existence? It is likely that a combination of factors was responsible. We have already noted that O_2 levels increased throughout the Proterozoic, and it is likely that high atmospheric and dissolved O_2 concentrations were needed to support large multicellular organisms. In addition, geologic evidence points to a series of intensely cold periods during the late Proterozoic, which would have resulted in seas that were largely covered by ice and continents that were covered by glaciers. The "snowball

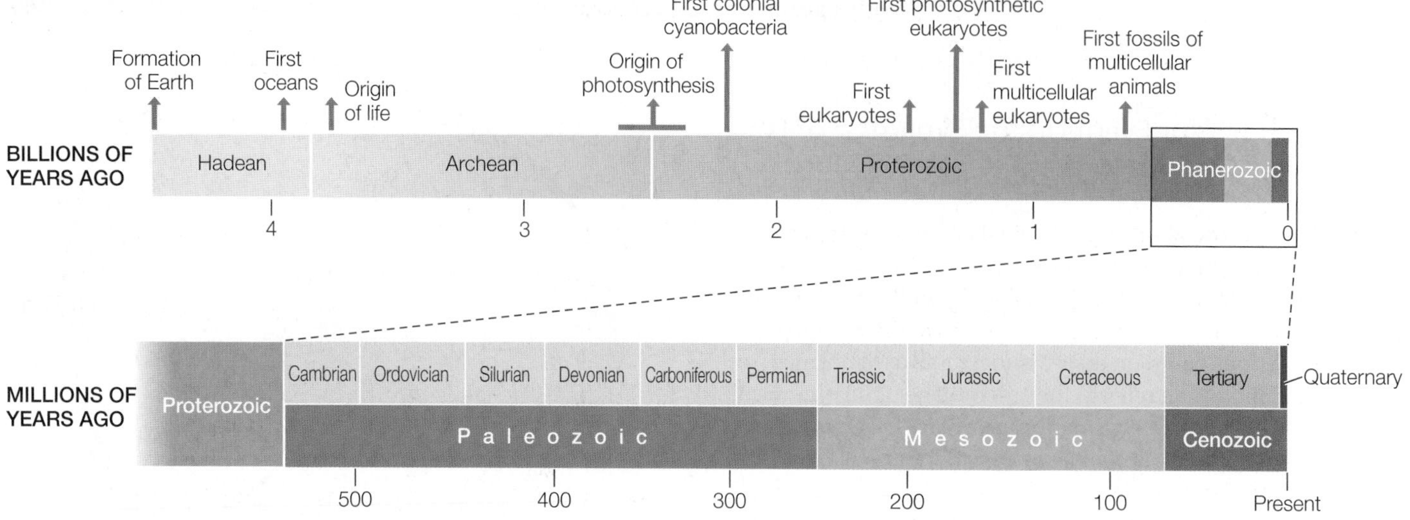

FIGURE 18.10 A Sense of Life's Time The top timeline shows the 4.5-billion-year history of Earth. Most of this history is accounted for by the Hadean, Archean, and Proterozoic eons, a 3.8-billion-year time span that saw the origin of life and the evolution of cells, photosynthesis, and multicellularity. The final 600 million years are expanded in the bottom timeline and detailed in Figure 18.12.

Earth" hypothesis suggests that cold conditions confined life to warm places such as hot springs, deep thermal vents, and perhaps a few equatorial oceans that avoided ice cover. The last of these Proterozoic glaciations ended about 580 million years ago, just before several major radiations of multicellular eukaryotes appear in the fossil record (**FIGURE 18.11**). Many of the multicellular organisms known from the late Proterozoic and early Phanerozoic were very different from any animals living today and may be members of groups that left no living descendants.

Life expanded rapidly during the Cambrian period

The Cambrian period (542–488 mya) marks the beginning of the Paleozoic, the first era of the Phanerozoic. The O_2 concentration in the Cambrian atmosphere was approaching its current level, and the glaciations of the late Proterozoic had ended nearly 40 million years earlier. A geologically rapid diversification of life took place that is often called the **Cambrian explosion**. This name is somewhat misleading, as the series of radiations it refers to actually began before the start of the Cambrian and continued for about 60 million years into the early Cambrian (see Figure 18.11). Nonetheless, 60 million years represents a relatively short amount of time, especially considering that the first eukaryotes had appeared about a billion (= 1,000 million) years earlier. Many of the major animal groups represented by species alive today first appeared during these evolutionary radiations. **FIGURE 18.12** provides an overview of the numerous

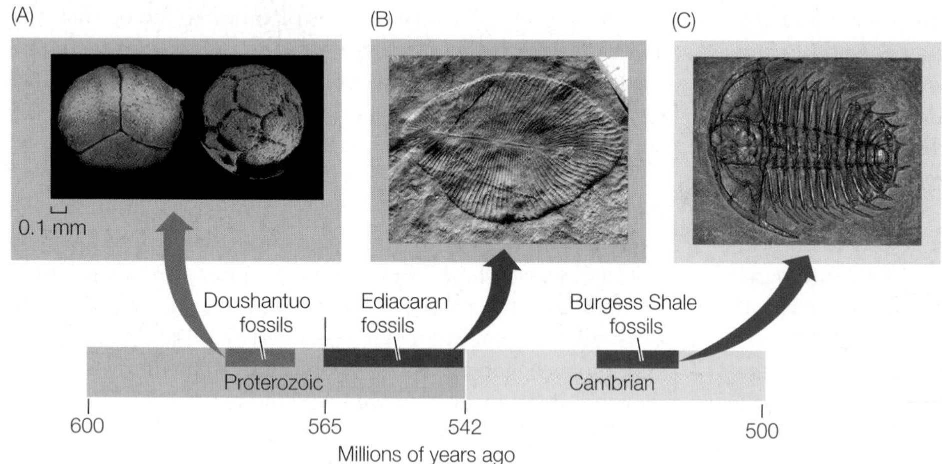

FIGURE 18.11 Diversification of Multicellular Organisms: The "Cambrian Explosion" Shortly after the end of Proterozoic glaciations (about 580 mya), several major radiations of multicellular organisms appear in the fossil record. (A) These microscopic fossils from the Doushantuo rock formation of China are the remains of tiny four- and eight-celled stages of multicellular organisms. (B) Unusual soft-bodied marine invertebrates, unlike any animals alive at present, characterize the fossilized fauna preserved at Ediacara is southern Australia. (C) By the early Phanerozoic, fossilized faunas such as those preserved in Canada's Burgess Shale include extinct representatives of some of the major animal groups alive today.

FIGURE 18.12 A Brief History of Multicellular Life on Earth The geologically rapid "explosion" of life before and during the Cambrian saw the rise of several animal groups that have representatives surviving today. The following three pages depict life's history through the Phanerozoic. The movements of the major continents during the past half-billion years are shown in the maps of Earth, and associated biotas for each time period are depicted. The artists' reconstructions are based on fossils such as those shown in the photographs. ▶

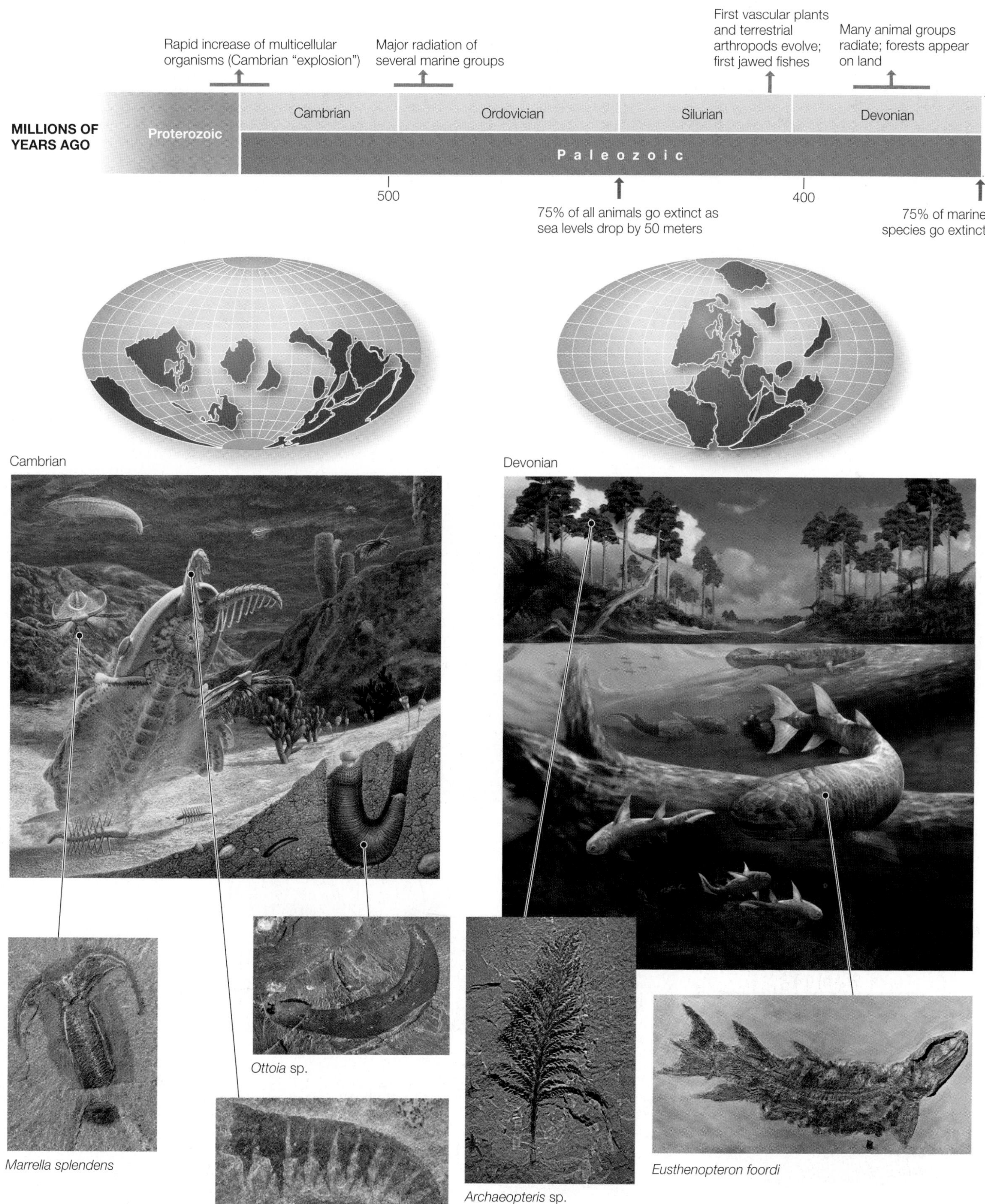
Rapid increase of multicellular organisms (Cambrian "explosion")
Major radiation of several marine groups
First vascular plants and terrestrial arthropods evolve; first jawed fishes
Many animal groups radiate; forests appear on land
MILLIONS OF YEARS AGO
Proterozoic
Cambrian
Ordovician
Silurian
Devonian
Paleozoic
500
400
75% of all animals go extinct as sea levels drop by 50 meters
75% of marine species go extinct
Cambrian
Devonian

Marrella splendens

Ottoia sp.

Anomalocaris canadensis (claw only)

Archaeopteris sp.

Eusthenopteron foordi

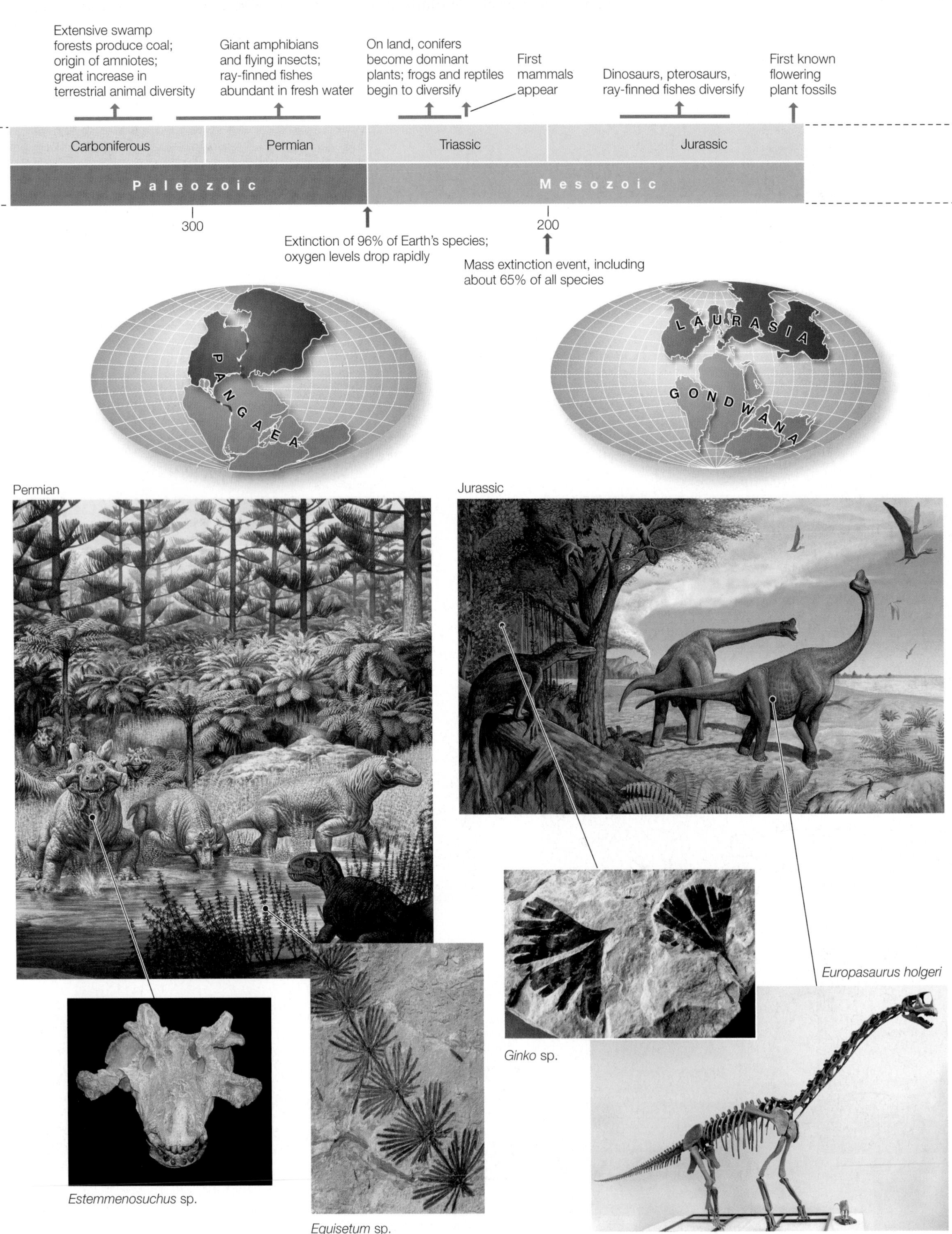

Extensive swamp forests produce coal; origin of amniotes; great increase in terrestrial animal diversity
Giant amphibians and flying insects; ray-finned fishes abundant in fresh water
On land, conifers become dominant plants; frogs and reptiles begin to diversify
First mammals appear
Dinosaurs, pterosaurs, ray-finned fishes diversify
First known flowering plant fossils
Carboniferous
Permian
Triassic
Jurassic
Paleozoic
Mesozoic
300
200
Extinction of 96% of Earth's species; oxygen levels drop rapidly
Mass extinction event, including about 65% of all species
PANGAEA
LAURASIA
GONDWANA
Permian
Jurassic
Europasaurus holgeri
Ginko sp.
Estemmenosuchus sp.
Equisetum sp.

Flowering plants diversify
Many radiations of animal groups, on both land and sea
Flowering plants dominate on land; rapid radiation of mammals
Grasslands spread as climates cool
Four major ice ages; evolution of *Homo*
Cretaceous
Tertiary
Quaternary
Mesozoic
Cenozoic
100
Present
Mass extinction event, including loss of most dinosaurs
Cretaceous
Tertiary
Hyracotherium leporinum
Sapindopsis belviderensis (leaves)
Chasmosaurus belli
Plesiadapis fodinatus (jaw)

continental and biotic innovations that have characterized the Phanerozoic.

For the most part, fossils tell us only about the hard parts of organisms, but in some well-studied Cambrian fossil beds, the soft parts of many animals were preserved. Multicellular life was largely or completely aquatic during the Cambrian. If there was life on land at this time, it was probably restricted to microorganisms.

Many groups of organisms that arose during the Cambrian later diversified

Geologists divide the remainder of the Paleozoic era into the Ordovician, Silurian, Devonian, Carboniferous, and Permian periods. Each period is characterized by the diversification of specific groups of organisms. Mass extinctions marked the ends of the Ordovician, Devonian, and Permian.

THE ORDOVICIAN (488–444 MYA) During the Ordovician period, the continents, which were located primarily in the Southern Hemisphere, still lacked multicellular life. Evolutionary radiation of marine organisms was spectacular during the early Ordovician, especially among animals, such as brachiopods and mollusks, that lived on the seafloor and filtered small prey from the water. At the end of the Ordovician, as massive glaciers formed over the southern continents, sea levels dropped about 50 meters, and ocean temperatures dropped. About 75 percent of all animal species became extinct, probably because of these major environmental changes.

THE SILURIAN (444–416 MYA) During the Silurian period, the continents began to merge together. Marine life rebounded from the mass extinction at the end of the Ordovician. Animals able to swim in open water and feed above the ocean floor appeared for the first time. Fishes diversified as bony armor gave way to the less rigid scales of modern fishes, and the first jawed fishes and the first fishes with supporting rays in their fins appeared. The tropical sea was uninterrupted by land barriers, and most marine organisms were widely distributed. On land, the first vascular plants evolved late in the Silurian (about 420 mya). The first terrestrial arthropods—scorpions and millipedes—evolved at about the same time.

THE DEVONIAN (416–359 MYA) Rates of evolutionary change accelerated in many groups of organisms during the Devonian period. The major land masses continued to move slowly toward each other. In the oceans there were great evolutionary radiations of corals and of shelled, squidlike cephalopod mollusks.

Terrestrial communities changed dramatically during the Devonian. Club mosses, horsetails, and tree ferns became common, and some attained the size of large trees. Their roots accelerated the weathering of rocks, resulting in the development of the first forest soils. The first plants to produce seeds appeared in the Devonian. The earliest fossil centipedes, spiders, mites, and insects date to this period, as do the earliest terrestrial vertebrates.

A massive extinction of about 75 percent of all marine species marked the end of the Devonian. Paleontologists are uncertain about its cause, but two large meteorites collided with Earth at about that time and may have been responsible, or at least a contributing factor. The continued merging of the continents, with the corresponding reduction in the area of continental shelves, also may have contributed to this mass extinction.

THE CARBONIFEROUS (359–299 MYA) Large glaciers formed over high-latitude portions of the southern land masses during the Carboniferous period, but extensive swamp forests grew on the tropical continents. These forests were dominated by giant tree ferns and horsetails with small leaves. Their fossilized remains formed the coal we now mine for energy. In the seas, crinoids (a group of echinoderms, related to sea stars and sea urchins) reached their greatest diversity, forming "meadows" on the seafloor.

The diversity of terrestrial animals increased greatly during the Carboniferous. Snails, scorpions, centipedes, and insects were abundant and diverse. Insects evolved wings, becoming the first animals to fly. Flight gave herbivorous insects easy access to tall plants, and plant fossils from this period show evidence of chewing by insects (**FIGURE 18.13**). The terrestrial vertebrates split into two lineages. The amphibians became larger and better adapted to terrestrial existence, while the sister lineage led to the amniotes—vertebrates with well-protected eggs that can be laid in dry places.

THE PERMIAN (299–251 MYA) During the Permian period, the continents merged into a single supercontinent called **Pangaea**. Permian rocks contain representatives of many of the major groups of insects we know today. By the end of the period the amniotes had split into two lineages: the reptiles, and a second lineage that would lead to the mammals. Ray-finned fishes became common in the freshwaters of Pangaea.

Toward the end of the Permian, conditions for life deteriorated. Massive volcanic eruptions resulted in outpourings of lava that covered large areas of Earth. The ash and gases produced by the volcanoes blocked sunlight and cooled the climate. The death and decay of the massive Permian forests rapidly used up atmospheric oxygen, and the loss of photosynthetic

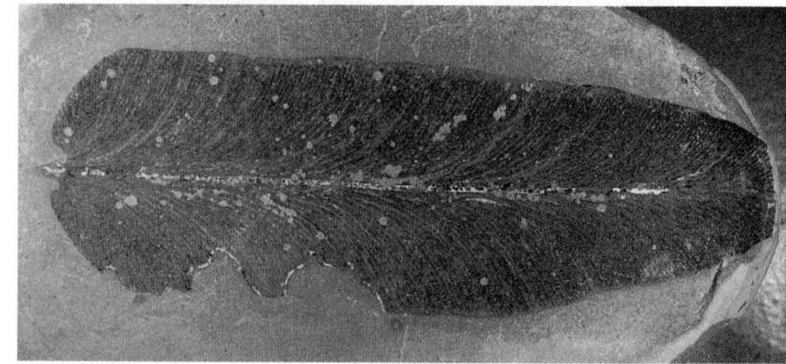

FIGURE 18.13 Evidence of Insect Diversification The margins of this fossil fern leaf from the Carboniferous have been chewed by insects.

organisms meant that relatively little new atmospheric oxygen was produced. In addition, much of Pangaea was located close to the South Pole by the end of the Permian. All of these factors combined to produce the most extensive continental glaciers since the "snowball Earth" times of the late Proterozoic. Atmospheric oxygen concentrations gradually dropped from about 30 percent to 15 percent. At such low concentrations, most animals would have been unable to survive at elevations above 500 meters, so about half of the land area would have been uninhabitable at the end of the Permian. The combination of these changes resulted in the most drastic mass extinction in Earth's history. Scientists estimate that about 96 percent of all multicellular species became extinct at the end of the Permian.

Geographic differentiation increased during the Mesozoic era

The few organisms that survived the Permian mass extinction found themselves in a relatively empty world at the start of the Mesozoic era (251 mya). As Pangaea slowly began to break apart in the Mesozoic, the biotas of the newly separated continents began to diverge. The oceans rose and once again flooded the continental shelves, forming huge, shallow inland seas. Atmospheric oxygen concentrations gradually rose. Life once again proliferated and diversified, but different groups of organisms came to the fore. The three groups of phytoplankton (floating photosynthetic organisms) that dominate today's oceans—dinoflagellates, coccolithophores, and diatoms—became ecologically important at this time, and their remains are the primary origin of the world's oil deposits. Seed-bearing plants replaced the trees that had ruled the Permian forests.

The Mesozoic era is divided into three periods: the Triassic, Jurassic, and Cretaceous. The Triassic and Cretaceous were terminated by mass extinctions, probably caused by meteorite impacts.

THE TRIASSIC (251–201.6 MYA) Pangaea remained largely intact through the Triassic. Many invertebrate groups diversified, and many burrowing animals evolved from groups living on the surfaces of seafloor sediments. On land, conifers and seed ferns were the dominant trees. The first frogs and turtles appeared. A great radiation of reptiles began, which eventually gave rise to crocodilians, dinosaurs, and birds. The first mammals appear. The end of the Triassic was marked by a mass extinction that eliminated about 65 percent of the species on Earth.

THE JURASSIC (201.6–145.5 MYA) Late in the Jurassic period, Pangaea became fully divided into two large continents: **Laurasia**, which drifted northward, and **Gondwana** in the south. Ray-finned fishes rapidly diversified in the oceans. The first lizards appeared, and flying reptiles (pterosaurs) evolved. Most of the large terrestrial predators and herbivores of the period were dinosaurs. Several groups of mammals made their first appearance, and the earliest known fossils of flowering plants are from late in this period.

THE CRETACEOUS (145.5–65.5 MYA) By the mid-Cretaceous period, Laurasia and Gondwana had largely broken apart into the continents we know today (although the Indian subcontinent was still separated from Asia). A continuous sea encircled the tropics. Sea levels were high, and Earth was warm and humid. Life proliferated both on land and in the oceans. Marine invertebrates increased in diversity. On land, the reptile radiation continued as dinosaurs diversified further and the first snakes appeared. Early in the Cretaceous, flowering plants began the radiation that led to their current dominance of the land. By the end of the period, many groups of mammals had appeared.

As described in Concept 18.2, another meteorite-caused mass extinction took place at the end of the Cretaceous. In the seas, many planktonic organisms and bottom-dwelling invertebrates became extinct. On land, almost all animals larger than about 25 kg in body weight became extinct. Many species of insects died out, perhaps because the growth of the plants they fed upon was greatly reduced following the impact. Some species in northern North America and Eurasia survived in areas that were not subjected to the devastating fires that engulfed most low-latitude regions.

Modern biotas evolved during the Cenozoic era

By the early Cenozoic era (65.5 mya), the continents were getting closer to their present positions, but the Indian subcontinent was still separated from Asia, and the Atlantic Ocean was much narrower. The Cenozoic was characterized by an extensive radiation of mammals, but other groups were also undergoing important changes.

Flowering plants diversified extensively and came to dominate world forests, except in the coolest regions, where the forests were composed primarily of gymnosperms. Mutations of two genes in one group of plants (the legumes) allowed them to use atmospheric nitrogen directly by forming symbioses with a few species of nitrogen-fixing bacteria. The evolution of this symbiosis was the first "green revolution" and dramatically increased the amount of nitrogen available for terrestrial plant growth. This symbiosis remains fundamental to the ecological base of life as we know it today.

LINK
The symbiosis between plants and nitrogen-fixing bacteria is covered in detail in **Concept 25.2**

The Cenozoic era is divided into the Tertiary and the Quaternary periods, which are commonly subdivided into **epochs** (**TABLE 18.2**).

THE TERTIARY (65.5–2.6 MYA) During the Tertiary period, the Indian subcontinent continued its northward drift. By about 55 mya it made initial contact with parts of southeastern Asia. By about 35 mya, the Indian Plate ran fully into the Eurasian Plate, and the Himalayas began to be pushed up as a result.

The early Tertiary was a hot and humid time, and the ranges of many plants shifted latitudinally. The tropics were probably too hot to support rainforest vegetation and instead were clothed in low-lying vegetation. In the middle of the Tertiary,

TABLE 18.2 Subdivisions of the Cenozoic Era

Period	Epoch	Onset (mya)
Quaternary	Holocene (Recent)	0.01 (~10,000 years ago)
	Pleistocene	2.6
Tertiary	Pliocene	5.3
	Miocene	23
	Oligocene	34
	Eocene	55.8
	Paleocene	65.5

however, Earth's climate became considerably cooler and drier. Many lineages of flowering plants evolved herbaceous (non-woody) forms, and grasslands spread over much of Earth.

By the start of the Cenozoic era, invertebrate faunas had already come to resemble those of today. Frogs, snakes, lizards, birds, and mammals all underwent extensive radiations during the Tertiary. Three waves of mammals dispersed from Asia to North America across one of the several land bridges that have intermittently connected the two continents during the past 55 million years. Rodents, marsupials, primates, and hoofed mammals appeared in North America for the first time.

THE QUATERNARY (2.6 MYA TO PRESENT) We are living in the Quaternary period. It is subdivided into two epochs, the Pleistocene and the Holocene (the Holocene is also known as the Recent).

The Pleistocene was a time of drastic cooling and climate fluctuations. During 4 major and about 20 minor "ice ages," massive glaciers spread across the continents, and the ranges of animal and plant populations shifted toward the equator. The last of these glaciers retreated from temperate latitudes less than 15,000 years ago. Organisms are still adjusting to this change. Many high-latitude ecological communities have occupied their current locations for no more than a few thousand years.

It was during the Pleistocene that divergence within one group of mammals, the primates, resulted in the evolution of the hominoid lineage. Subsequent hominoid radiation eventually led to the species *Homo sapiens*—modern humans. Many large bird and mammal species became extinct in Australia and in the Americas when *H. sapiens* arrived on those continents about 45,000 and 15,000 years ago, respectively. Many paleontologists believe these extinctions were the result of hunting and other influences of *Homo sapiens*.

LINK
The evolution of modern humans and their close relatives during the Pleistocene is discussed in **Concept 23.7**

The tree of life is used to reconstruct evolutionary events

The fossil record reveals broad patterns in life's evolution. To reconstruct major events in the history of life, biologists also rely on the phylogenetic information in the tree of life. We can use phylogeny, in combination with the fossil record, to reconstruct the timing of such major events as the acquisition of mitochondria in the ancestral eukaryotic cell, the several independent origins of multicellular organisms, and the movement of life onto dry land. We can also follow major changes in the genomes of organisms, and we can even reconstruct many gene sequences of species that are long extinct.

LINK
Concept 16.3 describes how biologists reconstruct the gene sequences of extinct organisms

Changes in Earth's physical environment have clearly influenced the diversity of organisms we see on the planet today. To study the evolution of that diversity, biologists examine the evolutionary relationships among species. Deciphering phylogenetic relationships is an important step in understanding how life has diversified on Earth. The next part of this book will explore the major groups of life and the different solutions these groups have evolved to major challenges such as reproduction, energy acquisition, dispersal, and escape from predation.

CHECKpoint CONCEPT 18.3

- ✓ Why have so few of the organisms that have existed over Earth's history become fossilized?
- ✓ What do we mean by the "Cambrian explosion"? How long did it last, and in what sense was it an "explosion"?
- ✓ What are some of the ways in which continental drift has affected the evolution of life?

Can modern experiments test hypotheses about the evolutionary impact of ancient environmental changes?

ANSWER Several experiments have been conducted to test the link between O_2 concentrations and evolution of body size in flying insects. One of these is discussed in Figure 18.8. The results of these experiments are consistent with the evolution of larger body size in flying insects in hyperoxic (high-oxygen) environments.

Experiments have also been conducted under hypoxic (low-oxygen) conditions, as existed at the end of the Permian. These experiments suggest that the evolution of body size is constrained under hypoxic conditions, even under strong artificial selection for larger body size. These latter results are consistent with the extinction of many large flying insects at the end of the Permian as a result of rapidly decreasing O_2 concentrations. Giant flying insects simply could not have survived the lower O_2 concentrations that existed at that time. The mass extinction at the end of the Permian is the only known mass extinction that involved considerable loss of insect diversity.

SUMMARY

CONCEPT 18.1 Events in Earth's History Can Be Dated

- The relative ages of organisms can be determined by the dating of fossils and the **strata** of **sedimentary rocks** in which they are found.
- **Radiometric dating** techniques use a variety of radioisotopes with different **half-lives** to date events in the remote past. **Review Figure 18.1**
- Geologists divide the history of life into eons, eras, and periods, based on major differences in the fossil assemblages found in successive strata. **Review Table 18.1**

CONCEPT 18.2 Changes in Earth's Physical Environment Have Affected the Evolution of Life

- Earth's crust consists of solid plates that float on fluid magma. **Continental drift** is caused by convection currents in the magma, which move the plates and the continents that lie on top of them. **Review Figure 18.2 and ANIMATED TUTORIAL 18.1**
- Major physical events on Earth, such as continental collisions and volcanic eruptions, have affected Earth's climate, atmosphere, and sea levels. In addition, extraterrestrial events such as meteorite strikes have created sudden and dramatic environmental shifts. All of these changes affected the history of life. **Review Figure 18.3 and Table 18.1**
- Oxygen-generating cyanobacteria liberated enough O_2 to open the door to oxidation reactions in metabolic pathways. Aerobic prokaryotes were able to harvest more energy than anaerobic organisms and began to proliferate. Increases in atmospheric O_2 concentrations supported the evolution of large eukaryotic cells and, eventually, multicellular organisms. **Review Figures 18.7 and 18.8**

CONCEPT 18.3 Major Events in the Evolution of Life Can Be Read in the Fossil Record

- Paleontologists use fossils and evidence of geological changes to determine what Earth and its **biota** may have looked like at different times. **Review Figures 18.11 and 18.12**
- Before the Phanerozoic, life was almost completely confined to the oceans. Multicellular life diversified extensively during the **Cambrian explosion**, a prime example of an evolutionary radiation.
- The periods of the Paleozoic era were each characterized by the diversification of specific groups of organisms. During the Mesozoic era, distinct terrestrial biotas evolved on each continent.
- Five episodes of **mass extinction** punctuated the history of life in the Paleozoic and Mesozoic eras.
- The Cenozoic era is divided into the Tertiary and the Quaternary periods, which in turn are subdivided into **epochs**. This era saw the emergence of the modern biotas as mammals radiated extensively and flowering plants became dominant. **Review Table 18.2**
- The tree of life can be used to reconstruct the timing of evolutionary events.

See **ACTIVITY 18.1** for a concept review of this chapter.

Go to the Interactive Summary to review key figures, Animated Tutorials, and Activities PoL2e.com/is18

Go to LaunchPad at **macmillanhighered.com/launchpad** for additional resources, including LearningCurve Quizzes, Flashcards, and many other study and review resources.

PART 4

Diversity

19 Bacteria, Archaea, and Viruses

KEY CONCEPTS

19.1 Life Consists of Three Domains That Share a Common Ancestor

19.2 Prokaryote Diversity Reflects the Ancient Origins of Life

19.3 Ecological Communities Depend on Prokaryotes

19.4 Viruses Have Evolved Many Times

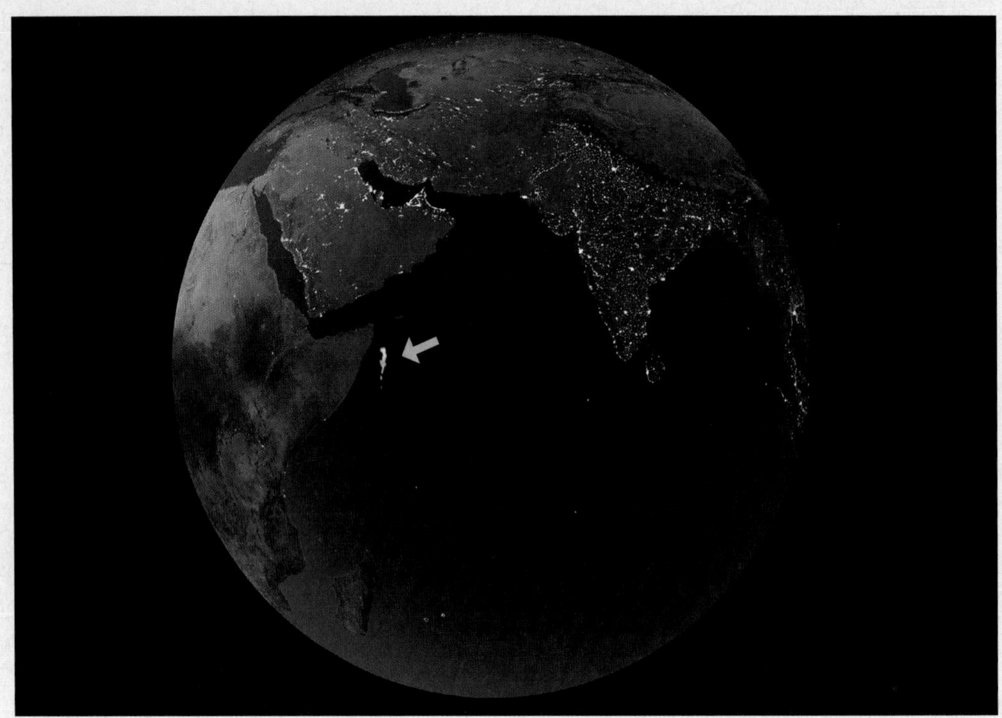

A satellite image reveals thousands of square kilometers of "milky seas" (arrow) in the Indian Ocean. This expanse of bioluminescence is produced by *Vibrio* bacteria.

On the night of January 25, 1995, the British merchant vessel *Lima* was off the coast of Somalia, near the Horn of Africa. This area is infamous for bands of pirates, so the crew was keeping a watchful eye on the seas. On the horizon, they spotted an eerie whitish glow. It was directly in their path, and there was no way to avoid it. Was the glow the result of some strange trick of piracy?

Within 15 minutes of first spotting the glow, the *Lima* was surrounded by glowing waters for as far as her crew could see. As the ship's log recorded, "it appeared as though the ship was sailing over a field of snow or gliding over the clouds." Fortunately for the crew, the glow had nothing to do with pirates.

For centuries, mariners in this part of the world had reported occasional "milky seas" in which the sea surface produced a strange glow at night, extending from horizon to horizon. Scientists to that point had never been able to confirm the reality or the cause of such phenomena. It was well established, however, that many organisms can emit light by bioluminescence—a complex, enzyme-catalyzed biochemical reaction that results in the emission of light but not heat.

What kind of organism could cause the vast expanse of bioluminescence observed by the *Lima*? Some marine organisms emit flashes of light when they are disturbed, but they could not produce the sustained and uniform glow seen in milky seas. The only organisms known to produce the quality of bioluminescence consistent with milky seas are prokaryotes, such as bacteria of the genus *Vibrio*. Using information supplied by the *Lima*, biologists scanned satellite images of the Indian Ocean for the specific light wavelengths emitted by *Vibrio*. The satellite images clearly showed thousands of square kilometers of *Vibrio*-produced milky seas.

Vibrio's bioluminescence requires a critical concentration of a specific chemical signal produced by the bacteria, so at low densities, free-living *Vibrio* populations do not glow. But as a colony establishes itself on phytoplankton, the bacteria's population density increases and concentrations of the luminescence signal build up. Eventually bacterial density (and concentration of the signal) becomes high enough for the huge colony to produce visible light. Such chemical-inducing action among bacterial cells is referred to as quorum sensing.

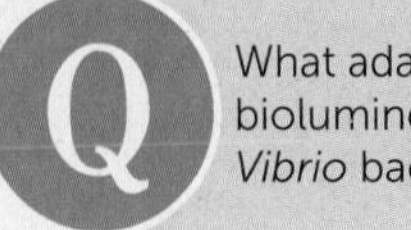

What adaptive advantage does bioluminescence provide to *Vibrio* bacteria?

You will find the answer to this question on page 399.

CONCEPT 19.1 Life Consists of Three Domains That Share a Common Ancestor

You may think that you have little in common with a bacterium. But all multicellular eukaryotes—including you—share many attributes with bacteria and archaea, together called **prokaryotes**. For example, all organisms, whether eukaryotes or prokaryotes,

- have cell membranes and ribosomes (see Chapter 4).
- have a common set of metabolic pathways, such as glycolysis (see Chapter 6).
- replicate DNA semiconservatively (see Chapter 9).
- use DNA as the genetic material to encode proteins, and use a similar genetic code to produce those proteins by transcription and translation (see Chapter 10).

These shared features support the conclusion that all living organisms are related. If life had multiple origins, there would be little reason to expect all organisms to use overwhelmingly similar genetic codes or to share structures as unique as ribosomes. Furthermore, similarities in the DNA sequences of genes that are shared by all organisms confirm the monophyly of life.

Despite these commonalities, major differences have also evolved across the diversity of life. Based on the deepest divisions of phylogenetic relationships, many biologists now recognize three **domains** (primary divisions) of life, two prokaryotic and one eukaryotic (**FIGURE 19.1**).

All prokaryotic organisms are unicellular, although they may form large coordinated colonies or biofilms consisting of many individuals. The domain Eukarya, by contrast, encompasses both unicellular and multicellular life forms. As was described in Chapter 4, prokaryotic cells differ from eukaryotic cells in some important ways:

- *Prokaryotic cells do not divide by mitosis.* Instead, after replicating their DNA, prokaryotic cells divide by their own method, binary fission (see Concept 7.2).
- *The organization of the genetic material differs.* The DNA of the prokaryotic cell is not organized within a membrane-enclosed nucleus. DNA molecules in prokaryotes are often circular. Many (but not all) prokaryotes have only one main chromosome and are effectively haploid, although many have additional smaller DNA molecules, called plasmids (see Concept 12.2).
- *Prokaryotes lack most of the membrane-enclosed cytoplasmic organelles—mitochondria, Golgi apparatus, and others—that are found in most eukaryotes.* However, the cytoplasm of a prokaryotic cell may contain a variety of infoldings of the cell membrane and photosynthetic membrane systems not found in eukaryotes.

Although the study and classification of eukaryotic organisms goes back centuries, much of our knowledge of the evolutionarily ancient prokaryotic domains is quite recent. Not until the final quarter of the twentieth century did advances in molecular genetics and biochemistry reach a point that enabled research that revealed deep-seated distinctions between the domains Bacteria and Archaea.

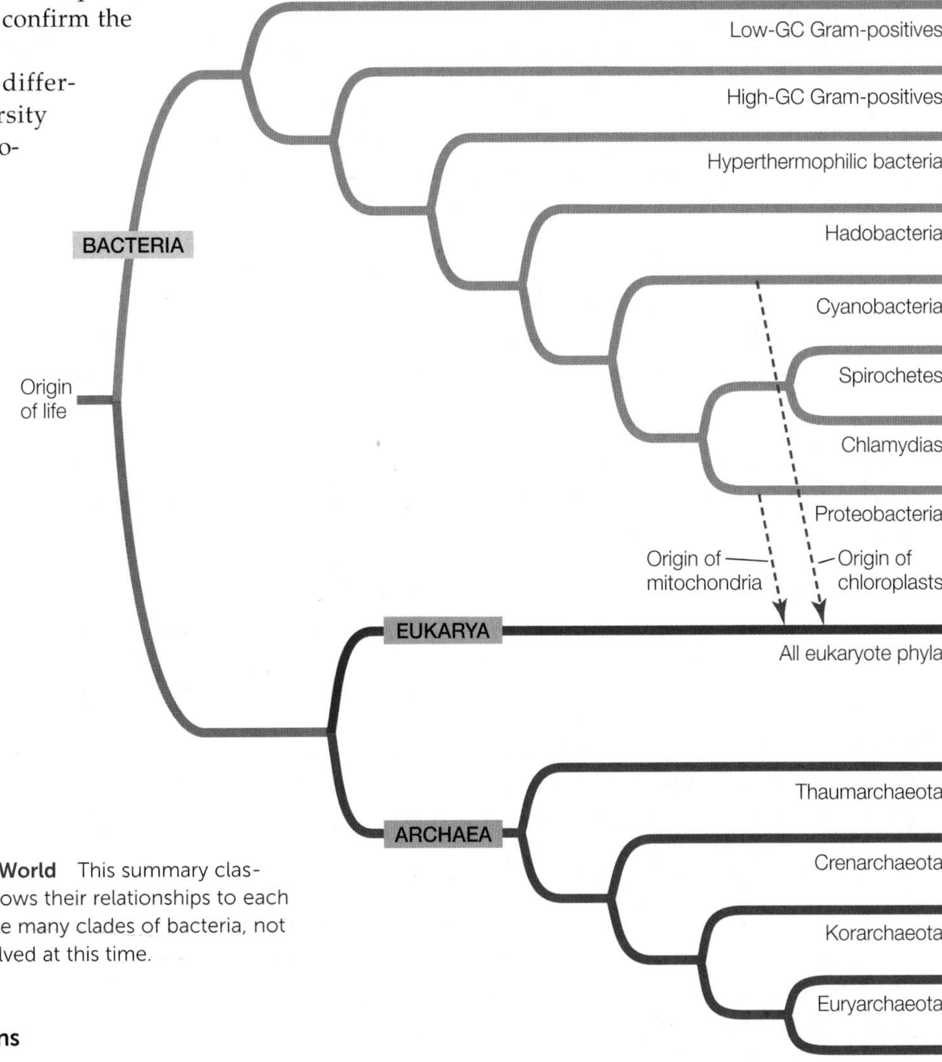

FIGURE 19.1 The Three Domains of the Living World This summary classification of the domains Bacteria and Archaea shows their relationships to each other and to Eukarya. The relationships among the many clades of bacteria, not all of which are listed here, are incompletely resolved at this time.

Go to ANIMATED TUTORIAL 19.1 The Evolution of the Three Domains
PoL2e.com/at19.1

TABLE 19.1 The Three Domains of Life on Earth

Characteristic	Domain: Bacteria	Domain: Archaea	Domain: Eukarya
Membrane-enclosed nucleus	Absent	Absent	**Present**
Membrane-enclosed organelles	Few	Absent	**Many**
Peptidoglycan in cell wall	**Present**	Absent	Absent
Membrane lipids	Ester-linked Unbranched	**Ether-linked** **Branched**	Ester-linked Unbranched
Ribosomes[a]	70S	70S	**80S**
Initiator tRNA	**Formylmethionine**	Methionine	Methionine
Operons	Yes	Yes	**Rare**
Plasmids	Yes	Yes	**Rare**
RNA polymerases	One	One[b]	**Three**
Ribosomes sensitive to chloramphenicol and streptomycin	**Yes**	No	No
Ribosomes sensitive to diphtheria toxin	**No**	Yes	Yes

[a]70S ribosomes are smaller than 80S ribosomes.
[b]Archaeal RNA polymerase is similar to eukaryotic polymerases.

The two prokaryotic domains differ in significant ways

A glance at **TABLE 19.1** will show you that there are major differences (most of which cannot be seen even under an electron microscope) between the two prokaryotic domains. In some ways archaea are more like eukaryotes; in other ways they are more like bacteria. (Note that we use lowercase when referring to the members of these domains and uppercase when referring to the domains themselves.) The basic unit of an archaeon (the term for a single archaeal organism) or bacterium (a single bacterial organism) is the prokaryotic cell. Each single-celled organism contains a full complement of genetic and protein-synthesizing systems, including DNA, RNA, and all the enzymes needed to transcribe and translate the genetic information into proteins. The prokaryotic cell also contains at least one system for generating the ATP it needs.

Genetic studies clearly indicate that all three domains had a single common ancestor. Across a major portion of their genome, eukaryotes share a more recent common ancestor with Archaea than they do with Bacteria (see Figure 19.1). However, the mitochondria of eukaryotes (as well as the chloroplasts of photosynthetic eukaryotes, such as plants) originated through endosymbiosis with a bacterium. Some biologists prefer to view the origin of eukaryotes as a fusion of two equal partners (one ancestor that was related to modern archaea, and another that was more closely related to modern bacteria). Others view the divergence of the early eukaryotes from the archaea as a separate and earlier event than the later endosymbioses. In either case, some eukaryote genes are most closely related to those of archaea, whereas others are most closely related to those of bacteria. The tree of life therefore contains some merging of lineages as well as the predominant divergence of lineages.

LINK

The origin of mitochondria and chloroplasts by endosymbiosis is described in **Concept 20.1**

Biologists estimate that the last common ancestor of the three domains lived about 3 billion years ago. We can deduce that it had DNA as its genetic material, and that its machinery for transcription and translation produced RNAs and proteins, respectively. This ancestor likely had a circular chromosome. Archaea, Bacteria, and Eukarya are all the products of billions of years of mutation, natural selection, and genetic drift, and they are all well adapted to their present-day environments. The earliest prokaryote fossils, which date back at least 3.5 billion years, indicate that there was considerable diversity among the prokaryotes even during those earliest days of life.

The small size of prokaryotes has hindered our study of their evolutionary relationships

Until about 300 years ago, nobody had even *seen* an individual prokaryote. Most prokaryotes remained invisible to humans until the invention of the first simple microscope. Prokaryotes are so small, however, that even the best light microscopes don't reveal much about them. It took advanced microscopic equipment and modern molecular techniques to open up the microbial world. (Microscopic organisms—both prokaryotes and eukaryotes—are often collectively referred to as "microbes.")

Before DNA sequencing became practical, taxonomists based prokaryote classification on observable characters such as shape, color, motility, nutritional requirements, and sensitivity to antibiotics. One of the characters most widely used to classify prokaryotes is the structure of their cell walls.

The cell walls of almost all bacteria contain **peptidoglycan**, a polymer that produces a meshlike structure around the cell. Peptidoglycan is a substance unique to bacteria. The absence of peptidoglycan from the cell walls of archaea is a key difference between the two prokaryotic domains. Peptidoglycan is also an excellent target for combating pathogenic (disease-causing) bacteria because it has no counterpart in eukaryotic cells. Antibiotics such as penicillin and ampicillin, as well as other agents that specifically interfere with the synthesis of

(A)

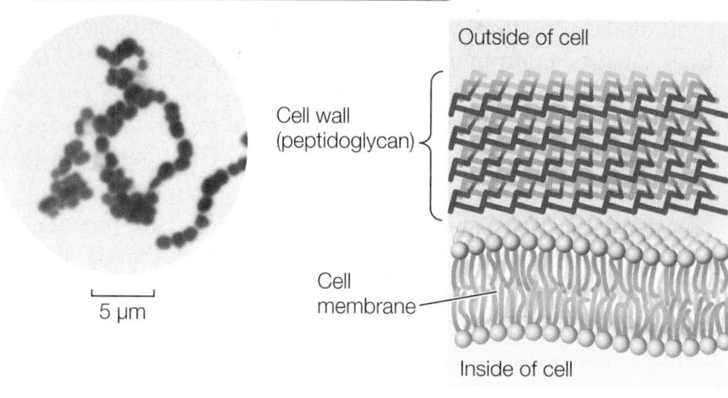

(B)

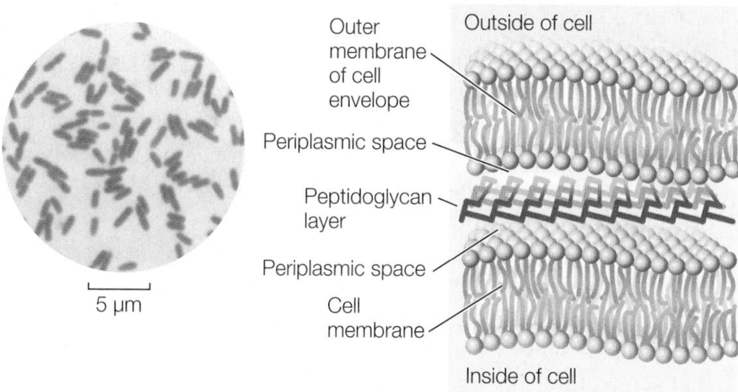

FIGURE 19.2 The Gram Stain and the Bacterial Cell Wall When treated with Gram stain, the cell walls of bacteria react in one of two ways. (A) Gram-positive bacteria have a thick peptidoglycan cell wall that retains the violet dye and appears deep blue or purple. (B) Gram-negative bacteria have a thin peptidoglycan layer that does not retain the violet dye, but picks up the counterstain and appears pink to red.

Go to ACTIVITY 19.1 Gram Stain and Bacteria PoL2e.com/ac19.1

peptidoglycan-containing cell walls, tend to have little, if any, effect on the cells of humans and other eukaryotes.

The **Gram stain** is a laboratory technique that can be used to classify most types of bacteria into two distinct groups, Gram-positive and Gram-negative. A smear of bacterial cells on a microscope slide is soaked in a violet dye and treated with iodine. Next it is washed with alcohol and counterstained with a red dye (safranin). **Gram-positive bacteria** retain the violet dye and appear blue to purple (**FIGURE 19.2A**). The alcohol washes the violet stain out of **Gram-negative bacteria**, which then pick up the safranin counterstain and appear pink to red (**FIGURE 19.2B**). For most bacteria, the effect of the Gram stain is determined by the chemical structure of the cell wall:

- *A Gram-negative cell wall* usually has a thin peptidoglycan layer, which is surrounded by a second, outer membrane quite distinct in chemical makeup from the cell membrane (see Figure 19.2B). The space between the cell membrane and the outer membrane (known as the periplasmic space) contains proteins that are important in digesting some materials, transporting others, and detecting chemical gradients in the environment.
- *A Gram-positive cell wall* usually has about five times as much peptidoglycan as a Gram-negative cell wall. Its thick peptidoglycan layer is a meshwork that may serve some of the same purposes as the periplasmic space of the Gram-negative cell envelope (cell wall plus outer membrane).

Shape is another phenotypic characteristic that is useful for the basic identification of bacteria. Three major shapes are common—spheres, rods, and spiral forms (**FIGURE 19.3**). Many bacterial names are based on these shapes. A spherical bacterium is called a **coccus** (plural *cocci*). Cocci may live singly or may associate in two- or three-dimensional arrays as chains, plates, blocks, or clusters of cells. A rod-shaped bacterium is called a **bacillus** (plural *bacilli*). A spiral bacterium (shaped like a corkscrew) is called a **spirillum** (plural *spirilla*). Bacilli and spirilla may be single, form chains, or gather in regular clusters. Among the other bacterial shapes are long filaments and branched filaments.

Less is known about the shapes of archaea because many of these organisms have never been seen. Many archaea are known only from samples of DNA from the environment. However, the species whose morphologies are known include cocci, bacilli, and even triangular and square-shaped species. Some flattened species grow on surfaces, arranged like sheets of postage stamps.

The nucleotide sequences of prokaryotes reveal their evolutionary relationships

Analyses of the nucleotide sequences of ribosomal RNA (rRNA) genes provided the first comprehensive evidence of evolutionary relationships among prokaryotes. For several

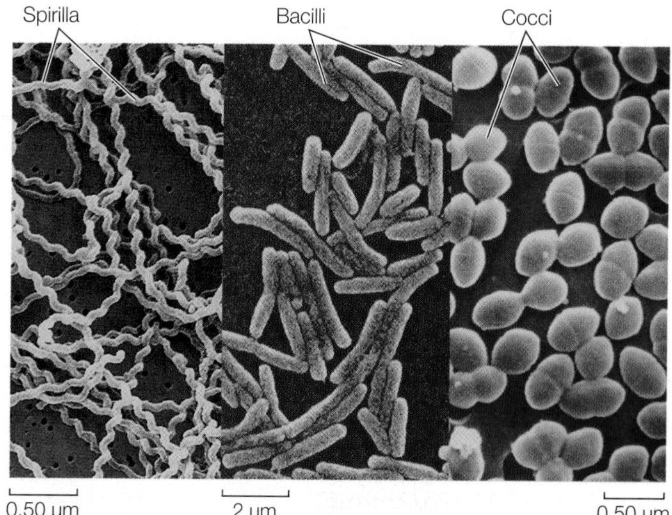

FIGURE 19.3 Bacterial Cell Shapes This composite colorized micrograph shows three common bacterial shapes. The spiral-shaped spirilla are *Leptospira* sp., a human pathogen. Rod-shaped cells are called bacilli; these *Legionella pneumophila* reside in the gut. Spherical cells are called cocci; these are a species of *Enterococcus*, also from the mammalian gut.

reasons, rRNA is particularly useful for phylogenetic studies of living organisms:

- rRNA was present in the common ancestor of all life and is therefore evolutionarily ancient.
- No free-living organism lacks rRNA, so rRNA genes can be compared throughout the tree of life.
- rRNA plays a critical role in translation in all organisms, so lateral transfer of rRNA genes among distantly related species is unlikely.
- rRNA has evolved slowly enough that gene sequences from even distantly related species can be aligned and analyzed.

Comparisons of rRNA genes from a great many organisms have revealed the probable phylogenetic relationships throughout the tree of life. Databases such as GenBank contain rRNA gene sequences from hundreds of thousands of species—more than any other type of gene sequence.

Although studies of rRNA genes reveal much about the evolutionary relationships of prokaryotes, they don't always reveal the entire evolutionary history of these organisms. In some groups of prokaryotes, analyses of multiple gene sequences have suggested several different phylogenetic patterns. How could such differences among different gene sequences arise? Studies of whole prokaryotic genomes have revealed that even distantly related prokaryotes sometimes exchange genetic material.

Lateral gene transfer can lead to discordant gene trees

As noted earlier, prokaryotes reproduce by binary fission. If we could follow these divisions back through evolutionary time, we would be tracing the path of the complete tree of life for bacteria and archaea. This underlying tree of relationships (represented in highly abbreviated form in Appendix A) is called the organismal (or species) tree. Because whole genomes are replicated during binary fission, we would expect phylogenetic trees constructed from most gene sequences (see Chapter 16) to reflect these same relationships.

Even though binary fission is an asexual process, there are other processes—including transformation, conjugation, and transduction—that allow the exchange of genetic information between some prokaryotes without reproduction. Thus prokaryotes can exchange and recombine their DNA with that of other individuals (this is sex in the genetic sense of the word), but this genetic exchange is not directly linked to reproduction, as it is in most eukaryotes.

LINK

Prokaryote exchange of genetic material by transformation, conjugation, and transduction is described in **Concept 8.4**

From early in evolution to the present day, some genes have been moving "sideways" from one prokaryote species to another, a phenomenon known as **lateral gene transfer**. Lateral gene transfers are well documented, especially among closely related species; some have been documented even across the domains of life.

Consider, for example, the genome of *Thermotoga maritima*, a bacterium that can survive extremely high temperatures. In comparing the 1,869 gene sequences of *T. maritima* against sequences encoding the same proteins in other species, investigators found that some of this bacterium's genes have their closest relationships not with the genes of other bacterial species, but with the genes of archaea that live in similar extreme environments.

When genes involved in lateral transfer events are sequenced and analyzed, the resulting individual gene trees will not match the organismal tree in every respect (**FIGURE 19.4**). The individual gene trees will vary because the history

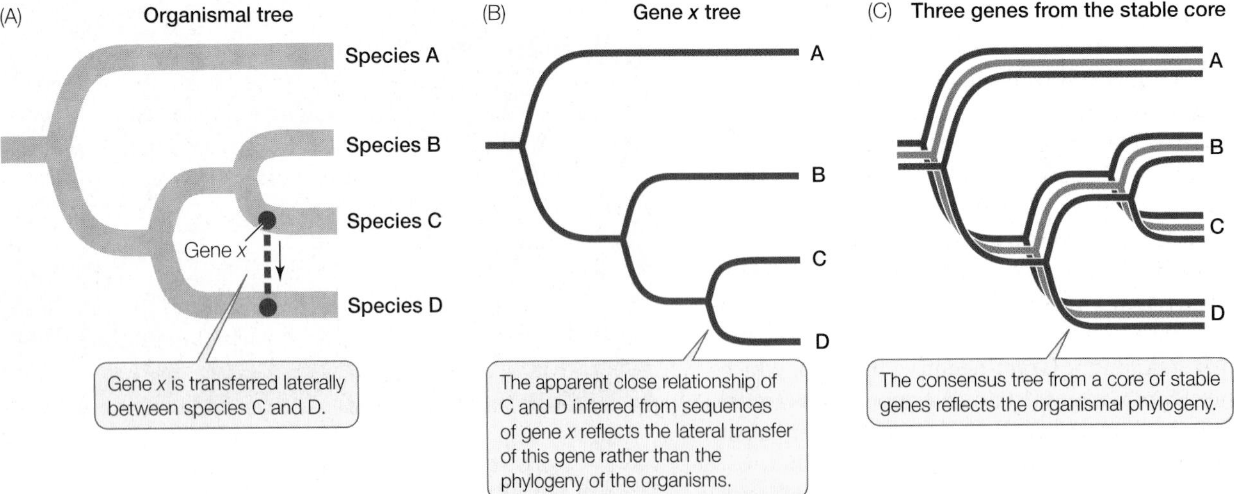
(A)
Organismal tree
Species A
Species B
Species C
Species D
Gene x
Gene x is transferred laterally between species C and D.
(B)
Gene x tree
A
B
C
D
The apparent close relationship of C and D inferred from sequences of gene x reflects the lateral transfer of this gene rather than the phylogeny of the organisms.
(C) Three genes from the stable core
A
B
C
D
The consensus tree from a core of stable genes reflects the organismal phylogeny.

FIGURE 19.4 Lateral Gene Transfer Complicates Phylogenetic Relationships (A) The phylogeny of four hypothetical prokaryote species, two of which have been involved in a lateral transfer of gene *x*. (B) A tree based only on gene *x* shows the phylogeny of the laterally transferred gene, rather than the organismal phylogeny. (C) In many cases, a consensus tree based on a "stable core" of genes accurately reflects the organismal phylogeny.

of lateral transfer events is different for each gene. Biologists can reconstruct the underlying organismal phylogeny by comparing multiple genes (to produce a consensus tree) or by concentrating on genes that are unlikely to be involved in lateral gene transfer events. For example, genes that are involved in fundamental cellular processes (such as the rRNA genes discussed above) are unlikely to be replaced by the same genes from other species because functional, locally adapted copies of these genes are already present.

What kinds of genes are most likely to be involved in lateral gene transfer? Genes that result in a new adaptation that confers higher fitness on a recipient species are most likely to be transferred repeatedly among species. For example, genes that produce antibiotic resistance are often transferred among bacterial species on plasmids, especially under the strong selection pressure such as that imposed by modern antibiotic medications. Improper or overly frequent use of antibiotics can select for resistant strains of bacteria that are much harder to treat. This selection for antibiotic resistance explains why informed physicians have become more careful in prescribing antibiotics.

It is debatable whether lateral gene transfer has seriously complicated our attempts to resolve the tree of prokaryotic life. Recent work suggests that it has not. Lateral gene transfer rarely creates problems at higher taxonomic levels, even though it may complicate our understanding of the relationships among individual species. Some species clearly obtain some of their genes from otherwise distantly related species, so evolutionary histories of individual genes may differ within a single organism. But it is now possible to make nucleotide sequence comparisons involving entire genomes, and these studies are revealing a stable core of crucial genes that are uncomplicated by lateral gene transfer. Gene trees based on this stable core more accurately reveal the organismal phylogeny (see Figure 19.4). The problem remains, however, that only a very small proportion of the prokaryotic world has been described and studied.

The great majority of prokaryote species have never been studied

Most prokaryotes have defied all attempts to grow them in pure culture, causing biologists to wonder how many species, and possibly even clades, we might be missing. A window onto this problem was opened with the introduction of a new way of examining nucleic acid sequences. When biologists are unable to work with the whole genome of a single prokaryote species, they can instead examine individual genes collected from a random sample of the environment (see Figure 12.6).

Norman Pace of the University of Colorado isolated individual rRNA gene sequences from extracts of environmental samples such as soil and seawater. Comparing such sequences with previously known ones revealed that an extraordinary number of the sequences were new, implying that they came from previously unrecognized species. Biologists have described only about 10,000 species of bacteria and only a few hundred species of archaea (see Figure 1.4). The results of Pace's and similar studies suggest that there may be millions—perhaps hundreds of millions—of prokaryote species on Earth. Other biologists put the estimate much lower, arguing that the high dispersal ability of many bacterial species greatly reduces local endemism (i.e., the number of species restricted to a small geographic area). Only the magnitude of these estimates differs, however; all sides agree that we have just begun to uncover Earth's bacterial and archaeal diversity.

CHECKpoint CONCEPT 19.1

- ✓ Why were all prokaryotes once considered members of a single clade, and what findings led to the establishment of Bacteria and Archaea as separate domains?
- ✓ How did biologists classify bacteria before it became possible to determine nucleotide sequences?
- ✓ Why are nucleotide sequences of rRNA genes particularly useful for evolutionary studies?
- ✓ How does lateral gene transfer complicate evolutionary studies?

Despite the challenges of reconstructing the phylogeny of prokaryotes, taxonomists are beginning to establish evolutionary classification systems for these organisms. With a full understanding that new information requires periodic revisions in these classifications, we will next apply a current system of classification to organize our survey of prokaryote diversity.

CONCEPT 19.2 Prokaryote Diversity Reflects the Ancient Origins of Life

The prokaryotes were alone on Earth for a very long time, adapting to new environments and to changes in existing environments. They have survived to this day, in massive numbers and incredible diversity, and they are found everywhere. In numbers of individuals, prokaryotes are far more abundant than eukaryotes are. Individual bacteria and archaea in the oceans number more than 3×10^{28}—about a million times more than the number of stars in the universe. Closer to home, the individual bacteria living in your intestinal tract outnumber all the humans who have ever lived.

Given our still fragmentary knowledge of prokaryote diversity, it is not surprising that there are several different hypotheses about the relationships of the major groups of prokaryotes. In this book we use a widely accepted classification system that has considerable support from nucleotide sequence data. We will discuss eight major bacterial groups that have the broadest phylogenetic support and have received the most study: the low-GC Gram-positives, high-GC Gram-positives, hyperthermophilic bacteria, hadobacteria, cyanobacteria, spirochetes, chlamydias, and proteobacteria (see Figure 19.1). We will then describe the archaea, whose diversity is even less well studied than that of the bacteria.

The low-GC Gram-positives include the smallest cellular organisms

The **low-GC Gram-positives**, also known as Firmicutes, derive the first part of their name from the relatively low ratio of G-C to A-T nucleotide base pairs in their DNA. The second part of their name is less accurate: some of the low-GC Gram-positives are in fact Gram-negative, and some have no cell wall at all. Despite these differences, phylogenetic analyses of DNA sequences support the monophyly of this bacterial clade.

One group of low-GC Gram-positives can produce heat-resistant resting structures called **endospores** (**FIGURE 19.5**). When a key nutrient such as nitrogen or carbon becomes scarce, the bacterium replicates its DNA and encapsulates one copy, along with some of its cytoplasm, in a tough cell wall heavily thickened with peptidoglycan and surrounded by a spore coat. The parent cell then breaks down, releasing the endospore. Endospore production is not a reproductive process, as the endospore merely replaces the parent cell. The endospore, however, can survive harsh environmental conditions that would kill the parent cell, such as high or low temperatures or drought, because it is dormant—its normal metabolic activity is suspended. Later, if it encounters favorable conditions, the endospore becomes metabolically active and divides, forming new cells that are like the parent cells. Members of this endospore-forming group of low-GC Gram-positives include the many species of *Clostridium* and *Bacillus*. Some of their endospores can be reactivated after more than 1,000 years of dormancy. There are even credible claims of reactivation of *Bacillus* endospores that are millions of years old.

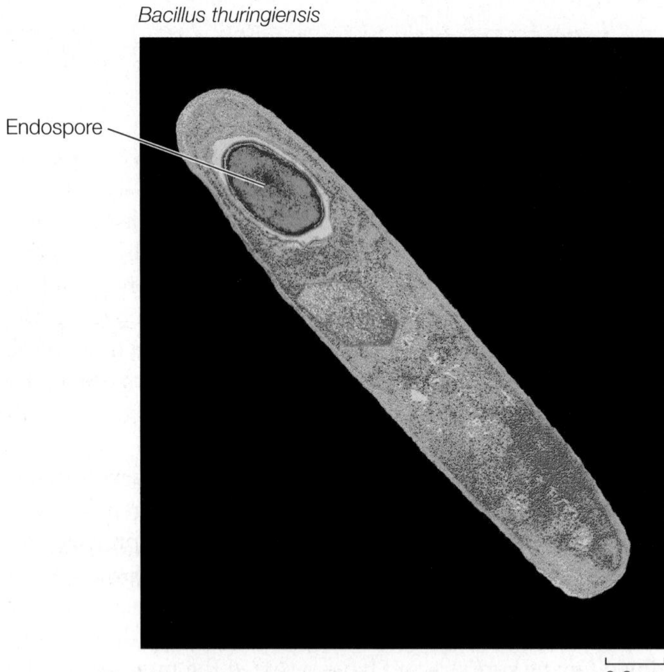

FIGURE 19.5 A Structure for Waiting Out Bad Times Some low-GC Gram-positive bacteria can replicate their DNA and encase it in a resistant endospore. In harsh conditions, the parent cell breaks down and the endospore survives in a dormant state until conditions improve.

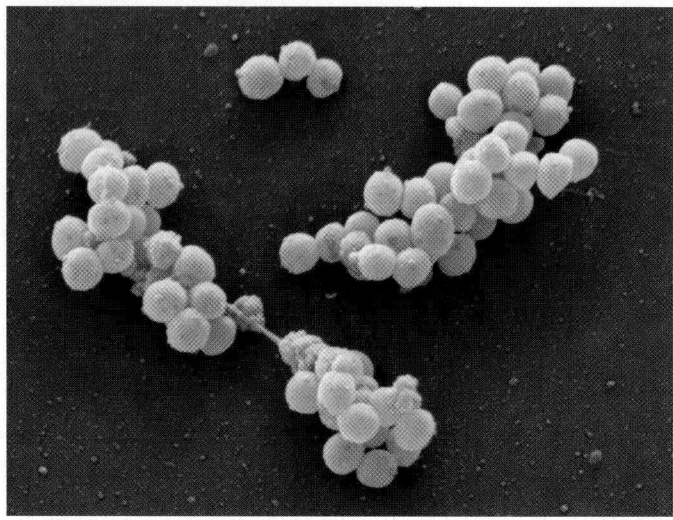

FIGURE 19.6 Staphylococci "Grape clusters" are the usual arrangement of these low-GC Gram-positive coccal bacteria, often the cause of skin or wound infections.

Endospores of *Bacillus anthracis* are the cause of anthrax. Anthrax is primarily a disease of grazing animals, but it can be fatal in humans if the infection starts in the respiratory system. When the endospores sense macrophages in mammalian blood, they reactivate and release toxins into the bloodstream. *Bacillus anthracis* has been used as a bioterrorism agent because large quantities of its endospores can be transported in a small space and then spread among human populations, where they may be inhaled or ingested.

Low-GC Gram-positives of the genus *Staphylococcus*—the **staphylococci** (**FIGURE 19.6**)—are abundant on the human body surface; they are responsible for boils and many other skin problems. *Staphylococcus aureus* is the best-known human pathogen in this genus; it is present in 20–40 percent of normal adults (and in 50–70 percent of hospitalized adults). In addition to skin diseases, *S. aureus* can cause respiratory, intestinal, and wound infections.

Another interesting group of low-GC Gram-positives, the **mycoplasmas**, lack cell walls, although some have a stiffening material outside the cell membrane. The mycoplasmas are among the smallest cellular organisms known (**FIGURE 19.7**). The smallest mycoplasmas capable of multiplying by binary fission have a diameter of about 0.2 µm. They are small in another crucial sense as well: they have less than half as much DNA as most other prokaryotes. It has been speculated that the DNA in a mycoplasma, which codes for fewer than 500 proteins, may be close to the minimum amount required to encode the essential properties of a living cell.

Some high-GC Gram-positives are valuable sources of antibiotics

High-GC Gram-positives, also known as actinobacteria, have a higher ratio of G-C to A-T nucleotide base pairs than do the low-GC Gram-positives. These bacteria develop an elaborately branched system of filaments (**FIGURE 19.8**) that resembles the

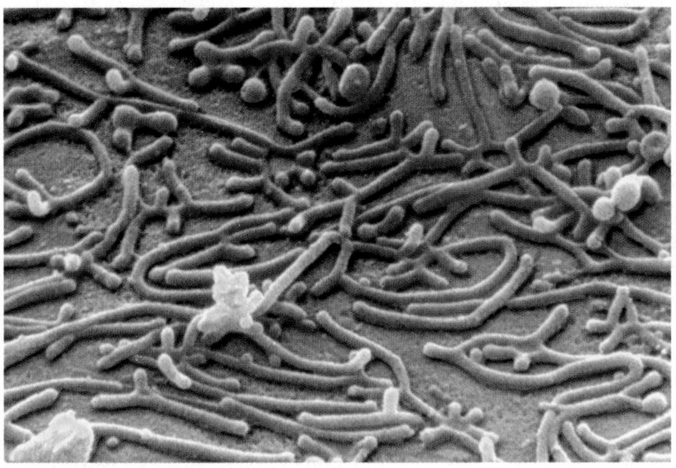

FIGURE 19.7 Tiny Cells With about one-fifth as much DNA as *E. coli*, mycoplasmas are among the smallest known bacteria.

filamentous growth habit of fungi, albeit at a reduced scale. Some high-GC Gram-positives reproduce by forming chains of spores at the tips of the filaments. In species that do not form spores, the branched, filamentous growth ceases and the structure breaks up into typical cocci or bacilli, which then reproduce by binary fission.

The high-GC Gram-positives include several medically important bacteria. *Mycobacterium tuberculosis* causes tuberculosis, which kills 3 million people each year. Genetic data suggest that this bacterium arose 3 million years ago in East Africa, making it the oldest known human bacterial pathogen. The genus *Streptomyces* produces streptomycin as well as hundreds of other antibiotics. We derive most of our antibiotics from members of the high-GC Gram-positives.

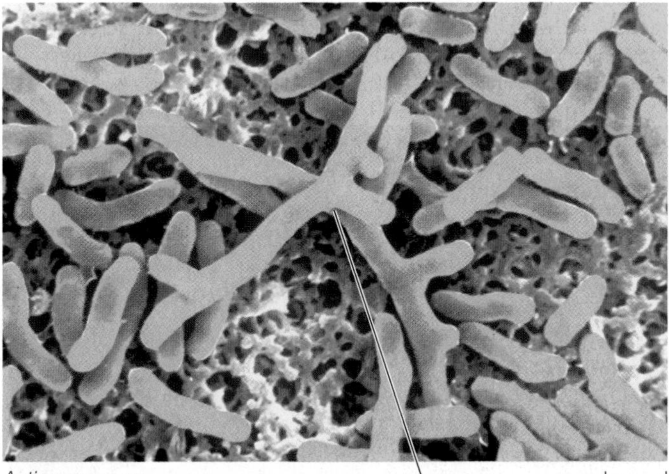

FIGURE 19.8 Actinobacteria Are High-GC Gram-Positives The tangled, branching filaments seen in this scanning electron micrograph are typical of this medically important bacterial group.

Hyperthermophilic bacteria live at very high temperatures

Several lineages of bacteria and archaea are **extremophiles**: they thrive under extreme conditions that would kill most other organisms. The **hyperthermophilic bacteria**, for example, are thermophiles (Greek, "heat-lovers"). Genera such as *Aquifex* live near volcanic vents and in hot springs, sometimes at temperatures near the boiling point of water. Some species of *Aquifex* need only hydrogen, oxygen, carbon dioxide, and mineral salts to live and grow. Species of the genus *Thermotoga* live deep underground in oil reservoirs, as well as in other high-temperature environments.

Biologists have hypothesized that high temperatures characterized the ancestral conditions for life on Earth, given that most environments on early Earth were much hotter than those of today. Reconstructions of ancestral bacterial genes have supported this hypothesis by showing that the ancestral sequences functioned best at elevated temperatures. The monophyly of the hyperthermophilic bacteria, however, is not well established.

Hadobacteria live in extreme environments

The **hadobacteria**, including such genera as *Deinococcus* and *Thermus*, are another group of thermophilic extremophiles. The group's name is derived from Hades, the ancient Greek name for the underworld. *Deinococcus* are resistant to radiation and can consume nuclear waste and other toxic materials. They can also survive extremes of cold as well as hot temperatures. Another member of this group, *Thermus aquaticus*, was the source of the thermally stable DNA polymerase that was critical for the development of the polymerase chain reaction. *Thermus aquaticus* was originally isolated from a hot spring, but it can be found wherever hot water occurs (including many residential hot water heaters).

Cyanobacteria were the first photosynthesizers

Cyanobacteria, sometimes called blue-green bacteria because of their pigmentation, are photosynthetic bacteria that require only water, nitrogen gas, oxygen, a few mineral elements, light, and carbon dioxide to survive. They use chlorophyll *a* for photosynthesis and release oxygen gas (O_2); many species also fix nitrogen. Photosynthesis by these bacteria was the basis of the "green revolution" that transformed Earth's atmosphere (see Concept 18.3).

Cyanobacteria carry out the same type of photosynthesis that is characteristic of eukaryotic photosynthesizers. They contain elaborate and highly organized internal membrane systems called **photosynthetic lamellae**. As mentioned in Concept 19.1, the chloroplasts of photosynthetic eukaryotes are derived from an endosymbiotic cyanobacterium.

Cyanobacteria may live free as single cells or associate in multicellular colonies. Depending on the species and on growth conditions, these colonies may range from flat sheets one cell thick to filaments to spherical balls of cells. Some filamentous colonies of cyanobacteria differentiate into three specialized cell types: vegetative cells, spores, and heterocysts. **Vegetative cells** photosynthesize, **spores** are resting stages that

(A) *Anabaena* sp.

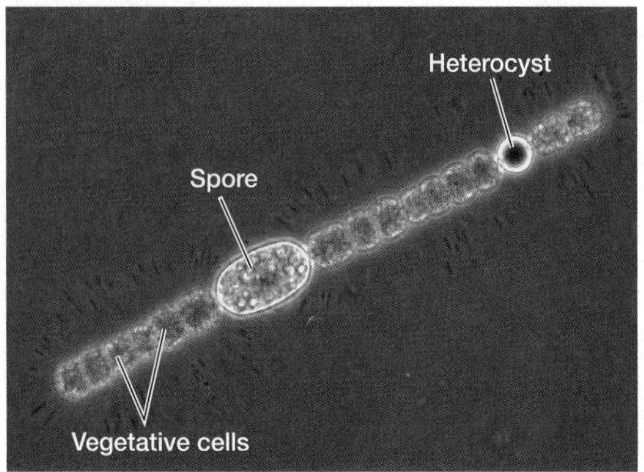

(B) *Nostoc punctiforme*

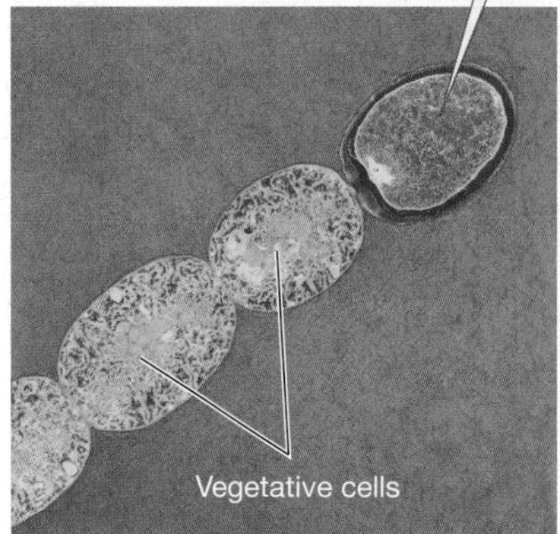

(C)

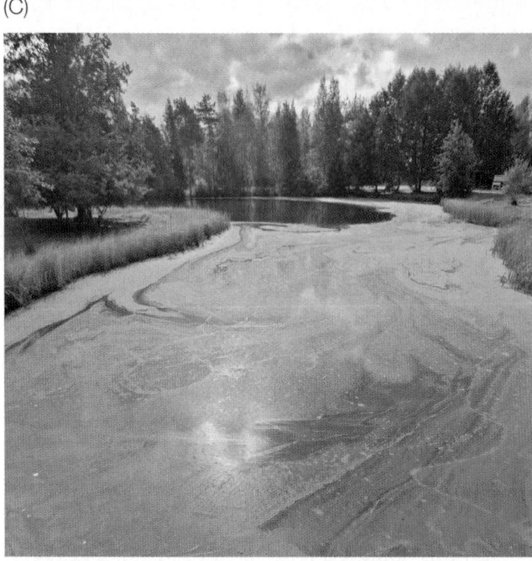

FIGURE 19.9 Cyanobacteria (A) Some cyanobacteria form filamentous colonies containing three cell types, including reproductive spores and photosynthesizing vegetative cells. (B) Heterocysts are specialized for nitrogen fixation and may serve as a breaking point when filaments reproduce. (C) This pond in Finland has experienced eutrophication: phosphorus and other nutrients generated by human activity have accumulated, feeding an immense green mat (commonly referred to as "pond scum") that is made up of several species of free-living cyanobacteria.

Go to MEDIA CLIP 19.1
Cyanobacteria
PoL2e.com/mc19.1

can survive harsh environmental conditions and eventually develop into new filaments, and **heterocysts** are cells specialized for nitrogen fixation (**FIGURE 19.9**). All of the known cyanobacteria with heterocysts fix nitrogen. Heterocysts also have a role in reproduction: when filaments break apart to reproduce, the heterocyst may serve as a breaking point.

Spirochetes move by means of axial filaments

Spirochetes are Gram-negative, motile bacteria characterized by unique structures called axial filaments (**FIGURE 19.10A**), which are modified flagella running through the periplasmic space. The cell body is a long cylinder coiled into a spiral (**FIGURE 19.10B**). The axial filaments begin at either end of the cell and overlap in the middle. Motor proteins connect the axial filaments to the cell wall, enabling the corkscrew-like movement of the bacterium. Many spirochetes are parasites of humans; a few are pathogens, including those that cause syphilis and Lyme disease. Others live free in mud or water.

Chlamydias are extremely small parasites

Chlamydias are among the smallest bacteria (0.2–1.5 µm in diameter). They can live only as parasites in the cells of other organisms. It was once believed that their obligate parasitism

FIGURE 19.10 Spirochetes Get Their Shape from Axial Filaments (A) A spirochete from the gut of a termite, seen in cross section, shows the axial filaments used to produce a corkscrew-like movement of these spiral prokaryotes. (B) This spirochete species causes syphilis in humans.

(A)

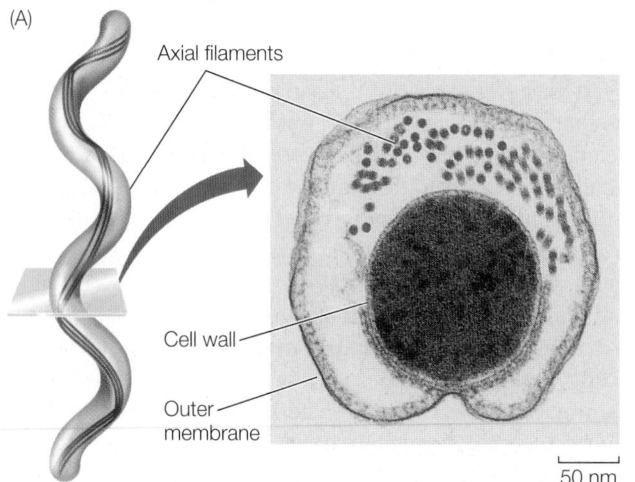

(B)

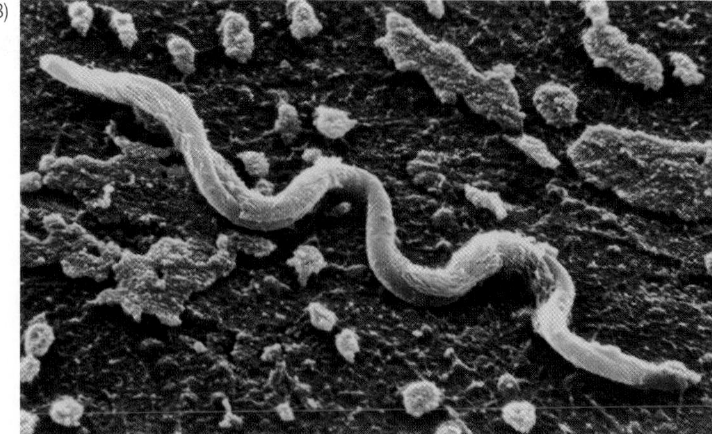

Cristispira sp.

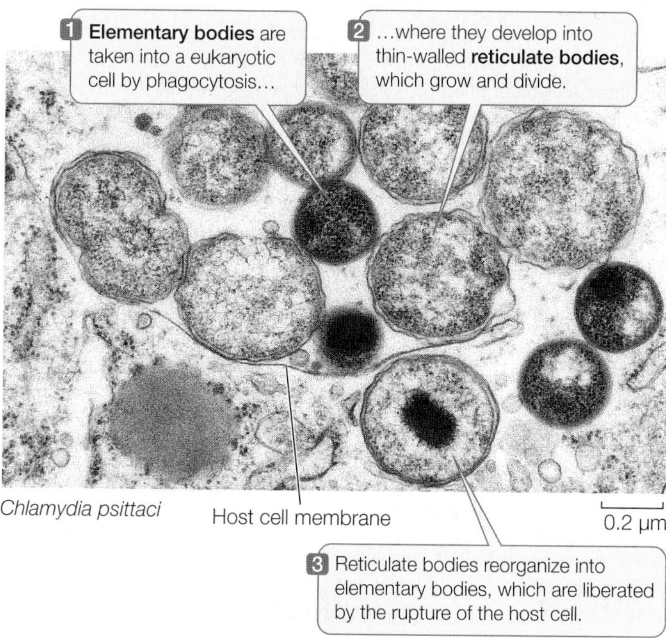

1 Elementary bodies are taken into a eukaryotic cell by phagocytosis...
2 ...where they develop into thin-walled reticulate bodies, which grow and divide.
Chlamydia psittaci
Host cell membrane
0.2 μm
3 Reticulate bodies reorganize into elementary bodies, which are liberated by the rupture of the host cell.

FIGURE 19.11 Chlamydias Change Form during Their Life Cycle Elementary bodies and reticulate bodies are the two major phases of the chlamydia life cycle.

resulted from an inability to produce ATP—that chlamydias were "energy parasites." However, genome sequencing indicates that chlamydias have the genetic capacity to produce at least some ATP. They can augment this capacity by using an enzyme called a translocase, which allows them to take up ATP from the cytoplasm of their host in exchange for ADP from their own cells.

These tiny, Gram-negative cocci are unique among prokaryotes because of a complex life cycle that involves two different forms of cells, elementary bodies and reticulate bodies (**FIGURE 19.11**). Various strains of chlamydias cause eye infections (especially trachoma), sexually transmitted diseases, and some forms of pneumonia in humans.

The proteobacteria are a large and diverse group

By far the largest group of bacteria, in terms of number of described species, is the **proteobacteria**. The proteobacteria include many species of Gram-negative photoautotrophs that use light-driven reactions to metabolize sulfur, as well as dramatically diverse bacteria that bear no phenotypic resemblance to the photoautotrophic species. Genetic and morphological evidence indicates that the mitochondria of eukaryotes were derived from a proteobacterium by endosymbiosis.

Among the proteobacteria are some nitrogen-fixing genera, such as *Rhizobium*, and other bacteria that contribute to the global nitrogen and sulfur cycles. *Escherichia coli*, one of the most studied organisms on Earth, is a proteobacterium. So, too, are many of the most famous human pathogens, such as *Yersinia pestis* (which causes bubonic plague), *Vibrio cholerae* (cholera), and *Salmonella typhimurium* (gastrointestinal disease; **FIGURE 19.12**).

The bioluminescent *Vibrio* we discussed at the opening of this chapter are also members of this group. There are many potential applications of the genes that encode bioluminescent proteins in bacteria. Already, these genes are being inserted into the genomes of other species in which the resulting bioluminescence is used as a marker of gene expression. Futuristic proposals for making use of bioluminescence in bioengineered organisms include crop plants that glow when they become water-stressed and need to be irrigated, and glowing trees that could light highways at night in place of electric lights.

Although fungi cause most plant diseases, and viruses cause others, about 200 known plant diseases are of bacterial origin. Crown gall, with its characteristic tumors (**FIGURE 19.13**), is one of the most striking. The causal agent of crown gall is *Agrobacterium tumefaciens*, a proteobacterium that harbors a plasmid used in recombinant DNA studies as a vehicle for inserting genes into new plant hosts.

Gene sequencing enabled biologists to differentiate the domain Archaea

The separation of Archaea from Bacteria and Eukarya was originally based on phylogenetic relationships determined from sequences of rRNA genes. It was supported when biologists sequenced the first archaeal genome. That genome consisted of 1,738 genes, more than half of which were unlike any genes ever found in the other two domains.

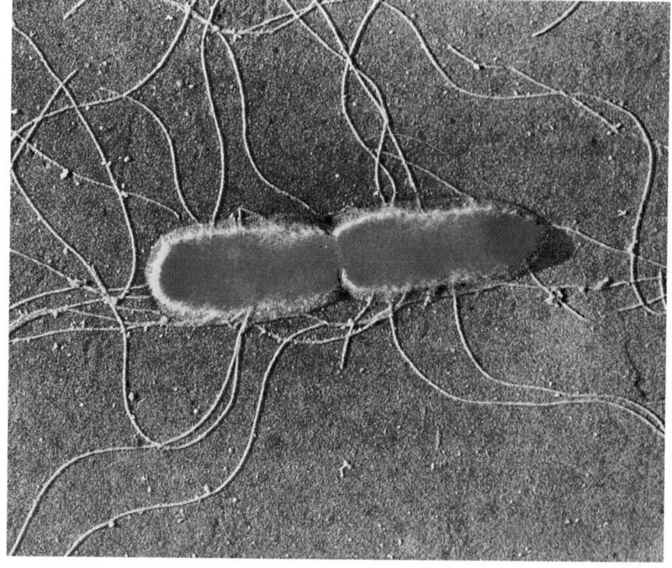

FIGURE 19.12 Proteobacteria Include Human Pathogens These conjugating cells of *Salmonella typhimurium* are exchanging genetic material (see Concept 8.4). Representing only one group of the highly diverse proteobacteria, this pathogen causes a wide range of gastrointestinal illnesses in humans.

Go to MEDIA CLIP 19.2
A Swarm of *Salmonella*
PoL2e.com/mc19.2

FIGURE 19.13 Crown Gall Crown gall, a type of tumor shown here growing on the trunk of a white oak, is caused by the proteobacterium *Agrobacterium tumefaciens*.

Archaea are well known for living in extreme habitats such as those with high salinity (salt content), low oxygen concentrations, high temperatures, or high or low pH (**FIGURE 19.14**). Many archaea are not extremophiles, however, but live in moderate habitats—they are common in soil, for example. Perhaps the largest numbers of archaea live in the ocean depths.

One current classification scheme divides Archaea into two principal groups, **Crenarchaeota** and **Euryarchaeota**. Less is known about three more recently discovered groups, **Korarchaeota**, **Nanoarchaeota**, and **Thaumarchaeota**. In fact, we know relatively little about the phylogeny of archaea, in part because the study of these prokaryotes is still in its early stages.

Two characteristics shared by all archaea are the absence of peptidoglycan in their cell walls and the presence of lipids of distinctive composition in their cell membranes (see Table 19.1). The unusual lipids in the membranes of archaea are found in all archaea and in no bacteria or eukaryotes. Most lipids in bacterial and eukaryotic membranes contain unbranched long-chain fatty acids connected to glycerol molecules by **ester linkages**:

$$\begin{array}{ccccc} & \mathrm{O} & & \mathrm{H} & \\ & \| & & | & \\ — & \mathrm{C} & —\mathrm{O}— & \mathrm{C} & — \\ & & & | & \\ & & & \mathrm{H} & \end{array}$$

In contrast, some lipids in archaeal membranes contain long-chain hydrocarbons connected to glycerol molecules by **ether linkages**:

$$\begin{array}{ccccc} & \mathrm{H} & & \mathrm{H} & \\ & | & & | & \\ — & \mathrm{C} & —\mathrm{O}— & \mathrm{C} & — \\ & | & & | & \\ & \mathrm{H} & & \mathrm{H} & \end{array}$$

INVESTIGATION

FIGURE 19.14 What Is the Highest Temperature Compatible with Life? Can any organism thrive at temperatures above 120°C? This is the temperature used for sterilization, known to destroy all previously described organisms. Kazem Kashefi and Derek Lovley isolated an unidentified prokaryote from water samples taken near a hydrothermal vent and found that it survived and even multiplied at 121°C. The organism was dubbed "Strain 121," and its gene sequencing results indicate that it is an archaeal species.[a]

HYPOTHESIS

Some prokaryotes can survive at temperatures above the 120°C threshold of sterilization.

METHOD

1. Seal samples of unidentified, iron-reducing, thermal vent prokaryotes in tubes with a medium containing Fe^{3+} as an electron acceptor. Control tubes contain Fe^{3+} but no organisms.
2. Hold both tubes in a sterilizer at 121°C for 10 hours. If the iron-reducing organisms are metabolically active, they will reduce the Fe^{3+} to Fe^{2+} (as magnetite, which can be detected with a magnet).

RESULTS

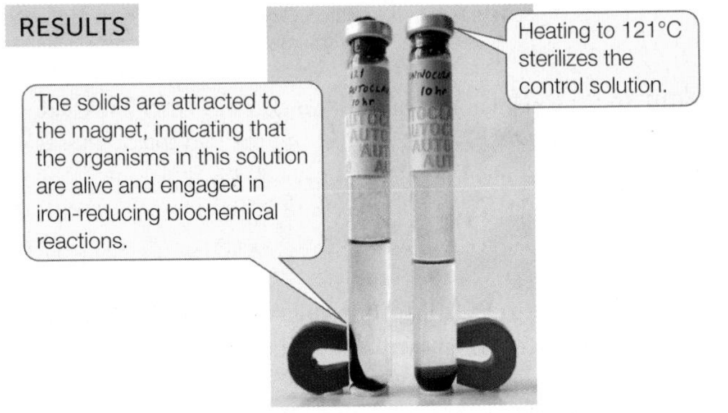

The solids are attracted to the magnet, indicating that the organisms in this solution are alive and engaged in iron-reducing biochemical reactions.
Heating to 121°C sterilizes the control solution.

CONCLUSION

This thermal vent organism (Strain 121) can survive at temperatures above the previously defined sterilization limit.

ANALYZE THE DATA

After Strain 121 was isolated, its growth was examined at various temperatures. The table shows generation time (the time between cell divisions) at nine temperatures.

Temperature (°C)	Generation time (hr)
85	10
90	4
95	3
100	2.5
105	2
110	4
115	6
120	20
130	No growth, but cells not killed

A. Make a graph showing generation time as a function of temperature.

B. Which temperature appears to be closest to the optimum for growth of Strain 121?

C. Note that no growth occurred at 130°C, but that the cells were not killed. How would you demonstrate that these cells were still alive?

Go to **LaunchPad** for discussion and relevant links for all **INVESTIGATION** figures.

[a]K. Kashefi and D. R. Lovley. 2003. *Science* 301: 934.

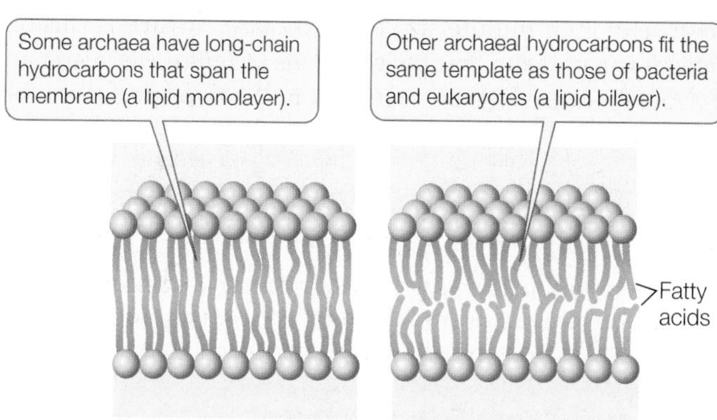
Some archaea have long-chain hydrocarbons that span the membrane (a lipid monolayer).
Other archaeal hydrocarbons fit the same template as those of bacteria and eukaryotes (a lipid bilayer).
Fatty acids

FIGURE 19.15 Membrane Architecture in Archaea The long-chain hydrocarbons of many archaeal lipids have glycerol molecules at both ends, so that the membranes they form consist of a lipid monolayer. In contrast, the membranes of other archaea, bacteria, and eukaryotes consist of a lipid bilayer.

FIGURE 19.16 Some Crenarchaeotes Like It Hot Thermophilic archaea can thrive in the intense heat of volcanic hot sulfur springs such as those in Wyoming's Yellowstone National Park.

These ether linkages are a synapomorphy (a shared, derived trait viewed as evidence of common ancestry) of archaea. In addition, the long-chain hydrocarbons of archaea are branched. One class of archaeal lipids, with hydrocarbon chains 40 carbon atoms in length, contains glycerol at *both* ends of the hydrocarbons (**FIGURE 19.15**). These lipids form a lipid monolayer structure that is unique to archaea. They still fit into a biological membrane because they are twice as long as the typical lipids in the bilayers of other membranes. Lipid monolayers and bilayers are both found among the archaea. The effects, if any, of these structural features on membrane performance are unknown. In spite of this striking difference in their lipids, the membranes of all three domains have similar overall structures, dimensions, and functions.

Most crenarchaeotes live in hot or acidic places

Most known crenarchaeotes are either thermophilic, acidophilic (acid loving), or both. Members of the genus *Sulfolobus* live in hot sulfur springs at temperatures of 70°C–75°C. They become metabolically inactive at 55°C (131°F). Hot sulfur springs are also extremely acidic. *Sulfolobus* grows best in the range of pH 2–pH 3, but some members of this genus readily tolerate pH values as low as 0.9. Most acidophilic thermophiles maintain an internal pH of 5.5–7 (close to neutral) in spite of their acidic environment. These and other thermophiles thrive where few other organisms can even survive (**FIGURE 19.16**).

Euryarchaeotes are found in surprising places

Some species of Euryarchaeota are **methanogens**: they produce methane (CH_4) by reducing carbon dioxide, and this is the key step in their energy metabolism. All of the methanogens are obligate anaerobes (see Concept 19.3). Comparison of their rRNA gene sequences has revealed a close evolutionary relationship among these methanogenic species, which were previously assigned to several different groups of bacteria.

Methanogenic euryarchaeotes release approximately 2 billion tons of methane gas into Earth's atmosphere each year, accounting for 80–90 percent of the methane that enters the atmosphere, including that produced in some mammalian digestive systems. Approximately a third of this methane comes from methanogens living in the guts of ruminants such as cattle, sheep, and deer, and another large fraction comes from methanogens living in the guts of termites and cockroaches. Methane is increasing in Earth's atmosphere by about 1 percent per year and contributes to the greenhouse effect (see Concept 45.4). Part of that increase is due to increases in cattle and rice farming and the methanogens associated with both.

Another group of euryarchaeotes, the **extreme halophiles** (salt lovers), lives exclusively in very salty environments. Because they contain pink carotenoid pigments, these archaea are sometimes easy to see (**FIGURE 19.17**). Extreme halophiles grow in the Dead Sea and in brines of all types. The reddish pink spots that can occur on pickled fish are colonies of halophilic archaea. Few other organisms can live in the saltiest of

FIGURE 19.17 Extreme Halophiles Highly saline environments such as these commercial seawater evaporating ponds in San Francisco Bay are home to extreme halophiles. The archaea are easily visible here because of the rich red coloration from their carotenoid pigments.

the homes that the extreme halophiles occupy. In these environments, most organisms would "dry" to death, losing too much water to the hypertonic environment. Extreme halophiles have been found in lakes with pH values as high as 11.5—the most alkaline environment inhabited by living organisms, and almost as alkaline as household ammonia.

Some of the extreme halophiles have a unique system for trapping light energy and using it to form ATP—without using any form of chlorophyll—when oxygen is in short supply. They use the pigment retinal (also found in the vertebrate eye) combined with a protein to form a light-absorbing molecule called microbial rhodopsin.

Another member of the Euryarchaeota, *Thermoplasma*, has no cell wall. It is thermophilic and acidophilic, its metabolism is aerobic, and it lives in coal deposits. Its genome of 1,100,000 base pairs is among the smallest (along with the mycoplasmas) of any free-living organism, although some parasitic organisms have even smaller genomes.

Several lineages of Archaea are poorly known

Most known archaea are crenarchaeotes or euryarchaeotes, but studies of extreme environments have identified several small lineages that are not closely related to either of these major groups. For example, the korarchaeotes and thaumarchaeotes are known only from DNA isolated directly from hot environments. Neither group has been successfully grown in pure culture. The thaumarchaeotes oxidize ammonia and may play an important role in the nitrogen cycle.

Another distinctive archaeal lineage has been discovered at a deep-sea hydrothermal vent off the coast of Iceland. It is the first representative of the group christened Nanoarchaeota because of their minute size (Greek *nanos*, "dwarf"). This organism is a parasite that lives on cells of *Ignicoccus*, a crenarchaeote. Because of their association, the two species can be grown together in culture (**FIGURE 19.18**).

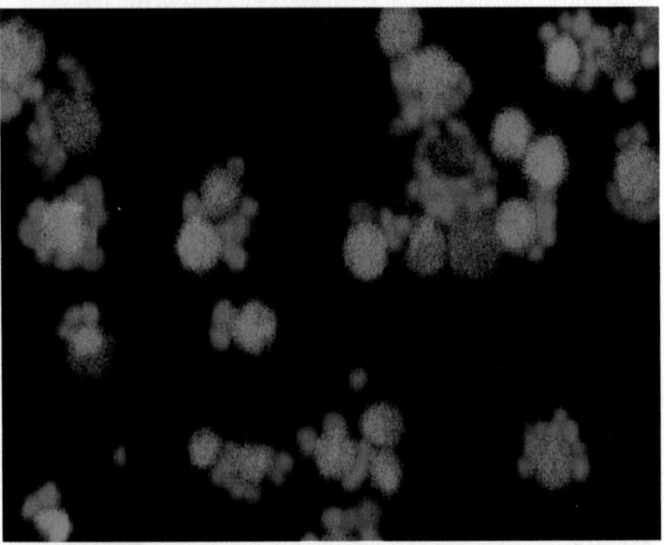

FIGURE 19.18 A Nanoarchaeote Growing in Mixed Culture with a Crenarchaeote *Nanoarchaeum equitans* (red), discovered living near deep-sea hydrothermal vents, is the only representative of the nanoarchaeote group so far identified. This tiny organism lives attached to cells of the crenarchaeote *Ignicoccus* (green). For this confocal laser micrograph, the two species were visually differentiated by fluorescent dye "tags" that are specific to their distinct gene sequences.

CHECKpoint CONCEPT 19.2

- ✓ How does the diversity of environments occupied by prokaryotes compare with the diversity of environments occupied by multicellular organisms you may be familiar with?
- ✓ Explain why Gram staining is of limited use in understanding the evolutionary relationships of bacteria.
- ✓ What makes the membranes of archaea unique?
- ✓ Given that all species of life have evolved for the same amount of time since their common origin, how would you respond to someone who characterizes prokaryotes as "primitive"?

Prokaryotes are found almost everywhere on Earth and live in a wide variety of ecosystems. Next we will examine the contributions of prokaryotes to the functioning of those ecosystems.

CONCEPT 19.3 Ecological Communities Depend on Prokaryotes

Prokaryotic cells and their associations do not usually live in isolation. Rather, they live in communities of many different species, often including microscopic eukaryotes. Whereas some microbial communities are harmful to humans, others provide important services. They help us digest our food, break down municipal waste, and recycle organic matter and chemical elements in the environment.

Many prokaryotes form complex communities

Some microbial communities form layers in sediments, and others form clumps a meter or more in diameter. Many microbial communities tend to form dense **biofilms**. Upon contacting a solid surface, the cells bind to that surface and secrete a sticky, gel-like polysaccharide matrix that traps other cells (**FIGURE 19.19**). Once a biofilm forms, the cells become more difficult to kill.

Biofilms are found in many places, and in some of those places they cause problems for humans. The material on our teeth that we call dental plaque is a biofilm. Pathogenic bacteria are difficult for the immune system—and modern medicine—to combat once they form a biofilm, which may be impermeable to antibiotics. Worse, some drugs stimulate the bacteria in a biofilm to lay down more matrix, making the film even more impermeable. Biofilms may form on just about any available surface, including contact lenses and artificial joint replacements. They foul metal pipes and cause corrosion, a major problem in steam-driven electricity generation plants. Fossil stromatolites—large, rocky structures made up

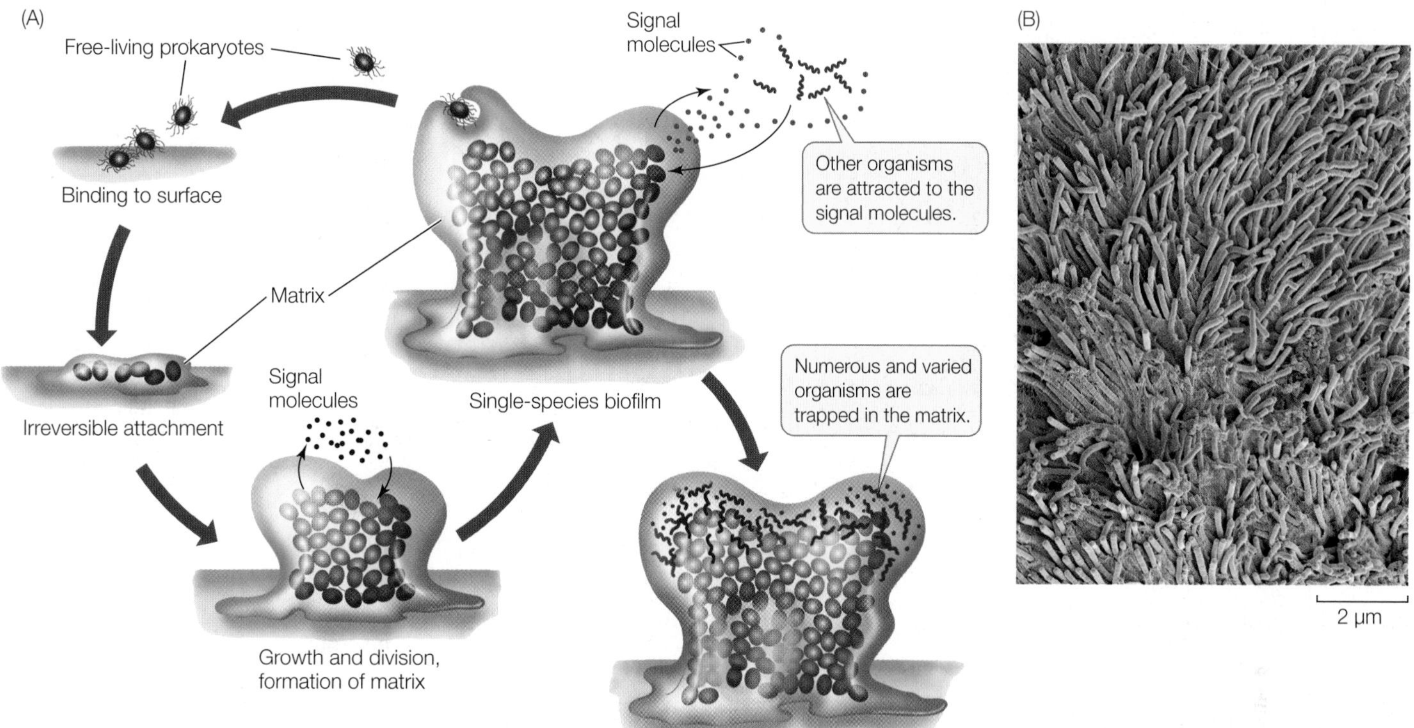

FIGURE 19.19 Forming a Biofilm (A) Free-living prokaryotes readily attach themselves to surfaces and form films that are stabilized and protected by a surrounding matrix. Once the population is large enough, the developing biofilm can send chemical signals that attract other microorganisms. (B) Scanning electron microscopy reveals a biofilm of dental plaque. The bacteria (red) are embedded in a matrix consisting of proteins from both bacterial secretions and saliva.

of alternating layers of fossilized biofilm and calcium carbonate—are among the oldest remnants of life on Earth (see Figure 18.6B).

Some biologists are studying the chemical signals that prokaryotes use to communicate with one another and that trigger density-linked activities such as biofilm formation. We saw one example of this type of communication—called **quorum sensing**—in the chapter-opening discussion of bioluminescent *Vibrio*. In the case of health-threatening bacteria, researchers hope to find ways to block the quorum-sensing signals that lead to the production of the matrix polysaccharides, thus preventing pathological biofilms from forming.

Microbiomes are critical to the health of many eukaryotes

Although only a few bacterial species are pathogens, popular notions of bacteria as "germs" and fear of the consequences of infection cause many people to assume that most bacteria are harmful. Increasingly, however, biologists are discovering that the health of humans (as well as that of other eukaryotes) depends in large part on the health of our **microbiomes**: the communities of bacteria and archaea that inhabit our bodies.

Every surface of your body is covered with diverse communities of bacteria. The Human Microbiome Project has identified more than 10,000 species living in and on humans. Inside your body, your digestive system teems with bacteria (**FIGURE 19.20**). If you count up all the cells in a human body, only about 10 percent of them are human cells. The rest are microbes—mostly bacteria, along with some archaea and microscopic eukaryotes. When these communities are disrupted, they must be restored before the body can function normally.

Biologists are discovering that many complex health problems are linked to the disruption of our microbiomes. These diverse microbial communities affect the expression of our genes and play a critical role in the development and maintenance of a healthy immune system. When our microbiomes contain an appropriate community of beneficial species, our bodies function normally. But these communities are strongly affected by our life experiences, by the food we eat, by the medicines we take, and by our exposure to various environmental toxins. The recent rapid increase in the rate of autoimmune diseases in humans—diseases in which the immune system begins to attack the body—has been linked to the changing diversity and composition of our microbiomes.

The early acquisition of an appropriate microbiome is critical for lifelong health. Normally, a human infant acquires much of its microbiome at birth, from the microbiome in its mother's vagina. Other components of the microbiome are also acquired from the mother, especially through breast feeding. Recent studies have shown that babies born by cesarean section, as well as babies that are bottle-fed on artificial milk formula, typically acquire microbes from a wider variety of sources. Many of the bacteria acquired in this way are not well suited for human health. Biologists have discovered that the incidence of many autoimmune diseases is higher in people who were born

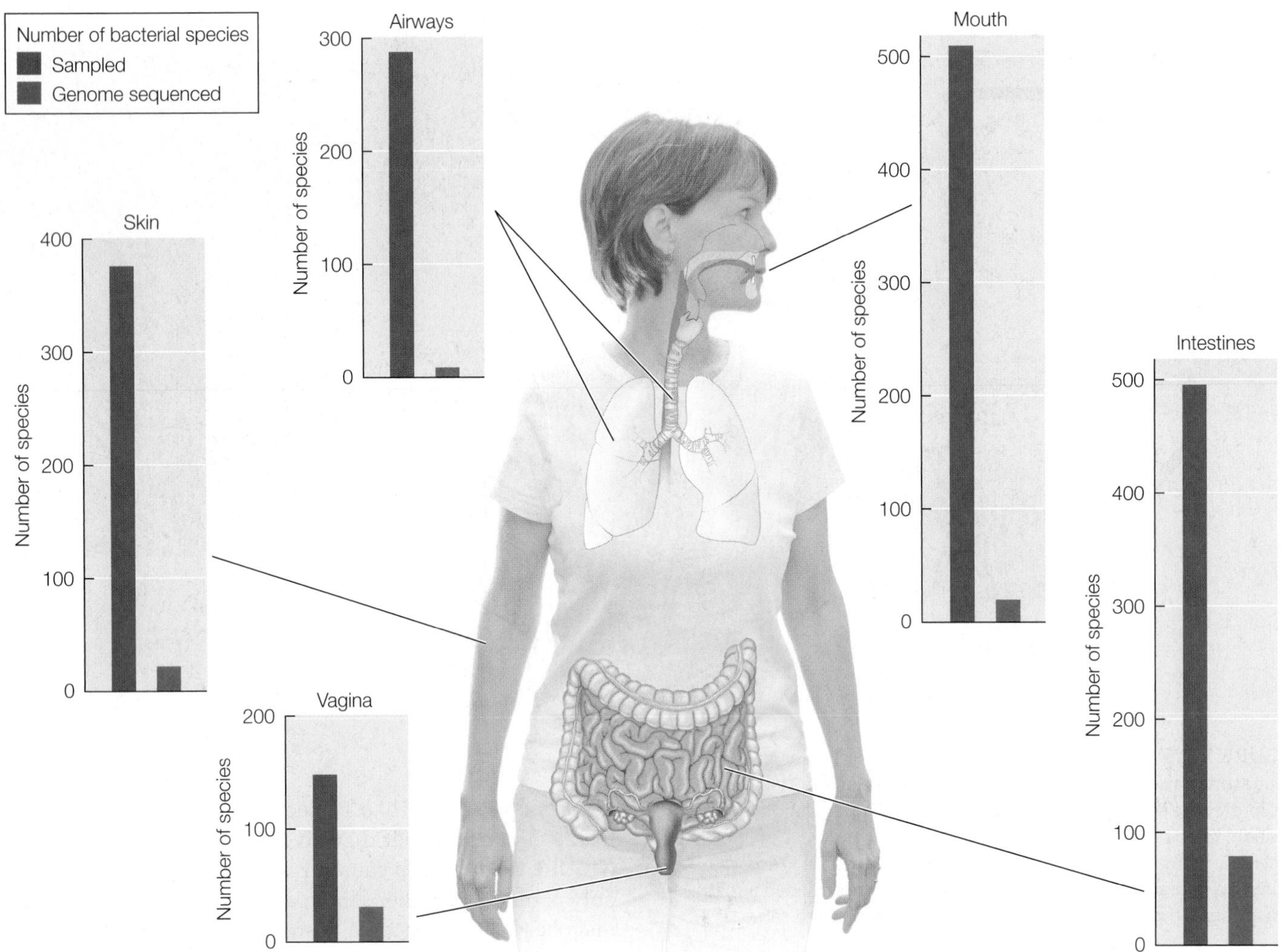

FIGURE 19.20 The Human Body's Microbiome Is Critical to the Maintenance of Health Surveys of the human microbiome have shown that this diverse community includes thousands of microbial species that are adapted to grow in or on various parts of the human body. Although it has become clear that the composition of this microbiome is closely associated with many aspects of human health, most of the component species are poorly characterized and remain largely unstudied by biologists. What has become clear is that, although the "subcommunities" in different parts of the body share similarities, each is a site-specific assemblage of many distinctive species.

by cesarean section and in those who were fed on formula as infants, compared with individuals who were born vaginally and breast-fed as infants. The difference appears to be related to the composition of the individual's original microbiome.

Our microbiomes may be related to many other health concerns as well. For example, physicians have long noted a connection between autism and gastrointestinal disorders in humans. In 2012, microbiologists discovered that children with autism have high levels of a genus of bacteria known as *Sutterella* adherent to their intestinal walls. These bacteria are absent or rare in children without autism. It is not yet known if these bacteria are causing the gastrointestinal problems, or if they are merely symptoms of the gastrointestinal disorders, but there is growing evidence that intestinal microbiomes are distinct in children with autism.

Humans require some of the metabolic products—especially vitamins B_{12} and K—produced by the microbiome living in the large intestine. Communities of bacteria line our intestines with a dense biofilm that is in intimate contact with the mucosal lining of the gut. This biofilm facilitates nutrient transfer from the intestine into the body, functioning like a specialized "tissue" that is essential to our health. This biofilm has a complex ecology that scientists have just begun to explore in detail—including the possibility that the species composition of an individual's gut microbiome may contribute to obesity (or the resistance to it).

Animals also harbor a variety of bacteria and archaea in their digestive tracts, many of which play important roles in digestion. Cattle depend on prokaryotes to break down plant material. Like most animals, cattle cannot produce cellulase, the enzyme needed to start the digestion of the cellulose that makes up the bulk of their plant food. However, bacteria living in a special section of the gut, called the rumen, produce enough cellulase to process the daily diet for the cattle (see Concept 30.2).

Marshall and Warren set out to satisfy Koch's postulates:

Test 1

The microorganism must be present in every case of the disease.

Results: Biopsies from the stomachs of many patients revealed that the bacterium was always present if the stomach was inflamed or ulcerated.

Test 2

The microorganism must be cultured from a sick host.

Results: The bacterium was isolated from biopsy material and eventually grown in culture media in the laboratory.

Test 3

The isolated and cultured bacteria must be able to induce the disease.

Results: Marshall was examined and found to be free of bacteria and inflammation in his stomach. After drinking a pure culture of the bacterium, he developed stomach inflammation (gastritis).

Test 4

The bacteria must be recoverable from newly infected individuals.

Results: Biopsy of Marshall's stomach 2 weeks after he ingested the bacteria revealed the presence of the bacterium, now christened *Helicobacter pylori*, in the inflamed tissue.

Conclusion

Antibiotic treatment eliminated the bacteria and the inflammation in Marshall. The experiment was repeated on healthy volunteers, and many patients with gastric ulcers were cured with antibiotics. Thus Marshall and Warren demonstrated that the stomach inflammation leading to ulcers is caused by *H. pylori* infections in the stomach.

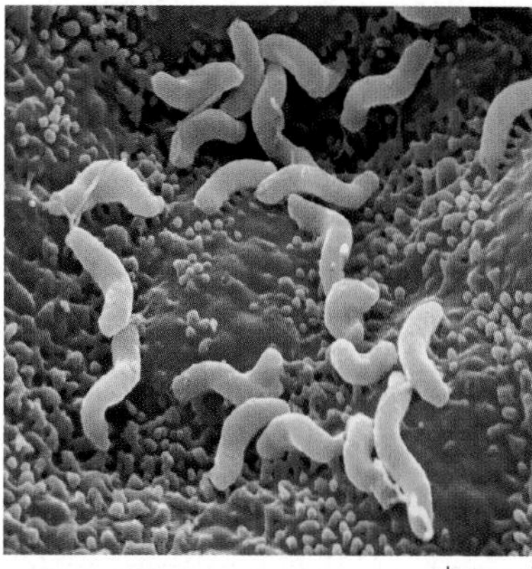

FIGURE 19.21 Satisfying Koch's Postulates Robin Warren and Barry Marshall of the University of Western Australia won the 2005 Nobel Prize in Medicine for showing that ulcers are caused not by the action of stomach acid but by infection with the bacterium *Helicobacter pylori*.

A small minority of bacteria are pathogens

The late nineteenth century was a productive era in the history of medicine—a time when bacteriologists, chemists, and physicians proved that many diseases are caused by microbial agents. During this time, the German physician Robert Koch laid down a set of four rules for establishing that a particular microorganism causes a particular disease:

1. The microorganism is always found in individuals with the disease.
2. The microorganism can be taken from the host and grown in pure culture.
3. A sample of the culture produces the same disease when injected into a new, healthy host.
4. The newly infected host yields a new, pure culture of microorganisms identical to those obtained in the second step.

These rules, called **Koch's postulates**, were important tools in a time when it was not widely understood that microorganisms cause disease. Although modern medical science has more powerful diagnostic tools, Koch's postulates remain useful. For example, physicians were taken aback in the 1990s when stomach ulcers—long accepted and treated as the result of excess stomach acid—were shown by Koch's postulates to be caused by the bacterium *Helicobacter pylori* (**FIGURE 19.21**).

For an organism to be a successful pathogen, it must:

- arrive at the body surface of a potential host;
- enter the host's body;
- evade the host's defenses;
- reproduce inside the host; and
- infect a new host.

Failure to complete any of these steps ends the reproductive career of a pathogenic organism. Yet in spite of the many defenses available to potential hosts (see Chapter 39), some bacteria are

APPLY THE CONCEPT

A small minority of bacteria are pathogens

Imagine you are in charge of maintaining a trout hatchery. Some trout are exhibiting loss of tissue at the tips of their fins, and you suspect a bacterial infection. You isolate and culture two species of bacteria from a trout with affected fins.

1. How would you satisfy the first of Koch's postulates?
2. Imagine that bacterium 1 satisfies the first of Koch's postulates, but bacterium 2 does not. What presence or absence of data from the cultures from many samples of infected fish would lead you to this conclusion?
3. You already know that bacterium 1 satisfies the second of Koch's postulates. You decide to conduct a test of the third postulate. To your surprise, the test animals are all healthy and show no sign of disease. What are some possible explanations? How would you test your hypotheses?

very successful pathogens. Pathogenic bacteria are often surprisingly difficult to combat, even with today's arsenal of antibiotics. One source of this difficulty is their ability to form biofilms.

For the host, the consequences of a bacterial infection depend on several factors. One is the **invasiveness** of the pathogen: its ability to multiply in the host's body. Another is its **toxigenicity**: its ability to produce toxins (chemical substances that are harmful to the host's tissues). *Corynebacterium diphtheriae*, the agent that causes diphtheria, has low invasiveness and multiplies only in the throat, but its toxigenicity is so great that the entire body is affected. In contrast, *Bacillus anthracis*, which causes anthrax, has low toxigenicity but is so invasive that the entire bloodstream ultimately teems with the bacteria.

There are two general types of bacterial toxins: endotoxins and exotoxins. **Endotoxins** are released when certain Gram-negative bacteria grow or lyse (burst). Endotoxins are lipopolysaccharides (complexes consisting of a polysaccharide and a lipid component) that form part of the outer bacterial membrane. Endotoxins are rarely fatal to the host; they normally cause fever, vomiting, and diarrhea. Among the endotoxin producers are some strains of the proteobacteria *Salmonella* and *Escherichia*.

Exotoxins are soluble proteins released by living, multiplying bacteria. They are highly toxic—sometimes fatal—to the host. Human diseases induced by bacterial exotoxins include tetanus (*Clostridium tetani*), cholera (*Vibrio cholerae*), and bubonic plague (*Yersinia pestis*). Anthrax is caused by three exotoxins produced by *Bacillus anthracis*. Botulism is caused by exotoxins produced by *Clostridium botulinum*, which are among the most poisonous ever discovered. The lethal dose for humans of one exotoxin of *C. botulinum* is about one-millionth of a gram. Nonetheless, much smaller doses of this exotoxin, marketed under various trade names (e.g., Botox), are used to treat muscle spasms and for cosmetic purposes (temporary wrinkle reduction in the skin).

Prokaryotes have amazingly diverse metabolic pathways

Bacteria and archaea outdo the eukaryotes in terms of metabolic diversity. Although they are much more diverse in size and shape, eukaryotes draw on fewer metabolic mechanisms for their energy needs. In fact, much of the eukaryotes' energy metabolism is carried out in organelles—mitochondria and chloroplasts—that are endosymbiotic descendants of bacteria. The long evolutionary history of bacteria and archaea, during which they have had time to explore a wide variety of habitats, has led to the extraordinary diversity of their metabolic "lifestyles"—their use or nonuse of oxygen, their energy sources, their sources of carbon atoms, and the materials they release as waste products.

ANAEROBIC VERSUS AEROBIC METABOLISM Some prokaryotes can live only by anaerobic metabolism because oxygen is poisonous to them. These oxygen-sensitive organisms are called **obligate anaerobes**. Other prokaryotes can shift their metabolism between anaerobic and aerobic modes and are thus called **facultative anaerobes**. Many facultative anaerobes alternate between anaerobic metabolism (such as fermentation) and cellular respiration as conditions dictate. **Aerotolerant anaerobes** cannot conduct aerobic cellular respiration, but they are not damaged by oxygen when it is present. By definition, an anaerobe does not use oxygen as an electron acceptor for its respiration.

At the other extreme from the obligate anaerobes, some prokaryotes are **obligate aerobes**, unable to survive for extended periods in the *absence* of oxygen. They require oxygen for cellular respiration.

TABLE 19.2 How Organisms Obtain Their Energy and Carbon

Nutritional category	Energy source	Carbon source
Photoautotrophs (some bacteria, some eukaryotes)	Light	Carbon dioxide
Photoheterotrophs (some bacteria)	Light	Organic compounds
Chemoautotrophs (some bacteria, many archaea)	Inorganic substances	Carbon dioxide
Chemoheterotrophs (found in all three domains)	Usually organic compounds; sometimes inorganic substances	Organic compounds

NUTRITIONAL CATEGORIES All living organisms face the same nutritional challenges: they must synthesize energy-rich compounds such as ATP to power their life-sustaining metabolic reactions, and they must obtain carbon atoms to build their own organic molecules. Biologists recognize four broad nutritional categories of organisms: photoautotrophs, photoheterotrophs, chemoautotrophs, and chemoheterotrophs. Prokaryotes are represented in all four groups (**TABLE 19.2**).

Photoautotrophs perform photosynthesis. They use light as their energy source and carbon dioxide (CO_2) as their carbon source. The cyanobacteria, like green plants and other photosynthetic eukaryotes, use chlorophyll *a* as their key photosynthetic pigment and produce oxygen gas (O_2) as a by-product of noncyclic electron transport.

There are other photoautotrophs among the bacteria, but these organisms use bacteriochlorophyll as their key photosynthetic pigment, and they do not produce O_2. Instead, some of these photosynthesizers produce particles of pure sulfur, because hydrogen sulfide (H_2S), rather than H_2O, is their electron donor for photophosphorylation. Many proteobacteria fit into this category. Bacteriochlorophyll molecules absorb light of longer wavelengths than the chlorophyll molecules used by other photosynthesizing organisms. As a result, bacteria using this pigment can grow in water under fairly dense layers of algae, using light of wavelengths that are not absorbed by the algae (**FIGURE 19.22**).

Photoheterotrophs use light as their energy source but must obtain their carbon atoms from organic compounds made by other organisms. Their "food" consists of organic compounds

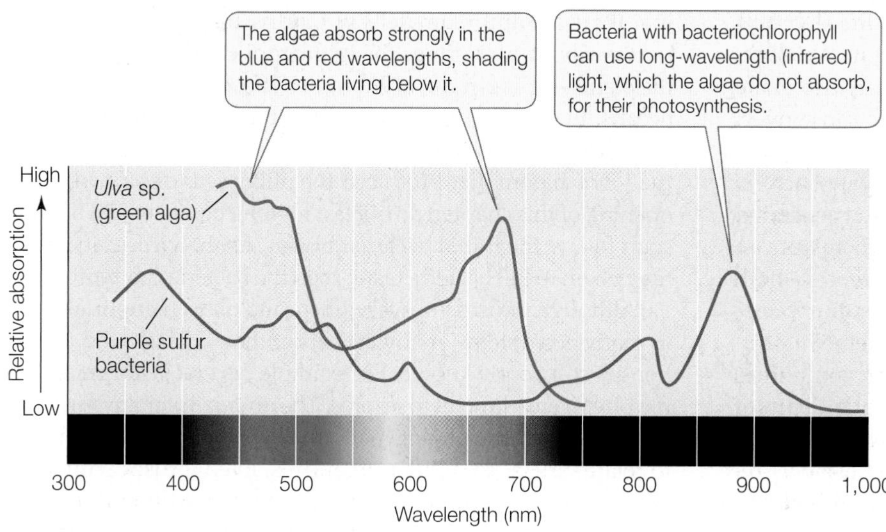
The algae absorb strongly in the blue and red wavelengths, shading the bacteria living below it.
Bacteria with bacteriochlorophyll can use long-wavelength (infrared) light, which the algae do not absorb, for their photosynthesis.
High
Low
Relative absorption
Ulva sp. (green alga)
Purple sulfur bacteria
300
400
500
600
700
800
900
1,000
Wavelength (nm)

FIGURE 19.22 Bacteriochlorophyll Absorbs Long-Wavelength Light The green alga *Ulva* contains chlorophyll, which absorbs no light of wavelengths longer than 750 nm. Purple sulfur bacteria, which contain bacteriochlorophyll, can conduct photosynthesis using longer wavelengths. As a result, these bacteria can grow under layers of algae.

such as carbohydrates, fatty acids, and alcohols. For example, compounds released from plant roots (as in rice paddies) or from decomposing photosynthetic bacteria in hot springs are taken up by photoheterotrophs and metabolized to form building blocks for other compounds. Sunlight provides the energy necessary for metabolism through photophosphorylation.

Chemoautotrophs obtain their energy by oxidizing inorganic substances, and they use some of that energy to fix carbon. Some chemoautotrophs use reactions identical to those of the typical photosynthetic cycle, but others use alternative pathways for carbon fixation. Some bacteria oxidize ammonia or nitrite ions to form nitrate ions. Others oxidize hydrogen gas, hydrogen sulfide, sulfur, and other materials. Many archaea are chemoautotrophs.

Finally, **chemoheterotrophs** obtain both energy and carbon atoms from one or more complex organic compounds that have been synthesized by other organisms. Most known bacteria and archaea are chemoheterotrophs—as are all animals and fungi and many protists.

Although most chemoheterotrophs rely on the breakdown of organic compounds for energy, some chemoheterotrophic prokaryotes obtain their energy by breaking down inorganic substances. Organisms that obtain energy from oxidizing inorganic substances (both chemoautotrophs as well as some chemoheterotrophs) are also known as lithotrophs (Greek, "rock consumers").

Prokaryotes play important roles in element cycling

The metabolic diversity of the prokaryotes makes them key players in the cycles that keep elements moving through ecosystems. Many prokaryotes are decomposers: organisms that metabolize organic compounds from dead organic material and return the products to the environment as inorganic substances. Prokaryotes, along with fungi, return tremendous quantities of carbon to the atmosphere as carbon dioxide, thus carrying out a key step in the carbon cycle.

The key metabolic reactions of many prokaryotes involve nitrogen or sulfur. For example, some bacteria carry out respiratory electron transport without using oxygen as an electron acceptor. These organisms use oxidized inorganic ions such as nitrate, nitrite, or sulfate as electron acceptors. Examples include the **denitrifiers**, which release nitrogen to the atmosphere as nitrogen gas (N_2). These normally aerobic bacteria, mostly species of the genera *Bacillus* and *Pseudomonas*, use nitrate (NO_3^-) as an electron acceptor in place of oxygen if they are kept under anaerobic conditions:

$$2\ NO_3^- + 10\ e^- + 12\ H^+ \rightarrow N_2 + 6\ H_2O$$

Denitrifiers play a key role in the cycling of nitrogen through ecosystems. Without denitrifiers, which convert nitrate ions back into nitrogen gas, all forms of nitrogen would leach from the soil and end up in lakes and oceans, making life on land much more difficult.

Nitrogen fixers convert atmospheric nitrogen gas into a chemical form (ammonia) that is usable by the nitrogen fixers themselves as well as by other organisms:

$$N_2 + 6\ H \rightarrow 2\ NH_3$$

All organisms require nitrogen in order to build proteins, nucleic acids, and other important compounds. Nitrogen fixation is thus vital to life as we know it. This all-important biochemical process is carried out by a wide variety of archaea and bacteria (including cyanobacteria) but by no other organisms, so we depend on these prokaryotes for our very existence.

LINK

For descriptions of the role of nitrogen in plant nutrition and of the global nitrogen cycle, see **Concepts 25.2 and 45.3**

Ammonia is oxidized to nitrate in soil and in seawater by chemoautotrophic bacteria called **nitrifiers**. Bacteria of two genera, *Nitrosomonas* and *Nitrosococcus*, convert ammonia (NH_3) to nitrite ions (NO_2^-), and *Nitrobacter* oxidize nitrite to nitrate (NO_3^-), the form of nitrogen most easily used by many plants. What do the nitrifiers get out of these reactions? Their metabolism is powered by the energy released by the oxidation of ammonia or nitrite. For example, by passing the electrons from nitrite through an electron transport system, *Nitrobacter* can make ATP and, using some of this ATP, can also make NADH. With this ATP and NADH, the bacterium can convert CO_2 and H_2O into glucose.

We have already seen the importance of the cyanobacteria in the cycling of oxygen: in ancient times, the oxygen generated by their photosynthesis converted Earth's atmosphere from an anaerobic to an aerobic environment. Other prokaryotes—both bacteria and archaea—contribute to the cycling of sulfur. Deep-sea hydrothermal vent ecosystems depend on chemoautotrophic prokaryotes that are incorporated into large communities of crabs, mollusks, and giant worms, all living at a depth of 2,500 meters—below any hint of sunlight. These bacteria obtain energy by oxidizing hydrogen sulfide and other substances released in the near-boiling water flowing from volcanic vents in the ocean floor.

CHECKpoint CONCEPT 19.3

✓ How do biofilms form, and why are they of special interest to researchers?

✓ What are three ways that modern lifestyles may be disrupting human microbiomes in ways that affect our health?

✓ How is nitrogen metabolism in the prokaryotes vital to other organisms? Given the roles of bacteria in the nitrogen cycle, how would you answer people who consider all bacteria to be "germs" and dangerous?

Before moving on to discuss the diversity of eukaryotic life, it is appropriate to consider another category of life that includes some pathogens: the viruses. Although they are not cellular, viruses are numerically among the most abundant forms of life on Earth. Their effects on other organisms are enormous. Where did viruses come from, and how do they fit into the tree of life? Biologists are still working to answer these questions.

CONCEPT 19.4 Viruses Have Evolved Many Times

Some biologists do not think of viruses as living organisms, primarily because they are not cellular and must depend on cellular organisms for basic life functions such as replication and metabolism. But viruses are derived from the cells of living organisms. They use the same essential forms of genetic information storage and transmission as do cellular organisms. Viruses infect all cellular forms of life—bacteria, archaea, and eukaryotes. They replicate, mutate, evolve, and interact with other organisms, often causing serious diseases in their hosts. Finally, viruses clearly evolve independently of other organisms, so it is almost impossible not to treat them as a part of life.

Viruses are abundant in many environments. In some freshwater and marine ecosystems, they can occur at densities of up to 10 million viruses per milliliter of water. Biologists estimate that there are approximately 10^{31} individual virus particles on Earth—about 1,000 times the number of cellular organisms on the planet. Viruses have an enormous effect on the ecology of the oceans. Every day, about one-half of the bacteria in the oceans are killed by viruses. Huge marine blooms of bacteria, such as the *Vibrio* bloom that produced the milky seas described at the opening of this chapter, do not last for long because viral blooms soon follow the initial bacterial bloom. As the viruses increase, they begin to kill bacteria faster than the bacteria can reproduce.

Although viruses are everywhere and play an important role in many ecosystems, many aspects of their ecology and evolution are still poorly known. For example, several factors make virus phylogeny difficult to resolve. The tiny size of many virus genomes restricts the phylogenetic analyses that can be conducted to relate viruses to cellular organisms. Even viruses with large genomes often contain many genes that are difficult to align with the genomes of cellular organisms. Their rapid mutation rate, which results in rapid evolution of virus genomes, tends to cloud evolutionary relationships over long periods. There are no known fossil viruses (viruses are too small and delicate to fossilize), so the paleontological record offers no clues to virus origins. Finally, viruses are highly diverse (**FIGURE 19.23**). Several lines of evidence support the hypothesis that viruses have evolved repeatedly within each of the major groups of life. The difficulty in resolving deep evolutionary relationships of viruses makes a phylogeny-based classification difficult. Instead, viruses are placed in one of several functionally similar groups on the basis of the structure of their genomes (for example, whether the genomes are composed of RNA or DNA, and are double- or single-stranded). Most of these defined groups are not thought to represent monophyletic taxa, however.

Many RNA viruses probably represent escaped genomic components

Although viruses are now obligate parasites of cellular species, many viruses may once have been cellular components involved in basic cellular functions—that is, they may be "escaped" components of cellular life that now evolve independently of their hosts.

NEGATIVE-SENSE SINGLE-STRANDED RNA VIRUSES A case in point is a class of viruses whose genome is composed of single-stranded **negative-sense RNA**: RNA that is the complement of the mRNA needed for protein translation. Many of these negative-sense single-stranded RNA viruses have only a few genes, including one for an RNA-dependent RNA polymerase that allows them to make mRNA from their negative-sense RNA genome. Modern cellular organisms cannot generate mRNA in this manner (at least in the absence of viral infections), but scientists speculate that single-stranded RNA genomes may have been common in the distant past, before DNA became the primary molecule for genetic information storage.

A self-replicating RNA polymerase gene that began to replicate independently of a cellular genome could conceivably acquire a few additional protein-coding genes through recombination with its host's DNA. If one or more of these genes were to foster the development of a protein coat, the virus

(A)

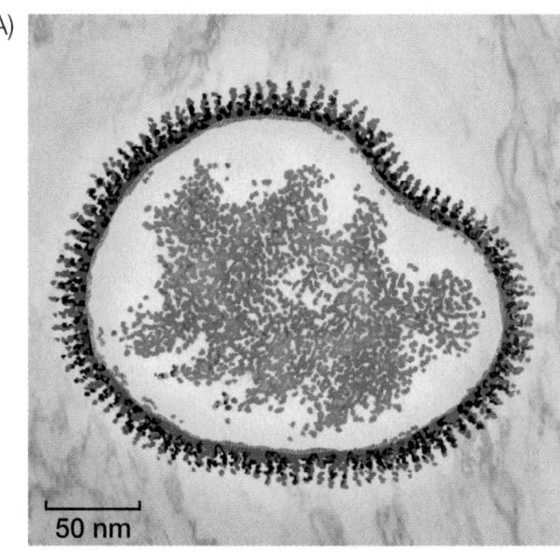

A negative-sense single-stranded RNA virus: The influenza A virus. This virus is responsible for seasonal influenza epidemics in humans. Surface view.

(B)

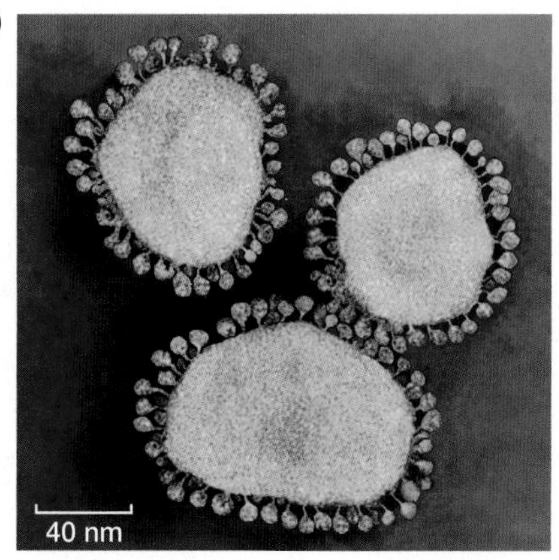

A positive-sense single-stranded RNA virus: Coronavirus of a type thought to be responsible for severe acute respiratory syndrome (SARS). Surface view.

(C)

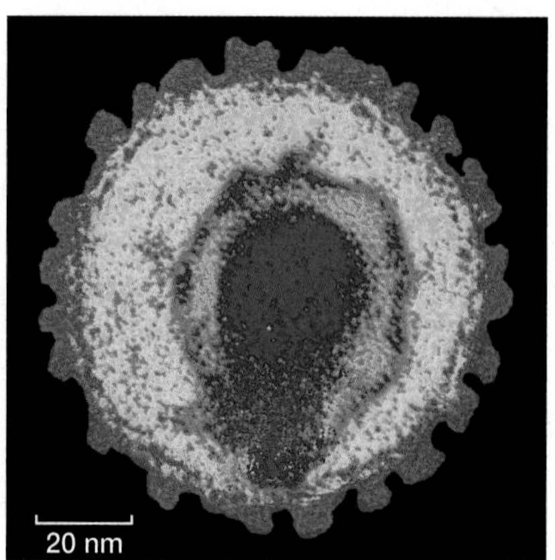

An RNA retrovirus: One of the human immunodeficiency viruses (HIV) that causes AIDS. Cutaway view.

(D)

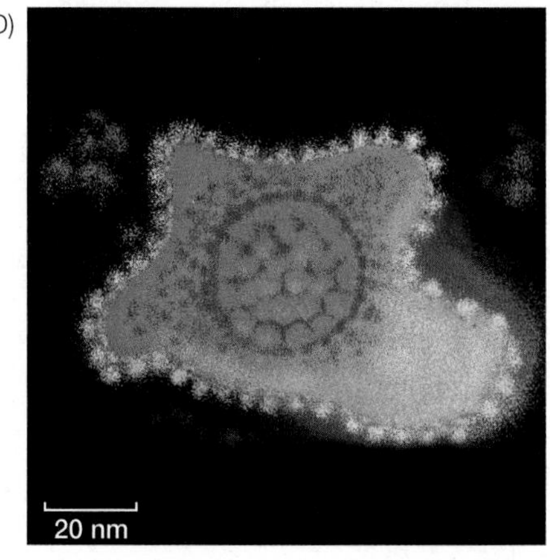

A double-stranded DNA virus: One of the many herpes viruses (Herpesviridae). Different herpes viruses are responsible for many human infections, including chicken pox, shingles, cold sores, and genital herpes (HSV1/2). Surface view.

(E)

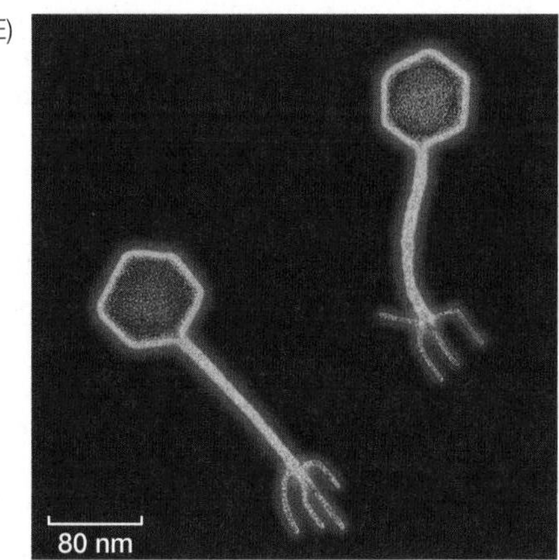

A double-stranded DNA virus: Bacteriophage T4. Viruses that infect bacteria are referred to as bacteriophage (or simply phage). T4 attaches leglike fibers to the outside of its host cell and injects its DNA into the cytoplasm through its "tail" (pink structure in this rendition).

(F)

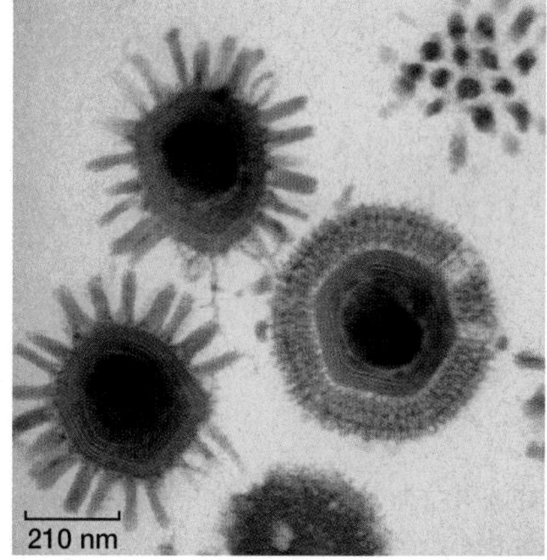

A double-stranded DNA mimivirus: This *Acanthamoeba polyphaga* mimivirus (APMV) has a genome larger than some prokaryote genomes. Cutaway view.

FIGURE 19.23 Viruses Are Diverse Relatively small genomes and rapid evolutionary rates make it difficult to reconstruct phylogenetic relationships among some classes of viruses. Instead, viruses are classified largely by general characteristics of their genomes. The images here are derived from cryoelectron micrographs.

might then survive outside the host and infect new hosts. It is believed that this scenario has been repeated many times independently across the tree of life, given that many of the negative-sense single-stranded RNA viruses that infect organisms from bacteria to humans are not closely related to one another. In other words, negative-sense single-stranded RNA viruses do not represent a distinct taxonomic group, but rather exemplify a particular process of cellular escape that probably happened many different times.

Familiar examples of negative-sense single-stranded RNA viruses include the viruses that cause measles, mumps, rabies, and influenza (see Figure 19.23A).

POSITIVE-SENSE SINGLE-STRANDED RNA VIRUSES The genome of another type of single-stranded RNA virus is composed of positive-sense RNA. Positive-sense genomes are already set for translation; no replication of the genome to form a complement strand is needed before protein translation can take place. Positive-sense single-stranded RNA viruses (see Figure 19.23B) are the most abundant and diverse class of viruses. Most of the viruses that cause diseases in crop plants are members of this group. These viruses kill patches of cells in the leaves or stems of plants, leaving live cells amid a patchwork of discolored dead tissue (giving them the name of mosaic or mottle viruses; **FIGURE 19.24**). Other viruses in this group infect bacteria, fungi, and animals. Human diseases caused by positive-sense single-stranded RNA viruses include polio, hepatitis C, and the common cold. As is true of the other functionally defined groups of viruses, these viruses appear to have evolved multiple times across the tree of life from different groups of cellular ancestors.

APPLY THE CONCEPT

Viruses have evolved many times

When Gram staining revealed the first mimivirus (discovered living inside an amoeba), it was mistakenly identified as a parasitic Gram-positive bacterium. It was soon discovered that this species was not cellular, however, and it was reclassified as a virus. The graph below shows the genome size and number of protein-coding genes for various species of bacteria. Use the graph to answer the following questions.

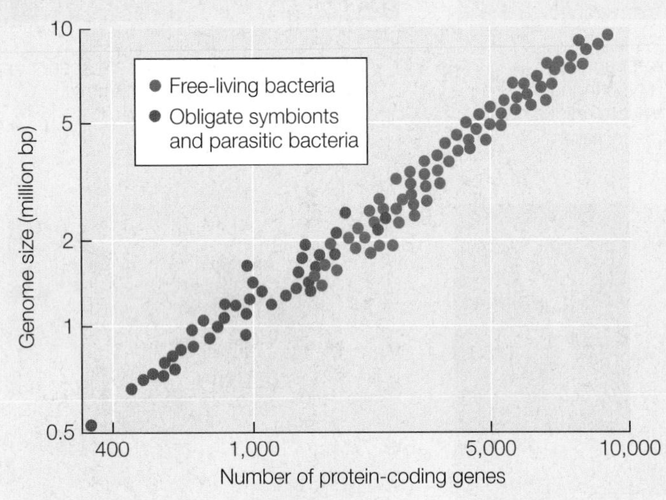

1. What is a likely explanation for the generally smaller genome size and smaller number of protein-coding genes in parasitic bacteria than in most free-living species? Can you form a hypothesis about which species of parasitic bacteria are likely to have existed for the longest time as parasites?
2. The genome of *Acanthamoeba polyphaga mimivirus* (the first discovered mimivirus) is 1,181,404 base pairs, encompassing 911 protein-coding genes. Plot the relevant data for this mimivirus on the graph. How does your result fit with the hypothesis that the mimivirus evolved from a parasitic bacterium?
3. Earlier in this chapter we described a minute, parasitic group of archaea, the Nanoarchaeota (see Figure 19.18). The genome of one species, *Nanoarchaeum equitans*, is 490,885 base pairs and contains 536 protein-coding genes. Plot the relevant data for this species on the graph. How does the genome of this parasitic archaea species differ from the genomes of the parasitic bacteria? From the genome of the mimivirus?

RNA RETROVIRUSES The RNA retroviruses are best known as the group that includes the human immunodeficiency viruses (HIV; see Figure 19.23C). Like the previous two categories of viruses, RNA retroviruses have genomes composed of single-stranded RNA and probably evolved as escaped cellular components.

Retroviruses are so named because they regenerate themselves by reverse transcription. When the retrovirus enters the nucleus of its vertebrate host, viral reverse transcriptase produces complementary DNA (cDNA) from the viral RNA genome and then replicates the single-stranded cDNA to produce double-stranded DNA. Another virally encoded enzyme called integrase catalyzes the integration of the new piece of double-stranded viral DNA into the host's genome. The viral genome is then replicated along with the host cell's DNA. The integrated retroviral DNA is known as a **provirus**.

Retroviruses are only known to infect vertebrates, although genomic elements that resemble portions of these viruses are a component of the genomes of a wide variety of organisms, including bacteria, plants, and many animals. Several retroviruses are associated with the development of various forms of cancer, as cells infected with these viruses are more likely to undergo uncontrolled replication.

As retroviruses become incorporated into the genomes of their hosts, many become nonfunctional copies that are no longer expressed as functional viruses. These sequences may provide a record of ancient viral infections that plagued our ancestors. Humans, for example, carry about 100,000 fragments of **endogenous retroviruses** in our genome. These fragments make up about 8 percent of our DNA—a considerably larger fraction of our genome than the fraction that comprises all our protein-coding genes (about 1.2 percent of our genome).

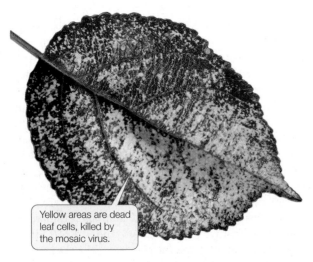

Yellow areas are dead leaf cells, killed by the mosaic virus.

FIGURE 19.24 Mosaic Viruses Are a Problem for Agriculture Mosaic, or "mottle," viruses are the most diverse class of viruses. This leaf is from an apple tree infected with mosaic virus.

DOUBLE-STRANDED RNA VIRUSES Double-stranded RNA viruses may have evolved repeatedly from single-stranded RNA ancestors—or perhaps vice versa. These viruses, which are not closely related to one another, infect organisms from throughout the tree of life. Many plant diseases are caused by double-stranded RNA viruses. Other viruses of this type cause many cases of infant diarrhea in humans.

Some DNA viruses may have evolved from reduced cellular organisms

Another class of viruses is composed of viruses that have a double-stranded DNA genome (see Figure 19.23D–F). This group is also almost certainly polyphyletic (with many independent origins). Many of the common phage that infect bacteria are double-stranded DNA viruses, as are the viruses that cause smallpox and herpes in humans.

Some biologists think that at least some of the DNA viruses may represent highly reduced parasitic organisms that have lost their cellular structure as well as their ability to survive as free-living species. For example, the mimiviruses (see Figure 19.23F) have a genome in excess of a million base pairs of DNA that encode more than 900 proteins. This genome is similar in size to the genomes of many parasitic bacteria and about twice as large as the genome of the smallest bacteria. The recently discovered pandoraviruses have even larger genomes (up to about 2.5 million base pairs)—larger than the smallest genomes of eukaryotes. Phylogenetic analyses of DNA viruses suggest that they have evolved repeatedly from cellular organisms, possibly including major branches of life that are now extinct. Furthermore, recombination among different viruses may have allowed the exchange of various genetic modules, further complicating the history and origins of these viruses.

Viruses can be used to fight bacterial infections

Although some viruses cause devastating diseases, other viruses have been used to fight disease. Most bacterial diseases are treated today with antibiotics. But antibiotics were first discovered in the 1930s, and they were not widely used to treat bacterial diseases until the 1940s. So antibiotics were not yet available during World War I, when bacterial infections plagued the battlefields. Battlefield wounds were often infected by bacteria, and in the absence of antibiotics, these infections often led to the loss of limbs and lives. While trying to find a way to combat this problem, a physician named Felix d'Herelle discovered the first evidence of viruses that attack bacteria. He named these viruses bacteriophage, or "eaters of bacteria." Herelle extracted bacteriophage from the stool of infected patients. He then used these extracts to treat patients with deadly bacterial infections, including dysentery, cholera, and bubonic plague. This practice became known as **phage therapy**. After the war, phage therapy was widely used among the general public to treat bacterial infections of the skin and intestines.

Go to MEDIA CLIP 19.3
Bacteriophages Attack *E. coli*
PoL2e.com/mc19.3

Phage therapy was mostly replaced by the use of antibiotics in the 1930s and 1940s as physicians grew concerned about treating patients with live viruses. Phage therapy continued to be used in the Soviet Union but largely disappeared from Western medical practice. Today, however, many antibiotics are losing their effectiveness as bacterial pathogens evolve resistance to these drugs. Phage therapy is once again an active area of research, and it is likely that bacteriophage will become increasingly important as weapons against bacterial diseases. One advantage that bacteriophage may have over antibiotics is that, like bacteria, bacteriophage can evolve. As bacteria evolve resistance to a strain of bacteriophage, biologists can select for new strains of bacteriophage that retain their effectiveness against the pathogens. In this way, biologists are using their understanding of evolution to combat the problem of antibiotic-resistant bacteria.

CHECKpoint CONCEPT 19.4

- ✓ Why is it difficult to place viruses precisely within the tree of life?
- ✓ What are the two main hypotheses of virus origins?
- ✓ How can viruses be used to treat some human diseases?

It appears that the enormous diversity of viruses is, at least in part, a result of their multiple origins from many different cellular organisms. It may be best to view viruses as spinoffs from the various branches on the tree of life—sometimes evolving independently of cellular genomes, sometimes recombining with them. One way to think of viruses is as the "bark" on the tree of life: certainly an important component all across the tree, but not quite like the main branches.

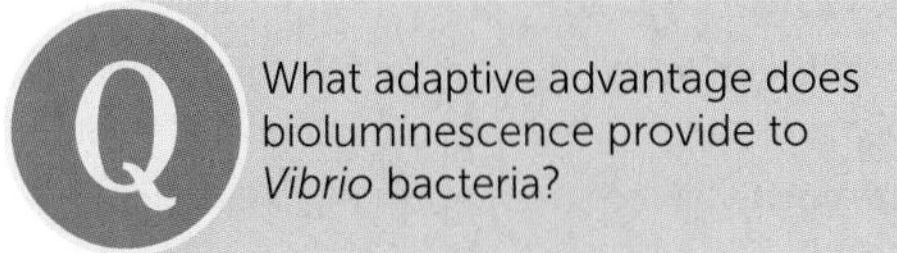

What adaptive advantage does bioluminescence provide to *Vibrio* bacteria?

ANSWER Although marine *Vibrio* can live independently, they thrive inside the guts of fish and other marine animals. Inside a fish, *Vibrio* cells attach themselves to food particles, including phytoplankton, and are eventually expelled into the ocean as waste. How can they get back into another fish gut? The bioluminescent glow produced by a dense colony of free-living *Vibrio* growing on phytoplankton attracts fish, which consume the phytoplankton and thus ingest the bacteria—which gets the bacteria into a new host fish.

SUMMARY

CONCEPT 19.1 Life Consists of Three Domains That Share a Common Ancestor

- Two of life's three **domains**, Bacteria and Archaea, are **prokaryotic**. They are distinguished from Eukarya in several ways, including their lack of a nucleus and of membrane-enclosed organelles. **Review Table 19.1**
- Eukaryotes are related to both Archaea and Bacteria and appear to have formed through endosymbiosis between members of these two lineages. The last common ancestor of all three domains probably lived about 3 billion years ago. **Review Figure 19.1 and ANIMATED TUTORIAL 19.1**
- Early attempts to classify prokaryotes were hampered by the organisms' small size and the difficulties of growing them in pure culture.
- The cell walls of almost all bacteria contain **peptidoglycan**, whereas the cell walls of archaea and eukaryotes lack peptidoglycan.
- Bacteria can be differentiated into two groups by the **Gram stain**. **Gram-negative bacteria** have a periplasmic space between the cell membrane and a distinct outer membrane. **Gram-positive bacteria** have a thick cell wall containing about five times as much peptidoglycan as a Gram-negative wall. **Review Figure 19.2 and ACTIVITY 19.1**
- The three most common bacterial morphologies are **spirilla** (spirals), **bacilli** (rods), and **cocci** (spheres). **Review Figure 19.3**
- The cells of some bacteria aggregate, forming multicellular colonies.
- Phylogenetic classification of prokaryotes is now based principally on the nucleotide sequences of rRNA and other genes involved in fundamental cellular processes.
- Prokaryotes reproduce asexually by binary fission, but many can exchange genetic material. Reproduction and genetic exchange are not directly linked in prokaryotes.
- Although **lateral gene transfer** has occurred throughout prokaryotic evolutionary history, elucidation of prokaryote phylogeny is still possible. **Review Figure 19.4**

CONCEPT 19.2 Prokaryote Diversity Reflects the Ancient Origins of Life

- Prokaryotes are the most numerous organisms on Earth, but only a small fraction of prokaryote diversity has been characterized to date.
- Some **high-GC Gram-positives** produce important antibiotics.
- The **low-GC Gram-positives** include the **mycoplasmas**, which are among the smallest cellular organisms ever discovered.
- The photosynthetic **cyanobacteria** release oxygen into the atmosphere. Cyanobacteria may live free as single cells or associate in multicellular colonies. **Review Figure 19.9**
- **Spirochetes** have unique structures called axial filaments that allow them to move in a corkscrew-like manner. **Review Figure 19.10**
- The **proteobacteria** embrace the largest number of known species of bacteria. Smaller groups include the **hyperthermophilic bacteria**, **hadobacteria**, and **chlamydias**.
- The best-studied groups of Archaea are **Crenarchaeota** and **Euryarchaeota**.
- Many archaea are **extremophiles**. Most known crenarchaeotes are thermophilic, acidophilic, or both. Some euryarchaeotes are **methanogens**; **extreme halophiles** are also found among the euryarchaeotes. **Review Figure 19.14**
- **Ether linkages** in the branched long-chain hydrocarbons of the lipids that make up their cell membranes are a synapomorphy of archaea. **Review Figure 19.15**

CONCEPT 19.3 Ecological Communities Depend on Prokaryotes

- Prokaryotes form complex communities, of which **biofilms** are one example. **Review Figure 19.19**
- Communities of bacteria can communicate information about their density using chemical signals in a process known as **quorum sensing**.
- **Microbiomes** are the communities of prokaryotes that live in and on the bodies of multicellular organisms. These communities are often important to the health of the hosts, and changes to the microbiome may lead to serious health consequences. **Review Figure 19.20**
- Prokaryotes inhabit the guts of many animals (including humans) and help them digest food.
- **Koch's postulates** establish the criteria by which an organism may be classified as a pathogen. Relatively few bacteria—and no archaea—are known to be pathogens. **Review Figure 19.21**
- Prokaryote metabolism is very diverse. Some prokaryotes are anaerobic, others are aerobic, and still others can shift between these modes.
- Prokaryotes fall into four broad nutritional categories: **photoautotrophs**, **photoheterotrophs**, **chemoautotrophs**, and **chemoheterotrophs**. **Review Table 19.2**
- Prokaryotes play key roles in the cycling of elements such as nitrogen, oxygen, sulfur, and carbon. One such role is as decomposers of dead organisms.
- Some prokaryotes metabolize sulfur or nitrogen. **Nitrogen fixers** convert nitrogen gas into a form that organisms can metabolize. **Nitrifiers** convert that nitrogen into forms that can be used by plants, and **denitrifiers** return nitrogen gas to the atmosphere.

CONCEPT 19.4 Viruses Have Evolved Many Times

- Viruses have evolved many times from many different groups of cellular organisms. They do not represent a single taxonomic group.
- Some viruses are probably derived from escaped genetic elements of cellular species; others are thought to have evolved as highly reduced parasites.
- Viruses are categorized by the nature of their genomes.
- Viruses can be used to fight bacterial infections in a process known as **phage therapy**.

Go to the Interactive Summary to review key figures, Animated Tutorials, and Activities
PoL2e.com/is19

Go to LaunchPad at **macmillanhighered.com/launchpad** for additional resources, including LearningCurve Quizzes, Flashcards, and many other study and review resources..

20 The Origin and Diversification of Eukaryotes

KEY CONCEPTS

20.1 Eukaryotes Acquired Features from Both Archaea and Bacteria

20.2 Major Lineages of Eukaryotes Diversified in the Precambrian

20.3 Protists Reproduce Sexually and Asexually

20.4 Protists Are Critical Components of Many Ecosystems

Blooms of dinoflagellates can cause toxic red tides, such as this one in Puget Sound in the U.S. state of Washington.

In summer 2005, a devastating red tide crippled the shellfish industry along the Atlantic coast of North America from Canada to Massachusetts. This red tide was produced by a bloom of dinoflagellates of the genus *Alexandrium*. These protists produce a powerful toxin that accumulates in clams, mussels, and oysters. A person who eats a mollusk contaminated with the toxin can experience a syndrome known as paralytic shellfish poisoning. Many people were sickened by eating mollusks that were harvested before the problem was diagnosed, and losses to the shellfish industry in 2005 were estimated at $50 million.

Several species of dinoflagellates produce toxic red tides in many parts of the world. Along the Gulf of Mexico, red tides caused by dinoflagellates of the genus *Karenia* produce a neurotoxin that affects the central nervous systems of fish, which become paralyzed and cannot respire effectively. Huge numbers of dead fish wash up on Gulf Coast beaches during a *Karenia* red tide. In addition, wave action can produce aerosols of the *Karenia* toxin, and these aerosols often cause asthma-like symptoms in humans on shore.

After the losses that resulted from the 2005 red tide, biologists at the Woods Hole Oceanographic Institution (WHOI) on Cape Cod began to monitor and model dinoflagellate populations off the New England coast. If biologists could accurately forecast future blooms, people in the area could be made aware of the problem in advance and adjust the shellfish harvest (and their eating habits) accordingly.

Biologists from WHOI monitored counts of dinoflagellates in the water and in seafloor sediments. They also monitored river runoff, water currents, water temperature and salinity, winds, and tides. An additional environmental factor was the "nor'easter" storms common along the New England coast. By correlating their measurements of these environmental factors with dinoflagellate counts, biologists produced a model that predicted growth of dinoflagellate populations.

In spring 2008, the WHOI team determined that all the factors were in place to produce another red tide like the one of 2005—if a nor'easter occurred to blow the dinoflagellates toward the coast. A nor'easter did occur at just the wrong time, and another red tide materialized in summer 2008, just as predicted. But this time, people were warned. Shellfish harvesters adjusted their harvest, and many fewer people were harmed by eating toxic mollusks.

Red tides are harmful, but can dinoflagellates also be beneficial to marine ecosystems?

You will find the answer to this question on page 418.

CONCEPT 20.1 Eukaryotes Acquired Features from Both Archaea and Bacteria

We easily recognize trees, mushrooms, and insects as plants, fungi, and animals, respectively. But there is a dazzling assortment of other eukaryotic organisms—mostly microscopic—that do not fit into these three groups. Eukaryotes that are not plants, animals, or fungi have traditionally been called **protists**. But phylogenetic analyses reveal that many of the groups we commonly refer to as protists are not, in fact, closely related. Thus the term "protist" does not describe a formal taxonomic group, but is a convenience term for "all the eukaryotes that are not plants, animals, or fungi."

The unique characteristics of the eukaryotic cell lead scientists to conclude that the eukaryotes are monophyletic, and that a single eukaryotic ancestor diversified into the many different protist lineages as well as giving rise to the plants, fungi, and animals. As we saw in Concept 19.1, eukaryotes are generally thought to be more closely related to archaea than to bacteria. The mitochondria and chloroplasts of eukaryotes, however, are clearly derived from bacterial lineages (see Figure 19.1).

Traditionally, biologists have hypothesized that the *split* of Eukarya from Archaea was followed by the endosymbioses with bacterial lineages that led to the origin of mitochondria and chloroplasts. Some biologists prefer to view the origin of eukaryotes as the *fusion* of lineages from the two prokaryote groups. This difference is largely a semantic one that hinges on the subjective point at which we deem the eukaryote lineage to have become definitively "eukaryotic." In either case, we can make some reasonable inferences about the events that led to the evolution of a new cell type, bearing in mind that the environment underwent an enormous change—from low to high availability of free atmospheric oxygen—during the course of these events.

LINK

You can compare eukaryotic and prokaryotic cells by reviewing **Concepts 4.2 and 4.3**

The modern eukaryotic cell arose in several steps

Several events were important in the origin of the modern eukaryotic cell (**FIGURE 20.1**):

- The origin of a flexible cell surface

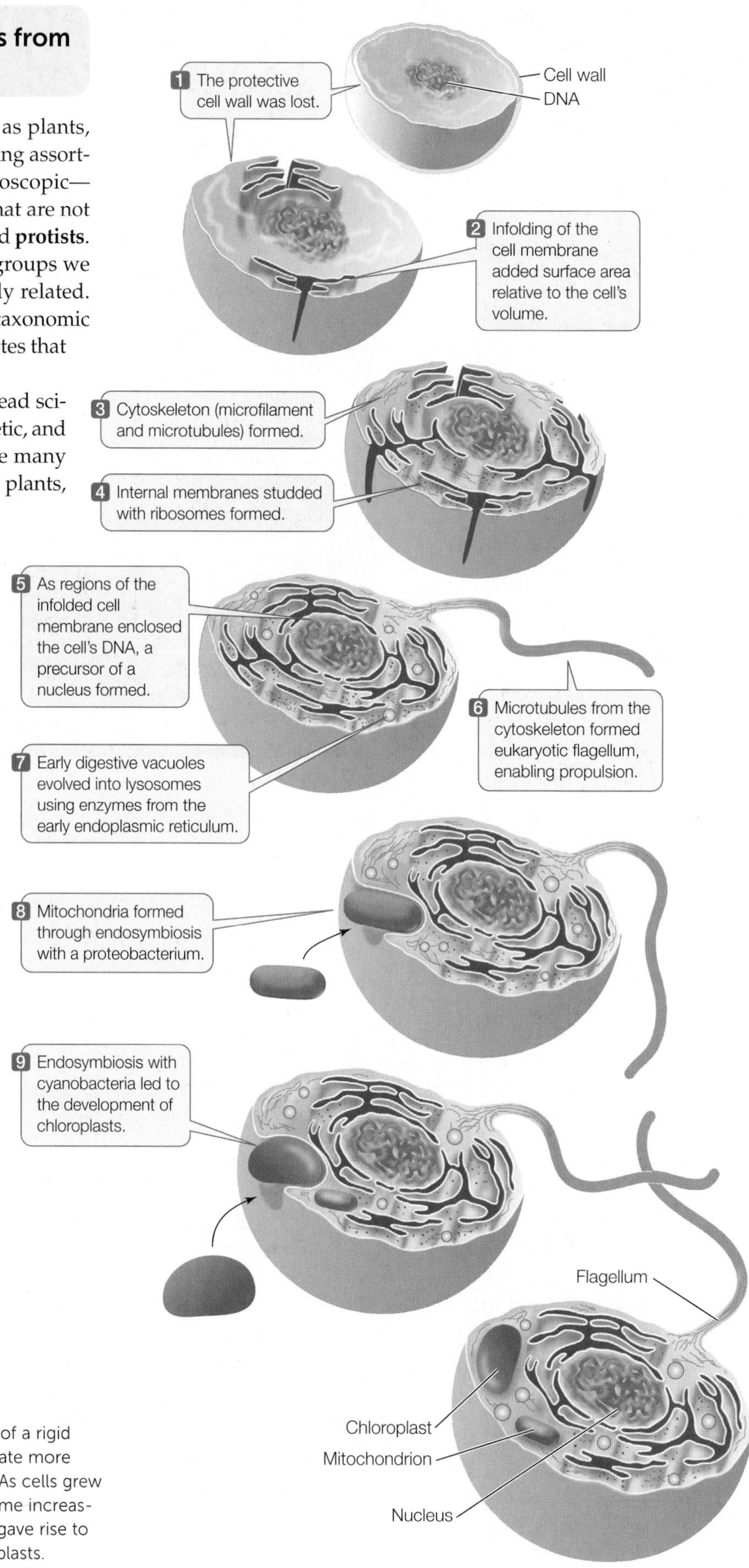

1 The protective cell wall was lost.
Cell wall
DNA
2 Infolding of the cell membrane added surface area relative to the cell's volume.
3 Cytoskeleton (microfilament and microtubules) formed.
4 Internal membranes studded with ribosomes formed.
5 As regions of the infolded cell membrane enclosed the cell's DNA, a precursor of a nucleus formed.
6 Microtubules from the cytoskeleton formed eukaryotic flagellum, enabling propulsion.
7 Early digestive vacuoles evolved into lysosomes using enzymes from the early endoplasmic reticulum.
8 Mitochondria formed through endosymbiosis with a proteobacterium.
9 Endosymbiosis with cyanobacteria led to the development of chloroplasts.
Flagellum
Chloroplast
Mitochondrion
Nucleus

FIGURE 20.1 Evolution of the Eukaryotic Cell The loss of a rigid cell wall allowed the cell membrane to fold inward and create more surface area, which facilitated the evolution of larger cells. As cells grew larger, cytoskeletal complexity increased, and the cell became increasing compartmentalized. Endosymbioses involving bacteria gave rise to mitochondria and (in photosynthetic eukaryotes) to chloroplasts.

- The origin of a cytoskeleton
- The origin of a nuclear envelope, which enclosed a genome organized into chromosomes
- The appearance of digestive vacuoles
- The acquisition of certain organelles via endosymbiosis

FLEXIBLE CELL SURFACE We presume that ancient prokaryotic organisms, like most present-day prokaryotic cells, had firm cell walls. The first step toward the eukaryotic condition was the loss of the cell wall by an ancestral prokaryotic cell. This wall-less condition occurs in some present-day prokaryotes.

Consider the possibilities open to a flexible cell without a firm cell wall, starting with cell size. As a cell grows larger, its surface area-to-volume ratio decreases (see Figure 4.2). Unless the surface area can be increased, the cell volume will reach an upper limit. If the cell's surface is flexible, however, it can fold inward and become more elaborate, creating more surface area for gas and nutrient exchange. With a surface flexible enough to allow infolding, the cell can exchange materials with its environment rapidly enough to sustain a larger volume and more rapid metabolism (see Figure 20.1, steps 1–2). Furthermore, a flexible surface can pinch off bits of the environment, bringing them into the cell by endocytosis. These infoldings of the cell surface, which also exist in some modern prokaryotes, were important for the evolution of large eukaryotic cells.

CHANGES IN CELL STRUCTURE AND FUNCTION Other early steps that were important for the evolution of the eukaryotic cell involved increased compartmentalization and complexity of the cell (see Figure 20.1, steps 3–7):

- The development of a more complex cytoskeleton
- The formation of ribosome-studded internal membranes, some of which surrounded the DNA
- The enclosure of the cell's DNA in a nucleus
- The formation of a flagellum from microtubules of the cytoskelton
- The evolution of digestive vacuoles

Until a few years ago, biologists thought that cytoskeletons were restricted to eukaryotes. Improved imaging technology and molecular analyses have now revealed homologs of many cytoskeletal proteins in prokaryotes, so simple cytoskeletons evolved before the origin of eukaryotes. The cytoskeleton of a eukaryote, however, is much more developed and complex than that of a prokaryote. This greater development of microfilaments and microtubules supports the eukaryotic cell and allows it to manage changes in shape, to distribute daughter chromosomes, and to move materials from one part of its larger cell to other parts. In addition, the presence of microtubules in the cytoskeleton allowed some cells to develop the characteristic eukaryotic flagellum.

The DNA of a prokaryotic cell is attached to a site on its cell membrane. If that region of the cell membrane were to fold into the cell, the first step would be taken toward the evolution of a nucleus, a primary feature of the eukaryotic cell. The nuclear envelope appeared early in the eukaryote lineage. The

APPLY THE CONCEPT

Eukaryotes acquired features from both archaea and bacteria

Ribosomal RNA (rRNA) genes are present in the nuclear genome of eukaryotes. There are also rRNA genes in the genomes of mitochondria and chloroplasts. Therefore, photosynthetic eukaryotes have three different sets of rRNA genes, which encode the structural RNA of separate ribosomes in the nucleus, mitochondria, and chloroplasts, respectively. Translation of each genome takes place on its own set of ribosomes.

The gene tree shows the evolutionary relationships of rRNA gene sequences isolated from the nuclear genomes of humans, yeast, and corn; from an archaeon (*Halobacterium*), a proteobacterium (*E. coli*), and a cyanobacterium (*Chlorobium*); and from the mitochondrial and chloroplast genomes of corn. Use the gene tree to answer the following questions.

1. Why aren't the three rRNA genes of corn one another's closest relatives?
2. How would you explain the closer relationship of the mitochondrial rRNA gene of corn to the rRNA gene of *E. coli* than to the nuclear rRNA genes of other eukaryotes? Can you explain the relationship of the rRNA gene from the chloroplast of corn to the rRNA gene of the cyanobacterium?

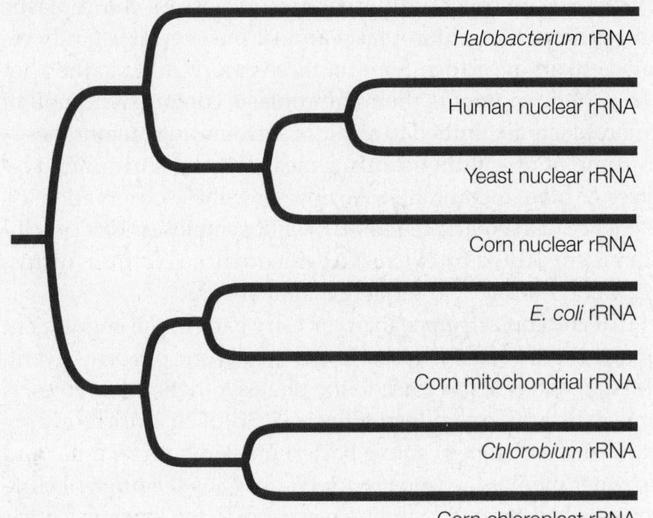

3. If you were to sequence the rRNA genes from human and yeast mitochondrial genomes, where would you expect these two sequences to fit on the gene tree?

next step was probably phagocytosis—the ability to engulf and digest other cells.

ENDOSYMBIOSIS At the same time the processes outlined above were taking place, cyanobacteria were generating O_2 as a product of photosynthesis. The increasing concentrations of O_2 in the oceans, and eventually in the atmosphere, had disastrous consequences for most organisms of the time, which were unable to tolerate the newly oxidizing environment. But some prokaryotes evolved strategies to use the increasing oxygen, and—fortunately for us—so did some of the new phagocytic eukaryotes.

At about this time, endosymbioses began to play a role in eukaryote evolution (see Figure 20.1, steps 8–9). The theory of endosymbiosis proposes that certain organelles are the descendants of prokaryotes engulfed, but not digested, by ancient eukaryotic cells. One crucial event in the history of eukaryotes was the incorporation of a proteobacterium that evolved into the mitochondrion. Initially the new organelle's primary function was probably to detoxify O_2 by reducing it to water. Later this reduction became coupled with the formation of ATP in cellular respiration. Upon completion of this step, the essential modern eukaryotic cell was complete.

LINK

You may wish to review the reactions of cellular respiration in **Concept 6.2**

Photosynthetic eukaryotes are the result of yet another endosymbiotic step: the incorporation of a prokaryote related to today's cyanobacteria. This endosymbiont evolved into the modern chloroplast.

Chloroplasts have been transferred among eukaryotes several times

Eukaryotes in several different groups possess chloroplasts, and groups with chloroplasts appear in several distantly related eukaryote clades. Some of these groups differ in the photosynthetic pigments their chloroplasts contain. And not all chloroplasts are limited to a pair of surrounding membranes—in some photosynthetic eukaryotes, they are surrounded by three or more membranes. We now view these observations as evidence of a remarkable series of endosymbioses. This conclusion is supported by extensive evidence from electron microscopy and nucleic acid sequence comparisons.

All chloroplasts trace their ancestry back to the engulfment of one cyanobacterium by a larger eukaryotic cell. This event, the step that first gave rise to the photosynthetic eukaryotes, is known as **primary endosymbiosis** (**FIGURE 20.2A**). The cyanobacterium, a Gram-negative bacterium, had both an inner and an outer membrane (see Figure 19.2B). Thus the original chloroplasts had two surrounding membranes: the inner and outer membranes of the cyanobacterium. Remnants of the peptidoglycan-containing cell wall of the bacterium are present in the form of a bit of peptidoglycan between the chloroplast membranes of glaucophytes, the first eukaryote group to branch off following primary endosymbiosis (as we will see in Chapter 21). Primary

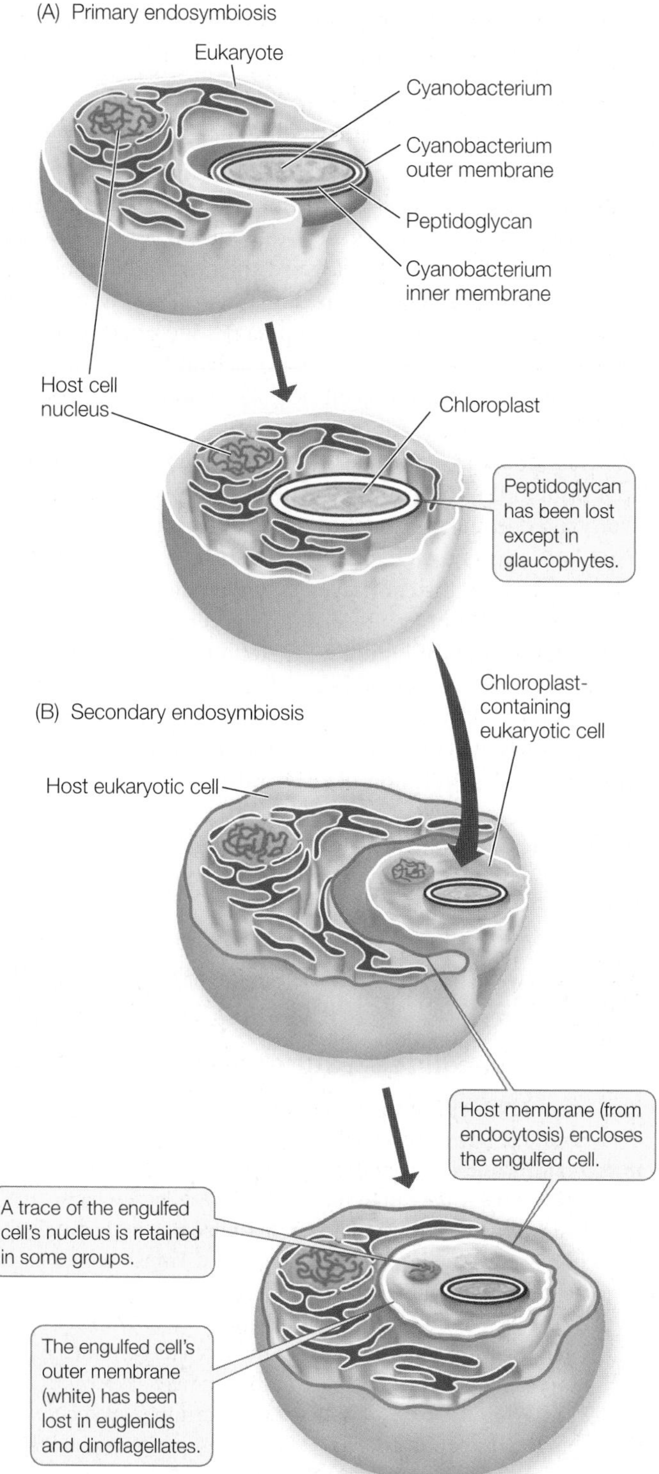
(A) Primary endosymbiosis
Eukaryote
Cyanobacterium
Cyanobacterium outer membrane
Peptidoglycan
Cyanobacterium inner membrane
Host cell nucleus
Chloroplast
Peptidoglycan has been lost except in glaucophytes.
(B) Secondary endosymbiosis
Chloroplast-containing eukaryotic cell
Host eukaryotic cell
Host membrane (from endocytosis) encloses the engulfed cell.
A trace of the engulfed cell's nucleus is retained in some groups.
The engulfed cell's outer membrane (white) has been lost in euglenids and dinoflagellates.

FIGURE 20.2 Endosymbiotic Events in the Evolution of Chloroplasts (A) A single instance of primary endosymbiosis ultimately gave rise to all of today's chloroplasts. (B) Secondary endosymbiosis—the uptake and retention of a chloroplast-containing cell by another eukaryotic cell—took place several times, independently.

Go to ANIMATED TUTORIAL 20.1
Family Tree of Chloroplasts
PoL2e.com/at20.1

endosymbiosis also gave rise to the chloroplasts of the red algae, green algae, and land plants. The red algal chloroplast retains certain pigments of the original cyanobacterial endosymbiont that are absent in green algal chloroplasts.

Almost all remaining photosynthetic eukaryotes are the result of additional rounds of endosymbiosis. For example, the photosynthetic euglenids derived their chloroplasts from **secondary endosymbiosis** (**FIGURE 20.2B**). Their ancestor took up a unicellular green alga, retaining its chloroplast and eventually losing the rest of the constituents of the alga. This history explains why the photosynthetic euglenids have the same photosynthetic pigments as the green algae and land plants. It also accounts for the third membrane of the euglenid chloroplast, which is derived from the euglenid's cell membrane (as a result of endocytosis). An additional round—**tertiary endosymbiosis**—occurred when a dinoflagellate apparently lost its chloroplast and took up another protist that had acquired its chloroplast through secondary endosymbiosis.

CHECKpoint CONCEPT 20.1

- ✓ Why was the development of a flexible cell surface a key event for eukaryote evolution?
- ✓ What do you consider the most critical events that led to the evolution of the eukaryotic cell? Why?
- ✓ Explain how increased availability of atmospheric oxygen (O_2) could have affected the evolution of the eukaryotic cell.

The features that eukaryotes gained from archaea and bacteria have allowed them to exploit many different environments. This led to the evolution of great diversity among eukaryotes, beginning with a radiation that started in the Precambrian.

CONCEPT 20.2 Major Lineages of Eukaryotes Diversified in the Precambrian

Most eukaryotes can be classified in one of eight major clades that began to diversify about 1.5 billion years ago: alveolates, excavates, stramenopiles, plants, rhizarians, amoebozoans, fungi, and animals (**FIGURE 20.3**). Plants, fungi, and animals each have close protist relatives (such as the choanoflagellate relatives of animals), which we will discuss along with those major multicellular eukaryote groups in Chapters 21–23.

Each of the five major groups of protist eukaryotes covered in this chapter consists of organisms with enormously diverse body forms and nutritional lifestyles. Some protists are motile, whereas others do not move. Some protists are photosynthetic, whereas others are heterotrophic. Most protists are unicellular, but some are multicellular. Most protists are microscopic, but a few are huge (giant kelps, for example, can grow to half the length of a football field). We refer to the unicellular species of protists as **microbial eukaryotes**, but keep in mind that there are large, multicellular protists as well.

Multicellularity has arisen dozens of times across the evolutionary history of eukaryotes. Four of the origins of multicellularity resulted in large organisms that are familiar to most

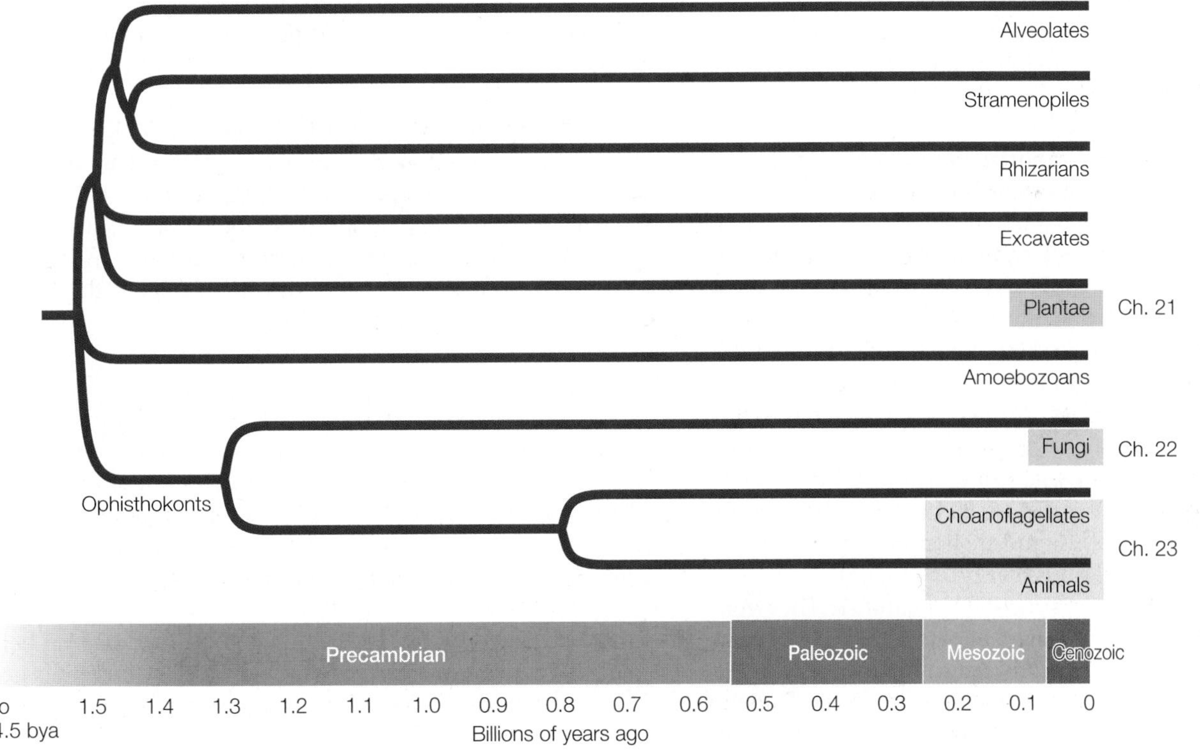

FIGURE 20.3 Precambrian Divergence of Major Eukaryote Groups A phylogenetic tree shows one current hypothesis and estimated time line for the origin of the major groups of eukaryotes. The rapid divergence of major lineages between 1.5 and 1.4 billion years ago makes reconstruction of their precise relationships difficult. The major multicellular groups (tinted boxes) will be covered in subsequent chapters.

people: plants, animals, fungi, and brown algae (the last are a group of stramenopiles). In addition, there are dozens of smaller and less familiar groups among the eukaryotes that include multicellular species. Recent experimental studies have shown that artificial selection for multicellularity can produce repeated, convergent evolution of multicellular forms over just a few months in some normally unicellular eukaryotic species. In addition, many unicellular species retain individual identities but nonetheless associate in large multicellular colonies. There is a near-continuum between fully integrated, multicellular organisms on the one hand and loosely integrated multicellular colonies of cells on the other. Biologists do not always agree on where to draw the line between the two.

Biologists used to classify protists largely on the basis of their life histories and reproductive features (see Concept 20.3). In recent years, however, electron microscopy and gene sequencing have revealed many new patterns of evolutionary relatedness among these groups. Analyses of slowly evolving gene sequences are making it possible to explore evolutionary relationships among eukaryotes in ever greater detail and with greater confidence. Nonetheless, some substantial areas of uncertainty remain, and lateral gene transfer may complicate efforts to reconstruct the evolutionary history of protists (as was also true for prokaryotes; see Concept 19.1). Today we recognize great diversity among the many distantly related protist clades.

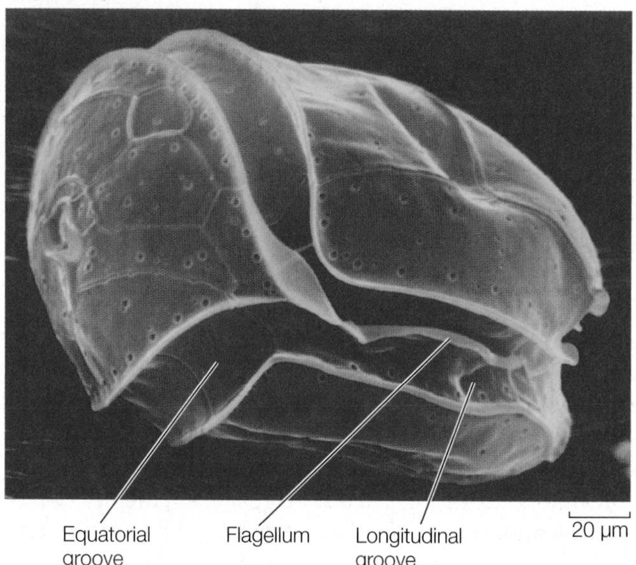

FIGURE 20.4 A Dinoflagellate The presence of two flagella is characteristic of many dinoflagellates, although only one of the two flagella is visible in this photograph. One flagellum originates within the equatorial groove and provides forward thrust and spin to the organism. The second flagellum originates in the longitudinal groove and acts like the rudder of a boat.

Go to MEDIA CLIP 20.1
A Dinoflagellate Shows Off Its Flagellum
PoL2e.com/mc20.1

Alveolates have sacs under their cell membranes

Alveolates are so named because they possess sacs, called alveoli, just beneath their cell membranes, which may play a role in supporting the cell surface. All alveolates are unicellular, and most are photosynthetic, but they are diverse in body form. The groups considered in detail here are the dinoflagellates, apicomplexans, and ciliates.

DINOFLAGELLATES Most **dinoflagellates** are marine and photosynthetic; they are important primary producers of organic matter in the oceans. Although fewer species of dinoflagellates live in fresh water, individuals can be abundant in freshwater environments. The dinoflagellates are of great ecological, evolutionary, and morphological interest. A distinctive mixture of photosynthetic and accessory pigments gives their chloroplasts a golden brown color. Some dinoflagellate species cause red tides, as discussed at the start of this chapter. Other species are photosynthetic endosymbionts that live within the cells of other organisms, including invertebrate animals (such as corals; see Figure 20.19) and other marine protists (see Figure 20.11A). Some are nonphotosynthetic and live as parasites within various marine organisms.

Dinoflagellates have a distinctive appearance. They generally have two flagella, one in an equatorial groove around the cell, the other starting near the same point as the first and passing down a longitudinal groove before extending into the surrounding medium (**FIGURE 20.4**). Some dinoflagellates can take on different forms, including amoeboid ones, depending on environmental conditions. It has been claimed that the dinoflagellate *Pfiesteria piscicida* can occur in at least two dozen distinct forms, although this claim is highly controversial. In any case, this remarkable dinoflagellate is harmful to fish and can, when present in great numbers, both stun and feed on them.

APICOMPLEXANS The exclusively parasitic **apicomplexans** derive their name from the apical complex, a mass of organelles contained in the apical end (the tip) of the cell. These organelles help the apicomplexan invade its host's tissues. For example, the apical complex enables *Plasmodium*, the causative agent of malaria, to enter its target cells in the human body after transmission by a mosquito.

Like many obligate parasites, apicomplexans have elaborate life cycles featuring asexual and sexual reproduction by a series of very dissimilar life stages. In many species, these life stages are associated with two different types of host organisms, as is the case with *Plasmodium*. Another apicomplexan, *Toxoplasma*, alternates between cats and rats to complete its life cycle. A rat infected with *Toxoplasma* loses its fear of cats, which makes it more likely to be eaten by, and thus transfer the parasite to, a cat.

CILIATES The **ciliates** are named for their numerous hairlike cilia, which are shorter than, but otherwise identical to, eukaryotic flagella. The ciliates are much more complex in body form than are most other unicellular eukaryotes (**FIGURE 20.5**). Their definitive characteristic is the possession of two types of nuclei (whose roles we will describe in Concept 20.3 when we discuss protist reproduction). Almost all ciliates

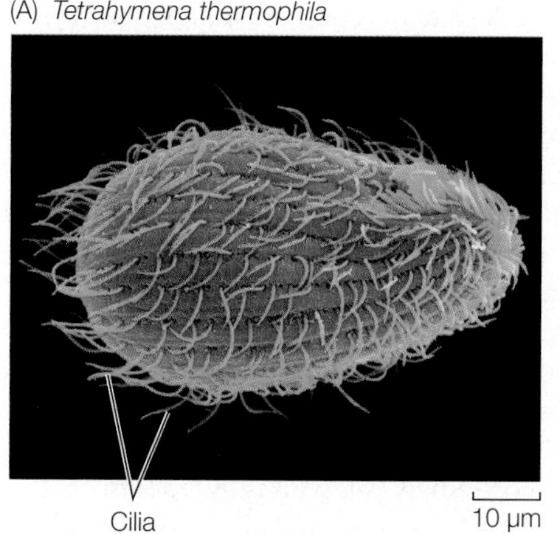

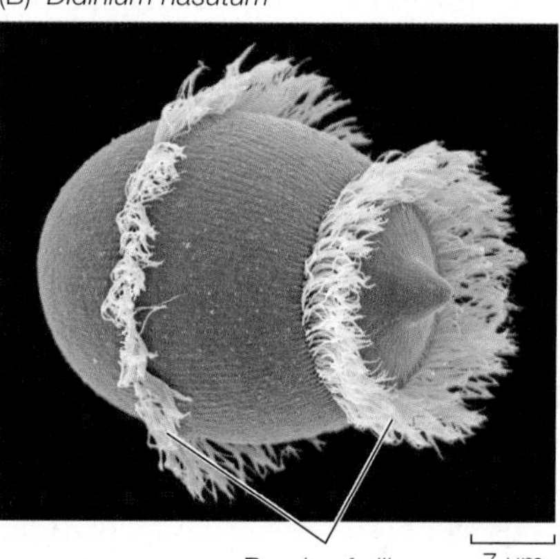

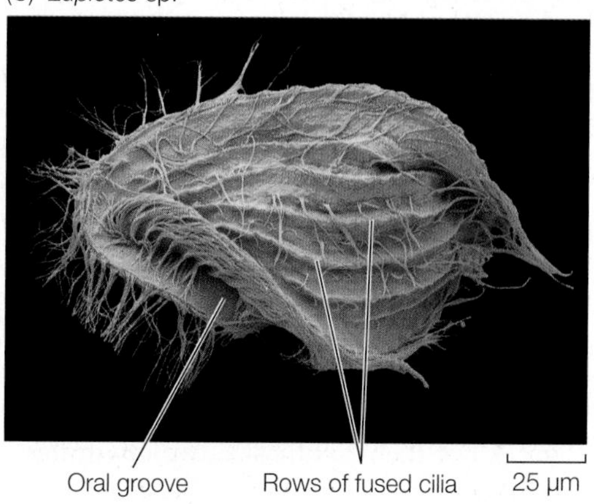

FIGURE 20.5 Diversity among the Ciliates (A) *Tetrahymena thermophila* is used as a model organism for research on gene expression as well as the structure and function of microtubule arrays. (B) The barrel-shaped *Didinium nasutum* feeds on other ciliates, including *Paramecium*. Their cilia occur in two separate bands. (C) Some of the cilia in *Euplotes* fuse into flat sheets that direct food particles into an oral groove.

are heterotrophic, although a few contain photosynthetic endosymbionts.

Paramecium, a frequently studied ciliate genus, exemplifies the complex structure and behavior of ciliates (**FIGURE 20.6**). The slipper-shaped cell is covered by an elaborate pellicle, a structure composed principally of an outer membrane and an inner layer of closely packed, membrane-enclosed sacs (the alveoli) that surround the bases of the cilia. Defensive organelles called trichocysts are also present in the pellicle. In response to a threat, a microscopic explosion expels the trichocysts in a few milliseconds, and they emerge as sharp darts, driven forward at the tip of a long, expanding filament.

The cilia provide *Paramecium* with a form of locomotion that is generally more precise than locomotion by flagella or pseudopods. A *Paramecium* can coordinate the beating of its cilia to propel itself either forward or backward in a spiraling manner. It can also back off swiftly when it encounters a barrier or a negative stimulus. The coordination of ciliary beating is probably the result of a differential distribution of ion channels in the cell membrane near the two ends of the cell.

Organisms living in fresh water are hypertonic to their environment. Many freshwater protists, including *Paramecium*, address this problem by means of specialized **contractile vacuoles** that excrete the excess water the organisms constantly take in by osmosis. The excess water collects in the contractile vacuoles, which then contract and expel the water from the cell.

Paramecium and many other protists engulf solid food by endocytosis, forming a **digestive vacuole** within which the food is digested. Smaller vesicles containing digested food pinch away from the digestive vacuole and enter the cytoplasm. These tiny vesicles provide a large surface area across which the products of digestion can be absorbed by the rest of the cell.

Go to ANIMATED TUTORIAL 20.2 Digestive Vacuoles
PoL2e.com/at20.2

LINK

The role of a pathogen's cell surface recognition molecules in the mammalian immune response is covered in **Chapter 39**

FIGURE 20.6 Anatomy of a *Paramecium* A *Paramecium*, with its many specialized organelles, exemplifies the complex body form of ciliates.

Go to ACTIVITY 20.1 Anatomy of a *Paramecium*
PoL2e.com/ac20.1

Stramenopiles typically have two unequal flagella, one with hairs

A morphological synapomorphy of most **stramenopiles** is the possession of rows of tubular hairs on the longer of their two flagella. Some stramenopiles lack flagella, but they are descended from ancestors that possessed flagella. The stramenopiles include the diatoms and the brown algae, which are photosynthetic, and the oomycetes, which are not.

DIATOMS All of the **diatoms** are unicellular, although some species associate in filaments. Many have sufficient carotenoids in their chloroplasts to give them a yellow or brownish color. All of them synthesize carbohydrates and oils as photosynthetic storage products. Diatoms lack flagella except in male gametes.

Architectural magnificence on a microscopic scale is the hallmark of the diatoms. Almost all diatoms deposit silica (hydrated silicon dioxide) in their cell walls. The cell wall of a diatom is constructed in two pieces, with the top overlapping the bottom like the top of a petri dish. The silica-impregnated walls have intricate patterns unique to each species (**FIGURE 20.7**). Despite their remarkable morphological diversity, all diatoms are symmetrical—either bilaterally (with "right" and "left" halves) or radially (with the type of symmetry possessed by a circle).

Diatoms reproduce both sexually and asexually. Asexual reproduction by binary fission is somewhat constrained by the stiff cell wall. Both the top and bottom of the "petri dish" become tops of new "dishes" without changing appreciably in size. As a result, the new cell made from the former bottom is smaller than the parent cell. If this process continued indefinitely, one cell line would simply vanish, but sexual reproduction largely solves this potential problem. Gametes are formed, shed their cell walls, and fuse. The resulting zygote then grows substantially in size before a new cell wall is laid down.

Diatoms are found in all the oceans and are frequently present in great numbers. They are major photosynthetic producers in coastal waters and are among the dominant organisms in the dense "blooms" of phytoplankton that occasionally appear in the open ocean (see Concept 20.4). Diatoms are also common in fresh water and even occur on the wet surfaces of terrestrial mosses.

...or bilateral (left-right) symmetry.

Diatoms display either radial (circular) symmetry...

FIGURE 20.7 Diatom Diversity This dark-field micrograph illustrates the variety of species-specific forms found among the diatoms.

Go to MEDIA CLIP 20.2
Diatoms in Action
PoL2e.com/mc20.2

BROWN ALGAE The **brown algae** obtain their namesake color from the carotenoid fucoxanthin, which is abundant in their chloroplasts. The combination of this yellow-orange pigment with the green of chlorophylls *a* and *c* yields a brownish tinge. All brown algae are multicellular, and some are extremely large. Giant kelps, such as those of the genus *Macrocystis*, may be up to 60 meters long.

Go to MEDIA CLIP 20.3
A Kelp Forest
PoL2e.com/mc20.3

The brown algae are almost exclusively marine. They are composed either of branched filaments (**FIGURE 20.8A**) or of leaflike growths (**FIGURE 20.8B**). Some float in the open ocean. The most famous example is the genus *Sargassum*, which forms dense mats in the Sargasso Sea in the mid-Atlantic. Most brown algae, however, attach themselves to rocks near the shore. A few thrive only where they are regularly exposed to heavy surf. All of the attached forms develop a specialized structure, called a holdfast, that literally glues them to the rocks. The "glue" of the holdfast is alginic acid, a gummy polymer of sugar acids found in the walls of many brown algal cells. In addition to its function in holdfasts, it cements algal cells and filaments together. It is harvested and used by humans as an emulsifier in ice cream, cosmetics, and other products.

OOMYCETES The **oomycetes** consist in large part of the water molds and their terrestrial relatives, such as the downy mildews. Water molds are filamentous and stationary. They are **absorptive heterotrophs**—that is, they secrete enzymes that digest large food molecules into smaller molecules that they can absorb. They are all aquatic and **saprobic**—meaning they feed on dead organic matter. If you have seen a whitish, cottony mold growing on dead fish or dead insects in water, it was probably a water mold of the common genus *Saprolegnia* (**FIGURE 20.9**).

(A) *Himanthalia elongata*

(B) *Postelsia palmaeformis*

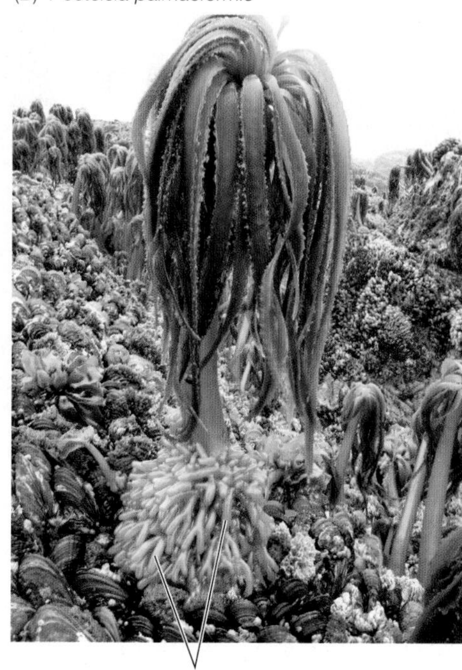

FIGURE 20.8 Brown Algae (A) This seaweed illustrates the filamentous growth form of the brown algae. (B) Sea palms exemplify the leaflike growth form of brown algae. Sea palms and many other brown algal species are "glued" to the rocks by tough, branched structures called holdfasts that can withstand the pounding of the surf.

Some other oomycetes, such as the downy mildews, are terrestrial. Although most of the terrestrial oomycetes are harmless or helpful decomposers of dead matter, a few are plant parasites that attack crops such as avocados, grapes, and potatoes.

Oomycetes were once classified as fungi. However, we now know that their similarity to fungi is only superficial, and that the oomycetes are more distantly related to the fungi than are many other eukaryote groups, including humans (see Figure 20.3). For example, the cell walls of oomycetes are typically made of cellulose, whereas those of fungi are made of chitin.

Saprolegnia sp.

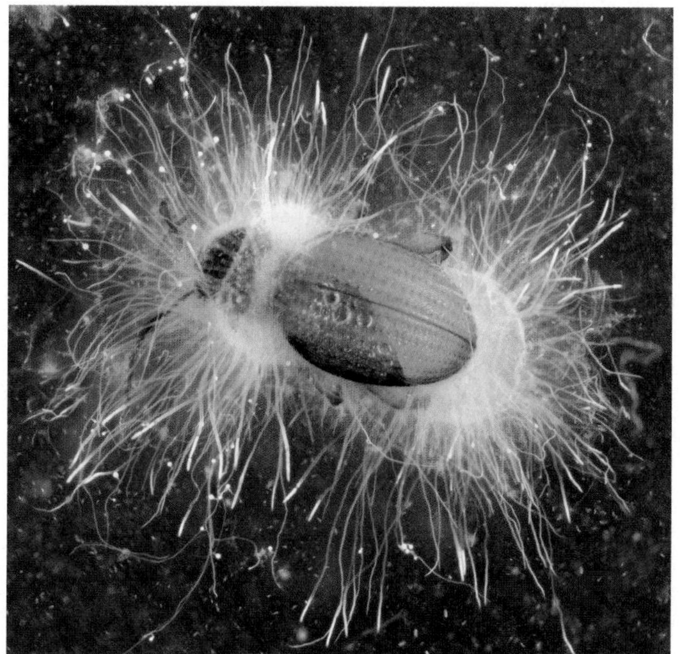

FIGURE 20.9 An Oomycete The filaments of a water mold radiate from the carcass of a beetle.

Rhizarians typically have long, thin pseudopods

The three primary groups of **rhizarians**—cercozoans, foraminiferans, and radiolarians—are unicellular and mostly aquatic. These organisms typically have long, thin pseudopods that contrast with the broader, lobelike pseudopods of the more familiar amoebozoans. The rhizarians have contributed to ocean sediments, some of which have become terrestrial features in the course of geological history.

CERCOZOANS The **cercozoans** are a diverse group with many forms and habitats. Some are aquatic, and others live in soil. One group of cercozoans possesses chloroplasts derived from a green alga by secondary endosymbiosis, and those chloroplasts contain a trace of the alga's nucleus.

FORAMINIFERANS Some **foraminiferans** secrete external shells of calcium carbonate (**FIGURE 20.10**). These shells have accumulated over time to produce much of the world's limestone. Foraminiferans live as plankton (drifing with the curent) or on the seafloor. Living foraminiferans have been found 10,896 meters down in the western Pacific's Challenger Deep—the deepest point in the world's oceans. At that depth, however, they cannot secrete normal shells because the surrounding water is too poor in calcium carbonate.

In living planktonic foraminiferans, long, threadlike, branched pseudopods extend through numerous microscopic openings in the shell and interconnect to create a sticky network, which the foraminiferans use to catch smaller

FIGURE 20.10 **Building Blocks of Limestone** Some foraminiferans secrete calcium carbonate to form shells. The shells of different species have distinctive shapes. Over millions of years, the shells of foraminiferans have accumulated to form limestone deposits.

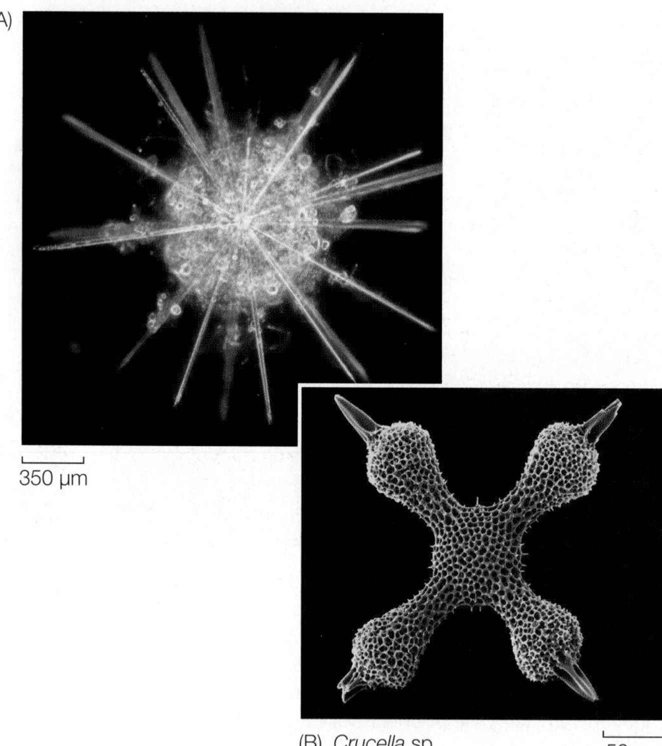

FIGURE 20.11 **Radiolarians Exhibit Distinctive Pseudopods and Radial Symmetry** (A) The radiolarians are distinguished by their thin, stiff pseudopods and by their radial symmetry. The pigmentation seen at the center of this radiolarian's glassy endoskeleton is imparted by endosymbiotic dinoflagellates. (B) The endoskeleton secreted by a radiolarian.

plankton. In some foraminiferan species, the pseudopods provide locomotion.

RADIOLARIANS **Radiolarians** are recognizable by their thin, stiff pseudopods, which are reinforced by microtubules (**FIGURE 20.11A**). These pseudopods greatly increase the surface area of the cell, and they help the cell stay afloat in its marine environment.

Radiolarians also are immediately recognizable by their distinctive radial symmetry. Almost all radiolarian species secrete glassy endoskeletons (internal skeletons). The skeletons of the different species are as varied as snowflakes, and many have elaborate geometric designs (**FIGURE 20.11B**). A few radiolarians are among the largest of the unicellular eukaryotes, measuring several millimeters in diameter.

Excavates began to diversify about 1.5 billion years ago

The **excavates** include a number of diverse groups that began to split from one another soon after the origin of eukaryotes. Several groups of excavates lack mitochondria, an absence that once led to the view that these groups might represent early-diverging eukaryotes that diversified before the evolution of mitochondria. However, the discovery of nuclear genes normally associated with mitochondria in these organisms suggests that the absence of mitochondria is a derived condition. In other words, ancestors of these excavate groups probably possessed mitochondria that were lost or reduced over the course of evolution. The existence of these organisms today shows that eukaryotic life is possible without mitochondria.

DIPLOMONADS AND PARABASALIDS The **diplomonads** and **parabasalids** are unicellular and lack mitochondria. *Giardia lamblia*, a diplomonad, is a familiar parasite that contaminates water supplies and causes the intestinal disease giardiasis. This tiny organism contains two nuclei bounded by nuclear envelopes, and it has a cytoskeleton and multiple flagella (**FIGURE 20.12A**).

In addition to flagella and a cytoskeleton, the parabasalids have undulating membranes that also contribute to the cell's locomotion. *Trichomonas vaginalis* (**FIGURE 20.12B**) is a parabasalid responsible for a sexually transmitted disease (trichomoniasis) in humans. Infection of the male urethra, where it may occur without symptoms, is less common than infection of the vagina.

HETEROLOBOSEANS The constantly changing amoeboid body form appears in several protist groups that are only distantly related to one another. The body forms of **heteroloboseans**, for example, resemble those of loboseans, an amoebozoan group that is not at all closely related to heteroloboseans (see p. 412). Amoebas of the free-living heterolobosean genus *Naegleria* usually have a two-stage life cycle in which one stage has amoeboid cells and the other flagellated cells. Some amoebas of this genus can enter the human body and cause a fatal disease of the nervous system.

EUGLENIDS AND KINETOPLASTIDS The **euglenids** and **kinetoplastids** together constitute a clade of unicellular excavates with flagella. Their mitochondria contain distinctive, disc-shaped cristae, and their flagella contain a crystalline rod not found in other organisms. They reproduce primarily asexually by binary fission.

The flagella of euglenids arise from a pocket at the anterior end of the cell. Spiraling strips of proteins under the

(A) *Giardia muris*

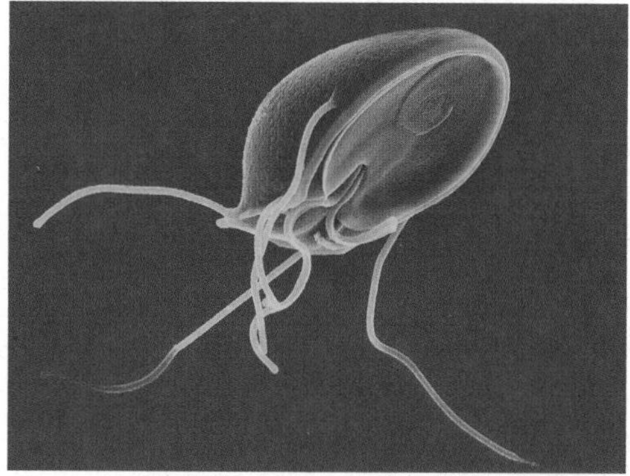

(B) *Trichomonas vaginalis*

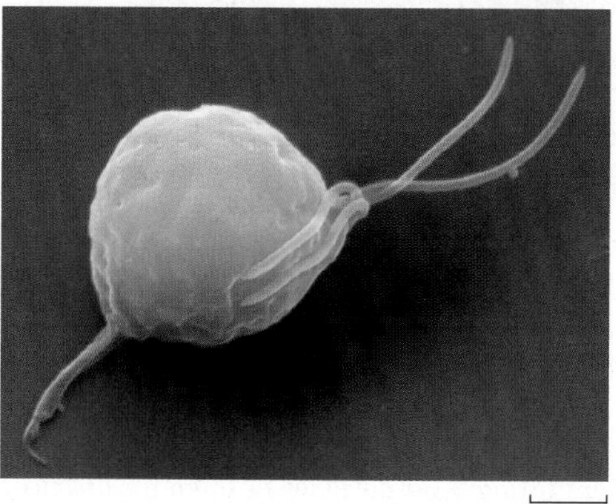

FIGURE 20.12 Some Excavate Groups Lack Mitochondria (A) *Giardia*, a diplomonad, has flagella and two nuclei. (B) *Trichomonas*, a parabasalid, has flagella and undulating membranes. Neither of these organisms possesses mitochondria.

cell membrane control the cell's shape. Some euglenids are photosynthetic.

FIGURE 20.13 depicts a cell of the genus *Euglena*. Like most other euglenids, this common freshwater organism has a complex cell structure. It propels itself through the water with the longer of its two flagella, which may also serve as an anchor to hold the organism in place. The second flagellum is often rudimentary.

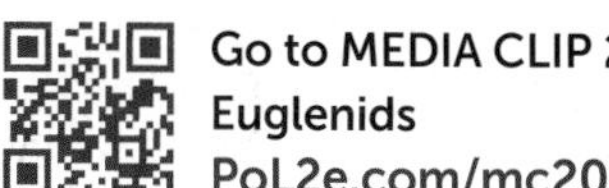

Go to MEDIA CLIP 20.4
Euglenids
PoL2e.com/mc20.4

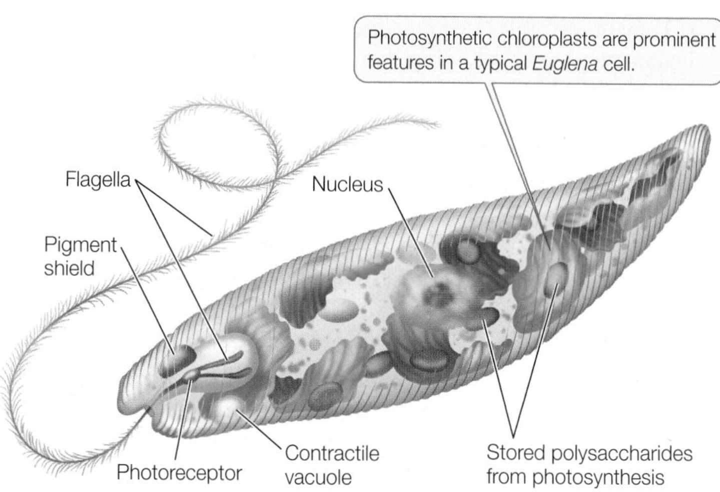

FIGURE 20.13 A Photosynthetic Euglenid In the *Euglena* species illustrated in this drawing, the second flagellum is rudimentary. Note that the primary flagellum originates at the anterior of the organism and trails toward its posterior.

The euglenids have diverse nutritional requirements. Many species are always heterotrophic. Other species are fully autotrophic in sunlight, using chloroplasts to synthesize organic compounds through photosynthesis. When kept in the dark, these euglenids lose their photosynthetic pigment and begin to feed exclusively on dissolved organic material in the water around them. Such a "bleached" *Euglena* resynthesizes its photosynthetic pigment when it is returned to the light and becomes autotrophic again. But *Euglena* cells treated with certain antibiotics or mutagens lose their photosynthetic pigment completely; neither they nor their descendants are ever autotrophs again. However, those descendants function well as heterotrophs.

The kinetoplastids are unicellular parasites with two flagella and a single, large mitochondrion. That mitochondrion contains a kinetoplast, a unique structure housing multiple circular DNA molecules and associated proteins. Some of these DNA molecules encode "guide proteins" that edit mRNA within the mitochondrion.

The kinetoplastids include several medically important species of pathogenic trypanosomes (**TABLE 20.1**). Some of these

TABLE 20.1 A Comparison of Three Kinetoplastid Trypanosomes

	Trypanosoma brucei	*Trypanosoma cruzi*	*Leishmania major*
Human disease	Sleeping sickness	Chagas' disease	Leishmaniasis
Insect vector	Tsetse fly	Assassin bug	Sand fly
Vaccine or effective cure	None	None	None
Strategy for survival	Changes surface recognition molecules frequently	Causes changes in surface recognition molecules on host cell	Reduces effectiveness of macrophage hosts
Site in human body	Bloodstream; attacks nerve tissue in final stages	Enters cells, especially muscle cells	Enters cells, primarily macrophages
Approximate number of deaths per year	50,000	43,000	60,000

organisms are able to change their cell surface recognition molecules frequently, allowing them to evade our best attempts to kill them and thus eradicate the diseases they cause.

Amoebozoans use lobe-shaped pseudopods for locomotion

Amoebozoans appear to have diverged from other eukaryotes about 1.5 billion years ago (see Figure 20.3). It is not yet clear if they are more closely related to opisthokonts (including fungi and animals) or to other major groups of eukaryotes.

The lobe-shaped pseudopods of amoebozoans (**FIGURE 20.14**) are a hallmark of the amoeboid body form. Amoebozoan pseudopods differ in form and function from the slender pseudopods of rhizarians. We consider three amoebozoan groups here: the loboseans and two groups known as slime molds.

LOBOSEANS **Loboseans** are small amoebozoans that feed on other small organisms and particles of organic matter by phagocytosis, engulfing them with pseudopods. Many loboseans are adapted for life on the bottoms of lakes, ponds, and other bodies of water. Their creeping locomotion and their manner of engulfing food particles fit them for life close to a relatively rich supply of sedentary organisms or organic particles. Most loboseans exist as predators, parasites, or scavengers. Members of one group of loboseans, the testate amoebas, live inside shells. Some of these amoebas produce casings by gluing sand grains together (**FIGURE 20.15**). Other testate amoebas have shells secreted by the organism itself.

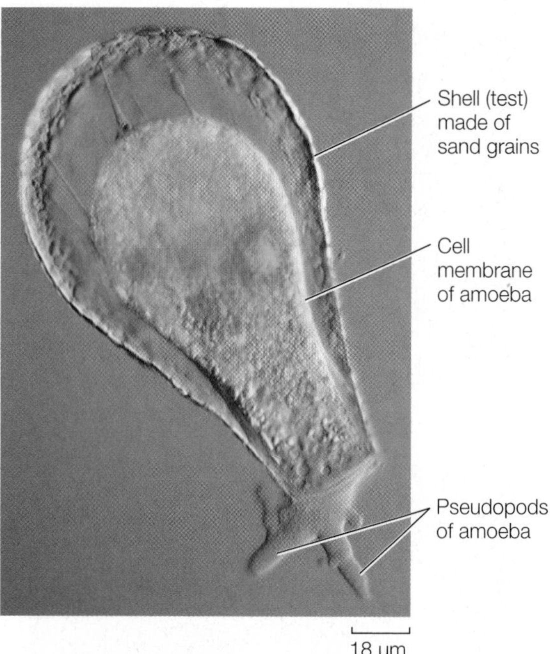

FIGURE 20.15 Life in a Glass House This testate amoeba has built a lightbulb-shaped shell, or test, by gluing sand grains together. Its pseudopods extend through the single aperture in the test.

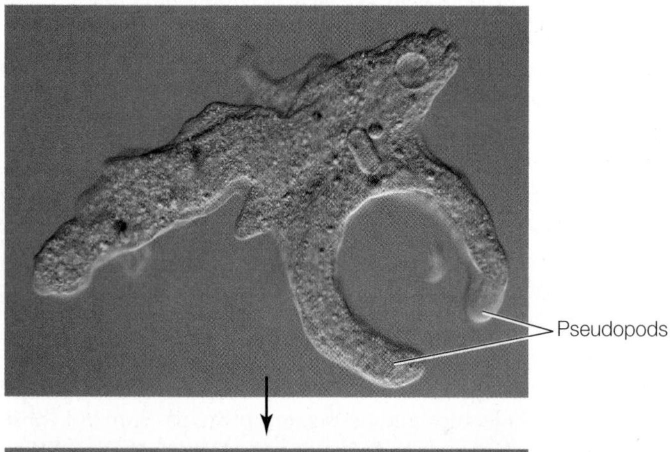

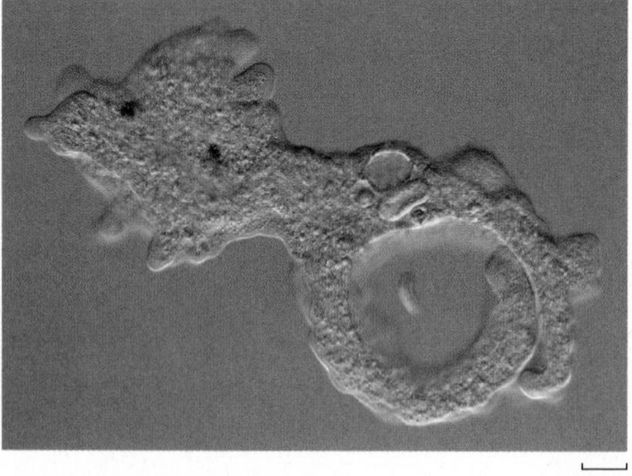

FIGURE 20.14 An Amoeba in Motion The flowing pseudopods of this "chaos amoeba" are constantly changing shape as it moves and feeds.

Go to MEDIA CLIP 20.5
Amoeboid Movement
PoL2e.com/mc20.5

PLASMODIAL SLIME MOLDS If the nucleus of an amoeba began rapid mitotic division, accompanied by a tremendous increase in cytoplasm and organelles but no cytokinesis (division of the cytoplasm), the resulting organism would resemble the multinucleate mass of a **plasmodial slime mold**. During its vegetative (feeding, nonreproductive) stage, a plasmodial slime mold is a wall-less mass of cytoplasm with numerous diploid nuclei. This mass streams very slowly over its substrate in a remarkable network of strands called a plasmodium (**FIGURE 20.16A**). The plasmodium of such a slime mold is an example of a **coenocyte**: many nuclei enclosed in a single cell membrane. The outer cytoplasm of the plasmodium (closest to the environment) is normally less fluid than the interior cytoplasm and thus provides some structural rigidity.

Plasmodial slime molds provide a dramatic example of movement by **cytoplasmic streaming**. The outer cytoplasmic region of the plasmodium becomes more fluid in places, and cytoplasm rushes into those areas, stretching the plasmodium. This streaming somehow reverses its direction every few minutes as cytoplasm rushes into a new area and drains away from an older one, moving the plasmodium over its substrate. Sometimes an entire wave of plasmodium moves across a surface, leaving strands behind. Microfilaments and a contractile protein interact to produce the streaming movement. As it

(A)

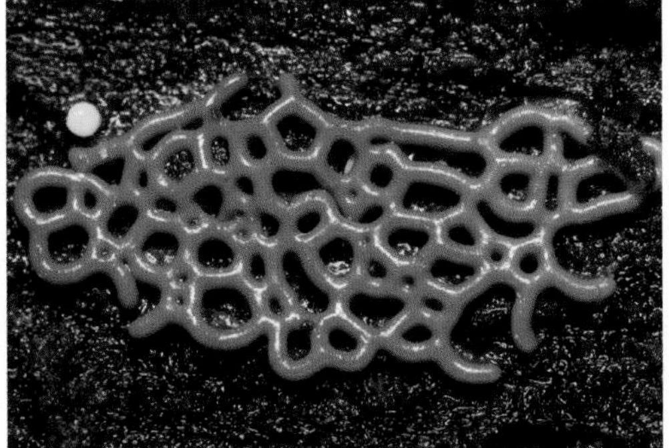

30 mm

FIGURE 20.16 Plasmodial Slime Molds (A) The plasmodial form of the slime mold *Hemitrichia serpula* covers rocks, decaying logs, and other objects as it engulfs bacteria and other food items; it is also responsible for its common name of "pretzel mold." (B) Fruiting structures of *Hemitrichia*.

Go to MEDIA CLIP 20.6
Plasmodial Slime Mold Growth
PoL2e.com/mc20.6

(B)

1.5 mm

moves, the plasmodium engulfs food particles by endocytosis—predominantly bacteria, yeasts, spores of fungi, and other small organisms as well as decaying animal and plant remains.

A plasmodial slime mold can grow almost indefinitely in its plasmodial stage as long as the food supply is adequate and other conditions, such as moisture and pH, are favorable. If conditions become unfavorable, however, one of two things can happen. First, the plasmodium can form an irregular mass of hardened cell-like components. This resting structure rapidly becomes a plasmodium again when favorable conditions are restored.

Alternatively, the plasmodium can transform itself into spore-bearing fruiting structures (**FIGURE 20.16B**). These stalked or branched structures rise from heaped masses of plasmodium. They derive their rigidity from walls that form and thicken between their nuclei. The diploid nuclei of the plasmodium divide by meiosis as the fruiting structure develops. One or more knobs, called sporangia, develop on the end of the stalk. Within a sporangium, haploid nuclei become surrounded by walls to form spores. Eventually, as the fruiting structure dries, it sheds its spores.

The spores germinate into wall-less, haploid cells called swarm cells, which can either divide mitotically to produce more haploid swarm cells or function as gametes. Swarm cells can live as separate individual cells that move by means of flagella or pseudopods, or they can become walled and resistant resting cysts when conditions are unfavorable. When conditions improve again, the cysts release swarm cells. Two swarm cells can also fuse to form a diploid zygote, which divides by mitosis (but without a wall forming between the nuclei) and thus forms a new, coenocytic plasmodium.

CELLULAR SLIME MOLDS Whereas the plasmodium is the basic vegetative unit of the plasmodial slime molds, an amoeboid cell is the vegetative unit of the **cellular slime molds** (**FIGURE 20.17**). Cells called myxamoebas, which have single haploid nuclei, engulf bacteria and other food particles by endocytosis and reproduce by mitosis and binary fission. This simple life cycle stage, consisting of swarms of independent, isolated cells, can persist indefinitely as long as food and moisture are available.

Dictyostelium discoideum

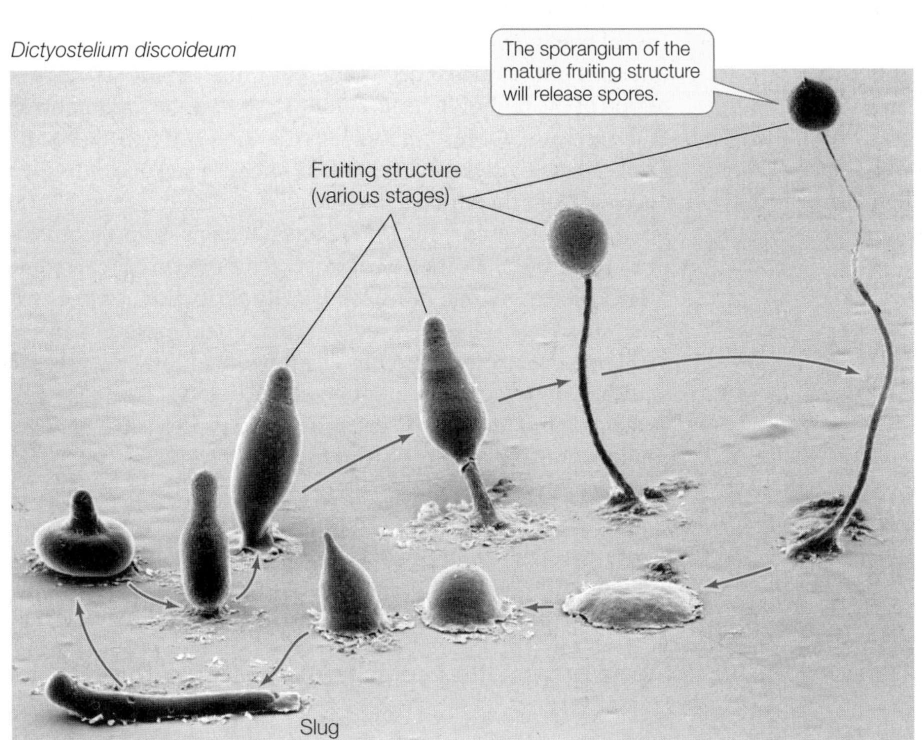
The sporangium of the mature fruiting structure will release spores.
Fruiting structure (various stages)
Slug
0.25 mm

FIGURE 20.17 A Cellular Slime Mold This composite micrograph shows the life cycle of the slime mold *Dictyostelium*.

Go to MEDIA CLIP 20.7
Cellular Slime Mold Aggregation
PoL2e.com/mc20.7

When conditions become unfavorable, the cellular slime molds aggregate and form fruiting structures, as do their plasmodial counterparts. The individual myxamoebas aggregate into a mass called a **slug** or **pseudoplasmodium**. Unlike the true plasmodium of the plasmodial slime molds, however, this structure is not simply a giant sheet of cytoplasm with many nuclei. Instead, the individual myxamoebas retain their cell membranes, and therefore their identity.

A slug may migrate over a substrate for several hours before becoming motionless and reorganizing to construct a delicate, stalked fruiting structure. Cells at the top of the fruiting structure develop into thick-walled spores, which are eventually released. Later, under favorable conditions, the spores germinate, releasing myxamoebas.

The cycle from myxamoebas through slug and spores to new myxamoebas is asexual. Cellular slime molds also have a sexual cycle, in which two myxamoebas fuse. The product of this fusion develops into a spherical structure that ultimately germinates, releasing new haploid myxamoebas.

CHECKpoint CONCEPT 20.2

- ✓ Explain why the term "protists" does not refer to a formal taxonomic group.
- ✓ Contrast the major distinctive features of alveolates, excavates, stramenopiles, rhizarians, and amoebozoans. Which of these groups is most closely related to fungi and animals? What morphological evidence supports this relationship?
- ✓ The fossil record of eukaryotes in the Precambrian is poor compared with those of the Cambrian and later geological periods, even though eukaryotes were diversifying for the last billion years of the Precambrian. Can you think of some possible reasons for the better fossil record beginning in the Cambrian?

The ancient origins of major eukaryote lineages, and the adaptation of these lineages to a wide variety of lifestyles and environments, resulted in enormous protist diversity. It is not surprising, then, that reproductive modes among protists are also highly diverse.

CONCEPT 20.3 Protists Reproduce Sexually and Asexually

Although most protists engage in both asexual and sexual reproduction, sexual reproduction has yet to be observed in some groups. In some protists, as in all prokaryotes, the acts of sex and reproduction are not directly linked.

Several asexual reproductive processes have been observed among the protists:

- The equal splitting of one cell into two by mitosis followed by cytokinesis
- The splitting of one cell into multiple (i.e., more than two) cells
- The outgrowth of a new cell from the surface of an old one (known as **budding**)
- The formation of specialized cells (spores) that are capable of developing into new individuals (known as **sporulation**)

Asexual reproduction results in offspring that are genetically nearly identical to their parents (they only differ by new mutations that may arise during DNA replication). Such asexually reproduced groups of nearly identical organisms are known as **clonal lineages**.

Sexual reproduction among the protists takes various forms. In some protists, as in animals, the gametes are the only haploid cells. In others, the zygote is the only diploid cell. In still others, both diploid and haploid cells undergo mitosis, giving rise to alternating multicellular diploid and haploid life stages.

LINK

Asexual and sexual reproduction are described in **Concept 7.1**

Some protists have reproduction without sex and sex without reproduction

As we noted in Concept 20.2, members of the genus *Paramecium* are ciliates, which commonly have two types of nuclei in a single cell (one macronucleus and from one to several micronuclei; see Figure 20.6). The micronuclei, which are typical eukaryotic nuclei, are essential for genetic recombination. Each macronucleus contains many copies of the genetic information, packaged in units containing very few genes each. The macronuclear DNA is transcribed and translated to regulate the life of the cell.

When paramecia reproduce asexually, all of the nuclei are copied before the cell divides. Paramecia also have an elaborate sexual behavior called **conjugation**, in which two individuals line up tightly against each other and fuse in the oral groove region of the body. Nuclear material is extensively reorganized and exchanged over the next several hours (**FIGURE 20.18**). Each cell ends up with two haploid micronuclei, one of its own and one from the other cell, which fuse to form a new diploid micronucleus. A new macronucleus develops from the micronucleus through a series of dramatic chromosomal rearrangements. The exchange of nuclei is fully reciprocal: each of the two paramecia gives and receives an equal amount of DNA. The two organisms then separate and go their own ways, each equipped with new combinations of alleles.

Conjugation in *Paramecium* is a sexual process of genetic recombination, but it is not a reproductive process. Two cells begin the process, and two cells are there at the end, so no new cells are created. As a rule, each asexual clone of paramecia must periodically conjugate. Experiments have shown that if some species are not permitted to conjugate, the clones can live through no more than approximately 350 cell divisions before they die out.

Some protist life cycles feature alternation of generations

Alternation of generations is a type of life cycle found in many multicellular protists, all land plants, and some fungi. A

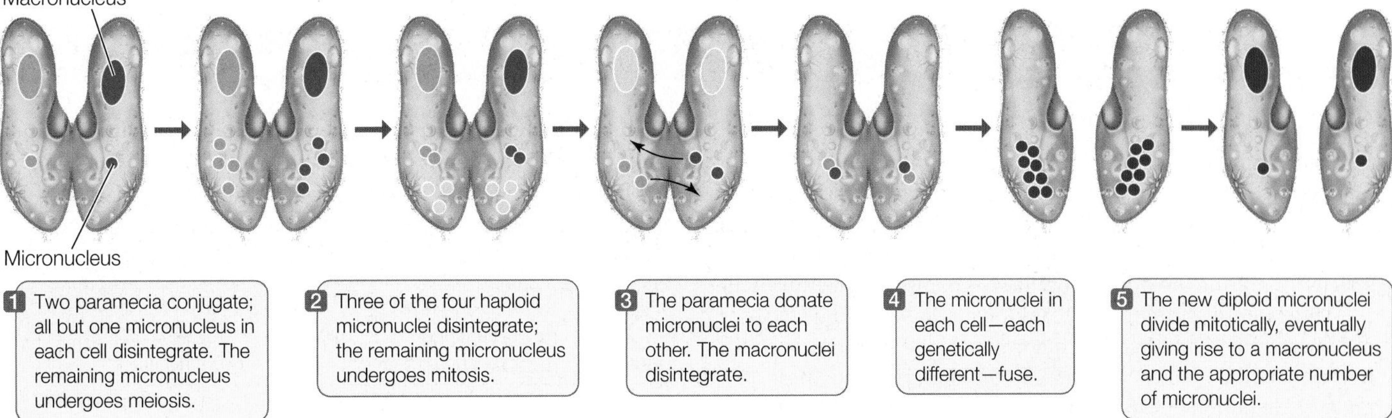

Macronucleus
Micronucleus
1 Two paramecia conjugate; all but one micronucleus in each cell disintegrate. The remaining micronucleus undergoes meiosis.
2 Three of the four haploid micronuclei disintegrate; the remaining micronucleus undergoes mitosis.
3 The paramecia donate micronuclei to each other. The macronuclei disintegrate.
4 The micronuclei in each cell—each genetically different—fuse.
5 The new diploid micronuclei divide mitotically, eventually giving rise to a macronucleus and the appropriate number of micronuclei.

FIGURE 20.18 Paramecia Achieve Genetic Recombination by Conjugating The exchange of micronuclei by two conjugating *Paramecium* individuals results in genetic recombination. After conjugation, the cells separate and continue their lives as two individuals.

multicellular, diploid, spore-producing organism gives rise to a multicellular, haploid, gamete-producing organism. When two haploid gametes fuse, a diploid organism is produced. The haploid organism, the diploid organism, or both may also reproduce asexually.

The two alternating (spore-producing and gamete-producing) generations differ genetically (one has diploid cells, the other haploid cells), but they may or may not differ morphologically. In **heteromorphic** alternation of generations, the two generations differ morphologically, but in **isomorphic** alternation of generations, they do not. Examples of both heteromorphic and isomorphic alternation of generations are found among the brown algae.

The gamete-producing generation does not produce gametes by meiosis because the gamete-producing organism is already haploid. Instead, specialized cells of the diploid spore-producing organism, called **sporocytes**, divide meiotically to produce four haploid spores. The spores may eventually germinate and divide mitotically to produce the multicellular haploid generation, which then produces gametes by mitosis and cytokinesis.

Gametes, unlike spores, can produce new organisms only by fusing with other gametes. The fusion of two gametes produces a diploid zygote, which then undergoes mitotic divisions to produce a diploid organism. The diploid organism's sporocytes then undergo meiosis and produce haploid spores, starting the cycle anew.

CHECKpoint CONCEPT 20.3

- ✓ Why is conjugation between paramecia considered a sexual process but not a reproductive process?
- ✓ Why do you think paramecia that are not allowed to conjugate begin to die out after about 350 rounds of asexual reproduction?
- ✓ Although most diploid animals have haploid stages (for example, eggs and sperm), their life cycles are not considered alternation of generations. Why not?

CONCEPT 20.4 Protists Are Critical Components of Many Ecosystems

Some protists are food for marine animals, while others poison those animals. Some are packaged as nutritional supplements for humans, and some are human pathogens. The remains of some form the sands of many modern beaches, and others are a major source of the oil that sometimes fouls those beaches.

Phytoplankton are primary producers

A single protist clade, the diatoms, performs about one-fifth of all photosynthetic carbon fixation on Earth—about the same amount as all of Earth's rainforests. These spectacular unicellular organisms (see Figure 20.7) are the predominant component of the phytoplankton, but the phytoplankton include many other protists that contribute heavily to global photosynthesis. Like green plants on land, these "floating photosynthesizers" are the gateway for energy from the sun into the rest of the living world. In other words, they are **primary producers**. These autotrophs are eaten by heterotrophs, including animals and many other protists. Those consumers are, in turn, eaten by other consumers. Most aquatic heterotrophs (with the exception of some species in the deep sea) depend on photosynthesis performed by phytoplankton.

LINK

Much of the energy contained in autotrophs is unavailable to the heterotrophs that eat them; see **Concept 44.3** for a discussion of energy flow through communities

Some microbial eukaryotes are deadly

Some microbial eukaryotes are pathogens that cause serious diseases in humans and other vertebrates. The best-known pathogenic protists are members of the genus *Plasmodium*, a highly specialized group of apicomplexans that spend part of their life cycle as parasites in human red blood cells, where they are the cause of malaria. In terms of the number of people affected, malaria is one of the world's three most serious infectious diseases: it infects more than 350 million people, and kills more than 1 million people, each year. On average, about two people die from malaria every minute of every day—most of

APPLY THE CONCEPT

Protists are critical components of many ecosystems

In most temperate regions of the oceans, there is a spring bloom of phytoplankton. Although the red tide blooms described in this chapter's opening story are harmful, phytoplankton blooms can also be beneficial for marine communities. In fact, many species of marine life depend on these blooms for their survival.

The dates of spring phytoplankton blooms near the coast of Nova Scotia, Canada, were determined by examining remote satellite images. The table to the right presents these dates as deviations from the mean date of the spring bloom in this region.[a] The table also gives the survival index for larval haddock (an important commercial fish) for the year after each bloom. The survival index is the ratio of the mass of juvenile fish to the mass of mature fish; higher values indicate better survival of larval fish.

1. Plot the survival index of larval haddock against the deviation in the date of the spring phytoplankton bloom. Calculate a correlation coefficient for their relationship (see Appendix B).
2. Formulate one or more hypotheses to explain your results. Keep in mind that larval haddock include phytoplankton in their diet, and that phytoplankton blooms also provide some cover in which larval fish can hide from potential predators.

YEAR	DEVIATION IN BLOOM DATE* (DAYS)	SURVIVAL INDEX
1	+5	1.9
2	+11	2.2
3	−15	6.8
4	+5	1.9
5	−4	4.9
6	−20	10.3
7	+6	2.1
8	+14	1.9

*Negative values indicate blooms occurring earlier than the mean date; positive values indicate later blooms.

[a]T. Platt et al. 2003. *Nature* 423: 398–399.

them in sub-Saharan Africa, although malaria occurs in more than 100 countries.

Mosquitoes of the genus *Anopheles* transmit *Plasmodium* to humans. The parasites enter the human circulatory system when an infected female *Anopheles* mosquito penetrates the skin in search of blood. The parasites find their way to cells in the liver and the lymphatic system, change their form, multiply, and re-enter the bloodstream, where they invade red blood cells.

The parasites multiply inside the red blood cells, which then burst, releasing new swarms of parasites. These episodes of bursting red blood cells coincide with the primary symptoms of malaria, which include fever, shivering, vomiting, joint pains, and convulsions. If another *Anopheles* bites the victim, the mosquito takes in *Plasmodium* cells along with blood. Some of the ingested cells develop into gametes that unite in the mosquito, forming zygotes. The zygotes lodge in the mosquito's gut, divide several times, and move into its salivary glands, from which they can be passed on to another human host. Thus *Plasmodium* is an extracellular parasite in the mosquito vector and an intracellular parasite in the human host.

Go to ANIMATED TUTORIAL 20.3
Life Cycle of the Malarial Parasite
PoL2e.com/at20.3

Plasmodium has proved to be a singularly difficult pathogen to attack. The complex *Plasmodium* life cycle is best broken by the removal of stagnant water, in which mosquitoes breed. Using insecticides to reduce the *Anopheles* population can also be effective, but the benefits must be weighed against the ecological, economic, and health risks posed by the insecticides themselves.

Even some of the phytoplankton that are such important primary producers can be deadly, as described in this chapter's opening story. Some diatoms and dinoflagellates reproduce in enormous numbers when environmental conditions are favorable for their growth. In the resulting red tides, the concentration of dinoflagellates may reach 60 million per liter of ocean water and produce potent nerve toxins that harm or kill many vertebrates, especially fish.

Some microbial eukaryotes are endosymbionts

Endosymbiosis is common among the microbial eukaryotes, many of which live within the cells of animals. Many radiolarians harbor photosynthetic endosymbionts (see Figure 20.11A). As a result, these radiolarians, which are not photosynthetic themselves, appear greenish or golden, depending on the type of endosymbiont they contain. This arrangement is often mutually beneficial: the radiolarian can make use of the carbon compounds produced by its photosynthetic guest, and the guest may in turn make use of metabolites made by the host or receive physical protection. In some cases, the guest is exploited for its photosynthetic products while receiving little or no benefit itself.

Dinoflagellates are also common endosymbionts and can be found in both animals and other protists. Most, but not all, dinoflagellate endosymbionts are photosynthetic. Some dinoflagellates live endosymbiotically in the cells of corals, contributing the products of their photosynthesis to the partnership. Their importance to the corals is demonstrated when the dinoflagellates die or are expelled by the corals as a result of changing environmental conditions such as rising water temperatures or increased water turbidity. This phenomenon is known as **coral bleaching**. Unless the corals can acquire new endosymbionts, they are ultimately damaged or destroyed as a result of their reduced food supply (**FIGURE 20.19**).

INVESTIGATION

FIGURE 20.19 Can Corals Reacquire Dinoflagellate Endosymbionts Lost to Bleaching? Some corals lose their chief nutritional source when their photosynthetic endosymbionts die, often as a result of changing environmental conditions. This experiment by Cynthia Lewis and Mary Alice Coffroth investigated the ability of corals to acquire new endosymbionts after bleaching.[a]

HYPOTHESIS

Bleached corals can acquire new photosynthetic endosymbionts from their environment.

METHOD

1. Count numbers of *Symbiodinium*, a photosynthetic dinoflagellate, living symbiotically in samples of a coral (*Briareum* sp.).
2. Stimulate bleaching by maintaining all *Briareum* colonies in darkness for 12 weeks.
3. After 12 weeks of darkness, count numbers of *Symbiodinium* in the coral samples; then return all colonies to light.
4. In some of the bleached colonies (the experimental group), introduce *Symbiodinium* strain B211—dinoflagellates that contain a unique molecular marker. A control group of bleached colonies is not exposed to strain B211. Maintain both groups in the light for 6 weeks.

RESULTS

Error bars indicate 95% confidence intervals for the mean (see Appendix B).

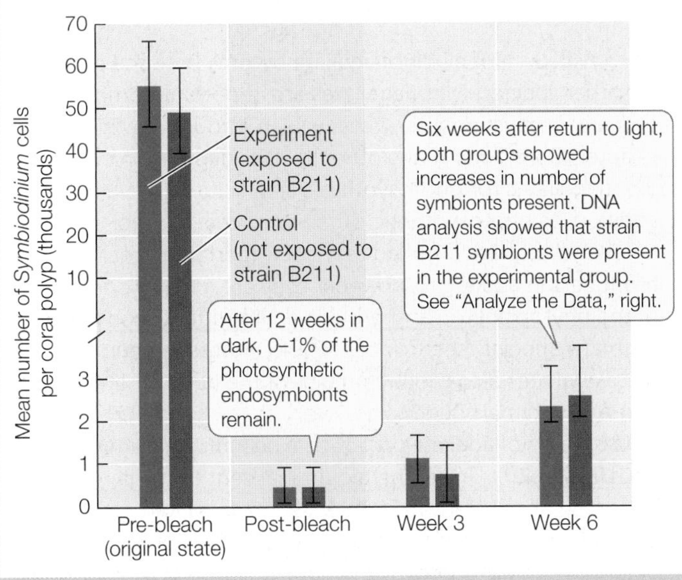
Mean number of *Symbiodinium* cells per coral polyp (thousands)
70
60
50
40
30
20
10
3
2
1
0
Experiment (exposed to strain B211)
Control (not exposed to strain B211)
After 12 weeks in dark, 0–1% of the photosynthetic endosymbionts remain.
Six weeks after return to light, both groups showed increases in number of symbionts present. DNA analysis showed that strain B211 symbionts were present in the experimental group. See "Analyze the Data," right.
Pre-bleach (original state)
Post-bleach
Week 3
Week 6

CONCLUSION

Corals can acquire new strains of endosymbionts from their environment following bleaching.

ANALYZE THE DATA

These data—the results of DNA analysis of the *Symbiodinium* endosymbionts—reveal that many of the experimental colonies took up strain B211 from their environment. The control colonies recovered their native *Symbiodinium*, except in colonies in which endosymbionts were completely lost. Use these data to answer the questions below.

	Symbiodinium strain present (% of colonies)			
	Non-B211	B211	None*	Colony died
Experimental colonies (strain B211 added)				
Pre-bleach	100	0	0	0
Post-bleach	58	0	42	0
Week 3	0	92	0	8
Week 6	8	58	8	25
Control colonies (no strain B211)				
Pre-bleach	100	0	0	0
Post-bleach	67	0	33	0
Week 3	67	0	33	0
Week 6	67	0	17	17

*Colonies remained alive but no *Symbiodinium* were detected.

A. Are new strains of *Symbiodinium* taken up only by coral colonies that have lost all their original endosymbionts?

B. Does the acquisition of a new *Symbiodinium* strain always result in survival of a recovering *Briareum* colony?

C. In week 3, only strain B211 was detected in the experimental colonies, but in week 6, non-B211 *Symbiodinium* were detected in 8% of the experimental colonies. Can you suggest an explanation for this observation?

Pre-bleach

Post-bleach

Go to **LaunchPad** for discussion and relevant links for all **INVESTIGATION** figures.

[a]C. L. Lewis and M. A. Coffroth. 2004. *Science* 304: 1490–1492.

We rely on the remains of ancient marine protists

Diatoms are lovely to look at, but their importance to us goes far beyond aesthetics, and even beyond their role as primary producers. Diatoms store oil as an energy reserve and to keep themselves afloat at the correct depth in the ocean. Over millions of years, diatoms have died and sunk to the ocean floor, where they have been compressed by sediments and undergone chemical changes. In this way, they have become a major source of petroleum and natural gas, two of our most important energy supplies and political concerns.

Because the silica-containing cell walls of dead diatoms resist decomposition, some sedimentary rocks are composed almost entirely of diatom skeletons that sank to the seafloor over time. Diatomaceous earth, which is obtained from such rocks, has many industrial uses, such as insulation, filtration, and metal polishing. It has also been used as an "Earth-friendly" insecticide that clogs the tracheae (breathing structures) of insects.

Other ancient marine protists have also contributed to today's world. Some foraminiferans, as we have seen, secrete shells of calcium carbonate. After they reproduce (by mitosis and cytokinesis), the daughter cells abandon the parent shell and make new shells of their own. The discarded shells of ancient foraminiferans make up extensive limestone deposits in various parts of the world, forming a layer hundreds to thousands of meters deep over millions of square kilometers of ocean bottom. Foraminiferan shells also make up much of the sand of some beaches. A single gram of such sand may contain as many as 50,000 foraminiferan shells and shell fragments.

The shells of individual foraminiferans are easily preserved as fossils in marine sediments. Each geological period has a distinctive assemblage of foraminiferan species. Because the shells of foraminiferan species have distinctive shapes (see Figure 20.10), and because they are so abundant, the remains of foraminiferans are especially valuable in classifying and dating sedimentary rocks. In addition, analyses of the chemical makeup of foraminiferan shells can be used to estimate the global temperatures prevalent at the time when the shells were formed.

CHECKpoint CONCEPT **20.4**

- ✓ What is the role of female *Anopheles* mosquitoes in the transmission of malaria?
- ✓ Explain the role of dinoflagellates in the two very different phenomena of coral bleaching and red tides.
- ✓ What are some of the ways in which diatoms are important to human society?

The next three chapters will explore the three major groups of multicellular eukaryotes, along with the protist ancestors from which they arose. Chapter 21 will describe the origin and diversification of the plants, Chapter 22 will present the fungi, and Chapter 23 will describe the animals.

FIGURE 20.20 Light Up the Sea Bioluminescent dinoflagellates flash as an outrigger disturbs the ocean surface off the island of Bali.

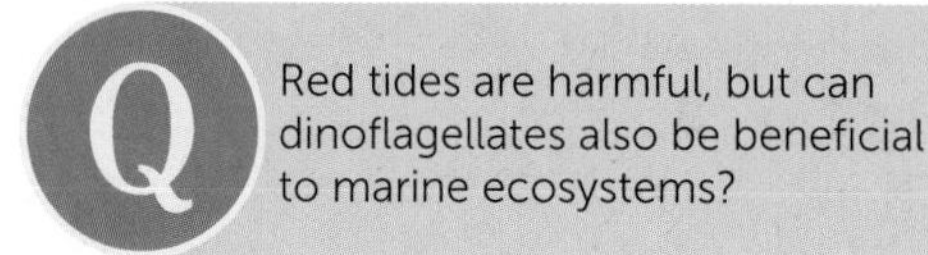

Red tides are harmful, but can dinoflagellates also be beneficial to marine ecosystems?

ANSWER Not all dinoflagellate blooms produce problems for other species. Dinoflagellates are important components of many ecosystems, as we have seen throughout this chapter. Corals and many other species depend on symbiotic dinoflagellates for photosynthesis (see Figure 20.19). In addition, as photosynthetic organisms, free-living planktonic dinoflagellates are among the most important primary producers in aquatic food webs. They are a major component of the phytoplankton and provide an important food source for many species (see Concept 20.4). Photosynthetic dinoflagellates also produce much of the atmospheric oxygen that most animals need to survive.

Some dinoflagellates produce a beautiful bioluminescence (**FIGURE 20.20**). Unlike the bioluminescent bacteria discussed at the start of Chapter 19, however, dinoflagellate bioluminescence is not steady. These protists produce flashes of light when disturbed, as people who swim in the ocean at night in certain regions often observe.

SUMMARY

CONCEPT 20.1 Eukaryotes Acquired Features from Both Archaea and Bacteria

- Early events in the evolution of the eukaryotic cell probably included the loss of the firm cell wall and infolding of the cell membrane. Such infolding probably led to segregation of the genetic material in a membrane-enclosed nucleus. **Review Figure 20.1**
- Some organelles were acquired by endosymbiosis. Mitochondria evolved by endosymbiosis with a proteobacterium.
- **Primary endosymbiosis** of a eukaryote and a cyanobacterium gave rise to the first chloroplasts. **Secondary endosymbiosis** and **tertiary endosymbiosis** between chloroplast-containing eukaryotes and other eukaryotes gave rise to the distinctive chloroplasts of euglenids, dinoflagellates, and other groups. **Review Figure 20.2 and ANIMATED TUTORIAL 20.1**

CONCEPT 20.2 Major Lineages of Eukaryotes Diversified in the Precambrian

- Most eukaryotes can be placed in one of eight major clades that originated in the Precambrian: alveolates, excavates, stramenopiles, rhizarians, amoebozoans, Plantae, fungi, and animals. The first five of these clades are collectively referred to as **protists**. **Review Figure 20.3**
- The term "protist" does not describe a formal taxonomic group, but rather is shorthand for "all eukaryotes that are not plants, animals, or fungi." Most, but not all, protists are unicellular.
- **Alveolates** are unicellular organisms with sacs (alveoli) beneath their cell membranes. Alveolate clades include the marine **dinoflagellates**, the parasitic **apicomplexans**, and the diverse, highly motile **ciliates**. **Review Figure 20.6, ACTIVITY 20.1, and ANIMATED TUTORIAL 20.2**
- **Stramenopiles** typically have two flagella of unequal length, the longer one bearing rows of tubular hairs. Among the stramenopiles are the unicellular **diatoms**, the multicellular **brown algae**, and the nonphotosynthetic **oomycetes**, many of which are **saprobic**.
- **Rhizarians** are unicellular and aquatic. They include the **foraminiferans**, whose shells have contributed to great limestone deposits; the **radiolarians**, which have thin, stiff pseudopods and glassy endoskeletons; and the **cercozoans**, which take many forms and live in diverse habitats.
- The **excavates** include a wide variety of symbiotic as well as free-living species. The **diplomonads** and **parabasalids** lack mitochondria, having apparently lost them during the course of their evolution. **Heteroloboseans** are amoebas with a two-stage life cycle. **Euglenids** are often photosynthetic and have anterior flagella and spiraling strips of protein that support their cell surface. The **kinetoplastids**, which include several human pathogens, have a single, large mitochondrion. **Review Figure 20.13**
- The **amoebozoans** move by means of lobe-shaped pseudopods. A **lobosean** consists of a single amoeboid cell. **Plasmodial slime molds** are amoebozoans whose vegetative stage is a **coenocyte** that moves by cytoplasmic streaming. In **cellular slime molds**, the individual cells maintain their identity at all times but aggregate to form fruiting structures. **Review Figure 20.17**

CONCEPT 20.3 Protists Reproduce Sexually and Asexually

- Asexual reproduction gives rise to **clonal lineages** of organisms.
- **Conjugation** in *Paramecium* is a sexual process but not a reproductive one. **Review Figure 20.18**
- **Alternation of generations**, which includes a multicellular diploid phase and a multicellular haploid phase, is a feature of the life cycle of many multicellular protists (as well as of some fungi and all land plants). The alternating generations may be **heteromorphic** or **isomorphic**.
- In alternation of generations, specialized cells of the diploid organism, called **sporocytes**, divide meiotically to produce haploid spores. Spores give rise to the multicellular, haploid, gamete-producing generation through mitosis. Gametes fuse and give rise to the diploid organism.

CONCEPT 20.4 Protists Are Critical Components of Many Ecosystems

- The diatoms are responsible for about one-fifth of the photo-synthetic carbon fixation on Earth. They and other members of the phytoplankton are the **primary producers** in the marine environment.
- Endosymbiotic relationships are common among microbial protists and typically benefit both the endosymbionts and their protist or animal partners. **Review Figure 20.19**
- Some protists are pathogens of humans and other vertebrates. **Review ANIMATED TUTORIAL 20.3**
- Ancient diatoms are the major source of today's petroleum and natural gas deposits.

Go to the Interactive Summary to review key figures, Animated Tutorials, and Activities PoL2e.com/is20

Go to LaunchPad at **macmillanhighered.com/launchpad** for additional resources, including LearningCurve Quizzes, Flashcards, and many other study and review resources.

21 The Evolution of Plants

KEY CONCEPTS

On examining a specimen of the orchid *Angraecum sesquipedale*, Charles Darwin predicted the existence of a pollinator with an exceptionally long proboscis. This pollinator, the sphinx moth *Xanthopan morgani*, was not discovered until after Darwin's death.

In the early 1860s, while the United States was entangled in its tragic civil war, much of middle- and upper-class England was caught up in an orchid frenzy. Amateur plant breeders and professional botanists alike were enchanted with raising the beautiful flowers. After *On the Origin of Species* appeared in 1859, Charles Darwin wrote his next book on this group of plants, publishing *Fertilisation of Orchids* in 1862.

There are more than 25,000 species of orchids, which makes them one of the most diverse plant groups. Darwin wanted to know why orchids had experienced such rapid diversification and was particularly impressed with the role that insect pollinators might have played in this process. He wanted examples to demonstrate the power of natural selection; he found such examples in abundance among the orchids.

Orchids show an impressive variety of specialized pollination mechanisms, many of which demonstrate that they have coevolved with their pollinators. For example, Darwin observed a South American orchid of the genus *Catasetum* shooting a packet of pollen at an insect that landed on its flower. When he was shown *Angraecum sesquipedale*, an orchid from Madagascar with a nectar tube over a foot long, Darwin hypothesized that there must be a moth with a proboscis of unprecedented length that fed from and pollinated that flower. Many people scoffed at his vision, but the moth he described was eventually discovered—21 years after his death.

In 1836 the explorer Robert Schomburgk shook the botanical world with a report that he had seen flowers described as belonging to three different genera of orchids—*Catasetum*, *Monachanthus*, and *Myanthus*—growing together on a single plant. The English botanist John Lindley remarked that this observation would "shake to the foundation all our ideas of the stability of genera and species." Orchid enthusiasts were befuddled by their efforts to grow specimens of *Myanthus*, only to have them flower with the more common blooms of *Catasetum*. Darwin knew that he needed to find the explanation for these odd observations, for otherwise he would have to conclude that individual plants were able to change their specific identity, something that did not fit with his explanations of the evolution of diversity.

What was Darwin's explanation for the three distinct flowers growing on a single orchid plant?

You will find the answer to this question on page 447.

CONCEPT 21.1 Primary Endosymbiosis Produced the First Photosynthetic Eukaryotes

More than a billion years ago, when a cyanobacterium was first engulfed by an early eukaryote, the history of life was altered radically. The chloroplasts that resulted from primary endosymbiosis of this cyanobacterium (see Figure 20.2) were obviously important for the evolution of plants and other photosynthetic eukaryotes, but they were also critical to the evolution of all life on land. Until photosynthetic plants were able to move onto land, there was very little on land to support multicellular animals or fungi, and almost all life was restricted to the oceans and fresh waters.

Primary endosymbiosis is a shared derived trait—a synapomorphy—of the group known as **Plantae** (**FIGURE 21.1**). Although *Plantae* is Latin for "plants," in everyday language—and throughout this book—the unmodified common name "plants" is usually used to refer only to the land plants. However, the first several clades to branch off the tree of life after primary endosymbiosis are all aquatic. Most aquatic photosynthetic eukaryotes (other than those secondarily derived from land plants) are known by the common name **algae**. This name, however, is just a convenient way to refer to all such groups, which are not all closely related.

Several distinct clades of algae were among the first photosynthetic eukaryotes

The ancestor of Plantae was unicellular and may have been similar in general form to the modern **glaucophytes** (**FIGURE 21.2A**). These microscopic freshwater algae are thought to be the sister group to the rest of Plantae (see Figure 21.1A). The chloroplast of glaucophytes is unique in containing a small amount of peptidoglycan between its inner and outer membranes—the same arrangement found in cyanobacteria. Peptidoglycan has been lost from the remaining photosynthetic eukaryotes.

In contrast to the glaucophytes, almost all **red algae** are multicellular (**FIGURE 21.2B**). Their characteristic color is a result of the accessory photosynthetic pigment **phycoerythrin**, which is found in relatively large amounts in the chloroplasts of many red algae. In addition to phycoerythrin, red algal chloroplasts contain chlorophyll *a* as well as several other accessory pigments.

The red algae include species that grow in the shallowest tide pools as well as the photosynthesizers found deepest in the ocean (as deep as 260 meters if nutrient conditions are right and the water is clear enough to permit light to penetrate). A few red algae inhabit fresh water. Most grow attached to a substrate by a structure known as a holdfast.

Despite their name, red algae don't always appear red in color. The ratio of two pigments—phycoerythrin (red) and chlorophyll *a* (green)—depends largely on the intensity of light that reaches the alga. In deep water, where light is dim, algae accumulate large amounts of phycoerythrin (which absorbs light at short wavelengths) and appear red. But many species growing near the surface contain a higher concentration of chlorophyll *a* and thus appear bright green.

The remaining algal groups in Plantae are the various "green algae." Like land plants, the green algae contain both chlorophylls *a* and *b* and store their reserve of photosynthetic products as starch in chloroplasts. All the groups that share these features are commonly called **green plants** because both of their photosynthetic pigments are green.

Three important clades of green algae are the **chlorophytes**, **coleochaetophytes**, and **stoneworts** (**FIGURE 21.3**). The chlorophytes are the sister group of all the other green plants, which are collectively called **streptophytes**. Among the streptophytes, the coleochaetophytes and stoneworts retain their eggs in the parental organism, as do land plants. They also share other cellular features with the land plants (see Figure 21.1A). Of these two groups, the stoneworts are thought to be the closest relatives of the land plants, based largely on similarities of their genes. The stoneworts also exhibit the branched apical growth that is typical of many land plants (see Figure 21.3C).

There are ten major groups of land plants

One of the key synapomorphies of the **land plants** is development from an embryo that is protected by tissues of the parent plant. For this reason, land plants are sometimes called **embryophytes** (*phyton*, "plant"). The green plants, the streptophytes, and the land plants each have been called "the plant kingdom" by different authorities, and others take an even broader view to include red algae and glaucophytes as "plants." To avoid confusion in this chapter, we will use modifying terms ("land plants" or "green plants," for example) to refer to the various clades of Plantae shown in Figure 21.1.

The land plants that exist today fall naturally into ten major clades (**TABLE 21.1**). Members of seven of those clades possess well-developed vascular systems that transport materials throughout the plant body. We call these seven groups, collectively, the **vascular plants**, or **tracheophytes**, because they all possess fluid-conducting cells called **tracheids**. The remaining three clades (liverworts, hornworts, and mosses) lack tracheids and are referred to collectively as **nonvascular land plants**. Note, however, that the three groups of nonvascular land plants do *not* form a clade (unlike the vascular plants, which *are* a clade).

CHECKpoint CONCEPT 21.1

- ✓ Explain the different possible uses of the term "plant."
- ✓ What are some of the key differences between glaucophytes, red algae, and the various clades of green algae?
- ✓ What evidence supports the phylogenetic relationship between land plants and green algae? Why doesn't the name "algae" designate a formal taxonomic group?

The green algal ancestors of the land plants lived at the margins of ponds or marshes, ringing them with a mat of dense green. It was from such a marginal habitat, which was sometimes wet and sometimes dry, that early plants made the transition onto land.

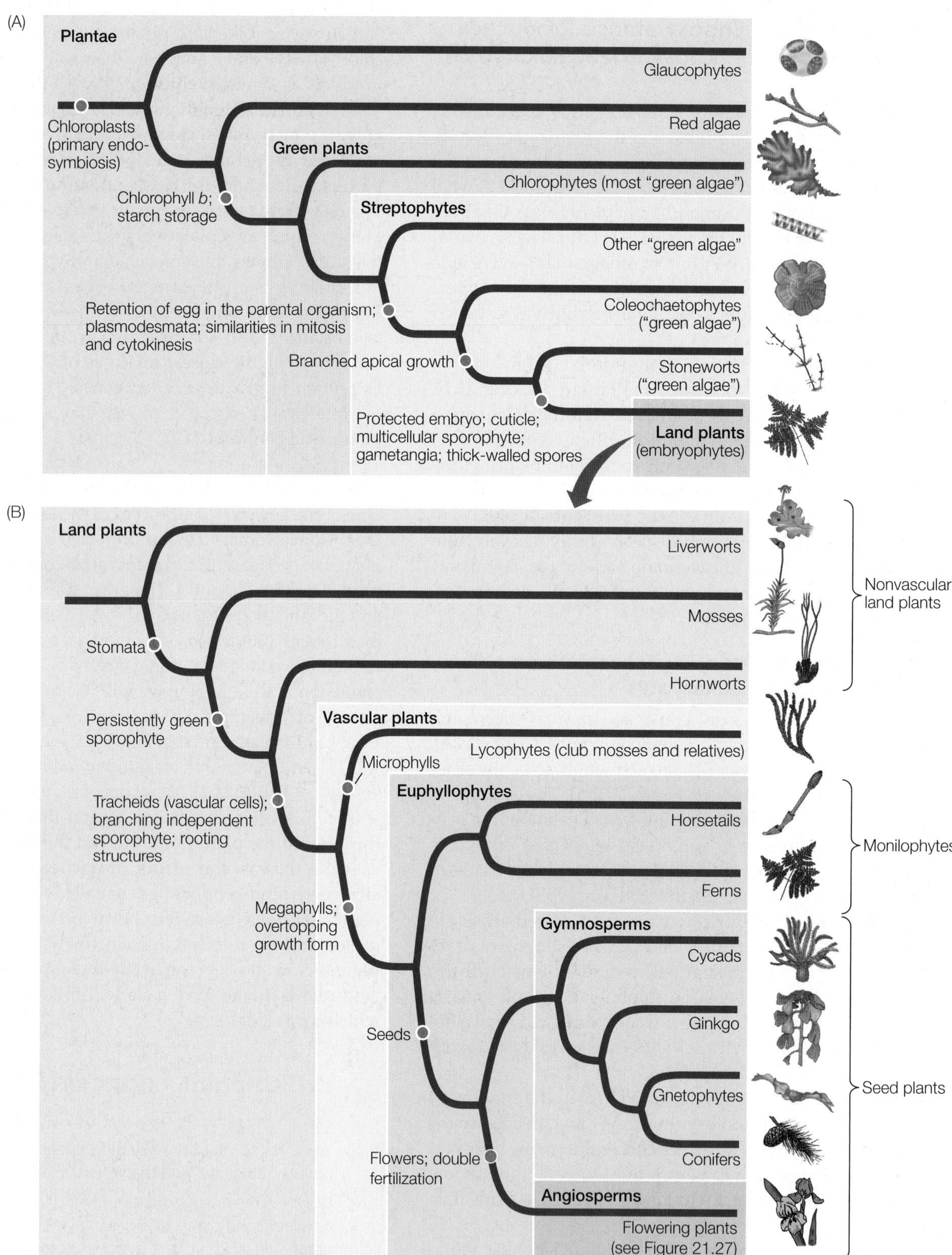

FIGURE 21.1 The Evolution of Plants (A) In its broadest definition, the term "plant" includes the glaucophytes, red algae, and green plants—all the groups descended from a common ancestor with chloroplasts. (B) Some biologists restrict the term "plant" to the green plants (those with chlorophyll *b*) or, even more narrowly, to the land plants. Three key characteristics that emerged during the evolution of land plants—protected embryos, vascular tissues, and seeds—led to their success in the terrestrial environment. (See Analyze the Data 21.1 at LaunchPad.)

(A) *Glaucocystis* sp.

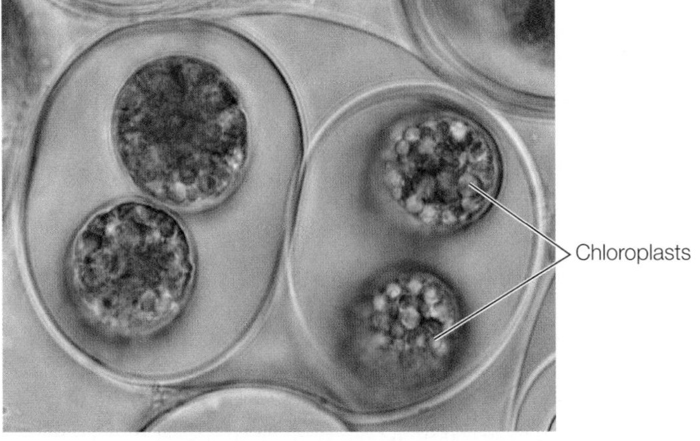

(B) *Mastocarpus papillatus*

FIGURE 21.2 Early Branching Groups of Plantae (A) The large chloroplasts of unicellular glaucophytes differ from chloroplasts of other Plantae in retaining a layer of peptidoglycan. This feature is thought to have been retained from the endosymbiotic cyanobacteria that gave rise to the chloroplasts of Plantae. This photograph shows a colony of two individuals, each with two chloroplasts. (B) The photosynthetic pigment phycoerythrin gives this red alga its rich hue, showing up vividly against a background of sand.

(A) *Ulva latuca*

(B) *Coleochaete* sp.

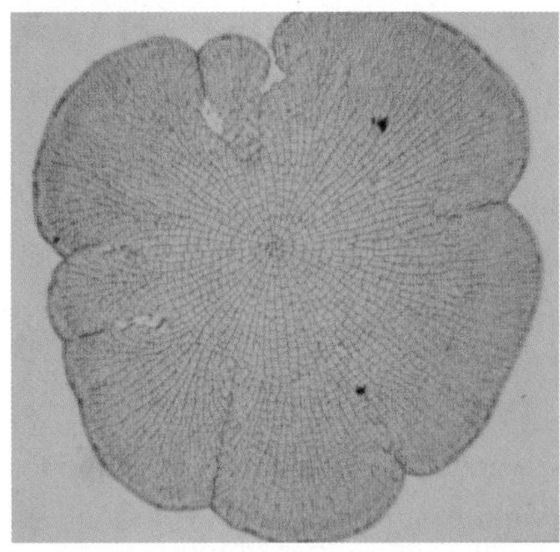

(C) *Chara vulgaris*

FIGURE 21.3 "Green Algae" Consist of Several Distantly Related Groups (A) Sea lettuce, a chlorophyte, grows in ocean tidewaters. (B) The coleochaetophytes are thought to be the sister group of stoneworts plus land plants. (C) The land plants probably evolved from a common ancestor shared with stoneworts, which display the branching pattern we otherwise associate with land plants. A stonewort in the common freshwater genus *Chara* is shown here. The orange structures are male sex organs.

Go to MEDIA CLIP 21.1
Reproductive Structures of *Chara*
PoL2e.com/mc21.1

CONCEPT 21.2 Key Adaptations Permitted Plants to Colonize Land

How did the land plants arise? To address this question, we can compare land plants with their closest relatives among the green algae. The features that differ between the two groups include the adaptations that allowed the first land plants to survive in the terrestrial environment.

Adaptations to life on land distinguish land plants from green algae

Land plants first appeared in the terrestrial environment between 450 and 500 million years ago. How did they survive in an environment that differed so dramatically from the aquatic environment of their ancestors? While the water essential for

TABLE 21.1 Classification of Land Plants

Group	Common name	Characteristics
NONVASCULAR LAND PLANTS		
Hepatophyta	Liverworts	No filamentous stage; gametophyte flat
Anthocerophyta	Hornworts	Embedded archegonia; sporophyte grows basally (from the ground)
Bryophyta	Mosses	Filamentous stage; sporophyte grows apically (from the tip)
VASCULAR PLANTS		
Lycopodiophyta	Lycophytes: Club mosses and allies	Microphylls in spirals; sporangia in leaf axils
Monilophyta	Horsetails, ferns	Megaphylls with a leaf gap (a space in the stem from which the leaf emerges)
SEED PLANTS		
Gymnosperms		
Cycadophyta	Cycads	Compound leaves; swimming sperm; seeds on modified leaves
Ginkgophyta	Ginkgo	Deciduous; fan-shaped leaves; swimming sperm
Gnetophyta	Gnetophytes	Vessels in vascular tissue; opposite, simple leaves
Coniferophyta	Conifers	Seeds in cones; needlelike or scalelike leaves
Angiosperms	Flowering plants	Endosperm; carpels; gametophytes much reduced; seeds within fruit

life is everywhere in the aquatic environment, water is difficult to obtain and retain in the terrestrial environment.

No longer bathed in fluid, organisms on land faced potentially lethal desiccation (drying). Large terrestrial organisms had to develop ways to transport water to body parts distant from the source of the water. And whereas water provides aquatic organisms with support against gravity, a plant living on land must either have some other support system or sprawl unsupported on the ground. A land plant must also use different mechanisms for dispersing its gametes and progeny than its aquatic relatives, which can simply release them into the water.

Survival on land was facilitated by the evolution among plants of numerous adaptations, including:

- The *cuticle*, a coating of waxy lipids that retards water loss
- *Stomata*, small closable openings in leaves and stems that are used to regulate gas exchange and water loss
- *Gametangia*, multicellular organs that enclose plant gametes and prevent them from drying out
- *Embryos*, young plants contained within a protective structure
- Certain *pigments* that afford protection against the mutagenic ultraviolet radiation that bathes the terrestrial environment
- Thick *spore walls* containing a *polymer* that protects the spores from desiccation and resists decay
- A *mutually beneficial association with fungi* (mycorrhizae) that promotes nutrient uptake from the soil

The cuticle may be the most important—and the earliest—of these features. Composed of several unique waxy lipids that coat the leaves and stems of land plants, the cuticle has several functions, the most obvious and important of which is to keep water from evaporating from the plant body.

As ancient plants colonized the land, they not only adapted to the terrestrial environment, they also modified it by contributing to the formation of soil. Acids secreted by plants help break down rock, and the organic compounds produced by the breakdown of dead plants contribute nutrients to the soil. Such effects are repeated today as plants grow in new areas.

Life cycles of land plants feature alternation of generations

A universal feature of the life cycles of land plants is alternation of generations. Recall from Concept 20.3 the two hallmarks of alternation of generations:

- The life cycle includes both a multicellular diploid stage and a multicellular haploid stage.
- Gametes are produced by mitosis, not by meiosis. Meiosis produces **spores** that develop into multicellular haploid organisms.

If we begin looking at the land plant life cycle at the single-cell stage—the diploid zygote—then the first phase of the cycle is the formation, by mitosis and cytokinesis, of a multicellular **embryo**, which eventually grows into a mature diploid plant. This multicellular diploid plant is called the **sporophyte** ("spore plant").

Cells contained within specialized reproductive organs of the sporophyte, called **sporangia** (singular *sporangium*), undergo meiosis to produce haploid, unicellular spores. By mitosis and cytokinesis, a spore develops into a haploid plant. This multicellular haploid plant, called the **gametophyte** ("gamete plant"), produces haploid gametes by mitosis. The fusion of two gametes (fertilization) forms a single diploid cell—the zygote—and the cycle is repeated (**FIGURE 21.4**).

The sporophyte generation extends from the zygote through the adult multicellular diploid plant and sporangium formation. In contrast, the gametophyte generation extends from the

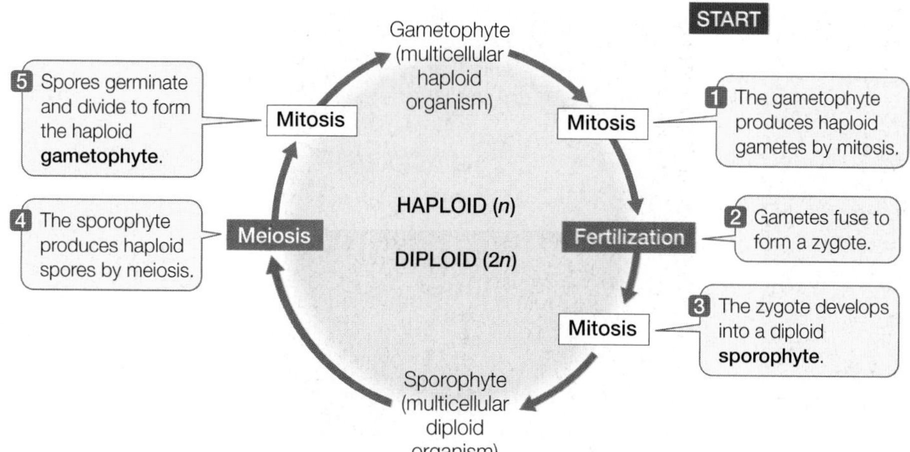

FIGURE 21.4 Alternation of Generations in Land Plants A multicellular diploid sporophyte generation that produces spores by meiosis alternates with a multicellular haploid gametophyte generation that produces gametes by mitosis.

spore through the adult multicellular haploid plant to the gametes. The transitions between the generations are accomplished by fertilization and by meiosis. In all land plants, the sporophyte and the gametophyte differ genetically: the sporophyte has diploid cells, and the gametophyte has haploid cells.

There is a trend toward reduction of the gametophyte generation in plant evolution. In the nonvascular land plants, the gametophyte is larger, longer-lived, and more self-sufficient than the sporophyte. In those groups that appeared later in plant evolution, however, the sporophyte is the larger, more conspicuous, longer-lived, and more self-sufficient generation.

Nonvascular land plants live where water is readily available

The living species of nonvascular land plants are the liverworts, mosses, and hornworts. These three groups are thought to be similar in many ways to the earliest land plants. Most of these plants grow in dense mats, usually in moist habitats. Even the largest of these species are only about half a meter tall, and most are only a few centimeters tall or long. Why have they not evolved to be taller? The probable answer is that they lack an efficient vascular system for transporting water and minerals from the soil to distant parts of the plant body.

The nonvascular land plants lack the true leaves, stems, and roots that characterize the vascular plants, although they have structures analogous to each. Their growth form allows water to move through the mats of plants by capillary action. They have leaflike structures that readily catch and hold any water that splashes onto them. They are small enough that minerals can be distributed throughout their bodies by diffusion. As in all land plants, layers of maternal tissue protect their embryos from desiccation. Nonvascular land plants also have a cuticle, although it is often very thin (or even absent in some species) and thus is not highly effective in retarding water loss.

Most nonvascular land plants live on the soil or on vascular plants, but some grow on bare rock, on dead and fallen tree trunks, and even on buildings. Their ability to grow on such marginal surfaces results from a mutualistic association with fungi. The earliest association of land plants with fungi dates back at least 460 million years. This mutualism probably facilitated the absorption of water and minerals, especially phosphorus, from the first soils.

LINK

Land plants of many groups have mutualistic associations with fungi, as described in **Concept 22.2**

Nonvascular land plants are widely distributed over six continents and even exist (albeit very locally) on the coast of the seventh, Antarctica. Most are terrestrial. Although a few species live in fresh water, these aquatic species are descended from terrestrial ones. None live in the oceans.

LIVERWORTS There are about 9,000 species of **liverworts**. Most liverworts have green, leaflike gametophytes that lie close to or flat on the ground (**FIGURE 21.5A**). The simplest liverworts are flat plates of cells a centimeter or so long with structures that produce sperm or eggs on their upper surfaces and rootlike filaments on their lower surfaces. The sporophyte remains attached to the larger gametophyte and rarely exceeds a few millimeters in length. Most liverworts can reproduce asexually (through simple division of the gametophyte) as well as sexually.

Go to MEDIA CLIP 21.2
Liverwort Life Cycle
PoL2e.com/mc21.2

MOSSES The most familiar of the nonvascular land plants are the **mosses** (**FIGURE 21.5B**). These hardy little plants, of which there are about 15,000 species, are found in almost every terrestrial environment. They are often found on damp, cool ground, where they form thick mats.

The mosses are the sister lineage to the vascular plants plus the hornworts (see Figure 21.1B). They share with those lineages an advance over the liverworts in their adaptation to life on land: they have stomata, which are important for both water retention and gas exchange.

Some moss gametophytes are so large that they cannot transport enough water throughout their bodies solely by diffusion. Gametophytes and sporophytes of many mosses

(A) *Marchantia polymorpha*

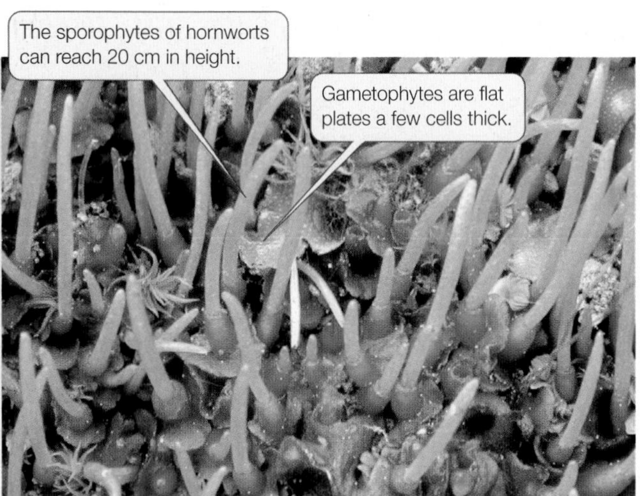
The sporophytes of hornworts can reach 20 cm in height.
Gametophytes are flat plates a few cells thick.

(C) *Anthoceros* sp.

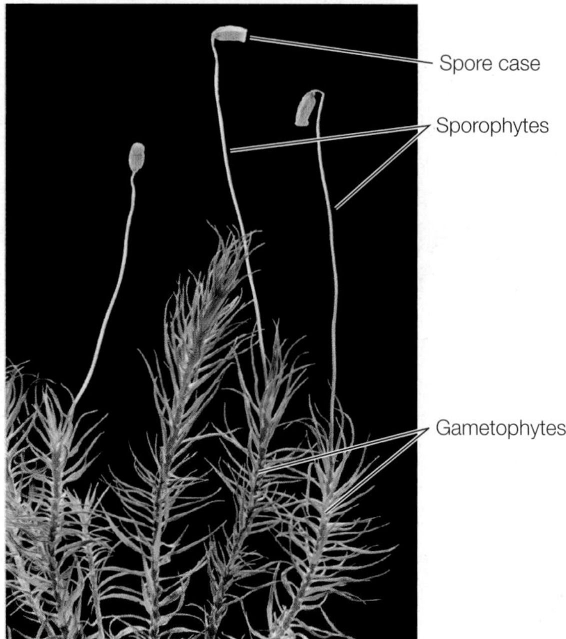
Spore case
Sporophytes
Gametophytes

(B) *Polytrichum commune*

FIGURE 21.5 Diversity among Nonvascular Land Plants (A) Gametophytes of a liverwort. (B) The sporophytes and gametophytes are easily distinguished in this moss. (C) The sporophytes of many hornworts resemble little horns.

contain a type of cell called a hydroid, which dies and leaves a tiny channel through which water can travel. The hydroid is functionally similar to the tracheid, the characteristic water-conducting cell of vascular plants, but it lacks the lignin and the cell-wall structure that characterize tracheids. The possession of hydroids shows that the term "nonvascular land plant" is somewhat misleading when applied to mosses. Despite their simple systems of internal transport, however, the mosses are not considered vascular plants because they lack tracheids and other vascular tissues.

HORNWORTS The approximately 100 species of **hornworts** are so named because their sporophytes often look like little horns (**FIGURE 21.5C**). Hornworts have two characteristics that distinguish them from liverworts and mosses. First, the cells of hornworts each contain a single large, platelike chloroplast, whereas the cells of the other two groups contain numerous small, lens-shaped chloroplasts. Second, of the sporophytes in all three groups, those of the hornworts come closest to being capable of growth without a set limit. Liverwort and moss sporophytes have a stalk that stops growing as the spore-producing structure matures, so elongation of the sporophyte is strictly limited. The hornwort sporophyte, however, has no stalk, and it is persistently green (a trait shared with vascular plants). A basal region of the sporangium remains capable of indefinite cell division, continuously producing new spore-bearing tissue above. The sporophytes of some hornworts growing in mild and continuously moist conditions can become as tall as 20 centimeters. Eventually, however, the sporophyte's growth is limited by the lack of a transport system.

Hornworts have evolved a symbiotic relationship that promotes their growth by providing them with greater access to nitrogen, which is often a limiting resource. Hornworts have internal cavities filled with mucilage, and the cavities are often populated by symbiotic cyanobacteria that convert atmospheric nitrogen gas into a form of nitrogen usable by their host plant.

The sporophytes of nonvascular land plants are dependent on the gametophytes

In the nonvascular land plants, the conspicuous green structure visible to the naked eye is the gametophyte. The gametophyte is photosynthetic and is therefore nutritionally independent. The sporophyte may or may not be photosynthetic, but it is always nutritionally dependent on the gametophyte and remains permanently attached to it.

FIGURE 21.6 illustrates a moss life cycle that is typical of the life cycles of nonvascular land plants. A sporophyte produces unicellular haploid spores as products of meiosis within a sporangium. When a spore germinates, it gives rise to a multicellular haploid gametophyte whose cells contain chloroplasts and are thus photosynthetic. Eventually gametes form within specialized

FIGURE 21.6 A Moss Life Cycle The life cycles of nonvascular land plants, exemplified here by that of a moss, are dependent on an external source of liquid water. The visible green structure of such plants is the gametophyte.

Go to ANIMATED TUTORIAL 21.1
Life Cycle of a Moss
PoL2e.com/at21.1

Go to MEDIA CLIP 21.3
Bryophyte Reproduction
PoL2e.com/mc21.3

sex organs, called the **gametangia**. The **archegonium** is a multicellular, flask-shaped female sex organ that produces a single egg. The **antheridium** is a male sex organ in which sperm, each bearing two flagella, are produced in large numbers. Both archegonia and antheridia are produced on the same individual, so each individual has both male and female reproductive structures. Adjacent individuals often fertilize one another's gametes, however, which helps maintain genetic diversity in the population.

Once released from the antheridium, the sperm must swim or be splashed by raindrops to a nearby archegonium on the same or a neighboring plant—a constraint that reflects the aquatic origins of the nonvascular land plants' ancestors. The sperm are aided on their journey by chemical attractants released by the egg or the archegonium. Before sperm can enter the archegonium, however, certain cells in the neck of the archegonium must break down, leaving a water-filled canal through which the sperm can swim to complete their journey. Notice that *all of these events require liquid water*.

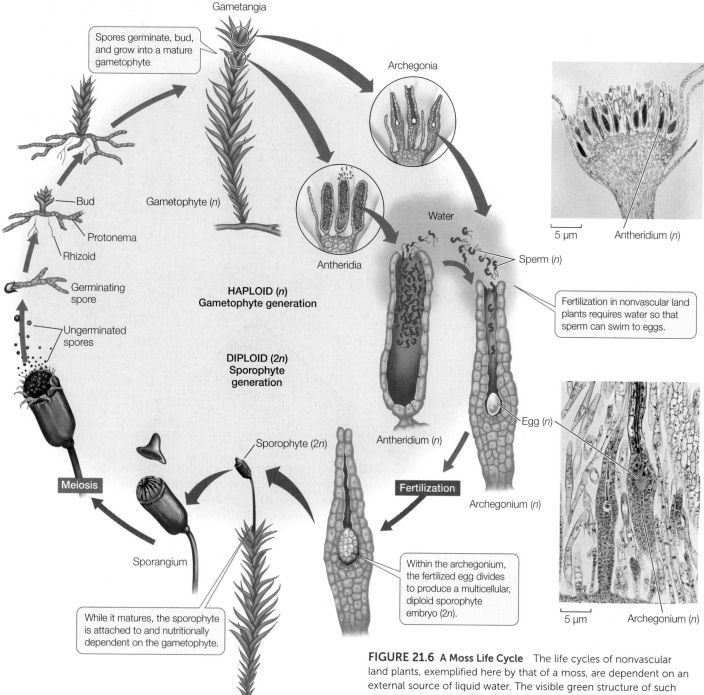

Once sperm arrive at an egg, the nucleus of a sperm fuses with the egg nucleus to form a diploid zygote. Mitotic divisions of the zygote produce a multicellular, diploid sporophyte embryo. The sporophyte matures and produces a sporangium, within which meiotic divisions produce spores and thus the next gametophyte generation.

CHECKpoint CONCEPT 21.2

- ✓ Describe several adaptations of plants to the terrestrial environment, and describe the distribution of those adaptations in liverworts, mosses, and hornworts.
- ✓ Explain what is meant by alternation of generations.
- ✓ What aspects of the reproductive cycle of nonvascular land plants appear to have been retained from their aquatic ancestors?

Further adaptations to the terrestrial environment appeared as plants continued to evolve. One of the most important of these later adaptations was the appearance of vascular tissues—the characteristic that defines the vascular plants.

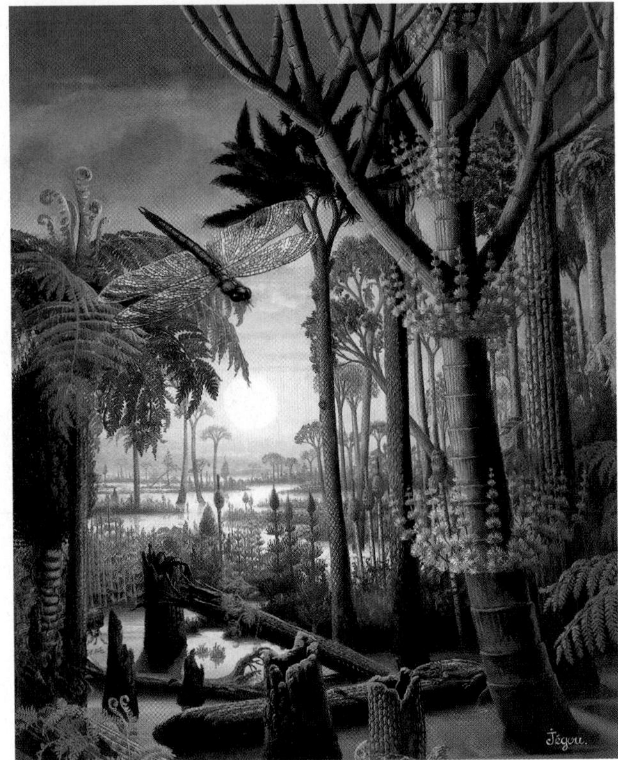

FIGURE 21.7 Reconstruction of an Ancient Forest Forests of the Carboniferous period were characterized by abundant vascular plants such as club mosses, ferns, and horsetails, some of which reached heights of 40 meters. Huge flying insects (see p. 357) thrived in these forests, which are the source of modern coal deposits.

CONCEPT 21.3 Vascular Tissues Led to Rapid Diversification of Land Plants

The first plants possessing vascular tissues did not arise until tens of millions of years after the earliest plants had colonized the land. But once vascular tissues arose, their ability to transport water and food throughout the plant body allowed vascular plants to spread to new terrestrial environments and to diversify rapidly.

Vascular tissues transport water and dissolved materials

Vascular plants differ from the other land plants in crucial ways, one of which is the possession of a well-developed vascular system consisting of tissues that are specialized for the transport of materials from one part of the plant to another. One type of vascular tissue, the **xylem**, conducts water and minerals from the soil to aerial parts of the plant. Because some of its cell walls contain a stiffening substance called lignin, xylem also provides support against gravity in the terrestrial environment. The other type of vascular tissue, the **phloem**, conducts the products of photosynthesis from sites where they are produced or released to sites where they are used or stored.

LINK

The vascular tissues of plants are described in detail in **Chapter 24**

Although the vascular plants are an extraordinarily large and diverse group, a particular event was critical to their evolution. Sometime during the Paleozoic era, probably in the mid-Silurian (430 mya), a new cell type—the tracheid—evolved in sporophytes of the earliest vascular plants. The tracheid is the principal water-conducting element of the xylem in all vascular plants except the angiosperms (flowering plants) and one small group of gymnosperms—and tracheids persist even in these groups, along with a more specialized and efficient system derived from them.

The evolution of tracheids set the stage for the complete and permanent invasion of land by plants. First, these cells provided a pathway for the transport of water and mineral nutrients from a source of supply to regions of need in the plant body. And second, the cell walls of tracheids, stiffened by lignin, provided rigid structural support. This support is a crucial factor in a terrestrial environment because it allows plants to grow upward and thus compete for sunlight. A taller plant can intercept more direct sunlight (and thus conduct photosynthesis more readily) than a shorter plant, which may be shaded by the taller one. Increased height also improves the dispersal of spores.

The vascular plants featured another evolutionary novelty: a branching, independent sporophyte. A branching sporophyte body can produce more spores than an unbranched body, and it can develop in complex ways. The sporophyte of a vascular plant is nutritionally independent of the gametophyte at maturity. Among the vascular plants, the sporophyte is the large and obvious plant that one normally pays attention to in nature, in contrast to the relatively small, dependent sporophytes typical of most nonvascular land plants.

The diversification of vascular plants made land more suitable for animals

The initial absence of herbivores (plant-eating animals) on land helped make the first vascular plants successful. By the late

(A) *Lycopodium annotinum*

(B) *Equisetum pratense*

(C) *Marsilea mutica*

(D) *Alsophilia spinulosa*

FIGURE 21.8 Lycophytes and Monilophytes (A) A strobilus is visible at the tip of this club moss. Club mosses have microphylls arranged spirally on their stems. (B) Horsetails have a distinctive growth pattern in which the stem grows in segments above each whorl of leaves. These are fertile shoots with sporangia-bearing structures at the apex. (C) The leaves of a species of water fern. (D) A tree fern on a mountain in India.

Silurian period (about 425 mya), the proliferation of land plants made the terrestrial environment more hospitable to animals. Arthropods, vertebrates, and other animals moved onto land only after vascular plants became established there.

Trees of various kinds appeared in the Devonian period and dominated the landscape of the Carboniferous period (359–299 mya). Forests of lycophytes (club mosses) up to 40 meters tall, along with horsetails and tree ferns, flourished in the tropical swamps of what would become North America and Europe (**FIGURE 21.7**). Plant parts from those forests sank into the swamps and were gradually covered by layers of sediment. Over millions of years, as the buried plant material was subjected to intense pressure and elevated temperatures, it was transformed into coal. Today that coal provides more than half of our electricity. The world's coal deposits, although huge, are not infinite, and humans are burning coal deposits at a far faster rate than they were produced.

In the subsequent Permian period, when the continents came together to form Pangaea, the continental interior became warmer and drier. The 200-million-year reign of the lycophyte–fern forests came to an end as they were replaced by forests of early gymnosperms.

The earliest vascular plants lacked roots

The earliest known vascular plants belonged to a now-extinct group called the **rhyniophytes**. The rhyniophytes were one of a very few types of vascular plants in the Silurian period. The landscape at that time probably consisted mostly of bare ground, with stands of rhyniophytes in low-lying moist areas. Early versions of the structural features of all the other vascular plant groups appeared in the rhyniophytes of that time. These shared features strengthen the case for the origin of all vascular plants from a common nonvascular land plant ancestor.

Rhyniophytes did not have roots. Like most modern ferns and lycophytes, they were apparently anchored in the soil by horizontal portions of stem, called **rhizomes**, which bore water-absorbing unicellular filaments called **rhizoids**. These plants also bore aerial branches, and sporangia—homologous to the sporangia of mosses—were found at the tips of those branches. Their branching pattern was **dichotomous**, meaning the apex (tip) of the shoot divided to produce two equivalent new branches, each pair diverging at approximately the same angle from the original stem.

The lycophytes are sister to the other vascular plants

The club mosses and their relatives, the spike mosses and quillworts, are collectively called **lycophytes**. The lycophytes are the sister group to the remaining vascular plants (see Figure 21.1B). There are relatively few (just over 1,200) surviving species of lycophytes.

The lycophytes have true roots that branch dichotomously. The arrangement of vascular tissue in their stems is simpler than that in other vascular plants. They bear simple leaflike structures called **microphylls**, which are arranged spirally on the stem. Growth in lycophytes comes entirely from apical cell division. Branching in the stems, which is also dichotomous, occurs by division of an apical cluster of dividing cells.

The sporangia of many club mosses are aggregated in cone-like structures called **strobili** (singular *strobilus*; **FIGURE 21.8A**). The strobilus of a club moss is a cluster of spore-bearing leaves inserted on an axis (a linear supporting structure). Other club mosses lack strobili and bear their sporangia on (or adjacent to) the upper surfaces of specialized leaves.

Horsetails and ferns constitute a clade

The horsetails and ferns were once thought to be only distantly related. As a result of genomic analyses, we now know that

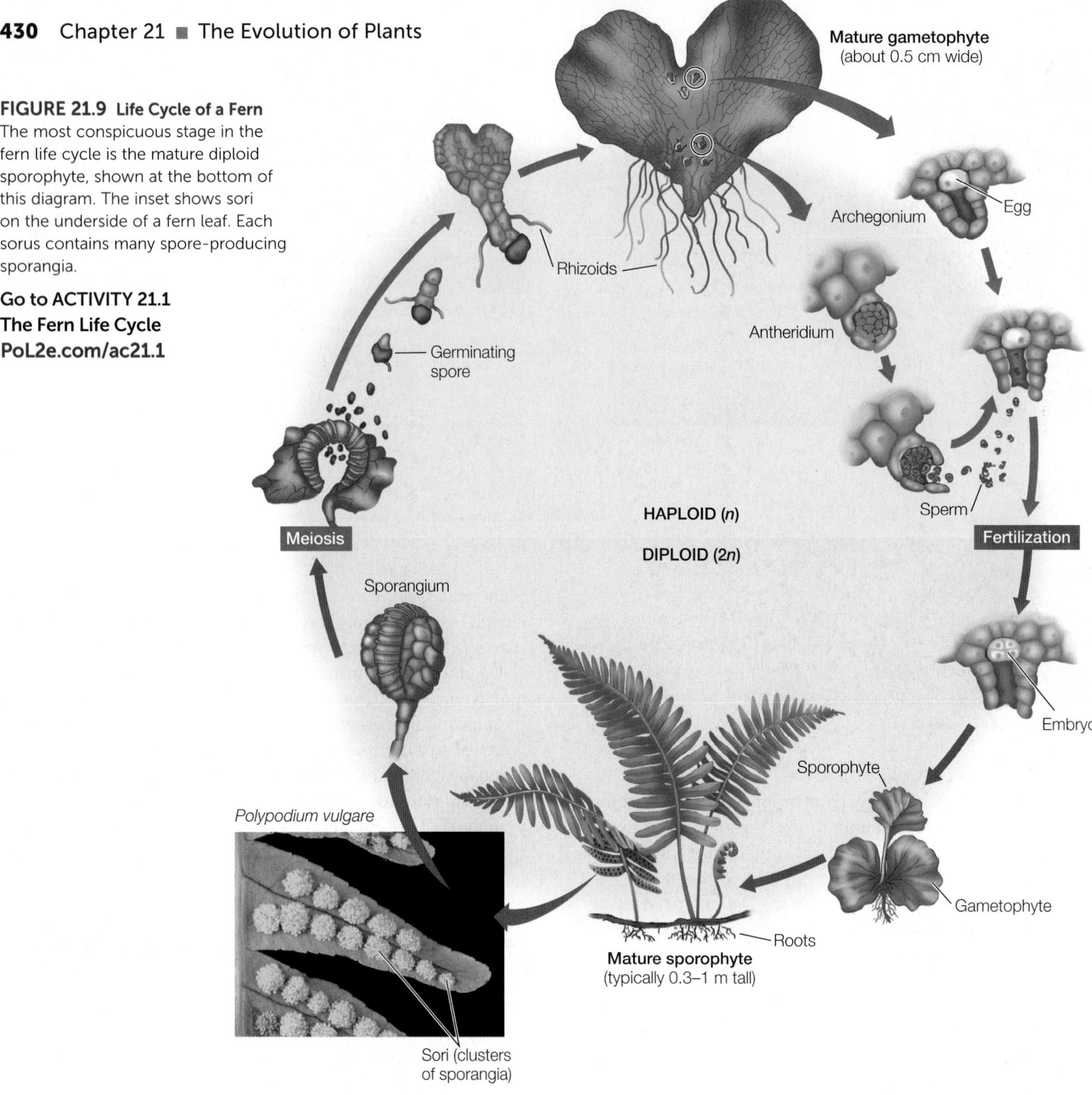

FIGURE 21.9 Life Cycle of a Fern The most conspicuous stage in the fern life cycle is the mature diploid sporophyte, shown at the bottom of this diagram. The inset shows sori on the underside of a fern leaf. Each sorus contains many spore-producing sporangia.

Go to ACTIVITY 21.1 The Fern Life Cycle PoL2e.com/ac21.1

they form a clade: the **monilophytes**. In the monilophytes—as in the seed plants, to which they are the sister group (see Figure 21.1B)—there is differentiation between a main stem and side branches. This pattern contrasts with the dichotomous branching characteristic of the lycophytes and rhyniophytes, in which each split gives rise to two branches of similar size.

Today there are only about 15 species of **horsetails**, all in the genus *Equisetum*. The horsetails have reduced true leaves that form in distinct whorls (circles) around the stem (**FIGURE 21.8B**). Horsetails are sometimes called "scouring rushes" because rough silica deposits found in their cell walls once made them useful for cleaning. They have true roots that branch irregularly. Horsetails have a large sporophyte and a small gametophyte, each independent of the other.

The first **ferns** appeared during the Devonian period. Today this group comprises more than 12,000 species. Analyses of gene sequences indicate that a few species traditionally allied with ferns may in fact be more closely related to horsetails than to ferns. Nonetheless, the majority of ferns form a monophyletic group.

Although most ferns are terrestrial, a few species live in shallow fresh water (**FIGURE 21.8C**). Terrestrial ferns are characterized by large leaves with branching vascular strands (**FIGURE 21.8D**). Some fern leaves become climbing organs and may grow to be as long as 30 meters.

In the alternating generations of a fern, the gametophyte is small, delicate, and short-lived, but the sporophyte can be very large and can sometimes survive for hundreds of years (**FIGURE 21.9**). Ferns require liquid water for the transport of the male gametes to the female gametes, so most ferns inhabit shaded, moist woodlands and swamps. The sporangia of ferns typically are borne on a stalk in clusters called **sori** (singular *sorus*). The sori are found on the undersurfaces of the leaves, sometimes covering the entire undersurface and sometimes located at the edges.

The vascular plants branched out

Several features that were new to the vascular plants evolved in lycophytes and monilophytes. Roots probably had their evolutionary origins as a branch, either of a rhizome or of the aboveground portion of a stem. That branch presumably penetrated the soil and branched further. The underground portion could anchor the plant firmly, and even in this primitive condition, it could absorb water and minerals.

The microphylls of lycophytes were probably the first leaflike structures to evolve among the vascular plants. Microphylls are usually small and only rarely have more than a single vascular strand, at least in existing species. Some biologists believe that microphylls had their evolutionary origins as sterile sporangia (**FIGURE 21.10A**). The principal characteristic of a microphyll is a vascular strand that departs from the vascular system of the stem in such a way that the structure of the stem's vascular system is scarcely disturbed. This pattern was evident even in the lycophyte trees of the Carboniferous period, many of which had microphylls many centimeters long.

The monilophytes and seed plants constitute a clade called the **euphyllophytes** (*eu*, "true"; *phyllos*, "leaf"). An important synapomorphy of the euphyllophytes is **overtopping**, a growth pattern in which one branch differentiates from and grows beyond the others (**FIGURE 21.10B**). Overtopping would have given these plants an advantage in the competition for light, enabling them to shade their dichotomously branching competitors. The overtopping growth of the euphyllophytes also allowed a new type of leaflike structure to evolve. This larger, more complex leaf is called a **megaphyll**. The megaphyll is thought to have arisen from the flattening of a portion of a branching stem system that exhibited overtopping growth. This change was followed by the development of photosynthetic tissue between the members of overtopped groups of branches, which had the advantage of increasing the photosynthetic surface area of those branches.

Heterospory appeared among the vascular plants

Among the earliest vascular plants, the gametophyte and the sporophyte were independent of one another, and both were photosynthetic. The spores produced by the sporophyte were of a single type and developed into a single type of gametophyte that bore both female and male reproductive organs. This is still true in some surviving lineages of vascular plants, such as horsetails and many ferns. Such plants, which bear a single type of spore, are said to be **homosporous** (**FIGURE 21.11A**).

A system with two distinct types of spores evolved several times among the monilophytes and in the ancestor of seed plants. Plants of this type are said to be **heterosporous** (**FIGURE 21.11B**). In heterospory, one type of spore—the **megaspore**—develops into a specifically female gametophyte (a **megagametophyte**) that produces only eggs. The other type, the **microspore**, is smaller and develops into a male gametophyte (a

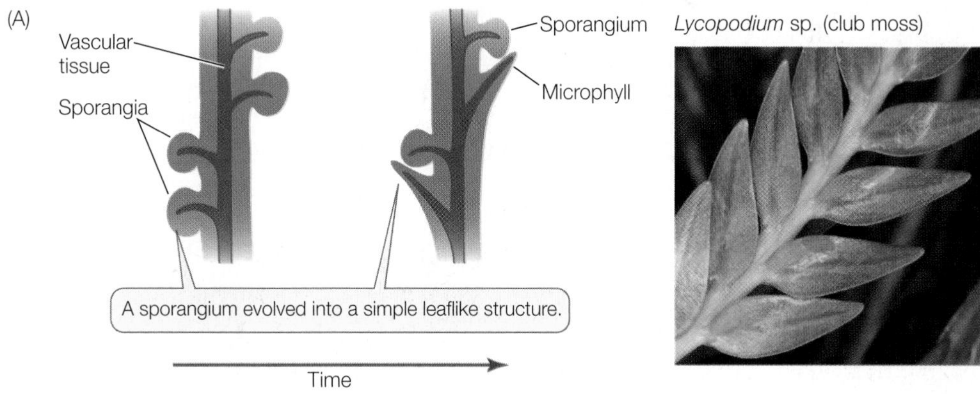
(A)
Vascular tissue
Sporangia
Sporangium
Microphyll
A sporangium evolved into a simple leaflike structure.
Time
Lycopodium sp. (club moss)

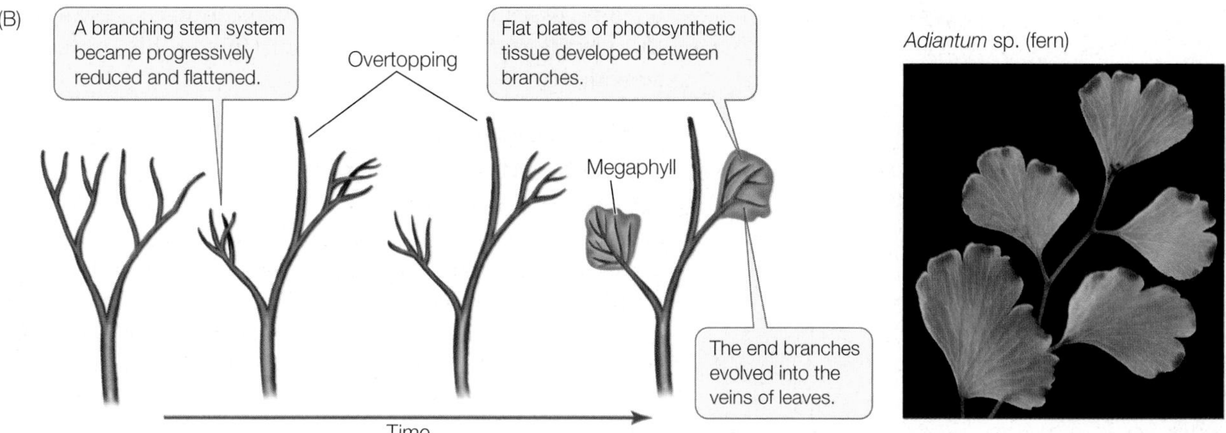
(B)
A branching stem system became progressively reduced and flattened.
Overtopping
Flat plates of photosynthetic tissue developed between branches.
Megaphyll
The end branches evolved into the veins of leaves.
Time
Adiantum sp. (fern)

FIGURE 21.10 Evolution of Leaves (A) Microphylls are thought to have evolved from sterile sporangia. (B) The megaphylls of monilophytes and seed plants may have arisen as photosynthetic tissue developed between branch pairs that were "left behind" as dominant branches overtopped them.

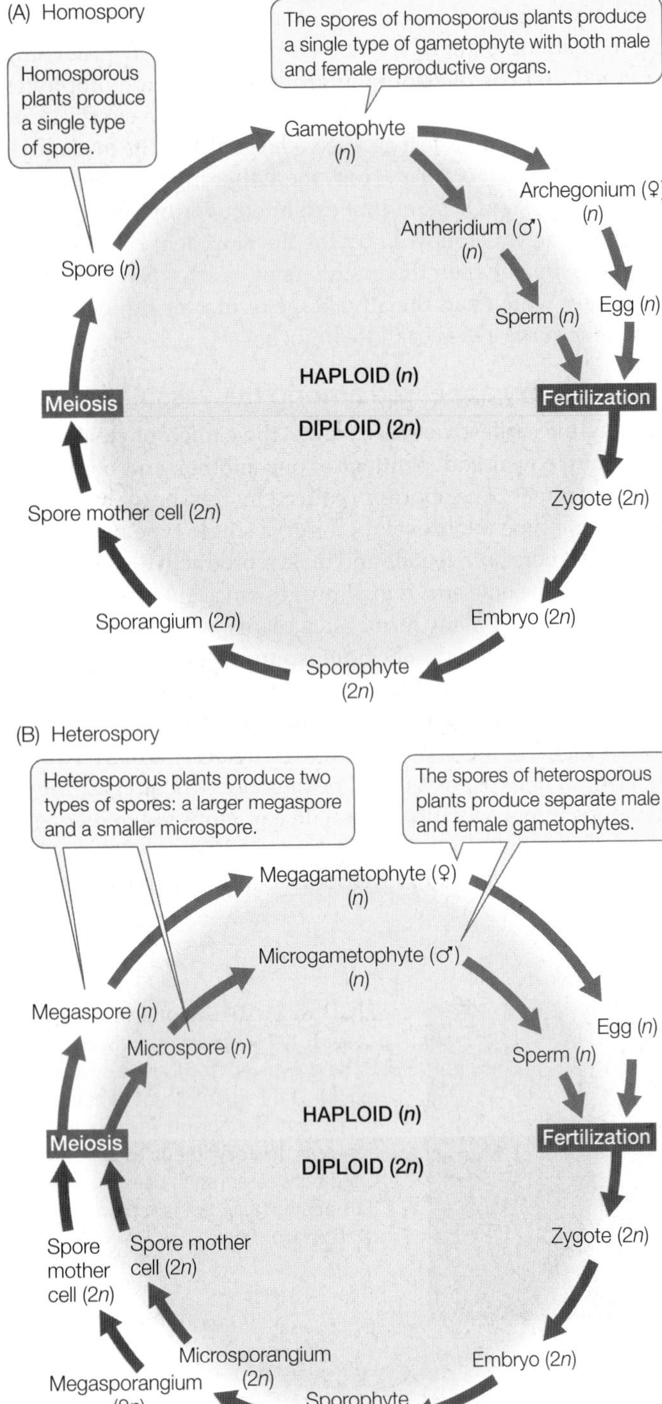

FIGURE 21.11 Homospory and Heterospory (A) Homosporous plants bear a single type of spore. Each gametophyte has two types of sex organs, antheridia (male) and archegonia (female). (B) Heterosporous plants bear two types of spores that develop into distinctly male and female gametophytes.

Go to ACTIVITY 21.2 Homospory
PoL2e.com/ac21.2

Go to ACTIVITY 21.3 Heterospory
PoL2e.com/ac21.3

microgametophyte) that produces only sperm. The sporophyte produces megaspores in small numbers in **megasporangia** and microspores in large numbers in **microsporangia**. Heterospory affects not only the spores and the gametophytes, but also the sporophyte plant itself, which must develop two types of sporangia.

The fact that heterospory evolved repeatedly suggests that it affords selective advantages. What advantages does heterospory provide? Heterospory allows the production of many small microspores, which are easily transported from plant to plant. Heterospory also results in the production of a few large megaspores in large megasporangia, which can provide nutrition and protection for the developing embryo. With this division, plants can increase the opportunities for long-distance cross-fertilization, and at the same time increase the chances for survival of the developing embryo.

CHECKpoint CONCEPT 21.3

- ✓ How do the vascular tissues xylem and phloem serve the vascular plants?
- ✓ Describe the evolution and distribution of different kinds of leaves and roots among the vascular plants.
- ✓ Explain the concept of heterospory. How does heterospory provide selective advantages over homospory?

All of the vascular plant groups we have discussed thus far disperse by means of spores. The embryos of these seedless vascular plants develop directly into sporophytes, which either survive or die, depending on environmental conditions. The spores of some seedless plants may remain dormant and viable for long periods, but the embryos of seedless plants are relatively unprotected (see Figure 21.9). Greater protection of the embryo evolved in the seed plants.

CONCEPT 21.4 Seeds Protect Plant Embryos

The **seed plants**—the gymnosperms and the angiosperms—provide a secure and lasting dormant stage for the embryo, known as a seed. In seeds, embryos may safely lie dormant, in some cases for many years (even centuries), until conditions are right for germination. How is this protection provided?

Features of the seed plant life cycle protect gametes and embryos

In Concept 21.2 we described a trend in plant evolution: the sporophyte became less dependent on the gametophyte, which became smaller in relation to the sporophyte. This trend continued with the seed plants, whose gametophyte generation is reduced even further than it is in the ferns (**FIGURE 21.12**). The haploid gametophyte develops partly or entirely while attached to and nutritionally dependent on the diploid sporophyte.

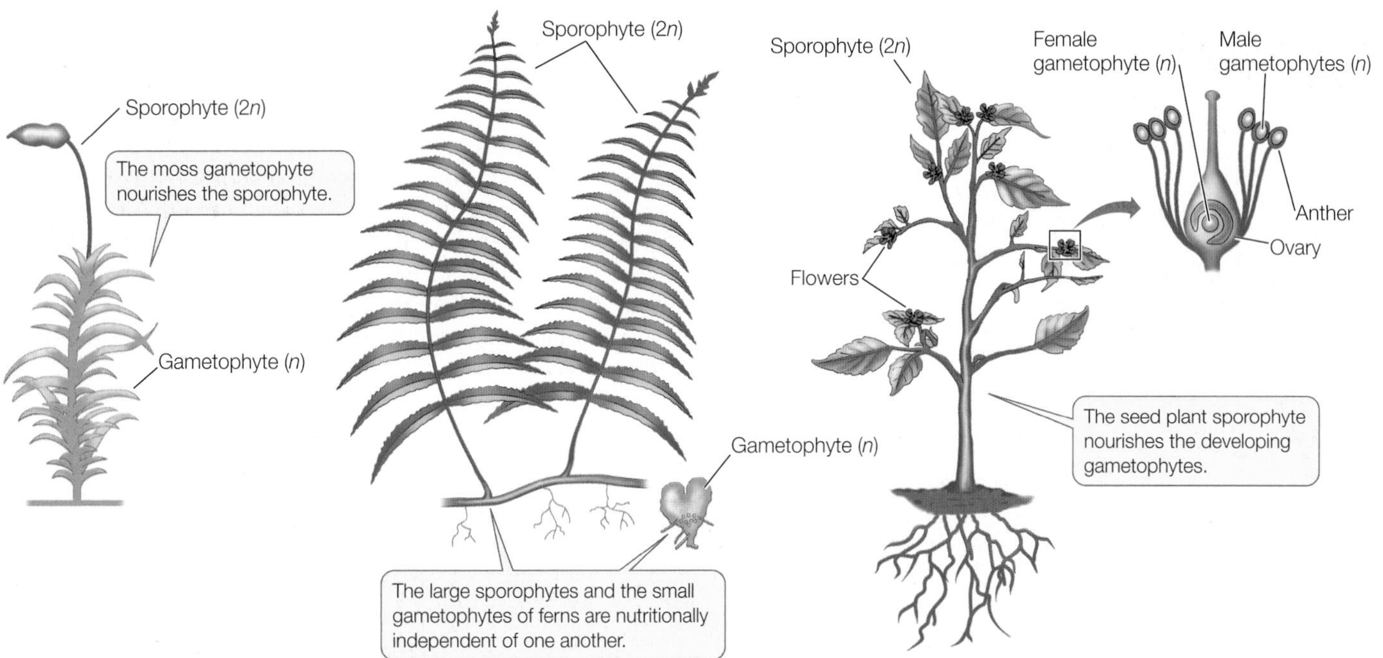

FIGURE 21.12 The Relationship between Sporophyte and Gametophyte In the course of plant evolution, the gametophyte (brown) has been reduced and the sporophyte (blue) has become more prominent.

Among the seed plants, only the earliest diverging groups of gymnosperms (including modern cycads and ginkgos) have swimming sperm. Even in these groups, sperm is transferred via pollen grains, so fertilization does not require liquid water outside the plant body. The evolution of pollen, along with the advent of seeds, gave seed plants the opportunity to colonize drier areas and spread over the terrestrial environment.

Seed plants are heterosporous (see Figure 21.11B)—that is, they produce two types of spores, one that becomes a microgametophyte (male gametophyte) and one that becomes a megagametophyte (female gametophyte). They form separate microsporangia and megasporangia on structures that are grouped on short stems, such as the stamens and carpels of an angiosperm flower.

Within the microsporangium, the meiotic products are microspores. Within its spore wall, a microspore divides mitotically one or a few times to form a multicellular male gametophyte called a **pollen grain**. Pollen grains are released from the microsporangium to be distributed by wind or by an animal pollinator (**FIGURE 21.13**). The spore wall that surrounds the pollen grain contains a substance called sporopollenin, the most chemically resistant biological compound known. Sporopollenin protects the pollen grain against dehydration and chemical damage—another advantage in terms of survival in the terrestrial environment.

In contrast to the microspores, the megaspores of seed plants are not shed. Instead they develop into female gametophytes within the megasporangia. These megagametophytes are dependent on the sporophyte for food and water.

In most seed plant species, only one of the meiotic products in a megasporangium survives. The surviving haploid nucleus divides mitotically, and the resulting cells divide again to produce a multicellular female gametophyte. The megasporangium is surrounded by sterile sporophytic structures, which form an **integument** that protects the megasporangium

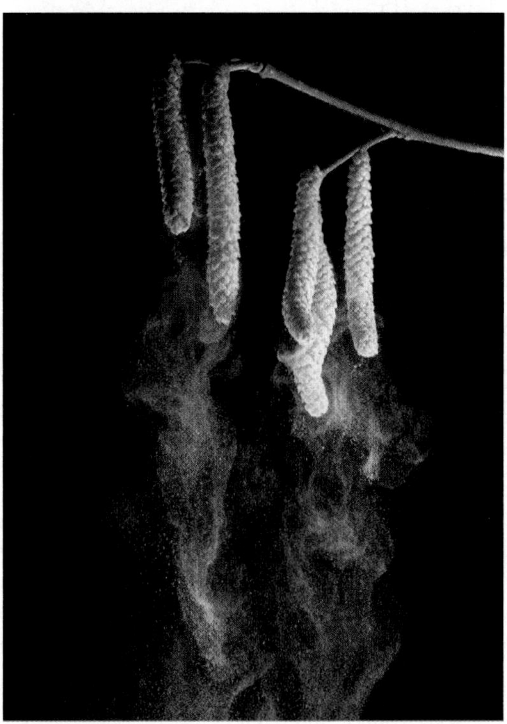

FIGURE 21.13 Pollen Grains Pollen grains are the male gametophytes of seed plants. The pollen of the hazel tree is dispersed by the wind. The grains may land near female gametophytes of the same or other hazel plants.

Go to MEDIA CLIP 21.4
Pollen Transfer by Wind
PoL2e.com/mc21.4

and its contents. Together, the megasporangium and integument constitute the **ovule**, which will develop into a seed after fertilization.

The arrival of a pollen grain at an appropriate landing point, close to a female gametophyte on a sporophyte of the same species, is called **pollination**. A pollen grain that reaches this point develops further. It produces a slender **pollen tube** that elongates and digests its way toward the megagametophyte (**FIGURE 21.14**). When the tip of the pollen tube reaches the megagametophyte, sperm are released from the tube and fertilization occurs.

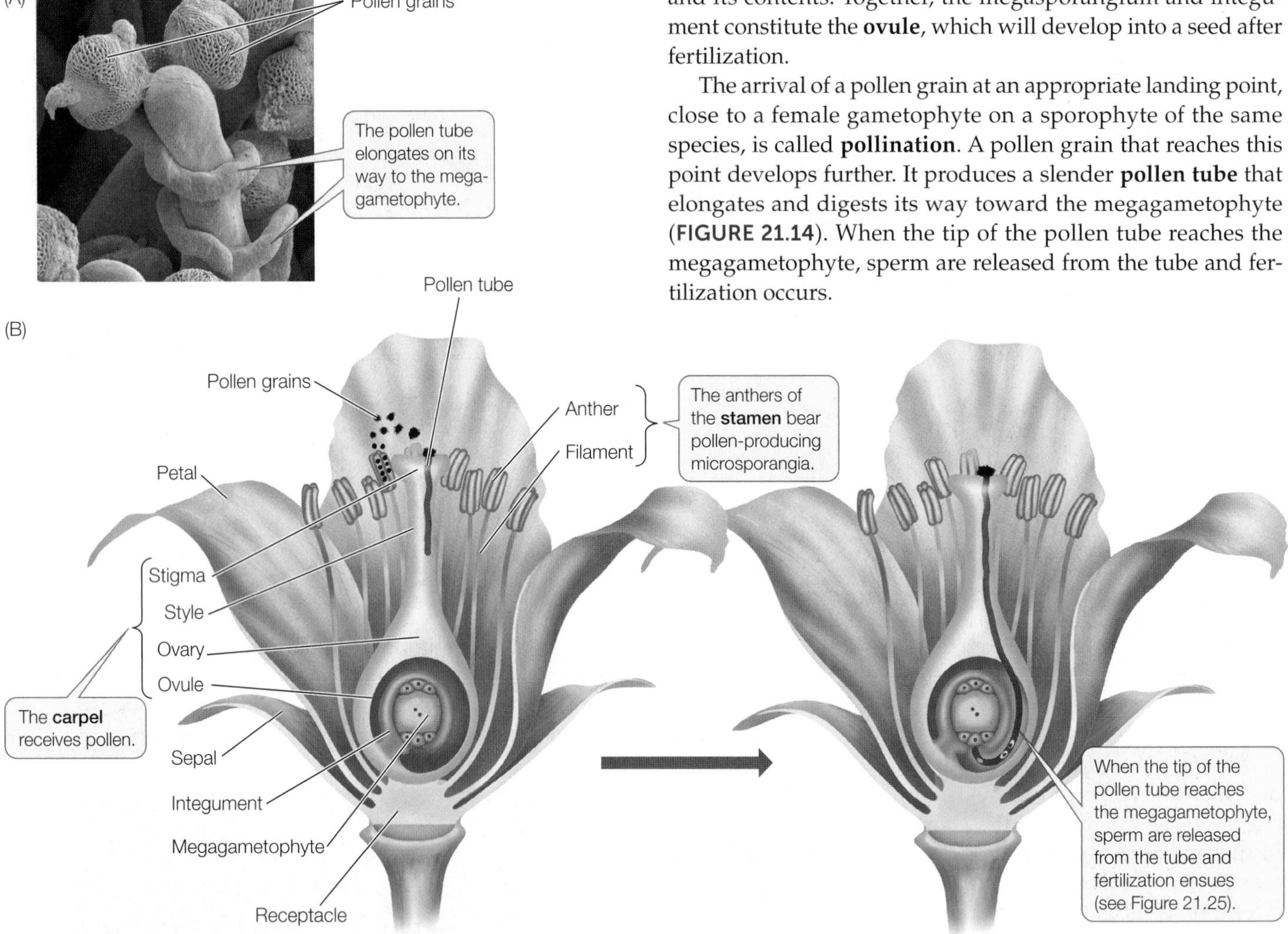
(A)
Pollen grains
The pollen tube elongates on its way to the megagametophyte.
(B)
Pollen tube
Pollen grains
Anther
Filament
The anthers of the **stamen** bear pollen-producing microsporangia.
Petal
Stigma
Style
Ovary
Ovule
The **carpel** receives pollen.
Sepal
Integument
Megagametophyte
Receptacle
When the tip of the pollen tube reaches the megagametophyte, sperm are released from the tube and fertilization ensues (see Figure 21.25).

FIGURE 21.14 Pollination in Seed Plants In all seed plants, a pollen tube grows from the pollen grain to the megagametophyte, where sperm are released. (A) Scanning electron micrograph of a pollen tube growing in a prairie gentian flower. (B) The process of pollination is diagrammed for a generalized angiosperm flower.

APPLY THE CONCEPT

Seeds protect plant embryos

In 1879 W. J. Beal began an experiment that he could not hope to finish in his lifetime. He prepared 20 lots of seeds for long-term storage. Each lot consisted of 50 seeds from each of 23 species. He mixed each lot of seeds with sand and placed the mixture in an uncapped bottle, then buried all the bottles on a sandy knoll. At regular intervals over the next century, several different biologists have excavated a bottle and checked the viability of its contents. The table shows the number of germinating seeds (of the original 50) from three species in years 50–100 of this ongoing experiment.

	YEARS AFTER BURIAL					
SPECIES	50	60	70	80	90	100
Oenothera biennis (Evening primrose)	19	12	7	5	0	0
Rumex crispus (Curly dock)	26	2	7	1	0	0
Verbascum blattaria (Moth mullein)	31	34	37	35	10	21

1. Calculate the percentage of surviving seeds for these three species in years 50–100 and graph seed survivorship as a function of time buried.
2. No seeds of the first two species were viable after 90 years of the experiment. Assume 100% seed viability at the start of the experiment (year 0), and predict from your graph the approximate year when you think the last of the *Verbascum blattaria* seeds will germinate.
3. What factors do you think might influence the differences in long-term seed viability among the species?

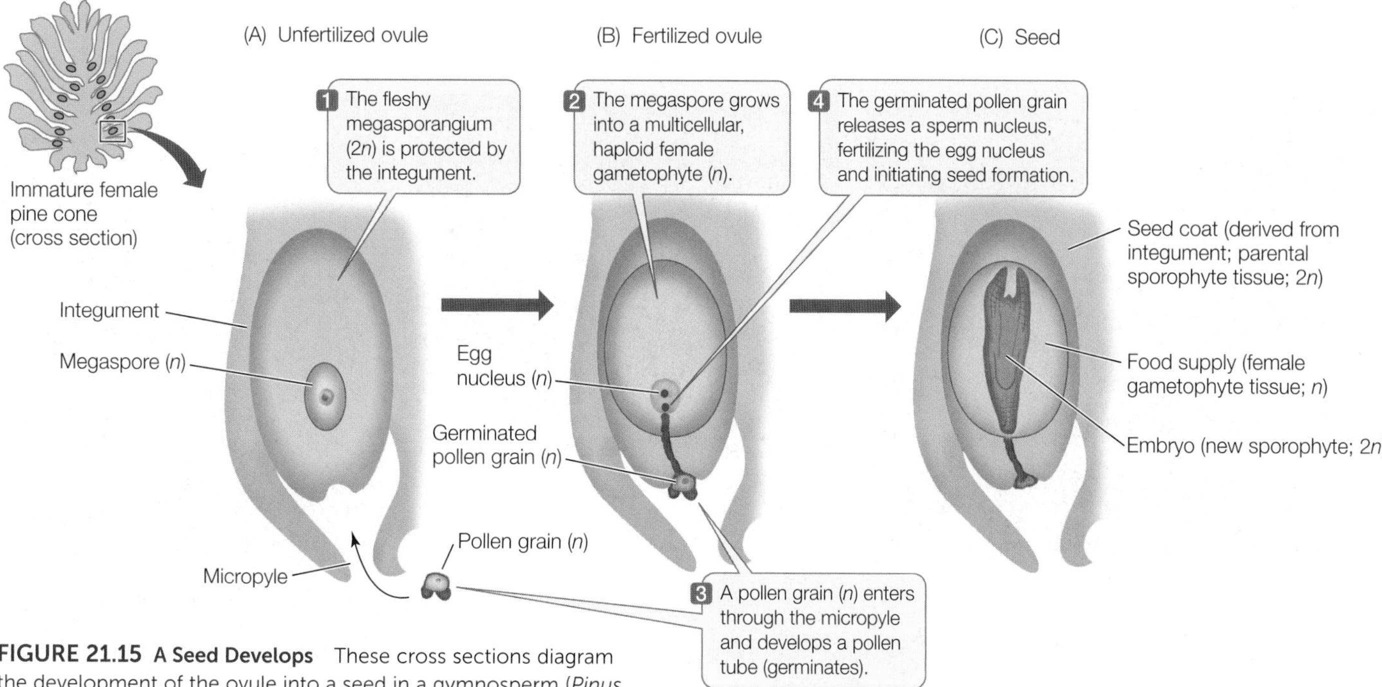
(A) Unfertilized ovule
(B) Fertilized ovule
(C) Seed
Immature female pine cone (cross section)
1 The fleshy megasporangium (2n) is protected by the integument.
2 The megaspore grows into a multicellular, haploid female gametophyte (n).
4 The germinated pollen grain releases a sperm nucleus, fertilizing the egg nucleus and initiating seed formation.
Integument
Megaspore (n)
Egg nucleus (n)
Germinated pollen grain (n)
Pollen grain (n)
Micropyle
Seed coat (derived from integument; parental sporophyte tissue; 2n)
Food supply (female gametophyte tissue; n)
Embryo (new sporophyte; 2n)
3 A pollen grain (n) enters through the micropyle and develops a pollen tube (germinates).

FIGURE 21.15 A Seed Develops These cross sections diagram the development of the ovule into a seed in a gymnosperm (*Pinus* sp.). Angiosperm seed development has differences (e.g., angiosperm integuments have two layers rather than one, and the angiosperm embryo is nourished by specialized tissue called endosperm) but follows the same general steps. (A) The haploid megaspore is nourished by tissues of the parental sporophyte (the diploid megasporangium). (B) The mature megaspore is fertilized by a pollen grain that penetrates the integument, germinates (grows a pollen tube), and releases a sperm nucleus. (C) Fertilization initiates production of a seed. A mature seed contains three generations: a diploid embryo (the new sporophyte), which is surrounded by haploid female gametophyte tissue that supplies nutrition, which is in turn surrounded by the seed coat (diploid parental sporophyte tissue).

The resulting diploid zygote divides repeatedly, forming an embryonic sporophyte. After a period of embryonic development, growth is temporarily suspended (the embryo enters a dormant stage). The end product at this stage is the multicellular seed.

The seed is a complex, well-protected package

A **seed** contains tissues from three generations (**FIGURE 21.15**). A seed coat develops from the integument—the tissues of the diploid sporophyte parent that surround the megasporangium. Within the megasporangium is haploid tissue from the female gametophyte, which contains a supply of nutrients for the developing embryo. (This tissue is fairly extensive in most gymnosperm seeds. In angiosperm seeds it is greatly reduced, and nutrition for the embryo is supplied instead by a tissue called endosperm.) In the center of the seed is the third generation, the embryo of the new diploid sporophyte.

The seed is a well-protected resting stage. The seeds of some species may remain dormant but stay viable (capable of growth and development) for many years, germinating only when conditions are favorable for the growth of the sporophyte. During the dormant stage, the seed coat protects the embryo from excessive drying and may also protect it against potential predators that would otherwise consume the embryo and its nutrient reserves. Many seeds have structural adaptations or are contained in fruits that promote their dispersal by wind or, more often, by animals. When the young sporophyte resumes growth, it draws on the food reserves in the seed. The possession of seeds is a major reason for the enormous evolutionary success of the seed plants, which are the dominant life forms of most modern terrestrial floras.

A change in stem anatomy enabled seed plants to grow to great heights

The earliest fossil seed plants have been found in late Devonian rocks. These plants had extensively thickened woody stems, which developed through the proliferation of xylem. This type of growth, which increases the diameter of stems and roots in some modern seed plants, is called **secondary growth**, and its product is called secondary xylem, or wood.

The younger portion of the wood produced by secondary growth is well adapted for water transport, but older wood becomes clogged with resins or other materials. Although no longer functional in transport, the older wood continues to provide support for the plant. This support allows woody plants to grow taller than other plants around them and thus capture more light for photosynthesis.

Not all seed plants are woody. In the course of seed plant evolution, many groups lost the woody growth habit. Nonetheless, other advantageous attributes helped them become established in an astonishing variety of places.

Gymnosperms have naked seeds

The two major groups of living seed plants are the **gymnosperms** (such as pines and cycads) and the **angiosperms** (flowering plants; see Figure 21.1B). We'll discuss the flowering plants in Concept 21.5 and examine the gymnosperms here.

The gymnosperms are seed plants that do not form flowers or fruits. Gymnosperms (which means "naked-seeded")

(A) *Encephalartos* sp.

(B) *Ginkgo biloba*

(C) *Welwitschia mirabilis*

(D) *Picea glauca* and *P. mariana*

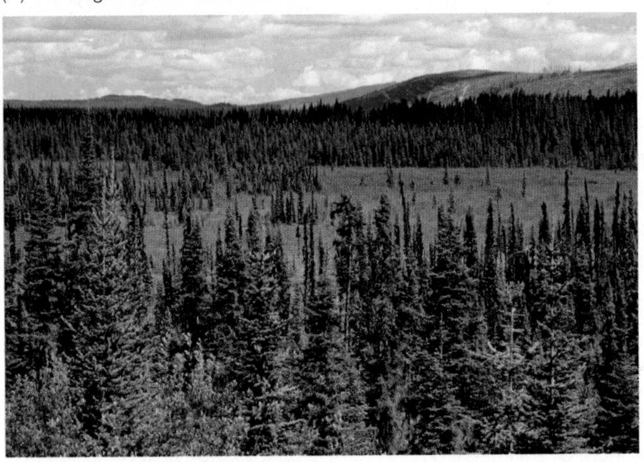

FIGURE 21.16 Diversity among the Gymnosperms (A) Many cycads have growth forms that resemble both ferns and palms, although cycads are not closely related to either group. (B) The characteristic broad leaves of the maidenhair tree. (C) The straplike leaves of *Welwitschia*, a gnetophyte, grow throughout the life of the plant, breaking and splitting as they grow. (D) Conifers dominate many types of landscapes in the Northern Hemisphere. Shown here are white and black spruce in Canada.

are so named because their ovules and seeds, unlike those of angiosperms, are not protected by ovary or fruit tissue. Although there are probably fewer than 1,200 living species of gymnosperms, these plants are second only to the angiosperms in their dominance of the terrestrial environment. The gymnosperms can be divided into four major groups:

- **Cycads** are palmlike plants of the tropics and subtropics (**FIGURE 21.16A**). Of the present-day gymnosperms, the cycads are probably the earliest-diverging clade. There are about 300 species, some of which grow as tall as 20 meters. The tissues of many species are highly toxic to humans if ingested.
- **Ginkgos**, common during the Mesozoic era, are represented today by a single genus and species: *Ginkgo biloba*, the maidenhair tree (**FIGURE 21.16B**). There are both male (microsporangiate) and female (megasporangiate) maidenhair trees. The difference is determined by X and Y sex chromosomes, as in humans. Few other plants have distinct sex chromosomes.
- **Gnetophytes** number about 90 species in three very different genera, which share certain characteristics analogous to ones found in the angiosperms. One of the gnetophytes is *Welwitschia* (**FIGURE 21.16C**), a long-lived desert plant with just two permanent straplike leaves, which split into many pieces that sprawl across the sand.
- **Conifers** are by far the most abundant of the gymnosperms. There are about 700 species of these cone-bearing plants, including the pines, spruce, and redwoods (**FIGURE 21.16D**).

With the exception of the gnetophytes, the living gymnosperm groups have only tracheids as water-conducting and support cells within the xylem. Most gymnosperms lack the vessel elements and fibers (cells specialized for water conduction and support, respectively) that are found in angiosperms. While the gymnosperm water-transport and support system may thus seem somewhat less efficient than that of the angiosperms, it serves some of the largest trees known. The coastal redwoods of California are the tallest gymnosperms, with some individuals growing to well over 100 meters tall.

During the Permian, as environments became warmer and dryer, the conifers and cycads flourished. Gymnosperm forests changed over time as the gymnosperm groups evolved. Gymnosperms dominated the Mesozoic era, during which the continents drifted apart and large dinosaurs lived. Gymnosperms were the principal trees in all forests until about 65 million years ago, and even today conifers are the dominant trees in many forests, especially at high latitudes and altitudes. The oldest living single organism on Earth today is a gymnosperm

FIGURE 21.17 Female and Male Cones (A) The scales of female cones (megastrobili) are modified branches. (B) The scales of male cones (microstrobili) are modified leaves.

in California—a bristlecone pine that germinated about 4,800 years ago, at about the time the ancient Egyptians were starting to develop writing.

Conifers have cones and lack swimming sperm

The great Douglas fir and cedar forests found in the northwestern United States and the massive boreal forests of pine, fir, and spruce of the northern regions of Eurasia and North America, as well as on the upper slopes of mountain ranges elsewhere, rank among the great forests of the world. All these trees belong to one group of gymnosperms: the conifers, or cone-bearers.

Male and female **cones** contain the reproductive structures of conifers. The female (seed-bearing) cone is known as a **megastrobilus** (plural *megastrobili*). An example of a familiar megastrobilus is the woody cone of pine trees. The seeds in a megastrobilus are protected by a tight cluster of woody scales, which are modifications of branches extending from a central axis (**FIGURE 21.17A**). The typically much smaller male (pollen-bearing) cone is known as a **microstrobilus**. The microstrobilus is typically herbaceous rather than woody, as its scales are composed of modified leaves, beneath which are the pollen-bearing microsporangia (**FIGURE 21.17B**).

The life cycle of a pine illustrates reproduction in gymnosperms (**FIGURE 21.18**). The production of male gametophytes in the form of pollen grains frees the plant completely from its dependence on liquid water for fertilization. Wind, rather than water, assists conifer pollen grains in their first stage of travel from the microstrobilus to the female gametophyte inside a cone. A pollen tube provides the sperm with the means for the last stage of travel by elongating through maternal sporophytic tissue. When the pollen tube reaches the female gametophyte, it releases two sperm, one of which degenerates after the other unites with an egg. Union of sperm and egg results in a zygote. Mitotic divisions and further development of the zygote result in an embryo.

The megasporangium, in which the female gametophyte will form, is enclosed in a layer of sporophytic tissue—the integument—that will eventually develop into the seed coat that protects the embryo. The integument, the megasporangium inside it, and the tissue attaching it to the maternal sporophyte constitute the ovule. The pollen grain enters through a small opening in the integument at the tip of the ovule, the **micropyle**.

Most conifer ovules (which will develop into seeds after fertilization) are borne exposed on the upper surfaces of the scales of the cone (megastrobilus). The only protection of the ovules comes from the scales, which are tightly pressed against one another within the cone. Some pines, such as the lodgepole pine, have such tightly closed cones that only fire suffices to split them open and release the seeds. These species are said to be fire-adapted, and fire is essential to their reproduction.

About half of all conifer species have soft, fleshy modifications of cones that envelop their seeds. Some of these are fleshy, fruitlike cones, as in junipers. Others are fruitlike

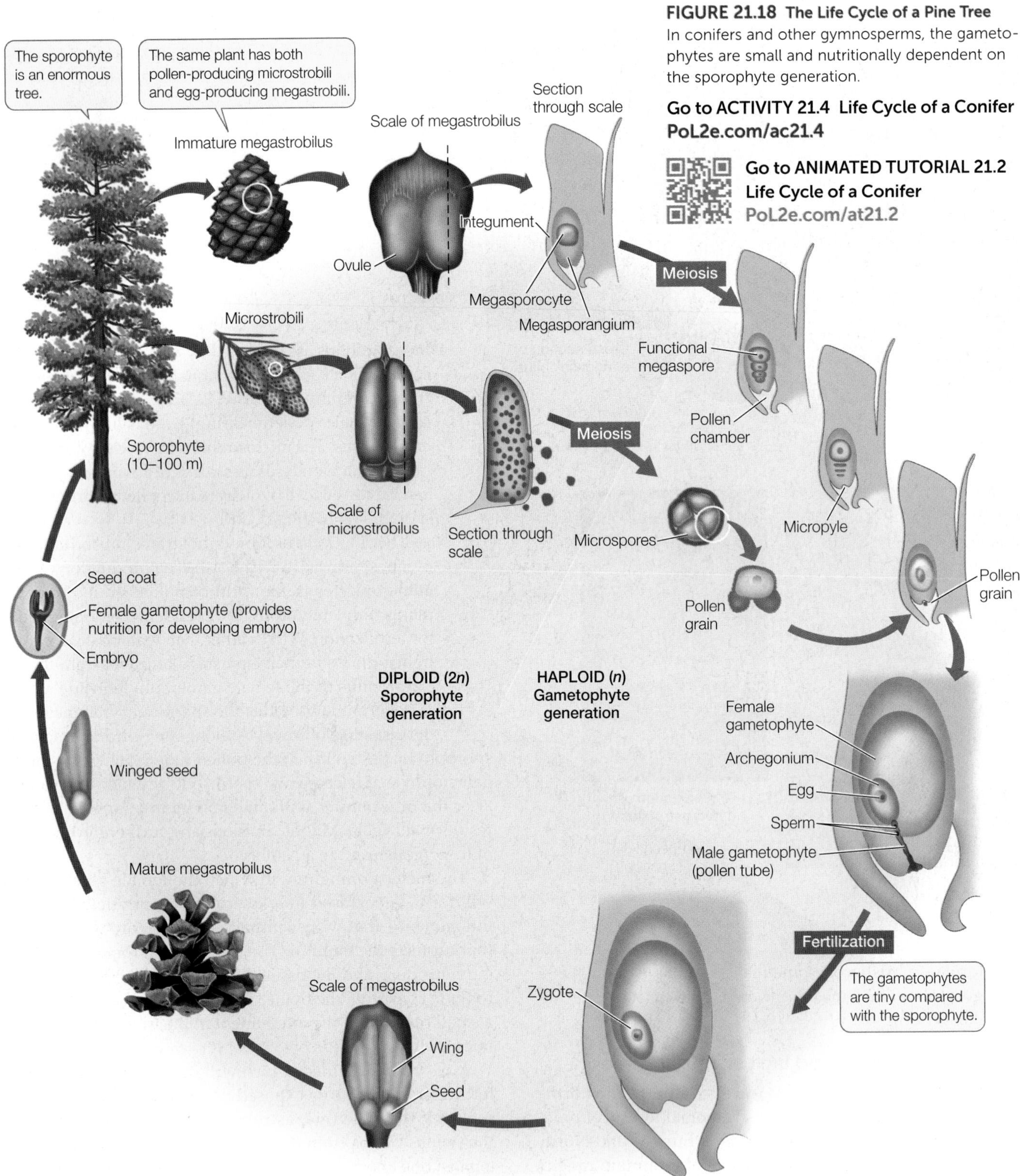

FIGURE 21.18 The Life Cycle of a Pine Tree In conifers and other gymnosperms, the gametophytes are small and nutritionally dependent on the sporophyte generation.

Go to ACTIVITY 21.4 Life Cycle of a Conifer PoL2e.com/ac21.4

Go to ANIMATED TUTORIAL 21.2 Life Cycle of a Conifer PoL2e.com/at21.2

extensions of the seeds, as in yews. These tissues, although often mistaken for "berries," are not true fruits. As we will see when we discuss angiosperms, true fruits are a plant's ripened ovaries, which are absent in gymnosperms. Nonetheless, the fleshy tissues that surround many conifer seeds serve a similar purpose as that of the fruits of flowering plants, acting as an enticement for seed-dispersing animals. Animals eat these fleshy tissues and disperse the seeds in their feces, often depositing the seeds considerable distances from the parent plant.

CHECKpoint CONCEPT 21.4

- ✓ Distinguish between the roles of the megagametophyte and the pollen grain.
- ✓ Explain the importance of pollen in freeing seed plants from dependence on liquid water.
- ✓ What are some of the advantages afforded by seeds? By wood?
- ✓ Explain how fire can be necessary for the survival of some plant species.

CONCEPT 21.5 Flowers and Fruits Increase the Reproductive Success of Angiosperms

The most obvious feature defining the angiosperms is the **flower**, which is their sexual structure. Production of **fruits** is also a shared derived trait of angiosperms. After fertilization, the ovary of a flower (together with the seeds it contains) develops into a fruit that protects the seeds and can promote seed dispersal. As we will see, both flowers and fruits afford major reproductive advantages to angiosperms.

Angiosperms have many shared derived traits

The name *angiosperm* ("enclosed seed") is derived from another distinctive trait of flowering plants that is related to the formation of fruits: the ovules and seeds are enclosed in a modified leaf called a **carpel**. Besides protecting the ovules and seeds, the carpel often interacts with incoming pollen to prevent self-fertilization, thus favoring cross-fertilization and increasing genetic diversity.

The female gametophyte of angiosperms is even more reduced than that of gymnosperms, usually consisting of only seven cells. Thus the angiosperms represent the current extreme of the trend we have traced throughout the evolution of vascular plants: the sporophyte generation becomes larger and more independent of the gametophyte, while the gametophyte becomes smaller and more dependent on the sporophyte.

The xylem of most angiosperms is distinguished by the presence of specialized water-transporting cells called **vessel elements**. These cells are larger in diameter than tracheids and connect with one another without obstruction, allowing easy water movement. A second distinctive cell type in angiosperm xylem is the **fiber**, which plays an important role in supporting the plant body. Angiosperm phloem possesses another unique cell type, called a companion cell. Like the gymnosperms, woody angiosperms show secondary growth, increasing in diameter by producing secondary xylem and secondary phloem.

A more comprehensive list of angiosperm synapomorphies, then, includes the following (some of these traits will be discussed later in this chapter):

- Flowers
- Fruits
- Ovules and seeds enclosed in a carpel
- Highly reduced gametophytes
- Germination of pollen on a stigma
- Double fertilization
- Endosperm (nutritive tissue for the embryo)
- Phloem with companion cells

The majority of these traits bear directly on angiosperm reproduction, which is a large factor in the success of this dominant plant group.

The sexual structures of angiosperms are flowers

Flowers come in an astonishing variety of forms—just think of some of the flowers you recognize. Flowers may be single, or they may be grouped together to form an **inflorescence**. Different families of flowering plants have characteristic types of inflorescences, such as the compound umbels of the carrot family (**FIGURE 21.19A**), the heads of the aster family (**FIGURE 21.19B**), and the spikes of many grasses (**FIGURE 21.19C**).

If you examine any familiar flower, you will notice that the outer parts look somewhat like leaves. In fact, all the parts of a flower *are* modified leaves. The diagram in Figure 21.14B represents a generalized flower (for which there is no exact counterpart in nature). The structures bearing microsporangia are called **stamens**. Each stamen is composed of a **filament** bearing an **anther** that contains the pollen-producing microsporangia. The structures bearing megasporangia are called carpels. The swollen base of the carpel, containing one or more ovules (each containing a megasporangium surrounded by two protective integuments), is called the **ovary**. The stalk at the top of the carpel is the **style**, and the terminal surface that receives pollen grains is the **stigma**. Two or more fused carpels, or a single carpel if only one is present, are also called a **pistil**.

Go to ACTIVITY 21.5 Flower Morphology
PoL2e.com/ac21.5

In addition, many flowers have specialized sterile (non-spore-bearing) leaves. The inner ones are called **petals** (collectively, the **corolla**) and the outer ones **sepals** (collectively, the calyx). The corolla and calyx can be quite showy and often play roles in attracting animal pollinators to the flower. The calyx more commonly protects the immature flower in bud. From base to apex, these floral organs—sepals, petals, stamens, and carpels—are usually positioned in circular arrangements or whorls and attached to a central stalk.

The generalized flower in Figure 21.14B has functional megasporangia and microsporangia. Such flowers are referred to as **perfect** (or hermaphroditic). Many angiosperms produce two types of flowers, one with only megasporangia and the other with only microsporangia. Consequently, either the stamens or the carpels are nonfunctional or absent in a given flower, and the flower is referred to as **imperfect**.

Species such as corn or birch, in which both megasporangiate (female) and microsporangiate (male) flowers occur on the same plant, are said to be **monoecious** ("one-housed"—but, it must be added, one house with separate rooms). Complete

(A) *Daucus carota*

Umbel

Flowers

Compound umbel

(B) *Zinnia* sp.

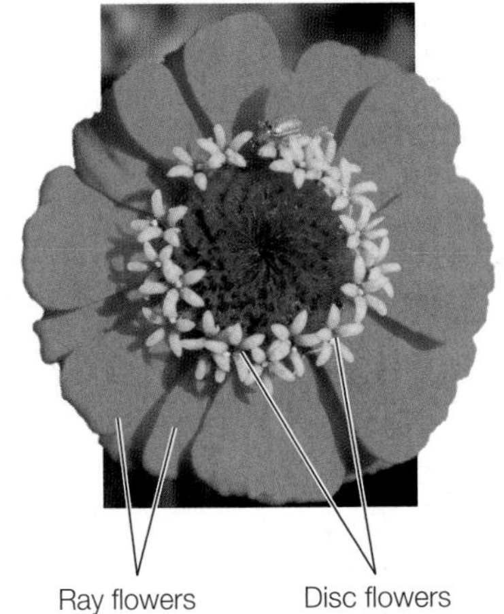

(C) *Agropyron repens*

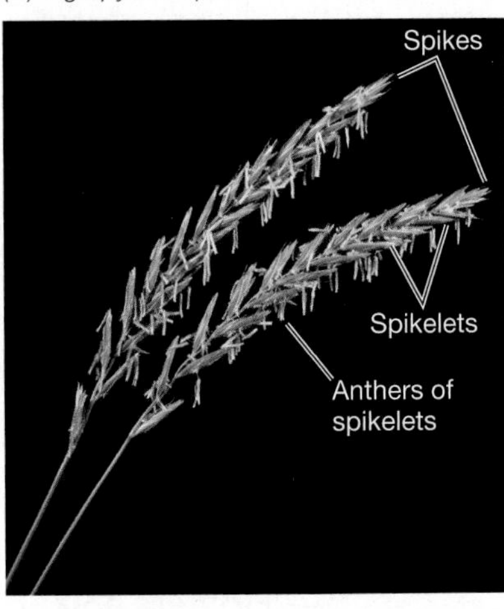

FIGURE 21.19 Inflorescences (A) The inflorescence of Queen Anne's lace, a member of the carrot family, is a compound umbel. Each umbel bears flowers on stalks that arise from a common center. (B) Zinnias are members of the aster family; their inflorescence is a head. Within the head, each of the long, petal-like structures is a ray flower; the central portion of the head consists of dozens to hundreds of disc flowers. (C) Some grasses, such as quack grass, have inflorescences called spikes, which are composed of many individual flowers, or spikelets.

separation of imperfect flowers occurs in some other angiosperm species, such as willows and date palms; in these species, an individual plant produces either flowers with stamens or flowers with carpels, but never both. Such species are said to be **dioecious** ("two-housed").

Flower structure has evolved over time

The flowers of the earliest-diverging clades of angiosperms have a large and variable number of tepals (undifferentiated sepals and petals), carpels, and stamens (**FIGURE 21.20A**). Evolutionary change within the angiosperms has included some striking modifications of this early condition: reductions in the number of each type of floral organ to a fixed number, differentiation of petals from sepals, and changes in symmetry from radial (as in a lily or magnolia) to bilateral (as in a sweet pea or orchid), often accompanied by an extensive fusion of parts (**FIGURE 21.20B**).

According to one hypothesis, the first carpels to evolve were leaves with marginal sporangia, folded but incompletely closed. Early in angiosperm evolution, the carpels fused with one another, forming a single, multichambered ovary (**FIGURE 21.21A**). In some flowers, the other floral organs are attached at the top of the ovary, rather than at the bottom as in Figure 21.14B. The stamens of the most ancient flowers may have been leaflike (**FIGURE 21.21B**), with little resemblance to the stamens of the generalized flower in Figure 21.14B.

A perfect flower represents a compromise of sorts. On the one hand, by attracting a pollinating bird or insect, the plant is attending to both its female and male functions with a single flower type, whereas plants with imperfect flowers must create that attraction twice—once for each type of flower. On the other hand, the perfect flower can favor self-pollination, which is usually disadvantageous. Another potential problem is that the female and male functions might interfere with each other—for example, the stigma might be so placed as to make it difficult for pollinators to reach the anthers, thus reducing the export of pollen to other flowers.

Might there be a way around these problems? One solution is seen in the bush monkeyflower (*Mimulus aurantiacus*), which is pollinated by hummingbirds. Its flower has a stigma that

(A) *Nymphaea* sp.

(B) *Paphiopedilum* sp.

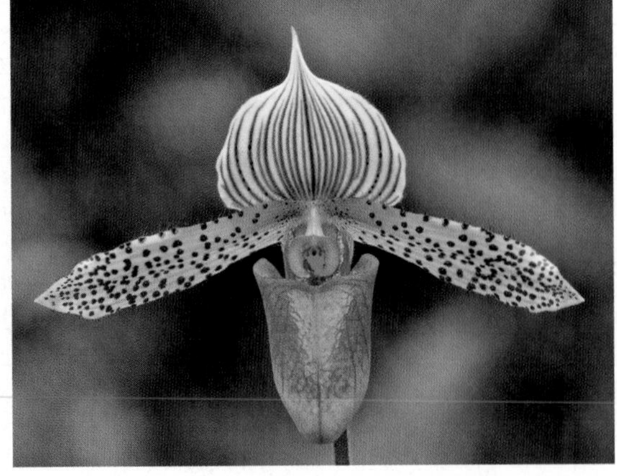

FIGURE 21.20 Flower Form and Evolution (A) A water lily shows the major features of early flowers: it is radially symmetrical, and the individual tepals, stamens, and carpels are separate, numerous, and attached at their bases. (B) Orchids such as this lady slipper have a bilaterally symmetrical structure that evolved much later than radial flower symmetry.

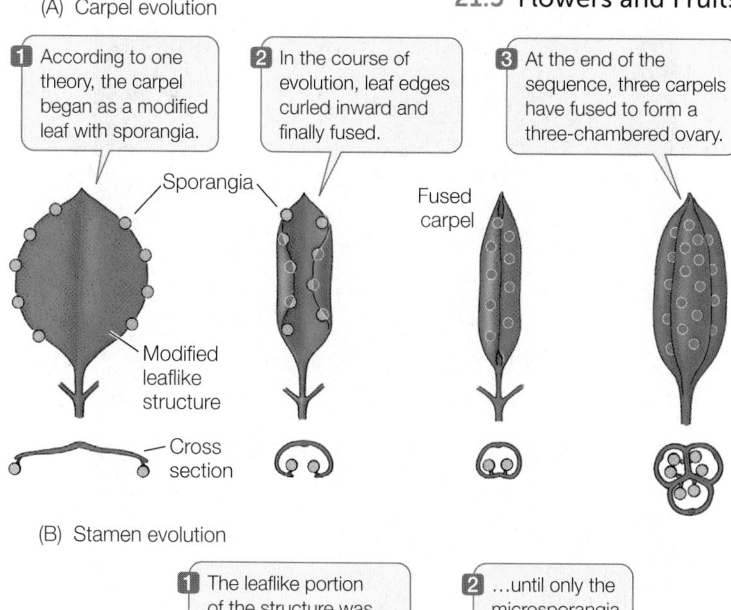

(A) Carpel evolution
1 According to one theory, the carpel began as a modified leaf with sporangia.
2 In the course of evolution, leaf edges curled inward and finally fused.
3 At the end of the sequence, three carpels have fused to form a three-chambered ovary.
Sporangia
Fused carpel
Modified leaflike structure
Cross section
(B) Stamen evolution
1 The leaflike portion of the structure was progressively reduced...
2 ...until only the microsporangia remained.
Austrobaileya sp.
Magnolia
Lily
Modified leaf
Anther
Filament
Sporangia
Cross section

FIGURE 21.21 Carpels and Stamens Evolved from Leaflike Structures (A) Possible stages in the evolution of a carpel from a more leaflike structure. (B) The stamens of three modern plants show three possible stages in the evolution of that organ. (It is *not* implied that these species evolved from one another; their structures simply illustrate the possible stages.)

initially serves as a screen, hiding the anthers (**FIGURE 21.22**). Once a hummingbird touches the stigma, one of the stigma's two lobes is retracted, so that subsequent hummingbird visitors pick up pollen from the previously screened anthers. Thus the first bird to visit the flower transfers pollen from another plant to the stigma, eventually leading to fertilization. Later visitors pick up pollen from the now-accessible anthers, fulfilling the flower's male function. **FIGURE 21.23** describes the experiment that revealed the function of this mechanism.

Angiosperms have coevolved with animals

Whereas most gymnosperms are pollinated by wind, most angiosperms are pollinated by animals. The many different mutualistic pollination relationships between plants and animals are vital to both parties. We mentioned coevolution between insects and orchids at the opening of this chapter, but we'll consider a few additional aspects of plant–pollinator coevolution here.

Many flowers entice animals to visit them by providing food rewards. Pollen grains themselves sometimes serve as food for animals. In addition, some flowers produce a sugary fluid called nectar as a pollinator attractant, and some of these flowers have specialized structures to store and distribute it, as we saw at the opening of this chapter. In the process of visiting flowers to obtain nectar or pollen, animals often carry pollen from one flower to another or from one plant to another. Thus, in their quest for food, the animals contribute to the genetic diversity of the plant population. Insects, especially bees, are among the most important pollinators. Other major pollinators include some species of birds and bats.

Go to MEDIA CLIP 21.5
Pollen Transfer by a Bat
PoL2e.com/mc21.5

For more than 150 million years, angiosperms and their animal pollinators have coevolved in the terrestrial environment. The animals have affected the evolution of the plants, and the plants have affected the evolution of the animals. Flower structure has become incredibly diverse under these selection pressures. Some of the products of coevolution are highly specific. For example, the flowers of some yucca species are pollinated by only one species of yucca moth, and that moth may exclusively pollinate just one species of yucca. Such specific relationships provide plants with a reliable mechanism for transferring pollen only to members of their own species.

Most plant–pollinator interactions are much less specific. In most cases, many different animal species pollinate the same plant species, and the same animal species pollinates many different plant species. However, even these less

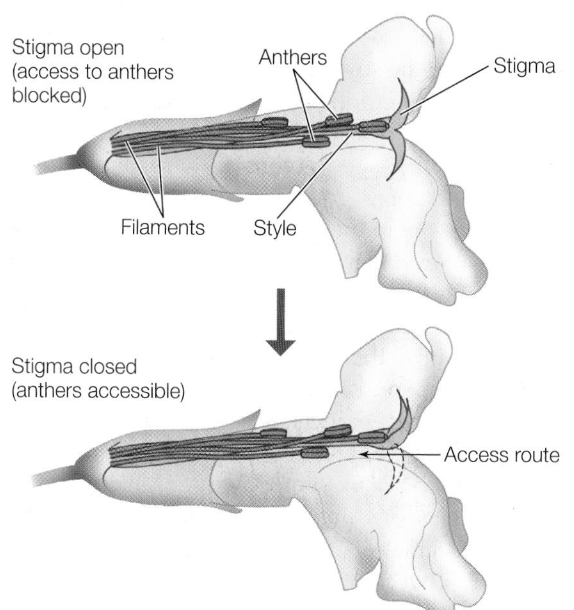

FIGURE 21.22 An Unusual Way to Prevent Self-Fertilization Both long stamens and long styles facilitate cross-pollination, but if these male and female structures are too close to each other, the likelihood of (disadvantageous) self-pollination increases. In *Mimulus aurantiacus*, the stigma is initially open, blocking access to the anthers. A hummingbird's touch as it deposits pollen on the stigma causes one lobe of the stigma to retract, creating a path to the anthers and allowing pollen dispersal.

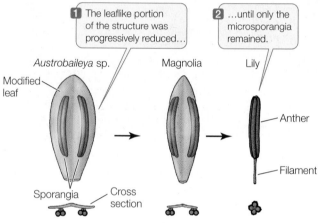

INVESTIGATION

FIGURE 21.23 The Effect of Stigma Retraction in Monkeyflowers Elizabeth Fetscher's experiments showed that the unusual stigma retraction response to pollination in monkeyflowers (illustrated in Figure 21.22) enhances the dispersal of pollen to other flowers.[a]

HYPOTHESIS

The stigma-retraction response in *M. aurantiacus* increases the likelihood than an individual flower's pollen will be exported to another flower once pollen from another flower has been deposited on its stigma.

METHOD

1. Set up three groups of monkeyflower arrays. Each array consists of one pollen-donor flower and multiple pollen-recipient flowers (with the anthers removed to prevent pollen donation).
2. In control arrays, the stigma of the pollen donor is allowed to function normally.
3. In one set of experimental arrays, the stigma of the pollen donor is permanently propped open.
4. In a second set of experimental arrays, the stigma of the pollen donor is artificially sealed closed.
5. Allow hummingbirds to visit the arrays, then count the pollen grains transferred from each donor flower to the recipient flowers in the same array.

RESULTS

Error bars show 1 standard error of the mean; the three groups are significantly different at $P < 0.05$ (see Appendix B).

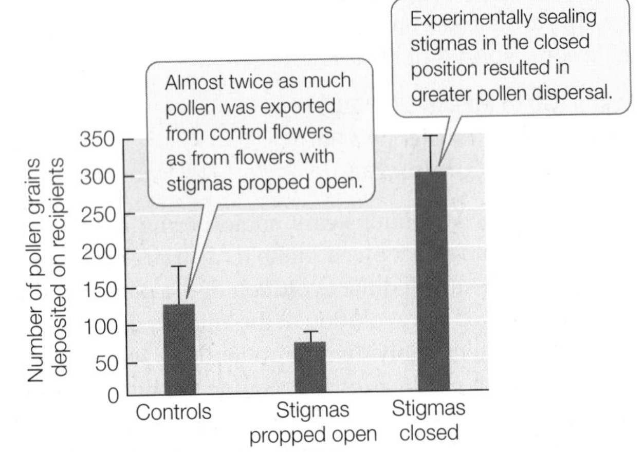
Experimentally sealing stigmas in the closed position resulted in greater pollen dispersal.
Almost twice as much pollen was exported from control flowers as from flowers with stigmas propped open.
Number of pollen grains deposited on recipients
350
300
250
200
150
100
50
0
Controls
Stigmas propped open
Stigmas closed

CONCLUSION

The stigma-retraction response enhances the male function of the flower (dispersal of pollen) once the female function (receipt of pollen) has been performed.

Go to **LaunchPad** for discussion and relevant links for all **INVESTIGATION** figures.

[a]A. E. Fetscher. 2001. *Proceedings of the Royal Society B* 268: 525–529.

specific interactions have developed some specialization. Bird-pollinated flowers are often red and odorless. Many insect-pollinated flowers have characteristic odors, and bee-pollinated flowers may have conspicuous markings, called

Taraxacum officinale

FIGURE 21.24 See Like a Bee To normal human vision (above), the petals of a dandelion appear solid yellow. Ultraviolet photography reveals patterns that attract bees to the central region, where pollen and nectar are located.

nectar guides, that are conspicuous only to animals, such as bees, that can see colors in the ultraviolet region of the spectrum (**FIGURE 21.24**).

The angiosperm life cycle produces diploid zygotes nourished by triploid endosperms

Like all seed plants, angiosperms are heterosporous. As we have seen, their ovules are contained within carpels rather than being exposed on the surfaces of scales, as in most gymnosperms. The male gametophytes, as in the gymnosperms, are pollen grains.

Pollination in the angiosperms consists of the arrival of a microgametophyte—a pollen grain—on a receptive surface in a flower (the stigma). As in the gymnosperms, pollination is the first in a series of events that results in the formation of a seed. The next event is the growth of a pollen tube extending to the megagametophyte. The third event is a fertilization process that, in detail, is unique to the angiosperms (**FIGURE 21.25**).

In nearly all angiosperms, *two* male gametes, contained in a single microgametophyte, participate in fertilization. The nucleus of one sperm combines with that of the egg to produce a diploid zygote, the first cell of the sporophyte generation. In most angiosperms, the other sperm nucleus combines with two other haploid nuclei of the female gametophyte to form a cell with a *triploid* ($3n$) nucleus. That cell, in turn, gives rise to triploid tissue, the **endosperm**, which nourishes the embryonic sporophyte

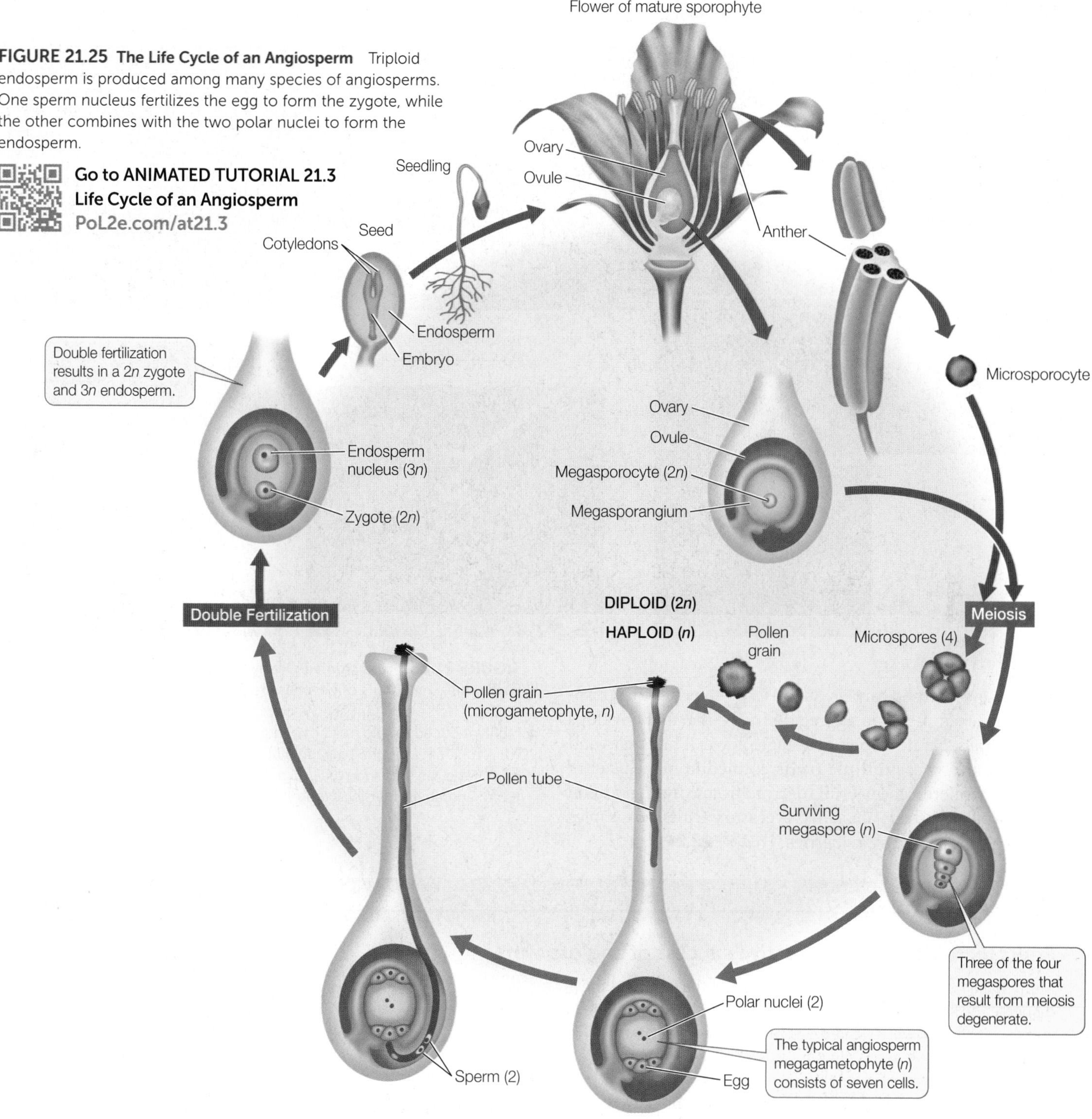

FIGURE 21.25 The Life Cycle of an Angiosperm Triploid endosperm is produced among many species of angiosperms. One sperm nucleus fertilizes the egg to form the zygote, while the other combines with the two polar nuclei to form the endosperm.

during its early development. This process, in which two fertilization events take place, is known as **double fertilization**.

As Figure 21.25 shows, the zygote develops into an embryo, which consists of an embryonic axis (the "backbone" that will become a stem and a root) and one or two **cotyledons**, or "seed leaves." The cotyledons have different fates in different plants. In many, they serve as absorptive organs that take up and digest the endosperm. In others, they enlarge and become photosynthetic when the seed germinates. Often they play both roles.

The ovule develops into a seed containing the products of the double fertilization that characterizes angiosperms: a diploid zygote and a triploid endosperm (see Figure 21.25). The endosperm serves as storage tissue for starch or lipids, proteins, and other substances that will be needed by the developing embryo.

Fruits aid angiosperm seed dispersal

Fruits typically aid in seed dispersal. Fruits may attach to or be eaten by an animal. The animal is then likely to move, after which the seeds may fall off or be defecated. Fruits are not necessarily fleshy. Fruits can also be hard and woody, or small and have modified structures that allow the seeds to be dispersed by wind or water.

A fruit may consist of only the mature ovary and its seeds, or it may include other parts of the flower or structures associated with it. A **simple fruit** is one that develops from a single

FIGURE 21.26 **Fruits Come in Many Forms** (A) The single seeds inside the simple fruits of peaches are dispersed by animals. (B) Each horse chestnut seed is covered by a spiny husk. (C) The highly reduced simple fruits of dandelions are dispersed by wind. (D) A multiple fruit, the pineapple (*Ananas comosus*), has become one of the most economically significant fruit crops of the tropics. (E) An aggregate fruit (blackberry). (F) An accessory fruit (pear).

carpel or several fused carpels, such as a plum or peach. A raspberry is an example of an **aggregate fruit**—one that develops from several separate carpels of a single flower. Pineapples and figs are examples of **multiple fruits**, formed from a cluster of flowers (an inflorescence). Fruits derived from parts in addition to the carpel and seeds are called **accessory fruits**—examples are apples, pears, and strawberries (**FIGURE 21.26**).

APPLY THE CONCEPT

Fruits increase the reproductive success of angiosperms

Many fleshy fruits attract animals, which eat the fruit and then disperse the seeds in their feces. If all other factors are equal, large seeds have a better chance of producing a successful seedling than small seeds. So why isn't there selection for larger seeds in the fruits of all plants?

In one study in Peru, the feces of the spider monkey *Ateles paniscus* were found to contain seeds from 71 species of plants. After eating fruit, the monkeys usually travel some distance before defecating, thus dispersing any undigested seeds.

If monkey feces are left undisturbed on the forest floor, rodents eat and destroy the vast majority of the seeds in the feces. To germinate successfully, the seeds in spider monkey feces need to be buried by dung beetles, which makes the discovery and destruction of seeds by rodents much less likely.

Ellen Andresen hypothesized that dung beetles were more likely to remove larger than smaller seeds from spider monkey dung before burying the dung. She added plastic beads of various diameters to spider monkey dung (to simulate seeds) and measured the percentage of beads buried with the dung by the beetles. Use her data to answer the questions at right.[a]

Bead diameter (mm)	2	4	6	8	10	12
Percentage buried	100	76	52	39	20	4

[a]Andresen, E. (1999). *Biotropica* 31(1): 145–158.

1. Plot bead size (the independent variable) versus percentage of beads buried by dung beetles (the dependent variable).
2. Calculate a regression line for the relationship shown in your graph (see Appendix B). Approximately what percentage of beads with a diameter of 5 mm would you predict would be buried by the beetles? What about beads 14 mm in diameter?
3. What other factors besides size might influence the probability of seed burial by dung beetles? Can you design an experiment to test your hypotheses?
4. Describe how changes in the population sizes of spider monkeys, rodents, and dung beetles would affect the reproductive success of various plant species.

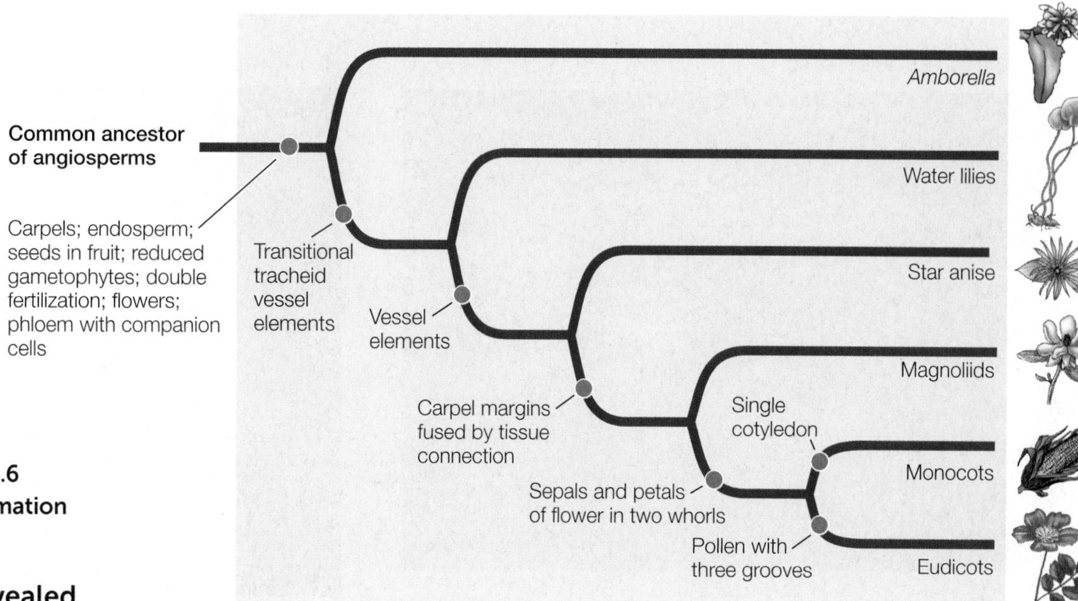

FIGURE 21.27 Evolutionary Relationships among the Angiosperms Recent analyses of many angiosperm genes have clarified the relationships among the major groups.

Go to MEDIA CLIP 21.6 Flower and Fruit Formation PoL2e.com/mc21.6

Recent analyses have revealed the phylogenetic relationships of angiosperms

FIGURE 21.27 shows the relationships among the major angiosperm clades. Recent molecular and morphological analyses have supported the hypothesis that the sister group of remaining flowering plants is a single species of the genus *Amborella* (see Figure 21.28A). This woody shrub, with cream-colored flowers, lives only on New Caledonia, an island in the South Pacific. Other early branching angiosperm groups include the water lilies, star anise and its relatives, and the **magnoliids** (**FIGURE 21.28**). The magnoliids include many familiar and useful plants, such as avocados, cinnamon, black pepper, and magnolias.

FIGURE 21.28 Monocots and Eudicots Are Not the Only Surviving Angiosperms (A) *Amborella*, a shrub, is sister to the remaining extant angiosperms. Notice the sterile stamens on this female flower, which may serve to lure insects that are searching for pollen. (B) The water lily clade was the next to diverge after *Amborella*. (C) Star anise and its relatives belong to another early-diverging angiosperm clade. (D,E) The largest clade other than the monocots and eudicots is the magnoliid complex, which includes magnolias and the group known as "Dutchman's pipe."

(A) *Amborella trichopoda*

(B) *Victoria cruziana*

(C) *Illicium floridanum*

(D) *Magnolia* sp.

(E) *Aristolochia littoralis*

(A) *Tulipa* sp. and *Narcissus* sp.

(B) *Triticum aestivum*

(C) *Posidonia* sp.

(D) *Phoenix dactylifera*

FIGURE 21.29 Monocots (A) Monocots include many popular garden flowers such as these tulips (pink and white), and daffodils (yellow). (B) Monocot grasses such as wheat feed the world; sugarcane, rice, and maize (corn) are also grasses. (C) Seagrasses form "meadows" in the shallow, sunlit waters of the world's oceans. (D) Palms are among the few monocot trees. Date palms like these are a major food source in some areas of the world.

The two largest clades—the **monocots** and the **eudicots**—include the great majority of angiosperm species. The monocots are so called because they have a single embryonic cotyledon, whereas the eudicots have two.

Representatives of the two largest angiosperm clades are everywhere. The monocots (**FIGURE 21.29**) include grasses, cattails, lilies, orchids, and palms. The eudicots (**FIGURE 21.30**) include the vast majority of familiar seed plants, including most herbs (i.e., nonwoody plants), vines, trees, and shrubs. Among the eudicots are such diverse plants as oaks, willows, beans, snapdragons, roses, and sunflowers.

CHECKpoint CONCEPT 21.5

- ✓ Explain the difference between pollination and fertilization.
- ✓ Give some examples of how animals have affected the evolution of the angiosperms.
- ✓ What are the respective roles of the two sperm in double fertilization?
- ✓ What are the different functions of flowers, fruits, and seeds?

Once life moved onto land, it was plants that shaped the terrestrial environment. Terrestrial ecosystems could not function without the foods and habitats provided by plants. Plants produce oxygen and remove carbon dioxide from the atmosphere. They play important roles in forming soils and renewing their fertility. Plant roots help hold soil in place, providing protection against erosion by wind and water. Plants also moderate local

(A) *Malus* sp.

(B) *Banksia coccinea*

(C) *Rafflesia arnoldii*

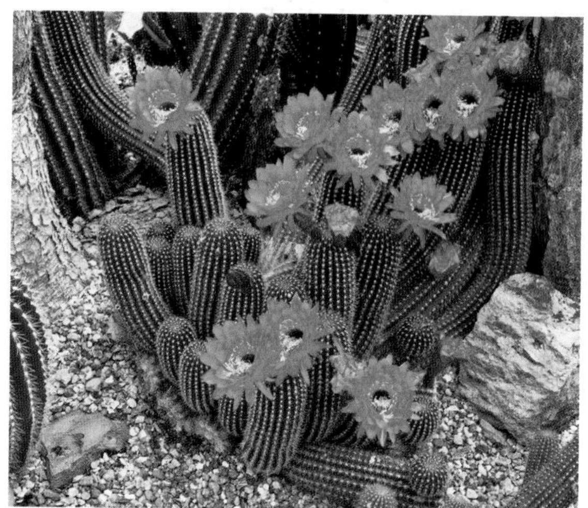

(D) *Echinopsis calochlora*

FIGURE 21.30 Eudicots (A) Eudicots include many trees, such as this crabapple tree. (B) Scarlet Banksia is a species of an Australian genus of eudicots that attracts a wide diversity of pollinators by producing large quantities of nectar. (C) The largest flower in the world is *Rafflesia arnoldii*, found in the rainforests of Indonesia. This plant is a root parasite of tropical vines and has lost its leaf, stem, and even root structures. The flower is the only part of the plant that can be easily seen. (D) Cacti comprise a large group of eudicots, with about 1,500 species in the Americas. Many bear large flowers for a brief period of each year.

climates in various ways, such as by increasing humidity, providing shade, and blocking wind. All of these ecosystem services permit a great diversity of fungi and animals to exist on land.

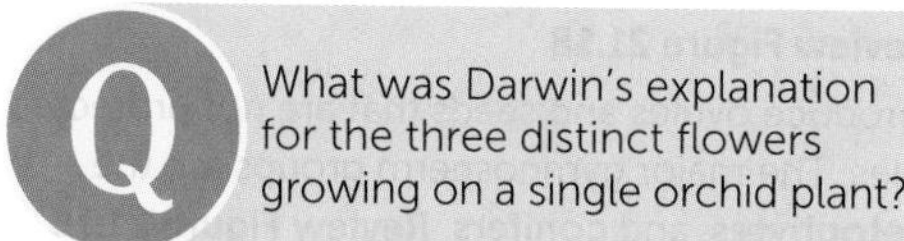

What was Darwin's explanation for the three distinct flowers growing on a single orchid plant?

ANSWER After obtaining specimens of the plant in question and dissecting its flowers, Darwin was able to demonstrate that the orchid was a single species (*Catasetum macrocarpum*) that bore three distinct types of flowers: megasporangiate (female), microsporangiate (male), and perfect (hermaphroditic). The three types of flowers were remarkable in their morphological differences, which had misled botanists into describing the different flower types as species in different genera. Most plants were either male (specimens identified as *Catasetum*) or female (specimens identified as *Monachanthus*), but some individuals that bore predominately male or female flowers also produced perfect flowers (specimens identified as *Myanthus*).

The case of *C. macrocarpum* demonstrates that some plants blur the lines between strict dioecy (male and female flowers in separate individuals), monoecy (male and female flowers in the same individual), and perfect flowers (flowers with both male and female parts). The flowers on a *C. macrocarpum* are either male or female at any one time, except that plants can bear some perfect flowers as well.

Why do the male and female flowers of *C. macrocarpum* look so different? Part of the explanation is their different roles in pollination. Recall Darwin's observation of a *Catasetum* flower shooting a packet of pollen at an insect that landed on its flower. The pollinia (pollen packets) and associated structures in male flowers of *Catasetum* are coiled like springs and are released suddenly when disturbed by an insect. This release forcefully propels the pollinia precisely into position on the

22 The Evolution and Diversity of Fungi

KEY CONCEPTS

22.1 Fungi Live by Absorptive Heterotrophy

22.2 Fungi Can Be Saprobic, Parasitic, Predatory, or Mutualistic

22.3 Major Groups of Fungi Differ in Their Life Cycles

22.4 Fungi Can Be Sensitive Indicators of Environmental Change

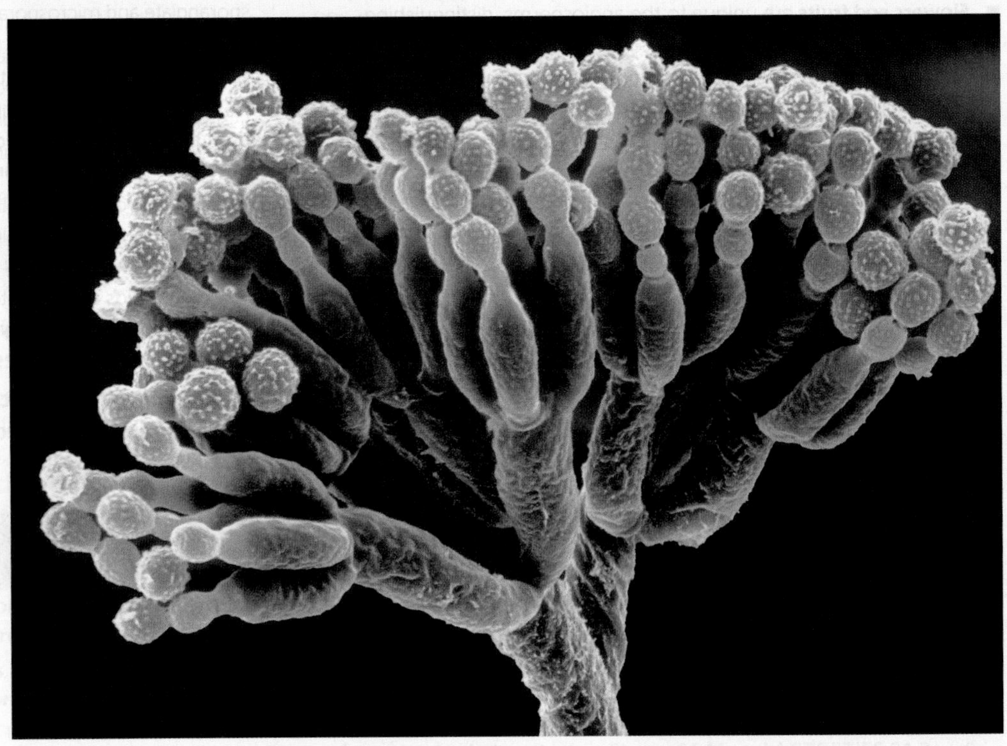

All species of the fungus *Penicillium* are recognizable by their dense spore-bearing structures (see Figure 22.18).

Alexander Fleming was already a famous scientist in 1928, but his laboratory was often a mess. That year he was studying the properties of *Staphylococcus* bacteria, the agents of dangerous staph infections. In August he took a long vacation with his family. When he returned in early September, he found that some of his petri dishes of *Staphylococcus* had become infested with a fungus that killed many of the bacteria.

Many scientists would have sighed at the loss, thrown out the petri dishes, and started new cultures of bacteria. But when Fleming looked at the dishes, he saw something exciting. Around each colony of fungi was a ring within which all the bacteria were dead.

Fleming hypothesized that the bacteria-free rings around the fungal colonies were produced by a substance excreted from the fungi, which he initially called "mould juice." He identified the fungi as members of the genus *Penicillium* and eventually named the antibacterial substance produced by these fungi penicillin. Fleming published his discovery in 1929, but initially the finding received very little attention.

Over the next decade, Fleming produced small quantities of penicillin for testing as an antibacterial agent. Some of the tests showed promise, but many were inconclusive, and eventually Fleming gave up on the research. But his tests had shown enough promise to attract the attention of several chemists, who worked out the practical problems of producing a stable form of the substance. Clinical trials of this stable form of penicillin were extremely successful, and by 1945 it was being produced and distributed as an antibiotic on a large scale. That same year, Fleming and two of the chemists, Howard Florey and Ernst Chain, won the Nobel Prize in Medicine for their work on penicillin.

The development of penicillin was one of the most important achievements in modern medicine. Until the introduction of modern antibiotics, the most widespread agents of human death included bacterial infections such as gangrene, tuberculosis, and syphilis. Penicillin proved to be highly effective in curing such infections, and its success led to the creation of the modern pharmaceutical industry. Soon many additional antibiotic compounds were isolated from other fungi or synthesized in the laboratory, leading to a "golden age" of human health.

Have antibiotics derived from fungi eliminated the danger of bacterial diseases in human populations?

You will find the answer to this question on page 467.

CONCEPT 22.1 Fungi Live by Absorptive Heterotrophy

Fungi are organisms that digest their food outside their bodies. They secrete digestive enzymes to break down large food molecules in the environment, then absorb the breakdown products through their cell membranes in a process known as **absorptive heterotrophy**. This mode of nutrition is successful in a wide variety of environments. Many fungi are **saprobes**, meaning they absorb nutrients from dead organic matter. Others are parasites, absorbing nutrients from living hosts. Still others are mutualists, living in intimate associations with other organisms that benefit both partners.

Modern fungi are believed to have evolved from a unicellular protist ancestor that had a flagellum. The probable common ancestor of the animals was also a flagellated protist much like the living choanoflagellates (see Figure 23.2). Current evidence, including the sequences of many genes, suggests that the fungi, choanoflagellates, and animals share a common ancestor not shared by other eukaryotes. These three lineages are often grouped together as the **opisthokonts** (**FIGURE 22.1**). A synapomorphy (shared derived trait) of the opisthokonts is a flagellum that, if present, is posterior, as in animal sperm. The flagella of all other eukaryotes are anterior.

Synapomorphies that distinguish the fungi as a group among the opisthokonts include absorptive heterotrophy and the presence of **chitin**, a structural polysaccharide, in their cell walls. The fungi represent one of four independent evolutionary origins of large multicellular organisms (plants, brown algae, and animals are the other three).

Unicellular yeasts absorb nutrients directly

Most fungi are multicellular, but single-celled species are found in most fungal groups. Unicellular, free-living fungi are referred to as **yeasts** (**FIGURE 22.2**). Some fungi that have yeast life stages also have multicellular life stages. Thus the term "yeast" does not refer to a single taxonomic group, but rather to a lifestyle that has evolved multiple times. Yeasts live in liquid or moist environments and absorb nutrients directly across their cell surfaces.

The ease with which many yeasts can be cultured, combined with their rapid growth rates, has made them ideal model organisms for study in the laboratory. They present many of the same advantages to laboratory investigators as do many bacteria, but because they are eukaryotes, their genome structures and cells are much more like those of humans and other eukaryotes than are those of bacteria.

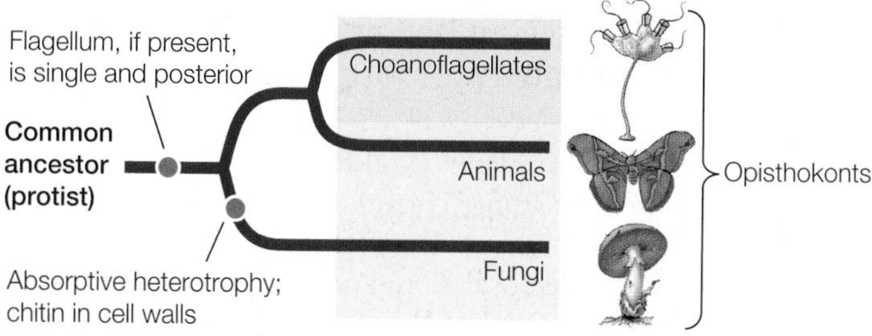

FIGURE 22.1 Fungi in Evolutionary Context Absorptive heterotrophy and the presence of chitin in their cell walls distinguish the fungi from other opisthokonts.

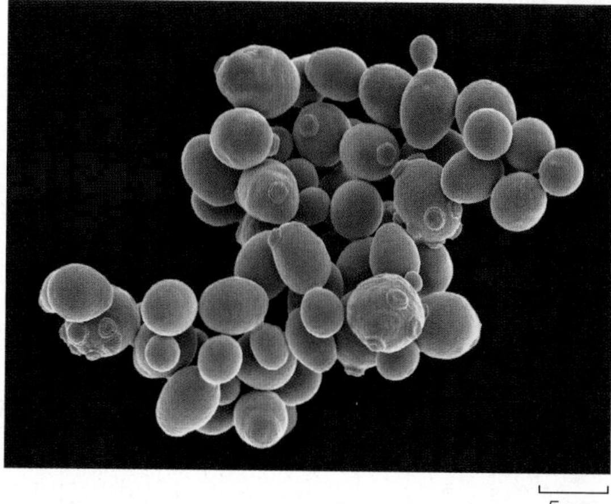

FIGURE 22.2 Yeasts Unicellular, free-living fungi are known as yeasts. Many yeasts reproduce by budding—mitosis followed by asymmetrical cell division—as those shown here are doing.

> **LINK**
> You can read about other model organisms in **Concept 12.3**

Multicellular fungi use hyphae to absorb nutrients

The body of a multicellular fungus is called a **mycelium** (plural *mycelia*). A mycelium is composed of a mass of individual tubular filaments called **hyphae** (singular *hypha*; **FIGURE 22.3A**), in which absorption of nutrients takes place. The cell walls of the hyphae are greatly strengthened by microscopic strands of a complex polysaccharide called chitin. In some species of fungi, the hyphae are subdivided into cell-like compartments by incomplete cross-walls called **septa** (singular *septum*). These subdivided hyphae are referred to as **septate hyphae**. Septa do not completely close off compartments in the hyphae. Pores at the centers of the septa allow organelles—sometimes even nuclei—to move in a controlled way between compartments (**FIGURE 22.3B**). In other species of fungi, the hyphae lack septa but may contain hundreds of nuclei. These multinucleate, undivided hyphae are referred to as **coenocytic**. The coenocytic condition results from repeated nuclear divisions without cytokinesis.

Certain modified hyphae, called **rhizoids**, anchor some fungi to their substrate (i.e., the dead organism or other matter on which they are feeding). These rhizoids are not homologous to the rhizoids of plants, and they are not specialized to absorb nutrients and water.

Fungi can grow very rapidly when conditions are favorable. In some species, the total hyphal growth of a fungal mycelium (not the growth of an individual hypha) may exceed 1 kilometer a day! The hyphae may be widely dispersed to forage for nutrients over a large area, or they may clump together in

(A)

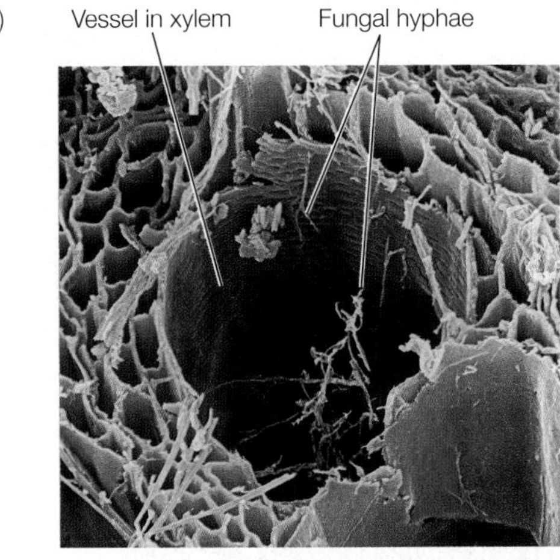
Vessel in xylem
Fungal hyphae
10 μm

(B)

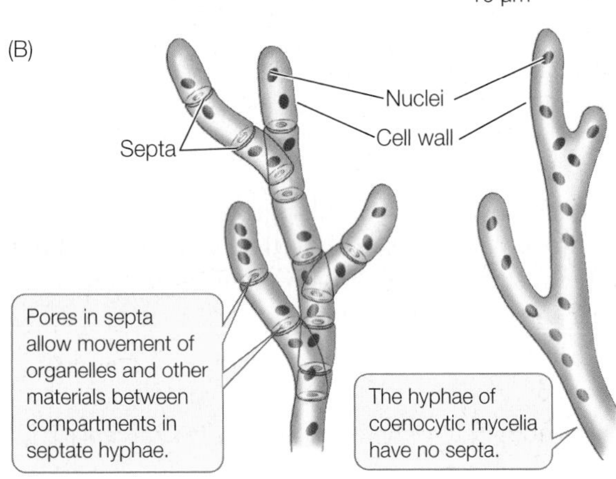
Septa
Nuclei
Cell wall
Pores in septa allow movement of organelles and other materials between compartments in septate hyphae.
The hyphae of coenocytic mycelia have no septa.

(C)

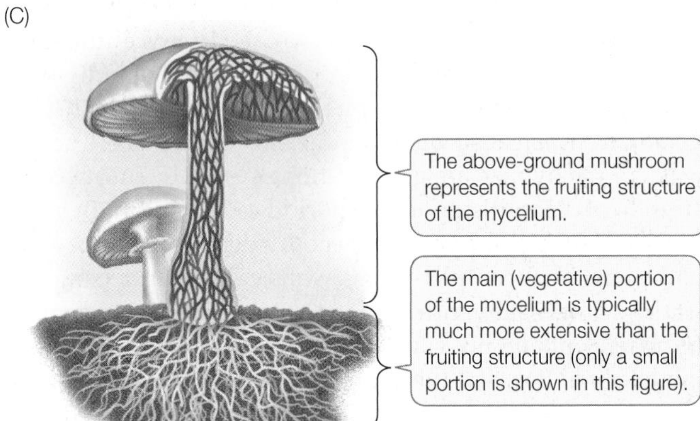
The above-ground mushroom represents the fruiting structure of the mycelium.
The main (vegetative) portion of the mycelium is typically much more extensive than the fruiting structure (only a small portion is shown in this figure).

FIGURE 22.3 Mycelia Are Made Up of Hyphae (A) The minute individual hyphae of fungal mycelia can penetrate small spaces. In this artificially colored micrograph, hyphae (yellow structures) of a dry-rot fungus are penetrating the xylem tissues of a log. (B) The hyphae of septate fungal species are divided into organelle-containing compartments by porous septa, whereas coenocytic hyphae have no septa. (C) The fruiting structure of a club fungus is short-lived, but the filamentous, nutrient-absorbing mycelium can be long-lived and cover a large area.

a cottony mass to exploit a rich nutrient source. The familiar mushrooms you may notice growing in moist areas are spore-producing fruiting structures (**FIGURE 22.3C**). In the fungal species that produce these structures, the mycelial mass is often far larger than the visible mushroom. The mycelium of one individual fungus discovered in Oregon covers almost 900 hectares underground and weighs considerably more than a blue whale (the largest animal). Aboveground, this individual is evident only as isolated clumps of mushrooms.

Fungi are in intimate contact with their environment

The filamentous hyphae of a fungus give it a unique relationship with its physical environment. The fungal mycelium has an enormous surface area-to-volume ratio compared with that of most large multicellular organisms. This large ratio is a marvelous adaptation for absorptive heterotrophy. Throughout the mycelium (except in fruiting structures), all of the hyphae are very close to their food source.

The downside of the large surface area-to-volume ratio of the mycelium is its tendency to lose water rapidly in a dry environment. Thus fungi are most common in moist environments. You have probably observed the tendency of molds, toadstools, and other fungi to appear in damp places.

Another characteristic of some fungi is a tolerance for highly hypertonic environments (those with a solute concentration higher than their own; see Concept 5.2). Many fungi are more resilient than bacteria in hypertonic surroundings. Jelly in the refrigerator, for example, will not become a growth medium for bacteria because it is too hypertonic to those organisms, but it may eventually harbor mold colonies. Mold in the refrigerator illustrates yet another trait of many fungi: tolerance of temperature extremes. Many fungi grow in temperatures as low as –6°C, and some tolerate temperatures higher than 50°C.

CHECKpoint CONCEPT 22.1

- ✓ Describe the relationship between fungal structure and absorptive heterotrophy.
- ✓ What are the advantages and disadvantages to multicellular fungi of the large surface-to-volume ratio of the mycelium?

Fungi are important components of healthy ecosystems. They interact with other organisms in many ways, some of which are harmful and some beneficial to those other organisms.

CONCEPT 22.2 Fungi Can Be Saprobic, Parasitic, Predatory, or Mutualistic

Without the fungi, our planet would be very different. Picture Earth with only a few stunted plants and watery environments choked with the remains of dead organisms. Fungi do much of Earth's garbage disposal. Fungi not only help clean up the landscape and form soil, but also play a key role in recycling mineral nutrients. Furthermore, the colonization of the terrestrial environment was made possible in large part by associations that fungi formed with land plants and other organisms.

Saprobic fungi are critical to the planetary carbon cycle

Saprobic fungi, along with bacteria, are the major decomposers on Earth, contributing to the decay of nonliving organic matter and thus to recycling of the elements used by living things. In forests, for example, the mycelia of fungi absorb nutrients from fallen trees, thus decomposing their wood. Fungi are the principal decomposers of cellulose and lignin, the main components of plant cell walls (most bacteria cannot break down these materials). Other fungi produce enzymes that decompose keratin and thus break down animal structures such as hair and nails.

Were it not for the fungal decomposers, Earth's carbon cycle would fail. Great quantities of carbon atoms would remain trapped forever on forest floors and elsewhere. Instead, those carbon atoms are returned to the atmosphere in the form of CO_2 by fungal respiration, where they are again available for photosynthesis by plants.

LINK

Earth's carbon cycle is described in **Concept 45.3**

There was a time in Earth's history when populations of saprobic fungi declined dramatically. Vast tropical swamps existed during the Carboniferous period, as we saw in Chapter 21. When plants in these swamps died, they began to form peat. Peat formation led to acidification of the swamps. That acidity, in turn, drastically reduced the fungal population. The result? With the decomposers largely absent, large quantities of peat remained on the swamp floor and over time were converted into coal.

In contrast to their decline during the Carboniferous, fungi did very well at the end of the Permian, a quarter of a billion years ago, when the aggregation of continents produced volcanic eruptions that triggered a global mass extinction (see Chapter 18). The fossil record shows that even as 96 percent of all multicellular species became extinct, fungi flourished—demonstrating both their hardiness and their role in recycling the elements in dead plants and animals.

Simple sugars and the breakdown products of complex polysaccharides are the favored source of carbon for saprobic fungi. Most fungi obtain nitrogen from proteins or the products of protein breakdown. Many fungi can use nitrate (NO_3^-) or ammonium (NH_4^+) ions as their sole source of nitrogen. No known fungus can get its nitrogen directly from inorganic nitrogen gas, however, as can some bacteria and plant–bacteria associations (that is, fungi cannot fix nitrogen; see Concept 19.3).

What happens when a fungus faces a dwindling food supply? A common strategy is to reproduce rapidly and abundantly. When conditions are good, fungi produce great quantities of spores, but the rate of spore production is commonly even higher when nutrient supplies go down. The spores may then remain dormant until conditions improve, or they may be dispersed to areas where nutrient supplies are higher.

Not only are fungal spores abundant in number, but they are extremely tiny and easily spread by wind or water (**FIGURE 22.4**). These attributes virtually ensure that they will be scattered over great distances, and that at least some of them will find conditions suitable for growth. The air we breathe contains as many as 10,000 fungal spores per cubic meter. No wonder we find fungi just about everywhere.

Lycoperdon perlatum

FIGURE 22.4 Spores Galore Puffballs (a type of club fungus) disperse trillions of spores in great bursts. Few of the spores travel very far, however; some 99 percent of them fall within 100 meters of the parent puffball.

Go to MEDIA CLIP 22.1
Fungal Decomposers
PoL2e.com/mc22.1

Some fungi engage in parasitic or predatory interactions

Whereas saprobic fungi obtain their energy, carbon, and nitrogen directly from dead organic matter, other species of fungi obtain their nutrition from parasitic—and even predatory—interactions.

PARASITIC FUNGI Mycologists (biologists who study fungi) distinguish between two classes of parasitic fungi based on their degree of dependence on their host. **Facultative parasites** can grow on living organisms but can also grow independently (including on artificial media). **Obligate parasites** can grow only on a living host. The fact that their growth depends on a living host shows that obligate parasites have specialized nutritional requirements.

Plants and insects are the most common hosts of parasitic fungi. The filamentous structure of fungal hyphae is especially well suited to a life of absorbing nutrients from living plants. The slender hyphae of a parasitic fungus can invade a plant through

FIGURE 22.5 Invading a Leaf (A) Hyphae of the mildew *Phyllactinia guttata* growing on the surface of a hazel leaf. (B) Haustoria are fungal hyphae that push into the living cells of plants, from which they absorb nutrients.

stomata, through wounds, or in some cases, by direct penetration of epidermal cell walls (**FIGURE 22.5A**). Once inside the plant, the hyphae branch out to expand the mycelium. Some hyphae produce **haustoria** (singular *haustorium*), branching projections that push through cell walls into living plant cells, absorbing the nutrients within those cells. The haustoria do not break through the cell membranes inside the cell walls, but instead push pockets into the membranes, so that the cell membrane fits them like a glove (**FIGURE 22.5B**). Fruiting structures may form, either within the plant body or on its surface.

Some parasitic fungi live in a close physical (symbiotic) relationship with their host that is usually not lethal to the plant. Others are pathogenic, sickening or even killing the host from which they derive nutrition.

Go to MEDIA CLIP 22.2
Mind-Control Killer Fungi
PoL2e.com/mc22.2

PATHOGENIC FUNGI Although most human diseases are caused by bacteria or viruses, fungal pathogens are a major cause of death among people with compromised immune systems. Many people with AIDS die of fungal diseases, such as the pneumonia caused by *Pneumocystis jirovecii*. Even *Candida albicans* and certain other yeasts that are normally a part of a healthy microbiome can cause severe diseases, such as esophagitis (which impairs swallowing), in individuals with AIDS and in individuals taking immunosuppressive drugs. Various fungi cause other, less threatening human diseases, such as ringworm and athlete's foot. Our limited understanding of the basic biology of these fungi still hampers our ability to treat the diseases they cause. As a result, fungal diseases are a growing international health problem.

The worldwide decline of amphibian species has been linked to the spread of a chytrid fungus, *Batrachochytrium dendrobatidis* (or *Bd* for simplicity). In some areas of the world where this fungus has been present for millennia, amphibian populations appear to be tolerant of *Bd* and it does not cause widespread die-offs. But in other areas, such as western North America and Australia, introductions of *Bd* are implicated in the loss or decline of many frog and salamander populations, or the extinction of entire species. Some strains of *Bd* are native to southern Africa, where they do not appear to cause widespread amphibian declines. Biologists have hypothesized that the spread of *Bd* around the world may have been initiated in the 1930s with exports of the African clawed frog (*Xenopus laevis*), which was once widely used in human pregnancy tests. Recent studies by Erica Rosenblum and her colleagues have shown that Africa may indeed have been one source of the introductions, but that strains of *Bd* appear to have been introduced multiple times from several geographic areas (**FIGURE 22.6**).

Fungi are by far the most important plant pathogens, causing crop losses amounting to billions of dollars. Bacteria and viruses are less important than fungi as plant pathogens. Major fungal diseases of crop plants include black stem rust of wheat and other diseases of wheat, corn, and oats. The agent of black stem rust is *Puccinia graminis*, which has a complicated life cycle that involves two plant hosts (wheat and barberry). In an epidemic in 1935, *P. graminis* was responsible for the loss of about one-fourth of the wheat crop in Canada and the United States.

Although many pathogenic fungi cause problems when they attack agricultural crops, pathogenic fungi of the genus *Fusarium* can benefit agriculture by killing certain weed species, such as witchweed, a serious pest of cereal crops. Another strain of *Fusarium* has been proposed as a tool in the war against cocaine. The fungus could be applied to kill coca plants, the source of the drug. The use of *Fusarium* to kill coca plants is highly controversial, however, as the fungus might infect neighboring plant species in addition to the coca plants.

PREDATORY FUNGI Some fungi have adaptations that enable them to function as active predators, trapping nearby microscopic protists or animals. The most common predatory

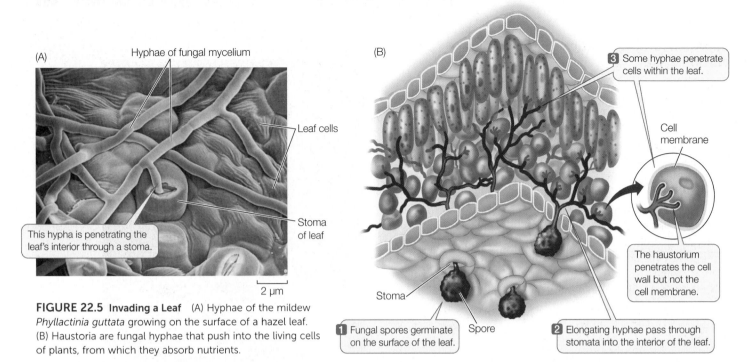

INVESTIGATION

FIGURE 22.6 What is the Origin of Amphibian-Killing Chytrids in Western North America? An investigation by Erica Rosenblum and colleagues of *Batrachochytrium dendrobatidis* samples from around the world suggests that this pathogenic fungus was introduced to western North America several times, from different geographic localities.[a]

HYPOTHESIS

Pathogenic forms of *Bd* that kill many amphibians were introduced into western North America from Africa, through importation of African clawed frogs beginning in the 1930s.

METHOD

1. Collect samples of *Bd* from throughout the world.
2. Sequence the genomes of the sampled chytrids.
3. Construct a phylogenetic tree by comparing the DNA sequences of the different samples of *Bd*.
4. Assess whether or not the resulting tree is consistent with the introduction of *Bd* from Africa to western North America.

RESULTS

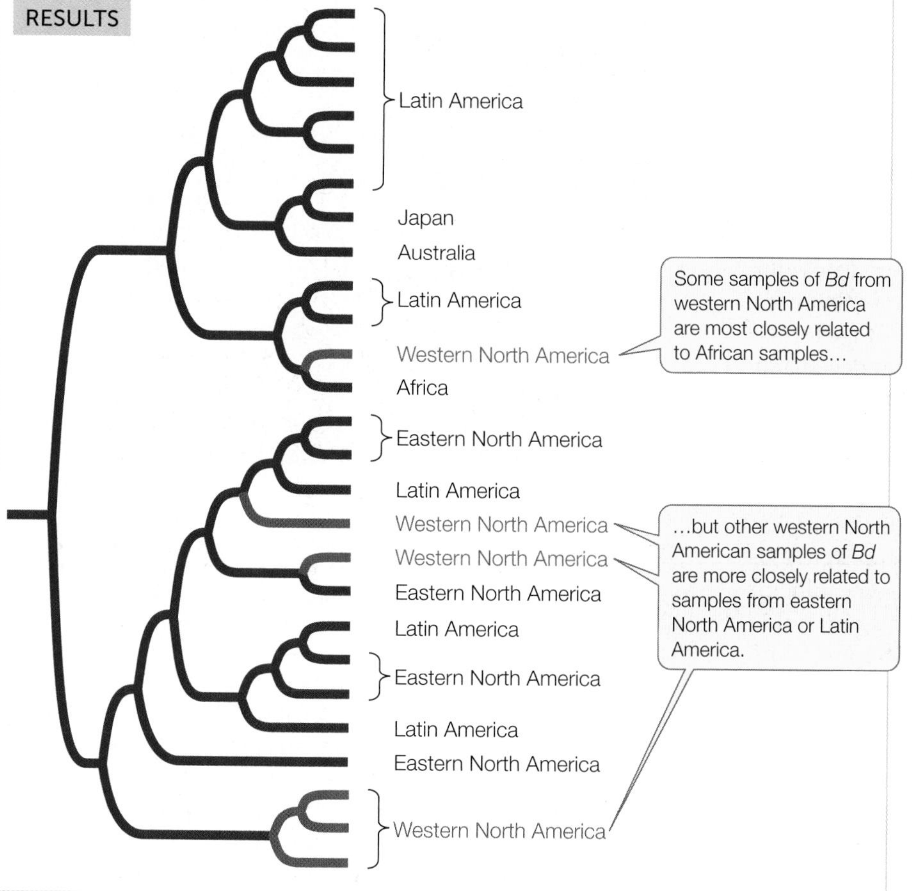

CONCLUSION

Pathogenic strains of *Bd* appear to have been introduced to western North America from several different sources. At least one strain of *Bd* is consistent with an introduction from Africa, but other strains were more likely introduced from eastern North America or Latin America.

Go to **LaunchPad** for discussion and relevant links for all **INVESTIGATION** figures.

[a]E. B. Rosenblum et al. 2013. *Proceedings of the National Academy of Sciences USA* 110: 9385–9390.

strategy seen in fungi is to secrete sticky substances from the hyphae so that passing organisms stick to them. The hyphae then quickly invade the trapped prey, growing and branching within it, spreading through its body, absorbing nutrients, and eventually killing it.

A more dramatic adaptation for predation is the constricting ring formed by some species of soil fungi (**FIGURE 22.7**). When nematodes (tiny roundworms) are present in the soil, these fungi form three-celled rings with a diameter that just fits a nematode. A nematode crawling through one of these rings stimulates the fungus, causing the cells of the ring to swell and trap the worm. Fungal hyphae quickly invade and digest the unlucky victim.

Mutualistic fungi engage in relationships beneficial to both partners

Certain kinds of relationships between fungi and other organisms have nutritional consequences for both partners. Two relationships of this type are highly specific and are **symbiotic** (the partners live in close, permanent contact with each other) as well as **mutualistic** (the relationship benefits both partners).

LINK

The obligate mutualistic relationship between leaf-cutter ants and the fungus they cultivate is necessary for the survival of each partner; see the opener of **Chapter 43 and Concepts 43.3 and 43.4**

LICHENS A **lichen** is not a single organism, but rather a symbiosis between two radically different species: a fungus and a photosynthetic microorganism. Together the organisms that constitute a lichen can survive some of the harshest environments on Earth (although they are sensitive to poor air quality; see Concept 22.4). The biota of Antarctica, for example, features more than 100 times as many species of lichens as of vascular plants. Relatively little experimental work has focused on lichens, perhaps because they grow so slowly—typically less than 1 centimeter in a year.

There are nearly 20,000 described "species" of lichens, each of which is assigned the name of its fungal component. These fungal components may constitute as many as 20 percent of all fungal species. Most of them are sac fungi (Ascomycota). Some of them are able to grow independently without a photosynthetic partner, but most have never been observed in nature other than in a lichen association. The photosynthetic component of a lichen is most often a unicellular green alga, but it can be a cyanobacterium, or may even include both.

FIGURE 22.7 Fungus as Predator A nematode is trapped by hyphal rings of the soil-dwelling fungus *Arthrobotrys dactyloides*.

FIGURE 22.8 Lichen Body Forms Lichens fall into three principal categories—crustose, fruticose, or foliose—based on their body form. (A) A crustose lichen growing on the surface of an exposed rock. (B) Fruticose lichens appear "shrubby," whereas foliose lichens look "leafy."

Lichens are found in all sorts of exposed habitats: on tree bark, on open soil, and on bare rock. Reindeer moss (not a moss at all, but the lichen *Cladonia subtenuis*) covers vast areas in Arctic, sub-Arctic, and boreal regions, where it is an important part of the diets of reindeer and other large mammals.

The body forms of lichens fall into three principal categories: **Crustose** (crustlike) lichens look like colored powder dusted over their substrate (**FIGURE 22.8A**). **Foliose** (leafy) lichens are more loosely attached and have large leaflike forms, whereas **fruticose** (shrubby) lichens are highly branched and grow upward like little shrubs, or hang in long strands from tree branches or rocks (**FIGURE 22.8B**).

A cross section of a typical foliose lichen reveals a tight upper region of fungal hyphae, a layer of photosynthetic cyanobacteria or algae, a looser hyphal layer, and finally hyphal rhizoids that attach the entire structure to its substrate (**FIGURE 22.9**). The meshwork of fungal hyphae takes up some mineral nutrients needed by the photosynthetic cells and also holds water tenaciously, providing a suitably moist environment. The fungi obtain fixed carbon from the photosynthetic products of the algal or cyanobacterial cells.

Within the lichen, the fungal hyphae are tightly pressed against the algal or cyanobacterial cells and sometimes even invade them without breaching the cell membrane (similar to the haustoria in parasitic fungi; see Figure 22.5). The photosynthetic cells not only survive these intrusions but continue to grow. Algal cells in a lichen "leak" photosynthetic products at a greater rate than do similar cells growing on their own, and photosynthetic cells taken from lichens grow more rapidly on their own than when associated with a fungus. On the basis of these observations, we could consider lichen fungi to be parasitic on their photosynthetic partners. In many places where lichens grow, however, the photosynthetic cells could not grow at all on their own.

Lichens can reproduce simply by fragmentation of the vegetative body (the **thallus**) or by means of specialized structures called **soredia** (singular *soredium*). Soredia consist of one or a few photosynthetic cells bound by fungal hyphae. They become detached from the lichen, are dispersed by air currents, and upon arriving at a favorable location, develop into a new lichen thallus. Alternatively, the fungal partner may go through its sexual reproductive cycle, producing haploid spores. When these spores are discharged, however, they disperse alone, unaccompanied by the photosynthetic partner.

Lichens are often the first colonists on new areas of bare rock. They get most of the mineral nutrients they need from the air and rainwater, augmented by minerals absorbed from dust. A lichen begins to grow shortly after a rain, as it begins to dry. As it grows, the lichen acidifies its environment slightly, and this acidity contributes to the slow breakdown of rocks, an early step in soil formation. With further drying, the lichen's photosynthesis ceases. The water content of the lichen may drop to less than 10 percent of its dry weight, at which point it becomes highly insensitive to extremes of temperature.

MYCORRHIZAE Many vascular plants depend on a symbiotic association with fungi. This ancient association between

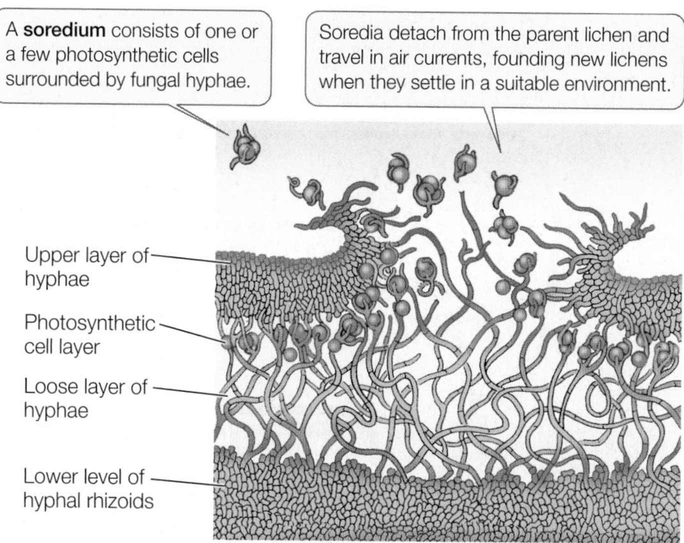

FIGURE 22.9 Lichen Anatomy Cross section showing the layers of a foliose lichen and the release of soredia.

plants and fungi was critical to the successful exploitation of the terrestrial environment by plants. Unassisted, the root hairs of many plants often do not take up enough water or minerals to sustain growth. However, the roots of such plants usually become infected with fungi, forming an association called a **mycorrhiza** (plural *mycorrhizae*). There are two types of mycorrhizae, distinguished by whether or not the fungal hyphae penetrate the plant cell walls.

In **ectomycorrhizae**, the fungus wraps around the root, and its mass is often as great as that of the root itself (**FIGURE 22.10A**). The fungal hyphae wrap around individual cells in the root but do not penetrate the cell walls. An extensive web of hyphae penetrates the soil in the area around the root, so that up to 25 percent of the soil volume near the root may be fungal hyphae. The hyphae attached to the root increase the surface area for the absorption of water and minerals, and the mass of hyphae in the soil acts like a sponge to hold water in the neighborhood of the root. Infected roots are short, swollen, and club-shaped, and they lack root hairs.

The fungal hyphae of **arbuscular mycorrhizae** enter the root and penetrate the cell walls of the root cells, forming arbuscular (treelike) structures inside the cell wall but outside the cell membrane. These structures, like the haustoria of parasitic fungi and the contact regions of fungal hyphae and photosynthetic cells in lichens, become the primary site of exchange between plant and fungus (**FIGURE 22.10B**). As in the ectomycorrhizae, the fungus forms a vast web of hyphae leading from the root surface into the surrounding soil.

LINK

Concept 25.2 contains a detailed description of the role of arbuscular mycorrhizae in plant nutrition

The mycorrhizal association is important to both partners. The fungus obtains needed organic compounds, such as sugars and amino acids, from the plant. In return, the fungus, because of its very high surface area-to-volume ratio and its ability to penetrate the fine structure of the soil, greatly increases the plant's ability to absorb water and minerals (especially phosphorus). The fungus may also provide the plant with certain growth hormones and may protect it against attack by disease-causing microorganisms.

Plants that have active arbuscular mycorrhizae typically are a deeper green and may resist drought and temperature extremes better than plants of the same species that have little mycorrhizal development. Attempts to introduce some plant species to new areas have failed until a bit of soil from the native area (presumably containing the fungus necessary to establish mycorrhizae) was provided. Trees without ectomycorrhizae do not grow well in the absence of abundant nutrients and water, so the health of our forests depends on the presence of ectomycorrhizal fungi. Many agricultural crops require inoculation of seeds with appropriate mycorrhizal fungi prior to planting. Without these fungi, the plants are unlikely to grow well, or in some cases at all. Certain plants that live in nitrogen-poor habitats, such as cranberry bushes and orchids, invariably have mycorrhizae. Orchid seeds will not germinate in nature unless they are already infected by the fungus that will form their mycorrhizae. Plants that lack chlorophyll always have mycorrhizae, which they often share with the roots of green, photosynthetic plants. In effect, these plants without chlorophyll are feeding on nearby green plants, using the fungus as a bridge.

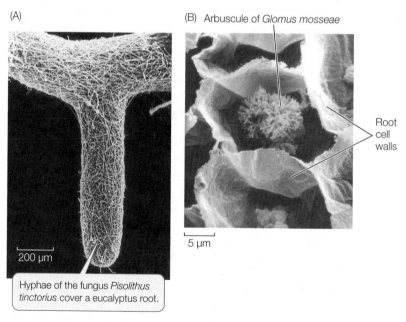

(A)

200 µm

Hyphae of the fungus *Pisolithus tinctorius* cover a eucalyptus root.

(B) Arbuscule of *Glomus mosseae*

Root cell walls

5 µm

FIGURE 22.10 Mycorrhizal Associations (A) Ectomycorrhizal fungi wrap themselves around a plant root, increasing the area available for absorption of water and minerals. (B) Hyphae of arbuscular mycorrhizal fungi infect the root internally and penetrate the root cell walls, branching within the cells and forming a treelike structure, the arbuscule. (For purposes of this scanning electron micrograph, the cell cytoplasm was removed to better visualize the arbuscule.)

Endophytic fungi protect some plants from pathogens, herbivores, and stress

In a tropical rainforest, 10,000 or more fungal spores may land on a single leaf each day. Some are plant pathogens, some do not affect the plant at all, and some invade the plant in a beneficial way. Fungi that live within aboveground parts of plants without causing obvious deleterious symptoms are called **endophytic fungi**. Recent research has shown that endophytic fungi are abundant in plants in all terrestrial environments.

Among the grasses, individual plants with endophytic fungi are more resistant to pathogens and to insect and mammalian herbivores than are plants lacking endophytes. The fungi produce alkaloids (nitrogen-containing compounds) that are toxic to animals. The alkaloids do not harm the host plant. In fact, some plants produce alkaloids (such as nicotine) themselves. The fungal alkaloids also increase the ability of grasses to resist stress of various types, including drought (water shortage) and salty soils. Such resistance is useful in agriculture.

The role, if any, of endophytic fungi in most broad-leaved plants is unclear. They may convey protection against pathogens, or they may simply occupy space within leaves without conferring any benefit, but also without doing harm. The benefit, in fact, might be all for the fungus.

CHECKpoint CONCEPT 22.2

- ✓ What is the role of fungi in Earth's carbon cycle?
- ✓ Describe the nature and benefits of the lichen association.
- ✓ Why do plants grow better when infected with mycorrhizal fungi?

Before molecular techniques clarified the phylogenetic relationships of fungi, one criterion used for assigning fungi to taxonomic groups was the nature of their life cycles—including the types of fruiting bodies they produce. The next section will take a closer look at life cycles in the six major groups of fungi.

CONCEPT 22.3 Major Groups of Fungi Differ in Their Life Cycles

Major fungal groups were originally defined by their structures and processes for sexual reproduction and also, to a lesser extent, by other morphological differences. Although fungal life cycles are even more diverse than was once realized, specific types of life cycles generally distinguish the six major groups of fungi: microsporidia, chytrids, zygospore fungi (Zygomycota), arbuscular mycorrhizal fungi (Glomeromycota), sac fungi (Ascomycota), and club fungi (Basidiomycota). **FIGURE 22.11** diagrams the evolutionary relationships of these groups as they are understood today.

The chytrids and the zygospore fungi may not represent monophyletic groups, as they each consist of several distantly related lineages that retain some ancestral features. The clades

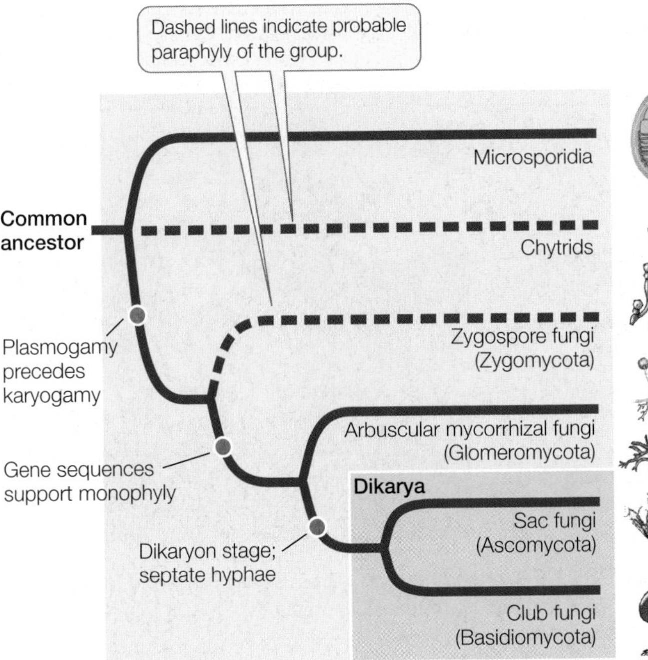

FIGURE 22.11 A Phylogeny of the Fungi Microsporidia are reduced, parasitic fungi whose relationships among the fungi are uncertain. They may be the sister group of most other fungi or more closely related to particular groups of chytrids or zygospore fungi. The dashed lines indicate that chytrids and zygospore fungi are thought to be paraphyletic; the relationships of the lineages within these two informal groups (see Table 22.1) are not yet well resolved. The sac fungi and club fungi together form the clade Dikarya.

Go to ACTIVITY 22.1 Fungal Phylogeny PoL2e.com/ac22.1

that are thought to be monophyletic within these two informal groupings are listed in **TABLE 22.1**. Recent evidence from DNA analyses has established the placement of the microsporidia among the fungi, the likely paraphyly of the chytrids and the zygospore fungi (in other words, members of these groups are probably not each other's closest relatives), the independence of arbuscular mycorrhizal fungi from the other fungal groups, and the monophyly of sac fungi and club fungi.

Fungi reproduce both sexually and asexually

Both asexual and sexual reproduction occur among the fungi (**FIGURE 22.12**). Asexual reproduction takes several forms:

- The production of (usually) haploid spores within sporangia
- The production of haploid spores (not enclosed in sporangia) at the tips of hyphae; such spores are called **conidia** (Greek *konis*, "dust")
- Cell division by unicellular fungi—either a relatively equal division of one cell into two (fission) or an asymmetrical division in which a smaller daughter cell is produced (budding)
- Simple breakage of the mycelium

Sexual reproduction is rare (or even unknown) in some groups of fungi but common in others. Sexual reproduction may not

TABLE 22.1 Classification of the Fungi

Group	Common name	Features
Microsporidia	Microsporidia	Intracellular parasites of animals; greatly reduced, among smallest eukaryotes known; polar tube used to infect hosts
Chytrids (paraphyletic)[a] Chytridiomycota Neocallimastigomycota Blastocladiomycota	Chytrids	Mostly aquatic and microscopic; flagellated gametes and zoospores
Zygomycota (paraphyletic)[a] Entomophthoromycotina Kickxellomycotina Mucoromycotina Zoopagomycotina	Zygospore fungi	Reproductive structure is a unicellular zygospore with many diploid nuclei in a zygosporangium; hyphae coenocytic; usually no fleshy fruiting body
Glomeromycota	Arbuscular mycorrhizal fungi	Form arbuscular mycorrhizae on plant roots; only asexual reproduction is known
Ascomycota	Sac fungi	Sexual reproductive saclike structure known as an ascus, which contains haploid ascospores; hyphae septate; dikaryon
Basidiomycota	Club fungi	Sexual reproductive structure is a basidium, a swollen cell at the tip of a specialized hypha that supports haploid basidiospores; hyphae septate; dikaryon

[a]The formally named groups within the chytrids and Zygomycota are each thought to be monophyletic, but their relationships to one another (and to microsporidia) are not yet well resolved.

occur in some species, or it may occur so rarely that biologists have never observed it. Species in which no sexual stage has been observed were once placed in a separate taxonomic group because knowledge of the sexual life cycle was considered necessary for classifying fungi. Now, however, these species can be related to other species of fungi through analysis of their DNA sequences.

Sexual reproduction in most fungi features an interesting twist: There is no morphological distinction between female and male structures, or between female and male individuals, in most groups of fungi. Rather, there is a genetically determined distinction between two *or more* **mating types**. Individuals of the same mating type cannot mate with each other, but they can mate with individuals of another mating type within the same species, thus avoiding self-fertilization. Individuals of different mating types differ genetically but are often visually and physiologically indistinguishable.

Microsporidia are highly reduced, parasitic fungi

Microsporidia are unicellular parasitic fungi. They are among the smallest eukaryotes known, with infective spores that are only 1–40 μm in diameter. About 1,500 species have been described, but many more species are thought to exist. Their relationships among the eukaryotes have puzzled biologists for many decades.

Microsporidia lack true mitochondria, although they have reduced structures, known as **mitosomes**, that are derived from mitochondria. Unlike mitochondria, however, mitosomes contain no DNA—the mitochondrial genome has been completely transferred to the nucleus. Because microsporidia lack mitochondria, biologists initially suspected that they represented an early lineage of eukaryotes that diverged before the endosymbiotic event from which mitochondria evolved. The presence of mitosomes, however, indicates that

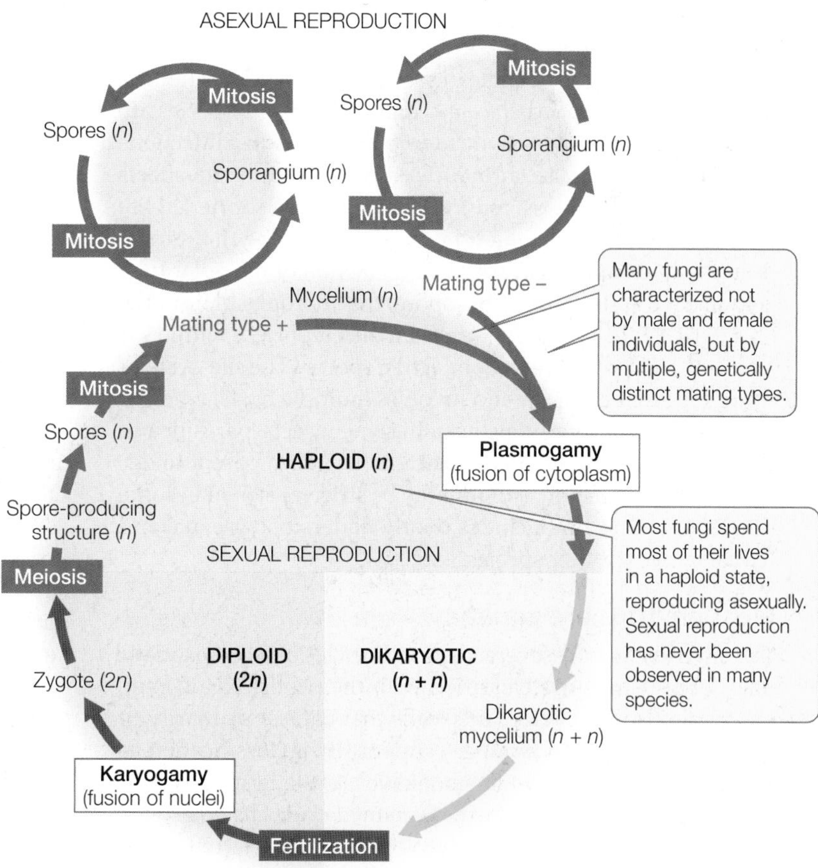

FIGURE 22.12 A Generalized Fungal Life Cycle Environmental conditions may determine which mode of reproduction—sexual or asexual—takes place at a given time.

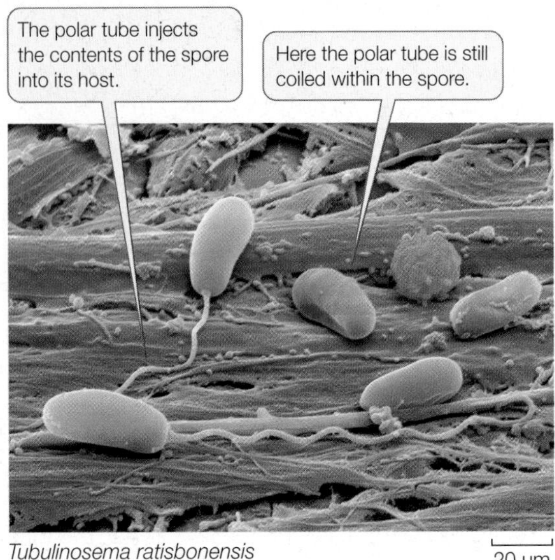
The polar tube injects the contents of the spore into its host.
Here the polar tube is still coiled within the spore.
Tubulinosema ratisbonensis
20 μm

FIGURE 22.13 Invasion of the Microsporidia The spores of microsporidia grow polar tubes that transfer the contents of the spores into the host's cells. The species shown here infects many animals, including humans.

this hypothesis is incorrect. DNA sequence analysis, along with the fact that their cell walls contain chitin, has confirmed that the microsporidia are in fact highly reduced, parasitic fungi, although their exact placement among the fungal lineages is still being investigated.

Microsporidia are obligate intracellular parasites of animals, especially of insects, crustaceans, and fishes. Some species are known to infect mammals, including humans. Most infections by microsporidia cause chronic disease in the host, with effects that include weight loss, reduced fertility, and a shortened life span. The host cell is penetrated by a polar tube that grows from the microsporidian spore. The function of the polar tube is to inject the contents of the spore, the sporoplasm, into the host (**FIGURE 22.13**). The sporoplasm then replicates within the host cell and produces new infective spores. The life cycle of some species is complex and involves multiple hosts, whereas other species infect a single host. In some insects, parasitic microsporidia are transmitted vertically (i.e., from parent to offspring). Reproduction is thought to be strictly asexual in some microsporidians, but includes poorly understood asexual and sexual cycles in other species.

Most chytrids are aquatic

The **chytrids** include several distinct lineages of mostly aquatic microorganisms once classified with the protists. However, morphological evidence (cell walls that consist primarily of chitin) and molecular evidence support their classification as early-diverging fungi. In this book we use the term "chytrid" to refer to all three of the formally named clades listed as chytrids in Table 22.1, but some mycologists use this term to refer to only one of those clades, the Chytridiomycota. There are fewer than 1,000 described species among the three groups of chytrids.

Chytrids reproduce both sexually and asexually. Like the animals, most chytrids that reproduce sexually possess flagellated gametes. The spores (called zoospores) are also flagellated (**FIGURE 22.14A**). The retention of flagella in the spores and gametes reflects the aquatic environment in which fungi first evolved. Chytrids are the only fungi that include species with flagella at any life cycle stage.

The chytrids are diverse in form. Some are unicellular, others have rhizoids, and still others have coenocytic hyphae. They may be parasitic (on organisms such as algae, mosquito larvae, nematodes, and amphibians) or saprobic. Some have complex mutualistic relationships with foregut-fermenting animals such as cattle and deer. Many chytrids live in freshwater habitats or in moist soil, but some are marine.

Some fungal life cycles feature separate fusion of cytoplasms and nuclei

Most members of the remaining four groups of fungi are terrestrial. Although the terrestrial fungi grow in moist places, they do not have motile gametes, and they do not release gametes into the environment, so liquid water is not required for fertilization. Instead, the cytoplasms of two individuals of different mating types fuse (a process called **plasmogamy**) before their nuclei fuse (a process called **karyogamy**). Sexual species of terrestrial fungi include some zygospore fungi, sac fungi, and club fungi.

Zygospore fungi reproduce sexually when adjacent hyphae of two different mating types release chemical signals that cause them to grow toward each other. These hyphae produce gametangia, which are specialized cells for reproduction that are retained as part of the hyphae. In the gametangia, nuclei replicate without cell division, resulting in multiple haploid nuclei in both gametangia. The two gametangia then fuse to form a zygosporangium that contains many haploid nuclei of each mating type (**FIGURE 22.14B**). Haploid nuclei of different mating types then pair up to form multiple diploid nuclei within the zygosporangium. A thick, multilayered cell wall develops around the zygosporangium to form a well-protected resting stage that can remain dormant for months. In harsh environmental conditions, this resting stage may be the only cell that survives as the surrounding cells of the hyphae die. At this stage the single surviving cell is known as a **zygospore**, which is the basis of the name of the zygospore fungi. When environmental conditions improve, the nuclei within the zygospore undergo meiosis and one or more stalked **sporangiophores** sprout, each bearing a sporangium. Each sporangium contains the products of meiosis: haploid nuclei that are incorporated into spores. These spores disperse and germinate to form a new generation of haploid hyphae.

The zygospore fungi include four major lineages of terrestrial fungi that live on soil as saprobes, as parasites of insects and spiders, or as mutualists of other fungi and invertebrate animals. They produce no cells with flagella, and only one diploid cell—the zygospore—appears in the entire life cycle. Their hyphae are coenocytic. Most species do not form a fleshy

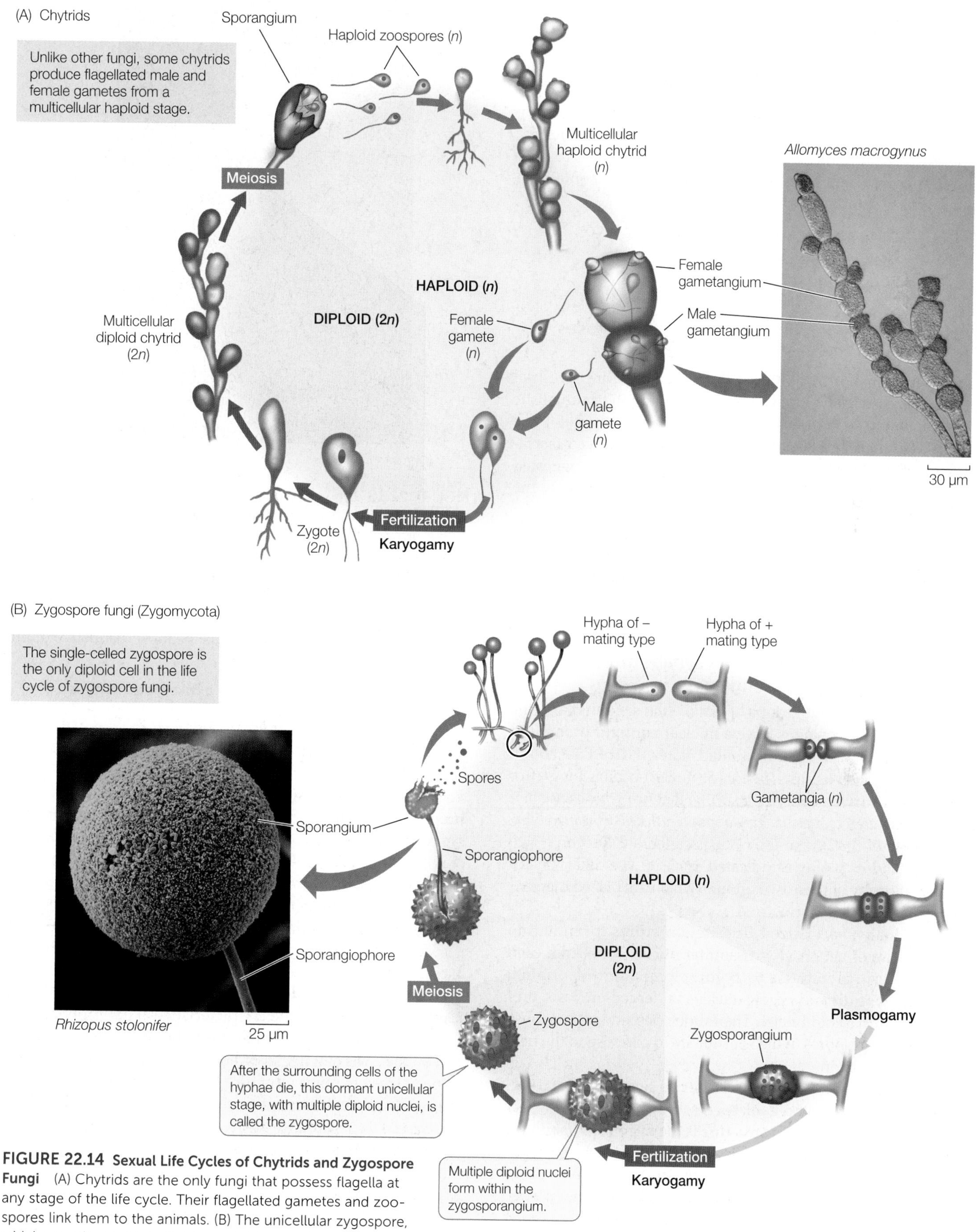

(A) Chytrids

Unlike other fungi, some chytrids produce flagellated male and female gametes from a multicellular haploid stage.

Sporangium
Haploid zoospores (n)
Multicellular haploid chytrid (n)
Allomyces macrogynus
Meiosis
HAPLOID (n)
DIPLOID ($2n$)
Female gametangium
Male gametangium
Multicellular diploid chytrid ($2n$)
Female gamete (n)
Male gamete (n)
30 μm
Zygote ($2n$)
Fertilization
Karyogamy

(B) Zygospore fungi (Zygomycota)

The single-celled zygospore is the only diploid cell in the life cycle of zygospore fungi.

Hypha of – mating type
Hypha of + mating type
Spores
Gametangia (n)
Sporangium
Sporangiophore
HAPLOID (n)
DIPLOID ($2n$)
Sporangiophore
Meiosis
Rhizopus stolonifer
25 μm
Zygospore
Plasmogamy
Zygosporangium

After the surrounding cells of the hyphae die, this dormant unicellular stage, with multiple diploid nuclei, is called the zygospore.

Multiple diploid nuclei form within the zygosporangium.

Fertilization
Karyogamy

FIGURE 22.14 Sexual Life Cycles of Chytrids and Zygospore Fungi (A) Chytrids are the only fungi that possess flagella at any stage of the life cycle. Their flagellated gametes and zoospores link them to the animals. (B) The unicellular zygospore, which contains multiple diploid nuclei, is a resting structure unique to the zygospore fungi.

fruiting structure. Instead, the hyphae spread in a radial pattern from the spore, with occasional stalked sporangiophores reaching up into the air (**FIGURE 22.15**).

More than 1,000 species of zygospore fungi have been described. One species you may have seen is *Rhizopus stolonifer*, the black bread mold. *Rhizopus* produces many stalked sporangiophores, each bearing a single sporangium containing hundreds of minute spores (see Figure 22.14B).

Go to ANIMATED TUTORIAL 22.1
Life Cycle of a Zygospore Fungus
PoL2e.com/at22.1

Arbuscular mycorrhizal fungi form symbioses with plants

Arbuscular mycorrhizal fungi (Glomeromycota) are terrestrial fungi that associate with plant roots in a symbiotic, mutualistic relationship (see Figure 22.10B). Fewer than 200 species have been described, but 80–90 percent of all plants have associations with them. Molecular systematic studies have suggested that arbuscular mycorrhizal fungi are the sister group to the Dikarya (sac fungi and club fungi; see Figure 22.11).

The hyphae of arbuscular mycorrhizal fungi are coenocytic. These fungi use glucose from their plant partners as their primary energy source, converting it into other, fungus-specific sugars that cannot return to the plant. Arbuscular mycorrhizal fungi are only known to reproduce asexually.

The dikaryotic condition is a synapomorphy of sac fungi and club fungi

In the two remaining groups of fungi—the sac fungi and the club fungi—some stages have a nuclear configuration other than the familiar haploid or diploid states (**FIGURE 22.16**). In these fungi, karyogamy (fusion of nuclei) occurs long after plasmogamy (fusion of cytoplasm), so that *two genetically different haploid nuclei coexist and divide within each cell of the mycelium*. This stage of the fungal life cycle is called a **dikaryon** ("two nuclei") and its ploidy is indicated as $n + n$. The dikaryon is a synapomorphy of these two groups, which are placed together in a clade called **Dikarya**.

Eventually, specialized fruiting structures form, within which pairs of genetically dissimilar nuclei—one from each parent—fuse, giving rise to zygotes long after the original "mating." The diploid zygote nucleus undergoes meiosis, producing four haploid nuclei. The mitotic descendants of those nuclei become spores, which germinate to give rise to the next haploid generation.

A life cycle with a dikaryon stage has several unusual features. First, there are no gamete *cells*, only gamete *nuclei*. Second, the only true diploid structure is the zygote, although for a long period the genes of both parents are present in the dikaryon and can be expressed. In effect, the dikaryon is neither diploid ($2n$) nor haploid (n). Rather, it is dikaryotic ($n + n$). Therefore a harmful recessive mutation in one nucleus may be compensated for by a normal allele on the same chromosome in the other nucleus, and dikaryotic hyphae often have characteristics that are different from their n or $2n$ products.

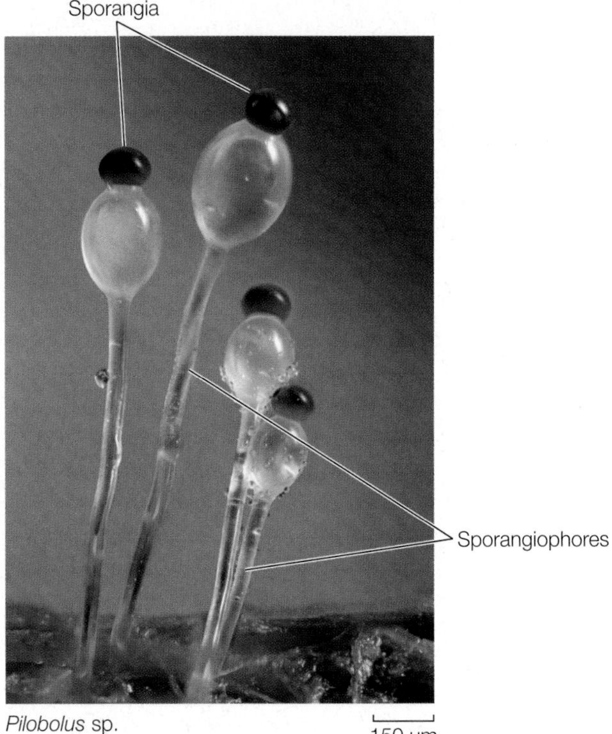

FIGURE 22.15 Zygospore Fungi Produce Sporangiophores These transparent structures are sporangiophores produced by a zygospore fungus growing on decomposing animal dung. The sporangiophores grow toward the light and end in tiny sporangia, which the stalked sporangiophores can eject as far as 2 meters. Animals ingest sporangia that land on grass and then disseminate the spores in their feces.

The dikaryotic condition is perhaps the most distinctive of the genetic peculiarities of the fungi.

Go to ACTIVITY 22.2 Life Cycle of a Dikaryotic Fungus
PoL2e.com/ac22.2

The sexual reproductive structure of sac fungi is the ascus

The **sac fungi** (Ascomycota) are a large and diverse group of fungi found in marine, freshwater, and terrestrial habitats. There are approximately 64,000 known species, almost a third of which are the fungal partners in lichens. The hyphae of sac fungi are segmented by more or less regularly spaced septa. A pore in each septum permits extensive movement of cytoplasm and organelles (including nuclei) from one segment to the next.

Sac fungi are distinguished by the production of sacs called **asci** (singular *ascus*), which after meiosis and spore cleavage contain sexually produced haploid **ascospores** (see Figure 22.16A). The ascus is the characteristic sexual reproductive structure of the sac fungi. In the past, the sac fungi were classified on the basis of whether or not the asci are contained within a specialized fruiting structure known as an **ascoma** (plural *ascomata*) and on differences in the morphology of that fruiting structure. DNA sequence analyses have resulted in a revision of these traditional groupings, however.

SAC FUNGUS YEASTS Some species of sac fungi are unicellular yeasts. The 1,000 or so species in this group are among the most important domesticated fungi. Perhaps the best known is baker's, or brewer's, yeast (*Saccharomyces cerevisiae*; see Figure 22.2), which metabolizes glucose obtained from its environment into

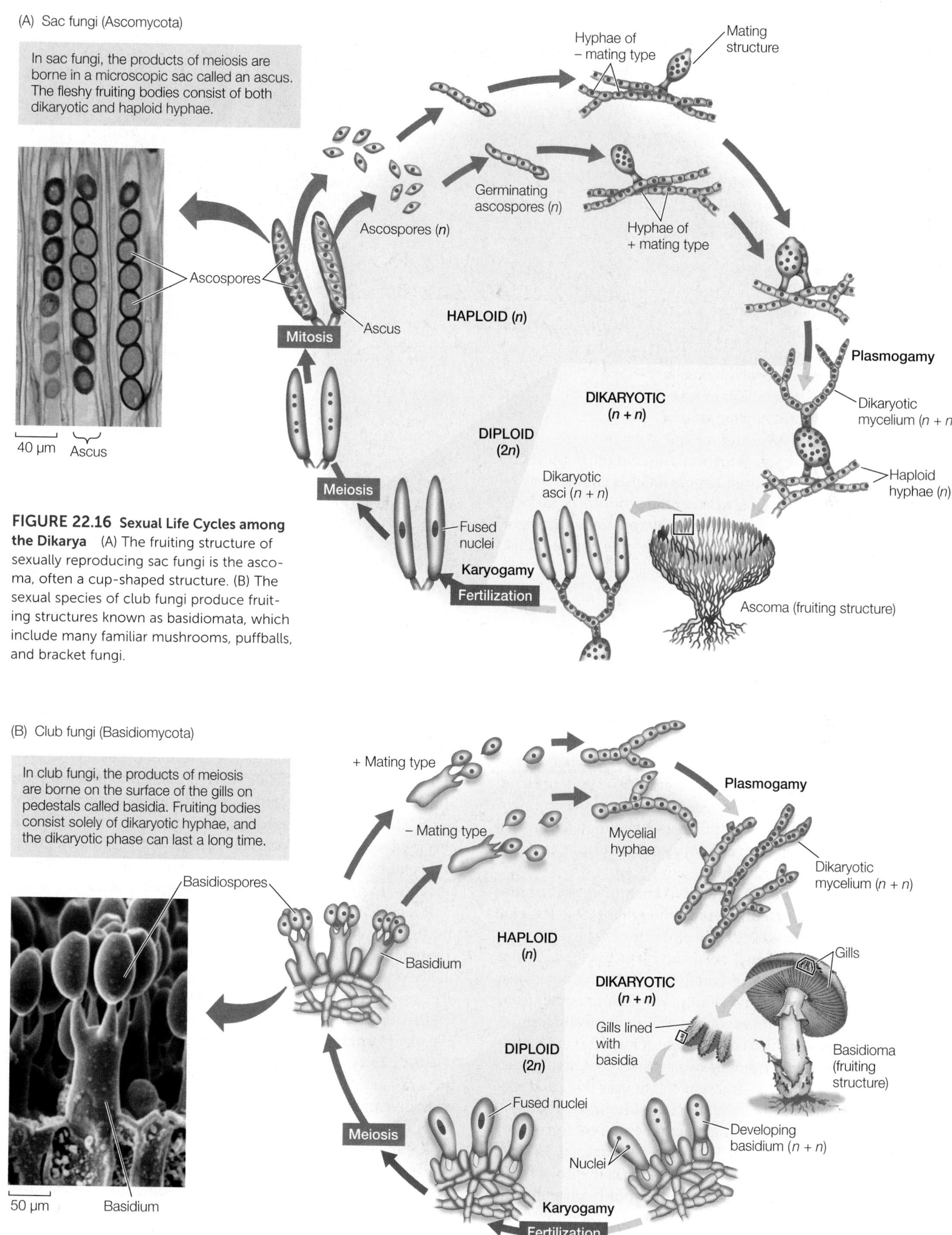

FIGURE 22.16 Sexual Life Cycles among the Dikarya (A) The fruiting structure of sexually reproducing sac fungi is the ascoma, often a cup-shaped structure. (B) The sexual species of club fungi produce fruiting structures known as basidiomata, which include many familiar mushrooms, puffballs, and bracket fungi.

ethanol and carbon dioxide by fermentation. It forms carbon dioxide bubbles in bread dough and gives baked bread its light texture. Although they are baked away in bread making (which produces the pleasant aroma of baking bread), the ethanol and carbon dioxide are both retained when the yeast ferments grain into beer. Other sac fungus yeasts live on fruits such as figs and grapes and play an important role in the making of wine. Many others are associated with insects. In the guts of some insects, they provide enzymes that break down materials that are otherwise difficult for the insects to digest, especially cellulose.

(A) *Cookeina tricholoma*

(B) *Morchella esculenta*

FIGURE 22.17 Sac Fungi (A) These brilliant red cups are the ascomata of a cup fungus. (B) Morels, which have a spongelike ascoma and a subtle flavor, are considered a culinary delicacy by humans.

Sac fungus yeasts reproduce asexually by budding. Sexual reproduction takes place when two adjacent haploid cells of opposite mating types fuse. In some species, the resulting zygote buds to form a diploid cell population. In others, the zygote nucleus undergoes meiosis immediately to form an ascus. Depending on whether the products of meiosis then undergo mitosis, a yeast ascus usually contains either eight or four ascospores, which germinate to become haploid cells. The sac fungus yeasts have lost the dikaryon stage.

FILAMENTOUS SAC FUNGI Most sac fungi are filamentous species, such as the cup fungi (**FIGURE 22.17A**), in which the ascomata are cup-shaped and can be as large as several centimeters across (although most are much smaller). The inner surfaces of the ascomata, which are covered with a mixture of specialized hyphae and asci, produce huge numbers of spores. The edible ascomata of some species, including morels and truffles (**FIGURE 22.17B**), are regarded by humans as gourmet delicacies (and can sell at prices higher than gold). The underground ascomata of truffles have a strong odor that attracts mammals such as pigs, which then eat and disperse the fungus. Humans sometimes take advantage of pigs' attraction to truffles and use pigs to locate truffles for human consumption.

The sexual reproductive cycle of filamentous sac fungi includes the formation of a dikaryon, although this stage is relatively brief compared with that in club fungi. Many filamentous sac fungi form multinucleate mating structures (see Figure 22.16A). Mating structures of two different mating types fuse and produce a dikaryotic mycelium, containing nuclei from both mating types. The dikaryotic mycelium typically forms a cup-shaped ascoma, which bears the asci. Only after the formation of asci do the nuclei from the two mating types finally fuse. Both nuclear fusion and the subsequent meiosis that produces haploid ascospores take place within individual asci. The ascospores are ultimately released (sometimes shot off forcefully) by the ascus to begin the new haploid generation.

The sac fungi also include many of the filamentous fungi known as molds. **Molds** consist of filamentous hyphae that do not form large ascomata, although they can still produce asci and ascospores. Many molds are parasites of flowering plants. Chestnut blight and Dutch elm disease are both caused by molds. The chestnut blight fungus, which was introduced to the United States in the 1890s, had destroyed the American chestnut as a commercial species by 1940. Before the blight, this species accounted for more than half the trees in the eastern North American forests. Another familiar story is that of the American elm. Sometime before 1930, the Dutch elm disease fungus (first discovered in the Netherlands, but native to Asia) was introduced into the United States on infected elm logs from Europe. Spreading rapidly—sometimes by way of connected root systems—the fungus destroyed great numbers of American elm trees.

Other plant pathogens among the sac fungi include the powdery mildews that infect cereal crops, lilacs, and roses, among many other plants. Mildews can be a serious problem to farmers and gardeners, and a great deal of research has focused on ways to control these agricultural pests.

Brown molds of the genus *Aspergillus* are important in some human diets. *A. tamarii* acts on soybeans in the production of soy sauce, and *A. oryzae* is used in brewing the Japanese alcoholic beverage sake. Some species of *Aspergillus* that grow on grains and on nuts such as peanuts and pecans produce extremely carcinogenic (cancer-inducing) compounds called aflatoxins.

Penicillium is a genus of green molds, of which some species produce the antibiotic penicillin, presumably for defense against competing bacteria. Two species, *P. camembertii* and *P. roqueforti*, are the organisms responsible for the characteristic strong flavors of Camembert and Roquefort cheeses, respectively.

The filamentous sac fungi can also reproduce asexually by means of conidia that form at the tips of specialized hyphae (**FIGURE 22.18**). Small chains of conidia are produced by the millions and can survive for weeks in nature. The conidia are what give molds their characteristic colors.

The sexual reproductive structure of club fungi is the basidium

Club fungi (Basidiomycota) produce some of the most spectacular fruiting structures found among the fungi. These fruiting structures, called **basidiomata** (singular *basidioma*), include

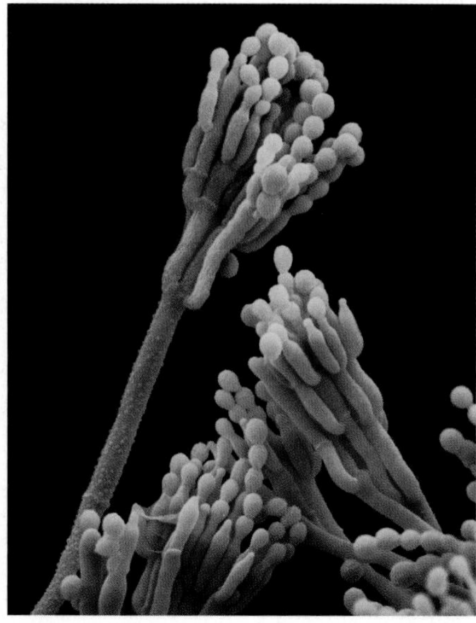

FIGURE 22.18 Conidia Chains of conidia (yellow) are developing at the tips of specialized hyphae arising from a *Penicillium* mold; compare the conidia of this species (which is used to produce "bleu" cheese) with the species shown on p. 450.

mushrooms of all kinds, puffballs (see Figure 22.4), and the bracket fungi often encountered on trees and fallen logs in a damp forest. About 30,000 species of club fungi have been described. They include about 4,000 species of mushrooms, including both poisonous and edible species (**FIGURE 22.19A**). Bracket fungi (**FIGURE 22.19B**) play an important role in the carbon cycle by breaking down wood. They also do great damage to both cut lumber and timber stands. Some of the most economically damaging plant pathogens are club fungi, including the rust fungi and smut fungi that parasitize cereal grains. In contrast, other club fungi contribute to the survival of plants as fungal partners in ectomycorrhizae.

The hyphae of club fungi characteristically have septa with small, distinctive pores. As these hyphae grow, haploid nuclei of different mating types meet and fuse, forming dikaryotic hyphae. The dikaryotic mycelium grows and eventually, when triggered by rain or another environmental cue, produces a basidioma. The dikaryon stage may persist for years, or even centuries. This pattern contrasts with the life cycle of the sac fungi, in which the dikaryon is found only in the stages leading up to formation of the asci.

The **basidium** (plural *basidia*), a swollen cell at the tip of a specialized hypha, is the characteristic sexual reproductive structure of the club fungi (see Figure 22.16B). In mushroom-forming club fungi, the basidia typically form on specialized structures known as gills. The basidium is the site of nuclear fusion and meiosis and thus plays the same role in the club fungi as the ascus does in the sac fungi and the zygosporangium does in the zygospore fungi.

After nuclei fuse in the basidium, the resulting diploid nucleus undergoes meiosis, and the four resulting haploid nuclei are incorporated into haploid **basidiospores**, which form on tiny stalks on the outside of the basidium. A single basidioma of the common bracket fungus *Ganoderma applanatum* can produce as many as 4.5 *trillion* basidiospores in one growing season. Basidiospores typically are forcibly discharged from their basidia and then germinate, giving rise to hyphae with haploid nuclei.

CHECKpoint CONCEPT 22.3

- ✓ Explain the concept of mating types. How are they different from male and female sexes?
- ✓ Explain how the microsporidia infect the cells of their animal hosts.
- ✓ What feature of the chytrids suggests that the fungi had an aquatic ancestor?
- ✓ What is the role of the zygospore in the life cycle of zygospore fungi?
- ✓ What distinguishes the fruiting structures of sac fungi from those of club fungi?

(A) *Amanita muscaria*

(B) *Trametes versicolor*

FIGURE 22.19 Club Fungus Basidiomata (A) Fly agaric is poisonous to humans, as are many species in the genus *Amanita*. Various species of *Amanita* account for about 95% of the human fatalities from mushroom poisoning. (B) Saprophytic bracket fungi are agents of decay on dead wood.

Fungi are of interest to biologists because of the roles they play in interactions with other organisms. But fungi can also be useful as tools for studying environmental problems and for finding solutions to those problems.

CONCEPT 22.4 Fungi Can Be Sensitive Indicators of Environmental Change

We've already noted the important roles that fungi play in ecosystems, from decomposers to pathogens to plant mutualists. These diverse ecological roles have led to the use of fungi in studies of environmental change and in environmental remediation.

Lichen diversity and abundance indicate air quality

Lichens can live in many harsh environments where few other species can survive, as we saw in Concept 22.2. In spite of their hardiness, however, lichens are highly sensitive to air pollution because they are unable to excrete any toxic substances they absorb. This sensitivity means that lichens are good biological indicators of air pollution levels. It also explains why they are not commonly found in heavily industrialized regions or in large cities.

Monitoring the diversity and abundance of lichens growing on trees is a practical and inexpensive system for gauging air quality around cities (**FIGURE 22.20**). Maps of lichen diversity provide environmental biologists with a tool for tracking the distribution of air pollutants and their effects. Sensitive biological indicators of pollution, such as lichen growth, allow biologists to monitor air quality without the use of specialized equipment. Lichens are naturally distributed across the environment, and they can also provide a long-term measure of the effects of air pollution across many seasons and years.

(A)

(B)

FIGURE 22.20 More Lichens, Better Air Lichen abundance and diversity are excellent indicators of air quality. (A) Many lichen species show luxuriant growth on trees in suitable environments with few pollutants in the air. (B) As air quality declines, so do the number and diversity of lichens.

Fungi record and help remediate environmental pollution

Each year, biologists deposit samples of many groups of organisms in the collections of natural history museums. These museum collections serve many purposes. Biologists borrow specimens from these museums to study many aspects of evolution and ecology, and the collections document changes in the biota of our planet over time.

Collections of fungi made over many decades or centuries provide a record of the environmental pollutants that were present when the fungi were growing. Biologists can analyze these historical samples to see how different sources of pollutants were affecting our environment before anyone thought to take direct measurements. These long-term records are also useful for analyzing the effectiveness of cleanup efforts and regulatory programs for controlling environmental pollutants.

We have already seen that fungi are critical to the planetary carbon cycle because of their role in breaking down dead organic matter. Fungi are also used in remediation efforts to help clean up sites that have been polluted by oil spills or contaminated with toxic petroleum-derived hydrocarbons. Many herbicides, pesticides, and other synthetic hydrocarbons are broken down primarily through the action of fungi.

Reforestation may depend on mycorrhizal fungi

When a forest is cut down, it is not just the trees that are lost. A forest is an ecosystem that depends on the interaction of many species. As we have discussed, many plants depend on close relationships with mycorrhizal fungal partners. When trees are removed from a site, the populations of mycorrhizal fungi there decline rapidly. If we wish to restore the forest on the site, we cannot simply replant it with trees and other plants and expect them to survive. The mycorrhizal fungal community must be reestablished as well. For large forest restoration projects, a planned succession of plant growth and soil improvement is often necessary before forest trees can be replanted. As the community of soil fungi gradually recovers, trees that have been inoculated with appropriate mycorrhizal fungi in tree nurseries can be planted to reintroduce greater diversity to the soil fungal community.

APPLY THE CONCEPT

Fungi can be sensitive indicators of environmental change

Biologists analyzed museum samples of lace lichens (*Ramalina menziesii*) collected near San Francisco, California, from 1892 to 2006 for evidence of lead contamination. They measured concentrations of lead (Pb) as well as the ratios of its isotopes ^{206}Pb and ^{207}Pb. The latter measurement was used to determine the source of lead contamination. Possible sources included a lead smelter that operated in the area from 1885 to 1971 (which produced emissions with a ^{206}Pb/^{207}Pb ratio of about 1.15–1.17); leaded gasoline in use from the 1930s to the early 1980s, peaking in 1970 (with a ^{206}Pb/^{207}Pb ratio of 1.18–1.23); and resuspension of historic lead contamination as atmospheric aerosols in recent decades (with an intermediate ^{206}Pb/^{207}Pb ratio of about 1.16–1.19).[a]

Before analyzing the data, use the information provided above to formulate hypotheses about these questions: What trends in atmospheric lead concentrations would you expect to see? What ^{206}Pb/^{207}Pb ratios would you expect to find at different times from the late 1800s to the early 2000s?

1. Plot lead concentration in the lichen samples against year of sample collection. Make a second plot of ^{206}Pb/^{207}Pb ratio against year of sample collection.
2. Do your analyses support the hypotheses you formulated? Are your hypotheses consistent with your analyses of both lead concentrations and ^{206}Pb/^{207}Pb ratios through time? If not, how would you modify your hypotheses, and what additional tests can you design to test your ideas?

SAMPLE	YEAR COLLECTED	LEAD CONCENTRATION (μg OF Pb/g OF LICHEN)	^{206}Pb/^{207}Pb RATIO
1	1892	11.9	1.165
2	1894	4.0	1.155
3	1906	13.7	1.154
4	1907	22.9	1.157
5	1945	49.9	1.187
6	1957	34.2	1.185
7	1978	50.9	1.221
8	1982	10.0	1.215
9	1983	4.6	1.224
10	1987	1.0	1.198
11	1988	1.3	1.199
12	1995	1.9	1.202
13	2000	0.4	1.184
14	2006	1.8	1.184

[a]A. R. Flegal et al. 2010. *Environmental Science and Technology* 44: 5613–5618.

CHECKpoint CONCEPT 22.4

- ✓ What are some advantages of using surveys of lichen diversity and museum collections of lichens to measure long-term changes in air quality, compared with direct measurements of atmospheric pollutants?
- ✓ Can you develop a strategy for tree harvest that would ease the difficulty and expense of reforestation projects by retaining viable communities of mycorrhizal fungi?

Whether living on their own or in symbiotic associations, fungi have spread successfully over much of Earth since their origin from a protist ancestor. An earlier ancestor of fungi also gave rise to the choanoflagellates and the animals, as we will describe in Chapter 23.

Have antibiotics derived from fungi eliminated the danger of bacterial diseases in human populations?

ANSWER Beginning in the 1940s, antibiotics derived from fungi ushered in a "golden age" of freedom from bacterial infections. Today, however, that golden age may be coming to an end. Many antibiotics are losing their effectiveness as pathogenic bacteria evolve resistance to these drugs (**FIGURE 22.21**). Some bacterial diseases, such as tuberculosis, are increasingly serious health problems because of the evolution of new strains that are resistant to most classes of antibiotics.

Why do bacteria evolve resistance to antibiotics? Mutations that allow bacteria to survive in the presence of an antibiotic are favored by selection in a bacterial population whenever an antibiotic is used. Such mutations often carry a cost to the bacteria, so they may be selected against in the absence of regular antibiotic use. To reduce the rate of evolution of antibiotic resistance, antibiotics should be used only for the

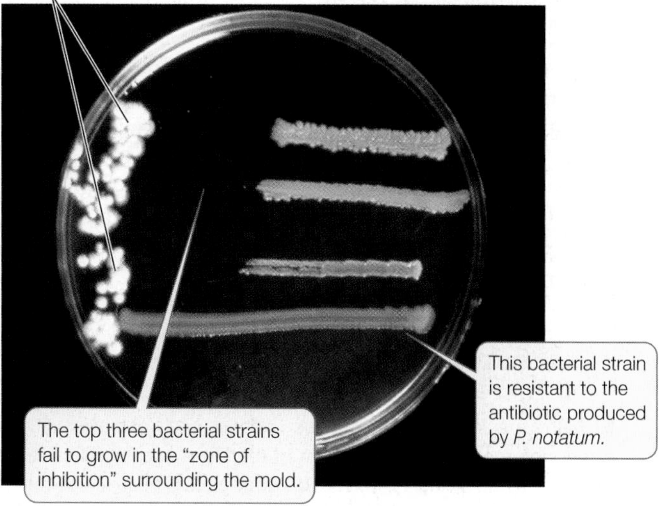

Penicillium notatum
This bacterial strain is resistant to the antibiotic produced by *P. notatum*.
The top three bacterial strains fail to grow in the "zone of inhibition" surrounding the mold.

FIGURE 22.21 Penicillin Resistance In a petri dish similar to those in Alexander Fleming's lab, four strains of a pathogenic bacterium have been cultured along with *Penicillium* mold. One strain is resistant to the mold's antibiotic substance, as is evidenced by its growth up to the mold.

treatment of appropriate bacterial diseases, and then used to completely clear the bacterial infection.

Most medical antibiotics are chemically modified forms of the substances that are found naturally in fungi and other organisms. Fungi naturally produce antibiotic compounds to defend themselves against bacterial growth and to reduce competition from bacteria for nutritional resources. These naturally occurring compounds are usually chemically modified to increase their stability, improve their effectiveness, and facilitate synthetic production.

From the late 1950s to the late 1990s, no new major classes of antibiotics were discovered. In recent years, however, three new classes of antibiotics have been synthesized based on the information learned from naturally occurring antibiotics, leading to improved treatment of some formerly resistant strains of bacteria.

Fungi have also been used to combat non-bacterial diseases. One of the more unusual applications of fungi is in the war against malaria, one of the biggest killers of humans in sub-Saharan Africa. Biologists have discovered that two species of fungi, *Beauveria bassiana* and *Metarhizium anisopliae*, can kill malaria-causing mosquitoes when applied to mosquito netting. Mosquitoes have not yet shown evidence of developing resistance to these biological pathogens, as they have to most chemical pesticides.

SUMMARY

CONCEPT 22.1 Fungi Live by Absorptive Heterotrophy

- Fungi are **opisthokonts** with **absorptive heterotrophy** and with **chitin** in their cell walls. Fungi have various nutritional modes: some are **saprobes**, others are parasites, and some are mutualists. **Yeasts** are unicellular, free-living fungi. **Review Figure 22.1**
- The body of a multicellular fungus is a **mycelium**—a meshwork of filaments called **hyphae**. Hyphae may be **septate** (having **septa**) or **coenocytic** (multinucleate). **Review Figure 22.3**
- Fungi are often tolerant of hypertonic environments, and many are tolerant of extreme temperatures.

CONCEPT 22.2 Fungi Can Be Saprobic, Parasitic, Predatory, or Mutualistic

- Saprobic fungi, as decomposers, make crucial contributions to the recycling of elements, especially carbon.
- Many fungi are parasites of plants, harvesting nutrients from plant cells by means of **haustoria**. **Review Figure 22.5**
- Certain fungi have relationships with other organisms that are both **symbiotic** and **mutualistic**.
- Some fungi associate with unicellular green algae, cyanobacteria, or both to form **lichens**, which live on exposed surfaces of rocks, trees, and soil. **Review Figure 22.9**
- **Mycorrhizae** are mutualistic associations of fungi with plant roots. They improve a plant's ability to take up nutrients and water. **Review Figure 22.10**
- **Endophytic fungi** live within plants and may protect their hosts from herbivores and pathogens.

CONCEPT 22.3 Major Groups of Fungi Differ in Their Life Cycles

- The microsporidia, chytrids, and zygospore fungi diversified early in fungal evolution. The mycorrhizal fungi, sac fungi, and club fungi form a monophyletic group, and the latter two groups form the clade **Dikarya**. **Review Figure 22.11, Table 22.1, and ACTIVITY 22.1**
- Many species of fungi reproduce both sexually and asexually, although most fungi spend most of their lives in a haploid, asexual state. When sexual reproduction does occur, it usually occurs between individuals of different **mating types**. **Review Figure 22.12**
- The **microsporidia** are highly reduced unicellular fungi. They are obligate intracellular parasites that infect several animal groups.
- The three distinct lineages of **chytrids** all have flagellated gametes. **Review Figure 22.14A**
- In the sexual reproduction of terrestrial fungi, hyphae fuse, allowing "gamete" nuclei to be transferred. **Plasmogamy** (fusion of cytoplasm) precedes **karyogamy** (fusion of nuclei).
- **Zygospore fungi** have a resting stage with many diploid nuclei, known as a **zygospore**. Their fruiting structures are simple stalked **sporangiophores**. **Review Figure 22.14B and ANIMATED TUTORIAL 22.1**
- **Arbuscular mycorrhizal fungi** form symbiotic associations with plant roots. They are only known to reproduce asexually. Their hyphae are coenocytic.
- In sac fungi and club fungi, a mycelium containing two genetically different haploid nuclei, called a **dikaryon**, is formed. The dikaryotic ($n + n$) condition is unique to the fungi. **Review Figure 22.16 and ACTIVITY 22.2**
- **Sac fungi** have septate hyphae; their sexual reproductive structures are **asci**. Many sac fungi are partners in lichens. Filamentous sac fungi produce fleshy fruiting structures called **ascomata**. The dikaryon stage in the sac fungus life cycle is relatively brief. **Review Figure 22.16A**
- **Club fungi** have septate hyphae. Many club fungi are plant pathogens, although mushroom-forming species are more familiar to most people. Their fruiting structures are called **basidiomata**, and their sexual reproductive structures are **basidia**. The dikaryon stage may last for years. **Review Figure 22.16B**

CONCEPT 22.4 Fungi Can Be Sensitive Indicators of Environmental Change

- The diversity and abundance of lichen growth on trees provide sensitive indications of air quality. **Review Figure 22.20**
- Museum collections of fungi provide a historical record of atmospheric pollutants that were present when the fungi were growing. Fungi are often useful in environmental remediation.
- Reforestation projects require restoration of the mycorrhizal fungal community.

Go to the Interactive Summary to review key figures, Animated Tutorials, and Activities
PoL2e.com/is22

Go to LaunchPad at **macmillanhighered.com/launchpad** for additional resources, including LearningCurve Quizzes, Flashcards, and many other study and review resources.

23 Animal Origins and Diversity

KEY CONCEPTS

Entomologists collecting insects from a tree canopy in the Yungas rainforest of Argentina. Such research illuminates the vast number of insect species that remain undescribed by science.

About 1.3 million species of animals have been discovered and named by biologists. One group of animals, the insects, accounts for more than 1 million of these species, or more than half of all known species of living organisms. Although these numbers may seem incredibly large, they represent a relatively small fraction of the total animal diversity that is thought to exist on Earth.

As recently as the 1980s, many biologists thought that about half of existing insect species had been described, but today they think that the number of described insect species may be a much smaller fraction of the total number of living species. Why did they change their minds?

A simple but important field study suggested that the number of existing insect species had been significantly underestimated. Knowing that the insects of tropical forests—the most species-rich habitat on Earth—were poorly known, entomologist Terry Erwin made a comprehensive sample of one group of insects, the beetles, in the canopies of a single species of tropical forest tree, *Luehea seemannii*, in Panama. Erwin fogged the canopies of 19 large *L. seemannii* trees with a pesticide and collected the insects that fell from the trees in collection nets. His sample contained about 1,200 species of beetles—many of them undescribed—from this one species of tree.

Erwin then used a set of assumptions to estimate the total number of insect species in tropical evergreen forests. His assumptions included estimates of the number of species of host trees in these forests, the proportion of beetles that specialized on a specific species of host tree, the relative proportion of beetles to other insect groups, and the proportion of beetles that live in trees versus leaf litter. From this and similar studies, Erwin estimated that there may be 30 million or more species of insects on Earth. Although recent tests of Erwin's assumptions suggest 30 million was an overestimate, it is clear that the vast majority of insect species remain to be discovered.

Erwin's pioneering study highlighted the fact that we live on a poorly known planet, most of whose species have yet to be named and described. Much of the undiscovered diversity occurs among several groups of physically small yet biologically diverse animal groups, although even many larger species of animals continue to be discovered.

Besides beetles, which other groups of animals are thought to contain many more species than are known at present?

You will find the answer to this question on page 517.

CONCEPT 23.1 Distinct Body Plans Evolved among the Animals

How do we recognize an animal? The answer may seem obvious for familiar animals, but consider the sponges, which were once thought to be plants, or the rotifers, which are smaller than some protists.

The general characteristics often used to recognize animals include multicellularity, heterotrophy (using nutrients produced by other organisms), internal digestion, and motility (independent movement). Although these features are useful, none is diagnostic for all animals. Some animals do not move independently (at least during certain life stages), and some plants and fungi do have limited movement. Many parasitic animals lack a gut for internal digestion, and many other organisms (notably the fungi) are multicellular and heterotrophic. So what evidence leads us to group the animals in a single clade?

Animal monophyly is supported by gene sequences and cellular morphology

The most convincing evidence that all the organisms considered to be animals share a common ancestor comes from phylogenetic analyses of their gene sequences. Relatively few complete animal genomes are available, but more are being sequenced each year. Analyses of these genomes and of many individual gene sequences have shown that all animals are indeed monophyletic. A currently well supported phylogenetic tree of the animals is shown in **FIGURE 23.1**.

Surprisingly few morphological features are shared across all animal species. The morphological synapomorphies (shared, derived traits) that are present are evident primarily at the cellular level. They include several unique features of the junctions between animal cells and a common set of extracellular matrix molecules, including collagen and proteoglycans. Although some animals in a few groups lack one or another of these cellular characteristics, it is believed that these traits were possessed by the ancestor of all animals and were subsequently lost in those groups.

FIGURE 23.1 Animal Phylogeny This tree presents the best-supported current hypotheses of the evolutionary relationships among major groups of animals. The traits highlighted by red circles will be explained as you read this chapter. (See Analyze the Data 23.1 at LaunchPad.)

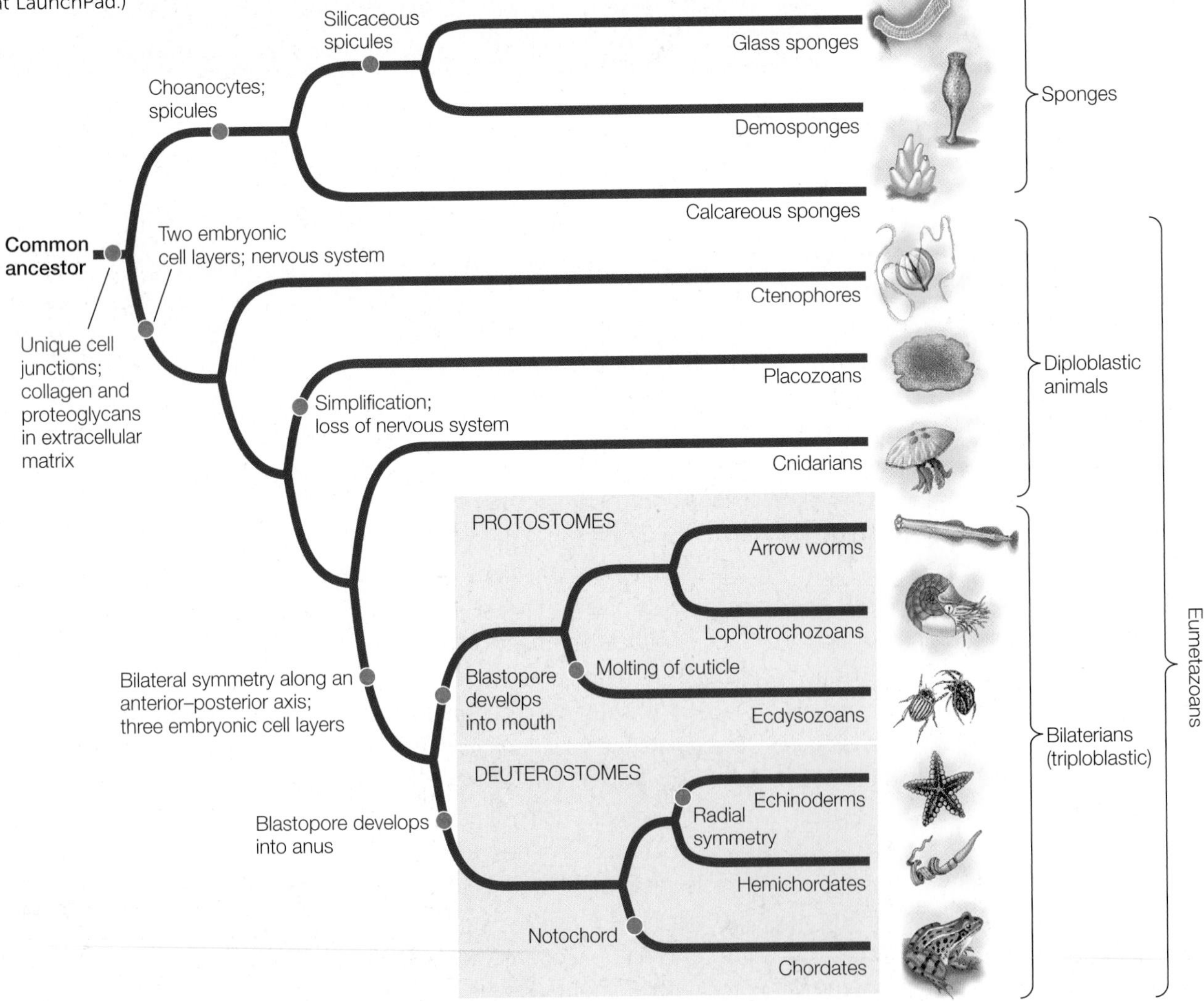

LINK
Animal cell junctions and extracellular matrix molecules are described in **Concept 4.5**

Similarities in the organization and function of Hox and other developmental genes provide additional evidence of developmental mechanisms shared by a common animal ancestor. The Hox genes specify body pattern and axis formation, leading to developmental similarities across animals (see Chapter 14).

The common ancestor of animals was probably a colonial flagellated protist similar to existing colonial choanoflagellates (**FIGURE 23.2A**), which have similarities to the multicellular sponges (**FIGURE 23.2B**). Why did these early animal ancestors begin to form multicellular colonies? One hypothesis is that multicellular colonies are more efficient than are single cells at capturing their prey. Experiments with living species of choanoflagellate show that they spontaneously form multicellular colonies in response to signaling compounds that are found on certain species of planktonic bacteria they eat (**FIGURE 23.3**).

Once multicellular colonies were formed in the ancestral animal lineage, it is likely that certain cells in the colony began to be specialized—some for movement, others for nutrition, others for reproduction, and so on. Once this functional specialization had begun, cells could continue to differentiate. Coordination among groups of cells could have improved by means of specific regulatory and signaling molecules that guided differentiation and migration of cells in developing embryos. Such coordinated groups of cells eventually evolved into the larger and more complex organisms that we call animals.

Nearly 80 percent of the 1.8 million named species of living organisms are animals, and millions of additional animal species await discovery. Evidence about the evolutionary relationships among animal groups can be found in fossils, in patterns of embryonic development, in the morphology and physiology of living animals, in the structure of animal proteins, and in gene sequences. Increasingly, studies of higher-level relationships have come to depend on genomic sequence comparisons, as genomes are ultimately the source of all inherited trait information.

Basic developmental patterns and body plans differentiate major animal groups

Distinct layers of cells form during the early development of most animals. These cell layers differentiate into specific organs and organ systems as development continues. The embryos of **diploblastic** animals have two cell layers: an outer **ectoderm** and

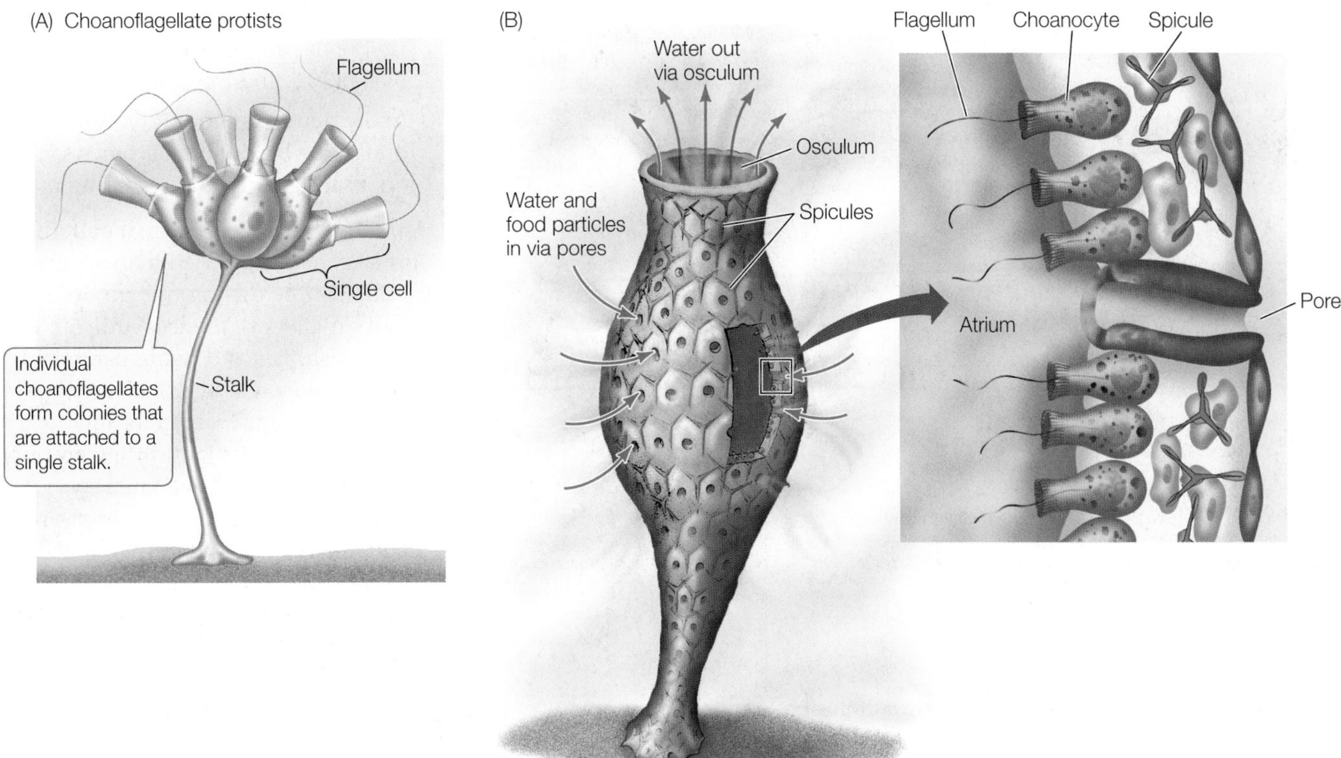

FIGURE 23.2 Choanocytes in Sponges Resemble Choanoflagellate Protists (A) The similarity of choanoflagellate protist colonies to sponge choanocytes supports an evolutionary link between this protist lineage and the animals. (B) A sponge moves food-containing water through its body by beating the flagella of its choanocytes—specialized feeding cells. Water enters through small pores and passes into water canals or an open atrium, where the choanocytes capture food particles from the water. The spicules are supportive, skeletal structures.

INVESTIGATION

FIGURE 23.3 What Induces Choanoflagellates to Form Multicellular Colonies?
The choanoflagellate *Salpingoeca rosetta* forms multicellular, rose-shaped colonies (rosettes) at low frequencies in natural conditions. The rosette colonies are more efficient than are single cells at capturing some prey species of planktonic bacteria. Rosanna Alegado and her collaborators investigated the chemical signals that induce the choanoflagellates to form multicellular colonies when environmental conditions favor the rosette formations.[a]

HYPOTHESIS

Multicellular colonies of *S. rosetta* are induced by chemical compounds found on specific species of planktonic bacteria.

METHOD

1. Isolate individual species of bacteria from an environmental sample of *S. rosetta* and its associated bacterial community.
2. Use a combination of antibiotics to produce a bacteria-reduced culture line of *S. rosetta* that does not spontaneously form rosette colonies (the RCA line, for "rosette colonies absent").
3. Feed the RCA line with each of the isolated species of bacteria from step 1. Record any rosette colony formation.
4. Isolate compounds that induce multicellular colony formation from rosette-producing bacteria identified in step 3.

RESULTS

Alegado and colleagues tested various compounds from *Algoriphagus* and found that RIF-1, a sulfonolipid produced by the bacteria, is the signal that induces formation of multicellular colonies in *S. rosetta*.

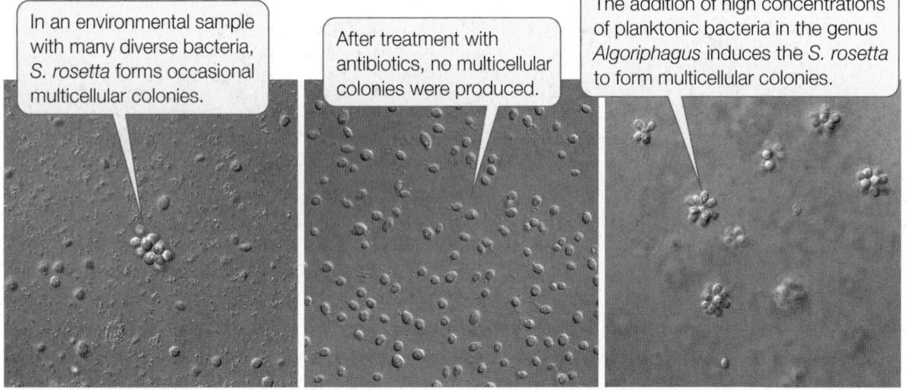

In an environmental sample with many diverse bacteria, *S. rosetta* forms occasional multicellular colonies.
After treatment with antibiotics, no multicellular colonies were produced.
The addition of high concentrations of planktonic bacteria in the genus *Algoriphagus* induces the *S. rosetta* to form multicellular colonies.

CONCLUSION

A chemical signal (RIF-1) from a particular species of prey bacteria can induce the formation of multicellular colonies in some species of choanoflagellates.

Go to **LaunchPad** for discussion and relevant links for all **INVESTIGATION** figures.

[a]R. A. Alegado et al. 2012. *eLife* 1:e00013.

an inner **endoderm**. Embryos of **triploblastic** animals have, in addition to ectoderm and endoderm, a third distinct cell layer, **mesoderm**, between the ectoderm and the endoderm. The existence of three cell layers in embryos is a synapomorphy of triploblastic animals, whereas the diploblastic animals (ctenophores, placozoans, and cnidarians) retain the ancestral condition. Some biologists consider sponges to be diploblastic, but since they do not have clearly differentiated tissue types or embryonic cell layers, the term is not usually applied to them.

During early development in many animals, in a process known as gastrulation, a hollow ball one cell thick indents to form a cup-shaped structure. The opening of the cavity formed by this indentation is called the blastopore:

Gastrulation will be covered in detail in Chapter 38. The point to remember here is that the overall pattern of gastrulation immediately after the blastopore stage divides the triploblastic animals into two major groups:

- In the **protostomes** (Greek, "mouth first"), the mouth arises from the blastopore, and the anus forms later.
- In the **deuterostomes** ("mouth second"), the blastopore becomes the anus, and the mouth forms later.

Although the developmental patterns of animals are more varied than suggested by this simple dichotomy, phylogenies based on DNA sequences indicate that the protostomes and deuterostomes are distinct clades. These groups are known as the **bilaterians** (named for their usual bilateral symmetry), and they account for the vast majority of animal species.

The general structure of an animal, the arrangement of its organ systems, and the integrated functioning of its parts are referred to as its **body plan**. As Chapter 14 described, the regulatory and signaling genes that govern the development of body symmetry, body cavities, segmentation, and appendages are widely shared among the different animal groups. Although the myriad animal body plans cover a wide range of morphologies, they can be seen as variations on four key features:

- The symmetry of the body
- The structure of the body cavity
- The segmentation of the body

- The existence and location of external appendages that are used for sensing, feeding, locomotion, mating, and other functions

Each of these features affects how an animal interacts with its environment.

Most animals are symmetrical

The overall shape of an animal can be described by its **symmetry**. An animal is said to be symmetrical if it can be divided along at least one plane into similar halves. Animals that have no plane of symmetry are asymmetrical. Placozoans and many sponges are asymmetrical, but most other animals have some kind of symmetry, which is governed by the expression of regulatory genes during development.

In organisms with **radial symmetry**, body parts are arranged around a single axis at the body's center:

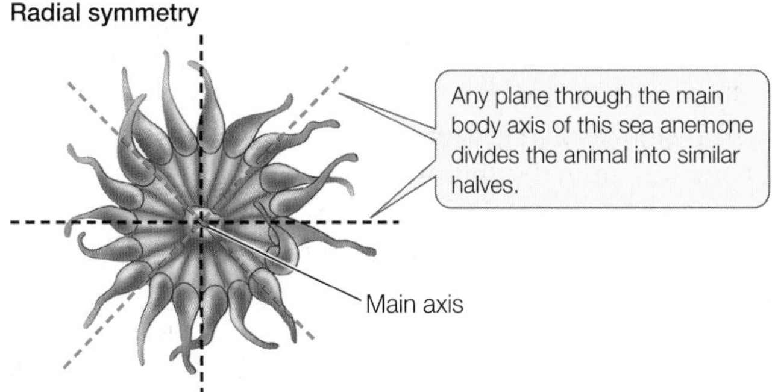

Radial symmetry
Any plane through the main body axis of this sea anemone divides the animal into similar halves.
Main axis

A perfectly radially symmetrical animal can be divided into similar halves by any plane that contains the central axis. However, most radially symmetrical animals are slightly modified, so fewer planes can divide them into identical halves. Some radially symmetrical animals are sessile (they remain fixed in one place) or drift with water currents. Others move slowly but can move equally well in any direction.

Bilateral symmetry is characteristic of animals that have a distinct front end, which typically precedes the rest of the body as the animal moves. A bilaterally symmetrical animal can be divided into mirror-image (left and right) halves by a single plane that passes through the midline of its body. This plane runs from the front, or **anterior**, end of the body, to the rear, or **posterior**, end:

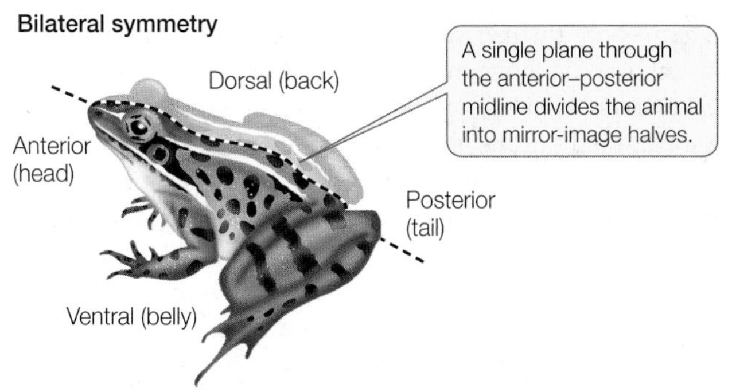

A plane at right angles to the midline divides the body into two dissimilar sides, its **dorsal** (top) and **ventral** (bottom) halves.

Bilateral symmetry is strongly correlated with **cephalization** (Greek *kephalos*, "head"), which is the concentration of sensory organs and nervous tissues at the anterior end of the animal. Cephalization has been evolutionarily favored because the anterior end of a bilaterally symmetrical animal typically encounters the environment first.

The structure of the body cavity influences movement

Animals with three embryonic cell layers (see Figure 23.1) can be divided into three types based on the presence and structure of an internal, fluid-filled **body cavity**.

- **Acoelomate** animals such as flatworms lack an enclosed, fluid-filled body cavity. Instead, the space between the gut (derived from endoderm) and the muscular body wall (derived from mesoderm) is filled with masses of cells called mesenchyme (**FIGURE 23.4A**). These animals typically move by beating cilia.
- **Pseudocoelomate** animals have a body cavity called a pseudocoel, a fluid-filled space in which many of the internal organs are suspended. A pseudocoel is enclosed by muscles (mesoderm) only on its outside; there is no inner layer of mesoderm surrounding the internal organs (**FIGURE 23.4B**).
- **Coelomate** animals have a body cavity, the coelom, that develops within the mesoderm. It is lined with a layer of muscular tissue called the peritoneum, which also surrounds the internal organs. The coelom is thus enclosed on both the inside and the outside by mesoderm (**FIGURE 23.4C**).

The structure of an animal's body cavity strongly influences the ways in which it can move. The body cavities of many animals function as **hydrostatic skeletons**. Fluids are relatively incompressible, so when the muscles surrounding a fluid-filled body cavity contract, fluids shift to another part of the cavity. If the body tissues around the cavity are flexible, fluids squeezed out of one region can cause another region to expand. The moving fluids can thus move specific body parts. (You can see how a hydrostatic skeleton works by watching a snail emerge from its shell.) A coelomate animal has better control over the movement of the fluids in its body cavity than a pseudocoelomate animal does. An animal that has longitudinal muscles (running along the length of the body) as well as circular muscles (encircling the body cavity) has even greater control over its movement.

In terrestrial environments, the hydrostatic function of fluid-filled body cavities applies mostly to relatively small, soft-bodied organisms. Most larger animals (as well as many smaller ones) have hard skeletons that provide protection and facilitate movement. Muscles are attached to those firm structures, which may be inside the animal or on its outer surface (in the form of a shell or cuticle).

Segmentation improves control of movement

Segmentation—the division of the body into segments—is seen in many animal groups. Segmentation facilitates specialization of different body regions. It also allows an animal to

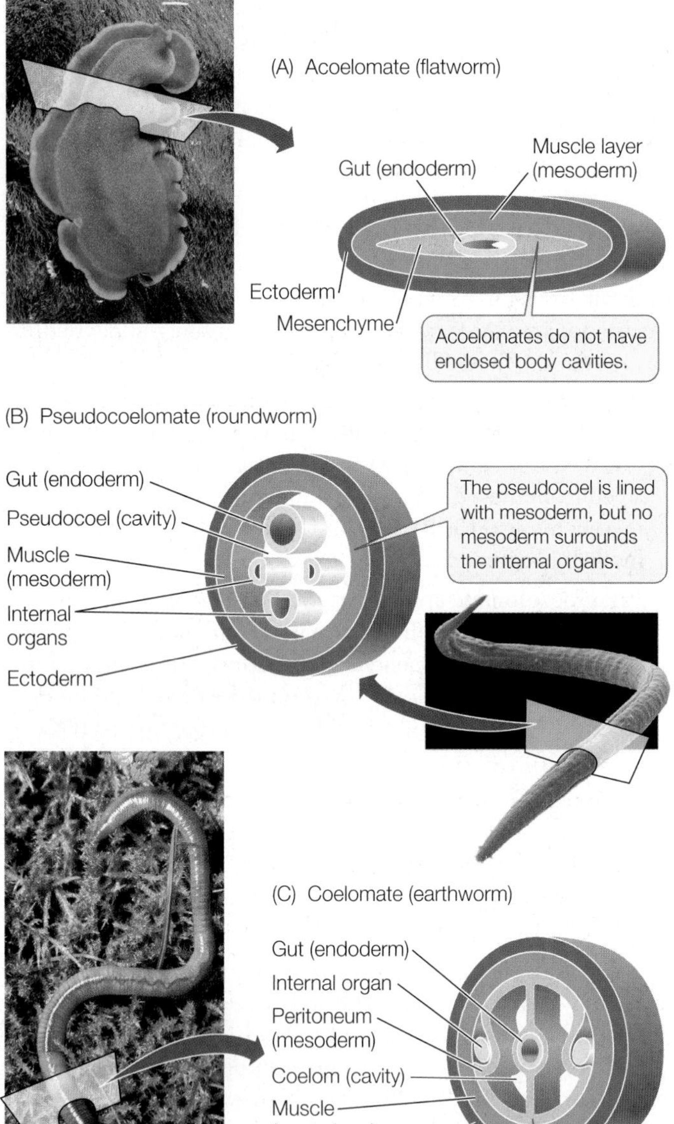

FIGURE 23.4 Animal Body Cavities (A) Acoelomates, like this marine flatworm, do not have enclosed body cavities. (B) Pseudocoelomates have a body cavity bounded by endoderm and mesoderm. (C) Coelomates have a peritoneum surrounding the internal organs in a region bounded by mesoderm.

Go to ACTIVITY 23.1 Animal Body Cavities PoL2e.com/ac23.1

alter the shape of its body in complex ways and to control its movements precisely. If an animal's body is segmented, muscles in each individual segment can change the shape of that segment independently of the others. In only a few segmented animals is the body cavity separated into discrete compartments, but even partly separated compartments allow better control of movement. Segmentation occurs in several groups of protostomes and deuterostomes.

In some animals, such as annelids (earthworms and their relatives), similar body segments are repeated many times (**FIGURE 23.5A**). In other animals, including most arthropods, segments differ strikingly from one another (**FIGURE 23.5B**). The dramatic evolutionary radiation of the arthropods (including the insects, spiders, centipedes, and crustaceans) was based on changes in a segmented body plan that features muscles attached to the inner surface of an external skeleton, including a variety of external appendages that move these animals. In some animals, distinct body segments are not apparent externally—for example, the segmented vertebrae of vertebrates, including humans. Nonetheless, muscular segmentation is clearly visible in humans with well-defined, muscular bodies (**FIGURE 23.5C**).

Appendages have many uses

Getting around under their own power allows animals to obtain food, to avoid predators, and to find mates. Even some sedentary species, such as sea anemones, have larval stages that use cilia to swim, thus increasing the animal's chances of finding a suitable habitat.

Appendages that project externally from the body greatly enhance an animal's ability to move around. Many echinoderms, including sea urchins and seas stars, have large numbers of tube feet that allow them to move slowly across the substrate. Animals whose appendages have become modified into limbs are capable of highly controlled, more rapid movement. The presence of jointed limbs has been a prominent factor in the evolutionary success of the arthropods and the vertebrates. In four independent instances—among the arthropod insects and among the vertebrate pterosaurs, birds, and bats—body plans emerged in which limbs were modified into wings, allowing these animals to use powered flight.

Appendages also include many structures that are not used for locomotion. Many animals have antennae, which are specialized appendages used for sensing the environment. Other appendages (such as claws and the mouthparts of many arthropods) are adaptations for capturing prey or chewing food. In some species, appendages are used for reproductive purposes, such as sperm transfer or egg incubation.

Nervous systems coordinate movement and allow sensory processing

The bilaterian animals have a well-coordinated central nervous system. More diffuse nervous systems, called **nerve nets**, are present in some other animals, such as jellyfishes and their relatives. Nervous systems appear to be completely absent in a few sessile animal groups, such as sponges.

The central nervous system of bilaterians coordinates the actions of muscles, which allows coordinated movement of appendages and body parts. This coordination of muscles permits highly effective and efficient movement on land, in water, or through the air. The central nervous system is also essential for the processing of sensory information gathered from a wide variety of sensory systems. Many animals have sensory systems for detecting light, for forming images of their environment (sight), for mechanical touch, for detecting

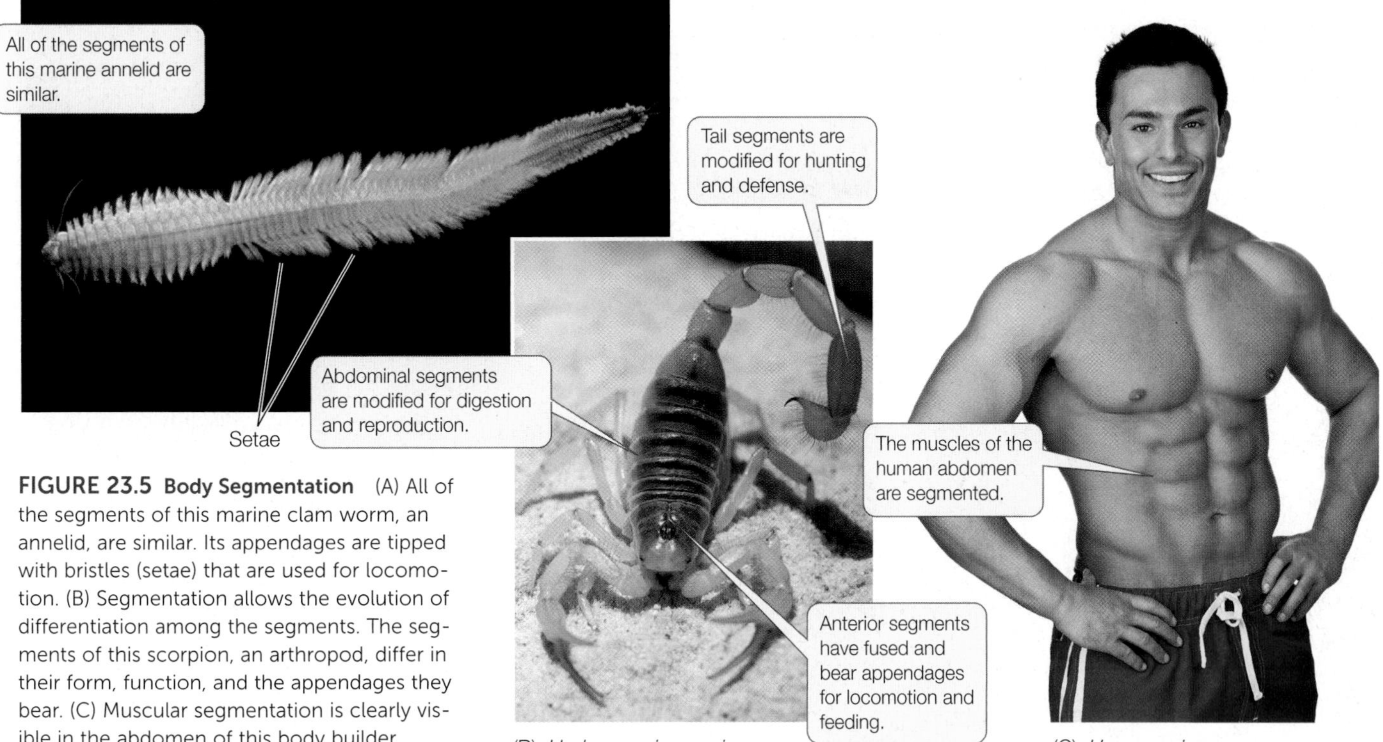

(A) Nereis succinea
All of the segments of this marine annelid are similar.
Setae
Abdominal segments are modified for digestion and reproduction.
Tail segments are modified for hunting and defense.
Anterior segments have fused and bear appendages for locomotion and feeding.
(B) Hadrurus arizonensis
The muscles of the human abdomen are segmented.
(C) Homo sapiens

FIGURE 23.5 Body Segmentation (A) All of the segments of this marine clam worm, an annelid, are similar. Its appendages are tipped with bristles (setae) that are used for locomotion. (B) Segmentation allows the evolution of differentiation among the segments. The segments of this scorpion, an arthropod, differ in their form, function, and the appendages they bear. (C) Muscular segmentation is clearly visible in the abdomen of this body builder.

movement, for detecting sounds (hearing), for detecting electrical fields, and for chemical detection (e.g., taste and smell). These sensory systems allow animals to find food, and the ability of animals to move allows them to capture or collect food from their environment. These same abilities also allow most animals to move to avoid potential predators or to search for suitable mates. Most animals can also assess the suitability of different environments and move appropriately in response to that information.

Go to MEDIA CLIP 23.1
Nervous Systems Lead to Efficient Predators
PoL2e.com/mc23.1

CHECKpoint CONCEPT 23.1

- ✓ Describe the difference between diploblastic and triploblastic embryos, and between protostomes and deuterostomes.
- ✓ Describe the main types of symmetry found in animals. How can an animal's symmetry influence the way it moves?
- ✓ Explain several ways in which body cavities and segmentation improve control over movement.

Variations in body symmetry, body cavity structure, life cycles, patterns of development, and survival strategies differentiate the many animal species. **TABLE 23.1** summarizes the living members of the major animal groups we will describe in the rest of this chapter.

CONCEPT 23.2 Some Animal Groups Fall outside the Bilateria

Looking at Figure 23.1, you can see that the protostome and deuterostome animals together comprise a monophyletic group known as the **Bilateria**. Some major traits that support the monophyly of bilaterians (in addition to genomic analyses) are strong bilateral symmetry, the presence of three distinct cell layers in embryos (triploblasty), and the presence of at least seven Hox genes (see Chapter 14). There are, however, four animal groups—the sponges, ctenophores, placozoans, and cnidarians—that are not bilaterians.

The simplest animals, the sponges, have no distinct tissue types. All other animals groups are usually known as **eumetazoans**. Two groups treated here as eumetazoans, the ctenophores and placozoans, have weakly differentiated tissue layers (placozoans also lack a nervous system), so some biologists exclude these two groups from Eumetazoa as well. Most eumetazoans have some form of body symmetry, a gut, a nervous system, and tissues organized into distinct organs (although there have been secondary losses of some of these structures in some eumetazoans).

Go to ACTIVITY 23.2 Sponge and Diploblast Classification
PoL2e.com/ac23.2

Sponges are loosely organized animals

Although they have some specialized cells, **sponges** have no distinct embryonic cell layers and no true organs. Early naturalists classified sponges as plants because they were sessile and lacked body symmetry.

TABLE 23.1 Summary of Living Members of the Major Animal Groups

	Approximate number of living species described	Major subgroups, other names, and notes
Sponges	8,500	Demosponges, glass sponges, calcareous sponges
Ctenophores	250	Comb jellies
Placozoans	2	Additional species have been discovered but not yet formally named
Cnidarians	12,500	Anthozoans: Corals, sea anemones Hydrozoans: Hydras and hydroids Scyphozoans: Jellyfishes
PROTOSTOMES		
Arrow worms	180	Glass worms
Lophotrochozoans		
Bryozoans	5,500	Moss animals
Flatworms	30,000	Free-living flatworms; flukes and tapeworms (all parasitic); monogeneans (ectoparasites of fishes)
Rotifers and relatives	3,800	Rotifers, hairy-backs, spiny-headed worms, and jaw worms
Ribbon worms	1,200	Proboscis worms
Brachiopods	450	Lampshells
Phoronids	10	Sessile marine filter feeders
Annelids	19,000	Polychaetes (generally marine; may not be monophyletic) Clitellates: Earthworms, freshwater worms, leeches
Mollusks	117,000	Monoplacophorans Chitons Bivalves: Clams, oysters, mussels Gastropods: Snails, slugs, limpets Cephalopods: Squids, octopuses, nautiloids
Ecdysozoans		
Kinorhynchs	180	Mud dragons
Loriciferans	30	Brush heads
Priapulids	20	Penis worms
Nematodes	25,000	Roundworms
Horsehair worms	350	Gordian worms
Tardigrades	1,200	Water bears
Onychophorans	180	Velvet worms
Arthropods		
Chelicerates	114,000	Horseshoe crabs, sea spiders, and arachnids (scorpions, harvestmen, spiders, mites, ticks)
Myriapods	12,000	Millipedes, centipedes
Crustaceans	67,000	Crabs, shrimps, lobsters and crayfishes, barnacles, copepods
Hexapods	1,020,000	Insects and their wingless relatives
DEUTEROSTOMES		
Echinoderms	7,500	Crinoids (sea lilies and feather stars); brittle stars; sea stars; sea urchins; sea cucumbers
Hemichordates	120	Acorn worms and pterobranchs
Chordates		
Tunicates	2,800	Sea squirts (ascidians), salps, and larvaceans
Lancelets	35	Cephalochordates
Vertebrates	65,000	Hagfishes, lampreys, chondrichthyans, ray-finned fishes, coelacanths, lungfishes, amphibians, reptiles (including birds), and mammals

Sponges have hard skeletal elements called **spicules**, which may be small and simple or large and complex. Three major groups of sponges, which separated soon after the split between sponges and the rest of the animals, are distinguished by their spicules. Members of two groups (demosponges and glass sponges) have skeletons composed of silicaceous spicules made of hydrated silicon dioxide (**FIGURE 23.6A,B**). These spicules have greater flexibility and toughness than synthetic glass rods of similar length. Members of the third group, the calcareous sponges, take their name from their calcium carbonate spicules (**FIGURE 23.6C**).

The body plan of all sponges—even large ones, which may reach a meter or more in length—is an aggregation of cells built around a water canal system. Sponges bring water into their bodies by beating the flagella of their specialized feeding cells, the **choanocytes** (see Figure 23.2B). Water, along with any food particles it contains, enters the sponge by way of small pores and passes into the water canals or a central atrium, where choanocytes capture food particles.

A skeleton of simple or branching spicules, often combined with a complex network of elastic fibers, supports the body of most sponges. Sponges also have an extracellular matrix, composed of collagen, adhesive glycoproteins, and other molecules, that holds the cells together. Most sponges are filter feeders. A few species are carnivores that trap prey on hook-shaped spicules that protrude from the body surface.

FIGURE 23.6 Sponge Diversity (A) The great majority of sponge species are demosponges, such as this barrel sponge. (B) As in the demosponges, the supporting structures of glass sponges are silicaceous spicules. The pores and water canals "typical" of the sponge body plan are apparent in both (A) and (B). (C) The spicules of calcareous sponges are made of calcium carbonate.

Most of the 8,500 species of sponges are marine animals. Only about 50 species of sponges live in fresh water. Sponges come in a wide variety of sizes and shapes that are adapted to different movement patterns of water. Sponges living in intertidal or shallow subtidal environments are subject to strong wave action and attach firmly to the substrate. Most sponges that live in slowly flowing water are flattened and are oriented at right angles to the direction of current flow, allowing them to intercept water and the food it contains as it flows past them.

Sponges reproduce both sexually and asexually. In most species, a single individual produces both eggs and sperm, but individuals do not self-fertilize. Water currents carry sperm from one individual to another. Sponges also reproduce asexually by budding and fragmentation.

Ctenophores are radially symmetrical and diploblastic

Ctenophores, also known as comb jellies, were until recently thought to be most closely related to the cnidarians (jellyfishes, corals, and their relatives). But ctenophores lack most of the Hox genes found in all other eumetazoans, and recent studies of their genomes have indicated that ctenophores were among the earliest lineages to split from the remaining animals.

Ctenophores have a radially symmetrical, diploblastic body plan. The two cell layers are separated by an inert, gelatinous extracellular matrix called **mesoglea**. Ctenophores, unlike sponges, have different openings for food entry and waste elimination: food enters through a mouth, and wastes are eliminated through two anal pores.

Ctenophores move by beating cilia rather than by muscular contractions. Most of the 250 known species have eight comblike rows of cilia-bearing plates, called **ctenes** (**FIGURE 23.7**).

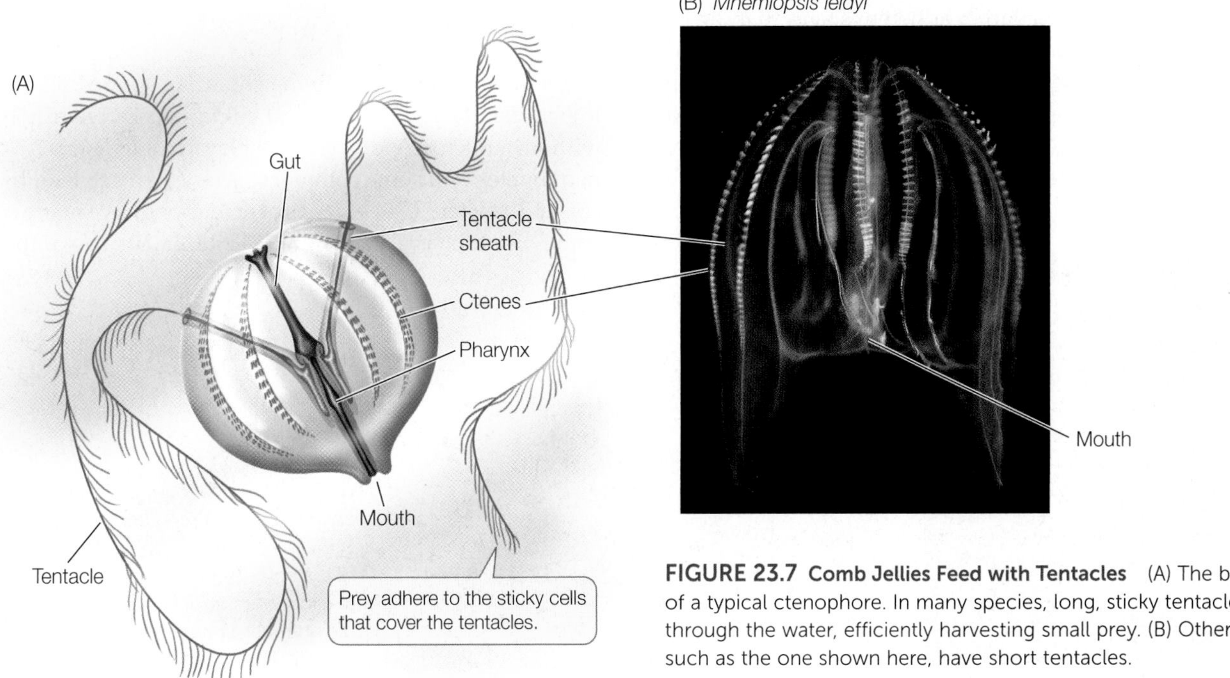

FIGURE 23.7 Comb Jellies Feed with Tentacles (A) The body plan of a typical ctenophore. In many species, long, sticky tentacles sweep through the water, efficiently harvesting small prey. (B) Other species, such as the one shown here, have short tentacles.

The feeding tentacles of ctenophores are covered with cells that discharge adhesive material when they contact prey. After capturing its prey, a ctenophore retracts its tentacles to bring the food to its mouth. In some species, the entire surface of the body is coated with sticky mucus that captures prey. Most ctenophores eat small planktonic organisms, although some eat other ctenophores. Ctenophores are common in open seas and can become abundant in protected bodies of water, where large populations may inhibit the growth of other organisms.

Ctenophore life cycles are uncomplicated. Gametes are released into the body cavity and then discharged through the mouth or the anal pores. Fertilization takes place in open seawater. In nearly all species, the fertilized egg develops directly into a miniature ctenophore, which gradually grows into an adult.

Go to MEDIA CLIP 23.2
Ctenophores
PoL2e.com/mc23.2

Placozoans are abundant but rarely observed

Placozoans are structurally very simple animals with only four distinct cell types (**FIGURE 23.8A**). Individuals in the mature life stage are usually observed adhering to surfaces (such as the glass of aquariums, where they were first discovered, or to rocks and other hard substrates in nature). Their structural simplicity—they have no mouth, gut, or nervous system—initially led biologists to suspect they might be the sister group of all other animals. Most phylogenetic analyses have not supported this hypothesis, however, and some aspects of the placozoans' structural simplicity may be secondarily derived. They are generally considered to have a diploblastic body plan, with upper and lower epithelial (surface) layers that sandwich a layer of contractile fiber cells.

Recent studies have found that placozoans have a pelagic (open-ocean) stage that is capable of swimming (**FIGURE 23.8B**), but the life history of placozoans is incompletely known. Most studies have focused on the larger adherent stages that can be found in aquariums, where they appear after being inadvertently collected with other marine organisms. The transparent nature and small size of placozoans make them very difficult to observe in nature. Nonetheless, it is known that placozoans can reproduce both asexually as well as sexually, although the details of their sexual reproduction are poorly understood. They are found in warm coastal seas around the world.

Cnidarians are specialized carnivores

The **cnidarians** (jellyfishes, sea anemones, corals, and hydrozoans) make up the largest and most diverse group of nonbilaterian animals. The mouth of a cnidarian is connected to a blind sac called the **gastrovascular cavity**. The gastrovascular cavity functions in digestion, circulation, and gas exchange, and it also acts as a hydrostatic skeleton. The single opening serves as both mouth and anus.

The life cycle of many cnidarians has two distinct stages, one sessile and the other motile (**FIGURE 23.9**), although one or the other of these stages is absent in some groups. In the sessile **polyp** stage, a cylindrical stalk is attached to the substrate. The motile **medusa** (plural *medusae*) is a free-swimming stage shaped like a bell or an umbrella. It typically floats with its mouth and feeding tentacles facing downward.

Mature polyps produce medusae by asexual budding. Medusae then reproduce sexually, producing eggs or sperm by meiosis and releasing the gametes into the water. A fertilized egg develops into a free-swimming, ciliated larva called a **planula**, which eventually settles to the bottom and develops into a polyp.

Cnidarians have epithelial cells with muscle fibers that contract and enable the animals to move, as well as simple nerve nets that integrate their body activities. They are specialized carnivores, using the toxin in harpoonlike structures called **nematocysts** to capture relatively large and complex prey. Some cnidarians, including many corals and anemones, gain additional nutrition from photosynthetic endosymbionts that live in their tissues.

LINK
The importance of protist endosymbionts in coral colonies is discussed in **Concept 20.4**

Of the roughly 12,500 living cnidarian species, all but a few live in the oceans. The smallest cnidarians can hardly be seen without a microscope. The largest known jellyfish is 2.5 meters in diameter, and some colonial species can reach lengths in excess of 30 meters. Three clades of cnidarians have many species, some of which are shown in **FIGURE 23.10**. The anthozoans

(A)
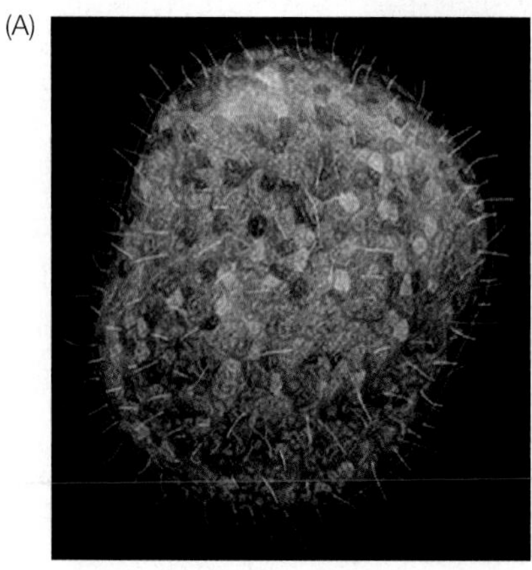
(B)
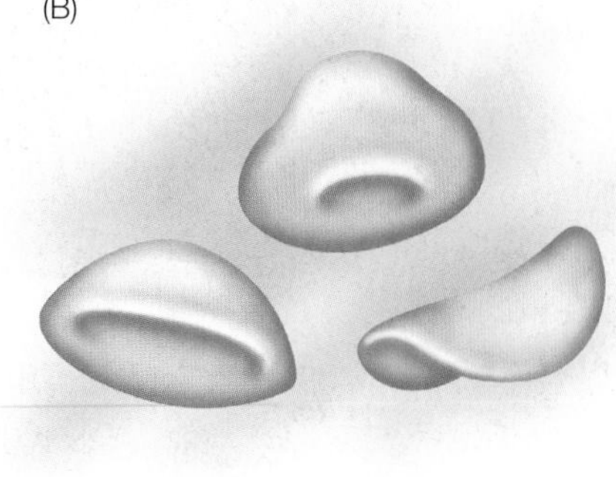

FIGURE 23.8 Placozoan Simplicity (A) As seen in this artist's rendition, adult placozoans are tiny (1–2 mm across), flattened, asymmetrical animals. (B) Recent studies have found a weakly swimming pelagic stage of placozoan to be abundant in many warm tropical and subtropical seas.

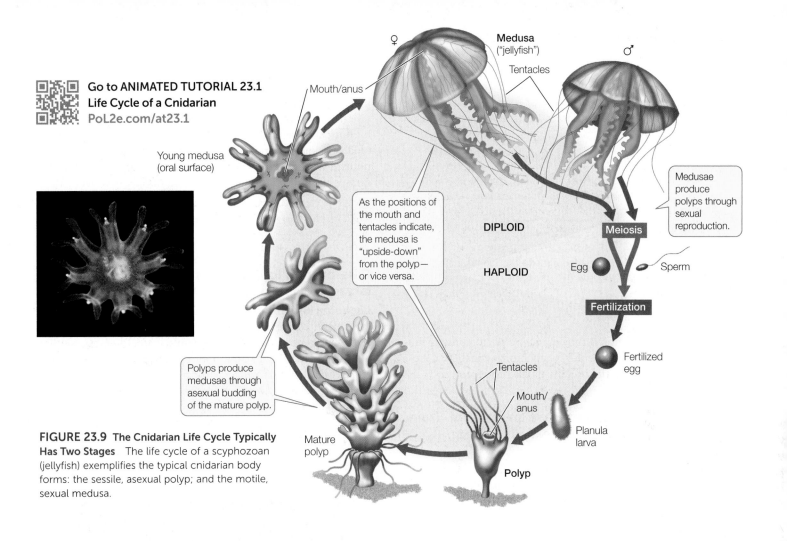

Go to ANIMATED TUTORIAL 23.1
Life Cycle of a Cnidarian
PoL2e.com/at23.1

♀
Medusa
("jellyfish")
♂
Tentacles
Mouth/anus
Young medusa
(oral surface)
Medusae produce polyps through sexual reproduction.
As the positions of the mouth and tentacles indicate, the medusa is "upside-down" from the polyp—or vice versa.
DIPLOID
HAPLOID
Meiosis
Egg
Sperm
Fertilization
Fertilized egg
Polyps produce medusae through asexual budding of the mature polyp.
Tentacles
Mouth/anus
Planula larva
Mature polyp
Polyp

FIGURE 23.9 The Cnidarian Life Cycle Typically Has Two Stages The life cycle of a scyphozoan (jellyfish) exemplifies the typical cnidarian body forms: the sessile, asexual polyp; and the motile, sexual medusa.

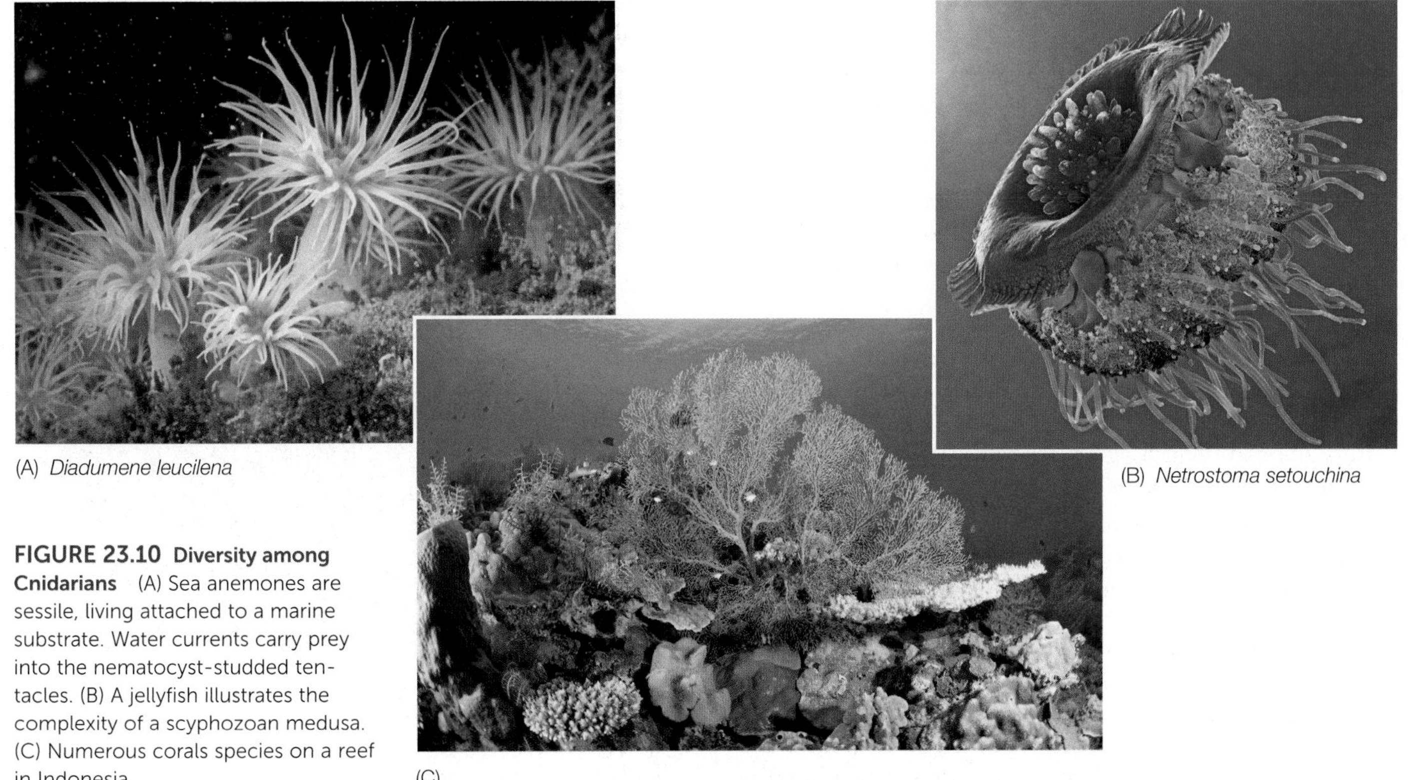

(A) *Diadumene leucilena*

(B) *Netrostoma setouchina*

(C)

FIGURE 23.10 Diversity among Cnidarians (A) Sea anemones are sessile, living attached to a marine substrate. Water currents carry prey into the nematocyst-studded tentacles. (B) A jellyfish illustrates the complexity of a scyphozoan medusa. (C) Numerous corals species on a reef in Indonesia.

include sea anemones, sea pens, and corals. The scyphozoans are commonly known as jellyfishes or sea jellies. The less familiar hydrozoans include both freshwater and marine species.

CHECKpoint CONCEPT 23.2

- ✓ Why are sponges and placozoans considered to be animals, even though they lack the complex body structures found among most other animal groups?
- ✓ How does feeding differ between ctenophores and cnidarians?
- ✓ Can you think of any advantages to the biphasic life cycle (sessile polyp and motile medusa) of many cnidarians?

In Concept 23.1 we noted that the name protostome means "mouth first," a developmental synapomorphy that links these animals. The vast majority of animals are protostomes.

CONCEPT 23.3 Protostomes Have an Anterior Brain and a Ventral Nervous System

The protostomes are a highly diverse group of animals. Although their body plans are extremely varied, they are all bilaterally symmetrical animals whose bodies exhibit two major derived traits:

- An anterior brain that surrounds the entrance to the digestive tract
- A ventral nervous system consisting of paired or fused longitudinal nerve cords

Other aspects of protostome body organization differ widely from group to group. Although the common ancestor of the protostomes likely had a coelom, subsequent modifications of the coelom distinguish many protostome lineages. In at least one protostome lineage (the flatworms), the coelom has been lost (that is, the flatworms reverted to an acoelomate state). Some lineages are characterized by a pseudocoel (see Figure 23.4B). In two of the most prominent protostome groups, the coelom has been highly modified:

- The arthropods lost the ancestral condition of the coelom over the course of evolution. Their internal body cavity has become a hemocoel, or "blood chamber," in which fluid from an open circulatory system bathes the internal organs before returning to blood vessels.
- Most mollusks have an open circulatory system with some of the attributes of the hemocoel, but they retain vestiges of an enclosed coelom around their major organs.

The evolutionary relationships of one small group of protostomes, the **arrow worms** (see Figure 23.1), have been debated for many years. Although recent gene sequence studies clearly identify arrow worms as protostomes, there remains some question as to their exact placement within the protostomes. The 180 living species of arrow worms are small (3 mm–12 cm) marine predators of planktonic protists and small fish.

The protostomes can be divided into two major clades—the lophotrochozoans and the ecdysozoans—largely on the basis of DNA sequence analysis. However, there are also some morphological characteristics that unite most members of these two groups.

Go to ANIMATED TUTORIAL 23.2
An Overview of the Protosomes
PoL2e.com/at23.2

Cilia-bearing lophophores and trochophore larvae evolved among the lophotrochozoans

Lophotrochozoans derive their name from two different features that involve cilia: a feeding structure known as a lophophore and a free-living larva known as a trochophore. Neither feature is universal for all lophotrochozoans, however.

Several distantly related groups of lophotrochozoans (including bryozoans, brachiopods, and phoronids) have a **lophophore**, a circular or U-shaped ring of ciliated, hollow tentacles around the mouth (**FIGURE 23.11A**). This complex structure is

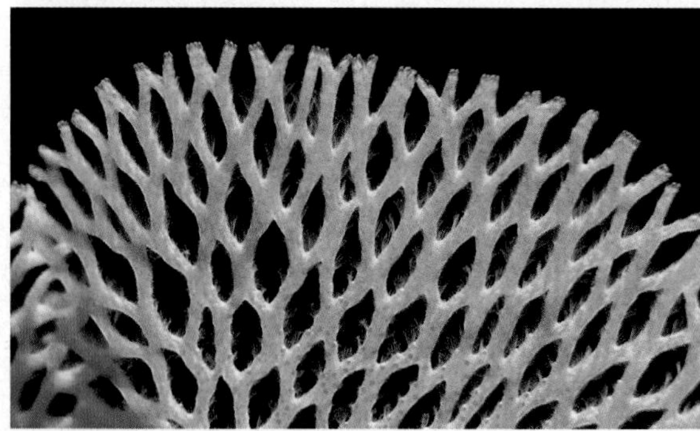

FIGURE 23.11 Bryozoans Form Colonies (A) The extended lophophores of its individual members dominate the anatomy of this colonial bryozoan. (B) The rigid orange tissue of this marine bryozoan colony connects and supplies nutrients to thousands of individual animals.

an organ for both food collection and gas exchange. The lophophore appears to have evolved independently several times, although it may be an ancestral feature that has been lost in many groups. Nearly all animals with a lophophore are sessile as adults, using the lophophore's tentacles and cilia to capture small floating organisms from the water.

Go to MEDIA CLIP 23.3
Feeding with a Lophophore
PoL2e.com/mc23.3

Some lophotrochozoans, especially in their larval form, use cilia for locomotion. The larval form known as a **trochophore** moves by beating a band of cilia:

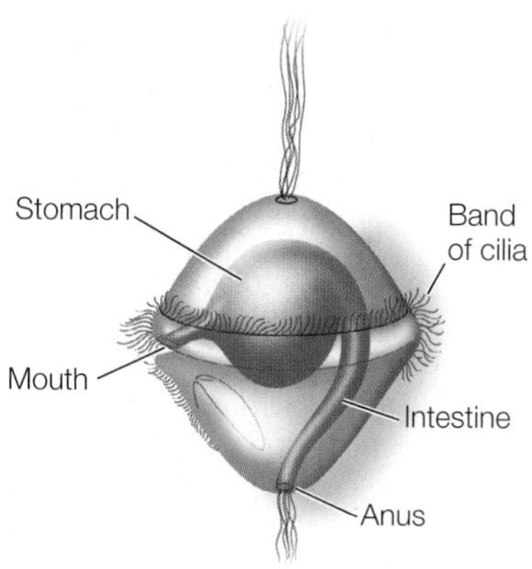

This movement of cilia also brings plankton closer to the larva, where the plankton can be captured and ingested (similar in function to the cilia of the lophophore). Trochophore larvae are found among many of the major groups of lophotrochozoans, including the mollusks, annelids, ribbon worms, and bryozoans. This larval form was probably present in the common ancestor of lophotrochozoans, although it has been subsequently lost in several lineages.

Lophotrochozoans range from relatively simple animals with a blind gut (that is, a gut with a single opening that both takes in food and expels wastes) and no internal transport system to animals with a complete gut (having separate entrance and exit openings) and a complex internal transport system. A number of these groups exhibit wormlike bodies, but the lophotrochozoans encompass a wide diversity of morphologies, including a few groups with external shells. Included among the lophotrochozoans are species-rich groups such as flatworms, annelids, and mollusks, along with many less well known groups, some of which have only recently been discovered.

BRYOZOANS Most of the 5,500 species of **bryozoans** ("moss animals") are colonial animals that live in a "house" made of material secreted by the external body wall. Almost all bryozoans are marine, although a few species occur in fresh or brackish water. A bryozoan colony consists of many small (1–2 mm) individuals connected by strands of tissue along which nutrients can be moved (**FIGURE 23.11B**). The colony is created by the asexual reproduction of its founding member, and a single colony may contain as many as 2 million individuals. Rocks in coastal regions in many parts of the world are covered with luxuriant growths of bryozoans. Some bryozoans create miniature reefs in shallow waters. In some species, the individual colony members are differentially specialized for feeding, reproduction, defense, or support.

Bryozoans can also reproduce sexually. Sperm are released into the water, which carries the sperm to other individuals. Eggs are fertilized internally; developing embryos are brooded before they exit as larvae to seek suitable sites for attachment to the substrate.

FLATWORMS Most of the 30,000 species of **flatworms** are tapeworms and flukes; members of these two groups are internal parasites, particularly of vertebrates (**FIGURE 23.12A**).

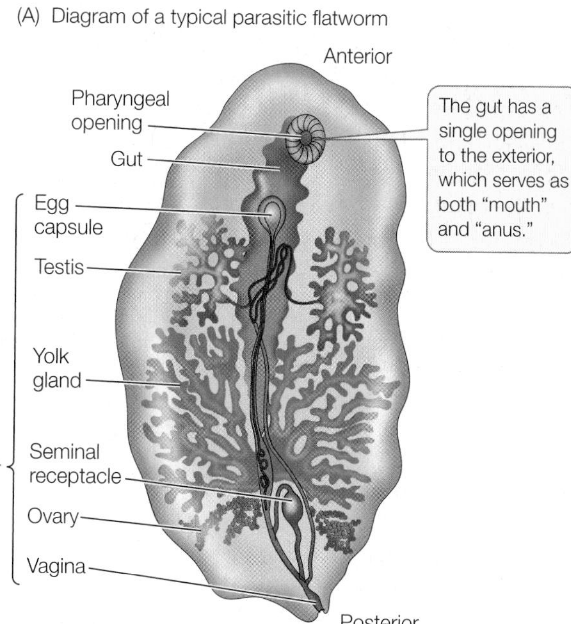

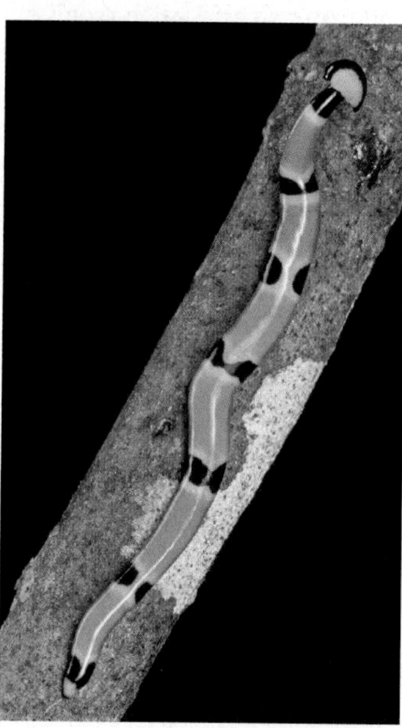

FIGURE 23.12 Flatworms (A) The fluke diagrammed here is representative of many parasitic flatworms. Absorbing nutrition from the host animal's gut, these internal parasites do not require elaborate feeding or digestive organs and can devote most of their bodies to reproduction. (B) The bright coloration of this terrestrial flatworm from Borneo resembles some species of venomous snakes, and may provide it protection from predation.

Because they absorb digested food from the guts of their hosts, many parasitic flatworms lack digestive tracts of their own. Some cause serious human diseases, such as schistosomiasis, which is common in parts of Asia, Africa, and South America. The species that causes this devastating disease has a complex life cycle involving both freshwater snails and mammals as hosts. Other flatworms are external parasites of fishes and other aquatic vertebrates. The turbellarians include most of the free-living species (**FIGURE 23.12B**).

Flatworms lack specialized organs for transporting oxygen to their internal tissues. Lacking a gas transport system, each cell must be near a body surface, a requirement met by the dorsoventrally flattened body form. In flatworms that have a digestive tract, this consists of a mouth opening into a blind sac. The sac is often highly branched, forming intricate patterns that increase the surface area available for the absorption of nutrients. Most free-living flatworms are cephalized, with a head bearing chemoreceptor organs, two simple eyes, and a small brain composed of anterior thickenings of the longitudinal nerve cords. Free-living flatworms glide over surfaces, powered by broad bands of cilia.

ROTIFERS Most species of **rotifers** are tiny—50–500 μm long, smaller than some ciliate protists—but they have specialized internal organs (**FIGURE 23.13**). A complete gut passes from an anterior mouth to a posterior anus. The body cavity is a pseudocoel that functions as a hydrostatic skeleton. Rotifers typically propel themselves through the water by means of rapidly beating cilia rather than by muscular contraction.

The most distinctive organ of rotifers is a conspicuous ciliated organ called the corona, which surmounts the head of many species. Coordinated beating of the cilia sweeps particles of organic matter from the water into the animal's mouth and down to a complicated structure called the mastax, in which food is ground into small pieces. By contracting muscles around the pseudocoel, a few rotifer species that prey on protists and small animals can protrude the mastax through their mouth and seize small objects with it.

Most species of rotifers live in fresh water. Some species rest on the surfaces of mosses or lichens in a desiccated, inactive state until it rains. When rain falls, they absorb water and become mobile, feeding in the films of water that temporarily cover the plants. Most rotifers live no longer than a few weeks.

Both males and females are found in some species of rotifers, but only females are known among the bdelloid rotifers (the b in "bdelloid" is silent). Biologists have concluded that the bdelloid rotifers may have existed for tens of millions of years without regular sexual reproduction. In general, lack of genetic recombination leads to the buildup of deleterious mutations, so long-term asexual reproduction typically leads to extinction (see Concept 15.6). However, recent studies indicate that bdelloid rotifers may avoid this problem by taking up fragments of genes directly from the environment during the desiccation–rehydration cycle. Such a mechanism allows genetic recombination among individuals in the absence of direct sexual exchange.

Go to MEDIA CLIP 23.4
Rotifer Feeding
PoL2e.com/mc23.4

RIBBON WORMS **Ribbon worms** (nemerteans) have simple nervous and excretory systems similar to those of flatworms. Unlike flatworms, however, they have a complete digestive tract with a mouth at one end and an anus at the other. Small ribbon worms move slowly by beating their cilia. Larger ones employ waves of muscle contraction to move over the surface of sediments or to burrow into them.

Within the body of nearly all of the 1,200 species of ribbon worms is a fluid-filled cavity called the rhynchocoel, within which lies a hollow, muscular **proboscis**. The proboscis, which is the worm's feeding organ, may extend much of the length of the body. Contraction of the muscles surrounding the rhynchocoel causes the proboscis to evert (turn inside out) explosively through an anterior pore (**FIGURE 23.14A**). The proboscis may be armed with sharp stylets that pierce prey and discharge paralytic toxins into the wound.

Ribbon worms are largely marine, although there are some freshwater and terrestrial species. Most ribbon worms are less than 20 centimeters long, but individuals of some species reach 20 meters or more. Some genera include species that are conspicuous and brightly colored (**FIGURE 23.14B**).

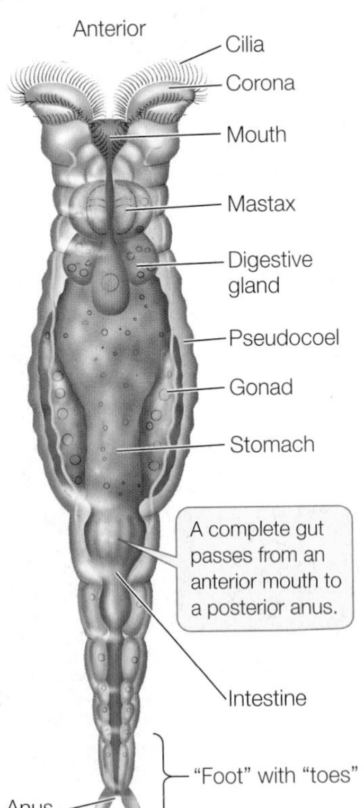

(B) *Philodina* sp.

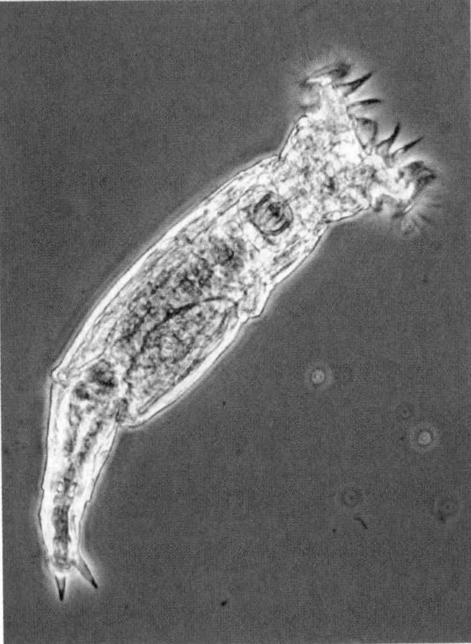

FIGURE 23.13 Rotifers (A) The individual diagrammed here reflects the general structure of many rotifers. (B) A light micrograph reveals the internal complexity of these tiny animals.

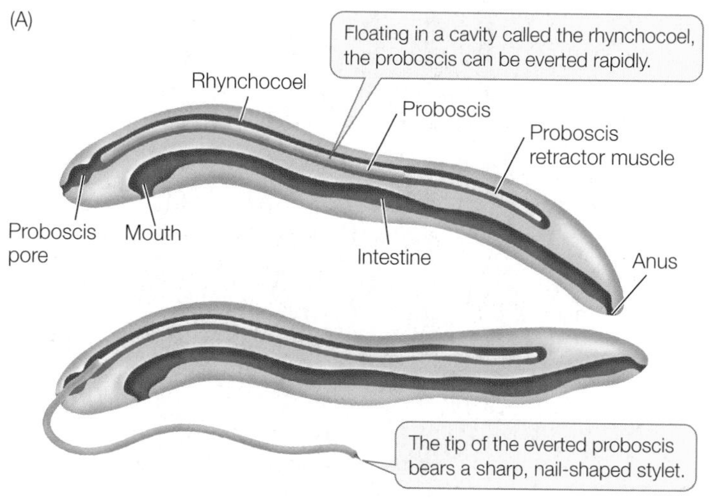

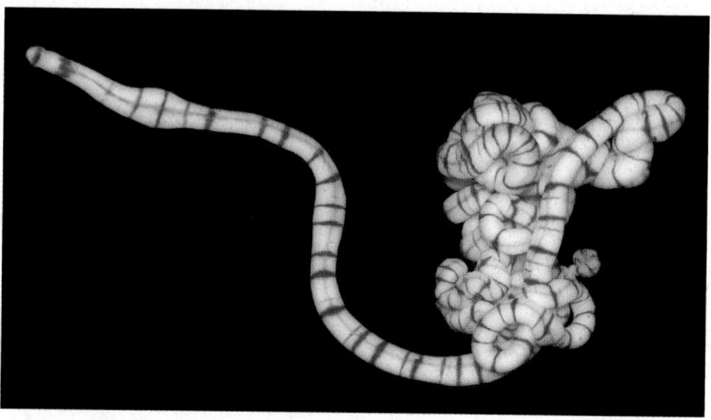

FIGURE 23.14 Ribbon Worms (A) The proboscis is the ribbon worm's feeding organ. (B) Although most nemerteans are small, some marine species such as this one can reach several meters in length.

Go to MEDIA CLIP 23.5
Explosive Extrusion of Ribbon Worm Proboscis
PoL2e.com/mc23.5

BRACHIOPODS AND PHORONIDS Like bryozoans (see Figure 23.11A), brachiopods and phoronids also feed using a lophophore, but this structure may have evolved independently in these groups. Although neither the brachiopods nor the phoronids are represented by many living species, the brachiopods—which have shells and thus leave an excellent fossil record—are known to have been much more abundant during the Paleozoic and Mesozoic eras.

Brachiopods (lampshells) are solitary marine animals with a rigid shell that is divided into two parts connected by a ligament (**FIGURE 23.15**). Although brachiopods superficially resemble bivalve mollusks, shells evolved independently in the two groups. The two halves of the brachiopod shell are dorsal and ventral, rather than lateral as in bivalves. The lophophore is located within the shell. Most brachiopods are 4–6 centimeters long. More than 26,000 fossil brachiopod species have been described, but only about 450 species survive.

The 10 known species of **phoronids** are small (5–25 cm long), sessile worms that live in muddy or sandy sediments or attached to rocky substrates. Phoronids are found in marine waters, from the intertidal zone to about 400 meters deep. They secrete tubes made of chitin, within which they live (**FIGURE 23.16**). Their cilia drive water into the top of the lophophore, and the water exits through the narrow spaces between the tentacles. Suspended food particles are caught and transported to the mouth by ciliary action. Eggs are fertilized internally, and the embryos are either released into the water or, in species

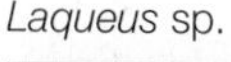

FIGURE 23.15 Brachiopods The lophophore of this North Pacific brachiopod can be seen between the valves of its shell. Although its shell resembles those of the bivalve mollusks (see Figure 23.19D), it evolved independently.

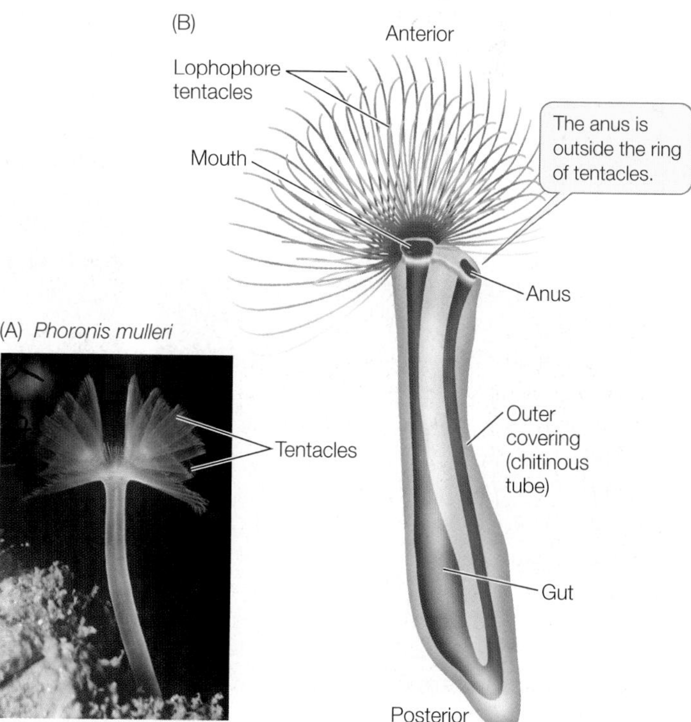

FIGURE 23.16 Phoronids (A) This phoronid has two lophophores, each composed of rings of tentacles. (B) The phoronid gut is U-shaped, as seen in this generalized diagram.

FIGURE 23.17 Annelids Have Many Body Segments The segmented structure of the annelids is apparent both externally and internally. Many organs of this earthworm, a common annelid, are repeated serially.

Cross section
Segments
Septum between segments
Circular muscle
Blood vessel
Longitudinal muscle
Clitellum
Setae (bristles)
Head
Tail
Excretory organ
Ring vessels of circulatory system
Intestine
Brain
Pharynx
Esophagus
Coelom
Nerve cord
Ventral nerve cord
Ovary
Seminal receptacle
Testes and sperm sacs
Crop
Gizzard
Intestine

with large embryos, retained in the parent's body, where they are brooded until they hatch.

ANNELIDS As discussed in Concept 23.1, segmentation allows an animal to move different parts of its body independently of one another, giving it much better control of its movement. A clear and obvious example of segmentation is seen in the body plan of the **annelids** (**FIGURE 23.17**; see also Figure 23.5A).

In most large annelids, the coelom in each segment is isolated from those in other segments. A separate nerve center called a ganglion (plural ganglia) controls each segment. Nerve cords that connect the ganglia coordinate their functioning. Most annelids lack a rigid external protective covering. Instead, they have a thin, permeable body wall that serves as a general surface for gas exchange. These animals are thus restricted to moist environments because they lose body water rapidly in dry air. The approximately 19,000 described annelid species live in marine, freshwater, and moist terrestrial environments.

Many annelids have one or more pairs of eyes and one or more pairs of tentacles (with which they capture prey or filter food from the surrounding water) at the anterior end of the body (**FIGURE 23.18A**). In some species, the body wall of most segments extends laterally as a series of thin outgrowths called parapodia. The parapodia function in gas exchange, and some species use them to move. Stiff bristles called setae protrude

(A) *Protula* sp.

(B) *Riftia* sp.

(C) *Placobdella parasitica*

FIGURE 23.18 Diversity among the Annelids (A) "Fan worms," or "feather duster worms," are sessile marine annelids that grow in chitinous tubes, from which their tentacles extend and filter food from the water. (B) Pogonophorans live around hydrothermal vents deep in the ocean. As in fan worms, their tentacles protrude from chitinous tubes. They do not possess a digestive tract, however, and obtain most of their nutrition from endosymbiotic bacteria. (C) This hermaphroditic freshwater leech is brooding a clutch of fertilized eggs.

from each parapodium, forming temporary contact with the substrate and preventing the animal from slipping backward when its muscles contract.

Some annelids, such as the pogonophorans, secrete tubes made of chitin and other substances, in which they live (**FIGURE 23.18B**). Pogonophorans have lost their digestive tract (they have no mouth or gut). So how do they obtain nutrition? Part of the answer is that pogonophorans can take up dissolved organic matter directly from the sediments in which they live or from the surrounding water. Much of their nutrition, however, is provided by endosymbiotic bacteria that live in a specialized organ known as the trophosome. These bacteria oxidize hydrogen sulfide and other sulfur-containing compounds, fixing carbon from methane in the process. The uptake of the hydrogen sulfide, methane, and oxygen used by the bacteria is facilitated by hemoglobin in the pogonophorans' tentacles. It is this hemoglobin that gives the tentacles their red color (see Figure 23.18B).

Oligochaetes have no parapodia or anterior tentacles, and they have only four pairs of setae bundles per segment. Earthworms—the most familiar oligochaetes—burrow in and ingest soil, from which they extract food particles. All oligochaetes are hermaphroditic; that is, each individual is both male and female. Sperm are exchanged simultaneously between two copulating individuals. Eggs and sperm are deposited outside the adult's body, in a cocoon secreted by the clitellum (see Figure 23.17). Fertilization occurs within the cocoon after it is shed, and when development is complete, miniature worms emerge and immediately begin independent life.

Leeches, like oligochaetes, lack parapodia and tentacles (**FIGURE 23.18C**). They live in freshwater, marine, and terrestrial habitats. The leech coelom is not divided into compartments, and consists of a large, fluid-filled cavity. Groups of segments at each end of the body are modified to form suckers, which serve as temporary anchors that help the leech move. With its posterior sucker attached to a substrate, the leech extends its body by contracting its circular muscles. The anterior sucker is then attached, the posterior one detached, and the leech shortens itself by contracting its longitudinal muscles.

Many leeches feed on vertebrate hosts by making an incision from which blood flows. A feeding leech secretes an anticoagulant into the wound to keep the blood flowing. For centuries, medical practitioners used leeches to treat diseases they believed were caused by an excess of blood or by "bad blood." Although most leeching practices (such as inserting a leech in a person's throat to alleviate swollen tonsils) have been abandoned, *Hirudo medicinalis* (the medicinal leech) is used medically even today to reduce fluid pressure and prevent blood clotting in damaged tissues, to eliminate pools of coagulated blood, and to prevent scarring. The anticoagulants of certain other leech species contain anesthetics and blood vessel dilators and are being studied for possible medical uses.

Go to MEDIA CLIP 23.6
Leeches Feeding on Blood
PoL2e.com/mc23.6

MOLLUSKS The most diverse group of lophotrochozoans are the **mollusks**, with about 117,000 species that inhabit a wide array of aquatic and terrestrial environments. There are four major clades of mollusks: chitons, gastropods, bivalves, and cephalopods. Although these groups differ dramatically in morphology, they all share the same three major body components: a foot, a visceral mass, and a mantle (**FIGURE 23.19A**).

- The molluscan **foot** is a muscular structure that originally was both an organ of locomotion and a support for internal organs. In cephalopods such as squids and octopuses, the foot has been modified to form arms and tentacles borne on a head with complex sensory organs. In other groups, such as clams, the foot is a burrowing organ. In some groups the foot is greatly reduced.
- The heart and the digestive, excretory, and reproductive organs are concentrated in a centralized, internal **visceral mass**.
- The **mantle** is a fold of tissue that covers the organs of the visceral mass. The mantle secretes the hard, calcareous shell that is typical of many mollusks.

In most mollusks, the mantle extends beyond the visceral mass to form a mantle cavity. Within this cavity lie gills that are used for gas exchange. When cilia on the gills beat, they create a current of water. The gill tissue, which is highly vascularized (contains many blood vessels), takes up oxygen from the water and releases carbon dioxide. Many mollusks use their gills as filter-feeding devices. Other mollusks feed using a rasping structure known as the radula to scrape algae from rocks. In some mollusks, such as the marine cone snails, the radula has been modified into a drill or poison dart.

Except in cephalopods, molluscan blood vessels do not form a closed circulatory system. Blood and other fluids empty into a large, fluid-filled hemocoel, through which fluids move around the animal and deliver oxygen to the internal organs. Eventually the fluids reenter the blood vessels and are moved by a heart.

The approximately 1,000 living species of **chitons** (**FIGURE 23.19B**) are characterized by eight overlapping calcareous plates, surrounded by a structure known as the girdle. These plates and girdle protect the chiton's internal organs and muscular foot. The chiton body is bilaterally symmetrical, and the internal organs, particularly the digestive and nervous systems, are relatively simple. Most chitons are marine omnivores that scrape algae, bryozoans, and other organisms from rocks with their sharp radula. An adult chiton spends most of its life clinging to rock surfaces with its large, muscular, mucus-covered foot. It moves slowly by means of rippling waves of muscular contraction in the foot. Fertilization in most chitons takes place in the water, but in a few species fertilization is internal and embryos are brooded within the body.

Gastropods are the most species-rich and widely distributed mollusks, with about 85,000 living species. Snails, whelks, limpets, slugs, nudibranchs (sea slugs), and abalones are all gastropods. Most species move by gliding on their muscular foot, but in a few species—the sea butterflies and heteropods—the foot is a swimming organ with which the animal moves through open ocean waters. The only mollusks that live in terrestrial

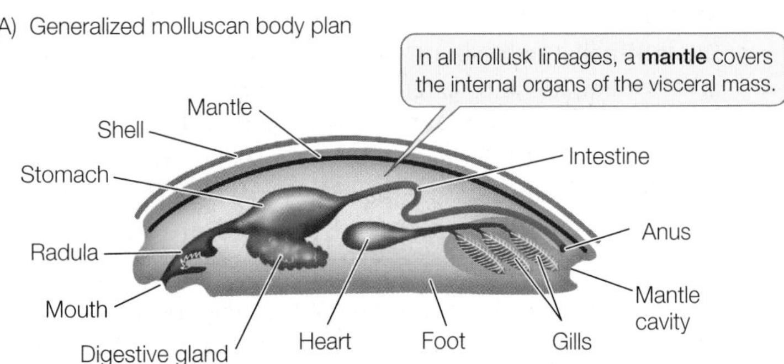

FIGURE 23.19 Organization and Diversity of Molluscan Bodies (A) The major molluscan groups display different variations on a general body plan that includes three major components: a foot, a visceral mass of internal organs, and a mantle. In many species, the mantle secretes a calcareous shell. (B) Chitons have eight overlapping calcareous plates surrounded by a girdle. (C) Most gastropods have a single dorsal shell, into which they can retreat for protection. (D) Bivalves get their name from their two hinged shells, which can be tightly closed. (E) Cephalopods are active predators; they use their arms and tentacles to capture prey.

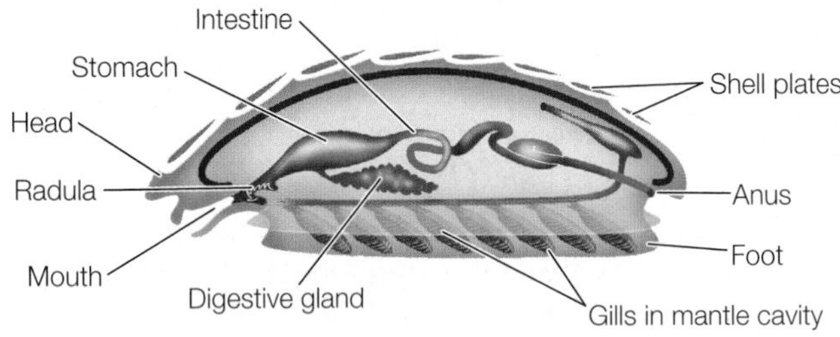

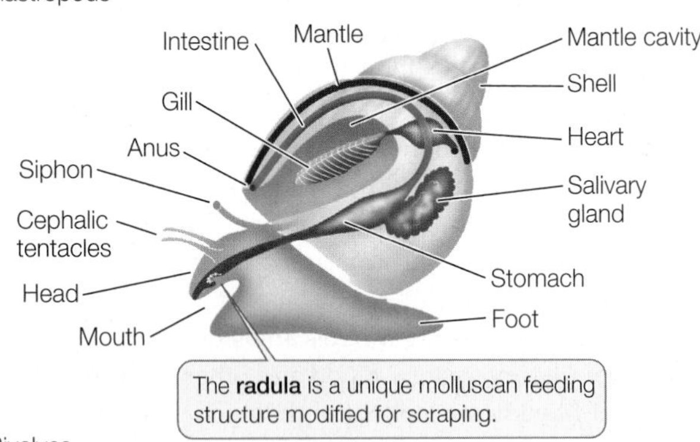

Achatina fulica

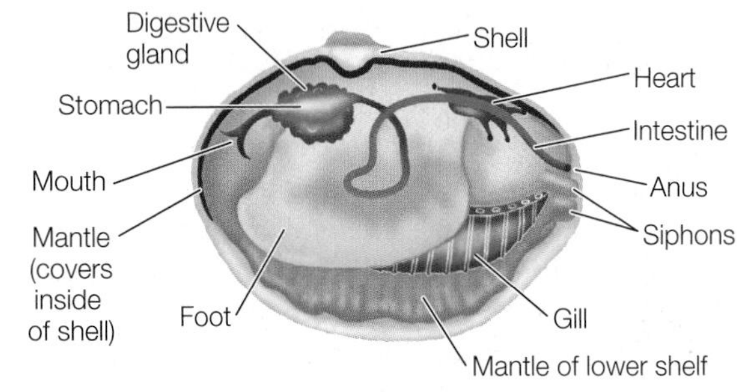

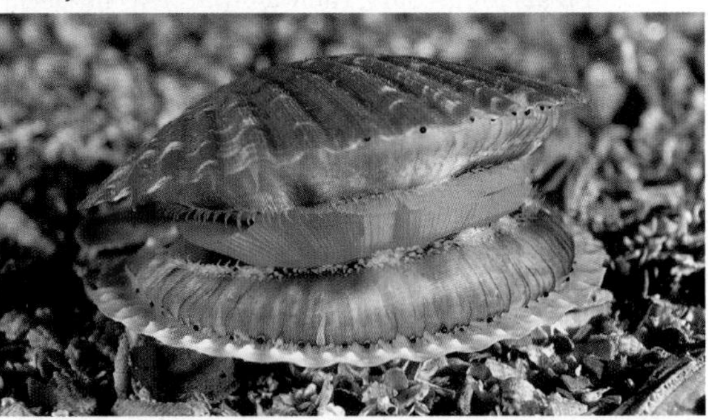

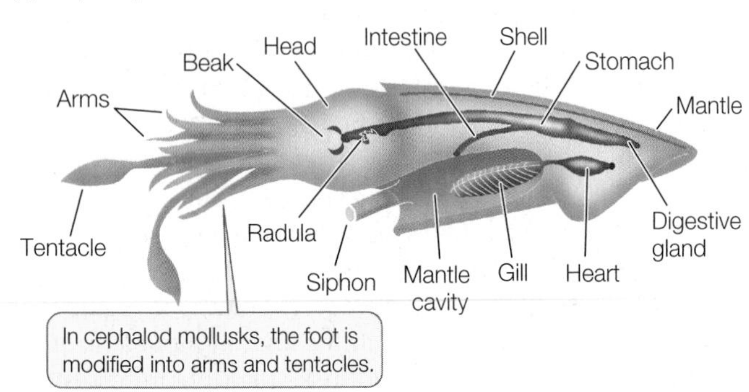

environments—land snails and slugs—are gastropods (**FIGURE 23.19C**). In these terrestrial species, the mantle tissue is modified into a highly vascularized lung.

Clams, oysters, scallops, and mussels are all familiar **bivalves**. The 30,000 living species are found in both marine and freshwater environments. Bivalves have a very small head and a hinged, two-part shell that extends over the sides of the body as well as the top (**FIGURE 23.19D**). Many clams use their foot to burrow into mud and sand. Bivalves feed by taking in water through an opening called an incurrent siphon and filtering food from the water with their large gills, which are also the main sites of gas exchange. Water and gametes exit through the excurrent siphon. Fertilization takes place in open water in most species.

There are about 800 living species of **cephalopods**—squids, octopuses, and nautiluses. Their excurrent siphon is modified to allow the animal to control the water content of the mantle cavity. The modification of the mantle into a device for forcibly ejecting water from the cavity through the siphon enables these animals to move rapidly by "jet propulsion" through the water. With their greatly enhanced mobility, cephalopods (which first appeared early in the Cambrian) became the major predators in the open waters of the Devonian oceans. They remain important marine predators today.

Cephalopods capture and subdue prey with their tentacles (see Figure 23.20B). As is typical of active, rapidly moving predators, cephalopods have a head with complex sensory organs, most notably eyes that are comparable to those of vertebrates in their ability to resolve images. The head is closely associated with a large, branched foot that bears the tentacles and a siphon (**FIGURE 23.19E**). The large, muscular mantle provides an external supporting structure. The gills hang in the mantle cavity. Many cephalopods have elaborate courtship behavior, which can involve striking color changes.

Many early cephalopods had a chambered external shell divided by partitions penetrated by tubes through which gases and liquids could be moved to control the animal's buoyancy. Nautiluses are the only surviving cephalopods that have such external chambered shells, although squids and cuttlefishes retain internal shells.

Shells have been lost several times among the mollusks, as in several groups of gastropods, including the slugs and nudibranchs (**FIGURE 23.20A**). These shell-less gastropods gain some protection from predation by being distasteful or toxic to many species. The often brilliant coloration of nudibranchs is aposematic, meaning it serves to warn potential predators of toxicity. Among the cephalopods, the octopuses have lost both external and internal shells (**FIGURE 23.20B**). Their lack of shells allows octopuses to escape predators by squeezing into small crevices.

Go to MEDIA CLIP 23.7
Octopuses Can Pass through Small Openings
PoL2e.com/mc23.7

Ecdysozoans must shed their cuticles

The distinguishing characteristic of **ecdysozoans** is their external covering, or **cuticle**, which is secreted by the underlying epidermis (the outermost cell layer). The cuticle provides these animals with both protection and support. Once formed, however, the cuticle cannot grow. How, then, can ecdysozoans increase in size? They do so by shedding, or **molting**, the cuticle and replacing it with a new, larger one. This molting process gives the clade its name (Greek *ecdysis*, "to get out of").

Go to MEDIA CLIP 23.8
Molting a Cuticle
PoL2e.com/mc23.8

A soft-bodied arthropod from the Cambrian, fossilized in the process of molting, shows that molting evolved more than 500 million years ago (**FIGURE 23.21A**). An increasingly rich array of molecular and genetic evidence, including a set of Hox genes shared by all ecdysozoans, suggests they have a single common ancestor. Thus molting of a cuticle is a trait that may have evolved only once during animal evolution.

Before an ecdysozoan molts, a new cuticle is already forming underneath the old one. Once the old cuticle is shed, the new one expands and hardens. Until it has hardened, though,

(A) *Chromodoris kuniei*

(B) *Octopus bimaculoides*

FIGURE 23.20 Mollusks in Some Groups Have Lost Their Shells (A) Nudibranchs ("naked gills"), also called sea slugs, are shell-less gastropods. This species is brightly colored, alerting potential predators of its toxicity. (B) Octopuses have lost both the external and internal shell, which allows these cephalopods to squeeze through tight spaces.

(A)

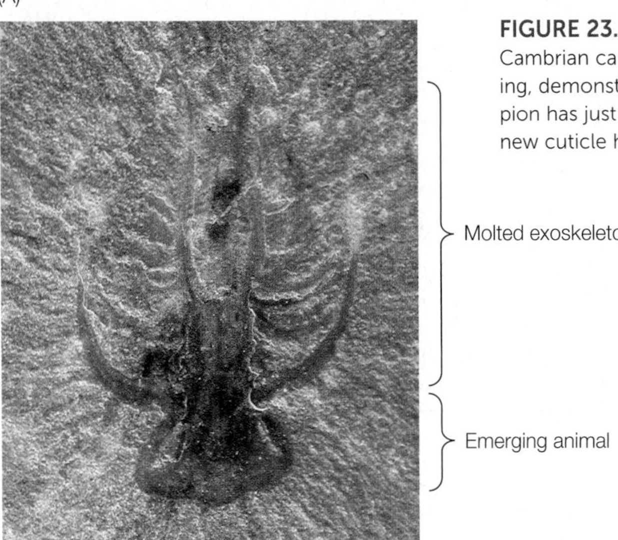
Molted exoskeleton
Emerging animal

(B) *Heterophrynus batesi*

Molted exoskeleton
The newly emerged whip scorpion's body is still soft and vulnerable.

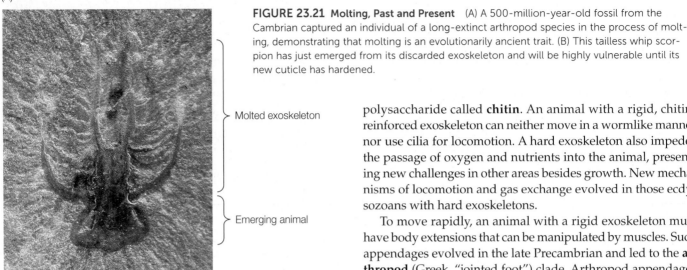

FIGURE 23.21 Molting, Past and Present (A) A 500-million-year-old fossil from the Cambrian captured an individual of a long-extinct arthropod species in the process of molting, demonstrating that molting is an evolutionarily ancient trait. (B) This tailless whip scorpion has just emerged from its discarded exoskeleton and will be highly vulnerable until its new cuticle has hardened.

the animal is vulnerable to its enemies, both because its outer surface is easy to penetrate and because an animal with a soft cuticle moves slowly or not at all (**FIGURE 23.21B**).

In many ecdysozoans that have wormlike bodies, the cuticle is relatively thin and flexible. Such a cuticle offers the animal some protection but provides only modest body support. A thin cuticle allows the exchange of gases, minerals, and water across the body surface, but it restricts the animal to moist habitats. Many species of ecdysozoans with thin cuticles live in marine sediments from which they obtain prey. Some freshwater species absorb nutrients directly through their thin cuticles, as do parasitic species that live within their hosts.

The cuticles of other ecdysozoans, notably the arthropods, function as external skeletons, or **exoskeletons**. These exoskeletons are thickened by layers of protein and a strong, waterproof polysaccharide called **chitin**. An animal with a rigid, chitin-reinforced exoskeleton can neither move in a wormlike manner nor use cilia for locomotion. A hard exoskeleton also impedes the passage of oxygen and nutrients into the animal, presenting new challenges in other areas besides growth. New mechanisms of locomotion and gas exchange evolved in those ecdysozoans with hard exoskeletons.

To move rapidly, an animal with a rigid exoskeleton must have body extensions that can be manipulated by muscles. Such appendages evolved in the late Precambrian and led to the **arthropod** (Greek, "jointed foot") clade. Arthropod appendages exist in an amazing variety of forms. They serve many functions, including walking and swimming, gas exchange, food capture and manipulation, copulation, and sensory perception. Arthropods grasp food with their mouths and associated appendages and usually digest it internally. Their muscles are attached to the inside of the exoskeleton.

The arthropod exoskeleton has had a profound influence on the evolution of these animals. Encasement within a rigid body covering provides support for walking on dry land, and the waterproofing provided by the cuticle keeps the animal from dehydrating in dry air. These features have allowed arthropods to invade terrestrial environments several times.

The evolution of the diverse arthropods and their relatives will be described in greater detail in Concept 23.4. Here we describe the other major groups of ecdysozoans.

NEMATODES **Nematodes** (roundworms) have a thick, multilayered cuticle that gives their unsegmented body its shape. As a nematode grows, it sheds its cuticle four times. Nematodes exchange oxygen and nutrients with their environment through both the cuticle and the gut, which is only one cell layer thick. Materials are moved through the gut by rhythmic contraction of a highly muscular organ, the pharynx, at the worm's anterior end. Nematodes move by contracting their longitudinal muscles.

Nematodes are probably the most abundant and universally distributed of all major animal groups. About 25,000 species have been described, but the actual number of living species may be in the millions. Many are microscopic, but the largest known nematode reaches a length of 9 meters (it is a parasite in the placentas of sperm whales). Countless nematodes live as scavengers in the upper layers of the soil, on the bottoms of lakes and streams, and in marine sediments. The topsoil of rich farmland may contain 3–9 billion nematodes per acre. A single rotting apple may contain as many as 90,000 individuals.

One soil-inhabiting nematode, *Caenorhabditis elegans*, serves as a model organism in the laboratories of many geneticists and developmental biologists. It is ideal for such research because it is easy to cultivate, matures in 3 days, and has a fixed number of body cells.

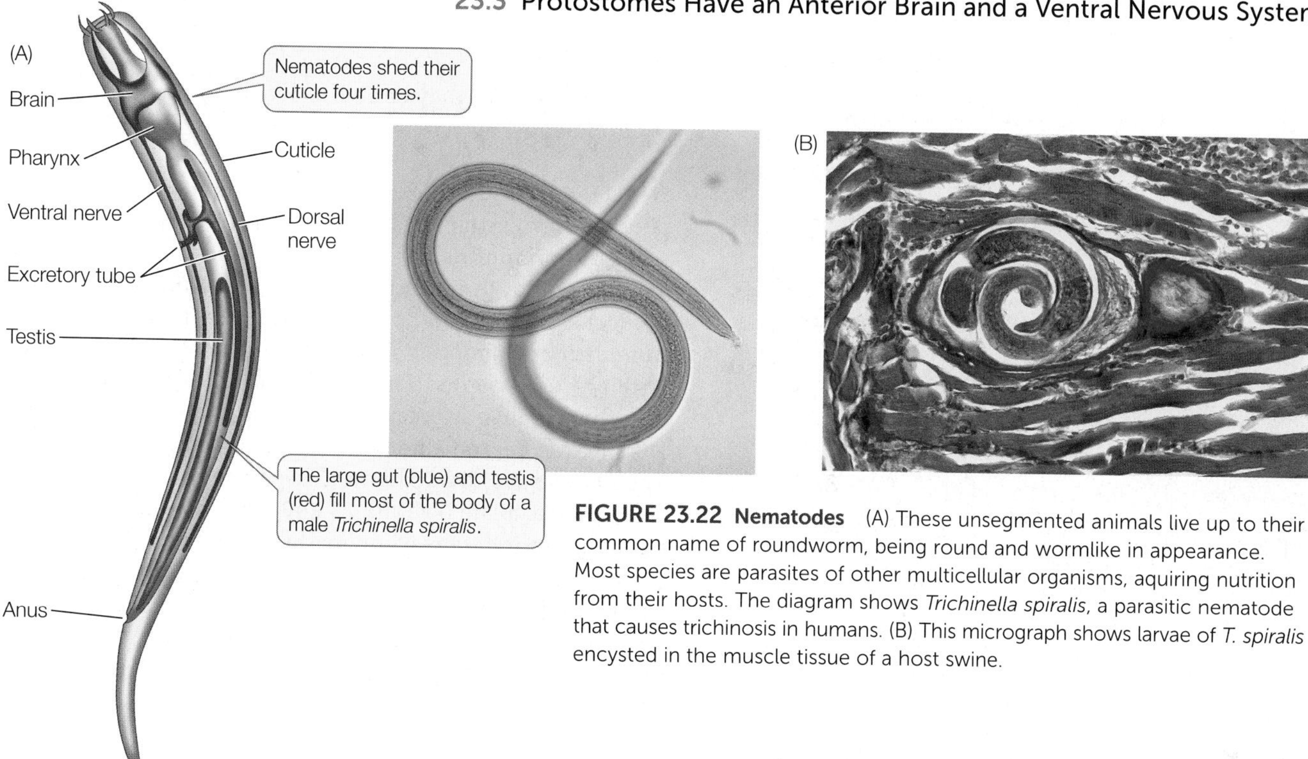
(A)
Brain
Pharynx
Ventral nerve
Excretory tube
Testis
Anus
Nematodes shed their cuticle four times.
Cuticle
Dorsal nerve
The large gut (blue) and testis (red) fill most of the body of a male *Trichinella spiralis*.
(B)

FIGURE 23.22 Nematodes (A) These unsegmented animals live up to their common name of roundworm, being round and wormlike in appearance. Most species are parasites of other multicellular organisms, aquiring nutrition from their hosts. The diagram shows *Trichinella spiralis*, a parasitic nematode that causes trichinosis in humans. (B) This micrograph shows larvae of *T. spiralis* encysted in the muscle tissue of a host swine.

Many nematodes are predators, feeding on protists and small animals (including other roundworms). Most significant to humans, however, are the many species that parasitize plants and animals (**FIGURE 23.22A**). The nematodes that parasitize humans (causing serious diseases such as trichinosis, filariasis, and elephantiasis), domesticated animals, and economically important plants have been studied intensively in an effort to find ways of controlling them.

The structure of parasitic nematodes is similar to that of free-living species, but the life cycles of many parasitic species have special stages that facilitate the transfer of individuals among hosts. *Trichinella spiralis*, the species that causes the human disease trichinosis, has a relatively simple life cycle. A person may become infected by eating the flesh of an animal (usually a pig) that has *Trichinella* larvae encysted in its muscles (**FIGURE 23.22B**). The larvae are activated in the person's digestive tract, emerge from their cysts, and attach to the intestinal wall, where they feed. Later they bore through the intestinal wall and are carried in the bloodstream to muscles, where they form new cysts. If present in great numbers, these cysts can cause severe pain or death.

HORSEHAIR WORMS About 350 species of the unsegmented **horsehair worms** have been described. As their name implies, these animals are extremely thin in diameter; they range from a few millimeters up to a meter in length. Most adult horsehair worms live in fresh water among the leaf litter and algal mats that accumulate near the shores of streams and ponds. A few species live in damp soil.

APPLY THE CONCEPT

There are two major groups of protostomes

The table shows a small sample of amino acid residues that have been used to reconstruct the relationships of some major animal groups. (The full dataset includes data on 11,234 amino acid positions across 77 species of animals.)

1. Construct a phylogenetic tree of these seven animals using the parsimony method (see Chapter 16). Use the sponge as your outgroup to root the tree. Assume that all changes among amino acids are equally likely.
2. How many changes (from one amino acid residue to another) occur along each branch on your tree?
3. Which branch on your tree represents the eumetazoans?
4. The protostomes? The lophotrochozoans? The ecdysozoans? The deuterostomes?

	AMINO ACID POSITION											
Animal	1	2	3	4	5	6	7	8	9	10	11	12
Sponge	P	P	M	S	V	R	Q	N	L	V	I	L
Clam	E	A	M	R	I	K	L	S	I	V	I	L
Earthworm	E	P	M	R	I	K	L	S	I	V	M	L
Tardigrade	E	K	M	R	I	R	L	N	L	V	L	L
Fruit fly	E	K	M	R	I	S	L	D	L	V	L	L
Sea urchin	E	P	M	R	V	R	Q	N	L	T	V	K
Human	E	P	I	R	V	R	Q	N	L	T	V	K

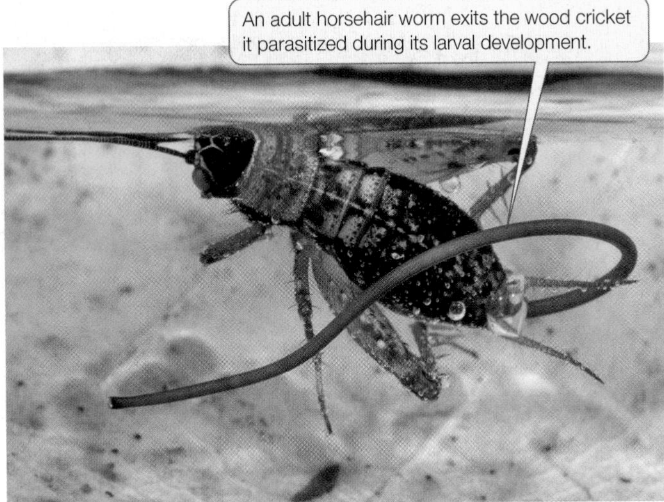

An adult horsehair worm exits the wood cricket it parasitized during its larval development.

FIGURE 23.23 Horsehair Worm Larvae Are Parasitic The larvae of this horsehair worm (*Paragordius tricuspidatus*) can manipulate its host's behavior. The hatching worm causes the cricket to jump into water, where the worm will continue its life cycle as a free-living adult. The insect, having delivered its parasitic burden, drowns.

Horsehair larvae are internal parasites of freshwater crayfishes and of terrestrial and aquatic insects (**FIGURE 23.23**). An adult horsehair worm has no mouth, and its gut is greatly reduced and probably nonfunctional. Some species feed only as larvae, absorbing nutrients from their hosts across the body wall. But other species continue to grow and shed their cuticles even after they have left their hosts, suggesting that some adult worms may be able to absorb nutrients from their environment.

SMALL MARINE CLADES There are several small, relatively poorly known groups of benthic marine ecdysozoans. The 20 species of **priapulids** are cylindrical, unsegmented, wormlike animals with a three-part body plan consisting of a proboscis, trunk, and caudal appendage ("tail"). It should be clear from their appearance why they were named after the Greek fertility god Priapus (**FIGURE 23.24A**). The 180 species of **kinorhynchs** live in marine sands and muds and are virtually microscopic; no kinorhynchs are longer than 1 millimeter. Their bodies are divided into 13 segments, each with a separate cuticular plate (**FIGURE 23.24B**). The minute (less than 1 mm long) **loriciferans** were not discovered until 1983. About 100 living species are known to exist, although only about 30 species have been described. The body is divided into a head, neck, thorax, and abdomen and is covered by six plates, from which the loriciferans get their name (Latin *lorica*, "corset"; **FIGURE 23.24C**).

CHECKpoint CONCEPT 23.3

- ✓ How does an animal's body covering influence the way it breathes, feeds, and moves?
- ✓ Why are most annelids restricted to moist environments?
- ✓ Briefly describe how the basic body organization of mollusks has been modified to yield a wide diversity of animals.
- ✓ What are three ways in which nematodes have a significant impact on humans?

We will turn now to the arthropods. Members of the four arthropod subgroups not only dominate the ecdysozoan clade but are also among the most diverse animals on Earth.

FIGURE 23.24 Benthic Marine Ecdysozoans Members of these groups are marine bottom-dwellers. (A) Most priapulid species live in burrows on the ocean floor, extending the proboscis for feeding. (B) Kinorhynchs are virtually microscopic. The cuticular plates that cover their bodies are molted periodically. (C) Six cuticular plates form a "corset" around the minute loriciferan body.

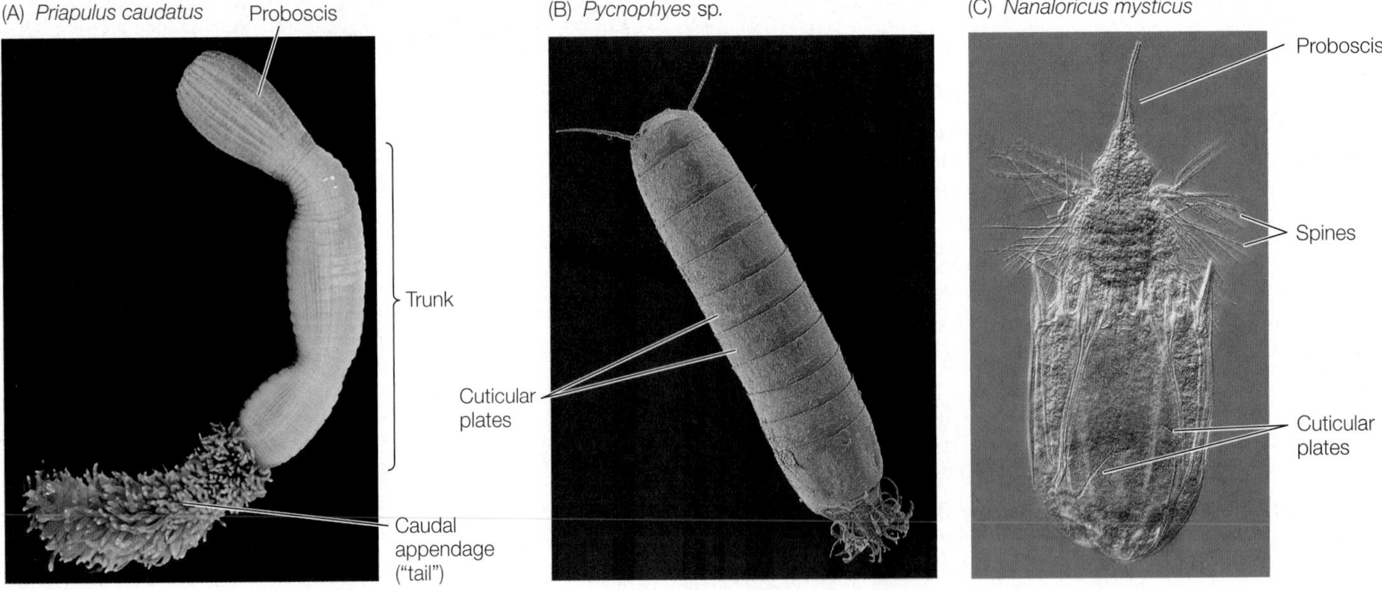

CONCEPT 23.4 Arthropods Are Diverse and Abundant Animals

Arthropods and their relatives are ecdysozoans with paired appendages. Arthropods are an extremely diverse group of animals in numbers of species. Furthermore, the number of individual arthropods alive at any one time is estimated to be about 10^{18}, or a billion billion. Among the animals, only the nematodes are thought to exist in greater numbers.

Several key features have contributed to the success of the arthropods. Their bodies are segmented, and their muscles are attached to the inside of their rigid exoskeletons. Each segment has muscles that operate that segment and the jointed appendages attached to it. Jointed appendages permit complex movements, and different appendages are specialized for different functions. Encasement of the body within a rigid exoskeleton provides the animal with support for walking in the water or on dry land and provides some protection against predators. The waterproofing provided by chitin keeps the animal from dehydrating in dry air.

Representatives of the four major arthropod groups living today are all species-rich: the chelicerates (including the arachnids—spiders, scorpions, mites, and their relatives), myriapods (millipedes and centipedes), crustaceans (including shrimps, crabs, and barnacles), and hexapods (insects and their relatives). Phylogenetic relationships among arthropod groups are currently being reexamined in light of a wealth of new information, much of it based on gene sequences. These studies suggest that the chelicerates are the sister group to the remaining arthropods, and that the crustaceans may be paraphyletic (some crustaceans are more closely related to hexapods than to other crustaceans). There is strong support for the monophyly of arthropods as a whole.

The jointed appendages of arthropods gave the clade its name, from the Greek words *arthron*, "joint," and *podos*, "foot" or "limb." Arthropods evolved from ancestors with simple, unjointed appendages. The exact forms of those ancestors are unknown, but some arthropod relatives with segmented bodies and unjointed appendages survive today. Before we describe the modern arthropods, we will discuss those arthropod relatives.

Arthropod relatives have fleshy, unjointed appendages

Until fairly recently, biologists debated whether the **velvet worms** (onychophorans) were more closely related to annelids or arthropods, but molecular evidence clearly links them to the arthropods. Indeed, with their soft, fleshy, unjointed, claw-bearing legs, velvet worms may be similar in appearance to the arthropod ancestor (**FIGURE 23.25A**). The 180 species of velvet worms live in leaf litter in humid tropical environments. Their soft, segmented bodies are covered by a thin, flexible cuticle that contains chitin. They use their fluid-filled body cavities as hydrostatic skeletons. Fertilization is internal, and large, yolky eggs are brooded within the female's body.

Tardigrades (water bears) also have fleshy, unjointed legs and use their fluid-filled body cavities as hydrostatic skeletons (**FIGURE 23.25B**). Tardigrades are tiny (0.5–1.5 mm long) and lack both a circulatory system and gas exchange organs. The 1,200 known species live in marine sands and on temporary water films on plants, as well as in ephemeral pools. When these films and shallow pools dry out, the animals also lose water and shrink to small, barrel-shaped objects that can survive for at least a decade in a dormant state. Tardigrades have been found in densities as high as 2 million per square meter of moss.

(A) *Macroperipatus torquatus*

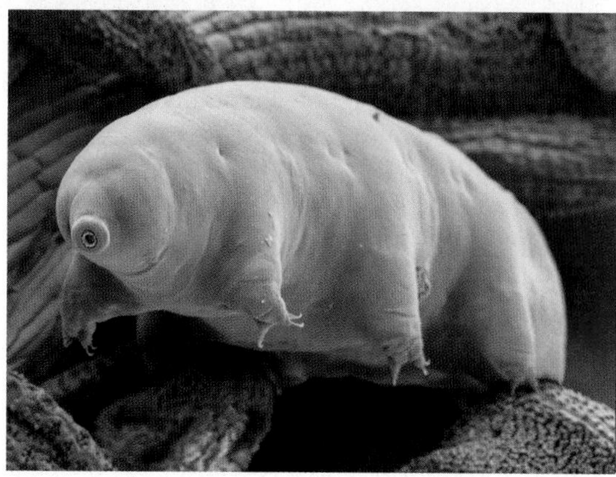

(B) *Macrobiotus sapiens*

FIGURE 23.25 Arthropod Relatives with Unjointed Appendages (A) Velvet worms have unjointed legs and use the body cavity as a hydrostatic skeleton. (B) Tardigrades can be abundant on the wet surfaces of mosses and plants and in temporary pools of water.

Chelicerates are characterized by pointed, nonchewing mouthparts

In the **chelicerates**, the head bears two pairs of pointed appendages modified to form mouthparts, called chelicerae, which are used to grasp (rather than chew) prey. In addition, many chelicerates have four pairs of walking legs. The 114,000 described species are placed in three major clades: sea spiders, horseshoe crabs, and arachnids.

The sea spiders (pycnogonids) are a poorly known group of about 1,000 marine species (**FIGURE 23.26A**). Most are small, with leg spans less than 1 cm, but some deep-sea species have leg spans up to 60 cm. A few sea spiders eat algae, but most are carnivorous, eating a variety of small invertebrates.

There are four living species of horseshoe crabs, but many close relatives are known from fossils. Horseshoe crabs, which have changed very little morphologically during their long

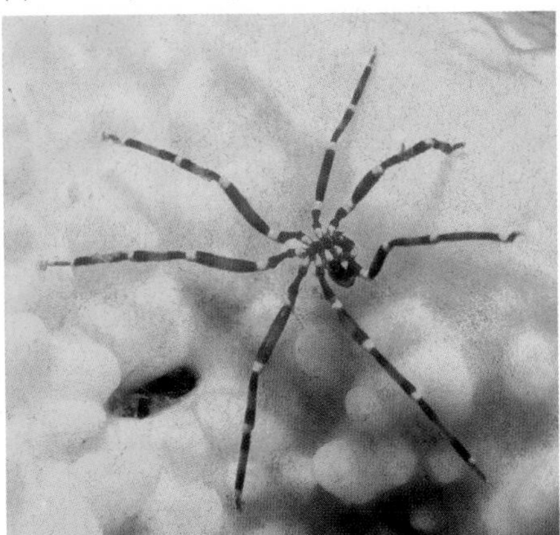
(A) *Pseudopallene* sp.

(B) *Limulus polyphemus*

FIGURE 23.26 Two Small Chelicerate Groups (A) Although they are not spiders, it is easy to see why sea spiders were given their common name. (B) Horseshoe crabs are an ancient group that has changed very little in morphology over time.

fossil history, have a large horseshoe-shaped covering over most of the body. They are common in shallow waters along the eastern coast of North America and the southern and eastern coasts of Asia, where they scavenge and prey on bottom-dwelling animals. Periodically they crawl into the intertidal zone in large numbers to mate and lay eggs (**FIGURE 23.26B**).

Arachnids are abundant in terrestrial environments. Most arachnids have a simple life cycle in which miniature adults hatch from internally fertilized eggs and begin independent lives almost immediately. Some arachnids retain their eggs during development and give birth to live young.

The most species-rich and abundant arachnids are the spiders, scorpions, harvestmen, mites, and ticks (**FIGURE 23.27**). More than 60,000 described species of mites and ticks live in soil, leaf litter, mosses, and lichens, under bark, and as parasites of plants and animals. Mites are vectors for wheat and rye mosaic viruses and cause mange in domestic animals as well as skin irritation in humans.

Spiders, of which about 50,000 species have been described, are important terrestrial predators with hollow chelicerae, which they use to inject venom into their prey. Some species have excellent vision that facilitates elaborate courtship displays and allows the spiders to hunt their prey. Others spin elaborate webs made of protein threads in which they snare prey. The threads are produced by modified abdominal appendages connected to internal glands that secrete the proteins, which solidify on contact with air. The webs of different groups of spiders are strikingly varied, and this variation enables the spiders to position their snares in many different environments for many different types of prey.

Mandibles and antennae characterize the remaining arthropod groups

The remaining three arthropod groups—the myriapods, crustaceans, and hexapods—have mouthparts composed of mandibles rather than chelicerae, so they are together called **mandibulates**.

(A) *Lycosa* sp.

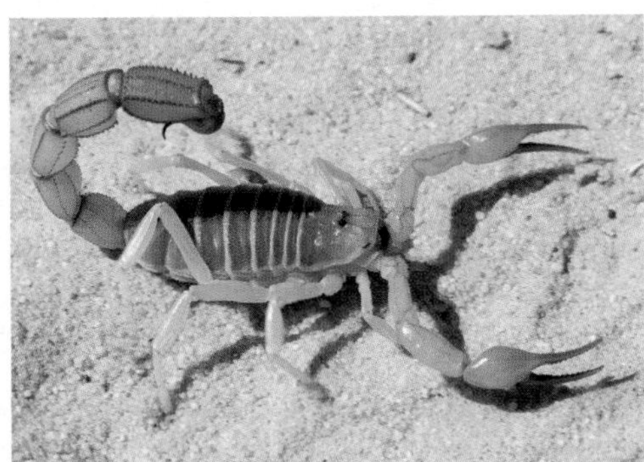
(B) *Androctonus* sp.

(C) *Leiobunum rotundum*

(D) *Brevipalpus phoenicis*

FIGURE 23.27 Arachnid Diversity (A) Known for their web-spinning abilities, spiders are expert predators. (B) A desert scorpion assumes the animal's aggressive attack position. These arachnids are fearsome predators, and their sting can be dangerous to humans. (C) Harvestmen, also called daddy longlegs, are scavengers. (D) Mites include many free-living species as well as blood-sucking, external parasites.

Go to MEDIA CLIP 23.9 Elaboration of Courtship Displays: Even Spiders Twerk PoL2e.com/mc23.9

(A) *Scolopendra hardwicki*

(B) *Polydesmus* sp.

FIGURE 23.28 Myriapods (A) Centipedes have modified appendages that function as poisonous fangs for capturing active prey. They have one pair of legs per segment. (B) Millipedes, which are scavengers and plant eaters, have smaller jaws and legs than centipedes do. They have two pairs of legs per segment.

Mandibles are often used for chewing as well as for biting and holding food. Another distinctive characteristic of the mandibulates is the presence of sensory antennae on the head.

The **myriapods** comprise the centipedes, millipedes, and their close relatives. Centipedes and millipedes have a well-formed head with the mandibles and antennae characteristic of mandibulates. Their distinguishing feature is a long, flexible, segmented trunk that bears many pairs of legs (**FIGURE 23.28**). Centipedes, which have one pair of legs per segment, prey on insects and other small animals. In millipedes, two adjacent segments are fused so that each fused segment has two pairs of legs. Millipedes are detritivores. More than 3,000 species of centipedes and 9,000 species of millipedes have been described, and many more species probably remain unknown. Although most myriapods are less than a few centimeters long, some tropical and subtropical species are ten times that size.

Crustaceans are the dominant marine arthropods today, and they are also common in freshwater and some terrestrial environments. The most familiar crustaceans are the shrimps, lobsters, crayfishes, and crabs (all decapods; **FIGURE 23.29A**) and the pillbugs (isopods; **FIGURE 23.29B**). Additional species-rich groups include the amphipods, ostracods, copepods

(A) *Graspus graspus*

(B) *Armadillidium vulgare*

(C) *Cyclops strenuus*

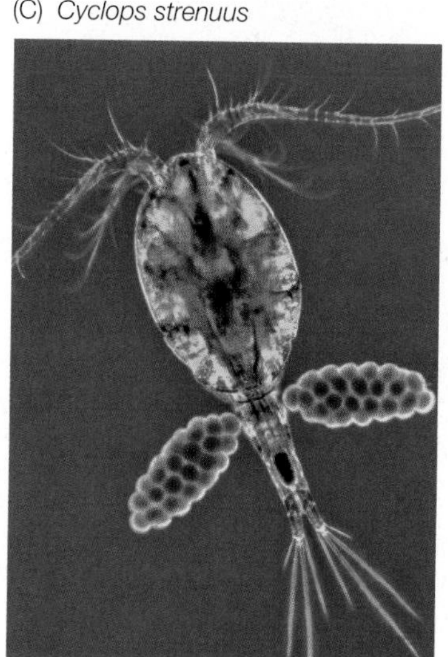

(D) *Triops longicaudatus*

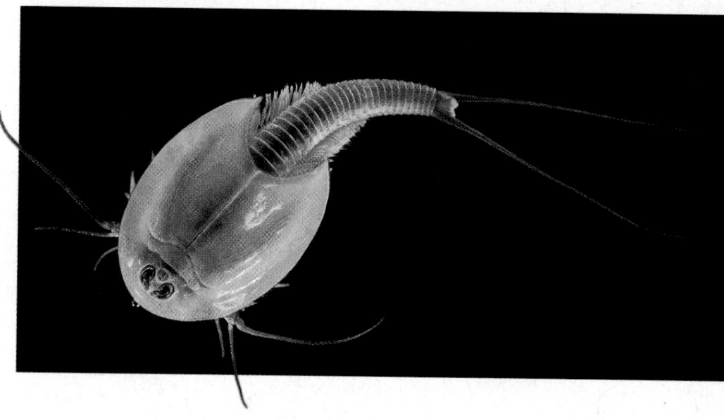

(E) *Lepas anatifera*

FIGURE 23.29 Crustacean Diversity (A) This decapod crustacean is a "Sally Lightfoot" crab from the Galápagos. (B) This pillbug, a terrestrial isopod, can roll into a tight ball when threatened. (C) Minute copepods such as this one are an important link in aquatic food chains. (D) Tadpole shrimp are branchiopods, not to be confused with the decapod crustaceans commonly called shrimp, or with the brachiopods seen in Figure 23.15. (E) Barnacles are sessile, attaching to a substrate by their muscular stalks and feeding with retractable feeding appendages (the delicate, tentacle-like structures seen here).

(**FIGURE 23.29C**), and branchiopods (**FIGURE 23.29D**), all of which are found in freshwater and marine environments.

Barnacles are unusual crustaceans that are sessile as adults (**FIGURE 23.29E**). Adult barnacles look more like mollusks than like other crustaceans, but as the zoologist Louis Agassiz remarked more than a century ago, a barnacle is "nothing more than a little shrimp-like animal, standing on its head in a limestone house and kicking food into its mouth."

Go to MEDIA CLIP 23.10
Barnacles Feeding
PoL2e.com/mc23.10

Most of the 67,000 described species of crustaceans have a body that is divided into three regions: head, thorax, and abdomen (**FIGURE 23.30A**). The segments of the head are fused together, and the head bears five pairs of appendages. Each of the multiple thoracic and abdominal segments usually bears one pair of appendages. The appendages on different parts of the body are specialized for different functions, such as gas exchange, chewing, capturing food, sensing, walking, and swimming. In many species, a fold of the exoskeleton, the carapace, extends dorsally and laterally back from the head to cover and protect some of the other segments.

The fertilized eggs of most crustacean species are attached to the outside of the female's body, where they remain during their early development. At hatching, the young of some species are released as larvae; those of other species are released as juveniles that are similar in form to the adults. Still other species release eggs into the water or attach them to an object in the environment.

More than half of all described species are insects

During the Devonian period, more than 400 million years ago, some mandibulates colonized terrestrial environments. Of the several groups (including some crustacean isopods and decapods) that successfully colonized the land, none is more prominent today than the six-legged **hexapods**: the insects and their relatives. Insects are abundant and diverse in terrestrial and freshwater environments. Only a few insects live in salt water.

The wingless relatives of the insects—the springtails, two-pronged bristletails, and proturans—are probably the most similar of living forms to insect ancestors. These insect relatives have a simple life cycle; they hatch from eggs as miniature adults. They differ from insects in having internal mouthparts. Springtails can be extremely abundant (up to 200,000 per square meter) in soil, leaf litter, and on vegetation and are the most abundant hexapods in the world in terms of number of individuals (rather than number of species).

More than 1 million of the 1.8 million described species of living organisms are insects. As we discussed at the beginning of this chapter, biologists have estimated that many millions of insect species remain to be discovered. Like crustaceans, insects have a body with three regions—head, thorax, and abdomen. They have a single pair of antennae on the head and three pairs of legs attached to the thorax. In most groups of insects, the thorax also bears two pairs of wings. Unlike other arthropods, insects have no appendages associated with their abdominal segments (**FIGURE 23.30B**).

Insects can be distinguished from springtails and other hexapods by their external mouthparts and by antennae that contain a motion-sensitive receptor called Johnston's organ. In addition, insects have a derived mechanism for gas exchange in air: a system of air sacs and tubular channels called tracheae (singular *trachea*) that extend from external openings called spiracles inward to tissues throughout the body (see Figure 31.10). Insects use nearly all species of plants and many species of animals as food.

TABLE 23.2 lists the major insect groups. Two groups—the jumping bristletails and silverfish—are wingless and have

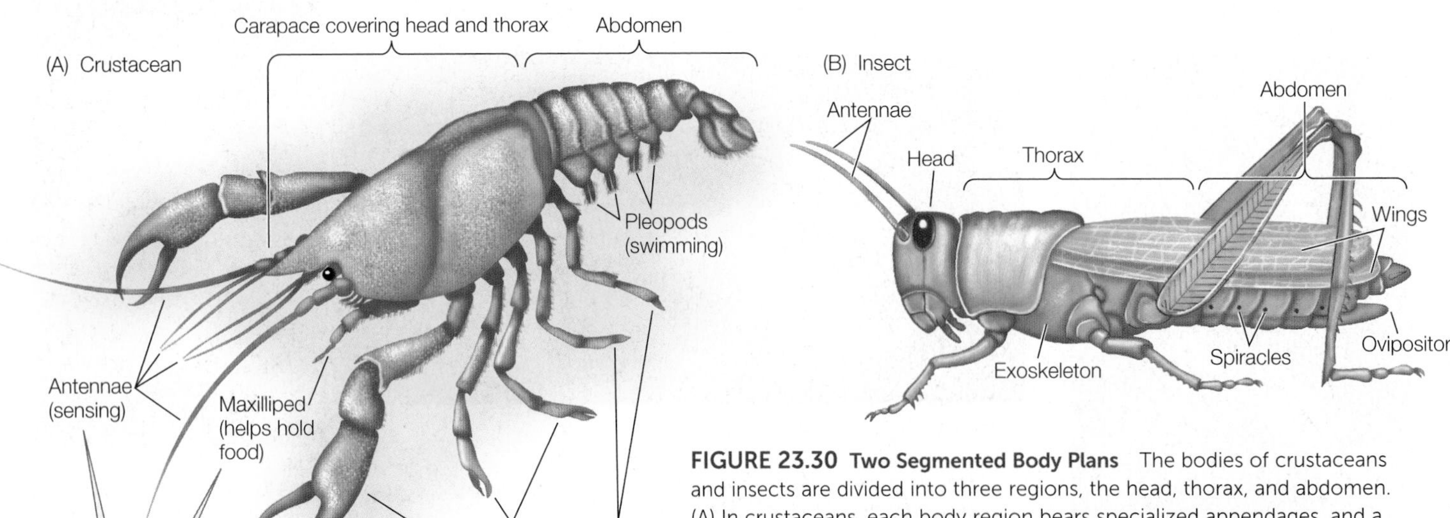

FIGURE 23.30 Two Segmented Body Plans The bodies of crustaceans and insects are divided into three regions, the head, thorax, and abdomen. (A) In crustaceans, each body region bears specialized appendages, and a shell-like carapace covers the head and thorax. (B) The thorax of insects bears three pairs of legs, and in most groups, two pairs of wings. Unlike other arthropods, insects have no appendages growing from their abdominal segments.

TABLE 23.2 The Major Insect Groups[a]

Group	Approximate number of described living species
Jumping bristletails (Archaeognatha)	515
Silverfish (Thysanura)	560
PTERYGOTE (WINGED) INSECTS (PTERYGOTA)	
Mayflies (Ephemeroptera)	3,250
Dragonflies and damselflies (Odonata)	5,900
Neopterans (Neoptera)[b]	
Ice-crawlers (Grylloblattodea)	35
Gladiators (Mantophasmatodea)	15
Stoneflies (Plecoptera)	3,750
Webspinners (Embioptera)	465
Angel insects (Zoraptera)	40
Earwigs (Dermaptera)	2,000
Grasshoppers and crickets (Orthoptera)	24,000
Stick insects (Phasmida)	3,000
Cockroaches (Blattodea)	4,650
Termites (Isoptera)	2,700
Mantids (Mantodea)	2,400
Booklice and barklice (Psocoptera)	5,750
Thrips (Thysanoptera)	6,000
Lice (Phthiraptera)	5,100
True bugs, cicadas, aphids, leafhoppers (Hemiptera)	104,000
Holometabolous neopterans (Holometabola)[c]	
Ants, bees, wasps (Hymenoptera)	117,000
Beetles (Coleoptera)	388,000
Twisted-wing parasites (Strepsiptera)	610
Lacewings, ant lions, mantidflies (Neuropterida)	5,900
Dobsonflies, alderflies, fishflies (Megaloptera)	350
Snakeflies (Raphidoptera)	250
Scorpionflies (Mecoptera)	760
Fleas (Siphonaptera)	2,100
True flies (Diptera)	155,000
Caddisflies (Trichoptera)	14,300
Butterflies and moths (Lepidoptera)	158,000

[a]The hexapod relatives of insects include the springtails (Collembola; 3,000 spp.), two-pronged bristletails (Diplura; 600 spp.), and proturans (Protura; 10 spp.). All are wingless and have internal mouthparts.
[b]Neopteran insects can tuck their wings close to their bodies.
[c]Holometabolous insects are neopterans that undergo complete metamorphosis.

simple life cycles, like the springtails and other close insect relatives. The remaining groups are all pterygote insects. **Pterygotes** have two pairs of wings, except in some groups where one or both pairs of wings have been secondarily lost. Secondarily wingless groups include the parasitic lice and fleas, some beetles, and the worker individuals in many ants.

Hatchling pterygotes do not look like adults, and they undergo substantial changes at each molt. The immature stages of insects between molts are called **instars**. A substantial change that occurs between one developmental stage and another is called **metamorphosis**. If the changes between its instars are gradual, an insect is said to have **incomplete metamorphosis**. If the change between at least some instars is dramatic, an insect is said to have **complete metamorphosis**. In many insects with complete metamorphosis, the different life stages are specialized for different environments and use different food sources. In many species, the larvae are adapted for feeding and growing, whereas the adults are specialized for reproduction and dispersal.

Go to MEDIA CLIP 23.11
Complete Metamorphosis
PoL2e.com/mc23.11

LINK

Insect metamorphosis is under the control of hormones, as described in **Concept 35.5**

Pterygote insects were the first animals in evolutionary history to achieve the ability to fly. Flight opened up many new lifestyles and feeding opportunities that only the insects could exploit, and it is almost certainly one of the reasons for the remarkable numbers of insect species and individuals, and for their unparalleled evolutionary success.

Molecular data suggest that insects began to diversify about 450 million years ago, about the time of the appearance of the first land plants (although earliest known fossil insects are only about 400 million years old). The earliest hexapods evolved in a terrestrial environment that lacked any similar organisms, which in part accounts for their remarkable success. But the success of the insects is also due to their wings. Homologous genes control the development of insect wings and crustacean appendages, suggesting that the insect wing evolved from a dorsal branch of a crustacean-like limb:

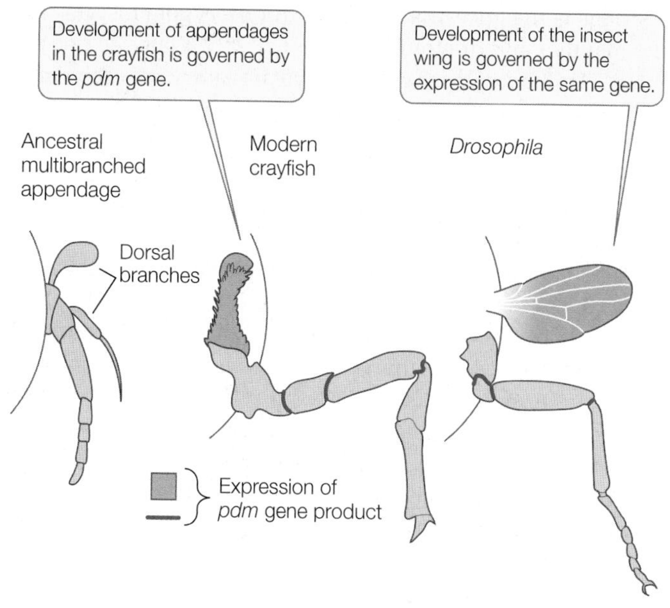

APPLY THE CONCEPT

More than half of all described species are insects

The following data were used by entomologist Terry Erwin to estimate the undescribed diversity of insects. Review the design of Erwin's experiment in the opening story of this chapter. Then use Erwin's data to answer the questions and consider how changes in the assumptions may affect the estimates.[a]

Approximate number of beetle species collected from *Luehea seemannii* trees:	1,200
Estimated number of host-specific beetles in this sample:	163
Number of species of canopy trees per hectare of forest:	70
Percentage of beetle species living in canopy (as opposed to ground-dwelling species):	75%
Percentage of beetles among all insect species:	40%

[a]T. L. Erwin. 1982. *Coleopterists Bulletin* 36(l): 74–75; T. L. Erwin and J. C. Scott. 1980. *Coleopterists Bulletin* 34(3): 305–322.

1. From the data above, estimate the number of insect species in an average hectare of Panamanian forest. Assume that the data for beetles on *L. seemannii* are representative of the other tree species, and that all the species of beetles that are *not* host-specific were collected in the original sample. Remember to sum your estimates of the number of (a) host-specific beetle species in the forest canopy; (b) non-host-specific beetle species in the forest canopy; (c) beetle species on the forest floor; and (d) species of all insects other than beetles.
2. Use the following information to estimate the number of insect species on Earth. There are about 50,000 species of tropical forest trees. Assume that the data for beetles on *L. seemannii* are representative for other species of tropical trees, and calculate the number of host-specific beetles found on these trees. Add an estimated 1 million species of beetles that are expected across different species of trees (including temperate regions). Estimate the number of ground-dwelling beetle species based on the percentage used in question 1. Now estimate the number of species of insects found worldwide, based on the percentage of beetles among all insect species.
3. The estimates in questions 1 and 2 are based on many assumptions. Do you think these assumptions are reasonable? Why or why not? Would you argue for a different set of assumptions? How do you think these changes in assumptions would affect your calculations? Can you think of ways to test your assumptions?

The dorsal limb branch of crustaceans is used for gas exchange. Thus the insect wing probably evolved from a gill-like structure that had a gas exchange function.

The adults of most flying insects have two pairs of stiff, membranous wings attached to the thorax. True flies, however, have one pair of wings and a pair of stabilizers called haltares. In winged beetles, one pair of wings—the forewings—forms heavy, hardened wing covers. Flying insects are important pollinators of flowering plants.

Two groups of pterygotes, the mayflies and dragonflies (**FIGURE 23.31A**), cannot fold their wings against their bodies. This is the ancestral condition for pterygote insects, and the mayflies and dragonflies are not closely related to one another. Members of these groups have predatory or herbivorous aquatic larvae that transform into flying adults after they crawl out of the water. Dragonflies (and their relatives the damselflies) are active predators as adults. In contrast, adult mayflies lack functional digestive tracts. Mayflies live only about a day, just long enough to mate and lay eggs.

All other pterygote insects—the **neopterans**—can tuck their wings out of the way upon landing and crawl into crevices and other tight places. Some neopteran groups undergo incomplete metamorphosis, so hatchlings of these insects are sufficiently similar in form to adults to be recognizable. Examples include the grasshoppers (**FIGURE 23.31B**), roaches, mantids, stick insects, termites, stoneflies, earwigs, thrips, true bugs (**FIGURE 23.31C**), aphids, cicadas, and leafhoppers. They acquire adult organ systems, such as wings and compound eyes, gradually through several juvenile instars.

More than 80 percent of all insects belong to a subgroup of the neopterans called the **holometabolous** insects (see Table 23.2), which undergo complete metamorphosis (**FIGURE 23.31D**). The many species of beetles account for almost half of this group (**FIGURE 23.31E**). Also included are lacewings and their relatives; caddisflies; butterflies and moths (**FIGURE 23.31F**); sawflies; true flies (**FIGURE 23.31G**); and bees, wasps, and ants, some species of which display unique and highly specialized social behaviors (**FIGURE 23.31H**).

CHECKpoint CONCEPT 23.4

- ✓ What features have contributed to making arthropods among the most abundant animals on Earth, both in number of species and number of individuals?
- ✓ Describe the difference between incomplete and complete metamorphosis.
- ✓ Can you think of some possible reasons why there are so many species of insects?

The majority of Earth's animal species are protostomes, so it is not surprising that the protostome groups display a huge variety of different body plans and other characteristics. We will now consider the diversity of the deuterostomes, a group that contains far fewer species than the protostomes but which is even more intensively studied and, not incidentally, is the group to which humans belong.

(A) *Libellula depressa*, Odonata

(B) *Brachystola magna*, Orthoptera

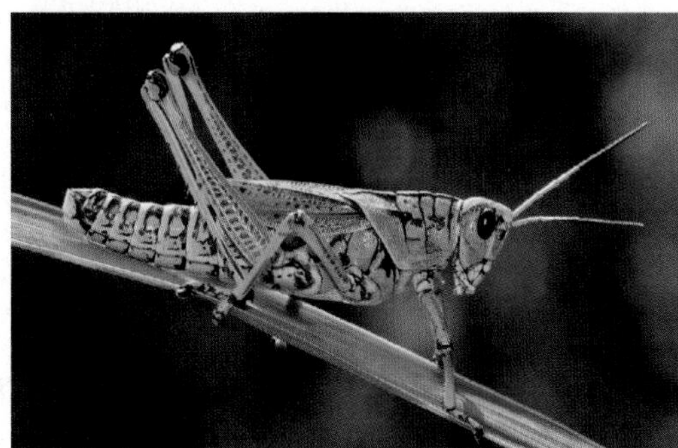

(C) *Graphosoma lineatum*, Hemiptera

(D) *Gumaga* sp., Trichoptera

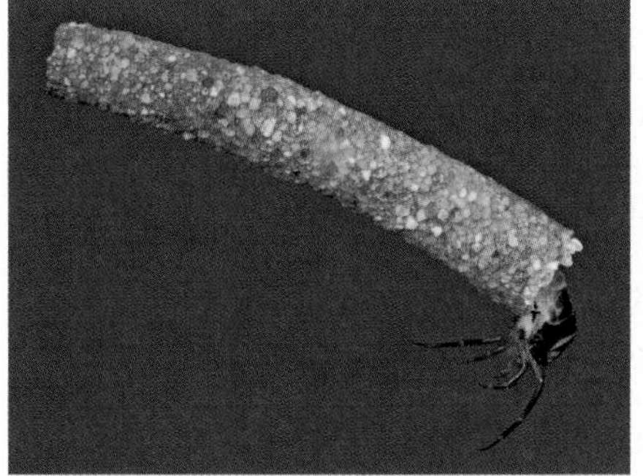

(E) *Cetonia aurata*, Coleoptera

(F) *Papilio machaon*, Lepidoptera

(G) *Poecilanthrax willistoni*, Diptera

(H) *Mischocyttarus flavitarsis*, Hymenoptera

FIGURE 23.31 Diverse Winged Insects (A) Unlike most flying insects, a dragonfly cannot fold its wings over its back. (B) Orthopteran insects such as grasshoppers undergo several larval molts (instars), but the juveniles resemble small adults (incomplete metamorphosis). (C) Hemipterans are "true" bugs. (D–H) Holometabolous insects undergo complete metamorphosis. (D) A larval caddisfly (right) emerges from its case constructed from silk and sand. (E) Coleoptera is the largest insect group; beetles such as this rose chafer account for more than half of all holometabolous species. (F) The swallowtail butterfly is found in most parts of North America. (G) This bee fly is a true fly. Although bee flies cannot sting, most potential predators avoid them because of their resemblance to bees. (H) This western paper wasp is a hymenopteran, a group in which most members display social behaviors.

CONCEPT 23.5 Deuterostomes Include Echinoderms, Hemichordates, and Chordates

It may surprise you to learn that you and a sea urchin are both deuterostomes. Adult sea stars, sea urchins, and sea cucumbers—the most familiar echinoderms—look so different from adult vertebrates (fishes, frogs, lizards, birds, and mammals) that it is difficult to believe all these animals are closely related. Two major pieces of evidence indicate that the deuterostomes share a common ancestor not shared with the protostomes:

- Deuterostomes share a pattern of early development in which the mouth forms at the opposite end of the embryo from the blastopore, and the blastopore develops into the anus (this is opposed to the protostomes, in which the blastopore becomes the mouth).
- Recent phylogenetic analyses of the DNA sequences of many different genes offer strong support for the shared evolutionary relationships of deuterostomes.

Note that neither of the above factors is apparent in the morphology of the adult animals.

Although there are far fewer species of deuterostomes than of protostomes (see Table 23.1), the deuterostomes are of special interest because they include many large animals—including humans—that strongly influence the characteristics of ecosystems. Many deuterostome species have been intensively studied in all fields of biology. Complex behaviors are especially well developed among some deuterostomes and are a vast and fascinating field of study in themselves (see Chapter 40).

There are three major clades of living deuterostomes:

- **Echinoderms**: sea stars, sea urchins, and their relatives
- **Hemichordates**: acorn worms and pterobranchs
- **Chordates**: sea squirts, lancelets, and vertebrates

All deuterostomes are triploblastic and coelomate (see Figure 23.4C). Skeletal support features, where present, are internal rather than external. Some species have segmented bodies, but the segments are less obvious than those of annelids and arthropods.

Go to ANIMATED TUTORIAL 23.3
An Overview of the Deuterostomes
PoL2e.com/at23.3

The earliest deuterostomes were bilaterally symmetrical, segmented animals with a pharynx that had slits through which water flowed. Echinoderms evolved their adult forms with unique symmetry (in which the body parts are arranged along five radial axes) much later, whereas other deuterostomes retained the ancestral bilateral symmetry.

The echinoderms and hemichordates (together known as **ambulacrarians**) have a bilaterally symmetrical, ciliated larva (**FIGURE 23.32A**). Adult hemichordates also are bilaterally symmetrical. Echinoderms, however, undergo a radical change in form as they develop into adults (**FIGURE 23.32B**), changing from a bilaterally symmetrical larva to an adult with **pentaradial symmetry** (symmetry in five or multiples of five). As is typical of animals with radial symmetry, echinoderms have no head, and they move equally well (but usually slowly) in many directions. Rather than having an anterior–posterior (head–tail) and dorsal–ventral (back–belly) body organization, echinoderms have an oral side (containing the mouth) and an opposite aboral side (containing the anus).

Echinoderms have unique structural features

About 13,000 species of echinoderms in 23 major groups have been described from fossil remains. They are probably only a small fraction of those that actually lived. Only 6 of the 23 major groups known from fossils are represented by species that survive today. Many clades of echinoderms became extinct during the periodic mass extinctions that have occurred throughout Earth's history (see Table 18.1). Nearly all of the 7,500 extant species of echinoderms live only in marine environments.

In addition to having pentaradial symmetry, adult echinoderms have two unique structural features. One is a system of calcified internal plates covered by thin layers of skin and some muscles. The calcified plates of most echinoderms are thick, and they fuse inside the entire body, forming an internal skeleton. The other unique feature is a **water vascular system**, a network of water-filled canals leading to extensions called **tube feet**. This system functions in gas exchange, locomotion, and feeding (see Figure 23.32B).

FIGURE 23.32 Bilaterally Symmetrical Echinoderm Larvae become Radially Symmetrical Adults (A) The ciliated larva of a sea star has bilateral symmetry. Hemichordates have a similar larval form. (B) The radially symmetrical adult sea star displays the canals and tube feet of the echinoderm water vascular system, as well as the calcified internal skeleton. The body's orientation is oral–aboral rather than anterior–posterior.

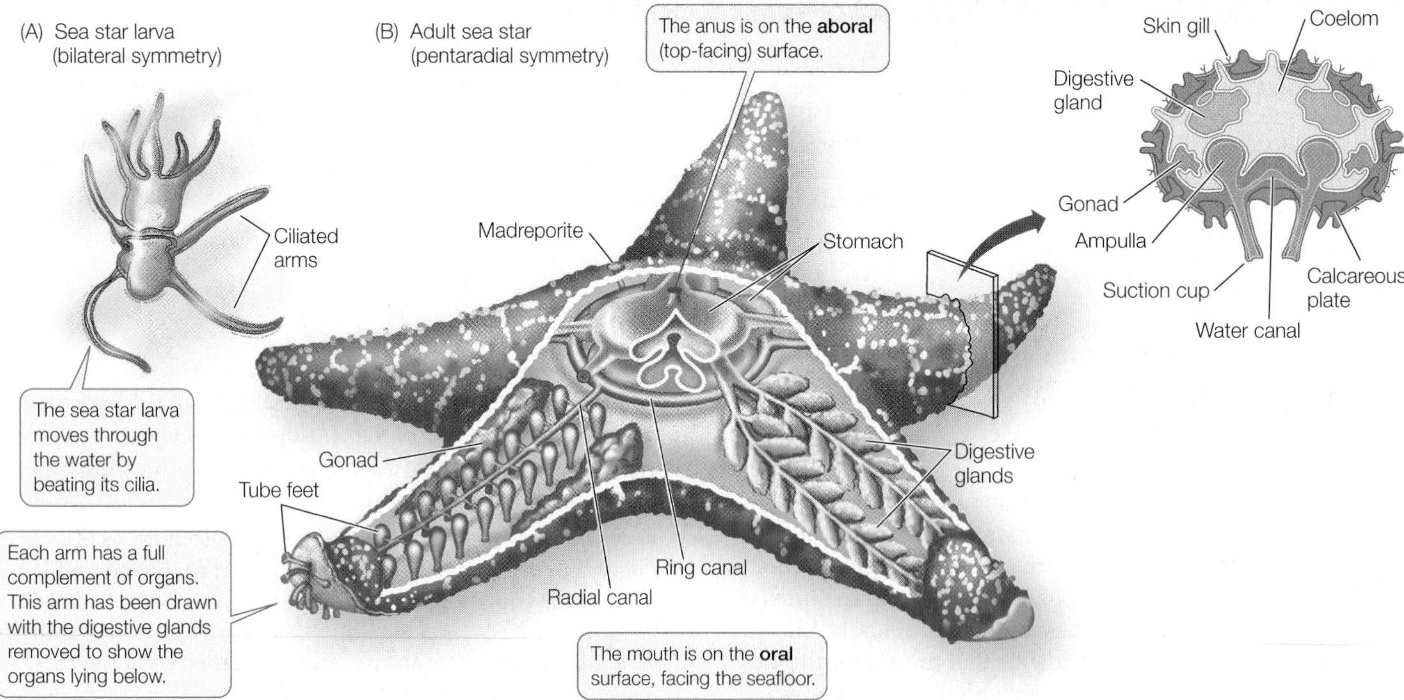

(A) *Comanthina schlegeli*

(B) *Strongylocentrotus purpuratus*

(C) *Pseudocolochirus* sp.

(D) *Asterias rubens*

(E) *Ophiothrix spiculata*

FIGURE 23.33 Echinoderm Diversity (A) The flexible arms of this feather star are clearly visible. (B) Sea urchins are important grazers on algae in the intertidal zones of the world's oceans. (C) Sea cucumbers are unique among echinderms in having an anterior–posterior rather than an oral–aboral orientation of the mouth and anus. (D) Sea stars are important predators on bivalve mollusks such as mussels and clams. Suction tips on its tube feet allow a sea star to grasp both shells of the bivalve and pull them open. (E) The arms of the brittle star are composed of hard but jointed plates.

Members of one major extant clade, the crinoids (sea lilies and feather stars), were more abundant and species-rich 300–500 million years ago than they are today. There are some 80 described living sea lily species, most of which are sessile organisms attached to a substrate by a stalk. Feather stars (**FIGURE 23.33A**) grasp the substrate with flexible appendages that allow for limited movement. About 600 living species of feather stars have been described.

Unlike the crinoids, most of the other surviving echinoderms are motile. The two main groups of motile echinoderms are the echinozoans (sea urchins and sea cucumbers; **FIGURE 23.33B,C**) and asterozoans (sea stars and brittle stars; **FIGURE 23.33D,E**).

The tube feet of the different echinoderm groups have been modified in a great variety of ways to capture prey. Sea lilies, for example, feed by orienting their arms in passing water currents. Food particles strike and stick to their tube feet, which are covered with mucus-secreting glands. The tube feet transfer food particles to grooves in the arms, where ciliary action carries the particles to the mouth. Sea cucumbers capture food with their anterior tube feet, which are modified into large, feathery, sticky tentacles that can be protruded from the mouth. Periodically, a sea cucumber withdraws the tentacles, wipes off the material that has adhered to them, and digests it.

Many sea stars use their tube feet to capture large prey such as annelids, gastropod and bivalve mollusks, small crustaceans such as crabs, and fishes. With hundreds of tube feet acting simultaneously, a sea star can grasp a bivalve in its arms, anchor the arms with its tube feet, and by steady contraction of the muscles in its arms, gradually exhaust the muscles the bivalve uses to keep its shell closed (see Figure 23.33D). To feed on a bivalve, a sea star can push its stomach out through its mouth and then through the narrow space between the two halves of the bivalve's shell. The sea star's stomach then secretes enzymes that digest the prey.

Go to MEDIA CLIP 23.12
Sea Star Hunting Bivalves
PoL2e.com/mc23.12

Most sea urchins eat algae, which they catch with their tube feet from the plankton or scrape from rocks with a complex rasping structure. Most of the 2,000 species of brittle stars ingest particles from the upper layers of sediments and assimilate the organic material from them, although some species filter suspended food particles from the water, and others capture small animals.

Hemichordates are wormlike marine deuterostomes

The roughly 120 species of hemichordates—acorn worms and pterobranchs—have a body organized in three major parts, consisting of a proboscis, a collar (which bears the mouth), and

(A)

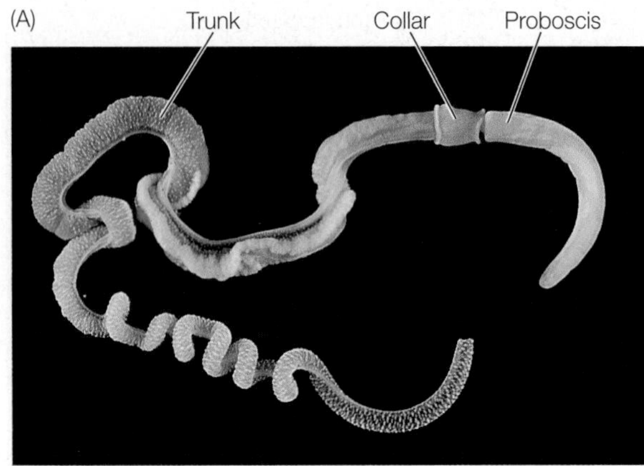

(B)

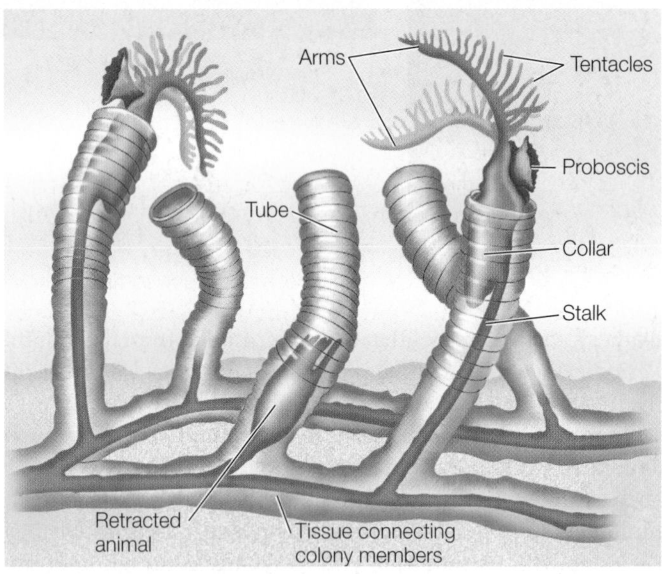

FIGURE 23.34 Hemichordates (A) The proboscis of an acorn worm is modified for burrowing. (B) The structure of a colonial pterobranch.

a trunk (which contains the other body parts). The 90 known species of acorn worms range up to 2 meters in length (**FIGURE 23.34A**). They live in burrows in muddy and sandy marine sediments. The digestive tract of an acorn worm consists of a mouth behind which are a muscular pharynx and an intestine. The pharynx opens to the outside through a number of pharyngeal slits through which water can exit. Highly vascularized tissue surrounding the pharyngeal slits serves as a gas exchange apparatus. Acorn worms breathe by pumping water into the mouth and out through the pharyngeal slits. They capture prey with the large proboscis, which is coated with sticky mucus to which small organisms in the sediment stick. The mucus and its attached prey are conveyed by cilia to the mouth. In the esophagus, the food-laden mucus is compacted into a ropelike mass that is moved through the digestive tract by ciliary action.

The 30 living species of pterobranchs are sedentary marine animals up to 12 millimeters long that live in a tube secreted by the proboscis. Some species are solitary; others form colonies of individuals joined together (**FIGURE 23.34B**). Behind the proboscis is a collar with anywhere from one to nine pairs of arms. The arms bear long tentacles that capture prey and function in gas exchange.

Chordate characteristics are most evident in larvae

As mentioned earlier, it is not obvious from examining the morphology of adult animals that echinoderms and chordates share a common ancestor. The evolutionary relationships among some chordate groups are not immediately apparent either. The features that reveal these evolutionary relationships are seen primarily in the larvae—in other words, it is during the early developmental stages that their evolutionary relationships are evident.

There are three principal chordate clades: the **lancelets** (also called cephalochordates), **tunicates** (also called urochordates), and **vertebrates**. Adult chordates vary greatly in form, but all chordates display the following derived structures at some stage in their development (**FIGURE 23.35**):

- A dorsal hollow nerve cord
- A tail that extends beyond the anus
- A dorsal supporting rod called the notochord

The **notochord** is the most distinctive derived chordate trait. It is composed of a core of large cells with turgid fluid-filled vacuoles, which make it rigid but flexible. In most tunicates the notochord is lost during metamorphosis to the adult stage. In most vertebrate species, it is replaced during development by vertebrae that provide support for the body.

The ancestral pharyngeal slits (not a derived feature of this group) are present at some developmental stage of chordates but are often lost in adults. The pharynx, which develops around the pharyngeal slits, functioned in chordate ancestors as the site for oxygen uptake and the elimination of carbon dioxide and water (as in acorn worms). The pharynx is much enlarged in some chordate species (as in the pharyngeal basket of the lancelet in Figure 23.35B).

Adults of most lancelets and tunicates are sessile

The 35 species of lancelets are small animals that rarely exceed 5 centimeters in length. The notochord, which provides body support, extends the entire length of the body throughout their lives (see Figure 23.35B). Lancelets are found in shallow marine and brackish waters worldwide. Most of the time they lie covered in sand with their head protruding above the sediment, but they can swim. They filter prey from the water with their pharyngeal basket.

All members of the three major tunicate groups—the sea squirts, salps, and larvaceans—live in marine environments. More than 90 percent of the 2,800 known species of tunicates are sea squirts (also called ascidians). Individual sea squirts range in length from less than 1 millimeter to 60 centimeters. Some sea squirts form colonies by asexual budding from a single founder. Colonies may measure several meters across. The baglike body of an adult sea squirt is enclosed in a tough tunic,

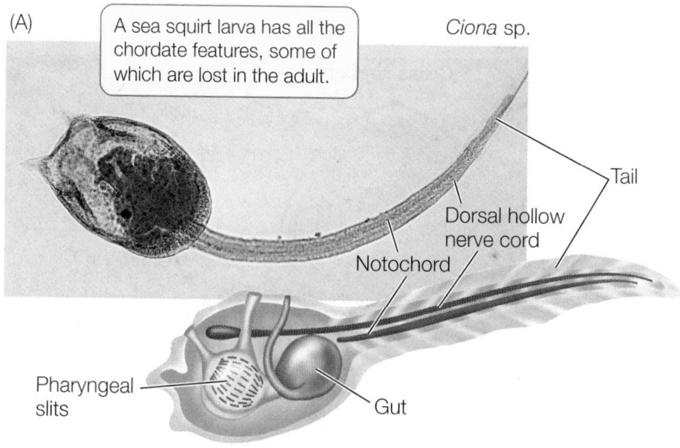

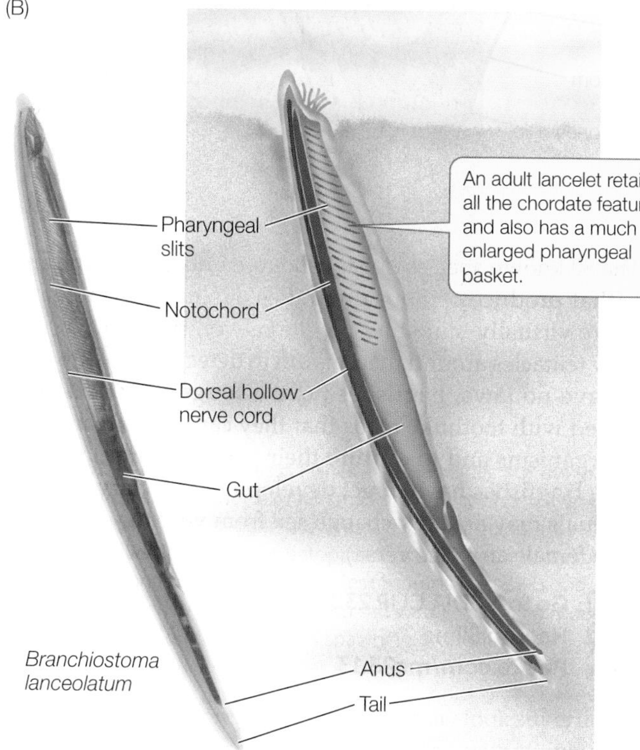

FIGURE 23.35 The Key Features of Chordates Are Most Apparent in Early Developmental Stages The pharyngeal slits of both the tunicate and the lancelet develop into a pharyngeal basket. (A) The sea squirt larva (but not the adult) has all three chordate features: a dorsal hollow nerve cord, postanal tail, and notochord. (B) All three chordate synapomorphies are retained in the adult lancelet.

FIGURE 23.36 An Adult Tunicate The transparent tunic and the pharyngeal basket are clearly visible in this sea squirt.

which is the basis for the name "tunicate" (**FIGURE 23.36**). The tunic is composed of proteins and a complex polysaccharide secreted by epidermal cells. The sea squirt pharynx is enlarged into a pharyngeal basket that filters prey from the water passing through it.

In addition to its pharyngeal slits, a sea squirt larva has a dorsal hollow nerve cord and a notochord that is restricted mostly to the tail region (see Figure 23.35A). Bands of muscle that surround the notochord provide support for the body. After a short time swimming in the plankton, the larvae of most species settle on the seafloor and transform into sessile adults. The swimming, tadpolelike larvae suggest a close evolutionary relationship between sea squirts and vertebrates (see Figure 16.4).

As we have discussed, the hemichordates, echinoderms, lancelets, and tunicates are marine animals. The lineage that led to the vertebrates is also thought to have evolved in the oceans, although probably in an estuarine environment (where fresh water meets salt water). Vertebrates have since radiated into marine, freshwater, terrestrial, and aerial environments worldwide.

The vertebrate body plan can support large, active animals

The vertebrates take their name from the unique, jointed, dorsal **vertebral column** that replaces the notochord during early development as their primary supporting structure. The individual elements in the vertebral column are called vertebrae. Four other key features characterize the vertebrates as well (**FIGURE 23.37**):

- An anterior skull with a large brain
- A rigid internal skeleton supported by the vertebral column, which also encloses the spinal cord
- Internal organs suspended in a coelom
- A well-developed circulatory system, driven by contractions of a ventral heart

The internal skeleton provides support for an extensive muscular system, which receives oxygen from the circulatory system and is controlled by the central nervous system. The evolution of these features allowed many vertebrates to become large, active predators, which in turn allowed the vertebrates to diversify widely (**FIGURE 23.38**).

There are two groups of living jawless fishes

The **hagfishes** are thought by many biologists to be the sister group to the remaining vertebrates (see Figure 23.38). Hagfishes (**FIGURE 23.39A**) have a weak circulatory system, with three small accessory hearts (rather than a single, large heart), a partial cranium or skull (containing no cerebrum or cerebellum,

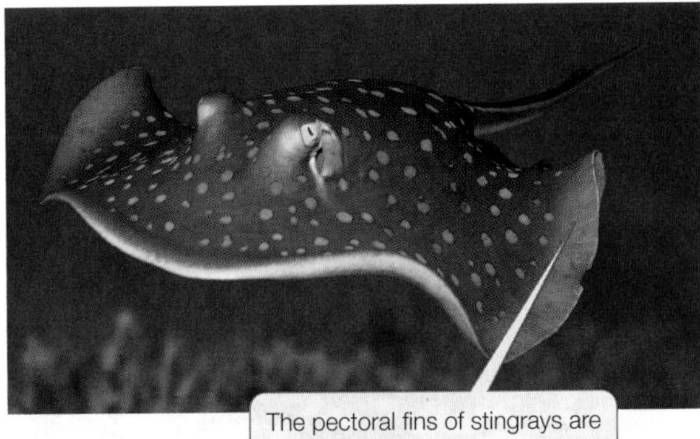

FIGURE 23.41 Chondrichthyans The fins of chondrichthyans lack supportive rays. (A) Most sharks, such as this great white shark, are active marine predators. (B) The fins of skates and rays, represented here by a ribbontail stingray, are modified for feeding on the ocean bottom. (C) A chimaera, or ratfish.

skates, and rays (about 1,000 living species) and chimaeras (40 living species). Like hagfishes and lampreys, these fishes have a skeleton composed entirely of firm but pliable cartilage. Their skin is flexible and leathery, sometimes bearing scales that give it the consistency of sandpaper. Sharks move forward by means of lateral undulations of their body and caudal fin (**FIGURE 23.41A**). Skates and rays propel themselves by means of vertical undulating movements of their greatly enlarged pectoral fins (**FIGURE 23.41B**).

Most sharks are predators, but some feed by straining plankton from the water. Most skates and rays live on the ocean floor, where they feed on mollusks and other animals buried in the sediments. Nearly all cartilaginous fishes live in the oceans, but a few are estuarine or migrate into lakes and rivers. One group of stingrays is found in river systems of South America. The less familiar chimaeras (**FIGURE 23.41C**) live in deep-sea or cold waters.

One lineage of aquatic gnathostomes gave rise to the bony vertebrates, which soon split into two main lineages—the **ray-finned fishes** and the **lobe-limbed vertebrates**. Bony vertebrates have internal skeletons of calcified, rigid bone rather than flexible cartilage. In early bony vertebrates, gas-filled sacs that extended from the digestive tract supplemented the gas exchange function of the gills by giving the animals access to atmospheric oxygen. These features enabled those fishes to live where oxygen was periodically in short supply, as it often is in freshwater environments. These lunglike sacs evolved into **swim bladders** (organs of buoyancy), as well as into the lungs of tetrapods. By adjusting the amount of gas in its swim bladder, a fish can control the depth at which it remains suspended in the water while expending very little energy to maintain its position.

The outer body surface of most species of ray-finned fishes is covered with flat, thin, lightweight scales that provide some protection or enhance movement through the water. The gills of ray-finned fishes open into a single chamber covered by a hard flap, called an operculum. Movement of the operculum improves the flow of water over the gills, where gas exchange takes place.

Ray-finned fishes began to radiate during the Mesozoic era and continued to radiate extensively throughout the Tertiary period. Today there are about 32,000 known living species, encompassing a remarkable variety of sizes, shapes, and lifestyles (**FIGURE 23.42**). The smallest are less than 1 centimeter long as adults; the largest weigh as much as 900 kilograms. Ray-finned fishes exploit nearly all types of aquatic food sources. In the oceans they filter plankton from the water, rasp algae from rocks, eat corals and other soft-bodied colonial animals, dig animals from soft sediments, and prey on virtually all kinds of other fishes. In fresh water they eat plankton, devour insects, eat fruits that fall into the water in flooded forests, and prey on other aquatic vertebrates and, occasionally, terrestrial vertebrates.

CHECKpoint CONCEPT 23.5

- ✓ What are three developmental patterns the earliest deuterostomes had in common?
- ✓ What are some of the ways that echinoderms use their tube feet to obtain food?
- ✓ What synapomorphies respectively characterize the chordates and the vertebrates?
- ✓ How do the hagfishes differ from the lampreys in morphology and life history? Why do some biologists not consider the hagfishes to be vertebrates?

(A)

A cleaner wrasse (Labroides phthirophagus) feeds on parasites off the body of a much larger brown surgeonfish (Acanthurus nigrofuscus).

(B) *Phyllopteryx taeniolatus*

FIGURE 23.42 Ray-Finned Fishes (A) The two ray-finned species shown here are mutualists in their coral reef ecosystem. The larger surgeonfish illustrates the body plan most commonly associated with this highly diverse clade. (B) This slow-moving weedy seadragon relies on its unusual morphology to provide camouflage in swaying seaweed.

(A) *Latimeria chalumnae*

(B) *Protopterus annectens*

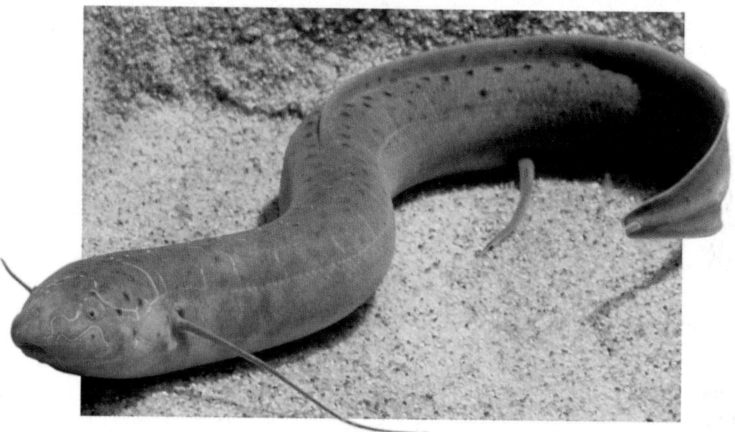

FIGURE 23.43 The Closest Relatives of Tetrapods (A) The African coelacanth, discovered in deep waters of the Indian Ocean, represents one of two surviving species of a group that was once thought to be extinct. (B) All surviving lungfish species, such as this African lungfish, live in the Southern Hemisphere.

In some fishes, the lunglike sacs that gave rise to swim bladders became specialized for another purpose: breathing air. That adaptation set the stage for the vertebrates to move onto the land.

CONCEPT 23.6 Life on Land Contributed to Vertebrate Diversification

The evolution of lunglike sacs in fishes set the stage for the invasion of the land. Some early ray-finned fishes probably used those sacs to supplement their gills when oxygen levels in the water were low, as lungfishes and many groups of ray-finned fishes do today. But with their unjointed fins, those fishes could only flop around on land. Changes in the structure of the fins first allowed some fishes to support themselves better in shallow water and, later, to move better on land.

Jointed fins enhanced support for fishes

In the lobe-limbed vertebrates, the paired pelvic and pectoral fins developed into more muscular fins that were joined to the body by a single enlarged bone. The modern representatives of these lobe-limbed vertebrates include the coelacanths, lungfishes, and tetrapods.

The coelacanths flourished from the Devonian until about 65 million years ago, when they were thought to have become extinct. However, in 1938 a commercial fisherman caught a living coelacanth off South Africa. Since that time, hundreds of individuals of this extraordinary fish, *Latimeria chalumnae*, have been collected. A second species, *L. menadoensis*, was discovered in 1998 off the Indonesian island of Sulawesi. *Latimeria*, a predator of other fishes, reaches a length of about 1.8 meters and weighs up to 82 kilograms (**FIGURE 23.43A**). Its skeleton is composed mostly of cartilage, not bone. The cartilaginous skeleton is a derived feature in this clade because it had bony ancestors.

Go to MEDIA CLIP 23.14
Coelacanths in the Deep Sea
PoL2e.com/mc23.14

Lungfishes were important predators in shallow-water habitats in the Devonian, but most lineages died out. The six surviving species live in stagnant swamps and muddy waters in South America, Africa, and Australia (**FIGURE 23.43B**).

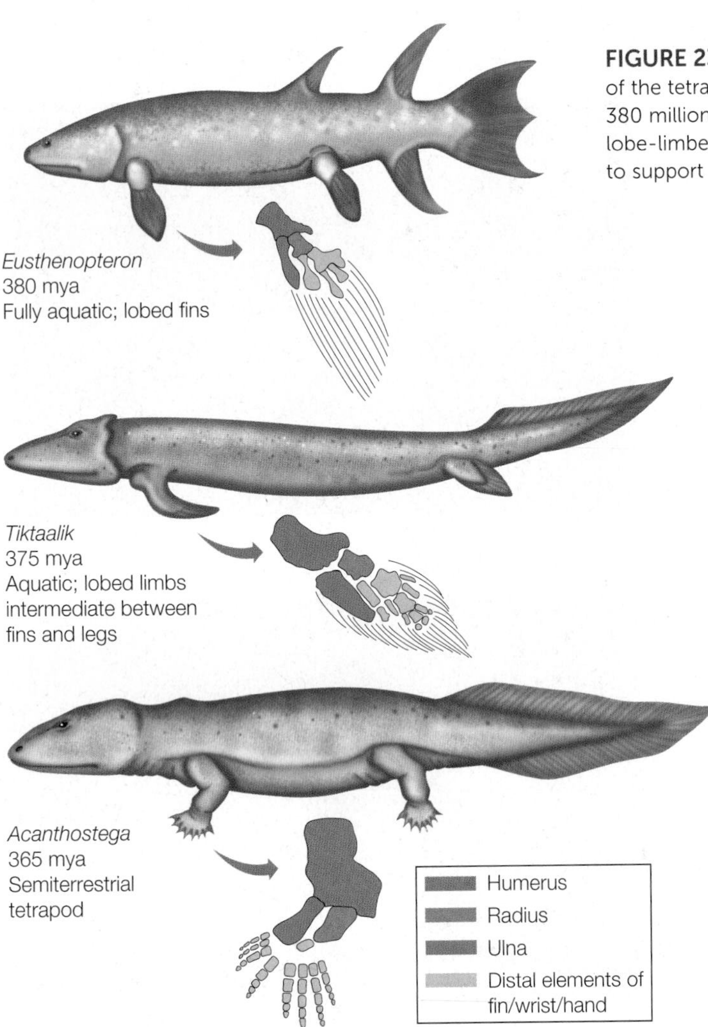

FIGURE 23.44 Tetrapod Limbs Are Modified Fins The major skeletal elements of the tetrapod limb were already present in aquatic lobe-limbed vertebrates some 380 million years ago. The relative sizes and positions of these elements changed as lobe-limbed vertebrates moved to a terrestrial environment, where limbs were used to support and move the animal's body on land.

Lungfishes have lungs derived from the lunglike sacs of their ancestors as well as gills. When ponds dry up, individuals of most species can burrow deep into the mud and survive for many months in an inactive state while breathing air.

Evidence suggests that some early aquatic lobe-limbed vertebrates began to use terrestrial food sources, became more fully adapted to life on land, and eventually evolved to become ancestral **tetrapods** ("four legs"). How was this transition from an animal that swam in water to one that walked on land accomplished? Fossil evidence suggests that limbs able to prop up a large fish and make the front-to-rear movements necessary for walking evolved while these animals still lived in water. These limbs appear to have functioned in holding the animals upright in shallow water, perhaps even allowing them to hold the head above the water's surface. These same structures were then co-opted for movement on land, at first probably for foraging on brief trips out of water (**FIGURE 23.44**). The basic skeletal elements of those limbs can be traced through major changes in limb form and function among the terrestrial vertebrates.

An early split in the tetrapod tree led to two main groups of terrestrial vertebrates: the **amphibians**, most of which remained tied to moist environments, and the **amniotes**, many of which adapted to much drier conditions.

Amphibians adapted to life on land

Most modern amphibians are confined to moist environments because they lose water rapidly through the skin when exposed to dry air. In addition, their eggs are enclosed within delicate membranous envelopes that cannot prevent water loss in dry conditions. In some amphibian species, adults live mostly on land but return to fresh water to lay and fertilize their eggs (**FIGURE 23.45**). The fertilized eggs give rise to larvae that live in water until they undergo metamorphosis to become terrestrial adults. However, many amphibians (especially those in tropical and subtropical areas) have evolved a wide diversity of additional reproductive modes and types of parental care. Internal fertilization, for example, evolved many times among the amphibians. Many species develop directly into adultlike forms from fertilized eggs laid on land or carried by the parents. Other species of amphibians are entirely aquatic, never leaving the water at any stage of their lives, and many of these species retain a larval-like morphology.

Go to ANIMATED TUTORIAL 23.4
Life Cycle of a Frog
PoL2e.com/at23.4

The more than 7,000 known species of amphibians living on Earth today belong to three major groups: the wormlike, limbless, tropical, burrowing or aquatic caecilians (**FIGURE 23.46A**), the tail-less frogs and toads (collectively called anurans; **FIGURE 23.46B**), and the tailed salamanders (**FIGURE 23.46C,D**).

Anurans are most diverse in wet tropical and warm temperate regions, although a few are found at very high latitudes. There are far more anurans than any other amphibians, with well over 6,000 described species and more being discovered every year. Some anurans have tough skins and other adaptations that enable them to live for long periods in very dry deserts, whereas others live in moist terrestrial and arboreal environments. Some species are completely aquatic as adults. All anurans have a very short vertebral column, with a strongly modified pelvic region that is adapted for leaping, hopping, or propelling the body through water by kicking the hind legs.

The more than 600 described species of salamanders are most diverse in temperate regions of the Northern Hemisphere, but many species are also found in cool, moist environments in the mountains of Central America, and a few species penetrate into tropical regions. Many salamanders live in rotting logs or moist soil. One major group has lost lungs, and these species exchange gases entirely through the skin and mouth lining—body parts that all amphibians use in addition to their lungs. A completely aquatic lifestyle has evolved several times among the salamanders (see Figure 23.46D). These aquatic species have arisen through a developmental process known as **neoteny**, or the retention of juvenile traits (such as gills) by delayed somatic development. Most species of salamanders

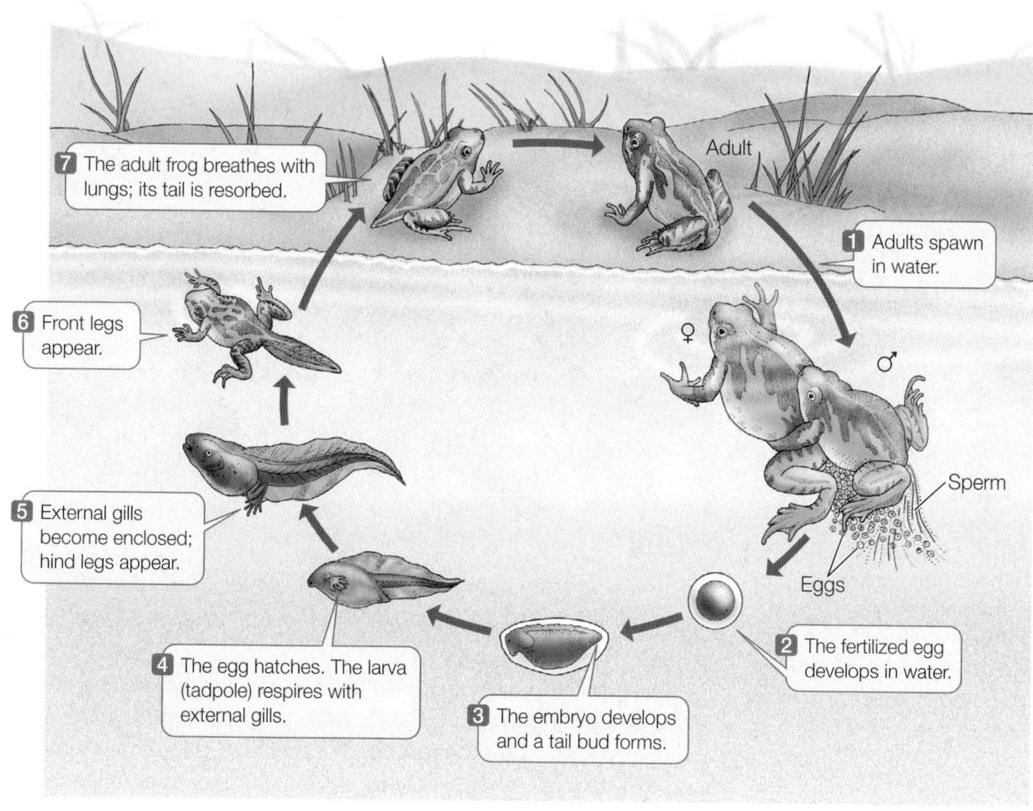

1 Adults spawn in water.
Adult
♀
♂
Sperm
Eggs
2 The fertilized egg develops in water.
3 The embryo develops and a tail bud forms.
4 The egg hatches. The larva (tadpole) respires with external gills.
5 External gills become enclosed; hind legs appear.
6 Front legs appear.
7 The adult frog breathes with lungs; its tail is resorbed.

FIGURE 23.45 In and Out of the Water Most early stages in the life cycle of many amphibians take place in water. The aquatic tadpole transforms into a terrestrial adult through metamorphosis. Some species of amphibians, however, have direct development (with no aquatic larval stage), and others are aquatic throughout life.

have internal fertilization, which is usually achieved through the transfer of a small jellylike, sperm-embedded capsule called a spermatophore.

Many amphibians have complex social behaviors. Most male anurans utter loud, species-specific calls to attract females of their own species (and sometimes to defend breeding territories), and they compete for access to females that arrive at the breeding sites. Many amphibians lay large numbers of eggs, which they abandon once they are deposited and fertilized. Some amphibians lay only a few eggs, which are fertilized

(A) *Siphonops annulatus*

(B) *Bufo periglenes*

(C) *Salamandra salamandra*

(D) *Eurycea waterlooensis*

FIGURE 23.46 Diversity among the Amphibians (A) Burrowing caecilians superficially look more like worms than amphibians. (B) A golden toad pair in Costa Rica. This species has recently become extinct, one of many amphibian species to do so in the past few decades. (C) A European fire salamander. (D) This Austin blind salamander's life cycle remains aquatic; it has no adult terrestrial stage. The eyes of this cave dweller have become greatly reduced.

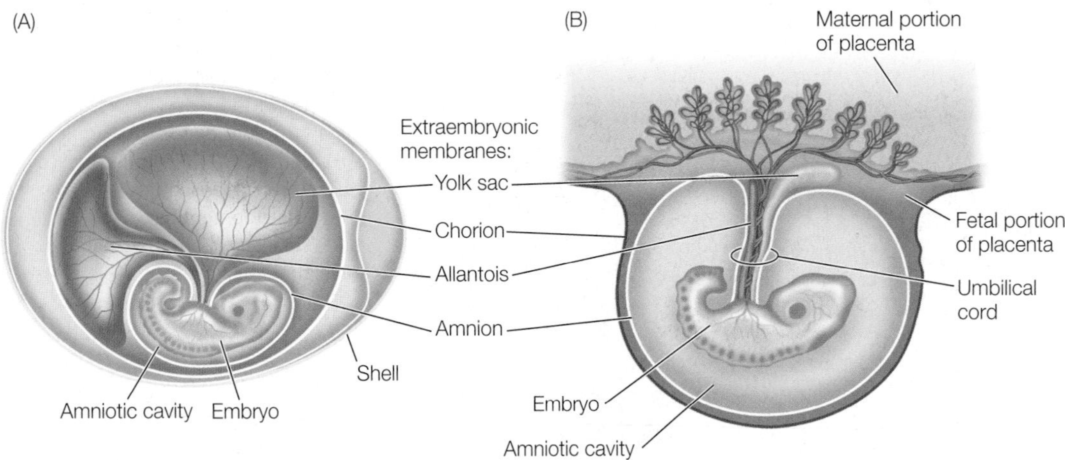

FIGURE 23.47 The Amniote Egg (A) The evolution of the amniote egg, with its water-retaining shell, four extraembryonic membranes, and embryo-nourishing yolk, was a major step in adaptation to the terrestrial environment. A chick egg is shown here. (B) In mammals, the developing embryo is retained inside the mother's body, with which it exchanges nutrients and wastes via the placenta. Note the correspondence between the various membranes in (A) and (B).

Go to ACTIVITY 23.3 The Amniote Egg
PoL2e.com/ac23.3

and then guarded in a nest or carried on the backs, in the vocal pouches, or even in the stomachs of one of the parents. A few species of frogs, salamanders, and caecilians are viviparous, meaning they give birth to well-developed young that have received nutrition from the female during gestation.

Go to MEDIA CLIP 23.15
Answering a Mating Call
PoL2e.com/mc23.15

Amphibians are the focus of much attention today because populations of many species are declining rapidly, especially in mountainous regions of western North America, Central and South America, and northeastern Australia. Worldwide, about one-third of amphibian species are now threatened with extinction or have disappeared completely in the last few decades. Scientists are investigating several hypotheses to account for these population declines, as described in Chapters 1 and 22.

Amniotes colonized dry environments

Several key innovations contributed to the ability of amniotes to exploit a wide range of terrestrial habitats. The **amniote egg** (which gives the group its name) is relatively impermeable to water and allows the embryo to develop in a contained aqueous environment (**FIGURE 23.47A**). The leathery or brittle, calcium-impregnated shell of the amniote egg retards evaporation of the fluids inside but permits passage of oxygen and carbon dioxide. The egg also stores large quantities of food in the form of yolk, allowing the embryo to attain a relatively advanced state of development before it hatches. Within the shell are extraembryonic membranes that protect the embryo from desiccation and assist its gas exchange and excretion of waste nitrogen.

LINK
The roles of the extraembryonic membranes of the amniote egg are discussed in **Concept 38.5**

In several different groups of amniotes, the amniote egg became modified, allowing the embryo to grow inside (and receive nutrition from) the mother. For instance, the mammalian egg lost its shell whereas the functions of the extraembryonic membranes were retained and expanded (**FIGURE 23.47B**).

Other innovations evolved in the organs of terrestrial adults. A tough, impermeable skin, covered with scales or modifications of scales such as hair and feathers, greatly reduced water loss. Adaptations of the vertebrate excretory organs, the kidneys, allowed amniotes to excrete concentrated urine, ridding the body of waste nitrogen without losing a large amount of water in the process.

During the Carboniferous, amniotes split into two major groups: the **reptiles** and the lineage that eventually led to the **mammals** (**FIGURE 23.48**).

Reptiles adapted to life in many habitats

The lineage leading to modern reptiles began to diverge from other amniotes more than 300 million years ago. More than 19,000 species of reptiles exist today, more than half of which are birds. Birds are the only living members of the otherwise extinct dinosaurs, the dominant terrestrial predators of the Mesozoic.

The **lepidosaurs** constitute the second-most species-rich clade of living reptiles. This group is composed of the **squamates** (lizards, snakes, and amphisbaenians—the last a group of mostly legless, wormlike, burrowing reptiles with greatly reduced eyes) and the **tuataras**, which superficially resemble lizards but differ from them in tooth attachment and several internal anatomical features. Many species related to the tuataras lived during the Mesozoic era, but today only two species, restricted to a few islands off New Zealand, survive (**FIGURE 23.49A**).

The skin of a lepidosaur is covered with horny scales that greatly reduce loss of water from the body surface. These scales, however, make the skin unavailable as an organ of gas

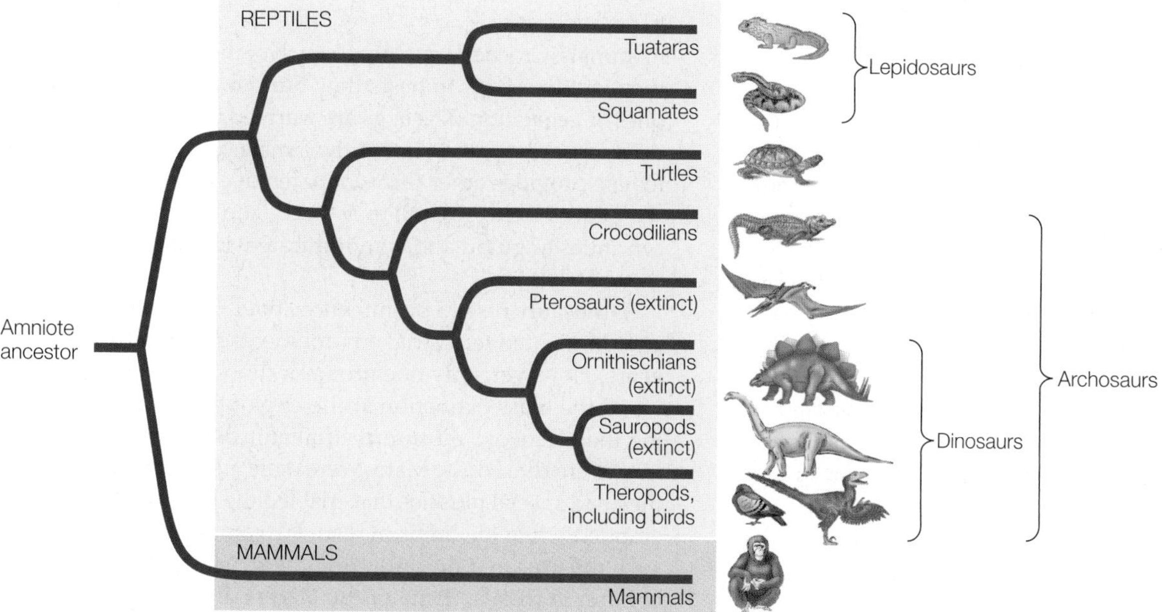

FIGURE 23.48 Phylogeny of Amniotes This tree of amniote relationships shows the primary split between mammals and reptiles. The reptile portion of the tree shows a lineage that led to the lepidosaurs (snakes, lizards, and tuataras), another to the turtles, and a third to the archosaurs (crocodiles, several extinct groups, and the birds).

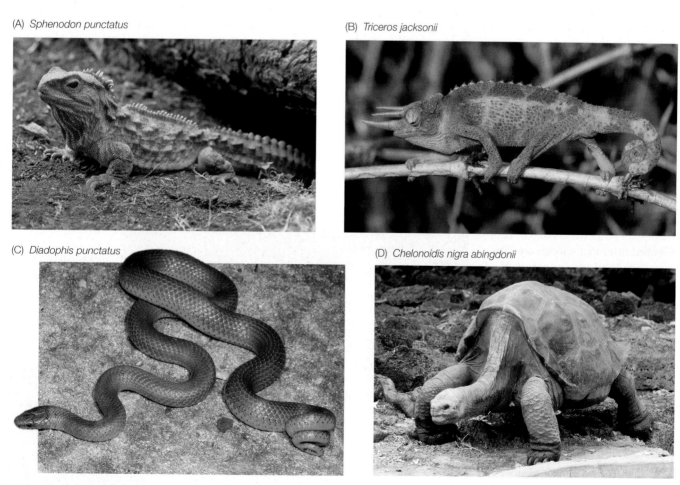

(A) *Sphenodon punctatus*

(B) *Triceros jacksonii*

(C) *Diadophis punctatus*

(D) *Chelonoidis nigra abingdonii*

FIGURE 23.49 Reptilian Diversity (A) The tuatara represents one of only two surviving species in a lineage that diverged long ago. (B) The Jackson's chameleon is native to east Africa but was introduced to Hawaii in 1972. (C) The ringneck snake of North America is harmless to humans. It coils its tail to reveal a bright orange underbelly, which distracts potential predators from the vital head region. (D) Galápagos tortoises are the largest turtles and among the largest reptiles. They have been documented to live for more than 100 years in the wild.

exchange. Gases are exchanged almost entirely via the lungs, which are proportionally much larger in surface area than those of amphibians. A lepidosaur forces air into and out of its lungs by bellowslike movements of its ribs. The three-chambered lepidosaur heart partially separates oxygenated blood from the lungs from deoxygenated blood returning from the body. With this type of heart, lepidosaurs can generate high blood pressure and can sustain a relatively high metabolism.

Most lizards are insectivores, but some are herbivores; a few prey on other vertebrates. Many lizards walk on four limbs (**FIGURE 23.49B**), although limblessness has evolved repeatedly in the group, especially in burrowing and grassland species. The largest lizard, which grows as long as 3 meters and can weigh more than 150 kilograms, is the predaceous Komodo dragon of the East Indies.

Go to MEDIA CLIP 23.16
Komodo Dragons Bring Down Prey
PoL2e.com/mc23.16

One major group of limbless squamates is the snakes (**FIGURE 23.49C**). All snakes are carnivores, and many can swallow objects much larger than themselves. Several snake groups evolved venom glands and the ability to inject venom rapidly into their prey.

The **turtles** comprise a reptilian group that has changed relatively little since the early Mesozoic. In these reptiles, dorsal and ventral bony plates form a shell into which the head and limbs can be withdrawn in many species (**FIGURE 23.49D**). The dorsal shell is a modification of the ribs. It is a mystery how the pectoral girdles evolved to be inside the ribs of turtles, making them unlike any other vertebrates. Most turtles live in aquatic environments, but several groups, such as tortoises and box turtles, are terrestrial. Sea turtles spend their entire lives at sea except when they come ashore to lay eggs. Human exploitation of sea turtles and their eggs has resulted in worldwide declines of these species, all of which are now endangered. A few species of turtles are strict herbivores or carnivores, but most species are omnivores that eat a variety of aquatic and terrestrial plants and animals.

Crocodilians and birds share their ancestry with the dinosaurs

Another reptilian clade, the **archosaurs**, includes the crocodilians, pterosaurs, dinosaurs, and birds. Only the crocodilians and birds are represented by living species today. Modern **crocodilians**—crocodiles, caimans, gharials, and alligators—are confined to tropical and warm temperate environments (**FIGURE 23.50A**). All crocodilians are carnivorous; they eat vertebrates of all kinds, including large mammals. Crocodilians spend much of their time in water, but they lay their eggs in nests they build on land or on floating piles of vegetation. The eggs are warmed by heat generated by decaying organic matter that the female places in the nest. The female provides other forms of parental care as well: typically, she guards the eggs until they hatch, and in some species, she continues to guard and communicate with her offspring after they hatch.

Dinosaurs rose to prominence about 215 million years ago and dominated terrestrial environments for about 150 million years. However, only one group of dinosaurs, the **birds**, survived the mass extinction at the end of the Cretaceous. During the Mesozoic, most terrestrial animals more than a meter long were dinosaurs. Many were agile and could run rapidly; they had special muscles that enabled the lungs to be filled and emptied while the limbs moved. We can infer the existence of such muscles in dinosaurs from the structure of the vertebral column in fossils. Some of the largest dinosaurs weighed as much as 70,000 kilograms.

Biologists have long accepted the phylogenetic position of birds among the reptiles, although birds clearly have many unique, derived morphological features. In addition to the strong morphological evidence for this placement, fossil and molecular data emerging over the last few decades have provided definitive supporting evidence. Birds are a specialized group of **theropods**, a clade of predatory dinosaurs that shared such traits as a bipedal stance, hollow bones, a furcula ("wishbone"), elongated metatarsals with three-fingered feet, elongated forelimbs with three fingers, and a pelvis that points backward. Modern birds are homeothermic, meaning they regulate their body temperatures at a relatively constant temperature through physiological control. Although we cannot directly assess this physiological trait in extinct species, many fossil theropods share morphological traits that suggest they may have been homeothermic as well.

(A) *Caiman crocodilus*

(B) *Struthio camelus*

FIGURE 23.50 Archosaurs (A) Caimans, alligators, and their relatives live in tropical and warm temperate climates. The spectacled caiman is the most common crocodilian throughout much of tropical Central and South America. (B) Birds are the other living archosaur group. This common ostrich is a palaeognath, a group that includes several flightless and weakly flying birds of the Southern Hemisphere. Flightless members of this early-diverging group lack the sternum keel structure that anchors the flight muscles of other bird groups.

(A)

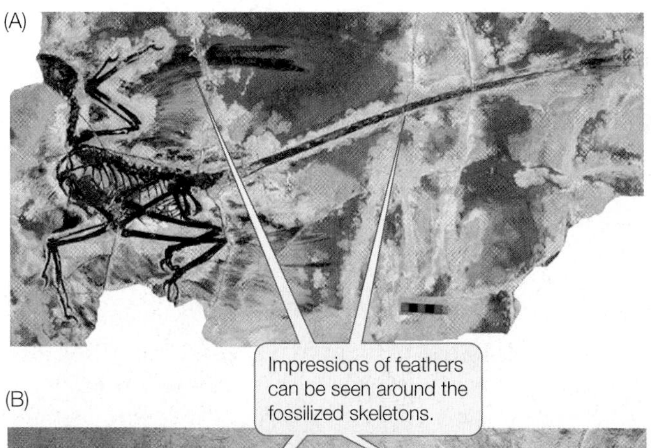

(B)

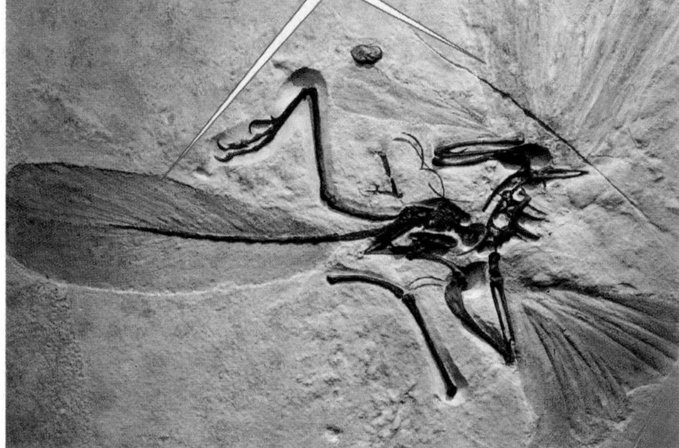

Impressions of feathers can be seen around the fossilized skeletons.

FIGURE 23.51 Mesozoic Bird Relatives The fossil record supports the evolution of birds from other dinosaurs. (A) *Microraptor gui* was a feathered dinosaur from the early Cretaceous (about 140 mya). (B) Dating from roughly the same time frame, *Archaeopteryx* is the oldest known birdlike fossil.

LINK

Homeothermy and its physiological control are described in **Concept 29.3**

The living bird species fall into two major groups that diverged about 80–90 million years ago from a flying ancestor. The few modern descendants of one lineage include a group of secondarily flightless and weakly flying birds, some of which are very large. This group, called the palaeognaths, includes the South and Central American tinamous and several large flightless birds of the southern continents—the rheas, emu, kiwis, cassowaries, and the world's largest bird, the ostrich (**FIGURE 23.50B**). The second lineage, the neognaths, has left a much larger number of descendants, most of which have retained the ability to fly.

The evolution of feathers allowed birds to fly

During the Mesozoic era, about 175 million years ago, a lineage of theropods gave rise to the birds. Recent fossil discoveries show that the scales of some theropod dinosaurs were modified to form feathers. The feathers of many of these dinosaurs were structurally similar to those of modern birds (**FIGURE 23.51**).

The evolution of feathers was a major force for diversification. Feathers are lightweight but are strong and structurally complex. The large quills of the flight feathers on the wings arise from the skin of the forelimbs to create the flying surfaces. Other strong feathers sprout like a fan from the shortened tail and serve as stabilizers during flight. The feathers that cover the body, along with an underlying layer of down feathers, provide birds with insulation that helps them survive in virtually all of Earth's climates.

Go to MEDIA CLIP 23.17
Falcons in Flight
PoL2e.com/mc23.17

The bones of theropod dinosaurs, including birds, are hollow with internal struts that increase their strength. Hollow bones would have made early theropods lighter and more mobile. Later, the hollow bones of theropods facilitated the evolution of flight. The sternum (breastbone) of flying birds forms a large, vertical keel to which the flight muscles are attached.

Flight is metabolically expensive. A flying bird consumes energy at a rate 15–20 times faster than a running lizard of the same weight. Because birds have such high metabolic rates, they generate large amounts of heat. They control the rate of heat loss using their feathers, which may be held close to the body or elevated to alter the amount of insulation they provide. The lungs of birds allow air to flow through unidirectionally rather than by pumping air in and out. This flow-through structure of the lungs increases the efficiency of gas exchange and thereby supports an increased metabolic rate.

LINK

You can read more about the effects of flying on metabolic rate in **Concept 29.2**. See **Concept 31.2**, pp. 652–653, for a description of the breathing system of birds

There are about 10,000 species of living birds, which range in size from the 150-kilogram ostrich to a hummingbird weighing only 2 grams (**FIGURE 23.52**). The teeth so prominent among other dinosaurs were secondarily lost in the ancestral birds, but birds nonetheless eat almost all types of animal and plant material. Insects and fruits are the most important dietary items for terrestrial species. Birds also eat seeds, nectar and pollen, leaves and buds, carrion, and fish and other vertebrates. By eating the fruits and seeds of plants, birds serve as major agents of seed dispersal.

Mammals radiated as non-avian dinosaurs declined in diversity

Small and medium-sized mammals coexisted with the large dinosaurs throughout most of the Mesozoic era, and most of the major groups that are alive today arose in the Cretaceous. After the non-avian dinosaurs disappeared during the mass extinction at the end of the Cretaceous, mammals increased dramatically in numbers, diversity, and size. Today mammals range in size from tiny shrews and bats weighing only about 2 grams to the blue whale, the largest animal on Earth, which measures up to 33 meters long and can weigh as much as 160,000 kilograms. Mammals have far fewer, but more highly differentiated, teeth than do fishes, amphibians, or reptiles. Differences among

(A) *Grus canadensis*

(B) *Aegolius acadicus*

(C) *Piranga olivacea*

FIGURE 23.52 Some Diverse Birds (A) Sandhill cranes feed on plants, insects, small reptiles, and even small mammals. As in other crane species, their mating and courtship behavior includes an elaborate dance display. (B) Owls such as this northern saw-whet owl are nighttime predators that can locate prey using their sensitive auditory systems. (C) Perching, or passeriform, birds such as this scarlet tanager comprise the most species-rich group of birds.

mammals in the number, type, and arrangement of teeth reflect their varied diets.

Four key features distinguish the mammals:

- Sweat glands, which secrete sweat that evaporates and thereby cools an animal
- Mammary glands, which in females secrete a nutritive fluid (milk) on which newborn individuals feed
- Hair, which provides a protective and insulating covering
- A four-chambered heart that completely separates the oxygenated blood coming from the lungs from the deoxygenated blood returning from the body (this last characteristic evolved independently in the archosaurs, including modern birds and crocodiles)

Mammalian eggs are fertilized within the female's body, and the embryos undergo a period of development in the female's body in an organ called the uterus. In the uterus, the embryo is contained in an amniotic sac that is homologous to one of four membranes found in the amniotic egg (see Figure 23.47). The embryo is connected to the wall of the uterus by an organ called a placenta. The placenta allows for nutrient and gas exchange, as well as waste elimination from the developing embryo, via the female's circulatory system. Most mammals have a covering of hair (fur), which is luxuriant in some species but has been greatly reduced in others, including the cetaceans (whales and dolphins) and humans. Thick layers of insulating fat (blubber) replace hair as a heat-retention mechanism in the cetaceans. Humans learned to use clothing for heat retention when they dispersed from warm tropical areas.

The approximately 5,700 species of living mammals are divided into two primary groups: the **prototherians** and the **therians**. Only five species of prototherians are known, and they are found only in Australia and New Guinea. These mammals, the duck-billed platypus and four species of echidnas, differ from other mammals in lacking a placenta, laying eggs, and having sprawling legs (**FIGURE 23.53A**). Prototherians supply milk for their young, but they have no nipples on their mammary glands. The milk simply oozes out and is lapped off the fur by the offspring.

Most mammals are viviparous

Members of the viviparous therian clade are further divided into the **marsupials** and the **eutherians**. Females of most marsupial species have a ventral pouch in which they carry and feed their offspring (**FIGURE 23.53B**). Gestation (pregnancy) in marsupials is brief; the young are born tiny but with well-developed forelimbs, with which they climb to the pouch. They attach to a nipple but cannot suck. The mother ejects milk into the tiny offspring until they grow large enough to suckle. Once her offspring have left the uterus, a female marsupial may become sexually receptive again. She can then carry fertilized eggs that are capable of initiating development and can replace the offspring in her pouch should something happen to them.

Eutherians include the majority of mammals. Eutherians are sometimes called placental mammals, but this name is inappropriate because some marsupials also have placentas. Eutherians are more developed at birth than are marsupials, and no external pouch houses them after they are born.

The more than 5,300 living species of morphologically diverse eutherians are divided into 20 major groups (**TABLE 23.3**). The relationships of the major groups of eutherians to one another have been difficult to determine because most of the major groups diverged in a short period of time during an explosive adaptive radiation. The two most diverse groups are the rodents (**FIGURE 23.53C**) and bats (**FIGURE 23.53D**), which together comprise about two-thirds of the species. Rodents are traditionally defined by the unique morphology of their teeth,

(A) *Ornithorhynchus anatinus*

(B) *Didelphis virginiana*

(C) *Erethizon dorsatum*

(D) *Leptonycteris curasoae*

(E)

(F) *Tursiops truncatus*

FIGURE 23.53 Mammalian Diversity (A) The duck-billed platypus is one of the five surviving species of prototherians—mammals that lay eggs. (B) Female marsupial mammals have a ventral pouch in which they nurture their offspring, which are extremely small at birth. The young of this opossum have grown large enough to leave her pouch. (C–F) Eutherian mammals. (C) Almost half of all eutherians are rodents, such as this North American porcupine. (D) Flight evolved in the common ancestor of bats. Virtually all bat species are nocturnal. (E) Many large mammals are important herbivores in terrestrial environments. Nowhere are these assemblages more spectacular than on the grass and brushlands of the African continent. (F) Bottlenose dolphins are cetaceans, a cetartiodactyl group that returned to the marine environment.

which are adapted for gnawing through substances such as wood. The bats probably owe much of their success to the evolution of flight, which allows them to exploit a variety of food sources and colonize remote locations with relative ease.

Go to MEDIA CLIP 23.18
Bats Feeding in Flight
PoL2e.com/mc23.18

Grazing and browsing by members of several eutherian groups helped transform the terrestrial landscape. Herds of grazing herbivores feed on open grasslands, whereas browsers feed on shrubs and trees. The effects of herbivores on plant life favored the evolution of the spines, tough leaves, and difficult-to-eat growth forms found in many plants. In turn, adaptations to the teeth and digestive systems of many herbivore lineages allowed these species to consume many plants despite such defenses—a striking example of coevolution. A large animal can survive on food of lower quality than a small animal can, and large size evolved in several groups of grazing and browsing mammals (**FIGURE 23.53E**). The evolution of large herbivores, in turn, favored the evolution of large carnivores able to attack and overpower them.

TABLE 23.3 Major Groups of Living Eutherian Mammals

Group	Number of described living species	Examples
Rodents (Rodentia)	2,337	Rats, mice, squirrels, woodchucks, ground squirrels, beaver, capybara
Bats (Chiroptera)	1,171	Fruit bats, echo-locating bats
Even-toed hoofed mammals and cetaceans (Cetartiodactyla)	469	Deer, sheep, goats, cattle, antelopes, giraffes, camels, swine, hippopotamus, whales, dolphins
Shrews, moles, and relatives (Soricomorpha)	428	Shrews, moles, solenodons
Primates (Primates)	396	Lemurs, monkeys, apes, humans
Carnivores (Carnivora)	284	Wolves, dogs, bears, cats, weasels, pinnipeds (seals, sea lions, walruses)
Rabbits and relatives (Lagomorpha)	92	Rabbits, hares, pikas
African insectivores (Afrosoricida)	50	Tenrecs, golden moles
Spiny insectivores (Erinaceomorpha)	24	Hedgehogs
Armadillos (Cingulata)	21	Armadillos
Tree shrews (Scandentia)	20	Tree shrews
Odd-toed hoofed mammals (Perissodactyla)	16	Horses, zebras, tapirs, rhinoceroses
Long-nosed insectivores (Macroscelidea)	15	Elephant shrews
Anteaters and sloths (Pilosa)	10	Anteaters, tamanduas, sloths
Pangolins (Pholidota)	8	Asian and African pangolins
Hyraxes and relatives (Hyracoidea)	5	Hyraxes, dassies
Sirenians (Sirenia)	4	Manatees, dugongs
Elephants (Proboscidea)	3	African and Indian elephants
Colugos (Dermoptera)	2	Flying lemurs
Aardvark (Tubulidentata)	1	Aardvark

Several lineages of terrestrial eutherians subsequently returned to the aquatic environments their ancestors had left behind (**FIGURE 23.53F**). The completely aquatic cetaceans—whales and dolphins—evolved from artiodactyl ancestors (whales are closely related to the hippopotamuses). The seals, sea lions, and walruses also returned to the marine environment, and their limbs became modified into flippers. Weasel-like otters retain their limbs but have also returned to aquatic environments, colonizing both fresh and salt water. The manatees and dugongs colonized estuaries and shallow seas.

CHECKpoint CONCEPT 23.6

- ✓ Many amphibians have a biphasic life cycle that includes an aquatic larva and a terrestrial adult. What are some exceptions to this common pattern?
- ✓ In the not-too-distant past, the idea that birds were reptiles met with skepticism. Explain how fossils, morphology, and molecular evidence now support the position of birds among the reptiles.
- ✓ Contrast the reproductive modes of prototherians, marsupials, and eutherians.

The biology of one eutherian group—the primates—has been the subject of extensive research. The behavior, ecology, physiology, and molecular biology of the primates are of special interest to us because this lineage includes humans.

CONCEPT 23.7 Humans Evolved among the Primates

The **primates** evolved from a lineage of relatively small, arboreal, insectivorous eutherians. Grasping limbs with opposable digits are one of the major adaptations to arboreal life that distinguish primates from other mammals. Primates underwent extensive evolutionary radiation throughout the Tertiary (**FIGURE 23.54**).

Two major lineages of primates split late in the Cretaceous

About 90 million years ago, late in the Cretaceous period, the primates split into two main clades: the prosimians and the anthropoids. **Prosimians**—lemurs, lorises, and galagos—once lived on all continents, but today they are restricted to Africa, Madagascar, and tropical Asia. Most prosimians are arboreal and noctural. There has been a remarkable radiation of lemurs on Madagasgar, where there are also diurnal and terrestrial species. Tarsiers were once considered prosimians as well, although today we know that they are more closely related to monkeys and apes than to lemurs, lorises, and galagos.

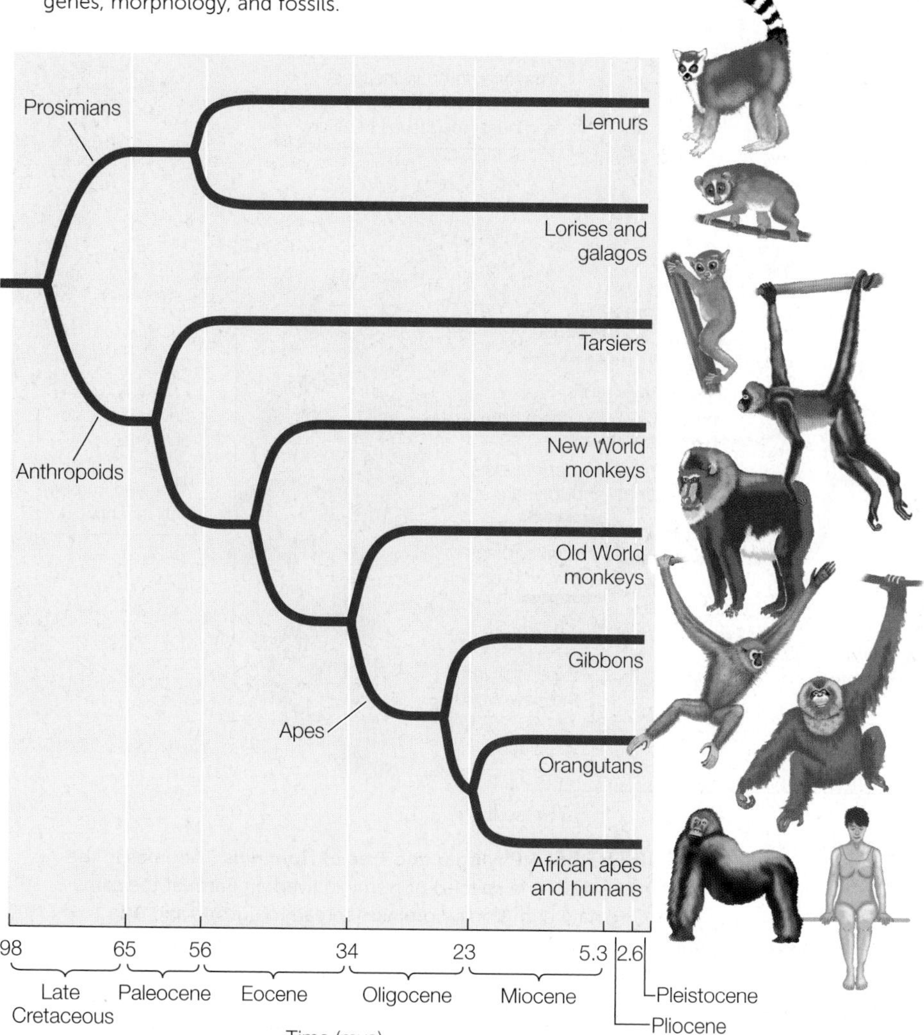

FIGURE 23.54 Phylogeny of the Primates
The phylogeny of primates is among the best studied of any major group of mammals. This tree is based on evidence from many genes, morphology, and fossils.

The second primate lineage, the **anthropoids**—tarsiers, New World monkeys, Old World monkeys, and apes—began to diversify shortly after the mass extinction event at the end of the Cretaceous, in Africa and Asia. New World monkeys diverged from Old World monkeys and apes slightly later, but early enough that they might have reached South America from Africa when those two continents were still close to each other. All New World monkeys now live only in South and Central America, and all of them are arboreal. Many of them have long, prehensile tails with which they can grasp branches. Many Old World monkeys are arboreal as well, but a number of species are terrestrial. No Old World primate has a prehensile (grasping) tail.

About 35 million years ago, a lineage that led to the apes separated from the Old World monkeys. Between 22 and 5.5 million years ago, dozens of species of apes lived in Europe, Asia, and Africa. The Asian apes (gibbons and orangutans), African apes (gorillas and chimpanzees), and humans are their modern descendants.

Bipedal locomotion evolved in human ancestors

About 6 million years ago in Africa, a lineage split occurred that would lead to the chimpanzees on the one hand and to the **hominid** clade, which includes modern humans and their extinct close relatives, on the other.

The earliest protohominids, known as ardipithecines, had distinct morphological adaptations for **bipedal locomotion** (walking on two legs). Bipedal locomotion frees the forelimbs to manipulate objects and to carry them while walking. It also elevates the eyes, enabling the animal to see over tall vegetation to spot predators and prey. Bipedal locomotion is also energetically more economical than quadrupedal locomotion (walking on four legs). All three advantages were probably important for the ardipithecines and their descendants, the australopithecines (**FIGURE 23.55**).

The first australopithecine skull was found in South Africa in 1924. Since then australopithecine fossils have been found at many sites in Africa. The most complete fossil skeleton yet found was discovered in Ethiopia in 1974. The skeleton, approximately 3.5 million years old, was that of a young female who has since become known to the world as "Lucy." Lucy was assigned to the species *Australopithecus afarensis*. Fossil remains of more than 100 *A. afarensis* individuals have since been discovered, and there have been recent discoveries of fossils of other australopithecine species that lived in Africa 4–5 million years ago.

Experts disagree over how many species are represented by australopithecine fossils, but it is clear that multiple species of hominids lived together over much of eastern Africa several million years ago. A lineage of larger species (weighing about 40 kilograms) is represented by *Paranthropus robustus* and *P. boisei*, both of which died out between 1 and 1.5 million years ago. A lineage of smaller australopithecines gave rise to the genus *Homo*.

Early members of the genus *Homo* lived contemporaneously with *Paranthropus* in Africa for about a million years. Some 2-million-year-old fossils of an extinct species called *H. habilis* were discovered in the Olduvai Gorge, Tanzania. Other fossils of *H. habilis* have been found in Kenya and Ethiopia. Associated with these fossils were tools that these early hominids used to obtain food.

Another extinct hominid species, *Homo erectus*, arose in Africa about 1.6 million years ago. Soon thereafter it had spread as far as eastern Asia, becoming the first hominid to leave Africa. Members of *H. erectus* were nearly as large as modern people, but their brains were smaller and they had comparatively thick

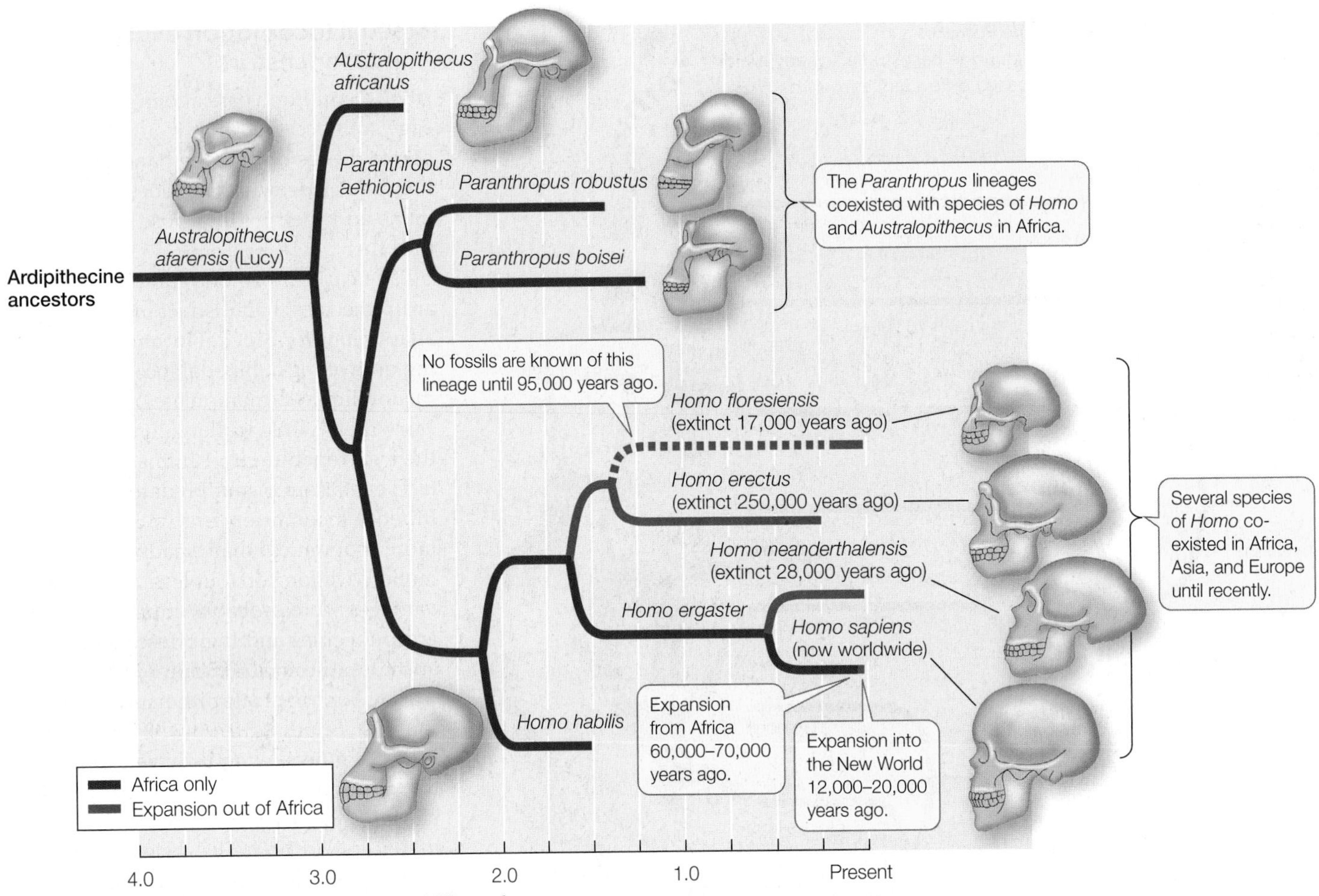

FIGURE 23.55 A Phylogenetic Tree of Hominids At times in the past, more than one species of hominid lived on Earth at the same time. Originating in Africa, hominids spread to Europe and Asia multiple times. All but one of these closely related species are now extinct; modern *Homo sapiens* have colonized nearly every corner of the planet.

skulls. The cranium, which had thick, bony walls, may have been an adaptation to protect the brain, ears, and eyes from impacts caused by a fall or a blow from a blunt object. What would have been the source of such blows? Fighting with other *H. erectus* individuals is a possible answer.

Homo erectus used fire for cooking and for hunting large animals, and made characteristic stone tools that have been found in many parts of Africa and Asia. Populations of *H. erectus* survived until at least 250,000 years ago, although more recent fossils may also be attributable to this species. In 2004 some 18,000-year-old fossil remains of a small *Homo* were found on the island of Flores in Indonesia. Since then, numerous additional fossils of this diminutive hominid have been found on Flores, dating from 95,000 to 17,000 years ago. Many anthropologists think that this small species, named *H. floresiensis*, was most closely related to *H. erectus*.

Human brains became larger as jaws became smaller

In another hominid lineage that diverged from *H. erectus* and *H. floresiensis*, the brain increased rapidly in size, and the jaw muscles, which were large and powerful in earlier hominids, dramatically decreased in size. These two changes were simultaneous, which suggests that they might have been developmentally linked. These changes are another example of evolution by neoteny (which, you may recall from our discussion of amphibians, is the retention of juvenile traits through delayed somatic development). Human and chimpanzee skulls are similar in shape at birth, but chimpanzee skulls undergo a dramatic change in shape as the animals mature (**FIGURE 23.56**). In particular, the jaw grows considerably in relation to the brain case. As human skulls grow, relative proportions much closer to those of the juvenile skull are retained, which results in a large brain case and small jaw compared with those of chimpanzees. A mutation in a regulatory gene that is expressed only in the head may have removed a barrier that had previously prevented this remodeling of the human cranium.

The striking enlargement of the brain relative to body size in the hominid lineage was probably favored by an increasingly complex social life. Any features that allowed group members to communicate more effectively with one another would have been valuable in cooperative hunting and gathering as well as for improving one's status in the complex social interactions that must have characterized early human societies, just as they do ours today.

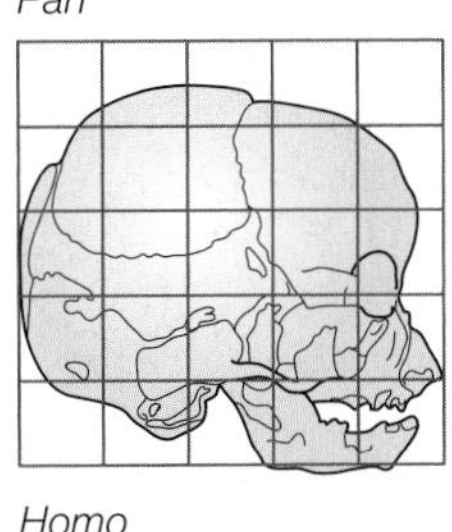

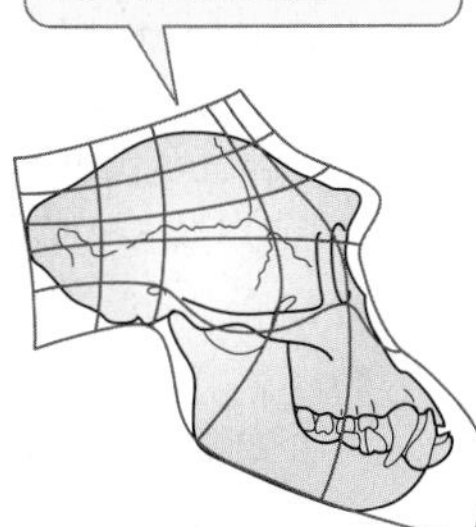

The chimp skull undergoes much more dramatic change in development than does the human skull.

The skulls of juvenile chimpanzees and humans are quite similar to each other.

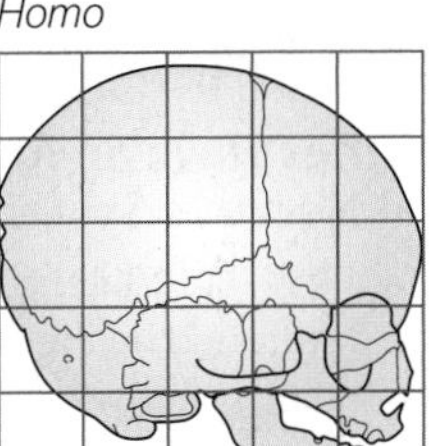

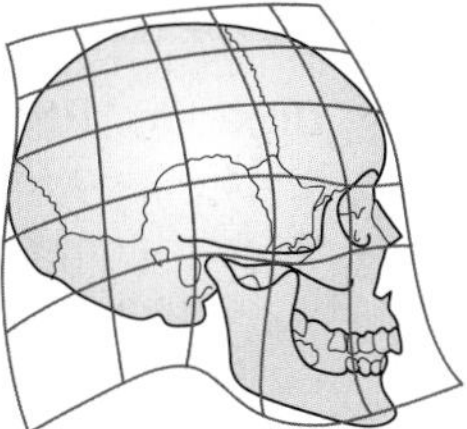

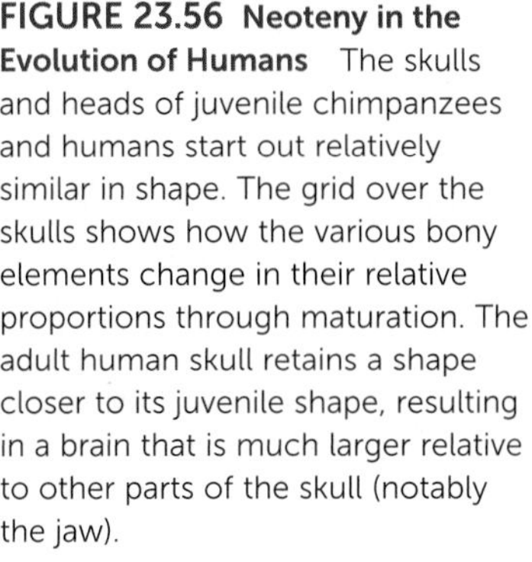

FIGURE 23.56 Neoteny in the Evolution of Humans The skulls and heads of juvenile chimpanzees and humans start out relatively similar in shape. The grid over the skulls shows how the various bony elements change in their relative proportions through maturation. The adult human skull retains a shape closer to its juvenile shape, resulting in a brain that is much larger relative to other parts of the skull (notably the jaw).

Several *Homo* species coexisted during the mid-Pleistocene, from about 1.5 million to about 250,000 years ago. All were skilled hunters of large mammals, but plants were important components of their diets as well. During this period another distinctly human trait emerged: rituals and a concept of life after death. Deceased individuals were buried with tools and clothing, supplies for their presumed existence in the next world.

One species, *Homo neanderthalensis*, was widespread in Europe and Asia between about 500,000 and 28,000 years ago. Neanderthals were short, stocky, and powerfully built. Their massive skull housed a brain somewhat larger than our own. They manufactured a variety of tools and hunted large mammals, which they probably ambushed and subdued in close combat. Early modern humans (*H. sapiens*) expanded out of Africa between 70,000 and 60,000 years ago. Then, about 35,000 years ago, *H. sapiens* moved into the range of *H. neanderthalensis* in Europe and western Asia, so the two species must have interacted with each other. Neanderthals abruptly disappeared about 28,000 years ago. Many anthropologists believe that Neanderthals were exterminated by those early modern humans. Scientists have been able to isolate large parts of the genome of *H. neanderthalensis* from recent fossils and to compare it with our own. These studies suggest that there was some limited interbreeding between the two species while they occupied the same range. As a result, in humans with Eurasian ancestry, 1–4 percent of the genes in their genomes may be derived from Neanderthal ancestors.

Early modern humans made and used a variety of sophisticated tools. They created the remarkable paintings of large mammals, many of them showing scenes of hunting, found in European caves. The animals they depicted were characteristic of the cold steppes and grasslands that occupied much of Europe during periods of glacial expansion. Early modern humans also spread across Asia, reaching North America perhaps as early as 20,000 years ago, although the date of their arrival in the Americas is still uncertain. Within a few thousand years, they had spread southward through North America to the southern tip of South America.

CHECKpoint CONCEPT 23.7

- ✓ Describe the differences between Old World and New World monkeys.
- ✓ What leads biologists to conclude that the evolution of brain size and jaw size may be functionally linked?
- ✓ What evidence leads biologists to think that limited interbreeding between *Homo sapiens* and *H. neanderthalensis* took place after *H. sapiens* expanded out of Africa?

Besides beetles, which other groups of animals are thought to contain many more species than are known at present?

ANSWER It is perhaps easier to list the groups of animals for which a nearly complete inventory of living species has been completed than to list all the groups for which many new species remain to be described. Among the insects, the best-studied group in terms of species is the butterflies, which are widely collected and studied. There are still many species of other lepidopterans (such as moths), however, remaining to be discovered. Most other major insect groups contain many undescribed species. Among the vertebrates, most species of birds and mammals have been described, although several new species of mammals are still discovered and named each year. Species discovery and description remains high for almost all other major groups of animals.

After insects, and perhaps even rivaling the insects in undiscovered diversity, are the nematodes. Although known nematode diversity is only about 1/40th of the known insect diversity (in terms of number of described species), the taxonomy of nematodes is much more poorly studied than that of insects. Some biologists think there are likely to be species-specific parasitic nematodes of most other species of multicellular organisms. If so, then there may be as many species of nematodes as there are of plants, fungi, and other animals combined.

Most of the other diverse groups of animals contain many as yet undetected species, judging from the rate of new species descriptions. In particular, flatworms (especially the parasitic flukes and tapeworms), marine annelids, mollusks, crustaceans, myriapods, and chelicerates all contain large numbers of undescribed species. Among the vertebrates, the rate of discovery of new species remains particularly high for the fishes and amphibians.

SUMMARY

CONCEPT 23.1 Distinct Body Plans Evolved among the Animals

- Animals share a set of derived traits not found in other groups of organisms, including similarities in the sequences of many of their genes, the structure of their cell junctions, and the components of their extracellular matrix. **Review Figure 23.1**
- Patterns of embryonic development provide clues to the evolutionary relationships among animals. **Diploblastic** animals develop two embryonic cell layers; **triploblastic** animals develop three cell layers.
- Differences in their patterns of early development characterize two major clades of triploblastic animals, the **protostomes** and the **deuterostomes**.
- Animal **body plans** can be described in terms of **symmetry**, **body cavity** structure, **segmentation**, and type of appendages.
- Most animals have either **radial symmetry** or **bilateral symmetry**. Many bilaterally symmetrical animals exhibit **cephalization**, with sensory and nervous tissues in an anterior head.
- On the basis of their body cavity structure, animals can be described as **acoelomates**, **pseudocoelomates**, or **coelomates**. **Review Figure 23.4 and ACTIVITY 23.1**
- Segmentation takes many forms and improves control of movement, especially if the animal also has appendages.

CONCEPT 23.2 Some Animal Groups Fall outside the Bilateria

- All animals other than **sponges**, **ctenophores**, **placozoans**, and **cnidarians** belong to a large, monophyletic group called the **Bilateria**. **Eumetazoans**, which have tissues organized into distinct organs, include all animals other than sponges. **Review ACTIVITY 23.2**
- Sponges are simple, asymmetrical animals that lack differentiated cell layers and true organs. Sponges have skeletons made up of silicaceous or calcareous **spicules**. They create water currents and capture food with flagellated feeding cells called **choanocytes**. Choanocytes are an evolutionary link between the animals and the choanoflagellate protists.
- The two cell layers of the radially symmetrical ctenophores are separated by an inert extracellular matrix called **mesoglea**. Ctenophores move by beating comblike plates of cilia called **ctenes**. **Review Figure 23.7**
- The life cycle of most cnidarians has two distinct stages: a sessile **polyp** stage and a motile **medusa**. A fertilized egg develops into a free-swimming larval **planula**, which settles to the bottom and develops into a polyp. **Review Figure 23.9 and ANIMATED TUTORIAL 23.1**

CONCEPT 23.3 Protostomes Have an Anterior Brain and a Ventral Nervous System

- Protostomes ("mouth first") are bilaterally symmetrical animals that have an anterior brain surrounding the entrance to the digestive tract and a ventral nervous system. Protostomes comprise two major clades, the lophotrochozoans and the ecdysozoans. **Review ANIMATED TUTORIAL 23.2**
- **Lophotrochozoans** include a wide diversity of animals. Within this group evolved **lophophores** (a complex organ for food collection and gas exchange) and free-living **trochophore** larvae. These features were subsequently lost in some lineages. The exact placement of **arrow worms** within the protostomes remains uncertain.
- Lophophores, wormlike body forms, and external shells are each found in many distantly related groups of lophotrochozoans. The most species-rich groups of lophotrochozoans are the **flatworms**, **annelids**, and **mollusks**.
- Annelids are a diverse group of segmented worms that live in moist terrestrial and aquatic environments. **Review Figures 23.17 and 23.18**
- Mollusks underwent a dramatic evolutionary radiation based on a body plan consisting of three major components: a **foot**, a **mantle**, and a **visceral mass**. The four major living molluscan clades—**chitons**, **bivalves**, **gastropods**, and **cephalopods**—demonstrate the diversity that evolved from this three-part body plan. **Review Figure 23.19**
- **Ecdysozoans** have a **cuticle** covering their body, which they must **molt** in order to grow. Some ecdysozoans, notably the **arthropods**, have a rigid cuticle reinforced with **chitin** that functions as an **exoskeleton**. New mechanisms of locomotion and gas exchange evolved among the arthropods.
- **Nematodes**, or roundworms, have a thick, multilayered cuticle. Nematodes are among the most abundant and universally distributed of all animal groups. **Review Figure 23.22**
- **Horsehair worms** are extremely thin; their larvae are internal parasites.
- Many ecdysozoan groups are wormlike in form. Members of several species-poor groups of marine ecdysozoans—**priapulids**, **kinorhynchs**, and **loriciferans**—have thin cuticles.
- One major ecdysozoan clade, the arthropods, has evolved jointed, paired appendages that have a wide diversity of functions.

CONCEPT 23.4 Arthropods Are Diverse and Abundant Animals

- Arthropods are the dominant animals on Earth in number of described species, and among the most abundant in number of individuals.

SUMMARY *(continued)*

- Encasement within a rigid exoskeleton provides arthropods with support for walking as well as some protection from predators. The waterproofing provided by chitin keeps arthropods from dehydrating in dry air.
- Jointed appendages permit complex movements. Each arthropod segment has muscles attached to the inside of the exoskeleton that operate that segment and the appendages attached to it.
- Two groups of arthropod relatives, the **velvet worms** and the **tardigrades**, have simple, unjointed appendages.
- **Chelicerates** have a two-part body and pointed mouthparts that grasp prey; most chelicerates have four pairs of walking legs.
- Mandibles and antennae are synapomorphies of the **mandibulates**, which include **myriapods**, crustaceans, and hexapods.
- **Crustaceans** are the dominant marine arthropods and are also found in many freshwater and some terrestrial environments. Their segmented bodies are divided into three regions (head, thorax, and abdomen) with different, specialized appendages in each region.
- **Hexapods**—insects and their relatives—are the dominant terrestrial arthropods. They have the same three body regions as crustaceans, but no appendages form in their abdominal segments. Wings and the ability to fly first evolved among the insects, allowing them to exploit new lifestyles. **Review Figure 23.30**

CONCEPT 23.5 Deuterostomes Include Echinoderms, Hemichordates, and Chordates

- Deuterostomes vary greatly in adult form, but based on the distinctive patterns of early development they share and on phylogenetic analyses of gene sequences, they represent a monophyletic group. There are far fewer species of deuterostomes than of protostomes, but many deuterostomes are large and ecologically important. **Review ANIMATED TUTORIAL 23.3**
- **Echinoderms** and **hemichordates** both have bilaterally symmetrical, ciliated larvae.
- Most adult echinoderms have **pentaradial symmetry**. Echinoderms have an internal skeleton of calcified plates and a unique **water vascular system** connected to extensions called **tube feet**. **Review Figure 23.32**
- Hemichordate adults are bilaterally symmetrical and have a three-part body that is divided into a proboscis, collar, and trunk. They include the acorn worms and the pterobranchs. **Review Figure 23.34**
- Chordates fall into three principal subgroups: **lancelets**, **tunicates**, and **vertebrates**.
- At some stage in their development, all chordates have a dorsal hollow nerve cord, a postanal tail, and a **notochord**. **Review Figure 23.35**
- Tunicates include the sea squirts, which are sessile filter feeders as adults. Lancelets live buried in the sand of shallow marine and brackish waters.
- The vertebrate body is characterized by a rigid internal skeleton, which is supported by a **vertebral column** that replaces the notochord, internal organs suspended in a coelom, a ventral heart, and an anterior skull with a large brain. **Review Figure 23.37**
- The evolution of jaws from gill arches enabled individuals to grasp large prey and, together with teeth, cut them into small pieces. **Review Figure 23.40**
- **Chondrichthyans** have skeletons of cartilage; almost all species are marine. The skeletons of bony vertebrates are made of bone. Among the bony vertebrates, **ray-finned fishes** have colonized most aquatic environments.

CONCEPT 23.6 Life on Land Contributed to Vertebrate Diversification

- Lungs and jointed appendages enabled one lineage of **lobe-limbed vertebrates** to colonize the land. This lineage gave rise to the **tetrapods**. **Review Figure 23.44**
- The earliest split in the tetrapod tree is between the **amphibians** and the **amniotes** (**reptiles** and **mammals**).
- Most modern amphibians are confined to moist environments because they and their eggs lose water rapidly. **Review Figure 23.45 and ANIMATED TUTORIAL 23.4**
- An impermeable skin, efficient kidneys, and an egg that could resist desiccation evolved in the amniotes. **Review Figure 23.47 and ACTIVITY 23.3**
- The major living reptile groups are the **lepidosaurs** (**tuataras**, lizards, snakes, and amphisbaenas), **turtles**, and **archosaurs** (**crocodilians** and **birds**). **Review Figure 23.48**
- **Mammals** are unique among animals in supplying their young with a nutritive fluid (milk) secreted by mammary glands. There are two primary mammalian clades: the **prototherians** (of which there are only five species) and the species-rich **therians**. The therian clade is subdivided into the **marsupials** and the **eutherians**. **Review Table 23.3**

CONCEPT 23.7 Humans Evolved among the Primates

- Grasping limbs with opposable digits distinguish **primates** from other mammals. The **prosimian** clade includes the lemurs, lorises, and galagos; the **anthropoid** clade includes tarsiers, monkeys, apes, and humans. **Review Figure 23.54**
- **Hominid** ancestors developed efficient **bipedal locomotion**. Several species of *Homo* coexisted in parts of the world until recently. **Review Figure 23.55**
- In the lineage leading to *Homo*, brains became larger as jaws became smaller. The two events appear to be developmentally linked and are an example of evolution via **neoteny**. **Review Figure 23.56**

See **ACTIVITY 23.4** for a concept review of the major groups of organisms.

Go to the Interactive Summary to review key figures, Animated Tutorials, and Activities PoL2e.com/is23

Go to LaunchPad at **macmillanhighered.com/launchpad** for additional resources, including LearningCurve Quizzes, Flashcards, and many other study and review resources.

PART 5

Plant Form and Function

24 The Plant Body

KEY CONCEPTS

24.1 The Plant Body Is Organized and Constructed in a Distinctive Way

24.2 Apical Meristems Build the Primary Plant Body

24.3 Many Eudicot Stems and Roots Undergo Secondary Growth

24.4 Domestication Has Altered Plant Form

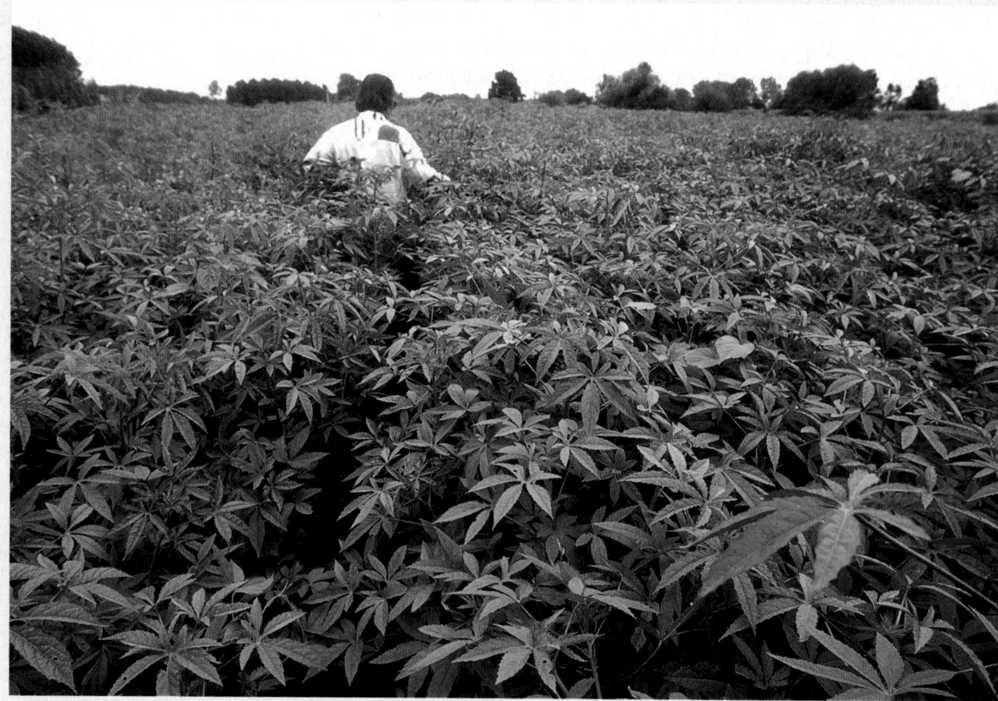

Kenaf is being grown as an alternative source of pulp for paper production.

People use plants and substances made from them not only for food but also for fiber, fuel, shelter, and medicines. The clothes you are wearing probably contain cotton, and the paper this book is printed on comes from wood pulp. About 5,000 years ago, Egyptians found that when stems of the papyrus plant (*Cyperus papyrus*) were peeled and dried, they could be formed into sheets that were easier to write on and store than parchment made from animal skins. About 3,000 years later, the Chinese invented a new method of papermaking that persists today: they mixed fibrous plant materials with water, mashed the mixture into a pulp, and pressed it into sheets, which they then dried.

With the advent of computers, economists predicted a lower demand for paper and therefore for the trees that provide the wood pulp used in papermaking. But this has not happened. Every year, U.S. residents use about 500 kilograms of paper per person, and paper constitutes about one-third of all solid waste. In both Europe and the United States, 90 percent of the forest cover that existed when papermaking was invented has disappeared because of the demand for fiber for paper as well as wood for energy production, heating, and construction. To regenerate these forests and to provide a sustainable resource for papermaking, there is considerable interest in growing fibrous crops that can substitute for trees.

Kenaf (*Hibiscus cannabinus*) is an angiosperm that grows readily and rapidly in Asia and parts of Africa. Although it does not produce wood, kenaf has stems rich in fiber. It has been cultivated for three millennia in Asia, where its fiber is used to make bags, rope, dry bedding for animals, and even sails. There is now interest in using kenaf fiber for papermaking. Compared with southern pine trees, the current main source of pulp for paper, kenaf grows more rapidly (it matures in months rather than years), produces more fiber per hectare of land, and requires less water and fewer chemicals when used in paper production. Newspapers and greeting cards made from kenaf paper are stronger, yet thinner, than products made from typical wood pulp. The future of this plant as a resource for papermaking looks bright.

What are the properties of the kenaf plant that make it suitable for papermaking?

You will find the answer to this question on page 535.

CONCEPT 24.1 The Plant Body Is Organized and Constructed in a Distinctive Way

Plants must harvest energy from sunlight and mineral nutrients from the soil. Because they are stationary (sessile), they cannot move to find a more favorable environment or to avoid predators. Their body plan and physiology allow plants to respond to these challenges:

- Stems, leaves, and roots have structural adaptations that enable plants anchored to one spot to capture scarce resources effectively, both above and below the ground.
- Their ability to grow throughout their lifetimes enables plants to respond to environmental cues by redirecting their growth to exploit environmental opportunities, as when roots grow toward a water supply. We will return to plant–environment relationships in Chapter 28.

In Chapter 21 we saw how modern plants arose from aquatic ancestors, giving rise to simple land plants and then vascular plants. Despite their obvious differences in size and form, all vascular plants have essentially the same simple structural organization. This chapter describes the basic architecture of the largest group of vascular plants, the angiosperms (flowering plants), and shows how so much diversity (there are more than 250,000 species of angiosperms) can literally grow out of such a simple basic form. In this chapter we focus on the angiosperms' three kinds of vegetative (nonsexual) organs: roots, stems, and leaves. We will cover reproductive organs in Chapter 27.

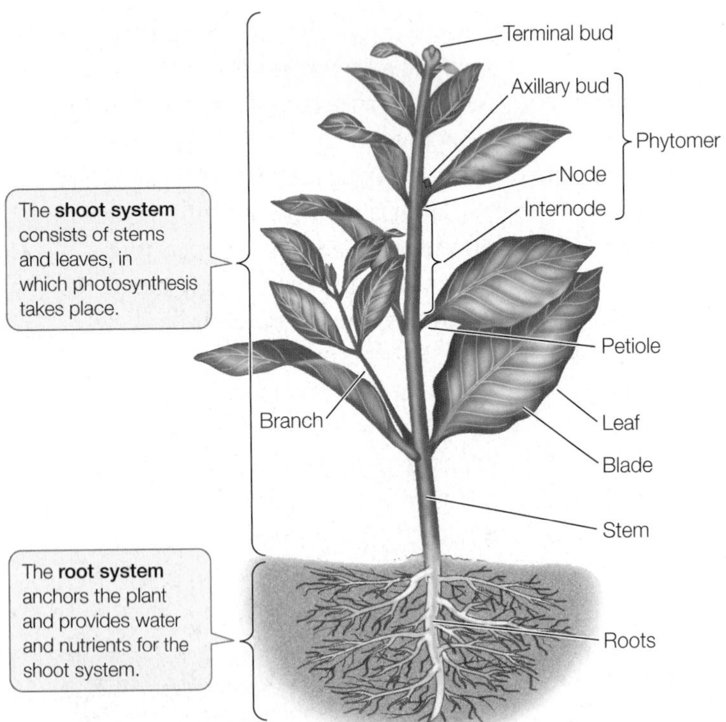

FIGURE 24.1 Vegetative Plant Organs and Systems The basic plant body plan, with root and shoot systems, and the principal vegetative organs.

Plant organs are organized into two systems (**FIGURE 24.1**):

- The **root system** anchors the plant in place. **Roots** absorb water and dissolved minerals and store the products of photosynthesis. The extreme branching of roots and their high surface area-to-volume ratios allow them to absorb water and mineral nutrients from the soil efficiently.
- The **shoot system** consists of the stems, leaves, and flowers. The **leaves** are the chief organs of photosynthesis. The **stems** hold and orient the leaves to the sun and provide connections for the transport of materials between roots and leaves. The flowers are shoots modified for reproduction.

Shoots are composed of repeating modules called **phytomers**. Each phytomer consists of a node carrying one or more leaves; an internode, which is the interval of stem between two nodes; and one or more axillary buds, each of which forms in the angle where the leaf meets the stem. A **bud** is an undeveloped shoot that can produce another leaf, a phytomer, or a flower.

As noted in Chapter 21, most angiosperms belong to one of two major clades (see Figure 21.27). **Monocots** are generally narrow-leaved plants such as grasses, lilies, orchids, and palms. **Eudicots** are generally broad-leaved plants such as soybeans, roses, sunflowers, and maples. These two clades, which account for 97 percent of angiosperm species, differ in several important structural characteristics (**FIGURE 24.2**).

Plants develop differently than animals

The four processes that govern the development of all organisms, whether plant or animal, are determination (the commitment of an embryonic cell to its ultimate fate in the organism); differentiation (the specialization of a cell); morphogenesis (the organization and spatial distribution of cells into tissues and organs); and growth (increase in body size). In plants, these processes are influenced by three properties unique to plants: meristems, cell walls, and the totipotency of most cells.

LINK
The processes of development are described in **Concept 14.1**

MERISTEMS Plants grow throughout their lifetimes. Although adult humans, for example, have stem cells to replace tissues lost through damage or apoptosis, these cells do not contribute to further growth. In contrast, plants have **meristems** that are capable of producing new roots, stems, leaves, and flowers throughout the plant's life, enabling the plant to continue growing.

CELL WALLS Plant cells are surrounded by a **cell wall**. This rigid extracellular matrix makes it impossible for cells to move from place to place as they do in animal development. Instead, plant morphogenesis occurs through alterations in the plane of cell division at cytokinesis, which in turn change the direction of cell expansion and thereby the direction in which a piece of tissue grows (**FIGURE 24.3**). Plant cytokinesis occurs along a cell plate laid down by Golgi vesicles (see Figure 7.7B). Unlike animal cells, in which the location of cytokinesis depends on the location of the middle of the mitotic spindle, the location of the plant cell plate is determined earlier—during interphase.

TOTIPOTENCY In animals, only early embryonic cells are totipotent, although some stem cell populations, such as those in bone marrow, are multipotent (capable of differentiating into a few different cell types). In contrast, in plants even differentiated cells are pluripotent, and most are totipotent (see Figure 14.3). This means a plant can readily repair damage wrought by the environment or herbivores.

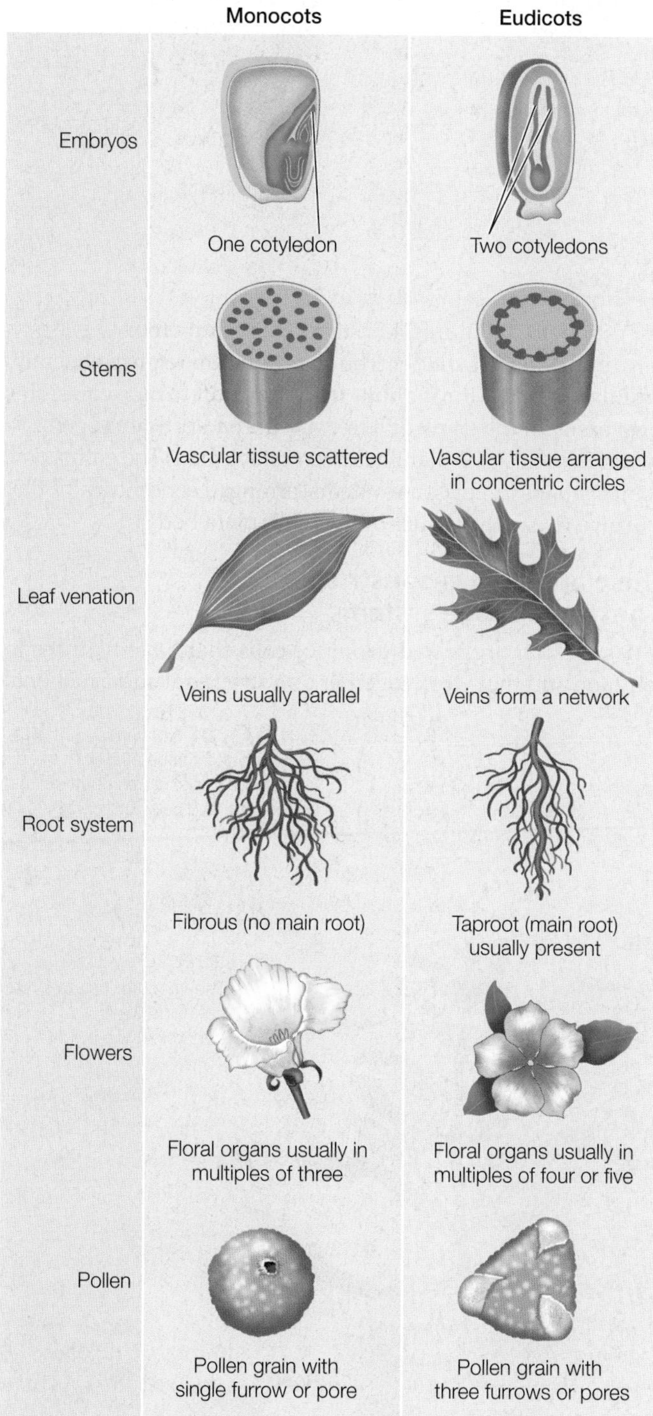

FIGURE 24.2 Comparing the Two Major Angiosperm Clades The figure shows the major structural differences between monocots and eudicots.

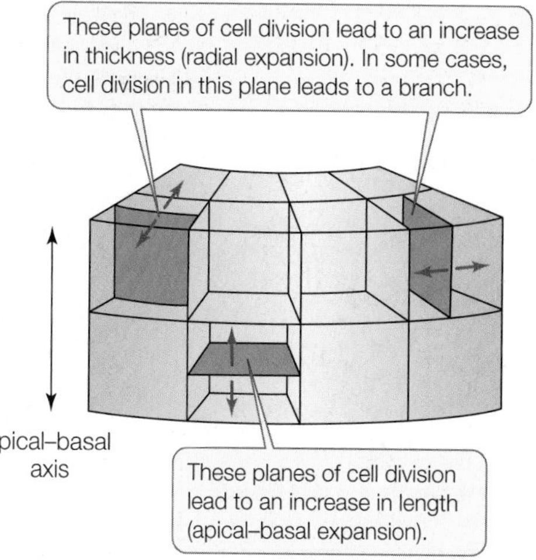

FIGURE 24.3 Cytokinesis and Morphogenesis The plane of cell division can determine the growth pattern of a plant organ, as in this section of a shoot.

LINK

See **Concept 13.4** for the use of totipotency in plant biotechnology

The plant body has an apical–basal axis and a radial axis

Two basic patterns are established early in plant embryogenesis (embryo formation) (**FIGURE 24.4**):

- The **apical–basal** axis: the arrangement of cells and tissues along the main axis from root to shoot
- The **radial** axis: the concentric (circular) arrangement of the tissue systems (which we will discuss in the next section)

Both axes are best understood in developmental terms. We focus here on embryogenesis in the model eudicot *Arabidopsis thaliana*, in which the process has been intensively studied.

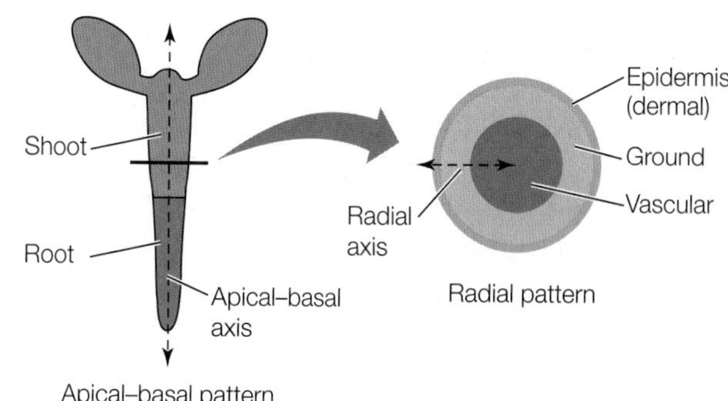

FIGURE 24.4 Two Patterns for Plant Morphogenesis (A) The apical–basal pattern is the arrangement of cells and tissues along the main axis from root to shoot. (B) The radial pattern determines the concentric arrangement of tissues.

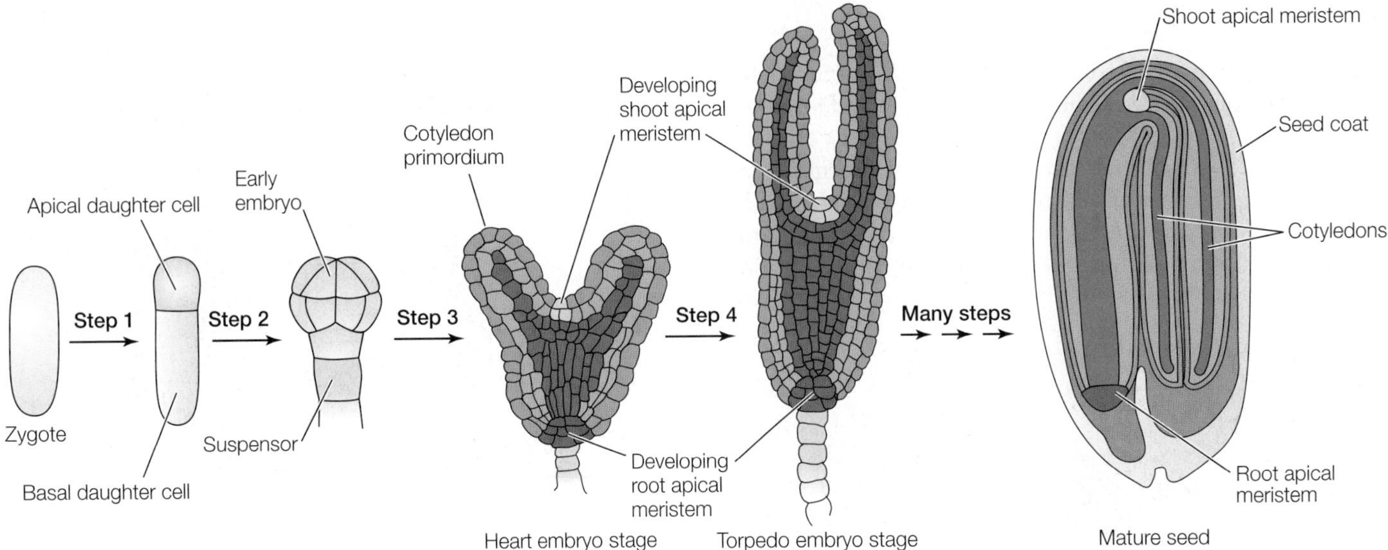

FIGURE 24.5 Plant Embryogenesis The basic body plan of the model eudicot *Arabidopsis thaliana* is established in several steps. By the heart embryo stage, the three tissue systems are established: the dermal (gold), ground (light green), and vascular (blue) tissue systems. These tissues are shown in the mature seed, which develops after many more stages.

The first step in the formation of a plant embryo is a mitotic division of the zygote that gives rise to two daughter cells (**FIGURE 24.5, STEP 1**). An asymmetrical plane of cell division results in an uneven distribution of cytoplasm between these two cells, which face different fates. Signals in the smaller, apical (upper) daughter cell induce it to produce the embryo proper, while the other, larger daughter cell produces a supporting structure, the **suspensor** (**FIGURE 24.5, STEP 2**). This division not only establishes the apical–basal axis of the new plant but also determines its polarity (which end is the tip, or apex, and which is the base). A longer, thinner suspensor and a more spherical or globular embryo are distinguishable after just four mitotic divisions (see Figure 14.1).

In eudicots, the initially globular embryo develops into the characteristic heart stage as the **cotyledons** ("seed leaves") start to grow (**FIGURE 24.5, STEP 3**). Further elongation of the cotyledons and of the apical–basal axis of the embryo gives rise to the torpedo stage, during which some of the internal tissues begin to differentiate (**FIGURE 24.5, STEP 4**). Between the cotyledons is the **shoot apical meristem**; at the other end of the axis is the **root apical meristem**. These meristems contain undifferentiated cells that will divide to give rise to the shoot and root systems.

As shown in Figure 24.5, step 2, the plant embryo is first a sphere and later a cylinder. The root and stem retain a generally cylindrical shape throughout the plant's life. You can see this most easily in the trunk of a tree. By the end of embryogenesis, the radial axis of the plant has been established. The embryonic plant contains three tissue systems, arranged concentrically, that will give rise to the tissues of the adult plant body.

The plant body is constructed from three tissue systems

A **tissue** is an organized group of cells that have features in common and that work together as a structural and functional

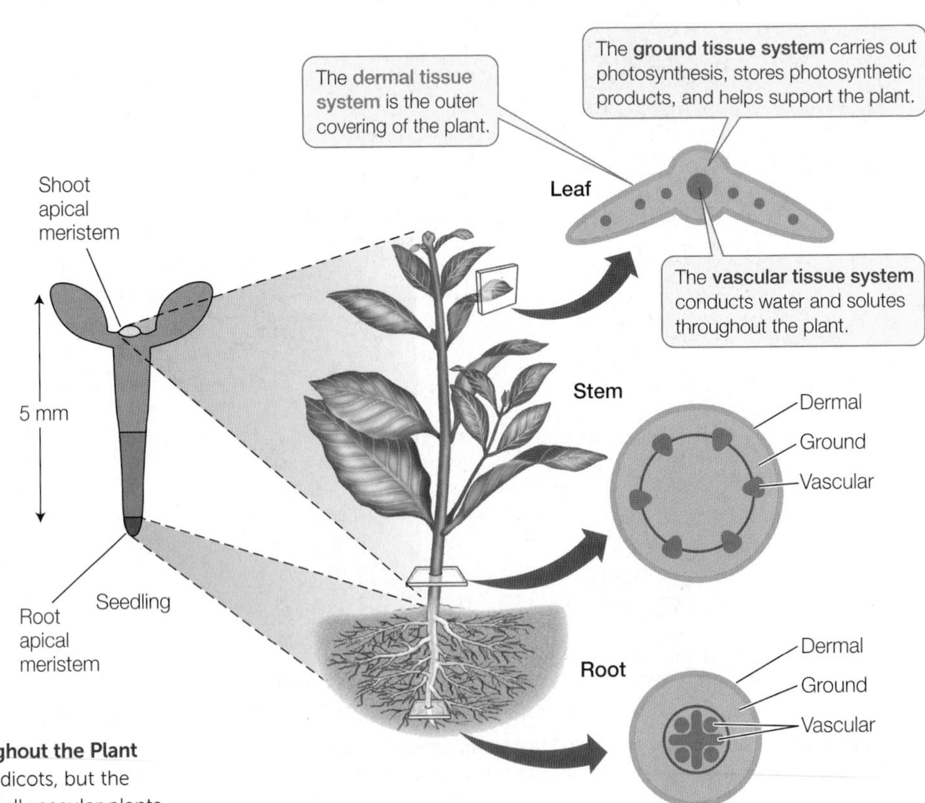

FIGURE 24.6 Three Tissue Systems Extend throughout the Plant Body The arrangement shown here is typical of eudicots, but the three tissue systems are continuous in the bodies of all vascular plants.

unit. In plants, tissues, in turn, are grouped into **tissue systems**. Despite their structural diversity, all vascular plants are constructed from three tissue systems: dermal, vascular, and ground. These three tissue systems are established during embryogenesis and ultimately extend throughout the plant body in a concentric arrangement (**FIGURE 24.6**). Each tissue system has distinct functions and is composed of different mixtures of cell types.

DERMAL TISSUE SYSTEM The **dermal tissue system** forms the **epidermis**, or outer covering, of a plant, which usually consists of a single cell layer. Epidermal cells are typically small. Some of them differentiate to form one of three specialized structures:

- *Stomata*, which are pores for gas exchange in leaves (see Figure 25.13)
- *Trichomes*, or leaf hairs, which provide protection against insects and damaging solar radiation (see Figure 28.7)
- *Root hairs*, which greatly increase root surface area, thus providing more surface for the uptake of water and mineral nutrients (see Figure 24.8)

Aboveground epidermal cells secrete a protective extracellular **cuticle** made of cutin (a polymer composed of long chains of fatty acids), a complex mixture of waxes, and cell wall polysaccharides. The cuticle limits water loss, reflects potentially damaging solar radiation, and serves as a barrier against pathogens.

As the stems and roots of woody plants expand, the epidermis is shed and is replaced by a dermal tissue called periderm. As we will see in Concept 24.3, this tissue includes an outer layer of cork that is impermeable to water and gases.

GROUND TISSUE SYSTEM Virtually all the tissue lying between dermal tissue and vascular tissue in both shoots and roots is part of the **ground tissue system**, which therefore makes up most of the plant body. Ground tissue contains three cell types.

Parenchyma cells are the most abundant ground tissue cells. They have large vacuoles and relatively thin cell walls. They perform photosynthesis (in the shoot) and store protein (in seeds) and starch (in roots).

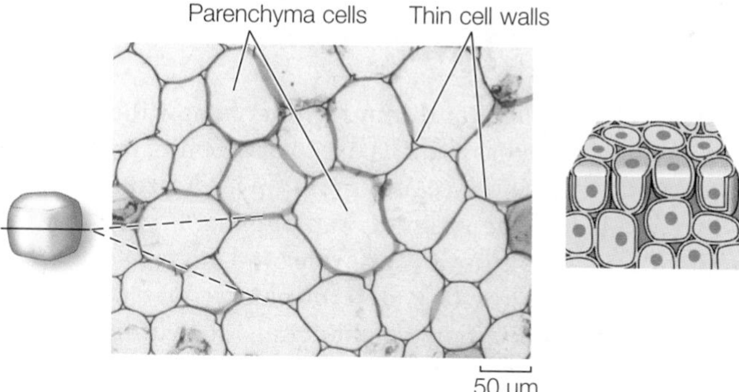

Collenchyma cells are elongated and have unevenly thickened primary cell walls. They provide support for growing tissues such as stems. The familiar "strings" in celery consist primarily of collenchyma cells.

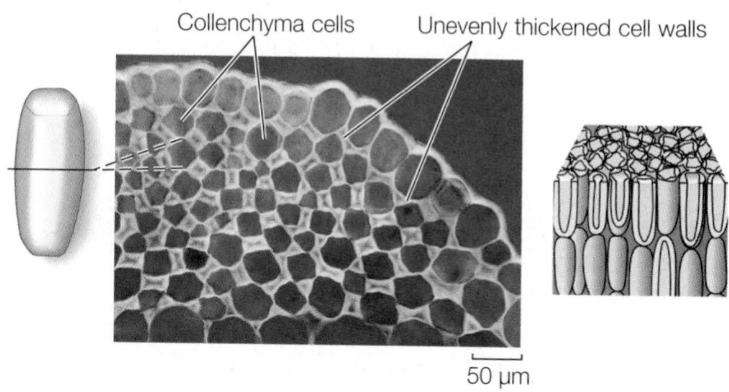

Sclerenchyma cells have very thick secondary walls reinforced with the polyphenol polymer **lignin**. Most sclerenchyma cells undergo programmed cell death (apoptosis; see Concept 7.5), but their strong cell walls remain to provide support for the plant. There are two types of sclerenchyma cells: fibers and sclereids. Elongated **fibers** provide relatively rigid support to wood and other parts of the plant, within which they are often organized into bundles. When wood is used to make high-quality paper, the wood pulp must be treated extensively to break down and remove the lignin in the fibers.

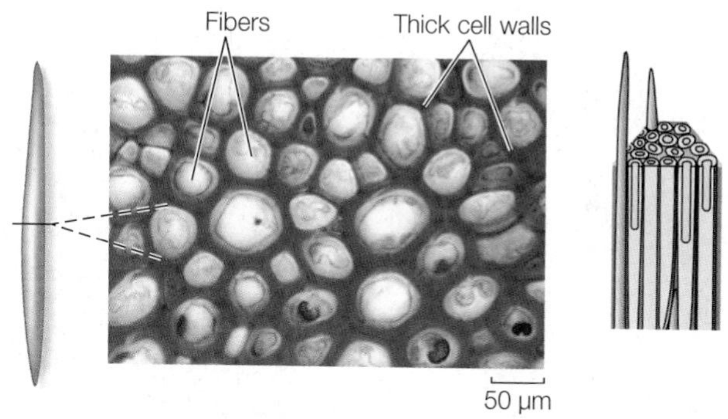

The bark of trees owes much of its mechanical strength to long fibers. **Sclereids** occur in various shapes and may pack together densely, as in a nut's shell or in some seed coats.

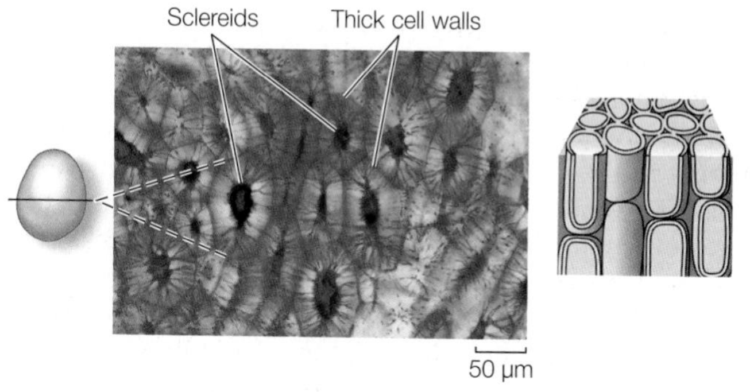

Isolated clumps of sclereids, called stone cells, in pears and some other fruits give them their characteristic gritty texture.

VASCULAR TISSUE SYSTEM The **vascular tissue system** is the plant's plumbing or transport system—the distinguishing feature of vascular plants. Its two constituent tissues, the xylem and phloem, distribute materials throughout the plant.

LINK

The functions of xylem and phloem are described in **Concepts 25.3 and 25.4**

Xylem distributes water and mineral ions taken up by the roots to all the cells of the roots and shoots. It contains two types of conducting cells—tracheids and vessel elements—which, like most sclerenchyma cells, are dead when mature. **Tracheids** are spindle-shaped cells, with thinner regions in the cell wall called pits through which water can move with little resistance from one tracheid to its neighbors. **Vessel elements** are larger in diameter than tracheids. They meet end-to-end and partially break down their end walls, forming an open pipeline for water conduction. The xylem also contains living parenchyma cells that regulate the movement of ions into the conducting cells.

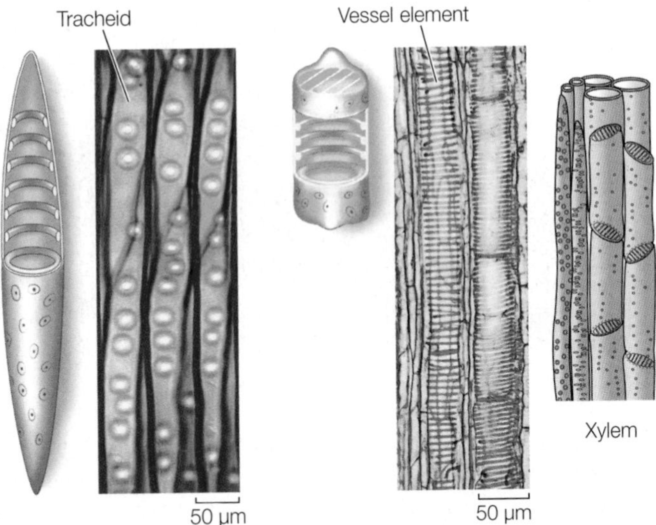

Phloem generally consists of living cells. It transports carbohydrates (primarily sugars) from sites where they are produced (called sources) to sites where they are used or stored (called sinks). Sources are often photosynthetic organs (such as leaves) but also include nonphotosynthetic storage organs (such as tubers and seeds) when they are drawn on to supply sugars to other parts of a plant. Sinks include growing tissues, roots, and developing flowers and fruit. The characteristic cells of the phloem are **sieve tube elements**, which, like vessel elements, meet end-to-end, forming sieve tubes. Instead of the ends of the cells breaking down, however, the sieve tube elements are connected by plasmodesmata (see Figure 4.7), which form a set of pores called a sieve plate. Although still alive, mature sieve tube elements have lost much of their cellular contents; adjacent, fully functional **companion cells** are connected to the sieve tube elements by plasmodesmata and perform many of the phloem's metabolic functions.

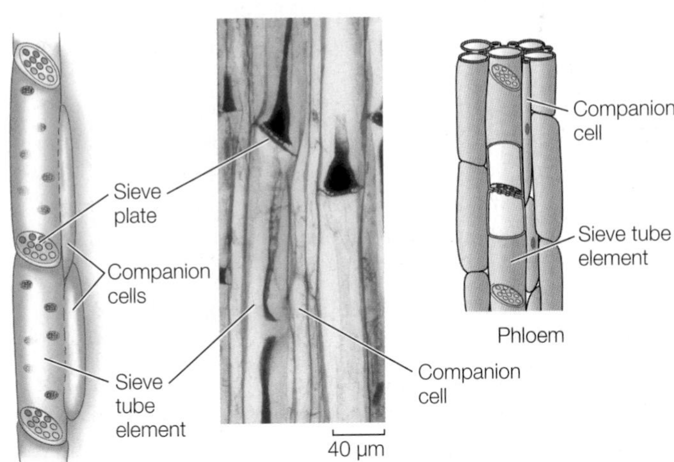

Xylem and phloem tissues may also contain fibers, which provide structural support.

CHECKpoint CONCEPT 24.1

- ✓ Make a table listing the three tissue systems of plants and their functions.
- ✓ What are the similarities and differences between plant and animal development?
- ✓ What are the similarities and differences between a root and a shoot?
- ✓ A mutant strain of *Arabidopsis thaliana* lacks apical–basal polarity in the embryo. What would you expect the phenotype of the plant to be?

We've discussed the basic body plan and tissue systems of plants, and how the apical–basal axis and radial axis arise during embryo development. Now let's see how the cells and tissues we have just described allow the embryo to build an adult plant body.

CONCEPT 24.2 Apical Meristems Build the Primary Plant Body

As noted above, plants and animals differ in how they develop and function. Whereas animals use their mobility to forage for food, plants stay in place and grow toward resources, both above and below the ground. So plant shoots grow toward sunlight, and roots grow toward water and minerals in the soil. In most animals, growth is **determinate**: it ceases when the adult state is reached. Some plant organs, such as most leaves and flowers, show this characteristic. By contrast, the growth of roots and shoots is **indeterminate**, an open-ended process which can be lifelong.

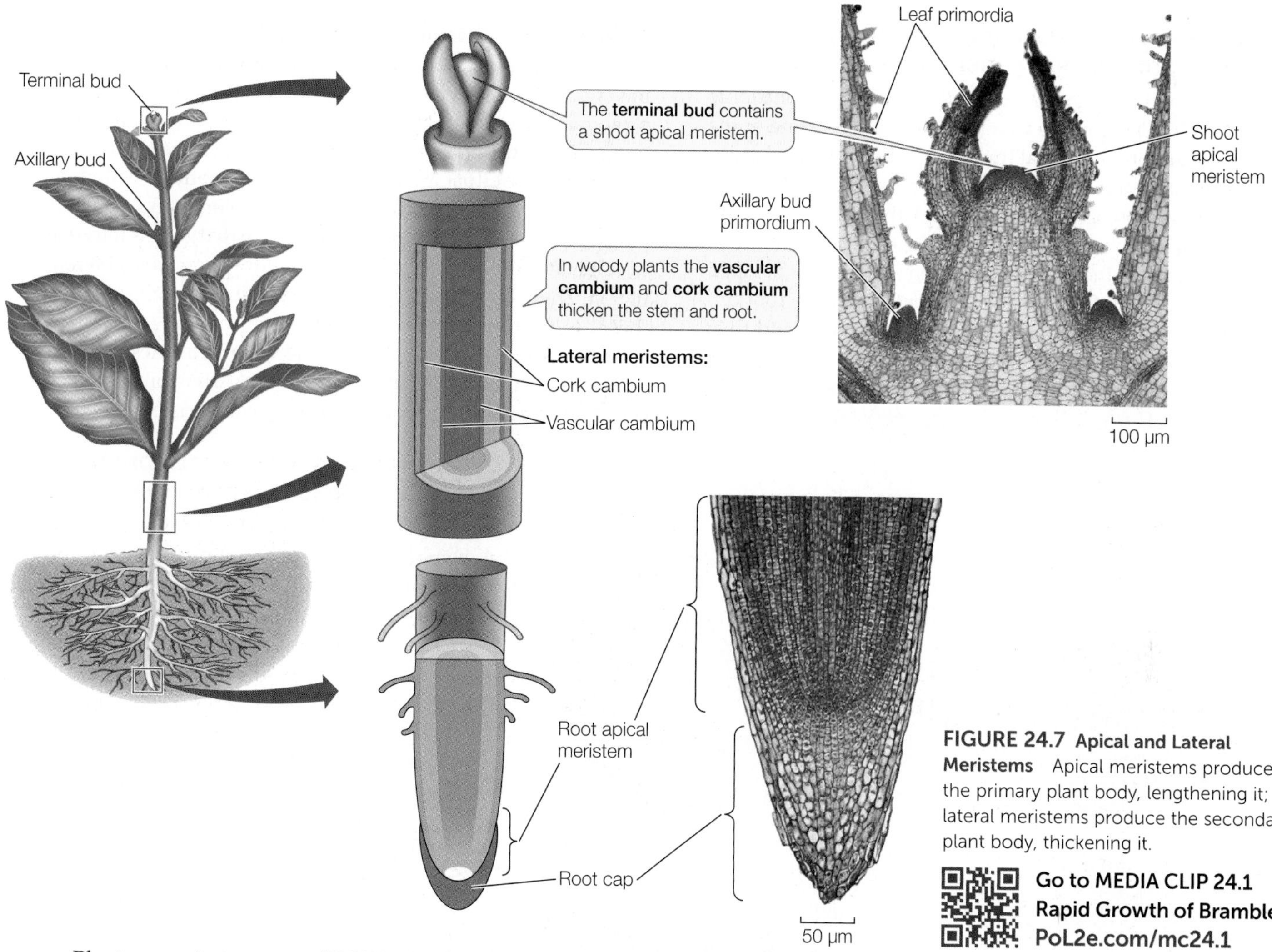

FIGURE 24.7 Apical and Lateral Meristems Apical meristems produce the primary plant body, lengthening it; lateral meristems produce the secondary plant body, thickening it.

Go to MEDIA CLIP 24.1 Rapid Growth of Brambles PoL2e.com/mc24.1

Plants grow in two ways (**FIGURE 24.7**):

- **Primary growth** results in the formation and lengthening of shoots and roots. **Apical meristems** produce primary growth, giving rise to the primary plant body. All seed plants have a primary plant body.
- **Secondary growth** increases plant thickness. **Lateral meristems** produce secondary growth, resulting in the secondary plant body. Secondary growth occurs in woody eudicots, such as trees and shrubs. Monocots typically do not undergo secondary growth.

The remainder of this section will focus on primary growth. We will return to secondary growth in Concept 24.3.

A hierarchy of meristems generates the plant body

Even before a seed germinates, the plant embryo inside the seed has two meristems: a shoot apical meristem at the end of the embryonic shoot and a root apical meristem near the end of the embryonic root (see Figure 24.5). The dividing cells of the meristems are comparable to animal stem cells: after their division, one daughter cell is capable of differentiating, and the other retains its undifferentiated phenotype.

The apical meristems form the root and shoot organs of the adult plant. They do so by producing three **primary meristems**. From the outside to the inside of the root or shoot, which are both cylindrical, the primary meristems are the **protoderm**, the **ground meristem**, and the **procambium**. These meristems, in turn, give rise to the three tissue systems:

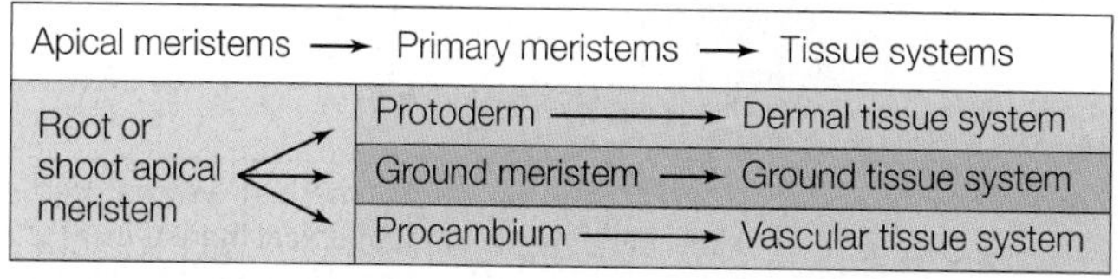

Apical meristems → Primary meristems → Tissue systems
Root or shoot apical meristem
Protoderm → Dermal tissue system
Ground meristem → Ground tissue system
Procambium → Vascular tissue system

In plants that have secondary growth, the lateral meristems represent an additional level in the meristem hierarchy; for this reason, lateral meristems are also referred to as secondary meristems.

Not all meristems fit within this framework. The totipotency of many plant cells means that even differentiated, non-meristematic cells can give rise to new meristems under certain circumstances. Examples include some forms of asexual reproduction and the propagation of plants from stem cuttings. The ability of non-meristematic cells to further divide is important in wound repair.

The root apical meristem gives rise to the root cap and the root primary meristems

The root apical meristem produces all the cells that contribute to growth in the length of a root (**FIGURE 24.8A**). Some of the daughter cells from the apical (tip) end of the root apical meristem contribute to a **root cap**, which protects the delicate growing region of the root as it pushes through the soil. The root cap secretes a mucopolysaccharide (slime) that acts as a lubricant. Even so, the cells of the root cap are often damaged or scraped away and must therefore be replaced constantly. The root cap is also the structure that detects the pull of gravity and thus controls the downward growth of roots.

Above the root cap, the cells in the root tip are arranged in several zones. The apical and primary meristems constitute the **zone of cell division**, the source of all the cells of the root's primary tissues. Daughter cells of the divisions of the apical meristem become the three cylindrical primary meristems: the protoderm, the ground meristem, and the procambium. These meristems in turn produce all the cells of the three tissues systems in the root. Just above this zone is the **zone of cell elongation**, where the newly formed cells are elongating and thus pushing the root farther into the soil. Above that zone is the **zone of cell maturation**, where the cells are completing their differentiation. These three zones grade imperceptibly into one another; there is no abrupt line of demarcation.

In the middle of the root apical meristem is a quiescent center, where cell divisions are rare. The rare divisions produce cells that ultimately become the root apical meristem.

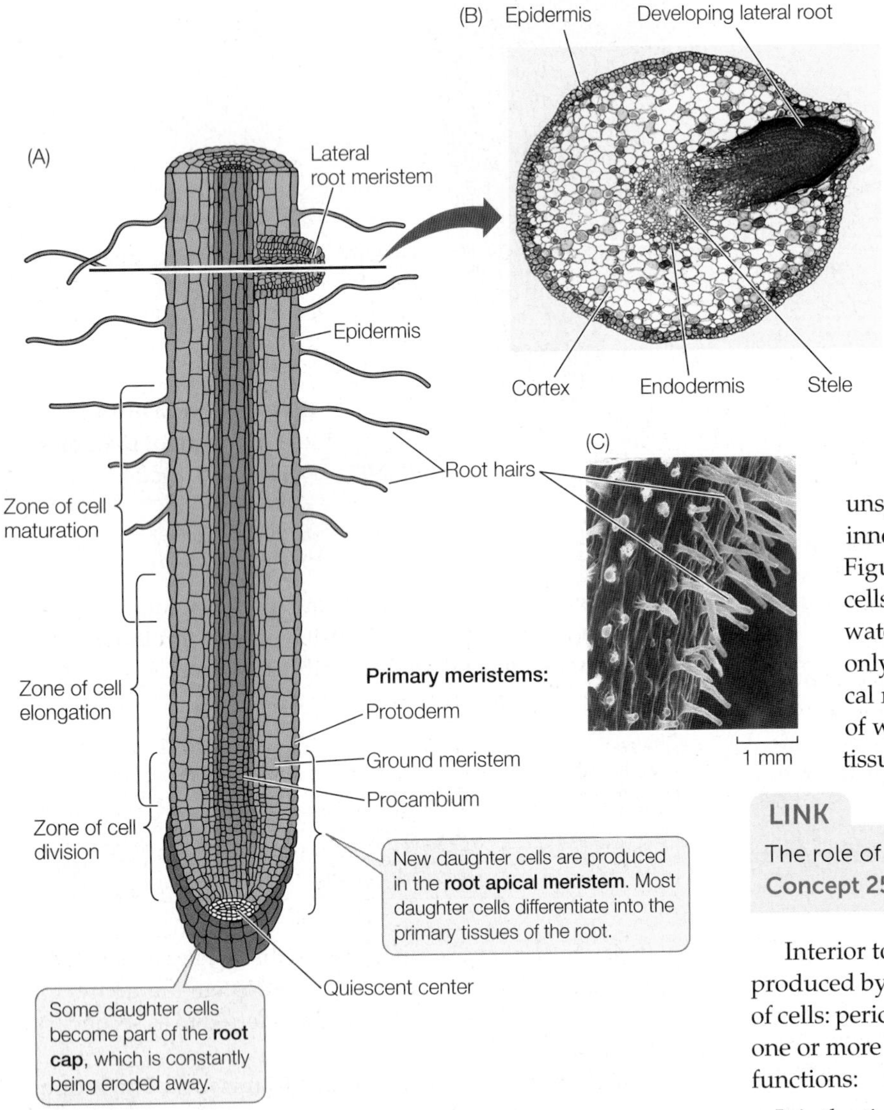

FIGURE 24.8 Tissues and Regions of the Root (A) Diagram of a longitudinal section of a root, showing how cell division creates the complex structure of the root. (B) A cross-section through a region of a willow root. Note the emerging lateral root, which arises from tissues deep within the root. (C) Root hairs, seen with a scanning electron microscope.

The products of the root's primary meristems become root tissues

The products of the three primary meristems (the protoderm, ground meristem, and procambium) are the tissue systems of the mature root: the epidermis (dermal tissue), cortex (ground tissue), and stele (vascular tissue) (**FIGURE 24.8B**).

The protoderm gives rise to the epidermis, an outer layer of cells that protect the root and absorb mineral ions and water. Many of the epidermal cells produce long, delicate root hairs (**FIGURE 24.8C**).

Internal to the protoderm, the ground meristem gives rise to a region of ground tissue that is many cells thick, called the **cortex**. The parenchyma cells of the cortex are relatively unspecialized and often serve as storage depots. The innermost layer of the cortex is the **endodermis** (see Figure 24.8B). Unlike the cell walls of other cortical cells, those of the endodermal cells contain suberin, a waterproof substance. Strategic placement of suberin in only certain parts of the cell wall enables the cylindrical ring of endodermal cells to control the movement of water and dissolved mineral ions into the vascular tissue system.

LINK
The role of the endodermis is described in more detail in **Concept 25.3**

Interior to the endodermis is the vascular cylinder, or **stele**, produced by the procambium. The stele consists of three types of cells: pericycle, xylem, and phloem. The **pericycle** consists of one or more layers of parenchyma cells. It has three important functions:

- It is the tissue from which lateral roots arise.
- It can contribute to secondary growth by giving rise to lateral meristems that thicken the root.
- Its cells contain membrane transport proteins that export nutrient ions into cells of the xylem.

APPLY THE CONCEPT

Meristems build roots

A research team investigated how long root cells continued to divide after being produced by the root apical meristem. They incubated germinating bean seeds for 2 hours with their roots suspended in water containing radioactive thymidine, which is taken up by dividing cells and incorporated into their DNA as it replicates (making the DNA radioactive).

After this incubation period, the researchers removed some of the seeds (time 0 in the table) and took longitudinal slices from the roots (see Figure 24.8A). They transferred the rest of the seeds to water containing nonradioactive thymidine and left them there for various times (12, 36, and 72 hours) before taking root slices. They tallied the number of radioactive cells found in each of the four regions of each root and compared those numbers with the total number of cells in each region. The table shows the results.

	RADIOACTIVE CELLS/TOTAL CELLS			
TIME AFTER TRANSFER (hr)	ROOT CAP	ROOT APICAL MERISTEM	ZONE OF CELL ELONGATION	ZONE OF CELL MATURATION
0	0/63	90/115	0/32	0/22
12	2/58	95/110	16/29	0/27
36	36/60	45/119	14/30	12/31
72	12/24	0/116	2/31	30/32

1. At 0 hours, why did only root apical meristem cells contain radioactive thymidine?
2. For each region, calculate the percentage of radioactive cells and plot that percentage over time.
3. Explain the rise and fall in the percentage of radioactive cells in each region.

At the very center of the root of a eudicot lies the xylem. Seen in cross section, it typically has the shape of a star with a variable number of points (**FIGURE 24.9A**). Between the points are bundles of phloem. In monocots, a region of parenchyma cells called the **pith** typically lies in the center of the root, surrounded by xylem and phloem (**FIGURE 24.9B**). Pith, which often stores carbohydrate reserves, is also found in the stems of eudicots.

The root system anchors the plant and takes up water and dissolved minerals

Water and mineral nutrients enter most plants through the root system, which is located in the soil. Although hidden from view, the root system is often larger than the visible shoot system. For example, the root system of a 4-month-old winter rye plant (*Secale cereale*) was found to be 130 times larger than the shoot system, with almost 13 million branches that had a cumulative length of over 500 kilometers! The large size of root systems gives them an enormous surface area for absorbing water and nutrients from the soil.

The root system of angiosperms originates in an embryonic root called the radicle. In most eudicots the radicle develops as a primary root (called the **taproot**) that extends downward by tip growth and outward by initiating **lateral roots** (**FIGURE 24.10A**).

In monocots, the radicle-derived initial root is short-lived. Monocots form a **fibrous root system** composed of numerous thin roots that are all roughly equal in diameter (**FIGURE 24.10B**). These roots are **adventitious**: they arise from stem tissues above the initial root. Fibrous root systems often form dense networks of roots near the soil surface, binding the soil together. The fibrous root systems of grasses, for example, may protect steep hillsides where runoff from rain would otherwise cause erosion.

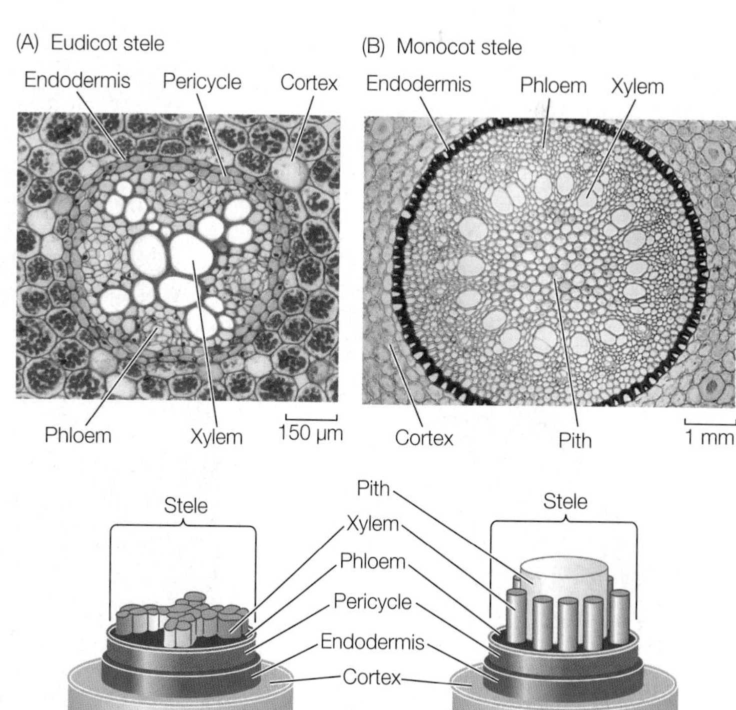

FIGURE 24.9 Products of the Root's Primary Meristems The protoderm gives rise to the epidermis, the outermost layer of cells. The ground meristem produces the cortex, the innermost layer of which is the endodermis. The vascular tissues of the root are found in the stele, which is the product of the procambium. The arrangement of tissues in the stele differs in the roots of (A) eudicots and (B) monocots.

Go to ACTIVITY 24.1 Eudicot Root
PoL2e.com/ac24.1

Go to ACTIVITY 24.2 Monocot Root
PoL2e.com/ac24.2

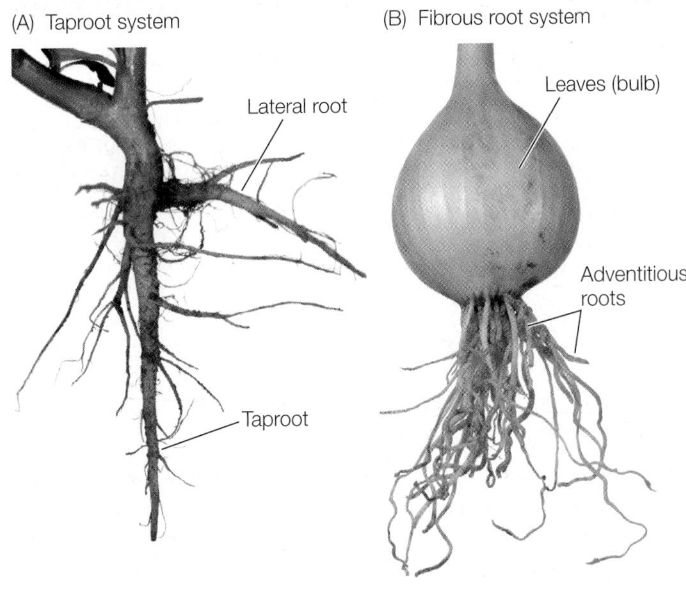

FIGURE 24.10 Root Systems of Eudicots and Monocots (A) Eudicots typically have taproot systems, as in the seedling of the herb epazote. (B) This garlic bulb shows the fibrous root system typical of monocots.

The products of the shoot's primary meristems become shoot tissues

The shoot apical meristem, like the root apical meristem, forms three primary meristems: protoderm, ground meristem, and procambium. These primary meristems, in turn, give rise to the three shoot tissue systems. The shoot apical meristem repetitively lays down the beginnings of leaves and axillary buds (corresponding to the phytomers in Figure 24.1). Leaves arise from bulges called leaf primordia, which form as cells divide on the sides of the shoot apical meristem (see Figure 24.7). Bud primordia form above the bases of the leaf primordia, where they may initiate new shoots. The growing stem has no protective structure analogous to the root cap, but the leaf primordia can act as a protective covering for the shoot apical meristem.

As in roots, the shoot protoderm, ground meristem, and procambium give rise, respectively, to the shoot epidermis, the shoot cortex, and the shoot vascular system. The vascular tissue of young stems is divided into discrete **vascular bundles** (**FIGURE 24.11**). Each vascular bundle contains both xylem and

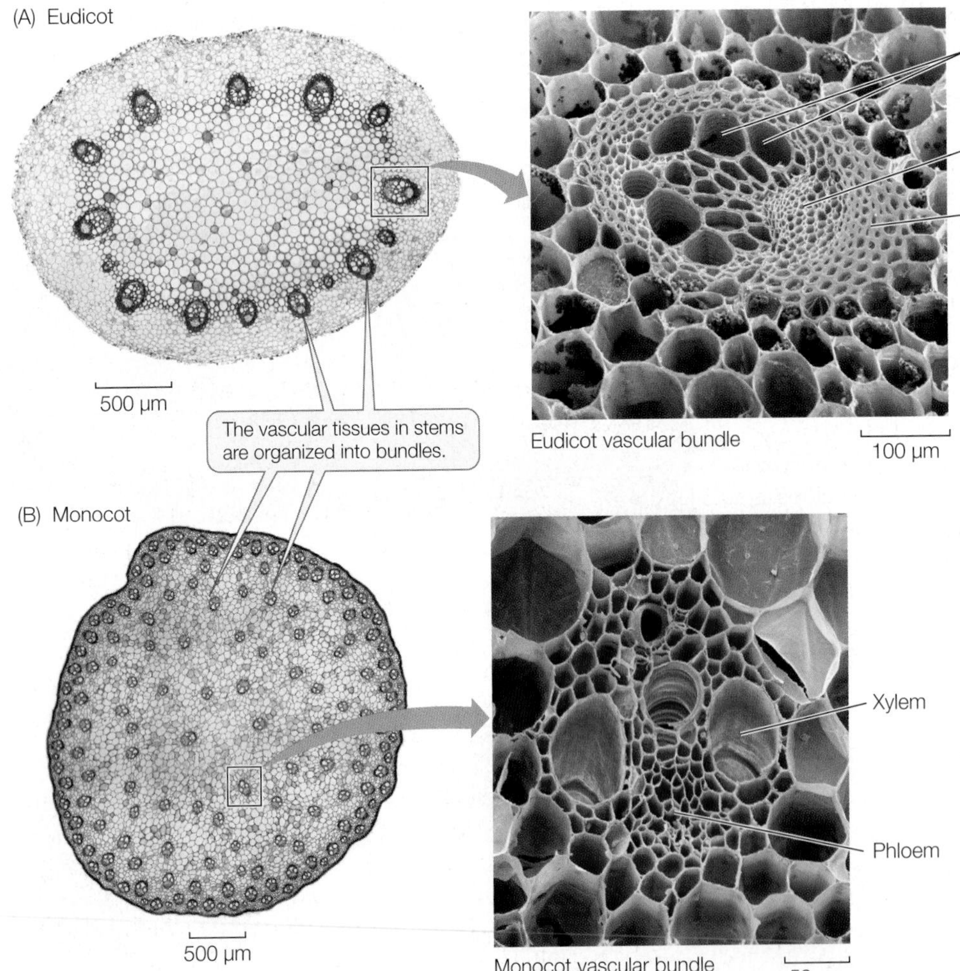

FIGURE 24.11 Vascular Bundles in Stems (A) In herbaceous and young woody eudicot stems, the vascular bundles are arranged in a cylinder, with pith in the center and the cortex outside the cylinder. (B) A scattered arrangement of vascular bundles is typical of monocot stems.

Go to ACTIVITY 24.3 Eudicot Stem PoL2e.com/ac24.3

Go to ACTIVITY 24.4 Monocot Stem PoL2e.com/ac24.4

phloem. In eudicots, the vascular bundles are arranged along the periphery and appear in a circle in cross section, but in monocots they are scattered throughout the stem.

Leaves are photosynthetic organs produced by shoot apical meristems

For most of its life, a plant produces leaves from shoot apical meristems. As shown in Figure 24.7, leaves originate from leaf primordia at the sides of the shoot apical meristem. Unlike the growth of roots and stems, the growth of leaves is finite, or determinate: leaves generally stop growing once they reach a predetermined mature size and shape.

Typically, the **blade** of a leaf is a thin, flat structure. In many eudicots, the blade is attached to the stem by a stalk called a **petiole** (see Figure 24.1). Often the petiole can adjust the orientation of the leaf blade relative to the rays of the sun, ranging from perpendicular to the light rays (to maximize photosynthesis) to parallel (to prevent overheating). Some leaves track the sun over the course of the day, moving so that they constantly face it. Monocots typically do not have petioles, and the base of the leaf may form a sheath that wraps around the stem.

Leaf anatomy is beautifully adapted to carry out photosynthesis and to support that process by exchanging the gases O_2 and CO_2 with the environment, limiting evaporative water loss, and exporting the products of photosynthesis to the rest of the plant. **FIGURE 24.12** shows a section of a typical eudicot leaf in three dimensions.

Most eudicot leaves have two zones of photosynthetic parenchyma tissue called **mesophyll** (which means "middle of the leaf"). The upper layer or layers of mesophyll, which consist of elongated cells, constitute a zone called palisade mesophyll. The lower layer or layers, which consist of irregularly shaped cells, constitute a zone called spongy mesophyll. Within the mesophyll is a network of air spaces through which CO_2 can diffuse to photosynthesizing cells. Vascular tissue branches extensively throughout the leaf, forming a network of veins that extend to within a few cell diameters of all its cells, ensuring that the mesophyll cells are well supplied with water and mineral nutrients by the xylem. The products of photosynthesis are loaded into the phloem of the veins for export to the rest of the plant.

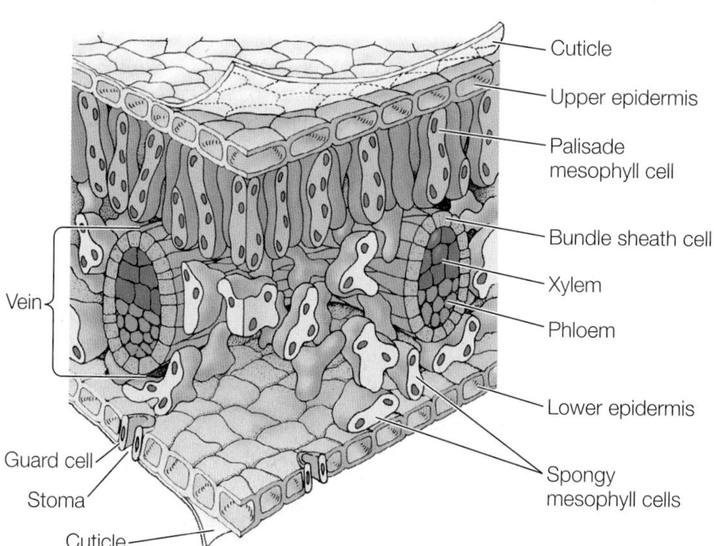

FIGURE 24.12 Eudicot Leaf Anatomy This three-dimensional diagram shows a section of a eudicot leaf.

Go to ACTIVITY 24.5 Eudicot Leaf
PoL2e.com/ac24.5

LINK

Photosynthesis is described in **Concepts 6.5 and 6.6**

Covering virtually the entire leaf on both its upper and lower surfaces is a layer of nonphotosynthetic epidermal cells. These cells secrete a cuticle that is impermeable to water. Although the epidermis keeps water in the leaf, it also keeps out CO_2—which is needed for photosynthesis. Therefore water and gases are exchanged through pores in the leaf called stomata (see Concept 25.3).

Plant organs can have alternative forms and functions

Flowering plants are incredibly diverse in form. Leaves, stems, and roots of many shapes can perform the basic functions of plant life: photosynthesis, the acquisition of water and nutrients, sensing the environment, and reproduction. These plant organs may also be modified to serve more specialized functions.

Taproots, for example, often function as nutrient storage organs, as in carrots, sugar beets, and sweet potato (**FIGURE 24.13A**). In some monocots—corn and some palms, for example—adventitious roots function as props to help support the shoot (**FIGURE 24.13B**). **Prop roots** are critical to these plants, which, unlike most eudicot tree species, are unable to support aboveground growth through the thickening of their stems.

Various modifications of stems are also seen in nature. The tuber of a potato, for example—the part of the plant eaten by humans—is not a root but rather an underground stem. The "eyes" of a potato are depressions containing axillary buds—in other words, a sprouting potato is just a branching stem (**FIGURE 24.13C**). Many desert plants have enlarged, water-retaining stems (**FIGURE 24.13D**). The runners of strawberry plants are horizontal stems from which roots grow at frequent intervals (**FIGURE 24.13E**). If the links between the rooted portions are broken, independent plants can develop on each side of the break—a form of vegetative (asexual) reproduction (see Concept 7.1).

Leaves are sometimes highly modified for special functions. For example, some leaves serve as storage depots for energy-rich molecules, as in the bulbs of onions. In succulents, the leaves store water. The protective spines of cacti are modified leaves (see Figure 24.13D). Other plants have modified portions of leaves called tendrils that support the plant by wrapping around other structures or plants (**FIGURE 24.13F**).

These are all examples of vegetative plant organs whose functions deviate from the norm. The evolution of these novel roles for established organs is an example of natural selection working with what is already present, as well as an example

FIGURE 24.13 Modified Roots, Stems, and Leaves (A) The taproots of some eudicots, such as carrots (*Daucus carota*) and sugar beets (*Beta vulgaris*), are nutrient storage organs. (B) The stems of this hala tree (*Pandanus tectorius*), a monocot, are supported by prop roots. (C) A potato (*Solanum tuberosum*) is a modified stem called a tuber; the sprouts that grow from its eyes are shoots, not roots. (D) The stem of this barrel cactus (*Echinocactus grusonii*) is enlarged to store water. Its highly modified leaves serve as thorny spines. Most of this plant's photosynthesis occurs in the stem. (E) The runners of strawberries (*Fragaria* sp.) are horizontal stems that produce roots and shoots at intervals. Rooted portions of the plant can live independently if the runner is cut. (F) The tips of some leaves, as in this red bryony vine (*Bryonia dioica*), are modified into tendrils.

of the interaction between evolution and development—the topic of evolutionary developmental biology (evo-devo).

LINK

See **Concepts 14.4 and 14.5** for a discussion of evolutionary developmental biology

In Concept 24.4 we will return to the modification of plant organs—in this case, the ways they have been manipulated by artificial selection to meet the needs of humans.

CHECKpoint CONCEPT 24.2

- ✓ What is the difference between determinate and indeterminate growth? What plant organs undergo each type of growth?
- ✓ What are the selective advantages of indeterminate growth?

Primary growth forms the plant organs and lengthens them. We now turn to secondary growth, in which some organs can grow in thickness.

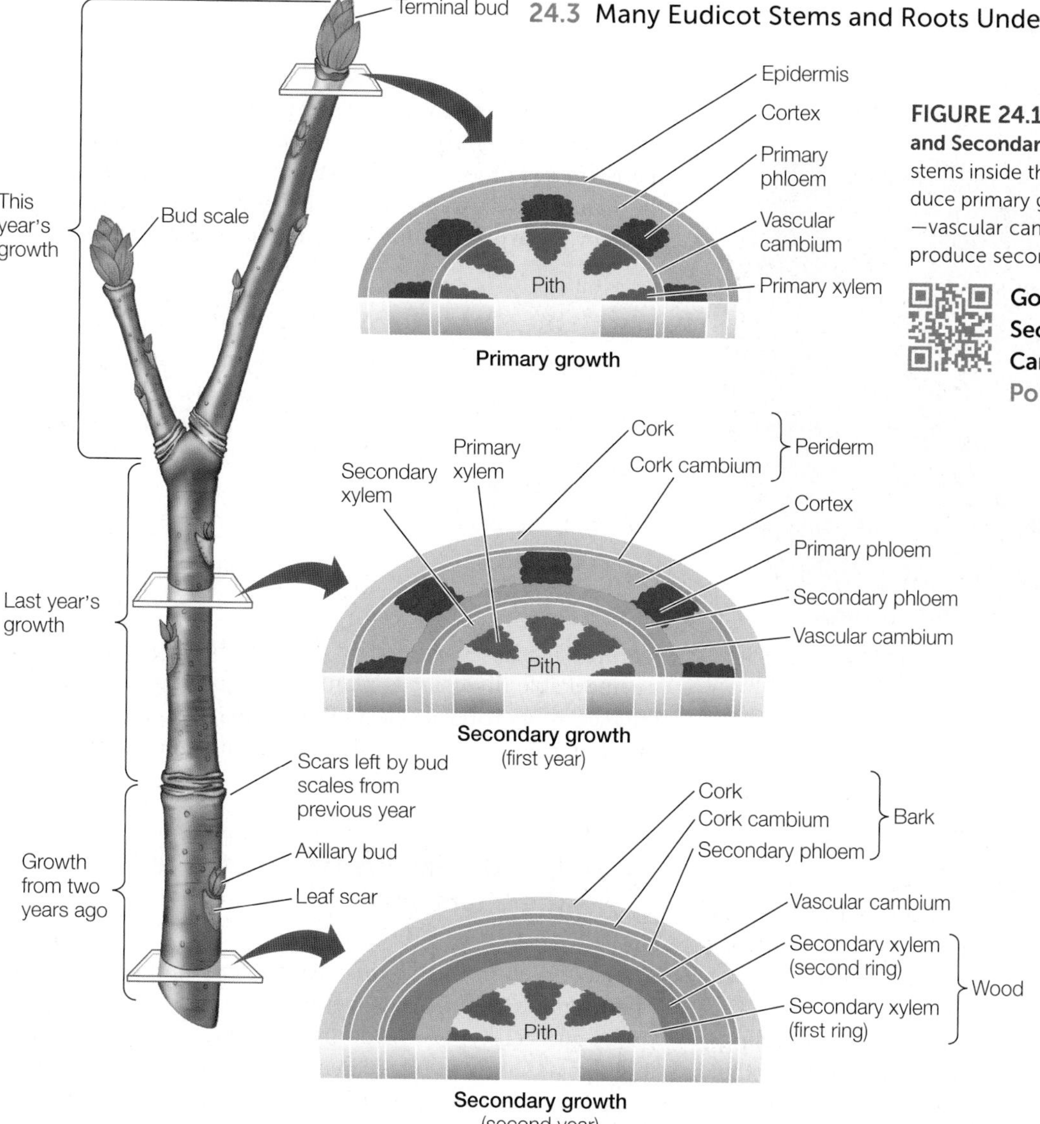

FIGURE 24.14 A Woody Twig Has Both Primary and Secondary Tissues The shoot apical meristems inside the buds of this dormant twig will produce primary growth in spring. Lateral meristems —vascular cambium and cork cambium—will produce secondary growth.

Go to ANIMATED TUTORIAL 24.1 Secondary Growth: The Vascular Cambium
PoL2e.com/at24.1

CONCEPT 24.3 Many Eudicot Stems and Roots Undergo Secondary Growth

As we mentioned earlier, the roots and stems of some eudicots develop a secondary plant body, the tissues of which we commonly refer to as wood and bark. These tissues are derived by secondary growth from the two lateral meristems:

- The **vascular cambium** produces cells that make up secondary xylem (wood) and secondary phloem (inner bark).
- The **cork cambium** produces mainly waxy-walled protective cells. It supplies some of the cells that become outer bark.

Although secondary growth occurs in broadly similar ways in both roots and stems, the details differ. We focus on stems in our discussion here.

Each year, deciduous trees lose their leaves, leaving bare branches and twigs in winter. These twigs illustrate both primary and secondary growth (**FIGURE 24.14**). The apical meristems of the twigs are enclosed in buds protected by bud scales. When the buds begin to grow in spring, the scales fall away, leaving scars, which show us where the bud was and allow us to identify each year's growth. The woody portions of the twig are the product of primary and secondary growth. Only the buds and the newest growth from the apical meristem consist entirely of primary tissues.

In the stem, the vascular cambium is initially a single layer of cells lying between the primary xylem and the primary phloem within the vascular bundles. A stem increases in diameter when the cells of the vascular cambium divide, producing secondary xylem cells toward the inside of the stem and secondary phloem cells toward the outside (see Figure 24.14). Each year the vascular cambium produces a new layer of secondary xylem and a new layer of secondary phloem, further increasing the diameter of the stem.

The successive layers of secondary xylem, which constitute the **wood**, all continue to conduct water and minerals for many years. A given year's secondary phloem, however, is active in transport of sugars for only that year; the sieve tube elements and companion cells are crushed as the next year's layer of secondary phloem is formed and pushes outward.

The cork cambium initially forms just inside the epidermis (see Figure 24.14). This meristematic tissue produces layers

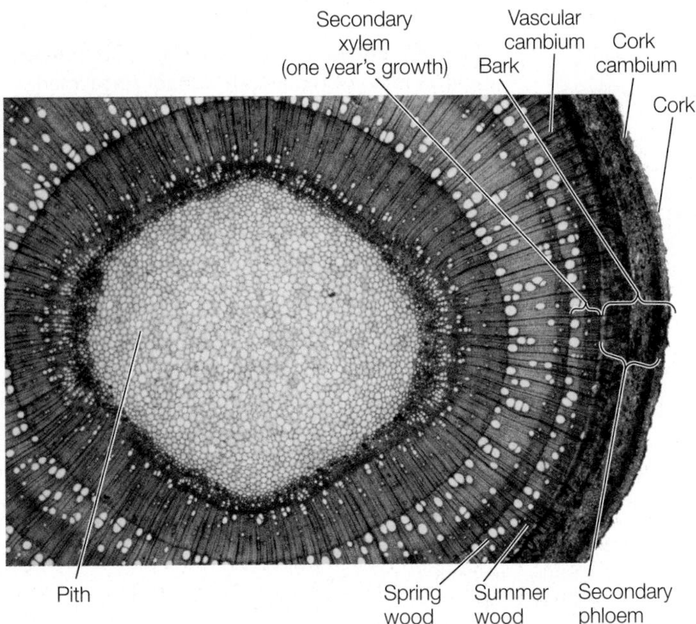

FIGURE 24.15 Annual Rings Rings of secondary xylem are a noticeable feature of this cross section from a tree trunk.

of **cork**, a protective tissue composed of cells with thick walls waterproofed with suberin. The cork cambium and the cork constitute a secondary dermal tissue called **periderm**. Once the periderm forms, the epidermis flakes off, leaving the cork as the outmost layer. The cork prevents excessive water loss and protects the plant from herbivores and pathogens.

As secondary growth of a stem continues, the expanding vascular tissue stretches and breaks the outer layers of the cortex and the initial periderm, which ultimately flake away. A new cork cambium arises from the underlying secondary phloem and produces a new periderm. The periderm and the secondary phloem—that is, all the tissues external to the vascular cambium—constitute the **bark**.

Cross sections of most tree trunks (which are mature stems) in temperate-zone forests reveal annual rings in the wood (**FIGURE 24.15**), which result from seasonal changes in environmental conditions. In spring, when water is relatively plentiful, the tracheids or vessel elements produced by the vascular cambium tend to be large in diameter and thin-walled. Such wood is well adapted for transporting large quantities of water and mineral nutrients. As water becomes less available in summer, narrower cells with thicker walls are produced, which make this summer wood darker and sometimes more dense than the wood formed in spring. Thus each growing season is recorded by a clearly visible annual ring. Trees in the moist tropics do not experience these seasonal changes, so they do not lay down such obvious regular rings. In temperate-zone trees, variations in temperature or water supply during the growing season can sometimes lead to the formation of more than one "annual" ring in a single year.

Only eudicots and other non-monocot angiosperms, along with many gymnosperms, have a vascular cambium and a cork cambium and thus undergo secondary growth. The few monocots that form thickened stems—palms, for example—do so without secondary growth. Palms have a very wide apical meristem that produces a wide stem, and dead leaf bases add to the diameter of the stem. Almost all monocots grow in essentially this way, as do other angiosperms that lack secondary growth.

CHECKpoint CONCEPT 24.3

- ✓ Compare primary and secondary growth, creating a table with the following headings: the plant groups that undergo them; the organs that are involved during the plant life cycle; the meristems that are involved; and the products of the meristems.
- ✓ Why are annual rings used to determine the age of a tree trunk?

Natural selection has acted on the plant body plan to produce a remarkable diversity of plant forms. In the next section we will see how artificial selection by humans on the plants that provide us with food and fiber has greatly altered plant form as well.

CONCEPT 24.4 Domestication Has Altered Plant Form

We have seen in this chapter that a very simple plant body plan—with roots, stems, leaves, meristems, and relatively few tissue and cell types—underlies the diversity of the flowering plants that cover our planet. Although differences in form among species are expected, members of the same plant species can be remarkably diverse in form as well. From a genetic perspective, this observation suggests that minor differences in genomes or gene regulation can underlie dramatic differences in plant form. (Nevertheless, some plant species do differ greatly in genome size, content, and organization.)

The form of the plant body is subject to natural selection. Phenotypes that can outcompete their peers for scarce resources have the highest fitness. Some plants have developed a viny, climbing phenotype that gives them access to light in crowded conditions. Others have a branched phenotype that puts neighboring plants in the shade. Still others have a deep and extensive root system that gives them access to water. Clearly, the vegetative body plan is as important in terms of evolution as are phenotypes relating directly to reproduction, such as flower morphology.

When humans domesticated crops and chose from nature those phenotypes most suitable for cultivation, they set in motion a long experiment in artificial selection of the plant body. For example, in the case of corn (maize), farmers selected plants with tall, compact growth and few branches, minimizing competition among individual plants. Corn was domesticated from the wild grass teosinte, which still grows in the hills of Mexico (**FIGURE 24.16**). One of the most conspicuous

FIGURE 24.16 Corn Was Domesticated from the Wild Grass Teosinte Beginning more than 8,000 years ago in Mexico, farmers favored plants with minimal branching. Reducing the number of branches results in fewer ears per plant but allows each ear to grow larger and produce more seeds.

differences between teosinte and domesticated corn is that teosinte, like other wild grasses, is highly branched, with many shoots, whereas domesticated corn has a single shoot. This morphological difference is due in large part to the activity of a single gene called *TEOSINTE BRANCHED 1* (*TB1*). The protein product of *TB1* regulates the growth of axillary buds. The allele of *TB1* in domesticated corn represses branching, whereas the allele in teosinte permits branching. Similarly, in the domestication of rice, farmers selected for a vertical orientation of the upper leaves so that sunlight could reach all the leaves and photosynthesis was maximized.

It may be hard to believe that a single species, *Brassica oleracea* (wild mustard), is the ancestor of so many familiar and morphologically diverse crops, including kale, broccoli, brussels sprouts, and cabbage (see Figure 15.5). An understanding of how the basic body plan of plants arises makes it possible to appreciate how each of these crops was domesticated. Starting with morphologically diverse populations of wild mustard, humans selected and planted the seeds from variants with the trait they found desirable. Many generations of such artificial selection produced the crops that fill the produce section of the supermarket and the stands of the farmers' market.

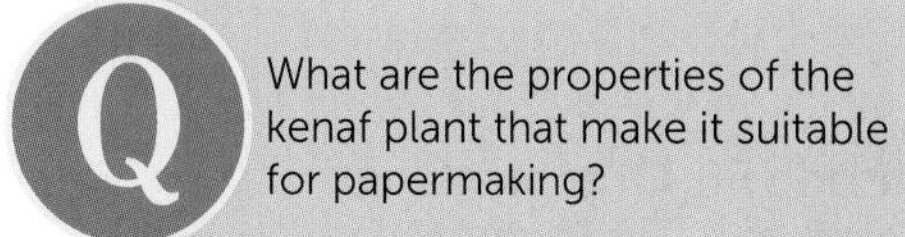

Q
What are the properties of the kenaf plant that make it suitable for papermaking?

ANSWER Kenaf is a biennial, herbaceous plant that grows rapidly from very active shoot apical meristems (Concept 24.1), reaching a height of 4–5 meters in 5 months. There are more than 500 known genetic strains of kenaf. Over the centuries since its domestication, kenaf has been selected to grow taller and to branch less (Concept 24.4). In addition, the plant's adventitious roots have become longer, at about 2 meters (Concept 24.2), and more numerous, better adapting the plant to soils that are waterlogged.

Kenaf stems are harvested for fiber (**FIGURE 24.17**). The phloem fibers are 3 millimeters long, somewhat longer than the fibers in wood; longer fibers are desirable for stronger products. In addition, the cell walls in kenaf stems are rich in cellulose but relatively poor in lignin. This makes kenaf easier to pulp than wood, which requires strong chemical treatments to hydrolyze the lignin in its cell walls. Furthermore, 1 hectare of kenaf produces 11–17 tons of fiber suitable for pulp and paper production; this is at least three times more than a hectare of southern pine trees produces. And bear in mind that it takes 5 months to grow the kenaf and 20 years to grow the pines!

The main uses of kenaf today are for the manufacture of rope, fiber-based bags, particleboard, and most recently, paper. The last use looks increasingly viable economically.

FIGURE 24.17 Kenaf Stems The stem of kenaf plants is rich in long phloem fibers. Their low lignin content makes the fibers attractive for making paper as well as other products. Paper derived from kenaf fibers has been used for printing newspapers.

SUMMARY

CONCEPT 24.1 The Plant Body Is Organized and Constructed in a Distinctive Way

- The vegetative organs of flowering plants are **roots**, which form a **root system**, and **stems** and **leaves**, which (together with flowers, which are sexual organs) form a **shoot system**. **Review Figure 24.1**
- Plant development is influenced by three unique properties of plants (compared with animals): **meristems**, the presence of **cell walls**, and the totipotency of most plant cells.
- During embryogenesis, the **apical–basal** axis and the **radial** axis of the plant body are established, as are the **shoot apical meristem** and the **root apical meristem**. **Review Figures 24.4 and 24.5**
- Three **tissue systems**, arranged concentrically, extend throughout the plant body: the **dermal tissue system**, **ground tissue system**, and **vascular tissue system**. **Review Figure 24.6**
- The vascular tissue system includes **xylem**, which conducts water and mineral ions absorbed by the roots to the shoot, and **phloem**, which conducts the products of photosynthesis throughout the plant body.

CONCEPT 24.2 Apical Meristems Build the Primary Plant Body

- **Primary growth** is characterized by the lengthening of roots and shoots and by the proliferation of new roots and shoots through branching. Some plants also undergo **secondary growth**, by which they increase in thickness.
- **Apical meristems** generate primary growth, and **lateral meristems** generate secondary growth. **Review Figure 24.7**
- Apical meristems at the tips of shoots and roots give rise to three **primary meristems** (**protoderm**, **ground meristem**, and **procambium**), which in turn produce the three tissue systems of the plant body.
- The root apical meristem gives rise to the **root cap** and the three primary meristems. The cells in the root tip are arranged in three zones that grade into one another: the **zone of cell division**, **zone of cell elongation**, and **zone of cell maturation**. **Review Figure 24.8**
- The vascular tissue of roots is contained within the **stele**. It is arranged differently in eudicot and monocot roots. **Review Figure 24.9 and ACTIVITIES 24.1 and 24.2**
- The root system of eudicots typically consists of a **taproot** and **lateral roots**. Monocots typically have a **fibrous root system** made up of **adventitious** roots. **Review Figure 24.10**
- In stems, the vascular tissue is divided into **vascular bundles**, each containing both xylem and phloem. **Review Figure 24.11 and ACTIVITIES 24.3 and 24.4**
- Eudicot leaves have two zones of photosynthetic **mesophyll** cells that are supplied by veins with water and minerals. **Review Figure 24.12 and ACTIVITY 24.5**
- The organs of the primary plant body may be modified to perform specialized functions. **Review Figure 24.13**

CONCEPT 24.3 Many Eudicot Stems and Roots Undergo Secondary Growth

- Two lateral meristems, the **vascular cambium** and **cork cambium**, are responsible for secondary growth. The vascular cambium produces secondary xylem (**wood**) and secondary phloem (inner **bark**). The cork cambium produces a protective tissue called **cork**. **Review Figures 24.14 and 24.15 and ANIMATED TUTORIAL 24.1**

CONCEPT 24.4 Domestication Has Altered Plant Form

- Although the plant body plan is simple, it can be changed dramatically by minor genetic differences, as evidenced by the natural diversity of wild plants.
- Crop domestication involves artificial selection of certain desirable traits found in wild populations. As a result of artificial selection over many generations, the body forms of crop plants are very different from those of their wild relatives. **Review Figure 24.16**

Go to the Interactive Summary to review key figures, Animated Tutorials, and Activities PoL2e.com/is24

Go to LaunchPad at **macmillanhighered.com/launchpad** for additional resources, including LearningCurve Quizzes, Flashcards, and many other study and review resources.

25 Plant Nutrition and Transport

KEY CONCEPTS

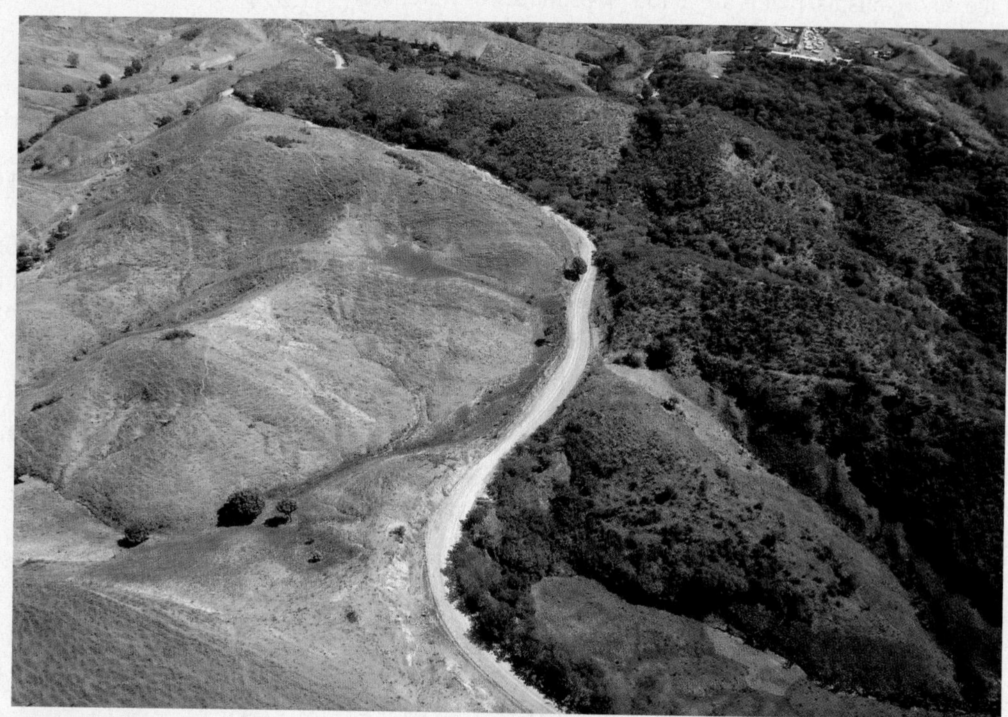

An aerial view of the border between Haiti (left) and the Dominican Republic (right) provides a dramatic illustration of the extent of soil degradation in Haiti.

The 2010 earthquake was a catastrophe for Haiti, the poorest country in the Western Hemisphere. More than 200,000 people died, and more than a million people were left homeless. The urban infrastructure in the capital city was destroyed. Amid the outpouring of help from the rest of the world came a realization that another disaster has been occurring in Haiti for decades: the destruction of its forests and farms. Plants need fertile soil to grow, and in Haiti that has been in increasingly short supply.

Haiti shares the island of Hispaniola with the Dominican Republic. The two countries have had dramatically different policies with regard to land conservation, with equally dramatic results. While the Dominicans have carefully managed their land, the story of Haiti is one of overexploitation. First, Spanish colonists planted sugarcane, a crop that depleted the soil. Then the French colonists that followed them cut down forests to plant coffee and tobacco for export to Europe. After the former slaves revolted and expelled the Europeans in the early 1800s, land speculators chased out indigenous farmers from fertile valleys and planted export crops such as corn and beans. Today only 4 percent of the forests that existed when the Europeans arrived in 1492 remain. As Haitians have said, "Te a fatige (The earth is tired)." Because there are few plant roots to stabilize the soil, especially in Haiti's many mountainous regions, rain washes the soil into the sea. From an airplane window it is easy to see the 120-mile-long border between the two countries. The Dominican side is verdant and forested, while the Haitian side is devoid of plant life and has lost most of its soil.

As human land use intensifies, ecological disasters like Haiti's are becoming all too common. Today, parts of sub-Saharan Africa's farmland are losing their topsoil as a result of poor land management, swelling populations, and a challenging climate. Crop failures, starvation, and large-scale human displacements are inevitable consequences. The need for a better understanding of how plants interact with the soil is a pressing human concern.

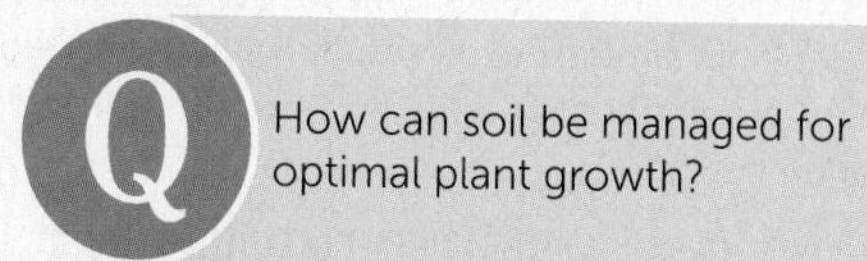

How can soil be managed for optimal plant growth?

You will find the answer to this question on page 553.

CONCEPT 25.1 Plants Acquire Mineral Nutrients from the Soil

Like all organisms, plants are made up primarily of water and molecules that contain the major elements of life. Nearly all plants are **autotrophs** that obtain carbon from atmospheric carbon dioxide through the carbon-fixation reactions of photosynthesis. They obtain carbon, hydrogen, and oxygen from air and water, so these elements are plentiful as long as plants have an adequate water supply.

The plant body contains other elements as well. Nitrogen, for example, is essential for the synthesis of proteins and nucleic acids. Plants obtain nitrogen, directly or indirectly, through the activities of bacteria and fungi, as we will see later in this chapter. In addition, proteins contain sulfur (S), nucleic acids contain phosphorus (P), and chlorophyll contains magnesium (Mg). Metal ions also play important biochemical roles in plants. For example, Fe^{2+} is used as a cofactor in electron transport in mitochondria and chloroplasts. Plants obtain these and other mineral nutrients mainly from the soil. Mineral nutrients dissolve in water in the soil as ions, forming a solution—called the **soil solution**—that surrounds the roots of plants.

LINK

The major elements of life are described in **Chapter 2**, and the chemical reactions of photosynthesis in **Concepts 6.5 and 6.6**

Nutrients can be defined by their deficiency

A plant nutrient is called an **essential element** if its absence causes severe disruption of normal plant growth and reproduction. An essential element cannot be replaced by any other element. There are two categories of essential elements:

- **Macronutrients** have concentrations of at least 1 gram per kilogram of the plant's dry matter. There are six macronutrients:

  Nitrogen (taken up as NO_3^- or NH_4^+)
  Phosphorus (taken up as PO_4^{3-})
  Potassium (taken up as K^+)
  Sulfur (taken up as SO_4^{2-})
  Calcium (taken up as Ca^{2+})
  Magnesium (taken up as Mg^{2+})

- **Micronutrients** have concentrations of less than 0.1g/kg of the plant's dry matter. They include iron (Fe^{2+}), chlorine (Cl^-), manganese (Mn^{2+}), zinc (Zn^{2+}), copper (Cu^{2+}), nickel (Ni^{2+}), boron (BO_3^{-3}), and molybdenum [$Mo(O_4)^{2-}$].

Plants show characteristic symptoms, most noticeable in leaves, that can be used to diagnose deficiencies of essential elements (**FIGURE 25.1**). With proper diagnosis, the missing nutrient(s) can be provided in the form of a supplement, or **fertilizer**.

Experiments using hydroponics have identified essential elements

The essential elements for plants were largely identified by growing plants **hydroponically**—that is, with their roots suspended in nutrient solutions instead of soil:

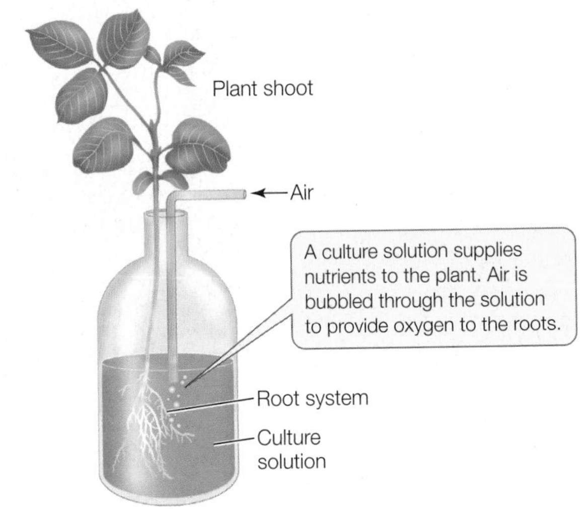

Plant shoot
Air
A culture solution supplies nutrients to the plant. Air is bubbled through the solution to provide oxygen to the roots.
Root system
Culture solution

Growing plants in this manner allows for greater control of nutrient availability than is possible in a complex medium like soil.

In the first successful experiments of this type, performed a century and a half ago, plants grew seemingly normally in solutions containing only calcium nitrate [$Ca(NO_3)_2$], magnesium sulfate ($MgSO_4$), and potassium phosphate (KH_2PO_4).

FIGURE 25.1 Mineral Nutrient Deficiency Symptoms The plants on the left were grown with a full complement of essential nutrients. The yellow leaves of the center plants are typical of iron deficiency, while the stunted growth of the plants on the right indicate nitrogen deficiency.

Go to ANIMATED TUTORIAL 25.1 Nitrogen and Iron Deficiencies
PoL2e.com/at25.1

INVESTIGATION

FIGURE 25.2 Nickel Is an Essential Element for Plants Using highly purified minerals in a hydroponic solution, Patrick Brown and his colleagues tested whether barley can complete its life cycle in the absence of nickel.[a] Other investigators showed that no other element could substitute for nickel.

HYPOTHESIS

Nickel is an essential element for plants.

METHOD

1. Grow barley plants for three generations in hydroponic solutions containing 0, 0.6, and 1.0 μM nickel sulfate ($NiSO_4$).
2. Harvest seeds from 5 or 6 third-generation plants in each of the groups.
3. Determine the nickel concentration in some of the seeds from each plant.
4. Germinate the other seeds from the same plants in a nickel-free hydroponic solution and plot the success of germination against nickel concentration.

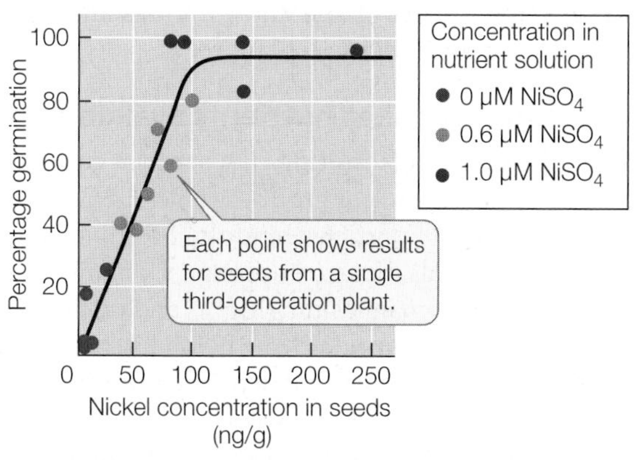

RESULTS

There was a positive correlation between seed germination and nickel concentration. There was significantly less germination at the lowest nickel concentrations.

CONCLUSION

Barley seeds require nickel in order to germinate and complete their life cycle.

ANALYZE THE DATA

A. What do the blue points have in common? Why are there five blue points instead of just one?

B. What is the biological meaning of the zero/zero point?

C. Why didn't the investigators examine only the plants grown without nickel in their hydroponic solution?

D. Plant biologists define the critical value for a mineral nutrient as that concentration in plant tissue that results in a 15 percent reduction in the optimal yield of the plant. Using maximum germination percentage in this experiment as the optimal yield, estimate the critical value for nickel.

Go to **LaunchPad** for discussion and relevant links for all **INVESTIGATION** figures.

[a]P. H. Brown et al. 1987. *Plant Physiology* 85: 801–803.

A solution missing any of these compounds could not support normal plant growth. Tests with other compounds that included various combinations of these same elements soon established the six macronutrients listed above.

Identifying micronutrients by this experimental approach proved to be more difficult because of the small amounts involved. Sufficient amounts of a micronutrient can be present as contaminants in the environment used to grow plants. Therefore nutrition experiments must be performed in tightly controlled laboratories with special air filters that exclude microscopic mineral particles, and they must use only the purest available chemicals. Furthermore, a seed may contain enough of a micronutrient to supply the embryo and the resultant second-generation plant throughout its lifetime. There might even be enough left over to pass on to the next generation of plants. Thus even if a plant is completely deprived of a micronutrient, it may take a long time for deficiency symptoms to appear. The most recent micronutrient to be listed as essential was nickel, in 1983 (**FIGURE 25.2**).

Soil provides nutrients for plants

Most terrestrial plants grow in soil, which provides plants with a number of benefits:

- Anchorage for mechanical support of the shoot
- Mineral nutrients and water from the soil solution
- O_2 for root respiration in root tissues from air spaces between soil particles
- The services of other soil organisms

Soils have both living and nonliving components. The living components include plant roots as well as populations of bacteria, fungi, protists, and animals such as earthworms and insects. The nonliving portions of the soil include rock fragments ranging in size from large stones to clay particles 2 μm or less in diameter. Soil also contains water and dissolved mineral nutrients, air spaces, and dead organic matter:

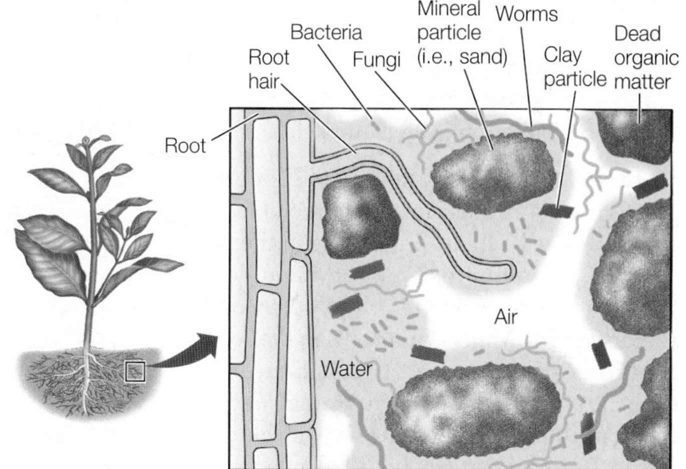

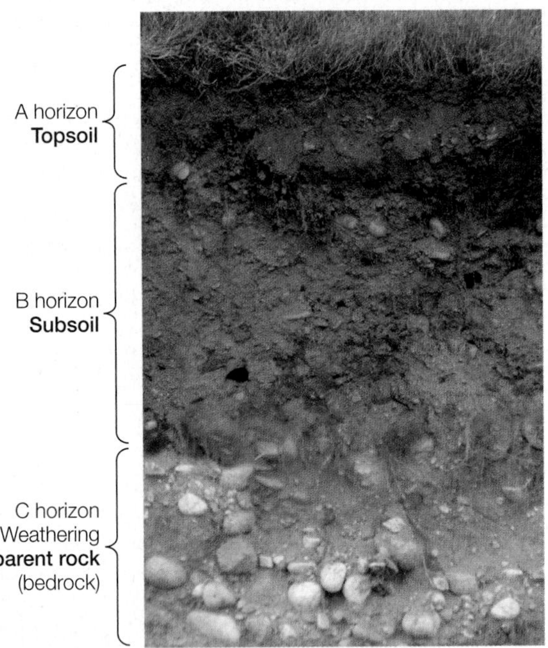

FIGURE 25.3 A Soil Profile The A, B, and C horizons of a soil can sometimes be seen at construction sites such as this one in Massachusetts. The darker upper layer (the A horizon) is home to most of the living organisms in the soil.

Although soils vary greatly, almost all of them have a soil profile consisting of several horizontal layers, called **horizons**. Soil scientists recognize three major horizons—termed A, B, and C—in a typical soil (**FIGURE 25.3**):

- The A horizon is the **topsoil** that supports the plant's mineral nutrient needs. It contains most of the soil's living and dead organic matter.
- The B horizon is the **subsoil**, which accumulates materials from the topsoil above it and the parent rock below.
- The C horizon is the **parent rock** from which the soil arises.

Parent rock is broken down into soil particles (weathered) in two ways:

- Mechanical weathering is the physical breakdown of materials by wetting, drying, and freezing.
- Chemical weathering is the alteration of the chemistry of the materials in the rocks by oxidation, water, and acids.

The soil's fertility (its ability to support plant growth) is determined by several factors. The size of the soil particles in the A horizon, for example, is important in determining nutrient retention and other soil characteristics. Mineral nutrients in the soil solution can be **leached** (washed away) from the A horizon downward by rain (or irrigation) and thereby become unavailable to plants. Tiny clay particles bind a lot of water, and mineral nutrients are not readily leached from them, but clay particles pack tightly and leave little room for air. Large sand particles have opposite properties; they provide plenty of air spaces but do not retain water or nutrients. A **loam** is a soil that is an optimal mixture of sand, clay, and silt (mineral particles intermediate in size between sand and clay) and thus has sufficient supplies of air, water, and nutrients for plants.

In addition to mineral particles, soils contain dead organic matter, largely from plants. Soil organisms break down dead leaves and other organic materials that fall to the ground into a substance called **humus**. This material is used as a food source by microbes that break down complex organic molecules and release simpler molecules into the soil solution. Humus also improves the texture of soil and provides air spaces that increase oxygen availability to plant roots.

Ion exchange makes nutrients available to plants

Chemical weathering results in clay particles covered with negatively charged chemical groups (**FIGURE 25.4**). These negatively charged clay particles bind the positively charged ions (cations) of minerals that are important for plant nutrition, such as potassium (K^+), magnesium (Mg^{2+}), and calcium (Ca^{2+}). This binding prevents the cations from being leached out of the soil, but to become available to plants, they must be released from the clay particles into the soil solution.

Recall that the root surface is covered with epidermal cells bearing root hairs (see Figure 24.8C). Proteins in the cell membrane of these cells actively pump protons (H^+) out of the cells, as we'll describe in Concept 25.3. In addition, cellular respiration in the roots releases CO_2, which dissolves in the soil water and reacts with it to form carbonic acid. This acid ionizes to form bicarbonate and free protons:

$$CO_2 + H_2O \rightleftharpoons H_2CO_3 \rightleftharpoons H^+ + HCO_3^-$$

Proton pumping by the root cells and ionization of carbonic acid both act to increase the proton concentration in the soil

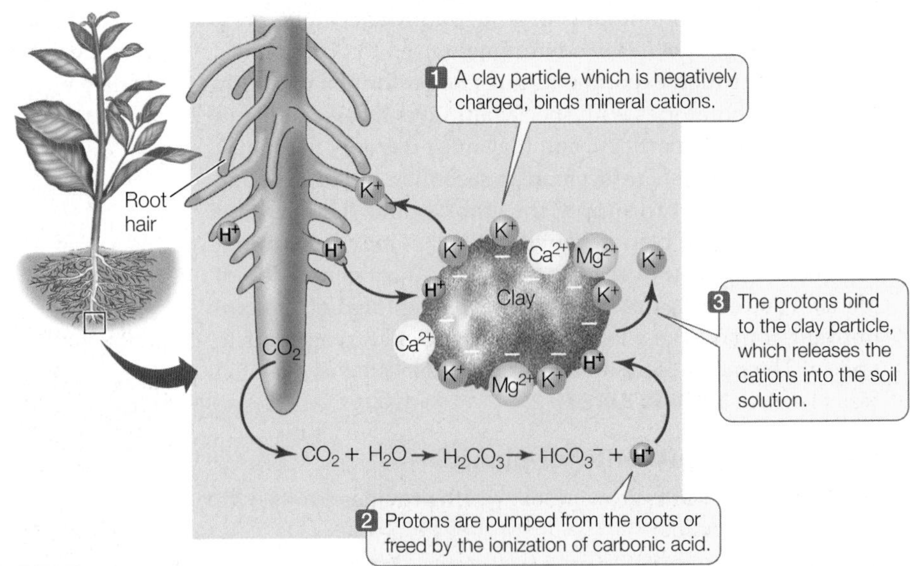

FIGURE 25.4 Ion Exchange Plants obtain some mineral nutrients from the soil in the form of positive ions bound to clay particles in the soil.

surrounding the root. The protons bind more strongly to clay particles than do mineral cations; in essence, the protons trade places with the cations in a process called **ion exchange** (see Figure 25.4). Ion exchange releases mineral nutrient cations into the soil solution, where they are available to be taken up by roots. Soil fertility is determined in part by the soil's ability to provide nutrients in this manner.

There is no comparable mechanism for binding and releasing negatively charged ions. As a result, important anions such as nitrate (NO_3^-) and sulfate (SO_4^{2-}) may be leached from the A horizon.

Fertilizers can be used to add nutrients to soil

Leaching and the harvesting of crops may deplete a soil of its nutrients, in which case the growth of subsequent crops on that soil will be reduced. Three ways to restore the nutrient content of soils are shifting agriculture, organic fertilizers, and inorganic (chemical) fertilizers.

SHIFTING AGRICULTURE In the past, when the soil could no longer support a level of plant growth sufficient for agricultural purposes, people simply moved to another location. This nomadic existence allowed the soil to replenish its nutrients naturally, by weathering of the parent rock and gradual accumulation and breakdown of organic matter. These processes take a long time, but that was not a problem as long as a lot of land was available. After many years in other locations, the people could return to their original land, which was again able to support farming. Today the food needs of a large human population are too great to allow land to be left unfarmed for a long time; in addition, most people live more settled lives now and do not want to move.

ORGANIC FERTILIZERS Humus is used as a food source by soil organisms, which in turn release simpler molecules to the soil solution. For example:

$$\text{Leaf proteins} \xrightarrow{\text{Bacteria}} NH_4^+ \xrightarrow{\text{Bacteria}} NO_3^-$$

These simpler molecules can dissolve in soil water and be taken up by plant roots.

Farmers can increase the nutrient content of soil by adding organic materials such as compost (partially decomposed plant material) or manure (the waste from farm animals), which is a particularly good source of nitrogen. In either case, these **organic fertilizers** add nutrients to the soil much more rapidly than natural processes, but still allow for a slow release of ions, with little leaching, as the materials decompose.

INORGANIC FERTILIZERS Because organic fertilizers may still act too slowly to restore fertility if a soil is to be farmed every year, **inorganic fertilizers** are used to supply mineral nutrients directly in forms that can be immediately taken up by plants. Inorganic fertilizers are characterized by their "N-P-K" percentages. A 5-10-10 fertilizer, for example, contains 5 percent nitrogen (as NH_4^+), 10 percent phosphate (P_2O_5), and 10 percent potash (K_2O) by weight (of the nutrient-containing compound, not as weights of the elements N, P, and K). Sulfur, in the form of ammonium sulfate [$(NH_4)_2SO_4$], is also occasionally added to soils.

To return to our example of nitrogen, inorganic fertilizers reach plants by this process:

$$\text{Chemical manufacture} \rightarrow NH_4^+ \xrightarrow{\text{Bacteria}} NO_3^-$$

The manufacture of nitrogen fertilizer begins with an energy-intensive industrial process in which N_2 is combined with H_2 at high temperatures and pressures in the presence of a catalyst to form ammonia (NH_3). This process accounts for about 2 percent of the world's energy use, mostly in the form of natural gas. Although organic and inorganic fertilizers have the same effects on crop nutrition, their costs are very different: manure and compost are free, whereas NH_4^+ made in a factory can be very expensive.

LINK

The cycling of nitrogen through biological systems is described in **Chapter 45**, and some of the prokaryotes that participate in that cycling are described in **Concept 19.3**

CHECKpoint CONCEPT 25.1

✓ What are the differences between:
- a. macronutrients and micronutrients?
- b. organic and inorganic fertilizers?
- c. the availability to plant roots of a cation and an anion?

✓ Outline an experimental method for determining whether an element is essential to a plant.

✓ A label on a bin in a market states "Organic apples, grown without chemical fertilizers." Are these apples different in terms of plant molecules than they would be if they were grown with chemical fertilizers?

We have just described some of the physical and chemical aspects of soils that are vital to plant nutrition. But many other organisms besides plants inhabit soils. We will turn now to the effects of those organisms on plants.

CONCEPT 25.2 Soil Organisms Contribute to Plant Nutrition

One gram of soil contains from 6,000 to 50,000 bacterial species and up to 200 meters of fungal hyphae, although both are largely invisible to the naked eye. Many of these soil organisms consume living and dead plant materials. Plants have an array of defenses they can use against potentially harmful soil organisms (which we will describe in Chapter 28). In many cases, however, plants *actively encourage* fungi and bacteria to infect their roots. Here we describe the resulting "intercellular trading posts," where products are exchanged to the mutual benefit of plants and a few very special soil microbes. In these interactions—called symbioses (singular *symbiosis*)—the plants usually "trade" the products of photosynthesis, which the microbes use as an energy source, for other essential elements, such as phosphorus and nitrogen, which the microbes can supply.

LINK

These plant–microbe interactions are examples of mutualism, one of five major types of interspecies interactions discussed in **Concept 43.1**

We will focus on two important plant–microbe interactions:

- Arbuscular mycorrhizae, a plant–fungus symbiosis (see Figure 22.10)
- Symbiotic nitrogen fixation, an association of nitrogen-fixing bacteria (see Concept 19.3) with plant roots

We'll begin by describing how these associations form. Then we'll see how they benefit the plants and their microbial partners.

Plants send signals for colonization

In the case of both arbuscular mycorrhizae and symbiotic nitrogen fixation, the roots of the potential host plant secrete signaling molecules that attract the soil organisms. These signals contribute to the specificity of the interactions, because only one or a few microbe species will respond to a particular signal molecule.

FORMATION OF MYCORRHIZAL ASSOCIATIONS In the case of arbuscular mycorrhizae, plant roots produce chemical signal molecules called **strigolactones** that stimulate rapid growth of fungal hyphae toward the root (**FIGURE 25.5A**). In response, the fungi signal the plant to form a prepenetration apparatus (PPA), which guides the growth of the fungal hyphae into the root cortex. The sites of nutrient exchange between fungus and plant are the arbuscules, which form inside root cortical cells. Despite the intimacy of this association, the plant and fungal cytoplasms never mix—they are separated by two membranes, the fungal cell membrane and the periarbuscular membrane (PAM), which is continuous with the plant cell membrane.

FORMATION OF NITROGEN-FIXING NODULES A group of plants called legumes (family Fabaceae) can form symbioses with several species of soil bacteria collectively known as rhizobia, which live in nodules that form on the legumes' roots. The roots of legumes release flavonoids and other chemical signals that attract the rhizobia (**FIGURE 25.5B**). The flavonoids also stimulate the bacteria to secrete Nod (nodulation) factors, which cause cells in the root cortex to divide and form the plant tissue that constitutes the root nodule. Bacteria enter the root via an infection thread, analogous to the PPA in mycorrhizal associations, and eventually reach cells inside the root nodule. There the bacteria are released into the cytoplasm of the nodule cells, enclosed in plant membrane–derived vesicles that are functionally equivalent to the PAM. The bacteria differentiate into **bacteroids**—the form of the bacteria that can fix nitrogen.

A COMMON MECHANISM There is increasing evidence that nodule formation depends on some of the same genes and mechanisms that allow mycorrhizal associations to develop. For example, both processes involve folding in of the cell membrane to allow entry of the fungal hyphae or rhizobia. The similarities in the structures formed (see Figure 25.5) are striking, considering that the symbioses involve members of two different groups of organisms (fungi and bacteria).

LINK

The fact that similar genes and developmental processes are used in two different types of symbiosis provides a striking demonstration of the principles of evolutionary developmental biology; see **Concept 14.4**

Mycorrhizae expand the root system

In most vascular plants, the roots alone cannot support an optimal growth rate—they simply cannot reach all the nutrients available in the soil. Mycorrhizae, which occur in more than 90 percent of terrestrial plants, expand the root surface area

APPLY THE CONCEPT

Soil organisms contribute to plant nutrition

Researchers grew lemon seedlings in soils that contained phosphate fertilizer, arbuscular mycorrhizal fungi, or both. The table shows the mean dry weight of the seedlings after 6 months.

1. What do these results tell us about the role of phosphate fertilizer in the growth of lemon plants?
2. Why didn't the mean weight of the seedlings increase when the amount of phosphate fertilizer, in the absence of mycorrhizae, was doubled?

PHOSPHATE FERTILIZER ADDED (g)	MYCORRHIZAE PRESENT	SEEDLING DRY WEIGHT (g)
0	No	1
0	Yes	10
12	No	28
12	Yes	166
24	No	20
24	Yes	210

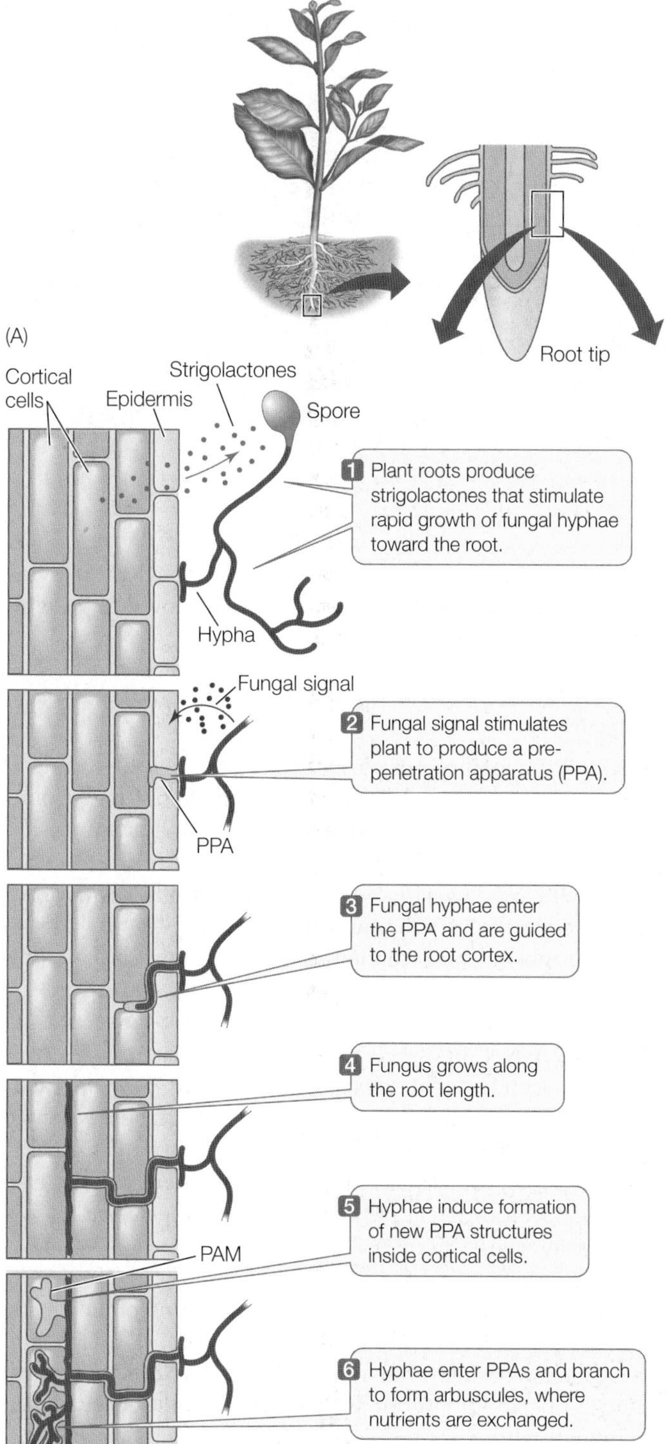

Root tip
(A)
Cortical cells
Strigolactones
Epidermis
Spore
1 Plant roots produce strigolactones that stimulate rapid growth of fungal hyphae toward the root.
Hypha
Fungal signal
2 Fungal signal stimulates plant to produce a pre-penetration apparatus (PPA).
PPA
3 Fungal hyphae enter the PPA and are guided to the root cortex.
4 Fungus grows along the root length.
PAM
5 Hyphae induce formation of new PPA structures inside cortical cells.
6 Hyphae enter PPAs and branch to form arbuscules, where nutrients are exchanged.

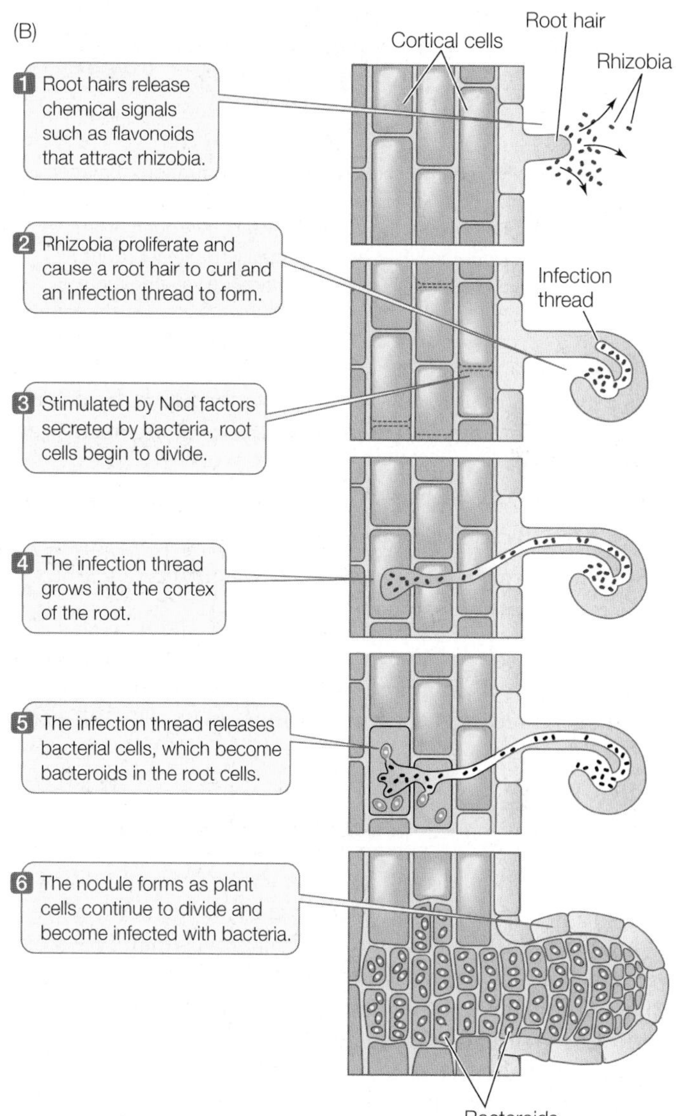

(B)
Cortical cells
Root hair
Rhizobia
1 Root hairs release chemical signals such as flavonoids that attract rhizobia.
2 Rhizobia proliferate and cause a root hair to curl and an infection thread to form.
Infection thread
3 Stimulated by Nod factors secreted by bacteria, root cells begin to divide.
4 The infection thread grows into the cortex of the root.
5 The infection thread releases bacterial cells, which become bacteroids in the root cells.
6 The nodule forms as plant cells continue to divide and become infected with bacteria.
Bacteroids

FIGURE 25.5 Roots Send Signals for Colonization Plant roots send chemical signals to arbuscular mycorrhizal fungi (A) and nitrogen-fixing bacteria (B) to stimulate colonization. Both processes do not occur in all plants.

10-fold to 1,000-fold (see Figure 22.10A). Fungal hyphae are also much finer than root hairs; they can get into pores in the soil that are inaccessible to roots. Mycorrhizae thus probe a vast expanse of soil for nutrients and deliver them into root cortical cells.

The primary nutrient that the plant obtains from a mycorrhizal interaction is phosphorus. In exchange, the fungus obtains an energy source: the products of photosynthesis. In fact, up to 20 percent of the photosynthate (carbohydrates produced by photosynthesis) of terrestrial plants is directed to and consumed by arbuscular mycorrhizal fungi.

Rhizobia capture nitrogen from the air and make it available to plant cells

Although nitrogen is essential for the survival of plants, they cannot use atmospheric N_2 because they lack the biochemical machinery to break the very stable triple bond that links the two N atoms in N_2. Some bacteria, however, have an enzyme called **nitrogenase** that enables them to convert N_2 into a more reactive and biologically useful form, ammonia (NH_3), by a process called **nitrogen fixation**:

$$N_2 + 6\,H \rightarrow 2\,NH_3$$

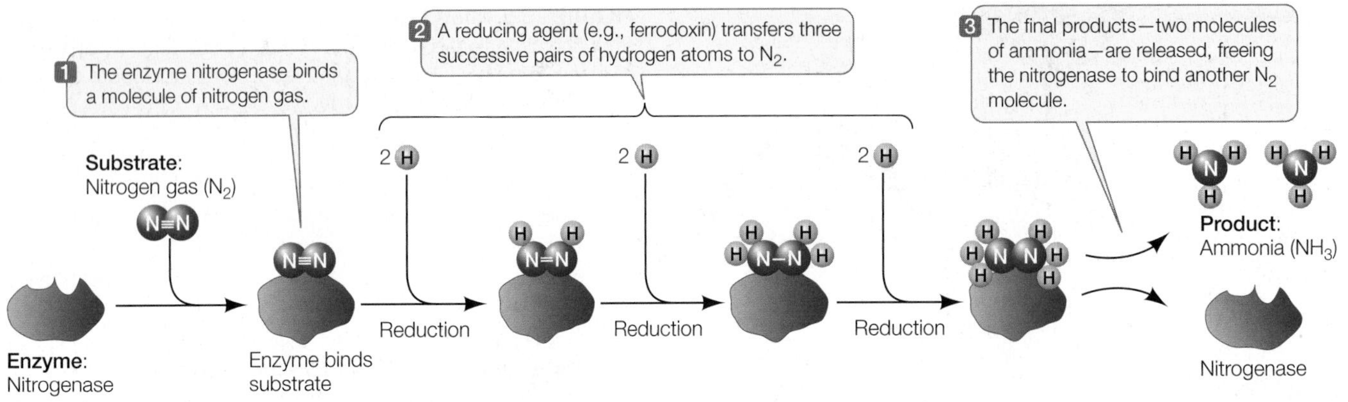

FIGURE 25.6 Nitrogenase Fixes Nitrogen Throughout the chemical reactions of nitrogen fixation, the reactants are bound to the enzyme nitrogenase. A reducing agent transfers hydrogen atoms to a molecule of nitrogen gas (N_2), and eventually the final product—ammonia (NH_3)—is released. This reaction requires a large input of energy: about 16 ATPs.

This process is by far the most important route by which nitrogen enters biological systems. Two types of organisms can fix nitrogen:

- Free-living (nonsymbiotic) organisms in soil and water (e.g., *Azotobacter* bacteria and *Nostoc* cyanobacteria)
- Symbiotic organisms living in other organisms (e.g., rhizobia in the roots of legumes)

Biological nitrogen fixation is the reduction of nitrogen gas, and it requires a great deal of energy. It proceeds by the stepwise addition of three pairs of hydrogen atoms to N_2 (**FIGURE 25.6**).

LINK

You can review reduction–oxidation reactions in **Concept 6.1**

Nitrogenase is strongly inhibited by oxygen. Many free-living nitrogen-fixing bacteria are anaerobes that live in environments that naturally contain little or no O_2. The environment in the plant root tissues is typically aerobic, but the nodules provide the rhizobia with the conditions necessary for nitrogen fixation (see Figure 25.5B). Within a nodule, O_2 is maintained at a low level that is sufficient to support aerobic respiration by the bacteria (which is necessary to supply energy for the fixation reaction) but not so high as to inactivate nitrogenase. The O_2 level is regulated by a plant-produced protein called **leghemoglobin**, which is an O_2 carrier. Leghemoglobin is a close relative of hemoglobin, the red, oxygen-carrying pigment of animals, and is thus an evolutionarily ancient molecule.

As with mycorrhizae, maintaining nitrogen-fixing bacteria is costly to the plant, consuming as much as 20 percent of all of the energy stored in photosynthate. Relatively few plant species are legumes, but those species are important to humans because they can be grown in nitrogen-poor soil. They include peas, soybeans, clover, alfalfa, and some tropical shrubs and trees. Most other plants depend on nitrogen fixed by free-living bacteria in the soil or released by the breakdown of proteins present in dead organic matter.

Some plants obtain nutrients directly from other organisms

Although the majority of plants obtain most of their mineral nutrients from the soil solution (with the help of fungi), some, such as carnivorous and parasitic plants, use other sources.

CARNIVOROUS PLANTS There are about 500 carnivorous plant species that obtain some of their nutrients by digesting arthropods. These plants typically grow in boggy soils where little nitrogen or phosphorus is available. Digestion (hydrolysis) of arthropod prey helps provide those missing nutrients. A well-known example of a carnivorous plant is the Venus flytrap (genus *Dionaea*; **FIGURE 25.7A**), which has a modified leaf with two halves that fold together. When an insect touches trigger hairs on the leaf, its two halves quickly come together, and their spiny margins interlock and trap the insect before it can escape. The leaf then secretes enzymes that digest the prey.

PARASITIC PLANTS Approximately 1 percent of flowering plant species derive some or all of their water, nutrients, and even photosynthate from other plants. **Hemiparasites** can photosynthesize but derive water and mineral nutrients from the living bodies of other plants. Mistletoes, for example, carry out some photosynthesis, but they parasitize other plants for water and mineral nutrients and may derive some photosynthetic products from them as well. Dwarf mistletoe (*Arceuthobium americanum*) is a serious parasite in forests of the western United States, destroying more than 3 billion board feet of lumber per year. **Holoparasites** are completely parasitic and do not perform photosynthesis. Witchweed (*Striga*) is a serious parasite of crops, causing more than $1 billion a year in corn losses in Africa (**FIGURE 25.7B**).

A COMMON ASSOCIATION Interestingly, strigolactone, the molecule made by plant roots that attracts arbuscular mycorrhizal fungi, is closely related to the molecule that *Striga* is attracted to as it seeks potential hosts. The mycorrhizal

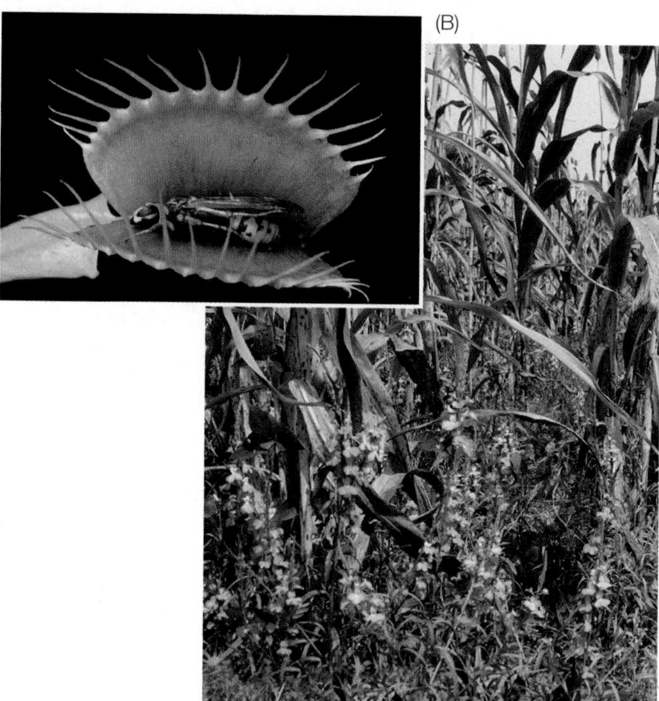

FIGURE 25.7 Nutrients from Other Organisms (A) The Venus flytrap (*Dionaea muscipula*) obtains nitrogen from the bodies of arthropods trapped inside the plant when its specialized leaves snap shut. (B) Purple witchweed (*Striga hermonthica*) parasitizes sorghum plants.

association is ancient (more than 400 million years old) and predates the evolution of parasitic plants. Scientists hypothesize that the ancestors of modern *Striga* evolved to take advantage of a compound that was already produced by plants to attract beneficial soil microbes. This is an example of "opportunistic evolution," the repurposing of preexisting processes rather than the development of new processes.

CHECKpoint CONCEPT **25.2**

- ✓ Compare the events that occur when arbuscular mycorrhizal fungi and rhizobia infect a plant root. What is the nutritional advantage to the plant in each case?
- ✓ How might you turn wheat, a plant that does not have symbiotic relationships with nitrogen-fixing bacteria, into a nitrogen-fixing plant? What biochemical challenges would you face?
- ✓ How do the nutritional needs of holoparasitic plants differ from those of carnivorous plants?
- ✓ What characteristics are shared among nonparasitic plant–parasitic plant, plant–mycorrhizal fungus, and plant–rhizobia associations?

Now that we have learned about plant nutrients in the soil, let's consider how soil water, with its dissolved nutrients, enters the plant root, and how it moves from the root to the rest of the plant body.

CONCEPT 25.3 Water and Solutes Are Transported in the Xylem by Transpiration–Cohesion–Tension

Plants require water to carry out photosynthesis, to transport solutes between plant organs, to cool their leaves by evaporation, and to maintain the internal pressure that supports their bodies. Here we focus on how water moves into and out of plant cells and between different parts of the plant.

Differences in water potential govern the direction of water movement

Water moves into and out of plant cells by osmosis. In plant cells, the direction of water movement is determined by **water potential** (**psi**, Ψ), defined as the tendency of a solution (water plus solutes) to take up water from pure water across a membrane. Whenever water moves across a selectively permeable membrane by osmosis, it moves toward the region of lower (more negative) water potential (**FIGURE 25.8A, LEFT**). Water potential is expressed in megapascals (MPa), a unit of pressure. Atmospheric pressure—"one atmosphere"—is about 0.1 MPa, or 14.7 pounds per square inch (a typical pressure in an automobile tire is about 0.2 MPa).

LINK

You can review the structure and function of the cell membrane in **Chapter 5**; **Concept 5.2** describes osmosis

Water potential has two components:

- **Solute potential** (Ψ_s): As solutes are added to pure water, the concentration of free water is reduced, the tendency to take up water increases, and water potential decreases. (In terms of cells, as solutes are added to the environment, it becomes hypertonic relative to the cells.)
- **Pressure potential** (Ψ_p): When any closed compartment takes up water, it tends to swell. The walls of the compartment, however, resist that swelling (think of blowing up a balloon). The result is an increase in internal pressure in the compartment, which decreases the tendency to take up more water (increases the water potential; **FIGURE 25.8A, RIGHT**).

We can express the effects of solute potential and pressure potential on water potential in the form of an equation:

$$\Psi = \Psi_s + \Psi_p$$

By definition the solute potential of pure water is zero; because added solutes decrease water potential, solute potential is usually negative. Pressure potential is defined as zero when it equals atmospheric pressure. Pressure potentials less than atmospheric pressure are negative, and those greater than atmospheric pressure are positive.

In a plant cell (**FIGURE 25.8B**), the cell wall resists the swelling that would otherwise occur as the cell takes up water (see

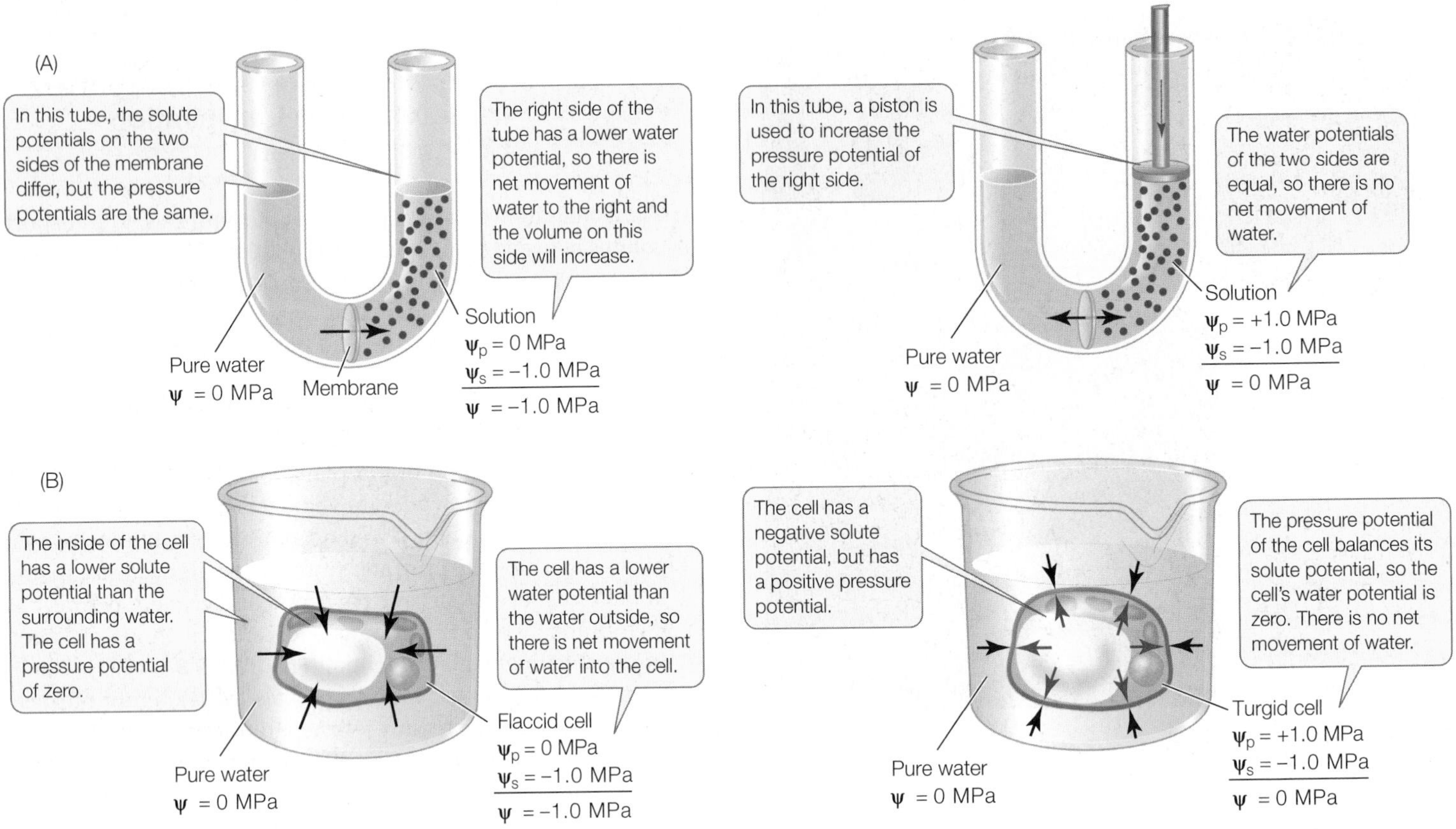

FIGURE 25.8 Water Potential, Solute Potential, and Pressure Potential (A) A theoretical illustration of water potential. (B) The effect of differences in water potential on a plant cell.

Go to ANIMATED TUTORIAL 25.2
Water Uptake in Plants
PoL2e.com/at25.2

also Figure 5.3C). The resulting **turgor pressure** is equivalent to the pressure potential exerted by the piston in Figure 25.8A. Water will enter plant cells by osmosis until the pressure potential exactly balances the solute potential. At this point the cell is said to be "turgid"—that is, it has a significant positive pressure potential. The physical structure of many plants is maintained by the (positive) pressure potential of their cells; if the pressure potential drops (for example, if the plant does not have enough water), the plant wilts (**FIGURE 25.9**).

Water and ions move across the root cell's cell membrane

The movement of a soil solution containing mineral ions across a root cell's cell membrane faces two major challenges:

- The membrane is hydrophobic, whereas water and mineral ions are polar.
- Some mineral ions must be moved against their concentration gradients.

These two challenges are met by the activities of membrane proteins.

AQUAPORINS Aquaporins are membrane channels through which water can diffuse (see Figure 5.5). Although aquaporins were first discovered in animal cells, they are also abundant in plant cell membranes. The abundance and activity of aquaporins are both regulated so that the rate of osmosis can be controlled. But the direction of osmosis is always the same: water moves to the region of more negative water potential.

ION CHANNELS AND PROTON PUMPS When the concentration of an ion in the soil solution is greater than that inside the

The cells in this plant have low turgor pressure and the plant is wilted.
The water potential of cells of this plant is zero because the negative solute potential is balanced by an equally positive pressure potential. The plant is upright because its cells are turgid.

FIGURE 25.9 A Wilted Plant A plant wilts when the pressure potential of its cells drops (its cells have low turgor pressure).

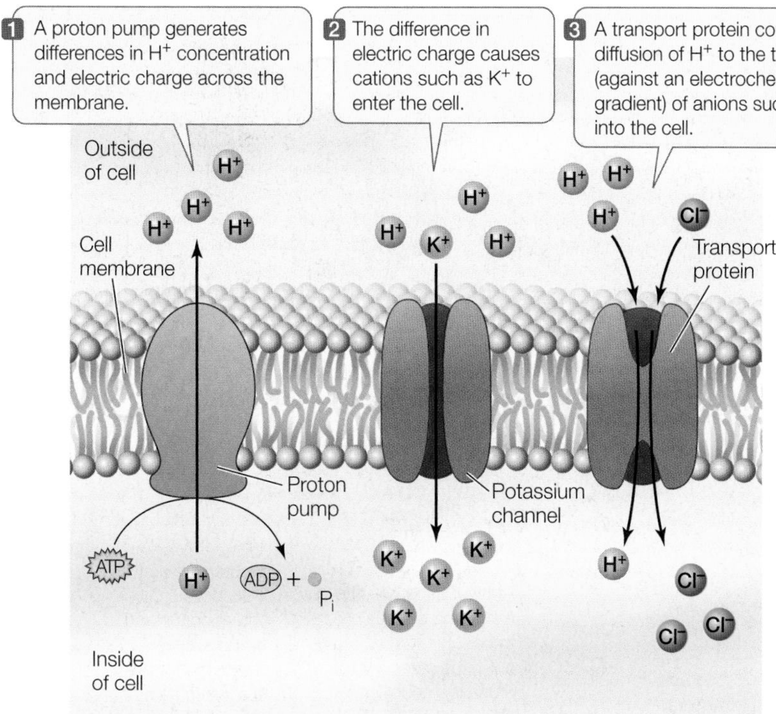

FIGURE 25.10 Ion Transport into Plant Cells The active transport of protons (H^+) out of the cell by a proton pump (step 1) drives the movement of both cations (step 2) and anions (step 3) into the cell.

root cells, transport proteins can move the ion into the root cells by facilitated diffusion. The concentrations of many mineral ions in the soil solution, however, are lower than those inside the plant. In addition, electric charge differences play a role in the uptake of mineral ions. For example, a negatively charged ion that moves into a negatively charged cellular compartment is moving against an electrical gradient, and this movement requires energy. Concentration and electrical gradients combine to form an electrochemical gradient. Uptake against an electrochemical gradient requires active transport.

LINK

Electrochemical gradients are discussed in more detail in **Concept 34.2**

Animals use a Na^+–K^+ pump to generate a Na^+ gradient that can be used to drive the uptake of other solutes (see Figure 5.7). Instead of a Na^+–K^+ pump, plants have a **proton pump**, which uses energy from ATP to move protons out of the cell against a proton concentration gradient (**FIGURE 25.10, STEP 1**). Because protons (H^+) are positively charged, their accumulation outside the cell has two results:

- An electrical gradient is created such that the region just outside the cell becomes more positively charged than the inside of the cell.
- A proton concentration gradient develops, with more protons just outside the cell than inside the cell.

With the inside of the cell more negative than the outside, cations such as potassium (K^+) move into the cell down the electrical gradient through specific membrane channels (**FIGURE 25.10, STEP 2**). Anions such as chloride (Cl^-) are moved into the cell against an electrochemical gradient by a membrane transport protein that couples their movement with that of H^+ (**FIGURE 25.10, STEP 3**). These processes are examples of secondary active transport.

Water and ions pass to the xylem by way of the apoplast and symplast

The journey of water and ions from the soil through the roots to the xylem follows two pathways—a fast lane (the apoplast) and a slower lane (the symplast)—either separately or in tandem:

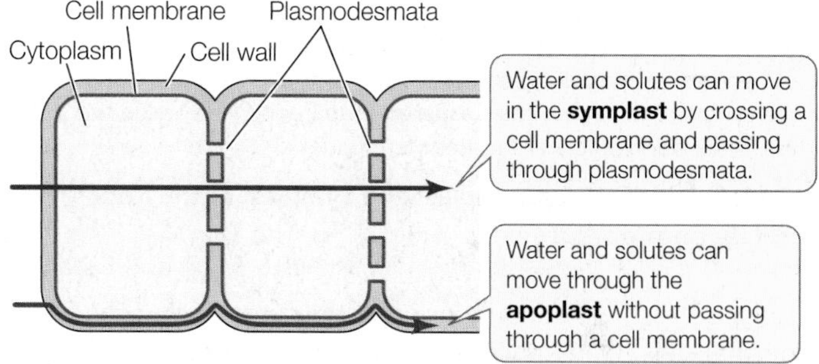

Cell membrane
Plasmodesmata
Cytoplasm
Cell wall
Water and solutes can move in the **symplast** by crossing a cell membrane and passing through plasmodesmata.
Water and solutes can move through the **apoplast** without passing through a cell membrane.

- The **apoplast** (Greek *apo*, "away from"; *plast*, "living material") consists of the cell walls, which lie outside the cell membranes, and intercellular spaces (spaces between cells), which are common in many plant tissues. The apoplast is a continuous meshwork through which water and solutes can flow without ever having to cross a membrane.
- The **symplast** (Greek *sym*, "together with") passes through the continuous cytoplasm of the living cells connected by plasmodesmata. The selectively permeable cell membranes of the root cells control access to the symplast, so movement of water and solutes into the symplast is tightly regulated.

Water and minerals that pass from the soil solution through the apoplast can travel freely as far as the endodermis, the innermost layer of the root cortex (**FIGURE 25.11**).

The endodermis is distinguished by the presence of the **Casparian strip**. This waxy, suberin-impregnated region of the endodermal cell wall forms a hydrophobic belt around each endodermal cell where it is in contact with other endodermal cells. The Casparian strip acts as a seal that prevents water and ions from moving through spaces between the endodermal

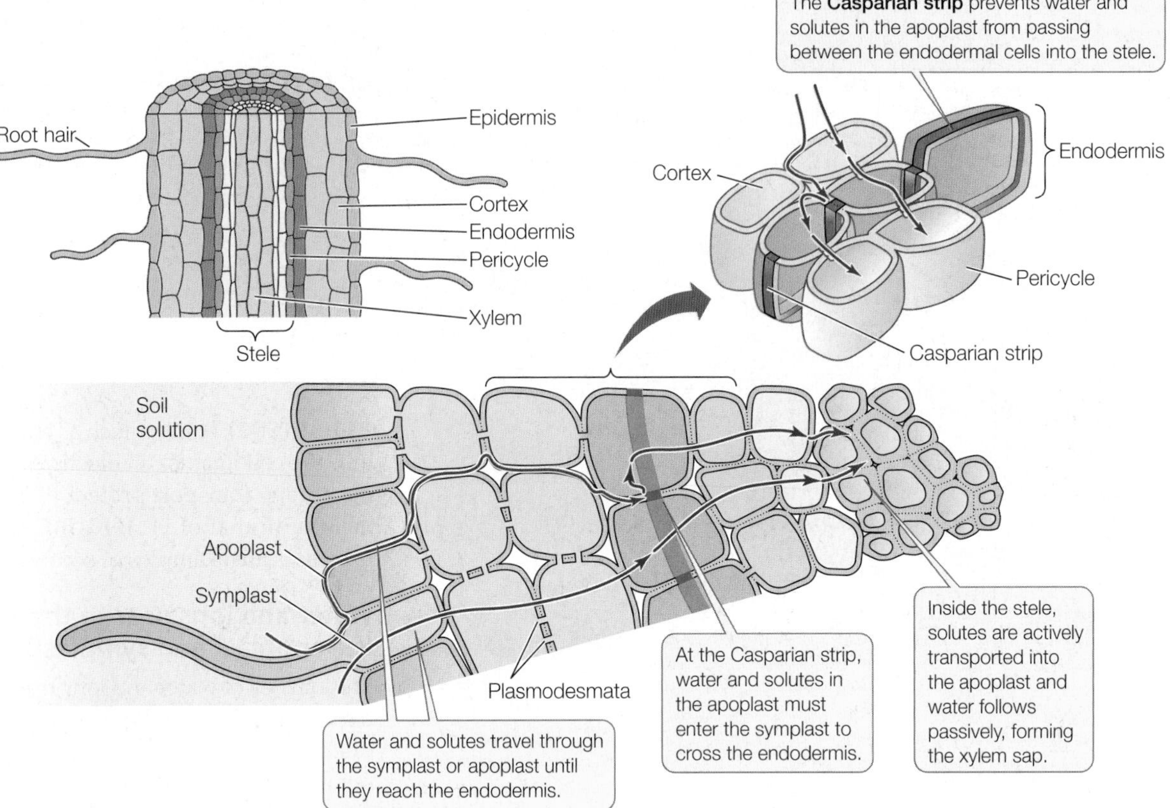

The **Casparian strip** prevents water and solutes in the apoplast from passing between the endodermal cells into the stele.
Root hair
Epidermis
Cortex
Endodermis
Pericycle
Xylem
Stele
Cortex
Endodermis
Pericycle
Casparian strip
Soil solution
Apoplast
Symplast
Plasmodesmata
Water and solutes travel through the symplast or apoplast until they reach the endodermis.
At the Casparian strip, water and solutes in the apoplast must enter the symplast to cross the endodermis.
Inside the stele, solutes are actively transported into the apoplast and water follows passively, forming the xylem sap.

FIGURE 25.11 Pathways to the Root Xylem Water and solutes can move into the root through the symplast and apoplast until they reach the endodermis (dark green). There the water and solutes must enter the symplast to bypass the Casparian strip (red). Once inside the stele, the water and solutes enter the xylem (blue).

Go to ACTIVITY 25.1 Apoplast and Symplast of the Root PoL2e.com/ac25.1

cells; instead, water and ions enter the cytoplasm of the endodermal cells. Water and ions already in the symplast can enter the endodermal cells through plasmodesmata.

LINK

The Casparian strip is functionally equivalent to the tight junctions found in some animal tissues; see **Figure 4.18**

Once they have passed the endodermal barrier, water and minerals remain in the symplast until they reach parenchyma cells in the pericycle or xylem. These cells then actively transport mineral ions into the apoplast of the stele. As ions are transported into the solution in the cell walls of the stele, water potential in the apoplast becomes more negative; consequently, water moves out of the cells and into the apoplast by osmosis. In other words, ions are transported actively, and water follows passively. The end result is that water and minerals end up in the xylem, where they constitute the xylem sap.

Water moves through the xylem by the transpiration–cohesion–tension mechanism

Once water has arrived in the xylem, it is all uphill from there! Consider the magnitude of what xylem accomplishes. A single maple tree 15 meters tall has been estimated to have some 177,000 leaves, with a total leaf surface area of 675 square meters—1.5 times the area of a basketball court. During a summer day, that tree loses 220 liters of water *per hour* to the atmosphere by evaporation from its leaves. To prevent wilting, the xylem must transport 220 liters of water 15 meters from the roots up to the leaves every hour.

How is the xylem able to move so much water to such great heights? Part of the answer lies in the structure of the xylem, which we described in Chapter 24. Recall that xylem vessels consist of a long "straw" of cell walls, which provide both structural support and the rigidity needed to maintain a pressure gradient within the xylem sap. Early experiments ruled out two possible ways that had been hypothesized for moving water up through these vessels from roots to leaves: upward pressure and capillary action.

Upward pressure would move water by "pushing the bottom." A simple experiment in 1893 ruled out this mechanism: A tree was cut at its base, and the sawed-off upper stem was placed in a vat of poison that killed living cells. (Recall that mature xylem is made up of dead cells.) The poison continued to rise through the trunk, even though cells were killed along the way. This observation showed that a living "pump" is not required to push the xylem sap up a tree. Furthermore, the roots were absent, yet upward flow in the trunk continued, demonstrating that the roots do not provide upward pressure. When the poison reached the leaves and killed them, however, all further movement up the trunk stopped, demonstrating that leaves must be alive for water to move in the xylem.

Capillary action was also ruled out as the mechanism that makes water rise in the xylem. Water molecules have strong cohesion, so a column of water can indeed rise a short distance

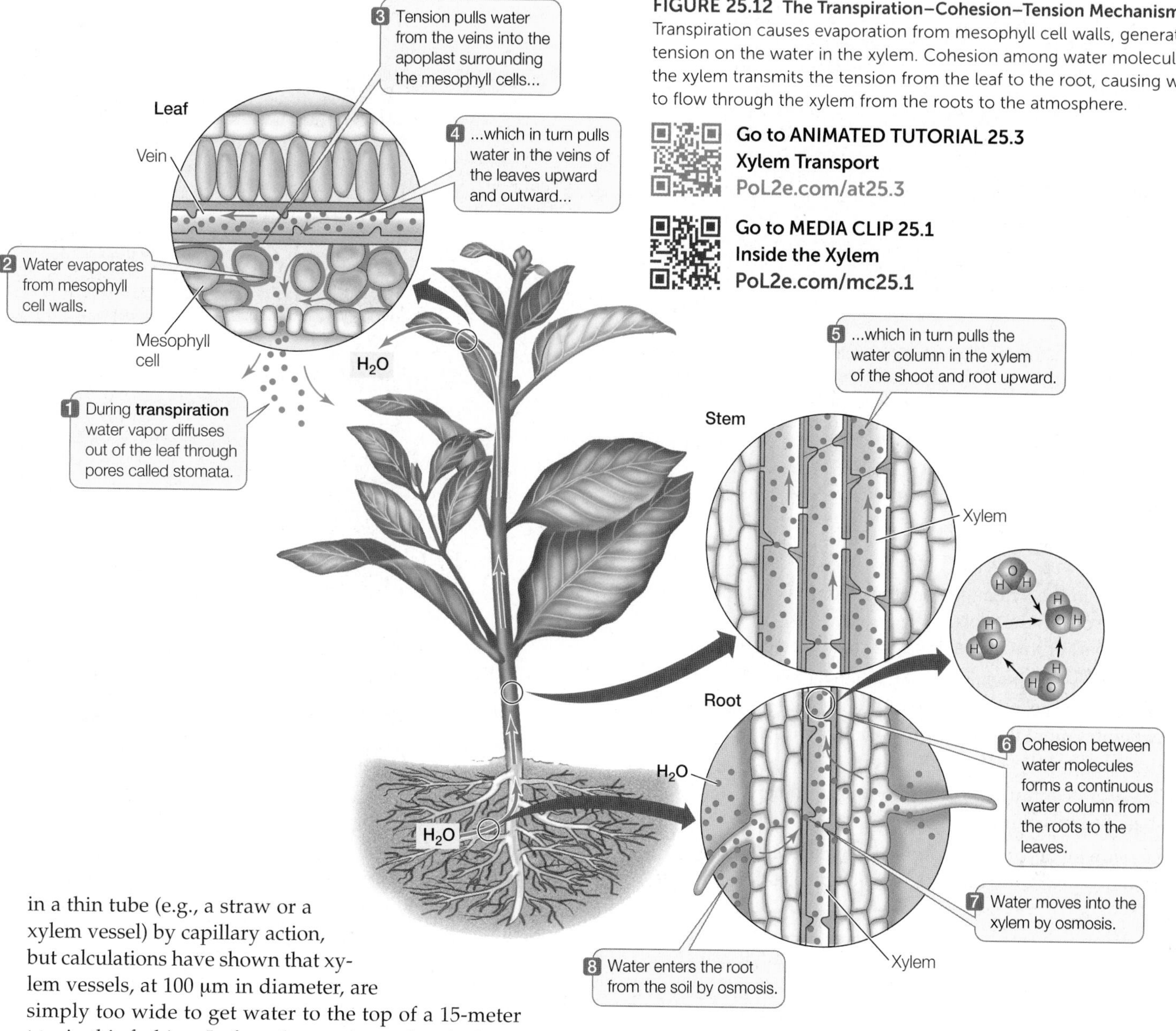

3 Tension pulls water from the veins into the apoplast surrounding the mesophyll cells...
Leaf
Vein
4 ...which in turn pulls water in the veins of the leaves upward and outward...
2 Water evaporates from mesophyll cell walls.
Mesophyll cell
H2O
1 During transpiration water vapor diffuses out of the leaf through pores called stomata.
5 ...which in turn pulls the water column in the xylem of the shoot and root upward.
Stem
Xylem
Root
H2O
6 Cohesion between water molecules forms a continuous water column from the roots to the leaves.
7 Water moves into the xylem by osmosis.
8 Water enters the root from the soil by osmosis.
Xylem

FIGURE 25.12 The Transpiration–Cohesion–Tension Mechanism Transpiration causes evaporation from mesophyll cell walls, generating tension on the water in the xylem. Cohesion among water molecules in the xylem transmits the tension from the leaf to the root, causing water to flow through the xylem from the roots to the atmosphere.

Go to ANIMATED TUTORIAL 25.3
Xylem Transport
PoL2e.com/at25.3

Go to MEDIA CLIP 25.1
Inside the Xylem
PoL2e.com/mc25.1

in a thin tube (e.g., a straw or a xylem vessel) by capillary action, but calculations have shown that xylem vessels, at 100 μm in diameter, are simply too wide to get water to the top of a 15-meter tree in this fashion. In fact, the maximum height for a water column raised by capillary action alone in a 100-μm-diameter vessel is only 0.15 meters.

LINK
You can review the properties of water in **Concept 2.2**

Almost a century ago, the **transpiration–cohesion–tension** theory was proposed to explain water transport in the xylem (**FIGURE 25.12**). According to this theory, the key elements of xylem transport are:

- *Transpiration*: evaporation of water from cells within the leaves
- *Cohesion* of water molecules in the xylem sap as a result of hydrogen bonding
- *Tension* on the xylem sap resulting from transpiration

The concentration of water vapor in the atmosphere is lower than that in the air spaces between the cells of a leaf. Because of this difference, water vapor diffuses out of the leaf in a process called **transpiration**. Within the leaf, water evaporates from the moist walls of the mesophyll cells and enters the intercellular air spaces. As water evaporates from the aqueous film coating each cell, the film shrinks back into tiny spaces in the cell walls, increasing the curvature of the water surface and thus increasing its surface tension. Because of hydrogen bonding, water molecules have cohesion, and therefore the increased tension (negative pressure potential) in the surface film draws more water into the cell walls, replacing that which was lost by evaporation. The resulting tension in the mesophyll draws water from the xylem of the nearest vein into the apoplast surrounding the mesophyll cells. The removal of water from the veins, in turn, establishes tension on the entire column of water

(A)

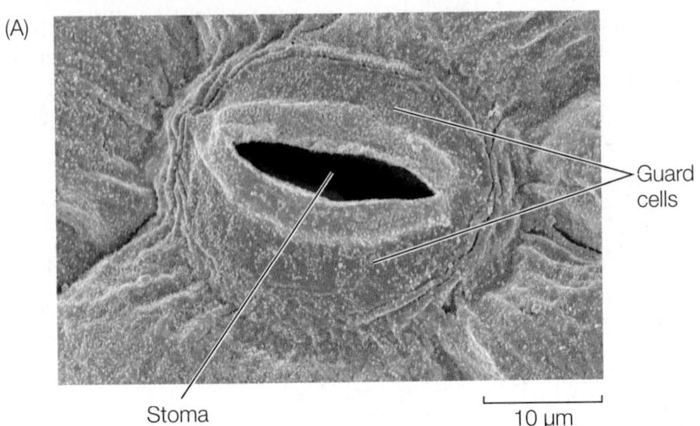
Guard cells
Stoma
10 μm

(B)

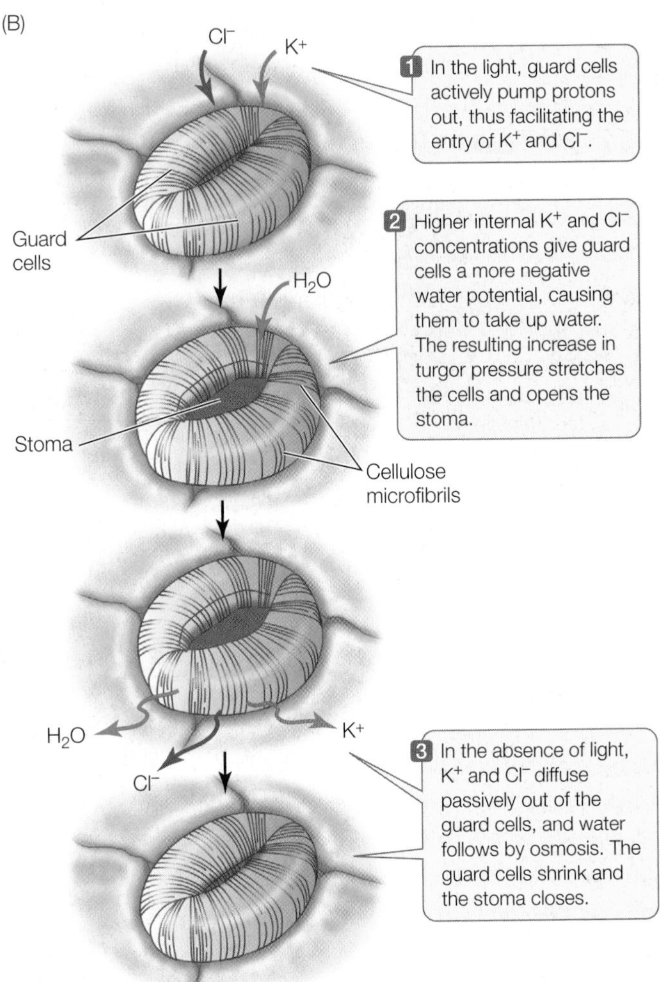
Cl⁻
K+
1 In the light, guard cells actively pump protons out, thus facilitating the entry of K+ and Cl⁻.
Guard cells
H2O
2 Higher internal K+ and Cl⁻ concentrations give guard cells a more negative water potential, causing them to take up water. The resulting increase in turgor pressure stretches the cells and opens the stoma.
Stoma
Cellulose microfibrils
H2O
K+
Cl⁻
3 In the absence of light, K+ and Cl⁻ diffuse passively out of the guard cells, and water follows by osmosis. The guard cells shrink and the stoma closes.

FIGURE 25.13 Stomata Regulate Gas Exchange and Transpiration (A) Scanning electron micrograph of an open stoma formed by two sausage-shaped guard cells. (B) Potassium ion (K^+) concentrations affect the water potential of the guard cells, controlling the opening and closing of stomata. Negatively charged ions (e.g., Cl^-) that accompany K^+ maintain electrical balance and contribute to the changes in water potential that open and close the stomata.

contained in the xylem, so that the column is drawn upward all the way from the roots.

Each part of this model is supported by evidence:

- The difference in water potential between the soil solution and air is huge, on the order of –100 MPa. This difference generates more than enough tension to pull a water column up the tallest tree.
- There is a continuous column of water in the xylem, which is caused by cohesion.
- Measurements of xylem pressures in cut stems show a negative pressure potential, which indicates considerable tension.

In addition to its role in xylem transport, transpiration has the added benefit of cooling a plant's leaves (much as humans sweat to cool off). The evaporation of water from mesophyll cells consumes heat, thereby decreasing the leaf temperature. A farmer can hold a leaf between thumb and forefinger to estimate its temperature; if the leaf doesn't feel cool, that means transpiration is not occurring and it must be time to water.

Although transpiration provides the driving force for the transport of water and minerals in the xylem, it also results in the loss of tremendous quantities of water from the plant. How do plants control this loss?

Stomata control water loss and gas exchange

In Chapter 6 we described photosynthesis, which has the general equation

$$CO_2 + H_2O \rightarrow \text{carbohydrate} + O_2$$

Leaves are the chief organs of photosynthesis, so they must have a large surface area across which to exchange the gases O_2 and CO_2. This surface area is provided by the mesophyll cells being surrounded by abundant air spaces inside the leaf (see Figure 25.12), which are also the location of the evaporation that drives water movement in the xylem. The epidermis of leaves and stems minimizes transpirational water loss by secreting a waxy cuticle, which is impermeable to water. However, the cuticle is also impermeable to CO_2 and O_2. How can the plant balance its need to retain water with its need to obtain CO_2 for photosynthesis?

Plants have met this challenge with the evolution of **stomata** (singular *stoma*): pores in the leaf epidermis, typically on the underside of the leaf (**FIGURE 25.13A**). There are up to 1,000 stomata per square millimeter in some leaves, constituting up to 3 percent of the total surface area of the shoot. Such an abundance of stomata could lead to excessive water loss. But stomata are not static structures. A pair of specialized epidermal cells, called **guard cells**, controls the opening and closing of each stoma.

When the stomata are open, CO_2 can enter the leaf by diffusion, but water vapor diffuses out of the leaf at the same time. Closed stomata prevent water loss but also exclude CO_2 from the leaf. Most plants open their stomata when the light intensity is sufficient to maintain a moderate rate of photosynthesis. At night, when darkness precludes photosynthesis, their stomata are closed; no CO_2 is needed at this time, and water is conserved. Even during the day, the stomata close if water is being lost at too great a rate.

APPLY THE CONCEPT

Water and solutes are transported in the xylem by transpiration–cohesion–tension

Suppose you measure water potential (Ψ) in a 100-m-tall tree and its surroundings and obtain the results in the table. Note that gravity exerts a force of –0.01 MPa per meter of height above ground.

1. Is the water potential in the leaf sufficiently low to draw water to the top of the tree?
2. Would transpiration continue if soil water potential decreased to –1.0 MPa?
3. What would you expect to happen to the xylem water potential if all of the stomata closed?

REGION	Ψ (MPa)
Soil water	–0.3
Xylem of root	–0.6
Xylem of trunk	–1.2
Inside of leaf	–2.0
Outside air	–58.5

Stomata can respond rapidly to changes in light, CO_2 concentration, and water loss. Plants can also change their total number of stomata in response to longer-term changes in environmental conditions.

LIGHT AND CO_2 CONCENTRATION Guard cells can respond to changes in light and CO_2 concentration in a matter of minutes by changing their solute potential. The absorption of light by a pigment in the guard cell's cell membrane activates a proton pump (see Figure 25.10), which actively transports H^+ out of the guard cells and into the apoplast of the surrounding epidermis. The resulting electrochemical gradient drives K^+ into the guard cells, where it accumulates (**FIGURE 25.13B**). Negatively charged chloride (Cl^-) ions and organic ions also move into and out of the guard cells along with the K^+ ions, maintaining electrical balance. The increased concentration of K^+ and other solutes inside the guard cells makes the solute potential of the guard cells more negative. Water then enters by osmosis, increasing the turgor pressure of the guard cells. The guard cells change their shape, becoming more turgid in response to the increase in pressure potential, so that a space—the stoma—appears between them. The stoma closes in the absence of light: the proton pump becomes less active, K^+ ions diffuse passively out of the guard cells, water follows by osmosis, the pressure potential decreases, and the guard cells sag together and seal off the stoma. Guard cell membranes are particularly rich in aquaporins, making guard cells well adapted for the rapid water movements involved in stomatal responses.

WATER Stomata also respond to water availability. Water stress is a common problem for plants, especially on a hot, windy day, when plants might close their stomata even when the sun is shining. The water potential of the leaf's mesophyll cells is the cue for this response. If the mesophyll becomes dehydrated, its cells release the hormone abscisic acid, which causes the stomata to close.

REGULATION BY STOMATA NUMBER A plant can regulate water loss not just by the opening or closing of stomata, but by changing the number of stomata. This occurs much more slowly, over a period of days to weeks. Trees can reduce stomata numbers by shedding leaves, or by making new leaves that have few stomata.

CHECKpoint CONCEPT 25.3

✓ What distinguishes water potential, solute potential, and pressure potential?

✓ What are the roles of transpiration, cohesion, and tension in xylem transport?

✓ If *Arabidopsis* is exposed to high CO_2 concentrations, the new leaves that form on the plant have fewer stomata than they would have had under normal conditions. Why do you think this might be advantageous?

Now that we understand how leaves obtain the water and CO_2 required for photosynthesis, let's examine how the products of photosynthesis are transported to other parts of the plant where they are needed.

CONCEPT 25.4 Solutes Are Transported in the Phloem by Pressure Flow

As photosynthesis occurs in the leaf, the carbohydrate products of photosynthesis (photosynthate; mainly sucrose) diffuse to the nearest small vein, where they are actively transported into sieve tube elements. The movement of carbohydrates and other solutes through the plant in the phloem is called **translocation**. These solutes are translocated from sources to sinks:

- A **source** is an organ (such as a mature leaf or a storage root) that *produces* (by photosynthesis or by digestion of stored reserves) more carbohydrates than it requires.
- A **sink** is an organ (such as a root, flower, developing fruit or tuber, or immature leaf) that *consumes* carbohydrates for its own growth and storage needs.

Sources and sinks can change roles. For example, storage roots (such as sweet potatoes) are sinks when they accumulate carbohydrates, but they are sources when their stored reserves are mobilized to nourish other organs in the plant.

LINK

The anatomy of sieve tube elements is described in **Concept 24.1**

Sucrose and other solutes are carried in the phloem

Evidence that the phloem carries sucrose and other solutes initially came in the 1600s, when Marcello Malpighi removed a ring of bark (containing the phloem) from the trunk of a tree, while leaving the xylem intact—that is, he "girdled" the tree. Over time, the bark in the region above the girdle swelled:

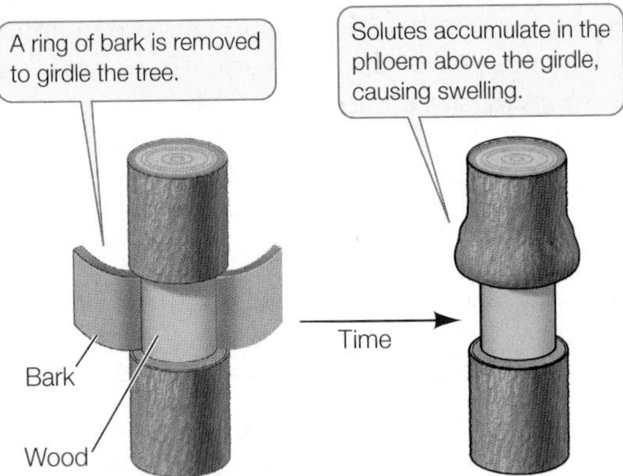

A ring of bark is removed to girdle the tree.
Solutes accumulate in the phloem above the girdle, causing swelling.
Time
Bark
Wood

Malpighi correctly concluded that the solution coming from the leaves above the girdle was trapped in the bark. Later, the bark below the girdle died because it was no longer receiving sugars from the leaves. Eventually the roots, and then the entire tree, died.

More recently plant biologists have analyzed the contents of phloem with the help of aphids. These insects feed on plants by drilling into sieve tube elements with a specialized organ, the stylet. The pressure potential in the sieve tube is higher than that outside the plant, so the phloem sap is forced through the stylet and into the aphid's digestive tract. So great is the pressure that sugary liquid is forced through the insect's body and out its anus:

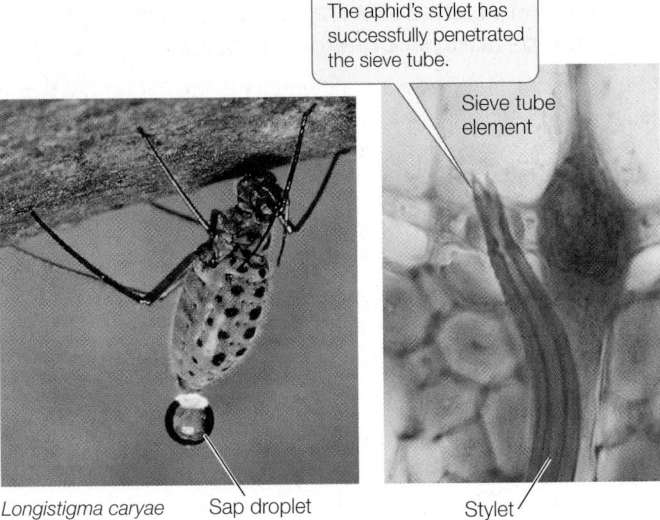

If an aphid is frozen in the act of feeding, its body can be broken off the plant stem, leaving the stylet intact. Phloem sap continues to flow out of the stylet for hours.

Several observations have come from analyzing phloem sap in this way, as well as from other experiments:

- Ninety percent of the phloem solutes consist of sucrose, with the rest hormones, other small molecules such as mineral nutrients and amino acids, and viruses.
- The flow rate can be very high, as much as 100 centimeters per hour.

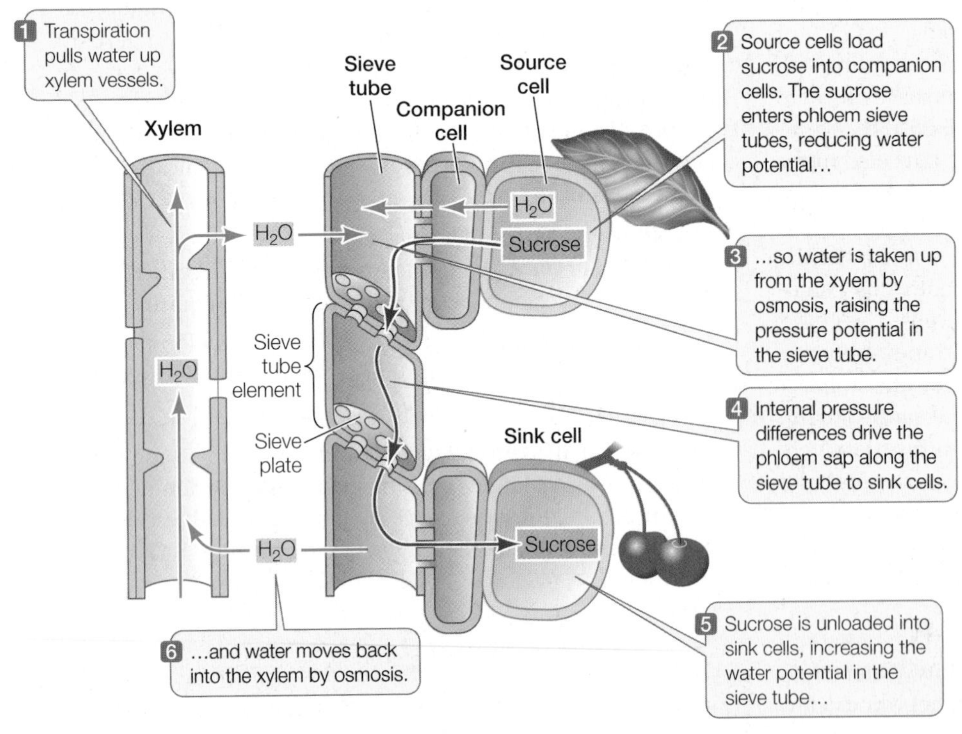

FIGURE 25.14 The Pressure Flow Model Differences in water potential produce a pressure gradient that moves phloem sap from sources to sinks.

Go to ANIMATED TUTORIAL 25.4 The Pressure Flow Model
PoL2e.com/at25.4

- Different sieve tube elements conduct their contents in different directions (up or down the stem, for example). Movement in the phloem as a whole is bidirectional, but it is unidirectional within an individual sieve tube.
- Movement of sap in the phloem requires living cells, in contrast to movement in the xylem.

The pressure flow model describes the movement of fluid in the phloem

The observations described above led to the **pressure flow model** as an explanation for translocation in the phloem (**FIGURE 25.14**). According to this model, sucrose is actively transported at a source into companion cells, from which it flows through plasmodesmata into the sieve tube elements. This gives those cells a higher sucrose concentration than the surrounding cells (a more negative solute potential), and water therefore enters the sieve tube elements from the xylem by osmosis. The entry of this water increases turgor pressure (causes a more positive pressure potential) at the source end of the sieve tube, so that the entire fluid content of the sieve tube is pushed toward the sink end of the tube—in other words, the sap moves in response to a pressure gradient. In the sink, the sucrose is unloaded both passively and by active transport, and water moves back into the xylem. In this way, the gradient of solute potential and pressure potential needed for the movement of phloem sap (translocation) is maintained.

Two steps in phloem translocation require metabolic energy:

- Transport of sucrose and other solutes from sources into companion cells and then into the sieve tubes; called "loading"
- Transport of solutes from the sieve tubes into sinks; called "unloading"

The need for metabolic energy is the reason why phloem transport, unlike xylem transport, requires living cells.

Two general routes can be taken by sucrose and other solutes as they move from the mesophyll cells into the phloem. In many plants, sucrose and other solutes follow an apoplastic pathway: they leave the mesophyll cells and enter the apoplast before they reach the sieve tube elements. Specific sugars and amino acids are then actively transported into cells of the phloem. Because the solutes cross at least one selectively permeable membrane in the apoplastic pathway, selective transport can be used to regulate which specific substances enter the phloem. In other plants, solutes follow a symplastic pathway: the solutes remain within the symplast all the way from the mesophyll cells to the sieve tube elements. Because no membranes are crossed in the symplastic pathway, a mechanism that does not involve membrane transport is used to load sucrose into the phloem.

In sink regions, the solutes are actively transported *out* of the sieve tube elements and into the surrounding tissues. This unloading serves several purposes: it helps maintain the gradient of solute potential, and hence of pressure potential, in the sieve tubes; it helps build up high concentrations of carbohydrates in storage organs, such as developing fruits and seeds; and it moves nutrients to parts of the plant that are growing rapidly, such as shoots that form after the winter.

CHECKpoint CONCEPT 25.4

- ✓ Deer sometimes graze on evergreen trees by eating bark during the winter when other food is scarce. In the spring, those trees may die. How does this happen?
- ✓ What would happen if the ends of sieve tubes were blocked and did not allow phloem sap to flow through them?
- ✓ Make a table that compares the flow of fluids in xylem and in phloem with respect to the driving force for movement; where flow occurs; whether the cells are alive; and whether the driving pressure is positive or negative.

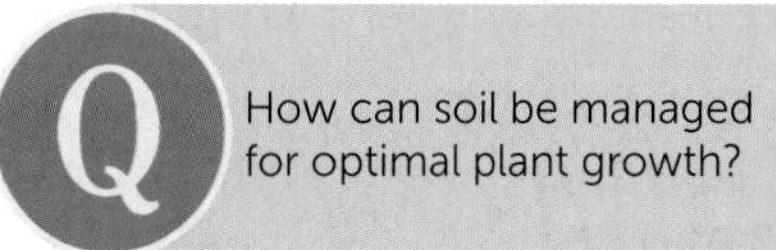

Q
How can soil be managed for optimal plant growth?

ANSWER Every year, 25 billion tons of topsoil (A horizon) are lost worldwide from potentially arable land. This loss reduces the nutrient content (Concept 25.1) and water-holding capacity (Concept 25.3) of the remaining soil. It therefore creates added demand for fertilizers and irrigation to restore the soils, but these inputs are too expensive in most areas of the world, and they can have adverse effects on the rest of the environment. For example, adding fertilizer to soils can pollute nearby waters or result in excess salts in the soil, and huge irrigation projects destroy natural habitats and can lead to flooding.

Faced with these challenges, scientists and farmers (often the same people!) have invented methods of sustainable farming. In the United States, for example, a branch of the federal government called the Natural Resource Conservation Service (formerly the Soil Conservation Service) was established in response to the country's experiences with soil degradation during the Dust Bowl of the 1930s.

There are several strategies to sustain the structure, nutrient content, and water-holding capacity of soil. One of these strategies is crop rotation. Different plants have different nutritional requirements and so remove different nutrients from the soil as they grow and are harvested. For example, corn requires large amounts of nitrogen, whereas soybeans harbor nitrogen-fixing bacteroids in their roots (Concept 25.2). In the midwestern United States, corn and soybeans are planted in alternate growing seasons.

Another strategy is conservation tillage. This refers to methods such as plowing that are used to prepare soil for planting. Conservation tillage can help maintain soil and reduce soil losses. If you fly over any heavily farmed area of the United States, look out the window and you will see contour plowing patterns; these patterns are used to control runoff of the topsoil during heavy rains. More recently, it has become increasingly common for farmers to leave behind on the field up to half of the stems and leaves of plants that are harvested for grain (**FIGURE 25.15**). The farmers plow these crop residues back into the soil, where they help maintain soil structure, bind water, and provide nutrients because they are organic matter that builds up humus. More than one-third of the farmland in the United States is now being farmed in this way.

Our increasing understanding of plants and their relationship to soils will doubtless lead to more sustainable practices in the future.

FIGURE 25.15 Conservation Tillage These soybean plants are growing amidst the residue of corn plants that were harvested for their grain. The crop residues help conserve the soil.

SUMMARY

CONCEPT 25.1 Plants Acquire Mineral Nutrients from the Soil

- Plants are photosynthetic **autotrophs** that require water and certain mineral nutrients to survive. They obtain most of these mineral nutrients as ions from the **soil solution**.
- The **essential elements** for plants include six **macronutrients** and several **micronutrients**. Plants that lack a particular nutrient show characteristic deficiency symptoms. **Review Figure 25.1 and ANIMATED TUTORIAL 25.1**
- The essential elements were discovered by growing plants **hydroponically** in solutions that lacked individual elements. **Review Figure 25.2**
- Soils supply plants with mechanical support, water and dissolved ions, air, and the services of other organisms. **Review Figure 25.3**
- Protons take the place of mineral nutrient cations bound to clay particles in soil in a process called **ion exchange**. **Review Figure 25.4**
- Farmers may use shifting agriculture or **fertilizer** to make up for nutrient deficiencies in soil.

CONCEPT 25.2 Soil Organisms Contribute to Plant Nutrition

- Signaling molecules called **strigolactones** induce the hyphae of arbuscular mycorrhizal fungi to invade root cortical cells and form arbuscules, which serve as sites of nutrient exchange between fungus and plant. **Review Figure 25.5A**
- Legumes signal nitrogen-fixing bacteria (rhizobia) to form **bacteroids** within nodules that form on their roots. **Review Figure 25.5B**
- In **nitrogen fixation**, nitrogen gas (N_2) is reduced to ammonia in a reaction catalyzed by **nitrogenase**. **Review Figure 25.6**
- Carnivorous plants supplement their nutrient supplies by trapping and digesting arthropods. Parasitic plants obtain minerals, water, or products of photosynthesis from other plants.

CONCEPT 25.3 Water and Solutes Are Transported in the Xylem by Transpiration–Cohesion–Tension

- Water moves through biological membranes by osmosis, always moving toward regions with a more negative water potential. The **water potential** (Ψ) of a cell or solution is the sum of its **solute potential** (Ψ_s) and its **pressure potential** (Ψ_p). **Review Figure 25.8 and ANIMATED TUTORIAL 25.2**
- The physical structure of many plants is maintained by the positive pressure potential of their cells (**turgor pressure**); if the pressure potential drops, the plant wilts.
- Water moves into root cells by osmosis through aquaporins. Mineral ions move into root cells through ion channels, by facilitated diffusion, and by secondary active transport. **Review Figure 25.10**
- Water and ions may pass from the soil into the root by way of the **apoplast** or the **symplast**, but they must pass through the symplast to cross the endodermis and enter the xylem. The **Casparian strip** in the endodermis blocks further movement of water and ions through the apoplast. **Review Figure 25.11 and ACTIVITY 25.1**
- Water is transported in the xylem by the **transpiration–cohesion–tension** mechanism. Evaporation from the leaf produces tension in the mesophyll, which pulls a column of water—held together by cohesion—up through the xylem from the root. **Review Figure 25.12 and ANIMATED TUTORIAL 25.3**
- **Stomata** allow a balance between water retention and CO_2 uptake. Their opening and closing is regulated by **guard cells**. **Review Figure 25.13**

CONCEPT 25.4 Solutes Are Transported in the Phloem by Pressure Flow

- **Translocation** is the movement of the products of photosynthesis, as well as some other small molecules, through sieve tubes in the phloem. The solutes move from **sources** to **sinks**.
- Translocation is explained by the **pressure flow model**: the difference in solute potential between sources and sinks creates a difference in pressure potential that pushes phloem sap along the sieve tubes. **Review Figure 25.14 and ANIMATED TUTORIAL 25.4**

Go to the Interactive Summary to review key figures, Animated Tutorials, and Activities PoL2e.com/is25

Go to LaunchPad at **macmillanhighered.com/launchpad** for additional resources, including LearningCurve Quizzes, Flashcards, and many other study and review resources.

26 Plant Growth and Development

KEY CONCEPTS

26.1 Plants Develop in Response to the Environment

26.2 Gibberellins and Auxin Have Diverse Effects but a Similar Mechanism of Action

26.3 Other Plant Hormones Have Diverse Effects on Plant Development

26.4 Photoreceptors Initiate Developmental Responses to Light

Plant geneticist Norman Borlaug, seen here in a field of semi-dwarf wheat, carried out a program of genetic crosses that led to high-yielding varieties of wheat and saved millions of people from starvation.

In their constant search for ways to help farmers produce more food for a growing population, biologists have developed cereal crops whose physiology allows them to produce more grain per plant (a higher yield). When a plant produces a lot of seeds, however, the sheer weight of the seeds may cause the stem to bend over or even break. This makes harvesting the seeds impossible: think of how hard it would be to get enough food for your family if you had to pick up seeds on the ground, some of which had already sprouted.

In 1945 the U.S. Army temporarily occupied Japan, which it had defeated in the Second World War. During the war, Japan, an island nation with a limited amount of land suitable for farming, was blockaded and could not import food. How had Japan been able to grow enough grain to feed its people? The answer lay in the fields: the Japanese had bred genetic strains of rice and wheat with short, strong stems that could bear a high yield of grain without bending or breaking. This innovation made an impression on an agricultural advisor who happened to be among the first wave of U.S. occupiers, and seeds of the Japanese strains were sent back to the United States.

A decade later, Norman Borlaug, a plant geneticist who was working in Mexico at the time, began genetic crosses of what were known as semi-dwarf wheat plants from Japan with wheat varieties that had genes conferring rapid growth, adaptability to varying climates, and resistance to fungal diseases. The results were new genetic strains of wheat that produced record yields, first in Mexico and then in India and Pakistan in the 1960s. At about the same time, and using a similar strategy, scientists in the Philippines developed new semi-dwarf strains of rice with equally spectacular results. People who had lived on the edge of starvation now produced enough food. Countries that had been relying on food aid from other countries were now growing so much grain that they could export the surplus. The development of these new semi-dwarf strains began what was called the "Green Revolution." Borlaug was awarded the Nobel Peace Prize for his research on wheat, which is estimated to have saved a billion lives.

What changes in their growth patterns made the new strains of cereal crops produced by the Green Revolution so successful?

You will find the answer to this question on page 570.

CONCEPT 26.1 Plants Develop in Response to the Environment

Plants are sessile (nonmobile) organisms that must seek out resources above and below the ground. Plants have several characteristics that distinguish them from animals and allow them to obtain the resources that they need to grow and reproduce:

- *Meristems*. Plants have permanent collections of stem cells (undifferentiated, constantly dividing cells) that allow them to continue growing throughout their lifetimes (see Concept 24.2).
- *Post-embryonic organ formation*. Unlike many animals, plants can initiate development of new organs such as leaves and flowers throughout their lifetimes.
- *Differential growth*. Plants can allocate their resources so that they grow more of the organs that will benefit them most—for example, more leaves to harvest sunlight or more roots to obtain water and nutrients.

The development of a plant—the series of progressive changes that take place throughout its life—is regulated in many ways. Several key factors are involved in regulating plant growth and development:

- *Environmental cues*, such as day length
- *Receptors* that allow a plant to sense environmental cues, such as photoreceptors that absorb light
- *Hormones*—chemical signals that mediate the plant's adaptations and effects of environmental cues, including those sensed by receptors
- The plant's *genome*, which encodes regulatory proteins and enzymes that catalyze the biochemical reactions of development

We will explore these factors in more detail later in this chapter. But first let's look at the initial steps of plant development—from seed to seedling—and the environmental cues and internal responses that guide those steps.

The seed germinates and forms a growing seedling

Concepts 24.1 and 27.1 describe the events of plant development and reproduction that lead to the formation of seeds. Here we begin with the seed, the structure that contains the plant embryo. Unlike most animal embryos, plant embryos may be held in "suspended animation" inside the seed, with development halted, for long periods. If development stops, even when external conditions (such as water supply) are adequate for growth, the seed is said to be **dormant**. Development restarts when the seed **germinates** (sprouts).

DORMANCY Seed dormancy may last for weeks, months, years, or even centuries: in 2005, a botanist was able to germinate a date palm seed recovered from a 2,000-year-old storage bin at Masada in Israel. Plants use several mechanisms to maintain dormancy:

- *Exclusion of water or oxygen* from the embryo by an impermeable seed coat
- *Mechanical restraint* of the embryo by a tough seed coat
- *Chemical inhibition* of germination
- *Photodormancy*: some seeds require a period of light or dark to germinate
- *Thermodormancy*: some seeds require a period of high or low temperature to germinate

Dormancy can be broken by factors that overcome these mechanisms. For example, passage through an animal's digestive system may damage the seed coat, or burial beneath soil (in darkness) might trigger germination.

Seed dormancy is a common phenomenon, so it must have selective advantages for plants. Dormancy ensures survival during unfavorable conditions and results in germination when environmental conditions are most favorable for growth and development of the young plant. For example, to avoid germination in the dry days of late summer, some seeds require exposure to a long cold period (winter) before they will germinate; this ensures that the plant has the entire growing season to mature. Dormancy also helps seeds survive long-distance dispersal, allowing plants to colonize new territory.

GERMINATION Seeds begin to germinate when dormancy is broken. For example, a seed may germinate after a germination inhibitor is washed away in heavy rain, which means the soils will have plenty of water and germination will be possible.

The first step in germination is the uptake of water, called **imbibition** (from *imbibe*, "to drink in"). A dormant seed contains very little water: only 5–15 percent of its weight is water, compared with 80–95 percent for most other plant parts. Seeds also contain polar macromolecules, such as cellulose and starch, that attract and bind polar water molecules. Consequently a seed has a very negative water potential (see Concept 25.3) and will take up water if the seed coat is permeable. The force exerted by imbibing seeds, which expand severalfold in volume, demonstrates the magnitude of their water potential. Imbibing cocklebur seeds can exert a pressure of up to 1,000 atmospheres (approximately 100 kilopascals, or 15,000 pounds per square inch).

As a seed takes up water, it undergoes metabolic changes: enzymes are activated, RNA and then proteins are synthesized, the rate of cellular respiration increases, and other metabolic pathways are activated. In many seeds, cell division is not initiated during the early stages of germination. Instead, growth results solely from the expansion of small, preformed cells.

As the seed begins to germinate, the growing embryo obtains chemical building blocks for its development—carbohydrates, amino acids, and lipid monomers—by hydrolyzing (see Figure 2.8) starch, proteins, and lipids stored in the seed. These reserves are stored in the **cotyledons**—embryonic leaves—or in the **endosperm**—a seed tissue specialized for storage. Germination is completed when the **radicle** (embryonic root) emerges from the seed coat. The plant is then called a **seedling**.

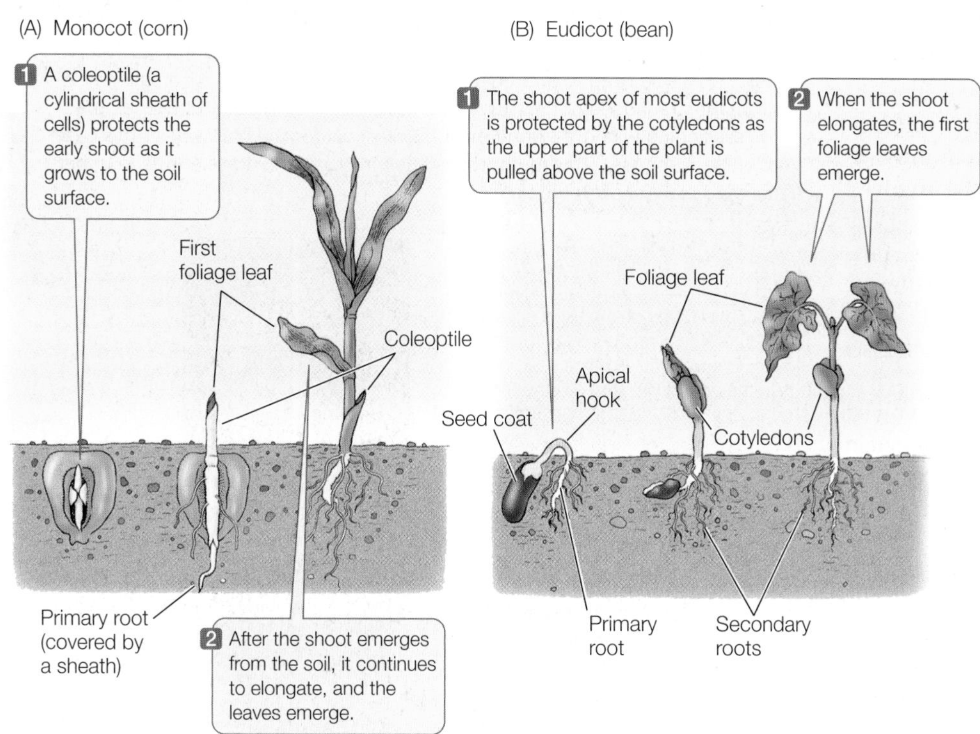
(A) Monocot (corn)
1 A coleoptile (a cylindrical sheath of cells) protects the early shoot as it grows to the soil surface.
First foliage leaf
Coleoptile
Primary root (covered by a sheath)
2 After the shoot emerges from the soil, it continues to elongate, and the leaves emerge.
(B) Eudicot (bean)
1 The shoot apex of most eudicots is protected by the cotyledons as the upper part of the plant is pulled above the soil surface.
2 When the shoot elongates, the first foliage leaves emerge.
Foliage leaf
Apical hook
Seed coat
Cotyledons
Primary root
Secondary roots

FIGURE 26.1 Patterns of Early Shoot Development (A) In grasses and some other monocots, growing shoots are protected by a coleoptile until they reach the soil surface. (B) In most eudicots, the growing point of the shoot is protected within the cotyledons.

Go to ACTIVITY 26.1 Monocot Shoot Development PoL2e.com/ac26.1

Go to ACTIVITY 26.2 Eudicot Shoot Development PoL2e.com/ac26.2

If the seed germinates underground, the new seedling must elongate rapidly (in the right direction!) and cope with a period of life in darkness or dim light. A series of photoreceptors direct this stage of development.

The pattern of early shoot development varies among the flowering plants. **FIGURE 26.1** shows the shoot development patterns of monocots and eudicots. In monocots, the growing shoot is protected by a cylindrical sheath of cells called the **coleoptile** as it pushes its way through the soil. In eudicots, the shoot is usually protected by the cotyledons.

Several hormones and photoreceptors help regulate plant growth

This survey of the early stages of plant development illustrates the many environmental cues that influence plant growth. A plant's responses to these cues are prompted and maintained by two types of regulators: hormones and photoreceptors. Both types of regulators act through signal transduction pathways (series of biochemical steps within a cell that occur between a stimulus and a response).

LINK

The general characteristics of signal transduction pathways are discussed in **Concepts 5.5 and 5.6**

Hormones are chemical signals that act at very low concentrations at sites often distant from where they are produced. Plant hormones are different from animal hormones (**TABLE 26.1**; see also Chapter 35). Each plant hormone plays multiple regulatory roles, and interactions among them can be complex. Several hormones regulate the growth and development of plants from seedling to adult (**TABLE 26.2**). Other hormones are involved in the plant's defenses against herbivores and microorganisms (which we will discuss in Chapter 28).

Photoreceptors are proteins with pigments associated with them. Light acts directly on photoreceptors, which in turn regulate developmental processes that need to be responsive to light, such as the many changes that occur as a seed germinates and a seedling emerges from the soil. Note that this effect of light as an information source is different from the effect of light as an energy source in photosynthesis.

Genetic screens have increased our understanding of plant signal transduction

In Chapter 14 we described how genetic studies can be used to identify the steps along a developmental pathway. You will recall the reasoning behind these experiments: if a mutation in a gene involved in a certain biochemical process disrupts a

TABLE 26.1 Comparison of Plant and Animal Hormones

Characteristic	Plant hormones	Animal hormones
Size, chemistry	Proteins, small organic molecules	Peptides, proteins, small molecules
Site of synthesis	Usually at many locations	Specialized glands or cells
Site of action	Local or distant	Usually distant, transported
Effects	Often diverse	Often specific
Regulation	By biochemical feedback	By central nervous system, ions, or feedback

developmental event, then that biochemical process must be essential for that developmental event. Similarly, genetic studies can be used to analyze pathways of receptor activation and signal transduction in plants: if proper signaling does not occur in a mutant strain, then the mutated gene must be involved in the signal transduction process. Mapping the mutated gene and identifying its function are starting points for understanding the signaling pathway. *Arabidopsis thaliana*, a member of the mustard family, has been a major model organism for plant biologists investigating signal transduction.

One technique for identifying the genes involved in a plant signal transduction pathway is illustrated in **FIGURE 26.2**. This technique, called a **genetic screen**, involves creating a large collection of randomly mutated plants and identifying those individuals that are likely to have a defect in the pathway of interest. Plant genes can be randomly mutated in two ways:

- By insertion of transposons (see Concepts 12.2 and 12.3) or T DNA (see Concept 13.2)
- By treatment of seeds with chemical mutagens, most commonly ethyl methane sulfonate or radiation

In both cases, the treated plants are grown and then examined for a specific phenotype, usually a characteristic that is known to be influenced by the pathway of interest. Once mutant plants have been selected, their genotypes are compared with those of wild-type plants. Mutant *Arabidopsis* plants with altered developmental patterns have provided a wealth of new information about the hormones present in plants and the mechanisms of hormone and photoreceptor action.

RESEARCH TOOLS

FIGURE 26.2 A Genetic Screen Genetics of the model plant *Arabidopsis thaliana* can be used to identify the steps of a signal transduction pathway. If a mutant strain does not respond to a hormone (in this case, ethylene), the corresponding wild-type gene must be essential for the pathway (in this case, ethylene response). This method has been instrumental to scientists in understanding plant growth regulation.

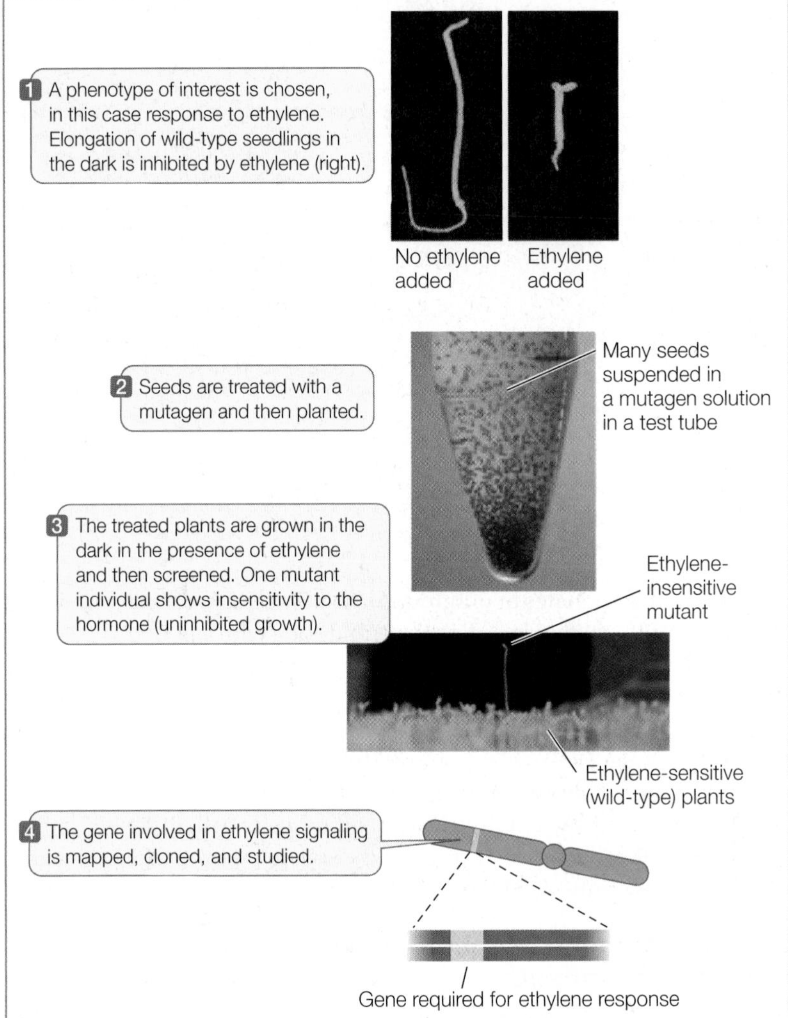

1 A phenotype of interest is chosen, in this case response to ethylene. Elongation of wild-type seedlings in the dark is inhibited by ethylene (right).
No ethylene added
Ethylene added
2 Seeds are treated with a mutagen and then planted.
Many seeds suspended in a mutagen solution in a test tube
3 The treated plants are grown in the dark in the presence of ethylene and then screened. One mutant individual shows insensitivity to the hormone (uninhibited growth).
Ethylene-insensitive mutant
Ethylene-sensitive (wild-type) plants
4 The gene involved in ethylene signaling is mapped, cloned, and studied.
Gene required for ethylene response

CHECKpoint CONCEPT 26.1

- ✓ Describe how monocots and eudicots differ in their early development.
- ✓ Seeds of the Australian shrub *Acacia myrtifolia* require extreme heat (fire) to germinate. How do you think fire might work as an environmental cue to break dormancy? What would be the selective advantage of this mechanism for breaking dormancy?
- ✓ Seedlings grown in the dark grow thin and tall, a phenomenon called etiolation. How would you set up a genetic screen using *Arabidopsis* to investigate the signaling pathway that controls etiolation?

With this overview of plant development, we will begin our examination of the specific regulatory mechanisms that underlie these events. Two hormones—gibberellins and auxin—have prominent roles in plant development.

CONCEPT 26.2 Gibberellins and Auxin Have Diverse Effects but a Similar Mechanism of Action

The discovery of two key plant hormones exemplifies the experimental approaches that plant biologists have used to investigate the mechanisms of plant development. **Gibberellins** (of which there are several active forms) and **auxin** were the first plant hormones to be identified, early in the twentieth century. In both cases, the discoveries came from observations of natural phenomena:

- *Gibberellins*: In rice plants, a disease caused by the fungus *Gibberella fujikuroi* resulted in plants that grew overly tall and spindly.

TABLE 26.2 Plant Growth Hormones

Hormone	Common structure	Typical activities
Abscisic acid		Maintains seed dormancy; closes stomata
Auxin (indole-3-acetic acid)		Promotes stem elongation, adventitious root initiation, and fruit development; inhibits axillary bud outgrowth, leaf abscission, and root elongation
Brassinosteroids		Promote stem and pollen tube elongation; promote vascular tissue differentiation
Cytokinins		Inhibit leaf senescence; promote cell division and axillary bud outgrowth; affect root growth
Ethylene		Promotes fruit ripening and leaf abscission; inhibits stem elongation and gravitropism
Gibberellins		Promote seed germination, stem growth, and ovule and fruit development; break winter dormancy; mobilize nutrient reserves in grass seeds

- *Auxin*: Charles Darwin and his son Francis noted that canary grass seedlings would bend toward the light when placed near a light source.

In both cases, a chemical substance was isolated that could cause the phenomenon:

- Gibberellic acid (see Table 26.2) made by the fungus caused rice plants to overgrow. Later it was found that plants also make gibberellic acid, and that applying it to plants caused growth.
- Auxin (indole-3-acetic acid) applied asymmetrically to the growing tips of seedlings caused cell expansion on the side away from the light, and this resulted in bending toward the light.

In each case, mutant plants that do not make the hormone exhibit a phenotype expected in the absence of the hormone, and adding the hormone reverses that phenotype (**FIGURE 26.3**):

- Tomato plants that do not make gibberellic acid are very short; supplying them with the hormone results in normal growth.
- *Arabidopsis thaliana* individuals that do not make auxin are also short; supplying them with that hormone reverses that phenotype.

Note that the phenotype involved—short stature, or dwarfism—is similar in both cases. This

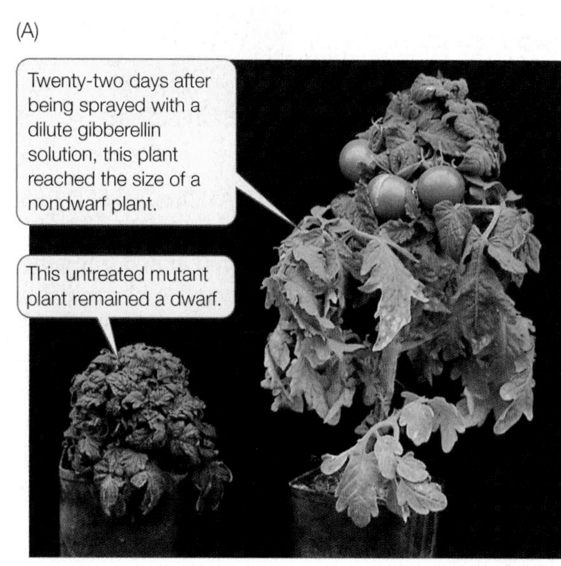

(A)
Twenty-two days after being sprayed with a dilute gibberellin solution, this plant reached the size of a nondwarf plant.
This untreated mutant plant remained a dwarf.

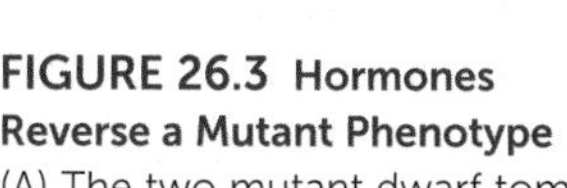

FIGURE 26.3 Hormones Reverse a Mutant Phenotype
(A) The two mutant dwarf tomato plants in this photograph were the same size when the one on the right was treated with gibberellins. (B) The short phenotype of this mutant *Arabidopsis* was reversed in the plant on the right by supplying auxin.

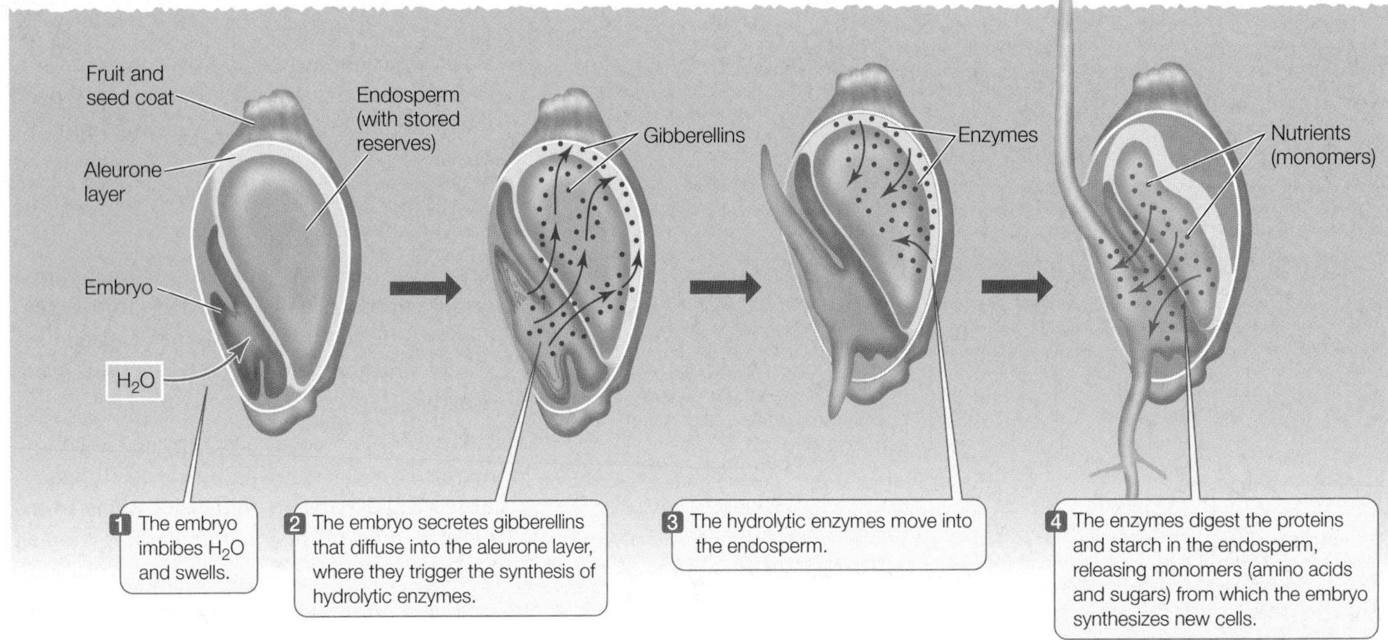

FIGURE 26.4 Gibberellins and Seed Germination During seed germination in cereal crops, gibberellins trigger a cascade of events that result in the conversion of starch and protein stores in the endosperm into monomers that can be used by the developing embryo.

Go to ACTIVITY 26.3 Events of Seed Germination PoL2e.com/ac26.3

observation exemplifies a concept that is important to keep in mind when studying plant hormones: their actions are not unique and specific, as is the case with animal hormones (see Table 26.1).

The three-part approach outlined above for gibberellins and auxin—observation, hormone isolation, and analysis of mutant plants—has been used to identify other plant hormones.

Gibberellins have many effects on plant growth and development

The functions of gibberellins can be inferred from the effects of experimentally decreasing concentrations of gibberellins or blocking their action at various times in plant development. Such experiments reveal that gibberellins have multiple roles in regulating plant growth.

STEM ELONGATION The effects of gibberellins on wild-type plants are not as dramatic as their effects on dwarf plants. We know, however, that gibberellins are indeed active in wild-type plants because inhibitors of gibberellin synthesis cause a reduction in stem elongation. Such inhibitors can be put to practical uses. For example, plants such as chrysanthemums that are grown in greenhouses tend to get tall and spindly. Because such plants do not appeal to consumers, flower growers spray them with gibberellin synthesis inhibitors to control their height.

FRUIT GROWTH Gibberellins can regulate the growth of fruit. Grapevines that produce seedless grapes develop smaller fruit than varieties that produce seed-bearing grapes. Biologists wanting to explain this phenomenon removed seeds from immature seeded grapes and found that this prevented normal fruit growth. This suggested that the seeds are sources of a growth regulator. Biochemical studies showed that developing seeds produce gibberellins, which diffuse out into the immature fruit tissue. Spraying young seedless grapes with a gibberellin solution, which causes them to grow as large as seeded ones, is now a standard commercial practice.

MOBILIZATION OF SEED RESERVES As noted in Concept 26.1, an important event early in seed germination is the hydrolysis of stored starch, proteins, and lipids. In germinating seeds of barley and other cereal crops, the embryo secretes gibberellins just after imbibition. The hormones diffuse through the endosperm to a surrounding tissue called the **aleurone layer**, which lies underneath the seed coat. The gibberellins trigger a cascade of events in the aleurone layer, causing it to synthesize and secrete enzymes that digest proteins and starch stored in the endosperm (**FIGURE 26.4**). These observations have practical importance: in the beer-brewing industry, gibberellins are used to enhance the "malting" (germination) of barley and the breakdown of its endosperm, producing sugars that are fermented into alcohol by yeast.

The transport of auxin mediates some of its effects

As we noted above, auxin was discovered in the context of **phototropism**: a response to light in which plant stems bend toward a light source. Auxin is made in the shoot apex (tip) and diffuses down the shoot in a polar (unidirectional) fashion, stimulating cell expansion. Note that this use of "polar" differs from its use in chemistry, where it means hydrophilic (see Concept 2.2). Unidirectional longitudinal movement of auxin occurs in other plant organs as well. For example, in a leaf petiole, which connects the leaf blade to the stem, auxin moves from the leaf blade end toward the stem. In roots, auxin moves unidirectionally toward the root tip.

Polar transport of auxin depends on four biochemical processes that may be familiar from earlier chapters (**FIGURE 26.5**):

- *Diffusion across a cell membrane.* Polar (hydrophilic) molecules diffuse across cell membranes less readily than nonpolar (hydrophobic) molecules (see Concept 5.2).
- *Membrane protein asymmetry.* Active transport, or efflux, carriers (see Concept 5.3) for auxin are located only in the portion of the cell membrane at the basal (bottom) end of the cell.
- *Proton pumping/chemiosmosis.* Proton pumps (see Concept 25.3) remove H^+ from the cell, thereby increasing the pH inside the cell and decreasing the pH in the cell wall. Proton pumping also sets up an electrochemical gradient, which provides potential energy to drive the transport of auxin by the carriers mentioned above.
- *Ionization of a weak acid.* Indole-3-acetic acid (the chemical name for auxin; see Table 26.2) is a weak acid; in solution it forms ions (H^+ and A^-, where A^- stands for indole acetate) that are in equilibrium with the non-ionized acid (HA):

$$A^- + H^+ \rightleftharpoons HA$$

When the pH is low, the increased H^+ concentration drives this reaction to the right, and HA (non-ionized auxin) is the predominant form. When the pH is higher, there is more A^- (ionized auxin).

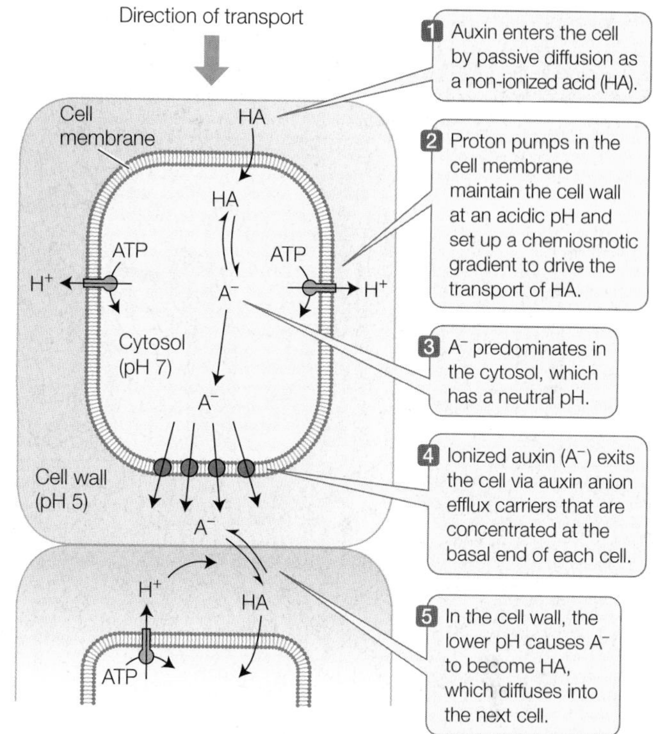

FIGURE 26.5 Polar Transport of Auxin Proton pumps set up a chemiosmotic gradient that drives the uptake of non-ionized auxin (HA) and the efflux of ionized auxin (auxin anion; A^-) through the basally placed auxin anion efflux carriers, leading to a net movement of auxin in a basal direction.

Whereas polar auxin transport distributes the hormone along the longitudinal axis of the plant, lateral (side-to-side) redistribution of auxin is responsible for directional plant growth. This redistribution is carried out by auxin efflux carriers that move from the base of the cell to one side; because of this, auxin exits the cell only on that side of the cell, rather than at the base, and moves sideways within the tissue.

This lateral movement of auxin explains the bending of canary grass seedlings toward light. When light strikes a canary grass coleoptile on one side, auxin at the tip moves laterally toward the shaded side. The asymmetry thus established is

APPLY THE CONCEPT

Gibberellins and auxin have diverse effects but a similar mechanism of action

Oat seedlings with coleoptiles 20 mm long were exposed to light from one side and checked for bending after 6 hours (see photo on p. 562). Various regions of the seedlings were covered with foil to block the light. The results varied depending on which region of the coleoptile was covered (see table).

1. Which part of the coleoptile senses the light?
2. For each treatment, what would happen if supplementary auxin were applied to the coleoptiles on the same side as the light? On the side away from the light?

REGION COVERED	BENDING
1. None	Yes
2. All	No
3. Apical 5 mm	No
4. Apical 10 mm	No
5. Apical 5–10 mm	Yes
6. Apical 10–15 mm	Yes

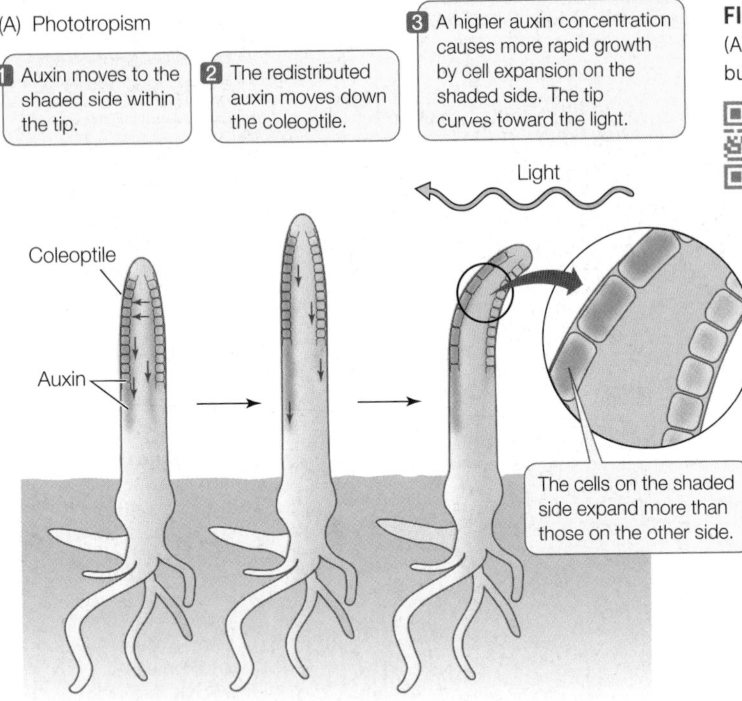
(A) Phototropism
1 Auxin moves to the shaded side within the tip.
2 The redistributed auxin moves down the coleoptile.
3 A higher auxin concentration causes more rapid growth by cell expansion on the shaded side. The tip curves toward the light.
Light
Coleoptile
Auxin
The cells on the shaded side expand more than those on the other side.

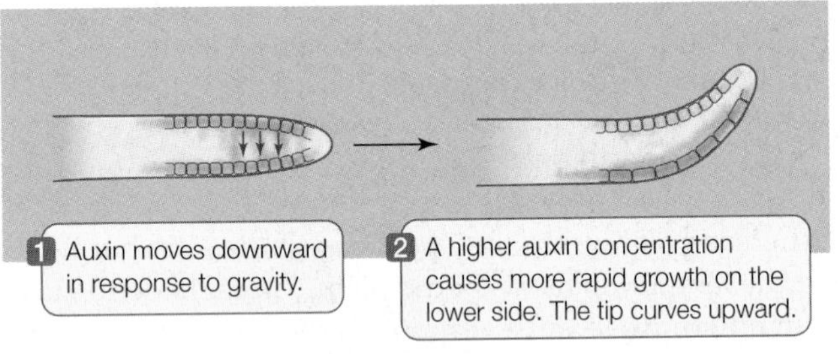
(B) Negative gravitropism of shoot
1 Auxin moves downward in response to gravity.
2 A higher auxin concentration causes more rapid growth on the lower side. The tip curves upward.

FIGURE 26.6 Plants Respond to Light and Gravity Phototropism (A) and gravitropism (B) occur in shoot apices in response to a redistribution of auxin.

Go to ANIMATED TUTORIAL 26.1
Tropisms
PoL2e.com/at26.1

maintained as polar transport moves auxin down the coleoptile, so that in the growing region below, the auxin concentration is highest on the shaded side. Cell expansion is thus increased on that side, causing the coleoptile to bend toward the light (**FIGURE 26.6A**). The same type of bending also occurs in eudicots:

Light

Light is not the only signal that can cause redistribution of auxin. Auxin moves to the lower side of a shoot that has been tipped sideways, causing more rapid growth in the lower side and hence an upward bending of the shoot. Such growth in a direction determined by gravity is called **gravitropism** (**FIGURE 26.6B**). The upward gravitropic response of shoots is defined as negative gravitropism; that of roots, which bend downward, is positive gravitropism.

Auxin plays many roles in plant growth and development

Like the gibberellins, auxin has multiple roles in plant growth and development.

ROOT INITIATION Cuttings from the shoots of some plants can produce roots and develop into entire new plants. For this to occur, cells in the interior of the shoot, originally destined to function only in food storage, must change their fate and become organized into the apical meristem of a new root (this is an example of the remarkable totipotency of plant cells; see Concept 14.1). Shoot cuttings of many species can be made to develop roots by dipping the cut surfaces into an auxin solution. These observations suggest that in an intact plant, the plant's own auxin plays a role in the initiation of lateral roots.

LEAF ABSCISSION In contrast to its stimulatory effect on root initiation, in leaves auxin inhibits the detachment of old leaves from stems. This detachment process, called **abscission**, is the cause of autumn leaf fall. Recall that the blade of a leaf produces auxin that moves toward the petiole. If the blade is cut off, the petiole falls from the plant more rapidly than if the leaf had remained intact. If the cut surface is treated with an auxin solution, however, the petiole remains attached to the plant, often longer than an intact leaf would have. The timing of leaf abscission in nature appears to be determined in part by a decrease in the movement of auxin through the petiole.

APICAL DOMINANCE Auxin helps maintain **apical dominance**, a phenomenon in which the apical bud (the very top/first bud that will form leaves) inhibits the growth of axillary buds (side buds; see Figure 24.1), resulting in the growth of a single main stem with minimal branching. A diffusion gradient of auxin from the apex of the shoot down the stem results in lower branches receiving less auxin and therefore branching more, while higher branches receive more auxin and branch less.

FRUIT DEVELOPMENT Fruit development normally depends on prior fertilization of the ovule (egg), but in many species, treatment of an unfertilized ovary with auxin or gibberellins causes **parthenocarpy**: fruit formation without fertilization. Parthenocarpic fruits form spontaneously in some cultivated varieties of plants, including seedless grapes, bananas, and some cucumbers.

CELL EXPANSION The expansion of plant cells is a key process in plant growth. Because the plant cell wall normally resists expansion of the cell, it plays a key role in controlling the rate and direction of plant cell growth. Auxin acts on the cell wall to regulate this process.

The expansion of a plant cell is driven primarily by the uptake of water, which enters the cell by osmosis. As it expands, the cell contents press against the cell wall, which resists this force (producing turgor pressure; see Figure 25.8). The cell wall is an extensively cross-linked network of polysaccharides and proteins, dominated by cellulose fibrils (see Figures 2.10 and 4.15). If the cell is to expand, some adjustments must be made in the wall structure to allow the wall to "give" under turgor pressure. Think of a balloon (the cell surrounded by a membrane) inside a box (the cell wall).

LINK

Water movement and turgor pressure in plant cells are discussed in **Concept 25.3**

The **acid growth hypothesis** explains auxin-induced cell expansion (**FIGURE 26.7**). This hypothesis proposes that protons (H^+) are pumped from the cytoplasm into the cell wall, lowering the pH of the wall and activating enzymes called expansins, which catalyze changes in the cell wall structure such that the polysaccharides adhere to each other less strongly. This change loosens the cell wall, allowing it to stretch as the cell expands. Auxin is believed to have two roles in this process: to increase the synthesis of proton pumps, and to guide their insertion into the cell membrane. Several lines of evidence support the acid growth hypothesis. For example, adding acid to the cell wall to lower the pH stimulates cell expansion even in the absence of auxin. Conversely, when a buffer is used to prevent the wall from becoming more acidic, auxin-induced cell expansion is blocked.

At the molecular level, auxin and gibberellins act similarly

The molecular mechanisms underlying both auxin and gibberellin action have been worked out with the help of genetic screens (see Figure 26.2). Biologists started by identifying mutant plants whose growth and development are insensitive to the hormones; that is, plants that are *not* affected by added hormone. Such mutant plants fall into two general categories:

- *Excessively tall plants*. These plants resemble wild-type plants given an excess of hormone, and they grow no taller when given extra hormone. They grow tall even when treated with inhibitors of hormone synthesis. Their hormone response is always "on," even in the absence of the hormone. It is presumed that the normal allele for the mutated gene codes for an inhibitor of the hormone signal transduction pathway. In wild-type plants, the signal transduction pathway is "off" unless stimulated by the hormone—therefore the plants are sensitive to added hormone. In the mutant plants the pathway is always "on"; the plant grows tall and is insensitive to added hormone.
- *Dwarf plants*. These plants resemble dwarf plants that are deficient in hormone synthesis (see Figure 26.3), but they do not respond to added hormone. In these mutant plants the hormone response is always "off," regardless of the presence of the hormone.

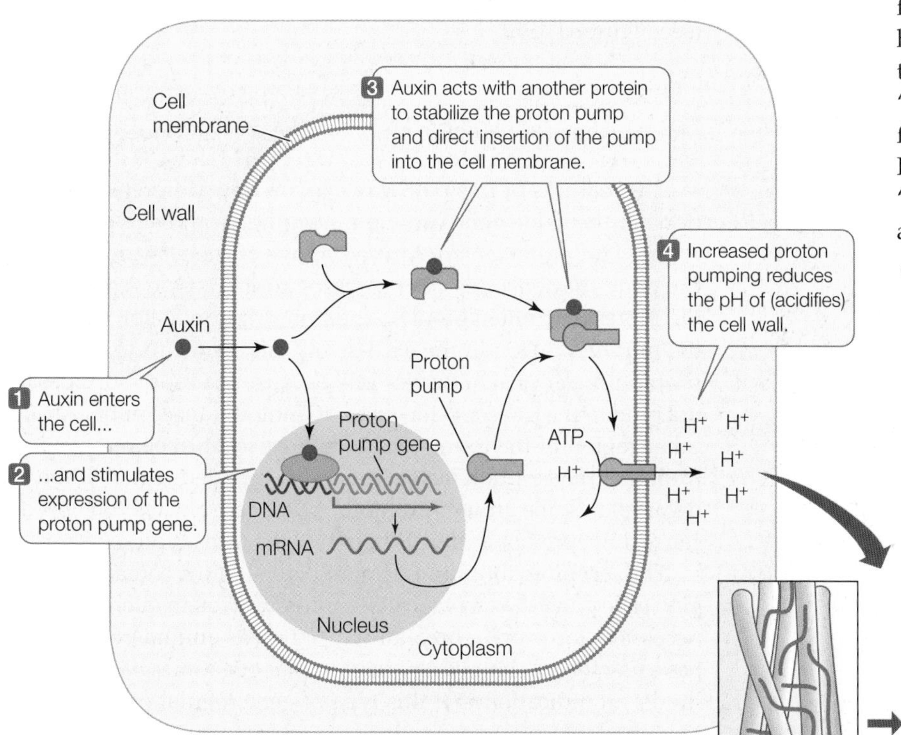
Cell membrane
3 Auxin acts with another protein to stabilize the proton pump and direct insertion of the pump into the cell membrane.
Cell wall
4 Increased proton pumping reduces the pH of (acidifies) the cell wall.
Auxin
Proton pump
1 Auxin enters the cell...
Proton pump gene
ATP
H+
2 ...and stimulates expression of the proton pump gene.
DNA
mRNA
Nucleus
Cytoplasm
H+ H+ H+ H+ H+ H+ H+

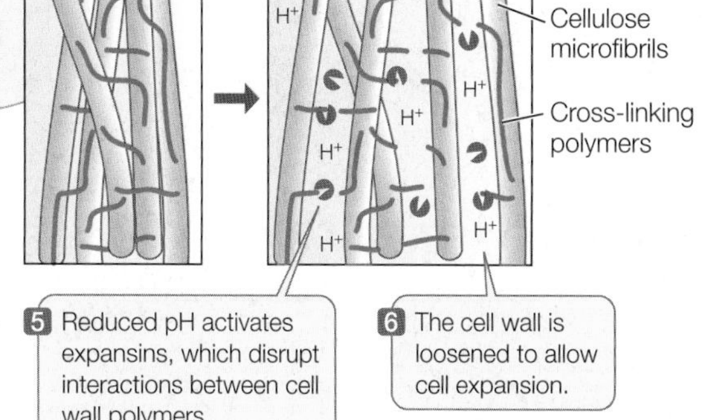
Expansin
Cellulose microfibrils
Cross-linking polymers
H+
5 Reduced pH activates expansins, which disrupt interactions between cell wall polymers.
6 The cell wall is loosened to allow cell expansion.

FIGURE 26.7 Auxin and Cell Expansion The plant cell wall is an extensive network of cross-linked polymers. Auxin induces loosening of the cell wall by activating proton pumps that reduce pH in the cell wall.

Go to ANIMATED TUTORIAL 26.2
Auxin Affects Cell Walls
PoL2e.com/at26.2

APPLY THE CONCEPT

Gibberellins and auxin have diverse effects but a similar mechanism of action

The aleurone layer of germinating barley can be isolated and studied for the induction of α-amylase, the enzyme that catalyzes starch hydrolysis. Predict the amount of α-amylase activity in aleurone layers subjected to the following treatments:

1. Incubation with gibberellic acid
2. Incubation with auxin
3. Incubation with gibberellic acid in the presence of an inhibitor of DNA transcription to mRNA
4. Incubation without gibberellic acid
5. Incubation without gibberellic acid in the presence of an inhibitor of transcription
6. Incubation with gibberellic acid in the presence of an inhibitor of the proteasome
7. Incubation without gibberellic acid in the presence of an inhibitor of the proteasome

Remarkably, some mutations of both types turned out to affect the same gene. How could this be? The mystery was solved once biologists figured out how auxin and gibberellins work in wild-type plants. Of course, the actual proteins involved with the two hormones are different, but the actions of both hormones are similar: they act by removing a repressor from a transcription factor that stimulates the expression of growth-promoting genes (**FIGURE 26.8**). The hormones bind to a receptor protein, which in turn binds to the repressor. Binding of the hormone–receptor complex stimulates polyubiquitination of the repressor, targeting it for breakdown in the proteasome (see Figure 11.19).

LINK

Review the process of transcription initiation in **Concept 11.1**

The features of this pathway explain how different mutations in the same gene can have seemingly opposite phenotypes. The mutations described above turned out to be in the gene that encodes the repressor protein. One region of the repressor protein binds to the transcription factor. This is the mutated region in the excessively tall plants: the growth-promoting genes are always "on" because the repressor does not bind to the transcription factor to inhibit transcription. Another region of the repressor protein causes it to be removed from the transcription factor. This is the mutated region in the dwarf plants: the growth-promoting genes are always "off" because the repressor is always bound to the transcription factor.

The receptors for both auxin and gibberellins contain a region called an **F-box** that facilitates protein–protein interactions necessary for protein breakdown. Whereas animal genomes have few F-box–containing proteins, plant genomes have hundreds, an indication that this type of gene regulation is common in plants.

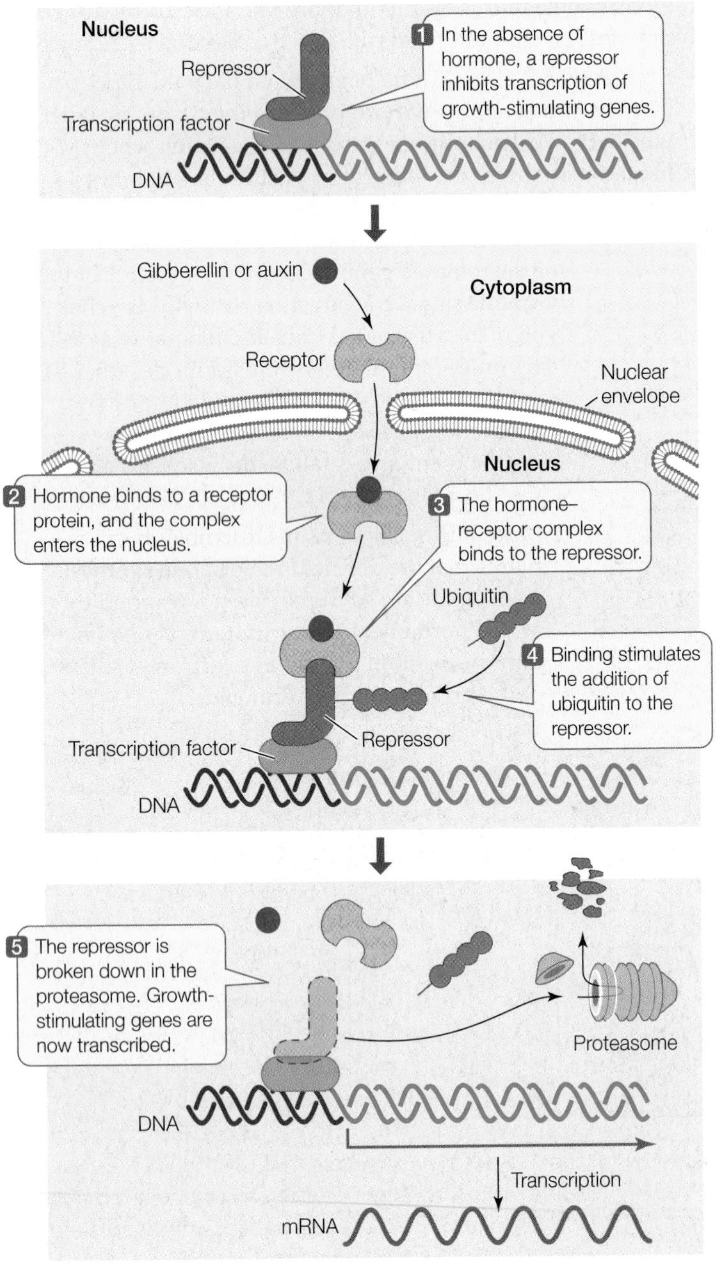

FIGURE 26.8 Gibberellins and Auxin Have Similar Signal Transduction Pathways Both hormones act to stimulate gene transcription by inactivating a repressor protein.

Go to MEDIA CLIP 26.1
Gibberellin Binding to Its Receptor
PoL2e.com/mc26.1

CHECKpoint CONCEPT 26.2

- ✓ Make a table for one auxin response and one gibberellin response with three column headings: site of synthesis, site of action, and effect.
- ✓ Explain why, even though auxin moves *away* from the lighted side of a coleoptile tip, the coleoptile bends *toward* the light.
- ✓ What would be the phenotype, in terms of seed germination, of a plant that has a mutation that disables the F-box region of the receptor for gibberellin?

The same approaches that led to the discovery of auxin and gibberellins have revealed several other classes of plant hormones. These hormones are diverse in their actions and affect many of the same developmental processes as auxin and gibberellins.

CONCEPT 26.3 Other Plant Hormones Have Diverse Effects on Plant Development

The discoveries of gibberellins and auxin have led to the isolation of additional classes of plant hormones (see Table 26.2). These hormones have multiple effects, and they interact with one another in many cases.

Ethylene is a gaseous hormone that promotes senescence

Whereas the cytokinins (see below) delay senescence (aging), another plant hormone promotes it: the gas **ethylene** (see Table 26.2), which is sometimes called the senescence hormone. When streets were lit by gas rather than by electricity, leaves on trees near street lamps dropped earlier in the fall than those on trees farther from the lamps. We now know why: ethylene, a combustion product of the illuminating gas, caused the early abscission. Whereas auxin inhibits leaf abscission, ethylene strongly promotes it; thus the ratio of auxin and ethylene controls abscission.

FRUIT RIPENING By promoting senescence, ethylene speeds the ripening of fruit. As a fruit ripens, it loses chlorophyll and its cell walls break down, making the fruit softer; ethylene promotes both of these processes. Ethylene also causes an increase in its own production. Thus once ripening begins, more and more ethylene forms, and because it is a gas, it diffuses readily throughout the fruit and even to neighboring fruits on the same or other plants. The old saying that "one rotten apple spoils the barrel" is true. That rotten apple is a rich source of ethylene, which speeds the ripening and subsequent rotting of the other fruit in a barrel or other confined space.

LINK

The action of ethylene in fruit ripening is an example of positive feedback; see **Figure 1.8**

Farmers in ancient times poked holes in developing figs to make them ripen faster. Wounding causes an increase in ethylene production by the fruit, and that increase promotes ripening. Today commercial shippers and storers of fruit hasten ripening by adding ethylene to storage chambers. This use of ethylene is the single most important use of a natural plant hormone in agriculture and commerce. Conversely, ripening can be delayed by the use of "scrubbers" and adsorbents (binding substances) that remove ethylene from the atmosphere in fruit storage chambers.

These stored tomatoes were treated with ethylene to promote ripening.

As flowers senesce, their petals may abscise (fall off). Flower growers and florists often immerse the cut stems of ethylene-sensitive flowers in dilute solutions of silver thiosulfate before selling them. Silver salts inhibit ethylene action by interacting directly with the ethylene receptor—thus they delay senescence, keeping flowers "fresh" for a longer time.

STEM GROWTH The stems of many eudicot seedlings (which lack protective coleoptiles) form an **apical hook** that protects the delicate shoot apex while the stem grows through the soil.

The apical hook is maintained through an asymmetrical production of ethylene, which inhibits the expansion of cells on the inner surface of the hook. Once the seedling breaks through the soil surface and is exposed to light, ethylene synthesis stops, and the cells of the inner surface are no longer inhibited. Those cells now expand, and the hook unfolds, raising the shoot apex and the expanding leaves into the sun. Ethylene also inhibits stem elongation in general, promotes lateral swelling of stems (as do the cytokinins), and decreases the sensitivity of stems to gravitropic stimulation. The short, thick, and exaggerated apical hook constitutes the "triple response," a well-characterized stunted growth habit observed when plants are treated with ethylene.

Cytokinins are active from seed to senescence

Like bacteria and yeasts, some plant cells, such as parenchyma cells, can be grown in a liquid or solidified medium containing sugars and salts. For continuous cell division, cultured plant cells also require a type of hormone called a **cytokinin**, first isolated from corn. More than 150 different cytokinins have been isolated, and most are derivatives of adenine.

Cytokinins have a number of different effects (see Table 26.2), in many cases by interacting with auxin:

- Adding an appropriate combination of auxin and cytokinins to a growth medium induces rapid proliferation of cultured plant cells.
- In cell cultures, a high cytokinin-to-auxin ratio (i.e., a medium containing more cytokinin than auxin) promotes the formation of shoots; a low ratio promotes the formation of roots.
- Cytokinins can cause certain light-requiring seeds to germinate even when they are kept in constant darkness.
- Cytokinins usually inhibit the elongation of stems, but they cause lateral swelling of stems and roots (the fleshy roots of radishes are an extreme example).
- Cytokinins stimulate axillary buds to grow into branches. The auxin-to-cytokinin ratio controls the extent of branching (bushiness) of a plant; a high ratio inhibits branching, a low ratio stimulates it.
- Cytokinins delay the senescence of leaves. If leaf blades are detached from a plant and floated on water or a nutrient solution, they gradually turn yellow and show other signs of senescence. If instead they are floated on a solution containing a cytokinin, they remain green and senesce much more slowly. Roots contain abundant cytokinins, and cytokinin transport to the leaves delays senescence:

This leaf was treated with cytokinin, which delays senescence.
This detached leaf was not treated.

Cytokinin appears to act through a pathway that includes proteins similar to those in two-component systems, a type of signal transduction pathway commonly found in bacteria. Indeed, this pathway was one of the first of its kind discovered in eukaryotes. The two components in such a system are:

- a *receptor* that can act as a protein kinase, phosphorylating itself as well as a target protein
- a *target protein*, generally a transcription factor, that regulates the response

Genetic screens in *Arabidopsis* for abnormalities in the response to cytokinin have identified the receptor (AHK; *Arabidopsis histidine kinase*) and the target protein (ARR; *Arabidopsis response regulator*). The target protein acts as a transcription factor when phosphorylated. The cytokinin signal transduction pathway also includes a third protein (AHP; *Arabidopsis histidine phosphotransfer protein*), which transfers phosphates from the receptor to the target protein (**FIGURE 26.9**). The plant genome has more than 20 genes that are expressed in response to this signaling pathway.

Brassinosteroids are plant steroid hormones

In animals, steroid hormones such as cortisol and estrogen, which are formed from cholesterol, have been well studied for decades (see Chapter 35). In the 1970s, biologists isolated a steroid from the pollen of rape (*Brassica napus*; sometimes called canola), a member of the mustard family. When applied to various plant tissues, this hormone stimulated cell expansion, pollen tube elongation, and vascular tissue differentiation, but it inhibited root expansion. Mutant plants with defects in brassinosteroid responses are stunted:

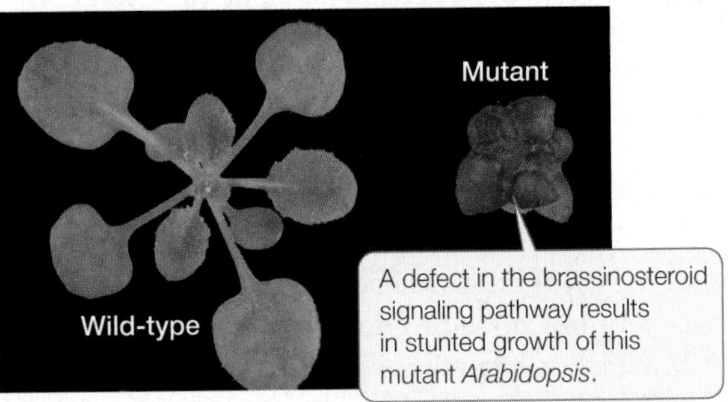

Dozens of chemically related, growth-affecting **brassinosteroids** have been found in plants. Like other plant hormones, brassinosteroids have diverse effects (see Table 26.2):

- They enhance cell expansion and cell division in shoots.
- They promote xylem differentiation.
- They promote growth of pollen tubes during reproduction.
- They promote seed germination.
- They promote apical dominance and leaf senescence.

The signaling pathway for these plant steroids differs sharply from those for steroid hormones in animals. In animals, steroids diffuse through the cell membrane and bind to receptors in the cytoplasm. In contrast, the receptor for brassinosteroids is an integral protein in the cell membrane.

Abscisic acid acts by inhibiting development

The hormone **abscisic acid** is found in most plant tissues. Many of its actions involve inhibition of other hormones' actions in the plant (see Table 26.2):

- *Prevention of seed germination*. Abscisic acid prevents seeds from germinating on the parent plant before the seeds are

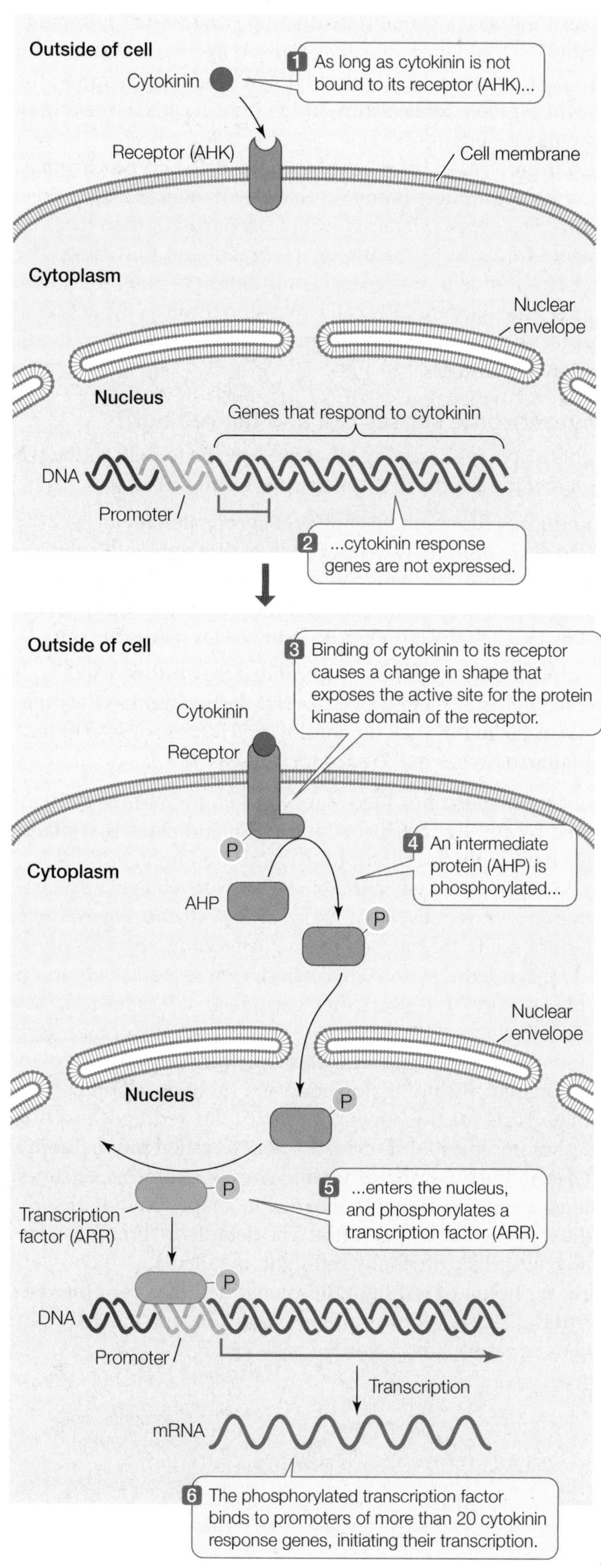

FIGURE 26.9 The Cytokinin Signal Transduction Pathway Plant cells respond to cytokinins through a two-component signal transduction pathway involving a receptor and a target protein.

dry. Premature germination, termed vivipary, is undesirable in cereal crops because the grain is ruined for human consumption if it has started to sprout. Viviparous seedlings are unlikely to survive if they remain attached to the parent plant and are unable to establish themselves in the soil.

- *Promotion of seed dormancy*. Abscisic acid promotes seed dormancy and inhibits the initiation of germination events, such as the hydrolysis of stored macromolecules, by gibberellins.
- *Reaction to stress*. Abscisic acid is a "stress hormone" that mediates a number of plant responses to pathogens and environmental stresses. We will describe some of these effects in Chapter 28. For now, consider that when the air is dry, abscisic acid stimulates the closure of stomata to prevent water loss by transpiration (see Concept 25.3).

Abscisic acid provides examples of the complex interactions between plant hormones. Whereas abscisic acid promotes seed dormancy, gibberellic acid inhibits it; whereas abscisic acid inhibits seed germination, both gibberellic acid and ethylene promote it. How the signaling pathways used by these hormones interact is under intense study by plant molecular biologists.

CHECKpoint CONCEPT 26.3

- ✓ Which plant hormones affect cell expansion, and what is the effect in each case?
- ✓ At the grocery store, you can buy plastic bags that keep fruit fresh for a much longer period than other containers. How do you think they work?
- ✓ A mutant strain of corn shows vivipary. Why is this a disadvantage for the plant and the crop, and what gene might be involved?

As we have seen, plant development is strongly influenced by the environment. For example, in Concept 26.1 we alluded to the role of photoreceptors whose responses to light guide the early growth of seedlings. We will now turn to the molecular nature of these and other plant photoreceptors.

CONCEPT 26.4 Photoreceptors Initiate Developmental Responses to Light

Plants are completely dependent on sunlight for energy, so it is not surprising that they have multiple ways to sense light and respond appropriately. A number of physiological and developmental events in plants are controlled by light, a process called **photomorphogenesis**. Plants respond to two aspects of light: its *quality*—that is, the wavelengths of light to which the plant is exposed; and its *quantity*—that is, the intensity and duration of light exposure.

Chapter 6 described how chlorophyll and other photosynthetic pigments absorb light of certain wavelengths (see Figure 6.17) and how plants use light energy to synthesize carbohydrates. Here we consider how light affects plant development. Light influences seed germination, phototropism, shoot elongation, initiation of flowering, and many other important aspects of plant development. Several photoreceptors take part in these processes.

Phototropin, cryptochromes, and zeaxanthin are blue-light receptors

An action spectrum for the phototropic response of a coleoptile (**FIGURE 26.10**) shows that blue light at a wavelength of 436 nm is the most effective in stimulating that response. To identify the blue-light-absorbing molecule involved, biologists have taken a genetic approach using the model plant *Arabidopsis*.

LINK

Action spectra are discussed in **Concept 6.5**

By obtaining blue-light-insensitive *Arabidopsis* mutant plants from a genetic screen, researchers were able to identify the gene for a blue-light receptor protein located in the cell membrane, called **phototropin**. This protein has a flavin mononucleotide associated with it that absorbs blue light at 436 nm. Light absorption leads to a change in the shape of the protein that exposes an active site for a protein kinase, which in turn initiates a signal transduction cascade that ultimately results in stimulation of cell expansion by auxin. Phototropin also participates with another type of blue-light receptor, the plastid pigment **zeaxanthin**, in the light-induced opening of stomata (see Figure 25.13).

A third class of blue-light receptors is the **cryptochromes**, proteins with yellow pigments that absorb blue and ultraviolet light. These pigments are located primarily in the plant cell nucleus and affect seedling development and flowering. The exact mechanism of cryptochrome action is not yet known. Strong blue light inhibits cell expansion through the action of cryptochromes, although the most rapid responses are mediated by phototropin.

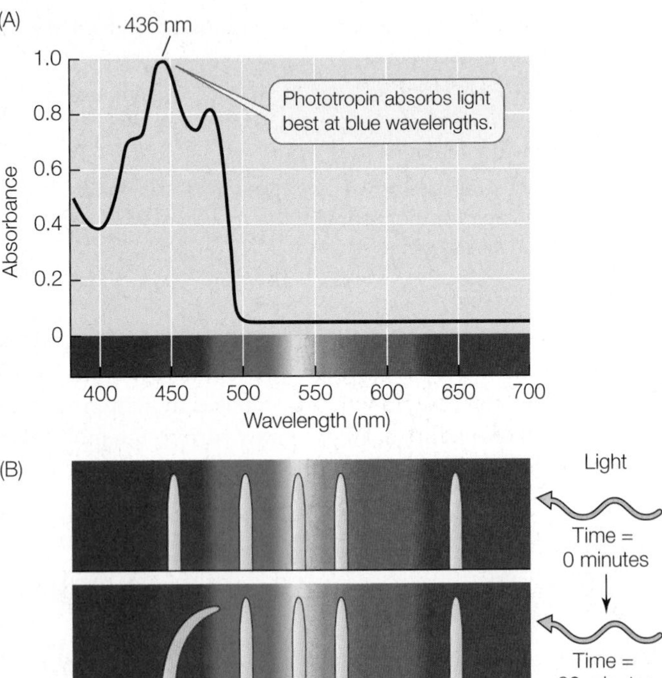

FIGURE 26.10 Action Spectrum for Phototropism The absorption spectrum for phototropin (A) is similar to the action spectrum for the bending of a coleoptile toward light (B). After 90 minutes, only the coleoptiles exposed to blue light bend.

Phytochrome senses red and far-red light

Plants also have responses that are specific to light in the red region of the spectrum. Here are two examples:

- Lettuce seeds spread on soil will germinate only in response to light. Even a brief flash of dim light will suffice. This response prevents the seeds from germinating when they are so deep in the soil that the seedlings would not be able to reach the light before their seed reserves ran out.
- Adult cocklebur plants flower when they are exposed to long nights. If there is even a brief flash of light in the middle of the night, they do not flower. This response enables plants to flower in the appropriate season.

The action spectra of these processes show that they are induced by red light (650–680 nm). This indicates that plants must have a photoreceptor that absorbs red light.

What is especially remarkable about these red light responses is that *they are reversible by far-red light* (710–740 nm). For example, if lettuce seeds are exposed to brief, alternating periods of red and far-red light in close succession, they respond only to the final exposure. If it is red, they germinate; if it is far-red, they remain dormant (**FIGURE 26.11**). This reversibility of the effects of red and far-red light regulates many other aspects of plant development, including flowering and shoot development.

The basis for the effects of red and far-red light is a bluish photoreceptor in the cytosol of plants called **phytochrome**. Phytochrome exists in two interconvertible isoforms, or states. The molecule undergoes a change in shape when it absorbs light at particular wavelengths. The default or "ground" state, which absorbs principally red light, is called $\mathbf{P_r}$. When P_r absorbs a photon of red light, it is converted into the other isoform, $\mathbf{P_{fr}}$. The P_{fr} isoform preferentially absorbs far-red light; when it does so, it is converted back into P_r:

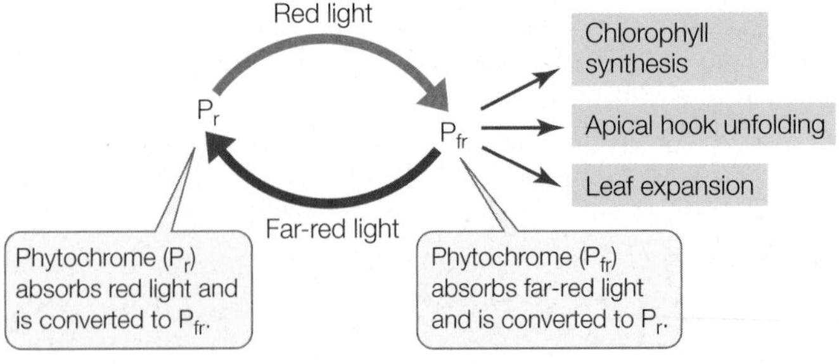

INVESTIGATION

FIGURE 26.11 Photomorphogenesis and Red Light Lettuce seeds will germinate if exposed to a brief period of light. The action spectrum for germination indicated that red light was most effective in promoting it, but far-red light would reverse the effect if presented right after the red light. Harry Borthwick and his colleagues asked what the effect of repeated alternating flashes of red and far-red light would be. In each case, the final exposure determined the germination response. This observation led to the conclusion that a single, photo-reversible molecule was involved. That molecule turned out to be phytochrome.[a]

HYPOTHESIS

The effects of red and far-red light on lettuce seed germination are mutually reversible.

METHOD

Expose lettuce seeds to alternating periods of red light R for 1 minute and far-red light FR for 4 minutes.

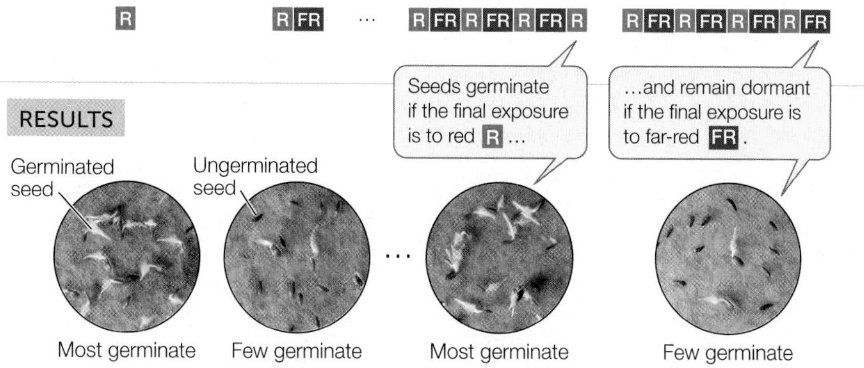

CONCLUSION

Red light and far-red light reverse each other's effects.

ANALYZE THE DATA

Seven groups of 200 lettuce seeds each were incubated in water for 16 hours in the dark. One group was then exposed to white light for 1 minute. A second group (controls) remained in the dark. Five other groups were exposed to red (R) and/or far-red (FR) light. All the seeds were then returned to darkness for 2 more days. The table shows the number of seeds that germinated in each group.

Condition	Seeds germinated
1. White light	199
2. Dark	17
3. R	196
4. R then FR	108
5. R then FR then R	200
6. R then FR then R then FR	86
7. R then FR then R then FR then R	198

A. Calculate the percentage of seeds that germinated in each case.

B. What can you conclude about the photoreceptors involved?

Go to **LaunchPad** for discussion and relevant links for all **INVESTIGATION** figures.

[a]H. Borthwick et al. 1952. *Proceedings of the National Academy of Sciences USA* 38: 662–666.

P_{fr}, not P_r, is the active isoform of phytochrome—the form that triggers important biological processes in various plants. These processes include chlorophyll synthesis, apical hook unfolding, and leaf expansion.

The ratio of red to far-red light determines whether a phytochrome-mediated response will occur. For example, in full daylight the ratio is about 1.2:1; because there is more red than far-red light, the P_{fr} isoform predominates. But for a plant growing in the shade of other plants, the ratio may be as low as 0.13:1, and most phytochrome is in the P_r isoform. The low ratio of red to far-red light in the shade results from absorption of red light by chlorophyll in the leaves overhead, so less of the red light gets through to the plants below. Shade-intolerant species respond by stimulating cell expansion in the stem and thus growing taller to escape the shade (similar to the behavior of etiolated seedlings grown in the dark). Similarly, shade cast by other plants prevents germination of seeds that require red light to germinate (see Figure 26.11).

Phytochrome stimulates gene transcription

How does phytochrome—or more specifically, P_{fr}—work? Phytochrome is a cytoplasmic protein composed of two subunits (**FIGURE 26.12**). Each subunit contains a protein chain and a nonprotein pigment called a chromophore. In *Arabidopsis*, there is a gene family that encodes five slightly different phytochromes, each functioning in different photomorphogenic responses.

The transcription of genes involved in phytochrome responses is changed when P_r is converted into the P_{fr} isoform. When P_r absorbs red light, the chromophore changes shape, which leads to a change in the conformation of the phytochrome protein from the P_r isoform to the P_{fr} isoform. Conversion to the P_{fr} isoform exposes two important regions of the protein (see Figure 26.12), both of which affect transcriptional activity:

- Exposure of a *nuclear localization signal sequence* (see Figure 10.20) results in movement of P_{fr} from the cytosol to the nucleus. Once in the nucleus, P_{fr} binds to transcription factors and thereby stimulates expression of genes involved in photomorphogenesis.
- Exposure of a *protein kinase* domain causes the P_{fr} protein to phosphorylate itself and other proteins involved in red-light signal transduction. Those proteins change the activity of other transcription factors.

The effect of this activation of transcription factors is quite large: in *Arabidopsis*, phytochrome affects an amazing 2,500 genes (10 percent of the entire genome) by either increasing or

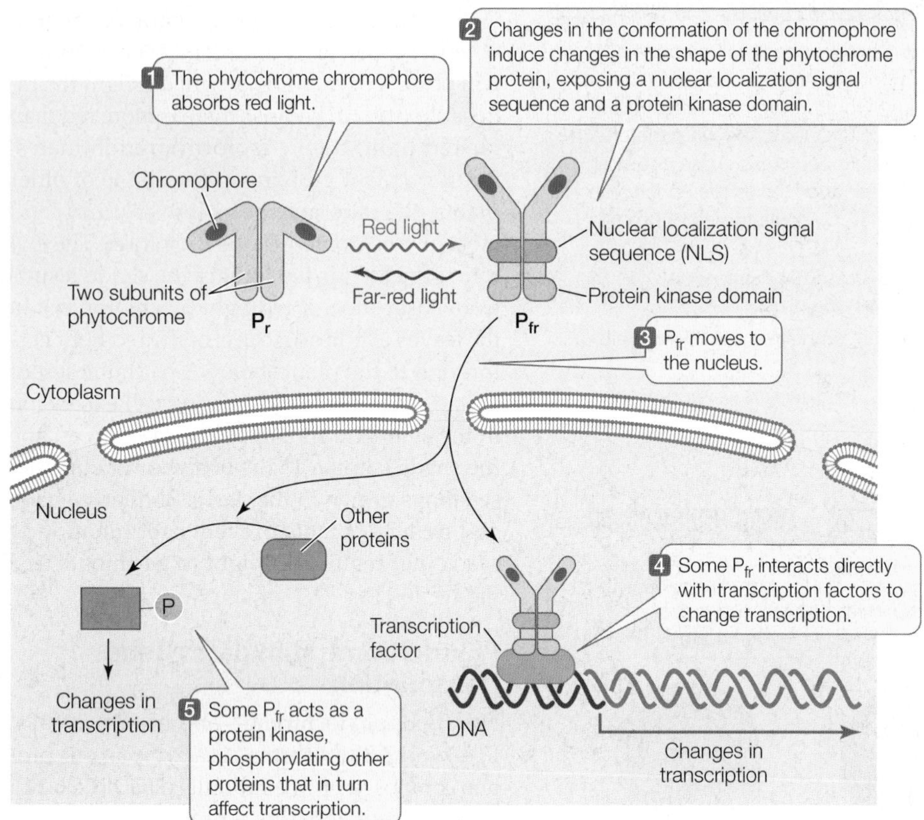

FIGURE 26.12 Phytochrome Stimulates Gene Transcription Phytochrome is composed of two subunits, each containing a protein chain and a chromophore. When the chromophore absorbs red light, phytochrome is converted into the P_{fr} isoform, which activates transcription of phytochrome-responsive genes.

decreasing their expression. Some of these genes are related to other hormones. For example, when P_{fr} is formed at the time of seed germination, genes for gibberellin synthesis are activated and genes for gibberellin breakdown are repressed. As a result, gibberellins accumulate and seed reserves are mobilized (see Figure 26.4).

Circadian rhythms are entrained by photoreceptors

The timing and duration of biological activities in living organisms are governed in all eukaryotes and some prokaryotes by what is commonly called a biological clock: an oscillator within cells that alternates back and forth between two states at roughly 12-hour intervals. The major outward manifestations of this clock are known as **circadian rhythms** (Latin *circa*, "about"; *dies*, "day"). In plants, circadian rhythms influence, for example, the opening (during the day) and closing (at night) of stomata in *Arabidopsis* and the elevation toward the sun (during the day) and lowering (at night) of leaves in bean plants.

LINK

For a description of the role of circadian rhythms in animal behavior, see **Concept 40.4**

The circadian rhythms of plants can be reset, or entrained, within limits, by light–dark cycles that change. Consider what happens when plants adapt to day length as the seasons change during the year. The action spectrum for plant entrainment indicates that phytochrome (and to a lesser extent, blue-light receptors) is very likely involved. At sundown, phytochrome is mostly in the active P_{fr} isoform. But as the night progresses, P_{fr} is gradually converted back into the inactive P_r isoform. By dawn, most phytochrome is in the P_r state, but as daylight begins, it is rapidly converted into P_{fr}. The switch to the P_{fr} isoform resets the plant's biological clock. However long the night, the clock is still reset at dawn every day. Thus while the total period measured by the clock does not change, the clock adjusts to changes in day length over the course of the year.

CHECKpoint CONCEPT 26.4

- ✓ Trace the events that occur in the phytochrome-mediated germination response of lettuce seeds, from the reception of light to alteration in gene expression.
- ✓ Compare the wavelengths of light absorbed by chlorophyll (see Figure 6.17) and by phototropin (see Figure 26.10). How does the difference influence a plant's responses to light when a plant grows beneath a forest canopy?
- ✓ How would you show that a physiological process in a plant is brought about by phytochrome?

What changes in their growth patterns made the new strains of cereal crops produced by the Green Revolution so successful?

ANSWER It is only recently that plant biologists have discovered why semi-dwarf wheat and rice have short stems (**FIGURE 26.13**). This knowledge has come from an understanding of the effects of various hormones on plant growth and development (Concept 26.1), the signal transduction pathways for response to these hormones (Concepts 26.2 and 26.3), and the biochemical pathways for the synthesis of these hormones. It turns out that the genes involved in wheat and in rice are different.

In semi-dwarf wheat, the mutant allele that produces short plants is involved in signal transduction in response to gibberellin. The gene involved is called *Rht* ("reduced height"). The mutant allele was first seen in Japanese semi-dwarf varieties. It is dominant, so its phenotypes are readily seen. Norman Borlaug used conventional plant breeding to transfer the *Rht* mutant allele into wheat strains in Mexico. In addition to having short stature and not falling over when bearing a high grain yield, *Rht* mutant plants put a greater proportion of their photosynthate into making seeds than wild-type plants do. This higher "harvest index" is also a major reason why most wheat varieties today carry an *Rht* mutation.

A hint that *Rht* might be involved in hormone regulation came when semi-dwarf plants harboring *Rht* mutations were found to be insensitive to added gibberellin and did not grow to normal height when the hormone was applied. *Rht* encodes the transcriptional repressor in the gibberellin signal transduction pathway (see Figure 26.8). Mutations in semi-dwarf plants are in the region that binds the gibberellin receptor and that signals polyubiquitination and degradation in the proteasome. Thus in the mutant plants, the repressor is always bound to the transcription factor, inhibiting its activity and keeping the gibberellin response "off." This mutation therefore results in shorter stature.

In semi-dwarf rice, the mutant allele that produces short plants is involved with gibberellin *synthesis*. The gene involved is called *sd-1* ("semi-dwarf"). In contrast to semi-dwarf wheat, rice plants with mutations in *sd-1* are sensitive to gibberellin—that is, the semi-dwarf phenotype can be corrected by supplying gibberellin. This observation indicates that the signal transduction machinery is intact in the mutant plants but that they lack the hormone. Indeed, *sd-1* encodes an enzyme that catalyzes one of the last steps in gibberellin synthesis.

FIGURE 26.13 Semi-Dwarf Rice A short variety of rice (left) produces higher yields of grain than its taller counterpart (right). The semi-dwarf rice produces less than normal amounts of gibberellins.

SUMMARY

CONCEPT 26.1 Plants Develop in Response to the Environment

- Plant development is regulated by environmental cues, receptors, hormones, and the plant's genome.
- Seed **dormancy**, which has adaptive advantages, is maintained by a variety of mechanisms. When dormancy ends, the seed **germinates** and develops into a **seedling**. **Review Figure 26.1 and ACTIVITIES 26.1 and 26.2**
- **Hormones** and **photoreceptors** act through signal transduction pathways to regulate plant growth and development.
- **Genetic screens** using the model organism *Arabidopsis thaliana* have contributed greatly to our understanding of signal transduction pathways in plants. **Review Figure 26.2**

CONCEPT 26.2 Gibberellins and Auxin Have Diverse Effects but a Similar Mechanism of Action

- **Gibberellins** stimulate growth of stems and fruits as well as mobilization of seed reserves in cereal crops. **Review Figure 26.4 and ACTIVITY 26.3**
- **Auxin** is made in cells at the shoot apex and moves down to the growing region in a polar manner. **Review Figure 26.5**
- Lateral movement of auxin, mediated by auxin efflux carriers, is responsible for **phototropism** and **gravitropism**. **Review Figure 26.6 and ANIMATED TUTORIAL 26.1**
- Auxin plays roles in lateral root formation, leaf **abscission**, and **apical dominance**.
- The **acid growth hypothesis** explains how auxin promotes cell expansion by increasing proton pumps in the cell membrane, which loosens the cell wall. **Review Figure 26.7 and ANIMATED TUTORIAL 26.2**
- Both auxin and gibberellins act by binding to their respective receptors, which then bind to a transcriptional repressor, leading to the repressor's breakdown in the proteasome. **Review Figure 26.8**

CONCEPT 26.3 Other Plant Hormones Have Diverse Effects on Plant Development

- **Cytokinins**, most of which are adenine derivatives, often interact with auxin. They promote plant cell division, promote seed germination in some species, and inhibit stem elongation, among other activities.
- Cytokinins act on plant cells through a two-component signal transduction pathway. **Review Figure 26.9**

(continued)

SUMMARY (continued)

- The ratio of auxin to **ethylene** controls leaf abscission. Ethylene promotes senescence and fruit ripening. It causes the stems of eudicot seedlings to form a protective **apical hook**. In stems, it inhibits elongation, promotes lateral swelling, and decreases sensitivity to gravitropic stimulation.
- **Brassinosteroids** promote cell expansion, pollen tube elongation, and vascular tissue differentiation but inhibit root elongation. Unlike animal steroids, these hormones act at a cell membrane receptor.
- **Abscisic acid** inhibits seed germination, promotes dormancy, and stimulates stomatal closing in response to dry conditions in the environment.

CONCEPT 26.4 Photoreceptors Initiate Developmental Responses to Light

- **Phototropin** is a blue-light receptor protein involved in phototropism. **Zeaxanthin** acts in conjunction with phototropin to mediate the light-induced opening of stomata. **Cryptochromes** are blue-light receptors that affect seedling development and flowering and inhibit cell expansion. **Review Figure 26.10**
- **Phytochrome** is a photoreceptor that exists in the cytosol in two interconvertible isoforms, $\mathbf{P_r}$ and $\mathbf{P_{fr}}$. The relative amounts of these two isoforms are a function of the ratio of red to far-red light. Phytochrome plays a number of roles in **photomorphogenesis**. **Review Figure 26.11**
- The phytochrome signal transduction pathway affects transcription in two ways: the P_{fr} isoform interacts directly with some transcription factors and influences transcription indirectly by phosphorylating other proteins. **Review Figure 26.12**
- **Circadian rhythms** are changes that occur on a daily cycle. Light can entrain circadian rhythms through photoreceptors such as phytochrome.

Go to the Interactive Summary to review key figures, Animated Tutorials, and Activities PoL2e.com/is26

Go to LaunchPad at **macmillanhighered.com/launchpad** for additional resources, including LearningCurve Quizzes, Flashcards, and many other study and review resources.

27 Reproduction of Flowering Plants

KEY CONCEPTS

27.1 Most Angiosperms Reproduce Sexually

27.2 Hormones and Signaling Determine the Transition from the Vegetative to the Reproductive State

27.3 Angiosperms Can Reproduce Asexually

The poinsettia is a popular Christmas flower. But the red "flowers" are not flowers at all; instead, they are leaves (bracts). The flowers are yellow structures in the center of a group of red leaves and are not visible in this photograph.

Dairy farmer Albert Ecke was fascinated by the red and green shrubs that grew all over southern California and were used by Mexican-Americans for red dye. The shrub, *Euphorbia pulcherrima*, got the name "poinsettia" from the man who first brought it to the United States: the first U.S. ambassador to Mexico, Joel Roberts Poinsett. In the early 1900s, Ecke started selling the plants at his farm in Hollywood. But two challenges stood in the way of his making this plant a commercial success.

First, although poinsettias generally bloomed in the fall and early winter in the mild climate of southern California, when Ecke grew them in fields in Hollywood, the formation of flowers was unreliable. Biologists later found that the flowering of poinsettias required at least a 14-hour night for several weeks. Any interruption in the long night—by passing cars or street lamps, for example—inhibited flowering. So Ecke's son Paul moved the growing operation south, to isolated fields far from Los Angeles. A second challenge was that the plants were tall and gangly. Although pretty, they were hard to transport and not attractive indoors. Paul Ecke found a variety of poinsettia that was much more compact. He propagated this variety and eventually developed it into the short potted plant that is so popular today. The Eckes decided that the time of flowering was just right for making the poinsettia a "Christmas flower," so they promoted this now portable plant as a holiday decoration, blanketing live television shows with free plants between Thanksgiving and Christmas. The campaign was successful: more than 100 million poinsettias are now sold in the United States during the winter holidays every year, making it the nation's best-selling potted plant.

Breeding more attractive flowering plants that are easier to grow is an ongoing part of floriculture, the industry involved with the production of floral crops. You may be surprised to learn that the brightly colored "flowers" of poinsettias are not flowers at all. The red parts of the plant that we most notice and appreciate are actually specialized leaves called bracts. The poinsettia has a single tiny yellow female flower, without petals, surrounded by male flowers.

How did an understanding of angiosperm reproduction allow floriculturists to develop a commercially successful poinsettia?

You will find the answer to this question on page 587.

CONCEPT 27.1 Most Angiosperms Reproduce Sexually

Most angiosperms (flowering plants) have evolved to reproduce sexually because this strategy results in the genetic diversity that is the raw material for evolution. Sexual reproduction in angiosperms involves mitosis, meiosis, and the alternation of haploid and diploid generations (see Figure 7.3). There are several important differences between sexual reproduction in angiosperms and in vertebrate animals (see Chapter 37):

- Meiosis in plants produces spores, which undergo mitosis to produce gametes; in animals, meiosis usually produces gametes directly.
- In most plants, there are multicellular diploid (sporophyte) and haploid (gametophyte) life stages (alternation of generations); in animals, there is no multicellular haploid stage.
- In plants, the cells that will form gametes (germline cells) develop in the adult organism, usually in response to environmental conditions; in animals, the germline cells are set aside before birth.

LINK

The life cycles of plants have many unique characteristics, which are detailed in **Chapter 21**

The flower is the reproductive organ of angiosperms

In angiosperms, flowers contain male or female gametophytes or often both. As we saw in Concept 21.5, a generalized flower consists of four concentric groups (whorls) of organs arising from modified leaves: the carpels, stamens, petals, and sepals.

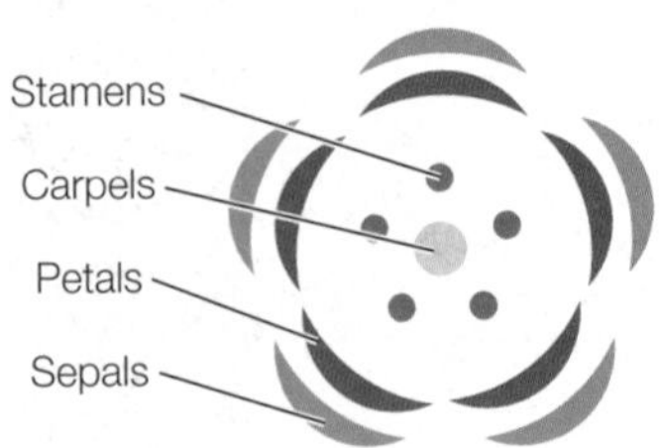

The parts of the flower are all derived from a modified shoot apical meristem.

- Carpels are the female sex organs that contain the developing female gametophytes (inside ovules).
- Stamens are the male sex organs that produce microspores, the first cells of the developing male gametophytes.

The differentiation of these organs is controlled by specific transcription factors (see Figure 14.12).

As you know, flowers come in many shapes, sizes, and forms. Some have both stamens (male sex organs) and carpels (female sex organs); such flowers are termed perfect (**FIGURE 27.1A**). Imperfect flowers, by contrast, are those with only male or only female sex organs. Male flowers have stamens but not

(A) Perfect: lily (*Lilium* sp.)

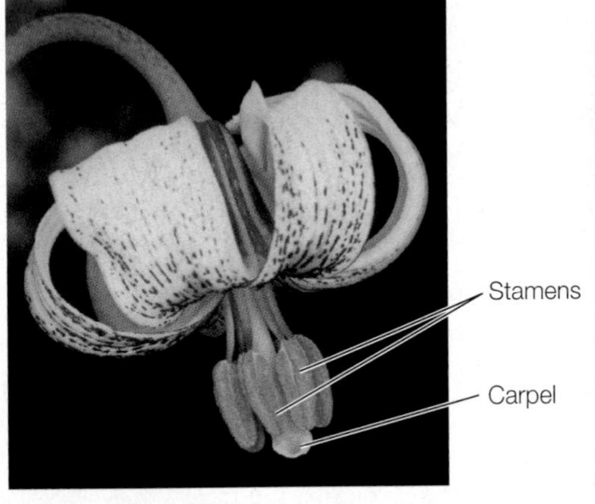

(B) Imperfect monoecious: corn (*Zea mays*)

(C) Imperfect dioecious: Red campion (*Silene dioica*)

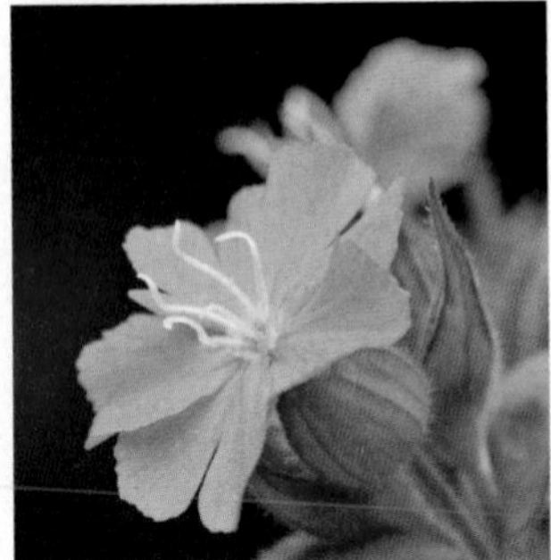

Female flowers with carpels

Male flower with stamens

FIGURE 27.1 Perfect and Imperfect Flowers (A) A lily is an example of a perfect flower, meaning one that has both male and female sex organs. (B) Imperfect flowers are either male or female. Corn is a monoecious species: both types of imperfect flowers are borne on the same plant. (C) Red campion is a dioecious species; some red campion plants bear male imperfect flowers, and others bear female imperfect flowers.

carpels, and female flowers have carpels but not functioning stamens. Some plants, such as corn, bear both male and female flowers on the same individual plant; such species are called **monoecious** ("one housed"; FIGURE 27.1B). In **dioecious** ("two-housed") species, individual plants bear either male-only or female-only flowers; an example is red campion (FIGURE 27.1C).

Angiosperms have microscopic gametophytes

FIGURE 27.2 shows the development of angiosperm gametophytes. These haploid, microscopic gamete-producing structures develop from haploid spores in the flower:

- Each female gametophyte (megagametophyte) is called an **embryo sac**, and it develops inside an ovule. One or more ovules are contained inside an ovary, which is the lower part of a carpel.
- Male gametophytes (microgametophytes), which are also called **pollen grains**, develop in the anther, which is part of the stamen.

FEMALE GAMETOPHYTE Of the four haploid megaspores resulting from meiosis, three undergo apoptosis (programmed cell death). Typically, the remaining megaspore undergoes three mitotic divisions without cytokinesis, producing eight haploid nuclei, all initially contained within a single cell—three nuclei at one end, three at the other end, and two in the middle. Subsequent cell wall formation leads to an elliptical, seven-celled megagametophyte with a total of eight nuclei:

- At one end of the megagametophyte are three small cells: the egg cell and two cells called synergids. The egg cell is the female gamete, and the synergids participate in fertilization by attracting the pollen tube. The tube enters one of the synergids before the sperm cells are released for fertilization.
- At the opposite end of the megagametophyte are three antipodal cells, which eventually degenerate by apoptosis.
- In the large central cell are two **polar nuclei**.

The megagametophyte, or embryo sac, is the entire seven-cell, eight-nucleus structure.

MALE GAMETOPHYTE The four haploid products of meiosis in the stamen each develop a cell wall and undergo a single mitotic division, producing four two-celled pollen grains (male gametophytes) that are released into the environment. The two cells in a pollen grain have different roles:

- The generative cell divides by mitosis to form two sperm cells that participate in double fertilization. (In some plants

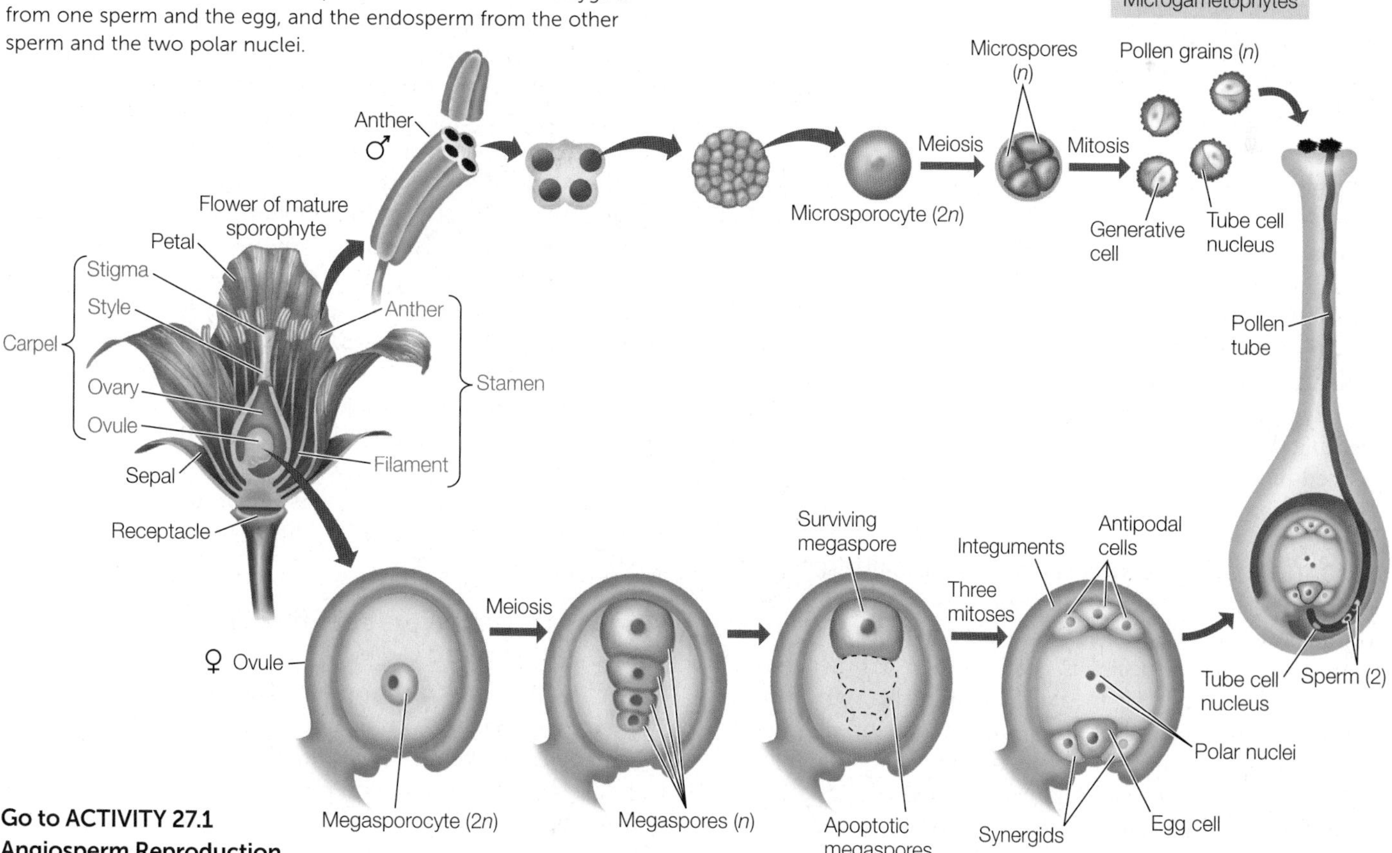

FIGURE 27.2 Sexual Reproduction in Angiosperms The embryo sac is the female gametophyte; the pollen grain is the male gametophyte. The male and female cells meet and fuse in the embryo sac. Angiosperms have double fertilization, in which a zygote and an endosperm nucleus form from separate fusion events—the zygote from one sperm and the egg, and the endosperm from the other sperm and the two polar nuclei.

Go to ACTIVITY 27.1 Angiosperm Reproduction PoL2e.com/ac27.1

this division takes place before pollen is released from the anther, resulting in three-celled pollen grains.)

- The tube cell forms the elongating pollen tube that delivers the sperm to the embryo sac after the pollen grain is transferred to a stigma (part of the female reproductive organ)—a process called **pollination**.

Go to MEDIA CLIP 27.1
Pollen Germination in Real Time
PoL2e.com/mc27.1

Angiosperms have mechanisms to prevent inbreeding

As we discussed in Concept 21.4, the evolution of pollen in the seed plants made it possible for male gametes to reach the female gametophyte in the absence of liquid water. The transfer of pollen from plant to plant presents its own challenges, however. In the first seed plants to evolve, tiny pollen grains were carried to stigmas by wind. Wind-pollinated flowers have sticky or featherlike stigmas, and they produce a great number of pollen grains because of the low probability of an individual grain encountering a suitable stigma. More recently in evolutionary time, most angiosperms have come to rely on animals such as insects and birds for pollination. Pollen transport by animals greatly increases the probability that pollen will get to a female gametophyte of the same species.

LINK

The evolution of plant–pollinator interactions and their effects on plant speciation are described in **Concepts 17.4 and 21.5**

Some plants self-pollinate: that is, they are fertilized by pollen that lands on the stigma of a flower on the same plant, or even the same flower. Mendel's garden peas, for example, reproduce in this way, so he had to remove stamens to control the parentage of his genetic crosses (see Figure 8.1). Self-pollination mitigates the difficulties of wind or animal pollination of a distant flower. However, "selfing" leads to homozygosity, which can reduce the reproductive fitness of offspring (known as inbreeding depression). Most plants have evolved mechanisms that prevent self-pollination and thus self-fertilization.

SEPARATION OF MALE AND FEMALE GAMETOPHYTES Self-fertilization is not possible in dioecious species, which bear only male or female flowers on an individual plant. Pollination in dioecious species is accomplished only when one plant pollinates another. In monoecious plants, which bear both male and female flowers on the same plant, the physical separation of male and female flowers is often sufficient to prevent self-fertilization. Some monoecious species prevent self-fertilization by staggering the development of male and female flowers so they do not bloom at the same time, making these species functionally dioecious.

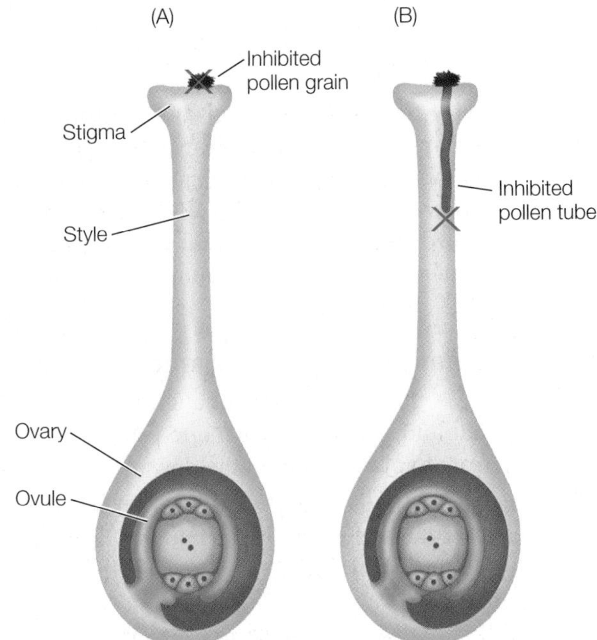

FIGURE 27.3 Self-incompatibility In a self-incompatible plant, pollen is rejected if it expresses an *S* allele that matches one of the *S* alleles of the recipient carpel. There are two mechanisms of rejection: (A) the pollen may fail to germinate, or (B) the pollen tube may die before reaching an ovule. Both mechanisms prevent the egg from being fertilized by a sperm from the same plant.

GENETIC SELF-INCOMPATIBILITY A pollen grain that lands on a stigma of the same plant will germinate and fertilize the female gamete only if the plant is self-compatible (capable of self-pollination). To prevent self-fertilization, many plants are genetically **self-incompatible**. This mechanism depends on the ability of a plant to determine whether pollen is genetically similar or genetically different from "self." Rejection of "same-as-self" pollen prevents self-fertilization.

Self-incompatibility in plants is controlled by a cluster of tightly linked genes called the *S* locus (for *s*elf-incompatibility). The *S* locus encodes proteins in the pollen and style that interact during the recognition process. A self-incompatible species typically has many alleles of the *S* locus. The pollen phenotype may be determined by its own haploid genotype or by the diploid genotype of its parent plant. In either case, if the pollen expresses an allele that matches one of the alleles of the recipient carpel, the pollen is rejected. Depending on the type of self-incompatibility, the rejected pollen either fails to germinate or the pollen tube is prevented from growing through the style (**FIGURE 27.3**); either way, self-fertilization is prevented.

A pollen tube delivers sperm cells to the embryo sac

When a genetically appropriate pollen grain lands on the stigma of a compatible carpel, it germinates. As with seed germination, a key event is water uptake by pollen, in this case from the stigma. Germination involves the development of a pollen tube. The pollen tube either traverses the spongy tissue of the style or, if the style is hollow, grows on the inner surface of the style until it reaches an ovule. The pollen tube typically grows at the rate of 1.5–3 millimeters per hour, taking just 1–2 hours to reach its destination, the female gametophyte.

The growth of the pollen tube may be guided in part by a chemical signal produced by the synergids in the female gametophyte. If one synergid is destroyed, the ovule still attracts pollen tubes, but destruction of both synergids renders the ovule unable to attract pollen tubes, and fertilization does not occur. The attractant appears to be species-specific: in some cases, isolated female gametophytes attract only pollen tubes of the same species.

Angiosperms perform double fertilization

The pollen tube grows down the style, under the direction of the haploid tube cell. Meanwhile the generative cell divides once (if it hasn't already done so) to form two haploid sperm cells. Once the pollen tube reaches the embryo sac, one of the two synergids degenerates, and the two sperm cells are released into its remains. *Two fertilization events occur* (**FIGURE 27.4**):

- One sperm cell fertilizes the egg cell, producing the diploid zygote, which forms the new sporophyte embryo.
- The other sperm cell fertilizes the central cell of the embryo sac, when the sperm fuses with the two polar nuclei. This fusion forms a triploid nucleus that undergoes rapid mitosis to form a specialized nutritive tissue, the **endosperm**. The endosperm's primary function is nourishment of the developing embryo. In some plants, notably cereals, the endosperm persists in the mature seed and contains reserves that provide chemical building blocks for the germinating seedling while it is underground and cannot perform photosynthesis.

This process of double fertilization is a characteristic feature of angiosperm reproduction. The remaining cells of the female gametophyte—the antipodal cells and the remaining synergid—eventually degenerate, as does the pollen tube nucleus.

Embryos develop within seeds contained in fruits

Fertilization initiates the highly coordinated growth and development of the embryo, endosperm, integuments, and carpel. The integuments—tissue layers immediately surrounding the ovule—develop into the seed coat, and the ovary becomes the fruit that encloses the seed (see Figure 27.2).

In Chapter 26 we described the events in plant embryonic development and its hormonal control. As seeds develop, they prepare for dispersal and dormancy by losing up to 95 percent of their water content. You can see this desiccation by comparing corn grains (e.g., popcorn) with ripe corn from the cob or a can. The embryo inside a dry seed contains very little water but is still alive; it has protective proteins that keep its cells in a viscous state.

In angiosperms, the ovary—together with the seeds it contains—develops into a fruit after fertilization has occurred. Fruits have two main functions:

- They protect the seed(s) from damage by animals and infection by microbial pathogens.
- They aid in seed dispersal.

A **fruit** may consist of only the mature ovary and seeds, or it may include other parts of the flower. Some species produce fleshy, edible fruits, such as peaches and tomatoes, whereas the fruits of other species are dry or inedible (**FIGURE 27.5**; see also Figure 21.26).

The diverse forms of fruits reflect the varied strategies plants use to disperse their progeny. Because plants cannot move, their progeny need some mechanism for separating

FIGURE 27.4 Double Fertilization Two sperm are involved in two nuclear fusion events, hence the term "double fertilization." One sperm is involved in the formation of the diploid zygote and the other in the formation of the triploid endosperm. The events shown here are for two-celled pollen grains; for three-celled pollen grains, the generative cell divides before the pollen is released from the anther.

Go to ANIMATED TUTORIAL 27.1
Double Fertilization
PoL2e.com/at27.1

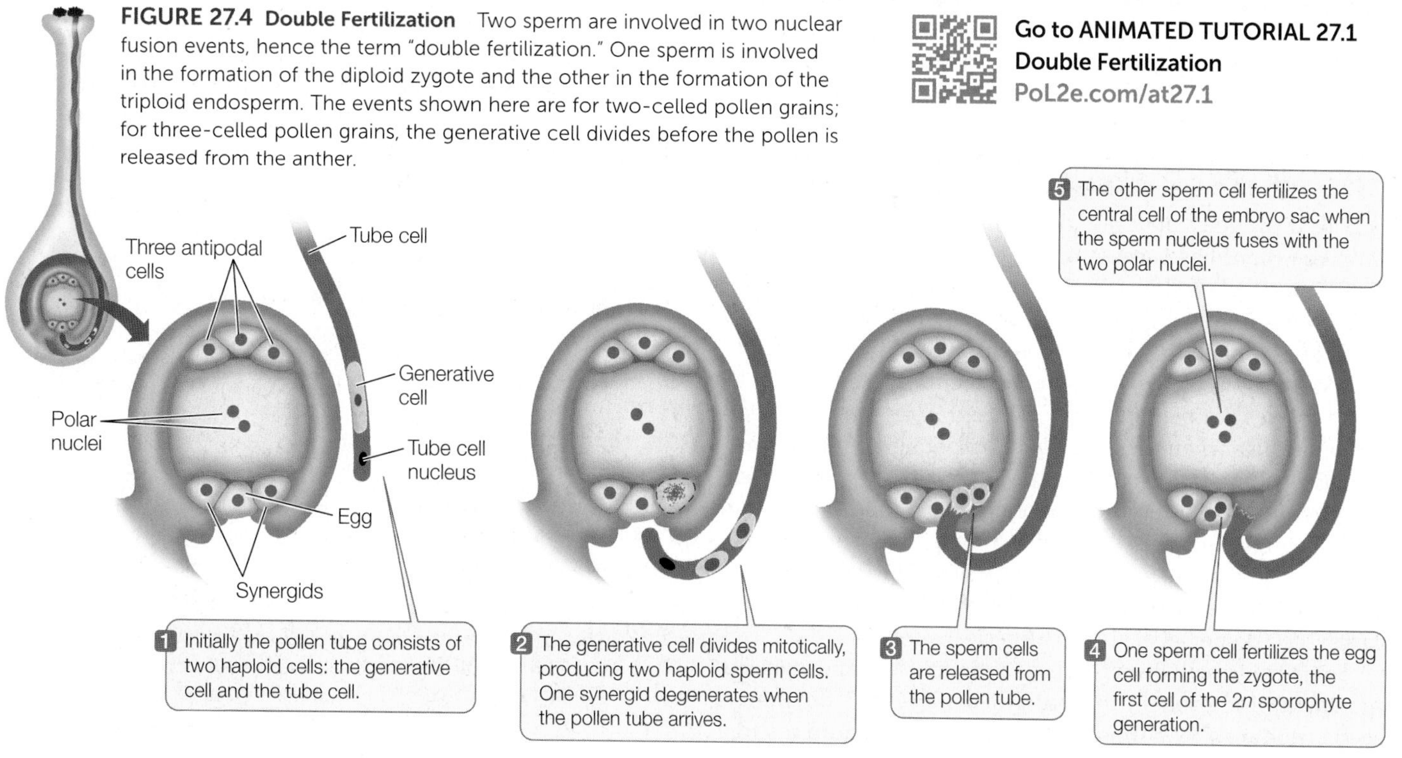
Three antipodal cells
Tube cell
Generative cell
Polar nuclei
Tube cell nucleus
Egg
Synergids
1 Initially the pollen tube consists of two haploid cells: the generative cell and the tube cell.
2 The generative cell divides mitotically, producing two haploid sperm cells. One synergid degenerates when the pollen tube arrives.
3 The sperm cells are released from the pollen tube.
4 One sperm cell fertilizes the egg cell forming the zygote, the first cell of the 2n sporophyte generation.
5 The other sperm cell fertilizes the central cell of the embryo sac when the sperm nucleus fuses with the two polar nuclei.

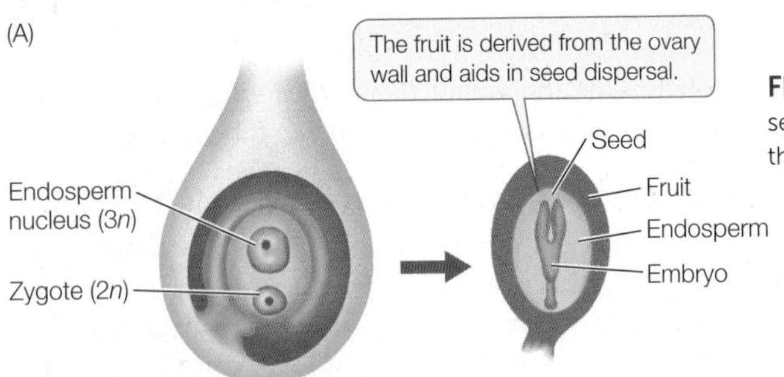

FIGURE 27.5 Angiosperm Fruits (A) There are a variety of fruits, but most have seed containing the embryo, surrounded by a fruit that derives from the wall of the ovary. Some examples include (B) garden pea, (C) tomato, and (D) corn.

themselves from their parent. Fruits are clearly an important way to move seeds away from the parent plant. Wide dispersal of progeny may not always be advantageous, however. If a plant has successfully grown and reproduced, its location is likely to be favorable for the next generation too. Some offspring do indeed stay near the parent, as is the case in many tree species, whose seeds simply fall to the ground. This strategy, however, has several potential disadvantages. If the species is a perennial (see Concept 27.2), offspring that germinate near their parent will be competing with their parent for resources, which may be too limited to support a dense population. Furthermore, even though local conditions were good enough for the parent to produce at least some seeds, there is no guarantee that conditions will still be good the next year, or that they won't be better elsewhere. Thus in many cases, seed dispersal is vital to a species' survival.

Some fruits help disperse seeds over substantial distances, increasing the probability that at least a few of the many seeds produced by a plant will find suitable conditions for germination and growth to sexual maturity. Some, like those of the thistle, are carried to new locations by wind:

Others attach themselves to animals (or to your clothes and shoes):

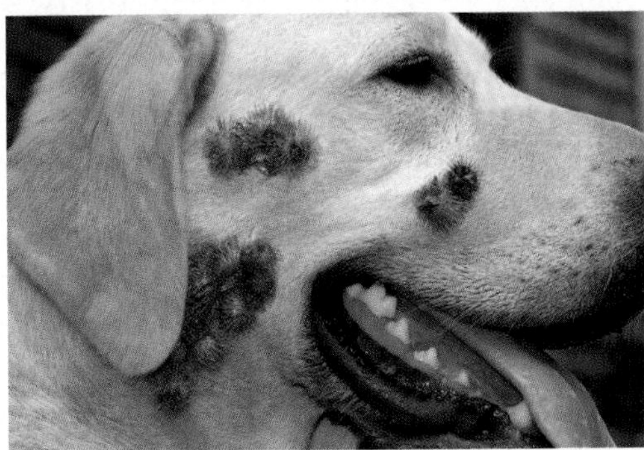

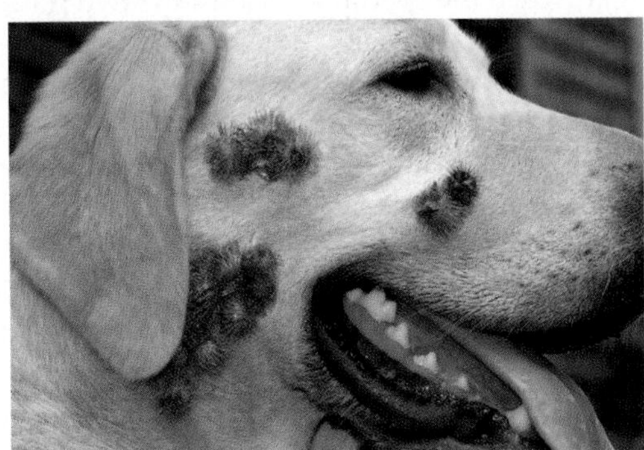

Water disperses some fruits; coconuts have been known to float thousands of miles between islands. Seeds swallowed whole by an animal along with fruits such as berries travel through the animal's digestive tract and are deposited some distance from the parent plant (see Apply the Concept in Chapter 21, p. 444).

CHECKpoint CONCEPT 27.1

- ✓ Make a table that compares the haploid and diploid life stages of a human and a corn plant with regard to (1) the relative sizes of the haploid and diploid stages, (2) how many cells are in the haploid stage, (3) the names of the haploid cells, (4) the time in the life cycle that the haploid stage occurs, (5) the location of the haploid cells, and (6) how the haploid stages develop in males and females. (Refer to Chapter 37 for information on human reproduction.)
- ✓ Describe the processes that angiosperms use to ensure a genetically diverse population.
- ✓ The two sperm that participate in plant double fertilization are genetically identical. Are the two products of fertilization genetically identical?

Now that we have described the cellular aspects of angiosperm reproduction, let's turn to the events that signal a transition from vegetative growth of the shoot to the formation of flowers.

CONCEPT 27.2 Hormones and Signaling Determine the Transition from the Vegetative to the Reproductive State

Flowering represents a reallocation of energy and materials away from making more roots, stems, and leaves (vegetative growth) to making flowers and gametes (reproductive growth). Flowering can happen at maturity as part of a predetermined developmental program (as in a dandelion plant in summer) or in response to environmental cues such as light or temperature (as with the poinsettias we described at the opening of this chapter). In either case, flowering is initiated through a cascade of changes in gene expression.

Herbaceous plants fall into three categories based on when they mature and initiate flowering and what happens after they flower:

- **Annuals** complete their lives within a year. Annuals include many crops important to humans, such as corn, wheat, rice, and soybeans. When the environment is suitable, annuals grow rapidly. After flowering, they channel most of their energy into the development of seeds and fruits. Then the rest of the plant withers and dies.
- **Biennials** take two years to complete their life cycle. They are much less common than annuals and include carrots, cabbage, and onions. Typically, biennials produce only vegetative growth during their first year and store carbohydrates in underground roots (carrots) or stems (onions). In their second year, they use most of the stored carbohydrates to produce flowers and seeds rather than vegetative growth, and the plant dies after seeds form.
- **Perennials** live three or more—sometimes many more—years. Maple trees, for example, can live up to 400 years. Perennials include many wildflowers as well as trees and shrubs. Typically these plants flower each year but stay alive and keep growing the next season.

Shoot apical meristems can become inflorescence meristems

The transition from purely vegetative growth to the flowering state requires a change in one or more apical meristems in the shoot system. As we described in Chapter 24, shoot apical meristems continually produce leaves, axillary buds, and stem tissues (**FIGURE 27.6A**) in a kind of unrestricted growth called indeterminate growth.

A shoot apical meristem becomes an **inflorescence meristem** (recall that an inflorescence is a cluster of flowers; see Figure 21.19) when it ceases production of leaves and stems and produces other structures: smaller leafy structures called bracts (such as the red leaves in poinsettias), as well as new meristems in the angles between the bracts and the stem (**FIGURE 27.6B**). These new meristems may also be inflorescence meristems, or they may be **floral meristems**, each of which gives rise to a single flower.

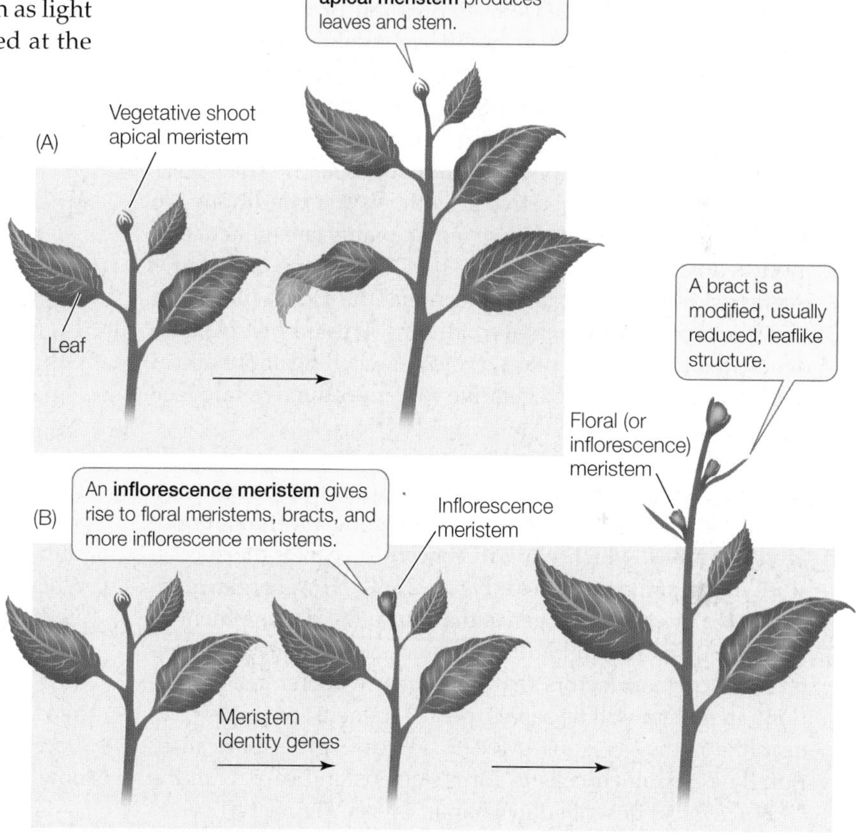

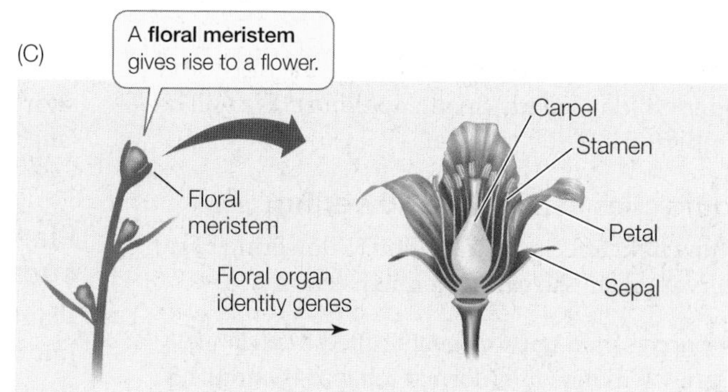

FIGURE 27.6 The Transition to Flowering (A) A shoot apical meristem produces vegetative organs without producing flowers. (B) A shoot apical meristem becomes an inflorescence meristem when it ceases production of leaves and stems and produces other structures. (C) A floral meristem becomes a single flower.

Each floral meristem typically produces four consecutive whorls, or spirals, of organs—the sepals, petals, stamens, and carpels—separated by very short internodes, which keep the flower compact (**FIGURE 27.6C**). In contrast to shoot apical meristems and some inflorescence meristems, floral meristems are responsible for determinate growth—growth of limited extent, like that of leaves.

A cascade of gene expression leads to flowering

The genes that determine the transition from shoot apical meristems to inflorescence meristems and from inflorescence meristems to floral meristems have been studied in model organisms such as *Arabidopsis thaliana*.

MERISTEM IDENTITY GENES Expression of two **meristem identity genes** initiates a cascade of further gene expression that leads to flower formation. Expression of the genes *LEAFY* and *APETALA1* is both necessary and sufficient for flowering. Evidence for the role of these genes is both genetic and molecular. For example, a mutant allele of *APETALA1* leads to continued vegetative growth, even if conditions are suitable for flowering. However, if plants are genetically modified so that the wild-type *APETALA1* gene is strongly expressed in the shoot apical meristem, the plant will flower regardless of environmental conditions. This is powerful evidence that *APETALA1* plays a role in switching shoot apical meristem cells from a vegetative to a reproductive fate (see Figure 27.6B).

FLORAL ORGAN IDENTITY GENES The products of the meristem identity genes trigger the expression of **floral organ identity genes**, which work in concert to specify the successive whorls of the flower (see Figure 27.6C). Floral organ identity genes are homeotic genes that cause the development of specific structures. In plants, the products of homeotic genes are transcription factors that determine whether cells in the floral meristem will be sepals, petals, stamens, or carpels. An example is the gene *AGAMOUS*, a class C gene that causes florally determined cells to form stamens and carpels in the "ABC" system described in Chapter 14 (see Figure 14.12).

Depending on the species, plants initiate these gene expression changes, and the events that follow, in response to either internal or external cues. Among external cues, the most studied are photoperiod (day length) and temperature. We will begin with photoperiod.

Photoperiodic cues can initiate flowering

The study of how light affects the transition to flowering began with two observations in the early twentieth century:

- A mutant tobacco strain appropriately called Maryland Mammoth grew 5 meters tall (normal tobacco is about 1.5 m tall). Instead of flowering in summer, it continued to grow until the late fall frost killed it. Farmers in Virginia were frustrated because they could not get seed of this luxuriant plant for the next year's crop.

- Because of improvements in agricultural techniques, soybean yields became so great that it was hard for farmers to harvest all the plants at once. So they tried planting the seeds in groups, several weeks apart. Unfortunately, the resulting plants all formed flowers and seeds at the same time.

The explanation for both of these observations was the same: the signal that set the plants' shoot apical meristems on the path to flowering was the length of daylight, or **photoperiod**. When soybeans experienced days of a certain length, they flowered, regardless of how "old" they were. Maryland Mammoth tobacco *could* flower, but it did not do so in Virginia because it died when the weather got cold. When the plant was grown in a greenhouse to prevent freezing, however, it flowered in December, when the days were short. Maryland Mammoth is now grown commercially in Florida.

Greenhouse experiments measured the day length required for different plant species to flower. Maryland Mammoth tobacco did not flower if exposed to more than 14 hours of light per day; flowering was only initiated once day length became shorter than 14 hours, as it does in December. Other plants (such as soybeans and henbane) flowered only when the days were long (**FIGURE 27.7**).

Plants vary in their responses to photoperiodic cues

Plants that flower in response to photoperiodic stimuli fall into two main classes, although there are variations on these patterns:

- **Short-day plants (SDPs)** flower only when the day is *shorter* than a critical maximum. They include poinsettias and chrysanthemums as well as Maryland Mammoth tobacco. Thus, for example, we see chrysanthemums in nurseries

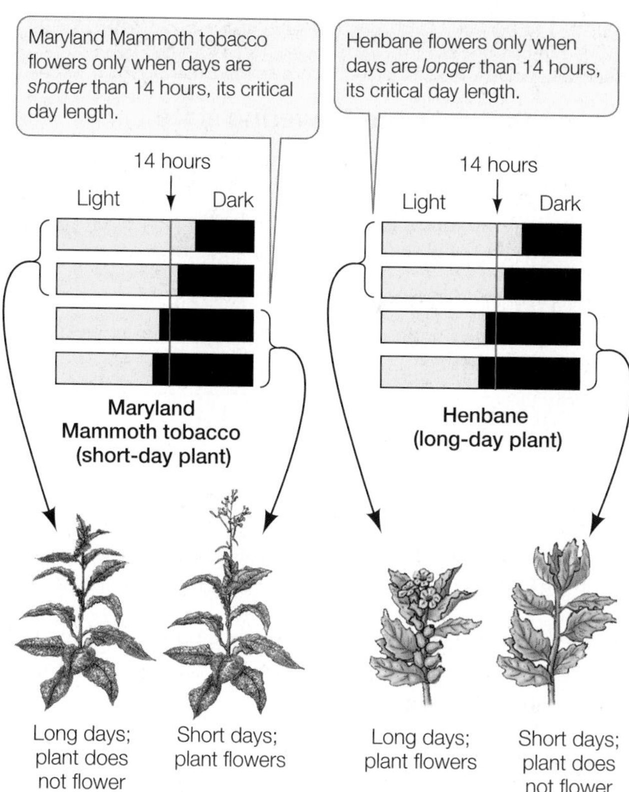

FIGURE 27.7 Photoperiod and Flowering By artificially varying the length of daylight in a 24-hour period, biologists showed that the flowering of Maryland Mammoth tobacco is initiated when the days become shorter than a critical length. Maryland Mammoth tobacco is thus called a short-day plant. Henbane, a long-day plant, shows an inverse pattern of flowering.

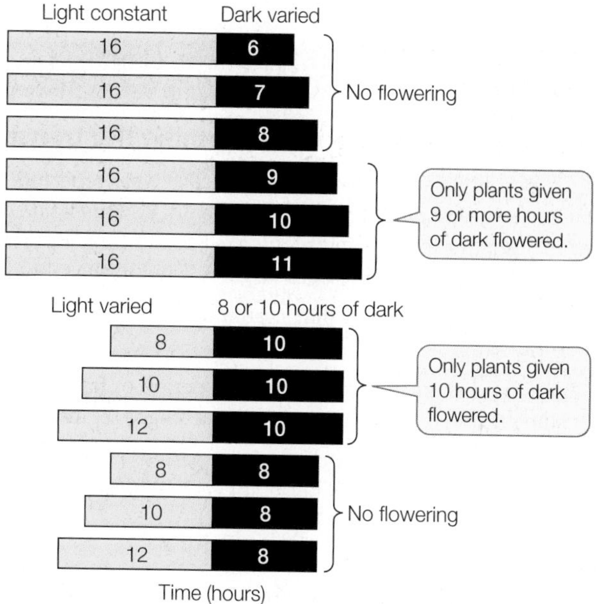

FIGURE 27.8 Night Length and Flowering Greenhouse experiments using cocklebur, an SDP, demonstrated that night length, not day length, is the environmental cue that initiates flowering.

in fall and poinsettias in winter (as noted at the opening of this chapter).

- **Long-day plants (LDPs)** flower only when the day is *longer* than a critical minimum. Spinach and clover are examples of LDPs. Spinach tends to flower and become bitter in the summer and is therefore normally planted in early spring.

Photoperiodic control of flowering serves an important role: it synchronizes the flowering of plants of the same species in a local population. This synchronization promotes cross-pollination and successful reproduction. It also means that floriculturists can vary light exposures in greenhouses to produce flowers at any time of year.

Night length is the key photoperiodic cue that determines flowering

The terms "short-day plant" and "long-day plant" assume that the key events that induce flowering occur in the daytime. What about the night? After all, short-day plants could just as easily be called "long-night plants," and long-day plants could be called "short-night plants." As it turns out, *night length is indeed the critical factor that induces flowering*, as a series of greenhouse experiments confirmed (**FIGURE 27.8**). In a greenhouse, the overall length of a day or night can be varied irrespective of the 24-hour natural cycle. For example, if cocklebur, an SDP, is exposed to several long periods of light (16 hours each), it will still flower as long as the dark period between them is 9 hours or longer. This 9-hour inductive dark period also induces flowering even if the light period varies from 8 hours to 12 hours.

Biologists noticed that when the inductive dark period was interrupted by a brief period of light, the flowering signal generated by the long night disappeared. It took several days of long nights for the plant to recover and initiate flowering. Interrupting the day with a dark period had no effect on flowering. A clue as to what occurred in the plant when the flash of light was given came when biologists determined the action spectrum for the wavelengths of light that were effective. As with lettuce seed germination (see Figure 26.11), red light was most effective at breaking the "night" stimulus, and its effect was reversible by far-red light. This indicated that the photoreceptor involved in regulating flowering is phytochrome.

LINK

The properties of phytochrome as a photoreversible photoreceptor of red and far-red light are described in **Concept 26.4**

Go to ANIMATED TUTORIAL 27.2
The Effect of Interrupted Days and Nights
PoL2e.com/at27.2

Plants sense night length by measuring the ratio of the P_{fr} and P_r isoforms of phytochrome. During the day there is more light in the red than in the far-red range. By the end of the day, much of the P_r has absorbed red light and been converted to P_{fr}. At night there is a gradual, spontaneous conversion of this P_{fr} back to P_r. The longer the night, the more P_r there is at the beginning of the next day. An SDP flowers when the ratio of P_{fr} to P_r is low at the end of the night, whereas an LDP flowers

APPLY THE CONCEPT

Hormones and signaling determine the transition from the vegetative to the flowering state

The experiments reported on in Figure 27.8 were repeated using a long-day plant (LDP) that normally flowers when days are 16 hours (or longer) and nights are 8 hours (or shorter). The plant was subjected to the light regimes listed in the table.

1. List the light regimes that should have resulted in flowering.
2. Predict the effect of adding a brief period of far-red light in the middle of the dark period of each light regime.
3. Predict the effect of adding a brief period of red light in the middle of the dark period of each light regime.

A. 14 hours light + 6 hours dark	G. 8 hours light + 8 hours dark
B. 14 hours light + 7 hours dark	H. 10 hours light + 8 hours dark
C. 14 hours light + 8 hours dark	I. 12 hours light + 8 hours dark
D. 14 hours light + 10 hours dark	J. 8 hours light + 10 hours dark
E. 14 hours light + 12 hours dark	K. 10 hours light + 10 hours dark
F. 14 hours light + 14 hours dark	L. 12 hours light + 10 hours dark

when this ratio is high. You can compare this mechanism to an hourglass egg timer. The sand in the top chamber is analogous to P_{fr} and gradually runs down (is converted to P_r) over the course of the night; the shorter the night, the more sand remains at the top at dawn. To "reset" an hourglass timer, you simply turn it over; this would be analogous to dawn of the next day, when P_r begins being converted back to P_{fr}.

The flowering stimulus originates in the leaf

The photoperiodic receptor for flowering, phytochrome, is located in the leaf, not in the shoot apical meristem. "Masking" experiments can show this: if a single leaf of an SDP growing under short-night conditions is covered to simulate a long-night exposure, the shoot apical meristems will transition to flowering as if the entire plant were exposed to long nights (**FIGURE 27.9**). Masking the buds (leaving the leaves exposed) does not induce flowering.

Because the receptor of the photoperiodic stimulus (phytochrome in the leaf) is physically separated from the tissue on which the stimulus acts (the shoot apical meristem), the inference can be drawn that a signal travels from the leaf through the plant's tissues to the shoot apical meristem. Unlike animals, plants do not have a nervous system, so the signal must be a diffusible chemical. Biologists have found additional evidence that a diffusible signal travels from the leaf to the shoot apical meristem:

- If a photoperiodically induced leaf is immediately removed from a plant after the inductive dark period, the plant does not flower. If the induced leaf remains attached to the plant for several hours, however, the plant flowers. This result suggests that something is synthesized in the leaf in response to the inductive dark period, and then moves out of the leaf to induce flowering.
- If two or more cocklebur plants are grafted together, and if one plant is exposed to inductive long nights and its graft partners are exposed to noninductive short nights, all of the plants flower.
- In several species, if an induced leaf from one species is grafted onto another, noninduced plant of a different species, the recipient plant flowers. This indicates that the same diffusible chemical signal is used by both species.

Although the diffusible signal was given a name, **florigen** (**FT**), meaning "flower inducing," decades ago, the nature of the signal was only recently explained.

Florigen is a small protein

The characterization of florigen was made possible by genetic and molecular studies of the model organism *Arabidopsis*, an LDP. Three genes are involved (**FIGURE 27.10**):

- *FT* (*FLOWERING LOCUS T*) codes for florigen. FT is a small protein (20 kDa molecular weight) that can travel through plasmodesmata. It is synthesized in phloem companion cells in the leaf and diffuses into the adjacent sieve tube elements, where it moves with the phloem to the shoot apical meristem.
- *CO* (*CONSTANS*) codes for a transcription factor that activates the synthesis of FT. Like *FT*, *CO* is expressed in phloem companion cells in the leaf. CO protein is expressed all the time but is unstable; an appropriate photoperiodic stimulus stabilizes CO so that there is enough to turn on FT synthesis.
- *FD* (*FLOWERING LOCUS D*) codes for a protein that binds to FT protein when it arrives in the shoot apical meristem. The FD protein is a transcription factor that forms a complex with FT protein; the complex activates promoters for meristem identity genes, such as *APETALA1* (*AP1*).

Genetic studies have shown that the *FT* gene is involved in photoperiod signaling in many species:

- Transgenic plants (e.g., tobacco and tomato) that express the *Arabidopsis FT* gene at high levels flower regardless of day length.
- Transgenic *Arabidopsis* plants that express high levels of *FT* genes from other plants (e.g., rice and tomato) flower regardless of day length.

Whereas photoperiod induces flowering via the FT–FD pathway in some plants, in many other species flowering is induced by other stimuli through other pathways.

INVESTIGATION

FIGURE 27.9 The Flowering Signal Moves from Leaf to Bud Phytochrome, the receptor for photoperiod, is in the leaf, but flowering occurs in the shoot apical meristem. To investigate whether there is a diffusible substance that travels from leaf to bud, James Knott exposed a single leaf of cocklebur plants to the inductive dark period.[a]

HYPOTHESIS

The leaves measure the photoperiod.

METHOD

Grow cocklebur plants under long days and short nights. Mask a leaf on some plants and see if flowering occurs.

RESULTS

If even one leaf is masked for part of the day—thus shifting that leaf to short days and long nights—the plant will flower.

CONCLUSION

The leaves measure the photoperiod. Therefore, some signal must move from the induced leaf to the flowering parts of the plant.

ANALYZE THE DATA

In related experiments, leaves were removed from plants before the plants were exposed to the inductive dark period. There were six plants in each condition.[b]

Condition	Number of plants that flowered
No inductive dark period, intact plant	0
Inductive dark period, intact plant	6
Inductive dark period, all leaves removed	0
Inductive dark period, all but one leaf removed	6

A. Do the data support the conclusion of the experiment by Knott described above? Explain your answer.

B. How would you modify the experiments shown in the table to find out how many days of the inductive dark period are required before the signal is produced? Would you expect the results to be different for the intact plants and plants with only one leaf? Why or why not?

Go to **LaunchPad** for discussion and relevant links for all **INVESTIGATION** figures.

[a]J. E. Knott. 1934. *Proceedings of the Society of Horticultural Science* 31: 152–154.
[b]K. C. Hamner and J. Bonner. 1938. *Botanical Gazette* 100: 388–431.

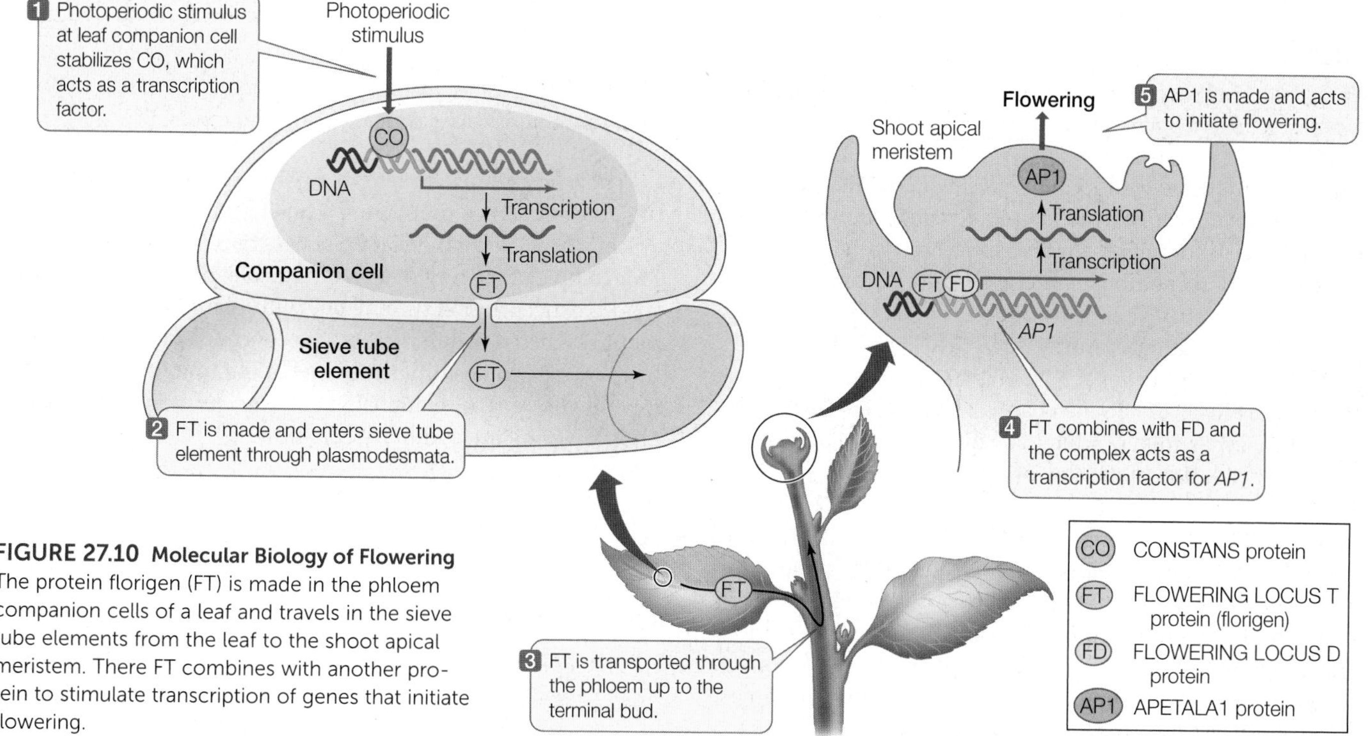

FIGURE 27.10 Molecular Biology of Flowering The protein florigen (FT) is made in the phloem companion cells of a leaf and travels in the sieve tube elements from the leaf to the shoot apical meristem. There FT combines with another protein to stimulate transcription of genes that initiate flowering.

Flowering can be induced by temperature or gibberellins

TEMPERATURE In some plant species, notably certain cereal crops, the environmental signal for flowering is exposure to cold temperatures, or **vernalization** (Latin *vernus*, "spring"). In both wheat and rye, we can distinguish two categories of flowering behavior. Spring wheat, for example, is a typical annual plant: it is sown in the spring and flowers in the same year. Winter wheat is sown in the fall, grows into a seedling, overwinters (often covered by snow), and flowers the following summer. If winter wheat is not exposed to cold, it will not flower normally in the next growing season. Vernalization evolved in plants adapted to grow in climates with harsh winters and a short growing season. It is a remarkable phenomenon:

- Plants seem to "remember" that they have been exposed to winter. When the weather gets warmer, exposed plants flower sooner than genetically identical plants that were not exposed to cold.
- Plants seem to "know" how long winter lasted, flowering sooner if the winter was long.

Some strains of *Arabidopsis* require vernalization, and those strains have been useful in figuring out the molecular pathway involved (**FIGURE 27.11**). The gene *FLC* (*FLOWERING LOCUS C*) encodes a transcription factor that blocks the FT–FD pathway by inhibiting expression of FT and FD. Cold temperatures inhibit the synthesis of FLC protein epigenetically: the chromatin is modified so that the *FLC* gene is not expressed (it is silenced). The longer the cold, the more cells there are that have the *FLC* gene silenced. This allows the FT and FD proteins to be expressed and flowering to proceed. Similar proteins control some steps in vernalization in cereals.

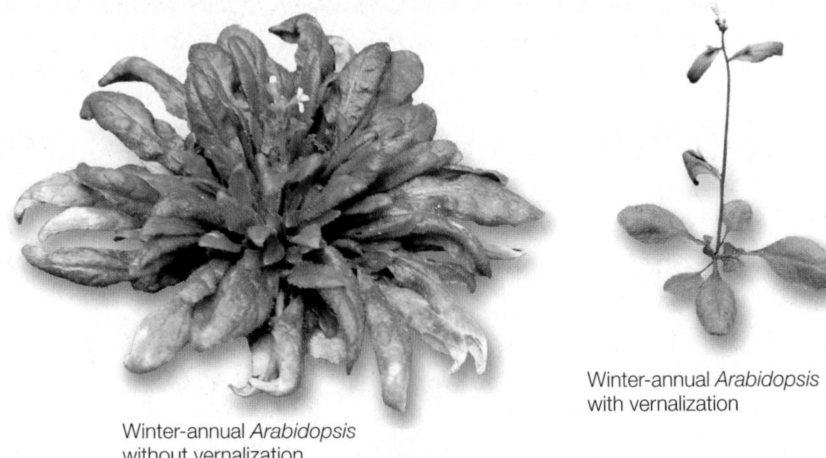

FIGURE 27.11 Vernalization A genetic strain of *Arabidopsis* (winter-annual *Arabidopsis*) requires vernalization for flowering. Without it, the plant is large and vegetative (left), but when exposed to a cold period, it is smaller and flowers (right).

LINK

Chromatin modification is a common way to regulate gene expression; see **Concept 11.3**

In addition to the well-known effects of cold temperature, there is also an effect of elevated temperature on flowering. Recently, biologists identified a transcription factor that increases the expression of the florigen gene *FT* and is more active at elevated temperatures (27°C as opposed to 22°C). The plants exposed to warm temperatures flower earlier, typically with smaller flowers and fewer seeds. This has important implications for the effects of climate change on crops. As Earth's climate gets warmer, food production may be reduced because of premature activation of the flowering pathway.

GIBBERELLINS *Arabidopsis* plants do not flower if they are genetically deficient in gibberellin synthesis or if they are treated with an inhibitor of gibberellin synthesis. These observations implicate gibberellins in flowering. Direct application of gibberellins to buds in *Arabidopsis* results in activation of the meristem identity gene *LEAFY*, which in turn promotes the transition to flowering (see Concept 14.3).

Some plants do not require an environmental cue to flower

Some plants flower in response to cues from an "internal clock." For example, flowering in some strains of tobacco is initiated in the terminal bud when the stem has grown four phytomers in length (recall that stems are composed of repeating units called phytomers; see Figure 24.1). If a terminal bud and a single adjacent phytomer are removed from a plant this size and planted, the cutting will flower because the bud has already received the cue for flowering. But the rest of the shoot below the bud that has been removed will not flower because it is only three phytomers long. After it grows an additional phytomer, it will flower. These results suggest that there is something about the *position* of the bud (atop four phytomers of stem) that determines its transition to flowering.

LINK

The role of positional information in morphogenesis in both plants and animals is described in **Concept 14.3**

The bud might "know" its position by the concentration of some substance that forms a positional gradient along the apical–basal axis of the plant. Such a gradient could be formed, for example, if the root makes a diffusible inhibitor of flowering whose concentration diminishes with plant height. When the plant reached a certain height, the concentration of the inhibitor would become sufficiently low at the tip of the shoot to allow flowering. The relationship between environmental factors and flowering are summarized below:

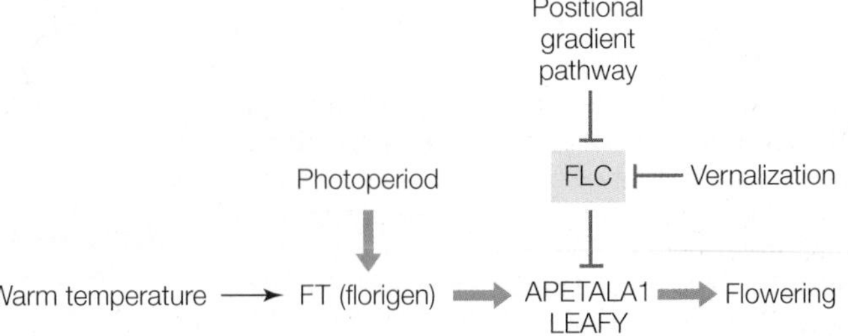

APPLY THE CONCEPT

Hormones and signaling determine the transition from the vegetative to the flowering state

Describe the genes and mutations that could be involved in each of the following observations:

1. A mutant plant flowers without its normal inductive dark period. When a leaf from the mutant plant is grafted onto an unexposed wild-type plant, the recipient plant flowers.
2. A mutant plant does not flower when exposed to the normal inductive dark period. When a leaf from a mutant plant that has been exposed to the inductive dark period is grafted onto an unexposed wild-type plant, the recipient plant flowers.
3. A plant that does not usually require vernalization flowers only after exposure to cold.
4. If a gene is coupled to an active promoter and expressed at high levels in the shoot apical meristem, flowering is induced even in the absence of an appropriate photoperiodic stimulus.
5. If a gene is experimentally overexpressed in the leaf, flowering is induced. Overexpression of the gene in the shoot apical meristem does not, however, induce flowering.

CHECKpoint CONCEPT 27.2

- ✓ How do shoot apical meristems, inflorescence meristems, and floral meristems differ, and what are the genes that control the transitions between them?
- ✓ Explain why "short-day plant" is a misleading term.
- ✓ If an LDP is exposed to its inductive dark period and one of its leaves is removed and grafted onto an SDP that has not been exposed to its inductive dark period, what will happen?
- ✓ How would you show that vernalization acts through the *FLC* gene?

Sexual reproduction is not the only way in which angiosperms give rise to more plants. We will now turn to some ways in which plants can reproduce asexually.

CONCEPT 27.3 Angiosperms Can Reproduce Asexually

A plant that reproduces asexually produces progeny genetically identical to itself (clones). This mode of reproduction has the disadvantage of not generating the diversity of offspring that sexual reproduction can produce. But if a plant is well adapted to its environment, the parent plant can pass on to all its progeny a combination of alleles that function well in that environment, and which might otherwise be separated by sexual recombination. Furthermore, the plant avoids the costs of producing flowers and the potentially unreliable processes of cross-pollination and seed germination. Many plants are capable of both sexual and asexual reproduction and can switch between the two depending on environmental conditions or the stage in their life cycle.

Angiosperms use many forms of asexual reproduction

Asexual reproduction is often accomplished through the modification of vegetative organs such as roots, stems, or leaves, which is why the term **vegetative reproduction** is sometimes used to describe asexual reproduction in plants. Another type of asexual reproduction, **apomixis**, involves flowers but no fertilization.

VEGETATIVE REPRODUCTION The totipotency of plant cells is dramatically shown when a cell from a fully differentiated organ gives rise to an entire new plant. Strawberries, for example, produce horizontal stems, called stolons or runners, that grow along the soil surface, form roots at intervals, and establish potentially independent plants (see Figure 24.13E). Another example is bamboo, which has underground stems called rhizomes that can form an entire forest (**FIGURE 27.12A**). Fleshy underground stems, such as potato tubers, produce multiple plants from "eyes" (see Figure 24.13C). Garlic can produce multiple plants from bulbs, which are also modified stems (**FIGURE 27.12B**). Even leaves can be the source of new plants, as in the succulent *Kalanchoe* (**FIGURE 27.12C**).

Plants that reproduce vegetatively often grow in environments such as eroding hillsides, or places where conditions for seed germination are unreliable. Plants with stolons or rhizomes, such as beach grasses, rushes, and sand verbena, are common on coastal sand dunes. Rapid vegetative reproduction enables these plants, once introduced, not only to multiply but also to stay ahead of the shifting sand; in addition, the dunes are stabilized by the extensive network of rhizomes or stolons that develops. Vegetative reproduction is also common in some deserts, where for much of the time the environment is unsuitable for seed germination and the establishment of seedlings.

Vegetative reproduction is efficient in an environment that is stable over the long term. A change in the environment, however, can leave an asexually reproducing species at a disadvantage. For example, the English elm tree (*Ulmus procera*) reproduces by suckers and is incapable of sexual reproduction. In 1967 a fungal pathogen called Dutch elm disease first struck English elms. The population lacked genetic diversity, and no individuals had genes that would confer resistance to the pathogen. Today the English elm is all but gone from England.

APOMIXIS Some plants produce flowers but use them to reproduce asexually rather than sexually. Dandelions, blackberries, some citrus trees, and some other plants reproduce by the asexual production of seeds, called apomixis. In sexual reproduction,

(A)

These bamboo shoots all arise from the same underground stem.

(B)

Each clove of garlic is a branch of a bulb and can give rise to a new plant.

(C)

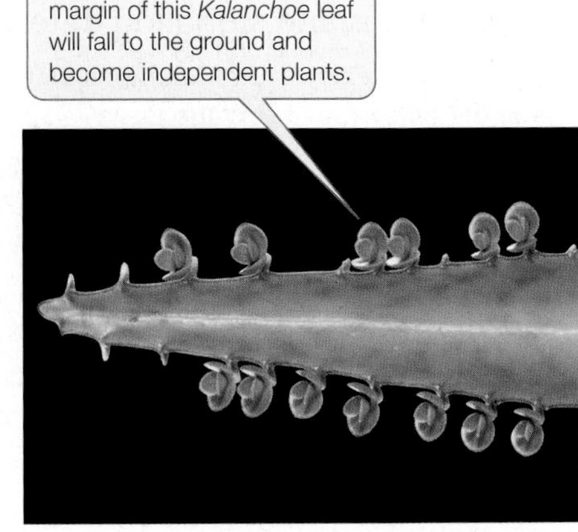

The plantlets forming on the margin of this *Kalanchoe* leaf will fall to the ground and become independent plants.

FIGURE 27.12 Vegetative Reproduction (A) The rhizomes of bamboo are underground stems that produce plants at intervals. (B) Bulbs are short stems with large leaves that store nutrients and can give rise to new plants. (C) In *Kalanchoe*, new plantlets can form on leaves.

seeds form from the union of haploid gametes in the embryo sac. But in apomixis, one of two other things happens:

- The cell in the ovule that is supposed to undergo meiosis fails to do so, resulting in a diploid egg cell, which then goes on to form an embryo and seed.
- Diploid cells from the integument surrounding the embryo sac form a diploid embryo sac, and the sac goes on to form an embryo and seed.

In both cases, seed and fruit development proceeds normally. But the genetic consequences are profound: *apomixis produces clones.*

Some agriculturally important plants are hybrids that are derived from the crossing of two genetically different species. Because these hybrids have two sets of chromosomes that are not homologous to one another, they cannot undergo meiosis and are sterile. This means they cannot form the seeds and fruits that are desired by people by ordinary means. In citrus, apomixis occurs naturally, which gets around this problem. Kentucky bluegrass, a mainstay of lawns, reproduces in this manner as well.

Many important crops, such as corn, are grown as hybrids because the offspring of a cross between two inbred, homozygous genetic strains are often superior to either of their parents, a phenomenon called hybrid vigor (see Concept 8.2). Unfortunately, once farmers have obtained a hybrid with desirable characteristics, they cannot use those plants for further crosses with themselves (selfing) to get more seeds for the next generation. You can imagine the genetic chaos when a hybrid, which is heterozygous at many of its loci (e.g., *AaBbCcDdEe*, etc.), is crossed with itself: there will be many new combinations of alleles (e.g., *AabbCCDdee*, etc.), resulting in highly variable progeny. The only efficient and reliable way to reproduce the hybrid is to maintain populations of the original parents to cross again each year. That is exactly what seed companies do. For farmers, it means buying new seeds every year—an expensive proposition.

If a hybrid carried a gene for apomixis, however, it could reproduce asexually, and its offspring would be genetically identical to itself (**FIGURE 27.13**). New hybrids could be developed that would be adapted to specific environments and could be propagated by the farmers on the spot. Investigators are searching for genes that promote apomixis; these could be introduced into desirable crops and allow them to be propagated indefinitely. Such a gene has been identified in corn, but the yield of the variety that contains it is low.

Vegetative reproduction is important in agriculture

One of the oldest methods of vegetative reproduction used in agriculture consists of simply making cuttings of stems, inserting them in soil, and waiting for them to form roots and become autonomous plants. The cuttings are usually encouraged to root by treatment with auxin, as described in Concept 26.2.

Woody plants can be propagated asexually by **grafting**: attaching a bud or a piece of stem from one plant to a root or root-bearing stem of another plant. The part of the resulting plant that comes from the root-bearing "host" is called the **stock**; the part grafted on is the **scion** (**FIGURE 27.14**). The vascular cambium of the scion associates with that of the stock, forming a continuous cambium that produces xylem and phloem. The cambium allows the transport of water and minerals to the scion and of photosynthate to the stock. Much of the fruit grown for market in the United States is produced on grafted trees, as are wine grapes. One advantage of this method is that a desirable scion (one that produces good fruit, for example) can be grafted onto a genetically different stock that has other desirable qualities, such as disease-resistant roots.

Another method widely used for asexual plant propagation is **meristem culture**, in which pieces of shoot apical meristem are cultured on growth media to generate plantlets, which can then be planted in the field. This strategy is vital when uniformity is desired, as in forestry, or when virus-free plants are the goal, as with strawberries and potatoes.

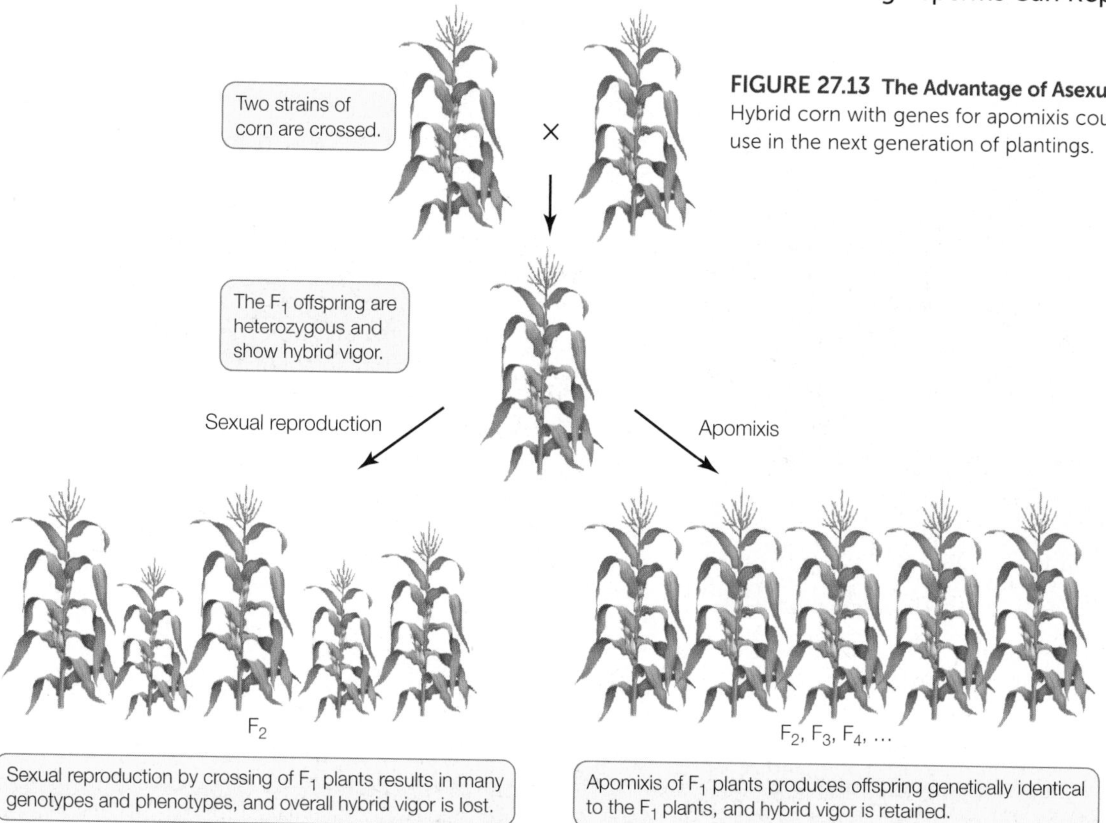

FIGURE 27.13 The Advantage of Asexual Reproduction by Apomixis Hybrid corn with genes for apomixis could be planted by farmers for use in the next generation of plantings.

CHECKpoint CONCEPT 27.3

- ✓ What are the advantages and disadvantages of asexual as compared with sexual reproduction in plants?
- ✓ What kinds of gene mutations would you search for that might promote apomixis?
- ✓ Most of the navel orange trees that once grew all over southern California are derived by grafting and apomixis from a single tree that still exists in Riverside, CA. How would you prove this?

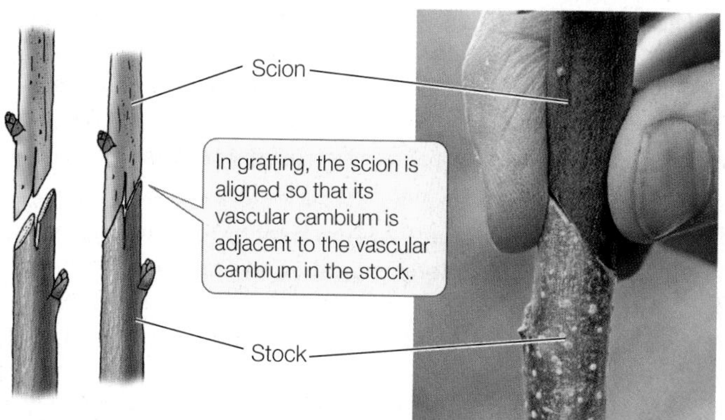

Scion
In grafting, the scion is aligned so that its vascular cambium is adjacent to the vascular cambium in the stock.
Stock

FIGURE 27.14 Grafting Grafting—attaching a piece of a plant to the root or root-bearing stem of another plant—is a common horticultural technique. The "host" root or stem is the stock; the upper grafted piece is the scion. In the photo, a scion of one apple variety is being grafted onto a stock of another variety.

How did an understanding of angiosperm reproduction allow floriculturists to develop a commercially successful poinsettia?

ANSWER The poinsettia illustrates well how the concepts of plant physiology and reproduction are used in floriculture. As Albert Ecke found, poinsettias are short-day plants, using the long nights of winter as a signal for flowering (Concept 27.2). In fact, the poinsettia's inductive dark period of 14 hours was among the first described in a flowering plant. The Eckes' plants did not flower consistently until they moved their growing operations from urban Los Angeles to an isolated rural region with less light pollution. Today the plants are grown in greenhouses, where the photoperiod is carefully regulated.

A second important factor in the poinsettia's commercial success was the plant's growth habit. In Mexico, the wild relatives of cultivated varieties grow up to 3 meters tall, with few branches (**FIGURE 27.15**; compare these plants with the 0.5 m tall plants in the photograph at the opening of this chapter). Paul Ecke found a variety that was much shorter and formed an attractive, branching plant with axillary shoots. These compact poinsettias were initially propagated asexually by grafting to native plants (Concept 27.3).

The cause of the compact phenotype was assumed to be a dwarfing gene, but several observations cast doubt on this assumption. First, heat treatment of short, branching plants changed their phenotype to the tall, less branched growth habit. Second, heat-treated plants would regain their short, branch-

FIGURE 27.15 A Wild Relative of Poinsettia Poinsettias growing in the wild in Mexico have an entirely different growth habit from the commercially propagated plants sold for holiday decorations.

ing growth habit if they were grafted onto untreated short, branching plants. These and other results pointed to a transmissible infective agent as the cause of the short, branching phenotype.

Dr. Ing-Ming Lee at the U.S. Department of Agriculture isolated a bacterium called *Phytoplasma* from the short, branching plants and proved by experiments that it was the causative agent of the compact growth habit. When young plants with the tall growth habit were inoculated with the bacteria, they grew up short and branching. It is suspected that the bacteria change cytokinin levels in the plants, since this hormone regulates branching (see Concept 26.3).

Many new varieties of poinsettias have been generated by conventional sexual reproduction (Concept 27.1). You can look for the phenotypes of these varieties next holiday season. They include stiff stems, new colors (such as white), and longer lifetimes for the colorful bracts.

SUMMARY

CONCEPT 27.1 Most Angiosperms Reproduce Sexually

- Sexual reproduction promotes genetic diversity in a population. The flower is an angiosperm's structure for sexual reproduction.
- Flowering plants have microscopic gametophytes. The megagametophyte is the **embryo sac**, which typically contains eight nuclei in seven cells. The microgametophyte is the two- or three-celled **pollen grain**. **Review Figure 27.2 and ACTIVITY 27.1**
- Following pollination, the pollen grain delivers sperm cells to the embryo sac by means of a pollen tube.
- Angiosperms exhibit double fertilization, forming a diploid zygote that becomes the embryo and a triploid **endosperm** that stores reserves. **Review Figure 27.4 and ANIMATED TUTORIAL 27.1**

CONCEPT 27.2 Hormones and Signaling Determine the Transition from the Vegetative to the Reproductive State

- In **annuals** and **biennials**, flowering and seed formation are followed by the death of the rest of the plant. **Perennials** live longer and reproduce repeatedly.
- For a vegetatively growing plant to flower, a shoot apical meristem must become an **inflorescence meristem**, which in turn must give rise to one or more **floral meristems**. These events are determined by specific genes. **Review Figure 27.6**
- Some plants flower in response to **photoperiod**. **Short-day plants** (**SDPs**) flower when nights are longer than a critical length specific to each species; **long-day plants** (**LDPs**) flower when nights are shorter than a critical length. **Review Figures 27.7 and 27.8 and ANIMATED TUTORIAL 27.2**
- The mechanism of photoperiodic control of flowering involves phytochromes and a diffusible protein signal, **florigen** (**FT**), which is formed in the leaf and is translocated to the shoot apical meristem. **Review Figures 27.9 and 27.10**
- In some angiosperms, exposure to cold—called **vernalization**—is required for flowering. In others, internal signals (such as gibberellin) induce flowering. All of these stimuli converge on the **meristem identity genes**.

CONCEPT 27.3 Angiosperms Can Reproduce Asexually

- Asexual reproduction allows rapid multiplication of organisms that are well suited to their environment.
- **Vegetative reproduction** involves the modification of a vegetative organ for reproduction. **Review Figure 27.12**
- Some plant species produce seeds asexually by **apomixis**. **Review Figure 27.13**
- Woody plants can be propagated asexually by **grafting**. **Review Figure 27.14**

Go to the Interactive Summary to review key figures, Animated Tutorials, and Activities PoL2e.com/is27

Go to LaunchPad at **macmillanhighered.com/launchpad** for additional resources, including LearningCurve Quizzes, Flashcards, and many other study and review resources.

28 Plants in the Environment

KEY CONCEPTS

A field of wheat infected with the wheat rust fungus (right) has a much reduced growth rate and produces much less grain than an uninfected field of wheat that is resistant to the fungus (left).

In late 1998, William Wagoire, a plant geneticist in Uganda, was astounded when he noticed some red blotches on the stems of wheat plants he was breeding. Wheat rust (*Puccinia graminis*), the fungus that causes the blotches, had supposedly been rendered almost extinct 25 years previously by the "Green Revolution," when a gene from rye plants had been crossed into wheat, making it resistant to this fungal disease. This landmark achievement had protected the crop that provides one-third of the human diet from a mold that can devastate it.

Two generations of farmers worldwide had never seen an epidemic of wheat rust, but its presence in history is vivid. Even the ancient Romans feared this disease, worshipping a god, "Robigus," who was thought to help their crop ward off the mold. In the seventeenth century, the early colonists in Massachusetts almost starved because the wheat they planted got infected with wheat rust. In 1917, an epidemic of wheat rust in the United States reduced the crop by one-third, leading to widespread panic.

Wagoire used DNA markers to identify the strain of wheat rust he had found (called Ug99 for Uganda 1999). When he compared it with known strains, which are stored in a few laboratories, he found that his strain was new and unique. Clearly, Ug99 had evolved a way to get around the resistance genes in modern wheat.

When the blotches of wheat rust on a stem burst, they release thousands of spores, any one of which can be carried by the wind to a susceptible plant. When you consider that 1 hectare of an infected wheat field can release more than 10 billion spores, the possibility of epidemic spread becomes apparent. Ug99 has begun a relentless path of infection, carried by prevailing winds. By 2001 it had infected wheat in Kenya; in 2003 it was in Ethiopia; in 2006 it crossed the Red Sea to Yemen; in 2009 it was in Iran; at the same time, winds carried the spores south to Zimbabwe and South Africa. Biologists fear that Ug99 could reach central Asia and Australia by prevailing winds, and even North America by the "747 route," accidentally dusting a traveler's clothes somewhere in the now widespread infective region. The Borlaug Global Rust Initiative has been set up to try to use knowledge of plant and fungal biology to stop the spread of this disease before it is too late.

How can knowledge of plant and fungal biology be used to prevent the spread of wheat rust?

You will find the answer to this question on page 602.

CONCEPT 28.1 Plants Have Constitutive and Induced Responses to Pathogens

Botanists know of dozens of diseases that can kill a wheat plant, each of them caused by a different pathogen, each of which in turn has many different genetic strains. Plant pathogens—which include fungi, bacteria, protists, and viruses—are part of nature, and for that reason alone they merit our study in biology. Because we humans depend on plants for our food, the stakes in our effort to understand plant pathology are especially high. That is why, just as medical schools have departments of pathology, universities in agricultural regions have departments of plant pathology.

Successful infection by a pathogen can have significant effects on a plant, reducing photosynthesis and causing massive cell and tissue death. Plants and pathogens have evolved together in a continuing "arms race": pathogens have evolved mechanisms with which to attack plants, and plants have evolved mechanisms for defending themselves against those attacks. Like the responses of the human immune system, the responses by which plants fight off infection can be either **constitutive**—always present in the plant—or **induced**—produced in reaction to the presence of a pathogen.

Physical barriers form constitutive defenses

As with humans and their skin, a plant's first line of defense is its outer surfaces, which can prevent the entry of pathogens. The parts of stems and leaves exposed to the outside environment are covered with cutin, suberin, and waxes. These substances not only prevent water loss by evaporation but can also prevent fungal spores and bacteria from entering the underlying tissues. The plant cell wall is also an important physical barrier to pathogens and can be strengthened further as an induced defense response (as we will see in the next section).

In addition to using physical barriers to prevent infection, plants use chemical warfare. Unlike animals, plants cannot flee from animals that eat them, so they make molecules that deter or destroy herbivores, as we will see in Concept 28.2. Some of these constitutively synthesized molecules also inhibit pathogens. For example, in plantains (*Plantago*), iridoid glycosides, which are glucose molecules covalently linked to short-chain fatty acids, inhibit the growth of fungal pathogens.

When constitutive defenses fail to deter a pathogen, plants initiate induced resistance mechanisms. As we discuss these mechanisms, refer to the overview in **FIGURE 28.1**.

Induced responses can be general or specific

Plant pathogens cause the host plant to activate various chemical defense responses. A wide range of molecules called **elicitors** have been identified that trigger these defenses. These molecules vary from peptides made by bacteria to cell wall fragments from fungi. Elicitors can also be derived from fragments of plant cell wall components broken down by pathogens. Plants have receptors that recognize elicitors and initiate the induced defense responses.

The responses of plants to elicitors can be described in terms of the "plant immune system." Two forms of immunity are recognized (see Figure 28.1):

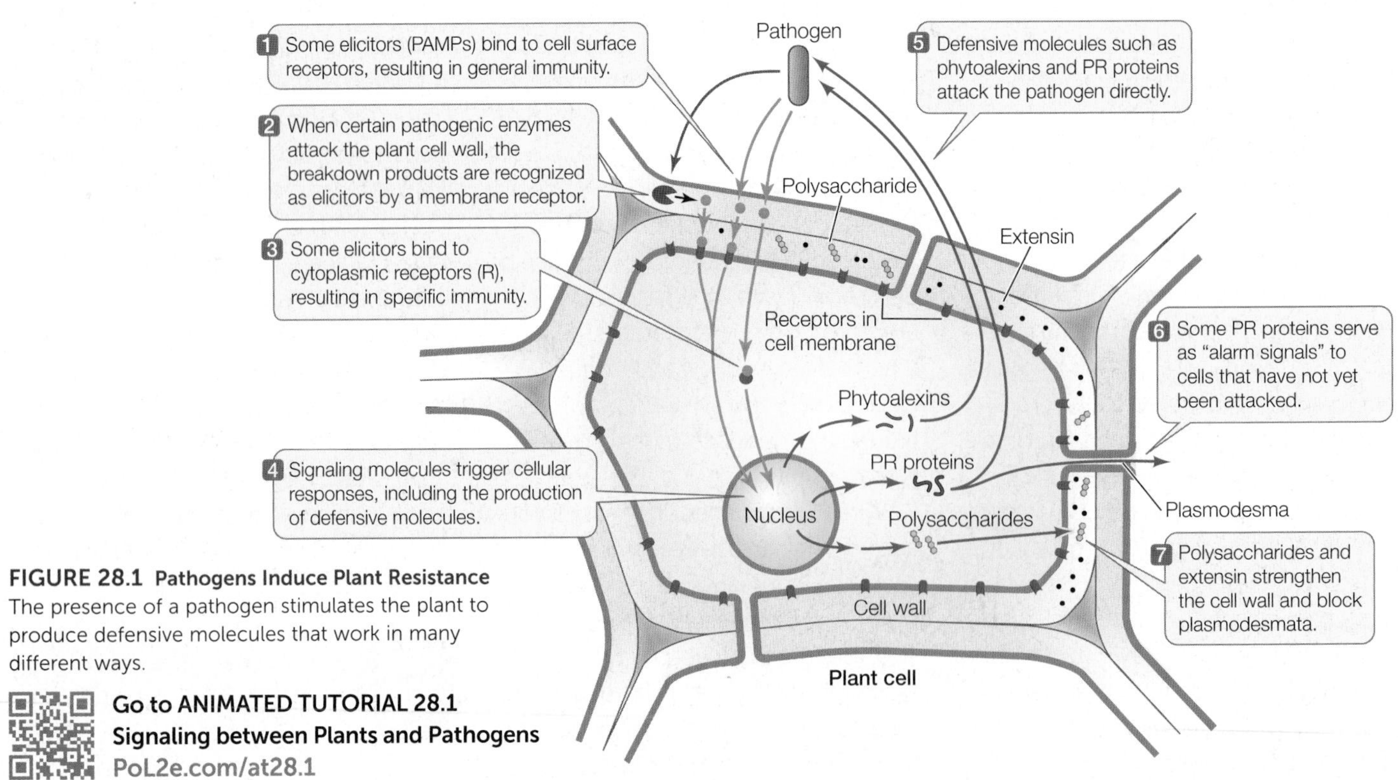
1 Some elicitors (PAMPs) bind to cell surface receptors, resulting in general immunity.
2 When certain pathogenic enzymes attack the plant cell wall, the breakdown products are recognized as elicitors by a membrane receptor.
3 Some elicitors bind to cytoplasmic receptors (R), resulting in specific immunity.
4 Signaling molecules trigger cellular responses, including the production of defensive molecules.
Pathogen
5 Defensive molecules such as phytoalexins and PR proteins attack the pathogen directly.
Polysaccharide
Extensin
Receptors in cell membrane
Phytoalexins
PR proteins
Nucleus
Polysaccharides
Cell wall
Plant cell
6 Some PR proteins serve as "alarm signals" to cells that have not yet been attacked.
Plasmodesma
7 Polysaccharides and extensin strengthen the cell wall and block plasmodesmata.

FIGURE 28.1 Pathogens Induce Plant Resistance The presence of a pathogen stimulates the plant to produce defensive molecules that work in many different ways.

Go to ANIMATED TUTORIAL 28.1
Signaling between Plants and Pathogens
PoL2e.com/at28.1

- General immunity is triggered by general elicitors called pathogen associated molecular patterns (PAMPs). PAMPs are usually molecules that are produced by entire classes of pathogens, such as the protein flagellin (found in bacterial flagella) or the polysaccharide chitin (found in fungal cell walls). Thus general immunity is an overall response rather than a response that is triggered by a specific pathogen in a particular plant. PAMPs are recognized by transmembrane receptors called pattern recognition receptors, which activate signaling pathways that lead to general immunity.
- Specific immunity is triggered by specific elicitors called effectors. Effectors include a wide variety of specific pathogen-produced molecules that enter the plant cell. Once inside the cell, effectors bind to cytoplasmic receptors called R proteins that trigger the specific immunity response.

LINK

The human immune system, with its general (innate) and specific (adaptive) defenses, is described in **Chapter 39**

General and specific immunity both involve multiple responses

Many of the signaling pathways associated with general and specific immunity are the same, although the latter is specific for particular pathogens in particular plants, and is much stronger than general immunity. Both forms of immunity involve signaling pathways that are triggered by binding between the elicitors (PAMPs or effectors) and their receptors. These pathways lead to various responses:

- *Formation of NO and reactive oxygen species*: Receptor binding triggers the rapid production of nitric oxide (NO) and reactive oxygen species such as hydrogen peroxide (H_2O_2). These reactive molecules are toxic to some pathogens, and they are components of signal transduction pathways leading to local and systemic (plant-wide) defenses.
- *Polymer deposition*: Polymers such as lignin are deposited on the inside of the cell wall to strengthen the wall. These macromolecules may also block the plasmodesmata, limiting the ability of pathogens to spread from cell to cell through these channels (see Figure 28.1, step 7).
- *Hormone signaling*: Some pathways result in the production of plant hormones, including salicylic acid and jasmonic acid. We will describe the roles of these hormones in immunity later in the chapter.
- *Changes in gene expression*: Signal transduction cascades lead to changes in gene expression. The upregulated genes include pathogenesis-related (PR) genes and genes encoding the production of antimicrobial substances called phytoalexins.

PHYTOALEXINS **Phytoalexins** are small molecules produced within hours of an infection by plant cells near the infection site. They are antibiotic-like molecules, toxic to pathogens. An example is camalexin, a phytoalexin made by the model organism *Arabidopsis thaliana* from the amino acid tryptophan:

COOH N S N H N H

Tryptophan → Camalexin

PATHOGENESIS-RELATED PROTEINS Plants produce several types of **pathogenesis-related**, or **PR, proteins**. Some are enzymes that break down the cell walls of pathogens. Chitinase, for example, is a PR protein that breaks down chitin, a distinctive component of fungal cell walls. In some cases the breakdown products of the pathogen's cell walls serve as elicitors that trigger further defensive responses. Other PR proteins may serve as alarm signals to plant cells that have not yet been attacked, so they can initiate defensive responses of their own (see Figure 28.1, step 6). PR proteins are not rapid-response weapons; rather, they act more slowly, perhaps after other mechanisms have blunted the pathogen's attack.

APPLY THE CONCEPT

Plants have constitutive and induced responses to pathogens

Because you are a biology student, your nonscientist neighbors are confident that you know all about plants. A neighbor brings you a potted plant he bought 3 weeks ago from a nursery. The leaves have brown spots. He bought another plant of the same species from a different nursery 3 weeks ago, and that plant has no spots.

1. You suspect that the spots are caused by either a disease or a nutritional deficiency. Given a large supply of the plants and soils involved, how would you distinguish between these possibilities?
2. Assuming that the spots are caused by a fungal disease, how would you investigate at the molecular level the mechanism for formation of the brown leaf spots?
3. You go to the nursery where the neighbor bought the infected plant and find several pots of plants with no brown spots on their leaves. What plant characteristics might account for the absence of spots?

Specific immunity is genetically determined

Pathogen genes that code for elicitors of specific immunity are called **avirulence (*Avr*) genes**; there are hundreds of such genes, and they vary among pathogen species and strains. When an elicitor enters a plant cell, it may encounter an R protein, a cytoplasmic receptor protein encoded by a **resistance (*R*) gene**. There are hundreds of *R* genes, each encoding a receptor that is specific for one or a few elicitors. If a plant receptor binds to an elicitor, a signal transduction pathway is set in motion that triggers the plant's specific immunity. This type of resistance is called **gene-for-gene resistance** because it involves a recognition event between two specific molecules: one determined by an *Avr* gene in the pathogen and the other determined by an *R* gene in the plant (**FIGURE 28.2**).

If the plant has no receptor to bind to the elicitor made by a particular species or strain of pathogen, the plant does not recognize the pathogen and does not turn on specific inducible defenses. The lack of a receptor makes the plant especially susceptible to that pathogen. There is an ongoing coevolutionary "arms race" between pathogens attempting to evade detection by plants and plants evolving ways to detect them. A major goal of plant breeders for the past 50 years has been to identify *R* and *Avr* genes and to breed new *R* genes into crops to make them more resistant to pathogens. This effort speeds up the arms race that occurs naturally.

Specific immunity usually leads to the hypersensitive response

As we described earlier, some induced defenses, such as polymer deposition, help prevent spread of the pathogen to the rest of the plant. Specific immunity often takes this a step further: cells around the site of infection undergo apoptosis (programmed cell death) in what is called the **hypersensitive response**. This prevents spread of the pathogen by depriving it of nutrients. Some of the cells also produce phytoalexins and other toxic chemicals before they die. Surrounding (still living) cells produce cell wall polysaccharides to seal off plasmodesmata and prevent the spread of the pathogen to healthy tissues. The dead tissue, called a necrotic lesion, contains and isolates what is left of the infection. The rest of the plant remains free of the infecting pathogen:

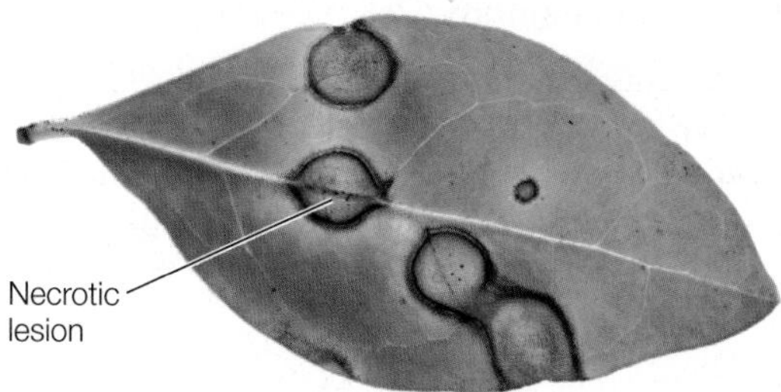

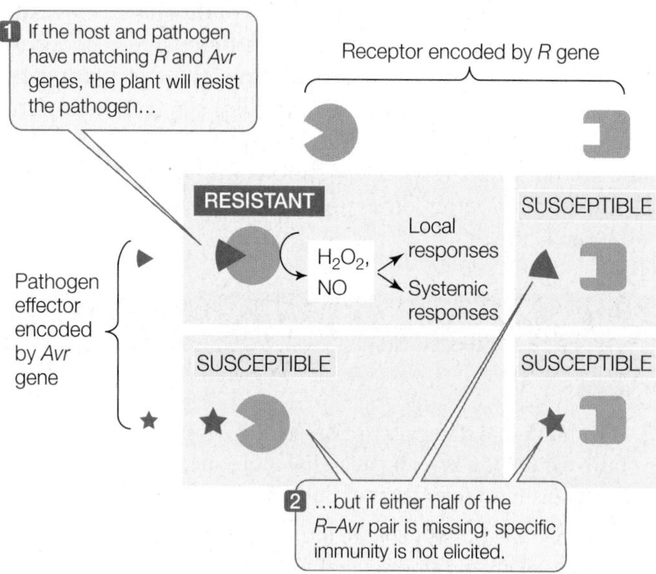

FIGURE 28.2 Genes and the Response to a Pathogen If an *Avr* gene in a pathogen that codes for an elicitor "matches" an *R* gene in a plant that codes for a receptor, the receptor binds the elicitor, and specific immunity results. In addition to the production of H_2O_2 and NO, responses may include gene activation and modifications of signaling proteins.

LINK

Apoptosis, or programmed cell death, is involved in many normal biological processes; see, for example, **Concept 14.3**. The cellular events of apoptosis are described in **Concept 7.5**

General and specific immunity can lead to systemic acquired resistance

The hypersensitive response contrasts with **systemic acquired resistance**, a general increase in the resistance of the entire plant to a wide range of pathogens. It is not localized and not limited to the pathogen that originally triggered it, and its effect may last as long as an entire growing season. This defensive response is initiated by salicylic acid:

Salicylic acid

Salicylic acid production is triggered by receptor binding in both general and specific immunity; salicylic acid then acts as a signal to turn on various defense responses.

Systemic acquired resistance is accompanied by the synthesis of PR proteins. Infection in one part of a plant can lead to the export of salicylic acid to other parts, where it triggers the production of PR proteins ahead of the spread of infection. Infected plant parts also produce the closely related compound methyl salicylate (also known as oil of wintergreen). This volatile (readily forms a gas) substance travels to other plant parts through the air and may trigger the production of PR proteins in neighboring plants that have not yet been infected.

Another type of systemic acquired resistance is a more specific defense against viruses whose genomes are made of RNA. The plant uses its own enzymes to convert some of the single-stranded RNA of the invading virus into double-stranded RNA (dsRNA) and to chop that dsRNA into small pieces called small interfering RNAs (siRNAs). The siRNAs interact with another cellular component to degrade viral mRNAs, blocking viral replication. This phenomenon is an example of RNA interference (RNAi) (see Figure 13.9). The siRNAs spread rapidly through the plant by way of plasmodesmata, providing systemic resistance.

CHECKpoint CONCEPT **28.1**

- ✓ Compare constitutive and induced responses to pathogens.
- ✓ Explain the inherited traits involved in both constitutive and induced responses to pathogens.
- ✓ If a group of plants is sprayed with oil of wintergreen, they all show resistance to a variety of pathogens. Explain how this occurs.

Pathogens are not the only foes of plants; plants must also defend themselves against herbivores, animals that eat plants. Because they are sessile, plants cannot run away from herbivores, but that does not mean they are defenseless against their would-be consumers, as we will see next.

CONCEPT 28.2 Plants Have Mechanical and Chemical Defenses against Herbivores

Herbivores depend on plants for energy and nutrients. Their foraging activities cause physical damage to plants, and they often spread pathogens among plants as well. While the majority of herbivores are insects (**FIGURE 28.3**), most other groups of animals also include some herbivores. Plants have both constitutive and induced mechanisms to protect themselves from herbivory.

Constitutive defenses are physical and chemical

Plants have a number of constitutive anatomical features, such as trichomes (leaf hairs), thorns, and spines that may be specialized for defense. Some plants have insoluble crystals of salts such as calcium oxalate that damage the tissues of insects that consume the plants. In addition, normal plant features, including thick cell walls and tree bark, may deter some herbivores.

Plants resort to "chemical warfare" to repel or inhibit other herbivores, using special chemicals called **secondary metabolites**. These substances are referred to as "secondary" because, unlike molecules such as sugars and amino acids, they are not used for basic cellular processes. The more than 10,000 known secondary metabolites range in molecular mass from about 70 to more than 390,000 daltons, but most are small (**TABLE 28.1**). Some are produced by only a single plant species, whereas others are characteristic of entire genera or even families. The effects of defensive secondary metabolites on animals are diverse. Some act on the nervous systems of herbivorous insects, mollusks, or mammals. Others mimic the natural hormones of insects, causing some larvae to fail to develop into adults. Still others damage the digestive tracts of herbivores.

An example of a secondary metabolite is canavanine, an amino acid that is similar to the amino acid arginine:

Arginine: NH_2–C(=NH)–NH–CH_2–CH_2–CH_2–CH(NH_2)–COOH

Canavanine: NH_2–C(=NH)–NH–O–CH_2–CH_2–CH(NH_2)–COOH

A seemingly slight chemical difference... (H—C—H in arginine) ...produces an inactive protein. (O in canavanine)

When an insect larva consumes canavanine-containing plant tissue, the canavanine is incorporated into the insect's proteins in some of the places where the insect's mRNA codes for arginine. The enzyme that charges the tRNA specific for arginine fails to discriminate accurately between the two amino acids. The structure of canavanine, however, is different enough from that of arginine that some of the resulting proteins end up with a modified tertiary structure and hence reduced biological activity. These defects in protein structure and function lead to developmental abnormalities that kill the insect.

(A) *Locusta migratoria*

(B) *Manduca sexta*

FIGURE 28.3 Insect Herbivores The great majority of herbivores are insects. (A) Some herbivores, such as this locust, are generalists that will attack nearly any plant. (B) Others are specialists, like this tobacco hornworm (the caterpillar of the Carolina sphinx moth), which feeds only on tobacco plants and a few closely related species.

TABLE 28.1 Secondary Metabolites Used in Plant Defense

Class	Type	Role	Example
Nitrogen-containing 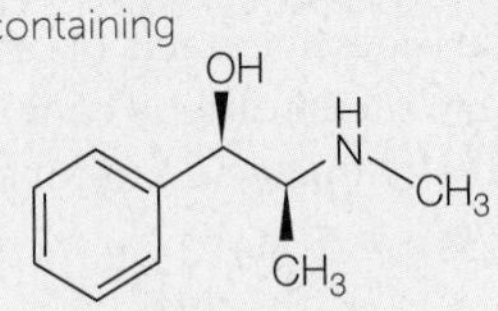 Ephedrine (an alkaloid)	Alkaloids	Neurotoxin	Nicotine in tobacco
	Glycosides	Inhibit electron transport	Dhurrin in sorghum
	Nonprotein amino acids	Disrupt protein structure	Canavanine in jack bean
Nitrogen- and sulfur-containing 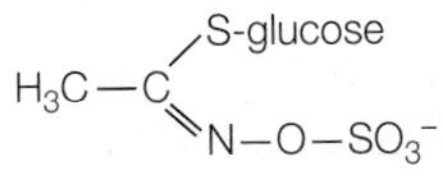 Methylglucosinolate	Glucosinolates	Inhibit respiration	Methylglucosinolate in cabbage
Phenolics 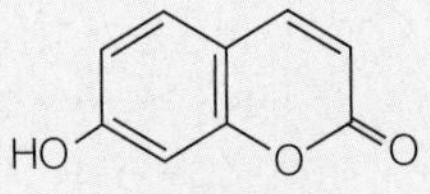 Umbelliferone	Coumarins	Block cell division	Umbelliferone in carrots
	Flavonoids	Phytoalexins	Capsaicin in peppers
	Tannins	Inhibit enzymes	Gallotannin in oak trees
Terpenes 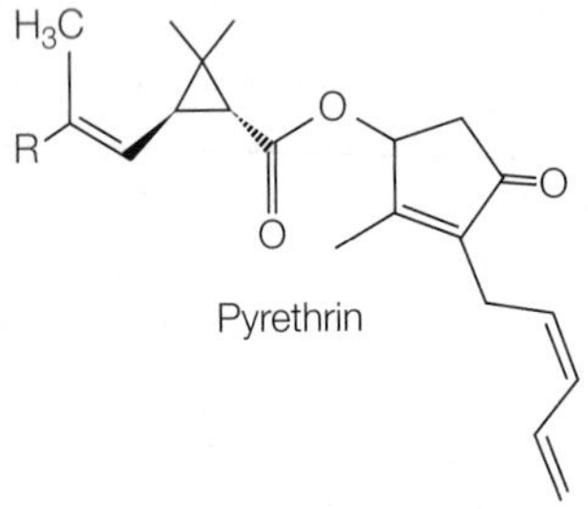 Pyrethrin	Monoterpenes	Neurotoxins	Pyrethrin in chrysanthemums
	Diterpenes	Disrupt reproduction and muscle function	Gossypol in cotton
	Triterpenes	Inhibit ion transport	Digitalis in foxglove
	Sterols	Block animal hormones	Spinasterol in spinach
	Polyterpenes	Deter feeding	Latex in *Euphorbia*

LINK
The charging of tRNA is described in **Concept 10.4**

While the mechanisms of action of secondary metabolites and their presence in plant tissues are circumstantial evidence of their role in defense, experiments can be done to provide cause-and-effect evidence. Nicotine, for example, kills insects by acting as an inhibitor of nervous system function, and its function in tobacco is presumed to be insecticidal. Yet commercial varieties of tobacco and related plants that produce nicotine are still attacked, with moderate damage, by pests such as the tobacco hornworm (see Figure 28.3B). Given that observation, does nicotine really deter herbivores? Biologists answered this question conclusively with a study that used tobacco plants in which an enzyme involved in nicotine biosynthesis had been inhibited, lowering the nicotine concentration in the plants by more than 95 percent. These low-nicotine plants suffered much more damage from insect herbivory than normal plants did (**FIGURE 28.4**).

Plants respond to herbivory with induced defenses

The first step in a plant's response to an attack by an herbivore is to sense the attack, either by means of changes in membrane potential or by means of chemical signals. The detection of an herbivore attack then triggers signal transduction pathways that induce plant defenses.

MEMBRANE SIGNALING The cell membrane is the part of the plant cell that is in contact with the environment. Within the first minute after an herbivore strikes, changes in the electric potential of the cell membrane occur in the damaged area. The continuity of the symplast (see Concept 25.3) ensures that the signal travels over much of the plant within 10 minutes.

LINK
In animals, direct electrical coupling between cells occurs only in specific tissues, such as cardiac muscle; see **Concept 33.1**

CHEMICAL SIGNALING When some insects (such as caterpillars) chew on a plant, substances in the insect's saliva combine with fatty acids derived from the consumed plant tissue. The resulting compounds act as elicitors to trigger both local and systemic responses to the herbivore. In corn, the elicitor produced by one particular moth larva has been named volicitin for its ability to induce production of volatile signals that can travel to other plant parts—and to neighboring corn plants—and stimulate their defensive responses.

INVESTIGATION

FIGURE 28.4 Nicotine Is a Defense against Herbivores The secondary metabolite nicotine, made by tobacco plants, is an insecticide, yet most commercial varieties of tobacco are susceptible to insect attack. Ian Baldwin demonstrated that a tobacco strain with a reduced nicotine concentration was much more susceptible to insect damage than native tobacco was.[a]

HYPOTHESIS

Nicotine helps protect tobacco plants against insects.

METHOD

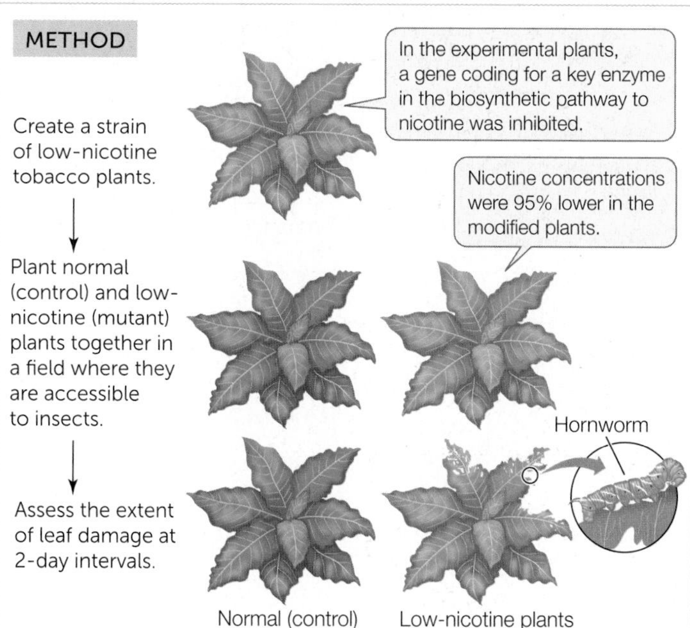

Create a strain of low-nicotine tobacco plants.
In the experimental plants, a gene coding for a key enzyme in the biosynthetic pathway to nicotine was inhibited.
Nicotine concentrations were 95% lower in the modified plants.
Plant normal (control) and low-nicotine (mutant) plants together in a field where they are accessible to insects.
Hornworm
Assess the extent of leaf damage at 2-day intervals.
Normal (control)
Low-nicotine plants

RESULTS

The low-nicotine plants suffered more than twice as much leaf damage as did the normal controls.

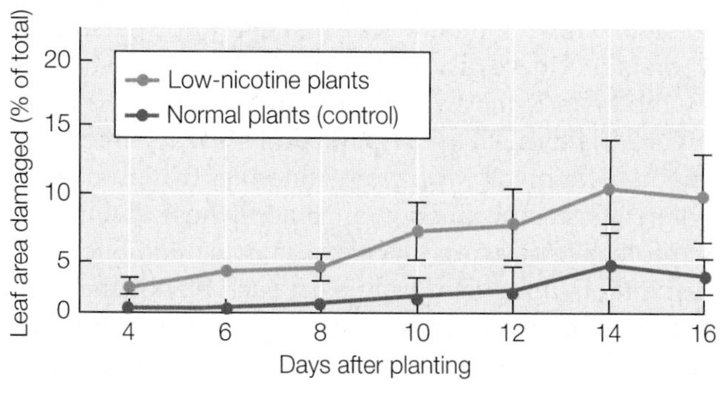

CONCLUSION

Nicotine provides tobacco plants with at least some protection against insects.

ANALYZE THE DATA

In a separate experiment, Baldwin and his colleagues showed that treatment with jasmonic acid (jasmonate) increased the concentration of nicotine in normal tobacco plants but not in the low-nicotine plants. The researchers planted a group of normal and low-nicotine plants and treated them with jasmonate 7 days after planting. The plants were assessed for herbivore damage every 2 days after being planted. The results are shown in the graph.

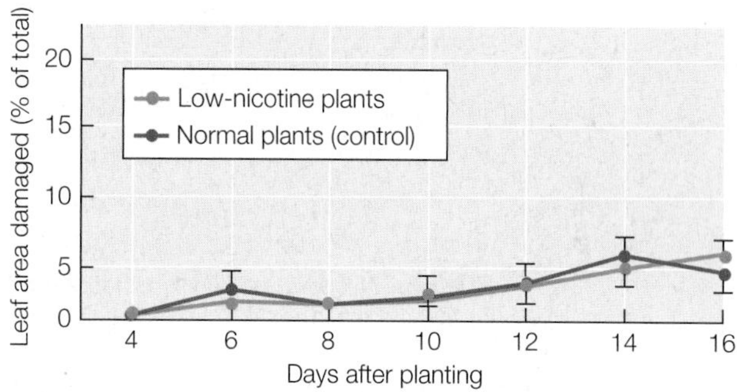

A. Compare these data to those at left for the untreated plants. What was the effect of jasmonate treatment on the resistance of normal plants to herbivore damage? What was the effect on low-nicotine plants?

B. What do these data reveal about the role of nicotine in preventing herbivore damage?

C. Explain how jasmonate could have had the effect it did on the low-nicotine plants, even though their nicotine levels were still low after jasmonate treatment.

D. The error bars at each data point are the standard error of the mean (SEM). What statistical test (see Appendix B) would you use to determine the possible significance of differences between the jasmonate-treated and untreated plants? At 10 days, the mean damage ± SEM for untreated low-nicotine plants was 6.0 ± 1.5 % (n = 36), and for treated low-nicotine plants it was 2.2 ± 0.6 % (n = 28). Run a statistical test comparing the two results, calculate the P-value, and comment on significance.

Go to **LaunchPad** for discussion and relevant links for all **INVESTIGATION** figures.

[a]P. Steppuhn et al. 2004. *PLOS Biology* 2: e217.

SIGNAL TRANSDUCTION PATHWAY The perception of damage from herbivory can initiate a signal transduction pathway in the plant that involves the plant hormone **jasmonic acid (jasmonate)** (**FIGURE 28.5**; also see Figure 28.4, Analyze the Data):

Jasmonic acid

When the plant senses an herbivore-produced elicitor, it makes jasmonate, which triggers many plant defenses, including the synthesis of protease inhibitors. The inhibitors, once in an insect's gut, interfere with the protease-catalyzed digestion of proteins and thus stunt the insect's growth. Jasmonate also "calls for help" by triggering the formation of volatile compounds that attract insects that prey on the herbivores attacking the plant. Similarly, the volatile compounds induced by volicitin in corn attract wasps that parasitize

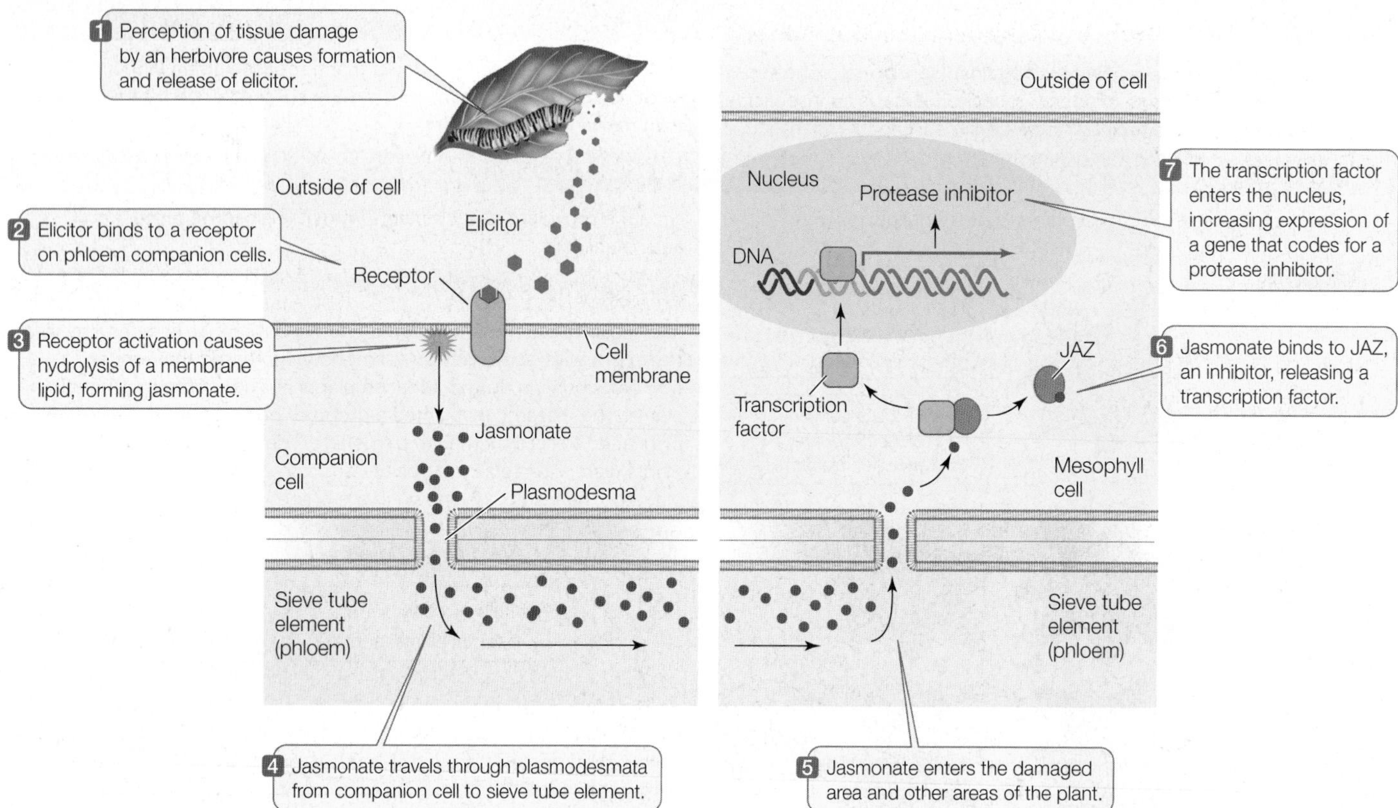

FIGURE 28.5 A Signaling Pathway for Induced Defenses against Herbivory The chain of events initiated by herbivory that leads to the production of a defensive chemical can consist of many steps. These steps may include the synthesis of one or two hormones, binding of receptors, gene activation, and finally, synthesis of the defensive chemical.

herbivorous caterpillars, so the "call for help" is a recurrent theme in plant–insect interactions.

Why don't plants poison themselves?

Why don't the constitutive defensive chemicals that are so toxic to herbivores and pathogens kill the plants that produce them? Plants that produce toxic defensive chemicals generally use one of several measures to protect themselves.

COMPARTMENTALIZATION Isolation of the toxic substance is the most common means of avoiding exposure. Plants store their toxins in vacuoles if the toxins are water-soluble. If they are not water-soluble, the toxins may be dissolved in latex (a complex mixture of dissolved molecules and suspended particles) and stored in a specialized compartment, or they may be dissolved in waxes on the epidermal surface. Such compartmentalized storage keeps the toxins away from mitochondria, chloroplasts, and other parts of the plant's metabolic machinery.

LINK

Both plants and animals use compartmentalization to isolate toxins and waste products that occur as by-products of some biochemical reactions; see **Concept 4.3**

STORAGE OF PRECURSORS Some plants store the precursors of toxic substances in one type of tissue, such as the epidermis, and store the enzymes that convert those precursors into the active toxin in another tissue, such as the mesophyll. When an herbivore chews part of the plant, cells are ruptured, the enzymes come into contact with the precursors, and the toxin is produced. The only part of the plant that is damaged by the toxin is that which was already damaged by the herbivore. Plants such as sorghum and some legumes, which respond to herbivory by having molecules that produce cyanide when ingested by animals, are among those that use this type of protective measure. Cyanide is a potent inhibitor of cellular respiration, and when people eat such plants, they first cook them or otherwise prepare them (e.g., soaking, fermenting, drying) to inactivate the cyanide-producing molecules.

MODIFIED PROTEINS Some plants have modified proteins that do not react with the toxin. As noted above, canavanine resembles arginine and therefore plays havoc with protein synthesis in insect larvae. In plants that make canavanine, the enzyme that catalyzes bond formation between arginine and its counterpart tRNA molecule discriminates correctly between arginine and canavanine, so canavanine is not incorporated into the plant's proteins.

Plants don't always mount a successful defense

Milkweeds such as *Asclepias syriaca* store their defensive chemicals in latex in specialized tubes called **laticifers**, which run alongside the veins in the leaves. When damaged, a milkweed releases copious amounts of toxic latex from its laticifers. Field studies have shown that most insects that feed on neighboring plants of other species do not attack laticiferous plants, but there are exceptions. One population of beetles that feeds on *A. syriaca* exhibits a remarkable prefeeding behavior: these beetles cut a few veins in the leaves before settling down to dine. Cutting the veins causes massive latex leakage from the adjacent laticifers and interrupts the latex supply to a downstream portion of the leaf. The beetles then move to the relatively latex-free portion and eat their fill.

This is just one of many ways that herbivores circumvent plant defenses. A successful plant defense exerts strong selection pressure on herbivores to get around it somehow; a successful herbivore, in turn, exerts strong selection pressure on plants to develop new defensive strategies. One can imagine, for example, that over time *A. syriaca* might evolve to have thicker walls around the base of its laticifers, such that the beetles can no longer cut them, or to produce a different toxin that does not depend on laticifers.

LINK

You can find more examples of the "evolutionary arms race" between plants and herbivores in **Concept 43.4**; see especially **Figures 43.10 and 43.11**

CHECKpoint CONCEPT 28.2

- ✓ Using a plant growing near you as an example, describe anatomical features that help plants avoid herbivores.
- ✓ A mutant strain of a plant makes jasmonate constitutively. What would be the herbivory defense phenotype of this plant?
- ✓ You have isolated a secondary metabolite from apple tree leaves. What circumstantial and experimental evidence would you use to show that the metabolite prevents the herbivorous larva of the moth *Spilonota ocellana* from damaging apple orchards?

In addition to coping with their biological enemies, plants must survive in their physical environment. Plants have a number of responses that allow them to cope with stressful environmental conditions.

CONCEPT 28.3 Plants Adapt to Environmental Stresses

In an ever-changing environment, plants face several potential stressors: drought, submersion, heat, cold, and high concentrations of salt and heavy metals in the soil. Here we will focus on the overall mechanisms by which plants deal with extremes of climate and soil conditions. You will see that in these cases, as with defenses against pathogens and herbivores, there are constitutive defenses as well as induced responses.

Go to MEDIA CLIP 28.1
Leaves for Every Environment
PoL2e.com/mc28.1

Some plants have special adaptations to live in very dry conditions

Many plants, especially those living in deserts, must cope with extremely limited water supplies. A variety of anatomical and life-cycle adaptations allow plants to survive under these conditions. Many of these adaptations are ways to avoid or reduce the inevitable water loss through transpiration that occurs during active photosynthesis (see Concept 25.3). Other adaptations help plants tolerate the high levels of light and heat that are often found in deserts.

DROUGHT AVOIDERS Some desert plants have no special structural adaptations for water conservation. Instead, these desert annuals, called drought avoiders, simply evade periods of drought. Drought avoiders carry out their entire life cycle—from seed to seed—during a brief period in which rainfall has made the surrounding desert soil sufficiently moist for growth and reproduction (**FIGURE 28.6**). A different drought avoidance strategy is seen in deciduous perennial plants, particularly in

FIGURE 28.6 Desert Annuals Avoid Drought The seeds of many desert annuals lie dormant for long periods, awaiting conditions appropriate for germination. When they do receive enough moisture to germinate, they grow and reproduce rapidly before the short wet season ends. During the long dry spells, only dormant seeds remain alive.

Africa and South America, that shed their leaves in response to drought as a way to conserve water. These plants remain dormant until conditions are again favorable for growth.

LEAF STRUCTURES Most desert plants are not drought avoiders, but rather grow in their dry environment year-round. Plants adapted to dry environments are called **xerophytes** (Greek *xeros*, "dry"). Three structural adaptations are found in the leaves of many xerophytes:

- Specialized leaf anatomy that reduces water loss
- A thick cuticle and a profusion of trichomes over the leaf epidermis, which retard water loss
- Trichomes that diffract and diffuse sunlight, thereby decreasing the intensity of light impinging on the leaves and the risk of damage to the photosynthetic apparatus by excess light

In some xerophytes, the stomata are strategically located in sunken cavities below the leaf surface (known as stomatal crypts), where they are sheltered from the drying effects of air currents (**FIGURE 28.7**). Trichomes surrounding the stomata slow air currents further. Cacti and similar plants have spines rather than typical leaves, and photosynthesis is confined to the fleshy stems (see Figure 24.13D). The spines may help the plants cope with desert conditions by reflecting solar radiation or by dissipating heat. The spines also deter herbivores.

WATER-STORING STRUCTURES **Succulence**—the possession of fleshy, water-storing leaves or stems—is another adaptation to dry environments (**FIGURE 28.8**). This adaptation allows plants to take up large amounts of water when it is available (such as after a brief thunderstorm) and then draw on the stored water during subsequent dry periods. Other adaptations of succulents include a reduced number of stomata and a variant form of photosynthesis, both of which reduce water loss.

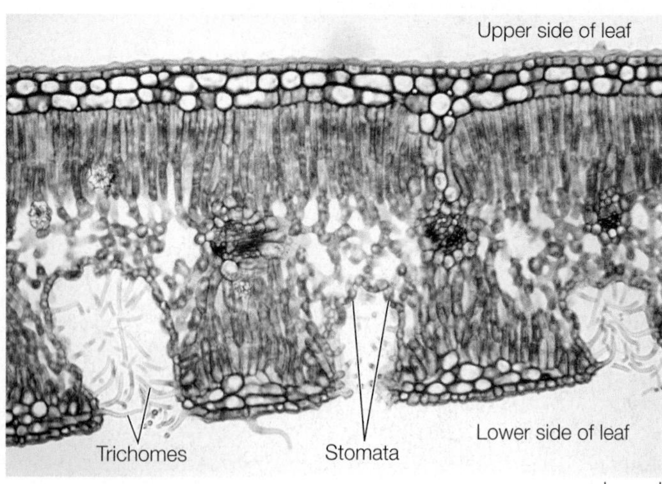

FIGURE 28.7 Stomatal Crypts Stomata in the leaves of some xerophytes, in this case oleander (*Nerium oleander*), are located in sunken cavities called stomatal crypts. The trichomes (hairs) covering these crypts trap moist air.

FIGURE 28.8 Succulence *Aloe vera* stores water in its fleshy leaves.

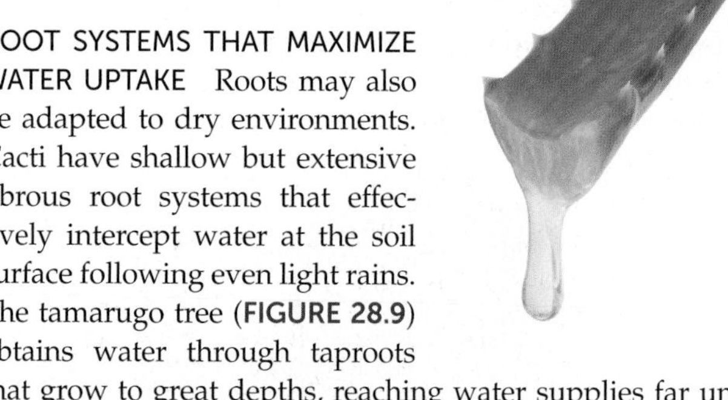

ROOT SYSTEMS THAT MAXIMIZE WATER UPTAKE Roots may also be adapted to dry environments. Cacti have shallow but extensive fibrous root systems that effectively intercept water at the soil surface following even light rains. The tamarugo tree (**FIGURE 28.9**) obtains water through taproots that grow to great depths, reaching water supplies far underground, as well as from condensation on its leaves. The Atacama Desert in northern Chile often goes several years without measurable rainfall, but the landscape there has many surprisingly large shrubs.

SOLUTE ACCUMULATION Xerophytes and other plants that must cope with inadequate water supplies may accumulate high concentrations of the amino acid proline or of secondary metabolites in their vacuoles. This solute accumulation lowers the water potential in the plant's cells below that in the soil, which allows the plant to take up water via osmosis. Plants living in saline environments share this and several other adaptations with xerophytes, as we will see shortly.

LINK

Review the principles of water potential and water movement in **Concept 25.3**

FIGURE 28.9 Mining Water with Deep Taproots In the Atacama Desert in Chile, the roots of this tamarugo tree (*Prosopis tamarugo*) must grow far beneath the soil surface to reach water.

Some plants grow in saturated soils

For some plants, the environmental challenge is the opposite of that faced by xerophytes: too much water. They live in environments so wet that the diffusion of oxygen to their roots is severely limited. These plants have shallow root systems that grow slowly; oxygen levels are likely to be highest near the surface of the soil, and slow growth decreases the roots' need for oxygen.

The root systems of some plants adapted to swampy environments, such as coastal mangrove habitats, have **pneumatophores**, which are extensions that grow out of the water and up into the air (**FIGURE 28.10A**). Pneumatophores contain lenticels (openings) that allow oxygen to diffuse through them, aerating the submerged parts of the root system.

Many submerged or partly submerged aquatic plants have large air spaces in the leaf and stem parenchyma and in the petioles. Tissue containing such air spaces is called **aerenchyma** (**FIGURE 28.10B**). Aerenchyma stores oxygen produced by photosynthesis and permits its ready diffusion to parts of the plant where it is needed for cellular respiration. Aerenchyma also imparts buoyancy. Furthermore, because aerenchyma contains far fewer cells than most other plant tissues, metabolism in aerenchyma proceeds at a lower rate, so the need for oxygen is much reduced.

Plants can respond to drought stress

The adaptations of xerophytes for coping with dry environments are generally constitutive—they are always present—and under normal conditions they prevent the plants from experiencing drought stress. When conditions become so dry that even xerophytes are stressed, the plants turn to inducible responses. The same responses are found in many other plants, including those that are not adapted to grow in dry climates.

When the weather is abnormally dry, the water content of the soil is reduced, and less water is available to plants. If a plant cannot take up water from the soil, it cannot grow. Indeed, inadequate water supply is the single most important factor that limits production of our most important food crops. Extreme water deficits in plant cells have two additional major biochemical effects: a reduction in membrane integrity as the polar–nonpolar forces that orient the lipid bilayer are reduced, and changes in the three-dimensional structures of proteins. A plant can suffer irreversible damage when the structure of its cells is compromised in these ways.

When plants sense a water deficit in their roots, a signaling pathway is set in motion that initiates several measures to conserve water and maintain cellular integrity. This pathway begins with the production of the hormone abscisic acid (see Concept 26.3) in the roots. This hormone travels from the roots to the shoot, where it causes stomatal closure and initiates gene transcription that leads to other physiological events that conserve water and cellular integrity (**FIGURE 28.11**).

Many plant genes whose expression is altered by drought stress have been identified, largely through research using DNA microarrays, proteomics, and other molecular approaches (see Chapters 12 and 13). One group of proteins in which production is upregulated during drought stress is the *l*ate *e*mbryogenesis *a*bundant (LEA; pronounced "lee-yuh") proteins. These hydrophobic proteins also accumulate in maturing seeds as they dry out (hence their name). The LEA proteins bind to membrane proteins and other cellular proteins to stabilize them, preventing them from clumping together as seeds dry out.

FIGURE 28.10 Plant Adaptations to Water-Saturated Habitats (A) The submerged roots of mangrove trees obtain oxygen through pneumatophores. (B) This cross section of a petiole of the yellow water lily (*Nymphaea mexicana*) shows the air-filled channels of aerenchyma tissue.

(A)

Pneumatophores are root extensions that grow out of the water, under which the rest of the roots are submerged.

(B)

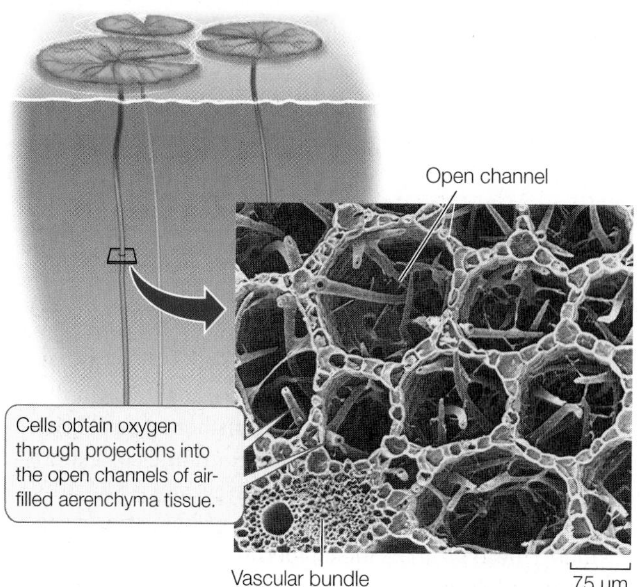

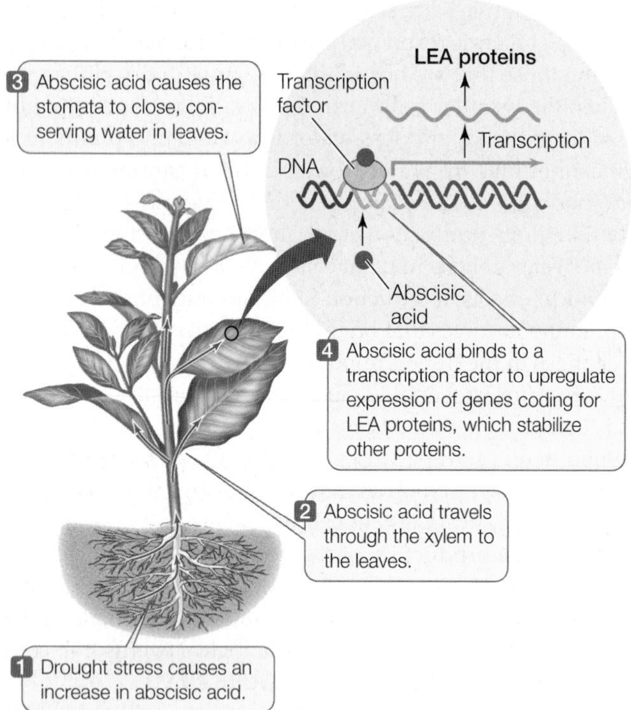

3 Abscisic acid causes the stomata to close, conserving water in leaves.
Transcription factor
LEA proteins
Transcription
DNA
Abscisic acid
4 Abscisic acid binds to a transcription factor to upregulate expression of genes coding for LEA proteins, which stabilize other proteins.
2 Abscisic acid travels through the xylem to the leaves.
1 Drought stress causes an increase in abscisic acid.

FIGURE 28.11 A Signaling Pathway in Response to Drought Stress Acclimation to drought stress begins in the root with the production of the hormone abscisic acid.

Plants can cope with temperature extremes

Temperatures that are too high or too low can stress plants and even kill them. Plant species differ in their sensitivity to heat and cold, but all plants have their limits. Any temperature extreme can damage cellular membranes:

- High temperatures destabilize membranes and denature many proteins, especially some of the enzymes of photosynthesis.
- Low temperatures cause membranes to lose their fluidity and alter their permeabilities to solutes.
- Freezing temperatures may cause ice crystals to form, damaging membranes.

Plants have both constitutive defenses and inducible responses for coping with temperature extremes.

ANATOMICAL ADAPTATIONS Many plants living in hot environments have constitutive defenses similar to those of xerophytes. These adaptations include hairs and spines that dissipate heat and leaf forms that intercept less direct sunlight.

HEAT SHOCK RESPONSE The plant inducible response to heat stress is similar to the response to drought stress in that new proteins are made, often under the direction of an abscisic acid–mediated signaling pathway. Within minutes of experimental exposure to raised temperatures (typically a 5°C–10°C increase), plants synthesize several kinds of **heat shock proteins**. Among these proteins are chaperonins, which help other proteins maintain their structures and avoid denaturation. Threshold temperatures for the production of heat shock proteins vary, but 39°C is sufficient to induce them in most plants.

COLD-HARDENING Low temperatures above the freezing point can cause chilling injury in many plants, including crops such as rice, corn, and cotton as well as tropical plants such as bananas. Many plant species can adjust to cooler temperatures through a process called **cold-hardening**, which requires repeated exposure to cool temperatures over many days. A key change during the hardening process is an increase in the proportion of unsaturated fatty acids in cell membranes, which allows them to retain their fluidity and function normally at cooler temperatures (see Figure 2.12). Plants have a greater ability to modify the degree of saturation of their membrane lipids than animals do. In addition, low temperatures induce the formation of proteins similar to the LEA proteins related to drought stress and heat shock, which protect against chilling injury (see Figure 28.11).

If ice crystals form within plant cells, they can kill the cells by puncturing organelles and cell membranes. Furthermore, the growth of ice crystals outside the cells can draw water from the cells and dehydrate them. Freeze-tolerant plants have a variety of adaptations to cope with these problems, including the production of antifreeze proteins that slow the growth of ice crystals.

Some plants can tolerate soils with high salt concentrations

Salty, or saline, environments (high in Na^+, K^+, Ca^{2+}, and Cl^-) are found in nature in diverse locales, from hot, dry deserts to moist, cool coastal marshes. Much of the land in Australia is naturally saline. In addition, agricultural land can become increasingly saline as a result of irrigation and the application of chemical fertilizers. This salinization, which can eventually make land unsuitable for farming, is an increasing problem worldwide (**FIGURE 28.12**).

FIGURE 28.12 Salty Soil Accumulation of salt from irrigation water in an area with inadequate drainage has caused this soil in central California (San Joaquin Valley) to become unsuitable for most plant growth.

APPLY THE CONCEPT

Plants adapt to environmental stresses

The *Arabidopsis* gene *cor15a* is under the control of a promoter that responds to environmental conditions. *Arabidopsis* plants were genetically modified with a copy of the *cor15a* promoter fused to a reporter gene that is expressed as an enzyme that is readily detected in leaf extracts (see Concept 13.2, p. 260 for information about the use of reporter genes to study gene expression). The genetically modified plants were transferred to environments kept at two different temperatures (19°C and 2°C), and the reporter enzyme activity in their leaves was measured at different times after the transfer. The results are shown in the table.[a]

	ENZYME ACTIVITY (UNITS/g LEAF PROTEIN)	
HOURS AFTER TRANSFER	19°C	2°C
12	0.35	3.21
24	0.51	5.66
36	0.45	9.65
48	0.60	10.10
72	0.49	11.33

1. Plot these data with hours after transfer on the x axis and enzyme activity on the y axis. What can you conclude about the role of *cor15a*?
2. In a separate experiment, plants were kept at 19°C for 36 hours in the presence of abscisic acid, and the enzyme activity was 8.99. What can you conclude from this result?
3. How would you investigate the possible role of *cor15a* in the plant's response to drought? Why would you expect that there might be such a role?

[a]S. S. Baker et al. 1994. *Plant Molecular Biology* 24: 701–713.

Because of its high salt concentration, a saline environment has a very negative water potential. To obtain water from such an environment, a plant must have an even more negative water potential (see Concept 25.3); otherwise water will diffuse out of its cells, and the plant will wilt and die. Plants in saline environments are also challenged by the potential toxicity of sodium ions, which inhibit enzymes and protein synthesis.

Plants that are adapted for survival in saline soils are called **halophytes**. Most halophytes take up Na^+, and many take up Cl^-, into their roots and transport those ions to their leaves, where they accumulate in the central vacuoles of leaf cells, away from more sensitive parts of the cells. The accumulated salts in the tissues of halophytes make their water potential more negative than the soil solution and allow them to take up water from their saline environment.

Some halophytes have **salt glands** in their leaves. These glands excrete salt, which collects on the leaf surface until it is removed by rain or wind (**FIGURE 28.13**). This adaptation, which reduces the danger of poisoning by accumulated salt, is found in some desert plants and in some plants growing in mangrove habitats. The negative water potential in the salt-laden leaves also promotes water flow from the roots up through the xylem by transpiration–cohesion–tension (see Concept 25.3).

Some plants can tolerate heavy metals

High concentrations of heavy metal ions are toxic to most plants. Some soils are naturally rich in heavy metals as a result of normal geological processes or acid rain. The mining of metallic ores also leaves localized areas with high concentrations of heavy metals and low concentrations of nutrients. Such sites are hostile to most plants, and seeds falling on them generally do not produce adult plants.

Some plants can survive in these environments, however, by accumulating concentrations of heavy metals that would kill most plants. More than 200 plant species have been identified as **hyperaccumulators** that store large quantities of metals such as arsenic (As), cadmium (Cd), nickel (Ni), aluminum (Al), and zinc (Zn).

Perhaps the best-studied hyperaccumulator is alpine pennycress (*Thlaspi caerulescens*). Before the advent of chemical analysis, miners used the presence of this plant as an indicator of mineral-rich deposits. A *Thlaspi* plant may accumulate as much as 30 g/kg dry weight Zn (most plants contain 0.1 g/kg dry weight) and 1.5 g/kg dry weight Cd (most plants contain 0.001 g/kg dry weight). Studies of *Thlaspi* and other hyperaccumulators have revealed the presence of several common adaptations:

- Increased ion transport into the roots
- Increased rates of translocation of ions to the leaves

FIGURE 28.13 Excreting Salt This saltwater mangrove plant has specialized glands that excrete salt, which appears here as crystals on the leaves.

FIGURE 28.14 Phytoremediation Plants that accumulate heavy metals can be used to clean up contaminated soils. Here, poplars are being used to remove contaminants from an air base.

- Accumulation of ions in vacuoles in the shoot
- Resistance to the ions' toxicity

Knowledge of these hyperaccumulation mechanisms and the genes underlying them has led to the emergence of **phytoremediation**, a form of bioremediation (see Chapter 13, p. 271) that uses plants to clean up environmental pollution in soils. Some phytoremediation projects use natural hyperaccumulators, whereas others use genes from hyperaccumulators to create transgenic plants that grow more rapidly in and are better adapted to a particular polluted environment. In either case, the plants are grown in the contaminated soil, where they act as natural "vacuum cleaners" by taking up the contaminants (**FIGURE 28.14**). The plants are then harvested and safely disposed of to remove the contaminants. Perhaps the most dramatic use of phytoremediation occurred after an accident at the nuclear power plant at Chernobyl, Ukraine (then part of the Soviet Union), in 1986, when sunflower plants were used to remove uranium from the nearby soil. Phytoremediation is now widely used to clean up land after strip mining.

CHECK**point** CONCEPT 28.3

- ✓ Compare the constitutive and induced mechanisms that allow plants to cope with drought conditions.
- ✓ Describe a mechanism shared by plants that tolerate high concentrations of salts and those that tolerate heavy metals.
- ✓ If an inhibitor of abscisic acid synthesis were sprayed on a plant and the plant was exposed to drought conditions, what would happen?

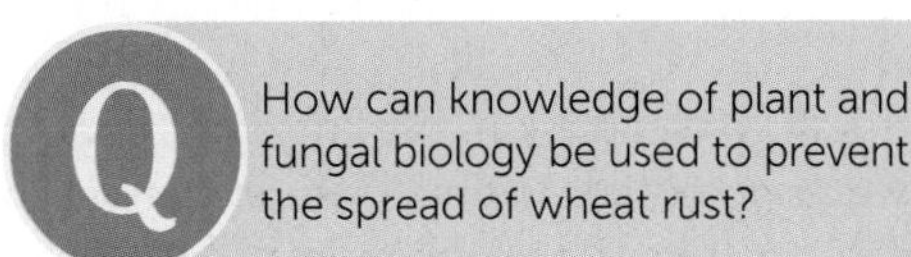

How can knowledge of plant and fungal biology be used to prevent the spread of wheat rust?

ANSWER Today, *Puccinia graminis* strain Ug99 continues its spread and threatens wheat crops worldwide. In the Green Revolution of the 1960s to 1980s, wheat plants were bred to have the gene complexes *Sr24* and *Sr31*, *R* genes that conferred resistance to existing strains of wheat rust fungus (Concept 28.1). The first strain of Ug99 found in Uganda had *Avr* genes that form elicitor proteins that do not bind to *Sr31* and therefore overcome part of the plant's resistance. Then, in 2006 a new genetic variant of Ug99 was found in Kenya with additional *Avr* genes that allow the pathogen to overcome the *Sr24* gene resistance as well (**FIGURE 28.15**).

Because 90 percent of the wheat grown in the world has no resistance to this new strain of Ug99, an intensive search is under way for additional wheat resistance genes. In the regions where Ug99 infections are now present, some wheat plants grow that are resistant. Samples of both wheat plants and fungus are sent to laboratories, where, under high security and sterility, they are tested and examined for *Avr* and *R* genes. In addition, seeds from thousands of varieties of wheat, collected from all over the world, are being grown into seedlings and examined for *R* genes.

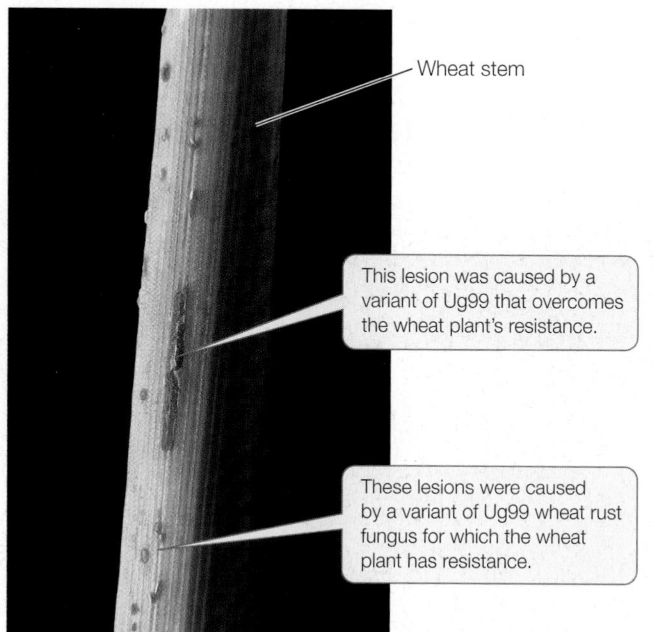

Wheat stem
This lesion was caused by a variant of Ug99 that overcomes the wheat plant's resistance.
These lesions were caused by a variant of Ug99 wheat rust fungus for which the wheat plant has resistance.

FIGURE 28.15 Overcoming Resistance to Pathogens Blotches caused by wheat rust are visible on the stem of a wheat plant. This plant has a gene called *Sr24* that confers resistance to the wheat rust pathogen, but it has also been infected by a variant of the pathogen, Ug99, that can overcome that resistance.

SUMMARY

CONCEPT 28.1 Plants Have Constitutive and Induced Responses to Pathogens

- Plants and pathogens have evolved together in a continuing "arms race": pathogens have evolved mechanisms for attacking plants, and plants have evolved mechanisms for defending themselves against those attacks.
- Some of the responses by which plants fight off pathogens are **constitutive**—always present in the plant—whereas others are **induced**—produced in reaction to the presence of a pathogen. **Review Figure 28.1 and ANIMATED TUTORIAL 28.1**
- Plants use physical barriers to block pathogen entry and seal off infected regions.
- **Gene-for-gene resistance** depends on a match between a plant's **resistance (*R*) genes** and a pathogen's **avirulence (*Avr*) genes**. **Review Figure 28.2**
- In the induced response to infection, cells produce **phytoalexins** and **pathogenesis-related (PR) proteins**, and the plant isolates the area of infection.
- The **hypersensitive response** (the formation of necrotic lesions) is rapid and localized.
- In **systemic acquired resistance**, salicylic acid activates defense responses throughout the plant.

CONCEPT 28.2 Plants Have Mechanical and Chemical Defenses against Herbivores

- Physical structures such as spines and thick cell walls deter some herbivores.
- Plants produce **secondary metabolites** as defenses against herbivores. **Review Table 28.1 and Figure 28.4**
- Hormones, including **jasmonic acid (jasmonate)**, participate in signaling pathways leading to the production of defensive compounds. **Review Figure 28.5**
- Plants protect themselves against their own toxic defensive chemicals by compartmentalizing those chemicals, by storing their precursors separately, or through modifications of their own proteins.

CONCEPT 28.3 Plants Adapt to Environmental Stresses

- **Xerophytes** are plants adapted to dry environments. Their structural adaptations include thickened cuticles, specialized trichomes, stomatal crypts, **succulence**, and long taproots. **Review Figure 28.7**
- Some plants accumulate solutes in their cells, which lowers their water potential so they can more easily take up water.
- Adaptations to water-saturated habitats include **pneumatophores**, extensions of roots that allow oxygen uptake from the air, and **aerenchyma**, tissue in which oxygen can be stored and can diffuse throughout the plant. **Review Figure 28.10**
- A signaling pathway involving abscisic acid initiates a plant's response to drought stress. **Review Figure 28.11**
- Plants respond to high temperatures by producing **heat shock proteins**. Low temperatures can result in **cold-hardening**.
- Plants that are adapted for survival in saline soils are called **halophytes**. Most halophytes accumulate salt. Some have **salt glands** that excrete salt to the leaf surface.
- Some plants living in soils that are rich in heavy metals are **hyperaccumulators** that take up and store large amounts of those metals in their tissues.
- **Phytoremediation** is the use of hyperaccumulating plants or their genes to clean up environmental pollution in soils.

See **ACTIVITY 28.1** for a concept review of this chapter.

Go to the Interactive Summary to review key figures, Animated Tutorials, and Activities PoL2e.com/is28

Go to LaunchPad at **macmillanhighered.com/launchpad** for additional resources, including LearningCurve Quizzes, Flashcards, and many other study and review resources.

PART 6

Animal Form and Function

29 Fundamentals of Animal Function

KEY CONCEPTS

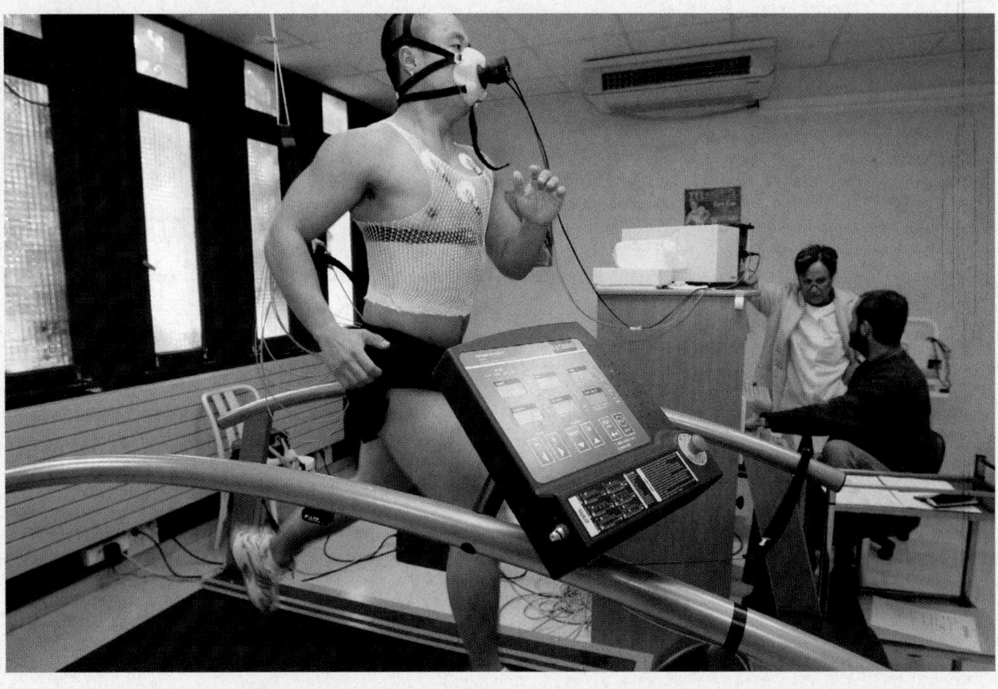

A man exercises on a treadmill wearing a device that measures his rate of O_2 consumption. His speed can be controlled by adjusting the speed of the treadmill.

Animals have muscles and are able to move. This man on a treadmill provides a dramatic illustration as he uses his leg muscles to propel himself forward. His ancient ancestors did much the same thing to pursue prey and escape dangers. Today investigators are studying the man's exercise performance on a treadmill, where his speed of running can be controlled by adjusting the speed of the treadmill. As the surface of the treadmill moves backward, driven by an electric motor, the man has to run forward at the same speed if he is to avoid being carried backward and dumped on the floor.

Treadmills have been used to study many questions in animal biology. Chimpanzees, for example, can be trained to run on a treadmill on four legs or on only two legs. Then their abilities to run in both ways can be compared to learn about advantages and disadvantages of two-legged and four-legged running. Studies of this kind help us better understand why humans became two-legged early in their evolution.

The man in the photo is breathing from a device that is measuring his rate of oxygen (O_2) consumption. From that information, investigators will be able to calculate the energy cost of his exercise.

One way to rate the efficiency of animal locomotion is to measure the energy cost to cover a certain distance. From measures of O_2 consumption on treadmills, for example, we know that a person expends about 300 kilojoules (70 kilocalories) per kilometer when jogging at a speed of 11 kilometers per hour.

The energy costs for birds to fly and for fish to swim can be measured in similar ways. Birds have been trained to fly in wind tunnels, and fish have been coaxed to swim in water tunnels. Then their O_2 consumption has been measured as they fly or swim.

Believe it or not, birds flying and fish swimming can cover distance much more cheaply—in terms of energy used—than mammals running. This may help explain why long-distance annual migrations have evolved principally in birds and fish rather than in mammals.

We humans have much higher costs to cover distance when swimming than fish do, in part because of the difference between swimming at the surface and swimming underwater. A person on a bicycle covers distance with roughly the same energy efficiency as a human-size fish swimming. So, to get a subjective sense of a fish's energy cost of covering distance, hop on a bicycle!

Why can we use a person's rate of O_2 consumption to measure the rate at which he or she consumes energy to run?

You will find the answer to this question on page 622.

CONCEPT 29.1 Animals Eat to Obtain Energy and Chemical Building Blocks

Animals are **heterotrophs**—organisms that require preformed organic molecules as food (see Concept 45.3). They obtain their energy by breaking the chemical bonds of organic compounds obtained from other organisms, and they build their tissues from preexisting organic compounds obtained from other organisms. Animals, in other words, eat organic matter, such as starch or fat, that other organisms synthesized, and they use that organic matter as their source of both energy and chemical building blocks.

In contrast, **autotrophs** obtain energy from inorganic sources and can synthesize organic compounds from inorganic precursors. Plants are examples of autotrophs. They obtain energy from sunlight and synthesize organic compounds, such as the sugar glucose, principally from H_2O and CO_2.

In most ecosystems, sunlight is the primary energy input. Plants and algae use some of the light energy in sunlight to carry out photosynthesis and build organic compounds from simple precursors. In the process, a small fraction of the light energy in sunlight is incorporated into the chemical bonds of the organic compounds that plants and algae make. Animals then ingest the organic compounds in the plant or algal material and use them as their sources of energy and chemical building blocks.

LINK

The flow of energy through ecological communities is discussed in **Concept 44.3**

Animals need chemical building blocks to grow and to replace chemical constituents throughout life

Why does an animal need the chemical building blocks it obtains by eating? One reason is growth. When an animal grows during its early development, it requires chemical building blocks—such as fatty acids and amino acids—to build each new cell it adds to its body.

Even after they are fully grown, animals need chemical building blocks throughout their lives. This important fact was discovered by using isotopic forms of carbon atoms and nitrogen atoms (see Concept 2.1 for a discussion of isotopes). Suppose you make some food in which the nitrogen atoms are nitrogen-15 instead of the usual isotope, nitrogen-14. Suppose, then, that you feed that food to a full-grown dog or other animal. After a few weeks you will find that many of the nitrogen-14 atoms in the dog's body have been replaced with nitrogen-15 atoms—even though the dog looks no different. If you then start to feed the dog regular food (with nitrogen-14 atoms) again, you will find that the nitrogen-15 atoms in the dog's body are gradually replaced with nitrogen-14 atoms.

Experiments such as this demonstrate that the atoms inside an adult animal's body are constantly exchanged with atoms in the animal's surrounding environment. For example, our red blood cells live for only about 4 months. When a red blood cell is at the end of its life span, it is broken down and replaced. As it breaks down, some of its constituents are excreted. For example, some of the iron from the cell is excreted. To replace the cell, our bodies need to get replacement constituents—such as replacement iron—from our environment.

Nearly all the cells in our body are engaged in the process of breakdown and replacement (although not necessarily on a 4-month schedule). As the process takes place, some atoms originally inside our body are lost, and some atoms are taken in from the environment. In humans, 2–3 percent of body protein is broken down and rebuilt every day! In the process, some of the amino acids in the initial proteins are lost and must be replaced. For this reason, we need to eat amino acid–containing foods all our lives, even after we are fully grown.

We have seen that individual atoms constantly come and go in an animal. Over time, therefore, an animal is not defined by its atomic building blocks. Instead it is defined by the organization of its body.

Animals need inputs of chemical-bond energy to maintain their organized state throughout life

Organization is the most essential attribute of life. Animals are organized at multiple levels. In individual cells, atoms are organized into molecules such as phospholipids and proteins. In turn, these individual molecules are organized into cellular structures such as the cell membrane and into cellular biochemical processes such as sequences of enzyme-catalyzed reactions.

Organization is also evident in animals at the level of the organism, in the ways cells relate to each other in tissues, organs, and multi-organ systems. Consider the structure of an individual tissue such as the retina of the vertebrate eye. In each part of the retina, individual nerve cells are highly organized in relation to each other (**FIGURE 29.1**), and this structural organization among cells is essential for vision. Another type of organization at the organismal level is the interaction of tissues and organs to form systems of organs that work together. A large set of organs—including the stomach, intestines, pancreas, and brain—must interact in an organized, coordinated way to carry out digestion, for instance (we will return to these topics in greater detail in Concept 29.4).

Recall the second law of thermodynamics, which tells us that, left to itself, any organized system tends to lose organization and become more random as time passes. Animals must combat this tendency because any loss of organization in an animal's body threatens the animal's well-being or even life itself.

LINK

You can review the laws of thermodynamics in **Concept 2.5**

For biologists, the best definition of **energy** is that it is the capacity to create or maintain organization. In essence, the reason animals need to obtain energy is so they can do work (of several types) to maintain their organization by combating the effects of the second law of thermodynamics.

To fully understand the need for energy, we need to recognize that energy exists in several forms, and that only some

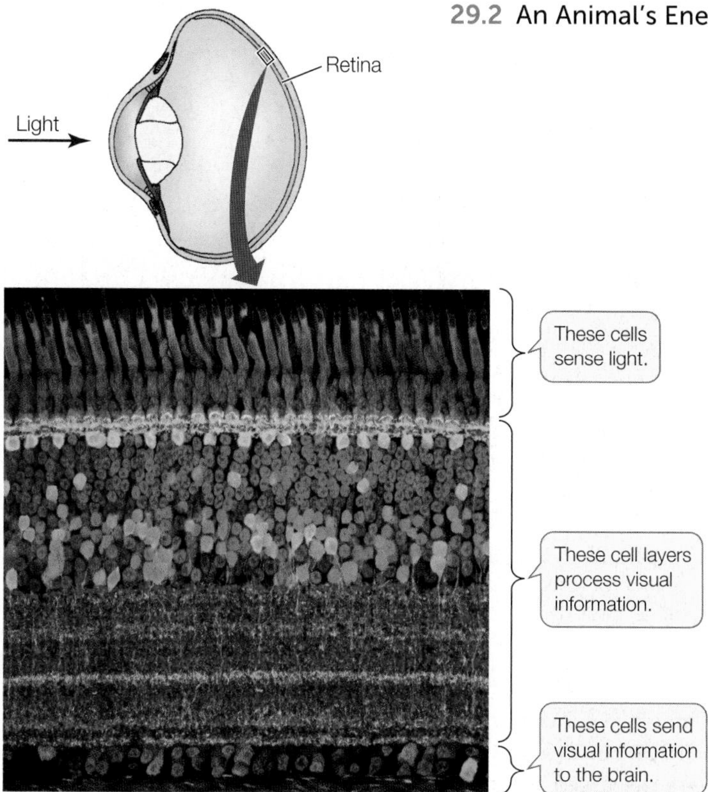
Light
Retina
These cells sense light.
These cell layers process visual information.
These cells send visual information to the brain.

FIGURE 29.1 Highly Organized Cells in the Retina of a Chick's Eye In this highly magnified image of a chick's retina, the differently colored layers are different types of cells. The cells in the top layer are the rods and cones responsible for light sensing, and the layers of cells beneath are responsible for successive steps of image processing in the retina before visual information is sent to the brain.

forms can do work. In living organisms, heat is a form of energy that is unable to do work. To do work, animals need chemical-bond energy, also known as chemical energy. Moreover, when animals use chemical energy to do work, they convert that energy to heat, which they are unable to convert back to chemical energy. These principles explain why animals need to eat to obtain new chemical energy from their foods throughout their lives. As animals use chemical energy to do work, they convert it to heat. Thus they continually need more chemical energy, which they get by ingesting organic molecules that they can break down.

CHECKpoint CONCEPT 29.1

- ✓ From what sources do autotrophs obtain chemical energy and chemical building blocks?
- ✓ Why is it necessary for animals to continue gathering chemical building blocks even after they've reached their full adult sizes?
- ✓ In what form does energy enter a heterotroph's body, and in what form does the majority of that energy inevitably leave?

Animals vary tremendously in how large their energy needs are. Among animals of any particular body size, some need 10 or even 50 times as much energy as others. Next we will discuss some of the reasons for this.

CONCEPT 29.2 An Animal's Energy Needs Depend on Physical Activity and Body Size

How much energy does an animal need? This is a complicated question because animals use energy for many different purposes. Every cell and tissue in an animal's body uses chemical energy every day to maintain its own internal organized state. Cells and tissues also use energy to do additional types of physiological work. When animals run, for example, their skeletal muscles use energy to generate mechanical forces for locomotion. An animal's heart muscle uses energy to pump blood. After a meal, the digestive tract uses energy to synthesize digestive enzymes and propel the ingested food through the gut. In this concept we look at how energy use is measured and at two major factors that determine an animal's total rate of energy use.

We quantify an animal's metabolic rate by measuring heat production or O_2 consumption

The word "consumed" is used in more than one way in studying animals. One of its most important uses is in discussing energy. When we say energy is "consumed" by an animal, we mean the energy is converted to heat. Energy cannot be destroyed. However, whereas chemical energy can do physiological work, heat cannot. For this reason, the conversion of chemical energy to heat has a huge consequence: the energy loses its ability to do work for the organism. This is why biologists say that energy converted to heat is "consumed" or used up.

An animal's **metabolic rate** is defined as its rate of energy consumption—in other words, the rate at which it consumes chemical energy and converts it to heat. Why are metabolic rates important? The main reason is that when energy is consumed, it needs to be replaced. An animal with a high metabolic rate needs to find and eat a lot of food per unit of time to replace the energy it is consuming. Animals with lower metabolic rates have lower food needs. All living animals have measurable metabolic rates and are found to produce heat when measurements are done with sensitive instruments. Even those animals that do not feel warm to the touch, such as frogs or earthworms, are producing heat. They do not warm up simply because they are losing heat as fast as they produce it.

How can we quantify an animal's metabolic rate? Because consumed energy becomes heat, one way is to measure an animal's rate of heat production. But measuring heat production directly is difficult, so biologists usually take advantage of a simple relationship. When organic matter is oxidized during aerobic metabolism, O_2 is used as heat is made (see Concept 6.2):

$$\text{Organic compound} + O_2 \rightarrow CO_2 + H_2O + \text{heat}$$

There is an approximately one-to-one relationship between the rate at which O_2 is used and the rate at which heat is produced. To quantify an animal's metabolic rate, therefore, we can measure its rate of O_2 use. Biologists nearly always use this method to quantify metabolic rates. The rate of O_2 use is usually called the rate of O_2 consumption. Note that in this case "consumption" refers to the combination of O_2 with other atoms. Thus

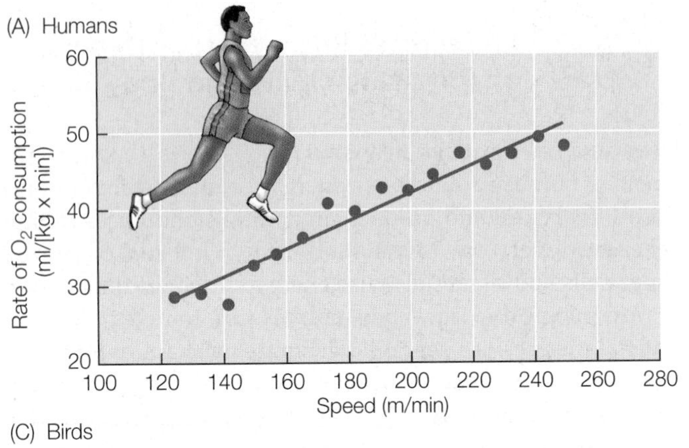

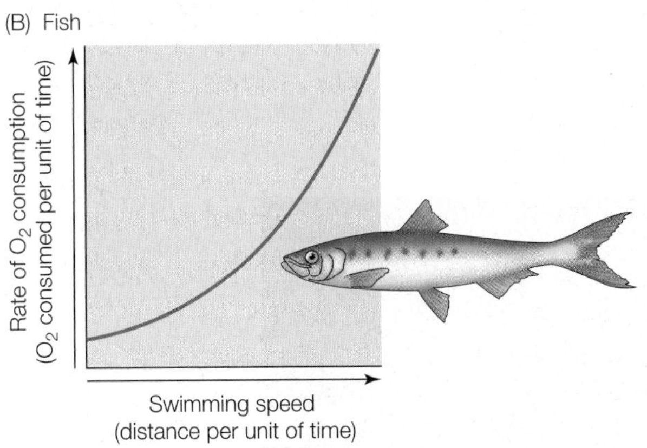

FIGURE 29.2 Cost of Exercise Rate of O_2 consumption as a function of forward speed for (A) people running, (B) fish swimming, and (C) birds flying.

"consumption" is being used in a different way than when we talk about energy consumption.

Physical activity increases an animal's metabolic rate

Physical activity often exerts a strong effect on an animal's metabolic rate. As a general rule, when people and other vertebrates engage in sustained exercise, their metabolic rate increases about tenfold between rest and all-out exertion. **FIGURE 29.2A** shows the results of a treadmill test—like the one we discussed at the chapter's start—on several people. When people run, their metabolic rate increases linearly with their speed. This is the usual relationship when animals cover distance by running or walking.

The relationship between speed and metabolic rate is different for animals that swim or fly. When fish swim, their metabolic rate increases exponentially with their speed because the resistance posed by the water increases exponentially with speed (**FIGURE 29.2B**). Birds show a U-shaped relationship between metabolic rate and speed (**FIGURE 29.2C**). At low speeds, a bird must maintain a high metabolic rate just to stay airborne. As it speeds up, its metabolic rate decreases because flight is

APPLY THE CONCEPT

An animal's energy needs depend on physical activity and body size

The graph shows the rate of energy consumption—energy cost per unit of time—of a flying parakeet as a function of its horizontal flight speed.[a] This graph can be used to quantify the energy cost per unit of distance. For assessing the efficiency of our cars, we often focus on miles per gallon. In other words, we focus on how far we can go per unit of energy used. In the graph for the bird, each dot shows the energy cost per unit of time (in kilojoules [kJ] per hour) at a particular speed (in kilometers per hour). Notice that if you divide speed in kilometers per hour by cost in kilojoules per hour, your result will be in kilometers per kilojoule:

$$\frac{\text{km}}{\text{hr}} \div \frac{\text{kJ}}{\text{hr}} = \frac{\text{km}}{\text{hr}} \times \frac{\text{hr}}{\text{kJ}} = \frac{\text{km}}{\text{kJ}}$$

In other words, your result will tell you how many kilometers a parakeet can travel for each kilojoule of energy it uses. This is a very similar concept to miles per gallon!

1. Calculate the value of kilometers per kilojoule for each of the eight dots on the graph.
2. Based on your answer to question 1, what is the best speed for the bird to fly at to cover as much distance as possible per unit of energy used?

[a]Tucker, V. A. 1968. *J. Exp. Biol.* 48: 67–87.

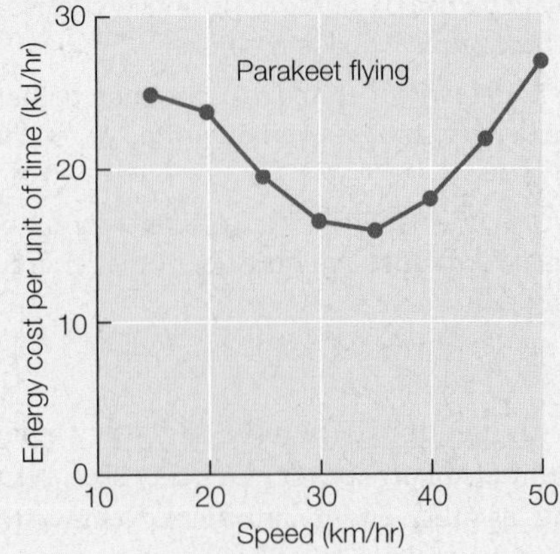

aided by the rapid flow of air over the wings. As the bird flies faster, however, its metabolic rate rises again, because flying at a high speed requires more effort than flying at a medium speed. Airplanes exhibit a similar type of relationship between their rate of energy consumption and speed.

Among related animals, metabolic rate usually varies in a regular way with body size

Species of terrestrial mammals vary in adult body size from about 20 grams in some mice (or even as small as 2 g in some shrews) to more than 5,000 kilograms (5,000,000 g) in elephants. When we look at related animals of widely different body sizes, do their metabolic rates vary in any particular way with their sizes?

To answer this question, we need to measure the metabolic rates of various species under standardized conditions. To standardize the conditions when studying mammals, investigators typically measure the metabolic rates of animals that are in a comfortable thermal environment and resting—and that have not eaten in the recent past. These measurements are called **basal metabolic rates** or **BMRs**.

The simplest way to compare the BMRs of mammals of different sizes is to express the rates per unit of body weight. To do this, we take the BMR for each animal and divide it by the animal's body weight to obtain BMR per gram, or BMR/g.

You might expect that BMR/g would simply be the same in all terrestrial mammals, from mice to elephants, but it isn't. If you start with mice and look at bigger and bigger species of mammals, BMR/g gets lower and lower. The "mouse-to-elephant" curve shows this (**FIGURE 29.3**). BMR/g is almost 20 times higher in a mouse than in an elephant. Relationships of this sort—in which animal characteristics are examined as functions of body size—are called **scaling relationships**.

What are the implications of the scaling relationship between BMR/g and body size in mammals? One thing we can deduce is that small mammals need much more food per gram of body weight than large mammals do—a fact that pet owners and zoo keepers need to keep in mind. Food requirements in relation to body size are also ecologically important. Suppose there are 100 mice and 1 rabbit living in a field or forest. If each mouse weighs 20 g and the rabbit weighs 2,000 g, the mice collectively weigh as much as the rabbit—2,000 g. You might therefore expect that both the mice and the rabbit would eat similar amounts of food each day. However, the mice will in fact eat about three times as much food as the rabbit, because the BMR/g is three times higher in a small mouse than in a medium-size rabbit.

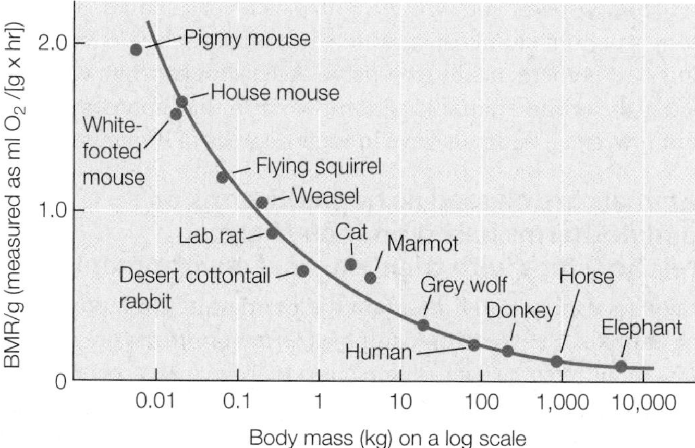

FIGURE 29.3 The Mouse-to-Elephant Curve in Basal Metabolic Rate Basal metabolic rate (measured as the rate of O_2 consumption) per gram of body weight in terrestrial mammals decreases dramatically as body size increases. The *x* axis is a logarithmic plot of body weight.

The same type of scaling relationship between metabolic rate per gram and body size (i.e., the same shape as seen in the mouse-to-elephant curve) is seen in almost all types of animals, including crabs, fish, lizards, and birds. Within each group of related species, metabolic rate per gram tends to decrease as body size increases.

CHECKpoint CONCEPT 29.2

- ✓ Define "consumption" as it relates to energy in biological systems.
- ✓ Which is likely to require more food over a given time period, a group of 60 skates, each weighing 1 kg, or a pair of 30-kg sharks? Skates and sharks both belong to the same group of animals, the cartilaginous fish.
- ✓ Biologists sometimes quantify metabolic rates by measuring heat production despite the technical challenges of using this method. Why can an animal's metabolic rate be quantified by measuring its rate of heat production?

Physical activity and body size are only two of the factors that influence metabolic rates. Now we will look at some of the others.

CONCEPT 29.3 Metabolic Rates Are Affected by Homeostasis and by Regulation and Conformity

The cells inside an animal's body are bathed with body fluids called **tissue fluids** or, more formally, **interstitial fluids**. Suppose we ask, what is the environment of cells inside the body? The tissue fluids, which exist around and between cells, are that environment. For this reason, an animal's tissue fluids are sometimes called the animal's **internal environment**, whereas the outside world is the animal's **external environment**.

To see this distinction more clearly, suppose you are standing outside on a cold, dry day. Your external environment will have a low temperature and little moisture. However, inside your body, your body fluids will be warm and wet. The environment of your cells—your body's internal environment—will therefore be warm and wet.

Animals are classed as regulators and conformers

What happens inside an animal when its external environment changes? A discussion of temperature provides an excellent starting point for understanding the answer.

Some types of animals, such as mammals and birds, maintain a constant internal temperature, much in the way that a house with a furnace and an air conditioner maintains a steady inside temperature. Humans, for example, maintain a constant internal temperature of about 37°C (98.6°F). In these types of animals, if the temperature of the external environment changes, the internal temperature does not change. Such animals are called **regulators** and are said to exhibit regulation. By definition, **regulation** occurs when the internal environment stays constant even as the external environment changes (**FIGURE 29.4A**).

In other types of animals, such as fish and frogs, if the temperature of the external environment changes, the internal temperature changes to match the new external temperature. These animals are called **conformers** and are said to exhibit conformity. By definition, **conformity** occurs when the internal environment varies so that it always matches the external environment (**FIGURE 29.4B**).

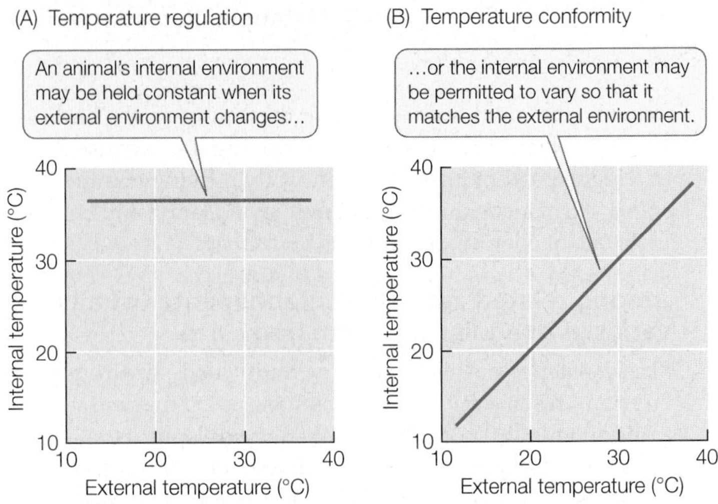

FIGURE 29.4 Regulation versus Conformity These graphs from the study of temperature illustrate the general principles of (A) regulation and (B) conformity.

The concepts of regulation and conformity can be applied to all characteristics of an animal's internal and external environments, not just temperature. Moreover, an animal that conforms in some characteristics may regulate in others. We've seen, for example, that fish are conformers in regard to temperature. They are regulators, however, in terms of the concentration of sodium (Na^+) ions, symbolized [Na^+], in their tissue fluids. If the sodium concentration of the external water changes, a fish keeps a constant [Na^+] in its tissue fluids, even though it lets its internal temperature change when the external temperature changes.

Regulation is more expensive than conformity

To think further about regulation and conformity, let's contrast a modern house with a simple log cabin without a stove or furnace. In terms of temperature, when we are inside the modern house, we are always comfortable. This happy state of affairs changes if we move to the cabin. In hot weather, we sometimes boil. In cold weather, we sometimes freeze. We also realize, however, that the cabin is not entirely bad. We have no fuel costs for the cabin, whereas we pay steep utility bills for the house. This contrast helps us understand the advantages and disadvantages of regulation and conformity in animals.

Regulation (such as we have in the modern house) provides stability but is energetically expensive. If an animal maintains a constant internal environment, its body cells enjoy constancy in their surroundings. This is the advantage of regulation. However, the animal must expend energy to maintain a constant internal environment when the outside environment varies. Regulation tends to increase an animal's metabolic rate. This is the disadvantage of regulation.

Conformity (such as we experience in the log cabin) does not provide stability but is energetically cheap. If an animal simply lets its internal environment change to match its external environment, the animal need invest no energy in the process. However, its cells must cope with considerable variation in their immediate surroundings.

Homeostasis is a key organizing concept

As the nineteenth century ended and the twentieth century began, physiologists were studying humans and "model" nonhuman animals such as dogs used in biomedical research. These investigators gradually realized for the first time that when people are healthy, their blood and tissue fluids are remarkably consistent in their characteristics from time to time. The investigators often focused their studies on the **blood plasma**, which is the blood solution in which the red blood cells and other blood cells are suspended. The investigators discovered that the concentration of glucose ("blood sugar") in a healthy person's blood plasma and tissue fluids is almost constant, whether a meal has just been eaten or not. They likewise discovered that temperature, pH, and the concentrations of major ions are approximately constant in the blood plasma and tissue fluids.

A famous early-twentieth-century physiologist, Walter Cannon, coined the term **homeostasis** to refer to this stability of the internal environment and the mechanisms that maintain the stability. To this day, homeostasis is an exceedingly important organizing concept in medicine, veterinary medicine, and many other fields of biology. The meaning of homeostasis has shifted a little over time; the term is sometimes used today to refer to any sort of stability or constancy in an animal.

Homeostasis, you will recognize, is a very similar concept to regulation. Homeostasis offers advantages, but as you can guess, it is energetically expensive. Accordingly, when we consider the entire animal kingdom, we find that homeostasis is not universal. Animals vary in their degrees of homeostasis.

Animals are classed as homeotherms or poikilotherms based on their thermal relationships with their external environment

Let's focus in greater detail on the temperature relationships of animals. These relationships have major effects on animal metabolic rates, as well as being important in many other ways.

Animals that maintain an approximately constant internal body temperature—such as mammals and birds—are said to exhibit **thermoregulation** and are called **homeotherms**. In most homeotherms, if a resting individual is exposed to a variety of external temperatures, its body temperature remains constant but its metabolic rate varies with external temperature as seen in **FIGURE 29.5**. Over a range of relatively mild external temperatures, called the **thermoneutral zone** (**TNZ**), the animal's metabolic rate is constant.

At temperatures below the TNZ, a homeotherm's metabolic rate increases as the external temperature falls. This happens for the same reason that the furnace in a house burns fuel faster as the weather outdoors becomes colder. A homeotherm must progressively increase its rate of metabolic heat production as the external temperature becomes progressively colder, in order to offset its progressively increasing rate of heat loss.

At temperatures above the TNZ, a homeotherm's metabolic rate increases as the external temperature rises. This happens for the same reason that the air conditioner in a house works harder as the weather outdoors becomes hotter. As the animal's external temperature becomes progressively warmer, the animal must work progressively harder to keep its body temperature from rising. Its metabolic rate reflects this harder work by rising.

Most animals are not homeotherms, and their relationship between metabolic rate and temperature is different than that of homeotherms. Fish, frogs, sea stars, crayfish, and many other kinds of animals permit their body temperature to match their external temperature. Such animals are called **poikilotherms** or **ectotherms**. These two names reflect two complementary characteristics of these animals: The animals have body temperatures that are variable and that are determined by the external temperature. The prefix *poikilo* means "variable," and *ecto* means "external." The name poikilotherm emphasizes the variability of body temperature in these animals, and the name ectotherm emphasizes the fact that their external temperature determines their body temperature. In this book we favor the term poikilotherm.

An important question about poikilotherms is how high their body temperature can rise—and how low it can fall. Poikilotherms vary widely in their body temperature limits:

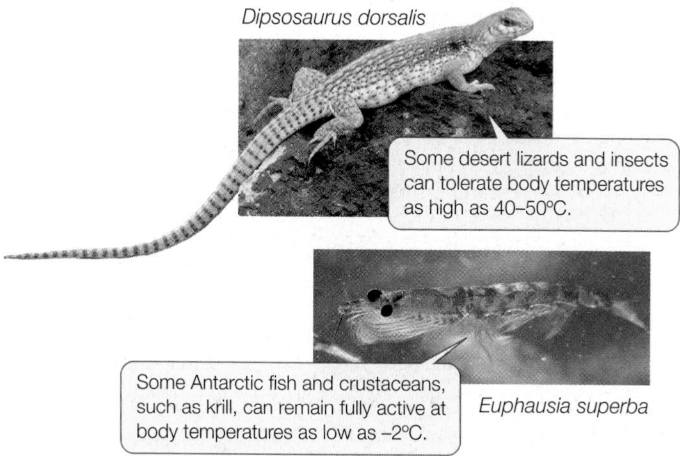

When the external temperature of a poikilotherm rises or falls, the animal's metabolic rate varies approximately exponentially (**FIGURE 29.6A**). The reason for this is that the animal's tissues become cooler or warmer as the external temperature varies. The tissues are cold in a cold environment and warm in a warm environment. Low tissue temperatures slow both biochemical processes (such as enzyme-catalyzed reactions; see Figure 3.22)

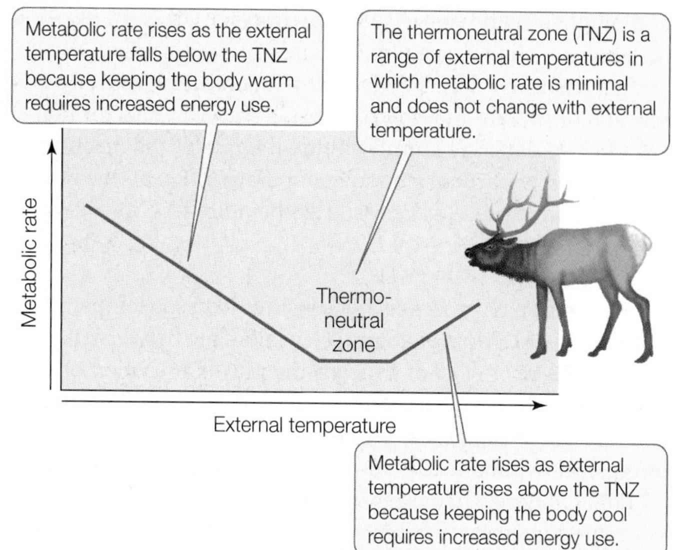

Metabolic rate rises as the external temperature falls below the TNZ because keeping the body warm requires increased energy use.
The thermoneutral zone (TNZ) is a range of external temperatures in which metabolic rate is minimal and does not change with external temperature.
Metabolic rate
Thermo-
neutral
zone
External temperature
Metabolic rate rises as external temperature rises above the TNZ because keeping the body cool requires increased energy use.

FIGURE 29.5 Homeothermy The metabolism–temperature relationship in mammals and birds.

Go to ACTIVITY 29.1 Thermoregulation
PoL2e.com/ac29.1

(A)

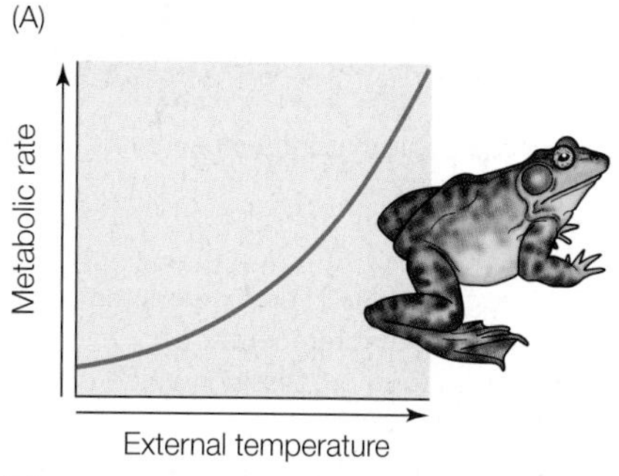

(B)

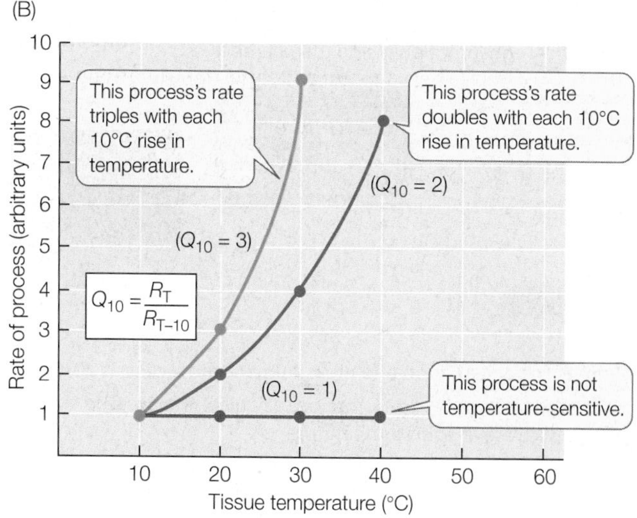

FIGURE 29.6 Poikilothermy and Q_{10} (A) The metabolism–temperature relationship in poikilothermic animals. Internal temperature increases as external temperature increases, and this speeds both biochemical and biophysical processes in the animal's body. (B) Q_{10} is a way to calculate the sensitivity of a rate, such as metabolic rate, to changes in tissue temperature.

and biophysical processes (such as diffusion), thereby slowing metabolism. High tissue temperatures do the opposite.

In a tissue, we can describe the temperature sensitivity of a reaction or process in numerical terms with Q_{10}, a factor that is easy to calculate. We simply divide the rate of the reaction or process at any particular tissue temperature, R_T, by the rate of that reaction or process at a tissue temperature 10°C lower, R_{T-10}:

$$Q_{10} = R_T/R_{T-10}$$

If a reaction or process is not sensitive to tissue temperature, it has a Q_{10} of 1. Most biological Q_{10} values are between 2 and 3 (**FIGURE 29.6B**). A Q_{10} of 2 means the rate of a reaction or process doubles as tissue temperature increases by 10°C; a Q_{10} of 3 means the rate triples. **FIGURE 29.7** uses the Q_{10} concept to help interpret how mud turtles respond to changes in temperature.

Poikilotherms often exert some control over their tissue temperatures by means of behavior. Because a poikilotherm's body temperature is determined by external temperature, the animal can adjust its body temperature by positioning itself in suitable environments (**FIGURE 29.8**). For example, if a poikilotherm's body temperature is too low, the animal can raise it by moving to a warmer place in its environment (if a warmer place is available).

Homeothermy is far more costly than poikilothermy

FIGURE 29.9 shows the metabolic-rate-versus-external-temperature relationship for a mammal or bird (homeotherm) on the same plot as that of a fish or lizard (poikilotherm) of the same body size. Notice that at each external temperature, the metabolic rate of the homeotherm is far higher than that of the poikilotherm. Notice especially, toward the left side of the graph, that in cold environments, as the external temperature falls, the metabolic rates of the animals become more different. With a falling external temperature, the homeotherm must make heat

INVESTIGATION

FIGURE 29.7 The Metabolic Rate of a Poikilotherm Slows Down as Temperature Decreases Jacqueline Litzgus and William Hopkins measured the standard metabolic rate of the mud turtle *Kinosternon subrubrum* at varying body temperatures.[a] They showed that temperature has a significant effect on standard metabolic rate in this species. The standard metabolic rate of a poikilotherm is its metabolic rate when resting and fasting. It can be measured at any body temperature of interest. (Poikilotherms do not have thermoneutral zones, explaining why the BMR concept does not apply to them.)

HYPOTHESIS

A mud turtle's standard rate of metabolism is influenced by its body temperature.

METHOD

1. Acclimate a group of wild-caught turtles to an external temperature of 30°C, by letting them live at that temperature for a long time. Deprive them of food so no food processing (which can increase metabolic rate) will occur during the experiment.
2. Place each turtle in a sealed chamber connected to a device that measures O_2 consumption for 24 hours. Periodically measure the rate of O_2 consumption inside the chamber.

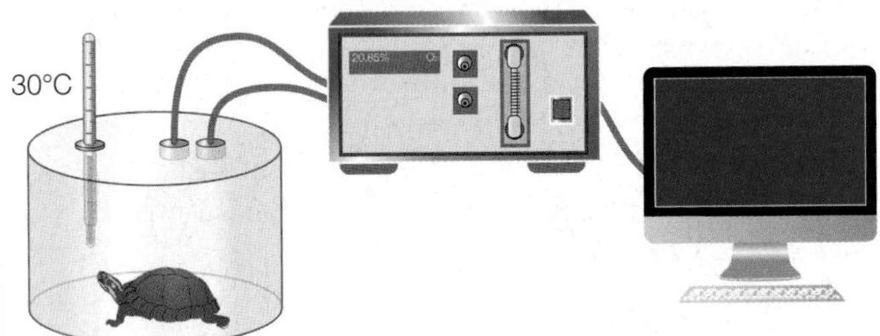

3. Repeat the acclimation and measurement processes at 20°C.
4. From the rates of O_2 consumption, calculate the average standard metabolic rate at each temperature.

RESULTS

The average standard metabolic rate at 20°C was 2.10 ml O_2 consumed/hour. The average standard metabolic rate at 30°C was 9.25 ml O_2 consumed/hour.

CONCLUSION

In turtles, which are poikilotherms, a lower external temperature results in a lower standard metabolic rate.

ANALYZE THE DATA

A. Calculate the value of Q_{10} for the mud turtle's standard metabolic rate. Based on the Q_{10} that you calculate, what would you expect the standard metabolic rate (expressed as rate of O_2 consumption) to be at 10°C?

B. The graph shows data for a single turtle recorded at 30°C. In calculating the turtle's standard metabolic rate, the researchers chose to use only the lowest 30% of O_2 consumption rates recorded. Why might the researchers have made this decision?

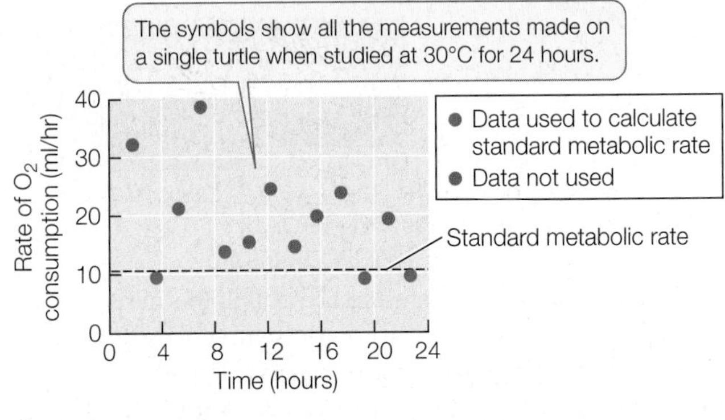

Go to **LaunchPad** for discussion and relevant links for all **INVESTIGATION** figures.

[a]J. D. Litzgus and W. A. Hopkins. 2003. *Journal of Thermal Biology* 28: 595–600.

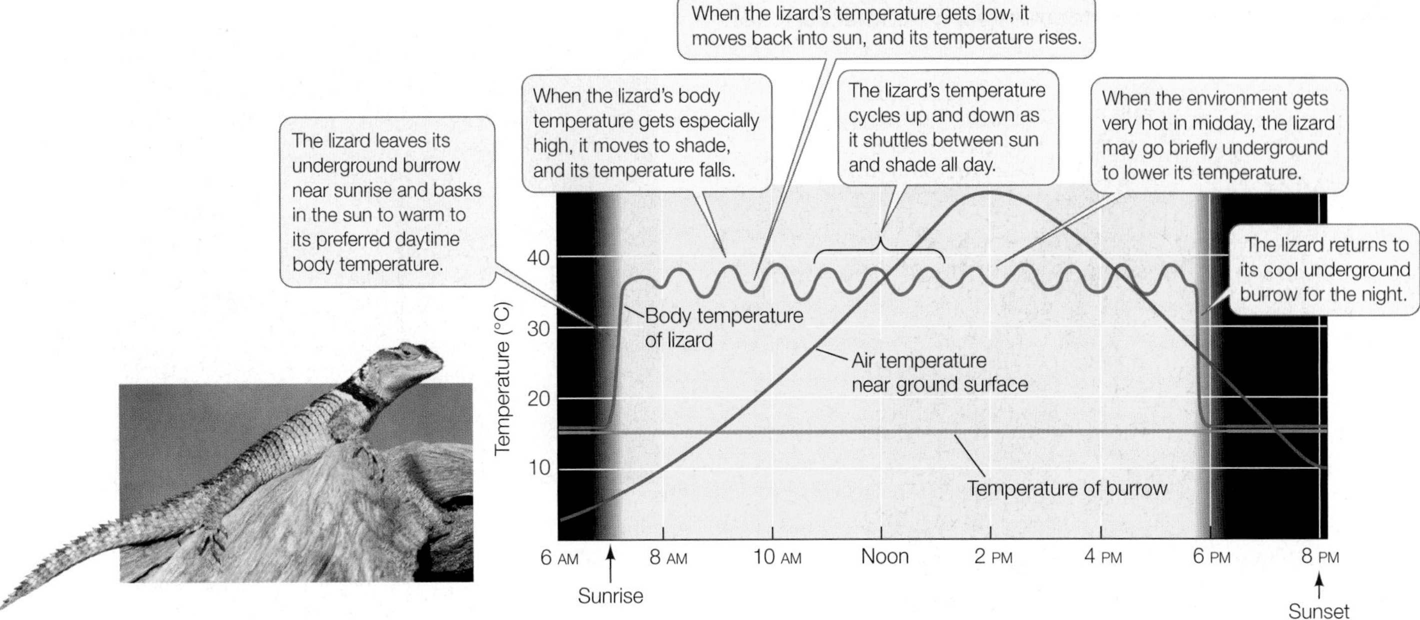
When the lizard's temperature gets low, it moves back into sun, and its temperature rises.
When the lizard's body temperature gets especially high, it moves to shade, and its temperature falls.
The lizard's temperature cycles up and down as it shuttles between sun and shade all day.
When the environment gets very hot in midday, the lizard may go briefly underground to lower its temperature.
The lizard leaves its underground burrow near sunrise and basks in the sun to warm to its preferred daytime body temperature.
The lizard returns to its cool underground burrow for the night.
Temperature (°C)
40
30
20
10
Body temperature of lizard
Air temperature near ground surface
Temperature of burrow
6 AM
8 AM
10 AM
Noon
2 PM
4 PM
6 PM
8 PM
Sunrise
Sunset

FIGURE 29.8 Poikilotherms Often Use Behavior to Regulate Body Temperature Behavioral regulation of body temperature is particularly well developed in lizards. The body temperature of a lizard depends on the temperature of its external environment, but the lizard can regulate its body temperature by moving from place to place within its environment. The behavior of a desert species, the spiny lizard *Sceloporus serrifer*, is illustrated here.

faster and faster to keep its tissues warm, whereas the poikilotherm's tissues become colder, slowing its metabolism.

Over the course of an entire year, the average metabolic rates of mammals and birds living in their natural environments are typically 15–30 times higher than the average metabolic rates of fish and lizards in their natural environments (comparing animals of equal body size). This contrast dramatizes the advantages and disadvantages of homeothermy. The cells inside the body of a homeotherm function at approximately one temperature whether the outside environment is cold or warm. The price of this regulation of body temperature is a 15- to 30-fold increase in the animal's food needs!

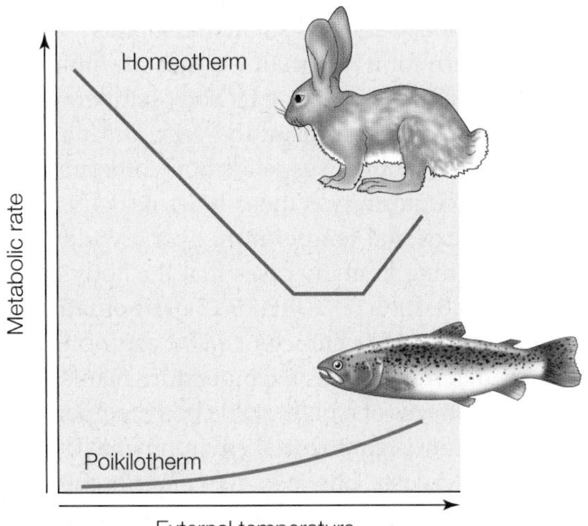
Homeotherm
Metabolic rate
Poikilotherm
External temperature

FIGURE 29.9 Metabolic Rate in a Vertebrate Homeotherm and a Vertebrate Poikilotherm of the Same Body Size This graph shows the typical relationships between metabolic rate and temperature for a homeotherm and a poikilotherm. Notice that at every temperature, the metabolic rate of the homeotherm is far higher than that of the poikilotherm.

LINK

Maintenance of homeostasis can affect life-history tradeoffs—the amount of resources an animal can allocate to reproduction, growth, and defense; see **Concept 42.3**

Homeotherms have evolved thermoregulatory mechanisms

Mammals, birds, and insects are the major groups of homeothermic animals. Placental mammals have core body temperatures of about 37°C whether they live in cold or warm climates. Birds have higher body temperatures, about 39°C. Among insects, only some exhibit homeothermy. Moreover, the insects that display homeothermy do so just at certain times, such as during flight. Many moths, butterflies, bees, and dragonflies maintain relatively constant, high tissue temperatures in their thorax—the location of their flight muscles—when they are flying. Crickets and katydids are similar when singing. Honeybee hives also display homeothermy. Collectively, the workers in a hive keep the brood (their eggs and larvae) at about 32–36°C even when the outside air temperature is far lower than freezing or as hot as 50°C.

Go to MEDIA CLIP 29.1
Thermoregulation in Animals
PoL2e.com/mc29.1

All homeothermic animals have mechanisms for increasing their metabolic rate—therefore increasing their heat production—when their external environment is cold. In some of these mechanisms, ATP (see Concept 6.1) is the source of energy to make heat. ATP is an energy-shuttle compound in cells. It stores energy when food molecules are broken down. It then can itself be broken down, releasing that energy.

Both mammals and birds **shiver** to increase heat production. During shivering, skeletal muscles contract in such a way

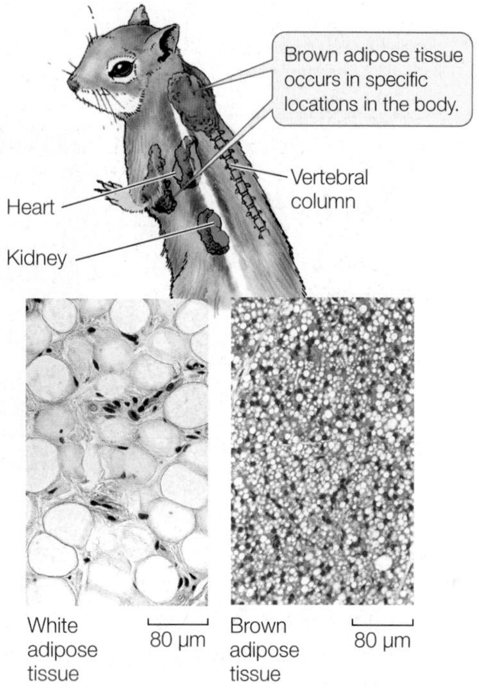

FIGURE 29.10 Brown Adipose Tissue In many mammals, specialized brown adipose tissue produces heat. Cells of white adipose tissue (left) contain large droplets of lipid but have a limited blood supply. Cells of brown adipose tissue (right) have many tiny lipid droplets, are packed with mitochondria, and are richly supplied with blood.

that they produce only subtle, quivering motions. The energy for these contractions comes from the chemical bonds of ATP. As molecules of ATP are used for contraction by the quivering muscles, chemical energy from ATP is converted to heat at a heightened rate.

Many mammals, but not birds, are also capable of **nonshivering thermogenesis**, in which thermogenesis (heat production) occurs without shivering. This process takes place in a specialized type of fatty tissue called **brown adipose tissue** (**BAT**) (**FIGURE 29.10**). The mechanism used by BAT to make heat is uncoupling of oxidative phosphorylation. In this process, no ATP is made. Instead, heat is made directly in the mitochondria because the proton gradient in the mitochondria is short-circuited.

LINK

Uncoupling during oxidative phosphorylation is discussed in **Concept 6.2**

When insects maintain a warm thorax during preparation for flight or during flight or singing, the flight muscles in their thorax provide the heat needed (**FIGURE 29.11**). In this case, the heat comes from ATP.

Besides heat production, body insulation is another important characteristic that allows homeotherms to maintain high tissue temperatures in cold environments. Fur and feathers provide high insulation. Heat is also often retained by specialized blood flow patterns (see Concept 32.5).

In hot environments, homeotherms often need to increase their rate of heat loss to keep from overheating. The most common way of doing this is to increase the rate of evaporation of water. When water turns from a liquid to a gas (water vapor), it absorbs a great deal of heat per gram (see Concept 2.2). This heat is removed from the animal when the vapor travels away in the atmosphere. Many large mammals—such as humans, horses, and camels—sweat to increase their rate of evaporation. Dogs pant. Honey bees in a hive spread water that they have collected on interior surfaces of the hive and then fan the moistened surfaces with their wings to speed evaporation.

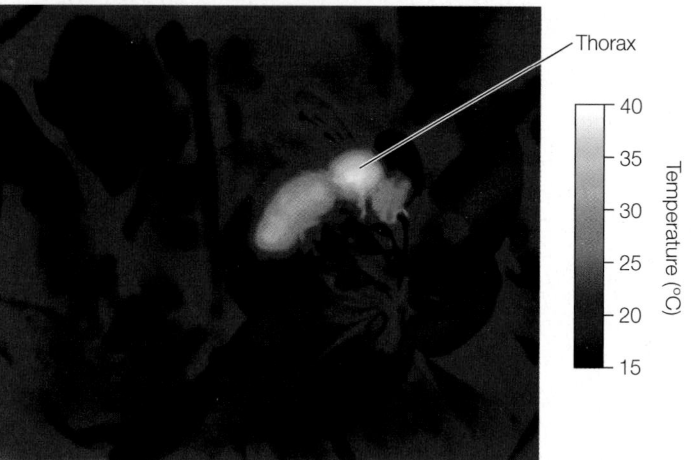

FIGURE 29.11 Thermoregulation in a Honey Bee This worker bee is foraging on a flower on a chilly day. During such behavior, the thorax is kept at a high, relatively stable temperature by heat produced by the flight muscles. The temperatures of the external surfaces of the bee, flower, and other plant parts are color-coded in this image, which was made by visualizing the surfaces at infrared wavelengths.

Hibernation allows mammals to reap the benefits of both regulation and conformity

Mammalian hibernation is remarkable in many ways—one being that hibernators are able to reap the benefits of both regulation and conformity. Most hibernating species are relatively small. They include certain mice, hamsters, ground squirrels, groundhogs, marmots, bats, and small marsupials.

The usual pattern for a hibernator is that the animal functions as a homeotherm—regulating its body temperature near 37°C—during the warm months of the year. During the cold months, however, the animal suspends homeothermy and exhibits temperature conformity. It then allows its body temperature to match the external temperature over a wide range of temperatures, meaning in many cases that the body temperature falls as low as 0–10°C (**FIGURE 29.12**). **Hibernation** is the term for this winter period characterized by conformity.

When an animal functions as a temperature regulator, it experiences the advantages of a high, stable body temperature: its cells function in a consistent thermal environment. The animal faces high metabolic costs, however, like other homeotherms. When the animal functions as a temperature conformer, it experiences the advantages of conformity, namely that its metabolic costs are very low. Hibernators essentially switch between two metabolism–temperature curves resembling those in Figure 29.9.

By reducing energy needs, hibernation allows an animal to spend the winter continuously in a relatively protective and safe place, such as an underground burrow. The amount of

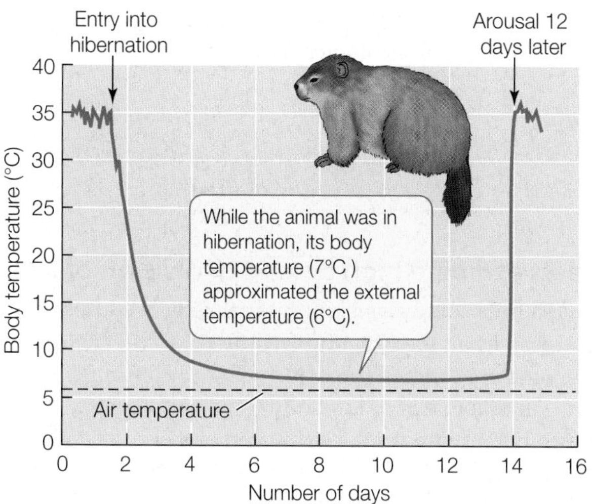

FIGURE 29.12 A Bout of Hibernation in a Groundhog This groundhog or woodchuck (*Marmota monax*) was living where the external temperature was 6°C. Its body temperature was over 35°C before and after the bout of hibernation, but during the 12 days of hibernation its body temperature was very close to the external temperature. Bouts of hibernation like the one shown here are interrupted by brief periods when the body temperature is high, but a hibernator stays continuously in its burrow throughout these events during the hibernating season.

metabolic fuel the animal requires is reduced to the point that the animal need not roam around its habitat in search of food. Instead, it can live entirely on body fat or on stored food it collected the previous summer.

CHECKpoint CONCEPT 29.3

- ✓ How would the pH of the interstitial fluids of a healthy, water-dwelling regulator be affected by a moderate increase in the pH of the external environment?
- ✓ Would a reaction that proceeds faster as temperature increases have a Q_{10} value higher or lower than 1?
- ✓ What is one advantage and one disadvantage of the conforming lifestyle?
- ✓ Does the metabolic rate of a homeotherm increase or decrease as external temperature decreases? Assume the temperature is below the homeotherm's thermoneutral zone. Would the answer differ for a poikilotherm?

Now that we've discussed how animals use energy, let's consider their body organization and the ways they make ATP.

CONCEPT 29.4 Animals Exhibit Division of Labor, but Each Cell Must Make Its Own ATP

Throughout the living world, structure and physiology—form and function—are intimately related. Often, form cannot be understood without understanding function, and vice versa.

Looking at the structure of animals, perhaps the first point worth noting is that animals typically consist mostly of water. Adult mammals (including us), for example, tend to be about 60 percent water. That water is said to be distributed in various body "compartments," which by definition are location categories rather than specific organs. Typically, most of the fluid in an animal's body is located within its cells. This fluid is called **intracellular fluid**, and the water in all the intracellular fluid throughout the body is said to be in the intracellular compartment. The rest of the fluid in the body is the **extracellular fluid**, said to be in the extracellular compartment. There are two subcategories of extracellular fluid, which we have already mentioned: the fluid portion of blood—termed the blood plasma—and the extracellular fluid found between cells in tissues throughout the body—termed the tissue fluid or, more formally, interstitial fluid.

Fluid compartments are separated from one another by physiologically active epithelia and cell membranes

How are the various fluid compartments kept separate in the body? There are two different scales of space we need to consider to answer this question, and they correspond to separation by epithelia and by cell membranes.

EPITHELIA Consider the following three examples in which fluids of two types lie close together:

- Blood plasma inside a blood capillary and tissue fluid surrounding the outside of the capillary
- Watery fluid inside the lumen (central cavity) of a vertebrate's intestines and tissue fluid in the intestinal wall
- Pond water on the outside of a submerged frog's skin and tissue fluid in the tissue of the frog's skin

In all three cases, the two fluids are separated by an **epithelium** (plural **epithelia**), which is a sheet of cells (epithelial cells) that covers a body surface or organ or lines a body cavity. The simplest and most common type of epithelium, called a **simple epithelium**, consists of just a single layer of cells that rests on a nonliving, highly permeable basement membrane (also called basal lamina; not to be confused with a cell membrane) that the cells help synthesize (**FIGURE 29.13**). Simple epithelia are exceedingly common. In our bodies, for example, a simple epithelium lines all our blood vessels and our intestines, as in the cases above. A simple epithelium also lines our kidney tubules, mammary glands, and sweat glands. In a frog, the skin consists entirely of living cells (unlike our skin, which has dead cells on the outside), and the outermost part of the skin is an epithelium.

Besides helping compartmentalize the body by separating fluids, an epithelium typically has numerous functional capabilities and plays major physiological roles. The cells of an epithelium, for example, commonly pump ions between the fluids on either side of the epithelium. Some epithelia secrete substances such as hormones, mucus, digestive enzymes, milk, or sweat. Some absorb nutrients from the gut. Other epithelia serve sensory functions, including smell and taste.

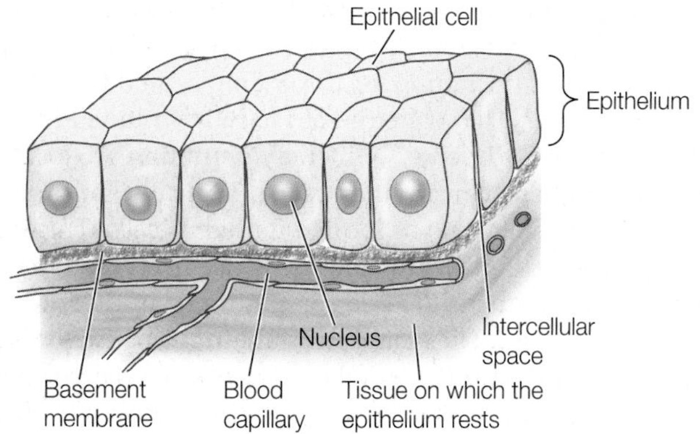

FIGURE 29.13 A Simple Epithelium A simple epithelium consists of a single layer of cells. The cells rest on a basement membrane that is nonliving but highly permeable to gases, water, and ions. Blood capillaries do not penetrate into the epithelium itself but instead are under the basement membrane.

CELL MEMBRANES At a much smaller scale, separation of fluid compartments is by cell membranes. Each cell has intracellular fluid inside separated from the extracellular fluid bathing the cell on the outside. The cell membrane separates the two. Cell membranes, like epithelia, have numerous functional capabilities and play major physiological roles. Cell membranes pump ions between the intracellular fluid and extracellular fluid. They often also play key roles in producing and receiving physiological signals.

LINK

You can review the properties of cell membranes and their transport and signaling functions in **Chapter 5**

Animals exhibit a high degree of division of labor

In multicellular organisms, individual cells need not perform for themselves all the functions required for life. Instead, cells can become specialized to perform just certain functions and depend on other cells in the body to perform others. Animals display a high degree of such division of labor.

For understanding the structure and function of an animal's body, we usually recognize cells, tissues, organs, and multi-organ systems as ever-more-complex levels in the hierarchy of organization of the animal body. A **tissue** is an assemblage of cells of similar type. An **organ** consists of two or more types of tissue with a defined structural relationship to each other. A **multi-organ system** consists of multiple organs working together.

As an example, **FIGURE 29.14** shows parts of the human digestive tract, including the detailed structure of the beginning of the small intestine. The small intestine itself is an organ. It consists of several tissue types: muscle layers, nervous tissue, connective tissue, an epithelium lining the lumen (central

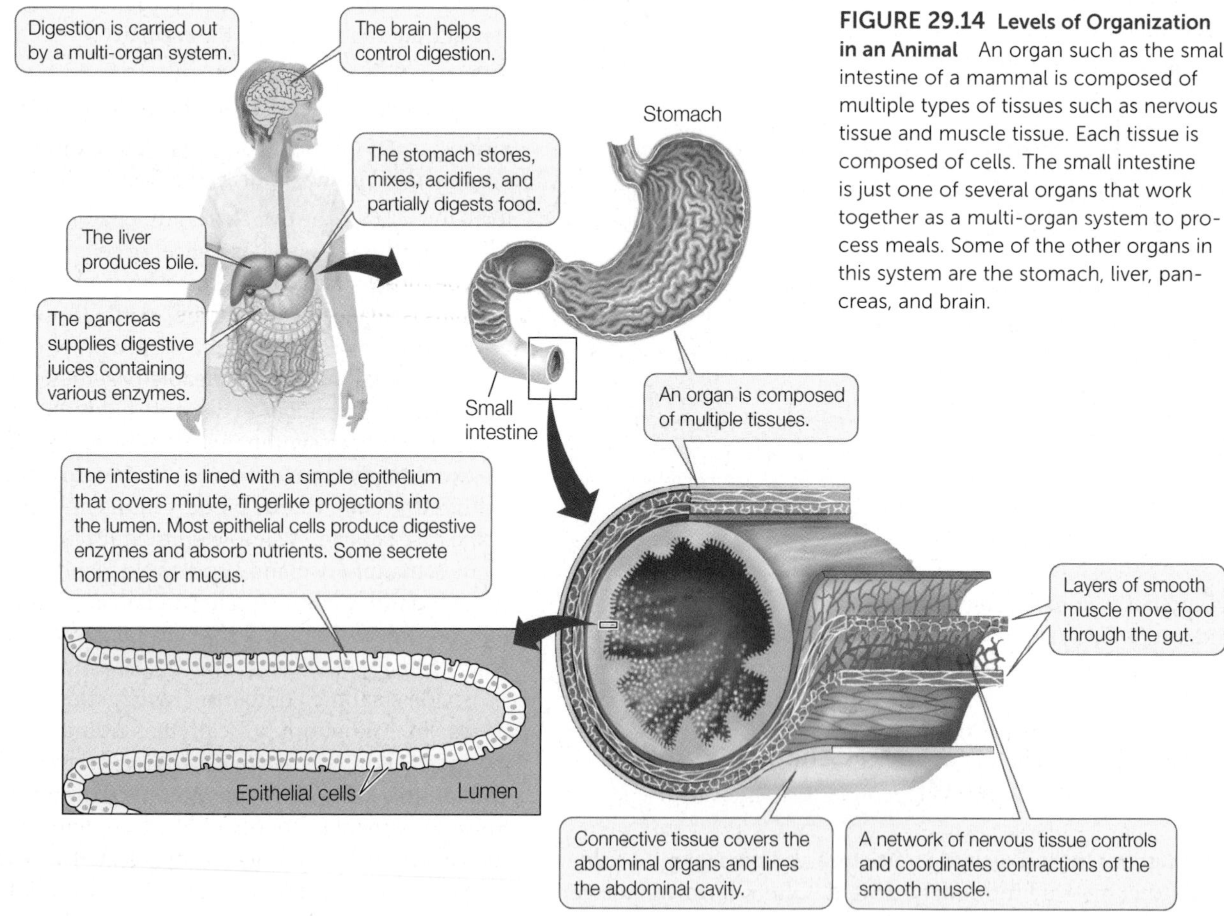

Digestion is carried out by a multi-organ system.
The brain helps control digestion.
The stomach stores, mixes, acidifies, and partially digests food.
Stomach
The liver produces bile.
The pancreas supplies digestive juices containing various enzymes.
Small intestine
An organ is composed of multiple tissues.
The intestine is lined with a simple epithelium that covers minute, fingerlike projections into the lumen. Most epithelial cells produce digestive enzymes and absorb nutrients. Some secrete hormones or mucus.
Layers of smooth muscle move food through the gut.
Epithelial cells
Lumen
Connective tissue covers the abdominal organs and lines the abdominal cavity.
A network of nervous tissue controls and coordinates contractions of the smooth muscle.

FIGURE 29.14 Levels of Organization in an Animal An organ such as the small intestine of a mammal is composed of multiple types of tissues such as nervous tissue and muscle tissue. Each tissue is composed of cells. The small intestine is just one of several organs that work together as a multi-organ system to process meals. Some of the other organs in this system are the stomach, liver, pancreas, and brain.

cavity), and so forth. Each of these tissue types consists principally of a single, specialized type of cell, such as smooth muscle cells in the muscle layers and nerve cells in the nervous tissue. During its normal function, the small intestine receives partially digested food from the stomach, digestive juices from the pancreas, and an emulsifying agent called bile from the liver, and it is controlled in part by the brain. The intestine, stomach, pancreas, liver, and brain are all organs, and they work together as a multi-organ system to process a meal.

Each organ in an animal's body is specialized to perform certain functions and is utterly dependent on other organs to perform other vital functions. The small intestine of a mammal, for example, requires O_2 for its survival and thus has a life-and-death dependency on the lungs, which take up O_2 from the atmosphere. The intestine also cannot carry out reproduction. Instead, an animal depends on its ovaries or testes to produce gametes. In turn, the lungs, ovaries, and testes depend on the intestine to supply them with nutrients.

Go to ACTIVITY 29.2 Tissues and Cell Types
PoL2e.com/ac29.2

Division of labor requires a rapid transport system

As a consequence of division of labor, organs often require inputs of materials from other organs, as exemplified by the small intestine needing O_2 from the lungs. Organs also need materials they make to be delivered to other organs. For example, wastes need to go to the kidneys and hormones need to be delivered to their targets.

Because of these needs, a rapid transport system is typically required for division of labor to be successful. Most types of animals have a circulatory system that serves this role. In many cases, the need for O_2 transport is the most pressing of all transport needs. For this reason, the sophistication of the circulatory system is usually correlated with metabolic rate (see Concept 32.1). Animals with high metabolic rates require high rates of O_2 transport and have evolved circulatory systems with exceptional abilities for rapid O_2 transport.

LINK

The diverse circulatory systems of animals are compared in Concepts 32.1 and 32.2

Each cell must make its own ATP

One of the most important principles in the study of animals is the fact that each cell must make its own ATP. ATP serves as the immediate energy source for most cellular functions, but it is not transported from cell to cell. Each cell must make its own ATP, and the rate of ATP supply in each cell depends on that particular cell's ability to synthesize ATP.

How, then, is energy transported in an animal's body? Energy is transported from organ to organ and from cell to cell in the form of food compounds, such as fatty acids or the sugar glucose, that can be broken down to release chemical energy. Each cell in the body takes up fatty acids or glucose from the blood and then breaks these compounds down internally to make ATP from the chemical energy released from chemical bonds.

Animal cells have aerobic and anaerobic processes for making ATP

Most animal cells make most of their ATP by processes that require O_2. Such processes are termed **aerobic**, or oxygen-requiring. In aerobic ATP production, there is, speaking roughly, a fixed relationship between ATP production and O_2 use. Thus the rate at which a cell can make ATP aerobically depends in part on the rate at which the cell is supplied with O_2 by the circulatory system or by other O_2-transport systems. Think back to the man running steadily on a treadmill in our opening photo. For a muscle cell in his legs to make ATP aerobically at a high rate—enabling the cell to have a high muscular power output—the cell must receive O_2 at a high rate.

Aerobic ATP production takes place mostly in the mitochondria of a cell, by the processes of electron transport and oxidative phosphorylation. During aerobic ATP production, all types of food molecules—sugars, lipids, and proteins—can be used as fuels, that is, as sources of chemical energy for ATP synthesis.

Some cells also have processes for making ATP without O_2. These mechanisms—termed **anaerobic**, meaning without oxygen—are not universally present. Brain cells in nearly all vertebrates, for example, have little ability to make ATP anaerobically. This explains why the vertebrate brain is quickly damaged by the absence of O_2 (as, for example, during a heart attack or stroke). However, many types of cells—including muscle, kidney, and gut cells—have well-developed mechanisms for making ATP anaerobically.

LINK

Aerobic and anaerobic metabolism are discussed in **Concepts 6.2 and 6.3**, respectively, and are compared in **Concept 33.3**

The most common process of anaerobic ATP production in animals is **anaerobic glycolysis**. In this mechanism, glycolysis (see Figure 6.7) is the only ATP-producing biochemical pathway that is operational (glycolysis is not itself either aerobic or anaerobic, but it is termed anaerobic glycolysis when it takes place where O_2 is lacking). The mitochondria, which require O_2, do not participate in anaerobic glycolysis. With only glycolysis operational, only sugars can be used as fuels. Anaerobic glycolysis produces just a small amount of ATP per sugar molecule, but it can process sugar molecules at exceedingly high rates and thus make ATP at extraordinary rates. However, ATP cannot be made for very long because anaerobic glycolysis produces lactic acid as its final product (see Figure 6.12A), and the lactic acid accumulates in the animal's body, eventually leading to imbalances that cause the process to stop. When people compete in mile or half-mile runs, much of the ATP used by their leg muscles is made anaerobically. This helps explain why the intensity of muscular work is high but the exercise cannot be continued for very long. Marathons must be run more slowly (mostly aerobically) to avoid this sort of fatigue.

CHECKpoint CONCEPT 29.4

- ✓ Could a single cell extracted from a multicellular organism survive on its own, similar to the way the lone cell of a unicellular organism does? Why or why not?
- ✓ In what form is energy transported from cell to cell?
- ✓ Why is permeability an essential attribute of compartmentalizing structures such as cell membranes and simple epithelia?

Before ending our introductory chapter on animals, we will now turn our spotlight on two additional topics of great importance: phenotypic plasticity and the ways animals control how they function.

CONCEPT 29.5 The Phenotypes of Individual Animals Can Change during Their Lifetimes

Although the genotype, or genetic constitution, of an individual animal is constant for the animal's entire life, countless examples are known of changes in an individual's phenotype, or observable characteristics. Change of this type is called **phenotypic plasticity**, which refers to an individual's ability to display two or more different phenotypes at different times during its life despite a constant genotype. When an individual's phenotype changes as a result of long-term exposure to a particular environment, the individual is said to **acclimate** or **acclimatize** to that environment.

Phenotypic plasticity is common at the biochemical level

Phenotypic plasticity occurs often at the biochemical level, and one of the most important ways it does so is by the action of **inducible enzymes**. Enzymes are called "inducible" if their concentrations in an animal's cells, rather than being constant, depend on the animal's environment or experiences. In other words, with inducible enzymes, the enzyme phenotype can change. Many fish and aquatic invertebrates, for example, have inducible enzymes that respond to long-term changes in body temperature. When winter comes and water temperature falls, these animals acclimatize over days or weeks by producing higher concentrations of enzymes that play key roles in energy metabolism. The fish and invertebrates thus have high concentrations of the enzymes in winter and low concentrations in summer. These enzyme changes make the metabolic functions of the animals less sensitive to long-term changes in temperature than they are to short-term changes.

Detoxification enzymes also provide outstanding examples of plasticity in enzyme phenotype. Human liver enzymes that detoxify ethanol (the alcohol in beer and wine) are usually almost absent in people who do not drink ethanol. However, if a person drinks ethanol regularly, higher concentrations of the enzymes are produced. This helps explain why people who regularly drink are more resistant to drunkenness than people who don't drink.

Enzymes known as P450 enzymes, found throughout the animal kingdom, play important roles in detoxifying a wide range of environmental toxins such as halogenated aromatic hydrocarbons (HAHs). Fish that live in polluted waters have higher levels of P450 enzymes than fish that live in pristine waters.

Phenotypic plasticity also occurs at the scales of tissues and organs

When individual small mammals—such as squirrels or mice—have been living in a warm environment, they have small amounts of brown adipose tissue (BAT; see Figure 29.10), the tissue responsible for nonshivering thermogenesis. Yet when the same individuals experience a cold environment over a period of days or weeks—and therefore become cold-acclimated—their BAT grows dramatically and becomes far more capable of producing heat at a high rate. Because of this plasticity, the animals need not invest in maintaining BAT when their environment is warm, but can have lots of BAT heat production when their environment is cold.

The tiny crustaceans known as water fleas (*Daphnia*), often abundant in freshwater ponds and lakes, display visibly dramatic phenotypic plasticity (**FIGURE 29.15**). They are pale-colored and contain almost no O_2-transporting hemoglobin when their environmental water has a high O_2 concentration. Their production of hemoglobin, being phenotypically plastic, changes in other circumstances, however. When individuals experience periods of low environmental O_2 concentration, they increase hemoglobin synthesis and turn bright red.

Phenotypic plasticity is under genetic control

How can alcohol detoxification enzymes be increased by the experience of drinking alcohol, and brown adipose tissue be increased by exposure to cold? More generally, how can mechanisms that help meet a stress be increased by that very stress?

The answer is that, over the course of evolution, natural selection can favor *patterns of response* to environmental

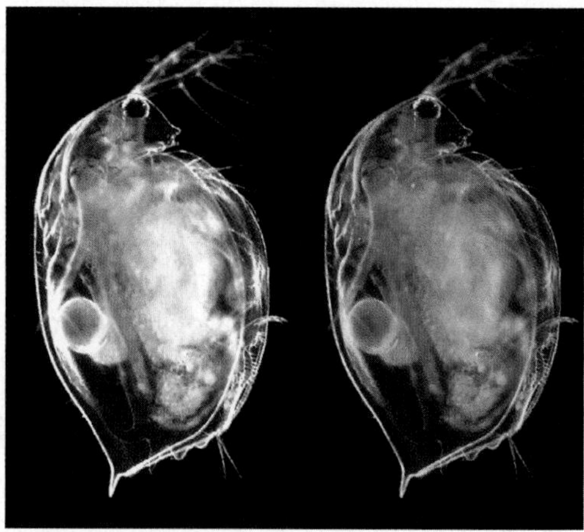

FIGURE 29.15 Phenotypic Plasticity During their lifetimes, individual water fleas (*Daphnia*) can be pale (left) or bright red (right), depending on whether they have been living in O_2-rich or O_2-poor water. The color change reflects how much hemoglobin the animals are synthesizing.

challenges. Phenotypic plasticity is a type of evolved genetic programming. During the evolution of phenotypic plasticity, natural selection—rather than favoring a fixed trait, such as sharp teeth in a carnivore—instead favors a genetic *program of responses* (such as programmed changes in gene expression) so that some environments lead to one set of responses and other environments lead to other sets of responses.

CHECKpoint CONCEPT 29.5

✓ The metabolic rate of a poikilotherm increases immediately when the temperature of its environment increases. Is this an example of phenotypic plasticity? Explain your answer.

✓ How might a scientist justify determining the relative purity of a stream by measuring the concentrations of P450 enzymes in the fish that inhabit it?

Phenotypic plasticity is one way animals can control the ways they function. To close this chapter, we will now take a much broader look at animal control systems.

CONCEPT 29.6 Animal Function Requires Control Mechanisms

The nervous and endocrine systems are the principal systems that control or regulate the ways tissues and organs interact harmoniously throughout an animal's body. In addition, there are control systems in individual cells that control cellular processes such as biochemical reactions and cell division. The importance of controls is emphasized dramatically when controls fail. Cancer in part reflects a failure of control of cell division. Lethal fevers reflect failure of control of body temperature.

LINK

Cancer and regulation of the cell cycle are discussed in **Chapter 7**; see **Concept 7.3** and **pp. 148–149**

Homeothermy exemplifies negative-feedback control

In discussing animal control systems, the control of body temperature in humans and other mammals is a helpful example because it takes place in ways analogous to the familiar engineered systems we use to regulate temperature in our homes, as we mentioned earlier. Not all animal control systems operate in the same way as the control of body temperature, but many do.

The four essential elements of a control system are the controlled variable, sensors, effectors, and control mechanism. The **controlled variable** is the characteristic of an animal that is being controlled—body temperature in this example. **Sensors** detect the current level of the controlled variable. In body temperature regulation, the sensors are neural structures that detect the existing temperature in various parts of the body (e.g., skin, brain, and scrotum) and send the information to the control mechanism in the brain. **Effectors** are tissues or organs that can change the level of the controlled variable. Shivering muscles, for example, are effectors because—by generating heat—they can change the body temperature. The **control mechanism** uses information from the sensors to determine which effectors to activate—and how intensely—to modify the controlled variable. The control mechanism for body temperature is composed of nerve cells located in and near the hypothalamus of the brain.

Go to ANIMATED TUTORIAL 29.1 The Hypothalamus
PoL2e.com/at29.1

In the regulation of mammalian body temperature, as shown in **FIGURE 29.16**, when the control mechanism receives information on the existing body temperature from the sensors, it compares the current body temperature with a set-point temperature—the temperature that is supposed to prevail, in this case 37°C. The control mechanism then activates effectors to reduce or remove the difference between actual body temperature and the set point. If the body temperature is lower than the set point, the control mechanism activates shivering or nonshivering thermogenesis, resulting in increased heat production, which tends to raise the body temperature until it equals 37°C. If the body temperature is higher than the set point, the control mechanism activates sweating or panting to reduce or remove the difference.

This control mechanism is an example of a **negative-feedback** system. The process is called "feedback" because after the control mechanism issues commands to effectors, it receives information on how its commands have altered the body temperature (the controlled variable). Information is "fed back" to

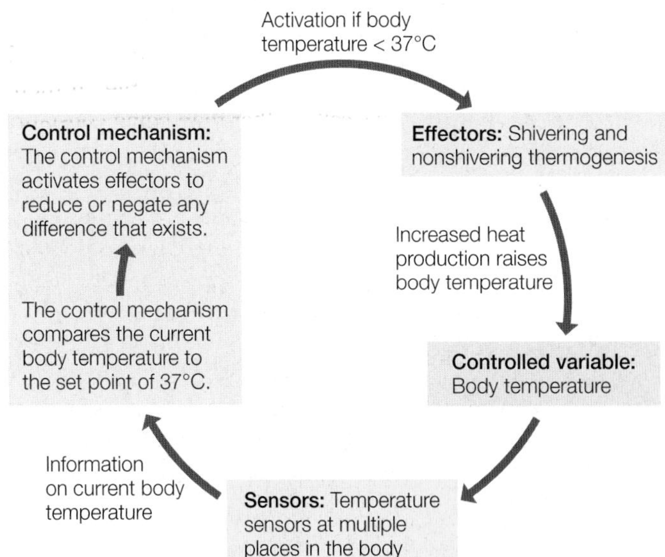

FIGURE 29.16 Negative Feedback: A Systems Diagram of the Temperature Control System of a Mammal The major components of this system are the controlled variable (body temperature), sensors, effectors, and the control mechanism. The system stabilizes body temperature by negative feedback.

the control mechanism. The process is described as "negative" because the control mechanism activates effectors in ways that reduce or negate any difference that exists between the body temperature and the set point. Negative feedback is a stabilizing influence and is often involved in maintaining homeostasis. In this case, it keeps the body temperature near 37°C.

Positive feedback occurs in some cases

Positive feedback occurs when deviations of a controlled variable from its existing level are increased or amplified by the action of the control mechanism. Positive feedback destabilizes a system. However, destabilization can sometimes be advantageous, provided it is ultimately brought under control. A nerve impulse is an example of positive feedback that lasts for a very brief time and then is terminated (see Figure 34.8).

The mammalian birth process depends on positive feedback. During the birth process, waves of contraction of a mother's uterine muscles are stimulated by the hormone oxytocin, which is secreted by neurons in the hypothalamus of the brain and released into the blood by the posterior pituitary gland. A control mechanism in the hypothalamus receives information from sensors in the uterus on the force of uterine contractions. When this control mechanism detects an increased force of contraction, it activates increased oxytocin secretion by the neurons in the hypothalamus, which are the effectors in this system. The heightened oxytocin secretion further increases the force of contraction in a self-reinforcing, positive-feedback loop (**FIGURE 29.17**). Ultimately the force becomes great enough to cause the fetus to move from the uterus into the outside world, ending further stimulus for oxytocin production and uterine contractions.

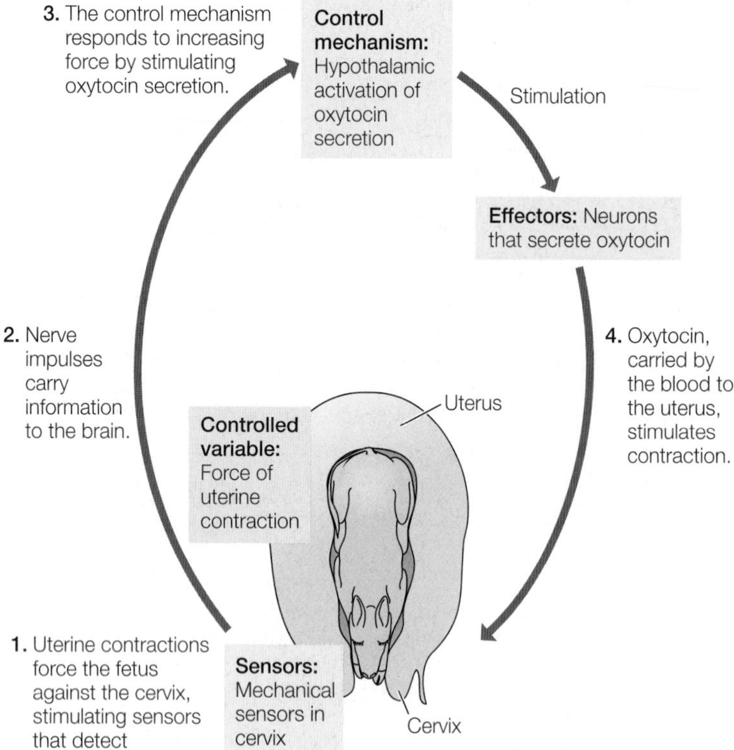
3. The control mechanism responds to increasing force by stimulating oxytocin secretion.
Control mechanism: Hypothalamic activation of oxytocin secretion
Stimulation
Effectors: Neurons that secrete oxytocin
4. Oxytocin, carried by the blood to the uterus, stimulates contraction.
2. Nerve impulses carry information to the brain.
Controlled variable: Force of uterine contraction
Uterus
1. Uterine contractions force the fetus against the cervix, stimulating sensors that detect mechanical forces.
Sensors: Mechanical sensors in cervix
Cervix

FIGURE 29.17 Positive Feedback: The Development of the Force Required for Birth to Occur in Mammals This system produces an escalating force by positive feedback of uterine muscle contractions.

Biological clocks make important contributions to control

Animals have self-contained, metabolic mechanisms of keeping track of time, known as **biological clocks**. These mechanisms do not require outside information to function. Instead, they have a self-contained ability to keep time. Biological clocks have several advantages. One of the most important is that they allow animals to anticipate future events, just as our watches let us anticipate when future events will occur.

The early experiments that convinced biologists of the existence of biological clocks focused on activities such as wheel-running in small mammals, such as mice and squirrels. Given a running wheel in a normal cage environment with day and night, nocturnal species of squirrrels tend to run on the wheel each night and to be inactive each day. They alternate between running and not running on a 24-hour cycle (**FIGURE 29.18A**). Investigators found that if they deprived these squirrels of an external day–night cycle, for example by keeping them in constant darkness, the squirrels still tended to alternate between periods of high and low running activity on an approximately 24-hour cycle (**FIGURE 29.18B**). This suggested that squirrels can keep time without external timing cues. To make sure that the squirrels were not receiving timing cues from their surroundings, investigators not only deprived the animals of an external day–night cycle but also isolated them from outside temperature variations, outside sounds, and any other outside cues imaginable. The squirrels still alternated between periods of high and low running activity. By now, based on revolutionary discoveries of the past 20 years, the mechanisms of biological clocks have been deciphered. Thus the existence of biological clocks has been confirmed by knowledge of the ways they work.

When an animal is prevented from receiving external timing cues, its biological clocks are said to free-run, and under these conditions they do not keep perfect time (see Figure 29.18B). They run fast or slow relative to actual time. Consider a biological clock that, in its free-running state, goes through an entire timing cycle in 23.8 hours. After 1 day of free-running, the clock will be 0.2 hour off, compared with the actual 24-hour solar-day rhythm. After 2 days it will be 0.4 hour off, and after 5 days it will be an entire hour off. In natural environments, this long-term drift is prevented because animals receive external timing cues and their biological clocks do not free-run. Each day, the clocks are reset to be synchronous with actual time—a process termed entrainment. Sunrise and sunset are the most commonly used cues for entrainment. In a natural environment, animals observe sunrise and sunset each day, and they use these indicators to synchronize their biological clocks with the actual 24-hour solar-day rhythm.

Biological clocks that have free-running timing cycles of about 24 hours are called **circadian clocks**. The word "circadian" means "about a day." It emphasizes that the self-contained, free-running periodicities of these clocks are not exactly 24 hours. By now we know of the existence of

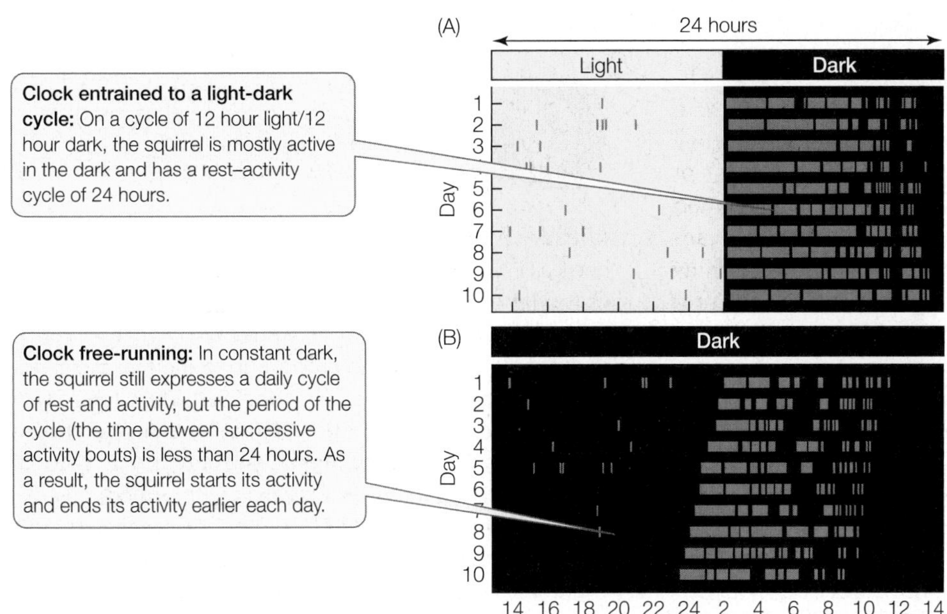
(A)
24 hours
Light
Dark
Clock entrained to a light-dark cycle: On a cycle of 12 hour light/12 hour dark, the squirrel is mostly active in the dark and has a rest–activity cycle of 24 hours.
Day
(B)
Dark
Clock free-running: In constant dark, the squirrel still expresses a daily cycle of rest and activity, but the period of the cycle (the time between successive activity bouts) is less than 24 hours. As a result, the squirrel starts its activity and ends its activity earlier each day.
14 16 18 20 22 24 2 4 6 8 10 12 14
Time of day (hours)

FIGURE 29.18 Control of Activity by a Circadian Biological Clock The rest–activity cycle of flying squirrels (*Glaucomys volans*) responds to the light–dark cycle under which the squirrels are kept. The squirrels are nocturnal. All the data shown are for a single individual studied (A) with a light-dark cycle and (B) in continuous darkness. Each horizontal line corresponds to one 24-hour day. Where a horizontal line is shaded gray, the squirrel was running. In each set of data, ten successive days, numbered 1 to 10, are portrayed from top to bottom.

Go to ANIMATED TUTORIAL 29.2
Circadian Rhythms
PoL2e.com/at29.2

biological clocks that have free-running timing cycles of about 1 year, called circannual clocks. Many animals in the ocean, where there are lunar tides, have circatidal clocks. All require entrainment to keep exact time.

Circadian clocks are believed to be virtually universal among animals. In an individual animal, most or all cells have their own circadian clocks. A "master clock" is present in an individual, however, and sends out signals to other cells in the individual's body, keeping the body as a whole synchronized. In mammals, the master clock consists of nerve cells in parts of the brain called the suprachiasmatic nuclei.

APPLY THE CONCEPT

Animal function requires control mechanisms

To better understand the physiological mechanism of circadian clocks, Martin Ralph and colleagues used artificial selection to produce two strains of hamsters: one with a short circadian period and one with a long circadian period.[a] (The circadian period is the length of time between two successive days as timed by the clock.) When the suprachiasmatic nuclei (SCN; parts of the brain involved in controlling circadian rhythms) in some of these adult hamsters were destroyed, the animals became arrhythmic. After several weeks, the scientists transplanted SCN tissue from fetal hamsters into the brains of the adult hamsters with destroyed SCNs. Long-period adult hamsters received tissue from short-period fetuses, and short-period adults received tissue from long-period fetuses. The effects of these treatments on the hamsters' circadian periods are shown in the table.

1. Why did destroying the SCN make the adult hamsters arrhythmic?
2. What result would you expect if SCN tissue from a long-period fetal hamster were transplanted into a long-period adult that had a destroyed SCN?
3. What do the results of the transplantation experiments tell you about where in the body the circadian rhythm phenotype (long versus short) is expressed?

[a]M. R. Ralph et al. 1990. *Science* 247: 975–978.

	TYPE OF RHYTHMICITY SHOWN AFTER VARIOUS TREATMENTS			
RECIPIENT	NO TREATMENT	SCN DESTROYED	SCN DESTROYED AND SHORT-PERIOD TRANSPLANT	SCN DESTROYED AND LONG-PERIOD TRANSPLANT
Short-period adult	Short period	Arrhythmic	Not done	Long period
Long-period adult	Long period	Arrhythmic	Short period	Not done

Biological clocks help control many animal functions. Circadian clocks can give animals an internal sense of when to wake up each day. They help control when animals are active and at rest. They also help control when the digestive system is most able to digest foods each day and when detoxification systems are most able to counteract toxic chemicals. Circannual clocks are known to control annual hibernation rhythms in some species of mammals, and they control annual reproductive rhythms in some animals, such as sheep.

LINK

Plants also have circadian clocks, which are entrained by light cues; see **Concept 26.4**

CHECKpoint CONCEPT 29.6

- ✓ Imagine a house with an air conditioner, furnace, and electric thermostat consisting of a thermometer and simple computer. For this control system, identify the controlled variable, sensor(s), effector(s), and control mechanism. Draw a negative-feedback loop for cooling a house using these components.
- ✓ What is the likely fate of a biological clock that is denied any chance for entrainment?
- ✓ Does negative feedback always have a diminishing effect on the controlled variable? That is, because it's "negative," does it always make the level of the controlled variable very low?

Why can we use a person's rate of O_2 consumption to measure the rate at which he or she consumes energy to run?

ANSWER To be able to do muscular work, muscle cells need ATP, which they must make themselves (Concept 29.4). As food molecules are broken down in muscle cells to provide energy to synthesize the ATP needed to power the work of muscles, O_2 is required in proportion to the energy released from the food molecules (Concept 29.2). Thus, logically, O_2 is required in proportion to the ATP that is made and broken down in metabolism. By measuring the rate at which O_2 is consumed, we can calculate the rate of breakdown of food molecules to release energy, the rate of ATP synthesis from that energy, and the rate of heat production as muscles liberate energy from ATP and use it to contract, converting it to heat.

The rate of oxygen consumption of a running horse can be determined remotely today with electronic devices that measure O_2 in inhaled and exhaled air by use of a mask. The data can then be stored in onboard digital memory or radioed.

SUMMARY

CONCEPT 29.1 Animals Eat to Obtain Energy and Chemical Building Blocks

- Animals build new molecules and cells from ingested organic compounds. A growing animal adds new cells to its body, and all animals are constantly replacing worn-out molecules and cells.
- An animal is characterized by the organization of its body's cells, tissues, and organs. **Energy** can be seen as the capacity to create or maintain this organization. **Review Figure 29.1**

CONCEPT 29.2 An Animal's Energy Needs Depend on Physical Activity and Body Size

- An animal's **metabolic rate** is defined as its rate of energy consumption—the rate at which the animal uses chemical energy, turning it into heat. Metabolic rate can be quantified by measuring the rate of O_2 consumption or (far less commonly) by measuring the rate of heat production.
- Physical activity increases an animal's metabolic rate, but the exact relationship between activity level and metabolic rate varies. **Review Figure 29.2**
- **Basal metabolic rate** (**BMR**) is the metabolic rate of a resting, fasting individual in a comfortable thermal environment. When BMR is expressed per unit of body weight in grams, it is called BMR/g.
- Smaller mammals consistently have higher BMR/g than larger mammals do. This is an example of a **scaling relationship**. An animal with a high BMR/g must consume more food relative to its body weight than an animal with a lower BMR/g. **Review Figure 29.3**

CONCEPT 29.3 Metabolic Rates Are Affected by Homeostasis and by Regulation and Conformity

- Within the animal body, cells exist in an **internal environment** of **tissue fluids**. Conditions outside the body are the **external environment**.
- Animals that maintain a consistent internal environment despite variation in the external environment are called **regulators**. **Conformers** are animals in which the internal environment varies to match the external environment. **Review Figure 29.4**
- Regulation of the internal environment is more energetically expensive than conformity.
- **Homeostasis** refers to the stability of an animal's internal environment and the mechanisms that maintain this stability.
- **Homeotherms** exhibit thermoregulation, maintaining a constant internal temperature by varying metabolic rate and insulation with external temperature. Above or below a specific temperature range called the **thermoneutral zone** (**TNZ**), metabolic rate increases to compensate for increased heat input or to offset heat loss, respectively. **Review Figure 29.5 and ACTIVITY 29.1**

SUMMARY *(continued)*

- Most animals are **poikilotherms** (**ectotherms**), with body temperatures that vary with, and match, external temperature. Metabolic rates decrease as temperature drops and increase as temperature rises. **Review Figures 29.6 and 29.7**
- The sensitivity of a reaction or process to changes of internal (tissue) temperature can be expressed as the numerical factor Q_{10}, which usually has a value of 2 to 3. **Review Figures 29.6B and 29.7**
- A poikilotherm can exert some control over its tissue temperature through behavior, by moving to a more favorable location. **Review Figure 29.8**
- Homeothermy expends more energy than poikilothermy. **Review Figure 29.9**
- Homeothermic heat-producing mechanisms used by mammals include **shivering**, the subtle contraction of the skeletal muscles to convert ATP to heat; and **nonshivering thermogenesis** (**NST**) in **brown adipose tissue**, which produces heat directly in the mitochondria by short-circuiting oxidative phosphorylation. Insects sometimes regulate their thoracic temperature using heat generated from ATP by the flight muscles in their thorax **Review Figures 29.10 and 29.11**
- In hot environments, homeothermic animals increase their rate of heat loss by sweating or panting, processes that increase heat transfer to the environment by evaporation of water.
- Mammalian **hibernation** refers to a strategy of reducing energy (and thus food) needs by entering a state of thermoconformity during extended cold periods. **Review Figure 29.12**

CONCEPT 29.4 Animals Exhibit Division of Labor, but Each Cell Must Make Its Own ATP

- The fluids in an animal's body are distributed among various body compartments. **Intracellular fluid** is inside the body's cells and is said to be in the intracellular compartment. **Extracellular fluid**, which includes blood plasma and interstitial fluid, is outside the cells and is said to be in the extracellular compartment.
- Different extracellular fluids are separated by **epithelia**, sheets of cells that cover a body surface or organ or line a body cavity. **Review Figure 29.13**
- The boundary between intracellular fluids and extracellular fluids is formed by cell membranes.
- Animals exhibit a high degree of division of labor. This division of labor typically necessitates a rapid transport system to distribute materials among parts of the body.
- Cells, **tissues**, **organs**, and **multi-organ systems** represent ever-more-complex levels in the hierarchy of organization of the animal body. **Review Figure 29.14 and ACTIVITY 29.2**
- Each cell must make its own ATP. Cells may use **aerobic** (with oxygen) or **anaerobic** (without oxygen) processes to produce ATP.

CONCEPT 29.5 The Phenotypes of Individual Animals Can Change during Their Lifetimes

- **Phenotypic plasticity** refers to an individual's ability to display two or more different phenotypes at different times during its life. Phenotypic plasticity is a way for an animal to **acclimate** or **acclimatize** to changes in its environment. **Review Figure 29.15**
- At a biochemical level, phenotypic plasticity often involves **inducible enzymes**, enzymes that change in abundance or type in response to an animal's environment.
- Phenotypic plasticity also occurs at the level of tissues and organs.
- The capacity for phenotypic plasticity is genetically based and is subject to natural selection.

CONCEPT 29.6 Animal Function Requires Control Mechanisms

- The four essential elements of a control system are the **controlled variable**, **sensors**, **effectors**, and **control mechanism**. The controlled variable is the property or characteristic of an animal that is being controlled. Sensors detect the current level of the controlled variable. Effectors are tissues or organs that can change the level of the controlled variable. The control mechanism uses information from the sensors to determine which effectors to activate—and how intensely—to modify the controlled variable.
- A control mechanism may involve **negative feedback**, in which the control mechanism activates effectors in ways that reduce or negate any difference that exists between the controlled variable's actual state and its set point. Negative feedback is stabilizing. **Review Figure 29.16 and ANIMATED TUTORIAL 29.1**
- **Positive feedback** occurs when deviations of a controlled variable from its existing level are increased or amplified by the action of the control mechanism. Positive feedback is destabilizing but still can be advantageous. **Review Figure 29.17**
- Animals have self-contained mechanisms of keeping track of time, known as **biological clocks**, that allow the animals to anticipate future events. Biological clocks that have free-running timing cycles of about 24 hours are called **circadian clocks**. Some animals also have circannual or circatidal clocks. **Review Figure 29.18 and ANIMATED TUTORIAL 29.2**

Go to the Interactive Summary to review key figures, Animated Tutorials, and Activities PoL2e.com/is29

Go to LaunchPad at **macmillanhighered.com/launchpad** for additional resources, including LearningCurve Quizzes, Flashcards, and many other study and review resources.

30 Nutrition, Feeding, and Digestion

KEY CONCEPTS

Herds of wildebeest move across a savanna in Kenya during the Serengeti migration.

Let's imagine we are lucky enough to travel in a hot-air balloon over the savannas of East Africa. Most of the animals below us will ignore the steady rushing sound of our hot-air burner. Thus as we float in a gentle breeze less than 100 feet above them, we will be able to observe their ordinary daily lives.

Textbooks tell us that Earth is dynamic. Let's use the wonders of human imagination to look at the scene below us dynamically. Let's pretend that we can see energy and matter traveling from place to place.

Light that has traveled 93,000,000 miles through cosmic space from the sun steadily strikes the grasses and thinly scattered trees. Through photosynthesis, these plants convert a small percentage of the light energy into chemical energy in the bonds of organic molecules they manufacture. The remainder of the sunlight energy—more than 99 percent of it—turns to heat as it strikes the savanna. As the plants synthesize the organic molecules of their bodies, they extract CO_2 from the air to provide the carbon atoms they need. Some plants—but not many—can even take nitrogen from the air to build their proteins.

The zebras, wildebeests, and other herbivores below us need the tissues of plants to sustain their lives because they depend on organic compounds for the energy and atoms they require. Dynamically, as the grasses and trees grow and make new organic matter, the animals eat a portion of the organic matter. Some of the organic compounds in the plants are particularly valuable because—while the animals require them—they can be difficult for the animals to obtain in adequate amounts. Proteins are often in this category.

We'll see many herbivores below us. The flux of energy from the sun into the plants is great enough for many herbivores to survive by exploiting it.

Only rarely, however, will we see a lion, cheetah, or hyena eat one of the herbivores. Organisms cannot live on heat. Yet the plants turn some of their chemical energy into heat, and the herbivores turn some to heat as well. Thus by the time the plants and herbivores have used the chemical energy from photosynthesis, only a small percentage of it is left. Only a few carnivores can get enough chemical energy to exist, so carnivore populations are smaller than herbivore populations.

In the midst of all these dynamic fluxes, each animal below us must somehow meet its needs for both energy and chemical substances. The study of how animals meet this challenge is the study of **nutrition**.

What are some of the specific types of organic molecules that animals must obtain in their foods, and why do they need them?

You will find the answer to this question on page 641.

CONCEPT 30.1 Food Provides Energy and Chemical Building Blocks

Animals—being heterotrophs (see Concept 29.1)—obtain their energy and chemical building blocks from eating other organisms. For animals in the wild, this process is frequently dangerous. For example, Thomas Ruf, Claudia Bieber, and Christopher Turbill recently studied various species of mammals that hibernate, asking whether individuals have different death rates when they are hibernating and not hibernating. When an individual is hibernating, it remains continually in an underground burrow, out of sight. When the same individual is awake, its metabolic rate is far higher and it has to venture out of its burrow to collect food. Ruf and his colleagues found that, on average, the odds of dying each month are five times higher when an individual is awake rather than hibernating. They attributed this difference to the dangers of getting food. As modern humans in an industrialized society, we may have a hard time understanding this. When we need food, we go to a supermarket, and maybe have an ice-cream treat on our way. For animals in the wild, getting food is risky. Plant eaters can be seen by predators and possibly attacked while they munch on grass or leaves. Meat eaters face the very real risk that animals they attack may fight back and injure them, possibly leading to blood loss or infection.

Eating must be of great importance for animals to risk their lives to get food. What exactly do animals require from their food? We discussed these needs briefly in Chapter 29. First, because animals are heterotrophs, they require chemical-bond energy to live: they need organic molecules that they can break down to release energy that they can then use to move about, pump blood, synthesize body parts, and carry out the other energy-demanding requirements of life. Second, animals eat because they require chemical building blocks. They need materials that they can use to build the parts of their bodies—such as amino acids to build proteins, fatty acids to build cell membranes, and calcium to build a skeleton. This need for chemical building blocks is far more complicated than you might imagine, because animals often cannot synthesize for themselves certain molecules they require for life. They need to get those specific molecules from food.

Food provides energy

We quantify the energy content of a food by measuring the amount of heat the food produces when it is completely burned in the presence of oxygen, producing CO_2, H_2O, and other fully oxidized chemical products (**FIGURE 30.1**). We can determine the energy content of a food this way because both burning and aerobic animal respiration follow the same chemical rules. Burning takes place by different chemical steps than respiration, but the products of the two processes are almost identical. Heat is one product. Therefore the amount of heat released by burning a food is close to the same as that produced when an animal metabolically breaks down the same food (and we can correct for any differences by use of chemical rules).

Food energy values in the United States are typically expressed in calories and kilocalories. A **calorie (cal)** is the amount of heat required to raise the temperature of 1 gram of water by 1°C, and a **kilocalorie (kcal)** is 1,000 times greater. We can use these units for any type of energy, not just heat, because all forms of energy are proportional to one another.

Confusion arises because people often refer to kilocalories as "calories" in everyday life. When you see the word "calories," you should check whether calories or kilocalories are being used. In the Nutrition Facts on food packages in the United States, for example, the energy values reported as "calories" are actually kilocalories (**FIGURE 30.2**). The **joule** is an alternative, metric unit for measuring energy (1 calorie = 4.2 joule). Most countries in the world use the joule, and scientists almost always use it in their scientific work.

Let's consider a meal of macaroni and cheese for which the Nutrition Facts are presented in Figure 30.2. As

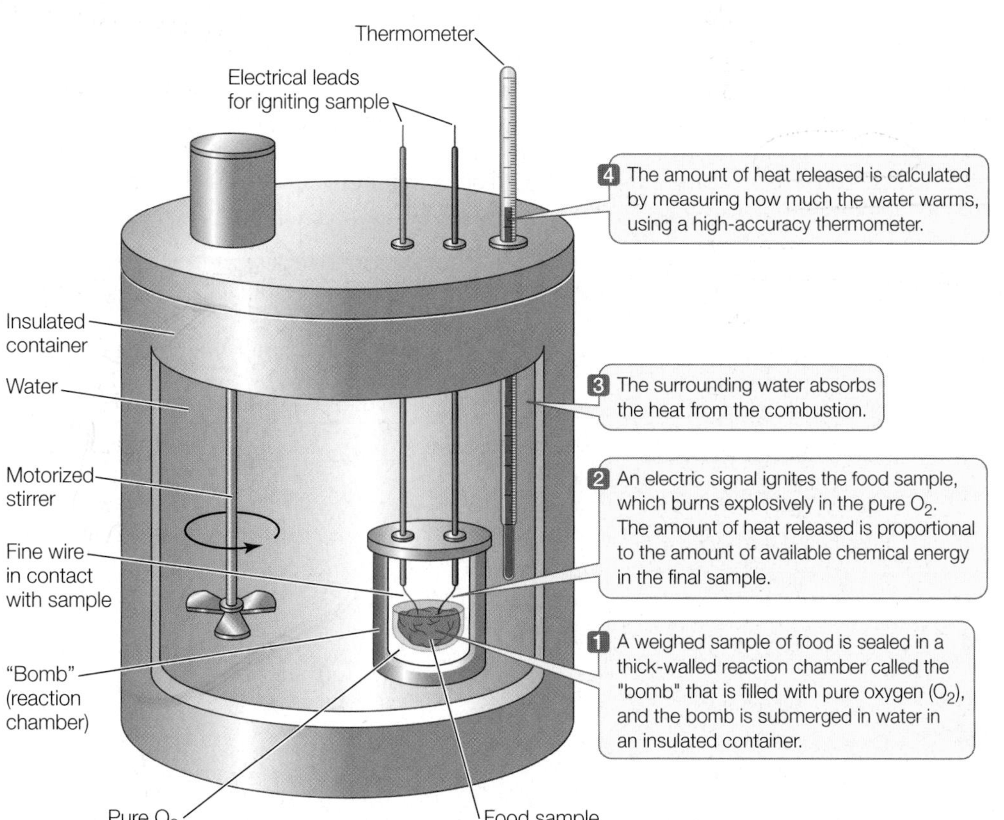
Thermometer
Electrical leads for igniting sample
4 The amount of heat released is calculated by measuring how much the water warms, using a high-accuracy thermometer.
Insulated container
Water
3 The surrounding water absorbs the heat from the combustion.
Motorized stirrer
2 An electric signal ignites the food sample, which burns explosively in the pure O_2. The amount of heat released is proportional to the amount of available chemical energy in the final sample.
Fine wire in contact with sample
"Bomb" (reaction chamber)
1 A weighed sample of food is sealed in a thick-walled reaction chamber called the "bomb" that is filled with pure oxygen (O_2), and the bomb is submerged in water in an insulated container.
Pure O_2
Food sample

FIGURE 30.1 Food Energy Values Are Measured by Explosive Combustion in a Bomb Calorimeter

Nutrition Facts

Serving Size 1 cup (228 g)
Servings Per Container 2

Amount Per Serving	
Calories 250	Calories from Fat 110
	% Daily Values*
Total Fat 12g	**18%**
Saturated Fat 3g	**15%**
Trans Fat 0g	
Cholesterol 30mg	**10%**
Sodium 470mg	**20%**
Total Carbohydrate 31g	**10%**
Dietary Fiber 0g	**0%**
Sugars 5g	
Protein 5g	**10%**
Vitamin A 4% •	Vitamin C 2%
Calcium 20% •	Iron 4%

*Percent Daily Values are based on a 2,000 calorie diet. Your Daily Values may be higher or lower depending on your calorie needs.

FIGURE 30.2 A Nutrition Label of the Sort Used in the United States This is the label for a particular brand of macaroni and cheese. Although this label refers to "calories," the values on it are in fact kilocalories.

you can see, each serving supplies 250 kcal. When the meal is sitting on your plate waiting to be eaten, this energy is in the form of chemical energy—in the chemical bonds of the starch, fats, and other molecules in the meal. After you've eaten the macaroni and cheese, if your body breaks down all the molecules in it to release their energy, your body will produce 250 kcal of heat as it uses the energy.

How much chemical energy does an animal consume each day, converting the chemical energy to heat? Metabolic rate provides the answer (see Concept 29.2). An animal's metabolic rate, expressed per day, is the amount of energy the animal converts to heat per day. What if we focus specifically on people? A rule of thumb is that adults in modern society use an average of 2,000–2,500 kcal per day in their daily mix of rest and activity. For general purposes, you can be sure you don't overeat and gain weight by adding up the energy values of all the foods you eat each day and making sure the total is no more than 2,000–2,500 kcal.

There are three major types of food molecules: lipids (fats and oils), carbohydrates, and proteins. Lipids are about twice as dense with energy as carbohydrates and proteins. Lipids yield about 9.4 kcal per gram when they are broken down, compared with 4.1 kcal/g for carbohydrates and 4.5 kcal/g for proteins. As you might expect, foods containing lots of lipids tend to provide lots of kilocalories per serving.

Chemical energy from food is sometimes stored for future use

When an animal eats foods containing more chemical energy than the animal needs, the animal often stores some or all of the extra chemical energy for future use. Animals store energy mostly in the form of lipids. This makes sense because lipids are particularly energy dense. Stored materials need to be carried around until they are used. For each gram of lipid that an animal stores and carries, it gets more than 9 kcal of energy—a large amount—when it finally uses the lipid. Animals do not need to eat lipids to store lipids. They can convert carbohydrates and proteins to lipids to store chemical energy.

Animals also store glycogen, a type of carbohydrate (specifically, polymerized glucose), but they do not store nearly as much energy as glycogen as they store as lipid. Glycogen is always stored with a lot of water bound to its molecules. When all this water is taken into account, the glycogen is heavy, and its energy density is only about 1 kcal/g.

Why do animals store glycogen at all if its energy value is only 1 kcal/g compared with more than 9 kcal/g for lipids? The answer is that some tissues require carbohydrates as their energy source. When glycogen is broken down, it yields the sugar glucose. The vertebrate brain gets its energy almost

APPLY THE CONCEPT

Food provides energy and chemical building blocks

MET, the *m*etabolic *e*quivalent of *t*ask, is an index of the intensity of physical activity. Exercise physiologists define the resting metabolic rate as 1 MET and increasingly energetic activities as multiples of 1 MET. Walking at 4 miles per hour has a value of about 3 METs (i.e., it uses 3 times as much energy per unit of time as resting quietly). Jogging (7 mph) requires about 8 METs, and vigorous running (10 mph) about 12 METs. Assuming that your resting metabolic rate is 80 kcal/hr, fill in the table with the time it would take you to burn off the calories from the listed foods while resting, jogging, or running.

		TIME REQUIRED TO BURN OFF CALORIES		
FOOD	KILOCALORIES	RESTING	JOGGING	RUNNING
Mixed carrots and peas, half-pound	136			
Beer, one 12-oz bottle	145			
Potato chips, kettle-cooked, 30 chips	225			
Sports energy bar (e.g., chocolate power bar)	250			
Double cheeseburger	440			
Chicago-style deep-dish pizza, half-pound	720			

entirely from glucose. Thus glycogen stores can directly meet the brain's energy needs. When skeletal muscles employ anaerobic glycolysis to make ATP for exercise (see Concept 29.4), they also need glucose as their energy source.

Proteins are not usually used as storage compounds. There are large masses of protein in an animal's body. Muscles are principally proteins, for example. However, these proteins are in use, rather than being stored for future use.

Food provides chemical building blocks

The molecules in an animal's body are made from the molecules in the animal's food. For carrying out this process, animals have enormous capacities to interconvert molecules. Lipid molecules can be converted to carbohydrate molecules, and certain amino acids can be converted to others, for example. These are just two of the many interconversion processes that occur in animals.

As an animal grows, its food must meet both its energy needs and its needs for chemical building blocks to form a larger and larger body. The need for chemical building blocks does not end with adulthood, however. As stressed in Concept 29.1, adult animals constantly rebuild their bodies. Mammals, for example, rebuild 2–3 percent of the protein molecules in their bodies every day. Much of this rebuilding is carried out by recycling building blocks within the body. Recycling is not perfect, however. For this reason, adults have a steady need throughout their lives for new building blocks that they obtain by eating.

LINK

For more on the reactions that interconvert macromolecules and their constituent monomers, see **Concepts 2.2–2.4, 3.1, and 3.2**

Some nutrients in foods are essential

The word "essential" has a particular meaning in the study of nutrition. An **essential nutrient** is one that an animal cannot synthesize for itself. We will start our discussion of essential nutrients by focusing on the **standard amino acids**: the 20 kinds of amino acids that animals use to build proteins (see Concept 3.2). These amino acids exemplify the distinction between essential and nonessential nutrients.

Biochemists have discovered that animals can synthesize some of the 20 standard amino acids but not others. Adult humans, for example, can synthesize 12 of the standard amino acids by rearranging the subparts (carbon skeletons and functional groups) of other molecules in their diet. Those 12 amino acids are *nonessential*. The remaining eight amino acids cannot be synthesized by adult humans. These **essential amino acids**—isoleucine, leucine, lysine, methionine, phenylalanine, threonine, tryptophan, and valine—must be obtained fully formed in the diet. The list of essential amino acids tends to be similar from one type of animal to another, but there are important differences. Infant humans, for example, require histidine in their diet as well as the eight amino acids that are essential for adults.

You might wonder: Where do the essential amino acids come from? Plants and algae can synthesize all 20 standard amino acids. For this reason, animals can obtain their essential amino acids by eating plants or algae. Different plants and algae vary in their proportions of amino acids. For an animal to get all of its essential amino acids in adequate quantities, it may therefore need to eat more than one kind of plant or alga. For adult humans, combinations of grain foods and legume foods—such as corn and beans—are often effective in providing all the essential amino acids (**FIGURE 30.3**).

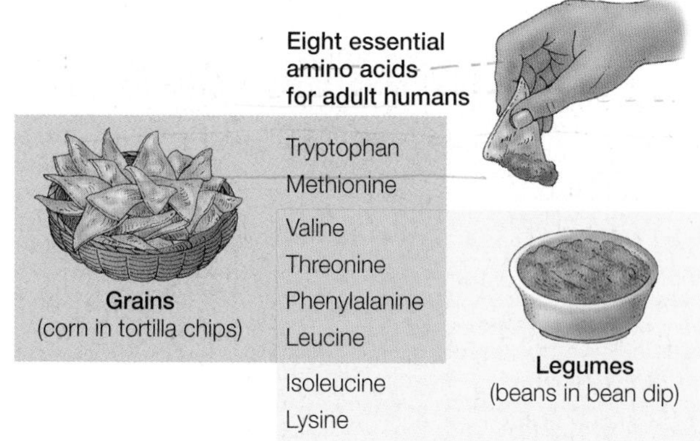

FIGURE 30.3 A Common Food Combination Provides All Eight Essential Amino Acids All around the world, cultures exist in which corn and beans are major parts of the daily diet. We mimic this diet when we snack on tortilla chips and bean dip. Eaten together, these two foods provide all the essential amino acids required by adult humans. (Corn by itself often does not provide enough lysine or isoleucine, and beans often do not provide enough methionine or tryptophan.) When two or more foods must be eaten to get all eight essential amino acids, it is important to eat the foods at about the same time.

Foods of animal origin—such as milk, eggs, and meat—tend to be good sources of all eight essential amino acids. Cows, chickens, and other farm animals get their essential amino acids by eating plants. They build those amino acids into proteins in their bodies or in the products they make, such as milk and eggs. We then get the amino acids when we eat them.

Animals also need certain **essential fatty acids**. There are a great many kinds of fatty acids, but in humans and many other animals only two are commonly essential: alpha-linolenic (α-linolenic) acid and linoleic acid. If α-linolenic acid is essential for an animal, the animal must either eat it in its diet or eat very similar fatty acids that can be converted to it by simple reactions. Alpha-linolenic acid and these other fatty acids are called the omega-3 fatty acids. Linoleic acid and other fatty acids that are closely similar to it are the omega-6 fatty acids.

Vitamins are a third group of essential nutrients. Vitamins are difficult to define chemically. This is true because needs for various vitamins have evolved independently or relatively independently, meaning that different vitamins often have unrelated chemical structures. Vitamin A provides an excellent illustration of the evolutionary history of a vitamin:

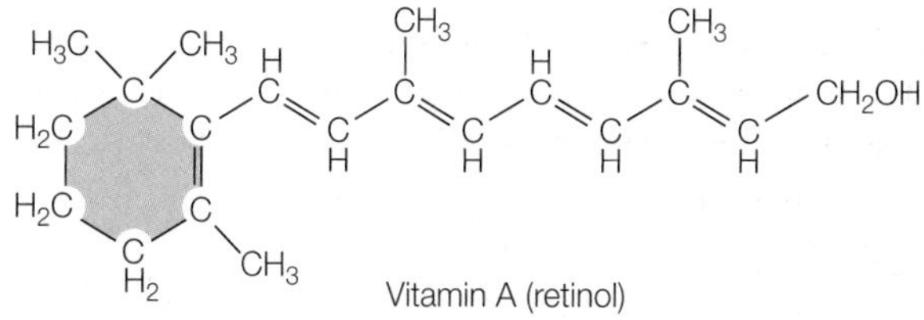

Vitamin A (retinol)

TABLE 30.1 A Full List of Vitamins Required by Humans

Vitamin	Some sources in human diet	Some functions	Deficiency symptoms
WATER-SOLUBLE			
B_1 (thiamin)	Liver, legumes, whole grains	Coenzyme in cellular respiration	Beriberi, loss of appetite, fatigue
B_2 (riboflavin)	Dairy foods, meat, eggs, green leafy vegetables	Coenzyme in FAD	Lesions in corners of mouth, eye irritation, skin disorders
B_3 (niacin)	Meat, fowl, liver, yeast	Coenzyme in NAD and NADP	Pellagra, skin disorders, diarrhea, mental disorders
B_5 (pantothenic acid)	Liver, eggs, yeast	Found in acetyl CoA	Adrenal problems, reproductive problems
B_6 (pyridoxine)	Liver, whole grains, dairy foods	Coenzyme in amino acid metabolism	Anemia, slow growth, skin problems, convulsions
Biotin	Liver, yeast, bacteria in gut	Found in coenzymes	Skin problems, loss of hair
B_{12} (cobalamin)	Liver, meat, dairy foods, eggs	Formation of nucleic acids, proteins	Pernicious anemia, psychological disorders
Folic acid	Vegetables, eggs, liver, whole grains	Coenzyme in formation of heme and nucleotides	Anemia
C (ascorbic acid)	Citrus fruits, tomatoes, potatoes	Formation of connective tissues; antioxidant	Scurvy, slow healing, poor bone growth
LIPID-SOLUBLE			
A (retinol)	Fruits, vegetables, liver, dairy foods	Key part of visual pigments; regulation of nervous system development	Night blindness
D (cholecalciferol)	Fortified milk, fish oils; also made in the body in presence of sunshine	Absorption of calcium and phosphate	Rickets
E (tocopherol)	Meat, dairy foods, whole grains	Muscle maintenance, antioxidant	Anemia
K (menadione)	Liver; also made by gut microbes	Blood clotting	Blood-clotting problems

When animals evolved vision, they used a slightly modified version of this complex structure for light collection in their eyes. But they did not evolve an ability to make the structure! Instead, they depended on getting the structure, already made, from plants and algae that synthesize the structure and incorporate it into pigments (e.g., β-carotene, see p. 121) that help collect light for photosynthesis. To this day, humans and many other animals must obtain vitamin A as a fully formed ("prefabricated") chemical structure in their diets because they cannot synthesize it themselves, yet they require it for vision and certain other functions.

All the vitamins are carbon compounds, and they are needed in just tiny amounts. Thus we can quickly define a **vitamin** as an essential carbon compound required in tiny amounts. **TABLE 30.1** summarizes the 13 vitamins commonly recognized as required by humans. Note that vitamins fall into two groups: those that are soluble in water and those that are soluble in lipids. The list of vitamins varies a bit among animals. For example, most mammals can make their own ascorbic acid, so it is not a vitamin for them. However, primates—including humans—cannot synthesize ascorbic acid, and for them it is a vitamin (vitamin C). Vitamin D is a special case because humans can synthesize it if their skin is exposed to sunlight. Vitamin D is essential for people only if they have inadequate sun exposure, as is often true in winter when people stay indoors or wear heavy clothing.

Go to ACTIVITY 30.1 Vitamins in the Human Diet
PoL2e.com/ac30.1

The final category of essential nutrients is the **essential minerals**. These are chemical elements that animals require in addition to carbon, oxygen, hydrogen, and nitrogen (the most abundant elements in organic molecules). Familiar examples of essential minerals are calcium, iodine, and iron (**TABLE 30.2**).

We need some essential minerals in substantial quantities. Calcium, for example, is the fifth most abundant element in the human body. A 70-kg person contains about 1.2 kg of calcium. Calcium is a component of calcium phosphate, the principal structural material in bones and teeth, and is also necessary for nerve function and muscular contraction (see Concepts 33.1 and 34.3). We lose calcium from our bodies in urine, sweat, and feces, so it must be replaced regularly. Humans require about 1 g of calcium per day in their diet.

We require some of the other essential minerals in only small amounts. Iron is one of these. It is the oxygen-binding atom in hemoglobin and myoglobin (see Concept 32.5) and is also a component of enzymes in the electron transport chain. Yet the total amount of iron in the tissues of a 70-kg person is only about 4 g. Only tiny amounts (10–20 mg) are required in the diet each day to replace iron lost from the tissues.

Go to ACTIVITY 30.2 Mineral Elements Required by Animals
PoL2e.com/ac30.2

Nutrient deficiencies result in diseases

The lack of any essential nutrient in an animal's diet produces a state of deficiency called **malnutrition**. When malnutrition is

TABLE 30.2 A Partial List of Minerals Required by Animals

Element	Some sources in human diet	Major functions
NEEDED IN LARGE AMOUNTS		
Calcium (Ca)	Dairy foods, eggs, green leafy vegetables, whole grains, legumes, nuts, meat	Found in bones and teeth; blood clotting; nerve and muscle action; enzyme activation
Magnesium (Mg)	Green vegetables, meat, whole grains, nuts, milk	Required by many enzymes; found in bones and teeth
Phosphorus (P)	Dairy foods, eggs, meat, whole grains, legumes, nuts	Found in nucleic acids, ATP, and phospholipids; bone formation; found in buffers; metabolism of sugars
Sodium (Na)	Table salt, dairy foods, meat, eggs	Nerve and muscle action; water balance; principal positive ion in extracellular fluids
NEEDED IN SMALL AMOUNTS		
Chromium (Cr)	Meat, dairy foods, whole grains, legumes, yeast	Glucose metabolism
Cobalt (Co)	Meat, tap water	Found in vitamin B_{12}; formation of red blood cells
Copper (Cu)	Liver, meat, fish, shellfish, legumes, whole grains, nuts	Found in active site of many redox enzymes and electron carriers; production of hemoglobin; bone formation
Iodine (I)	Fish, shellfish, iodized salt	Found in thyroid hormones
Iron (Fe)	Liver, meat, green vegetables, eggs, whole grains, legumes, nuts	Found in active sites of many redox enzymes and electron carriers, hemoglobin, and myoglobin
Manganese (Mn)	Organ meats, whole grains, legumes, nuts, tea, coffee	Activates many enzymes
Zinc (Zn)	Liver, fish, shellfish, and many other foods	Found in some enzymes and some transcription factors; insulin physiology

chronic, it leads to a specific **deficiency disease** related to the missing nutrient. Scurvy, for example, is a deficiency disease caused by inadequate amounts of a vitamin (see Table 30.1). This complex illness has many debilitating symptoms, including bleeding gums, and is ultimately fatal if not cured. Scurvy was a major problem for sailors at sea, who often had no fresh foods during long trips, until physicians learned in the eighteenth century that it is caused by deficiency of a nutrient, later identified as vitamin C. The disease disappeared when sailors made a point of eating limes or other citrus fruits that provide ample vitamin C.

LINK

Deficiencies of essential minerals—for example, iodine—can also cause disease; see **Concept 35.4**

CHECKpoint CONCEPT 30.1

- Imagine that a slice of pizza contains 300 kcal of energy. Using only the energy stored in the pizza, what is the maximum amount of water you could heat by 1°C? (Hint: Start by converting kilocalories to calories and express your final answer in grams.)
- Which of these food molecules will produce the most heat per gram if burned in the presence of oxygen: lipids, carbohydrates, or proteins?
- Do all organisms require the same essential amino acids?

Now that we have discussed why animals require food, let's turn to the methods they use to get food.

CONCEPT 30.2 Animals Get Food in Three Major Ways

If we look at animals in the broadest way possible, we realize that they get their food in three major ways. Some target easily visible, individual food items. Some collect tiny—often invisibly tiny—food particles, which they must gather in great numbers to obtain a sufficient amount. Finally, some depend on symbioses with microbes to obtain critical food needs.

These three methods are not mutually exclusive. Many animals employ two methods. For example, as we note below, there is a growing realization that we humans are probably far more dependent on microbes than previously thought.

Some animals feed by targeting easily visible, individual food items

An osprey (fish hawk; *Pandion haliaetus*) catching a fish provides an iconic example of feeding by targeting easily visible, individual food items (**FIGURE 30.4**). The osprey spots an individual fish from the air while flying. Then it plunges to the water to seize the fish with its talons in a targeted way. This type of feeding also occurs when a lion kills a zebra with a bite to its neck, or when a crab grasps a clam with its claws and crushes the clam's shell to get to the meat inside.

Animals that eat leaves and grasses also exemplify this type of feeding. As a deer plucks leaves from a tree, it targets each leaf. Many grazing animals are remarkably selective, targeting some types of plants but avoiding others. Bees target individual flowers to collect their nectar and pollen. Squid catch fish one at a time.

FIGURE 30.4 An Osprey Eats by Targeting Individual Food Items

Suspension feeders collect tiny food particles in great numbers

The water in a lake or ocean often contains great numbers of suspended organic particles, many of them too tiny to be seen with the naked eye. For example, photosynthesis in bodies of water is typically carried out by single-celled algae suspended in the water's sunlit, upper layers. Each milliliter of this water can contain hundreds or thousands of individual algal cells, each microscopically small. Tiny animals that eat the algae are often also abundant. Dead algal cells, dead microscopic animals, and the feces of tiny animals are additional types of common, suspended organic particles. Here we use the term "particles" to refer to small organic objects that may be plant or animal, living or dead.

Clams, many fish, many aquatic insects, and numerous other aquatic animals have evolved ways of collecting tiny particulate foods. Such animals are called **suspension feeders** in recognition of the fact that they feed on suspended particles. These animals were originally called filter feeders because people assumed they use sievelike or filterlike body parts to collect particles. However, we realize today that many of these animals do not sieve or filter the water, and specialists now prefer the more general term "suspension feeders."

Collecting tiny particles is not the only challenge suspension feeders face. To obtain sufficent amounts of food, these animals also must be able to collect the particles in huge numbers. A single clam eats millions of individual algal cells each day. Suspension feeders usually do not target individual food particles, but instead collect huge numbers indiscriminately.

A common definition of suspension feeding is that it is a process by which an animal collects individual food particles that are only 1/100th the size of the animal or smaller. By this definition, many whales are classed as suspension feeders. About half the living species of whales have no teeth. Instead, they have closely spaced plates (made of a material like our fingernails) that allow them to collect small organisms from the water (**FIGURE 30.5A**). Some of the biggest whales, measuring 20–30 meters long, feed principally on crustaceans that are each about the size of a rice grain. These whales are suspension feeders by the definition we are using. The two biggest fish alive today, the whale shark (*Rhincodon typus*) and basking shark (*Cetorhinus maximus*), are also suspension feeders, as are the huge devil (manta) rays (*Manta*).

Most suspension feeders, however, are less than 30 centimeters long and feed on particles that are microscopic or nearly microscopic in size. These small suspension feeders include sardines, anchovies, menhadens, and many other commercially important fish. They also include scallops, mussels, clams, and many other aquatic invertebrates (**FIGURE 30.5B**).

Many animals live symbiotically with microbes of nutritional importance

A **symbiosis** is an intimate, long-term association between two different types of organisms. Many animals have evolved

(A)
The plates in the mouth of this baleen whale can collect food objects the size of rice grains.
Eschrichtius robustus

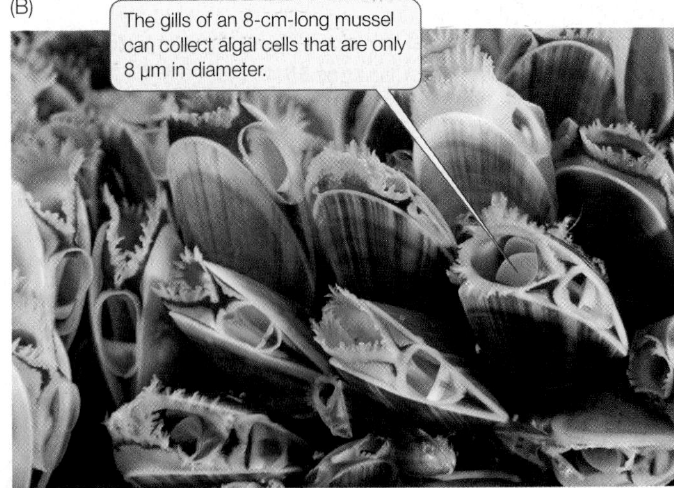

(B)
The gills of an 8-cm-long mussel can collect algal cells that are only 8 μm in diameter.
Mytilus edulis

FIGURE 30.5 Two Suspension Feeders (A) A toothless grey whale showing the array of plates in its mouth. The plates, made of a fingernail-like material, are called baleen, and the toothless whales are often called baleen whales. (B) Mussels (a type of mollusk) feed by pumping environmental water over and through their gills, which remove tiny food particles from the water.

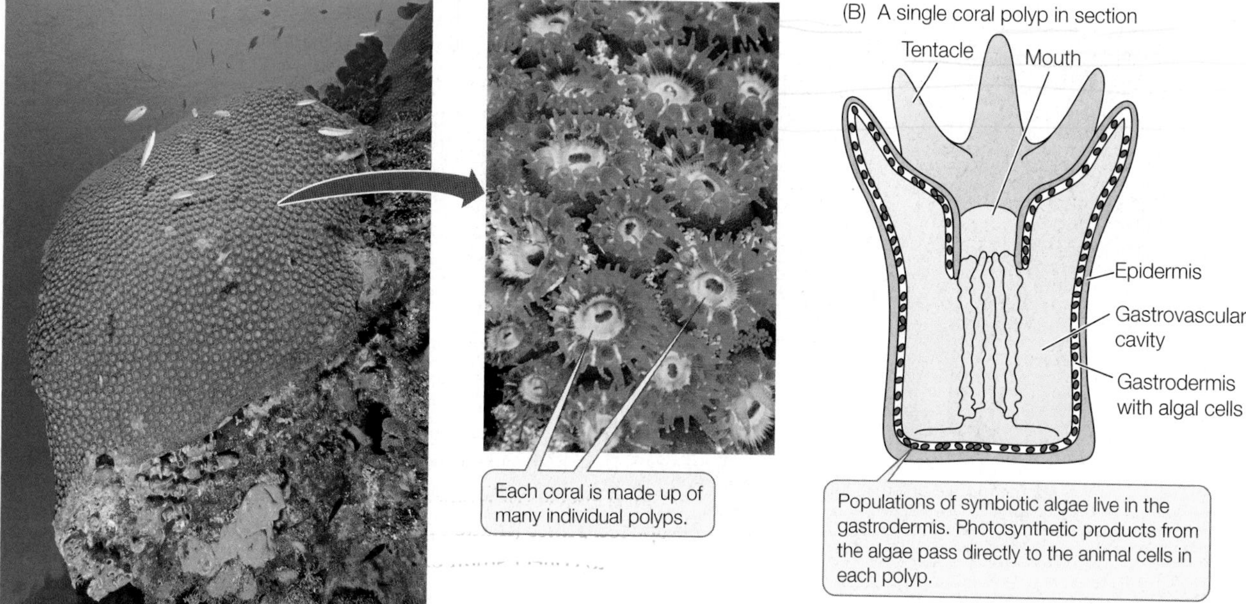
(A) A tropical coral-reef ecosystem
Each coral is made up of many individual polyps.
(B) A single coral polyp in section
Tentacle
Mouth
Epidermis
Gastrovascular cavity
Gastrodermis with algal cells
Populations of symbiotic algae live in the gastrodermis. Photosynthetic products from the algae pass directly to the animal cells in each polyp.

FIGURE 30.6 Reef-Building Corals Live Symbiotically with Algal Cells Corals, unlike most animals, are photosynthetic and require sunlight. (A) Coral reefs consist of a diverse array of corals. Each coral is a large colony of hydra-like polyps, which build the coral's skeleton. (B) Algal cells (often called zooxanthellae) live inside the cells of a polyp's gastrodermal cell layer. The algae carry out photosynthesis, synthesizing glucose and other nutritionally important molecules for the polyp. Polyps also feed heterotrophically by using stinging cells on their tentacles to capture tiny animals that swim by.

symbioses with microbes, such as bacteria or algae, in which the microbes synthesize nutritionally important molecules for the animals. Because the microbes live inside the animals, the animals do not eat them. However, molecules transported directly from the microbes to the animals serve the same functions as foods, and thus we will call such molecules foods.

All reef-building corals contain algae (**FIGURE 30.6**). The cells of these algae live inside the cells of the corals. A reef-building coral is an animal–algal symbiosis composed of a cnidarian animal (resembling a hydra or anemone) and a specific species of alga in the genus *Symbiodinium* (see Concept 20.4). The algal cells contain photosynthetic pigments that are responsible for some of a coral's coloration. By carrying out photosynthesis, the algal cells produce food molecules such as glucose. Some of these food molecules then diffuse out of the algal cells into the animal cells and become food for the heterotrophic animal cells. This symbiosis is of extreme importance nutritionally. Corals can lose their algal cells when stressed (a process called bleaching). If that happens, the corals deteriorate and will die unless they obtain new algal cells.

Animals have also evolved nutritionally important symbioses with many types of microbes that are heterotrophic rather than photosynthetic. The ruminant mammals—including cows, sheep, moose, antelopes, deer, and giraffes—are the most famous examples. All ruminants are herbivores and eat foods such as grasses and leaves. They have evolved a complex stomach that consists of four chambers through which food passes sequentially (**FIGURE 30.7**). The first of these chambers, the rumen, is huge, occupying more than 10 percent of the animal's body. It is filled with a microbial community consisting of a complex and changeable mix of bacteria, protozoa, yeasts, and fungi living in a nonacidic, saliva-like fluid.

When a ruminant eats, its food (after chewing) goes directly to the rumen, which acts like a fermentation vat. Here the resident microbes attack the food, breaking it down and transforming it in ways the animals cannot. The processes carried out by the microbes are called **fermentations** because they take place without O_2 (the rumen is anoxic). Three functions of the rumen microbes are particularly important:

- They break down cellulose and hemicelluloses (cellulose-like compounds; see Figure 2.10). Vertebrate animals by themselves cannot break these compounds down and thus cannot obtain nutritional value from cellulose, one of the most abundant organic materials on land (see Concept 30.3). By breaking down cellulose, the microbes turn it into small molecules (short-chain fatty acids, such as acetic acid) that the ruminant can absorb into its blood and use for chemical energy.
- The microbes make several types of B vitamins and amino acids that are essential nutrients for the ruminants.
- The microbes enable ruminants to recycle nitrogen, one of the chemical elements that is often in short supply. The microbes take up waste nitrogen from the ruminant's metabolism and convert it to microbial protein. The animal then digests this protein, obtaining amino acids from which it can synthesize animal proteins. Animals without this recycling process simply excrete their waste nitrogen, and then need to replace it by finding nitrogen-rich foods.

One of the breaking stories in the study of animal nutrition is the recent realization that heterotrophic microbes probably play important nutritional roles in most animals, not just in animals such as ruminants that have evolved large fermentation

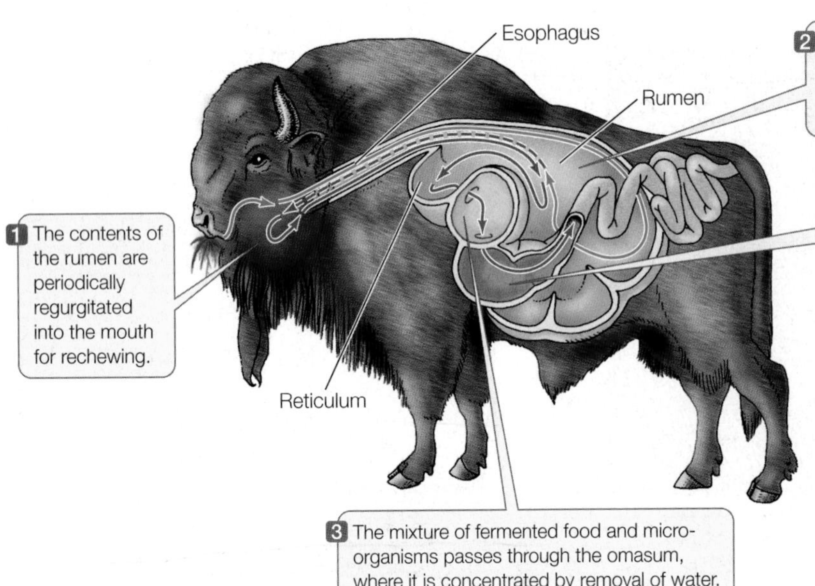

Esophagus
Rumen
Reticulum
1 The contents of the rumen are periodically regurgitated into the mouth for rechewing.
2 The rumen and reticulum serve as fermentation vats for the mixed microbial community. The microbes ferment cellulose, forming short-chain fatty acids that diffuse into the blood.
3 The mixture of fermented food and microorganisms passes through the omasum, where it is concentrated by removal of water.
4 The abomasum functions similarly to our stomach. It secretes hydrochloric acid and protein-digesting enzymes. The microbes are killed and partly digested here, then passed on to the small intestine where digestion is completed and absorption occurs.

FIGURE 30.7 A Ruminant's Stomach Bison, like all ruminants, have a specialized stomach with four compartments that enables them to obtain energy and nutrients from coarse, otherwise indigestible plant material through fermentation. The fermentation is carried out by a mixed microbial community of bacteria, protists, yeasts, and fungi that is housed primary in the rumen, which is not acidified.

vats. The microbes of greatest interest are those that live in the lumen—the central cavity—of the intestines and other parts of the gut and collectively are called the **gut microbiome**.

Humans have a large gut microbiome, consisting mostly of bacteria, which we acquire during birth and after we are born. People differ in the composition of their gut microbiome, and increasing evidence indicates that the gut microbiome has important effects on individual digestion, nutrition, and metabolism.

LINK

For more on the diverse microbes in the gut microbiome and their possible effects on animal nutrition and human health, see **Concepts 19.3 and 41.1**

CHECKpoint CONCEPT 30.2

- ✓ A newly hatched, nearly microscopic octopus larva and a fist-sized oyster may both feed on the same type of microscopic animals suspended in the water column. Why is the oyster considered a suspension feeder whereas the octopus larva is not?
- ✓ Reef corals are animals. However, they can be severely damaged if the water in which they live becomes cloudy, blocking sunlight from reaching them. Why?
- ✓ Rabbits have large fermentation chambers like ruminants do. However, unlike ruminants, rabbits are *hind*gut fermenters. Ingested food does not pass through their fermentation chamber (their cecum; see Figure 30.8) until *after* it leaves their stomach and small intestine. With the functional significance of symbiotic fermentation in mind, how might you explain the common sight of a rabbit eating its own feces?
- ✓ Certain large worms, called hydrothermal-vent worms, living on the ocean floor have no mouth, gut, or anus. Does this imply that they are not receiving any nutrients from the surrounding environment? Which of the three major ways of getting food might these worms employ?

Now that we have surveyed the ways in which animals get foods, let's turn to the ways they process their foods after ingestion. Processing is as important as the chemical composition of food in determining the nutrients that animals obtain.

CONCEPT 30.3 The Digestive System Plays a Key Role in Determining the Nutritional Value of Foods

After animals ingest foods, the foods usually pass through a tubular digestive tract—a tubular gut—in most cases. The foods enter at the mouth, and later their indigestible components exit as feces at the other end of the tube, the anus. Exceptions to this pattern exist. For example, in animals with symbiotic microbes (see Concept 30.2), sugars or other nutritional compounds synthesized by the microbes enter the animal tissues directly without needing to pass through the digestive tract. For the rest of this chapter we will focus mainly on food processing by the digestive tract.

The inner walls of the digestive tract are lined, from mouth to anus, with an epithelium (usually a simple epithelium; see Figure 29.13):

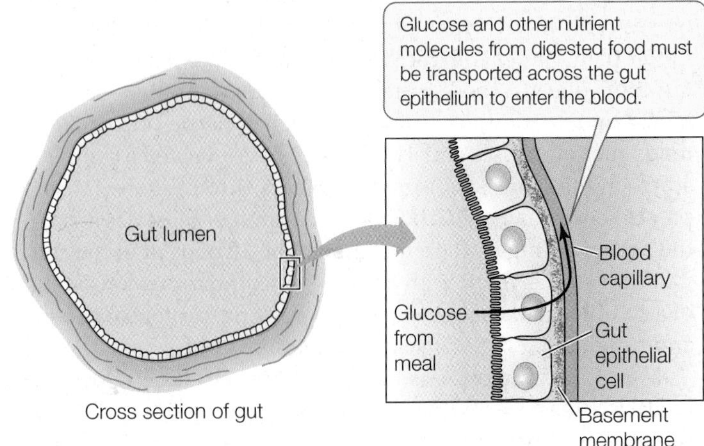

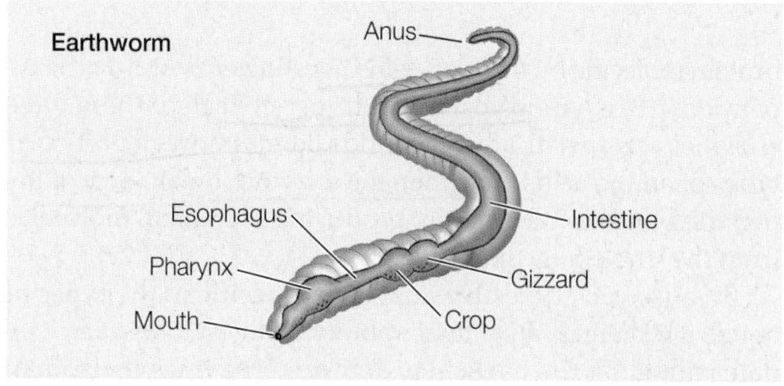

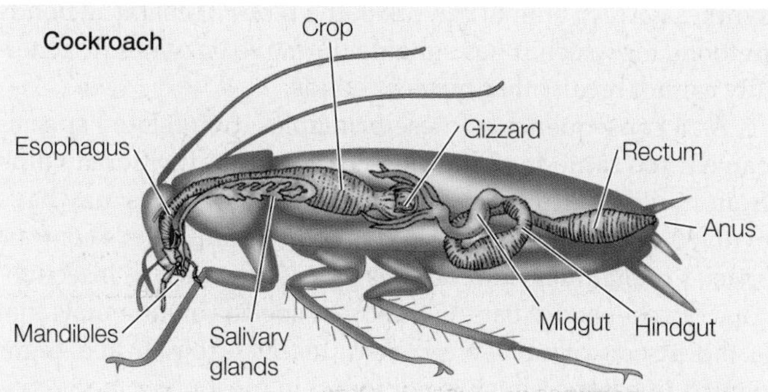

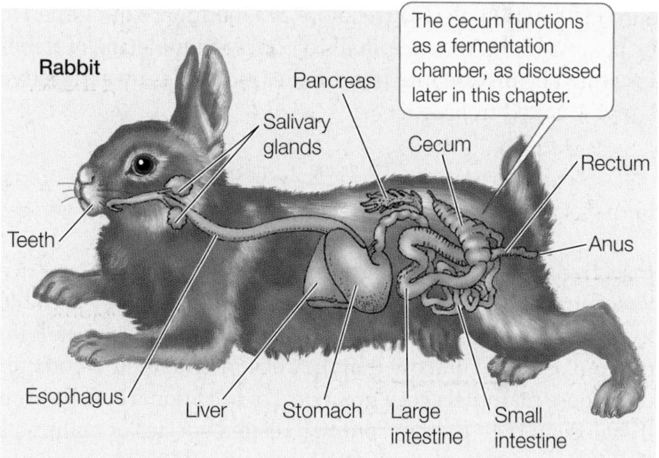

FIGURE 30.8 Digestive Systems Are Diverse, but They All Carry Out Digestion and Absorption In each group of animals, specific regions of the digestive tract vary from species to species, in part because different species are adapted to different diets.

After being ingested, food materials are in the gut lumen on the inside of this epithelium (i.e., enclosed by the epithelium). The blood vessels of the gut are located in the tissues in the walls of the gut, surrounding the epithelium. Thus for food materials to enter the blood for transport to cells throughout the body, they must cross the gut epithelium. Most materials in the foods animals ingest cannot do that. The molecules of carbohydrates, lipids, and proteins in foods are far too large and complex to cross the gut epithelium.

The digestive system's principal function is to perform two processes that together permit food materials to be taken into the blood and used: digestion and absorption. Here we define them in ways that apply to most animals. **Digestion** is the breakdown of food molecules in the gut lumen by enzyme action to produce smaller molecules that can cross the gut epithelium to enter the blood. **Absorption** is the process of transporting molecules from the gut lumen into the blood. The mechanisms of both digestion and absorption are specific for certain types of molecules. For the most part, the molecules that an animal can absorb are the very ones that the animal's digestion produces.

The digestive systems of all animals carry out digestion and absorption. This is true even though digestive systems are diverse in their detailed anatomy and physiology (**FIGURE 30.8**).

Digestive abilities determine which foods have nutritional value

To see a critically important aspect of animal nutrition, let's consider what happens if a person eats cellulose, the principal material in the cell walls of plants (see Concept 2.3). Cellulose molecules are very large, and they are unable to pass through the gut epithelium from the gut lumen into the blood. Moreover, humans have no way to digest cellulose: we have no cellulose-digesting enzymes, and unlike ruminants, we don't have a rumen stocked with symbiotic microbes that can break down cellulose molecules into smaller pieces. As a consequence, we get no nutritional benefit from cellulose. Even though cellulose molecules contain a lot of chemical energy that would be useful for us, we cannot access that energy. If we eat cellulose, it simply stays inside our gut lumen and is eliminated in our feces.

This example illustrates that the nutritional value of a food depends on more than its chemical composition. It also depends on an animal's ability to process the food so its components can be absorbed into the blood and delivered to the cells of the animal's body.

Animals are diverse in the foods they can digest

Each digestive enzyme is specific in the types of chemical bonds it can break (although some enzymes are more specific than others). **FIGURE 30.9** shows the digestion of a short

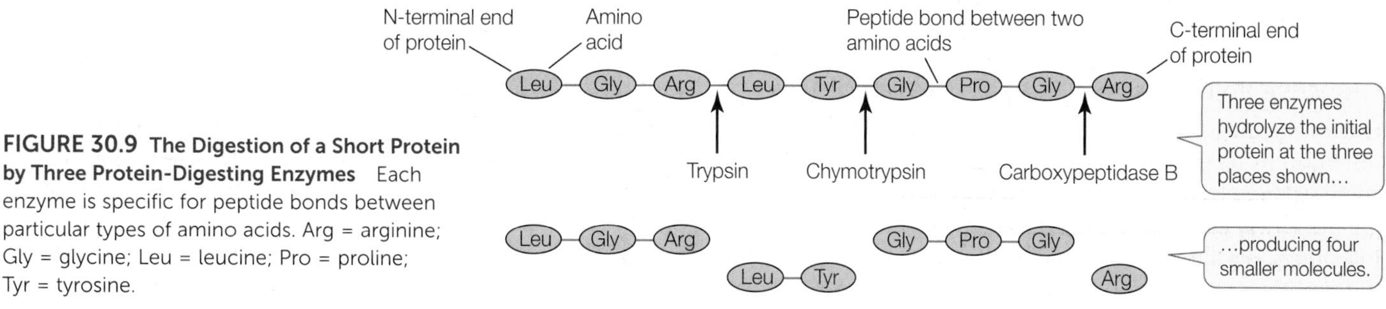

FIGURE 30.9 The Digestion of a Short Protein by Three Protein-Digesting Enzymes Each enzyme is specific for peptide bonds between particular types of amino acids. Arg = arginine; Gly = glycine; Leu = leucine; Pro = proline; Tyr = tyrosine.

protein molecule by three protein-digesting enzymes (carboxypeptidase B, chymotrypsin, and trypsin). Each of these three enzymes acts specifically on peptide bonds between particular types of amino acids. Together, the enzymes break three of the peptide bonds in the protein, producing four small molecules from the single large molecule.

Because each digestive enzyme is specific in the types of bonds it can break, an animal's ability to digest foods is strictly determined by the particular digestive enzymes the animal synthesizes. An animal can break the types of chemical bonds in foods for which it has suitable digestive enzymes. It generally cannot break other types of bonds.

As a consequence of these principles, two related species can eat the same food but derive different nutritional value from it (because different sets of digestive enzymes may have evolved in the two species). For a vivid example, let's contrast species of mammals that eat lots of insects—such as insectivorous (insect-eating) bats—with other species of mammals that avoid insects or eat few of them. Insects are rich in a sugar named trehalose, which is not found in most other foods consumed by mammals. For trehalose in a mammal's gut lumen to be absorbed into the mammal's blood, each molecule of trehalose must be broken into two smaller molecules by a digestive enzyme called trehalase:

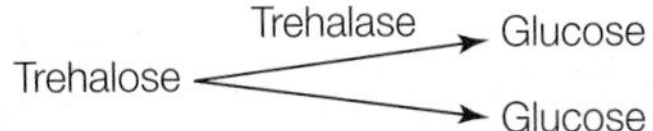

Insectivorous mammals have evolved the ability to produce lots of trehalase. Thus when they ingest insects, they are able to digest the trehalose in them, breaking each molecule into two molecules of glucose that they absorb into their blood and use. These mammals get a great deal of nutritional benefit from trehalose in their diet. In contrast, related species of mammals that do not eat insects typically have not evolved the ability to produce much trehalase. If they eat trehalose, they get almost no nutritional value from it. Thus two related mammals could eat identical meals of insects and obtain radically different nutritional benefits.

As we have seen, cellulose provides another example of these same points. Humans get no nutritional value from cellulose because we cannot digest it, but ruminants obtain a great deal of nutritional value from it because their symbiotic microbes break down cellulose. Among insects, some species are like people and get no nutritional value from cellulose. By contrast, certain wood-eating insects produce cellulose-digesting enzymes that allow them to digest cellulose and absorb it. And other wood-eating insects, such as some termites, maintain symbioses with microbes that break down cellulose for them.

Digestive abilities sometimes evolve rapidly

One of the most revealing discoveries in the study of human nutrition in the past decade concerns the digestion of milk sugar, lactose, by various human populations. When humans are first born, we produce the digestive enzyme lactase in our intestines. For this reason, babies can readily digest lactose in their mother's milk, breaking each lactose molecule into two

FIGURE 30.10 Rapid Evolution Occurring in Cattle-Raising Tribes Has Enabled People in the Tribes to Digest Milk Sugar throughout Life Most African people have little or no ability to digest lactose (milk sugar) in adulthood. However, tribes that raise cattle, such as the Maasai, often exhibit high frequencies of distinctive genes that code for synthesis of lactase (an enzyme that digests lactose) throughout life.

parts (the monosaccharides galactose and glucose) that the infants can absorb and use. In most human populations, however, people stop making lactase as they grow older; by the time they are adults, they make little or no lactase and thus are unable to digest and use milk sugar. Most African tribes follow this pattern: the people in the tribes make lactase as infants but lose that ability in adulthood.

Researchers went to Africa to answer the question, What happens in African tribes that raise cattle? Certain tribes maintain herds of cattle that have enormous cultural significance. Tribal life is centered on caring for the herds. The cattle are used as special gifts, and the people use the blood and milk of the cattle as important foods (**FIGURE 30.10**).

The researchers discovered that in many cattle-raising tribes, the people have evolved the ability to synthesize lactase in adulthood. Thus, unlike most adult Africans, adults in the cattle-raising tribes can digest milk sugar all their lives. This ability to synthesize lactase in adulthood has evolved independently several times. We know this because the gene mutations responsible for lactase synthesis differ among tribes.

From this research came another important discovery—that nutritional processes can evolve rapidly in human populations. Some of the mutant genes responsible for adult lactase synthesis in African tribes first appeared, through the process of mutation, only about 5,000 years ago (a short time on an evolutionary time scale). Today these genes are widespread in the tribes where they occur.

Almost 100 percent of white-skinned people in northern Europe, such as white-skinned Dutch and Danish people, synthesize lactase as adults. How can this be? A large body of evidence indicates that the relevant mutant genes in northern European populations first appeared after their ancient ancestors migrated out of Africa and into Europe. Thus mutations that maintain lactase production in adulthood occurred in the European populations independently from the mutations in the African tribes we just discussed. Possibly the mutant genes in northern Europe were favored by natural selection because people there eat large amounts of milk, cheese, and other dairy products.

When people from northern Europe immigrated to North America centuries ago, all of them probably had genes enabling them to synthesize lactase in adulthood. However, they intermarried with people who immigrated from other parts of the world. Today about 75 percent of white-skinned people in North America produce lactase in adulthood, but 25 percent do not and may have sufficient problems with digesting dairy products to be lactose intolerant.

Digestive abilities are phenotypically plastic

Individual animals—from fish to mammals—often modify their production of digestive enzymes during their lifetimes in response to the types of foods they are eating. In this way individuals can adaptively modify the nutritional values of foods.

If a lab rat has been eating a low-protein diet but is switched to a high-protein diet, its digestive tract starts to increase production of protein-digesting enzymes within 24 hours. A week later, when the response is fully developed, the rat will secrete protein-digesting enzymes at rates that are five times higher than before the switch in diet. By means of this digestive phenotypic plasticity, a rat that starts to eat a lot of protein soon develops a greatly enhanced ability to digest protein and thus tap dietary proteins for energy and amino acids. Conversely, when a rat that has been eating lots of protein is suddenly deprived of protein, it will greatly reduce production of protein-digesting enzymes.

Similarly, if a rat is given increased amounts of carbohydrates in its diet, it will increase production of carbohydrate-digesting enzymes. If it is given lots of lipids, it will increase lipid-digesting enzymes.

Parallel changes take place in absorption mechanisms. For example, researchers have observed that when people start to eat lots of sugar after a period of eating little sugar, their absorption mechanisms for taking up the products of sugar digestion become more effective.

CHECKpoint CONCEPT **30.3**

- ✓ Are the terms "digestion" and "absorption" interchangeable? Compare the two.
- ✓ Humans derive no energy from the cellulose molecules in brussels sprouts. Does this imply that cellulose molecules, and brussels sprouts in general, are low in chemical energy?
- ✓ Why might the ability to produce lactase into adulthood be beneficial to people living in cultures heavily dependent on dairy products?

At this point we have discussed the fundamentals of nutrition, food gathering, and extraction of energy and nutrients from food. With this background, let's now take a closer look at the digestive systems of vertebrate animals.

CONCEPT 30.4 The Vertebrate Digestive System Is a Tubular Gut with Accessory Glands

The vertebrate digestive system consists of a tube, extending from mouth to anus and called the gut, that receives secretions from certain nearby glands, most notably the liver and pancreas (**FIGURE 30.11**; also see the bottom panel of Figure 30.8). The tissues of the gut wall are arranged in concentric layers (**FIGURE 30.12**). Surrounding the gut lumen is the gut epithelium, a functionally dynamic cell layer. Most cells in the epithelium are digestive–absorptive cells, which produce distinctive proteins, such as digestive enzymes, that function in digestion and absorption. In addition to the dominant digestive–absorptive cells, the gut epithelium also contains scattered mucus-secreting cells, plus scattered endocrine cells that secrete hormones involved in controlling the digestive tract (as you will see in Concept 30.5). Surrounding the epithelium and its underlying connective tissue (together called the mucosa) is a tissue layer called the submucosa, which contains blood and lymph vessels that carry absorbed nutrients to the rest of the body. The submucosa also contains a network of neurons that help control the secretory functions of the gut.

Two layers of smooth muscle surround the submucosa. The inner layer, composed of circular muscles, encircles the gut and constricts it when it contracts. The outer layer, composed of longitudinal muscles, shortens the gut when it contracts. Between the two muscle layers is a second network of neurons, which coordinates contractions of the muscle layers so that food materials in the gut are pushed along by waves of contraction called **peristalsis** or, at times, are pushed back and forth. The passage of materials through the gut is further controlled by sphincter muscles located at several places, such as the entrances to the stomach and small intestine. Sphincter muscles encircle the gut, and they contract so tightly that they seal off one part of the gut from another except when they are opened briefly in a controlled manner to let food materials pass through.

The neuron networks in the gut wall make up the enteric nervous system. This system is part of the autonomic nervous system, which controls many body functions without any need of conscious input and, in general, without voluntary control (see Concept 34.5).

An interesting aspect of the open central cavity of the gut tube, the gut lumen, is that it is anoxic, meaning it is devoid of O_2. Tapeworms and other gut parasites thus must have the ability to live without O_2, a very unusual ability for animals.

Several classes of digestive enzymes take part in digestion

Like most enzymes, the digestive enzymes are proteins. They break chemical bonds by hydrolysis, a reaction that splits a bond by adding a water molecule (see Figure 2.8). The digestive enzymes are classified in two ways. One is based on the type of food molecules the enzymes digest. Lipases hydrolyze bonds of lipids; proteases and peptidases hydrolyze the bonds between adjacent amino acids in proteins; and carbohydrases hydrolyze the bonds between adjacent sugar units of

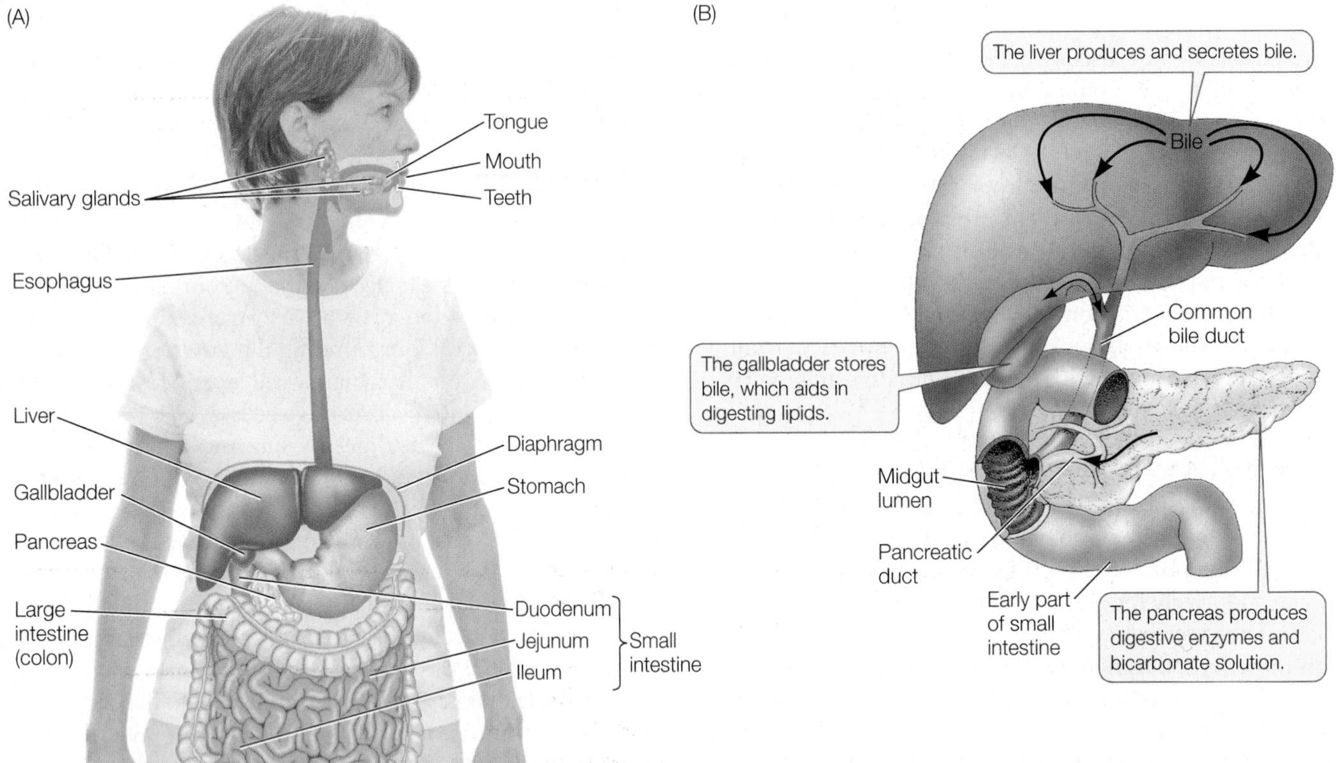

FIGURE 30.11 The Human Digestive System (A) The mouth, esophagus, and stomach are parts of the foregut. The small intestine corresponds to the midgut, and the large intestine corresponds to the hindgut. Three accessory glands shown are the salivary glands, liver, and pancreas. (B) The ducts via which secretions of the liver and pancreas enter the midgut lumen.

Go to ACTIVITY 30.3 The Human Digestive System
PoL2e.com/ac30.3

carbohydrates. Besides these general names, specific names are used for specific enzyme proteins.

Digestive enzymes are also classed by where they act:

- *Intraluminal enzymes* are secreted into the gut lumen and digest food molecules in the lumen.
- *Membrane-associated enzymes* are attached to the epithelial cells lining the gut lumen. They act when food molecules in the lumen come in contact with the gut epithelium.
- *Intracellular enzymes* are inside some of the gut epithelial cells and are involved in protein digestion.

Digestive enzymes typically act in sequences. Food proteins, for example, are first attacked by protein-digesting enzymes in the stomach that break the protein molecules into relatively large pieces. Other protein-digesting enzymes in the intestines then break these pieces into smaller pieces.

Processing of food starts in the foregut

The digestive tract begins with the **foregut**, which consists of the mouth, esophagus, and stomach—and in some animals, an expanded food-storage chamber (crop) or food-grinding chamber (gizzard) ahead of the stomach. When most vertebrates eat, they simply seize food with their mouth, then swallow it whole. Fish, amphibians, birds, and non-avian reptiles eat this way. Mammals, however, chew their food by using their teeth and jaw muscles, and as they chew, they start the process of carbohydrate digestion by secreting saliva from their salivary glands into their mouth cavity. Mammalian saliva contains a starch-digesting enzyme, amylase.

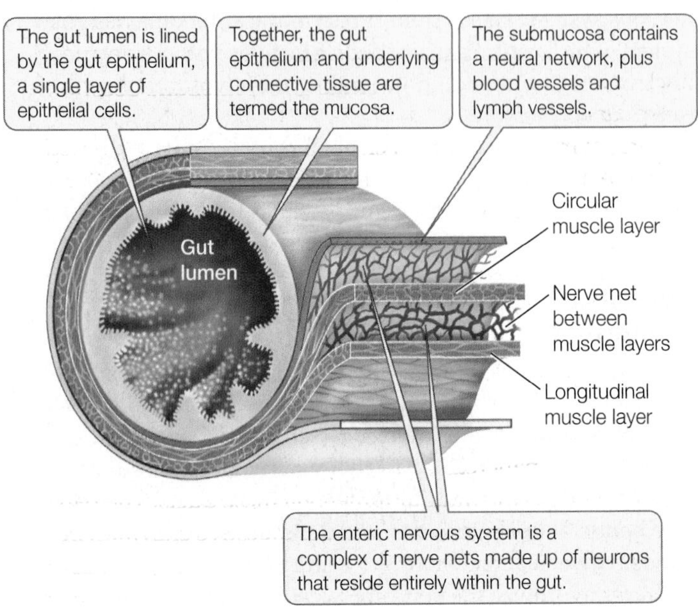

FIGURE 30.12 Tissue Layers of the Vertebrate Gut The organization of tissue layers is the same in all regions of the gut, but specialized adaptations of specific tissues characterize different regions.

Go to MEDIA CLIP 30.1
Following Food from Mouth to Gut
PoL2e.com/mc30.1

The esophagus typically serves simply as a passageway for food to reach the stomach from the mouth. Each time we swallow, a sphincter at the start of our esophagus opens briefly (under control of the enteric nervous system) to let food pass from our mouth into our esophagus. Then, after peristalsis has moved the food from the start of the esophagus to the end, the sphincter at the entrance to the stomach briefly opens to permit the food to enter the stomach.

The stomach is an expanded part of the foregut that, in most vertebrates, performs four major functions. First, the stomach serves as a storage chamber for newly ingested food. It can hold a large meal that is awaiting processing. Second, the stomach secretes hydrochloric acid (HCl), which mixes with food in the stomach lumen, helping cause food disintegration. When you think about the pHs that are sometimes achieved in the stomach, you will probably be amazed that living cells are capable of secreting the acid needed. In humans and dogs, the contents of the stomach lumen during meal processing can reach a pH of 0.8—the same as that of battery acid and ten times more acidic than lemon juice! Acid-secreting cells in the stomach epithelium secrete protons (H^+ ions) into the lumen, using energy from ATP, to achieve this astounding pH. The third major function of the stomach is to begin protein digestion. Certain cells in the stomach epithelium secrete protein-digesting enzymes called pepsins into the stomach lumen, where these enzymes start to break up proteins. Finally, the stomach mechanically squeezes food and mixes it with the acid and digestive enzymes. This is achieved by contractions of the smooth muscles in the stomach wall.

A fascinating challenge that had to be met during evolution arises from the fact that protein-digesting enzymes could potentially digest the very cells that synthesize them, killing the cells! The evolutionary solution to this challenge is that these enzymes are synthesized and secreted in inactive forms. Stomach cells synthesize and secrete pepsins as inactive pepsinogens, for example. Only after the pepsinogens arrive in the stomach lumen are they activated (by exposure to HCl) to form pepsins:

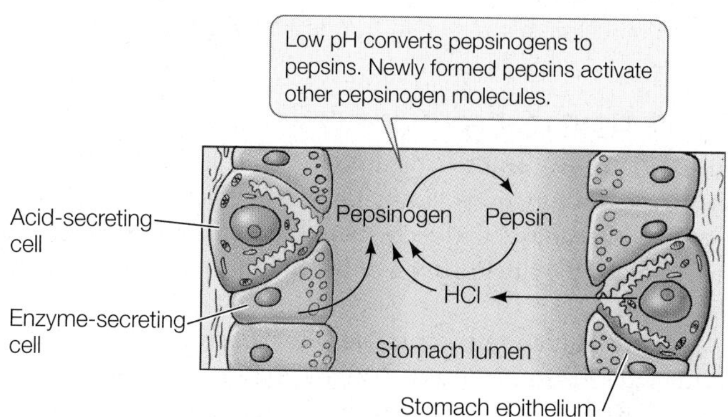
Low pH converts pepsinogens to pepsins. Newly formed pepsins activate other pepsinogen molecules.
Pepsinogen
Pepsin
HCl
Acid-secreting cell
Enzyme-secreting cell
Stomach lumen
Stomach epithelium

As we have noted, some vertebrates have evolved a fermentation vat for microbial symbionts in specialized, nonacidic parts of the foregut. All such animals are classified as **foregut fermenters**. The ruminants are one such group (see Figure 30.7). Kangaroos, hippos, sloths, colobus monkeys, and a few ground-dwelling, herbivorous birds are also foregut fermenters.

Food processing continues in the midgut and hindgut

After leaving the stomach, food travels through the **midgut** and **hindgut**, which are the first and second parts of the intestines. In humans and other mammals, the two parts differ in diameter: the midgut has a smaller diameter and is termed the small intestine; the hindgut has a larger diameter and is called the large intestine. The distinction between "small" and "large" does not apply in many vertebrates, however, because the two parts—although they differ in function—are the same in diameter. This is often the case in fish, for instance.

Fermentation vats for heterotrophic microbial symbionts have evolved as parts of the midgut or hindgut in some vertebrates. These animals, like the foregut fermenters, are all herbivores. Some fish, including catfish, carp, and tilapia, are **midgut fermenters**. Horses, zebras, rabbits (see Figure 30.8), rhinos, and apes are **hindgut fermenters**. So are geese, chickens, and ostriches. All profit from being able to ferment cellulose and hemicelluloses, greatly enhancing the energy value of plant foods.

The midgut is the principal site of digestion and absorption

The midgut is the principal site of digestion of all three major food categories: proteins, carbohydrates, and lipids. It is also typically the principal site of nutrient absorption. The products of digestion of all three categories of food are absorbed in the midgut, as are vitamins, minerals, and water.

In many vertebrates the inner midgut wall is highly folded, and the folds bear fingerlike projections called **villi** (singular *villus*) (**FIGURE 30.13**). In turn, each cell of the gut epithelium bears many microscopic projections called **microvilli**. Together, the midgut wall foldings, the villi, and the microvilli vastly enlarge the area of the luminal surface of the midgut epithelium (luminal surface = the surface facing toward the gut lumen). This huge surface area is of great functional importance for two reasons. First, many membrane-associated digestive enzymes are attached to the luminal surface of the gut epithelium. Second, absorption occurs across this surface.

Material from the stomach enters the midgut in a regulated manner. A sphincter muscle, the pyloric sphincter, seals off the stomach from the midgut when it is contracted. Under control of hormones and the enteric nervous system, this sphincter opens briefly to allow controlled, small squirts of digesting material to enter the midgut from the stomach. This measured introduction of material into the midgut ensures that the midgut digestive and absorptive processes are not overwhelmed and can keep up.

Both the pancreas and the liver are connected by ducts to the early part of the midgut (see Figure 30.11B). Digestive juices secreted by the pancreas and emulsifying agents secreted by the liver flow into the midgut lumen via these ducts. The pancreatic digestive juices contain a host of digestive enzymes, including ones that hydrolyze proteins, carbohydrates, and lipids. They also contain bicarbonate ions (HCO_3^-) that neutralize

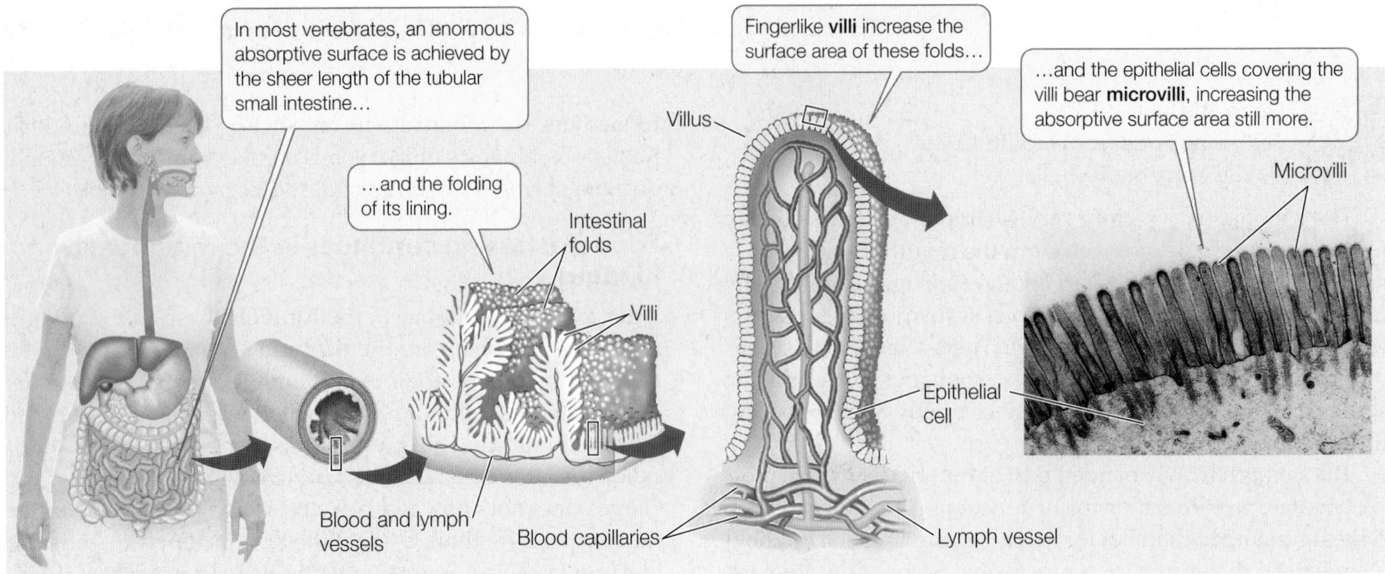

In most vertebrates, an enormous absorptive surface is achieved by the sheer length of the tubular small intestine...
...and the folding of its lining.
Intestinal folds
Villi
Blood and lymph vessels
Fingerlike villi increase the surface area of these folds...
Villus
Epithelial cell
Blood capillaries
Lymph vessel
...and the epithelial cells covering the villi bear microvilli, increasing the absorptive surface area still more.
Microvilli

FIGURE 30.13 Folds, Villi, and Microvilli Make the Surface Area of the Midgut Epithelium Enormous The large surface area increases the midgut's ability to absorb nutrients. It also increases the area of the surface to which membrane-associated digestive enzymes are attached.

the highly acidic pH of the material that enters the midgut from the stomach. The pancreas synthesizes and secretes protein-digesting enzymes as inactive molecules termed proenzymes. Only after they reach the midgut lumen are they enzymatically activated to hydrolyze food proteins.

Large food molecules are broken down into their smallest molecular subunits during digestion in the midgut. Proteins are broken down into amino acids; carbohydrates into simple monosaccharides (see Concept 2.3) such as glucose, galactose, and fructose; and lipids into fatty acids and glycerol (glycerin) (see Concept 2.4). These small molecules are then absorbed.

Let's now look briefly at the digestion and absorption of each of the major food categories. Recall that protein digestion starts in the stomach. It continues in the midgut lumen, where the products of stomach digestion are broken down further by protein-digesting enzymes made by the pancreas and gut epithelium (see Figure 30.9). These processes produce both single amino acids and short chains of two or three amino acids, which are absorbed into midgut epithelial cells. Inside these cells, the short chains are further hydrolyzed, so that the final product of protein digestion consists of just single amino acids. These are then transported into the blood vessels in the submuscosa, completing absorption.

In mammals, carbohydrate digestion starts in the mouth with the action of saliva enzymes that help break down starches. It is completed in the midgut. Often the final step in carbohydrate digestion in the midgut is the hydrolysis of a disaccharide—consisting of two linked simple sugar molecules—into two monosaccharide (simple sugar) molecules. The enzymes responsible are called disaccharidases. (Two of the midgut disaccharidases—lactase and trehalase—have been discussed earlier in this chapter.) The monosaccharides are then absorbed across the cells of the midgut epithelium to enter the submucosal blood vessels.

Lipid digestion starts and ends primarily in the midgut. Lipids present a special challenge for digestion because they are not water-soluble. The liver meets this challenge by secreting bile into the midgut. Bile salts and other bile components act as detergents, emulsifying food lipids in the midgut lumen into great numbers of tiny droplets. The lipids then have a high surface area, allowing lipid-digesting enzymes to break the lipids into fatty acids and glycerol. The bile salts then continue to associate with the fatty acids, keeping them in solution until the fatty acids are absorbed into midgut epithelial cells. An interesting characteristic of bile salts is that they are not modified by this process, and in fact are reabsorbed and then reused in processing one meal after another. After fatty acids and glycerol are absorbed into midgut epithelial cells, they are reassembled into simple lipid molecules, which form aggregates with proteins synthesized by the cells. These lipoprotein aggregates—in which the lipids are emulsified by the proteins—then pass into lymph vessels in the submucosa, and later they enter the bloodstream.

The cellular mechanisms by which digestive products enter and leave the gut epithelial cells during absorption include diffusion, facilitated diffusion, and secondary active transport. Of these, the most important is secondary active transport (see Concept 5.3), which requires ATP that the midgut epithelial cells synthesize.

LINK

The various mechanisms for moving molecules across membranes are discussed in **Concepts 5.2 and 5.3**

The hindgut reabsorbs water and salts

The hindgut—called the large intestine or colon in mammals—has two main functions. First, it stores indigestible wastes, in the form of accumulated feces, between defecations. The feces are stored in the final portion of the hindgut, the rectum, until eliminated.

Second, the hindgut completes the reabsorption of needed water and salts from the gut contents prior to elimination of the feces. This may sound like a humble function, but it is vitally important. In humans, about 7 liters of a watery solution containing Na^+, Cl^-, and other ions are secreted from the blood into the gut per day to facilitate processing of food. This large volume needs to be returned to the blood. Diseases such as cholera and dysentery can block reabsorption of water and salts by the hindgut, and in such cases a person can rapidly lose such massive quantities of water and salts through diarrhea that death quickly results.

CHECKpoint CONCEPT 30.4

- ✓ Digestive enzymes fall into three classes based on where they act. What are these three classes, and where do they act?
- ✓ Using correct terminology, explain how the cells that synthesize protein-digesting enzymes avoid being digested by the enzymes.
- ✓ Catfish, carp, and tilapia are some of the fish that are most useful for producing fish biomass for human consumption in fish farming (aquaculture). One reason for this is that they can be fed plant foods—in contrast, for example, to salmon, which are carnivores and therefore must be fed fish. In this context, why is it important that catfish, carp, and tilapia are fermenters?

We have seen how the vertebrate digestive system digests and absorbs a single meal. Finally we will look more closely at nutrition from meal to meal.

CONCEPT 30.5 The Processing of Meals Is Regulated

Most animals do not eat continuously. Instead, they alternate between an absorptive state (food in the gut) and a postabsorptive state (no food in the gut). This alternation poses two important challenges. First, for blood homeostasis to be maintained, the concentrations of glucose and other nutrient molecules in the blood plasma must be stabilized, so they do not skyrocket during meals and plummet between meals. Second, nutrients must be stored while they are abundant during the absorptive state and then released from those stores between meals.

Hormones help regulate appetite and the processing of a meal

Vertebrates have more than ten different types of scattered endocrine (hormone-producing) cells in the gut epithelium. The endocrine cells in one region secrete different hormones than the endocrine cells in other regions. This means that location-specific endocrine signals can be sent between gut regions to help coordinate the processing of a meal, complementing the coordination provided by the enteric nervous system.

When an animal eats and food arrives in its stomach, endocrine cells in the stomach epithelium secrete the hormone gastrin into the blood. Gastrin then travels in the blood to all the other stomach epithelial cells, some of which are specialized to synthesize acid (HCl) or pepsins. Gastrin stimulates these cells to secrete acid and pepsins, which help break down the food (**FIGURE 30.14**).

The entry of acidified, partly digested food material into the midgut initiates several important endocrine controls. Recall that the material is released into the midgut in small bursts. Each burst stimulates two distinct sets of endocrine cells in the midgut epithelium (actually there are more than two sets, but we will discuss two here). One set releases secretin (the first hormone ever identified) into the blood, and the other releases cholecystokinin (CCK). When these two hormones reach the stomach, they tend to inhibit stomach acid secretion and muscle contraction (see Figure 30.14). This inhibition is maintained until the midgut has processed the current burst of material from the stomach. The two hormones also affect the pancreas and liver. Secretin stimulates the pancreas to secrete bicarbonate, which flows in the pancreatic juice to the midgut, where it neutralizes the acidity of the recently arrived stomach material. Secretin and CCK, acting synergistically, also stimulate the pancreas to secrete its digestive enzymes—and the liver to secrete bile—into the midgut.

Appetite is also controlled in part by hormones. For example, when the stomach is empty, it secretes ghrelin (pronounced gre-lin), a hormone that stimulates appetite. Another hormone, leptin (named after a Greek word for "thin"),

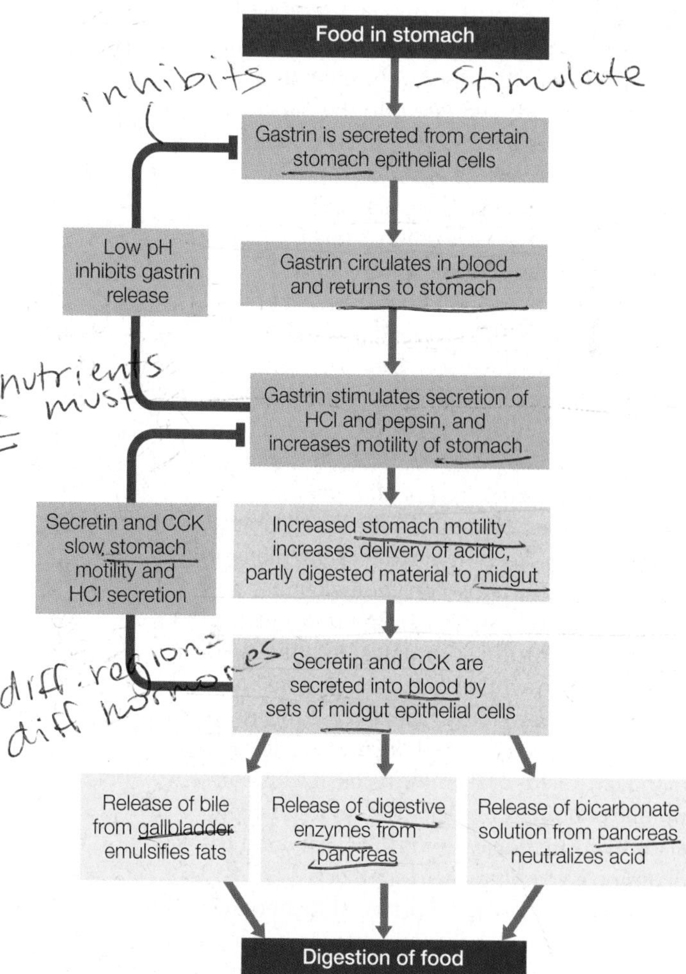

FIGURE 30.14 Hormones Control the Processing of a Meal The hormones gastrin, cholecystokinin (CCK), and secretin are involved in feedback loops that control the processing of food in the digestive tract. Red lines indicate inhibitory actions; green lines indicate stimulatory actions.

provides information to the brain to suppress appetite. Leptin is secreted by fat cells. High leptin levels indicate that there are many fat cells, and thus there is not a great need to eat. Leptin was discovered, and its properties are still being explored, through studies of mutant mice (**FIGURE 30.15**).

Go to ANIMATED TUTORIAL 30.1
Parabiotic Mice Simulation
PoL2e.com/at30.1

Insulin and glucagon regulate processing of absorbed food materials from meal to meal

Insulin and glucagon are two of the most important hormones that regulate blood nutrient levels—and the storage and release of nutrients in the body—as animals switch back and forth between the absorptive and postabsorptive states. Both hormones originate in the pancreas. The pancreas is composed of two major parts. The cells in one part perform the exocrine functions we have already discussed—secreting digestive enzymes and bicarbonate into the pancreatic ducts to be carried to the midgut. The other part of the pancreas consists of structures termed the islets of Langerhans. The cells here are endocrine cells that secrete hormones into the blood. One subset of the endocrine cells, called beta (β) cells, secretes insulin. Another subset, the alpha (α) cells, secretes glucagon.

As a meal is processed, glucose, fatty acids, and other products of digestion enter the blood in abundance. Insulin is secreted at this time. It stimulates many types of cells in the body to take up glucose. In this way insulin plays two major roles. It stabilizes the blood glucose concentration, keeping it from rising too high (**FIGURE 30.16**). It also promotes the use and storage of glucose when glucose is available in abundance. Insulin stimulates both liver and muscle cells to synthesize glycogen from blood glucose, thereby storing the glucose for future use. Insulin also has roughly parallel effects on blood fatty acids and amino acids. It stimulates uptake of both from the blood, storage of fatty acids by formation of lipids in fat cells, and use of amino acids for protein synthesis.

During the postabsorptive state between meals, insulin secretion stops, and the pancreas secretes glucagon. In many ways, glucagon's effects are the opposite of insulin's. Glucagon promotes both breakdown of glycogen into glucose molecules and release of glucose into the blood. In this way it helps stabilize the blood glucose concentration by preventing the

INVESTIGATION

FIGURE 30.15 A Single-Gene Mutation Leads to Obesity in Mice In mice the *Ob* gene codes for the protein leptin, a satiety (feeling of fullness) factor that signals the brain when enough food has been consumed. The recessive *ob* allele is a loss-of-function allele. Accordingly, *ob/ob* mice do not produce leptin; they do not experience satiety and thus become obese. The receptor protein in the brain that responds to leptin is coded by the *Db* gene. Mice homozygous for the recessive loss-of-function allele *db* cannot respond to leptin, even if they produce it, and they become obese.[a]

HYPOTHESIS

Mice that cannot produce the protein leptin will not become obese if they are able to obtain leptin from an outside source.

METHOD

1. Create two strains of genetically obese laboratory mice, one of which lacks functional leptin (genotype *ob/ob*) and one which lacks the receptor for leptin (genotype *db/db*).
2. Create parabiotic pairs by surgically joining the circulatory systems of a non-obese (wild-type) mouse with a partner from one of the obese strains.
3. Allow mice to feed at will.

Parabiotic pair

Wild-type mouse (*Ob/–* and *Db/–*) Genetically obese mouse (either *ob/ob* or *db/db*)

RESULTS

Parabiotic *ob/ob* mice obtain leptin from the wild-type partner and lose fat. Parabiotic *db/db* mice remain obese because they lack the leptin receptor and thus the leptin they obtain from their partner has no effect.

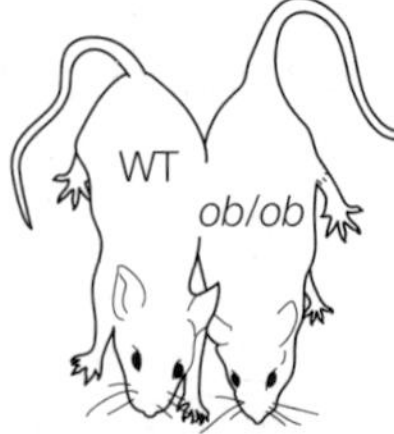

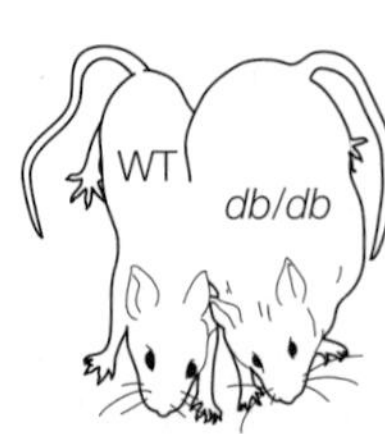

CONCLUSION

The protein leptin is a satiety signal that acts to prevent overeating and resultant obesity.

ANALYZE THE DATA

In a further experiment, leptin was purified and injected into *ob/ob* and *Ob/Ob* (wild-type) mice daily. The data in the table were collected before the injections began (baseline) and 10 days later.

	Baseline		Day 10	
Characteristic measured	*ob/ob*	*Ob/Ob*	*ob/ob*	*Ob/Ob*
Food intake (g/day)	12.0	5.5	5.0	6.0
Body mass (g)	64	35	50	38
Metabolic rate [ml O_2/(kg × hr)]	900	1,150	1,100	1,150
Body temperature (°C)	34.8	37.0	37.0	37.0

A. Do the data support the hypothesis that leptin is a satiety signal? Why or why not?

B. Discuss what factors might explain the loss of body mass in the *ob/ob* mice.

Go to **LaunchPad** for discussion and relevant links for all **INVESTIGATION** figures.

[a]D. L. Coleman and K. P. Hummel. 1969. *American Journal of Physiology* 217: 1298–1304.
D. L. Coleman. 1973. *Diabetologia* 9: 294–297.

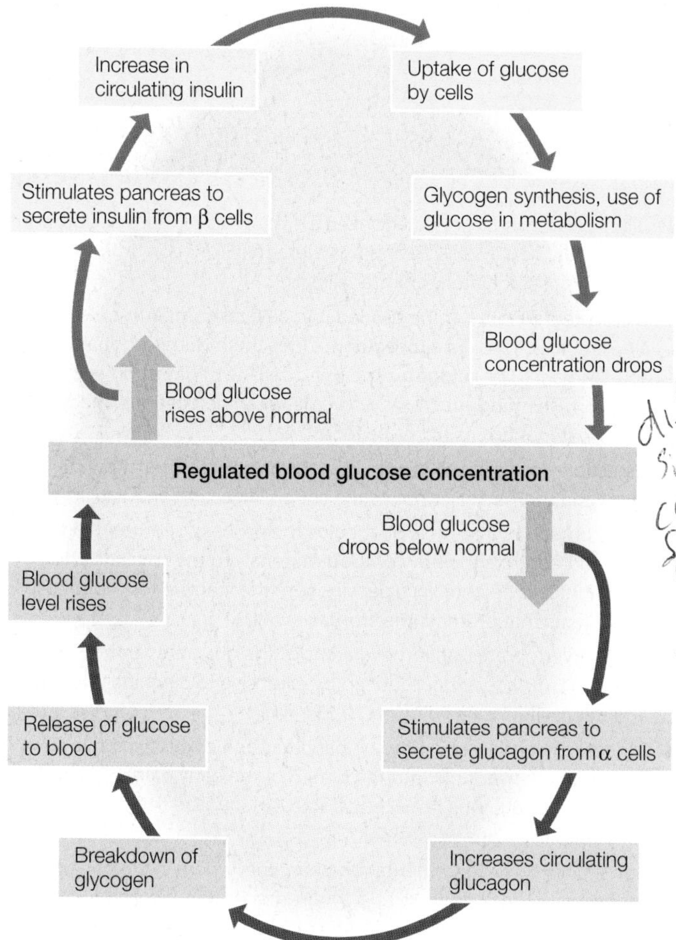

FIGURE 30.16 Regulating Glucose Concentration in the Blood Insulin (blue) and glucagon (brown) together maintain homeostasis in blood glucose concentration by negative feedback. Stability of blood glucose concentration is important for several reasons. One is that glucose, obtained from the blood, is the brain's essential source of metabolic energy.

Go to ANIMATED TUTORIAL 30.2
Insulin and Glucose Regulation
PoL2e.com/at30.2

concentration from falling too low (see Figure 30.16). It also ensures that cells throughout the body have a glucose supply even when foods are not being digested. In addition, glucagon stimulates breakdown of stored lipids and release of fatty acids into the blood.

CHECKpoint CONCEPT 30.5

- ✓ Blood glucose concentration is often said to be subject to negative feedback control by insulin. Explain.
- ✓ Why is it important that the endocrine cells in the gut epithelium do not all secrete the same hormone?
- ✓ Glucose is not normally found in a person's urine. Particularly in earlier decades, glucose in the urine was used as a diagnostic indication of diabetes. In one form of diabetes, termed Type I, the β cells in the islets of Langerhans are defective. Why would this condition lead to glucose in the urine?

Q What are some of the specific types of organic molecules that animals must obtain in their foods, and why do they need them?

ANSWER Animals must obtain all their essential amino acids in their foods because they are not able to synthesize those amino acids and will develop deficiency diseases if they have inadequate amounts (Concept 30.1). Many types of mammals require eight amino acids, called the essential amino acids: isoleucine, leucine, lysine, methionine, phenylalanine, threonine, tryptophan, and valine. Similarly, animals are unable to synthesize vitamins and essential fatty acids. To get adequate amounts, they must obtain these molecules in their foods. Two points should be stressed for fully understanding these concepts. First, the essential nutrients are not the same for all animals (Concept 30.1). Second, microbial symbionts sometimes provide essential nutrients, in which case animals do not literally need to eat the nutrients (Concept 30.2). Ruminants, for example, do not need to ingest certain required B vitamins—even though they are unable to synthesize them—because the microbial symbionts in their rumen synthesize the vitamins.

SUMMARY

CONCEPT 30.1 Food Provides Energy and Chemical Building Blocks

- Animals are heterotrophs that derive their energy and chemical building blocks from eating other organisms.
- A measure of the energy content of food is the **calorie** (**cal**), the amount of heat required to raise the temperature of 1 gram of water by 1°C. What we typically refer to as "calories" in our food are actually **kilocalories** (**kcal**). The metric unit for measuring energy is the **joule**. **Review Figures 30.1 and 30.2**
- There are three major types of food molecules: lipids (fats and oils), carbohydrates, and proteins. Lipids contain about twice as many calories per unit of weight as carbohydrates and proteins. Animals store extra energy as lipids and glycogen.
- Food provides carbon skeletons, termed **essential**, that animals cannot synthesize themselves.
- Adult humans require eight **essential amino acids** and at least two **essential fatty acids**. **Review Figure 30.3**
- **Vitamins** are essential carbon molecules needed in tiny amounts. They are either water-soluble or lipid-soluble. **Review Table 30.1 and ACTIVITY 30.1**
- **Essential minerals** are chemical elements that are required in the diet in addition to carbon, hydrogen, oxygen, and nitrogen. **Review Table 30.2 and ACTIVITY 30.2**
- **Malnutrition** results when any essential nutrient is lacking from the diet. Chronic malnutrition causes **deficiency diseases**.

(continued)

SUMMARY *(continued)*

CONCEPT 30.2 Animals Get Food in Three Major Ways

- Animals can be characterized by how they acquire their food. Some animals feed on easily visible food items that they eat by targeting them individually. **Suspension feeders** (or filter feeders) are aquatic animals that feed indiscriminately on large numbers of tiny food particles that they collect from the surrounding water. **Review Figures 30.4 and 30.5**
- Some animals rely on **symbiosis** with microbes to obtain at least some of their nutrients. Reef-building corals, for example, contain photosynthetic algae that supply carbon compounds to the coral. Microbes in the rumen of ruminants break down cellulose that the animals are unable to digest. **Review Figures 30.6 and 30.7**

CONCEPT 30.3 The Digestive System Plays a Key Role in Determining the Nutritional Value of Foods

- In most animals, digestion takes place in a tubular gut that has two openings: the mouth and the anus. Food is processed within the gut lumen. **Review Figure 30.8**
- **Digestion** is the enzymatic breakdown of food molecules into smaller molecules that can cross the gut epithelium. The process of transporting molecules from the gut lumen into the blood is called **absorption**.
- Each digestive enzyme can break only specific types of chemical bonds in food molecules. For example, the sugars trehalose and lactose are broken down by trehalase and lactase, respectively. **Review Figure 30.9**
- The nutritional value of a food depends on the ability of an animal's digestive tract to process the food in such a way that nutrients in the food can be absorbed.
- Animal species vary in the digestive enzymes they produce and thus in their ability to gain nutritional value from specific food molecules.
- An individual animal can adjust its digestive and absorptive capabilities in response to changes in its diet.

CONCEPT 30.4 The Vertebrate Digestive System Is a Tubular Gut with Accessory Glands

- The vertebrate digestive system consists of a tubular gut and several secretory organs—notably the liver and pancreas—that aid in digestion. **Review Figure 30.11 and ACTIVITY 30.3**
- The gut consists of an inner gut epithelium that secretes mucus, enzymes, and hormones and absorbs nutrients; a submucosa containing blood and lymph vessels; two layers of smooth muscle; and a nerve network called the enteric nervous system, which is part of the autonomic nervous system. **Review Figure 30.12**
- **Peristalsis** moves food along the length of the gut. Sphincters block food passage at certain locations but relax at controlled moments to let food through.
- Digestive enzymes are categorized on the basis of the types of food molecules they hydrolyze and by where they function (intraluminal, membrane-associated, or intracellular).
- The vertebrate gut can be divided into several compartments with different functions. The **foregut** includes the mouth, esophagus, and stomach. The **midgut** is the principal site of digestion and absorption. The **hindgut** stores waste between defecations and reabsorbs water and salts from the feces.
- Digestion begins in the mouth in mammals. Mammals chew their food, and their saliva contains the starch-digesting enzyme amylase.
- The stomach breaks up food, begins the process of protein digestion, and controls the flow of food materials to the small intestine. Stomach cells secrete HCl, pepsinogen (the inactive form of pepsin), and mucus that protects the stomach wall.
- In most vertebrates, absorptive areas of the gut are characterized by a large surface area produced by extensive folding, including **villi** and **microvilli**. **Review Figure 30.13**
- Digestive juices secreted by the pancreas and emulsifying bile secreted by the liver flow into the lumen of the early part of the midgut. Bicarbonate ions from the pancreas neutralize the pH of the food entering from the stomach.
- Some animals have fermentation chambers where symbiotic microbes help break down foods, synthesize B vitamins, and recycle nitrogen. The host animal is classified as a **foregut fermenter**, **midgut fermenter**, or **hindgut fermenter**, based on the location of the fermentation compartment within the digestive tract.

CONCEPT 30.5 The Processing of Meals Is Regulated

- Animals alternate between an absorptive state (food in the gut) and a postabsorptive state (no food in the gut).
- The actions of the stomach and small intestine are controlled by hormones such as gastrin, secretin, and cholecystokinin (CCK). **Review Figure 30.14**
- Food intake is governed by sensations of hunger and satiety, which are determined by brain mechanisms responding to feedback signals provided in part by the hormones ghrelin and leptin. **Review Figure 30.15 and ANIMATED TUTORIAL 30.1**
- Insulin and glucagon from the pancreas control the glucose concentration of the blood. Insulin stimulates the uptake and use of glucose by many cells of the body. If blood glucose concentration falls, glucagon secretion increases, stimulating the liver to break down glycogen and release glucose to the blood. **Review Figure 30.16 and ANIMATED TUTORIAL 30.2**

Go to the Interactive Summary to review key figures, Animated Tutorials, and Activities PoL2e.com/is30

Go to LaunchPad at **macmillanhighered.com/launchpad** for additional resources, including LearningCurve Quizzes, Flashcards, and many other study and review resources.

31 Breathing

KEY CONCEPTS

31.1 Respiratory Gas Exchange Depends on Diffusion and Bulk Flow

31.2 Animals Have Evolved Diverse Types of Breathing Organs

31.3 The Mammalian Breathing System Is Anatomically and Functionally Elaborate

Reinhold Messner on the summit of Mount Everest after ascending without supplemental oxygen.

Who knows when someone first tried to reach the top of Mount Everest? People have been living in the Himalayan highlands for 10,000–20,000 years, and it seems inevitable that some would have been intrepid explorers who climbed the mountains. We can be sure those prehistoric explorers had no tanks of compressed O_2. Whatever they accomplished, they did while breathing the natural air.

In the modern era of mountaineering, E. F. Norton was one of the first to try to reach Everest's summit while breathing natural air. In 1924 he reached 8,500 meters (28,000 feet)—just 300 meters (1,000 feet) from the top. Biologists at the time knew little about the challenges of extreme elevations. The overwhelming difficulty of climbing at 8,500 meters was a surprise. Norton tried to take charge of his situation by counting steps. Attempting to take 20 consecutive uphill steps, he never could and had to turn back.

When Reinhold Messner and Peter Habeler finally made the first recorded climb to the summit of Everest while breathing natural air in 1978, the extreme demands of the final 1,000 meters were well known. Messner is one of the most gifted mountaineers in history. By now he has climbed all 14 peaks in the world that are higher than 8,000 meters—all while breathing natural air. Yet he and Habeler required an entire hour to climb Everest's final 100 meters.

The most fundamental problem at high elevations is that the atmospheric pressure is far lower than at sea level. Of the molecules in the air, 21 percent are O_2 at all elevations. But because of the low pressure at high elevation, the air is dilute: the number of molecules per liter is reduced. An unusual aspect of this stress is that it is inescapable. People or other animals can escape the cold temperatures of high elevation by crawling into tents or burrows. The air they breathe, however, remains dilute everywhere. The breathing organs—the lungs—face an unavoidable challenge to take in enough O_2 to meet the body's needs. At the peak of Everest, the air has only a third as many molecules per unit of volume as at sea level.

The highest permanent human settlements are at 3,500–4,500 meters, where the air is about 60 percent as concentrated as at sea level. Prehistoric people established such settlements at three places on Earth: the Andes, the Tibetan plateau, and the Ethiopian highlands. Biologists have recently discovered that genetic adaptations to high elevation have started to evolve in these populations even though they have been living there for only a few hundred generations. Some species of birds and mammals, such as guanacos and vicuñas, are highly adapted because of their long evolutionary histories in the high mountains.

Why is climbing uphill at 1.7 meters per minute (100 meters/hour) the most intense work that is possible for a human near Everest's summit?

You will find the answer to this question on page 659.

CONCEPT 31.1 Respiratory Gas Exchange Depends on Diffusion and Bulk Flow

The **respiratory gases** that animals must exchange with their environment are oxygen (O_2) and carbon dioxide (CO_2). Cells require O_2 from the environment so they can produce ATP by cellular respiration (see Concept 6.2). Specifically, O_2 is needed in the mitochondria of each cell because electrons removed from food molecules must have a final resting place after they pass through the mitochondrial electron transport chain, releasing the energy that the mitochondria use to make ATP. Their final resting place is in combination with O_2 to make water (see Figure 6.9). This water simply mingles with the rest of the water in an animal's body.

Cellular respiration also produces CO_2 because CO_2 is one of the products of the citric acid cycle (see Figure 6.8). Unlike the H_2O produced by cellular respiration, CO_2 cannot be allowed simply to accumulate in an animal's body water. CO_2 is often called a "gaseous acid" because it tends to lower the pH of the body fluids if it accumulates in them. It can also have other disadvantageous effects. For these reasons, as an animal takes in O_2 from its environment and delivers it to cells for use in cellular respiration, the animal must rid itself of the CO_2 made by cellular respiration by releasing the CO_2 into its environment.

LINK

You can review aerobic cellular respiration and the production of ATP in **Concept 6.2**

Think about the path an O_2 molecule must travel from the atmosphere to a mitochondrion inside a muscle cell in your calf. First the O_2 molecule must travel deep into your lungs. Then it must cross two simple epithelia to enter the blood: first the epithelium forming your lung wall and then the epithelium forming the wall of one of your blood capillaries. The molecule must then travel in the blood from your thorax to your leg to reach a blood capillary next to the muscle cell. Then, to enter the cell, the molecule must cross the wall of the blood capillary (an epithelium) and pass through the cell membrane of the muscle cell. Finally, the molecule must travel through the cell cytoplasm to reach and enter a mitochondrion. CO_2 must make essentially the same journey in reverse.

Notice that some of the steps in an O_2 molecule's trip from atmosphere to mitochondrion are very short (**FIGURE 31.1**). The distance across each epithelium is about 1 micrometer (μm) or less. The distance across a cell membrane is roughly 0.01 μm. The distance from the cell membrane of a muscle cell to the deepest mitochondrion inside the cell is likely to be only 5–50 μm.

By contrast, some of the other steps in an O_2 molecule's trip are extraordinarily long. To pass deep into your lungs from your nostrils, an O_2 molecule needs to travel about 500,000 μm (0.5 meter). To travel in your blood from your lungs to a muscle cell in your calf, the molecule needs to travel about 1,000,000 μm!

Distance is therefore one of the factors we need to take into account as we consider the travels of O_2 and CO_2 molecules between the atmosphere and an animal's cells.

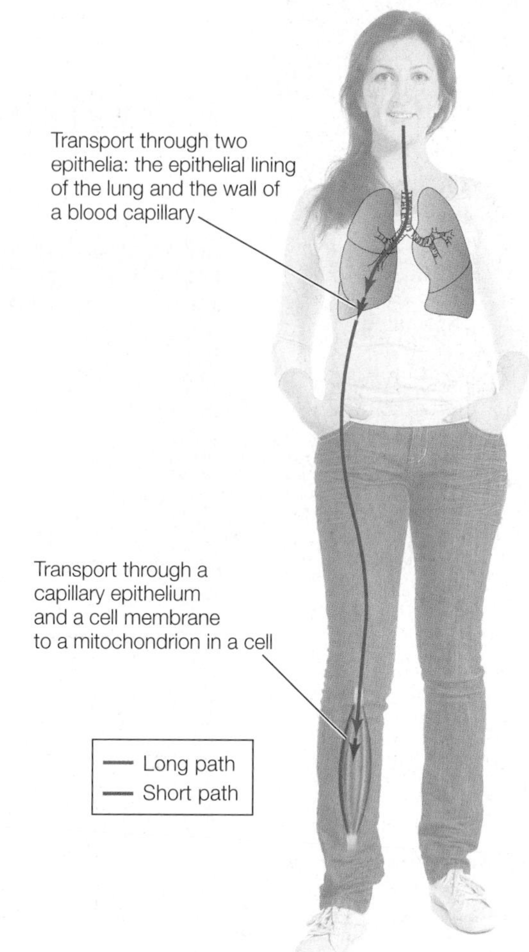

FIGURE 31.1 The Path of O_2 Molecules into an Animal's Body Consists of Long and Short Steps

Also note this important point: active transport mechanisms for O_2 do not exist (see Concept 5.3). This means that energy supplied by ATP cannot be used to drive O_2 through an epithelium or cell membrane. Instead, O_2 always crosses epithelia and cell membranes by diffusion. Much the same can be said for CO_2, although active transport of CO_2 is known in certain cases in the animal kingdom.

The diffusion of gases depends on their partial pressures

Atoms and molecules undergo ceaseless random motions at atomic–molecular scales of space. Diffusion results from these random motions (see Concept 5.2). If a certain type of molecule is more concentrated in one region of a solution than another, it is a simple statistical fact that random motions will carry more molecules away from the concentrated region than they will carry into it. In this way, the random motions cause a net transport of molecules from the concentrated region to neighboring dilute regions. This is the principle by which O_2 and CO_2 diffuse in a solution.

LINK

See **Concept 5.2** for further information about the diffusion of molecules in solution

Complexities arise where liquids and gases are in contact. When O_2 is present in a gas mixture, it is simply one of the types of gas molecules. For example, in air 21 percent of the molecules are O_2. When O_2 is present in a liquid, it is in solution. You have seen that when you dissolve sugar in water, the sugar molecules become invisible. O_2 does exactly the same thing. When it dissolves in water, the individual O_2 molecules fit in between water molecules and become invisible (bubbles, regardless of how tiny, are not in solution).

What happens when O_2 diffuses from a gas into an aqueous (watery) solution—so that the O_2 molecules switch from being in a gaseous to a dissolved state—or vice versa? How can we predict whether this diffusion will be from the gas to the liquid or in the opposite direction? Chemists have discovered that if you measure a property called **partial pressure**, you can solve this problem. O_2 always diffuses from where its partial pressure is high to where its partial pressure is low. Partial pressure is easily defined in a gas phase (not so easily in a liquid phase): it's simply the portion of the total pressure exerted by any particular gas, such as O_2, in a mixture of gases.

To clarify the relationship between concentration and partial pressure, suppose we have a gas phase in contact with a liquid phase, as in **FIGURE 31.2**. Within a gas phase, the concentration and partial pressure of O_2 are proportional to each other. O_2 always diffuses from where its concentration is high to where its concentration is low, for the reasons already discussed. Because concentration and partial pressure are proportional, O_2 also diffuses from where its partial pressure is high to where its partial pressure is low. Within a liquid phase, the same statements are true. Concentration and partial pressure are proportional, and O_2 always diffuses from places where both are high to places where both are low.

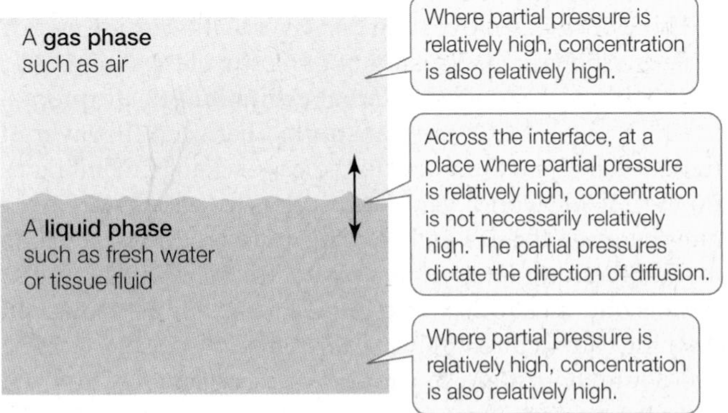

A gas phase
such as air
Where partial pressure is relatively high, concentration is also relatively high.
Across the interface, at a place where partial pressure is relatively high, concentration is not necessarily relatively high. The partial pressures dictate the direction of diffusion.
A liquid phase
such as fresh water
or tissue fluid
Where partial pressure is relatively high, concentration is also relatively high.

FIGURE 31.2 A Gas Always Diffuses from High Partial Pressure to Low Partial Pressure The concepts illustrated here apply to any gas. The gas is assumed to be in solution in the liquid phase. *Within* a gas phase or *within* a liquid phase, the partial pressure and concentration of a gas are proportional. Thus as the gas diffuses from high to low partial pressure, it also diffuses from high to low concentration. Concentration, however, is not a guide to diffusion across an interface *between* a gas and a liquid phase. Diffusion occurs from high to low partial pressure, but this does not necessarily mean the gas moves from where its concentration is high to where its concentration is low.

However, when O_2 diffuses *between* a gas phase and a liquid phase, its direction of diffusion cannot be predicted from its concentrations in the two phases. When diffusing from one phase to the other, O_2 moves from where its partial pressure is high to where it is low, even though this direction may be opposite to what concentrations would suggest (the partial pressures have the advantage that they take into account the effects of the solubility of O_2 in water, which the concentrations do not take properly into account). The same principles apply to all other gases, such as CO_2 and N_2.

The **law of diffusion for gases**—often called **Fick's law**—is expressed by the equation

$$\text{Rate of diffusion between two locations} = DA\frac{P_1 - P_2}{L}$$

where

- D, the diffusion coefficient, depends on the diffusion medium (air or water), the temperature, and the particular gas that's diffusing
- A is the cross-sectional area across which the gas is diffusing
- P_1 and P_2 are the partial pressures of the gas at the two locations
- L is the distance (diffusion path length) between the two locations

This equation applies to the diffusion of all gases, and it applies within a gas phase, within a liquid phase, and between gas and liquid phases. Partial pressures are expressed in units of pressure. Thus, the standard unit for partial pressure is the pascal (Pa). Alternative units in common use are atmospheres (atm) and millimeters of mercury (mm Hg), the latter referring to the height of a mercury column the pressure can support (1 atm = 760 mm Hg = 101.3 kilopascals).

Diffusion can be very effective but only over short distances

Let's ask the simple question, Can diffusion move lots of O_2 (or CO_2) from one place to another in a short period of time? The answer is both yes and no. This distinction is one of the most important concepts to understand about diffusion. As we'll see, the answer depends on the distance involved.

Notice that the distance between two locations, L, appears in the denominator of the diffusion equation. The rate of diffusion between two locations increases as the distance between the locations decreases.

Suppose O_2 on the outside of a cell is diffusing across the cell membrane into the cell. The distance, L, for this diffusion is very short—about 0.01 μm. Thus this diffusion will be very fast. In fact, if you were to arrange to have a high O_2 concentration outside a cell and a low concentration inside, diffusion would move enough O_2 into the cell in just *0.1 microsecond* to halve the concentration difference between outside and inside! This explains why all cells depend simply on diffusion for O_2 to pass into them through their cell membranes.

By contrast, suppose O_2 present in body fluids near your lungs is diffusing through your body fluids to a muscle cell in your calf. For this diffusion, L is much longer—about 1 meter. This diffusion therefore will be slow—far slower, in fact, than you might ever have imagined. If you had a high O_2 concentration near your lungs and a low concentration in your calf muscle, diffusion would require more than *30 years* to halve the concentration difference! Diffusion would obviously be totally useless for getting O_2 to your calf muscle.

We see, then, that diffusion can be very effective for moving O_2 over short distances but not over long distances. The founder of modern respiratory physiology, August Krogh, calculated an important rule of thumb for understanding diffusion. He analyzed a system in which O_2 had to diffuse through tissue to reach mitochondria that were carrying out cellular respiration at an average rate for animals, and he asked how thick the tissue could be. His answer was about 0.5 millimeter. Today, physiologists still use Krogh's rule of thumb and view 0.5 mm as the greatest distance over which O_2 can diffuse fast enough through tissue or tissue fluids to meet metabolic needs.

Gas transport in animals often occurs by alternating diffusion and bulk flow

Returning to Figure 31.1, we can see that diffusion is suitable for moving O_2 through the short steps in the path from the atmosphere to a muscle cell. It is not suitable, however, for the long steps.

This limitation explains why animals have often evolved processes of bulk flow to transport O_2 over long distances. **Bulk flow** refers to a flow of matter from one place to another. Winds and water currents are forms of bulk flow in the physical world. When you breathe, you create a bulk flow of air into and out of your lungs. The beating of your heart causes a bulk flow of blood through your blood vessels. Bulk flow can carry O_2 over long distances at high rates.

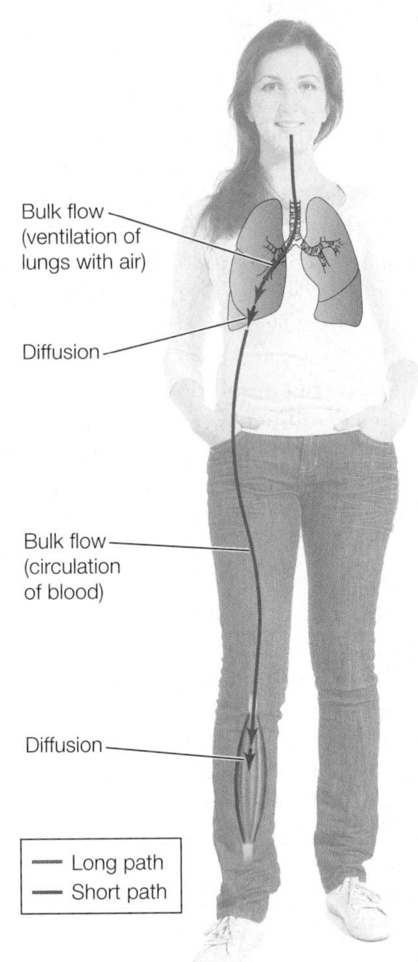

FIGURE 31.3 O_2 Transport into an Animal Often Occurs by Alternating Bulk Flow and Diffusion This alternation is clear if you read from top to bottom. Compare this figure with Figure 31.1.

LINK

The systems that circulate blood throughout the bodies of animals are described in **Concepts 32.1 and 32.2**

Diffusion and bulk flow alternate in humans and other mammals (**FIGURE 31.3**). O_2 is carried by bulk flow over the long distance from the atmosphere to the depths of the lungs. Then it diffuses across the thin epithelia in the lungs to pass into the blood. O_2 is then again carried by bulk flow, in the circulating blood, to reach the legs. Finally, O_2 diffuses out of blood in the legs to cross a blood-capillary epithelium, cross a cell membrane, and finally reach a mitochondrion.

Breathing is the transport of O_2 and CO_2 between the outside environment and gas exchange membranes

The **gas exchange membranes** of an animal are thin layers of tissue—usually comprising only one or two simple epithelia—where the respiratory gases move between the animal's environmental medium—air or water—and its internal tissues. Our own gas exchange membranes, for example, are the membranes deep in our lungs where O_2 is removed from the air and enters our blood and where CO_2 leaves our blood and enters the air. All animals have gas exchange membranes of some type. O_2 moves into the internal tissues from the layer of environmental medium immediately next to the gas exchange membranes, and CO_2 moves from the internal tissues into that layer.

How does O_2 from the outside world get to the layer of environmental medium next to the gas exchange membranes? And how does CO_2 move from that layer to the outside world? These questions are the subject of the study of breathing.

Breathing—also called **external expiration**—is the process by which O_2 and CO_2 are transported between the layer of environmental medium next to the gas exchange membranes and the outside world. **Ventilation** refers to bulk flow of air or water between the gas exchange membranes and the outside world. Breathing often occurs by ventilation. This is true in humans and most other vertebrates, for example. However, breathing can occur entirely by diffusion if the distances to be covered are short. Many tiny insects, such as gnats, are believed to breathe by diffusion.

Physiologists categorize breathing organs into two fundamental types: lungs and gills (**FIGURE 31.4**). **Lungs** are invaginated (folded inward) into the body, and their passages *contain* the environmental medium. **Gills** are evaginated (folded outward) from the body and are *surrounded* by the environmental medium. In general, air-breathing animals have lungs and water-breathing animals have gills, although exceptions occur.

When lungs are ventilated, air usually moves into them along lung passages during inhalation and then moves out

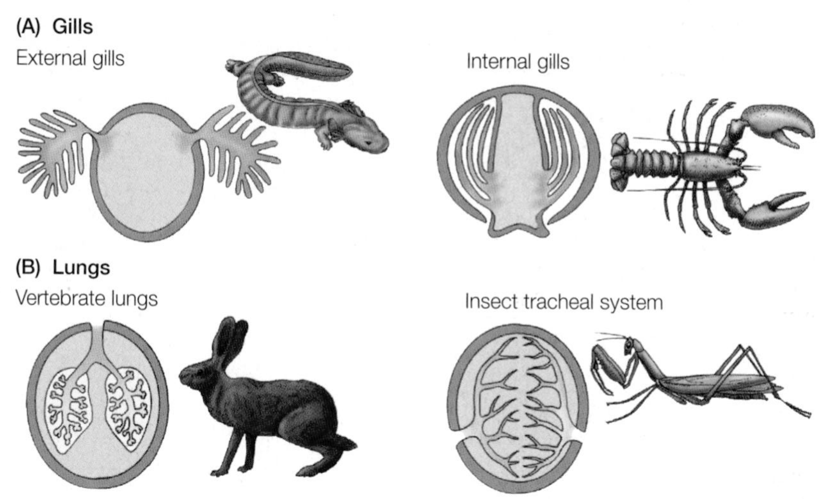
(A) Gills
External gills
Internal gills
(B) Lungs
Vertebrate lungs
Insect tracheal system

FIGURE 31.4 Gills and Lungs Physiologists classify all specialized breathing organs (blue in these diagrams) as either gills or lungs, depending on whether they are (A) evaginated structures (folded outward from the body) that are surrounded by the environmental medium or (B) invaginated structures (folded inward into the body) that contain the environmental medium. The use of the word "lungs" does not always match everyday usage. For example, the tracheal breathing system of an insect (discussed later in this chapter) is a lung because it is invaginated and contains air, but it is rarely called a lung in everyday language. Gills are subdivided into external gills (which project directly into the animal's environment) and internal gills (which project into a protective gill chamber that is ventilated with water).

through the same passages during exhalation. This ventilation is described as **tidal** because air flow occurs first in one direction and then in the opposite direction in the same airways—in and out like the ocean tides. Humans exemplify tidal ventilation.

In animals with gills, breathing usually involves pumping water in a one-way stream over the gill surfaces. This water movement is the ventilation of the gills, and the ventilation is described as **unidirectional**.

Air and water are very different respiratory environments

Diffusion takes place much faster through air than water (a fact reflected by the value of *D* in the diffusion equation). If all factors that affect diffusion are kept constant except that diffusion may be through air or water, the rate of diffusion of O_2 will be about 200,000 times faster through air!

Sea turtle nests that become flooded provide a striking illustration of the difference in diffusion rates through air and water (**FIGURE 31.5**). Loggerhead sea turtles lay their eggs about 40 centimeters deep on ocean beaches and cover them with sand. Nathan Miller outfitted nests with tiny probes that continually monitored the partial pressure of O_2 in the nests. He found that the eggs received plenty of O_2 by diffusion through the sand from the atmosphere above, provided the spaces between sand grains were full of air (Krogh's rule does not apply in air). However, if the sand became flooded with water for a long period—so the spaces between sand grains were water-filled—the O_2 partial pressure in the nest could fall to zero, causing the death of many eggs.

In addition to its slow rate of diffusion through water, O_2 also is not very soluble in water. Suppose we have some fresh water in an aquarium and we bubble it continuously with air so that the water and air are at equilibrium with each other (in this state they have the same O_2 partial pressure). At 12°C the air will contain about 200 milliliters (ml) of O_2 in each liter. The water, however, will contain only about 7.7 ml of O_2 in each liter! The solubility of O_2 in water is so low that the amount present per liter is only about 4 percent as high as in air—even when the water is steadily bubbled with air.

The solubility of O_2 in water decreases as water temperature rises. Thus the concentration of O_2 in water tends to decrease as water warms. Fresh water at 24°C, for example, dissolves 6.2 ml of O_2 in each liter, substantially less than the 7.7 ml per liter dissolved at 12°C. Most aquatic animals are poikilothermic (see Concept 29.3). Thus as water warms, their body temperatures and rates of O_2 consumption increase (see Figure 29.6). At the same time, the amount of dissolved O_2 available from

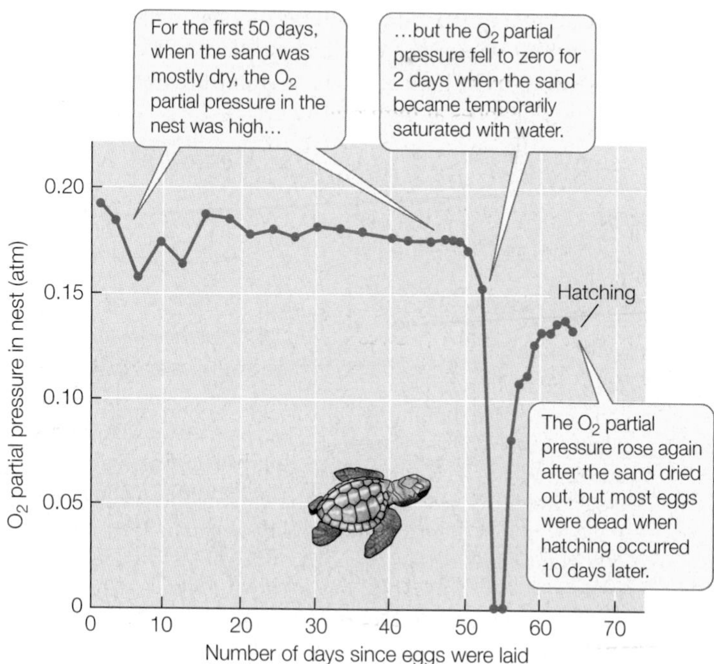
For the first 50 days, when the sand was mostly dry, the O_2 partial pressure in the nest was high...
...but the O_2 partial pressure fell to zero for 2 days when the sand became temporarily saturated with water.
Hatching
The O_2 partial pressure rose again after the sand dried out, but most eggs were dead when hatching occurred 10 days later.
O_2 partial pressure in nest (atm)
0.20
0.15
0.10
0.05
0
0
10
20
30
40
50
60
70
Number of days since eggs were laid

FIGURE 31.5 The O_2 Supply for Sea Turtle Eggs Depends on whether the Sand in which They Are Buried Is Dry or Wet The O_2 partial pressure in a nest of sea turtle eggs depends on how fast the developing turtles use O_2 and how fast O_2 is resupplied by diffusion from the atmosphere. During the first 50 days of development in the nest studied here, the sand was dry and O_2 diffused into the nest as fast as the developing turtles used O_2—keeping the O_2 partial pressure in the nest high. When the sand became soaked with water for 2 days, as the developing turtles used O_2, the O_2 they used was not replaced, and their metabolism reduced the O_2 partial pressure to zero.

APPLY THE CONCEPT

Respiratory gas exchange depends on diffusion and bulk flow

The need for O_2 and the availability of O_2 can clash in fish as water temperatures rise. Water contains much less O_2 than an equal volume of air, and it requires more effort to be pumped over respiratory gas exchange surfaces. Another factor that affects respiration in water-breathing animals is temperature, which affects both their O_2 needs and the solubility of O_2 in water. Answer the following questions by constructing a graph from the data in the table. Use water temperature as your *x* axis.

WATER TEMPERATURE (°C)	O_2 CONCENTRATION OF WATER (ml/L)	AVERAGE METABOLIC RATE OF INACTIVE FISH [ml O_2/(kg × hr)]	AVERAGE METABOLIC RATE OF ACTIVE FISH [ml O_2/(kg × hr)]
5	9.0	8	30
15	7.0	50	110
25	5.8	140	255
35	5.0	225	285

1. How does water temperature influence the O_2 available to a water-breathing animal?
2. Why does water temperature affect the O_2 demands of a fish?
3. From the data, what is the evidence that O_2 availability is a limiting factor for metabolism in warm water?
4. From the data, what general argument would you make about the effect of environmental warming on the viability of aquatic species?

each liter of water decreases. Water-breathing animals can find themselves caught in a respiratory trap, in which their O_2 needs increase as O_2 availability decreases. This can weaken them—interfering with growth or reproduction—and even kill them if the mismatch between O_2 needs and O_2 availability becomes too great.

All the animals alive today with the highest metabolic rates are air-breathers. Water is a challenging place to live in terms of getting enough O_2, if an animal is a water-breather.

CHECKpoint CONCEPT 31.1

✓ Where in the human body do inhaled O_2 molecules first cross a cell membrane?

✓ Review the equation for Fick's law on p. 645. What happens to the rate of diffusion as *A* increases? As *L* increases?

✓ Many ocean fish depend on the phenomenon of upwelling, where cold, nutrient-rich water rises up from the depths to mix with warm, nutrient-poor water near the surface. Which has the greater potential to store O_2, warm water or cold water? And how might the concentration of O_2 influence a fish's ability to metabolize nutrients available in its environment?

Now that we have examined the basics of respiratory gas exchange, let's look at the diversity of structures and processes by which animals obtain O_2 and get rid of CO_2.

CONCEPT 31.2 Animals Have Evolved Diverse Types of Breathing Organs

Some types of animals do not have specialized breathing organs. Instead, they have body plans in which most of their cells are close to a body surface. Planaria and other flatworms provide an example:

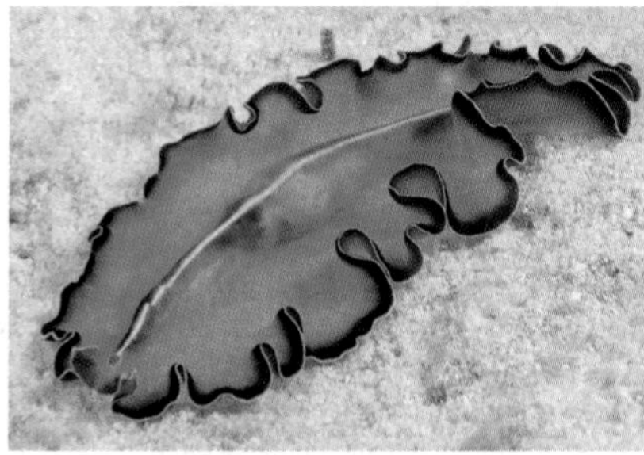

A flatworm (*Pseudobiceros* sp.)

Because of a flatworm's flat and thin body plan, O_2 can diffuse directly to each cell at a rate high enough to meet the cells' metabolic needs, simply by entering the body across an external surface. Likewise, CO_2 can exit the cells and the body by diffusion.

Sponges also lack specialized breathing organs:

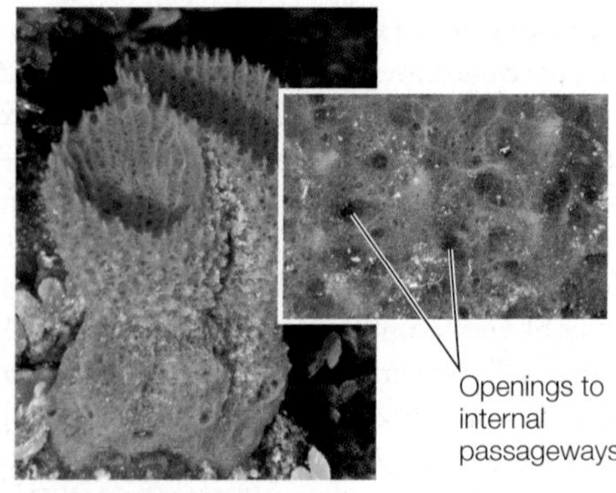

A sponge (*Niphates digitalis*)

The bodies of sponges are organized around an extensive system of internal passageways through which water flows continually, driven by beating flagella. All of a sponge's cells are close enough to one of these passageways to receive O_2 by diffusion from the environmental water flowing through. CO_2 leaves the cells in the same way.

Most types of animals, however, have evolved specialized breathing organs. The reason is that in most animal body plans, the great majority of cells are far from an external body surface. If we look at a snail, a lobster, a fish, or a bird, we quickly see that most cells are much farther than 0.5 millimeters from a body surface. Even if the external body surface were freely permeable to O_2 and CO_2, Krogh's rule of thumb tells us that diffusion through body fluids, directly from the outside environment, could not possibly meet the needs of most cells for respiratory gas exchange.

Specialized breathing organs have large surface areas of thin membranes

In what ways are lungs and gills specialized for breathing? One of their most important features is they have large surface areas of thin gas exchange membranes. Both of these properties—large surface area and thinness—make perfect sense when we look at the diffusion equation. As you'll recall, O_2 and CO_2 cross the gas exchange membrane by diffusion. The equation tells us that a large area, *A*, helps ensure a high rate of diffusion. The equation also tells us that a short diffusion path, *L*, as exists with a thin membrane, aids rapid diffusion.

Whether the breathing organs are lungs or gills, a large surface area is achieved by complex patterns of tissue folding and branching. These foldings and branchings are invaginated into the animal's body in the case of lungs, and evaginated in the case of gills (see Figure 31.4).

Consider your own lungs. The area of the gas exchange membranes in them is about 130 square meters! That's the same as the floor area of a classroom for 80 students. All that surface area can fit into the small volume of your lungs because of branching and folding. The branching resembles that of a tree. Thus, to understand it better, you can think of your lungs as an upside-down tree, where your trachea, or windpipe, corresponds to the tree trunk. Your trachea branches to form two tubes, then each of those tubes branches, and those branches also branch—and so on until the final branches are 23 steps removed from the trunk—the trachea. Each final branch ends in a closed sac of gas exchange membrane that is itself highly folded to increase its surface area. The result is 130 square meters of thin membrane with air on one side and blood on the other side. By virtue of both its large area and thinness, this membrane is ideally suited for O_2 to diffuse rapidly into your blood and for CO_2 to diffuse rapidly out.

How else are lungs and gills specialized for breathing? Another way is the presence of a ventilation mechanism. Without ventilation, our lungs would not be useful. Muscles in our thorax and abdomen expand our lungs when we inhale and sometimes help to contract our lungs when we exhale, causing a bulk flow of air in and out. This bulk flow is as essential as the gas exchange membranes for our lungs to work.

Finally, most breathing organs need circulatory specializations because cells throughout the body depend on blood circulation to receive O_2 taken up in the lungs or gills. **Perfusion** refers to blood flow through the capillaries or other small blood vessels of a tissue. Perfusion of the gas exchange membranes in a breathing organ must be rapid enough for the blood to pick up O_2 as fast as it is delivered by breathing. Then blood must circulate throughout the body at a high enough rate to deliver O_2 by bulk flow from the lungs or gills to meet the O_2 requirements of all the cells.

In short, most specialized breathing organs have three elements:

- A ventilation system that brings air or water rapidly to the gas exchange membrane
- A thin gas exchange membrane of large surface area, with air or water on one side and blood on the other side, for rapid diffusion
- A high rate of blood perfusion provided by the circulatory system, which also pumps blood rapidly between the lungs or gills and the rest of the body

The directions of ventilation and perfusion can greatly affect the efficiency of gas exchange

When ventilation is unidirectional, a key question is whether blood perfusion of the gas exchange membrane is in the same, or opposite, direction. Let's consider a water-breathing animal, although the principles we'll discuss apply whether the environmental medium is water or air. We can think of blood flow along the gas exchange membrane as being a fluid stream, and of water flow along the membrane as a second fluid stream. The entire system of blood and water flow is a **cocurrent system** (sometimes called a **concurrent system**) if the two fluid streams flow in the same direction. It is a **countercurrent system** if the two fluid streams flow in opposite directions. Gas exchange between blood and water is called **cocurrent gas exchange** in a cocurrent system and **countercurrent gas exchange** in a countercurrent system.

Keep in mind that as blood circulates in the body of a water-breathing animal, it flows back and forth between the gills and the systemic tissues. The **systemic tissues** in any animal are all the tissues other than those of the breathing organ. In the systemic tissues, the blood releases O_2 to the cells and picks up CO_2. Then the blood returns to the gills to unload CO_2 and pick up more O_2.

Also keep in mind how partial pressure affects the rate of gas diffusion. The diffusion equation shows this effect. For our present purposes, the partial pressures of interest are those in the water and blood at the gas exchange membrane. Thus we can think of P_1 and P_2 in the diffusion equation as being the O_2 partial pressures in the water and blood. The rate of diffusion is proportional to the difference between these two partial pressures.

Now let's follow the blood from the moment it arrives at the gas exchange membrane. At that moment, the blood has a low O_2 partial pressure because it has just returned from the systemic tissues. What is the O_2 partial pressure of the water the blood first encounters? In a cocurrent system (**FIGURE 31.6A**), this water is fresh and has a high O_2 partial pressure. Thus there is a large difference in O_2 partial pressure between the water and blood, and O_2 diffuses rapidly into the blood. However, as the blood and water flow in the same direction along the

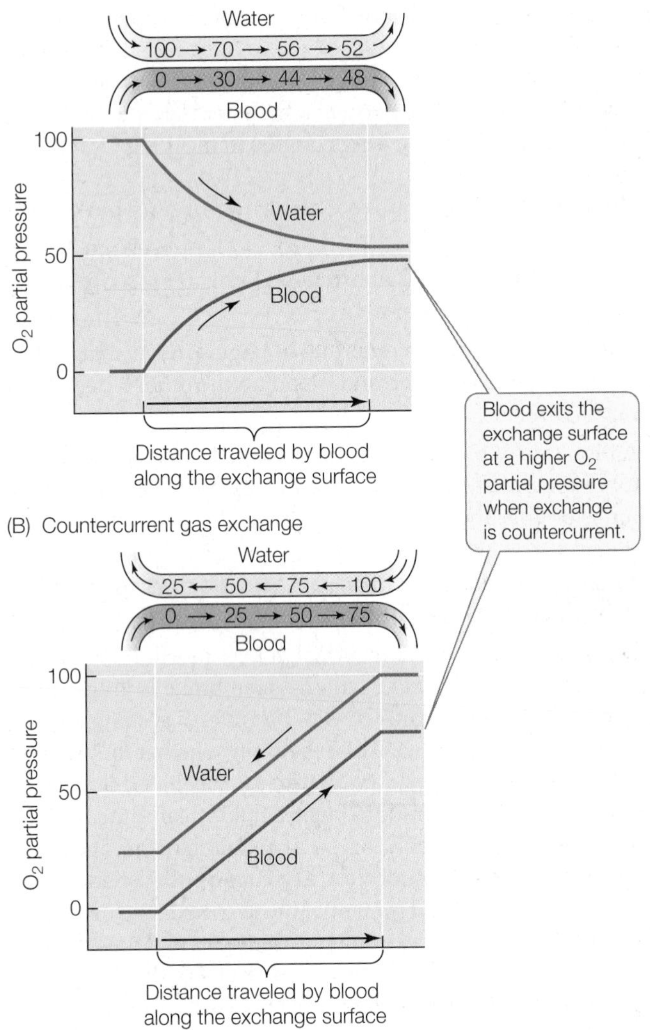

FIGURE 31.6 Countercurrent Gas Exchange Is More Efficient Than Cocurrent (A) Cocurrent gas exchange. (B) Countercurrent gas exchange. The diagram above each graph depicts the flow of blood and environmental water along the gas exchange membrane. Numbers are O_2 partial pressures in arbitrary units. The blood reaches a higher O_2 partial pressure when gas exchange is countercurrent.

gas exchange membrane in a cocurrent system, the O_2 partial pressure of the blood rises whereas that of the water falls, so the difference in partial pressure becomes smaller and smaller. Accordingly, the rate of O_2 diffusion into the blood becomes lower and lower.

A key question to ask is, When the blood finally leaves the gas exchange membrane, what is the O_2 partial pressure of the water with which it last undergoes gas exchange? As Figure 31.6A shows, that O_2 partial pressure is far lower than that in fresh environmental water. The O_2 partial pressure of the blood cannot be any higher. Thus the blood O_2 partial pressure is doomed to being far lower than the O_2 partial pressure of fresh environmental water.

A countercurrent system works in a very different way (**FIGURE 31.6B**). In this system, What is the O_2 partial pressure of the water that the blood first encounters? This water is stale and has a low O_2 partial pressure in comparison with fresh environmental water. Thus at the spot where blood first arrives at the gas exchange membrane, there is just a moderate difference in O_2 partial pressure between the water and blood in comparison to the difference in the cocurrent system, and O_2 diffuses into the blood at only a moderate rate by comparison. However, this moderate difference is maintained in the countercurrent system rather than becoming smaller as it does in the cocurrent system. As the blood flows along the gas exchange membrane and its O_2 partial pressure rises, the blood keeps encountering fresher and fresher water with a higher and higher O_2 partial pressure. For this reason, at every step of the way the O_2 partial pressure of the water is substantially higher than that of the blood. As a consequence, O_2 keeps diffusing into the blood at a substantial rate, even as the blood becomes richer and richer in O_2.

Now we come back to the same key question we asked earlier, When the blood finally leaves the gas exchange membrane, what is the O_2 partial pressure of the water with which it last undergoes gas exchange? In the countercurrent system, this water is entirely fresh environmental water and has a high O_2 partial pressure. The O_2 partial pressure of the blood can accordingly rise high enough to be relatively close to that of fresh environmental water.

In conclusion, blood leaves the gas exchange membrane with a higher O_2 partial pressure when exchange is countercurrent than when it is cocurrent. This difference occurs because, as blood flows through the gills, the blood takes up more O_2 from the water in a countercurrent system, where uptake continues at a substantial rate every step of the way along the whole length of the gas exchange membrane. A countercurrent system is more efficient in transferring O_2 from the environmental medium to the blood.

LINK

Countercurrent exchange is also important for thermoregulation in some animals; see **Figure 32.16**

Many aquatic animals with gills use countercurrent exchange

The gills of many aquatic invertebrates, such as clams and aquatic snails, employ countercurrent gas exchange. The gills of fish do also, and here we will look at those in detail.

The gills of a fish are supported by bone-reinforced gill arches positioned between the mouth cavity and the protective opercular flaps on the sides of the fish just behind its eyes (**FIGURE 31.7A**). The gill arches resemble vertical pillars with spaces—termed gill slits—in between. A fish pumps water unidirectionally into its mouth, through its mouth cavity, through its gill slits, over its gills, and out from under the opercular flaps.

Fish gills have an enormous surface area for gas exchange because they are extensively branched and folded. Each gill consists of two rows of long, flat filaments called gill filaments (**FIGURE 31.7B**). The upper and lower surfaces of each gill filament are folded into many flat folds—called secondary lamellae—that run perpendicular to the long axis of the filament

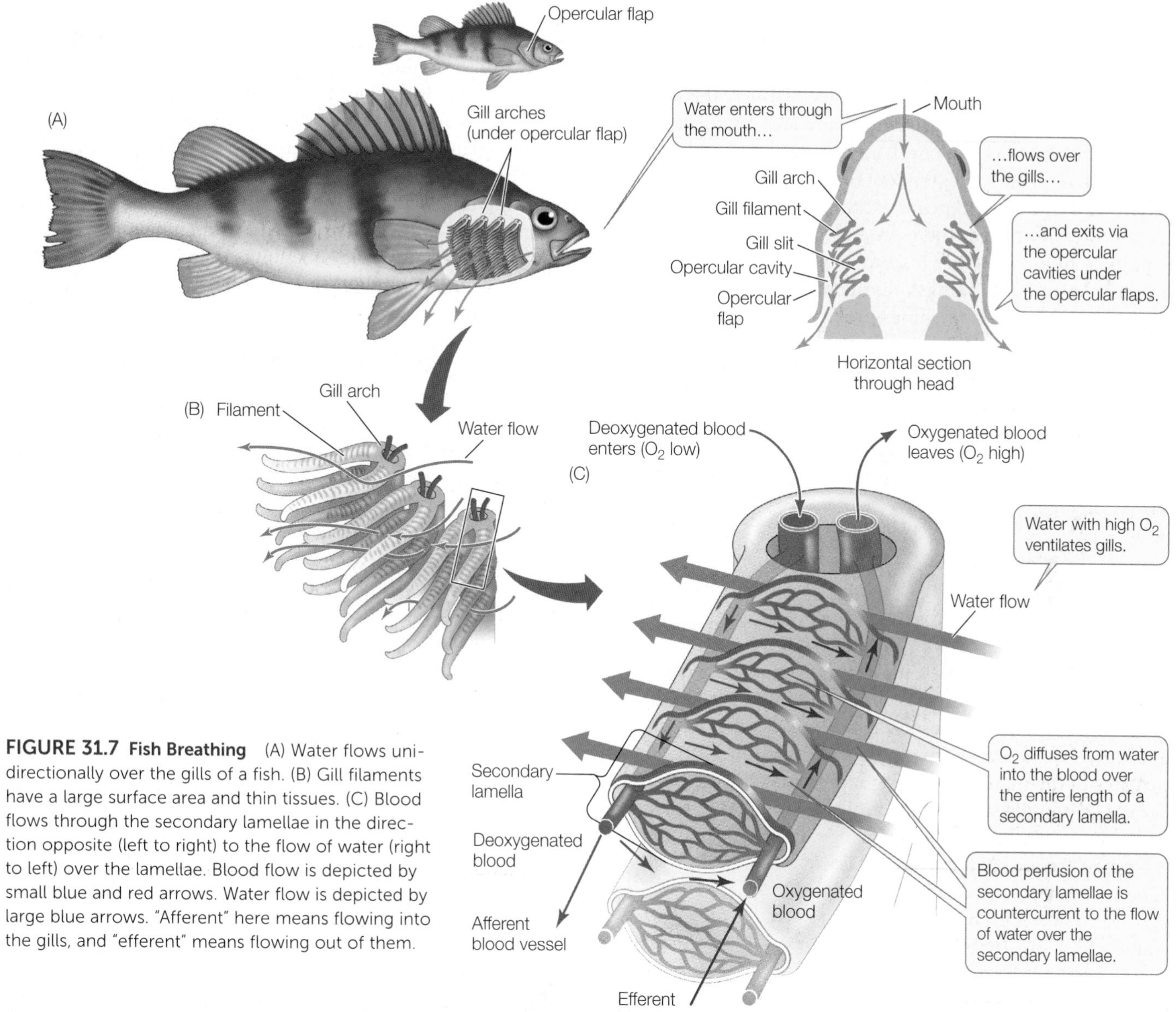

FIGURE 31.7 Fish Breathing (A) Water flows unidirectionally over the gills of a fish. (B) Gill filaments have a large surface area and thin tissues. (C) Blood flows through the secondary lamellae in the direction opposite (left to right) to the flow of water (right to left) over the lamellae. Blood flow is depicted by small blue and red arrows. Water flow is depicted by large blue arrows. "Afferent" here means flowing into the gills, and "efferent" means flowing out of them.

(**FIGURE 31.7C**). The secondary lamellae are the principal sites of respiratory gas exchange. Blood flows through small blood passages within each lamella, and these blood passages are separated from the water flowing over the gills by just a thin epithelium. This epithelium is the gas exchange membrane. Its total surface area is enormous because of the great numbers of filaments and secondary lamellae.

Blood flows unidirectionally through each secondary lamella, as Figure 31.7C shows, and the direction of water flow over the lamella is opposite to the direction of blood flow. Thus efficient, countercurrent gas exchange takes place. As blood enters a lamella, it is deoxygenated, or low in O_2 (see Concept 32.5). It gradually becomes oxygenated by taking up O_2 by diffusion from the water. If we focus on the spot where blood leaves a lamella, we see that the water on the outside of the lamella at that spot is water that has just started to flow across the gills. That water is accordingly high in O_2, ensuring that the blood is brought to the highest possible O_2 partial pressure before exiting the lamella and traveling to the systemic tissues.

Most fish ventilate their gills by means of a two-pump mechanism based on contractions of breathing muscles. One pump occurs in the mouth cavity. Muscles first expand the mouth cavity so it fills with water through the mouth. Muscles then contract the mouth cavity with the mouth closed, driving the water through the gill slits. The second pump occurs in the opercular cavities—the spaces under the opercular flaps. Muscles pull the opercular flaps away from the body wall, creating a sucking force that draws water through the gill slits into the opercular cavities. Muscles then pull the opercular flaps toward the body wall, discharging water.

Some highly active fish—such as tunas and some sharks—ventilate their gills by driving water across them as they swim. A swimming tuna holds its mouth open so that water is rammed into its mouth and across its gills by the fish's forward motion. Fish of this type can die if prevented from swimming.

Most terrestrial vertebrates have tidally ventilated lungs

The lungs of terrestrial vertebrates evolved as saclike outpocketings of the digestive tract. This ancient heritage is still reflected today in the fact that, except in birds and some crocodilians, the finest branches of the tubular parts of the lungs end in closed, dead-end sacs of gas exchange membrane. Because of this structure, ventilation of the gas exchange membranes cannot be unidirectional but must be tidal: fresh air flows in and exhaled gas flows out by the same route. Without unidirectional ventilation, cocurrent and countercurrent gas exchange are not possible.

As we already noted, there are 23 levels of branching in the lungs of humans. Other mammals have similar levels of complexity. Amphibian lungs are much simpler and help us understand what vertebrate lungs probably were like early in their evolution. An amphibian lung has an open central cavity, which is surrounded by intricately folded, well-perfused membranes:

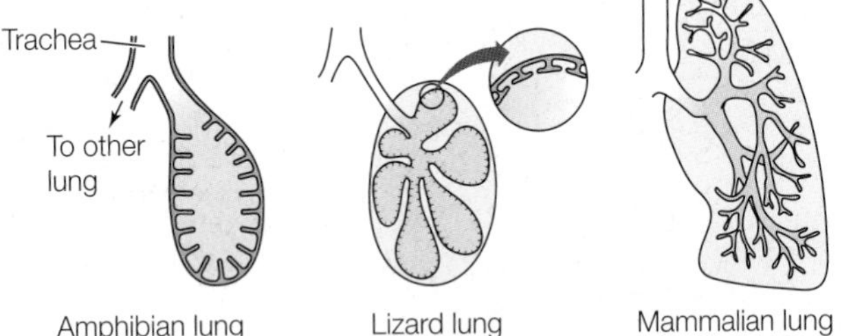

Compared with amphibians, some lizards exemplify increasing complexity in that each lung is subdivided into many separate parts served by separate airway tubes, instead of having a single, open central cavity (see above). The intricate branching of airways in the mammalian lung represents a further, more elaborate step in the same direction. The area of gas exchange membranes in the lungs of a mouse or rat is about ten times greater than the area in a mouse-size or rat-size lizard.

The inside surfaces of vertebrate lungs are coated with a thin layer of metabolically produced lipids and proteins called **lung surfactant**. By reducing surface tension at the surfaces of the lung epithelium, this material helps stabilize the complex three-dimensional structure of the gas exchange membranes. Infants born without adequate lung surfactant (a problem most often seen in premature infants) are in danger of dying unless they receive medical care because the interior surfaces of their lung membranes tend to stick together, requiring lots of effort to pull them apart with each inhalation.

In amphibians, the lungs are inflated by positive pressure. During each inhalation, the mouth cavity is first filled with air from the atmosphere. The mouth and nostrils are then sealed, and muscles contract the mouth cavity, raising the pressure high enough to drive air into the lungs. This mechanism is the same as that used by air-breathing fish (such as lungfish) to fill their air-breathing organs, suggesting that amphibians carry over the mechanism from their fishlike ancestors.

With the evolution of reptiles, a major transition in ventilation took place. Air is drawn into the breathing system by suction. This is true in today's reptiles, including birds. It is also true in mammals. In most non-avian (non-bird) reptiles and in mammals, the lungs are expanded during inhalation, creating a lower air pressure inside the lungs than in the atmosphere, thereby sucking air into the lungs (see Concept 31.3 for greater detail in humans).

Birds have rigid lungs ventilated unidirectionally by air sacs

Birds and mammals both have high metabolic rates and thus require sophisticated lungs that can supply O_2 at high rates to the rest of the body. The lungs of birds and mammals, moreover, are both believed to have evolved from lungs similar to those seen in today's non-avian reptiles. The lungs of birds, however, are radically different from those of mammals.

The lungs of birds are often described as "rigid." This does not mean they are solid or hard, but rather that they do not change much in volume as a bird inhales and exhales.

The breathing system of a bird includes air sacs as well as lungs (**FIGURE 31.8A**); the air sacs are the parts that undergo

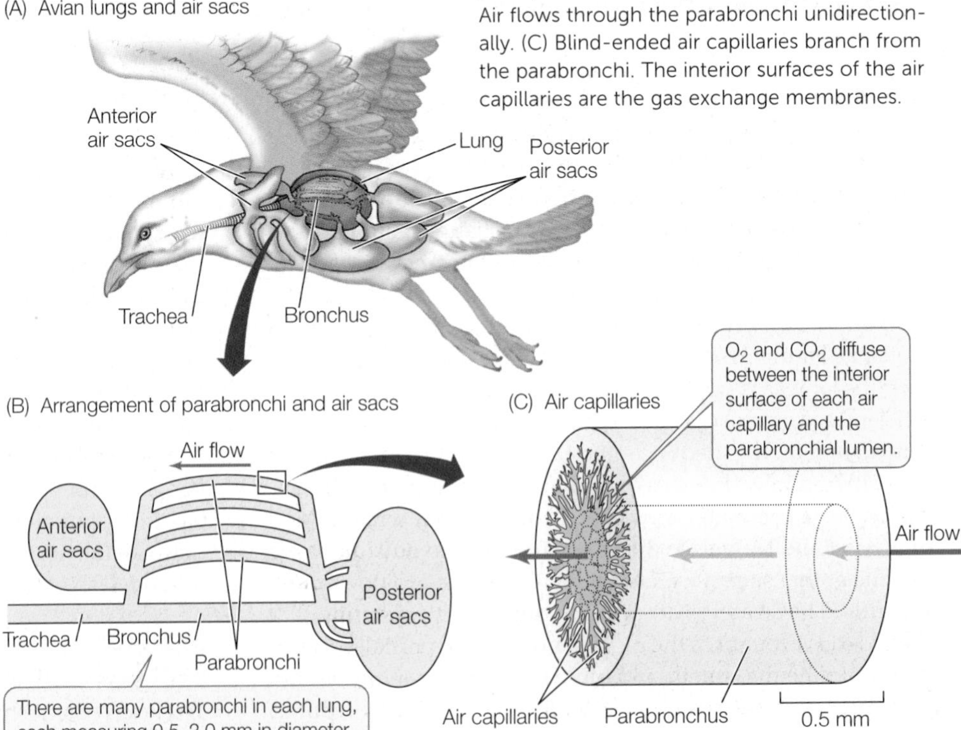

FIGURE 31.8 The Breathing System of a Bird (A) The lungs in the thorax are relatively rigid. The air sacs—of which there are usually nine—are located in the thorax and abdomen and serve as bellows to ventilate the lungs. (B) Arrays of numerous parabronchi in the lungs are the principal sites where gas exchange takes place. Air flows through the parabronchi unidirectionally. (C) Blind-ended air capillaries branch from the parabronchi. The interior surfaces of the air capillaries are the gas exchange membranes.

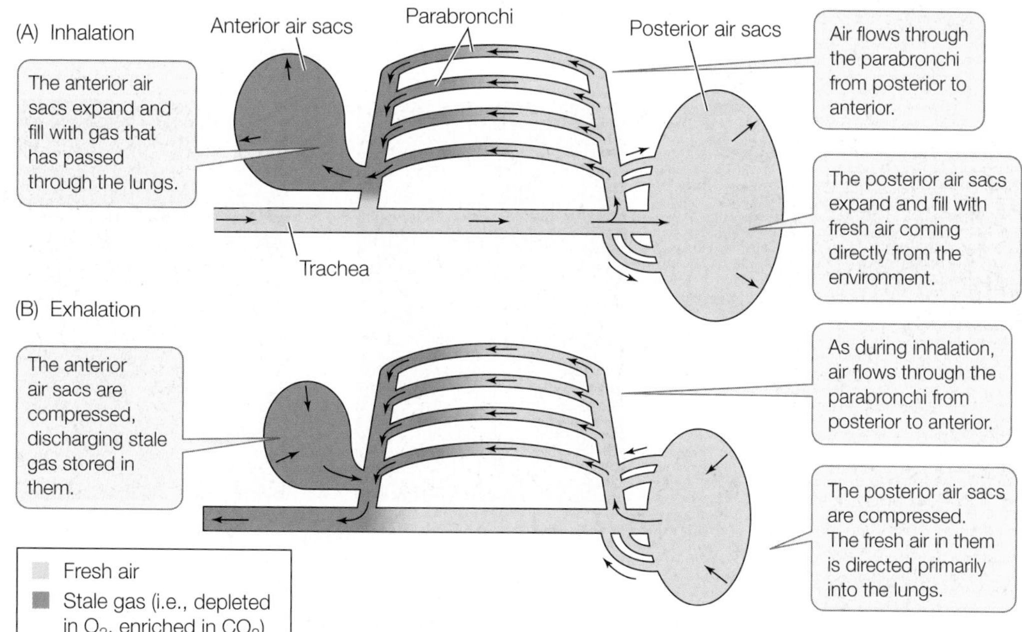
(A) Inhalation
Anterior air sacs
Parabronchi
Posterior air sacs
Air flows through the parabronchi from posterior to anterior.
The anterior air sacs expand and fill with gas that has passed through the lungs.
The posterior air sacs expand and fill with fresh air coming directly from the environment.
Trachea
(B) Exhalation
The anterior air sacs are compressed, discharging stale gas stored in them.
As during inhalation, air flows through the parabronchi from posterior to anterior.
The posterior air sacs are compressed. The fresh air in them is directed primarily into the lungs.
Fresh air
Stale gas (i.e., depleted in O_2, enriched in CO_2)

FIGURE 31.9 Airflow in the Lungs and Air Sacs of Birds Airflow during (A) inhalation, when air sacs expand, and (B) exhalation, when air sacs contract. Air flows from posterior to anterior through the parabronchi during both inhalation and exhalation.

large changes in volume as a bird inhales and exhales. These thin-walled sacs connect to airways in the lungs and fill much of the internal space in a bird's body. The air sacs do not participate in exchange of O_2 and CO_2 between the air and blood. Instead, their function is ventilation. They expand and contract during inhalation and exhalation, acting like bellows and forcing air to flow past the gas exchange membranes in the rigid lungs.

Long, tiny tubes called parabronchi (singular parabronchus) are the most important airways for gas exchange in the lungs of a bird (**FIGURE 31.8B**). As astounding as it may sound, air flows through most of a bird's parabronchi *unidirectionally*, from posterior to anterior, during all phases of its breathing cycle, whether the bird is inhaling or exhaling! Air flows unidirectionally through the parabronchi because of coordinated actions of thoracic and abdominal muscles that expand and contract the anterior and posterior sets of air sacs (**FIGURE 31.9**). When a bird inhales, some of the incoming fresh air travels to the posterior of each lung and enters the parabronchi. That air then flows through the parabronchi from posterior to anterior and exits into anterior air sacs. Some of the incoming fresh air also inflates the posterior air sacs. Then, when the bird exhales, this fresh air in the posterior air sacs is driven into the parabronchi, again flowing from posterior to anterior. As this occurs, stale air exiting the parabronchi at the anterior end, plus stale air from the anterior air sacs, is exhaled into the atmosphere.

Go to ANIMATED TUTORIAL 31.1
Airflow in Birds
PoL2e.com/at31.1

Tiny air capillaries extend into the walls of each parabronchus from the parabronchial lumen (**FIGURE 31.8C**). These air capillaries are so minute that if the openings of 1,000 of them were lined up in a row, they would cover only 1 centimeter or less. The walls of the air capillaries are richly perfused with blood and serve as the gas exchange membranes. As air moves by bulk flow through the parabronchi, O_2 and CO_2 are thought to move in and out of the tiny air capillaries by diffusion. Gas exchange with the blood is termed "cross-current"—a type of exchange intermediate in efficiency between cocurrent and countercurrent.

Insects have airways throughout their bodies

In insects (and spiders), a breathing system has evolved that is unlike that of any other animals (**FIGURE 31.10**). This breathing system is a type of lung because it consists of gas-filled tubules that are invaginated into the body from the body surface. The tubules branch everywhere in an insect's body. In fact they branch in such fantastically intricate ways that a gas-filled tubule comes close to every cell. In some cases tubules grow into cells, indenting the cell membrane inward. The tubules connect to the atmosphere at several openings, called spiracles, on the outside surface of an insect's body. O_2 makes its way directly from the atmosphere, through the gas-filled tubules, to each cell. CO_2 moves directly from all cells to the atmosphere. Because of this unique breathing system, blood flow is generally unimportant in delivering O_2 and getting rid of CO_2.

Most of the gas-filled tubules are called **tracheae**, and for this reason, insects are said to have a **tracheal breathing system**. The tracheae are lined with a thin, gas-permeable layer of chitin, which provides support and keeps them from collapsing shut. The finest tubules are called tracheoles. Air sacs are connected to the tracheae in many insects.

The air inside the tracheal breathing system is believed to be completely still in some insects. In these cases O_2 and CO_2 travel everywhere by diffusion. Keep in mind that Krogh's rule of thumb refers just to diffusion through water. Diffusion in air is much faster (200,000 times faster in the case of O_2) because random molecular movements are more vigorous in gases than in water. Because of the fast diffusion that can occur in air, insects can be moderately large (e.g., 0.5 cm long) and still have their gas exchange needs met by diffusion in the tracheal system, provided they are not very active.

In insects that are large or active, mechanisms have evolved to create bulk flow of air inside the tracheal system. Grasshoppers, for example, pump air through their tracheae

(A)

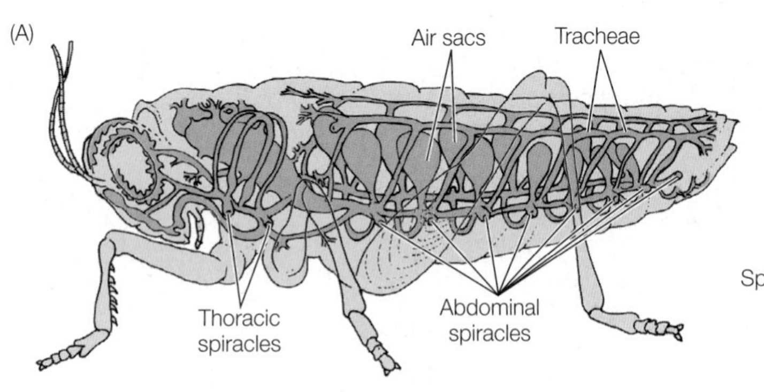

(B)

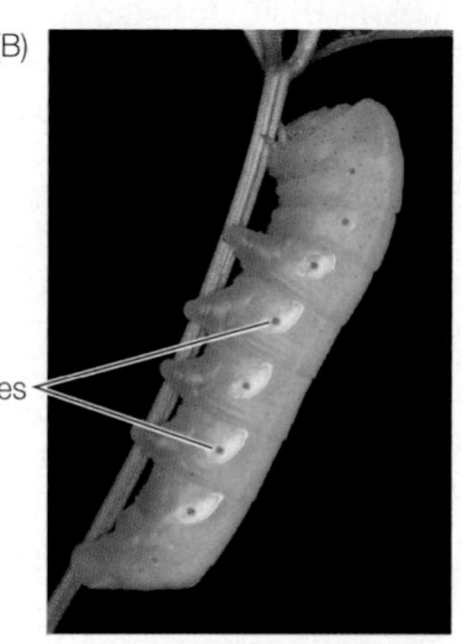

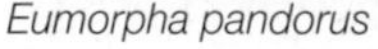

(C)

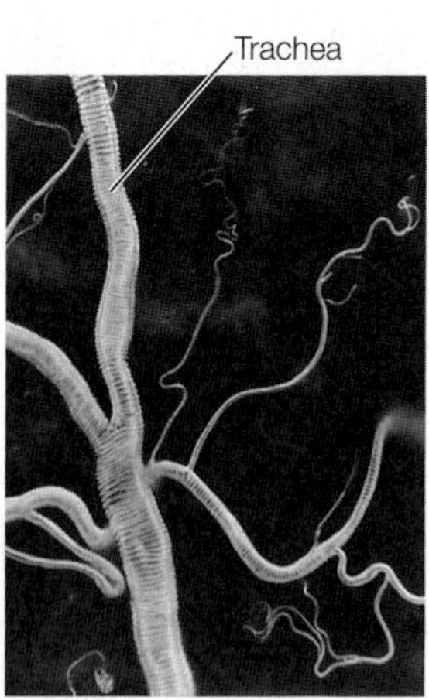

FIGURE 31.10 The Tracheal Breathing System of Insects
(A) The insect breathing system consists of elaborate arrays of branching, gas-filled tubes called tracheae, which open to the atmosphere through pores called spiracles. In many insects, air sacs connect to the tracheae and can act as bellows to cause tracheal ventilation. (B) The spiracles of a Pandorus sphinx moth larva are visible on the sides of its body. (C) This scanning electron micrograph shows an insect trachea dividing into smaller tracheae and tracheoles. All cells in an insect's body are close enough to one of these branches to receive O_2 directly from the tracheal system by diffusion.

by contracting and elongating their bodies. Bees use air sacs as bellows to drive air through their tracheae. The tracheae throughout an insect's body sometimes undergo steady, rhythmic contractions.

CHECKpoint CONCEPT 31.2

- ✓ What process is responsible for the movement of O_2 into a flatworm's body?
- ✓ As blood flows along the gas exchange membrane, how does the difference in O_2 partial pressure between the environmental medium and blood vary in countercurrent and cocurrent systems?
- ✓ Why can't diffusion alone supply fresh O_2 molecules to the breathing systems of many organisms?
- ✓ Mammals, birds, and insects are the three groups of animals in which species with the highest known metabolic rates occur. Of these three groups, two—mammals and birds—have high-performance circulatory systems that can circulate blood rapidly throughout the body and perfuse tissues at high rates. Insects do not have high-performance circulatory systems. Explain.

Now let's take a closer look at the anatomy and function of mammalian lungs, focusing on the human breathing system.

CONCEPT 31.3 The Mammalian Breathing System Is Anatomically and Functionally Elaborate

In mammals, air enters the lungs through the oral cavity and the nasal passages, which join together in the pharynx, or throat (**FIGURE 31.11A**). Below the pharynx, the esophagus, or food tube, directs food to the stomach, and the trachea, or windpipe, leads to the lungs. At the beginning of the trachea is the larynx, or voice box, which houses the vocal cords. The walls of the trachea and many further branches of the airways contain rings of cartilage, which help prevent airway collapse.

The trachea branches into two primary bronchi, one leading to each lung. Each primary bronchus then branches more than 20 times to generate the treelike structure of progressively smaller airways we discussed earlier (**FIGURE 31.11B**). At the farthest reaches of the branching airways, each tiny branch ends blindly in an **alveolar sac** of epithelial tissue (**FIGURE 31.11C**). The wall of each alveolar sac consists of numerous semispherical pocket-like structures called **alveoli** (singular *alveolus*). These alveoli number about 300 million in the two lungs of an adult person, creating an enormous surface area (**FIGURE 31.11D**) where gas exchange occurs (**FIGURE 31.11E**).

Most branches of the airway system in a mammal are not involved in the exchange of O_2 and CO_2 between the air and blood. These branches—including the trachea, primary bronchi, and most of the succeeding branches—are called the **conducting airways** because their principal function is to conduct flowing air in and out of the lungs during inhalation and exhalation.

Exchange of O_2 and CO_2 between the air and blood occurs in the **respiratory airways**. These are the parts of the lungs where the membranes between air and blood are thin and perfused with blood at high rates. The respiratory airways include the last few branches of the branching airway system—termed respiratory bronchioles—and the alveolar sacs, including the alveoli. In healthy human lungs, blood and air are separated by only about 0.5 μm in the alveoli. Bulk air flow—inflow and outflow—takes place in the respiratory bronchioles as a person inhales and exhales. But the air inside the alveoli is motionless. This means that as O_2 and CO_2 travel to and from the gas exchange membranes, they must diffuse across the central cavities of the alveoli. Moreover, as we already discussed, O_2 and CO_2 must diffuse across the two thin epithelia that line the alveoli and the surrounding blood capillaries. Diffusion occurs rapidly because the distances to be covered are minute.

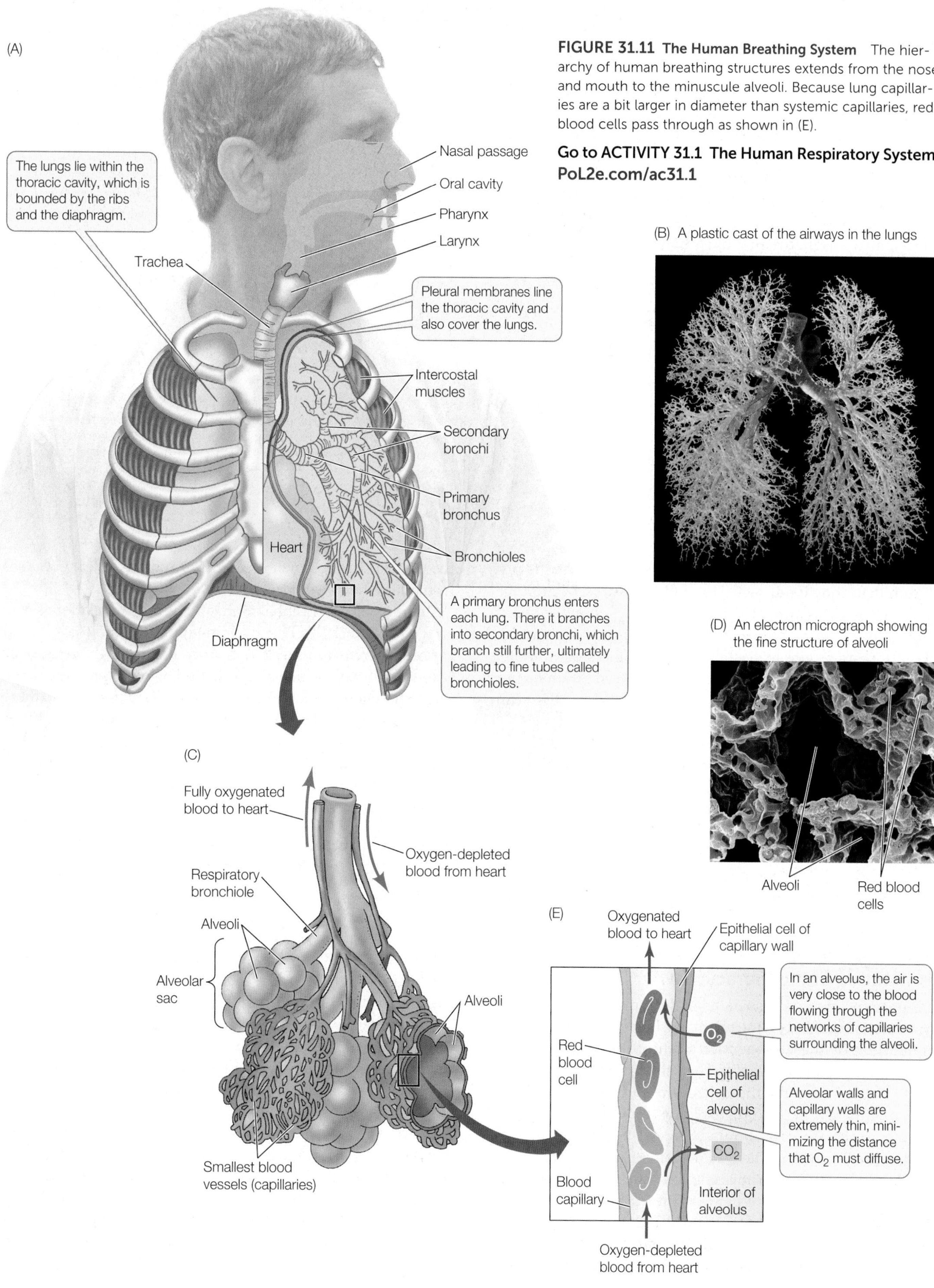

FIGURE 31.11 The Human Breathing System The hierarchy of human breathing structures extends from the nose and mouth to the minuscule alveoli. Because lung capillaries are a bit larger in diameter than systemic capillaries, red blood cells pass through as shown in (E).

Go to ACTIVITY 31.1 The Human Respiratory System PoL2e.com/ac31.1

At rest, only a small portion of the lung volume is exchanged

The amount of air that moves in and out of the lungs per breath is called the **tidal volume** (**FIGURE 31.12**). When a person is at rest, it is called the resting tidal volume and amounts to about 0.5 liter. The maximal tidal volume—also called vital capacity—is the amount of air that moves in and out per breath when we inhale and exhale as vigorously as possible. It is about 5 liters. We use only about 10 percent of our maximal tidal volume when we breathe at rest.

RESEARCH TOOLS

FIGURE 31.12 Measuring Lung Ventilation A spirometer is a device that measures tidal volume: the volume of air a person breathes in and out per breath.

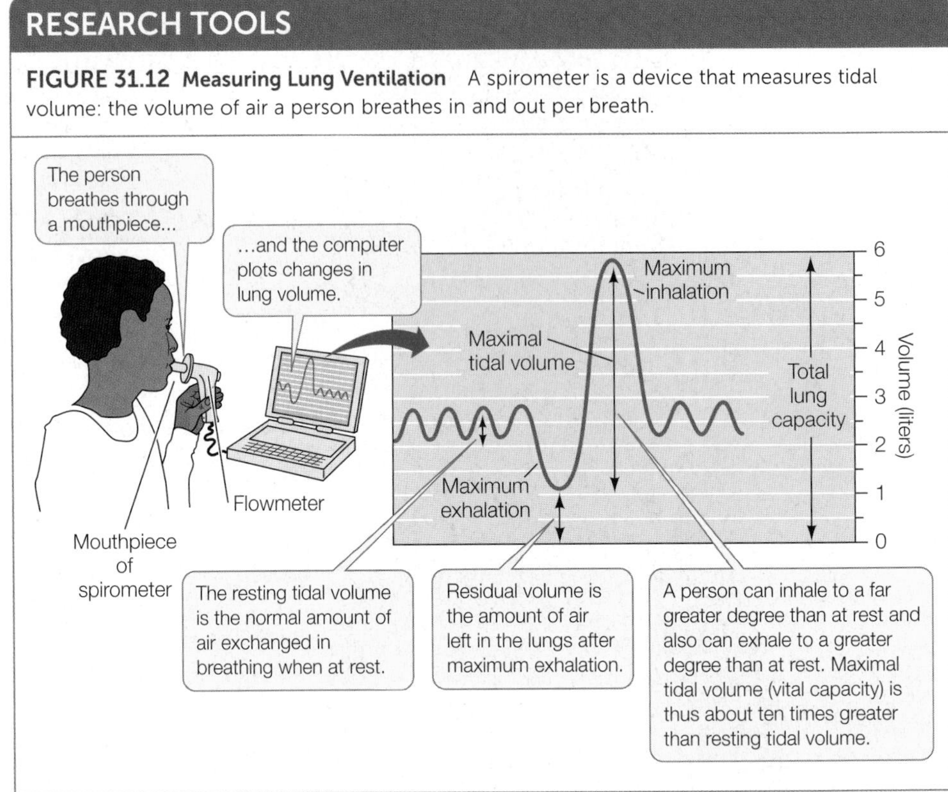

The person breathes through a mouthpiece...
...and the computer plots changes in lung volume.
Flowmeter
Mouthpiece of spirometer
Maximum inhalation
Maximal tidal volume
Total lung capacity
Maximum exhalation
Volume (liters)
6
5
4
3
2
1
0
The resting tidal volume is the normal amount of air exchanged in breathing when at rest.
Residual volume is the amount of air left in the lungs after maximum exhalation.
A person can inhale to a far greater degree than at rest and also can exhale to a greater degree than at rest. Maximal tidal volume (vital capacity) is thus about ten times greater than resting tidal volume.

A person's **respiratory minute volume** is the total volume of air that he or she inhales and exhales per minute. We calculate the minute volume by multiplying tidal volume (the volume exchanged per breath) by the number of breaths per minute. When people exercise and "breathe harder," they increase their tidal volume above the resting tidal volume, and they also take more breaths per minute than they do at rest. These two factors can produce a large increase in minute volume. At rest we breathe about 12 times per minute. With a resting tidal volume of about 0.5 liter, our resting minute volume is about 6 liters per minute. Well-conditioned athletes can breathe at least 30 times per minute at a tidal volume of over 3 liters, meaning they can increase their minute volume by a factor of more than 15!

Mammals and other animals with tidally ventilated lungs do not empty their lungs completely when they exhale. Some air is always left behind in the lungs. The amount left after a maximum exhalation is called the residual volume (see Figure 31.12). During normal breathing, whenever a mammal inhales, the fresh air mixes with stale air from the previous breath that was not exhaled. This mixing lowers the O_2 partial pressure at the alveolar gas exchange membranes but ordinarily presents no problems.

The lungs are ventilated by expansion and contraction of the thoracic cavity

Mammalian lungs are suspended in the thoracic, or chest, cavity, which is bounded on the top by the bones of the shoulders, on the sides by the rib cage, and on the bottom by a sheet of muscle, unique to mammals, called the **diaphragm**. Any increase in volume of the thoracic cavity pulls on the walls of the lungs, thereby expanding the lungs and causing inflation.

Two muscles or sets of muscles are critical for breathing: the diaphragm and the intercostal (from the Latin *costa*, "rib") muscles that run between adjacent ribs (see Figure 31.11A). The diaphragm is dome-shaped at rest, protruding upward from the abdomen into the thoracic cavity. It flattens when it contracts, so it protrudes less into the thoracic cavity, expanding it. There are two sets of intercostal muscles. One set enlarges the thoracic cavity when it contracts, and the other set makes it smaller when it contracts.

In humans at rest, inhalation is active but exhalation is passive (**FIGURE 31.13**). During inhalation, the diaphragm and certain intercostal muscles contract, expanding the thoracic cavity and lungs, which fill by suction (a lower pressure in the lungs than in the atmosphere). During exhalation at rest, these muscles simply relax, allowing the elastic recoil of the lungs and thoracic cavity to restore the relaxed lung volume. During exercise, additional breathing muscles come into play. Inhalation is more forceful, and muscle contractions are employed to make exhalation faster and more forceful than it is at rest.

The breathing rhythm depends on nervous stimulation of the breathing muscles

All the muscles involved in breathing require nervous stimulation to contract. The nerve impulses originate in the brain. These facts explain why a broken neck can kill a person rapidly. If the nerves carrying the breathing impulses from the brain to the breathing muscles are severed, the muscles stop contracting. Without artificial respiration, the person suffocates.

Physiologists have long sought the places in the brain where the breathing rhythm originates. Investigators recently identified these rhythm-generating (rhythmogenic) centers by studying thin slices of brain tissue taken from nestling rodents. Although these slices are cut off from all their normal connections with other parts of the nervous system, they remain functional for some time. Investigators have shown that two neural complexes in the medulla oblongata of the brain (a part of the brainstem) generate a breathing rhythm under these circumstances (**FIGURE 31.14**). These rhythm-generating centers, called the pre-Bötzinger complexes, are believed to be the beginning points for the nerve impulses that activate the muscles we use in breathing.

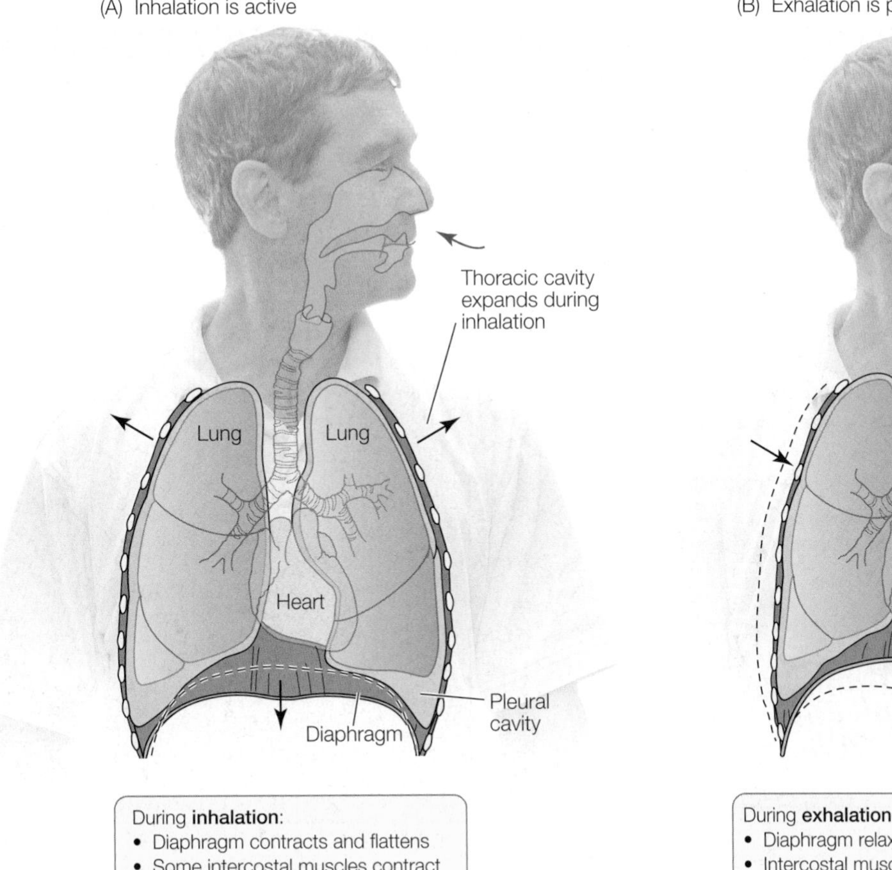
(A) Inhalation is active
Thoracic cavity expands during inhalation
Lung
Lung
Heart
Pleural cavity
Diaphragm
During **inhalation**:
• Diaphragm contracts and flattens
• Some intercostal muscles contract
• Thoracic cavity expands because of these muscle actions
• Lungs expand
• Air is sucked in

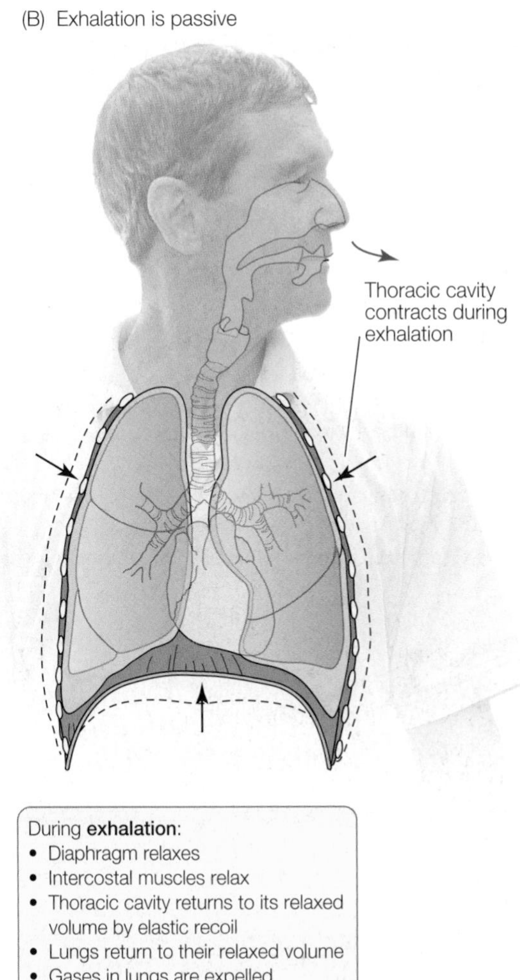
(B) Exhalation is passive
Thoracic cavity contracts during exhalation
During **exhalation**:
• Diaphragm relaxes
• Intercostal muscles relax
• Thoracic cavity returns to its relaxed volume by elastic recoil
• Lungs return to their relaxed volume
• Gases in lungs are expelled

FIGURE 31.13 Resting Human Ventilation When a person is at rest, inhalation (A) is an active process driven by contraction of the diaphragm and some of the intercostal muscles. However, exhalation (B) is a passive process that occurs by elastic recoil; the thoracic cavity and lungs elastically return to their relaxed volume as soon as muscle contractions stop maintaining them in an expanded state.

During exertion, when tidal volume and breathing rate increase to meet a person's increased O_2 needs, muscle contractions aid both exhalation and inhalation.

Go to ANIMATED TUTORIAL 31.2
Airflow in Mammals
PoL2e.com/at31.2

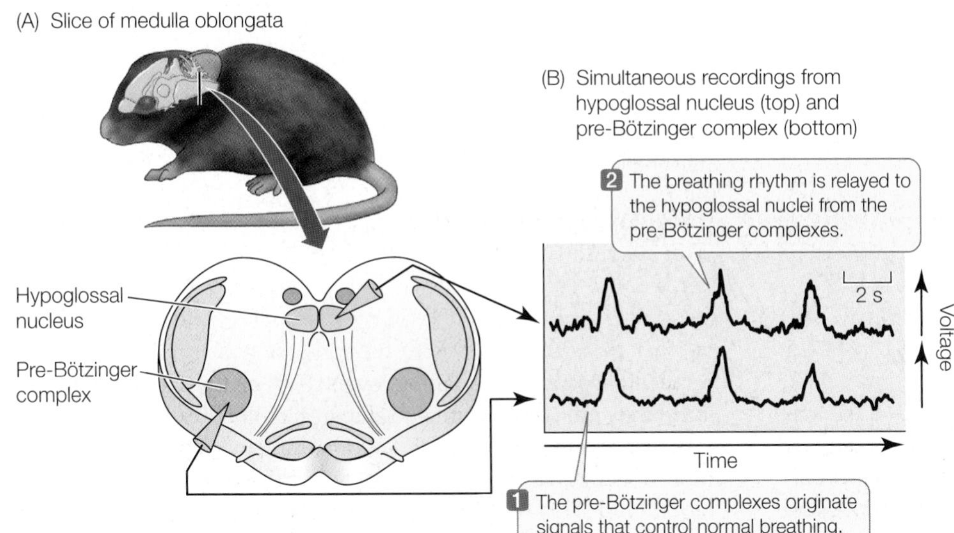
(A) Slice of medulla oblongata
Hypoglossal nucleus
Pre-Bötzinger complex
(B) Simultaneous recordings from hypoglossal nucleus (top) and pre-Bötzinger complex (bottom)
2 The breathing rhythm is relayed to the hypoglossal nuclei from the pre-Bötzinger complexes.
2 s
Voltage
Time
1 The pre-Bötzinger complexes originate signals that control normal breathing.

FIGURE 31.14 Where Does the Breathing Rhythm Originate? To find the neural tissue where the breathing rhythm starts, physiologists studied thin slices of living brain tissue from nestling mice. (A) This slice, cut from the medulla oblongata, has no connections to other parts of the nervous system. Its neural activity is therefore completely self-contained. (B) Electrical recordings were made from one of two clusters of nerve cells called the pre-Bötzinger complexes. The investigators found that these complexes generate electrical signals that control normal breathing. From simultaneous recordings from one of two other nerve cell clusters called the hypoglossal nuclei, the investigators found that the breathing rhythm that starts in the pre-Bötzinger complexes is relayed to the hypoglossal nuclei. In an intact mouse, the breathing rhythm would then be relayed to the breathing muscles via the hypoglossal nerves.

INVESTIGATION

FIGURE 31.15 The Respiratory Control System Detects Exercise What is the metabolic feedback signal that controls ventilation during exercise? In experiments with dogs running on a treadmill at a constant speed but at varying upward slopes, the respiratory minute volume increased as the CO_2 partial pressure in the arterial blood increased.[a]

HYPOTHESIS

Arterial CO_2 partial pressure is a feedback stimulus controlling ventilation.

METHOD

1. Dogs are trained to run on a treadmill.
2. The dogs are equipped with instruments that measure respiratory minute volume and with arterial catheters that enable sampling of blood.
3. As a dog runs at a constant speed, the slope of the treadmill is changed to modify the metabolic workload.
4. Minute volume is plotted as a function of arterial CO_2 partial pressure.

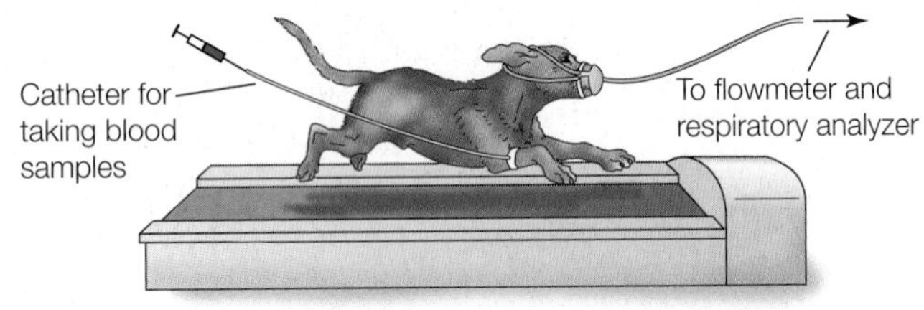

RESULTS

When the workload is altered by slowly changing the slope of the treadmill (no change in speed), the respiratory minute volume is a function of arterial CO_2 partial pressure.

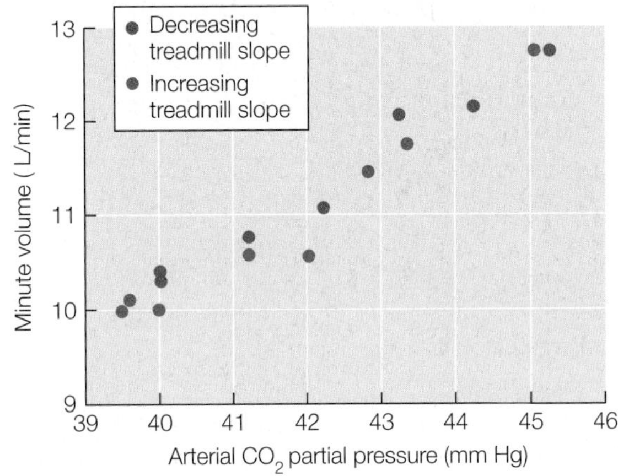

CONCLUSION

The data support the hypothesis that arterial CO_2 partial pressure is a metabolic feedback signal controlling ventilation in response to changes in workload.

ANALYZE THE DATA

Additional experiments were done in which the speed of the treadmill was changed instead of the slope to modify the metabolic workload. Again the respiratory minute volume (L/min) and arterial CO_2 partial pressure (P_{CO_2}, in units of mm Hg) were measured. The average values when the dogs were running at 3 mph were CO_2 partial pressure = 39.0 mm Hg and L/min = 9. The data for the first seven breaths after the treadmill speed was increased to 6 mph are shown in the table. After the dogs had run at 6 mph for several minutes, the values averaged CO_2 partial pressure = 41.0 mm Hg and L/min = 15.0. Plot these data and use the information to answer the questions.

Breath	P_{CO_2}	L/min
1	38.0	13.0
2	37.0	14.0
3	36.5	11.2
4	37.2	11.2

Breath	P_{CO_2}	L/min
5	37.2	12.0
6	36.8	13.0
7	37.2	13.0

A. Do these data support the hypothesis that CO_2 partial pressure is the feedback signal controlling ventilation? Why or why not?

B. How do you explain the average values after the dogs had been running at the higher speed for a few minutes?

C. Relate these results to those obtained in the experiment in which the slope of the treadmill was varied. Explain the differences between the results.

Go to **LaunchPad** for discussion and relevant links for all **INVESTIGATION** figures.

[a]C. R. Bainton. 1972. *Journal of Applied Physiology* 33: 778–787.

Breathing is under negative-feedback control by CO_2

You might expect that a mammal's breathing would be regulated by the blood's O_2 content. In fish, blood O_2 is indeed the primary feedback stimulus for gill ventilation. However, mammals are relatively insensitive to falling levels of O_2 in arterial blood. Instead they are very sensitive to changes in blood CO_2.

To fully understand control of breathing by CO_2, it is important to recognize that CO_2 dissolved in the blood reacts with blood H_2O. This reaction forms carbonic acid (H_2CO_3), which dissociates almost completely into H^+ ions and bicarbonate ions (HCO_3^-).

$$CO_2 + H_2O \rightleftharpoons H_2CO_3 \rightleftharpoons H^+ + HCO_3^-$$

What happens when the cells of your body increase their metabolism and therefore add CO_2 to the blood at an increased rate? There are two immediate consequences. First, the partial pressure of CO_2 in the blood rises. Second, the concentration of H^+ in the blood also rises, meaning the blood becomes more acidic and its pH falls. Chemosensitive (chemical-sensing) neural centers located on the ventral surface of the medulla oblongata in the brain (see Figure 31.14) independently monitor the blood CO_2 partial pressure and H^+ concentration. These centers detect the increases in CO_2 partial pressure and H^+ concentration. This information is then used to control breathing in a negative-feedback manner. Breathing is stimulated so that the respiratory minute volume increases. This increased ventilation increases exhalation of CO_2, helping the body get rid of CO_2 at an accelerated rate when the cells have increased their rate of CO_2 production. The increased ventilation also tends to return both the blood CO_2 partial pressure and H^+ concentration back toward their values that existed before CO_2 production increased.

What if, under different circumstances, your cells decrease the rate at which they add CO_2 to the blood? In this case, the blood CO_2 partial pressure and H^+ concentration initially tend to fall. The chemosensitive centers detect these changes, leading again to negative-feedback control of breathing. In this instance, breathing is inhibited, lowering the rate of CO_2 exhalation and causing the blood CO_2 partial pressure and H^+ concentration to rise.

These controls serve to adjust the respiratory minute volume so that the rate of CO_2 exhalation approximately matches the rate at which the cells of the body produce CO_2. The controls also stabilize the blood CO_2 partial pressure and H^+ concentration, by tending to bring them down if they start to rise and tending to bring them up if they start to fall.

In the experiment in **FIGURE 31.15**, dogs ran on a treadmill at a constant speed. The slope of the treadmill could be tilted up at various angles, allowing the researchers to vary the dogs' metabolic workload and thus their rates of CO_2 production. Negative-feedback systems rarely lead to complete stability. In this case, the CO_2 partial pressure in a dog's arterial blood tended to vary a little (it would have varied much more without negative feedback). The results demonstrate that a dog's respiratory minute volume is closely correlated with its blood CO_2 partial pressure. This evidence supports the idea that breathing is stimulated when the blood CO_2 partial pressure increases.

Go to MEDIA CLIP 31.1
Hunting on the Sea Bed with One Breath
PoL2e.com/mc31.1

Breathing is also under control of factors in addition to CO_2

You may have noticed that when you start running, your breathing rate increases almost immediately—far sooner than your blood CO_2 partial pressure would be expected to change. This everyday observation suggests that the simple act of moving muscles and joints may serve to accelerate ventilation during exercise. Investigators studied this possibility in the experiments on dogs we've been discussing (see Analyze the Data in Figure 31.15). For this purpose the dogs ran on a horizontal treadmill, and investigators altered the dogs' intensity of work by varying how fast they ran. When the dogs worked harder, respiratory minute volume sometimes increased to a greater extent than CO_2 partial pressure by itself would explain. According to the results of this study and other studies, information from receptors that detect motion in muscles and joints is used during exercise to help control the intensity of ventilation.

Blood O_2 partial pressure is also sometimes used in controlling the intensity of ventilation. O_2 partial pressure is not as influential as CO_2 partial pressure, and minor fluctuations in the O_2 partial pressure have little effect on ventilation. However, a large drop in blood O_2 partial pressure—such as might occur at high elevation—strongly stimulates ventilation.

In mammals, two sets of oxygen-sensing chemosensitive bodies detect the blood O_2 partial pressure. They are called peripheral chemoreceptors because they are not in the brain. They are positioned along the large arteries leaving the heart (see Figure 32.6). One set (by far the most important set in humans) is a pair of carotid bodies found along the carotid arteries in the neck. The second set consists of aortic bodies associated with the aorta. These bodies detect blood O_2 partial pressure and send information on it by way of nerves to the breathing control centers in the medulla oblongata.

CHECKpoint CONCEPT 31.3

- ✓ The alveoli become filled with tissue fluid in several types of diseases, such as heart weakening that causes pulmonary edema. This condition is a health emergency. Why?
- ✓ In a healthy individual, how would the respiratory minute volume change in response to a decrease in blood pH?
- ✓ How could a neck injury that leaves the trachea and other air passageways intact result in an inability to breathe?

Why is climbing uphill at 1.7 meters per minute (100 meters/hour) the most intense work that is possible for a human near Everest's summit?

ANSWER Any particular type of sustained, physical work demands a certain amount of sustained metabolic effort and thus a certain, fixed rate of O_2 consumption (regardless of elevation). Consequently a person's maximal rate of O_2 consumption determines the maximal sustained work intensity of which he or she is capable. A person's maximal rate of O_2 consumption decreases with elevation because O_2 becomes more dilute in the atmosphere as elevation increases (**FIGURE 31.16**). For this reason, maximal work intensity decreases with elevation. At the top of Mount Everest, hardly any work is possible. The maximal intensity of work is a slow walk.

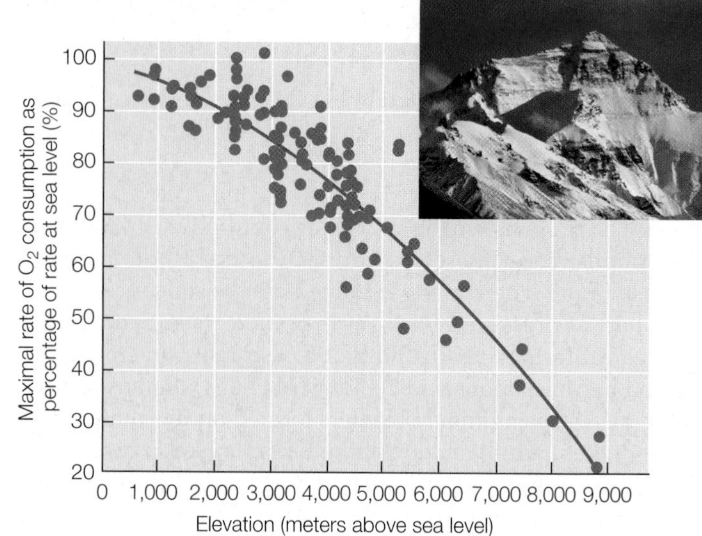

FIGURE 31.16 A Person's Maximal Rate of O_2 Consumption Decreases as Elevation Increases Each symbol represents one measurement on one person. Data from many studies were compiled to create this plot.

SUMMARY

CONCEPT 31.1 Respiratory Gas Exchange Depends on Diffusion and Bulk Flow

- Most cells require a constant supply of O_2 and continuous removal of CO_2. These **respiratory gases** are exchanged between an animal's cells and the animal's environment by a combination of diffusion and **bulk flow**. **Review Figures 31.1 and 31.3**
- A gas diffuses from where its **partial pressure** is high to where its partial pressure is low. The **law of diffusion for gases (Fick's law)** shows how various physical factors influence the diffusion rate of gases. Adaptations to maximize respiratory gas exchange influence one or more variables of Fick's law. **Review Figure 31.2**
- As they move between an animal's internal tissues and the environment, respiratory gases diffuse across **gas exchange membranes**.
- Breathing organs are classified based on their structure: **lungs** fold inward into the body; **gills** fold outward from the body. **Review Figure 31.4**
- In water, gas exchange is limited by the low diffusion rate and low solubility of O_2 in water. O_2 becomes less soluble in water as water temperature increases. **Review Figure 31.5**

CONCEPT 31.2 Animals Have Evolved Diverse Types of Breathing Organs

- Adaptations to maximize gas exchange include increasing the surface area for gas exchange and maximizing both **ventilation** with the respiratory medium and **perfusion** with blood. The gas exchange membranes are also very thin, which minimizes the distance gases must diffuse.
- **Cocurrent gas exchange** occurs when perfusion flows in the same direction as ventilation; flows in opposite directions result in **countercurrent gas exchange**. The most efficient transfer of gases is achieved by countercurrent systems, such as those found in fish. **Review Figures 31.6 and 31.7**
- Internal partitioning of the lungs is greater in mammals than in nonavian reptiles like lizards, and greater in these reptiles than in amphibians. Ventilation in all these groups is **tidal**.
- The breathing system of birds includes air sacs that are filled and emptied tidally and act as bellows to ventilate the lungs. Air flows unidirectionally and continuously through the parabronchi (the principal gas exchange parts) of bird lungs. **Review Figures 31.8 and 31.9, ANIMATED TUTORIAL 31.1**
- Insects distribute air throughout their bodies in a system of **tracheae** and tracheoles. The circulatory system is not involved in gas exchange because all cells are supplied directly by the tracheal system. **Review Figure 31.10**

CONCEPT 31.3 The Mammalian Breathing System Is Anatomically and Functionally Elaborate

- In mammalian lungs, the gas exchange surface area provided by the millions of **alveoli** is enormous, and the diffusion path length is short. **Review Figure 31.11 and ACTIVITY 31.1**
- The **tidal volume** is the amount of air that moves in and out of the lungs per breath. The **respiratory minute volume** is the total volume of air that is inhaled and exhaled per minute. **Review Figure 31.12**
- Inhalation occurs when contractions of the **diaphragm** and some of the intercostal muscles expand the thoracic cavity. In a resting person, when the diaphragm and intercostal muscles relax, the thoracic cavity contacts by elastic recoil, resulting in exhalation. During exercise, greater muscular forces are used, increasing the volume of air inhaled and exhaled per breath and the rate of breathing. **Review Figure 31.13 and ANIMATED TUTORIAL 31.2**
- The breathing rhythm is generated by neurons in rhythm-generating centers in the medulla oblongata. **Review Figure 31.14**
- The most important feedback stimulus for breathing is the level of CO_2 in the blood, which is detected by chemosensitive neural centers on the surface of the medulla oblongata. These centers detect both the blood CO_2 partial pressure and the blood pH, which indirectly reflects the level of CO_2. **Review Figure 31.15**
- Breathing rate can also be affected by the onset of exercise (sensed by receptors that detect motion in muscles and joints) and by large changes in blood O_2 partial pressure (detected by chemosensitive carotid and aortic bodies on the large vessels leaving the heart). **See ACTIVITY 31.2 for a concept review of this chapter.**

Go to the Interactive Summary to review key figures, Animated Tutorials, and Activities PoL2e.com/is31

Go to LaunchPad at **macmillanhighered.com/launchpad** for additional resources, including LearningCurve Quizzes, Flashcards, and many other study and review resources.

32 Circulation

KEY CONCEPTS

Bluefin tuna are vigorous swimmers. They need more than strong swimming muscles to swim, because the muscles require ATP at high rates and thus need O_2 at high rates. The circulatory system bears responsibility for moving O_2 rapidly to the muscles from the gills.

If tuna, such as these bluefin tuna, lived on land, they would be as famous as wolves. Like wolves, bluefin tuna are carnivores that hunt in groups and depend on speed, strength, and awesome teeth to attack prey: in their case, other fish. Tuna are strong, mobile predators at the top of the food chain.

Judging by the length of time spent in motion, tuna are actually more mobile than wolves. Using their red swimming muscles, they swim continuously, day and night, at speeds of one to two body lengths per second. For bluefin, this means traveling continuously at speeds of 2–6 meters per second, interrupted at times by anaerobically fueled bursts of attack swimming at far higher speeds. Tuna rank with the elite endurance athletes of the sea.

To meet the oxygen (O_2) demands of their vigorous lifestyle, tuna have evolved some of the most elaborate gills known in fish: gills that can take up O_2 at high rates from seawater. However, the swimming muscles that must receive O_2 rapidly for endurance swimming are located in the abdomen, far from the gills.

Rapid circulation of blood is thus as important as elaborate gills for the steady, endurance swimming of tuna. As stressed in Concept 31.1, the flow of blood through an animal's circulatory system is a form of bulk flow. As such, it can move substances at high rates over long distances within the animal's body. Tuna researcher John Magnuson once commented that tuna are "astounding bundles of adaptations for efficient and rapid swimming." One of those adaptations is a high-performance circulatory system that can transport O_2 at high rates from the gills to the swimming muscles.

Tuna belong to the small set of fish that have warm bodies, and in this regard their circulatory system plays a second important role in swimming. In more than 99 percent of fish, body temperature equals water temperature. Only a tiny minority of fish are warm-bodied as tuna are. The entire body of a tuna is not warm, however. Just certain parts are. The red swimming muscles represent the largest mass of warm tissue. In bluefin tuna, those muscles remain at 25–35°C even when the ocean is as cold as 5°C. This warming of the red swimming muscles is believed to help the muscles maintain high power outputs. A specialized arrangement of blood vessels helps the swimming muscles retain heat and stay warm.

How can the arrangement of blood vessels help warm the swimming muscles in warm-bodied fish such as tuna?

You will find the answer to this question on page 678.

CONCEPT 32.1 Circulatory Systems Can Be Closed or Open

A **circulatory system** typically consists of a muscular pump (heart), a fluid (blood), and a series of conduits (blood vessels or channels) through which the blood is pumped. Because blood must flow close to all the cells of the body, the profusion of blood vessels is often spectacular when visualized (**FIGURE 32.1**).

For an animal to have a high degree of division of labor among its organs, a circulatory system is essential because it allows each organ to export its products promptly to other parts of the body and to receive promptly the products of other organs. With a circulatory system, O_2 taken up by gills or lungs can be delivered rapidly to all other organs (see Figure 31.3); hormones secreted by endocrine glands can be distributed quickly throughout the body; lipid molecules stored in body fat can be delivered rapidly to the cells of any organ where they are needed; and waste products of cell metabolism can be carried promptly to the kidneys for excretion.

The sophistication of the circulatory system in a species tends to be correlated with the species' need for circulatory O_2 transport. This is because O_2 transport is typically the most urgent circulatory function. Many minutes can go by without any transport of hormones, food molecules, or wastes—animals can tolerate such interruptions. However, animals typically cannot tolerate long interruptions in O_2 transport. For instance, if a person's blood circulation stops during a cardiac arrest, depriving brain cells of O_2, the person becomes unconscious within 20 seconds! When we consider how rapidly blood must flow in a circulatory system, the need for O_2 transport is typically the most important consideration because—of all the substances carried by blood flow—O_2 must be transported the fastest. This explains why animals that have evolved high metabolic rates (and therefore have high O_2 needs) also usually have evolved high-performance circulatory systems.

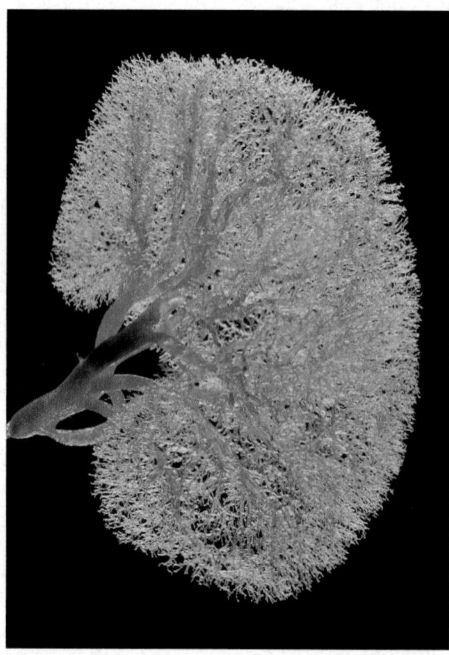

FIGURE 32.1 The Arteries of Visible Size That Carry Blood to the Human Kidney To obtain this image, the arteries were injected with a liquid plastic that hardened after injection, and then the tissues were removed, leaving just this plastic cast. Keep in mind that, regardless of the complexity, the vessels visualized here are only the ones that can be seen with the naked eye. The microscopic vessels, which are even more profuse, are not visualized with this technique.

The vertebrate animals nicely illustrate the relationship between metabolic rate and circulatory function. Mammals and birds typically have far higher rates of blood flow and higher blood pressures than lizards, frogs, or fish of the same body size. This difference in circulatory performance correlates with differences in metabolic rates. Mammals and birds have far higher metabolic rates than lizards, frogs, or fish (see Concept 29.3) and therefore require circulatory systems that can deliver O_2 far faster. The same relationship is seen among types of fish. Fish species with high metabolic rates—such as the tuna in our opening story—have circulatory systems capable of higher performance than those of sluggish fish species. Similarly, among the mollusks, squid—which are fast-swimming predators—have high-performance circulatory systems compared with those of sedentary mollusks such as snails and clams.

Insects are the exception that proves the rule. Some insects, when flying, have very high O_2 needs per gram of body weight. Yet insects have modest, low-performance circulatory systems. The tracheal breathing system of insects accounts for this paradox. O_2 is brought to each cell of an insect by a branch of the animal's breathing system, not by blood circulation. The circulatory system, freed of responsibility to transport O_2, need not perform at a high level, despite the high metabolic rates of insects.

LINK

The breathing system of insects is described in **Concept 31.2**; see especially **Figure 31.10**

Closed circulatory systems move blood through blood vessels

A circulatory system is **closed** if the blood is always contained within blood vessels as it circulates. The vertebrates again provide excellent examples and will be our focus here. The heart of a vertebrate pumps blood into large blood vessels called arteries (which we will discuss in more detail in Concept 32.2). The arteries branch like trees into smaller and smaller blood vessels as they carry the blood to tissues and organs throughout the body. The branches ultimately become so small in diameter that we cannot see them with the naked eye. Nonetheless, the blood always remains within vessels. As blood leaves tissues and organs, it flows through vessels that merge into larger and larger vessels—the veins—that carry it back to the heart:

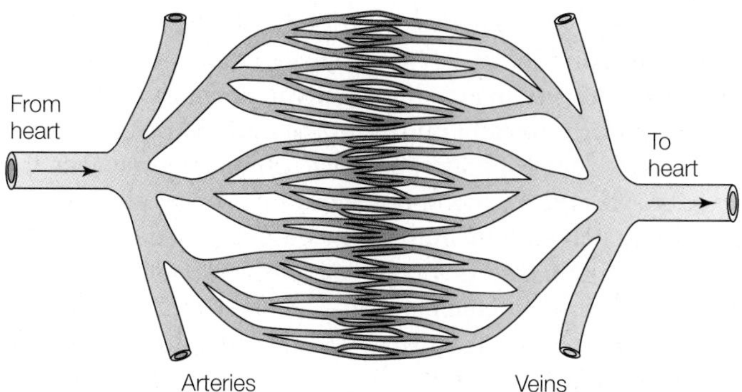

The blood vessels of smallest diameter—which we can see only with a microscope—are collectively called the vessels of the **microcirculation** and consist of arterioles, capillaries, and venules. The arterioles and capillaries are particularly critical for the functioning of a closed circulatory system. As we discuss them, an important point to recognize is that in the walls of all blood vessels, the layer of cells that immediately surrounds the blood is a simple epithelium (see Figure 29.13). For historical reasons, this layer of cells is called the **vascular endothelium**.

The walls of **capillaries** consist only of vascular endothelium. In other words, they consist only of a single layer of thin, flattened epithelial cells, less than 1 μm thick (**FIGURE 32.2A**). Moreover, capillaries are of tiny diameter. The open passage in a blood vessel is called the vessel lumen. In capillaries, the lumen is often just slightly larger in diameter than individual red blood cells. For this reason, the cells (~7–8 μm in diameter in humans) must pass through capillaries in single file (**FIGURE 32.2B**). In each tissue of the body, capillaries are found near every tissue cell. For instance, capillaries are positioned near every brain cell and every muscle cell. The number of capillaries is so great as to be almost unimaginable. Added together, for example, the capillaries in a single cubic centimeter of mammalian skeletal muscle or heart muscle are 10–30 meters long!

Capillaries are the principal sites in a closed circulatory system where blood releases O_2, nutrient molecules, hormones, and other substances to the tissues. They are also the principal sites where blood takes up CO_2 and other waste products from the tissues. Capillaries play these roles in part because they are so numerous. They are also suited to these roles because their walls are thin and often porous, allowing materials to move rapidly across their walls by diffusion and other means.

Arterioles are as important for the function of the microcirculation as capillaries are. The walls of an arteriole consist of an inner layer of vascular endothelium surrounded by an outer layer of smooth muscle cells that are under the control of the autonomic nervous system (described in Concept 34.5) and other control mechanisms. The muscle cells in the wall can adjust the diameter of an arteriole's lumen. Contraction of the muscle cells causes vasoconstriction, reducing the lumen diameter. Relaxation of the muscle cells allows the lumen diameter to increase, a process called vasodilation. Control of the arteriole lumen diameter in these ways is called **vasomotor control**.

Why do changes in the lumen diameter matter? They are of enormous importance because they control where blood flows. The famous French physiologist Jean Poiseuille (pronounced *Pwa-sul*) demonstrated in the mid-19th century that the rate of fluid flow through a tube is extraordinarily sensitive to the diameter of the tube lumen because this rate depends directly on the 4th power of the diameter. In other words, if the diameter of the lumen of a tube is doubled, the flow rate through the tube increases by a factor of 16! Conversely, halving the diameter reduces flow to 1/16 of its previous rate. An arteriole precedes each capillary bed (set of interconnected capillaries) and controls the flow of blood into it (**FIGURE 32.3**). By controlling the diameters of the arterioles, the autonomic nervous system and other systems exert control over which capillary beds receive blood flow.

Several everyday examples illustrate arteriolar control of blood flow. The skin of people with fair complexions is often more pinkish in warm than in cold weather. This change reflects vasomotor control. Skin arterioles dilate in warm weather, permitting vigorous blood flow to the capillary beds near the skin

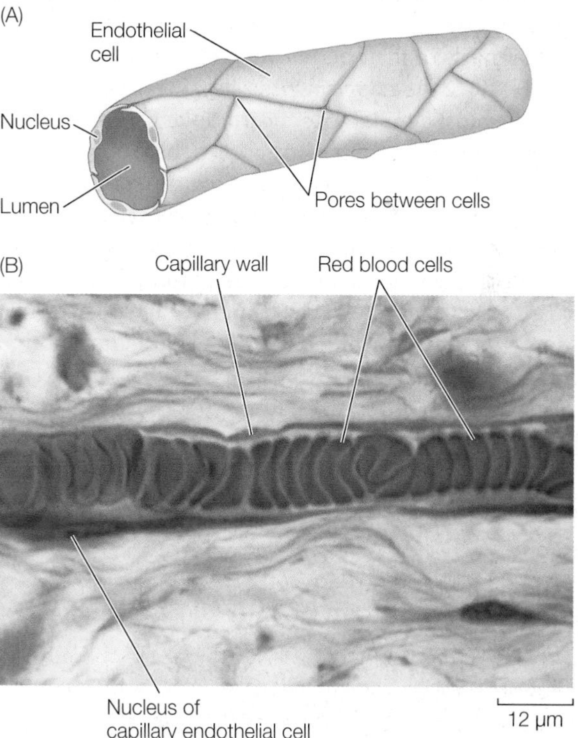

FIGURE 32.2 Blood Capillaries The capillaries may be small, but they are where blood flow accomplishes most of its missions. (A) The wall of a capillary consists of a single layer of highly flattened epithelial cells (called endothelial cells in this context). Pores (often termed fenestrations) are found between cells in the capillaries of some tissues but not others. (B) The diameter of the lumen of capillaries is small enough that red blood cells must move through in single file.

Go to MEDIA CLIP 32.1
Capillary Flow: A Tight Squeeze
PoL2e.com/mc32.1

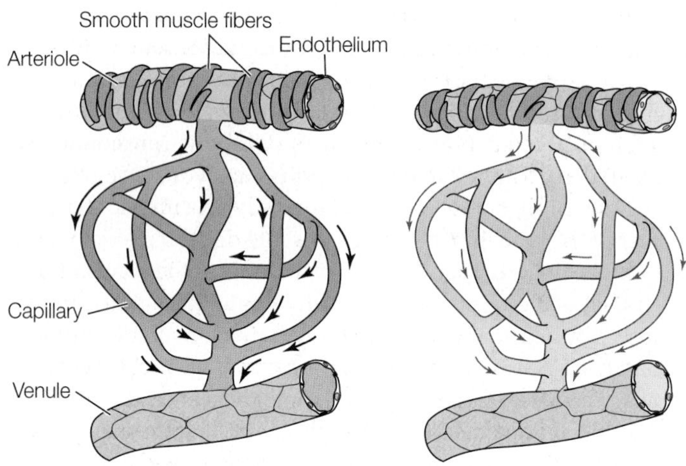

FIGURE 32.3 Arterioles Control the Flow of Blood into Capillary Beds (A) An arteriole with a large diameter allows for a high rate of blood flow. (B) The rate of blood flow drops dramatically when the arteriole diameter decreases.

surface—a response that makes the skin turn pink and helps metabolic heat leave the body. When people exercise, arterioles in their exercising skeletal muscles dilate, and arterioles in many other parts of their body constrict. These changes increase blood flow to the exercising muscles. Erection of the penis is one of the most familiar cases of arteriolar control. During sexual arousal, dilation occurs in arterioles that supply blood to the spongy tissue of the penis, permitting the spongy tissue to inflate with blood under high pressure (see Concept 37.2).

In general, closed circulatory systems can support higher levels of metabolic activity than the open systems we will discuss next, in part because they have a higher capacity to direct blood flow to the tissues that need it the most. In addition to vertebrates, other animals with closed circulatory systems are the annelid worms, such as earthworms, and the cephalopod mollusks, such as squid and octopuses.

In open circulatory systems, blood leaves blood vessels

A circulatory system is **open** if the blood exits blood vessels as it flows through the body. Arthropods—such as insects, spiders, crabs, and lobsters—have open circulatory systems, as do most mollusks. In an animal with an open circulatory system, the heart typically pumps blood into arteries that carry the blood for at least a short distance. Ultimately, however, the vessels end, and the blood pours out into spaces surrounded not by vascular endothelium but by ordinary tissue cells. After this, the blood makes its way back to the heart, not in veins but by flowing through channels called sinuses and lacunae. The sinuses and lacunae are (respectively) small and large spaces among ordinary tissue cells.

In an open circulatory system, no distinction exists between blood (the fluid pumped by the heart and flowing in the blood vessels) and interstitial fluid (the fluid found between cells in the ordinary tissues—also called tissue fluid; see Concept 29.4). Debate exists about the correct name to use for the fluid in open circulatory systems. Here we call it blood. Some biologists, however, prefer to call the fluid hemolymph, referring to the fact that it has attributes of both blood (*hemo*, "blood") and tissue fluid (lymph). Tissue cells are directly bathed by the blood or hemolymph. For this reason, the cells can readily take up O_2 and other materials from the blood, or release wastes into it.

Open circulatory systems are not very well understood, although we know they sometimes have very high rates of blood flow. They are thought to be limited in how well they can control where the blood goes. Circulation in a crayfish or lobster illustrates the pattern of flow in an open circulatory system (**FIGURE 32.4**). The heart pumps blood into arteries that carry the blood to most major parts of the body before ending. After the blood leaves the vessels, it makes its way through sinuses and lacunae to ventral blood spaces along the bases of the walking legs, where the gills arise. Sucking forces developed by the heart then pull the blood through the gills and back into the heart to be pumped again into the arteries.

APPLY THE CONCEPT

Circulatory systems can be closed or open

Atherosclerosis ("hardening of the arteries") is a potentially deadly chronic disease. The immediate problem is that fatty deposits—composed of lipids such as cholesterol and triacylglycerols—build up in blood vessels. Take out a ruler, and using the principles developed by Poiseuille (principles known today as Poiseuille's Law), estimate how much more slowly blood will pass through the atherosclerotic artery at the right, compared with the healthy artery to the left, assuming everything is the same except the fatty deposits. People start their lives with arteries like the one at the left.

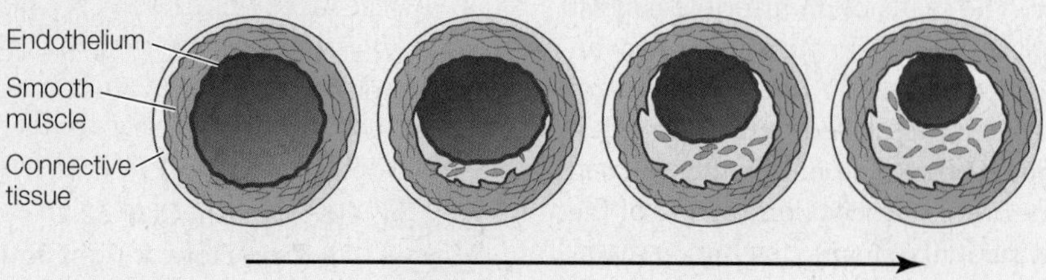

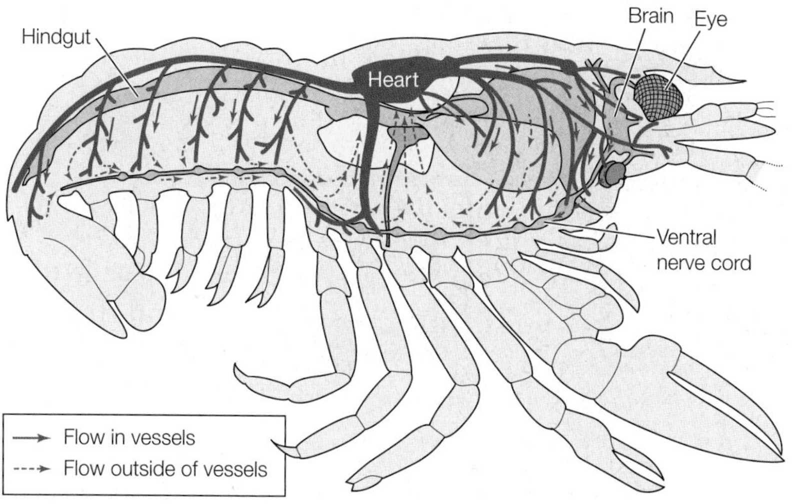

FIGURE 32.4 Blood Flow in the Open Circulatory System of a Crayfish or Lobster Crayfish and lobsters have many arteries, which extend forward and backward from the heart. The heart pumps blood into these arteries, which transport the blood to most parts of the body. The arteries end, however, and then the blood is released from vessels into sinuses and lacunae that are lined with ordinary systemic tissue cells rather than with vascular endothelium. After release from the arteries, blood from the anterior parts of the body flows posteriorly, and blood from the posterior parts of the body flows anteriorly. These flows converge on a ventral sinus (cavity) that runs along the bases of the principal legs. From there the blood returns to the heart. Although not shown here, the blood passes through the gills (which are attached at the bases of the principal legs) in this final step.

CHECKpoint CONCEPT 32.1

- ✓ What is the most fundamental distinction between open and closed circulatory systems?
- ✓ How do arterioles and capillaries differ structurally and functionally?
- ✓ A grasshopper and a clam both have open circulatory systems. What enables the grasshopper to be more active and sustain a higher metabolic rate?

For the rest of this chapter, we will focus mainly on the closed circulatory systems of vertebrates. One reason for this focus is that we ourselves have a closed circulatory system. We'll start by discussing the major patterns of blood flow throughout the vertebrate body.

CONCEPT 32.2 The Breathing Organs and Systemic Tissues Are Usually, but Not Always, in Series

For understanding circulation, we need to recognize two fundamentally different types of tissues or organs in the body of an animal: the systemic tissues and the breathing organs. The **systemic tissues** are all the tissues and organs of an animal's body other than the breathing organs (lungs or gills). They take up O_2 from the blood and add CO_2 to the blood. Conversely, the **breathing organs** add O_2 to the blood and take up CO_2 from the blood. With these definitions in mind, we can also define the systemic circuit and the breathing-organ circuit.

The **systemic circuit** consists of the blood vessels (or blood channels in the case of an open circulatory system) that carry blood to and from the systemic tissues. The **breathing-organ circuit** consists of the blood vessels that carry blood to and from the breathing organs. This circuit can be called the pulmonary circuit or lung circuit in animals that breathe with lungs, or it can be called the gill circuit in those that breathe with gills.

A final set of important definitions applies to the visible blood vessels, the arteries and veins. These vessels are defined by direction of blood flow, not by the O_2 content of the blood they carry. **Arteries** are vessels that carry blood *away from* the heart. **Veins** carry blood *toward* the heart. Recall from Chapter 31 that **perfusion** refers to flow of blood through a tissue or organ.

Most fish have the systemic circuit and gill circuit connected in series

The most common circulatory plan in animals is for the systemic and breathing-organ circuits to be connected in series, meaning that blood flows through the two circuits sequentially: first through one circuit, then through the other, then back through the first.

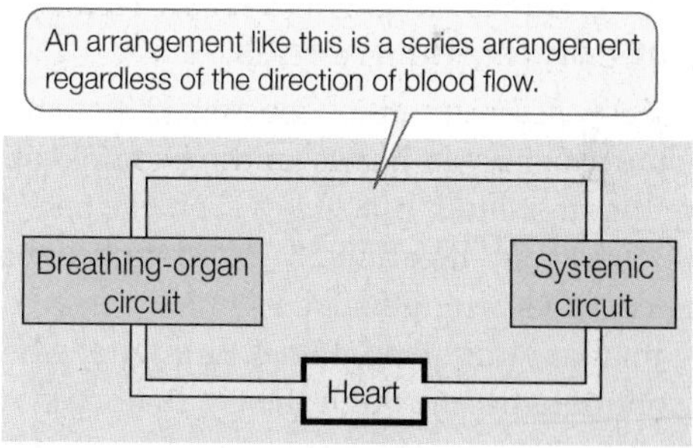

Fish display such a series arrangement (**FIGURE 32.5A**). The only exceptions are air-breathing species. In a typical fish, the heart pumps blood into arteries that lead directly to the gills. After the blood perfuses the gills and exits them, it is collected into additional arteries, which branch to supply blood to all the systemic tissues, from the brain in the head to the kidneys and swimming muscles in the abdomen. After perfusing the systemic tissues, the blood enters veins that take it back to the heart.

A series arrangement of this sort is considered to be the most efficient for delivering O_2. It is efficient because all the blood pumped to the systemic tissues is O_2-rich, having just passed through the gills. Moreover, all the blood pumped to the gills is O_2-depleted, having just exited the systemic tissues. Energy is not wasted in sending O_2-rich blood to the gills or O_2-depleted blood to the systemic tissues.

Natural selection has favored the evolution of a series arrangement in many types of animals. We have already seen,

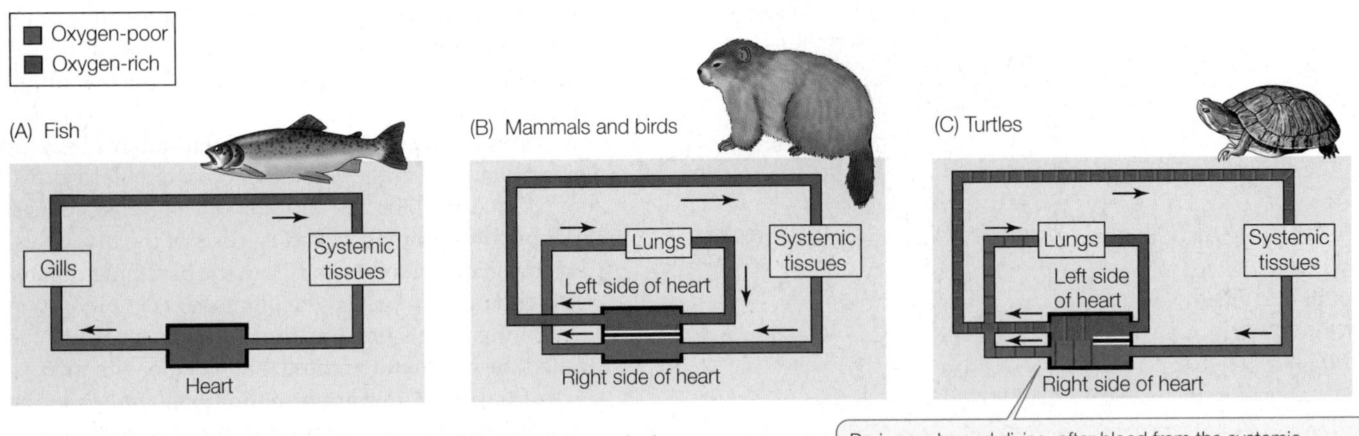

FIGURE 32.5 Vertebrate Circulatory Plans Shown here are typical circulatory plans for (A) fish, (B) mammals and birds, and (C) turtles and most other non-avian reptiles.

Go to ACTIVITY 32.1 Vertebrate Circulatory Systems PoL2e.com/ac32.1

for example, that crustaceans have series circulation in their open circulatory systems (see Figure 32.4). After the blood of a crayfish or crab passes through the gills, the heart pumps it through the systemic circuit, and then the blood returns to the gills. Squid and octopuses, which have closed circulatory systems, also have the gill circuit and systemic circuit connected in series.

Mammals and birds have the systemic circuit and lung circuit connected in series

As we will see in more detail in Concept 32.3, mammals and birds have four-chambered hearts, adding complexities to understanding the circulatory plan. However, like fish, they have the systemic circuit and breathing-organ circuit connected in a high-efficiency series arrangement.

In a mammal or bird, blood passes through the heart twice to make a complete trip through the circulatory system. The heart is divided into right and left halves. Moreover, it is *completely divided*, meaning that sheets of tissue completely separate the right and left halves, so that blood flowing through the right and left sides of the heart cannot mix. As blood travels from the systemic circuit to the lung circuit, it passes through the right half of the heart. As it travels from the lung circuit to the systemic circuit, it passes through the left half of the heart. **FIGURE 32.5B** shows the circulatory plan. If you trace the flow of blood from any particular point back to the same point, you will see that blood passes from the systemic circuit to the lung circuit, then back to the systemic circuit, and so forth: a series circulation.

A limitation that fish face is that, after their heart pumps their blood, the blood must pass through both the gill and systemic circuits without receiving additional propulsive energy (an added push) from the heart. Mammals and birds have evolved circulatory plans without this limitation. In a mammal or bird, the heart pumps the blood just before it travels to the lungs, and it pumps it again just before it travels to the systemic circuit.

Amphibians and most non-avian reptiles have circulatory plans that do not anatomically guarantee series flow

What about circulation in the amphibians and non-avian reptiles such as turtles and lizards? Let's start with a few words about blood flow through the **ventricle**, the heart's largest and most muscular chamber. In fish, only O_2-depleted blood (from the systemic circuit) enters the ventricle. Thus when the ventricle contracts and ejects blood to flow to the gills, it ejects only O_2-depleted blood. In mammals and birds, the ventricle is divided into two anatomically separate chambers (as we will see in Figure 32.6). Only O_2-depleted blood enters the right chamber, which then ejects only O_2-depleted blood when it contracts. Only O_2-rich blood from the lungs enters the left chamber, which therefore pumps exclusively O_2-rich blood.

Adult amphibians and most non-avian reptiles have a single ventricle like fish, but blood flow in and out is far more complicated. The lumen of the single ventricle is not a wide-open space because it is occupied by crisscrossing strands of heart muscle and by sheets of connective tissue. Yet the lumen is not fully divided into subparts. O_2-rich blood from the lungs enters the ventricle at one point, while simultaneously, O_2-depleted blood from the systemic circuit enters it at another point. The ventricle pumps blood to both the lungs and the systemic circuit. In theory, therefore, O_2-rich and O_2-depleted blood can mix in the ventricle or be pumped to the wrong destination. Decades ago, when biologists knew only about anatomy, the hearts of these animals were often viewed erroneously as being defective intermediate stages in the evolution of vertebrates.

With data from functional studies, we now know that O_2-rich and O_2-depleted blood are often kept separate as they flow through the ventricle. Investigators have added fluids that can be seen with X-rays to blood of one oxygenation state or the other, for example, and often have observed that little mixing occurs between the O_2-rich and O_2-depleted blood. In these cases, a series circulation is maintained, with just a small amount of inefficiency as a result of mixing in the ventricle (**FIGURE 32.5C**). O_2-rich blood from the lungs is pumped by the ventricle principally to the systemic circuit, and O_2-depleted blood from the systemic circuit is sent primarily to the lungs.

We also now realize from functional studies that sometimes an incompletely divided ventricle can have advantages for certain of these animals. When turtles dive underwater for long periods, for example, their lungs become useless. The turtles, however, have evolved control mechanisms whereby they can stop circulating blood to their lungs. Taking advantage of the fact that they have just a single, undivided ventricle, they direct blood that enters the ventricle from the systemic circuit so that the ventricle pumps it back into the systemic circuit, rather than sending it to the lungs. In this way, the turtles can pump blood round and round in the systemic circuit—permitting the circulatory system to perform all its functions other than gas transport from the lungs—during the time their lungs are useless.

CHECKpoint CONCEPT 32.2

- ✓ What is the systemic circuit?
- ✓ Why is a series arrangement of the systemic and breathing-organ circuits efficient for O_2 transport?
- ✓ In a turtle that has been diving underwater for a long time, does blood pass through the systemic and lung circuits in series? Explain your answer.

By now you have learned a lot about hearts in a general way. But how do hearts work? What controls their regular beating? How do the cells of the heart muscle get the O_2 they need? We will address questions such as these in the next concept.

CONCEPT 32.3 A Beating Heart Propels the Blood

A **heart** is a discrete, localized pumping structure. Some animals with a circulatory system lack a heart. In earthworms and other annelid worms, for example, the blood is propelled entirely by peristaltic waves of contraction in muscular blood vessels. Some animals, by contrast, have more than one heart. Crayfish and lobsters, for instance, have a small accessory heart that helps perfuse their brain, in addition to their large, principal heart.

The muscle tissue of a heart is termed the **myocardium** (*myo*, "muscle"; *cardi*, "heart"). The cells of the myocardium often have distinctive properties, compared with skeletal muscle cells or smooth muscle cells. They are thus distinguished as cardiac muscle cells (see Concept 33.4).

The volume of blood pumped per beat by a heart is called the heart's stroke volume. A heart's cardiac output is the volume of blood the heart pumps per minute. We calculate cardiac output by multiplying stroke volume by the number of beats per minute, the heart rate. For example, when you are at rest, your heart has a stroke volume of about 70 milliliters per beat (volume pumped into the systemic circuit) and beats about 70 times per minute. Your resting cardiac output is thus about 5 liters per minute (70 ml/beat × 70 beats/min = 4,900 ml/min, or about 5 L/min).

During exercise, we increase both stroke volume and heart rate. An average young person can increase cardiac output about fourfold. Even at rest, however, the amazing performance of the heart is evident. At rest the heart pumps about 300 liters each hour, corresponding to 2.6 million liters per year and almost 200 million liters in an average lifetime.

Vertebrate hearts are myogenic and multi-chambered

Hearts are classified based on the answers to two questions: how is each beat initiated, and how many chambers are present?

Vertebrate hearts do not require nervous stimulation to beat. Instead, each beat is initiated within the myocardium in modified muscle cells, and the muscle of a vertebrate heart beats on its own, even when all connections to other tissues of the body are severed. Vertebrate hearts are classed as **myogenic**, meaning "beginning in muscle," to recognize that heartbeats originate in the muscle tissue. Mollusks, too, have myogenic hearts. Crustaceans do not, however, as we will see later.

Vertebrate hearts are also classed as multi-chambered because they consist of two or more distinct chambers. Fish hearts have two muscular chambers: an **atrium** (plural *atria*)—which is a thin-walled chamber through which blood passes on its way to the principal pumping chamber—and a thick-walled, highly muscular ventricle. Amphibian hearts have three chambers (two atria that empty into one partially divided ventricle), and the hearts of mammals and birds have four chambers. One-way valves are present where one heart chamber opens into another, ensuring than blood flows through the chambers in the correct order.

Let's take a closer look at the mammalian heart (**FIGURE 32.6**). It consists of four chambers: two highly muscular ventricular chambers, which contract together, and two less-muscular atria, which also contract together. O_2-depleted blood returning to the heart from the systemic tissues enters the right atrium and then flows through a one-way valve into the right ventricle. When the ventricles contract, the O_2-depleted blood is pumped into the pulmonary arteries leading to the lungs. O_2-rich blood delivered to the heart from the lungs enters the left atrium and then passes through a one-way valve into the left ventricle. When the ventricles contract, this O_2-rich blood is pumped into a massive artery, the **aorta**, from which it flows through branching arteries to the systemic tissues.

The cycle of contraction and relaxation of the heart is called the **cardiac cycle**. This cycle is divided into two phases: **systole** (pronounced sís-toll-ee), when the ventricles contract, and **diastole** (die-ás-toll-ee), when the ventricles relax (**FIGURE 32.7**).

Go to ANIMATED TUTORIAL 32.1
The Cardiac Cycle
PoL2e.com/at32.1

The ventricles pressurize the blood inside them when they contract, and this positive pressure—developed by contraction of the ventricular myocardium—is the principal force that drives blood through the blood vessels. The positive pressure developed by the right ventricle forces blood to flow to the

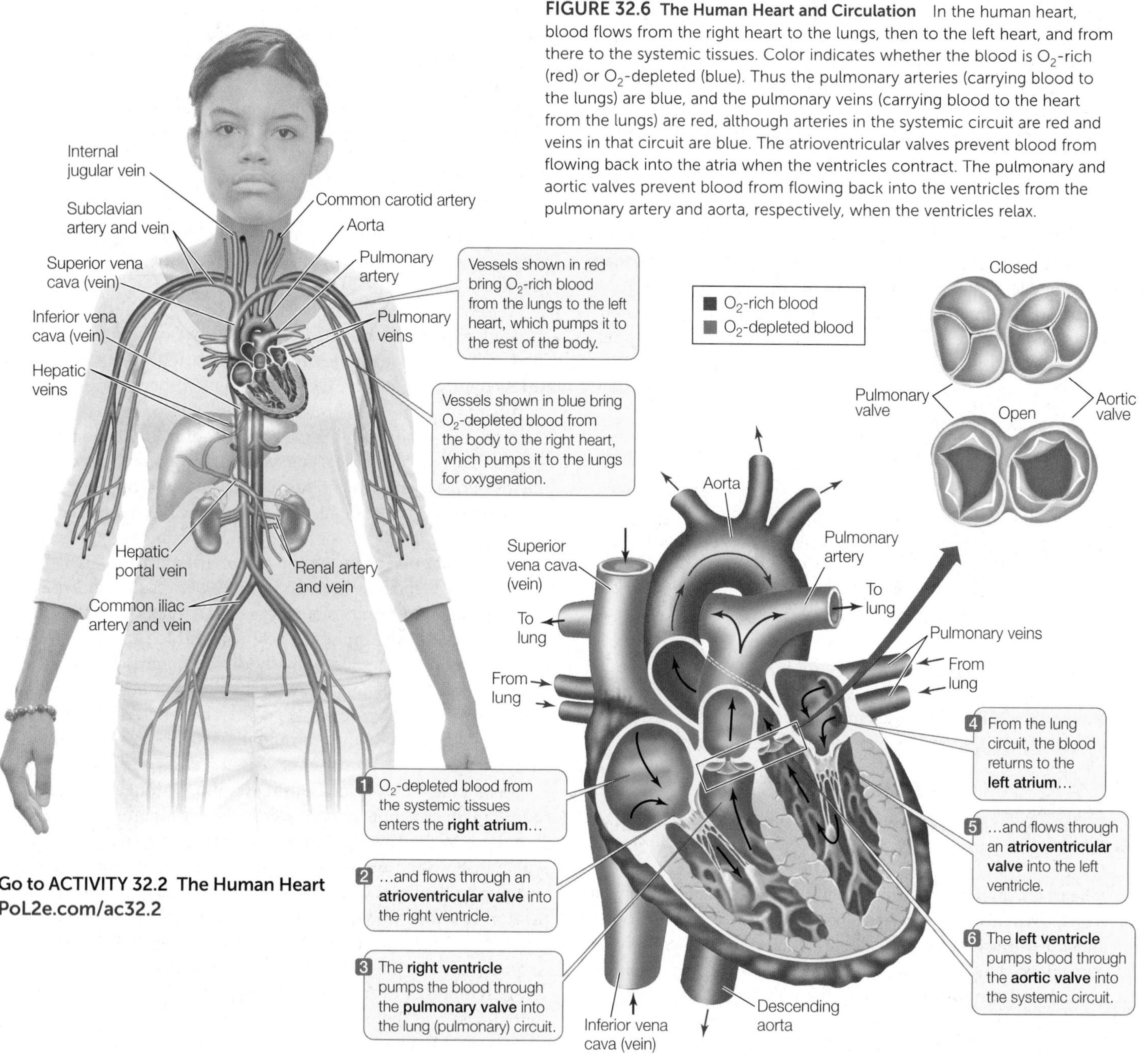
Internal jugular vein
Subclavian artery and vein
Superior vena cava (vein)
Inferior vena cava (vein)
Hepatic veins
Common carotid artery
Aorta
Pulmonary artery
Pulmonary veins
Vessels shown in red bring O_2-rich blood from the lungs to the left heart, which pumps it to the rest of the body.
Vessels shown in blue bring O_2-depleted blood from the body to the right heart, which pumps it to the lungs for oxygenation.
Hepatic portal vein
Renal artery and vein
Common iliac artery and vein
O_2-rich blood
O_2-depleted blood
Closed
Open
Pulmonary valve
Aortic valve
Aorta
Superior vena cava (vein)
Pulmonary artery
To lung
To lung
From lung
Pulmonary veins
From lung
1 O_2-depleted blood from the systemic tissues enters the **right atrium**...
2 ...and flows through an **atrioventricular valve** into the right ventricle.
3 The **right ventricle** pumps the blood through the **pulmonary valve** into the lung (pulmonary) circuit.
4 From the lung circuit, the blood returns to the **left atrium**...
5 ...and flows through an **atrioventricular valve** into the left ventricle.
6 The **left ventricle** pumps blood through the **aortic valve** into the systemic circuit.
Inferior vena cava (vein)
Descending aorta

FIGURE 32.6 The Human Heart and Circulation In the human heart, blood flows from the right heart to the lungs, then to the left heart, and from there to the systemic tissues. Color indicates whether the blood is O_2-rich (red) or O_2-depleted (blue). Thus the pulmonary arteries (carrying blood to the lungs) are blue, and the pulmonary veins (carrying blood to the heart from the lungs) are red, although arteries in the systemic circuit are red and veins in that circuit are blue. The atrioventricular valves prevent blood from flowing back into the atria when the ventricles contract. The pulmonary and aortic valves prevent blood from flowing back into the ventricles from the pulmonary artery and aorta, respectively, when the ventricles relax.

Go to ACTIVITY 32.2 The Human Heart
PoL2e.com/ac32.2

lungs and back to the heart. The positive pressure developed by the left ventricle forces blood to flow to all the systemic tissues and back. The left ventricle is more muscular than the right (see Figure 32.6) and develops higher pressures, which help ensure flow through the larger systemic circuit.

The **blood pressure** at any point in the circulatory system is the extent to which the pressure in the blood exceeds the pressure in the environment of the animal. As you might imagine, the blood pressure in the aorta rises and falls during the cardiac cycle, even though the aortic walls are elastic and therefore tend to dampen pressure changes. In healthy young people, the aortic pressure during systole is about 120 mm Hg (millimeters of mercury, referring to the height of a vertical column of mercury that the pressure can support). The aortic pressure during diastole is about 80 mm Hg. These pressures undergo little change as blood flows through major arteries and can therefore be measured in the arm, as is usually done. They are typically quoted as a ratio: "120 over 80." **Hypertension** ("high blood pressure") is a disease in which the pressures are significantly higher than this norm.

Each heartbeat is initiated by pacemaker cells, which are modified muscle cells found in a localized part of the right atrial wall, the sinoatrial (S-A) node (**FIGURE 32.8A**). Cardiac muscle cells in the vertebrate myocardium are electrically coupled, meaning that when a particular cell undergoes electrical depolarization (a reversal of the electric charge difference across the

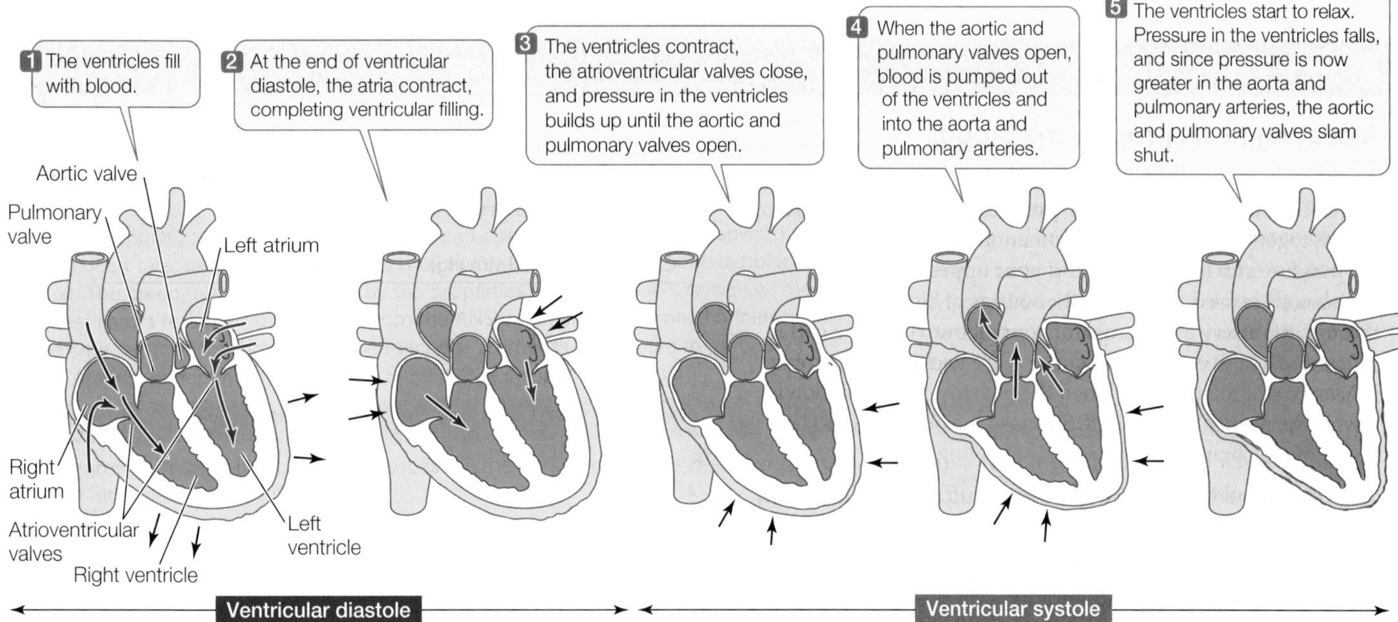

1 The ventricles fill with blood.
2 At the end of ventricular diastole, the atria contract, completing ventricular filling.
3 The ventricles contract, the atrioventricular valves close, and pressure in the ventricles builds up until the aortic and pulmonary valves open.
4 When the aortic and pulmonary valves open, blood is pumped out of the ventricles and into the aorta and pulmonary arteries.
5 The ventricles start to relax. Pressure in the ventricles falls, and since pressure is now greater in the aorta and pulmonary arteries, the aortic and pulmonary valves slam shut.
Aortic valve
Pulmonary valve
Left atrium
Right atrium
Atrioventricular valves
Right ventricle
Left ventricle
Ventricular diastole
Ventricular systole

FIGURE 32.7 The Cardiac Cycle The rhythmic relaxation (diastole) and contraction (systole) of the ventricles is called the cardiac cycle.

cell membrane) and consequently contracts, it initiates electrical depolarization and contraction in the cells next to it. Accordingly, when electrical depolarization and contraction are initiated at any one point in a mass of heart muscle, they spread rapidly throughout the muscle by moving from cell to cell. This process ensures that all the cells within a large mass of muscle depolarize and contract almost simultaneously. After the pacemaker cells in the S-A node initiate a wave of depolarization and contraction, the depolarization and contraction rapidly spread throughout the atria (**FIGURE 32.8B**, steps 1 and 2). Depolarization cannot spread directly into the ventricles, however, because a sheet of fibrous connective tissue lies between the atrial and ventricular muscle masses.

(A) The conducting system and sinoatrial node

Sinoatrial (S-A) node
Right atrium
Atrioventricular (A-V) node
Right ventricle
Septum between ventricles
Left atrium
Fibrous connective tissue
Conducting system
Left ventricle

(B) The initiation and spread of depolarization during a heartbeat

Depolarized
Not depolarized

1 Depolarization begins in the S-A node and spreads outward through atrial muscle.

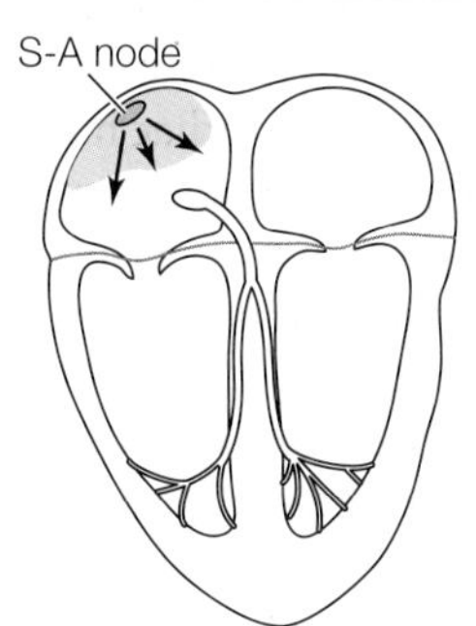

S-A node

2 Although depolarization spreads rapidly throughout the atrial muscle, its spread into the A-V node is delayed. The depolarized atria start to contract.

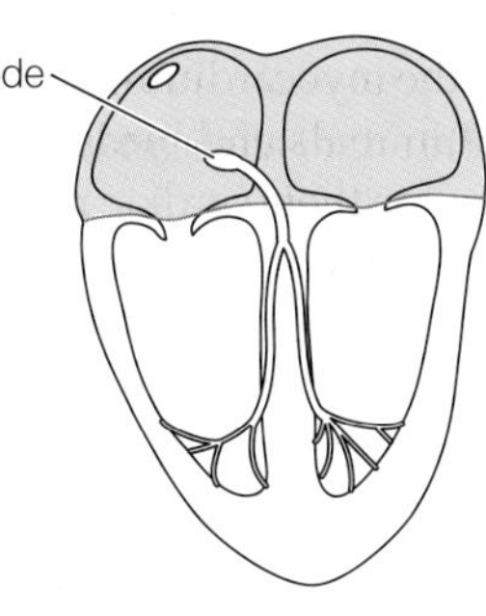

A-V node

3 Once the A-V node becomes depolarized, the depolarization spreads very rapidly into the ventricles along the conducting system. Atrial muscle starts to repolarize.

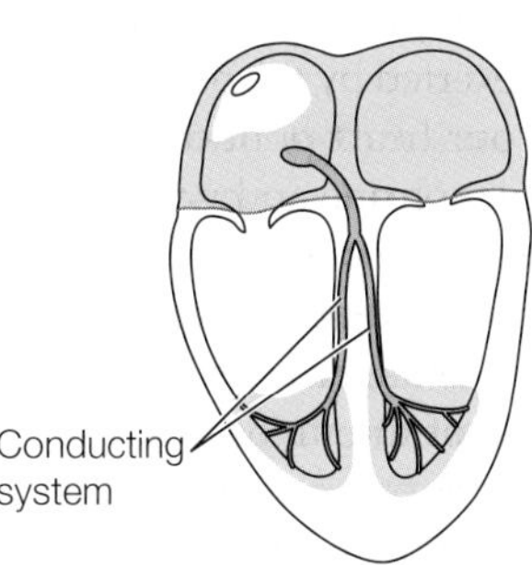

Conducting system

4 The nearly simultaneous depolarization of cells throughout the ventricular myocardium leads to forceful ventricular contraction.

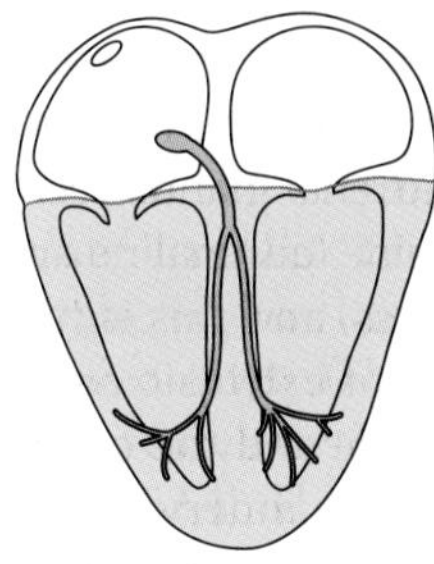

FIGURE 32.8 The Process of Activation of the Mammalian Heart during Each Heartbeat The muscle cells of the heart, like virtually all muscle cells, are activated to contract by electrical depolarization of their cell membrane. (A) Structures composed of modified muscle cells—the S-A node, A-V node, and conducting system—initiate and coordinate the spread of depolarization through the heart during each heartbeat. (B) Muscle cells in pink areas are depolarized during the successive stages shown. Cells in white areas are not depolarized; they are either awaiting depolarization or have repolarized following depolarization.

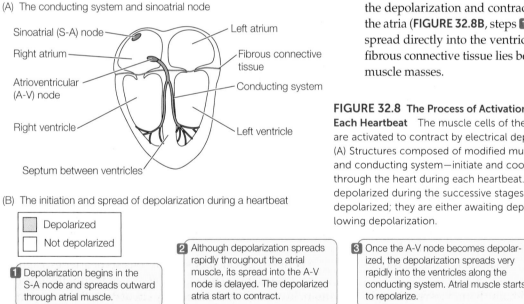

Hg because of the control of breathing we discussed in Concept 31.3. This means that the O_2 partial pressure in blood oxygenated in the alveoli is also kept near 100 mm Hg. As shown by the "Loading" point on the graph in Figure 32.20, human hemoglobin combines with a maximum amount of O_2—about 100 percent of the O_2 it can possibly carry—when the O_2 partial pressure is near 100 mm Hg. In short, human hemoglobin becomes highly oxygenated in the lungs.

In the systemic tissues of a person at rest, the O_2 partial pressure, on average, is about 40 mm Hg. The "Unloading at rest" point on the graph shows the consequences. O_2 binding is only about 75 percent of maximum when human hemoglobin is at 40 mm Hg. When highly oxygenated blood from the lungs first flows into the capillaries of the systemic tissues, its oxygen binding is about 100 percent, but as soon as it is exposed to the partial pressure of 40 mm Hg that prevails in the systemic tissues, it releases some O_2 so that its oxygen binding falls to 75 percent. This is the process by which hemoglobin releases O_2 to the tissues. Later, when the blood flows back into the lungs, the hemoglobin is exposed once more to a partial pressure of about 100 mm Hg and therefore takes up O_2, so that its oxygen binding is once again about 100 percent.

During exercise, hemoglobin can greatly increase the amount of O_2 it releases in the exercising muscles. The O_2 partial pressure in exercising muscles often falls to about 20 mm Hg during moderately intense exercise because the muscle mitochondria are using O_2 rapidly. When hemoglobin is exposed to this partial pressure, its oxygen binding falls to only 35 percent (see "Unloading during exercise" in Figure 32.20). The hemoglobin thus releases much more of its O_2 than it does in resting tissues.

In summary, there are two major ways that the circulatory system increases O_2 delivery when we exercise. First, we circulate blood more rapidly than at rest (by increasing heart rate and stroke volume). Second, our hemoglobin releases more than twice as much O_2 each time it flows from the lungs to the exercising muscles.

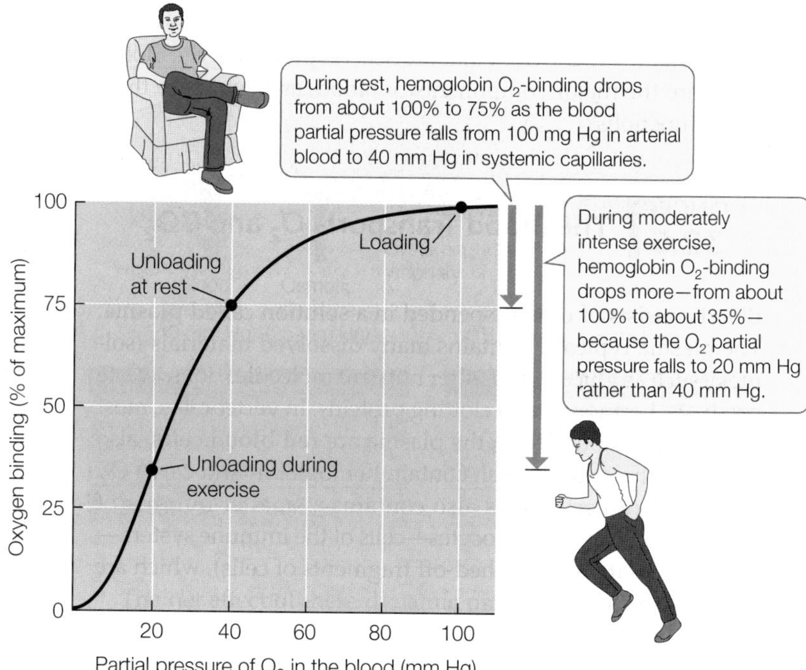

FIGURE 32.20 Loading and Unloading of O_2 by Human Hemoglobin This curve—known as an oxygen equilibrium curve—shows the chemical behavior of human hemoglobin under conditions in arterial blood. Hemoglobin almost instantly adjusts its O_2 binding to match changes in the prevailing O_2 partial pressure, taking up or releasing O_2 as needed. The vertical green arrows show how much O_2 is unloaded by hemoglobin as it flows from the lungs through the systemic tissues during rest and exercise. The blood O_2 partial pressure is expressed in millimeters of mercury (mm Hg) of pressure.

Go to ACTIVITY 32.4 Oxygen-Binding Curves
PoL2e.com/ac32.4

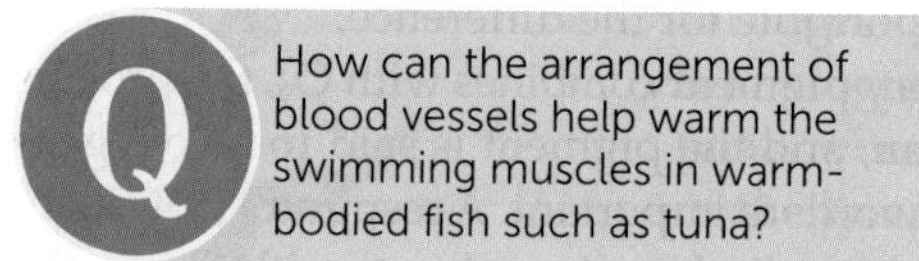

Q
How can the arrangement of blood vessels help warm the swimming muscles in warm-bodied fish such as tuna?

ANSWER The red swimming muscles produce considerable heat from their metabolism during steady, long-duration swimming in high-performance species of fish. In most species, however, this heat is lost to the environmental water as fast as it is produced, and the swimming muscles are not warmed by the heat they produce. This is true regardless of body size because heat moves readily through water, and in the gills it merely needs to move from one watery fluid, the blood, to another, the environmental water, to escape from the fish.

The key to keeping heat in the swimming muscles, so that they are warmed, is to prevent the heat they produce from leaving them. This is not possible in most fish because of the arrangement of the arteries and veins in the vasculature of the swimming muscles (**FIGURE 32.21**). Tuna (and certain sharks, such as great whites) have evolved a unique arrangement of arteries and veins to carry blood to and from their red swimming muscles, which are deep in the body (see Figures 32.17

CHECKpoint CONCEPT 32.5

- ✓ By what factor does hemoglobin increase the capacity of human blood to carry O_2?
- ✓ What do we mean when we say that hemoglobin and hemocyanin combine reversibly with O_2?
- ✓ Myoglobin, the red pigment inside muscle cells that are specialized for endurance exercise (see Concept 33.3)—giving the cells a red color—is actually a type of hemoglobin. There's a property of respiratory pigments called oxygen affinity. It is a measure of how strongly a respiratory pigment and O_2 are chemically attracted. Red muscles need O_2 because they make ATP aerobically. Keeping this in mind, would you expect myoglobin to have a higher or lower oxygen affinity than the hemoglobin in the blood? Explain.

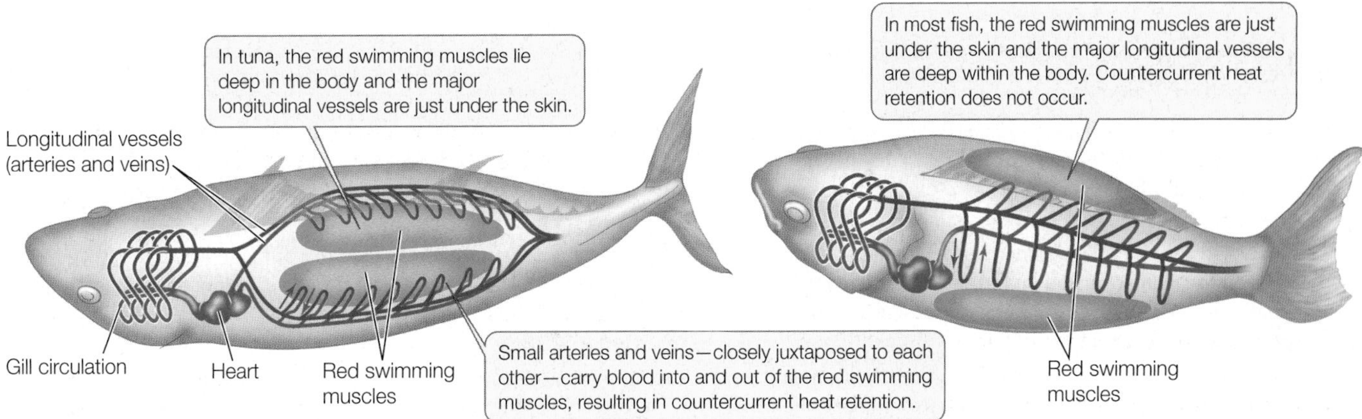

In tuna, the red swimming muscles lie deep in the body and the major longitudinal vessels are just under the skin.
In most fish, the red swimming muscles are just under the skin and the major longitudinal vessels are deep within the body. Countercurrent heat retention does not occur.
Longitudinal vessels (arteries and veins)
Gill circulation
Heart
Red swimming muscles
Small arteries and veins—closely juxtaposed to each other—carry blood into and out of the red swimming muscles, resulting in countercurrent heat retention.
Red swimming muscles

FIGURE 32.21 Tuna and Other Warm-Bodied Fish Have a Unique Arrangement of Arteries and Veins

and 33.14). In this arrangement, as veins leave the swimming muscles—carrying heat with them—they come into close contact with the arteries carrying blood into the muscles. This arrangement allows countercurrent heat exchange to take place. Heat moves out of the venous blood into the arterial blood, and the arterial blood carries the heat back into the swimming muscles. Heat made by the muscles thus has a hard time exiting them (very unlike the case in most fish), and the muscles are warmed to temperatures higher than that of the environmental water.

SUMMARY

CONCEPT 32.1 Circulatory Systems Can Be Open or Closed

- Animals with a high degree of division of labor need a **circulatory system** that transports respiratory gases, nutrients, hormones, and metabolic wastes throughout the body at high rates.
- O_2 transport is the most urgent function of the circulatory system in most types of animals. Because of this, animals with relatively high rates of O_2 consumption tend to have relatively high-performance circulatory systems. Insects are the exception that proves the rule. Although many insects have high rates of O_2 consumption, insects do not have high-performance circulatory systems because their tracheal breathing system, not their circulatory system, meets needs for O_2 transport.
- In **closed** circulatory systems—found in vertebrates, annelid worms, squid, and octopuses—the blood never leaves a system of vessels. In **open** circulatory systems—found in arthropods and most mollusks—the blood leaves vessels and bathes tissue cells directly. One advantage of closed circulatory systems is their ability to direct blood selectively to specific tissues.
- The blood in a closed circulatory system flows from the heart, through arteries, to the microscopic vessels of the **microcirculation**, which consists of arterioles, capillaries, and venules. Veins then carry blood back to the heart.
- **Capillaries** are the smallest vessels and are the principal sites of exchange of gases, nutrients, and other substances between the blood and the tissues. **Arterioles** control the rate of blood flow into the capillaries. **Review Figures 32.2 and 32.3**
- In open circulatory systems, the heart typically pumps blood into arteries that carry the blood for at least a short distance. These vessels then end and the blood flows out into the animal's tissues. The blood makes its way back to the heart through channels between the tissue cells called sinuses and lacunae. **Review Figure 32.4**
- In open circulatory systems there is no distinction between blood and interstitial fluid.

CONCEPT 32.2 The Breathing Organs and Systemic Tissues Are Usually, but Not Always, in Series

- The **systemic circuit** consists of the blood vessels (or blood channels in the case of an open circulatory system) that carry blood to and from the systemic tissues. The **breathing-organ circuit** consists of the blood vessels that carry blood to and from the breathing organs (lungs or gills).
- **Arteries** are vessels that carry blood away from the heart. **Veins** carry blood toward the heart.
- In most animals the systemic circuit and breathing-organ circuit are connected in series: the blood flows from one circuit to the other sequentially. Circulation in series is the most efficient way to deliver O_2 to the systemic tissues. **Review Figure 32.5A and B, ACTIVITY 32.1**
- Adult amphibians and non-avian reptiles have a partially divided ventricle that does not ensure series circulation. Some of these animals, such as turtles, can shunt blood away from the breathing-organ circuit at times when doing so may be advantageous, such as during prolonged diving. **Review Figure 32.5C**

CONCEPT 32.3 A Beating Heart Propels the Blood

- Vertebrate hearts are multi-chambered and **myogenic**; heartbeats are initiated by specialized muscle cells within the heart.
- The simplest vertebrate heart, found in fish, is two-chambered. It has an **atrium** that receives blood from the body and a **ventricle** that pumps blood out of the heart. Amphibians and non-avian reptiles have a three-chambered heart with two atria and one partially divided ventricle.
- The heart of mammals and birds has four chambers—two atria and two ventricles—with valves to prevent backflow. Blood flows from the right atrium and ventricle to the lungs, then to the left atrium and ventricle, and then to the rest of the body. **Review Figure 32.6 and ACTIVITY 32.2**

(continued)

SUMMARY (continued)

- The **cardiac cycle** has two phases: **systole**, when the ventricles contract, and **diastole**, when the ventricles relax. **Review Figure 32.7 and ANIMATED TUTORIAL 32.1**
- The heartbeat in mammals is initiated by pacemaker cells in the sinoatrial node that spontaneously depolarize, triggering a wave of depolarization that spreads through the atria and then spreads with a brief delay into the ventricles via the atrioventricular node and the **conducting system**. **Review Figure 32.8**
- The autonomic nervous system controls heart rate by changing the rate of depolarization of the pacemaker cells. Norepinephrine from sympathetic nerves increases heart rate, and acetylcholine from parasympathetic nerves decreases it. **Review Figure 32.9**
- In mammals and many other vertebrates, a **coronary circulatory system** supplies O_2 to the cells of the heart muscle (myocardium).
- The depolarization of heart muscle cells can be detected on the surface of the body and recorded as an **electrocardiogram** (**ECG** or **EKG**). **Review Figure 32.10**
- The hearts of crustaceans have one chamber and are **neurogenic**; the muscle cells of the heart require nervous stimulation to beat. **Review Figure 32.11**

CONCEPT 32.4 Many Key Processes Occur in the Vascular System

- Arteries are composed of muscle fibers and elastic fibers that enable them to dampen the surge in pressure when the heart beats and store in their stretched state some of the energy imparted to the blood by the heart. Veins have thinner walls and have a high capacity for storing blood. **Review Figure 32.12 and ACTIVITY 32.3**
- The pressure and linear velocity of blood vary greatly as it flows through the vascular system. Blood pressure is high in arteries and low in veins. Blood velocity is lowest in capillary beds. **Review Figure 32.13**
- Veins have one-way valves that assist in returning blood to the heart under low pressure. **Review Figure 32.14**
- Heat loss from certain regions of an animal's body—such as the extremities of the arctic fox and the red swimming muscles of tuna—can be reduced by countercurrent heat exchange. **Review Figures 32.15–32.17**
- High blood pressure forces fluid out of the blood as it travels through the initial part of a capillary bed; osmotic pressure draws some of the fluid back into the blood as blood pressure falls. **Review Figure 32.18**
- The excess fluid that does not reenter the blood as it travels through capillary beds is called lymph. The **lymphatic system** collects this fluid and returns it to the blood. **Review Figure 32.19**

CONCEPT 32.5 The Blood Transports O_2 and CO_2

- In vertebrates, blood consists of a liquid **plasma** and cellular components (**red blood cells**, or erythrocytes; **white blood cells**, or leukocytes; and **platelets**). Red blood cells are produced in the bone marrow. **Hemoglobin**, the **respiratory pigment** in red blood cells, binds O_2 reversibly. Each hemoglobin molecule can carry a maximum of four O_2 molecules.
- Hemoglobin binds O_2 when the partial pressure of O_2 is high and releases it when the partial pressure is low. Therefore hemoglobin picks up (loads) O_2 as it flows through a breathing organ and gives up (unloads) O_2 in the systemic tissues. This chemical behavior can be summarized in an **oxygen equilibrium curve**. **Review Figure 32.20, ACTIVITY 32.4, and ANIMATED TUTORIAL 32.2**
- **Hemocyanin** is the respiratory pigment in mollusks and arthropods.
- CO_2 reacts with blood water to form bicarbonate ions (HCO_3^-) and is transported mostly in the form of bicarbonate.

Go to the Interactive Summary to review key figures, Animated Tutorials, and Activities PoL2e.com/is32

Go to LaunchPad at **macmillanhighered.com/launchpad** for additional resources, including LearningCurve Quizzes, Flashcards, and many other study and review resources.

33 Muscle and Movement

KEY CONCEPTS

A sphinx moth (*Hemaris thysbe*) uses its flight muscles to hover at a flower, collecting nectar with its long proboscis.

This moth, a type of sphinx or hawk moth, is sometimes mistaken for a hummingbird, accounting for its name, the hummingbird clearwing moth (*Hemaris thysbe*). Common in several parts of the United States and Canada, it feeds during daylight by hovering at flowers using rapid wingstrokes driven by the flight muscles in its thorax. It has a strikingly long proboscis that it can insert deeply into flowers to collect nectar. Other species of sphinx moth also have a striking proboscis, the most amazing being that of the famous Madagascar sphinx moth we mentioned at the start of Chapter 21. The hummingbird clearwing moth buzzes as it hovers because of the rapid movements of its flight apparatus.

Insects probably first took to the air about 350 million years ago. Over that long evolutionary time, diverse forms of flight have evolved, from the relatively slow and silent up-and-down wingstrokes of many butterflies, to the high-frequency buzzing wingstrokes of sphinx moths, to the very high frequency whining wingstrokes of mosquitoes. Today, engineers are studying insects to learn more about the aerodynamics of flight. Sphinx moths are of particular interest because they are powerful fliers that can quickly alternate between hovering and flying straight ahead at high speeds.

Muscle cells—one of the defining properties of animals—develop forces by means of interactions among fiber-like, contractile proteins within the cells. These forces typically set matter in motion. For example, the forces developed by the flight muscles of an insect set the wings in motion. For matter to be set in motion, an energy source is required. Insect flight-muscle proteins draw the necessary energy from adenosine triphosphate (ATP) made by the mitochondria in the muscle cells.

The great majority of muscle cells in the animal kingdom require a nerve impulse to activate each contraction: contractions and nerve impulses occur in a 1:1 ratio. Sphinx moths draw the interest of muscle physiologists because they are among the animals that have the highest frequencies of muscle contraction during flight while retaining this 1:1 ratio. Their wingstrokes can number more than 30 per second (30 Hz).

Per gram, insect flight muscle stands out as one of the tissues that attain the highest known rates of oxygen consumption. Physiologists reason that if high mechanical power generation is critical for ecological success—as it certainly is in powerfully flying insects—natural selection will have favored the evolution of muscle cells that have exceptional capacities to use O_2 to produce ATP to power muscle contraction. Electron micrographs of insect flight muscle show cells that are literally packed with contractile proteins and mitochondria.

Q Why is it likely that available space inside cells has limited the contents of contractile proteins and mitochondria in high-performance muscles?

You will find the answer to this question on page 697.

CONCEPT 33.1 Muscle Cells Develop Forces by Means of Cycles of Protein–Protein Interaction

Muscle tissue makes up a large portion of body mass in vertebrates and many other animals. In humans, for example, muscles account for almost half of body mass (**FIGURE 33.1**). Muscles are as important functionally as they are anatomically. They are the basis for virtually all behavior. They permit animals to run, fly, or swim to acquire food or escape dangers. In humans and some other mammals, muscles of the face account for changing facial expressions. Besides behavior, many other physiological actions also depend on muscle contraction, such as the circulation of blood by the beating of the heart.

There are several types of muscle tissue in the animal kingdom. The contractile mechanism is basically the same in most muscle types, however. We will begin our study of muscle function by focusing on the contractile mechanism in vertebrate **skeletal muscle**—the type of muscle that is attached to the bones of the skeleton and that provides the power for walking, flying, swimming, and other forms of locomotion.

Contraction occurs by a sliding-filament mechanism

Muscle tissue develops mechanical forces. This is its defining feature. In the scientific study of muscle, **contraction** refers to force development. A muscle does not necessarily *shorten* when it undergoes contraction. To see the distinction between contraction and shortening, imagine grabbing a bar of steel attached to a concrete wall with your hand and pulling as hard as you can with your arm muscles. Your muscles will develop great forces, but they will not shorten because the bar of steel is immovable. For a muscle to shorten during contraction, the force it exerts must be greater than opposing forces. When you use your arm to pull on a drawer, your arm muscles shorten during contraction because the force they exert exceeds the opposing forces, and the drawer slides toward you.

The current theory of how muscle contracts is often known as the **sliding-filament theory**. This name describes what the process *looks* like and is useful in that way. When muscle cells contract in circumstances in which they shorten, tiny filaments in the cells look like they slide past each other during contraction. However, the name "sliding" can be misleading, because "slide" suggests effortless motion, like a sled sliding down a snow-covered hill. In fact, the motion of the filaments in muscle cells is not at all effortless. It is quite the opposite: during contraction, the filaments use energy from ATP to develop mechanical forces and pull on each other. The "sliding" of the filaments is merely a consequence of this energy-using, force-generating process.

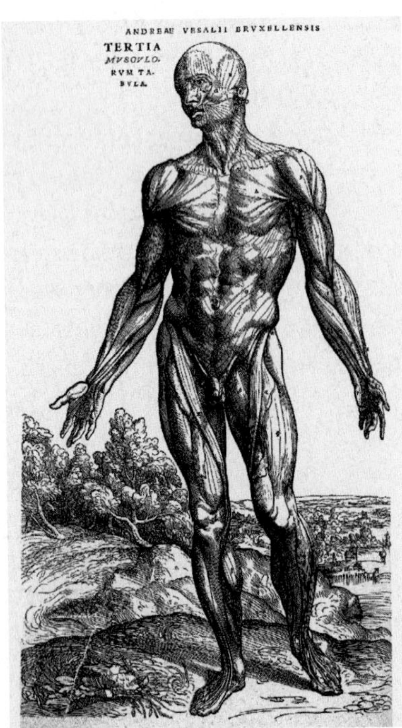

FIGURE 33.1 Muscles Account for a Large Portion of Body Mass Andrea Vesalius produced this wood block print, the first accurate depiction of the skeletal muscles of the human body, in 1543. He was noted for posing his subjects as they might have stood in life.

Actin and myosin filaments slide in relation to each other during muscle contraction

The cells of skeletal muscle are sometimes called "cells" and sometimes "fibers." These two terms are synonyms: a **muscle fiber** is a muscle cell. The cells in vertebrate skeletal muscle are large and have many nuclei, making them multinucleate. During development they form through the fusion of many individual embryonic muscle cells called myoblasts. A muscle such as your biceps (the muscle that pulls your forearm up toward your upper arm) is composed of hundreds or thousands of muscle fibers that are organized in bundles surrounded by connective tissue (**FIGURE 33.2A**).

Muscle contraction results from the interaction of two contractile proteins that are abundant in each muscle fiber: **actin** and **myosin**. Molecules of actin are organized into **actin filaments**, also called thin filaments. Molecules of myosin are organized into **myosin filaments**, or thick filaments. These two kinds of filaments interdigitate in parallel arrays as seen in **FIGURE 33.2B**.

> **LINK**
> Actin and myosin are also part of the cytoskeleton, which is responsible for other types of movement in cells; see **Concept 4.4**

Let's look now in more detail at the anatomical relationships among a muscle, its muscle fibers, and the actin and myosin filaments inside each fiber. Looking at Figure 33.2A and moving outward from the person's arm, you will see that a muscle is made up of bundles of long, parallel muscle fibers. Similarly, each muscle fiber is composed internally of long, parallel **myofibrils**, each of which is composed of arrays of actin

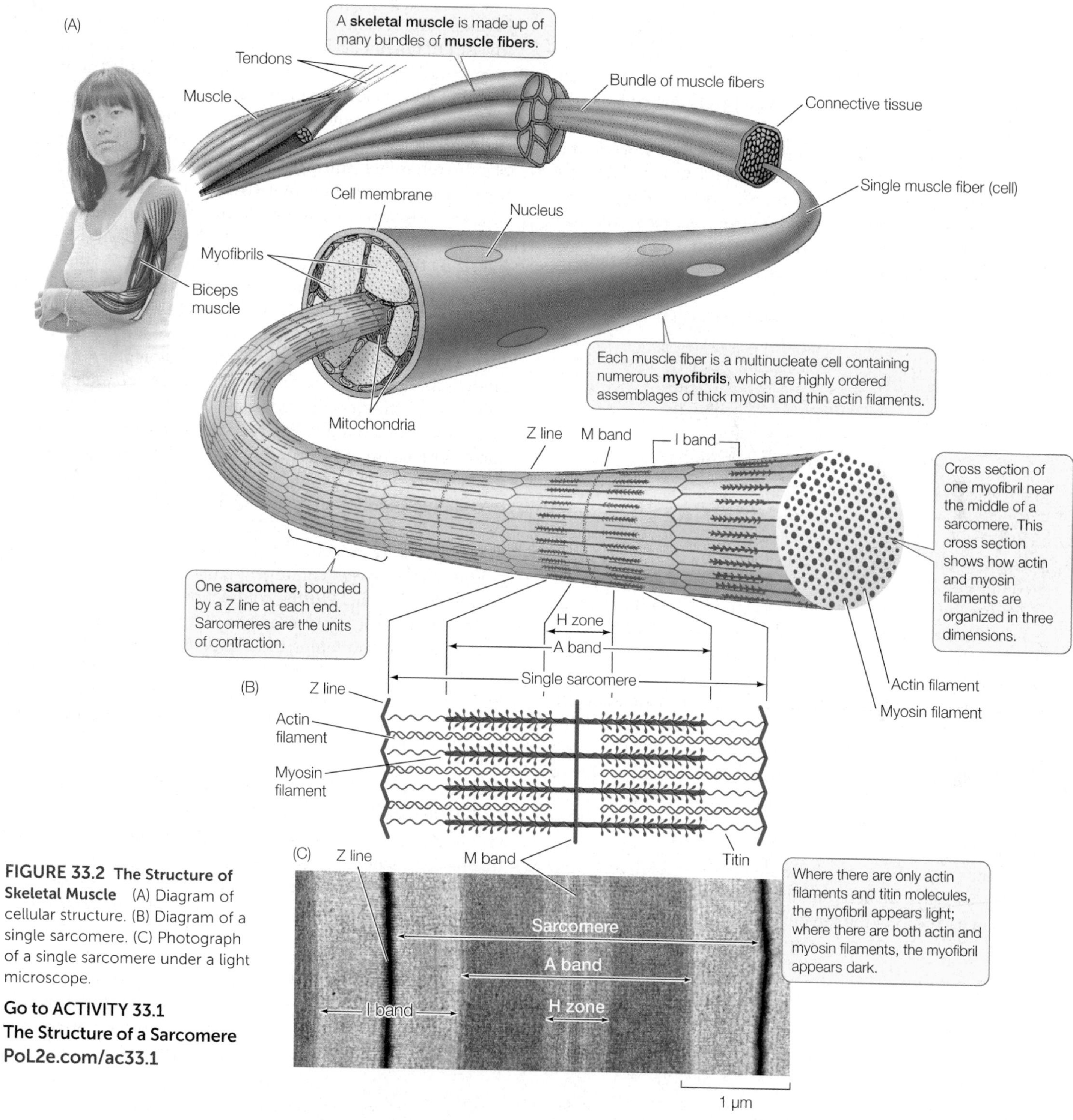

FIGURE 33.2 The Structure of Skeletal Muscle (A) Diagram of cellular structure. (B) Diagram of a single sarcomere. (C) Photograph of a single sarcomere under a light microscope.

Go to ACTIVITY 33.1 The Structure of a Sarcomere PoL2e.com/ac33.1

and myosin filaments oriented parallel to the long axis of the myofibril.

Understanding the structure of a myofibril is critical. Figure 33.2A is drawn so you can view a myofibril from the side. From this view, you can see that a myofibril consists of repeating units called **sarcomeres**. Each sarcomere starts and ends with a Z line. Internally, as shown in Figure 33.2B, each sarcomere contains overlapping arrays of actin and myosin filaments, as we have already noted. At the far right of Figure 33.2A, the myofibril is shown in cross section so you can see the three-dimensional arrangement of its actin and myosin filaments. In most regions of a myofibril, as the cross section shows, each thick myosin filament is surrounded by six thin actin filaments, and each actin filament sits within a triangle of three myosin filaments.

In side view, the structures we have been discussing give skeletal muscle a distinctly banded appearance when the muscle is examined under a simple light microscope. To see individual actin and myosin filaments, you would need to use an electron microscope because they are far too thin to be seen with a light microscope. When you look at muscle with a light microscope, however, you see patterns that result from the organization of large sets of actin and myosin filaments—much as you can see black and white lines on a computer screen even though you cannot see the individual pixels that collectively give rise to the patterns. **FIGURE 33.2C** shows the banding pattern in a single muscle fiber as seen with a light microscope at fairly high magnification. By comparing Figures 33.2B and 33.2C, you can see how the bands seen with the light microscope line up with the actin and myosin filaments in a sarcomere. At lower magnification under a light microscope, the banding in a region of skeletal muscle looks like this:

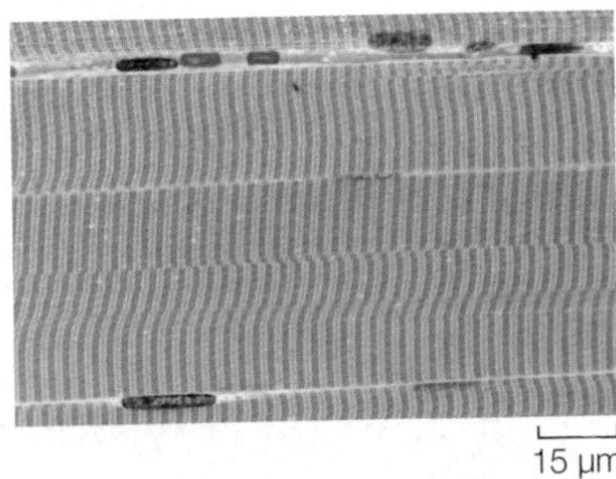

Because of this banding pattern, skeletal muscle is called striated muscle.

Before the molecular nature of the muscle banding pattern was known, the bands were given names that are still used today (see Figure 33.2B,C). Each sarcomere is bounded by Z lines, as already mentioned, which represent structures to which the actin filaments are attached. At the center of each sarcomere is a structure called the M band (for "middle"). It contains proteins that help hold the myosin filaments in position. Each myosin filament extends out symmetrically on either side of the M band. Collectively, the myosin filaments in each sarcomere give rise to a dark region—called the A band—centered on the M band.

Finally, in each sarcomere there are three relatively light bands: an H zone centered on the M band, and one half of an I band at each end of the sarcomere. These bands appear light because, unless a myofibril is fully contracted, the actin and myosin filaments do not overlap in them: only myosin filaments are found in the H zone, and only actin filaments are found in the I bands (see Figure 33.2B,C).

Within a sarcomere, the thick myosin filaments are held in their correct positions in three-dimensional space by molecules of a protein called **titin** (**FIGURE 33.3A**). This giant molecule is the largest known protein. A myosin filament consists of about 200–400 individual myosin molecules in a bundle. Each titin molecule in a sarcomere runs between the M band and a Z line, directly through a myosin filament along the filament's long axis. Between the end of the filament closest to the Z line and the Z line itself, the titin molecule is elastic, like a bungee cord. Collectively, the titin molecules do two things. They hold the myosin filaments in position, and if a sarcomere becomes stretched, their elasticity returns the sarcomere to its shorter length when the stretching forces relax.

A myofibril looks like Figure 33.3A when it is in its relaxed state. When a myofibril contracts in circumstances in which it can shorten, its sarcomeres shorten and the banding pattern changes to look like **FIGURE 33.3B**. The light-colored H zone and I bands become much narrower, and the dark Z lines move toward the A band. These changes are easy to see under a light microscope as Figure 33.3 shows. Why do they occur? At a molecular level, during contraction the actin and myosin filaments slide past each other, so that they overlap to an increased degree, shortening the muscle fiber. In each sarcomere the actin filaments slide into the H zone, the region occupied by only myosin filaments in the relaxed state. As they do so, the Z lines at opposite ends of the sarcomere are pulled closer to each other.

ATP-requiring actin–myosin interactions are responsible for contraction

To understand how a sarcomere develops mechanical forces during contraction, it is critical to understand the structures of myosin and actin molecules (**FIGURE 33.4**). A myosin molecule consists of two long polypeptide chains coiled together, each ending in a large globular head. A myosin filament is made up of many myosin molecules, the heads of which project sideways near the ends of the filament.

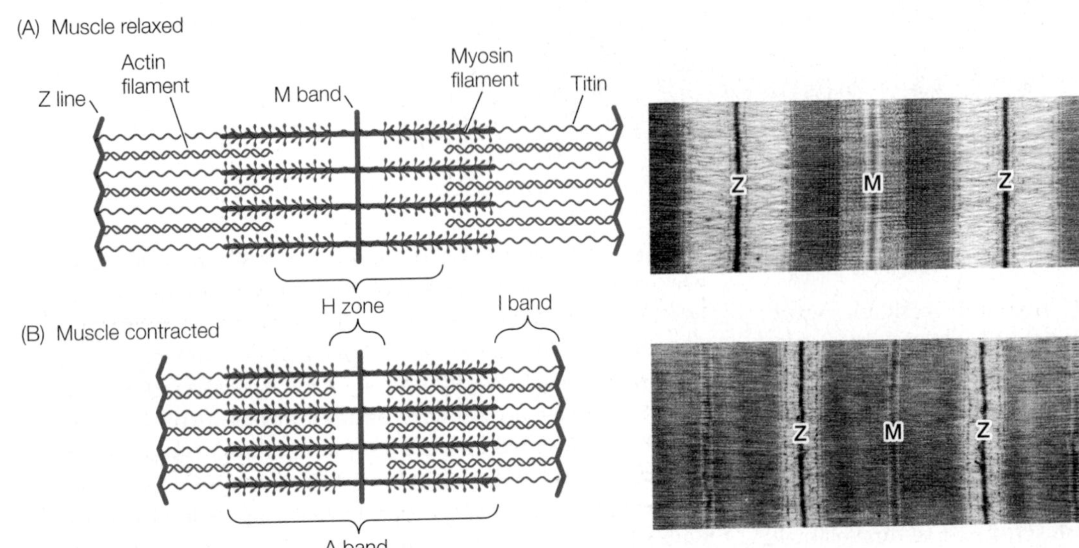

FIGURE 33.3 Sliding Filaments Diagrams and light-microscope images of (A) a relaxed sarcomere and (B) the same sarcomere shortened during contraction. The banding pattern of the sarcomere changes as it shortens. This shortening results from the sliding-filament contractile mechanism of muscle contraction.

FIGURE 33.4 The Structure of Actin and Myosin Filaments Myosin filaments are bundles of myosin molecules with globular heads and polypeptide tails. Actin filaments consist of two chains of actin monomers twisted together, along with two other proteins: tropomyosin and troponin.

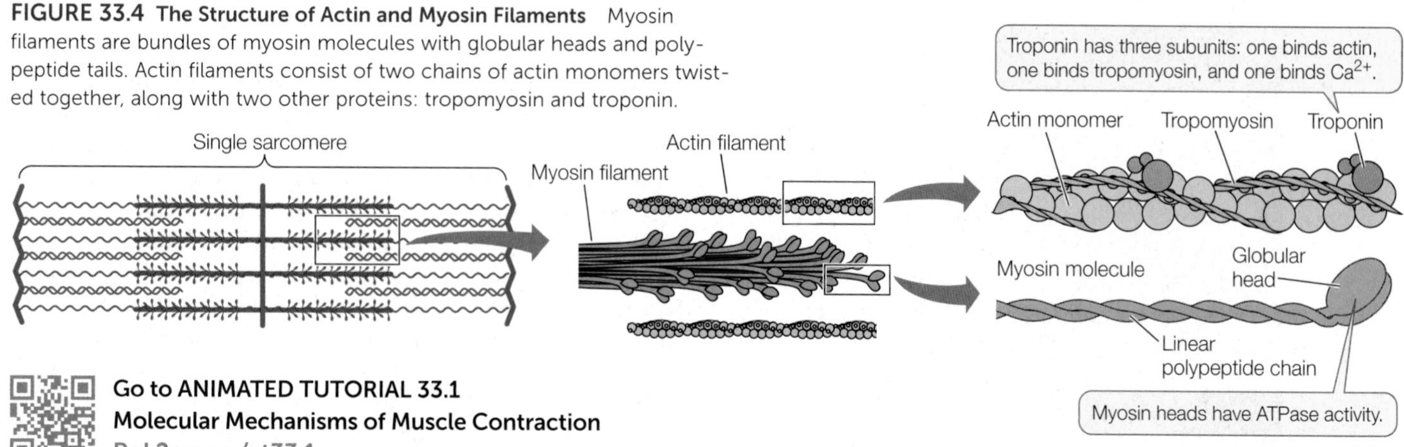

Go to ANIMATED TUTORIAL 33.1
Molecular Mechanisms of Muscle Contraction
PoL2e.com/at33.1

An actin filament consists of small actin monomers polymerized into two long chains that look like two strands of pearls twisted together. Twisting around each actin chain is another protein, **tropomyosin**, and attached to tropomyosin at intervals are molecules of still another protein, **troponin**. These molecules control contraction and relaxation, as we will see in the next section.

How do myosin and actin interact? The globular heads of myosin can bind to specific sites on actin, forming bridge-like links, called **cross-bridges**, between the myosin and actin filaments. Two functional properties of the myosin heads are exceedingly important. First, when a myosin head binds to an actin filament, the head's molecular shape (conformation) changes in such a way that the head bends and exerts a tiny force on the actin filament. This force causes the actin filament to be pulled a distance of 5–10 nanometers toward the M band along the myosin filament (see the transition from step 3 to step 4 in Figure 33.7). Second, when a myosin head is bound to actin, it has ATPase activity: that is, it binds to and hydrolyzes ATP, releasing energy. The energy released changes the molecular shape of the myosin head, causing it to disconnect from actin and unbend, returning to its extended position. At that point, the myosin head can again bind to actin and undergo another change of molecular shape that causes a tiny force to be exerted on the actin filament, moving the actin filament relative to the myosin filament.

Together, these details explain how the actin and myosin filaments in a sarcomere interact to develop mechanical forces and slide past each other during contraction. The details also help explain why the force a muscle fiber can exert during contraction depends on how many actin–myosin cross-bridges can form.

Rigor mortis—the stiffening of muscles soon after death—is still another muscle property that can be explained by the details of muscle contraction we are discussing. ATP is needed to break actin–myosin bonds. When ATP production ceases with death, actin–myosin bonds cannot be broken and muscles stiffen.

To fully understand contraction, remember that each myosin filament has many myosin heads near both ends and is surrounded by six actin filaments. As contraction takes place, numerous myosin heads go through cycles of interaction with the actin filaments. Their cycles of binding and release are not exactly synchronous, however. Accordingly, when a single myosin head breaks its contact with actin, other cross-bridges are still connected and flexing. As a result, the actin filaments do not slip backward.

Excitation leads to contraction, mediated by calcium ions

Muscle cells, as well as nerve cells (see Chapter 34), are excitable. A cell is **excitable** if its cell membrane can generate and conduct electrical impulses (action potentials). Cell membranes ordinarily have an electrical polarity in which the outside of the membrane is more positive than the inside. An **impulse**, or **action potential**, is a region of reversed polarity, also called depolarization (see Concept 34.2). In an excitable cell, after an impulse—a region of reversed polarity, or depolarization—is initiated at one point in a cell membrane, it travels along the full length of the cell membrane:

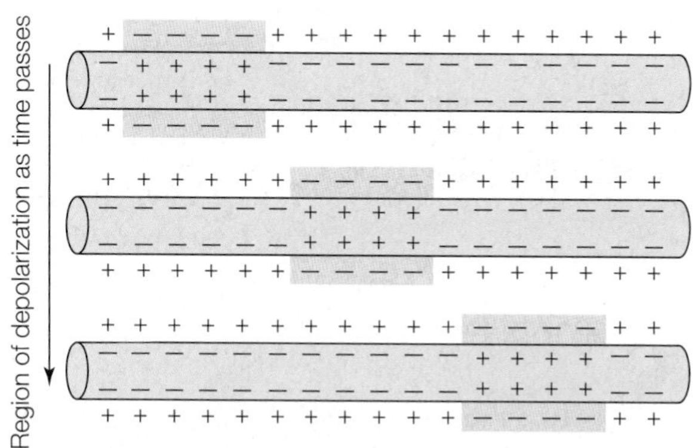

What triggers muscle cells to contract? In skeletal muscle, nerve impulses stimulate muscle fibers to contract. The nerve cells that are involved have long, threadlike cell processes (cell extensions) called axons (see Figure 34.2). Each muscle fiber is in physical contact with an axon of a nerve cell. The nerve cell is said to **innervate** the muscle fiber. The point of contact between the nerve cell and muscle fiber is known as a **neuromuscular junction** (**FIGURE 33.5**). When a nerve impulse arrives at the neuromuscular junction, it is said to excite the muscle fiber, and the event is called **excitation**. During excitation (see Concept 34.3), an impulse or action potential is initiated in the cell membrane of the muscle fiber and then travels the length of that cell membrane.

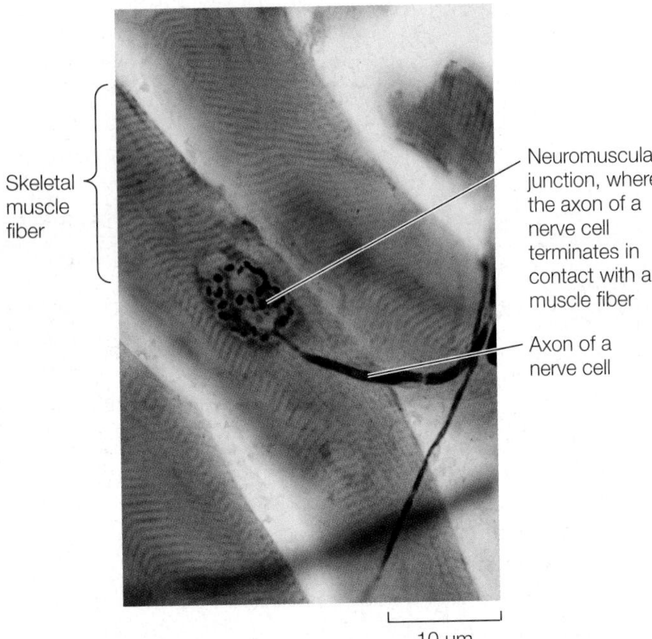

Skeletal muscle fiber
Neuromuscular junction, where the axon of a nerve cell terminates in contact with a muscle fiber
Axon of a nerve cell
10 μm

FIGURE 33.5 Innervation of a Muscle Fiber In this view of several muscle fibers, we see a nerve cell process (axon) from a far-away nerve cell that controls the contraction of one of the muscle fibers. The point of contact of the nerve cell process with the muscle fiber is called a neuromuscular junction. When the nerve cell is stimulated, an impulse travels along its axon to the neuromuscular junction and excites the muscle fiber. It is possible for one nerve cell to send a process to more than one muscle fiber. In such cases, all the muscle fibers that the nerve cell innervates are simultaneously excited, and contract together, when the nerve cell is stimulated.

Excitation–contraction coupling is the process by which the excitation of a muscle fiber leads to contraction of the muscle fiber. Several decades ago, when excitation–contraction coupling was first being investigated, researchers injected muscle fibers in a frog's leg with the bioluminescent protein aequorin, which is activated to emit light by calcium ions (Ca^{2+}). The researchers found that these injected fibers emitted a flash of light each time they were stimulated to contract. This elegant experiment announced to the world that Ca^{2+} plays a role in excitation–contraction coupling. After a muscle fiber is excited, Ca^{2+} is released into the general cytoplasm of the fiber from internal Ca^{2+}-storage locations. Ca^{2+} in the cytoplasm of an injected cell activates the aequorin, producing a flash of light.

Where is the Ca^{2+} stored in muscle fibers, and how is it released? A distinctive feature of muscle cells is that each cell has many spots where the cell membrane is indented inward into the cell cytoplasm to form a **transverse tubule**, or **T tubule** (**FIGURE 33.6**). The membrane that forms the walls of each T tubule is continuous with the outer cell membrane. Thus when an impulse or action potential travels along the outer cell membrane, it also travels inward along each T tubule, thereby spreading into the interior of the cell at multiple places via the system of multiple T tubules.

The T tubules of a muscle cell are physically very close to the endoplasmic reticulum of the cell. In a muscle cell, the cytoplasm is called the **sarcoplasm**, and the endoplasmic reticulum is called the **sarcoplasmic reticulum**. The sarcoplasmic reticulum is a closed compartment that surrounds every myofibril in a muscle cell (see Figure 33.6). Calcium pumps in the sarcoplasmic reticulum membrane take up Ca^{2+} from the sarcoplasm. Thus when a muscle cell is at rest there is a high

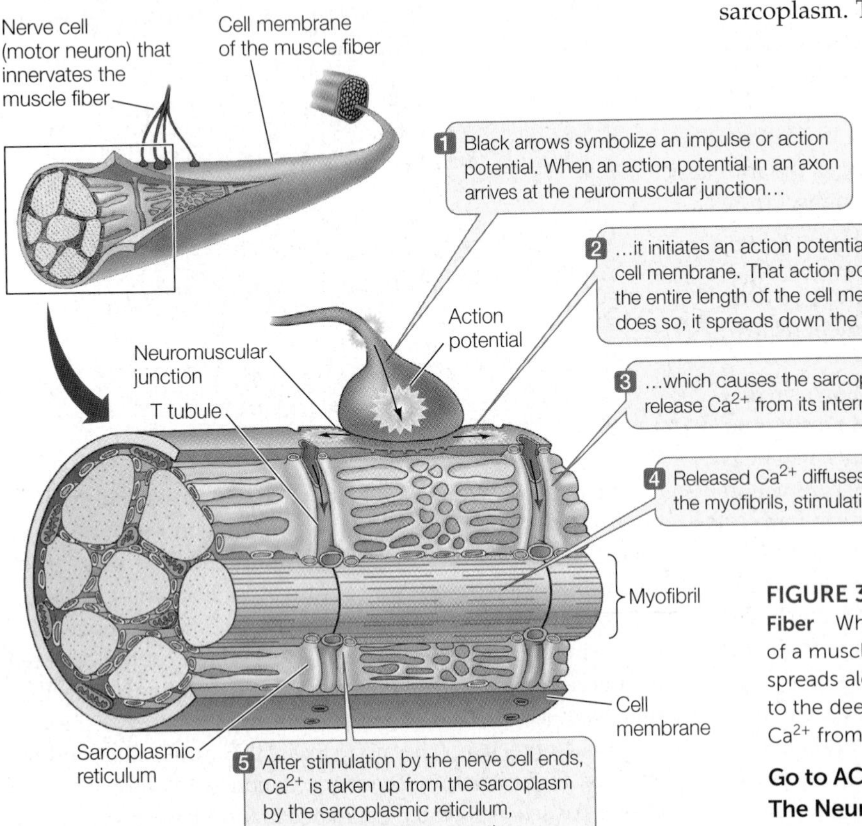

Nerve cell (motor neuron) that innervates the muscle fiber
Cell membrane of the muscle fiber
1 Black arrows symbolize an impulse or action potential. When an action potential in an axon arrives at the neuromuscular junction...
2 ...it initiates an action potential in the muscle fiber cell membrane. That action potential spreads along the entire length of the cell membrane, and, as it does so, it spreads down the T tubules...
Action potential
Neuromuscular junction
T tubule
3 ...which causes the sarcoplasmic reticulum to release Ca^{2+} from its internal stores of Ca^{2+}.
4 Released Ca^{2+} diffuses into sarcoplasm bathing the myofibrils, stimulating muscle contraction.
Myofibril
Cell membrane
Sarcoplasmic reticulum
5 After stimulation by the nerve cell ends, Ca^{2+} is taken up from the sarcoplasm by the sarcoplasmic reticulum, terminating muscle contraction.

FIGURE 33.6 T Tubules Spread Action Potentials into a Muscle Fiber When an action potential is initiated in the cell membrane of a muscle fiber at a neuromuscular junction, the action potential spreads along the cell membrane. It also spreads via the T tubules to the deep interior of the muscle fiber, triggering the release of Ca^{2+} from the sarcoplasmic reticulum.

Go to ACTIVITY 33.2
The Neuromuscular Junction
PoL2e.com/ac33.2

concentration of Ca^{2+} inside the sarcoplasmic reticulum and a low concentration in the sarcoplasm. Essentially, Ca^{2+} is stored inside the sarcoplasmic reticulum when a muscle cell is resting.

Spanning the space between the membranes of the T tubules and the membranes of the sarcoplasmic reticulum are two proteins that are physically connected. One of these, the dihydropyridine (DHP) receptor, is located in the T tubule membrane. It changes its conformation—its molecular shape—when an action potential reaches it. The other protein, the ryanodine receptor, is a Ca^{2+} channel located in the sarcoplasmic reticulum membrane.

The two receptor proteins act to release Ca^{2+} into the sarcoplasm when a muscle fiber is excited. When the DHP receptor is activated by an action potential traveling down a T tubule, both the DHP and ryanodine receptor proteins change conformation. As a consequence of this change, the ryanodine receptor (a Ca^{2+} channel) allows Ca^{2+} to leave the sarcoplasmic reticulum. Ca^{2+} then floods into the sarcoplasm surrounding the actin and myosin filaments (**FIGURE 33.7**, step 1).

The Ca^{2+} ions in the sarcoplasm then trigger actin and myosin to interact so that the muscle fiber contracts. How do the Ca^{2+} ions do this? Recall that strands of the protein tropomyosin are twisted around the chainlike actin polymers (see Figure 33.4), and molecules of the globular protein troponin occur at regular intervals. Each troponin molecule has three subunits: one binds actin, one binds tropomyosin, and one binds Ca^{2+}.

When a muscle fiber is at rest, the tropomyosin strands are positioned so they block the sites on actin filaments where myosin heads can bind. When Ca^{2+} is released into the sarcoplasm, it binds to troponin, changing troponin's conformation. Because troponin is bound to tropomyosin, this troponin change affects the conformation of tropomyosin, which twists enough to expose the actin–myosin binding sites (see Figure 33.7, step 2). Myosin heads can then form cross-bridges with actin (step 3). When a myosin head forms a cross-bridge, it is in a conformation that poises it to undergo its "power stroke,"

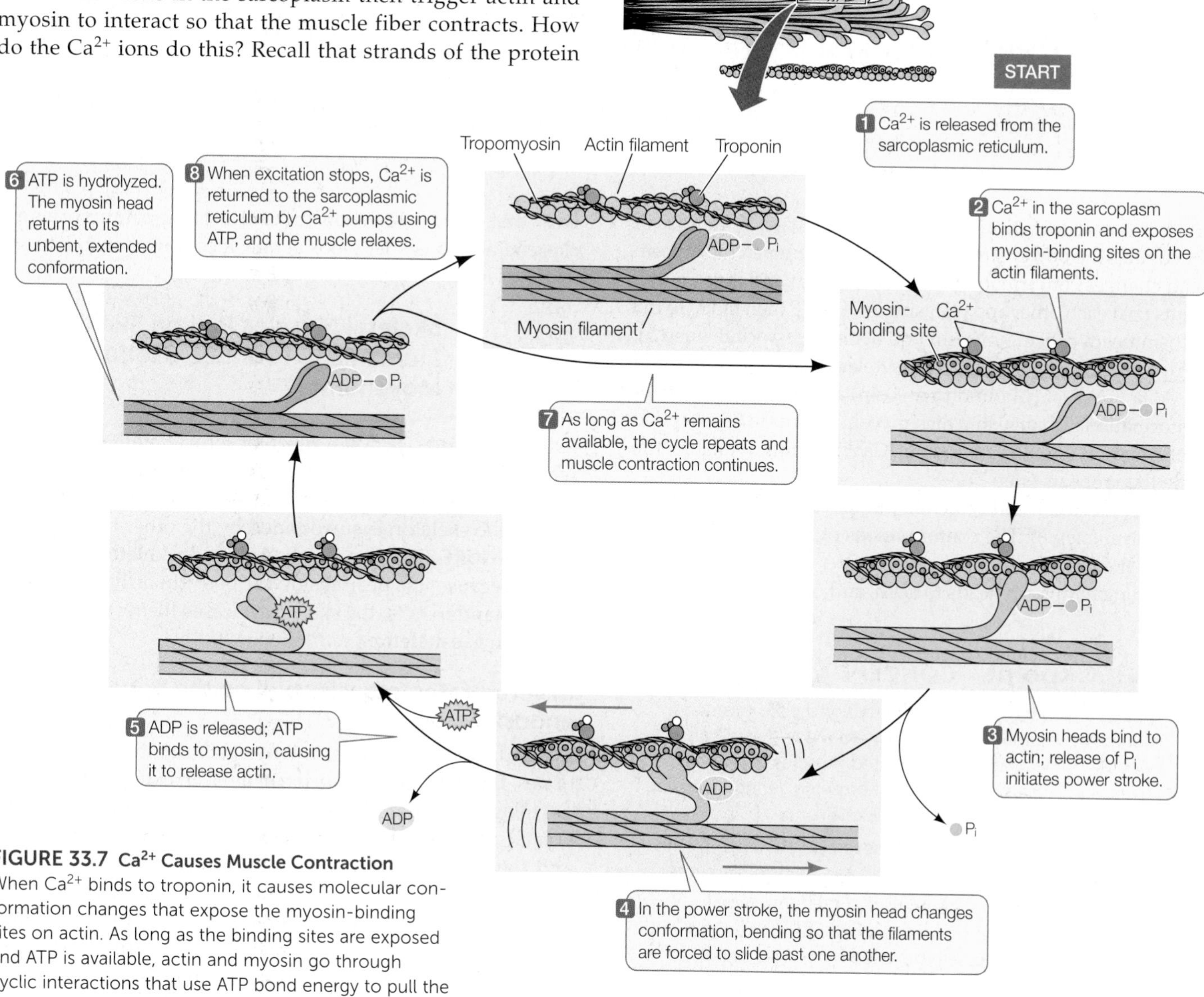

FIGURE 33.7 Ca^{2+} Causes Muscle Contraction When Ca^{2+} binds to troponin, it causes molecular conformation changes that expose the myosin-binding sites on actin. As long as the binding sites are exposed and ATP is available, actin and myosin go through cyclic interactions that use ATP bond energy to pull the actin and myosin filaments past each other, developing mechanical forces. P_i = inorganic phosphate.

APPLY THE CONCEPT

Muscle cells develop forces by means of cycles of protein–protein interaction

The force that a muscle fiber can generate depends on how many actin–myosin cross-bridges can form. This number in turn depends on the degree of overlap between the actin and myosin filaments in the sarcomeres. This overlap can vary considerably, as shown in Figure 33.3. When a muscle is stretched from its resting length, there is less overlap between the filaments. We can hypothesize that the force a muscle can produce during a given contraction will be a function of its length at the beginning of the contraction relative to its resting length. In an experiment, a bundle of muscle fibers can be attached at either end to an apparatus that can stretch the fibers and measure the force they generate when they are stimulated to contract. The results are shown in the table.

Plot these results with force generated on the *y* axis, and length of fibers at the beginning of contraction on the *x* axis. In answering the questions here, keep in mind that "length" refers to a percentage of resting length.

LENGTH OF FIBERS AT BEGINNING OF CONTRACTION AS PERCENTAGE OF RESTING LENGTH	FORCE GENERATED AS PERCENTAGE OF MAXIMUM FORCE
35%	10%
75%	50%
100%	100%
115%	100%
130%	80%
150%	50%
175%	10%

1. At what fiber length is a muscle's force-generating capacity greatest?
2. When a muscle is stretched to longer than its resting length, at what percentage of the resting length do you predict that the muscle fibers will not be able to generate any force?
3. Why does the force-generating capacity of a muscle decrease at exceptionally great muscle lengths? At exceptionally short lengths?
4. When you do a pull-up, which stages of the pull-up are most difficult, and why?

and it is bound to ADP and inorganic phosphate, P_i. Release of P_i activates the power stroke (step 4), during which the myosin head changes conformation, pulling the actin and myosin filaments past each other and releasing ADP. ATP then binds to the myosin head, causing myosin and actin to dissociate (step 5). Hydrolysis of the ATP then occurs, releasing energy (although the ADP and P_i stay bound to myosin) and leading to another conformational change in which myosin returns to its extended conformation (step 6). As long as Ca^{2+} remains available, the cycle then repeats (step 7).

When excitation ends, calcium pumps remove Ca^{2+} from the sarcoplasm (step 8). The conformations of troponin and tropomyosin then return to their original state, tropomyosin blocks the binding of myosin heads to actin, and contraction stops.

CHECKpoint CONCEPT 33.1

- ✓ Imagine planting your feet and trying to push through a concrete wall that's far too heavy to move. As you push, would you describe the associated muscles in your back, arms, and legs as contracting, shortening, lengthening, or a combination of these words? Explain.
- ✓ In a muscle fiber, how is force development aided by the overlapping arrangement of actin and myosin filaments?
- ✓ Describe how the concentration of Ca^{2+} in the sarcoplasmic reticulum of a muscle cell changes before, during, and after the cell is excited by a nerve impulse (action potential).

Now that we have seen how the contractile apparatus of a muscle generates forces, let's look at how these forces are put to use.

CONCEPT 33.2 Skeletal Muscles Pull on Skeletal Elements to Produce Useful Movements

Skeletal systems provide rigid structural elements against which the skeletal muscles pull to create directed movements. Two major types of skeletal systems exist in animals. In vertebrates, the skeleton is surrounded by the other tissues of the body (**FIGURE 33.8**) and is called an **endoskeleton** (*endo*, "inside") because it is deep inside the body. In arthropods, such as crabs and insects, the skeleton encases the rest of the body and is an **exoskeleton** (*exo*, "outside").

In vertebrates, muscles pull on the bones of the endoskeleton

The endoskeleton consists mainly of bone in most adult vertebrates. **Bone** is composed principally of an extracellular matrix of collagen protein fibers with insoluble calcium phosphate crystals deposited among them, giving the bone hardness and rigidity. Bone also contains several types of living cells throughout life. The activities of these cells make bone a dynamic tissue. It is not inert, as many people tend to assume. Bone is remodeled—that is, repaired or altered in shape—throughout life by the activities of its living cells. Because of this process, a bone can undergo modest changes in shape or structure to adjust to the forces placed on it. Bone also serves

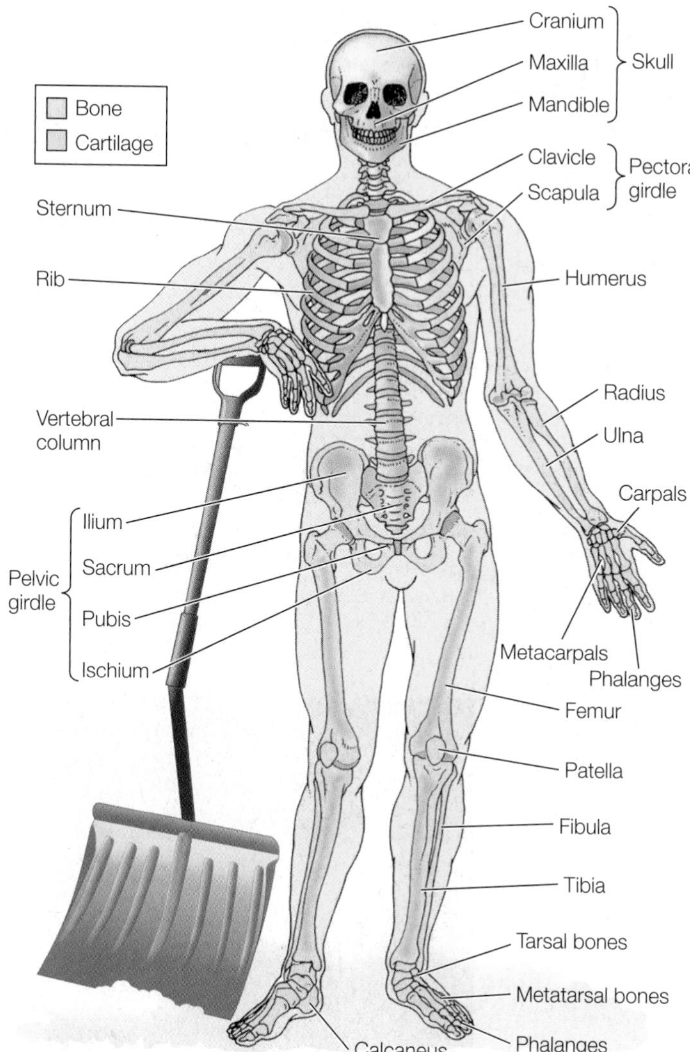

FIGURE 33.8 The Human Endoskeleton Bone and cartilage make up the internal skeleton of a human being. The human adult skeleton consists of more than 200 named bones, many of which are attached to other bones by moveable joints.

as a reservoir of calcium for the rest of the body, being in dynamic exchange with soluble calcium in the extracellular body fluids. This exchange is under the control of hormones, such as parathyroid hormone (see Figure 35.10), and vitamin D. The endoskeleton also includes cartilage in its structure. **Cartilage** is a flexible, rather than hard, form of skeletal tissue that plays several roles, such as imparting flexibility to skeletal elements (e.g., the rib cage).

To produce movements, muscles and bones work together around **joints**, the structures where two or more bones come together. Muscles attach to bones by bands of flexible connective tissue called **tendons**. Tendons often extend *across* joints. In such cases, muscles on one side of a joint exert forces on bones on the opposite side of the joint. Tunas, which are among the most powerful swimmers of all fish, provide dramatic examples. A tuna swims by beating its tail fin, but there are no muscles of any size in its tail. The massive swimming muscles of a tuna are in the middle of its body (**FIGURE 33.9**). Forces

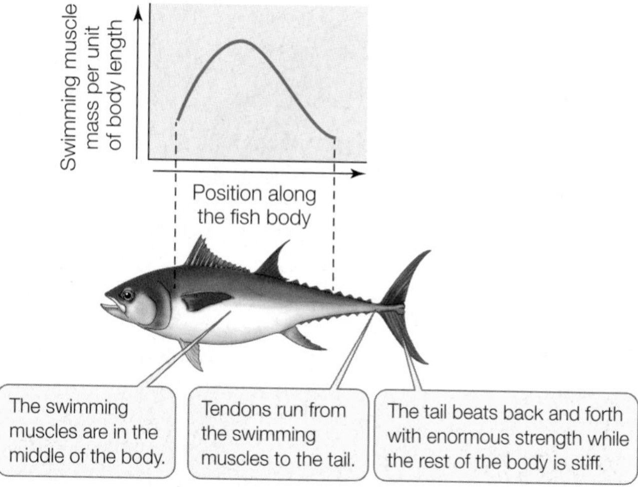

FIGURE 33.9 Tuna Swimming The blue line is a graph of swimming muscle mass. Where the line is highest, the muscle mass is greatest.

developed by these muscles are transmitted to the tail by tendons that run from the swimming muscles to the tail fin across the joint at the base of the tail.

A muscle can exert force in only one direction. Thus for the bones at a joint to move in opposite directions, two muscles are required. These muscles work as an **antagonistic pair** in which the two have opposite actions: when one contracts, the other relaxes. You can see this action in the motions of your forearm (**FIGURE 33.10A**). When your biceps contracts, your triceps relaxes, and your forearm moves up because the tendons of the biceps cross the elbow joint to attach to the forearm. When your triceps contracts, your biceps relaxes, and your forearm moves down.

Different sets of muscles, coordinated by the nervous system, typically must work together to control complex sets of movements. For example, consider the large muscle on the back of your calf, the gastrocnemius muscle. When it contracts, the movement created depends on the state of contraction of the muscle on the front of your thigh, the quadriceps femoris (**FIGURE 33.10B**). If your quadriceps is not contracting strongly, contraction of your gastrocnemius bends your knee joint and lifts your calf as in walking. If your quadriceps is contracting strongly—an action that keeps your knee joint from bending—contraction of your gastrocnemius extends your foot as in jumping.

In arthropods, muscles pull on interior extensions of the exoskeleton

Crabs, lobsters, insects, and other arthropods have an exoskeleton that covers every part of their body. It is composed of the polymerized carbohydrate chitin, and in crabs and lobsters it is extensively hardened by deposition of calcium-containing minerals. An exoskeleton functions like armor, protecting the soft tissues of the body inside. However, the exoskeleton of an arthropod cannot grow after it is formed. As an arthropod grows, it must therefore replace its exoskeleton periodically with a new, larger version.

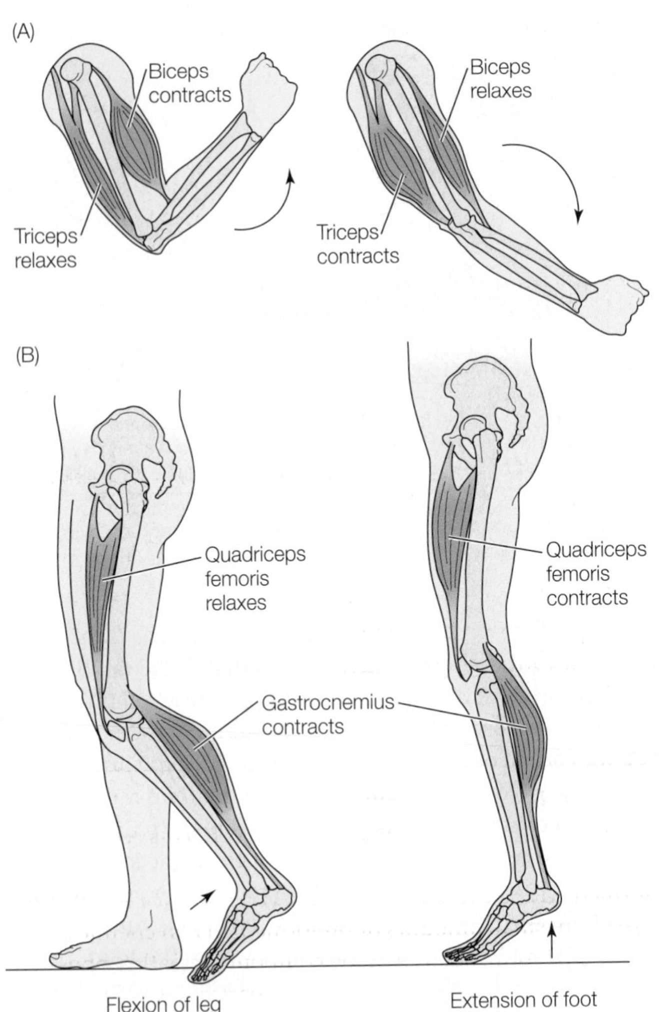

FIGURE 33.10 Movements Result from Interactions of Muscles with the Skeletal System (A) Antagonistic muscles are coordinated to allow the bones at a joint to move in opposite directions. (B) Different sets of muscles, coordinated by the nervous system, typically work together to control complex movements.

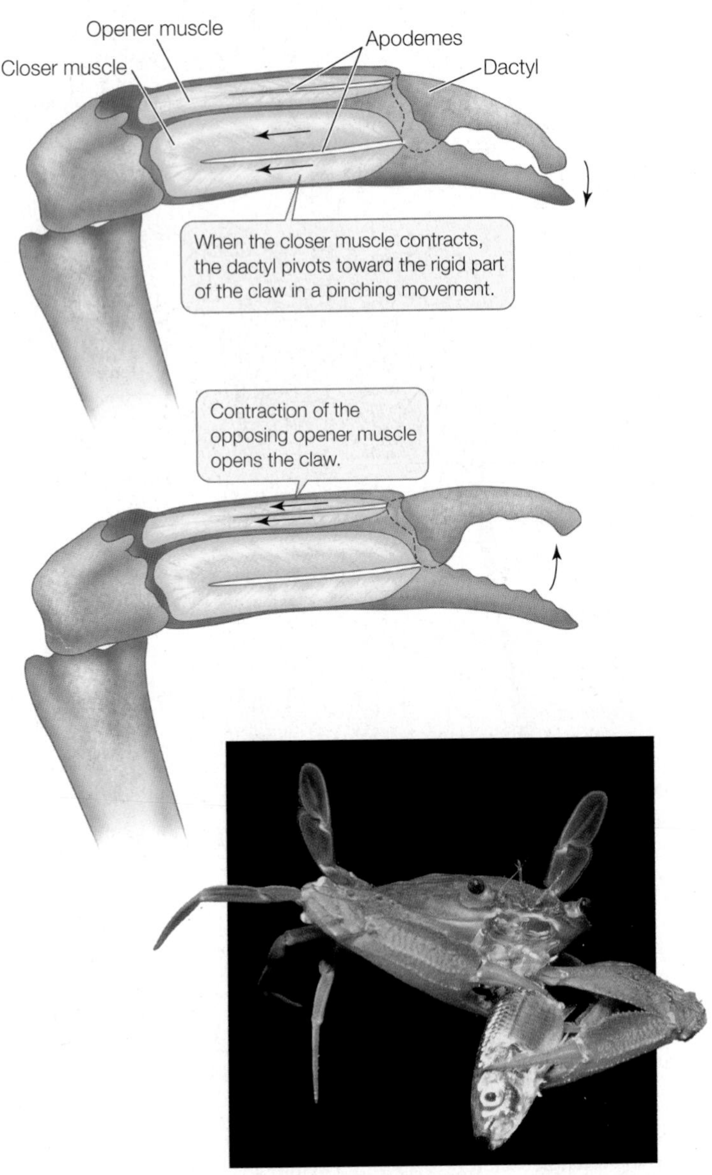

Opener muscle
Apodemes
Closer muscle
Dactyl
When the closer muscle contracts, the dactyl pivots toward the rigid part of the claw in a pinching movement.
Contraction of the opposing opener muscle opens the claw.

FIGURE 33.11 In Animals with an Exoskeleton, Muscles Pull on Interior Projections of the Exoskeleton This principle is illustrated here by the muscles that operate the claw of a crab. When the elements of the claw are pulled together, they exert high forces to gather food or pinch attackers. The arrows indicate the direction of movement of the apodemes and dactyl.

LINK

The process of shedding an exoskeleton—called molting—is controlled by hormones, as described in **Concept 35.5**

The skeletal muscles have a different relationship to the skeleton in arthropods than they do in animals with endoskeletons. Typically, where there is a joint between two parts of the exoskeleton, the exoskeleton projects inward, forming structures called apodemes to which the muscles inside the body attach. The large "pinching" claws at the end of a crab's front legs provide an excellent illustration. One side of each claw does not move, but the other side (the dactyl) connects to the leg by a joint and is moveable. At the joint, the moveable element of the claw has two apodemes to which two muscles attach (**FIGURE 33.11**). These muscles act as an antagonistic pair. When the larger of the two contracts, it pulls the claw closed with great force (leading to a pinching movement and a loud scream if your finger is caught!). The smaller of the two opens the claw with less force when it contracts.

LINK

The diverse structures and compositions of exoskeletons are detailed in **Concepts 23.3 and 23.4**

Hydrostatic skeletons have important relationships with muscle

An animal is said to have a **hydrostatic skeleton** if its body or a part of its body becomes stiff and skeleton-like because of a high fluid pressure inside. The most familiar hydrostatic skeleton is the human erected penis; blood pressure, produced by

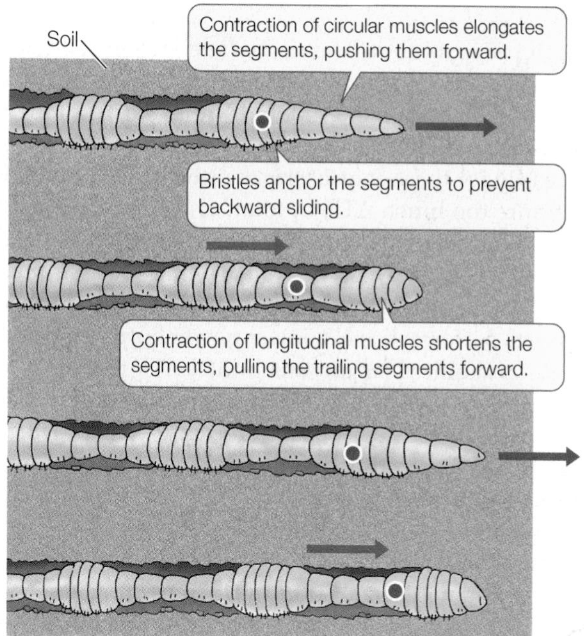

Soil
Contraction of circular muscles elongates the segments, pushing them forward.
Bristles anchor the segments to prevent backward sliding.
Contraction of longitudinal muscles shortens the segments, pulling the trailing segments forward.

FIGURE 33.12 A Hydrostatic Skeleton Alternating waves of muscle contraction move the earthworm through the soil. The red dot enables you to follow changes in the position of one segment as the worm moves forward. Red arrows show forward extension of the body when contraction of circular muscles pressurizes the worm and the interior pressure forces the anterior end to extend forward. Blue arrows show trailing parts of the body being pulled forward while the anterior is anchored.

the beating of the heart muscle, gives it its rigid, skeleton-like character (see Concept 37.2).

An earthworm digs through soil by using muscles in its body wall to create a hydrostatic skeleton (**FIGURE 33.12**). Each time the worm moves forward (red arrows in the figure), contractions of certain muscles raise the pressure inside the worm to a high enough level that the front end of the worm's body expands forward through the soil, much as a ribbon-shaped balloon expands when inflated. The worm then anchors its front end and pulls its back end toward its front end (blue arrows in the figure) before repeating the process. Clams do much the same thing to dig into sand. Through muscular contractions, they cyclically inflate their foot under pressure, so that the foot shoves deeper into the sand, and then pull the rest of their body down toward the foot.

CHECKpoint CONCEPT 33.2

- ✓ What are some biological structures, apart from muscle, that allow animals with endoskeletons to perform complex movements?
- ✓ Can endoskeletal components continue to change shape and structure after an animal is done growing?
- ✓ Are all muscles that function in locomotion necessarily attached to hard skeletal elements?

Now that we have looked at how skeletal muscles work together with different skeletal systems, let's turn to the factors that affect the performance of the skeletal muscles.

CONCEPT 33.3 Skeletal Muscle Performance Depends on ATP Supply, Cell Type, and Training

Skeletal muscles vary in performance from one muscle to another and from time to time. Our postural and finger muscles illustrate differences among muscles. Our postural muscles sustain loads steadily over long periods of time, whereas our finger muscles typically contract quickly for brief periods.

Muscle power output depends on a muscle's current rate of ATP supply

Muscles perform work, by which we mean that they apply forces to bones or other structures—forces that often make the structures move. For understanding the performance of work by a muscle cell, the cell's fuel supply for work is often as important as the properties of its contractile apparatus. Muscle cells depend on ATP as their source of energy to do work. For this reason, their *rate* of doing work—their power output—at a given time can be only as high as the rate at which they supply their contractile apparatus with ATP.

Muscle cells have three systems for supplying the ATP they need for contraction. Here we term these the immediate, glycolytic, and oxidative systems:

- The *immediate system* uses preformed ATP. This ATP is already present in a cell when contraction begins. Some preformed ATP is present literally as ATP, but muscle cells can also store small amounts of preformed ATP in chemically combined form, such as in the compound creatine phosphate found in vertebrate skeletal muscles.
- The *glycolytic system* synthesizes ATP during the process of contraction by anaerobically breaking down carbohydrates (see Concepts 6.3 and 29.4). In vertebrates, crustaceans, and some other animals, this system produces lactic acid. It does not require O_2.
- The *oxidative system* also synthesizes ATP during the process of contraction, but unlike the glycolytic system, it employs aerobic breakdown of food molecules to synthesize ATP (see Concept 6.2). It requires O_2 because it depends on the citric acid cycle and electron transport, fully oxidizing carbohydrates and lipids to H_2O and CO_2.

These three systems vary greatly in how rapidly they can produce ATP and in how long they can continue to make ATP at a high rate. Accordingly, the power output and endurance of a working muscle cell vary with which system the cell is using to make ATP (**FIGURE 33.13**).

When a muscle cell starts to do work, the immediate system can supply ATP extremely rapidly, enabling the cell to achieve a very high power output. The immediate system can do this because it uses ATP made at an earlier time and already present in a cell when work starts. Only a small amount of preformed ATP can be present in a cell, however. Thus the immediate system can function for only a brief period of time.

The glycolytic system can synthesize ATP at a very high rate, although not as high a rate as that at which the immediate

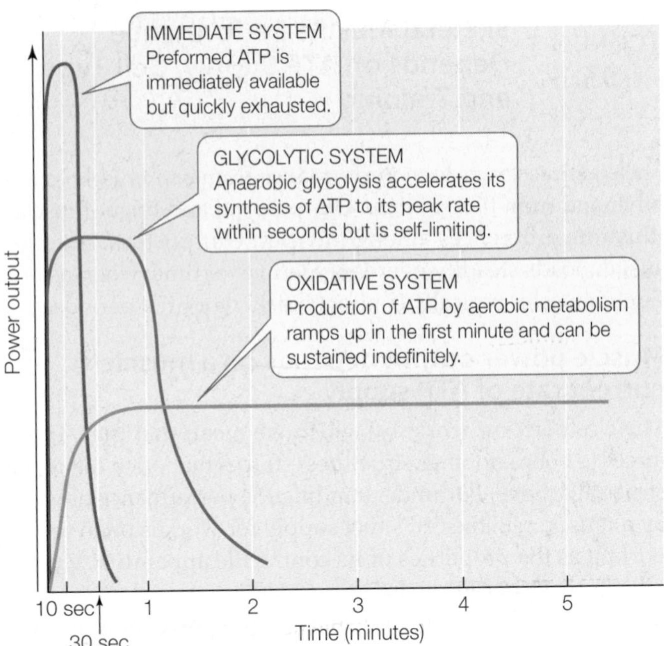

FIGURE 33.13 Supplying Fuel for High Performance Power output during all-out exertion is shown as a function of time when muscles are using the immediate, glycolytic, and oxidative systems for obtaining the ATP they need for contraction. This diagram refers to human performance.

system can supply ATP. However, the glycolytic system is self-limiting. For poorly understood reasons that may be related to the lactic acid it produces, the glycolytic system cannot keep making ATP indefinitely. As the glycolytic system makes ATP, it reaches a point at which it can no longer function, and then it must rest for a long time (tens of minutes or hours) before it can be used again to supply ATP.

The oxidative system makes ATP at a slower rate than the other two systems. However, its products—besides ATP—are simply H_2O and CO_2, which healthy animals can deal with effectively. Thus the oxidative system is not self-limiting and can continue making ATP for very long periods of time if a cell's supply of O_2 is sufficient for it to function.

A muscle cell that is capable of using all three systems of ATP supply varies in its performance, depending on which system it is using. To explore this important point, let's return to Figure 33.13. Keeping in mind that the power output of a contracting muscle cell depends directly on its rate of ATP supply, we see that *if we focus on power output*, the three systems of ATP supply rank as follows: the immediate system permits a muscle cell to reach highest power output, the glycolytic system ranks next, and the oxidative system permits the lowest power output. However, *if we focus on endurance*, the three systems rank in the opposite order: the immediate system permits the lowest endurance, the glycolytic system ranks next, and the oxidative system permits the greatest endurance.

These principles explain why, in human athletic competition, speed and endurance are inversely related. Sprints, such as the 100-yard dash, are run at a very high speed. They depend greatly on the immediate system for ATP, which permits very high power output. The immediate system cannot supply ATP for very long, however, meaning sprints must be of short duration. Events like the half-mile run and mile run are performed at lower speeds and last longer than sprints. A key reason for these contrasts is that the half-mile and mile runs require too much ATP, in total, to be run with just the immediate system as their source of ATP. Thus they depend to a great extent on the glycolytic system, which permits muscle cells to have greater endurance than the immediate system but restricts performance to a lower power output and speed. A marathon requires far too much total ATP to be run with just the immediate and glycolytic systems as ATP sources. Only the oxidative system—based on O_2 use—can provide the amount of ATP a marathon requires. The speed of running in a marathon must therefore be lower than in races fueled by the immediate and glycolytic systems, although long endurance is possible.

Long-endurance casual sports such as jogging and swimming laps depend almost entirely on the oxidative system. This is true because the oxidative system can accelerate its ATP production quickly at the start, and then can meet all the ATP needs of these forms of exercise on a minute-by-minute basis, indefinitely. This is why these sports are described as aerobic.

Muscle cell types affect power output and endurance

All muscle cells are not equal in their capability to use the glycolytic and oxidative systems to synthesize ATP. Some cells principally use the glycolytic system, and others principally use the oxidative system. There are no cells, however, that depend principally on the immediate system, which can supply only small amounts of ATP.

Multiple muscle cell types typically occur within a single species of animal. Two major types are often recognized: **slow oxidative cells** (sometimes called slow-twitch cells) and **fast glycolytic cells** (fast-twitch cells). In vertebrates, these cell types are also called "red" and "white." Cells of the two types typically occur mixed together in the skeletal muscles of mammals. In some other animals, however, certain skeletal muscles are composed almost entirely of one muscle cell type or the other. For example, many fish have distinct masses of muscle that are visibly red or white (**FIGURE 33.14**). The red muscles are composed mostly of slow oxidative cells. The white muscles consist mostly of fast glycolytic cells.

Slow oxidative and fast glycolytic muscle cells differ in three main ways:

- Their principal mechanisms for making ATP
- Their mechanisms for taking up O_2 from the blood
- Their molecular forms of myosin; the two types of cells express different genes for myosin, and their molecular forms of myosin differ in ATPase (ATP-splitting) activity

Slow oxidative cells are poised to make most of their ATP aerobically using the oxidative system of ATP synthesis. These cells need lots of O_2. This explains why, in vertebrates, these cells have evolved high levels of the red, hemoglobin-like compound **myoglobin** in their sarcoplasm (making them red).

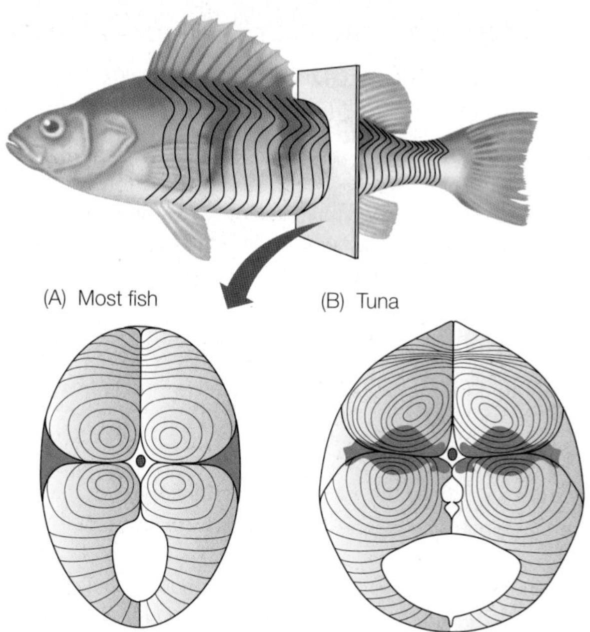

FIGURE 33.14 Red and White Swimming Muscles in Fish The fish shown is a perch. The cross sections, made approximately where indicated on the fish, show the red (dark) and white (light) swimming muscles in (A) perch and most other fish and (B) tuna.

Myoglobin speeds entry of O_2 into cells from the cell surfaces. Slow oxidative cells also have large numbers of mitochondria and are rich in mitochondrial enzymes, notably the enzymes of the citric acid cycle and electron transport system (see Concept 6.2). They are not well endowed with the enzymes required for the glycolytic system to make ATP anerobically, however. Slow oxidative cells thus make most of their ATP aerobically. Accordingly, they can sustain ATP production for long, continuous periods of time, but their peak rates of ATP production are only moderately high (see Figure 33.13). Slow oxidative muscle cells also have molecular forms of myosin with relatively low ATPase activity, meaning they recycle actin–myosin cross-bridges at relatively low rates.

Because of their limited rates of ATP production and myosin properties, slow oxidative cells tend to contract and develop tension slowly (accounting for the word "slow" in their name). They are well suited for maintaining relatively low power output for sustained periods.

Fast glycolytic cells contrast with slow oxidative cells in all the ways we've mentioned. They have high levels of the enzymes needed for glycolytic, anaerobic ATP synthesis and thus can produce ATP very rapidly. They also have molecular forms of myosin with relatively high ATPase activity, meaning they can recycle actin–myosin cross-bridges rapidly. Because of their high rates of ATP production and myosin properties, fast glycolytic cells tend to contract and develop tension rapidly (thus they are "fast"). They also fatigue rapidly, however. They have relatively few mitochondria, and they lack myoglobin, so they are not red.

Fish use principally their red muscles, composed mainly of slow oxidative cells, during routine cruising, when they move slowly for long periods. They use principally their white muscles, composed mainly of fast glycolytic cells, during burst swimming, when they swim at high speeds to evade danger or catch prey. Their white muscles enable them to achieve very high power outputs for brief periods.

Individual people vary greatly in the proportions of red and white cells in their major muscles. These differences generally do not matter in sports done for fun. At the level of national or international competition, however, researchers know from many studies that champion long-distance runners, swimmers, and cyclists—who require endurance—tend to differ from champion weight-lifters, wrestlers, and sprinters—who require high power outputs for brief periods. The endurance athletes tend to have leg and arm muscles consisting mostly of slow oxidative muscle cells (**FIGURE 33.15**). The other athletes tend to have muscles with a high percent of fast glycolytic cells.

Training modifies muscle performance

Skeletal muscle has a remarkable ability to change its properties with use. In this way, muscle provides dramatic examples of phenotypic plasticity (see Concept 29.5). When a muscle

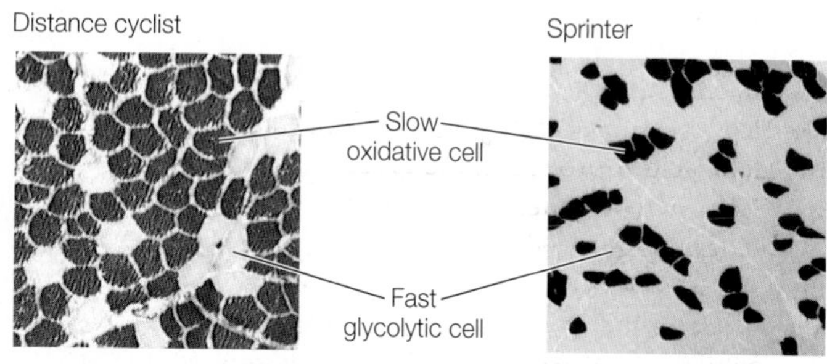

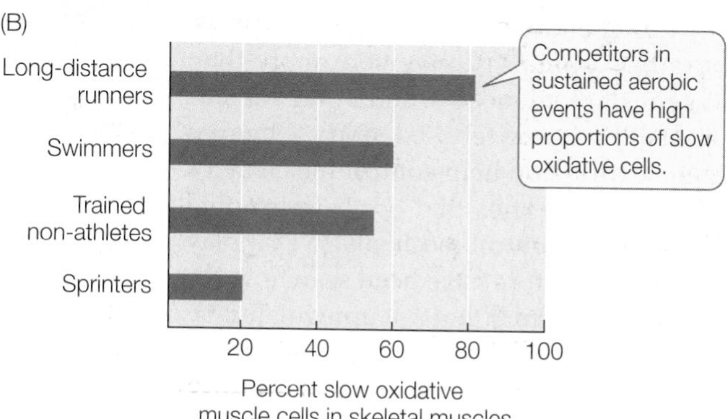

FIGURE 33.15 Competitive Athletes Tend to Have Muscle Fiber Types Corresponding to Their Sports (A) People vary in the cellular composition of their skeletal muscles. In these microscopic images, cells were stained with a reagent that shows slow oxidative cells as dark and fast glycolytic cells as light. (B) At the level of international or national competition, athletes successful in sustained aerobic events tend to have a high percent of slow oxidative cells, whereas successful sprinters tend to have a high percent of fast glycolytic cells. These differences are caused by heredity, early development, and training. Differences among individuals in muscle cell composition do not matter in sports done for fun.

changes in response to use, the change is often termed a "training effect." We observe a familiar training effect when we lift weights or do other similar forms of exercise. Our muscles gradually become visibly larger. The number of muscle cells in a muscle does not change under these conditions. Instead, a muscle becomes larger because its already-existing cells increase in size (termed hypertrophy) by increasing their content of actin and myosin.

To fully understand training effects, two types of exercise need to be distinguished. **Endurance exercise** (endurance training) refers to steady, long-duration exercise such as cross-country running or long-distance cycling. **Resistance exercise** (resistance training) refers to exercise that generates particularly large forces, typically in relatively few repetitions spread over a short period of time. Weight-lifting and wrestling are forms of resistance exercise.

When people or research animals engage in training programs, endurance exercise causes their muscle cells to increase their numbers of mitochondria. It also causes some cells to transform from being fast glycolytic to being slow oxidative cells. It stimulates growth of additional blood capillaries in muscles—an effect that reduces the average distance between each muscle cell and O_2-carrying blood in its nearest capillary. All of these training effects enhance capabilities for steady, sustained forms of exercise. Recent studies have shown that the expression of more than 100 genes is dramatically increased in muscle cells during endurance training in people or laboratory animals. Conversely, more than 100 genes are silenced when a person's or animal's life is entirely sedentary. Changes in gene expression help control the changes in muscle properties that occur (hormonal and metabolic control mechanisms also play roles). Investigators have been searching for transcription factors that are upregulated by endurance exercise and that may help regulate expression of genes that control transformation of cells from fast glycolytic to slow oxidative (**FIGURE 33.16**).

When people or research animals engage in training with resistance exercise, as we have already seen, the training causes their muscle cells to build a bigger contractile apparatus, enabling the cells to better develop the forces required for weight-lifting or other resistance activities. Resistance exercise also causes some muscle cells to transform from slow oxidative to fast glycolytic.

INVESTIGATION

FIGURE 33.16 A Regulator of Muscle Fiber Type In response to training with endurance exercise, skeletal muscles undergo changes that make the muscles more resistant to fatigue. How are these responses regulated? Some evidence indicated that a transcription factor called PPARδ ("PPAR delta"), which is present in skeletal muscle and is activated in response to endurance exercise, might stimulate the formation of slow oxidative cells, a key response to endurance training. Yong-Xu Wang and colleagues used transgenic mice to investigate whether artificially increasing the amount of active PPARδ would initiate exercise-associated responses in muscles even in the absence of endurance training.[a]

HYPOTHESIS

Activation of PPARδ is sufficient to drive formation of slow oxidative muscle fibers.

METHOD

1. Engineer a gene that encodes a permanently activated form of PPARδ.
2. Produce transgenic mice that synthesize the permanently activated form of PPARδ in skeletal muscle.
3. Test the transgenic mice for (a) muscle characteristics and (b) endurance. Error bars represent 1 standard error of the mean; the t-test was used to produce P-values (see Appendix B).

RESULTS

Muscle tissue from the transgenic mice when examined by light microscopy had more slow oxidative muscle fibers (cells stained dark blue) than did muscle tissue from wild-type mice.

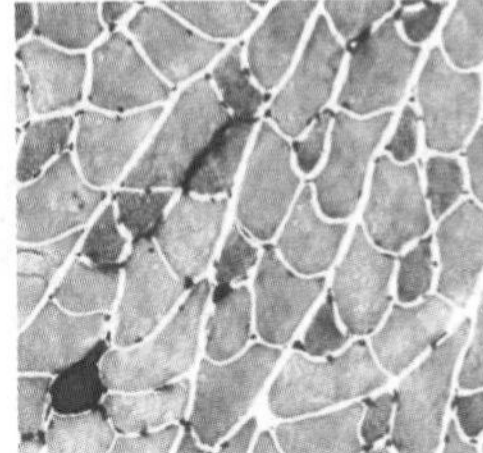

Wild-type mouse Transgenic mouse

When trained to run on a treadmill, transgenic mice were able to run for a longer time before becoming exhausted than wild-type mice were.

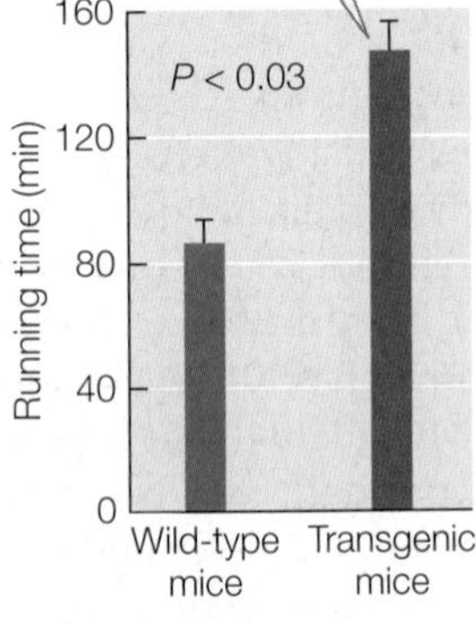

CONCLUSION

Activated PPARδ induces formation of slow oxidative muscle fibers and enhances endurance.

ANALYZE THE DATA

In a related experiment, using different sets of individual mice, the researchers tested the endurance of mutant mice that had no PPARδ transcription factor (see graph).

A. Based on these data, what can you conclude about the effect of PPARδ on fatigue resistance of muscle?

B. What would you expect to find if you examined cross sections of skeletal muscle tissue from the mutant mice?

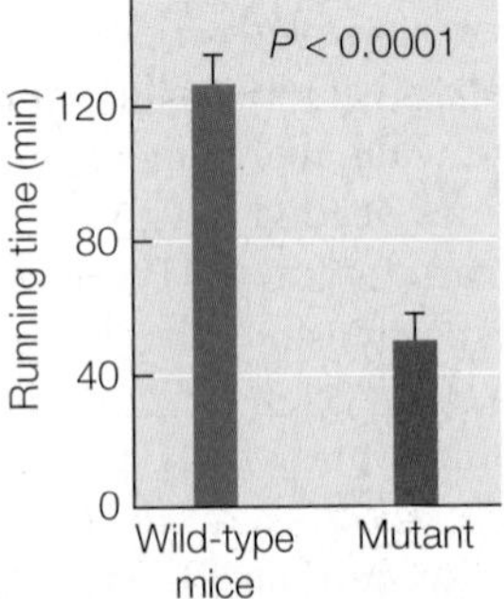

Go to **LaunchPad** for discussion and relevant links for all **INVESTIGATION** figures.

[a]Y.-X. Wang et al. 2004. *PLoS Biology* 2(10): e294.

CHECKpoint CONCEPT 33.3

- ✓ What factors prevent muscle cells from continuously using ATP provided by the immediate system to power contractions?
- ✓ Consider a muscle cell that contains myosin proteins specialized for rapid recycling of actin–myosin cross-bridges. Is the cell likely to make most of its ATP by using the glycolytic system or the oxidative system?
- ✓ Weekly measurements of an athlete's gastrocnemius (calf) muscle reveal that the density of capillaries in the muscle and the number of mitochondria per muscle cell are increasing. Would you predict that the athlete has been training for a lengthy marathon or lifting weights at the gym?

Skeletal muscles, as we have seen, are enormously important. We breathe using skeletal muscles, and the skeletal muscles are responsible for most behaviors, including athletic performance. Skeletal muscle is not the only type of muscle, however, and next we will look at the other types.

CONCEPT 33.4 Many Distinctive Types of Muscle Have Evolved

Besides skeletal muscle, several other distinctive types of muscle have evolved. Vertebrate **cardiac muscle**—heart muscle—is one type. During each of our lives, our cardiac muscle beats regularly from our early embryonic development until we die.

Vertebrate cardiac muscle is both similar to and different from skeletal muscle

Like skeletal muscle, cardiac muscle appears striated when viewed under a light microscope. This is because the actin and myosin filaments in cardiac muscle are arranged in regular patterns in sarcomeres. In other ways, however, cardiac muscle differs morphologically from skeletal muscle. Cardiac muscle cells are much smaller than skeletal muscle cells, and unlike skeletal muscle cells, they have only one nucleus.

A key way that cardiac muscle differs from skeletal muscle is in the relationship of adjacent cells. In cardiac muscle, adjacent cells are electrically coupled by gap junctions that occur in structures called intercalated discs. At a gap junction, the sarcoplasms (cytoplasms) of two adjacent cardiac muscle cells are continuous with each other via tiny pores:

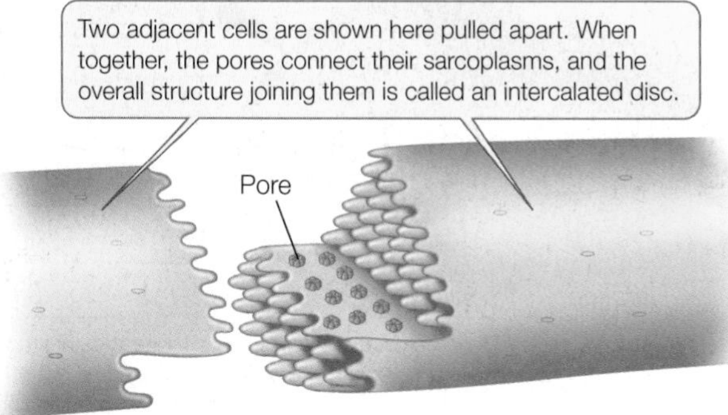

Two adjacent cells are shown here pulled apart. When together, the pores connect their sarcoplasms, and the overall structure joining them is called an intercalated disc.
Pore

Because of this continuity, if one of the two cells is excited and an impulse spreads over its cell membrane, the impulse spreads directly to the second cell (see Figure 4.18 and Concept 4.5). Accordingly, when an impulse is initiated at any one point in a mass of heart muscle, it spreads rapidly throughout the muscle by moving directly from one cell to another. This process ensures that all the cells within a large mass of muscle are excited at approximately the same time. Thus all the cells contract almost simultaneously, raising the blood pressure in the heart to a high level and propelling the blood out of the heart and through the blood vessels.

In a vertebrate heart, certain of the cells in the heart muscle itself generate the heartbeat rhythm (see Concept 32.3 and Figure 32.8). The heart rhythm is thus very different from the breathing rhythm. The breathing muscles of vertebrates are skeletal muscles, and like all skeletal muscles, they contract only if stimulated by nerves. If a person suffers a neck fracture, he or she will stop breathing if the nerves that stimulate the breathing muscles are damaged. His or her heart will keep beating, however, because of the properties of certain of the cardiac muscle cells.

Go to MEDIA CLIP 33.1
Be Still My Beating Stem Cell Heart
PoL2e.com/mc33.1

Vertebrate smooth muscle powers slow contractions of many internal organs

In addition to skeletal and cardiac muscle, vertebrate animals have a third type of muscle known as **smooth muscle**. Structurally, smooth muscle cells are smaller than skeletal muscle cells. They are usually spindle-shaped, and each has a single nucleus. They are termed "smooth" because their actin and myosin filaments are not as regularly arranged as those in skeletal and cardiac muscle, and for this reason they do not have a striated appearance when viewed under a light microscope:

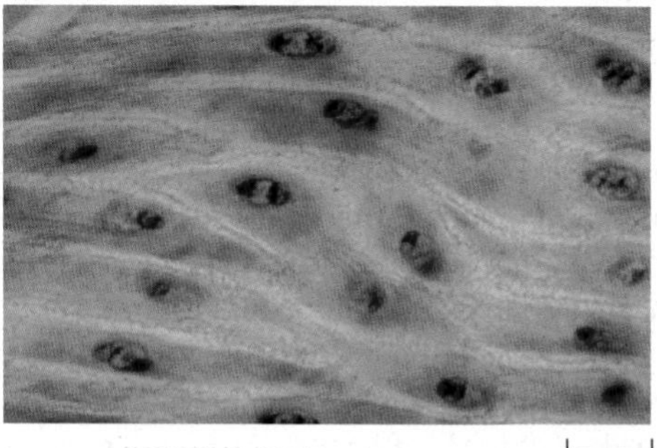

30 μm

Smooth muscles provide the contractile forces for most of our internal organs, under the control of the nervous system. Smooth muscles in the walls of the digestive tract move food through the digestive tract. Smooth muscles in the walls of small arteries and arterioles adjust the diameters of these blood vessels, helping control blood flow through them. During sexual climax, smooth muscles in the ejaculatory passages of the male propel the ejaculation of semen, and in the female smooth muscles contribute to the rhythmic contractions of orgasm. Smooth muscles in the walls of the bladder produce forces for bladder emptying.

Some smooth muscle tissue, such as that in the walls of the digestive tract, consists of sheets of cells in which adjacent cells are in electrical contact through gap junctions. As a result, excitation of one cell can spread directly to adjacent cells, helping coordinate the contractions of the cells. This occurs during **peristalsis**, a pattern of contraction in which a wave of contraction moves along the gut tube from one end to the other.

LINK

The structure and function of the vertebrate gut are detailed in **Concept 30.4**

In both cardiac muscle and smooth muscle, excitation–contraction coupling depends on Ca^{2+} influx into the sarcoplasm, as it does in skeletal muscle. Details of excitation–contraction coupling differ among the three types of vertebrate muscle, however.

Go to ANIMATED TUTORIAL 33.2
Smooth Muscle Action
PoL2e.com/at33.2

Some insect flight muscle has evolved unique excitation–contraction coupling

Invertebrates have evolved a variety of specialized types of muscle. Here we discuss just three of them. Two of these, asynchronous flight muscle and catch muscle, are specialized to contract in unusual ways that are highly adaptive for the animals in which they occur. Another specialized type does not contract at all but instead has become modified to produce electricity!

The muscles that drive the wings during insect flight are of two types in terms of their excitation–contraction coupling. In about 25 percent of insect species, each contraction requires excitation. The sphinx moth in our opening photograph is one of these. Each excitation of a muscle cell brings about one contraction, meaning the ratio of excitations to contractions is 1:1. For muscle of this type to contract at a higher and higher frequency, all the processes of excitation–contraction coupling must take place faster and faster. At high frequencies, contraction becomes severely compromised in several ways as a result. Both muscular efficiency and power output diminish at high frequencies.

A different and unique form of muscle that has evolved only in insects solves these problems. It is termed **asynchronous muscle** because each excitation results in many contractions, meaning contractions are not synchronized in a 1:1 ratio with excitations. At high frequencies of contraction, asynchronous muscle maintains higher efficiency and greater power output than synchronous muscle. The reason asynchronous muscle has evolved in insects seems to be that it is well adapted in these ways to meeting the particular challenges of insect flight. Asynchronous muscle has evolved independently in insects about ten times and is found today in about 75 percent of insect species.

Catch muscle in clams and scallops stays contracted with little ATP use

A scallop or clam has a shell composed of two half-shells. One or two powerful adductor muscles connect these half-shells. Contraction of the adductor muscles pulls the two half-shells tightly together so that the animal becomes fully enclosed in its hard shell—a form of armor:

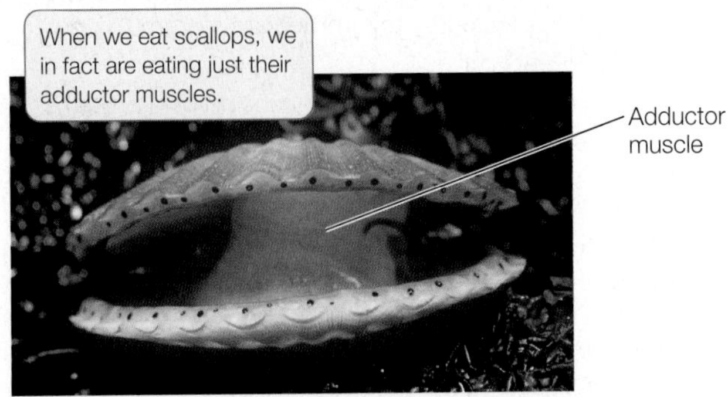

When we eat scallops, we in fact are eating just their adductor muscles.
Adductor muscle

A scallop or clam's adductor muscles sometimes need to remain contracted for long periods of time when predators are nearby.

The adductor muscles (and certain other muscles of mollusks) are able to enter a specialized state termed **catch** in which they maintain high contractile force for tens of minutes or even for hours with almost no use of ATP. The way catch works is not well understood. However, measurements show that catch muscles maintain contraction at a far lower ATP cost than most muscles. Catch is a highly controlled state that can be quickly ended by signals from the nervous system. Its energy-saving feature is adaptive for animals that have body armor that depends on muscular contraction to be effective.

Fish electric organs are composed of modified muscle

Electric eels (*Electrophorus*), which live in tropical rivers, can produce external voltage pulses as large as 700 volts. Electric rays (*Torpedo*) in the ocean also produce large external electric pulses, although they produce high electric currents (up to 20 amperes) rather than high voltages. Fish such as these use their electric pulses to stun prey. Many other fish produce small external electric pulses that they use to detect their surroundings for orientation.

The electric organs of nearly all electric fish have evolved from skeletal muscle and consist of modified muscle cells. These cells cannot contract because the contractile apparatus (e.g., actin and myosin) in each cell is poorly developed. However, the cell membranes of the muscle cells are unique in that the tiny voltage differences across the cell membranes of many cells all add together when the cells are excited during an electric pulse:

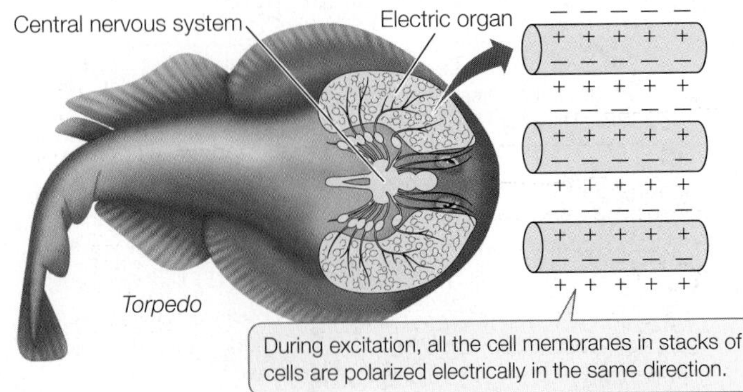

Central nervous system
Electric organ
Torpedo
During excitation, all the cell membranes in stacks of cells are polarized electrically in the same direction.

The voltage difference across each cell membrane is only about 0.1 V. In an electric organ, the voltage differences of many membranes add up, in the same way that the voltages of batteries add up when the batteries are connected in series. For this reason, an electric organ with 1,000 cell membranes can produce 100 V.

CHECKpoint CONCEPT 33.4

- ✓ Why does the heart keep beating in victims of neck fractures, although the victims stop breathing?
- ✓ What type of muscle would be used in each of the following activities: the extension of a lizard's forelimbs during walking; the pumping of O_2-poor blood from a person's heart to his or her lungs; the expulsion of food during vomiting; the holding shut of a clam's shell?
- ✓ If needed, could the modified muscle cells of fish electric organs be used to generate forces for swimming? Why or why not?

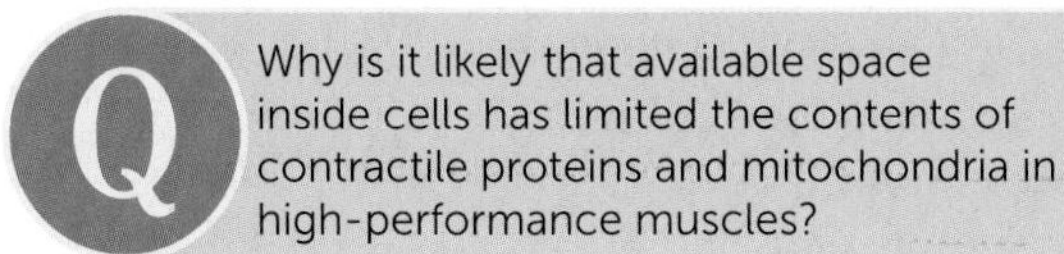

Why is it likely that available space inside cells has limited the contents of contractile proteins and mitochondria in high-performance muscles?

ANSWER **FIGURE 33.17** is a highly magnified image of an insect flight muscle cell, obtained by electron microscopy. The inside of the cell is filled almost completely by mitochondria and contractile proteins. Open space is thus a scarce resource. Put simply, a high-performance muscle cell needs as large a set of contractile proteins as possible and as many mitochondria as possible—meaning there is a sort of "competition" for space in which the amounts of contractile proteins and mitochondria are each limited by space shortage. If, over evolutionary time, natural selection started to favor larger numbers of contractile protein molecules, the contractile proteins could edge out some of the mitochondria—jeopardizing the ability of the contractile proteins to get enough ATP. If natural selection started to favor more mitochondria, the mitochondria would edge out contractile proteins—jeopardizing the ability of the cell to use the ATP it could produce. The fact that space is limited has resulted in a sort of compromise in the use of space inside a high-performance muscle cell.

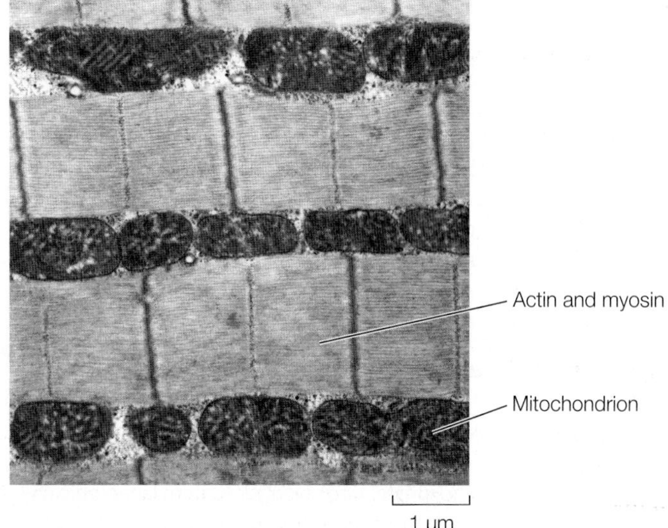

FIGURE 33.17 An Electron Micrograph Showing the Fine Structure of an Insect Flight Muscle Cell

SUMMARY

CONCEPT 33.1 Muscle Cells Develop Forces by Means of Cycles of Protein–Protein Interaction

- **Skeletal muscle** consists of bundles of **muscle fibers**. Each skeletal muscle fiber is a large, elongated cell containing multiple nuclei. **Review Figure 33.2 and ACTIVITY 33.1**
- A skeletal muscle fiber contains numerous **myofibrils**, which contain bundles of **actin** and **myosin** filaments. The regular, overlapping arrangement of the actin and myosin filaments into **sarcomeres** gives skeletal muscle its striated appearance. **Review Figure 33.2**
- **Contraction** is the development of force by a muscle. The molecular mechanism of contraction is described by the **sliding-filament theory** and involves the binding of the globular heads of myosin molecules to actin molecules to form **cross-bridges**. Upon binding to actin, a myosin head changes its conformation, causing the two filaments to move past each other. Release of the myosin heads from actin and their return to their original conformation requires ATP. **Review Figures 33.3 and 33.4 and ANIMATED TUTORIAL 33.1**
- Nerve cells make contact with skeletal muscle fibers at **neuromuscular junctions**. In general, each muscle fiber in skeletal muscle is innervated by a single nerve cell. **Review Figure 33.5**
- **Excitation** by a nerve impulse (action potential) stimulates a muscle fiber to contract by **excitation–contraction coupling**. An action potential spreads across the muscle fiber's cell membrane and through the **transverse (T) tubules**, causing Ca^{2+} to be released from the **sarcoplasmic reticulum**. **Review Figure 33.6 and ACTIVITY 33.2**
- Ca^{2+} binds to **troponin** and changes its conformation, pulling the **tropomyosin** strands away from the myosin-binding sites on the actin filaments. The muscle fiber continues to contract until the Ca^{2+} is returned to the sarcoplasmic reticulum. **Review Figure 33.7**

CONCEPT 33.2 Skeletal Muscles Pull on Skeletal Elements to Produce Useful Movements

- Skeletal systems provide structures against which skeletal muscles can pull to produce useful movements. **Endoskeletons** are internal systems of rigid supports, consisting of **bone** and **cartilage**, to which muscles are attached. **Review Figure 33.8**
- **Tendons** connect muscles to bones. Muscles and bones work together around **joints** to produce movement. **Review Figures 33.9 and 33.10**

(*continued*)

SUMMARY (continued)

- **Exoskeletons** are skeletons that enclose an animal, notably the hardened outer surfaces of arthropods. In arthropods muscles attach to apodemes, internal projections at the joints of the exoskeleton. **Review Figure 33.11**
- **A hydrostatic skeleton** is said to exist if an animal's body or a part of its body becomes stiff and skeleton-like because of a high fluid pressure inside. **Review Figure 33.12**

CONCEPT 33.3 Skeletal Muscle Performance Depends on ATP Supply, Cell Type, and Training

- Muscle contraction depends on a supply of ATP. Muscle cells have three systems for supplying ATP: the immediate system (preexisting ATP and creatine phosphate); the glycolytic system (anaerobic glycolysis); and the oxidative system (aerobic metabolism). The immediate system can fuel high muscle power output instantaneously but is exhausted within seconds. Glycolysis can regenerate ATP rapidly but is self-limiting. Oxidative metabolism delivers ATP more slowly but can continue to do so for a long time. **Review Figure 33.13**
- **Slow oxidative muscle cells** have cellular properties (e.g., abundant mitochondria) that facilitate extended, aerobic work; **fast glycolytic muscle cells** generate great forces for short periods of time. In vertebrates, slow oxidative cells contain the hemoglobin-like compound **myoglobin**, making these cells red. **Review Figure 33.14**
- Skeletal muscles contain varying proportions of slow oxidative cells and fast glycolytic cells, depending on genetic controls during muscle development and the demands placed on the muscles. The properties of muscles can be modified by training. **Review Figures 33.15 and 33.16**

CONCEPT 33.4 Many Distinctive Types of Muscle Have Evolved

- **Cardiac muscle** cells are striated, uninucleate, and electrically connected by gap junctions, so that action potentials spread rapidly throughout masses of cardiac muscle and cause coordinated contractions. In vertebrates some modified cardiac muscle cells serve as the pacemaker cells for rhythmic beating of the heart.
- **Smooth muscle** provides contractile force for internal organs such as the gut, blood vessels, and reproductive ducts. Some smooth muscle tissue (e.g., in the digestive tract) consists of sheets of cells that are electrically coupled through gap junctions, helping coordinate the contractions of adjacent cells. **Review ANIMATED TUTORIAL 33.2**
- Some insects drive flapping of their wings with **asynchronous muscles**, which unlike most muscles undergo multiple contractions with each excitation.
- **Catch** muscles, such as the adductor muscles of clams and scallops, can sustain strong contractions for long periods with little ATP.
- The electric organs of nearly all electric fish evolved from skeletal muscle and consist of modified, noncontractile muscle cells.

Go to the Interactive Summary to review key figures, Animated Tutorials, and Activities
PoL2e.com/is33

Go to LaunchPad at **macmillanhighered.com/launchpad** for additional resources, including LearningCurve Quizzes, Flashcards, and many other study and review resources.

34 Neurons, Sense Organs, and Nervous Systems

The star-nosed mole has tiny eyes but large, specialized nose tentacles—correlated with the fact that it hunts in darkness for tiny prey that it locates by touch.

You might imagine that this strange-looking mammal exists only in some remote forest in one of Earth's most inaccessible places. Actually its home is in the eastern United States and Canada. If you live in that part of the world and like to walk in the wild outdoors, you have probably walked right by one. This mammal lives mostly underground, however, explaining why most people have never seen one. It's called the star-nosed mole (*Condylura cristata*) because its nose bears two symmetrical arrays of tentacles that surround the nostrils in a starlike pattern.

Star-nosed moles live in burrows in wet, marshy soils where tiny invertebrates are abundant. The moles are specialized to eat these minute animals. They eat lots of tiny worms and insects. Almost as amazing as their tentacled nose is the fact that the moles forage underwater to a great extent, as well as in their burrows. Most of what biologists know of them comes from studies of burrow foraging.

Why did their strange star-shaped nose evolve? Biologists have deduced that if star-nosed moles eat tiny animals, they must find and eat great numbers of them each day. The current hypothesis regarding the evolution of this strange nose is that it is an adaptation for rapid foraging on tiny prey in darkness.

The star-shaped nose is clearly a sensory organ. All you need to do to be convinced of that is to watch a mole for a few minutes. The moles constantly explore their environment with their tentacles. People at first hypothesized that the moles detect tastes or odors with their tentacles. Researchers have since learned that the sensory receptors in the tentacles are touch receptors.

The nose is no ordinary touch organ, however. The array of tentacles is about 1 centimeter in diameter—not very large. Yet there are about 25,000 tiny touch receptors (called mechanoreceptors) in the skin of the tentacles. And about 100,000 nerve processes (axons) travel from the tentacles to the mole's brain, carrying signals from the receptors. The star-nosed mole is believed to have the highest resolution touch perception found in a mammal. The brain uses this information to perceive potential food items.

The moles are active explorers. As they move from place to place in darkness, they investigate each potential food item in an intense flurry of moving tentacles, providing touch sensations to their brain. With this flow of information, they typically decide in less than a second whether to eat, allowing them to move on promptly in their ceaseless quest to find enough tiny animals to meet their daily food need.

How might the star-nosed mole's brain be specialized to process information from its nose?

You will find the answer to this question on page 730.

CONCEPT 34.1 Nervous Systems Are Composed of Neurons and Glial Cells

Animals need a way to transmit signals at high speeds from place to place within their bodies. Suppose, for example, that you are crowded with other people in the kitchen during a party and, being distracted, you put your hand behind you and lean on a hot stove. You respond immediately—pulling your hand away—because of nerve cells, also called neurons. Mammalian neurons routinely transmit signals at 20–100 meters per second! Thus a danger signal travels the length of your arm to reach your spinal column in less than 0.02 seconds, the first step in allowing you to respond. The speed of signal transmission by neurons is also evident in more ordinary actions. During banjo plucking, each pluck of a string takes place because of neuronal signals that originate by way of interactions among many neurons in the brain and travel to the player's fingers (**FIGURE 34.1**). These signals must move at enormous speeds for notes to burst from the instrument as fast as they do.

GENERATION
RETROSPEKTIVE

FIGURE 34.1 Banjo Plucking Dramatizes the Speed of Neuronal Signaling For each finger pluck, nerve impulses are generated in the brain and travel down the spinal cord and out to the finger that plucks the string. The entire process occurs several times per second.

A neuron is a type of **excitable** cell, meaning its cell membrane can generate and conduct signals called impulses, or action potentials. The excitable nature of neurons is one of their key specializations. Most types of cells in an animal's body are not excitable. The two principal types that are excitable are neurons and muscle cells (see Concept 33.1). What do we mean by an impulse, or action potential? Cell membranes ordinarily have an electrical polarity in which the outside of the membrane is more positive than the inside. An impulse, or action potential, is a state of reversed polarity. In a region of the cell membrane where an impulse or action potential is present, the cell membrane is said to be depolarized because its electrical polarity is outside-negative instead of outside-positive. In an excitable cell, by definition, after an action potential is generated at one point in the cell membrane, it propagates over the whole membrane. During this process, the region of depolarization moves along the cell membrane, and the membrane is said to "conduct" the impulse:

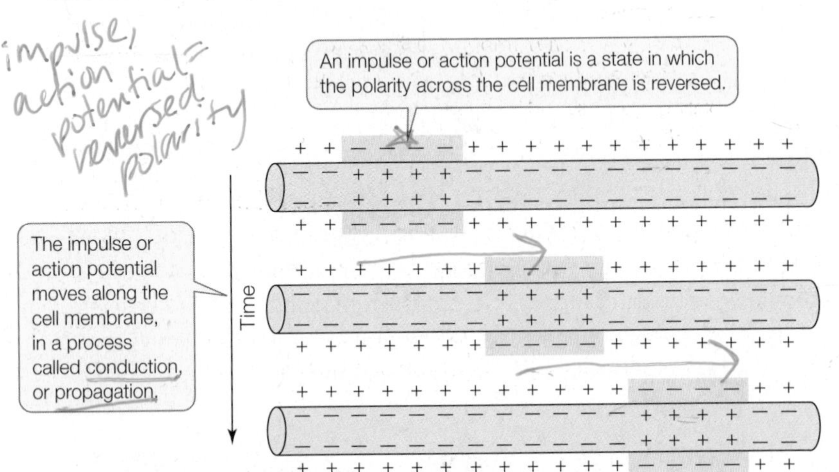
An impulse or action potential is a state in which the polarity across the cell membrane is reversed.
The impulse or action potential moves along the cell membrane, in a process called conduction, or propagation.
Time

This conduction, or propagation, is what happens when a neuron carries a signal from the brain to the fingers, telling the fingers to pull away from danger or to pluck a banjo string. We will discuss action potentials in more detail in Concept 34.2.

Nervous systems are composed of two types of cells, neurons and glial cells. The neurons are excitable, as we have said, but the glial cells typically are not.

Neurons are specialized to produce electric signals

A **neuron**, or nerve cell, is a cell that is specially adapted to generate electric signals, typically in the form of action potentials. Neurons are very diverse in structure. Here we describe just their most common features. Neurons often are highly elongated. Because of this, they can act like telephone wires ("land lines"), carrying signals (action potentials) rapidly over long distances.

Another way in which neurons resemble telephone wires is that their signals travel to specific destinations—that is, to anatomically defined destinations. A neuron typically must make contact with a target cell for signals in the neuron to affect the target cell directly. Thus, in most cases, cells must be contacted by a neuron to receive the neuron's signals. The signals, in essence, are addressed: they are received only at defined locations. In short, signal transmission by neurons is *fast* and *addressed*.

Places where neurons make functionally relevant contact with other cells are called synapses. More specifically, a **synapse** is a cell-to-cell contact point that is specialized for signal transmission from one cell to another. One of the two cells at a synapse is a neuron. The other can be another neuron, or it can be some other type of cell such as a sensory cell or muscle cell. Signal transmission at a synapse is usually one-way. A signal arrives at the synapse by way of one cell, called the **presynaptic cell**, and leaves by way of another, the **postsynaptic cell**.

Most neurons have four anatomical regions:

- a set of dendrites
- a cell body
- an axon
- a set of presynaptic axon terminals

We can also think of a neuron as having four functional regions, which generally correspond to the anatomical regions (**FIGURE 34.2**).

The **dendrites** (from the Greek *dendron*, "tree") are relatively short cell processes (extensions) that tend to branch from the cell body like twigs on a shrub. They are typically the principal sites where incoming signals arrive from other cells. Hundreds or thousands of other cells—neurons or sensory cells—may bring signals to a single neuron. These cells typically make synaptic contact with the dendrites or cell body.

The **cell body** contains the nucleus and most of a neuron's other organelles. The most important function of the cell body is to combine and integrate incoming signals. Earlier we stressed that neurons achieve fast and addressed signal transmission. Of equal importance, neurons integrate signals. This means that a neuron can receive many inputs and combine them to produce a single output, and it can do this in ways that are adaptive and help promote harmonious function of the animal.

In most neurons, one cell process—the **axon**—is much longer than the others. The axon may extend a long distance. For instance, single axons extend from your spinal cord to your fingers. The axon is the part of a neuron that is anatomically specialized for long-distance signal conduction. Usually the signals (action potentials) conducted by the axon originate in or near the cell body and travel outward from the cell body.

Where the axon ends, it branches, and each of these branches terminates with a small swelling called a **presynaptic axon terminal**. These terminals make synaptic contact with other cells, enabling signals generated by the neuron to initiate signals in other neurons or in muscle cells. A neuron is said to **innervate** the cells with which its presynaptic axon terminals make synaptic contact.

A neuron, as we have stressed, is a single cell. Often the axons of many neurons travel together in a bundle from place to place, like a telephone cable composed of many wires. These bundles are called **nerves**. Individual neurons are far too small to be seen with the naked eye, but nerves consisting of many axons are often easily visible. Biologists use the term "nerve" only for axon bundles outside the brain and spinal cord. A bundle of axons in the brain or spinal cord is called a **tract** (or sometimes a commissure or connective).

Glial cells support, nourish, and insulate neurons

Glial cells, also called glia or neuroglia, are the second major type of cell in the nervous system. Unlike neurons, they typically are not excitable and do not conduct action potentials. Nonetheless, they are critically important. They are particularly numerous in mammals. Half the volume of the mammalian brain is glial cells, which are smaller than neurons and ten times more numerous than neurons.

Several distinct types of glial cells are present in the nervous system and play distinct roles. Certain glial cells help orient developing neurons toward their target cells during embryonic development. In adult vertebrates, glial cells in the brain and spinal cord provide metabolic support for neurons, help regulate the composition of the extracellular fluids bathing the neurons, and perform immune functions. Sometimes glial cells also assist signal transmission across synapses.

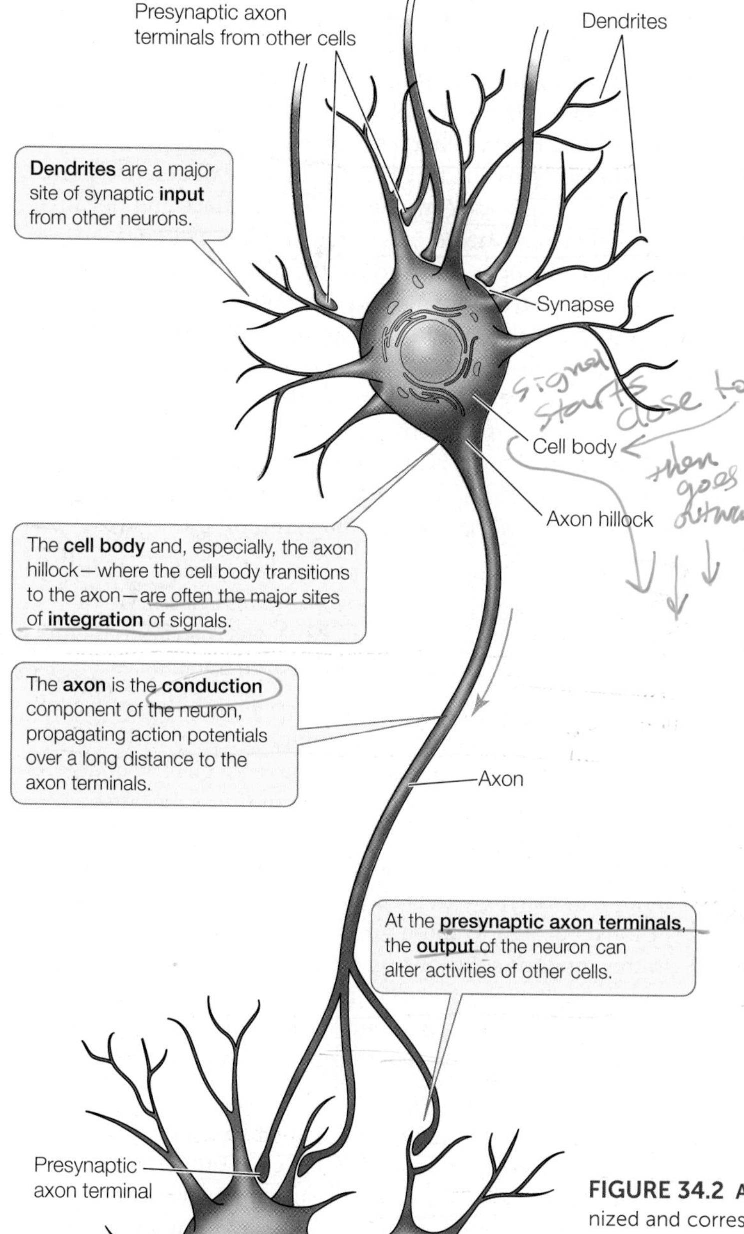
Presynaptic axon terminals from other cells
Dendrites
Dendrites are a major site of synaptic input from other neurons.
Synapse
Cell body
Axon hillock
The cell body and, especially, the axon hillock—where the cell body transitions to the axon—are often the major sites of integration of signals.
The axon is the conduction component of the neuron, propagating action potentials over a long distance to the axon terminals.
Axon
At the presynaptic axon terminals, the output of the neuron can alter activities of other cells.
Presynaptic axon terminal
Postsynaptic cells

FIGURE 34.2 A Generalized Neuron Four *functional* regions of the cell are recognized and correspond approximately to the four major *anatomical* regions. The drawing shows the features typical of most neurons. The form of these features, including axon length and the density and branching pattern of dendrites, can vary greatly among the many different types of neurons.

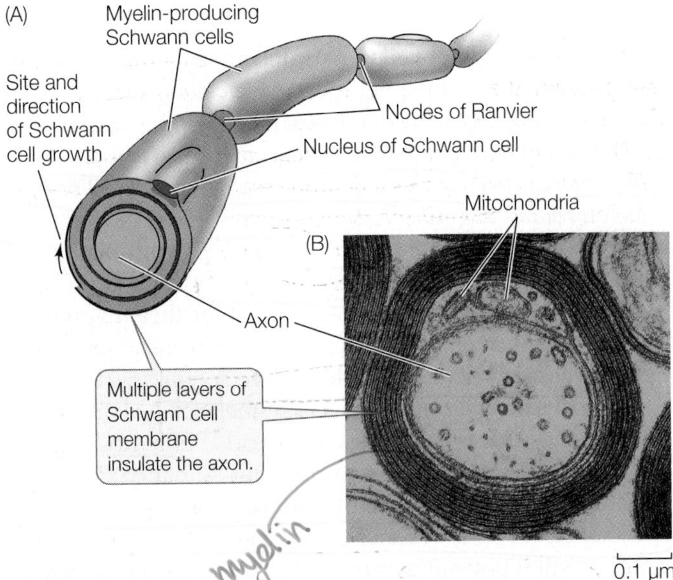

FIGURE 34.3 Electrically Insulating an Axon (A) Schwann cells wrap around the axon, enclosing the axon in many layers of lipid-rich Schwann cell membrane, called myelin, without intervening cytoplasm. Myelin provides electrical insulation to the axon. At the intervals between Schwann cells—the nodes of Ranvier—the axon is exposed. (B) A myelinated axon, seen in cross section at high magnification by use of an electron microscope.

In vertebrates, but generally not in invertebrates, certain glial cells insulate axons electrically by wrapping around them, covering them with concentric layers of lipid-rich cell membrane. In the brain and spinal cord, the cell membranes of glial cells called **oligodendrocytes** wrap around axons. In neurons outside the brain and spinal cord, glial cells of a different type, called **Schwann cells**, perform this function (**FIGURE 34.3**). This multilayered wrap of cell membranes on axons forms a lipid-rich, electrically nonconductive sheath called **myelin**. Parts of the nervous system that consist mostly of myelinated axons have a glistening white appearance and are sometimes called **white matter**. Not all vertebrate axons are myelinated, but those that are can conduct action potentials more rapidly than unmyelinated axons, as we will discuss in Concept 34.2.

CHECKpoint CONCEPT 34.1

- ✓ What is the most dramatic way in which typical neurons and typical glial cells differ functionally?
- ✓ Use a tape measure to measure the distance from your brain to the part of your spine at the same level as your shoulders, then measure the distance from that part of your spine to the tip of one of your index fingers. Estimate how fast action potentials would have to travel for you to pluck four strings on a guitar or banjo per second with your index finger. For simplification, assume that the action potentials merely need to travel the distance you've measured when your brain orders your finger to contract.
- ✓ What is meant when we say signals transmitted by a neuron are "addressed"?

We've now seen that nervous systems are composed of neurons and glial cells. In the next concept we will focus on how the neurons generate and transmit action potentials.

CONCEPT 34.2 Neurons Generate Electric Signals by Controlling Ion Distributions

The one feature common to all nervous systems is that they encode and transmit information in the form of action potentials. Some basic electrical principles help us understand how neurons produce action potentials and other electric signals. **Current** is a flow of electric charges from place to place. In a wire, electrons flow, and current is based on the flow of electrons. However, in the cells of organisms, current is based on the flow of ions in solution (see Concept 2.2), such as sodium ions (Na^+), potassium ions (K^+), or chloride ions (Cl^-). Each Na^+ ion, for example, bears a single positive charge. When Na^+ ions move from one place to another, positive charges flow, and this flow of charges is an electric current.

A **voltage**, also called an electric **potential difference**, exists if positive charges are concentrated in one place and negative charges are concentrated in a different place. Voltages *produce currents* because opposite charges attract and will move toward one another if given a chance. To illustrate, suppose positively charged ions are concentrated in one area and negatively charged ions are concentrated in another. Positive ions such as Na^+ will be attracted to and tend to flow toward the negative area, giving rise to a current.

Of all the background information that is helpful for understanding the electrical function of neurons, the locations where voltage differences can exist are most important.

- *No voltage differences exist from one place to another within open solutions such as the intracellular fluids of a cell.* In an open solution, positive and negative ions distribute themselves evenly so that there are no charge imbalances.
- *Voltage differences exist only across membranes such as the cell membrane.* A cell membrane is composed primarily of lipids (see Figure 5.1). Because ions have an electric charge, they tend to associate with water, not lipids, because water molecules are polar (see Concept 2.2). Moreover, the lipid molecules in a membrane, being hydrophobic, tend to associate with each other rather than with water or ions. Because of these principles, ions tend to remain in the aqueous solutions on either side of a membrane and do not readily cross the membrane. Consequently, positive ions may be concentrated on one side of a cell membrane while negative charges are concentrated on the opposite side.

Solutions that are not immediately in contact with a membrane are termed bulk solutions. We can summarize by saying that a bulk solution is electrically neutral and does not show charge differences from place to place, but a membrane may have a charge difference across it, caused by ion imbalances in the water immediately next to the membrane surfaces:

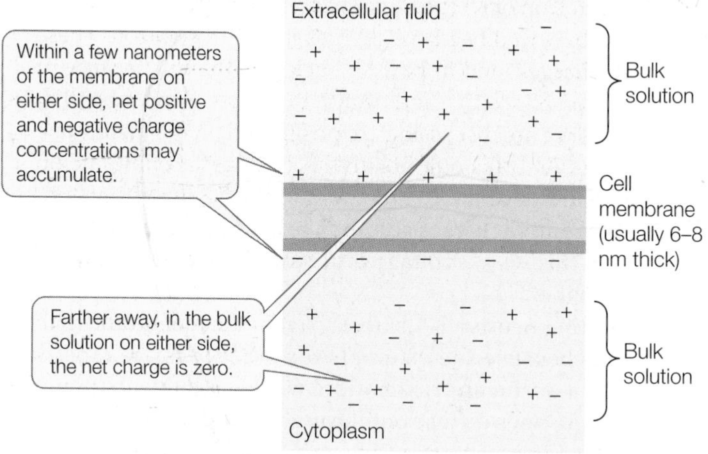

Extracellular fluid
Within a few nanometers of the membrane on either side, net positive and negative charge concentrations may accumulate.
Bulk solution
Cell membrane (usually 6–8 nm thick)
Farther away, in the bulk solution on either side, the net charge is zero.
Bulk solution
Cytoplasm

A voltage exists across a membrane that has a charge difference from one side to the other. This voltage (potential difference) is called the **membrane potential** and can be readily measured (**FIGURE 34.4**). A neuron is described as "resting" when it's not carrying a signal. In a resting neuron, the membrane potential is referred to as the **resting potential** and is typically between –60 and –70 millivolts (mV). The minus sign means that the inside of the cell is electrically negative relative to the outside. Membrane potentials are always written in this way: the sign (negative or positive) always refers to the charge inside relative to the charge outside.

Like other cell membranes, neuron cell membranes contain proteins that serve as ion transporters and channels (see Concept 5.2). Ions cannot pass through the membrane lipid bilayer, as we have just emphasized, and this permits charge differences to exist across a neuron cell membrane. However, ions can pass through the transporters and channels in the membrane. The transporter proteins and channel proteins determine the magnitudes and directions of the voltages and currents across the membrane.

Only small shifts of ions are required for rapid changes in membrane potential

As we talk in this chapter about the functions of neurons, we will often encounter cases in which rapid changes of the membrane potential are of critical importance. We need, therefore, to focus briefly on some key properties of these changes.

The membrane potential changes when ions are redistributed. During the rapid events of routine electrical signaling such as action potentials, ion redistribution occurs through membrane channel proteins. Channels are proteins that permit ions to cross the membrane by simple diffusion (see Concept 5.2).

Suppose that a membrane shifts from being negative on the inside to being positive on the inside. Suppose also that this change occurs because channels in the membrane allow a current of positive charges (ions) to flow inward across the membrane.

ion redistribution occurs through membrane channel proteins

RESEARCH TOOLS

FIGURE 34.4 Measuring the Membrane Potential An electrode can be made from a fine glass tube with a sharp tip filled with a solution that conducts electric charges. When such an electrode is inserted into an axon and another electrode is placed in the extracellular fluid just outside the axon, the difference in voltage across the axon cell membrane—called the membrane potential—can be measured.

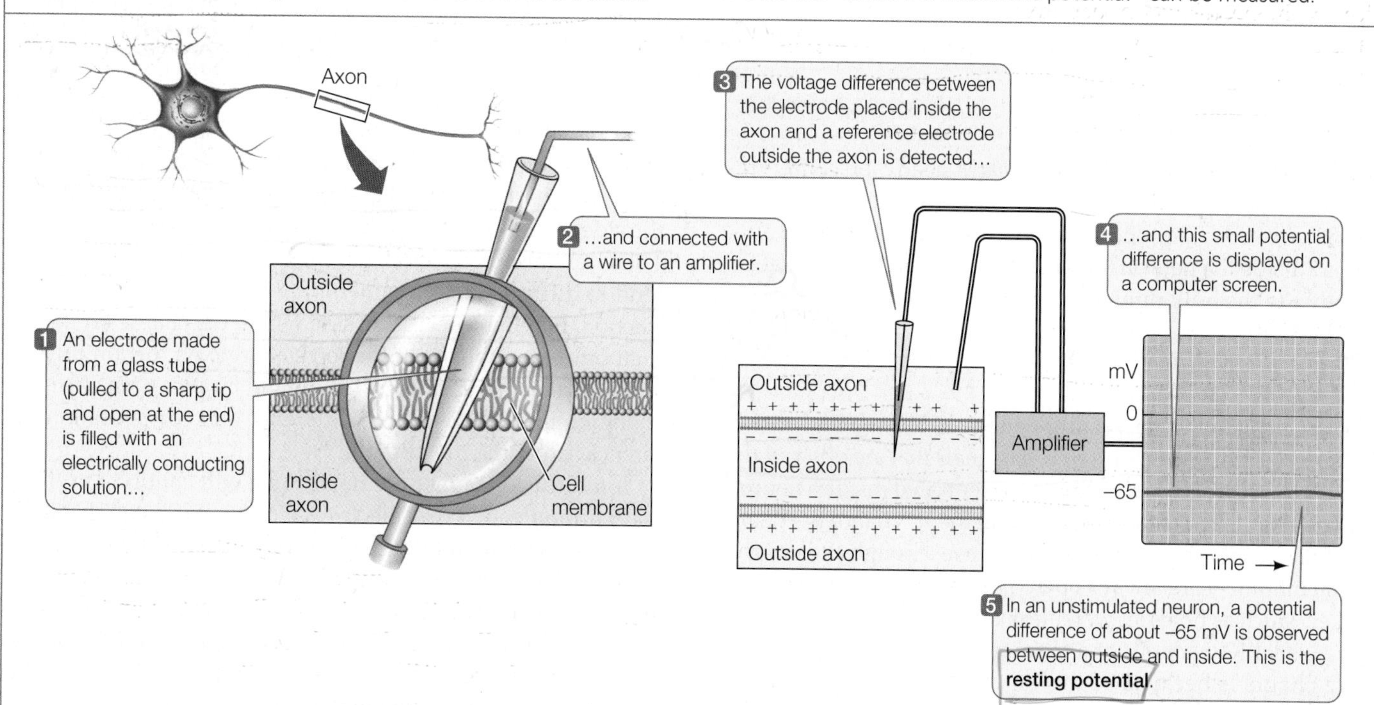

A critically important point is that only relatively small numbers of positive charges need to move through the membrane for this change of membrane potential to occur. This is true because only the charges within a few nanometers of the membrane surfaces are changing. Robert Schmidt and his colleagues once asked, Of all the ions within 1,000 nanometers of a membrane surface, what proportion are responsible for the membrane potential? Their answer was less than 100 out of every 1 million. Just a relatively few ions—positioned within a few nanometers of the membrane surfaces—are involved in rapid changes of membrane potential. The other ions, located in the bulk solutions on either side, are not directly involved.

Similarly, the current through the membrane moves only relatively small numbers of ions during rapid changes of membrane potential. The current is not changing the compositions of the bulk solutions: the compositions of the intracellular and extracellular fluids. A small current is sufficient to redistribute ions on the inner and outer membrane surfaces and lead to a large change in membrane potential—a phenomenon that is fundamental to how neurons generate signals.

Go to ANIMATED TUTORIAL 34.1
The Resting Membrane Potential
PoL2e.com/at34.1

The sodium–potassium pump sets up concentration gradients of Na^+ and K^+

One of the key membrane proteins is a transporter called Na^+-K^+-ATPase or the **sodium–potassium pump**. The protein goes through repeating cycles in which it actively transports ions. During each cycle, it uses the energy of an ATP molecule to expel three Na^+ ions from inside the neuron and to pump two K^+ ions into the neuron from the outside. Na^+ and K^+ are the principal positive ions in both the intracellular and extracellular fluids.

LINK
For more on the sodium–potassium pump see **Concept 5.3**

The sodium–potassium pump creates steady ion concentration differences between the intracellular and extracellular fluids. Because of it, the concentration of Na^+ is higher in the bulk fluids outside the neuron than inside, and the K^+ concentration is higher in the bulk fluids inside than outside.

The resting potential is mainly a consequence of K^+ leak channels

What is the cause of the resting potential? This voltage difference is principally the result of K^+ channels that are open all the time. As we have seen, K^+ is more concentrated inside a neuron than outside. Some (fewer than half) of the K^+ channels—channels that specifically permit K^+ to diffuse through the neuron cell membrane—are always open. They are like little holes in a bucket. Much as holes let water leak out of a bucket, these open channels let K^+ leak out of a neuron. K^+ diffuses out through these channels because of the K^+ concentration gradient (a higher K^+ concentration inside than outside).

This diffusion of K^+ (positive charges) out of a neuron leaves behind unbalanced negative charges inside the neuron. These negative charges tend to pull K^+ back into the cell because of the attraction of positive and negative charges. Remember that the charges we are discussing are positioned on the membrane surfaces. As K^+ leaks out, positive K^+ charges accumulate on the outer surface of the cell membrane. At the same time, unbalanced negative charges accumulate on the inner membrane surface. In this way, a voltage difference is created across the cell membrane.

In a resting neuron, K^+ is freer to cross the membrane than other ions because the channels for other ions are mostly closed. As a useful approximation, we can assume that only K^+ crosses. K^+ ions diffuse out through K^+ leak channels until the voltage difference—negative on the inside—pulls K^+ back inside with exactly the same force as the concentration gradient drives K^+ out. At this equilibrium point, K^+ stops moving through the membrane. There is no further net movement of K^+. The membrane potential in this equilibrium state is called the **equilibrium potential** of K^+. Simply put, if K^+ is the only ion that is free to diffuse across the cell membrane in a resting neuron, its diffusion drives the cell's membrane potential toward the K^+ equilibrium potential, which is the voltage difference that will draw K^+ into the cell to the same degree as the high inside K^+ concentration makes K^+ diffuse out.

The Nernst equation predicts an ion's equilibrium potential

Let's now look at the general principles that are exemplified by K^+ diffusion in a resting neuron. These are the principles of electrochemical equilibrium. The principles are easiest to understand when the channels for only one ion are open and, as a result, only one ion can diffuse across the membrane. When these conditions exist, we can say that the membrane is permeable to just the one ion.

Suppose the concentration of an ion is different on one side of a membrane than the other, and suppose also that one side is more positive than the other. In this case there will be two effects on ion diffusion. There will be a *concentration effect* on diffusion: the ion will tend to diffuse across the membrane from the side of high concentration to the side of low concentration. There will also be an *electrical effect* on diffusion: the ion will tend to diffuse across the membrane because of electrical attraction, moving toward the side where the charge on the cell membrane is opposite to the ion's charge. As the ion diffuses across the membrane, its diffusion will modify both effects by changing the electric charge difference and concentration difference across the membrane.

Ion diffusion across the membrane will continue until the concentration effect and electrical effect become equal and opposite. This is the state of **electrochemical equilibrium**: a state in which the electrical effect causes diffusion in one direction and the concentration (chemical) effect causes diffusion of equal magnitude in the opposite direction. When an ion has reached electrochemical equilibrium, there is no net diffusion of the ion across the membrane in either direction:

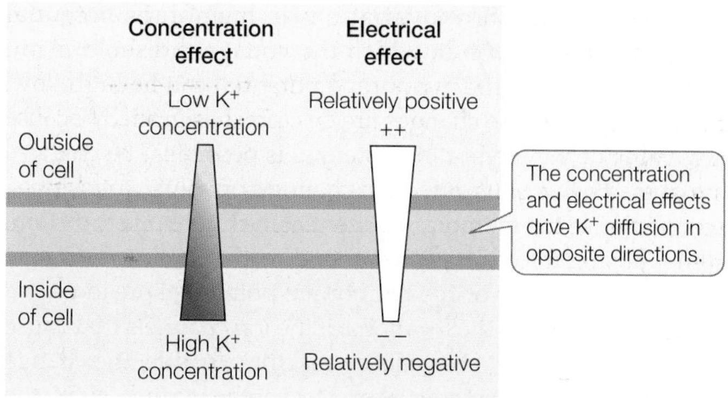

In studies of neurons, the electric potential difference at electrochemical equilibrium is usually called the equilibrium potential of the ion and is symbolized E_{ion}. This potential can be calculated from the ion concentrations inside and outside the cell membrane by use of a famous equation called the **Nernst equation**.

Expressions with brackets represent concentrations. Thus "[ion]" represents the concentration of the ion of interest.

$$\text{Potential difference at electrochemical equilibrium} = E_{ion} = 2.3\frac{RT}{zF}\log\frac{[\text{ion}]_{outside}}{[\text{ion}]_{inside}}$$

In this equation, z is the number of charges per ion, T is the absolute temperature, R is a constant (the universal gas constant), and F is another constant (the Faraday constant).

Go to ACTIVITY 34.1 The Nernst Equation PoL2e.com/ac34.1

The Nernst equation can be used whenever we can assume that only a single ion is able to diffuse across a membrane. In such cases, the equation predicts the equilibrium membrane potential. For example, we have already said that in a resting neuron, just one ion, K^+, can readily cross the cell membrane because the open channels are mostly K^+ channels. Under these circumstances, K^+ is expected to diffuse to electrochemical equilibrium, and this diffusion is expected to determine the membrane potential. Taking account of the concentrations of K^+ inside and outside the cell, the K^+ equilibrium potential, E_K, calculated with the Nernst equation is about –85 mV, depending on temperature. The actual neuron resting potential, as we have said, is close to this, about –65 mV. The principal reason the equation's prediction is a little off is that resting neurons are slightly permeable to other ions besides K^+.

As we will see, the Nernst equation can be applied to any ion, not just K^+, and helps explain each of the major phases of an action potential. It can be used, in an approximate way, to calculate the equilibrium membrane potential whenever just one ion accounts for most diffusion across the membrane and other ions are diffusing to just a minor extent.

Ion distributions across nerve cell membranes are not always at equilibrium. In fact, they sometimes are far from equilibrium. In such cases, ions will move toward equilibrium. The principles of electrochemical equilibrium are then useful for predicting the direction of ion diffusion.

Gated ion channels can alter the membrane potential

Instead of being open all the time like the K^+ leak channels, most ion channels behave as if they contain gates: they are open under some conditions and closed under others. The channels for ions other than K^+ are gated channels for the most part. And even most K^+ channels are gated. Most gated channels are closed in a resting neuron, explaining why the K^+ leak channels determine the resting membrane potential.

We categorize gated channels on the basis of the agent that causes them to open or close (**FIGURE 34.5**):

- **Voltage-gated channels** open or close in response to local changes in the membrane potential.

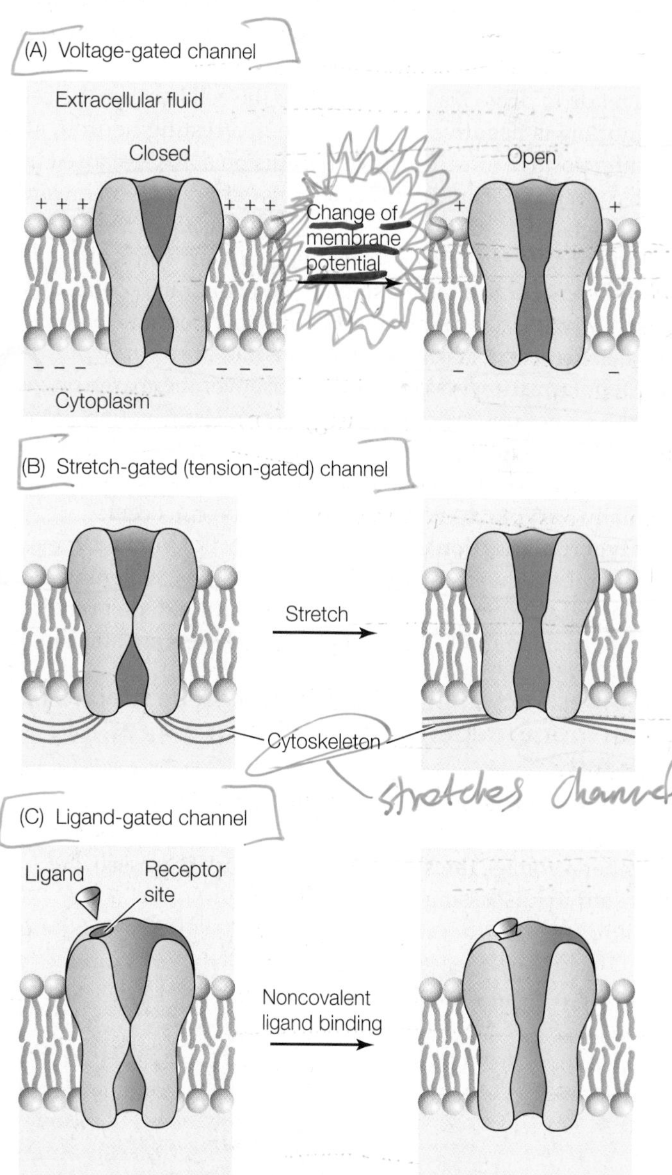

FIGURE 34.5 Three Types of Gated Ion Channels

- **Stretch-gated channels** open or close in response to stretch or tension applied to the cell membrane.
- **Ligand-gated channels** have binding sites where they bind noncovalently with specific chemical compounds that control them. A compound that controls a channel is termed a ligand of that channel. A ligand-gated channel opens and closes depending on whether it is bound to its ligand. Although Figure 34.5C shows a ligand-binding site on the outside of the cell membrane, binding sites (e.g., for second messengers) can also be on the inside.

Openings and closings of gated channels alter the membrane potential. For example, let's consider what happens in a resting neuron if some of the Na^+ channels in the cell membrane open (Na^+ channels are ordinarily closed in a resting neuron). If Na^+ channels open, Na^+ ions diffuse into the neuron from the outside. This is true because Na^+ is far from electrochemical equilibrium in a resting neuron. Na^+ is far more concentrated outside the cell than inside, and this concentration difference tends to drive diffusion into the cell. The electrical difference also tends to drive Na^+ diffusion into the cell because the cell membrane is negative on the inside in a resting neuron, attracting positive ions. The inward diffusion of Na^+—caused by both concentration and electrical effects—represents an inward flow of positive charges which makes the inside of the neuron cell membrane become less negative. In other words, the voltage difference across the membrane (the membrane potential)—which is ordinarily near –65 mV in a resting neuron—becomes smaller. Perhaps it becomes –60 mV instead of –65 mV.

Depolarization is said to occur whenever the charge on the inside of a neuron cell membrane becomes less negative—relative to the charge on the outside—than it is at rest. In Concept 34.1 we implied that depolarization is either present or absent. Actually, varying degrees of depolarization can occur.

Hyperpolarization can also take place. It occurs whenever the charge on the inside becomes *more* negative. Opening and closing of gated ion channels can cause either depolarization or hyperpolarization, depending on which ions are affected.

Changes in membrane potential can be graded or all-or-none, depending on whether a threshold is crossed

Two types of changes in membrane potential occur: graded and all-or-none. The type of change that occurs depends on whether a voltage threshold of about –50 mV is crossed. We will start by discussing graded changes.

FIGURE 34.6A shows variations in membrane potential over time in a neuron in which the voltage threshold is not crossed. Note that the membrane potential can be high, low, or in between. These changes are described as **graded** because any value of the membrane potential is possible. The changes are caused by various gated ion channels opening or closing—some causing the membrane potential to shift in one direction, some causing it to shift in the other direction.

Graded changes of the membrane potential spread only a short distance from the locations of the ion channels that cause them. If a set of Na^+ channels opens, for example, the change in membrane potential will be greatest at the location along the cell membrane where the channels are found. The change will spread on either side. But it will become smaller and smaller at greater and greater distances from the channels:

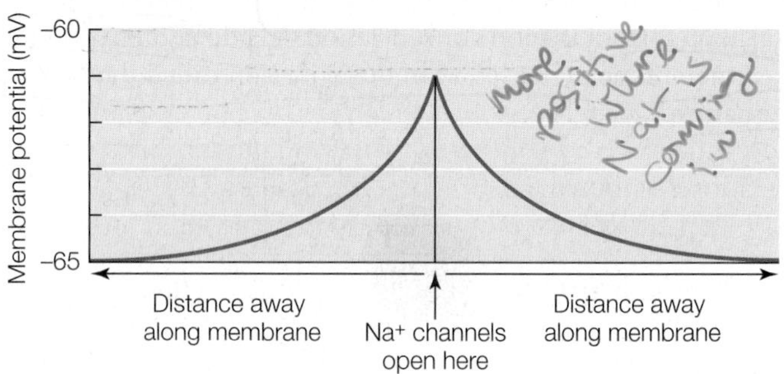

Let's now discuss all-or none changes in the membrane potential. If a neuron depolarizes to the point that its inside charge is within about 50 mV of the outside charge, a massive, all-or-none event occurs: an impulse, or action potential, is triggered (**FIGURE 34.6B**). A membrane potential of about –50 mV is thus the voltage threshold for an action potential. An action potential is described as all-or-none because, when it occurs, it is always the same size. Action potentials are not graded. Moreover, as we'll soon discuss, an action potential does not become smaller and smaller as it moves away from the location on a cell membrane where it originates. It stays the same in size as it propagates along the cell membrane of the axon of the neuron.

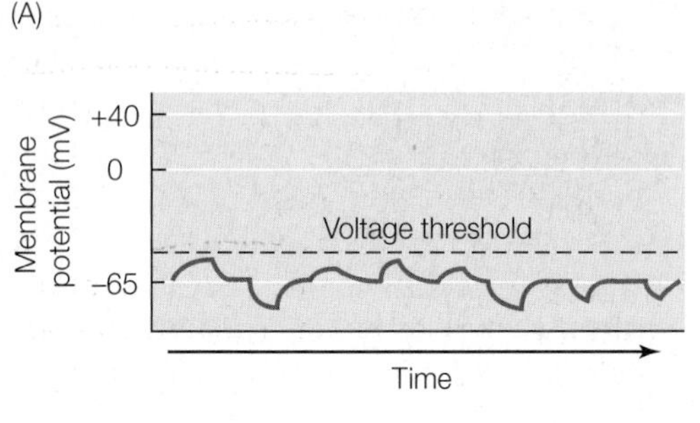

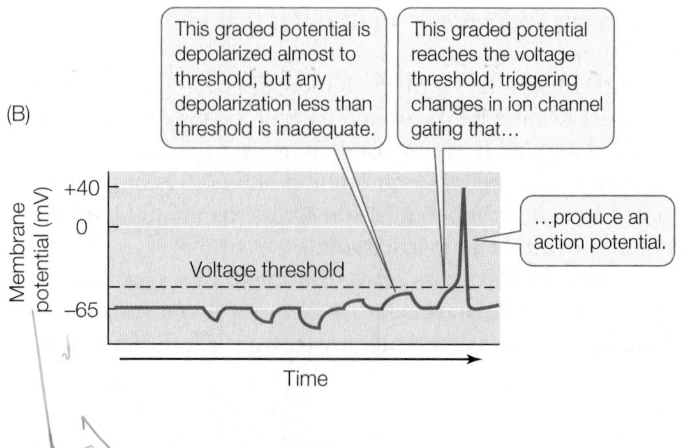

FIGURE 34.6 Graded and All-or-None Changes in Membrane Potential (A) Graded changes in membrane potential may not reach the voltage threshold. (B) When graded changes in membrane potential reach the voltage threshold, the membrane generates a large all-or-none change in membrane potential—the action potential.

Graded changes give rise to all-or-none changes when the voltage threshold is crossed, and this interaction between the two types of changes provides one of the most important mechanisms by which a neuron carries out integration. Graded changes in membrane potential at a single location in the cell membrane add together. Suppose, for example, that a neuron simultaneously receives four signals—A, B, C, and D—that affect different types and numbers of gated ion channels in the neuron cell membrane, producing graded changes in the membrane potential. Suppose that at a particular spot on the cell membrane, signal A causes a 2.5-mV depolarization, B causes a 1.5-mV depolarization, C causes a 1-mV hyperpolarization, and D causes a 1-mV depolarization. All these graded changes in the membrane potential sum together. Thus at this particular spot, the membrane potential will be depolarized by 4 mV. It will be –61 mV instead of –65 mV as at rest.

The spot of greatest importance for this type of integration in a neuron is the axon hillock, where the axon arises from the cell body (see Figure 34.2). This is the spot where action potentials are most commonly initiated. A neuron may be receiving hundreds or thousands of signals at its dendrites and cell body. These signals bring about graded changes in membrane potential that spread to the axon hillock, where all the depolarizations and hyperpolarizations sum.

As this summation occurs, if the membrane potential at the axon hillock varies only in the range below about –50 mV, as in Figure 34.6A, no action potentials are produced. Moreover, no electric signals reach the axon terminals because the graded potentials disappear over the distance between the axon hillock and the axon terminals.

However, as the summation of graded potentials occurs, if the membrane potential at the axon hillock crosses the threshold as at the right in Figure 34.6B—that is, if it becomes –50 mV or even more depolarized—the axon hillock initiates an action potential. This action potential travels to the axon terminals, and the axon terminals then send signals to the other cells they innervate.

These processes are the major means by which a single neuron integrates its inputs to determine its outputs. In short, all the inputs create small, graded changes of membrane potential that sum together as they spread. If their integrated effect is a membrane potential that is not as depolarized as about –50 mV, the neuron produces no action potentials. But if their integrated effect is a membrane potential that is depolarized to at least –50 mV, the neuron generates one or more action potentials that propagate long distances down the axon.

An action potential is a large depolarization that propagates with no loss of size

An **action potential**, or **nerve impulse**, is a large, brief, localized change in the membrane potential—so large that the polarity reverses from inside-negative to inside-positive. During an action potential, the membrane depolarizes from –65 mV at rest to about +40 mV: a change of more than 100 mV. This event is extremely brief. It lasts only about 1 millisecond. It is also very localized: it is restricted to just a small region of the cell membrane.

An action potential propagates with no loss of size, however. Propagation occurs because an action potential at one location in the cell membrane of a neuron causes currents to flow that depolarize neighboring regions to the voltage threshold. When an action potential occurs in the axon hillock, currents flow into the neighboring region of the axon that is closer to the axon terminals, causing a graded depolarization there that is great enough that threshold is reached. That region then produces an action potential. Currents from that action potential then spread into the axon region still closer to the axon terminals. That region then produces an action potential. Each region of the axon stimulates the next region to produce an action potential until the action potential reaches the axon terminals. This process is what is meant by "propagation" in discussions of neuron function. Recall that action potentials are all-or-none and always the same size. Thus there is no loss of magnitude as an action potential propagates along the length of an axon.

Why does the cell membrane depolarize so profoundly during an action potential? And why does this depolarization end so quickly—within a millisecond? The properties of voltage-gated K^+ and Na^+ channels in the neuron cell membrane are chiefly responsible.

When the membrane potential in any local region of an axon reaches threshold, massive numbers of voltage-gated Na^+ channels in that region quickly open. Suddenly, therefore, the principal ion that can cross the membrane becomes Na^+, instead of K^+. This results in a sudden, rapid influx of Na^+ ions into that part of the axon because, as we discussed earlier, both the Na^+ concentration difference and the electrical difference across the cell membrane favor inward Na^+ diffusion. If Na^+ were the only ion that could cross the membrane at this stage and if enough time were available to reach electrochemical equilibrium, the membrane potential would become E_{Na}—the equilibrium potential for Na^+ (about +50 mV according to the Nernst equation). The membrane potential in fact almost reaches E_{Na}: it typically becomes about +40 mV (**FIGURE 34.7**, transition from A to B). Quickly, however, the Na^+ channels close, and voltage-gated K^+ channels, which respond in a delayed fashion to the local membrane potential, open in massive numbers. With the Na^+ channels closed and the voltage-gated K^+ channels open, the principal ion that can cross the membrane becomes K^+. K^+ diffuses out of the cell because K^+ is more concentrated inside than outside and also because, at this moment (with the membrane potential at +40 mV), the outside of the cell membrane is relatively negative, attracting K^+. The movement of K^+ out of the cell drives the membrane potential back toward –65 mV, which is close to E_K, the K^+ equilibrium potential (Figure 34.7C).

Positive feedback (see Concept 29.6) plays a major role in the depolarization phase of an action potential (**FIGURE 34.8**). Positive feedback occurs because the Na^+ channels are voltage-gated and the voltage across the neuron cell membrane is rapidly changing. When the cell membrane becomes partly depolarized, some voltage-gated Na^+ channels open. Na^+ ions then cross the cell membrane faster (diffusing from outside to inside) and depolarize the membrane even more. This greater depolarization causes more voltage-gated Na^+ channels to

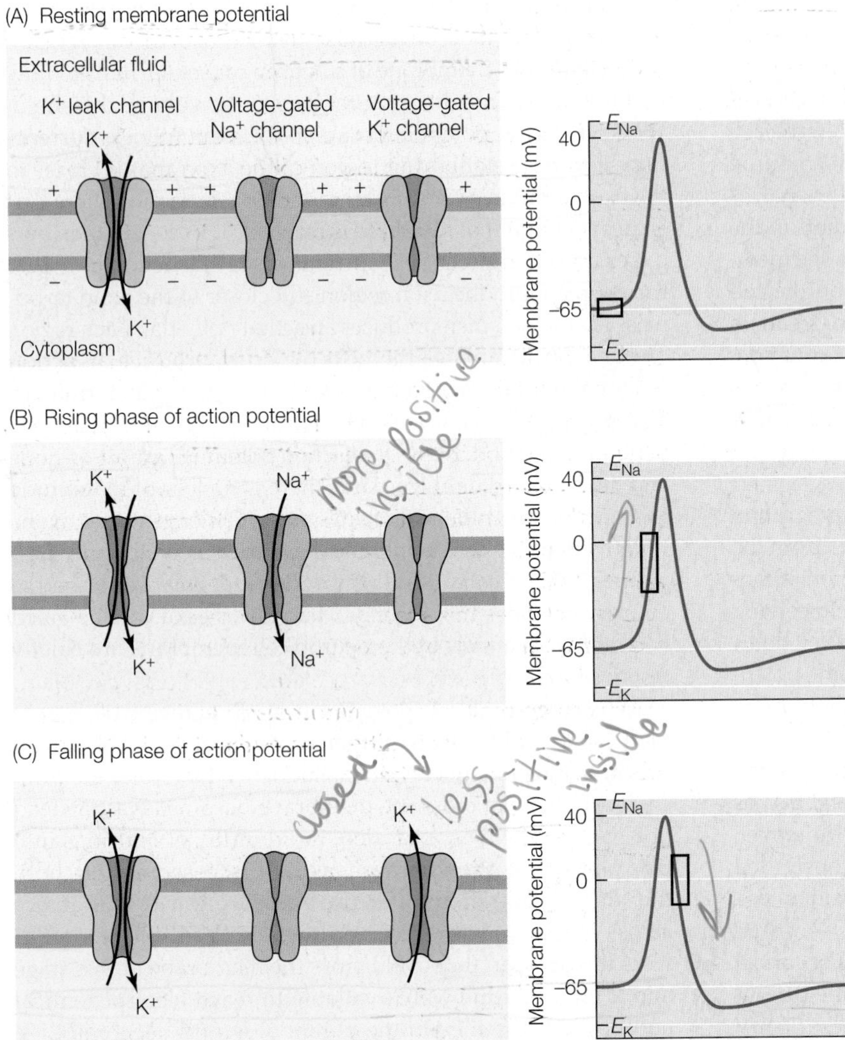

FIGURE 34.7 Production of an Action Potential The diagrams to the left show changes in voltage-gated Na^+ and K^+ channels, and those on the right show changes in membrane potential during (A) rest, (B) the rising phase of an action potential, and (C) the falling phase. The black rectangles on the membrane potential curves indicate the part of the curve during which the events diagrammed at left occur. The closing of the Na^+ channels in (C) is of a specialized type (called inactivation) that cannot be reversed (i.e., the channels cannot reopen) for a brief period. K^+ leak channels are always open.

Go to ANIMATED TUTORIAL 34.2
The Action Potential
PoL2e.com/at34.2

open. This positive-feedback loop continues until all the voltage-gated Na^+ channels open and maximum depolarization (approaching E_{Na}) occurs.

Why does an action potential travel in just one direction along an axon? After any particular local part of the axon membrane has produced an action potential, it undergoes a brief refractory period during which it cannot produce another action potential. Thus as an action propagates along an axon, the region of the cell membrane just behind it is in its refractory period. Currents are able to induce an action potential only in the neighboring region in the direction of the axon terminals.

Action potentials travel particularly fast in large axons and in myelinated axons

Action potentials do not travel along all axons at the same speed. Because of physical laws, they tend to travel faster in large-diameter than in small-diameter axons. This principle has been particularly important in the evolution of invertebrate nervous systems. Many invertebrates have evolved axons of large diameter that increase the speed of conduction of action potentials to muscles on which the animals depend for

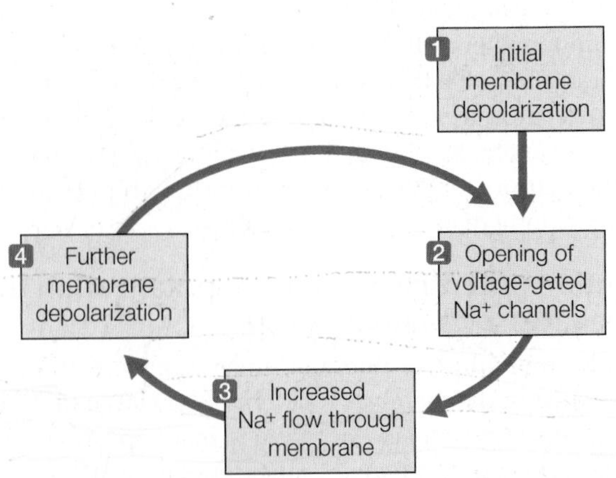

FIGURE 34.8 Positive Feedback Plays a Key Role in the Generation of an Action Potential During the rising phase of an action potential, positive feedback progressively drives all voltage-gated Na^+ channels to open.

rapid escape from danger. In squid, for example, the muscle cells used for escape swimming are innervated by giant axons that in some cases are 1 millimeter in diameter. These giant axons are relatively easy to manipulate and study. Many of the basic principles of neuron function were discovered by studying them.

Vertebrates have evolved an additional means of increasing the speed of conduction: myelination. When glial cells wrap around axons, covering them with myelin, they leave uncovered spaces, called **nodes of Ranvier**, in between (see Figure 34.3). In a myelinated axon, action potentials are generated only at the nodes of Ranvier, because voltage-gated ion channels are clustered at these nodes and also because the myelin-covered axon segments are too well insulated electrically to permit rapid ion fluxes (flows) across the cell membrane. As an action potential travels along a myelinated axon, it jumps from one node of Ranvier to the next, a process called saltatory ("jumping") conduction. For an axon of any particular diameter, an action potential travels far faster by saltatory conduction than it would by continuously traveling the entire length of the axon membrane:

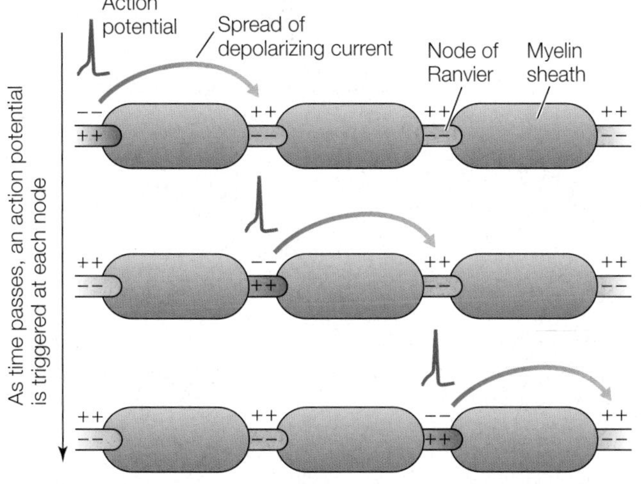

CHECKpoint CONCEPT 34.2

- What are the major differences in size and transmission between a graded potential and an action potential?
- What are two different evolutionary adaptations to speed the conduction of action potentials?
- Suppose the inner surface of a neuron cell membrane changes from being 60 mV more negative than the outer surface to being 50 mV more negative. Is the neuron closer to producing an action potential or farther from doing so? Explain.

Now that we have described how an action potential is generated and propagated along an axon, we will address the question of what happens when the action potential arrives at the axon terminal. How is the information communicated to the next cell?

CONCEPT 34.3 Neurons Communicate with Other Cells at Synapses

Recall that synapses are cell-to-cell contact points that are specialized for signal transmission from one cell to another. They play extremely important roles in information processing and integration. They also play central roles in learning and memory. A great deal of modern research on the function of the nervous system focuses on synapses.

Chemical synapses are most common, but electrical synapses also exist

The most common type of synapse is termed a **chemical synapse** because signals pass from cell to cell by means of molecules ("chemicals") released by the presynaptic cell. At a chemical synapse, the cell membranes of the pre- and postsynaptic cells are actually separated by an extremely narrow space, about 20–30 nm wide, termed the synaptic cleft. Action potentials cannot cross this space (although the space is bridged by occasional trans-synaptic protein molecules). When an action potential arrives at the end of the presynaptic cell, that cell releases a compound called a **neurotransmitter** into the synaptic cleft, and the neurotransmitter crosses the cleft. Acetylcholine is an example of a neurotransmitter:

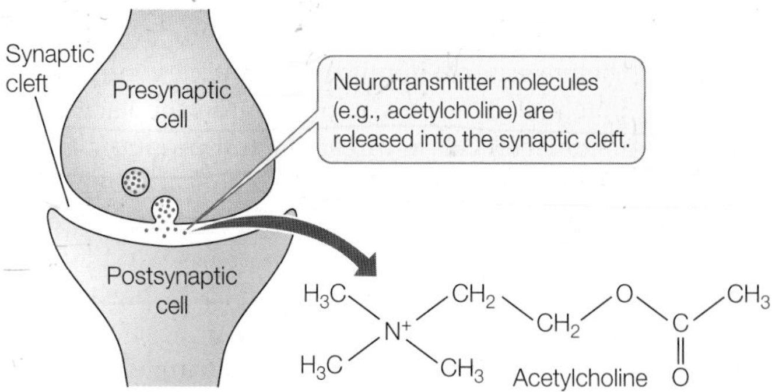

Neurotransmitter molecules diffuse across the synaptic cleft very rapidly because the distance is extremely short. The molecules then bind to receptor proteins in the cell membrane of the postsynaptic cell. These receptor molecules are specialized to generate an electrical change or some other sort of change when they bind to neurotransmitter molecules—completing signal transmission across the synapse. In many cases, the entire process of synaptic signal transmission takes just a few milliseconds.

Neurotransmitter molecules are rapidly removed from the synaptic cleft after they have been released into it by the presynaptic cell. Removal terminates the signal and is essential for the proper function of chemical synapses. Removal may occur by two principal mechanisms. The neurotransmitter molecules may be enzymatically broken down, or they may be taken up by nearby neurons or glial cells. In some well-known cases, neurotransmitter molecules are removed from the synapse by reuptake into the presynaptic cell.

In an **electrical synapse**, signal transmission is "all electrical" instead of employing a neurotransmitter. At such a synapse, the cell membranes of the pre- and postsynaptic cells are joined by gap junctions at which the cytoplasms of the two cells are continuous (see Figure 4.18). Thus when an action potential arrives in the presynaptic cell, it can spread easily and directly to the postsynaptic cell. Signals cross electrical synapses with essentially no delay.

Electrical synapses have evolved principally where very fast, invariant signal transmission is advantageous. They are found in neurons that control escape swimming in some fish and crustaceans. They are also found in some tissues where great numbers of cells must be stimulated synchronously to act together, such as fish electric organs (see Concept 33.4).

The vertebrate neuromuscular junction is a model chemical synapse

Each muscle cell (muscle fiber) in a vertebrate skeletal muscle is typically innervated by one neuron, called a motor neuron. A muscle such as your biceps contains thousands of muscle cells, and many neurons are involved in innervating the whole muscle. However, each muscle cell is typically innervated by just one neuron, and the muscle cell will contract only if stimulated to do so by that neuron.

A **neuromuscular junction**—a type of chemical synapse—is a synapse between a motor neuron and a skeletal muscle cell (**FIGURE 34.9A**). This type of synapse, sometimes called a motor end-plate, has been studied intensively to understand how chemical synapses work. It is not entirely representative of all chemical synapses because it is specialized for a one-to-one relationship between nerve impulses and muscle contractions: each action potential in the presynaptic neuron depolarizes the postsynaptic (muscle) cell to threshold and produces an action potential in the postsynaptic cell. Chemical synapses in the brain do not operate this way, as we will discuss later in this chapter.

At a neuromuscular junction, the axon of the presynaptic neuron branches close to the muscle cell, creating several buttonlike axon terminals (called boutons) that are in synaptic contact with the muscle cell (**FIGURE 34.9B**). Each of these axon terminals contains spherical synaptic vesicles filled with molecules of acetylcholine (ACh), the neurotransmitter used by all vertebrate neuromuscular junctions (**FIGURE 34.9C**).

When an action potential arrives at an axon terminal, the terminal releases ACh into the synaptic cleft by a mechanism that depends on voltage-gated calcium channels. The action potential causes Ca^{2+} channels in the presynaptic membrane to open (see Figure 34.9C, step 1). Ca^{2+} diffuses into the axon terminal through the open channels because the concentration of Ca^{2+} is greater outside the cell than inside (step 2). Increased Ca^{2+} inside the terminal induces synaptic vesicles to fuse with the presynaptic membrane, releasing ACh into the synaptic cleft by exocytosis (step 3; see Figure 5.8B).

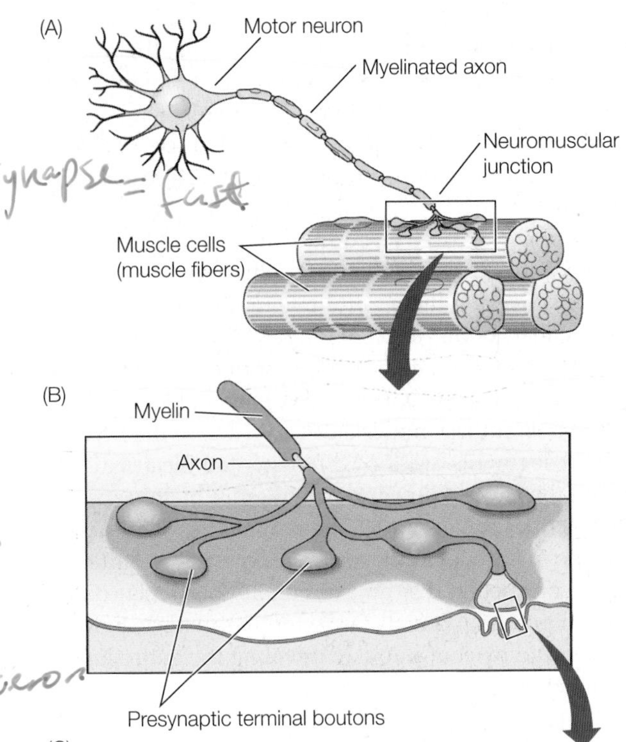

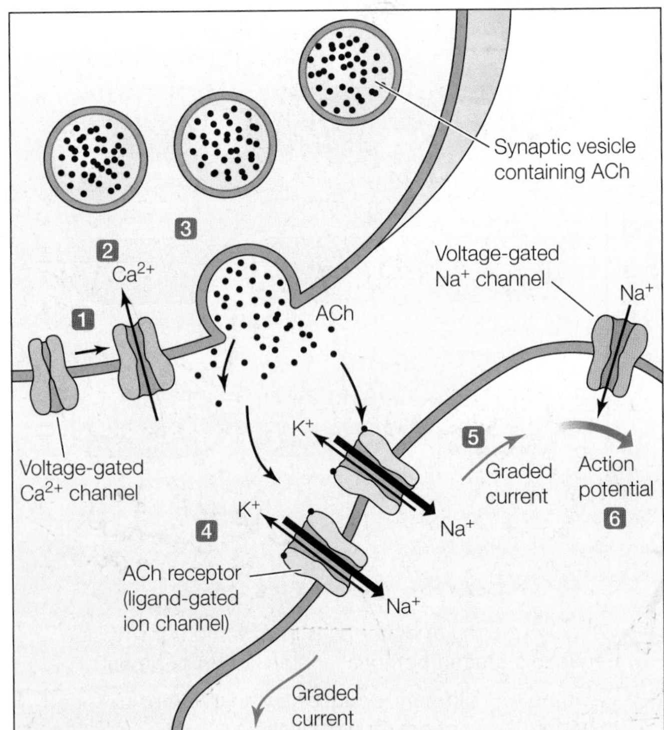

FIGURE 34.9 Synaptic Transmission at a Vertebrate Neuromuscular Junction (A, B) A motor neuron communicates with a skeletal muscle cell at a neuromuscular junction. (C) The motor neuron releases acetylcholine (ACh) into the synaptic cleft, stimulating an action potential in the muscle cell.

Go to ANIMATED TUTORIAL 34.3
Synaptic Transmission
PoL2e.com/at34.3

Go to ANIMATED TUTORIAL 34.4
Neurons and Synapses
PoL2e.com/at34.4

Go to MEDIA CLIP 34.1
Put Some ACh Into It!
PoL2e.com/mc34.1

APPLY THE CONCEPT

Neurons communicate with other cells at synapses

How do we know that Ca^{2+} influx into the presynaptic neuron ending causes the release of neurotransmitter? The squid giant axon and its neuron endings are a convenient system for experiments because they are so large, making it possible to inject substances into the presynaptic cell at a synapse. Some of the substances that can be injected are Ca^{2+} ions and BAPTA, a substance that binds Ca^{2+} ions. Shown below are the results of a series of experiments using these substances.

1. What is happening during the delay between the pre- and postsynaptic membrane events in the control condition?
2. Explain the postsynaptic response in the absence of a presynaptic electric signal in experiment 1.
3. Explain why there is a presynaptic but no postsynaptic response in experiment 2.

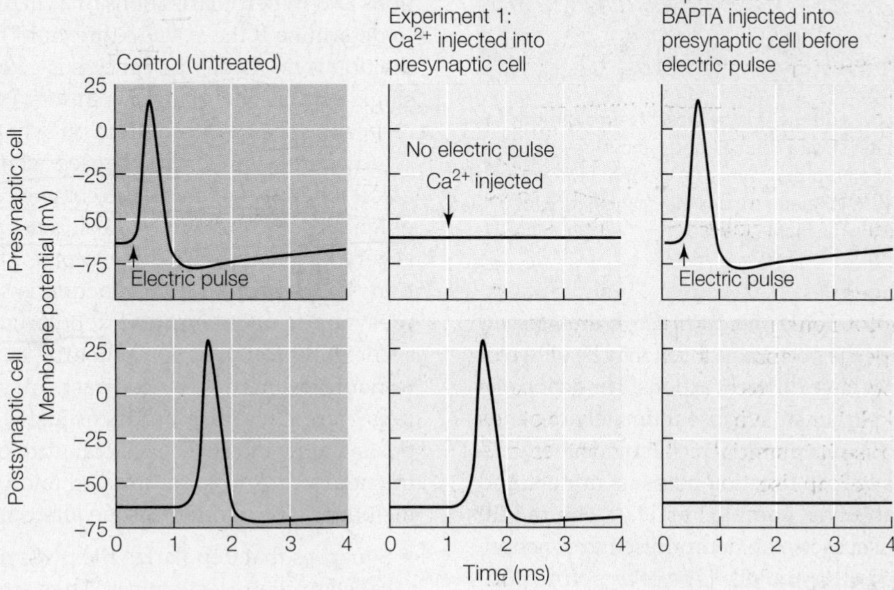

The response of the muscle cell depends on ligand-gated receptors in the postsynaptic cell membrane at the synapse (see Figure 34.5C). These receptors, which are gated by ACh, act as ion channels when they are open. When ACh is released into the synaptic cleft, it binds to the ACh-gated receptor proteins (step 4). These receptors open, allowing Na^+ and K^+ to diffuse through. Influx of Na^+ is dominant because of the locally prevailing ion and electrical gradients. The rapid influx of Na^+ causes a large, graded depolarization, which induces graded ionic currents that spread to neighboring regions of the postsynaptic membrane (step 5) where there are voltage-gated Na^+ and K^+ channels. The depolarization in these neighboring regions is great enough to exceed the threshold of the voltage-gated ion channels, setting off an action potential (step 6). This action potential is propagated throughout the cell membrane of the muscle cell, and it propagates into the cell's transverse tubules (T tubules), activating contraction.

LINK

The physical and molecular interactions of muscle contraction are described in **Concept 33.1**

Many neurotransmitters are known

Several dozen neurotransmitters are now known, and more are being discovered all the time. They fall into three major chemical categories:

- *Amino acids* that serve as neurotransmitters include glutamate, glycine, and γ-aminobutyric acid (GABA; γ = gamma).
- *Biogenic amines* include acetylcholine and also dopamine, norepinephrine, and serotonin.
- A variety of *peptides* (strings of amino acids) serve as neurotransmitters.

Each presynaptic neuron produces specific neurotransmitters. Some produce only one, and others produce two or more, but all presynaptic neurons are specific in the neurotransmitters they release. For simplicity, we'll assume as we go forward that each neuron releases one neurotransmitter.

Let's now consider a neuron that is functioning as the postsynaptic cell. In the brain, a postsynaptic neuron may have chemical synapses with hundreds or thousands of presynaptic neurons (**FIGURE 34.10**). Each of these neurons will use one neurotransmitter. However, different neurons may use different neurotransmitters. Collectively, the hundreds or thousands of presynaptic cells may use dozens of neurotransmitters.

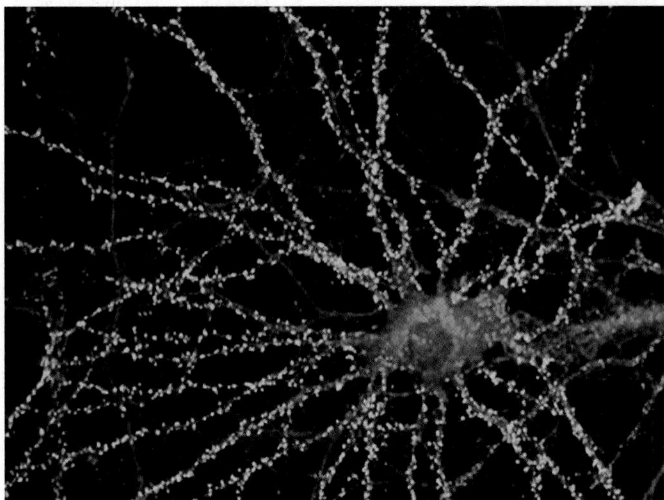

FIGURE 34.10 Synapses on a Single Postsynaptic Neuron in the Brain of a Mouse This image shows the cell body and many dendrites of a single neuron, labeled green. Synapses on the neuron are yellow. The synapses were labeled red, but the combination of green and red produces yellow at most synapses.

Moreover, the receptor proteins *for any particular neurotransmitter* in the cell membrane of the postsynaptic cell may be of two or more molecular types that have different actions. The action of a neurotransmitter at any particular synapse ultimately depends on the action of the specific receptor to which it binds there.

This complexity in synapse function helps us understand how brain function can be as complex as it is. The model chemical synapse we discussed, the neuromuscular junction, employs one neurotransmitter and one type of receptor. Modern brain scientists, however, seek to understand neurons that have great numbers of synapses that function with a variety of neurotransmitters, and often with a variety of receptor proteins for any given neurotransmitter.

Synapses can be fast or slow depending on the nature of receptors

Two broad classes of neurotransmitter receptors are recognized. These are fast, ionotropic receptors and slow, metabotropic receptors.

Ionotropic receptors are ligand-gated ion channels similar to the ACh receptor in the vertebrate neuromuscular junction. Neurotransmitter binding to an ionotropic receptor directly causes a fast, short-lived change in ion diffusion across the cell membrane of the postsynaptic cell. In this way it can directly cause an immediate change in the local membrane potential. This change may in turn initiate other changes in the postsynaptic cell.

Metabotropic receptors are not ion channels. Instead, they are commonly G protein–linked receptors (see Figure 5.14). Such receptors often exert their effects by controlling production of second messengers such as cyclic adenosine monophosphate (see Figure 5.17). Recall that second messengers carry messages to the interior of a cell, where they can exert controlling effects on a wide variety of cell mechanisms, and even on gene expression. Postsynaptic cell responses mediated by metabotropic receptors are very diverse and are often slow and long-lived.

LINK

The structure and properties of G protein–linked receptors and the cellular responses they activate are described in **Concepts 5.5 and 5.6**

Fast synapses produce postsynaptic potentials that sum to determine action potential production

In Concept 34.2 we discussed the fact that graded depolarizations and hyperpolarizations of a cell membrane sum together to determine if the voltage threshold for action potential production is reached. This process is a key mechanism by which single neurons integrate their inputs. Let's now review the process with specific information on synaptic function.

At neuron-to-neuron synapses, when a postsynaptic ligand-gated ion channel opens, the change caused in the membrane potential of the postsynaptic cell is usually less than 1 mV! This is very different from what happens at a neuromuscular junction. The neuromuscular junction is specialized so that each presynaptic nerve impulse depolarizes the postsynaptic cell sufficiently for an action potential to be generated. Neuron-to-neuron synapses, by contrast, are specialized so that each presynaptic impulse produces just a fast, small, short-lived postsynaptic effect. The exact nature of the effect depends on the neurotransmitter employed and on the receptor protein, including which ion fluxes are altered.

- Synapses that depolarize the postsynaptic membrane are called **excitatory synapses**. They are so named because depolarization shifts the membrane potential toward the threshold for action potential production, increasing the likelihood that an action potential will be triggered. Excitatory synapses produce graded membrane depolarizations called **excitatory postsynaptic potentials** (**EPSPs**).
- Synapses that shift the membrane potential away from threshold are called **inhibitory synapses** because they lower the likelihood of action potential production. Inhibitory synapses produce graded membrane hyperpolarizations called **inhibitory postsynaptic potentials** (**IPSPs**).

Each individual EPSP or IPSP, produced at a single time by a single synapse, is usually less than 1 mV, as we have said, and disappears in 10–20 milliseconds. The EPSPs and IPSPs are typically produced at dendrites and at the cell body (the dendrites and cell body are the sites of the synapses).

EPSPs and IPSPs, being graded potentials, fade as they spread along the cell membrane of a postsynaptic neuron. When they spread, however, they affect the membrane potential in the part of the neuron near the axon hillock where action potentials are initiated. Membrane potentials must be present at a single time to sum together. They may originate from many spatially separate synapses (**FIGURE 34.11**). Summation is thus both temporal (dependent on time) and spatial. Keep in mind that each individual EPSP or IPSP lasts only 10–20 milliseconds. Thus the sum of the graded potentials changes from moment to moment.

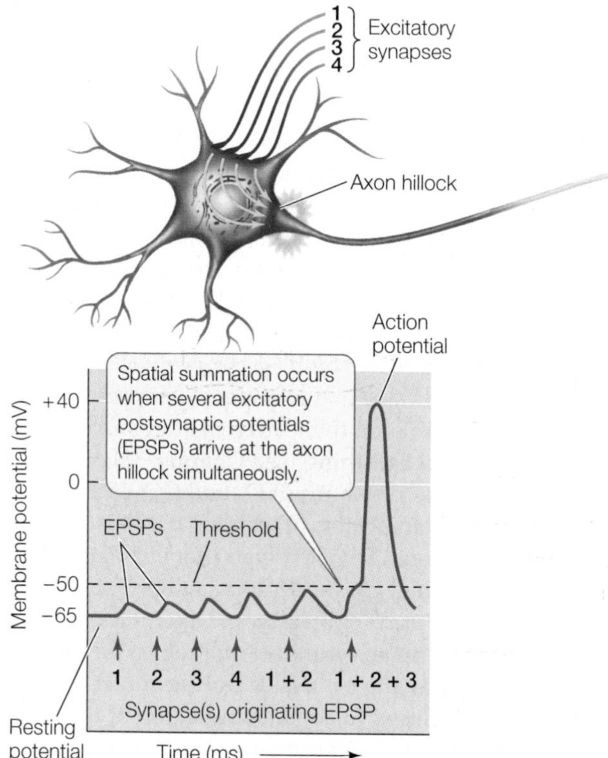

FIGURE 34.11 A Postsynaptic Neuron Carries Out Spatial Summation An individual neuron sums excitatory and inhibitory postsynaptic potentials received at spatially separate synapses. The graph shows changes in the membrane potential at the axon hillock. When the sum of the potentials present at one time depolarizes the axon hillock to threshold, the neuron generates an action potential.

Summation determines whether the postsynaptic cell produces action potentials. If the sum of EPSPs and IPSPs at the axon hillock is a depolarization great enough to reach threshold, the cell produces an action potential. If the depolarization is greater than threshold after the refractory period, the cell produces another action potential. In this way, a long period of depolarization above threshold results in a long sequence of closely spaced action potentials.

Synaptic plasticity is a mechanism of learning and memory

Synaptic plasticity is a phenomenon of extreme importance that is a principal focus of modern synapse research. **Synaptic plasticity** refers to the fact that synapses in the nervous system of an individual animal can undergo long-term changes in their functional properties and even their physical shape during the individual's lifetime. Such changes are currently hypothesized to be one of the major mechanisms of learning and memory.

Learning refers to the ability of an individual animal to modify its behavior as a consequence of individual experiences that took place at earlier times. For instance, if we have been accustomed to nibbling mushrooms while walking in the woods but one day we become deathly ill after doing so, we will change our behavior in the future: seeing mushrooms, we will no longer grab one for a snack. What is the mechanism by which such learning might occur? Synaptic plasticity is one hypothesis. According to this hypothesis, an individual's experiences at one time in its life can produce long-term changes in the individual's synapses, so that future experiences are then processed by the nervous system in altered ways.

LINK

Some of the many ways animal behavior is modified by learning and early experiences are described in **Concept 40.2**

Two model examples of synaptic plasticity illustrate its mechanisms and importance. Sea hares (marine mollusks in the genus *Aplysia*) pull their gills inside when certain parts of their body are touched:

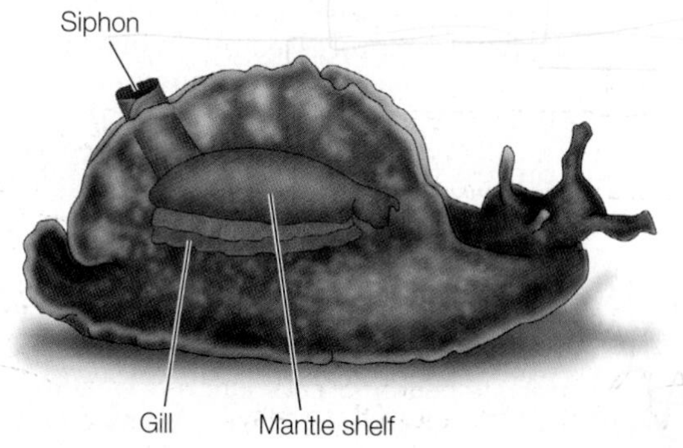

The sea hares learn from past experience how vigorously to do this. They withdraw their gills more vigorously if they have previously been exposed to a very noxious agent—a type of learning called sensitization. Sensitization occurs because of synaptic plasticity in synapses between the sensory neurons that detect touching and the motor neurons that stimulate the gill-withdrawal muscles. These synapses are said to be functionally strengthened when a sea hare is subjected to a noxious stimulus: the synapses become altered so that the amount of neurotransmitter released per presynaptic signal is increased, causing the postsynaptic cell to be excited to a greater degree by an identical presynaptic input. Long-term changes brought about by a second-messenger system are responsible.

Studies of the mammalian hippocampus provide a vivid example of synaptic plasticity in which synapses change not only in function but also in physical characteristics. The hippocampus lies deep within the brain:

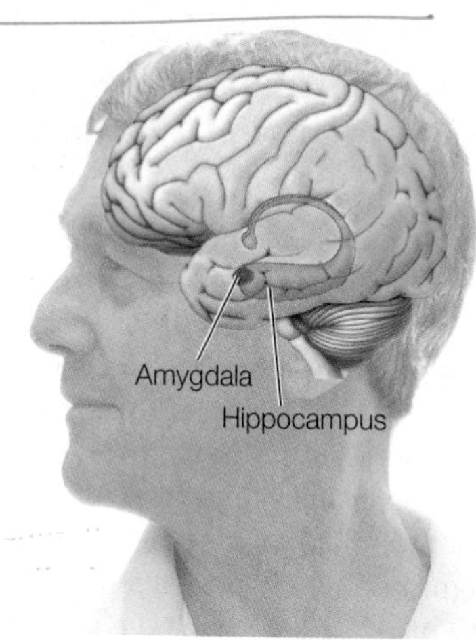

It is involved in spatial learning and memory formation. Studies have been carried out on living hippocampal tissue slices isolated from fully mature mice. Certain neuronal circuits can be stimulated in these preparations while neighboring circuits are not. Investigators have shown that when a circuit is repeatedly stimulated, certain of the postsynaptic structures in that particular circuit physically grow. Along with this physical response, the synapses also strengthen functionally. In this case, the number of active postsynaptic receptor molecules in each postsynaptic cell increases. Consequently, the cell responds more strongly to synaptic input from the presynaptic cell. Synaptic plasticity in this case has been shown to depend on second messengers, altered protein synthesis, and altered gene transcription in the postsynaptic cells. If you imagine some information being processed by circuits that are identical except that some have received a lot of prior stimulation and others have not, it is clear that the previously stimulated circuits will process the information differently, providing a basis for both learning and memory.

CHECKpoint CONCEPT 34.3

- At a vertebrate neuromuscular junction, what would happen if a poison blocked the mechanism that removes acetylcholine from the synapse after it has had its usual effect?
- Consider the arrival of a single excitatory presynaptic signal at a single chemical synapse. What is the difference between the magnitude of the postsynaptic graded potential at a vertebrate neuromuscular junction and at a neuron-to-neuron synapse, and why does it matter?
- What is meant by synaptic plasticity, and speaking very generally, how could it be a mechanism of learning?

We have seen how neurons use ion pumps and ion channels to generate action potentials, and how action potentials trigger neurotransmitter release, allowing neurons to communicate with each other and with other cells. Next we'll look at how systems of neurons can receive sensory information.

CONCEPT 34.4 Sensory Processes Provide Information on an Animal's External Environment and Internal Status

Animals require information on their external environments to function successfully. They need to perceive where they are going, and they need information to locate food, find mates, and avoid danger. Animals have evolved an enormous variety of sensory processes that enable them to gather information on their external environments. Imagine, for example, a male moth, an insect-eating bat, and a migrating bird moving across the landscape on a moonless, pitch-black night in late summer. The moth might be able to detect a female of his species, hundreds of yards away, by using sense organs that are extremely sensitive to odors that females release into the air. The bat might be able to zero in on him as he flies toward her by using echolocation. In this process, a bat emits bursts of vocal sound, at frequencies humans can't hear, and senses the echoes of these sounds that bounce off the moth. The moth, in turn, might use its own sense of hearing to detect the bat's vocalizations and take evasive action. Meanwhile the bird migrating overhead might use its eyes to locate a key star, which it can use as a navigational aid to fly south (see Concept 40.2).

Sensory processes are equally important for providing animals with information on their internal status. For instance, when we discussed breathing (see Chapter 31), we emphasized that the partial pressures of O_2 and CO_2 in an animal's blood are important indicators of whether it is breathing properly. Sensory processes have evolved that enable animals to measure their internal O_2 and CO_2 partial pressures so that they can adjust their rate of breathing. Other internal sensory processes provide animals with information on the amount of tension produced by their muscles during contraction. This information can be used to adjust stimulation of the muscles by the nervous system so that the tension is neither too high nor too low.

Sensory receptor cells transform stimuli into electric signals

Sensory processes depend on **sensory receptor cells**, which are cells, typically neurons, specialized to transform the energy of a stimulus—called stimulus energy—into an electric signal. Energy transduction is the process of changing one type of energy into another. A sensory receptor cell is said to carry out **sensory transduction** when it produces an electric signal from stimulus energy.

The electric signal that is produced then directly or indirectly causes action potentials to be generated in a neuron. The action potentials—which encode the initial stimulus—convey the sensory information to the brain or other integrative parts of the nervous system for processing and interpretation. In this way, the animal acquires a sensory perception of its external environment and its internal status.

Sensory receptor cells are highly specific in the stimuli to which they ordinarily respond. A receptor cell that responds to vibration does not respond to light, and vice versa. A cell that responds to sweet tastes responds to neither vibration nor light.

Some stimuli—such as light, heat, and mechanical motion—are clearly forms of energy. Thus when we speak of stimulus energy, our meaning is straightforward. But what do we mean when we speak of stimulus energy when the stimulus is a chemical compound such as a sweet-tasting molecule? In these cases we are not referring to the bond energy of the molecule (the energy that would be released by burning or respiration). Instead, in the case of chemical stimuli, when the stimulus molecule binds with a protein during sensory reception, a change takes place in the protein's conformation (molecular shape), and the energy involved in this process represents the stimulus energy.

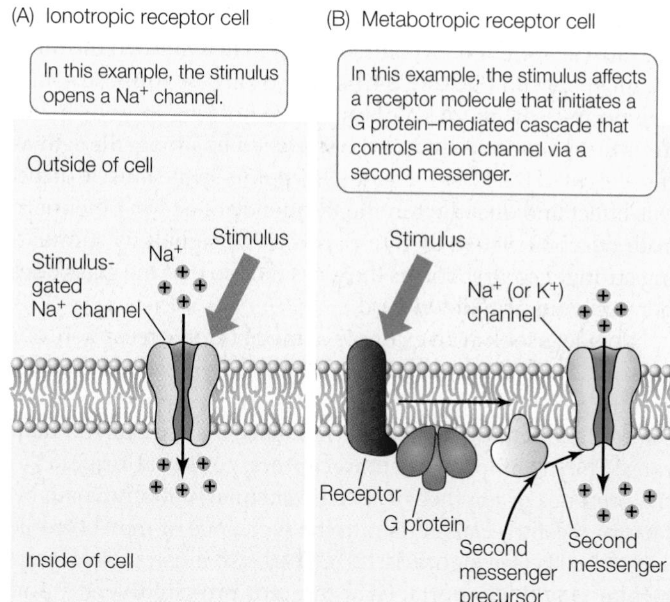

FIGURE 34.12 Sensory Receptor Cells Are Ionotropic or Metabotropic A metabotropic cell that affects an ion channel is illustrated here, but metabotropic cells do not always work in this way.

Sensory receptor cells depend on specific receptor proteins and are ionotropic or metabotropic

A **sensory receptor protein** is a membrane protein in a sensory receptor cell that initially detects a stimulus and directly or indirectly produces a graded change, called a receptor potential, in the cell's membrane potential. There are many types of sensory receptor proteins, and they respond to different kinds of stimuli. For example, some sensory receptor proteins respond to stretch. Others respond to particular chemicals. Still others respond to light. A receptor cell is specific in the sensory receptor protein it expresses, and this is one of the major reasons that receptor cells are highly specific in the stimuli to which they respond.

Receptor cells often have modified cell membranes with exceptionally large surface areas. The cell membrane, for example, may be greatly enlarged by the presence of microvilli, cilia, or membrane foldings. The enlarged cell membrane enables a receptor cell to have great numbers of receptor molecules, making the cell highly sensitive to stimuli:

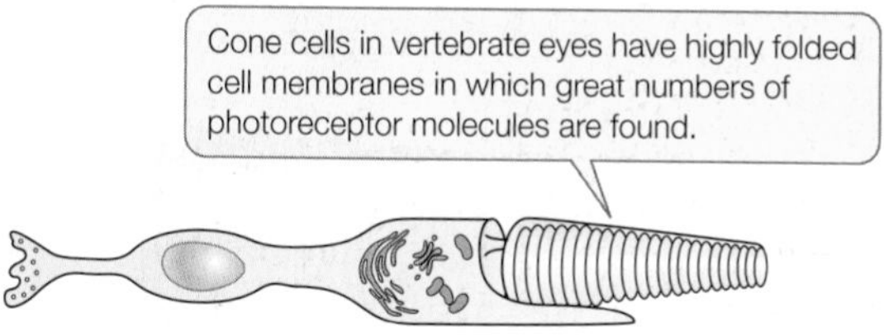

Cone cells in vertebrate eyes have highly folded cell membranes in which great numbers of photoreceptor molecules are found.

Sensory receptor cells may be ionotropic or metabotropic. The meanings of these terms in this context are similar to their meanings in the study of synapses (see Concept 34.3).

An **ionotropic receptor cell** typically has a receptor protein that is a stimulus-gated Na^+ channel (**FIGURE 34.12A**). What happens during stimulus reception? First, the stimulus acts on the receptor protein. Second, the protein changes conformation so that it operates as an open Na^+ channel, and Na^+ diffuses in through the open channel, creating a receptor potential: a graded change in the cell membrane potential. Third, the graded change of membrane potential directly or indirectly results in action potentials.

A **metabotropic receptor cell** typically has a receptor protein that activates a G protein when exposed to the stimulus (**FIGURE 34.12B**; see also Figure 5.14). In most cases, the receptor *is* a G protein–linked receptor. Activation of the G protein typically leads to a change in the concentration of a second messenger in the receptor cell. This change can directly or indirectly alter ion fluxes across the cell membrane, producing a receptor potential. The receptor cell itself may then generate action potentials, or it may secrete a neurotransmitter that affects another cell that produces action potentials.

Sensation depends on which neurons in the brain receive action potentials from sensory cells

All sensory systems convey information to the brain or other integrative parts of the nervous system in the form of action potentials. Accordingly, the brain receives action potentials from all types of sensory systems. How, then, do we perceive some of these action potentials as representing light, others as representing smells, and still others as representing sounds? How do we decode the action-potential codes?

Much of the answer lies in physical separation in the brain, between areas that receive axons from various sensory systems. In vertebrates, axons from the eyes travel to a specific brain region called the visual cortex. In turn, action potentials arriving in the visual cortex are interpreted as representing light or other visual sensations. Axons from the inner ear travel to another brain region called the auditory cortex, where arriving action potentials are interpreted as representing sounds. The intensity of the sensations (e.g., the brightness of a light or loudness of a sound) is coded by the frequencies of the action potentials.

We can see the importance of these principles on occasions when action potentials are generated by the wrong stimulus. People sometimes see flashes of light when they are poked in the eye. A strong poke can produce action potentials in axons from the retina that go to the visual cortex. Because the action potentials arrive in the visual cortex, the brain interprets them as representing light.

The parts of the brain that receive particular types of sensory information are determined during development, at least in part by interaction with the axons conveying the information. In a person blind from birth, for example, the visual cortex does not simply occupy space in the brain, unused, for life. Instead, much of the brain region is taken over to serve other sensory functions.

Sensation of stretch and smell exemplify ionotropic and metabotropic reception

Animals commonly have stretch receptor cells associated with their muscles. These are types of **mechanoreceptors**: cells that

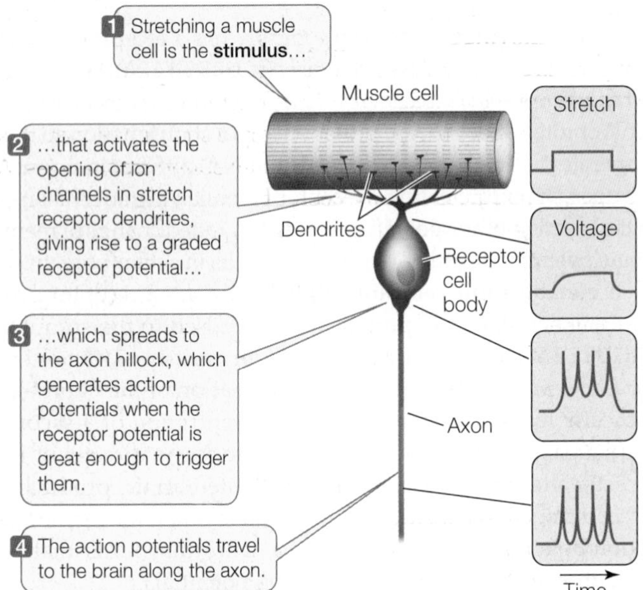

1 Stretching a muscle cell is the stimulus...
Muscle cell
Stretch
2 ...that activates the opening of ion channels in stretch receptor dendrites, giving rise to a graded receptor potential...
Dendrites
Voltage
Receptor cell body
3 ...which spreads to the axon hillock, which generates action potentials when the receptor potential is great enough to trigger them.
Axon
4 The action potentials travel to the brain along the axon.
Time

FIGURE 34.13 Stimulating a Crayfish Stretch Receptor Cell Produces Electric Signals Signal transduction in a stretch receptor cell of a crayfish can be investigated by measuring the membrane potential (voltage) at various places in the stretch receptor neuron (purple).

respond specifically to mechanical distortion of their cell membrane. When a muscle contracts, its stretching forces are applied to its stretch receptor cells, which provide the nervous system with information on the effectiveness of the contraction. Mechanoreceptors are typically ionotropic, as are thermoreceptors, which detect heat and cold.

A stretch receptor cell of a crayfish provides a model of how an ionotropic receptor cell generates action potentials. The receptor cell is a specialized neuron. When the cell membranes of its dendrites are stretched, stretch-gated Na^+ channels in the membranes open (see Figure 34.5B). Consequently, a graded, depolarizing receptor potential is produced and spreads to the axon hillock, and action potentials are generated there if the graded depolarization is above threshold (**FIGURE 34.13**). The rate at which action potentials are generated depends on the magnitude of the receptor potential, which depends on how much stretch the receptor cell detects.

Go to ANIMATED TUTORIAL 34.5
Mechanoreceptor Simulation
PoL2e.com/at34.5

Stretch receptors associated with our biceps help us adjust the muscle's strength of contraction to match the load the muscle must sustain (**FIGURE 34.14**). For example, suppose someone pours a drink into a glass you're holding in your hand. Your biceps muscle tends to be stretched to a longer length as the weight of the glass increases. Receptors in the muscle detect this effect and encode it in the frequency of action potentials. Your nervous system then responds to the signals by stimulating stronger contraction as the glass fills, so that the glass does not pull your arm downward.

Now let's look at an example of metabotropic reception, our sense of smell. Chemicals detected by smell are called odorants, and the sense of smell is called olfaction. The sensory cells for odorants, called olfactory receptor cells, are specialized neurons. They exemplify **chemoreceptors**: cells that respond to the presence or absence of specific chemicals. Mammalian olfactory receptor cells illustrate the workings of metabotropic sensory cells (see Figure 34.12B). They also illustrate in a spectacular way the importance of receptor protein diversity and specificity.

The olfactory receptor cells of a mammal are found in the olfactory epithelium, a part of the lining of the internal nasal cavity. One end of each receptor cell projects into the mucus that covers the surface of this epithelium. The other end of the cell is an axon that extends into a part of the brain called the olfactory bulb (see Figure 34.25). Odorants must dissolve in the liquid of the mucus to be detected.

The receptor protein of a mammalian olfactory receptor cell is a G protein–linked receptor found in the part of the cell that projects into the mucus. When an odorant molecule binds to the receptor, the receptor activates a G protein. The G protein in turn activates an enzyme that causes an increase of a second messenger in the cytoplasm. The second messenger opens ion channels in the cell membrane, causing a graded

FIGURE 34.14 Negative Feedback: Sensory Signals from Stretch Receptors Enable the Nervous System to Control Muscle Contraction In vertebrates, tiny sensory organs called muscle spindles send information to the nervous system about how much a muscle is being lengthened by stretching. This information is used here to keep limb position stable by increasing the contractile force as the downward force increases.

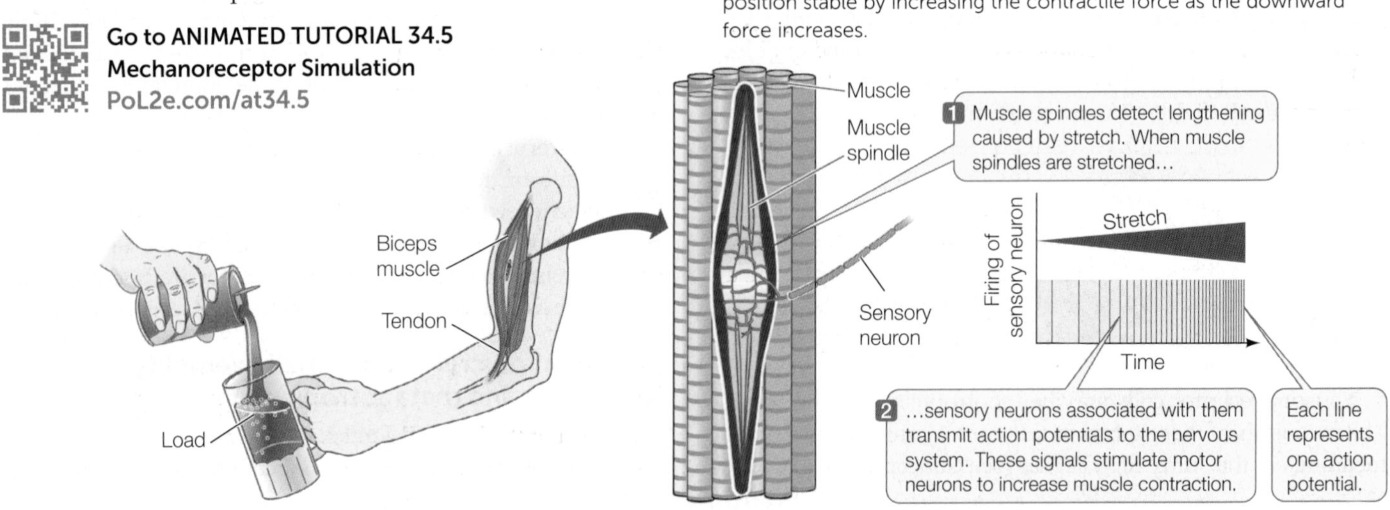

depolarization, which in turn causes action potentials to be generated in the cell. These propagate into the brain by way of the cell axon.

One of the amazing aspects of these cells is the number of different G protein–linked receptor proteins (GPLRs) they can express. Each cell expresses just one type of GPLR and can detect only the odorant molecules that bind to that GPLR. However, the genome codes for many options. About 1,000 different GPLRs are found in mice, for example. Collectively, the tens of thousands of olfactory receptor cells—each with one type of GPLR—can detect a very wide range of odorants. Each specific type of GPLR typically can bind to a range of odorants. However, different types of GPLRs have different sets of odorants they bind. The combination of olfactory cells that are stimulated by any particular odorant is unique to that odorant. Consequently, the brain can interpret the pattern of signals from the olfactory cells as pointing to a particular smell.

Some animals have olfactory receptor cells of extraordinary specificity. Because these cells effectively respond to only one ordinary odorant, they can be extremely sensitive to it. The feathery antennae of some male moths have huge numbers of odorant receptor cells that respond in a highly specific way to pheromone molecules emitted by the females of their species (see Concept 40.4). These moths can detect a female when her pheromone molecules represent only 0.000000000000001 percent of the molecules in the air:

A male moth can detect the female pheromone at very low concentations in the air passing over his antennae.

Auditory systems use mechanoreceptors to sense sound pressure waves

Sounds consist of waves of compression of the air. An easy way to understand the nature of sound is to envision a loudspeaker. A close look reveals that the membrane of a loudspeaker moves back and forth. When the membrane moves toward you, it compresses the air, creating a region of relatively high pressure that moves toward you like a wave (at the speed of sound). When the speaker membrane moves away from you, it creates a region of relatively low pressure that also moves toward you between the regions of high pressure. You thus experience a sound pressure wave of alternating high and low pressure. The frequency of these alternations accounts for tone or pitch. We perceive high frequencies as high-pitch, treble tones and low frequencies as low-pitch, bass tones.

The hearing organs of many different types of animals are based on a single principle. The hearing organ contains a thin membrane that is free to move when a sound pressure wave strikes it—moving in when high pressure strikes and out when low pressure strikes. Specialized mechanoreceptors are linked to the membrane and detect the movements.

In an insect, the membrane is a simple structure in which a single area of membrane responds to all sound frequencies. A system of this sort is mainly useful for detecting sound presence or absence (and loudness), rather than discriminating one frequency from another.

The hearing system of mammals, by contrast, can separately detect many different frequencies. When we listen to someone playing the piano, our brain receives specific information on each frequency in the music, allowing us to perceive high notes as distinct from low notes.

In the ear of a mammal, the membrane that initially responds to sound pressure waves is the tympanic membrane (ear drum). On the inside of the tympanic membrane is an air-filled cavity called the middle ear (**FIGURE 34.15A**). Because the tympanic membrane has air on both sides, it is free to vibrate when a sound pressure wave strikes it. Three interconnected, delicate bones—called ossicles and named the malleus, incus, and stapes—are found in the middle ear (**FIGURE 34.15B**). They act as a lever system that turns vibrations of the tympanic membrane into more forceful vibrations of another flexible membrane called the oval window.

The oval window is part of the inner ear. Specifically, it is part of the **cochlea**, a coiled, fluid-filled tube where sound energy is transduced into electric signals. A cross section of the cochlea reveals that it is composed of three parallel canals separated by two membranes. Of these the **basilar membrane** is of particular importance (**FIGURE 34.15C**).

The basilar membrane gradually changes in width and stiffness from one end to the other. At the end nearest the oval window, it is only about 0.04 millimeters wide in the human ear and relatively stiff. It becomes progressively wider and more flexible until, at its far end, it is 0.5 millimeters wide and not as stiff.

When the sound pressure wave is transmitted into the cochlear fluid at the oval window, different frequencies in the sound pressure wave cause different, local regions of the basilar membrane to oscillate. Different regions move in response to different frequencies because of the gradations of membrane width and stiffness. Low-frequency vibrations (representing low pitches) have their greatest effect on the wide end of the membrane farthest from the oval window, setting that part into motion to a far greater degree than other parts (**FIGURE 34.16A**). Other frequencies affect other parts of the membrane (**FIGURE 34.16B**).

How is information on these oscillations sent to the brain? Sitting on the basilar membrane is a structure, the organ of Corti, that houses distinctive mechanoreceptor cells called hair cells (**FIGURE 34.15D**). Each hair cell bears extensions called stereocilia (the "hairs," which are actually microvilli, not cilia) that project into a semirigid membrane, the tectorial membrane. When a particular part of the basilar membrane oscillates, the stereocilia of the hair cells in that region are bent because they are effectively anchored at one end in the tectorial membrane but moving at the other end.

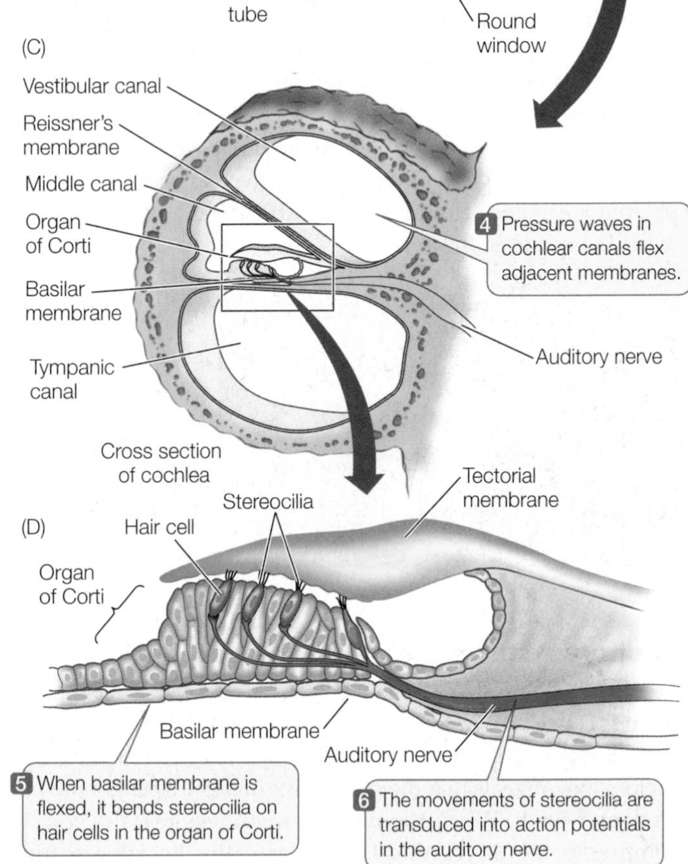

FIGURE 34.15 The Human Ear The ear produces action potentials that are sent to the brain, permitting environmental sound pressure waves to be perceived.

Go to ACTIVITY 34.2 Structures of the Human Ear PoL2e.com/ac34.2

The hair cells transduce the bending motions of their stereocilia into electric signals. They do not produce action potentials. Instead, they synapse with neurons, and when they are stimulated by the bending of their stereocilia, they release a neurotransmitter that excites the neurons to produce action potentials, which travel to the brain. In this way, the frequency composition of a sound pressure wave is encoded by the particular hair cells that trigger action potentials as the wave exerts its effect in the inner ear (see Figure 34.16A).

The photoreceptors involved in vision detect light using rhodopsins

Both simple and complex animals have visual systems by which they sense and respond to light. The simplest visual systems give animals the ability to orient to the sun and sky. Complex visual systems provide animals with rapid and extremely detailed visual information about objects in their environments. Remarkably, the basic process of detecting light in visual systems has been conserved across the entire animal kingdom. **Photoreceptors** are sensory receptor cells that are sensitive to light. In all photoreceptors employed in vision, the sensory receptor protein is a member of a family of closely similar membrane pigments called **rhodopsins**. The rhodopsins act as G protein–linked receptors, and the photoreceptors are metabotropic receptor cells.

Detection of light depends on the ability of rhodopsins to absorb photons of light and undergo a change in molecular conformation (shape) as a consequence. A rhodopsin molecule consists of two parts: a protein, opsin (which alone is not photosensitive), and a nonprotein light-absorbing structure, 11-*cis*-retinal. The retinal part of each molecule is derived from vitamin A, explaining why sufficient vitamin A is essential for proper vision.

A rhodopsin molecule changes to a different isomer (arrangement of atoms in three-dimensional space) when it

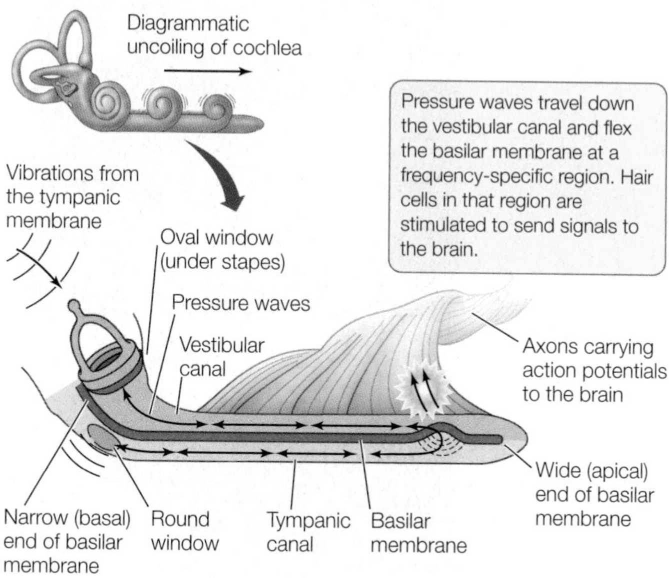
(A) Oscillation of the basilar membrane in response to a low-frequency sound
Diagrammatic uncoiling of cochlea
Pressure waves travel down the vestibular canal and flex the basilar membrane at a frequency-specific region. Hair cells in that region are stimulated to send signals to the brain.
Vibrations from the tympanic membrane
Oval window (under stapes)
Pressure waves
Vestibular canal
Axons carrying action potentials to the brain
Wide (apical) end of basilar membrane
Narrow (basal) end of basilar membrane
Round window
Tympanic canal
Basilar membrane

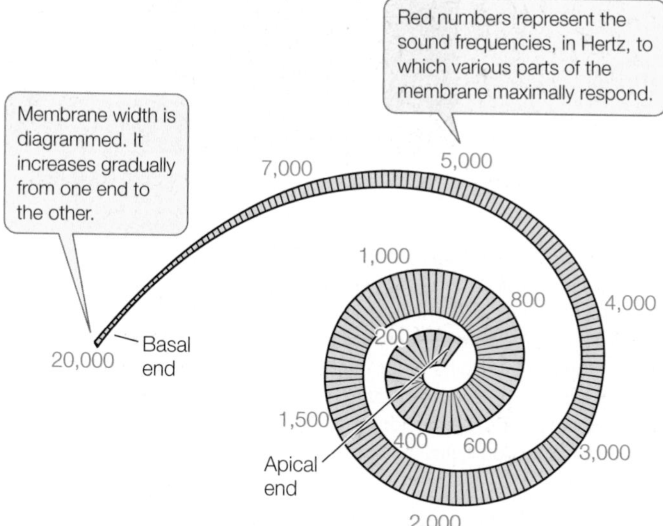
(B) The human basilar membrane in its coiled form
Red numbers represent the sound frequencies, in Hertz, to which various parts of the membrane maximally respond.
Membrane width is diagrammed. It increases gradually from one end to the other.
7,000
5,000
1,000
800
4,000
200
20,000
Basal end
1,500
400
600
3,000
Apical end
2,000

FIGURE 34.16 Frequencies in Sound Pressure Waves Are Transduced into Signals from Distinctive Sets of Hair Cells (A) Each frequency causes oscillations at a particular region of the basilar membrane. Here low-frequency sound pressure waves are shown flexing a region of the basilar membrane far from the oval window. (B) Different regions of the basilar membrane move in response to different frequencies because of the gradations in the membrane's width and stiffness. Hair cells positioned along the basilar membrane produce signals that lead to generation of action potentials that encode the regions that are oscillating.

Go to ANIMATED TUTORIAL 34.6
Sound Transduction in the Human Ear
PoL2e.com/at34.6

absorbs light. When the 11-*cis*-retinal in a molecule of rhodopsin absorbs light, it undergoes a twist around one of its bonds, so its atoms become reoriented in space to the all-*trans*-retinal isomer:

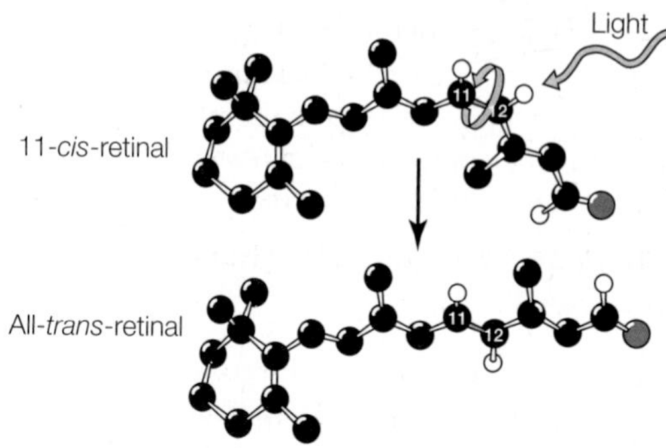
Light
11-*cis*-retinal
All-*trans*-retinal
11
12

This change places strains on bonds in the opsin part of the molecule, changing the conformation of the opsin. The rhodopsin is thus activated. It in turn activates a G protein–mediated signal transduction cascade, in which a second messenger exerts effects on voltage-gated Na^+ channels in the photoreceptor cell membrane.

Go to ANIMATED TUTORIAL 34.7
Photosensitivity
PoL2e.com/at34.7

Metabotropic control has a huge potential advantage in that it can greatly amplify signals. With large numbers of rhodopsin molecules and strong amplification, some photoreceptor cells undergo a measurable change in membrane potential in response to just a single photon of light!

LINK
Signal amplification by metabotropic control is described in **Concept 5.6**

The vertebrate retina is a developmental outgrowth of the brain and consists of specialized neurons

Vertebrates have image-forming eyes. Like a camera, an image-forming eye has a lens that focuses images on an internal surface that is sensitive to light. The eye is a fluid-filled structure bounded by a tough connective tissue layer, as exemplified by the human eye:

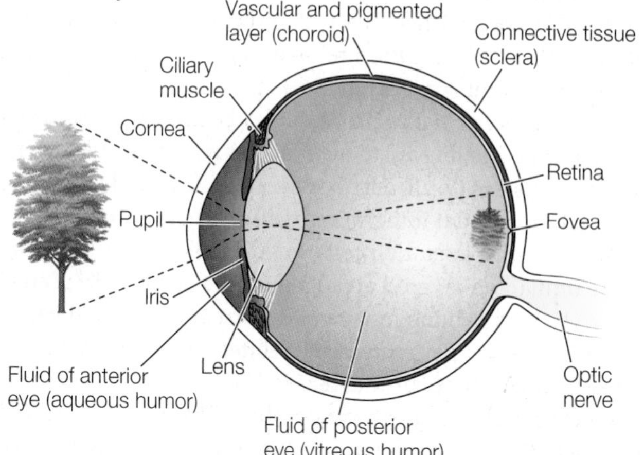
Vascular and pigmented layer (choroid)
Connective tissue (sclera)
Ciliary muscle
Cornea
Retina
Fovea
Pupil
Iris
Lens
Fluid of anterior eye (aqueous humor)
Fluid of posterior eye (vitreous humor)
Optic nerve

At the front of the eye, the connective tissue forms the transparent **cornea**. Just inside the cornea is the colored **iris**. Light enters the light-sensing part of the eye through the **pupil**, the central opening of the iris. The iris is equipped with small muscles that enlarge the diameter of the pupil in dim light and reduce it in bright light. Behind the iris is the **lens**, made of crystal-clear proteins called crystallins. In mammals, muscle cells pulling on the lens produce fine adjustments in its shape, focusing images on the photosensitive layer—the **retina**—at the back of the eye.

Go to ACTIVITY 34.3 Structure of the Human Eye
PoL2e.com/ac34.3

The retina is a developmental outgrowth of the brain and consists of several types of specialized neurons. Two types, the **rod** cells and **cone** cells, are the photoreceptor cells responsible for vision (**FIGURE 34.17A**). Both are specialized to have a large membrane area and large numbers of rhodopsin molecules. All rhodopsin molecules do not respond to the same wavelengths of light, even though their photoreceptive molecular subpart, the retinal, is the same. They differ in their opsins.

Rods outnumber cones in the human retina: 100 million rods versus 5 million cones. The rods are far more sensitive to light than the cones and are especially important for vision in dim light. Cones provide high-acuity color vision. Humans and some other primates have three types of cones, which are distinguished by having three different molecular forms of rhodopsin. These forms differ from one another in their sensitivities to various wavelengths of light, and all three differ from the single form of rhodopsin in rod cells (**FIGURE 34.17B**). As a group, the cone cells are able to detect and encode color because each wavelength of light has a unique collective effect on the three types.

Rod and cone cells do not produce action potentials. They only produce graded membrane potentials. Compared with other sensory receptor cells, they are unusual in that they hyperpolarize in response to their stimulus. How does this work? When in darkness, they have *open* Na^+ channels in their cell membranes. Na^+ ions entering by way of these channels depolarize the cells to some extent. When the cells are stimulated by light, the second messenger produced under G protein control *closes* Na^+ channels. This closing lowers the Na^+ current flowing into the cell, permitting the charge on the inside of the cell membrane to become more negative relative to the charge on the outside. This effect is graded, depending on the light intensity (**FIGURE 34.18**). The rod and cone cells synapse with other neurons in the retina and relay signals to them by way of graded neurotransmitter release.

A great deal of signal processing occurs in the retina before visual signals are relayed to the brain by way of axons in the optic nerve. Besides the visual photoreceptor cells (rods and cones), four types of visual integrating neurons—called bipolar cells, horizontal cells, amacrine cells, and ganglion cells—are found in the retina in a layered arrangement (**FIGURE 34.19**; see also Figure 29.1). A striking feature of the arrangement of these cells is that light must pass through the integrating cells (which are transparent) to stimulate the rods and cones! This unexpected arrangement is a by-product of the embryonic developmental path by which the eyes form from the brain. The photoreceptor cells

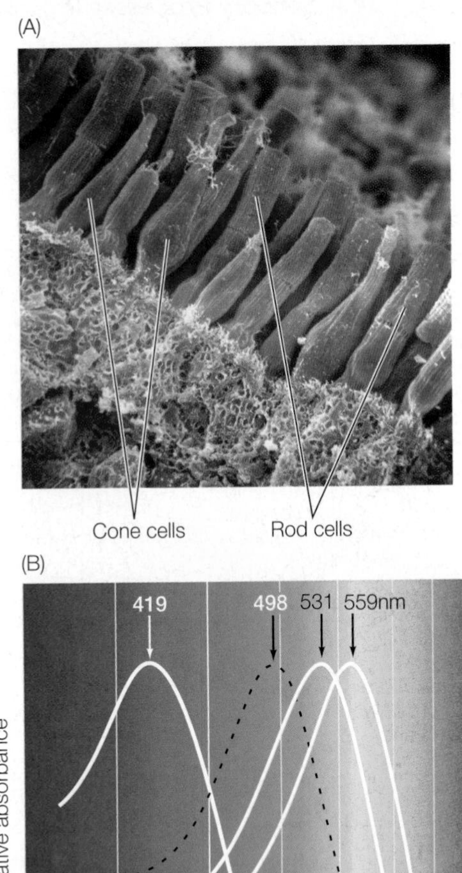

FIGURE 34.17 Rods and Cones (A) This scanning electron micrograph of photoreceptor cells in the retina of a mud puppy (an amphibian) shows cylindrical rods and tapered cones. The cells are artificially colored to distinguish them. (B) The three types of cone cells in the human retina differ in their absorption of the various wavelengths in the spectrum of light, and they therefore differ in their sensitivities to various wavelengths. The dashed line shows the spectral response of rod cells.

(rods and cones) synapse with bipolar cells, which in turn synapse with ganglion cells. In contrast to the cells in these straight-through connections from the back to the front of the retina, the horizontal cells and amacrine cells communicate laterally within the retina and in this way are able to extract information on the relative signal strengths in neighboring photoreceptor cells and neighboring bipolar cells. Horizontal cells help sharpen the perception of contrast between light and dark. Amacrine cells have several functions, including detecting motion. Ultimately, all of this information converges onto the ganglion cells, which are the only cells that produce action potentials. Axons of these cells converge to form the optic nerve and carry the action potentials to the visual cortex in the brain.

Each ganglion cell is said to have a defined, circular receptive field formed by the cells from which it receives signals. Some ganglion cells are excited by light falling on photoreceptor cells at the center of their receptive field and inhibited by

INVESTIGATION

FIGURE 34.18 A Rod Cell Responds to Light The cell membrane of a rod cell hyperpolarizes—develops a more negative membrane potential—in response to a flash of light. Rod cells develop graded receptor potentials in response to the absorption of light energy.[a]

HYPOTHESIS

When a rod cell absorbs photons (light energy), its membrane potential changes in proportion to the strength of the light stimulus.

METHOD

1. Record membrane potentials from the inner segment of a rod cell.
2. Stimulate the rod cells with light flashes of varying intensity and record the results.

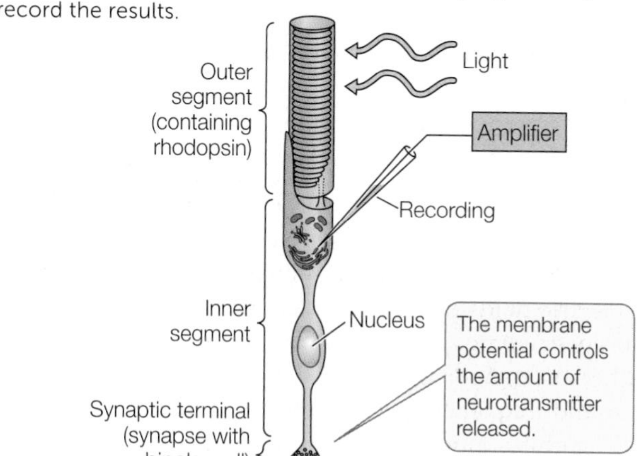

RESULTS

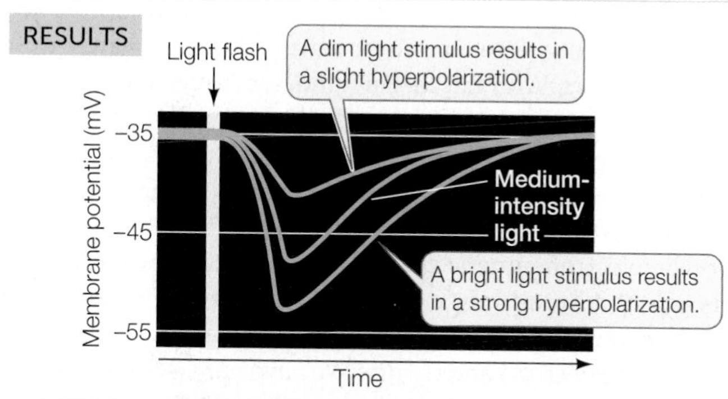

CONCLUSION

The membrane potential of rod cells is depolarized in the dark and hyperpolarizes (becomes more negative) in response to light.

Go to **LaunchPad** for discussion and relevant links for all **INVESTIGATION** figures.

[a]D. A. Baylor et al. 1979. *Journal of Physiology* 288: 589–611.

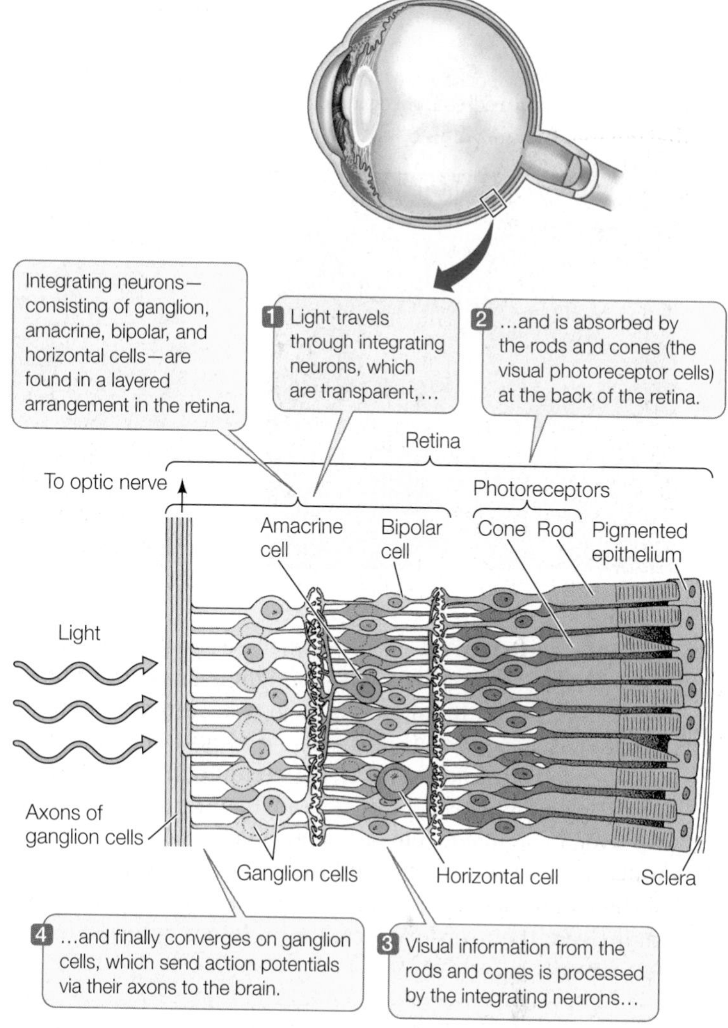

FIGURE 34.19 The Human Retina The human retina consists of the visual photoreceptor cells (rods and cones) and four types of integrating neurons found, in a layered arrangement, at the back of the eye.

Go to ACTIVITY 34.4 Structure of the Human Retina PoL2e.com/ac34.4

Go to MEDIA CLIP 34.2 Into the Eye PoL2e.com/mc34.2

light falling on photoreceptor cells in the periphery. Other ganglion cells are inhibited by light falling in the center of their receptive field and stimulated by light falling on the periphery. The result is that the ganglion cells can communicate information to the brain about visual patterns such as spots, edges, and areas of contrast.

The huge numbers of cells involved in vision say a lot about the amount of processing of visual images that takes place in the retina and brain. Each human retina sends a million axons to the brain. The visual cortex consists of hundreds of millions of neurons that receive and process the information from those million axons.

Some retinal ganglion cells are photoreceptive and interact with the circadian clock

One of the most revolutionary, recent discoveries about the mammalian eye is that 1–2 percent of the retinal ganglion cells are photoreceptive. Most ganglion cells (the other 98–99 percent) cannot directly detect light and function only to help integrate visual information that originates in the rods and cones.

The photoreceptive ganglion cells are not employed in vision. They have a relatively low sensitivity to light and do not produce images.

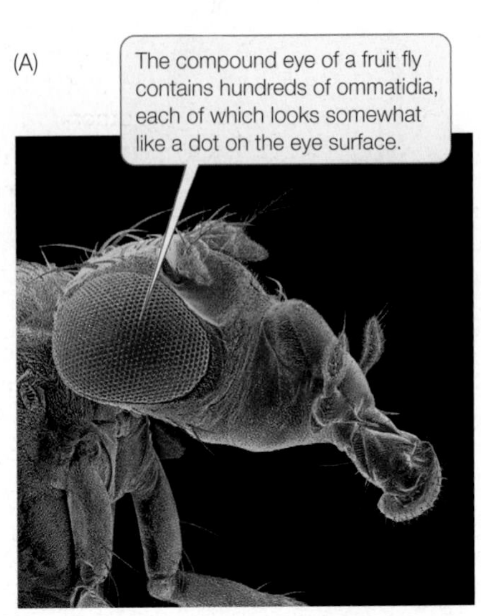

(A)
The compound eye of a fruit fly contains hundreds of ommatidia, each of which looks somewhat like a dot on the eye surface.

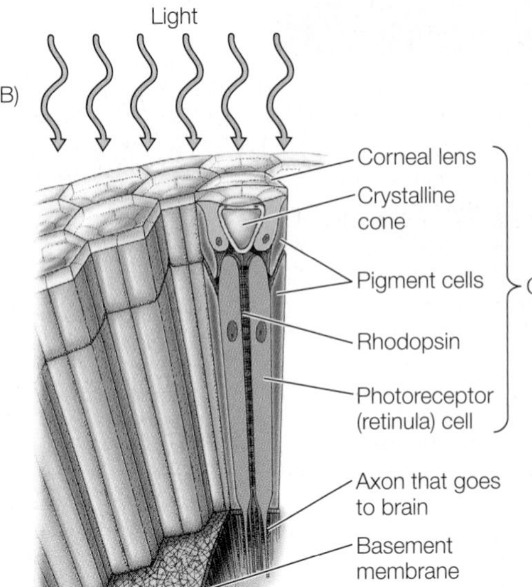

FIGURE 34.20 Ommatidia: The Functional Units of Arthropod Eyes (A) The micrograph shows the compound eye of a fruit fly. (B) The rhodopsin-containing retinula cells are the photoreceptors in ommatidia.

Rather than contributing to vision, these ganglion cells provide information simply on whether light is present and how bright it is. This information is used in two ways. It is relayed to the suprachiasmatic nuclei, where it is used to entrain the master circadian biological clock to the environmental rhythm of day and night (see Concept 29.6). The information is also employed in regulating the size of the pupils of the eyes, so the pupils are large in dim light and small in bright light.

The photoreceptive ganglion cells do not use rhodopsin. Instead, their sensory receptor protein is believed to be melanopsin.

Arthropods have compound eyes

Invertebrates have evolved a variety of types of eyes that differ from vertebrate eyes. Even the eyes of octopuses and squid, which look very similar to vertebrate eyes on the outside, have very different properties than vertebrate eyes.

The eyes of arthropods (insects, crayfish, crabs, and lobsters) are called **compound eyes** because each eye consists of many optical units, called **ommatidia** (singular *ommatidium*), each with its own lens. Light entering an ommatidium through its lens strikes photoreceptor (retinula) cells that contain rhodopsin and generate action potentials (**FIGURE 34.20**).

Each ommatidium of a compound eye is directed at a slightly different part of the visual world. The more ommatidia in the eye, the higher the resolution of the image compiled by the brain, rather like the pixel density in a computer monitor. The number of ommatidia in a compound eye varies from only a few in some ants to 800 in fruit flies and to 30,000 in some fast-flying predators such as dragonflies.

Animals have evolved a remarkable diversity of sensory abilities

Vision, hearing, olfaction, and touch are almost universal in animals. Animals use these sensory abilities to gather information about their external environment so they can move about effectively, avoid dangers, and exploit opportunities. Some animals, in addition, have evolved unusual sensory abilities that let them make use of additional information sources, such as electric fields or Earth's magnetic field (see Figure 40.12). Moreover, some animals are able to use vision or hearing to gather information from parts of the electromagnetic spectrum or sound-frequency spectrum that are unfamiliar to us.

The philosopher Bertrand Russell titled one of his most famous books *Our Knowledge of the External World.* In this book he stressed that, as humans, we can only know about those aspects of our external environment that our human sensory systems detect. With this fact in mind, we realize that some animals that live around us perceive our external environment in very different ways than we do. They can see electromagnetic wavelengths to which we are blind or hear sound frequencies to which we are deaf.

The electromagnetic spectrum consists of electromagnetic energy at a broad range of wavelengths from 0 to over 100 micrometers (μm) (see Figure 6.16). Our eyes can see the wavelengths between 0.4 and 0.7 μm. We have names for the parts of the spectrum we cannot see: the wavelengths that are invisible to us even when our eyes are healthy. The wavelengths that are too long for our eyes to see (i.e., longer than 0.7 μm) are called **infrared wavelengths**. The wavelengths too short for our eyes to see (i.e., shorter than 0.4 μm) are called **ultraviolet wavelengths**.

Some snakes obtain vision-like images of their surroundings using infrared wavelengths. They are therefore able to see patterns of warmth and coolness in darkness—an ability that helps them find prey. The best-understood snakes with this ability are the pit vipers, such as rattlesnakes, which have specialized sense organs in pits near their eyes. The organ in each pit focuses infrared electromagnetic energy onto a membrane that can detect it, much like our eyes focus visible light onto our retina. Unlike the retina, the pit organ does not employ rhodopsins or similar molecules. Biologists still do not fully understand the sensory mechanism by which infrared electromagnetic energy is detected.

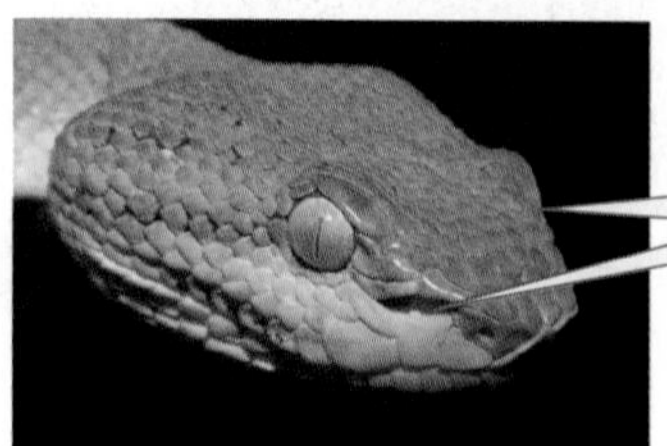

This bamboo pit viper (*Trimeresurus gramineus*) has a sense organ that detects infrared wavelengths, located in a pit on each side of its head.

At the other end of the electromagnetic spectrum, many animals have eyes that can see ultraviolet wavelengths to which we are blind. Honey bees, for example, see flowers in different ways than we see them because they see in the ultraviolet. Many flowers that look uniform in color to our eyes have bold color patterns at the ultraviolet wavelengths bees see (see Figure 21.24). Many species of birds also see ultraviolet wavelengths. The eyes of these birds look much like ours. However, unlike the case in our eyes, their cornea and lens are transparent to ultraviolet wavelengths, permitting ultraviolet light to enter the eye, and the eye has specialized cone cells receptive to ultraviolet wavelengths.

The discovery of ultraviolet vision in birds led researchers to wonder, If the males and females of a species look identical to us, do they also look identical to the birds? Cedar waxwings (*Bombycilla cedrorum*), found almost everywhere in North America, are one species studied to answer this question. Male and female waxwings look the same to us. But we see only at 0.4–0.7 μm. When we use instruments to learn what the waxwings look like at shorter, ultraviolet wavelengths, we find that males and females look different. Thus the birds themselves can tell males and females apart by sight even though we cannot. This is equally true for many other species.

Some animals can sense electric fields—fields of voltage and current—enabling them to perceive their surroundings using information that's entirely foreign to us. The weakly electric fish provide examples. These fish produce weak electric fields in their surroundings by use of modified muscles (see Concept 33.4). Electric currents in these self-produced environmental fields pass through the outer skin layers of the fish, where sensory receptor cells detect the currents. The fish detect distortions in these fields caused by objects in their external environment; by analyzing these distortions, the fish can perceive the objects even in complete darkness. Fish in the genus *Gnathonemus* use this sensory ability to locate and capture stream insects at night.

African fish of the species *Gnathonemus petersii* are nocturnal. They find food using self-produced electric fields.

CHECKpoint CONCEPT 34.4

- ✓ As a mosquito lands on your arm, a cell in the mosquito's leg responds to altered stretch within the leg. Another cell in its antennae detects the CO_2 exhaled by you and other animals. What types of sensory receptor proteins do these cells probably contain?
- ✓ Referring to the structure of the vertebrate retina, explain how it is possible for visual signals to be processed extensively in the retina itself.
- ✓ A range of sound frequencies is changed to a range of locations in the vertebrate cochlea. How does this happen?

Now that we have discussed cellular processes from action potentials to receptor function, let's examine how cells are organized to form the nervous systems of animals.

CONCEPT 34.5 Neurons Are Organized into Nervous Systems

A nervous system consists of many neurons, interacting with each other by way of synapses. When we say "many," we mean a huge number. The human brain is currently estimated to contain about 86 billion neurons!

The simplest nervous systems seen in animals alive today occur in anemones, corals, jellyfish, and other cnidarians. In these animals, the individual neurons are widely and relatively randomly dispersed throughout the body. Together they form a nerve net:

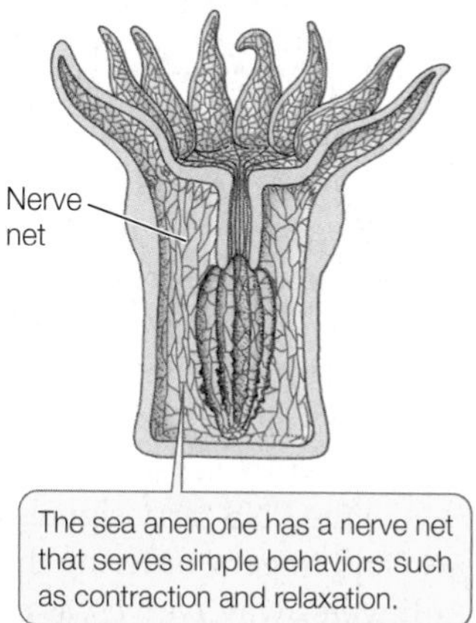

The sea anemone has a nerve net that serves simple behaviors such as contraction and relaxation.

Biologists generally assume that when nervous systems first evolved they were similar to nerve nets.

As nervous systems evolved further, they followed two major trends, centralization and cephalization. Some neurons became specialized to integrate the signals of other neurons. Most of these integrating neurons became clustered together in centralized integrating organs, rather than being distributed randomly throughout the body. **Centralization** refers to this evolutionary trend toward clustering. At the same time, the major integrating areas became concentrated toward the anterior end of the animal's body, a trend called **cephalization** (from a Greek word for "head"). Mammals vividly exemplify the outcomes of these evolutionary trends. Most of the integrating neurons of a mammal are centralized in the brain and spinal cord, and the major integrating organ—the brain—is at the anterior end of the body in the head. Why has cephalization occurred? Most kinds of animals have particularly important sets of sense organs—such as eyes, ears, and sensory tentacles—at their anterior end. This is easy to understand because when an animal moves forward in its environment, its anterior end is the first to make contact with new dangers or opportunities. The concentration of integrating parts of the nervous system at the anterior end facilitates having a concentration of sense organs there.

Here we focus on the groups of animals with high degrees of centralization and cephalization. The nervous system in such animals consists of two major parts, the central and peripheral systems. The **central nervous system** (**CNS**) consists of relatively large structures that are composed principally of integrating neurons and their associated glial cells. The largest part of the CNS—containing the greatest number of neurons—is the **brain**.

The CNS functions as a control system, and like all control systems it needs to interact with sensors and effectors to function (see Concept 29.6). The sensors are the sensory cells and sense organs that we've just discussed. They provide the CNS with information on the animal's external environment and internal status. **Effectors** are cells or tissues that perform actions, that "carry out orders." Muscle cells are effectors because they can contract and produce motion under orders from the nervous system. Other types of effectors include glands, some of which secrete materials in response to signals from the nervous system, and specialized structures such as fish electric organs (which produce high currents or voltages when ordered to do so by the nervous system).

The **peripheral nervous system** (**PNS**) consists of neurons and parts of neurons that are located outside the CNS. These neurons are mostly engaged in two functions related to sensors and effectors. They bring sensory information from sense organs to the CNS. They also carry orders from the CNS to muscle cells or other effectors. For instance, if you touch a hot object with your fingers, neurons in the PNS carry sensory information from temperature sensors in your fingers to the CNS. Other neurons in the PNS carry signals from the CNS to muscles in your arm and fingers, ordering the muscles to contract and pull your fingers away. Nerves, as mentioned before, are bundles of axons grouped together in the PNS, analogous to telephone cables containing many wires.

Neurons are classed as interneurons, sensory neurons, or efferent (including motor) neurons.

- **Interneurons** are neurons that are confined to the CNS. Most neurons in the CNS are interneurons. For the most part, interneurons engage in integrative and information-storage functions.
- **Sensory neurons** function as sensory receptor cells or are neurons (also called afferent neurons) that carry signals to the CNS from sensory cells or organs.

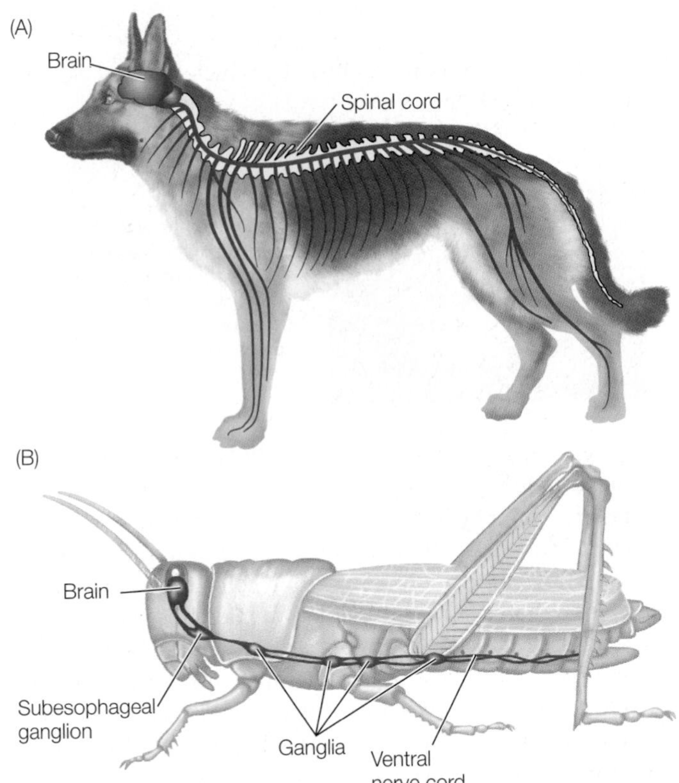

FIGURE 34.21 The Organization of Vertebrate and Arthropod Nervous Systems Only a few of the ganglia are shown in (B).

- **Efferent neurons** convey signals from the CNS to muscles or other effectors such as glands. Among these neurons, those that carry signals to skeletal muscles are called **motor neurons**.

The cellular components of nervous systems—the neurons, synapses, and sensory cells—are remarkably similar in the fundamental ways they function in all groups of animals. However, some animals differ substantially from others in the ways their neurons are organized to form their nervous systems. Let's briefly compare vertebrates and arthropods to see this point.

In vertebrates, the central nervous system is positioned in the dorsal part of the body (**FIGURE 34.21A**). It consists of two major parts: the brain in the head and the spinal cord that extends for most of the length of the body behind the brain. Both of these parts are to some degree hollow inside. In short, the vertebrate CNS is dorsal and hollow. The neurons and glial cells in the spinal cord form a continuous column of tissue, in the sense that the cord does not exhibit sharply differentiated subparts along its length.

In arthropods, such as insects and crayfish, the CNS is mostly positioned in the ventral part of the body (**FIGURE 34.21B**; see also Figure 32.4). It too consists of two major parts: the brain, and a ventral nerve cord that extends for most of the length of the body behind the brain. These structures are not hollow. In other words, the arthropod CNS is mostly ventral and solid. Moreover, instead of being a continuous column of tissue, the ventral nerve cord is composed of a series of alternating subparts. Arthropods are segmented (see Concept 23.1), and there is an enlarged structure called a ganglion (plural, ganglia) in each segment of the animal's body. Two strands of nervous tissue, called connectives, run from each ganglion to the next. (A word of warning: the

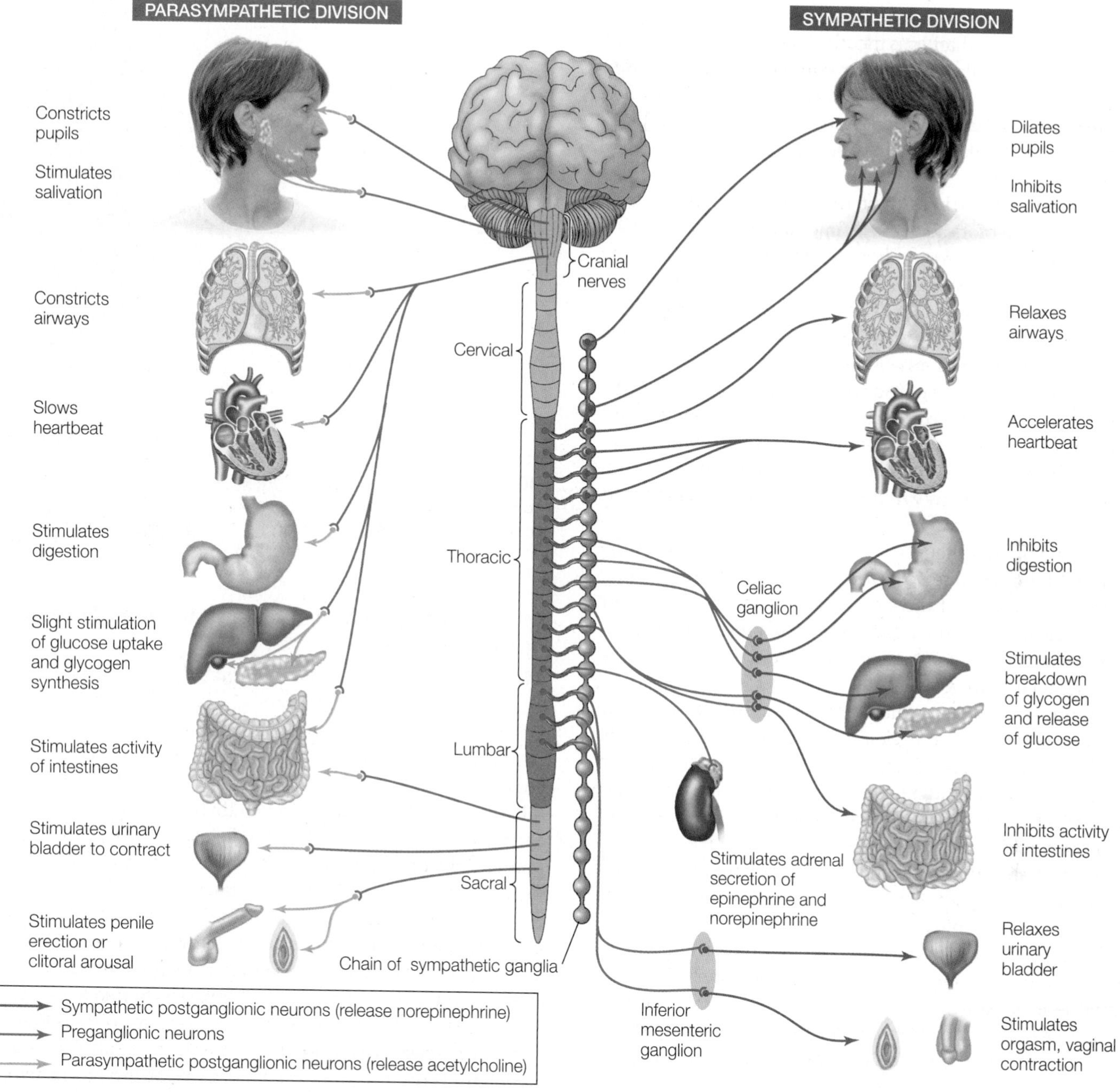

FIGURE 34.22 The Autonomic Nervous System The sympathetic and parasympathetic divisions work in opposition to each other in their effects on most organs. One stimulates an increase in activity, and the other causes a decrease in activity. The preganglionic neurons of both divisions mostly release acetylcholine.

word "ganglion" is also used in describing vertebrate nervous systems, but it has a very different meaning in discussing vertebrates and arthropods, as we'll soon see.)

The autonomic nervous system controls involuntary functions

The nervous system as a whole controls several types of effectors: skeletal muscles, smooth muscles, glands, and so forth. An important distinction is recognized between the part of the nervous system that controls skeletal muscles and the part that controls all other effectors. This distinction is entirely different from the one between the CNS and PNS.

The **autonomic nervous system** (**ANS**) is the part of the nervous system (both CNS and PNS) that controls effectors other than the skeletal muscles. These effectors are called **autonomic** (or internal) **effectors**. We generally are unaware that they are being controlled, and we have little or no ability to influence them by conscious, voluntary effort. ("Autonomic" means involuntary or unconscious.) The ANS exists in all groups of vertebrates. Some invertebrates are also considered to have an ANS.

The ANS of a human or other mammal controls a wide variety of smooth muscles, such as those in the gut, blood vessels, bladder, eyes (e.g., iris muscles), airways of the lungs, and reproductive organs (e.g., during orgasm) (**FIGURE 34.22**). It controls a variety of exocrine glands (glands that secrete their products elsewhere than into the blood) such as the tear and sweat glands. It also controls a few endocrine glands (glands that secrete hormones into the blood), most notably the adrenal medullary glands. And it controls color-change cells in fish, the acid-secreting cells of the stomach, and several other effectors.

The vertebrate ANS has three divisions. The **enteric division** is composed of the nerve cells internal to the gut wall (see Figure 29.14). Here we focus on the **sympathetic** and **parasympathetic divisions** (see Figure 34.22).

We can think of these two divisions as consisting mainly of efferent neurons that carry orders from the CNS to autonomic effectors. Each efferent pathway is composed of two neurons that synapse at a ganglion. (As we discuss this topic, remember that "ganglion" has different meanings in vertebrates and arthropods.) A vertebrate **ganglion** is a discrete, anatomically clustered set of neuron cell bodies in the PNS. The two neurons of each efferent pathway are described as preganglionic and postganglionic.

The preganglionic neurons of the parasympathetic division exit the CNS from two regions: a cranial region where the brain and spinal cord connect and a sacral region, the most posterior region of the spinal cord. The axons of these preganglionic neurons travel almost to their target cells before synapsing with the postganglionic neurons.

In contrast, the preganglionic neurons of the sympathetic division exit the CNS from the thoracic and lumbar regions of the spinal cord. Most of the ganglia of the sympathetic division lie next to the spinal cord, lined up in two chains on either side. Accordingly, the postganglionic neurons need to cover most of the distance to the target cells.

The most important neuronal difference between the sympathetic and parasympathetic divisions is that their postganglionic neurons employ different neurotransmitters where they synapse with target cells. The sympathetic neurons release norepinephrine as their neurotransmitter. The parasympathetic neurons release acetylcholine (ACh). In organs that receive both sympathetic and parasympathetic input, the target cells typically respond in opposite ways to norepinephrine and to ACh.

The sympathetic and parasympathetic divisions thus often work in opposition to each other, as emphasized in Figure 34.22. In this way, the two divisions—acting together—can control autonomic effectors in fine-tuned ways, adjusting their functions up or down as needed. For example, both divisions innervate the pacemaker of the heart. Sympathetic stimulation causes the heart rate to increase, and parasympathetic stimulation causes it to decrease.

The fight-or-flight response is the most famous effect of the sympathetic division. When the sympathetic division is strongly activated, it greatly increases the heart rate, force of heart contraction, and cardiac output. It dilates the passageways of the lungs and increases release of glucose into the blood from the liver. In all these ways, it prepares the body to meet an emergency. At the same time, it reduces less urgent activities such as digestion.

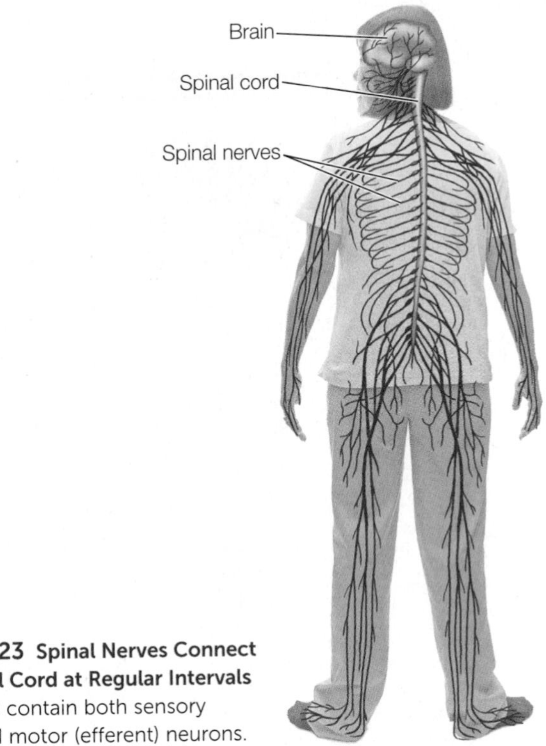

FIGURE 34.23 Spinal Nerves Connect to the Spinal Cord at Regular Intervals These nerves contain both sensory (afferent) and motor (efferent) neurons.

Spinal reflexes represent a simple type of skeletal muscle control

Now let's turn to control of the skeletal muscles. These effectors are controlled by the parts of the CNS and PNS other than the autonomic nervous system. Many of the neurons that control the skeletal muscles enter or leave the CNS in spinal nerves that connect with the spinal cord at regular intervals on each side (**FIGURE 34.23**). These nerves contain both motor neurons that carry orders from the CNS to the skeletal muscles and sensory neurons that bring information to the CNS, such as information from stretch receptors in the muscles.

As we all know, we can consciously or voluntarily control our skeletal muscles. This is possible because neurons in our brain can control signals in the motor neurons to the muscles. Also, however, some signals to the muscles originate by way of **spinal reflexes**. These are neuron-mediated responses that do not involve participation of the brain. Instead, all the neuron interactions occur in the spinal cord, requiring no conscious attention. Because spinal reflex systems are relatively simple, they are among the best-known vertebrate neural circuits.

We are all familiar with the spinal reflex known as the knee-jerk reflex (**FIGURE 34.24**). It occurs when a physician or nurse taps your knee with a little hammer and—if your nervous system is healthy—your lower leg jerks forward. The tap stretches the patellar tendon that connects your quadriceps muscle—the muscle on the forward-facing side of your thigh—to your tibia bone in your lower leg. The stretch caused by the tap stimulates stretch receptors in the muscle. These receptors initiate action potentials in sensory neurons, which propagate these action potentials along their axons to the spinal cord via a spinal nerve.

In the spinal cord, each sensory axon synapses with a motor neuron that serves the same muscle, the quadriceps. This synapse is excitatory and causes the motor neuron to generate action potentials. These signals travel in the axon of the motor

1 A hammer tap stretches the tendon in the knee, stretching receptors in the quadriceps.
2 Stretch receptors fire action potentials.
3 The sensory neuron synapses with a motor neuron in the spinal cord.
4 The motor neuron conducts action potentials to the quadriceps, causing contraction.
5 Simultaneously a spinal interneuron inhibits firing in the motor neuron for the antagonistic muscle.
6 The leg jerks forward.
Gray matter
White matter
Dorsal root (sensory neurons)
Dorsal horn
Ventral horn
Ventral root (motor neurons)
Spinal interneuron
Sensory neuron
Motor neurons
Spinal nerve
Quadriceps muscle

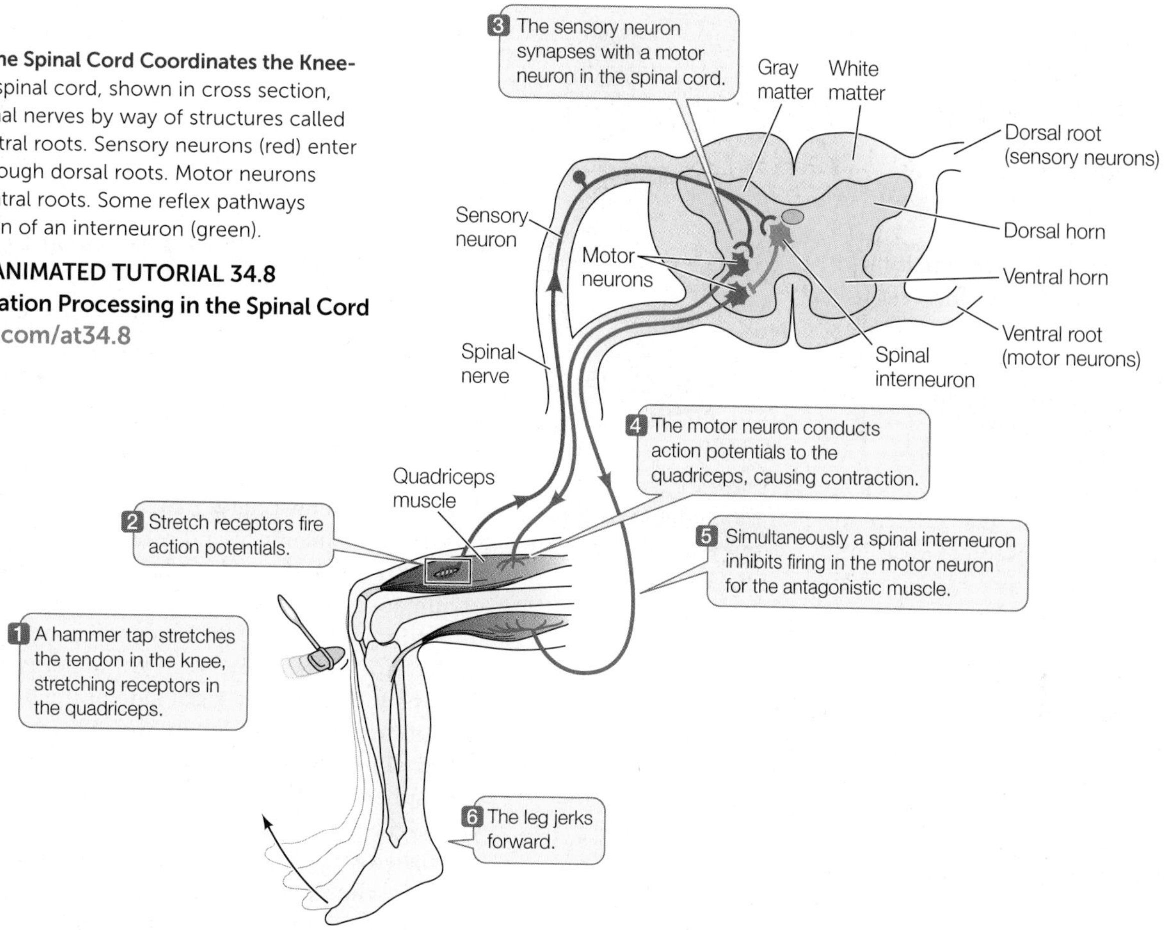

FIGURE 34.24 The Spinal Cord Coordinates the Knee-Jerk Reflex The spinal cord, shown in cross section, connects with spinal nerves by way of structures called the dorsal and ventral roots. Sensory neurons (red) enter the spinal cord through dorsal roots. Motor neurons (blue) leave via ventral roots. Some reflex pathways involve participation of an interneuron (green).

Go to ANIMATED TUTORIAL 34.8 Information Processing in the Spinal Cord
PoL2e.com/at34.8

neuron via the spinal nerve to the quadriceps. There the action potentials excite the muscle at neuromuscular junctions. The muscle contracts, and your lower leg kicks forward.

There's more to the story. An additional important factor is that limb muscles are organized into opposing (antagonistic) pairs (see Concept 33.2). For contraction of the quadriceps to move your lower leg forward easily, the opposing muscle on the back of your thigh needs to relax. It does so because the sensory neurons coming from the stretch receptors also synapse onto interneurons in the spinal cord (see Figure 34.24, step 5). Those interneurons make inhibitory synapses onto the motor neurons that control the opposing muscle. The motor neurons are inhibited from stimulating that muscle to contract, allowing your lower leg to move forward when the quadriceps contracts.

These interactions among cells are unusually simple, but they nicely illustrate the interaction of sensory and motor neurons, interneurons, and the PNS and CNS in controlling contractions of skeletal muscles.

The most dramatic changes in vertebrate brain evolution have been in the forebrain

All vertebrate brains consist of three major regions: the forebrain (the most anterior part), midbrain, and hindbrain (the most posterior part). The most posterior part of the hindbrain is the medulla oblongata. All information traveling between the brain and spinal cord must pass through the medulla oblongata. During the course of vertebrate brain evolution, the overall shape and relative size of the medulla oblongata have changed little (**FIGURE 34.25**).

In contrast, the cerebral hemispheres, which are a major component of the forebrain, have undergone dramatic changes during vertebrate evolution. Both mammals and birds have evolved especially large cerebral hemispheres (see Figure 34.25). This is noteworthy because the cerebral hemispheres are particularly important in carrying out high-order sensory, motor, and integrative functions. In humans, for example, the cerebral hemispheres are responsible for language and reasoning.

Mammals and birds also have brains that are about 10 times larger than the brains of lizards or other nonavian reptiles that are the same in body size. A mouse, for example, has a brain about 10 times larger than that of a mouse-sized lizard, and a parrot has a brain about 10 times larger than that of a parrot-sized lizard. The cerebral hemispheres account for much of this difference.

There are two take-home messages from these comparisons. First, mammals and birds have evolved strikingly large cerebral hemispheres, the neurons and glia of which are of enormous importance in high-order brain function. Second, brain size matters. The evolution of enhanced functionality in mammals and birds has gone hand in hand with large increases in the numbers of neurons. From this vantage, it is significant that primates and dolphins have the largest brains, for their body sizes, of all mammals.

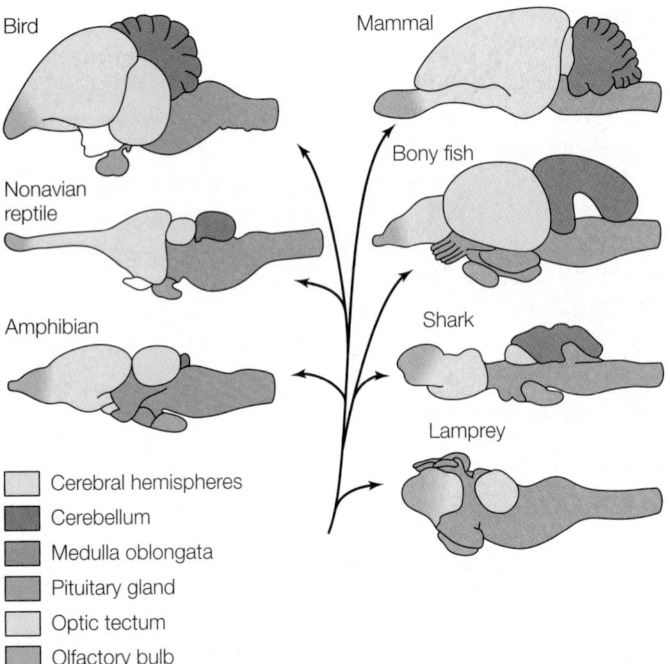

FIGURE 34.25 Evolution of the Vertebrate Brain A reasonably representative brain of a modern species in each group is shown. In the mammal, the large cerebral hemispheres keep us from seeing the pituitary, which is found in a position similar to other groups. The optic tectum is involved in processing visual information. Medulla oblongata, a term used with several variations in exact meaning, is here used in its most general sense.

We need to be careful not to overextrapolate the significance of these broad trends, however. Brain size matters in a general sense, but it does not matter when we look at all particulars. Some animals with small brains exhibit stunning behavioral capacities (see Figure 40. 8). Moreover, among human beings—where the subject has been studied for well over a century—all the evidence available indicates that individual intelligence is not correlated with individual brain size. We know too little about brain function to explain how all these pieces of the puzzle fit together.

Location specificity is an important property of the mammalian cerebral hemispheres

In humans, the cerebral hemispheres (which together make up the cerebrum) are so enlarged that they cover most other parts of the brain. The outermost layer of the cerebral hemispheres is the **cerebral cortex**, a thin, important layer rich in cell bodies. The cerebral cortex is only about 4 millimeters thick, but it is folded into ridges (convolutions) that increase its size. The cerebral hemispheres play major roles in sensory perception, learning, memory, and conscious behavior. Indeed, they provide the neural resources for the great intellectual capacity of humans.

A curious feature of the mammalian nervous system is that the left side of the body is served (in both sensory and motor aspects) mostly by the right side of the brain, and the right side of the body is served by the left side of the brain. Thus sensory input from your right hand goes to your left cerebral hemisphere, and sensory input from your left hand goes to your right cerebral hemisphere.

Earlier we noted that all action potentials arriving in the brain are the same, even though some represent visual images, others represent sounds, and still others represent odors. Some action potentials from a person's touch receptors represent touches to the fingers, whereas others represent touches to the abdomen or feet. How does the brain process a simple, invariant electrical event—an action potential—and develop an accurate perception of the type of stimulus (sight, sound, or

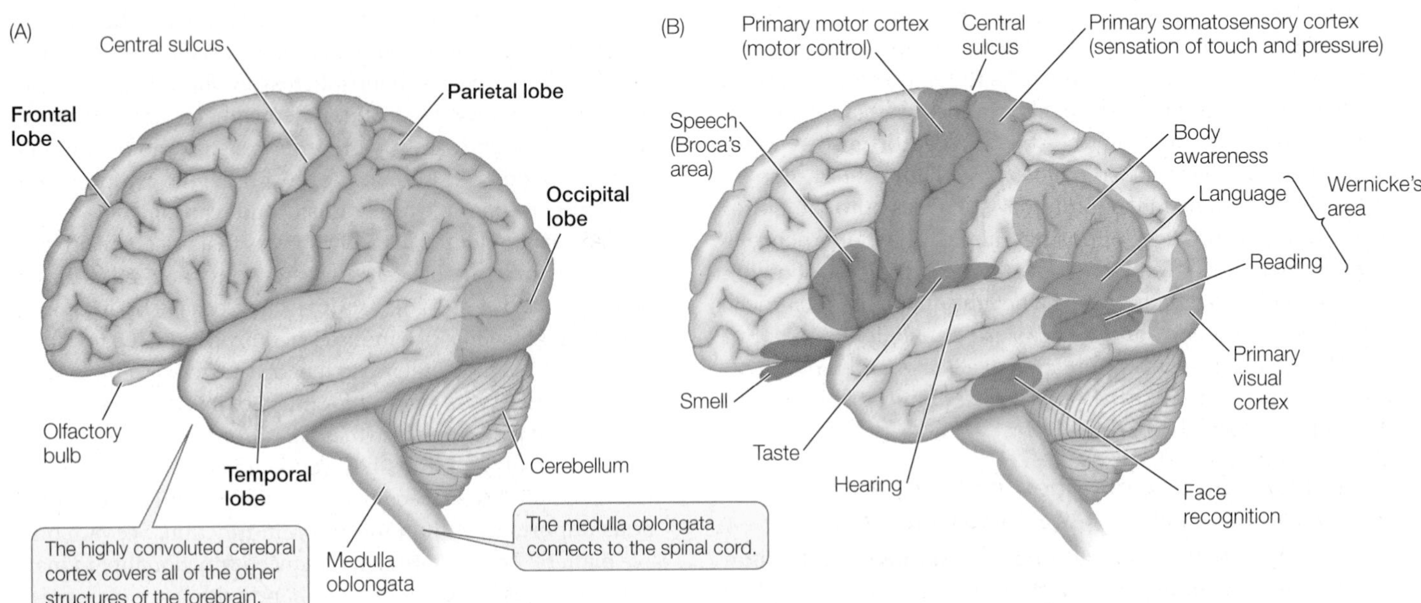

FIGURE 34.26 The Human Left Cerebral Hemisphere (A) The major anatomical regions of the cerebral cortex. (B) Different areas of the cerebral cortex have specific, localized functions.

Go to ACTIVITY 34.5 Structures of the Human Brain
PoL2e.com/ac34.5

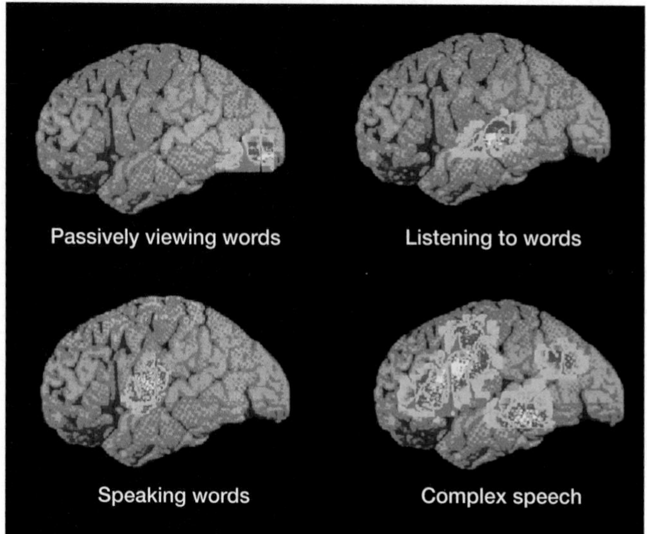

FIGURE 34.27 Imaging Techniques Reveal Active Parts of the Brain Positron emission tomography (PET) scanning reveals the brain regions activated by various aspects of language use. Radioactively labeled glucose is given to the subject. Brain areas take up radioactivity in proportion to their metabolic use of glucose. The PET scan visualizes levels of radioactivity in specific brain regions when a particular activity is performed. The red and yellow areas are the most active, green and blue the least active.

odor) that caused it, or an accurate perception of the place of origin of the stimulus?

Location specificity is one of the major mechanisms the brain uses to interpret the action potentials it receives. In each cerebral hemisphere, various specific regions are specialized in adults to carry out specific sensory and motor functions. Distinct regions are responsible for vision, hearing, and smell, for example (**FIGURE 34.26**). Axons from the retinas connect to the visual cortex, the region responsible for vision. Similarly, axons from the cochleas in the ears and axons from the odorant receptor cells in the nasal epithelium connect to the regions responsible for hearing and smell, respectively. Because of these physical connections, axons are effectively labeled in regards to the meaning of the action potentials they bring to the brain.

Motor functions are also localized. Moreover, combined sensory and motor functions often occur in certain, relatively localized brain areas. An explosive area of research at present is the exploration of sensory, motor, and combined sensory and motor localization by use of imaging methods that allow us to visualize where neurons exhibit evidence of increased electrical activity. For instance, the positron emission tomography (PET) images in **FIGURE 34.27** show brain regions where there is evidence of increased neuronal activity when several language functions are carried out. Functional magnetic resonance imaging (fMRI) is another technique used for similar research.

Localization of function is not limited to the cerebral hemispheres. **FIGURE 34.28**, for example, is a fMRI image of the brain of a person experiencing fear. The image indicates increased neuronal activity in the amygdala, a part of the forebrain (outside the cerebral hemispheres) that is well known to participate in fear responses. Abundant evidence points to the hippocampus, still another part of the forebrain, as being critical for spatial learning and memory (we discussed synaptic plasticity in the hippocampus in Concept 34.3). Damage to the hippocampus renders mice and rats unable to solve maze problems that depend on learning the relationships of objects in space.

Brain regions have maps in some cases. In such regions, the parts of the brain that serve various anatomical regions of the body are physically related to each other in ways that mirror the physical relationships of the rest of the body. A well-known example is seen in the somatosensory ("body sensing") part of the cerebral cortex. Regions of the cerebral cortex that process sensory information from touch and pressure receptors on the body surface in the leg, hip, trunk, neck, and head are lined up (**FIGURE 34.29A**). Similarly, regions processing such information from the eyes, nose, lips, and jaw are lined up. In the illustration, the size of each body part in the drawing reflects the amount of cortical area devoted to the part. The large size of the lips, for example, reflects the fact that a large area is devoted to processing sensory signals from the lips.

Near the mapped somatosensory area of the cortex is a parallel, mapped motor area. You can see from the size of drawings of body parts in the motor area (**FIGURE 34.29B**) that large areas of motor cortex are devoted to sending motor signals to the hands and face.

Not all parts of the brain fit easily into the concepts of localized function and mapped brain regions. Thus we need to recognize that these concepts are both important and limited in their ability to clarify brain function.

A curious aspect of language processing in humans is that it resides principally in one cerebral hemisphere—which in about 85 percent of people is the left hemisphere. The two hemispheres share responsibilities for other functions. Individuals with damage to the left hemisphere frequently experience deficits in their ability to use or understand

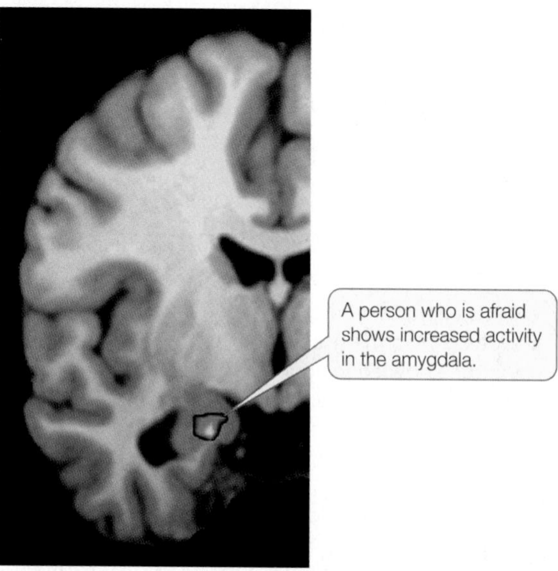

A person who is afraid shows increased activity in the amygdala.

FIGURE 34.28 Source of the Fear Response Frightening situations—or even memories of such situations—activate the amygdala, as shown in this functional magnetic resonance image (fMRI) of the left side of the brain of a person experiencing fear.

FIGURE 34.29 The Body Is Represented in Maplike Ways in the Primary Somatosensory and Primary Motor Parts of the Cerebral Cortex The amount of cortical tissue devoted to various body parts is not even. More cortical tissue is devoted to some body parts than others, as reflected in the sizes of the parts in the drawings. (A) The primary somatosensory cortex. (B) The primary motor cortex.

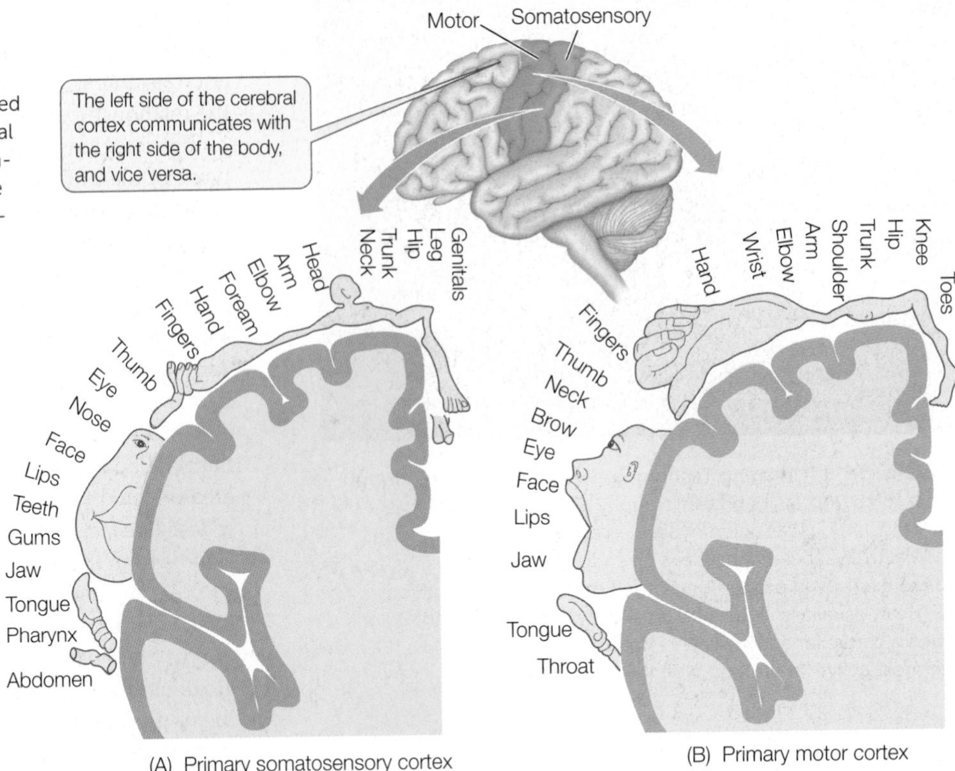

words. The deficit depends on the specific part of the left hemisphere that is damaged. Damage to Broca's area, for example, interferes with speech and writing, whereas damage to Wernicke's area interferes with understanding language (see Figure 34.26B).

The development of a full understanding of the brain is one of the great frontiers of twenty-first century biology. Many major unanswered questions remain. "What is consciousness?" is one of the most compelling.

CHECKpoint CONCEPT 34.5

- ✓ Give two specific examples in which the sympathetic and parasympathetic divisions exert opposite effects on the function of an organ.
- ✓ Why is localization of function in the brain critical for sensory perception?
- ✓ What is cephalization of the nervous system, and why has it evolved in so many animals? Try to think this through using the knowledge you have gained about the relationships between the brain and sense organs.

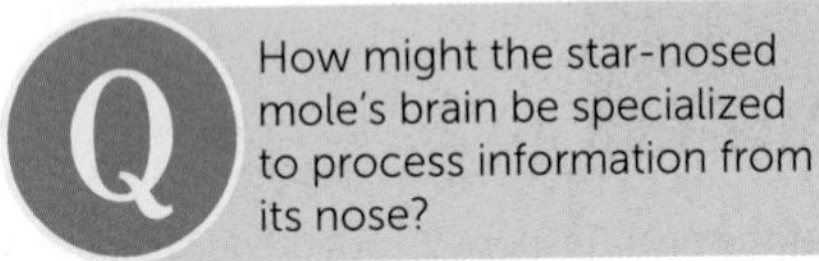

How might the star-nosed mole's brain be specialized to process information from its nose?

ANSWER Just as large areas of cerebral cortex are devoted to processing sensory information from a person's lips and fingers (see Figure 34.29A), large areas of a star-nosed mole's cerebral cortex are devoted to processing sensory information from the tentacles of its nose. The drawings in **FIGURE 34.30A** show a mole in two ways: first, proportioned as it actually is and, second, proportioned in a way that reflects how much cortical tissue is devoted to processing sensory information from its specific body parts, such as its nose, eyes, whiskers, and anterior digging feet. Note that the amount of cortical tissue devoted to processing signals from the nose is larger than that devoted to the rest of the mole's body.

Additionally, a localized part of the cerebral cortex is devoted to receiving sensory information from the nose, and this part of the cortex is highly mapped. The tentacles surrounding each nostril are numbered from 1 to 11 in order from the most dorsal-pointing tentacle to the most ventral-pointing one. **FIGURE 34.30B** shows the relevant part of the cortex, including the places in the cortex that receive axons from the numbered tentacles. Note that the map of the cortical regions exactly mirrors the arrangement of the tentacles. By means of such localization in the cortex, the brain can readily determine the tentacle that originates any action potential that arrives by way of any particular axon from the nose.

The concepts of localized function and mapped brain regions apply to the star-nosed mole as much as they do to people, reflecting the fact that the mole's entire nervous system—including its sense organs—is highly integrated.

(A) Cortical magnification in star-nosed mole

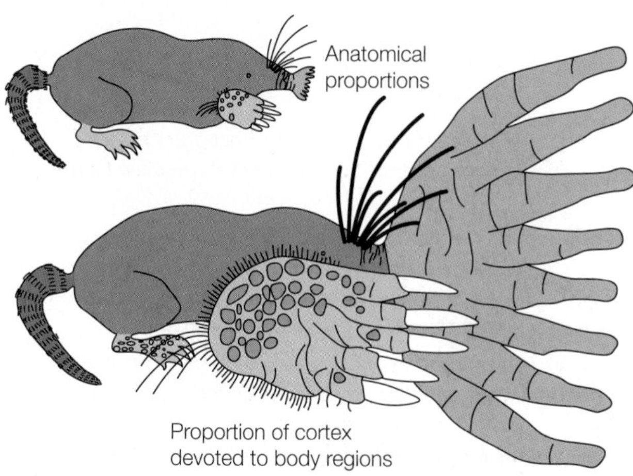

(B)

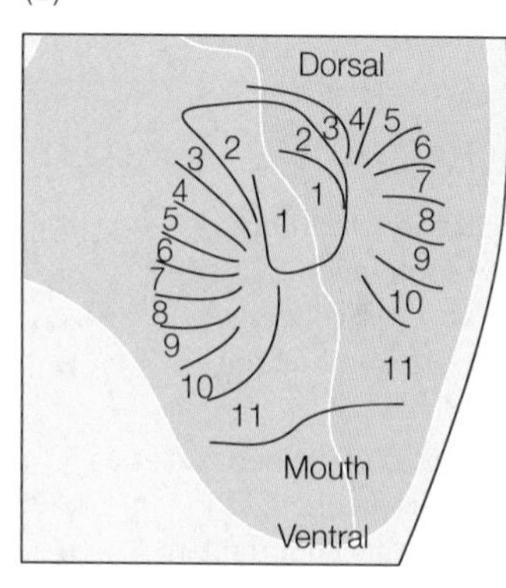

FIGURE 34.30 Representation of the Nose of the Star-nosed Mole in the Cerebral Cortex (A) Proportions of the body, and proportions of the cortex devoted to processing information from various parts of the body. (B) A part of the cerebral cortex, showing where sensory information arrives from the numbered tentacles of the nose.

SUMMARY

CONCEPT 34.1 Nervous Systems Are Composed of Neurons and Glial Cells

- A **neuron** (nerve cell) is **excitable**; it is specially adapted to generate and propagate electric signals, typically in the form of action potentials. Neurons make functionally relevant contact with other cells at **synapses**.
- Neurons usually have four anatomical regions: a set of **dendrites**, a **cell body**, an **axon**, and a set of **presynaptic axon terminals**. Dendrites receive signals from other neurons, and presynaptic axon terminals send signals to other cells. **Review Figure 34.2**
- Unlike neurons, **glial cells** are usually not excitable. In vertebrates, they include **oligodendrocytes**, which form insulating **myelin** sheaths around axons in the brain and spinal cord, and **Schwann cells**, which perform the same function outside the brain and spinal cord. **Review Figure 34.3**

CONCEPT 34.2 Neurons Generate Electric Signals by Controlling Ion Distributions

- Neurons have an electric charge difference across their cell membranes, called the **membrane potential**. The membrane potential is created by ion transporters and channels. The membrane potential is considered to be negative if the inside of the cell membrane is negative in relation to the outside. In inactive neurons, the membrane potential is called the **resting potential** and is negative. **Review Figure 34.4 and ANIMATED TUTORIAL 34.1**
- The **sodium–potassium pump** concentrates K^+ on the inside of a neuron and Na^+ on the outside. In a resting neuron, K^+ leak channels allow K^+ to diffuse through the cell membrane. The resting potential is negative because a negative charge on the inside of the cell membrane is needed to balance the tendency for K^+ to diffuse outward because of its high internal concentration.
- The **Nernst equation** can be used to predict a neuron's membrane potential. **Review ACTIVITY 34.1**
- Most ion channels are gated: they open only under certain conditions. **Voltage-gated channels** open or close in response to local changes in the membrane potential. **Stretch-gated channels** open or close in response to stretch or tension applied to the cell membrane. **Ligand-gated channels** have binding sites where they bind noncovalently with specific chemical compounds that control them. **Review Figure 34.5**
- When gated ion channels in the cell membrane of a neuron open or close, changes of the membrane potential called **depolarization** or **hyperpolarization** can occur. Depolarization occurs when the inside of the cell membrane becomes less negative in relation to the outside.
- When depolarization occurs, a key question is whether it is great enough for a voltage threshold to be reached. If the threshold is not reached, changes in the membrane potential are **graded**. **Review Figure 34.6A**
- An **action potential**, or **nerve impulse**, is a rapid reversal in membrane potential across a portion of the cell membrane resulting from the opening and closing of voltage-gated Na^+ channels and K^+ channels. The voltage-gated Na^+ channels open when the cell membrane depolarizes to the voltage threshold. **Review Figures 34.6B, 34.7, and 34.8, and ANIMATED TUTORIAL 34.2**
- An action potential is an all-or-none event that is conducted (propagated) along the entire length of a neuron's axon without any loss of magnitude. It is conducted along the axon because local current flow depolarizes adjacent regions of membrane so they reach voltage threshold. In myelinated axons, action potentials jump between **nodes of Ranvier**, patches of membrane that are not covered by myelin.

CONCEPT 34.3 Neurons Communicate with Other Cells at Synapses

- A **synapse** is a cell-to-cell contact point specialized for signal transmission from one cell to another. Most synapses are **chemical synapses** (with **neurotransmitters**). Some are **electrical synapses**.
- At a chemical synapse, when an action potential reaches the axon terminal of the presynaptic neuron, it causes the release of a neurotransmitter, which diffuses across the synaptic cleft and binds to receptors on the cell membrane of the postsynaptic cell. The postsynaptic cell is usually another neuron or a muscle cell. **Review ANIMATED TUTORIALS 34.3 and 34.4**
- The vertebrate **neuromuscular junction** is a well-studied chemical synapse between a motor neuron and a skeletal muscle cell. Its neurotransmitter is acetylcholine (ACh). **Review Figure 34.9**
- Synapses can be fast or slow, depending on the nature of their receptors. **Ionotropic receptors** are ion channels and generate fast, short-lived responses. **Metabotropic receptors** initiate second-messenger cascades that lead to slower, more sustained responses.
- There are many different neurotransmitters and types of receptors. The action of a neurotransmitter depends on the receptor to which it binds.

(continued)

SUMMARY *(continued)*

- Synapses between neurons can be either **excitatory** or **inhibitory**. A postsynaptic neuron integrates information by summation of graded postsynaptic potentials (which may be excitatory or inhibitory), in both space (spatial summation) and time (temporal summation). **Review Figure 34.11**
- **Synaptic plasticity** is the process by which synapses in the nervous system of an individual animal can undergo long-term changes in their functional properties. Such changes are probably important in learning and memory.

CONCEPT 34.4 Sensory Processes Provide Information on an Animal's External Environment and Internal Status

- **Sensory receptor cells** transduce information about an animal's external and internal environment into action potentials.
- Sensory receptor cells have **sensory receptor proteins** that respond to sensory input, causing a graded change of membrane potential called a receptor potential.
- An **ionotropic receptor cell** typically has a receptor protein that is a stimulus-gated Na^+ channel. A **metabotropic receptor cell** typically has a receptor protein that activates a G protein when exposed to the stimulus. **Review Figure 34.12**
- The action potentials sent to the brain are interpreted as particular sensations based on which neurons in the brain receive them.
- **Mechanoreceptors** are cells that respond specifically to mechanical distortion of their cell membrane and are typically ionotropic. Stretch receptors are mechanoreceptors. **Review Figures 34.13 and 34.14 and ANIMATED TUTORIAL 34.5**
- **Chemoreceptors** are metabotropic receptor cells that respond to the presence or absence of specific chemicals. The sense of smell depends on chemoreceptors.
- In the mammalian auditory system, sound pressure waves enter the auditory canal, where the tympanic membrane vibrates in response to them. The movements of the tympanic membrane are relayed to the oval window. Movements of the oval window create sound pressure waves in the fluid-filled **cochlea**. **Review Figure 34.15 and ACTIVITY 34.2**
- The **basilar membrane** running down the center of the cochlea is distorted by sound pressure waves at specific locations that depend on the frequency of the waves. These distortions cause the localized bending of stereocilia of hair cells, mechanoreceptors in the organ of Corti. Pitch or tone is encoded by the specific locations along the basilar membrane where action potentials are generated. **Review Figure 34.16 and ANIMATED TUTORIAL 34.6**
- **Photoreceptors** are sensory receptor cells that are sensitive to light. The photosensitivity of photoreceptor cells involved in vision depends on the absorption of photons of light by sensory receptor proteins called **rhodopsins**. Vertebrates have two types of visual photoreceptor cells, **rods** and **cones**. Color vision in humans arises from three types of cone cells with different spectral absorption properties. **Review Figure 34.17 and ANIMATED TUTORIAL 34.7**
- When excited by light, vertebrate visual photoreceptor cells hyperpolarize. They do not produce action potentials. **Review Figure 34.18**
- The vertebrate **retina** consists of four types of integrating neurons and the photoreceptor cells lining the back of the eye. Extensive processing of visual information occurs in the retina and also in the brain. **Review Figure 34.19 and ACTIVITIES 34.3 and 34.4**
- Arthropods have **compound eyes** consisting of many optical units called **ommatidia**, each with its own lens. **Review Figure 34.20**

CONCEPT 34.5 Neurons Are Organized into Nervous Systems

- Nerve nets are the simplest nervous systems. As nervous systems evolved further, they followed two major trends: **centralization**—the clustering of neurons into centralized integrating organs—and **cephalization**—the concentration of major integrating centers at the anterior end of the animal's body.
- The **brain** and spinal cord make up the **central nervous system** (**CNS**). Neurons that extend or reside outside the brain and spinal cord make up the **peripheral nervous system** (**PNS**).
- Neurons are classed as **interneurons**, **sensory neurons**, or **motor neurons**.
- In vertebrates, the central nervous system is positioned in the dorsal part of the body. In arthropods, such as insects and crayfish, the CNS is primarily positioned in the ventral part of the body. **Review Figure 34.21**
- The **autonomic nervous system** (**ANS**) is the part of the nervous system (both CNS and PNS) that controls involuntary functions. Its **enteric**, **sympathetic**, and **parasympathetic divisions** differ in anatomy, neurotransmitters, and effects on target tissues. The sympathetic and parasympathetic divisions usually exert opposite effects on an organ. **Review Figure 34.22 and ANIMATED TUTORIAL 34.8**
- The vertebrate brain consists of a forebrain, midbrain, and hindbrain. During the course of vertebrate evolution, some parts of the brain (e.g., the medulla oblongata) have remained relatively unchanged, whereas other parts (e.g., the cerebral hemispheres) have changed dramatically. **Review Figure 34.25**
- The cerebral hemispheres, including the **cerebral cortex** of each hemisphere, are the dominant structures of the human brain. **Review Figure 34.26**
- In each cerebral hemisphere, specific regions are specialized to carry out specific sensory and motor functions, for example language functions and fear. **Review Figures 34.26–34.28 and ACTIVITY 34.5**
- Some brain regions have maps: the parts of the brain that serve various anatomical regions of the body are physically related to each other in ways that mirror the physical relationships of the rest of the body. Examples are the somatosensory ("body sensing") and motor areas of the cerebral cortex. **Review Figures 34.29 and 34.30**

Go to the Interactive Summary to review key figures, Animated Tutorials, and Activities PoL2e.com/is34

Go to LaunchPad at **macmillanhighered.com/launchpad** for additional resources, including LearningCurve Quizzes, Flashcards, and many other study and review resources.

35 Control by the Endocrine and Nervous Systems

KEY CONCEPTS

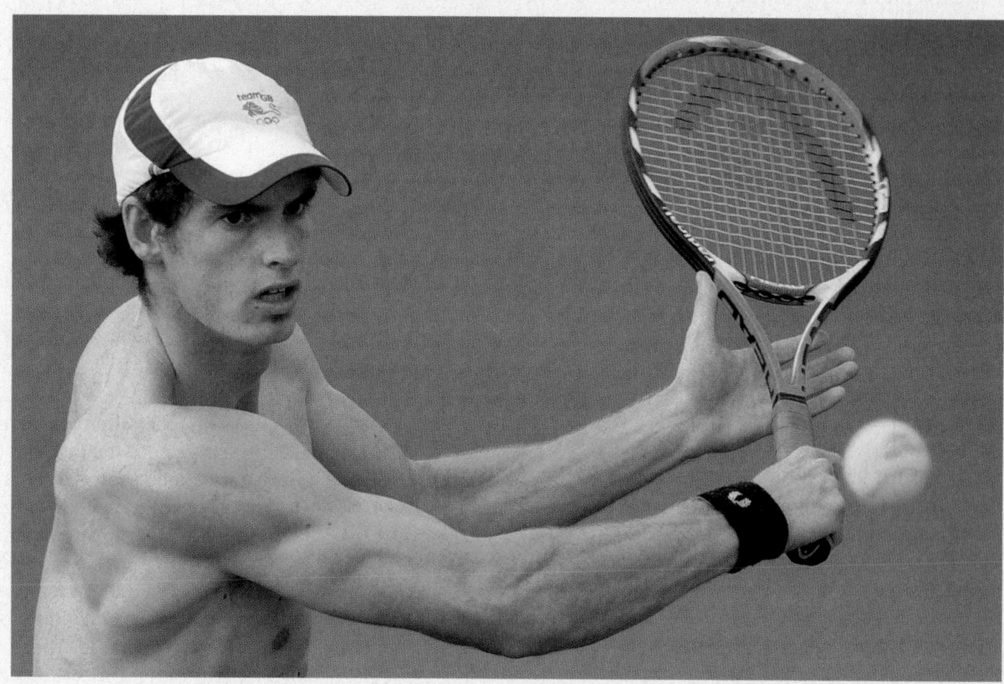

This young man's muscles are being controlled simultaneously by the nervous and endocrine systems.

Two types of control of skeletal muscle are evident in this scene. First are the quick, moment-by-moment moves of the game. To return the ball with a backhand shot, this player runs to a good position to execute his return, plants his feet, pulls back the racket, focuses his eyes on the ball, and swings—all in a few seconds. His nervous system controls the muscle contractions responsible for these quick moves. In addition, this player's muscles are far different than they were before he entered puberty. His muscles have become more highly developed, larger, and stronger since puberty began—a process requiring years. His endocrine system has controlled these long-term muscle changes.

Castration revealed long ago that the testes—although they are just one part of the body—play a critical role throughout the body in male maturation. No one knows how the practice of castration began, because it started in prehistorical times, before written historical records were kept. Presumably, farmers discovered that removal of the testes rendered male farm animals more docile, and then the practice was extended to people such as harem servants. Boys castrated prior to puberty do not undergo normal adolescent muscular and skeletal development, and their voices do not deepen because their vocal cords do not increase in length. By 1700, several thousand prepubescent boys were castrated each year in Italy to create adult male singers with high voices. Some of these castrati, as they were called, are legendary for being among the greatest opera singers the world has ever known.

The effects of castration pointed to the involvement of the testes in male maturation for thousands of years, but not until 1849 did someone carry out scientific experiments to understand why the testes matter. The German physiologist Arnold Berthold knew from farmers that when male chickens are castrated prior to sexual maturation, they do not develop adult male characteristics. The fleshy crest (comb) on the top of a castrated male's head remains immature, for example. After Berthold castrated young male chickens, he transplanted surgically isolated testes into their bodies at abnormal locations. These chickens then matured normally, complete with adult combs.

These experiments started the modern study of endocrinology. Berthold concluded that the testes secrete a substance into the blood that circulates with the blood to all parts of the body and controls the development of other tissues. Substances of this type were named hormones about 50 years later, and in 1935, the key substance secreted by the testes was chemically identified and named testosterone.

Why are the skeletal muscles under control of both the nervous system and the endocrine system?

You will find the answer to this question on page 749.

CONCEPT 35.1 The Endocrine and Nervous Systems Play Distinct, Interacting Roles

All the cells in an animal's body need to act together harmoniously. During muscular exercise such as a game of tennis, for example, cells in the breathing muscles and heart need to work harder to ensure that the cells of the exercising skeletal muscles receive O_2 at an accelerated rate. Cells in the liver also need to contribute, by mobilizing stored glucose for use by the exercising muscles. When we survey all the tissues of the body, we quickly recognize that various cells are specialized to produce forces, store nutrients, make gametes, or secrete digestive juices. Some cells must also be specialized to carry out control and coordination throughout the body.

We saw in Chapter 34 that nerve cells are one type of cell specialized for control and coordination. Endocrine cells are the second principal type of cell specialized in this way. We will define endocrine cells and hormones in a formal way at the beginning of Concept 35.2. Here it's sufficient to say simply that endocrine cells secrete hormones into the blood. Nerve and endocrine cells work together to ensure that an animal functions in ways that are harmonious and coordinated, rather than clashing and disjointed.

Nerve and endocrine cells need to communicate with other cells to carry out their functions of control and coordination. Most such intercellular communication takes place by means of chemical signals that are released from a nerve cell (neuron) or an endocrine cell and travel to another cell, called the **target cell** (**FIGURE 35.1**). The signal molecules bind to receptors on the target cell, triggering a response in it. These responses can affect an animal's function, anatomy, and behavior.

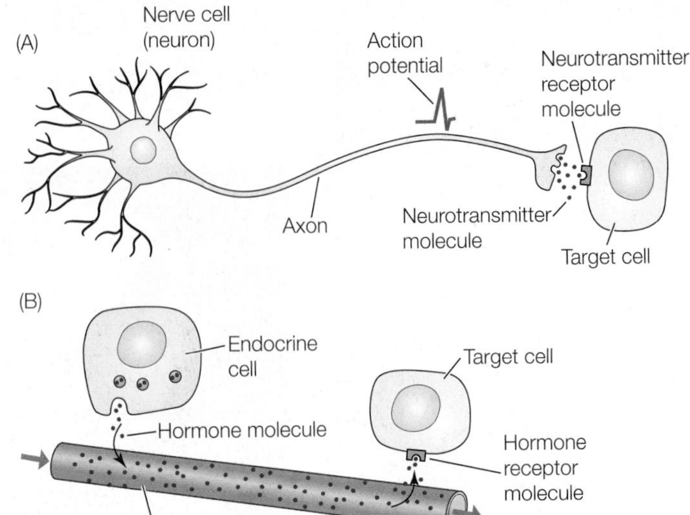

FIGURE 35.1 Chemical Signaling by Nerve and Endocrine Cells Both nerve cells (neurons) and endocrine cells release chemical signaling molecules (red dots) that affect target cells. (A) A nerve cell makes synaptic contact with a target cell at the end of its axon. Action potentials in the axon trigger release of neurotransmitter molecules at the synapse. These molecules then bind with neurotransmitter receptor molecules in the postsynaptic cell, initiating a response in the postsynaptic cell. (B) An endocrine cell releases hormone molecules into the blood. These molecules circulate throughout the body and bind with hormone receptor molecules on target cells, initiating responses.

The nervous and endocrine systems work in different ways

Most tissues and organs in an animal's body are affected by both the nervous and endocrine systems. The heart, for example, is subject to both nervous and endocrine controls. Why have two systems evolved rather than just one? The answer is undoubtedly complicated. One reason is that the nervous and endocrine systems work in different ways. The two systems are specialized to carry out different types of control and coordination.

As we saw in Chapter 34, neurons typically operate in ways similar to wires in a telephone system or computer network. A neuron is a cell specially adapted to generate brief signals—the nerve impulses, or action potentials—that travel at high speeds along its axon (see Figure 35.1A). The axon makes functional contacts—synapses—with only certain target cells, and only those cells are affected by the neuron. Each axon terminal releases a chemical signal—neurotransmitter—that affects the cell with which it has synaptic contact. This target cell responds quickly to the signal, and the signal ends rapidly because the neurotransmitter molecules released into a synapse are immediately either broken down or taken up into cells.

We see that nervous control has two essential features: it is *fast* and *addressed*. Neuronal signals are *fast* in that they travel rapidly and begin and end abruptly. Like a letter or an e-mail, they are *addressed* because they are delivered to highly defined target cells.

Unlike nerve cells, endocrine cells release their chemical signals—the hormones—into the blood. These chemical signals are then carried throughout the body by the circulation of the blood, potentially reaching all the cells present in most or all tissues and organs (see Figure 35.1B).

Endocrine control has two essential features, both of which distinguish it from nervous control: it is *slow* and *broadcast*. Individual hormonal signals are relatively *slow* because they operate on much longer time scales than individual neuronal signals. Initiation of hormonal effects requires at least several seconds or minutes because a hormone, once it is released into the blood, must circulate to target tissues and diffuse to effective concentrations within the tissues before it can elicit responses. After a hormone has entered the blood, it may act on target cells for a substantial amount of time—minutes, hours, or even days—before its blood concentration is reduced to ineffective levels by metabolic destruction or excretion. A final reason that endocrine signals can be slow is that they sometimes act on target cells by altering gene transcription and protein synthesis—processes that require at least many minutes to have effects on cell function.

Endocrine control is said to be *broadcast* because after a hormone is released into the blood, all cells in the body are potentially exposed to it. Those that respond are the ones that express a receptor protein for the hormone. That is, hormone action has specificity, but its specificity depends not on addressed delivery of the chemical signal but on which cells have receptor molecules for the signal. In any one tissue, it is common for all cells to express receptor molecules for a particular hormone, meaning all cells in the tissue respond to a single release of the hormone. Moreover, cells in more than one tissue may respond, perhaps with different cell types responding in different ways. In principle, hormones may exert either limited or widespread effects, but in practice they commonly affect at least an entire tissue, and often multiple tissues.

Nervous systems and endocrine systems tend to control different processes

Lines of communication in the nervous system are capable of much finer control—in both time and space—than is possible for endocrine systems. Not surprisingly, the two systems tend to be used to control different functions in the body. The nervous system controls predominantly the fine, rapid movements of discrete skeletal muscles. The endocrine system typically controls more widespread, prolonged activities such as developmental or metabolic changes.

Consider, for example, the tennis player in our opening photo. Tennis requires rapid body movements that anticipate ball movements and an opponent's possible responses. It also requires specific control of multiple skeletal muscles in split-second time. These functions could only be coordinated by the nervous system. In contrast, the coordination of adolescent development requires adjusting the activities of many tissues over a prolonged period. In principle, the nervous system *could* carry out a coordination task of this sort. To do so, however, the nervous system would require tens of thousands of discrete axons between integrating centers (such as the brain) and target cells, and would need to send trains of impulses along all these axons for years. In contrast, endocrine glands can accomplish this task with greater economy, by secreting a relatively small number of long-lasting chemicals into the blood.

Most tissues in an animal's body are under dual control of the nervous and endocrine systems, as noted earlier. The tennis player's skeletal muscles illustrate this dual control. A single muscle, such as the biceps, typically contains thousands of muscle cells innervated by more than 100 motor neurons. The nervous system can selectively activate a few, many, or all of the motor neurons to control rapidly and precisely the amount of force the player's biceps generates. And it can separately control each of the other skeletal muscles in his body. Simultaneously—as days and months go by—testosterone and other hormones secreted steadily by endocrine glands during adolescence facilitate widespread muscle growth and development.

The nervous and endocrine systems work together

In spite of the differences between the nervous and endocrine systems, these two systems do not operate in mutually exclusive ways. On the contrary, they work together closely, and each system can affect the other.

The nervous system exerts control over the endocrine system in many well known ways. The brain initiates endocrine development of the gonads during puberty in mammals, for example. Moreover, throughout life the brain helps control the secretion of gonadal hormones, thyroid hormones, and other hormones. We will discuss these controls in Concepts 35.3 and 35.4.

Conversely, the endocrine system sometimes exerts control over the nervous system. Sex hormones from the testes and ovaries affect brain development during mammalian puberty, for example. Studies on rats, hamsters, and other research animals indicate that the sex hormones have an organizing effect on the development of certain neural circuits in the brain so that the circuits develop some different specific properties in females and males.

Chemical signaling operates over a broad range of distances

Animals use chemical signaling over a very broad range of spatial scales (**FIGURE 35.2**). In this concept we have discussed chemical signaling on two spatial scales—the two different scales of distance seen in nervous signaling and in endocrine signaling. However, chemical signaling also takes place on even shorter and even longer scales of distance.

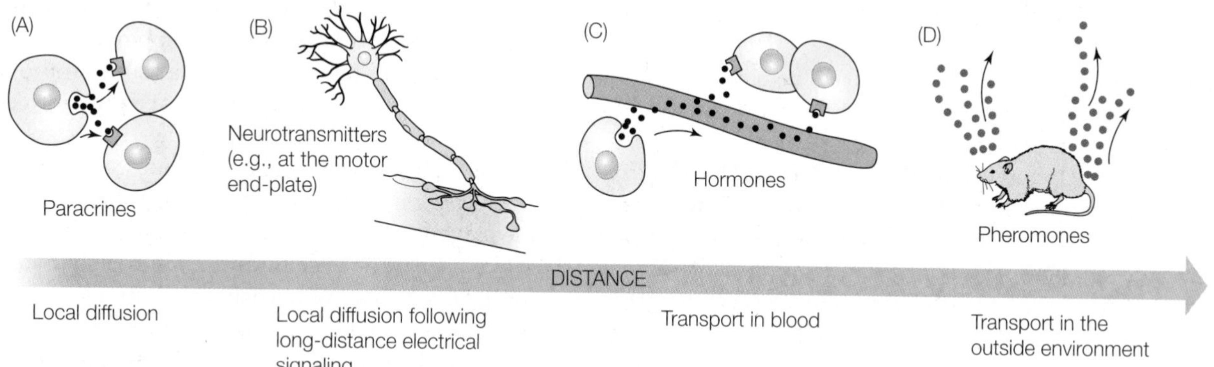

FIGURE 35.2 Chemical Signaling Occurs over a Broad Range of Spatial Scales (A) Cells communicate with neighboring cells in a tissue via molecules called paracrines. (B) Neurons have long processes (axons), at the ends of which they release molecules called neurotransmitters to communicate across synapses with target cells. (C) Endocrine cells secrete molecules called hormones that can travel long distances by circulating in the blood. (D) Animals give off molecules called pheromones that can communicate specific information (such as their sex) over very long distances to other members of their species.

Many chemical signals diffuse from cell to cell in a tissue without entering the blood. Two such types of signals are paracrines and autocrines. Paracrines are chemicals that are secreted by one cell and affect the functions of other, neighboring cells in a tissue by binding to receptors in or on the neighboring cells (see Figure 35.2A). Autocrines are chemicals that are secreted by a cell into the surrounding intercellular fluids and then diffuse to receptors on that very same cell and affect its functions (see Figure 5.10).

Neurotransmitters and hormones work at intermediate distances. Neurotransmitters resemble paracrines in that they move just short distances from cell to cell by diffusion, because only the width of a synaptic cleft separates a presynaptic neuron from its target cell. However, the secretion of neurotransmitters is controlled from farther away by electrical signals that travel the length of the presynaptic cell (see Figure 35.2B). Hormones are carried to the farthest reaches of an animal's body by the circulation of the blood (see Figure 35.2C).

Pheromones are chemical signals that an individual animal releases into its external environment and that exert specific effects (e.g., behavioral effects) on other individuals of the same species (see Figure 35.2D). Some pheromones travel hundreds of meters before reaching their targets.

CHECKpoint CONCEPT 35.1

- ✓ Why is endocrine control described as slow and broadcast?
- ✓ Why is it correct to say that both nervous and endocrine control depend on chemical signaling?
- ✓ If all cells in the body are bathed by a hormone, why are some affected while others are not?

Now that we've compared and contrasted the nervous and endocrine systems, let's turn our spotlight on the endocrine system.

CONCEPT 35.2 Hormones Are Chemical Messengers Distributed by the Blood

In Chapter 34 we discussed the nervous system in detail. Now we will turn to a detailed discussion of the endocrine system, starting by defining its fundamental properties.

Animals have two types of glands, which are termed exocrine and endocrine. All glands produce and secrete materials. The exocrine glands typically have outflow tubes, called ducts, that carry their secretions away. Examples include the mammary, salivary, and tear glands. In each case, the gland secretes a material (milk, saliva, or tears) that flows out by way of tubular ducts. The prostate gland, which contributes fluid to the semen, is another exocrine gland.

The **endocrine glands** do not have ducts (and therefore are sometimes called ductless glands). **Endocrine cells**, by definition, secrete their products into the blood flowing through nearby blood capillaries or other blood passages (keep in mind that in this book we use the term blood to refer to the fluid circulating in the circulatory system in all animals; see Concept 32.1). Because endocrine cells secrete into the blood, endocrine glands require only a blood supply, not ducts, for their secretions to be carried away.

In some cases, endocrine cells exist as single cells scattered within a tissue composed mostly of other cells, as we have observed in the gut epithelium (see Concept 30.5). In other cases, endocrine cells are aggregated (grouped) together to form discrete tissues or organs. These are the endocrine glands. Examples include the thyroid gland and adrenal glands.

The testes, ovaries, and pancreas illustrate a complexity that sometimes arises: these organs are composed of a mix of endocrine and non-endocrine tissues. A vertebrate testis, for example, consists mostly of tiny, non-endocrine tubules in which sperm are produced. In between these tubules are small aggregations of endocrine cells called the Leydig cells or interstitial cells. The testes are sometimes called endocrine glands. Technically, however, they are only partially endocrine glands.

A **hormone** is defined to be a chemical substance that is secreted into the blood by endocrine cells and that regulates the function of other cells that it reaches by blood circulation. An additional defining feature of hormones is that they act at very low blood concentrations: sometimes as low as 10^{-12} molar (moles per liter). Still another defining feature is that the effects of hormones on target cells are initiated by noncovalent binding of the hormone molecules to receptor protein molecules synthesized by the target cells. This feature gives hormones their specificity of action, as we stressed before. The specific action of a hormone on a target cell depends on the receptor proteins that the cell synthesizes and on the biochemical pathways the receptor proteins activate.

Growth, development, reproductive cycles, water balance, and long-term stress responses are some of the processes that typically are under primarily hormonal control in animals. This list reflects the fact that, as we discussed in Concept 35.1, hormones are well suited to control processes that involve many tissues and that occur on time scales of hours, days, months, or years.

Endocrine cells are neurosecretory or nonneural

Two broad classes of endocrine cells exist (**FIGURE 35.3**). Some, called **neurosecretory cells** or neuroendocrine cells, resemble neurons in that they are excitable cells that propagate action potentials. Others, called **nonneural endocrine cells** or epithelial endocrine cells, are not excitable.

The cell bodies of neurosecretory cells (see Figure 35.3A) are located in the central nervous system (CNS). These cells have axons, which typically extend outside the CNS. Neurosecretory cells release hormones into the blood at their axon terminals, instead of releasing neurotransmitters at synapses as neurons do. Often, groups of neurosecretory cells have their axon terminals positioned within a **neurohemal organ** (*neuro*, "related to neurons"; *hemal*, "related to blood"), a specialized part of the circulatory system in which the terminals are closely juxtaposed with the blood.

Neurosecretory cells provide a direct interface between the nervous and endocrine systems. Ordinary neurons make synaptic

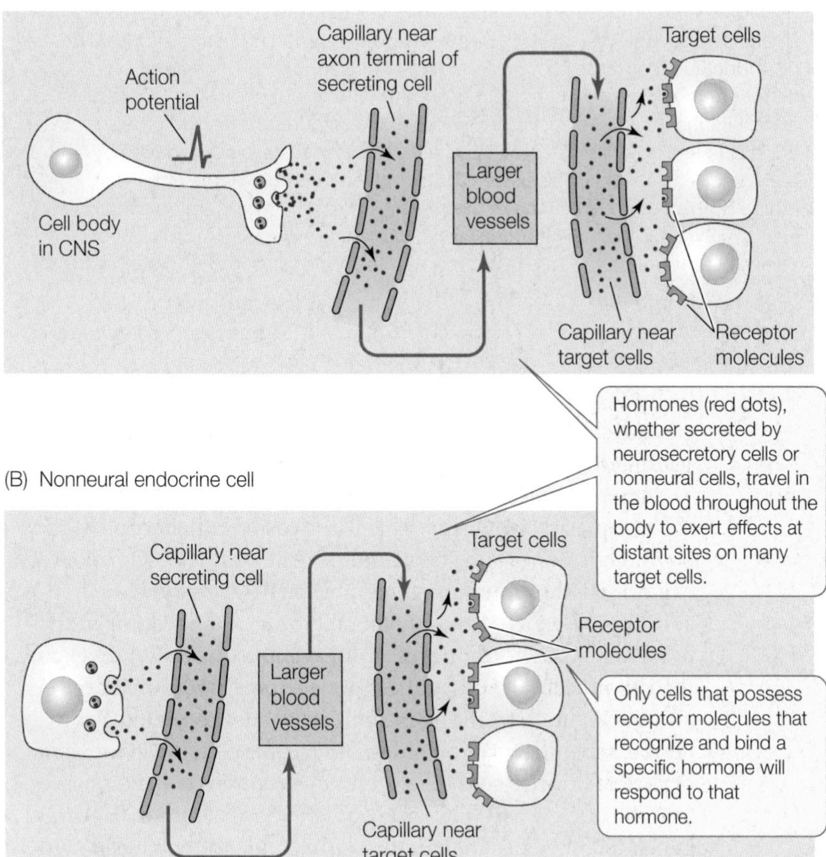

FIGURE 35.3 Two Types of Endocrine Cells The two principal types of endocrine cells are (A) neurosecretory cells and (B) nonneural endocrine cells.

contact with the dendrites and cell body of a neurosecretory cell in the CNS. The neurosecretory cell responds to neural activity in the CNS by secreting a hormone into the blood. The typical order of events in the function of a neurosecretory cell is that neurons in the CNS activate the cell by exciting it at synapses. The neurosecretory cell, in response, initiates and propagates action potentials. These action potentials activate hormone release into the blood at the cell's axon terminals. An example is provided by the cells that produce the hormone oxytocin in mammals. Oxytocin plays critical roles during the birth process and during nursing. It is secreted by neurosecretory cells in the hypothalamus of the brain.

The secretions of neurosecretory cells can quite properly be called hormones. Sometimes, however, scientists distinguish these secretions by calling them neurohormones or neurosecretions.

Nonneural endocrine cells (see Figure 35.3B) do not employ action potentials. They are typically stimulated to secrete their hormones by other hormones, for which they express receptor proteins. The typical order of events in the function of a nonneural endocrine cell is that the cell receives a hormonal signal from another gland. This hormone binds to receptor proteins, which initiate secretion by the nonneural endocrine cell. The cells of the anterior pituitary gland and thyroid gland provide examples.

Most hormones belong to one of three chemical groups

There is enormous diversity in the chemical structure of hormones, but we can classify most hormones into one of three groups:

- **Peptide and protein hormones** are molecules composed of chains of amino acids. They are water-soluble and therefore easily transported in the blood. Most hormones are in this group. Some are large molecules that may contain nearly 200 amino acid units (**FIGURE 35.4A**). Others are small molecules containing as few as three amino acid units. For simplicity, these hormones are often all referred to as peptide hormones (disregarding the fact that the larger ones are technically proteins), and we follow that practice here. During their synthesis, the peptide hormones are packaged in vesicles within the cells that make them. They are released by exocytosis (see Concept 5.4). Because water-soluble molecules cannot easily cross cell membranes, the receptors for peptide hormones are on the exterior of a target cell. There are dozens, perhaps hundreds, of peptide hormones, with diverse structures and effects.
- **Steroid hormones** are synthesized from cholesterol and share a similar structure of four interlinked rings (**FIGURE 35.4B**). Steroid hormones are lipid-soluble and readily pass through the membranes of the cells that synthesize them, but they do not dissolve well in blood plasma. Thus they are usually bound to carrier proteins for transport in the blood. Once they reach a target cell, they typically diffuse directly through its cell membrane. Most steroid receptors are found inside target cells, in the cytoplasm or nucleus.
- **Amine hormones** are small molecules that are modified amino acids. For example, both the thyroid hormone thyroxine and the fight-or-flight hormone epinephrine are made from the amino acid tyrosine (**FIGURE 35.4C**), and the hormone melatonin (which affects an animal's daily rhythms) is made from tryptophan. Depending on whether the modified amino acid is polar or nonpolar, an amine hormone may be water-soluble (can't readily cross membranes) or lipid-soluble (can readily cross membranes).

Receptor proteins can be on the cell surface or inside a cell

Some receptor proteins are positioned in the cell membrane of a target cell, as in the case of the receptors for peptide hormones and some amine hormones. The hormone binds to a part of the receptor protein that projects outside the cell membrane, and then the receptor molecule initiates processes that alter cell function. Many receptors of this type are G protein–linked receptors (see Concept 5.5) that, after binding to a hormone, initiate second messenger cascades inside the target cell. These

(A) Peptide hormones

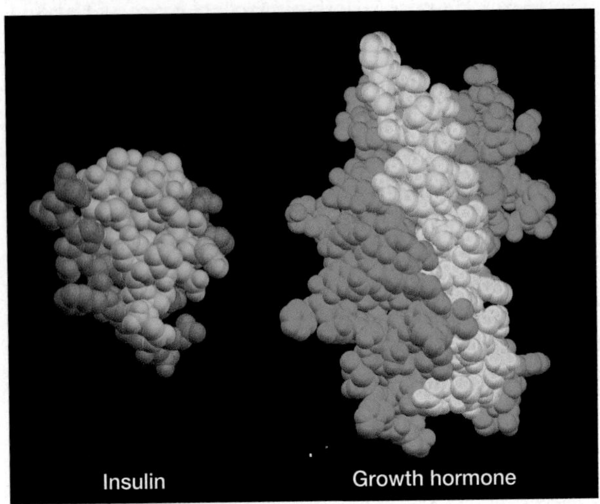

(B) Steroid hormones

Sterol backbone

Cholesterol

Corticosteroids

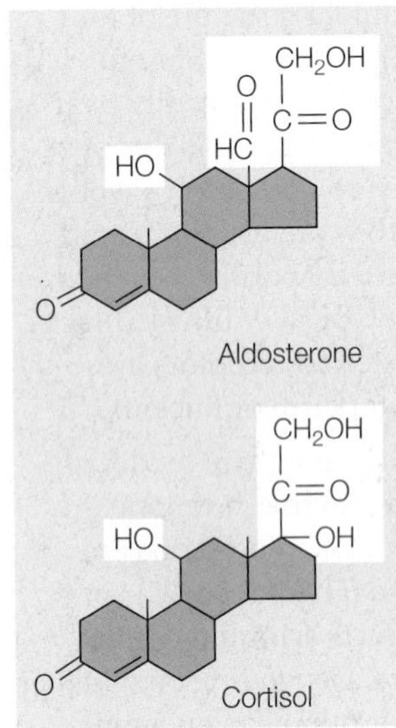

Sex steroids

Testosterone

Estradiol

(C) Amine hormones

Tyrosine

Epinephrine

Thyroxine

FIGURE 35.4 The Three Principal Chemical Classes of Hormones (A) Peptide hormones are composed of strings of amino acids. They include proteins and small polypeptide molecules. Shown here are two proteins, insulin and growth hormone. Each of these proteins consists of two subunits, which are shown in different colors. (B) Steroid hormones are modified from cholesterol molecules and all include a characteristic set of four ring structures, here called the sterol backbone and colored green. They include the corticosteroids produced by the adrenal cortex and the sex steroids produced primarily by the gonads. (C) Amine hormones are small molecules synthesized from single amino acid molecules. Thyroxine and epinephrine are both made from the amino acid tyrosine. The structures in B and C are in abbreviated form; not all carbon (C) atoms are shown.

cascades activate or inactivate enzymes in the cell cytoplasm, leading to the cell's response (see Figure 5.14).

Other receptor proteins are located inside the target cell, in the cytoplasm or nucleus. This is the case for the receptors for steroid hormones and some amine hormones. These hormones can diffuse through the cell membrane because they are lipid-soluble. When a steroid hormone binds to its receptor in the cytoplasm, the hormone–receptor complex usually moves to the cell nucleus. There it alters gene expression (transcription and translation), resulting in synthesis of new proteins. For example, when testosterone binds to receptors in the cytoplasm of skeletal muscle cells, testosterone–receptor complexes move into the nuclei of the cells and activate transcription of several genes. These genes include those that code for enhanced synthesis of the contractile proteins actin and myosin (see Concept 33.1). In this way, testosterone tends to bring about an increase in muscle mass.

LINK

You can learn more about the various types of receptor proteins in **Concept 5.5**

A target cell is not limited to expressing only a single kind of hormone receptor protein. Cells often express two or more receptor proteins, which make them target cells for two or more hormones.

The number of receptor molecules that a target cell has for a particular hormone can vary. In some cases, negative or positive feedback affects the number of receptor molecules that a target cell expresses. For example, continuous high concentrations of a hormone sometimes cause a target cell to decrease its number of receptors for the hormone. This decrease makes the cell less sensitive to the hormone, tending to offset the effect of the continuous high concentration. For example, in type II diabetes mellitus, chronically high levels of the hormone insulin (usually caused by excessive carbohydrate intake) lead to decreased production of insulin receptors throughout the body. As a result, cells become less sensitive to insulin.

LINK

The crucial role of insulin in regulating glucose metabolism is detailed in **Concept 30.5**

Hormone action depends on the nature of the target cells and their receptors

The action of a hormone on a target cell depends on characteristics of the target cell—in particular, the receptor protein the cell expresses and the cellular processes activated by the receptor protein. For this reason, one hormone can have dramatically different effects on different cells.

In the course of evolution, many hormones have been conserved for long periods of evolutionary time, but their functions have changed as various types of animals have evolved. This pattern—in which a single hormone molecule evolves changing functions—arises because the actions of a hormone depend on the receptor systems, which can change. The peptide hormone prolactin provides a dramatic example.

Prolactin was first discovered in mammals, where it stimulates milk synthesis in the mammary glands (breasts). Prolactin has since been found in all other vertebrate groups—from fish to birds—even though these other phyla have neither mammary glands nor milk. In salmon, prolactin is involved in stabilizing blood ion composition as the fish migrate between salt and fresh water. In birds, prolactin can stimulate nest building and parental care. The hormone—in one molecular form or another—has existed in vertebrates throughout their evolutionary history, but prolactin has taken on radically different functions in various vertebrate groups by evolution of the target cells it affects and their receptor systems.

Within one individual animal, some target cells for a hormone can have different receptor proteins and receptor systems than other target cells do. The hormone then elicits different responses from the different types of target cells. For example, in a sudden emergency, our fight-or-flight response is activated to help us confront the emergency. As part of this response, our adrenal medullary glands (**FIGURE 35.5**) secrete epinephrine and norepinephrine (sometimes known as adrenaline and noradrenaline), while at the same time our sympathetic nervous system (see Concept 34.5) exerts its effects. Within seconds, these hormones are circulating in our blood. We have at least five types of receptors for the hormones, all of them G protein–linked receptors. These receptors fall into two major categories—called α-adrenergic and β-adrenergic receptors—that bind epinephrine and norepinephrine with different degrees of ease. Cells of different tissues have different receptors. Moreover, even when different types of cells have the same receptor, they may respond very differently because the receptor activates different processes in different cell types. For these reasons, the hormones bring about a large range of responses in various tissues: faster and stronger beating of the heart, constriction of blood vessels in the skin, dilation of blood vessels in skeletal muscles, breakdown of glycogen to release glucose in the liver (see Figure 5.17), and a decrease in both blood flow and secretion of digestive enzymes in the gut.

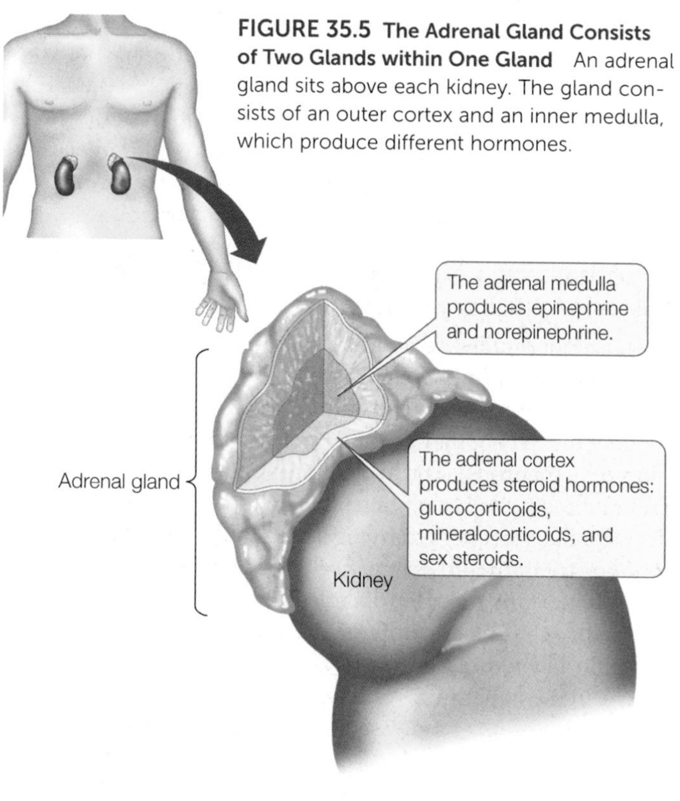

FIGURE 35.5 The Adrenal Gland Consists of Two Glands within One Gland An adrenal gland sits above each kidney. The gland consists of an outer cortex and an inner medulla, which produce different hormones.

The adrenal medulla produces epinephrine and norepinephrine.
The adrenal cortex produces steroid hormones: glucocorticoids, mineralocorticoids, and sex steroids.
Adrenal gland
Kidney

LINK

Signal transduction pathways—the cellular processes activated by receptors—are discussed in **Concept 5.6**

A hormonal signal is initiated, has its effect, and is terminated

When endocrine cells are activated to secrete a hormone, they vary in how quickly they start hormone release. Peptide and amine hormones are usually synthesized prior to use. They can be secreted very rapidly because they are present in stored form in endocrine cells. In contrast, steroid hormones are typically synthesized on demand. Initiation of secretion of steroid hormones is relatively slow because of the need to synthesize the hormones before they can be released.

Just as there are processes for secreting a hormone into the blood, there are ways to remove it. Removal takes place in several ways. Target cells sometimes enzymatically degrade hormones. Certain organs, such as the liver and kidneys in vertebrates, also enzymatically degrade hormones. Hormones may also be excreted.

The removal processes ultimately terminate a hormone signal. The half-life of a hormone in the blood is the time required for half of a group of simultaneously secreted hormone molecules to be removed from the blood. Some hormones have half-lives measured in minutes, and in these cases an individual hormonal signal does not last long. Epinephrine, for example, has a half-life of 1–2 minutes, explaining why the jolt of a fight-or-flight response disappears quickly if we decide no emergency exists and we are safe. Other hormones have far longer half-lives. In the human bloodstream, for example, the hormones antidiuretic hormone (involved in water conservation), cortisol (involved in long-term responses to stress), and thyroxine (affects cellular metabolism) display average half-lives of about 15 minutes, 1 hour, and nearly 1 week, respectively. They may therefore have prolonged effects on target tissues.

When a hormone is secreted steadily over a period of time, the blood hormone concentration is determined by an interaction between the rate of secretion (adding hormone to the

APPLY THE CONCEPT

Hormones are chemical messengers distributed by the blood

One way to characterize the time course of a hormone is to measure its half-life in the blood. A hormone's half-life can be measured in the laboratory by injecting some of the hormone and then determining the length of time required for the blood level of the hormone to fall from its maximum level halfway back to its baseline value. The table gives blood concentrations of the thyroid hormone thyroxine (T_4) following a 600-µg injection in a particular experiment. Plot these data.

1. Use the data in the table to estimate the half-life of 600 µg of T_4 in the bloodstream in this experiment. (The concentration at time 0 is the baseline.)
2. If you were trying to correct a hormone deficiency by administering hormone therapy and needed to keep the hormone level in the blood from falling too low, how would your dosing frequency differ for hormones with different half-lives?

TIME (hr) AFTER INJECTION	BLOOD CONCENTRATION (µg/100ml)
0 (just prior to injection)	7.5
6	13.7
12	12.3
24	11.1
36	10.7
48	10.3
60	9.9
72	9.5
84	9.3
96	9.1

blood) and the rate of removal. Removal of any particular hormone tends to take place at a fairly steady rate. Thus changes in the blood concentration of a hormone depend chiefly on changes in the rate at which the hormone is secreted.

The blood concentration is stabilized by negative feedback in many cases, making negative feedback a major theme in the study of endocrinology. During negative feedback, if the blood concentration rises above a set-point level, secretion of the hormone is inhibited.

LINK

Many physiological processes in animals have set points and are regulated by negative feedback; **Concept 29.6** describes the components of negative-feedback systems

When they are in the blood, some hormones are bound to carrier proteins. For example, steroid hormones must be bound with water-soluble proteins to be in solution in the blood. Carrier proteins often extend the half-lives of hormones.

Some hormones are converted to more active forms after they are secreted, a process called peripheral activation. A classic example is provided by the hormones of the thyroid gland. The thyroid secretes a molecule that contains four iodine atoms. This molecule is called thyroxine, T_4, or tetraiodothyronine (see Figure 35.4C). The hormone becomes more active when peripheral tissues enzymatically remove one of the iodine atoms, forming T_3, triiodothyronine.

CHECKpoint CONCEPT 35.2

- ✓ Growth hormone is a large, water-soluble peptide. Given these characteristics, how would you expect this hormone to interact with its target cells?
- ✓ Are sweat glands correctly described as endocrine glands?
- ✓ Name one major similarity and one major difference between an ordinary neuron and a neurosecretory cell.

For understanding the vertebrate endocrine system, one of the most important themes is that the brain often controls the secretion of hormones by the endocrine system. We will start with that theme as we now turn our attention to the vertebrate endocrine system.

CONCEPT 35.3 The Vertebrate Hypothalamus and Pituitary Gland Link the Nervous and Endocrine Systems

The **pituitary gland**, found at the base of the brain (**FIGURE 35.6A**; see also Figure 34.25), is sometimes called the "master gland" because it secretes hormones that control the functions of many other glands. It is attached by a stalk to a part of the brain called the **hypothalamus**. The pituitary has two parts—the **anterior pituitary** (adenohypophysis) and **posterior pituitary** (neurohypophysis). The two have different developmental origins, which helps account for marked differences in the way they function. Both parts of the pituitary have close functional links with the brain.

Hypothalamic neurosecretory cells produce the posterior pituitary hormones

The posterior pituitary is basically an extension of the brain (**FIGURE 35.6B**). Its hormones are synthesized and secreted by brain neurosecretory cells. The cell bodies of these cells are in the hypothalamus, where they receive synapses from other, ordinary brain cells (neurons). The axons of the neurosecretory cells extend into the posterior pituitary, where they terminate in a rich blood capillary network, forming a neurohemal organ where the hormones of the neurosecretory cells are secreted into the blood.

An important general principle is that different sets of neurosecretory cells secrete different hormones. We can apply this principle as we focus on the hypothalamus and posterior pituitary. In most mammals and also in many other vertebrates, there are two sets of neurosecretory cells, which secrete two

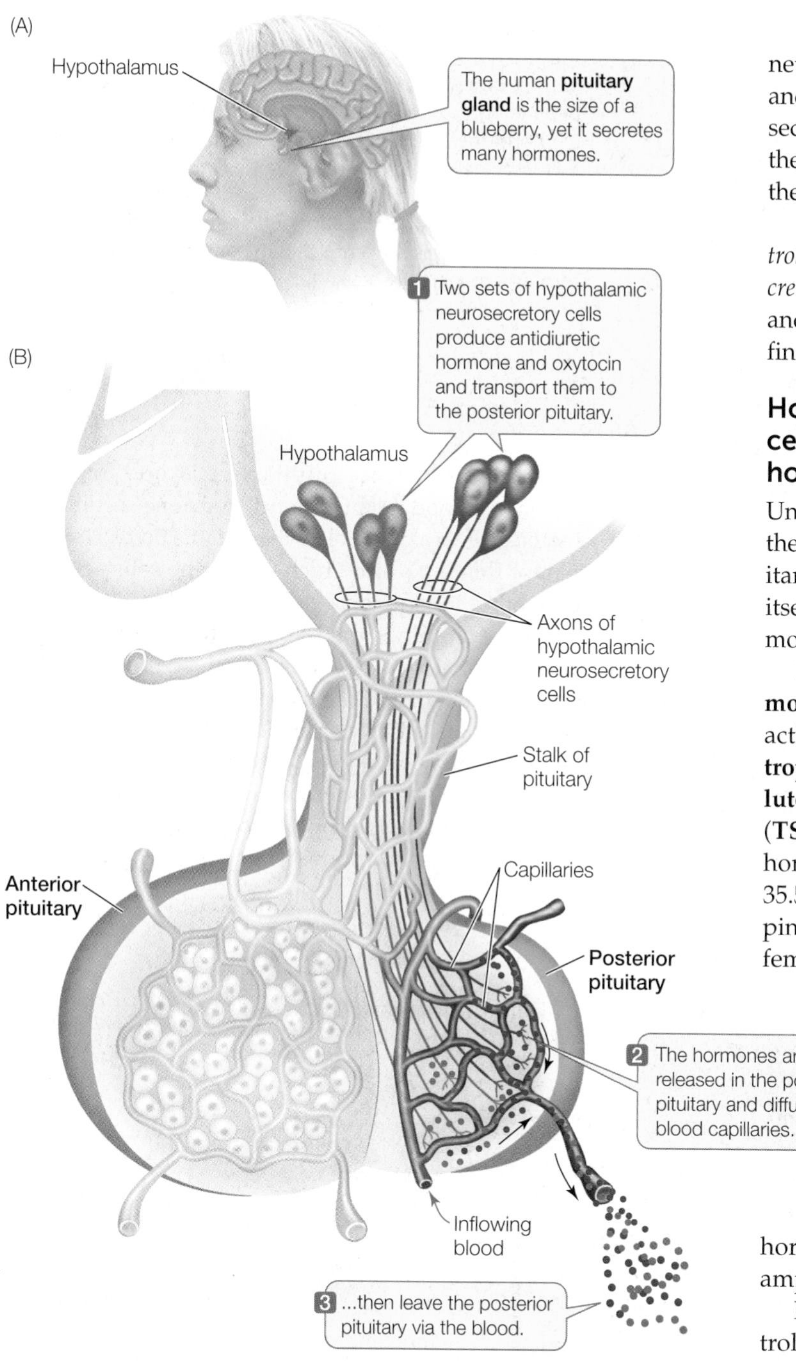
(A)
Hypothalamus
The human **pituitary gland** is the size of a blueberry, yet it secretes many hormones.
(B)
1 Two sets of hypothalamic neurosecretory cells produce antidiuretic hormone and oxytocin and transport them to the posterior pituitary.
Hypothalamus
Axons of hypothalamic neurosecretory cells
Stalk of pituitary
Capillaries
Anterior pituitary
Posterior pituitary
2 The hormones are released in the posterior pituitary and diffuse into blood capillaries...
Inflowing blood
3 ...then leave the posterior pituitary via the blood.

FIGURE 35.6 The Posterior Pituitary Gland (A) The human pituitary gland is located just below the hypothalamus of the brain. (B) The posterior pituitary is a neurohemal organ for neurosecretory cells in the brain. Two sets of neurosecretory cells in the hypothalamus produce two peptide hormones that are released in the posterior pituitary.

hormones, in the posterior pituitary system. Both hormones are short peptides containing nine amino acid units. In mammals these two hormones are **antiduiretic hormone** (**ADH**; also called vasopressin) and **oxytocin**. ADH helps control water excretion by the kidneys, as we will discuss in Chapter 36. Oxytocin controls milk ejection from the mammary glands and helps control contractions of the uterus during the birth process (see Figure 29.17).

Signals produced by the brain control secretion by these neurosecretory cells. In this way, the enormous computational and integrative powers of the brain can be marshaled to control secretion of the posterior pituitary hormones. When activated, the neurosecretory cells generate action potentials that travel to their axon terminals, leading to hormone release into the blood.

In short, in the posterior pituitary system, *brain neurons control the secretion of hormones into the general circulation by neurosecretory cells*. This is one type of interface between the nervous and endocrine systems—a type of interface that we commonly find in both vertebrates and invertebrates.

Hormones from hypothalamic neurosecretory cells control production of the anterior pituitary hormones

Unlike the posterior pituitary (which is a neurohemal organ), the anterior pituitary is an endocrine gland. The anterior pituitary hormones are synthesized in the anterior pituitary gland itself by nonneural endocrine cells (**FIGURE 35.7**). All the hormones are peptides or closely related to peptides.

Four of the hormones are known as **tropins**, or **tropic hormones**, meaning their principal functions are to control the activities of other endocrine glands. They are **adrenocorticotropic hormone** (**ACTH**), **follicle-stimulating hormone** (**FSH**), **luteinizing hormone** (**LH**), and **thyroid-stimulating hormone** (**TSH**). A different set of pituitary cells produces each tropic hormone. ACTH helps control the adrenal cortex (see Figure 35.5). Both FSH and LH, which together are called gonadotropins, control the gonads: the testes in males and the ovaries in females. TSH controls the thyroid gland.

The anterior pituitary gland also produces several other hormones. These hormones act on non-endocrine targets and therefore are not tropic hormones. They include **growth hormone** (**GH**; see Figure 35.4A), which acts on a wide variety of tissues to promote growth. They also include prolactin (which we've already mentioned), and melanocyte-stimulating hormone, which is named for its action on skin-color cells in amphibians.

How is secretion of the anterior pituitary hormones controlled? A critical role is played by a specialized vascular network, the **hypothalamo–hypophysial portal system** ("hypophysis" is an alternative name for pituitary). This vascular system consists of two capillary beds that are connected so that blood flows through them sequentially. Blood flows first through a capillary bed in the hypothalamus. Neurosecretory cells in the hypothalamus secrete hormones into the blood in this capillary bed. When the blood flows out of the hypothalamus, it travels a short distance to the second capillary bed, which is in the anterior pituitary. In this way, the hormones secreted by the hypothalamic neurosecretory cells are brought to the pituitary nonneural endocrine cells. These hormones control secretion by the endocrine cells in the pituitary. A typical sequence of action begins when ordinary brain neurons that synapse on the hypothalamic neurosecretory cells stimulate those cells to secrete hormones. Those hormones then travel to the anterior

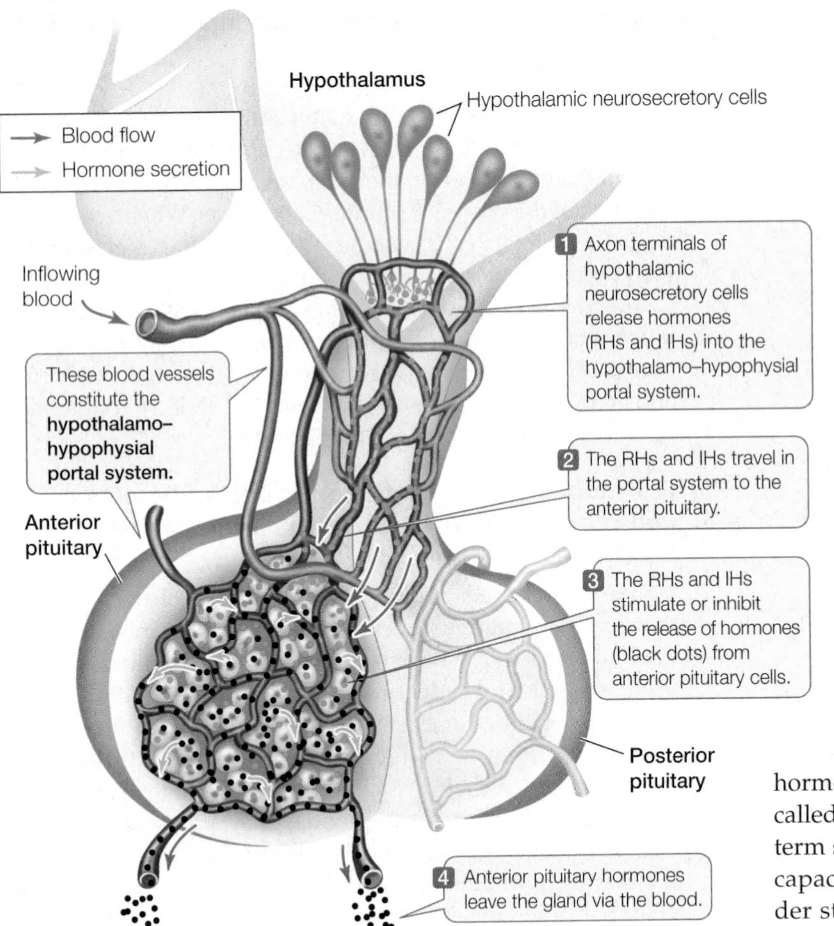

Hypothalamus
Hypothalamic neurosecretory cells
Blood flow
Hormone secretion
Inflowing blood
These blood vessels constitute the **hypothalamo–hypophysial portal system.**
Anterior pituitary
1 Axon terminals of hypothalamic neurosecretory cells release hormones (RHs and IHs) into the hypothalamo–hypophysial portal system.
2 The RHs and IHs travel in the portal system to the anterior pituitary.
3 The RHs and IHs stimulate or inhibit the release of hormones (black dots) from anterior pituitary cells.
Posterior pituitary
4 Anterior pituitary hormones leave the gland via the blood.

FIGURE 35.7 The Anterior Pituitary Gland Cells of the anterior pituitary produce four tropic hormones that control other endocrine glands, as well as several other peptide hormones. These cells are controlled by releasing hormones (RHs) and inhibiting hormones (IHs) produced by neurosecretory cells in the hypothalamus. The RHs and IHs are delivered to cells in the anterior pituitary by blood flowing through portal blood vessels that run between the hypothalamus and the anterior pituitary through the pituitary stalk.

pituitary and control secretion by the anterior pituitary non-neural endocrine cells.

The hormones added to the blood by the neurosecretory cells in the hypothalamus are called **releasing hormones (RHs)** and **inhibiting hormones (IHs**; also called release-inhibiting hormones). Each RH or IH is specific in its actions. For example, there are separate RHs for growth hormone and thyroid-stimulating hormone. When the RH for growth hormone is secreted in the hypothalamus, it affects the specific cells in the anterior pituitary that secrete growth hormone, stimulating them to secrete (release) their hormone into the general circulation. The RH for thyroid-stimulating hormone, in contrast, stimulates the specific pituitary cells that secrete TSH.

In summary, in the anterior pituitary system, *neurosecretory cells in the brain control the secretion of hormones into the general circulation by other endocrine cells*. This is a second type of interface between the nervous and endocrine systems. It too is a type of interface that we commonly find in both vertebrates and invertebrates.

Endocrine cells are organized into control axes

A typical sequence in vertebrates is for a hormone secreted by brain neurosecretory cells to control the secretion of a tropic hormone by anterior pituitary cells, and for the tropic hormone then to control hormone secretion by the cells in a gland elsewhere in the body. A system of this type—in which endocrine cells act on each other in sequence—is called an **axis**.

One example is the hypothalamus–pituitary–adrenal cortex (HPA) axis. Some of the steroid hormones secreted by the adrenal cortex, including a class called glucocorticoids, are important in responding to long-term stress. When the brain, using its great computational capacity to integrate inputs, detects that an animal is under stress, some of the cells in the hypothalamus secrete corticotropin-releasing hormone. This releasing hormone

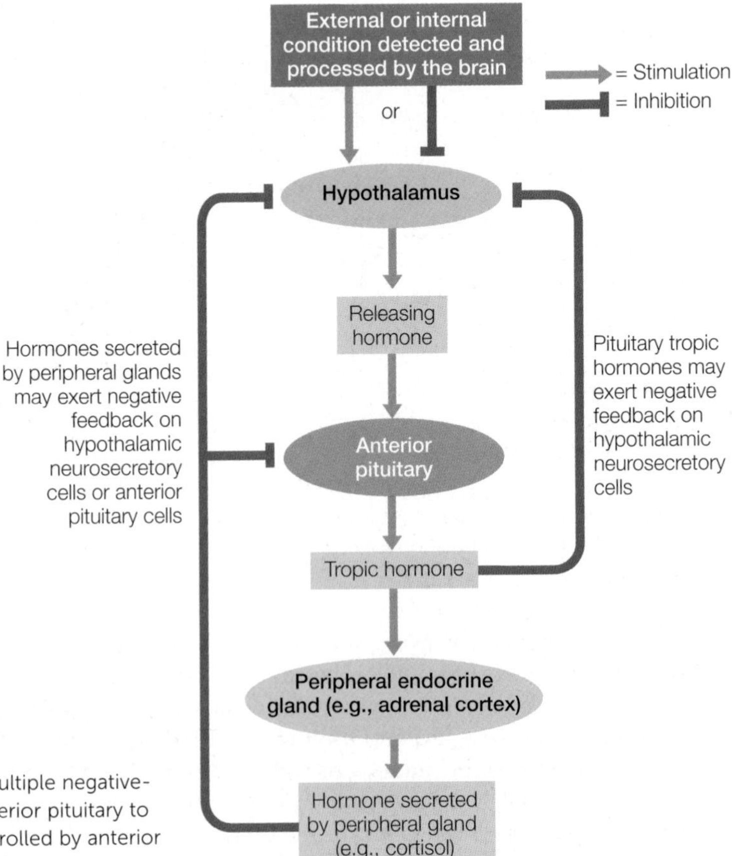

External or internal condition detected and processed by the brain
or
= Stimulation
= Inhibition
Hypothalamus
Releasing hormone
Anterior pituitary
Tropic hormone
Peripheral endocrine gland (e.g., adrenal cortex)
Hormone secreted by peripheral gland (e.g., cortisol)
Hormones secreted by peripheral glands may exert negative feedback on hypothalamic neurosecretory cells or anterior pituitary cells
Pituitary tropic hormones may exert negative feedback on hypothalamic neurosecretory cells

FIGURE 35.8 Multiple Feedback Loops Control Hormone Secretion Multiple negative-feedback loops regulate the chain of command from hypothalamus to anterior pituitary to peripheral endocrine glands. For purposes of this diagram, the glands controlled by anterior pituitary tropic hormones are termed peripheral glands.

travels in the portal system to cells in the anterior pituitary gland that respond by increasing secretion of the tropic hormone adrenocorticotropic hormone (ACTH). ACTH then travels in the general circulation to the adrenal cortex, which responds by increasing secretion of glucocorticoids into the blood. In primates, the principal glucocorticoid is cortisol (see Figure 35.4B).

Negative feedback often takes place within an axis. How does it work? One way is that the hypothalamic neurosecretory cells and the endocrine cells of the anterior pituitary are often under negative-feedback control by the hormones of their target glands (**FIGURE 35.8**). For example, when the blood concentration of cortisol (or other glucocorticoids) increases following stimulation of the adrenal cortex by the HPA axis, blood cortisol tends to cause cells in the hypothalamus to reduce release of corticotropin-releasing hormone and cells in the anterior pituitary to reduce secretion of ACTH. This negative feedback helps stabilize the blood cortisol concentration at a relatively constant level.

Go to ANIMATED TUTORIAL 35.1
The Hypothalamus and Negative Feedback
PoL2e.com/at35.1

Hypothalamic and anterior pituitary hormones are often secreted in pulses

In many cases the brain stimulates the neurosecretory cells in the hypothalamus to secrete in pulses instead of continuously. Corticotropin-releasing hormone, for example, is typically secreted in mammals in two or three pulses per hour. Pulsed secretion of RHs or IHs often proves to be necessary for these hormones to act (**FIGURE 35.9**). Continuous secretion can lead target cells in the pituitary to reduce their numbers of RH or IH receptor molecules and thus reduce their sensitivity. Pulsed secretion is hypothesized to prevent this loss of sensitivity. At the next step in the axis, the anterior pituitary hormones are often also secreted in pulses.

INVESTIGATION

FIGURE 35.9 Effectiveness of Gonadotropin-Releasing Hormone (GnRH) Depends on a Pulsatile Pattern of Release In adult monkeys, GnRH is released from the hypothalamus in short pulses, about once every 1–3 hours. When the supply of GnRH is stopped (for example, by removing or destroying the hypothalamus), the anterior pituitary no longer secretes FSH or LH, and levels of these gonad-regulating hormones decline. Early studies in rhesus monkeys showed that in such cases FSH and LH production could be restored by injections of GnRH, but only if GnRH was delivered in pulses rather than continuously. The two patterns of GnRH delivery, however, also resulted in very different *total* doses of GnRH, complicating the interpretation of these initial results. The researchers did further experiments to determine whether delivery of GnRH in pulses is critical to its function in stimulating the secretion of FSH and LH.[a]

HYPOTHESIS

The effectiveness of GnRH depends on the dose, not on the pattern of delivery.

METHOD

1. Administer GnRH to rhesus monkeys in which the hypothalamic cells responsible for GnRH secretion have been destroyed. Administer GnRH either continuously at 0.001, 0.01, 0.1, or 1.0 μg/min, or in pulses (one 6-min pulse per hr) at 1.0 μg/min.
2. Measure the concentrations of LH and FSH in the blood.

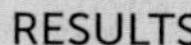

RESULTS

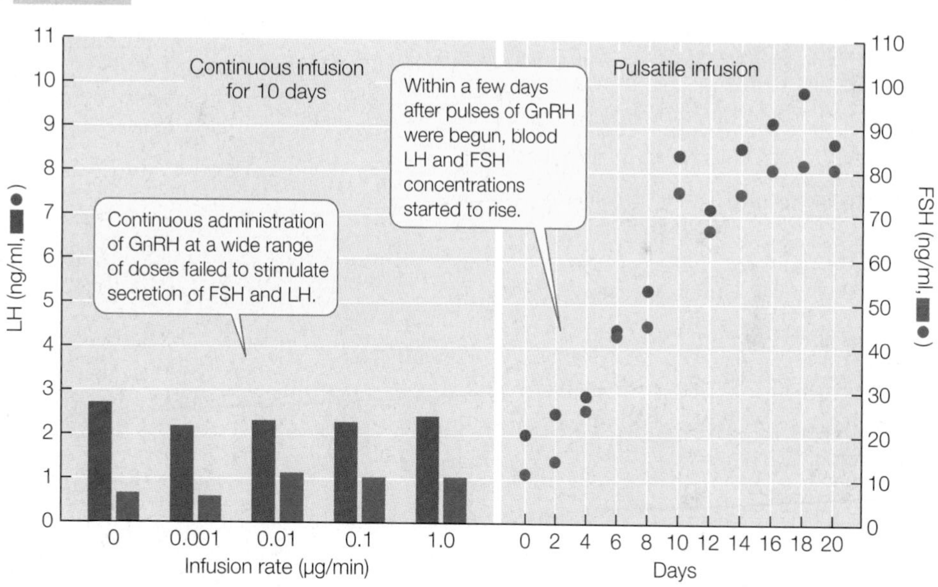

CONCLUSION

The hypothesis is not supported: the effectiveness of the GnRH depends on the pattern of delivery.

ANALYZE THE DATA

A. For delivery in pulses, GnRH was administered at 1 μg/min for 6 minutes once per hour. Which continuous rate delivered the same amount of GnRH per hour as the delivery in pulses? How much GnRH was delivered per hour at the lowest continuous rate? How much lower was this than the dose from pulses?

B. For interpreting this experiment, why is it important that one of the continuous delivery rates provides the same total dose per hour as the pulsatile delivery?

C. We now know that the anterior pituitary stops responding to GnRH if GnRH is always present. What does this tell you about how fast each pulse of GnRH disappeared from the bloodstream? Be as quantitative as possible, taking account of the GnRH dose that you calculated for the lowest continuous rate above.

Go to **LaunchPad** for discussion and relevant links for all **INVESTIGATION** figures.

[a]P. E. Belchetz et al. 1978. *Science* 202: 631–633.

CHECKpoint CONCEPT 35.3

- ✓ Considering the hierarchy of control, why might it be misleading to call the pituitary gland the "master gland"?
- ✓ Why are many anterior pituitary hormones called tropic hormones?
- ✓ Neurosecretory cells in the hypothalamus and cells in the anterior pituitary gland both have cortisol receptors. If they were to lose their cortisol receptors, these cells would immediately begin secreting unusually large amounts of corticotropin-releasing hormone and ACTH. Why?

The hypothalamus and anterior pituitary gland control the secretion of hormones from several other endocrine glands. We will now look at the hormones that some of these other glands release, and the functions of these hormones.

CONCEPT 35.4 Hormones Regulate Mammalian Physiological Systems

Hormones are involved in controlling and coordinating a wide range of mammalian physiological systems, and these regulatory processes in mammals often have parallels in other vertebrates. Here we take a brief look at the some of the major roles of hormones in the lives of mammals.

FIGURE 35.10 shows the locations of the principal mammalian endocrine glands and summarizes their major hormones and actions. In addition to the principal glands, many organs—such as the heart and stomach—are not named as glands but nonetheless contain endocrine cells. The box in the figure lists some of these. We discuss several endocrine control systems in other chapters, as **FIGURE 35.11** indicates.

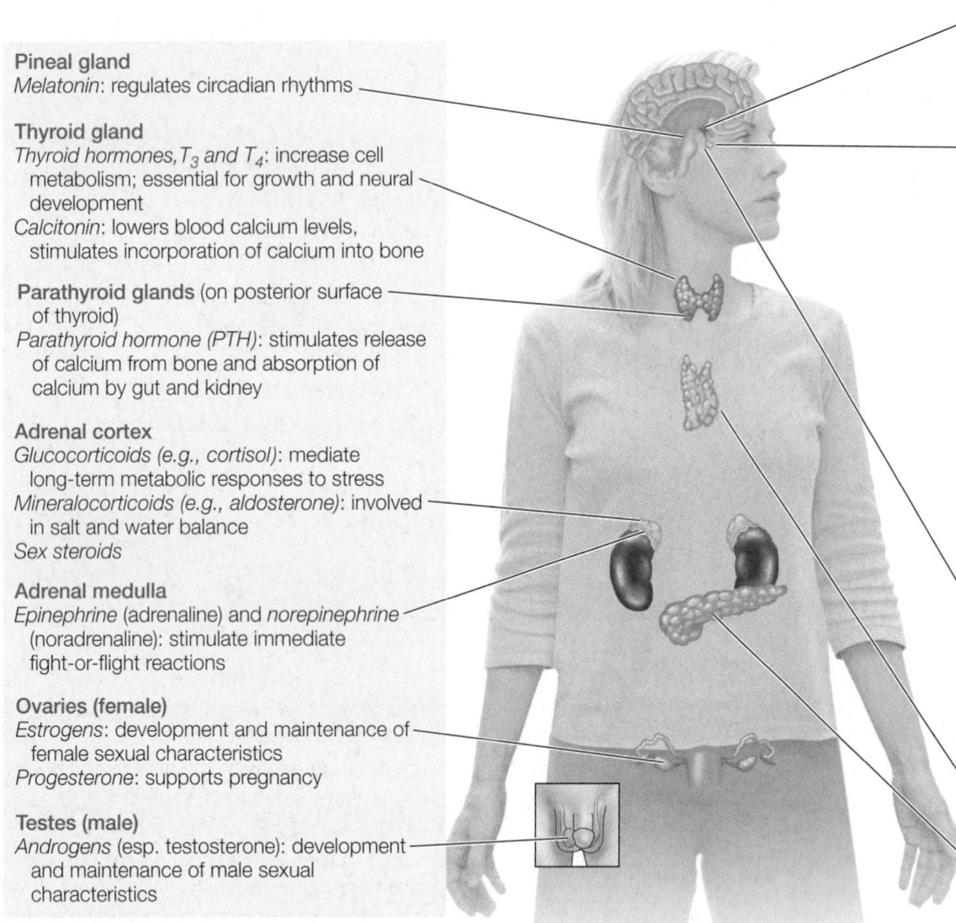
Pineal gland
Melatonin: regulates circadian rhythms
Thyroid gland
Thyroid hormones, T_3 and T_4: increase cell metabolism; essential for growth and neural development
Calcitonin: lowers blood calcium levels, stimulates incorporation of calcium into bone
Parathyroid glands (on posterior surface of thyroid)
Parathyroid hormone (PTH): stimulates release of calcium from bone and absorption of calcium by gut and kidney
Adrenal cortex
Glucocorticoids (e.g., cortisol): mediate long-term metabolic responses to stress
Mineralocorticoids (e.g., aldosterone): involved in salt and water balance
Sex steroids
Adrenal medulla
Epinephrine (adrenaline) and *norepinephrine* (noradrenaline): stimulate immediate fight-or-flight reactions
Ovaries (female)
Estrogens: development and maintenance of female sexual characteristics
Progesterone: supports pregnancy
Testes (male)
Androgens (esp. testosterone): development and maintenance of male sexual characteristics
Hypothalamus
Releasing and inhibiting hormones: control the anterior pituitary
ADH and *oxytocin*: transported to and released from the posterior pituitary (below)
Anterior pituitary gland
Thyroid-stimulating hormone (TSH): stimulates the thyroid gland
Follicle-stimulating hormone (FSH): in females, stimulates maturation of ovarian follicles; in males, stimulates spermatogenesis
Luteinizing hormone (LH): in females, triggers ovulation and ovarian production of estrogens and progesterone; in males, stimulates production of testosterone
Adrenocorticotropic hormone (ACTH): stimulates secretory activity of the adrenal cortex
Growth hormone (GH): stimulates protein synthesis and growth
Prolactin: stimulates milk production
Melanocyte-stimulating hormone (MSH): controls skin pigmentation
Endorphins and *enkephalins*: pain control
Posterior pituitary gland (site of release of two hormones made by hypothalamic neurosecretory cells)
Oxytocin: stimulates contraction of uterus, flow of milk
Antidiuretic hormone (ADH): promotes water conservation by kidneys
Thymus (diminishes in adults)
Thymosins: play roles in the immune system
Pancreas (only certain parts, the islets of Langerhans, are endocrine)
Insulin: stimulates cells to take up and use glucose
Glucagon: stimulates liver to release glucose

Other organs include cells that produce and secrete hormones:

Organ	Hormone
White adipose tissue	Leptin
Heart	Atrial natriuretic peptide
Kidney	Erythropoietin
Stomach	Gastrin, ghrelin
Intestine	Secretin, cholecystokinin
Liver	Insulin-like growth factors

FIGURE 35.10 The Human Endocrine System

Go to ACTIVITY 35.1 The Human Endocrine Glands
PoL2e.com/ac35.1

Chapter 29
Fundamentals of Animal Function

Oxytocin (from the posterior pituitary) plays a critical role in controlling uterine muscle contractions essential for a baby to be born (see Figure 29.17).

Chapter 30
Nutrition, Feeding, and Digestion

Secretin and cholecystokinin (both from scattered endocrine cells in the midgut epithelium) help coordinate parts of the gut during meal processing (see Concept 30.5; Figure 30.14).

Ghrelin (from the stomach) and **leptin** (from white adipose tissue) help control appetite (see Concept 30.5; Figure 30.15).

Insulin and glucagon (both from the pancreas) regulate processing of absorbed food materials between meals (see Concept 30.5; Figure 30.16).

Chapter 36
Water and Salt Balance

Antidiuretic hormone (from the hypothalamus and the posterior pituitary) controls excretion of water by the kidneys (see Concept 36.5; Figure 36.9).

Chapter 37
Animal Reproduction

Follicle-stimulating hormone and luteinizing hormone (from the anterior pituitary) help control the menstrual and estrous cycles and are essential for male fertility (see Concept 37.2; Figure 37.11).

Estrogens (from the ovaries) help control the menstrual and estrous cycles (see Concept 37.2; Figure 37.11).

Progesterone and estrogens (from the ovaries and placenta) help control uterine development that is required for pregnancy (see Concept 37.2; Figure 37.11).

FIGURE 35.11 Several Endocrine Control Systems Are Discussed in Other Chapters Endocrine control is of such widespread importance that it features in many chapters in Part 6.

The thyroid gland is essential for normal development and provides an example of hormone deficiency disease

The thyroid gland is found in the neck, wrapped around the front of the trachea (windpipe) (see Figure 35.10). Two different cell types in the thyroid gland produce two different hormones. One hormone is thyroxine (T_4), described earlier, and the other is calcitonin (involved in calcium metabolism). Here we focus on T_4, an amine hormone that contains four iodine atoms (see Figure 35.4C). The presence of iodine in the T_4 structure is the reason for most of our dietary iodine requirement. The thyroid also releases a relatively small amount of T_3 (triiodothyronine)—a molecule formed by removal of an iodine from T_4—that has an iodine atom at only three of the four positions where T_4 does. T_3 is more active in affecting target cells, and after T_4 has been secreted into the blood, peripheral tissues convert it to T_3, as noted earlier. When biologists speak of "thyroid hormones," they mean T_4 and T_3 unless stated otherwise.

The thyroid hormones are vital during development and growth. They promote cellular amino acid uptake and protein synthesis. They enter cells and bind to an intracellular receptor that stimulates transcription of numerous genes. Insufficiencies of the thyroid hormones in a human fetus or growing child greatly retard mental and physical development. The thyroid hormones also are noted for elevating metabolic rate in mammals and birds.

FIGURE 35.12 Goiter

Soils deficient in iodine occur in many parts of the world. Plants that grow in these soils tend to be low in iodine, and people who depend on local foods in these areas often suffer from iodine deficiency. Iodized table salt (salt containing traces of iodine) was invented to solve this problem. If people purchase iodized salt and use it in a customary way to salt their food, they get enough iodine regardless of whether or not they live in an iodine-deficient part of the world.

Impaired mental development caused by childhood deficiency of thyroid hormones—resulting from iodine deficiency—remains a massive worldwide problem because in many poverty-stricken regions, iodized salt is not readily available. Global human mental performance would be raised in a generation if iodine deficiency during pregnancy and childhood could be eliminated.

Goiter is another hormone deficiency disease that (in its most common form) results from inadequate dietary iodine. It illustrates that the proper operation of feedbacks in endocrine control systems can be critical for health. Goiter refers to an enlarged thyroid gland (**FIGURE 35.12**). It was common in many parts of the United States until use of iodized salt became the norm.

Adults develop goiter when they get too little iodine in their diet and, as a result, their thyroid produces inadequate amounts of thyroid hormones. When thyroid hormones are present in sufficient amounts in the blood, they exert negative feedback on the production of thyroid-stimulating hormone (TSH) by the anterior pituitary. This feedback does not take place in people producing only small amounts of thyroid hormones. Their blood levels of thyroid hormones are low, and accordingly such people produce high levels of TSH due to a lack of negative feedback on TSH production. Their steady, high production of TSH stimulates their thyroid gland to grow to excessive size.

Sex steroids control reproductive development

Many hormones, including the thyroid hormones, play critical roles in mammalian development. In certain ways, however, the hormones of greatest importance for development are the sex steroids and the hypothalamic and pituitary hormones that control them.

The gonads—the testes of males and the ovaries of females—produce steroid hormones as well as gametes (sperm and ova). Many of these hormones are described as androgens or estrogens. Masculinizing steroid hormones are called **androgens** (Greek, "male-makers"). Testosterone is an androgen and is the principal hormone produced by the testes. It was named testosterone when it was discovered in 1935 for the two-fold reason that it was collected from testes and it belongs chemically to the family of cholesterol-like compounds (see Figure 35.4B).

Go to MEDIA CLIP 35.1
The Testosterone Factor
PoL2e.com/mc35.1

Feminizing steroid hormones are called **estrogens**. Two or more types of estrogen molecules are typically secreted, one often being estradiol (see Figure 35.4B). The ovaries produce estrogens and another female sex steroid, **progesterone**, that is involved principally in coordinating processes associated with pregnancy. During pregnancy, the placenta also secretes estrogens and progesterone. We will discuss these hormones—and the hormones that control their secretion—at length in Chapter 37.

Androgens are not exclusive to males, and estrogens are not exclusive to females. Both sexes make both androgens and estrogens. However, the relative blood concentrations differ in the two sexes. In some cases androgens are simply precursors in the synthesis of estrogens, to which androgens are converted by a process called aromatization.

During embryonic development, the sex steroids play a key role in controlling whether a human embryo develops phenotypically into a male or female. Androgens are required for male phenotypic development, and in their absence an embryo develops as a female. The sex of an individual mammal is determined genetically at the time of conception. Females inherit two X chromosomes, whereas males inherit an X and a Y chromosome (see Concept 37.1). At the start of embryonic development, the gonads and external genital structures are the same whether an individual is XX or XY. In males, early in development a gene on the Y chromosome, the *SRY* gene, causes the gonads to differentiate as testes (otherwise they become ovaries). As the differentiation into testes proceeds, the gonads of a male embryo begin to produce testosterone. In humans, starting 2–3 months after conception, testosterone increases from a low blood concentration to a high blood concentration, remains high for about 2 months, and then falls to a low concentration again because the gonads stop producing it rapidly. This period of high blood testosterone causes the external genital structures

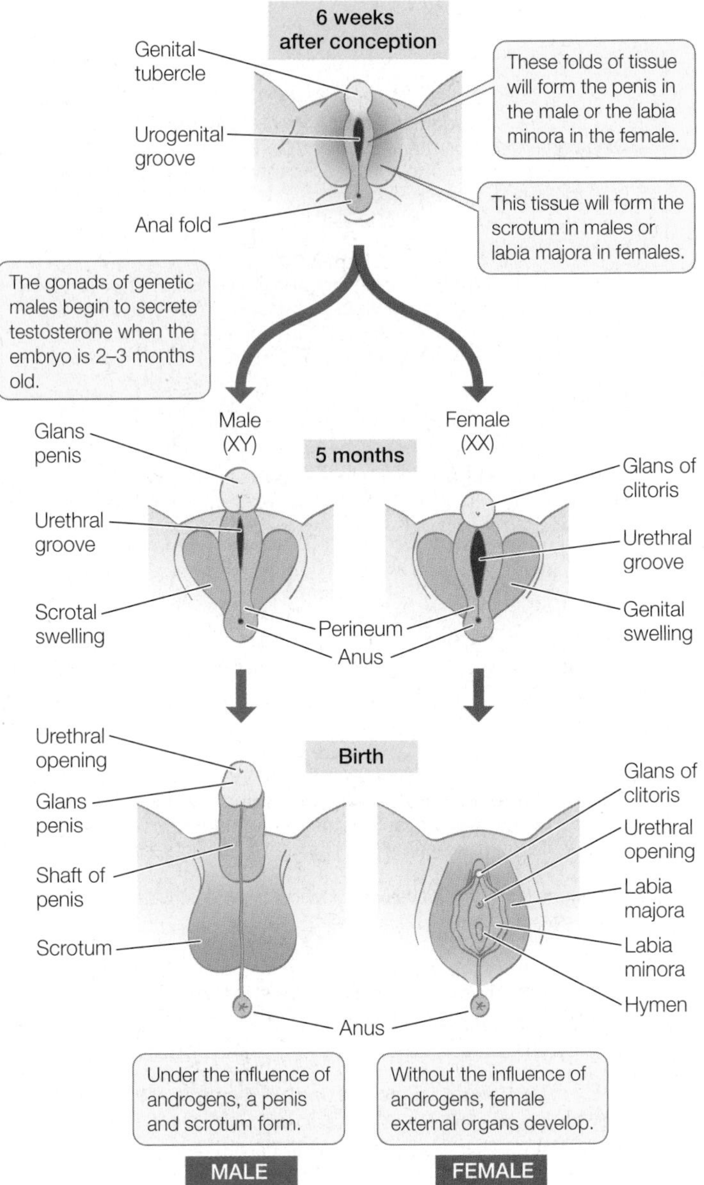

FIGURE 35.13 Sex Steroids Direct the Prenatal Development of Human Sex Organs The external sex organs of early human embryos are undifferentiated. Testosterone promotes the development of male external sex organs. In its absence, female sex organs form.

to differentiate into the male form. Without testosterone, the structures differentiate into female form.

Particular external genital structures diverge into male or female form during this phenotypic sexual differentiation. The scrotum of males and certain labial folds of females develop from the same early structures (**FIGURE 35.13**). The head (glans) of the penis and the head of the clitoris also have the same embryonic origins. The gonads are initially located in the abdomen in both sexes. In males, the testes migrate out of the abdomen into the scrotum. The ovaries remain in the abdomen in females.

Just as hormones play a crucial role in embryonic development, they again play a crucial role at puberty. Recall that luteinizing hormone (LH) and follicle-stimulating hormone (FSH)—both anterior pituitary tropic hormones that act on the gonads—are together called the **gonadotropins**. Their secretion by the anterior pituitary is under the control of hypothalamic neurosecretory cells that produce **gonadotropin-releasing hormone (GnRH)**. As we have been doing, let's focus on humans as we discuss the process of puberty.

When a person is between about 1 year of age and the start of puberty, the GnRH neurosecretory cells in the hypothalamus are quiet, producing a GnRH pulse only once every few hours. Puberty begins when these cells suddenly shift to producing GnRH pulses at a far higher rate, a change that results in a sustained increase in the average concentration of GnRH in blood flowing into the anterior pituitary. Despite many theories, researchers remain uncertain about the causes of the change in activity of the GnRH neurosecretory cells. The consequences, however, are well known and dramatic.

The anterior pituitary responds to the increase in the pulse rate of GnRH secretion with a large increase in FSH and LH secretion. In males, the elevated blood concentration of LH stimulates cells in the testes to produce testosterone vigorously. A large increase occurs in the blood concentration of testosterone, which (along with other androgens) brings about many changes. Testosterone enters cells, binds to intracellular receptors, and alters gene expression throughout the body. As a consequence, a boy's voice deepens, hair begins to grow on his face and body, his skeletal muscles increase in mass, and his testes and penis grow larger. The increased blood concentration of FSH stimulates the production of sperm.

In females, increased blood levels of LH stimulate the ovaries to increase production of estrogens. Concentrations of estrogens in the blood rise and initiate development of many traits: a girl's breasts, vagina, and uterus become larger; her hips broaden; she develops increased subcutaneous fat; and her menstrual cycles begin. Increased FSH stimulates the maturation of ovarian follicles, which are necessary for production of mature eggs (see Concept 37.1).

Sex steroids continue to play essential roles in reproductive cycles and sexual functioning throughout adulthood. We will discuss these roles in Chapter 37.

CHECKpoint CONCEPT 35.4

- ✓ Boys are occasionally born with a testicle missing from the scrotum. The missing testicle is nearly always in the abdomen, and its position can be corrected surgically. But why might a testicle be in the abdomen at birth?
- ✓ Explain why the thyroid gland can grow to great size when it secretes inadequate amounts of thyroid hormones because of iodine deficiency.
- ✓ What endocrine event in the hypothalamus initiates puberty?

Now that we've examined vertebrate endocrine systems, let's conclude with a look at endocrine systems in invertebrates. Of all the invertebrates, the insects are the best understood in this regard. A striking parallel is that endocrine controls are of enormous importance in insect development, just as they are in vertebrate development.

CONCEPT 35.5 The Insect Endocrine System Is Crucial for Development

Insects and other arthropods, such as crayfish and crabs, have elaborate endocrine systems. For example, many insects have antidiuretic and diuretic hormones that control excretion of water by the insect organs that serve kidney functions. Diuretic hormones promote excretion of a high volume of water. Some of the blood-sucking insects secrete diuretic hormones immediately after each blood meal. These hormones promote rapid excretion of much of the water in the blood, thereby concentrating the nutritious part of the meal (e.g., the blood proteins) in the gut.

The best-understood endocrine systems in insects control growth and development. A key feature of insects is that they have exoskeletons. For an insect to grow, it must periodically shed its exoskeleton and produce a larger one. This shedding process is called molting or ecdysis. The life of an insect consists of a series of discrete steps: growing followed by molting, then more growing followed by molting, and so on. Insects typically undergo changes in their external shape as they develop from young individuals to adults. However, the rigid exoskeleton cannot change shape between molts. This means that all the changes of external shape in the life of an insect must occur during molting. In this way, the molting process is intimately connected to both growth and morphological development (development of form).

Many hormones are involved in controlling the molting process, but three are of principal importance (**FIGURE 35.14**). We will start by discussing two of them and then address the

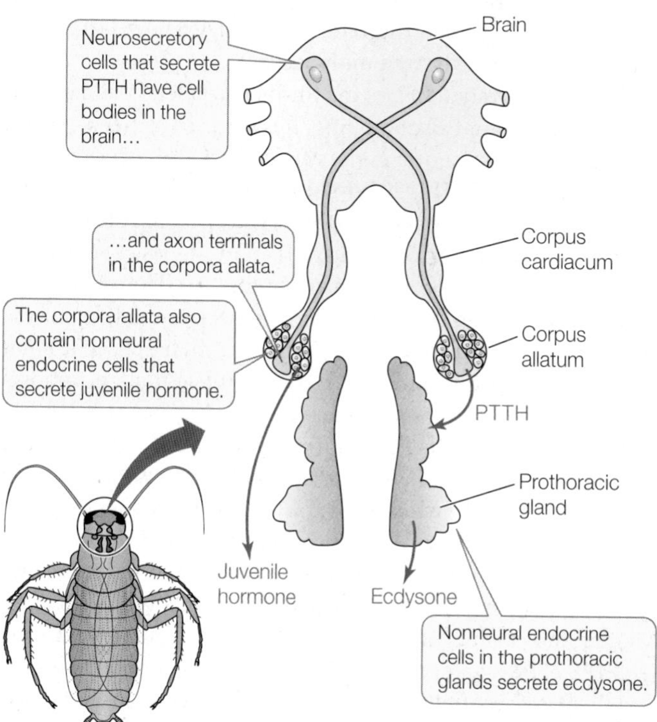

FIGURE 35.14 Key Structures and Hormones in the Control of Insect Development Hormones are in red. PTTH = prothoracicotropic hormone

third. One is prothoracicotropic hormone (PTTH), a peptide tropic hormone secreted by brain neurosecretory cells. The second is ecdysone, a steroid secreted by nonneural endocrine cells in the prothoracic glands. PTTH controls secretion of ecdysone. In the regulation of molting, *neurosecretory cells in the brain control secretion of hormones into the general circulation by other endocrine cells*. This is one of the two types of interface between the nervous and endocrine systems that we identified in vertebrates. Invertebrates and vertebrates, we see, sometimes exhibit similar major organizational features.

The brain uses its integrative and computational powers to determine when each molt will occur. The neurosecretory cells that secrete PTTH receive synaptic inputs from other, ordinary brain neurons that control when the neurosecretory cells will secrete their hormone. PTTH is not released directly from the brain, however. In an arrangement reminiscent of that in the posterior pituitary gland in vertebrates, axons of the brain neurosecretory cells extend out of the insect brain and end in or near organs called the corpora allata (singular, corpus allatum). These are neurohemal organs, where PTTH is released into the blood from the axon terminals of the brain neurosecretory cells.

During each episode of PTTH secretion, the flow of blood carries the PTTH to the two prothoracic glands, where it stimulates an episode of ecdysone secretion. As the ecdysone circulates, it undergoes peripheral activation to form 20-hydroxyecdysone, a hormone that primarily affects the insect's epidermis, the layer of living tissue just inside the exoskeleton. The epidermis secretes the exoskeleton. The hormone, being a steroid, enters the epidermal cells. There it combines with intracellular receptors, which alter gene transcription patterns. As a result, the epidermal cells secrete enzymes that loosen their connection with the old exoskeleton, allowing the old exoskeleton to be shed. Then the epidermal cells synthesize a new, larger exoskeleton. In other words, a molt occurs.

Juvenile hormone (JH) is the third hormone of primary importance in the control of molting. It is secreted into the general circulation by nonneural endocrine cells in the corpora allata (see Figure 35.14). It is lipid-soluble and enters target cells (chemically it is a terpene and thus of a different sort than the other hormones we have discussed). Many types of cells in an insect's body have intracellular receptors for JH and respond to it by undergoing changes in gene transcription. JH controls an insect's body form as the insect molts. Some insects, such as moths and butterflies, undergo a complete metamorphosis during their development. Each individual starts life as a larva (caterpillar), retains its larval form for several molts, and then metamorphoses into an adult of radically different form (see Concept 23.4). Here we discuss the role of JH in such insects.

When a larva molts, it retains its larval form if JH is at high concentration in the blood during the molting process. During the early life of an individual, the blood concentration of JH is high. Molting thus results in a series of larger and larger larval forms (**FIGURE 35.15**). Later in life, however, JH secretion is reduced, and the JH concentration in the blood falls to a low level. At that point, when molting occurs, the juvenile form is not retained. Instead, the insect enters a resting stage of distinctive body form, called a pupa. During the inactive pupal stage, the insect's body is extensively remodeled inside the pupal exoskeleton. Then, when the insect molts yet again without a high concentration of JH in the blood, the individual emerges as an adult.

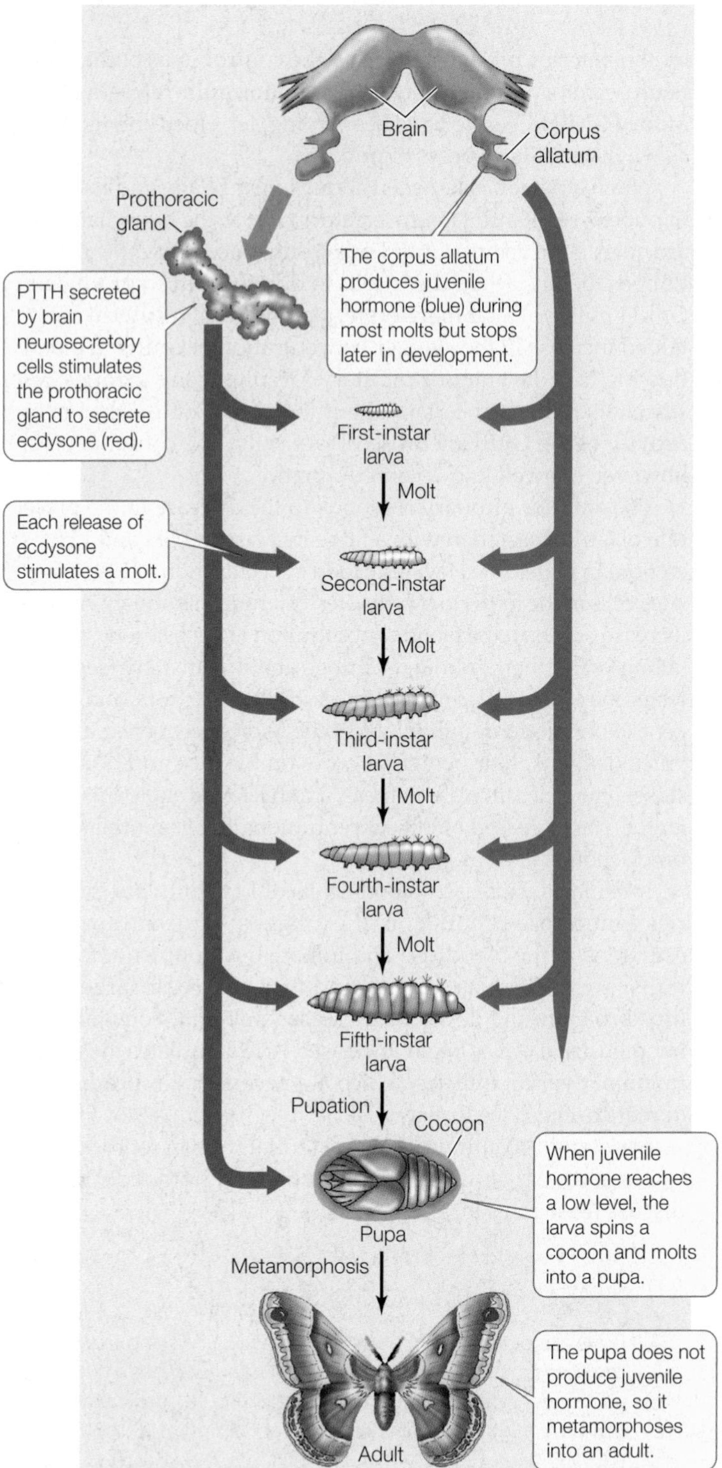

FIGURE 35.15 Hormonal Control of Molting and Metamorphosis in a Moth The silkworm moth *Hyalophora cecropia* is illustrated. An instar is a stage between molts. The ever-larger larval forms are thus referred to as successive larval instars.

Go to ANIMATED TUTORIAL 35.2
Complete Metamorphosis
PoL2e.com/at35.2

CHECKpoint CONCEPT 35.5

✓ Why is juvenile hormone called "juvenile" hormone?

✓ What two actions does ecdysone have on insect epithelial cells?

✓ In what way is there a fundamental similarity between (1) the control of ecdysone secretion by PTTH in insects and (2) the control of FSH secretion by GnRH in mammals?

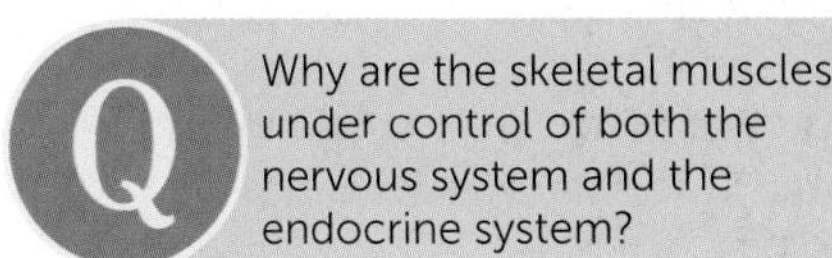

Why are the skeletal muscles under control of both the nervous system and the endocrine system?

ANSWER The muscles need to be controlled on two scales of spatial definition and two scales of time (Concept 35.1). The nervous and endocrine systems are suited to exerting control on these different scales—explaining why evolution has favored control by both systems. During a game of tennis, for example, the skeletal muscles of the arms, legs, and other parts of the body need to be controlled individually and on a moment-by-moment basis. Nervous control is suited to this task because it is addressed and fast. During adolescent development, muscles throughout the body need to be controlled together and on a sustained basis for many months as they grow and gain strength (Concept 35.4). Endocrine control is suited to this task because it is broadcast and slow.

SUMMARY

CONCEPT 35.1 The Endocrine and Nervous Systems Play Distinct, Interacting Roles

- Nerve and endocrine cells control and coordinate the functions of the body by releasing chemical signals that travel to another cell, called the **target cell**. **Review Figure 35.1**
- Neuronal signals are fast and addressed, whereas endocrine signals are slow and broadcast. The nervous system controls predominantly the fine, rapid movements of discrete skeletal muscles. The endocrine system typically controls more widespread, prolonged activities such as developmental or metabolic changes.
- Animals use chemical signaling over a very broad range of spatial scales. Autocrines and paracrines act on, respectively, the cells producing the signals and on their immediate neighbors. Neurotransmitters and hormones work at intermediate distances. **Pheromones** are released into the environment and can affect targets hundreds of meters away. **Review Figure 35.2**

CONCEPT 35.2 Hormones Are Chemical Messengers Distributed by the Blood

- Animals have two types of secretory glands: exocrine glands, which have ducts to carry away their secretions, and **endocrine glands**, which do not have ducts. **Endocrine cells** secrete chemical signals into the blood.
- A **hormone** is a chemical substance that is secreted into the blood by endocrine cells and that regulates the function of other cells that it reaches by blood circulation.
- Some endocrine cells are **neurosecretory cells**; they propagate action potentials and secrete hormones into the blood from their axon terminals. Other endocrine cells are **nonneural endocrine cells**; they are non-excitable cells that are typically stimulated to secrete their hormones by other hormones. **Review Figure 35.3**
- Most hormones are **peptide hormones** (polypeptides or proteins), **steroid hormones**, or **amine hormones**. Peptide hormones and some amine hormones are water-soluble; steroids and some amine hormones are lipid-soluble. **Review Figure 35.4**
- Receptors for water-soluble hormones are located on the cell surface of a target cell. Receptors for most lipid-soluble hormones are inside the cell.
- Hormones cause different responses in different target cells, depending on each target cell's type of receptor and the processes activated inside the cell by binding of hormone to that receptor.
- Each hormone has a characteristic half-life, the time required for half of a group of simultaneously secreted hormone molecules to be removed from the blood. Typical half-lives range from minutes to as long as a week.

CONCEPT 35.3 The Vertebrate Hypothalamus and Pituitary Gland Link the Nervous and Endocrine Systems

- The **pituitary gland** has two parts—the **anterior pituitary** and **posterior pituitary**—which have different developmental origins and function in different ways. Both parts of the pituitary have close functional links with the brain.
- The posterior pituitary is a neurohemal organ where hormones produced by hypothalamic neurosecretory cells are released into the blood. In mammals it secretes two peptide hormones: **antidiuretic hormone** (**ADH**) and **oxytocin**. **Review Figure 35.6**
- The anterior pituitary is a nonneural endocrine gland that secretes four **tropic hormones** as well as **growth hormone** (**GH**), prolactin, and a few other hormones. It is controlled by **releasing hormones** (**RH**s) and **inhibiting hormones** (**IH**s; also called release-inhibiting hormones) produced by neurosecretory cells in the **hypothalamus**. **Review Figure 35.7**
- Endocrine cells often act on each other in sequence, a system known as an **axis**. For example, the hypothalamus–pituitary–adrenal cortex (HPA) axis controls the adrenal secretion of glucocorticoids. **Review ANIMATED TUTORIAL 35.1**
- Hormone release often is controlled by negative-feedback loops. **Review Figure 35.8**
- Releasing hormones and inhibiting hormones are often released from the hypothalamus in pulses. Pulsed release is hypothesized to prevent loss of sensitivity in the target cells in the pituitary. **Review Figure 35.9**

(continued)

SUMMARY *(continued)*

CONCEPT 35.4 Hormones Regulate Mammalian Physiological Systems

- The thyroid gland is controlled by **thyroid-stimulating hormone** (**TSH**) and secretes the thyroid hormones thyroxine (T_4) and triiodothyronine (T_3), which control cellular metabolism. Iodine deficiency impairs thyroid hormone production and can lead to impaired mental development in children and to goiter in adults. **Review Figures 35.10–35.12 and ACTIVITY 35.1**
- Sex steroids (**androgens** in males, **estrogens** and **progesterone** in females) are produced by the gonads under control of tropic hormones, called gonadotropins, secreted by the anterior pituitary gland. Sex steroids control prenatal sexual development, puberty, and adult reproductive functions. **Review Figure 35.13**

CONCEPT 35.5 The Insect Endocrine System Is Crucial for Development

- Two hormones, prothoracicotropic hormone (PTTH) and ecdysone, control molting in insects. A third hormone, juvenile hormone (JH), prevents maturation. When an insect stops producing juvenile hormone, it molts into an adult. **Review Figures 35.14 and 35.15 and ANIMATED TUTORIAL 35.2**

See **ACTIVITY** 35.2 for a concept review of this chapter.

Go to the Interactive Summary to review key figures, Animated Tutorials, and Activities
PoL2e.com/is35

Go to LaunchPad at **macmillanhighered.com/launchpad** for additional resources, including LearningCurve Quizzes, Flashcards, and many other study and review resources.

36 Water and Salt Balance

KEY CONCEPTS

Salmon migrate between fresh water and the ocean. These Atlantic salmon spent the last few years in the ocean. To spawn, they now are swimming upstream to reach suitable spawning areas in the very river where they were conceived.

Salmon are amazing as they jump up waterfalls during their travels upstream. But even more amazing is something we don't see: the concentration of salts and water in their blood plasma stays almost constant as they migrate between the salty oceans and freshwater rivers. We humans can swim in the ocean or fresh water without upsetting our internal water and salt balance, because our skin does not allow water or salts to diffuse through it. However, fish are different. Some of their outer body surfaces—such as their gills and mouth membranes—are permeable to water and salts. To get oxygen, fish need gills that are highly permeable so that O_2 can diffuse easily into their blood from the water in which they swim. Gills that are highly permeable to O_2 are also highly permeable to water and salts. The blood plasma of fish can therefore gain water from or lose water to the surrounding environmental water by osmosis, and it can gain or lose ions by diffusion. For these reasons, fish do not resemble us in the way they relate to the water and salts in their surroundings.

The blood plasma of salmon is about one-third as concentrated with salts as seawater. Each individual salmon starts life in fresh water, where its blood is far more concentrated than the environmental water in which it swims. During this stage of life, the salmon gains water by osmosis from its surroundings and loses salts by diffusion to its surroundings—processes that tend to dilute its blood. Later, the salmon migrates into the ocean, where its blood is far less concentrated with salts than its environment. During the years it lives in the ocean, it gains salts by diffusion from its surroundings and loses water by osmosis to its surroundings—processes that tend to make its blood become more concentrated.

Osmosis is relentless, whatever its direction. It carries water into or out of a salmon during every hour of the animal's life. Similarly, whatever the direction of salt diffusion, it too is relentless.

How, then, can the concentration of the blood plasma remain almost constant throughout a salmon's life? The salmon has mechanisms that allow it to control the concentration of its blood plasma. These cost energy, and they must be properly regulated by the nervous and endocrine systems.

Salmon dramatically illustrate homeostasis: stability of the internal environment (see Concept 29.3). For most cells in a salmon's body, the type of water in the fish's external environment—whether fresh water or seawater—is of almost no consequence. This is true because the cells are bathed by the tissue fluids, which are similar to blood plasma and therefore almost constant in composition. Because of this stable internal environment, the cells enjoy independence of the external environment.

Q How can a salmon keep the composition of its blood plasma stable regardless of whether it is swimming in seawater or fresh water?

You will find the answer to this question on page 766.

CONCEPT 36.1 Kidneys Regulate the Composition of the Body Fluids

The body fluids of an animal are crucial for its survival because they provide the context in which all living functions take place. An animal's cells are bathed by its tissue fluids (intercellular fluids), which make up the environment of the cells: the animal's internal environment (see Concept 29.3). The organelles inside cells, such as the mitochondria and nucleus, are bathed by the intracellular fluids.

The body fluids have three properties: osmotic pressure, ionic composition, and volume. These properties are interrelated, but often it is helpful to think of them as three distinct characteristics.

- *Osmotic pressure*. The osmotic pressure is basically a measure of the total concentration of solutes (dissolved matter). Each individual dissolved entity makes an approximately equal contribution to the osmotic pressure. For example, if a body fluid has glucose, Na^+, and Cl^- dissolved in it, the individual dissolved entities are the individual glucose molecules, Na^+ ions, and Cl^- ions. Each—regardless of its type—makes a roughly equal contribution to the osmotic pressure. Osmotic pressures are important because they determine the direction of osmotic water movement. Water moves by osmosis from regions where the osmotic pressure (total solute concentration) is low to regions where the osmotic pressure (total solute concentration) is high.

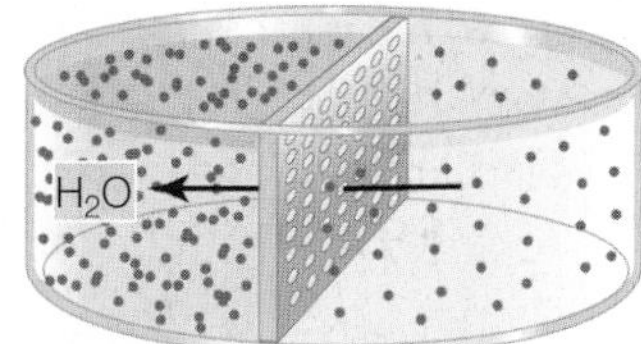

Water moves by osmosis from a region where the osmotic pressure is low to one where it is high. Dots symbolize individual dissolved entities.

  A good way to remember this is to recognize that a solution low in solutes is high in water, and vice versa. Water moves by osmosis from where it is high to where it is low: from low osmotic pressure to high osmotic pressure.

- *Ionic composition*. The principal solutes in most biological solutions are Na^+, K^+, and Cl^-. Other ions, such as Ca^{2+}, also occur. Two solutions of the same osmotic pressure are often very different in ion concentrations. One solution, for example, might be rich in Na^+ and the other rich in K^+. Thus it is important to consider the concentrations of particular ions, not just the osmotic pressure. Ion concentrations are important for several reasons. One is that they help determine the direction of ion diffusion. Each particular type of ion (such as Na^+ or K^+) diffuses in response to its particular concentration differences from place to place.
- *Volume*. We experience the importance of body fluid volume every time we get a bruise that causes tissue swelling. A body fluid may be correct in its osmotic pressure and ion concentrations but may still cause problems if its volume is wrong. Tissue swelling happens because of an accumulation of excess body fluids. Volume is always important even if we don't notice it. For example, consider a 100-gram goldfish in a goldfish bowl (**FIGURE 36.1**). The goldfish absorbs so much water from its surroundings by osmosis each day that if the water simply accumulated in its body, the fish would be one-third bigger each morning than it was the morning before. This swelling is prevented because the goldfish regulates its volume and voids the water as fast as it comes in. Large volumes of water are coming and going all the time, nonetheless.

FIGURE 36.1 Invisible Dynamism The water and salt exchanges between an animal and its environment often occur at high rates despite being invisible. A goldfish in its bowl may look almost inert. In fact, water is moving rapidly by osmosis into the fish's body from its environment, and the fish is producing urine equally rapidly to keep the volume of its internal body fluids constant.

"Water and salt balance" refers to homeostasis of the characteristics of the body fluids: their osmotic pressures, ionic compositions, and volumes. Such homeostasis is important because many of the functions of an animal depend on the characteristics of its body fluids.

LINK

You can review the processes of osmosis and diffusion in **Concept 5.2**

Kidneys make urine from the blood plasma

The kidneys—sometimes called the renal or excretory organs—of animals are exceptionally diverse, as we will see. In discussing the kidneys, a good starting point is to ask what similarities exist among the kidneys of various types of animals.

All kidneys consist of tubular structures that discharge a fluid directly or indirectly to the outside environment. The fluid that's excreted into the outside environment is the **urine**. Sometimes a single kidney consists of a single tubule. This is the case in crayfish and lobsters, for example. In contrast, a kidney may consist of great numbers of tubules. A vertebrate kidney, for example, consists of many microscopically tiny tubules called **nephrons**. There are about 1 million nephrons in a single human kidney.

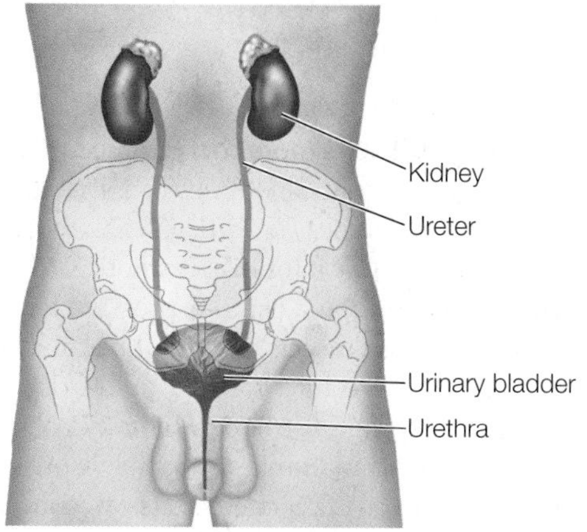

FIGURE 36.2 **Major Parts of the Human Excretory System**
The organs shown in yellow at the top of the kidneys are the adrenal glands, which are not part of the excretory system.

Go to ACTIVITY 36.1 The Human Excretory System
PoL2e.com/ac36.1

Each nephron (see Figure 36.11) produces urine. The nephrons discharge into a system of tubes that collects the urine from all of them and discharges it into a single tube, the ureter, that carries the fluid out of the kidney. In mammals, the two ureters carry the urine to the urinary bladder, and from there the fluid is expelled through the urethra during urination (**FIGURE 36.2**).

Go to MEDIA CLIP 36.1
Inside the Bladder
PoL2e.com/mc36.1

Kidneys typically make urine from the blood plasma. Recall that the blood plasma is the blood solution: the part of the blood other than the blood cells (see Concept 29.3).

In essence, a kidney tubule starts with one fluid—the blood plasma—and produces another fluid—the urine. We can analyze kidney function by thinking of the kidney as an input–output organ. How does its output (urine) differ from its input (plasma), and how and why does this transformation take place?

Kidneys regulate the composition and volume of the blood plasma

The most important characteristic of kidneys is that they regulate the composition and volume of the blood plasma. This is their primary function.

The kidneys regulate the composition of the blood plasma by selectively removing materials from the plasma. Think for a moment of a bottle containing a salty solution composed of water, Na^+, Cl^-, and other ions. Suppose there is a target osmotic pressure for the solution and that its current osmotic pressure is too low. We could raise the osmotic pressure to the target level by removing water from the solution. To be more realistic, however, suppose we can't remove only water. Suppose that when we remove water, we can't avoid also removing some ions. In this case, we could raise the osmotic pressure to the target level by unequally, or disproportionately, removing water and ions. If we remove lots of water but only small amounts of ions, the solution's osmotic pressure will go up. This exemplifies the way that kidneys usually work.

At this point we can define a **kidney** as an organ composed of tubular structures that produces an aqueous (watery) solution—urine—for excretion. This solution is derived from the blood plasma. The kidney's primary function is to regulate the composition and volume of the blood plasma by means of controlled removal of solutes and water from the plasma.

Urine/plasma (U/P) ratios are essential tools for understanding kidney function

Suppose the osmotic pressure of an animal's blood plasma is 100 (we'll simply use numbers here and not worry about units of measure). Suppose that, at the same time, the animal's kidneys are excreting urine that has an osmotic pressure of 25. Notice that the urine (25) in this case is more watery (less concentrated with solutes) than the plasma (100).

One way to understand kidney function is to express the composition of the urine as a ratio of the composition of the plasma. Such a ratio is called a urine/plasma, or U/P, ratio. The U/P ratio is 0.25 in the case we're discussing (25/100 = 0.25). Specifically, this is an *osmotic* U/P ratio because it's the ratio of urine and plasma osmotic pressures.

Now recall that the function of making urine is to regulate the plasma. If the osmotic U/P ratio is 0.25, what is the effect on the composition of the plasma? *Both the water and the solutes in the urine come from the plasma.* We see then that when the kidneys make this watery urine, they take a lot of water out of the plasma but only a little solute. In this way the kidneys modify the composition of the plasma, causing it to become more concentrated. They are raising the osmotic pressure of the plasma.

Suppose, by contrast, that the osmotic U/P ratio is 2.0. In this case the urine is less watery than the plasma. When the kidneys make this urine, they take a lot of solute out of the plasma but only a little water. In this way they cause the plasma to become less concentrated.

These principles are of great importance. An animal's kidneys typically adjust the U/P ratio up and down to regulate the blood plasma. The principles allow us to interpret the effects of these adjustments:

- If the urine is less concentrated than the plasma (U/P ratio < 1), the kidneys are making the plasma become more concentrated.
- If the urine is more concentrated than the plasma (U/P ratio > 1), the kidneys are making the plasma become more dilute.

These principles can be applied to osmotic pressures (total solute concentrations), as we have done here. They can also be applied to individual ions. For example, if the urine is more concentrated in Na^+ than the plasma is, the effect will be to lower the plasma Na^+ concentration.

Our day-to-day urine concentrations illustrate these principles

Our urine osmotic pressure is constantly being adjusted up and down by our kidneys to regulate the osmotic pressure of our blood plasma. Human kidneys, at one extreme, can produce urine only one-tenth as concentrated as blood plasma: osmotic U/P = 0.1. At the other extreme, our urine can be four times more concentrated than our plasma: osmotic U/P = 4.0. Our kidneys use this range of variation to perform their regulatory functions.

Suppose you drink a great deal of water one day. Your blood plasma might then become too dilute. Sensory structures detect that the plasma osmotic pressure is tending to go down. In response, your kidneys start to make very dilute urine. The osmotic U/P, as we've said, could be as low as 0.1. To make this dilute urine, the kidneys unequally remove water and solutes from the plasma: lots of water, little solutes. This process keeps the plasma osmotic pressure from falling.

Suppose that at a different time you go out on a hike and run out of water, so you go 2 days without drinking. Your blood plasma might then become too concentrated. Your kidneys respond to sensory input in this case by making concentrated urine. The osmotic U/P could be as high as 4.0. To make this concentrated urine, the kidneys unequally remove water and solutes from the plasma: lots of solutes, little water. This process keeps the plasma osmotic pressure from rising.

Simultaneously, the kidneys adjust the volume of urine. They do this in ways that help keep plasma volume stable.

To summarize, our kidneys adjust both the composition and the volume of the urine to maintain homeostasis in the composition and volume of the blood plasma. We produce a high volume of dilute urine when we drink lots of water. We produce a low volume of concentrated urine when we are dehydrated. The kidneys, by regulating the composition and volume of the blood plasma in these ways, also regulate the composition and volume of the tissue fluids that bathe the cells of the body.

The range of action of the kidneys varies from one animal group to another

We have seen that the human kidney can adjust the osmotic U/P ratio to be between 0.1 and 4.0. Various types of animals differ in this regard. Why does the range of U/P ratios matter? It matters because the regulatory actions that the kidneys can perform depend on it.

The kidneys of most animals can produce urine that is more dilute than the blood plasma (osmotic $U/P < 1$). This means that the kidneys of most animals can help with problems of excess water in the body.

Animals vary a lot, however, in how concentrated their urine can be. In the great majority of animal groups, the osmotic U/P ratio cannot be higher than 1.0 because of the ways the kidneys function. Fish, amphibians, lizards, snakes, and turtles cannot produce urine that is more concentrated than their blood plasma. Nor can most aquatic invertebrates. *This means that if the blood plasma of these animals is becoming too concentrated, their kidneys cannot correct the problem* (only an osmotic U/P ratio higher than 1.0 will cause the plasma osmotic pressure to decrease).

Mammals, birds, and insects stand out because they are the only animals that can produce concentrated urine with an osmotic U/P ratio exceeding 1.0. Mammals have evolved the greatest abilities. Some desert mammals can achieve osmotic U/P ratios as high as 10 or even 20. Some insects can concentrate their urine to an osmotic U/P ratio of 8. Birds are more limited than mammals and insects but can have U/P ratios of 2–3. Notice that all three of these groups of animals live on land. The ability of their kidneys to make concentrated urine is a great asset as these animals try to keep their plasma osmotic pressure normal while facing risks of rapid dehydration in terrestrial environments.

Extrarenal salt excretion sometimes provides abilities the kidneys cannot provide

In some groups of animals, tissues or organs other than the kidneys have evolved the ability to excrete ions at high total concentrations—a process called **extrarenal salt excretion** (*extrarenal*, "outside the kidneys"). All the animals with extrarenal salt excretion exhibit two characteristics. First, they live in environments where they either face a high risk of dehydration or tend to gain excess salts during their routine daily lives. Second, they have kidneys that are too limited in their concentrating ability to meet these challenges fully.

Many birds and non-avian reptiles that live in close association with the ocean or in deserts have **salt glands**. These glands are located in the head and excrete highly concentrated salt solutions. Seabirds and sea turtles, for example, get overloaded with salts from eating marine (ocean) invertebrates, plants, and algae. They depend on their salt glands (**FIGURE 36.3**) to excrete excess

(A)

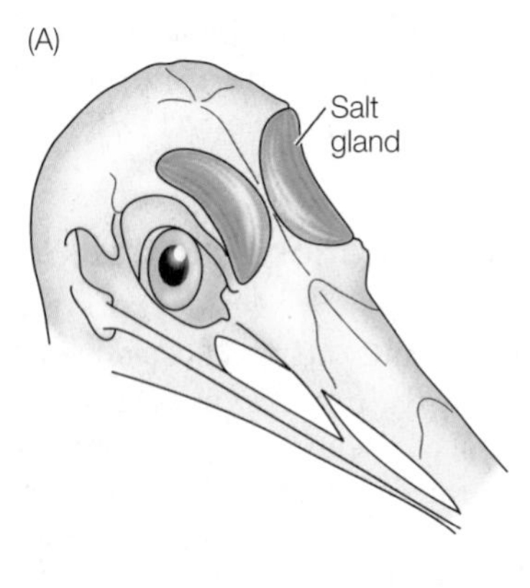
Salt
gland

(B)

A salty fluid drips
from the bird's bill.

(C)

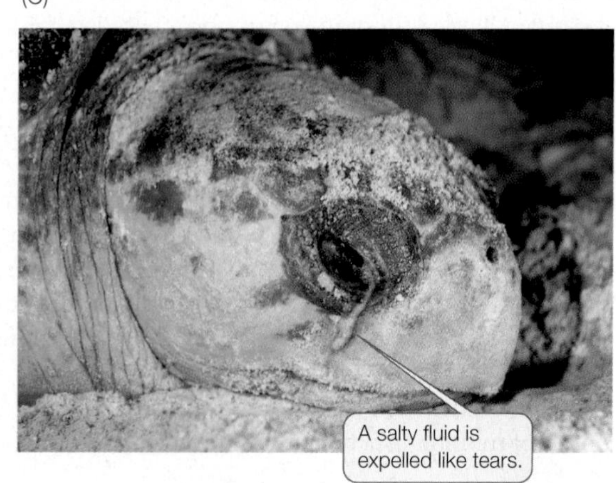
A salty fluid is
expelled like tears.

FIGURE 36.3 Salt Glands in Marine Birds and Sea Turtles (A) The position of the salt glands in a gull. In a bird, the fluid secreted by each gland enters a duct that empties into the nasal passages. The fluid then exits by way of the nostrils. (B) Fluid secreted by the salt glands exits the external tubular nostrils of a northern fulmar and then drips off the tip of the bird's bill. (C) Sea turtles, such as this leatherback sea turtle (*Dermochelys coriacea*), have salt glands, which secrete fluids that drip like tears from the area around the eyes.

salts. Saltwater fish have specialized salt-excreting cells in their gills, as we will see in Concept 36.3. Mammals—including those living in the oceans—do not have extrarenal salt excretion. Biologists think this is because mammals can depend on their high-performance kidneys to meet all their needs for salt excretion.

CHECKpoint CONCEPT 36.1

- ✓ Explain how production of urine with an osmotic U/P ratio of 0.3 helps raise the osmotic pressure of the blood plasma.
- ✓ Suppose a person eats lots of salty foods. What effect could this have on the composition of the person's blood plasma, and how do the kidneys combat it?
- ✓ The marine iguanas of the Galápagos Islands are unique among lizards because they principally eat ocean seaweeds that are as salty as seawater. Periodically a marine iguana "sneezes" a salty solution out of its nostrils. The solution is made by salt glands in the iguana's head. Why do these animals have salt glands instead of relying entirely on their kidneys for excretion?

We have seen how kidneys regulate the composition and volume of the blood plasma by producing urine of varying osmotic pressure and volume. Next we will discuss the special challenges for excretion posed by proteins and other nitrogen-containing molecules when they are broken down.

CONCEPT 36.2 Nitrogenous Wastes Need to Be Excreted

When carbohydrates and fats are fully broken down during metabolism, the end products of the process are H_2O and CO_2 (see Concept 6.2), which are not difficult to eliminate. Proteins and nucleic acids, however, contain nitrogen atoms. Their breakdown therefore produces not only H_2O and CO_2 but also nitrogen-containing waste molecules termed **nitrogenous wastes**. These molecules can be toxic, creating special challenges.

Kidneys are occasionally defined as the organs of nitrogen excretion. This is a misleading definition because in fact the chief function of kidneys is regulation of the blood plasma. Moreover, whereas the kidneys are important in nitrogen excretion in some types of animals, they play little or no role in nitrogen excretion in many other types.

Most water-breathing aquatic animals excrete ammonia

During the breakdown of proteins and nucleic acids, nitrogen atoms are removed from the molecules in the form of amino groups (—NH_2), as shown near the top of **FIGURE 36.4**. The —NH_2 groups are easily converted to ammonia (NH_3) or ammonium ions (NH_4^+) without any investment of energy. Ammonia therefore is the simplest and energetically cheapest nitrogenous waste to produce. The problem with ammonia is that it is highly toxic, and its concentration must be kept very low in the body fluids.

Crayfish, squid, most fish, and most other water-breathing animals excrete nitrogen principally as ammonia and for this reason are called **ammonotelic**. They are able to use ammonia as their principal waste because they can lose it rapidly and directly into their aquatic environment across their outer body surfaces. Ammonia is highly soluble in water and diffuses rapidly, so it can be lost readily across the gills and other permeable body surfaces as fast as it is produced, keeping internal concentrations low.

Most terrestrial animals excrete urea, uric acid, or compounds related to uric acid

Terrestrial animals typically must void nitrogenous wastes in their urine. Ammonia is so toxic that great volumes of urine would be required for it to be lost in the urine. The loss of water in urine would in fact be great enough to be dehydrating. Because of all these constraints, terrestrial animals almost never excrete nitrogenous wastes primarily as ammonia.

Instead, terrestrial animals incorporate waste —NH_2 groups into compounds that are far less toxic than ammonia. Producing these compounds requires ATP and thus has a metabolic cost. This is a disadvantage. However, the compounds have the advantage that they do not need to be excreted immediately. Their toxicity is low enough for them to be concentrated in the

FIGURE 36.4 Nitrogenous Wastes The breakdown of proteins and nucleic acids produces nitrogenous wastes. Most water-breathing aquatic animals, including most fish, excrete nitrogenous wastes as ammonia. Most terrestrial animals and some aquatic animals excrete either urea or uric acid.

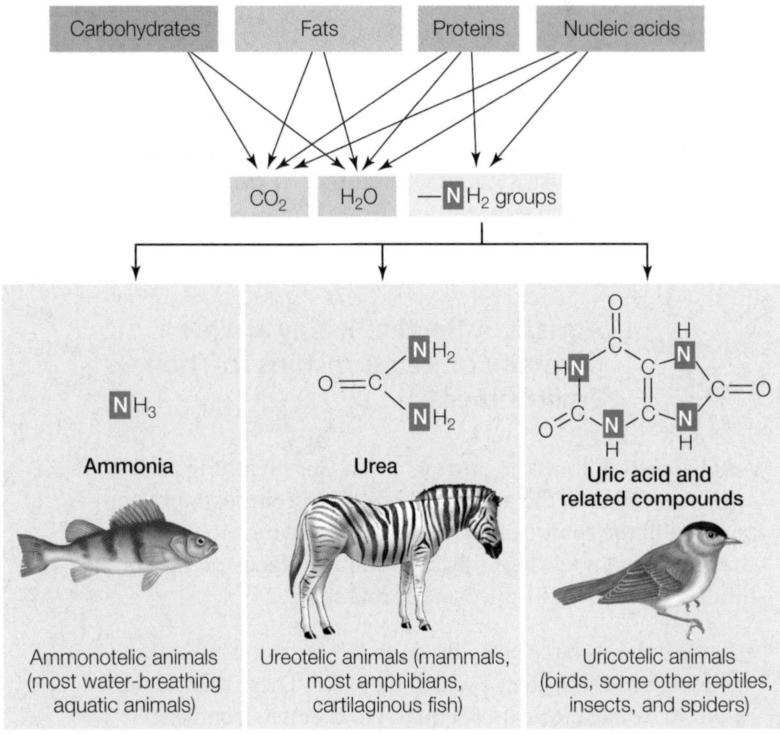

urine before they are excreted, reducing the amount of water required to excrete them:

- Mammals and most amphibians excrete waste nitrogen principally as urea and therefore are called **ureotelic** (see Figure 36.4). Urea is very soluble and relatively low in toxicity.
- Insects, spiders, and many reptiles (including birds) are **uricotelic**, meaning they excrete waste nitrogen mainly as uric acid or compounds related to uric acid. These nitrogenous wastes are poorly soluble in water and thus are often excreted in precipitated (solid or semisolid) form. Precipitation removes the compounds from solution, reducing their toxicity and the amount of water needed to excrete them. The whitish paste in bird droppings is uric acid.

Animals that live in different habitats at different developmental stages may change their primary nitrogenous waste when they change habitat. For example, the aquatic tadpoles of frogs and toads excrete ammonia across their gill membranes, whereas adult frogs and toads generally excrete urea.

CHECKpoint CONCEPT 36.2

- ✓ Name the three major nitrogenous wastes. Which one is the most toxic?
- ✓ Desert-adapted species of mice and rats survive in deserts by minimizing their water losses. Most are herbivores: they eat mostly plant foods. Why might a desert mouse or rat tend to be dehydrated by eating a large meal of animal flesh?
- ✓ Some species of amphibians that live in arid or semiarid habitats excrete uric acid as adults, instead of urea as other adult amphibians do. What selective advantage would this adaptation have had for the early ancestors of these species?

Now that we have reviewed the three forms in which nitrogenous wastes can be excreted, let's focus on the challenges that animals face in Earth's major environments. We'll start with the aquatic environments: the ocean and bodies of fresh water such as lakes and rivers.

CONCEPT 36.3 Aquatic Animals Display a Wide Diversity of Relationships to Their Environment

A key question to ask regarding any water-breathing aquatic animal is, What is the animal's body fluid composition compared with the composition of the water in which it lives? The simplest way to address this question is to focus on the total solute concentration—the osmotic pressure:

- Some animals have body fluids that have the same osmotic pressure as the water in which they live. These animals are said to be **isosmotic** (*iso*, "equal") to their environment.
- Other animals have body fluids that have a higher osmotic pressure than the environmental water. These animals are termed **hyperosmotic**, (*hyper*, "higher") to their environment.
- Still other animals have body fluids that have a lower osmotic pressure than the environmental water. They are described as **hyposmotic** (*hypo*, "lower") to their environment.

We quantify osmotic pressure in terms of **osmolarity**. Recall that each separate dissolved entity in a solution—whether it is a glucose molecule, protein molecule, or Na^+ ion—contributes approximately equally to osmotic pressure. A solution is defined to be 1 osmolar—symbolized 1 *Osm*—if it has 6×10^{23} dissolved entities per liter (6×10^{23} is Avogadro's number). A solution that is 1 osmolar is 1,000 milliosmolar, symbolized m*Osm*.

Most invertebrates in the ocean are isosmotic with seawater

The simplest and energetically cheapest way for an animal to live in the ocean is for its body fluids to be the same in their total solute concentration as seawater. Animal life first appeared in the ocean, and we think the earliest animals were simply isosmotic with the seawater. Today nearly all invertebrates in the ocean—including lobsters, squid, corals, and sea stars—are isosmotic with seawater. Seawater has an osmotic pressure of almost exactly 1 *Osm*. The blood and other body fluids of most marine invertebrates thus have an osmotic pressure of about 1 *Osm*.

Ocean bony fish are strongly hyposmotic to seawater

There are more than 20,000 species of fish with bony skeletons in the ocean. Their body fluids are far more dilute, in terms of both osmotic pressure and the concentrations of major ions, than the seawater in which they swim. The osmotic pressure of their blood plasma and other body fluids is 0.3–0.5 *Osm* (300–500 m*Osm*), whereas that of seawater is 1 *Osm*. The fish maintain this difference by use of regulatory mechanisms and thus are classed as **hyposmotic regulators**.

Recall from our opening story about salmon that the gills, mouth membranes, and some of the other body surfaces of fish are permeable to water and salts. Because of this permeability and their dilute body fluids, ocean fish face two relentless, never-ending problems:

- *They tend to lose water* by osmosis from their dilute body fluids into the surrounding, more concentrated seawater (**FIGURE 36.5A**). Their steady loss of water tends to dehydrate them (living in the ocean is like living in a desert for these fish).
- *They also tend to gain ions*, such as Na^+ and Cl^-, by diffusion from the seawater into their more dilute body fluids. Their steady losses of water and gains of ions tend to make their blood and other body fluids become too concentrated.

Ocean bony fish expend about 8–17 percent of their metabolic energy each day to fix these problems. To replace the water they are losing, they drink the seawater and expend energy

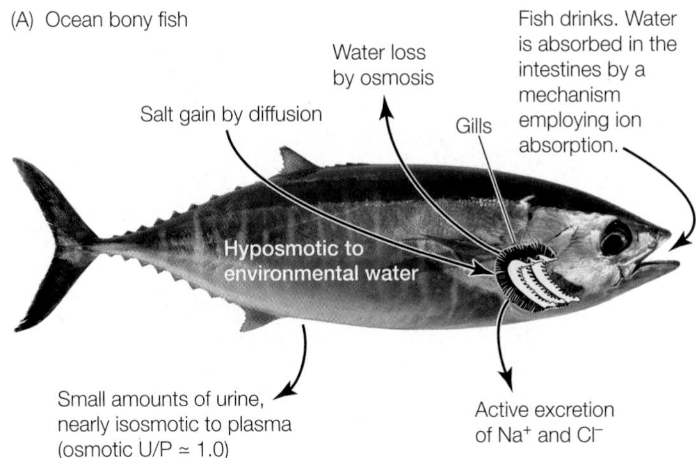

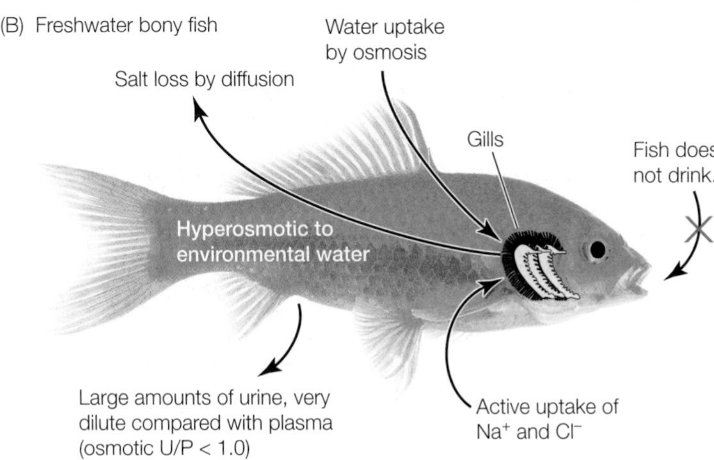

FIGURE 36.5 Principal Water and Salt Exchanges in Bony Fish The water and salt exchanges in ocean fish (A) are in many ways the opposite of those exchanges in freshwater fish (B).

in pumping ions across their intestinal wall so that water will enter their blood from their gut. They also use energy to excrete excess ions. The most concentrated urine their kidneys can produce (in terms of osmotic pressure) is U/P = 1.0. This is not sufficient to keep their body fluids more dilute than seawater. With inadequate kidneys, these fish have evolved vigorous extrarenal salt excretion. Specialized cells in their gills, known as mitochondria-rich cells or chloride cells, use ATP for active ion transport. These cells pump Na^+ and Cl^- out of the blood directly into the more concentrated seawater.

Why are the body fluids of these fish more dilute than seawater? Biologists do not know for certain. The preferred hypothesis today is that all the ocean bony fish evolved from freshwater ancestors, and that their dilute body fluids are a vestige of that earlier stage in their evolution.

All freshwater animals are hyperosmotic to fresh water

Animal life began in the oceans, as we've noted. During the long history of life, however, many phyla invaded rivers, streams, and lakes. Today's freshwater invertebrates include crayfish, shrimp, clams, snails, and cnidarians such as hydra. In all these cases, the freshwater species are descended from ocean ancestors that had internal osmotic pressures near 1 *Osm*. The freshwater forms alive today, however, have internal osmotic pressures that are less than half that. Why?

An important general principle is that when the body fluids of an animal differ from the animal's environment in some way, the cost of maintaining this difference tends to be greater when the difference is large than when it is small. (We are familiar with an analogy in our apartments and homes. The cost of keeping them warm in winter depends on how warm we keep them. Maintaining a big difference between the temperature inside and outside requires more energy than keeping the inside just slightly warmer than the outside.) When invertebrates first invaded fresh water from the ocean, they had a very large difference between their internal osmotic pressure, 1 *Osm*, and the external osmotic pressure, which is about 0 *Osm* in fresh water. Maintaining this difference required a high rate of energy use. By reducing their internal osmotic pressure over the course of evolutionary time, the animals lowered their rate of energy use. All are hyperosmotic to freshwater and are **hyperosmotic regulators**. Their cost of regulation is lower, however, than it would be if they had retained their ancestral internal concentrations.

Vertebrates have undergone similar changes. Biologists generally believe that the ancestors of today's vertebrates were jawless fish, isosmotic with seawater, living in the ocean. They invaded fresh water, and over evolutionary time, their body fluids became more dilute. Probably they evolved an internal osmotic pressure of about 300–350 m*Osm*. At the same time, they evolved jaws. These early jawed fish with internal osmotic pressures near 300–350 m*Osm* diversified to give rise to today's freshwater fish. They also invaded two other major habitats. Some reinvaded the oceans, giving rise to nearly all the ocean fish alive today, as noted in the last section. Others invaded the land.

LINK

For more on the evolutionary origins of terrestrial animals, see **Concept 23.6**

Because their body fluids are more concentrated than fresh water, all of today's water-breathing freshwater animals—including fish, other water-breathing vertebrates, and invertebrates—face two relentless, never-ending problems that are the opposite of those faced by ocean fish (**FIGURE 36.5B**):

- *They tend to gain water* by osmosis from the dilute freshwater environment. Their steady gain of water tends to bloat them (recall that a goldfish gains water equal to about one-third of its body weight per day).
- *They also tend to lose ions*, such as Na^+ and Cl^-, by diffusion from their body fluids into their environment. Their steady gains of water and losses of ions tend to make their body fluids become too dilute.

Freshwater animals must invest energy to maintain their hyperosmotic state. To void the water they are gaining osmotically, their kidneys produce a large volume of urine per day. In most cases, the kidneys also produce urine that is far more dilute than the blood plasma (osmotic U/P < 1.0). Producing a large volume of dilute urine costs energy, but it means the kidneys help keep the blood at its correct osmotic pressure. Lost salts have to be replaced. Freshwater animals have specialized active-transport cells that use ATP to pump Na^+ and Cl^- directly from their freshwater environment into their blood. These cells are in the gills of fish, crayfish, and clams. They are in the skin of adult frogs. The pumps are so effective that they take up a lot of Na^+ and Cl^- per day even though these ions are exceedingly dilute in fresh water.

LINK

You can review active transport processes in **Concept 5.3**

Some aquatic animals face varying environmental salinities

Some fish—notably salmon, sturgeon, and many eels—have life cycles in which they migrate between rivers and the ocean. Here we'll focus on salmon. They are conceived and undergo their early development in fresh water, complete much of their growth in the ocean, and finally return to fresh water to spawn. They switch between being typical hyperosmotic regulators when in fresh water to being typical hyposmotic regulators when in seawater (see Figure 36.5):

- A salmon in fresh water does not drink because it is already overloaded with water from osmosis. Upon entering the ocean, it starts to drink and activates intestinal ion transport mechanisms required to take water into its blood from the gut.
- When a salmon is in fresh water, its gills use ATP to pump ions into its blood from the surrounding water. In the ocean, its gills use ATP to pump ions out of its blood into the seawater.
- In fresh water, a salmon's kidneys produce abundant urine that is very dilute compared with the blood plasma (osmotic U/P < 1.0). In seawater, the kidneys produce just a small volume of urine that has about the same osmotic pressure as the blood plasma (osmotic U/P = 1.0).

These changes are controlled by hormones such as prolactin.

Many nonmigratory aquatic animals face a range of environmental options. For example, consider ocean invertebrates. They might move into coastal waters where seawater mixes with fresh water and salinities are intermediate. They might even move into freshwater streams.

Most marine invertebrates are **osmotic conformers** (osmoconformers), but a minority are **osmotic regulators** (osmoregulators). In osmotic conformers, the osmotic pressure of the body fluids is always the same as that of the environmental water. Thus when osmotic conformers enter more dilute water, their body fluids become more dilute to the same extent. We know this is stressful because most of these animals do not survive in water that is much more dilute than seawater. Marine mussels are an exception. They prosper at a wide range of salinities even though they are osmotic conformers (**FIGURE 36.6**).

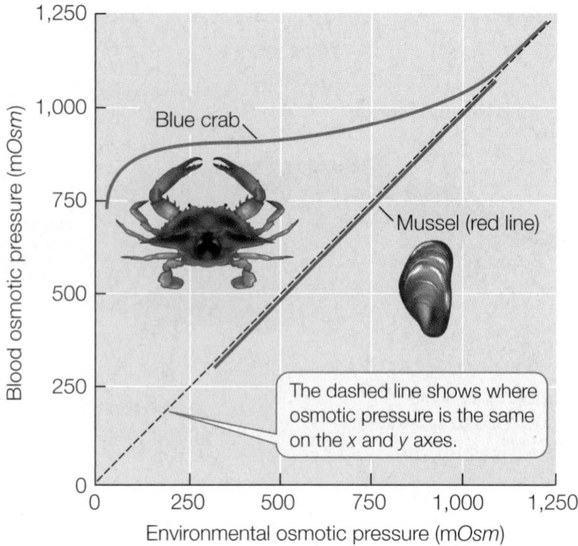

FIGURE 36.6 Osmotic Regulation and Conformity When ocean invertebrates enter waters where the salinities are lower than in the ocean, some (such as mussels) are osmotic conformers and allow their blood osmotic pressure to match the environmental osmotic pressure. Others (such as many crabs) are osmotic regulators. They keep the osmotic pressure of their blood from falling as low as the osmotic pressure in the environmental water, and their internal osmotic pressure is therefore more stable than the environmental osmotic pressure. The osmotic pressure of seawater is 1,000 m*Osm*.

Most marine invertebrates that prosper in dilute waters, including the blue crabs (*Callinectes sapidus*) that are abundant in the coastal waters of our east and Gulf coasts, are osmotic regulators. When living in the ocean, they are nearly isosmotic with seawater, but they are hyperosmotic regulators when in dilute waters, keeping their blood osmotic pressure higher than the environmental osmotic pressure (see Figure 36.6). They thus invest energy in osmotic regulation. In return, they have a more stable internal osmotic pressure than conformers, and this stability allows them to be successful over a broad range of external salinities.

LINK

The general principles of regulation versus conformity are discussed in **Concept 29.3**, along with specific coverage of thermal relationships of animals with their environment

CHECKpoint CONCEPT 36.3

- ✓ Some marine mussels are of exceptional commercial importance, being considered gourmet foods. From the viewpoint of water and salt balance, in what way are mussels exceptional?
- ✓ Do bony fish living in the ocean steadily gain or lose water by osmosis from the seawater, or do they neither gain nor lose in a steady manner? Explain.
- ✓ In what way do the bony fish living in the ocean exhibit extrarenal salt excretion?

Now that we've seen how various animals living in aquatic environments deal with fresh water and with seawater, let's turn to animals living on land.

CONCEPT 36.4 Dehydration Is the Principal Challenge for Terrestrial Animals

Water balance and the risk of dehydration make up the single greatest challenge that terrestrial animals face. We all know that water spilled on a table will gradually disappear by evaporation as it forms gaseous water vapor in the air. Likewise, the internal body fluids of animals on land also tend to evaporate—a process that can concentrate the body fluids and reduce their volume.

The normal blood osmotic pressure of terrestrial vertebrates is typically 300–350 m*Osm*. The early ancestors of terrestrial vertebrates were freshwater fish with similar osmotic pressures, as noted in Concept 36.3. Amphibians, reptiles (including birds), and mammals—including us—retain osmotic pressures in this range.

Humidic terrestrial animals have rapid rates of water loss that limit their behavioral options

During the early stages of animal evolution, the body coverings of animals were relatively permeable to water. This is understandable, because animals spent their early evolution in aquatic environments. Because of their aquatic history, animals had body coverings that were highly water-permeable when they emerged onto land.

Today, some terrestrial animals—classed as **humidic**—continue to have outer body coverings that are highly permeable to water. Examples include millipedes, centipedes, land snails (except when they are withdrawn into their shells), and most amphibians. Water evaporates across the skin of a leopard frog at a rate similar to that of evaporation from an open water surface!

Although humidic animals live on land, they are not free to expose themselves continually to the open air. Their behavioral options are limited because they lose water so rapidly in the open air that if they are forced to remain there, they die within tens of minutes or a few hours. Humidic animals have two options. One is to stay in places where the humidity of the air is high, such as in thick grass or under logs. Their other option is to limit the time they spend in dehydrating situations and return often to places where they can rehydrate. A frog, for instance, might hop out into an open, sunlit area where there are lots of insects to eat, and it might stay there for a short while. A frog loses water rapidly in such a situation, however, and must soon return to a place, such as a pond edge, where it can replenish its body water.

Xeric terrestrial animals have low rates of water loss, giving them enhanced freedom of action

The terrestrial animals we're accustomed to seeing out in the open typically have body coverings that protect them from high rates of evaporative water loss. We see birds, snakes, lizards, mammals, insects, and spiders. All have evolved body coverings in which layers of lipids—waxy or greasy materials—prevent water in their body fluids from freely evaporating into the air. With this protection, these animals—termed **xeric**—can spend indefinite periods in the open air.

Mammals, birds, and other xeric vertebrates have lipids in the outermost cellular layers of their skin. Insects and spiders have a thin layer of lipids on the outside of their exoskeleton.

> **LINK**
> See **Concept 23.6** for more on the adaptations that allowed animals to colonize dry environments

Some xeric animals are adapted to live in deserts

A desert, by definition, is a terrestrial environment that receives so little rain that the availability of water exerts a dominant, controlling influence on the organisms living there. Animals adapted to live in deserts exploit multiple mechanisms for reducing their rates of water loss, so that they have very low water requirements.

Kangaroo rats (**FIGURE 36.7**) and a variety of other desert rodents have such low rates of water loss that when individuals finish nursing from their mother, most never drink again for the rest of their lives. How is this possible? These animals have multiple adaptations that reduce their rate of water loss:

- They live in deep, cool burrows during the heat of the day.
- They have exceptionally low rates of evaporation through their skin and produce dry feces.
- Their kidneys can produce urine 10–20 times more concentrated than their blood plasma (osmotic U/P = 10–20), the highest values known among all animals. By concentrating their urine to such a great extent, the animals can excrete their soluble wastes with very little water loss.

FIGURE 36.7 A Desert Kangaroo Rat Distinctive rodents—many of which hop about on their hind legs like kangaroos—have evolved in all the world's deserts. Kangaroo rats—such as this Ord's kangaroo rat (*Dipodomys ordii*)—occur in the deserts of the American Southwest.

Desert rodents do need some water. Where does it come from? Kangaroo rats typically eat air-dried plant parts such as seeds. These provide a little water. The most important source of water for kangaroo rats, however, is the H_2O produced by the oxidation of organic molecules during metabolism (see Concept 6.2). This water is called **metabolic water** or oxidation water. All animals produce metabolic water, but it is important in the water budget only in animals that conserve water exceptionally well.

Many insects are also adapted outstandingly well to living in deserts. Some birds and mammals survive in deserts by eating the juicy bodies of desert insects or the leaves of juicy plants.

LINK

Plants that live in deserts also face numerous challenges; their adaptations to desert life are described in **Concept 28.3**

CHECKpoint CONCEPT 36.4

- ✓ What is metabolic water?
- ✓ Dromedary camels have kidneys that can produce urine eight times more concentrated than their blood plasma. The average concentrating ability of mammals of similar body size is about half as great. Why would camels benefit from having a higher concentrating ability than average?
- ✓ Humidic and xeric terrestrial animals differ in behavior because of their contrasting abilities to conserve body water. Speaking in broad terms, how do they differ in behavior, and why?

In the last two concepts we have considered ways in which aquatic and terrestrial animals meet the challenges they face in their environments. Earlier we stressed that the concentration and volume of an animal's urine are major considerations for understanding water and salt balance. To close the chapter, we will now examine how kidneys work and the characteristics of the urine they produce.

CONCEPT 36.5 Kidneys Adjust Water Excretion to Help Animals Maintain Homeostasis

Kidney tubules—such as the nephrons of vertebrates—typically make urine in two steps. First, fluid derived from the blood plasma enters a kidney tubule at one end. This fluid is called the primary urine. Next, the volume and composition of the fluid are modified as it flows through the remainder of the tubular system in the kidney, forming the final product, called the definitive urine, which is excreted. During this modification step, materials may be *reabsorbed* from the urine. Materials may also be *secreted* into it. Here we focus on these processes in vertebrate kidneys.

Fluid enters a nephron by ultrafiltration driven by blood pressure

A vertebrate nephron is open at one end and closed at the other. Urine formation starts at the closed end, which consists of a cup-shaped structure called a **Bowman's capsule** that encloses a dense cluster of blood capillaries called a **glomerulus** (plural *glomeruli*). The glomerulus sits within the Bowman's capsule much like a fist pushed into an inflated balloon (**FIGURE 36.8A,B**).

(A) The general form of a vertebrate nephron at the end where primary urine is formed

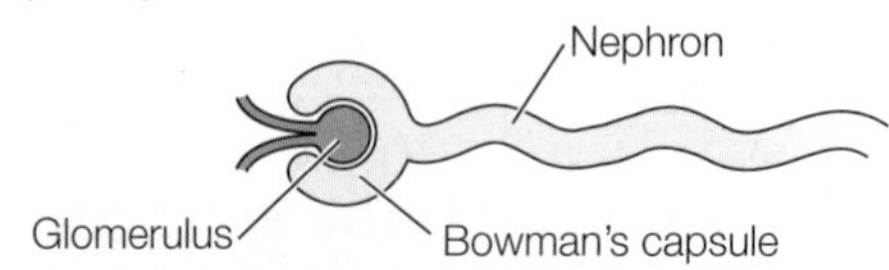

(B)

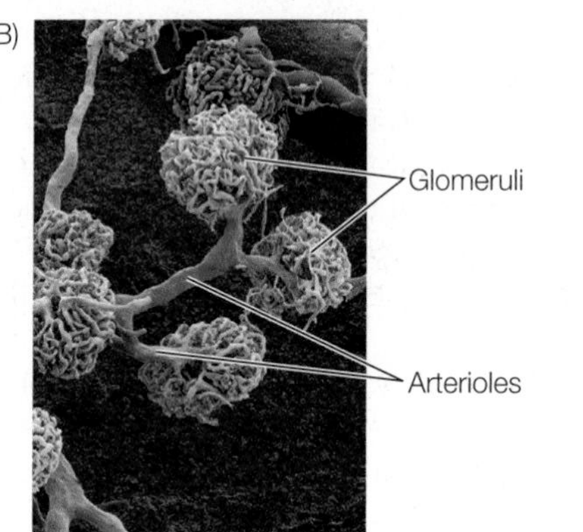

(C) A human glomerulus positioned in a Bowman's capsule

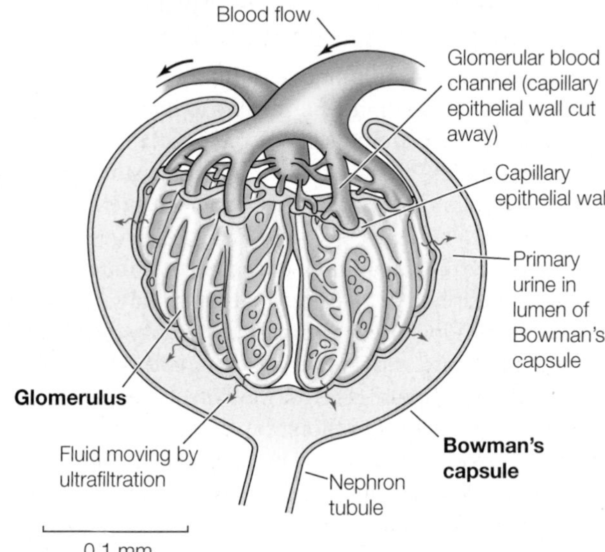

FIGURE 36.8 Primary Urine Is Produced in the Bowman's Capsules and Glomeruli (A) At its closed end, each vertebrate nephron begins with a Bowman's capsule containing a glomerulus. (B) The glomeruli look like balls of capillaries served by arterioles in this electron micrograph; only blood vessels are seen here because other tissue has been digested away. (C) This drawing of a human Bowman's capsule with glomerulus shows the formation of primary urine by ultrafiltration.

Go to ACTIVITY 36.2 The Vertebrate Nephron PoL2e.com/ac36.2

APPLY THE CONCEPT

Kidneys adjust water excretion to help animals maintain homeostasis

In analyzing kidney function, an important question is whether the nephrons reabsorb substances from the tubular fluid passing through them or secrete substances into the tubular fluid. A classic way to study this question starts with measuring the glomerular filtration rate (GFR). The GFR—the volume of fluid that passes into the Bowman's capsules of the kidneys per unit of time—can be calculated by using a marker substance (a substance artificially added to the blood) that is freely filtered out of the blood plasma (so that its concentration in the ultrafiltrate equals its plasma concentration) but *not* reabsorbed or secreted by the nephrons or other renal tubules. The amount of such a marker substance entering the Bowman's capsules equals the amount leaving the body in the definitive urine. Inulin, a polysaccharide, is the most commonly used marker substance. If a different substance *is* reabsorbed or secreted by the nephrons or other renal tubules, the amount of that substance leaving the body in the definitive urine will be either more than or less than the amount filtered. Consider the data in the table. Assume that all three substances are freely filtered, and assume a urine flow rate of 1 milliliter (ml) per minute. As you address the questions asked, keep in mind that the amount of a substance in a solution is equal to its concentration in the solution times the volume of the solution. The concentrations of Na^+ are in milliequivalents (mEq) per liter (L).

1. What is the glomerular filtration rate (in ml/min)?
2. Is PAH secreted or reabsorbed? Justify your answer.
3. Is Na^+ secreted or reabsorbed? Justify your answer.

SUBSTANCE	CONCENTRATION IN BLOOD PLASMA	CONCENTRATION IN URINE
Inulin	1 mg/ml	125 mg/ml
Para-aminohippuric acid (PAH)	0.01 mg/ml	5.85 mg/ml
Na^+	140 mEq/L	70 mEq/L

Blood is delivered to each glomerulus at a relatively high pressure that's sufficient to force fluid from the blood plasma to pass through minute, porelike openings in the walls of the glomerular blood capillaries and then through the interior wall of the Bowman's capsule. This fluid enters the lumen (the open central cavity) of the Bowman's capsule and is the primary urine (**FIGURE 36.8C**).

The walls of the capillaries and of the Bowman's capsule do not allow all components of the blood to pass through. In this way, the blood is filtered as it moves through these structures. Blood cells and dissolved proteins in the plasma are filtered out and remain in the plasma. By contrast, water, salts, and small organic molecules such as glucose and amino acids pass through freely, entering the nephron. The process of forming primary urine is called **ultrafiltration**. The rate at which the primary urine is formed by all the nephrons functioning collectively is the **glomerular filtration rate** (**GFR**).

The primary urine is similar to the blood plasma in most ways except for lacking blood cells and proteins. The concentrations of ions (e.g., Na^+ and Cl^-) and of small molecules (e.g., glucose) in the primary urine are the same as those in the blood plasma. Moreover, the osmotic pressure of the primary urine is essentially equal to that of the plasma. In other words, the primary urine is isosmotic with the plasma.

As the primary urine flows through a nephron to form the definitive urine, more than half of the water and ions are typically reabsorbed into the blood plasma. More than 99 percent are reabsorbed in a person producing concentrated urine. This sounds inefficient at first. Why produce a large volume of primary urine and reabsorb most of it? The process gives the kidneys intimate access to the plasma. Humans, for example, have a GFR of about 120 milliliters per minute. At this rate the equivalent of all the plasma water in the body enters the nephrons every 30 minutes! This rate enables the kidneys to remove wastes or toxins quickly and to make rapid adjustments in the plasma volume and composition.

Many types of animals are believed to employ ultrafiltration to produce primary urine. Besides vertebrates, these animals include squid and octopuses, and crustaceans such as crayfish.

The processing of the primary urine in amphibians reveals fundamental principles of nephron function

The amphibian nephron provides a model for understanding all vertebrate nephrons. The nephrons of freshwater fish and non-avian reptiles, and most of the nephrons of birds, are similar. The parts of a vertebrate nephron are termed "early" or "proximal" if they are relatively close to the Bowman's capsule. They are called "late" or "distal" if they are relatively far away. Along the entire length of a nephron, the nephron walls consist of a single layer of epithelial cells. However, these cells differ greatly in their structure and in the proteins they express from one part of a nephron to another.

An amphibian nephron consists of two highly convoluted segments, the proximal and distal convoluted tubules:

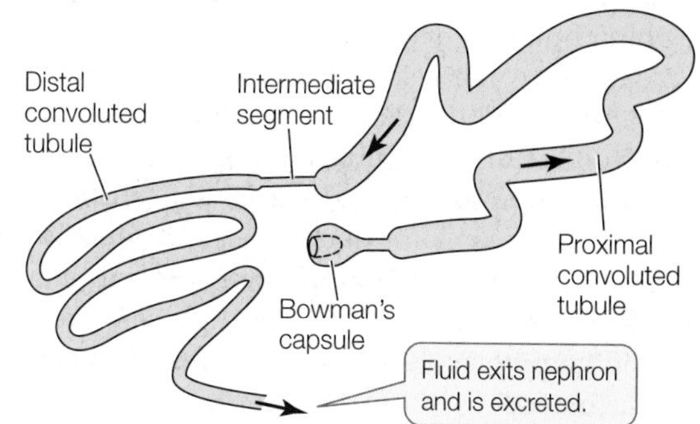

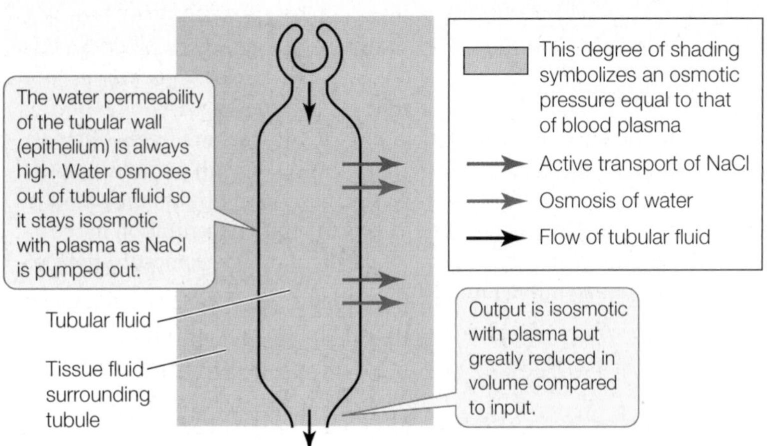

FIGURE 36.9 Amphibian and Mammalian Kidney Tubules Operate in Fundamentally Similar Ways The lightness or darkness of the shading represents the osmotic pressure in the fluids inside and outside the tubules, with darker shading corresponding to higher osmotic pressures. In both amphibians and mammals, the tubules can switch between producing a high volume of relatively dilute urine (diuresis) and a low volume of relatively concentrated urine (antidiuresis), under control of antidiuretic hormone (ADH). (A) The functioning of the early tubule is similar in amphibians and mammals. (B, C) The greatest difference between amphibians and mammals is that in amphibians the tissue fluids surrounding the late tubules are isosmotic with blood plasma (B), whereas in mammals these tissue fluids reach a far higher concentration than blood plasma (C). This difference explains why amphibians and mammals differ in their maximum urine concentrations.

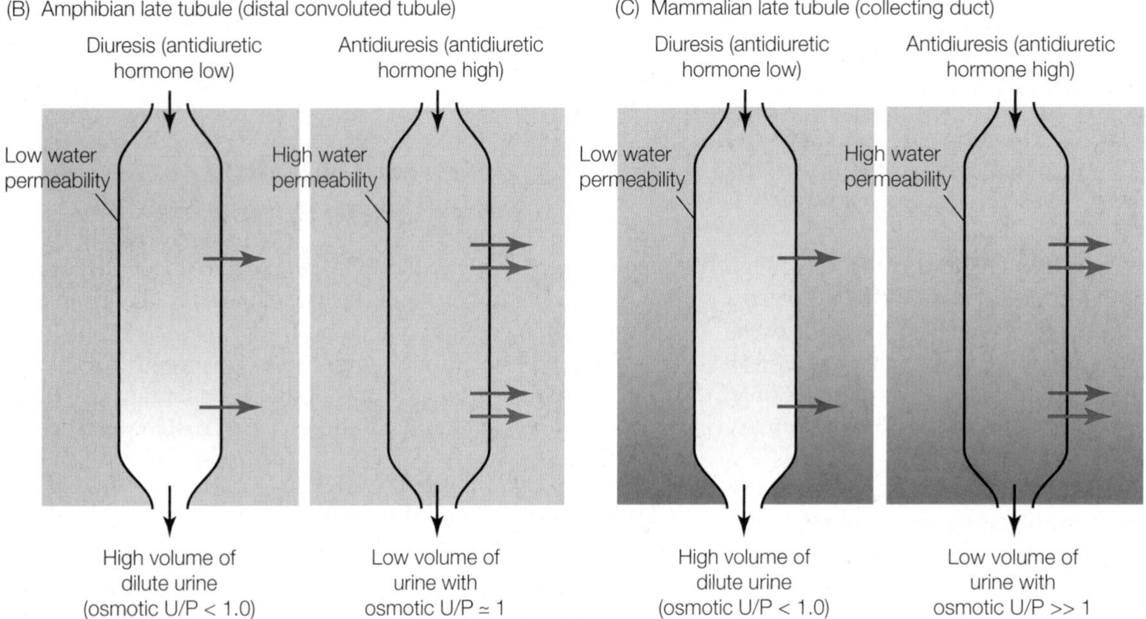

When primary urine flows out of the Bowman's capsule, it passes first through the proximal convoluted tubule (the early part of an amphibian nephron). The nephron wall in this segment is always highly permeable to water because **aquaporins** (water channel proteins; see Concept 5.2) are present at all times in the cell membranes of the epithelial cells composing the wall.

As the fluid in the tubular lumen—called **tubular fluid**—flows through the proximal tubule, Na^+ and Cl^- are removed by active transport and returned to the blood. This removal of Na^+ and Cl^- tends to make the tubular fluid more dilute than the blood plasma. However, as the tubular fluid tends to become dilute, water also moves out of it into the nearby, more concentrated blood plasma by osmosis. The end result is that the volume of the tubular fluid is greatly reduced as the fluid flows through the proximal tubule, but the osmotic pressure of the fluid remains unchanged and equal to that of the blood plasma (**FIGURE 36.9A**). The epithelial cells of the proximal tubule also actively transport glucose and amino acids out of the tubular fluid, returning these valuable molecules to the blood.

Farther along the nephron, the tubular fluid enters the distal convoluted tubule. The most important difference between the proximal and distal convoluted tubules is that the wall of the distal convoluted tubule has a variable permeability to water. This is believed to be controlled by insertion and retrieval of aquaporin proteins. Aquaporin molecules in the epithelial cells of the distal tubule can either be inserted in the cell membranes or not. If they are inserted, the cells are highly permeable to water. If they are not inserted, the cells are poorly permeable to water. Like the proximal tubule, the distal tubule pumps Na^+ and Cl^- out of the tubular fluid by active transport.

Insertion of aquaporins is controlled by **antidiuretic hormone** (**ADH**) secreted by the hypothalamus of the brain (see Concept 35.3). Variations in ADH secretion control whether the kidney produces abundant urine—a state called diuresis—or a small amount of urine—termed antidiuresis. ADH secretion also controls the concentration of the definitive urine (**FIGURE 36.9B**).

When the level of ADH is high, the wall of the distal convoluted tubule (the late part of the amphibian nephron) is highly permeable to water because aquaporin molecules are inserted in the epithelial cell membranes. The distal tubule then acts much like the proximal tubule. Na^+ and Cl^- are pumped out of the tubular fluid, and water follows by osmosis, so the osmotic pressure in the tubular fluid remains similar to that in the blood plasma. Accordingly, the kidney is in a state of antidiuresis. It produces a low volume of urine that is approximately isosmotic with the blood plasma (see Figure 36.9B, antidiuresis).

By contrast, when the level of ADH is low, aquaporin molecules are not inserted in the epithelial cell membranes, and the wall of the distal tubule is poorly permeable to water. Under these conditions, water in the tubular fluid cannot readily leave it, and most of the water that enters the distal tubule passes through and is excreted. Moreover, as Na^+ and Cl^- are pumped out of the tubular fluid, the fluid becomes more and more dilute. The kidney is in a state of diuresis and produces a high volume of urine that is far more dilute that the blood plasma (see Figure 36.9B, diuresis).

ADH secretion is controlled in ways that help maintain homeostasis of the blood plasma. If the plasma is tending to become too dilute and full of water, sensors inform the brain, and ADH is not secreted—resulting in a high volume of dilute urine. In contrast, if the plasma is tending to become too concentrated and low in volume, ADH is secreted, and a low volume of concentrated urine is excreted.

Mammalian kidneys produce exceptionally high urine concentrations

The kidneys of amphibians regulate the urine so that the osmotic U/P ratio varies between values far less than 1.0 (urine more dilute than the blood plasma) and equal to 1.0 (urine isosmotic with the plasma). How do mammals produce exceptionally concentrated urine with a U/P ratio much higher than 1.0?

Mammalian nephrons function in ways that are fundamentally similar to those of amphibian nephrons. For example, the proximal convoluted tubule of a mammalian nephron (the early part of the nephron) is always highly permeable to water and reduces the volume of the primary urine without changing its osmotic pressure (see Figure 36.9A).

The most important difference between amphibian and mammalian kidneys is that the late tubules of the mammalian kidney—which are called **collecting ducts**—are surrounded with tissue fluids that in places are far more concentrated than the blood plasma. This difference explains how the mammalian kidney can produce urine that is more concentrated than the plasma. **FIGURE 36.9C** illustrates the mechanism. Tubular fluid travels through a collecting duct just prior to leaving the kidney as definitive urine. The tissue fluids surrounding the beginning of the collecting duct are similar to blood plasma in osmotic pressure. Along the length of the duct, however, the tissue fluids surrounding the duct have an ever-higher osmotic pressure, reaching an osmotic pressure far higher than that of the blood plasma, as Figure 36.9C shows. Na^+, Cl^-, and other salts cannot diffuse readily across the wall of a collecting duct, although they can cross by means of ion pumping when pumps are active.

When the ADH level is high, the wall of the collecting duct is highly permeable to water because aquaporins are inserted in the cell membranes of the epithelial cells composing the wall. As tubular fluid flows through the collecting duct, it loses water by osmosis into the concentrated tissue fluids surrounding the duct (and the blood then picks up this water). This process of water removal concentrates the salts in the tubular fluid (see Figure 36.9C, antidiuresis). Just before the tubular fluid leaves the collecting duct to enter the ureter and travel to the bladder, the tubular fluid comes to osmotic equilibrium with the tissue fluids of highest concentration. The osmotic pressure of the urine is thus far higher than that of the blood plasma. A low volume of highly concentrated urine is produced.

When the ADH level is low, the wall of the collecting duct has a low permeability to water because aquaporins are removed from the cell membranes of the epithelial cells composing the wall (see Figure 36.9C, diuresis). As tubular fluid flows through the collecting duct, it does not lose much water by osmosis into the surrounding tissue fluid. Moreover, Na^+ and Cl^- are actively transported out of the tubular fluid, making the fluid more and more dilute as it moves through the collecting duct. A high volume of dilute urine is produced.

AQP-2 is the specific molecular form of aquaporin that is of greatest importance in controlling the water permeability of the collecting duct. Experiments have documented its important role (**FIGURE 36.10**).

How are the tissue fluids bathing the collecting ducts deep inside the kidney made highly concentrated? In a mammalian kidney, a portion of each nephron functions to increase the osmotic pressure of the tissue fluids surrounding it. This portion is a long, hairpin-shaped segment called the **loop of Henle** (pronounced Hen-lee), positioned immediately after the proximal convoluted tubule (**FIGURE 36.11**). All the nephrons in a mammalian kidney are organized so that their loops of Henle are parallel. This microscopic structure gives the entire kidney a visible structure. Each kidney has an outer layer, the renal cortex, where the Bowman's capsules and the convoluted tubules of the nephrons are found. Inside this cortical layer, the kidney has an inner layer called the renal medulla. This inner layer is composed of the loops of Henle and collecting ducts (see Figure 36.11). Parts of the medulla are said to be "deeper," the farther they are from the cortex.

The loops of Henle in a kidney work together to increase the osmotic pressure of the tissue fluids deep in the medulla. In this way, the loops create a gradient in the osmotic pressure of the medullary tissue fluids, so that the tissue fluids near the cortex are isosmotic with the blood plasma, but those deep in the medulla are far more concentrated. For example, in the human kidney, the osmotic pressure of the tissue fluids increases from about 300 m*Osm* near the cortex to about 1,200 m*Osm*—four times the blood osmotic pressure—in the deepest medullary tissue. To create this gradient, the loops of Henle use a mechanism called **countercurrent multiplication** (not to be confused with countercurrent exchange) in which the flow of tubular fluid in opposite directions in each loop (first into the medulla and then out) multiplies effects of active ion transport. The loops of Henle do not themselves concentrate the urine, but they set the stage for the urine to be concentrated in the collecting ducts.

INVESTIGATION

FIGURE 36.10 ADH Activates Insertion of Aquaporins into Cell Membranes Aquaporin proteins make some regions of renal tubules permeable to water. One type of aquaporin, AQP-2, is responsible for the permeability of the collecting duct cells in a mammal. How does antidiuretic hormone (ADH) act on molecules of AQP-2 to control the level of water permeability in renal cells?[a]

HYPOTHESIS

ADH controls water permeability by changing the cellular location of aquaporin molecules.

METHOD

1. Isolate collecting ducts from rat kidney.
2. Use immunochemical staining to localize AQP-2 molecules in collecting duct cells without ADH, then with ADH, and—finally—after the applied ADH is washed away.
3. Measure the water permeability of the collecting duct cells under the same three conditions.

RESULTS

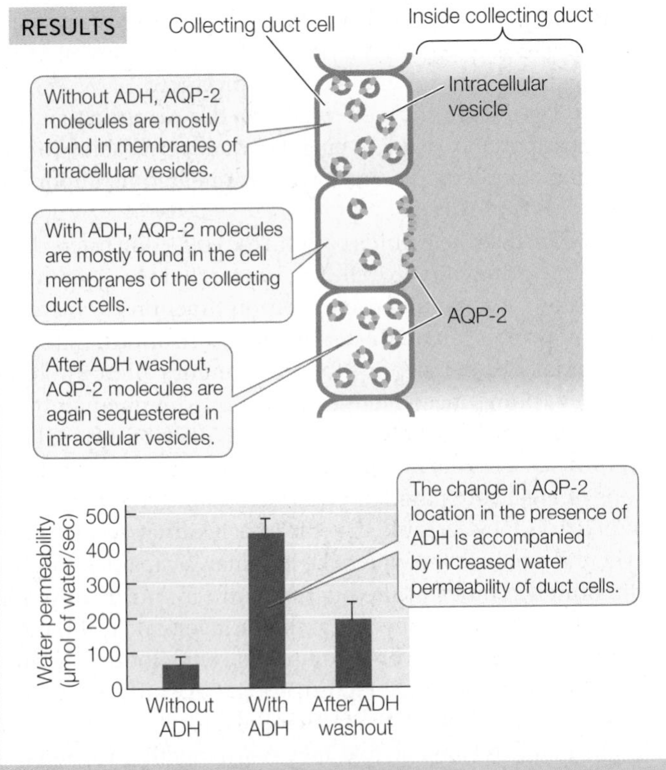
Collecting duct cell
Inside collecting duct
Intracellular vesicle
Without ADH, AQP-2 molecules are mostly found in membranes of intracellular vesicles.
With ADH, AQP-2 molecules are mostly found in the cell membranes of the collecting duct cells.
AQP-2
After ADH washout, AQP-2 molecules are again sequestered in intracellular vesicles.
The change in AQP-2 location in the presence of ADH is accompanied by increased water permeability of duct cells.
Water permeability (μmol of water/sec)
500
400
300
200
100
0
Without ADH
With ADH
After ADH washout

CONCLUSION

In the absence of ADH, AQP-2 water channel molecules are sequestered intracellularly. When ADH is present, these water channels are inserted into the cell membranes, making the cells more permeable to water.

ANALYZE THE DATA

The results showed that ADH controls the location of AQP-2 molecules in collecting duct cells. Does ADH also affect the amount of AQP-2? The Brattleboro strain of rats does not produce ADH, and the DI strain of mice has a defect in the signaling pathway of the ADH receptor. The amount of AQP-2 produced in response to ADH and to an ADH receptor antagonist was studied in these animals. Values were normalized to baseline amounts in normal animals.

	Amount of AQP-2			
Treatment	Normal rat	Brattleboro rat	Normal mouse	DI mouse
Baseline	1.0	0.5	1.0	0.1
ADH	1.0	1.5	1.0	0.1
ADH receptor antagonist	0.5	0.5	0.5	0.1

A. How does ADH affect the amount of AQP-2 in normal and Brattleboro rats? Explain the difference in the responses of the two rat strains.

B. Why does ADH not have an effect on the amount of AQP-2 in DI mice?

C. Considering these data as a whole, construct a hypothesis to explain why Brattleboro rats can maintain AQP-2 levels at 50% of normal under baseline conditions. How would you test your hypothesis?

Go to **LaunchPad** for discussion and relevant links for all **INVESTIGATION** figures.

[a]K. Fushimi et al. 1993. *Nature* 361: 549–552; S. Nielsen et al. 1995. *Proceedings of the National Academy of Sciences* 92: 1013–1017.

Each nephron empties into a collecting duct, and the fluid from the nephron then flows through the collecting duct from the cortex to deep in the medulla (see Figure 36.11). As the fluid flows through the collecting duct, it becomes definitive urine, which exits the collecting duct to flow to the bladder via the ureter draining the kidney (see Figure 36.2). Figure 36.9C, which we have already discussed, shows how the volume and concentration of the definitive urine are determined in the collecting duct, both when the ADH level is high and when it is low.

Birds differ from mammals in that only 10–30 percent of their nephrons have loops of Henle. This limits the concentrating ability of bird kidneys relative to mammalian kidneys.

The Malpighian tubules of insects employ a secretory mechanism of producing primary urine

The primary urine of insects is produced in blind-ended tubules, the **Malpighian tubules**, that open into the gut between the midgut and hindgut (**FIGURE 36.12**). Insects have a low-pressure, open circulatory system (see Concept 32.1) and do not use high blood pressure to cause ultrafiltration into the tubules. Instead, the epithelial cells in the tubule walls employ active transport to secrete KCl (or in some cases other salts) at a high rate into the tubular fluid, giving the tubular fluid a high osmotic pressure. The walls of the tubules are permeable to water. Water therefore enters the tubules by osmosis from the more dilute blood

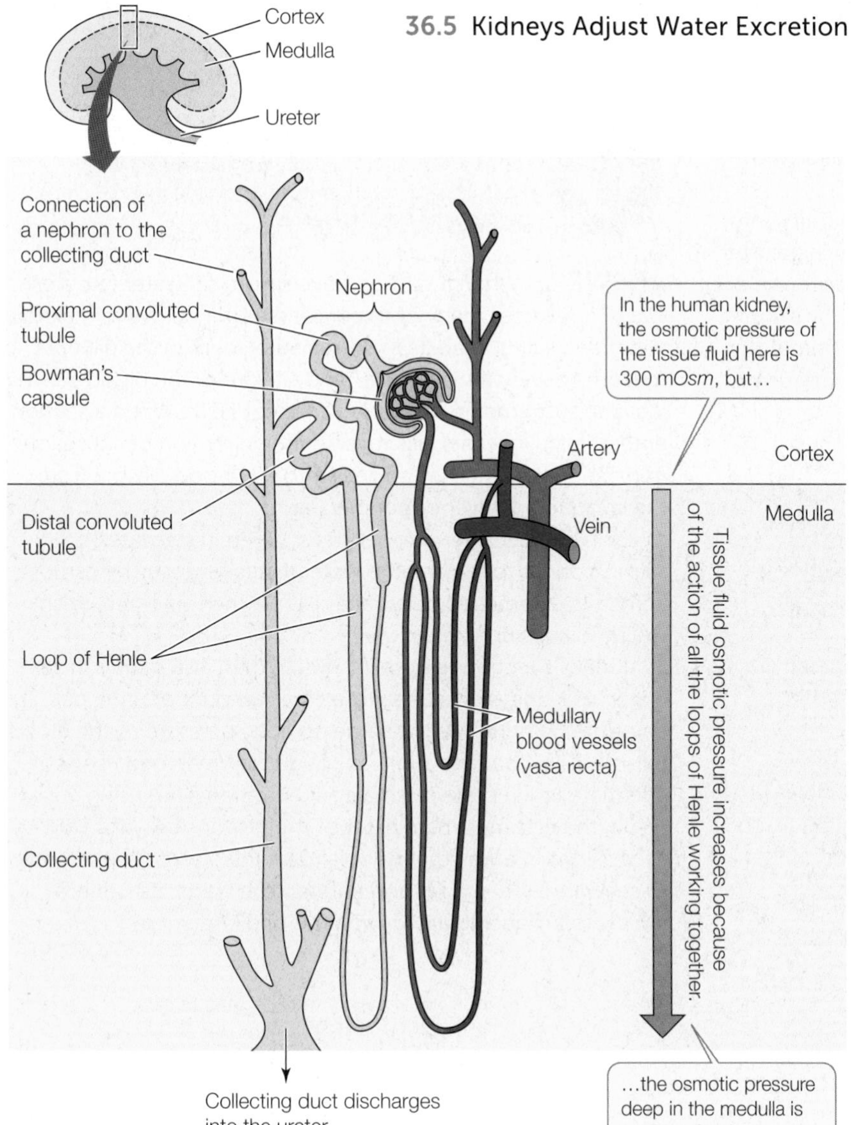

FIGURE 36.11 The Mammalian Kidney
The Bowman's capsules and convoluted tubules of the nephrons are in the kidney's outer region, the cortex. Other tubules are in the internal region, the medulla. The loops of Henle run parallel to one another in the medulla, where they create a steep gradient of osmotic pressure in the medullary tissue fluid. After tubular fluid leaves each nephron, it flows through a collecting duct to exit the kidney. As the tubular fluid within the collecting duct moves deeper and deeper into the medulla during this process, it passes by tissue fluids that are more and more concentrated.

Go to ANIMATED TUTORIAL 36.1
The Mammalian Kidney

PoL2e.com/at36.1

Go to ANIMATED TUTORIAL 36.2
Kidney Regulation Simulation
PoL2e.com/at36.2

Abdomen of ant

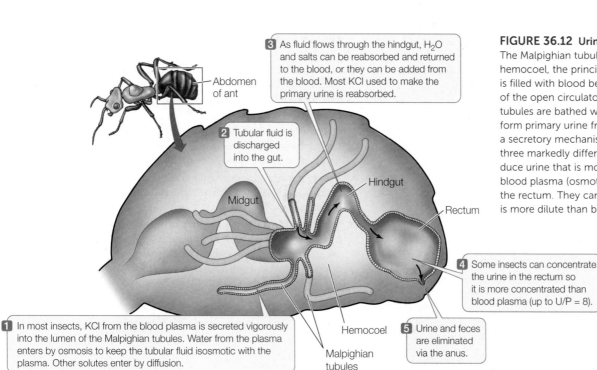

3 As fluid flows through the hindgut, H_2O and salts can be reabsorbed and returned to the blood, or they can be added from the blood. Most KCl used to make the primary urine is reabsorbed.

2 Tubular fluid is discharged into the gut.

Hindgut

Midgut

Rectum

4 Some insects can concentrate the urine in the rectum so it is more concentrated than blood plasma (up to U/P = 8).

1 In most insects, KCl from the blood plasma is secreted vigorously into the lumen of the Malpighian tubules. Water from the plasma enters by osmosis to keep the tubular fluid isosmotic with the plasma. Other solutes enter by diffusion.

Hemocoel

5 Urine and feces are eliminated via the anus.

Malpighian tubules

FIGURE 36.12 Urine Production in Insects
The Malpighian tubules project into the hemocoel, the principal body cavity, which is filled with blood because of the structure of the open circulatory system. There the tubules are bathed with blood. The tubules form primary urine from the blood plasma by a secretory mechanism. Insects use at least three markedly different mechanisms to produce urine that is more concentrated than blood plasma (osmotic U/P > 1.0), usually in the rectum. They can also produce urine that is more dilute than blood plasma (U/P < 1.0).

surrounding them, creating the primary urine. Other salts besides KCl enter by diffusion. The process is described as a "secretory mechanism" for producing primary urine.

The primary urine flows out of the Malpighian tubules into the hindgut and then flows through the hindgut, which greatly modifies the volume and composition of the fluid before it is excreted. Thus the excretory system consists of both the Malpighian tubules and the hindgut. One important function of the hindgut is to reabsorb the KCl used to produce primary urine, so the KCl can be used again in urine production.

CHECKpoint CONCEPT 36.5

- ✓ Human patients suffering from heart failure often become unable to produce adequate amounts of urine, a condition known as kidney failure. How could heart failure lead to kidney failure?
- ✓ What is the difference between an ultrafiltration-based mechanism of producing primary urine and a secretion-based mechanism?
- ✓ Antidiuretic hormone (ADH) has the same effect on the late tubules of both amphibians and mammals: it increases the water permeability of epithelial cells that are otherwise poorly permeable to water. If the effect is the same, why do amphibian kidneys produce urine of the same osmotic pressure as the blood plasma when the ADH concentration is high, whereas mammals produce urine more concentrated than blood plasma?

How can a salmon keep the composition of its blood plasma stable regardless of whether it is swimming in seawater or fresh water?

ANSWER When a salmon travels from seawater into fresh water, it reverses many of the processes it employs to maintain water and salt homeostasis. In seawater, cells in the gill epithelium actively transport Na^+ and Cl^- out of the blood plasma into the surrounding seawater (Concept 36.3). When a salmon enters fresh water, its gills reverse their direction of active ion transport, pumping Na^+ and Cl^- into the blood plasma from the surrounding river or pond water.

Drinking behavior also reverses. When in seawater, a salmon drinks to replace the water it steadily loses by osmosis across its external body surfaces. When the fish enters fresh water, it stops drinking.

Kidney function reverses as well. When a salmon is in seawater, its kidneys produce just a small amount of urine per day, and the urine has the same osmotic pressure as the blood plasma. This concentration, although maximal for a fish, is inadequate to keep the plasma concentration from rising. As a result, the gills must bear most of the responsibility for voiding excess ions in a way that will maintain the plasma hyposmotic to seawater. When a salmon enters fresh water, its kidneys switch to producing a large volume of dilute urine.

SUMMARY

CONCEPT 36.1 Kidneys Regulate the Composition of the Body Fluids

- The body fluids of an animal have three characteristics: osmotic pressure, ionic composition, and volume.
- **Kidneys** are organs composed of tubular structures that produce **urine**—an aqueous solution derived from the blood plasma—for excretion. The primary function of kidneys is to regulate the composition and volume of the blood plasma by means of controlled removal of solutes and water from the plasma. **Review Figure 36.2 and ACTIVITY 36.1**
- Kidney function can be expressed in terms of the composition of the urine as a ratio of the composition of the blood plasma. Such a ratio is called a urine/plasma, or U/P, ratio. If the urine is less concentrated in total solutes than the plasma (osmotic U/P < 1), the kidneys are making the plasma become more concentrated. If the urine is more concentrated than the plasma (U/P > 1), the kidneys are making the plasma become more dilute.
- Animals vary in the U/P ratios that can be achieved by their kidneys and thus in how concentrated their urine can be. Mammals, birds, and insects are the only animals that can produce concentrated urine with an osmotic U/P ratio exceeding 1.0.
- Some groups of animals have tissues or organs other than the kidneys that excrete ions at high total concentrations—a process called **extrarenal salt excretion**. **Review Figure 36.3**

CONCEPT 36.2 Nitrogenous Wastes Need to Be Excreted

- Metabolism of proteins and nucleic acids produces toxic **nitrogenous wastes**, which must be eliminated from the body. **Review Figure 36.4**
- **Ammonotelic** animals produce ammonia as their primary nitrogenous waste. They are typically water-breathing aquatic animals that eliminate ammonia by diffusion across their gills or other permeable body surfaces.
- Terrestrial animals nearly always excrete nitrogenous wastes that are far less toxic than ammonia. Synthesizing these alternative nitrogenous wastes requires investment of energy.
- **Ureotelic** animals detoxify ammonia by converting it mostly to urea before excretion. These animals include mammals and most amphibians.
- **Uricotelic** animals convert ammonia mostly to uric acid or other compounds closely related to uric acid. They include insects, spiders, and birds and some other reptiles. Uric acid and the compounds related to it can be excreted in precipitated form, resulting in little loss of water.

SUMMARY *(continued)*

CONCEPT 36.3 Aquatic Animals Display a Wide Diversity of Relationships to Their Environment

- **Osmolarity** is a measure of the overall solute concentration (osmotic pressure) of a fluid. An animal may have body fluids that have the same osmotic pressure as the animal's environment (**isosmotic**) or that have a higher (**hyperosmotic**) or lower (**hyposmotic**) osmotic pressure than the animal's environment.
- Ocean bony fish are hyposmotic to their environment; they tend to lose water by osmosis and gain ions by diffusion. Freshwater fish are hyperosmotic to their environment; they tend to gain water by osmosis and lose ions by diffusion. To regulate the composition of their body fluids, both ocean and freshwater fish employ their kidneys, ion pumping mechanisms in their gills, and drinking behavior. **Review Figure 36.5**
- Migratory bony fish such as salmon switch between being **hyperosmotic regulators** in fresh water and **hyposmotic regulators** in seawater.
- Some marine invertebrates face varying environmental osmolarities. These animals can be **osmotic conformers** or **osmotic regulators**. Most marine invertebrates are osmotic conformers and do not survive well in dilute waters. **Review Figure 36.6**

CONCEPT 36.4 Dehydration is the Principal Challenge for Terrestrial Animals

- Terrestrial animals can readily become dehydrated because of evaporative loss of water from their body fluids.
- **Humidic** terrestrial animals have outer body coverings (i.e., skin, exoskeleton, or the like) that are highly permeable to water. Such animals lose water so rapidly by dehydration that they can tolerate only limited exposure to dehydrating conditions.
- **Xeric** terrestrial animals have outer body coverings that highly limit evaporation of their body fluids. These animals can therefore spend indefinite periods in the open air. The low water permeability of their body coverings results from the presence of lipids.
- **Metabolic water** is the water produced by the oxidation of food materials during metabolism. It sometimes serves as a major source of water for desert animals that have very low total water needs because they conserve water exceptionally well.

CONCEPT 36.5 Kidneys Adjust Water Excretion to Help Animals Maintain Homeostasis

- Vertebrate kidneys consist of many tubules called **nephrons**.
- The fluid first introduced into a nephron is the primary urine. This fluid is modified as it flows through the nephron by processes of reabsorption and secretion. The fluid that is excreted into the outside environment is the definitive urine.
- Urine formation in a vertebrate nephron begins at the closed end of the nephron, which consists of a **Bowman's capsule** that surrounds a **glomerulus**. The primary urine is formed by **ultrafiltration**, during which fluid moves out of the blood plasma in the glomerulus and into the lumen of the Bowman's capsule, driven by the force of blood pressure. The primary urine is similar in composition to the blood plasma in most ways except for lacking blood cells and proteins. **Review Figure 36.8 and ACTIVITY 36.2**
- The rate at which primary urine is introduced into all the Bowman's capsules collectively is the **glomerular filtration rate** (**GFR**).
- The proximal convoluted tubule of a nephron reabsorbs water, Na^+, Cl^-, and certain other solutes, reducing the volume of the **tubular fluid** without changing its osmolarity.
- **Antidiuretic hormone** (**ADH**) controls the function of the late kidney tubules: the distal convoluted tubules in amphibians and the **collecting ducts** in mammals. When the ADH concentration is high, **aquaporin** molecules are inserted in the cell membranes of the epithelial cells of the late tubules, making the cells water permeable. Consequently, the tubular fluid loses water by osmosis into the surrounding tissue fluids, reducing the volume of the tubular fluid and raising its concentration (an antidiuretic state). When the ADH concentration is low, the epithelial cells are poorly permeable to water. Thus, the volume of the tubular fluid is not reduced as much by osmosis, and the fluid is rendered dilute by removal of solutes (a diuretic state). **Review Figures 36.9 and 36.10**
- When the concentration of ADH is high, an amphibian makes urine that is isosmotic with the blood plasma because the tissue fluids surrounding the distal convoluted tubules are isosmotic with plasma. A mammal, however, makes urine that has an osmotic pressure higher than the plasma because the tissue fluids surrounding the collecting ducts are hyperosmotic to the plasma.
- In a mammal, the **loops of Henle** create a concentration gradient in the tissue fluids of the renal medulla by **countercurrent multiplication**. This is the mechanism by which the tissue fluids surrounding the collecting ducts are made hyperosmotic to the plasma. **Review Figure 36.11 and ANIMATED TUTORIALS 36.1 and 36.2**
- The excretory system of an insect is composed of the **Malpighian tubules** (which join the gut at the junction of midgut and hindgut) and the hindgut. Primary urine is formed by a secretory mechanism, often driven by active transport of KCl into the Malpighian tubules. **Review Figure 36.12**

Go to the Interactive Summary to review key figures, Animated Tutorials, and Activities PoL2e.com/is36

Go to LaunchPad at **macmillanhighered.com/launchpad** for additional resources, including LearningCurve Quizzes, Flashcards, and many other study and review resources.

37 Animal Reproduction

KEY CONCEPTS

The ability of rabbit populations to expand rapidly is based on distinctive features of their reproductive biology. European rabbits (*Oryctolagus cuniculus*) such as these are noted for their reproductive potential.

Rabbits are famed for their reproductive potential. Some of this fame is of a shocking kind. European rabbits (*Oryctolagus cuniculus*) were introduced into Australia around 1860 so they could be hunted for sport. Populations quickly grew into the millions, and today rabbits are blamed for dramatic habitat destruction in much of the country.

Reproductive biology is not the only factor that affects whether a species becomes a pest or an asset in the wild. The presence or absence of natural predators, for example, is also important.

Nonetheless, any analysis of rabbits must take into consideration that, for mammals of their body size, their reproductive biology endows them with an extraordinary capacity to multiply. One pair of European rabbits has the biological potential to produce 15 pairs in a year and (by the principles of exponential growth) more than 200 pairs in 2 years.

Two aspects of the reproductive biology of rabbits stand out in explaining this reproductive potential. First, rabbits exhibit postpartum estrus. Estrus, or "heat," is a behavioral state in which a female expresses readiness to mate. Females of most nonhuman mammals, including rabbits, avoid or repel males except when in estrus. Postpartum estrus is a specialized type of estrus in which a female enters heat immediately after she gives birth (*postpartum* means "following birth"). When a female rabbit gives birth, she of course has a litter of young to nurse. If a male is present, however, she also goes immediately into estrus and mates. Not uncommonly, she therefore supports two litters at once! As she nurses the just-born litter, she simultaneously nurtures a new litter in her uterus.

The second key reproductive characteristic of rabbits is induced ovulation. Ovulation is the release of eggs (ova) from the ovaries into the reproductive tract so they can potentially meet sperm. In most mammals, including humans, ovulation occurs independently of mating. Pregnancy thus has a large element of chance in whether it occurs. A female might mate and get lots of sperm, yet not become pregnant because her ovaries happen to release eggs at a different time. In a species with induced ovulation, by contrast, the act of mating directly causes ovulation to occur. When a female rabbit copulates, signals are immediately sent to her brain, which promptly sends signals to the ovaries, resulting in the release of eggs. The process guarantees that eggs will be present for fertilization in the female's reproductive tract at the same time that sperm are there to fertilize them.

What differences and similarities exist between mammals that exhibit induced ovulation and those that ovulate independently of mating?

You will find the answer to this question on page 784.

CONCEPT 37.1 Sexual Reproduction Depends on Gamete Formation and Fertilization

In casual conversation, "having sex" refers to human sexual activity. Biologists, however, mean something far broader when speaking of sex. As applied to animals, the process of **sex** is a mechanism by which the genes of two individuals are combined to produce offspring. In the case of animal **sexual reproduction**, each parent produces specialized reproductive cells called **gametes** by meiosis. Because the gametes are products of meiosis, each gamete cell has only half as many chromosomes as the other cells in the body, and the gametes are diversified in their chromosomes and genes by independent assortment and crossing over (see Concept 7.4). The gametes of the female parent are relatively large, nonmotile cells called **eggs** or **ova** (singular, *ovum*). The gametes of the male, called **spermatozoa** or **sperm**, are small cells that swim, typically using flagella (**FIGURE 37.1A**). A sperm and egg fuse to produce a single-celled zygote that has a full set of chromosomes and develops into a sexually produced offspring (**FIGURE 37.1B**; see Chapter 38). In this way, chromosomes and genes from two parents are combined to produce each offspring, and the new individual is not genetically identical to either parent.

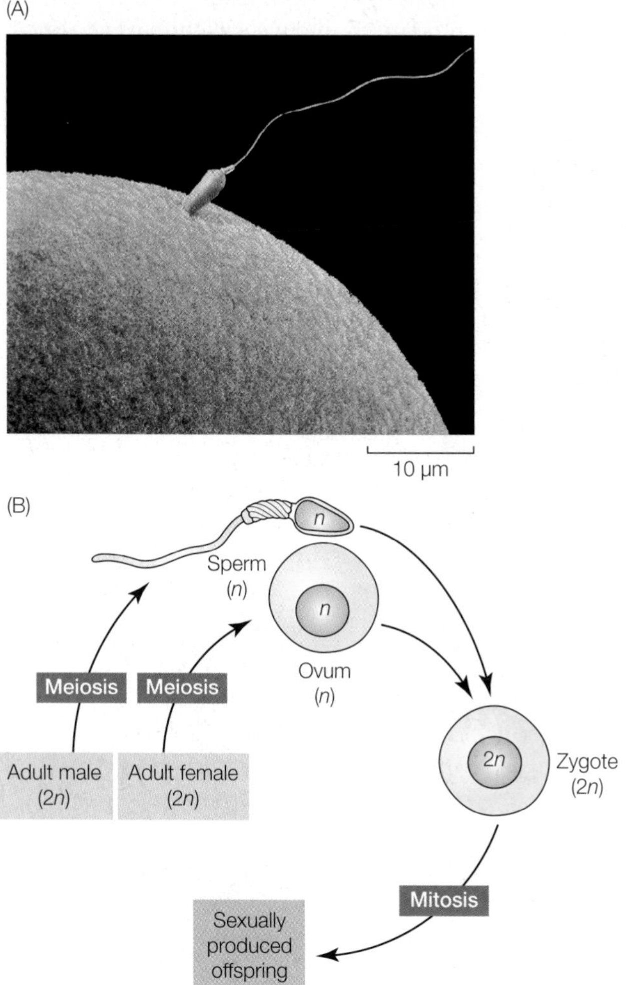

FIGURE 37.1 Sexual Reproduction in Animals (A) The sperm and egg of a human. (B) The essence of animal sexual reproduction is the combination of genes from two individuals. Animals have diplontic life cycles (see Figure 7.3). The body cells that are not specialized for reproduction (called somatic cells) have two copies of each chromosome and thus are diploid (2*n*). The gametes, by contrast, have only a single copy of each chromosome and thus are haploid (*n*). When a sperm and egg combine, they form a single-celled zygote that is 2*n* (diploid), and the zygote subsequently multiplies by mitosis to develop into the adult.

LINK

The genetic aspects of sexual reproduction and the diversity it generates are discussed in **Concepts 7.1 and 7.4**; the evolutionary consequences are discussed in **Concept 15.6**

During **asexual reproduction**, by contrast, the process of sex (combination of genetic material from two parents) does not occur. Instead, offspring are produced by just a single parent, usually by mitosis: parental cells with a full set of chromosomes divide to produce new offspring cells that each have the same full set of chromosomes. Offspring are thus genetically identical to their single parent. A variety of animals, mostly invertebrates, reproduce asexually. Usually they can also reproduce sexually. Budding and fission are the two most common types of asexual reproduction. In budding, new individuals form as outgrowths, or buds, from the bodies of other individuals. In fission, an individual splits into two or more pieces that grow into new individuals. Reef corals exemplify animals in which budding is a major mechanism of reproduction. Corals occasionally reproduce sexually with eggs and sperm. However, when a new colony is produced sexually, it starts with just one small polyp. The large reef structures with which we are familiar are produced by budding. The initial polyp produces other polyps by budding, resulting in a large colony of hundreds or thousands of genetically identical polyps (**FIGURE 37.2**).

Asexual reproduction has the advantage that it maintains favorable combinations of genes. An individual with a genotype that is highly suited to its environment can rapidly produce offspring with the same advantageous genotype by budding, fission, or other asexual means.

Most animals reproduce sexually

Most vertebrates reproduce only sexually, and the same can be said of many invertebrates. Almost all animals have the capacity to reproduce sexually even if, like corals, they also reproduce asexually. We will focus on sexual reproduction for the rest of this chapter.

Sexual reproduction can seem to us to be the obvious best option for producing offspring. Why would reproduction occur by any other means? However, sexual reproduction can have some disadvantages that asexual reproduction does not (see Concept 15.6). For example, potential mates have to find each other and succeed in mating despite the presence of competitors and predators. Possibly the greatest downside of sexual reproduction is that it breaks up favorable combinations of

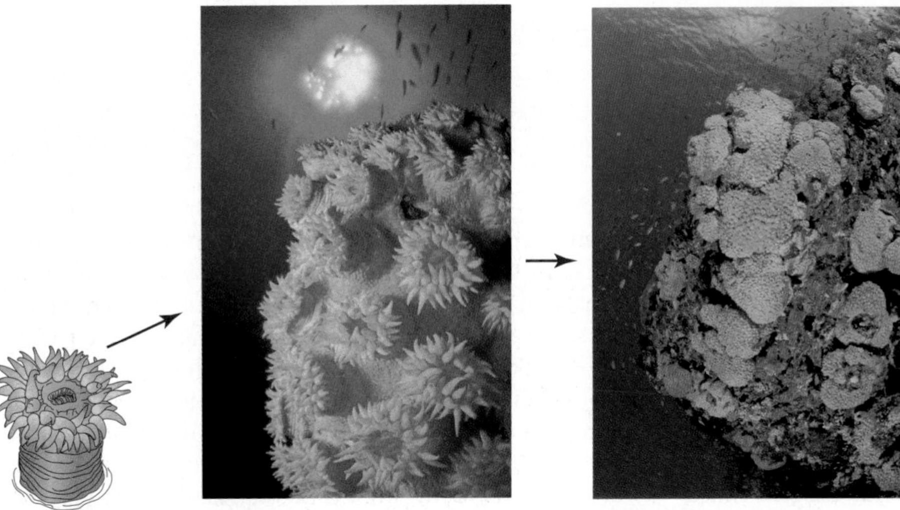

Sexual reproduction produces a polyp.
That polyp multiplies asexually by budding to produce a colony of genetically identical polyps.
Continued asexual reproduction produces massive colonies that form the key structures of coral reef ecosystems.

FIGURE 37.2 Asexual Reproduction in Animals Sexual reproduction in reef corals produces a single polyp, which then reproduces asexually by budding to produce a colony. Each polyp secretes skeletal material. The skeletal material of all the polyps in the colony fuses into a single structure, creating the massive skeletal structures that typify coral reefs.

genes. An individual with a genotype that is virtually perfect for its environment cannot simply pass that genotype along to its offspring. Biologists have theorized a great deal about why sex is so common despite this limitation (see Concept 15.6). According to one argument, the explanation lies in the changeable nature of environments. Environments, it is argued, change sufficiently—from time to time and from place to place—that no single genotype is likely to be highly successful for very long. Sexual reproduction has an enormous potential for producing genetic diversity in offspring, maximizing the odds that at least some offspring will succeed in a changing environment.

Among sexually reproducing animals, the two sexes, male and female, differ in three ways:

- Primary reproductive organs
- Accessory reproductive organs
- Secondary sexual characteristics

The **primary reproductive organs** (primary sex organs) are the **gonads**: the **ovaries** in females and **testes** (singular, *testis*)—sometimes called testicles—in males (**FIGURE 37.3**). These organs are "primary" because they produce the cells, the gametes, that will combine to produce a new individual.

The **accessory reproductive organs** (accessory sex organs) are the reproductive organs other than the gonads. In human males, for example, the ducts (e.g., the vas deferens) that carry sperm out of the testes to the urethra in the penis are accessory reproductive organs, as is the penis itself. Often, the accessory reproductive organs are more elaborate in females than males. The females of some animals, such as mammals and many sharks, nurture offspring internally for a long time before giving birth. In these cases, the uterus and other structures responsible for nurturing the young are among the accessory reproductive organs. In other animals, such as birds, the females lay eggs, and their accessory organs include the glands that add the shell or capsule to each egg. (Note that the term "egg" has two meanings in discussing reproduction. The female gamete (ovum) is called an "egg," but a structure like a hen's egg is also called an "egg.")

The **secondary sexual characteristics** are properties of nonreproductive tissues and organs that are distinctive in each sex. These characteristics often result from the actions of sex hormones on nonreproductive structures. Facial hair, chest hair, and a deep voice are some of the secondary sexual

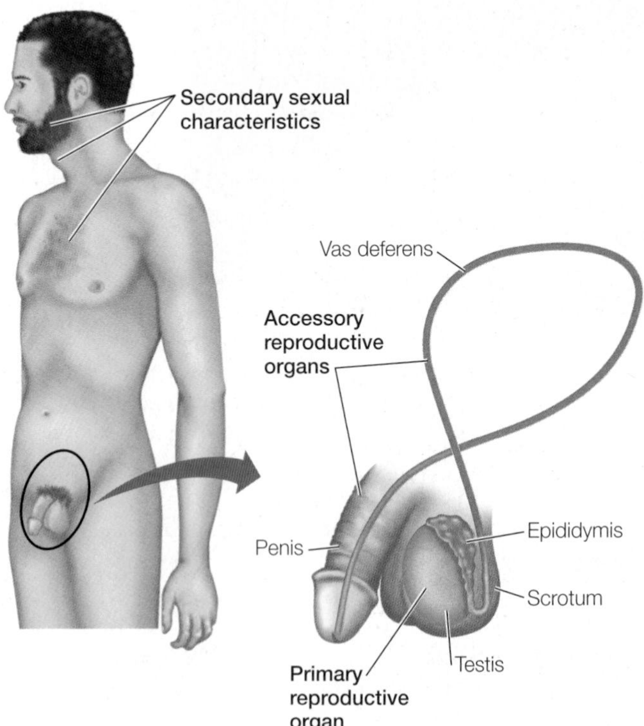

FIGURE 37.3 Sexually Reproducing Animals Have Three Types of Sex-Specific Characteristics Individuals of each sex have sex-specific primary reproductive organs (gonads) and accessory reproductive organs. They may also have sex-specific secondary sexual characteristics.

characteristics of human males (see Figure 37.3). The bright feather colors of many male birds and the enlarged horns of male caliper beetles (*Golofa porteri*) are also secondary sexual characteristics:

Female mallard ducks are mostly brown, and female caliper beetles do not have enlarged horns.

Secondary sexual characteristics are frequently subject to sexual selection.

LINK

Sexual selection results from nonrandom mating and is discussed in **Concept 15.2**; see especially **Figures 15.9 and 15.10**

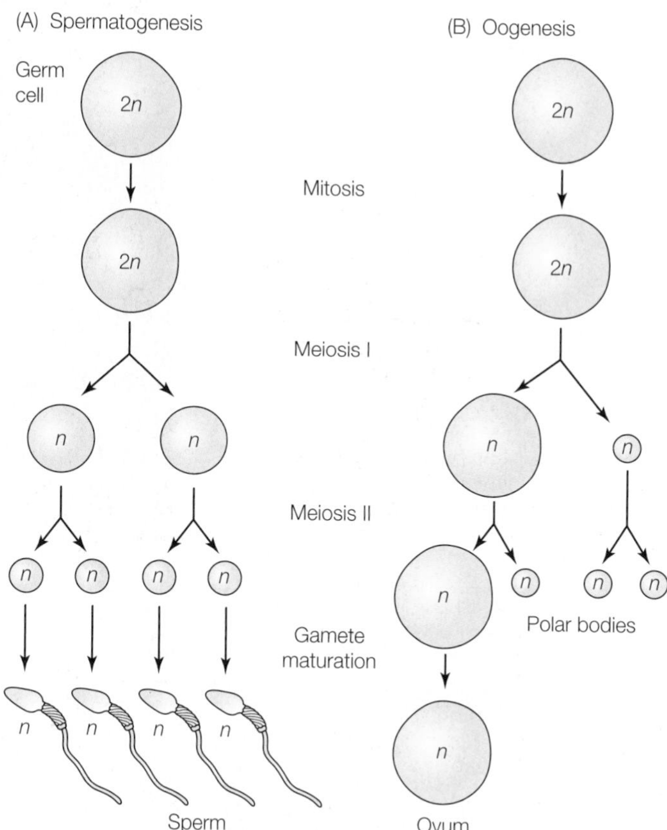

FIGURE 37.4 Gametogenesis Male and female germ cells proliferate by mitosis to produce an abundant supply of diploid cells that initiate and complete meiosis to produce sperm or ova. (A) Spermatogenesis. (B) Oogenesis. During oogenesis in many animals, the first polar body to be produced does not divide, so only two polar bodies are produced in total. $2n$ is the diploid number of chromosomes per cell; n is the haploid number. See Concept 7.4 for details of meiosis.

Gametogenesis in the gonads produces the haploid gametes

Gametogenesis is the process by which gametes are produced. More specifically, it is called spermatogenesis in males and oogenesis in females. The process begins with diploid **germ cells** (diploid, denoted $2n$, means two copies of each chromosome per cell; see Figure 37.1). These are cells in the gonads that are capable of undergoing meiosis to produce gametes. The diploid germ cells divide by mitosis until they reach a stage at which they switch to meiosis. These mitotic divisions increase the numbers of cells that are available to undergo meiosis and thereby increase the number of haploid gametes produced (haploid, $1n$, means one copy of each chromosome per cell). Recall that meiosis consists of two cell divisions, meiosis I and meiosis II, during which the DNA is replicated only once (see Figure 7.11).

In adult males, sperm are typically produced continuously during the time of year when reproduction occurs (i.e., all year in human males). During this time, the populations of diploid germ cells in the testes multiply continuously, producing cells that go through the two meiotic divisions, resulting in four, equal-sized haploid cells from each diploid cell (**FIGURE 37.4A**). The haploid cells then mature into fully differentiated sperm.

Gametogenesis in females differs in two ways from that in males. First, meiosis typically produces only one gamete from each diploid germ cell because the cytoplasm of the diploid cell is divided in highly unequal ways during each of the two meiotic divisions (**FIGURE 37.4B**). One of the cells produced by meiosis is a large cell, the ovum. The others are tiny cells, called polar bodies, that degenerate.

The second difference that often occurs is cellular developmental arrest during meiosis in females. When a cell destined to become an ovum is in arrest, it can continue to accumulate resources and grow, but its meiosis stops. Mammals exemplify developmental arrest during oogenesis. According to the prevailing view of most researchers, all the diploid germ cells in a young female placental mammal stop undergoing mitosis, begin meiosis (completing the very first step in meiosis I), and go into developmental arrest while the individual is a fetus in the uterus or shortly thereafter. The cells then remain in arrest for months or years until the individual matures to reproductive age. Then some of the arrested cells resume meiosis to become ova that are released from the ovaries. In humans, one of the arrested cells typically develops fully and is released from the ovaries during each month between the time a girl enters puberty (near age 10) and the time she goes through menopause (near age 50).

The gonads, in addition to having germ cells and producing gametes, typically contain somatic cells that contribute to reproduction. Somatic cells by definition do not produce gametes. If they divide, they do so strictly by mitosis. Certain somatic cells in the gonads provide metabolic support to the

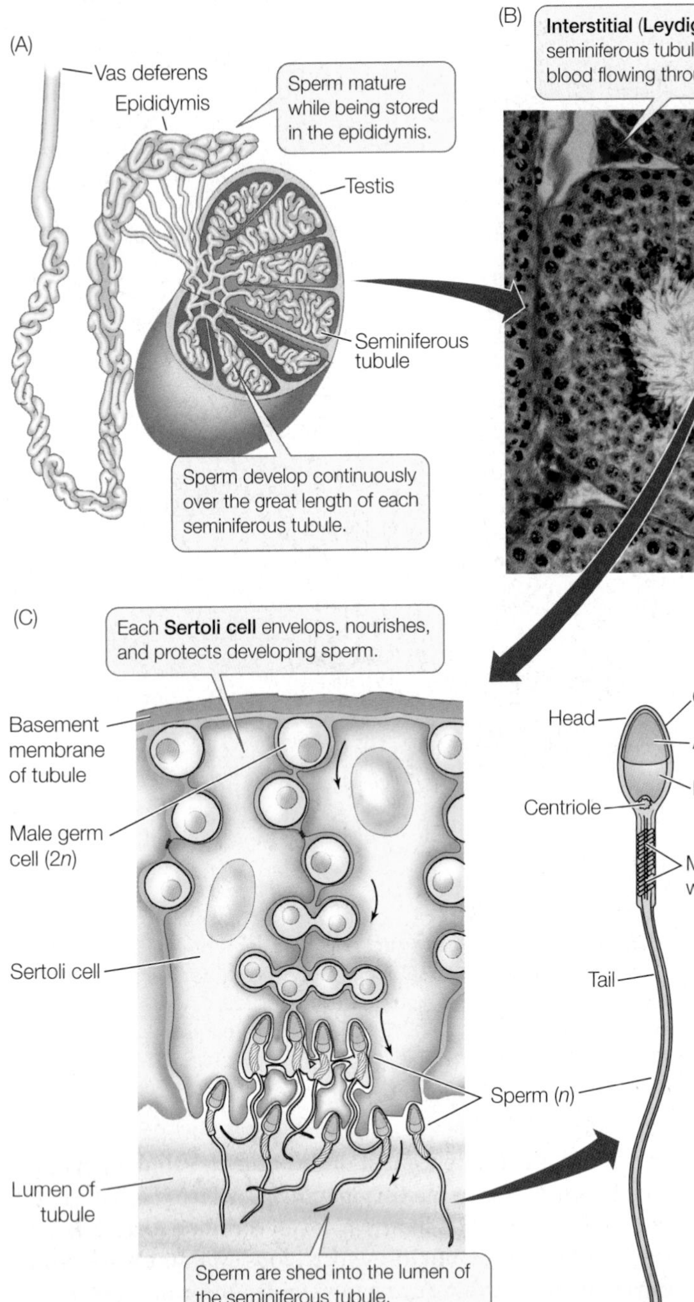
(A)
Vas deferens
Epididymis
Sperm mature while being stored in the epididymis.
Testis
Seminiferous tubule
Sperm develop continuously over the great length of each seminiferous tubule.
(B)
Interstitial (Leydig) cells in the tissue between seminiferous tubules secrete testosterone into blood flowing through the testis.
Lumen of seminiferous tubule (sperm project into it)
(C)
Each Sertoli cell envelops, nourishes, and protects developing sperm.
Basement membrane of tubule
Male germ cell (2n)
Sertoli cell
Lumen of tubule
Sperm (n)
Sperm are shed into the lumen of the seminiferous tubule.
Head
Cell membrane
Acrosome
Nucleus
Centriole
Mitochondria within midpiece
Tail

FIGURE 37.5 The Mammalian Testis (A) Seminiferous tubules fill the testis and continuously produce millions of sperm during the reproductive season. (B) This cross section of a testis shows both a seminiferous tubule and groups of interstitial (Leydig) cells in the spaces between this tubule and adjacent tubules. (C) As germ cells undergo gametogenesis, the cells move from the outer region of the seminiferous tubule toward its center (shown by arrows; compare with Figure 37.4A), where sperm are shed into the lumen of the tubule.

Go to ACTIVITY 37.1 Spermatogenesis
PoL2e.com/ac37.1

germ cells during gametogenesis. Somatic cells may also secrete hormones. The mammalian testis illustrates these points (**FIGURE 37.5**).

There are three principal cell types in a testis:

- Germ cells that produce the sperm
- Somatic **Sertoli cells** that assist sperm production
- Somatic **interstitial cells** (**Leydig cells**) that produce the steroid sex hormone **testosterone** (see Concept 35.4)

The testis is principally composed of long, coiling tubules, the **seminiferous tubules**. The lining of each tubule consists of germ cells in all the stages of gametogenesis and Sertoli cells. Sperm are shed into the lumen (open central cavity) of the tubule and make their way to the epididymis, a highly convoluted tubular structure attached to the outside of the testis, where the sperm undergo maturation. The epididymis connects to the **vas deferens**, a tube that carries sperm out of the testis during ejaculation (see Figure 37.3). The endocrine interstitial cells are located in the spaces between adjacent seminiferous tubules.

Let's focus more closely on the somatic cells that support the germ cells during gametogenesis. In male vertebrates, spermatogenesis occurs in the midst of Sertoli cells (see Figure 37.5C). The Sertoli cells envelop the developing germ cells, provide nutrients, and help regulate sperm production. A striking fact is that the Sertoli cells constitute well over half the volume of each testis in a human or other mammal. This testifies to their importance.

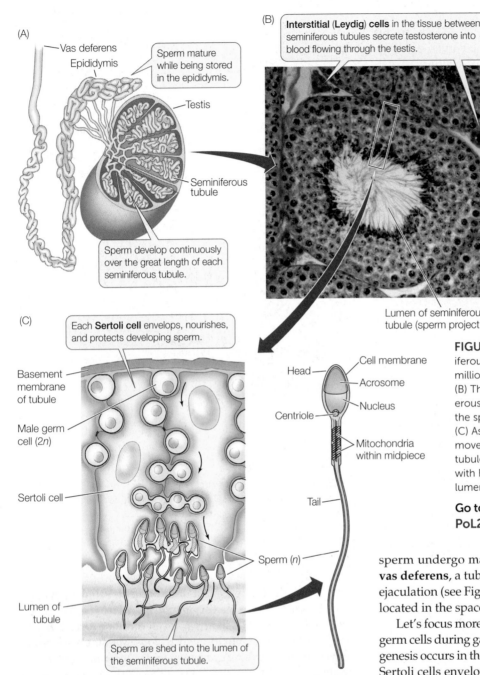

Ova often have a more defined relationship with somatic support cells, in the sense that each ovum is associated with a specific set of support cells during its development. This is true in vertebrates and some invertebrates (e.g., insects). A developing ovum and its support cells are together called an **ovarian follicle**. In human females, as we've said, one cell that entered developmental arrest during fetal life typically develops to become an ovum released from the ovaries during each month of adult life. The developing ovum, termed an **oocyte**, is surrounded by somatic support cells—termed granulosa and theca cells—that multiply as development takes place (**FIGURE 37.6**). As the support cells multiply, they provide increased metabolic support, and the follicle grows enormously from a

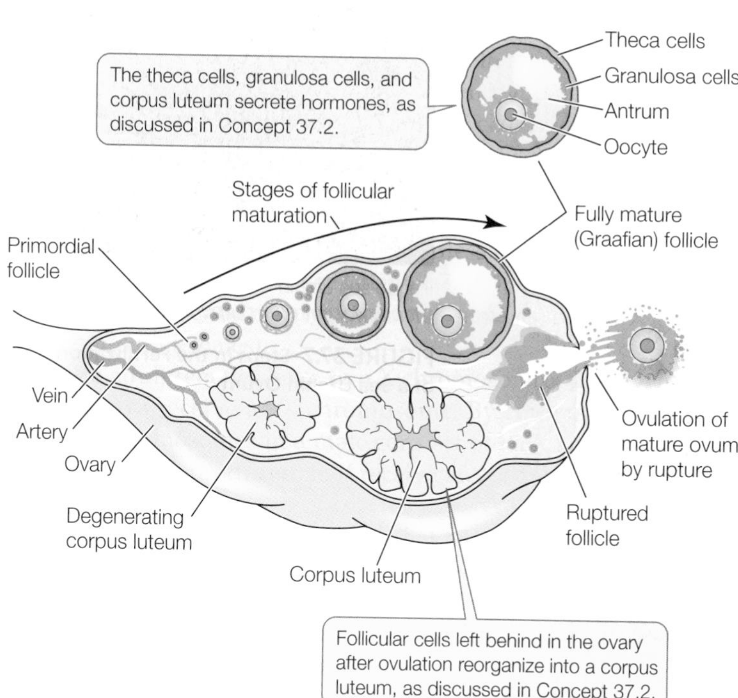

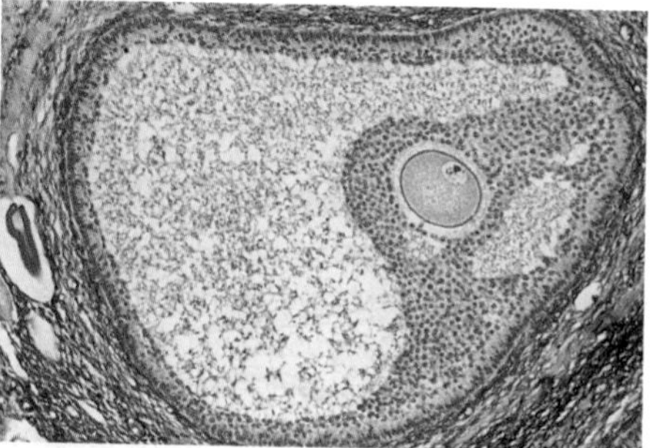

A mature follicle seen in cross section in a light microscope

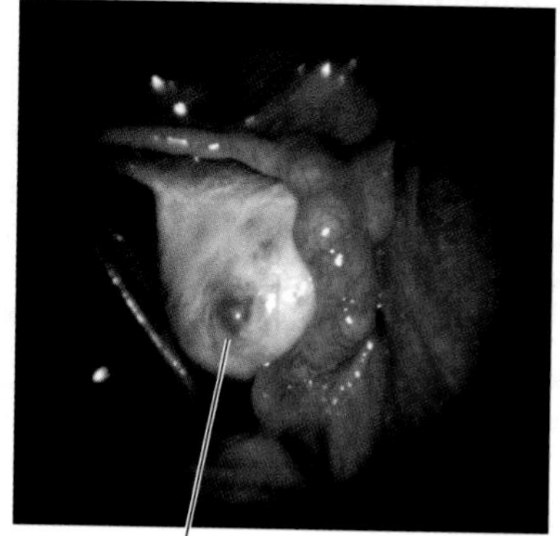

Mature follicle bulging on surface of ovary

FIGURE 37.6 The Mammalian Ovary Showing Stages of Follicle Development The stages of follicle development are shown in a clockwise series, starting at the upper left with a primordial follicle. In a human ovary, a primordial follicle progresses through all these stages in the course of approximately 28 days (one menstrual cycle). A new primordial follicle starts the next month. The clockwise arrangement is for learning purposes only. Follicles do not actually progress clockwise through an ovary.

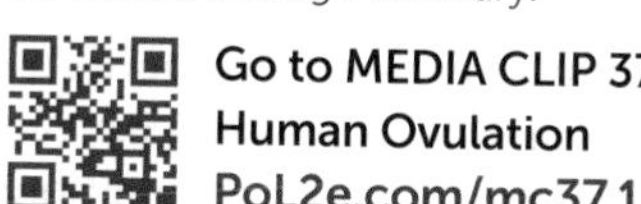

Go to MEDIA CLIP 37.1
Human Ovulation
PoL2e.com/mc37.1

primodial follicle to a mature (Graafian) follicle. Along the way, a fluid-filled cavity (antrum) opens up within it. Moreover, the follicle migrates to be at the very surface of the ovary, where at full maturity it bulges out like a large blister, 1.5–2 centimeters in diameter in a human. The release of an ovum from the ovary is called **ovulation**. In vertebrates, including humans, it occurs by rupture of the mature follicle.

Fertilization may be external or internal

Fertilization is the fusion of sperm and ovum. It can be external or internal. The product is a single diploid cell, the zygote, which develops into an embryo (see Chapter 38).

To achieve fertilization, most aquatic animals simply release their gametes into the water, a process termed **spawning**. Fertilization then occurs in the water outside the female's body and is described as **external fertilization**. In spawning species, the ducts and openings of the reproductive system can be similar in males and females because both sexes are simply releasing gametes into the environment. The zygotes typically do not receive parental care. They must develop into feeding larvae and find food before the yolk or other nutrients provided in the ovum run out. Mass deaths of commercially important fish larvae sometimes occur when the larvae are only 2–3 weeks old because at that time the larvae exhaust the yolk provided in the eggs but fail to find food. With the larvae dead, adults are not produced.

Internal fertilization occurs when a male inserts sperm inside the reproductive tract of a female, permitting fertilization to occur inside the female's body. Most terrestrial animals employ internal fertilization—not surprising, because it ensures that the gametes and zygotes are kept moist. Some aquatic animals also employ internal fertilization.

Internal fertilization can have several advantages. It permits parental defense of the new zygotes. Some animals enclose each fertilized egg in a protective shell or capsule before laying the egg. This is true in birds (e.g., chicken eggs), insects, turtles, skates, and some sharks. Internal fertilization is essential for this process because sperm must reach the ovum before the shell is formed:

Skates enclose each egg in an egg case like this.

Some animals—notably mammals and certain sharks (e.g., requiem and hammerhead sharks)—keep fertilized zygotes in the female reproductive tract for an extended period of development during which nutrients are steadily provided, so the

INVESTIGATION

FIGURE 37.9 Investigating the Mechanism of Sex Change In contrast to the anemonefish, females of the blackeye goby (*Coryphopterus nicholsii*) sometimes become males. Investigators have been studying the mechanism by which this occurs. By simply describing blood hormones, the investigators have found that the female-to-male transition is accompanied by a decline in the blood concentration of the female sex steroid 17β-estradiol and an increase in the concentration of the male sex steroid 11-ketotestosterone (11-KT). But do these changes in hormones *cause* the sex change? Are they *sufficient* to make a female change into a male? To find out, the investigators artificially manipulated the hormone levels.[a]

HYPOTHESIS

An increase in the concentration of 11-ketotestosterone (11-KT) is sufficient to cause female-to-male transition in blackeye gobies.

METHOD

1. Capture female blackeye gobies in the wild and hold in individual tanks.
2. Insert in each fish a hormone-releasing implant containing 11-KT or an implant with no hormone (control treatment). The implant is designed to release hormone gradually.
3. After 43 days of treatment, examine the gonads of the fish to determine the sex.

RESULTS

	Sex at end of treatment		
Treatment	Female	Transitional	Male
Control	13	0	0
11-KT or 11-ketoadrenosterone (converted to 11-KT in body)	0	1	13

Administration of 11-KT causes female blackeye gobies to become male.

CONCLUSION

The results support the hypothesis that an increase in the concentration of 11-KT is sufficient to cause female-to-male transition.

ANALYZE THE DATA

In a related experiment, 8 female fish were treated with fadrozole, a chemical that inhibits the synthesis of 17β-estradiol and thus should lower the concentration of this female sex hormone. At the end of treatment, 1 fish was still female, 5 fish were transitional, and 2 fish were male.

A. Was this treatment as effective as 11-KT (or 11-ketoadrenosterone) administration in causing females to become males? What data support your conclusion?

B. The researchers hypothesized that a decrease in 17β-estradiol causes sex change. Do the data support the hypothesis? Why or why not? What additional information would be helpful in deciding whether the hypothesis is supported?

Go to **LaunchPad** for discussion and relevant links for all **INVESTIGATION** figures.

[a]F. J. Kroon and N. R. Liley. 2000. *General and Comparative Endocrinology* 118: 273–283.

CHECKpoint CONCEPT 37.1

✓ What are some of the advantages and disadvantages of internal and external fertilization?

✓ Vasectomy is a procedure for birth control in which the principal ducts carrying sperm out of the testes are tied off. Basing your answer on the cellular structure of the human testis, explain why a man's testosterone level is not affected by vasectomy.

✓ Rarely, men are found who have two X chromosomes. The *SRY* gene is responsible for this syndrome. What type of unusual gene placement could account for the syndrome? Explain.

Having surveyed the general aspects of animal gametogenesis and fertilization, we will next consider the male and female reproductive systems in mammals, using our own species as the primary example.

CONCEPT 37.2 The Mammalian Reproductive System Is Hormonally Controlled

Hormones play critical roles in almost every aspect of mammalian sexual reproduction. Follicle-stimulating hormone (FSH) and luteinizing hormone (LH), secreted by the anterior pituitary gland, are key players in both sexes (see Concepts 35.3 and 35.4). They are named for their roles in females, but the same hormones are secreted in males and exert major effects on tissues of the male reproductive system. Secretion of FSH and LH is controlled by neurosecretory cells in the hypothalamus of the brain. These cells secrete gonadotropin-releasing hormone (GnRH) into blood that travels to the anterior pituitary via the hypothalamo–hypophysial portal system (see Figure 35.7). In the pituitary, GnRH regulates the cells that secrete FSH and LH.

Somatic cells in the gonads produce steroid sex hormones. We often say that "the ovaries" or "the testes" secrete hormones, but it's important to recognize that only certain cells in the gonads are responsible. The principal sex steroid in males is testosterone, which we've already seen is secreted by the interstitial (Leydig) cells in the testes. In females, the somatic cells of the ovarian follicles secrete feminizing steroids called **estrogens**. Typically, two or more specific types of estrogen molecules are secreted, but here we refer to them simply as "estrogens." After follicles have ruptured during ovulation, the somatic cells remaining in the ovary reorganize and—in addition to estrogens—secrete another type of female sex steroid, **progesterone**,

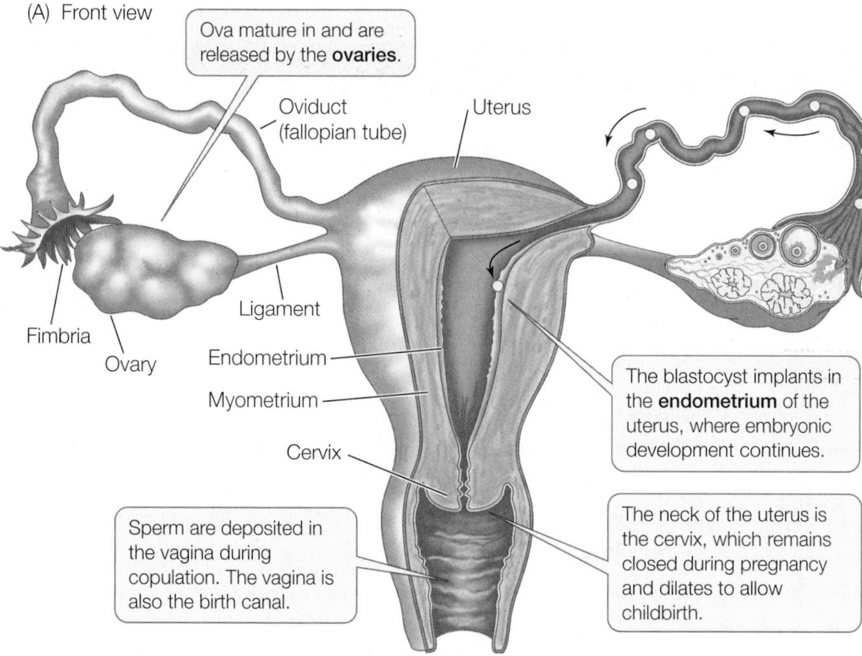

(A) Front view
Ova mature in and are released by the **ovaries**.
Oviduct (fallopian tube)
Uterus
Ova released from the ovaries are taken into the **oviducts**, where they travel to the uterus. Fertilization occurs in the upper regions of the oviduct, where development of the zygote begins.
Fimbria
Ovary
Ligament
Endometrium
Myometrium
Cervix
The blastocyst implants in the **endometrium** of the uterus, where embryonic development continues.
Sperm are deposited in the vagina during copulation. The vagina is also the birth canal.
The neck of the uterus is the cervix, which remains closed during pregnancy and dilates to allow childbirth.

FIGURE 37.10 Reproductive Organs of the Human Female The female reproductive organs are shown (A) from the front and (B) from the side.

Go to ACTIVITY 37.2
The Human Female Reproductive Tract
PoL2e.com/ac37.2

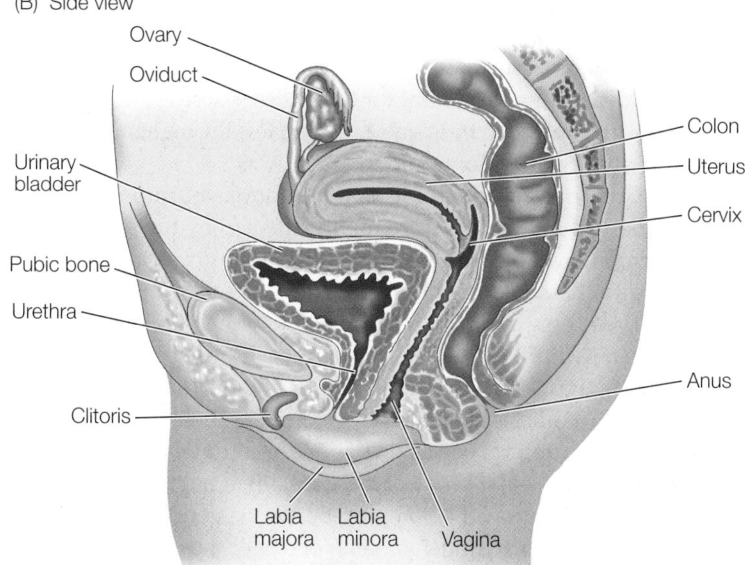

which is involved principally in coordinating processes associated with pregnancy. The placenta also secretes estrogens and progesterone during pregnancy.

> **LINK**
> For information on the roles of sex steroids during embryonic development, see **Concept 35.4**

Ova mature in the ovaries and move to the uterus

The ovaries are positioned in the lower abdomen, as seen in **FIGURE 37.10** (Figure 35.10 provides a larger perspective). Ova released from the ovaries make their way to the uterus by way of tubes called the **oviducts** or fallopian tubes. Each ovary is positioned near the funnel-shaped opening of an oviduct. When an ovum is discharged from a follicle at the surface of the ovary during ovulation, it is drawn into the oviduct, and then it is propelled toward the uterus by beating cilia and contractions of smooth muscles.

Sperm are deposited in the vagina—a tubular structure that opens to the outside—during copulation (sexual intercourse). Fertilization of an ovum, if it occurs, takes place in the upper third of the oviduct. Sperm in the vagina enter the uterus at the cervix, travel through the uterus, and then travel up the oviduct to achieve fertilization. Sperm are often in the upper oviduct within 30 minutes, indicating that they are probably transported not only by their swimming powers but also by uterine contractions.

Passage of an ovum or zygote down the oviduct is relatively slow. It requires about 4 days in humans. If fertilization has occurred, the zygote undergoes its early development during this time and arrives in the uterus as an early embryo called a blastocyst (see Chapter 38). The uterus is a thick-walled organ consisting of two major parts: an outer layer of smooth muscle, the myometrium, and an inner, nurturing layer, the **endometrium**. The endometrium is highly vascular and composed of nutrient-providing cells, and it is where the zygote will develop further. After the blastocyst arrives in the uterus, it remains in the lumen for a few days and then buries itself in the endometrium by means of enzyme-catalyzed processes. The entry of the blastocyst into the tissue of the endometrium is termed **implantation**.

Ovulation is either induced or spontaneous

The immediate stimulus for ovulation is a surge in the blood concentration of LH. However, the cause of this surge differs between species that undergo induced and spontaneous ovulation.

Induced ovulation, described in our opening discussion, is ovulation triggered by copulation. Rabbits and certain cats, shrews, and camelids are induced ovulators. During copulation in an induced ovulator, the copulatory act stimulates key sensory neurons in the cervix. These neurons send action potentials to the brain, where the GnRH-secreting neurosecretory cells in the hypothalamus are stimulated. These cells release GnRH, resulting in a surge in LH secretion from the anterior pituitary into the general circulation. In rabbits that have been

APPLY THE CONCEPT

The mammalian reproductive system is hormonally controlled

In a famous and much-cited paper, M. C. Chang investigated the question: Are mammalian sperm altered by being in the female reproductive tract? He studied this question in rabbits. Because of the design of Chang's studies, the unusual properties of rabbits we've mentioned, such as induced ovulation, were not of any relevance. Chang collected freshly ejaculated semen from male rabbits. He diluted the semen, and then he injected it directly into the upper third of the oviduct of fertile females at various times relative to the time that ovulated ova arrived there. The results he obtained are shown in the table.[a]

TIMING OF SEMEN INJECTION	PERCENTAGE OF OVA FERTILIZED
Semen injected 6 hr prior to ovulation	78%
Semen injected 2 hr prior to ovulation	0%
Semen injected 2 hr after ovulation	0%
Semen held for 5 hr in the uterus of another rabbit and then injected 2 hr prior to ovulation	75%

1. What do these results indicate about the question Chang asked? Justify your answer.
2. In vitro fertilization is a procedure by which ova and sperm are collected and brought together in a dish. Are Chang's results relevant to in vitro fertilization?

[a]M. C. Chang. 1951. *Nature* 168: 697–698.

studied, the blood LH concentration increases in 1–2 hours to more than five times higher than usual. This LH surge causes the mature follicle to rupture.

Most mammals, including humans and other primates, are spontaneous ovulators. During **spontaneous ovulation**, follicle rupture and release of the mature ovum are again triggered by an LH surge. However, the timing of the surge is under control of endogenous hormonal cycles in the female. "Endogenous" refers to processes that originate within an animal. The hormonal cycles in a spontaneous ovulator occur on their own rhythm, more or less independently of whether copulation takes place. We will focus on spontaneous ovulation for the rest of our discussion of females.

In most species of mammals that display spontaneous ovulation, a female ovulates in cycles, as already implied. The duration of each cycle varies among species. For instance, cycles in rats and mice are typically 4–6 days, whereas the human cycle averages 28 days. The cycles repeat, over and over, during reproductive seasons (i.e., all year in humans) unless pregnancy occurs. If pregnancy occurs, the cycles stop.

Hormones have the remarkable feature that they can affect many tissues at once, as we stressed in Chapter 35. In female reproductive cycles, hormones coordinate the ovaries, uterus, and in some cases, behavior.

Primates menstruate in each cycle that does not result in pregnancy. **Menstruation** is a discharge of bloody material from the uterus via the vagina. During each cycle, the endometrium undergoes preparation for pregnancy. It develops glands and abundant vascularization, and it thickens (in humans, by a factor of three to five). If pregnancy fails to occur, much of the extra growth is sloughed off, creating the bloody discharge. The cycles of the reproductive system are called **menstrual cycles** because of this highly visible aspect.

Mammals other than primates do not menstruate or have menstrual cycles. Often, however, they undergo dramatic changes in behavior: they cyclically enter estrus. This is the most highly visible aspect of their cycles, which accordingly are called **estrous cycles**. (Note the cycles are "estrous" cycles, whereas the period of sexual receptivity is "estrus.") The endometrium cyclically prepares for pregnancy and regresses during estrous cycles, but the regression occurs by reabsorption, not sloughing off, of tissue.

Hormones secreted by the brain, anterior pituitary gland, and ovaries are important for the cycles. These organs interact and affect each other to create the cycles. If pregnancy occurs, the embryo and uterus join in.

Hormone receptors are as important as hormones. Tissue responses to hormones are highly affected by the numbers of hormone receptors present, and hormones can cause cells to modify their receptor numbers.

To see how cycling occurs, let's look at the human menstrual cycle (**FIGURE 37.11**). By convention, each cycle is considered to start on the day menstruation begins, and that day is numbered day "0." Follicles start their development then, and ovulation (follicular rupture) occurs on day 14. Accordingly days 0–14 are called the follicular phase of the cycle. What happens over those 14 days? The answer is complex, but by mastering it you will be rewarded with an understanding of one of the most fascinating and important processes in the human body. Follow the steps listed here by referring to the circled numbers in Figure 37.11A–D:

① At the start, FSH and LH are secreted by the anterior pituitary at fairly high rates compared with the rates seen following ovulation during the previous cycle (compare with step 10). The FSH and LH stimulate development of follicles, including the one follicle that will mature and release its ovum. Specifically, they stimulate the somatic cells of that follicle—the granulosa and thecal cells—to secrete estrogen.

② Estrogen secretion gradually increases as the follicle develops.

③ The increasing estrogen secretion affects the endometrium, stimulating endometrial development that prepares the uterus to accept a blastocyst if fertilization occurs. Estrogen also simulates the endometrial cells to express progesterone receptors, preparing these cells to respond to progesterone later on.

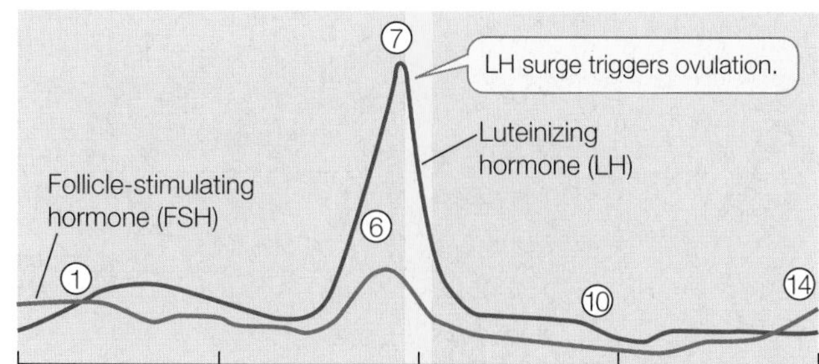

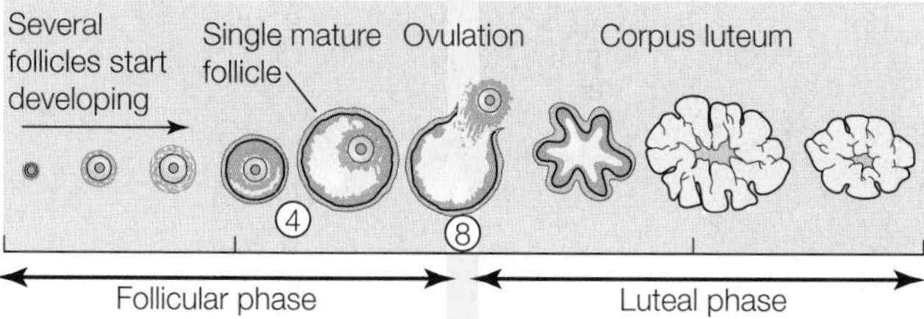

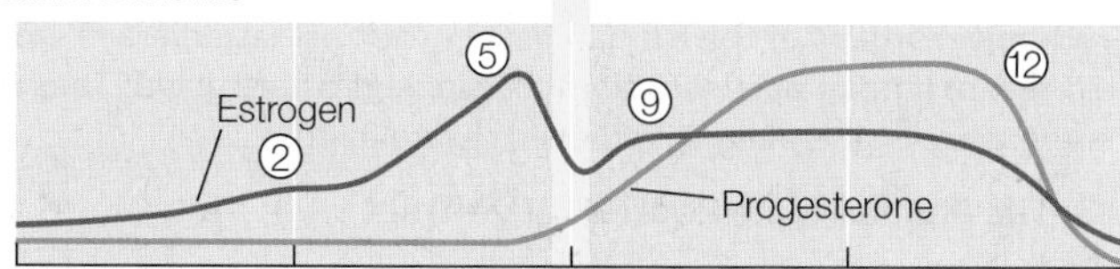

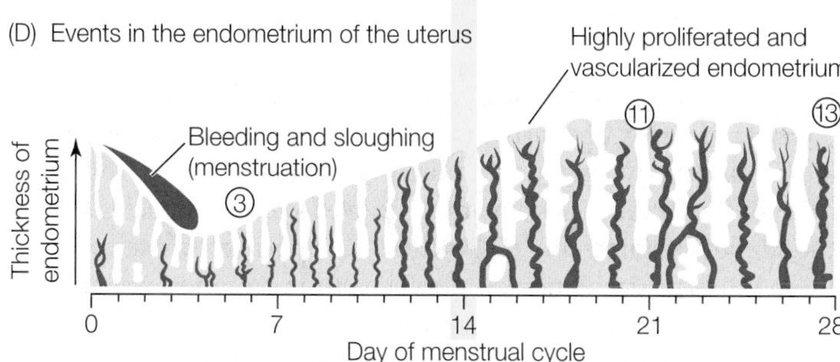

FIGURE 37.11 Events in the Human Female Reproductive Cycle Circled numbers correlate with the sequence of events described in the text.

Go to ANIMATED TUTORIAL 37.2
The Menstrual Cycle
PoL2e.com/at37.2

④ In the second half of the follicular phase, the granulosa cells of the maturing follicle increase their numbers of LH receptors.

⑤ As day 14 approaches, the blood estrogen concentration is high.

⑥ At this stage, estrogen stimulates the anterior pituitary and possibly the brain to bring about increasing secretion of LH and FSH.

⑦ Finally, estrogen stimulation of the anterior pituitary causes the surge in LH secretion.

⑧ The LH surge has a strong effect on the follicle because of the abundance of LH receptors in the granulosa cells (recall step 4). The follicle ruptures, and ovulation occurs. In mammals that (unlike humans) undergo estrus, the peak estrogen secretion at this time stimulates the behaviors of estrus. Estrus thus coincides with ovulation.

When the follicle ruptures, most of the somatic cells of the follicle are left behind in the ovary. LH (which, you'll recall, is *luteinizing* hormone) stimulates those cells to reorganize to form a new hormone-producing structure called the **corpus luteum** (see Figure 37.11B). If pregnancy has not occurred, the corpus luteum, at this stage known as a **corpus luteum of the cycle**, secretes for a finite length of time: about 10 days in humans. It then starts to undergo programmed degeneration, ceasing to function entirely on day 28 of the menstrual cycle. Days 14–28 are the luteal phase of the cycle. What happens during it? Again, follow the steps by referring to the circled numbers in Figure 37.11A–D:

⑨ The corpus luteum secretes progesterone (its principal secretion), estrogen, and a third hormone, inhibin.

⑩ These hormones, acting together, suppress FSH and LH secretion by the pituitary. In primates, this suppression is so strong that it prevents new follicles from starting to develop at this stage.

⑪ Progesterone secretion from the corpus luteum has a major effect on the endometrium. Exocrine glands in the endometrium are stimulated to secrete nutrient materials, and the endometrium becomes more vascularized. These changes fully prepare the endometrium to accept a blastocyst.

⑫ If fertilization has not occurred, the corpus luteum starts to undergo its programmed degeneration after about 10 days. This leads to a steep decline in progesterone and estrogen secretion.

⑬ Deprived of the luteal hormones (hormones secreted by the corpus luteum), the endometrium cannot maintain its highly developed state. It starts to regress, with menstruation beginning on day 0 of the next cycle.

⑭ The disappearance of luteal hormones also sets the stage for a new cycle to begin because the luteal hormones no longer act on the pituitary or brain to suppress FSH and LH secretion. FSH and LH increase, returning the cycle to step 1, in which development of a new follicle is stimulated.

Pregnancy is a specialized hormonal state

If fertilization takes place, the blastocyst burrows into the tissue of the endometrium within a few days after arriving in the uterus, and there is an urgent need to maintain the endometrium in its highly developed state. In humans and many other mammals, the endometrium is prevented from deteriorating by rescue of the corpus luteum. A signal of the existence of pregnancy is sent to the corpus luteum. In response, the corpus luteum does not undergo programmed degeneration and instead becomes a **corpus luteum of pregnancy**, which continues to grow and secrete hormones. The signal in primates is a hormone, chorionic gonadotropin (CG), secreted by the blastocyst. In other words, the embryo emits a signal that tells the mother's tissues to maintain the uterus in a state suitable for pregnancy. The concentration of CG rises high enough in the mother's blood and urine to be easily detected, and many pregnancy tests work by detecting it.

The uterus requires progesterone to remain in a state suitable for pregnancy. In humans, after the corpus luteum is rescued, it provides adequate progesterone for 7–10 weeks. By then, the placenta—which also secretes progesterone—has become sufficiently well developed to take over.

The **placenta** is a structure in which maternal blood vessels and embryonic (or fetal) blood vessels are closely juxtaposed (see Figure 38.19). Thus in addition to its hormonal role, the placenta enables exchange of materials between mother and embryo. The blood of the mother and that of the embryo do not mix. However, nutrients and O_2 move across vessel walls from the maternal blood into the blood of the embryo, and CO_2 and other wastes move from the embryo into the maternal blood.

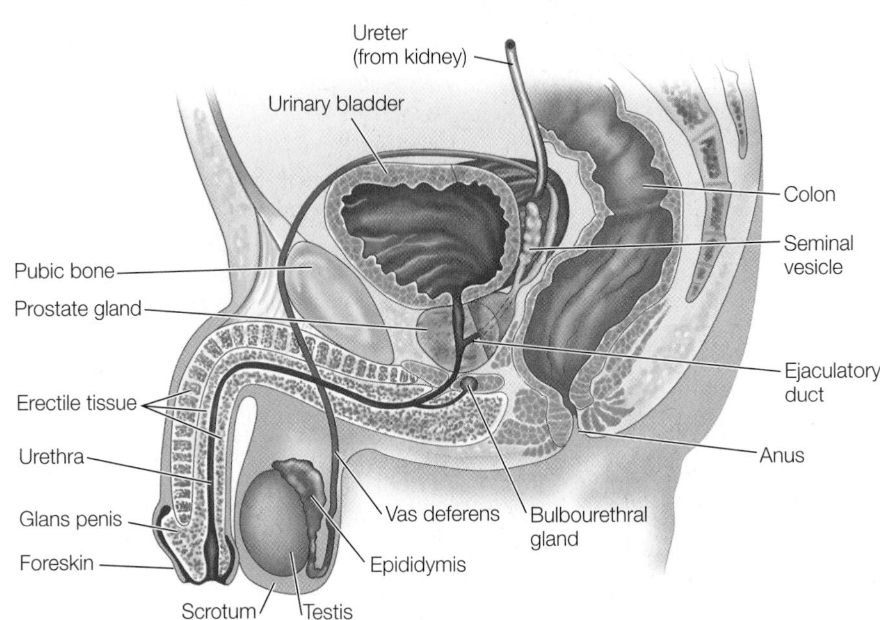

FIGURE 37.12 Reproductive Organs of the Human Male The two vas deferens, from the right and left testes, join at a point upstream from the single prostate gland. The foreskin, the skin covering the glans of the penis, is removed during circumcision.

Go to ACTIVITY 37.3 The Human Male Reproductive Tract
PoL2e.com/ac37.3

At the end of pregnancy, still another hormone, oxytocin (secreted by the hypothalamus and posterior pituitary gland; see Concept 35.3 and Figure 35.6) plays an essential role. As we have already discussed (see Figure 29.17), oxytocin stimulates the uterine muscle layer (the myometrium) to contract with ever-increasing force during the birth process so that the baby is expelled from the uterus. Following birth, oxytocin and the anterior pituitary hormone prolactin stimulate the mammary glands (breasts) to produce and eject milk.

Male sex organs produce and deliver semen

The testes of most mammals are located outside the body cavity in a sac of skin called the **scrotum** (see Figure 37.3). During early development they are inside the abdomen in the same positions as the ovaries, but they migrate into the scrotum as development proceeds (see Concept 35.4). In human males, the testes are normally in the scrotum all the time. In many other species, such as certain mice, the testes are in the scrotum only during the reproductive season, being withdrawn into the abdomen at other times. The testes of most mammals need to be about 2°C cooler than ordinary abdominal temperature (37°C) for normal production of viable sperm, and being in the scrotum permits this.

After sperm are produced in a testis, they are stored in the epididymis and vas deferens. The epididymis is a convoluted tubular structure located next to the testis in the scrotum (see Figure 37.5). The vas deferens is the tube that leads from the epididymis to the urethra, the passage in the penis through which urine and semen flow to the outside (see Figure 37.3). The vas deferens and urethra are endowed with smooth muscle, which contracts to propel semen during ejaculation.

Semen, the fluid expelled from the penis during ejaculation, consists of a mix of sperm and fluids secreted by male accessory reproductive glands. In humans, the glands that provide most of the fluids in semen are the paired seminal vesicles and the single prostate gland (**FIGURE 37.12**). Sperm are mixed with these fluids as ejaculation occurs. The fluids provide a supportive medium for the sperm, including fructose and other energy sources. The paired bulbourethral glands secrete a clear mucus prior to ejaculation. This fluid often emerges from the penis and may contain sperm even though it is not semen.

The shaft of the human penis is filled almost entirely by three compartments of spongy tissue that can be expanded with blood (see Figure 37.12). Erection occurs when these compartments are inflated with blood at high pressure until the penis is stiff. Nitric oxide (NO) is the agent that immediately controls erection. When NO is released from parasympathetic nerve endings in the penis, it leads to dilation of blood vessels that permit blood to fill the spongy tissues. NO is short-lived. It acts by increasing synthesis of cyclic GMP (cyclic guanosine monophosphate), which in turn acts as a second messenger (see Concept 5.6). Medications for erectile dysfunction (impotence) inhibit an enzyme that breaks down cyclic GMP. This permits the second messenger to act longer.

Many mammals have a bone in the penis that helps stiffen it for copulation. Included are dogs, seals, rodents, and some primates (but not humans).

Hormones play as crucial a role in male reproductive function as in female. Reproductive processes occur in far more steady ways in males than females, however, and the controlling hormones do not cycle in the dramatic ways seen in females.

Testosterone is required for spermatogenesis. In human males, the interstitial cells secrete an abundance of testosterone during a period of fetal life and again during the first year after birth. After that, testosterone levels remain low until puberty begins. At puberty, the blood testosterone level increases dramatically, and thereafter it stays high for the rest of life, although it declines gradually after middle age. Spermatogenesis begins at puberty when the testes mature and thereafter continues without interruption throughout life. Many nonhuman mammals exhibit strongly seasonal reproduction. In these animals, testosterone secretion and spermatogenesis often occur only in reproductive seasons. In those seasons, however, the animals exhibit the same steadiness as seen in humans in testosterone secretion and spermatogenesis.

The pituitary hormones FSH and LH play central roles in males. LH stimulates testosterone secretion by the interstitial cells. In this way it plays a mandatory role in spermatogenesis. FSH and testosterone, acting together, stimulate the Sertoli cells to support spermatogenesis.

Many contraceptive methods are available

Useful contraceptive methods are based on an accurate understanding of the biology of reproduction. They fall into a several categories (**TABLE 37.1**). The simplest to understand is abstinence.

TABLE 37.1 Methods of Routine Contraception

Method	How is the method used?	Annual failure rate[a] (pregnancies/100 women)		Helps prevent infection with STDs and HIV?[b]
		Typical use	Perfect use	
Unprotected	No form of birth control	85	85	No
NON-TECHNOLOGICAL METHODS				
Abstinence	Not engaging in sexual intercourse	0	0	Yes
Fertility awareness methods (e.g., rhythm method)	Engaging in intercourse only on days of the ovarian cycle when fertility is judged to be low based on cervical mucus or standard cycle properties	24	3–5	No
Withdrawal (pulling out, coitus interruptus)	The man withdraws his penis prior to ejaculating.	22	4	No
Barrier methods				
Male condom	A sheath of impermeable material (usually latex) is fitted over the erect penis.	18	2	Yes, latex condoms
Female condom; diaphragm	A physical device is inserted in the vagina to block entry of sperm into the cervix and uterus.	12–21	5–6	Yes, condom; no, diaphragm
Spermicides	Chemical compounds that kill or immobilize sperm are inserted in the vagina.	28	18	No
Sponge	A sponge with spermicides is inserted in the vagina to absorb and kill sperm and block entry to cervix.	12–24	9–20	No
HORMONE-BASED CONTRACEPTIVES FOR WOMEN				
Oral hormones ("the pill")	A hormonal pill that prevents ovulation is taken daily; contains either a combination of estrogens and progestin (progesterone) or just progestin.	9	0.3	No
Hormones administered non-orally on long-term basis	Based on the same hormonal actions as the pill; hormones are delivered by long-acting injection, skin patch that releases them through the skin, vaginal ring, etc.	6–9	0.3	No
INTRAUTERINE DEVICES				
Intrauterine devices (IUD)	A small plastic or metal device is inserted semipermanently into the uterus. Some contain copper, others hormones. Device impairs sperm or prevents implantation.	0.2–0.8	0.2–0.6	No
STERILIZATION				
Vasectomy (male sterilization)	The vas deferens on each side is cut and tied so that sperm can no longer pass into the urethra and be ejaculated. Does not affect male hormone levels or sexual response.	0.15	0.10	No
Tubal ligation (female sterilization)	The oviducts are cut, tied, or clamped so that eggs cannot reach the uterus. Does not affect female hormone levels or sexual response.	0.5	0.5	No

[a]"Typical use" failure rates are based on real-world statistics for people who sometimes apply methods incorrectly or skip using them. "Perfect use" refers to using methods according to directions without fail, every time intercourse takes place. Methods of short-term contraception (e.g., "morning-after pill") are not included. *Source*: J. Trussell. 2011. *Contraception* 83: 397–404.

[b]STD = sexually transmitted disease such as gonorrhea; HIV = human immunodeficiency virus.

Without sperm in the female reproductive tract, pregnancy cannot occur. Several methods are based on blocking sperm from entering the uterus during and after sexual intercourse. This is the rationale for condoms (male and female) and diaphragms. Some methods, such as birth control pills, manipulate a woman's hormone cycle so that ovulation does not occur.

Table 37.1 presents contraceptive failure rates in the United States according to a recent assessment. In many ways, the most informative aspect of the data is the comparison of failure rates for "typical" and "perfect" applications of the methods. Notice the large differences for some methods. For example, latex male condoms are estimated to have an annual failure rate of 2 pregnancies per 100 women (whose partners use condoms) when the condoms are used exactly according to directions. However, among people in the general population who say they use male condoms as their method of contraception, the failure rate is almost 10 times greater. This difference reflects deficiencies in how and when condoms are put on and removed, and it also reflects instances when avowed condom users skip using them. The message is clear: many methods are not very effective unless the users follow the directions in detail and without fail.

CHECKpoint CONCEPT 37.2

- ✓ What is the role of chorionic gonadotropin (CG) in human reproduction, and why is it a target for pregnancy tests?
- ✓ Where do the fluids come from that are expelled from a man's penis before and during ejaculation?
- ✓ Taking specific account of FSH, explain the relationship between the cycling of estrogen secretion and the cycling of follicle development during the human menstrual cycle.

In countless ways, reproduction must be integrated with the rest of an animal's life cycle. If reproduction requires nutrients, for example, it must be integrated with processes that provide nutrients. Thus a final aspect of reproduction is how it relates to other aspects of the life cycles of animals.

CONCEPT 37.3 Reproduction Is Integrated with the Life Cycle

Reproduction is an essential part of an animal's life cycle. Because of this, a full understanding of reproduction often involves asking questions about the life cycle. Does reproduction place limits on the rest of the life cycle? Does the life cycle compel reproduction to take place in certain ways? These are just some of the questions that arise.

Animals often gain flexibility by having mechanisms to decouple the steps in reproduction

In humans, the individual steps in the reproductive process are rigidly linked. Mating leads promptly to fertilization, fertilization leads promptly to embryonic development, and development adheres to a relatively rigid schedule, culminating in birth at a relatively fixed time after fertilization. Some other animals also show this sort of rigid sequencing. In cases like this, the individual steps in the reproductive process cannot be separately coordinated with conditions in the environment.

Many animal species, however, have evolved mechanisms of decoupling successive steps in the reproductive process, so that the time that elapses between one step and the next is flexible. Such mechanisms increase options for certain steps to be coordinated with environmental conditions independently of other steps.

Sperm storage in the female reproductive tract is a common mechanism that provides for flexible timing between copulation and fertilization. With sperm storage, fertilization can occur long after copulation.

Female blue crabs (*Callinectes sapidus*) illustrate the advantages of sperm storage. They typically copulate during only one short period in their lives. However, they store the sperm they acquire. Thus they can make and fertilize new masses of eggs for over a year. Queen honey bees store sperm for years. Sperm storage also occurs in certain other crustaceans and insects, and in certain bats, sharks, nonavian reptiles, and birds.

Embryonic diapause—a programmed state of arrested or profoundly slowed embryonic development—occurs in many animals. In cases of embryonic diapause, embryos start to develop but then stop for a while before continuing. This permits adjustment of the time between fertilization and completion of embryonic development.

Silkworm moths (*Bombyx*) use embryonic diapause to ensure that their offspring do not hatch out of their eggs in winter. The adult moths copulate, fertilization occurs, and the eggs are laid in the autumn of the year. These autumn-laid eggs are programmed to undergo a complete arrest of development (cessation of mitosis) early in embryonic development. After the eggs have entered this arrest, they must be exposed to a low temperature (5°C or lower) for about 2 months before they can emerge from the arrested state. These interacting processes—the programmed appearance of diapause and the need for extended cold exposure to terminate it—ensure that eggs laid in the autumn do not hatch into hungry caterpillars during winter. The eggs need winter cold to emerge from developmental arrest. They accordingly hatch in the spring.

LINK

Plants exhibit a similar strategy by using a variety of environmental cues to ensure that their seeds germinate when conditions are favorable for seedling growth; see **Concept 26.1**

Embryonic diapause in placental mammals is usually called **delayed implantation**. The reason for this name is that the developmental arrest involves postponing the implantation of the blastocyst into the endometrium.

One of the most astounding examples of embryonic diapause is in Antarctic fur seals (*Arctocephalus gazella*). This species uses

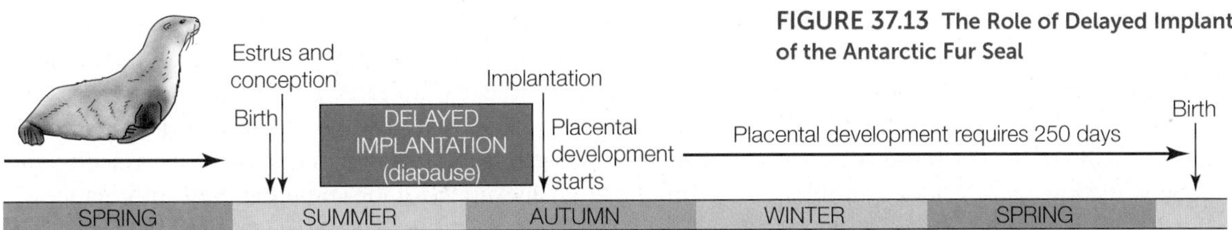

FIGURE 37.13 The Role of Delayed Implantation in the Life Cycle of the Antarctic Fur Seal

delayed implantation to achieve an exact match between the length of its reproductive cycle and the length of the calendar year. The challenge these seals face is that they have only a narrow window of time, early in the Antarctic summer, to give birth. If they give birth much earlier, their offspring can be killed by late winter storms. If they give birth much later, their offspring cannot grow as much as they need to before the next winter starts.

These fur seals have evolved a reproductive cycle in which the time between copulation and birth is almost exactly 365 days. Thus they give birth and copulate in the early summer one year, and they give birth to the offspring from that copulation 365 days later in the early summer of the next year.

However, the placental development of a young Antarctic fur seal (measured from implantation to birth) lasts only 250 days. If the seals were like people—with all the steps rigidly linked—a youngster conceived in the early summer would be born around March 1 in the cold of the Antarctic winter. The seals solve this problem with delayed implantation (**FIGURE 37.13**). After fertilization, the blastocyst enters the uterus, but it does not implant. Instead it goes into arrest for more than 3 months and then implants. At that point, 250 days remain before the proper birthing moment.

Some animals can reproduce only once, but most can reproduce more than once

One of the most important reproductive characteristics of a species is the number of times that individuals are physiologically capable of reproducing during their lives. In some animals, termed **semelparous**, each individual is physiologically programmed to reproduce only once. Octopuses, some fish, many insects, and many marine worms are semelparous.

Octopuses and Pacific salmon provide two dramatic examples. A species of octopus that is particularly well studied is *Octopus vulgaris*. Many other species are believed to resemble it. Males place spermatophores (packets of sperm) in the reproductive tracts of females during mating, and the females lay fertilized eggs soon after. At that point, a female stops eating and guards her eggs for 1–2 months (**FIGURE 37.14A**). Guarding her eggs is her final achievement, because she dies soon after her eggs hatch. Males also die at about the same time. Both the females and males reproduce only once in their lives.

The same is true of Pacific salmon such as sockeye salmon (*Oncorhynchus nerka*). After sockeye salmon grow to adulthood over a period of 1–4 years in the Pacific Ocean (see Concept 36.3), they migrate up rivers—sometimes hundreds of miles—before spawning. As soon as they enter the rivers from the sea, both sexes stop eating. Accordingly, they power their upriver migration entirely by breaking down their own body substance. After reaching a suitable spot to spawn, they release their gametes and die (**FIGURE 37.14B**).

When a species is semelparous, parents often sacrifice their own well-being to an extraordinary extent to provide the resources they need to produce their offspring. This is understandable because the parents have no future reproductive potential

(A)

(B)

FIGURE 37.14 Semelparity (A) In many species of octopuses, a female produces eggs just once in her life and guards them. She does not eat while guarding her eggs, and she dies after they hatch. (B) Sockeye salmon (*Oncorhynchus nerka*) expend enormous resources to reach freshwater spawning areas, then die after spawning.

after their single period of reproduction. By the time sockeye salmon spawn, they are often capable of little else but spawning because they have so thoroughly depleted their bodies.

Most animal species are **iteroparous**, meaning that individuals are physiologically capable of two or more separate periods of reproductive activity during their lives (i.e., multiple "iterations" of reproduction). In many species, for example, individuals reproduce at least once per year for as many years as they live.

The question of parental investment in offspring is far more complex in iteroparous animals than in semelparous ones. This is because when iteroparous parents produce offspring at any one time, they remain able to produce more offspring at future times. To gain the advantages of future reproduction, they must survive: as they reproduce, they must keep enough resources for themselves to preserve their own health. Their use of resources over their lives thus follows a different pattern than in semelparous species.

Seasonal reproductive cycles are common

In iteroparous animals that live in environments with regular seasonal cycles, the reproductive cycle is nearly always timed to coordinate with the environmental seasonal cycle. We humans are unusual in not showing such seasonality.

The most common pattern is for animals to reproduce in months when temperature, food supply, and other conditions are favorable—such as spring and summer—and suspend reproduction during unfavorable months such as those of winter. Some species of rabbits and mice, for example, shut down their gonads during winter. The testes of the males stop making sperm, become small, and sometimes are withdrawn from the scrotum into the abdomen. The testes then regrow, move into the scrotum, and resume sperm production in the spring. Females also suspend reproductive function and then restore it.

To describe life cycles like this, biologists often speak of animals as having a reproductive season and a nonreproductive season. In this context, "season" does not refer to the defined four seasons of the year. Instead, it refers to the months in which the animals are reproductive and those in which they are nonreproductive.

Some animals breed only once in each reproductive season. The red fox (*Vulpes vulpes*) of North America, for example, enters estrus for a single 1- to 6-day period—and ovulates once—each year during its reproductive season. Some other species of foxes and wolves are similar. If a female fails to mate at this time or loses her offspring, she does not enter estrus again until a full year later. Many birds also breed once per year.

Mammals more commonly superimpose an estrous cycle on their seasonal cycle. For example, during each month of her reproductive season each year, a female rat or mouse cycles in and out of estrus every 4–6 days unless she gets pregnant, as we've mentioned. Males produce sperm continuously during the months of the reproductive season.

Seasonal breeders require mechanisms to time their seasonal cycle. Often they employ environmental cues. Of these, the most important, in both invertebrates and vertebrates, is the amount of daylight per 24-hour day—termed the **photoperiod**. For example, many mice enter reproductive readiness when the photoperiod is increasing or long (i.e., during "the long days" of late spring and early summer). In this case, photoperiod is being used as a signal that the time of year is spring or summer. Because photoperiod varies in a mathematically exacting way with the time of year, it provides unambiguous information.

LINK

Plants also use the photoperiod to time reproduction; see **Concept 27.2**

Circannual biological clocks are a second important mechanism for timing seasonal cycles. Sheep have internal timing mechanisms that time events for an entire year. Some other animals do as well. They use these circannual clocks to control their seasonal reproductive cycles.

CHECKpoint CONCEPT 37.3

- ✓ If red foxes (*Vulpes vulpes*) enter estrus and ovulate only once a year, why are they not classed as semelparous?
- ✓ How does sperm storage provide flexibility in timing the parts of the reproductive cycle?
- ✓ Why, specifically, is embryonic diapause in mammals called delayed implantation?

What differences and similarities exist between mammals that exhibit induced ovulation and those that ovulate independently of mating?

ANSWER There is one important similarity. Ovulation occurs in response to a surge in the blood concentration of luteinizing hormone (LH) in both groups of mammals. Mammals that display induced and spontaneous ovulation differ, however, in how the LH surge is produced. In a species with induced ovulation, the LH surge is a direct response to copulation. The physical act of copulation stimulates sensory neurons that send signals to the brain, and the brain responds by promptly initiating the LH surge. In this way there is a deterministic connection between the presence of sperm in the female reproductive tract and the presence of ova. When sperm are provided by copulation, ova are provided by ovulation almost simultaneously.

In species with spontaneous ovulation, the timing of the LH surge depends on endogenous hormonal cycles in the female. These cycles occur on their own rhythm, more or less independently of whether copulation takes place. Accordingly, chance plays a greater role in determining whether sperm and ova will be present in the female reproductive tract at the same time. The degree of chance is reduced by estrus in many species of mammals with spontaneous ovulation. Females enter estrus approximately when they ovulate. Nonetheless, in a species with spontaneous ovulation, the timing of copulation and that of ovulation may be sufficiently different (because the two are not directly causally linked) for sperm to be present in the appropriate parts of the oviducts when ova are not, or vice versa.

SUMMARY

CONCEPT 37.1 Sexual Reproduction Depends on Gamete Formation and Fertilization

- In **sexual reproduction** the parents produce **gametes** that have only half as many chromosomes as the other cells in the body. The gametes of the female parent are relatively large, nonmotile cells called **eggs** or **ova**. Those of the male are small cells that swim, typically using flagella, and are called **spermatozoa** or **sperm**. The gametes fuse to produce a single-celled zygote that has a full set of chromosomes and develops into a sexually produced offspring. **Review Figure 37.1**
- Gametes are produced by meiosis and therefore are diversified in their chromosomes and genes by independent assortment and crossing over. The new individual that is formed when gametes fuse is not genetically identical to either parent.
- In **asexual reproduction** offspring are produced by mitosis and are genetically identical to their parent and to one another. Asexual reproduction produces no genetic diversity but has the advantage of maintaining favorable combinations of genes. **Review Figure 37.2**
- In sexually reproducing animals, the male and female sexes differ in their **primary reproductive organs**, **accessory reproductive organs**, and **secondary sexual characteristics**. The primary reproductive organs are the **gonads**: the **ovaries** in females and **testes** in males. The accessory reproductive organs are the reproductive organs other than the gonads. The secondary sexual characteristics are sex-specific properties of nonreproductive tissues and organs, such as facial hair, chest hair, and a deep voice in human males. **Review Figure 37.3**
- **Gametogenesis** occurs in testes and ovaries, where the **germ cells** proliferate mitotically, then undergo meiosis to produce gametes. **Review Figure 37.4**
- The gonads, in addition to having germ cells and producing gametes, typically contain somatic cells that contribute to reproduction. In testes, for example, **Sertoli cells** assist sperm production and **interstitial cells** (**Leydig cells**) produce the sex hormone **testosterone**. **Review Figure 37.5 and ACTIVITY 37.1**
- An **oocyte** (developing ovum) and its support cells are together called an **ovarian follicle**. The release of an ovum from the ovary is called **ovulation**. **Review Figure 37.6**
- **Fertilization**—the fusion of sperm and ovum—can be external or internal. **External fertilization** is common among aquatic animals. To achieve fertilization, these animals simply release their gametes into the water, a process termed **spawning**. Fertilization then occurs in the water outside the female's body. **Internal fertilization** occurs when a male inserts sperm inside the reproductive tract of a female, permitting fertilization to occur inside the female's body.
- For fertilization to occur, several steps must occur. In many animals these steps include species-specific binding of sperm to protective material surrounding the ovum, the acrosomal reaction, the sperm's passage through the protective layers covering the ovum, and the fusion of the sperm and ovum cell membranes. **Review Figure 37.7 and ANIMATED TUTORIAL 37.1**
- The entry of a sperm into the ovum triggers blocks to polyspermy, which prevent additional sperm from entering the ovum.
- **Sex determination** is the process by which the sex of an individual becomes fixed. In humans and other placental mammals, sex is determined at fertilization and depends on which **sex chromosomes** the zygote receives from its parents.
- In some animals, sex is not determined at fertilization. Some of these species exhibit **environmental sex determination**, in which the sex of an individual is determined by the temperature it experiences during its embryonic development.
- Some species are **sequential hermaphrodites**; they can be males at one time and females at another during their lives. **Review Figures 37.8, 37.9**

CONCEPT 37.2 The Mammalian Reproductive System Is Hormonally Controlled

- Somatic cells in the gonads produce steroid sex hormones. The principal sex steroid in males is testosterone, which is secreted by the interstitial (Leydig) cells in the testes. In females, the somatic cells of the ovarian follicles secrete feminizing steroids called **estrogens**. After ovulation, the follicular somatic cells remaining in the ovary reorganize to form a **corpus luteum**, which secretes another type of female sex steroid, **progesterone**, which is involved principally in coordinating processes associated with pregnancy.
- Ova mature in the female's ovaries, and after release (ovulation), they enter the **oviducts** (fallopian tubes). Sperm deposited in the vagina during copulation move through the cervix and uterus into the oviducts, where fertilization occurs. **Review Figure 37.10 and ACTIVITY 37.2**
- If fertilization has occurred, the zygote undergoes its early development as it travels the length of the oviduct and arrives in the uterus as an early embryo called a blastocyst. The blastocyst buries itself in the **endometrium** of the uterus—a process called **implantation**.
- Ovulation is either induced or spontaneous. **Induced ovulation** is ovulation triggered by copulation. In **spontaneous ovulation**, the timing of ovulation is under control of endogenous hormonal cycles in the female.
- In most species of mammals that display spontaneous ovulation, a female ovulates in cycles. In primates these cycles are called **menstrual cycles**, in recognition of the fact that **menstruation**—the sloughing off of part of the endometrium—occurs in each cycle that does not result in pregnancy.
- Mammals other than primates do not menstruate. Often, however, they undergo dramatic changes in behavior: they cyclically enter estrus (heat). Their cycles accordingly are called **estrous cycles**.
- The human menstrual cycle is under the control of hormones secreted by the hypothalamus, anterior pituitary gland, somatic cells in the ovarian follicle, and the corpus luteum. **Review Figure 37.11 and ANIMATED TUTORIAL 37.2**
- The uterus requires progesterone to support a pregnancy. In humans, chorionic gonadotropin, secreted by the blastocyst, signals the corpus luteum not to degenerate and instead to provide adequate progesterone for the first 7–10 weeks of pregnancy. By then, the **placenta**—which also secretes progesterone—is sufficiently well developed to take on this task.
- The male reproductive organs produce and deliver **semen**. Semen consists of sperm suspended in a fluid that nourishes the sperm and facilitates fertilization. This fluid is produced by accessory reproductive glands, such as the prostate gland. **Review Figure 37.12 and ACTIVITY 37.3**
- Spermatogenesis depends on testosterone secreted by the interstitial (Leydig) cells of the testes, which themselves are under the control of hormones produced in the anterior pituitary gland.

(continued)

SUMMARY (continued)

- Methods of contraception (prevention of pregnancy) include abstention from intercourse and the use of technologies that decrease the probability of fertilization and implantation. **Review Table 37.1**

CONCEPT 37.3 Reproduction Is Integrated with the Life Cycle

- Many animal species have evolved mechanisms of decoupling successive steps in the reproductive process. Such mechanisms increase options for certain steps to be coordinated with environmental conditions independently of other steps. These mechanisms include sperm storage and **embryonic diapause**, which in mammals is called **delayed implantation**. **Review Figure 37.13**
- Some animals are **semelparous**: each individual is physiologically programmed to reproduce only once in its lifetime. Examples include octopuses and Pacific salmon. **Review Figure 37.14**
- Most animal species are **iteroparous**, meaning that individuals are physiologically capable of two or more separate periods of reproductive activity during their lives. In iteroparous animals that live in environments with regular seasonal cycles, the reproductive cycle is nearly always timed to coordinate with the environmental seasonal cycle.

Go to the Interactive Summary to review key figures, Animated Tutorials, and Activities
PoL2e.com/is37

Go to LaunchPad at **macmillanhighered.com/launchpad** for additional resources, including LearningCurve Quizzes, Flashcards, and many other study and review resources.

38 Animal Development

KEY CONCEPTS

Vertebrates have two lateral eyes, clearly seen as this hatchling Greek tortoise (*Testudo hermanni*) takes its first peek at its new environment.

In Homer's *Odyssey,* the mythic hero Odysseus visits the island of the Cyclopes, a race of one-eyed giants. He and his men come across an enormous cave inhabited by one of these beings. After feasting on the cyclops's stored food, they hide as the monster returns: "The cyclops ... blotted out the light in the doorway. He was as tall and rugged as an alp. One huge eye glared out of the center of his forehead."

The one-eyed monster of Greek mythology has a basis in reality: a birth defect in humans that is known as holoprosencephaly (Greek, "one forebrain"). The defect varies in severity. In its extreme form, as the name implies, the forebrain does not develop as two hemispheres but rather as one, and the face can develop with a single, centrally located eye. These extreme cases invariably lead to death of the fetus and a miscarriage. In less extreme cases, infants are born with cleft lips and cleft palates.

Several factors have been implicated in holoprosencephaly, most of them genetic. Mutations in several genes that are part of a family of signaling factors called the hedgehog family can cause this condition. The first *hedgehog* gene was discovered in fruit flies, where it plays a role in determining the differences in the body segments of the fly. The name arises from the fact that flies with a defect in this gene are hunched up and have a continuous covering of tiny projections, so they resemble a hedgehog. The homologous vertebrate gene, *Sonic hedgehog* (named for the video-game character), is critical in central nervous system development.

Environmental factors can also induce holoprosencephaly. A particularly powerful mutagen is a molecule (an alkaloid) found in certain plants, notably the corn lily (*Veratrum californicum*) and its relative in the eastern North America, false hellebore (*V. viride*). Cows and sheep grazing in fields where these plants grow have high incidences of stillbirths with extreme forms of holoprosencephaly. In fact, when the molecule was extracted and purified from corn lily, it was named cyclopamine. In humans, the ingestion early in pregnancy of certain drugs may also be associated with these birth defects. Research into how these environmental factors cause holoprosencephalic development is helping explain why Sonic hedgehog is critical in limb and nervous system development.

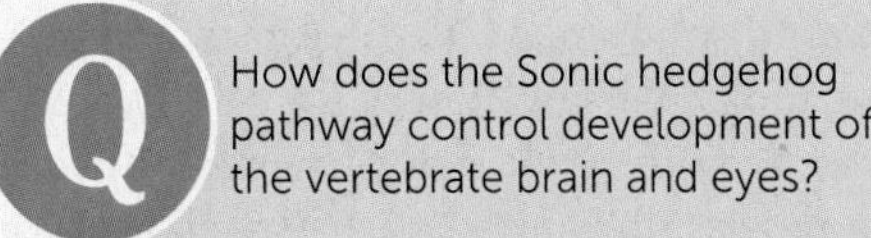

Q How does the Sonic hedgehog pathway control development of the vertebrate brain and eyes?

You will find the answer to this question on page 806

This chapter is contributed by Dr. Susan Douglas Hill, Michigan State University.

CONCEPT 38.1 Fertilization Activates Development

The development of an animal, a plant, or a fungus is one of the most breathtaking processes you will encounter in your study of biology. In sexually reproducing animals, one cell, the fertilized egg, gives rise to an entire organism. Early investigators assumed that a tiny person was encapsulated in a human egg or sperm and that this "homunculus" required only the appropriate stimulus and nutrients for growth:

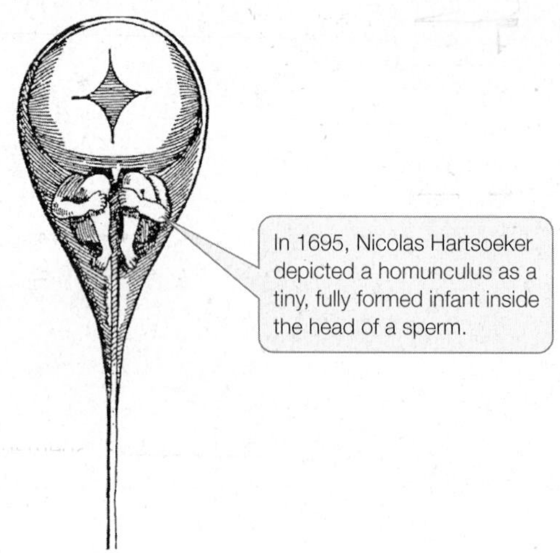

In 1695, Nicolas Hartsoeker depicted a homunculus as a tiny, fully formed infant inside the head of a sperm.

Improvement in microscopes allowed scientists to see that no tiny animal resides in either sperm or egg. That left the mystery—how does an organism achieve its final form?

You saw earlier that scientists obtain much information by studying "model" organisms (see Concept 1.1). Nowhere is this more true than in developmental biology. It is important to realize that certain animals are used as models, not because they represent the majority of the group under investigation, but because of particular characteristics that make them desirable for research, such as short developmental times and easy maintenance in a laboratory. Remember, too, that there are probably millions of animal species—many not yet discovered. Each has tweaked its development to adapt to its environment. The doors are open for further investigations in developmental biology.

Egg and sperm make different contributions to the zygote

Development usually begins with the union of egg and sperm to form a fertilized egg, or **zygote**. Egg cells are very large compared with other cells—and compared with sperm, they are gigantic. Even the non-yolky eggs of sea urchins and mammals are large compared with other cells, because they contain the materials and information needed to initiate and maintain early development. The cytoplasm of an egg contains most of the molecular machinery for cell division, energy production, and growth (yolk that provides nutrients for growth, mitochondria, ribosomes, etc.), as well as informational molecules such as transcription factors and messenger RNAs (mRNAs). These play important roles in setting up the signaling cascades that orchestrate the major processes of development discussed in Concept 14.1: determination, differentiation, morphogenesis, and growth.

The union of egg and sperm is complex. Many sperm are attracted to an egg. However, one sperm, and only one, must convey its nuclear material to the egg cytoplasm:

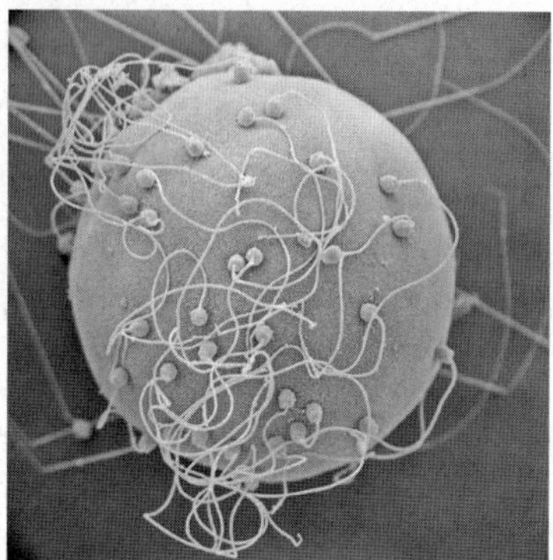
A sea urchin egg surrounded by sperm

All of the egg cell membrane may be available for sperm entry, as in sea urchins, or entry may be restricted to certain parts. In amphibians, for example, sperm can penetrate only half the egg surface. In *Drosophila*, sperm must enter through a tunnel whose position is fixed prior to fertilization.

Activation of an egg by a sperm brings about changes in the egg cell membrane and the release of ions from stores within the egg cytoplasm. These ion changes trigger a series of crucial events. Entry of additional sperm is blocked, metabolic rate rises, protein synthesis is initiated, and cytoplasm is reorganized.

Most of the sperm cytoplasm is discarded when the egg and sperm membranes fuse, but the nuclear material and usually the centrosome (the major microtubule organizing center of an animal cell) are transferred from the sperm to the egg. The centrosome then plays a major role in forming the zygote centrosome and organizing the zygote's microtubules. When egg and sperm nuclei join, the fertilized egg becomes a single-celled, diploid zygote that is ready for its first mitotic division.

LINK

Concept 37.1 explains sperm–egg recognition and blocks to polyspermy (fertilization by more than one sperm)

Polarity is established early in development

Most animals have a defined head and tail, back and belly, and right and left sides. These differences are known as polarities and are usually established early in development. In most animals, polarities are established during early cell divisions as materials deposited in the egg cytoplasm by the mother are redistributed and segregated. In some animals, however, polarities are set in place before or at fertilization.

In the fruit fly *Drosophila melanogaster*, a favorite model organism, the position of the egg in the ovary determines both anterior–posterior (head–tail) and dorsal–ventral (back–belly) polarity. Cells that surround the egg, called follicle cells, pass mRNAs from the mother to specific regions of the egg cytoplasm. After the egg is fertilized and development is underway, these mRNAs encode regulatory proteins that determine polarity.

In amphibians, another group of favorite model organisms, anterior–posterior and dorsal–ventral polarity are determined at fertilization by the position of the yolk and the point of sperm penetration. Cytoplasmic rearrangement following fertilization is easily observed in some frog species. In unfertilized eggs of these species, dense yolk granules are concentrated by gravity in the lower half of each egg, the **vegetal hemisphere**. This yolky vegetal hemisphere is unpigmented. In the top half of each egg, or **animal hemisphere**, the outermost (cortical) cytoplasm is heavily pigmented, while the pigment in the underlying cytoplasm is more diffuse. Sperm-binding sites occur only on the animal hemisphere. When a sperm enters the egg, the cortical cytoplasm rotates toward the site of sperm entry, producing a band of lightly pigmented cytoplasm called the **gray crescent** on the opposite side (**FIGURE 38.1**).

As we will soon see, the anus later forms within the gray crescent. Thus with the formation of the gray crescent, the zygote already has an anterior–posterior axis for a future head and tail, as well as a dorsal–ventral axis. The cytoplasmic rearrangement is initiated by the sperm centriole, which organizes microtubules in the egg cytoplasm in a parallel array that extends through the vegetal hemisphere. This cytoskeleton guides the shift of the cortical cytoplasm (see Concept 14.2), and the gray crescent appears.

Much of our understanding of developmental mechanisms builds on the insightful experiments of scientists in the late 1800s and early 1900s who used techniques that may seem primitive by today's standards but were very revealing. This point is beautifully illustrated by German biologist Hans Spemann's work, which also had a profound influence on future amphibian research. Similar experiments were carried out on many animals, and today's work is based on the results of these creative early investigators.

In the early 1900s, Spemann was studying the development of salamander eggs. With great patience and dexterity, he formed loops from single hairs taken from his baby daughter (baby hair is finer than adult hair) and tied them around fertilized eggs, effectively dividing the eggs in half (**FIGURE 38.2**). When his loops bisected the gray crescent, both halves of the zygote developed into a complete embryo. When he tied the loops so the gray crescent was on only one side of the constriction, only the half with the gray crescent developed normally. The half lacking gray crescent material became an unorganized clump of cells, which, because of its characteristics, he referred to as the "belly piece." Spemann concluded that important cytoplasmic factors—now called **cytoplasmic determinants**—necessary for the establishment of polarity and development of a normal salamander are unequally distributed in the fertilized egg and reside in the gray crescent.

Experiments using amphibian eggs have contributed much to our understanding of development. Keep in mind, though, that although amphibians are useful models, many animals develop differently, and in many, polarities are not established until cell divisions are underway.

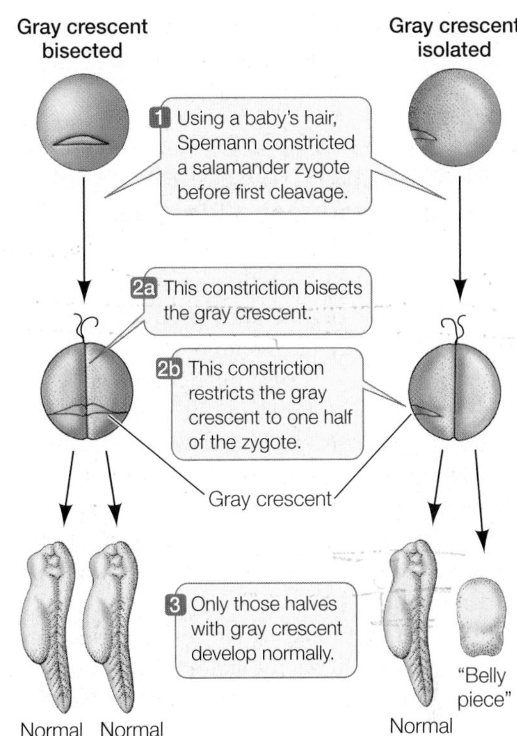

FIGURE 38.2 Importance of the Gray Crescent Spemann's research revealed that normal development in salamanders depends on cytoplasmic determinants localized in the gray crescent.

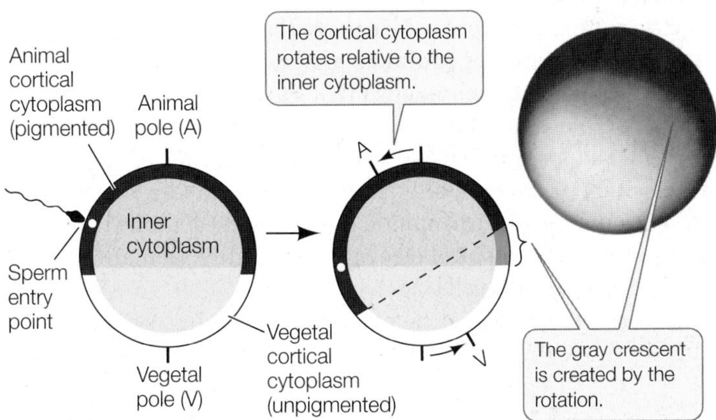

FIGURE 38.1 The Gray Crescent In frog eggs, rearrangement of the cytoplasm after fertilization creates the gray crescent and imposes anterior–posterior and dorsal–ventral axes on the egg.

CHECKpoint CONCEPT 38.1

- ✓ What normally initiates development?
- ✓ What does the sperm contribute to the zygote?
- ✓ What usually determines initial polarity in developing embryos?
- ✓ How are anterior–posterior and dorsal–ventral polarities established in fruit flies? In frogs?

We have now seen how the zygote forms and how polarities may be established. We will turn next to the early series of cell divisions that produce a multicellular embryo.

CONCEPT 38.2 Cleavage Creates Building Blocks and Produces a Blastula

Cleavage is the initial process that creates cellular building blocks to form the new organism by producing many smaller cells from one. Ultimately, as a result of cleavage, these smaller cells will differ in their cytoplasmic determinants. To understand cleavage, it is important to understand that no growth occurs at this stage. Cleavage divisions differ from other mitotic divisions in that there is no "gap" or growth phase between divisions, so no new cytoplasm is added. Thus as cells increase in number, they become smaller and smaller. The cells produced are called **blastomeres**. The cluster of cells produced is termed a **blastula**.

Early investigators thought that, through cleavage, genes were divided up so that each blastomere would later form its specific tissue or organ according to its restricted genetic makeup. We now know that this is not the case. Each blastomere receives the entire genome of the animal. Now we have the next mystery: How can cells that all have the same genetic information give rise to the diversity of cell types that we see in an adult? The answer depends on several mechanisms (see Concept 14.2).

During early stages of development, events are controlled by specific information passed on by the mother. Proteins and mRNAs, referred to as cytoplasmic determinants, are deposited in the cytoplasm of the egg as it matures. Early cleavages occur rapidly, and there is little or no transcription of mRNA in the zygote nucleus. Rather, mRNA already placed by the mother in the egg is used to direct protein synthesis. In most animals, it is not until the mid- to late blastula stage that the genome of the developing embryo takes over. Prior to that time, maternal factors control development. As we saw in the formation of the gray crescent, the cytoskeleton may redistribute such materials from one point to another. Also, at some point, as mitotic divisions continue, new cell membranes separate cytoplasmic determinants so they become distributed differentially among blastomeres (see Concept 14.2, especially Figure 14.7). These processes plus cell–cell communication play a role in the determination of blastomere fates. All of these processes are involved in the development of most animals, but as you will see, the stage at which each process comes into play differs from group to group. At some point, the fate of a blastomere becomes restricted or determined and it can no longer form an entire embryo.

Scientists have learned much about the developmental potential of blastomeres by separating them at different times of development. In such experiments, blastomeres are isolated to learn how they will develop on their own. It is also possible to leave all the cells together, label a specific cell (e.g., with a dye), and then identify the tissues and organs that form from that cell. A new technique called light sheet microscopy allows individual cells to be followed through many cleavages without the

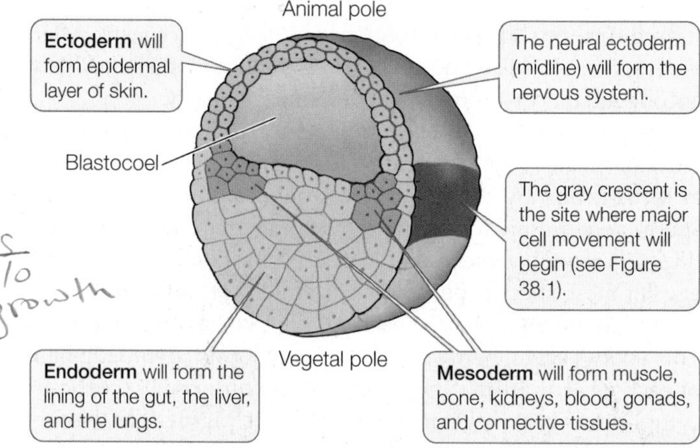

FIGURE 38.3 Fate Map of a Frog Blastula Colors indicate the portions of the early-stage frog embryo (blastula) that will form the three germ layers—ectoderm, mesoderm, and endoderm—and subsequently the frog's tissues and organs.

addition of a dye. Experiments of all types—isolation experiments and tracing experiments—have been used to generate **fate maps** for many species (**FIGURE 38.3**). To understand this, we need to explore different cleavage patterns.

Cleavage is easiest to understand in animals that produce eggs with little yolk, such as sea urchins. Blastomeres may be the same size or different. They will be the same size if the egg cytoplasm is divided equally among all the blastomeres. Unequal cytoplasmic divisions will result in large blastomeres and small blastomeres. As cleavage continues, the zygote becomes divided into progressively smaller blastomeres:

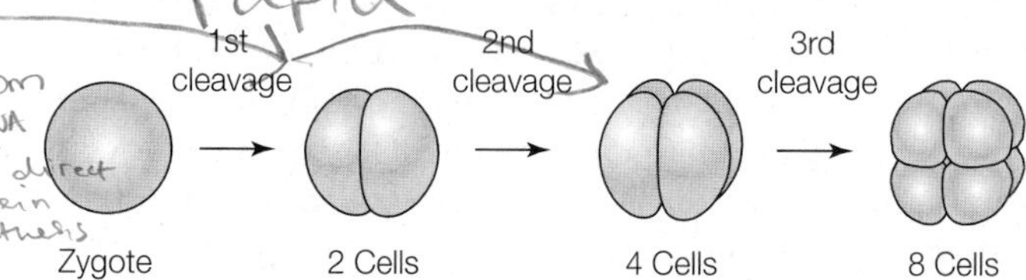

If the cytoplasm is completely divided by each cell division, cleavage is termed **complete cleavage**. If it is not, as you will see is the case in birds, insects, and some other animals, it is termed **incomplete cleavage**.

Two different patterns of cleavage that seem to have deep evolutionary roots are **radial cleavage** and **spiral cleavage**. These patterns are seen in eggs that have little yolk and undergo complete cleavage, such as those of snails and sea urchins. In all types of cleavage, the plane of cleavage (shown by the position of the new cell membrane that divides the cells) is determined by the orientation of the mitotic spindle. Differences in radial and spiral cleavage are readily seen after the third cleavage (the 8-cell stage).

At this time, in radial cleavage the four daughter cells lie directly above the four parent cells:

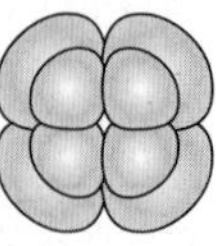

During spiral cleavage, mitotic spindles align so that, after the third and following cleavages, the four daughter cells lie in the grooves between the four parent cells, creating the appearance of a spiral:

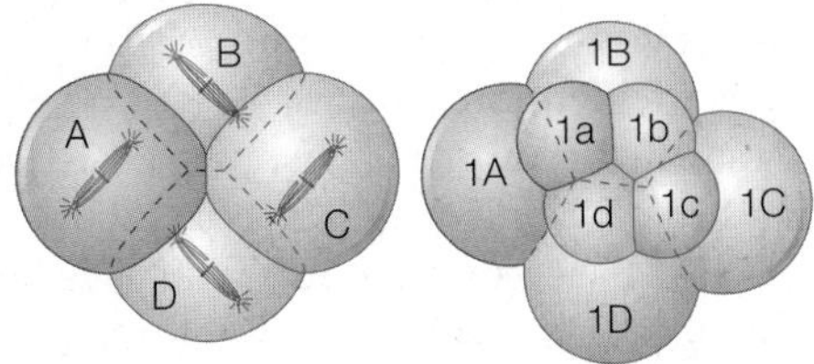

A spirally cleaving egg has the most thermodynamically stable shape possible. Appropriately sized soap bubbles assume the same configuration as a spirally cleaving egg.

You are probably most familiar with deuterostome animals, the large clade (group) that includes the vertebrates and all other chordates, echinoderms, and a few other groups. Echinoderms and several other deuterostomes display radial cleavage. Radial cleavage is considered to be the basic deuterostome pattern, although it is often highly modified in the presence of yolk.

The other large group of animals, the protostomes, includes most of the invertebrates. Protostomes do not have radial cleavage. Here we focus on the lophotrochozoans, a clade that includes about half of all animal phyla, including the annelids and mollusks. In the lophotrochozoans, spiral cleavage predominates.

LINK

Concept 23.1 shows how early development determines body plans that reflect phylogenetic relationships among the animals; **Concept 23.3** discusses the protostomes, explaining that there are two major clades; **Concept 23.5** discusses the deuterostomes and some of their developmental characteristics

Specific blastomeres generate specific tissues and organs

Blastomeres become determined—irreversibly committed to specific fates—at different times in different species. In many protostomes with spirally cleaving eggs and in a few deuterostomes, blastomere fates are determined as early as the 2-cell stage. If one of these blastomeres is experimentally removed, a particular portion of the embryo will not form. This type of development has been called **mosaic development** because each blastomere appears to contribute a specific set of "tiles" to the final "mosaic" that is the adult animal (**FIGURE 38.4**).

A feature of many spirally cleaving eggs that is not seen in radially cleaving eggs is the production of a polar lobe, at the vegatal pole, during early cleavages. Cytoplasmic determinants as well as yolk are sequestered in this "out-pouching" during cell division and are then returned to a specific blastomere after cell membranes have re-formed. If the polar lobe is removed experimentally, one abnormal larva develops that typically lacks specific structures (see Figure 38.4B). If the two blastomeres are experimentally separated, two abnormal larvae develop, each lacking very specific structures (see Figure 38.4C). These experiments clearly demonstrate one way in which information for further development is distributed to specific blastomeres.

In contrast to mosaic development, in **regulative development** the loss of some cells during cleavage does not affect the developing embryo. Instead, the remaining cells compensate for the loss. If blastomeres of a sea urchin egg are separated at the 2-cell stage, two half-sized, but complete, larvae form. Even at the 4-cell stage, separation produces four small larvae (**FIGURE 38.5**). We now know that cytoplasmic determinants are found in the cytoplasm of eggs showing both mosaic and regulative development. In animals that display regulative development, however, the distribution of these determinants happens later as cell divisions continue (see Concept 14.2; Figure 14.7). Regulative development is typical of vertebrate species, including humans.

As noted earlier, cleavage leads to the production of a ball of cells termed a blastula. Blastulas produced by radially cleaving eggs, and by those whose cleavage is derived from this basic pattern, have an inner fluid-filled cavity called a blastocoel (see Figure 38.3). The blastula of a sea urchin, for example, is

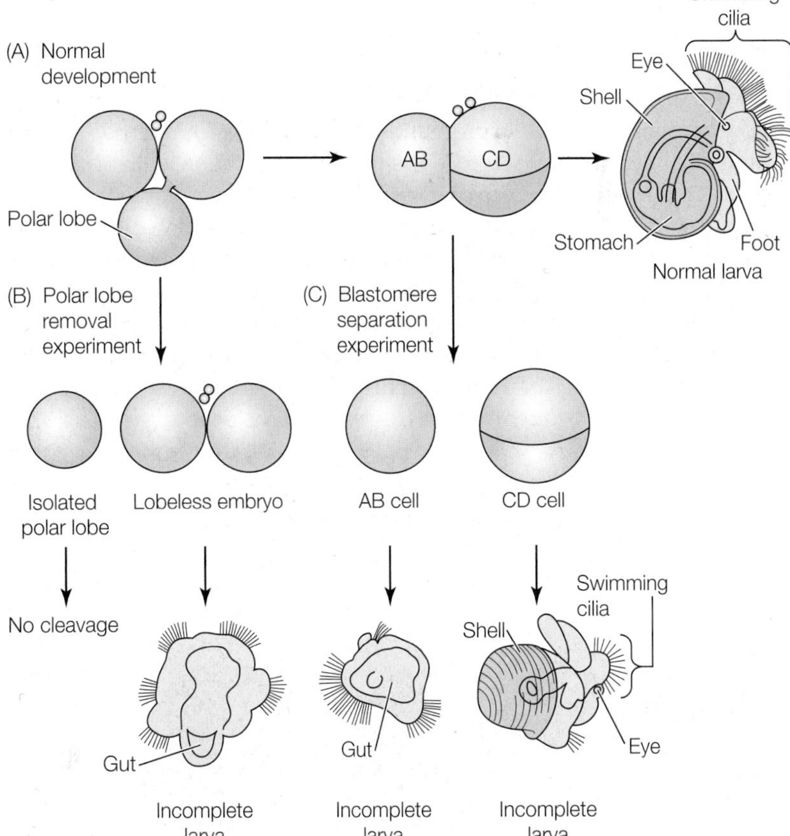

FIGURE 38.4 Mosaic Development of a Snail Egg (A) Normal development. (B) Removal of the polar lobe results in an abnormal, incomplete larva. (C) Separation of the two blastomeres results in two abnormal larvae, incomplete in different ways.

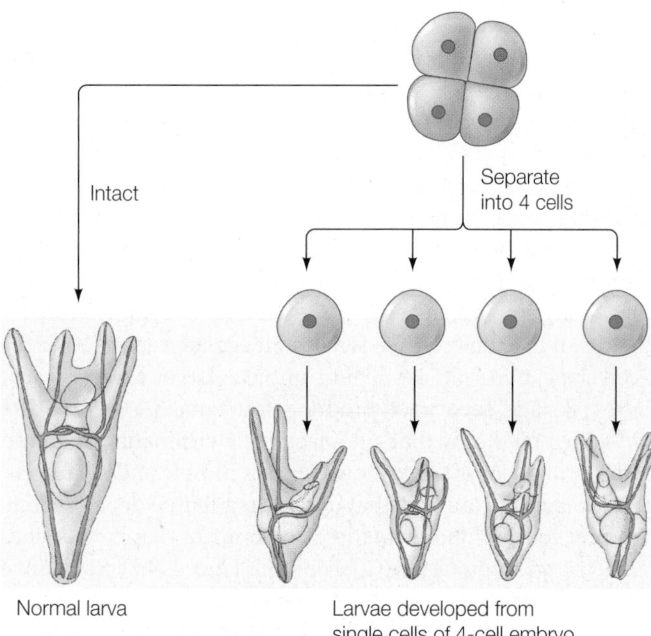

FIGURE 38.5 Regulative Development of a Sea Urchin Egg Separation of four blastomeres results in four small but almost normal larvae.

a hollow sphere. Blastulas produced by spirally cleaving eggs may have a blastocoel, but often they are a solid ball of cells, without the fluid-filled space.

The amount of yolk affects cleavage

The amount of yolk has a profound effect on cleavage patterns and other developmental processes such as gastrulation, which we will discuss in Concept 38.3. Some eggs, such as those of sea urchins, snails, and placental mammals, have little yolk. Some, such as those of frogs and salamanders, have a moderate amount of yolk. Others, including those of birds and egg-laying mammals, are extremely yolky. As you will see, it is as if the presence of yolk slows down or impedes cell divisions. This means that the eggs of many animals do not follow the radial

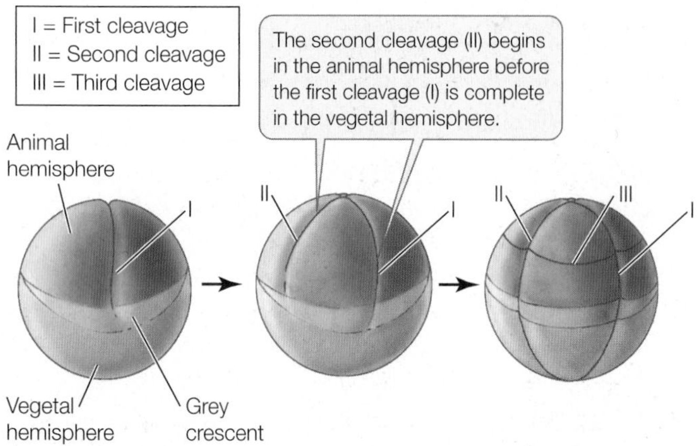

FIGURE 38.6 Complete Cleavage in a Frog Egg Cleavage proceeds more rapidly in the animal hemisphere that in the vegetal hemisphere.

or spiral cleavage patterns that we consider basic in deuterostomes and protostomes.

In eggs with little yolk, like those of sea urchins and snails, cleavage usually proceeds as described above and is complete.

In moderately yolky eggs like those of a frog, cleavage is complete and cell membranes form around every nucleus. However, cells divide more slowly in the yolk-filled vegetal hemisphere than in the animal hemisphere (**FIGURES 38.6** and **38.7A**). In the vegetal hemisphere, the yolk retards the process of division so that the second round of cleavage has begun in the animal hemisphere before the first round is completed in the vegetal hemisphere. In the example shown in Figure 38.6, the third cleavage is horizontal and occurs in the animal hemisphere. Because cells divide faster in the animal hemisphere, more cells are produced in the dorsal portion of the egg than in the ventral, resulting in an animal hemisphere consisting of small blastomeres and a vegetal hemisphere consisting of fewer, larger blastomeres. The blastocoel forms within the small blastomeres of the animal hemisphere.

In very yolky eggs such as those of a squid (a molluscan protostome), fish, bird, or egg-laying mammal (e.g., the platypus), the yolk remains undivided and the embryo forms as a small **blastodisc** of cleaving cells on the yolk surface (**FIGURE 38.7B**). The yolk never divides, and so the cleavage is incomplete. The disc of cells on top of the yolk enlarges and becomes several cells thick. For example, the blastula of a bird, or of the zebrafish shown in Figure 38.7B, is a flattened structure that sits atop the large undivided yolk and consists of an upper layer called epiblast, which will give rise to the embryo proper, and a lower hypoblast, with a fluid-filled space between the two. (Epiblast and hypoblast in the bird egg are shown in Figure 38.11.)

A different type of incomplete cleavage is seen in insects. In *Drosophila* eggs, for example, the original zygote nucleus undergoes a series of divisions unaccompanied by divisions of the cytoplasm. This early embryo, then, is a syncytium: a mass of cytoplasm in which several nuclei are surrounded by the same cell membrane. As development proceeds, nuclei migrate to the periphery of the egg, and cell membranes form (**FIGURE 38.7C**).

Cleavage in placental mammals is unique

Placental mammals have non-yolky eggs (which undergo development within the mother's body after fertilization), but their cleavage pattern differs from that of other non-yolky deuterostome eggs. You might expect that their eggs would show radial cleavage like sea urchin eggs. They do not. Remember that mammals evolved from a reptilian ancestor. As a result, several aspects of their development—such as the formation of epiblast and hypoblast, as well as of the yolk sac and allantois (which we will discuss in Concept 38.5)—were derived from processes in yolky eggs of reptiles, including birds. Other features, however, are unique. The two groups of placental mammals—the marsupials and the eutherians (see Concept 23.6) have evolved slightly different developmental steps. In this chapter, we will consider only eutherian development. Early cell divisions in eutherians do not follow the radial cleavage pattern. For example, at the second cleavage division, the two blastomeres divide in different

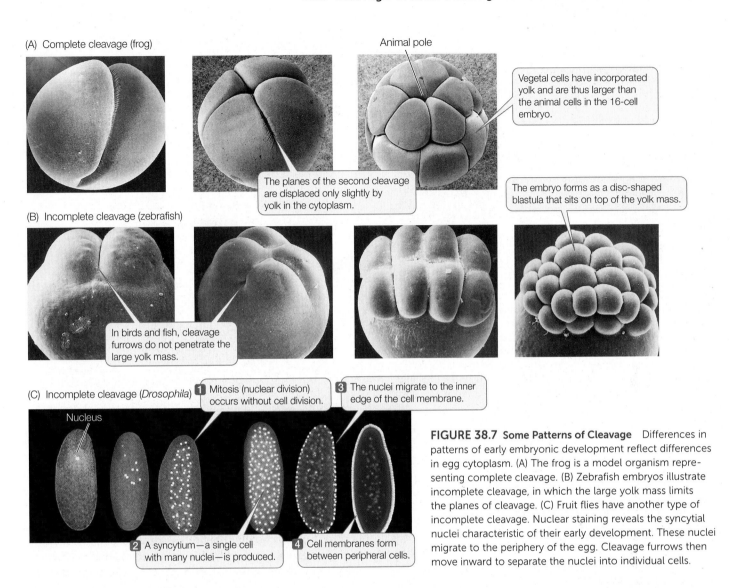

(A) Complete cleavage (frog)

Animal pole

Vegetal cells have incorporated yolk and are thus larger than the animal cells in the 16-cell embryo.

The planes of the second cleavage are displaced only slightly by yolk in the cytoplasm.

(B) Incomplete cleavage (zebrafish)

The embryo forms as a disc-shaped blastula that sits on top of the yolk mass.

In birds and fish, cleavage furrows do not penetrate the large yolk mass.

(C) Incomplete cleavage (*Drosophila*)

1 Mitosis (nuclear division) occurs without cell division.

3 The nuclei migrate to the inner edge of the cell membrane.

Nucleus

2 A syncytium—a single cell with many nuclei—is produced.

4 Cell membranes form between peripheral cells.

FIGURE 38.7 Some Patterns of Cleavage Differences in patterns of early embryonic development reflect differences in egg cytoplasm. (A) The frog is a model organism representing complete cleavage. (B) Zebrafish embryos illustrate incomplete cleavage, in which the large yolk mass limits the planes of cleavage. (C) Fruit flies have another type of incomplete cleavage. Nuclear staining reveals the syncytial nuclei characteristic of their early development. These nuclei migrate to the periphery of the egg. Cleavage furrows then move inward to separate the nuclei into individual cells.

planes, so that one blastomere cleaves vertically and the other horizontally, in a pattern called **rotational cleavage**:

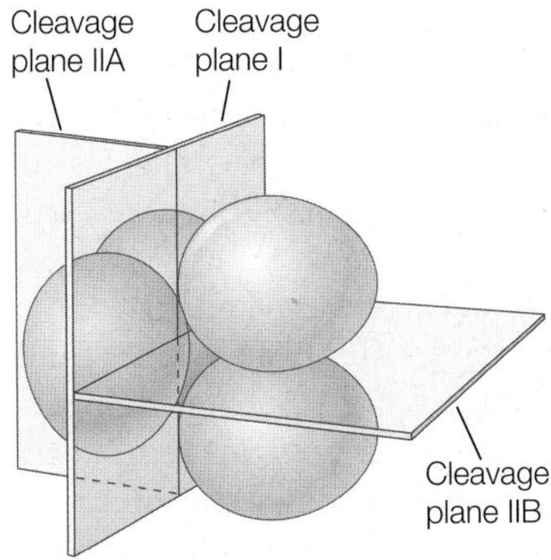

Blastomeres within the same egg divide at different times, so the eutherian egg may contain an odd number of cells. This is not usually the case during early cleavage in other groups. Finally, the mammalian genome is activated very early, so cleavage and development are governed by the new nuclei rather than by cytoplasmic determinants passed on from the mother. In most deuterostomes, this transition happens toward the end of the blastula stage.

Another unique feature of mammalian development occurs during the fourth cleavage, when the cells separate into two groups (**FIGURE 38.8A**). The **inner cell mass** will become the embryo, while the surrounding outer cells become an encompassing sac called the **trophoblast**. Trophoblast cells secrete fluid, creating a blastocoel with the inner cell mass at one end. At this stage, the mammalian embryo is called a **blastocyst**. Later, trophoblast cells will contribute to the formation of a unique mammalian organ, the placenta, which we will discuss in Concept 38.5.

In a mammal, fertilization occurs in the upper reaches of the mother's oviduct, and cleavage takes place as the zygote travels down the oviduct. When the blastocyst arrives in the uterus, it adheres to the lining of the uterus, which is called the endometrium. This starts the process of **implantation** in which the early embryo burrows into the uterine wall. In humans, implantation begins about 6 days after fertilization (**FIGURE 38.8B**).

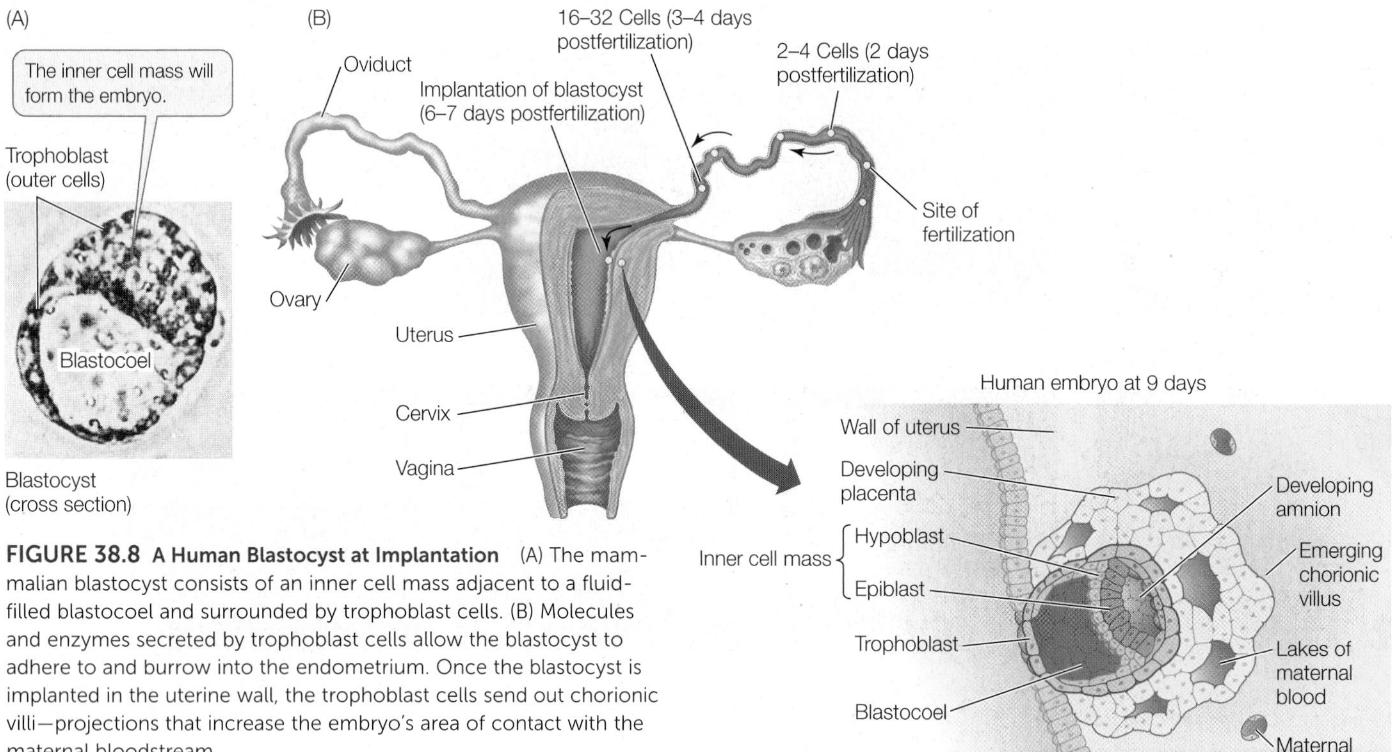

FIGURE 38.8 A Human Blastocyst at Implantation (A) The mammalian blastocyst consists of an inner cell mass adjacent to a fluid-filled blastocoel and surrounded by trophoblast cells. (B) Molecules and enzymes secreted by trophoblast cells allow the blastocyst to adhere to and burrow into the endometrium. Once the blastocyst is implanted in the uterine wall, the trophoblast cells send out chorionic villi—projections that increase the embryo's area of contact with the maternal bloodstream.

CHECKpoint CONCEPT 38.2

- ✓ What two major functions does cleavage perform?
- ✓ Compare radial and spiral cleavage. What determines the direction of the cleavage plane?
- ✓ What does experimental removal of the polar lobe of a snail egg demonstrate?
- ✓ Describe complete and incomplete cleavage and explain where mammals fit into this scheme.
- ✓ What type of development must a species have for identical twins to occur? What do you think leads to the development of such twins? How do you think fraternal twins are produced?

After formation of the blastula, the next stage of development is gastrulation. The developmental biologist Louis Wolpert once said, "It is not birth, marriage, or death, but gastrulation which is the most important time in your life."

CONCEPT 38.3 Gastrulation Produces a Second, then a Third Germ Layer

Gastrulation involves major movements of cells from the external surface to the interior of the embryo. These cell movements create a three-layered body plan and set the stage for development of the first organs. Triploblastic animals (see Concept 23.1) create all of their organs and tissues from three basic germ layers—ectoderm, endoderm, and mesoderm. These layers are formed by the process of gastrulation. The resulting embryo is known as a **gastrula**.

Just as the amount of yolk affects cleavage, it also affects how gastrulation occurs. The process is easiest to understand in radially cleaving eggs with little yolk, such as the much-studied eggs of sea urchins (**FIGURE 38.9**). One wall of the blastula (the vegetal pole) bulges inward, or invaginates, as if someone were poking a finger into a hollow ball. The new space that forms is the archenteron, or "first gut." The opening into the archenteron is the blastopore. Recall that in deuterostomes, the blastopore becomes the anus and the mouth (the second opening) forms opposite it. The outer layer of cells is now ectoderm, and the wall of the archenteron is endoderm and future mesoderm. The archenteron elongates, assisted by contractions of wandering cells called mesenchyme cells. The archenteron will eventually squeeze the blastocoel out of existence.

In eggs such as those of amphibians, which contain a moderate amount of yolk, cleavage continues to produce many ever-smaller cells in the animal hemisphere and larger cells in the vegetal hemisphere. To initiate gastrulation, some cells in the gray crescent invaginate to form a slitlike blastopore (**FIGURE 38.10**). Small blastomeres of the animal pole then begin to roll like a sheet over the dorsal lip of the blastopore and push into the blastocoel. At the same time, cell divisions continue to increase the number of cells. Because the small cells of the animal hemisphere continue to divide more rapidly than the larger cells of the vegetal hemisphere, these smaller cells grow over and surround the larger cells of the vegetal hemisphere. As these smaller cells reach the blastopore, they too

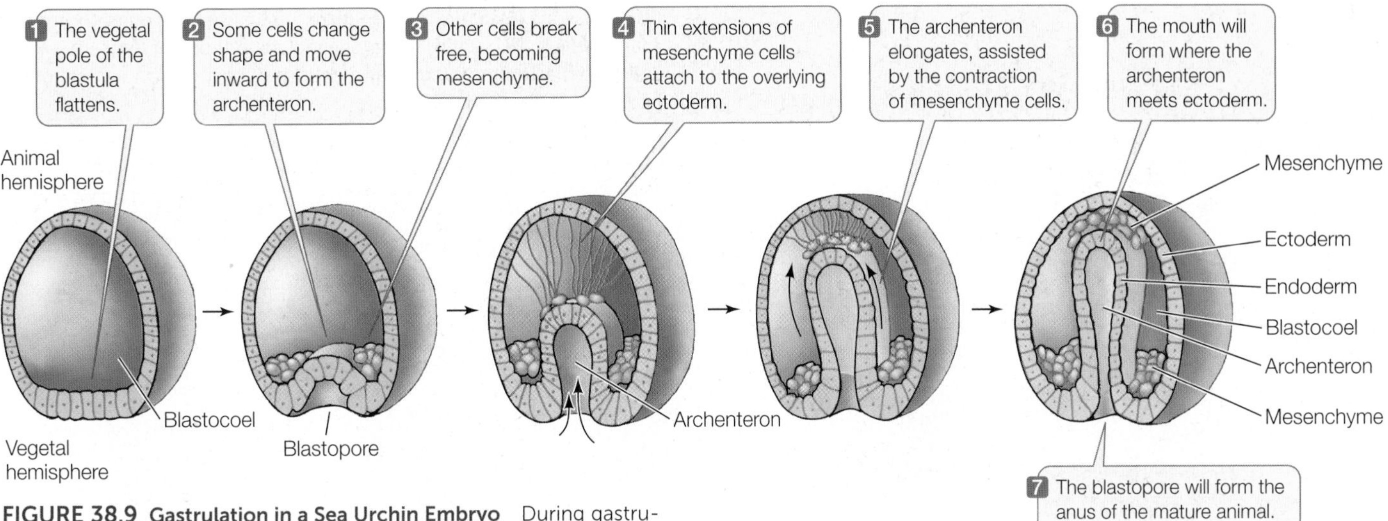

1 The vegetal pole of the blastula flattens.
2 Some cells change shape and move inward to form the archenteron.
3 Other cells break free, becoming mesenchyme.
4 Thin extensions of mesenchyme cells attach to the overlying ectoderm.
5 The archenteron elongates, assisted by the contraction of mesenchyme cells.
6 The mouth will form where the archenteron meets ectoderm.
Animal hemisphere
Vegetal hemisphere
Blastocoel
Blastopore
Archenteron
Mesenchyme
Ectoderm
Endoderm
Blastocoel
Archenteron
Mesenchyme
7 The blastopore will form the anus of the mature animal.

FIGURE 38.9 Gastrulation in a Sea Urchin Embryo During gastrulation, cells move to new positions. Invagination to form endoderm in this non-yolky egg is illustrated. The archenteron, or "first gut," remains connected to the exterior via the blastopore. Mesenchyme cells assist with gastrulation and will also form the temporary larval skeleton.

move inward so that the inward migration occurs all around the blastopore. These cells will form much of the endoderm and mesoderm of the developing embryo. The blastopore becomes the anus.

In very yolky eggs like those of birds, in which the blastodisc forms on top of the yolk, the blastopore forms as a groove in the primitive streak, a collection of cells marked at the anterior end by Hensen's node (**FIGURE 38.11**). Cells that will become mesoderm and endoderm migrate inward through Hensen's node and all along the groove of the primitive streak.

Placental mammals retain the avian/reptilian gastrulation pattern, even though they lack yolk. Mammals and reptiles, including birds, are all amniotes (i.e., they produce eggs that develop extraembryonic membranes), so it is not surprising that they share certain patterns of early development. In Concept 38.2 we discussed the development of the mammalian inner cell mass and the outer trophoblast. As in avian development, the inner cell mass of placental mammals splits into an upper layer called the epiblast and a lower layer called the hypoblast

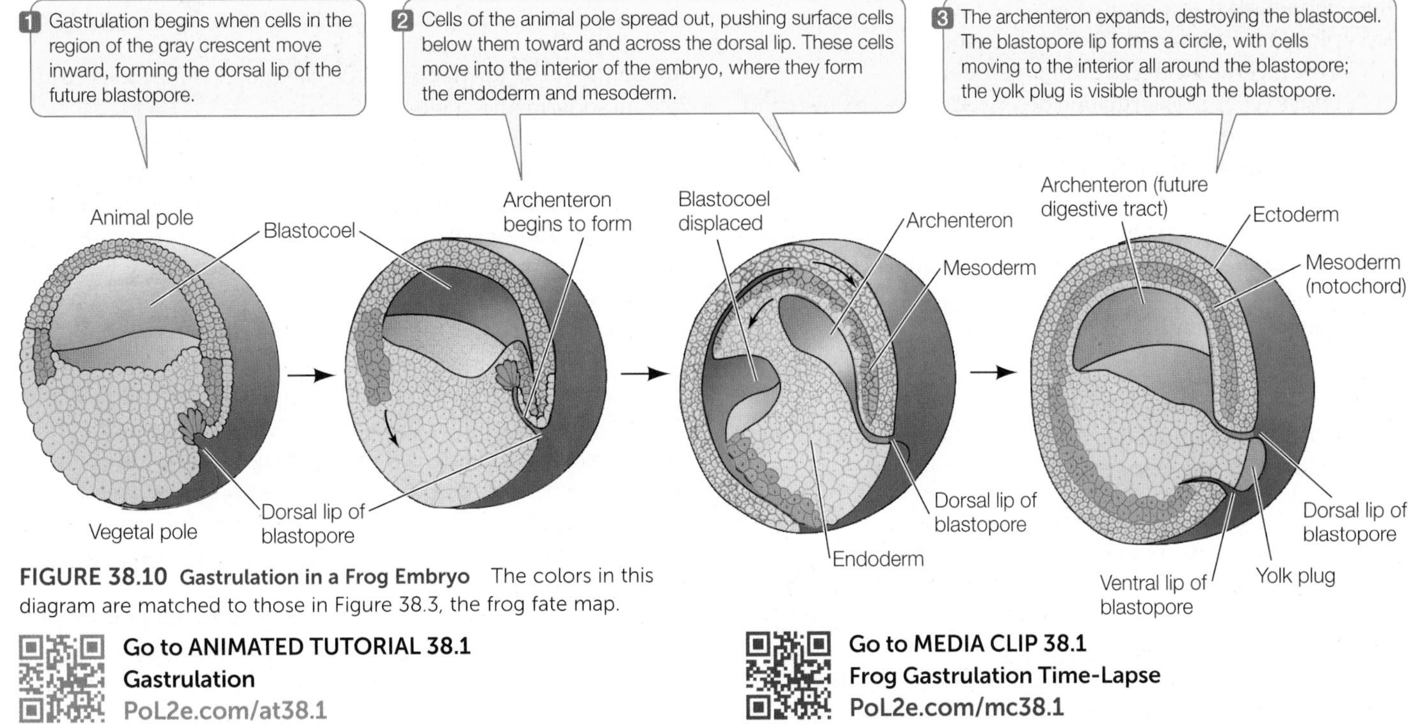

FIGURE 38.10 Gastrulation in a Frog Embryo The colors in this diagram are matched to those in Figure 38.3, the frog fate map.

Go to ANIMATED TUTORIAL 38.1
Gastrulation
PoL2e.com/at38.1

Go to MEDIA CLIP 38.1
Frog Gastrulation Time-Lapse
PoL2e.com/mc38.1

Chick embryo viewed from above
Flattened blastodisc
Yolk
1 Posterior epiblast cells change shape and thicken, forming the primitive streak.
Anterior
Midline
Embryo
Yolk
Posterior
2 Cells migrate, converging at the primitive streak and causing it to elongate.
Primitive streak
3 The primitive streak narrows and lengthens...
4 ...forming the primitive groove—the chick blastopore. Cells migrate inward through the primitive groove and Hensen's node.
Hensen's node
5 Cells generated in Hensen's node and passing into the gastrula migrate anteriorly and form head structures and notochord.
Hensen's node
Primitive groove
Surface cells move toward the groove and into the gastrula.
Primitive groove
Hensen's node
Epiblast
Blastocoel
Yolk
Hypoblast
Endoderm
Cells moving over the sides of the primitive groove form mesoderm and endoderm.
The hypoblast is displaced by spreading endoderm.
Cross section through chick embryo

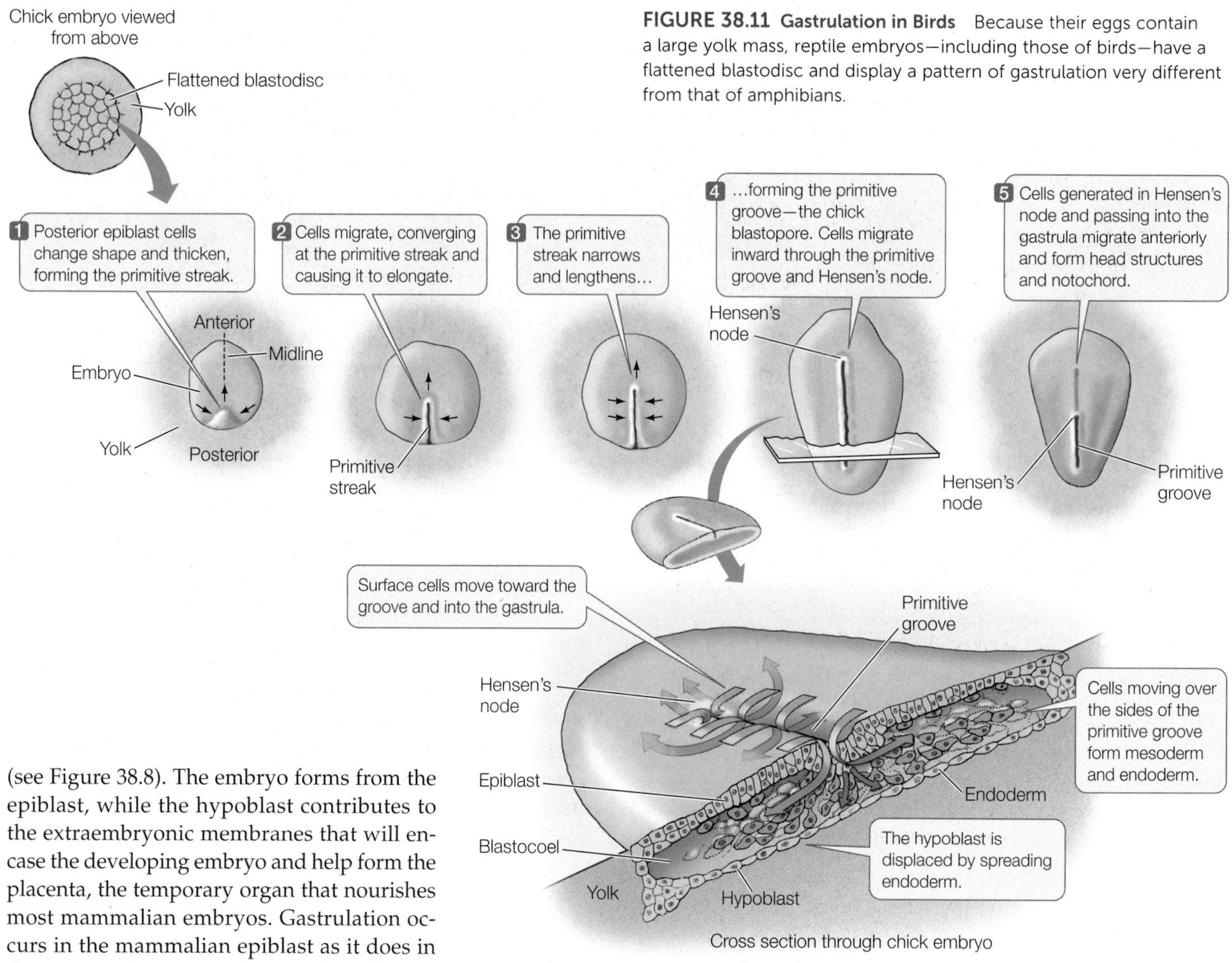

FIGURE 38.11 Gastrulation in Birds Because their eggs contain a large yolk mass, reptile embryos—including those of birds—have a flattened blastodisc and display a pattern of gastrulation very different from that of amphibians.

(see Figure 38.8). The embryo forms from the epiblast, while the hypoblast contributes to the extraembryonic membranes that will encase the developing embryo and help form the placenta, the temporary organ that nourishes most mammalian embryos. Gastrulation occurs in the mammalian epiblast as it does in the avian epiblast. A primitive groove forms, and epiblast cells migrate through the groove to become layers of endoderm and mesoderm.

All three types of gastrulation described above can be found among the protostomes, but recall that there is a major difference in the fate of the blastopore. In protostomes, the blastopore (the first opening) becomes the mouth and the anus forms at some distance away; in deuterostomes, the blastopore becomes the anus, and the mouth breaks through secondarily. This is why animals are called protostomes ("mouth first") or deuterostomes ("mouth second").

LINK

For more on the differences between protostomes and deuterostomes, see **Concept 23.1**

Gastrulation results in the production of the three germ layers and sets the stage for the formation of the **coelom**, a space completely enclosed in mesoderm. Your body cavity—the space between the gut and body wall—is a coelom (see Concept 23.1). Because of the coelom, gut contractions and heartbeats are independent of an animal's outer body wall. To understand the formation of the coelom, it is best to consider the mesoderm and coelom together.

In deuterostomes, the mesoderm often forms from the archenteron wall, and the coelom develops within in it, sometimes as a direct out-pouching of the archenteron (**FIGURE 38.12A**). This type of mesoderm and coelom formation is believed to be characteristic of deuterostomes, although it may be highly modified by yolk and other factors.

Protostomes show a different pattern that fits with the fact that specific blastomeres usually give rise to each tissue and organ (mosaic development). For example, in the snail *Ilyanassa*, one recognizable small blastomere, 4d, forms the bulk of the mesoderm. This single cell (4d) received information directing production of mesoderm from the polar lobe (see Figure 38.4), undergoes divisions and forms most of the mesoderm. The coelom forms as a split in the prospective mesoderm cells (**FIGURE 38.12B**).

Whatever the mechanism of gastrulation, the process produces an embryo with three germ layers—ectoderm, endoderm and mesoderm—that are positioned to influence one another through tissue interactions.

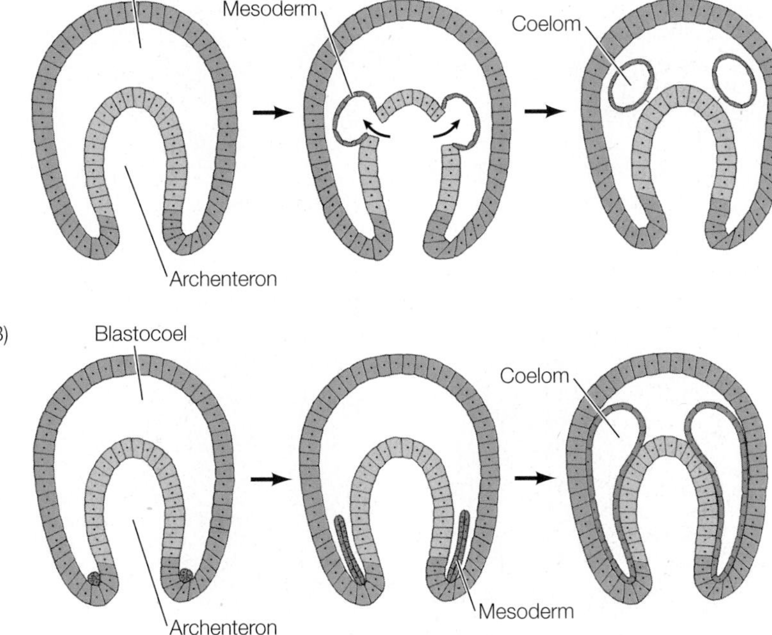

FIGURE 38.12 Formation of the Coelom Non-yolky eggs of deuterostomes and protostomes form their coeloms in different ways. (A) In deuterostomes, the coelom often forms as an out-pouching of the archenteron. (B) In protostomes, it forms as a split in the mesoderm.

CHECKpoint CONCEPT 38.3

- ✓ What does the blastopore become in protostomes? What does it become in deuterostomes?
- ✓ What are the major similarities and differences between sea urchin gastrulation and frog gastrulation?
- ✓ Why is the primitive streak in the chick embryo the equivalent of the blastopore?
- ✓ Contrast mesoderm formation in the non-yolky eggs of a deuterostome such as a starfish with those of a protostome such as a snail. Compare coelom formation in these two groups.

Following gastrulation and the subsequent formation of three germ layers, an embryo is ready to start forming the organs of the new individual. Let's now look at that key process.

CONCEPT 38.4 Gastrulation Sets the Stage for Organogenesis and Neurulation in Chordates

Organogenesis refers to the phase in development in which all of the organs and organ systems form. We will discuss the organs sequentially, but keep in mind that in an actual embryo, the processes of organogenesis occur simultaneously. Here we focus on organogenesis principally in the chordates, the large clade that includes the vertebrates and all other animals that possess a notochord. These animals also possess a dorsal nerve cord with a central cavity—a defining chordate characteristic (see Figure 16.4). The nerve cord and brain form by a process known as **neurulation.**

Each of the embryo's three germ layers gives rise to a defined set of tissues and organs:

- Ectoderm, the outer germ layer, forms the epidermis of the skin and structures such as hair, claws, and sweat glands, the brain and all of the nervous system, and pigment cells that provide color to skin, hair, feathers, and more.
- Endoderm, the innermost germ layer, produces the lining of the digestive tract and the organs that arise from it, among them the lungs, liver, pancreas, and gall bladder.
- Mesoderm, the middle layer that surrounds the coelom, gives rise to the notochord, heart, blood and blood vessels, urogenital system, muscles, bones, and the dermis (the inner layer of the skin). In vertebrates, the dermis interacts with the overlying ectoderm (the epidermis) to produce many structures, including scales, teeth, hair, feathers, horns, sweat glands, and mammary glands.

Taking a big-picture view, the basic body plan can be seen as a tube within a tube. The outer tube, arising from ectoderm and mesoderm, is the body wall. The inner tube, arising from endoderm and mesoderm, is the gut. In between is the coelom. The major organs and organ systems develop as elaborations of this basic plan.

As development proceeds, cells become recognizable as belonging to a particular type (neurons, blood cells, liver cells and so on). When a cell displays its final specialized characteristics, it is said to be **differentiated.** Differentiation takes place gradually throughout development and often into adult life. For example, mammalian red blood cells contain hemoglobin and lack nuclei. Old red blood cells are replaced continuously throughout the animal's life as new cells are produced and differentiate.

The notochord induces formation of the neural tube

In many chordates, some of the cells that move through the blastopore during gastrulation separate from the roof of the archenteron (**FIGURE 38.13**) and form the notochord (see Figure 23.35). In amphibians, for example, these are cells from the dorsal blastopore lip. In addition to providing structural support, the notochord plays a critical role during development because it induces the overlying ectoderm to form the dorsal nerve cord. **Induction** refers to the ability of one tissue to direct or lead another along a specific pathway of development. We now know that this is done by chemical signaling. Induction is one of the most important mechanisms in biology. It probably takes place in all animals and plays a role in the formation of most organ systems.

LINK

Some of the molecular mechanisms of induction are discussed in **Concept 14.2**; see especially **Figure 14.9**

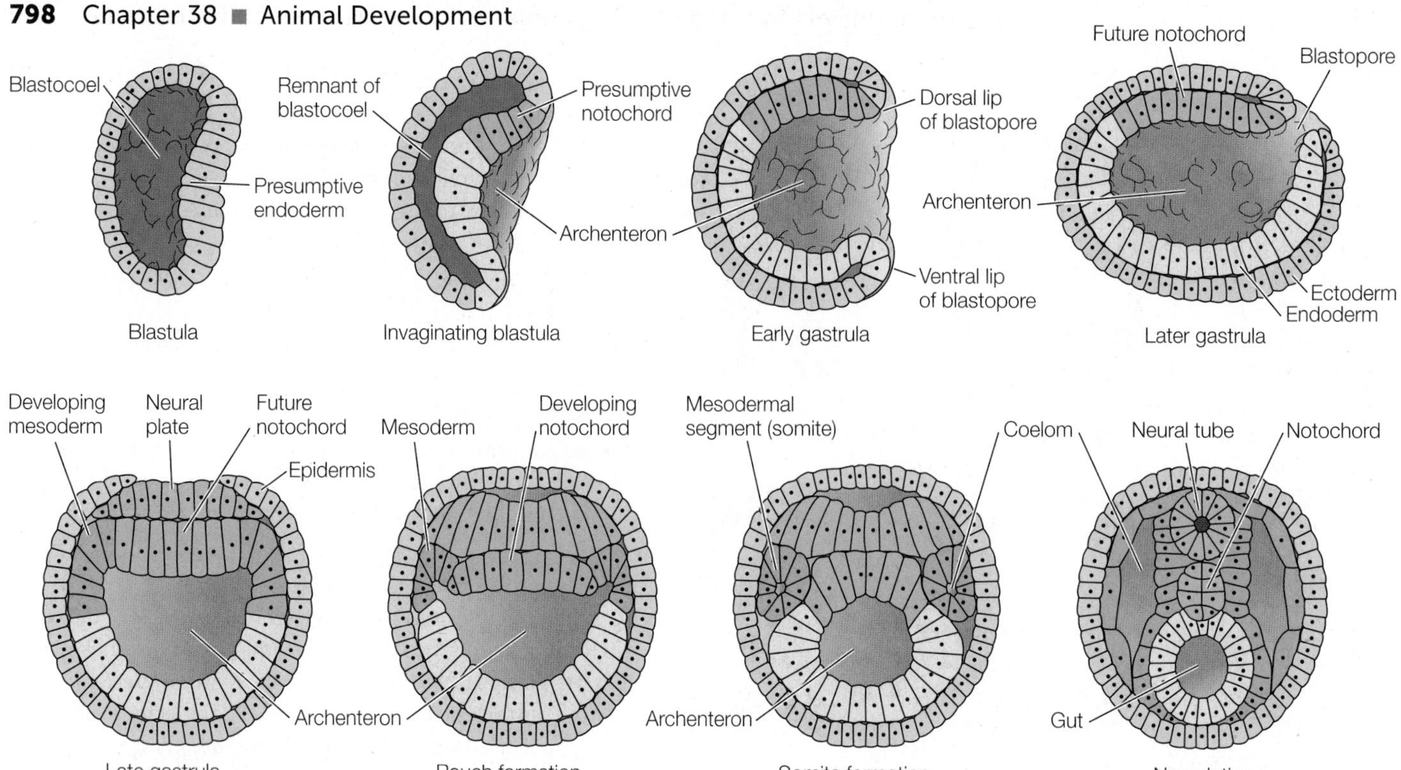

FIGURE 38.13 Gastrulation, Mesoderm Formation, and Neurulation in the Lancelet Amphioxus, a Non-vertebrate Chordate In these classic descriptions (by Edwin Grant Conklin and Berthold Hatschek), the formation of tissue layers in a non-yolky chordate egg is clearly seen. Bottom row depicts sections of embryos cut at right angles to those in the top row.

Under the influence of the developing notochord, a portion of the dorsal ectoderm flattens and begins to roll up, forming first a neural plate and then a neural groove. When the edges of the groove grow together, they fuse to form the neural tube, and ectoderm closes over them. At its anterior end, the vertebrate neural tube forms three swellings, which will become the major divisions of the adult brain: the **hindbrain**, **midbrain**, and **forebrain**. The rest of the neural tube will become the spinal cord. If the neural tube does not close properly, serious problems result. Examples in humans include spina bifida (an opening in the spine that allows the spinal cord to protrude) and anencephaly (absence of a major portion of the brain).

APPLY THE CONCEPT

Gastrulation sets the stage for organogenesis and neurulation in chordates

What about the amphibian dorsal blastopore lip makes it so crucial to "organizing" the three germ layers? The distribution of β-catenin in the late blastula corresponds to the location of this organizer in the early gastrula. A series of experiments endeavored to ascertain whether β-catenin is "necessary and sufficient" for subsequent functions of the organizer. Review the results of these experiments in the table and answer the questions.

EXPERIMENT	RESULT
A. "Knock out" β-catenin by injecting β-catenin antisense mRNA into frog egg (see Figure 13.9).	Dorsal lip of blastopore does not form.
B. Repeat experiment A, but follow injection of antisense mRNA with an injection of exogenous (i.e., from an external source) β-catenin protein into the egg.	Gastrulation is normal.
C. Inject exogenous β-catenin into ventral region of an early gastrula (where β-catenin does not normally occur).	A second site of gastrulation is stimulated, producing results similar to those of Spemann and Mangold (see Figure 38.14).
D. Repeat experiment C, but follow exogenous β-catenin with β-catenin antisense mRNA injected into same area.	Gastrulation is normal, occurring only at original site

1. Which experiment shows that β-catenin is *necessary* for induction of the organizer? Explain.
2. Which experiment shows that β-catenin is *sufficient* (enough) for induction of the organizer? Explain.
3. What was the purpose of experiment B? Of experiment D?
4. This series of experiments demonstrates what is sometimes called the "find it, lose it, move it" approach to understanding a gene's function. Explain why this catchy phrase is appropriate.

Much of our understanding of induction comes from work that began on amphibian embryos in the early twentieth century and has continued since. When Hans Spemann and his student Hilde Mangold were studying the dorsal lip of the blastopore in frog eggs, they transplanted this region to another embryo at the same stage of development. The results were momentous. This small piece of tissue stimulated a second site of gastrulation, and a second complete embryo formed (**FIGURE 38.14**). Because the dorsal blastopore lip of amphibians was apparently capable of inducing host tissue to form an entire embryo, Spemann and Mangold dubbed it the **organizer**, and its action became known as primary embryonic induction. We now know that the dorsal blastopore lip is not responsible for the first inductive event in development, but because of its extreme importance, it is still referred to as the organizer. For more than 80 years, the organizer has been an active area of research. Hensen's node, which we discussed in Concept 38.3, acts as an organizer in birds, and a homologous region does so in mammals.

Molecular biologists are making great strides in working out the molecular mechanisms of the organizer. Thanks to a series of experiments in the 1980s and 1990s, we now know that the protein β-catenin plays a key role in determining which cells become the organizer. It does this by triggering a complex series of interactions between various transcription factors and growth factors that control gene expression. Several of these transcription factors have been identified.

During the last century, clever experiments revealed the source and approximate molecular size of many inductive factors. Now, modern techniques are enabling scientists to determine the identity of many such molecules. One of the best-studied examples is the induction of the dorsal nerve cord in amphibians.

The development of the central nervous system depends on the activation of a specific genetic cascade within certain ectodermal cells that will become neuronal instead of epidermal. Work on amphibian embryos has shown that specific polypeptides released from the developing notochord act as inducers that control this activation. Two of these polypeptides that induce the formation of the neural tube are appropriately named Noggin and Chordin. They act, not by direct activation of genes, but rather by inhibiting the activity of another substance, the TGF-β bone morphogenetic protein BMP4 (**FIGURE 38.15**). Among its many actions, BMP4 induces the ectoderm to form epidermis. In the absence of BMP4, ectoderm forms neural tissue. By blocking the effect of BMP4, Chordin and Noggin allow the ectoderm to carry out its "default" differentiation of becoming neural.

INVESTIGATION

FIGURE 38.14 The Dorsal Lip Induces Embryonic Organization In a classic experiment, Hans Spemann and Hilde Mangold transplanted the dorsal blastopore lip mesoderm of an early gastrula stage salamander embryo. The results showed that the cells of this embryonic region, which they dubbed "the organizer," could direct the formation of an entire embryo.[a]

HYPOTHESIS

Cytoplasmic factors in the early dorsal blastopore lip organize cell differentiation in amphibian embryos.

METHOD

1. Excise a patch of tissue from above the dorsal blastopore lip of an early gastrula stage salamander embryo (the donor).
2. Transplant the donor tissue onto a recipient embryo at the same stage. The donor tissue is transplanted onto a region of ectoderm that should become epidermis (skin).

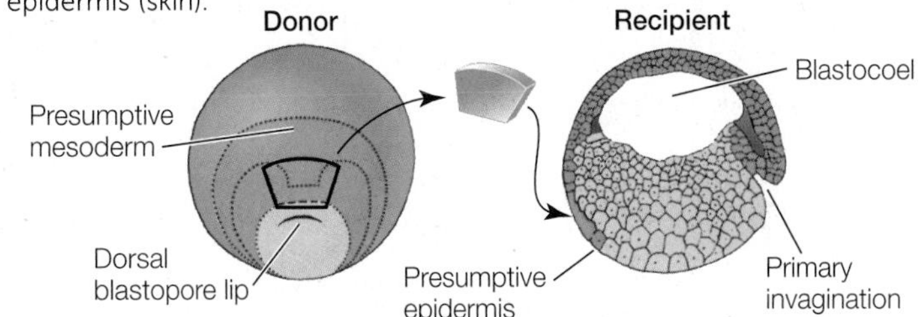
Donor
Recipient
Blastocoel
Presumptive mesoderm
Dorsal blastopore lip
Presumptive epidermis
Primary invagination

RESULTS

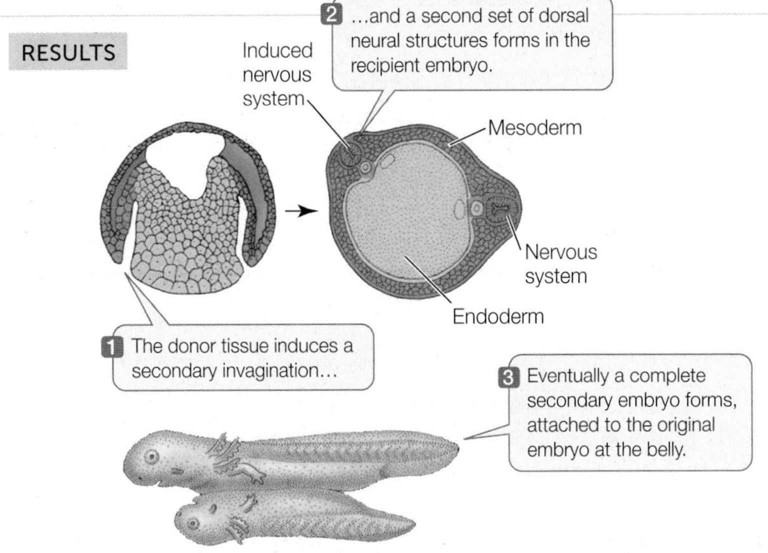
2 ...and a second set of dorsal neural structures forms in the recipient embryo.
Induced nervous system
Mesoderm
Nervous system
Endoderm
1 The donor tissue induces a secondary invagination...
3 Eventually a complete secondary embryo forms, attached to the original embryo at the belly.

CONCLUSION

The cells of the dorsal blastopore lip can induce other cells to change their developmental fates.

Go to **LaunchPad** for discussion and relevant links for all **INVESTIGATION** figures.

[a]H. Spemann and H. Mangold. 1924. *Roux' Arch. Entw. Mech.* 100: 599–638. Viktor Hamburger's translation appeared in *Foundations of Experimental Embryology*, 1964, (B.H. Willier and J.M. Oppenheimer, eds.), pp. 146–184.

Go to ANIMATED TUTORIAL 38.2
Tissue Transplants Reveal the Process of Determination
PoL2e.com/at38.2

INVESTIGATION

FIGURE 38.15 Differentiation Can Be Due to Inhibition of Transcription Factors When organizer cells involute to underlie dorsal ectoderm along the embryo midline, that overlying ectoderm becomes neural tissue rather than skin (epidermis). But do the organizer cells cause dorsal ectoderm to become neural tissue, or do they *prevent* this ectoderm from becoming skin?[a]

HYPOTHESIS

The default state of amphibian dorsal ectoderm is neural; it is induced by underlying mesoderm to become epidermis.

METHOD

1. Excise animal caps (presumptive ectoderm) from amphibian blastulas. Culture presumptive mesodermal cells from early gastrulas and extract BMP4. BMP4 is a growth factor released from the notochord that induces overlying ectoderm to become skin.

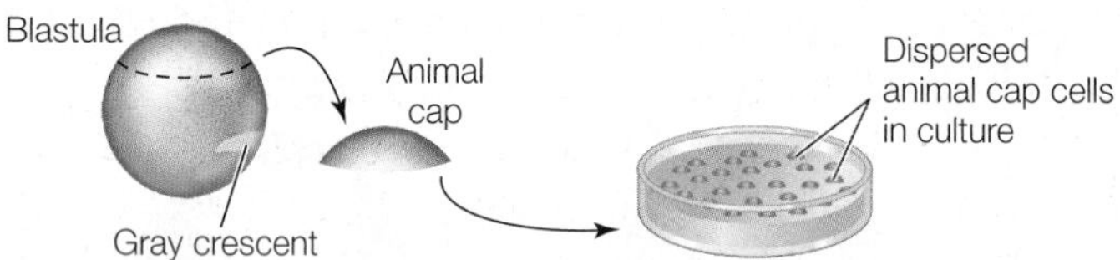

2. Prepare four separate cultures of embryonic ectodermal cells. Incubate with no additions (control); with BMP4 from step 1; with a BMP4 inhibitor; and with both molecules.

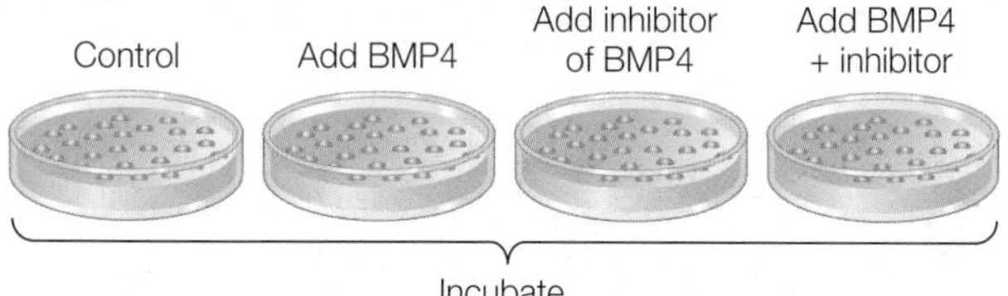

3. After incubation, extract mRNAs from the cultured ectodermal cells and run on gels to reveal expression of NCAM (a neural tissue marker) and keratin (an epidermal tissue marker).

RESULTS

Gels reveal that different treatments with BMP4 and its inhibitor alter patterns of gene expression in cultured ectodermal cells.

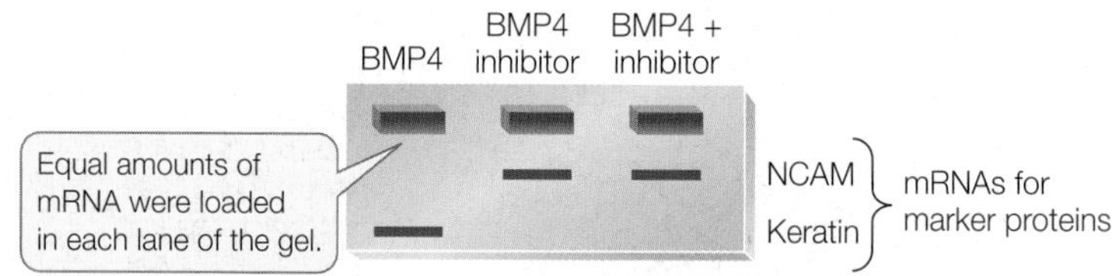

CONCLUSION

The organizer cells secrete an inhibitor of BMP4.

ANALYZE THE DATA

Use the gel results shown above to answer the questions.

A. Does BMP4 induce expression of neural-specific or epidermal-specific genes?

B. Does BMP4 block any gene expression in the ectodermal cells?

C. What is the evidence that BMP4 has an inductive and/or an inhibitory effect on gene expression in ectodermal cells?

D. Do these results support the hypothesis?

Go to **LaunchPad** for discussion and relevant links for all **INVESTIGATION** figures.

[a]P. A. Wilson and A. Hemmati-Brivanlou. 1995. *Nature* 376: 331–333.

Another important factor controlling differentiation of the neural tube is Sonic hedgehog, symbolized Shh, the transcription factor mentioned at the start of this chapter. Shh is released by the notochord and diffuses into the ventral region of the neural tube, where it directs the development of this region into the ventral structures and circuits of the spinal cord. These circuits will carry the major output from the motor pathways of the nervous system.

LINK

Sonic hedgehog also plays an important role in the development of vertebrate limbs; see **Concept 14.3**

As the neural tube is closing, certain cells dissociate from it and come to lie between the neural tube and the overlying epidermis. These are known as **neural crest cells** (**FIGURE 38.16**). They are star-shaped cells that are termed "pluripotent" because they can differentiate in many different ways. Neural crest cells migrate extensively and have amazing developmental capacities. Not only do they form sensory neurons and major parts of the autonomic nervous system, but they are also responsible for most of the skull bones, the pigment cells, and many other structures. Without neural crest cells, vertebrates would not have heads, as we recognize them.

Mesoderm forms tissues of the middle layer

On either side of the forming notochord, three regions of mesoderm can be recognized—**somites**, **intermediate mesoderm**, and **lateral plate mesoderm** (see Figure 38.16). The vertebrate body plan, like that of annelids and arthropods, consists of repeating segments that are modified during development. These segments are first seen as the blocks of cells known as somites (**FIGURE 38.17**; also see Figure 38.16), then later as the repeating patterns of vertebrae, ribs, nerves, and muscles along the anterior–posterior axis as differentiation proceeds.

Somites form bones, cartilage, skeletal muscle, and the dermis of the skin.

Intermediate mesoderm forms much of our urinary and reproductive system.

Lateral plate mesoderm surrounds the coelom and lines the vertebrate body cavity (the peritoneal cavity). The peritoneal cavity makes up much of the coelom. The thin tissue layer that surrounds the peritoneal cavity is the peritoneum. (You may have heard of peritonitis, a severe inflammation of the covering of the internal organs. A major cause is the entry of bacteria into the peritoneal cavity, as occurs when an appendix ruptures.) The inner layer of

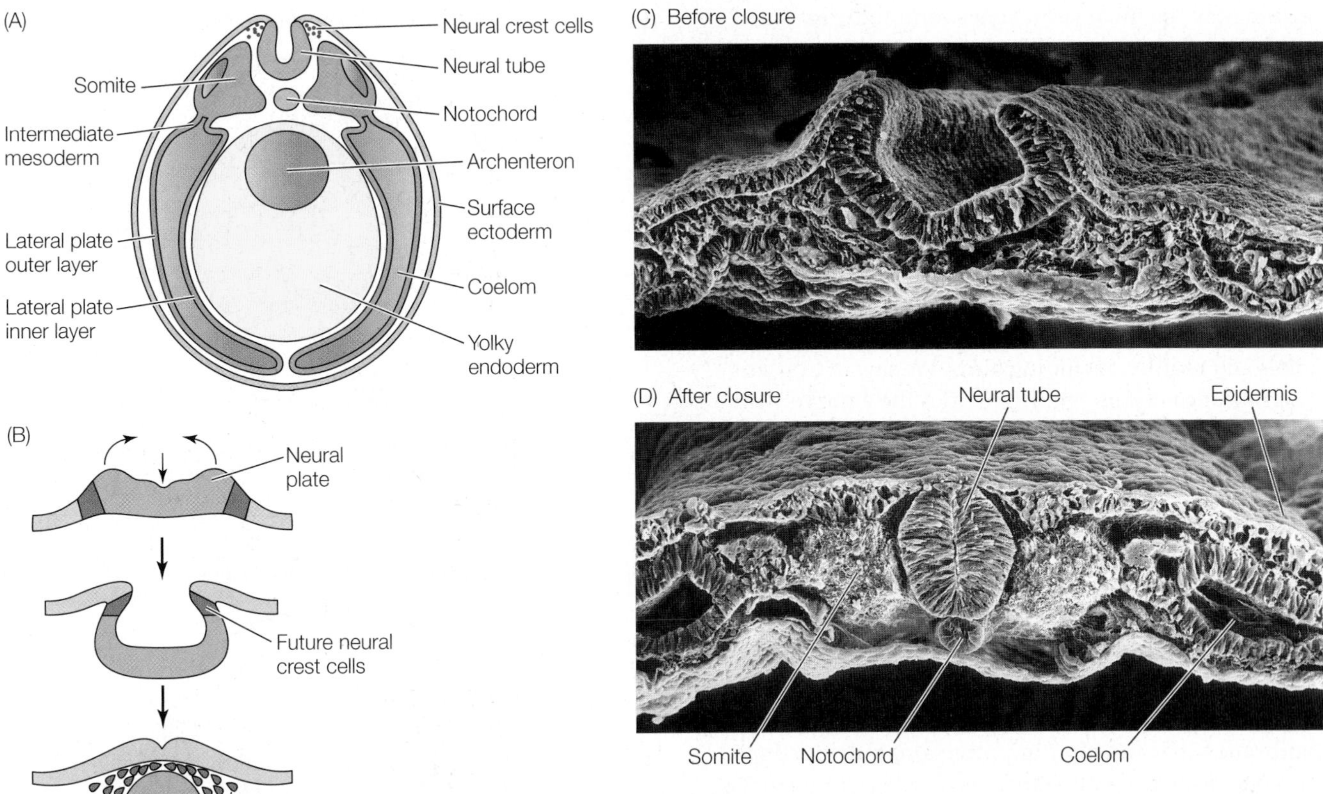

FIGURE 38.16 Neurulation and Differentiation of Mesoderm in Vertebrates (A) This illustration of a frog egg shows how the neural folds draw together and close, releasing neural crest cells. The mesoderm extends between the ectoderm and endoderm. The three divisions of trunk mesoderm—somite, intermediate mesoderm, and lateral plate mesoderm—are apparent. (B) At the start of vertebrate neurulation, the ectoderm of the neural plate is flat. The neural plate invaginates and folds, forming a tube. Neural crest cells are released. (C,D) Scanning electron micrograph of the neural tube in a chick, before and after closure.

lateral plate mesoderm lies close to the endoderm. It forms part of the peritoneum, the muscles of the digestive tract, and most of the circulatory system, including the heart. The outer layer of lateral plate mesoderm lies adjacent to the ectoderm and forms part of the peritoneum and some of the muscle of the body wall (see Figure 38.16).

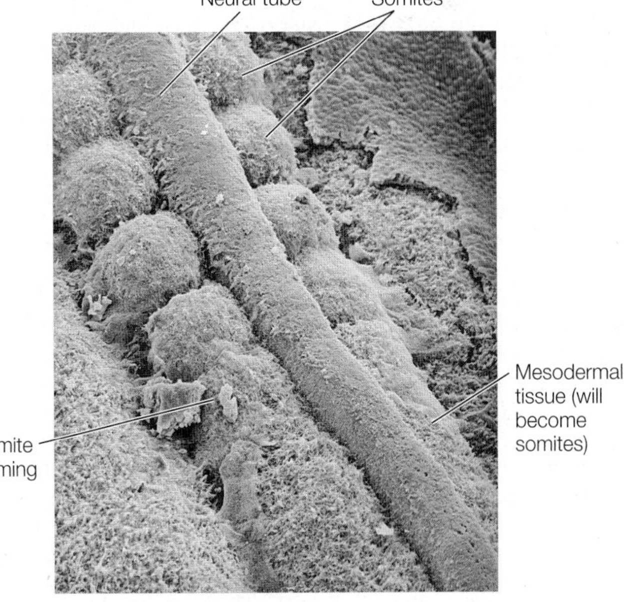

FIGURE 38.17 Body Segmentation In this scanning electron micrograph of somite formation in a chick embryo, the overlying ectoderm has been removed and the neural tube and somites are seen from above. Skeletal muscle, cartilage, bone, and the lower layer of the skin form from the somites.

CHECKpoint CONCEPT 38.4

- ✓ What two major mechanisms are responsible for determining the differentiated state of a cell?
- ✓ What is the role of the notochord in neurulation?
- ✓ A chordate synapomorphy (a shared derived trait) is a hollow dorsal nerve chord (see Concept 23.5). How does the process of neurulation explain why the spinal cord has a central cavity?
- ✓ What major tissues does ectoderm form? What three major mesodermal regions form in chordates, and what does each form?

We have seen how the three germ layers form the organs and tissues of the vertebrate embryo. Next we will focus on how they also protect and nourish the embryo in many of the animals with which you are most familiar.

CONCEPT 38.5 Extraembryonic Membranes Protect and Nourish the Embryo

A major step in the evolution of tetrapods (four-limbed vertebrates) was the formation of the amniotic egg, seen today in mammals and reptiles, including birds. We saw in Concept 23.6 the significance of this egg. It provides the embryo with a contained aqueous environment, freeing the processes of reproduction and development from dependence on an external water supply. The three germ layers that form the embryo also create the membranes that provide this protective environment.

Extraembryonic membranes form with contributions from all germ layers

Whether they develop inside or outside the mother's body, embryos of amniotes are surrounded by four **extraembryonic membranes** that function in protection, nutrition, gas exchange, and waste removal. The chicken embryo provides a good example, but the process is similar in other reptiles and in egg-laying mammals. (We will discuss placental mammals in the next section.) The four membranes are the yolk sac, allantois, amnion, and chorion (**FIGURE 38.18**).

- The **yolk sac** forms first, by extension of the endoderm and the inner layer of lateral plate mesoderm (see Figure 38.18A,B). These two layers grow around the yolk and enclose it. The inner layer of lateral plate mesoderm is special because it can form blood cells and blood vessels. The first blood cells of the embryo and the vessels that carry them—the vitelline artery and its branches—form in the mesoderm of the yolk sac. Stored yolk is not transferred directly to the gut of the embryo through the yolk stalk. Rather, it is

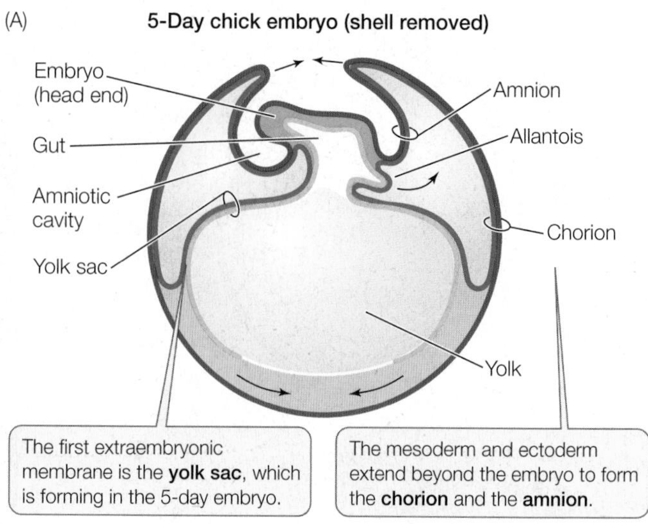

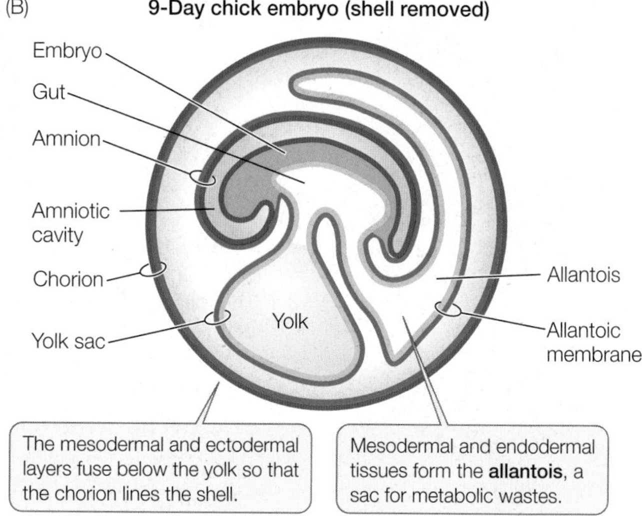

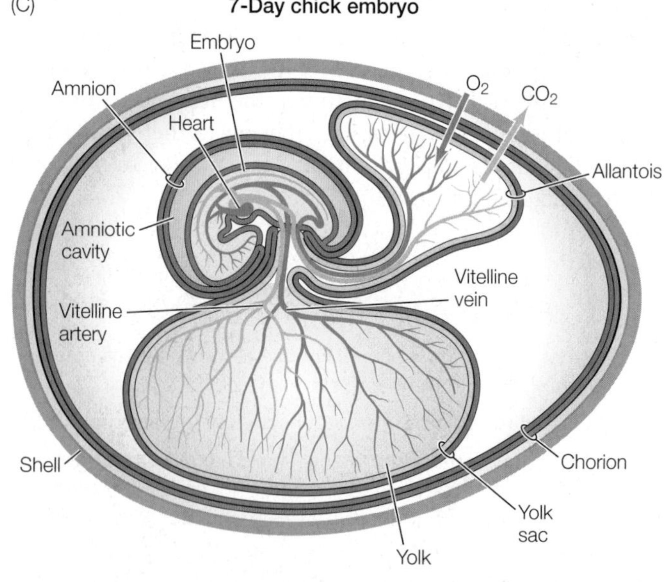

FIGURE 38.18 The Extraembryonic Membranes of Amniotes In birds, other reptiles, and mammals, the embryo constructs four extraembryonic membranes. (A,B) Membrane outgrowth and fusion in 5- and 9-day embryos. (A) Developing amnion and chorion form from ectoderm and endoderm as seen in this 5-day chick embryo. The membranes will fuse when they meet dorsally (arrows), enclosing the embryo in 2 envelopes. The yolk sac and allantois form from endoderm and mesoderm. The yolk sac grows around the yolk (arrows) and the allantois forms a pouch. (B) Membranes surround the embryo in this 9-day embryo. Some will continue to expand as the embryo grows, but the yolk sac shrinks as yolk is used by the embryo. Fluids secreted by the amnion fill the amniotic cavity, providing an aqueous environment for the embryo. The allantois stores the embryo's waste products. The allantoic membrane will become pressed against the chorion and shell and functions in gas exchange between the embryo and its environment. (C) Developing chick in shell showing development of the circulatory system, which is well underway by day 7. Only the yolk sac and the allantois are capable of making blood vessels. As a result, they are involved in nutrient transport and gas exchange.

Go to ACTIVITY 38.1 Extraembryonic Membranes
PoL2e.com/ac38.1

digested by the endoderm of the yolk sac, and the products are transported to the embryo by the developing blood and blood vessels in the yolk sac wall (Figure 38.18C).

- The **allantois**, a sac for storage of metabolic wastes, is also an outgrowth of endoderm and the inner layer of lateral plate mesoderm (Figure 38.18A–C). Because the yolk sac and the allantois are the extraembyronic membranes that contain lateral plate mesoderm, they are the only two that develop blood vessels (Figure 38.18C).
- Ectoderm and the outer layer of the lateral plate mesoderm also grow out from the embryo and surround it (Figure 38.18A). Where they meet, they fuse, forming two membranes, an inner amnion and outer chorion (Figure 38.18B). The **amnion** surrounds the embryo and secretes fluid into the enclosed cavity, providing a protective, shock-absorbing environment.
- The **chorion**, located just beneath the egg's shell, forms a continuous membrane that limits water loss. The chorion cannot make blood vessels, but it fuses with the blood vessel–forming allantoic membrane, forming the chorioallantoic membrane that provides for exchange of O_2 and CO_2 between the embryo and the outside world (Figure 38.18C).

Extraembryonic membranes in mammals form the placenta

In placental mammals, the entire trophoblast becomes embedded in the endometrium of the uterus (see Figure 38.8). Hypoblast cells proliferate to form what in birds would be the yolk sac, even though there is little yolk in the eggs of placental mammals. In eutherian placental mammals, the allantois and chorion combine, forming the chorioallantoic placenta. These embryonic tissues, along with maternal tissue of the uterine wall (the endometrium), produce the placenta. Note that the placenta is a unique organ because it is composed of tissues from two organisms—the mother and her offspring. Interestingly, the yolk sac in mammals continues to be an early site of blood cell production as it is in birds, even though no or very little yolk is present. Meanwhile, the amnion grows around the embryo, enclosing it in a fluid-filled amniotic cavity (**FIGURE 38.19**). (A pregnant woman's "water breaks" when the amnion bursts during labor and releases the amniotic fluid.)

Human gestation is traditionally divided into three periods of roughly 12 weeks each, called trimesters. The first trimester is a time of rapid cell division and tissue differentiation, and the embryo is very sensitive during this time to damage from radiation, drugs, chemicals, and pathogens that can cause birth defects such as those referred to in the opening story of the chapter. By the end of the first trimester, most organs have started to form, and the embryo is about 8 centimeters long. At about this time, the human embryo is medically and legally referred to as a **fetus**.

Fish also make yolk sacs

Many fish also have a very yolky egg (see Figure 38.7B). They are not amniotes, however, so how do they obtain nutrients from this stored material? Embryonic fish produce a yolk sac, but it differs from that of amniotes. As the fish embryo forms, all three germ layers—ectoderm, mesoderm and endoderm—grow around the yolk. The yolk sac becomes vascularized as in amniotes, and materials are carried in the blood vessels to the embryo (**FIGURE 38.20**).

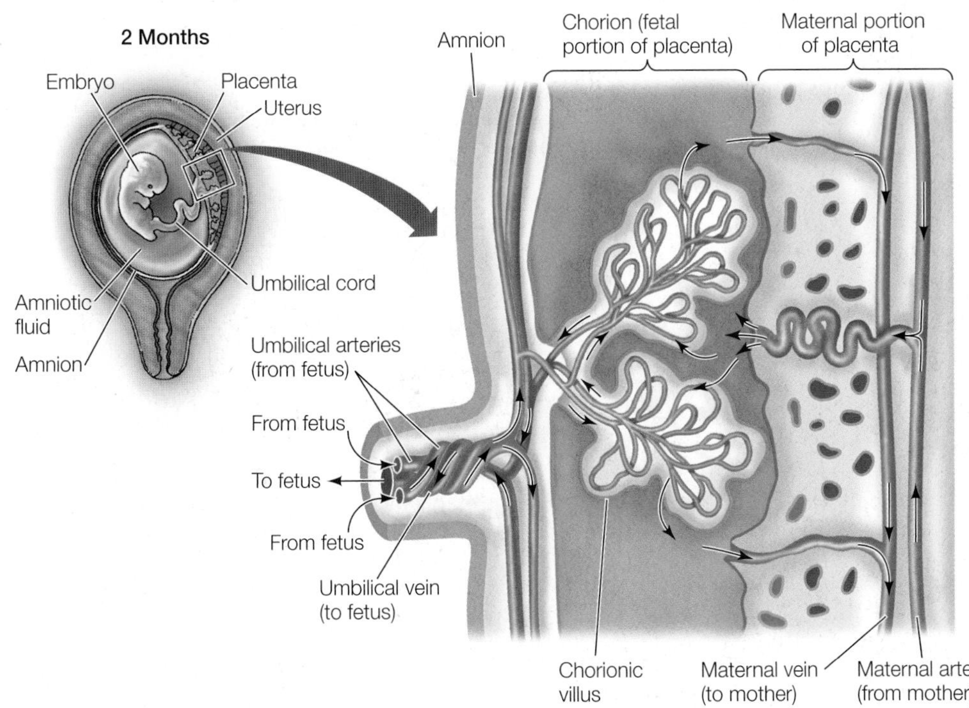

FIGURE 38.19 The Mammalian Placenta In humans and other placental mammals, nutrients and wastes are exchanged between maternal and fetal blood in the placenta. This organ forms from both fetal tissue and tissues of the mother's uterine wall. The embryo is attached to the placenta by the umbilical cord. Embryonic blood vessels invade the placental tissue to form fingerlike chorionic villi. Maternal blood flows into the spaces surrounding the villi, allowing nutrients and respiratory gases to be exchanged between the maternal and fetal blood.

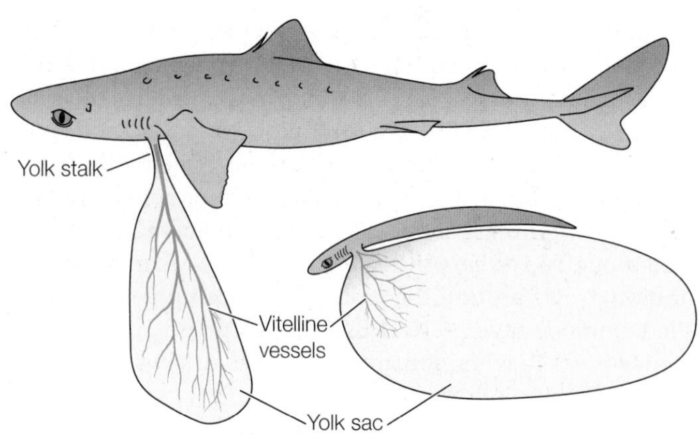

FIGURE 38.20 Fish Yolk Sac The yolk sac in this developing shark is trilaminar because it is composed of all three tissue layers. Contrast this with the yolk sac of mammals, which forms from just two layers and thus is bilaminar.

CHECKpoint CONCEPT 38.5

- ✓ What are the four extraembryonic membranes of a chick embryo? Describe their functions.
- ✓ Would you expect a frog embryo to have the same extra-embryonic membranes as a lizard? Why or why not?
- ✓ What function does the yolk sac have in both egg-laying and placental mammals?
- ✓ The yolk sac of fish is referred to as being trilaminar. Why is this term appropriate? How does a fish yolk sac differ from a chick yolk sac?
- ✓ What outstanding feature of the placenta makes it a unique organ?

We have now seen how a single-celled zygote becomes a complex organism, but development does not stop at birth or hatching. Animals must grow. For some, growth stops when they reach adulthood. Others grow throughout their lives. In either case, this growth is part of development. We will consider post-embryonic growth next.

CONCEPT 38.6 Development Continues throughout Life

No sexually produced animal is as large as its parent at birth or hatching. All must obtain nourishment and grow. Some animals immediately look like their parents and undergo **direct development** to reach full size. Others appear very different and spend part of their lives as one or more different forms, as they undergo **indirect development**.

A change in the proportions of various parts of an organism relative to one another as a consequence of unequal growth is known as **allometry**. Isometric growth is a 1:1 size increase; allometric growth is any change from this. Allometry is often seen in direct developers as parts of their body change in relative size and shape with growth. For example, at birth a human baby is not a miniature adult. Not only are many of its systems immature, but its overall morphology (form and shape) is also quite different than that of an adult. One difference is the ratio of head to trunk size. A baby's head is much larger in comparison with the rest of its body than an adult's head is, and the head stops growing much earlier in life than the rest of the body (**FIGURE 38.21**). It has been suggested that the large head of infant mammals, especially humans, corresponds to the large brain and development of the cerebral cortex. Many examples of allometry exist. Indeed, nearly all direct developers experience allometric growth of some sort.

Indirect development involves changes in morphology that can be dramatic. Many animals, both protostome and deuterostome, have a larval stage that may be quite unlike the adult.

FIGURE 38.21 Allometric Growth of a Human Proportions of body parts change as an individual develops from fetus to adult.

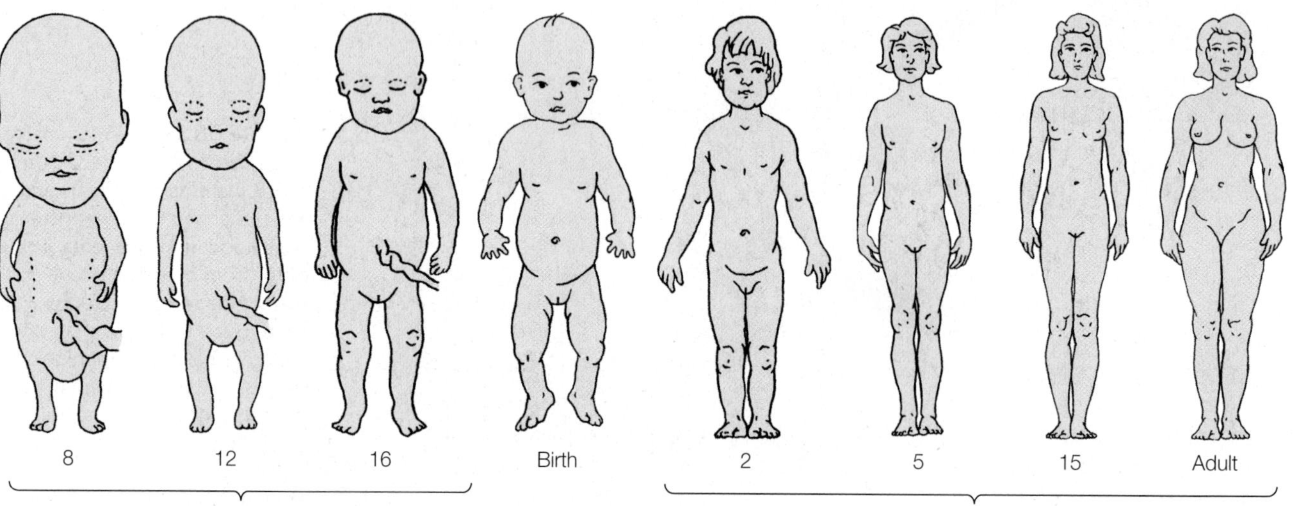

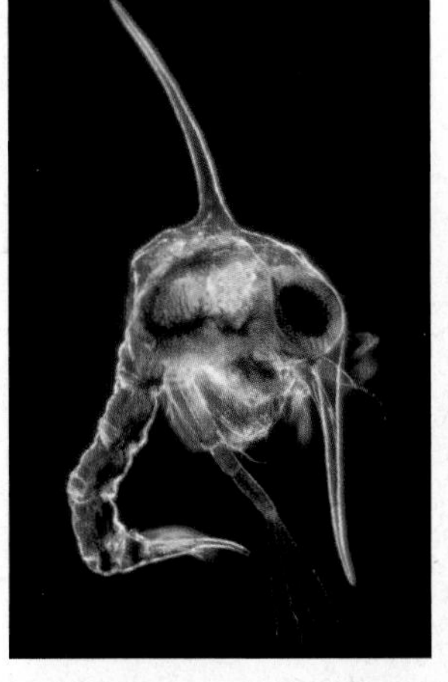

(A) Shore crab (*Carsinus maenas*)

(B) Sea star (*Asterias rubens*)

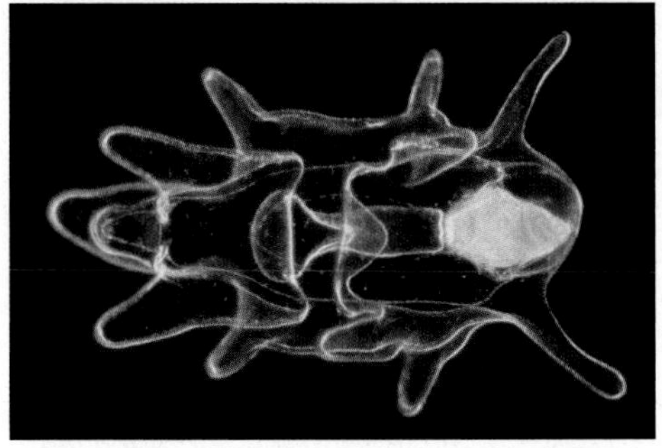

FIGURE 38.22 Larval Forms Are Often Quite Unlike Their Parents Here we show larval (left) and adult (right) forms of (A) a crab (a protostome) and (B) a sea star (a deuterostome).

A familiar example is the monarch butterfly, which has a caterpillar that feeds on the milkweed plant on which its egg was placed, and a winged adult that can migrate from Canada to Mexico. **FIGURE 38.22** shows two larvae with which you might not be familiar.

A larval stage usually performs one of two important roles—or both. Like the caterpillar, it may be primarily a feeding stage. This is seen in animals that provision their eggs with little or moderate yolk resources. Insects, fish, and amphibians are prime examples of this lifestyle. Some adult insects do not feed at all. Ephemeridae—mayflies—last only 1 day as adults. (They are ephemeral, as their name tells us.) They molt from the pupal to adult stage, seek a mate, deposit eggs, and die, relying solely on energy stored by the larva.

Larvae may also play an important role in species distribution. This is especially true in animals in which the adult is sedentary or relatively so. Corals, for example, can reproduce by fragmentation, but new sites are colonized by swimming larvae. Some polychaete worms (capitellids) are rapid colonizers of freshly disturbed, organically rich areas (substrates of polluted harbors and beneath fish farms, oil spills, and whale falls—decaying whale carcasses on the ocean floor) that are widely distributed around the world. Larvae, swimming only with ciliary bands, are carried by currents and then settle in these areas.

When a larva transforms into an adult at metamorphosis, changes can be dramatic, as in the monarch and other butterflies and moths. Major portions of the body may be remodeled or even discarded (see Concept 23.4). Many amphibians lay their eggs in spring, and after quite rapid development, small tadpoles emerge. A tadpole feeds, and its size increases. At metamorphosis, the tadpole shows changes in virtually every system. Tail and gills are resorbed, and limbs grow. In many species, the tadpole stage is herbivorous whereas the adult is carnivorous. Accompanying this change is an alteration of gut length. The gut of a tadpole is actually several times longer than the gut of the small froglet it will become! A tunicate (sea squirt; see Figure 23.35A) tadpole settles and rapidly resorbs its tail, including the notochord and nerve cord. Since the three major chordate characteristics (notochord, hollow dorsal nerve chord, and post-anal tail) are no longer present, the adult tunicate does not resembles a chordate! The larva of the lamprey is so different from the adult that it was once considered a different species. It is still called by the earlier name of "ammocoetes."

CHECKpoint CONCEPT 38.6

- ✓ Prior to the Renaissance, painters often depicted infants as miniature adults. Why do these babies not appear natural?
- ✓ In addition to head size, give another example of allometric growth in humans.
- ✓ What are two major functions of larvae?

As we have seen, the wonderful process that we recognize as development begins with the fusion of a single egg and a single sperm and continues throughout the entire life of the organism. Although not considered here, the processes of regeneration, wound healing, and aging are all developmental mechanisms. Medical interests in stem cell technology, in wound repair and regeneration, in the prevention of birth defects, and in many other areas, as well as a strong interest in the overlapping field of evolutionary biology, are adding impetus to our desire to understand developmental processes in many organisms. With all of the exciting techniques now available, the door for future studies is wide open.

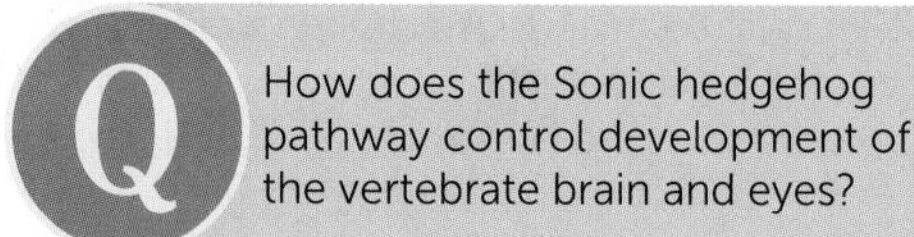

Q
How does the Sonic hedgehog pathway control development of the vertebrate brain and eyes?

ANSWER Sonic hedgehog (Shh) is a transcription factor secreted by the notochord (Concept 38.4). It diffuses to the overlying neural tube, where it induces differentiation of the ventral neural tube into motor neurons and other features of the ventral regions of the spinal cord. Shh also plays different inductive roles at other locations. At the anterior end of the neural tube where the forebrain will develop, the closed neural tube undergoes reciprocal inductive tissue interactions with the overlying ectoderm. Signaling molecules from the neural tissue induce the ectoderm to form the lens of the eye, and the developing lens produces signaling molecules that induce the underlying neural tissue to form an optic cup that will differentiate into the retina.

But why do these events happen bilaterally? Mesoderm that lies anterior to the notochord produces and releases Shh, as does the notochord. Shh induces the ventral region of the anterior neural tube to differentiate into the structures of the forebrain. One of the actions of Shh is to inhibit a transcription factor called Pax6 that is expressed by the neural tissue. Pax6 is essential for determination of the eye field (i.e., tissues that can form eyes; see Figure 14.22). The midline expression of Shh splits the eye field into two and also enables the forebrain to develop into left and right hemispheres.

When environmental factors affect the action of Shh, holoprosencephaly can result. When pregnant cows or sheep graze on *Veratrum californicum* at certain critical points in gestation, the embryo is exposed to cyclopamine and the Shh signaling pathway is blocked. Depending on when the exposure occurs, bilateral development of the forebrain may not take place, resulting in holoprosencephaly and cyclopia (**FIGURE 38.23**). Another area of great interest for humans is the observation that Shh requires cholesterol to be an effective signal. Abnormally low levels of cholesterol during pregnancy (such as might occur with the use of certain statin drugs) could cause some aspects of holoprosenecephaly.

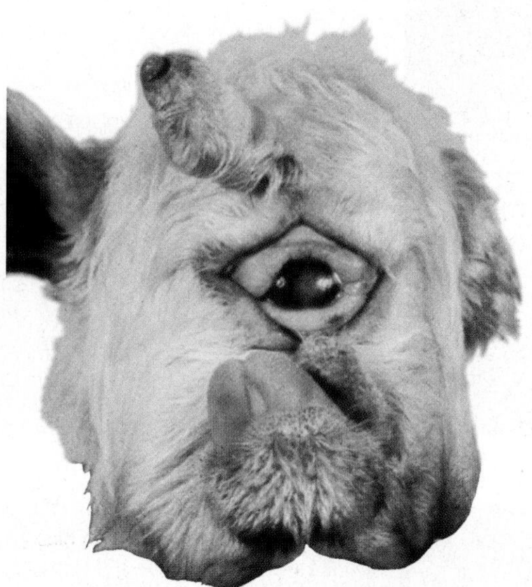

FIGURE 38.23 Environmentally Induced Holoprosencephaly A stillborn lamb born of a ewe that grazed on corn lily leaves. Alkaloids in the plant interfered with the Sonic hedgehog signaling pathway in the embryo, resulting in fusion of the cerebral hemispheres and the formation of a single eye.

SUMMARY

CONCEPT 38.1 Fertilization Activates Development

- One sperm and only one sperm must fertilize an egg. Fertilization activates the egg.
- Egg and sperm make different contributions to the **zygote**. The egg contributes a haploid nucleus, most of the molecular machinery for cell division, energy production, and growth (yolk, mitochondria, ribosomes, etc.), as well as informational molecules such as transcription factors and mRNAs. The sperm contributes its nuclear material and usually its centrosome, the major microtubule-organizing center.
- Cytoplasmic contents of the egg often are not distributed uniformly. Nutrient molecules, usually in yolk granules, are often found in the **vegetal hemisphere**. In amphibians, rearrangement of **cytoplasmic determinants** at fertilization establishes the major axes of the future embryo. **Review Figures 38.1 and 38.2**
- Fusion of egg and sperm nuclei produces a single-celled, diploid zygote.

CONCEPT 38.2 Cleavage Creates Building Blocks and Produces a Blastula

- **Cleavage** is a period of rapid cell division without cell growth. Usually, little if any gene expression occurs during cleavage. Cleavage can be complete or incomplete. It results in a ball or mass of cells called a **blastula**. **Review Figures 38.3 and 38.4**
- Cell membranes that form during cleavage segregate cytoplasmic determinants into different blastomeres.
- Some species undergo **mosaic development**, in which the fate of each cell is already determined at first division. Other species, including vertebrates, undergo **regulative development**, in which cell fates become determined later.
- **Radial cleavage** and **spiral cleavage** patterns are different. Both describe complete cleavage. Radial cleavage is the basal deuterostome pattern. Spiral cleavage occurs only in protostomes.
- The amount of yolk influences cleavage. Eggs with little yolk undergo complete cleavage. Very yolky eggs often undergo incomplete cleavage, so that part of the cytoplasm remains undivided.

SUMMARY (continued)

- Early cell divisions in eutherian mammals are unique. These cell divisions produce a **blastocyst** composed of an **inner cell mass** that will become the embryo and an outer layer of cells called the **trophoblast**. The trophoblast helps the embryo implant into the uterine wall. **Review Figure 38.8**

CONCEPT 38.3 Gastrulation Produces a Second, then a Third Germ Layer

- **Gastrulation** involves massive cell movements during which cells move from the outside to the inside of the embryo. It produces three germ layers and places cells from various regions of the blastula into new associations with one another. **Review Figure 38.9**
- The initial step of gastrulation is inward movement of certain blastomeres. The site of inward movement becomes the blastopore. Cells that move into the blastula become the endoderm and mesoderm; cells remaining on the outside become the ectoderm. **Review Figures 38.9 and 38.10 and ANIMATED TUTORIAL 38.1**
- Gastrulation in reptiles, including birds, differs from that in sea urchins and frogs because the blastula forms as a flattened disc of cells atop the yolk. **Review Figure 38.11**
- Although their eggs have very little yolk, placental mammals have a pattern of gastrulation similar to that of birds and other reptiles.

CONCEPT 38.4 Gastrulation Sets the Stage for Organogenesis and Neurulation in Chordates

- **Organogenesis** is the process whereby tissues interact to form organs and organ systems.
- The dorsal lip of the amphibian blastopore is a critical site for cell determination and has been called the **organizer**. **Review Figures 38.14 and 38.15 and ANIMATED TUTORIAL 38.2**
- In the formation of the chordate nervous system, one group of cells that migrates through the blastopore becomes the notochord. The notochord induces the overlying ectoderm to form the neural tube. **Review Figure 38.16**
- Preventing the differentiation of ectoderm into epidermis allows ectoderm to become neural tissue.
- **Neural crest cells** are released from the closing neural tube. Neural crest cells migrate to different regions and form many parts of the vertebrate body—the sensory nervous system, bones of the head, pigment cells, and other structures. **Review Figure 38.16**
- Mesoderm on either side of notochord forms **somites** that will give rise to muscle, bone, dermis and other tissues, **intermediate mesoderm** that will form urinary and reproductive tissues, and **lateral plate mesoderm** that encloses the coelom and gives rise to the heart, muscles of the digestive system, and other tissues. **Review Figures 38.16 and 38.17**

CONCEPT 38.5 Extraembryonic Membranes Protect and Nourish the Growing Embryo

- Amniotic eggs contain four **extraembryonic membranes**. In reptiles, including birds, the **yolk sac** surrounds the yolk and provides nutrients to the embryo. The **chorion** lines the eggshell and participates with the allantois in gas exchange. The **amnion** surrounds the embryo and encloses it in an aqueous environment. The **allantois** is a storage sac for metabolic wastes and participates with the chorion in gas exchange (as part of the chorioallantoic membrane).
- Only the yolk sac and the allantois can develop blood vessels.
- The first blood cells and blood vessels appear in the yolk sac. **Review Figure 38.18 and ACTIVITY 38.1**
- In eutherian mammals, the chorion and allantois interact with maternal uterine tissues to form a placenta, which provides the embryo with nutrients and gas exchange. The amnion encloses the embryo in an aqueous environment. **Review Figure 38.19**
- The yolk sac of fish is trilaminar, meaning it is composed of all three germ layers.

CONCEPT 38.6 Development Continues throughout Life

- Development continues throughout life. To attain adult size, organisms undergo **direct** or **indirect development**.
- Not all parts of a body grow at the same rate. Unequal growth that results in differential growth of one body part relative to another is called **allometry**. Most direct developers display allometric growth.
- Morphologies of adult and larval forms of the same species may be very different. Larval forms serve two major functions—feeding and/or species distribution. **Review Figure 38.22**

Go to the Interactive Summary to review key figures, Animated Tutorials, and Activities
PoL2e.com/is38

Go to LaunchPad at **macmillanhighered.com/launchpad** for additional resources, including LearningCurve Quizzes, Flashcards, and many other study and review resources.

39 Immunology: Animal Defense Systems

KEY CONCEPTS

39.1 Animals Use Innate and Adaptive Mechanisms to Defend Themselves against Pathogens

39.2 Innate Defenses Are Nonspecific

39.3 The Adaptive Immune Response Is Specific

39.4 The Adaptive Humoral Immune Response Involves Specific Antibodies

39.5 The Adaptive Cellular Immune Response Involves T Cells and Their Receptors

The siege of Athens by Sparta and its Peloponnesian allies made the city crowded and vulnerable to the spread of an infectious disease.

> As a rule, however, there was no ostensible cause; but people in good health were all of a sudden attacked by violent heats in the head, and redness and inflammation in the eyes, the inward parts, such as the throat or tongue, becoming bloody and emitting an unnatural and fetid breath.

The Athenian historian Thucydides wrote these words in 430 BCE. He was describing a rapidly spreading infectious disease, or plague, that was sweeping the city of Athens in ancient Greece, and would ultimately kill about one-third of its inhabitants.

One year earlier, in 431 BCE, the Spartans had surrounded Athens. Sparta was a rival state with a highly disciplined army, while Athens was near a port and had a powerful navy. The Athenian leader Pericles decided to take advantage of his naval supremacy. He instructed his soldiers to abandon the countryside to the Spartans, relying on the navy to keep the enemies out of the city and to keep the Athenians supplied with food. People living in the countryside poured into Athens.

As Athens became more crowded, deteriorating sanitation and close living conditions resulted in the city becoming an incubator for infectious diseases. In his famous *History of the Peloponnesian War*, Thucydides speculated that this disease came to Athens from Africa. His precise clinical description has provoked great debate among medical historians regarding its nature. In 1994 a burial ground was found near Athens that apparently contained the remains of several dozen people who died of the disease. Samples of dental pulp from the skeletons were examined using the polymerase chain reaction (PCR), and the results revealed the presence of DNA sequences from the bacterium that causes typhoid fever. Based on these results, it was proposed that the disease that killed so many Athenians was typhoid fever. Others disagree, using arguments based on a clinical description of typhoid fever today.

Whatever the cause, another passage from Thucydides (who was one of the few lucky survivors of the disease) is an interesting observation about immunity—the subject of this chapter:

> Yet it was with those who had recovered from the disease that the sick and the dying found most compassion. These knew what it was from experience and now had no fear for themselves; for the same man was never attacked twice, never at least fatally.

How can a person survive an infection and be resistant to further infection?

You will find the answer to this question on page 825.

CONCEPT 39.1 Animals Use Innate and Adaptive Mechanisms to Defend Themselves against Pathogens

Animals have a number of ways of defending themselves against **pathogens**—harmful organisms and viruses that can cause disease. These defense systems are based on the distinction between self—the animal's own molecules—and nonself, or foreign, molecules. Some defensive mechanisms are present all the time. For example, the skin is always present to protect a mammal from invaders. Other defenses are activated in response to invaders. These responses involve three phases:

- *Recognition phase*: The organism must be able to recognize pathogens and discriminate between self and nonself.
- *Activation phase*: The recognition event leads to a mobilization of cells and molecules to fight the invader.
- *Effector phase*: The mobilized cells and molecules destroy the invader.

There are two general types of defense mechanisms:

- **Innate defenses**, or nonspecific defenses, are inherited mechanisms that provide the first line of defense against pathogens. They include physical barriers such as the skin, molecules that are toxic to invaders, and phagocytic cells that ingest invaders. This system recognizes broad classes of organisms or molecules and gives a quick response, within minutes or hours. Some innate defenses are present all the time, whereas others are rapidly activated in response to an injury or invasion by a pathogen. All animals (and plants—see Chapter 28) have innate defenses.

LINK

Phagocytosis is a form of endocytosis in which a cell engulfs a large particle or another cell; see **Concept 5.4**

- **Adaptive defenses** are aimed at specific pathogens and are activated by the innate immune system. For example, cells in the adaptive defense system can make **antibodies**—proteins that will recognize, bind to, and aid in the destruction of specific pathogens, if the pathogens enter the body. Adaptive defenses are typically slow to develop and long-lasting. Adaptive defenses evolved in vertebrate animals.

Immunity occurs when an organism has sufficient defenses to successfully avoid biological invasion by a pathogen.

Innate defenses evolved before adaptive defenses

All animals have innate defenses against their enemies. For example, the crustacean *Tachypleus tridentatus*—the Japanese horseshoe crab—first appeared in the fossil record about 400 million years ago. It relies only on innate defenses. These defenses include barriers, defensive cells, and defensive molecules.

- Barriers include physical, chemical, and biological mechanisms for resisting infections (see Concept 39.2). The horseshoe crab has the hard exoskeleton that is characteristic of arthropods. This shell acts to protect the crab from invasion by pathogens.

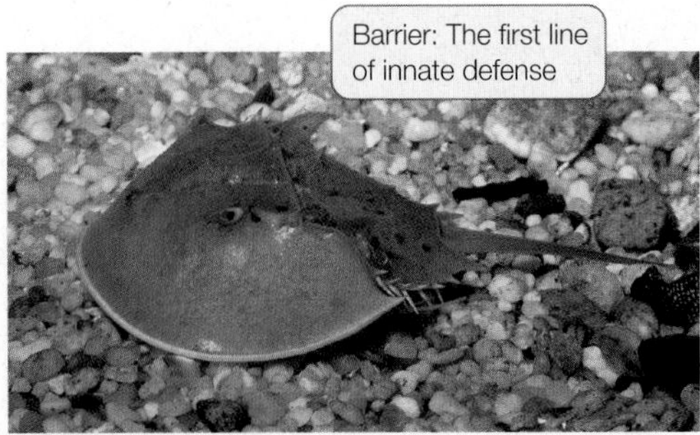

- Cells involved in innate defenses include phagocytes that bind to microbial pathogens, ingest them by endocytosis, and destroy their large molecules by hydrolysis. Amebocytes in the blood of the horseshoe crab fulfill this defensive role.

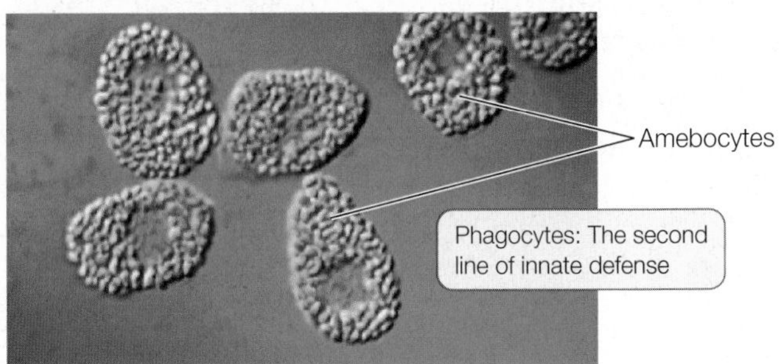

- Molecules that are toxic to invading pathogens are important in innate defense. The horseshoe crab has a wide array of such molecules that are released from cells in its blood. These molecules include peptides that disrupt the bacterial cell membrane, rendering it permeable; and peptides that bind to bacterial surfaces and cross-link them.

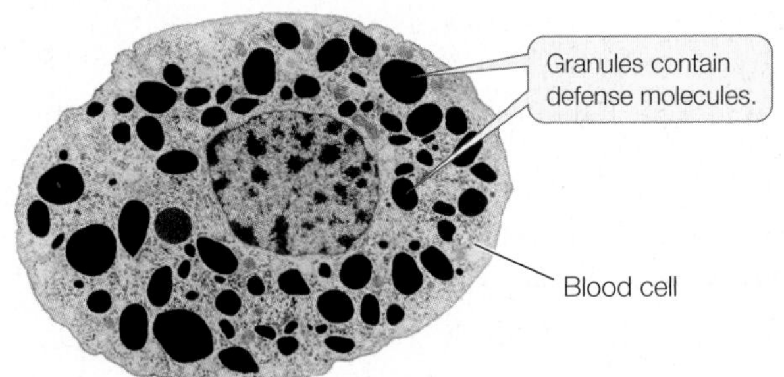

Studies of innate immunity, along with genome sequencing, have revealed that the recognition and activation phases of innate immunity evolved very early in animals. For example, animals as diverse as humans and fruit flies share a class of receptors, called Toll-like receptors, that participate in innate defense responses. In vertebrates, each Toll-like receptor recognizes and binds to a specific molecule that is found in a broad class of pathogens, such as a component of the bacterial cell wall. Binding sets off a signal transduction pathway that ends with the expression of genes for anti-pathogen molecules (**FIGURE 39.1**). This pathway exists in some form in many animal groups, including humans.

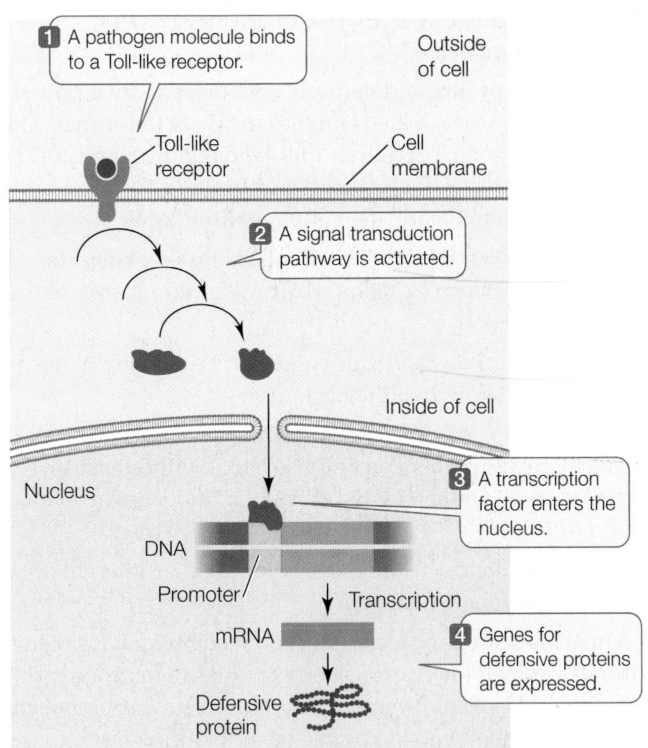

FIGURE 39.1 Cell Signaling and Innate Defense Binding of a pathogen molecule to a Toll-like receptor initiates a signal transduction pathway that results in the expression of genes involved in innate defense.

Mammals have both innate and adaptive defenses

In mammals and other vertebrates, the innate and adaptive defenses operate together, usually in sequence, as a coordinated defense system. We will focus on these defenses for the rest of the chapter. **TABLE 39.1** gives an overview of innate and adaptive defenses during the course of an infection. Innate defenses are the body's first line of defense; adaptive defenses often require days or even weeks to become effective.

The major players in adaptive immunity are specific cells and proteins. These are produced in the blood and lymphoid tissues and are circulated throughout the body, where they interact with almost all the other tissues and organs.

One milliliter of human blood typically contains about *5 billion* red blood cells and *7 million* white blood cells. Whereas the main function of red blood cells is to carry oxygen throughout the body, **white blood cells** (also called leukocytes) are specialized for various functions in the immune system (**FIGURE 39.2**). There are two major groups of white blood cells: phagocytes and lymphocytes. **Phagocytes** are large cells that engulf pathogens and other substances by phagocytosis. Some phagocytes are involved in both innate and adaptive immunity. In particular, macrophages and dendritic cells play key roles in communicating between the innate and adaptive immune systems. **Lymphocytes** include B cells and T cells, which are involved in adaptive immunity; and natural killer cells, which are involved in both innate and adaptive immunity.

TABLE 39.1 Innate and Adaptive Immune Defenses

Response (time after infection by a pathogen)	System	Mechanisms
Early (0–4 hr)	Innate, nonspecific (first line)	Barrier (skin and lining of organs) Dryness, low pH Mucus Lysozyme, defensins
Middle (4–96 hr)	Innate, nonspecific (second line)	Inflammation Phagocytosis Natural killer cells Complement system Interferons
Late (>96 hr)	Adaptive, specific	Humoral immunity (B cells, antibodies) Cellular immunity (T cells)

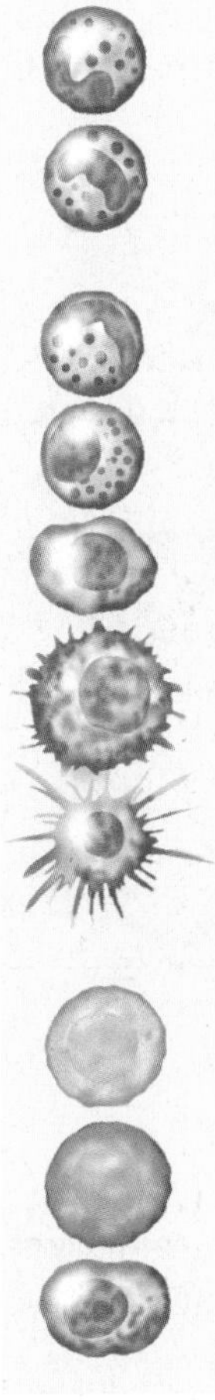

TYPE OF CELL	FUNCTION
Basophils (I)	Release histamine and other molecules involved in inflammation
Eosinophils (A)	Kill antibody-coated parasites
Phagocytes:	
Neutrophils (I)	Stimulate inflammation
Mast cells (I)	Release histamine
Monocytes (I, A)	Develop into macrophages and dendritic cells
Macrophages (I, A)	Antigen presentation
Dendritic cells (I, A)	Present antigens to T cells
Lymphocytes:	
B lymphocytes (A)	Differentiate to form antibody-producing cells and memory cells
T lymphocytes (A)	Kill pathogen-infected cells; regulate activities of other white blood cells
Natural killer cells (I, A)	Attack and lyse virus-infected or cancerous body cells

FIGURE 39.2 White Blood Cells White blood cells have key roles in both innate (I) and adaptive (A) immunity.

Go to ANIMATED TUTORIAL 39.1
Cells of the Immune System
PoL2e.com/at39.1

CHECKpoint CONCEPT 39.1

✓ Make a table that compares innate and adaptive immunity.

✓ If you compared the genomes of an invertebrate (e.g., a crab) and a human with regard to the presence of genes for immunity, what might you find?

Now that we've seen a brief overview of innate and adaptive immunity, let's look at some of the innate defenses that mammals have against invading organisms.

CONCEPT 39.2 Innate Defenses Are Nonspecific

Innate, nonspecific defenses are general protection mechanisms that attempt to either stop pathogens from invading the body or quickly eliminate pathogens that do manage to invade. They are genetically programmed (innate) and "ready to go," in contrast to adaptive, specific defenses, which take time to develop after a pathogen or toxin has been recognized as nonself. In mammals, innate defenses include physical barriers as well as cellular and chemical defenses (**FIGURE 39.3**).

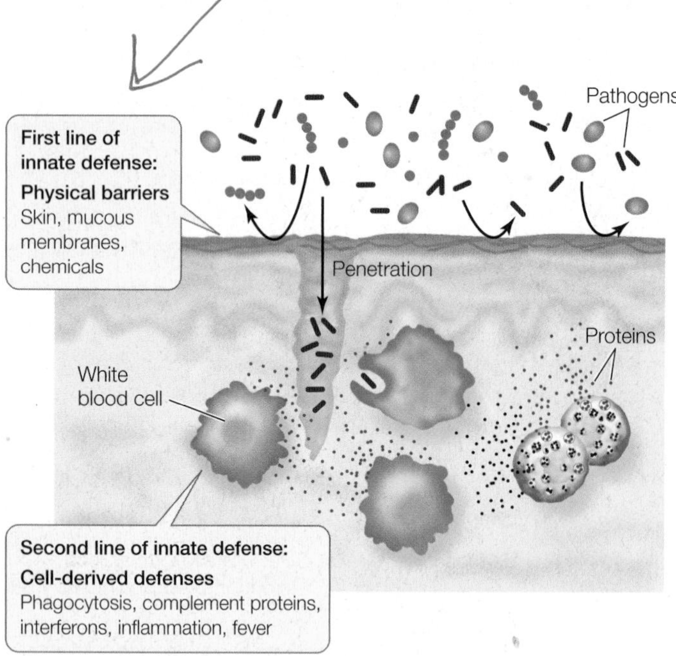

FIGURE 39.3 Innate Defenses Physical barriers, cells, and proteins (complement and interferons) provide nonspecific defenses against invading pathogens.

Go to MEDIA CLIP 39.1
The Chase Is On: Phagocyte versus Bacteria
PoL2e.com/mc39.1

Barriers and local agents defend the body against invaders

The first line of innate defense is encountered by a potential pathogen as soon as it lands on the surface of the animal. Consider a pathogenic bacterium that lands on human skin. The challenges (physical, chemical, and biological) faced by the bacterium just to invade the body are formidable:

- *The physical barrier of the skin*: Bacteria rarely penetrate intact skin (which explains why broken skin increases the risk of infection).
- *The saltiness of the skin*: This condition is usually not hospitable to the growth of the bacterium.
- *The presence of normal flora*: Bacteria and fungi that normally live in great numbers on body surfaces without causing disease will compete with potential pathogens for space and nutrients.

If a pathogen lands inside the nose or another internal organ, it faces other innate defenses:

- **Mucus** is a slippery secretion produced by mucous membranes, which line various body cavities that are exposed to the external environment. Mucus traps microorganisms so they can be removed by the beating of cilia (see Concept 4.4, pp. 74–75), which continuously move the mucus and its trapped debris toward the outside of the body.
- **Lysozyme**, an enzyme made by mucous membranes, cleaves bonds in the cell walls of many bacteria, causing them to lyse (burst open).
- **Defensins**, also made by mucous membranes, are peptides of 18–45 amino acids. They contain hydrophobic domains and are toxic to a wide range of pathogens, including bacteria, microbial eukaryotes, and enveloped (membrane-enclosed) viruses. Defensins insert themselves into the cell membranes of these organisms and make the membranes freely permeable to water and all solutes, thus killing the invaders:

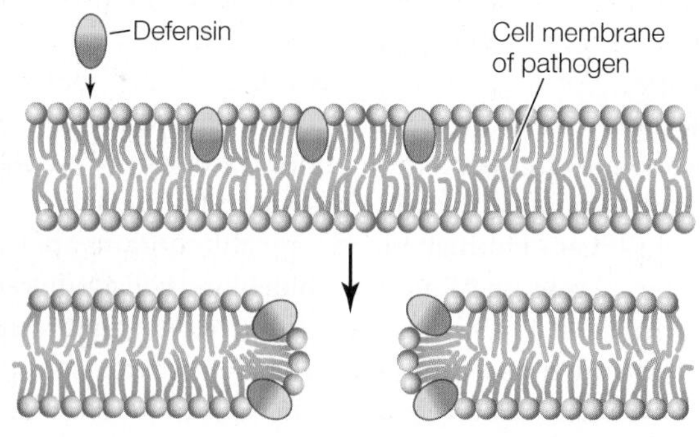

Defensins are also produced inside phagocytes, where they kill pathogens trapped by phagocytosis. Plants also produce molecules that kill pathogens (see Concept 28.1).

- Harsh conditions in the internal environment can kill pathogens. For example, the gastric juice in the stomach is a deadly environment for many bacteria because of the hydrochloric acid and proteases that are secreted into it.

Cell signaling pathways stimulate additional innate defenses

Pathogens that are able to penetrate the body's outer and inner surfaces encounter a second line of innate defenses. These include the activation of defensive cells. In some animals and plants, this activation involves recognition of nonself molecules by a class of receptors called **pattern recognition receptors** (**PRRs**). In mammals, these receptors are mainly present on phagocytes and natural killer cells. The Toll-like receptor shown in Figure 39.1 is an example. The molecules recognized by PRRs are called **pathogen associated molecular patterns** (**PAMPs**). These are molecules that are unique to large classes of microbes, such as bacterial flagellin and fungal chitin. Other well-known PAMPs include bacterial lipopolysaccharide, found in bacterial cell membranes, and nucleic acid variants that are unique to viruses, such as double-stranded RNA. Binding of PAMPs to PRRs stimulates signal transduction pathways that lead to a variety of responses. One response is the production of antimicrobial peptides such as defensins (see above). Other responses include phagocytosis of invading organisms, activation of natural killer cells, activation of the complement system, and production of cytokines.

PHAGOCYTOSIS Pathogenic cells, viruses, and virus-infected cells can be recognized by phagocytes, which then ingest the invaders by phagocytosis:

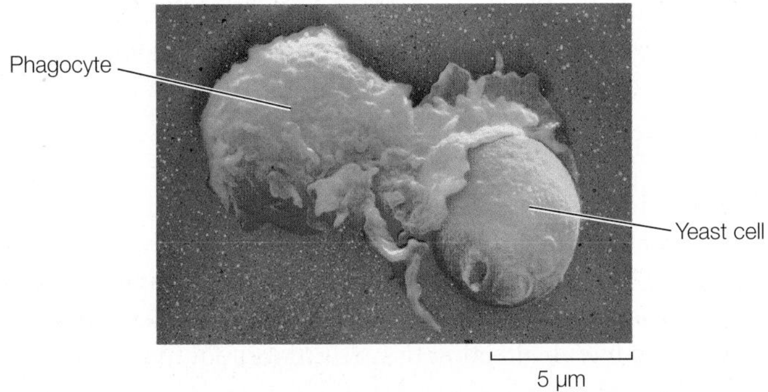

Once inside the phagocyte, the invader is killed either by hydrolysis within the lysosomes (see Figure 4.9) or by defensins.

NATURAL KILLER CELLS One class of lymphocytes, known as **natural killer cells**, can distinguish between healthy body cells and those that are infected by viruses or have become cancerous. In the latter cases they initiate apoptosis (programmed cell death; see Concept 7.5) in the target cells. In addition to this nonspecific action, natural killer cells interact with the adaptive defense mechanisms by lysing target cells that have been tagged with antibodies.

COMPLEMENT PROTEINS Vertebrate blood contains more than 20 different proteins that make up the antimicrobial complement system. Once this system has been activated, the proteins function in a characteristic sequence, or cascade, with each protein activating the next:

1. One complement protein binds to components on the surface of an invading cell. This binding helps phagocytes recognize and destroy the invading cell.
2. Another protein activates the inflammation response (see below) and attracts phagocytes to the site of infection.
3. Finally, other proteins lyse the invading cell.

The complement system can be activated by various mechanisms, including the innate and adaptive immune systems. In the latter case, the complement cascade is initiated by the binding of a complement protein to an antibody bound to the surface of the invading cell.

CYTOKINES **Cytokines** are signaling proteins that are released by many cell types. Cytokines that are produced in response to PRR activation include inflammatory cytokines (some interleukins and tumor necrosis factor; see below) and **interferons**, which help increase the resistance of neighboring cells to infection. Interferons are found in many vertebrates, and their production is induced by a variety of molecules. One such inducer is double-stranded viral RNA; thus interferons are particularly important in defense against viruses. Interferons bind to receptors on the cell membranes of uninfected cells, stimulating a signaling pathway that inhibits viral reproduction if the cells are subsequently infected. In addition, interferons stimulate the cells to hydrolyze bacterial or viral proteins to peptides, an initial step in adaptive immunity.

Inflammation is a coordinated response to infection or injury

When a tissue is damaged because of infection or injury, the body responds with **inflammation**: redness, swelling, and heat near the damaged site, which can then become painful. This response can happen almost anywhere in the body, internally as well as on the outer surface. Inflammation is an important phenomenon: it isolates the area to stop the spread of the damage; it recruits cells and molecules to the damaged location to kill any pathogens that might be present; and it promotes healing.

Among the first responders to tissue damage are **mast cells**, which adhere to the skin and the linings of organs and release numerous chemical signals, including:

- **Tumor necrosis factor**, a cytokine protein that kills target cells and activates immune cells
- **Prostaglandins**, fatty acid derivatives that play roles in various responses, including the initiation of inflammation in nearby tissues
- **Histamine**, an amino acid derivative that increases the permeability of blood vessels to white blood cells and molecules so they can act in nearby tissues

The redness and heat of inflammation result from the dilation of blood vessels in the infected or injured area (**FIGURE 39.4**). Phagocytes enter the inflamed area, where they engulf pathogens and dead tissue cells. Phagocytes are responsible for most of the healing associated with inflammation. They produce several cytokines, which (among other functions) can signal the brain to produce a fever. This rise in body temperature accelerates lymphocyte production and phagocytosis, thereby speeding the immune response. In some cases pathogens are temperature-sensitive, and their growth is inhibited by the

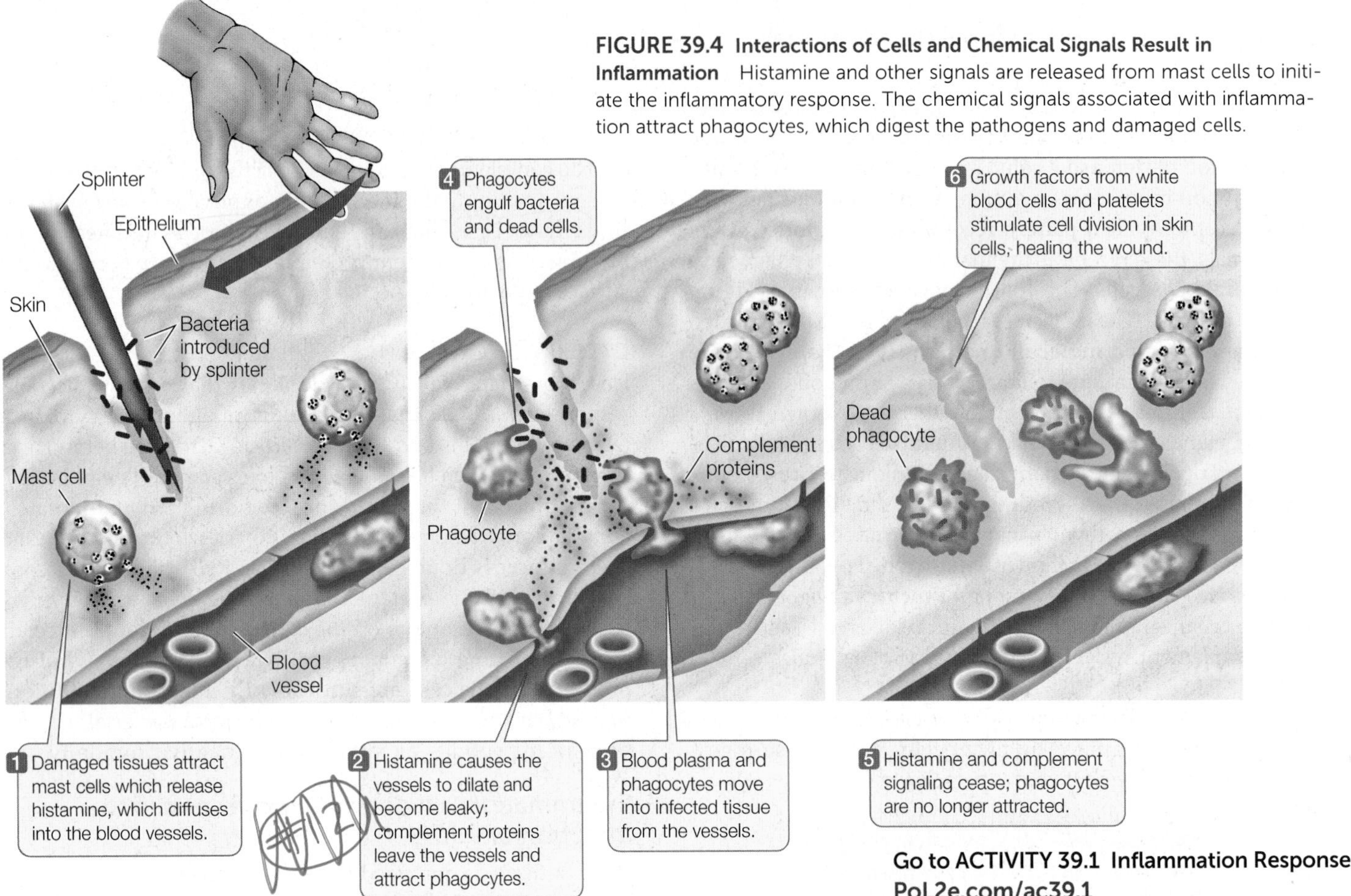

Splinter
Epithelium
Skin
Bacteria introduced by splinter
Mast cell
Blood vessel
4 Phagocytes engulf bacteria and dead cells.
Phagocyte
Complement proteins
6 Growth factors from white blood cells and platelets stimulate cell division in skin cells, healing the wound.
Dead phagocyte
1 Damaged tissues attract mast cells which release histamine, which diffuses into the blood vessels.
2 Histamine causes the vessels to dilate and become leaky; complement proteins leave the vessels and attract phagocytes.
3 Blood plasma and phagocytes move into infected tissue from the vessels.
5 Histamine and complement signaling cease; phagocytes are no longer attracted.

FIGURE 39.4 Interactions of Cells and Chemical Signals Result in Inflammation Histamine and other signals are released from mast cells to initiate the inflammatory response. The chemical signals associated with inflammation attract phagocytes, which digest the pathogens and damaged cells.

Go to ACTIVITY 39.1 Inflammation Response
PoL2e.com/ac39.1

fever. The pain of inflammation results from increased pressure due to swelling, the action of leaked enzymes on nerve endings, and the action of prostaglandins, which increase the sensitivity of the nerve endings to pain. (Recall that in Chapter 3 we described aspirin and how it alleviates pain by blocking the synthesis of prostaglandins.)

Following inflammation, pus may accumulate. Pus is a mixture of leaked fluid and dead cells: bacteria, white blood cells, and damaged body cells. Pus is a normal result of inflammation, and is gradually consumed and further digested by macrophages. If there is a wound, platelets—small, irregularly shaped cell fragments that are present in the blood—aggregate at the wound site. The platelets produce growth factors that stimulate nearby skin cells to divide and heal the wound.

Inflammation can cause medical problems

Although inflammation is generally a good thing, sometimes the inflammatory response is inappropriately strong:

- In an **allergic** reaction, a nonself molecule that is normally harmless binds to mast cells, causing the release of histamine and subsequent inflammation (along with itchy, watery eyes, and rashes in some cases). The nonself molecule may come from food or from the environment—for example, from the surface of a plant pollen grain (as in hay fever) or a particle in dust.
- In **autoimmune diseases** such as rheumatoid arthritis, the immune system fails to distinguish between self and nonself, and attacks tissues in the organism's own body. In these cases the inflammation can be local, attacking an organ, or general, affecting tissues throughout the body.
- In **sepsis**, the inflammation that is due to a bacterial infection does not remain local. Instead it becomes widespread, with the dilation of blood vessels throughout the body. The resulting drop in blood pressure is a medical emergency and can be lethal.

CHECKpoint CONCEPT 39.2

✓ Outline the sequence of innate defenses encountered by a pathogenic bacterial cell if it is ingested in food.

✓ Antihistamines are used to treat the symptoms, such as sneezing, that are due to inflammation caused by irritants in the airways. How do you think antihistamines might work?

✓ A massive inflammation caused by a food allergy can be treated with an injection of epinephrine. Refer to the description of epinephrine's effects in Concept 35.2. How do you think this hormone relieves the inflammation symptoms?

In most instances innate immunity is sufficient to block a pathogen from affecting the body. But many pathogens are present in huge numbers (think of a viral infection), and some may escape the innate defenses and begin to proliferate in the body. In these cases, adaptive immunity takes over.

CONCEPT 39.3 The Adaptive Immune Response Is Specific

Long before the twentieth century, scientists suspected that blood was somehow involved in immunity against pathogens, but they did not have definitive experimental evidence for this. More than a century ago, Emil von Behring and Shibasaburo Kitasato at the University of Marburg in Germany performed a key experiment that suggested that blood contained important factors for adaptive immunity (**FIGURE 39.5**). They showed that guinea pigs injected with a sublethal dose of diphtheria toxin developed in their blood serum (the noncellular fluid that remains after blood is clotted) a factor that, when injected into other guinea pigs, was able to protect them from a lethal dose of the same toxin. In other words, the serum recipients had somehow acquired immunity. The response of the donor guinea pigs to the diphtheria toxin is an example of adaptive immunity: after exposure to the toxin, they made a protective factor that was not present before their exposure. Moreover, the immunity was specific: the factor made by the first guinea pig protected the others only against the specific toxin produced by the strain of bacteria with which the first guinea pig had been injected.

Based on the animal model, Behring realized that serum protection might work for human diseases as well. It did, and he won the Nobel Prize for his efforts in protecting children against diphtheria. Later, the agent of this immunity was identified as an antibody protein, and the process of acquiring immunity from antibodies received from another individual was called passive immunity.

LINK

Many bacteria produce toxins of various types; see **Concept 19.3**

The injection of antibodies to produce a passive immune response is an artificial system that allows an organism to mount a defense against a single pathogenic strain. Vertebrates have evolved systems of adaptive immunity that have the potential to recognize and eliminate

INVESTIGATION

FIGURE 39.5 The Discovery of Specific Immunity Until the twentieth century, most people did not survive an attack of diphtheria, but a few did. Emil von Behring and Shibasaburo Kitasato performed a key experiment using an animal model. They demonstrated that the factor(s) responsible for immunity against diphtheria was in the blood serum.[a]

HYPOTHESIS

Serum from guinea pigs injected with a sublethal dose of diphtheria toxin protects other guinea pigs that are exposed to a lethal dose of the same toxin.

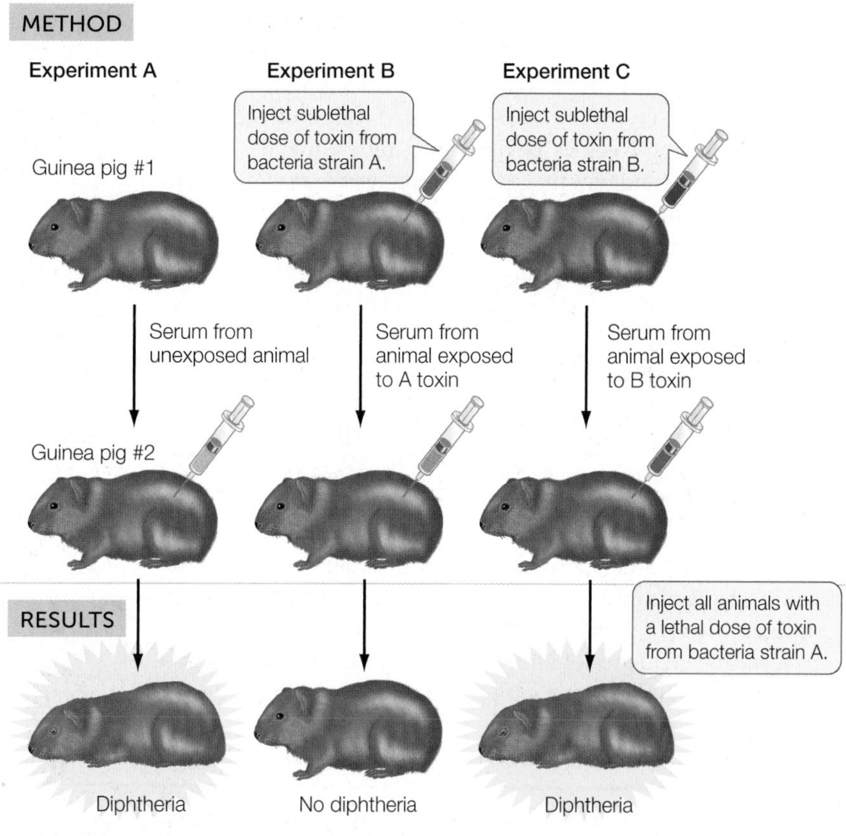

CONCLUSION

Serum of toxin-exposed guinea pigs is protective against later exposure to a lethal dose of toxin from the same genetic strain of bacteria, but not a different strain.

ANALYZE THE DATA

In their experiments, Behring and Kitasato used different doses of toxin to test the immunity of guinea pig #2 in each experiment. The results are shown in the table.

	Symptoms	
Experiment	0.5 ml dose	10 ml dose
A	Diphtheria	Diphtheria
B	No diphtheria	No diphtheria
C	Diphtheria	Diphtheria

A. Explain these data in terms of the level of protection afforded by the serum.

B. These experiments could be performed with either intact bacteria that cause diphtheria, or with a bacteria-free filtrate of a 10-day-old culture of the bacteria. Explain.

Go to **LaunchPad** for discussion and relevant links for all **INVESTIGATION** figures.

[a]E. von Behring and S. Kitasato. 1890. *Deutsche medizinisch Wochenschrift* 16: 1113–1114.

virtually any pathogen that may infect the body. We will now turn to some key features of adaptive immunity.

Adaptive immunity has four key features

Four important features of the adaptive immune system are: specificity; the ability to distinguish self from nonself; the ability to respond to an enormous diversity of nonself molecules; and immunological memory.

SPECIFICITY B and T lymphocytes are crucial for the specificity of adaptive immune responses. T cell receptors and the antibodies produced by B cells recognize and bind to specific nonself substances called **antigens**, and this interaction initiates a specific immune response. Each T cell and each antibody-producing B cell is specific for a single antigen.

The sites on antigens that the immune system recognizes are called **antigenic determinants** or **epitopes**:

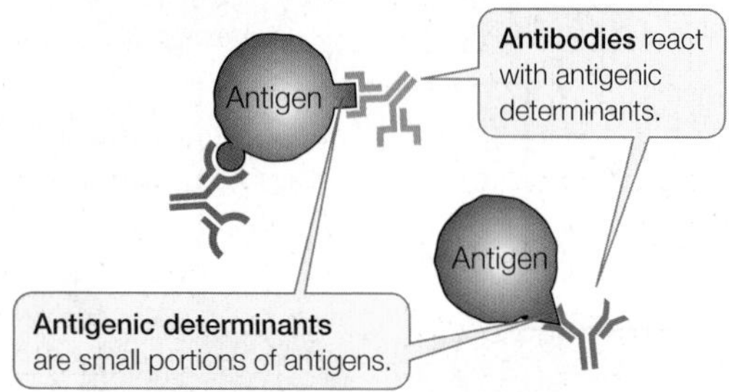

Antigens are usually proteins or polysaccharides, and there can be multiple antigens on a single invading bacterium. An antigenic determinant is a specific portion of an antigen, such as a certain sequence of amino acids that may be present in a protein. A single antigenic molecule can have multiple different antigenic determinants. For the remainder of the chapter, we will often refer to antigenic determinants simply as "antigens."

DISTINGUISHING SELF FROM NONSELF As we saw in Concept 39.2, the innate immune system can recognize certain nonself molecules (PAMPs), and mount defensive responses to the pathogens from which these nonself molecules were derived. The adaptive immune system recognizes both self and non-self molecules. The human body contains tens of thousands of different molecules, each with a specific three-dimensional structure capable of generating an immune response. Thus every cell in the body bears a tremendous number of antigens. A crucial requirement of an individual's immune system is that it recognize the body's own antigens and not attack them.

Normally, the body is tolerant of its own molecules—the same molecules that would generate an immune response in another individual. This occurs primarily during the early differentiation of T and B cells, when they first encounter self antigens. Any immature B or T cell that shows the potential to mount a strong immune response against self antigens undergoes apoptosis within a short time. This process is called **clonal deletion**.

A failure of clonal deletion can lead to an immune response within an individual to self antigens, or **autoimmunity**. Autoimmunity does not always result in disease, but a number of autoimmune diseases are common:

- People with systemic lupus erythematosis (SLE) have antibodies to many cellular components. These antibodies can cause serious damage when they bind to normal tissue antigens and form large circulating antigen–antibody complexes. These can become stuck in tissues and provoke inflammation.
- Hashimoto's thyroiditis is the most common autoimmune disease in women over age 50. Immune cells attack thyroid tissue, resulting in fatigue, depression, weight gain, and other symptoms associated with impaired thyroid function.

DIVERSITY Pathogens take many forms: viruses, bacteria, protists, fungi, and multicellular parasites. Furthermore, each pathogenic species usually exists as many subtly different genetic strains, and each strain possesses multiple surface features. Estimates vary, but a reasonable guess is that humans can respond specifically to 10 million different antigens. Upon recognizing an antigen, the immune system responds by activating lymphocytes of the appropriate specificity.

To have the ability to respond to the large number of potential pathogens, the body needs to generate a vast diversity of lymphocytes that are specific for different antigens. These lymphocytes represent a pool from which specific cells are selected when needed.

- *Diversity is generated primarily by DNA changes*—chromosomal rearrangements and other mutations—that occur just after the B and T cells are formed in the bone marrow. Each B cell is able to produce only one kind of antibody; thus there are millions of different B cells. Similarly, there are millions of different T cells with specific T cell receptors. The adaptive immune system is "predeveloped"—*all of the machinery available to respond to an immense diversity of antigens is already there, even before the antigens are encountered.*
- *Antigen binding "selects" a particular B or T cell for proliferation*. For example, when an antigen fits the surface receptor on a B cell and binds to it, that B cell is activated. It divides to form a clone of cells (a genetically identical group derived from a single cell), all of which produce and/or secrete antibodies with the same specificity as the receptor (**FIGURE 39.6**). Binding, activation, and proliferation also apply to T lymphocytes. A particular lymphocyte is selected via binding and activation, and then it proliferates to generate a clone—hence the name **clonal selection** for this mechanism of producing an immune response.

IMMUNOLOGICAL MEMORY After responding to a particular type of pathogen once, the immune system "remembers" that pathogen and can usually respond more rapidly and powerfully to the same threat in the future. This **immunological memory** usually saves us from repeats of childhood infectious diseases.

The first time a vertebrate animal is exposed to a particular antigen, it takes several days before the adaptive immune system produces antibodies and T cells specific for the antigen. This is the **primary immune response**. Immunological memory arises from the fact that activated lymphocytes divide and

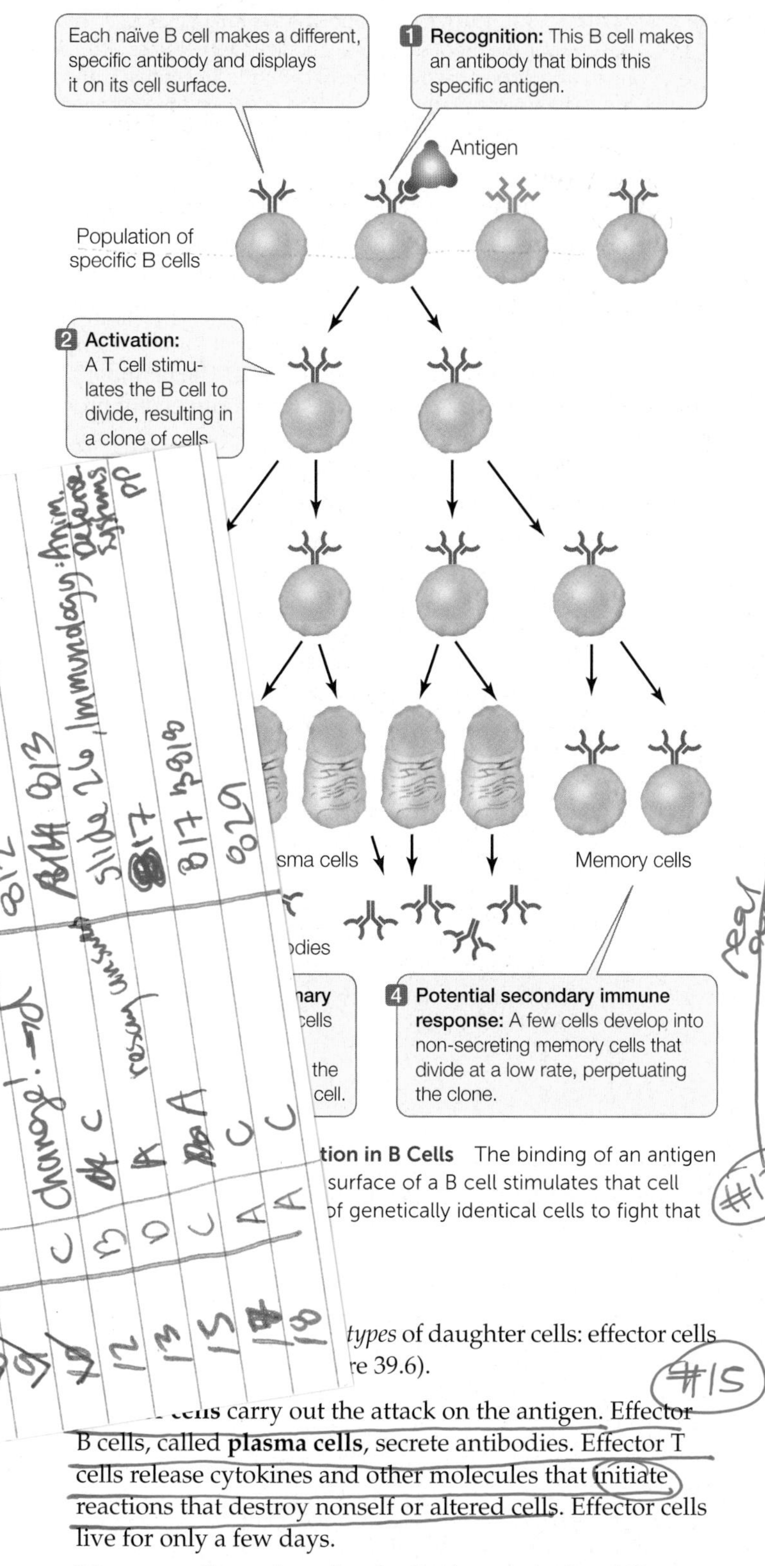
Each naïve B cell makes a different, specific antibody and displays it on its cell surface.
1 Recognition: This B cell makes an antibody that binds this specific antigen.
Antigen
Population of specific B cells
2 Activation: A T cell stimulates the B cell to divide, resulting in a clone of cells
Memory cells
4 Potential secondary immune response: A few cells develop into non-secreting memory cells that divide at a low rate, perpetuating the clone.

…tion in B Cells The binding of an antigen … surface of a B cell stimulates that cell … of genetically identical cells to fight that …

… *types* of daughter cells: effector cells … re 39.6).

- … cells carry out the attack on the antigen. Effector B cells, called **plasma cells**, secrete antibodies. Effector T cells release cytokines and other molecules that initiate reactions that destroy nonself or altered cells. Effector cells live for only a few days.
- **Memory cells** are long-lived cells that retain the ability to start dividing on short notice to produce more effector and more memory cells. Memory B and T cells may survive in the body for decades, rarely dividing.

After a primary immune response to a particular antigen, subsequent encounters with the same antigen will trigger a much more rapid and powerful **secondary immune response**. The memory cells that bind with that antigen proliferate, launching a huge army of plasma cells and effector T cells. The principle behind vaccination is to trigger a primary immune response that prepares the body to mount a stronger, quicker secondary response if it encounters the actual pathogen.

Macrophages and dendritic cells play a key role in activating the adaptive immune system

We have just seen how the adaptive immune system uses antigens to distinguish between self and nonself, and how it uses clonal selection to produce large numbers of specific B and T cells. In Concepts 39.4 and 39.5 we will focus on how these cells function to eliminate specific pathogens from the body. But how is the adaptive immune system activated in the first place?

One mechanism for eliminating pathogens in the innate immune system is phagocytosis. After ingestion of a pathogenic organism or infected host cell, phagocytic cells display fragments of the pathogen on their cell surfaces. These fragments function as antigens, and **antigen presentation** is one way that components of the innate immune system communicate with the adaptive immune system. Macrophages and dendritic cells play a key role in activating the adaptive immune system. After engulfing pathogens or infected host cells, these cells migrate to lymph nodes, where they present antigen to immature T cells. In addition, the antigen-presenting cells secrete cytokines and other signals that stimulate the activation and differentiation of the T cells.

Two types of adaptive immune responses interact

The adaptive immune system mounts two types of responses against invaders: the **humoral immune response** and the **cellular immune response**. These two responses work simultaneously and cooperatively, sharing many mechanisms. We will use the example of a viral infection in an overview of these two types of responses (**FIGURE 39.7**). These responses also occur when bacteria or other pathogens infect and grow inside host cells.

B cells that make antibodies are the workhorses of the humoral immune response, and **cytotoxic T (T_C) cells** are the workhorses of the cellular immune response. There are three phases in both types of immune response:

- *Recognition phase:* In both cellular and humoral immunity, recognition occurs when an antigen is inserted into the cell membrane of an antigen-presenting cell, with the unique antigen structure protruding from the cell membrane. In addition to dendritic cells and macrophages, developing B cells can also perform phagocytosis and function as antigen-presenting cells. Figure 39.7 shows a virus-infected host cell being engulfed by an antigen-presenting cell; in other cases, the free viral particles may be engulfed. The antigen on the surface of the antigen-presenting cell is recognized by a **T-helper (T_H) cell** bearing a T cell receptor protein that is specific for the antigen. In both humoral and cellular immunity, binding initiates the activation phase.
- *Activation phase:* When the T_H cell recognizes an antigen on an antigen-presenting cell, it propagates and releases cytokines that stimulate B cells and T_C cells bearing receptors

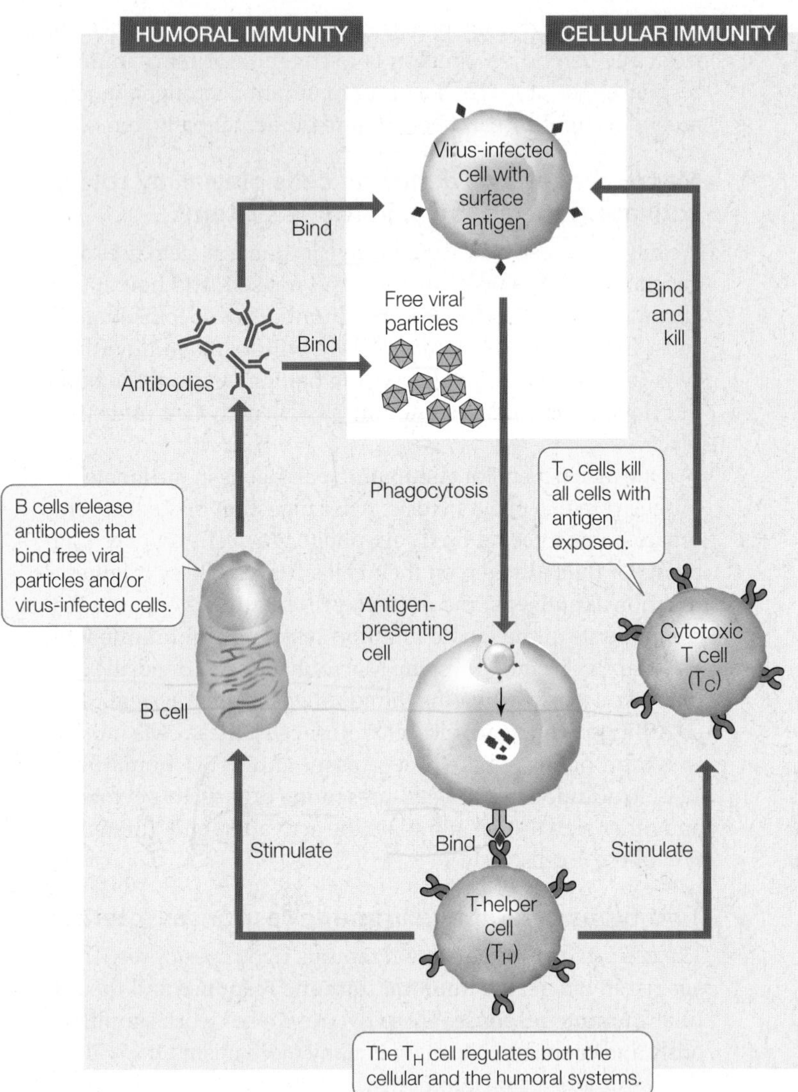

FIGURE 39.7 The Adaptive Immune System Humoral immunity involves the production of antibodies by B cells. Cellular immunity involves the activation of cytotoxic T cells that bind to and destroy self cells that are mutated or infected by pathogens. Details of these processes are described later in this chapter.

to the same antigen to divide. The results are a clone of B cells that function in humoral immunity and a clone of T_C cells that function in cellular immunity.

- *Effector phase:* In the humoral immune response, cells of the B clone produce antibodies that bind to viral particles and/or virus-infected cells. The bound antibodies attract phagocytes and complement proteins that engulf and destroy the virus and the virus-infected cells. In cellular immunity, cells of the T_C clone bind to virus-infected cells and destroy them.

CHECKpoint CONCEPT 39.3

- ✓ List the four key features of the adaptive immune system.
- ✓ In 2009, the H1N1 strain of influenza was a worldwide epidemic. Notably, very old people, who had been alive during the 1918 flu outbreak, had low rates of H1N1 infection. Explain this in terms of immunological memory.
- ✓ Insulin-dependent (type 1) diabetes results from a destruction of the cells in the pancreas that make the hormone insulin (see Concept 30.5). One hypothesis for this disease is that it is caused by an autoimmune reaction. Explain how this might happen and how you would investigate your hypothesis.

APPLY THE CONCEPT

The adaptive immune response is specific

In humans, the dominant gene *D* codes for the RhD (Rhesus) protein on the surfaces of red blood cells. The recessive allele *d* does not encode a cell surface protein. If cells from an individual who is *DD* or *Dd* (Rh^+) enter the bloodstream of an individual who is *dd* (Rh^-), the Rh^- individual makes antibodies that react with the Rh^+ cells. During development, the blood vessels of a mother and her fetus are near each other but do not mix. However, antibodies can pass through capillary walls between the two blood supplies.

MOTHER		FETUS		
GENOTYPE	Rh PHENOTYPE	GENOTYPE	Rh PHENOTYPE	RESULT
Dd		*Dd*		
dd		*Dd*		

1. Hemolytic disease of the newborn (sometimes resulting in stillbirth) can occur when antibodies from the mother pass into the fetal circulatory system. Fill in the table, which predicts RhD incompatibility.
2. During birth there is mixing of the blood supplies of mother and fetus. In a situation of RhD incompatibility, the first birth does not result in a clinical problem. But a second RhD incompatible birth with this mother results in a hemolytic disease. Explain how this can happen (hint: immunological memory).
3. Today, RhD incompatibility does not result in stillbirths. The mother is treated with serum containing anti-RhD antibodies late in the first and subsequent pregnancies. How can this prevent the immunological reactions?

With this overview of the basic characteristics of adaptive immunity and the events of the immune response, we will now turn to some of the cellular and molecular details of how these events occur.

CONCEPT 39.4 The Adaptive Humoral Immune Response Involves Specific Antibodies

Every day in the human body, billions of B cells survive the test of clonal deletion and are released from the bone marrow into the circulatory system. B cells are the basis for the humoral immune response.

Go to ANIMATED TUTORIAL 39.2 Humoral Immune Response
PoL2e.com/at39.2

Plasma cells produce antibodies that share a common overall structure

A B cell begins as a "naïve" B cell with a receptor protein on its cell surface that is specific for a particular antigen. The cell is activated by antigen binding to this receptor, and after stimulation by a T_H cell, it gives rise to a clone of plasma cells that make antibodies as well as to a smaller number of memory cells (see Figure 39.6). The stimulation occurs after the B cell presents antigen to a T_H cell with a receptor that can recognize the antigen. The T_H cell then secretes cytokines that stimulate the B cell to divide. This process is described in more detail in Concept 39.5.

All the plasma cells arising from a given B cell produce antibodies that are specific for the antigen that originally bound to the parent B cell. Thus antibody specificity is maintained as B cells proliferate.

There are several classes of antibodies, also called **immunoglobulins**, but all contain a tetramer with four polypeptide chains (**FIGURE 39.8**). In each immunoglobulin molecule, two of the polypeptides are identical light chains, and two are identical heavy chains. Disulfide bridges (see Chapter 3, p. 44) hold the chains together. Each polypeptide chain has a constant region and a variable region:

- The amino acid sequence of the **constant region** determines the general structure and function (the class) of an immunoglobulin. All immunoglobulins in a particular class have a similar constant region. When an antibody acts as a B cell receptor, the constant region is inserted into the cell membrane, anchoring the antibody to the cell surface.
- The amino acid sequence of the **variable region** is different for each specific immunoglobulin. The three-dimensional shape of the variable region's antigen-binding site is responsible for antibody specificity.

The two antigen-binding sites on each immunoglobulin molecule are identical, making the antibody bivalent (*bi*, "two"; *valent*, "binding"). This ability to bind two antigen molecules at once, along with the existence of multiple epitopes (antigenic determinants) on each antigen, permits antibodies to form large complexes with the antigens. For example, one antibody might bind two molecules of an antigen. Another antibody might bind the same antigen at a different epitope or to an antigen that is already bound to the first antibody, along with a third antigen molecule:

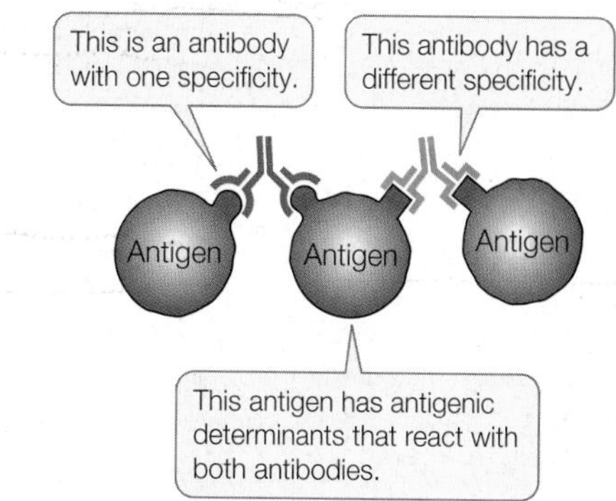

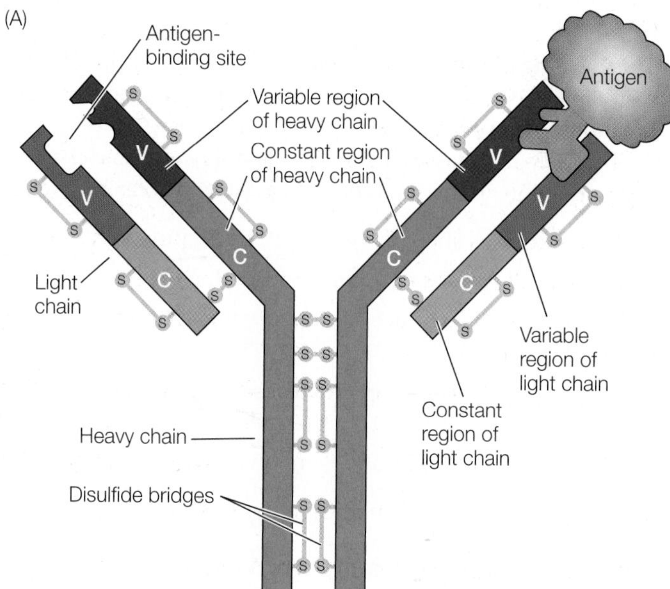

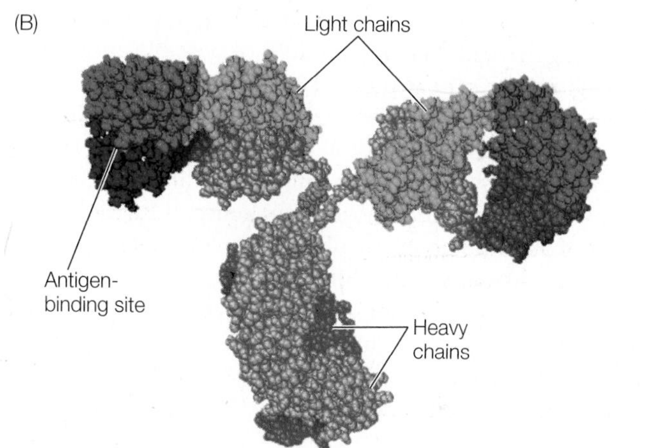

FIGURE 39.8 The Structure of an Immunoglobulin Four polypeptide chains (two light, two heavy) make up an immunoglobulin molecule. Here we show both diagrammatic (A) and space-filling (B) representations of an immunoglobulin.

Go to ACTIVITY 39.2 Immunoglobulin Structure
PoL2e.com/ac39.2

This binding of multiple antigens and multiple antibodies can result in large complexes that are easy targets for ingestion and breakdown by phagocytes.

There are five classes of immunoglobulins (Ig), and they differ in function and in the type of heavy chain constant region:

- IgG constitutes about 80 percent of circulating antibodies.
- IgD is a cell surface receptor on a B cell.
- IgM is the initial surface and circulating antibody released by a B cell.
- IgA protects mucous membranes exposed to the environment.
- IgE binds to mast cells and is involved with inflammation.

We will focus on IgG antibodies in this chapter.

Antibody diversity results from DNA rearrangements and other mutations

Each mature B cell makes antibodies targeted to only one single antigen. And there are millions of possible antigens to which a human can be exposed. How can the genome encode enough different antibodies to protect the body against all the possible pathogens? Although there are millions of possible amino acid sequences in immunoglobulins, there are not millions of different immunoglobulin genes. It turns out that instead of a single gene encoding each complete immunoglobulin, the genome of the differentiating B cell has multiple different coding regions for each domain of the protein, and diversity is generated by putting together different combinations of these regions. Shuffling of this genetic deck generates the enormous immunological diversity that characterizes each individual mammal.

Each gene encoding an immunoglobulin chain is in reality a "supergene" assembled by means of genetic recombination from several clusters of smaller genes scattered along part of a chromosome (**FIGURE 39.9**). Every cell in the body has hundreds of immunoglobulin genes located in separate clusters that are potentially capable of participating in the synthesis of both the variable and constant regions of immunoglobulin chains. In most body cells and tissues, these genes remain intact and separated from one another. But during B cell development, these genes are cut out, rearranged, and joined together in DNA recombination events. One gene from each cluster is chosen randomly for joining, and the others are deleted. In the case of one multigene set, the *J* genes, some of the extra sequences are removed by RNA splicing (**FIGURE 39.10**).

In this manner, a unique immunoglobulin supergene is assembled from randomly selected "parts." Each B cell precursor assembles two supergenes, one for a specific heavy chain and the other, assembled independently, for a specific light chain. This remarkable example of irreversible cell differentiation generates an enormous diversity of immunoglobulins from the same genome. It is a major exception to the generalization that all somatic cells derived from the fertilized egg have identical DNA.

Figure 39.9 illustrates the gene families that encode the constant and variable regions of the heavy chain in mice. There are multiple genes that encode each of the three parts of the variable region: 100 *V*, 30 *D*, and 6 *J* genes. Each B cell randomly selects one gene from each of these clusters to make the final coding sequence (*VDJ*) of the heavy-chain variable region. So the number of different heavy chains that can be made through this random recombination process is quite large:

100 *V* × 30 *D* × 6 *J* = 18,000 possible combinations

Now consider that the light chains are similarly constructed, with a similar amount of diversity made possible by random recombination. If we assume that the degree of potential light-chain diversity is the same as that for heavy-chain diversity, the number of possible combinations of light- and heavy-chain variable regions is:

18,000 different light chains × 18,000 different heavy chains = 324 million possibilities!

Other mechanisms generate even more diversity:

- When the DNA sequences that encode the *V*, *D*, and *J* regions are rearranged so that they are next to one another, the recombination event is not precise, and errors occur at the junctions. This imprecise recombination can create frameshift mutations, generating new codons at the junctions, with resulting amino acid changes.
- After the DNA sequences are cut and before they are rejoined, the enzyme terminal transferase often adds some nucleotides to the free ends of the DNA pieces. These additional bases create insertion mutations.

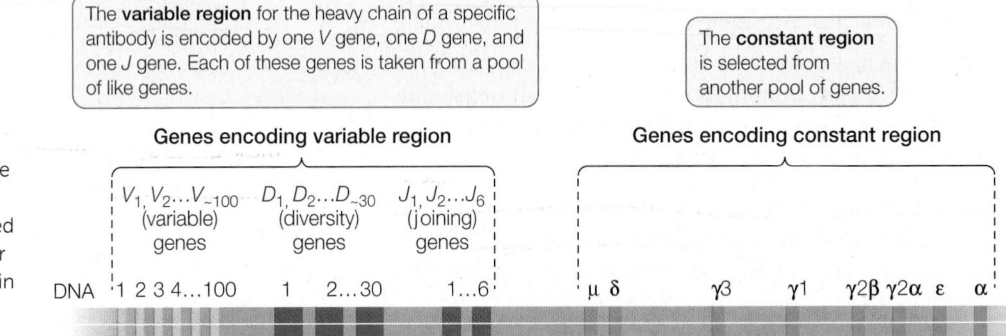

FIGURE 39.9 Super Genes Mouse immunoglobulin heavy chains have four domains, each of which is coded for by a gene selected from a cluster of similar genes. The immunoglobulin protein has one domain from each cluster.

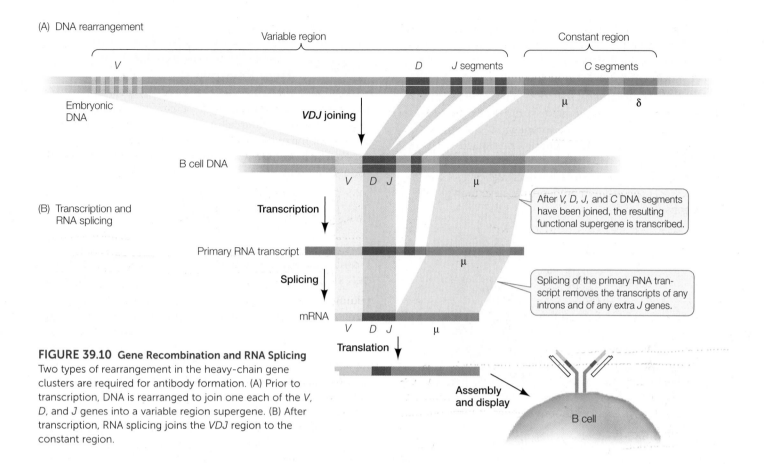

FIGURE 39.10 Gene Recombination and RNA Splicing Two types of rearrangement in the heavy-chain gene clusters are required for antibody formation. (A) Prior to transcription, DNA is rearranged to join one each of the *V*, *D*, and *J* genes into a variable region supergene. (B) After transcription, RNA splicing joins the *VDJ* region to the constant region.

- There is a relatively high spontaneous mutation rate in immunoglobulin genes. Once again, this process creates many new alleles and adds to antibody diversity.

LINK

The effects of DNA mutations on the amino acid sequences of proteins are described in **Concept 10.3**

When we include these possibilities with the millions of combinations that can be made by random DNA rearrangements, it is not surprising that the immune system can mount a response to almost any natural or artificial substance.

Once the DNA rearrangements are completed, each supergene is transcribed and then translated to produce an immunoglobulin light chain or heavy chain. These chains combine to form an active immunoglobulin protein.

Go to ANIMATED TUTORIAL 39.3
A B Cell Builds an Antibody
PoL2e.com/at39.3

Antibodies bind to antigens and activate defense mechanisms

Recall that antibodies have two roles in B cells after they undergo DNA rearrangements and RNA splicing. First, by being expressed on the cell surface, a unique antibody can act as a receptor for an antigen in the recognition phase of the humoral response (see Figures 39.6 and 39.7). Second, in the effector

APPLY THE CONCEPT

The adaptive humoral immune response involves specific antibodies

Mammals have two kinds of light chains: the kappa (κ) and lambda (λ) chains, and each variable region consists of a V domain and a J domain. Imagine a strain of mouse that has a family of 350 genes encoding the κ V domain and 10 encoding the κ J domain, while the λ variable regions are encoded by 300 *V* and 5 *J* genes. The IgG heavy chain is encoded by 350 *V*, 8 *J*, and 5 *D* genes.

1. How many different types of antibodies can be made from these genes by recombination?
2. What are two other ways to generate antibody diversity from these genes?

phase of the humoral response, specific antibodies are produced in large amounts by a clone of B cells. These antibodies are secreted from the B cells and enter the bloodstream where they act in either of two ways:

- *Activation of the complement system*: When antibodies bind to antigens that are present on the surfaces of pathogens, they attract components of the complement system (see Concept 39.2). Some complement proteins attack and lyse the pathogenic cells. In addition, the binding of antibody and complement proteins stimulates phagocytes (monocytes, macrophages, or dendritic cells) to ingest the pathogens.

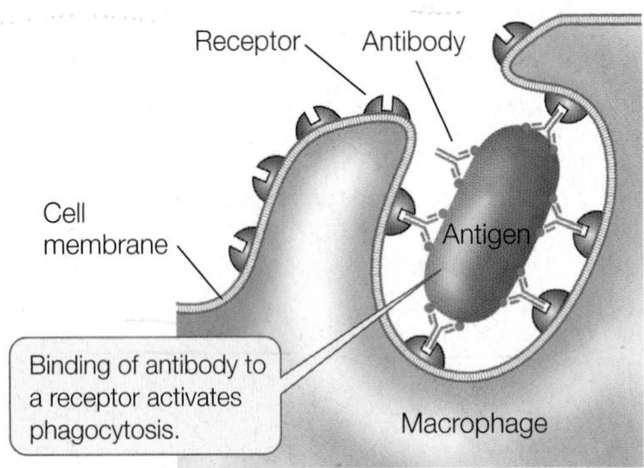

- *Activation of effector cells:* Because of its cross-linking function (see p. 819), antibody binding can result in large, insoluble antibody–pathogen or antibody–antigen complexes. These attract effector cells such as phagocytes and natural killer cells.

CHECKpoint CONCEPT 39.4

- ✓ Sketch an IgG antibody, identifying the variable and constant regions, light and heavy chains, and antigen-binding sites.
- ✓ Assume that one IgG heavy chain is encoded by 1,500 base pairs of DNA, and that one IgG light chain is encoded by 600 base pairs of DNA. If each antibody were encoded by a different gene, how many base pairs of DNA would be needed for 10 million antibody genes? Is it possible for the human genome to contain all these genes? Explain your answer.
- ✓ The bacterium that causes diphtheria (see Figure 39.5) synthesizes a toxic protein. You have probably not been exposed to this bacterium or its toxin. At the present time, are you making B cells and antibodies that bind specifically to diphtheria toxin? Explain your answer.
- ✓ When an antigenic protein such as diphtheria toxin (see above) enters the bloodstream, clones of numerous B cells are produced. Explain.

The humoral immune response works in concert with the cellular immune response in adaptive immunity. Now let's turn to a closer examination of the cellular response.

CONCEPT 39.5 The Adaptive Cellular Immune Response Involves T Cells and Their Receptors

Two types of effector T cells (T-helper cells and cytotoxic T cells) are involved in the cellular immune response, along with proteins of the major histocompatibility complex (MHC proteins), which underlie the immune system's tolerance for the body's own cells.

Go to ANIMATED TUTORIAL 39.4
Cellular Immune Response
PoL2e.com/at39.4

T cell receptors specifically bind to antigens on cell surfaces

Like B cells, T cells possess specific membrane receptors. The T cell receptor is not an immunoglobulin, however, but a glycoprotein with a molecular weight of about half that of an IgG. It is made up of two polypeptide chains, each encoded by a separate gene (**FIGURE 39.11**). The two chains have distinct regions with constant and variable amino acid sequences. As in the immunoglobulins, the variable regions provide the site for specific binding to antigens. However, there is an important difference between immunoglobulin binding and T cell receptor binding: whereas an immunoglobulin can bind to an antigen whether it is present on the surface of a cell or not, a T cell receptor binds only to an antigen displayed by an MHC protein on the surface of an antigen-presenting cell.

MHC proteins present antigens to T cells and result in recognition

Recall that so far we have described two types of T cells: T_H and T_C cells. Both cell types express T cell receptors that bind to antigen on the cell surface. But the response of each cell type to binding is quite different. T_H binding results in activation of the adaptive immune response, whereas T_C binding results in death of the cell carrying the antigen. MHC proteins form complexes with antigens on cell surfaces and assist with recognition by the T cells, so that the appropriate type of T cell binds.

The MHC proteins are cell membrane glycoproteins. Two types of MHC proteins function to present antigens to the two different types of T lymphocytes:

- **Class I MHC** proteins are present on the surface of every nucleated cell in the mammalian body. They present antigens to T_C cells. These antigens can be fragments of virus proteins in virus-infected cells or abnormal proteins made by cancer cells as a result of somatic mutations.
- **Class II MHC** proteins are on the surfaces of macrophages, B cells, and dendritic cells. They present antigens to T_H cells. The three cell types ingest antigens and break them down; one of the fragments then binds to MHC II for presentation (**FIGURE 39.12**).

In humans, there are three genetic loci for class I MHC proteins and three for class II MHC proteins. Each of these six loci

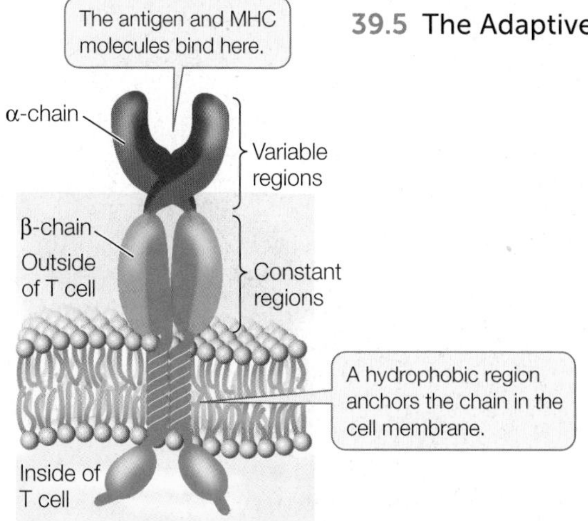

FIGURE 39.11 A T Cell Receptor The receptors on T lymphocytes are smaller than those on B lymphocytes, but their two polypeptides contain both variable and constant regions. As with the B cell receptors, the constant regions fix the receptor in the cell membrane, while the variable regions establish the specificity for binding to antigen.

has as many as 1,000 different alleles. With so many possible allele combinations, it is not surprising that different people are very likely to have different MHC genotypes. This is why it can be difficult to find a good "match" for organ donations. MHC proteins are "self" markers. To accomplish its role in antigen presentation, an MHC protein has an antigen-binding site that can hold a peptide of about 10–20 amino acids. The T cell receptor recognizes not just the antigenic fragment but also the class I or II MHC molecule to which the fragment is bound.

FIGURE 39.12 Macrophages Are Antigen-Presenting Cells A fragment of an antigen is displayed by MHC II protein on the surface of a macrophage. T cell receptors on a specific T-helper cell can then bind to and interact further with the antigen–MHC II protein complex.

Information on MHC proteins, the cellular origins of antigens, and T lymphocytes is summarized in **TABLE 39.2**.

T_H cells contribute to the humoral and cellular immune responses

A T_H cell is activated when it has receptors that can bind a specific antigen on the surface of an antigen-presenting cell. This results in the rapid clonal propagation of the T_H cell. The T_H cells then release cytokines that stimulate other immune cells, including T_C cells, to propagate.

A T_H cell can also directly interact with a B cell that recognizes the same antigen. Like dendritic cells and macrophages, B cells are antigen-presenting cells. B cells take up antigens bound to their surface immunoglobulin receptors by endocytosis, break them down, and display antigenic fragments on class II MHC proteins. When a T_H cell binds to the displayed antigen–MHC II complex, it stimulates the B cell to propagate and to secrete antibodies.

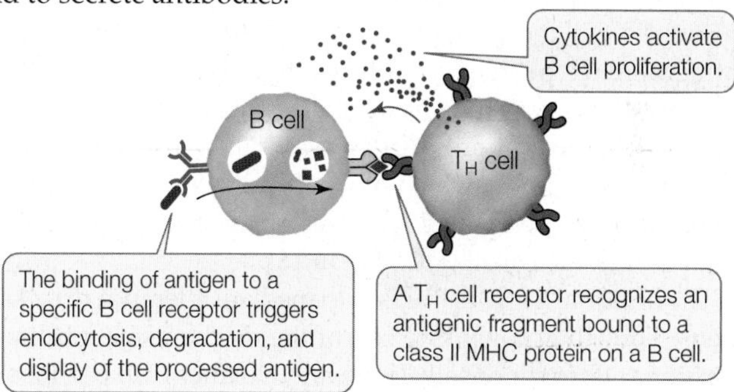

Activation of the cellular response results in death of the targeted cell

T_C cells are activated when they bind to cells carrying an antigen–MHC I protein complex, such as a virus-infected cell. This activation, along with signals from T_H cells, results in the production of a clone of T_C cells with the same specific receptor. When bound, the T_C cells do two things to eliminate the antigen-carrying cell:

- They produce perforin, which lyses the bound target cell.
- They stimulate apoptosis in the target cell.

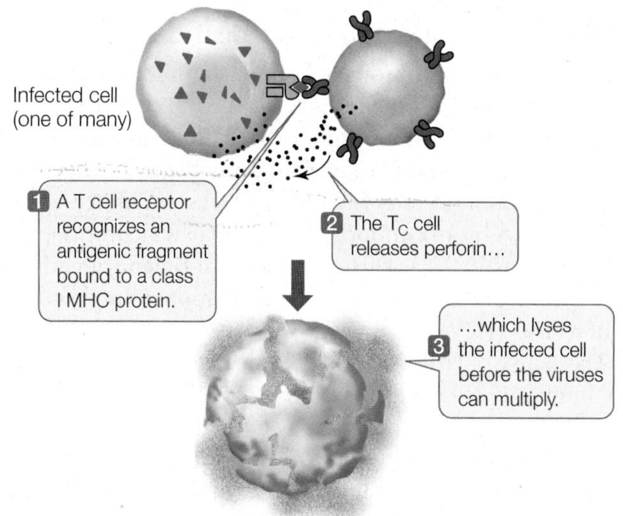

TABLE 39.2 The Interaction between T Cells and Antigen-Presenting Cells

Presenting cell type	Antigen presented	MHC class	T cell type
Any cell	Intracellular protein fragment	Class I	Cytotoxic T cell (T_C)
Macrophages, dendritic cells, and B cells	Fragments from extracellular proteins	Class II	Helper T cell (T_H)

Regulatory T cells suppress the humoral and cellular immune responses

A third class of T cells called **regulatory T cells (Tregs)** ensures that the immune response does not spiral out of control. Like T_H and T_C cells, Tregs are made in the thymus gland, express the T cell receptor, and become activated if they bind to antigen–MHC protein complexes. But Tregs are different in one important way: the antigens that Tregs recognize are self antigens. The activation of Tregs causes them to secrete the cytokine interleukin-10, which blocks T cell activation and leads to apoptosis of the T_C and T_H cells that are bound to the same antigen-presenting cell (**FIGURE 39.13**).

The important role of Tregs is to mediate tolerance to self antigens. Thus they constitute one of the mechanisms for distinguishing self from nonself. There are two lines of experimental evidence for the role of Tregs:

- If Tregs are experimentally destroyed during development in the thymus of a mouse, the mouse grows up with an out-of-control immune system, mounting strong immune responses to self antigens—autoimmunity.
- In humans, a rare X-linked inherited disease occurs when a gene critical to Treg function is mutated. An infant with this disease, called IPEX (*i*mmune dysregulation, *p*olyendocrinopathy and *e*nteropathy, *X*-linked), mounts an immune response that attacks the pancreas, thyroid, and intestine. Most affected individuals die within the first few years of life.

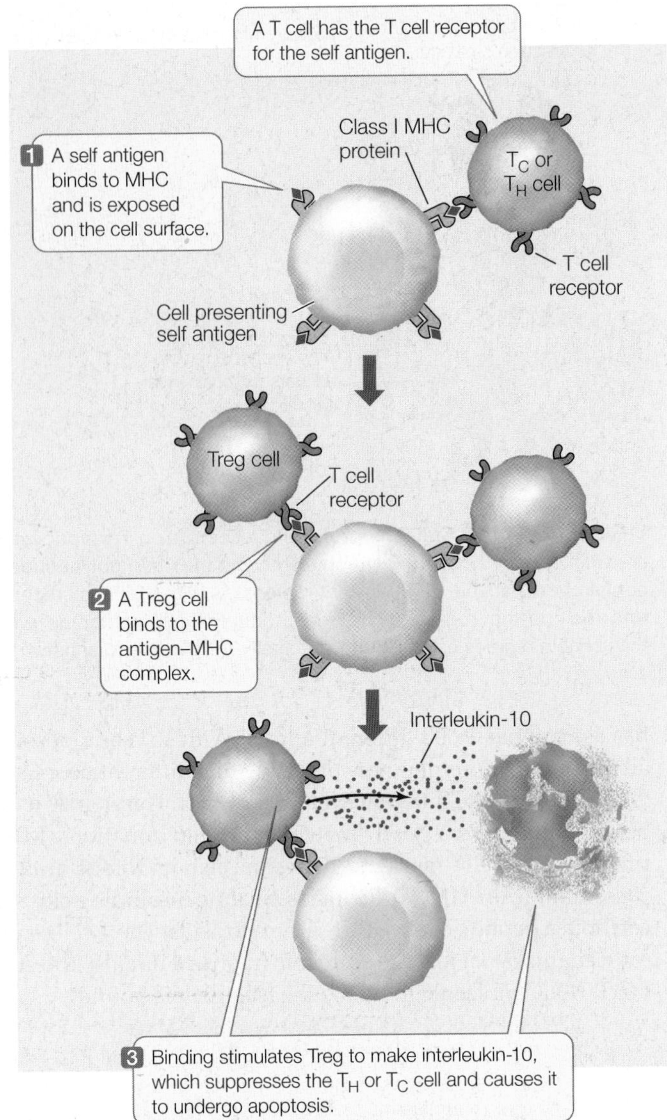

A T cell has the T cell receptor for the self antigen.
Class I MHC protein
T_C or T_H cell
1 A self antigen binds to MHC and is exposed on the cell surface.
T cell receptor
Cell presenting self antigen
Treg cell
T cell receptor
2 A Treg cell binds to the antigen–MHC complex.
Interleukin-10
3 Binding stimulates Treg to make interleukin-10, which suppresses the T_H or T_C cell and causes it to undergo apoptosis.

FIGURE 39.13 Tregs and Tolerance A special class of T cells called regulatory T cells (Tregs) inhibits the activation of the immune system in response to self antigens.

AIDS is an immune deficiency disorder

There are a number of inherited and acquired immune deficiency disorders. In some individuals, T or B cells never form; in others, B cells lose the ability to give rise to plasma cells. In either case, the affected individual is unable to mount an immune response and thus lacks a major line of defense against pathogens. A more common disease that was first detected in the early 1980s is **acquired immune deficiency syndrome (AIDS)**, which results from infection by **human immunodeficiency virus (HIV)**.

The course of HIV infection provides a good review of adaptive immunity (see Figure 39.7). HIV is transmitted from person to person in blood, semen, vaginal fluid, or breast milk; the recipient tissue is either blood (by transfusion or injection) or a mucous membrane lining an organ. HIV initially infects macrophages, T_H cells, and antigen-presenting dendritic cells in the blood and tissues. At first there is an immune response to the viral infection, and some T_H cells are activated. But because HIV infects the T_H cells, they are killed both by HIV itself and by T_C cells that lyse infected T_H cells. Consequently, T_H cell numbers decline after the first month or so of infection. Meanwhile, the extensive production of HIV by infected cells activates the humoral immune system. Antibodies bind to HIV and the complexes are removed by phagocytes. The HIV level in blood goes down. There is still a low level of infection, however, mostly because of the depletion of T_H cells (**FIGURE 39.14**).

During this dormant period, people carrying HIV generally feel fine, and their T_H cell levels are adequate for them to mount immune responses against other infections. Eventually, however, the virus destroys the T_H cells, and their numbers fall to the point where the infected person is susceptible to infections that the T_H cells would normally eliminate. These infections result in conditions such as Kaposi's sarcoma, a skin tumor caused by a herpes virus; pneumonia caused by the fungus *Pneumocystis jirovecii*; and lymphoma tumors caused by the Epstein–Barr virus. These conditions result from opportunistic

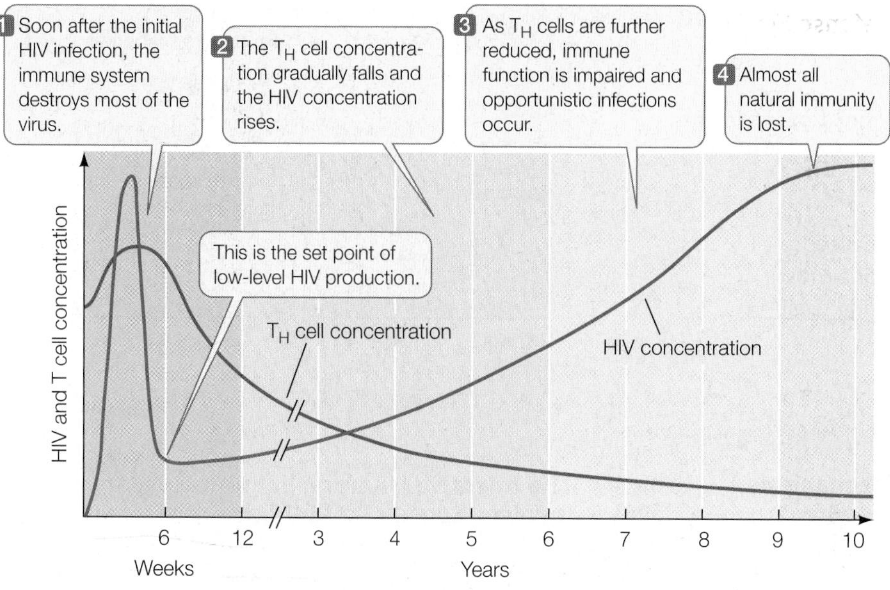

FIGURE 39.14 The Course of an HIV Infection An HIV infection may be carried, unsuspected, for many years before the onset of symptoms.

infections because the pathogens take advantage of the crippled immune system of the host. In addition, HIV infection somehow stimulates Treg cells, and these also downregulate the immune response. The combination of infections and a weakened immune response can lead to death within a year or two of development of full-blown AIDS.

The molecular biology of HIV and its life cycle have been intensively studied. Drug treatments are focused on inhibiting processes necessary for viral entry, assembly, and replication, such as the synthesis of cDNA from viral RNA by reverse transcriptase, and the cutting of the large precursor viral protein into mature, active proteins by viral protease. Combinations of such drugs result in long-term survival, a major achievement of molecular biology applied to medicine. Unfortunately, like many medical treatments, HIV drugs are not available to all who need them—particularly in poor regions of the world where AIDS is prevalent. As a result, there are about 1.7 million deaths per year worldwide from AIDS.

LINK

For more on the life cycle of HIV, see **Concept 11.2**

CHECKpoint CONCEPT 39.5

- ✓ Compare the T cell receptor and B cell receptor in terms of structure, diversity, and function.
- ✓ What are the similarities and differences in function between class I and II MHC proteins?
- ✓ What are the roles of T_H cells in cellular and humoral immunity?
- ✓ Since MHC proteins are highly variable and almost always differ among unrelated people, an organ transplant from one person to another will generally provoke a cellular immune response, and the organ will be rejected. Patients receiving organ transplants are treated with cyclosporin, a drug that inhibits T cell development. How do you think cyclosporin prevents rejection? What side effects might you expect in treated people?

Q How can a person survive an infection and be resistant to further infection?

ANSWER Almost 2,500 years after Thucydides's observations, we have some answers to this question. During the Athenian plague, innate immunity (Concept 39.2) kept most of the disease agent out of the body. Any agent that evaded the innate defenses was attacked by the adaptive immune system (Concepts 39.3–39.5). The agent was engulfed by macrophages or dendritic cells, which broke up the agent and presented fragments on its cell surface, in complexes with class II MHC proteins. The antigen-presenting cell migrated to a lymph node, where the MHC–antigen complex was recognized by T_H cells with the appropriate T cell receptors. Clones of these T cells developed, and some of these cells bound to B cells that displayed the same antigen on their surfaces. The B cells then formed clones of plasma cells, which made antibodies that bound to the agent. The antibody-bound agent was then destroyed by the complement system and phagocytes. Meanwhile, if cells were infected by the agent, a parallel series of events occurred in the cellular immune response, resulting in a clone of T_C cells that killed the infected cells (Concept 39.5). These two systems allowed Thucydides to survive the infection that killed so many of his fellow Athenians.

The T and B cells activated by the two adaptive immune responses also formed smaller clones of memory cells. When Thucydides took care of others in Athens who had the disease, the reentry of the pathogen into his body provoked a rapid, massive T and B cell response. This response prevented the pathogen from proliferating and making him sick again.

An important application of immunological memory is the use of **vaccines**. Exposure to an antigen in a form that does not cause disease can still initiate a primary immune response, generating memory cells without making the person ill. Later, if a pathogen carrying the same antigen attacks, specific memory cells already exist. They recognize the antigen and quickly overwhelm the invaders with a massive production of lymphocytes and antibodies (**FIGURE 39.15**).

Because the antigens used for immunization or vaccination are derived from pathogenic organisms, they must be altered so they cannot cause disease. This is achieved in a variety of ways, including inactivation of the pathogen by chemicals or heat, mutations that render the pathogen inactive, or recombinant DNA technology to make nonharmful peptides derived from the pathogen. The use of vaccines, especially in children,

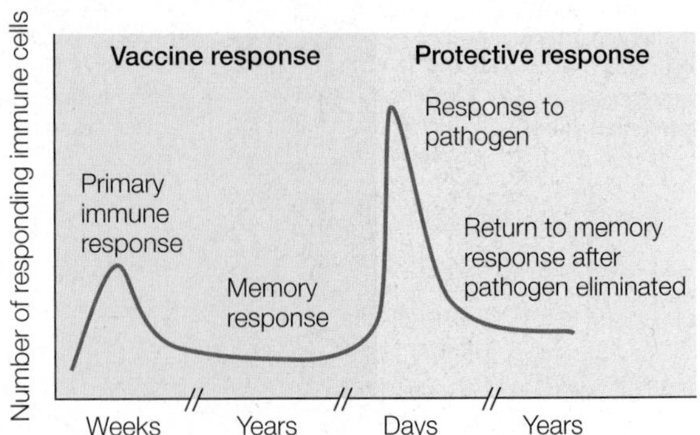

FIGURE 39.15 Vaccination Immunological memory from exposure to an antigen that does not cause disease can result in a massive response to the disease agent when it appears later.

has been a great success in public health. In industrialized countries, widespread vaccination has completely or almost completely wiped out some deadly diseases, including smallpox, diphtheria, and polio.

SUMMARY

CONCEPT 39.1 Animals Use Innate and Adaptive Mechanisms to Defend Themselves against Pathogens

- Animal defenses against **pathogens** are based on the body's ability to distinguish between self and nonself.
- **Innate defenses** are nonspecific, inherited mechanisms that protect the body from many kinds of pathogens. These defenses typically act rapidly.
- **Adaptive defenses** are specific mechanisms that respond to specific pathogens. Adaptive defenses develop more slowly than innate defenses but are long-lasting.
- Innate defenses evolved before adaptive defenses and include barriers, phagocytic cells, molecules that are toxic to pathogens, and a common recognition-signaling pathway. **Review Figure 39.1**
- Many innate and adaptive defenses are implemented by cells carried in the circulatory and lymphatic systems. **Review Figure 39.2 and ANIMATED TUTORIAL 39.1**

CONCEPT 39.2 Innate Defenses Are Nonspecific

- Innate defenses include physical barriers such as the skin, mucous membranes, and competing resident microorganisms. **Review Figure 39.3**
- Circulating defensive cells, such as **phagocytes** and **natural killer cells**, work to eliminate invaders.
- The vertebrate complement system consists of more than 20 different antimicrobial proteins that act to alter membrane permeability and kill targeted cells.
- The **inflammation** response activates several types of cells and proteins that act against invading pathogens. **Mast cells** release **histamine**, which increases the permeability of blood vessels to aid in inflammation. **Review Figure 39.4 and ACTIVITY 39.1**

CONCEPT 39.3 The Adaptive Immune Response Is Specific

- The adaptive immune system recognizes specific **antigens**, responds to an enormous diversity of **antigenic determinants** (also called **epitopes**), distinguishes self from nonself, and remembers the antigens it has encountered.
- **Clonal selection** accounts for the specificity and diversity of the immune response and for **immunological memory**. **Review Figure 39.6**
- Adaptive immunity includes the **humoral immune response**, which involves antibody production by B cells, and the **cellular immune response**, mediated by T cells. Both require specific receptors to bind to each antigen. **Review Figure 39.7**

CONCEPT 39.4 The Adaptive Humoral Immune Response Involves Specific Antibodies

See **ANIMATED TUTORIAL** 39.2

- Naïve B cells are activated by binding of the antigen and by stimulation from T_H cells with the same specificity, and then form **plasma cells**. These cells synthesize and secrete specific antibodies.
- The basic unit of an **immunoglobulin** is a tetramer of four polypeptides: two identical light chains and two identical heavy chains, each consisting of a **constant region** and a **variable region**. The variable regions determine the specificity of an immunoglobulin, and the constant regions of the heavy chain determine its class. **Review Figure 39.8 and ACTIVITY 39.2**
- There are five classes of immunoglobulins, differing in function and in the constant region of the heavy chain.
- B cell genomes undergo recombination events in which the genes that encode specific domains of the immunoglobulin variable regions are randomly selected from large clusters of genes. This DNA rearrangement yields millions of different immunoglobulin proteins. **Review Figures 39.9 and 39.10 and ANIMATED TUTORIAL 39.3**

CONCEPT 39.5 The Adaptive Cellular Immune Response Involves T Cells and Their Receptors

See **ANIMATED TUTORIAL** 39.4

- T cell receptors are somewhat similar in structure to the immunoglobulins, having variable and constant regions. **Review Figure 39.11**
- There are several types of T cells. **Cytotoxic T (T_C) cells** recognize and kill virus-infected cells or mutated cells. **T-helper (T_H) cells** direct both the cellular and humoral immune responses.
- The genes of the major histocompatibility complex (MHC) encode membrane proteins that bind antigenic fragments and present them to T cells. **Review Figure 39.12**
- **Regulatory T cells (Tregs)** inhibit the other T cells from mounting an immune response to self antigens. **Review Figure 39.13**
- **Acquired immune deficiency syndrome (AIDS)** arises from depletion of the T_H cells as a result of infection with **human immunodeficiency virus (HIV)**. **Review Figure 39.14**

See **ACTIVITY 39.3** for a review of the major human organ systems.

Go to the Interactive Summary to review key figures, Animated Tutorials, and Activities PoL2e.com/is39

Go to LaunchPad at **macmillanhighered.com/launchpad** for additional resources, including LearningCurve Quizzes, Flashcards, and many other study and review resources.

40 Animal Behavior

KEY CONCEPTS

Within a coral reef ecosystem, individuals of some species of fish live in schools while individuals of other species move around in pairs. The behaviors shown by both types of fish help give structure to the fish community. Here we see a loose school of orange anthias fish at the left and a pair of golden butterflyfish (*Chaetodon semilarvatus*).

For scuba divers, the chance to see the diversity of fish in coral reef ecosystems is one of the greatest thrills of diving. Fish are by far the most diverse group of vertebrate animals. The number of described species of fish is only slightly lower than the number of all other types of vertebrates combined. Coral reefs are hotbeds of fish diversification. Fifteen percent of fish species—many of them dramatically colored—are found nowhere else and thus are endemic to coral reefs.

Without actually visiting reefs and looking hard at the fish living there, you might envision chaos. On any one reef, you might expect thousands of individuals, representing dozens of species, in endless random motion, mixing with each other in a riot of color.

In reality, however, the behaviors of the fish give a great deal of regularity and structure to these fish communities. If you are able to visit a reef and watch closely, you will notice that individuals of some species occur almost entirely in single-species groups—schools—that often stay close to particular coral structures. Individuals of some other species—such as the butterflyfish seen here—swim about in pairs. Virtually every time you see them, they are in pairs. Individuals of still other species are always associated with sea anemones, often living among the anemones' tentacles. The ecological community is structured by the behaviors of the fish.

We can observe the same phenomenon in other types of communities, such as those of antelopes in the savannas of East Africa. Dozens of antelope species coexist in these grasslands—but not in chaos. Often when you see impala antelopes (*Aepyceros melampus*), for example, many of the individuals are in groups in which only one animal has large antlers. This structuring of the ecological community results from behavior. In this case, a dominant male—the animal with antlers—defends an area in which he gathers females around him and repels other males that attempt to join the group.

"Social behavior" refers to the behavioral relationships of individuals with other individuals of their own species. On the coral reef in the photo, we see two types of social behavior: school formation and pair formation. Throughout the study of behavior, a central question is why the behaviors we observe have evolved. Regarding social behaviors, what advantages might they have for the animals that participate in them?

In what ways might schooling behavior and pairing behavior be advantageous for the individuals involved?

You will find the answer to this question on page 843.

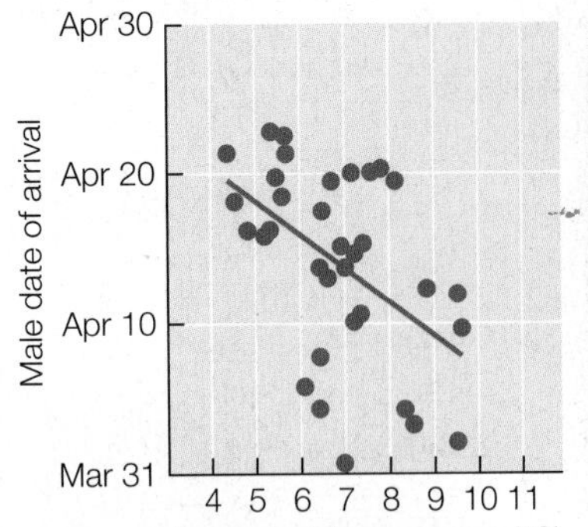

FIGURE 40.1 Migratory Birds Have Returned Earlier to the Netherlands in the Spring as Temperatures Have Risen Pied flycatchers (*Ficedula hypoleuca*) overwinter in Africa and migrate back to the Netherlands in the spring. Over a 34-year period, air temperature in the Netherlands has increased because of climate change, and males have arrived earlier. There has been a statistically significant correlation between average arrival date and average air temperature, as this graph shows. Each symbol corresponds to one year. Air temperatures were measured each year between March 16 and April 15, and all the measurements for each year were averaged to get the temperature plotted for the year.

CONCEPT 40.1 Behavior Is Controlled by the Nervous System but Is Not Necessarily Deterministic

Animals are distinctive in that behaviors play central roles in their interactions with each other and with their environments. They can rapidly approach each other, flee from each other, vocalize, seek shelter, and migrate—all by using their muscles. Of all animal characteristics, adjustments of behavior are often the most visible responses to environmental change. Today, for example, many migratory animals are changing the timing of their migrations in response to climate change (**FIGURE 40.1**). Behavioral shifts of populations into new habitats closer to the poles may be the greatest hope of many to survive if conditions where they currently live become too warm.

LINK

The effects of climate change on ecosystems and communities are complex, and changes in the behavior of a species may not be sufficient to ensure its survival; see **Concept 45.5**

When an animal engages in behaviors, biologists today agree that the animal's nervous system activates and coordinates the behaviors. When lizards run, for instance, neurons stimulate their leg motions. Similarly, when people take part in animated conversations, their brains activate their hand movements and facial expressions. According to this viewpoint, although behaviors—such as leg motions and smiles—are fleeting and not in themselves material objects, they have a material basis: they arise from brain tissue, neurons, and the movements of ions that give rise to nerve impulses (see Concept 34.2).

Many types of evidence point to the neural basis of behavior

We know that particular types of human behavior depend on the function of particular brain regions. For example, if the part of a person's brain known as Broca's area (see Figure 34.26B) is badly damaged by accident or disease, the person may be perfectly fine in almost all ways but will have extreme difficulty with talking or writing. This sort of "negative" evidence (what goes wrong when a brain region is missing?) is complemented by positive evidence from brain-imaging studies in normal people. As seen in Figure 34.27, when a person speaks, specific brain regions show evidence of increased metabolism. These types of evidence point to a neural basis for complex behaviors involved in human communication.

Additional evidence for the neural basis of behavior comes from a very different source: the study of certain highly stereotyped animal behaviors, often called **fixed action patterns**, which are expressed by animals without prior learning and are often resistant to modification by learning. These behaviors point to control by the nervous system because they are sometimes elaborate and yet they depend solely on the simple presence of a healthy nervous system to be performed. A classic example is the begging behavior of gull chicks. In some species of gulls, adults have a red dot on their bills. The chicks, in a fixed action pattern, peck at the red dot when a parent returns to the nest from foraging, and this pecking stimulates the parent to regurgitate food. Pecking at a red dot is so stereotyped that newly hatched, naive gull chicks peck at models of gull heads with a red dot (although they respond only slightly to models without a dot) and even at red-dotted models of a beak without a head (**FIGURE 40.2**). Web spinning by some spiders is another example of a highly stereotyped, unlearned behavior. The spinning of a web involves hundreds or thousands of movements performed in sequence, and each time a spider builds a web, the sequence is largely the same. Yet in many species there is no opportunity for a spider to learn how to build a web from another spider because the adults die before their eggs hatch.

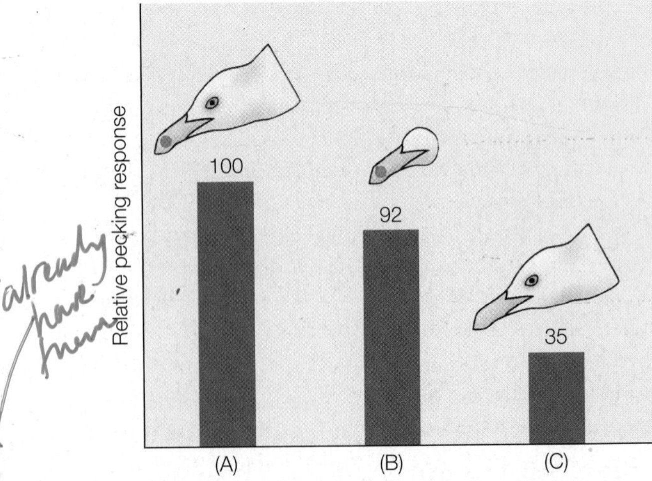

FIGURE 40.2 Expression of a Fixed Action Pattern (A) Herring gull (*Larus argentatus*) chicks instinctively peck at a cardboard model of an adult gull head with a red dot on the bill and (B) even at a model of just a bill with a red dot. (C) If presented with a model of a gull head without a red dot, the chicks respond to only a small extent.

Behaviors evolve

Recognizing that neural tissue gives rise to behaviors, we can expect behaviors to evolve. As is true of all tissues, the properties of neural tissue are in part encoded by genes, which are mutable and passed from generation to generation. If certain gene variants, or alleles, produce more adaptive behaviors than others, natural selection can favor those alleles. In this way the frequencies of alleles in a population can change so that the properties of the neural tissue—and the behaviors coded by the tissue—adaptively evolve.

Many studies establish that genes can exert important effects on behavior. Perhaps the most graphic evidence comes from studies of mutants in fruit flies (*Drosophila*). One of the most important examples involves mutations in a fruit fly gene named *per* (for *period*). (We now know that mammals have homologs of this same gene.) Around 1970, Seymour Benzer and his colleagues reported mutations in this gene that radically altered the flies' daily rhythm of activity under control of the circadian clock. To investigate these mutant flies, the investigators studied them in constant darkness, where their alternating periods of inactivity and activity were timed by their circadian clock, not by observations of day and night. In flies with ordinary, non-mutant *per* genes, each episode of activity started about 24 hours after the previous episode. One of the mutant alleles of *per* caused episodes of activity to be only about 19 hours apart. Another caused them to be about 29 hours apart. Studies of *per* thus demonstrated that single gene mutations sometimes alter behavior.

LINK
You can review the properties of circadian clocks in Concept 29.6

House mice (*Mus musculus*) in North America provide an elegant illustration of adaptive evolution of behavior in free-living wild populations. House mice did not live in North America prior to European colonization. They arrived as stowaways on European ships, presumably in random groups. Carol Lynch trapped wild house mice at five locations, from warm Florida to cold Maine. Within the mice from each location, she then paired males and females in her lab to obtain lab-born, lab-reared offspring, which she studied. All the mice studied had been born and grew up in exactly the same lab environment. When Lynch measured the nests they made, she found a trend: under identical conditions, the mice from progressively more northern populations tended to build bigger (and thus warmer) nests (**FIGURE 40.3**). This result points to the evolution by natural selection of a genetically controlled, behavioral propensity to build bigger nests in wild mouse populations that settled in locations where big nests are particularly advantageous.

Current studies with artificial selection indicate that behavior can evolve remarkably rapidly. In artificial selection (as opposed to natural selection), human researchers choose which combinations of animals have the most offspring, mimicking natural selection but in a highly controlled way. In one such experiment, researchers gave each mouse in a large population a running wheel and then monitored the number of revolutions

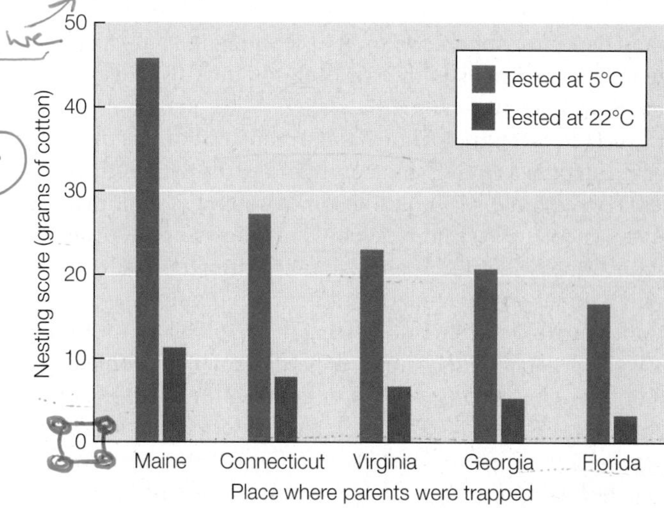

FIGURE 40.3 Nest Building by House Mice from Five Locations All mice were born in the lab and tested under identical conditions. Their parents were trapped at the five locations shown. Mice were tested at two different temperatures. Nest building by each mouse was measured by summing the weight of cotton the mouse used to build four nests. The graph shows average nest building by the mice tested at each temperature.

each mouse voluntarily ran per day. Males that ran the highest numbers of revolutions were then mated with females that ran the highest numbers to produce a new generation of mice. When those mice grew to adulthood, they were put through the same process. This selection procedure was repeated generation after generation. After only 13 generations, the selected mice on average ran 12.1 kilometers per day, more than twice as far as control mice (5.5 kilometers per day). Primarily, the difference reflected faster running by the selected mice, rather than a longer time spent running. Studies revealed critical changes in the brains of the selected mice, indicating that a difference had evolved in the neural control of running behavior.

Despite its neural basis, behavior is not necessarily simplistically deterministic

Earlier we noted that behaviors are often fleeting and not in themselves material objects, but they have a material basis because they arise from neural tissue. Regarding human behavior, this dichotomy—nonmaterial behavior arising from material substance—attracts the attention of virtually all branches of scholarship. Biologists, psychologists, anthropologists, sociologists, theologians, and criminologists are all concerned with the dichotomy and ponder it.

Ever since the modern study of biology began, some scientists have argued for **biological determinism**. Determinism, as applied to behavior, means that the behaviors of animals are hardwired by genetics. In this view, an individual's genes orchestrate his or her neural properties in fixed ways that in turn affect his or her behavior in fixed ways, leading to predictable behaviors. Also, in this view, populations of individuals, such as different racial groups, can differ genetically in ways that produce inflexible and predictable differences in behavior among the populations.

Almost all biologists are convinced that some simple animals exhibit determinism. Clams, for example, are inflexible in

many of their responses to their environment. Here, however, we focus on animals with advanced nervous systems. In particular, we focus on people.

Belief in human biological determinism flourished among some biologists in the nineteenth and early twentieth centuries. For example, at that time some prominent biologists believed (incorrectly) that a person's intelligence is determined by his or her inherited brain size. According to this discarded view, the mental capacities of racial groups could be predicted by measuring relative brain sizes. In the mid-twentieth century, the Second World War and the slaughter of people in gas chambers—based on a theory of genetic inferiority—greatly reduced enthusiasm for the biological determinist view of human behavior. Later, people who argue for the deterministic view became more vocal as the twentieth century drew to a close, a trend that has continued into the twenty-first century.

Students of biology need to be aware of the two sides in this debate. Arguments for and against determinism are too important to be lost in a fog. Many arguments for determinism found in books and magazine articles marketed to the general public are not well supported by objective, statistically supported, and relevant data of the sort required for ideas to be deemed scientific.

In what ways could the behavior of animals with advanced nervous systems not be simplistically deterministic despite its neural origins? Behavior is dramatically more flexible than any other biological trait. This is true in part because learning modifies behavior. In addition, recent discoveries (which we will shortly discuss) point to important epigenetic effects on behavior. These are non-genomic effects that can exert lifelong influences and may be transmissible from one generation to the next (see Concept 11.3). As animals develop and live their lives, each individual can interact in unique ways with its environment, and—because of learning and epigenetic effects—these interactions can result in dramatic behavioral differences among genetically similar individuals.

The greatest unresolved question in all of biology is probably the question of how the human mind emerges from the human brain. The unresolved nature of this question, despite its profound importance, emphasizes that scientists have much to learn about mental phenomena and the behaviors they unleash.

CHECKpoint CONCEPT 40.1

✓ What physical objects provide the basis for animal behaviors?

✓ When a male robin is in reproductive condition, it has bright red breast feathers and will aggressively attack other male robins. It will also attack a simple tuft of red feathers on a stick. How does this seemingly maladaptive behavior relate to the concept of a fixed action pattern?

✓ Considering behaviors classified as fixed action patterns, in what way are they passed from one generation of a species to the next?

✓ How do experiments on artificial selection of wheel running in mice provide insight into the rate of evolution?

We have already noted that development and learning often affect an animal's behavior. Now we will focus on these important processes.

CONCEPT 40.2 Behavior Is Influenced by Development and Learning

The particulars of an individual animal's life—such as the exact place it lives and the exact appearances of its parents—are typically not predictable. Information about these particulars therefore cannot be inherited. Biologists think this is the fundamental reason why animals have evolved the ability to **learn**—by which we mean the ability of an individual animal to modify its behaviors as a consequence of individual experiences that took place earlier in its life. Suppose that a mouse living in a forest is less likely to be caught by a predator if it can rapidly return to its burrow. Prior to the mouse's birth, there would be no way to predict the location of that mouse's burrow or landmarks useful for finding the burrow. For this reason, natural selection could not provide inherited information on these particulars. However, natural selection has favored the evolution of learning abilities. A mouse inherits mechanisms by which it can learn locations. Then, during its life, it uses those mechanisms to incorporate specific information on its actual burrow location and useful landmarks into its escape behavior.

Lee Metzgar did experiments on white-footed mice (*Peromyscus leucopus*) to test the validity of the idea that animals learn about their surroundings in ways that aid their survival. An important background point for understanding Metzgar's experiments is that when these mice move around on the forest floor, they commonly run next to logs or sticks. An adult mouse was allowed to live for several days in a seminatural arena with a simulated forest floor, consisting of small logs, sticks, and dead leaves. There was no burrow in this room, but there were logs big enough to provide protection from aerial attack. Then the mouse was temporarily removed from the room and a hungry screech owl was introduced. Soon afterward, the mouse was returned to the room, along with a new mouse that lacked prior experience with the room. The arena was then observed until the screech owl caught one of the mice (**FIGURE 40.4**). To a strongly significant extent, the new mouse was the one more likely to be caught. However, if the logs, sticks, and leaves were replaced with new but similar items in different locations immediately before the test with the owl, the two mice in the room were equally likely to be caught. These results indicate that the experienced mouse learned the locations of logs and sticks during the days it lived in the room. It knew escape routes and places where it was safe, provided the logs and sticks were where the mouse had learned them to be.

Specific information of critical survival value is often learned during early postnatal development

Konrad Lorenz, one of the pioneers of the modern study of behavior, discovered that if he associated with young greylag geese (*Anser anser*) soon after they hatched, they thereafter treated him as if he were their parent. They associated with him instead of

FIGURE 40.4 Screech Owls Prey on Mice Screech owls watch from perches above the forest floor. When an owl sees or hears a mouse, it drops from its perch and grasps the mouse with its talons unless the mouse behaviorally evades capture. The species shown here is an eastern screech owl (*Megascops asio*).

other geese and followed him everywhere. This attachment bordered on impossible to change. Lorenz named this phenomenon **behavioral imprinting** (not to be confused with genomic imprinting; see Concept 11.3). By now we know that imprinting occurs in many other species of birds and some other animals. Today behavioral imprinting is defined to be a type of learning that is distinctive because the learning takes place within a relatively narrow window of time early in postnatal life and, after that, is inflexible. Imprinting can have lifelong consequences. When male geese imprint on a person at the start of their lives, they later, in adulthood, prefer trying to mate with people. During normal development, geese imprint on—and establish a strong attraction to—their true parents and species.

Indigo buntings (*Passerina cyanea*) and some other bird species are noteworthy for navigating by the stars. As adults, the birds migrate at night, using the stars to determine the direction in which they fly. After this amazing phenomenon was first discovered, Stephen Emlen studied young buntings in a planetarium to learn more about the mechanism they use. Emlen established that the birds need to know the North Star to navigate correctly. But how do they know which star is the North Star? His experiments revealed that each young bird has to learn the correct star during its first few weeks after hatching. In the planetarium, Emlen could make the artificial sky rotate around any star he chose. Young birds learned to identify that star—the one the sky rotated around—and later, when they were under a natural sky, they treated that star as the North Star for the rest of their lives. When buntings develop outdoors in natural environments in the Northern Hemisphere, the sky appears to rotate around the true North Star (**FIGURE 40.5**). In this way, buntings in nature learn the true North Star.

Another type of early learning with enormous consequences—seen in many birds—is learning of species-specific songs. Darwin's finches in the Galápagos Islands provide just one of many examples (see Concept 17.3). In adulthood, the males of each finch species sing a species-specific song that attracts females (which do not sing) in a species-specific way. The songs are not inherited, however. Instead, each male learns the song he will sing for the rest of his life during a period of about 30 days, 10–40 days after he hatches. During that period a young male is in or near his nest, and his father sings nearby. The young male learns his song by an imprinting-like process by listening to (and later duplicating) his father's song. Particular brain regions, larger in males than in females, are required for this learning. If a young male's father dies and he learns the song of a different species by hearing an adult of a different species sing during his critical learning period, the young male will later sing an incorrect song and attract females of the incorrect species.

Early experience also has other, more global effects on behavior

A large body of knowledge points to early experience as having multiple, intricate lifelong effects. These effects are specific to each individual, depending on that individual's particular experiences. For example, an ongoing series of experiments on lab rats is bringing to light new revelations regarding behavior and early development. When they are carefully observed, rat mothers are found to exhibit different parenting behaviors. Two categories of mothers are recognized. Here we will call them "high-caring" and "low-caring." Mothers in both categories are statistically the same in how often they nurse their young and in the absolute amount of contact they have with their suckling offspring. They differ, however, in how much they lick and groom their young during nursing and in how much they adopt a favorable posture for easy suckling. Mothers in the high-caring category lick and groom a lot and adopt a favorable posture. Low-caring mothers lick and groom to a lesser extent and adopt a less favorable posture.

FIGURE 40.5 The Stars in the Northern Hemisphere Seem to Make Circles around the North Star The stars seem to make circles around the northern end of Earth's axis—the position of the North Star—because of Earth's rotation. This apparent movement of the stars is evident in a long exposure of the northern sky, as in this photograph, in which the camera shutter was kept open for several hours.

Offspring of the two categories of mothers have been raised to adulthood and then studied in two potentially threatening environments that are new to them. In both environments, the adult offspring of low-caring mothers are far more likely to exhibit fear. For example, rats deprived of food for 24 hours were placed, one at a time, in a novel, brightly lit arena in which food was provided at only one place in the very center. This environment is intimidating because rats prefer low light and prefer not being out in the open. The behavioral differences between the two groups of rats were dramatic. Hungry adult offspring of low-caring mothers waited much longer than did hungry adult offspring of high-caring mothers to go to the food and start eating, and they spent far less time eating.

How can a young rat's early experiences change its behavior throughout its life? In the case we are discussing, epigenetic effects are in part responsible and play a pivotal role (see Concept 11.3). Key regulatory genes in stress-response biochemical and hormonal pathways are tagged with epigenetic marks in early life. The tagging of these genes is different in offspring of low- and high-caring mothers. Such differences in epigenetic marking (which by definition do not alter the DNA itself) are maintained throughout the life of an individual, permanently altering the individual's stress responses.

Malnutrition in early life is also known to affect epigenetic tagging in rats, as is abandonment. Marks from these early experiences persist into adulthood, altering gene expression and behavior throughout life.

In Africa, massive swarms of millions of migratory locusts (*Locusta*) sometimes form and fly across the landscape, causing devastation to crops wherever they land to eat. An amazing feature of this phenomenon is that each individual locust is capable of having two radically different behavioral phenotypes. One phenotype is that the individual avoids other individuals. A population of individuals expressing this phenotype is spread out and inconspicuous. Alternatively, each individual may be highly gregarious—meaning the population forms a swarm. Individual experience determines the phenotype. Individuals that are never forced to live closely together retain the solitary phenotype. The same individuals become a swarm if forced into close contact—for instance, if they are forced to feed next to each other, in the same places, because of food shortage.

CHECKpoint CONCEPT 40.2

- In captive breeding programs for whooping cranes, an endangered species of bird, the young birds must be prevented from seeing people because otherwise they will not mate with cranes. How can this be?
- Do epigenetic marks alter the sequence of base pairs in an animal's genome?
- Penguins often breed in large groups in which nesting females are very close to each other, in contrast to some other species of birds in which nesting females are widely separated. Why might penguins have evolved particularly strong imprinting on their mothers?

Behavior is integrated with the other characteristics of animals. Next we will discuss a few ways in which behavior interacts with other aspects of animal function.

CONCEPT 40.3 Behavior Is Integrated with the Rest of Function

Escape running by pronghorn (*Antilocapra americana*) is one of the greatest spectacles of animal behavior on Earth (**FIGURE 40.6**). Although cheetahs (*Acinonyx jubatus*) achieve the highest speeds of all running animals (about 110 kilometers per hour), they can sustain those speeds for only a minute or two. Pronghorn can run at more than 50 kilometers per hour for tens of minutes—the highest *sustained* speeds known in running animals.

Pronghorn strikingly illustrate that an animal's behavior often depends on—and is integrated with—the animal's other characteristics. For pronghorn to *behave* as they do, their muscles must be able to produce enormous forces over long periods. For exercise to be sustained in this way, it has to be based on aerobic, oxygen-based ATP production (see Concept 33.3). Investigators have discovered that almost all aspects of pronghorn are specialized so that the animals can deliver O_2 at very high rates to their muscles, use the O_2 at very high rates to make ATP in the muscle cells, and use the ATP at very high rates to perform intense muscular work. For example, compared with other mammals of their size, pronghorn have exceptionally large lungs and skeletal muscles, and their muscle cells are exceptionally tightly packed with mitochondria.

Toads and frogs have evolved contrasting behavioral specializations that depend on their biochemistry of ATP synthesis

When toads such as the western toad (*Bufo boreas*) are chased, they hop away at modest speeds that they can maintain for many minutes. In contrast, when leopard frogs (*Rana pipiens*)

FIGURE 40.6 Pronghorn Run at the Fastest Sustained Speeds of Any Running Animals Their running behavior is possible only because they have also evolved systems of oxygen transport, ATP production, and muscular contraction that are up to the task. Pronghorn occur in grasslands from northern Mexico to southern Alberta and Saskatchewan.

APPLY THE CONCEPT

Behavior is integrated with the rest of function

When chased, both frogs and toads respond by hopping away until fatigue prevents them from going any farther—a behavior dependent on the synthesis and subsequent use of ATP during muscle contraction. Frogs hop away at high speeds using anaerobic glycolysis, a biochemical process that produces large amounts of ATP very rapidly but that can't be sustained for long periods of time. Toads use aerobic respiration, a process that produces less ATP per unit of time but that can operate for a much longer period. Suppose you make records of speed and duration for two cases of escape behavior as in the figure (these data, although modeled on actual data, are hypothetical).

1. Which animal fatigued more quickly?
2. Which strategy of ATP synthesis allowed for the achievement of a higher maximum speed?
3. Which animal traveled a greater distance? (Hint: Determining the number of squares under each graph of speed versus time will allow you to integrate the information and estimate distance traveled.)

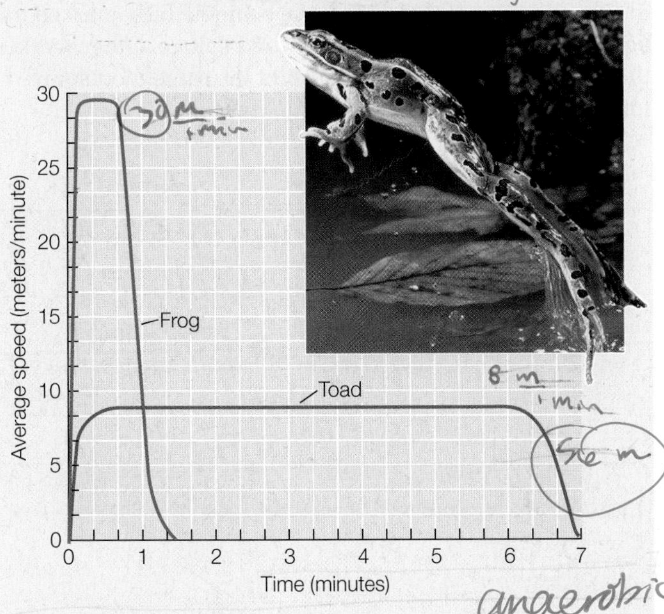

are chased, they leap away at lightning speeds—far faster than toads—but they become fatigued and stop leaping after a minute or two. These contrasting escape behaviors in toads and frogs reflect differences in the biochemical mechanisms that their leg muscles use to make ATP for escape running. Toads and frogs thus illustrate that there are often intimate links between escape behavior and the biochemistry of ATP synthesis.

The usual rule, when comparing related animals, is that aerobic ATP production is slow and resistant to fatigue, whereas anaerobic ATP production is fast and subject to rapid fatigue (see Concept 33.3). Enzyme studies show that toads have evolved high levels of enzymes required for *aerobic* ATP production in their leg muscles. Frogs, by contrast, have an enzyme profile that emphasizes high rates of *anaerobic* ATP production. These evolved enzyme differences explain the differences in their hopping behavior.

Behaviors are often integrated with body size and growth

Many insects, such as crickets and katydids, sing to communicate with other members of their species. Among these insects there is a broad tendency for the tonal frequencies of their songs to vary regularly with body size. In a pipe organ, the large pipes produce the lowest frequencies of sound. For much the same reason, insect species of large body size tend to produce songs of lower frequency than species of small size. This same trend is observed in singing frogs. Body size helps determine how communication occurs in these animals.

In many species, individuals must grow to a large adult size before they can successfully consummate reproductive behaviors. Male elk (*Cervus elaphus*) in Scotland, for example, rarely father offspring before they are 5 years old, even though they are reproductively mature at 2 years of age. Being reproductively mature is not sufficient for successful reproductive behavior. A male must also be big enough and experienced enough to dominate other males.

When spotted hyenas (*Crocuta crocuta*) in a pack capture a zebra or antelope, they are famous for eating everything, including the skeleton, within minutes in a frenzy of feeding. For young hyenas to get food during this group feeding behavior, they need to be as effective as possible in grabbing bites to eat. Being able to bite through bone is a great advantage because it permits much faster feeding than trying to strip flesh off bone. A young hyena's feeding behavior is transformed—in ways that raise its odds of survival—as its jaws, jaw muscles, and teeth mature, permitting faster and faster bone consumption (**FIGURE 40.7**).

CHECKpoint CONCEPT 40.3

- ✓ Prior to mating, European eels swim for months at a time as they travel from their adult habitats in Europe to spawning grounds in the ocean near Bermuda, a distance of about 5,500 kilometers. This behavior depends on ATP production for the swimming muscles. Would you predict that European eels mostly use aerobic or anaerobic means of synthesizing ATP during this migration?
- ✓ Simply because of laws of physics, if an animal falls from a substantial height, its risk of injury from whole-body impact on the ground depends on its body size. For a mouse-size animal, the odds of being hurt by the whole-body impact of a fall are near zero. The risk, however, increases with body size to the point that for a horse, death is inevitable. How would this size-dependent trend affect the evolution of behaviors that involve leaping from branch to branch within trees?
- ✓ What aspects of pronghorn muscle cell physiology allow the animals to maintain high escape running speeds?

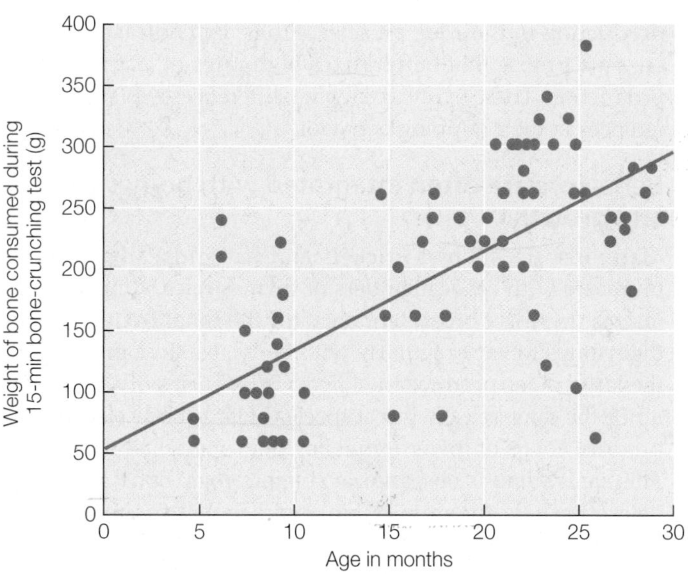

FIGURE 40.7 Young Hyenas Must Mature to Feed on Bone Rapidly Spotted hyenas of various ages were given a standardized "bone-crunching test." The amount of bone they could consume in 15 minutes increased dramatically as they aged and their jaws and jaw muscles became stronger. Young animals are limited in their ability to compete with adults during group feeding behavior at a kill. The line was fitted by linear least-squares regression (see Appendix B).

Let's now turn our attention to travel behavior. Most animals move from place to place over the surface of Earth during their lives. Travel behavior of this sort can help animals find things they need, but also entails risks, including the risks of getting lost. Travel behavior is one of the important topics in the study of animal behavior.

CONCEPT 40.4 Moving through Space Presents Distinctive Challenges

For most animals, movement through space is critical for success. The distances covered may be short, as when animals search for food each day. Some animals, by contrast, travel almost halfway around the globe during seasonal migrations. Moving through space requires muscle function. It also requires that animals meet the challenges of navigation and orientation, ensuring that they can find their goals and return home. **Navigation** refers to the act of moving toward a particular destination or along a particular course. **Orientation** refers to adopting a position, or a path of locomotion, relative to an environmental cue such as the sun.

Trail following and path integration are two mechanisms of navigation

To navigate, animals sometimes simply follow trails. Foraging workers in many species of ants use this method. If certain workers find a food source, they use a pheromone to mark their path as they return to their colony's home. A **pheromone** is a chemical compound or mix of compounds that is emitted into the outside environment by individuals of a species and that elicits specific behavioral responses from other members of the same species. One sex sometimes uses pheromones to attract the other sex. In the case of the ants, as worker ants return home from a profitable food source, they deposit minute quantities of their species-specific trail pheromone along the way. In this manner they leave behind a fully marked trail between the colony nest and the food source. Other workers then detect the pheromone and find the same food source by following the trail.

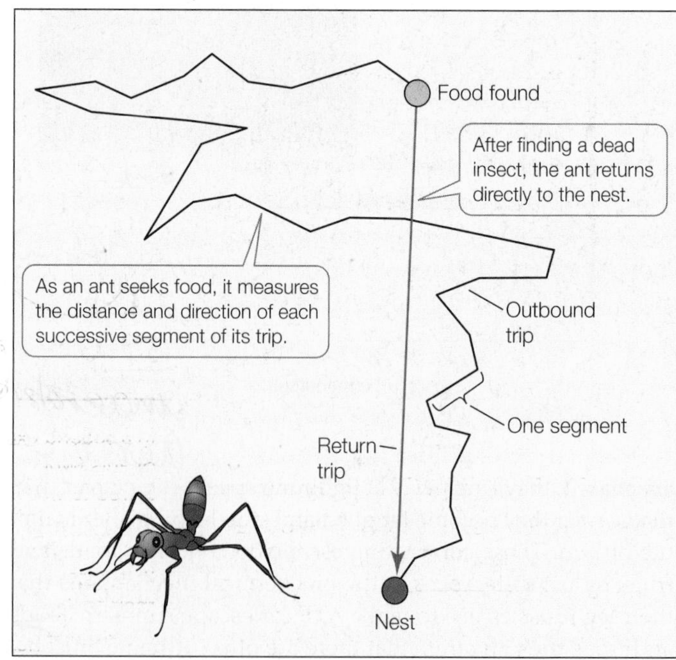

FIGURE 40.8 A Foraging Trip by a Worker Ant of a Desert Ant Species (*Cataglyphis*) The worker's outbound trip consists of a series of segments as the ant wanders the desert surface seeking dead bodies of heat-killed insects. At any point, however, the ant can return to its home nest quickly in a straight line. The ants accomplish this task with a brain that weighs 0.1 milligram!

Ants of the genus *Cataglyphis* are specialized to live in hot, dry deserts. Instead of using trail pheromones, they use path integration to navigate during foraging. Colonies of *Cataglyphis* live underground, where temperatures are mild. Worker ants emerge to collect food during the daytime, when the desert surface is fiercely hot. Their principal food is the dead bodies of other desert insects killed by overheating. Why do the workers forage during the heat of the day? The answer seems to be that this behavior enables them to find heat-killed insects promptly after death, before the bodies of the dead insects dry out. In this way the workers acquire water—which is extremely valuable in the desert—as well as food. Experiments have shown that a *Cataglyphis* worker that is searching far and wide for a dead body always knows how to run straight back to its underground nest. A worker does not retrace its path to get home. Instead, if it finds a dead body or starts to overheat, it returns home by the shortest possible, straight path. This behavior enables a worker to get underground quickly.

How does a foraging *Cataglyphis* worker know how to find its home nest without retracing its outbound travels? Let's think of the worker's outbound path as a series of segments (**FIGURE 40.8**). The worker monitors the length and the compass direction of each segment. Then, by a process called path integration, it integrates, or puts together, all this information on segment lengths and directions to know where it is relative to its nest. We know this sort of process can work because it closely resembles the process (called dead reckoning) that sailors often used 150 years ago to know where they were at sea on an hour-by-hour basis.

How does a *Cataglyphis* worker measure the length and direction of each segment of its foraging path? It measures the length of each segment by counting steps (**FIGURE 40.9**). It can determine the direction by several means, as we will soon discuss.

INVESTIGATION

FIGURE 40.9 An Insect Odometer Desert ants navigate by path integration, which requires information about both direction and distance. How do they measure distance? Matthias Wittlinger and his colleagues investigated the possibility that Saharan desert ants (*Cataglyphis fortis*) "count" steps to keep track of how far they've walked.[a] Because every stride is a certain length, the number of steps can theoretically measure the distance traveled.

HYPOTHESIS

Desert ants measure distance by keeping track of the number of steps they take.

METHOD

1. Train ants to walk from their nest to a feeder 10 meters away and then back to their nest.
2. Capture trained ants that have walked from the nest to the feeder and modify them before they start back to their nest. Separate them into three groups, with 25 ants in each group. Shorten the legs of one group ("Stumps") by removing their lower leg segments. Lengthen the legs of another group ("Stilts") by gluing pig bristle extensions to their legs. Leave the legs of the third group ("Normal") unchanged as a control.
3. Release the ants at the feeder, so that they can return to their nest. Measure how far they walk before they stop to search for the nest entrance (indicating that they've arrived where they expect the nest to be).

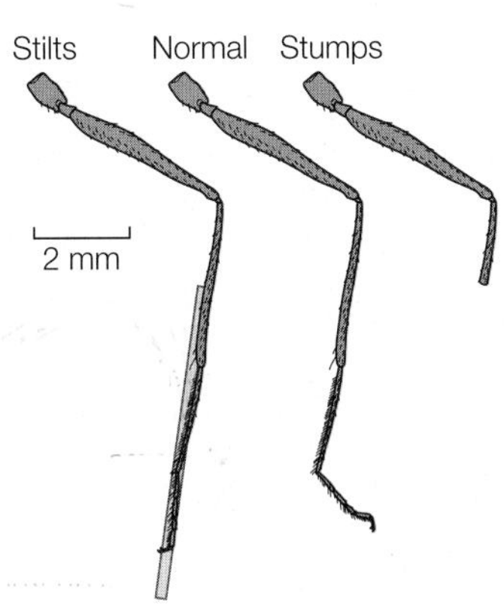

Stilts
Normal
Stumps
2 mm

RESULTS

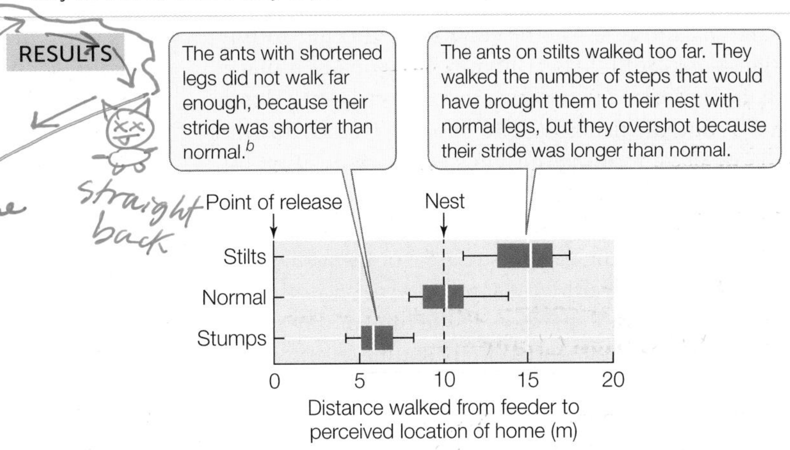

The ants with shortened legs did not walk far enough, because their stride was shorter than normal.[b]
The ants on stilts walked too far. They walked the number of steps that would have brought them to their nest with normal legs, but they overshot because their stride was longer than normal.
Point of release
Nest
Stilts
Normal
Stumps
0
5
10
15
20
Distance walked from feeder to perceived location of home (m)

CONCLUSION

Desert ants behave as if they keep track of how far they travel by "counting" their steps.

ANALYZE THE DATA

In a second experiment, the same groups of ants (which had altered legs) were allowed to walk from their nest to the feeder on their altered legs and then return to their nest. The graph shows the distances the three groups of ants walked to get back to their nest.

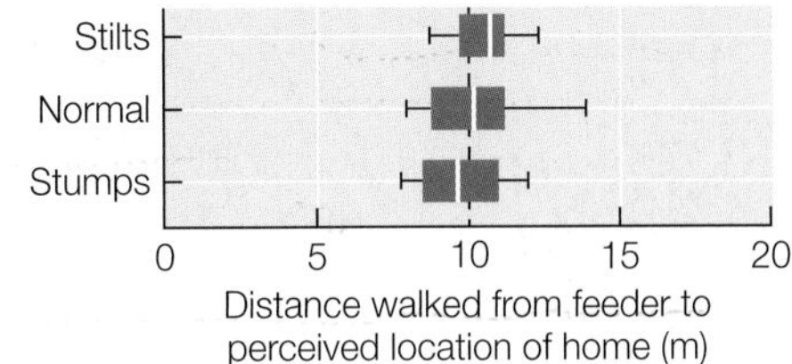

Stilts
Normal
Stumps
0
5
10
15
20
Distance walked from feeder to perceived location of home (m)

1. How do these results differ from those obtained in the first experiment? Explain why they differ.
2. Why is this experiment an important control?

Go to **LaunchPad** for discussion and relevant links for all **INVESTIGATION** figures.

[a]M. Wittlinger et al. 2006. *Science* 312: 1965–1967.
[b]The data are shown as a box plot. The vertical line inside the box is the median, the box indicates the middle 50% of the data, and the horizontal line that extends outside the box indicates the middle 90% of the data.

Animals have multiple ways of determining direction

Much of the earliest research on animal orientation focused on homing pigeons. These pigeons fly back to their home nests even after having been transported tens of kilometers away. For instance, during both World Wars, homing pigeons were extensively used to carry messages. They were reared in home nests near command centers. They then were carried into battle and released with critical messages that they transported back to the command centers.

How can homing pigeons determine direction so they can fly home? One mechanism they use is a **sun compass**. They do not possess a physical compass. We say they have a sun compass because, in essence, they can determine where north, south, east, and west are by observing the sun. To understand the challenge of using the sun as a compass, suppose for a moment that you have no way to know the time of day and that you are suddenly dropped at an unfamiliar location. Suppose also that you see the sun just above the horizon. With the sun in that position, you would reason that when you face the sun, you are facing either east or west. Without knowing the time of day, however, you would not know which: east or west? From thought exercises like this, researchers hypothesized that when pigeons use the sun as a compass, the birds not only must observe the position of the sun but also must know the time of day so that they can calculate compass directions—north, south, east, and west—from the sun's position. The researchers hypothesized that pigeons use their circadian clock (see Concept 29.6) to provide the necessary information on time of day.

To test this hypothesis, the researchers (working in the Northern Hemisphere) released homing pigeons at a place from which the birds needed to fly south to return home. Think now about the compass position of the sun as it crosses the sky each day. (You can determine the sun's compass position by dropping an imaginary vertical line from the sun to the horizon and noting where the line intersects the horizon between east and west). The sun is in the east at sunrise, southeast at midmorning, south at noon, southwest at midafternoon, and west at sunset (**FIGURE 40.10**). Regardless of the time of day when pigeons were released, they tended to fly south. As shown in Figure 40.10A, in midmorning they flew at angle α to the right of the sun's position. Later in the day, in midafternoon, they flew at angle β to the left of the sun's position.

The pigeons were then placed in rooms where the lights were turned on 6 hours before sunrise and turned off 6 hours before sunset. After 2 weeks or so, their circadian clocks became entrained (see Concept 29.6) to the artificial light cycle. When the birds were then taken outdoors again and tested for their ability to return home, they did not fly south. For example, when released in midmorning, they flew east! This result provided strong evidence that the birds were using their circadian clocks to determine compass directions from the sun's position. In midmorning, the time-shifted birds thought the time was midafternoon because their internal clocks had been artificially reset. They flew at angle β to the left of the sun's position, traveling east (see Figure 40.10B).

By now, a wide variety of animals have been shown to have a sun compass. In most cases, their circadian clocks play an essential role.

Redundancy in mechanisms of orientation was one of the great discoveries to come from the early study of homing pigeons. Investigators found that instead of having a single orientation mechanism, the pigeons have multiple mechanisms. We now know that this sort of redundancy is observed often in the animal kingdom. Homing pigeons are able to find home on cloudy days when they cannot see the sun. A key orientation method they use on cloudy days is to detect Earth's magnetic field and orient to it. Homing pigeons also sometimes use landmarks such as hills to orient, and they may use odors, low-frequency environmental sounds, and learning from other pigeons.

Go to ANIMATED TUTORIAL 40.2
Homing Simulation
PoL2e.com/at40.2

Many insects and birds are able to determine compass directions by detecting patterns of polarized light in the sky. This

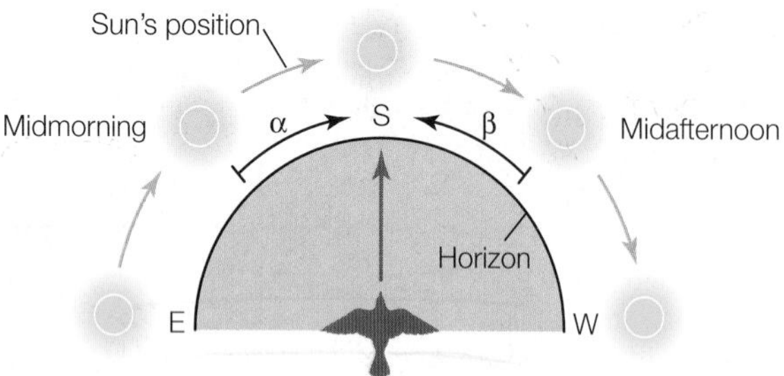

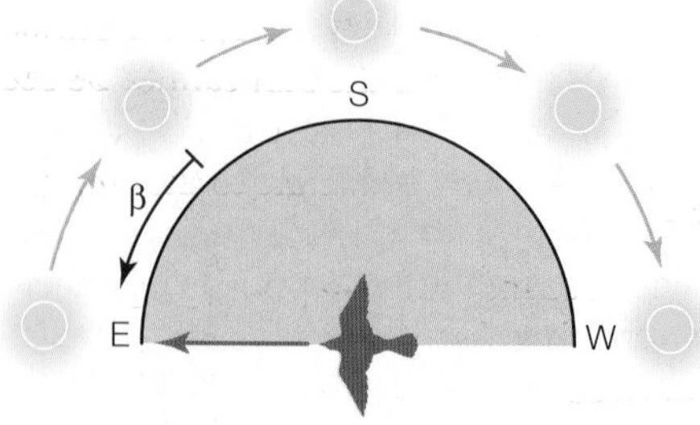

FIGURE 40.10 The Sun Compass of a Homing Pigeon Requires That the Pigeon Interpret the Sun's Position in the Sky Based on Knowledge of Time of Day A pigeon trained to fly south to find home was tested in the Northern Hemisphere. Here we see a bird's eye view of the southern horizon, from east to west, with the sun passing overhead between dawn and dusk. (A) The pigeon with its circadian clock functioning normally. Using time-of-day information from its circadian clock, the bird flies at angle α to the right of the sun in midmorning but at angle β to the left of the sun in midafternoon. (B) The pigeon with its circadian clock entrained so that the clock says the time is midafternoon when the actual time is midmorning. The pigeon flies at angle β to the left of the sun and thus flies east instead of south.

Go to ANIMATED TUTORIAL 40.1
Time-Compensated Solar Compass
PoL2e.com/at40.1

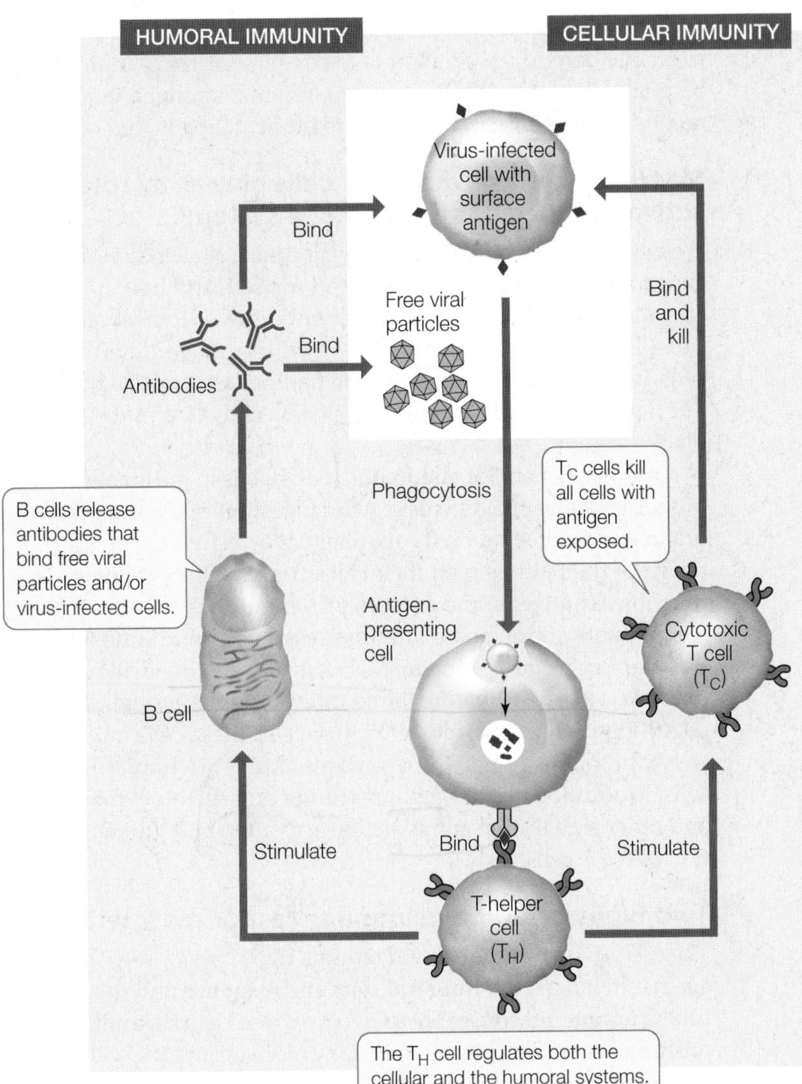

FIGURE 39.7 The Adaptive Immune System Humoral immunity involves the production of antibodies by B cells. Cellular immunity involves the activation of cytotoxic T cells that bind to and destroy self cells that are mutated or infected by pathogens. Details of these processes are described later in this chapter.

to the same antigen to divide. The results are a clone of B cells that function in humoral immunity and a clone of T_C cells that function in cellular immunity.

- *Effector phase:* In the humoral immune response, cells of the B clone produce antibodies that bind to viral particles and/or virus-infected cells. The bound antibodies attract phagocytes and complement proteins that engulf and destroy the virus and the virus-infected cells. In cellular immunity, cells of the T_C clone bind to virus-infected cells and destroy them.

CHECKpoint CONCEPT 39.3

- ✓ List the four key features of the adaptive immune system.
- ✓ In 2009, the H1N1 strain of influenza was a worldwide epidemic. Notably, very old people, who had been alive during the 1918 flu outbreak, had low rates of H1N1 infection. Explain this in terms of immunological memory.
- ✓ Insulin-dependent (type 1) diabetes results from a destruction of the cells in the pancreas that make the hormone insulin (see Concept 30.5). One hypothesis for this disease is that it is caused by an autoimmune reaction. Explain how this might happen and how you would investigate your hypothesis.

APPLY THE CONCEPT

The adaptive immune response is specific

In humans, the dominant gene *D* codes for the RhD (Rhesus) protein on the surfaces of red blood cells. The recessive allele *d* does not encode a cell surface protein. If cells from an individual who is *DD* or *Dd* (Rh^+) enter the bloodstream of an individual who is *dd* (Rh^-), the Rh^- individual makes antibodies that react with the Rh^+ cells. During development, the blood vessels of a mother and her fetus are near each other but do not mix. However, antibodies can pass through capillary walls between the two blood supplies.

MOTHER		FETUS		
GENOTYPE	Rh PHENOTYPE	GENOTYPE	Rh PHENOTYPE	RESULT
Dd		*Dd*		
dd		*Dd*		

1. Hemolytic disease of the newborn (sometimes resulting in stillbirth) can occur when antibodies from the mother pass into the fetal circulatory system. Fill in the table, which predicts RhD incompatibility.
2. During birth there is mixing of the blood supplies of mother and fetus. In a situation of RhD incompatibility, the first birth does not result in a clinical problem. But a second RhD incompatible birth with this mother results in a hemolytic disease. Explain how this can happen (hint: immunological memory).
3. Today, RhD incompatibility does not result in stillbirths. The mother is treated with serum containing anti-RhD antibodies late in the first and subsequent pregnancies. How can this prevent the immunological reactions?

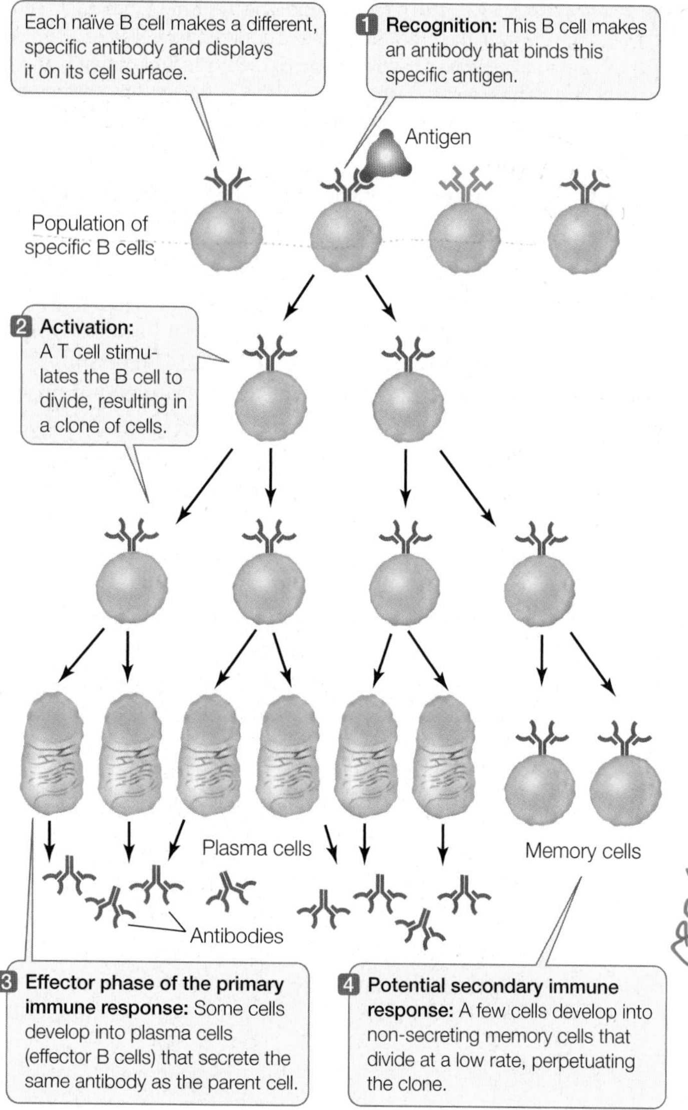

Each naïve B cell makes a different, specific antibody and displays it on its cell surface.
1 Recognition: This B cell makes an antibody that binds this specific antigen.
Antigen
Population of specific B cells
2 Activation: A T cell stimulates the B cell to divide, resulting in a clone of cells.
Plasma cells
Memory cells
Antibodies
3 Effector phase of the primary immune response: Some cells develop into plasma cells (effector B cells) that secrete the same antibody as the parent cell.
4 Potential secondary immune response: A few cells develop into non-secreting memory cells that divide at a low rate, perpetuating the clone.

FIGURE 39.6 Clonal Selection in B Cells The binding of an antigen to a specific receptor on the surface of a B cell stimulates that cell to divide, producing a clone of genetically identical cells to fight that invader.

differentiate to produce *two types* of daughter cells: effector cells and memory cells (see Figure 39.6).

- **Effector cells** carry out the attack on the antigen. Effector B cells, called **plasma cells**, secrete antibodies. Effector T cells release cytokines and other molecules that initiate reactions that destroy nonself or altered cells. Effector cells live for only a few days.
- **Memory cells** are long-lived cells that retain the ability to start dividing on short notice to produce more effector and more memory cells. Memory B and T cells may survive in the body for decades, rarely dividing.

After a primary immune response to a particular antigen, subsequent encounters with the same antigen will trigger a much more rapid and powerful **secondary immune response**. The memory cells that bind with that antigen proliferate, launching a huge army of plasma cells and effector T cells. The principle behind vaccination is to trigger a primary immune response that prepares the body to mount a stronger, quicker secondary response if it encounters the actual pathogen.

Macrophages and dendritic cells play a key role in activating the adaptive immune system

We have just seen how the adaptive immune system uses antigens to distinguish between self and nonself, and how it uses clonal selection to produce large numbers of specific B and T cells. In Concepts 39.4 and 39.5 we will focus on how these cells function to eliminate specific pathogens from the body. But how is the adaptive immune system activated in the first place?

One mechanism for eliminating pathogens in the innate immune system is phagocytosis. After ingestion of a pathogenic organism or infected host cell, phagocytic cells display fragments of the pathogen on their cell surfaces. These fragments function as antigens, and **antigen presentation** is one way that components of the innate immune system communicate with the adaptive immune system. Macrophages and dendritic cells play a key role in activating the adaptive immune system. After engulfing pathogens or infected host cells, these cells migrate to lymph nodes, where they present antigen to immature T cells. In addition, the antigen-presenting cells secrete cytokines and other signals that stimulate the activation and differentiation of the T cells.

Two types of adaptive immune responses interact

The adaptive immune system mounts two types of responses against invaders: the **humoral immune response** and the **cellular immune response**. These two responses work simultaneously and cooperatively, sharing many mechanisms. We will use the example of a viral infection in an overview of these two types of responses (**FIGURE 39.7**). These responses also occur when bacteria or other pathogens infect and grow inside host cells.

B cells that make antibodies are the workhorses of the humoral immune response, and **cytotoxic T (T_C) cells** are the workhorses of the cellular immune response. There are three phases in both types of immune response:

- *Recognition phase:* In both cellular and humoral immunity, recognition occurs when an antigen is inserted into the cell membrane of an antigen-presenting cell, with the unique antigen structure protruding from the cell membrane. In addition to dendritic cells and macrophages, developing B cells can also perform phagocytosis and function as antigen-presenting cells. Figure 39.7 shows a virus-infected host cell being engulfed by an antigen-presenting cell; in other cases, the free viral particles may be engulfed. The antigen on the surface of the antigen-presenting cell is recognized by a **T-helper (T_H) cell** bearing a T cell receptor protein that is specific for the antigen. In both humoral and cellular immunity, binding initiates the activation phase.
- *Activation phase:* When the T_H cell recognizes an antigen on an antigen-presenting cell, it propagates and releases cytokines that stimulate B cells and T_C cells bearing receptors

mechanism is particularly common in insects, where it depends on specialized photoreceptors. The atmosphere is filled with tiny particles, such as dust, water droplets, and ice crystals. When sunlight strikes such particles, the reflected light rays become polarized, or aligned parallel to one another, to some degree, giving rise to polarization patterns. These patterns are present even if there are only small patches of clear sky. For this reason, the "polarization compass" is often functional on cloudy days.

We discussed earlier how *Cataglyphis* ants in the desert employ path integration by measuring the distances and directions of all segments of a trip they make away from their nest. If a prominent landmark is present, the ants determine their direction of travel relative to the landmark. However, they function perfectly well without any landmarks. Under those circumstances the principal mechanism they use to determine direction is a compass based on polarized light detection. They also have a sun compass.

Birds that migrate at night can use the stars to determine their directions of travel (see Figure 40.5). A great many animals besides birds are now known to be able to orient by using Earth's magnetic field.

Honey bee workers communicate distance and direction by a waggle dance

Honey bees, like ants, live in colonies in which worker individuals perform services for the colony as a whole. When a worker bee goes on a foraging flight to gather food and finds flowers containing pollen and nectar, she returns to the hive and communicates her discovery to other workers by a specialized behavior called the waggle dance. During her foraging flight, she measures the distance and direction to the flowers. Her dance communicates both of those pieces of information to the other workers so that they also can fly to the flowers.

A bee that discovers a new patch of flowers measures the distance to the patch by monitoring the rate at which she flies past local landmarks (she measures this rate visually). To measure the direction to the flower patch, she monitors the angle of her flight relative to the compass position of the sun. Of course, as she measures these critical pieces of information, she is flying *horizontally* outdoors. After she returns to the hive, she performs her waggle dance on *vertical* honeycomb surfaces inside the hive, where it is dark and the sun cannot be seen. The marvel of the waggle dance is that it communicates the information on horizontal distance and angle to the sun while the bee is running ("dancing") on vertical surfaces in the dark.

During the dance, the bee moves in a repeating figure-eight pattern. She alternates half-circles to the left and right with a short, straight run between each half-circle (**FIGURE 40.11**). During each straight run, she waggles her abdomen. The duration of the straight, waggle run communicates the distance to the food source. The angle of the straight run relative to vertical communicates the direction to the food source. That angle matches the angle a bee needs to fly relative to the sun's compass position to find the food source. What happens if the sun is hidden by clouds? Honey bees can infer the position of the sun from atmospheric polarization patterns.

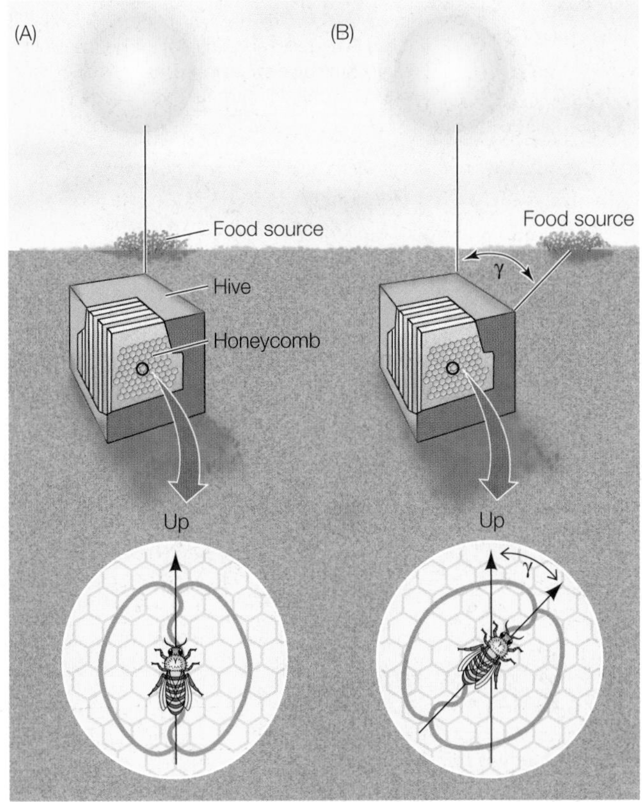

FIGURE 40.11 The Honey Bee Waggle Dance (A) A honey bee (*Apis mellifera*) worker runs straight up on the vertical surface of the honeycomb in a dark hive while wagging her abdomen to tell other workers in her hive that there is a food source in the direction of the sun. (B) When the waggle runs are at angle γ from the vertical, the other bees know that the same angle, γ, separates the direction of the food source from the direction of the sun.

Go to ACTIVITY 40.1 Honey Bee Dance Communication PoL2e.com/ac40.1

Migration: Many animals have evolved periodic movements between locations

Today biologists are better able to track moving animals than ever before, because of technological innovations. Consequently, a revolution is underway in our understanding of migrations. For example, natural historians have long known that shorebirds called bar-tailed godwits (*Limosa lapponica*) spend their summers in Alaska and winters in New Zealand:

Bar-tailed godwit (*Limosa lapponica*)

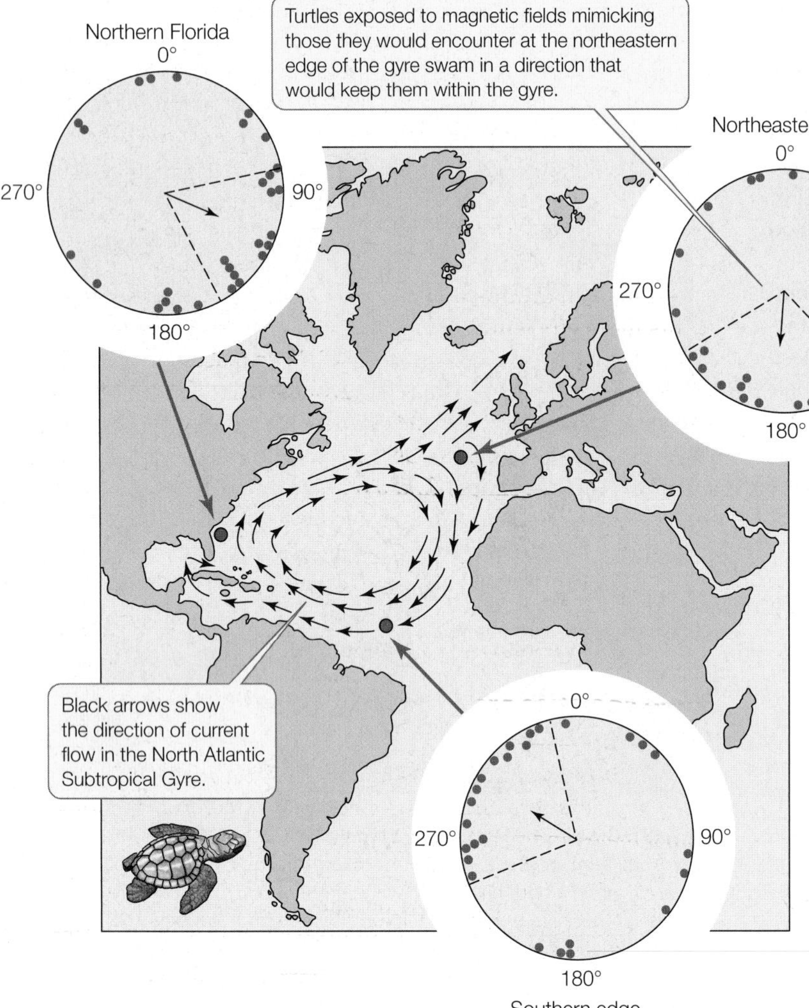
Northern Florida
0°
270°
90°
180°
Turtles exposed to magnetic fields mimicking those they would encounter at the northeastern edge of the gyre swam in a direction that would keep them within the gyre.
Northeastern edge
0°
270°
90°
180°
Black arrows show the direction of current flow in the North Atlantic Subtropical Gyre.
0°
270°
90°
180°
Southern edge

FIGURE 40.12 Young Loggerhead Sea Turtles Orient to Magnetic Fields as They Travel in a Giant Closed Circle across the Atlantic Ocean and Back Newly hatched loggerhead sea turtles, studied in a pool of water represented by the large tan circles, were exposed to magnetic fields that mimicked Earth's magnetic field at each of the three red-dotted locations in the ocean. Within each diagram of the pool (each large tan circle), each small dot shows the direction a turtle tried to swim, the solid arrow shows the average swimming direction, and the dashed lines show the 95% confidence limits for swimming direction. The data indicate that as sea turtles travel in the currents of the North Atlantic Subtropical Gyre, they orient relative to local magnetic fields to swim in directions that keep them within the closed circle of current flow.

Some of the mind-boggling details of their migrations could not be known, however, until individual birds could be outfitted with miniaturized electronics to track their flights. Researchers have just discovered that when the birds leave Alaska, they fly *nonstop* to New Zealand over the open water of the central Pacific Ocean—a trip of more than 10,000 kilometers. For the 6–9 days of the trip, the birds cannot eat or drink. They presumably do not sleep. Their flight muscles must engage in powering or controlling flight continuously day and night—not to mention the necessity of navigating accurately over endless expanses of water. As often happens in the practice of science, the tracking studies of godwits seem to raise more questions than they answer.

Specialists vary in the way they define migration. A useful general definition is that animals **migrate** when they periodically move from one location to a different location, remain for a substantial length of time, and later return.

One of the long-distance migrations that is currently best understood is that of young loggerhead sea turtles (*Caretta caretta*). Several decades ago, Archie Carr coined the term "the missing year" in describing these turtles, which hatch out of eggs on beaches in places such as Florida. For a while, the little turtles can be found in the ocean along the Florida coast, but then they disappear. The next time they are seen, about a year later, they are much bigger. Where have they been during their missing year, and how did they find their way back?

Recent genetic studies have been crucial for determining where the turtles go. The turtles that hatch in a particular region of Florida display a distinctive genetic signature in terms of allele frequencies. Researchers discovered that turtles in the ocean west of Africa have the same signature. This and other evidence indicate that after hatching in Florida, young loggerhead sea turtles migrate across the Atlantic Ocean to African waters. Then, growing as they go, they migrate back to the waters off Florida. The turtles do not need to power their migration entirely by swimming because a large circular water current (the North Atlantic Subtropical Gyre) helps them travel all along the way. As shown in **FIGURE 40.12**, the turtles use Earth's magnetic field to help them complete their journey and ensure that they do not get lost at sea. The tan circles in the figure show the results of experiments demonstrating that the turtles are able to orient their direction of swimming in relation to magnetic fields so that they stay in the circular water current.

CHECKpoint CONCEPT 40.4

- ✓ Why is an accurate circadian clock crucial to the sun compass method of orientation?
- ✓ In some species of ants, when workers employ a trail pheromone to mark a trail to a food source, each worker that goes to the food source adds more pheromone to the trail on its way home if it readily found food at the source. How could this process act as a positive feedback loop to draw large numbers of workers to a distinctively valuable food source?
- ✓ Why might natural selection favor the emergence of redundant methods of orientation?

Now we will turn to one of the themes with which we began this chapter, namely that animals often do not function alone. When animals interact with each other, behavior is the paramount way in which they do so.

CONCEPT 40.5 Social Behavior Is Widespread

A moment's reflection reveals that many animal species live in groups. Flocks of birds, schools of fish, and herds of antelopes are just a few examples. Many types of groups are considered societies. Specialists debate the exact meaning of a "society," but a broad, commonly used definition is that a **society** is a group of individuals of a single species, organized to some degree in a cooperative manner. The subject of **social behavior** includes both the behaviors of individuals that integrate the individuals into societies and the group behaviors of entire societies.

Living in a group has downsides. Grouping tends to make animals more visible. Diseases can spread more easily within a group, and a group may rapidly deplete food or other resources in the place where it is living.

Biologists presume that group living has evolved only if it provides advantages that exceed its disadvantages. Much of the research being done on societies today focuses on testing hypotheses regarding how individuals in a society benefit from living together in a group. Physiological advantages prevail in some cases. Penguins, for example, reduce their thermoregulatory costs in the cold of winter by huddling. Our focus here will be on potential behavioral advantages of group living.

Some societies consist of individuals of equal status

Sometimes the individuals in a flock, school, or herd seem to play similar, approximately equal roles in the function of the group. In a group of this sort, one advantage hypothesized for group living is that a group may enjoy enhanced sensory awareness of its environment because the members of the group can pool their individual sensory capabilities. A group of 50 vertebrate animals has 100 eyes instead of just 2. Groups may therefore have greater awareness of danger than individuals. One investigator found that a trained goshawk's success in capturing a pigeon in a flock decreased as the number of pigeons in the flock increased (**FIGURE 40.13A**). The larger the flock, the sooner one of the individuals in the flock spotted the goshawk and started to fly away, stimulating the other individuals to take flight as well. Belding's ground squirrels, which live in large colonies in open meadows, use alarm calls to reduce predation risk. When an aerial predator such as a hawk arrives overhead, one of the squirrels that first spots the hawk gives a loud whistle (**FIGURE 40.13B**). Immediately, other individuals also whistle, and the animals run to their burrows or other shelter. Aerial predators almost never capture a ground squirrel after the alarm whistles have begun.

Animals in groups may also discover preferred environments more efficiently. In a recent experiment, investigators studied a species of fish that prefers to be in places where light levels are low. They studied schools of the fish of various sizes in a large aquarium that had both darkened and well-lit areas of water. The darkened areas moved from place to place, so the fish could not simply find a dark spot and stay there. The efficiency with which a school could find dark areas increased dramatically as the number of fish in the school increased (**FIGURE 40.14**). There is good reason to think that results would be similar if schools were seeking other desirable regions in the water, such as regions with abundant food.

Some societies are composed of individuals of differing status

A group of vervet monkeys (*Chlorocebus*) typically contains multiple subadult and adult males (**FIGURE 40.15**), along with females and youngsters. One of the adult males typically **dominates** all the other adult and subadult males in the group, meaning it "wins" during one-on-one behavioral contests with each of the others. This dominant male has by far the greatest chances of mating with the adult females in the group.

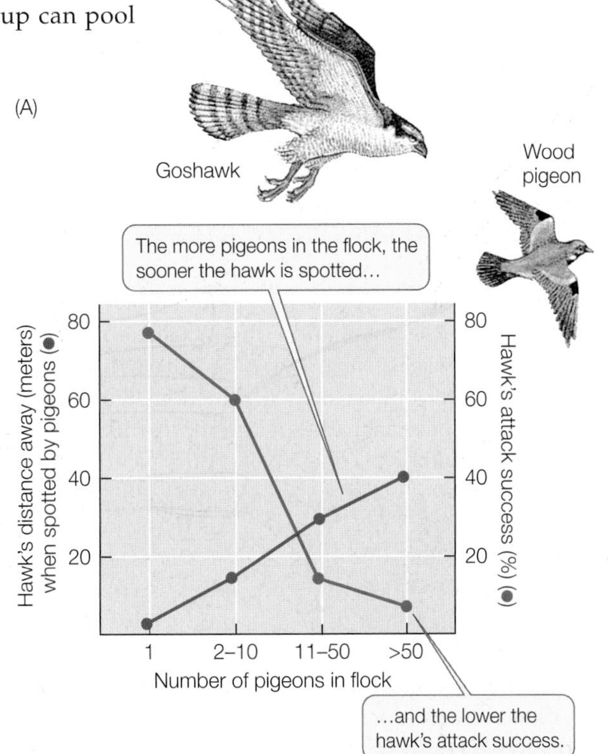

FIGURE 40.13 Group Living Provides Protection from Predators Animals that live in groups can spread the cost of looking out for predators. (A) The larger the number of wood pigeons in a flock, the greater the chances that one of the pigeons will spot a goshawk before it attacks, and the lower the chances that the goshawk will capture one of the pigeons. (B) A young male Belding's ground squirrel gives an alarm call upon spotting a predator.

(A)

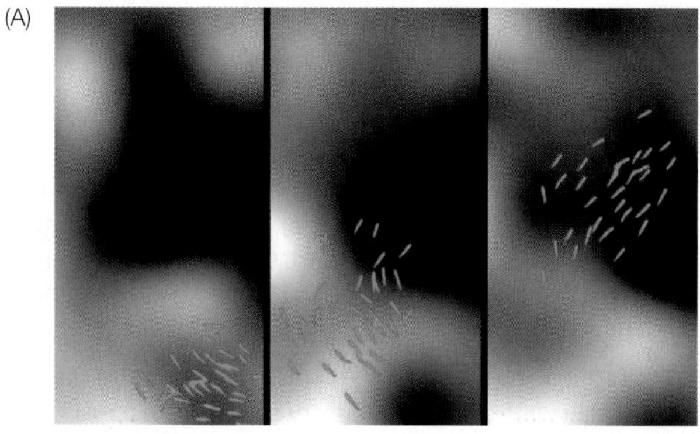

(B)

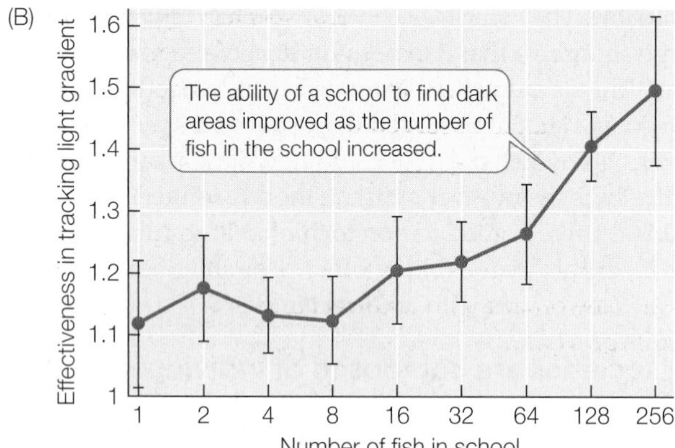

FIGURE 40.14 A School of Darkness-Preferring Fish Finds Dark Patches Faster as the Number of Fish in the School Increases Juvenile golden shiners (*Notemigonus*) prefer areas in the water where the light intensity is low. Schools were placed in an aquarium where dark areas were present but moved from place to place. (A) Three successive images of a school, taken 2 seconds apart, as the school finds and enters a dark patch. (B) An index of a school's effectiveness in finding dark patches. The index is plotted against the number of fish in the school. Vertical bars are 95% confidence intervals (see Appendix B).

FIGURE 40.15 An Adult Male Vervet Monkey Note his colorful genitalia, which signal his adult status.

Dominance by a particular male is even more evident in impala antelopes (*Aepyceros melampus*). If you ever go on safari in Africa and encounter a group of impala during the breeding season, you'll notice the existence of two subgroups. One subgroup consists entirely of males. By marking the animals in this type of subgroup, researchers established that these males do not get to mate with females. The second subgroup consists of a single male and a large number of females. This single male does essentially all of the mating! As you might guess, the dominant male—the single male in the group of females—must expend extraordinary effort to repel intruding males and intercept females that try to leave. This activity is exhausting, and dominant males tend to retain their dominance for only a few months.

LINK

Dominance behaviors may result in sexual selection, which would tend to favor increases in these behaviors in subsequent generations; see **Concept 15.2**

Within groups in which certain individuals have distinctive status—such as groups of vervet monkeys or antelopes with dominant males—individuals are able to benefit from group living in the same ways we discussed in the last section. In addition, biologists hypothesize that the evolution of social systems based on distinctive status has taken place because such systems can provide additional benefits. In cases of dominant males, researchers currently hypothesize that the process of becoming dominant may often serve as a test of a male's strength, endurance, and other properties critical for success. Females that mate with the dominant male may thereby ensure that their offspring are genetically well endowed. In this way, both the individual males and the individual females that mate in such systems of social behavior realize advantages.

Eusociality represents an extreme type of differing status

In most animals, all the adult members of a social group are physiologically capable of reproducing, even though some may be behaviorally excluded from reproducing at certain times. An entirely different situation exists in many social insects, such as honey bees.

In a honey bee colony, a single female—the queen—is reproductive and lays eggs. Occasionally she produces a few male offspring (drones) that are fertile, producing sperm. However, most of the thousands of other individuals in the colony—all of which are her offspring—are sterile female worker bees. Similar colony structures occur in many other types of bees, wasps, ants, and termites.

Eusociality refers to a social structure in which some members of a social group are nonreproductive and assist the reproduction of fertile members of the group, typically their mother. It represents an extreme form of differing status within a social group. Nearly all known cases of eusociality occur within the insects, although eusociality is also documented in naked mole rats (*Heterocephalus glaber*) and certain coral-reef shrimps.

All cases of structured, unequal reproductive status among adults in social groups raise the question of why evolution has produced systems in which the inequality exists. This question reaches its extreme manifestation in the eusocial animals. In discussing such questions, an important concept is altruism. As used in discussions of evolution, **altruism** refers to any characteristic of an individual that imposes a cost on that individual while aiding another individual. Workers in eusocial colonies exemplify altruism because, whereas they themselves are unable to breed, they help another individual breed. Altruism also occurs if an individual—when seeing a predator—gives warning calls that draw the predator's attention to itself (endangering its life) while giving neighboring individuals a better chance to hide. The reasons why altruism exists have been hotly debated in evolutionary biology for a half century.

Go to MEDIA CLIP 40.1
Social Shrimps
PoL2e.com/mc40.1

CHECKpoint CONCEPT **40.5**

- Name one advantage and one disadvantage of group living.
- Consider the phenomenon of mating exclusion, in which certain adult individuals in a group are excluded from mating. How does mating exclusion differ in groups with a dominant individual and in eusocial groups?
- What is altruism, and why might a biologist have difficulty reconciling it with the theory of evolution by natural selection?

Isolated individuals and entire ecological communities are at opposite ends of a spectrum of complexity in biological systems. Having started this chapter with a focus on individuals, we will now end by looking at some of the ways that behavior is important in communities.

CONCEPT 40.6 Behavior Helps Structure Ecological Communities and Processes

In ecological communities, the behaviors of animals often give critically important structure to the use of time and space and to interrelationships among species. One way that behavior has this effect is the structuring of the 24-hour day into activity periods. Some animals are active only during daylight hours. Others are active only at night. These behavioral differences can determine whether two species encounter each other. For example, because insect-eating bats are typically nocturnal, spending their days in secluded places such as caves and attics, diurnal insects are not threatened by the bats even though the two live in the same ecological community.

Behavior helps maintain species

In much of North America, the principal native mice are members of the genus *Peromyscus*, and it is common for two species in this genus to be found living in one general region. White-footed mice (*P. leucopus*) and cotton mice (*P. gossypinus*) are one example. When males and females of these two species are housed together in cages, they breed readily and produce fertile offspring. In nature, however, hybrids are rarely found. To a large degree, behavior explains this difference. In nature the two species prefer different types of woodlands. They thus live in different places and probably rarely meet. We can also hypothesize that if they were to meet, they would detect oddities in each other's interactive behavior and separate before mating. Many similar cases are known of species that can be demonstrated to be interfertile yet remain distinct in undisturbed natural habitats. Their behaviors keep them apart in such habitats. Behavior is thus a key factor in allowing species to maintain their species distinctions.

The species of Darwin's finches in the Galápagos Islands are of particular interest because intense study of the finches has provided unusually detailed insight into hybridization in the wild. Peter and Rosemary Grant have established by analyses of DNA that, among the several species of ground-living Darwin's finches, about 1 percent of birds that grow to adulthood in the wild are fertile hybrids of two distinct species. The hybridization that is observed demonstrates that the species of Darwin's finches are not genetically walled off from each other. Instead, behavior ordinarily keeps the species distinct. Males ordinarily learn their songs from their fathers (see Concept 40.2), and a female ordinarily will not mate with a male singing a song other than the correct one for her species. These behaviors prevent most cross-breeding.

LINK

Behaviors that prevent cross-breeding are an important mechanism of reproductive isolation and speciation; see **Concept 17.4**

Animals often behaviorally partition space into territories or home ranges

In many types of animals—including ones as diverse as fish, dragonflies, and birds—individuals within a single population restrict their movements to limited areas. For example, **FIGURE 40.16** shows use of space by six adult male white-footed mice within a 4-hectare forest. Each mouse roamed within a defined area on the forest floor, with almost no overlap among the areas occupied by the different mice.

In cases like this, behavior acts to structure the space available. Rather than interacting ecologically with the entire forest,

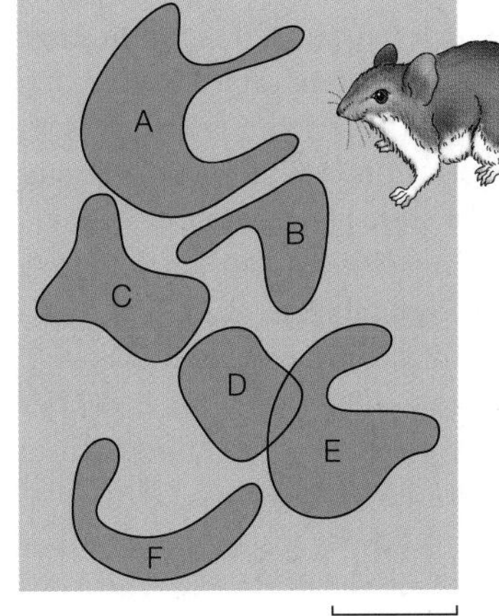

FIGURE 40.16 Home Ranges of Male White-footed Mice in a Woodland At the time of study, six male *Peromyscus leucopus*, designated A–F, were living in this approximately 4-hectare area within a large woodland. Each mouse moved around just within the circumscribed home range shown.

each individual interacts with the ecological community in just a small part, which is more or less distinct from the parts with which other, neighboring individuals are engaged.

The region occupied by an individual is known as its **territory** if the individual actively keeps out other individuals of the same species. A region is known as an individual's **home range** if other individuals are not excluded. Home ranges often overlap considerably. Both territories and home ranges provide familiarity. An individual occupying a territory or a home range learns all the locations within its living area where it can escape attacks or find resources or shelter. When an individual defends a territory, it enjoys the additional benefit of having sole use of any resources present.

Go to ANIMATED TUTORIAL 40.3
The Costs of Defending a Territory
PoL2e.com/at40.3

Behavior helps structure relationships among species

For understanding behavioral relationships among species in ecosystems, ecologists often use a **cost–benefit approach**. Such an approach assumes that an animal has only a limited amount of time and energy, and therefore—in its relationships with other species—cannot afford to engage in behaviors that cost more to perform than they bring in benefits. In cost–benefit analysis, ecologists quantify both costs and benefits. Then, by calculating relative costs and benefits, they make—and test—predictions about which behaviors to expect.

A specific case studied in detail by Bernd Heinrich is the choice of flowers in the foraging ecology of bumblebees (*Bombus*). As a bumblebee forages in a flower patch, it flies from flower to flower, landing on flowers to collect nectar (**FIGURE 40.17**). During this process, the bee must keep its flight muscles warm so it will be able to fly. Keeping them warm during flight is no problem because the working muscles produce heat at a high rate. Keeping them warm when the bee has landed on a flower requires extra heat production when the air is cold for the same reason a house furnace needs to make heat at an increased rate in cold weather. Adding all these considerations together, Heinrich calculated the energy cost per minute of foraging and determined that the cost is considerably greater in cold air than in warm air.

Heinrich also calculated the benefit of foraging. To do this, he measured how much nectar could be obtained from each flower and the sugar concentration of the nectar. Then he calculated the energy a bee could obtain from each flower by metabolically oxidizing the sugar in the flower's nectar.

One common flowering plant in Heinrich's study area was a rhododendron with large flowers. A bumblebee got 1.7 joules (J) of energy by taking the nectar from each rhododendron flower. In cold air at 0°C, a bee's cost of foraging was about 12.5 J per minute. Thus a bumblebee needed to harvest the nectar from only about 7 flowers each minute for its foraging benefit to equal its foraging cost. Actually, a bee could harvest 20 flowers per minute, receiving a benefit that far exceeded its cost—and enabling the bee to take energy back to its hive. Heinrich predicted that bumblebees would forage on rhododendron flowers in both cold and warm weather, and this is what the bees do.

FIGURE 40.17 A Bumblebee Landing on a Flower to Collect Nectar

Another common plant was wild cherry. Heinrich noticed that bumblebees foraged on the cherry flowers only in warm weather. By applying his cost–benefit analysis, he discovered why. Cherry flowers are smaller than rhododendron flowers and provide only 0.2 J of energy each. Accordingly, at 0°C a bee would have to harvest more than 60 cherry flowers per minute just to get enough nectar to equal its cost of foraging. Bumblebees cannot harvest that many flowers per minute. Thus they would have to go into an energy deficit—spending more energy than they received—to forage on cherry flowers in 0°C air.

With this study, Heinrich showed that bumblebee foraging behavior is structured in ways that keep a bee in energy balance. This fact, in turn, structures the relationship between bumblebees and the plants that flower in their ecosystems.

Go to ANIMATED TUTORIAL 40.4
Foraging Behavior
PoL2e.com/at40.4

LINK

A cost–benefit approach can also be used to analyze life history trade-offs during evolution; see **Concept 42.3**

CHECKpoint CONCEPT 40.6

- ✓ How might Lee Metzgar's study (see Concept 40.2, Figure 40.4) on the survival of certain white-footed mice in the presence of screech owls help explain the existence of home range behavior?
- ✓ Imagine that on a cold day a bee requires 15 joules of energy per minute to forage continuously. If the bee can harvest 20 flowers of a certain species every minute, and gets 0.8 J of energy for each flower it visits, would it be to the bee's energetic benefit to forage on those flowers that day?
- ✓ Animals that actively defend a territory gain sole access to all the energy sources available in that territory, a major energetic benefit. What might some of the energetic costs of territory maintenance be?

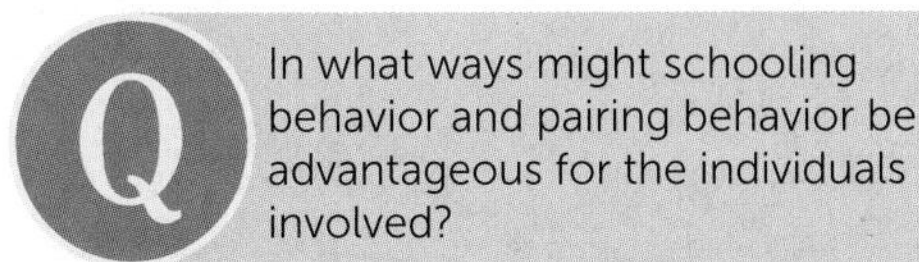

In what ways might schooling behavior and pairing behavior be advantageous for the individuals involved?

ANSWER Schooling behavior can be advantageous in several ways. For example, a school has more sense organs than a single individual. For this reason, a school may detect predators sooner than an individual so that evasive action can be taken. Similarly, a school may be able to find valuable resources in the environment more rapidly or efficiently. Pairing behavior could have several advantages—one being that in male–female pairs, each individual is assured of a mate.

SUMMARY

CONCEPT 40.1 Behavior Is Controlled by the Nervous System but Is Not Necessarily Deterministic

- Changes in behavior are an important way for animals to respond to changes in their environment. **Review Figure 40.1**
- An animal's behaviors are activated and coordinated by the animal's nervous system.
- Some animal behaviors, called **fixed action patterns**, are highly stereotyped. They do not require learning and in fact are often resistant to modification by learning. **Review Figure 40.2**
- Animal behavior has a genetic basis and is subject to natural and artificial selection. Behavior therefore evolves. **Review Figure 40.3**
- Behavior is often flexible; it can be modified by an animal's experiences.

CONCEPT 40.2 Behavior Is Influenced by Development and Learning

- Many animals are able to **learn**—meaning that individuals can alter their behaviors on the basis of previous individual experiences.
- **Behavioral imprinting** is a process by which an animal learns to respond to a specific set of stimuli during a limited period early in postnatal life and thereafter the behavioral responses are fixed.
- Early experiences can affect the behavior of an animal for the remainder of its life. Some of these effects are consequences of epigenetic effects on gene expression.

CONCEPT 40.3 Behavior Is Integrated with the Rest of Function

- Behavior and other aspects of animal function are interdependent; the behavior of an animal is constrained by its physiological capacity to perform the required actions.
- Body structure, size, and growth rate can influence behavior, as can cellular biochemistry and metabolism. **Review Figure 40.7**

CONCEPT 40.4 Moving through Space Presents Distinctive Challenges

- Animals find their way in the environment by **navigation** (moving toward a particular destination or along a particular course) and **orientation** (adopting a position or trajectory relative to an environmental cue).
- Animals sometimes navigate by following trails, which in many cases are marked with **pheromones**. Navigation may also involve path integration. During this process, an animal integrates the distances and directions it travels to maintain an awareness of its position relative to a particular site. **Review Figures 40.8 and 40.9**
- Animals may sense direction by using information on the sun, magnetic fields, landmarks, and atmospheric polarization patterns. A sun compass involves not only observing the sun's position but also knowing the time of day, information that can be provided by a circadian biological clock. Animals often have multiple, redundant mechanisms of orientation. **Review Figure 40.10 and ANIMATED TUTORIALS 40.1 and 40.2**
- Honey bees use a waggle dance to communicate the position of a food source relative to the hive. **Review Figure 40.11 and ACTIVITY 40.1**
- Some animals display remarkable feats of navigation when they **migrate**, traveling as far as thousands of kilometers between locations. **Review Figure 40.12**

CONCEPT 40.5 Social Behavior Is Widespread

- A **society** is a group of individuals of a single species that exhibits some degree of cooperative organization.
- Group living confers benefits such as greater foraging efficiency and protection from predators, but it also has costs, such as increased competition for food and ease of transmission of diseases. **Review Figures 40.13 and 40.14**
- Individuals in a society can be of equal status—as in a school of fish—or of unequal status—as in a group of impala antelopes. In some cases a single male **dominates** the other males in a group; this usually confers a reproductive advantage on the dominant male.
- An extreme form of unequal status is seen in cases of **eusociality**, in which some individuals in the group are infertile and assist the reproduction of fertile individuals, usually their mother and a few male siblings. Eusocial animals include the social insects (e.g., honey bees) and naked mole rats.

CONCEPT 40.6 Behavior Helps Structure Ecological Communities and Processes

- Behavioral differences help maintain species by decreasing the likelihood of interspecies breeding.
- An individual within a population often restricts its movements to a limited portion of the area occupied by the population as a whole. The smaller area is known as a **territory** if the individual actively keeps out other individuals of the same species, or as a **home range** if other individuals are not excluded. **Review Figure 40.16 and ANIMATED TUTORIAL 40.3**
- A **cost–benefit approach** can be used to investigate the adaptive value of specific behaviors. **Review ANIMATED TUTORIAL 40.4**

See **ACTIVITY 40.2** for a concept review of this chapter.

Go to the Interactive Summary to review key figures, Animated Tutorials, and Activities PoL2e.com/is40

Go to LaunchPad at **macmillanhighered.com/launchpad** for additional resources, including LearningCurve Quizzes, Flashcards, and many other study and review resources.

PART 7

Ecology

41 The Distribution of Earth's Ecological Systems

KEY CONCEPTS

41.1 Ecological Systems Vary over Space and Time

41.2 Solar Energy Input and Topography Shape Earth's Physical Environments

41.3 Biogeography Reflects Physical Geography

41.4 Biogeography Also Reflects Geological History

41.5 Human Activities Affect Ecological Systems on a Global Scale

A cow grazes in a mesquite-dominated former grassland in southeastern Arizona. Reducing grazing has failed to restore this region's grassland ecosystem.

The U.S.–Mexico Borderlands of southeastern Arizona have a colorful "Wild West" history peopled by Native Americans, Spanish explorers, cowboys, buffalo soldiers, outlaws and sheriffs, miners and ranchers, and railroad tycoons. This history has faded with time—except for one lasting legacy that dates from the cattle boom of the 1880s.

The first herds of cattle were brought to the Borderlands in 1687, when Spanish missions and settlements were established, but livestock numbers remained below 40,000 head until the 1880s. At that time, the end of the Apache Wars, the completion of transcontinental railroads (which opened huge markets for beef), and an influx of investment capital combined to bring 1.5 million animals to fatten on the free grass of Arizona's "open range." The sea of grass that had once grown "as high as the belly of a horse" was soon grazed to the ground.

The boom did not last long. A drought in 1892 and 1893 killed 75 percent of the cattle, leaving so many carcasses that one could walk across the San Simon Valley, it is said, without touching the ground. Prickly shrubs such as mesquite (*Prosopis* species) gained a toehold in the denuded landscape when torrential rains returned in the late 1890s, washing away soil and cutting deep gullies in the landscape. The once rich grasslands had been turned into shrublands that could support few cattle.

Soon thereafter, state and federal governments closed the "open range," substituting a lease system that would allow them to better manage public lands. They tried to restore the grasses using a method that had worked for thousands of years in the grassy meadows of Europe: adjust the number of cattle up or down depending on meadow conditions. This method is based on the assumption that grazing depletes grasses that cattle prefer and thereby favors the growth of less desirable plants. If that is the case, then reducing the number of cattle should allow the desirable species to recover. Alas, "resting the range" has not worked in the arid Borderlands: shrubs have continued to spread, and grasses have not increased measurably.

Can basic ecological principles suggest why removing cattle has not restored grasses to the Borderlands?

You will find the answer to this question on page 862.

CONCEPT 41.1 Ecological Systems Vary over Space and Time

During the Age of Exploration of the eighteenth and nineteenth centuries, the prospect of visiting unknown lands filled with unfamiliar plants and animals lured many European scientists—including Charles Darwin—to travel the world. They described and took careful measurements of a tremendous variety of environments and organisms on their journeys, thus initiating the study of two new topics: **physical geography**, the spatial distributions of Earth's climates and surface features; and **biogeography**, the spatial distributions of species. It soon became apparent that the distributions of species and environments are linked.

Organisms and their environments are ecological systems

The link between physical geography and biogeography is not surprising, because all organisms, humans included, are surrounded by and interact with an external environment that is made up of nonliving **(abiotic)** and living **(biotic)** components. These components—one or more organisms and the environment with which they exchange energy and materials—make up an **ecological system**.

The term "ecological" derives from **ecology**, a word that the German biologist Ernst Haeckel constructed in 1866 from the Greek roots *oikos*, "household," and *logia*, "study of." Haeckel meant this word to describe "the entire science of the relations of the organism to its external environment." Ecological knowledge is very old, for even the earliest humans understood a great deal about the plants and animals they depended on for food, the animals that competed with them for food, and the organisms and physical circumstances that presented natural hazards. Indeed, the earliest available religious texts and scholarly writings are full of ecological topics. By coining a formal term, however, Haeckel not only helped establish the study of "all those complicated interrelations described by Darwin as the conditions of the struggle for existence" as a formal science, but also underscored ecology's relevance to Darwin's theory of evolution by natural selection.

Ecological systems can be small or large

A system is defined by the interacting parts it contains. But scientists have a great deal of freedom to choose which set of interacting parts to study. In the case of ecological systems, we can draw a boundary around any part of the biological hierarchy from the individual organism on up (see Figure 1.5). Each successive level of this ecological hierarchy brings in new interacting parts at progressively larger scales of space and time.

At the smallest scale is the fundamental ecological unit: the individual organism and its immediate environment. Individuals remove materials and energy from the environment, convert those materials and energy into forms that can be used by other organisms, and by their presence and activities, modify the environment. In aggregate, these individual-level exchanges determine the nature and magnitude of interactions at higher levels of the hierarchy.

The individual organism is part of a **population**: a group of individuals of the same species that live, interact, and reproduce in a particular geographic area. The assemblage of interacting populations of different species within a particular geographic area forms a biological **community**. As we expand the geographic area included within the boundary of our ecological system, we include multiple communities in **landscapes** until, at the largest scale, we include all the organisms of Earth and the environments they occupy—the entire **biosphere**. Ecologists tend to replace the term "ecological system" with **ecosystem** when they are explicitly including the abiotic components of the environment, and in particular when they are considering communities and their environmental context.

Large ecological systems tend to be more complex than small ones, not only because they contain more interacting parts, but also because those parts interact over a greater span of space and time. The components of the biosphere, for example, are linked not only by the rapid and localized exchanges between individual organisms and their surroundings, but also by slower and more extensive migrations of organisms, great cycles of air and water movement, and even the much slower geological churning of Earth's crust.

Even when very small, however, ecological systems can be complex. Consider an ecosystem within each of us—the human intestine. It is one of the most densely populated ecosystems on Earth, hosting hundreds of species of bacteria, archaea, and yeasts, and on the order of 10^{12} individual microbes per gram of fecal material. The microbial cells in our bodies vastly outnumber our own trillion or so human cells, and their combined metabolism rivals that of an organ such as the human liver.

It's no surprise that microbes thrive in the mammalian gut. Mammals regulate the gut's physical environment within narrow limits of temperature and chemistry, and they supply it with a steady stream of ingested nutrients. Gut microbes metabolize food materials, including some that the host cannot digest, and excrete waste products that provide nutrition to the host or to other members of the microbial community. Microbial species interact with one another and with host cells by forming layers (biofilms) that coat the gut lining.

As with any system, these interactions determine the properties of the whole and its components: they affect host nutrition and immune function as well as the makeup of the microbial community itself. In fact, a number of human diseases—including infection by pathogenic gut microbes, irritable bowel syndrome, obesity, and wasting diseases such as kwashiorkor—have been linked to disruption of the gut's microbial community.

LINK

Human health is linked to the health of our microbial communities. Review the description of the communities of microbes that live on and in our bodies in **Concept 19.3**

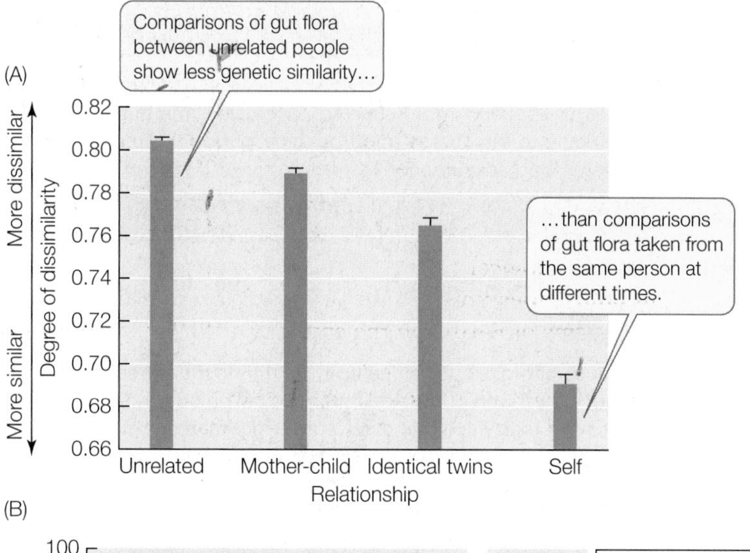

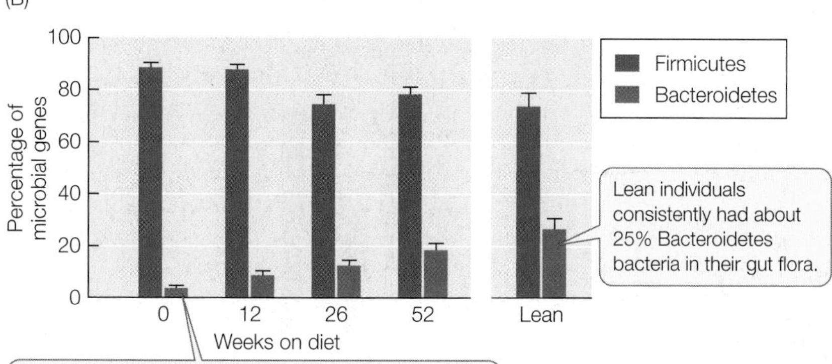

FIGURE 41.1 Genetics and Diet Affect the Composition of the Gut Microbial Community (A) The closer the genetic relationship between two people, the more similar their gut microbial communities are. (B) Obese people have proportionally more bacteria in the phylum Firmicutes, and proportionally fewer bacteria in the phylum Bacteroidetes, than do lean people. However, the microbial community of obese people approaches that of lean people as they lose weight on a calorie-restricted diet. Error bars indicate 1 standard error from the mean in both (A) and (B) (see Appendix B).

Ecological systems vary, but in ways that can be understood with scientific methods

The biotic and abiotic components of ecosystems are distributed unevenly in space, as European scientists discovered during the Age of Exploration. Individual ecosystems also change over time, as cattle ranchers in the Borderlands discovered to their dismay. Such variations may be bewildering at first, but further study often reveals repeated patterns that provide clues to their causes.

The ecosystem of the human gut illustrates this spatial and temporal variation, and scientists are using natural history observations, experiments, and modeling to understand it. With modern molecular methods (see Figure 12.6), we can now characterize the phylogenetic and functional properties of entire gut microbial communities, which is impossible with traditional culture techniques because many gut microbes won't grow outside the gut. The first surveys revealed bewildering variation: each person, it turns out, has a unique gut community. More recent work, however, has demonstrated that there is some order to this variation. The gut communities of genetic relatives, for instance, are more similar than are those of unrelated people (**FIGURE 41.1A**), suggesting that a host's genotype favors some microbes over others. In addition, the gut communities of obese and lean people differ in their ratios of two bacterial phyla. When obese people lose weight on experimental calorie-restricted diets, the ratios of these two bacterial phyla become more like those of lean people (**FIGURE 41.1B**), suggesting that diet also affects the gut community. The relationship between gut community and obesity appears to be a two-way street, because systems models of nutrient processing by different microbial groups indicate that obesity might be related to how much energy the community can extract from food. This conclusion is supported by transplant experiments done with mice, whose gut ecology is similar to ours (**FIGURE 41.2**). Exciting results such as these open the door to new therapies that use ecological principles to restore human health by manipulating our "ecosystem within."

CHECKpoint CONCEPT 41.1

- ✓ What two components make up the external environment of organisms?
- ✓ Explain why ecological systems at large scales tend to be more complex than smaller-scale systems.
- ✓ Humans often experience digestive problems after taking a course of antibiotics to treat a bacterial infection. Why might this be so?

Variation over space and time can be seen in ecosystems at all scales, from the human gut to the entire biosphere. We will now turn to some of the physical factors that underlie this heterogeneity at the biosphere scale: Earth's climates and topography.

INVESTIGATION

FIGURE 41.2 The Microbial Communities of Genetically Obese Mice Contribute to Their Obesity Mice (*Mus musculus*) that are homozygous (*ob*/*ob*) for a particular gene, *Lep*^{ob}, are obese, but their *ob*/+ and +/+ littermates are lean. Like obese humans, *ob*/*ob* mice host a gut microbial community that is relatively rich in bacteria of the phylum Firmicutes and poor in Bacteroidetes. Does this gut community contribute to the obesity of *ob*/*ob* mice, or does it simply reflect their greater food consumption?[a]

HYPOTHESIS

The gut microbial community of obese mice contributes to their obesity.

METHOD

1. Rear +/+ mice to adulthood in a germ-free environment so that their guts contain no microbes.
2. Select 9 mice at random to have their guts inoculated with fecal material obtained from obese *ob*/*ob* donors, and select 10 mice at random to be inoculated with material from lean +/+ donors.
3. Verify that the recipient mice have developed microbial communities similar to those of the donors.
4. Feed both groups of recipient mice on mouse chow containing 3.7 kcal/g. Measure the food intake of the mice over the next 2 weeks.
5. Use a noninvasive body-scanning method to measure changes in the body fat content of the two groups of recipient mice over the same 2-week period.

RESULTS

Both groups of recipients gained body fat during the 2 weeks, but those inoculated with gut material from an *ob*/*ob* donor gained more body fat than mice that received +/+ gut material, even though they ate the same amount of food. Error bars indicate 1 standard error from the mean; bars with different letters are statistically different from one another.

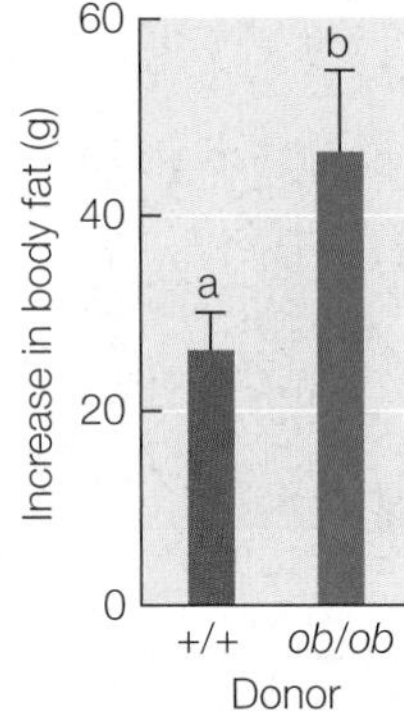

CONCLUSION

The Firmicutes-dominated gut community of obese mice contributes to their obesity.

ANALYZE THE DATA

Bacteria in the phylum Firmicutes are particularly adept at breaking indigestible dietary polysaccharides into monosaccharides and fatty acids that mammals can absorb. The researchers therefore hypothesized that the reason a Firmicutes-dominated gut community contributes to obesity is that it is more efficient at extracting energy from food. To evaluate this hypothesis, the researchers collected all feces produced in 24 hours by 13 *ob*/*ob* mice and 9 +/+ or *ob*/+ littermates, then burned them in a closed chamber and used the temperature change to measure the energy content of each gram of feces. Error bars indicate 1 standard error from the mean; bars with different letters are statistically different from one another.

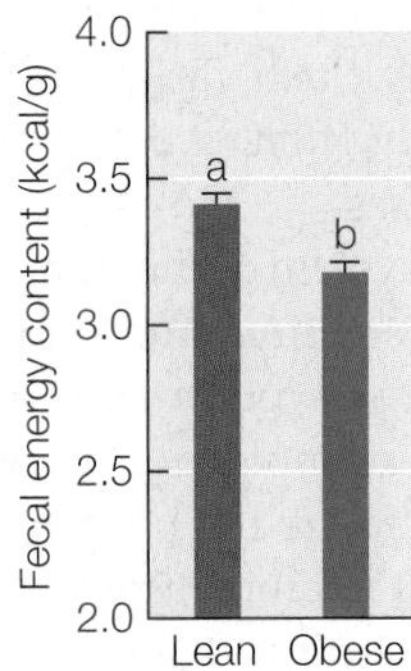

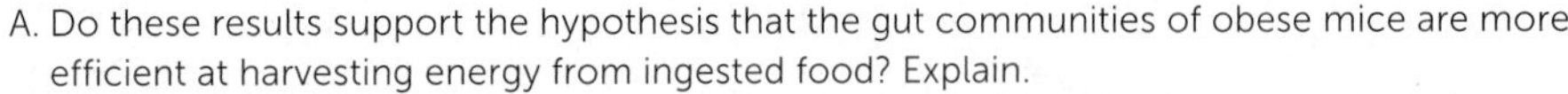

A. Do these results support the hypothesis that the gut communities of obese mice are more efficient at harvesting energy from ingested food? Explain.

B. Does it matter to your conclusion that *ob*/*ob* mice typically eat more chow than +/+ or *ob*/+ mice?

C. Assume that mice store all the excess energy they obtain in the form of body fat, and that mouse fat contains 9.3 kcal/g. If the *ob*/*ob* recipient mice added 0.44 g more body fat than the +/+ recipients in 24 hours, how much more energy did they obtain from their food in this time period?

D. If both groups of recipients consumed 55 g of chow, what percentage of the energy consumed in chow does this extra energy gain represent? (Hint: Recall that the energy content of chow is 3.7 kcal/g.)

Go to **LaunchPad** for discussion and relevant links for all **INVESTIGATION** figures.

[a]P. J. Turnbaugh et al. 2006. *Nature* 444: 1027–1031.

CONCEPT 41.2 Solar Energy Input and Topography Shape Earth's Physical Environments

Earth's physical environments vary enormously—from hot to cold, wet to dry, aquatic to terrestrial, salt water to fresh water. To this variation across space we must add variation over time, on scales from the very slow coming and going of oceans, mountains, or ice ages to the very fast shifts in temperature as clouds pass overhead. On the short time scales that we humans experience, physical conditions depend largely on solar energy input, which drives the circulation of the atmosphere and the oceans. Patterns of input and circulation determine whether a particular place at a particular time of year tends to be hot or cold, wet or dry—and as a consequence, the types of organisms found in that place.

LINK

Review the effects that changes in Earth's physical environments have had on the evolution of life in **Concept 18.2**

Variation in solar energy input drives patterns of weather and climate

All of us are familiar with **weather**, the atmospheric conditions—temperature, humidity, precipitation (rainfall or snowfall), and wind direction and speed—that we experience at a particular place and time. Averaged over long periods, the state of these atmospheric conditions and their pattern of variation over time constitute the **climate** of a place. In other words, climate is what you expect; weather is what you get. We and other organisms respond to both weather and climate. Responses to weather are usually short-term: animals may seek shelter from a sudden rainstorm (human animals open their umbrellas), and plants may close their stomata (small openings in the leaf epidermis) on a hot summer day to reduce water loss from their tissues. Responses to climate, by contrast, tend to involve adaptations that prepare organisms for expected conditions. Organisms in hot climates can tolerate heat; those that inhabit seasonal climates use cues such as day length to prepare for an impending hot, cold, or dry season.

Some global patterns of climate stem from variation in solar energy input over space. Because Earth is spherical, the sun's rays strike the planet at a shallower angle near the poles than near the equator (**FIGURE 41.3**). This difference results in less solar energy input at high latitudes (near the poles) than at low latitudes (near the equator), for two reasons. First, the energy of rays of sunlight is spread over a larger surface area at high latitudes. Second, more sunlight is reflected back into space by the atmosphere and Earth's surface when sunlight strikes at a shallow angle. A consequence is that air temperatures decrease from low latitudes to high latitudes. Averaged over the course of a year, air temperature drops by about 0.8°C with every degree of latitude (about 110 km) north or south of the equator.

At any given place, the sun's input also varies through the year, causing seasonal variation in climate. This **seasonality** is a consequence of the 23.5° tilt of Earth's axis of rotation relative to the plane of its yearly orbit around the sun. The tilt causes different latitudes to receive their greatest solar energy input at different times of the year (**FIGURE 41.4**). At the solstice in late June, the North Pole tilts toward the sun, and the sun is directly above the Tropic of Cancer (latitude 23.5° N): the Northern Hemisphere experiences summer, with its long days and warm temperatures. At the solstice in late December, the North Pole points away from the sun, the sun is directly above the Tropic of Capricorn (latitude 23.5° S), and the climate pattern is reversed: it is summer in the Southern Hemisphere and winter in the Northern Hemisphere. At the equinoxes in late March and late September, the sun is directly above the equator, and both hemispheres experience equal lengths of day and night. These seasonal variations in solar input are relatively small at the equator and become more pronounced toward the poles.

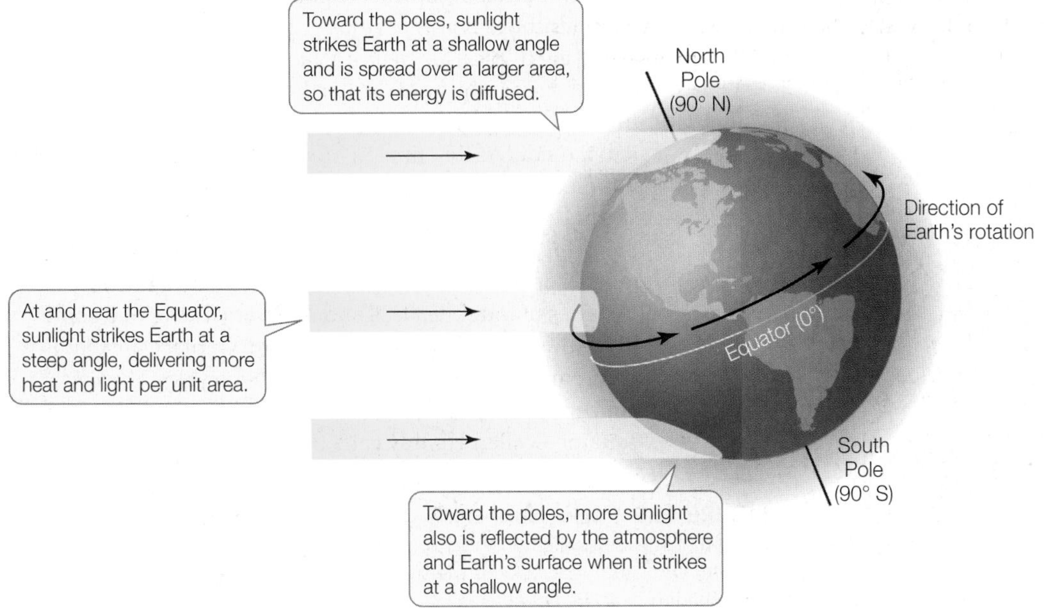
Toward the poles, sunlight strikes Earth at a shallow angle and is spread over a larger area, so that its energy is diffused.
North Pole (90° N)
Direction of Earth's rotation
At and near the Equator, sunlight strikes Earth at a steep angle, delivering more heat and light per unit area.
Equator (0°)
South Pole (90° S)
Toward the poles, more sunlight also is reflected by the atmosphere and Earth's surface when it strikes at a shallow angle.

FIGURE 41.3 Solar Energy Input Varies with Latitude The angle of incoming sunlight affects the amount of solar energy that reaches a given area of Earth's surface. On average, sunlight strikes Earth's surface at a steeper angle at low than at high latitudes. This results in greater input of energy at low (i.e., equatorial) latitudes because the energy is spread over a smaller area and because less sunlight is reflected back into space by the atmosphere and land surface.

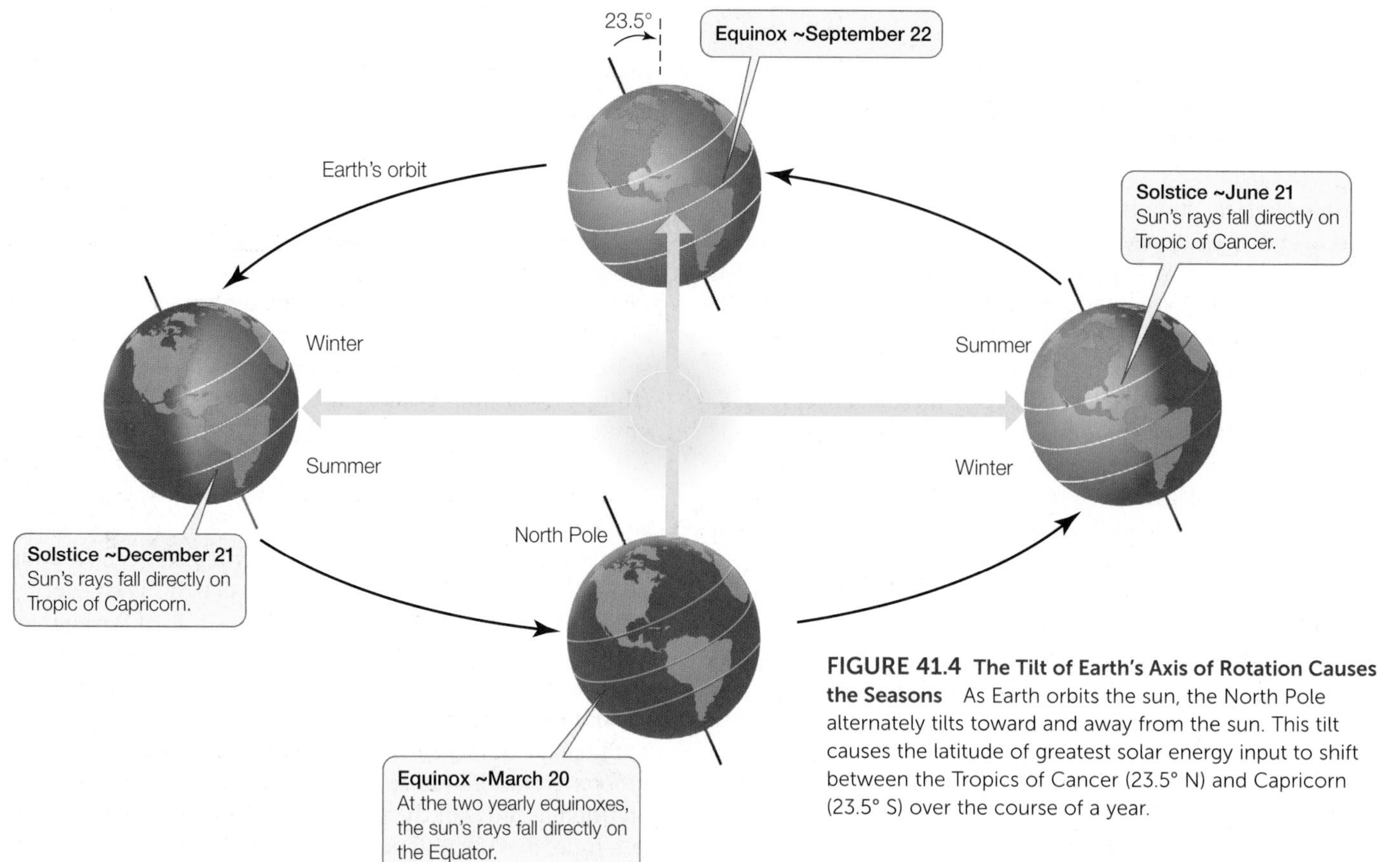

FIGURE 41.4 The Tilt of Earth's Axis of Rotation Causes the Seasons As Earth orbits the sun, the North Pole alternately tilts toward and away from the sun. This tilt causes the latitude of greatest solar energy input to shift between the Tropics of Cancer (23.5° N) and Capricorn (23.5° S) over the course of a year.

The circulation of Earth's atmosphere redistributes heat energy

The yearly input of solar energy is greatest in the **tropics** (the latitudes between the Tropics of Cancer and Capricorn) and decreases poleward. This latitudinal gradient in solar input drives the vast circulation patterns of Earth's gaseous atmosphere.

As the sun beats down on the tropics, the air there heats up—its molecules move faster (temperature is a measure of disordered molecular motion), colliding and pushing one another farther apart (see Concept 2.5). The heated air expands, becoming less dense and more buoyant, and rises. As it rises, it continues to expand and cool **adiabatically** (this is the principle on which an air conditioner works: when compressed gas is allowed to expand, it cools). The air rises to a level at which the surrounding air has its same temperature and density. Because tropical air gets so warm, it ascends to great altitudes before it stops rising.

Rising tropical air creates a vacuum that pulls in cooler, denser surface air from the north and south. As that air in turn heats, expands, and rises, it pushes the cool air above it to the side. As this air aloft moves away from the tropics, it radiates heat to outer space, becoming more dense and less buoyant. It sinks back toward Earth, is compressed by the overlying atmosphere, and warms adiabatically (the reverse of adiabatic cooling). It reaches Earth's surface at latitudes of roughly 30° N and 30° S, at which point some of it flows back toward the equator to replace air that is rising in the tropics. Thus it forms two cycles of vertical atmospheric circulation—one north and one south of the equator—called **Hadley cells** (**FIGURE 41.5**).

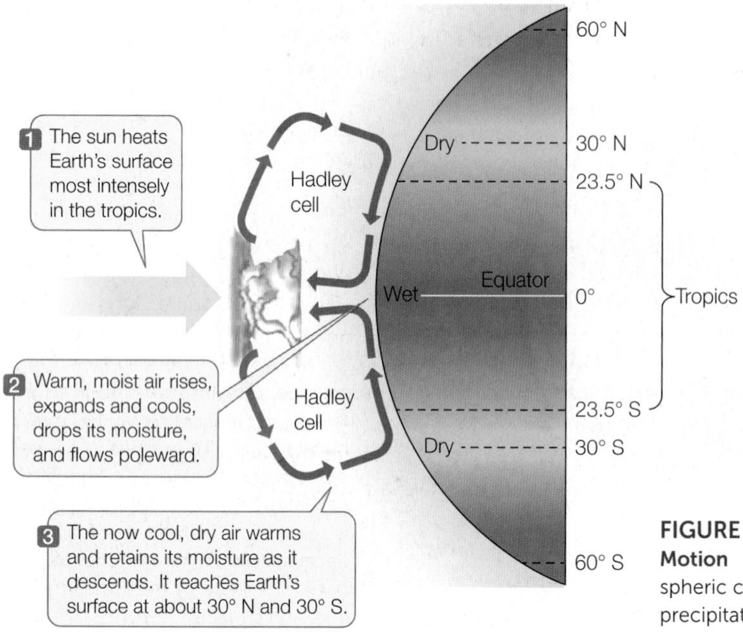

FIGURE 41.5 Tropical Solar Energy Input Sets the Atmosphere in Motion Intense sunlight in the tropics drives a vertical pattern of atmospheric circulation that reaches the subtropics and that results in high precipitation near the equator.

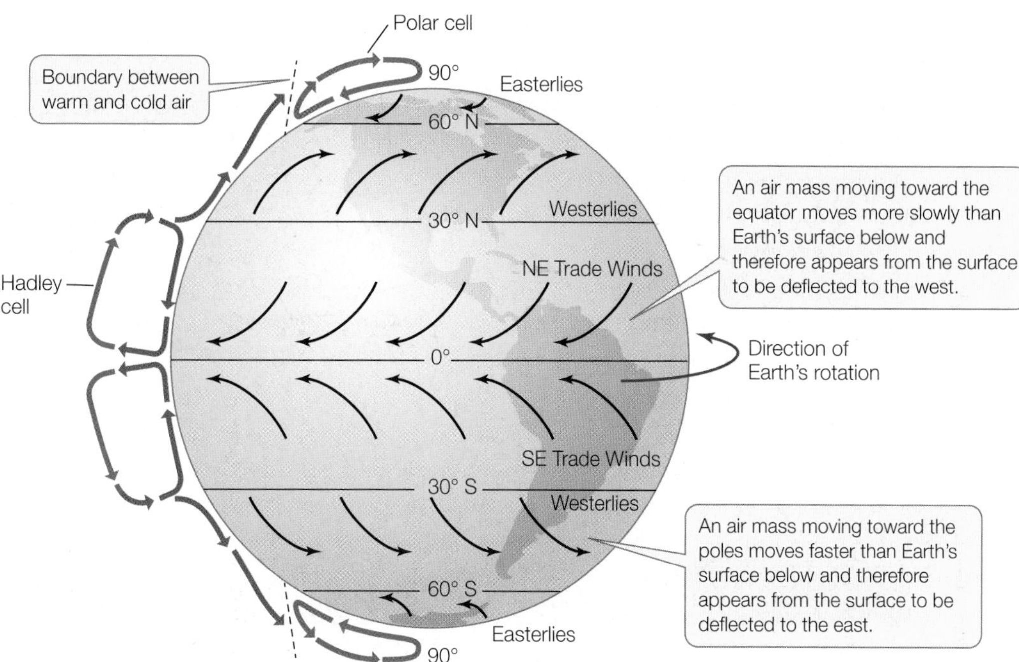

FIGURE 41.6 Global Atmospheric Circulation and Prevailing Winds Because the speed of rotation of Earth's surface decreases with latitude, air flowing toward the equator lags behind the surface beneath it, whereas air flowing toward the poles moves faster than the surface beneath it. The resulting Coriolis effect explains patterns of prevailing surface winds.

The Hadley cell circulation produces strong latitudinal patterns of precipitation. As water molecules in the tropics (most notably the oceans) absorb the intense sunlight there, they evaporate, so the air that rises in the tropics is rich in water vapor. As this warm, moist air expands and cools, fewer water molecules are moving fast enough to bounce apart after they collide. Instead, they form droplets of liquid water (or ice crystals), which grow and—when massive enough—fall to the ground as precipitation. Thus the warm, rising air of the Hadley cells drops most of its moisture near the equator as rain. The high-altitude air that eventually descends near 30° N and 30° S is largely depleted of water; moreover, because it warms as it descends and is compressed, the little water that remains stays in a gaseous state. Earth's great deserts are located around these latitudes.

Not all the descending air in the Hadley cells returns immediately to the tropics. Some of it continues to flow toward the poles, encountering and overriding frigid, dense polar air that is flowing equatorward as part of polar circulation cells that are smaller and weaker than the Hadley cells. The interaction of these warm and cold air masses generates the spinning winter storms that sweep from west to east through the middle latitudes between the tropics and the polar regions.

To better understand these storms and other patterns of horizontal air movement, we must consider one more thing—Earth's rotation on its axis from west to east (see Figure 41.3). We have already seen that the latitudinal gradient in solar input creates a latitudinal, or north–south, movement of air. The planet's rotation adds an east–west component to air movement. Because Earth is a sphere, the rotation of its surface is fastest at the equator, where its circumference is greatest, and slower toward the poles. Surface air that is not moving north or south is carried eastward at the same speed as the surface beneath it. But air that is moving toward the equator lags behind the faster-moving surface, and therefore appears from the surface to be deflected to the west. Conversely, air that is moving poleward races ahead of the surface and appears to be deflected to the east. This **Coriolis effect** causes prevailing surface winds to blow from east to west in the tropics (the northeast or southeast trade winds, so called because of their importance during the era of trade by sailing ships), from west to east in the middle latitudes (the westerlies), and from east to west again above 60° N and 60° S latitude (the easterlies; **FIGURE 41.6**). The Coriolis effect also leads to the rotation of air characteristic of middle-latitude storms.

These aspects of atmospheric circulation combine to affect climate patterns by transferring an immense amount of heat energy from the hot tropics to the cold poles. Without this transfer, the poles would sink toward absolute zero in winter, and the equator would reach fantastically high temperatures throughout the year.

Ocean circulation also influences climate

The prevailing surface winds and Coriolis effect drive massive circulation patterns in the waters of the oceans, called currents, which carry materials, organisms, and heat with them. In the northern tropics, the trade winds drag surface water toward the west. When this water nears the western edge of an ocean basin, much of it flows northward, only to be turned to the east by the Coriolis effect and, at higher latitudes, by drag from the westerlies of the northern middle latitudes. This eastward flow turns south along the western edge of the continents and then westward again, completing a clockwise circuit, or **gyre**. The pattern in the Southern Hemisphere is essentially a mirror image of that just described, with surface circulation moving in counterclockwise rather than clockwise rotation (see Figure 41.7). The westward-flowing tropical water that is not caught up in these gyres flows back to the east in an equatorial countercurrent.

As they move poleward, tropical surface waters transfer heat from low to high latitudes, adding to the heat transfer that accompanies atmospheric circulation. The Gulf Stream and North Atlantic Drift, for example, bring warm water from the tropical

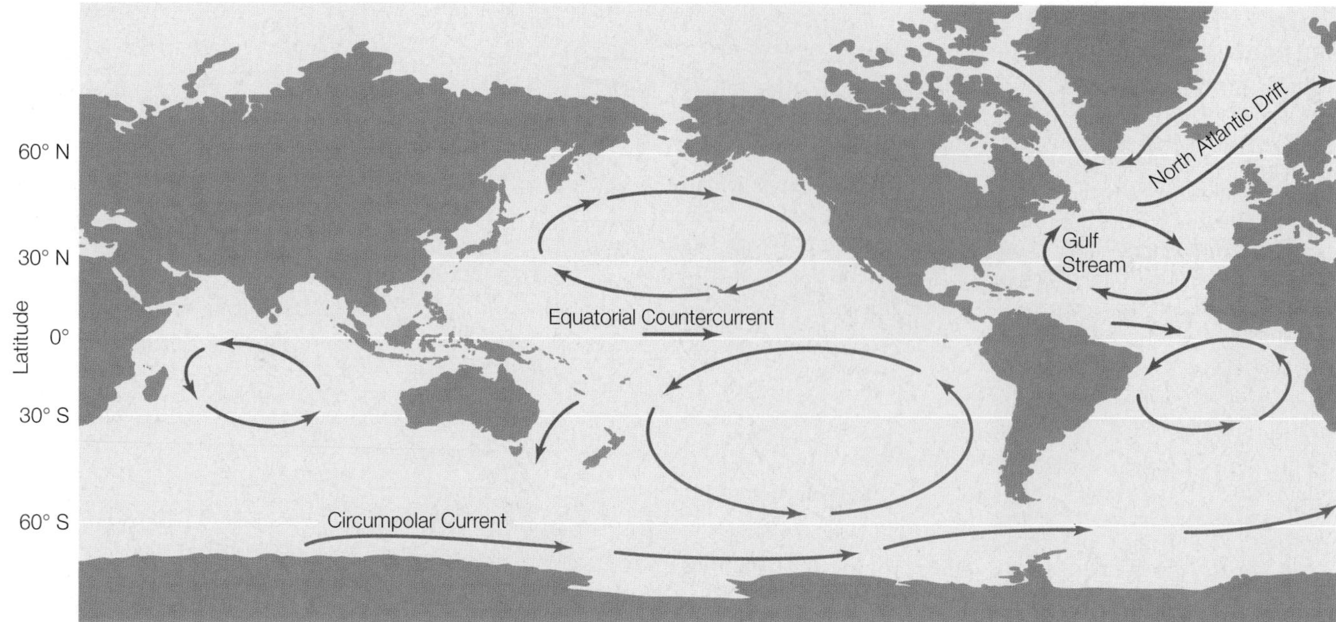

FIGURE 41.7 Ocean Surface Currents Ocean surface currents are characterized by giant patterns (gyres) of clockwise rotation in the Northern Hemisphere and counterclockwise rotation in the Southern Hemisphere, caused by frictional drag from prevailing winds (see Figure 41.6), the Coriolis effect, and deflection at the edges of continents.

Go to MEDIA CLIP 41.1
Perpetual Ocean Currents
PoL2e.com/mc41.1

Atlantic Ocean and the Gulf of Mexico north and east across the Atlantic (**FIGURE 41.7**), which warms the air above the North Atlantic. Westerlies then carry this air across northern Europe, moderating temperatures there. The effect on organisms is dramatic: polar bears roam the shores of Hudson Bay in Canada but are missing from comparable latitudes in Scotland or Sweden.

Surface currents are not the entire story, however. Gradients in water density caused by variation in temperature and salinity cause surface waters to sink in certain regions, forming deep currents. These deep currents return to the surface in areas of upwelling, completing a vertical ocean circulation. Thus oceanic circulation, like atmospheric circulation, is three-dimensional.

Oceans and large lakes moderate Earth's terrestrial climates because water has a high heat capacity, as we saw in Concept 2.2. This means that the temperature of water changes relatively slowly as it exchanges heat with the air it contacts and as it absorbs or emits radiant energy. As a result, water temperatures fluctuate less with the seasons and with the day–night cycle than do land temperatures, and the air over land close to oceans or lakes also shows less seasonal and daily temperature fluctuation.

Topography produces additional environmental heterogeneity

Earth's surface is not flat; it is crumpled into mountains, valleys, and ocean basins. This variation in the elevation of Earth's surface, known as **topography**, affects the physical environment.

Mountains, for example, are "weather formers." Because rising air expands and cools adiabatically, temperatures become cooler as elevation (distance above sea level) increases. If you climb a mountain, you will find that the temperature drops by about 1°C for each 220 meters of elevation gained. Where prevailing winds encounter a mountain range, air flows up the windward side of the range (the side facing into the wind). As the air cools, water vapor may condense or crystallize and fall as rain or snow. The leeward side of the mountain range (the side away from the wind) receives little precipitation because air descending on that side warms and retains any remaining

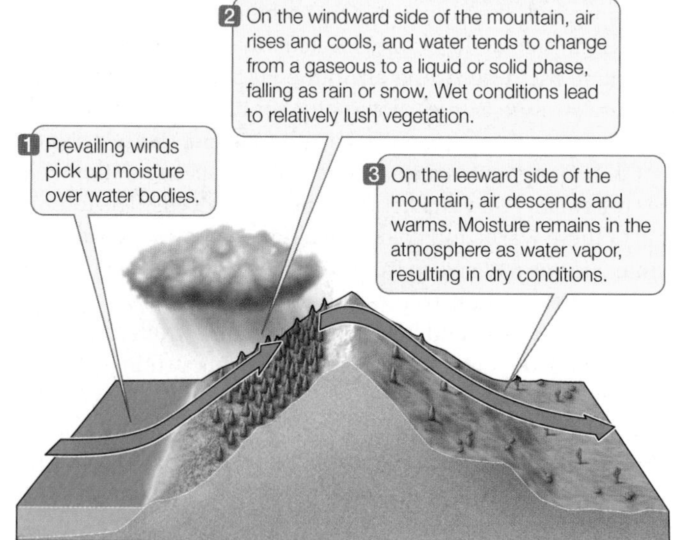

1 Prevailing winds pick up moisture over water bodies.
2 On the windward side of the mountain, air rises and cools, and water tends to change from a gaseous to a liquid or solid phase, falling as rain or snow. Wet conditions lead to relatively lush vegetation.
3 On the leeward side of the mountain, air descends and warms. Moisture remains in the atmosphere as water vapor, resulting in dry conditions.

FIGURE 41.8 A Rain Shadow Precipitation tends to be lower on the leeward side of a mountain range than on the windward side.

Go to ANIMATED TUTORIAL 41.1
Rain Shadow
PoL2e.com/at41.1

APPLY THE CONCEPT

Solar energy input and topography shape Earth's physical environments

Shown below are climate diagrams for three different localities. (Note the discontinuity in the precipitation axis in graph C.) Using the information in these diagrams and what you've learned about factors that determine climate, answer the following questions and explain the logic that led to your answer.

1. Which locality is near the equator?
2. Which locality has the greatest seasonality in temperature?
3. Which locality has the longest growing season?
4. Which locality is in the Northern Hemisphere?

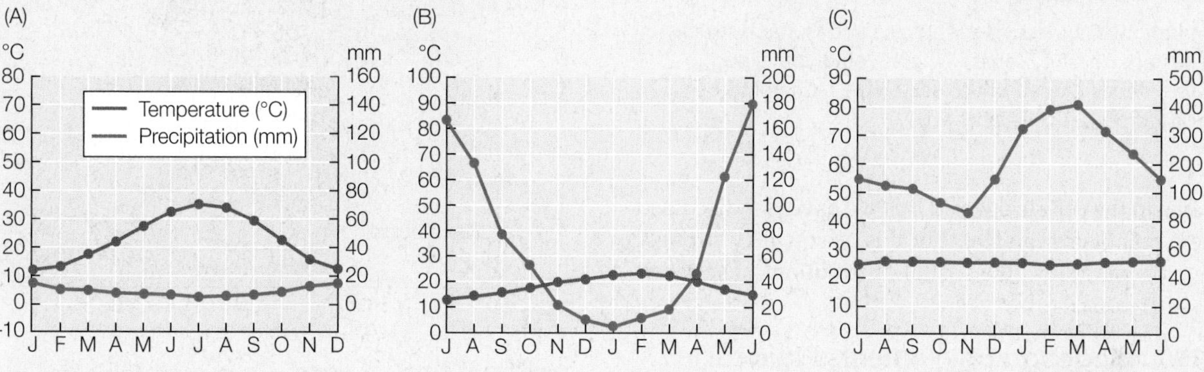

water in a gaseous state. Thus the mountain range is said to form a **rain shadow** (**FIGURE 41.8**). The difference in precipitation fosters more luxuriant growth of plants on the windward side than on the leeward side.

Topography also influences physical conditions in aquatic environments. The velocity of water flow, for example, is dictated by topography: water flows rapidly down steep slopes, more slowly down gradual slopes, and pools as lakes, ponds, or oceans in depressions. The depth of a water-filled depression determines the gradients of many abiotic factors, including temperature, pressure, light penetration, and water movement (see Figure 41.13).

Climate diagrams summarize climates in an ecologically relevant way

The climate at any location can be summarized in a **climate diagram**. Recognizing that ecological processes are strongly influenced by temperature and moisture, the German biogeographer Heinrich Walter superimposed graphs of average monthly temperature and average monthly precipitation through the year. He scaled the axes of the two graphs to incorporate the rule of thumb that plant growth requires at least 20 millimeters of precipitation per 10°C of temperature above freezing (0°C). This scaling makes it easy to see when conditions allow terrestrial plant growth—when temperature is greater than 0°C and the precipitation line is above the temperature line. This period is known as the growing season (**FIGURE 41.9**).

FIGURE 41.9 Climate Diagrams Summarize Climate in an Ecologically Relevant Way Climate diagrams traditionally combine graphs of average temperature (left axis) and average precipitation (right axis) throughout the year, as shown here for San Simon Valley near the Arizona–New Mexico border. The summer solstice at the location in question (late June in the Northern Hemisphere, late December in the Southern hemisphere) is placed at the center of the diagram. The axes are scaled such that precipitation favors plant growth when the precipitation line is above the temperature line. The months highlighted in green are typically favorable for plant growth.

CHECKpoint CONCEPT 41.2

- ✓ What is the difference between weather and climate?
- ✓ Outside the tropics, in the Northern Hemisphere we find plants adapted to warmer and drier conditions on the south-facing slopes of hills and mountains. In the Southern Hemisphere, however, such plants are found on north-facing slopes. Why?
- ✓ How would Earth's climates be different if Earth's axis of rotation were not tilted?
- ✓ In March 2011 a massive earthquake and tsunami struck the eastern coast of Japan. Where would you expect floating debris carried out to sea by the tsunami to be deposited on beaches, and why?

The patterns in the physical world that we have just discussed limit where organisms can live. For this reason, they strongly influence biogeography—the spatial distributions of species.

CONCEPT 41.3 Biogeography Reflects Physical Geography

An organism's physiology, morphology, and behavior—all aspects of its phenotype—affect how well it can tolerate a particular physical environment. For this reason, the physical environment of a place greatly influences what species can live there. Furthermore, we expect species that occur in similar environments to have evolved similar phenotypic adaptations to those conditions.

Similarities in terrestrial vegetation led to the biome concept

The European scientist–explorers who traveled the globe noticed that the vegetation found in climatically similar regions on different continents was strikingly convergent. This observation led to the concept of a **biome**: a distinct physical environment that is inhabited by ecologically similar organisms with similar adaptations. The species that occupy the same biome in geographically separate regions are often not closely related. Their similarities therefore reflect convergent evolution by natural selection (see Concept 16.1).

Terrestrial biomes are generally distinguished by the characteristics of their vegetation—whether the dominant plants are woody trees or shrubs, grasses, or broad-leaved herbaceous plants, and whether their leaves are big or small, deciduous (dropped seasonally) or evergreen. Characteristics of organisms other than plants also distinguish the biomes, since these characteristics represent adaptations to the vegetation as well as the climate. The distribution of terrestrial biomes is broadly determined by annual patterns of temperature and precipitation (**FIGURE 41.10**). Latitudinal gradients in average temperature and differences in its seasonal fluctuation separate tropical, subtropical, temperate, boreal, and polar biomes. Within those broad temperature zones, average annual precipitation and its pattern of seasonal fluctuation predict what biome will be found. In tropical biomes, for example, the length of the dry season determines

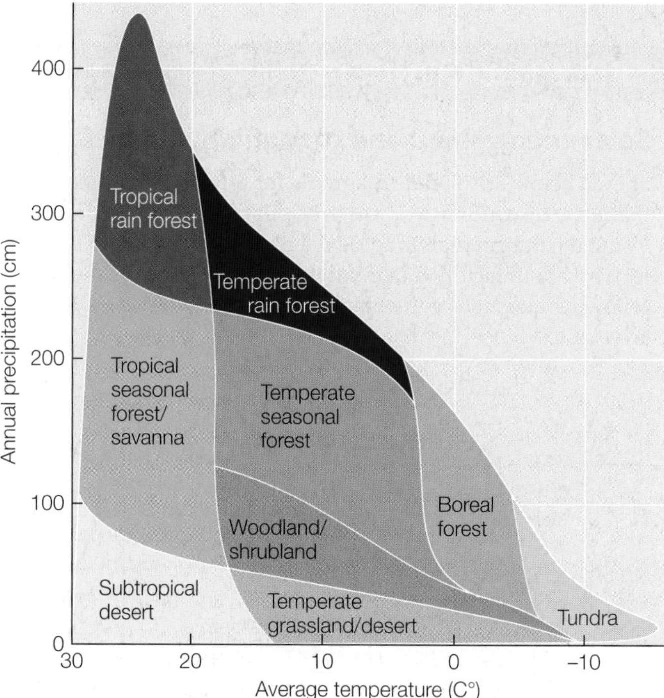

FIGURE 41.10 Terrestrial Biomes Reflect Average Annual Temperature and Precipitation Average annual temperature and precipitation strongly influence the structure of terrestrial vegetation.

whether tropical rainforest, tropical seasonal forest, or savanna develops. Elevation-associated gradients in temperature also determine the distribution of biomes; for example, tundra can be found on mountaintops, even in the tropics (**FIGURE 41.11**).

LINK

Some plant species can tolerate the cold temperatures of the tundra. You can find out more about plant adaptations to various environments in **Concept 28.3**

Climate is not the only factor that molds terrestrial biomes

Other features of the physical environment interact with climate to influence the character of terrestrial vegetation. Southwestern Australia, for example, has a "Mediterranean" climate with hot, dry summers and cool, moist winters. This climate generally supports woodland/shrubland vegetation. Succulent plants, which are well adapted to summer drought, are common in Mediterranean climates—but not in Australia, which has nutrient-poor soils. The lack of nitrogen in these soils makes it expensive for plants to construct the nitrogen-rich photosynthetic machinery of leaves. As a result, many plants produce long-lived leaves that are defended against herbivores in ways that do not require nitrogen (see Concept 28.2). These defenses include tough, indigestible cell walls, compounds that reduce the nutritional quality of leaves, and toxic oils and resins—features that also make the vegetation highly flammable. Plant material produced during the winter rains dries out during the summer and fuels intense fires that periodically sweep across the landscape. As a result, succulent plants, which are easily killed by fires, are rare in southwestern Australia.

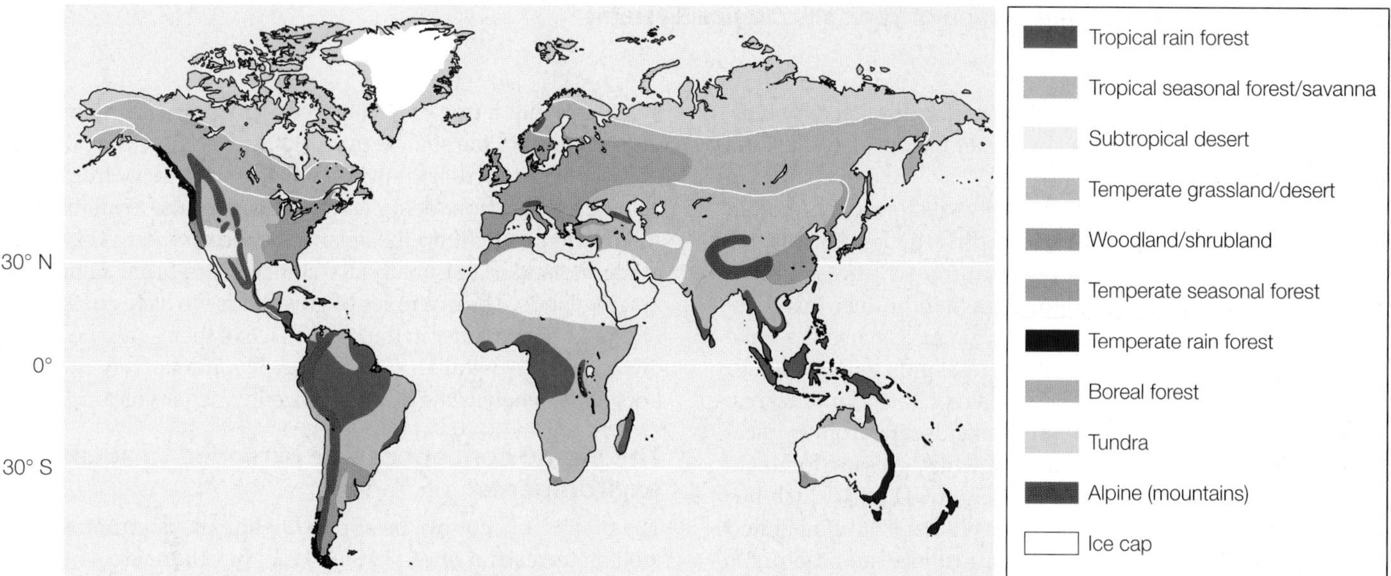

FIGURE 41.11 Global Terrestrial Biomes The global distribution of terrestrial biomes reflects latitudinal and elevational gradients in temperature and precipitation. The same biomes can be found in widely separated locations. (Note that tundra occurs on the tops of high mountains, even in the tropics, but in patches too small to see on this map.)

Go to ANIMATED TUTORIAL 41.2
Terrestrial Biomes
PoL2e.com/at41.2

Soil nutrients and fire bring us back to the desert grasslands of the U.S.–Mexico Borderlands. Grasslands—areas dominated by grasses and herbs rather than trees or shrubs—normally occur where precipitation is too scanty, or too sporadic, to support forests but is more plentiful than is typical of deserts. Fully one-fourth of Earth's land surface experiences such conditions, and much of this area supports grasslands. But grasslands are also found in some places where we do not expect them (**FIGURE 41.12**). Their presence in these places demonstrates that biome

FIGURE 41.12 Similar Vegetation Types, Different Conditions Landscapes dominated by grasses, rather than trees or shrubs, are found throughout the world, sometimes under very different conditions. (A) The desert grassland of the U.S.–Mexico Borderlands, as it still appears in some places that have not been transformed into shrubland. (B) Grassland savanna of the Serengeti Plain, Tanzania, East Africa. (C) "Spinifex" grassland (genus *Triodia*) in Northern Territory, Australia. (D) Low-nutrient grassy meadows ("Magerwiesen") in the Swiss Alps.

boundaries are fuzzy and are not perfectly predicted by temperature and precipitation because other factors also affect the character of vegetation.

One such factor is fire, which rarely kills grasses but often kills shrubs and trees. Before humans arrived, the U.S.–Mexico Borderlands experienced periodic natural wildfires, ignited by lightning, that produced desert grassland rather than desert shrubland (see Figure 41.12A). Humans themselves have contributed to the establishment of grasslands in many areas. Fires lit by humans in the Serengeti Plain of northern Tanzania, Africa, may have helped convert the landscape from tropical seasonal forest to savanna, a biome in which grasses dominate and trees are widely scattered (see Figure 41.12B). Such human effects are not necessarily recent. Humans have inhabited Australia for at least 50,000 years, for instance, and their fires appear to have transformed some of the original woodlands in the arid "red center" of the continent into grasslands dominated by "spinifex" grasses (genus *Triodia*; see Figure 41.12C). Ecosystems in which grasses dominate are also found in other places where we might expect to find forests. When humans brought domesticated livestock into temperate regions such as the European Alps, they often converted forests into grassy meadows to feed their animals (see Figure 41.12D).

The properties of soils interact with climate, fire, or grazing to affect the character of vegetation. Trees and shrubs grow particularly slowly in nutrient-poor soils. Lack of sufficient nutrients might prevent a tree or shrub seedling from sending its roots deep enough in the soil to survive the next dry season, or from growing large enough to survive the next wildfire, or from being able to recover from being eaten. If so, then the availability of soil nutrients might tip the balance between woody vegetation and grassland in seasonally dry climates, fire-prone regions, or grazing lands. The examples of grasslands we have chosen (see Figure 41.12) do differ in their soils. Two of them—alpine meadows and the Australian spinifex—grow on relatively nutrient-poor soils, whereas the other two occupy richer volcanic soils.

The biome concept can be extended to aquatic environments

The biome concept can be applied to aquatic environments as well as terrestrial ones (**TABLE 41.1**), but there are some differences. In aquatic environments, there is no single group of organisms, like terrestrial plants, that is useful for distinguishing biomes. Furthermore, climate (atmospheric conditions of temperature, precipitation, and the like) is less important in determining what organisms grow in a region than are such things as water depth and movement, temperature, pressure, salinity, oxygen content, and the characteristics of the substrate.

Go to ANIMATED TUTORIAL 41.3
Aquatic Biomes
PoL2e.com/at41.3

TABLE 41.1 Major Aquatic Biomes

Biome[a]	Description
FRESHWATER	
Rivers and streams	Flowing water. Many small, fast-flowing source streams form on high ground, feeding into networks of ever larger, slower-flowing streams and rivers. Biota adapted to constantly moving water.
Wetlands	Glades, swamps, and marshes. Biota adapted to water-saturated soil or standing fresh water.
Ponds and lakes	Significant bodies of standing fresh water. Ponds are smaller and shallower, subject to drying. Biotic zones determined by distance from shore and light penetration (see Figure 41.13A).
ESTUARINE	
Salt marshes	Cool-temperate stands of salt-tolerant grasses, herbaceous plants, and low-growing shrubs. Crucial to nutrient cycling and coastal protection. Habitat supports diverse aquatic and terrestrial life.
Mangrove forests	Tropical and warm subtropical coasts and river deltas. Dominated by mangrove trees with aerial roots (see Figure 28.11A). Rich in animal life; protect against coastal erosion.
MARINE	
Intertidal	Sandy or rocky coastlines subject to rising and falling tides; organisms adapted to withstand both submerged and dry conditions, as well as the force of waves and moving water.
Kelp forests	Found in shallow coastal waters of temperate and cold regions. Dominated by large, leaflike brown algae (kelp) that support a wide variety of marine life.
Seagrass beds	"Meadows" of grasses (see Figure 21.29C) found in shallow, light-filled temperate and tropical waters.
Coral reefs	Species-rich, highly endangered ecosystems of shallow tropical waters. Dependent on corals (see Figure 23.9C) and their photosynthetic endosymbionts (see Concept 20.4).
Open ocean	The pelagic zone (see Figure 41.13B) hosts photosynthetic planktonic (free-floating) organisms that support a rich array of marine animals. At the greatest depths, below the level of light penetration, the abyssal zone supports a fauna largely dependent on detritus that sinks down from pelagic regions.
Hydrothermal vents	Abyssal ecosystems warmed by volcanic emissions. Chemoautotrophic prokaryotes (see Concept 19.3) nourish large annelid worms (see Figure 23.17B) and other invertebrates.

[a]A benthic region—a bottom zone containing silt, sand, or some other substrate and the organisms that occur there—is part of all three biome types.

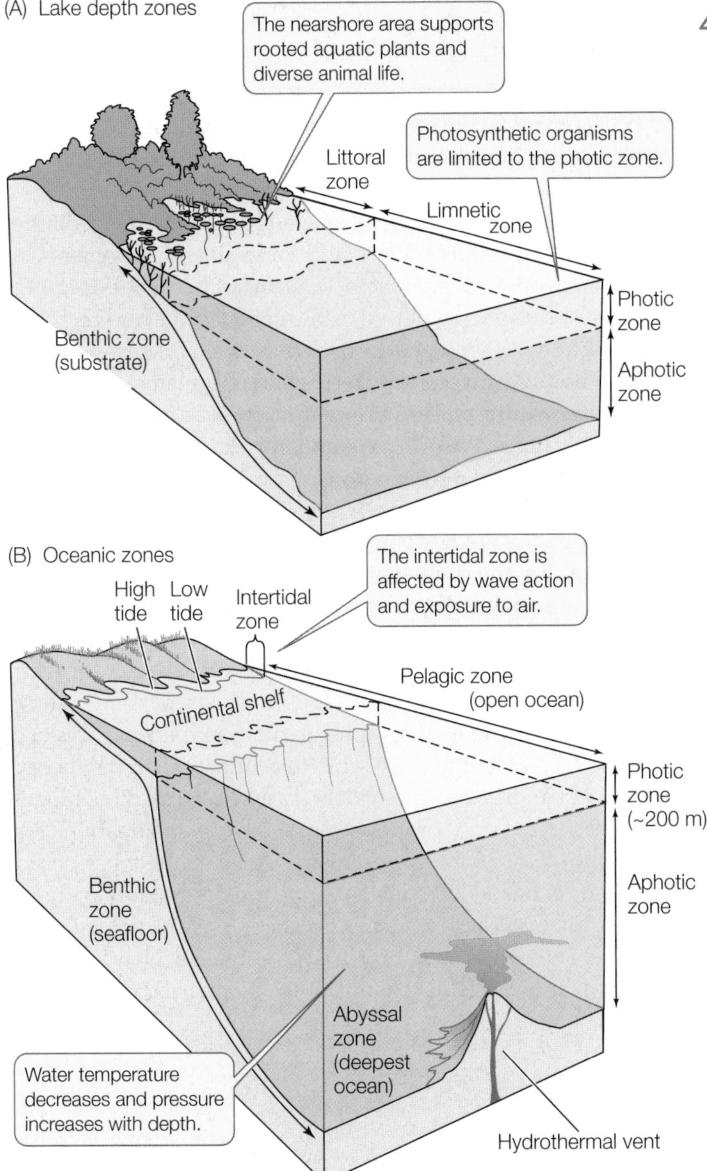

FIGURE 41.13 Water-Depth Zones Freshwater (A) and marine (B) environments can be divided into water-depth zones.

Salinity distinguishes freshwater biomes (streams, ponds, and lakes), saltwater biomes (salt lakes and oceans), and estuarine biomes (at river mouths where fresh and salt water mix). Salt concentration strongly affects the ability of aquatic organisms to regulate the water content of their bodies (see Concept 36.1) and thus influences which species are found where.

Freshwater biomes can be distinguished by water movement. Streams form wherever precipitation exceeds the amount of water lost to the atmosphere or soil, and the excess water flows downhill. Streams start out small but grow as they join together, forming rivers that eventually reach a lake or ocean in most cases. Within streams, the velocity of the water's flow determines the strength of the current against which fish must swim and the amount of force that can dislodge bottom-dwelling insects and algae from the streambed. Water flow also determines the presence or absence of sediment. Rapid water flow scours away sediment, exposing rocky surfaces to which organisms can cling. As water slows, it deposits sediment, forming a soft bottom in which organisms can burrow.

Nonflowing biomes such as lakes and oceans can be divided into water-depth zones (**FIGURE 41.13**). The nearshore regions of lakes (**littoral zone**) and oceans (littoral or **intertidal zone**) are shallow, affected by wave action, and periodically exposed to air by fluctuations in water level. Here, different species live in different depth zones according to their tolerance of heat, desiccation, and wave energy.

Because only the surface water is in contact with air, dissolved oxygen concentrations are highest in surface waters. Light, too, penetrates only a short distance into water. Photosynthetic organisms are confined to this zone of light penetration, called the **photic zone**. Rooted aquatic vascular plants and their multicellular algal equivalents (as well as corals with their photosynthetic algal symbionts) grow where the photic zone extends to the bottom, sometimes forming communities with a complex vertical structure much like that of terrestrial plant communities. In the open-water **limnetic zone** of lakes and the **pelagic zone** of oceans beyond the continental shelf, the prominent photosynthesizers are **phytoplankton** (free-floating photosynthetic organisms).

No photosynthetic organisms inhabit the **aphotic zone**, below the reach of light, which as a consequence is sparsely populated. The lake bottom or ocean floor is called the **benthic zone**. Water is heavy, so pressure increases with water depth. Temperature generally decreases with depth because cold water is denser than warm water and sinks. Organisms that dwell in the deepest **abyssal zone** of the oceans experience very high pressures, low oxygen levels, and (except near hydrothermal vents) cold temperatures.

CHECKpoint CONCEPT 41.3

- ✓ What is meant by a biome?
- ✓ What conditions limit the distribution of photosynthetic organisms in aquatic biomes? In terrestrial biomes?
- ✓ Is it legitimate to consider the human gut a biome? Explain your answer.

We have just seen why physical geography and biogeography are linked. But the distributions of species also are contingent on the movements of continents over Earth's history.

CONCEPT 41.4 Biogeography Also Reflects Geological History

During the Age of Exploration it became clear that spatial patterns in the physical environment were not sufficient by themselves to explain the distributions of organisms. The observations of one scientist in particular suggested an important role for geological history.

Barriers to dispersal affect biogeography

Alfred Russel Wallace—who, along with Charles Darwin, advanced the idea that natural selection could account for the evolution of life's diversity—first noticed an odd pattern of species distributions during a seven-year exploration of the

Malay Archipelago. He observed that dramatically different bird faunas inhabited two neighboring islands, Bali and Lombok. Wallace pointed out that these differences could not be explained by climate or by soil characteristics, because in those respects the two islands are similar. Instead, he suggested that the Malay Archipelago was divided into two distinct halves by a line (now known as Wallace's line; see Figure 41.14) that follows a deep-water channel separating Bali and Lombok. This channel is so deep that it would have remained full of water—and thus would have been a barrier to the movement of terrestrial animals—even during the glaciations of the Pleistocene era, when sea level dropped more than 100 meters and Bali and the islands to its west were connected to the Asian mainland. As a consequence, the terrestrial faunas on either side of Wallace's line evolved mostly in isolation over a long period.

LINK

The development of the theory of natural selection, and Darwin's and Wallace's contributions to it, are discussed in **Concept 15.1**

The movements of continents account for biogeographic regions

Wallace's observations of animal distributions worldwide led him to divide Earth's land masses into six continental-scale areas called **biogeographic regions**. Each biogeographic region encompasses multiple biomes and contains a distinct assemblage of species, many of which are phylogenetically related. Although Wallace based his regions' boundaries on animal distributions, schemes based on plants have similar boundaries, and the biogeographic regions as we define them today have changed little from those first proposed by Wallace, despite improvements in our understanding of phylogenetic relationships (**FIGURE 41.14**).

Many boundaries between biogeographic regions correspond to geographic barriers to movement. These barriers include bodies of water (for terrestrial organisms) and areas with extreme physical environments, such as deserts or high mountain ranges. At Wallace's line, for example, the Oriental biogeographic region is separated from the Australasian biogeographic region by a deep-water channel.

But not all biogeographic regions are bounded by obvious barriers, and some of them span several continents. It's not at all clear at first why there are major biogeographic boundaries in Mexico, northern Africa, and Asia; or why the Antarctic region includes the southern tips of South America and Africa as well as New Zealand.

These aspects of biogeographic patterns remained a mystery until the second half of the twentieth century, when the theory of continental drift became widely accepted. Until then, biogeographers could only propose unsatisfactory explanations for the distributions of many organisms, such as rare cross-ocean colonization events or transient appearances of land bridges. Take the distribution of southern beeches (genus *Nothofagus*), for example. These trees are found in South America, New Zealand, Australia, and on certain islands of the southern Pacific Ocean

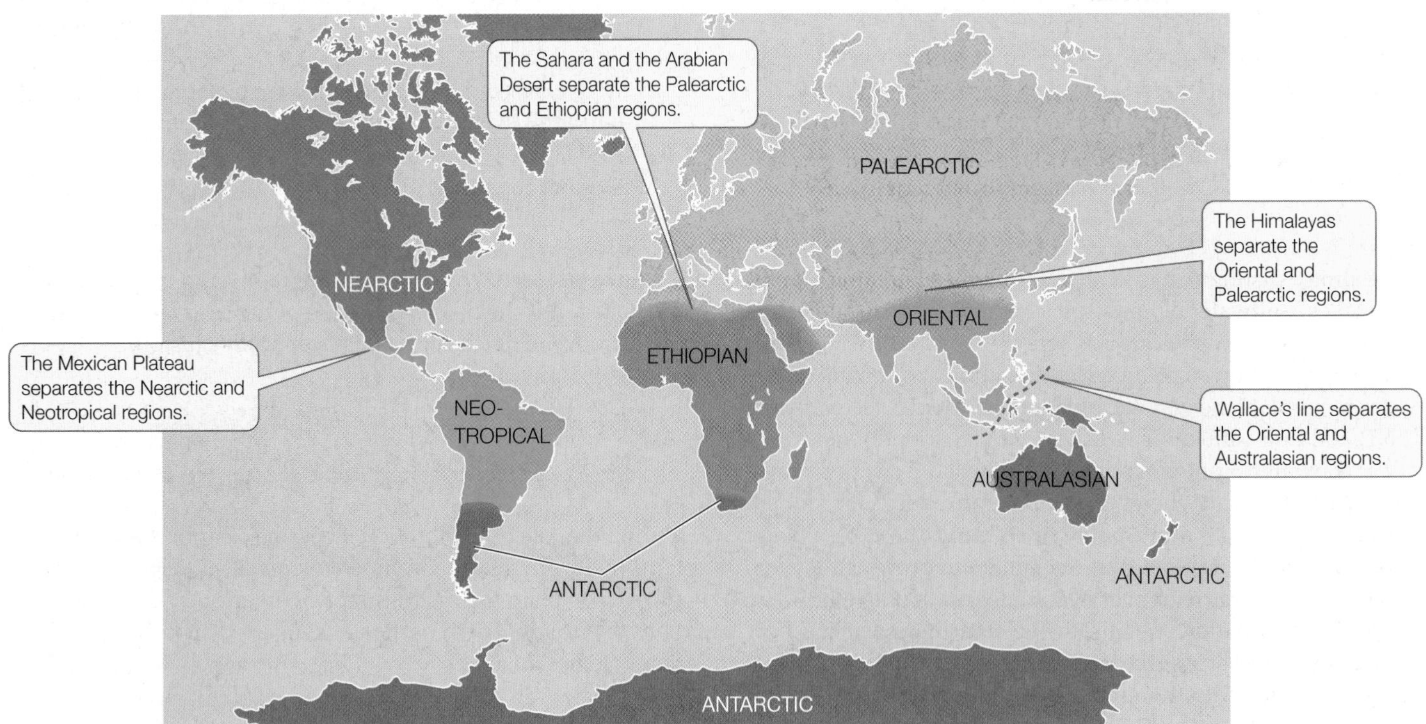

FIGURE 41.14 Earth's Terrestrial Biogeographic Regions The terrestrial biogeographic regions we recognize today are essentially the six regions Wallace proposed based on their distinctive assemblages of animals; modern biogeographers add an Antarctic region. Today we know that the boundaries delineating these regions are largely the result of continental drift.

Go to ACTIVITY 41.1 Major Biogeographic Regions PoL2e.com/ac41.1

Nothofagus sp.

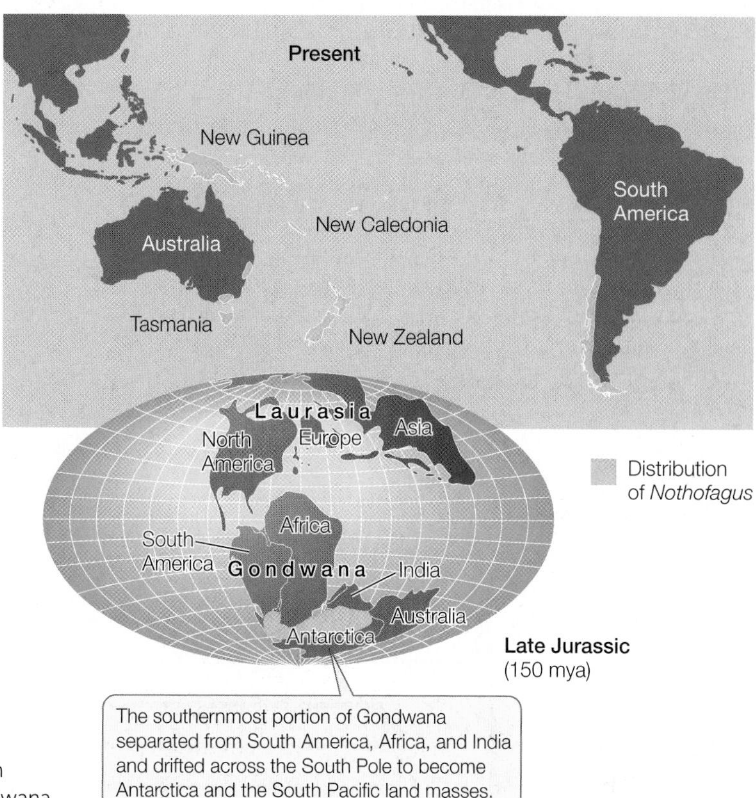

FIGURE 41.15 Distribution of *Nothofagus* The modern distribution of southern beeches is best explained by their origin on Gondwana during the late Jurassic period. The breakup of Gondwana and subsequent continental drift resulted in the modern distribution of *Nothofagus* in South America, Australia, New Zealand, and some islands of the South Pacific.

(**FIGURE 41.15**). How could their distribution be explained, if not by several separate transoceanic colonization events?

Once the movements of the continents were understood, it became clear that they have shaped the distributions of organisms by creating and disrupting dispersal routes. The genus *Nothofagus* is now thought to have originated on the southern supercontinent Gondwana and to have been carried along when Gondwana broke apart and its fragments drifted to their present locations. Paleontologists have found fossilized *Nothofagus* pollen dated at over 80 mya in Australia, New Zealand, Antarctica, and South America. This fossil evidence pre-dates the breakup of these portions of Gondwana, suggesting that *Nothofagus* was once widely distributed across that supercontinent.

LINK

The movements of the continents over geological time are depicted in **Figure 18.12** and **Animated Tutorial 18.1**, and described in **Table 18.1**; see also **Apply the Concept** on p. 860

Our understanding of continental drift makes it clear that the major biogeographic regions occupy land masses that have been isolated from one another long enough to allow the organisms present at the time of isolation to undergo independent evolutionary radiations. The northern biogeographic regions (Nearctic and Palearctic) became isolated from the southern regions (Neotropical, Ethiopian, Oriental, Australasian, and Antarctic) during the Jurassic period, about 150 mya, when Pangaea broke into two great land masses, Laurasia to the north and Gondwana to the south (see Figure 41.15). Southern biogeographic regions subsequently became isolated from one another when Gondwana began to break apart, about 130 mya. Neotropical, Ethiopian, Oriental, and Australasian regions were separated approximately 110 mya, and Antarctica became isolated from Australia around 60 mya. At about the same time, the Nearctic and Palearctic regions became isolated by the opening of the Atlantic Ocean, except for a periodic connection via the Bering land bridge between Asia and North America. Thus the biotas of the seven biogeographic regions developed largely in isolation throughout the Tertiary period (about 65 to 2.6 mya), when extensive evolutionary radiations of flowering plants and vertebrates took place.

More recently, continued movements of continents have eliminated some barriers to dispersal and have caused mixing of species that had previously been isolated, referred to as **biotic interchange**. The collision of India with Asia, for example, formed the Oriental biogeographic region and at the same time caused uplift of the Himalayas. These mountains separate the Oriental and Palearctic regions. Exchange of species between the Ethiopian and Palearctic regions became possible when Africa collided with Eurasia, but the Sahara and Arabian deserts present a climate-based barrier to such exchange. Similarly, mixing of species between the Neotropical and Nearctic regions began when a land bridge formed between South America and North America. During this Great American Interchange, many North American species displaced South American lineages and drove them to extinction. Currently the Mexican Plateau provides a climate-based barrier to the northward dispersal of tropical species that slows their mixing with the communities to the north.

Phylogenetic methods contribute to our understanding of biogeography

As we saw in Chapter 16, taxonomists have developed powerful methods of reconstructing phylogenetic relationships

APPLY THE CONCEPT

Biogeography also reflects geological history

Below is a phylogenetic tree of mammals based on molecular data, accompanied by the original and current distribution of each taxon.[a] The time course of major continental drift events over the last 150 million years is shown below the tree. Use these data to answer the following questions.

[a]J. W. Murphy et al. 2007. *Genome Research* 17: 413–421.

1. Which one node (split) in the tree most clearly corresponds to an event of continental separation?
2. Are there groups of mammals that appear to have dispersed to new areas following removal of a barrier to dispersal?

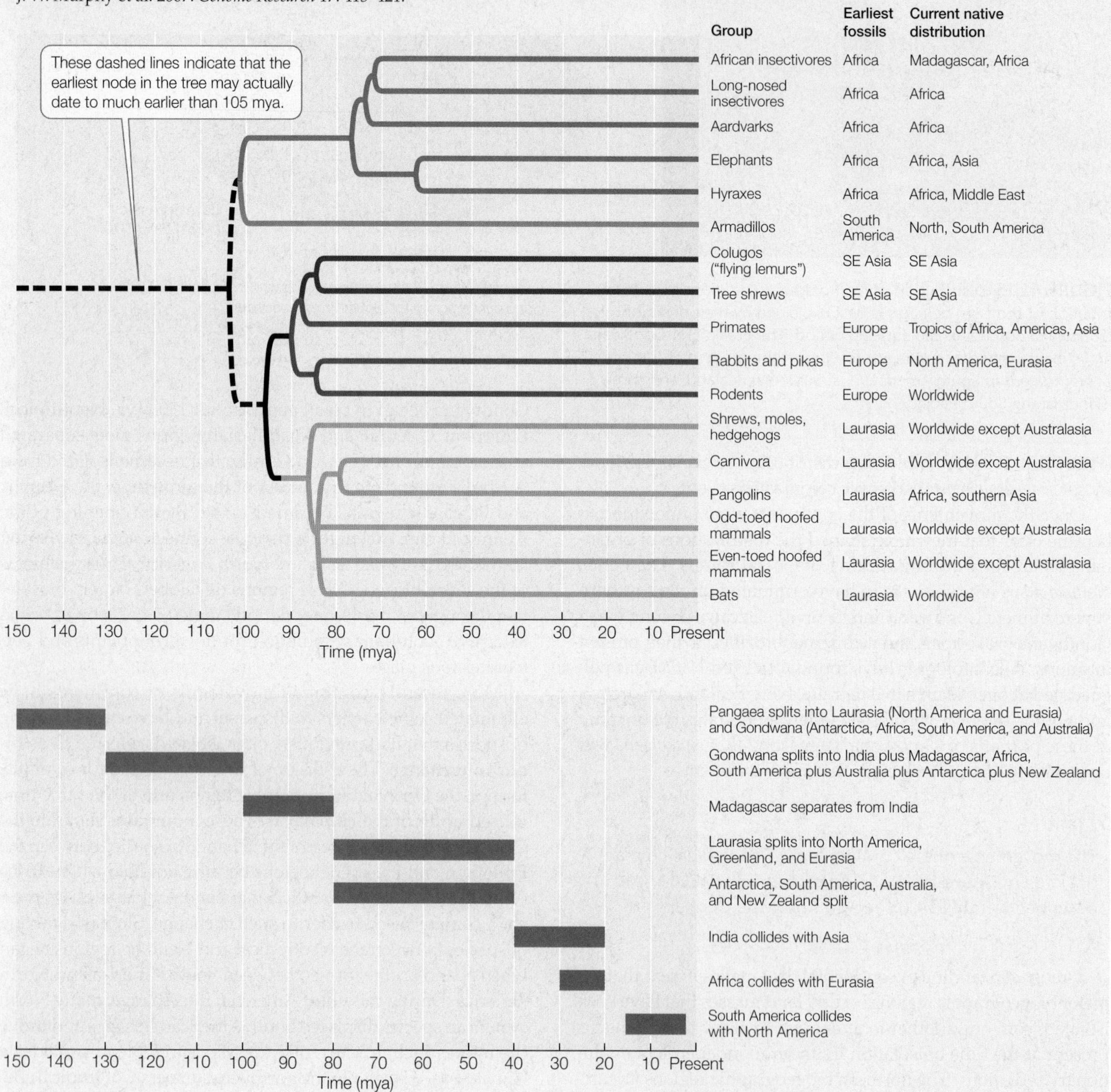

among organisms. Biogeographers use phylogenetic information, in conjunction with the fossil record and geological history, to address questions about how the modern distributions of organisms came about. They begin by superimposing on a phylogenetic tree the geographic area that is or has been occupied by each group. By comparing the sequence and timing of splits (nodes) in the phylogenetic tree with the sequence of separation or connection of geographic areas, biogeographers can determine where a lineage originated and reconstruct the history of its diversification and dispersal. For example, if a phylogenetic split coincides with the appearance of a barrier to dispersal, it is reasonable to conclude that the barrier caused the split by isolating portions of the original range of the ancestral species. If the phylogenetic split comes long before or after a barrier arises, then something else caused portions of the ancestral population to evolve in isolation from one another.

CHECKpoint CONCEPT 41.4

- ✓ What is Wallace's line?
- ✓ What kinds of barriers might limit the dispersal of terrestrial species? Of aquatic species?
- ✓ Biomes and biogeographic regions are two different concepts that relate to two different fundamental factors that mold the distributions of organisms. Explain.

As we have seen, the makeup of ecological systems on Earth is enormously variable over time and space. Recently, however, this variation has been greatly influenced by the activities of one species: *Homo sapiens*. How are our activities affecting ecological systems?

CONCEPT 41.5 Human Activities Affect Ecological Systems on a Global Scale

Some say that we are entering a new geological period, dubbed the "Anthropocene," or Age of Humans, because our activities are altering ecological systems on a global scale. Others say that the new age should be called the "Homogenocene," or Homogeneous Age, because the net effect of our activities is to make ecological systems less complex and more uniform. We affect ecological systems by using species for food or timber, by altering landscapes, by spreading organisms across the globe without regard to natural barriers to dispersal, and as Concept 45.4 will describe, by changing Earth's climates.

FIGURE 41.16 Human Agricultural Practices Produce a Uniform Landscape The cultivation of crop plants for food and other uses diminishes the diversity of ecosystems. This monoculture is a field of soybean plants (*Glycine* sp.) in Mississippi.

We are altering natural ecosystems as we use them

When we use natural ecosystems—for example, through hunting, fishing, grazing, or logging—we remove individuals of particular species and thereby change their abundances. If we remove too many, we can even cause some species to disappear entirely—to go extinct. The resulting shifts in the relative abundances of components of natural ecosystems change the patterns of interaction among species, and thereby change how entire ecosystems function. The decreased abundance of grasses in relation to shrubs in the U.S.–Mexico Borderlands, for example, has reduced the ability of these lands to support grass-eating animals, whether domesticated or not.

We are replacing natural ecosystems with human-dominated ones

We have now converted almost half of Earth's land area into human-dominated ecosystems. Cities occupy only about 1 percent of the land area (about 1.5 million km^2), but an area roughly the size of South America (about 18 million km^2) is devoted to agriculture, and another roughly the size of Africa (about 35 million km^2) is managed as pasture or used as rangeland for cattle and other livestock.

Human-dominated ecosystems contain fewer interacting species than natural ecosystems and are therefore less complex. Pastures or heavily grazed rangelands, for example, often contain fewer species than do natural grasslands or forests. And when an area is converted to agriculture, monocultures (plantings of single crops) often replace species-rich natural plant communities (**FIGURE 41.16**). Furthermore, agricultural systems worldwide are dominated by a relatively few plant species. Some 19 species comprise 95 percent of total global crop production, and overall crop diversity is a tiny fraction of the approximately 350,000 named plant species on Earth.

Human activities are similarly reducing the complexity of natural landscapes. For example, when rivers are dammed to control flooding or tapped to irrigate crops, species-rich natural streamside ecosystems disappear. Similarly, conversion of land to human uses not only reduces the total area of natural vegetation that remains, but also breaks it into smaller fragments. Such changes, and others that we will discuss in the following chapters, simplify natural ecosystems and change the way they function.

We are blurring biogeographic boundaries

Humans are also moving organisms around on a global scale, sometimes deliberately—as when Spanish settlers brought domesticated cattle to the U.S.–Mexico Borderlands—and sometimes inadvertently. Only about 1 percent of inadvertent introductions result in self-sustaining populations in the new locality, but their cumulative effect on geographic distributions is huge. Of the insects found in both Europe and North America, for example, half are estimated to have been transported between the two continents by humans. Similarly, more than half of the plant species on many oceanic islands are not native, and in many continental areas the figure is 20 percent or more. The pace of introductions is astonishing: for example, in California's San Francisco Bay, one new species on average became established every 12 weeks during the 1990s. Introducing new species to communities changes their species composition and adds novel interactions to the original system that may precipitate further changes in species abundances—even the complete loss of some species from the system.

This human-assisted biotic interchange is homogenizing the biota of the planet, blurring the differences among communities that evolved during long periods of continental isolation. One wonders how Alfred Wallace would draw the boundaries of biogeographic regions if he were using present-day data!

Science provides tools for conserving and restoring ecological systems

With the dawning of the Anthropocene, ecologists are called upon as never before to understand how humans are influencing Earth's ecological systems and how we can preserve their ability to sustain life on our planet. Accordingly, several new subdisciplines of ecology have emerged. The goal of **conservation ecology** is to understand the process of extinction and devise ways to prevent the extinction of vulnerable species. Often this requires that we protect the ecosystems of which these species are a part. The goal of **restoration ecology** is to restore the health of damaged ecosystems. These two fields are related because natural systems are sometimes altered so strongly that extinction will occur unless the systems are restored.

To achieve their goals, conservation and restoration ecologists deploy the same tools of scientific inquiry that all other biologists (and scientists) use: observation, questioning, logic, and experimentation (see Concept 1.5). They often begin with **natural history**—the observation of nature outside of a formal, hypothesis-testing investigation—because knowledge of nature is fundamental to all stages of ecological inquiry. Natural history observations are the source of new questions and hypotheses and are critical for the design of good experiments—ones that do not mislead us because they use unnatural conditions or leave out an important feature of the system. A lack of natural history knowledge often limits our ability to answer important questions. For example, we cannot fully answer the question "Has the introduction of honey bees caused declines in native bee species?" without knowing what bee species were present before honey bees were introduced and how abundant they were. It turns out that such information is available for only a handful of places on Earth.

As we saw in Chapter 1, scientists test hypotheses by seeing if their predictions are true. Sometimes we can use simple rules of logic to develop predictions. But with systems as complex as ecological systems, we often need to use a mathematical model, or perhaps even a computer simulation. What is critical is that we base the model on natural history knowledge about the system. In the case of the degraded rangeland system of the Borderlands, for example, predicting the conditions that would restore desirable grasses might require a complex model of the many factors that affect the growth of grasses relative to shrubs.

Conservation and restoration ecology are only two examples of the important practical applications of ecology in the Anthropocene. The chapters that follow will highlight numerous additional examples.

CHECKpoint CONCEPT 41.5

✓ Identify two human activities that alter the distributions of species.

✓ In what ways do agricultural ecosystems differ from the natural ecosystems they replace?

✓ Suppose that your job is to design an experiment to see whether allowing natural wildfires, rather than suppressing them, can restore a certain shrubland in the U.S.–Mexico Borderlands to a grassland. What natural history knowledge would you want to have about fire in the Borderlands, and how would this knowledge influence the design of your experiment?

Now that we have explored some of the properties of ecological systems, we are ready to take a closer look at them, beginning with the smallest scales in the ecological hierarchy: individual organisms and populations.

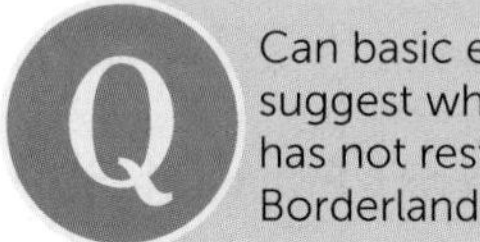

Can basic ecological principles suggest why removing cattle has not restored grasses to the Borderlands?

ANSWER The physical environment of the Borderlands is strongly influenced by Earth's climate patterns: its position near 32° N latitude, where the warm, dry air of the Hadley cells descends, makes it an arid place (Concept 41.2). Aridity, combined with other features of the physical environment, determines the characteristics of organisms that live in the Borderlands (Concept 41.3). Ecological communities of the Borderlands are also affected by the region's history of geographic isolation from or connection with other regions (Concept 41.4). All of these principles prepare us to realize that grasslands in different locations may look similar but may respond very differently to human activities.

As we've seen, communities dominated by grasses do differ from one another (Concept 41.1, Figure 41.12). Some occupy moist and cool environments, whereas others, like the Borderlands, occupy dry and hot ones. Some, including many meadows and pastures in Europe and eastern North America, were once forests that were cleared to make way for grasses; managing the size of cattle herds in these places can conserve or restore the grassland ecosystem. Others, such as the tropical savannas of Africa, have long histories of evolution with grazing mammals, and so tolerate grazing well. Some grasslands in more arid regions, such as the Borderlands, originally experienced natural wildfires and may require periodic burning for their maintenance. Still others, such as the spinifex grasslands of Australia, grow on nutrient-poor soils that support sparse growth of grasses at best.

Armed with this understanding, you may see several possible reasons why the grasslands in the Borderlands were lost and why "resting the range" by itself has not restored them. Perhaps periodic burning is needed. Perhaps the original landscape, without deep gullies that carry water away, must be re-created to return to the soil the water and nutrients that grasses require. Perhaps an interaction between fire and nutrients influences which vegetation type will be favored. And another thought may occur to you: perhaps it will be necessary to begin with drastic restoration measures—for example, removal of the existing shrubs—before other measures will have a chance of success. Ecologists are now using natural history knowledge, experiments in natural grassland plots, and modeling as scientific tools (Concept 41.5) to explore all of these possibilities.

SUMMARY

CONCEPT 41.1 Ecological Systems Vary over Space and Time

- An **ecological system**, or **ecosystem**, consists of one or more organisms and the **biotic** and **abiotic** components of the environment with which they interact.
- Ecological interactions lead to evolution by natural selection.
- Ecological systems can be studied at scales in the biological hierarchy ranging from an individual organism and its immediate surroundings to **populations**, **communities**, **landscapes**, or the entire **biosphere**.
- Ecological systems are highly complex and variable but can be understood with scientific methods.

CONCEPT 41.2 Solar Energy Input and Topography Shape Earth's Physical Environments

- **Weather** is the state of atmospheric conditions in a particular place at a particular time, and **climate** is their average state and pattern of variation over longer periods.
- Latitudinal differences in solar energy input are caused by differences in the angle of the sun's incoming rays. **Seasonality** results from the tilt of Earth's axis of rotation relative to its orbit around the sun. **Review Figures 41.3 and 41.4**
- Latitudinal differences in solar energy input drive north–south patterns of atmospheric circulation. These patterns, in turn, influence latitudinal patterns of temperature and precipitation and distribute heat from low to high latitudes. **Review Figure 41.5**
- Prevailing winds result from the interaction of north–south atmospheric circulation with Earth's eastward rotation, which adds an east–west component to air movement because of the **Coriolis effect**. **Review Figure 41.6**
- Oceanic circulation patterns called **gyres** are caused by prevailing surface winds and by the Coriolis effect. These patterns redistribute heat from low to high latitudes and influence patterns of temperature and precipitation. **Review Figure 41.7**
- Variation in the elevation of Earth's surface, known as **topography**, affects the physical environment. **Review Figure 41.8 and ANIMATED TUTORIAL 41.1**
- **Climate diagrams** summarize climate at a particular location in an ecologically relevant way. **Review Figure 41.9**

CONCEPT 41.3 Biogeography Reflects Physical Geography

- A **biome** is a distinct physical environment that is inhabited by ecologically similar organisms with similar adaptations. The species that occupy the same biome in geographically separate regions may not be closely related phylogenetically but often show convergent adaptations.
- The distribution of terrestrial biomes is broadly determined by annual patterns of temperature and precipitation. **Review Figures 41.10 and 41.11 and ANIMATED TUTORIAL 41.2**
- Other features of the physical environment—particularly soil characteristics—interact with climate to influence the character of terrestrial vegetation. **Review Figure 41.12**
- Climate is less important in distinguishing aquatic biomes than are water depth and flow, temperature, pressure, salinity, and characteristics of the substrate. **Review Table 41.1, Figure 41.13, and ANIMATED TUTORIAL 41.3**

CONCEPT 41.4 Biogeography Also Reflects Geological History

- Alfred Russel Wallace first noticed a boundary between two distinct assemblages of terrestrial species, now known as Wallace's line, that reflected a barrier to dispersal, rather than a boundary between climates or soils.
- The world can be divided into **biogeographic regions**, each of which contains a distinct assemblage of species, many of which are phylogenetically related. **Review Figure 41.14 and ACTIVITY 41.1**
- The boundaries of biogeographic regions generally correspond to present or past barriers to dispersal and can be explained by continental drift. **Review Figure 41.15**
- Biogeographers use phylogenetic information, in conjunction with the fossil record and geological history, to determine how the modern distributions of organisms came about.

CONCEPT 41.5 Human Activities Affect Ecological Systems on a Global Scale

- Humans have converted almost half of Earth's land area into human-dominated ecosystems, which are much more homogenous and less complex than natural ecosystems. Remaining natural ecosystems occur in small, isolated fragments.
- Humans move species around the globe without regard to natural barriers to dispersal. These movements are homogenizing the biota of the planet.
- **Conservation ecology** and **restoration ecology** are new subdisciplines of ecology whose goals are to prevent extinction of species and to restore damaged ecosystems.
- Knowledge of **natural history** is essential to these subdisciplines, as it is to all ecological inquiry.

Go to the Interactive Summary to review key figures, Animated Tutorials, and Activities PoL2e.com/is41

Go to LaunchPad at **macmillanhighered.com/launchpad** for additional resources, including LearningCurve Quizzes, Flashcards, and many other study and review resources.

42 Populations

KEY CONCEPTS

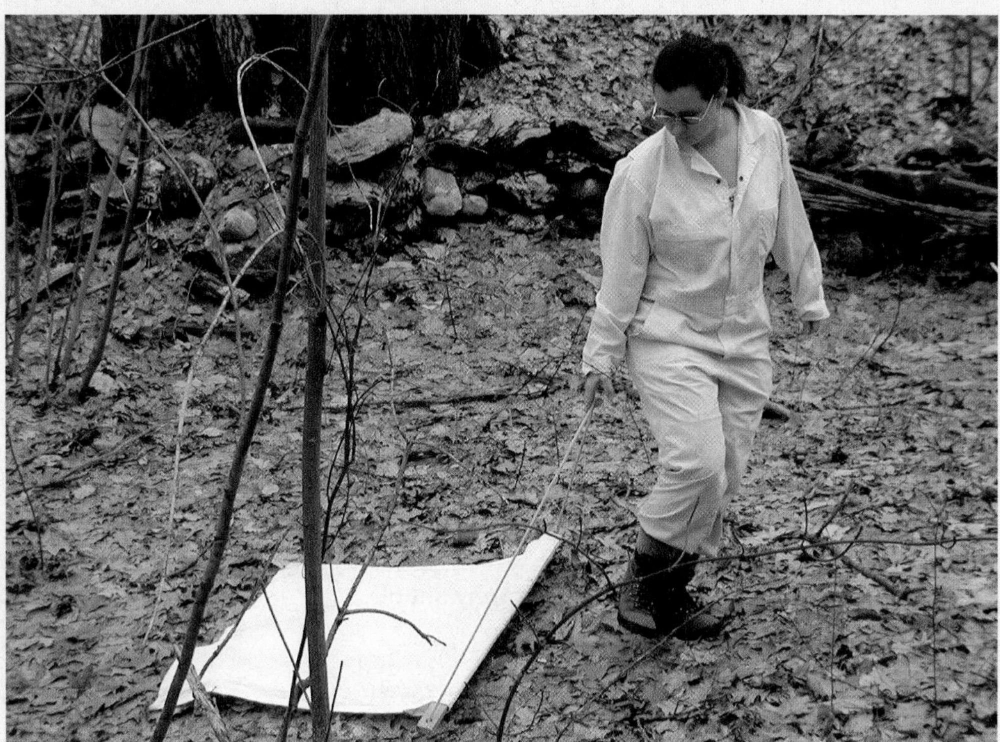

A field worker sweeps for ticks in the forests near Millbrook, New York.

Ecologist Rick Ostfeld was studying white-footed mice (*Peromyscus leucopus*) and eastern chipmunks (*Tamias striatus*) at the Cary Institute of Ecosystem Studies in Millbrook, New York, when an unexpected event took his work in a new direction. Ostfeld and his colleague Clive Jones were working on these rodents in order to understand what triggers the dramatic once-in-a-decade outbreaks of invasive gypsy moths (*Lymantria dispar*) that defoliate oak forests in North America. They suspected that the outbreaks start when the moths escape control by mice and chipmunks, which eat moth pupae.

In 1991, soon after the study began, one of the field assistants discovered that half his face was paralyzed and that he could not close one eyelid. (He devised a protective eye patch so that he could continue his fieldwork.) The eventual diagnosis was Lyme disease. Scientists at the time were only beginning to understand this disease, which was not described until 1975 and whose cause—infection by bacteria of the genus *Borrelia*—was not identified until 1982. By 1991, epidemiologists had discovered that Lyme disease is transmitted by *Borrelia*-infected black-legged ticks (*Ixodes scapularis*) when they bite hosts, and that Lyme disease risk increases with the abundance of infected ticks. Tick hosts include white-tailed deer, birds, lizards, and small mammals as well as humans.

Millbrook turned out to be at the epicenter of the emerging Lyme disease epidemic, and Ostfeld turned his attention to black-legged ticks as the disease vectors (organisms that transmit pathogens). He had already been counting ticks on live-trapped mice and chipmunks. Now he began to study how the abundance of rodents and other vertebrate hosts affects tick abundance and infection rate. He discovered that small mammals are the primary reservoir for *Borrelia* (they harbor it without getting sick). Ticks are not infected when they hatch from eggs as larvae; they become infected when they bite an infected host as a larva or nymph. It turns out that birds and lizards do not harbor *Borrelia*, and that most ticks are already infected by the time they are adults able to climb high enough into the vegetation to encounter large hosts such as deer.

From his previous work, Ostfeld already knew that rodent abundance soars when oaks produce a big acorn crop. Now he found that the number of tick larvae that survive to become nymphs, and of nymphs that survive to become adults, increases with rodent numbers. His next goal was to use this information about rodent, oak, and tick population sizes to determine how humans can slow the spread of Lyme disease.

How does understanding the population ecology of disease vectors help us combat infectious diseases?

You will find the answer to this question on page 880.

CONCEPT 42.1 Populations Are Patchy in Space and Dynamic over Time

In Chapter 41 we saw that individual organisms are the fundamental units of all larger ecological systems. Many important insights are gained by focusing on these fundamental units, but additional insights arise by studying **populations**—groups of individuals of the same species. Populations have properties that individuals do not have, and we cannot understand larger ecological systems, such as communities, without understanding their component populations. One important property of populations is abundance because, together with the average properties of individuals, abundance determines a species' ecological role in the larger community or ecosystem.

Humans have long had a practical interest in understanding what determines abundance because we want to manage our ecological interactions with other species. We want to know how to increase the abundance of organisms that provide us with resources such as food or fiber for clothing, how to conserve wild organisms that benefit us by acting as agents of pollination or pest control, and how to decrease the abundance of undesirable organisms such as crop pests, weeds, and pathogens. Understanding population ecology is a key to achieving these goals.

Population density and population size are two measures of abundance

"Abundance" has several meanings in ecology. We might consider a species abundant because there are many individuals in an area. If that same species occurs only in a small geographic region, however, we might consider it rare because the total number of individuals of that species is small. Depending on why they want to measure abundance, ecologists use one of two approaches. If they are interested in the causes or consequences of local abundance, they usually measure **population density**: the number of individuals per unit of area (for surface-dwelling organisms) or volume (for organisms that live in air, soil, or water). Alternatively, if ecologists are interested in the population as a whole, they usually measure **population size**: the number of individuals in the population.

Counting all the individuals in a population is practical only for very small populations in which individuals can be distinguished. For this reason, ecologists usually measure population density first. Then they multiply population density by the area or volume occupied by the entire population to estimate population size.

Abundance varies in space and over time

The abundance of organisms varies on several spatial scales. At the largest scale, as we saw in Chapter 41, many species are found only in a particular region, and their densities are zero elsewhere. Within that region, which is called the species' **geographic range**, the species may be restricted to particular kinds of environments, often called **habitats**, which may themselves be patchily distributed. For example, Edith's checkerspot butterfly (*Euphydryas editha*) is found from southern British Columbia and Alberta to Baja California, Nevada, Utah, and Colorado (**FIGURE 42.1A**). Within this geographic range, it occurs in the open habitats (coastal chaparral, meadows, grasslands, or open woodlands) where the plants that its caterpillars eat are found. In the San Francisco Bay area, suitable food plants grow only in soils derived from a chemically unique rock called serpentine (**FIGURE 42.1B**). Outcrops of serpentine rocks constitute habitat "islands" for the butterflies—that is, patches of suitable habitat separated by areas of unsuitable habitat. Habitat patchiness has consequences for abundance that we will explore in Concept 42.5.

In any given locality, population densities are **dynamic**—they change over time. For example, abundances of rodents, black-legged ticks, and oak acorns vary greatly from year to year in the forests described in the opening story (**FIGURE 42.2**).

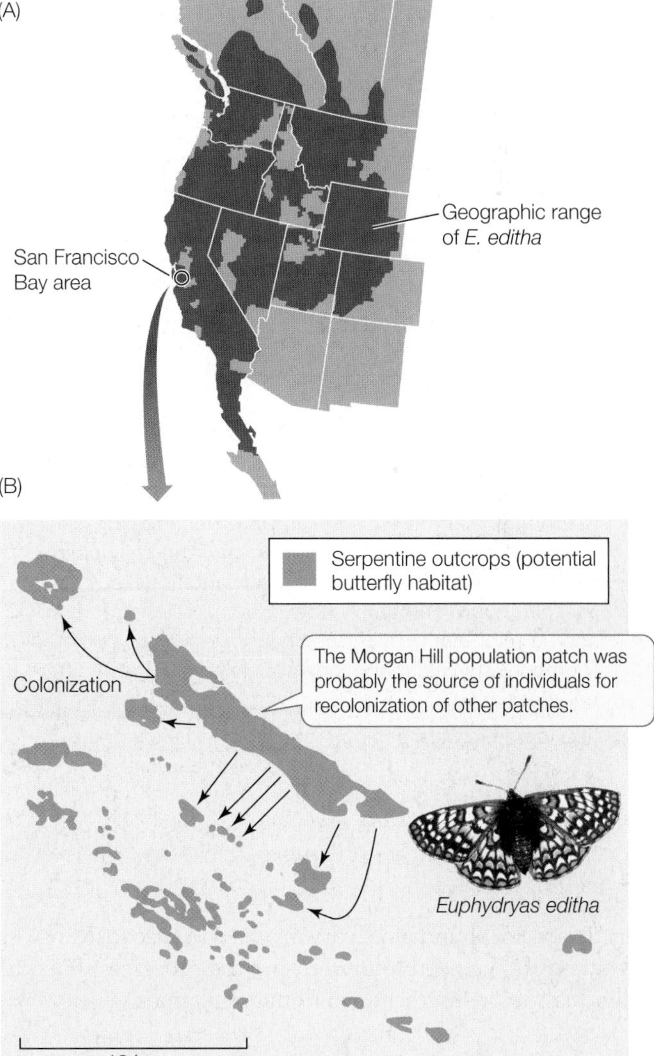

FIGURE 42.1 Species Are Patchily Distributed on Several Spatial Scales (A) The geographic range of Edith's checkerspot butterfly (*Euphydryas editha*) extends from British Columbia and Alberta to Baja California. (B) Within the geographic range, the butterflies occur in patches of suitable habitat which, in the San Francisco Bay area, are on serpentine soils. Arrows indicate colonization events that are discussed in Concept 42.5.

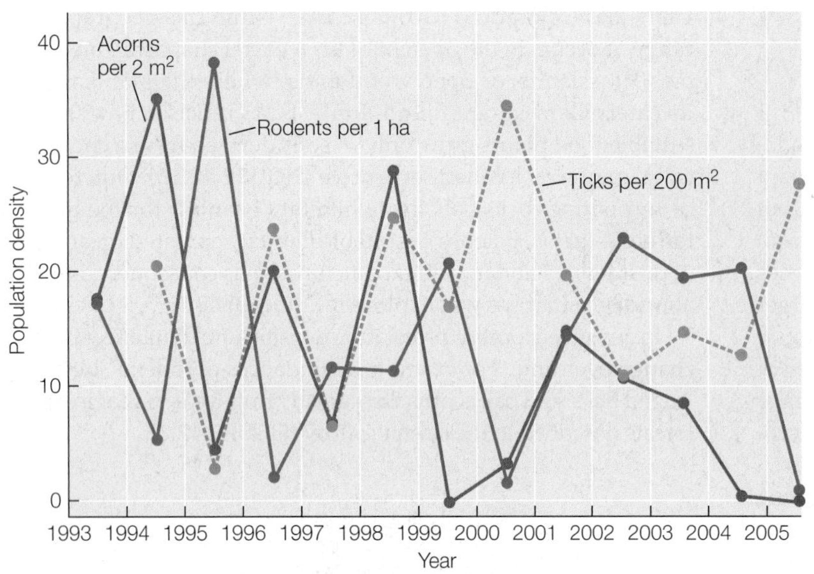

FIGURE 42.2 Population Densities Are Dynamic Densities of acorns, rodents, and nymphal black-legged ticks varied over time in an oak forest near Millbrook, New York. (Notice that the population density of each organism is measured on a spatial scale appropriate for that organism's average density; see Appendix C for definitions of units of area.)

CHECKpoint CONCEPT 42.1

- ✓ What aspects of abundance do population density and population size measure, respectively?
- ✓ Big-leaf mahogany (*Swietenia macrophylla*) is a valuable timber tree native to South American tropical rainforests. In Brazil, it is a species of conservation concern because natural populations are shrinking due to habitat destruction and illegal logging. In Sri Lanka and the Philippines, mahogany is grown in commercial plantations, but it has escaped cultivation and is now invading wild rainforest and harming native species. What goals would you set for managing wild mahogany populations in these different areas?
- ✓ The image in the opening story shows a field worker "sweeping" the forest floor with a cloth to which ticks will attach themselves, as if the cloth were a host animal. How would you use this method to estimate tick population size in a given area of forest?

Why do species abundances vary in time and space? To answer this question, we need to understand the processes that add individuals to or subtract them from populations.

CONCEPT 42.2 Births Increase and Deaths Decrease Population Size

The size of anything, whether a bank account or a population, changes over time through additions and subtractions. Consider the size of something in the future. The size of your bank account, for example, will be equal to the amount of money you have now, plus the amount that is added, minus the amount that is taken out. What differs between bank balances and population sizes are the processes that add or subtract. In the case of your bank account, deposits and interest add to your balance, while withdrawals and bank fees subtract from it. In the case of populations, births (or the production of new individuals via seeds in the case of seed plants) add individuals, while deaths remove individuals. To spell this out,

the number of individuals in a population at some time in the future =
the number now + the number that are born –
the number that die

If we turn this word equation into symbols, we have the most basic mathematical model of population size:

$$N_{t+1} = N_t + B - D \qquad \text{Equation 42.1}$$

where N is the population size, B is the number of births in the time interval from time t to time $t + 1$, and D is the number of deaths in that same interval. Ecologists call this equation the "birth–death" or **BD model** of population size.

We can use Equation 42.1 to understand population dynamics. Most often, the size of a population does not remain constant over time—the population grows or shrinks. Ecologists call the amount of change over a specific amount of time the growth rate of the population (a somewhat confusing term until we realize that when a population shrinks, its growth rate is negative). The population growth rate can be obtained from Equation 42.1 in two steps. First, we subtract N_t (the present population size) from both sides of the equation to obtain the change in population size:

$$N_{t+1} - N_t = \Delta N = (N_t - N_t) + B - D = B - D$$

The term ΔN is shorthand for the change in population size (using the convention that the Greek letter delta means "change in"). Next, we divide both sides by ΔT, the time interval from t to $t + 1$, to obtain the change per unit of time:

$$\frac{\Delta N}{\Delta T} = \frac{B-D}{\Delta T} = \frac{B-D}{(t+1)-t} = \frac{B-D}{1} = B - D \qquad \text{Equation 42.2}$$

This shows, logically enough, that the change in size over one time interval is the number of births minus the number of deaths during that interval.

Change in the size of a population can be measured directly only for very small, countable populations, such as a zoo population. Ecologists can get around this difficulty by keeping track of a sample of individuals over time. From this sample, they can estimate the number of offspring that the average individual produces, called the **per capita birth rate** (b), and the average individual's chance of dying, called the **per capita death rate** (d), in some interval of time. Per capita (literally, "per head") means "per individual." Total birth and death rates in the population can then be estimated by multiplying per capita birth and death rates by population size, so that $B = bN_t$ and $D = dN_t$. Substituting into Equation 42.1, we have

APPLY THE CONCEPT

Births increase and deaths decrease population size

Jane is a rancher who is considering shifting her operation from cattle to American bison (*Bison bison*), to take advantage of the demand for lean, healthy bison meat. To decide, she needs to know how well bison will do on her ranch. She buys 50 female bison that are already inseminated and places 10 of them, picked at random, into their own pasture. These 10 females serve as a sample from which Jane collects demographic data over one year. Use her data to answer the questions below.

1. How many total births and how many total deaths occurred in this sample population?
2. What were the estimated average per capita birth and death rates (b and d) for the entire bison herd, based on this sample?
3. What was the estimated per capita growth rate (r)?
4. Based on these estimates, what was the size of Jane's entire bison herd at the end of the year? (Hint: Use Equation 42.3.)

FEMALE #	ALIVE AT END OF YEAR?	NUMBER OF OFFSPRING
1	Yes	1
2	Yes	0
3	Yes	1
4	Yes	0
5	No	0
6	Yes	1
7	Yes	1
8	No	0
9	Yes	1
10	Yes	0

$$N_{t+1} = N_t + bN_t - dN_t$$

or equivalently,

$$N_{t+1} = N_t + (b - d)N_t$$

The value $(b - d)$—the difference between per capita birth rate and per capita death rate—represents the average individual's contribution to total population growth rate. This value is the **per capita growth rate**, which ecologists symbolize as r. Substituting r for $(b - d)$, we have

$$N_{t+1} = N_t + rN_t \qquad \text{Equation 42.3}$$

By converting Equation 42.3 into an equation of growth rate analogous to Equation 42.2, we get

$$\frac{\Delta N}{\Delta T} = rN \qquad \text{Equation 42.4}$$

What happens if per capita birth rate is greater than per capita death rate, or if death rate is greater than birth rate, or if they are equal? If $b > d$, then $r > 0$, and the population grows. If $b < d$, then $r < 0$, and the population shrinks. If $b = d$, then $r = 0$, and the population size does not change.

We are now in a better position to understand why population sizes or densities change over time: they *have to* unless the number of births exactly equals the number of deaths. We are also in a better position to understand why population densities vary in space. We saw in Chapter 41 that barriers to dispersal can cause species to be absent from some places. Another possible reason for a species' absence from a place is a negative population growth rate. A species will not persist in a given location if its per capita growth rate, r, is negative there. Any population that decreases steadily in size over time eventually reaches a density of zero—that is, it goes extinct.

CHECKpoint CONCEPT 42.2

- ✓ What is the relationship between per capita birth rate and the total number of births in a population during a time period?
- ✓ What do we need in addition to estimates of per capita birth and death rates in order to predict population size at some future time?
- ✓ What two strategies could Jane the rancher (see Apply the Concept on this page) use to expand her bison herd faster?

We can understand the dynamics of whole populations, and even some aspects of spatial variation in abundance, if we can estimate the average birth and death rates of the individuals that constitute those populations. What determines these variables?

CONCEPT 42.3 Life Histories Determine Population Growth Rates

The study of processes that influence birth, death, and population growth rates is known as **demography** (Greek *demos*, "population"; *graphia*, "description"). Ecologists who wish to understand such processes for a species usually begin by characterizing its **life history**: the sequence of key events, such as growth and development, reproduction, and death, that occur during an average individual's life. To describe the life history of the black-legged tick, for example, we begin with the first life stage—one of the thousands of eggs laid by an adult female in spring (**FIGURE 42.3**). If the egg survives, it hatches

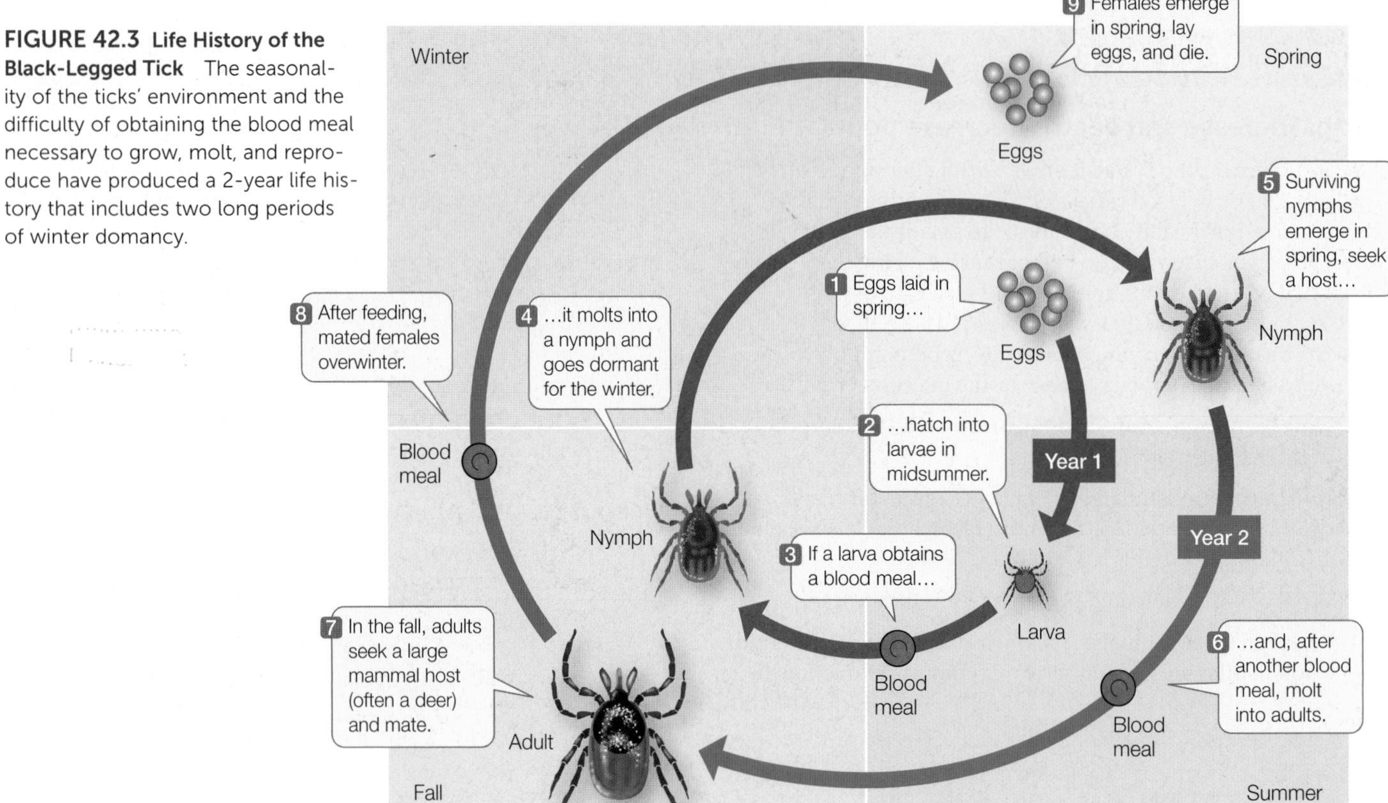

FIGURE 42.3 Life History of the Black-Legged Tick The seasonality of the ticks' environment and the difficulty of obtaining the blood meal necessary to grow, molt, and reproduce have produced a 2-year life history that includes two long periods of winter domancy.

into a larva in midsummer. If the larva can obtain a blood meal from a mammal, bird, or lizard, it molts into a nymph and goes dormant for the winter. If it survives the winter, the nymph becomes active again the following summer and seeks another blood meal from a vertebrate host. If it is successful, it molts into an adult in the fall and seeks a large mammal host, such as a deer, raccoon, opossum—or human. If successful, it mates on this final host, and if it is a female, goes dormant for another winter, lays eggs in the spring, and dies.

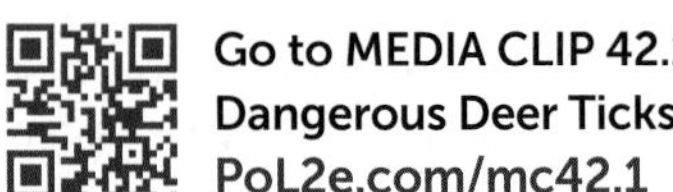
Go to MEDIA CLIP 42.1
Dangerous Deer Ticks
PoL2e.com/mc42.1

Even this qualitative description of the tick's life history gives us important insights into its ecological interactions. But to estimate birth and death rates, we need more information about the timing of life events and the number of offspring produced. A quantitative tick life history would indicate at what ages individuals make transitions between life stages and how many do so successfully: the fraction of eggs that hatch successfully, of larvae that find a host and molt successfully into nymphs, and of nymphs that live through the winter, find a host, and molt into adults. In addition, a quantitative life history would indicate the number of eggs that the average adult female lays.

Obtaining such information is not easy, and ecologists still don't have all of it for black-legged ticks. We do have quantitative life histories, however, for other species, such as the common cactus finch (*Geospiza scandens*) in the Galápagos archipelago. **TABLE 42.1** shows the life history, in table form, of a sample of 210 finches born on a single island in 1978. The table quantifies two important aspects of any life history: the fraction of individuals that survive from birth to different life stages (which are ages in the case of the finches), called **survivorship**, and the average number of offspring each surviving individual produces at those life stages, called **fecundity**. Survivorship can also be expressed as its opposite, **mortality**, which is the fraction of individuals that do not survive from birth to a given stage or age (mortality = 1 – survivorship).

Information on survivorship, fecundity, and the duration of life stages can be used to calculate per capita growth rates (r) for any life history. The calculations involved are complex, but it is not hard to see that survivorship and fecundity will affect r. All else being equal, the higher the fecundity and the higher the survivorship, the higher r will be. If reproduction shifts to earlier ages, r will increase as well.

Life histories are diverse

Species vary considerably in how many and what types of life stages they go through, how old they are when they begin to reproduce, how often they reproduce, how many offspring they produce, and how long they live. Most black-legged ticks spin out their lives over 2 years, whereas the life history of a periodical cicada spans nearly 2 decades. Other arthropods have life spans of days or weeks. Some species, like the tick, go through a discrete series of life stages of variable duration, each separated by a molt. Other species, such as common cactus finches and humans, develop continuously as they age. Some plants germinate, grow, flower, produce seeds, and die

TABLE 42.1 Quantitative Life History for 210 Common Cactus Finches Born in 1978 on Isla Daphne

Calendar year	Age of bird (years)	Survivorship[a]	Fecundity[b]
1978	0	1.00	0.00
1979	1	0.43	0.05
1980	2	0.37	0.67
1981	3	0.33	1.50
1982	4	0.31	0.66
1983 Increased rain	5	0.30	5.50
1984	6	0.20	0.69
1985 Drought	7	0.11	0.00
1986	8	0.07	0.00
1987 Increased rain	9	0.07	2.20

[a]Survivorship = the proportion of the 210 birds that survived from birth to a given age.
[b]Fecundity = the average number of young born per female of a given age.

all in one growing season—they are annuals (see Concept 27.2). Other plants live for centuries. Some species spend long periods in a dormant state, whereas others are continuously active. Some organisms, including the tick, reproduce only once and then die; others, such as finches and humans, can reproduce multiple times.

Life histories vary not only among species but within species as well, including among human populations. For example, **life expectancy** (the age to which an average person survives) of the Aeta people of the Philippines, who hunt and gather wild food, is 16.5 years, and girls reach puberty at age 12. The Turkana, herders of East Africa, experience food shortages as frequently as the Aeta, but they have a life expectancy of 47.5 years, and Turkana women reach sexual maturity at age 15. In well-nourished United States populations, life expectancy is 78.2 years—much greater than for either the Aeta or the Turkana—yet girls in the U.S. reach puberty at age 12.5 years on average.

Resources and physical conditions shape life histories

Individual organisms must acquire materials and energy to maintain themselves and to fuel metabolism, growth, activity, defense, and reproduction. Materials and energy—and the time available to acquire them—constitute **resources** for organisms. Organisms also need physical conditions that they are able to tolerate. The primary distinction between resources and conditions is that resources can be used up (e.g., mineral nutrients, photons of light, time, space, or food items), whereas conditions are experienced but not consumed.

The rate at which an organism can acquire a resource increases with the availability of that resource in its environment, up to the point at which the organism's capacity to take in and process the resource is saturated. A plant's net photosynthetic rate, for example, increases with sunlight intensity (**FIGURE 42.4A**), and an animal's rate of food intake increases with the density of food (**FIGURE 42.4B**)—but in both cases we see a leveling off of intake rate at high resource availability.

Organisms use the resources they obtain for functions such as maintenance, growth (see Concept 29.1), defense (see Chapter 39), and

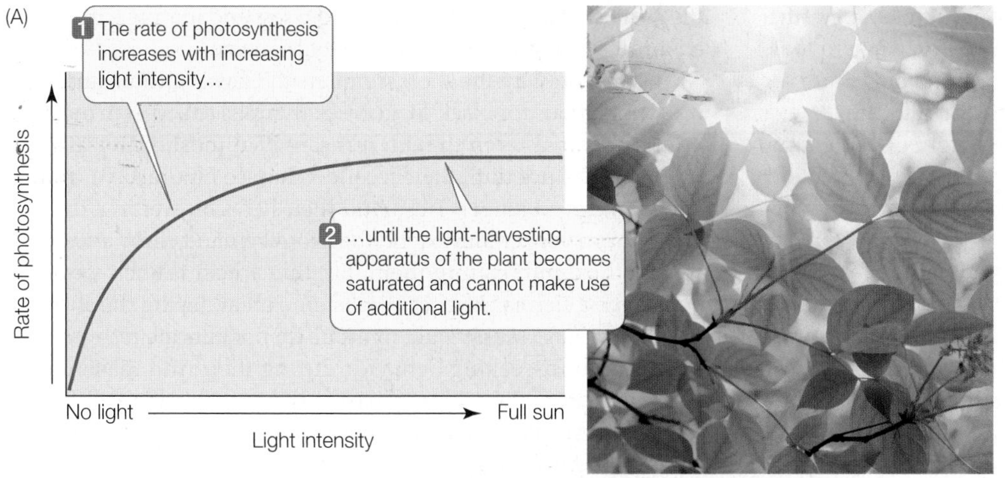

FIGURE 42.4 Resource Acquisition Increases with Resource Availability—Up to a Point (A) A plant's photosynthetic rate increases with light intensity. (B) Kangaroo rats (*Dipodomys* spp.) are able to harvest seeds faster as seed density increases. (Note the cheek pouch bulging with seeds.) Error bars indicate 95 percent confidence intervals around the mean. See Appendix B for discussion of statistical concepts.

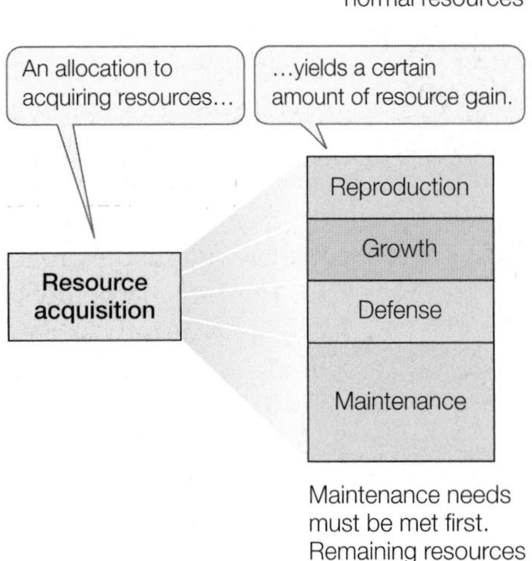
(A) Typical conditions, normal resources
An allocation to acquiring resources…
…yields a certain amount of resource gain.
Resource acquisition
Reproduction
Growth
Defense
Maintenance
Maintenance needs must be met first. Remaining resources are divided among other activities.

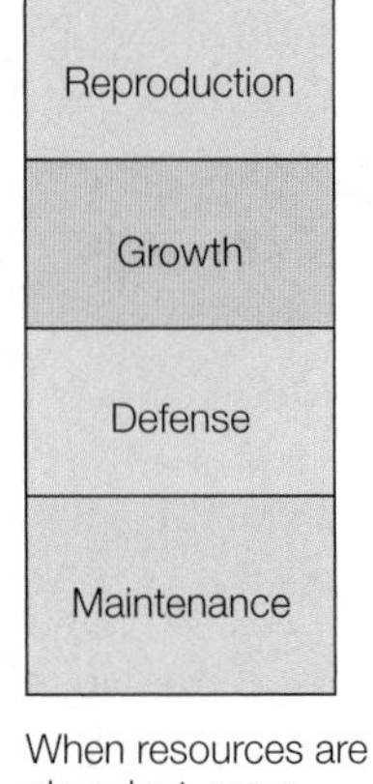
(B) Typical conditions, abundant resources
Reproduction
Growth
Defense
Maintenance
When resources are abundant, more resources are gained and more are available after maintenance needs are met.

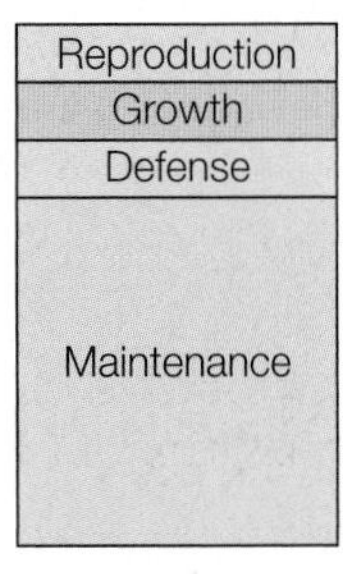
(C) Stressful conditions, normal resources
Reproduction
Growth
Defense
Maintenance
Stressful conditions mean more resources must be expended on maintenance and fewer are available for other purposes.

FIGURE 42.5 The Principle of Allocation Organisms use resources (energy and time) to acquire more resources. The organism must then allocate acquired resources among competing life functions. In general, an organism's first priority is to maintain the integrity of its body. This figure assumes that the same amount of energy or time is allocated to resource acquisition in all three conditions.

reproduction (see Chapter 37) (**FIGURE 42.5**). In addition, they must devote some resources to activities involved in obtaining more resources. The **principle of allocation** states that once an organism has acquired a unit of some resource (such as food), that bit of resource cannot be used for multiple functions at once.

In general, an organism's first priority is to maintain the integrity of its body. Investment in maintenance is greater in stressful than in unstressful environments because maintenance is more difficult under stressful physical conditions (as was illustrated for temperature in Concept 29.3). For example, the hotter their desert environment, the more time day-active lizards must spend in a cool retreat to keep body temperature below a lethal level, and thus the less time they have to look for food (see Figure 29.8). Once an organism obtains more resources than it needs for maintenance, it can allocate the excess to other functions, such as growing bigger or producing offspring. In general, as the average individual in a population acquires more resources, the average fecundity and survivorship increase.

Life-history variation among species and populations often reflects the principle of allocation. A species that invests heavily in growth early in life, for example, cannot simultaneously invest heavily in defense (e.g., protective structures or chemical defenses). As a consequence, it reaches adult size quickly, but at the cost of lower survivorship than a species that invests more in defense. A species that starts investing in reproduction early in life grows more slowly, and matures at a smaller adult size, than a species that waits to reproduce. Species that invest heavily in reproduction often do so at the expense of adult survivorship. They have high fecundity but short life spans. Such negative relationships among growth, reproduction, and survival are called **life-history trade-offs**.

The environment shapes much of this life-history variation because, together with a species' way of life, it determines the relative costs and benefits, in terms of survivorship and fecundity, of any particular allocation pattern. The Aeta people mentioned earlier, for example, live in an environment that imposes higher mortality than does that of the Turkana. High mortality means that the benefit of early reproduction (more individuals survive to reproduce at least once) outweighs the fecundity benefit of growing to a larger adult size (in general, human females produce 0.24 more offspring per additional centimeter of height). Most Aeta women reach puberty and stop growing at about 12 years of age, at an adult height averaging 140 cm. For the Turkana, the lower mortality rate shifts the balance—the advantage of early reproduction is outweighed by the fecundity advantage of attaining a larger adult size. Turkana women mature at age 15 and at a height averaging 166 cm. In the United States, plentiful food and modern medicine reduce the cost of maturing early—girls grow fast and mature early, at an adult size equivalent to that of the Turkana.

In a similar way, the short summers in New York State make it advantageous for black-legged tick nymphs to invest in overwinter survival, even though it means that most individuals do not reproduce until their second year (see Figure 42.3), and a few take even longer. Ticks find their hosts by waiting: they climb up on vegetation, extend their forelegs, and climb aboard when a host bumps into them. Finding a host is such a rare event that a tick can expect at most one meal during the short growing season. As a result, nymphs do not allocate resources to further host-seeking behavior during their first season of life, but instead allocate resources to overwinter survival. The fact that ticks take 2 years to complete their life history is thus a consequence of the difficulty of encountering a host and the seasonality of the environment.

Species' distributions reflect the effects of environment on per capita growth rates

A population cannot persist in an environment where its per capita growth rate is negative, because its size would inevitably shrink to zero—it would go extinct. We can predict where populations are likely to be found if we know how resource availability and physical conditions affect survivorship and fecundity. Obtaining complete knowledge of life histories is often impossible for species in the wild, but even incomplete knowledge helps us understand species' distributions. **FIGURE 42.6** illustrates this

INVESTIGATION

FIGURE 42.6 Climate Warming Stresses Spiny Lizards Barry Sinervo and colleagues wanted to know whether climate warming will stress Mexican blue spiny lizards (*Sceloporus serrifer*) by reducing the number of hours they can remain outside their burrows without overheating. These lizards feed only during daytime and retreat to their cool burrows when their body temperature gets too high (see Figure 29.8). In 2008, the researchers conducted experiments in four locations in Mexico where these lizards had been found in 1975. Climate records indicated that the four locations did not warm at the same rate between 1975 and 2008, perhaps because they differ in such things as proximity to the ocean, elevation above sea level, and latitude.[a]

HYPOTHESIS

Spiny lizards have fewer hours to forage on hotter days.

METHOD

1. Construct model "lizards" that have the same thermal properties as real lizards.

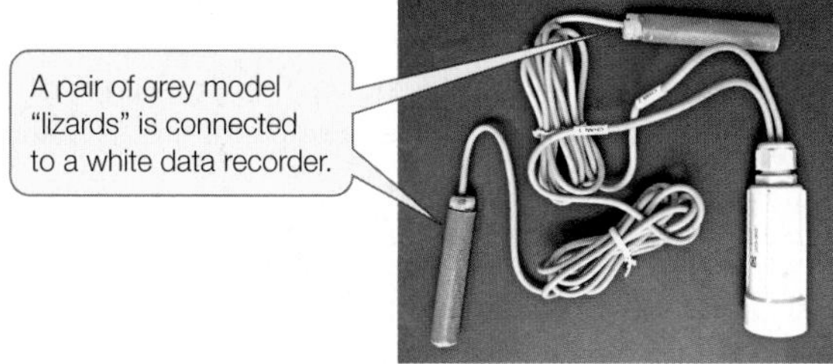

A pair of grey model "lizards" is connected to a white data recorder.

2. At four sites in the Yucatán where lizards occurred in 1975, place replicate model lizards in various perches that real lizards use when they forage. Monitor body temperature T_b of the models each hour and record maximum daily air temperatures during the breeding season (March and April).
3. For each day, calculate the number of hours that the body temperature T_b of the model lizards exceeded 31°C—the temperature at which *S. serrifer* are known to stop foraging and retreat to their burrows—to arrive at a predicted number of inactive hours for real lizards.
4. Determine whether the predicted number of inactive hours increased with maximum daily air temperature.

RESULTS

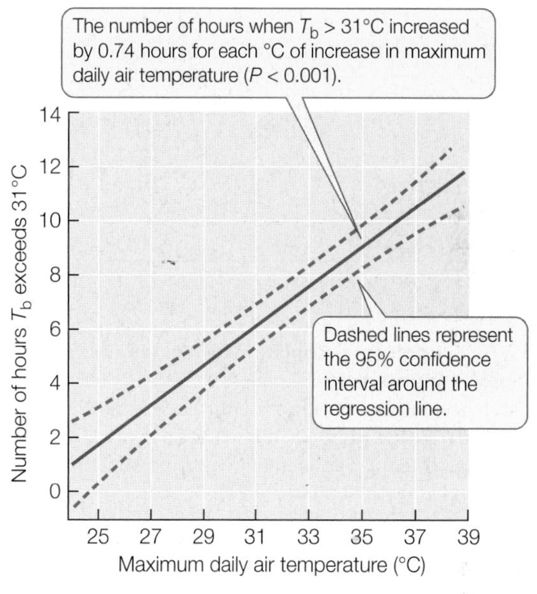

CONCLUSION

The hypothesis is supported by the data: *S. serrifer* can forage without overheating for fewer hours on days when maximum air temperature is higher.

ANALYZE THE DATA

Between 1975 and 2008, *S. serrifer* populations went extinct at two of the four sites monitored by Sinervo and colleagues. The graph at right shows March and April averages in 2008 for the number of hours (in some cases including hours after sundown!) when T_b > 31°C at the two "extinct" sites and at the two "persistent" sites. Error bars are 1 standard error of the mean (see Appendix B); sample size for each point is 240 observations (2 sites × 30 days × 4 lizard models). Use this graph to answer the questions below.

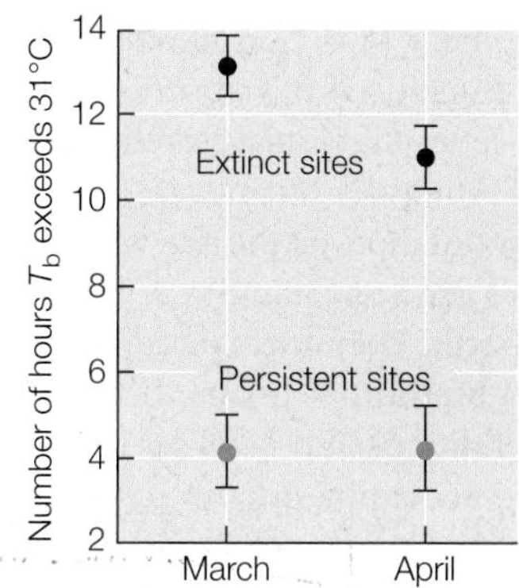

A. How did the hours available for foraging differ between the "extinct" sites (black symbols) and the "persistent" sites (green symbols)?

B. How might the availability of foraging time influence lizard fecundity, survivorship, and per capita growth rates?

C. Sinervo and colleagues knew that climate warming had taken place between 1975 and 2008 (for example, see Figure 45.13), the period during which some of the lizard populations had gone extinct. Is the extinction of some, but not all, lizard populations consistent with the conclusion that climate warming was one of the causes of extinction? Why or why not?

Go to **LaunchPad** for discussion and relevant links for all **INVESTIGATION** figures.

[a]B. Sinervo et al. 2010. *Science* 328: 894–899.

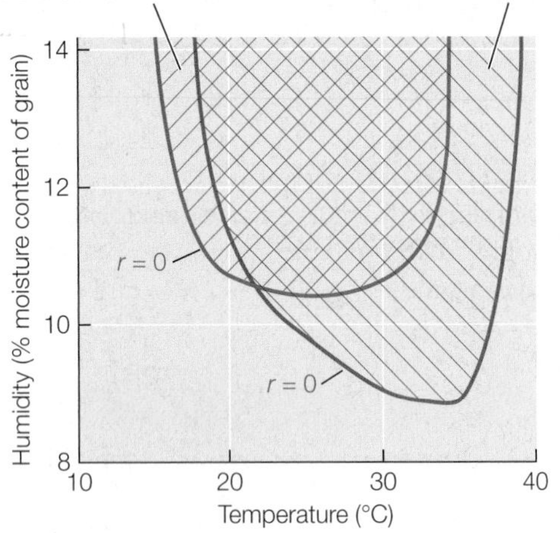

FIGURE 42.7 Environmental Conditions Affect Per Capita Growth Rates and Species Distributions Charles Birch's laboratory research revealed that *R. dominica* populations performed better under warmer conditions than did *S. oryzae* populations. The lines marked $r = 0$ indicate the limit of the conditions under which populations of each species can grow; within the hatched areas, per capita growth rates are positive for the species.

point for the lizard *Sceloporus serrifer*. By combining measurements of conditions in the lizard's natural environment with knowledge of its physiology and behavior, researchers were able to draw conclusions about how climate change is affecting the lizard's survivorship, fecundity, and distribution—with important implications for its conservation.

Sometimes it is possible to explore the links between environmental conditions, life histories, and species distributions with the help of experiments carried out in the laboratory. Charles Birch used this approach to understand why two species of beetles that infest stored grain are serious pests in some parts of Australia but not in others. In his laboratory, Birch reared populations of the rice weevil (Sitophilus oryzae, known in Birch's time as Calandra oryzae) and the lesser grain borer (Rhyzopertha dominica) under various conditions of temperature and humidity. He quantified beetle life histories for each set of conditions and used survivorship and fecundity data to calculate per capita growth rates for each species under those conditions. The results, summarized in **FIGURE 42.7**, explain the more tropical geographic range of R. dominica. In addition, both species had negative per capita growth rates in cool, dry environments. This finding suggested a strategy for managing stores of grain to minimize losses: keep the grain cool and dry.

CHECKpoint CONCEPT 42.3

- ✓ When kangaroo rats harvest seeds, they grab them with their front paws and put them into "grocery bags"—cheek pouches (see Figure 42.4B). Based on the graph in that figure, what do you think the harvest rate would be if seed density increased to 5 seeds/cm^2? Explain.
- ✓ Female hummingbirds in the Rocky Mountains usually live long enough to reproduce for several summer seasons. If they still have young in the nest when the supply of food (nectar produced by flowers) declines toward zero at the end of the summer, they abandon the nest and the young. Can you explain this behavior in terms of the principle of allocation?
- ✓ In an experiment, ecologists added predatory fish to some streams, leaving other streams predator-free as controls. How would you expect size and age of first reproduction of prey fish to evolve in the presence of predators? (Hint: Consider the example of the Aeta and Turkana.)
- ✓ Explain how the results shown in Figure 42.7 are consistent with the grain borer *Rhyzopertha dominica* having a more tropical geographic range than the rice weevil *Sitophilus oryzae*. (Hint: See Concept 41.2.)

Knowing whether population growth rates are negative or positive in particular environments tells us why species are absent or present in those environments. But this knowledge does not help us understand why population densities vary among different locations where population growth rates are positive. To do so, we must take a closer look at the dynamics of population growth.

CONCEPT 42.4 Populations Grow Multiplicatively, but the Multiplier Can Change

We saw in Concept 42.2 that a population will grow as long as the per capita growth rate, r, is greater than zero. Recall also that r refers to a specific period of time (it might be a day, a week, or a year, depending on the species) over which we have estimated per capita birth and death rates. During this period the population will add a number of individuals that is precisely r times its initial size. In other words, growth is multiplicative. We saw this in Equation 42.3, which states that $N_{t+1} = N_t + rN_t$. What may not be obvious at first is that this pattern continues in the next time period: the population will again add a multiple r of its numbers, since $N_{t+2} = N_{t+1} + rN_{t+1}$. Because N_{t+1} was already larger than the initial population N_t, this means that an ever-larger number of individuals is added in each successive period, as long as the multiplier r remains constant.

Multiplicative growth with a constant r differs dramatically in its form from **additive growth**, in which a certain *number* (rather than a certain *multiple*) is added in each period:

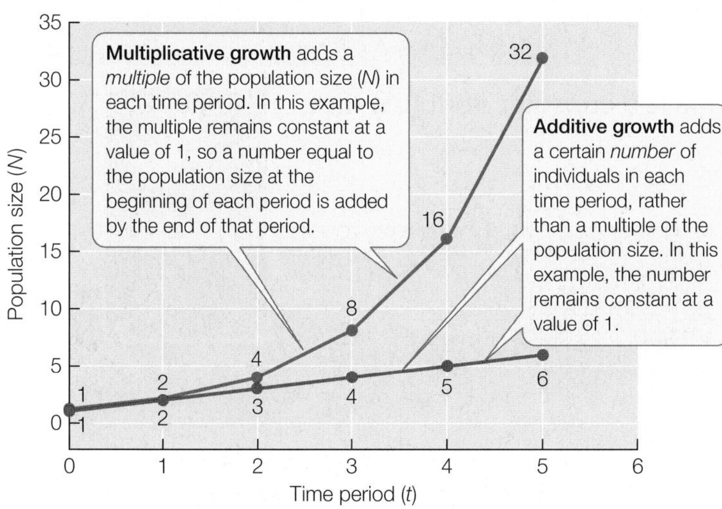

Multiplicative growth adds a multiple of the population size (N) in each time period. In this example, the multiple remains constant at a value of 1, so a number equal to the population size at the beginning of each period is added by the end of that period.
Additive growth adds a certain number of individuals in each time period, rather than a multiple of the population size. In this example, the number remains constant at a value of 1.
Population size (N)
Time period (t)

Multiplicative growth with constant *r* can generate large numbers very quickly

We naturally tend to think in terms of additive growth, so multiplicative growth with an unchanging, positive multiplier astonishes us with the phenomenal numbers it can generate in a short time. For example, in 1911 L. O. Howard, then chief entomologist (insect scientist) of the U.S. Department of Agriculture, estimated that a pair of flies reproducing for the first time on April 15 would give rise to a population of 5,598,720,000,000 adults by September 10 if all the offspring survived and reproduced. Other entomologists disagreed with Howard's calculation—they pegged the number much *higher*!

Charles Darwin was well aware of the power of multiplicative growth:

> Every organic being naturally increases at so high a rate, that, if not destroyed, the earth would soon be covered by the progeny of a single pair. ...As more individuals are produced than can possibly survive, there must in every case be a struggle for existence.

This ecological struggle for existence, fueled by multiplicative growth, is an essential part of natural selection and adaptation.

LINK

Darwin's theory of evolution by natural selection is described in **Concepts 15.1 and 15.2**

Populations growing multiplicatively with constant *r* have a constant doubling time

Multiplicative growth characterizes many phenomena that affect our everyday lives, such as interest-bearing bank accounts, fossil-fuel consumption, and radioactive decay (see Figure 18.1). Our tendency to think in additive terms impedes our ability to comprehend these phenomena and to understand their practical implications. For example, how much more money will we have in 5 years if we invest $1,000 in a bank account that earns 2 percent per year as opposed to 1 percent per year? How does the drop in the per capita growth rate from 2.20 percent per year in 1963 to the current 1.14 percent change our estimates of the future size of the global human population? Answering such questions requires an understanding of multiplicative growth.

Multiplicative growth with a constant *r* has a very striking property: a constant **doubling time**. A population growing in this way adds an ever-increasing number of individuals as time goes on. At some specific time, the number added exactly equals the initial population, and the population has doubled. As long as *r* does not change, the time the population takes to double will also remain constant. The doubling time depends on *r*—it gets shorter as *r* increases.

It is easy to see the implications of multiplicative growth with a constant, positive *r* if we think in terms of doubling times. For example, by doing so a municipality would see that even a seemingly small but constant growth rate of 3 percent per year would mean that its population would double every 23 years—and that its sewage treatment capacity and other such municipal services would need to do the same.

Go to ANIMATED TUTORIAL 42.1
Multiplicative Population Growth Simulation
PoL2e.com/at42.1

Density dependence prevents populations from growing indefinitely

If populations grow multiplicatively, why isn't Earth covered with flies, or any other organism? It turns out that populations do not grow multiplicatively with constant *r* for very long. Instead of continuing to increase (as in the red J-shaped curve in the graph below), population growth eventually slows, and population size usually reaches a more or less steady size (as in the blue S-shaped curve):

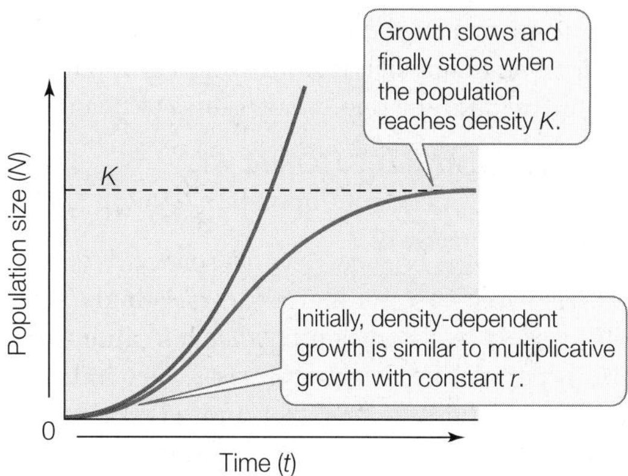

Go to ACTIVITY 42.1 Population Growth
PoL2e.com/ac42.1

Why do populations stop growing? Upon examination, we see that even though the population still adds a multiple of its previous size in each time period, the multiplier *r*, which determines what is added to the population, changes, rather than

APPLY THE CONCEPT

Populations growing multiplicatively with constant *r* have a constant doubling time

Yellow star-thistle (*Centaurea solstitialis*) is a spiny annual plant native to the Mediterranean region. The species has invaded several regions of the United States. It is a noxious weed that is unpalatable to livestock (including American bison). Jane, the rancher, discovers that 1 hectare of her 128-hectare pasture has been invaded by star-thistle. A year later, she finds that the weed population has grown to cover 2 hectares.

1. Based on the information above, how many hectares do you predict the star-thistle population will cover in 1, 2, and 3 more years if the population is growing additively by a constant number? How many hectares if the population is growing multiplicatively with constant *r*?
2. Imagine that Jane discovers the star-thistle population only after it has already covered 32 hectares of her pasture. How many years does she have until the weed completely covers the pasture if its population is growing additively by a constant number each year? How many years does she have if it is growing multiplicatively with constant *r*?

The invasive yellow star-thistle is noxious to grazing livestock.

remaining constant. As the population becomes more crowded, *r* decreases—in other words, *r* is **density dependent** (**FIGURE 42.8A**). At low population densities, per capita growth rates are at their highest. As the population grows and becomes more crowded, per capita birth rate tends to decrease and death rate to increase, and so *r* decreases toward zero. When $r = 0$, the population stops changing in size—in other words, it reaches an **equilibrium** size. That size is called the **carrying capacity**, or *K* (**FIGURE 42.8B**). *K* can be thought of as the number of individuals that a given environment can support indefinitely. If individuals are added to the population so that it exceeds the carrying capacity, the per capita growth rate becomes negative, and the population shrinks. If the population falls below *K*, the per capita growth rate becomes positive, and the population grows (see Figure 42.8B).

Go to ANIMATED TUTORIAL 42.2
Density-Dependent Population Growth Simulation
PoL2e.com/at42.2

Why does *r* decrease with crowding? When used by one organism, a resource becomes unavailable to other organisms. Therefore, as population density increases, each individual gets a smaller piece of the resource "pie" and has less to allocate to its life functions (see Figure 42.5). Per capita birth rate decreases, per capita death rate increases, and therefore *r* decreases with population density. Once again it is useful to spell this out:

Per capita growth rate =
the maximum possible value in uncrowded conditions –
an amount that is a function of density

When population density reaches *K*, the carrying capacity, the average individual has just the amount of resources it needs in order to produce enough offspring to have exactly replaced itself by the time it dies. When density is less than *K*, the average individual can more than replace itself; when density is greater than *K*, the average individual has fewer resources than it needs to replace itself.

LINK

Density dependence results from intraspecific (within species) competition—mutually detrimental interactions among individuals of the same species—as we shall see in **Concept 43.2**

Now we are prepared to return to the question posed at the end of Concept 42.1: Why do species abundances vary in time and space?

Changing environmental conditions cause the carrying capacity to change

We have learned that resource availability in the environment affects an individual's success in acquiring resources (see Figure 42.4), and that physical conditions affect the costs of maintenance (see Figure 42.5). If these factors vary over space, then the carrying capacity will vary over space as well.

Similarly, if resource availability or environmental conditions are good in one year but bad in another, the population will find itself alternately above and below the current carrying capacity. As a result, populations will fluctuate around an average *K* over time. Figure 42.2 illustrates this dynamic for rodents and ticks in Millbrook, New York. Rodent densities tend to increase following years of high acorn production because the rodents have enough food to produce many offspring, and they tend to decrease following poor acorn years. Similarly, nymphal tick densities go up and down following high and low rodent years, respectively, because host abundance determines the success of larval ticks in obtaining a blood meal.

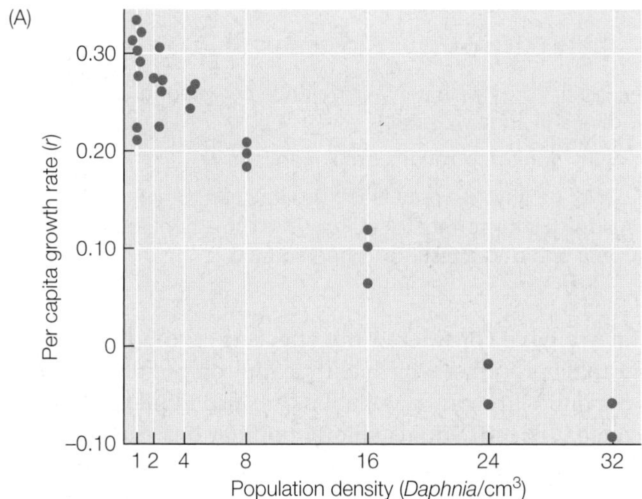

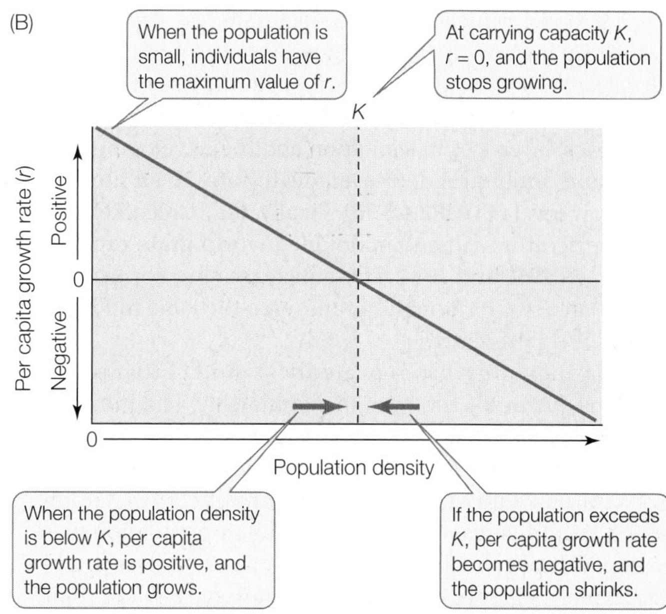

FIGURE 42.8 Per Capita Growth Rate Decreases with Population Density (A) The decline in *r* as population density increases, as measured in a population of a small crustacean, *Daphnia pulex* (see Figure 42.10). (B) Schematic representation of density-dependent growth, in which *r* declines linearly with population density. This graph is an idealization—few species follow it exactly. For example, close inspection of (A) suggests that the decline in *r* for *Daphnia pulex* is not precisely linear.

These increases and decreases are not immediate—there is a delay while acorns are converted into newborn rodents, and while newly fed larval ticks molt into nymphs.

A look back at Table 42.1 shows how temporal changes in the environment affected the fecundity of a group of 210 common cactus finches born on Isla Daphne in 1978. Notice that females produced no surviving young when they were 7 and 8 years old. These years followed a severe drought in 1985 that limited production of the seeds the finches eat. Conversely, fecundity was unusually high when the finches were 5 and 9 years old. These were the unusually wet years of 1983 and 1987, when food was plentiful.

Technology has increased Earth's carrying capacity for humans

The human population is unique among populations of large animals. Not only has it continued to grow, it has grown at an ever-faster per capita rate until very recently, as is indicated by its steadily decreasing doubling times (**FIGURE 42.9A**). The current population multiplier, or per capita growth rate (*r*), is somewhat above 1.1 percent per year, lower than the peak of 2.2 percent reached in 1963, because birth rates have declined faster than have death rates over the past 50 years. Even if this recent trend in *r* continues, however, the human population is projected to grow to 8 billion by 2025.

The human population growth rate has jumped with every technological advance that has raised carrying capacity by

FIGURE 42.9 Human Population Growth Advances in technology have allowed the human population to grow at an ever-faster per capita rate.

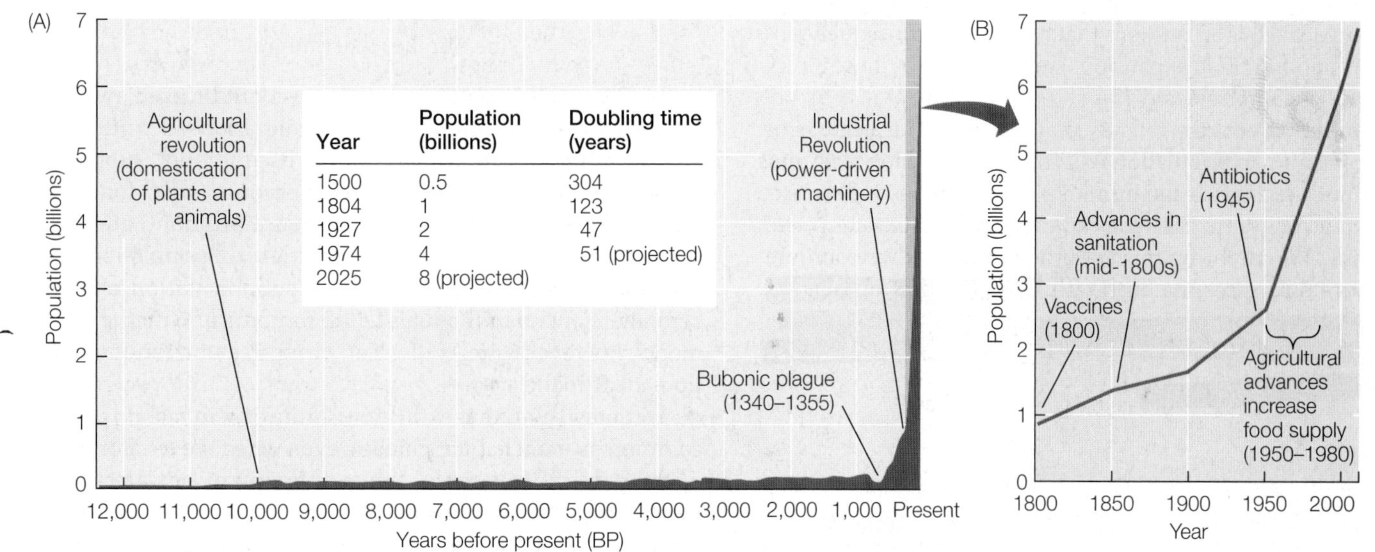

Year	Population (billions)	Doubling time (years)
1500	0.5	304
1804	1	123
1927	2	47
1974	4	51 (projected)
2025	8 (projected)	

increasing food production (thereby increasing per capita birth rate, b) or by improving health (thereby reducing per capita death rate, d). The first jump occurred with the rise of agriculture (about 10,000 years ago). In the nineteenth and twentieth centuries, advances in sanitation and the development of vaccines and antibiotics decreased death rates to an unprecedented low level (**FIGURE 42.9B**). Finally, the "Green Revolution" in agriculture more than doubled world grain production between 1950 and 1984. These increases in r correspond to shorter and shorter doubling times (see the table in Figure 42.9A) until very recently.

Human ingenuity has repeatedly stymied attempts to predict the future of the human population. The most famous attempt was that of Thomas Malthus, whose *Essay on the Principle of Population*, first published in 1798, provided Charles Darwin with a critical insight for his theory of natural selection. Malthus reasoned that the human population was growing multiplicatively, whereas its food supply was growing additively. He predicted that food shortages would limit human population growth, which he estimated would peak by the end of the nineteenth century. Malthus could not, of course, have anticipated the effects of medical discoveries and the Green Revolution.

Many believe that the human population has now overshot its long-term carrying capacity. They give two reasons. First, most twentieth-century technological advances relied on abundant fossil fuels instead of renewable biomass and muscle power, and fossil fuels are not renewable. Maintaining current levels of agricultural production depends heavily on fertilizers derived from natural gas, pesticides derived from oil, and petroleum-fueled engines that pump irrigation water, till the soil, and process and transport food. Similarly, the manufacture and distribution of medicines and diagnostic machinery depends on fossil fuels. World production of crude oil has continued to rise but cannot do so indefinitely because reserves are finite, and the same is true of coal, natural gas, and other fossil fuels. Eventually these fuels will be depleted.

A second reason many believe that our carrying capacity will decrease in the near future is the climate change and degradation of Earth's ecosystems that have been a consequence of our recent population expansion. We will consider these topics further in Chapters 44 and 45.

Population ecology tells us that the growth trajectory of the global human population will change only when birth and death rates change. If the human population has indeed exceeded its long-term carrying capacity, it will ultimately have to shrink. We can bring this about voluntarily if we continue to reduce our per capita birth rate. The less pleasant alternative is to wait for the Four Horsemen of the Apocalypse—famine, pestilence, war, and death—to achieve a sustainable human population size.

LINK

Voluntary reduction of birth rates can be achieved by the use of the contraceptive methods described in **Table 37.1**

CHECKpoint CONCEPT 42.4

- ✓ Explain the fundamental difference between multiplicative and additive growth.
- ✓ Why don't populations grow indefinitely?
- ✓ Why do limits to population growth lead to natural selection? (Hint: Consider that individuals in a population vary in b and d as a result of inherited traits.)

So far we have considered the effects of two universal demographic processes—birth and death—on populations. But the dynamics of many populations are also influenced by the movement of individuals among populations. Such "open" populations grow when immigrants enter from the outside world and shrink when emigrants leave. Immigration and emigration affect populations in important ways.

CONCEPT 42.5 Immigration and Emigration Affect Population Dynamics

Many species occupy habitat patches that are separated from other patches by unsuitable environments (**FIGURE 42.10**; see also Figure 42.1). Each patch of suitable habitat occupied by the species can be thought of as a **subpopulation**, and the set of subpopulations in a region is known as a **metapopulation**. When subpopulations are not too isolated from one another, individuals occasionally move, or disperse, among them.

When dispersal is possible, each subpopulation changes as all populations do—through births and deaths—but with the added input of individuals that move *into* the subpopulation (immigrants) and loss of individuals that move *away* (emigrants). Adding the number of immigrants (I) and emigrants (E) to the BD growth model, we get the **BIDE model**:

The number of individuals in a population
at some time in the future =
the number now + the number that are born +
the number that immigrate –
the number that die –
the number that emigrate

In the BD model, populations are considered **closed systems** because there is no immigration or emigration—no individuals move into or out of them. In the BIDE model, subpopulations are considered **open systems** because individuals can move among them. This difference has a profound effect on population dynamics. In the BD model, a subpopulation that goes extinct stays extinct: once the subpopulation reaches zero, there are no individuals left to reproduce. With the BIDE model, however, immigration can bring an extinct subpopulation "back to life."

You may think that extinction is rare, but in fact it is fairly common in small habitat patches, even when the environment of the patch is favorable. When subpopulations are small, there is a reasonable possibility during any time period that by chance no female will give birth and that all adult individuals

FIGURE 42.10 **A Metapopulation Has Many Subpopulations** The coast of Scandinavia is characterized by natural freshwater rock pools, which constitute suitable habitat for small crustaceans of the genus *Daphnia* that feed by filtering algae from the water. Each rock pool that is occupied by *Daphnia* constitutes a subpopulation within a larger metapopulation.

will die—in which case the subpopulation will be extinct. Small populations are also more likely than large ones to go extinct following environmental disturbances.

Once a subpopulation goes extinct, it will remain at zero density—and its patch of habitat will remain unoccupied—unless individuals move into, and thus recolonize, the patch. Thus if there is no dispersal among patches, the subpopulations will "wink off" one by one. The time to extinction of the entire metapopulation is longer the larger each subpopulation

Go to ANIMATED TUTORIAL 42.3
Metapopulation Simulation
PoL2e.com/at42.3

is, and the more subpopulations there are, but extinction will occur nonetheless. If subpopulations are connected by dispersal, however, the metapopulation as a whole will persist much longer, for two reasons. First, immigration can repopulate unoccupied patches, and "rescue" declining subpopulations from extinction. The repopulated patches, in turn, can colonize others, causing them to "wink on." Second, immigrants contribute to the genetic diversity within subpopulations by bringing new alleles with them. This gene flow combats the erosion of genetic diversity through genetic drift that can reduce r and compromise a species' evolutionary potential.

LINK

Gene flow and genetic drift are discussed in **Concept 15.2**

We can see the recolonization process at work in the metapopulation of Edith's checkerspot butterfly in the San Francisco Bay area (see Figure 42.1). Many subpopulations of this endangered butterfly went extinct during a severe drought that gripped California between 1975 and 1977. The only subpopulation that did not go extinct was the largest one, on Morgan Hill. The extinct subpopulations remained at zero density until 1986, when nine suitable habitat patches were recolonized from the Morgan Hill source population. Patches closest to Morgan Hill were most likely to be recolonized because adult butterflies do not fly very far.

APPLY THE CONCEPT

Immigration and emigration affect population dynamics

In a study of subpopulations of *Daphnia* in Swedish rock pools (see Figure 42.10), ecologist Jan Bengtsson established a number of artificial rock pools that held either 12, 50, or 300 liters of water. He introduced the same number of *Daphnia* to each pool and then covered the pools with mesh, which prevented immigration or emigration. Over the next 4 years, Bengtsson returned repeatedly to count the *Daphnia*, thereby obtaining the average size of the subpopulation in each pool. If a subpopulation went extinct, he included only the nonzero values for the size of that subpopulation measured *before* extinction in calculating its average size. The table shows the grand average of these time-based averages for pools in which the subpopulations went extinct, and for those in which they persisted. Use the table to explore how extinction risk changed with the volume of a pool and the number of *Daphnia* in it.[a]

1. Graph the relationship between pool size and percentage of subpopulations that went extinct, using size as the x axis. What does your graph show you?
2. Within each pool size, which subpopulations were at greatest risk of extinction?
3. How are these two results related to each other?
4. What might have happened if Bengtsson had not placed mesh over his experimental pools?

POOL SIZE (L)	NUMBER OF REPLICATE SUBPOPULATIONS	PERCENTAGE OF SUBPOPULATIONS GOING EXTINCT	GRAND AVERAGE SIZE OF PERSISTENT SUBPOPULATIONS	GRAND AVERAGE SIZE OF EXTINCT SUBPOPULATIONS
12	27	18.5	1,500	65
50	25	16	2,705	530
300	11	9.1	9,690	1,800

[a]J. Bengtsson. 1989. *Nature* 340: 713–715.

CHECKpoint CONCEPT 42.5

- ✓ What is the fundamental difference between the BD (see Concept 42.2) and BIDE models of population growth?
- ✓ Barriers to dispersal and environmental tolerance both influence where a species is found. Which of these factors is more important in determining the geographic range of a species, and which is more important in determining which habitats a species occupies within that geographic range?
- ✓ Human-caused changes in land use often break up the original habitat of a species into spatially separated fragments. Why is this likely to increase the risk of extinction?

In Chapter 41 we saw that ecological knowledge has many important applications in our everyday lives—in medicine, agriculture, and other areas. Our understanding of the dynamics of populations and metapopulations provides tools that allow us to influence their fates, as several examples will illustrate.

CONCEPT 42.6 Ecology Provides Tools for Conserving and Managing Populations

For millennia, humans have tried to maintain or increase populations of desirable or useful species and reduce populations of species they consider undesirable. Such efforts are most successful if they are based on knowledge of how those populations grow and what determines their densities.

Knowledge of life histories helps us manage populations

Knowing the life history of a species helps us identify those life stages that are most important for the species' reproduction and survival, and hence for its population growth rate. Here are a few examples of how this knowledge can be applied.

CONSERVING ENDANGERED SPECIES Populations of Edith's checkerspot butterfly in the San Francisco Bay area are threatened with extinction. Larval survival is critical to the population dynamics of these butterflies (see Figure 42.1). Temporal and spatial variation in butterfly density is tied to the availability of two larval food plants, California plantain (*Plantago erecta*) and purple owl's clover (*Orthocarpus densiflorus*), which are found only on serpentine soils. Maintaining healthy populations of these plants is critical for conservation of the butterflies, but the low-growing *Plantago* and *Orthocarpus* are suppressed by tall introduced grasses. Grazing by cattle can control the invasive grasses, and grazing is an important strategy for conserving *Euphydryas editha*. This example shows that achieving conservation goals is often compatible with human use of the environment.

MANAGING FISHERIES The black rockfish (*Sebastes melanops*) is an important game fish that lives off the Pacific coast of North America. Rockfish grow continually throughout their lives. As is true of many animals, larger females produce more eggs than smaller females. Larger females also provision their eggs with larger oil droplets. Larger droplets provide more energy to the newly hatched larvae, allowing them to grow faster and survive better than larvae with smaller oil droplets.

These life-history characteristics have important implications for rockfish populations. Because fishermen prefer to catch big fish, intensive fishing off the Oregon coast between 1996 and 1999 reduced the average age of female rockfish from 9.5 to 6.5 years. Thus in 1999, females were, on average, smaller than in 1996. This change decreased the average number of eggs produced by females in the population and reduced the average growth rate of larvae by about 50 percent, causing a rapid decline in population density. Because a relatively small number of large females can produce enough eggs to maintain the population, one strategy for maintaining rockfish without shutting fishing down completely is to set aside a few no-fishing areas where some females are protected and can grow to large sizes.

REDUCING DISEASE RISK Because adult black-legged ticks feed and mate primarily on large mammals—white-tailed deer for example—it seems logical that controlling deer densities would reduce the abundance of tick nymphs, which present the greatest risk of transmitting Lyme disease to humans. Surprisingly, experimental reductions in deer density have had little effect on subsequent nymph abundance. Studies of the tick's life history (see Figure 42.3) indicate that the success of larval ticks in obtaining a blood meal, not the number of them that hatch from eggs, has the greatest effect on the subsequent abundance of nymphs. As is true of rockfish, a few adult female ticks can produce enough eggs to maintain populations. Hence controlling the abundance of larval hosts (small mammals) is a more effective strategy for reducing tick populations than controlling the abundance of adult hosts (deer and other large mammals).

Knowledge of metapopulation dynamics helps us conserve species

Conservation biology strives to avoid the extinction of species. From studies of extinction rates, we know that the risk of extinction for a species with a metapopulation structure is affected by the number and average size of its subpopulations, and by rates of dispersal among them. Conservation planners therefore begin with an inventory of remaining areas of natural habitat and an evaluation of the risks to subpopulations in those areas. They then devise ways to protect as many habitat patches as possible, giving priority to those with the largest area because large patches potentially support the largest and most genetically diverse subpopulations. Planners also evaluate the quality of the patches, as measured by their carrying capacity for the species (often estimated from population densities), and develop ways to maintain their quality.

Finally, planners must consider opportunities for individuals to move among subpopulations (**FIGURE 42.11**). For some species (such as Edith's checkerspot butterfly, which can fly through

INVESTIGATION

FIGURE 42.11 Habitat Corridors Can "Rescue" Subpopulations from Extinction Data from the experiments summarized here suggest that corridors between patches of habitat for small arthropod species increase the chances of dispersal, and thus of subpopulation persistence.[a]

HYPOTHESIS

Subpopulations of a metapopulation are more likely to persist if there are no barriers to dispersal.

METHOD

1. On seven replicate moss-covered boulders, scrape off the continuous cover of moss to create a cluster of 13 moss patches surrrounded by bare rock. A large central 50 × 50 cm moss "mainland" (M) is surrounded by 12 small patches of moss, each 10 cm^2 in area. In the "insular" treatment (I), the patches are surrounded by bare rock (which is inhospitable to small moss-dwelling arthropods and is thus a barrier to dispersal). In the "corridor" treatment (C), the patches are connected to the mainland by a 7- by 2-cm strip of live moss. In the "broken-corridor" treatment (B), the configuration is the same as in the "corridor" treatment, except that the moss strip is cut by a 2-cm strip of bare rock.

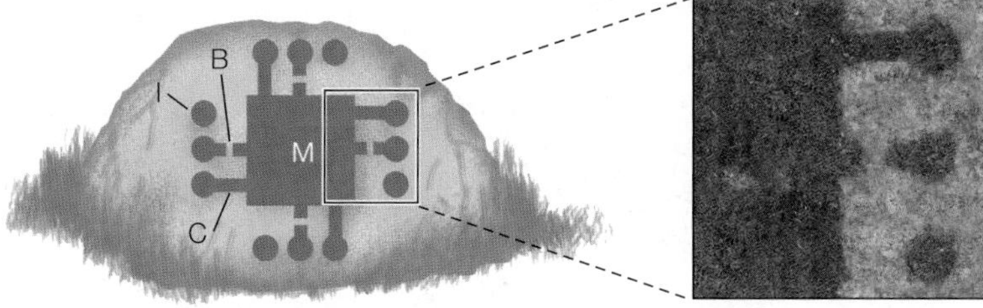

2. After 6 months, determine the number of small arthropod species present in each of the patches.

RESULTS

Patches connected to the mainland by corridors retained as many species as did the larger mainland patch to which they were connected. Significantly fewer species remained in the broken-corridor and insular treatments. Error bars indicate 1 standard error of the mean; bars with different letters are statistically different from one another.

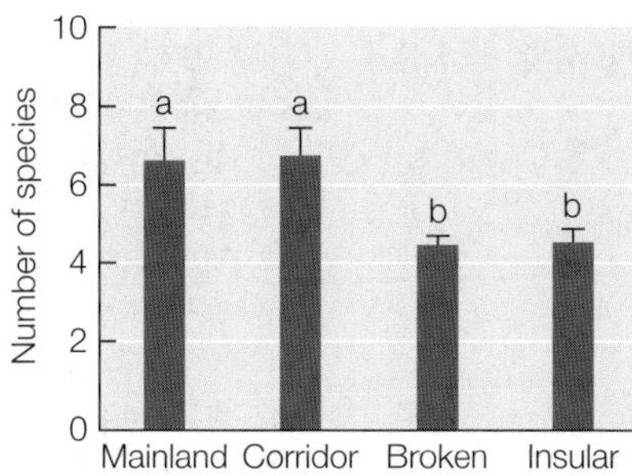

CONCLUSION

Barriers to dispersal reduce the number of subpopulations that persist in a metapopulation.

ANALYZE THE DATA

A. What percentage of the species present in the mainland patch was lost, on average, in the corridor patches?

B. What percentage of the species present in the mainland patch was lost, on average, in the insular and broken-corridor patches? What do these percentages tell you about the average risk of extinction for a subpopulation in a small patch?

C. The insular patches were two-thirds the size of the broken-corridor patches, because the latter contained 5 cm^2 of habitat in the corridor in addition to the 10 cm^2 in the patch itself. Did this difference in patch size affect subpopulation persistence?

D. What effect is patch size by itself expected to have on the persistence of a given species in a patch? Do you see evidence for this effect in the data?

Go to **LaunchPad** for discussion and relevant links for all **INVESTIGATION** figures.

[a]A. Gonzalez and E. J. Chaneton. 2002. *Journal of Animal Ecology* 71: 594–602.

Go to ANIMATED TUTORIAL 42.4 Habitat Corridors
PoL2e.com/at42.4

FIGURE 42.12 A Corridor for Large Mammals An overpass above the Trans-Canada Highway in Banff National Park, Alberta. Bears, elk, deer, and other large mammals use such overpasses as corridors to move between areas that would otherwise be effectively isolated from one another by the highway with its speeding traffic.

unsuitable habitat to reach a new serpentine outcrop), simple proximity of patches determines rates of dispersal and recolonization. For those species, patches that are distant from one another can be connected by a series of intervening patches that serve as "stepping stones." For other species, however, a continuous area of habitat through which individuals can move, called a **dispersal corridor**, is needed to connect subpopulations (see Figure 42.11). Dispersal corridors can sometimes be created by maintaining vegetation along roadsides, fence lines, or streams, or by creating bridges or underpasses that allow individuals to circumvent roads or other barriers to movement (**FIGURE 42.12**).

LINK

Strategies for conserving communities of interacting species are discussed in **Concept 44.5**

CHECKpoint CONCEPT 42.6

- ✓ Describe three elements of a conservation plan for a species that has a metapopulation structure.
- ✓ What similarities do you see between the rockfish and black-legged tick examples summarized above, in terms of the factors that will determine whether a population grows or shrinks?
- ✓ Conservation plans for marine species such as rockfish often lack a corridor component. Why?

You may have noticed that this chapter has focused on single-species populations, but it has also mentioned interactions among species—rodents interacting with acorns and ticks, ticks with hosts and bacteria, butterflies with food plants and invasive plants. These ubiquitous interactions between species, which play important roles in their population dynamics, will be the subject of the next two chapters.

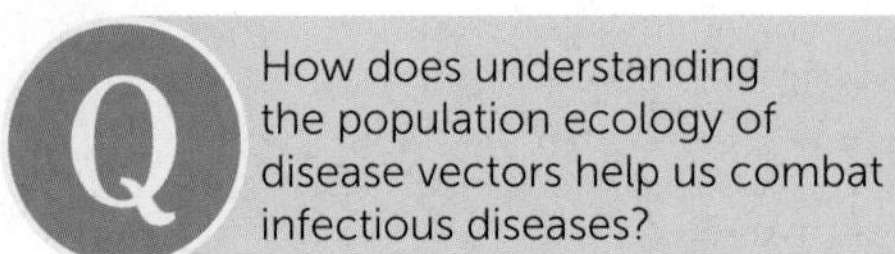

How does understanding the population ecology of disease vectors help us combat infectious diseases?

ANSWER How frequently people come into contact with a dangerous pathogen depends on its abundance and distribution and on those of any disease vectors that transmit it to humans. Population ecology helps us understand what determines the abundance and distribution of pathogens and their vectors, and hence allows us to devise ways of controlling their abundance or avoiding contact with them.

By studying the survivorship and fecundity of the black-legged tick throughout its life history (Concept 42.3), Rick Ostfeld learned that it is the abundance of hosts for larvae and nymphs, rather than weather or the abundance of hosts for adult ticks, that determines tick abundance. Furthermore, Ostfeld ascertained that ticks become infected with *Borrelia* when they feed on infected hosts, and that small mammals are the most effective reservoirs for *Borrelia*. Long-term studies of rodent population dynamics had already shown that rodent abundance is positively correlated with acorn availability (Concept 42.4).

This knowledge suggests a simple and quick way to reduce the probability of contracting Lyme disease (Concept 42.6). Acorn production can be used to predict areas that are likely to become infested with disease-carrying ticks, and various measures can be taken to reduce the risk of human contact with those ticks. These measures include boosting public education efforts for several years after a big acorn crop, posting warning signs in high-risk areas, and alerting health-care specialists whose patients live or work in high-risk areas. A longer-term possibility that has yet to be assessed is to reduce the fraction of ticks that carry *Borrelia* by augmenting populations of hosts that do not harbor *Borrelia* (such as lizards and birds) in the host community. Ticks that bite these hosts will not become infected with *Borrelia*.

SUMMARY

CONCEPT 42.1 Populations Are Patchy in Space and Dynamic over Time

- **Population density** is the number of individuals of a species per unit of area (or volume). **Population size** is the total number of individuals in a population.
- The region within which a species lives is called the species' **geographic range**. Within that range, the species may be restricted to particular suitable **habitats**; these habitats may occur in patches separated by areas of unsuitable habitat. **Review Figure 42.1**
- Population densities vary over time as well as over space. **Review Figure 42.2**

CONCEPT 42.2 Births Increase and Deaths Decrease Population Size

- According to the **BD model**, the size of a population at some future time is the current size plus the number of births minus the number of deaths that occur until that time. Equations based on the BD model can be used to calculate a population's growth rate.
- Ecologists can estimate the **per capita growth rate (*r*)** by tracking births and deaths in a sample of individuals.
- Unless the per capita growth rate is zero, the size of a population must be changing. If *r* is positive, the population is growing. If *r* is negative, the population is shrinking.

CONCEPT 42.3 Life Histories Determine Population Growth Rates

- A species' **life history** describes the sequence of key events of growth and development, reproduction, and death that occur during an average individual's life. **Review Figure 42.3**
- The fraction of individuals that survive to particular life stages or ages is **survivorship**, and the average number of offspring they produce at those life stages or ages is **fecundity**. Survivorship and fecundity determine the per capita growth rate, *r*. **Review Table 42.1**
- Life histories are diverse; they vary both within and among species.
- To survive and reproduce, organisms need materials and energy and the time to acquire them—all of which constitute **resources**. Organisms also need physical conditions they are able to tolerate. Physical conditions differ from resources in that they are experienced, but not used up.
- The rate at which an organism can acquire resources increases with the availability of resources in its environment, up to a point. **Review Figure 42.4**
- The **principle of allocation** states that organisms must divide the resources they obtain among competing life functions. **Review Figure 42.5**
- The environment shapes life histories because it determines the relative costs and benefits of any particular pattern of allocation to different life functions.
- Species can persist only in environments in which their per capita growth rate is positive. **Review Figures 42.6 and 42.7**

CONCEPT 42.4 Populations Grow Multiplicatively, but the Multiplier Can Change

- Populations grow **multiplicatively**, increasing or decreasing by a multiple of their current size in each time period. As a result, populations that are increasing in size by a constant multiple have a constant **doubling time**. **Review ACTIVITY 42.1 and ANIMATED TUTORIAL 42.1**
- As a population becomes denser and the average individual's share of the resource "pie" shrinks, birth rates decline and death rates increase. Thus the per capita growth rate is said to be **density-dependent**. **Review ANIMATED TUTORIAL 42.2**
- When $r = 0$, the population reaches an **equilibrium** size. That size, called the **carrying capacity**, or ***K***, is the number of individuals the environment can support indefinitely. **Review Figure 42.8**
- Changes in resource availability or physical conditions cause the carrying capacity to change.
- Technology has increased Earth's carrying capacity for humans, and the human population has responded by growing rapidly. **Review Figure 42.9**

CONCEPT 42.5 Immigration and Emigration Affect Population Dynamics

- The regional distribution of a species often takes the form of a **metapopulation**, a cluster of distinct **subpopulations** in separate habitat patches. **Review ANIMATED TUTORIAL 42.3**
- The **BIDE model** adds immigration and emigration to the BD model.
- Immigration can repopulate an area where a subpopulation has gone extinct. Immigration also corresponds to gene flow among subpopulations, which combats erosion of genetic diversity through genetic drift.

CONCEPT 42.6 Ecology Provides Tools for Conserving and Managing Populations

- Knowing the life history of a species makes it possible to identify those life stages that are most important for the species' population growth rate. That knowledge can be applied to reduce populations of species considered undesirable and to maintain or increase populations of desirable or useful species.
- Conservation efforts strive to protect as many large habitat patches as possible, because large patches host large populations with low risk of extinction, and a metapopulation with many subpopulations persists longer than one with few subpopulations.
- Providing continuous **dispersal corridors** of habitat through which individuals can move between patches helps maintain the larger metapopulation and so is important for many conservation efforts. **Review Figure 42.11 and ANIMATED TUTORIAL 42.4**

Go to the Interactive Summary to review key figures, Animated Tutorials, and Activities PoL2e.com/is42

Go to LaunchPad at **macmillanhighered.com/launchpad** for additional resources, including LearningCurve Quizzes, Flashcards, and many other study and review resources.

43 Ecological and Evolutionary Consequences of Interactions within and among Species

KEY CONCEPTS

Leaf-cutter ants of the tropical species *Atta cephalotes* carry coca leaves through the rainforest. The ants riding on the leaves act as "bodyguards," warding off parasitic flies that might otherwise lay their eggs on the larger worker ants.

On a morning walk in the Arizona desert, it's hard to resist the temptation to stop and watch the parades of colorful flower and leaf fragments that march across your path. The fragments aren't moving on their own, of course—they're being carried by *Acromyrmex versicolor*, one of the 47 described species of leaf-cutter ants. Leaf-cutter ants are tropical insects whose geographic range barely extends into North America. If you travel south through Mexico and into the Neotropics, however, you will see more and more ant highways and more and more different leaf-cutter species. Leaf-cutter ants originated in South America about 10 million years ago, crossing into North America only after the Panamanian land bridge connected the two continents.

Leaf-cutter ants are important herbivores of the Neotropics. A single underground nest can contain 8 million workers, occupy 20 cubic meters of soil, and consume more than 2 kilograms of plant material per day—enough to strip a large area of vegetation. Leaf-cutters aren't ordinary herbivores, however: they're fungus farmers! They cultivate a fungus in the nest that grows nowhere else on Earth, feeding it with the leaf fragments they harvest. Different leaf-cutter species cultivate different fungus species, all in the family Lepiotaceae.

The interaction between fungus and ant is an obligate one—neither species can live without the other—and it involves a remarkable array of specialized adaptations on the part of both partners. A new queen carries a pellet of fungus with her when she leaves her mother's nest and uses it to start her own small fungus garden, which she nourishes with her feces. She and the larvae that develop from her eggs feed on special nutrient-rich fungal structures. After the larvae pupate and metamorphose into worker ants, they collect leaf material to feed the growing fungus garden. As the colony expands, the workers dig new chambers, transport fragments of the fungus to those chambers, and take on other tasks as scouts, gardeners, nurses, garbage collectors, undertakers—and even bodyguards for the leaf harvesters.

Leaf-cutters associate with other species that influence their interaction with the fungus. A bacterium in the genus *Pseudonocardia*, carried on the ants' exoskeletons, manufactures antibiotics that suppress the crop-killing green mold *Escovopsis* but do not harm the cultivated fungus. And just recently, researchers discovered a nitrogen-fixing bacterium, *Klebsiella*, in the fungus gardens. Much remains to be learned about leaf-cutter interactions!

How could the intricate ecological relationship between leaf-cutter ants and fungus have evolved?

You will find the answer to this question on page 896.

CONCEPT 43.1 Interactions between Species May Increase, Decrease, or Have No Effect on Fitness

The interactions between leaf-cutter ants and their nest associates represent only a small fraction of the ants' interactions with other species. As they harvest leaves, the ants interact with many plant species. The ants themselves are consumed by parasitic flies, beetles, and pathogenic fungi, and the ant corpses that are hauled to the colony's refuse dump are consumed by many species of microbes. The leaf-cutter example illustrates one of life's certainties: at some point between birth and death, every organism will encounter and interact with individuals of other species. **Interspecific** (Latin, "between species") **interactions** affect each individual's life history, and thereby its success in surviving and reproducing, which in turn determines the individual's contribution to total population growth rate (see Concept 42.3; keep in mind that males contribute to the production of new individuals by inseminating females). This contribution to population growth rate is a measure of that individual's **fitness** (see Concept 15.4). By affecting survival and reproduction, interspecific interactions influence the dynamics and densities of populations, alter the distributions of species, and lead to evolutionary change in one or both of the interacting species.

Interspecific interactions are classified by their effects on fitness

The details of interspecific interactions are bewilderingly diverse, but we can cut through all those details by asking, "Does the interaction help or harm individuals of the participating species?" Or more specifically, "Does the interaction increase, decrease, or have no effect on survival and reproduction, and thus on fitness?" If we keep these questions in mind, we will see that there are only five broad categories of interspecific interactions (**FIGURE 43.1A**): competition (–/–), consumer–resource

(A)

Major Types of Species Interactions

Type of interaction	Effect on species 1	Effect on species 2
Competition	–	–
Consumer–resource: Predation, herbivory, parasitism	+	–
Mutualism	+	+
Commensalism	+	0
Amensalism	–	0

(C)

Consumer–resource
The American bison feeds on the grasses of the Great Plains.
Amensalism, Commensalism
The large mammal unwittingly destroys insects and their nests. The cowbirds feed on insects disturbed by the bison's passage.

(B)

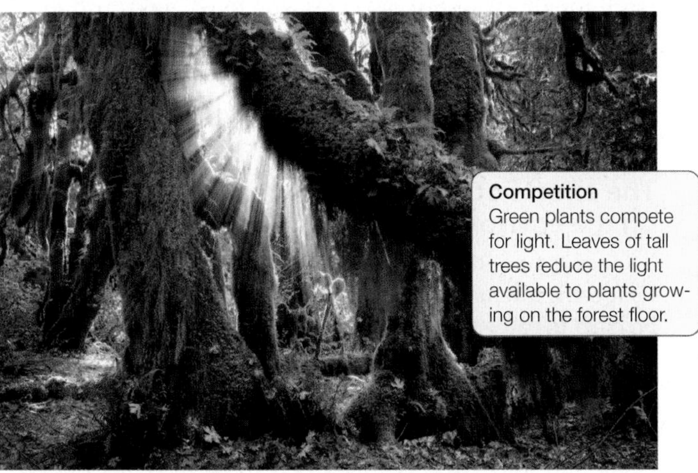

Competition
Green plants compete for light. Leaves of tall trees reduce the light available to plants growing on the forest floor.

(D)

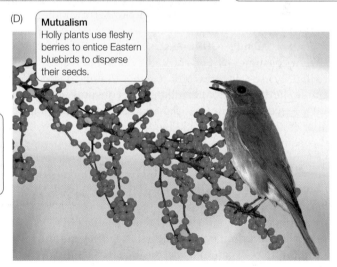

Mutualism
Holly plants use fleshy berries to entice Eastern bluebirds to disperse their seeds.

FIGURE 43.1 Types of Interspecific Interactions (A) Interactions between species can be grouped into categories based on whether their influence on fitness is positive (+), negative (–), or neutral (0). (B) The characteristics of canopy (treetop) and understory (low-growing) vegetation in a forest are largely a product of competition for light. (C) Commensalisms between large grazing ungulates (hoofed animals) and insect-eating birds are found in grassland environments around the world. (D) The interaction between berry-producing plants and berry-eating birds is a mutualism. Here an Eastern bluebird (*Sialia sialis*) gains nutrition by eating holly berries (*Ilex opaca*). The holly's seeds pass through the bird's gut unharmed, and are deposited in nutrient-rich feces away from the parent plant.

Go to ACTIVITY 43.1 Ecological Interactions
PoL2e.com/ac43.1

(+/−), mutualism (+/+), commensalism (+/0), and amensalism (−/0).

Interspecific competition refers to −/− interactions in which members of different species decrease one another's fitness because they require some of the same resources. Recall from Concept 42.4 that when an organism uses a resource, that resource becomes unavailable to other organisms, potentially reducing their survival and reproduction. We say "potentially" because not all resources are in short enough supply to limit these components of fitness. However, at any one time there is generally some **limiting resource** that is in the shortest supply relative to demand. Species that share limiting resources are likely to compete.

Competition may occur among organisms that consume the same food, whether they are African lions and spotted hyenas that attack wildebeest, or aphids and grasshoppers that eat the same plant. Even the interaction between the mold *Escovopsis* and leaf-cutter ants can be considered competitive, because both derive energy and nutrients from the fungus that the ants cultivate. But the limiting resource need not be food. Plants may be limited by availability of sunlight (**FIGURE 43.1B**) or inorganic nutrients in the soil. Some organisms are limited by water or space in which to grow (think of barnacles encrusting a rocky shore; see Figure 43.4) or in which to build a nest. Plants may even compete for the attention of animal partners that pollinate flowers or disperse seeds and thereby contribute to the plants' reproduction.

Consumer–resource interactions are those in which organisms gain their nutrition by consuming other living organisms or are themselves consumed. It's obvious why these are +/− interactions: the consumer benefits while the consumed organism—the resource—loses. Consumer–resource interactions include **predation**, in which an individual of one species (the predator) kills and consumes individuals of another species (the prey); **herbivory**, in which an herbivore consumes part or all of a plant (**FIGURE 43.1C**); and **parasitism**, in which a parasite consumes part of a host individual but usually does not kill it. These categories blur together to some degree. For example, herbivorous caterpillars that consume some of a plant's leaves (see Figures 28.4 and 28.5) are functionally similar to the parasitic black-legged ticks (see Figure 42.3) that consume some of a rodent's blood. And some parasites are functionally similar to predators because they consume so much of the host's body, or otherwise damage it so severely (e.g., some pathogenetic organisms; see pp. 393–394 and 454–455), that the host dies.

A **mutualism** is an interaction that benefits both species (+/+). The interaction between leaf-cutter ants and the fungus they cultivate is mutualistic: the ants feed, cultivate, and disperse the fungus; the fungus, in turn, converts inedible leaf fragments into special fungal structures that the ants can eat. The interactions between plants and pollinating or seed-dispersing animals (**FIGURE 43.1D**), or between humans and some of their gut bacteria (see Concept 41.1), are similarly beneficial to both partners. Mutualisms take many forms and involve many kinds of organisms. They also vary in how essential the interaction is to the partners.

LINK

We have seen several examples of mutualisms in this book, including interactions between mycorrhizal fungi and plants (see **Concepts 22.2 and 25.2**); between fungi, algae, and cyanobacteria in lichens (see **Concept 22.2**); and between corals and dinoflagellates (see **Concept 20.4**).

Competition, consumer–resource interactions, and mutualism all affect the fitness of both participants. The other two defined types of interactions affect only one of the participants.

In **commensalism**, one participant benefits while the other is unaffected (+/0). Some examples involve species whose feeding behavior makes food more accessible for another. For example, the brown-headed cowbird (*Molothrus ater*) owes its common name to its habit of following herds of grazing cattle, foraging on insects flushed from the vegetation by their hooves and teeth. (*M. ater* has also been called the buffalo bird because it followed the bison herds that were once abundant across North America; see Figure 43.1C). The ungulate–cowbird interaction is commensal because the birds have no effect on ungulate fitness. In other cases, the feeding of one species benefits a second species by converting food into a form it can use. Cattle, for example, convert plants into dung, which dung beetles (see Figure 44.4) can use, but the dung beetles have no effect on the cattle. Dung beetles, in turn, are on the giving end of another commensalism: dung-dwelling organisms that cannot fly, such as mites, nematodes, and even fungi, attach themselves to the bodies of the beetles, which not only can fly but are very good at locating fresh dung. The hitchhikers have no apparent effect on dung beetle survival or reproduction.

Amensalism refers to interactions in which one participant is harmed while the other is unaffected (−/0). Elephants moving through a forest or bison grazing the plains crush insects and plants with each step (see Figure 43.1C), but the large mammals are unaffected by this carnage. Amensal interactions tend to be more accidental in nature than the other interactions, as these examples suggest.

The effects of many interactions are contingent on the environment

Ecologists find it useful to distinguish the five categories of interspecific interactions just described, but real interactions can be hard to categorize. Many interactions have both beneficial and harmful aspects, and the net effect of one species on another can be positive or negative depending on environmental conditions.

Sea anemones, for example, sting and consume small animals, including fish. But a select few fish species (mostly anemonefishes; genus *Amphiprion*) are protected by a special mucus coat and thus can live among the anemones' stinging tentacles (**FIGURE 43.2**). Although the benefit of this association to the anemonefishes is clear—they escape their own predators by hiding behind the anemones' nematocysts and can scavenge food caught by the anemones—more research is needed to clarify the consequences for the anemones. Are they nourished by

APPLY THE CONCEPT

Interactions between species may increase, decrease, or have no effect on fitness

Many animals that face uncertain food availability have evolved a tendency to collect and store more food than they can eat right away. Desert seed-eating rodents such as Merriam's kangaroo rat (*Dipodomys merriami*; see Figure 42.4B), for example, will harvest as many seeds as you put in front of them and bury the seeds in shallow deposits scattered around their territories. In some years, the plants in the desert environment produce few seeds, and the kangaroo rats eat all the seeds they have stored. In other years, the kangaroo rats store more seeds than they are able to eat, and the uneaten seeds germinate—they have been "planted." In those years, the plant species whose seeds kangaroo rats favor increase in abundance.

Researchers in southern Arizona monitored the effect of kangaroo rats on the large-seeded native bunchgrasses that the animals prefer. They fenced several plots of land with rodent-proof fences and removed kangaroo rats from half the plots. Over the next 15 years they compared the area covered by the grasses (a measure of their abundance) in plots with and without kangaroo rats.[a]

From 1931 to 1935, southern Arizona experienced a severe drought, and grass populations declined. Rains returned in 1935, and the year 1941 saw particularly high rainfall (for the desert). Examine the graph showing differences in grass abundance (relative to 1931 values) between the two types of plots and answer the following questions.

1. During the drought, did kangaroo rats have a positive, negative, or neutral effect on the grasses? Explain your reasoning.
2. Did the effect of kangaroo rats on grasses change after rains returned? Explain your reasoning.
3. What aspects of kangaroo rat behavior are detrimental to grasses? What aspects might help grasses?
4. Imagine a hypothetical kangaroo rat species that finds and eats 50% of the seeds in its environment without storing any. Redraw the blue line in the figure for this hypothetical animal.

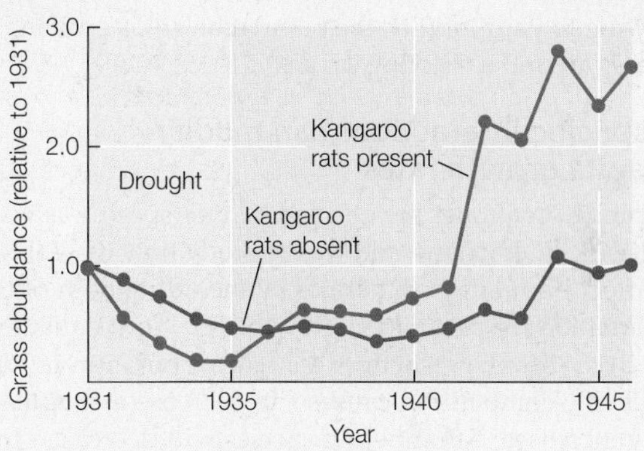

[a]H. G. Reynolds. 1950. *Ecology* 31: 456–463.

the fishes' nitrogen-rich feces? Do they suffer because the fish steal some of their prey? The net effect of the interaction for anemones may change from beneficial when food is scarce to detrimental when food is abundant.

Amphiprion ocellaris

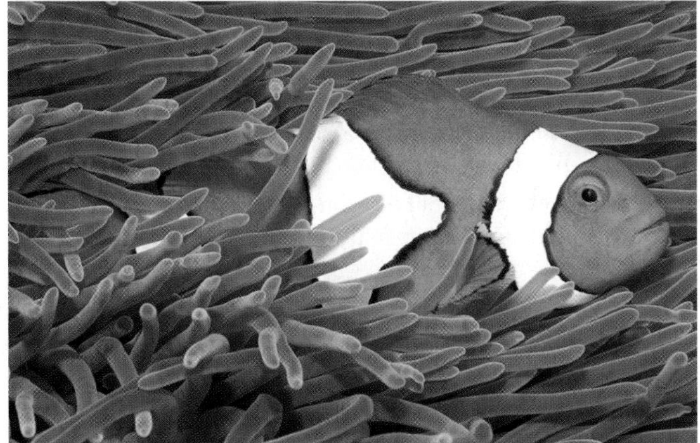

FIGURE 43.2 Interactions between Species Are Not Always Clear-Cut Ecologists long believed that the relationship between sea anemones and anemonefishes was a commensalism: that the fish, by living among the anemones' stinging tentacles, gained protection from predators. But could it also be considered a mutualism—if the fishes' feces provide nutrients that are important for the anemones—or competition—if the fish occasionally steal the anemones' prey?

CHECKpoint CONCEPT 43.1

- ✓ What is the criterion for classifying interspecific interactions?
- ✓ What type of interspecific interaction occurs between humans and their crop plants?
- ✓ Describe an experiment that would allow you to determine whether the effect of anemonefishes on sea anemones is positive, negative, or neutral.

Interspecific interactions affect population dynamics because they influence the contributions of individuals to population growth rate. How exactly are population dynamics affected?

CONCEPT 43.2 Interactions within and among Species Affect Population Dynamics and Species Distributions

Individual members of a population usually vary in their survival and reproduction, a point we will return to in Concept 43.4. Even so, we can sum up the contributions of all members to obtain the total population growth rate. The resulting value is exactly equivalent to what we obtained by using averages in Concept 42.2. There we showed that the population growth rate equals the average individual's contribution (the per capita growth rate, r) multiplied by the number of individuals in the population (N). If r does not change, population size either grows at an ever-faster rate (if r is positive) or gradually shrinks to zero (if r is negative).

The value of r is not fixed, however, because of **intraspecific** (Latin, "within species") **competition**: mutually detrimental interactions among members of the same species that occur because they use the same limiting resources. As population density increases, per capita resource availability shrinks, the per capita growth rate decreases, and overall population growth slows (see Concept 42.4). This negative feedback (see Concept 1.2) dramatically alters the growth trajectory: instead of growing toward infinity, the population stops growing and fluctuates around a carrying capacity (K).

Interspecific interactions can modify per capita growth rates

Per capita growth rates are affected by interspecific as well as intraspecific interactions, and we can study how these interactions affect population dynamics by including both of them in an equation for population growth. The word equation on p. 874 describes the growth of a single population in which intraspecific competition causes r to decrease as population density increases. A competing species will decrease r by an additional amount that depends on its density:

per capita growth rate (r) of species A =

{maximum possible r for species A in uncrowded conditions –

an amount that is a function of A's own population density} –

{an amount that is a function of the population density of competing species B}

Notice that to describe this interaction, we need to write a pair of equations: one for species A and one for species B. Each equation consists of an expression that describes the growth of one species in the absence of the other, minus an expression that describes the effect of the other species.

Interspecific interactions other than competition can be described in a similar way. Depending on the type of interaction, we must either subtract or add the effect of the other species (see Figure 43.1A). In consumer–resource interactions, we subtract the effect of the consumer species (e.g., the predator) in the equation for the resource species (e.g., the prey), since the consumer increases the mortality of the resource species. Conversely, we add an effect of the resource species in the equation for the consumer, since the consumer benefits from the presence of the resource. In mutualistic interactions, we add an effect of each mutualist to the equation for the other. In commensal interactions, we add a term in the equation of the species that benefits, and in amensal interactions, we subtract a term in the equation of the species that is harmed. (In these two types of interactions, equations for the partner species contain no reciprocal terms, because, by definition, the partner is unaffected by the interaction.)

Go to ANIMATED TUTORIAL 43.1
Species Interactions and Population Dynamics
PoL2e.com/at43.1

Interspecific interactions affect population dynamics and can lead to extinction

Because interspecific interactions modify per capita growth rates, populations show different dynamics in the presence and absence of other species. The Russian ecologist Georgii F. Gause explored the effects of interspecific interactions on populations of several species of the protist *Paramecium* (see Figure 20.6). He grew these unicellular organisms in single-species

APPLY THE CONCEPT

Interactions within and among species affect population dynamics and species distributions

A leaf-cutter ant nest can be considered a community—an ecological system (see Concepts 1.2 and 41.1) in which the species are components that interact with one another. Major components of the system are shown as labeled boxes, and their interactions as arrows between those boxes.

Use the description of interactions within leaf-cutter ant nests in the opening story of this chapter to answer the following questions:

1. What is the sign of the following direct effects of each species on another?

Ants on fungus	Fungus on ants
Fungus on mold	Mold on fungus
Mold on bacteria	Bacteria on mold
Bacteria on ants	Ants on bacteria

2. Explain for each interaction the mechanism by which the fitness of interacting individuals is affected.
3. To which of the five categories of interspecific interactions does each pairwise interaction belong?
4. Explain how the spatial distribution of the green mold *Escovopsis* might affect the spatial distribution of leaf-cutter ant colonies.

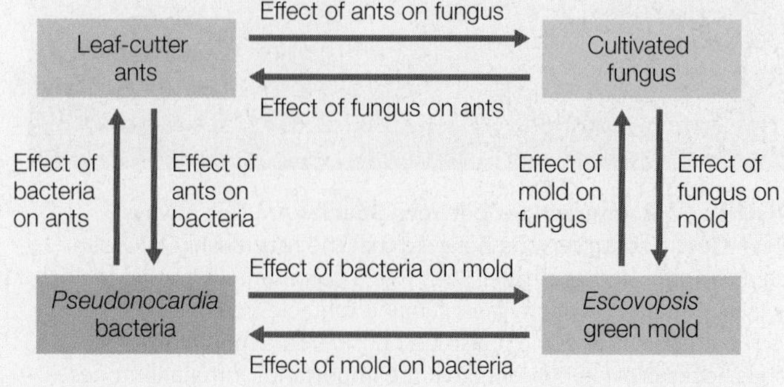

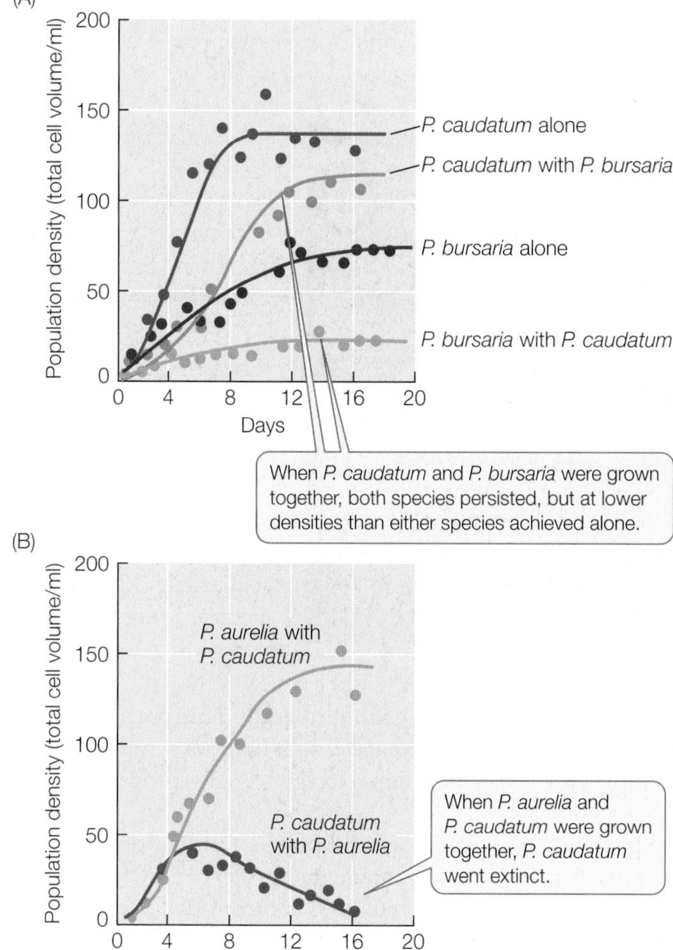

FIGURE 43.3 Interspecific Competition Affects Population Growth (A) Some species of the unicellular protist *Paramecium* are able to coexist, but their populations grow more slowly, and reach lower equilibrium densities, when a competitor is present. (B) Competition between some *Paramecium* species results in local extinction of one species.

cultures and in cultures containing either a competing species of *Paramecium* or a protist predator, *Didinium*. To understand the results, Gause used both word equations, such as the one we gave earlier, and mathematical expressions of the same interactions. In the decades since Gause's work, mathematical models have proved to be important for understanding how interspecific interactions affect population dynamics. In the case of competition, all models lead to the following conclusions:

- The presence of a competitor always reduces population growth rate. In **FIGURE 43.3A**, notice that the slope of the curve of population density versus time is shallower in the presence of a competitor, because the competitor reduces the per capita growth rate from what it would be with intraspecific density dependence alone.
- When two competitors coexist, as in Figure 43.3A, they achieve lower equilibrium population densities than either would achieve alone.
- In some cases, as in **FIGURE 43.3B**, one competitor drives the other extinct.

Other types of interspecific interactions also affect population growth rates and average densities:

- In all interspecific interactions, the per capita growth rate of at least one species is modified by the presence of the other, sometimes positively and sometimes negatively.
- In all interspecific interactions, the average population density of at least one species differs in the presence and in the absence of the other species. Population densities are increased in species for which the interaction is positive and are decreased in species for which the interaction is negative.
- In interactions that have negative effects—that is, consumer–resource interactions, competition, and amensalism—local extinction of one or both of the interacting species is possible.

Interspecific interactions can affect species distributions

We have seen that a species may be absent from a particular location because barriers to dispersal have prevented the species from colonizing it (see Concept 41.4), or because the environment is unfavorable for the species' population growth (see Concept 42.3). We can now see that other species—not just physical conditions—may determine whether a population can persist. A site may be unfavorable because it lacks suitable prey or because it harbors a competitor or predator; conversely, an otherwise unfavorable site may become favorable if a mutualist is present.

FIGURE 43.4 shows an example of how competitive interactions can restrict the habitats in which species occur. The rock barnacle (*Semibalanus balanoides*) and Poll's stellate barnacle (*Chthamalus stellatus*) compete for space on rocky shorelines of the North Atlantic Ocean. Adult barnacles attach themselves permanently to rocks and filter food particles from the water that flows past them; therefore, occupying a space amounts to having the opportunity to feed. The planktonic (swimming) larvae of both *S. balanoides* and *C. stellatus* settle in the intertidal zone and metamorphose into immobile adults. There is little overlap between the areas occupied by adults; the smaller stellate barnacles generally live higher in the intertidal zone, where they face longer periods of exposure to air and desiccation than do rock barnacles.

In a famous experiment conducted more than 50 years ago, Joseph Connell removed each of the species from its characteristic zone and observed the response of the other species. Stellate barnacle larvae settled in large numbers throughout much of the intertidal zone, including the lower levels where rock barnacles are normally found, but they thrived at those levels only when rock barnacles were absent. The rock barnacles grew so fast when present that they smothered, crushed, or undercut the stellate barnacles. By outcompeting them for space, rock barnacles made the lower intertidal zone unsuitable for stellate barnacles. In contrast, removing stellate barnacles from higher in the intertidal zone did not lead to their replacement by rock barnacles; the rock barnacles were less tolerant of desiccation and so failed to thrive there even when stellate barnacles were absent.

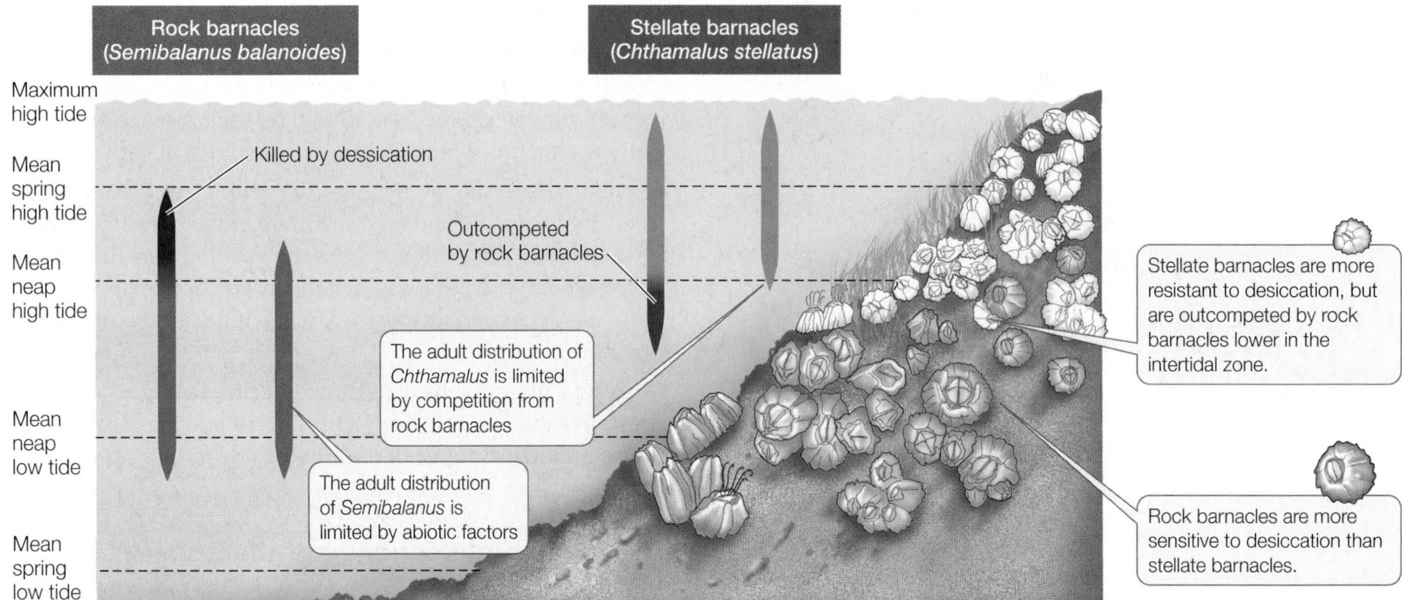

FIGURE 43.4 Interspecific Competition Can Restrict Distributions The left-hand bar for each species shows the zone over which larvae settle, and thus the potential adult distribution. The right-hand bar shows the actual adult distribution. Competition with rock barnacles restricts stellate barnacles to a smaller portion of the intertidal zone than they could otherwise occupy.

Rarity advantage promotes species coexistence

Mathematical models of interspecific interactions have been useful not only in helping us understand how these interactions affect population dynamics and species distributions, but also by giving us insights into the conditions that allow interacting species to coexist.

Models of interspecific competition indicate that two competitors will coexist when individuals of each species suppress their own per capita growth rate more than they suppress the per capita growth rate of their competitor. In other words, intraspecific competition must be stronger than interspecific competition: *you harm yourself more than you harm your competitor*. When this is the case, a species gains a growth advantage when it is rare—that is, when it is at a low density and its competitor is at a high density—and this **rarity advantage** prevents the population from decreasing to zero. The result is coexistence.

One mechanism that can cause intraspecific competition to be stronger than interspecific competition is **resource partitioning**: differences between competing species in resource use. Individuals of the same species, which use very similar resources, will suppress their own per capita resource availability more than they suppress that of the competitor, unless the competitor uses identical resources. But this is unlikely, and if differences in resource use are sufficiently large, competing species can coexist (**FIGURE 43.5**).

Paramecium aurelia, for instance, outcompetes both *P. caudatum* and *P. bursaria*, but the latter two species can coexist (see Figure 43.3A) because *P. bursaria* can feed on bacteria in the low-oxygen sediment layer at the bottom of culture flasks, a habitat that *P. caudatum* cannot tolerate. This partitioning of habitat is possible because *P. bursaria* harbors symbiotic algae that provide it with additional oxygen as a by-product of algal photosynthesis.

In the case of predators and prey, a number of processes can result in a rarity advantage that protects prey from being driven extinct by their predators. For example, prey may become harder to find as they become rarer because they can all hide in the best refuges; or they may be able to invest more in their defenses when they are at low densities and have more resources available per capita. Predators may also stop looking for rare prey and switch to a more abundant prey type.

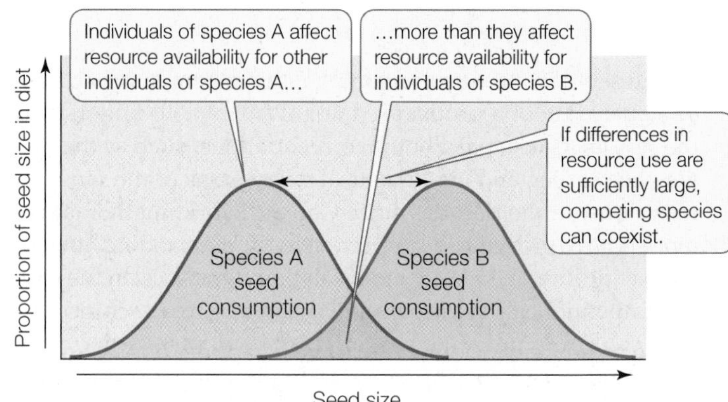

FIGURE 43.5 Resource Partitioning Can Cause Intraspecific Competition to Be Greater Than Interspecific Competition When two species differ in their use of resources, individuals will have a greater effect on the availability of resources to individuals of their own species than to individuals of the other species. In this hypothetical scenario, birds of species A eat smaller seeds on average than birds of species B. Individuals of species A affect the food supply for their own species (smaller seeds) more than they affect the food supply for species B (larger seeds), and vice versa. In other words, the two species partition, or divide up, the seed resource.

Alternatively, some other limiting factor (such as availability of nest sites) may prevent the predators from becoming numerous enough to eat all the prey. The effect of such processes is that per capita growth rates may rebound when prey become rare, thus making their extinction less likely.

CHECKpoint CONCEPT 43.2

- ✓ Describe the components of an equation for the per capita growth rate of a species that interacts with another species.
- ✓ Explain, in terms of resource availability, why the equilibrium population size of one species is reduced by the presence of a competing species.
- ✓ What are the possible outcomes of a predator–prey interaction in terms of the final population sizes of the two interacting species?

The examples of two-species interspecific interactions that we have discussed do not occur in isolation. As may already be apparent, they form parts of a larger web of interactions that connects any given species to many others living in the same area.

CONCEPT 43.3 Species Are Embedded in Complex Interaction Webs

Some species are specialized in their interspecific interactions, competing with, consuming, or being consumed by few other species. Examples include leaf-cutter ants, whose entire diet consists of fungal structures; yucca moths, which pollinate and oviposit only in flowers of specific *Yucca* species; and many parasitic insects that attack only a single host species. However, many species are involved in interactions of many types with diverse other species. For example, most animal-pollinated flowering plants are pollinated by more than one pollinator species, and few pollinators visit only one plant species. Furthermore, each species in a web of pollination interactions is connected to a much larger set of species that act as its competitors, hosts, consumers, and so forth. *Prosopis* shrubs (mesquite) are pollinated by many species of native bees and other insects, and their flowers and leaves are eaten, or harvested, by dozens of species of herbivorous insects, including leaf-cutter ants. Their seeds are eaten or dispersed by rodents, beetles, coyotes, foxes, and hoofed mammals, including cattle (see the opening story of Chapter 41). The roots of a *Prosopis* shrub interact with an entire community of soil fungi and bacteria as well as with roots of other plants. Each of the species that interacts with *Prosopis* in turn interacts with a constellation of other species.

Consumer–resource interactions form the core of interaction webs

These extensive webs of interspecific interactions may seem bewildering, but they make sense once we realize that they are organized around consumer–resource interactions—after all, everybody has to eat, or (to be more accurate) everybody needs energy and nutrients. Every species therefore participates in consumer–resource interactions as the consumer, the resource, or both. In addition, competition commonly occurs between consumers that use the same sources of energy and nutrients—for example, sunlight, chemical nutrients, or food—or (in the case of immobile organisms such as barnacles or plants) the space that gives them access to these resources. As we will see in Concept 43.4, even mutualisms usually involve consumers and resources. For example, leaf-cutter ants both feed and eat their fungus; algae that live inside the cells of sea anemones (see Figure 43.2) photosynthesize, producing carbohydrates that the animals consume, and in return, the anemones provide the algae with nitrogen. To be sure, energy and nutrients are not the only types of resources involved in interspecific interactions. Birds, for example, compete for nest sites that are safe for their offspring. Nevertheless, feeding, or **trophic** (Greek *trophes*, "nourishment"), interactions form the core of interaction webs.

Trophic interactions determine the flow of resources through communities. Nutrients and energy enter communities via organisms that obtain the *materials* that form their bodies from inorganic sources and that obtain the *energy* that fuels their metabolism from solar radiation or inorganic chemicals. These organisms are called **primary producers** because they convert energy and inorganic materials into organic compounds that can be used by the rest of the community. They are also called **autotrophs** (Greek, "self-feeders") because they create their own "food" from inorganic sources (see Concept 6.6).

Species that obtain energy by breaking apart organic compounds that have been assembled by other organisms are called **heterotrophs** (Greek, "other-feeders"). Heterotrophs that dine on primary producers are called **primary consumers**, or herbivores (Latin, "plant eaters"). Those that eat herbivores are called **secondary consumers**, or carnivores (Latin, "flesh eaters"); those that eat the flesh of secondary consumers are called **tertiary consumers**, and so on. These feeding positions—primary producers and primary, secondary, and tertiary consumers—are called **trophic levels**. Some organisms, called **omnivores**, feed from multiple trophic levels. **Decomposers** feed on waste products or dead bodies of organisms (**TABLE 43.1**).

Trophic interactions can be diagrammed as a **food web** that shows who eats whom. Arrows from resource to consumer species show how energy and materials flow, and species are arranged vertically according to trophic level (**FIGURE 43.6**). Food web diagrams do not show all the interactions within a community, but they do show the major ones, and many others can be inferred from them. Competition, for example, can be inferred whenever multiple consumers use the same resource: elk (*Cervus canadensis*) and American bison (*Bison bison*) compete for grass, and plants compete for sunlight and soil nutrients.

Losses or additions of species can cascade through communities

As the naturalist and conservationist John Muir noted in his 1911 book *My First Summer in the Sierra*, "When we try to

TABLE 43.1 The Major Trophic Levels

Trophic level	Source of energy	Examples
Primary producers (photosynthesizers, chemoautotrophs; see Table 19.2)	Solar energy, inorganic chemicals	Green plants, protists, bacteria, archaea
Primary consumers (herbivores)	Tissues of primary producers	Elk, grasshoppers, gypsy moth larvae, pollinating bees, geese
Secondary consumers (carnivores)	Tissues of herbivores	Spiders, great tits, cheetahs, parasites
Tertiary consumers (carnivores)	Tissues of carnivores	Tuna, killer whales, parasites
Omnivores	Several trophic levels	Coyotes, opossums, crabs, robins, white-footed mice
Decomposers	Dead bodies and waste products of other organisms	Fungi, many bacteria, vultures, earthworms, dung beetles

Go to ACTIVITY 43.2 The Major Trophic Levels
PoL2e.com/ac43.2

pick out anything by itself, we find it hitched to everything else in the Universe." This poetic statement encapsulates an ecological fact: adding, removing, or changing the abundance of any species has effects that reverberate throughout an interaction web.

An example of what we might call the "Muir Effect" comes from Yellowstone National Park in Wyoming. Wolves (*Canis lupus*) in the park prey on large mammals, including elk (see Figure 43.6). Unrestricted hunting eliminated wolves from the park by 1926, and they were reintroduced in 1995. This shift in trophic interactions has had cascading effects on the entire ecosystem.

Park managers initiated annual censuses of elk in 1920. Concerned by the rapid increase in elk numbers after wolves disappeared, they deliberately kept elk numbers in check for many years. But the removal of elk ceased in 1968, and elk densities increased rapidly after that (**FIGURE 43.7A**). The elk browsed aspen trees (*Populus tremuloides*) so heavily that no young trees were added to the population after 1960 (**FIGURE 43.7B**). The elk also browsed streamside willows (*Salix* spp.), with the result that beavers (*Castor canadensis*), which depend on willows for food, nearly went extinct locally.

After park managers reintroduced wolves in 1995, elk populations decreased, and as a consequence, young aspen now survive, willows regrow, and species associated with these trees (beavers as well as birds, insects, and many others) have increased. This history of removal and reintroduction illuminates a cascade of effects across trophic levels—a so-called **trophic cascade**. What happens to predators is felt not only by their prey, but also by the prey of their prey, and by many other species connected with all of them.

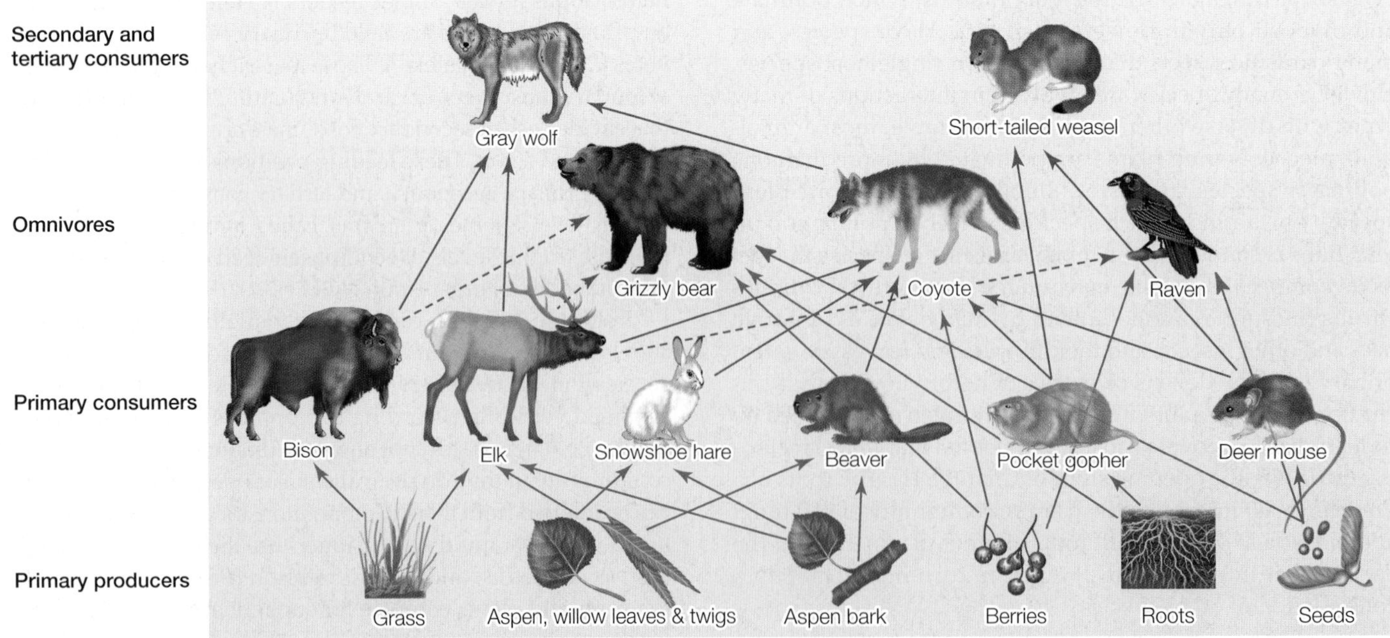

FIGURE 43.6 A Food Web in the Grasslands of Yellowstone National Park The arrows point from resource to consumer. Even this highly simplified food web is complex. Primary consumers eat plant material (blue arrows). Secondary and tertiary consumers are carnivores that kill and eat live animals (red arrows). Omnivores such as grizzly bears, coyotes, and ravens eat both plant and animal tissues; ravens and grizzlies also eat carrion (flesh of dead animals that they did not kill; dashed red arrows).

(A)

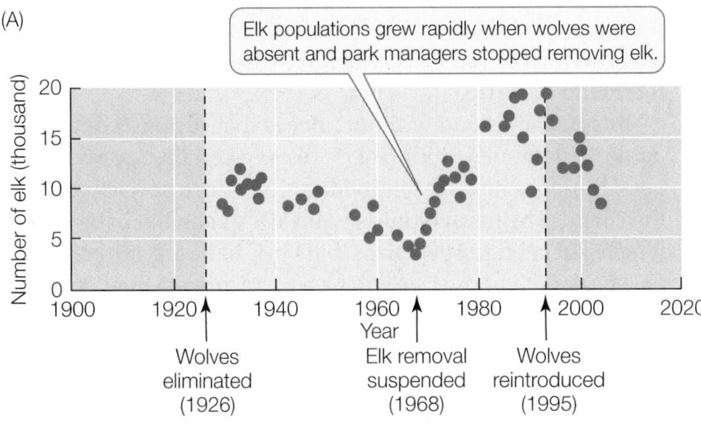

(B)

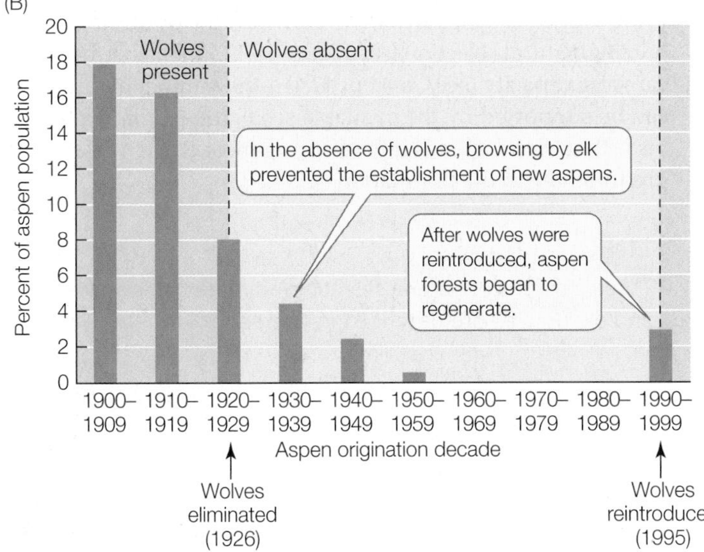

(C) *Populus tremuloides* inside elk-exclusion fence

FIGURE 43.7 Removing Wolves Initiated a Trophic Cascade (A) The elk population in Yellowstone National Park increased rapidly after wolves were eliminated and the park stopped removing elk. (B) Aspens did not become established when wolves were absent and abundant elk browsed the trees. (C) Researchers constructed an "elk exclusion" fence to demonstrate that aspen regenerated only inside the fence, where elk were absent.

Go to MEDIA CLIP 43.1
Wolf Cascade
PoL2e.com/mc43.1

The cascading effects of ecological interactions have implications for conservation

The Yellowstone study exemplifies the new scientific field of **conservation ecology** (one of many practical applications of ecology), and more specifically, the subfield of **restoration ecology**. Conservation ecology attempts to avoid the loss of elements or functions of an existing web of interactions, whereas restoration ecology provides scientific guidance for restoring lost elements or functions of a web. Reintroducing predators that humans once eliminated (naturally enough: a predator that eats our own prey or domestic animals is our competitor!) is a bold attempt to restore ancestral ecosystems. It also illustrates a bigger point: if we wish to restore or conserve ecological systems, we must consider the entire web of ecological interactions. And if we strive to conserve a particular species that is in danger of extinction, we must consider the web from the standpoint of that species, looking outward at all of its ecological interactions.

Nothing illustrates these points better than the problem of species introductions. As we saw in Concept 41.5, humans are introducing many species to new places, both intentionally and unintentionally. These introductions not only blur biogeographic boundaries but also have the potential to alter interactions among native species. Species that are introduced into a region where their natural enemies are absent may reproduce rapidly and spread widely. Such **invasive species** are likely to have negative effects on native species that lack adaptations to compete with or defend themselves against the newcomers. Alternatively, introduced species may benefit some native species if they provide a resource that was not available previously. In all cases, they alter interaction webs, as we will see in Chapter 44.

Introduced invasive species can harm native species in several ways. Introduced yellow star-thistle (*Centaurea solstitialis*), for example, grows rapidly in grasslands and physically crowds out native species, covering entire areas with its spiny flowering heads (see p. 874). Other flowering plants alter relationships between native plants and their mutualists. Purple loosestrife (*Lythrum salicaria*), introduced into North America in the early 1800s, not only crowds out natives but also—adding insult to injury—competes with them for the attention of insect pollinators. The native species *Lythrum alatum*, for example, receives fewer visits from bumble bees, and produces fewer seeds as a result, when purple loosestrife is present.

Some invaders alter ecological interactions by causing extinctions of native species. Concept 22.3 described the case of an introduced sac fungus, a pathogen that drove the once-dominant American chestnut (*Castanea dentata*) to extinction. The chestnut has been replaced by oaks throughout the forests of the northeastern United States. Because acorn production varies greatly from year to year, whereas chestnuts produced consistent nut crops every year, the loss of chestnuts has precipitated dramatic year-to-year fluctuations in rodents, ticks, and Lyme disease (see Figure 42.2).

Organisms that are deliberately introduced to control pests can alter interactions among native species if they do not confine their attention to the pest species. *Rhinocyllus conicus*, for example, is a Eurasian weevil whose larvae eat developing seeds. It was introduced into North America in 1968 to control the invasive musk thistle (*Carduus nutans*). But when the abundance of musk thistle declined, the weevil moved on to the native Platte thistle (*Cirsium canescens*) and then to wavyleaf thistle (*C. undulatum*). The introduced weevil has become not only a consumer of native thistles but also a competitor of the native insects that eat thistles.

CHECKpoint CONCEPT 43.3

- ✓ Why do some introduced species become invasive?
- ✓ At what trophic level would you place humans?
- ✓ What would a terrestrial landscape look like if its communities contained no decomposers?

The effects of interspecific and intraspecific interactions on individual members of a population depend on those individuals' characteristics. For instance, if individuals vary in heritable attributes that determine how much a competitor affects their access to resources, or that determine their risk of being eaten by a predator, then the interaction will result in evolution by natural selection.

CONCEPT 43.4 Interactions within and among Species Can Result in Evolution

Both intraspecific and interspecific interactions can affect the success of an average individual in surviving and reproducing. But averages tell us only part of the story, because no individual is exactly average. Interactions will therefore affect individuals to different extents, causing them to vary in their contribution to population growth rate—that is, in their fitness. Heritable traits that cause greater benefit from positive interactions, or less harm from negative interactions, will become more common in the population. The dynamics of this evolutionary change (see Chapter 15) will depend on the mix of positive and negative effects that characterize an interaction. Let's consider some examples, starting with intraspecific competition.

Intraspecific competition can increase the carrying capacity

Equations for density-dependent population growth, such as the word equation in Concept 42.4, assume that all the individuals in a population are equally affected by its density. However, individuals often vary in phenotypic traits that affect how well they can use available resources. Consider the seed-eating finches of the Galápagos archipelago off the west coast of Ecuador (see Figure 17.7). The average beak size for any given finch species varies from island to island, and the beak size on any given island is close to the size that is best for processing one of the types of seeds available on the island (**FIGURE 43.8**). Such patterns suggest that beak size has evolved in response to the available seeds. To visualize this process, consider the following scenario.

Imagine an island without finches that supports one abundant plant species producing seeds of a certain size and hardness. Suppose next that individuals of a finch species colonize the island. The new colonists will vary somewhat in beak size, which is related to body size—big birds have big, strong beaks. One beak (and body) size will be best: relative to this "optimal" size, smaller birds will have more difficulty cracking the seeds open, whereas larger birds will have to allocate more energy to maintenance of their larger bodies, thus leaving less for reproduction (see concept 42.3). Individuals with beaks closest to the optimal size will leave the most offspring. As a consequence, the average beak size will evolve toward the optimum by stabilizing natural selection (see Figure 15.13A). Furthermore, the carrying capacity (see Concept 42.4)—the number of birds that can be supported by the available seed crop—will increase as birds that convert seeds into offspring most efficiently come to predominate in the population.

LINK

Three patterns of natural selection—directional, stabilizing, and disruptive—are described in **Concept 15.4**

Interspecific competition can lead to resource partitioning and coexistence

Interspecific competition, like intraspecific competition, can lead to evolution, as is illustrated by another interaction in the Galápagos archipelago, between a finch and an insect. The small ground finch (*Geospiza fuliginosa*) consumes flower nectar as well as seeds. Small individuals are better at obtaining nectar from the islands' small flowers, but large individuals have larger beaks that are better at cracking seeds. Carpenter bees (*Xylocopa darwinii*) compete with *G. fuliginosa* for nectar on some islands but are absent from others. Small finches are affected more strongly by competition from bees than are larger individuals. As a consequence, the finches are larger in size, and drink less nectar, on islands where carpenter bees are present (**FIGURE 43.9**). This evolutionary shift means that resource use by *G. fuliginosa* has diverged from that of its bee competitors on islands where the two species coexist—an example of resource partitioning (see Concept 43.2) that resulted from natural selection.

Consumer–resource interactions can lead to an evolutionary arms race

Predators, parasites, and herbivores benefit if they can acquire lots of nutritious food quickly and inexpensively, whereas resource species benefit if they can evade or deter these natural enemies at little cost. The interests of consumer and resource species are obviously at odds. These opposing interests can lead to an "evolutionary arms race," in which prey continually evolve better defenses, predators continually evolve better offenses, and neither gains any lasting advantage over the other. The Red Queen summed up this situation in Lewis Carroll's

INVESTIGATION

FIGURE 43.8 Intra- and Interspecific Competition Influence the Morphology of Coexisting Species In the Galápagos archipelago, seed-eating finches (*Geospiza* spp.) use their beaks to crack open seeds, which are often in short supply. Individuals with big beaks can crack large, hard seeds that individuals with small beaks cannot eat, whereas small-beaked birds gain more fitness than do large-beaked birds from eating small, soft seeds. Dolph Schluter and Peter Grant documented how the islands differed both in their seed resources and in the morphology of *Geospiza* species present.[a]

HYPOTHESIS

Differences in the morphology of finches among islands reflect interisland differences in the availability of seeds of different sizes.

METHOD

1. Determine which species of seeds can be eaten by *Geospiza* species with different beak sizes, and determine the species' food requirements.
2. On each island, measure the abundances of seeds of different plant species and calculate the availability of food for finches of different beak sizes.
3. From these measurements, calculate for each island how many finches of a given beak size could be supported if no other finch species were present (this is an estimate of carrying capacity). Do this for hypothetical finches with beak sizes that span the possible range for seed-eating *Geospiza* species.
4. Display the calculated results as a graph for each island, with beak size on the *x* axis and estimated carrying capacity on the *y* axis. A peak in the graph predicts a high carrying capacity for a finch species with that beak size. A low point means the opposite—a species with that beak size is predicted to achieve only a low equilibrium density.
5. On each graph, place a dot showing the mean beak size of each *Geospiza* species actually present on the island.

RESULTS

1. Each finch species present on an island had a beak size that matched a peak quite closely.
2. The species present differed in beak size, so that only one species "occupied" a peak.
3. The number of finch species present on an island increased with the number of distinct peaks in predicted carrying capacity.

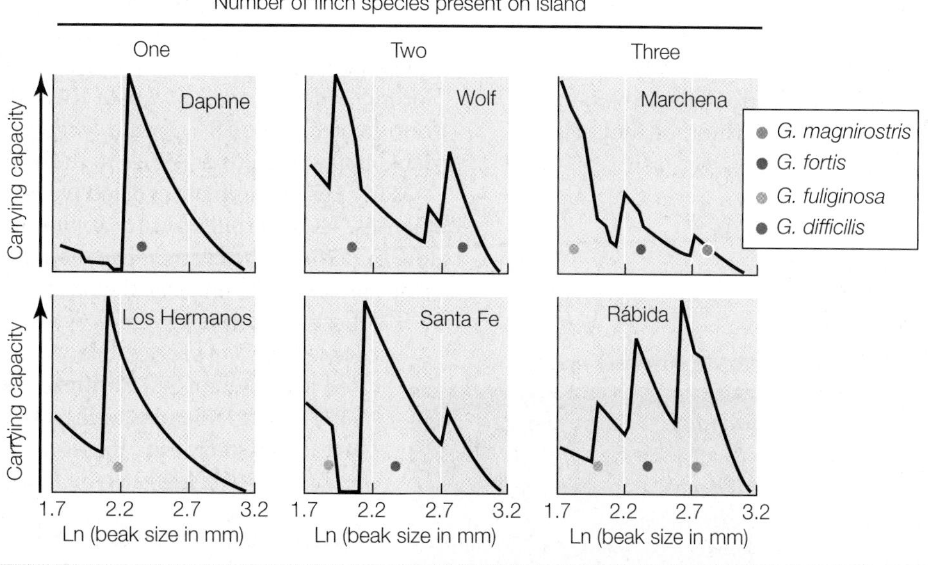

CONCLUSION

The availability of seeds of different sizes determines how many species of *Geospiza* occur on an island and what their beak sizes are.

ANALYZE THE DATA

A. Explain how intraspecific competition could cause the average beak size of a hypothetical finch species that colonizes an island to evolve to match a peak in carrying capacity.

B. Suppose the *G. fuliginosa* population on Marchena goes extinct, and the island is subsequently colonized by *G. fuliginosa* individuals from Rábida. Do you expect the average beak size of the colonizing population to increase or decrease through time? Explain your reasoning.

C. Explain how interspecific competition could cause peaks in seed availability to have no more than one associated finch species.

Go to **LaunchPad** for discussion and relevant links for all **INVESTIGATION** figures.

[a]D. Schluter and P. R. Grant. 1984. *American Naturalist* 123: 175–196.

Geospiza fuliginosa

Nectar Use and Size of *G. fuliginosa*

Island	Time spent feeding on flower nectar (%)	Mean size (wingspan, mm)
BEES ABSENT		
Pinta	10	59.8
Marchena	28	58.2
BEES PRESENT		
Fernandina	1	64.8
Santa Cruz	14	64.0
San Salvador	0	63.8
Española	0	64.7
Isabela	7	64.5

Xylocopa darwinii

FIGURE 43.9 Finch Morphology Evolves in Response to Competition with Carpenter Bees On those Galápagos islands where small ground finches are the sole pollinators of small flowers, the birds are small (as measured by their short wingspans). Small size lowers their energy requirements and may increase their ability to negotiate the flowers. On islands where carpenter bees compete with these finches for nectar, the birds are larger and have bigger, stronger beaks (which allows them to include seeds in their diet).

Through the Looking-Glass with the words "It takes all the running you can do, to keep in the same place."

Resource species have various defensive strategies available to them. Some mobile animals use speed, size, or weapons to thwart predators. Others hide or use camouflage to avoid being detected, or mimic unpalatable species (**FIGURE 43.10A**). Immobile organisms have other tricks up their sleeves; for example, they may have evolved thick armor, or they may be poisonous or not nutritious. In response to these strategies, natural selection in populations of consumers favors greater speed, size, or strength; keen senses; armor-piercing or crushing tools (**FIGURE 43.10B**); or means of detoxifying poisons.

Plants produce a great variety of defensive chemicals, as we saw in Chapter 28. Used in small quantities, many of these chemicals, including the caffeine of coffee and tea, the mustard oils of cabbages, the capsaicins of chilis, and the piperine of black pepper, spice up our own lives. But the production of such compounds evolved because they are toxic or repellant to some herbivores and pathogens. Defensive compounds explain in part why the leaf-cutter ant *Atta cephalotes* harvests leaves from only 17 of 332 plant species that grow in the Florencia Norte forest of Costa Rica: the terpene-like compounds (see Table 28.1) found in the leaves of many of the plants make them toxic either to the ants or to their fungus.

Many herbivores have evolved ways around plant chemical defenses. The caterpillars of *Heliconius* butterflies, for example, store or detoxify the cyanide-containing defensive compounds of the passionflower (*Passiflora*) vines they feed on, and even use these poisons as defenses against their own predators. Some *Passiflora* species, conversely, have evolved modified leaf structures that mimic butterfly eggs (**FIGURE 43.11**). Since female butterflies will not lay eggs on leaves that already carry eggs, these structures reduce the plant's probability of being eaten by *Heliconius* caterpillars.

Mutualisms involve conflict of interest

Mutualisms are often misunderstood. We sometimes hear that "bees visit flowers to pollinate them," or that "flowers provide food for their pollinators," but such statements imply an impossible evolutionary process

(A)

Episyrphus balteatus

Vespula vulgaris

(B) Anodorhyncus hyacinthinus

FIGURE 43.10 Defense Mechanisms and Evolutionary Arms Races (A) The harmless hoverfly (above) gains protection by mimicking a dangerous stinging wasp (below) that predators have learned to avoid. (B) Brazil nuts (the seeds of *Bertholletia excelsa*) are protected by extremely hard shells that most birds cannot penetrate. The evolution of beak strength in the hyacinth macaw, however, has kept pace with the nut's hardness.

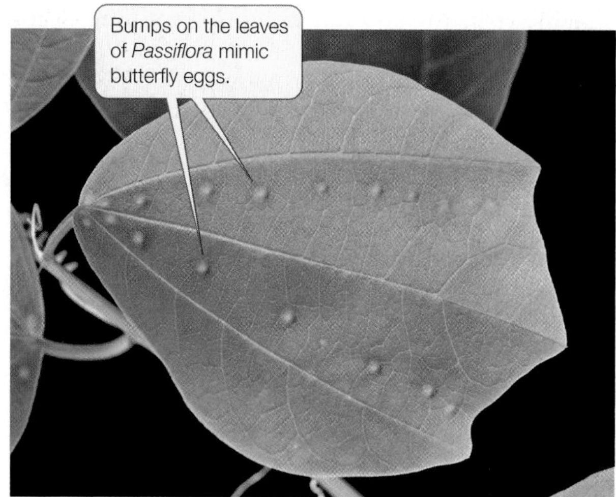
Bumps on the leaves of *Passiflora* mimic butterfly eggs.

FIGURE 43.11 Using Mimicry to Avoid Being Eaten The raised yellow bumps on these passionflower leaves resemble the eggs of the plant's principal herbivores, zebra butterflies (*Heliconius* spp.). Because female butterflies will not lay their eggs on leaves that already carry eggs, these "false eggs" protect the plant from being eaten by caterpillars.

in which species exhibit traits that evolved to benefit another species. As Charles Darwin said about his theory of natural selection (see Concept 15.1),

> If it could be proved that any part of the structure of any one species had been formed for the exclusive good of another species, it would annihilate my theory, for such could not have been produced through natural selection.

In truth, most pollinators visit flowers to get food and just happen to pollinate the flowers in the process; flowers provide food (usually as little as possible) to lure animals, not to fatten them up; birds and bees respond to any conspicuous color that signals the presence of food; and most flowers are visited by any animal that gains economically by doing so, whether it pollinates the plant or not. In the rough-and-tumble world of ecological interactions, species benefit other species not out of **altruism** but because acting in their own self-interest happens to benefit others.

LINK

The coevolution of plants and pollinators is discussed in **Concept 21.5**

APPLY THE CONCEPT

Interactions within and among species can result in evolution

Crabs feed on mussels, and individual young mussels can sometimes detect the presence of crabs and respond by developing a thicker protective shell as they grow to adult size. The abilities to detect and to respond to crabs are inherited traits.

On the Atlantic coast of North America, blue mussels (*Mytilus edulis*) face two non-native crab predators. The European green crab (*Carcinus maenas*) was introduced to the United States almost 200 years ago, and the Asian shore crab (*Hemigrapsus sanguineus*) arrived only 20 years ago. *C. maenas* has spread all along the Atlantic coast of North America, whereas *H. sanguineus* has not yet reached northern Maine. Have mussels evolved the ability to detect this new enemy and respond to it?

Researchers collected very young *M. edulis* from northern populations, which have experienced predation by *C. maenas* but not by *H. sanguineus*, and from southern populations, which have experienced both predator species. They grew mussels from each population on floating docks under three different conditions: no crab nearby (controls); a hungry *C. maenas* caged nearby; or a hungry *H. sanguineus* caged nearby. After 3 months, researchers measured mussel shell thickness.[a]

Use the graph to answer the following questions. Error bars indicate 95% confidence intervals of mean shell thickness; bars with different letters are statistically different from one another ($P < 0.05$). See Appendix B for discussion of statistical concepts.

1. Have mussel populations exposed to predation by non-native crabs evolved the ability to detect and respond to them by thickening their shells? Explain your answer.
2. Which comparison is critical to your conclusion?
3. Do mussels in southern populations have thinner shells than mussels in northern populations? Would a difference in average shell thickness between the two populations invalidate your conclusion? Why or why not?
4. What do these data suggest about how quickly evolution can take place?

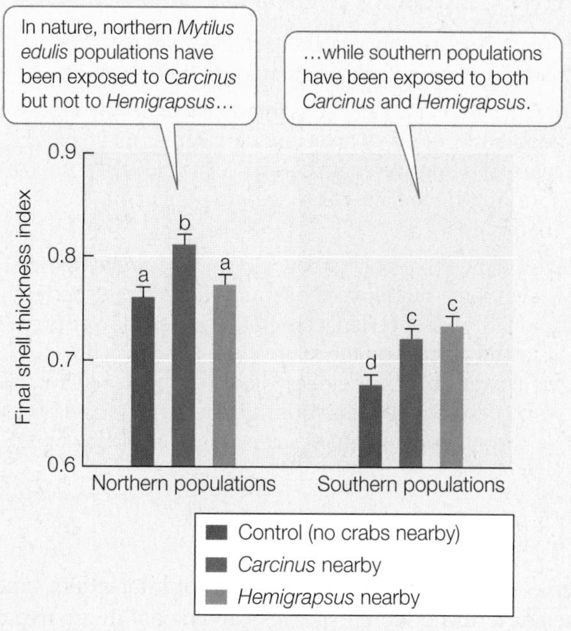
In nature, northern *Mytilus edulis* populations have been exposed to *Carcinus* but not to *Hemigrapsus*…
…while southern populations have been exposed to both *Carcinus* and *Hemigrapsus*.
Final shell thickness index
0.9
0.8
0.7
0.6
a
b
a
d
c
c
Northern populations
Southern populations
Control (no crabs nearby)
Carcinus nearby
Hemigrapsus nearby

[a]A. S. Freeman and J. E. Byers. 2006. *Science* 313: 831–833.

All mutualisms involve "biological barter": the exchange of resources and services. Plants and pollinators exchange food (usually nectar or pollen) for pollen transport. Plants and frugivores exchange food (fruit) for seed dispersal. Some mutualists (such as ants and plants) exchange defense or parasite and pathogen control for food or shelter. Others (such as plants and mycorrhizae, or mammals and gut bacteria) exchange one kind of nutrient for another. In such "barter" systems, each species is selected to obtain the greatest benefit for the least cost. Therefore the net effect of the mutualism on fitness may vary depending on environmental conditions that influence the value and cost of the goods and services that are exchanged. Mycorrhizae, for example, benefit plants in nutrient-poor soils, but they can be a liability to plants in nutrient-rich soils, where the cost of feeding the mycorrhizae outweighs their value in nutrient uptake. Accordingly, plants in nutrient-rich soils excrete fewer carbohydrate-rich compounds from their roots, thus discouraging colonization by mycorrhizal fungi.

Go to ANIMATED TUTORIAL 43.2
An Ant–Plant Mutualism
PoL2e.com/at43.2

Because mutualisms are diverse and often complex, it is difficult to predict their consequences for the evolution of the species involved. Some mutualisms evolve into one-on-one relationships with ever-tighter reciprocal adaptation—leaf-cutter ants and fungus are a case in point. Others evolve into many-species relationships that rely on generalized adaptations; many mutualisms between plants and the animals that disperse their seeds fall into this category. Some mutualisms are unstable and dissolve into consumer–resource relationships. Understanding the conditions under which mutualisms follow different evolutionary trajectories is an active area of current research.

CHECKpoint CONCEPT 43.4

- ✓ What goods or services are exchanged in the "biological barter" between leaf-cutter ants and their fungus?
- ✓ Why might individual plants in populations within the geographic range of an insect herbivore contain high levels of defensive chemicals while individuals in populations outside the insect's range produce no defensive chemicals?
- ✓ We claimed that pollinators do not visit flowers "to pollinate them," but females of some yucca moth species carefully collect pollen, carry it to the next flower they visit, and place it on the stigma before depositing eggs in the flower's ovary. The eggs will hatch into larvae that eat some developing seeds. Does this behavior "annihilate" Darwin's theory of natural selection? Explain why or why not.

The ecological and evolutionary effects of interactions among species determine which species can coexist in an area and what their abundances are. The number of species present, their phenotypic attributes, and their relative abundances determine the structure and function of communities—the topic of the next chapter.

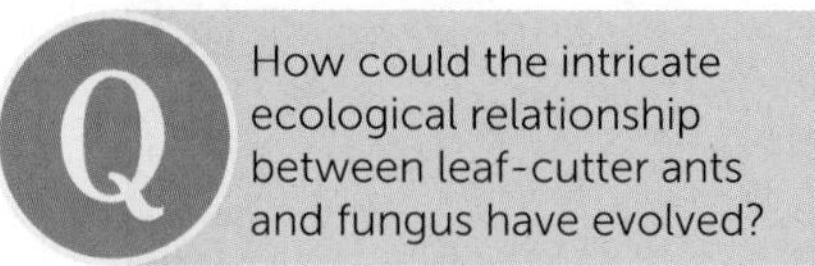

ANSWER We have seen that interactions between species can be mutually beneficial, positive for one species and negative or neutral for the other, negative for one species and neutral for the other, or mutually detrimental (Concept 43.1). These interspecific interactions affect the fitness of the individuals involved (Concept 43.1), and any traits that increase the benefit of a positive interaction, or decrease the cost of a negative interaction, will therefore spread by natural selection (Concept 43.4).

Although more details remain to be discovered, we now can outline a scenario for the evolution of the remarkable ecological interaction between leaf-cutter ants and their cultivated fungus. Phylogenetic analysis (see Chapter 16) suggests that fungus cultivation originated when ancestors of the leaf-cutters began to feed on fungi growing on the refuse in their nests. Ant colonies that provided better growing conditions for these fungi—by providing more refuse and, eventually, by providing fresh leaf material (which the ants cannot themselves digest)—and that chose productive fungus varieties to propagate (by moving fungal masses into newly founded nests and into newly dug chambers in established nests)—had more food to eat, and thus higher fitness, than colonies that did not do these things (**FIGURE 43.12**). Thus fungus-tending behaviors evolved by natural selection. Conversely, fungi that provided better ant food ben-

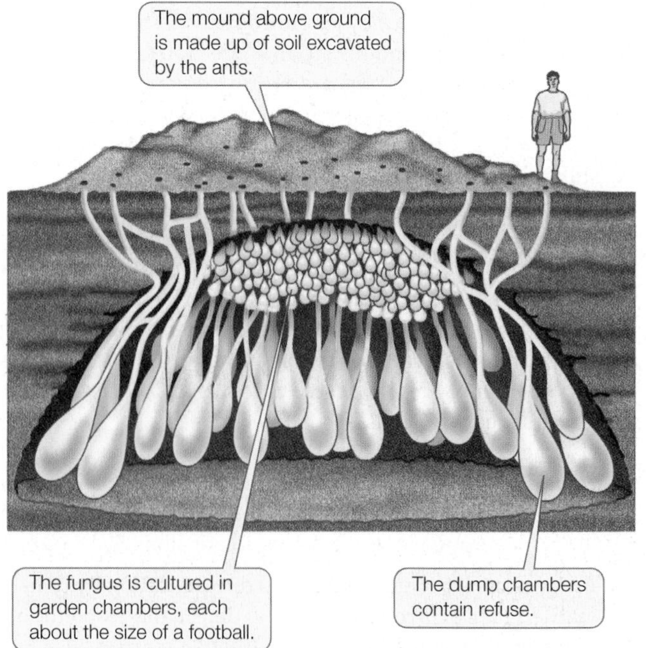

The mound above ground is made up of soil excavated by the ants.
The fungus is cultured in garden chambers, each about the size of a football.
The dump chambers contain refuse.

FIGURE 43.12 A Fungal Garden Diagram of the nest chamber of a large *Atta* leaf-cutter ant colony.

efited from faster ant colony growth and propagation, and these fungal adaptations also evolved by natural selection.

This history of reciprocal adaptation led to the great ecological success of leaf-cutter ants and their fungus. Leaf-cutter ants are not only the most important herbivores in the Neotropics, but also have expanded into dry environments, such as the Arizona desert, that are normally hostile to fungi.

SUMMARY

CONCEPT 43.1 Interactions between Species May Increase, Decrease, or Have No Effect on Fitness

- **Interspecific interactions** have consequences that affect the **fitness** of individuals and thus the dynamics and densities of populations and the distributions of species.
- Interspecific interactions can be divided into five broad categories depending on whether their effects on fitness are positive, negative, or neutral. **Review Figure 43.1 and ACTIVITY 43.1**
- **Interspecific competition** refers to mutually detrimental interactions in which members of two or more species use the same resource. Competition is generally strongest for the **limiting resource** that is in the shortest supply relative to demand.
- In **consumer–resource interactions**, the consumer gains nutrition and the resource species is killed or harmed. Consumer–resource interactions include **predation**, **herbivory**, and **parasitism**.
- A **mutualism** is a mutually beneficial interspecific interaction.
- **Commensalism**, in which one participant benefits while the other is unaffected, and **amensalism**, in which one participant is harmed while the other is unaffected, differ from other types of interspecific interactions in affecting only one of the participants.
- Many interspecific interactions have both beneficial and harmful aspects, and their effects may be contingent on environmental conditions. **Review Figure 43.2**

CONCEPT 43.2 Interactions within and among Species Affect Population Dynamics and Species Distributions

- The equation for density-dependent population growth, which describes **intraspecific competition**, can be extended to include the effects of interspecific competitors on per capita growth rates. Consumer–resource interactions and mutualisms can be described in a similar way by equations. **Review ANIMATED TUTORIAL 43.1**
- The per capita growth rate and the average population size of each interacting species are modified by the presence of the other in a way that depends on the type of interaction.
- In interactions with negative effects on one species (consumer–resource interactions, competition, and amensalism), local extinction of one or both of the populations of the interacting species is possible. **Review Figure 43.3**
- The presence or absence of other species can make an environment favorable or unfavorable for a species and so affect its distribution. **Review Figure 43.4**
- Coexistence of species competing for resources, and persistence of prey species and their predators, is possible if each species gains a growth advantage when it is rare. Differences between competing species in their use of resources, known as **resource partitioning**, are often great enough to generate such a **rarity advantage**. **Review Figure 43.5**

CONCEPT 43.3 Species Are Embedded in Complex Interaction Webs

- Most species are involved in webs of interaction with many other species.
- **Trophic** interactions, which determine the flow of energy and materials through communities, form the core of these interaction webs.
- Energy and materials enter communities through **primary producers**, which produce organic compounds from inorganic sources.
- Species that feed on primary producers are called **primary consumers**. Those that consume primary consumers are **secondary consumers**, and so on. These positions in the feeding hierarchy are called **trophic levels**. **Review Table 43.1 and ACTIVITY 43.2**
- Trophic interactions can be diagrammed as **food webs**. **Review Figure 43.6**
- Changes in species composition or abundance at higher trophic levels may cause a **trophic cascade** of changes at lower trophic levels. **Review Figure 43.7**
- Species introduced by humans to a new region where their natural enemies are absent may become **invasive species** that reproduce rapidly, spread widely, and alter interactions among the native species of the region.

CONCEPT 43.4 Interactions within and among Species Can Result in Evolution

- Intraspecific competition can increase the carrying capacity by causing selection for traits that allow individuals to use resources more efficiently.
- Interspecific competition can lead to resource partitioning and coexistence. It does so through selection for traits that allow individuals of one or both competing species to decrease use of resources that the competitor species uses. **Review Figures 43.8 and 43.9**
- The opposing interests of consumer species and resource species can lead to an "evolutionary arms race" in which resource species continually evolve better defenses and consumer species continually evolve better offenses.
- Mutualisms involve mutual exploitation and the exchange of goods and services. The fitness effects of mutualisms may vary with environmental conditions that affect the value and cost of the goods and services that are exchanged. **Review ANIMATED TUTORIAL 43.2**

Go to the Interactive Summary to review key figures, Animated Tutorials, and Activities PoL2e.com/is43

Go to LaunchPad at **macmillanhighered.com/launchpad** for additional resources, including LearningCurve Quizzes, Flashcards, and many other study and review resources.

44 Ecological Communities

KEY CONCEPTS

- 44.1 Communities Contain Species That Colonize and Persist
- 44.2 Communities Change over Space and Time
- 44.3 Community Structure Affects Community Function
- 44.4 Diversity Patterns Provide Clues to What Determines Diversity
- 44.5 Community Ecology Suggests Strategies for Conserving Community Function

This coffee plantation uses high-intensity sun cultivation techniques that increase crop yield at the expense of species diversity.

As we sip our morning cup of coffee, we can send silent thanks to *Coffea arabica*, the plant that yields 70 percent of our coffee beans. This understory shrub or small tree of tropical forests originated in the Ethiopian highlands as a natural hybrid between *C. canephora* ("robusta" coffee, from which we get the other 30 percent) and *C. eugenioides*, both of which hail from western and central Africa.

Coffea arabica has probably been cultivated in Ethiopia for 1,500 years, although the earliest written description of the plant dates to a tenth-century set of physician's notes. Molecular genetic evidence indicates that its cultivation spread to Yemen about 575 CE. Coffee was widely traded within the Muslim world, in spite of controversy there about the use of stimulants (see the opening story of Chapter 5), and the first coffeehouses were in Arabia. Arabs tried to corner the coffee market by banning exports of plants and fertile seeds, but the Dutch managed to smuggle some live plants out of Yemen. The beverage reached Italy in the 1600s and later spread through Europe.

By the 1700s, coffee cultivation had reached Southeast Asia and the Americas. Coffee is now one of the most valuable export commodities of the developing world, second only to oil. As demand has grown, worldwide coffee production has risen from about 5 million metric tons per year in the 1970s to about 7 million metric tons today, and coffee production employs some 25 million people in developing countries.

Coffee is grown mainly in tropical regions, using a variety of cultivation methods. The least intensive, traditional method is to plant coffee in the shade of undisturbed forests. This practice embeds coffee in natural forest communities, conserving most of their original plant and animal species. Medium-intensity shade cultivation replaces the forest with a low-diversity plantation of trees, under which the coffee is grown. High-intensity sun cultivation uses no shade trees; modern genetic cultivars are planted at high densities, heavily pruned, and subsidized with fertilizer, herbicides, and pesticides. Thus the more intensive methods increasingly replace natural communities with human-constructed communities that contain far fewer species. High-intensity methods increase coffee yields, but they sacrifice the greater biological diversity that traditional methods maintain, cause greater pollution, and require costly labor and chemicals.

Q
Can we use principles of community ecology to improve methods of coffee cultivation?

You will find the answer to this question on page 914.

CONCEPT 44.1 Communities Contain Species That Colonize and Persist

We saw in Chapter 43 that the abundances, distributions, and even phenotypes of species are affected by their interactions with other species. These interactions occur in the context of an ecological **community**—the group of species that occur together in a geographic area. Ecologists have practical reasons for wanting to know why communities contain the species that they do—we depend on them for all manner of natural resources and services. And like other dynamic systems, a community's components (species) and their interactions determine how well it functions. Understanding how communities are put together and how they work is essential to conserving them, as we will see.

Despite the simple definition of community, ecologists must make some practical decisions about how to draw the boundaries of a community, or about which species to include in a study. Sometimes there are obvious changes in habitat that indicate a community boundary. A pond, for example, defines a group of aquatic species that interact much more with one another than with terrestrial species outside the pond. Often ecologists choose to study a representative portion of a single habitat, such as a 10- × 10-meter area of forest. But even a small area may contain so many species that ecologists decide to focus their attention on subsets of species. Those subsets may be defined taxonomically, by a habitat or location, by some character traits, or by the types of interactions the species engage in. Hence biologists may speak of the bird community of a certain island, the *Daphnia* community of rock pools (see Figure 42.10), the microbial community of the human gut, or the community of grazing herbivores in the marine rocky intertidal zone.

Regardless of how we delineate them, communities can be characterized in terms of which species they contain, or their **species composition**. Communities are also characterized by how many species they contain, and by how abundant each species is. These three attributes of communities— species composition, the number of species, and how abundant each species is—are components of **community structure**.

The mix of species in a community is determined by the same factors that explain the distributions of individual species; as we saw in Concept 42.1, a species can occur in a location only if it is able to colonize that location and persist there. Species may fail to colonize a community, or be lost from it, for any of several reasons. Individuals of a species may simply never reach the location—consider how few dispersers from a mainland are likely to reach a remote island before they perish! Once colonists have arrived, they may be unable to tolerate the environmental conditions, or a resource or an essential mutualistic partner may be missing. Species may be excluded by competitors, predators, or pathogens. Finally, some species may go locally extinct because their populations are so small that, by chance, all the individuals die at some point without reproducing (see Concept 42.5).

LINK

The abundances of the species that persist in a community are determined by the population-level processes discussed in **Chapters 42 and 43**

We can think of communities as being assembled through such gains and losses of individual species. The processes of community assembly are well illustrated by Krakatau, a small (17 km^2) volcanic island in the Sunda Strait of Indonesia (see Figure 44.10). The volcano exploded in 1883, sterilizing what was left of the island and covering it with a thick layer of ash. Scientists quickly mounted expeditions to observe the return of life to the island, and they have surveyed it periodically ever since. By 1886, seeds of 10 plant species that grow on nearby tropical beaches had floated to Krakatau (**FIGURE 44.1A**), and wind had brought seeds or spores of 14 additional species of grasses and ferns. Tree seeds had arrived on the wind by the 1920s. As the forest canopy closed in, some pioneering plant species that require high levels of light disappeared from Krakatau's now-shady interior. Once forests developed, fruit-eating birds and bats began to be attracted to the island, bringing new animal-dispersed seeds with them.

FIGURE 44.1 Vegetation Recolonized Krakatau (A) Beach-adapted plants such as *Ipomoea pes-caprae* (beach morning glory) were among the first organisms to take hold on the denuded island. (B) One century after a massive eruption left Krakatau an ash-covered shell, vegetation once again covers much of the island.

(A)

(B)

Now, more than a century after Krakatau exploded, the island appears from afar to be covered by stable plant communities (**FIGURE 44.1B**). But a closer look reveals that their species compositions continue to change as new species colonize and earlier colonists go extinct.

CHECKpoint CONCEPT 44.1

✓ Describe at least two ways in which ecologists decide which species to include in a community study.

✓ Describe the processes that are involved in community assembly.

✓ As new species have colonized Krakatau, the densities of some existing species have changed, even though these species have not gone extinct. Why? (Hint: see Concept 43.2.)

Krakatau provides a dramatic illustration of a general pattern: the species composition of communities changes over time. Species composition also changes over space. In spite of this turnover of species in space and time, are there general rules that govern which species we find where and when?

CONCEPT 44.2 Communities Change over Space and Time

Ecologists have noted repeated patterns of spatial and temporal change, or **turnover**, in species composition. These patterns demonstrate that the species we find in one place and time are not a random subset of those that potentially could be there. Let's take a closer look at these patterns and what they mean.

Species composition varies along environmental gradients

We saw in Chapter 41 that physical conditions on Earth change over long distances—consider how temperature changes with latitude, or pressure with ocean depth. Physical conditions can also change on small spatial scales (see Concept 41.1). If we climb a steep tropical mountain, for example, we traverse gradients in temperature and moisture, and because different species have different tolerance limits, we encounter constantly changing plant communities. In coffee-growing regions, we would find coffee plantations between about 1,300 and 1,500 meters in elevation, where coffee grows best—reflecting its evolutionary origin in African highlands.

Ecologists have characterized how plant communities change along environmental gradients on Krakatau. Environments there range from shoreline habitat that is buffeted by wind and waves to inland habitat that varies in elevation from sea level to 800 meters at the highest point, and from gentle ash-covered slopes to steep rocky cliffs. The composition of Krakatau's plant communities is similar in locations with similar conditions and different in locations with different conditions. These spatial changes are easy to see from afar because the species in different habitats tend to look different (compare Figure 44.1A,B), just as vegetation in the tropical rainforest biome differs from vegetation in tropical savanna (see Concept 41.3). Sandy beaches are dominated by low-growing herbs and vines, bordered inland by a fringe of short trees and shrubs; gentle inland slopes are dominated by taller trees with an understory of vines and ferns, and the summit, often in the clouds, supports a thick growth of short, moss-covered trees.

Spatial turnover in the species composition of communities often signals hard-to-see gradients. The rock outcrops that provide habitat for Edith's checkerspot butterfly near San Francisco (see Figure 42.1), for example, contain soils derived from serpentine rock, which is unusually high in heavy metals such as chromium and nickel. If we measure off a transect—a straight line used for ecological surveys—running from a non-serpentine to a serpentine area and identify the plants that occur in evenly spaced samples along it, we will find that several species drop out and new ones appear partway along the transect, where heavy-metal concentrations jump (**FIGURE 44.2**).

The early naturalist–explorers who documented associations of climate and vegetation—the biomes described in Concept 41.3—likewise documented associations of particular animal species with particular vegetation. Some animal species, such as Edith's checkerspot butterfly, are found in certain plant communities because the plants they eat grow there. In other cases, animals are associated with particular plant communities because plants modify physical conditions such as temperature and humidity and because plants contribute to **habitat structure**: the horizontal and vertical distribution of objects in the habitat. The vertical distribution of leaves and the density, size, and surface texture of plant stems influence which species are present by determining the effectiveness of their methods for obtaining food, avoiding predators, building nests, and so on. Morphological, physiological, and behavioral traits of animals adapt them for the structure of the habitats with which they are associated (**FIGURE 44.3**).

Several processes cause communities to change over time

Species composition changes not only over space, but also over time. Krakatau is not unique in this regard—all communities, not just those in the early stages of assembly, are dynamic. Over longer periods, of course, the species composition of communities reflects evolution. But species turnover also occurs over much shorter periods. Three major processes contribute to this dynamism: extinction and colonization, disturbance, and climate change.

EXTINCTION AND COLONIZATION Ecological communities are likely to change even in a constant environment because species may go locally extinct and be replaced by new species that arrive from time to time. As we saw in Concept 44.1, there are several reasons why a species may go locally extinct. Dispersal is an ongoing process that delivers a constant influx of new individuals to all but the most isolated locations. When individuals arrive at a location that already contains a population of their species, they add to its size (and genetic diversity); when those individuals are members of a new species,

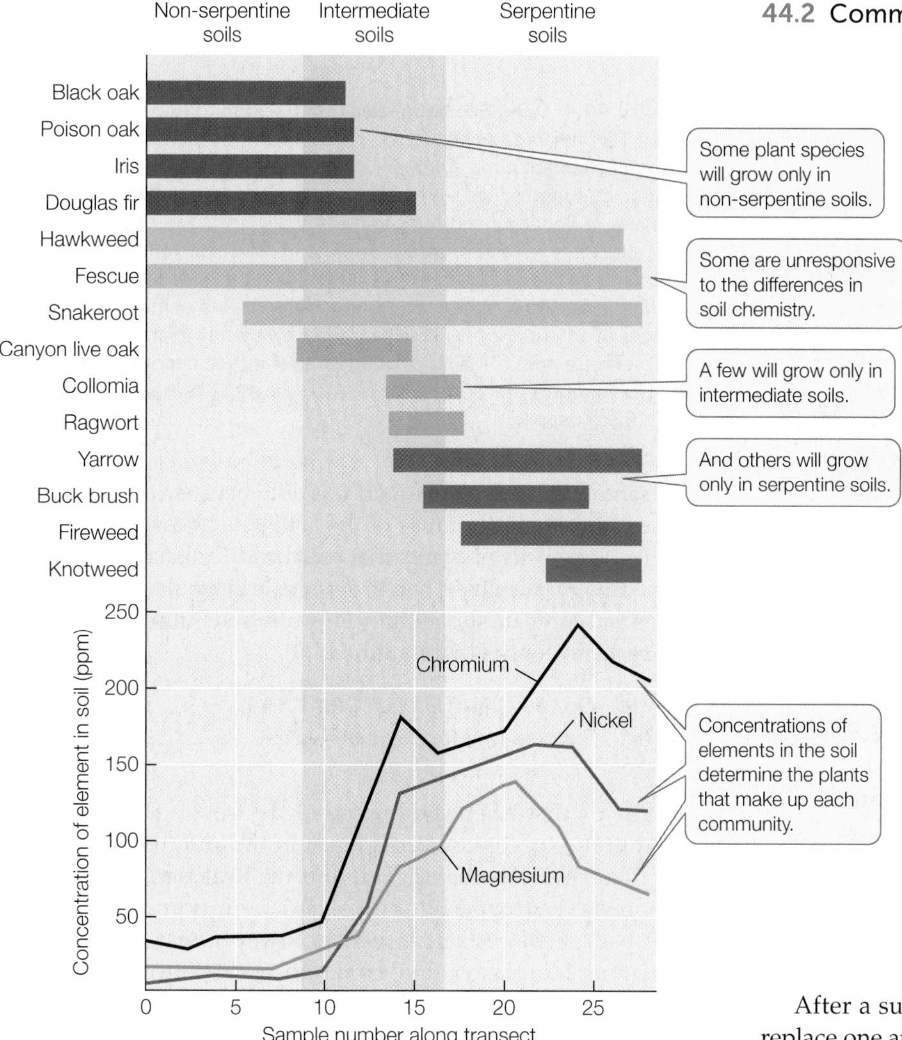

FIGURE 44.2 Change in Species Composition along an Environmental Gradient Ecologists sampled plant species along a transect running from non-serpentine to serpentine soils. They found that as concentrations of heavy metals in the soil changed, some species dropped out of the plant community and new ones appeared. Bars indicate samples in which a given species was present.

they can establish a new population and add another species to the community.

DISTURBANCE An event that causes sudden environmental changes—referred to as a **disturbance**—can change the species composition of a community. Examples of such events are volcanic eruptions, wildfires, hurricanes, landslides, and human activities. Such disturbances can wipe out some or all of the species in communities and alter the physical environment. Sudden change can also occur without the destruction of a resident community. New environments can appear, for example, when a glacier melts away, a depression fills with rainwater—or a large mammal deposits dung.

After a sudden change in the environment, species often replace one another in a relatively predictable sequence called **succession**. A fresh pile of elephant dung, for example, is colonized by a predictable sequence of dung beetle species (family Scarabidae). These animals feed on the dung and lay their eggs in it. Some species typically arrive early and are then replaced by later arrivals (**FIGURE 44.4**).

Go to MEDIA CLIP 44.1
Dung Beetles
PoL2e.com/mc44.1

Several factors are responsible for successional sequence. One key factor is variation in the colonizing ability of species. The first species to arrive in newly available habitat are good dispersers. The early-arriving dung beetles tend to be strong fliers with excellent senses of smell, or are "hitchhikers" that ride on dung-producing animals and colonize dung as it is produced. Similarly, the first plant species on Krakatau were those with seeds that are readily dispersed by water or wind. A second factor is that environmental conditions in the disturbed site change over time. Early-arriving dung beetles must cope

(A)

(B)

FIGURE 44.3 Many Animals Are Associated with Habitats of a Particular Structure (A) The legs and feet of the red-breasted nuthatch (*Sitta canadensis*) allow it to cling to the trunks, branches, and twigs of trees as it gleans small arthropods and seeds. It prefers conifer forests. (B) Pronghorn (*Antilocapra americana*), the fastest North American land mammal, uses speed to escape predators. It prefers open, flat habitats without obstacles to movement.

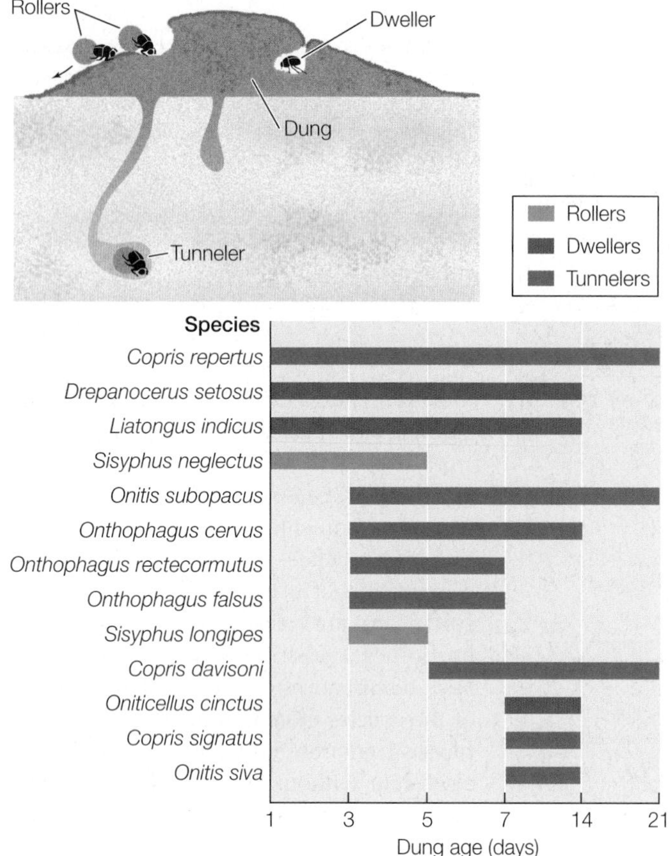

FIGURE 44.4 Dung Beetle Species Composition Changes over Time The Indian dung beetles represented in this diagram all consume elephant dung. Each bar indicates the time during which the species it represents is present. Whereas *Copris repertus* (top) colonizes newly produced dung and is still present 3 weeks later, other species enter and drop out of the community at various times, depending on their adaptations for finding dung and for using dung of different ages. Thus the composition of the beetle community changes over time. Notice also the different ways in which dung is used: some species roll balls of dung away and then bury them, some dwell in the dung, and some take bits of dung into chambers they dig under the dung pile.

with wet conditions, whereas late-arriving species face a drier environment; pioneering species on Krakatau had to tolerate full sun and volcanic ash. Later-arriving species experience an environment that has changed not only because of physical processes, but also because of the activities of early-arriving species. The seeds of plants that colonized Krakatau after the forest canopy had closed had to germinate in the shade of other plants and compete successfully for water and nutrients in soils that earlier colonists had modified.

Go to ANIMATED TUTORIAL 44.1
Succession after Glacial Retreat
PoL2e.com/at44.1

When a disturbance damages or destroys a preexisting community, succession often leads in the end to a community that resembles the original one. On Krakatau, the tropical forests destroyed by the volcanic eruption eventually came back. Ecologists now realize, however, that a return to something like the original community is far from guaranteed. Instead, some disturbances may push the system past a threshold, or tipping point, so that there is an abrupt **ecological transition** to a distinctly different community. The

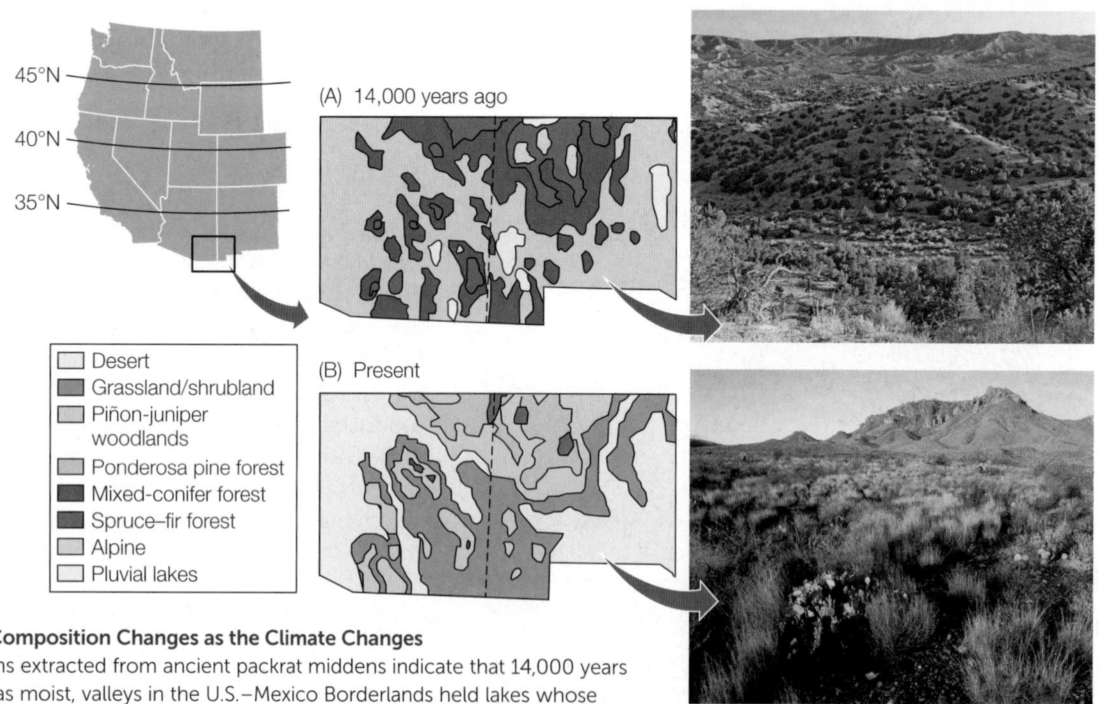

FIGURE 44.5 Species Composition Changes as the Climate Changes (A) Fossilized plant remains extracted from ancient packrat middens indicate that 14,000 years ago, when the climate was moist, valleys in the U.S.–Mexico Borderlands held lakes whose shores were clothed in piñon–juniper woodland. (B) In today's more arid climate, the lakes have dried up and the woodland has given way to desert grassland—and more recently to shrubland, following human disturbance (see the opening story of Chapter 41).

system may then stay in this alternative state and not return to its original state, even when the factor that caused the disturbance is removed. For example, intensive cattle grazing led to an abrupt conversion of grasslands to shrublands in the U.S.–Mexico Borderlands (see the opening story of Chapter 41). Grasslands have not returned, even in places where cattle have been removed.

CLIMATE CHANGE A final cause of temporal variation in communities is climate change. As physical conditions change, the geographic ranges of species necessarily change with them. One way in which ecologists have been able to reconstruct the history of such changes is by analyzing fossilized remains of plants found in middens—refuse piles—left by packrats (*Neotoma* spp.). These rodents bring plant material into their nests. When the material is preserved, it provides a record of the vegetation that was present when the animals were alive. By using carbon-14 (see Figure 18.1) to date packrat-midden fossils, biologists have been able to show how plant communities of the U.S.–Mexico Borderlands changed over the last 14,000 years as the climate became drier following the end of the last ice age (**FIGURE 44.5**).

CHECKpoint CONCEPT 44.2

- ✓ Why are animal species often associated with particular plant communities?
- ✓ How does succession differ from species turnover that results from ongoing colonization and extinction?
- ✓ Give an example of an environmental gradient that you have observed yourself, explaining what factors change along it and what visible changes you see in the species present.

We saw in Concept 44.1 that species composition, species number, and species abundances all are aspects of community structure. How do these attributes affect the way a community functions?

CONCEPT 44.3 Community Structure Affects Community Function

It is useful to think of an ecological community as a dynamic system (see Concept 1.2) similar to a human-made machine, with inputs, "internal workings," and outputs. The inputs to a community are energy and materials from the surrounding abiotic environment. The "internal workings" of the community—the metabolism of its individuals, the dynamics of its populations, and the interactions among its species—produce an output of transformed energy and materials.

Energy flux is a critical aspect of community function

If we think of a community as a machine, we can measure **community function** as we would the function of a machine, in terms of its inputs and outputs. Community function is measured by the amount of energy or matter that moves into and out of the community per unit of time. This **flux**, or flow rate, reflects an exchange between the community and its environment. We can also measure function for particular parts of the community. Wolves, for example, convert elk into new wolves—a conversion of primary consumers (an input) into secondary consumers (an output). The output of seeds and fruits by most flowering plants, including *Coffea arabica*, responds to visits by insect pollinators, a type of input. Retention of nutrients within an ecosystem is a function of the community's decomposers—the recyclers. Many community outputs affect not only species in the community, but also organisms in other ecosystems, including humans. These outputs represent "goods" and "services," which we will discuss in Concept 44.5.

LINK

The role of decomposers in recycling nutrients—particularly nitrogen—is described in **Concept 45.3**

To develop this idea of community function, let's first focus on how energy moves through communities. (The flux of matter through communities is coupled with that of energy, as will become more apparent in Chapter 45.) **FIGURE 44.6** shows, in simplified form, the flux of energy through the most common type of food web, in which the primary producers are photosynthetic organisms that use the sun's energy. The pattern of energy flux is similar in those rare communities (such as those in deep-sea hydrothermal vents) in which primary producers are chemoautotrophs.

Energy enters communities through primary producers; the total amount of energy that they capture and convert to chemical energy per unit of time is called **gross primary productivity** (**GPP**). Not all of the captured energy is available to primary consumers because primary producers use some of it to fuel cellular respiration and other catabolic processes. Ultimately this energy is lost as heat. The amount of energy stored in primary producer tissues per unit of time is called **net primary productivity** (**NPP**). GPP and NPP reflect two aspects of primary producer function: the rate at which primary producers capture energy, and the rate at which they convert it to forms available to other trophic levels. GPP can be measured from the rate with which primary producers take up carbon dioxide during photosynthesis, and NPP from the rate of CO_2 uptake minus its rate of release during respiration. New technologies allow us to measure such gas exchanges in nature, but it is often more convenient to use change in the **biomass** of primary producers (the dry mass of their tissues) per unit of time as an approximation for NPP.

The percentage of the energy contained in the biomass of one trophic level that is incorporated into biomass in the next trophic level is the **ecological efficiency** of energy transfer between those levels. Ecological efficiency of energy transfer averages only about 10 percent—the biomass of one trophic level is about 10 percent of that in the next lower level.

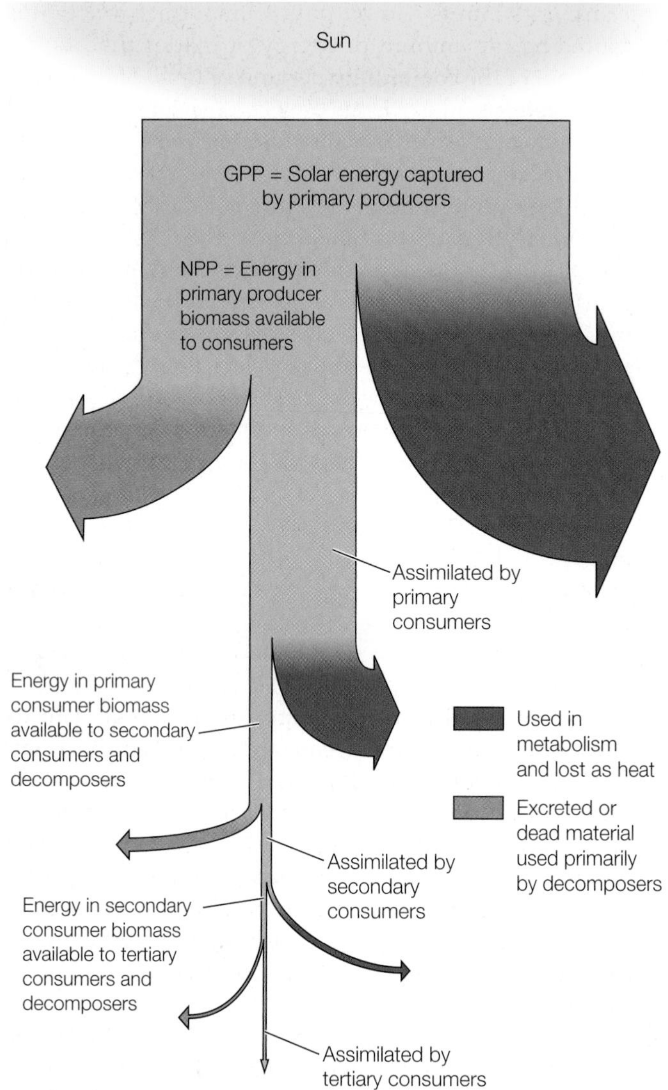

FIGURE 44.6 Energy Flow through Ecological Communities On average, the total amount of energy incorporated into the biomass of a trophic level per unit of time is about one-tenth that of the level it consumes. Of the 90 percent of energy *not* available to the next trophic level, some is waste or dead material used by decomposers, and some is lost during metabolism as heat.

Go to ACTIVITY 44.1
Energy Flow through an Ecological Community
PoL2e.com/ac44.1

LINK

Food-web structure is discussed in **Concept 43.3**; see especially **Table 43.1** and **Figure 43.6**. You can review photosynthesis and cellular respiration in **Chapter 6**; see especially **Concepts 6.2 and 6.5** and **Figure 6.14**

Just as NPP measures primary producer function, the biomass produced by primary consumers measures how well they function to convert primary producer tissue into their own tissue, and so on for secondary and tertiary consumers. The low efficiency of these energy transfers limits the number of trophic levels found in a community—few contain more than four—because fewer and fewer individuals can be supported as one moves to higher trophic levels. This is one reason we sometimes are urged to "eat low on the food chain": more humans can be fed if we act as primary consumers than if we act as secondary consumers.

There are several reasons for low ecological efficiency. First, not all the biomass at one trophic level is ingested by the next one; for example, herbivores routinely miss hard-to-get plant parts or avoid eating plants that are chemically well defended. Second, some ingested matter is indigestible and is excreted as waste. Tree bark, for example, contains lignin and cellulose, which most herbivores cannot digest. Finally, and most important, consumers as well as primary producers use much of the energy they assimilate (the digested portion of what they ingest) to fuel aspects of their own metabolism that do not add new biomass through growth or reproduction. Energy used in metabolism does the work of assembling and disassembling biological molecules, moving the body, and so on, and heat is generated in the process. Except for the energy stored in the chemical bonds of assembled molecules, none of the energy used in metabolism is available to the next trophic level.

Community function is affected by species diversity

The structure of a community influences how much energy enters it and how much is transferred through the food web because each species has a unique **niche** that determines its function in the community. The niche is a central ecological concept with two important meanings. The first meaning refers to the physical and biological environments in which a species has a positive per capita growth rate—these are the environments where we expect to find that species. Figure 42.7 shows, for example, that the niche of the beetle *Rhyzopertha dominica* includes drier and warmer conditions than that of *Sitophilus oryzae*. Physical tolerances do not completely define a species' niche, however, because the biological environment—the presence of competitors, predators, and so on—also affects population growth. Figure 43.4 shows, for example, that low intertidal zones are included in the niche of stellate barnacles only where rock barnacles are absent.

A species' niche also refers to what the species does in the community—its functional role, or "profession." This role is largely defined by how it affects other species—what resources it uses and makes unavailable to other species, when and where it obtains them, how efficiently it does so, what it produces from them that other species can use, and whether it affects the physical environment. Many of these effects, but not all of them, are determined by the species' trophic interactions.

Once we think of functional roles, we can begin to see that community structure affects community function. Loss of wolves from Yellowstone National Park in the 1920s, for example, decreased the flux of energy from elk to secondary consumers and increased energy flux from aspen and willow to primary consumers (see Figures 43.6 and 43.7). But how do we go from specific examples like this to a more general understanding of how structure affects function? Ecologists have begun to tackle this problem by looking for broad patterns in structure–function

Community A
Community A is less diverse than community B because it contains three equally abundant species rather than four.

Community B
With four equally abundant species, community B is the most diverse.

Community C
Community C is less diverse than community B because it has an uneven distribution of the four species.

FIGURE 44.7 Species Richness and Species Evenness Contribute to Diversity These hypothetical communities of fungi (mushrooms) are all the same size (12 individuals), but they differ in species richness (3 versus 4 species) and species evenness, both of which affect diversity.

Go to ACTIVITY 44.2 Measures of Species Diversity PoL2e.com/ac44.2

relationships across communities. They have discovered that one important aspect of community structure—**species diversity**—influences community function.

Species diversity has two components, illustrated in **FIGURE 44.7**. One is simply the number of species in the community, called **species richness**. A community that contains more species is more diverse, all else being equal. The second component is how similar the species are in their abundances, a property called **species evenness**. A community that has four equally abundant species is more diverse than one in which 75 percent of the individuals belong to one species and the remaining 25 percent are spread among three other species. This second component may not be intuitive until we understand how it affects communities: the properties of a community with a few abundant species will be defined mostly by those species, whereas the properties of a community with equally abundant species will reflect the influence of all of them. Ecologists sometimes use species richness as a simple measure of diversity, but often prefer to use a mathematical diversity index that incorporates both richness and evenness.

Community outputs often vary with species diversity. In a long-term study of prairie plant communities, for instance, David Tilman and his colleagues at the University of Minnesota cleared outdoor plots into which they planted up to 32 native prairie species. After the plantings had reached maturity, they harvested the plants and measured their aboveground biomass. They found that this measure of NPP increased as the species richness of a plot increased (**FIGURE 44.8A**).

Two hypotheses have been proposed for such an increase. First, the relationship may come from "sampling": communities with more species may be more likely, just by chance, to include some "superspecies" that are exceptionally good at predation, or at photosynthesizing, or at whatever they do, and so have a strong influence on total community output.

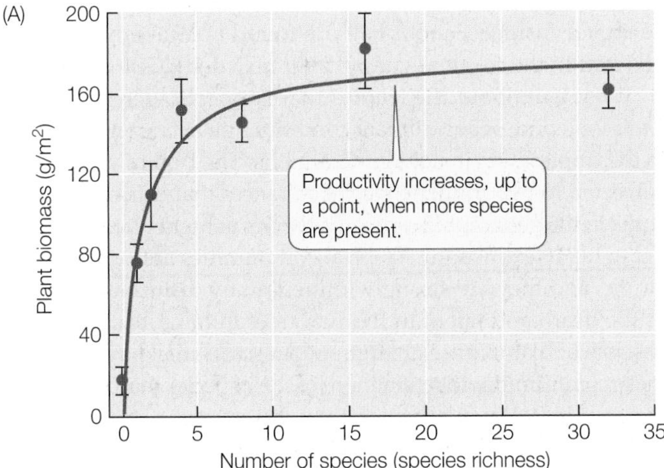

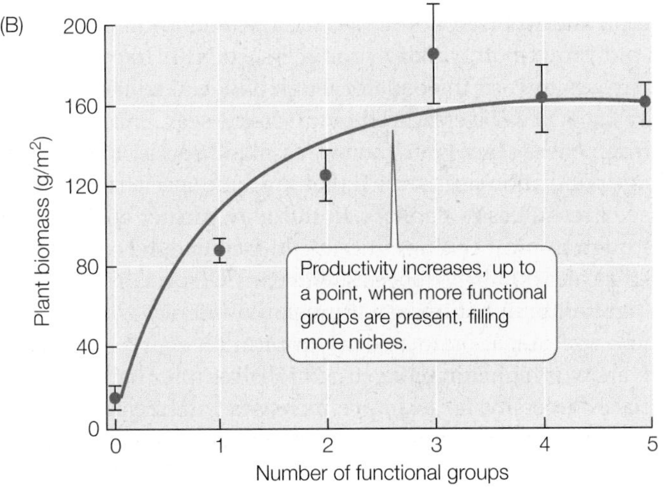

FIGURE 44.8 Species Richness and Number of Functional Groups Affect Primary Productivity (A) In 289 experimental prairie plots, NPP (measured as aboveground plant biomass produced per square meter per year) increased with species richness. (B) When the researchers separated out the effects of the number of functional groups from those of species richness, they found that the number of functional groups was the key influence on production. These results suggest that primary productivity is enhanced by a greater variety of species with complementary niches. Error bars indicate $\pm$ 1 standard error of the mean. See Appendix B for review of statistical concepts.

Alternatively, the relationship may come from "niche complementarity": differences (often subtle) among species in how efficiently they use alternative resources under different conditions. (This idea is related to resource partitioning; see Figure 43.5). For example, "C_4" grasses, which have a special type of photosynthesis, grow better than "C_3" grasses in warm conditions but worse in cool conditions; symbiotic nitrogen-fixing bacteria give legumes a boost in nutrient-poor soils but are a drain in nutrient-rich soils; shrubs and trees can tap the nutrients and water in deeper soil layers than can herbaceous species. Species-rich communities may be more likely than species-poor communities to contain representatives of several such complementary "functional groups" that together make more complete use of available resources. (To see this idea, consider how total consumption of seed resources in Figure 43.5 would change if Species B were removed from the hypothetical community.) Tilman and colleagues could test these two hypotheses because they planted some species-rich plots with multiple species from only one functional group, and others with species from several functional groups. They found that NPP increased with the number of functional groups, and that the number of species within functional groups had no additional effect on NPP (**FIGURE 44.8B**). This result suggested that niche complementarity caused the relationship between species richness and NPP.

LINK

Some ways in which differences in resource use evolve and promote species coexistence are described in **Concepts 43.2 and 43.4**

NPP is not the only measure of community function that increases with species diversity. Control of agricultural pests by natural predators and parasites is enhanced when multiple crops are planted together or when a diverse natural plant community borders agricultural fields. And as we'll see at the end of this chapter, two recent studies of coffee pollination found that the production of seeds was higher, and pollination rates were less variable between years, in plantations that were close to natural tropical forests, which harbor a great diversity of coffee-visiting bee species.

CHECKpoint CONCEPT 44.3

- ✓ What is the major source of energy for communities, and how does it enter them?
- ✓ What is ecological efficiency, and how does it influence the number of trophic levels in a community?
- ✓ Discuss parallels between the two components of species diversity and components of cultural and ethnic diversity in human societies.
- ✓ What would Tilman and colleagues have seen if "sampling" explained the relationship between species richness and NPP?

Species diversity is an important characteristic of communities because it affects how those communities function as ecological systems. But what factors determine species diversity in a particular community?

CONCEPT 44.4 Diversity Patterns Provide Clues to What Determines Diversity

Species diversity is not uniform around the world. Geographic patterns in diversity shed light on factors that affect diversity by modifying the balance between processes that add species to and remove species from communities.

Species richness varies with latitude

When the German naturalist Alexander von Humboldt traveled around Central and South America, Europe, and Asia two centuries ago, he noted that "the nearer we approach the tropics, the greater the ... variety of structure ... of organic life." This recognition that species richness increases toward the equator has been amply confirmed since then. A 2-hectare plot of tropical forest in Malaysia, for example, can contain 227 tree species, whereas an equivalent plot of temperate forest in Michigan contains only 10 to 15. There are 56 bird species that breed in Greenland, 105 in New York State, 469 in Guatemala, and 1,395 in Colombia. And although some taxa appear not to follow the pattern—of some 20,000 known bee species on Earth, for instance, about half are found in temperate regions and half in the tropics—many other taxa do (**FIGURE 44.9**).

What is it about the tropics that supports such diversity? Over long time spans, climate conditions have been more stable in the tropics than in temperate regions. The tropics were never subjected to the dramatic glacial advances that caused massive shifts in the geographic ranges of species at higher latitudes and led to many extinctions. Tropical communities are certainly dynamic, and they are strongly influenced by disturbances (such as the death and fall of individual trees in rainforests), but the absence of disturbance at large spatial scales may have allowed these communities to retain more of their species than temperate communities, leaving them with higher species richness today.

We also know that climate changes with latitude. The tropics receive abundant solar energy input, which makes them warm and wet (see Concept 41.2). These conditions promote rapid growth of primary producers, so NPP, like diversity, increases toward the equator. High NPP can enhance diversity (as well as be enhanced by diversity; see Concept 44.3) if greater energy input into food webs allows species to maintain large population sizes, even if they are more specialized in their use of resources or habitats. In this way, greater energy flow through tropical communities could facilitate the coexistence of a greater number of species with narrow, specialized niches.

Variation in habitat structure could interact with greater niche specialization to amplify tropical diversity. In general, diversity is higher in more structurally complex habitats. Bird species diversity, for example, increases with the diversity of foliage height in deciduous forests. If tropical species have

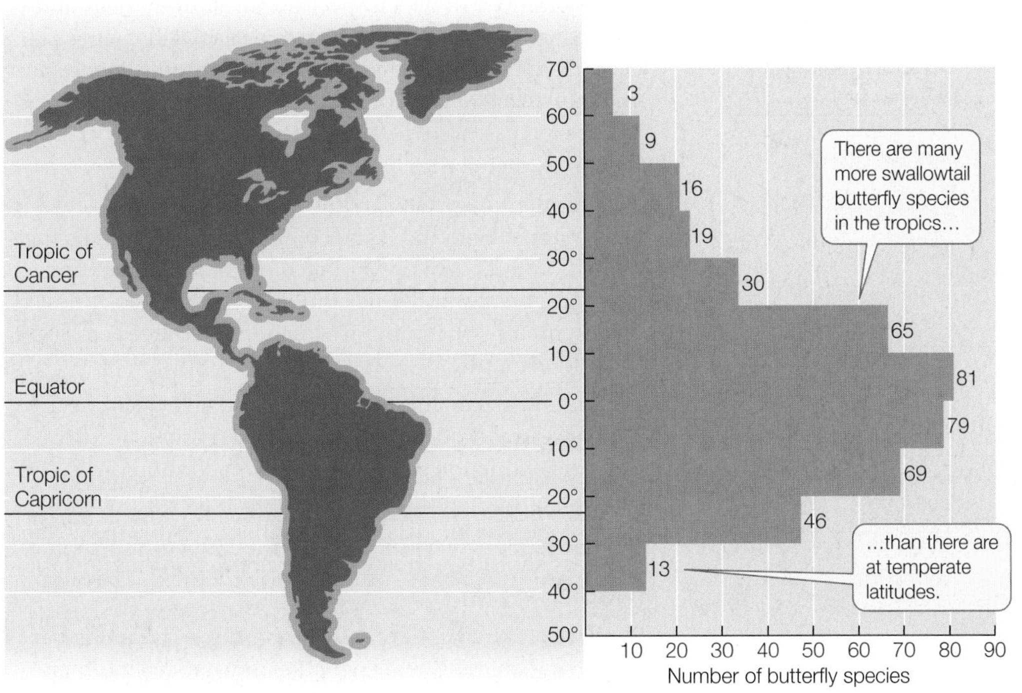

FIGURE 44.9 Species Richness Increases toward the Equator Among swallowtail butterflies (Papilionidae), species richness decreases with latitude both north and south of the equator. Similar latitudinal gradients of species richness have been observed in many other taxa.

more specialized habitat requirements, and if tropical vegetation is more structurally complex by virtue of greater plant species richness, then complex habitat structure and productivity may interact to increase tropical diversity.

Ecologists still have much to learn about the relative contributions of these various factors to the puzzle of tropical diversity, but the research stimulated by that puzzle has demonstrated that disturbance patterns, productivity, and habitat structure all affect species diversity.

Species richness varies with the size and isolation of oceanic islands

The numbers of species found on oceanic islands exhibit regular patterns that have long fascinated ecologists (**FIGURE 44.10**). First of all, species richness is greater on large than on small islands. And second, species richness is greater on islands near a mainland than on more distant islands. What can explain these differences?

As we saw in the case of Krakatau, communities are assembled by colonization, along with occasional extinctions. Just as a population reaches an equilibrium size when additions (births) and losses (deaths) are equal, it seems plausible that species richness on an island might ultimately represent an equilibrium between the rate at which new species colonize

FIGURE 44.10 Area and Isolation Influence Species Richness on Islands On islands in the tropical eastern Pacific (left), species richness increases with island area and decreases with distance from New Guinea, the major source of colonists (right).

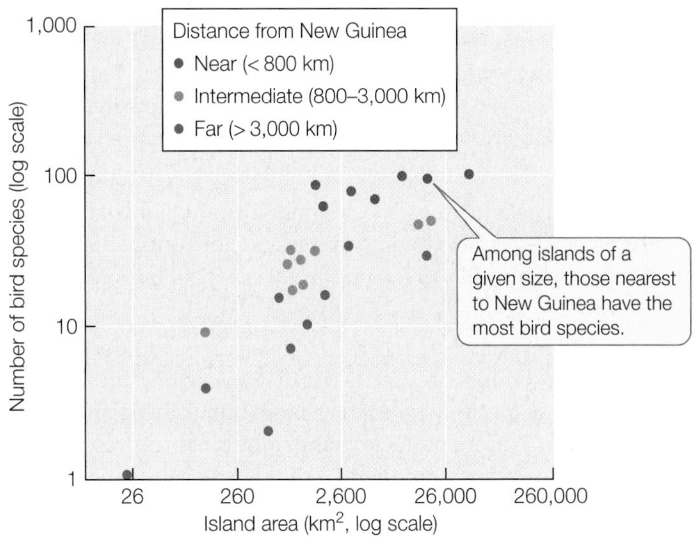

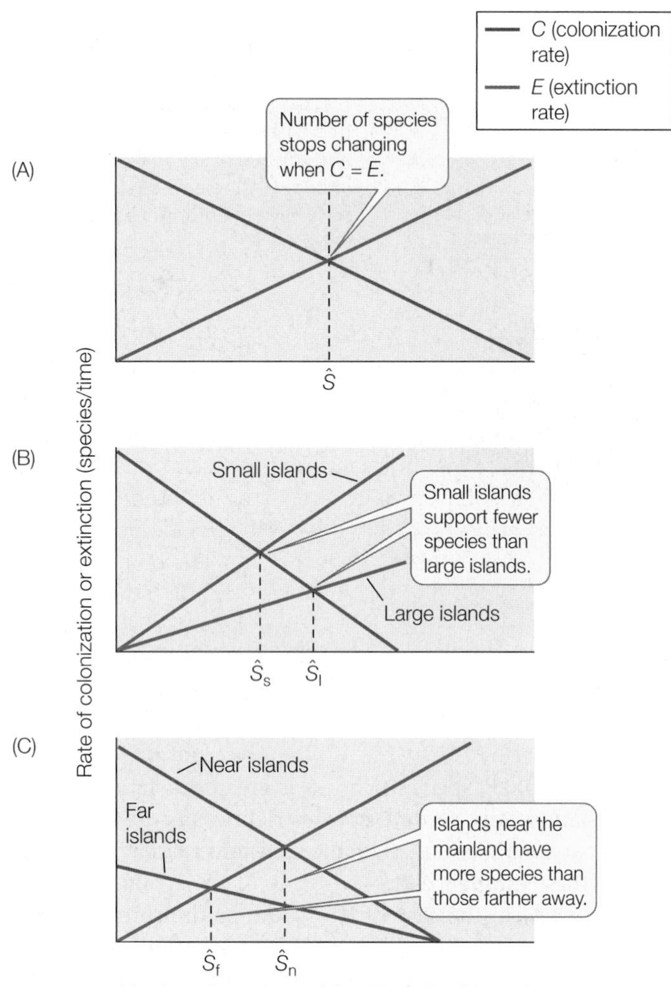

FIGURE 44.11 MacArthur and Wilson's Theory of Island Biogeography (A) Species richness reaches a stable equilibrium ($\hat{S}$) when the colonization rate equals the extinction rate. (B) Smaller islands have larger per-species extinction rates, and hence lower equilibrium species richness. (C) Islands more distant from the mainland have lower colonization rates, and hence lower equilibrium species richness.

and the rate at which resident species go extinct. This idea forms the basis for the theory of **island biogeography**, proposed in 1963 by Robert H. MacArthur and Edward O. Wilson. The theory can be summarized as follows:

- An oceanic island gains species only if they colonize from elsewhere—usually from nearby islands or continents, whose communities make up the species pool of potential colonists.
- The rate at which new species arrive on an island declines as the island fills with species. At first, every immigrant represents a species new to the island, but as the island fills up, more and more of the arriving individuals represent species already resident there. (If all species in the mainland species pool were present on the island, the colonization rate would be zero.)
- The overall rate at which species are lost from the island—the extinction rate—increases as the island fills with species. In any time period, each species has some chance of going extinct from various causes, and as the number of species increases, so does the overall number of species that go extinct per unit of time.
- The number of species on an island—its species richness—stops changing if the colonization rate equals the extinction rate. When we plot colonization and extinction rates on a graph, this equilibrium species richness is represented by the point at which the two lines cross, $\hat{S}$ (**FIGURE 44.11A**).
- Population sizes will decline as island size decreases, and since small populations are more at risk of extinction, the extinction rate for small islands should rise more steeply than that of large islands as a function of total species on the islands. Thus the equilibrium species richness should be greater on large than on small islands (**FIGURE 44.11B**).
- Similarly, fewer wind- and water-borne seeds, and fewer dispersing animals, will encounter a distant island than one closer to the mainland species pool. As a result, the colonization rate, and thus the equilibrium species richness, will be higher on close islands than on distant ones (**FIGURE 44.11C**).

The theory of island biogeography successfully explains the patterns of species richness found not only on oceanic islands, but also on "islands" of one habitat type surrounded by a "sea" of a different habitat type. It has proved to be one of the most successful explanatory theories in ecology (**FIGURE 44.12**).

Go to ANIMATED TUTORIAL 44.2
Island Biogeography Simulation
PoL2e.com/at44.2

CHECKpoint CONCEPT 44.4

- ✓ Describe three factors that may contribute to the high species richness of tropical rainforests.
- ✓ Tropical coral reef communities are among the most species-rich of any on Earth. Speculate on factors that might explain their diversity.
- ✓ What do Figures 44.11B and C tell you about the predicted rate of change in species composition (species turnover rate) at equilibrium on small versus large islands and on islands distant from versus close to a mainland? (Hint: think of turnover rate as the number of new species that replace previous species in a period of time.)

Quantitative studies of species richness and the theory of island biogeography have contributed greatly to our knowledge of the structure and function of ecological communities. We will now turn to some ways in which this knowledge can be applied to conserving and restoring these communities.

INVESTIGATION

FIGURE 44.12 The Theory of Island Biogeography Can Be Tested
By experimentally defaunating (removing all the arthropods from) four small mangrove islands of equal size but at different distances from the mainland, Simberloff and Wilson were able to observe the process of recolonization and compare the results with the predictions of island biogeography theory.[a] See Apply the Concept on p. 910 for a test of a possible alternative explanation of the species diversity on the mangrove islands.

HYPOTHESIS

The number of species on an island before it is experimentally defaunated will decrease with distance from a mainland source of colonists, and this number will be reestablished after the island has been recolonized.

METHOD

1. Take a census of the terrestrial arthropod species on four small mangrove islands of equal size (11–12 m diameter) but at different distances from a mainland source of colonists.
2. Erect scaffolding and tent the islands. Fumigate with methyl bromide (a chemical that kills arthropods but does not harm plants).

3. Remove tenting. Monitor recolonization for the following 2 years, periodically taking a census of arthropod species.

RESULTS

The number of species present before defaunation decreased with distance from the mainland. Two years after defaunation, each island had about the same number of species it had before the experiment.

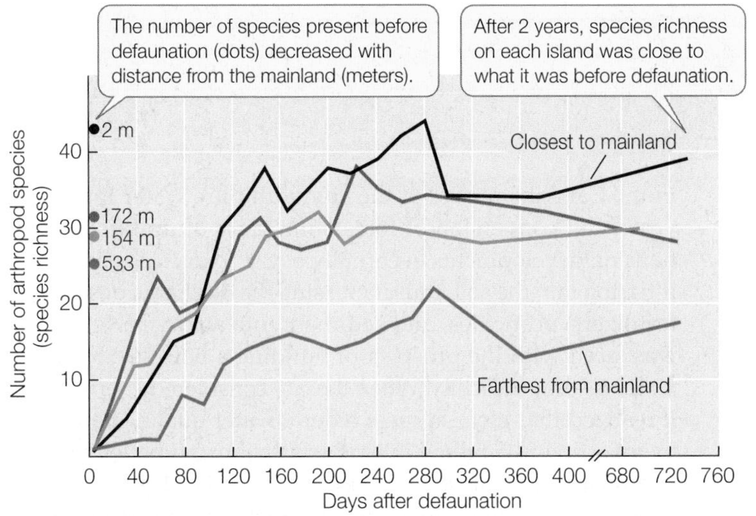

CONCLUSION

The results support predictions of island biogeography theory.

Go to **LaunchPad** for discussion and relevant links for all **INVESTIGATION** figures.

[a]D. S. Simberloff and E. O. Wilson. 1969. *Ecology* 51: 934–937.

CONCEPT 44.5

Community Ecology Suggests Strategies for Conserving Community Function

The ability of an ecological community to provide humans with critical goods and services often depends on its diversity. How can we use this knowledge to conserve community function?

Ecological communities provide humans with goods and services

Humans, like all species, depend on the ecological interactions that occur in communities and the interactions between communities and their physical environment. Although this seems obvious today, it was only in 1970 that ecosystems were first said to provide people with "goods and services." The goods include food, clean water, clean air, building materials, and fuel; the services include flood control, soil stabilization, pollination of plants, and climate regulation. Because many of these goods and services involve exchanges between communities and the abiotic environment, they are called **ecosystem services**; a short list of examples is given in **TABLE 44.1**.

Goods that are of direct biological origin are obvious: everybody knows that we get food from plants and animals, wood from trees, wool from sheep, and aesthetic enjoyment from flowers and birds. The roles of properly functioning communities in providing ecosystem services are less obvious, however, and we often take them for granted—until something goes wrong.

European settlers of Australia learned the hard way about the importance of the services provided by decomposers. When the first settlers arrived in 1788, they brought cattle with them. From the beginning, however, the Australian cattle industry ran into problems: native dung beetles, adapted as they were to the dry, fibrous dung of forest marsupials, shunned the wet manure produced by cattle in open pastures. As a consequence, manure piled up higher and deeper. The accumulated dung not only reduced pasture productivity (because the nitrogen in the dung was not being recycled, and because cattle do not eat fouled plant material), but also provided a perfect nursery for native bush flies and introduced blood-sucking buffalo flies. Populations of these pests exploded.

In 1964, Australia's national scientific research organization, CSIRO, finally tackled the problem. Insect biologists traveled the world to find dung beetles that could process cattle dung in Australian environments without disrupting native beetle communities—an important precaution, since many well-intended introductions have had unforeseen, detrimental impacts (see Concept 43.3). Between 1968 and 1984, CSIRO introduced more than 50 species of beetles. The project was successful: soil nitrogen content improved, pasture productivity

APPLY THE CONCEPT

Diversity patterns provide clues to what determines diversity

An alternative to MacArthur and Wilson's explanation for patterns of species richness on islands has been proposed: that any island has a particular mix of niches that support an island-specific community. For Simberloff and Wilson's mangrove islands (see Figure 44.12), this hypothesis would mean that each island gradually reassembled its original arthropod community, and that the number of species stopped changing when the full complement of species was restored. Review two additional results from the Simberloff and Wilson study shown below and answer the associated questions.[a]

1. Figure A shows the original community and post-defaunation census results for individual arthropod species on one mangrove island. Do these census results support or refute the alternative hypothesis? Explain your answer.
2. Figure B shows the accumulation of arthropod species in three functional groups—herbivores, carnivores, and parasites—for the same island. Did species in these three groups accumulate at the same rates? Why or why not?

[a]D. S. Simberloff and E. O. Wilson. 1969. *Ecology* 50(2): 278–296; J. W. Glasser. 1982. *American Naturalist* 119(3): 375–390.

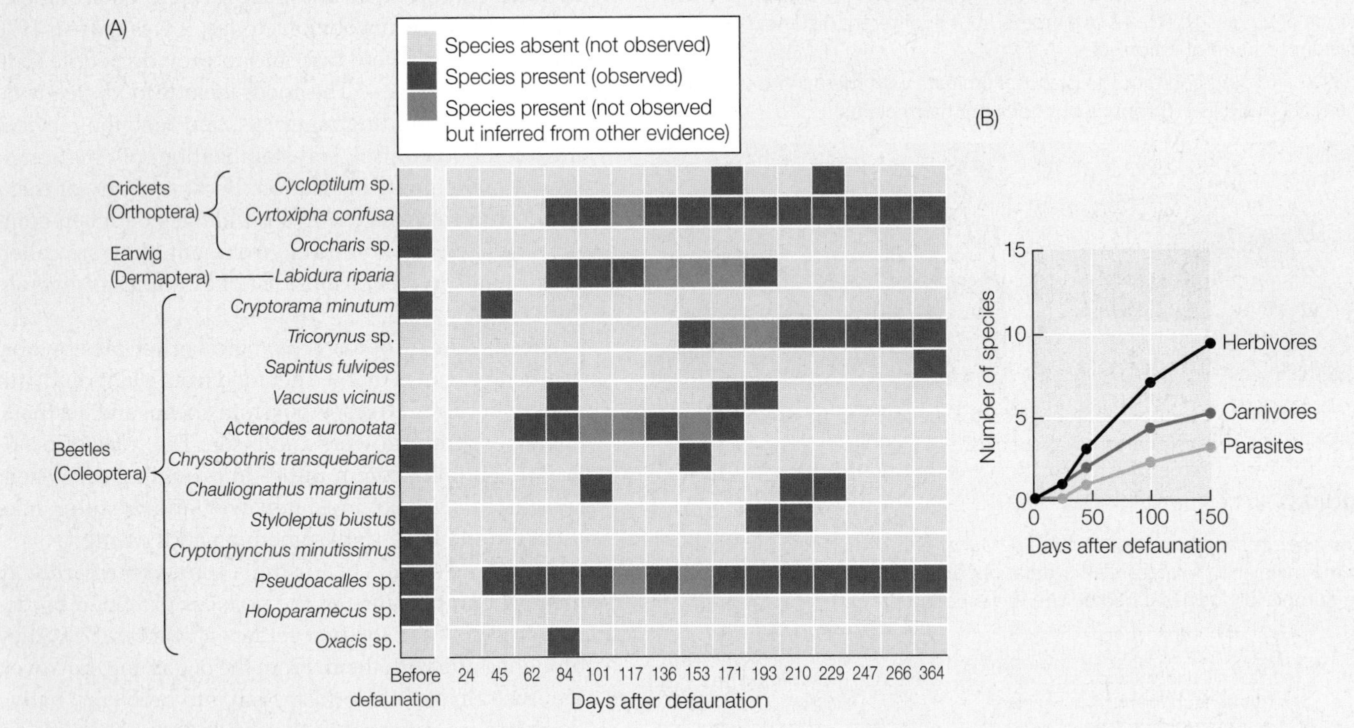

increased, fly numbers plummeted, cattle became healthier, and native dung beetles were unharmed.

Ecosystem services have economic value

Ecosystem services have economic value, as the Australian Dung Beetle Project shows. Some services related to food production are relatively straightforward to value because they are included in the price of the food. In the United States, for instance, wild native pollinating insects (not including domesticated honey bees) contribute $3 billion annually to crop production. Many other services—such as greenhouse gas regulation—are more difficult to value because they are not included in commercial markets, but that does not mean they lack value.

Consider the case of New York City's water, which comes from reservoirs in the Catskill Mountains. Natural ecosystems in the Catskills provided the city with pure water for many years. But conversion of forests into agricultural fields and housing developments eventually began to decrease water infiltration into the soil, thus increasing the discharge of sewage, fertilizers, pesticides, and sediment into streams. New York was faced with the prospect of building a new $6–$8 billion water treatment facility. When the city considered alternatives, it realized that it could meet federal water quality standards by investing $1.5 billion in land protection and better sewage treatment in the Catskills. Although expensive, this natural alternative cost a lot less than the technological alternative! New York's Long-Term Watershed Protection Plan, completed in 2006, has been a complete success.

The New York example is not unique. Many ecosystems can provide more goods and services, at lower cost, than artificial substitutes can. But as we have seen, the community functions

TABLE 44.1 Some Major Ecosystem Goods and Services

Good or service	Community process	Examples
Food production	Trophic interactions	Production of wild food; production of crops and livestock
Materials production	Trophic interactions	Production of lumber, fuel, fiber
Pollination and seed dispersal	Plant–pollinator and plant–disperser interactions	Plant reproduction and dispersal
Maintenance of fertile soil	Decomposition; composition of vegetation; plant- and animal-microbe mutualisms	Nitrogen fixation, nutrient recycling, erosion control
Waste treatment	Decomposition	Breakdown of toxins and wastes
Pest control	Predator–prey, host–parasite, and competitive interactions	Decreased pest abundance
Water supply	Regulation of water infiltration and runoff by vegetation structure and animal activity	Retention of water in watersheds, reservoirs, and aquifers
Climate regulation	Metabolic gas exchange	Regulation of greenhouse gases and cloud formation
Disturbance control	Composition and biomass of vegetation	Damping of storm winds and wave surges; flood control
Recreational and cultural opportunities	Ecological processes; species diversity	Ecotourism, outdoor recreation; aesthetic, educational, spiritual, scientific values

that deliver those goods and services depend on species diversity. Ecological communities can best meet our needs if we do not manage them so intensively, or disturb them so severely, that we lower their diversity.

Island biogeography suggests strategies for conserving community diversity

Concept 41.5 described how humans are converting natural communities into less diverse, human-managed communities such as croplands, pastures, and urban settlements. As a result, once-continuous areas of natural habitat are being reduced to scattered fragments—habitat islands embedded in human-modified landscapes (**FIGURE 44.13**). We expect such **fragmentation** of habitat to cause losses of species from communities for several reasons: the total amount of habitat decreases, the size of habitat patches decreases, and the patches become more isolated from one another. Given what you have learned about metapopulations (see Concept 42.5) and island biogeography, you now can predict the results. Subpopulations become smaller as habitat area shrinks and are therefore more prone to extinction. The human-dominated landscape surrounding the fragments may serve as a barrier to dispersal, decreasing colonization rates. The net consequence is lower species richness in the fragments than in the original habitat.

The insight that species richness reflects a balance between colonization and extinction suggests that the detrimental effects of fragmentation can be minimized by enhancing colonization and reducing extinction. Colonization rates can be

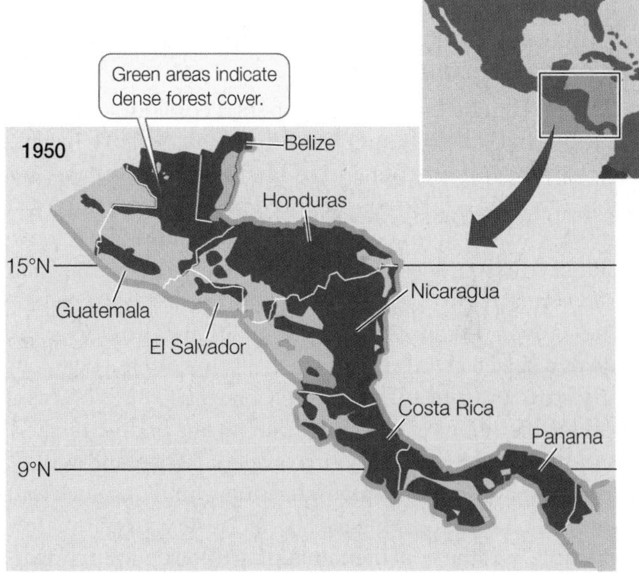

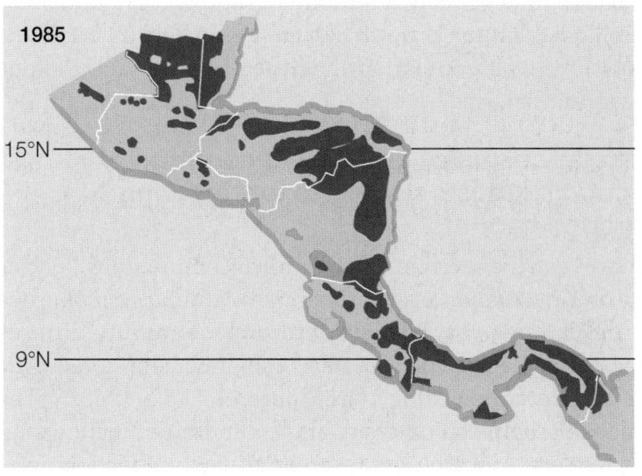

FIGURE 44.13 Habitat Fragmentation in Tropical Forests Less than half of the tropical forest that existed in Central America in 1950 (top) remained by 1985 (bottom), most of it in small patches. Although the rate of deforestation slowed somewhat after 1985, an additional 19 percent of the forest had disappeared by 2005.

increased by clustering habitat fragments close to one another or to a large "mainland" patch, or by connecting fragments with dispersal corridors (see Concept 42.6). Extinction rates can be minimized by retaining at least some large patches of the original habitat and by preserving the capacity of remaining patches to support healthy populations.

Several of these predictions have been tested in a remarkable large-scale experiment. Landowners in Brazil agreed to preserve tropical forest patches of certain sizes and configurations laid out by biologists (**FIGURE 44.14**), and they allowed the biologists to survey the parcels before and after the intervening forest was cleared for pasture. Species soon began to disappear from the patches. The first species lost were monkeys that travel over large areas of forest. Army ants and the birds that follow their columns also disappeared quickly. Species were lost more rapidly in isolated fragments than in fragments of equal area that were connected to nearby unfragmented forest, and small isolated fragments lost species more rapidly than larger isolates.

Results of studies such as these now guide land-use planning worldwide. They suggest that dispersal corridors and large habitat patches are necessary to maintain healthy, diverse communities.

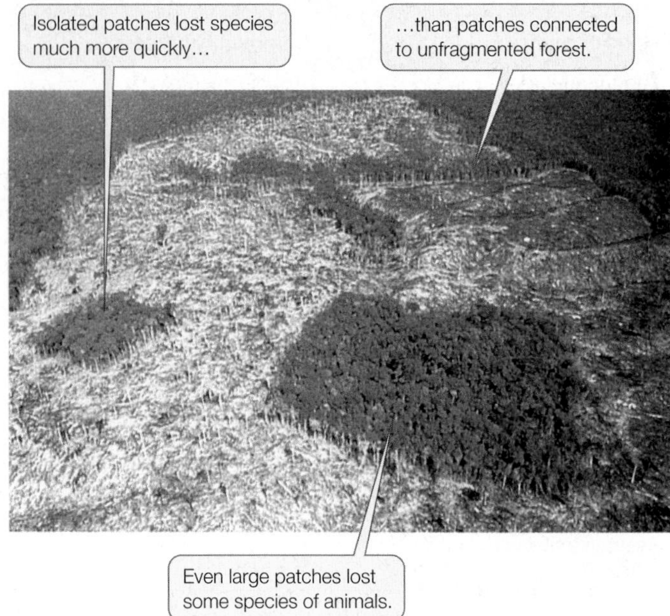

FIGURE 44.14 A Large-Scale Study of Habitat Fragmentation Biologists studied patches of tropical evergreen forest near Manaus, Brazil, before and after the parcels were isolated by forest clearing. The results of the study demonstrated that small, isolated habitat fragments support fewer species than larger or less isolated patches of the same habitat.

Go to ANIMATED TUTORIAL 44.3 Fragmentation Effects
PoL2e.com/at44.3

Trophic cascades suggest the importance of conserving certain species

It is well and good to plan for dispersal corridors and large patches of habitat, but species differ considerably in their habitat area requirements and in the kinds of dispersal corridors they will use. How big should natural areas be? What constitutes an effective corridor? These decisions can often be facilitated if they target species that play particularly important roles in communities.

The wolves of Yellowstone National Park provide an example of a species whose presence has a major effect on community structure and function (see Concept 43.3). Wolves are crucial for maintaining healthy aspen forests and watersheds via trophic cascades that also involve elk, aspen and willow, and beavers. Conservation planning is now targeting such critical species. The Yellowstone to Yukon Conservation Initiative, for example, has the goal of maintaining a continuous corridor of wolf habitat between Yellowstone National Park and areas of similar habitat to the north, which should help maintain the Yellowstone wolf population and hence a healthy Yellowstone ecosystem.

The relationship of diversity to community function suggests strategies for restoring degraded habitats

We have seen that even highly disturbed communities, such as those on Krakatau or on experimentally defaunated mangrove islands, can recover. This insight from community ecology suggests that degraded ecosystems can often be restored—the goal of restoration ecology (see Concepts 41.5 and 43.3). The relationship between diversity and community function suggests that one of the best ways to improve the functioning of a damaged system is to restore species diversity. To do so, restoration ecologists draw on their knowledge of the factors that shape diversity (**FIGURE 44.15**).

As we have also seen, however, disturbance sometimes results in an ecological transition to a community that is very different from the original community (see Concept 44.2). In such cases it may be difficult to find a way to reverse the transition and restore the function of the original community. This possibility is a sobering reminder that restoration is challenging and requires deep understanding of individual ecosystems.

CHECKpoint CONCEPT 44.5

- ✓ Give two examples each of goods and services provided to humans by natural ecosystems and by ecosystems constructed and managed by humans.
- ✓ If you were working as a restoration ecologist to restore a certain community, what measures might you use to assess whether your efforts were successful?

In this chapter we have discussed factors that influence the structure of communities and the interactions that shape the flow of energy through them. We have seen how community

INVESTIGATION

FIGURE 44.15 Species Richness Can Enhance Wetland Restoration In one large-scale field experiment, ecologists compared different methods for restoring denuded areas of the Tijuana Estuary, a wetlands environment near San Diego, California. They found that several measures of community function improved more rapidly in species-rich than in species-poor plantings.[a]

HYPOTHESIS

Faster progress toward restoring the community's original function will be made by planting mixtures of species than by planting single species alone.

METHOD

1. In an area of wetland denuded of vegetation, mark off replicate small experimental plots, all of the same size.
2. Choose 8 native species typical of the region. Plant some plots with each of the 8 species by itself, others with different 3-species subsets, and others with different 6-species subsets.
3. Plant the same total number of seedlings per plot. Leave some plots unplanted as controls.
4. Return over the next 18 months to measure the vegetation and ecosystem nitrogen levels.

RESULTS

Vegetation covered the bare ground more quickly in those plots with higher species richness. Those same plots developed complex vertical structure more quickly and accumulated more nitrogen in plant roots per m² of area.

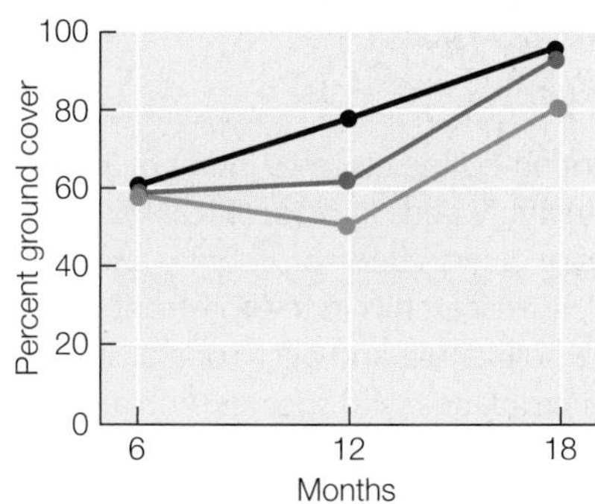

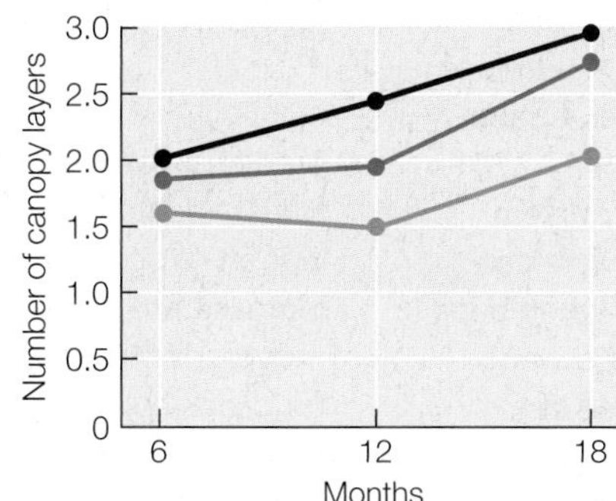

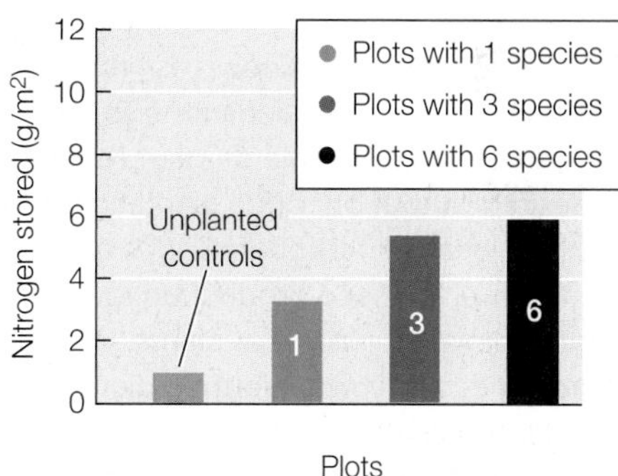

CONCLUSION

Planting more species into a community leads to more rapid restoration of wetland function.

ANALYZE THE DATA

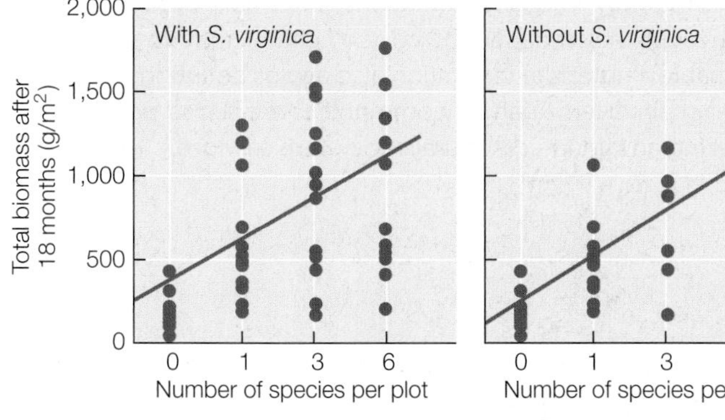

A. The unplanted control plots (above) showed nonzero values of stored nitrogen after 18 months. Is this a surprise? Why or why not?

B. One of the 8 species in the study, *Salicornia virginica*, contributed a disproportionate share of the biomass, cover, and nitrogen in mixtures in which it was included. In the two graphs at left, each circle represents the performance of a particular plot; the lines indicate best-fit regressions. Based on these two graphs, what do you conclude about the mechanism for the higher success of species-rich plantings? Is the greater chance of including *S. virginica* in species-rich plantings the likely explanation?

Go to **LaunchPad** for discussion and relevant links for all **INVESTIGATION** figures.

[a] G. H. Keer and J. B. Zedler. 2002. *Ecological Applications* 12: 456–473; J. C. Callaway et al. 2003. *Ecological Applications* 13: 1626–1639.

structure affects a community's function, including its outputs of goods and services. But outputs are limited by inputs, and to understand what determines inputs to communities, we need to consider their abiotic environments. The next chapter will apply this ecosystem perspective and consider how humans are influencing the global ecosystem.

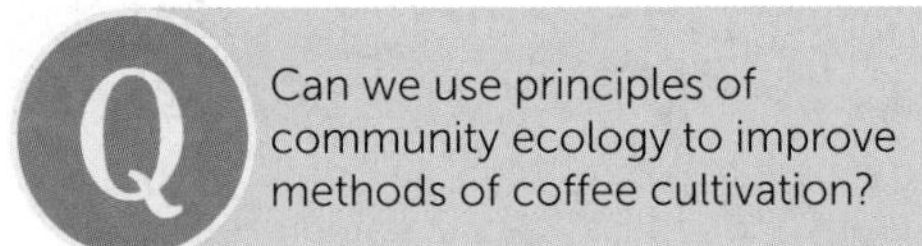

Q
Can we use principles of community ecology to improve methods of coffee cultivation?

ANSWER Croplands are ecological communities that are governed by the same factors that govern natural communities. We can therefore use our understanding of how natural communities function to improve the functioning of agricultural communities. For example, we know from studying natural communities that species diversity often enhances community function (Concept 44.3)—whether we are talking about pest control, pollination, or crop yield. We also know that species diversity is increased when colonization rates are high and extinction rates are low, and when complex habitat structure provides opportunities for species with diverse niches to coexist (Concept 44.4). We can use these ecological principles in many situations to produce more food at less cost and with fewer negative effects on the environment. For example, we can enhance crop production by growing crops in complementary functional groups together, a practice known as "intercropping." Planting coffee, which is shade-tolerant, in the shade of taller, timber-producing trees is an example of intercropping.

The desired output of coffee plantations is, of course, seeds (coffee "beans"). Although *Coffea arabica* flowers can pollinate themselves, seed production is enhanced when bees move pollen between individual plants. From our knowledge of community ecology, we might predict that diverse bee communities would provide the best pollination overall, because different bee species may pollinate in complementary ways. Recent research in Indonesia confirms this prediction: seed production was 50 percent higher, and varied less from year to year, in coffee plantations that hosted 20 wild bee species than in those that hosted 3 bee species.

We also know that wild species of insects are important for pollinating the flowers of most species of flowering plants. This knowledge suggests that embedding coffee plants in diverse natural communities of plants and animals might be an effective way to reap the benefits of better pollination. Research in both Costa Rica and Indonesia confirmed this prediction as well: planting coffee close to intact patches of forest and leaving flowering weeds in place (**FIGURE 44.16**) was all it took to attract a diverse bee community and thus increase production. This research is at the forefront of a new movement to enhance pollination and pest control services for crops by bordering croplands with species-rich and structurally diverse communities.

FIGURE 44.16 Traditional Coffee Cultivation and Community Diversity Low-intensity, high-diversity coffee cultivation (the product of which is often marketed as "shade-grown organic coffee") results in lower yields, but can net a higher profit by avoiding the costs of fertilizers and other chemicals—with the added benefit of maintaining the services provided by natural forests.

Traditional low-intensity cultivation methods therefore seem to be a cost-effective way to increase coffee production. Although traditional methods may not provide the highest yields per hectare, they may nevertheless be very profitable. A study of nine coffee plantations in Mexico reported that high-intensity sun cultivation produced more coffee per hectare, but that low-intensity plantations were more profitable because they reduced the costs of chemicals and labor. That low-intensity cultivation also avoids pollution and helps maintain diverse natural communities are added benefits of taking advantage of natural ecosystem services.

SUMMARY

CONCEPT 44.1 Communities Contain Species That Colonize and Persist

- A **community** is a group of species that coexist and interact with one another within a defined geographic area.
- A community is characterized by the particular mix of species it contains (**species composition**), by the number of species it contains, and by their abundances. Together these attributes of communities constitute **community structure**.
- A community contains those species that are able to colonize an area and persist there.

CONCEPT 44.2 Communities Change over Space and Time

- Spatial variation in physical conditions and in **habitat structure** is associated with predictable spatial variation in species composition. **Review Figure 44.2**
- Even in a constant environment, ongoing extinction and colonization result in ongoing change, or **turnover**, in the species composition of communities over time.
- After a sudden **disturbance**, species sometimes replace one another in a more or less predictable sequence called **succession**. Succession following a disturbance that removes the original community often results in a community that resembles the original one. **Review Figure 44.4 and ANIMATED TUTORIAL 44.1**
- Some disturbances cause an **ecological transition** to a different type of community, and removal of the disturbance factor does not restore the original type of community.
- Climate change can cause the species composition of a community to change over time. **Review Figure 44.5**

CONCEPT 44.3 Community Structure Affects Community Function

- An ecological community is a dynamic system with inputs, internal workings, and outputs. A fundamental aspect of this **community function** is the flux of energy into and out of the community.
- The total amount of energy that primary producers capture and convert into chemical energy per unit of time is called **gross primary productivity** (**GPP**). The amount of GPP that becomes available to consumers per unit of time is called **net primary productivity** (**NPP**). **Review Figure 44.6 and ACTIVITY 44.1**
- The **ecological efficiency** of energy transfer from one trophic level to the next is only about 10 percent.
- The term **niche** refers to the physical and biological environments in which a species has a positive per capita growth rate, as well as to the species' functional role in the community.
- **Species diversity** has two components: the number of species in the community, called **species richness**, and the distribution of those species' abundances, called **species evenness**. **Review Figure 44.7 and ACTIVITY 44.2**
- The outputs of a community are measures of its function. Community outputs increase with the species diversity of the community and the number of functional roles those species fill. **Review Figure 44.8**

CONCEPT 44.4 Diversity Patterns Provide Clues to What Determines Diversity

- Species richness is greatest in the tropics and decreases with increasing latitude. Possible causes of this pattern include greater climate stability, greater energy input, and greater habitat complexity at lower latitudes. **Review Figure 44.9**
- Species richness also varies among oceanic islands. Large islands support more species than small islands, and islands close to a mainland support more species than distant islands. **Review Figure 44.10**
- The theory of **island biogeography** explains these island patterns as a balance between the rate at which new species colonize and the rate at which resident species go extinct. **Review Figures 44.11 and 44.12 and ANIMATED TUTORIAL 44.2**

CONCEPT 44.5 Community Ecology Suggests Strategies for Conserving Community Function

- Ecosystems provide humans with a variety goods and services. These **ecosystem services** have economic value. **Review Table 44.1**
- Production of ecosystem services is threatened by human activities, including the **fragmentation** of natural habitats, that reduce species diversity. The detrimental effects of habitat fragmentation can be mitigated by land-use management that minimizes extinction in each habitat "island" and maximizes dispersal among them. **Review Figures 44.13 and 44.14 and ANIMATED TUTORIAL 44.3**
- Ecosystem function often can be preserved or restored by focusing on particular species that play especially important roles in the community, and by maintaining or restoring overall species diversity. **Review Figure 44.15**

Go to the Interactive Summary to review key figures, Animated Tutorials, and Activities PoL2e.com/is44

Go to LaunchPad at **macmillanhighered.com/launchpad** for additional resources, including LearningCurve Quizzes, Flashcards, and many other study and review resources.

45 The Global Ecosystem

KEY CONCEPTS

The Mauna Loa observatory on the "Big Island" of Hawaii provides scientists with accurate data on the composition of Earth's atmosphere.

The nineteenth century was a golden age for Earth science. Geologists not only deduced the astonishing age of the planet, but also discovered that massive sheets of ice had covered much of Europe and North America only 15,000 years ago—Earth's climate was once much colder than it is today.

What could cause such dramatic climate change? One possibility, proposed in 1861, pointed to atmospheric gases such as carbon dioxide and water vapor, which by then were known to absorb infrared (heat) radiation. Scientists hypothesized that these gases might influence Earth's climate by trapping energy in the lower atmosphere. By the end of the nineteenth century, they had outlined a "carbon dioxide theory" to explain how fluctuating atmospheric CO_2 concentrations could change the strength of this "greenhouse effect" and explain past ice ages.

Scientists pursuing the carbon dioxide theory had an additional insight. They realized that coal and petroleum contained fossilized carbon from dead plants and animals. Since the Industrial Revolution, humans had been burning these fossil fuels, releasing their carbon as CO_2. Might fossil fuel combustion be increasing CO_2 in the atmosphere, thus magnifying the greenhouse effect? Some early skeptics felt that any effect of CO_2 would be insignificant because it constituted only about 300 parts per million (0.03 percent) of all molecules in the atmosphere. In the 1930s, however, physicists and chemists demonstrated that changes in atmospheric CO_2 could indeed affect Earth's climate. In the 1950s the American scientist Gilbert Plass estimated the rate at which atmospheric CO_2 was increasing and used one of the first electronic computers to predict that increased CO_2 concentrations could significantly raise the average global temperature.

The challenge now facing scientists was to see if the predicted rise in atmospheric CO_2 could be detected. But nobody had yet been able to measure atmospheric CO_2 with precision. A young American chemist, Dave Keeling, developed sensitive new instruments and set them up atop Mauna Loa in Hawaii, 4,000 meters above sea level and far from human-generated pollution. Pledging himself to a long-term study, he began taking measurements in 1957 as part of the International Geophysical Year (IGY), one of the first government-supported international scientific collaborations. But the IGY money soon dried up. Undaunted by periods of insufficient funding, Keeling continued his research on atmospheric CO_2 and the carbon cycle over the next four decades.

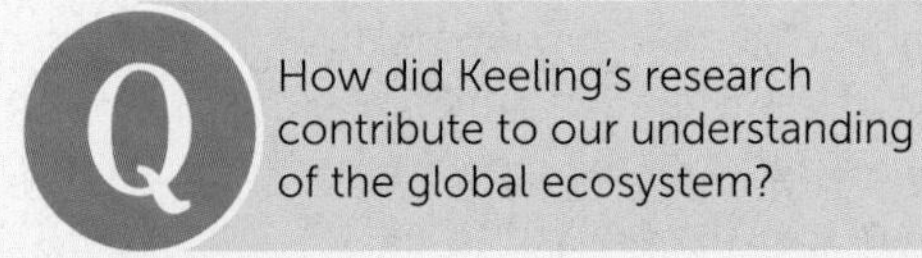

How did Keeling's research contribute to our understanding of the global ecosystem?

You will find the answer to this question on page 932.

CONCEPT 45.1 Climate and Nutrients Affect Ecosystem Function

As we learned in Concept 41.1, the term **ecosystem** is commonly applied to ecological communities plus the abiotic environments with which they exchange energy and materials. Most of the interactions among components of ecosystems happen at a local scale. A plant, for example, can absorb only those nitrate molecules that are in physical contact with its root hairs; an elk in Alberta cannot fall prey to a wolf living in Wyoming's Yellowstone National Park; and a raindrop interacts directly only with the soil particles it contacts. For this reason, most studies of ecosystems are done at relatively small spatial scales, such as a single river drainage, lake, or patch of forest.

Local ecosystems are linked, however, by slower exchanges of organisms, energy, and materials that occur on spatial scales ranging from tens of kilometers within regional landscapes to thousands of kilometers around the planet. These long-distance interactions occur when organisms move among communities and when energy and materials are transported by the physical circulation of the atmosphere and the oceans, by flowing water, and by geological processes that occur below Earth's crust. Large-scale patterns ultimately influence inputs to and losses from local ecosystems, making it impossible to understand a local ecosystem completely without considering it in the context of the larger systems of which it is a part. This chapter therefore takes both local and global perspectives.

NPP is a measure of ecosystem function

Ecosystems, like other biological systems (see Concept 1.2), can be characterized by their components and by patterns of interaction among those components. In ecosystem studies, the community as a whole is often treated as one component, and distinct portions of the abiotic environment, such as soil, atmosphere, or water, are treated as other components. The dynamic processes by which these components exchange and transform energy and materials are aspects of **ecosystem function**.

Ultimately, movements of materials between the abiotic and biotic components of ecosystems are tied to carbon. Primary producers combine inorganic carbon dioxide with inorganic water to make organic carbohydrates. Energy stored in the chemical bonds of those molecules fuels the synthesis of all the biomass in the bodies of organisms at every trophic level (see Concept 6.2). Most of the elemental constituents of that biomass enter communities through the primary producers as well. Calcium in the bones of a wolf, for example, came from the elk it ingested; the elk obtained calcium from the plants it ate; and the plants used energy from the breakdown of carbohydrates to extract inorganic calcium from the soil.

The rate at which a community makes primary producer biomass—its net primary productivity, or NPP (see Concept 44.3)—is therefore a measure of the flow of materials, as well as energy, into the biotic component of its ecosystem. This measure of ecosystem function does not incorporate all exchanges of materials between organisms and the abiotic environment: organisms, for example, also take in or excrete inorganic water or salts to maintain homeostasis. Such exchanges are nonetheless indirectly coupled with primary productivity because they are fueled by the energy that enters the community via primary producers.

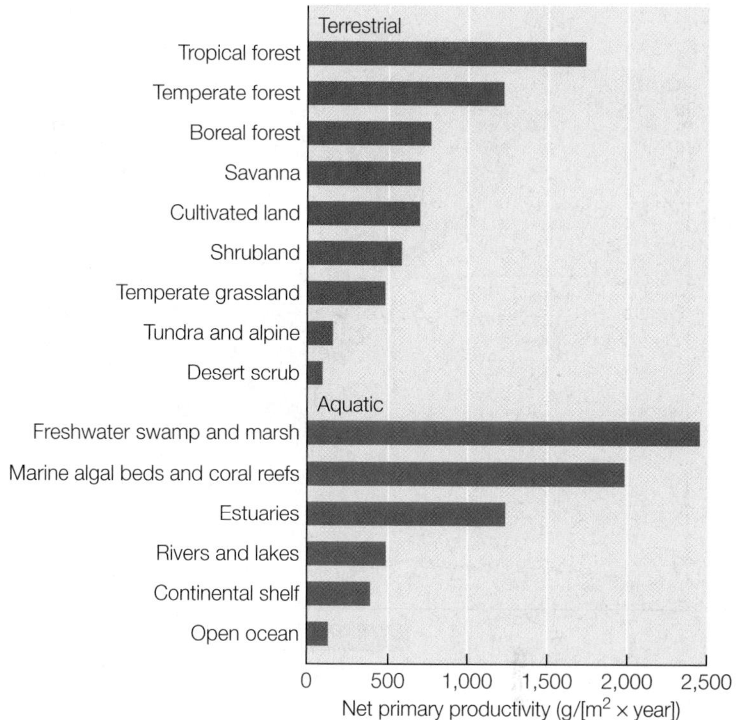

FIGURE 45.1 NPP Varies among Ecosystem Types Net primary productivity is expressed here as grams of biomass produced per square meter of area per year.

Scientists are now able to estimate NPP with instruments, carried by orbiting satellites, that measure the reflection of different wavelengths of light by Earth's surface. This "view from space" allows them to calculate how much sunlight is absorbed by chlorophyll in different regions and thus to map the distribution of photosynthetic biomass—which is correlated with NPP—across the planet. These maps show that NPP is not the same everywhere. As we will see next, patterns of variation in NPP indicate factors that limit ecosystem function.

NPP varies predictably with temperature, precipitation, and nutrients

Net primary productivity varies considerably among ecosystem types. Among terrestrial ecosystems, tropical forests are the most productive per unit of area, followed by temperate forests, whereas tundra and deserts are the least productive (**FIGURE 45.1**). Among aquatic ecosystems, freshwater swamps and marshes and marine algal beds and coral reefs are the most productive, followed by estuarine habitats, whereas the open ocean is the least productive. Much of this pattern derives from variation in climate and nutrient availability.

Satellite images of terrestrial ecosystems show that NPP varies with latitude: highly productive areas are concentrated in the tropics, and higher latitudes are less productive (**FIGURE**

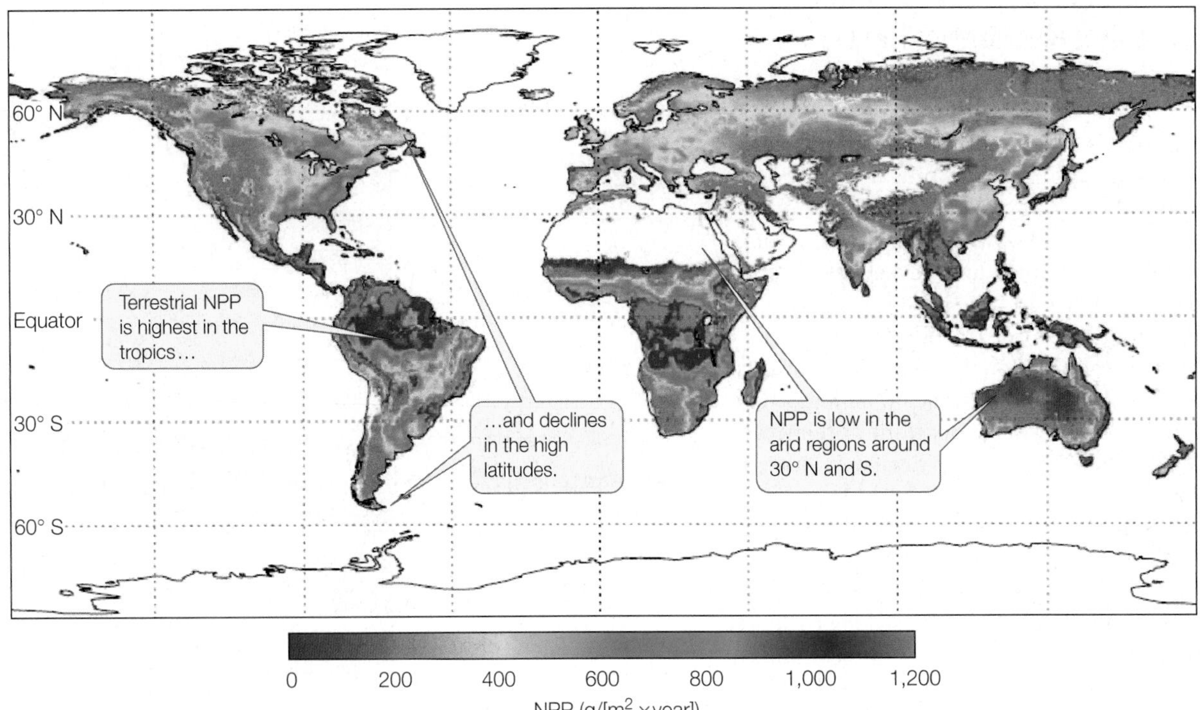

FIGURE 45.2 Terrestrial NPP Corresponds to Climate This map of estimated terrestrial NPP is based on satellite sensor data accumulated over the period 2000–2005. The white spaces on land represent unvegetated areas, including barren deserts and ice caps.

45.2). This pattern parallels latitudinal gradients in solar energy input (see Figure 41.3) and temperature, suggesting that NPP increases with temperature. Indeed, terrestrial NPP generally does increase with temperature (**FIGURE 45.3A**), which makes sense when we consider that the activity of photosynthetic enzymes, like that of other enzymes, increases up to the critical temperature at which enzymes denature (see Concept 3.4). This critical temperature varies among plant species that are adapted to different environments but is usually about 40°C–50°C.

The satellite images also show areas of low terrestrial NPP at 30° N and S latitude (see Figure 45.2), where deserts occur (see Figures 41.5 and 41.11). This pattern parallels gradients in precipitation, suggesting that moisture also affects NPP. Terrestrial NPP does increase with precipitation up to about 2,400 millimeters per year, then decreases at higher levels (**FIGURE**

FIGURE 45.3 Terrestrial NPP Varies with Temperature and Precipitation NPP is given in megagrams (1 Mg equals 10^6 g or 1 metric ton) per hectare per year.

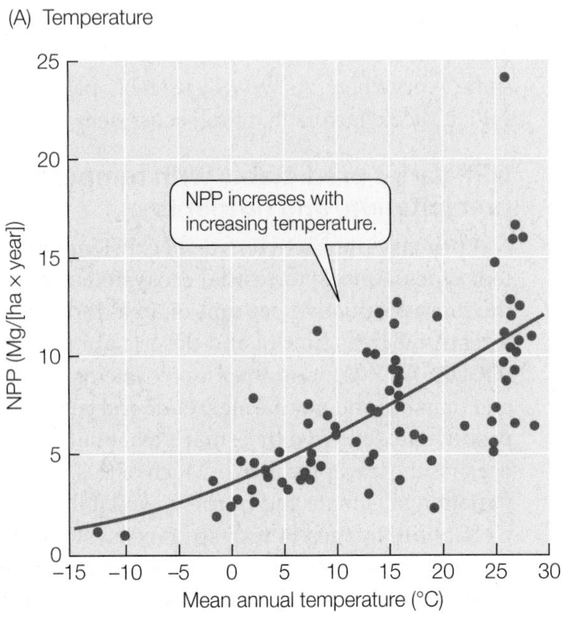

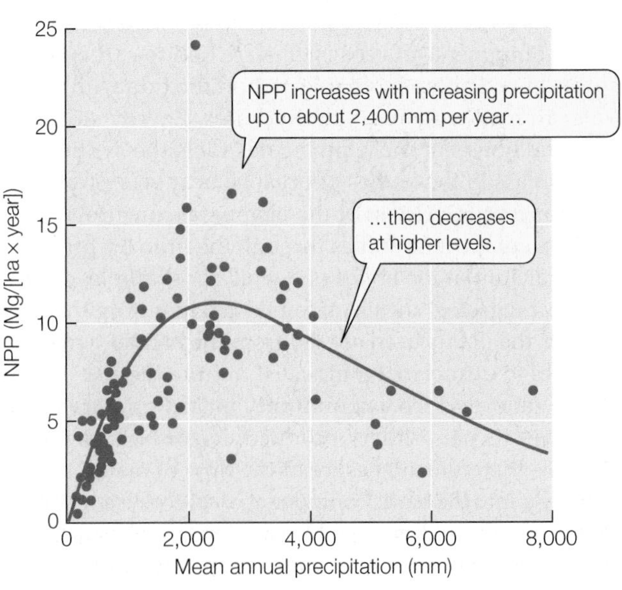

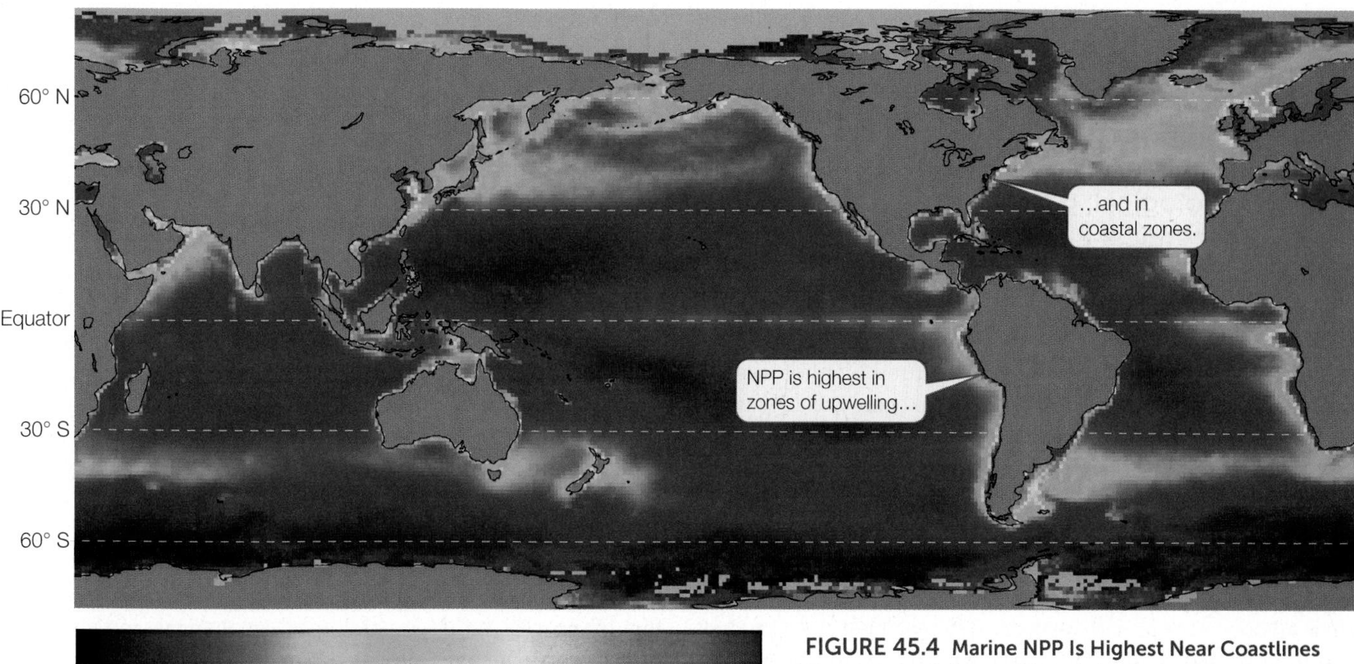

FIGURE 45.4 Marine NPP Is Highest Near Coastlines Photosynthesis is possible only in the top layers of oceans, where sunlight penetrates. The availability of nutrients determines how much NPP occurs in any given part of this photic zone. Highest NPP occurs where runoff from land brings nutrients into shallow coastal waters, and where upwelling of deep waters brings nutrients from the benthic zone to the surface.

45.3B). This also makes physiological sense. Moist soil conditions facilitate water and nutrient uptake by roots and allow plants to maintain pressure within their cells without closing their stomata (leaf pores), thus facilitating CO_2 uptake (see Concept 25.3). The decrease in NPP in extremely moist climates may result from increased cloud cover and lower solar input or from a lack of oxygen in water-saturated soils.

Finally, we also know that elements such as nitrogen, phosphorus, and potassium, which are critical for plant growth (see Figure 25.1), can be in short enough supply to limit NPP in both natural and agricultural ecosystems. Soils in southwestern Australia, for example, are nutrient-poor because most nutrients have been washed (leached) out of the soil over hundreds of millions of years, during which there has been no nutrient-renewing geological uplift. As we saw in Concept 41.3, many plants in this region do not allocate their scarce nitrogen to defensive compounds; instead, they allocate it to proteins that are essential for metabolism and growth.

LINK

The principle of allocation of resources is discussed in **Concept 42.3**

In aquatic ecosystems, NPP is affected more strongly by light and nutrients than by temperature. Photosynthesis is restricted to surface waters where light can penetrate (the photic zone; see Figure 41.13). As a result, NPP is generally higher in surface than in deep waters; the only productive deep-water areas are hydrothermal vents, where chemoautotrophs (see Concept 19.3) generate new biomass by using chemical energy rather than sunlight. But not all surface waters are productive. The open ocean, for example (see Figure 45.1), is unproductive because dissolved nutrients are often limited there—phytoplankton take up surface nutrients quickly and move them to deep waters when they (or the consumers that eat them) die and sink. Aquatic NPP is highest where nutrients are abundant—where rivers and streams discharge nutrients leached from terrestrial ecosystems into coastal marine areas, near-shore zones of lakes, and shallow freshwater wetlands (**FIGURE 45.4**). NPP is also high in areas of upwelling, where deep water from the bottom of an ocean or lake rises to the surface, bringing with it nutrients from benthic sediments.

CHECKpoint CONCEPT 45.1

- ✓ NPP is generally higher in freshwater wetlands (swamps and marshes) than in shallow marine waters (algal beds, coral reefs, and estuaries; see Figure 45.1). Why might this be so?
- ✓ Why do you think NPP in human-cultivated land is lower than that in several natural ecosystem types such as temperate and tropical forest? (See Figure 45.1. Hint: See Concept 44.3.)
- ✓ Outline an experiment that would allow you to determine whether NPP in a grassland ecosystem is limited by the availability of water in the soil.

We now have an overview of global patterns of ecosystem function, as measured by NPP, and some of the factors that cause them. We have seen that nutrient availability plays an important role in these patterns, which makes sense because nutrients are chemical elements that organisms require for their metabolism and as materials to build their bodies. What are the processes that affect the availability of nutrients?

CONCEPT 45.2 Biological, Geological, and Chemical Processes Move Materials through Ecosystems

The sun provides a steady input of radiant energy to Earth. A small amount of this radiant energy is captured by photosynthetic organisms and converted into chemical energy, as we saw in Concept 44.3. This energy fuels the metabolism of all organisms in the community and is ultimately transformed into waste heat. In contrast, the chemical elements that make up organisms come from within the Earth system itself. There is essentially a fixed amount of each element because little matter escapes Earth's gravity or enters the Earth system from space. Whereas Earth is an open system with respect to energy, it is a closed system with respect to matter. This does not mean that the distribution of matter is static, however: energy from the sun and heat from Earth's interior drive biological, geological, and chemical processes that transform matter and move it around the global ecosystem.

The forms and locations of elements determine their accessibility to organisms

Imagine an atom. At any time, this atom occurs in a particular molecular form and occupies a particular location in Earth's closed system. It might be part of a living organism or of dead organic matter. It might be part of a molecule in the atmosphere, in a body of fresh or salt water, or in soil, sediment, or rock. If in rock, it might be part of the surface crust of Earth or more deeply buried. These alternative forms and locations of matter can be thought of as **compartments**. Atoms cycle repeatedly among compartments when they are chemically transformed or physically moved. The atoms in Earth's crust, for example, have gone through repeated cycles of being buried, compressed, heated, uplifted, eroded, and deposited again in sediments (see Figure 18.2). The compartments of the global ecosystem, and the processes that cycle materials among them, can be depicted in a systems diagram (**FIGURE 45.5**; see also Figure 1.6).

The chemical form in which an atom exists determines whether it is accessible to life.

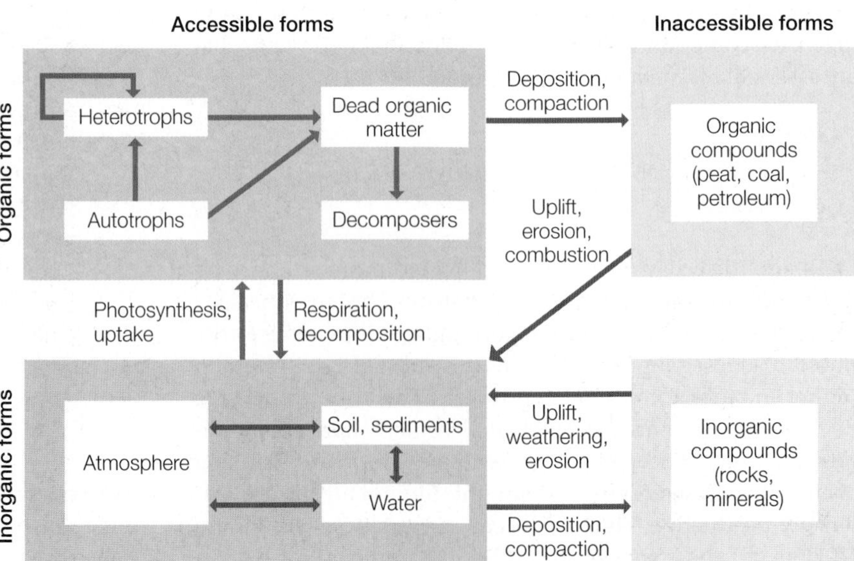

FIGURE 45.5 Chemical Elements and Compounds Cycle among Compartments of the Global Ecosystem The different chemical forms and locations of elements determine whether or not they are accessible to living organisms. These different forms and locations of matter represent compartments (boxes) in the diagram shown here. Biological (green arrows), geological, and chemical processes (brown arrows) cycle matter among the compartments.

With rare exceptions, autotrophs take up carbon in the form of CO_2, and they absorb nutrients such as nitrogen, phosphorus, and potassium in the form of ions dissolved in water (see Concept 25.1). Heterotrophs (including decomposers) extract most of the materials they need from living or dead organic matter. But even when an atom is in the right chemical form, organisms cannot use it if it is in the wrong place. Matter that is buried too deeply, or that occurs in places that are too hot, too cold, or otherwise too hostile, is inaccessible to life.

The processes that supply matter to organisms occur in the **biosphere**, the thin skin at Earth's surface where atmosphere, land, and water are in contact with one another and where organisms live. The biosphere is only about 23 kilometers thick—a mere 1/277 of Earth's radius—extending up to the stratosphere and down to the abyssal zone of the oceans.

Movement of matter is driven by biogeochemical processes

Matter is moved among the compartments of the global ecosystem by processes that are biological, geological, and chemical; thus these movements are called **biogeochemical cycles**. Biological and abiotic chemical reactions combine elements with other elements and compounds, converting them between inorganic and organic forms or between oxidized and reduced states. Many of these reactions involve the abundant elements oxygen and hydrogen. Chemical reactions are also involved in the weathering of rock during soil formation. Physical processes—the circulation of air and water and convective flows within Earth's mantle—move matter through the Earth system, alternately exposing it to air, water, and solar radiation or to heat and pressure.

LINK

Oxidation–reduction ("redox") reactions are described in **Concept 6.1**, weathering and soil formation in **Concept 25.1**, plate tectonic processes in **Concept 18.2**, and circulation of the atmosphere and oceans in **Concept 41.2**

The total amount, or **pool**, of an element or molecule in a given compartment depends on its rate of movement into and out of that compartment. The rate of movement, or **flux**, from one compartment to another is measured in amount per unit of time. If the fluxes into and out of a compartment are not identical, the pool will grow or shrink. Oxidation of organic matter by respiration, wildfires, or fossil fuel combustion, for example, will increase the atmospheric pool of carbon unless photosynthesis and other biogeochemical processes remove carbon from the atmosphere at an identical or greater rate.

All of the materials contained in the bodies of living organisms are ultimately derived from abiotic sources, either atmospheric gases or materials dissolved in water. Primary producers (see Concept 44.3) take up elements from these inorganic compartments, accumulating them as biomass. Trophic interactions (see Concept 43.3) pass those elements on to heterotrophs, which also take in other inorganic materials such as oxygen and water as part of their metabolism. Eventually all of these materials enter the nonliving organic matter compartment. Decomposers break down the dead organic matter into simpler chemical compounds and elements that can be taken up by primary producers. Much of this recycling of materials occurs locally, but some occurs on a larger scale when materials diffuse away in gaseous form or are carried out of the local ecosystem by wind or water.

CHECKpoint CONCEPT 45.2

✓ What is the distinction between the concepts of pool and flux with regard to a given chemical element or compound?

✓ Explain the meaning of "closed system" in the statement that Earth is a closed system with respect to matter.

✓ Draw a systems diagram to show how energy from sunlight enters and flows through the global ecosystem. (Hint: See Concept 6.5 and Figures 44.6 and 45.5.)

We have now seen that the materials that make up living tissue—chemical elements and compounds—cycle through ecosystems. Let's next consider the cycles of three materials that are especially critical for ecosystem function.

CONCEPT 45.3 Certain Biogeochemical Cycles Are Especially Critical for Ecosystems

All biogeochemical cycles can be described in terms of compartments containing the material involved, amounts (pools) of the material contained in each compartment, and rates (fluxes) at which the material enters and leaves those compartments (see Figure 45.5). At the same time, the cycles of different materials vary in their spatial scales and in the processes that drive them. We will illustrate this variation by tracing the cycles for three essential materials: water, nitrogen, and carbon. We will point out how each cycle affects the global ecosystem, particularly the atmosphere, and is in turn affected by human activities.

Water transports materials among compartments

Water is a most remarkable molecule. It makes up more than 70 percent of living biomass. It is the medium for metabolism and the solvent in which biologically accessible forms of many nutrients are dissolved. Precipitation of water as rain or snow transports materials from the atmosphere to Earth's surface. Water weathers rock into soil. And as the primary agent of erosion and transport of dissolved ions and sediment, water is responsible for much of the physical movement of materials around the planet. By virtue of its high heat capacity and its ability to change between solid, liquid, and gaseous states at normal Earth temperatures, water redistributes heat around the planet as it circulates through the oceans and atmosphere. And, as we will soon see, water in the atmosphere is an important player in the global radiation balance.

LINK

The molecular nature of water and its crucial role in the evolution of life as we know it are covered in **Chapter 2**

The global ecosystem contains almost 1.4 billion cubic kilometers of water. About 96.5 percent of this water is in the oceans, which cover 71 percent of the planet's surface—making it obvious why Earth is sometimes called the "blue planet." Other compartments of water include ice and snow (1.76%), groundwater (1.70%), the atmosphere (0.001%), and fresh water in lakes and rivers (surface fresh water; 0.013%). Even though biomass is mostly water, only about 0.0001 percent of Earth's water is found in living organisms.

Some fluxes among these compartments (**FIGURE 45.6**) involve gravity-driven flows of water from the atmosphere to Earth's surface (precipitation) and from land to the oceans and other bodies of water (runoff). Other fluxes, however, are associated with changes in the physical state of water—from liquid to gas (evaporation), from gas to liquid (condensation), between liquid and solid (freezing and melting), and between solid and gas (sublimation and deposition).

The driving force of the water cycle is solar-powered evaporation, which transforms water from liquid to gas and, in the process, moves it into the atmosphere from oceans and other bodies of water, from the soil, and from plants (via transpiration; see Concept 25.3) and other organisms. It takes considerable energy—2.24 kJ (about the energy stored in one AA battery)—to evaporate 1 gram of water. On a global scale, the rate of evaporation is enormous, taking up about one-third of the total solar energy that reaches Earth's surface. This energy is released again as heat when the water vapor condenses to form precipitation. Fluxes of water from Earth's surface to the atmosphere and from the atmosphere to the surface are approximately equal on a global basis, although precipitation exceeds evaporation over land, and evaporation exceeds precipitation over the oceans.

Humans affect the global water cycle when they change the way land is used. Removing vegetation reduces the amount of precipitation that is retained by soil or groundwater and recycled locally, thereby increasing the amount that leaves the local ecosystem as surface runoff. As a result, deforestation, grazing, and crop cultivation all tend to dry out local

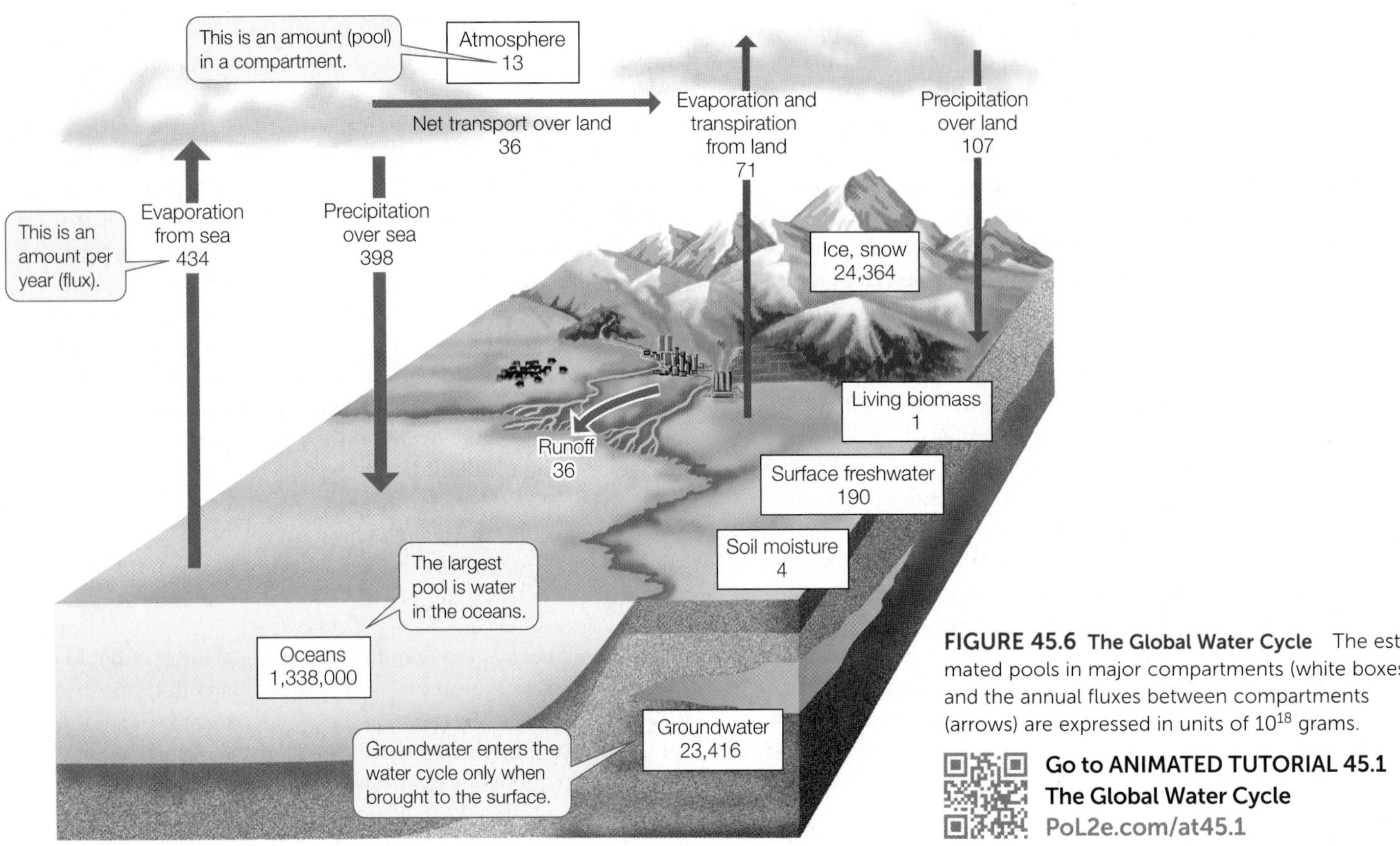

FIGURE 45.6 The Global Water Cycle The estimated pools in major compartments (white boxes) and the annual fluxes between compartments (arrows) are expressed in units of 10^{18} grams.

Go to ANIMATED TUTORIAL 45.1 The Global Water Cycle
PoL2e.com/at45.1

ecosystems. Pumping depletes groundwater, moving it to the surface, where it either evaporates or runs off into streams and, eventually, into oceans.

The water cycle also affects and is affected by the global climate. Warming of the climate is melting polar ice caps and glaciers, increasing the total amount of water in the oceans and causing sea level to rise. With more liquid water comes more evaporation, and thus more water entering the atmosphere and more precipitation globally. Water vapor absorbs infrared radiation—and thus contributes to the greenhouse effect—but it also forms clouds, which reflect incoming sunlight back into space. Scientists are working to predict the net effects on climate of these counterbalancing changes in the water cycle.

Within-ecosystem recycling dominates the global nitrogen cycle

Nitrogen is abundant on Earth, yet it is often in short supply in biological communities. Unlike water—whose availability to organisms depends only on where it occurs and its physical state—nitrogen's accessibility to life depends not only on where it occurs, but also on what other elements it is attached to. As a result, the nitrogen cycle, unlike the water cycle, involves chemical transformations (**FIGURE 45.7**).

The gas N_2 constitutes about 78 percent of the molecules in Earth's atmosphere, but most organisms cannot use nitrogen in this form because they cannot break the strong triple bond between the two nitrogen atoms. The exceptions are certain microbes that break the triple bond and attach hydrogen to make ammonium (NH_4^+)—which is usable by other organisms—through a metabolic process called **nitrogen fixation**. In terrestrial ecosystems, nitrogen fixation is carried out mostly by free-living bacteria in the soil and by symbiotic bacteria associated with plant roots, but it also occurs in other places, such as the fungal gardens of leaf-cutter ants described in the opening story of Chapter 43 (**FIGURE 45.8**). Analogous biochemical processes in aquatic ecosystems are carried out by free-living microorganisms and by microbial symbionts of phytoplankton and other aquatic organisms. A small amount of nitrogen is fixed in the atmosphere by lightning. Diverse other microbial species in terrestrial and aquatic ecosystems convert ammonium into other accessible forms of nitrogen such as nitrate (NO_3^-). Nitrogen fixation is reversed by another group of microbes in a process called **denitrification**, which ultimately completes the cycle by returning N_2 gas to the atmosphere.

LINK

The metabolic processes by which microbes transform nitrogen are described in **Concepts 19.3 and 25.2**

Ammonium and nitrate are accessible to autotrophs because they dissolve in water and thus are easily taken up through cell membranes. Heterotrophs obtain nitrogen by breaking proteins down into amino acids, which can also be transported through cell membranes. Much of the nitrogen in terrestrial ecosystems is recycled locally as decomposition of dead organic matter releases accessible nitrogen into the soil. Some nitrogen, however, is lost to the atmosphere when biomass burns, or is leached from soils and carried away in streams, ultimately into lakes

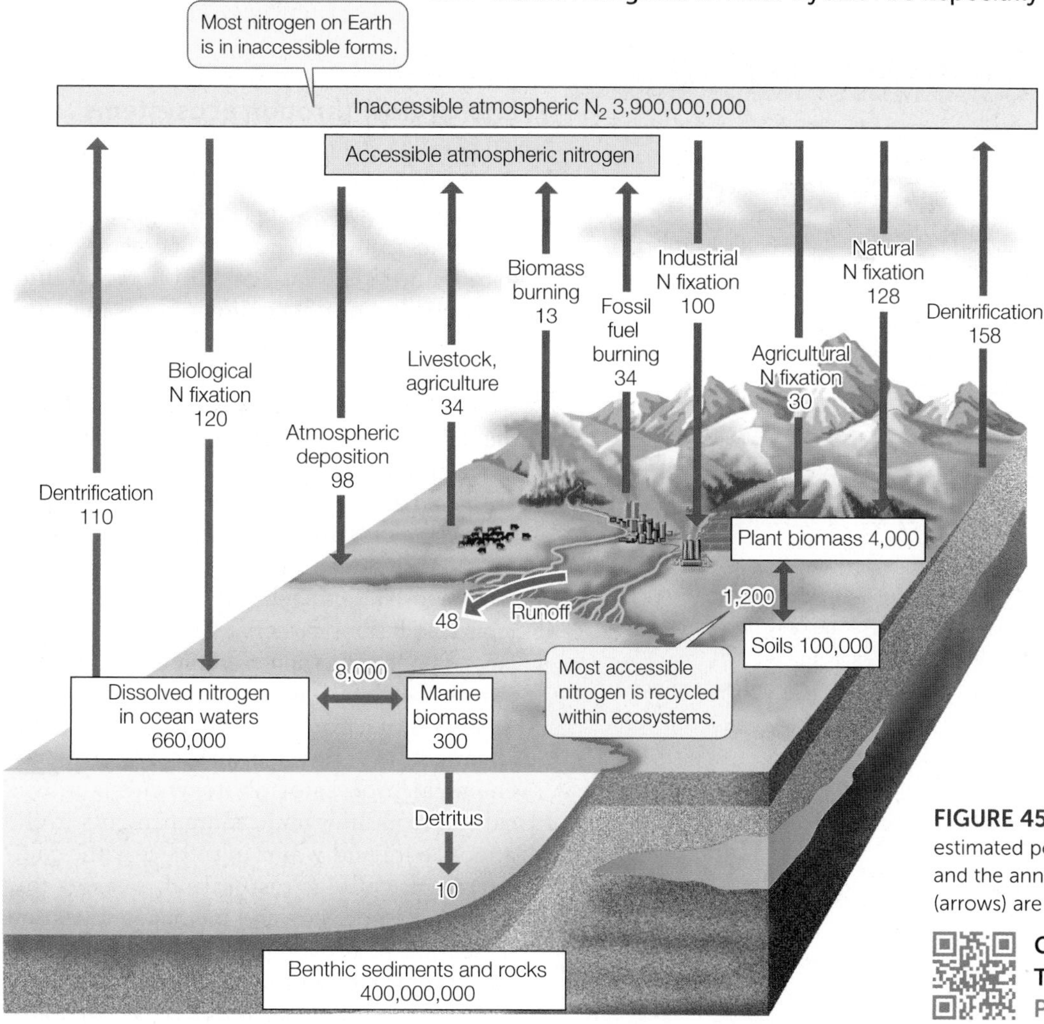

FIGURE 45.7 The Global Nitrogen Cycle The estimated pools in compartments (white boxes) and the annual fluxes between compartments (arrows) are expressed in units of 10^{12} grams.

Go to ANIMATED TUTORIAL 45.2
The Global Nitrogen Cycle
PoL2e.com/at45.2

and oceans. In aquatic systems, primary production occurs in surface waters, and some nitrogen is recycled there. However, much of it is recycled at greater depths in the water column as organisms in deep waters intercept and consume sinking organic matter. The nitrogen in this organic matter eventually accumulates in sediments. Some of it is returned to the surface by upwelling or, eventually, by tectonic processes followed by weathering of exposed rock.

Human activities are affecting nitrogen fluxes and pools. Cultivation of wetland crops (such as rice), livestock raising, and burning of plant material, coal, and petroleum release oxides of nitrogen and ammonia into the atmosphere. These molecules contribute to atmospheric smog and acid rain, and one of them, nitrous oxide (N_2O), absorbs infrared radiation. When they are eventually deposited on Earth's surface, sometimes far from their source, they can add as much nitrogen "fertilizer" as farmers place on their crops. Humans also alter the atmospheric pool of N_2 through the industrial process that fixes nitrogen to manufacture fertilizer and explosives. The rate of industrial fixation (see Concept 25.1) plus the rate of fixation by legume crops such as soybeans now adds up to more than the rate of natural terrestrial nitrogen fixation (see Figure 45.7). Transport of topsoil and dissolved nitrogen from fertilized croplands and deforested areas by wind or water runoff is increasingly exporting nitrogen from terrestrial to aquatic ecosystems.

Increased nitrogen inputs can increase primary production, but too much of a good thing can be a bad thing. In aquatic systems, nutrient-stimulated increases in primary productivity—a phenomenon called **eutrophication**—can result in rapid growth of phytoplankton (including cyanobacteria; see Figure 19.9C) and algae. Respiration by these primary producers and by the decomposers that process their dead bodies can reduce oxygen levels below the tolerances of many aquatic organisms, including fish and crustaceans. Eutrophication can lead to "dead zones" devoid of oxygen-requiring aquatic life, like that found where the Mississippi River discharges nutrient-rich agricultural runoff into the Gulf of Mexico (**FIGURE 45.9**). In nutrient-poor terrestrial ecosystems, excess nitrogen can change the species composition of plant communities. Species that are adapted to low nutrient levels grow slowly, even when fertilized, and thus are displaced by faster-growing species that can take advantage of the additional nutrient supply. In the Netherlands, nitrogen deposition from upwind industrial and agricultural regions has caused species-rich plant communities to give way to species-poor communities, contributing measurably to a recent loss of plant species diversity there.

Go to MEDIA CLIP 45.1
Tracking Dead Zones from Space
PoL2e.com/mc45.1

INVESTIGATION

FIGURE 45.8 Where Does the Extra Nitrogen Come From? Leaf-cutter ants require more nitrogen than is contained in the fresh leaves that they supply to their fungal gardens. Perhaps the fungus concentrates nitrogen from the leaves into the special fungal structures that the ants eat. If so, then spent leaf material in the ants' refuse dumps (see Figure 43.12) should be lower in nitrogen than fresh leaves—but the opposite turns out to be true. One possible explanation for this enrichment of the discarded leaf material is that the ants eat protein-rich insects as well as the fungus and fertilize the fungal garden with their feces. Another is that the fungus absorbs nitrogen from the soil. A third possibility is that the fungus absorbs nitrogen from nitrogen-fixing organisms in the nest. Pinto-Tomás and colleagues tested this third hypothesis.[a]

HYPOTHESIS

Nitrogen-fixing organisms in leaf-cutter ant nests supply nitrogen to the ants.

METHOD

1. Bring ant colonies into the laboratory and allow them to function in an environment with no other insects and no soil (that is, where the only non-atmospheric source of N is fresh leaves supplied by the researchers).
2. Measure the nitrogen content of leaves, fungus garden, ants, and leaf refuse.

RESULTS

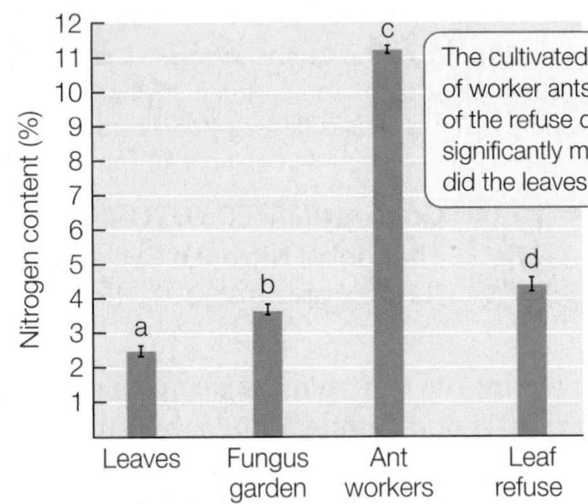

Error bars indicate 1 standard error from the mean; bars with different letters are statistically different from one another. (See Appendix B for discussion of statistical concepts.)

CONCLUSION

Nitrogen is being fixed within the ant nest.

ANALYZE THE DATA

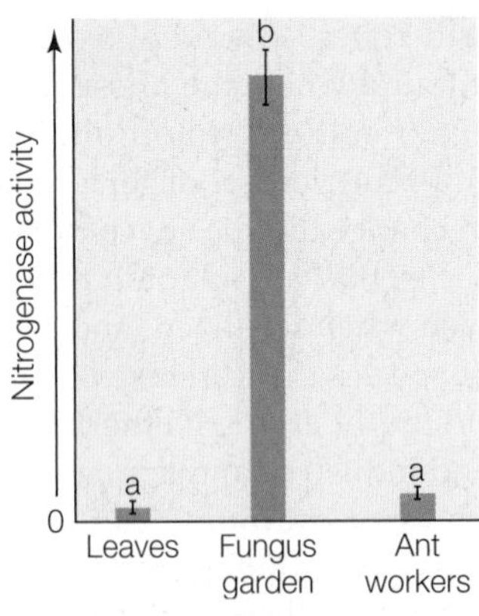

The researchers measured activity of nitrogenase, an enzyme in the bacterial metabolic pathway that fixes atmospheric nitrogen. The results are shown in the figure to the left. Error bars indicate 1 standard error from the mean; different letters indicate significant differences in nitrogenase activity.

A. Where in the nest is nitrogen fixation occurring?

B. How can the ant workers have a higher nitrogen content than the fungus garden (see figure above)?

C. Can the researchers now conclude that the extra nitrogen comes from nitrogen-fixing bacteria living in the fungus gardens?

Go to **LaunchPad** for discussion and relevant links for all **INVESTIGATION** figures.

[a]A. A. Pinto-Tomás et al. 2009. *Science* 326: 1120–1123.

Movement of carbon is linked to energy flow through ecosystems

All the macromolecules that make up living organisms contain carbon, and much of the energy that organisms use to fuel their metabolic activities comes from the oxidation of organic carbon compounds. The movement of carbon into, through, and out of communities is therefore intimately linked to energy flow through ecosystems, and biomass is an important compartment of the carbon cycle (**FIGURE 45.10**). Most of the carbon in the atmosphere occurs as the gases CO_2 (carbon dioxide) and CH_4 (methane), which are mixed on a global scale by the solar-powered circulation of Earth's atmosphere (see Concept 41.2). These same forms are found dissolved in the oceans and in fresh water, and aquatic systems also harbor carbonate (CO_3^{2-}). By far the largest pools of carbon occur in fossil fuels and in rocks containing carbonate compounds.

Fluxes of carbon are driven by biological, chemical, and physical processes. The biochemical process of photosynthesis moves carbon from inorganic compartments in the atmosphere and water into the organic compartment; cellular respiration reverses this flux. Carbon dioxide and methane move from their atmospheric compartments into the much larger compartments in the oceans when they dissolve, and they are returned to the atmosphere by outgassing; both of these movements are purely physical processes. The rate at which CO_2 dissolves is slightly greater than the rate at which it is outgassed, for two reasons. First, some of the dissolved CO_2 is converted into organic compounds by marine primary production. Most of this carbon is then recycled through the trophic interactions of aquatic organisms, but gravity moves a steady rain of nonliving organic matter into sediments in the benthic zone, where geological processes transform it eventually into **fossil fuels** such as coal, natural gas, and petroleum. Second, some of the dissolved CO_2 is incorporated into relatively insoluble carbonate compounds, especially calcium carbonate. These also ultimately reach the sediments and form carbonate rocks such as limestone. The chemical reactions involved in carbonate formation are inorganic, but they are orchestrated by marine organisms when they form their shells (see Figure 20.10).

Human activities influence the carbon cycle in a number of ways. Anything that changes primary productivity, such as nitrogen deposition or altered land use, alters the movement of carbon between inorganic and organic compartments. Any activity that affects water

Mississippi
Alabama
Oxygen level
Texas
Louisiana
Low
High
Anoxia—low oxygen levels—results when eutrophication occurs and algal populations increase dramatically. Their decompositon uses up all available oxygen.

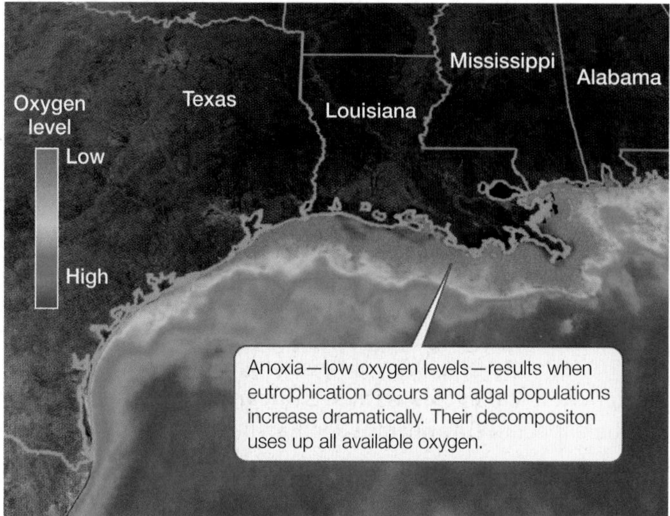

FIGURE 45.9 High Nutrient Input Creates Dead Zones Runoff from agricultural lands in the midwestern United States carries nitrogen and phosphorus down the Mississippi River and into the Gulf of Mexico. The resulting eutrophication creates an oxygen-poor "dead zone" that few aquatic organisms other than microbes can tolerate.

APPLY THE CONCEPT

Certain biogeochemical cycles are especially critical for ecosystems

Figure 45.10 depicts the global carbon cycle, showing organic and inorganic pools of carbon in various compartments as well as fluxes. Refer to that figure to answer the following questions.

1. What percentage of the flux of carbon from terrestrial ecosystems into the atmosphere is due to human activities?
2. What other flux of carbon from terrestrial ecosystems is being influenced by humans, and how?
3. Given the biogeochemical cycles we have discussed so far, which additional two cycles of materials would you wish to explore next in order to understand ecosystem function, and why? (Hint: Consider the major chemical constituents of living tissue, discussed in Concept 2.1.)

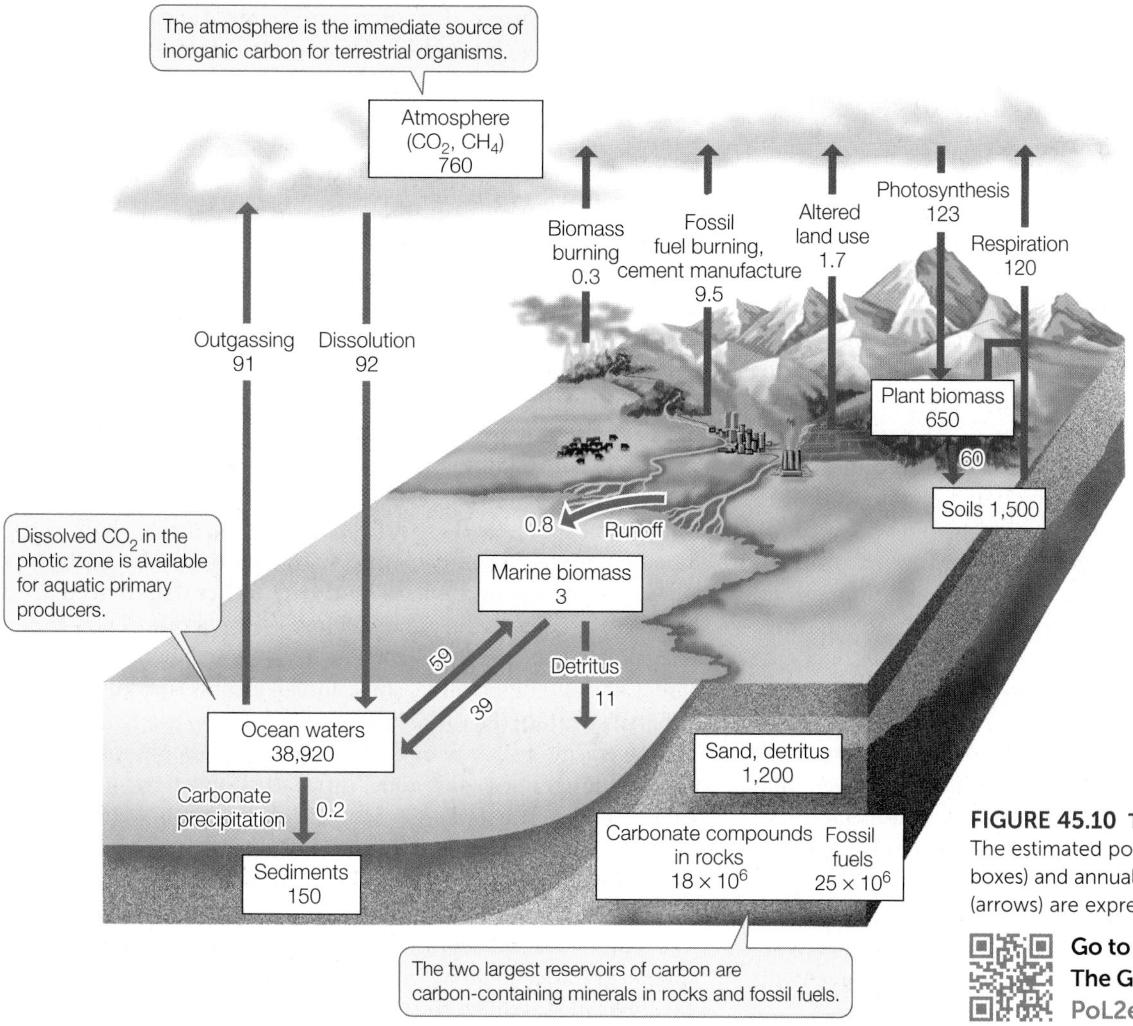

The atmosphere is the immediate source of inorganic carbon for terrestrial organisms.
Atmosphere (CO_2, CH_4) 760
Biomass burning 0.3
Fossil fuel burning, cement manufacture 9.5
Altered land use 1.7
Photosynthesis 123
Respiration 120
Outgassing 91
Dissolution 92
Plant biomass 650
60
Soils 1,500
0.8
Runoff
Dissolved CO_2 in the photic zone is available for aquatic primary producers.
Marine biomass 3
59
39
Detritus
11
Ocean waters 38,920
Sand, detritus 1,200
Carbonate precipitation
0.2
Sediments 150
Carbonate compounds in rocks 18×10^6
Fossil fuels 25×10^6
The two largest reservoirs of carbon are carbon-containing minerals in rocks and fossil fuels.

FIGURE 45.10 The Global Carbon Cycle The estimated pools in compartments (white boxes) and annual fluxes between compartments (arrows) are expressed in units of 10^{15} grams.

Go to ANIMATED TUTORIAL 45.3 The Global Carbon Cycle
PoL2e.com/at45.3

runoff, such as deforestation or impoundment or alteration of river flows, affects the movement of carbon between the terrestrial and aquatic compartments. Burning of organic matter, whether biomass or fossil fuels, increases the atmospheric pool of CO_2, as does the heating of calcium carbonate in the manufacture of cement. The atmospheric pool of methane (CH_4) is increased by livestock production, rice cultivation, and water storage in reservoirs, because microbes in the guts of cattle and in waterlogged sediments break down organic compounds anaerobically to produce methane gas. Although the atmospheric pool of CH_4 is far smaller than that of CO_2, both are potent absorbers of infrared radiation and affect Earth's radiation balance, as we will soon see.

Biogeochemical cycles are not independent

The water, nitrogen, and carbon cycles illustrate the diversity of processes, relative sizes of pools and fluxes, and spatial and temporal scales that are involved in the cycling of materials within the Earth system. As you can imagine, there is even more variation if we consider other materials with different chemical and physical properties. For example, few forms of phosphorus are gases, and the phosphorus cycle lacks a significant atmospheric component. Similarly, sulfur released into the atmosphere by volcanic eruptions reacts with other gases to form tiny, airborne particles that affect Earth's radiation balance (See Concept 45.4).

Go to ACTIVITY 45.1 The Phosphorus and Sulfur Cycles PoL2e.com/at45.1

This diversity does not mean that different biogeochemical cycles are entirely independent of one another, however. The cycles of many materials are at least partly linked, for several reasons.

First, the fluxes of different materials are linked if those fluxes share common physical or chemical processes. For example, many elements, including carbon, nitrogen, phosphorus, and sulfur, have water-soluble forms, which are transported together in precipitation and runoff. If runoff increases, the fluxes of all water-soluble materials from terrestrial to aquatic compartments will increase.

Second, the fluxes of different materials through communities are linked because the functional macromolecules in living tissue—proteins, nucleic acids, polysaccharides, and so forth—have a precise structure with fixed proportions of elemental building blocks. Thus when biomass is produced, there is coordinated movement of these building blocks into living organic compartments, and when biomass is decomposed or burned, they all move back into inorganic compartments.

Fluxes can also interact in more subtle ways. Increased atmospheric concentrations of CO_2, for example, can increase the water-use efficiency of many terrestrial plants. These plants obtain the CO_2 they need for photosynthesis by opening their stomata, which also leads to transpiration—the loss of water vapor from their tissues. In a high-CO_2 environment, these plants can leave their stomata open for a shorter time and therefore transpire less water, which slows the rate of water movement from soil to the atmosphere.

CHECKpoint CONCEPT 45.3

- ✓ How is the cycling of water important to the cycling of other materials?
- ✓ Why is nitrogen often a limiting nutrient for plant growth, even though it is the most abundant element in Earth's atmosphere?
- ✓ Explain why the rate with which CO_2 outgasses from oceans and fresh waters is lower than the rate with which it dissolves into those waters.

Biogeochemical cycles with significant atmospheric pools can be especially important for Earth's climate, as we will see next.

CONCEPT 45.4 Biogeochemical Cycles Affect Global Climate

Concept 41.2 discussed how latitudinal and seasonal variations in the input of solar radiation drive global climate patterns. Earth's climate is also influenced by the particular mix of gases that make up the atmosphere. As we saw in the opening story of this chapter, scientists have long known that some of these gases—particularly water vapor and carbon dioxide—absorb infrared radiation. Interactions of infrared-absorbing molecules with solar radiation have a profound effect on the radiation budget of the planet.

Earth's surface is warm because of the atmosphere

All objects that are warmer than absolute zero—our sun included—emit electromagnetic radiation (see Figure 6.16). The wavelengths emitted by an object depend on its temperature: hotter objects radiate at shorter, more energetic wavelengths. Because the surface of the sun is so hot—about 6,000°C—its radiation distribution peaks at wavelengths visible to the human eye. What happens to the incoming photons? Approximately 23 percent of them hit aerosols (tiny airborne particles and water droplets or ice crystals in clouds) and gas molecules in the atmosphere and are reflected back into space. Another 8 percent are reflected back into space when they hit Earth's surface. About 20 percent are absorbed, rather than reflected, by gas molecules in the atmosphere. Finally, about 49 percent of incoming solar energy (168 watts per square meter, averaged across day and night, all seasons, and all locations) is absorbed at Earth's surface (**FIGURE 45.11**).

The warmed surface of Earth in turn reemits photons, but at much longer, less energetic, infrared wavelengths (wavelengths beyond the red end of the visible spectrum, which we perceive as heat). Some of this infrared radiation is absorbed by greenhouse gas molecules rather than escaping immediately into space. Those molecules reradiate infrared photons back to Earth's surface, keeping the energy within the lower atmosphere. The additional warming of Earth caused by greenhouse gas molecules is called the **greenhouse effect**.

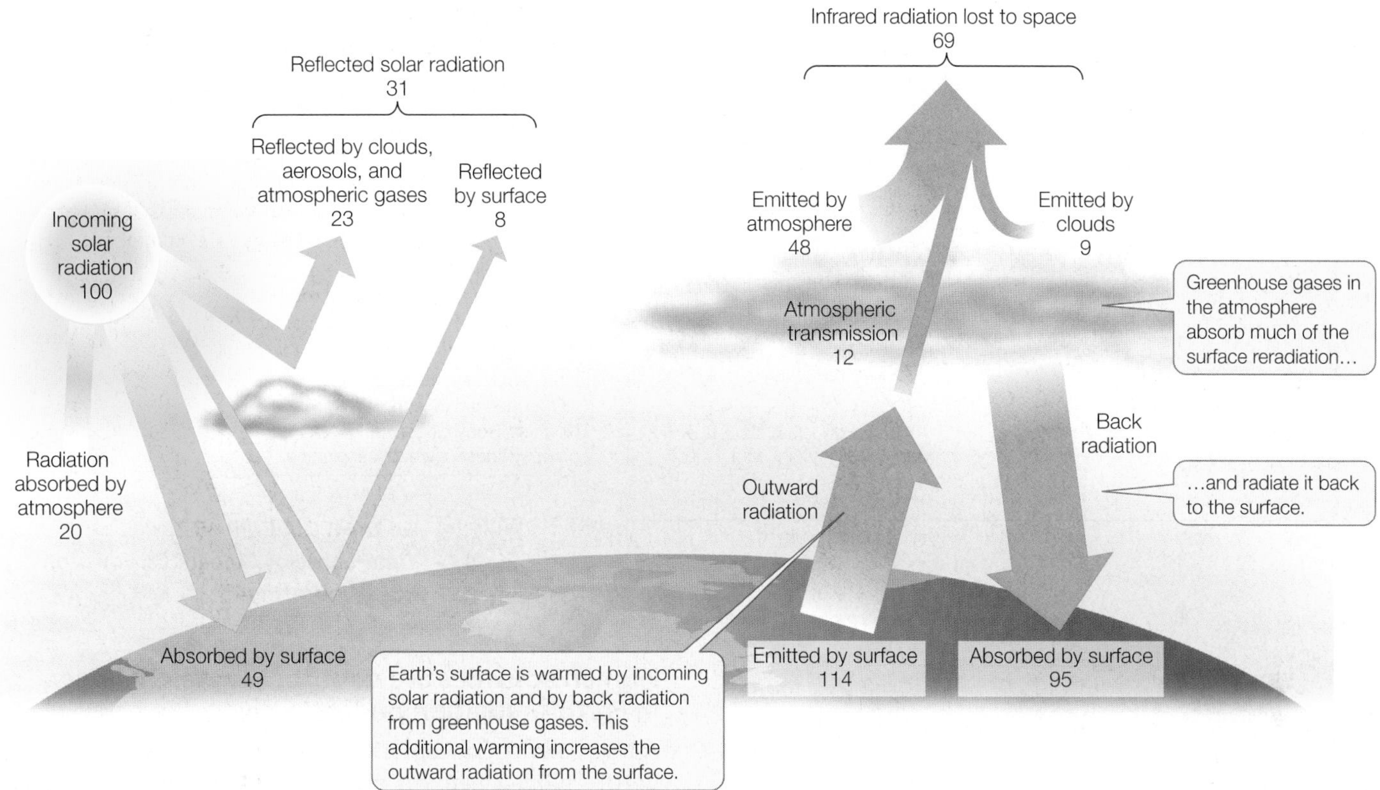

FIGURE 45.11 Earth's Radiation Budget The average annual energy gain from incoming short-wave solar radiation that is absorbed by Earth's surface and atmosphere (left-hand portion of figure) is balanced by the average annual loss to outer space in the form of long-wave (infrared) radiation (right-hand portion of figure). Greenhouse gases in the lower atmosphere absorb some of the infrared radiation emitted by Earth's surface and radiate it back to the surface, adding to the warming caused by solar radiation. This "greenhouse effect" is sufficiently strong that the total infrared radiation emitted by the surface exceeds the average flux of incoming solar radiation at the top of the atmosphere (342 watts/m^2 averaged across day and night at all locations and over all seasons). All fluxes (arrows) are given as percentages of this average.

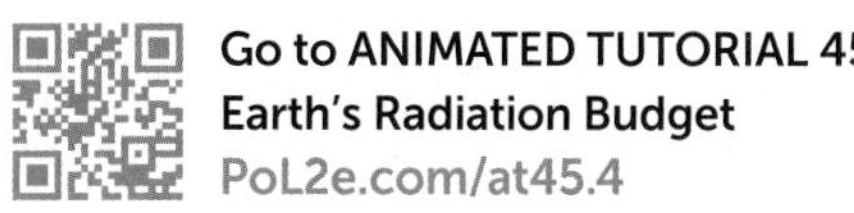
Go to ANIMATED TUTORIAL 45.4
Earth's Radiation Budget
PoL2e.com/at45.4

Water vapor, carbon dioxide, methane, nitrous oxide, and other molecules that absorb infrared radiation are referred to as **greenhouse gases**. The pools of these molecules in the atmosphere influence how much energy is retained in the lower atmosphere. Without its atmosphere, Earth's average surface temperature would be a frigid –18°C, about 34°C colder than it is at present. The greenhouse effect influences any planet with an atmosphere, but its magnitude depends on the amounts and types of materials in the atmosphere. Mars's thin atmosphere warms its surface by only 3°C, whereas the thick atmosphere of Venus warms that planet's surface by a whopping 468°C.

Recent increases in greenhouse gases are warming Earth's surface

Improved methods for measuring the content of air samples accurately—such as Dave Keeling's precise methods for measuring CO_2—show us that the composition of the atmosphere is changing. The measurements initiated by Keeling reveal a steady increase in atmospheric CO_2 over the last half century (**FIGURE 45.12A**). Analyses of air trapped in glacial ice demonstrate that CO_2 and other greenhouse gases have been increasing even longer—since about 1880 (**FIGURE 45.12B**). Average annual global temperatures have followed suit (**FIGURE 45.13**).

This **global warming** is affecting patterns of climate worldwide. A warmer Earth means not only hotter air temperatures, but also more intense fluxes of water, with greater overall evaporation and precipitation. Hadley cells (see Concept 41.2) are expected to expand poleward because warmer tropical air will rise higher in the atmosphere and expand farther toward the poles before sinking. This expansion should cause precipitation to increase near the equator and at high latitudes and to decrease at middle latitudes. Global warming is uneven across Earth's surface—for instance, the tropical Pacific and western Indian Oceans are warming more than other waters, and high latitudes are warming faster than low latitudes. As a result, changes in precipitation will be season- and region-specific, but in general, wet regions are expected to get wetter and dry regions drier. Precipitation trends over the twentieth century appear to support these expectations (**FIGURE 45.14**). Global warming is also predicted to increase storm intensity and weather variability. Recent analyses indicate that hurricane numbers have not increased, but strong hurricanes (category 4 and 5) and "megastorms" such as 2012's Sandy have become more frequent since the 1970s.

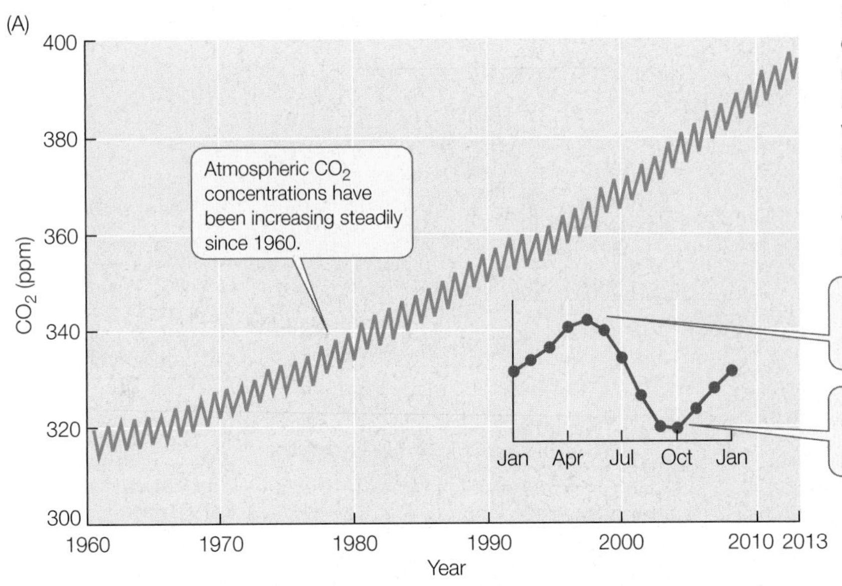

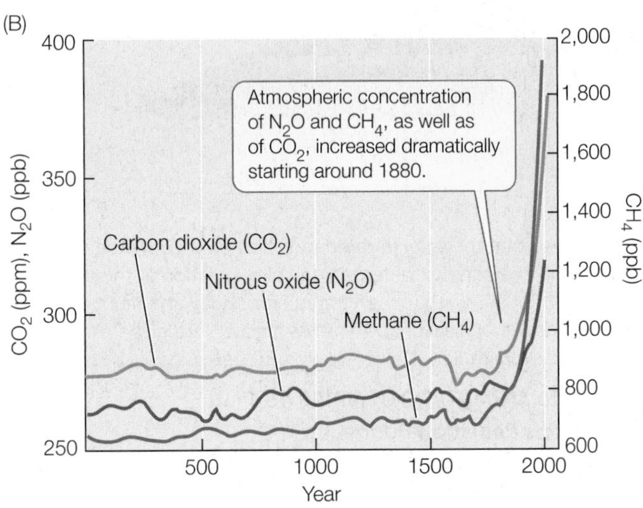

FIGURE 45.12 Atmospheric Greenhouse Gas Concentrations Are Increasing (A) The Keeling curve shows measurements of atmospheric CO_2 concentrations atop Mauna Loa, Hawaii, expressed as parts per million (ppm) by volume of dry air. (B) Measurements of air trapped in glacial ice have allowed researchers to extend our record of greenhouse gas concentrations back 2,000 years. Concentrations are expressed as parts per million (ppm) for CO_2 and parts per billion (ppb) for nitrous oxide (N_2O) and methane (CH_4).

Go to MEDIA CLIP 45.2
Time-Lapse of Global CO_2 Emissions
PoL2e.com/mc45.2

Human activities are contributing to changes in Earth's radiation budget

As we saw in Concept 45.3, human activities are affecting Earth's biogeochemical cycles. Fossil fuel burning, cement manufacture, and forest clearing are adding CO_2 to the atmosphere; expanded livestock production, water impoundment, and wetland crop cultivation are adding CH_4 and N_2O through anaerobic microbial metabolism. Other human activities are also influencing Earth's radiation budget. Expansion of livestock grazing, agriculture, and offroad vehicle traffic in arid regions is increasing airborne dust. Dust and dark-colored soot particles ("black carbon") from fossil fuel burning that are deposited on snow and ice increase absorption of incoming solar energy. The result is faster melting of persistent ice fields such as the Greenland ice sheet and Arctic sea ice, as well as advanced timing of spring snowmelt in temperate regions.

Other human activities have the potential to counteract these effects. Increasing amounts of aerosols in the atmosphere and land clearing, for example, increase the amount of solar radiation that is reflected. When climate scientists incorporate all of these effects into quantitative computer models of Earth's climate, however, they conclude that human activities are contributing significantly to global warming and all the aspects of climate change that go with it (see Apply the Concept, p. 930).

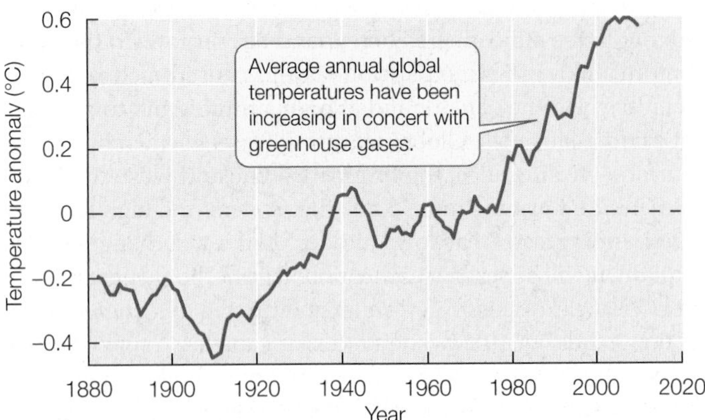

FIGURE 45.13 Average Annual Global Temperatures Are Increasing The curve shows the deviation from the average global temperature for the period 1951–1980 (indicated by 0).

CHECKpoint CONCEPT 45.4

- ✓ What factors caused the dramatic increases in carbon dioxide and methane in the atmosphere that became apparent by the late 1800s? (See Figure 45.12B. Hint: See also Figure 42.9.)
- ✓ Is the greenhouse effect something that occurs without humans, or is it caused by humans?
- ✓ Why is global warming linked to a greater intensity of storms? (Hint: Consider Concept 41.2.)

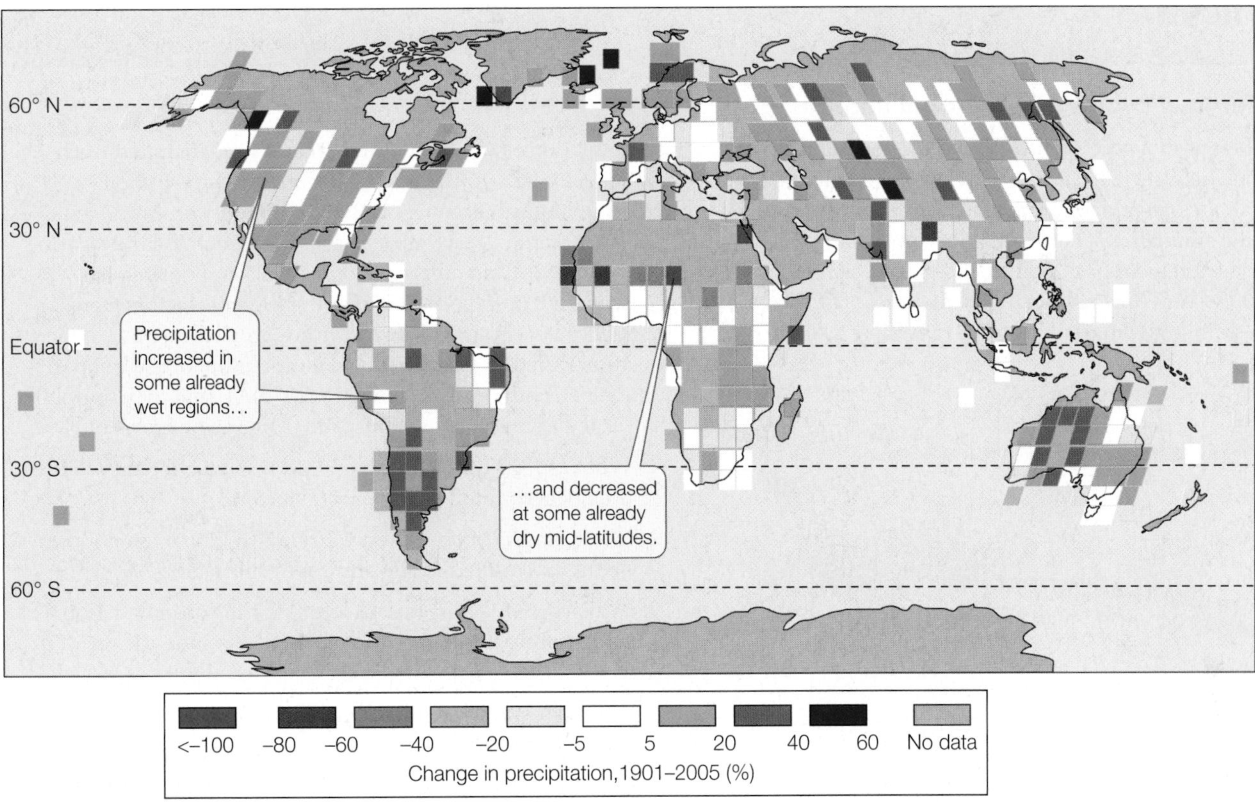

FIGURE 45.14 Global Precipitation Patterns Have Changed Over the twentieth century, precipitation increased at high latitudes and in some tropical regions, but decreased at some already dry mid-latitudes. Note, however, that much regional variation is apparent.

Now that we have seen how human activities are changing Earth's radiation budget and climate, we can consider how ecological systems are responding.

CONCEPT 45.5 Rapid Climate Change Affects Species and Communities

Climate change is not expected to affect all ecosystems in the same way because the changes are not distributed evenly around the globe or across the seasons. As we have seen, some parts of the planet are getting drier while others are getting wetter (see Figure 45.14), and some parts are warming much faster than others. In all cases, however, recent climate change has been far more rapid than what organisms have experienced in their evolutionary histories.

Rapid climate change presents ecological challenges

The life histories of organisms are shaped by their environment, as we saw in Concept 42.3. Thus organisms that live in seasonal environments have adaptations to ensure that crucial life events occur at favorable times of the year. Alpine plants, for example, start growing when the snow melts; seeds of desert plants germinate in response to certain combinations of soil temperature and moisture; changes in day length often trigger plant flowering, bird migration, and onset or cessation of insect dormancy. But climate change is altering the relative timing of many such environmental cues: in a warmer world, the day length that once signaled the start of good growing conditions may instead indicate the middle of the growing season. Although species' responses can evolve by natural selection, the rate of such evolution may be too slow, or available genetic variation may be too small, to keep up with an environment that changes too rapidly or that continues to change directionally for too long.

Over the short term, we have evidence that many species are continuing to track environmental changes so that events in their life histories remain within appropriate periods. For example, biological events that mark the end of the winter season in temperate ecosystems—such as the leafing out of trees—have generally been taking place at earlier dates as the climate has warmed. In Britain, the first flowers of a sample of 385 plant species have appeared, on average, 4.5 days earlier per 1°C increase in average temperature. However, this is an *average* value for advances in flowering, and not all species have responded in the same way—some did not shift their time of first flowering, and some now flower later in the spring rather than earlier. Thus some species may not be responding adaptively to climate change at all. Even for those that do, it is unclear that they can continue to track climate change over the long term.

Changes in seasonal timing can disrupt interspecific interactions

Not all environmental cues respond to the same degree, or in the same direction, to climate change, so relationships among

APPLY THE CONCEPT

Biogeochemical cycles affect global climate

The graphs below show the results of computer simulations of Earth's climate between 1850 and 2000. The blue-shaded regions indicate the range of average global temperatures produced by running the simulations many times (encompassing 95% of the outputs from the replicate simulations). The actual observed temperatures are shown by the red lines. The three graphs represent three different sets of assumptions used in the simulations: (A) only natural factors were included, such as solar radiation input, aerosols from volcanic eruptions, and oscillations in oceanic circulation; (B) only anthropogenic (human-generated) factors were included, such as land clearing, soot deposition, and emissions of aerosols and greenhouse gases; and (C) both natural and anthropogenic factors were included. Average global temperatures are expressed as deviation from the average global temperature of the last part of the nineteenth century (indicated by 0).[a] Using these graphs, answer the following questions, and explain your reasoning.

1. What aspects of the observed trends in global temperature are not accounted for by natural factors?
2. What aspects of the observed trends in global temperature are not accounted for by anthropogenic factors?
3. Which set of factors best accounts for the observed global temperature trends?
4. The arrows in panel (C) indicate five major volcanic eruptions. What effect did these natural events have on global temperatures?

[a]www.ipcc.ch/publications_and_data/ar4/wg1/en/contents.html

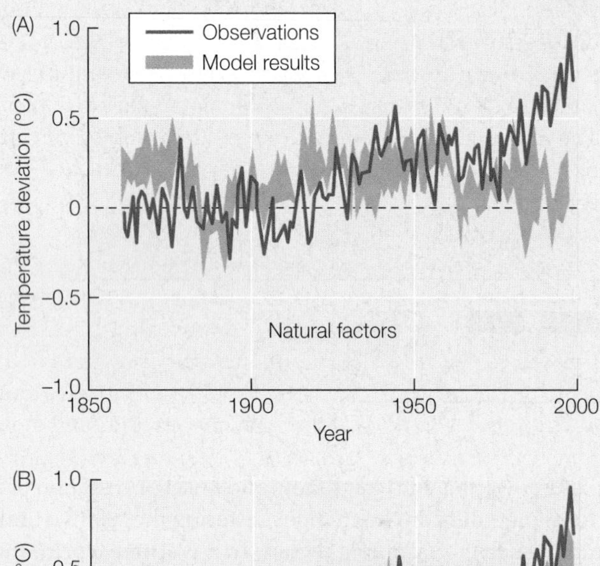

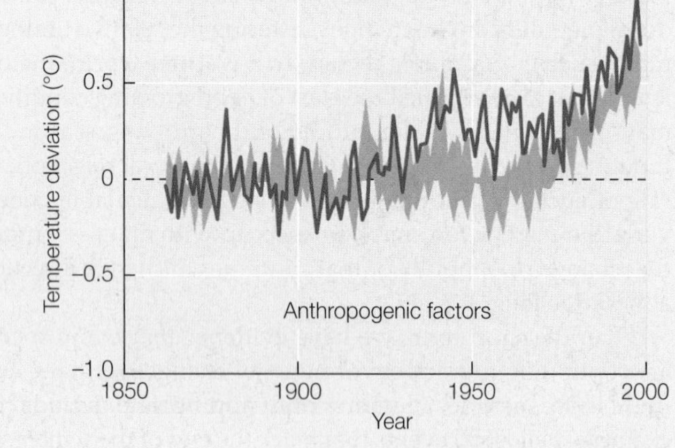

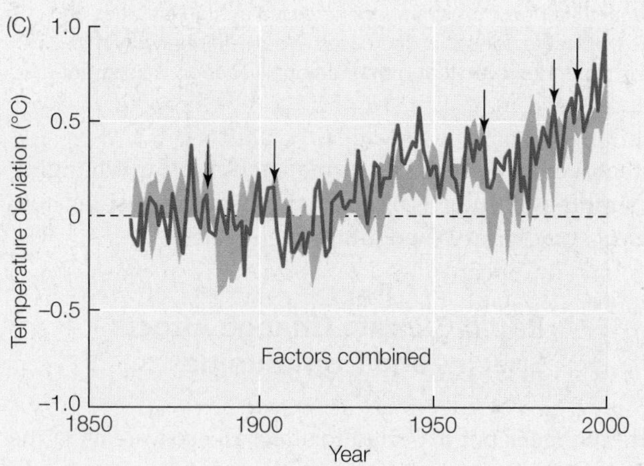

cues are shifting. Because different species make use of different cues, they may respond differently to climate change, as the different British plant species have done. This variation raises the possibility of mismatches in the timing of life history events among species that interact with one another.

Such mismatches are now becoming apparent in nature. In the Netherlands, for example, eggs of the winter moth (*Operophtera brumata*) hatch when temperatures warm after a fixed number of frosts, whereas the leaves of the oak trees on which the winter moth caterpillars feed (*Quercus robur*) emerge when warm days follow a certain number of cool days. The eggs have been hatching earlier, and oak leaves have been appearing earlier, in response to the earlier onset of warm temperatures in spring. The caterpillars do not use the same cues as the oaks, however, and they have been hatching too early and then starving. Great Tits (*Parus major*), which feed on winter moth caterpillars, are not nesting earlier because they use day length as the cue for egg laying (**FIGURE 45.15**). These birds now have less food for their nestlings not only because there are fewer caterpillars overall, but also because they are now feeding their nestlings after the peak in caterpillar abundance has passed. They are experiencing lower fecundity as a result.

Climate change can alter community composition by several mechanisms

If local populations cannot accommodate changing environmental conditions, they will go extinct, and their extinctions will change the species composition of their communities. As we saw in Figure 42.6, *Sceloporus* lizards in the Yucatán of Mexico are experiencing hotter temperatures during the spring

FIGURE 45.15 Climate Change Affects Life Histories A long-term study in the Netherlands shows that warming temperatures have affected the seasonal cycle of the winter moth *Operophtera brumata*, whose caterpillars are now emerging before the leaves on which they feed have expanded, causing more caterpillars to starve. Great Tits (*Parus major*) are not laying eggs earlier. They now have difficulty finding enough caterpillars to feed their nestlings because they have missed the time of peak caterpillar abundance, and there are fewer caterpillars overall.

breeding season, and this warming restricts the time during which they can forage. The lizards have gone locally extinct where they have insufficient feeding time to support reproduction. Extrapolating to lizards worldwide, researchers predict that continued global warming will place most lizard populations at risk of extinction by the year 2080. Losses of lizards may affect populations of their insect prey, of the plants those insects feed on, and of other animals that also eat the insects. As we saw in Concept 44.3, loss of diversity in communities can lead to overall loss of community function.

Altered relationships between environmental cues can also change communities even if species do not go locally extinct. In southwestern North America, the winter rainy season now arrives later in the fall, after soil temperatures have cooled. This shift favors plant species whose seeds use cool-season moisture as the cue to germinate, at the expense of species whose germination is cued by warm-season moisture. Thus the relative abundance of the cool-season species is increasing in these communities.

Climate-caused shifts in environmental gradients can also lead to the assembly of novel communities (see Concept 44.2). Some species have moved up mountains as their low-elevation habitats become too warm and higher-elevation habitats become more favorable. Others have moved toward higher latitudes for similar reasons; such poleward shifts have been documented for numerous taxonomic groups. Different species shift to different degrees or not at all (as is true of shifts in timing of life history events), so the mixture of species at any place changes. These distributional shifts also influence species interactions. Warmer temperatures in western North America, for example, are speeding up the life history of the mountain bark beetle (*Dendroctonus ponderosae*) and decreasing its winter mortality rate. As a result, beetle populations have increased and expanded, causing widespread tree mortality in pine and spruce forests from New Mexico to Alaska. Loss of the conifers has cascading effects on the consumers that depend on them, as well as on water runoff, wildfire frequency, and NPP.

Extreme weather events also have an impact

One of the predictions of climate models is that the frequency of extreme weather events will increase. Severe storms, flooding, droughts, heat waves, and other such events can elicit sudden shifts in species distributions and in community composition. For example, a brief but intense drought in the 1950s caused a ponderosa pine forest in northern New Mexico to shrink abruptly, and drought-adapted piñon–juniper woodland to expand, by more than 2 kilometers in less than 5 years. The new community boundary remained stable even after the drought ended. This is the most rapid shift in the boundary between two terrestrial plant communities so far documented at the scale of an entire landscape.

Such a response to an out-of-the-ordinary event might suggest that species are always able to change their geographic distributions rapidly in response to climate change. But as is true with shifts in timing, it remains virtually certain that many species will be left behind by continued shifts in environmental conditions, and that the number of species, their relative abundances, and their interactions will be altered in communities worldwide.

CHECKpoint CONCEPT 45.5

- ✓ Extrapolating from the British plant data, what change might you anticipate in average first flowering date by the middle of this century, when atmospheric CO_2 is predicted to reach double its preindustrial concentration and the average temperature in Britain is expected to rise by an additional 3°C?
- ✓ How might changes in the seasonal timing of some events (such as the onset of warm weather at the end of winter) affect mutualistic interactions between plants and insect pollinators?

The changes that human activities are causing in Earth's biogeochemical cycles and in its climate have serious ecological implications. What can we do to address them?

CONCEPT 45.6 Ecological Challenges Can Be Addressed through Science and International Cooperation

The current period of climate change is not unprecedented. Throughout the history of life on Earth, wobble in the planet's orbit around the sun, continental drift, volcanic activity, sunspots, and even asteroid impacts have caused Earth's climates to change, precipitating five major episodes of mass extinction (see Figure 18.12). There is even precedent for organism-caused change in the atmosphere. The first photosynthetic microbes increased atmospheric oxygen concentrations to a level that was toxic to the anaerobic prokaryotes that inhabited Earth at the time, and the first land plants raised oxygen concentrations even more about 250 million years ago. What is unprecedented about the present climate change is that it has been precipitated by diverse activities of a single species: *Homo sapiens*.

It is sobering to realize that one species can wield such power. However, it is encouraging to realize that, of all species on Earth, humans are uniquely able to address the problems they have caused—not only because science equips us to understand the natural world and to solve problems, but also because *Homo sapiens* has a remarkable capacity for cooperative action. We certainly are capable of selfishness and conflict too, but cooperative interactions are central features of all functional human societies. Other group-living animals cooperate as well, but scientific studies show that cooperation with unrelated individuals is especially highly developed in humans—even as infants we routinely offer help to another individual in need. We support cooperative societal ventures of all sorts, we police "bad" behaviors that can undermine cooperation—and we are unique, as far as we now know, in caring about the other species with which we share planet Earth.

The scale of cooperative human groups has gradually expanded through human history from prehistoric hunter–gatherer family groups to the United Nations, which spans the planet. It is encouraging that governments of separate nations have been cooperating in several global-scale initiatives to tackle complex environmental issues, from the IGY, which supported the first years of Dave Keeling's work, to the United Nations' Intergovernmental Panel on Climate Change and World Meteorological Organization, which operate today. Earth's nations have also negotiated international agreements to achieve environmental goals, including the Montreal Protocol to prevent depletion of UV-absorbing atmospheric ozone, the Kyoto Protocol to reduce emissions of greenhouse gases, and the Convention on International Trade in Endangered Species of Wild Fauna and Flora (CITES), which seeks to conserve species by eliminating the economic benefits of exploiting them.

Go to ACTIVITY 45.2 The Benefits of Cooperation PoL2e.com/ac45.2

Nonetheless, humans face huge challenges to achieving effective cooperation on a global scale. A major challenge is that the economic policies of virtually every nation are structured to achieve continual economic growth—ever-increasing production and consumption of goods and services—despite the fact that Earth has a finite capacity to provide those goods and services. We will need to make the transition to sustainable, steady-state economies, and that will require an overhaul of economic models and institutions. Another, related, challenge is the continued multiplicative growth of the human population (see Concept 42.4). On a crowded planet, with competition among societies for limited resources, cooperation inevitably becomes more difficult. Addressing both challenges will require us to devise international systems for establishing—and enforcing—rules of acceptable behavior among groups and nations.

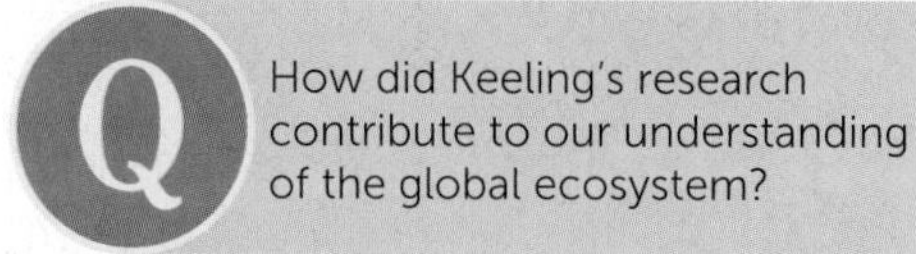

Q
How did Keeling's research contribute to our understanding of the global ecosystem?

ANSWER Dave Keeling's four decades of research left an important legacy, the Keeling curve (see Figure 45.12A), which documents the ongoing increase in atmospheric CO_2 (Concept 45.4). Almost immediately, Keeling discovered that CO_2 concentrations do not vary erratically, as earlier crude methods of measurement suggested, but instead change seasonally as summer warmth, and the primary productivity that goes with it, shifts from the Northern Hemisphere (where most of Earth's land mass lies) to the Southern Hemisphere (see Figure 45.12A inset).

Keeling's measurements quickly contributed to a better understanding of the pools and fluxes of the global carbon cycle (Concept 45.3), including the influence of fossil fuel burning and cement production (which together added 9.5×10^{15} g of carbon—and 34.7×10^{15} g of carbon dioxide—to the atmosphere in 2011). Measurements on Mauna Loa and elsewhere continue to the present day, and they show that CO_2 increased from 315 parts per million (ppm) in the atmosphere in 1958 to 399 ppm in July 2013. Better understanding of the carbon cycle has contributed to greatly improved global climate models.

Keeling's results were noticed almost immediately. In 1965 the U.S. President's Science Advisory Committee warned of the increasing greenhouse effect, and in 1966 the National Academy of Sciences issued a scholarly report on the topic. The first World Climate Conference was held in Geneva, Switzerland, in 1979, at a time when there was growing scientific consensus that climate change poses a critical environmental challenge (Concept 45.5). The number of scientists studying climate change, and the sophistication of their experiments and models, grew rapidly. At the end of the twentieth century, measurements showed that the average temperature of the planet had increased by 0.7°C since 1900—an increase very close to the predictions of global climate models.

The Keeling curve forms a crucial part of our understanding of the Earth system. It is an example of the carefully documented scientific information used by the Intergovernmental Panel on Climate Change (IPCC). The IPCC was formed in 1988 by another scientific collaboration of governments—a sort of "grandchild" of IGY—and its many scientists continue to summarize and report on the latest evidence for natural and human-caused climate change. The 2009 IPCC report predicted a further human-caused increase in the global average temperature of between 1.8°C and 4.0°C by the year 2100.

> We have a civilization based on science and technology, and we've cleverly arranged things so that almost nobody understands science and technology. That is as clear a prescription for disaster as you can imagine. While we might get away with this combustible mixture of ignorance and power for a while, sooner or later it's going to blow up in our faces. The powers of modern technology are so formidable that it's insufficient just to say, 'Well, those in charge, I'm sure, are doing a good job.' This is a democracy, and for us to make sure that the powers of science and technology are used properly and prudently, we ourselves must understand science and technology. We must be involved in the decision-making process.
>
> *Carl Sagan*

SUMMARY

CONCEPT 45.1 Climate and Nutrients Affect Ecosystem Function

- The term **ecosystem** is commonly applied to ecological communities plus the abiotic environments with which they exchange energy and materials.
- A community's net primary productivity (NPP) reflects the rate at which it exchanges energy and materials with its abiotic surroundings, and is thus a measure of **ecosystem function**.
- NPP varies considerably among ecosystem types. **Review Figure 45.1**
- The global pattern of variation in terrestrial NPP shows that it is influenced by temperature and precipitation. Soil nutrients also play a role in terrestrial NPP. **Review Figures 45.2 and 45.3**
- Aquatic NPP varies most strongly with availability of light and nutrients. **Review Figure 45.4**

CONCEPT 45.2 Biological, Geological, and Chemical Processes Move Materials through Ecosystems

- The chemical form in which an element exists and where it occurs on Earth determine whether the element is accessible to life. Alternative forms and locations of matter can be thought of as **compartments** of the global ecosystem
- The chemical transformations and physical transport of elements or molecules that move them among compartments are called **biogeochemical cycles**. **Review Figure 45.5**
- The **pool**, or amount, of an element or molecule in a compartment depends on its flows, or **fluxes**, into and out of that compartment.

CONCEPT 45.3 Certain Biogeochemical Cycles Are Especially Critical for Ecosystems

- Solar-powered evaporation drives the water cycle, moving water into the atmosphere in gaseous form. Condensation and precipitation return water to Earth's surface. Gravity-driven flows move water from land to lakes and oceans. **Review Figure 45.6 and ANIMATED TUTORIAL 45.1**
- The availability of nitrogen to living organisms depends on **nitrogen fixation** and other biochemical processes carried out by microbes. These processes convert gaseous N_2 into forms that other organisms can use. Other microbial processes return N_2 to the atmosphere. **Review Figure 45.7 and ANIMATED TUTORIAL 45.2**
- Human activities add various chemical forms of nitrogen to the atmosphere, where they contribute to smog and acid rain, to terrestrial systems, where they act as fertilizers, and to aquatic systems, where they cause **eutrophication**. **Review Figure 45.9**
- Photosynthesis and respiration move carbon between inorganic and organic compartments. As a result, energy flow and the flux of carbon through biological communities are intimately linked.
- Some carbon absorbed from the atmosphere by ocean waters is taken up through photosynthesis, some precipitates as carbonate compounds, and some sinks to the benthic zone as dead organic material. Carbonate compounds and organic material that accumulate in sediments and soils are transformed into carbonate rocks and **fossil fuels**.
- Humans are increasing the carbon pool in the atmosphere by burning fossil fuels, manufacturing cement, raising livestock, and raising wetland crops. **Review Figure 45.10 and ANIMATED TUTORIAL 45.3**
- The biogeochemical cycles of different materials are linked when the same physical, chemical, or biological processes drive their fluxes. **Review ACTIVITY 45.1**

CONCEPT 45.4 Biogeochemical Cycles Affect Global Climate

- Earth absorbs incoming solar radiation and reemits it at infrared wavelengths. Some of this infrared radiation is absorbed by atmospheric gases and is reradiated back to Earth's surface. The resulting retention of heat energy within the Earth system is called the **greenhouse effect**. **Review Figure 45.11 and ANIMATED TUTORIAL 45.4**
- Gases such as water vapor, carbon dioxide, methane, and nitrous oxide that absorb infrared radiation are called **greenhouse gases**. The pools of these molecules in the atmosphere influence Earth's radiation budget.
- Atmospheric concentrations of CO_2 and other greenhouse gases have been increasing rapidly since about 1880, and average annual global temperatures have followed suit. **Review Figures 45.12 and 45.13**

(continued)

SUMMARY (continued)

- **Global warming** is changing Earth's climates. High latitudes are warming more than low latitudes, precipitation patterns are changing, and storm intensities are increasing. **Review Figure 45.14**
- Computer models of the Earth system show that human activities have contributed significantly to the recent warming of the climate.

CONCEPT 45.5 Rapid Climate Change Affects Species and Communities

- Climate change is altering the timing of some seasonal environmental cues but not others. Although species sometimes can evolve new responses to such cues, the rate of evolution will not keep up with an environment that changes too rapidly.
- Because different species respond to different seasonal cues, altered cues result in timing mismatches among species in a community and thus disrupt their interactions.
- Climate change is altering the distributions and abundances of species, resulting in the assembly of novel communities.
- Climate change is increasing the frequency of extreme weather events, which can cause sudden shifts in species distributions and in community composition.

CONCEPT 45.6 Ecological Challenges Can Be Addressed through Science and International Cooperation

- Humans are causing major changes in the biosphere and in other aspects of the Earth system. However, we are also uniquely equipped to address these changes, not only because science enables us to understand the natural world and to devise solutions to problems, but also because of our capacity for cooperative action. **Review ACTIVITY 45.2**

See **ACTIVITY 45.3** for a concept review of this chapter.

Go to the Interactive Summary to review key figures, Animated Tutorials, and Activities PoL2e.com/is45

Go to LaunchPad at **macmillanhighered.com/launchpad** for additional resources, including LearningCurve Quizzes, Flashcards, and many other study and review resources.

Appendix A The Tree of Life

Phylogeny is the organizing principle of modern biological taxonomy. A guiding principle of modern phylogeny is monophyly. A monophyletic group is one that contains an ancestral lineage and all of its descendants. Any such group can be extracted from a phylogenetic tree with a single "cut" of the tree. The cut lineage, with all of its descendant taxa, is a monophyletic group. Biologists give names to the major monophyletic groups of life so that we may refer to them easily.

The tree shown here provides a guide to the relationships among the major groups of extant (living) organisms in the tree of life as we have presented them throughout this book. The position of the branching "splits" indicates the relative branching order of the lineages of life, but the time scale is not meant to be uniform. In addition, the groups appearing at the branch tips do not necessarily carry equal phylogenetic "weight." For example, the ginkgo [75] is indeed at the apex of its lineage; this gymnosperm group consists of a single living species. In contrast, a phylogeny of the eudicots [83] could continue on from this point to fill many more trees the size of this one.

The glossary entries that follow are informal descriptions of some major features of the organisms described in Part Four of this book. We show greater detail about the relationships and content of some groups than we have presented in this book so that this appendix may serve as a reference. Each entry gives the group's common name, followed by the formal scientific name of the group (in parentheses). Numbers in square brackets reference the location of the respective groups on the tree.

It is sometimes convenient to use an informal name to refer to a collection of organisms that are not monophyletic but nonetheless all share (or all lack) some common attribute. We call these "convenience terms"; such groups are indicated in these entries by quotation marks, and we do not give them formal scientific names. Examples include "prokaryotes," "protists," and "algae." Note that these groups cannot be removed with a single cut; they represent a collection of distantly related groups that appear in different parts of the tree.

Go to LaunchPad at **macmillanhighered.com/launchpad** for an interactive version of this tree, with links to photos, distribution maps, species lists, and identification keys.

A

acorn worms (*Enteropneusta*) Benthic marine hemichordates [119] with an acorn-shaped proboscis, a short collar (neck), and a long trunk.

"algae" Convenience term encompassing various distantly related groups of aquatic, photosynthetic eukaryotes [4].

alveolates (*Alveolata*) [5] Unicellular eukaryotes with a layer of flattened vesicles (alveoli) supporting the plasma membrane. Major groups include the dinoflagellates [51], apicomplexans [50], and ciliates [49].

amborella (*Amborella*) [78] An understory shrub or small tree found only on the South Pacific island of New Caledonia. Thought to be the sister group of the remaining living angiosperms [15].

ambulacrarians (*Ambulacraria*) [29] The echinoderms [118] and hemichordates [119].

amniotes (*Amniota*) [36] Mammals, reptiles, and their extinct close relatives. Characterized by many adaptations to terrestrial life, including an amniotic egg (with a unique set of membranes—the amnion, chorion, and allantois), a water-repellant epidermis (with epidermal scales, hair, or feathers), and, in males, a penis that allows internal fertilization.

amoebozoans (*Amoebozoa*) [84] A group of eukaryotes [4] that use lobe-shaped pseudopods for locomotion and to engulf food. Major amoebozoan groups include the loboseans, plasmodial slime molds, and cellular slime molds.

amphibians (*Amphibia*) [128] Tetrapods [35] with glandular skin that lacks epidermal scales, feathers, or hair. Many amphibian species undergo a complete metamorphosis from an aquatic larval form to a terrestrial adult form, although direct development is also common. Major amphibian groups include frogs and toads (anurans), salamanders, and caecilians.

amphipods (*Amphipoda*) Small crustaceans [116] that are abundant in many marine and freshwater habitats. They are important herbivores, scavengers, and micropredators, and are an important food source for many aquatic organisms.

angiosperms (*Anthophyta* or *Magnoliophyta*) [15] The flowering plants. Major angiosperm groups include the monocots [82], eudicots [83], and magnoliids [81].

animals (*Animalia* or *Metazoa*) [19] Multicellular heterotrophic eukaryotes. The majority of animals are bilaterians [22]. Other groups of animals include the sponges [20], ctenophores [95], placozoans [96], and cnidarians [97]. The closest living relatives of the animals are the choanoflagellates [91].

annelids (*Annelida*) [105] Segmented worms, including earthworms, leeches, and polychaetes. One of the major groups of lophotrochozoans [24].

anthozoans (*Anthozoa*) One of the major groups of cnidarians [97]. Includes the sea anemones, sea pens, and corals.

anthropoids (*Haplorhini*) A group of primate mammals [129] that consists of tarsiers, monkeys, and apes.

anurans (*Anura*) Comprising the frogs and toads, this is the largest group of living amphibians [128]. They are tailless, with a shortened vertebral column and elongate hind legs modified for jumping. Many species have an aquatic larval form known as a tadpole.

apicomplexans (*Apicomplexa*) [50] Parasitic alveolates [5] characterized by the possession of an apical complex at some stage in the life cycle.

arachnids (*Arachnida*) Chelicerates [114] with a body divided into two parts: a cephalothorax that bears six pairs of appendages (four pairs of which are usually used as legs) and an abdomen that bears the genital opening. Familiar arachnids include spiders, scorpions, mites and ticks, and harvestmen.

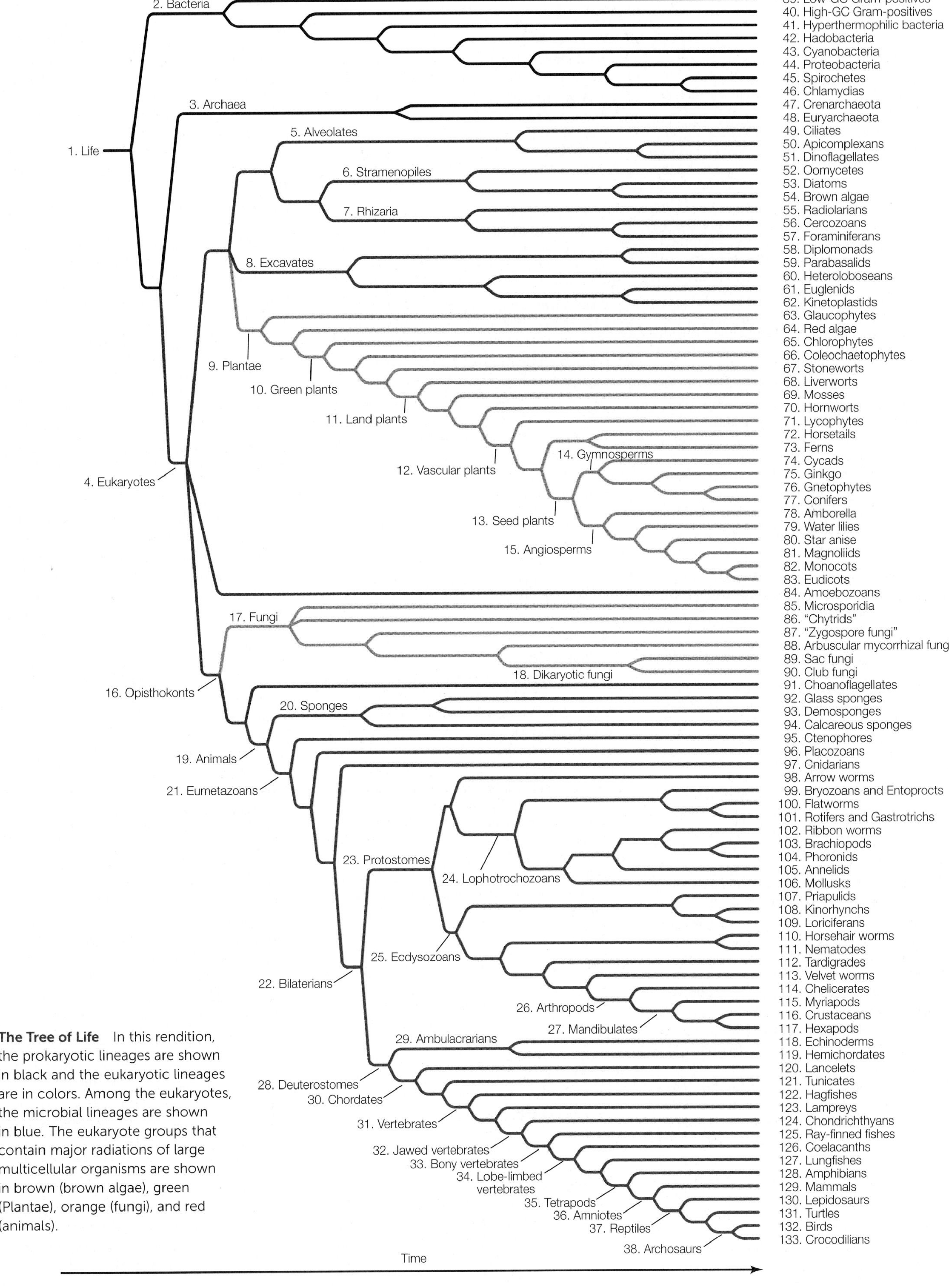

The Tree of Life In this rendition, the prokaryotic lineages are shown in black and the eukaryotic lineages are in colors. Among the eukaryotes, the microbial lineages are shown in blue. The eukaryote groups that contain major radiations of large multicellular organisms are shown in brown (brown algae), green (Plantae), orange (fungi), and red (animals).

arbuscular mycorrhizal fungi (*Glomeromycota*) [88] A group of fungi [17] that associate with plant roots in a close symbiotic relationship.

archaeans (*Archaea*) [3] Unicellular organisms lacking a nucleus and lacking peptidoglycan in the cell wall. Once grouped with the bacteria, archaeans possess distinctive membrane lipids.

archosaurs (*Archosauria*) [38] A group of reptiles [37] that includes dinosaurs and crocodilians [133]. Most dinosaur groups became extinct at the end of the Cretaceous; birds [132] are the only surviving dinosaurs.

arrow worms (*Chaetognatha*) [98] Small planktonic or benthic predatory marine worms with fins and a pair of hooked, prey-grasping spines on each side of the head.

arthropods (*Arthropoda*) The largest group of ecdysozoans [25]. Arthropods are characterized by a stiff exoskeleton, segmented bodies, and jointed appendages. Includes the chelicerates [114], myriapods [115], crustaceans [116], and hexapods (insects and their relatives) [117].

ascidians (*Ascidiacea*) "Sea squirts"; the largest group of tunicates [121]. They are sessile (as adults), marine, saclike filter feeders.

B

bacteria (*Eubacteria*) [2] Unicellular organisms lacking a nucleus, possessing distinctive ribosomes and initiator tRNA, and generally containing peptidoglycan in the cell wall. Different bacterial groups are distinguished primarily on nucleotide sequence data.

barnacles (*Cirripedia*) Crustaceans [116] that undergo two metamorphoses—first from a feeding planktonic larva to a nonfeeding swimming larva, and then to a sessile adult that forms a "shell" composed of four to eight plates cemented to a hard substrate.

bilaterians (*Bilateria*) [22] Those animal groups characterized by bilateral symmetry and three distinct tissue types (endoderm, ectoderm, and mesoderm). Includes the protostomes [23] and deuterostomes [28].

birds (*Aves*) [132] Feathered, flying (or secondarily flightless) tetrapods [35].

bivalves (*Bivalvia*) Major mollusk [106] group; clams and mussels. Bivalves typically have two similar hinged shells that are each asymmetrical across the midline.

bony vertebrates (*Osteichthyes*) [33] Vertebrates [31] in which the skeleton is usually ossified to form bone. Includes the ray-finned fishes [125], coelacanths [126], lungfishes [127], and tetrapods [35].

brachiopods (*Brachiopoda*) [103] Lophotrochozoans [24] with two similar hinged shells that are each symmetrical across the midline. Superficially resemble bivalve mollusks, except for the shell symmetry.

brittle stars (*Ophiuroidea*) Echinoderms [118] with five long, whip-like arms radiating from a distinct central disk that contains the reproductive and digestive organs.

brown algae (*Phaeophyta*) [54] Multicellular, almost exclusively marine stramenopiles [6] generally containing the pigment fucoxanthin as well as chlorophylls *a* and *c* in their chloroplasts.

bryozoans (*Ectoprocta* or *Bryozoa*) [99] A group of marine and freshwater lophotrochozoans [24] that live in colonies attached to substrates; also known as ectoprocts or moss animals. They are the sister group of entoprocts.

C

caecilians (*Gymnophiona*) A group of burrowing or aquatic amphibians [128]. They are elongate, legless, with a short tail (or none at all), reduced eyes covered with skin or bone, and a pair of sensory tentacles on the head.

calcareous sponges (*Calcarea*) [94] Filter-feeding marine sponges with spicules composed of calcium carbonate.

cellular slime molds (*Dictyostelida*) Amoebozoans [84] in which individual amoebas aggregate under stress to form a multicellular pseudoplasmodium.

cephalochordates (*Cephalochordata*) [120] *See* lancelets.

cephalopods (*Cephalopoda*) Active, predatory mollusks [106] in which the molluscan foot has been modified into muscular hydrostatic arms or tentacles. Includes octopuses, squids, and nautiluses.

cercozoans (*Cercozoa*) [56] Unicellular eukaryotes [4] that feed by means of threadlike pseudopods. Group together with foraminiferans [57] and radiolarians [55] to comprise the rhizaria [7].

charophytes (*Charales*) [67] *See* stoneworts.

chelicerates (*Chelicerata*) [114] A major group of arthropods [26] with pointed appendages (chelicerae) used to grasp food (as opposed to the chewing mandibles of most other arthropods). Includes the arachnids, horseshoe crabs, pycnogonids, and extinct sea scorpions.

chimaeras (*Holocephali*) A group of bottom-dwelling, marine, scaleless chondrichthyan fishes [124] with large, permanent, grinding tooth plates (rather than the replaceable teeth found in other chondrichthyans).

chitons (*Polyplacophora*) Flattened, slow-moving mollusks [106] with a dorsal protective calcareous covering made up of eight articulating plates.

chlamydias (*Chlamydiae*) [46] A group of very small Gram-negative bacteria; they live as intracellular parasites of other organisms.

chlorophytes (*Chlorophyta*) [65] The most abundant and diverse group of green algae, including freshwater, marine, and terrestrial forms; some are unicellular, others colonial, and still others multicellular. Chlorophytes use chlorophylls *a* and *c* in their photosynthesis.

choanoflagellates (*Choanozoa*) [91] Unicellular eukaryotes [4] with a single flagellum surrounded by a collar. Most are sessile, some are colonial. The closest living relatives of the animals [19].

chondrichthyans (*Chondrichthyes*) [124] One of the two main groups of jawed vertebrates [32]; includes sharks, rays, and chimaeras. They have cartilaginous skeletons and paired fins.

chordates (*Chordata*) [30] One of the two major groups of deuterostomes [28], characterized by the presence (at some point in development) of a notochord, a hollow dorsal nerve cord, and a post-anal tail. Includes the lancelets [120], tunicates [121], and vertebrates [31].

"chytrids" [86] Convenience term used for a paraphyletic group of mostly aquatic, microscopic fungi [17] with flagellated gametes. Some exhibit alternation of generations.

ciliates (*Ciliophora*) [49] Alveolates [5] with numerous cilia and two types of nuclei (micronuclei and macronuclei).

clitellates (*Clitellata*) Annelids [105] with gonads contained in a swelling (called a clitellum) toward the head of the animal. Includes earthworms (oligochaetes) and leeches.

club fungi (*Basidiomycota*) [90] Fungi [17] that, if multicellular, bear the products of meiosis on club-shaped basidia and possess a long-lasting dikaryotic stage. Some are unicellular.

club mosses (*Lycopodiophyta*) [71] Vascular plants [12] characterized by microphylls. *See* lycophytes.

cnidarians (*Cnidaria*) [97] Aquatic, mostly marine eumetazoans [21] with specialized stinging organelles (nematocysts) used for prey capture and defense, and a blind gastrovascular cavity. The sister group of the bilaterians [22].

coelacanths (*Actinista*) [126] A group of marine lobe-limbed vertebrates [34] that was diverse from the Middle Devonian to the Cretaceous, but is now known from just two living species. The pectoral and anal fins are on fleshy stalks supported by skeletal elements, so they are also called lobe-finned fishes.

coleochaetophytes (*Coleochaetales*) [66] Multicellular green algae characterized by flattened growth form composed of thin-walled cells. Thought to be the sister-group to the stoneworts [67] plus land plants [11].

conifers (*Pinophyta* or *Coniferophyta*) [77] Cone-bearing, woody seed plants [13].

copepods (*Copepoda*) Small, abundant crustaceans [116] found in marine, freshwater, or wet terrestrial habitats. They have a single eye, long antennae, and a body shaped like a teardrop.

craniates (*Craniata*) Some biologist exclude the hagfishes [122] from the vertebrates [31], and use the term craniates to refer to the two groups combined.

crenarchaeotes (*Crenarchaeota*) [47] A major and diverse group of archaeans [3], defined on the basis of rRNA base sequences. Many are extremophiles (inhabit extreme environments), but the group may also be the most abundant archaeans in the marine environment.

crinoids (*Crinoidea*) Echinoderms [118] with a mouth surrounded by feeding arms, and a U-shaped gut with the mouth next to the anus. They attach to the substratum by a stalk or are free-swimming. Crinoids were abundant in the middle and late Paleozoic, but only a few hundred species have survived to the present. Includes the sea lilies and feather stars.

crocodilians (*Crocodylia*) [133] A group of large, predatory, aquatic archosaurs [38]. The closest living relatives of birds [132]. Includes alligators, caimans, crocodiles, and gharials.

crustaceans (*Crustacea*) [116] Major group of marine, freshwater, and terrestrial arthropods [26] with a head, thorax, and abdomen (although the head and thorax may be fused), covered with a thick exoskeleton, and with two-part appendages. Crustaceans undergo metamorphosis from a nauplius larva. Includes decapods, isopods, krill, barnacles, amphipods, copepods, and ostracods.

ctenophores (*Ctenophora*) [95] Radially symmetrical, diploblastic marine animals [19], with a complete gut and eight rows of fused plates of cilia (called ctenes).

cyanobacteria (*Cyanobacteria*) [43] A group of unicellular, colonial, or filamentous bacteria that conduct photosynthesis using chlorophyll *a*.

cycads (*Cycadophyta*) [74] Palmlike gymnosperms with large, compound leaves.

cyclostomes (*Cyclostomata*) This term refers to the possibly monophyletic group of lampreys [123] and hagfishes [122]. Molecular data support this group, but morphological data suggest that lampreys are more closely related to jawed vertebrates [32] than to hagfishes.

D

decapods (*Decapoda*) A group of marine, freshwater, and semiterrestrial crustaceans [116] in which five of the eight pairs of thoracic appendages function as legs (the other three pairs, called maxillipeds, function as mouthparts). Includes crabs, lobsters, crayfishes, and shrimps.

demosponges (*Demospongiae*) [93] The largest of the three groups of sponges [20], accounting for 90 percent of all sponge species. Demosponges have spicules made of silica, spongin fiber (a protein), or both.

deuterostomes (*Deuterostomia*) [28] One of the two major groups of bilaterians [22], in which the mouth forms at the opposite end of the embryo from the blastopore in early development (contrast with protostomes). Includes the ambulacrarians [29] and chordates [30].

diatoms (*Bacillariophyta*) [53] Unicellular, photosynthetic stramenopiles [6] with glassy cell walls in two parts.

dikaryotic fungi (*Dikarya*) [18] A group of fungi [17] in which two genetically different haploid nuclei coexist and divide within the same hypha; includes club fungi [90] and sac fungi [89].

dinoflagellates (*Dinoflagellata*) [51] A group of alveolates [5] usually possessing two flagella, one in an equatorial groove and the other in a longitudinal groove; many are photosynthetic.

dinosaurs (*Dinosauria*) A group of archosaurs [38] that includes birds [132] as well as many extinct groups from the Mesozoic era. Extinct Mesozoic dinosaurs included some of the largest terrestrial vertebrates that have ever lived. Informally, many people use the term to refer only to the extinct Mesozoic species.

diplomonads (*Diplomonadida*) [58] A group of eukaryotes [4] lacking mitochondria; most have two nuclei, each with four associated flagella.

E

ecdysozoans (*Ecdysozoa*) [25] One of the two major groups of protostomes [23], characterized by periodic molting of their cuticles. Nematodes [111] and arthropods [26] are the largest ecdysozoan groups.

echinoderms (*Echinodermata*) [118] A major group of marine deuterostomes [28] with five-fold radial symmetry (at some stage of life) and an endoskeleton made of calcified plates and spines. Includes sea stars, crinoids, sea urchins, sea cucumbers, and brittle stars.

elasmobranchs (*Elasmobranchii*) The largest group of chondrichthyan fishes [124]. Includes sharks, skates, and rays. In contrast to the other group of living chondrichthyans (the chimaeras), they have replaceable teeth.

embryophytes *See* land plants [11].

entoprocts (*Entoprocta*) [99] A group of marine and freshwater lophotrochozoans [24] that live as single individuals or in colonies attached to substrates. They are the sister group of bryozoans, from which they differ in having both their mouth and anus inside the lophophore (the anus is outside the lophophore in bryozoans).

eudicots (*Eudicotyledones*) [83] A group of angiosperms [15] with pollen grains possessing three openings. Typically with two cotyledons, net-veined leaves, taproots, and floral organs typically in multiples of four or five.

euglenids (*Euglenida*) [61] Flagellate excavates characterized by a pellicle composed of spiraling strips of protein under the plasma membrane; the mitochondria have disk-shaped cristae. Some are photosynthetic.

eukaryotes (*Eukarya*) [4] Organisms made up of one or more complex cells in which the genetic material is contained in nuclei. Contrast with archaeans [3] and bacteria [2].

eumetazoans (*Eumetazoa*) [21] Those animals [19] characterized by body symmetry, a gut, a nervous system, specialized types of cell junctions, and well-organized tissues in distinct cell layers (although there have been secondary losses of some or most of these characteristics in a few eumetazoan lineages).

euphyllophytes (*Euphyllophyta*) The group of vascular plants [12] that is sister to the lycophytes [71] and which includes all plants with megaphylls.

euryarchaeotes (*Euryachaeota*) [48] A major group of archaeans [3], diagnosed on the basis of rRNA sequences. Includes many methanogens, extreme halophiles, and thermophiles.

eutherians (*Eutheria*) A group of viviparous mammals [129], eutherians are well developed at birth (contrast to prototherians and marsupials, the other two groups of mammals). Most familiar mammals outside the Australian and South American regions are eutherians (see Table 23.3).

excavates (*Excavata*) [8] Diverse group of unicellular, flagellate eukaryotes, many of which possess a feeding groove; some lack mitochondria.

F

ferns (*Pteridopsida* or *Polypodiopsida*) [73] Vascular plants [12] usually possessing large, frondlike leaves that unfold from a "fiddlehead." As used in this book, this name refers to a monophyletic clade, also known as the leptosporangiate ferns.

flatworms (*Platyhelminthes*) [100] A group of dorsoventrally flattened and generally elongate soft-bodied lophotrochozoans [24]. May be free-living or parasitic, found in marine, freshwater, or damp terrestrial environments. Major flatworm groups include the tapeworms, flukes, monogeneans, and turbellarians.

flowering plants *See* angiosperms [15].

flukes (*Trematoda*) A group of wormlike parasitic flatworms [100] with complex life cycles that involve several different host species. May be paraphyletic with respect to tapeworms.

foraminiferans (*Foraminifera*) [57] Amoeboid organisms with fine, branched pseudopods that form a food-trapping net. Most produce external shells of calcium carbonate.

fungi (*Fungi*) [17] Eukaryotic heterotrophs with absorptive nutrition based on extracellular digestion; cell walls contain chitin. Major fungal groups include the microsporidia [85], "chytrids" [86], "zygospore fungi" [87], arbuscular mycorrhizal fungi [88], sac fungi [89], and club fungi [90].

G

gastropods (*Gastropoda*) The largest group of mollusks [106]. Gastropods possess a well-defined head with two or four sensory tentacles (often terminating in eyes) and a ventral foot. Most species have a single coiled or spiraled shell. Common in marine, freshwater, and terrestrial environments.

gastrotrichs (*Gastrotricha*) [101] Tiny (0.06–3.0 mm), elongate acoelomate lophotrochozoans [24] that are covered in cilia. They live in marine, freshwater, and wet terrestrial habitats. They are simultaneous hermaphrodites.

ginkgo (*Ginkgophyta*) [75] A gymnosperm [14] group with only one living species. The ginkgo seed is surrounded by a fleshy tissue not derived from an ovary wall and hence not a fruit.

glass sponges (*Hexactinellida*) [92] Sponges [20] with a skeleton composed of four- and/or six-pointed spicules made of silica.

glaucophytes (*Glaucophyta*) [63] Unicellular freshwater algae with chloroplasts containing traces of peptidoglycan, the characteristic cell wall material of bacteria.

gnathostomes (*Gnathostomata*) *See* jawed vertebrates [32].

gnetophytes (*Gnetophyta*) [76] A gymnosperm [14] group with three very different lineages; all have wood with vessels, unlike other gymnosperms.

green plants (*Viridiplantae*) [10] Organisms with chlorophylls *a* and *b*, cellulose-containing cell walls, starch as a carbohydrate storage product, and chloroplasts surrounded by two membranes.

gymnosperms (*Gymnospermae*) [14] Seed plants [13] with seeds "naked" (i.e., not enclosed in carpels). Probably monophyletic, but status still in doubt. Includes the conifers [77], gnetophytes [76], ginkgo [75], and cycads [74].

H

hadobacteria (*Hadobacteria*) [42] A group of extremophilic bacteria [2] that includes the genera *Deinococus* and *Thermus*.

hagfishes (*Myxini*) [122] Elongate, slimy-skinned vertebrates [31] with three small accessory hearts, a partial cranium, and no stomach or paired fins. *See also* craniates; cyclostomes.

hemichordates (*Hemichordata*) [119] One of the two primary groups of ambulacrarians [29]; marine wormlike organisms with a three-part body plan.

heteroloboseans (*Heterolobosea*) [60] Colorless excavates [8] that can transform among amoeboid, flagellate, and encysted stages.

hexapods (*Hexapoda*) [117] Major group of arthropods [26] characterized by a reduction (from the ancestral arthropod condition) to six walking appendages, and the consolidation of three body segments to form a thorax. Includes insects and their relatives (see Table 23.2).

high-GC Gram-positives (*Actinobacteria*) [40] Gram-positive bacteria with a relatively high (G+C)/(A+T) ratio of their DNA, with a filamentous growth habit.

hominid (*Hominidae*) A group of primate mammals [129] that consists of the great apes (humans, chimpanzees, gorillas, orangutans, and their extinct close relatives).

hornworts (*Anthocerophyta*) [70] Nonvascular plants with sporophytes that grow from the base. Cells contain a single large, platelike chloroplast.

horsehair worms (*Nematomorpha*) [110] A group of very thin, elongate, wormlike freshwater ecdysozoans [25]. Largely nonfeeding as adults, they are parasites of insects and crayfish as larvae.

horseshoe crabs (*Xiphosura*) Marine chelicerates [114] with a large outer shell in three parts: a carapace, an abdomen, and a tail-like telson. There are only five living species, but many additional species are known from fossils.

horsetails (*Sphenophyta* or *Equisetophyta*) [72] Vascular plants [12] with reduced megaphylls in whorls.

hydrozoans (*Hydrozoa*) A group of cnidarians [97]. Most species go through both polyp and mesuda stages, although one stage or the other is eliminated in some species.

hyperthermophilic bacteria [41] A group of thermophilic bacteria [2] that live in volcanic vents, hot springs, and in underground oil reservoirs; includes the genera *Aquifex* and *Thermotoga*.

I

insects (*Insecta*) The largest group within the hexapods [117]. Insects are characterized by exposed mouthparts and one pair of antennae containing a sensory receptor called a Johnston's organ. Most have two pairs of wings as adults. There are more described species of insects than all other groups of life [1] combined, and many species remain to be discovered. The major insect groups are described in Table 23.2.

"invertebrates" Convenience term encompassing any animal [19] that is not a vertebrate [31].

isopods (*Isopoda*) Crustaceans [116] characterized by a compact head, unstalked compound eyes, and mouthparts consisting of four pairs of appendages. Isopods are abundant and widespread in salt, fresh, and brackish water, although some species (the sow bugs) are terrestrial.

J

jawed vertebrates (*Gnathostomata*) [32] A major group of vertebrates [31] with jawed mouths. Includes chondrichthyans [124], ray-finned fishes [125], and lobe-limbed vertebrates [34].

K

kinetoplastids (*Kinetoplastida*) [62] Unicellular, flagellate organisms characterized by the presence in their single mitochondrion of a kinetoplast (a structure containing multiple, circular DNA molecules).

kinorhynchs (*Kinorhyncha*) [108] Small (< 1 mm) marine ecdysozoans [25] with bodies in 13 segments and a retractable proboscis.

korarchaeotes (*Korarchaeota*) A poorly known group of Archaea [3] known only from DNA isolated from hot environments.

krill (*Euphausiacea*) A group of shrimplike marine crustaceans [116] that are important components of the zooplankton.

L

lampreys (*Petromyzontiformes*) [123] Elongate, eel-like vertebrates [31] that often have rasping and sucking disks for mouths.

lancelets (*Cephalochordata*) [120] A group of weakly swimming, eel-like benthic marine chordates [30].

land plants (*Embryophyta*) [11] Plants with embryos that develop within protective structures; also called embryophytes. Sporophytes and gametophytes are multicellular. Land plants possess a cuticle. Major groups are the liverworts [68], mosses [69], hornworts [70], and vascular plants [12].

larvaceans (*Larvacea*) Solitary, planktonic tunicates [121] that retain both notochords and nerve cords throughout their lives.

lepidosaurs (*Lepidosauria*) [130] Reptiles [37] with overlapping scales. Includes tuataras and squamates (lizards, snakes, and amphisbaenians).

leptosporangiate ferns (*Pteridopsida* or *Polypodiopsida*) [73] Vascular plants [12] usually possessing large, frondlike leaves that unfold from a "fiddlehead," and possessing thin-walled sporangia.

life (*Life*) [1] The monophyletic group that includes all known living organisms. Characterized by a nucleic-acid based genetic system (DNA or RNA), metabolism, and cellular structure. Some parasitic forms, such as viruses, have secondarily lost some of these features and rely on the cellular environment of their host.

liverworts (*Hepatophyta*) [68] Nonvascular plants lacking stomata; stalk of sporophyte elongates along its entire length.

lobe-limbed vertebrates (*Sarcopterygii*) [34] One of the two major groups of bony vertebrates [33], characterized by jointed appendages (paired fins or limbs).

loboseans (*Lobosea*) A group of unicellular amoebozoans [84]; includes the most familiar amoebas (e.g., *Amoeba proteus*).

"lophophorates" Convenience term used to describe several groups of lophotrochozoans [24] that have a feeding structure called a lophophore (a circular or U-shaped ridge around the mouth that bears one or two rows of ciliated, hollow tentacles). Not a monophyletic group.

lophotrochozoans (*Lophotrochozoa*) [24] One of the two main groups of protostomes [23]. This group is morphologically diverse, and is supported primarily on information from gene sequences. Includes bryozoans and entoprocts [99], flatworms [100], rotifers and gastrotrichs [101], ribbon worms [102], brachiopods [103], phoronids [104], annelids [105], and mollusks [106].

loriciferans (*Loricifera*) [109] Small (< 1 mm) ecdysozoans [25] with bodies in four parts, covered with six plates.

low-GC Gram-positives (*Firmicutes*) [39] A diverse group of bacteria [2] with a relatively low (G+C)/(A+T) ratio of their DNA, often but not always Gram-positive, some producing endospores.

lungfishes (*Dipnoi*) [127] A group of aquatic lobe-limbed vertebrates [34] that are the closest living relatives of the tetrapods [35]. They have a modified swim bladder used to absorb oxygen from air, so some species can survive the temporary drying of their habitat.

lycophytes (*Lycopodiophyta*) [71] Vascular plants [12] characterized by microphylls; includes club mosses, spike mosses, and quillworts.

M

magnoliids (*Magnoliidae*) [81] A major group of angiosperms [15] possessing two cotyledons and pollen grains with a single opening. The group is defined primarily by nucleotide sequence data; it is more closely related to the eudicots and monocots than to three other small angiosperm groups.

mammals (*Mammalia*) [129] A group of tetrapods [35] with hair covering all or part of their skin; females produce milk to feed their developing young. Includes the prototherians, marsupials, and eutherians.

mandibulates (*Mandibulata*) [27] Arthropods [26] that include mandibles as mouth parts. Includes myriapods [115], crustaceans [116], and hexapods [117].

marsupials (*Marsupialia*) Mammals [129] in which the female typically has a marsupium (a pouch for rearing young, which are born at an extremely early stage in development). Includes such familiar mammals as opossums, koalas, and kangaroos.

metazoans (*Metazoa*) *See* animals [19].

"methanogens" A convenience term for any of several groups of anaerobic Archaea [3] that produce methane as a metabolic biproduct.

microbial eukaryotes *See* "protists."

microsporidia (*Microsporidia*) [85] A group of parasitic unicellular fungi [17] that lack mitochondria and have walls that contain chitin.

mollusks (*Mollusca*) [106] One of the major groups of lophotrochozoans [24], mollusks have bodies composed of a foot, a mantle (which often secretes a hard, calcareous shell), and a visceral mass. Includes monoplacophorans, chitons, bivalves, gastropods, and cephalopods.

monilophytes (*Monilophyta*) A group of vascular plants [12], sister to the seed plants [13], characterized by overtopping and possession of megaphylls; includes the horsetails [72] and ferns [73].

monocots (*Monocotyledones*) [82] Angiosperms [15] characterized by possession of a single cotyledon, usually parallel leaf veins, a fibrous root system, pollen grains with a single opening, and floral organs usually in multiples of three.

monogeneans (*Monogenea*) A group of ectoparasitic flatworms [100].

monoplacophorans (*Monoplacophora*) Mollusks [106] with segmented body parts and a single, thin, flat, rounded, bilateral shell.

mosses (*Bryophyta*) [69] Nonvascular plants with true stomata and erect, "leafy" gametophytes; sporophytes elongate by apical cell division.

moss animals *See* bryozoans [99].

mycoplasmas (*Mycoplasma*) A group of tiny, low-GC Gram-positive bacteria [39] that lack a cell wall.

myriapods (*Myriapoda*) [115] Arthropods [26] characterized by an elongate, segmented trunk with many legs. Includes centipedes and millipedes.

N

nanoarchaeotes (*Nanoarchaeota*) Minute archaeans [3] that parasitize cells of crenarchaeotes.

nematodes (*Nematoda*) [111] A very large group of elongate, unsegmented ecdysozoans [25] with thick, multilayer cuticles. They are among the most abundant and diverse animals, although most species have not yet been described. Include free-living predators and scavengers, as well as parasites of most species of land plants [11] and animals [19].

neognaths (*Neognathae*) The main group of birds [132], including all living species except the ratites (ostrich, emu, rheas, kiwis, cassowaries) and tinamous. *See* palaeognaths.

neopterans (*Neoptera*) A group of hexapods [117] that includes all the winged insects that can flex their wings over their abdomens.

O

oligochaetes (*Oligochaeta*) Annelid [105] group whose members lack parapodia, eyes, and anterior tentacles, and have few setae. Earthworms are the most familiar oligochaetes.

onychophorans *See* velvet worms.

oomycetes (*Oomycota*) [52] Water molds and relatives; absorptive heterotrophs with nutrient-absorbing, filamentous hyphae.

opisthokonts (*Opisthokonta*) [16] A group of eukaryotes [4] in which the flagellum on motile cells, if present, is posterior. The opisthokonts include the fungi [17], animals [19], and choanoflagellates [91].

ostracods (*Ostracoda*) Marine and freshwater crustaceans [116] that are laterally compressed and protected by two clamlike calcareous or chitinous shells.

P

palaeognaths (*Palaeognathae*) A group of secondarily flightless or weakly flying birds [132]. Includes the flightless ratites (ostrich, emu, rheas, kiwis, cassowaries) and the weakly flying tinamous.

parabasalids (*Parabasalia*) [59] A group of unicellular eukaryotes [4] that lack mitochondria; they possess flagella in clusters near the anterior of the cell.

phoronids (*Phoronida*) [104] A small group of sessile, wormlike marine lophotrochozoans [24] that secrete chitinous tubes and feed using a lophophore.

placoderms (*Placodermi*) An extinct group of jawed vertebrates [32] that lacked teeth. Placoderms were the dominant predators in Devonian oceans.

placozoans (*Placozoa*) [96] A poorly known group of structurally simple, asymmetrical, flattened, transparent animals found in coastal marine tropical and subtropical seas. Most evidence suggests that placozoans are secondarily simplified eumetazoans [21].

Plantae (*Plantae* or *Archaeplastida*) [9] The most broadly defined plant group. In most parts of this book, we use the word "plant" as synonymous with "land plant" [11], a more restrictive definition.

plasmodial slime molds (*Myxogastrida*) Amoebozoans [84] that in their feeding stage consist of a coenocyte called a plasmodium.

pogonophorans (*Pogonophora*) Deep-sea annelids [105] that lack a mouth or digestive tract; they feed by taking up dissolved organic matter, facilitated by endosymbiotic bacteria in a specialized organ (the trophosome).

polychaetes (*Polychaeta*) A group of mostly marine annelids [105] with one or more pairs of eyes and one or more pairs of feeding tentacles; parapodia and setae extend from most body segments. May be paraphyletic with respect to the clitellates.

priapulids (*Priapulida*) [107] A small group of cylindrical, unsegmented, wormlike marine ecdysozoans [25] that takes its name from its phallic appearance.

primates (*Primates*) A group of mammals [129] that includes prosimians and anthropoids.

"prokaryotes" Not a monophyletic group; as commonly used, includes the bacteria [2] and archaeans [3]. A term of convenience encompassing all cellular organisms that are not eukaryotes.

prosimians (*Strepsirrhini*) A group of primate mammals [129] that includes lemurs, lorises, and galagos.

proteobacteria (*Proteobacteria*) [44] A large and extremely diverse group of Gram-negative bacteria that includes many pathogens, nitrogen fixers, and photosynthesizers. Includes the alpha, beta, gamma, delta, and epsilon proteobacteria.

"protists" This term of convenience is used to encompass a large number of distinct and distantly related groups of eukaryotes, many but far from all of which are microbial and unicellular. Essentially a "catch-all" term for any eukaryote group not contained within the land plants [11], fungi [17], or animals [19].

protostomes (*Protostomia*) [23] One of the two major groups of bilaterians [22]. In protostomes, the mouth typically forms from the blastopore (if present) in early development (contrast with deuterostomes). The major protostome groups are the lophotrochozoans [24] and ecdysozoans [25].

prototherians (*Prototheria*) A mostly extinct group of mammals [129], common during the Cretaceous and early Cenozoic. The five living species—four echidnas and the duck-billed platypus—are the only extant egg-laying mammals.

pterobranchs (*Pterobranchia*) A small group of sedentary marine hemichordates [119] that live in tubes secreted by the proboscis. They have one to nine pairs of arms, each bearing long tentacles that capture prey and function in gas exchange.

pterygotes (*Pterygota*) A group of hexapods [117] that includes all the winged insects.

pycnogonids (*Pycnogonida*) Treated in this book as a group of chelicerates [114], but sometimes considered an independent group of arthropods [26]. Pycnogonids have reduced bodies and very long, slender legs. Also called sea spiders.

R

radiolarians (*Radiolaria*) [55] Amoeboid organisms with needlelike pseudopods supported by microtubules. Most have glassy internal skeletons.

ray-finned fishes (*Actinopterygii*) [125] A highly diverse group of freshwater and marine bony vertebrates [33]. They have reduced swim bladders that often function as hydrostatic organs and fins supported by soft rays (lepidotrichia). Includes most familiar fishes.

red algae (*Rhodophyta*) [64] Mostly multicellular, marine and freshwater algae characterized by the presence of phycoerythrin in their chloroplasts.

reptiles (*Reptilia*) [37] One of the two major groups of extant amniotes [36], supported on the basis of similar skull structure and gene sequences. The term "reptiles" traditionally excluded the birds [132], but the resulting group is then clearly paraphyletic. As used in this book, the reptiles include turtles [131], lepidosaurs [130], birds [132], and crocodilians [133].

rhizaria (*Rhizaria*) [7] Mostly amoeboid unicellular eukaryotes with pseudopods, many with external or internal shells. Includes the foraminiferans [57], cercozoans [56], and radiolarians [55].

rhyniophytes (*Rhyniophyta*) A group of early vascular plants [12] that appeared in the Silurian and became extinct in the Devonian. Possessed dichotomously branching stems with terminal sporangia but no true leaves or roots.

ribbon worms (*Nemertea*) [102] A group of unsegmented lophotrochozoans [24] with an eversible proboscis used to capture prey. Mostly marine, but some species live in fresh water or on land.

rotifers (*Rotifera*) [101] Tiny (< 0.5 mm) lophotrochozoans [24] with a pseudocoelomic body cavity that functions as a hydrostatic organ, and a ciliated feeding organ called the corona that surrounds the head. Rotifers live in freshwater and wet terrestrial habitats.

roundworms (*Nematoda*) [111] *See* nematodes.

S

sac fungi (*Ascomycota*) [89] Fungi that bear the products of meiosis within sacs (asci) if the organism is multicellular. Some are unicellular.

salamanders (*Caudata*) A group of amphibians [128] with distinct tails in both larvae and adults and limbs set at right angles to the body.

salps *See* thaliaceans.

sarcopterygians (*Sarcopterygii*) [34] *See* lobe-limbed vertebrates.

scyphozoans (*Scyphozoa*) Marine cnidarians [97] in which the medusa stage dominates the life cycle. Commonly known as jellyfish.

sea cucumbers (*Holothuroidea*) Echinoderms [118] with an elongate, cucumber-shaped body and leathery skin. They are scavengers on the ocean floor.

sea spiders *See* pycnogonids.

sea squirts *See* ascidians.

sea stars (*Asteroidea*) Echinoderms [118] with five (or more) fleshy "arms" radiating from an indistinct central disk. Also called starfishes.

sea urchins (*Echinoidea*) Echinoderms [118] with a test (shell) that is covered in spines. Most are globular in shape, although some groups (such as the sand dollars) are flattened.

"seed ferns" A paraphyletic group of loosely related, extinct seed plants that flourished in the Devonian and Carboniferous. Characterized by large, frondlike leaves that bore seeds.

seed plants (*Spermatophyta*) [13] Heterosporous vascular plants [12] that produce seeds; most produce wood; branching is axillary (not dichotomous). The major seed plant groups are gymnosperms [14] and angiosperms [15].

sow bugs *See* isopods.

spirochetes (*Spirochaetes*) [45] Motile, Gram-negative bacteria with a helically coiled structure and characterized by axial filaments.

sponges (*Porifera*) [20] A group of relatively asymmetric, filter-feeding animals that lack a gut or nervous system and generally lack differentiated tissues. Includes glass sponges [92], demosponges [93], and calcareous sponges [94].

springtails (*Collembola*) Wingless hexapods [117] with springing structures on the third and fourth segments of their bodies. Springtails are extremely abundant in some environments (especially in soil, leaf litter, and vegetation).

squamates (*Squamata*) The major group of lepidosaurs [130], characterized by the possession of movable quadrate bones (which allow the upper jaw to move independently of the rest of the skull) and hemipenes (a paired set of eversible penises, or penes) in males. Includes the lizards (a paraphyletic group), snakes, and amphisbaenians.

star anise (*Austrobaileyales*) [80] A group of woody angiosperms [15] thought to be the sister-group of the clade of flowering plants that includes eudicots [83], monocots [82], and magnoliids [81].

starfish (*Asteroidea*) *See* sea stars.

stoneworts (*Charales*) [67] Multicellular green algae with branching, apical growth and plasmodesmata between adjacent cells. The closest living relatives of the land plants [11], they retain the egg in the parent organism.

stramenopiles (*Heterokonta* or *Stramenopila*) [6] Organisms having, at some stage in their life cycle, two unequal flagella, the longer possessing rows of tubular hairs. Chloroplasts, when present, surrounded by four membranes. Major stramenopile groups include the brown algae [54], diatoms [53], and oomycetes [52].

streptophytes (*Streptophyta*) All of the green plants [10] other than chlorophytes.

T

tapeworms (*Cestoda*) Parasitic flatworms [100] that live in the digestive tracts of vertebrates as adults, and usually in various other species of animals as juveniles.

tardigrades (*Tardigrada*) [112] Small (< 0.5 mm) ecdysozoans [25] with fleshy, unjointed legs and no circulatory or gas exchange organs. They live in marine sands, in temporary freshwater pools, and on the water films of plants. Also called water bears.

tetrapods (*Tetrapoda*) [35] The major group of lobe-limbed vertebrates [34]; includes the amphibians [128] and the amniotes [36]. Named for the presence of four jointed limbs (although limbs have been secondarily reduced or lost completely in several tetrapod groups).

thaliaceans (*Thaliacea*) A group of solitary or colonial planktonic marine tunicates [121]. Also called salps.

thaumarchaeotes (*Thaumarchaeota*) A poorly known group of Archaea [3] that oxidize ammonia; known only from DNA isolated from hot environments.

therians (*Theria*) Mammals [129] characterized by viviparity (live birth). Includes eutherians and marsupials.

theropods (*Theropoda*) Archosaurs [38] with bipedal stance, hollow bones, a furcula ("wishbone"), elongated metatarsals with three-fingered feet, and a pelvis that points backwards. Includes many well-known extinct dinosaurs (such as *Tyrannosaurus rex*), as well as the living birds [132].

tracheophytes *See* vascular plants [12].

trilobites (*Trilobita*) An extinct group of arthropods [26] related to the chelicerates [114]. Trilobites flourished from the Cambrian through the Permian.

tuataras (*Rhyncocephalia*) A group of lepidosaurs [130] known mostly from fossils; there are only two living tuatara species. The quadrate bone of the upper jaw is fixed firmly to the skull. Sister group of the squamates.

tunicates (*Tunicata*) [121] A group of chordates [30] that are mostly saclike filter feeders as adults, with motile larval stages that resemble tadpoles.

turbellarians (*Turbellaria*) A group of free-living, generally carnivorous flatworms [100]. Their monophyly is questionable.

turtles (*Testudines*) [131] A group of reptiles [37] with a bony carapace (upper shell) and plastron (lower shell) that encase the body in a fashion unique among the vertebrates.

U

urochordates (*Tunicata*) [121] *See* tunicates.

V

vascular plants (*Tracheophyta*) [12] Plants with xylem and phloem. Major groups include the lycophytes [71] and euphyllophytes.

velvet worms (*Onychophora*) [113] Elongate, segmented ecdysozoans [25] with many pairs of soft, unjointed, claw-bearing legs. Also known as onychophorans.

vertebrates (*Vertebrata*) [31] The largest group of chordates [30], characterized by a rigid endoskeleton supported by the vertebral column and an anterior skull encasing a brain. Includes hagfishes [122], lampreys [123], and the jawed vertebrates [32], although some biologists exclude the hagfishes from this group. *See also* craniates.

W

water bears *See* tardigrades.

water lilies (*Nymphaeaceae*) [79] A group of aquatic, freshwater angiosperms [15] that are rooted in soil in shallow water, with round floating leaves and flowers that extend above the water's surface. They are the sister-group to most of the remaining flowering plants, with the exception of the genus *Amborella* [78].

Y

"yeasts" Convenience term for several distantly related groups of unicellular fungi [17].

Z

"zygospore fungi" (*Zygomycota*, if monophyletic) [87] A convenience term for a probably paraphyletic group of fungi [17] in which hyphae of differing mating types conjugate to form a zygosporangium.

Appendix B Making Sense of Data: A Statistics Primer

This appendix is designed help you understand the analysis of biological data in a statistical context. We outline major concepts involved in the collection and analysis of data, and we present basic statistical analyses that will help you interpret, understand, and complete the Apply the Concept and Analyze the Data problems throughout this book. Understanding these concepts will also help you understand and interpret scientific studies in general. We present formulas for some common statistical tests as examples, but the main purpose of this appendix is to help you understand the purpose and reasoning behind these tests. Once you understand the basis of an analysis, you may wish to use one of many free, online websites for conducting the tests (such as http://vassarstats.net).

Why Do We Do Statistics?

ALMOST EVERYTHING VARIES We live in a variable world, but within the variation we see among biological organisms there are predictable patterns. We use statistics to find and analyze these patterns. Consider any group of common things in nature—all women aged 22, all the cells in your liver, or all the blades of grass in your yard. Although they will have many similar characteristics, they will also have important differences. Men aged 22 tend to be taller than women aged 22, but of course not every man will be taller than every woman in this age group.

Natural variation can make it difficult to find general patterns. For example, scientists have determined that smoking increases the risk of getting lung cancer. But we know that not all smokers will develop lung cancer and not all nonsmokers will remain cancer-free. If we compare just one smoker with just one nonsmoker, we may end up drawing the wrong conclusion. So how did scientists discover this general pattern? How many smokers and nonsmokers did they examine before they felt confident about the risk of smoking?

Statistics helps us find and describe general patterns in nature, and draw conclusions from those patterns.

AVOIDING FALSE POSITIVES AND FALSE NEGATIVES When a woman takes a pregnancy test, there is some chance that it will be positive even if she is not pregnant, and there is some chance that it will be negative even if she is pregnant. We call these kinds of mistakes "false positives" and "false negatives."

Doing science is a bit like taking a medical test. We observe patterns in the world, and we try to draw conclusions about how the world works from those observations. Sometimes our observations lead us to draw the wrong conclusions. We might conclude that a phenomenon occurs, when it actually does not; or we might conclude that a phenomenon does not occur, when it actually does.

For example, planet Earth has been warming over the past century (see Concept 45.4). Ecologists are interested in whether plant and animal populations have been affected by global warming. If we have long-term information about the locations of species and about temperatures in certain areas, we can determine whether shifts in species distributions coincide with temperature changes. Such information, however, can be very complicated. Without proper statistical methods, one may not be able to detect the true impact of temperature, or instead may think a pattern exists when it does not.

Statistics helps us avoid drawing the wrong conclusions.

How Does Statistics Help Us Understand the Natural World?

Statistics is essential to scientific discovery. Most biological studies involve five basic steps, each of which requires statistics:

- **Step 1: Experimental Design**
  Clearly define the scientific question and the methods necessary to tackle the question.
- **Step 2: Data Collection**
  Gather information about the natural world through observations and experiments.
- **Step 3: Organize and Visualize the Data**
  Use tables, graphs, and other useful representations to gain intuition about the data.
- **Step 4: Summarize the Data**
  Summarize the data with a few key statistical calculations.
- **Step 5: Inferential Statistics**
  Use statistical methods to draw general conclusions from the data about the world and the ways it works.

Step 1: Experimental Design

We make observations and conduct experiments to gain knowledge about the world. Scientists come up with scientific ideas based on prior research and their own observations. These ideas start as a question such as "Does smoking cause cancer?" From prior experience, scientists then propose possible answers to the question in the form of hypotheses such as "Smoking increases the risk of cancer." Experimental design then involves devising comparisons that can test predictions of the hypotheses. For example, if we are interested in evaluating the hypothesis that smoking increases cancer risk, we might decide to compare the incidence of cancer in nonsmokers and smokers. Our hypothesis would predict that the cancer incidence is higher in smokers than in nonsmokers. There are various ways we could make that comparison. We could compare the smoking history of people newly diagnosed with cancer with that of people whose tests came out negative for cancer. Alternatively, we could assess the current smoking habits of a sample of people and see whether those who smoke more heavily are more likely to develop cancer in, say, 5 years. Statistics provides us with tools for assessing which approach will provide us with an answer with smaller costs in time and effort in data collection and analysis.

We use statistics to guide us in planning exactly how to make our comparisons—what kinds of data and how many observations we will need to collect.

Step 2: Data Collection

TAKING SAMPLES When biologists gather information about the natural world, they typically collect representative pieces of information, called **data** or observations. For example, when evaluating the efficacy of a candidate drug for medulloblastoma brain cancer, scientists may test the drug on tens or hundreds of patients, and then draw conclusions about its efficacy for all patients with these tumors. Similarly, scientists studying the relationship between body weight and clutch size (number of eggs) for female spiders of a particular species may examine tens to hundreds of spiders to make their conclusions.

We use the expression "sampling from a population" to describe this general method of taking representative pieces of information from the system under investigation (**FIGURE B1**). The pieces of information together make up a **sample** of the larger system, or **population**. In the cancer therapy example, each observation was the change in a patient's tumor size 6 months after initiating treatment, and the population of interest was all individuals with medulloblastoma tumors. In the spider example, each observation was a pair of measurements—body weight and clutch size—for a single female spider, and the population of interest was all female spiders of this species.

Sampling is a matter of necessity, not laziness. We cannot hope (and would not want) to collect and weigh *all* of the female spiders of the species of interest on Earth! Instead, we use statistics to determine how many spiders we must collect to confidently infer something about the general population and then use statistics again to make such inferences.

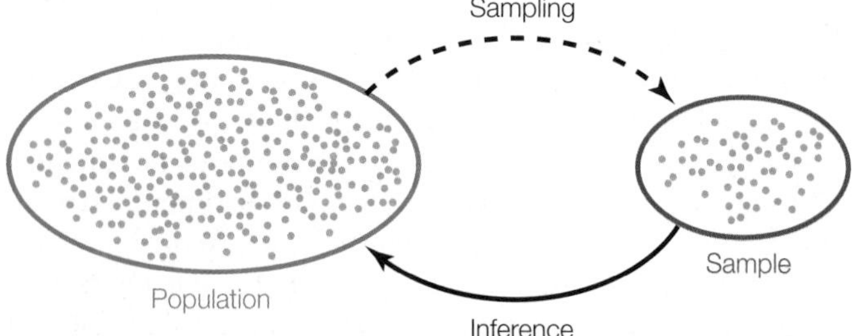

FIGURE B1 Sampling from a Population Biologists take representative samples from a population, use descriptive statistics to characterize their samples, and then use inferential statistics to draw conclusions about the original population.

TABLE B1 Weight and Length Measurements for a Sample of *Abramis brama* from Lake Laengelmavesi

Individual number	Fish weight (g)	Fish length (cm)
1	242	30.0
2	290	31.2
3	340	31.1
4	363	33.5
5	390	35.0
6	430	34.0
7	450	34.7
8	450	35.1
9	475	36.2
10	500	34.5
11	500	36.2
12	500	36.2
13	500	36.4
14	575	38.7
15	600	37.3
16	600	37.2
17	610	38.5
18	620	39.2
19	650	38.6
20	680	39.7
21	685	39.5
22	700	37.2
23	700	38.3
24	700	40.6
25	714	40.6
26	720	40.9
27	725	40.5
28	850	41.5
29	920	42.6
30	925	44.0
31	950	45.9
32	955	44.1
33	975	45.3
34	1,000	41.6

DATA COME IN ALL SHAPES AND SIZES In statistics we use the word "variable" to mean a measurable characteristic of an individual component of a system. Some variables are on a numerical scale, such as the daily high temperature or the clutch size of a spider. We call these **quantitative variables.** Quantitative variables that take on only whole number values (such as spider clutch size) are called **discrete variables,** whereas variables that can also take on a fractional value (such as temperature) are called **continuous variables.**

TABLE B2 Summary of Fish Weights of *Abramis brama* from Lake Laengelmavesi

Weight (g)	Frequency	Relative frequency
201–300	2	0.06
301–400	3	0.09
401–500	8	0.24
501–600	3	0.09
601–700	8	0.24
701–800	3	0.09
801–900	1	0.03
901–1,000	6	0.18
TOTAL	34	1.0

Other variables take categories as values, such as a human blood type (A, B, AB, or O) or an ant caste (queen, worker, or male). We call these **categorical variables**. Categorical variables with a natural ordering, such as a final grade in Introductory Biology (A, B, C, D, or F), are called **ordinal variables**.

Each class of variables comes with its own set of statistical methods. We will introduce a few common methods in this appendix that will help you work on the problems presented in this book, but you should consult a biostatistics textbook for more advanced tests and analyses for other data sets and problems.

Step 3: Organize and Visualize the Data

Your data consist of a series of values of the variable or variables of interest, each from a separate observation. For example, **TABLE B1** shows the weight and length of 34 fish (*Abramis brama*) from Lake Laengelmavesi in Finland. From this list of numbers, it's hard to get a sense of how big the fish in the lake are, or how variable they are in size. It's much easier to gain intuition about

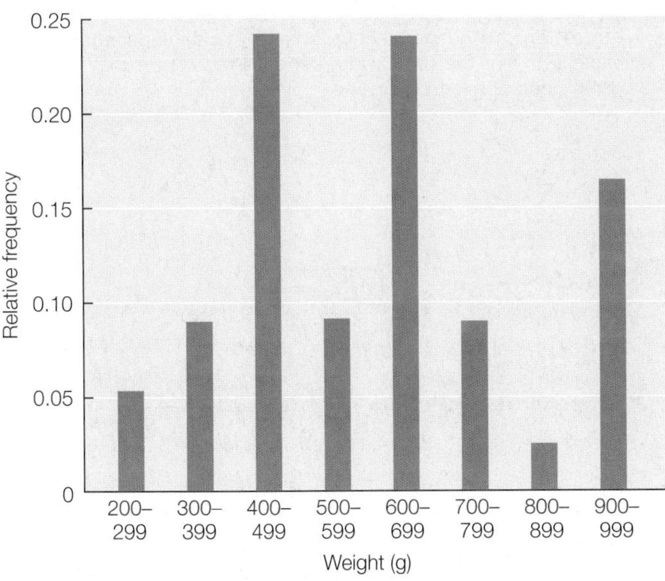

FIGURE B2 Histograms Depict Frequency Distributions of Quantitative Data This histogram shows the relative frequency of different weight classes of fish (*Abramis brama*).

TABLE B3 Poinsettia Colors

Color	Frequency	Proportion
Red	108	0.59
Pink	34	0.19
White	40	0.22
TOTAL	182	1.0

your data if you organize them. One way to do this is to group (or bin) your data into **classes**, and count up the number of observations that fall into each class. The result is a **frequency distribution**. **TABLE B2** shows the fish weight data as a frequency distribution. For each 100-gram weight class, Table B2 shows the number, or frequency, of observations in that weight class, as well as the relative frequency (proportion of the total) of observations in that weight class. Notice that the data take up much less space when organized in this fashion. Also notice that we can now see that most of the fish fall in the middle of the weight range, with relatively few very small or very large fish.

Go to MEDIA CLIP B1
Interpreting Frequency Distributions
PoL2e.com/mcB1

It is even easier to visualize the frequency distribution of fish weights if we graph them in the form of a **histogram** such as the one in **FIGURE B2**. When grouping quantitative data, it is necessary to decide how many classes to include. It is often useful to look at multiple histograms before deciding which grouping offers the best representation of the data.

Frequency distributions are also useful ways of summarizing categorical data. **TABLE B3** shows a frequency distribution of the colors of 182 poinsettia plants (red, pink, or white) resulting from an experimental cross between two parent plants. Notice that, as with the fish example, the table is a much more compact way to present the data than a list of 182 color observations would be. For categorical data, the possible values of the variable are the categories themselves, and the frequencies are the number of observations in each category. We can visualize frequency distributions of categorical data like this by constructing a **bar chart**. The heights of the bars indicate the number of observations in each category (**FIGURE B3**). Another way to display the

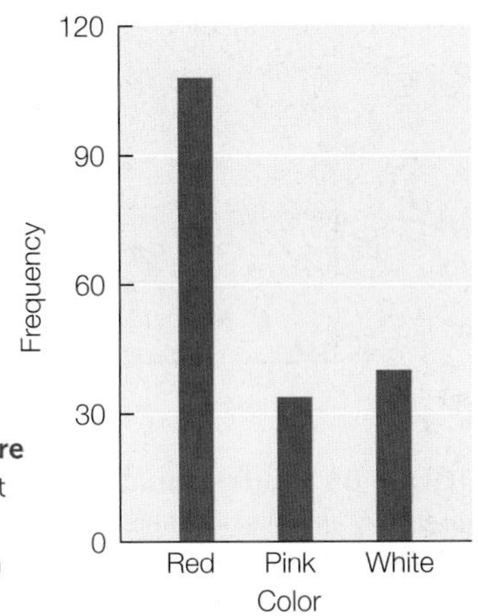

FIGURE B3 Bar Charts Compare Categorical Data This bar chart shows the frequency of three poinsettia colors that result from an experimental cross.

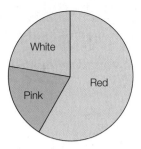

FIGURE B4 Pie Charts Show Proportions of Categories This pie chart shows the proportions of the three poinsettia colors presented in Table B3.

same data is in a **pie chart**, which shows the proportion of each category represented like pieces of a pie (**FIGURE B4**).

Sometimes we wish to compare two quantitative variables. For example, the researchers at Lake Laengelmavesi investigated the relationship between fish weight and length, from the data presented in Table B1. We can visualize this relationship using a **scatter plot** in which the weight and length of each fish is represented as a single point (**FIGURE B5**). These two variables have a **positive relationship** since the slope of a line drawn through the points is positive. As the length of a fish increases, its weight tends to increase in an approximately linear manner.

Tables and graphs are critical to interpreting and communicating data, and thus should be as self-contained and understandable as possible. Their content should be easily understood simply by looking at them. Axes, captions, and units should be clearly labeled, statistical terms should be defined, and appropriate groupings should be used when tabulating or graphing quantitative data.

Step 4: Summarize the Data

A **statistic** is a numerical quantity calculated from data, whereas **descriptive statistics** are quantities that describe general patterns in data. Descriptive statistics allow us to make straightforward comparisons among different data sets and concisely communicate basic features of our data.

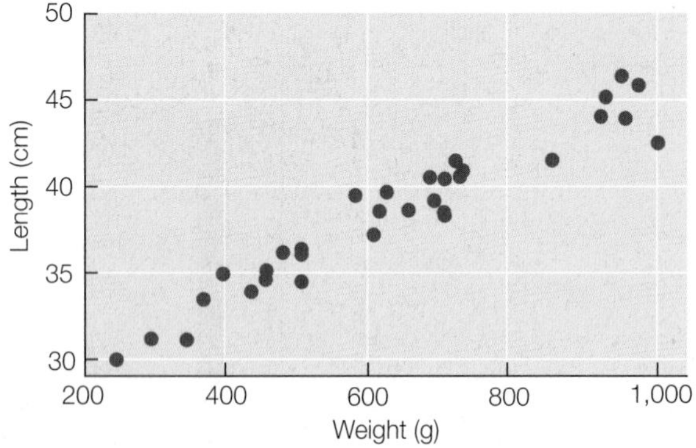

FIGURE B5 Scatter Plots Contrast Two Variables Scatter plot of *Abramis brama* weights and lengths (measured from nose to end of tail). These two variables have a positive relationship since the slope of a line drawn through the points is positive.

DESCRIBING CATEGORICAL DATA For categorical variables, we typically use proportions to describe our data. That is, we construct tables containing the proportions of observations in each category. For example, the third column in Table B3 provides the proportions of poinsettia plants in each color category, and the pie chart in Figure B4 provides a visual representation of those proportions.

DESCRIBING QUANTITATIVE DATA For quantitative data, we often start by calculating **measures of center**, quantities that roughly tell us where the center of our data lies. There are three commonly used measures of center:

- The **mean**, or average value, of our sample is simply the sum of all the values in the sample divided by the number of observations in our sample (**FIGURE B6**).
- The **median** is the value at which there are equal numbers of smaller and larger observations.
- The **mode** is the most frequent value in the sample.

It is often just as important to quantify the variation in the data as it is to calculate its center. There are several statistics

RESEARCH TOOLS

FIGURE B6 Descriptive Statistics for Quantitative Data

Below are the equations used to calculate the descriptive statistics we discuss in this appendix. You can calculate these statistics yourself, or use free internet resources to help you make your calculations.

Notation:

$x_1, x_2, x_3, \ldots x_n$ are the n observations of variable X in the sample.

$\sum_{i=1}^{n} x_i = x_1 + x_2 + x_3, \ldots + x_n$ is the sum of all of the observations. (The Greek letter sigma, Σ, is used to denote "sum of.")

In regression, the independent variable is X, and the dependent variable is Y. b_0 is the vertical intercept of a regression line. b_1 is the slope of a regression line.

Equations

1. Mean: $\bar{x} = \dfrac{\sum_{i=1}^{n} x_i}{n}$
2. Standard deviation: $s = \sqrt{\dfrac{\Sigma(x_i - \bar{x})^2}{n-1}}$
3. Correlation coefficient: $r = \dfrac{\Sigma(x_i - \bar{x})(y_i - \bar{y})}{\sqrt{\Sigma(x_i - \bar{x})^2 (y_i - \bar{y})^2}}$
4. Least-squares regression line: $Y = b_0 + b_1 X$ where $b_1 = \dfrac{\Sigma(x_i - \bar{x})(y_i - \bar{y})}{\Sigma(x_i - \bar{x})^2}$ and $b_0 = \bar{y} - b_1 \bar{x}$
5. Standard error of the mean: $SE_{\bar{x}} = \dfrac{s}{\sqrt{n}}$

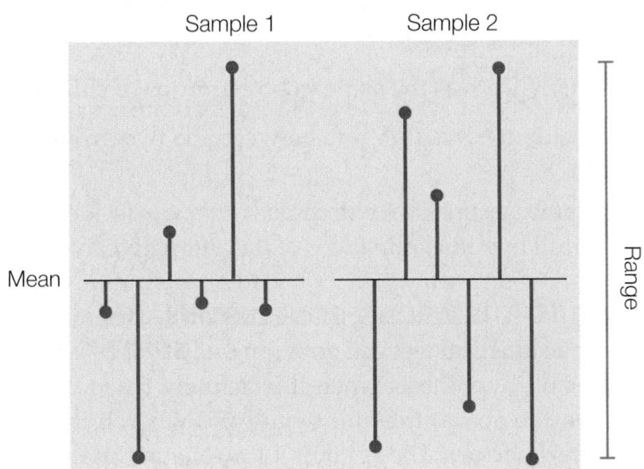

FIGURE B7 Measures of Dispersion Two samples with the same mean (black horizontal lines) and range (blue vertical line). Red lines show the deviations of each observation from the mean. Samples with large deviations have large standard deviations. Sample 1 has a smaller standard deviation than sample 2.

that tell us how much the values differ from one another. We call these **measures of dispersion**. The easiest one to understand and calculate is the **range**, which is simply the largest value in the sample minus the smallest value. The most commonly used measure of dispersion is the **standard deviation**, which calculates the extent to which the data are spread out from the mean. A deviation is the difference between an observation and the mean of the sample. The standard deviation is a measure of how far the average observation in the sample is from the sample mean. Two samples can have the same range, but very different standard deviations if observations in one are clustered closer to the mean than in the other. In **FIGURE B7**, for example, sample 1 has a smaller standard deviation (s = 2.6) than does sample 2 (s = 3.6), even though the two samples have the same means and ranges.

To demonstrate these descriptive statistics, let's return to the Lake Laengelmavesi study (see the data in Table B1). The mean weight of the 34 fish (see equation 1 in Figure B6) is:

$$\bar{x} = \frac{\text{sum of the weight of all fish in sample}}{\text{number of fish in sample}} = \frac{21{,}284\text{ g}}{34} = 626\text{ g}$$

Since there is an even number of observations in the sample, then the median weight is the value halfway between the two middle values:

$$\frac{610\text{ g} + 620\text{ g}}{2} = 615\text{ g}$$

The mode of the sample is 500 g, which appears 4 times. The standard deviation (see equation 2 in Figure B6) is:

$$s = \sqrt{\frac{\sum (x_i - \bar{x})^2}{n-1}} = 206.6\text{ g}$$

and the range is 1,000 g – 242 g = 758 g.

DESCRIBING THE RELATIONSHIP BETWEEN TWO QUANTITATIVE VARIABLES Biologists are often interested in understanding the relationship between two different quantitative variables: How does the height of an organism relate to its weight? How does air pollution relate to the prevalence of asthma? How does lichen abundance relate to levels of air pollution? Recall that scatter plots visually represent such relationships.

We can quantify the strength of the relationship between two quantitative variables using a single value called the Pearson product–moment **correlation coefficient** (see equation 3 in Figure B6). This statistic ranges between –1 and 1 and tells us how closely the points in a scatter plot conform to a straight line. A negative correlation coefficient indicates that one variable decreases as the other increases; a positive correlation coefficient indicates that the two variables increase together; and a correlation coefficient of zero indicates that there is no linear relationship between the two variables (**FIGURE B8**).

One must always keep in mind that *correlation does not mean causation*. Two variables can be closely related without one causing the other. For example, the number of cavities in a child's mouth correlates positively with the size of his or her feet. Clearly cavities do not enhance foot growth; nor does foot growth cause tooth decay. Instead the correlation exists because both quantities tend to increase with age.

Intuitively, the straight line that tracks the cluster of points on a scatter plot tells us something about the typical relationship between the two variables. Statisticians do not, however, simply eyeball the data and draw a line by hand. They often use a method called least-squares **linear regression** to fit a straight line to the data (see equation 4 in Figure B6). This method calculates the line that minimizes the overall vertical distances between the points in the scatter plot and the line itself. These distances are called **residuals** (**FIGURE B9**). Two parameters describe the regression line: b_0 (the vertical intercept of the line, or the expected value of variable Y when X = 0), and b_1 (the slope of the line, or how much values of Y are expected to change with changes in values of X).

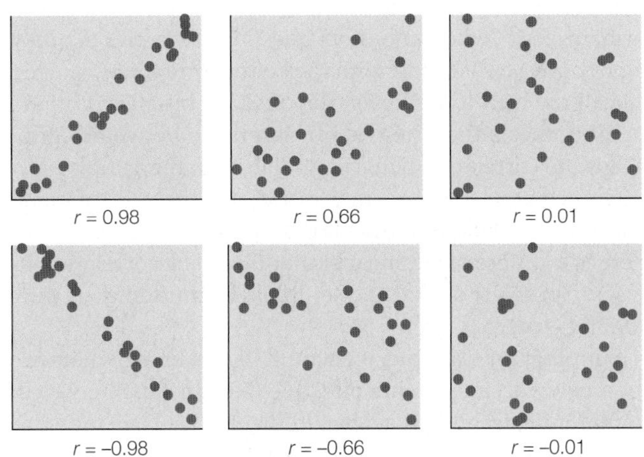

FIGURE B8 Correlation Coefficients The correlation coefficient (r) indicates both the strength and the direction of the relationship.

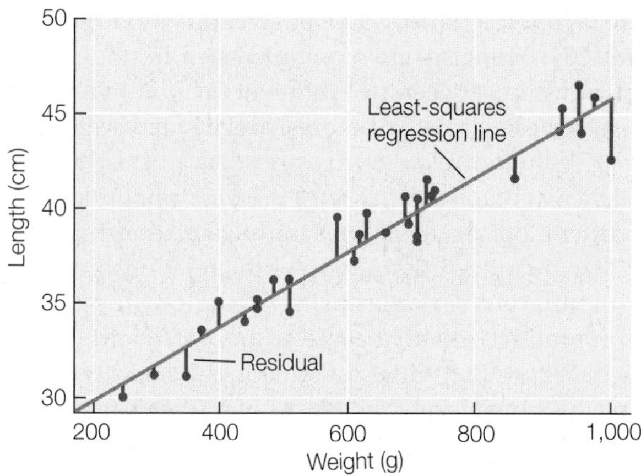

FIGURE B9 Linear Regression Estimates the Typical Relationship between Two Variables Least-squares linear regression line for *Abramis brama* weights and lengths (measured from nose to end of tail). The regression line (blue line) is given by the equation $Y = 26.1 + 0.02X$. It is the line that minimizes the sum of the squares of the residuals (red lines).

Step 5: Inferential Statistics

Data analysis often culminates with statistical inference—an attempt to draw general conclusions about the system under investigation—the larger population—from observations of a representative sample (see Figure B1). When we test a new medulloblastoma brain cancer drug on ten patients, we do not simply want to know the fate of those ten individuals; rather, we hope to predict the drug's efficacy on the much larger population of all medulloblastoma patients.

STATISTICAL HYPOTHESES Once we have collected our data, we want to evaluate whether or not they fit the predictions of our hypothesis. For example, we want to know whether or not cancer incidence is greater in our sample of smokers than in our sample of nonsmokers, whether or not clutch size increases with spider body weight in our sample of spiders, or whether or not growth of a sample of fertilized plants is greater than that of a sample of unfertilized plants.

Before making statistical inferences from data, we must formalize our "whether or not" question into a pair of opposing hypotheses. We start with the "or not" hypothesis, which we call the **null hypothesis** (denoted H_0) because it often is the hypothesis that there is no difference between sample means, no correlation between variables in the sample, or no difference between our sample and an expected frequency distribution. The **alternative hypothesis** (denoted H_A) is that there is a difference in means, that there is a correlation between variables, or that the sample distribution does differ from the expected one.

Suppose, for example, we would like to know whether or not a new vaccine is more effective than an existing vaccine at immunizing children against influenza. We have measured flu incidence in a group of children who received the new vaccine and want to compare it with flu incidence in a group of children who received the old vaccine. Our statistical hypotheses would be as follows:

H_0: Flu incidence was the same in both groups of children.

H_A: Flu incidence was different between the two groups of children.

In the next few sections we will discuss how we decide when to reject the null hypothesis in favor of the alternative hypothesis.

JUMPING TO THE WRONG CONCLUSIONS There are two ways that a statistical test can go wrong (**FIGURE B10**). We can reject the null hypothesis when it is actually true (**Type I error**), or we can accept the null hypothesis when it is actually false (**Type II error**). These kinds of errors are analogous to false positives and false negatives in medical testing, respectively. If we mistakenly reject the null hypothesis when it is actually true, then we falsely endorse the incorrect hypothesis. If we are unable to reject the null hypothesis when it is actually false, then we fail to realize a yet undiscovered truth.

Suppose we would like to know whether there are more females than males in a population of 10,000 individuals. To determine the makeup of the population, we choose 20 individuals randomly and record their sex. Our null hypothesis is that there are not more females than males; and our alternative hypothesis is that there are. The following scenarios illustrate the possible mistakes we might make:

- *Scenario 1*: The population actually has 40% females and 60% males. Although our random sample of 20 people is likely to be dominated by males, it is certainly possible that, by chance, we will end up choosing more females than males. If this occurs, and we mistakenly reject the null hypothesis (that there are not more females than males), then we make a Type I error.
- *Scenario 2*: The population actually has 60% females and 40% males. If, by chance, we end up with a majority of males in our sample and thus fail to reject the null hypothesis, then we make a Type II error.

Fortunately, statistics has been developed precisely to avoid these kinds of errors and inform us about the reliability of our conclusions. The methods are based on calculating the probabilities of different possible outcomes. Although you may have heard or even used the word "probability" on multiple

Our conclusion	The real world: Null hypothesis true (*not more females*)	The real world: Null hypothesis false (*more females*)
Null hypothesis true (*not more females*)	√	Type II error (*false negative*)
Null hypothesis false (*more females*)	Type I error (*false positive*)	√

FIGURE B10 Two Types of Errors Possible outcomes of a statistical test. Statistical inference can result in correct and incorrect conclusions about the population of interest.

occasions, it is important that you understand its mathematical meaning. A **probability** is a numerical quantity that expresses the likelihood of some event. It ranges between zero and one; zero means there is no chance the event will occur, and one means the event is guaranteed to occur. This only makes sense if there is an element of chance, that is, if it is possible the event will occur and possible that it will not occur. For example, when we flip a fair coin, it will land on heads with probability 0.5 and land on tails with probability 0.5. When we select individuals randomly from a population with 60% females and 40% males, we will encounter a female with probability 0.6 and a male with probability 0.4.

Probability plays a very important role in statistics. To draw conclusions about the real world (the population) from our sample, we first calculate the probability of obtaining our sample if the null hypothesis is true. Specifically, statistical inference is based on answering the following question:

Suppose the null hypothesis is true. What is the probability that a random sample would, by chance, differ from the null hypothesis as much as our sample differs from the null hypothesis?

If our sample is highly improbable under the null hypothesis, then we rule it out in favor of our alternative hypothesis. If, instead, our sample has a reasonable probability of occurring under the null hypothesis, then we conclude that our data are consistent with the null hypothesis and we do not reject it.

Returning to the sex ratio example, let's consider two new scenarios:

- *Scenario 3*: Suppose we want to infer whether or not females constitute the majority of the population (our alternative hypothesis) based on a random sample containing 12 females and 8 males. We would calculate the probability that a random sample of 20 people includes at least 12 females assuming that the population, in fact, has a 50:50 sex ratio (our null hypothesis). This probability is 0.13, which is too high to rule out the null hypothesis.
- *Scenario 4*: Suppose now that our sample contains 17 females and 3 males. If our population is truly evenly divided, then this sample is much less likely than the sample in scenario 3. The probability of such an extreme sample is 0.0002, and would lead us to rule out the null hypothesis and conclude that there are more females than males.

This agrees with our intuition. When choosing 20 people randomly from an evenly divided population, we would be surprised if almost all of them were female, but would not be surprised at all if we ended up with a few more females than males (or a few more males than females). Exactly how many females do we need in our sample before we can confidently infer that they make up the majority of the population? And how confident are we when we reach that conclusion? Statistics allows us to answer these questions precisely.

STATISTICAL SIGNIFICANCE: AVOIDING FALSE POSITIVES Whenever we test hypotheses, we calculate the probability just discussed, and refer to this value as the ***P*-value** of our test. Specifically, the *P*-value is the probability of getting data as extreme as our data (just by chance) if the null hypothesis is in fact true. In other words, it is the likelihood that chance alone would produce data that differ from the null hypothesis as much as our data differ from the null hypothesis. How we measure the difference between our data and the null hypothesis depends on the kind of data in our sample (categorical or quantitative) and the nature of the null hypothesis (assertions about proportions, single variables, multiple variables, differences between variables, correlations between variables, etc.).

For many statistical tests, *P*-values can be calculated mathematically. One option is to quantify the extent to which the data depart from the null hypothesis and then use look-up tables (available in most statistics textbooks or on the internet) to find the probability that chance alone would produce a difference of that magnitude. Most scientists, however, find *P*-values primarily by using statistical software rather than hand calculations combined with look-up tables. Regardless of the technology, the most important steps of the statistical analysis are still left to the researcher: constructing appropriate null and alternative hypotheses, choosing the correct statistical test, and drawing correct conclusions.

After we calculate a *P*-value from our data, we have to decide whether it is small enough to conclude that our data are inconsistent with the null hypothesis. This is decided by comparing the *P*-value to a threshold called the **significance level**, which is often chosen even before making any calculations. We reject the null hypothesis only when the *P*-value is less than or equal to the significance level, denoted α. This ensures that, if the null hypothesis is true, we have at most a probability α of accidentally rejecting it. Therefore the lower the value of α, the less likely you are to make a Type I error (see the lower left cell of Figure B10). The most commonly used significance level is $\alpha = 0.05$, which limits the probability of a Type I error to 5%.

If our statistical test yields a *P*-value that is less than our significance level α, then we conclude that the effect described by our alternative hypothesis is statistically significant at the level α and we reject the null hypothesis. If our *P*-value is greater than α, then we conclude that we are unable to reject the null hypothesis. In this case, we do not actually reject the alternative hypothesis, rather we conclude that we do not yet have enough evidence to support it.

POWER: AVOIDING FALSE NEGATIVES The **power** of a statistical test is the probability that we will correctly reject the null hypothesis when it is false (see the lower right cell of Figure B10). Therefore the higher the power of the test, the less likely we are to make a Type II error (see the upper right cell of Figure B10). The power of a test can be calculated, and such calculations can be used to improve your methodology. Generally, there are several steps that can be taken to increase power and thereby avoid false negatives:

- **Decrease the significance level, α.** The higher the value of α, the harder it is to reject the null hypothesis, even if it is actually false.

- **Increase the sample size.** The more data one has, the more likely one is to find evidence against the null hypothesis, if it is actually false.
- **Decrease variability in the sample.** The more variation there is in the sample, the harder it is to discern a clear effect (the alternative hypothesis) when it actually exists.

It is always a good idea to design your experiment to reduce any variability that may obscure the pattern you seek to detect. For example, it is possible that the chance of a child contracting influenza varies depending on whether he or she lives in a crowded (e.g., urban) environment or one that is less so (e.g., rural). To reduce variability, a scientist might choose to test a new influenza vaccine only on children from one environment or the other. After you have minimized such extraneous variation, you can use power calculations to choose the right combination of α and sample size to reduce the risks of Type I and Type II errors to desirable levels.

There is a trade-off between Type I and Type II errors: as α increases, the risk of a Type I error decreases but the risk of a Type II error increases. As discussed above, scientists tend to be more concerned about Type I than Type II errors. That is, they believe it is worse to mistakenly believe a false hypothesis than it is to fail to make a new discovery. Thus they prefer to use low values of α. However, there are many real-world scenarios in which it would be worse to make a Type II error than a Type I error. For example, suppose a new cold medication is being tested for dangerous (life-threatening) side effects. The null hypothesis is that there are no such side effects. A Type II error might lead regulatory agencies to approve a harmful medication that could cost human lives. In contrast, a Type I error would simply mean one less cold medication among the many that already line pharmacy shelves. In such cases, policymakers take steps to avoid a Type II error, even if, in doing so, they increase the risk of a Type I error.

STATISTICAL INFERENCE WITH QUANTITATIVE DATA Statistics that describe patterns in our samples are used to estimate properties of the larger population. Earlier we calculated the mean weight of a sample of *Abramis brama* in Lake Laengelmavesi, which provided us with an estimate of the mean weight of all the *Abramis brama* in the lake. But how close is our estimate to the true value in the larger population? Our estimate from the sample is unlikely to exactly equal the true population value. For example, our sample of *Abramis brama* may, by chance, have included an excess of large individuals. In this case, our sample would overestimate the true mean weight of fish in the population.

The **standard error** of a sample statistic (such as the mean) is a measure of how close it is likely to be to the true population value. The **standard error of the mean**, for example, provides an estimate of how far we might expect a sample mean to deviate from the true population mean. It is a function of how much individual observations vary within samples (the standard deviation) and the size of the sample (n). Standard errors increase with the degree of variation within samples, and they decrease with the number of observations in a sample (because large samples provide better estimates about the underlying population than do small samples). For our sample of 34 *Abramis brama*, we would calculate the standard error of the mean using equation 5 in Figure B6:

$$SE_{\bar{x}} = \frac{s}{\sqrt{n}} = \frac{206.6\text{ g}}{\sqrt{34}} = 35.4\text{ g}$$

The standard error of a statistic is related to the **confidence interval**—a range around the sample statistic that has a specified probability of including the true population value. The formula we use to calculate confidence intervals depends on the characteristics of the data, and the particular statistic. For many types of continuous data, the bounds of the **95% confidence interval** of the mean can be calculated by taking the mean and adding and subtracting 1.96 times the standard error of the mean. Consider our sample of 34 *Abramis brama*, for example. We would say that the mean of 626 grams has a 95% confidence interval from 556.6 grams to 695.4 grams (626 ± 69.4 grams). If all our assumptions about our data have been met, we would expect the true population mean to fall in this confidence interval 95% of the time.

Researchers typically use graphs or tables to report sample statistics (such as the mean) as well as some measure of confidence in them (such as their standard error or a confidence interval). This book is full of examples. If you understand the concepts of sample statistics, standard errors, and confidence intervals, you can see for yourself the major patterns in the data, without waiting for the authors to tell you what they are. For example, if samples from two groups have 95% confidence intervals of the mean that do not overlap, then you can conclude that it is unlikely that the groups have the same true mean.

Go to ANIMATED TUTORIAL B1
Standard Deviations, Standard Errors, and Confidence Intervals
PoL2e.com/atB1

Biologists conduct statistical tests to obtain more precise estimates of the probability of observing a given difference between samples if the null hypothesis that there is no difference in the populations is true. The appropriate test depends on the nature of the data and the experimental design. For example, we might want to calculate the probability that the mean weights of two different fish species in Lake Laengelmavesi, *Abramis brama* and *Leusiscus idus*, are the same. A simple method for comparing the means of two groups is the *t*-test, described in **FIGURE B11**. We looked earlier at data for *Abramis brama*; the researchers who collected these data also collected weights for six individuals of *Leusiscus idus*: 270, 270, 306, 540, 800, and 1,000 grams. We begin by stating our hypotheses and choosing a significance level:

H_0: *Abramis brama* and *Leusiscus idus* have the same mean weight.

H_A: *Abramis brama* and *Leusiscus idus* have different mean weights.

$\alpha = 0.05$

RESEARCH TOOLS

FIGURE B11 The *t*-test

What is the *t*-test? It is a standard method for assessing whether the means of two groups are statistically different from each another.

Step 1: State the null and alternative *hypotheses*:

H_0: The two populations have the same mean.
H_A: The two populations have different means.

Step 2: Choose a significance level, α, to limit the risk of a Type I error.

Step 3: Calculate the *test statistic*: $t_s = \frac{\bar{y}_1 - \bar{y}_2}{\sqrt{\frac{s_1^2}{n_1} + \frac{s_2^2}{n_2}}}$

Notation: $\bar{y}_1$ and $\bar{y}_2$ are the sample means; s_1 and s_2 are the sample standard deviations; and n_1 and n_2 are the sample sizes.

Step 4: Use the test statistic to assess whether the data are consistent with the null hypothesis:

Calculate the *P-value* (*P*) using statistical software or by hand using statistical tables.

Step 5: Draw conclusions from the test:

If $P \le \alpha$, then reject H_0, and conclude that the population distribution is significantly different.

If $P > \alpha$, then we do not have sufficient evidence to conclude that the means differ.

The test statistic is calculated using the means, standard deviations, and sizes of the two samples:

$$t_s = \frac{626 - 531}{\sqrt{\frac{207^2}{34} + \frac{310^2}{6}}} = 0.724$$

We can use statistical software or one of the free statistical sites on the internet to find that the *P*-value for this result is $P = 0.497$. Since *P* is considerably greater than α, we fail to reject the null hypothesis and conclude that our study does not provide evidence that the two species have different mean weights.

You may want to consult an introductory statistics textbook to learn more about confidence intervals, *t*-tests, and other basic statistical tests for quantitative data.

STATISTICAL INFERENCE WITH CATEGORICAL DATA With categorical data, we often wish to ask whether the frequencies of observations in different categories are consistent with a hypothesized frequency distribution. We can use a **chi-square goodness-of-fit** test to answer this question.

FIGURE B12 outlines the steps of a chi-square goodness-of-fit-test. As an example, consider the data described in Table B3. Many plant species have simple Mendelian genetic systems in which parent plants produce progeny with three different colors of flowers in a ratio of 2:1:1. However, a botanist believes that these particular poinsettia plants have a different genetic system that does not produce a 2:1:1 ratio of red, pink, and white plants. A chi-square goodness-of-fit can be used to assess whether or not the data are consistent with this ratio, and thus whether or not this simple genetic explanation is valid. We start by stating our hypotheses and significance level:

H_0: The progeny of this type of cross have the following probabilities of each flower color:

Pr{Red} = .50, Pr{Pink} = .25, Pr{White} = .25

H_A: At least one of the probabilities of H_0 is incorrect.

$\alpha = 0.05$

We next use the probabilities in H_0 and the sample size to calculate the expected frequencies:

	Red	Pink	White
Observed	108	34	40
Expected	(.50)(182) = 91	(.25)(182) = 45.5	(.25)(182) = 45.5

Based on these quantities, we calculate the chi-square test statistic:

$$\chi_s^2 = \sum_{i=1}^{C} \frac{(O_i - E_i)^2}{E_i} = \frac{(108-91)^2}{91} + \frac{(34-45.5)^2}{45.5} + \frac{(40-45.5)^2}{45.5} = 6.747$$

RESEARCH TOOLS

FIGURE B12 The Chi-Square Goodness-of-Fit Test

What is the chi-square goodness-of-fit test? It is a standard method for assessing whether a sample came from a population with a specific distribution.

Step 1: State the null and alternative *hypotheses*:

H_0: The population has the specified distribution.
H_A: The population does not have the specified distribution.

Step 2: Choose a significance level, α, to limit the risk of a Type I error.

Step 3: Determine the *observed frequency* and *expected frequency* for each category:

The observed frequency of a category is simply the number of observations in the sample of that type.

The expected frequency of a category is the probability of the category specified in H_0 multiplied by the overall sample size.

Step 4: Calculate the *test statistic*: $\chi_s^2 = \sum_{i=1}^{C} \frac{(O_i - E_i)^2}{E_i}$

Notation: *C* is the total number of categories, O_i is the observed frequency of category *i*, and E_1 is the expected frequency of category *i*.

Step 5: Use the test statistic to assess whether the data are consistent with the null hypothesis:

Calculate the *P-value* (*P*) using statistical software or by hand using statistical tables.

Step 6: Draw conclusions from the test:

If $P \le \alpha$, then reject H_0, and conclude that the population distribution is significantly different than the distribution specified by H_0.

If $P > \alpha$, then we do not have sufficient evidence to conclude that population has a different distribution.

Glossary

A

A horizon *See* topsoil.

abiotic (a′ bye ah tick) [Gk. *a*: not + *bios*: life] Nonliving. In ecology, the nonliving (physical and chemical) components of the environment of organisms. (Contrast with biotic.)

abscisic acid (ab sighs′ ik) A plant growth substance with growth-inhibiting action. Causes stomata to close; involved in a plant's response to salt and drought stress.

abscission (ab sizh′ un) [L. *abscissio*: break off] The process by which leaves, petals, and fruits separate from a plant.

absorption (1) Of light: complete retention (in contrast to reflection or transmission). (2) In the study of nutrition, the process of transporting molecules from the gut lumen into the blood.

absorption spectrum A graph of light absorption versus wavelength of light; shows how much light is absorbed at each wavelength.

absorptive heterotroph An organism (usually a fungus) that obtains its food by **absorptive heterotrophy**, i.e., by secreting digestive enzymes into the environment to break down large food molecules, then absorbing the breakdown products.

abyssal zone (uh biss′ ul) [Gk. *abyssos*: bottomless] The deepest portions of the oceans; characterized by extremely high pressures, low dissolved oxygen levels, and cold temperatures.

accessory fruit A fruit of a flowering plant in which the some of the flesh is derived from tissues other than the ovary; examples include apples and pears.

accessory reproductive organs Reproductive organs other than the gonads.

acclimation The change in an individual's phenotype that results from long-term exposure to a new environment. A form of phenotypic plasticity. Also called acclimatization.

acid growth hypothesis The hypothesis that auxin increases proton pumping, thereby lowering the pH of the cell wall and activating enzymes that loosen polysaccharides. Proposed to explain auxin-induced cell expansion in plants.

acoelomate An animal that does not have a coelom.

acquired immune deficiency syndrome (AIDS) A condition caused by human immunodeficiency virus (HIV) in which the body's T-helper cells are reduced, leaving the victim subject to opportunistic diseases.

ACTH *See* adrenocorticotropic hormone.

actin [Gk. *aktis*: ray] A protein that makes up the cytoskeletal microfilaments in eukaryotic cells and is one of the two contractile proteins in muscle. (*See also* myosin.)

actin filaments Chains of linked actin molecules. Also called thin filaments. (*See also* myosin filaments.)

action potential An impulse in a neuron taking the form of a brief, local, high-amplitude depolarization of the cell membrane. (Contrast with resting potential.)

action spectrum A graph of a biological process versus light wavelength; shows which wavelengths are involved in the process.

activation energy (E_a) The energy barrier that blocks the tendency for a chemical reaction to occur.

activator A transcription factor that stimulates transcription when it binds to a gene's promoter. (Contrast with repressor.)

active site The region on the surface of an enzyme or ribozyme where the substrate binds, and where catalysis occurs.

active transport The energy-dependent transport of a substance across a biological membrane against a concentration gradient—that is, from a region of low concentration (of that substance) to one of high concentration. (*See also* primary active transport, secondary active transport; contrast with facilitated diffusion, passive transport.)

adaptation (a dap tay′ shun) (1) In evolutionary biology, a particular structure, physiological process, or behavior that makes an organism better able to survive and reproduce. Also, the evolutionary process that leads to the development or persistence of such a trait. (2) In sensory neurophysiology, a sensory cell's loss of sensitivity as a result of repeated stimulation.

adaptive defenses In animals, one of two general types of defenses against pathogens. Involves antibody proteins and other proteins that recognize, bind to, and aid in the destruction of specific viruses and bacteria. Present only in vertebrate animals. (Contrast with innate defenses.)

additive growth Population growth in which a constant number of individuals is added to the population during successive time intervals. (Contrast with multiplicative growth.)

adenine (A) (a′ den een) A nitrogen-containing base found in nucleic acids, ATP, NAD, and other compounds.

adenohypophysis *See* anterior pituitary gland.

adenosine triphosphate *See* ATP.

ADH *See* antidiuretic hormone.

adhesion Binding of one substance or structure to another, as in a cell binding to its extracellular matrix.

adiabatic (ad′ e uh bat′ ik) [Gk. *adiabatos*: impassable] Referring to a system that exchanges no heat with its surroundings. When such a "heat insulated" system—for example a parcel of air—contracts or expands, its temperature increases or decreases, respectively. This thermodynamic principle applies to the functioning of air conditioners and refrigerators.

adrenocorticotropic hormone (ACTH) A tropic hormone produced by the anterior pituitary gland that stimulates release of cortisol and certain other hormones from the adrenal cortex.

adventitious roots (ad ven ti′ shus) [L. *adventitius*: arriving from outside] Roots originating from the stem at ground level or below; typical of the fibrous root system of monocots.

aerenchyma In plants, parenchymal tissue containing air spaces.

aerobic (air oh′ bic) [Gk. *aer*: air + *bios*: life] Requiring molecular oxygen, O_2. (Contrast with anaerobic.)

aerotolerant anaerobes Organisms that tolerate the presence of oxygen, but do not use it for metabolism.

afferent neuron (af′ ur unt) [L. *ad*: toward + *ferre*: to carry] *See* sensory neuron.

aggregate fruit A fruit of a flowering plant that develops from the merger of several different ovaries that were separate in the flower.

alcohol fermentation *See* fermentation.

aleurone layer In some seeds, a tissue that lies beneath the seed coat and surrounds the endosperm. Secretes digestive enzymes that break down macromolecules stored in the endosperm.

allantois (al′ lun toh is) [Gk. *allant*: sausage] An extraembryonic membrane sac that stores the embryo's nitrogenous wastes; because it contains blood vessels, it provides the means for gas exchange. In eutherian mammals, forms part of the placenta.

allele (a leel′) [Gk. *allos*: other] A specific form of a gene at a given locus on a chromosome, among multiple possible forms.

allergic reaction [Ger. *allergie*: altered] An overreaction of the immune system to amounts of an antigen that do not affect most people; often involves IgE antibodies.

allometry (1) In studies of animals of different sizes, nonproportional variation in relation to body size. (2) In studies of growth, a change in the proportions of various parts of an organism relative to one another as a consequence of unequal growth.

allopatric speciation (al′ lo pat′ rick) [Gk. *allos*: other + *patria*: homeland] The formation of two species from one when reproductive isolation occurs because of the interposition of (or crossing of) a physical geographic barrier such as a river. Also called geographic speciation. (Contrast with sympatric speciation.)

allopolyploidy The possession of more than two chromosome sets that are derived from more than one species.

allosteric regulation (al lo steer′ ik) [Gk. *allos*: other + *stereos*: structure] Regulation of the activity of a protein (usually an enzyme) by the binding of an effector molecule to a site other than the active site.

α (alpha) helix A prevalent type of secondary protein structure; a right-handed spiral.

alternation of generations The succession of multicellular haploid and diploid phases in some sexually reproducing organisms, notably plants.

alternative hypothesis In statistical inference, the hypothesis that contrasts with the null hypothesis; typically the hypothesis of primary interest.

alternative splicing A process for generating different mature mRNAs from a single gene by splicing together different sets of exons during RNA processing.

altruism Behavior that benefits another individual at a cost of reduced fitness to the individual performing the behavior.

alveolar sac A blind sac found at the end of a branch of the lung airways in a mammal; consists of many alveoli

alveolus (al ve′ o lus) (plural: alveoli) [L. *alveus*: cavity] A small, baglike cavity, especially a semispherical subpart of the wall of an alveolar sac in the mammalian lung

amensalism (a men′ sul ism) A "–/0" interaction between individuals of two different species that harms one individual (i.e., it suffers reduced fitness) but does not affect the other. (Contrast with commensalism, mutualism.)

amine hormones Small hormone molecules synthesized from single amino acids (e.g., thyroxine and epinephrine). (Compare to peptide hormones, steroid hormones.)

amino acid An organic compound containing both NH_2 and COOH groups. Proteins are polymers of amino acids.

ammonotelic (am moan′ o teel′ ic) [Gk. *telos*: end] Pertaining to an organism (typically water-breathing aquatic animals) in which ammonia is the principal final product of breakdown of nitrogen-containing compounds (primarily proteins). (Contrast with ureotelic, uricotelic.)

amnion (am′ nee on) The fluid-filled sac within which the embryos of reptiles (including birds) and mammals develop.

amniote egg A shelled egg surrounding four extraembryonic membranes and embryo-nourishing yolk. This evolutionary adaptation permitted mammals and reptiles to live and reproduce in drier environments than can most amphibians.

amphipathic (am′ fi path′ ic) [Gk. *amphi*: both + *pathos*: emotion] Of a molecule, having both hydrophilic and hydrophobic regions.

anabolic reaction (an uh bah′ lik) [Gk. *ana*: upward + *ballein*: to throw] A synthetic reaction in which simple molecules are linked to form more complex ones; requires an input of energy and captures it in the chemical bonds that are formed. (Contrast with catabolic reaction.)

anaerobic (an ur row′ bic) [Gk. *an*: not + *aer*: air + *bios*: life] Occurring without the use of molecular oxygen, O_2. (Contrast with aerobic.)

anaerobic glycolysis Glycolysis taking place in the absence of molecular oxygen, O_2. In animals, the most common process of anaerobic ATP production. In this process glucose is converted to lactic acid.

anaphase (an′ a phase) [Gk. *ana*: upward] The stage in cell nuclear division at which the first separation of sister chromatids (or, in the first meiotic division, of paired homologs) occurs.

ancestral trait The trait originally present in the ancestor of a given group; may be retained or changed in the descendants of that ancestor.

androgen (an′ dro jen) [Gk. *andros*: man + *genein*: to produce] Any of the several masculinizing steroid hormones (most notably testosterone).

aneuploidy (an′ you ploy dee) A condition in which one or more chromosomes or pieces of chromosomes are either lacking or present in excess.

angiosperms Flowering plants; one of the two major groups of living seed plants. (*See also* gymnosperms.)

animal hemisphere The non-yolky upper portion of some animal eggs. (Contrast with vegetal hemisphere.)

anion (an′ eye on) [Gk. *ana*: upward] A negatively charged ion. (Contrast with cation.)

annual A plant whose life cycle is completed in one growing season. (Contrast with biennial, perennial.)

ANS *See* autonomic nervous system.

antagonistic muscle pair Two muscles, or groups of muscles, that perform coordinated, opposing actions, thus allowing the bones at a joint to move in opposite directions. When one muscle contracts (to close the angle of a joint), its antagonist relaxes. To open the joint, the formerly relaxed muscle contracts and the formerly contracted muscle relaxes.

antenna system *See* light-harvesting complex.

anterior Toward or pertaining to the tip or headward region of the body axis. (Contrast with posterior.)

anterior pituitary gland The anterior portion of the vertebrate pituitary gland that is composed of nonneural endocrine cells and produces several hormones, including tropic hormones. Also called the adenohypophysis. (*See also* pituitary gland, posterior pituitary gland.)

anther (an′ thur) [Gk. *anthos*: flower] A pollen-bearing portion of the stamen of a flower.

antheridium (an′ thur id′ ee um) [Gk. *antheros*: blooming] The multicellular structure that produces the sperm in nonvascular land plants and ferns.

antibody One of the myriad proteins produced by the immune system that specifically binds to a foreign substance in blood or other tissue fluids and initiates its removal from the body.

anticodon The three nucleotides in transfer RNA that pair with a complementary triplet (a codon) in messenger RNA.

antidiuretic hormone (ADH) A hormone that promotes water reabsorption by the kidney. ADH is produced by neurons in the hypothalamus and released from nerve terminals in the posterior pituitary gland. Also called vasopressin in mammals.

antigen (an′ ti jun) Any substance that stimulates the production of an antibody or antibodies in the body of a vertebrate.

antigen presentation In cellular immunity, the process in which a cell ingests and digests an antigen, and then exposes fragments of that antigen, bound to a major histocompatibility complex (MHC) molecule, to the outside of the cell, where the antigen can be recognized by T cells.

antigenic determinant The specific region of an antigen that is recognized and bound by a specific antibody. Also called an epitope.

antiparallel Pertaining to molecular orientation in which a molecule or parts of a molecule have opposing directions.

antisense RNA A single-stranded RNA molecule complementary to, and thus targeted against, an mRNA of interest to block its translation.

aphotic zone The parts of lakes and oceans below the photic zone, where no sunlight penetrates.

apical–basal axis Longitudinal definition of function in an adult organism or future function in an embryo, as in the root tip (apex).

apical dominance In plants, inhibition by the apical bud of the growth of axillary buds.

apical hook A form taken by the stems of many eudicot seedlings that protects the delicate shoot apex while the stem grows through the soil.

apical meristem The meristem at the tip of a shoot or root; responsible for a plant's primary growth.

apomixis (ap oh mix′ is) [Gk. *apo*: away from + *mixis*: sexual intercourse] The asexual production of seeds.

apoplast (ap′ oh plast) In plants, the continuous meshwork of cell walls and extracellular spaces through which material can pass without crossing a plasma membrane. (Contrast with symplast.)

apoptosis (ap uh toh′ sis) A series of genetically programmed events leading to cell death.

aquaporin [L. *aqua*: water + *porus*: pore] A transport protein in plant and animal cell membranes through which water passes by osmosis. Also called a water channel protein.

archegonium (ar′ ke go′ nee um) The multicellular structure that produces eggs in nonvascular land plants, ferns, and gymnosperms.

arteriole A small blood vessel with muscular walls that carries blood from an artery into a capillary bed.

artery A muscular blood vessel carrying blood away from the heart to other parts of the body. (Contrast with vein.)

ascoma The fruiting structure of sexually reproducing sac fungi; often a cup-shaped structure.

ascospores Haploid spores contained with the sexual reproductive structure (the ascus) of sac fungi.

ascus (ass′ cus) (plural: asci) [Gk. *askos*: bladder] In sac fungi, the club-shaped sporangium within which spores (ascospores) are produced by meiosis.

asexual reproduction The formation of new individuals without the union of genetic material from two different parents. Often takes place by budding or fragmentation. The offspring are genetically identical to their parent.

asynchronous muscle A muscle in which each excitation results in multiple contractions.

atom [Gk. *atomos*: indivisible] The smallest unit of a chemical element. Consists of a nucleus and one or more electrons.

atomic number The number of protons in the nucleus of an atom; also equals the number of electrons around the neutral atom. Determines the chemical properties of the atom.

ATP (adenosine triphosphate) An energy-storage compound containing adenine, ribose, and three phosphate groups. When it is formed from ADP, useful energy is stored; when it is broken down (to ADP or AMP), energy is released to drive endergonic reactions.

ATP synthase An integral membrane protein that couples the transport of protons with the formation of ATP.

atrium (a′ tree um) [L. *atrium*: central hall] An internal chamber. In the hearts of vertebrates, the thin-walled chamber(s) entered by blood on its way to the ventricle(s).

autoimmune diseases Diseases (e.g., rheumatoid arthritis) that result from failure of the immune system to distinguish between self and nonself, causing it to attack tissues in the organism's own body.

autoimmunity An immune response by an organism to its own molecules or cells.

autonomic effector An effector other than skeletal muscle; includes smooth and cardiac muscles, exocrine glands, and endocrine glands. Also called an internal effector.

autonomic nervous system (ANS) The portion of the nervous system that controls autonomic effectors (effectors other than the skeletal muscles). Also called the involuntary nervous system.

autophagy The programmed destruction of a cell's components.

autopolyploidy The possession of more than two entire chromosomes sets that are derived from a single species.

autosome Any chromosome (in a eukaryote) other than a sex chromosome.

autotroph (au′ tow trowf′) [Gk. *autos:* self + *trophe*: food] An organism that is capable of living exclusively on inorganic materials, water, and an energy source other than the chemical bonds of organic compounds. Some autotrophs (photoautotrophs) use sunlight as their energy source. Others (chemoautotrophs) use oxidation of inorganic compounds. (Contrast with heterotroph.)

auxin (awk′ sin) [Gk. *auxein*: to grow] In plants, a substance (the most common being indoleacetic acid) that regulates growth and various aspects of development.

avirulence (*Avr*) genes Genes in a pathogen that may trigger defenses in plants. (*See* gene-for-gene resistance.)

axis In the study of endocrinology, a system in which several types of endocrine cells act on each other in sequence.

axon [Gk. axle] The part of a neuron that conducts action potentials away from the cell body.

B

B horizon *See* subsoil.

bacillus (bah sil′ us) [L: little rod] Any of various rod-shaped bacteria.

bacterial conjugation *See* conjugation.

bacteriophage (bak teer′ ee o fayj) [Gk. *bakterion*: little rod + *phagein*: to eat] Any of a group of viruses that infect bacteria. Also called phage.

bacteroids Nitrogen-fixing organelles that develop from endosymbiotic bacteria.

bar chart A figure that displays frequency distributions of categorical data using bar lengths to represent relative frequency.

bark All tissues external to the vascular cambium of a plant.

basal metabolic rate (BMR) The minimum rate of energy use in an awake (but resting) bird or mammal. For measuring the BMR, the animal is resting, fasting, and in its thermoneutral zone (i.e., its metabolic rate is not elevated for thermoregulation).

base In nucleic acids, the purine or pyrimidine that is attached to each sugar in the sugar–phosphate backbone.

base pair (bp) In double-stranded DNA, a pair of nucleotides formed by the complementary base pairing of a purine on one strand and a pyrimidine on the other.

basidioma (plural: basidiomata) A fruiting structure produced by club fungi.

basidium (bass id' ee yum) In club fungi, the characteristic sporangium in which four **basidiospores** are formed by meiosis and then borne externally before being shed.

basilar membrane A membrane in the human inner ear that flexes in response to sound waves and thereby activates hair cells; flexes at different locations in response to different pitches of sound.

BD model A mathematical model that describes how Birth and Death rates affect the size and change in size (dynamics) of ecological populations.

bedrock *See* parent rock.

behavioral imprinting In animal behavior, a rapid form of learning in which an animal learns, during a brief critical period, to make a particular response, which is maintained for life, to some object or other organism.

benthic zone [Gk. *benthos*: bottom] The substrate at the bottom of an ocean, lake, stream, or other body of water.

β (beta) pleated sheet A type of protein secondary structure; results from hydrogen bonding between polypeptide regions running antiparallel to each other.

BIDE model A mathematical model that describes how birth, immigration (movement of individuals into populations), death, and emigration (movement of individuals out of populations) affect the size and change in size (dynamics) of ecological populations.

biennial A plant whose life cycle includes vegetative growth in the first year and flowering and senescence in the second year. (Contrast with annual, perennial.)

bilateral symmetry The condition in which only the right and left sides of an organism, divided by a single plane through the midline, are mirror images of each other.

bilayer A structure that is two layers in thickness. In biology, most often refers to the phospholipid bilayer of membranes. (*See* phospholipid bilayer.)

binary fission Reproduction of a prokaryote by division of a cell into two comparable progeny cells.

binomial nomenclature A taxonomic naming system in which each species is given two names (a genus name followed by a species name).

biofilm A community of microorganisms embedded in a polysaccharide matrix, forming a highly resistant coating on almost any moist surface.

biogeochemical cycle A description of the cyclical pathways by which biological, geological, and chemical processes move a chemical element or molecule through biotic and abiotic compartments of planet Earth.

biogeographic region A large, continental-scale region of planet Earth, encompassing multiple biomes, that contains a distinct assemblage of species, many of which are phylogenetically related. (Contrast with biome.)

biogeography The spatial distribution of Earth's species.

bioinformatics The use of computers and/or mathematics to analyze complex biological information, such as DNA sequences.

biological clock A self-contained, metabolic mechanism that keeps track of time.

biological determination As applied to behavior, the belief that genes completely determine an organism's behavior.

biological species concept The definition of a species as a group of actually or potentially interbreeding natural populations that are reproductively isolated from other such groups. (Contrast with lineage species concept, morphological species concept.)

biomass The mass of a sample of biological tissue, living or dead. The term usually refers to the total mass of a designated group of organisms (or of certain of their tissues), and may be measured as the mass of their intact bodies or of their dried bodies.

biome (bye' ome) A distinct abiotic environment that is found at different places on the planet and is inhabited in each of these locations by ecologically-similar organisms with convergent adaptations. (Contrast with biogeographic region.)

bioremediation The use by humans of other organisms to remove contaminants from the environment.

biosphere (bye' oh sphere) The region that supports living organisms on Earth, extending about 23 km from the abyssal depths of the oceans into the stratosphere (an upper portion of the atmosphere).

biota (bye oh' tah) All of the organisms—animals, plants, fungi, and microorganisms—found in a given area. (Contrast with flora, fauna.)

biotechnology The use of cells or living organisms to produce materials useful to humans.

biotic (bye ah' tick) [Gk. *bios*: life] Living. In ecology, the living components of the environment of organisms. (Contrast with abiotic.)

biotic interchange The mixing of distinct groups of species that were isolated by physical or climatic barriers, after the barrier is removed.

bipedal locomotion The act of walking or running on two appendages.

blade The thin, flat portion of a leaf.

blastocyst (blass' toe cist) A mammalian blastula. The sphere of cells formed by early cell divisions and composed of an inner cell mass that becomes the embryo and an outer layer of cells called the trophoblast.

blastodisc (blass' toe disk) An embryo that forms as a disk of cells on the surface of a large yolk mass; comparable to a blastula, but occurring in animals such as birds and reptiles, in which the massive yolk restricts complete cleavage.

blastomere Any of the cells produced by the early divisions of a fertilized animal egg.

blastula (blass' chu luh) An early stage of the animal embryo; in many species, a hollow sphere of cells surrounding a central cavity, the blastocoel. (Compare with blastodisc.)

blood plasma The watery solution in which red blood cells and/or other blood cells are suspended. The part of the blood other than cells.

blood pressure The extent to which the pressure of the blood exceeds the pressure of the surrounding external environment.

BMR *See* basal metabolic rate.

body cavity An internal, fluid-filled space, enclosed by mesoderm in animals.

body plan The general structure of an animal, the arrangement of its organ systems, and the integrated functioning of its parts.

Bohr model A model for atomic structure that depicts the atom as largely empty space, with a central nucleus surrounded by electrons in orbits, or electron shells, at various distances from the nucleus.

bone A rigid component of vertebrate skeletal systems that contains an extracellular matrix of insoluble calcium phosphate crystals as well as collagen fibers.

bottleneck *See* population bottleneck.

Bowman's capsule An elaboration of the renal tubule, composed of podocytes, that surrounds and collects the filtrate from the glomerulus.

brain The part of the central nervous system of greatest size.

brassinosteroids Plant steroid hormones that mediate light effects promoting the elongation of stems and pollen tubes.

breathing The exchange of O_2 and CO_2 between the environmental medium next to the gas exchange membranes and the outside world; often occurs by ventilation, but can occur entirely by diffusion. Also called external respiration.

breathing-organ circuit In a circulatory system, the blood vessels that carry blood to and from the breathing organs. (Contrast with systemic circuit.)

breathing organs Specialized organs involved in the exchange of O_2 and CO_2 between an organism and the external environment. Classified as either lungs or gills.

brown adipose tissue (BAT) A specialized fatty tissue, known only in mammals, that serves as the primary site of nonshivering thermogenesis.

bud In plants, a small protuberance in the shoot that contains an undeveloped leaf, flower or shoot.

budding Asexual reproduction in which a more or less complete new organism grows from the body of the parent organism, eventually detaching itself.

bulk flow Mass movement of a fluid (gas or liquid) driven by a difference of pressure, from a region of high pressure to a region of low pressure.

C

C horizon *See* parent rock.

calorie (cal) [L. *calor*: heat] A unit of energy equal to the amount of heat required to raise the temperature of 1 gram of water by 1°C. Physiologists commonly use the kilocalorie (kcal) as a unit of energy (1 kcal = 1,000 calories). Nutritionists also use the kilocalorie, but refer to it as the *Calorie* (capital C).

Calvin cycle The stage of photosynthesis in which CO_2 reacts with RuBP to form 3PG, 3PG is reduced to a sugar, and RuBP is regenerated, while other products are released to the rest of the plant. Also known as the Calvin–Benson cycle.

Cambrian explosion The rapid diversification of multicellular life that took place during the Cambrian period.

cAMP (cyclic AMP) A compound formed from ATP that acts as a second messenger.

5′ cap *See* G cap.

capillary [L. *capillaris*: hair] In the circulatory system, a microscopic blood vessel with walls composed of just a simple epithelium (called the capillary endothelium), one cell layer thick. Capillaries are the principal sites of exchange between blood and other tissues in animals with a closed circulatory system.

carbohydrates Organic compounds containing carbon, hydrogen, and oxygen in the ratio 1:2:1 (i.e., with the general formula $C_nH_{2n}O_n$). Common examples are sugars, starch, and cellulose.

carbon-fixation reactions The phase of photosynthesis in which chemical energy captured in the light reactions is used to drive the reduction of CO_2 to form carbohydrates.

cardiac cycle Alternating contraction and relaxation of the heart. In a mammal, contraction of the two atria of the heart, followed by contraction of the two ventricles and then relaxation.

cardiac muscle A type of muscle tissue that makes up the wall of the heart and is responsible for the beating of the heart. Characterized by branching, electrically coupled cells with single nuclei and a striated (striped) appearance. (Contrast with smooth muscle, skeletal muscle.)

cardiovascular system [Gk. *kardia*: heart + L. *vasculum*: small vessel] The heart, and vessels of the circulatory system. Also called circulatory system.

carnivore [L. *carn*: flesh + *vorare*: to devour] An organism that eats animal tissues. (*See* secondary consumer; contrast with omnivore, detritivore, herbivore.)

carpel (kar′ pel) [Gk. *karpos*: fruit] The organ of the flower that contains one or more ovules.

carrier proteins Membrane proteins that bind specific molecules and transport them through the membrane.

carrying capacity (*K*) The maximum population size or density of a given species that a given unchanging environment can support indefinitely.

cartilage In vertebrates, a tough connective tissue found in joints, the outer ear, part of the nose, and elsewhere. Forms the entire skeleton in some animal groups, notably sharks and their relatives.

Casparian strip A band of cell wall containing suberin and lignin, found in the endodermis. Restricts the movement of water across the endodermis.

caspase One of a group of proteases that catalyze cleavage of target proteins and are active in apoptosis.

catabolic reaction (kat uh bah′ lik) [Gk. *kata*: to break down + *ballein*: to throw] A synthetic reaction in which complex molecules are broken down into simpler ones and energy is released. (Contrast with anabolic reaction.)

catalyst (kat′ a list) [Gk. *kata*: to break down] A chemical substance that accelerates a reaction without itself being consumed in the overall course of the reaction. Catalysts lower the activation energy of a reaction. Enzymes are biological catalysts.

catch A state in which some invertebrate muscles maintain a state of contraction over extended periods of time with minimal energy expenditure.

categorical variables Variables that take on qualitative categories as values, such as human blood types (A, AB, B, or O).

cation (cat′ eye on) An ion with one or more positive charges. (Contrast with anion.)

Cdk *See* cyclin-dependent kinase.

cDNA library A collection of complementary DNAs derived from mRNAs of a particular tissue at a particular time in the life cycle of an organism.

cDNA *See* complementary DNA.

cell body In a neuron, portion of the neuron that contains the cell nucleus and most of the neuron's other organelles.

cell cycle The stages through which a cell passes between one division and the next. Includes all stages of interphase and mitosis.

cell cycle checkpoints Points of transition between different phases of the cell cycle, which are regulated by cyclins and cyclin-dependent kinases (Cdk's).

cell division The reproduction of a cell to produce two new cells. In eukaryotes, this process involves nuclear division (mitosis) and cytoplasmic division (cytokinesis).

cell fate The type of cell that an undifferentiated cell in an embryo will become in the adult.

cell junctions Specialized structures associated with the plasma membranes of epithelial cells. Some contribute to cell adhesion, others to intercellular communication.

cell membrane Lipid bilayer also containing proteins and other molecules that encloses the cytoplasm of the cell and separates it from the surrounding environment.

cell potency In multicellular organisms, an undifferentiated cell's potential to become a cell of a specific type. (*See* multipotent, pluripotent, totipotent.)

cell signaling Communication between cells or within cells as they respond to specific cues.

cell theory States that cells are the basic structural and physiological units of all living organisms, and that all cells come from preexisting cells.

cell wall A relatively rigid structure that encloses cells of plants, fungi, many protists, and most prokaryotes, and which gives these cells their shape and limits their expansion in hypotonic media.

cellular immune response Immune system response mediated by T cells and directed against parasites, fungi, intracellular viruses, and foreign tissues (grafts). (Contrast with humoral immune response.)

cellular respiration The catabolic pathways by which electrons are removed from various molecules and passed through intermediate electron carriers to O_2, generating H_2O and releasing energy.

central nervous system (CNS) Large structures composed principally of integrating neurons and associated glial cells, forming the site of most information processing, storage, and retrieval; in vertebrates, the brain and spinal cord. (Contrast with peripheral nervous system.)

centralization (of the nervous system) Over the course of evolution, the tendency of neurons to become clustered into centralized, integrating organs.

centriole (sen′ tree ole) A paired organelle that helps organize the microtubules in animal and protist cells during nuclear division.

centromere (sen′ tro meer) [Gk. *centron*: center + *meros*: part] The region where sister chromatids join.

centrosome (sen′ tro soam) The major microtubule organizing center of an animal cell.

cephalization (sef ah luh zay′ shun) [Gk. *kephale*: head] The evolutionary trend toward increasing concentration of nervous tissue (e.g., the brain) and sensory organs at the anterior end of the animal.

cerebral cortex The outer layer of the cerebral hemispheres in the vertebrate brain. Plays key roles in sensory perception, learning, memory, and motor control.

channel protein An integral membrane protein that forms an aqueous passageway across the membrane in which it is inserted and through which specific solutes may pass.

character In genetics, an observable feature, such as eye color. (Contrast with trait.)

chemical bond An attractive force stably linking two atoms.

chemical reaction The change in the composition or distribution of atoms of a substance with consequent alterations in properties.

chemical synapse A point of contact between two neurons at which the neurons are separated by only a very narrow space and neurotransmitter molecules released from the presynaptic cell induce changes in ion flux and membrane potential in the postsynaptic cell. (Contrast with electrical synapse.)

chemiosmosis Formation of ATP in mitochondria and chloroplasts, resulting from a pumping of protons across a membrane (against a gradient of electrical charge and of pH), followed by the return of the protons through a protein channel with ATP synthase activity.

chemoautotroph An organism that uses carbon dioxide as a carbon source and obtains energy by oxidizing inorganic substances from its environment; also called chemolithotroph. (Contrast with chemoheterotroph, photoautotroph, photoheterotroph.)

chemoheterotroph An organism that must obtain both carbon and energy from organic substances. (Contrast with chemoautotroph, photoautotroph, photoheterotroph.)

chemolithotroph [Gk. *lithos*: stone, rock] *See* chemoautotroph.

chemoreceptor A sensory receptor cell that senses specific molecules (such as CO_2 or O_2 internally or environmental odorant molecules externally).

chi-square goodness-of-fit A statistical test used to assess whether the frequencies of observations in different categories are consistent with an hypothesized frequency distribution.

chitin (kye′ tin) [Gk. *kiton*: tunic] The characteristic tough but flexible organic component of the exoskeleton of arthropods, consisting of a complex, nitrogen-containing polysaccharide. Also found in cell walls of fungi.

chlorophyll (klor′ o fill) [Gk. *kloros*: green + *phyllon*: leaf] Any of several green pigments associated with chloroplasts or with certain bacterial membranes; responsible for trapping light energy for photosynthesis.

chloroplast [Gk. *kloros*: green + *plast*: a particle] An organelle bounded by a double membrane containing the enzymes and pigments that perform photosynthesis. Chloroplasts occur only in eukaryotes.

choanocyte (ko′ an uh site) The collared, flagellated feeding cells of sponges.

chorion (kor′ ee on) [Gk. *khorion*: afterbirth] The outermost of the membranes protecting mammal and reptile (including bird) embryos; in mammals it forms part of the placenta.

chromatin The nucleic acid–protein complex that makes up eukaryotic chromosomes.

chromatin remodeling A mechanism for epigenetic gene regulation by the alteration of chromatin structure.

chromosomal mutation Loss of or changes in position/direction of a DNA segment on a chromosome.

chromosome (krome′ o sowm) [Gk. *kroma*: color + *soma*: body] In bacteria and viruses, the DNA molecule that contains most or all of the genetic information of the cell or virus. In eukaryotes, a structure composed of DNA and proteins that bears part of the genetic information of the cell.

cilium (sil′ ee um) (plural: cilia) [L. eyelash] Hairlike organelle used for locomotion by many unicellular organisms and for moving water and mucus by many multicellular organisms. Generally shorter than a flagellum.

circadian clocks A biological clock that has a free-running cycle of about 24 hours.

circadian rhythm (sir kade′ ee an) [L. *circa*: approximately + *dies*: day] A rhythm of growth or activity that recurs about every 24 hours.

circulatory system A system consisting of a muscular pump (heart), a fluid (blood), and a series of conduits (e.g., blood vessels) that transports materials around the body by mass flow.

citric acid cycle In cellular respiration, a set of chemical reactions whereby acetyl CoA is oxidized to carbon dioxide and hydrogen atoms are stored as NADH and $FADH_2$. Also called the Krebs cycle.

clade [Gk. *klados*: branch] A monophyletic group made up of an ancestor and all of its descendants.

class (1) In biological classification: A group of related orders. (2) Generally: a group of things defined by particular attributes.

class I MHC molecules Cell surface proteins that are present on the surface of all nucleated cells. They present antigens to cytotoxic T (T_C) cells.

class II MHC molecules Cell surface proteins that are present on the surface of macrophages, B cells, and dendritic cells. They present antigens to T-helper (T_H) cells.

cleavage The first few cell divisions of an animal zygote. Cleavage divides the embryo without increasing its mass. (*See* complete cleavage, incomplete cleavage.)

climate The long-term average state of the weather at a given place over a yearly cycle. (Contrast with weather.)

climate diagram A way of summarizing the climate of a given location by superimposing graphs of average monthly temperature and precipitation through a year in a way that indicates periods that are favorable or unfavorable for plant growth.

clonal deletion Inactivation or destruction of lymphocyte clones that would produce immune reactions against the animal's own body.

clonal lineages Asexually reproduced groups of nearly identical organisms.

clonal selection Mechanism by which exposure to antigen results in the activation of selected T- or B-cell clones, resulting in an immune response.

clone [Gk. *klon*: twig, shoot] (1) Genetically identical cells or organisms produced from a common ancestor by asexual means. (2) To produce many identical copies of a DNA sequence by its introduction into, and subsequent asexual reproduction of, a cell or organism.

closed circulatory system Circulatory system in which the circulating fluid is contained within a continuous system of vessels. (Contrast with open circulatory system.)

closed system A system that isolated from, and does not interact with, its surroundings. (Contrast with open system.)

coccus (kock' us) (plural: cocci) [Gk. *kokkos*: berry, pit] Any of various spherical or spheroidal bacteria.

cochlea (kock' lee uh) [Gk. *kokhlos*: snail] A spiral tube in the inner ear of vertebrates; it contains the sensory cells involved in hearing.

cocurrent gas exchange The exchange of gas between blood and water in a **cocurrent system** (a system in which the blood and water flow in the same direction along the gas exchange membrane of the animal; sometimes called a concurrent system).

coding region Nucleotide sequences in a gene that directly specify amino acids in a protein.

coding strand One of the two strand of DNA that for a particular gene specifies the amino acids in a protein.

codominance A condition in which two alleles at a locus produce different phenotypic effects and both effects appear in heterozygotes.

codon Three nucleotides in messenger RNA that direct the placement of a particular amino acid into a polypeptide chain. (Contrast with anticodon.)

coelom (see' loam) [Gk. *koiloma*: cavity] An animal body cavity completely enclosed by mesoderm.

coelomate (see' lo mate) An animal that contains a body cavity, the coelom, that develops within the mesoderm.

coenocyte (see' no site) An organism that contains many nuclei within each cell membrane.

coenocytic (seen' a sit ik) [Gk. *koinos*: common + *kytos*: container] Referring to the condition, found in some fungal hyphae, of "cells" containing many nuclei but enclosed by a single plasma membrane. Results from nuclear division without cytokinesis.

coenzyme A (CoA) A coenzyme used in various biochemical reactions as a carrier of acyl groups.

cofactor An inorganic ion that is weakly bound to an enzyme and required for its activity.

cohesion The tendency of molecules (or any substances) to stick together.

cold-hardening A process by which plants can acclimate to cooler temperatures; requires repeated exposure to cool temperatures over many days.

coleoptile A sheath that surrounds and protects the shoot apical meristem and young primary leaves of a grass seedling as they move through the soil.

collagen [Gk. *kolla*: glue] A fibrous protein found extensively in bone and connective tissue.

collecting duct In mammals and birds, a tubule that receives urine produced in the nephrons of the kidney and delivers that fluid to the ureter for excretion. Concentration of the urine occurs in the collecting duct.

collenchyma (cull eng' kyma) [Gk. *kolla*: glue + *enchyma*: infusion] A type of plant cell, living at functional maturity, which lends flexible support by virtue of primary cell walls thickened at the corners. (Contrast with parenchyma, sclerenchyma.)

commensalism [L. *com*: together + *mensa*: table] A "+/0" interaction between individuals of two different species that benefits one individual (i.e., increases its fitness) but does not affect the other. (Contrast with amensalism, mutualism.)

community The assemblage of interacting individuals of different species within a particular geographic area.

community function Any aspect of the interactions that occur within an ecological community that affects the rate of input of energy and matter into the community and its output of transformed matter and energy.

community structure The attributes used to characterize an ecological community, which usually include the number of species in the community, their identities, and their relative abundances. (*See* species composition, species diversity.)

companion cell In angiosperms, a specialized cell found adjacent to a sieve tube element.

comparative genomics Computer-aided comparison of DNA sequences between different organisms to reveal genes with related functions.

compartment One of the interacting components of a system. Applied to the global ecosystem, the different forms and physical locations in which chemical elements or molecules may occur.

competition In ecology, a "–/–" interaction between individuals of the same or different species that causes both individuals to lose fitness.

competitive inhibitor A nonsubstrate that binds to the active site of an enzyme and thereby inhibits binding of its substrate. (Contrast with noncompetitive inhibitor.)

complementary base pairing The AT (or AU), TA (or UA), CG, and GC pairing of bases in double-stranded DNA, in transcription, and between tRNA and mRNA.

complementary DNA (cDNA) DNA formed by reverse transcriptase acting with an RNA template; essential intermediate in the reproduction of retroviruses; used as a tool in recombinant DNA technology; lacks introns.

complete cleavage Pattern of cleavage that occurs in eggs that have little yolk. Early cleavage furrows divide the egg completely. (*See also* radial cleavage, spiral cleavage; contrast with incomplete cleavage.)

complete metamorphosis A change of state during the life cycle of an organism in which the body is almost completely rebuilt to produce an individual with a very different body form. Characteristic of insects such as butterflies, moths, beetles, ants, wasps, and flies.

component The interacting parts of a biological system.

compound eye An eye characteristic of arthropods, composed of many individual optical units called ommatidia.

computational model A system in which the interactions among components are expressed as mathematical functions.

condensation reaction A chemical reaction in which two molecules become connected by a covalent bond and a molecule of water is released (AH + BOH → AB + H_2O.) (Contrast with hydrolysis reaction.)

conditional mutation A mutation that results in a characteristic phenotype only under certain environmental conditions.

conducting airways In the lungs of animals, the airways that do not participate in the exchange of respiratory gases between air and blood, but rather simply conduct air in and out of the lungs.

conducting system In the vertebrate heart, a system of specialized muscle cells by which depolarization rapidly spreads throughout the heart muscle.

cone (1) In conifers, a reproductive structure consisting of spore-bearing scales extending from a central axis. (Contrast with strobilus.) (2) In the vertebrate retina, a type of photoreceptor cell responsible for color vision.

confidence interval A numerical interval that is calculated to contain the true value of some parameter of interest at a stated probability level.

conformer An animal in which an internal condition (e.g., body temperature) adjusts to match an external condition (e.g., environmental temperature). (Contrast with regulators.)

conformity The state in which an internal condition of an animal (e.g., body temperature) changes to match an external condition (e.g., environmental temperature). (Contrast with regulation.)

conidium (ko nid′ ee um) (plural: conidia) [Gk. *konis*: dust] A type of haploid fungal spore borne at the tips of hyphae, not enclosed in sporangia.

conjugation (kon ju gay′ shun) [L. *conjugare*: yoke together] (1) A process by which DNA is passed from one cell to another through a conjugation tube, as in bacteria. (2) A nonreproductive sexual process by which *Paramecium* and other ciliates exchange genetic material.

consensus sequences Short stretches of DNA that appear, with little variation, in many different genes.

conservation ecology The subdiscipline of ecology whose goal is to understand the processes of extinction of populations and species, and to devise ways to conserve species and the ecological systems of which they are a part.

constant region The portion of an immunoglobulin molecule whose amino acid composition determines its class and does not vary among immunoglobulins in that class. (Contrast with variable region.)

constitutive Always present; produced continually at a constant rate. (Contrast with inducible.)

constitutive genes Genes that are expressed all the time. (Contrast with inducible genes.)

consumer–resource interactions In ecology, a "+/–" interaction between individuals of different species in which one individual gains fitness by consuming part or all of another individual, which loses fitness from being consumed. Examples are predation, herbivory, and parasitism.

continental drift The gradual movements of the world's continents that have occurred over billions of years.

continuous variables Variables that can take on a continuous range of values.

contractile vacuole (kon trak′ tul) A specialized vacuole that collects excess water taken in by osmosis, then contracts to expel the water from the cell.

contraction The condition in which a muscle produces force.

control mechanism One of four essential elements of a control system; modifies a controlled variable by using information from sensors to determine which effectors to activate and how intense the activation needs to be.

controlled variable One of four essential elements of a control system; the characteristic of a biochemical system, cell or animal that is being controlled.

convergent evolution Independent evolution of similar features from different ancestral traits.

coral bleaching A phenomenon in which corals lose their endosymbionts and become whitened as a result; leads to weakening, and sometimes death, of the coral.

corepressor A molecule that binds to a bacterial repressor protein, altering its shape so that it binds to the operator sequence in DNA and prevents its transcription.

Coriolis effect The apparent deflection of a moving body as viewed from a rotating frame of reference. In the system of the rotating Earth, an object (such as a packet of atmospheric gases) moving toward the Equator will appear to be deflected toward the west from the perspective of an observer on the surface; one moving toward a Pole will appear to be deflected toward the east.

cork In plants, a protective outermost tissue layer composed of cells with thick walls waterproofed with suberin.

cork cambium [L. *cambiare*: to exchange] In plants, a lateral meristem that produces secondary growth, mainly in the form of waxy-walled protective cells, including some of the cells that become bark.

cornea In a vertebrate, the clear, transparent tissue that covers the forward-facing surface of the eye and allows light to pass through to the retina.

correlation coefficient A measure of the strength of relationship between two quantitative variables, ranging from –1 (a perfect negative relationship) to 1 (a perfect positive relationship).

corolla (ko role′ lah) [L. *corolla*: a small crown] All of the petals of a flower, collectively.

coronary circulatory system A system of blood vessels that carries blood to and from the heart muscle

corpus luteum (kor′ pus loo′ tee um) (plural: corpora lutea) [L. yellow body] A hormone-producing (endocrine) structure formed from reorganization of the cells of an ovarian follicle that has undergone ovulation. In mammals, it secretes progesterone, estrogen, and inhibin. If fertilization occurs, it remains active during pregnancy; if not, it degenerates.

corpus luteum of pregnancy A corpus luteum that functions for an extended period of time during pregnancy. Secretes hormones that are important for maintenance of pregnancy.

corpus luteum of the cycle A corpus luteum that functions for a short time (e.g., about 10 days in humans) and then undergoes internally programmed degeneration.

cortex [L. *cortex*: covering, rind] (1) In plants, the tissue between the epidermis and the vascular tissue of a stem or root. (2) In animals, the outer tissue of certain organs, such as the adrenal gland (adrenal cortex) and the brain (cerebral cortex).

cost–benefit approach An approach to ecological and evolutionary analysis that assumes an animal has a limited amount of time and energy to devote to its activities, and accordingly activities will occur only if benefits exceed costs. (*See also* trade-off.)

cotyledon (kot′ ul lee′ dun) [Gk. *kotyledon*: hollow space] A "seed leaf." An embryonic organ that stores and digests reserve materials; may expand when seed germinates.

countercurrent heat exchange The exchange of heat that occurs between two closely juxtaposed fluid streams that are flowing in opposite directions (e.g. blood flowing in opposite directions in two juxtaposed blood vessels).

countercurrent gas exchange The exchange of gas between blood and water in a **countercurrent system** (a system in which the blood and water flow in opposite directions along the gas exchange membrane of the animal).

countercurrent multiplication The mechanism that increases the concentration of the interstitial fluids in the medulla of the mammalian kidney, permitting a mammal to produce concentrated urine. During countercurrent multiplication, countercurrent flow in the loops of Henle multiplies effects of active ion transport.

covalent bond Chemical bond based on the sharing of electrons between two atoms.

CpG islands DNA regions rich in C residues adjacent to G residues. Especially abundant in promoters, these regions are where methylation of cytosine usually occurs.

cross-bridges During muscle contraction, the links formed when the globular heads of myosin filaments bind to specific sites on actin filaments. The heads of a myosin molecule interact with actin molecules to produce contraction.

crossing over The mechanism by which linked genes undergo recombination. In general, the term refers to the reciprocal exchange of corresponding segments between two homologous chromatids.

crustose A growth form of organisms, such as lichens, in which the organism forms a thin, close, tight bond with the surface of a rock, tree, or other object.

cryptochromes [Gk. *kryptos*: hidden + *kroma*: color] Photoreceptors mediating some blue-light effects in plants and animals.

ctene (teen) [Gk. *cteis*: comb] In ctenophores, a comblike row of cilia-bearing plates. Ctenophores move by beating the cilia on their eight ctenes.

current (1) Flow of a fluid (gas or liquid), for example in ocean waters or in the atmosphere. (2) The flow of electrical charge.

cuticle (1) In plants, a waxy layer on the outer body surface that retards water loss. (2) In ecdysozoans, an outer body covering that provides protection and support and is periodically molted.

cyclic AMP *See* cAMP.

cyclic electron transport In photosynthetic light reactions, the flow of electrons that produces ATP but no NADPH or O_2.

cyclin A protein that activates a cyclin-dependent kinase, bringing about transitions in the cell cycle.

cyclin-dependent kinase (Cdk) A protein kinase whose target proteins are involved in transitions in the cell cycle and which is active only when complexed with additional protein subunits, called cyclins.

cytokine A regulatory protein made by immune system cells that affects other target cells in the immune system.

cytokinesis (sy′ toe kine ee′ sis) [Gk. *kytos*: container + *kinein*: to move] The division of the cytoplasm of a dividing cell. (Contrast with mitosis.)

cytokinin (sy′ toe kine′ in) A member of a class of plant growth substances that plays roles in senescence, cell division, and other phenomena.

cytoplasm The contents of the cell, excluding the nucleus.

cytoplasmic determinants In animal development, materials in the cytoplasm, the spatial distribution of which may determine such things as embryonic axes and cell determination.

cytoplasmic segregation The asymmetrical distribution of cytoplasmic determinants in a developing animal embryo.

cytoplasmic streaming A form of amoeboid movement in which a fluid cytoplasm moves and stretches the organism's body in new directions.

cytosine (C) (site′ oh seen) A nitrogen-containing base found in DNA and RNA.

cytoskeleton The network of microtubules and microfilaments that gives a eukaryotic cell its shape and its capacity to arrange its organelles and to move.

cytosol The fluid portion of the cytoplasm, excluding organelles and other solids.

cytotoxic T cells (T_C) Cells of the cellular immune system that recognize and directly eliminate virus-infected cells. (Contrast with T-helper cells.)

D

data Quantified observations about a system under study.

daughter chromosomes During mitosis, the separated chromatids from the beginning of anaphase onward.

decomposer An organism that obtains the energy and materials it needs to survive, grow, and reproduce from the waste products or dead bodies of other organisms.

defensin A type of protein made by phagocytes that kills bacteria and enveloped viruses by insertion into their plasma membranes.

deficiency disease A condition (e.g., scurvy) caused by chronic lack of any essential nutrient.

delayed implantation Embryonic diapause in placental mammals. A state of arrested embryonic development that occurs after an embryo has arrived in the uterus but before it implants in the uterine wall.

deletion A mutation resulting from the loss of a continuous segment of a gene or chromosome. Such mutations almost never revert to wild type. (Contrast with duplication, point mutation.)

demethylase An enzyme that catalyzes the removal of the methyl group from cytosine, reversing DNA methylation.

demography (Gk. *demos*: population + *graphia*: description) The quantitative study of the processes that determine the structure of populations (e.g., the relative proportions of individuals of different sizes, ages, stages, and genders), the size of populations, and the changes in these attributes.

denaturation Loss of activity of an enzyme or a nucleic acid molecule as a result of structural changes induced by heat or other means.

dendrite [Gk. *dendron*: tree] The receptive element of most neurons, which receives synaptic input from other neurons. Usually much branched and relatively short compared with the axon.

denitrification The chemical process by which "fixed" (biologically-useable) nitrogen is converted back into N_2 gas. (Contrast with nitrogen fixation.)

denitrifiers Bacteria that release nitrogen to the atmosphere as nitrogen gas (N_2).

density-dependent Pertaining to any factor or interaction that varies with density. For example, individual survival or reproductive success often changes as population density changes.

deoxygenation The release of O_2 from a combined state with a respiratory pigment such as hemoglobin.

deoxyribonucleic acid *See* DNA.

deoxyribonucleoside triphosphates (dNTPs) The raw materials for DNA synthesis: deoxyadenosine triphosphate (dATP), deoxythymidine triphosphate (dTTP), deoxycytidine triphosphate (dCTP), and deoxyguanosine triphosphate (dGTP). Also called deoxyribonucleotides.

deoxyribose A five-carbon sugar found in nucleotides and DNA.

depolarization A change in the membrane potential across a cell membrane so that the inside of the membrane becomes less negative, or even positive, compared with the outside of the membrane. (Contrast with hyperpolarization.)

derived trait A trait that differs from the ancestral trait. (Contrast with synapomorphy.)

dermal tissue system The outer covering of a plant, consisting of epidermis in the young plant and periderm in a plant with extensive secondary growth. (Contrast with ground tissue system, vascular tissue system.)

descent with modification Darwin's premise that all species share a common ancestor and have diverged from one another gradually over time.

descriptive statistics Quantitative measures that describe general patterns in data.

determinate growth A growth pattern in which the growth of an organism or organ ceases when an adult state is reached; characteristic of most animals and some plant organs. (Contrast with indeterminate growth.)

determination In development, the process whereby the fate of an embryonic cell or group of cells (e.g., to become epidermal cells or neurons) is set.

detritivore (di try' ti vore) [L. *detritus*: worn away + *vorare*: to devour] An organism that obtains its energy from the dead bodies or waste products of other organisms. (Contrast with carnivore, herbivore, omnivore.)

development The process by which a multicellular organism, beginning with a single cell, goes through a series of changes, taking on the successive forms that characterize its life cycle.

diaphragm (dye' uh fram) [Gk. *diaphrassein*: barricade] (1) A sheet of muscle that separates the thoracic and abdominal cavities, found only in mammals; helps drive air flow during breathing. (2) A birth control device in which a sheet of rubber is fitted over the woman's cervix, blocking the entry of sperm.

diastole (dye ass' toll ee) [Gk. dilation] The portion of the cardiac cycle when the heart muscle relaxes. (Contrast with systole.)

dichotomous (dye cot' oh mus) [Gk. *dichot*: split in two; *tomia*: removed] A branching pattern in which the shoot divides at the apex producing two equivalent branches that subsequently never overlap.

differentiation The process whereby originally similar cells follow different developmental pathways; the actual expression of determination.

diffusion Random movement of molecules or other particles, resulting in even distribution of the particles when no barriers are present.

digestion The breakdown (by hydrolysis) of ingested food molecules into smaller chemical components that an animal is capable of distributing to the tissues of its body.

digestive vacuole In protists, an organelle specialized for digesting food ingested by endocytosis.

dihybrid cross A mating in which the parents differ with respect to the alleles of two loci of interest.

dikaryon (di care' ee ahn) [Gk. *di*: two + *karyon*: kernel] A cell or organism carrying two genetically distinguishable nuclei. Common in fungi.

dioecious (die eesh' us) [Gk. *di*: two + *oikos*: house] Pertaining to organisms in which the two sexes are "housed" in two different individuals, so that eggs and sperm are not produced in the same individuals. Examples: humans, fruit flies, date palms. (Contrast with monoecious.)

diploblastic Having two cell layers. (Contrast with triploblastic.)

diploid (dip' loid) [Gk. *diplos*: double] Having a chromosome complement consisting of two copies (homologs) of each chromosome. Designated $2n$. (Contrast with haploid.)

direct development Pattern of development in which the young resemble the adults. (Contrast with indirect development.)

directional selection Selection in which phenotypes at one extreme of the population distribution are favored. (Contrast with disruptive selection, stabilizing selection.)

disaccharide A carbohydrate made up of two monosaccharides (simple sugars).

discrete variables Quantitative variables that take on only whole number values.

dispersal corridor A continuous strip of habitat that connects larger pieces of habitat, thus facilitating movement of organisms between them.

disruptive selection Selection in which phenotypes at both extremes of the population distribution are favored. (Contrast with stabilizing selection.)

disturbance Events that cause sudden change in the environment and in ecological systems as a result. Examples include fire, landslides, floods, strong storms, and volcanic eruptions—as well as less obvious events such as the death and fall of a large tree that opens up the forest canopy.

disulfide bridge The covalent bond between two sulfur atoms (–S—S–) linking two molecules or remote parts of the same molecule.

DNA (deoxyribonucleic acid) The fundamental hereditary material of all living organisms. In eukaryotes, stored primarily in the cell nucleus. A nucleic acid using deoxyribose rather than ribose.

DNA fingerprint An individual's unique pattern of allele sequences, commonly short tandem repeats and single nucleotide polymorphisms.

DNA ligase Enzyme that unites broken DNA strands during replication and recombination.

DNA methyltransferase An enzyme that catalyzes the methylation of DNA.

DNA microarray A small glass or plastic square onto which thousands of single-stranded DNA sequences are fixed so that hybridization of cell-derived RNA or DNA to the target sequences can be performed.

DNA polymerase Any of a group of enzymes that catalyze the formation of DNA strands from a DNA template.

DNA replication The creation of a new strand of DNA in which DNA polymerase catalyzes the exact reproduction of an existing (template) strand of DNA.

DNA segregation The separation of two DNAs formed by replication into two new cells during cell division.

DNA transposons Mobile genetic elements that move without making an RNA intermediate. (Contrast with retrotransposons.)

dNTPs *See* deoxyribonucleoside triphosphates.

domain (1) An independent structural element within a protein. Encoded by recognizable nucleotide sequences, a domain often folds separately from the rest of the protein. Similar domains can appear in a variety of different proteins across phylogenetic groups (e.g., "homeobox domain"; "calcium-binding domain"). (2) In phylogenetics, the three monophyletic branches of life (Bacteria, Archaea, and Eukarya).

dominance In genetics, the ability of one allelic form of a gene to determine the phenotype of a heterozygous individual in which the homologous chromosomes carry both it and a different (recessive) allele. (Contrast with recessive.)

dominate In animal behavior, to have control over resources (e.g., food or mates) as a result of defeating challengers in one-on-one behavioral contests.

dormancy A condition in which normal activity is suspended, as in some spores, seeds, and buds.

dorsal [L. *dorsum*: back] Toward or pertaining to the back or upper surface. (Contrast with ventral.)

double fertilization In angiosperms, a process in which the nuclei of two sperm fertilize one egg. One sperm's nucleus combines with the egg nucleus to produce a zygote, while the other combines with the same egg's two polar nuclei to produce the first cell of the triploid endosperm (the tissue that will nourish the growing plant embryo).

doubling time The time over which an ecological population doubles in size (number of individuals). Populations growing multiplicatively with a constant value of r have a constant doubling time, no matter what their current size is.

duplication A mutation in which a segment of a chromosome is duplicated, often by the attachment of a segment lost from its homolog. (Contrast with deletion.)

dynamic (Gk. *dunamikos*: powerful) Characterized by activity or change.

E

ecological efficiency The transfer of energy from one trophic level to the next, often estimated as the ratio of dry biomass at the higher level to dry biomass at the lower level. Note that this use of "efficiency" as the ratio of energy output to input is similar to the measure of efficiency used for many human-made machines.

ecological system One or more organisms and the abiotic and biotic environment with which they interact.

ecological transition A transition from one ecological community to a different community following a disturbance. The transition is not easily reversible, and the new community persists even if original environmental conditions are restored; thus the two ecological communities are considered to be alternative stable states of the system. (Contrast with succession.)

ecology [Gk. *oikos*: house] Defined by the German biologist Ernst Haeckel in 1866 as "the entire science of the relations of the organism to its external environment," by which he (and we) include both the biotic and abiotic environment.

ecosystem (eek' oh sis tum) In general, shorthand for "ecological system." Often used more specifically to refer to an ecological community and its abiotic environmental context.

ecosystem function Any aspect of the interactions that occur within an ecosystem that affects the rate of input of energy and matter into the ecosystem and its output of transformed matter and energy.

ecosystem services Ecosystem outputs that benefit humans.

ectoderm [Gk. *ektos*: outside + *derma*: skin] The outermost of the three embryonic germ layers first delineated during gastrulation. Gives rise to the skin, sense organs, and nervous system.

ectomycorrhizae Mutualistic fungi that cover the roots of plants and assist in the uptake of water and minerals from the soil by the plant.

ectotherm [Gk. *ektos*: outside + *thermos*: heat] An animal in which body temperature matches—and varies with—the external temperature. (*See also* poikilotherm.)

effector cells In cellular immunity, B cells and T cells that attack an antigen, either by secreting antibodies that bind to the antigen or by releasing molecules that destroy any cell bearing the antigen.

effectors One of four essential elements of a control system. A tissue, organ, or cell that performs actions ("carries out orders") under the direction of the nervous system or another physiological control system (e.g., the endocrine system).

efferent neurons In a nervous system, nerve cells that carry commands from the central nervous system to physiological and behavioral effectors such as muscles and glands.

egg In all sexually reproducing organisms, the female gamete (also called an ovum); in birds, reptiles, and some other animals, a structure within which the early embryonic development of a fertilized ovum occurs. (*See also* amniote egg.)

electrical synapse A type of synapse at which there is direct continuity between the cytoplasms of the presynaptic and postsynaptic neurons and action potentials spread directly from presynaptic cell to postsynaptic cell. (Contrast with chemical synapse.)

electrocardiogram (ECG or EKG) A graphic recording of electrical potentials that are measured on the body surface but originate from the heart.

electromagnetic radiation A self-propagating wave that travels though space and has both electrical and magnetic properties.

electron A subatomic particle outside the nucleus carrying a negative charge and very little mass.

electron shell The region surrounding the atomic nucleus at a fixed energy level in which electrons orbit.

electron transport The passage of electrons through a series of proteins with a release of energy which may be captured in a concentration gradient or in chemical form such as NADH or ATP.

electronegativity The tendency of an atom to attract electrons when it occurs as part of a compound.

element A substance that cannot be converted to simpler substances by ordinary chemical means.

elicitor A substance that induces the synthesis of defensive molecules such as phytoalexins in higher plants.

elongation (1) In molecular biology, the addition of monomers to make a longer RNA or protein during transcription or translation. (2) Growth of a plant axis or cell primarily in the longitudinal direction.

embryo [Gk. *en*: within + *bryein*: to grow] A young animal, or young plant sporophyte, while it is still contained within a protective structure such as a seed, egg, or uterus.

embryo sac In angiosperms, the female gametophyte. Found within the ovule, it consists of eight or fewer cells, membrane bounded, but without cellulose walls between them.

embryonic diapause A programmed state of arrested or profoundly slowed embryonic development. (*See also* delayed implantation.)

embryonic stem cell (ESC) A pluripotent cell in the blastocyst.

emigration Movement of individual organisms out of a population.

3′ end (3 prime) The end of a DNA or RNA strand that has a free hydroxyl group at the 3′ carbon of the sugar (deoxyribose or ribose).

5′ end (5 prime) The end of a DNA or RNA strand that has a free phosphate group at the 5′ carbon of the sugar (deoxyribose or ribose).

endocrine cells Cells that secrete hormones into the blood. (*See also* endocrine gland.)

endocrine gland (en' doh krin) [Gk. *endo*: within + *krinein*: to separate] A gland or tissue , lacking ducts, in which endocrine cells secrete hormones into the blood. (Contrast with exocrine gland.)

endocytosis A process by which liquids or solid particles are taken up by a cell through invagination of the plasma membrane. (Contrast with exocytosis.)

endoderm [Gk. *endo*: within + *derma*: skin] The innermost of the three embryonic germ layers delineated during gastrulation. Gives rise to the digestive and respiratory tracts and structures associated with them.

endodermis In plants, a specialized cell layer marking the inside of the cortex in roots and some stems. Frequently a barrier to free diffusion of solutes.

endogenous retroviruses A DNA sequence derived from a retrovirus that persists in and is transmitted as part of the genome of a non-viral organism. Some endogenous retroviruses can become active and produce new viral infections.

endomembrane system A system of intracellular membranes that exchange material with one another, consisting of the Golgi apparatus, endoplasmic reticulum, and lysosomes when present.

endometrium The inner tissue layer of the uterus of a mammal. Embryos implant in the endometrium, which later helps form the placenta.

endophytic fungi Fungi that live within the aboveground portions of plants without causing obvious harm to the host plant.

endoplasmic reticulum (ER) [Gk. *endo*: within + L. *reticulum*: net] A system of membranous tubes and flattened sacs found in the cytoplasm of eukaryotes. Exists in two forms: rough ER, studded with ribosomes; and smooth ER, lacking ribosomes.

endoskeleton [Gk. *endo*: within + *skleros*: hard] An internal skeleton covered by other, soft body tissues. (Contrast with exoskeleton.)

endosome Membrane-enclosed compartment in a cell that is formed after endocytosis.

endosperm [Gk. *endo*: within + *sperma*: seed] A specialized triploid seed tissue found only in angiosperms; contains stored nutrients for the developing embryo.

endospore [Gk. *endo*: within + *spora*: to sow] In some bacteria, a resting structure that can survive harsh environmental conditions.

endotoxin A lipopolysaccharide that forms part of the outer membrane of certain Gram-negative bacteria that is released when the bacteria grow or lyse. (Contrast with exotoxin.)

end-product inhibition *See* feedback inhibition.

endurance exercise Exercise that consists of many repetitions of relatively low-intensity muscular actions over long periods of time. Also termed endurance training. (Contrast with resistance exercise.)

energy The capacity to create and maintain organization. More narrowly defined, the capacity do work or move matter against an opposing force.

enteric division One of three divisions of the vertebrate autonomic nervous system; composed of neurons located between the smooth muscle layers of the gut.

entropy (S) (en′ tro pee) [Gk. *tropein*: to change] A measure of the degree of disorder in any system. Spontaneous reactions in a closed system are always accompanied by an increase in entropy.

environmental sex determination Mechanism in which the sex of an individual is determined by the temperature it experiences during embryonic development.

enzyme (en′ zime) [Gk. *zyme*: to leaven (as in yeast bread)] A catalytic protein that speeds up a biochemical reaction.

enzyme–substrate complex (ES) An intermediate in an enzyme-catalyzed reaction; consists of the enzyme bound to its substrate(s).

eons The primary division of geologic time. The history of Earth is divided into four eons: Hadeon, Archean, Proterozoic, and Phanerozoic.

epi- [Gk. upon, over] A prefix used to designate a structure located on top of another; for example, epidermis, epiphyte.

epiblast The upper or overlying portion of the avian blastula which is joined to the hypoblast at the margins of the blastodisc.

epidermis [Gk. *epi*: over + *derma*: skin] In plants and animals, the outermost cell layers. (Only one cell layer thick in plants.)

epigenetics The scientific study of changes in the expression of a gene or set of genes that occur without change in the DNA sequence.

epistasis Interaction between genes in which the presence of a particular allele of one gene determines how another gene will be expressed.

epithelium (plural: epithelia) A sheet of cells that lines a body cavity or covers an organ or body surface; one of the major tissue types in multicellular animals.

epitope *See* antigenic determinant.

epoch A subdivision of geologic time that divides geologic periods.

EPSP *See* excitatory postsynaptic potential.

equilibrium The state of a system in which there is no net change through time.

equilibrium potential The membrane potential at which an ion is at electrochemical equilibrium, i.e., at which the concentration and electrical effects on diffusion of the ion are balanced so there is no net diffusion of the ion across the membrane.

ER *See* endoplasmic reticulum.

erythrocyte (ur rith′ row site) [Gk. *erythros*: red + *kytos*: container] *See* red blood cell.

ES *See* enzyme–substrate complex.

ESC *See* embryonic stem cell.

essential amino acids Amino acids that an animal cannot synthesize for itself and must obtain from its food.

essential element A mineral nutrient required for normal growth and reproduction in plants and animals.

essential fatty acids Fatty acids that an animal cannot synthesize for itself and must obtain from its food.

essential mineral *See* essential nutrient.

essential nutrient A nutrient (e.g., amino acid, fatty acid, vitamin, mineral) that an animal cannot synthesize for itself and must obtain from its food.

ester linkage A condensation (water-releasing) reaction in which the carboxyl group of a fatty acid reacts with the hydroxyl group of an alcohol. Lipids, including most membrane lipids, are formed in this way. (Contrast with ether linkage.)

estrogens Any of several feminizing steroid sex hormones; produced chiefly by the ovaries in mammals, and by the placenta during pregnancy.

estrous cycle A cycle of behavioral readiness to copulate (called estrus and correlated with ovulation) in most female mammals.

estrus (es′ trus) [L. *oestrus*: frenzy] The period of heat, or maximum sexual receptivity, observed in most female mammals at times correlated with ovulation.

ether linkage The linkage of two hydrocarbons by an oxygen atom (HC—O—CH). Ether linkages are characteristic of the membrane lipids of the Archaea. (Contrast with ester linkage.)

ethylene One of the plant growth hormones, the gas $H_2C{=}CH_2$. Involved in fruit ripening and other growth and developmental responses.

euchromatin Diffuse, uncondensed chromatin. Contains active genes that will be transcribed into mRNA. (Contrast with heterochromatin.)

eudicots Angiosperms with two embryonic cotyledons; one of the two largest clades of angiosperms. (*See also* monocots.)

eukaryotes (yew car′ ree oats) [Gk. *eu*: true + *karyon*: kernel or nucleus] Organisms whose cells contain their genetic material inside a nucleus. Includes all life other than the viruses, archaea, and bacteria. (Contrast with prokaryotes.)

eusociality A type of social group in which some adults are infertile and provide services to other, fertile adults.

eutrophication (yoo trofe′ ik ay′ shun) [Gk. *eu*: truly + *trephein*: to flourish] Responses of aquatic ecosystems to excessive nutrient input, which include greatly increased primary productivity and resulting depletion of dissolved oxygen.

evolution Any gradual change. Most often refers to organic or Darwinian evolution, which is the genetic and resulting phenotypic change in populations of organisms from generation to generation. (Contrast with speciation.)

evolutionary developmental biology (evo-devo) The study of the interplay between evolutionary and developmental processes, with a focus on the genetic changes that give rise to novel morphology. Key concepts of evo-devo include modularity, genetic toolkits, genetic switches, and heterochrony.

evolutionary reversal The reappearance of an ancestral trait in a group that had previously acquired a derived trait.

evolutionary theory The understanding and application of the mechanisms of evolutionary change to biological problems.

excitable cell A cell that can generate and propagate action potentials because its cell membrane contains voltage-gated channels, notably neurons and muscle cells.

excitation In the study of muscle, the event in which a nerve impulse arrives at a neuromuscular junction and initiates an action potential in the cell membrane of a muscle cell.

excitation–contraction coupling In a muscle cell, the process by which electrical excitation of the cell membrane leads to contractile activity by the contractile proteins (actin and myosin) in the cell. Key events occur at the transverse tubules and sarcoplasmic reticulum, and result in release of Ca^{2+} into the cytoplasm.

excitatory postsynaptic potential (EPSP) At a chemical synapse (specifically, an excitatory synapse), a graded membrane depolarization produced in the postsynaptic cell by arrival of an impulse in the presynaptic cell. (Contrast with inhibitory postsynaptic potential.)

excitatory synapse A synapse at which impulses in the presynaptic cell act to depolarize the postsynaptic cell.

exocrine gland (eks' oh krin) [Gk. *exo*: outside + *krinein*: to separate] Any gland, such as a salivary gland, that secretes to the outside of the body or into the gut. (Contrast with endocrine gland.)

exocytosis A process by which a vesicle within a cell fuses with the plasma membrane and releases its contents to the outside. (Contrast with endocytosis.)

exon A portion of a DNA molecule, in eukaryotes, that codes for part of a polypeptide. (Contrast with intron.)

exoskeleton (eks' oh skel' e ton) [Gk. *exos*: outside + *skleros*: hard] A hard covering on the outside of the body to which muscles are attached. (Contrast with endoskeleton.)

exotoxin A highly toxic, usually soluble protein released by living, multiplying bacteria. (Contrast with endotoxin.)

expression vector A DNA vector, such as a plasmid, that carries a DNA sequence for the expression of an inserted gene into mRNA and protein in a host cell.

expressivity The degree to which a genotype is expressed in the phenotype; may be affected by the environment.

external environment The conditions of the surroundings outside an animal's body.

external fertilization Fertilization that occurs after release of sperm and eggs into the external environment; typical of aquatic animals. Also called spawning. (Contrast with internal fertilization.)

external respiration *See* breathing.

extracellular fluid The aqueous solution (body fluid) outside cells (in the extracellular compartment). In animals with closed circulatory systems, subdivided into blood plasma and interstitial fluid.

extracellular matrix A material of heterogeneous composition surrounding cells and performing many functions including adhesion of cells.

extraembryonic membranes Four membranes that support but are not part of the developing embryos of reptiles, birds, and mammals, defining these groups phylogenetically as amniotes. (*See* amnion, allantois, chorion, and yolk sac.)

extrarenal salt excretion Excretion of inorganic ions in concentrated solutions by structures other than the kidneys, such as the gills in marine fish and salt glands in marine birds and sea turtles.

extreme halophiles A group of euryarchaeotes that live exclusively in very salty environments.

extremophiles Archaea and bacteria that live and thrive under conditions (e.g., extremely high temperatures) that would kill most organisms.

F

F-box A protein domain that facilitates protein–protein interactions necessary for protein degradation. Often involved in the regulation of gene expression (through the breakdown of regulatory proteins such as repressors), especially in plants.

F_1 The first filial generation; the immediate progeny of a parental (P) mating.

F_2 The second filial generation; the immediate progeny of a mating between members of the F_1 generation.

facilitated diffusion Passive movement through a membrane involving a specific carrier protein; does not proceed against a concentration gradient. (Contrast with active transport, diffusion.)

facultative anaerobe A prokaryote that can shift its metabolism between anaerobic and aerobic modes depending on the presence or absence of O_2. (Alternatively, facultative aerobe.)

facultative parasite A organism that can parasitize other living organisms but is also capable of growing independently.

fallopian tube *See* oviduct.

family A group of related genera.

fast glycolytic cells Muscle cells that are highly dependent on ATP production by anaerobic glycolysis and that contract and develop high tension rapidly, but fatigue quickly. Also called fast-twitch cells. (Compare to slow oxidative cells.)

fate map A diagram of the blastula showing which cells (blastomeres) are "fated" to contribute to specific tissues and organs in the mature body.

fatty acid A molecule made up of a long nonpolar hydrocarbon chain and a polar carboxyl group. Found in many lipids.

fauna (faw' nah) All the animals found in a given area. (Contrast with flora.)

fecundity The average number of offspring produced by an individual female during a particular life stage, age, or period of time.

feedback inhibition A mechanism for regulating a metabolic pathway in which the end product of the pathway can bind to and inhibit the enzyme that catalyzes the first committed step in the pathway. Also called end-product inhibition.

fermentation (fur men tay' shun) [L. *fermentum*: yeast] (1) Speaking specifically about energy metabolism, the anaerobic degradation of a substance such as glucose to smaller molecules such as lactic acid or alcohol with the extraction of energy. (2) Speaking generally, metabolic processes that occur in the absence of O_2.

fertilization Union of gametes.

fertilizer Any of a number of substances added to soil to improve the soil's capacity to support plant growth. May be organic or inorganic.

fetus Medical and legal term for the stages of a developing human embryo from about the eighth week of pregnancy (the point at which all major organ systems have formed) to the moment of birth.

fiber In angiosperms, an elongated, tapering sclerenchyma cell, usually with a thick cell wall, that serves as a support function in xylem. (*See also* muscle fiber.)

fibrous root system A root system typical of monocots composed of numerous thin adventitious roots that are all roughly equal in diameter. (Contrast with taproot system.)

Fick's law of diffusion *See* law of diffusion for gases.

filament In flowers, the part of a stamen that supports the anther.

filter feeder (1) Speaking generally, an earlier term for a suspension feeder. (2) Speaking specifically, an organism that feeds by means of a straining or filtering device on organisms much smaller than itself that are suspended in water or air.

first filial generation *See* F_1.

first law of thermodynamics The principle that energy can be neither created nor destroyed.

fission *See* binary fission.

fitness An individual's contribution of genes to the next generation, as a consequence of its success in surviving and reproducing.

fixed The condition in which a particular allele is the only allele present at a genetic locus within a population (i.e., there is no genetic variation at the locus).

fixed action pattern In the study of animal behavior, a genetically determined behavior that is performed without learning, stereotypic (performed the same way each time), and not modifiable by learning.

flagellum (fla jell' um) (plural: flagella) [L. *flagellum*: whip] Long, whiplike appendage that propels cells. Prokaryotic flagella differ sharply from those found in eukaryotes.

flora (flore' ah) All of the plants found in a given area. (Contrast with fauna.)

floral meristem In angiosperms, a meristem that forms the floral organs (sepals, petals, stamens, and carpels).

floral organ identity genes In angiosperms, genes that determine the fates of floral meristem cells; their expression is triggered by the products of meristem identity genes.

florigen (FT) A plant hormone involved in the conversion of a vegetative shoot apex to a flower.

flower The sexual structure of an angiosperm.

fluid mosaic model A molecular model for the structure of biological membranes consisting of a fluid phospholipid bilayer in which suspended proteins are free to move in the plane of the bilayer.

flux The rate of movement of energy, or of matter (an element or a molecular compound), between two compartments of an ecosystem, or between the Earth system and space.

foliose Having a leafy growth form.

follicle-stimulating hormone (FSH) A hormone, produced by the anterior pituitary gland, that helps control secretion of hormones by the gonads (i.e., a gonadotropin).

food web A diagrammatic representation of which organisms consume which other organisms in an ecological community, i.e., a diagram of trophic relationships.

foot One of the three main body parts of a mollusk.

forebrain The region of the vertebrate brain that comprises the cerebrum, thalamus, and hypothalamus.

foregut The beginning of the digestive tract. Consists of the mouth, esophagus, and stomach, and in some animals, a crop or gizzard.

foregut fermenter An animal that has a specialized, nonacidic foregut chamber housing a mixed microbial community composed of fermenting microbes that assist with the breakdown of food materials.

fossil fuels Combustible materials, such as coal, natural gas, petroleum, and peat that are formed by geological processes over long periods of time from the remains of living organisms.

founder effect Random changes in allele frequencies resulting from establishment of a population by a very small number of individuals.

fragmentation In ecology, the subdivision of a larger piece of habitat into smaller pieces that are isolated from one another by a different habitat.

frame-shift mutation The addition or deletion of a single or two adjacent nucleotides in a gene's sequence. Results in the misreading of mRNA during translation and the production of a nonfunctional protein. (Contrast with missense mutation, nonsense mutation, silent mutation.)

frequency distribution A figure that displays the frequency of different classes of data.

free energy (*G*) Energy that is available for doing useful work, after allowance has been made for the increase or decrease of disorder.

fruit In angiosperms, a ripened and mature ovary (or group of ovaries) containing the seeds. Sometimes applied to reproductive structures of other groups of plants.

fruticose Having a shrubby growth form.

FSH *See* follicle-stimulating hormone.

FT *See* florigen.

functional genomics The assignment of functional roles to the proteins encoded by genes identified by sequencing entire genomes.

functional group A characteristic combination of atoms that contribute specific properties when attached to larger molecules.

G

G cap A chemically modified GTP added to the 5′ end of mRNA; facilitates binding of mRNA to ribosome and prevents mRNA breakdown. Also called 5′ cap.

G protein A membrane protein involved in signal transduction; characterized by binding GDP or GTP.

G protein–linked receptors A class of receptors that change configuration upon ligand binding such that a G protein binding site is exposed on the cytoplasmic domain of the receptor, initiating a signal transduction pathway.

G1 In the cell cycle, the gap between the end of mitosis and the onset of the S phase.

G1–S transition In the cell cycle, the point at which G1 ends and the S phase begins.

G2 In the cell cycle, the gap between the S (synthesis) phase and the onset of mitosis.

gain of function mutation A mutation that results in a protein with a new function. (Contrast with loss of function mutation.)

gametangium (gam uh tan' gee um) (plural: gametangia) [Gk. *gamos*: marriage + *angeion*: vessel] Any plant or fungal structure within which a gamete is formed.

gamete (gam' eet) [Gk. *gamete/gametes*: wife, husband] The mature sexual reproductive cell: the egg or the sperm.

gametogenesis (ga meet' oh jen' e sis) The specialized series of cellular divisions, including meiosis, that leads to the formation of gametes.

gametophyte (ga meet' oh fyte) In plants and photosynthetic protists with alternation of generations, the multicellular haploid phase that produces the gametes. (Contrast with sporophyte.)

ganglion (gang' glee un) (plural: ganglia) [Gk. tumor] (1) In a vertebrate, a discrete cluster of neuron cell bodies in the peripheral nervous system. (2) In an arthropod, an enlarged part of the ventral nerve cord found in each body segment.

gap genes In *Drosophila* (fruit fly) development, segmentation genes that define broad areas along the anterior–posterior axis of the early embryo. Part of a developmental cascade that includes maternal effect genes, pair rule genes, segment polarity genes, and Hox genes.

gas exchange membranes Thin layers of tissue (one or two cells thick) where respiratory gases move between the internal tissues of an animal and the animal's environmental medium.

gastrovascular cavity Serving for both digestion (gastro) and circulation (vascular); in particular, the central cavity of the body of jellyfish and other cnidarians.

gastrula The end product of gastrulation. At this point the embryo contains the forerunners of the three germ layers that will generate the entire body.

gastrulation Development of a blastula into a gastrula. In embryonic development, the process by which a blastula is transformed by massive movements of cells, from the exterior to the interior, into a gastrula.

gated channel A membrane protein that changes its three-dimensional shape, and therefore its ion conductance, in response to a stimulus. When open, it allows specific ions to move across the membrane.

gel electrophoresis (e lek' tro fo ree' sis) [L. *electrum*: amber + Gk. *phorein*: to bear] A technique for separating molecules (such as DNA fragments) from one another on the basis of their electric charges and molecular weights by applying an electric field to a gel.

gene [Gk. *genes*: to produce] A unit of heredity. Used here as the unit of genetic function which carries the information for a polypeptide or RNA.

gene family A set of similar genes derived from a single parent gene; need not be on the same chromosomes. The vertebrate globin genes constitute a classic example of a gene family.

gene flow Exchange of genes between populations through migration of individuals or movements of gametes.

gene-for-gene resistance In plants, a mechanism of resistance to pathogens in which resistance is triggered by the specific interaction of the products of a pathogen's *Avr* genes and a plant's *R* genes.

general transcription factors In eukaryotes, transcription factors that bind to the promoters of most protein-coding genes and are required for their expression. Distinct from transcription factors that have specific regulatory effects only at certain promoters or classes of promoters.

genetic drift Changes in gene frequencies from generation to generation as a result of random (chance) processes.

genetic linkage Association between genes on the same chromosome such that they do not show random assortment and seldom recombine; the closer the genes, the lower the frequency of recombination.

genetic screen A technique for identifying genes involved in a biological process of interest. Involves creating a large collection of randomly mutated organisms and identifying those individuals that are likely to have a defect in the pathway of interest. The mutated gene(s) in those individuals can then be isolated for further study.

genetic structure The frequencies of different alleles at each locus and the frequencies of different genotypes in a Mendelian population.

genetic switches Mechanisms that control how the genetic toolkit is used, such as promoters and the transcription factors that bind them. The signal cascades that converge on and operate these switches determine when and where genes will be turned on and off.

genetic toolkit A set of developmental genes and proteins that is common to most animals and is hypothesized to be responsible for the evolution of their differing developmental pathways.

genome (jee' nome) The complete DNA sequence for a particular organism or individual.

genomic equivalence The principle that no information is lost from the nuclei of cells as they pass through the early stages of embryonic development.

genomic imprinting The form of a gene's expression is determined by parental source (i.e., whether the gene is inherited from the male or female parent).

genomic library All of the cloned DNA fragments generated by the breakdown of genomic DNA into smaller segments.

genotype (jean' oh type) [Gk. *gen*: to produce + *typos*: impression] An exact description of the genetic constitution of an individual, either with respect to a single trait or with respect to a larger set of traits. (Contrast with phenotype.)

genus (jean' us) (plural: genera) [Gk. *genos*: stock, kind] A group of related, similar species recognized by taxonomists with a distinct name used in binomial nomenclature.

geographic range The entire region in which a species occurs.

geographic speciation *See* allopatric speciation.

geological time scale The standardized scale accepted by geologists to mark the major divisions in the history of the Earth.

germ cells [L. *germen*: to beget] Cells in the gonads that multiply mitotically and are capable of undergoing meiosis to produce gametes. (Contrast with somatic cell.)

germline mutation Mutation in a cell that produces gametes (i.e., a germline cell). (Contrast with somatic mutation.)

germination Sprouting of a seed or spore.

GFR *See* glomerular filtration rate.

gibberellin (jib er el' lin) A class of plant growth hormones playing roles in stem elongation, seed germination, flowering of certain plants, etc.

gill A breathing organ that is evaginated (folded outward from the body) and surrounded by the environmental medium. The type of breathing organ found most commonly in water-breathers. (Contrast with lung.)

glial cells One of the two major types of cells in the nervous system (the other being neurons); unlike neurons, glial cells are typically not excitable and do not conduct action potentials. Also called glia or neuroglia.

global warming An increase through time in Earth's average annual surface temperature.

glomerular filtration rate (GFR) The rate at which all the nephrons in the kidneys collectively produce primary urine by ultrafiltration of blood plasma.

glomerulus (glo mare' yew lus) (plural: glomeruli) [L. *glomus*: ball] Vascular structure in the kidney where blood filtration takes place, forming primary urine. Each glomerulus consists of a knot of capillaries served by arterioles.

gluconeogenesis The biochemical synthesis of glucose from other substances, such as amino acids, lactate, and glycerol.

glyceraldehyde 3-phosphate (G3P) A phosphorylated three-carbon sugar; an intermediate in glycolysis and photosynthetic carbon fixation.

glycerol (gliss' er ole) A three-carbon alcohol with three hydroxyl groups; a component of phospholipids and triglycerides.

glycosidic linkage Bond between carbohydrate (sugar) molecules through an intervening oxygen atom (—O—).

glycolipid A lipid to which sugars are attached.

glycolysis (gly kol' li sis) [Gk. *gleukos*: sugar + *lysis*: break apart] The enzymatic breakdown of glucose to pyruvic acid.

glycoprotein A protein to which sugars are attached.

glyoxysome (gly ox′ ee soam) An organelle found in plants, in which stored lipids are converted to carbohydrates.

Golgi apparatus (goal′ jee) A system of concentrically folded membranes found in the cytoplasm of eukaryotic cells; functions in secretion from the cell by exocytosis.

gonad (go′ nad) [Gk. *gone*: seed] An organ that produces gametes in animals: either an ovary (female gonad) or testis (male gonad). Certain of the tissues in a gonad secrete hormones.

gonadotropin A type of tropic hormone that controls hormone secretion by gonadal endocrine cells, and that supports and maintains gonadal tissue, including gamete production. The two gonadotropins in vertebrates are follicle-stimulating hormone and luteinizing hormone, both secreted by the anterior pituitary gland.

gonadotropin-releasing hormone (GnRH) In vertebrates, hormone produced by the hypothalamus that stimulates the anterior pituitary to secrete ("release") gonadotropins.

Gondwana The large southern land mass that existed from the Cambrian (540 mya) to the Jurassic (138 mya). Present-day remnants are South America, Africa, India, Australia, and Antarctica.

GPP *See* gross primary productivity.

graded membrane potentials Small local changes in membrane potential caused by opening or closing of ion channels. They can be of any size (thus "graded").

grafting Artificial transplantation of tissue from one organism to another. In horticulture, the transfer of a bud or stem segment from one plant onto the root of another as a form of asexual reproduction.

Gram-negative bacteria Bacteria that appear red when stained using the Gram-staining technique. These bacteria have an outer membrane outside the relatively thin peptidoglycan layer of the cell wall.

Gram-positive bacteria Bacteria that appear blue to purple when stained using the Gram-staining technique. These bacteria have an outer cell wall consisting of a thick layer of peptidoglycan.

Gram stain A differential purple stain useful in characterizing bacteria. The peptidoglycan-rich cell walls of Gram-positive bacteria stain purple; cell walls of Gram-negative bacteria generally stain orange.

gravitropism [Gk. *tropos*: to turn] A directed plant growth response to gravity.

gray crescent In frog development, a band of diffusely pigmented cytoplasm on the side of the egg opposite the site of sperm entry. Arises as a result of cytoplasmic rearrangements that establish the anterior–posterior axis of the zygote.

greenhouse effect The retention of heat energy near Earth's surface as a consequence of absorption and re-radiation of infrared radiation by certain gases (such as carbon dioxide and methane) in the atmosphere.

greenhouse gases Gases in the atmosphere, such as carbon dioxide and methane, that absorb infrared radiation and thereby retain heat energy within Earth's lower atmosphere.

gross primary productivity (GPP) The rate at which primary producers in a given area (or volume) capture energy from sunlight or other abiotic sources and convert it to chemical energy.

ground meristem That part of an apical meristem that gives rise to the ground tissue system of the primary plant body.

ground tissue system Those parts of the plant body not included in the dermal or vascular tissue systems. Ground tissues function in storage, photosynthesis, and support. (Contrast with dermal tissue system, vascular tissue system.)

growth An increase in the size of the body and its organs by cell division and cell expansion. In ecology, change in size of the population of a species.

growth factor A chemical signal that stimulates cells to divide.

growth hormone A peptide hormone released by the anterior pituitary gland that stimulates many processes involved in tissue formation and growth.

guanine (G) (gwan′ een) A nitrogen-containing base found in DNA, RNA, and GTP.

guard cells In plants, specialized, paired epidermal cells that surround and control the opening of a stoma (pore).

gut microbiome Populations of microbes, consisting of many species of bacteria and other heterotrophic microbes, living in the gut lumen of an animal.

gymnosperms Seed plants with seeds not enclosed in carpels. (*See also* angiosperms.)

gyre Large-scale circular ocean currents caused by prevailing winds and Earth's rotation.

H

habitat patches Areas of suitable habitat for a species that are separated by areas of unsuitable habitat.

habitat The characteristic environment(s) occupied by a species.

habitat structure The physical structure of a habitat, for example the horizontal and vertical distribution of objects (such as the stems and leaves of plants) in a habitat.

Hadley cells Large-scale vertical patterns of atmospheric circulation in which warm rises near the Equator, moves toward the poles, and cools and sinks around 30 degrees N or S latitude.

half-life The time required for half of a sample of a radioactive isotope to decay to its stable, nonradioactive form, or for a drug or other substance to reach half its initial dosage.

halophyte (hal′ oh fyte) [Gk. *halos*: salt + *phyton*: plant] A plant that grows in a saline (salty) environment.

haploid (hap′ loid) [Gk. *haploeides*: single] Having a chromosome complement consisting of just one copy of each chromosome; designated 1*n* or *n*. (Contrast with diploid.)

haplotype Linked nucleotide sequences that are usually inherited as a unit (as a "sentence" rather than as individual "words").

Hardy–Weinberg equilibrium The unchanging frequencies of alleles and genotypes in a population that is expected with random combination of gametes, in the absence of natural selection, mutation, migration, and genetic drift. The expectations of Hardy–Weinberg equilibrium can be used to calculate the expected frequencies of genotypes given the frequency of alleles, under the stated assumptions.

haustorium (haw stor′ ee um) (plural: haustoria) [L. *haustus*: draw up] A specialized hypha or other structure by which fungi and some parasitic plants draw nutrients from a host plant.

heart In a circulatory system, a muscular pump that drives flow of blood around the body.

heat capacity The ratio of energy absorbed by a substance the increase in temperature of that substance.

heat of vaporization The energy that must be supplied to convert a molecule from a liquid to a gas at its boiling point.

heat shock proteins Chaperone proteins expressed in cells exposed to high or low temperatures or other forms of environmental stress.

helical Shaped like a screw or spring (helix); this shape occurs in DNA and proteins.

helicase An enzyme that catalyzes the unwinding of a nucleic acid double helix.

helper T cell *See* T-helper cell.

hemiparasite A parasitic plant that can photosynthesize, but derives water and mineral nutrients from the living body of another plant. (Contrast with holoparasite.)

hemizygous (hem' ee zie' gus) [Gk. *hemi*: half + *zygotos*: joined] In a diploid organism, having only one allele for a given trait, typically the case for X-linked genes in male mammals and Z-linked genes in female birds. (Contrast with homozygous, heterozygous.)

hemocyanin A type of respiratory pigment found in arthropods and mollusks, consisting of a copper-based protein. Undergoes reversible combination with O_2 at copper-containing loci.

hemoglobin (hee' mo glow bin) [Gk. *heaema*: blood + L. *globus*: globe] A type of respiratory pigment consisting of an iron-containing protein. Undergoes reversible combination with O_2 at iron-containing loci called heme sites.

herbivore (ur' bi vore) [L. *herba*: plant + *vorare*: to devour] An animal that consumes plant tissues. *See* primary consumer. (Contrast with carnivore, detritivore, omnivore.)

herbivory An interaction between individuals of two different species in which one individual, the herbivore, consumes all or part of the other individual, which is a plant.

heritability The extent to which the value or expression of a trait of an organism is determined by genes transmitted from its parents.

hetero- [Gk.: *heteros*: other, different] A prefix indicating two or more different conditions, structures, or processes. (Contrast with homo-.)

heterochromatin Densely packed, dark-staining chromatin; any genes it contains are usually not transcribed. (Contrast with euchromatin.)

heterochrony Alteration in the timing of developmental events, leading to different results in the adult organism.

heterocyst A large, thick-walled cell type in the filaments of certain cyanobacteria that performs nitrogen fixation.

heteromorphic (het' er oh more' fik) [Gk. *heteros*: different + *morphe*: form] Having a different form or appearance, as two heteromorphic life stages of a plant. (Contrast with isomorphic.)

heterosis *See* hybrid vigor.

heterosporous (het' er os' por us) Producing two types of spores, one of which gives rise to a female megaspore and the other to a male microspore. (Contrast with homosporous.)

heterotopy [Gk. different place] Spatial differences in gene expression during development, controlled by developmental regulatory genes and contributing to the evolution of distinctive adult phenotypes.

heterotroph (het' er oh trof) [Gk. *heteros*: different + *trophe*: feed] An organism that requires preformed organic molecules as sources of energy and chemical building blocks. (Contrast with autotroph.)

heterotypy [Gk. different kind] Alteration in a developmental regulatory gene itself rather than the expression of the genes it controls. (Contrast with heterotopy.)

heterozygous (het' er oh zie' gus) [Gk. *heteros*: different + *zygotos*: joined] In diploid organisms, having different alleles of a given gene on the pair of homologs carrying that gene. (Contrast with homozygous.)

heterozygous carrier An individual that carries a recessive allele for a phenotype of interest (e.g., a genetic disease); the individual does not show the phenotype, but may have progeny with the phenotype if the other parent also carries the recessive allele.

hibernation [L. *hibernum*: winter] In mammals, the state of inactivity of some species during winter, during which body temperature falls low enough to match ambient temperature and metabolic rate is prodoundly depressed.

high-throughput sequencing Rapid DNA sequencing on a micro scale in which many fragments of DNA are sequenced in parallel.

highly repetitive sequences Short (less than 100 bp), nontranscribed DNA sequences, repeated thousands of times in tandem arrangements.

hindbrain The region of the developing vertebrate brain that gives rise to the medulla, pons, and cerebellum.

hindgut The second part of the intestine; in mammals, the hindgut is also called the large intestine.

hindgut fermenters An animal that has a specialized hindgut chamber housing mixed communities of fermenting microbes that assist with the breakdown of food materials.

histamine (hiss' tah meen) A substance released by damaged tissue, or by mast cells in response to allergens. Histamine increases vascular permeability, leading to edema (swelling).

histogram A figure that displays frequencies of classes of quantitative data binned by ranges of a particular variable.

histone Any one of a group of proteins forming the core of a nucleosome, the structural unit of a eukaryotic chromosome.

histone acetyltransferases Enzymes involved in chromatin remodeling. Add acetyl groups to the tail regions of histone proteins. (Contrast with histone deacetylase.)

histone deacetylase In chromatin remodeling, an enzyme that removes acetyl groups from the tails of histone proteins. (Contrast with histone acetyltransferases.)

HIV *See* human immunodeficiency virus.

holometabolous Undergoing complete metamorphosis.

holoparasite A fully parasitic plant (i.e., one that does not perform photosynthesis).

home range A region inhabited by an individual in which other individuals are not excluded. (Contrast with territory.)

homeobox 180-base-pair segment of DNA found in certain homeotic genes; regulates the expression of other genes and thus controls large-scale developmental processes.

homeostasis (home' ee o sta' sis) [Gk. *homos*: same + *stasis*: position] Stability of the internal environment of an individual, such as a constant body temperature, and the physiological or behavioral feedback responses that maintain that stability. In the words of Walter Cannon, who coined the term, "the coordinated physiological processes which maintain most of the [constant] states in the organism."

homeotherm An animal that maintains a relatively constant body temperature by physiological means, (e.g., changes in metabolic rate) rather than simply by behavior.

homeotic mutation Mutation in a homeotic gene that results in the formation of a different organ than that normally made by a region of the embryo.

homo- [Gk. *homos*: same] A prefix indicating two or more similar conditions, structures, or processes. (Contrast with hetero-.)

homologous pair A pair of matching chromosomes made up of a chromosome from each of the two sets of chromosomes in a diploid organism.

homologous recombination Exchange of segments between two DNA molecules based on sequence similarity between the two molecules. The similar sequences align and crossover. Used to create knockout mutants in mice and other organisms.

homology (ho mol' o jee) [Gk. *homologia*: of one mind; agreement] A similarity between two or more features that is due to inheritance from a common ancestor. The structures are said to be **homologous**, and each is a *homolog* of the others.

homoplasy (home′ uh play zee) [Gk. *homos*: same + *plastikos*: shape, mold] The presence in multiple groups of a trait that is not inherited from the common ancestor of those groups. Can result from convergent evolution, evolutionary reversal, or parallel evolution.

homosporous Producing a single type of spore that gives rise to a single type of gametophyte, bearing both female and male reproductive organs. (Contrast with heterosporous.)

homozygous (home′ oh zie′ gus) [Gk. *homos*: same + *zygotos*: joined] In diploid organisms, having identical alleles of a given gene on both homologous chromosomes. An individual may be a homozygote with respect to one gene and a heterozygote with respect to another. (Contrast with heterozygous.)

horizons The horizontal layers of a soil profile, including the topsoil (A horizon), subsoil (B horizon) and parent rock or bedrock (C horizon).

hormone (hore′ mone) [Gk. *hormon*: to excite, stimulate] A chemical substance produced in minute amounts by endocrine cells and transported in the blood to distant target cells, where it exerts regulatory influences on their function.

Human Genome Project A publicly and privately funded research effort, successfully completed in 2003, to produce a complete DNA sequence for the entire human genome.

human immunodeficiency virus (HIV) The retrovirus that causes acquired immune deficiency syndrome (AIDS).

humidic Referring to terrestrial animals, those that are restricted to humid, water-rich microenvironments; unable to live steadily in the open air.

humoral immune response The response of the immune system mediated by B cells that produces circulating antibodies active against extracellular bacterial and viral infections. (Contrast with cellular immune response.)

humus (hew′ mus) The partly decomposed remains of plants and animals on the surface of a soil.

hybrid vigor The superior fitness of heterozygous offspring as compared with that of their dissimilar homozygous parents. Also called heterosis.

hybrid zone A region of overlap in the ranges of two closely related species where the species may hybridize.

hydrogen bond A weak electrostatic bond which arises from the attraction between the slight positive charge on a hydrogen atom and a slight negative charge on a nearby oxygen or nitrogen atom.

hydrological cycle The movement of water from the oceans to the atmosphere, to the soil, and back to the oceans.

hydrolysis reaction (high drol′ uh sis) [Gk. *hydro*: water + *lysis*: break apart] A chemical reaction that breaks a bond by inserting the components of water ($AB + H_2O \rightarrow AH + BOH$). (Contrast with condensation reaction.)

hydrophilic (high dro fill′ ik) [Gk. *hydro*: water + *philia*: love] Having an affinity for water. (Contrast with hydrophobic.)

hydrophobic (high dro foe′ bik) [Gk. *hydro*: water + *phobia*: fear] Having no affinity for water. Uncharged and nonpolar groups of atoms are hydrophobic. (Contrast with hydrophilic.)

hydroponic Pertaining to a method of growing plants with their roots suspended in nutrient solutions instead of soil.

hydrostatic skeleton Stiffness imparted to the body or part of the body of an animal by high blood pressure inside

hyper- [Gk. *hyper*: above, over] Prefix indicating above, higher, more. (Contrast with hypo-.)

hyperaccumulators Species of plants that store large quantities of heavy metals such as arsenic, cadmium, nickel, aluminum, and zinc.

hyperosmotic In comparing two solutions, refers to the one having a higher osmotic pressure. The term is relative and has meaning only in comparing two specified solutions. (Compare with hyposmotic.)

hyperosmotic regulator An aquatic animal that maintains a stable blood osmotic pressure higher than the osmotic pressure of the water in which it lives.

hyperpolarization A change in the membrane potential across a cell membrane so that the inside of the membrane becomes more negative compared with the outside of the membrane. (Contrast with depolarization.)

hypersensitive response A defensive response of plants to microbial infection in which phytoalexins and pathogenesis-related proteins are produced and the infected tissue undergoes apoptosis to isolate the pathogen from the rest of the plant.

hypertension High blood pressure.

hypertonic Having a greater solute concentration. Said of one solution compared with another. (Contrast with hypotonic, isotonic.)

hypha (high′ fuh) (plural: hyphae) [Gk. *hyphe*: web] In the fungi and oomycetes, any single filament.

hypo- [Gk. *hypo*: beneath, under] Prefix indicating underneath, below, less. (Contrast with hyper-.)

hypophysis *See* pituitary gland.

hyposmotic In comparing two solutions, refers to the one having a lower osmotic pressure. The term is relative and has meaning only in comparing two specified solutions. The word is a contraction of *hypo-osmotic*. (Compare with hyperosmotic.)

hyposmotic regulator An aquatic animal that maintains a stable blood osmotic pressure lower than the osmotic pressure of the water in which it lives.

hypothalamo–hypophysial portal system In vertebrates, a system of blood vessels that connects capillaries in the hypothalamus to capillaries in the anterior pituitary gland; provides a direct pathway by which hypothalamic hormones can reach specific populations of cells in the anterior pituitary.

hypothalamus A part of the forebrain lying below the thalamus and relatively near the base of the brain. It plays roles in learning, memory, spatial orientation, and control of water balance, reproduction, and temperature regulation.

hypotonic Having a lesser solute concentration. Said of one solution in comparing it to another. (Contrast with hypertonic, isotonic.)

I

igneous Rocks formed from the cooling and hardening of molten magma.

imbibition Water uptake by a seed; first step in germination.

immunity In animals, the ability to avoid disease when invaded by a pathogen by deploying various defense mechanisms.

immunoglobulins A class of proteins containing a tetramer consisting of four polypeptide chains—two identical light chains and two identical heavy chains—held together by disulfide bonds; active as receptors and effectors in the immune system.

immunological memory The capacity to more rapidly and massively respond to a second exposure to an antigen than occurred on first exposure.

imperfect flower A flower lacking either functional stamens or functional carpels. (Contrast with perfect flower.)

implantation The process by which the early mammalian embryo becomes attached to and embedded in the endometrium (lining of the uterus).

imprinting *See* behavioral imprinting.

impulse *See* action potential.

in vitro evolution A method based on natural molecular evolution that uses artificial selection in the laboratory to rapidly produce molecules with novel enzymatic and binding functions.

inclusive fitness The sum of an individual's genetic contribution to subsequent generations both via production of its own offspring and via its influence on the survival of relatives who are not direct descendants.

incomplete cleavage A pattern of cleavage that occurs in many very yolky eggs, in which the cleavage furrows do not penetrate all of the cytoplasm. (Contrast with complete cleavage.)

incomplete dominance Condition in which the heterozygous phenotype is intermediate between the two homozygous phenotypes.

incomplete metamorphosis Insect development in which changes between instars are gradual.

independent assortment *See* law of independent assortment.

indeterminate growth A open-ended growth pattern in which an organism or organ continues to grow as long as it lives; characteristic of some animals and of plant shoots and roots. (Contrast with determinate growth.)

indirect development Development in which the young are morphologically very different than the adult. (Contrast with direct development.)

induced mutation A mutation resulting from exposure to a mutagen from outside the cell. (Contrast with spontaneous mutation.)

induced ovulation Ovulation (release of an egg from the ovaries of a female) triggered by the act of copulation. (Contrast with spontaneous ovulation.)

induced pluripotent stem cells (iPS cells) Multipotent or pluripotent animal stem cells produced from differentiated cells in vitro by the addition of several genes that are expressed.

induced responses Defensive responses that a plant produces only in the presence of a pathogen, in contrast to constitutive defenses, which are always present.

inducer (1) A compound that stimulates the synthesis of a protein. (2) In embryonic development, a substance that causes a group of target cells to differentiate in a particular way.

inducible enzyme An enzyme that is present in variable amounts, depending on the conditions to which a cell, tissue, or organism is exposed. Expression of an inducible enzyme depends on levels of specific inducing agents.

inducible genes Genes that are expressed only when their products are needed. (Contrast with constitutive genes.)

inducible Produced only in the presence of a particular compound or under particular circumstances. (Contrast with constitutive.)

induction In embryonic development, the process by which one cell population influences the fate of another cell population.

inflammation A nonspecific defense against pathogens; characterized by redness, swelling, pain, and increased temperature.

inflorescence A structure composed of several to many flowers.

inflorescence meristem A meristem that produces floral meristems as well as other small leafy structures (bracts).

infrared wavelengths Wavelengths in the electromagnetic spectrum that are too long to be seen by the human eye (i.e., greater than 0.7 μm).

ingroup In a phylogenetic study, the group of organisms of primary interest. (Contrast with outgroup.)

inhibiting hormones (IHs) Hormones, released by neurosecretory cells in the hypothalamus, that inhibit the secretion of hormones by the anterior pituitary gland. Each IH is specific for particular pituitary hormones. Also called release-inhibiting hormones.

inhibitory postsynaptic potential (IPSP) At a chemical synapse (specifically, an inhibitory synapse), a graded membrane hyperpolarization produced in the postsynaptic cell by arrival of an impulse in the presynaptic cell. (Contrast with excitatory postsynaptic potential.)

inhibitory synapse A synapse at which impulses in the presynaptic cell act to hyperpolarize the postsynaptic cell, thereby shifting the membrane potential of the postsynaptic cell away from the threshold for action potential production.

initiation complex In protein translation, a combination of a small ribosomal subunit, an mRNA molecule, and the tRNA charged with the first amino acid coded for by the mRNA; formed at the onset of translation.

innate defenses In animals, one of two general types of defenses against pathogens. Nonspecific and present in most animals. (Contrast with adaptive defenses.)

inner cell mass Part of the mammalian blastula (blastocyst), the inner cell mass will give rise to the embryo (via the epiblast) and the yolk sac, allantois, and amnion (via the hypoblast). (Contrast with trophoblast.)

innervate To provide neural input.

inorganic fertilizer A chemical or combination of chemicals applied to soil or plants to make up for a plant nutrient deficiency. Often contains the macronutrients nitrogen, phosphorus, and potassium (N-P-K).

instar (in' star) An immature stage of an insect between molts.

integral membrane proteins Proteins that are at least partially embedded in the plasma membrane. (Contrast with peripheral membrane proteins.)

integument [L. *integumentum*: covering] A protective surface structure. In gymnosperms and angiosperms, a layer of tissue around the ovule which will become the seed coat.

interference RNA (RNAi) *See* RNA interference.

interferon A glycoprotein produced by virus-infected animal cells; increases the resistance of neighboring cells to the virus.

intermediate filaments Components of the cytoskeleton whose diameters fall between those of the larger microtubules and those of the smaller microfilaments.

intermediate mesoderm The mesoderm that forms urinary and reproductive tissues.

internal effector *See* autonomic effector.

internal environment In multicellular organisms, the extracellular body fluids (interstitial fluids or tissue fluids), which bathe the cells of the body and therefore constitute the immediate environment of the cells.

internal fertilization Fertilization inside the body of the female that occurs because sperm are deposited into the female reproductive tract and encounter ova there (Contrast with external fertilization.)

internal membrane An intracellular membrane.

interneuron A neuron that communicates information between two other neurons.

interphase In the cell cycle, the period between successive nuclear divisions during which the chromosomes are diffuse and the nuclear envelope is intact. During interphase the cell is most active in transcribing and translating genetic information.

interspecific competition A mutually detrimental (–/–) interaction between individuals of two different species that causes both to suffer reduced fitness. (Contrast with intraspecific competition.)

interspecific interactions An interaction between individuals of different species that affects the per capita birth and/or death rate (or equivalently the fitness) of at least some of the individuals involved.

interstitial cells Testosterone-secreting cells located in the tissue between the seminiferous tubules of the vertebrate testis. Also called Leydig cells.

interstitial fluid Extracellular fluid that is found bathing the cells in tissues throughout the body. Also called tissue fluid.

intertidal zone The nearshore region of oceans that lies between the level of the high and low tides, so that it is alternately inundated with water and exposed to air.

intracellular fluid The aqueous solution inside a cell.

intraspecific competition A mutually detrimental (–/–) interaction between individuals of the same species that causes both to suffer reduced fitness. (Contrast with interspecific competition.)

intron Portion of a of a gene within the coding region that is transcribed into pre-mRNA but is spliced out prior to translation. (Contrast with exon.)

invasive species A species that increases in abundance and spreads widely, often to the detriment of other species, when it is introduced to a new location.

invasiveness The ability of a pathogen to multiply in a host's body. (Contrast with toxigenicity).

inversion A rare 180° reversal of the order of genes within a segment of a chromosome.

involuntary nervous system *See* autonomic nervous system.

ion (eye' on) [Gk. *ion*: wanderer] An electrically charged particle that forms when an atom gains or loses one or more electrons.

ion channel An integral membrane protein that allows ions to diffuse across the membrane in which it is embedded.

ion exchange A process by which protons produced by a plant root displace mineral cations from clay particles in the surrounding soil.

ionic attraction An electrostatic attraction between positively and negatively ions.

ionotropic receptor A receptor protein that is a ligand-gated ion channel and that therefore directly alters membrane ion permeability when it combines with its ligand.

ionotropic receptor cell A sensory receptor cell that typically has an ionotropic receptor protein that is a stimulus-gated Na^+ channel. *See* ionotropic receptor.

iPS cells *See* induced pluripotent stem cells.

iris (eye' ris) [Gk. *iris*: rainbow] The round, pigmented membrane that surrounds the pupil of the eye and adjusts its aperture to regulate the amount of light entering the eye.

island biogeography Patterns in the spatial distribution of species among oceanic islands or "islands" of one kind of habitat surrounded by an "ocean" of a different habitat.

iso- [Gk. *iso*: equal] Prefix used for two separate entities that share some element of identity.

isomorphic (eye so more' fik) [Gk. *isos*: equal + *morphe*: form] Having the same form or appearance, as when the haploid and diploid life stages of an organism appear identical. (Contrast with heteromorphic.)

isosmotic Having the same osmotic pressure; said of a solution in comparison to another solution. The term is relative and has meaning only in comparing two specified solutions.

isotonic Having the same solute concentration; said of two solutions. (Contrast with hypertonic, hypotonic.)

isotope (eye' so tope) [Gk. *isos*: equal + *topos*: place] Isotopes of a given chemical element have the same number of protons in their nuclei (and thus are in the same position on the periodic table), but differ in the number of neutrons.

iteroparity [L. *itero*: to repeat + *pario*: to beget] A type of reproductive life history in which individuals are physiologically capable of reproducing multiple times in a lifetime. (Contrast with semelparity.)

iteroparous Characterized by iteroparity.

J

jasmonate Also called jasmonic acid, a plant hormone involved in triggering responses to pathogen attack as well as other processes.

jasmonic acid *See* jasmonate.

jelly coat The outer protective layer of a sea urchin egg, which triggers an acrosomal reaction in sperm.

joint In skeletal systems, a junction between two or more bones.

joule (J) A metric unit for measuring energy: 1 J = 0.24 calorie; 1 calorie = 4.2 joule.

juxtacrine signal A type of cell signal that requires that the signaling and responding cells are in direct contact. Usually involves interaction between signaling molecules bound to the surfaces of the two cells.

K

karyogamy The fusion of nuclei of two cells. (Contrast with plasmogamy.)

karyotype The number, forms, and types of chromosomes in a cell.

kidney An organ that regulates the composition and volume of the blood plasma by producing and eliminating from the body an aqueous solution (urine) derived from the blood plasma or other extracellular fluids.

kilocalorie (kcal) *See* calorie.

kinase *See* protein kinase.

kinetic energy (kuh-net' ik) [Gk. *kinetos*: moving] The energy associated with movement. (Contrast with potential energy.)

kinetochore (kuh net' oh core) Specialized structure on a centromere to which microtubules attach.

kingdom A group of related phyla.

Koch's postulates A set of rules for establishing that a particular microorganism causes a particular disease.

Krebs cycle *See* citric acid cycle.

L

lactic acid fermentation Anaerobic series of reactions that convert glucose to lactic acid, in some bacteria and animal cells.

lactifers Plant cell or vessel that contains latex.

lagging strand In DNA replication, the daughter strand that is synthesized in discontinuous stretches. *See* Okazaki fragments.

landscape An ecological system consisting of multiple ecological communities within a geographical area larger than the area occupied by a single community.

lateral gene transfer The transfer of genes from one species to another, common among bacteria and archaea.

lateral meristem Either of the two meristems, the vascular cambium and the cork cambium, that give rise to a plant's secondary growth.

lateral plate mesoderm One of three regions of mesoderm on either side of the notochord; encloses the coelom and gives rise to the heart, muscles of the digestive system, muscles of the outer body wall, and other tissues.

lateral root A root extending outward from the taproot in a taproot system; typical of eudicots.

Laurasia The northernmost of the two large continents produced by the breakup of Pangaea.

law of diffusion for gases An equation that describes the factors that determine the rate of diffusion of gas molecules within a gas phase, liquid phase, or across the boundary between a gas phase and liquid phase; often called Fick's law.

law of independent assortment During meiosis, the random separation of genes carried on nonhomologous chromosomes into gametes so that inheritance of these genes is random. This principle was articulated by Mendel as his second law

law of segregation In genetics, the separation of alleles, or of homologous chromosomes, from each other during meiosis so that each of the haploid daughter nuclei produced contains one or the other member of the pair found in the diploid parent cell, but never both. This principle was articulated by Mendel as his first law.

laws of thermodynamics [Gk. *thermos*: heat + *dynamis*: power] Laws derived from studies of the physical properties of energy and the ways energy interacts with matter. (*See also* first law of thermodynamics, second law of thermodynamics.)

LDP *See* long-day plant.

leaching In soils, a process by which mineral nutrients in upper soil horizons are dissolved in water and carried to deeper horizons, where they are unavailable to plant roots.

leading strand In DNA replication, the daughter strand that is synthesized continuously. (Contrast with lagging strand.)

leaf (plural: leaves) In plants, the chief organ of photosynthesis.

learning The process of modifying an individual's behavior based on prior experiences of the individual.

leghemoglobin In nitrogen-fixing plants, an oxygen-carrying protein in the cytoplasm of nodule cells that transports enough oxygen to the nitrogen-fixing bacteria to support their respiration, while keeping free oxygen concentrations low enough to protect nitrogenase.

lens In an eye, a structure composed of transparent proteins that focuses images on the retina or other light-sensing structures.

leukocytes *See* white blood cells.

Leydig cells *See* interstitial cells.

LH *See* luteinizing hormone.

lichen (lie' kun) An organism resulting from the symbiotic association of a fungus and either a cyanobacterium or a unicellular alga.

life expectancy The average time from birth to death of an individual organism.

life history The sequence of key events, such as growth and development, reproduction, and death, that occur during the life of the average individual of a given species.

life-history tradeoffs Negative relationships among growth, reproduction, and survival that occur because an individual cannot simultaneously allocate resources to such competing life functions as resource acquisition, maintenance, growth, defense, or reproduction.

ligand (lig' and) Any molecule that binds to a receptor site of another (usually larger) molecule.

ligand-gated channel A channel protein that can open and close and that opens to allow diffusion of a solute as a result of binding by a neurotransmitter (or other specific signaling molecule) to a receptor site on the channel protein.

light reactions The initial phase of photosynthesis, in which light energy is converted into chemical energy.

light-harvesting complex In photosynthesis, a group of different molecules that cooperate to absorb light energy and transfer it to a reaction center. Also called antenna system.

lignin A complex, hydrophobic polyphenolic polymer in plant cell walls that crosslinks other wall polymers, strengthening the walls, especially in wood.

limiting resource A resource that constrains the fitness of individuals at a given time and place because it is in short supply relative to demand.

limnetic zone [Gk. *limn*☒: pool] The "open water" zone of freshwater lakes.

lineage A series of populations, species, or genes descended from a single ancestor over evolutionary time.

lineage species concept The definition of a species as a branch on the tree of life, which has a history that starts at a speciation event and ends either at extinction or at another speciation event. (Contrast with biological species concept, morphological species concept.)

linear regression A statistical method of fitting a straight line to describe the relationship between two variables in a scatter plot.

linkage *See* genetic linkage.

lipid (lip' id) [Gk. *lipos*: fat] Nonpolar, hydrophobic molecules that include fats, oils, waxes, steroids, and the phospholipids that make up biological membranes.

lipid bilayer *See* phospholipid bilayer.

littoral zone The nearshore regions of lakes and oceans. In oceans this includes the intertidal zone.

loam A type of soil consisting of a mixture of sand, silt, clay, and organic matter. One of the best soil types for agriculture.

locus (low' kus) (plural: loci, low' sigh) In genetics, a specific location on a chromosome. May be considered synonymous with *gene*.

long-day plant (LDP) A plant that requires long days (actually, short nights) in order to flower. (Compare to short-day plant.)

loop of Henle (hen' lee) Long, hairpin loop in a mammalian nephron. The loop runs from the cortex of the kidney down into the medulla of the kidney and back to the cortex; creates a concentration gradient in the interstitial fluids in the medulla. Some nephrons in birds kidneys also have loops of Henle.

lophophore A U-shaped fold of the body wall with hollow, ciliated tentacles that encircles the mouth of animals in several different groups. Used for filtering prey from the surrounding water.

loss of function mutation A mutation that results in the loss of a functional protein. (Contrast with gain of function mutation.)

lung A breathing organ that is invaginated (folded inward) into the body and contains the environmental medium. The most common type of breathing organ in terrestrial animals. (Contrast with gill.)

lung surfactant A thin layer of lipids and proteins that coats the inside of a lung; decreases surface tension at the surfaces of the lung epithelium, which reduces the amount of work necessary to inflate the lung.

luteinizing hormone (LH) A hormone, produced by the anterior pituitary gland, that helps control secretion of hormones by the gonads (i.e., a gonadotropin).

lymphatic system An elaborate system of vessels, separate from the blood vascular system, that picks up excess interstitial fluid (lymph) and returns it to the blood.

lymphocyte One of the two major classes of white blood cells; includes T cells, B cells, and other cell types important in the immune system.

lysogeny A form of viral replication in which the virus becomes incorporated into the host chromosome and remains inactive. Also called a lysogenic cycle. (Contrast with lytic cycle.)

lysosome (lie′ so soam) [Gk. *lysis*: break away + *soma*: body] A membrane-enclosed organelle originating from the Golgi apparatus and containing hydrolytic enzymes. (Contrast with secondary lysosome.)

lysozyme (lie′ so zyme) An enzyme in saliva, tears, and nasal secretions that hydrolyzes bacterial cell walls.

lytic cycle A viral reproductive cycle in which the virus takes over a host cell's synthetic machinery to replicate itself, then bursts (lyses) the host cell, releasing the new viruses. (Contrast with lysogeny.)

M

M phase The portion of the cell cycle in which mitosis takes place.

macromolecule A giant (molecular weight > 1,000) polymeric molecule. The macromolecules are the proteins, polysaccharides, and nucleic acids.

macronutrient In plants, a mineral element required in concentrations of at least 1 milligram per gram of plant dry matter; in animals, a mineral element required in large amounts. (Contrast with micronutrient.)

maintenance methyltransferase An enzyme that transfers methyl groups to DNA after DNA replication.

major histocompatibility complex (MHC) A complex of linked genes, with multiple alleles, that control a number of cell surface antigens that identify self and can lead to graft rejection.

malnutrition A condition caused by lack of adequate amounts of an essential nutrient.

Malpighian tubules (mal pee′ gy un) Fine tubules that initiate urine formation in insects. They empty their product into the gut at the junction of the midgut and hindgut.

mantle (1) In mollusks, a fold of tissue that covers the organs of the visceral mass and secretes the hard shell that is typical of many mollusks. (2) In geology, the Earth's crust below the solid lithospheric plates.

mass extinction A period of evolutionary history during which rates of extinction are much higher than during intervening times.

mass number The sum of the number of protons and neutrons in an atom's nucleus.

mast cells Cells, typically found in connective tissue, that release histamine in response to tissue damage.

maternal effect genes Genes coding for morphogens that determine the polarity of the egg and larva in fruit flies. Part of a developmental cascade that includes gap genes, pair rule genes, segment polarity genes, and Hox genes.

mating type A particular strain of a species that is incapable of sexual reproduction with another member of the same strain but capable of sexual reproduction with members of other strains of the same species.

maximum likelihood A statistical method of determining which of two or more hypotheses (such as phylogenetic trees) best fit the observed data, given an explicit model of how the data were generated.

mean The sum of all values in a sample divided by the number of observations in the sample.

measures of center Quantities that describe various aspects of the center of a group of observations.

measures of dispersion Measures that quantify the dispersion of observations in a sample of observations.

mechanoreceptor A sensory receptor cell that is sensitive to physical movement or physical distortion and generates action potentials in response.

median The value at which there are equal numbers of larger and smaller observations in a sample.

medusa (plural: medusae) In cnidarians, a free-swimming, sexual life cycle stage shaped like a bell or an umbrella.

megagametophyte In heterosporous plants, the female gametophyte; produces eggs. (Contrast with microgametophyte.)

megaphyll The generally large leaf of a fern, horsetail, or seed plant, with several to many veins. (Contrast with microphyll.)

megasporangia The structures on a heterosporous plant that produce a few large megaspores (which develop into female gametophytes).

megaspore [Gk. *megas*: large + *spora*: to sow] In plants, a haploid spore that produces a female gametophyte.

megastrobilus In conifers, the female (seed-bearing) cone. (Contrast with microstrobilus.)

meiosis (my oh′ sis) [Gk. *meiosis*: diminution] Division of a diploid nucleus to produce four haploid daughter cells. The process consists of two successive nuclear divisions with only one cycle of chromosome replication. In **meiosis I**, homologous chromosomes separate but retain their chromatids. The second division **meiosis II**, is similar to mitosis, in which chromatids separate.

membrane potential The difference in electrical charge (voltage) between the inside and the outside of a cell membrane, typically expressed in millivolts. The name derives from the fact that potential difference is a synonym for voltage.

memory cells In the immune system, long-lived lymphocytes produced after exposure to antigen. They persist in the body and are able to mount a rapid response to subsequent exposures to the antigen.

Mendel's laws *See* law of independent assortment, law of segregation.

menstrual cycle The cycle in which oocytes mature and are ovulated periodically in females of primates, including humans; one phase of each cycle is characterized by menstruation, the shedding of the uterine lining in a blood-tinged discharge from the vagina.

menstruation In humans and other primates, a process by which the lining of the uterus (the endometrium) breaks down (after a failure of pregnancy to occur), and the sloughed-off tissue, including blood, flows from the body.

meristem [Gk. *meristos*: divided] Plant tissue made up of undifferentiated actively dividing cells.

meristem culture A method for the asexual propagation of plants, in which pieces of shoot apical meristem are cultured to produce plantlets.

meristem identity genes In angiosperms, a group of genes whose expression initiates flower formation, probably by switching meristem cells from a vegetative to a reproductive fate.

mesoderm [Gk. *mesos*: middle + *derma*: skin] The middle of the three embryonic germ layers first delineated during gastrulation. Gives rise to the skeleton, circulatory system, muscles, excretory system, and most of the reproductive system.

mesoglea (mez′ uh glee uh) [Gk. *mesos*: middle + *gloia*, glue] A thick, gelatinous noncellular layer that separates the two cellular tissue layers of ctenophores, cnidarians, and scyphozoans.

mesophyll (mez′ uh fill) [Gk. *mesos*: middle + *phyllon*: leaf] Chloroplast-containing, photosynthetic cells in the interior of leaves.

messenger RNA (mRNA) Transcript of a region of one of the strands of DNA; carries information (as a sequence of codons) for the synthesis of one or more proteins.

meta- [Gk.: between, along with, beyond] Prefix denoting a change or a shift to a new form or level; for example, as used in metamorphosis.

metabolic pathway A series of enzyme-catalyzed reactions so arranged that the product of one reaction is the substrate of the next.

metabolic rate (MR) An animal's rate of energy consumption; the rate at which it converts chemical-bond energy to heat and external work.

metabolic water Water that is formed by oxidation of food during metabolism. For example, when glucose is oxidized, one of the products is H_2O that did not previously exist; this H_2O is metabolic water. Also called oxidation water.

metabolism (meh tab' a lizm) [Gk. *metabole*: change] The sum total of the chemical reactions that occur in an organism, or some subset of that total (as in respiratory metabolism).

metabolome The quantitative description of all the small molecules in a cell or organism.

metabolomics The study of the metabolome as it relates to the physiological state of a cell or organism.

metabotropic receptor A receptor protein that is not an ion channel. Typically a G protein–linked receptor protein. (*See* G protein–linked receptors.)

metabotropic receptor cell A sensory receptor cell in which the sensory receptor protein is not an ion channel but typically instead is a G protein–linked receptor.

metagenomics The practice of analyzing DNA from environmental samples without isolating intact organisms.

metamorphosis (met' a mor' fo sis) [Gk. *meta*: between + *morphe*: form, shape] A change occurring between one developmental stage and another, as for example from a tadpole to a frog. (*See* complete metamorphosis, incomplete metamorphosis.)

metaphase (met' a phase) The stage in nuclear division at which the centromeres of the highly supercoiled chromosomes are all lying on a plane (the metaphase plane or plate) perpendicular to a line connecting the division poles.

metapopulation A "population of populations"—the group of spatially-separated subpopulations that occurs within a defined geographic area.

MHC *See* major histocompatibility complex.

microbiomes The diverse communities of bacteria that live on or within the body and are essential to bodily function.

microcirculation The part of a closed circulatory system that consists of the smallest diameter blood vessels (i.e., arterioles, capillaries, and venules.)

microfilament In eukaryotic cells, a fibrous structure made up of actin monomers. Microfilaments play roles in the cytoskeleton, in cell movement, and in muscle contraction.

microgametophyte In heterosporous plants, the male gametophyte; produces sperm. (Contrast with megagametophyte.)

micronutrient In plants, a mineral element required in concentrations of less than 100 micrograms per gram of plant dry matter; in animals, a mineral element required in concentrations of less than 100 micrograms per day. (Contrast with macronutrient.)

microphyll A small leaf with a single vein, found in club mosses and their relatives. (Contrast with megaphyll.)

micropyle (mike' roh pile) [Gk. *mikros*: small + *pylon*: gate] Opening in the integument(s) of a seed plant ovule through which pollen grows to reach the female gametophyte within.

microRNA A small, noncoding RNA molecule, typically about 21 bases long, that binds to mRNA to inhibit its translation.

microsporangia The structures on a heterosporous plant that produce many small microspores (which develop into male gametophytes).

microspore [Gk. *mikros*: small + *spora*: to sow] In plants, a haploid spore that produces a male gametophyte.

microstrobilus In conifers, male pollen-bearing cone. (Contrast with megastrobilus.)

microtubules Tubular structures found in centrioles, spindle apparatus, cilia, flagella, and cytoskeleton of eukaryotic cells. These tubules play roles in the motion and maintenance of shape of eukaryotic cells.

microvilli (sing.: microvillus) Minute, rod-shaped projections of epithelial cells, such as the cells lining the small intestine, that increase their surface area.

midbrain One of the three regions of the vertebrate brain. Part of the brainstem, it serves as a relay station for sensory signals sent to the cerebral hemispheres, in addition to other functions.

midgut The first part of the intestine. In mammals, the midgut is also called the small intestine.

midgut fermenter An animal that has a specialized midgut chamber housing mixed communities of fermenting microbes that assist with the breakdown of food materials

migrate To periodically move from one location to another, then, after a period of time, return to the original location.

missense mutation A change in a gene's sequence that changes the amino acid at that site in the encoded protein. (Contrast with frame-shift mutation, nonsense mutation, silent mutation.)

mitochondrion (my' toe kon' dree un) (plural: mitochondria) [Gk. *mitos*: thread + *chondros*: grain] An organelle in eukaryotic cells that contains the enzymes of the citric acid cycle, the respiratory chain, and oxidative phosphorylation.

mitosis (my toe' sis) [Gk. *mitos*: thread] Nuclear division in eukaryotes leading to the formation of two daughter nuclei, each with a chromosome complement identical to that of the original nucleus.

mitosomes Reduced structures derived from mitochondria found in some organisms.

mode The most frequent value in a sample of observations.

model organism A species that is widely studied experimentally to understand general principles about biology.

moderately repetitive sequences DNA sequences repeated 10–1,000 times in the eukaryotic genome. They include the genes that code for rRNAs and tRNAs, as well as the DNA in telomeres.

molds Sac fungi composed of filamentous hyphae that do not form large ascomata.

molecular biology Study of the formation, structure and functions of nucleic acids and proteins.

molecular clock The approximately constant rate of divergence of macromolecules from one another over evolutionary time; used to date past events in evolutionary history.

molecular toolkit *See* genetic toolkit.

molecule A chemical substance made up of two or more atoms joined by covalent bonds or ionic attractions.

molting The process of shedding part or all of an outer covering, as the shedding of feathers by birds or of the entire exoskeleton by arthropods.

monocots Angiosperms with a single embryonic cotyledon; one of the two largest clades of angiosperms. (*See also* eudicots.)

monoecious (mo nee' shus) [Gk. *mono*: one + *oikos*: house] Pertaining to organisms in which both sexes are "housed" in a single individual that produces both eggs and sperm. (In some plants, these are found in different flowers within the same plant.) Examples include corn, peas, earthworms, hydras. (Contrast with dioecious.)

monohybrid cross A mating in which the parents differ with respect to the alleles of only one locus of interest.

monomer [Gk. *mono*: one + *meros*: unit] A small molecule, two or more of which can be combined to form oligomers (consisting of a few monomers) or polymers (consisting of many monomers).

monophyletic (mon′ oh fih leht′ ik) [Gk. *mono*: one + *phylon*: tribe] Pertaining to a group that consists of an ancestor and all of its descendants. (Contrast with paraphyletic, polyphyletic.)

monosaccharide A simple sugar. Oligosaccharides and polysaccharides are made up of monosaccharides.

morphogen A diffusible substance whose concentration gradient determines a developmental pattern in animals and plants.

morphogenesis (more′ fo jen′ e sis) [Gk. *morphe*: form + *genesis*: origin] The development of form; the overall consequence of determination, differentiation, and growth.

morphological species concept The definition of a species as a group of individuals that look alike. (Contrast with biological species concept, lineage species concept.)

morphology (more fol′ o jee) [Gk. *morphe*: form + *logos*: study, discourse] The scientific study of organic form, including both its development and function.

mortality The average individual's chance of dying during a specified set of life stages, ages, or period of time. Mortality = 1 - survivorship (Contrast with survivorship.)

mosaic development Pattern of animal embryonic development in which each blastomere contributes a specific part of the adult body. (Contrast with regulative development.)

motor neuron A neuron that carries signals from the central nervous system to a skeletal muscle, stimulating the muscle to contract. A type of efferent neuron.

motor proteins Specialized proteins that use energy to change shape and move cells or structures within cells.

mRNA *See* messenger RNA.

mucus A slippery substance secreted by mucous membranes (e.g., mucosal epithelium). A barrier defense against pathogens in innate immunity in animals.

Muller's ratchet The accumulation—"ratcheting up"—of deleterious mutations in the nonrecombining genomes of asexual species.

multi-organ system A system in which multiple organs work together.

multiple fruit A fruit derived from carpels of several flowers. An example is a pineapple.

multiplicative growth In ecology, population growth in which a multiple of the current population size is added to the population during each successive time interval. The multiple of growth may be a constant, or it may change. (Contrast with additive growth.)

multipotent Having the ability to differentiate into a limited number of cell types. (Contrast with pluripotent, totipotent.)

muscle fiber A single muscle cell. In the case of skeletal muscle, a multinucleate cell.

mutagen (mute′ ah jen) [L. *mutare*: change + Gk. *genesis*: source] Any agent (e.g., a chemical, radiation) that increases the mutation rate.

mutation A change in the genetic material not caused by recombination.

mutualism A mutually beneficial (+/+) interaction between individuals of two different species that causes both to benefit, in terms of increased fitness. (Contrast with amensalism, commensalism.)

mutualistic A pair of symbiotic organisms in which both partners benefit from the interaction.

mycelium (my seel′ ee yum) [Gk. *mykes*: fungus] In the fungi, a mass of hyphae.

mycorrhiza (my′ ko rye′ za) (plural: mycorrhizae) [Gk. *mykes*: fungus + *rhiza*: root] An association of the root of a plant with the mycelium of a fungus.

myelin (my′ a lin) Concentric layers of cell membrane wrapped around some axons, forming a sheath around the axons; myelin provides an axon with electrical insulation and increases the rate of transmission of action potentials.

myocardium The muscle tissue of a heart.

MyoD The protein encoded by the *myo*blast *d*etermining gene. A transcription factor involved in the differentiation of myoblasts (muscle precursor cells).

myofibril (my′ oh fy′ bril) [Gk. *mys*: muscle + L. *fibrilla*: small fiber] A long, longitudinally oriented component of a striated muscle cell that consists of a series of sarcomeres and extends the length of the cell. A myofibril is composed principally of actin and myosin filaments. In cross section, a muscle cell consists of multiple myofibrils.

myogenic heart A heart in which the electrical impulse to contract during each beating cycle originates in muscle cells or modified muscle cells. (Contrast with neurogenic heart.)

myoglobin (my′ oh globe′ in) [Gk. *mys*: muscle + L. *globus*: sphere] An oxygen-binding molecule found in red muscle cells. A type of hemoglobin, consisting of a single heme unit and a single globin chain per molecule

myosin One of the two contractile proteins of muscle. (*See also* actin.)

myosin filaments Bundles of linked myosin molecules. Also called thick filaments. (*See also* actin filaments.)

N

natural history An observation about nature, obtained outside of a formal hypothesis-testing context, that provides the basis for further investigation.

natural killer cell A type of lymphocyte that attacks virus-infected cells and some tumor cells as well as antibody-labeled target cells.

natural selection The differential contribution of offspring to the next generation by various genetic types belonging to the same population. The mechanism of evolution proposed by Charles Darwin.

navigation The act of moving on a particular course or toward a specific destination using sensory cues to determine direction and position.

necrosis (nec roh′ sis) [Gk. *nekros*: death] Premature cell death caused by external agents such as toxins.

negative feedback In a regulatory system, a type of control that acts to reduce differences that arise between the level of a controlled variable and its set-point level. It tends to stabilize the controlled variable at a level close to the set-point level. (Contrast with positive feedback.)

negative-sense RNA RNA that is complementary to mRNA. Before it can be translated, it must be converted to positive-sense RNA by an RNA polymerase.

nematocyst (ne mat′ o sist) [Gk. *nema*: thread + *kystis*: cell] An elaborate, threadlike structure produced by cells of jellyfishes and other cnidarians, used chiefly to paralyze and capture prey.

neoteny (knee ot′ enny) [Gk. *neo*: new, recent; *tenein*: to extend] The retention of juvenile or larval traits by the fully developed adult organism.

nephron (nef′ ron) [Gk. *nephros*: kidney] The functional unit of a vertebrate kidney, consisting of a structure for receiving a filtrate of blood plasma and a long tubule that modifies the filtrate by secretion and reabsorption.

Nernst equation A mathematical equation that calculates the electrical potential difference required for electrochemical equilibrium across a membrane that is permeable to a single type of ion, when the ion differs in concentration on the two sides of the membrane.

nerve A structure consisting of many neuronal axons bound together like wires in a telephone cable.

nerve impulse *See* action potential.

nerve nets Diffuse, loosely connected aggregations of nervous tissues in certain non-bilatarian animals such as cnidarians.

nervous tissue Tissue specialized for processing and communicating information; one of the four major tissue types in multicellular animals.

net primary productivity (NPP) The rate at which primary producers in a given area (or volume) convert energy they have captured from sunlight or other abiotic sources into energy stored in the molecules that make up their tissues, and which therefore is available for use by primary consumers. NPP is often estimated as the gain in biomass of primary producers in a unit of time.

neural crest cells During vertebrate neurulation, cells that migrate outward from the neural plate; they are pluripotent cells that form the sensory nervous system, autonomic nervous system, many bones of the skull, pigment cells, and other tissues

neuroendocrine cells *See* neurosecretory cells.

neurogenic heart A heart in which the electrical impulse to contract during each beating cycle originate in neurons. (Contrast with myogenic heart.)

neuroglia *See* glial cells.

neurohemal organ A specialized organ in the circulatory system that is made up of axon terminals of neurosecretory cells in association with a well-developed bed of capillaries.

neuromuscular junction Synapse (point of contact) where a motor neuron axon stimulates a muscle fiber (muscle cell).

neuron (noor′ on) [Gk. *neuron*: nerve] A nervous system cell that can generate and conduct action potentials along an axon to a synapse with another cell.

neurosecretory cells Neurons that synthesize and release hormones into the blood. Also called neuroendocrine cells.

neurotransmitter At a synapse, a substance produced and released by the presynaptic cell that diffuses across the synapse and excites or inhibits the postsynaptic cell.

neurulation Stage in vertebrate development during which the nervous system begins to form.

neutron (new′ tron) One of the three fundamental particles of matter (along with protons and electrons), with mass slightly larger than that of a proton and no electrical charge.

niche (nitch) [L. *nidus*: nest] In ecology, the abiotic and biotic conditions under which a given species can persist, and the functional role of the species in its community.

nitrifiers Chemoautotrophic bacteria that oxidize ammonia to nitrate in soil and in seawater.

nitrogen fixation Conversion of atmospheric nitrogen gas (N_2) to a water-soluble, biologically-usable form (usually ammonium, NH_4^+), in nature by free-living or symbiotic microbes, but also by industrial processes. (Compare with denitrification.)

nitrogen fixers Organisms that convert atmospheric nitrogen gas into a chemical form (ammonia) that is usable by the nitrogen fixers themselves as well as by other organisms.

nitrogenase An enzyme complex found in nitrogen-fixing bacteria that mediates the stepwise reduction of atmospheric N_2 to ammonia and which is strongly inhibited by oxygen.

nitrogenous wastes The potentially toxic nitrogen-containing end products resulting from the break down of proteins and nucleic acids. They are usually eliminated from the body by excretion.

node [L. *nodus*: knob, knot] (1) In plants, a (sometimes enlarged) point on a stem where a leaf is or was attached. (2) In phylogenetics, a split in a phylogenetic tree when one lineage diverges into two.

node of Ranvier In a myelinated axon, a gap in the myelin sheath where the axon cell membrane is not covered with myelin.

noncompetitive inhibitor A nonsubstrate that inhibits the activity of an enzyme by binding to a site other than its active site. (Contrast with competitive inhibitor.)

nondisjunction Failure of sister chromatids to separate in meiosis II or mitosis, or failure of homologous chromosomes to separate in meiosis I. Results in aneuploidy.

nonneural endocrine cells Cells that secrete hormones into the blood and that are not neurons or derived from neurons. (*See* neurosecretory cells.)

nonsense mutation Change in a gene's sequence that prematurely terminates translation by changing one of its codons to a stop codon.

nonshivering thermogenesis (NST) In mammals and some birds, elevation of heat production for thermoregulation by means other than shivering.

nonsynonymous substitution A change in a gene from one nucleotide to another that changes the amino acid specified by the corresponding codon (i.e., AGC → AGA, or serine → arginine). (Contrast with synonymous substitution.)

nonvascular land plants Land plants that lack specialized vascular tissues for the conduction of water or nutrients through the plant body. There are three living species of land plants: the liverworts, hornworts, and mosses.

notochord (no′ tow kord) [Gk. *notos*: back + *chorde*: string] A flexible rod of gelatinous material serving as a support in the embryos of all chordates and in the adults of tunicates and lancelets.

NPP *See* net primary productivity.

NST *See* nonshivering thermogenesis.

nucleic acid (new klay′ ik) A polymer made up of nucleotides, specialized for the storage, transmission, and expression of genetic information. DNA and RNA are nucleic acids.

nucleic acid hybridization A technique in which a single-stranded nucleic acid probe is made that is complementary to, and binds to, a target sequence, either DNA or RNA. The resulting double-stranded molecule is a hybrid.

nucleoid (new′ klee oid) The region that harbors the chromosomes of a prokaryotic cell. Unlike the eukaryotic nucleus, it is not bounded by a membrane.

nucleolus (new klee′ oh lus) A small, generally spherical body found within the nucleus of eukaryotic cells. The site of synthesis of ribosomal RNA.

nucleotide The basic chemical unit in nucleic acids, consisting of a pentose sugar, a phosphate group, and a nitrogen-containing base.

nucleus (new′ klee us) [L. *nux*: kernel or nut] (1) In cells, the centrally located compartment of eukaryotic cells that is bounded by a double membrane and contains the chromosomes. (2) In the brain, an identifiable group of neurons that share common characteristics or functions.

null hypothesis In statistics, the premise that any differences observed in an experiment are simply the result of random differences that arise from drawing two finite samples from the same population.

O

obligate aerobe Organisms that require oxygen for metabolism.

obligate anaerobe An anaerobic prokaryote that cannot survive exposure to O_2.

obligate parasites Organisms that can only survive and grow in or on other living organisms, to the detriment of the host.

octet rule Description of processes that atoms undergo whereby they obtain, give up or share electrons such that their outer (valence) shell contains eight electrons.

Okazaki fragments Newly formed DNA making up the lagging strand in DNA replication. DNA ligase links Okazaki fragments together to give a continuous strand.

oligodendrocyte A type of glial cell that forms a myelin sheath on axons in the central nervous system.

oligopeptide Peptide made up of less than 20 amino acids.

oligosaccharide A polymer containing a small number of monosaccharides.

ommatidium (plural: ommatidia) [Gk. *omma*: eye] A single visual unit in the compound eye of an arthropod.

omnivore [L. *omnis*: everything + *vorare*: to devour] An organism that obtains the energy and materials it needs to survive, grow, and reproduce from a variety of trophic levels. (Contrast with carnivore, detritivore, herbivore.)

oncogene [Gk. *onkos*: mass, tumor + *genes*: born] A gene that codes for a protein product that stimulates cell proliferation. Mutations in oncogenes that result in excessive cell proliferation can give rise to cancer. (Contrast with tumor suppressor.)

one gene–one polypeptide The idea, since shown to be an oversimplification, that each gene in the genome encodes only a single polypeptide—that there is a one-to-one correspondence between genes and polypeptides.

oocyte The developing ovum.

open circulatory system Circulatory system in which the blood leaves blood vessels and travels through spaces (sinuses and lacunae) bounded by ordinary tissue cells as it flows through the body (Contrast with closed circulatory system.)

open system A system that is not isolated from, and interacts with, its surroundings. (Compare with closed system.)

open reading frames Sequences of DNA within genes that begin with an initiation codon and end with a stop codon.

operator The region of an operon that acts as the binding site for the repressor.

operon A genetic unit of transcription, typically consisting of several structural genes that are transcribed together; the operon contains at least two control regions: the promoter and the operator.

order A group of related families.

ordinal variables Categorical variables with a natural ordering, such as the grades A, B, C, D, and F.

organ [Gk. *organon*: tool] A body part, such as the heart, liver, brain, root, or leaf, that is composed of two or more tissues integrated to perform a distinct function.

organ identity genes In angiosperms, genes that specify the different organs of the flower.

organelle (or gan el′) Any of the membrane-enclosed structures within a eukaryotic cell. Examples include the nucleus, endoplasmic reticulum, and mitochondria.

organic fertilizers Substances added to soil to improve the soil's fertility; derived from partially decomposed plant material (compost) or animal waste (manure).

organizer Region of the early amphibian embryo (the dorsal lip of the blastopore) that develops into the notochord and thus establishes the basic body plan. Also known as the primary embryonic inducer.

organogenesis The formation of organs and organ systems during development.

orientation A behavioral process in which an organism positions itself or moves in relation to environmental cues such as the sun.

origin of replication (*ori*) DNA sequence at which helicase unwinds the DNA double helix and DNA polymerase binds to initiate DNA replication.

orthologs Genes that are related by orthology.

osmoconformer *See* osmotic conformer.

osmolar (*Osm*) A unit of measure of osmotic pressure. A 1-osmolar solution is defined to be a solution that behaves osmotically as if it has one Avogadro's number (6×10^{23}) of independent dissolved entities per liter.

osmolarity The osmotic pressure of a solution expressed in osmolar units.

osmoregulator *See* osmotic regulator.

osmosis (oz mo′ sis) [Gk. *osmos*: to push] Movement of water across a differentially permeable membrane, from one region to another region where the water potential is more negative.

osmotic conformer An aquatic animal in which the osmotic pressure of the blood and other extracellular fluids matches the osmotic pressure of the external environment. Also called osmoconformer. (Contrast with osmotic regulator.)

osmotic regulator An aquatic animal that maintains a relatively constant osmotic pressure in its blood and other extracellular fluids regardless of changes in the osmotic pressure of its external environment. Also called osmoregulator. (Contrast with osmotic conformer.)

osmotic pressure Pressure exerted by the flow of water through a semipermeable membrane separating two solutions with different concentrations of solute.

outgroup In phylogenetics, a group of organisms used as a point of reference for comparison with the groups of primary interest (the ingroup).

ovarian follicle During gamete development in the ovary of a female animal, an oocyte (developing ovum) together with its support cells (theca and granulosa cells).

ovary (oh′ var ee) [L. *ovum*: egg] A female organ, in plants or animals, that produces ova (eggs).

overtopping Plant growth pattern in which one branch differentiates from and grows beyond the others.

oviduct In mammals, the tube serving to transport eggs from an ovary to the uterus. Also called a fallopian tube.

ovulation The process of releasing an ovum (egg) from the ovary.

ovule (oh′ vule) In plants, a structure comprising the megasporangium and the integument, which develops into a seed after fertilization.

ovum (plural: ova) *See* egg.

oxidation (ox i day′ shun) Relative loss of electrons in a chemical reaction; either outright removal to form an ion, or the sharing of electrons with substances having a greater affinity for them, such as oxygen. Most oxidations, including biological ones, are associated with the liberation of energy. (Contrast with reduction.)

oxidation water *See* metabolic water.

oxidative phosphorylation ATP formation in the mitochondrion, associated with flow of electrons through the respiratory chain.

oxidized Increase in positive charge of an element by removing electron(s); formation of an oxide by adding oxygen.

oxygenase An enzyme that catalyzes the addition of oxygen to a substrate from O_2.

oxygenation (1) Referring to water, the dissolution of O_2 in that water. (2) The combination of respiratory pigments with O_2; oxygenation of this sort is reversible and not equivalent to oxidation.

oxygen equilibrium curve A graph of the amount of O_2 per unit of blood volume as a function of the O_2 partial pressure of the blood (a measure of the concentration of O_2 dissolved in the blood plasma.)

oxytocin In mammals, a hormone released by the posterior pituitary gland;

its major functions are to stimulate contraction of the uterus and the flow of milk from mammary glands.

P

***P*-value** The calculated probability of observing a given result by chance sampling, given the null hypothesis is true.

pair rule genes In *Drosophila* (fruit fly) development, segmentation genes that divide the early embryo into units of two segments each. Part of a developmental cascade that includes maternal effect genes, gap genes, segment polarity genes, and Hox genes.

paleomagnetic dating A method for determining the age of rocks based on properties relating to changes in the patterns of Earth's magnetism over time.

Pangaea (pan jee' uh) [Gk. pan: all, every] The single land mass formed when all the continents came together in the Permian period.

para- [Gk. *para*: akin to, beside] Prefix indicating association in being along side or accessory to.

parallel phenotypic evolution The repeated evolution of similar traits, especially among closely related species; facilitated by conserved developmental genes.

paraphyletic (par' a fih leht' ik) [Gk. *para*: beside + *phylon*: tribe] Pertaining to a group that consists of an ancestor and some, but not all, of its descendants. (Contrast with monophyletic, polyphyletic.)

parasite An organism that consumes parts of an organism much larger than itself (known as its host). Parasites sometimes, but not always, kill their host.

parasitism A +/– interaction between individuals of two different species that causes one individual (the parasite) to benefit (it gains fitness) from consuming part or all of the other (the host), which suffers reduced fitness as a consequence of being consumed.

parasympathetic division A division of the vertebrate autonomic nervous system that is connected to the central nervous system via cranial and sacral nerves; the parasympathetic and sympathetic divisions tend to exert opposing controls on autonomic effectors. (Contrast with sympathetic division.)

parenchyma (pair eng' kyma) A plant tissue composed of relatively unspecialized cells without secondary walls.

parent rock The soil horizon consisting of the rock that is breaking down to form the soil. Also called bedrock, or the C horizon.

parental (P) generation The individuals that mate in a genetic cross. Their offspring are the first filial (F_1) generation.

parsimony principle Principle that states that the preferred explanation of observed data is the simplest explanation.

parthenocarpy Formation of fruit from a flower without fertilization.

passive transport Diffusion across a membrane; may or may not require a channel or carrier protein. (Contrast with active transport.)

pathogen (path' o jen) [Gk. *pathos*: suffering + *genesis:* source] An organism that causes disease.

pathogen associated molecular patterns (PAMPs) The molecules recognized by pattern recognition receptors.

pathogenesis-related (PR) proteins A component of the hypersensitive response of plants to pathogens. Some PR proteins function as direct defenses against pathogens, others initiate other defensive responses.

pattern formation In animal embryonic development, the organization of differentiated tissues into specific structures such as wings.

pattern recognition receptors (PRRs) Proteins made by cells that recognize molecular patterns on pathogens; part of innate immunity.

PCR *See* polymerase chain reaction.

pedigree The pattern of transmission of a genetic trait within a family.

pelagic zone [Gk. *pelagos*: open sea] The "open water" zone of oceans that lies beyond the continental shelves.

penetrance The proportion of individuals with a particular genotype that show the expected phenotype.

pentaradial symmetry Symmetry in five or multiples of five; a feature of adult echinoderms.

peptidyl transferase A catalytic function of the large ribosomal subunit that consists of two reactions: breaking the bond between an amino acid and its tRNA in the P site, and forming a peptide bond between that amino acid and the amino acid attached to the tRNA in the A site.

peptides Molecule containing two or more amino acids.

peptide hormones Hormone molecules made up of strings of amino acids. They are water soluble. (Compare to steroid hormones, amine hormones.)

peptide bond The bond between amino acids in a protein; formed between a carboxyl group and amino group (—CO—NH—) with the loss of water molecules.

peptidoglycan The cell wall material of many bacteria, consisting of a single enormous molecule that surrounds the entire cell.

per capita birth rate (*b*) The number of offspring that the average individual produces in some specified interval of time.

per capita death rate (*d*) The average individual's chance of dying in some specified interval of time.

per capita growth rate (*r*) The average individual's contribution to total population growth rate in some specified interval of time, which equals per capita birth rate (*b*) minus per-capita death rate (*d*).

perennial (per ren' ee al) [L. *per*: throughout + *annus*: year] A plant that survives from year to year. (Contrast with annual, biennial.)

perfect flower A flower with both stamens and carpels; a hermaphroditic flower. (Contrast with imperfect flower.)

perfusion The flow of blood through a tissue or organ.

pericycle [Gk. *peri*: around + *kyklos*: ring or circle] In plant roots, tissue just within the endodermis, but outside of the root vascular tissue. Meristematic activity of pericycle cells produces lateral root primordia.

periderm The outer tissue of the secondary plant body, consisting primarily of cork.

peripheral membrane proteins Proteins associated with but not embedded within the plasma membrane. (Contrast with integral membrane proteins.)

peripheral nervous system (PNS) The portion of the nervous system other than the central nervous system (CNS). It consists of neurons and parts of neurons located outside the CNS. (Contrast with central nervous system.)

peristalsis (pair' i stall' sis) Wavelike muscular contractions proceeding along a tubular organ, propelling the contents along the tube.

peroxisome An organelle that houses reactions in which toxic peroxides are formed and then converted to water.

petal [Gk. *petalon*: spread out] In an angiosperm flower, a sterile modified leaf, nonphotosynthetic, frequently brightly colored, and often serving to attract pollinating insects.

petiole (pet' ee ole) [L. *petiolus*: small foot] The stalk of a leaf.

P_{fr} *See* phytochrome.

phage (fayj) *See* bacteriophage.

phage therapy A strategy for treating bacterial infections using bacteriophage.

phagocyte [Gk. *phagein*: to eat + *kystos*: sac] One of two major classes of white blood cells; one of the nonspecific defenses of animals; ingests invading microorganisms by phagocytosis.

phagocytosis Endocytosis by a cell of another cell or large particle.

phagosome Membrane-enclosed vesicle inside a cell that results from infolding of the cell membrane and enclosing a particle to be taken into the cell.

pharmacogenomics The study of how an individual's genetic makeup affects his or her response to drugs or other agents, with the goal of predicting the effectiveness of different treatment options.

pharming The use of genetically modified animals to produce medically useful products in their milk.

phenotype (fee' no type) [Gk. *phanein*: to show] The observable properties of an individual resulting from both genetic and environmental factors. (Contrast with genotype.)

phenotypic plasticity The ability of an individual animal to express two or more phenotypes during its life. (*See also* acclimation.)

pheromone (feer' o mone) [Gk. *pheros*: carry + *hormon*: excite, arouse] A chemical substance that is released into the external environment by an individual and that brings about specific behavioral responses in other individuals of the same species.

phloem (flo' um) [Gk. *phloos*: bark] In vascular plants, the vascular tissue that transports sugars and other solutes from sources to sinks.

phospholipid A lipid containing a phosphate group; an important constituent of cellular membranes. (*See* lipid.)

phospholipid bilayer The basic structural unit of biological membranes; a sheet of phospholipids two molecules thick in which the phospholipids are lined up with their hydrophobic "tails" packed tightly together and their hydrophilic, phosphate-containing "heads" facing outward. Also called lipid bilayer.

photic zone The surface zone of lakes or oceans that is penetrated by sunlight.

photoautotroph An organism that obtains energy from light and carbon from carbon dioxide. (Contrast with chemoautotroph, chemoheterotroph, photoheterotroph.)

photoheterotroph An organism that obtains energy from light but must obtain its carbon from organic compounds. (Contrast with chemoautotroph, chemoheterotroph, photoautotroph.)

photomorphogenesis In plants, a process by which physiological and developmental events are controlled by light.

photon (foe' ton) [Gk. *photos*: light] A quantum of visible radiation; a "packet" of light energy.

photoperiod Day length; the number of hours of daylight in a 24-hour day.

photoreceptor (1) In plants, a pigment that triggers a physiological response when it absorbs a photon. (2) In animals, a sensory receptor cell that senses and responds to light energy.

photosynthesis (foe tow sin' the sis) [literally, "synthesis from light"] Metabolic processes carried out by green plants and cyanobacteria, by which visible light is trapped and the energy used to convert CO_2 into organic compounds.

photosynthetic lamellae Elaborate internal membrane systems found in cyanobacteria that are used for photosynthesis.

photosystem [Gk. *phos*: light + *systema*: assembly] A light-harvesting complex in the chloroplast thylakoid composed of pigments and proteins.

photosystem I In photosynthesis, the complex that absorbs light at 700 nm, passing electrons to ferrodoxin and thence to NADPH.

photosystem II In photosynthesis, the complex that absorbs light at 680 nm, passing electrons to the electron transport chain in the chloroplast.

phototropins A class of blue light receptors that mediate phototropism and other plant responses.

phototropism [Gk. *photos*: light + *trope*: turning] A directed plant growth response to light.

phycoerythrin A red accessory photosynthetic pigment found in red algae.

phylogenetic tree A graphic representation of lines of descent among organisms or their genes.

phylogeny (fy loj' e nee) [Gk. *phylon*: tribe, race + *genesis*: source] The evolutionary history of a particular group of organisms or their genes.

phylum (plural: phyla) A group of related classes in biological classification.

physical geography The spatial distribution of Earth's climates and physical features.

phytoalexins Substances toxic to pathogens, produced by plants in response to fungal or bacterial infection.

phytochrome (fy' tow krome) [Gk. *phyton*: plant + *chroma*: color] A plant pigment regulating a large number of developmental and other phenomena in plants. It has two isomers: P_r, which absorbs red light, and P_{fr}, which absorbs far red light. P_{fr} is the active form.

phytomers In plants, the repeating modules that compose a shoot, each consisting of one or more leaves, attached to the stem at a node; an internode; and one or more axillary buds.

phytoplankton Photosynthetic plankton.

phytoremediation A form of bioremediation that uses plants to clean up environmental pollution.

pie chart A circular figure that displays proportions of different classes of data in an observed sample.

pigment A substance that absorbs visible light.

pinocytosis Endocytosis by a cell of liquid containing dissolved substances.

pistil [L. *pistillum*: pestle] The structure of an angiosperm flower within which the ovules are borne. May consist of a single carpel, or of several carpels fused into a single structure. Usually differentiated into ovary, style, and stigma.

pith In plants, relatively unspecialized tissue found within a cylinder of vascular tissue.

pituitary gland A small endocrine gland attached to the base of the brain in vertebrates. Many of its hormones control the activities of other endocrine glands. Also known as the hypophysis. (*See also* posterior pituitary gland, anterior pituitary gland.)

placenta (pla sen' ta) The organ in female mammals that provides for the nourishment of the fetus and elimination of the fetal waste products.

plankton Aquatic organisms that drift with the current. Photosynthetic members of the plankton are referred to as phytoplankton.

planula (plan' yew la) [L. *planum*: flat] A free-swimming, ciliated larval form typical of the cnidarians.

plasma (plaz' muh) The liquid portion of blood, in which blood cells are suspended. The part of the blood that is left after the blood cells are removed.

plasma cell An antibody-secreting cell that develops from a B cell; the effector cell of the humoral immune system.

plasmid A DNA molecule distinct from the chromosome(s); that is, an extrachromosomal element; found in many bacteria. May replicate independently of the chromosome.

plasmodesma (plural: plasmodesmata) [Gk. *plassein*: to mold + *desmos*: band]

A cytoplasmic strand connecting two adjacent plant cells.

plasmogamy The fusion of the cytoplasm of two cells. (Contrast with karyogamy.)

plate tectonics [Gk. *tekton*: builder] The scientific study of the structure and movements of Earth's lithospheric plates, which are the cause of continental drift.

platelet In blood, a membrane-bounded body without a nucleus, arising as a fragment of a cell in the bone marrow of mammals. Important to blood-clotting action.

pluripotent [L. *pluri*: many + *potens*: powerful] Having the ability to form all of the cells in the body. (Contrast with multipotent, totipotent.)

pneumatophores Root, sometimes submerged, that function in gas exchange with the environment.

poikilotherm An animal in which the body temperature matches the temperature of the external environment and varies as the external temperature varies. Also called an ectotherm.

point mutation A mutation that results from the gain, loss, or substitution of a single nucleotide.

polar covalent bond A covalent bond in which the electrons are drawn to one nucleus more than the other, resulting in an unequal distribution of charge.

polar nuclei In angiosperms, the two nuclei in the central cell of the megagametophyte; following fertilization they give rise to the endosperm.

polarity (1) In chemistry, the property of unequal electron sharing in a covalent bond that defines a polar molecule. (2) In development, the difference between one end of an organism or structure and the other.

pollen [L. *pollin*: fine flour] In seed plants, microscopic grains that contain the male gametophyte (microgametophyte) and gamete (microspore).

pollen grain *See* pollen.

pollen tube A structure that develops from a pollen grain through which sperm are released into the megagametophyte.

pollination The process of transferring pollen from an anther to the stigma of a pistil in an angiosperm or from a strobilus to an ovule in a gymnosperm.

poly A tail A long sequence of adenine nucleotides (50–250) added after transcription to the 3′ end of most eukaryotic mRNAs.

poly- [Gk. *poly*: many] A prefix denoting multiple entities.

polymer [Gk. *poly*: many + *meros*: unit] A large molecule made up of similar or identical subunits called monomers. (Contrast with monomer.)

polymerase chain reaction (PCR) An enzymatic technique for the rapid production of millions of copies of a particular stretch of DNA where only a small amount of the parent molecule is available.

polymorphic (pol′ lee mor′ fik) [Gk. *poly*: many + *morphe*: form, shape] Coexistence in a population of two or more distinct traits.

polyp (pah′ lip) [Gk. *poly*: many + *pous*: foot] In cnidarians, a sessile, asexual life cycle stage.

polypeptide A large molecule made up of many amino acids joined by peptide linkages. Large polypeptides are called proteins.

polyphyletic (pol′ lee fih leht′ ik) [Gk. *poly*: many + *phylon*: tribe] Pertaining to a group that consists of multiple distantly related organisms, and does not include the common ancestor of the group. (Contrast with monophyletic, paraphyletic.)

polyploidy (pol′ lee ploid ee) The possession of more than two entire sets of chromosomes.

polyribosome (polysome) A complex consisting of a threadlike molecule of messenger RNA and several (or many) ribosomes. The ribosomes move along the mRNA, synthesizing polypeptide chains as they proceed.

polysaccharide A macromolecule composed of many monosaccharides (simple sugars). Common examples are cellulose and starch.

pool In ecology, the total amount of any form of matter—of an element or a molecular compound—in a given compartment of an ecosystem.

population A group of individuals of the same species that live, interact, and reproduce together in a particular geographic area.

population bottleneck A period during which only a few individuals of a normally large population survive.

population density The number of individuals in a population per unit of area (for organisms that live in two-dimensional habitats such as the land surface) or volume (for organisms that live in three-dimensional habitats such as air or water).

population size The total number of individuals in an ecological population.

positional information In development, the basis of the spatial sense that induces cells to differentiate as appropriate for their location within the developing organism; often comes in the form of a morphogen gradient.

positive feedback In a regulatory system, a type of control that acts to increase differences that arise between the level of a controlled variable and its set-point level. The period of amplifying deviation is followed by a period in which stabilization is restored in most biological systems. (Contrast with negative feedback.)

positive relationship A relationship in which two variables tend to vary among observations in the same direction.

positive selection Natural selection that acts to establish a trait that enhances survival in a population. (Contrast with purifying selection.)

post- [L. *postere*: behind, following after] Prefix denoting something that comes after.

posterior Toward or pertaining to the rear. (Contrast with anterior.)

posterior pituitary gland A portion of the pituitary gland that is derived from neural tissue. In mammals, it is involved in the release of antidiuretic hormone and oxytocin, both of which are synthesized by hypothalamic neurosecretory cells. (*See also* pituitary gland, anterior pituitary gland.)

postsynaptic cell A neuron or effector cell that receives a signal (chemical or electrical) from a presynaptic cell at a synapse.

postzygotic isolating mechanisms Barriers to the reproductive process that occur after the union of the nuclei of two gametes. (Contrast with prezygotic isolating mechanisms.)

potential difference *See* voltage.

potential energy Energy not doing work, such as the energy stored in chemical bonds. (Contrast with kinetic energy.)

power In reference to statistical tests: refers to the probability of correctly rejecting a null hypothesis when it is false.

P_r *See* phytochrome.

PR proteins *See* pathogenesis-related proteins.

PRRs *See* pattern recognition receptors.

Precambrian The first and longest period of geological time, during which life originated.

precursor RNA (pre-mRNA) Initial gene transcript before it is modified to produce functional mRNA. Also known as the primary transcript.

predator An organism that kills and eats other organisms.

predation A +/– interaction between individuals of two different species in which one individual (the predator) gains fitness by consuming the other individual (the prey), which of course loses fitness by being consumed.

pressure flow model An effective model for phloem transport in angiosperms. It holds that sieve element transport is driven by an osmotically generated pressure gradient between source and sink.

pressure potential (Ψ_p) The hydrostatic pressure of an enclosed solution in excess of the surrounding atmospheric pressure. (Contrast with solute potential, water potential.)

presynaptic axon terminal One of the end processes of an axon, where the axon terminates at a synapse.

presynaptic cell A neuron or other cell that transmits a signal to a postsynaptic cell at a synapse. (Contrast with postsynaptic cell.)

prey [L. *praeda*: booty] An organism consumed by a predator as an energy source.

prezygotic isolating mechanisms Barriers to the reproductive process that occur before the union of the nuclei of two gametes (Contrast with postzygotic isolating mechanisms.)

primary active transport Active transport in which ATP is hydrolyzed, yielding the energy required to transport an ion or molecule against its concentration gradient. (Contrast with secondary active transport.)

primary consumer Organisms that consume primary producers.

primary embryonic inducer *See* organizer.

primary endosymbiosis The engulfment of a cyanobacterium by a larger eukaryotic cell that gave rise to the first photosynthetic eukaryotes with chloroplasts.

primary growth In plants, growth that is characterized by the lengthening of roots and shoots and by the proliferation of new roots and shoots through branching. (Contrast with secondary growth.)

primary immune response The first response of the immune system to an antigen, involving recognition by lymphocytes and the production of effector cells and memory cells. (Contrast with secondary immune response.)

primary lysosome *See* lysosome.

primary meristem Meristem that produces the tissues of the primary plant body.

primary producer A photosynthetic or chemosynthetic organism that synthesizes complex organic molecules from simple inorganic ones.

primary reproductive organs The gonads: the ovaries in females and the testes (or testicles) in males.

primary structure The specific sequence of amino acids in a protein. (Contrast with secondary, tertiary, quaternary structure.)

primary transcript *See* precursor RNA.

primase An enzyme that catalyzes the synthesis of a primer for DNA replication.

primer Strand of nucleic acid, usually RNA, that is the necessary starting material for the synthesis of a new DNA strand, which is synthesized from the 3′ end of the primer.

principle of allocation The principle that a unit of resource (such as food) cannot be used simultaneously for multiple functions (such as growth and reproduction), but instead must be allocated to one or another function.

pro- [L.: first, before, favoring] A prefix often used in biology to denote a developmental stage that comes first or an evolutionary form that appeared earlier than another. For example, prokaryote, prophase.

probability A numerical quantity that expresses the likelihood of an event occurring on a scale from 0 (no chance of the event) to 1 (certainty of the event).

probe A segment of single-stranded nucleic acid used to identify DNA molecules containing the complementary sequence.

proboscis A hollow, muscular feeding organ.

procambium Primary meristem that produces the vascular tissue.

processes The ways in which the components of a biological system interact (e.g., protein synthesis, nutrient metabolism, grazing).

processive Pertaining to an enzyme that catalyzes many reactions each time it binds to a substrate, as DNA polymerase does during DNA replication.

products The molecules that result from the completion of a chemical reaction.

progesterone [L. *pro*: favoring + *gestare*: to bear] A female sex hormone that maintains pregnancy.

prokaryotes Unicellular organisms that do not have nuclei. (Contrast with eukaryotes.)

prometaphase The phase of nuclear division that begins with the disintegration of the nuclear envelope.

promoter A DNA sequence to which RNA polymerase binds to initiate transcription.

prop roots Adventitious roots in some monocots that function as supports for the shoot.

prophase (pro′ phase) The first stage of nuclear division, during which chromosomes condense from diffuse, threadlike material to discrete, compact bodies.

prostaglandin Any one of a group of specialized lipids with hormone-like functions. It is not clear that they act at any considerable distance from the site of their production.

proteasome In the eukaryotic cytoplasm, a huge protein structure that binds to and digests cellular proteins that have been tagged by ubiquitin.

protein (pro′ teen) [Gk. *protos*: first] Long-chain polymer of amino acids with twenty different common side chains. Occurs with its polymer chain extended in fibrous proteins, or coiled into a compact macromolecule in enzymes and other globular proteins.

protein kinase (kye′ nase) An enzyme that catalyzes the addition of a phosphate group from ATP to a target protein.

proteoglycan A glycoprotein containing a protein core with attached long, linear carbohydrate chains.

proteome The set of proteins that can be made by an organism. Because of alternative splicing of pre-mRNA, the number of proteins that can be made is usually much larger than the number of protein-coding genes present in the organism's genome.

proteomics The study of the proteome—the complete complement of proteins produced by an organism.

protocells Ordered structures enclosed within a membrane that can be formed in the laboratory and are hypothesized to resemble cells when life originated.

protoderm Primary meristem that gives rise to the plant epidermis.

proton (pro′ ton) [Gk. *protos*: first, before] (1) A subatomic particle with a single positive charge. The number of protons in the nucleus of an atom determine its element. (2) A hydrogen ion, H^+.

proton pump An active transport system that uses ATP energy to move hydrogen ions across a membrane, generating an electric potential.

provirus Double-stranded DNA made by a virus that is integrated into the host's chromosome and contains promoters that are recognized by the host cell's transcription apparatus.

proximate explanation The immediate genetic, physiological, neurological, and developmental explanations for the advantages of an adaptation.

pseudocoelomate (soo′ do see′ low mate) [Gk. *pseudes*: false + *koiloma*: cavity] Having a body cavity, called a pseudocoel, consisting of a fluid-filled space in which many of the internal organs are suspended, but which is enclosed by mesoderm only on its outside.

pseudogene [Gk. *pseudes*: false] A DNA segment that is homologous to a functional gene but is not expressed because of changes to its sequence or changes to its location in the genome.

pseudoplasmodium An aggregate of individuals in cellular slime molds that act in a coordinated fashion to form a fruiting structure.

Punnett square Method of predicting the results of a genetic cross by arranging the gametes of each parent at the edges of a square.

pupil The opening in the vertebrate eye through which light passes from the outside to the retina.

purifying selection The elimination by natural selection of detrimental characters from a population. (Contrast with positive selection.)

purine (pure′ een) One of the two types of nitrogenous bases in nucleic acids. Each of the purines—adenine and guanine—pairs with a specific pyrimidine.

pyrimidine (per im′ a deen) One of the two types of nitrogenous bases in nucleic acids. Each of the pyrimidines—cytosine, thymine, and uracil—pairs with a specific purine.

pyruvate oxidation Conversion of pyruvate to acetyl CoA and CO_2 that occurs in the mitochondrial matrix in the presence of O_2.

Q

qualitative traits Traits that differ from one another by discrete qualities.

quantitative trait Phenotype determined by multiple genes.

quantitative variables Variables that can take on values along a numerical scale.

quaternary structure The specific three-dimensional arrangement of protein subunits. (Contrast with primary, secondary, tertiary structure.)

quorum sensing The use of chemical communication signals to trigger density-linked activities such as biofilm formation in prokaryotes.

R

R group The distinguishing group of atoms of a particular amino acid. Also known as a side chain.

radial axis Radial definition of function in an adult organism or future function in an embryo, as in the developing flower with organs in whorls.

radial cleavage Embryonic development in some deuterostomes in which the planes of cell division are parallel and perpendicular to the animal-vegetal axis of the embryo. Considered the basic deuterostome cleavage pattern. (Compare to spiral cleavage.)

radial symmetry The condition in which any two halves of a body are mirror images of each other, providing the cut passes through the center; a cylinder cut lengthwise down its center displays this form of symmetry.

radicle An embryonic root.

radiometric dating A method for determining the age of objects such as fossils and rocks based on the decay rates of radioactive isotopes.

rain shadow The relatively dry area on the down-wind side of a mountain range.

range The largest minus the smallest observed value for a variable in a sample.

rarity advantage Any situation in which individuals of a species lose less fitness from interspecific competition when they are rare than when they are common.

reactant A chemical substance that enters into a chemical reaction with another substance.

reaction center A group of electron transfer proteins that receive energy from light-absorbing pigments and convert it to chemical energy by redox reactions.

receptor *See* receptor protein, sensory receptor cell.

receptor endocytosis Endocytosis initiated by macromolecular binding to a specific membrane receptor. Also called receptor-mediated endocytosis.

receptor protein A protein that can bind to a specific molecule, or detect a specific stimulus, within the cell or in the cell's external environment.

recessive In genetics, an allele that does not determine phenotype in the presence of a dominant allele. (Contrast with dominance.)

recombinant Pertaining to an individual, meiotic product, or chromosome in which genetic materials originally present in two individuals end up in the same haploid complement of genes.

recombinant DNA A DNA molecule made in the laboratory that is derived from two or more genetic sources.

recombination frequency The proportion of offspring of a genetic cross that have phenotypes different from the parental phenotypes due to crossing over between linked genes during gamete formation.

red blood cell (RBC) A cell in the blood of an animal that contains hemoglobin and transports O_2. Also called an erythrocyte.

redox reaction A chemical reaction in which one reactant becomes oxidized and the other becomes reduced. Short for reduction–oxidation reaction.

reduction Gain of electrons by a chemical reactant; any reduction is accompanied by an oxidation. (Contrast with oxidation.)

regulation The maintenance of internal conditions at an approximately constant level while external conditions vary. (Contrast with conformity.)

regulative development A pattern of animal embryonic development in which the fates of the first blastomeres are not absolutely fixed. (Contrast with mosaic development.)

regulators Animals that maintain relatively constant internal conditions in the presence of changing external conditions (i.e., the animals exhibit regulation.) (Contrast with conformers.)

regulatory T cells (T_{reg}) The class of T cells that mediates tolerance to self antigens.

reinforcement The evolution of enhanced reproductive isolation between populations due to natural selection for greater isolation.

release-inhibiting hormones *See* inhibiting hormones.

releasing hormone (RH) A hormone secreted by neuroendocrine cells in the hypothalamus of a vertebrate that travels to the anterior pituitary gland through the hypothalamo–hypophysial portal system and stimulates the secretion of a hormone by a specific population of anterior pituitary endocrine cells.

replication fork A point at which a DNA molecule is replicating. The fork forms by the unwinding of the parent molecule.

reporter gene A genetic marker included in recombinant DNA to indicate the presence of the recombinant DNA in a host cell.

repressor A protein encoded by a regulatory gene that can bind to a promoter and prevent transcription of the associated gene. (Contrast with activator.)

reproductive isolation Condition in which two divergent populations are no longer exchanging genes. Can lead to speciation.

reproductive signals External cues that stimulate cells to divide or organisms to reproduce.

RER *See* rough endoplasmic reticulum.

residuals The deviations, along the y-axis from the linear regression line, of individual observations in a bivariate scatter plot.

resistance exercise Exercise that consists of relatively short periods of high-intensity muscular actions against a large load, often repeated with intervening interruptions. Also called resistance training. (Contrast with endurance exercise.)

resistance (*R*) genes Plant genes that confer resistance to specific strains of pathogens.

resource partitioning A situation in which competing organisms (usually of different species) differ in their use of resources.

resource In ecology, the materials (e.g., food, mineral nutrients), energy (e.g., photons of light), or favorable physical conditions, that organisms need to survive and reproduce, and that they make unavailable to other organisms when they use these resources.

respiratory airways Airways in the lungs where O_2 and CO_2 are exchanged between the air and blood.

respiratory chain The terminal reactions of cellular respiration, in which electrons are passed from NAD or FAD, through a series of intermediate carriers, to molecular oxygen, with the concomitant production of ATP.

respiratory gases Oxygen (O_2) and carbon dioxide (CO_2).

respiratory minute volume In an animal with lungs, the total volume of air inhaled and exhaled per minute.

respiratory pigments Any of the pigments that undergo reversible combination with O_2 and thus are able to pick up O_2 in certain places in an animal's body (e.g., the breathing organs) and release it in other places (e.g., systemic tissues). Respiratory pigments include hemoglobin and hemocyanin.

resting potential The potential difference (voltage) across the cell membrane of a living cell at rest. In cells at rest, the interior is negative to the exterior. (Contrast with action potential.)

restoration ecology The applied subdiscipline of ecology whose goal is to restore the function of damaged ecosystems.

restriction enzyme Any of a type of enzyme that cleaves double-stranded DNA at specific sites; extensively used in recombinant DNA technology. Also called a restriction endonuclease.

restriction site A specific DNA base sequence that is recognized and acted on by a restriction endonuclease.

retina (rett' in uh) [L. *rete*: net] The light-sensitive layer of cells in the vertebrate eye and some invertebrate eyes. In the vertebrate eye, it consists of photoreceptor cells and cells that process signals from the photoreceptor cells.

retrotransposons Mobile genetic elements that are reverse transcribed into RNA as part of their transfer mechanism. (Contrast with DNA transposons.)

retrovirus An RNA virus that contains reverse transcriptase. Its RNA serves as a template for cDNA production, and the cDNA is integrated into a chromosome of the host cell.

reverse transcriptase An enzyme that catalyzes the production of DNA (cDNA), using RNA as a template; essential to the reproduction of retroviruses.

rhizoids (rye' zoids) [Gk. root] Hairlike extensions of cells in mosses, liverworts, and a few vascular plants that serve the same function as roots and root hairs in vascular plants. The term is also applied to branched, rootlike extensions of some fungi and algae.

rhizome (rye' zome) An underground stem (as opposed to a root) that runs horizontally beneath the ground.

rhodopsin A visual pigment, found in many molecular forms in the animal kingdom, used in the process of transducing energy of photons of light into changes in the membrane potential of photoreceptor cells.

ribonucleic acid *See* RNA.

ribose A five-carbon sugar in nucleotides and RNA.

ribosomal RNA (rRNA) Several species of RNA that are incorporated into the ribosome. Involved in peptide bond formation.

ribosome A small particle in the cell that is the site of protein synthesis.

ribozyme An RNA molecule with catalytic activity.

ribulose bisphosphate carboxylase/ oxygenase *See* rubisco.

RNA (ribonucleic acid) An often single-stranded nucleic acid whose nucleotides use ribose rather than deoxyribose and in which the base uracil replaces thymine found in DNA. Serves as genome from some viruses. (*See* ribosomal RNA, transfer RNA, messenger RNA, ribozyme.)

RNA interference (RNAi) A mechanism for reducing mRNA translation whereby a double-stranded RNA, made by the cell or synthetically, is processed into a small, single-stranded RNA, whose binding to a target mRNA results in the latter's breakdown.

RNA polymerase An enzyme that catalyzes the formation of RNA from a DNA template.

RNA splicing The last stage of RNA processing in eukaryotes, in which the transcripts of introns are excised through the action of small nuclear ribonucleoprotein particles (snRNP).

rod cells Light-sensitive cells in the vertebrate retina; these sensory receptor cells are sensitive to extremely dim light and are responsible for dim-light, black and white vision.

root (1) In reference to phylogenetic trees: the base (oldest) part of the tree. (2) In reference to plants: the organ responsible for anchoring the plant in the soil, absorbing water and minerals, and producing certain hormones. Some roots are storage organs.

root apical meristem Undifferentiated tissue at theof the root that gives rise to the organs of the root.

root cap A thimble-shaped mass of cells, produced by the root apical meristem, that protects the meristem; the organ that perceives the gravitational stimulus in root gravitropism.

root system The organ system that anchors a plant in place, absorbs water and dissolved minerals, and may store products of photosynthesis from the shoot system.

rotational cleavage Cleavage pattern typical of eutherian mammals in which the first cleavage is meridional whereas in the second cleavage, one blastomere divides meridionally and the other equatorially.

rough endoplasmic reticulum (RER) The portion of the endoplasmic reticulum whose outer surface has attached ribosomes. (Contrast with smooth endoplasmic reticulum.)

rRNA *See* ribosomal RNA.

RT-PCR Determination of the quantity of an RNA by first converting to it to DNA by reverse transcriptase, and then amplifying the DNA by the polymerase chain reaction.

rubisco Contraction of ribulose bisphosphate carboxylase/oxygenase, the enzyme that combines carbon dioxide or oxygen with ribulose bisphosphate to catalyze the first step of photosynthetic carbon fixation or photorespiration, respectively.

S

S phase In the cell cycle, the stage of interphase during which DNA is replicated. (Contrast with G1, G2, M phase.)

salt glands (1) In plants, glands on the leaves of some halophytic (salt-loving) plants that secrete salts onto the outer plant surfaces, thereby ridding the tissue fluids of the plants of excess salts. (2) In animals, organs other than kidneys that excrete concentrated salt solutions, thereby ridding the body fluids of excess salts.

sample A set of observations made from a population.

saprobe [Gk. *sapros*: rotten] An organism (usually a bacterium or fungus) that obtains its carbon and energy by absorbing nutrients from dead organic matter.

saprobic Feeding on dead organic matter.

sarcomere (sark' o meer) [Gk. *sark*: flesh + *meros*: unit] The contractile unit of a skeletal muscle cell. Each myofibril in a muscle cell consists of a series of sarcomeres.

sarcoplasm The cytoplasm of a muscle cell.

sarcoplasmic reticulum The endoplasmic reticulum of a muscle cell.

saturated fatty acid A fatty acid in which all the bonds between carbon atoms in the hydrocarbon chain are single bonds—that is, all the bonds are saturated with hydrogen atoms. (Contrast with unsaturated fatty acid.)

scaling relationships The study of the relations between physiological (or morphological) characteristics and body size within sets of phylogenetically related species (e.g., the study of metabolism-weight relations among mammals of different sizes.)

scatter plot A figure that displays the values of observations for two variables along perpendicular axes.

Schwann cell A type of glial cell that forms a myelin sheath on axons in the peripheral nervous system.

scion In horticulture, the bud or stem from one plant that is grafted to a root or root-bearing stem of another plant (the stock).

sclereid One of the principle types of cells in sclerenchyma.

sclerenchyma (skler eng' kyma) [Gk. *skleros*: hard + *kymus*: juice] A plant tissue composed of cells with heavily thickened cell walls. The cells are dead at functional maturity. The principal types of sclerenchyma cells are fibers and sclereids.

scrotum In most mammals, a pouch outside the body cavity that contains the testes (testicles).

SDP *See* short-day plant.

seasonality Predictable variation in the climate of a given place during the course of an annual cycle.

second filial generation *See* F_2.

second law of thermodynamics The principle that when energy is converted from one form to another, some of that energy becomes unavailable for doing work.

secondary active transport A form of active transport that does not use ATP as an energy source; rather, transport is coupled to ion diffusion down a concentration gradient established by primary active transport. (Contrast with primary active transport.)

secondary consumer An organism that consumes primary consumers.

secondary endosymbiosis The engulfment of a photosynthetic eukaryote by another eukaryotic cell that gave rise to certain groups of photosynthetic eukaryotes (e.g., euglenids).

secondary growth In plants, growth that contributes to an increase in girth. (Contrast with primary growth.)

secondary immune response A rapid and intense response to a second or subsequent exposure to an antigen, initiated by memory cells. (Contrast with primary immune response.)

secondary lysosome Membrane-enclosed organelle formed by the fusion of a primary lysosome with a phagosome, in which macromolecules taken up by phagocytosis are hydrolyzed into their monomers. (Contrast with lysosome.)

secondary metabolite A compound synthesized by a plant that is not needed for basic cellular metabolism. Typically has an antiherbivore or antiparasite function.

secondary sexual characteristics Sex-specific properties of nonreproductive tissues and organs.

secondary structure Of a protein, localized regularities of structure, such as the α helix and the β pleated sheet. (Contrast with primary, tertiary, quaternary structure.)

secrete To discharge a substance from a cell or gland.

sedimentary rock Rock formed by the accumulation of sediment grains on the bottom of a body of water.

seed A fertilized, ripened ovule of a gymnosperm or angiosperm. Consists of the embryo, nutritive tissue, and a seed coat.

seedling A plant that has just completed the process of germination.

segment polarity genes In *Drosophila* (fruit fly) development, segmentation genes that determine the boundaries and anterior–posterior organization of individual segments. Part of a developmental cascade that includes maternal effect genes, gap genes, pair rule genes, and Hox genes.

segmentation Division of an animal body into segments.

segregation *See* law of segregation.

selectable marker A gene, such as one encoding resistance to an antibiotic, that can be used to identify (select) cells that contain recombinant DNA from among a large population of untransformed cells.

selective permeability Allowing certain substances to pass through while other substances are excluded; a characteristic of membranes.

self-incompatability In plants, the possession of mechanisms that prevent self-fertilization.

semelparity A type of reproductive life history in which individuals are physiologically capable of reproducing only one time during their lives. In **semelparous** species, individuals are often programmed to die after reproducing once. (Contrast with iteroparity.)

semen (see' men) [L. *semin*: seed] The thick, whitish liquid produced by the male reproductive system in mammals, containing the sperm.

semiconservative replication The way in which DNA is synthesized. Each of the two partner strands in a double helix acts as a template for a new partner strand. Hence, after replication, each double helix consists of one old and one new strand.

seminiferous tubules The tubules within the testes (testicles) within which sperm production occurs.

sensor One of four essential elements of a control system; an organ or cell that detects the current level of the controlled variable.

sensory neuron A neuron that functions as a sensory receptor cell or that carries a signal from sensory cells or organs to the central nervous system. Also called an afferent neuron.

sensory receptor cell A cell, typically a neuron, that is specialized to transform the energy of a stimulus into an electric signal. Each type is highly specific in the stimuli to which it ordinarily responds.

sensory receptor proteins Proteins in sensory receptor cells that respond to sensory input, causing a receptor potential (a graded change of membrane potential) in the receptor cell.

sensory transduction The transformation of environmental stimuli into neural electric signals in a sensory receptor cell.

sepal (see′ pul) [L. *sepalum*: covering] One of the outermost structures of the flower, usually protective in function and enclosing the rest of the flower in the bud stage.

sepsis Generalized inflammation caused by bacterial infection. Can cause a dangerous drop in blood pressure.

septate [L. wall] Divided, as by walls or partitions.

septum (plural: septa) (1) A partition or cross-wall appearing in the hyphae of some fungi. (2) The bony structure dividing the nasal passages.

sequential hermaphrodites Species in which an individual can be male at some times and female at other times during its lifespan

SER *See* smooth endoplasmic reticulum.

Sertoli cells Cells in the seminiferous tubules that nurture the developing sperm.

sex A process by which the genes of two individuals are combined to produce offspring.

sex chromosome In organisms with a chromosomal mechanism of sex determination, one of the chromosomes involved in sex determination.

sex determination The process by which the sex (male or female) of an individual becomes established during the conception and development of the individual.

sex pilus A thin connection between two bacteria through which genetic material passes during conjugation.

sex-linked inheritance Inheritance of a gene that is carried on a sex chromosome. Also called sex linkage.

sexual reproduction Reproduction in which the genes of two individuals are combined to produce offspring, typically involving the union of male and female gametes.

sexual selection Selection by one sex of characteristics in individuals of the opposite sex. Also, the favoring of characteristics in one sex as a result of competition among individuals of that sex for mates.

shared derived trait *See* synapomorphy.

shiver (1) In a mammal or bird, the subtle contraction and relaxation of skeletal muscles that produces heat rather than motion as the primary product. (2) In an insect, contraction of the flight muscles in a nonflying mode to generate heat rather than flight.

shoot apical meristem Undifferentiated tissue at the apex of the shoot that gives rise to the organs of the shoot.

shoot system In plants, the organ system consisting of the leaves, stem(s), and flowers.

short-day plant (SDP) A plant that flowers when nights are longer than a critical length specific for that plant's species. (Compare to long-day plant.)

short tandem repeat (STR) A short (1–5 base pairs), moderately repetitive sequence of DNA. The number of copies of an STR at a particular location varies between individuals and is inherited.

side chain *See* R group.

sieve tube element The characteristic cell of the phloem in angiosperms, which contains cytoplasm but relatively few organelles, and whose end walls (sieve plates) contain pores that form connections with neighboring cells.

sigma factor In prokaryotes, a protein that binds to RNA polymerase, allowing the complex to bind to and stimulate the transcription of a specific class of genes (e.g., those involved in sporulation).

signal sequence The sequence within a protein that directs the protein to a particular organelle.

signal transduction pathway The series of biochemical steps whereby a stimulus to a cell (such as a hormone or neurotransmitter binding to a receptor) is translated into a response of the cell.

significance level A particular threshold of making a Type I error (incorrectly rejecting a true null hypothesis) in a statistical test, selected a priori by the investigator.

silent mutation A change in a gene's sequence that has no effect on the amino acid sequence of a protein because it occurs in noncoding DNA or because it does not change the amino acid specified by the corresponding codon. (Contrast with frame-shift mutation, missense mutation, nonsense mutation.)

silent substitution *See* synonymous substitution.

simple diffusion Diffusion that doesn't involve a direct input of energy or assistance by carrier proteins.

simple epithelium An epithelium (sheet of cells that lines a body cavity or covers an organ or body surface) consisting of a single cell layer.

simple fruit A fruit derived from a single ovary; examples include grapes and tomatoes.

single nucleotide polymorphisms (SNPs) Inherited variations in a single nucleotide base in DNA that differ between individuals.

sink In plants, any organ that imports the products of photosynthesis, such as roots, developing fruits, and immature leaves. (Contrast with source.)

sister chromatid Each of a pair of newly replicated chromatids.

sister clades Two phylogenetic groups that are each other's closest relatives.

sister species Two species that are each other's closest relatives.

skeletal muscle A type of muscle tissue characterized by multinucleated cells containing highly ordered arrangements of actin and myosin microfilaments that give the tissue a striated appearance under a light microscope. Also called striated muscle. (Contrast with cardiac muscle, smooth muscle.)

sliding filament theory Mechanism of muscle contraction (use of ATP to produce contractile forces) based on the formation and breaking of crossbridges between actin and myosin filaments, causing the filaments to slide past each other.

slow oxidative cells Skeletal muscle cells specialized for sustained, relatively low-intensity, aerobic work; contain myoglobin and abundant mitochondria. Also called slow-twitch cells. (Compare to fast glycolytic cells.)

slug A worm-like organism or colony of organisms; used to refer to unshelled terrestrial mollusks as well as to the mobile, colonial stage of cellular slime molds.

small intestine *See* midgut.

small nuclear ribonucleoprotein particle (snRNP) A complex of an enzyme and a small nuclear RNA molecule, functioning in RNA splicing.

smooth endoplasmic reticulum (SER) Portion of the endoplasmic reticulum that lacks ribosomes and has a tubular appearance. (Contrast with rough endoplasmic reticulum.)

smooth muscle Muscle tissue consisting of small, mononucleated muscle cells innervated by the autonomic nervous system. (Contrast with cardiac muscle, skeletal muscle.)

social behavior Behavior that helps integrate individuals into social groups (societies), and the group behaviors of social groups. (*See also* society.)

society A group of individuals of a single species that exhibits some degree of cooperative action.

sodium–potassium (Na^+–K^+) pump A membrane protein (anti-porter) that carries out primary active transport of ions; it uses energy from ATP to pump sodium ions out of a cell and potassium ions into the cell. Also called a sodium–potassium ATPase.

soil horizon *See* horizons.

soil solution The aqueous portion of soil, from which plants take up dissolved mineral nutrients.

solute potential (Ψ_s) A property of any solution, resulting from its solute contents; it may be zero or have a negative value. The more negative the solute potential, the greater the tendency of the solution to take up water through a differentially permeable membrane. (Contrast with pressure potential, water potential.)

solvent Liquid in which a substance (solute) is dissolved to form a solution.

somatic cell [Gk. *soma*: body] All the cells of the body that are not specialized for reproduction. (Contrast with germ cell.)

somatic mutation Permanent genetic change in a somatic cell. These mutations affect the individual only; they are not passed on to offspring. (Contrast with germline mutation.)

somite (so′ might) A segmental block of mesoderm; differentiates to form dermis, axial skeleton (vertebrae, ribs), and skeletal muscle.

soredia (sing.: soredium) Propagules of lichens consisting of one or a few photosynthetic cells bound by fungal hyphae.

sori Clusters of stalked sporangia that occur on the underside of fern fronds.

source In plants, any organ that exports the products of photosynthesis in excess of its own needs, such as a mature leaf or storage organ. (Contrast with sink.)

spawning *See* external fertilization.

speciation (spee′ see ay′ shun) The process of splitting one biological lineage into two biological lineages that evolve independently from one another.

species (spee′ sees) [L. kind] The base unit of taxonomic classification, consisting of an ancestor–descendant group of populations of evolutionarily closely related, similar organisms. The more narrowly defined "biological species" consists of individuals capable of interbreeding with each other but not with members of other species.

species composition The identities of all the species that make up a given ecological community.

species concepts Ways that biologists think about the existence of species.

species diversity In ecology, a measure of the chance that two individuals in a community are of the same species. Measures of species diversity include the number of distinct species in a community (species richness), and/or the relative abundances of different species (species evenness).

species evenness A measure of the similarity in the abundances of species in a community. If species are equally abundant, evenness is at a maximum.

species richness The total number of species in an ecological community.

sperm [Gk. *sperma*: seed] *See* spermatozoa.

spermatozoa Haploid gametes that are capable of swimming (usually by use of flagella), produced by spermatogenesis in the testis of a male animal. Also called sperm.

spicule [L. arrowhead] A hard, calcareous skeletal element typical of sponges.

spinal reflex In a vertebrate, the conversion of sensory neuronal signals to motor neuronal signals in the spinal cord without participation of the brain.

spindle Array of microtubules emanating from both poles of a dividing cell during mitosis and playing a role in the movement of chromosomes at nuclear division. Named for its shape.

spiral cleavage Embryonic development in many protostomes in which the plane of cell division is diagonal to the vertical axis of the embryo. (Compare to radial cleavage.)

spirillum A bacterium that is shaped like a corkscrew.

spleen Organ that serves as a reservoir for venous blood and eliminates old, damaged red blood cells from the circulation.

splicing *See* RNA splicing.

spliceosome RNA–protein complex that splices out introns from eukaryotic pre-mRNAs.

spontaneous mutation A genetic change caused by internal cellular mechanisms, such as an error in DNA replication. (Contrast with induced mutation.)

spontaneous ovulation Ovulation (release of an egg from the ovaries of a female) that results from endogenous processes (processes that originate within the animal), more or less independent of copulation. (Contrast with induced ovulation.)

sporangiophore A stalked reproductive structure produced by zygospore fungi that extends from a hypha and bears one or many sporangia.

sporangium (spor an′ gee um) (plural: sporangia) [Gk. *spora*: seed + *angeion*: vessel or reservoir] In plants and fungi, any specialized structure within which one or more spores are formed.

spore [Gk. *spora*: seed] (1) Any asexual reproductive cell capable of developing into an adult organism without gametic fusion. In plants, haploid spores develop into gametophytes, diploid spores into sporophytes. (2) In prokaryotes, a resistant cell capable of surviving unfavorable periods.

sporocyte Specialized cells of the diploid sporophyte that will divide by meiosis to produce four haploid spores. Germination of these spores produces the haploid gametophyte.

sporulation The formation of specialized cells (spores) that are capable of developing into new individuals.

sporophyte (spor′ o fyte) [Gk. *spora*: seed + *phyton*: plant] In plants and protists with alternation of generations, the diploid phase that produces the spores. (Contrast with gametophyte.)

stabilizing selection Selection against the extreme phenotypes in a population, so that the intermediate types are favored. (Contrast with disruptive selection.)

stamen (stay′ men) [L. *stamen*: thread] A male (pollen-producing) unit of a flower, usually composed of an anther, which bears the pollen, and a filament, which is a stalk supporting the anther.

standard amino acid One of the 20 amino acids that animals use to synthesize proteins.

standard deviation A measure of the spread of observations in a sample. See Appendix B for the mathematical formula.

standard error A measure of how close a sample statistic (such as the mean) is likely to be to the true population value.

standard error of the mean A measure of how close a sample mean is likely to be to the true population value. Calculated by dividing the standard deviation of a sample by the square root of the sample size.

staphylococci Bacteria of the genus *Staphylococcus*.

start codon The mRNA triplet (AUG) that acts as a signal for the beginning of translation at the ribosome. (Contrast with stop codon.)

statistic A numerical quantity calculated from data.

stele (steel) [Gk. *stylos*: pillar] The central cylinder of vascular tissue in a plant stem.

stem In plants, the organ that holds leaves and/or flowers and transports and distributes materials among the other organs of the plant.

stem cell In animals, an undifferentiated cell that is capable of continuous proliferation. A stem cell generates more stem cells and a large clone of differentiated progeny cells. (*See also* embryonic stem cell.)

steroid hormones Lipid-soluble hormones synthesized from cholesterol. The chemical structure of a steroid hormone includes a distinctive subpart consisting of four interlinked rings. (Compare to peptide hormones, amine hormones.)

stigma [L. *stigma*: mark, brand] The part of the pistil at the apex of the style that is receptive to pollen, and on which pollen germinates.

stock In horticulture, the root or root-bearing stem to which a bud or piece of stem from another plant (the scion) is grafted.

stoma (plural: stomata) [Gk. *stoma*: mouth, opening] Small opening in the plant epidermis that permits gas exchange; bounded by a pair of guard cells whose osmotic status regulates the size of the opening.

stop codon Any of the three mRNA codons that signal the end of protein translation at the ribosome: UAG, UGA, UAA. (Contrast with start codon.)

STR *See* short tandem repeat.

stratigraphy The study of geological strata, or layers.

stratum (plural: strata) [L. *stratos*: layer] A layer of sedimentary rock laid down at a particular time in the past.

stretch-gated channel A membrane ion channel that opens and closes in response to local stretching or pulling forces applied to the cell membrane.

striated muscle *See* skeletal muscle.

strigolactones Signaling molecules produced by plant roots that attract the hyphae of mycorrhizal fungi.

strobilus (plural: strobili) One of several conelike structures in various groups of plants (including club mosses, horsetails, and conifers) associated with the production and dispersal of reproductive products.

structural gene A gene that encodes the primary structure of a protein not involved in the regulation of gene expression.

style [Gk. *stylos*: pillar or column] In the angiosperm flower, a column of tissue extending from the tip of the ovary, and bearing the stigma or receptive surface for pollen at its apex.

sub- [L. under] A prefix used to designate a structure that lies beneath another or is less than another. For example, subcutaneous (beneath the skin); subspecies.

subduction In plate tectonics, the movement of one plate under another.

subpopulation In ecology, the portion of a larger metapopulation that occupies a given piece of suitable habitat. *See* metapopulation.

subsoil The soil horizon lying below the topsoil and above the parent rock (bedrock); the zone of infiltration and accumulation of materials leached from the topsoil. Also called the B horizon.

substrate (sub′ strayte) (1) The molecule or molecules on which an enzyme exerts catalytic action. (2) The base material on which a sessile organism lives.

succession A relatively predictable series of changes in the species composition of a community following a disturbance, that leads ultimately to a community that resembles the original, predisturbance community. (Contrast with ecological transition.)

succulence In plants, possession of fleshy, water-storing leaves or stems; an adaptation to dry environments.

sun compass A mechanism by which an animal uses the position of the sun and an internal clock to determine compass direction in navigation.

surface area-to-volume ratio For any cell, organism, or geometrical solid, the ratio of surface area to volume; this is an important factor in setting an upper limit on the size a cell or organism can attain.

surface tension The attractive intermolecular forces at the surface of liquid; an especially important property of water.

survivorship The average individual's chance of surviving over a specified set of life stages, ages, or period of time—often from birth to a given stage or age. (Contrast with mortality.)

suspensor In the embryos of seed plants, the stalk of cells that pushes the embryo into the endosperm and is a source of nutrient transport to the embryo.

suspension feeder An aquatic animal that feeds on objects much smaller than itself that are suspended in water. (*See also* filter feeder.)

swim bladders Organs used to regulate buoyancy in fish.

symbiosis (sim′ bee oh′ sis) [Gk. *sym*: together + *bios*: living] The living together of two or more species in a prolonged, interactive relationship.

symbiotic A relationship in which two or more organisms live in close association with one another.

symmetry Pertaining to an attribute of an animal body in which at least one plane can divide the body into similar, mirror-image halves. (*See* bilateral symmetry, radial symmetry.)

sympathetic division A division of the vertebrate autonomic nervous system that is connected to the central nervous system via thoracic and lumbar spinal nerves; the sympathetic and parasympathetic divisions tend to exert opposing controls over autonomic effectors. (Contrast with parasympathetic division.)

sympatric speciation (sim pat′ rik) [Gk. *sym*: same + *patria*: homeland] Speciation due to reproductive isolation without any physical separation of the subpopulation. (Contrast with allopatric speciation.)

symplast The continuous meshwork of the interiors of living cells in the plant body, resulting from the presence of plasmodesmata. (Contrast with apoplast.)

synapomorphy A trait that arose in the ancestor of a phylogenetic group and is present (sometimes in modified form) in all of its members, thus helping to delimit and identify that group. Also called a shared derived trait.

synapse (sin′ aps) [Gk. *syn*: together + *haptein*: to fasten] A specialized type of junction where a neuron meets its target cell (which can be another neuron or some other type of cell) and neurotransmitter molecules serve to carry signals from the presynaptic neuron to the postsynaptic cell across the synaptic cleft.

synaptic plasticity The process by which synapses in the nervous system of an individual animal can undergo long-term changes in their functional properties.

synonymous (silent) substitution A change of one nucleotide in a sequence to another when that change does not affect the amino acid specified (i.e., UUA → UUG, both specifying leucine). (Contrast with nonsynonymous substitution, missense mutation, nonsense mutation.)

system A set of interacting parts in which neither the parts nor the whole can be understood without taking into account the interactions among the parts.

systematics The scientific study of the diversity and relationships among organisms.

systemic acquired resistance A general resistance to many plant pathogens following infection by a single agent.

systemic circuit In a circulatory system, the blood vessels that take blood to and from the systemic tissues. (Contrast with breathing-organ circuit.)

systemic tissues All tissues other than those of the breathing organs (lungs or gills).

systems analysis The understanding of a biological system through evaluation of its components and their interactions.

systems biology The scientific study of an organism as an integrated and interacting system of genes, proteins, and biochemical reactions.

systole (sis' tuh lee) [Gk. *systole*: contraction] During the cardiac cycle, the period of contraction of the chambers of the heart, which drives blood forward in the circulatory system. (Contrast with diastole.)

T

T tubules *See* transverse tubules.

T-helper (T_H) cell Type of T cell that stimulates events in both the cellular and humoral immune responses by binding to the antigen on an antigen-presenting cell; target of the HIV-I virus, the agent of AIDS. (Contrast with cytotoxic T cell.)

taproot The main root of a plant that is usually thick and grows down into the soil.

taproot system A root system typical of eudicots consisting of a primary root (taproot) that extends downward by tip growth and outward by initiating lateral roots. (Contrast with fibrous root system.)

target cell A cell with the appropriate receptor proteins to bind and respond to a particular hormone or other chemical mediator.

TATA box An eight-base-pair sequence, found about 25 base pairs before the starting point for transcription in many eukaryotic promoters, that binds a transcription factor and thus helps initiate transcription.

taxon (plural: taxa) [Gk. *taxis*: arrange, put in order] A biological group (typically a species or a clade) that is given a name.

telomerase An enzyme that catalyzes the addition of telomeric sequences lost from chromosomes during DNA replication.

telomeres (tee' lo merz) [Gk. *telos*: end + *meros*: units, segments] Repeated DNA sequences at the ends of eukaryotic chromosomes.

telophase (tee' lo phase) [Gk. *telos*: end] The final phase of mitosis or meiosis during which chromosomes become diffuse, nuclear envelopes re-form, and nucleoli begin to reappear in the daughter nuclei.

template A molecule or surface on which another molecule is synthesized in complementary fashion, as in the replication of DNA.

template strand In double-stranded DNA, the strand that is transcribed to create an RNA transcript that will be processed into a protein. Also refers to a strand of RNA that is used to create a complementary RNA.

tendon A collagen-containing band of tissue that connects a muscle with a bone.

termination In molecular biology, the end of transcription or translation.

territory A region that an individual occupies and defends against occupation by other individuals of the same species. (Compare with home range.)

tertiary consumers An organism that consumes secondary consumers.

tertiary endosymbiosis The mechanism by which some eukaryotes acquired the capacity for photosynthesis; for example, a dinoflagellate that apparently lost its chloroplast became photosynthetic by engulfing another protist that had acquired a chloroplast through secondary endosymbiosis.

tertiary structure In reference to a protein, the relative locations in three-dimensional space of all the atoms in the molecule. The overall shape of a protein. (Contrast with primary, secondary, quaternary structures.)

test cross Mating of a dominant-phenotype individual (who may be either heterozygous or homozygous) with a homozygous-recessive individual.

testis (tes' tis) (plural: testes) [L. *testis*: witness] The male gonad; the organ that produces the male gametes. Also sometimes called a testicle.

testosterone An androgen (sex steroid hormone), produced by the Leydig cells of the testes.

tetrad [Gk. *tettares*: four] During prophase I of meiosis, the association of a pair of homologous chromosomes or four chromatids.

thallus The vegetative body of a lichen.

thermoneutral zone (TNZ) [Gk. *thermos*: temperature] In a homeotherm, the range of ambient temperatures over which the animal is under neither heat nor cold stress and the metabolic rate is constant.

thermoregulation The maintenance of a relatively constant body temperature.

thick filaments *See* myosin filaments.

thin filaments *See* actin filaments.

thylakoid (thigh la koid) [Gk. *thylakos*: sack or pouch] A flattened sac within a chloroplast. Thylakoid membranes contain all of the chlorophyll in a plant, in addition to the electron carriers of photophosphorylation. Thylakoids stack to form grana.

thymine (T) Nitrogen-containing base found in DNA.

thyroid-stimulating hormone (TSH) A hormone, released by the anterior pituitary gland, that signals the thyroid gland to secrete its hormones.

tidal ventilation Bidirectional pumping of air in and out of lungs, during which air enters and leaves the lungs by the same route. (Contrast with unidirectional ventilation.)

tidal volume The volume of air that is breathed in and out of the lungs during each breath.

tissue A group of similar cells organized into a functional unit; usually integrated with other tissues to form part of an organ.

tissue fluid *See* interstitial fluid.

tissue system In plants, any of three organized groups of tissues—dermal tissue, vascular tissue, and ground tissue—that are established during embryogenesis and have distinct functions.

titin A giant, elongated, elastic protein molecule that in a striated muscle cell spans an entire half-sarcomere from the M band to the Z line.

TNZ *See* thermoneutral zone.

topography Variation in the elevation of Earth's surface, in the form of mountains, valleys, ocean basins, and so on.

topsoil The uppermost soil horizon; contains most of the organic matter of soil, but may be depleted of most mineral nutrients by leaching. Also called the A horizon.

totipotent [L. *toto*: whole, entire + *potens*: powerful] Possessing all the genetic information and other capacities necessary to form an entire individual. (Contrast with multipotent, pluripotent.)

toxigenicity The ability of some pathogenic bacteria to produce chemical substances that harm the host.

trachea (tray' kee ah) [Gk. *trakhoia*: tube] (1) In a vertebrate lung, the principal tube leading into the lungs from the mouth cavity. (2) One of the airways in the tracheal breathing system of an insect.

tracheal breathing system The type of breathing system found in insects, in which airways (called tracheae and tracheoles) branch throughout the animal's body so that airways are nearby all cells.

tracheid (tray' kee id) A type of tracheary element found in the xylem of nearly all vascular plants, characterized by tapering ends and walls that are pitted but not perforated. (Contrast with vessel element.)

tract A bundle of axons within a vertebrate central nervous system.

trade-off Any situation in which a gain in one function, outcome, aspect, or amount comes only at the cost of a loss in other functions, outcomes, aspects, or amounts.

trait In genetics, a specific form of a character: eye color is a character; brown eyes and blue eyes are traits. (Contrast with character.)

transcription The synthesis of RNA using one strand of DNA as a template.

transcription factors Proteins that assemble on a eukaryotic chromosome, allowing RNA polymerase II to perform transcription.

transcription initiation site The part of a gene's promoter where synthesis of the gene's RNA transcript begins.

transfection Insertion of recombinant DNA into animal cells.

transfer RNA (tRNA) A family of folded RNA molecules. Each tRNA carries a specific amino acid and anticodon that will pair with the complementary codon in mRNA during translation.

transformation (1) A mechanism for transfer of genetic information in bacteria in which pure DNA from a bacterium of one genotype is taken in through the cell surface of a bacterium of a different genotype and incorporated into the chromosome of the recipient cell. (2) Insertion of recombinant DNA into a host cell.

transgenic Containing recombinant DNA incorporated into the genetic material.

transition state In an enzyme-catalyzed reaction, the reactive condition of the substrate after there has been sufficient input of energy (activation energy) to initiate the reaction.

translation The synthesis of a protein (polypeptide). Takes place on ribosomes, using the information encoded in messenger RNA.

translational repressor A protein that blocks translation by binding to mRNAs and preventing their attachment to the ribosome. In mammals, the production of ferritin protein is regulated by a translational repressor.

translocation (1) In genetics, a rare mutational event that moves a portion of a chromosome to a new location, generally on a nonhomologous chromosome. (2) In vascular plants, movement of solutes in the phloem.

transmembrane protein An integral membrane protein that spans the phospholipid bilayer.

transpiration [L. *spirare*: to breathe] The evaporation of water from plant leaves and stem, driven by heat from the sun, and providing the motive force to raise water (plus mineral nutrients) from the roots.

transpiration–cohesion–tension mechanism Theoretical basis for water movement in plants: evaporation of water from cells within leaves (transpiration) causes an increase in surface tension, pulling water up through the xylem. Cohesion of water occurs because of hydrogen bonding.

transposon Mobile DNA segment that can insert into a chromosome and cause genetic change.

transverse tubules (T tubules) Indentations of the cell membrane at regular intervals over the entire surface of a muscle cell. Transverse tubules conduct electrical excitation into the interior of the cell and thus play a key role in linking electrical excitation with contraction.

triglyceride A simple lipid in which three fatty acids are combined with one molecule of glycerol.

triploblastic Having three cell layers. (Contrast with diploblastic.)

tRNA *See* transfer RNA.

trochophore (troke′ o fore) [Gk. *trochos*: wheel + *phoreus*: bearer] A radially symmetrical larval form typical of annelids and mollusks, distinguished by a wheel-like band of cilia around the middle.

trophic [Gk *trophes*: nourishment] Referring to feeding.

trophic cascade Effects of abundances of species at one trophic level on species at other trophic levels. For example, introduction of a tertiary consumer to a community may reduce the population sizes of secondary consumers which it eats, which in turn may increase the population sizes of primary consumers that the secondary consumers eat, which may reduce the population sizes of primary producers that the primary consumers eat.

trophic interactions The consumer–resource relationships among species in a community.

trophic level A group of organisms united by obtaining their energy from the same part of the food web of a biological community.

trophoblast [Gk *trophes*: nourishment + *blastos*: sprout] In mammalian development, the outer group of cells of the blastocyst; will form the chorion and thus become part of the placenta, thereby helping to nourish the growing embryo. (Contrast with inner cell mass.)

tropic hormone A hormone produced by one endocrine gland that controls the secretion of hormones by another endocrine gland. Also called a tropin.

tropics The latitudes between the Tropic of Cancer and the Tropic of Capricorn (23.5° north and south, respectively), defined by the region of the planet in which the sun is directly overhead at least once during each annual cycle.

tropin *See* tropic hormone.

tropomyosin [troe poe my′ oh sin] A muscle protein that controls the interactions of actin and myosin necessary for muscle contraction.

troponin A muscle protein that controls the interactions of actin and myosin necessary for muscle contraction.

tube feet A unique feature of echinoderms; extensions of the water vascular system, which functions in gas exchange, locomotion, and feeding.

tubular fluid In a kidney, fluid flowing through the lumen of one of the tubular structures (e.g., a nephron) of which the kidney is composed.

tubulin A protein that polymerizes to form microtubules.

tumor necrosis factor A family of cytokines (growth factors) that causes cell death and is involved in inflammation.

tumor suppressor A gene that codes for a protein product that inhibits cell proliferation; inactive in cancer cells. (Contrast with oncogene.)

turgor pressure [L. *turgidus*: swollen] *See* pressure potential.

turnover In ecology, the change through time or over space in the composition of an ecological community.

Type I error The incorrect rejection of a true null hypothesis.

Type II error The incorrect acceptance of a false null hypothesis.

U

ubiquitin A small protein that is covalently linked to other cellular proteins identified for breakdown by the proteosome.

ultimate explanation The historical explanations of the processes that led to the evolution of an adaptation.

ultrafiltration A process of primary urine formation during which fluid moves out of the blood plasma into tubules in the kidney, driven by the force of blood pressure.

ultraviolet wavelengths Wavelengths in the electromagnetic spectrum that are too short to be seen by the human eye (i.e., shorter than 0.4 μm).

unidirectional ventilation Forced flow (of water or air) over or through a breathing organ in a single direction from the point of intake to the point of outflow. (Contrast with tidal ventilation.)

unipotent An undifferentiated cell that is capable of becoming only one type of mature cell. (Contrast with totipotent, multipotent, pluripotent.)

unsaturated fatty acid A fatty acid whose hydrocarbon chain contains one or more double bonds. (Contrast with saturated fatty acid.)

uracil (U) A pyrimidine base found in nucleotides of RNA.

ureotelic Pertaining to an organism in which the principal final product of the breakdown of nitrogen-containing compounds (primarily proteins) is urea. Mammals and most amphibians are ureotelic. (Contrast with ammonotelic, uricotelic.)

uricotelic Pertaining to an organism in which the principal final product of the breakdown of nitrogen-containing compounds (primarily proteins) is uric acid or other compounds closely related to uric acid. Insects, spiders, birds, and some other reptiles are uricotelic. (Contrast with ammonotelic, ureotelic.)

urine (you' rin) The fluid that the kidney produces and excretes from the body.

V

vacuole (vac' yew ole) Membrane-enclosed organelle in plant cells that can function for storage, water concentration for turgor, or hydrolysis of stored macromolecules.

van der Waals interactions Weak attractions between atoms resulting from the interaction of the electrons of one atom with the nucleus of another. This type of attraction is about one-fourth as strong as a hydrogen bond.

variable region The portion of an immunoglobulin molecule or T cell receptor that includes the antigen-binding site and is responsible for its specificity. (Contrast with constant region.)

vas deferens (plural: vasa deferentia) In a male mammal, the duct that transfers sperm from the testis and epididymis to the urethra for ejaculation.

vascular bundle In vascular plants, a strand of vascular tissue, including xylem and phloem as well as thick-walled fibers.

vascular cambium (kam' bee um) [L. *cambiare*: to exchange] In plants, a lateral meristem that gives rise to secondary xylem and phloem.

vascular endothelium In the walls of all blood vessels, the layer of cells—a simple epithelium—that immediately surrounds the blood.

vascular system All the blood vessels of the circulatory system. (*See also* cardiovascular system.)

vascular tissue system The transport system of a vascular plant, consisting primarily of xylem and phloem. (Contrast with dermal tissue system, ground tissue system.)

vasomotor control Related to changes in the inside (luminal) diameters of blood vessels mediated by contraction and relaxation of smooth muscles in the blood vessel walls. Changes in inside diameter have large effects on rate of blood flow through a vessel.

vasopressin *See* antidiuretic hormone.

vector (1) An agent, such as an insect, that carries a pathogen affecting another species. (2) A plasmid or virus that carries an inserted piece of DNA into a bacterium for cloning purposes in recombinant DNA technology.

vegetal hemisphere The yolky portion of some animal eggs. (Contrast with animal hemisphere.)

vegetative cells (1) Generally: non-reproductive cells. (2) The photosynthesizing cells of a cyanobacterial colony.

vegetative reproduction Asexual reproduction through the modification of stems, leaves, or roots.

vein [L. *vena*: channel] A blood vessel that returns blood to the heart. (Contrast with artery.)

ventilation Forced flow of air or water over or through structures used for breathing (external respiration) or over body surfaces used for external respiration.

ventral [L. *venter*: belly, womb] Toward or pertaining to the belly or lower side. (Contrast with dorsal.)

ventricle In reference to a heart, the principal pumping chamber. The walls of the ventricle are highly muscular, composed of cardiac muscle.

vernalization [L. *vernalis*: spring] Events occurring during a required chilling period, leading eventually to flowering.

vertebral column [L. *vertere*: to turn] The jointed, dorsal column that is the primary support structure of vertebrates.

vesicle Within the cytoplasm, a membrane-enclosed compartment that is associated with other organelles; the Golgi complex is one example.

vessel element A type of tracheary element with perforated end walls; found only in angiosperms. (Contrast with tracheid.)

villus (vil' lus) (plural: villi) [L. *villus*: shaggy hair or beard] A hairlike projection from a membrane; for example, the gut wall is often covered with villi that increase its surface area.

virus Any of a group of ultramicroscopic particles constructed of nucleic acid and protein (and, sometimes, lipid) that require living cells in order to reproduce. Viruses evolved multiple times from different cellular species.

visceral mass A centralized, internal portion of a mollusk's body that contains the heart as well as digestive, excretory, and reproductive organs.

vitamin [L. *vita*: life] A carbon compound that an organism cannot synthesize, but nevertheless requires in small quantities for normal growth and health.

voltage A measure of the difference in electrical charge between two points. Also called potential difference.

voltage-gated channel A type of gated channel protein, located in a cell membrane, that opens or closes in response to changes in the local membrane potential.

W

water channel protein *See* aquaporin.

water potential (psi, Ψ) In osmosis, the tendency for a system (a cell or solution) to take up water from pure water through a differentially permeable membrane. Water flows toward the system with a more negative water potential. (Contrast with solute potential, pressure potential.)

water vascular system In echinoderms, a network of water-filled canals that functions in gas exchange, locomotion, and feeding.

wavelength The distance between successive peaks of a wave train, such as electromagnetic radiation.

weather The atmospheric conditions of temperature, humidity, precipitation, and wind speed and direction experienced at a given place and time. (Contrast with climate.)

white blood cells Cells in the blood plasma that play defensive roles in the immune system. Also called leukocytes.

white matter In the nervous system, tissue that is rich in myelinated axons and therefore has a glistening white appearance.

wild type Geneticists' term for standard or reference type. Deviants from this standard, even if the deviants are found in the wild, are usually referred to as mutant. (Note that this terminology is not usually applied to human genes.)

wood Secondary xylem tissue.

X

xeric Referring to terrestrial animals, those that are able to live steadily in the open air and thus face the full drying power of the terrestrial environment.

xerophyte (zee' row fyte) [Gk. *xerox*: dry + *phyton*: plant] A plant adapted to an environment with limited water supply.

xylem (zy' lum) [Gk. *xylon*: wood] In vascular plants, the tissue that conducts water and minerals; xylem consists, in various plants, of tracheids, vessel elements, fibers, and other highly specialized cells.

Y

yolk sac In reptiles (including birds) and mammals, the extraembryonic membrane that forms from the endoderm and lateral plate mesoderm and surrounds the yolk. First source of blood and blood vessels. During development, the yolk sac digests the yolk, and its blood vessels carry products to the embryo.

Z

zeaxanthin A blue-light receptor involved in the opening of plant stomata.

zone of cell division The apical and primary meristems of a plant root; the source of all cells of the root's primary tissues.

zone of cell elongation The part of a plant root, generally above the zone of cell division, where cells are expanding (growing), primarily in the longitudinal direction.

zone of cell maturation The part of a plant root, generally above the zone of cell elongation, where cells are differentiating.

zygospore Multinucleate, diploid cell that is a resting stage in the life cycle of zygospore fungi.

zygote (zye′ gote) [Gk. *zygotos*: yoked] The fertilized egg. The cell created by the union of two gametes, in which the gamete nuclei are fused. The earliest stage of the diploid generation.

Illustration Credits

Table of Contents Page xvi: © Pat O'Hara/Corbis. Page xvii: © David M. Phillips/Science Source. Page xix: David McIntyre. Page xx: © Greg Dale/National Geographic Society/Corbis. Page xxii: © PjrStudio/Alamy. Page xxiii: © Eye of Science/Science Source. Pages xxiv, xxvi: David McIntyre. Page xxvii: © Konrad Wothe/Minden Pictures. Page xxx: © Robert Pickett/Visuals Unlimited, Inc. Page xxxii: © Andy Rouse/Corbis. Page xxxiii: © Wild Wonders of Europe/O. Haarberg/Naturepl.com.

Part 1 Opener (page 17): © Biophoto Associates/Science Source.

Part 2 Opener (page 128): © corfield/Alamy.

Part 3 Opener (page 297): © George Grall/National Geographic Society/Corbis.

Part 4 Opener (page 377): © David Fleetham/Alamy.

Part 5 Opener (page 520): © Titti Soldati/Alamy.

Part 6 Opener (page 604): © blickwinkel/Alamy.

Part 7 Opener (page 844): © Piotr Naskrecki/Minden Pictures/Corbis.

Chapter 1 *Opener*: © Lee Grismer. 1.2: © Eye of Science/Science Source. 1.3A: © Armands Pharyos/Alamy. 1.3B: © Jane Gould/Alamy. 1.5 *Biosphere*: NASA images by Reto Stöckli, based on data from NASA and NOAA. 1.10A: © Life on white/Alamy. 1.10B: © Stockbroker xtra/AGE Fotostock. 1.10C: © Michael Doolittle/Alamy. 1.10D: © Stephen Dalton/Minden Pictures/Corbis. 1.12: From T. Hayes et al., 2003. *Environ. Health Perspect.* 111: 568.

Chapter 2 *Opener*: Courtesy of NASA/JPL. 2.10C *left*: © Biophoto Associates/Science Source. 2.10C *center*: © Dennis Kunkel Microscopy, Inc. 2.10C *right*: © Barry King/Biological Photo Service. Page 23: © EpicStockMedia/istock. Page 24: © Kuttelvaserova Stuchelova/Shutterstock. Page 30: David McIntyre.

Chapter 3 *Opener*: © blickwinkel/Alamy. 3.9: Data from PDB 3LPO. L. Sim et al., 2010. *J. Biol. Chem.* 285: 17763. 3.15: Data from PDB 1IG8 (P. R. Kuser et al., 2000. *J. Biol. Chem.* 275: 20814) and 1BDG (A. M. Mulichak et al., 1998 *Nat. Struct. Biol.* 5: 555). 3.24: Data from PDB 1PTH. P. J. Loll et al., 1995. *Nat. Struct. Biol.* 2: 637.

Chapter 4 *Opener*: Courtesy of Tom Deerinck and Mark Ellisman of the National Center for Microscopy and Imaging Research at the University of California at San Diego. 4.1: After N. Campbell, 1990. *Biology*, 2nd Ed., Benjamin Cummings. 4.1 *Protein*: Data from PDB 1IVM. T. Obita, T. Ueda, & T. Imoto, 2003. *Cell. Mol. Life Sci.* 60: 176. 4.1 *T4*: © Dept. of Microbiology, Biozentrum/SPL/Science Source. 4.1 *Organelle*: © E.H. Newcomb & W.P. Wergin/Biological Photo Service. 4.1 *Bacterium*: © Jim Biddle/Centers for Disease Control. 4.1 *Plant cells*: © Carolina K. Smith/Shutterstock. 4.1 *Butterfly egg, Butterfly*: David McIntyre. 4.1 *Human*: © East/Shutterstock. 4.1 *Redwood*: © urosr/Shutterstock. 4.3A *Light microscope*: © Olaf Doering/Alamy. 4.3A *Light micrograph*: © Carolina K. Smith/Shutterstock. 4.3A *Electron microscope*: © Enzo/AGE Fotostock. 4.3A *Electron micrograph*: © SERCOMI/AGE Fotostock. 4.4 *ER*: From Y. Mizutani et al., 2001. *J. Cell Sci.* 114: 3727. 4.4 *Mitochondria*: From L. Argaud et al., 2004. *Cardiovasc. Res.* 61: 115. 4.4 *Nuclei*: Reproduced with permission from Y. Mizutani et al., 2001. *J. Cell Sci.* 114: 3727. 4.5: © J. J. Cardamone Jr. & B. K. Pugashetti/Biological Photo Service. 4.6A: © Dennis Kunkel Microscopy, Inc. 4.6B: Courtesy of David DeRosier, Brandeis U. 4.7 *Mitochondrion*: © K. Porter, D. Fawcett/Visuals Unlimited, Inc. 4.7 *Cytoskeleton*: © Don Fawcett, John Heuser/Science Source. 4.7 *Nucleolus*: © Richard Rodewald/Biological Photo Service. 4.7 *Peroxisome*: © E. H. Newcomb & S. E. Frederick/Biological Photo Service. 4.7 *Cell wall*: © Biophoto Associates/Science Source. 4.7 *Ribosome*: From M. Boublik et al., 1990. *The Ribosome*, p. 177. Courtesy of American Society for Microbiology. 4.7 *Centrioles*: © Conly L. Rieder/Biological Photo Service. 4.7 *Plasma membrane*: Courtesy of J. David Robertson, Duke U. Medical Center. 4.7 *Rough ER*: © Don Fawcett/Science Source. 4.7 *Smooth ER*: © Don Fawcett, D. Friend/Science Source. 4.7 *Chloroplast*: © E.H. Newcomb & W.P. Wergin/Biological Photo Service. 4.7 *Golgi apparatus*: Courtesy of L. Andrew Staehelin, U. Colorado. 4.9: © Sanders/Biological Photo Service. 4.10: Courtesy of Vic Small, Austrian Academy of Sciences, Salzburg, Austria. 4.11A *upper*: © Dennis Kunkel Microscopy, Inc. 4.11A *lower*, 4.11B: © W. L. Dentler/Biological Photo Service. 4.13: From N. Pollock et al., 1999. *J. Cell Biol.* 147: 493. Courtesy of R. D. Vale. 4.14: © micro_photo/istock. 4.15: © Biophoto Associates/Science Source. 4.16 *left*: © Dr. Donald Fawcett & D. Friend/Visuals Unlimited, Inc. 4.16 *upper right*: From J. A. Buckwalter & L. Rosenberg, 1983. *Coll. Rel. Res.* 3: 489. Courtesy of L. Rosenberg. 4.16 *lower right*: © Biophoto Associates/Science Source. 4.18 *Tight junction*: © Barry F. King/Biological Photo Service. 4.18 *Desmosome*: Courtesy of Darcy E. Kelly, U. Washington. 4.18 *Gap junction*: Courtesy of C. Peracchia. 4.19: Courtesy of Janet Iwasa, Szostak group, MGH/Harvard. Page 65: From Y.-L. Shih, T. Le, and L. Rothfield. 2003. *PNAS USA* 100: 7865. Courtesy of L. Rothfield. Page 72 *Daffodils*: © SilentWolf/istock. Page 72 *Chromoplasts*: From Gunning, B. E. S. *Plant Cell Biology on DVD: Information for students and a resource for teachers*. Springer, 2009 (www.springer.com/life+sciences/plant+sciences/book/978-3-642-03690-3). Page 72 *Potatoes*: © John Shepherd/istock. Page 72 *Leucoplasts*: Courtesy of R. R. Dute. Page 72 *Vacuole*: © E. H. Newcomb & S. E. Frederick/Biological Photo Service.

Chapter 5 *Opener*: © blickwinkel/Alamy. 5.3A *left*: © Stanley Flegler/Visuals Unlimited, Inc. 5.3A *center, right*: © David M. Phillips/Science Source. 5.3B: © Ed Reschke/Getty Images. 5.5: From G. M. Preston et al., 1992. *Science* 256: 385. 5.9: From M. M. Perry, 1979. *J. Cell Sci.* 39: 26. 5.19: Data from PDB 3EML. V. P. Jaakola et al., 2008. *Science* 322: 1211.

Chapter 6 *Opener*: © Thanassis Stavrakis/AP/Corbis. 6.17: © Martin Shields/Alamy. 6.24: David McIntyre.

Chapter 7 *Opener*: © Dr. Gopal Murti/Science Source. 7.1A: © Dr. Stanley Flegler/Visuals Unlimited, Inc. 7.1B: David McIntyre. 7.1C: © Stefan Sollfors/Alamy. 7.2: © Egmont Strigl/Alamy. 7.3 *left*: © Dr. Jeremy Burgess/SPL/Science Source. 7.3 *center*: David McIntyre. 7.3 *right*: Courtesy of Andrew D. Sinauer. 7.5 *Chromosome*: © Dr. Klaus Boller/Science Source. 7.5 *Nucleus*: © Richard Rodewald/Biological Photo Service. 7.6: © Conly L. Rieder/Biological Photo Service. 7.7A: © Dr. Gopal Murti/Science Source. 7.7B: © B. A. Palevitz, E. H. Newcomb/Biological Photo Service. 7.12: © C. A. Hasenkampf/Biological Photo Service. 7.15A: © Dr. Gopal Murti/Visuals Unlimited, Inc. Page 134 *upper*: © Power and Syred/Science Source. Page 134 *lower*: Courtesy of Dr. Thomas Ried and Dr. Evelin Schröck, NIH.

Chapter 8 *Opener*: Courtesy of Chesdovi/Wikipedia. 8.9 *Dark*: © Marina Golskaya/istock. 8.9 *Chinchilla*: © purelook/istock. 8.9 *Point*: © Carolyn A. McKeone/Science Source. 8.9 *Albino*: © ZTS/Shutterstock. 8.12 *Black*: © Silense/istock. 8.12 *Chocolate*: © Erik Lam/istock. 8.12 *Yellow*: © Gelpi

JM/Shutterstock. 8.19A: © David Scharf/Getty Images. Page 160: Courtesy of the Plant and Soil Sciences eLibrary (plantandsoil.unl.edu); used with permission from the Institute of Agriculture and Natural Resources at the University of Nebraska. Page 161 *Cat and rabbit*: © Mark Taylor/Naturepl.com. Page 162: © Peter Morenus/U. of Connecticut.

Chapter 9 *Opener*: © Universal Studios/AF archive/Alamy. 9.2: © Lee D. Simon/Science Source. 9.4 *X-ray crystallograph*: Courtesy of Prof. M. H. F. Wilkins, Dept. of Biophysics, King's College, U. London. 9.4B: © NLM/Science Source/Getty Images. 9.5A: © A. Barrington Brown/Science Source. 9.5B: Data from S. Arnott & D. W. Hukins, 1972. *Biochem. Biophys. Res. Commun.* 47(6): 1504. 9.10A: Data from PDB 1SKW. Y. Li et al., 2001. *Nat. Struct. Mol. Biol.* 11: 784. 9.13B: © Peter Lansdorp/Visuals Unlimited, Inc. 9.13C: David McIntyre. 9.20: Courtesy of Christoph P. E. Zollikofer, Marcia S. Ponce de León, and Elisabeth Daynès. Page 173: © Carolina Biological Supply, Co./Visuals Unlimited, Inc. Page 188: © Stanley Flegler/Visuals Unlimited, Inc. Page 190: Data from PDB 1AGU. I. S. Zegar et al., 1996. *Biochemistry* 35: 6212.

Chapter 10 *Opener*: © Dennis Kunkel Microscopy, Inc. 10.4: Data from PDB 1MSW. Y. W. Yin & T. A. Steitz, 2002. *Science* 298: 1387. 10.8: From D. C. Tiemeier et al., 1978. *Cell* 14: 237. 10.13: Data from PDB 1EHZ. H. Shi & P. B. Moore, 2000. *RNA* 6: 1091. 10.14: Data from PDB 1GIX and 1G1Y. M. M. Yusupov et al., 2001. *Science* 292: 883. 10.18B: Courtesy of J. E. Edström and *EMBO J.* 10.22: Data from PDB 1HNW. D. E. Brodersen et al. 2000. *Cell* 103: 1143.

Chapter 11 *Opener*: © Lenartowski/AGE Fotostock. 11.9: Data from PDB 1A66. P. Zhou et al., 1998. *Cell* 92: 687. 11.20: Data from PDB 1T2K. D. Panne et al., 2004. *EMBO J.* 23: 4384. Page 220: © Lee D. Simon/Science Source. Page 228 *Cat*: © Ocean/Corbis. Page 228 *Barr*: Courtesy of Irina Solovei, U. Munich (LMU), Germany.

Chapter 12 *Opener*: © Don Mason/Corbis. 12.4: From P. H. O'Farrell, 1975. High resolution two-dimensional electrophoresis of proteins. *J. Biol. Chem.* 250: 4007. Courtesy of Patrick H. O'Farrell. 12.15B: Bettmann/Corbis. 12.16A *left*: © Bruce Stotesbury/PostMedia News/Zuma Press. 12.16A *right*: © kostudio/Shutterstock. 12.16B: © Yann Arthus-Bertrand/Corbis. Page 244 *Yeast*: © Mediscan/Corbis. Page 244 *Nematode*: © Sinclair Stammers/Science Source. Page 245 *Drosophila*: David McIntyre. Page 245 *Arabidopsis*: David McIntyre (from seed provided by Tobias Baskin).

Chapter 13 *Opener*: © Ted Spiegel/Corbis. 13.2: © Explorer/Science Source. 13.5: © Dr. Jack Bostrack/Visuals Unlimited, Inc. 13.6: © Stephen Sewell/istock. 13.12: David McIntyre. 13.15: Courtesy of the Golden Rice Humanitarian Board, www.goldenrice.org. 13.16: Courtesy of Eduardo Blumwald. 13.17: © Peter Jordan/Alamy. Page 254: Data from PDB 1ERI. Y. C. Kim et al., 1990. *Science* 249: 1307. Page 269: © Dr. George Chapman/Visuals Unlimited, Inc.

Chapter 14 *Opener, left*: © RGB Ventures LLC dba SuperStock/Alamy. *Opener, right*: © David Scharf/Visuals Unlimited, Inc. 14.4: © Roddy Field, the Roslin Institute. 14.7B: © Cold Spring Harbor Press; courtesy of Dustin Updike and Susan Strome. 14.8: From J. E. Sulston & H. R. Horvitz, 1977. *Dev. Bio.* 56: 100. 14.12A: David McIntyre. 14.13 *upper*: Courtesy of C. Rushlow and M. Levine. 14.13 *center*: Courtesy of T. Karr. 14.13 *lower*: Courtesy of S. Carroll and S. Paddock. 14.14A: © Eye of Science/Science Source. 14.14B: © Science VU/Dr. F. Rudolph Turner/Visuals Unlimited, Inc. 14.16: David McIntyre. 14.17: © Bone Clones, www.boneclones.com. 14.18: Courtesy of J. Hurle and E. Laufer. 14.19 *Insect*: © flavijus piliponis/istock. 14.19 *Centipede*: © jeridu/istock. 14.21: Courtesy of Mike Shapiro and David Kingsley. 14.22: From Y. Yamamoto et al., 2004. *Nature* 431: 844. © Macmillan Publishers Ltd. Courtesy of Karl O. Stetter. Page 286: Courtesy of D. Daily and W. Sullivan.

Chapter 15 *Opener*: Courtesy of the National Museum of Health and Medicine, Armed Forces Institute of Pathology, Washington, D.C. 15.6 *Wild type*: © Luis César Tejo/Shutterstock. 15.6 *Fancy pigeons*: Photos by David McIntyre; pigeons bred by Matthew Sodoski, Cynthia Archambault, and Frank J. D'Alessandro. 15.9: © Richard Du Toit/Minden Pictures/Corbis. 15.15: Courtesy of David Hillis. 15.19A *Langur*: © Jean-Pierre Zwaenepoel/Naturepl.com. 15.19A *Longhorn*: Courtesy of David Hillis. 15.19B: © Marshall Bruce/istock. 15.25: © Pasieka/Science Source. Page 299 *Darwin*: © The Art Gallery Collection/Alamy. Page 299 *Beagle*: Painting by Ronald Dean, reproduced by permission of the artist and Richard Johnson, Esquire.

Chapter 16 *Opener*: Courtesy of Misha Matz. 16.4 *Tunicate larva*: Courtesy of William Jeffery. 16.4 *Tunicate adult*: © WaterFrame/Alamy. 16.4 *Frog larva*: © Michael & Patricia Fogden/Minden Pictures. 16.4 *Frog adult*: © George Grall/National Geographic Society/Corbis. 16.8 *Xiphophorus*: David McIntyre, courtesy of Exotic Fish and Pet World, Southampton, MA. 16.8 *Priapella*: Courtesy of Dbr/Wikipedia. 16.9 *L. bicolor*: Courtesy of Steve Matson. 16.9 *L. liniflorus*: Courtesy of Anthony Valois/National Park Service. 16.13A: © Frans Lanting Studio/Alamy. 16.13B: © blickwinkel/Alamy. 16.13C: © Clem Haagner/Bruce Coleman, Inc./Alamy. 16.14: Courtesy of Misha Matz.

Chapter 17 *Opener*: Photos by David McIntyre (courtesy of Exotic Fish and Pet World, Southampton, MA) and Nantawat Chotsuwan/Shutterstock. 17.1A *left*: © Donald Enright/Alamy. 17.1A *right*: © James Hager/Robert Harding World Imagery/Corbis. 17.1B: © Papilio/Alamy. 17.2A: David McIntyre. 17.2B: © Jack Glisson/Alamy. 17.4: © Barry Mansell/Naturepl.com. 17.8: © Tim Gainey/Alamy. 17.10 *G. olivacea*: © Phil A. Dotson/Science Source. 17.10 *G. carolinensis*: © Suzanne L. Collins/Science Source. 17.11A: © Gustav Verderber/Visuals Unlimited, Inc. 17.11B: © Nathan Derieg. 17.11C: David McIntyre. 17.11D: © Daniel L. Geiger/SNAP/Alamy. 17.12: Courtesy of Donald A. Levin. 17.13 *upper*: © Gerhard Schulz/AGE Fotostock. 17.13 *lower*: © Thomas Marent/Visuals Unlimited, Inc. 17.14: Courtesy of William R. Rice.

Chapter 18 *Opener*: © Graham Cripps/NHMPL. *Opener inset*: © Natasha Litova/istock. 18.2B: © Ric Ergenbright/Corbis. 18.4: Courtesy of Dave Harlow, U.S. Geological Survey. 18.5: © Martin Bond/SPL/Science Source. 18.6A: David McIntyre. 18.6B: © Walter Geiersperger/Corbis. 18.6C: © Mint Images/Frans Lan/AGE Fotostock. 18.9: © PjrStudio/Alamy. 18.11A: From S. Xiao et al., 1998. *Nature* 391: 553. © Macmillan Publishers Ltd. 18.11B: Courtesy of Martin Smith. 18.11C: © Albert J. Copley/Visuals Unlimited, Inc. 18.12 *Cambrian*: © John Sibbick/NHMPL. 18.12 *Marella*: Courtesy of the Amherst College Museum of Natural History, The Trustees of Amherst College. 18.12 *Ottoia*: © Alan Sirulnikoff/Science Source. 18.12 *Anomalocaris*: © Kevin Schafer/Alamy. 18.12 *Devonian*: © The Field Museum, #GEO86500_125d. 18.12 *Archaeopteris*: © John Cancalosi/Getty Images. 18.12 *Eusthenopteron*: © Gerald & Buff Corsi/Visuals Unlimited, Inc. 18.12 *Permian*: © John Sibbick/The Natural History Museum, London. 18.12 *Estemmenosuchus*: Courtesy of Gondwana Studios. 18.12 *Equisetum*: © John Cancalosi/Alamy. 18.12 *Jurassic*: © Sergey Krasovskiy/Stocktrek Images/Corbis. 18.12 *Ginkgo*: © Biophoto Associates/Science Source. 18.12 *Europasaurus*: © Holger Hollemann/dpa/Corbis. 18.12 *Cretaceous*: © Anness Publishing/NHMPL. 18.12 *Chasmosaurus*: © Oleksiy Maksymenko/Alamy. 18.12 *Sapindopsis*: © Barbara J. Miller/Biological Photo Service. 18.12 *Tertiary*: © Publiphoto/Science Source. 18.12 *Hyracotherium*: Courtesy of the Amherst College Museum of Natural History, The Trustees of Amherst College. 18.12 *Plesiadapis*: © The Natural History Museum, London. 18.13: Courtesy of Conrad C. Labandeira, Department of Paleobiology, National Museum of Natural History, Smithsonian Institution.

Chapter 19 *Opener*: "Milky seas" data prepared by Steven Miller of the Naval Research Laboratory, Monterey, CA. Background data (Earth at Night image) courtesy Marc Imhoff (NASA/GSFC) and Christopher Elvidge (NOAA/NGDC);

image by Craig Mayhew (NASA/GSFC) and Robert Simmon (NASA/GSFC). 19.2A: © Michael Abbey/Visuals Unlimited, Inc. 19.2B: Courtesy of the Centers for Disease Control. 19.3A: Courtesy of Rob Weyant/CDC. 19.3B, C: Courtesy of Janice Haney Carr/CDC. 19.5: © Dr. George Chapman/Visuals Unlimited, Inc. 19.6: © Eye of Science/Science Source. 19.7: © Don W. Fawcett/Science Source. 19.8: © BSIP SA/Alamy. 19.9A: © Paul W. Johnson/Biological Photo Service. 19.9B: © Dr. Terry Beveridge/Visuals Unlimited, Inc. 19.9C: © RJH_CATALOG/Alamy. 19.10A: © J. A. Breznak & H. S. Pankratz/Biological Photo Service. 19.10B: © Paul W. Johnson/Biological Photo Service. 19.11: Courtesy of Randall C. Cutlip. 19.12: © CNRI/Science Source. 19.13: © Dean A. Glawe/Biological Photo Service. 19.14: From K. Kashefi & D. R. Lovley, 2003. *Science* 301: 934. Courtesy of Kazem Kashefi. 19.16: © eye35.pix/Alamy. 19.17: © Nancy Nehring/istock. 19.18: From H. Huber et al., 2002. *Nature* 417: 63. © Macmillan Publishers Ltd. Courtesy of Karl O. Stetter. 19.19B: © Science Photo Library RF/Photolibrary.com. 19.21: © Eye of Science/Science Source. 19.23A: © Chris Bjornberg/Science Source. 19.23B: © Dr. Linda M. Stannard, University of Cape Town/Science Source. 19.23C: © Dr. Hans Gelderblom/Visuals Unlimited, Inc. 19.23D: © BSIP SA/Alamy. 19.23E: © Dr. Harold Fisher/Visuals Unlimited, Inc. 19.23F: © Dr. Raoult/Science Source. 19.24: © Nigel Cattlin/Alamy.

Chapter 20 *Opener*: © Don Paulson/PureStock/AGE Fotostock. 20.4: © Biophoto Associates/Science Source. 20.5A: © Aaron J. Bell/Science Source. 20.5B: © Aaron Bell/Visuals Unlimited, Inc. 20.5C: © Steve Gschmeissner/Science Source. 20.7: © Scenics & Science/Alamy. 20.8A: © John N. A. Lott/Biological Photo Service. 20.8B: © Carl W. May/Biological Photo Service. 20.9: David McIntyre. 20.10: © Robert Brons/Biological Photo Service. 20.11A: © Robert Brons/Biological Photo Service. 20.11B: © Manfred Kage/Science Source. 20.12A: Courtesy of Dr. Stan Erlandsen/CDC. 20.12B: © Eye of Science/Science Source. 20.14: © Wim van Egmond/Visuals Unlimited, Inc. 20.15: © Wim van Egmond/Visuals Unlimited, Inc. 20.16A: © Matt Meadows/Getty Images. 20.16B: © FLPA/Alamy. 20.17: Courtesy of R. Blanton and M. Grimson. 20.19: Courtesy of M.A. Coffroth and Cindy Lewis, University at Buffalo. 20.20: © Per-Andre Hoffmann/Alamy.

Chapter 21 *Opener*: © Mitsuhiko Imamori/Minden. 21.2A: © Dr. Peter Siver/Visuals Unlimited, Inc. 21.2B: © Robert & Jean Pollock/Visuals Unlimited, Inc. 21.3A: © Wolfgang Pölzer/Alamy. 21.3B: © Larry Mellichamp/Visuals Unlimited, Inc. 21.3C: © Bob Gibbons/Alamy. 21.5A: © Wildlife GmbH/Alamy. 21.5B: David McIntyre. 21.5C: © Daniel Vega/AGE Fotostock. 21.6 *top*: David McIntyre. 21.6 *bottom*: © Ed Reschke/Getty Images. 21.7: © Publiphoto/Science Source. 21.8A: David McIntyre. 21.8B: © Jolanta Dbrowska/AGE Fotostock. 21.8C: © Andreas Altenburger/Alamy. 21.8D: © India View/Alamy. 21.9 *inset*: © Chris Mattison/Alamy. 21.10A: Courtesy of the Biology Department Greenhouses, U. Massachusetts, Amherst. 21.10B: David McIntyre. 21.13: © Wildlife GmbH/Alamy. 21.14A: © Susumu Nishinaga/Science Source. 21.16A: Courtesy of Andrew D. Sinauer. 21.16B: David McIntyre. 21.16C: © Juan Carlos Muñoz/AGE Fotostock. 21.16D: © Gunter Marx/Alamy. 21.17: David McIntyre. 21.19A: © Barbara J. Miller/Biological Photo Service. 21.19B: © John N. A. Lott/Biological Photo Service. 21.19C: © Nigel Cattlin/Alamy. 21.20A: © Zena Elea/Alamy. 21.20B: © blickwinkel/Alamy. 21.24: © Bjorn Rorslett/Science Source. 21.26A: Courtesy of Keith Weller/USDA ARS. 21.26B: © Derek Middleton/ FLPA/Minden Pictures/Corbis. 21.26C: © Chris Lishman/Alamy. 21.26D: © Stefan Sollfors/Alamy. 21.26E: © Andrey Khrobostov/Alamy. 21.26F: Courtesy of Keith Weller/USDA ARS. 21.28A: Photo by David McIntyre, courtesy of the U. Massachusetts Biology Department Greenhouses. 21.28B: © Juergen Ritterbach/Alamy. 21.28C: © Rob & Ann Simpson/Visuals Unlimited, Inc. 21.28D: © Anthony Green/Alamy. 21.28E: © Florapix/Alamy. 21.29A: © Craig Roberts/Alamy. 21.29B: © Alfred Owczarzak/Biological Photo Service. 21.29C: © Wildlife GmbH/Alamy. 21.29D: © BSIP SA/Alamy. 21.30A: © John N. A. Lott/Biological Photo Service. 21.30B: © Science Photo Library/Alamy. 21.30C: © A & J Visage/Alamy. 21.30D: © DK Limited/Corbis.

Chapter 22 *Opener*: © Biophoto Associates/Science Source. 22.2: © Steve Gschmeissner/Science Photo Library/Corbis. 22.3A: © Dr. Jeremy Burgess/Science Source. 22.4: © Chris Howes/Wild Places Photography/Alamy. 22.5A: © Biophoto Associates/Science Source. 22.7: © N. Allin & G. L. Barron/Biological Photo Service. 22.8A: © Wolstenholme Images/Alamy. 22.8B: David McIntyre. 22.10A: © R. L. Peterson/Biological Photo Service. 22.10B: © M. F. Brown/Biological Photo Service. 22.13: © Eye of Science/Science Source. 22.14A: © Carolina Biological Supply, Co/Visuals Unlimited, Inc. 22.14B: © Dr. Jeremy Burgess/Science Source. 22.15: Photo by David McIntyre; manure courtesy of Myrtle Jackson. 22.16A: © Dr. Cecil H. Fox/Science Source. 22.16B: © Biophoto Associates/Science Source. 22.17A: © Thomas Marent/Minden Pictures/Corbis. 22.17B: © sgoodwin/AGE Fotostock. 22.18: © Dennis Kunkel Microscopy, Inc. 22.19A: © blickwinkel/Alamy. 22.19B: © George L. Barron/Biological Photo Service. 22.20: Courtesy of David Hillis. 22.21: © Biophoto Associates/Science Source.

Chapter 23 *Opener*: © Philippe Psaila/Science Source. 23.3: From Alegado et al. 2012, eLife 1:e00013. 23.4A: © Ed Robinson/Getty Images. 23.4B: © Steve Gschmeissner/Science Source. 23.4C: © Konrad Wothe/Minden Pictures/Corbis. 23.5A: © Charles R. Wyttenbach/Biological Photo Service. 23.5B: © Joseph Ruggeri/Alamy. 23.5C: © Stockphoto4u/istock. 23.6A: © Peter Szekely/Alamy. 23.6B: David McIntyre. 23.6C: © Robert Brons/Biological Photo Service. 23.7B: © Mint Images/Frans Lanting/AGE Fotostock. 23.8A: Courtesy of Wim van Egmond. 23.9: Adapted from F. M. Bayerand & H. B. Owre, 1968. *The Free-Living Lower Invertebrates*, Macmillan Publishing Co. 23.9 *inset*: © Larry Jon Friesen. 23.10A: © H. Wes Pratt/Biological Photo Service. 23.10B, C: © WaterFrame/Alamy. 23.11A: © blickwinkel/Alamy. 23.11B: © WaterFrame/Alamy. 23.12B: © Mark Moffett/Minden Pictures. 23.13B: © Roland Birke/Getty Images. 23.14B: © Larry Jon Friesen. 23.15: © David Wrobel/Visuals Unlimited, Inc. 23.16A: © Roger Klocek/Visuals Unlimited, Inc. 23.18A: © WaterFrame/Alamy. 23.18B: Courtesy of Cindy Lee Van Dover. 23.18C: © Larry Jon Friesen. 23.19B: © Martin Almqvist/Alamy. 23.19C: © guichaoua/Alamy. 23.19D: © Marevision/AGE Fotostock. 23.19E: © David Fleetham/Alamy. 23.20A: © Photoshot Holdings Ltd/Alamy. 23.20B: © Larry Jon Friesen. 23.21A: From D. C. García-Bellido & D. H. Collins, 2004. *Nature* 429: 40. Courtesy of Diego García-Bellido Capdevila. 23.21B: © Nature's Images/Science Source. 23.22A: Courtesy of Dr. Mae Melvin/CDC. 23.22B: © Carolina Biological Supply, Co/Visuals Unlimited, Inc. 23.23: © Pascal Goetgheluck/Science Source. 23.24A: Courtesy of Jen Grenier and Sean Carroll, U. Wisconsin. 23.24B: © David Scharf/Visuals Unlimited, Inc. 23.24C: Courtesy of Reinhardt Møbjerg Kristensen. 23.25A: © George Grall/National Geographic Society/Corbis. 23.25B: © Eye of Science/Science Source. 23.26A: © cbimages/Alamy. 23.26B: © Barbara J. Miller/Biological Photo Service. 23.27A: © Peter J. Bryant/Biological Photo Service. 23.27B: © Oliver Gerhard/Alamy. 23.27C: © Nigel Cattlin/Alamy. 23.27D: SEM by Eric Erbe; colorization by Chris Pooley/USDA ARS. 23.28A: © Rod Williams/Naturepl.com. 23.28B: © National Geographic Image Collection/Alamy. 23.29A: © F1online digitale Bildagentur GmbH/Alamy. 23.29B: © Larry Jon Friesen. 23.29C: © Robert Brons/Biological Photo Service. 23.29D: © Solvin Zankl/Naturepl.com. 23.29E: © Charles R. Wyttenbach/Biological Photo Service. 23.31A: © Jonathan Ashton/Alamy. 23.31B: © imagebroker/Alamy. 23.31C: © Jupe/Alamy. 23.31D: © Larry Jon Friesen. 23.31E: © Juniors Bildarchiv GmbH/Alamy. 23.31F: © blickwinkel/Alamy. 23.31G: © Peter J. Bryant/Biological Photo Service. 23.31H: © Larry Jon Friesen. 23.33A: © imagebroker/Alamy. 23.33B: © SERDAR/Alamy. 23.33C: © Bruce Coleman, Inc./Alamy. 23.33D: © Fred Winner/Science Source. 23.33E: © WaterFrame/Alamy. 23.34A: © C. R. Wyttenbach/Biological Photo Service.

23.35A: © Stan Elems/Visuals Unlimited, Inc. 23.35B: David McIntyre. 23.36: © Marevision/AGE Fotostock. 23.39A: © Ken Lucas/Biological Photo Service. 23.39B *left*: © Natural Visions/Alamy. 23.39B *right*: © anne de Haas/istock. 23.41A: © Wayne Lynch/AGE Fotostock. 23.41B: © blickwinkel/Alamy. 23.41C: © RGB Ventures LLC dba SuperStock/Alamy. 23.42: © Larry Jon Friesen. 23.43A: © Hoberman Collection/Corbis. 23.43B: © Tom McHugh/Science Source. 23.46A: © Dante Fenolio/Science Source. 23.46B: © Bill Coster/Alamy. 23.46C: © blickwinkel/Alamy. 23.46D: Courtesy of David Hillis. 23.49A: © Bruce Coleman, Inc./Alamy. 23.49B, C: © Larry Jon Friesen. 23.49D: © Gordon Chambers/Alamy. 23.50A: © blickwinkel/Alamy. 23.50B: Courtesy of Andrew D. Sinauer. 23.51A: From X. Xu et al., 2003. *Nature* 421: 335. © Macmillan Publishers Ltd. 23.51B: © James L. Amos/Science Source. 23.52A: © georgesanker.com/Alamy. 23.52B: © Rick & Nora Bowers/Alamy. 23.52C: © All Canada Photos/Alamy. 23.53A: © Dave Watts/Alamy. 23.53B: © Rick & Nora Bowers/Alamy. 23.53C: © Design Pics Inc/Photolibrary.com. 23.53D: © Rick & Nora Bowers/Alamy. 23.53E: © Danita Delimont/Alamy. 23.53F: © Mike Hill/Alamy. 23.56A: © Cyril Ruoso/Minden Pictures. 23.56B: Courtesy of Andrew D. Sinauer. 23.56C: © Arco Images GmbH/Alamy. 23.56D: David McIntyre. Page 472: Courtesy of J. B. Morrill.

Chapter 24 *Opener*: © paolo negri/Alamy. 24.7 *upper*: David McIntyre. 24.7 *lower*: © James Solliday/Biological Photo Service. 24.8B: © Larry Jon Friesen. 24.8C: © John N. A. Lott/Biological Photo Service. 24.9A: David McIntyre. 24.9B: © John N. A. Lott/Biological Photo Service. 24.10: David McIntyre. 24.11A *left*: David McIntyre. 24.11A *right*: © Andrew Syred/Science Source. 24.11B *left*: © RGB Ventures LLC dba SuperStock/Alamy. 24.11B *right*: © Steve Gschmeissner/Science Source. 24.13A: © Kim Karpeles/Alamy. 24.13B: © dk/Alamy. 24.13C: © Icon Digital Featurepix/Alamy. 24.13D: © Purple Marbles/Alamy. 24.13E: © Robert & Jean Pollock/Visuals Unlimited, Inc. 24.13F: © Rafael Campillo/AGE Fotostock. 24.15: © Phil Gates/Biological Photo Service. 24.17: Courtesy of Scott Bauer/USDA. Page 525 *Parenchyma*: David McIntyre. Page 525 *Collenchyma*: © Phil Gates/Biological Photo Service. Page 525 *Sclerenchyma*: © Biophoto Associates/Science Source. Page 525 *Sclereids*: © Dr. Dennis Drenner/Visuals Unlimited, Inc. Page 526 *Tracheids*: © Dr. John D. Cunningham/Visuals Unlimited, Inc. Page 526 *Vessel elements*: © J. Robert Waaland/Biological Photo Service. Page 526 *Sieve tube elements*: David McIntyre.

Chapter 25 *Opener*: Courtesy of the United Nations Environment Programme, www.unep.org. 25.1, 25.3: David McIntyre. 25.7A: © blickwinkel/Alamy. 25.7B: Courtesy of USDA APHIS PPQ Archive, USDA APHIS PPQ, Bugwood.org. 25.9: © Nigel Cattlin/Alamy. 25.13A: © Susumu Nishinaga/Science Source. 25.15: Courtesy of Lynn Betts/USDA NRCS. Page 552: © M. H. Zimmermann.

Chapter 26 *Opener*: © Micheline Pelletier/Sygma/Corbis. 26.2: From J. M. Alonso and J. R. Ecker, 2006. *Nature Reviews Genetics* 7: 524. 26.3A: Courtesy of J. A. D. Zeevaart, Michigan State U. 26.3B: From W. M. Gray, 2004. *PLoS Biol* 2(9): e311. 26.11: David McIntyre. 26.13: Courtesy of Drs. Matsuoka and Ashikari. Page 560: © Sylvan Wittwer/Visuals Unlimited, Inc. Page 562: © Robert J. Erwin/Science Source. Page 565 *Tomatoes*: Courtesy of Adel A. Kader. Page 565 *Seedling*: © Nigel Cattlin/Alamy. Page 566 *Cytokinin*: David McIntyre. Page 566 *Brassinosteroid*: Courtesy of Eugenia Russinova, VIB Department of Plant Systems Biology, Ghent University, Belgium.

Chapter 27 *Opener*: © Geoffrey Robinson/Alamy. 27.1A: © Phil Gates/Biological Photo Service. 27.1B *Male*: © Nigel Cattlin/Alamy. 27.1B *Female*: © Owaki - Kulla/Corbis. 27.1C *left*: © Adrian Davies/Alamy. 27.1C *right*: © Frank Young; Papilio/Corbis. 27.5B: © Leonid Nyshko/Alamy. 27.5C: © Maximilian Stock Ltd/PhotoCuisine/Corbis. 27.5D: © Scenics & Science/Alamy. 27.11: Courtesy of Richard Amasino and Colleen Bizzell. 27.12A: © Deco Images/Alamy. 27.12B: © Nigel Cattlin/Alamy. 27.12C: © Jerome Wexler/Visuals Unlimited, Inc. 27.14: David McIntyre. 27.15: © Jaroslaw Pyrih/Alamy. Page 578 *Thistle*: © Wolstenholme Images/Alamy. Page 578 *Dog*: © Scott Camazine/Alamy. Page 580: Courtesy of Richard Amasino.

Chapter 28 *Opener*: Courtesy of Robert Bowden/Kansas State University. 28.3A: © Arco Images GmbH/Alamy. 28.3B: © Daniel Borzynski/Alamy. 28.4: After A. Steppuhn et al., 2004. *PLoS Biology* 2: 1074. 28.6: © Charles Krebs/Corbis. 28.7: © John N. A. Lott/Biological Photo Service. 28.8: © D. Hurst/Alamy. 28.9: © blickwinkel/Alamy. 28.10: © John N. A. Lott/Biological Photo Service. 28.12: Courtesy of Scott Bauer/USDA. 28.13: © Jurgen Freund/Naturepl.com. 28.14: Courtesy of Ryan Somma. 28.15: From M. Ayliffe, R. Singh, & E. Lagudah. 2008. *Curr. Opin. Plant Biol.* 11: 187. Page 592: David McIntyre. Page 597 *Beetle*: Courtesy of Thomas Eisner, Cornell U.

Chapter 29 *Opener*: © BSIP SA/Alamy. 29.1: Courtesy of Andy J. Fischer. 29.8: © Rolf Nussbaumer Photography/Alamy. 29.10: From M. H. Ross, W. Pawlina, & T. A. Barnash, 2009. *Atlas of Descriptive Histology*. Sinauer Associates: Sunderland, MA. 29.11: © Custom Life Science Images. 29.15: Courtesy of Shinichi Tokishita. 29.18: © Carl Hanninen/Alamy. Page 611 *Dipsosaurus*: Courtesy of Mark A. Wilson/Wikipedia. Page 611 *Euphausia*: © Gerald & Buff Corsi/Visuals Unlimited, Inc. Page 622: Courtesy of Marco Brugnoli and Cosmed S.r.l.

Chapter 30 *Opener*: © Arco Images GmbH/Alamy. 30.4: © Wild Wonders of Europe/Cairns/Naturepl.com. 30.5A: © Christopher Swann/Science Source. 30.5B: © Alexander Semenov/Science Source. 30.6A *left*: © Masa Ushioda/Alamy. 30.6A *right*: © WaterFrame/Alamy. 30.10: © Curt Wiler/Alamy. 30.13 *right*: © Biophoto Associates/Science Source. 30.15: © Science VU/Jackson/Visuals Unlimited, Inc.

Chapter 31 *Opener*: Photo by Peter Habeler, courtesy of Leo Dickinson. 31.1, 31.3: © Daniel Kaesler/Alamy. 31.10B: © Scott Camazine/Alamy. 31.10C: Courtesy of Thomas Eisner, Cornell U. 31.11B: © Ralph Hutchings/Visuals Unlimited, Inc. 31.11D: © Dr. David Phillips/Visuals Unlimited, Inc. 31.15: After C. R. Bainton, 1972. *J. Appl. Physiol.* 33: 775. 31.16: © Dmitry Pichugin/istock. Page 648 *Flatworm*: © Ross Armstrong/AGE Fotostock. Page 648 *Sponge*: © Pete Oxford/Minden Pictures/Corbis.

Chapter 32 *Opener*: © Richard Herrmann/Visuals Unlimited, Inc. 32.1: © Manfred Kage/Science Source. 32.2: © Alfred Owczarzak/Biological Photo Service. 32.15: © Steven J. Kazlowski/Alamy.

Chapter 33 *Opener*: © Don Johnston/All Canada Photos/Corbis. 33.1: From Vesalius, A., 1543. *De humani corporis fabrica*. 33.2 *micrograph*: © Don W. Fawcett/Science Source. 33.3: © James Dennis/Phototake. 33.5: © Dr. Fred Hossler/Visuals Unlimited, Inc. 33.11: © Reinhard Dirscherl/Visuals Unlimited, Inc. 33.15: Courtesy of Jesper L. Andersen. 33.16: From Y-X Wang et al., 2004. *PLoS Biology* 2: e294. 33.17: © Biophoto Associates/Science Source. Page 684 *top*: From Ross, Pawlina, & Barnash, 2009. *Atlas of Descriptive Histology*. Sinauer Associates: Sunderland, MA. Page 695: © Biophoto Associates/Science Source. Page 696 *top*: © Premaphotos/Alamy.

Chapter 34 *Opener*: © Ken Catania/Visuals Unlimited, Inc. 34.1: © Eckehard Schulz/AP/Corbis. 34.3B: © C. Raines/Visuals Unlimited, Inc. 34.10: Courtesy of the Kennedy Lab, Division of Biology, Caltech. 34.17A: © Omikron/Science Source. 34.20A: © RGB Ventures LLC dba SuperStock/Alamy. 34.27: © Wellcome Dept. of Cognitive Neurology/SPL/Science Source. 34.28: Courtesy of Dr. Kevin LaBar, Duke U. Page 717: © Photoshot Holdings Ltd/Alamy. Page 723 *Viper*: © ephotocorp/Alamy. Page 723 *Fish*: © Jane Burton/Naturepl.com.

Chapter 35 *Opener*: © Toby Melville/Reuters/Corbis. 35.4A *Insulin*: Data from PDB 2HIU. Q. X. Hua et al., 1995. *Nat. Struct. Biol.* 2: 129. 35.4A *HGH*: Data from PDB 1HGU. L. Chantalat et al., 1995. *Protein Pept. Lett.* 2: 333. 35.12: © Alison Wright/Corbis.

Chapter 36 *Opener*: © blickwinkel/Alamy. 36.1: © David Cook/blueshiftstudios/Alamy. 36.3B: © A. Held/blickwinkel/AGE Fotostock. 36.3C: © Danita Delimont/Alamy. 36.5A: © Lunamarina/istock. 36.5B: © Cristian Baitg/istock. 36.7: © Rick & Nora Bowers/Alamy. 36.8B: © Steve Gschmeissner/Science Photo Library/Alamy.

Chapter 37 *Opener*: © blickwinkel/Alamy. 37.1: © Francis Leroy, Biocosmos/Science Source. 37.2A: © Aurora Photos/Alamy. 37.2B: © WaterFrame/Alamy. 37.5B: © Ed Reschke/Getty Images. 37.6 *top*: © Biophoto Associates/Science Source. 37.6 *bottom*: © CNRI/SPL/Science Source. 37.7 *inset*: © SERDAR/Alamy. 37.8: © Images & Stories/Alamy. 37.9: © Amar and Isabelle Guillen/Guillen Photography/Alamy. 37.14A: © Fred Bavendam/Minden Pictures. 37.14B: © Michel Roggo/Naturepl.com. Page 771 *Ducks*: © sasimoto/istock. Page 771 *Beetles*: © Pete Oxford/Minden Pictures/Corbis. Page 773 *bottom*: © Jeff Rotman/Alamy.

Chapter 38 *Opener*: © Wildlife GmbH/Alamy. 38.7A *left*: From H. W. Beams & R. G. Kessel, 1976. *American Scientist* 64: 279. 38.7A *center, right*: © Dr. Lloyd M. Beidler/Science Source. 38.7B: From H. W. Beams & R. G. Kessel, 1976. *American Scientist* 64: 279. 38.7C: Courtesy of D. Daily and W. Sullivan. 38.8A: From J. G. Mulnard, 1967. *Arch. Biol.* (Liege) 78: 107. Courtesy of J. G. Mulnard. 38.16C, D: Courtesy of K. W. Tosney and G. Schoenwolf. 38.17: Courtesy of K. W. Tosney. 38.22A *larva*: © FLPA/Alamy. 38.22A *adult*: © Sabena Jane Blackbird/Alamy. 38.22B *larva*: © D.P. Wilson/FLPA/Science Source. 38.22B *adult*: © Arterra Picture Library/Alamy. 38.23: Courtesy of Dr. Lynn James, USDA Poisonous Plant Research Laboratory. Page 786 *Sperm*: © Dr. David Phillips/Visuals Unlimited, Inc.

Chapter 39 *Opener*: © Christie's Images/Corbis. Page 810 *Crab*: © Alex Kerstitch/Visuals Unlimited, Inc. Page 810 *Amebocytes*: From P. Armstrong and M. Conrad, 2008. *Journal of Visualized Experiments* 20: e958. Page 810 *Granules*: From Iwanaga, 2002. *Current Opinion in Immunology* 14: 87. Page 813: © Biology Pics/Science Source.

Chapter 40 *Opener*: © Jane Gould/Alamy. 40.1: © Richard Becker/Alamy. 40.4: © Norbert Wu/Minden Pictures/Corbis. 40.5: © Scott Smith/Corbis. 40.6: © John E Marriott/All Canada Photos/Corbis. 40.7: © Andrew Parkinson/Naturepl.com. 40.13B: © FLPA/Tom and Pam Gard/AGE Fotostock. 40.14A: From Berdahl et al., 2013. Science 339: 574. 40.15: © Matteo Volpi/Shutterstock. 40.17: © Natural Visions/Alamy. Page 833: © Oxford Scientific/Getty Images. Page 837: © Markus Varesvuo/Naturepl.com.

Chapter 41 *Opener*: © Robert Shantz/Alamy. 41.2: © epa european press-photo agency b.v./Alamy. 41.12A: © Sunpix Travel/Alamy. 41.12B: © Ulrich Doering/Alamy. 41.12C: © Paul Mayall/Imagebroker/AGE Fotostock. 41.12D: © Niels van Kampenhout/Alamy. 41.15: © Nico Stengert/Novarc/Corbis. 41.16: © AgStock Images, Inc./Alamy.

Chapter 42 *Opener*: Courtesy of Richard S. Ostfeld Laboratory. 42.1 *inset*: © T. W. Davies/California Academy of Sciences. 42.4A: © Ozerov Alexander/Shutterstock. 42.4B: © Mary McDonald/Naturepl.com. 42.6: Courtesy of Barry Sinervo. 42.7 *left*: Courtesy of Joseph Berger, Bugwood.org. 42.7 *right*: Courtesy of Clemson University/USDA Cooperative Extension Slide Series/Bugwood.org. 42.10: © David Paterson/Alamy. 42.10 *inset*: © Nancy Nehring/istock. 42.11: David McIntyre. 42.12: © Joel Sartore/National Geographic Stock. Page 874 *Field*: Courtesy of Steve Dewey, Utah State University, Bugwood.org. Page 874 *inset*: Courtesy of Peggy Greb, USDA Agricultural Research Service, Bugwood.org.

Chapter 43 *Opener*: © Kim Taylor/Naturepl.com. 43.1B: © Inge Johnsson/Alamy. 43.1C: © Werner Bollmann/AGE Fotostock. 43.1D: © Gay Bumgarner/Alamy. 43.2: © cbpix/Shutterstock. 43.7C: Courtesy of Robert Beschta, from R. L. Beschta & W. J. Ripple, 2009. *Biological Conservation* 142: 2401. 43.9 *left*: © Tui De Roy/Minden Pictures/Corbis. 43.9 *right*: © Photoshot/Alamy. 43.10A *upper*: © Juniors Bildarchiv/Alamy. 43.10A *lower*: © ran-plett/istock. 43.10B: © Pete Oxford/Minden Pictures/Corbis. 43.11: David McIntyre.

Chapter 44 *Opener*: © Nigel Cattlin/Alamy. 44.1A: © Prisma Bildagentur AG/Alamy. 44.1B: Courtesy of Robert Whittaker. 44.3A: © Ed Reschke/Getty Images. 44.3B: © Mcmaster Studio/Alamy. 44.5 *upper*: David McIntyre. 44.5 *lower*: © William Manning/Corbis. 44.12: Courtesy of E. O. Wilson. 44.14: Richard Bierregaard, Courtesy of the Smithsonian Institution, Office of Environmental Awareness. 44.16: © John Mitchell/Alamy.

Chapter 45 *Opener*: © Jonathan Kingston/Getty Images. 45.2: From Zhao et al. 2006. *J. Geophys. Res.–Biogeosciences* 111: G01002. 45.4: Image by Robert Simmon, NASA GSFC Earth Observatory, based on data provided by Watson Gregg, NASA GSFC. 45.9: Courtesy of Goddard SVS/NASA. 45.14: After IPCC. 2007. *Climate Change 2007*. Cambridge University Press, Cambridge. 45.15: © blickwinkel/Alamy.

Index

Numbers in *italic* indicate that the information will be found in an illustration or table.

F

M

N

T

Z

ONLINE RESOURCES

The following online resources are referenced throughout the textbook, on the pages indicated. Each is available via the instant access code (QR code) and/or direct Web address that appears with the in-text reference. All of these resources (along with many additional resources) are also available within **LaunchPad** at highschool.bfwpub.com/launchpad/pol2e and on the **Companion Website** at bcs.whfreeman.com/pol2e.

MEDIA CLIPS

ANIMATED TUTORIALS

ONLINE RESOURCES continued

The following online resources are referenced throughout the textbook, on the pages indicated. Each is available via the instant access code (QR code) and/or direct Web address that appears with the in-text reference. All of these resources (along with many additional resources) are also available within **LaunchPad** at highschool.bfwpub.com/launchpad/pol2e and on the **Companion Website** at bcs.whfreeman.com/pol2e.

ACTIVITIES